CARDANO 1501–1576

Scholar, poet, physician, astrologer, and amateur mathematician, Geronimo Cardano gambled or played chess daily for 40 years, he says, and cast the horoscope of Christ. The execution of one son and the profligacy of another saddened his later years, yet did not prevent his writing *Ars Magna,* the first great Latin treatise on algebra.

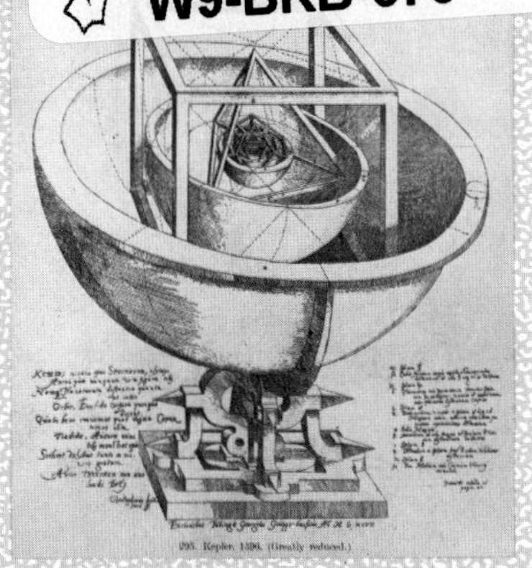

Kepler's model of the universe — a fantastic concept laboriously worked out. It was all wrong, but the same rich and fruitful mind later developed the three laws of planetary motion.

KEPLER 1571–1630

Born prematurely and a smallpox victim at the age of four, Johannes Kepler faced religious persecution, political upheavals, and in 1615 the loss of his wife and child. He was sustained by his deep faith in the mathematical harmony of nature—a harmony of the spheres with the sun as sole auditor. The concept of continuity was his focus.

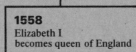

Euclid: The Elements of Geometrie, London, 1570

1536 Lodovico Ferrari, 14, became Cardano's secretary. A few years later he showed his master how to solve the general quartic equation.

1600 Guido del Monte wrote his important book on perspective, which actually contained Euclidean proofs. The book was later considered one of the jewels of Italian literature.

NAPIER 1550–1617

John Napier wrote his first book—a theological polemic—in English in order that "hereby the simple of this iland may be instructed." He designed burning mirrors for destroying objects at a distance, proclaimed a means for "sayling under the water," and unveiled the concept of logarithms. Some accused him of black art, others lauded his ingenuity.

GALILEO 1564–1642

A skillful musician who was attracted to painting as a profession, young Galileo Galilei studied medicine at his father's insistence until seduced by the charms of geometry. His superb lectures on physics, astronomy, and projectiles required a hall capacity of 2,000. Galileo's defense of the Copernican doctrine led to his persecution by Aristotelians of the church.

1500 — 1600

1509 Henry VIII becomes king of England

1513 Balboa discovers Pacific

1533 Spaniards conquer Aztecs and Incas

1558 Elizabeth I becomes queen of England

1564 Shakespeare born

DESCARTES 1596–1650

René Descartes founded his philosophy on systematic doubt: "I think, therefore I am" was ostensibly his sole premise. He had no mathematics teacher and few students, but his work on analytic geometry will long be remembered.

NEWTON 1642–1727

Isaac Newton was born on a farm on Christmas Day. He showed early ingenuity with mechanical devices and such an interest in books that he was sent to Cambridge. During the plague the university closed, and Newton returned to Woolsthorpe. "I was in the prime of my age for invention, and minded mathematics and philosophy more than at any time since," he said; and here, in isolation, the 23-year-old developed calculus and conceived his laws of motion and gravitation. Knighted and rich, he died after a lifetime devoted to alchemy, theology, public administration, and science.

FERMAT 1601–1665

Pierre-Simon de Fermat, son of a leather merchant, was educated at home. His profession (law) left time for his hobby (mathematics), which he started very late in life (at about 30). Noted for his honesty and fairness, but not his modesty, he addressed sharp challenges to those who questioned his problem-solving techniques. Both he and Descartes are universally credited as the inventors of analytic geometry.

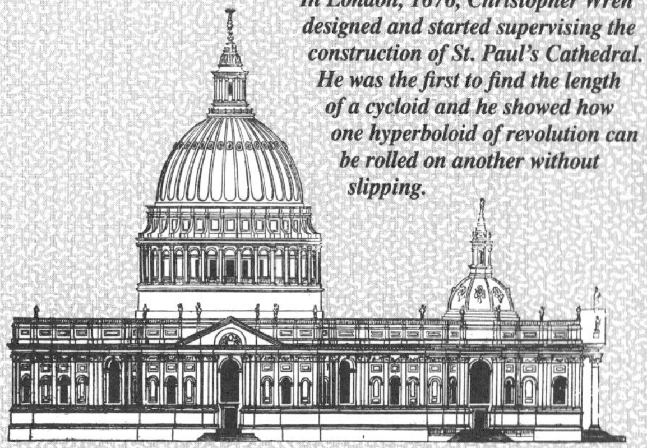

In London, 1676, Christopher Wren designed and started supervising the construction of St. Paul's Cathedral. He was the first to find the length of a cycloid and he showed how one hyperboloid of revolution can be rolled on another without slipping.

Pascal's Triangle was first printed in 1527 as a title page to the arithmetic of Petrus Apianus.

PASCAL 1623–1662

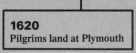

Attending no school, Blaise Pascal learned the classics from his father. Geometry was withheld until Blaise at age 12 discovered a theorem on his own. His "Provincial Letters," composed after he left science for theology, are a glory of French prose.

1600 **1700**

1620
Pilgrims land at Plymouth

1642
Isaac Newton born

1660
Charles II becomes king of England

1685
Bach born

CALCULUS

THE PRINDLE, WEBER & SCHMIDT SERIES IN MATHEMATICS

Althoen and Bumcrot, *Introduction to Discrete Mathematics*
Boye, Kavanaugh, and Williams, *Elementary Algebra*
Boye, Kavanaugh, and Williams, *Intermediate Algebra*
Burden and Faires, *Numerical Analysis, Fourth Edition*
Cass and O'Connor, *Fundamentals with Elements of Algebra*
Cullen, *Linear Algebra and Differential Equations, Second Edition*
Dick and Patton, *Calculus, Volume I*
Dick and Patton, *Calculus, Volume II*
Dick and Patton, *Technology in Calculus: A Sourcebook of Activities*
Eves, *In Mathematical Circles*
Eves, *Mathematical Circles Adieu*
Eves, *Mathematical Circles Squared*
Eves, *Return to Mathematical Circles*
Fletcher, Hoyle, and Patty, *Foundations of Discrete Mathematics*
Fletcher and Patty, *Foundations of Higher Mathematics, Second Edition*
Gantner and Gantner, *Trigonometry*
Geltner and Peterson, *Geometry for College Students, Second Edition*
Gilbert and Gilbert, *Elements of Modern Algebra, Third Edition*
Gobran, *Beginning Algebra, Fifth Edition*
Gobran, *Intermediate Algebra, Fourth Edition*
Gordon, *Calculus and the Computer*
Hall, *Algebra for College Students*
Hall, *Beginning Algebra*
Hall, *College Algebra with Applications, Third Edition*
Hall, *Intermediate Algebra*
Hartfiel and Hobbs, *Elementary Linear Algebra*
Humi and Miller, *Boundary-Value Problems and Partial Differential Equations*
Kaufmann, *Algebra for College Students, Fourth Edition*
Kaufmann, *Algebra with Trigonometry for College Students, Third Edition*
Kaufmann, *College Algebra, Second Edition*
Kaufmann, *College Algebra and Trigonometry, Second Edition*
Kaufmann, *Elementary Algebra for College Students, Fourth Edition*
Kaufmann, *Intermediate Algebra for College Students, Fourth Edition*
Kaufmann, *Precalculus, Second Edition*
Kaufmann, *Trigonometry*
Kennedy and Green, *Prealgebra for College Students*
Laufer, *Discrete Mathematics and Applied Modern Algebra*
Nicholson, *Elementary Linear Algebra with Applications, Second Edition*
Pence, *Calculus Activities for Graphic Calculators*
Pence, *Calculus Activities for the TI-81 Graphic Calculator*
Plybon, *An Introduction to Applied Numerical Analysis*
Powers, *Elementary Differential Equations*
Powers, *Elementary Differential Equations with Boundary-Value Problems*
Proga, *Arithmetic and Algebra, Third Edition*
Proga, *Basic Mathematics, Third Edition*
Rice and Strange, *Plane Trigonometry, Sixth Edition*
Schelin and Bange, *Mathematical Analysis for Business and Economics, Second Edition*
Strnad, *Introductory Algebra*
Swokowski, *Algebra and Trigonometry with Analytic Geometry, Seventh Edition*
Swokowski, *Calculus, Fifth Edition*
Swokowski, *Calculus, Fifth Edition (Late Trigonometry Version)*
Swokowski, *Calculus of a Single Variable*
Swokowski, *Fundamentals of College Algebra, Seventh Edition*
Swokowski, *Fundamentals of College Algebra and Trigonometry, Seventh Edition*
Swokowski, *Fundamentals of Trigonometry, Seventh Edition*
Swokowski, *Precalculus: Functions and Graphs, Sixth Edition*
Tan, *Applied Calculus, Second Edition*
Tan, *Applied Finite Mathematics, Third Edition*
Tan, *Calculus for the Managerial, Life, and Social Sciences, Second Edition*
Tan, *College Mathematics, Second Edition*
Trim, *Applied Partial Differential Equations*
Venit and Bishop, *Elementary Linear Algebra, Alternate Second Edition*
Venit and Bishop, *Elementary Linear Algebra, Third Edition*
Wiggins, *Problem Solver for Finite Mathematics and Calculus*
Willard, *Calculus and Its Applications, Second Edition*
Wood and Capell, *Arithmetic*
Wood and Capell, *Intermediate Algebra*
Wood, Capell, and Hall, *Developmental Mathematics, Fourth Edition*
Zill, *A First Course in Differential Equations with Applications, Fourth Edition*
Zill and Cullen, *Advanced Engineering Mathematics*
Zill, *Calculus, Third Edition*
Zill, *Differential Equations with Boundary-Value Problems, Second Edition*

THE PRINDLE, WEBER & SCHMIDT SERIES IN ADVANCED MATHEMATICS

Brabenec, *Introduction to Real Analysis*
Ehrlich, *Fundamental Concepts of Abstract Algebra*
Eves, *Foundations and Fundamental Concepts of Mathematics, Third Edition*
Keisler, *Elementary Calculus: An Infinitesimal Approach, Second Edition*
Kirkwood, *An Introduction to Real Analysis*
Ruckle, *Modern Analysis: Measure Theory and Functional Analysis with Applications*
Sieradski, *An Introduction to Topology and Homotopy*

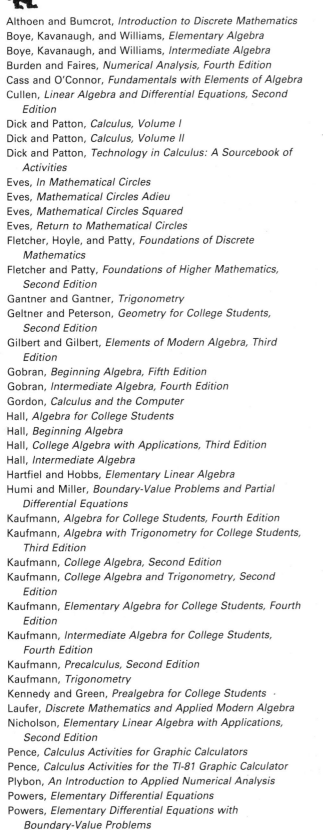

CALCULUS
THIRD EDITION

DENNIS G. ZILL
Loyola Marymount University

PWS-KENT
PUBLISHING COMPANY
Boston

PWS-KENT
Publishing Company

20 Park Plaza
Boston, Massachusetts 02116

PWS-KENT Publishing Company is a division of Wadsworth, Inc.

Library of Congress Cataloging-in-Publication Data

Zill, Dennis G.
Calculus / Dennis G. Zill. — 3rd ed.
p. cm. — (The Prindle, Weber & Schmidt series in mathematics)
Includes index.
ISBN 0-534-92793-9
1. Calculus. I. Title. II. Series.
QA303.Z52 1992 91-17494
515—dc20 CIP

About the Time Line (appearing on the endsheets):
Sections of the Time Line have been reproduced through the courtesy of the International Business Machines Corporation. Illustration credits include *The Bettman Archive* for da Vinci's Proportions of Man, St. Remi at Rheims (p. 1); Kepler's Model of the Universe, Euclid's The Elements of Geometrie (p. 2); Pascal's Triangle, Wren's St. Paul's Cathedral (p. 3); Karlskirche (p. 4); IBM corporate headquarters, Wright's "Falling Water," Empire State Building (p. 6); *Historical Pictures Service* for Hilbert, Poincaré (p. 5); Noether (p. 6); and *Smithsonian Institution photo* for Kepler (p. 2); Pascal, Newton (p. 3); Euler, Laplace, Gauss (p. 4); Birkhoff, Von Neumann (p. 6).

Sponsoring Editor: Steve Quigley
Developmental Editor: Barbara Lovenvirth
Production Coordinator and Cover Designer: Robine Andrau
Production: Cece Munson/The Cooper Company
Manufacturing Coordinators: Margaret Sullivan Higgins, Peter D. Leatherwood
Interior Designer: Julia Gecha
Copy Editor: Carol Reitz
Interior Illustrator: Tech-Graphics
Cover Photo: Slide Graphics of New England, Inc.
Cover Printer: Henry N. Sawyer Co., Inc.
Typesetter: Polyglot Pte Ltd
Printer and Binder: R.R. Donnelley & Sons

Printed in the United States of America
91 92 93 94 95 — 10 9 8 7 6 5 4 3 2 1

CONTENTS

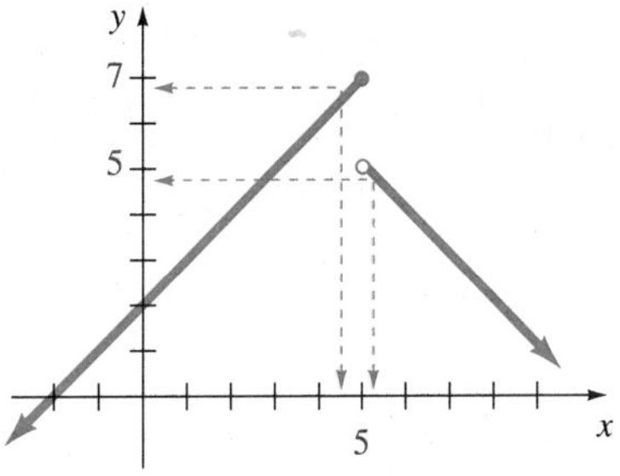

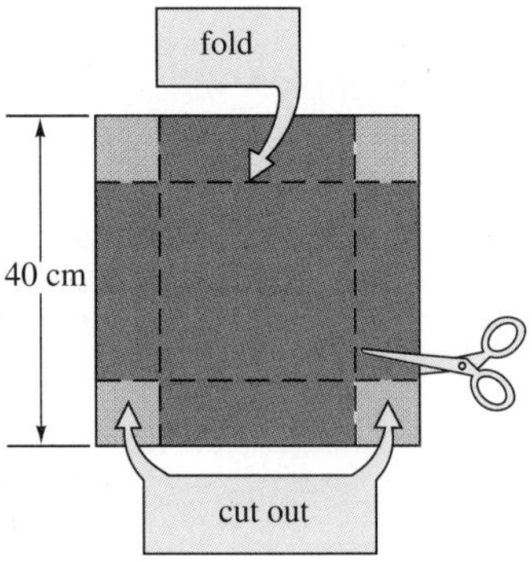
fold
40 cm
cut out

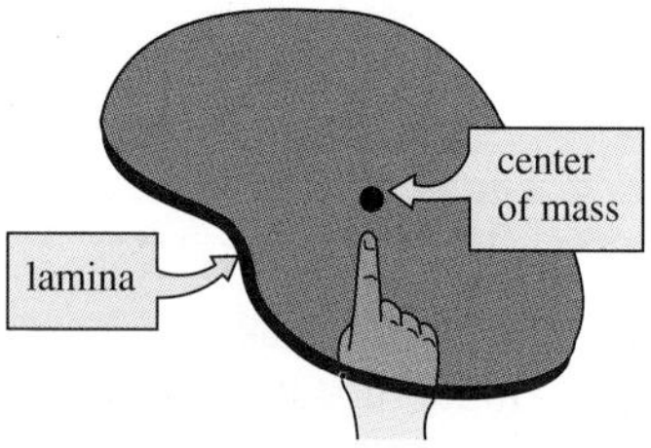
center
of mass
lamina

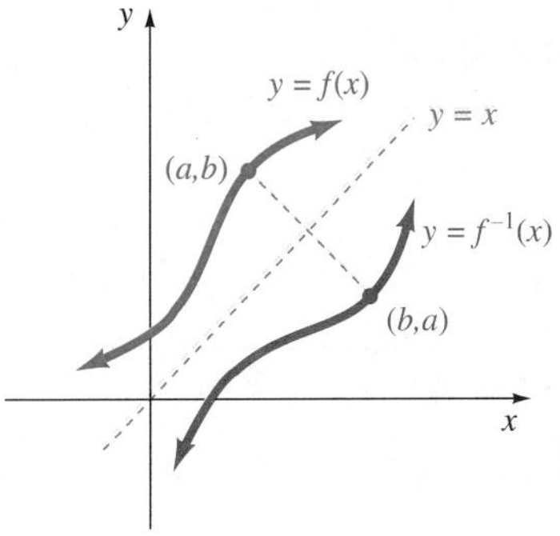
y
y = f(x)
y = x
(a,b)
y = f^{-1}(x)
(b,a)
x

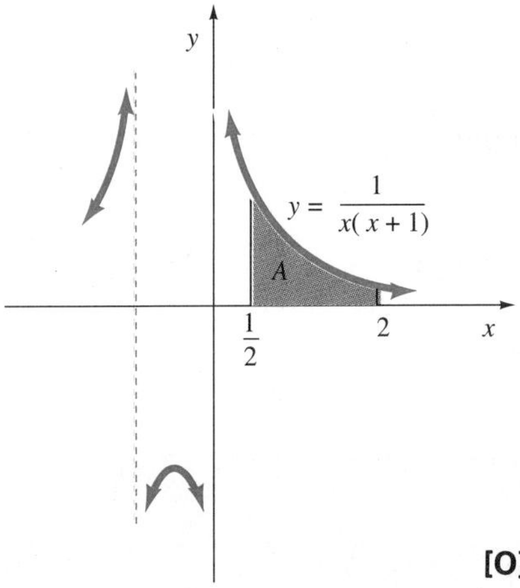
y
y = 1/(x(x + 1))
A
1/2
2
x

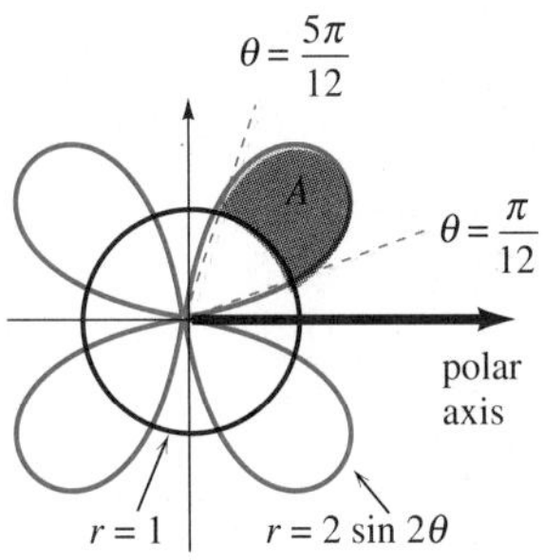

$\theta = \frac{5\pi}{12}$
A
$\theta = \frac{\pi}{12}$
polar axis
$r = 1$
$r = 2 \sin 2\theta$

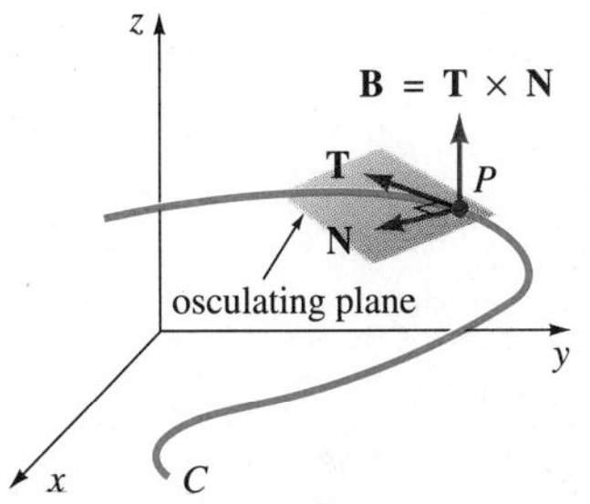
z
B = T × N
T
P
N
osculating plane
y
x
C

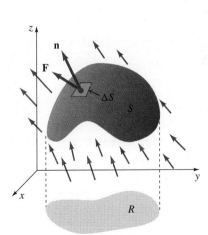

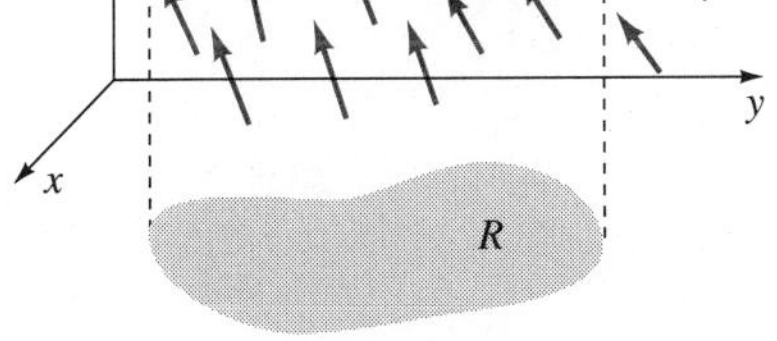
z
n
F
ΔS
S
y
x
R

m

PREFACE

The third edition of *Calculus* represents a substantial revision over the last edition. Although there is much new in this edition, I have striven to keep intact my original goal of compiling a calculus text that is not just a collection of definitions, theorems, and problems, but rather a book that communicates with its primary audience, the students. It is hoped that the changes make the text more relevant and interesting for both student and instructor.

One of my objectives in this revision was to modify the book in line with some of the refinements that I am currently making in my own course in calculus. Other changes reflect both the requests made by users (students and instructors) of the previous two editions and, of course, the suggestions of the reviewers.

Changes in the Third Edition

- The precalculus material on real numbers, the Cartesian plane, lines, and circles, has been moved to Appendix I.
- Chapter 1 now begins with the function concept, and the material on limits has been merged with that on functions. Limits are still presented intuitively and a few additional examples were added to that discussion (specifically, for quadratic functions).
- In Section 1.3, "Trigonometric Functions," a brief discussion on using calculators to evaluate inverse trigonometric functions has been added, enabling the student to utilize the calculator more fully in the exercise sets.
- Section 1.4, entitled "Setting Up Functions," was added to expose the student early on to the type of problems—namely, interpreting words and geometry to construct a function—that will be encountered in the subsequent sections dealing with related rates and applied max/min problems.
- The method of interval bisection for approximating roots of equations has been added to the discussion of continuity in Section 1.8. This topic is introduced as a direct consequence of the Intermediate Value Theorem.

- In Chapter 4, the Midpoint Rule has been added to Section 4.8, "Approximate Integration."
- The discussions on the disk method and the shell method have been combined into Section 5.3, "Solids of Revolution."
- The concepts of the natural logarithm and the natural exponential function now appear in Chapter 6. This change enables the instructor to include these functions in the first semester if desired. Section 6.1 of Chapter 6 is now "Inverse Functions," previously in Chapter 7.
- Another feature of Chapter 6 is the inclusion of linear first-order differential equations and their applications. This change was made to accommodate those students who encounter this type of differential equation early in their physics, engineering, and biology courses. However, as some instructors may legitimately feel that there is no time to cover this topic in the first semester, the material is marked optional.
- The discussion of logarithmic differentiation has been incorporated into Section 6.5.
- The material on inverse hyperbolic functions has been combined with the material on inverse trigonometric functions and appears now in Chapter 7.
- The "ladder method" or "tabular integration by parts" has been added to "Integration by Parts" in Chapter 8 and the "cover-up method" is briefly discussed as a means of evaluating coefficients in the section on "Partial Fractions."
- "Integral Tables, Calculators, and Computers" appears as Section 8.6. The problems from "Integration of Rational Functions of Sine and Cosine," now deleted, have been incorporated into this section.
- "A Review of Applications," previously in Chapter 9, has been deleted. These problems, which utilized techniques of integration, have been placed in various appropriate sections.
- In Chapter 11, "Analytic Geometry," the concept of eccentricity has been added to the first three sections.
- The material on curvature in Section 14.3 has been rewritten and appears before the discussion of the tangential and normal components of acceleration.
- In Chapter 15, "Method of Least Squares" has been added and is presented as an application of the second partials test for relative extrema.
- In Chapter 16, a new section, "Change of Variables in Multiple Integrals," has been added. In this discussion, the Jacobian of a transformation is introduced.
- In Chapter 17, "Green's Theorem" has been moved forward and now appears directly after the discussion of line integrals. The discussion of Stokes' Theorem and the Divergence Theorem has been expanded. These theorems now appear in separate sections, and are followed by proofs. Line integrals that are path independent in 3-space are now fully discussed in Section 17.2.
- SI notation is used throughout the text.

- There are many new examples, graphs, computer graphics, and computer programs interspersed throughout the text.
- New material on matrices has been added to Appendix I; and the table of integrals now appears as Appendix IV.

Exercises

Many new problems, especially applications and problems dealing with graphics calculators and computers, have been added to the exercise sets. For the most part, the added applications are "real life" in that they have been thoroughly researched, using original sources. Some problems have been marked with the phrase "This problem could present a challenge" to indicate that these should not be assigned casually by the instructor.

Problems dealing with interpretation of graphs have also been added. Moreover, there is an increased emphasis on the trigonometric functions, both in the examples and in the exercise sets throughout the text. There are over 7800 problems in this edition.

Design Features

The boxes around the examples have been eliminated to give the text a more open appearance. However, theorems and definitions appear in a colored outline box and when possible are labeled with their appropriate titles. Selected figures have been enhanced using the four-color format. I have tried to be consistent in the use of coloring; for example, vectors are red, curves are blue, and so on. Each chapter opens with its own table of contents, an introduction to the material covered in that chapter, a list of important concepts, and a brief motivational discussion on some application of the mathematics within that chapter.

Special Features

Two features new to this edition are "To the Student" and "Test Yourself." The latter is a self-test consisting of 40 questions on four broad areas of precalculus mathematics. This test is intended to encourage students to review, perhaps on their own, some of the more basic subjects, such as absolute values, the Cartesian plane, equations of lines, circles, and so on, that are used throughout the text as a matter of course. Answers to all questions in the test are given in the answer section.

Three new sections, entitled "Calculator/Computer Activities," are placed after Chapter 6, Chapter 12, and Chapter 18. These sections contain exercises requiring the use of calculators/graphics calculators/computers and reflect the type of problems contained in the preceding text. In total, there are 18 sets of activities keyed to the corresponding 18 chapters. These problems are in addition to the many new calculator/computer problems added to the regular exercise sets.

Discussions on computer algebra systems such as Mathematica and Maple have been added in appropriate places throughout the text.

Following the text material, there is a discussion of the relationship of some of the mathematics studied in calculus and the exciting new field of fractal geometry. This material was specially written at a level that a typical student of calculus can understand. One reason for including this discussion is to give the student some appreciation that mathematics is not a collection of dry facts and rules discovered hundreds of years ago but, rather, is a dynamic and constantly evolving subject.

For the Student

Users of the previous two editions have been very receptive to the "Remarks" that often conclude a discussion. As a consequence their number has been increased. The "Remarks" are intended to be informal discussions that are aimed directly at the student. These discussions range from warnings about common algebraic, procedural, and notational errors; to misinterpretations of theorems; to advice; to questions asking the student to think about and possibly extend the ideas just presented.

At the request of users, the number of guidance boxes in the examples has been increased. Also, this edition includes more of the review inserts prior to the start of a section (for example, see Section 5.2).

Supplements

FOR INSTRUCTORS

Complete Solutions Manual, Warren S. Wright, Loyola Marymount University. Contains complete worked-out solutions to every exercise found in the text.

Printed Test Bank. Contains all questions found in the ExpTest with answers, for each chapter.

FOR STUDENTS

Study Guide for Calculus, David C. Arney, David H. Olwell, Kathleen G. Snook, and Richard D. West, all of the United States Military Academy, West Point. This is a self-study workbook designed for use with any standard calculus text. Students can reinforce their understanding of all major topics through drills, review, and self-tests.

Student Solutions Manual, Warren S. Wright. Provides solutions to every third exercise in the text.

Calculus Activities for Graphic Calculators, Dennis Pence, Western Michigan University. Calculus topics using the Sharp, Casio, and HP-28S, HP-28C graphing calculators.

Calculus Activities for the TI-81 Graphing Calculator, Dennis Pence.

Technology in Calculus: A Sourcebook of Activities, Thomas Dick and Charles Patton, Oregon State University. Over 40 projects designed for discovery in calculus.

Instructors Resource Guide to MPP, Howard Penn, US Naval Academy. The guide provides instructors with notes on teaching key concepts using MPP. Example files containing examples and problems from the text are also included for the IBM PC and compatibles.

COMPUTERIZED TESTING

ExpTest. A testing program for the IBM-PCs and compatibles, this test bank allows users to view and edit all tests, adding to, deleting from, and modifying existing questions. Any number and variety of tests can be created. A graphics importation feature permits the display and printing of graphs, diagrams, and maps provided with the test banks. A demo disk is available.

ExamBuilder. A testing program for the Macintosh that allows users to create, edit, and print tests. It contains multiple-choice and open-ended questions. A demo disk is available. (Users must have HyperCard to run the demo disk.)

SOFTWARE

Mathematics Plotting Package (*MPP*). Available for IBM-PCs and compatibles. This public-domain program developed at the U.S. Naval Academy is accompanied by a text-specific Instructors Resource Guide to MPP by Howard Penn. Also included are Example Files for MPP—disks containing examples and problems from the Resource Guide.

Grapher, Steve Scarborough. For the Macintosh, this is a flexible program that can be used to generate several types of graphs. In addition to plotting rectangular and polar curves and interpolating polynomials, it handles parametric equations, numerical solutions of differential equations, series, and direction fields.

TrueBASIC CALCULUS. Available for IBM-PCs and compatibles. This is a disk and manual package for self-study, exploration of problems, and solutions.

Acknowledgments

Compiling a textbook of this complexity is a monumental task. Besides myself, many people put much time and energy into this project. I would like to recognize and express my appreciation to the production staff at PWS-KENT Publishing Company, Cece Munson for coordinating the phases of production, Carol Rcitz for an excellent job in copyediting the final manuscript, and the following reviewers who contributed many suggestions, valid criticisms, and even an occasional word of support:

Steven Blasberg
West Valley College

H. Edward Donley
Indiana University of Pennsylvania

John W. Dulin
GMI Engineering & Management Institute

Patrick J. Enright
Arapahoe Community College

Peter Frisk
Rock Valley College

David Green, Jr.
GMI Engineering & Management Institute

Christopher E. Hee
Eastern Michigan University

Rahim G. Karimpour
Southern Illinois University

William J. Keane
Boston College

Carlon A. Krantz
Kean College of New Jersey

John C. Lawlor
University of Vermont
Jill McKenney
Lane Community College
Antonio Magliaro
Southern Connecticut State University
Walter Fred Martens
University of Alabama at Birmingham
William E. Mastrocola
Colgate University
Edward T. Migliore
Monterey Peninsula College
James Osterburg
University of Cincinnati
Marvin C. Papenfuss
Loras College
Don Poulson
Mesa Community College
James J. Reynolds
Pennsylvania State University—Beaver Campus
Donald E. Rossi
De Anza College
Nedra Shunk
Santa Clara University
Phil R. Smith
American River College
Richard Werner
Santa Rosa Junior College
Loyd V. Wilcox
Golden West College

Finally, I would like to single out the following individuals for special recognition:

> Scott and Carol Wright for their excellent work in producing the answer supplements for the text and for their generous help in the final stages of the manuscript preparation,
>
> Barry A. Cipra for contributing many fine ideas, which form the basis for many of the new applied problems,
>
> Dennis Pence of Western Michigan University for his work on the new supplemental Calculator/Computer Activities,
>
> Terence H. Perciante of Wheaton College for his contributions to the new appendix on Fractals,
>
> Barbara Lovenvirth, Developmental Editor at PWS-KENT Publishing Company, for her many suggestions, overall help, and humor, which at times kept us both sane.

Even with all this help, the accuracy of every letter, word, symbol, equation, and figure contained in this final product is the responsibility of the author. I would be very grateful to have any errors or "typos" called to my attention.

Dennis G. Zill
Los Angeles

TO THE STUDENT

You are enrolled in one of the most interesting courses in mathematics. Many years ago when I was enrolled in Calculus I, I was amazed (and still am) at the power and beauty of the material. It was unlike any mathematics that I had studied up to that point. It was fun, it was exciting, it was a challenge.

After teaching mathematics for over 25 years, I have seen almost every type of student in calculus, from a budding genius who invented his own calculus, to students who could not or would not master the most rudimentary mechanics of the subject. Over these years I have also witnessed a sad phenomenon: some students fail calculus, not because they find the subject matter impossibly difficult, but because they have weak algebra skills and an inadequate working knowledge of trigonometry. This prelude is being written on an examination day in my calculus classes. An unacceptable number of students did poorly on one particular problem since, in the process of its solution, they wrote $(\tan x)^{1/2}(\tan x)^2 = \tan x$. From that point on, their calculus was flawless but meaningless. You all have most probably had four years of high school mathematics, and some of you have just finished a college-level course in precalculus mathematics. There is much new ground to be covered in calculus and consequently there is very little time to review basics in the formal classroom setting. Calculus builds on your prior knowledge. So we teachers must assume that you can factor, simplify and solve equations, solve inequalities, use a calculator, apply the laws of exponents (correctly), find equations of lines, plot points in the Cartesian plane, and know some trigonometric identities. I know too that you have also encountered the notion of a function, but because calculus is all about functions it is logical to start this text with that concept.

Following this note there is a list of 40 questions in a section called "Test Yourself." This "test" is an opportunity for you to check your knowledge on some of the topics that will be assumed in this text. Relax, take your time, read and work every question, and then compare your answers with those given on page A-63. Regardless of your "score" in this test, you are strongly encouraged to review the material in Appendix I.

TEST YOURSELF

Basic Mathematics

1. (True/False) $\sqrt{a^2 + b^2} = a + b$. ______
2. (True/False) For $a > 0$, $(a^{4/3})^{3/4} = a$. ______
3. (True/False) For $x \neq 0$, $x^{-3/2} = \dfrac{1}{x^{2/3}}$. ______
4. (True/False) $\dfrac{2^n}{4^n} = \dfrac{1}{2^n}$. ______
5. (Fill in the blank) In the expansion of $(1 - 2x)^3$ the coefficient of x^2 is ______.
6. Without the aid of a calculator, evaluate $(-27)^{5/3}$.
7. Write as one expression without negative exponents:

$$x^2 \frac{1}{2}(x^2 + 4)^{-1/2} 2x + 2x\sqrt{x^2 + 4}$$

8. Solve the equations: **(a)** $x^2 = 7x$ **(b)** $x^2 + 2x = 5$
9. Complete the square: $2x^2 + 6x + 5$
10. Factor completely: **(a)** $x^4 - 2x^3 - 15x^2$ **(b)** $x^4 - 16$ **(c)** $x^3 - 27$

Real Numbers

11. (True/False) If $a < b$, then $a^2 < b^2$. ______
12. (True/False) $\sqrt{(-9)^2} = -9$. ______
13. (True/False) If $a < 0$, then $\dfrac{-a}{a} < 0$. ______

14. (Fill in the blanks) If $|3x| = 18$, then $x =$ ______ or $x =$ ______.

15. (Fill in the blank) If $a - 5$ is a negative number, then $|a - 5| =$ ______.

16. Which of the following real numbers are rational numbers?

(a) 0.25	**(b)** $8.131313\ldots$	**(c)** π	**(d)** $\dfrac{22}{7}$
(e) $\sqrt{16}$	**(f)** $\sqrt{2}$	**(g)** 0	**(h)** -9
(i) $1\frac{1}{2}$	**(j)** $\dfrac{\sqrt{5}}{\sqrt{2}}$	**(k)** $\dfrac{\sqrt{3}}{2}$	**(l)** $\dfrac{-2}{11}$

17. Match the given interval with the appropriate inequality.

(*i*) $(2, 4]$	(*ii*) $[2, 4)$
(*iii*) $(2, 4)$	(*iv*) $[2, 4]$
(a) $\lvert x - 3 \rvert < 1$	**(b)** $\lvert x - 3 \rvert \le 1$
(c) $0 \le x - 2 < 2$	**(d)** $1 < x - 1 \le 3$

18. Express the interval $(-2, 2)$ as **(a)** an inequality and **(b)** an inequality involving absolute values.

19. Sketch the graph of $(-\infty, -1] \cup [3, \infty)$ on the number line.

20. Find all real numbers x that satisfy the inequality $|3x - 1| > 7$. Write your solution using interval notation.

Cartesian Plane

21. (Fill in the blank) If (a, b) is a point in the third quadrant, then $(-a, b)$ is a point in the ______ quadrant.

22. (Fill in the blank) The midpoint of the line segment from $P_1(2, -5)$ to $P_2(8, -9)$ is ______.

23. (Fill in the blanks) If $(-2, 6)$ is the midpoint of the line segment from $P_1(x_1, 3)$ to $P_2(8, y_2)$, then $x_1 =$ ______ and $y_2 =$ ______.

24. (Fill in the blanks) The point $(1, 5)$ is on a graph. Give the coordinates of another point on the graph if the graph is:
   (a) symmetric with respect to the x-axis. ______
   (b) symmetric with respect to the y-axis. ______
   (c) symmetric with respect to the origin. ______

25. (Fill in the blanks) The x- and y-intercepts of the graph of $|y| = 2x + 4$ are, respectively, ______ and ______.

26. In which quadrants of the Cartesian plane is the quotient x/y negative?

27. The y-coordinate of a point is 2. Find the x-coordinate of the point if the distance from the point to $(1, 3)$ is $\sqrt{26}$.

28. Find an equation of the circle for which $(-3, -4)$ and $(3, 4)$ are endpoints of a diameter.

29. If the points P_1, P_2, and P_3 are collinear as shown in the figure, then find an equation relating the distances $d(P_1, P_2)$, $d(P_2, P_3)$, and $d(P_1, P_3)$.

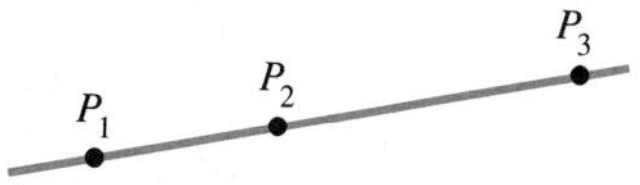

30. Which of the following equations best describes the circle given in the accompanying figure? The symbols a, b, c, d, and e stand for nonzero constants.

(a) $ax^2 + by^2 + cx + dy + e = 0$
(b) $ax^2 + ay^2 + cx + dy + e = 0$
(c) $ax^2 + ay^2 + cx + dy = 0$
(d) $ax^2 + ay^2 + e = 0$
(e) $ax^2 + ay^2 + cx + e = 0$

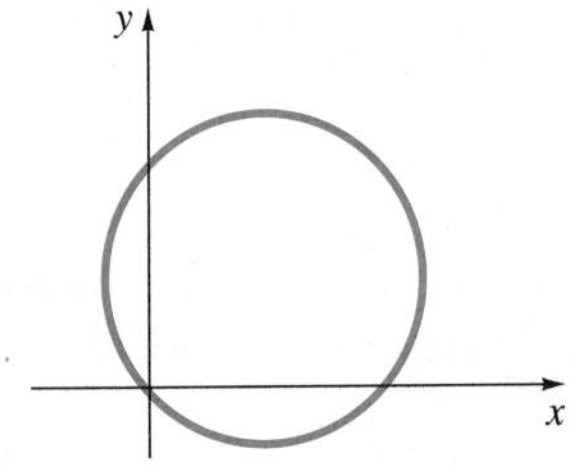

Lines

31. (True/False) The lines $2x + 3y = 5$ and $-2x + 3y = 1$ are perpendicular. ________

32. (Fill in the blank) The lines $6x + 2y = 1$ and $kx - 9y = 5$ are parallel if $k =$ ________.

33. (Fill in the blank) A line with x-intercept -4 and y-intercept 32 has slope ________.

34. (Fill in the blanks) The slope and the x- and y-intercepts of the line $2x - 3y + 18 = 0$ are, respectively, ________, ________, and ________.

35. (Fill in the blank) An equation of the line with slope -5 and y-intercept 3 is ________.

36. Find an equation of the line that passes through $(3, -8)$ and is parallel to the line $2x - y = -7$.

37. Find an equation of the line through the points $(-3, 4)$ and $(6, 1)$.

38. Find an equation of the line that passes through the origin and through the point of intersection of the graphs of $x + y = 1$ and $2x - y = 7$.

39. A tangent line to a circle at a point P on the circle is a line through P that is perpendicular to the line through P and the center of the circle. Find an equation of the tangent line L indicated in the figure.

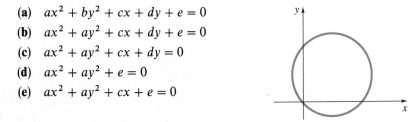

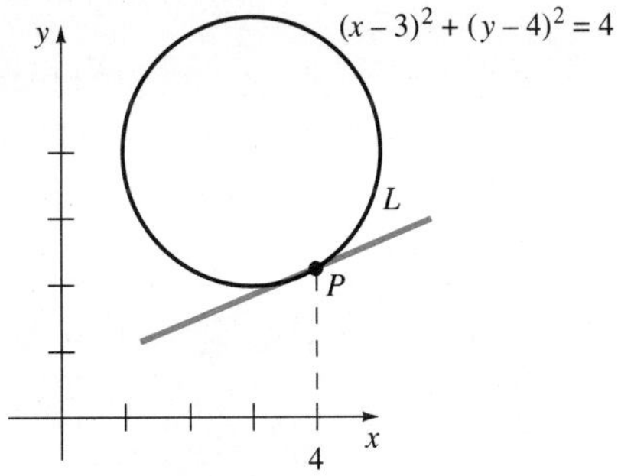

40. Match the given equation with the appropriate graph.

(*i*) $x + y - 1 = 0$ (*ii*) $x + y = 0$

(*iii*) $x - 1 = 0$ (*iv*) $y - 1 = 0$

(*v*) $10x + y - 10 = 0$ (*vi*) $-10x + y + 10 = 0$

(*vii*) $x + 10y - 10 = 0$ (*viii*) $-x + 10y - 10 = 0$

(a)
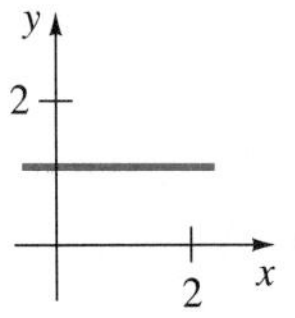
(b)
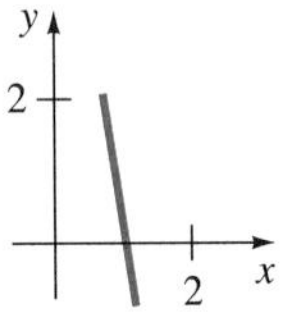
(c)
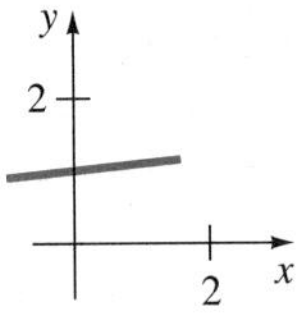
(d)
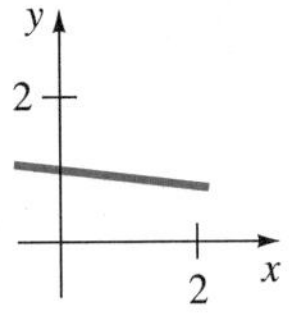
(e)
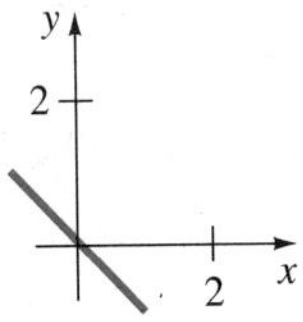
(f)
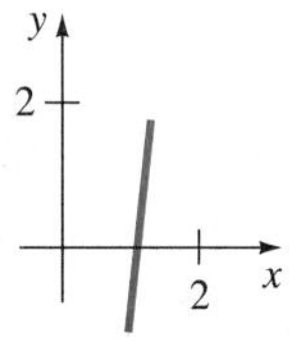
(g)
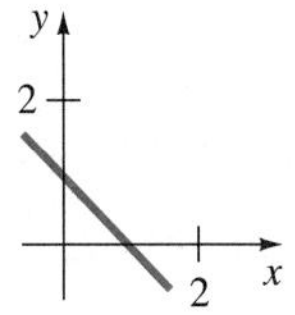
(h)
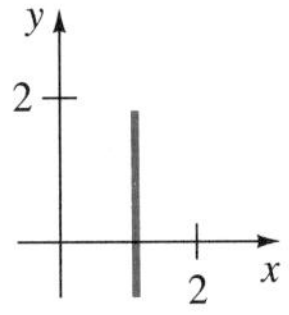

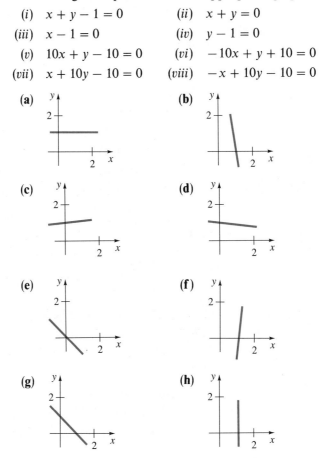

1

FUNCTIONS AND LIMITS

INTRODUCTION

Two of the most fundamental concepts in the study of calculus are the **function** and the **limit of a function**. After reviewing some basic facts about functions, we shall focus our attention on the seemingly simple, but extremely important, problem of determining whether the values $f(x)$ of a function f approach a number L as x approaches a number a. Using the symbol $\to$ for the word *approach*, we ask:

$$\text{Does} \quad f(x) \to L \quad \text{as} \quad x \to a?$$

Historically, two problems are used to introduce the basic tenets of calculus. These are the *tangent problem* and the *area problem.* We shall examine both these problems in detail in Chapters 2 and 4 and see that their solutions involve the limit concept.

We know that the area of a circle of radius r is $A = \pi r^2$. How was this formula discovered? Here is one way. Suppose, as shown in the figure, that regular polygons are inscribed in a

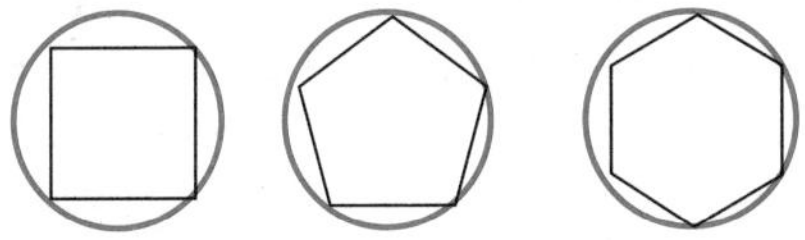

circle of radius r. As the number of the sides of the polygons increases, it seems reasonable to assume that the areas of the polygons approximate or *approach* the area of the circle. Of course, to see this we need to find a formula for the area of a regular n-sided polygon. You will be asked to find this formula and thereby derive the area of a circle in Exercises 1.7.

IMPORTANT CONCEPTS

function
independent variable
dependent variable
domain
range
polynomial function
rational function
power function
composition of functions
even function
odd function
horizontal asymptote
vertical asymptote
limit at a point
infinite limits
continuity at a point
discontinuity
continuity on an interval
intermediate value theorem
bisection method

1.1 FUNCTIONS AND GRAPHS

Have you ever heard remarks such as "Success is a function of hard work" and "Demand is a function of price"? The word *function* is often used to suggest a relationship or a dependence of one quantity on another. In mathematics the function concept has a similar, but slightly more specialized, interpretation. Before giving a precise definition, let us consider an example that uses the word *function* in a more restrictive sense.

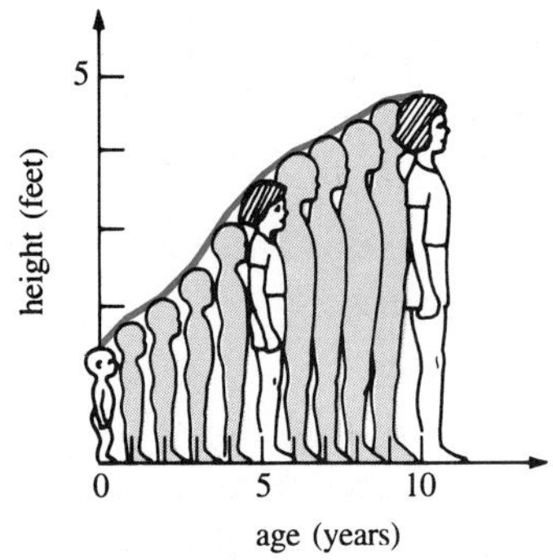

FIGURE 1.1

EXAMPLE 1

(*a*) The area of a circle is a function of its radius.

(*b*) As shown in Figure 1.1, the height of a child, when measured at yearly intervals, is a function of the child's age.

(*c*) The first-class postage for a letter is a function of its weight.

(*d*) The intensity of sound is a function of the distance from its source.

(*e*) The volume of a cubical box is a function of the length of one of its sides.

(*f*) The force between two particles of opposite charge is a function of the distance between them. □

Rule or Correspondence A function is a **rule**, or a **correspondence**, relating two sets in such a manner that each element in the first set corresponds to *one and only one* element in the second set. In other words, a functional relationship is a single-valued relationship. Thus, in Example 1, a circle of a given radius has only one area; at a specified instant in time a child can have only one height; and so on. As a further example, suppose four people are asked, first, to write their names and then their ages, and second, to write their names followed by the names of the cars that they own. They respond:

Ages	Cars
Jackie—44	Jackie—Ford, Porsche
Bill —55	Bill —Plymouth
Dena —46	Dena —Honda
Scott —50	Scott —Chevrolet, Oldsmobile, Buick

The first correspondence is a function, since there is only one age associated with each name. The second correspondence is not a function because two elements in the first set of names (Jackie and Scott) are associated with more than one car name.

We summarize the preceding discussion with a formal definition.

DEFINITION 1.1 Function

A **function** f from a set X to set Y is a rule that assigns to each element x in X a unique element y in Y. The set X is called the **domain** of f. The set of corresponding elements y in Y is called the **range** of f.

Unless stated to the contrary, we shall assume hereafter that the sets X and Y consist of real numbers.

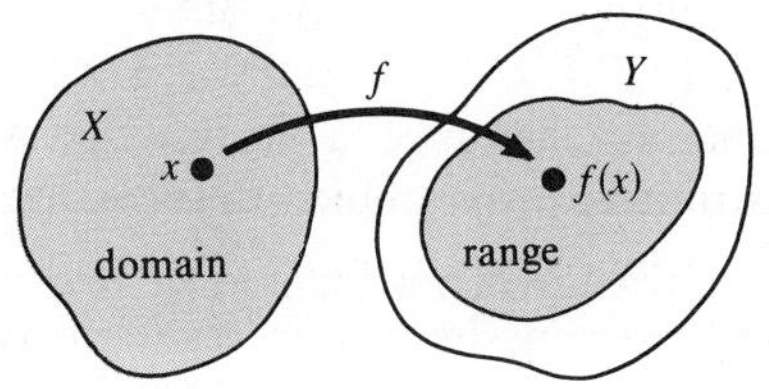

FIGURE 1.2

Value of a Function Let f be a function. The number y in the range that corresponds to a selected number x in the domain is said to be the *value* of the function at x, or the **image** of x, and is written $f(x)$. The latter symbol is read "f of x" or "f at x," and we write $y = f(x)$. See Figure 1.2. Since the value of y depends on the choice for x, it is called the **dependent variable**; x is called the **independent variable**.

A function is frequently defined by means of a formula or equation with the domain a subset of the set R of real numbers.

EXAMPLE 2 The rule for squaring a real number x is given by the equation

$$y = x^2 \quad \text{or} \quad f(x) = x^2$$

The values of f at, say, $x = -5$ and $x = \sqrt{7}$ are obtained by replacing x, in turn, by -5 and $\sqrt{7}$:

$$f(-5) = (-5)^2 = 25$$
$$f(\sqrt{7}) = (\sqrt{7})^2 = 7$$

□

EXAMPLE 3 In Example 2, since any real number can be squared, the domain of the function $f(x) = x^2$ is the set R of real numbers. Using interval notation, we also write the domain of f as $(-\infty, \infty)$. Since $x^2 \geq 0$ for all x, it follows that the range of f is $[0, \infty)$. □

For emphasis we could write the function in Example 2 as

$$f(\quad) = (\quad)^2 \tag{1}$$

This illustrates the fact that x is a *place holder* for any number in the domain of the function. Thus, if we wish to evaluate (1) at $3 + h$, where h is a real number, we put $3 + h$ into the parentheses:

$$f(3 + h) = (3 + h)^2 = 9 + 6h + h^2$$

Note that an inequality such as $y < x^2$ does not define a function. For any real number x there is no unique real number y that is less than x^2. For example, if $x = 4$, then $y = 3$, $y = 9$, and $y = 15.5$ are just some of the numbers that satisfy $y < 4^2$.

As a matter of course, the domain of a function defined by an equation is usually not specified. Unless stated to the contrary, it is understood that:

*the domain of a function f is the largest subset of the set R of real numbers for which the rule makes sense.**

For example, for $f(x) = 1/x$ we cannot compute $f(0)$, since $1/0$ is not defined. Thus, the domain of $f(x) = 1/x$ is the set of all real numbers except zero. By the same reasoning, we see that the domain of $f(x) = x/(x^2 - 4)$ is the set of all real numbers except -2 and 2. Similarly, $f(x) = \sqrt{x}$ requires that $x \geq 0$, and so the domain of the latter function is $[0, \infty)$.

A function is often compared to a computing machine. The "input" x is transformed by the "machine" f into the "output" $f(x)$. See Figure 1.3. The set of allowable inputs is the domain of the function; the range of the function is the set of outputs. Computers and calculators are programmed to recognize when a number is not in the set of allowable inputs of a function. For example, on most simple calculators entering -4 and pressing the $\sqrt{}$ key result in an error message.

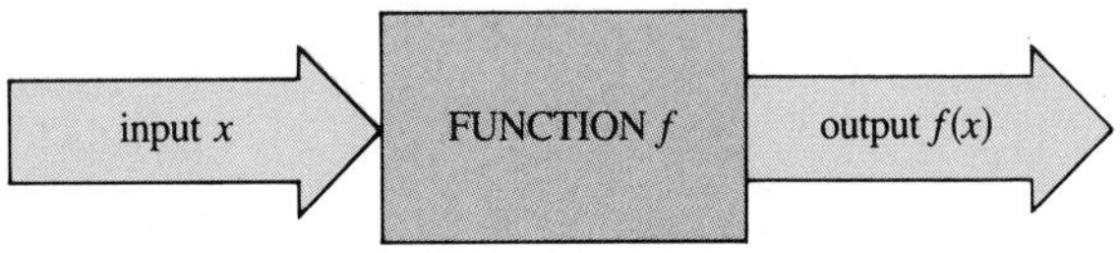

FIGURE 1.3

EXAMPLE 4 Determine the domain and range of the function

$$f(x) = 7 + \sqrt{3x - 6}$$

Solution The radicand $3x - 6$ must be nonnegative. Solving $3x - 6 \geq 0$ gives $x \geq 2$, and so the domain of f is $[2, \infty)$. Now, by the definition of the square root symbol, $\sqrt{3x - 6} \geq 0$ for $x \geq 2$ and consequently, $y = 7 + \sqrt{3x - 6} \geq 7$. Since $3x - 6$ and $\sqrt{3x - 6}$ increase as x increases, we conclude that the range of f is $[7, \infty)$. □

Note: Finding the range of a function by inspection is generally not an easy task.

Other Symbols The use of f or $f(x)$ to represent a function is a natural notation. However, in different contexts such as mathematics, science, engineering, and business, functions are denoted by diverse symbols such as F, G, H, g, h, p, and q. Different letters such as r, s, t, u, v, w, and z are often used for both the independent and dependent variables. Thus, a function could be written $w = G(z)$ or $v = h(t)$. For example, the area of a circle is $A = \pi r^2$; that is, $A = f(r)$ or $f(r) = \pi r^2$.

*This is sometimes referred to as the **natural domain** or **implicit domain** of the function.

EXAMPLE 5

(*a*) The distance s that a freely falling body will travel is a function of time t. See Figure 1.4(a).

(*b*) The minimum flying speed v of a bird is a function of its length L. See Figure 1.4(b).

(*c*) In the analysis of walking, the period T of oscillation of a leg is a function of its length L. See Figure 1.4(c).

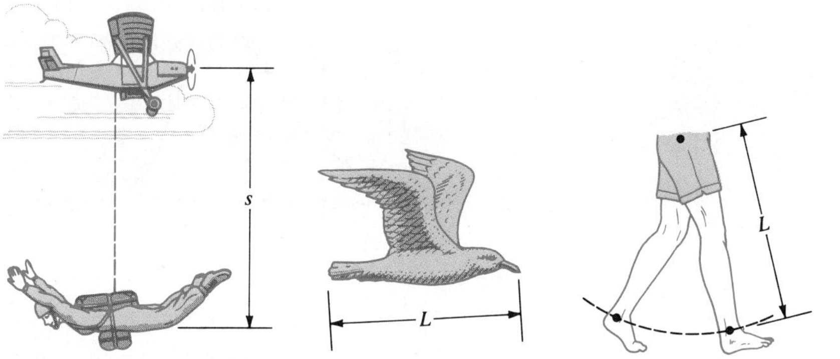

(a) $s = -\frac{1}{2}gt^2$, $g = \text{constant}$ (b) $v = k\sqrt{L}$, $k = \text{constant}$ (c) $T = 2\pi\sqrt{\frac{2}{3}\frac{L}{g}}$, $g = \text{constant}$

FIGURE 1.4 □

Graphs The **graph** of a function f is the set of points

$$\{(x, y) \mid y = f(x),\ x \text{ in the domain of } f\}$$

in the Cartesian plane. As a consequence of Definition 1.1, a function is characterized geometrically by the fact that *any* vertical line intersecting its graph does so *in exactly one point*. See Figure 1.5(a) and (b). Equivalently, if a vertical line intersects a graph in more than one point, the graph is *not* a graph of a function. See Figure 1.5(c) and (d).

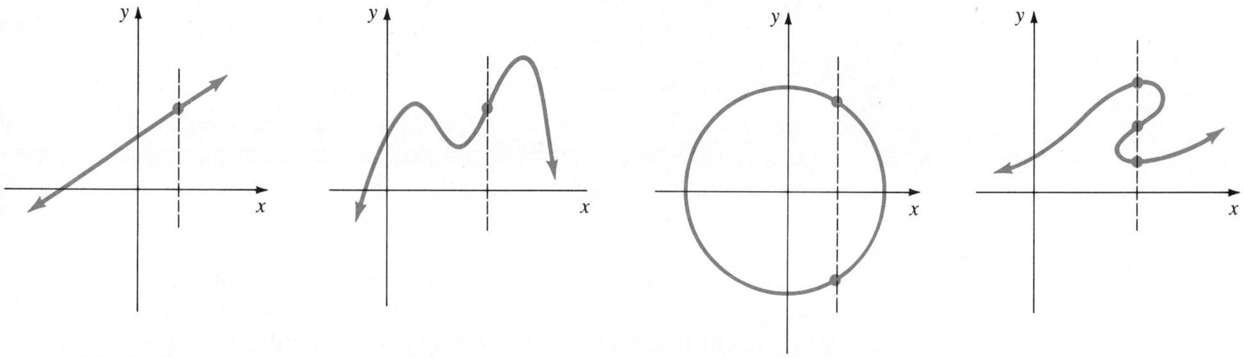

(a) a function (b) a function (c) not a function (d) not a function

FIGURE 1.5

If a point (a, b) is on the graph of a function f, then the y-coordinate b is the value of the function at a; that is, $b = f(a)$. As we see in Figure 1.6, the value $f(a)$ is a *directed distance* from the x-axis to the point. Moreover, it is often possible to discern the domain and range of a function from its graph. Figure 1.7 shows that the domain of f is an interval on the x-axis, and the range of f is an interval on the y-axis.

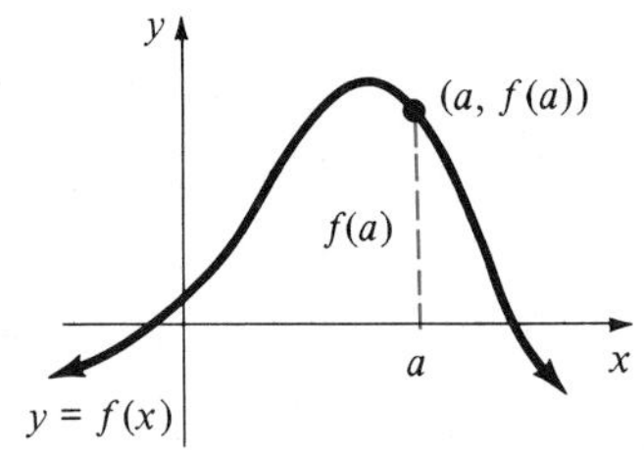

FIGURE 1.6

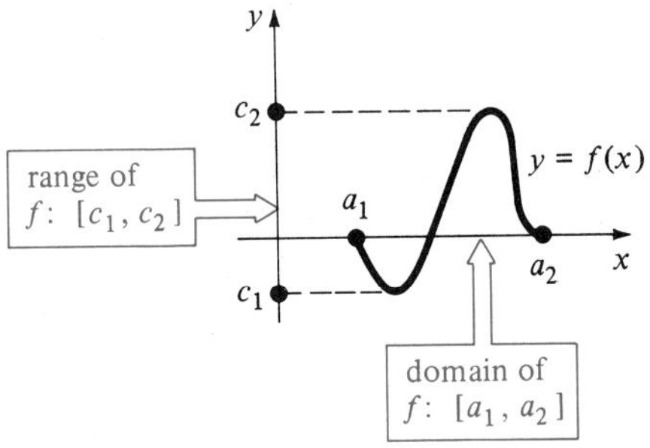

FIGURE 1.7

Throughout this text we shall encounter many different types of functions. The following table summarizes four types of functions and their domains.

Function		Domain
Polynomial of degree n	$f(x) = a_n x^n + a_{n-1}x^{n-1} + \cdots + a_1 x + a_0$, n a nonnegative integer, $a_n \neq 0$, coefficients a_i $(i = 0, 1, \ldots, n)$ are real numbers	The set R of real numbers
Rational	$f(x) = P(x)/Q(x)$, where P and Q are polynomial functions	The set R of real numbers except those numbers for which $Q(x) = 0$
Power	$f(x) = kx^n$, k a nonzero constant, n a real number	The domain of a power function depends on the value of n.
Piecewise-defined	It is one function but consists of two or more formulas defined over different intervals.	The domain of a piecewise-defined function is specified.

Polynomial functions of degrees 0, 1, and 2 are, respectively,

$$f(x) = a_0 \qquad \textbf{constant} \text{ function} \tag{2}$$

$$f(x) = a_1 x + a_0, \quad a_1 \neq 0 \qquad \textbf{linear} \text{ function} \tag{3}$$

$$f(x) = a_2 x^2 + a_1 x + a_0, \quad a_2 \neq 0 \qquad \textbf{quadratic} \text{ function} \tag{4}$$

By comparing (3) with the slope–intercept form of a line, $y = mx + b$, and making the obvious identifications, we see that the graph of a linear function is a **straight line**. Of course, the graph of constant function is a horizontal line. You may already know that the graph of quadratic function is called a **parabola**.

EXAMPLE 6 Graph the polynomial functions $f(x) = x^2$ and $f(x) = x^3$.

Solution In Figure 1.8(a) and (b) we plotted the points that correspond to the values of x and $f(x)$ given in the accompanying tables. Since the domain of each function is the set of real numbers, we connect the points with a smooth and continuous curve.

x	$f(x)$
-2	4
-1	1
$-\frac{1}{2}$	$\frac{1}{4}$
0	0
$\frac{1}{2}$	$\frac{1}{4}$
1	1
2	4

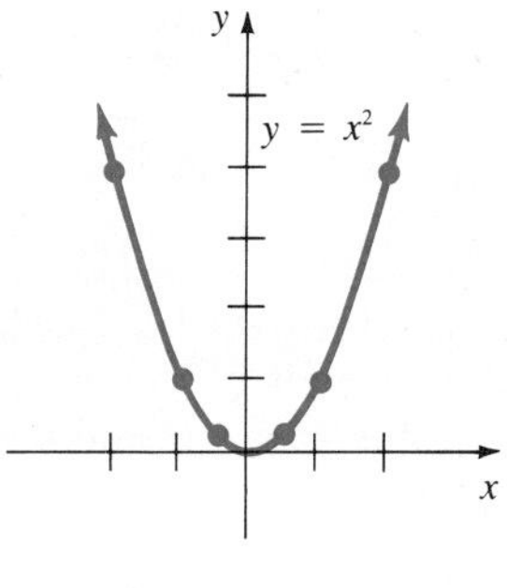

(a)

x	$f(x)$
-2	-8
-1	-1
$-\frac{1}{2}$	$-\frac{1}{8}$
0	0
$\frac{1}{2}$	$\frac{1}{8}$
1	1
2	8

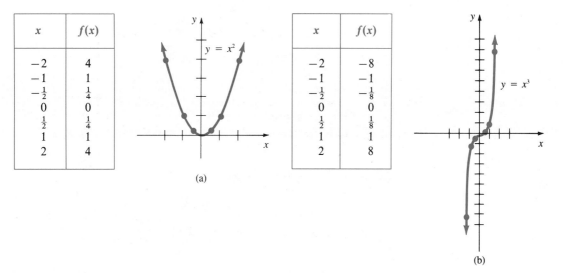

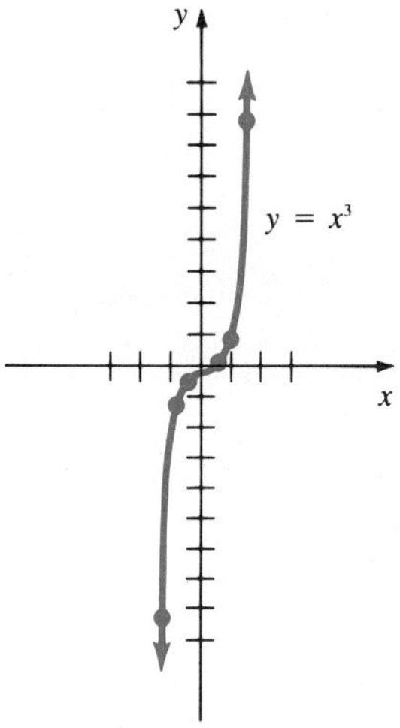

(b) □

FIGURE 1.8

Figure 1.8(a) shows the typical shape of a parabola. Figure 1.8(b) shows one of several possible graphs of a third-degree, or **cubic**, polynomial function.

EXAMPLE 7 $f(x) = \dfrac{x^3 + x + 5}{x^2 - 3x - 4}$ is a rational function. Since the denominator $x^2 - 3x - 4 = (x + 1)(x - 4)$ and $(x + 1)(x - 4) = 0$ for -1 and 4, the domain of f is the set of all real numbers except -1 and 4. □

The simple polynomial functions $y = x$, $y = x^2$, and $y = x^3$ are also examples of power functions. Although we shall not prove it, $f(x) = kx^n$ defines a function for any real-number exponent n. A power function such as $x^{\sqrt{2}}$ does indeed make sense, and the rule $y = x^{\sqrt{2}}$ gives a single value of y for each nonnegative value of x. As pointed out in the foregoing table, the domain of a power function depends on the exponent n. For example, when $k = 1$ and $n = \frac{1}{2}$, $y = x^{1/2} = \sqrt{x}$ is a function whose domain we have already seen to be $[0, \infty)$, whereas when $k = 1$ and $n = 4$, the domain of the power function $y = x^4$ is the set R of real numbers. Power functions occur often in the physical sciences; each of the functions given in Example 5 is a power function. In biology the pulse rate p of an animal of mass m is modeled by the power function $p = km^{-1/4}$. Thus, an animal with large mass, such as an elephant, has a slow pulse rate.

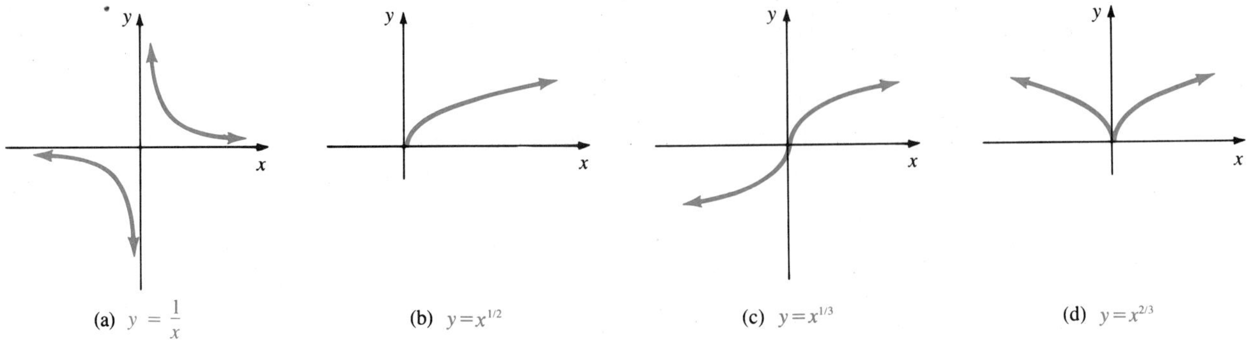

FIGURE 1.9

Figure 1.9 shows the graphs of the power functions $y = kx^n$ that correspond to $k = 1$, $n = -1$, $n = \frac{1}{2}$, $n = \frac{1}{3}$, and $n = \frac{2}{3}$, respectively.

EXAMPLE 8 Graph the piecewise-defined function

$$f(x) = \begin{cases} -1, & x < 0 \\ 0, & x = 0 \\ x + 2, & x > 0 \end{cases}$$

Solution Note that f is *not* three functions but rather one function with three pieces. The domain of f consists of the union of the intervals defined by $x < 0$ and $x > 0$ and the point $x = 0$. In this case the domain is the set R of real numbers. The graph of f is obtained by drawing, in turn,

the graph of the horizontal line $y = -1$ for $x < 0$,

the point (0, 0), and

the graph of the line $y = x + 2$ for $x > 0$.

The graph of the function f is given in Figure 1.10. □

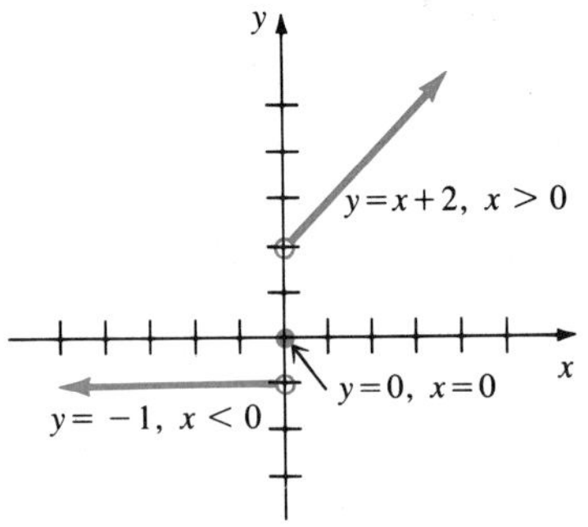

FIGURE 1.10

Combining Functions A function f can be combined with another function g by means of arithmetic operations to form other functions. Suppose f is a function with domain X_1 and g is a function with domain X_2. Then the sum, differences, product, and quotients are:

Sum $(f + g)(x) = f(x) + g(x)$

Differences $(f - g)(x) = f(x) - g(x), \quad (g - f)(x) = g(x) - f(x)$

Product $(fg)(x) = f(x)g(x)$ (5)

Quotients $(f/g)(x) = \dfrac{f(x)}{g(x)}, \quad (g/f)(x) = \dfrac{g(x)}{f(x)}$

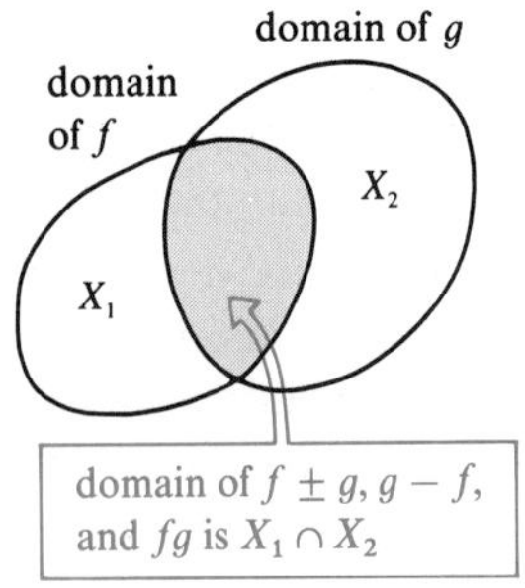

FIGURE 1.11

As illustrated in Figure 1.11, the domain of each of the functions $f + g$, $f - g$, $g - f$, and fg is the intersection of the domain of f with the domain of g.

The domain of f/g is the intersection of the domains of f and g without the numbers for which $g(x) = 0$. In the case of g/f, we exclude from the intersection the numbers for which $f(x) = 0$.

EXAMPLE 9 If $f(x) = 2x^2 - 5$ and $g(x) = 3x + 4$, find $f + g$, $f - g$, fg, and f/g.

Solution From (5) we obtain

$$(f + g)(x) = f(x) + g(x) = (2x^2 - 5) + (3x + 4) = 2x^2 + 3x - 1$$
$$(f - g)(x) = f(x) - g(x) = (2x^2 - 5) - (3x + 4) = 2x^2 - 3x - 9$$
$$(fg)(x) = f(x)g(x) = (2x^2 - 5)(3x + 4) = 6x^3 + 8x^2 - 15x - 20$$
$$(f/g)(x) = \frac{f(x)}{g(x)} = \frac{2x^2 - 5}{3x + 4}, \qquad x \neq -\frac{4}{3}$$ □

EXAMPLE 10 If $f(x) = \sqrt{x - 1}$ and $g(x) = \sqrt{2 - x}$, find the domains of fg and f/g.

Solution The domains of f and g are $[1, \infty)$ and $(-\infty, 2]$, respectively. Consequently, the domain of the product

$$(fg)(x) = \sqrt{x - 1}\,\sqrt{2 - x} = \sqrt{(x - 1)(2 - x)}$$

is the intersection of the domains: $[1, 2]$. However, the domain of the quotient

$$(f/g)(x) = \frac{\sqrt{x - 1}}{\sqrt{2 - x}} = \sqrt{\frac{x - 1}{2 - x}}$$

is $[1, 2)$, since $g(x) = 0$ at $x = 2$. □

If f is a constant function, say $f = c$, then the product fg is the function cg defined by

$$(cg)(x) = c \cdot g(x)$$

The domain of cg is the domain of g; for example, if $g(x) = 4x^3 - 3x$, then the function $5g$ is simply $(5g)(x) = 20x^3 - 15x$.

Composition In conclusion, we examine an entirely different way of combining two functions f and g. If the numbers represented by $g(x)$ are in the domain of f, we can form a new function by evaluating f at $g(x)$. This function whose outputs are $f(g(x))$ is called the **composition**, or **composite**, of f and g is denoted by the symbol $f \circ g$. Similarly, if the numbers represented by $f(x)$ are in the domain of g, we can form the composition of g and f. We summarize:

Composition of f and g $\quad (f \circ g)(x) = f(g(x))$

Composition of g and f $\quad (g \circ f)(x) = g(f(x))$ $\qquad (6)$

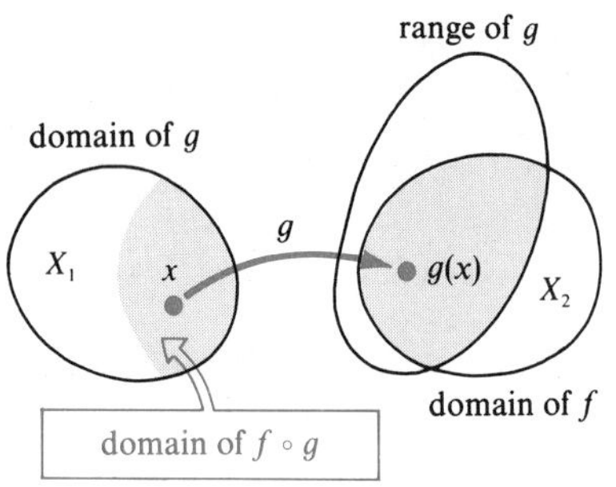

FIGURE 1.12

In Figure 1.12 we have illustrated that the domain of $f \circ g$ is that subset of the domain of g for which $g(x)$ is in the domain of f.

EXAMPLE 11 If $f(x) = x^2$ and $g(x) = x^2 + 1$, find $f \circ g$ and $g \circ f$.

Solution It follows from (6) that

$$(f \circ g)(x) = f(g(x)) = f(x^2 + 1) = (x^2 + 1)^2 = x^4 + 2x^2 + 1$$

and $$(g \circ f)(x) = g(f(x)) = g(x^2) = (x^2)^2 + 1 = x^4 + 1$$ □

Example 11 illustrates that, in general, $f \circ g \neq g \circ f$.

EXAMPLE 12 If $f(x) = 3x - \sqrt{x}$ and $g(x) = 2x + 1$, find the domain of $f \circ g$.

Solution $$(f \circ g)(x) = f(g(x)) = f(2x + 1)$$
$$= 3(2x + 1) - \sqrt{2x + 1} = 6x + 3 - \sqrt{2x + 1}$$

For $f \circ g$ to be defined, we must demand that $g(x) \geq 0$ or $2x + 1 \geq 0$; that is, the domain of $f \circ g$ is given by $[-\frac{1}{2}, \infty)$. □

▲ ***Remark*** In this beginning discussion we focused on only three types of functions. Of course, there are many other kinds. Throughout the subsequent sections of this text we shall consider functions involving absolute values, trigonometric functions, inverse trigonometric functions, logarithmic functions, and exponential functions, to name a few more.

EXERCISES 1.1 *Answers to odd-numbered problems begin on page A-63.*

1. Given that $f(x) = x^2 - x$, find:
   (a) $f(-3)$ (b) $f(1)$ (c) $f(\sqrt{3})$ (d) $f(1 + a)$

2. Given that $f(x) = \sqrt{x + 4}$, find:
   (a) $f(-3)$ (b) $f(0)$ (c) $f(1)$ (d) $f(5)$

3. Given that

$$F(x) = \begin{cases} |x|, & x < -1 \\ -2x + 1, & x \geq -1 \end{cases}$$

find:
   (a) $F(-2)$ (b) $F(-1.5)$ (c) $F(-1)$
   (d) $F(0)$ (e) $F(\frac{1}{2})$ (f) $F(a)$

4. Given that

$$g(t) = \begin{cases} t^2, & -1 < t \leq 1 \\ 2t, & t \leq -1 \text{ or } t > 1 \end{cases}$$

find for $0 < a < 1$:
   (a) $g(1 + a)$ (b) $g(1 - a)$ (c) $g(1.5 - a)$
   (d) $g(a)$ (e) $g(-a)$ (f) $g(2a)$

In Problems 5–10 find $\dfrac{f(a + h) - f(a)}{h}$, $h \neq 0$, and simplify.

5. $f(x) = 6x - 9$
6. $f(x) = -\frac{1}{4}x + 6$
7. $f(x) = -5x^2 + 9x$
8. $f(x) = x^2 + 2x - 4$

9. $f(x) = x^3$

10. $f(x) = \dfrac{5}{x}$

In Problems 11–24 find the domain of the given function.

11. $f(x) = \sqrt{x+1}$

12. $f(x) = x\sqrt{2x-3}$

13. $f(x) = \dfrac{1+x}{\sqrt{x}}$

14. $f(x) = \sqrt{x} + \sqrt{x-2}$

15. $g(x) = \sqrt{25 - x^2}$

16. $g(x) = \sqrt{x^2 - 5x + 4}$

17. $F(x) = \dfrac{x^2 - 16}{x - 4}$

18. $G(x) = \dfrac{1}{x^2 + x - 6}$

19. $Q(x) = \dfrac{x}{2 - 1/x}$

20. $H(x) = 7x^3 - x^{-2} + 8$

21. $f(x) = x^{3/2}$

22. $f(x) = 1 + x^{2/3}$

23. $f(x) = \begin{cases} x, & x < 2 \\ x^2, & 2 \le x < 3 \end{cases}$

24. $g(x) = \begin{cases} x+1, & x \le 0 \\ x-6, & x \ge 1 \end{cases}$

In Problems 25–28 find the range of the given function.

25. $f(x) = 1 + x^2$

26. $g(x) = (2x+1)^2$

27. $g(x) = 4 - \sqrt{x}$

28. $f(x) = 3 + \sqrt{4 - x^2}$

In Problems 29–32 determine whether the given graph is the graph of a function.

29.

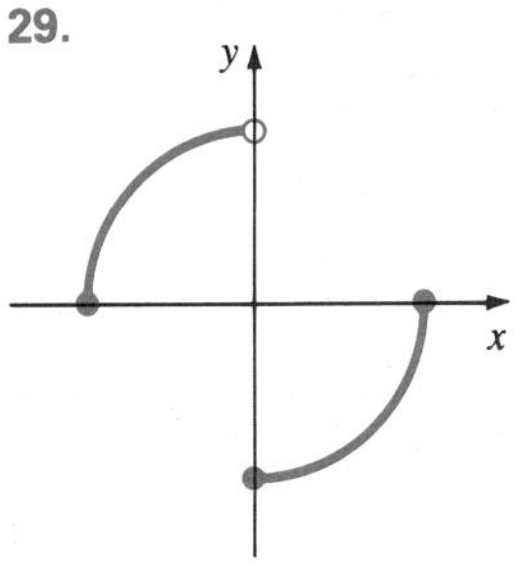

FIGURE 1.13

30.

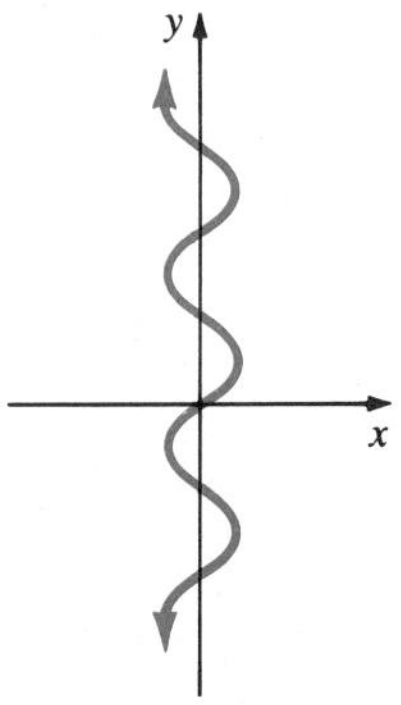

FIGURE 1.14

31.

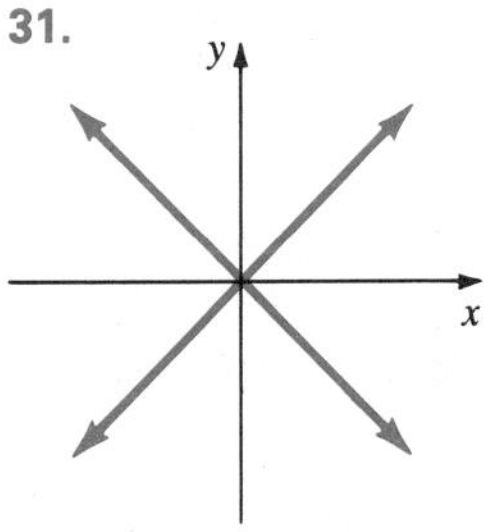

FIGURE 1.15

32.

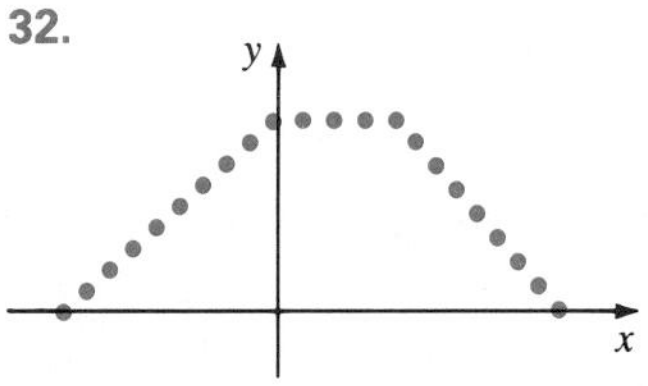

FIGURE 1.16

In Problems 33–36 determine the domain and range of the function whose graph is given.

33.

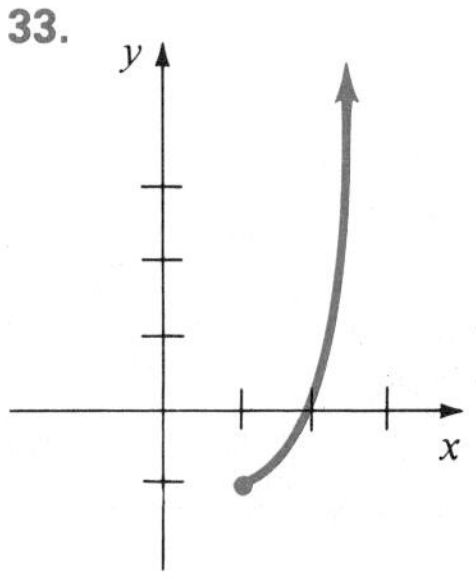

FIGURE 1.17

34.

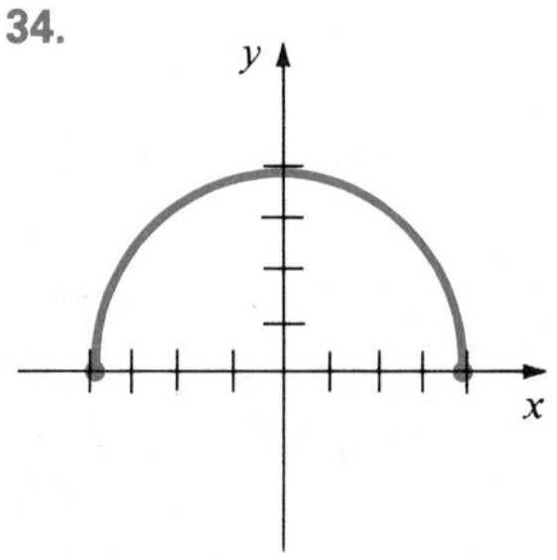

FIGURE 1.18

35.

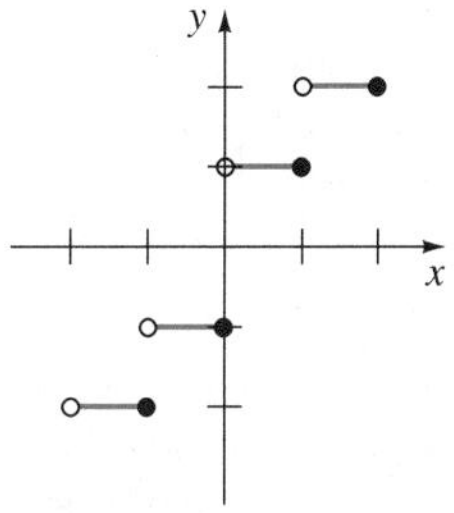

FIGURE 1.19

36.

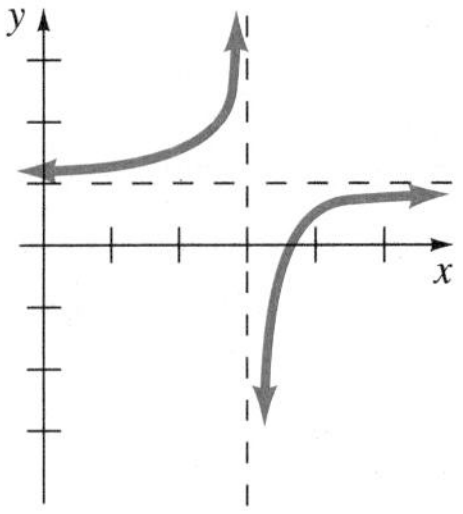

FIGURE 1.20

37. For what values of x is

$$f(x) = \begin{cases} x^3, & x < 0 \\ x^2, & x \ge 0 \end{cases}$$

equal to 25? to -64?

38. Determine whether the numbers -1 and 2 are in the range of the rational function $f(x) = (2x-1)/(x+4)$.

In Problems 39–54 graph the given function.

39. $f(x) = -x^2$

40. $f(x) = x^2 + 1$

41. $f(x) = (x+1)^2$

42. $f(x) = 1 - x^2$

43. $f(x) = x^3 + 2$

44. $f(x) = -x^3$

45. $f(x) = x^4$

46. $f(x) = x^5$

47. $f(x) = \dfrac{1}{x^2}$

48. $f(x) = x^{-1/2}$

49. $f(x) = \dfrac{2}{1 + x^2}$

50. $f(x) = \dfrac{1}{x - 1}$

51. $f(x) = \begin{cases} 3, & x \le 1 \\ -3, & x > 1 \end{cases}$

52. $f(x) = \begin{cases} x^2, & x < 0 \\ x, & x \ge 0 \end{cases}$

53. $f(x) = \begin{cases} x^2, & x < -1 \\ 1, & -1 \le x \le 1 \\ 2 - x^2, & x > 1 \end{cases}$

54. $f(x) = \begin{cases} x + 2, & x < 0 \\ 2 - x, & 0 \le x < 2 \\ x - 2, & x \ge 2 \end{cases}$

55. The International Whaling Commission has decreed that the weight W (in long tons) of a mature blue whale is given by the linear function $W = (3.51)L - 192$, where L is length and $L \ge 70$ ft.* Find the weight of a 90-ft blue whale.

56. The expected length (in cm) of a human fetus is given by the linear function $L = 1.53t - 6.7$, where $t \ge 12$ weeks. What is the expected length of a fetus at 36 weeks? What are the weekly increases in length? When is a fetus 1 ft (30.48 cm) in length?

57. The functional relationship between degrees Celsius T_c and degrees Fahrenheit T_f is linear. Express T_f as a function of T_c if (0°C, 32°F) and (60°C, 140°F) are on the graph of T_f. Show that 100°C is equivalent to the Fahrenheit boiling point 212°F. See Figure 1.21.

58. The functional relationship between degrees Celsius T_c and degrees Kelvin T_k is linear. Express T_k as a function of T_c given that (0°C, 273°K) and (27°C, 300°K) are on the graph of T_k. Express the boiling point 100°C in degrees Kelvin. Absolute zero is defined as 0°K. What is 0°K in degrees Celsius? Express T_k as a linear function of T_f. What is 0°K in degrees Fahrenheit? See Figure 1.21.

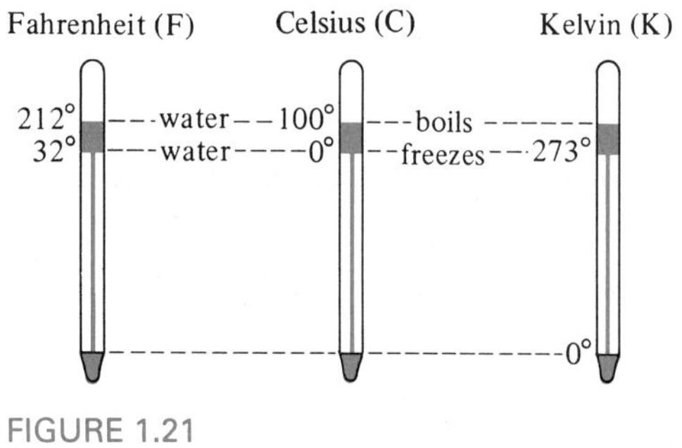

FIGURE 1.21

59. A ball is thrown upward from ground level with an initial velocity of 96 ft/s. The height of the ball from the ground is given by the quadratic function $s(t) = -16t^2 + 96t$. At what times is the ball on the ground? Graph s over the time interval for which $s(t) \ge 0$.

*For $0 < L < 70$ the whales are immature, and the formula is not used. A blue whale at birth is approximately 24 ft long.

60. In Problem 59 at what times is the ball 80 ft above the ground? How high does the ball go?

61. It is often assumed that the weight W of a body is a function of its length L. This assumption is given by the power function $W = kL^3$, where k is a constant. For a certain lizard $k = 400$ g/m^3. Determine the weight of a 0.5-m-long lizard.

62. A relationship between the weight W and length L of a sperm whale is given by the power function

$$W = 0.000137L^{3.18}$$

where W is measured in tons and L in feet. Determine the weight of a 100-ft-long sperm whale.

If x changes from x_1 to x_2, the amount of change is $\Delta x = x_2 - x_1$ and the corresponding change in the functional value is $\Delta y = f(x_2) - f(x_1)$. In Problems 63 and 64 find the values of Δx and Δy.

63. $f(x) = x^2 + 4x, x_1 = 4, x_2 = 7$

64. $f(x) = \sqrt{x + 1}, x_1 = 3, x_2 = 8$

In Problems 65–72 find $f + g$, $f - g$, fg, and f/g.

65. $f(x) = 2x + 5, g(x) = -4x + 8$

66. $f(x) = 5x^2, g(x) = 7x - 9$

67. $f(x) = 3x^2, g(x) = 4x^3$

68. $f(x) = x^2 - 3x, g(x) = x + 1$

69. $f(x) = \dfrac{x}{x + 1}, g(x) = \dfrac{1}{x}$

70. $f(x) = \dfrac{2x - 1}{x + 3}, g(x) = \dfrac{x - 3}{4x + 2}$

71. $f(x) = x^2 + 2x - 3, g(x) = x^2 + 3x - 4$

72. $f(x) = x^2, g(x) = \sqrt{x}$

In Problems 73–80 find $f \circ g$ and $g \circ f$.

73. $f(x) = 3x - 2, g(x) = x + 6$

74. $f(x) = 2x + 10, g(x) = \frac{1}{2}x - 5$

75. $f(x) = 4x + 1, g(x) = x^2$

76. $f(x) = x^2, g(x) = x^3 + x^2$

77. $f(x) = \dfrac{3}{x}, g(x) = \dfrac{x}{x + 1}$

78. $f(x) = 2x + 4, g(x) = \dfrac{1}{2x + 4}$

79. $f(x) = x^3 - 5, g(x) = \sqrt[3]{x + 5}$

80. $f(x) = x^2 + \sqrt{x}, g(x) = x^2$

In Problems 81 and 82 find $f \circ (2f)$ and $f \circ (1/f)$.

81. $f(x) = 2x^3$

82. $f(x) = \dfrac{1}{x - 1}$

In Problems 83 and 84 find $f(g(0))$, $f(g(\frac{1}{2}))$, $g(f(-2))$, and $g(f(g(1)))$.

83. $f(x) = x^2 + 1$, $g(x) = 2x^4 - 4x^2 + 3$

84. $f(x) = \dfrac{x}{x + 1}$, $g(x) = \dfrac{1}{2x + 1}$

In Problems 85 and 86 express the given function F as a composition $f \circ g$ of two functions f and g.

85. $F(x) = 2x^4 - x^2$

86. $F(x) = \dfrac{1}{x^2 + 9}$

The composition of three functions f, g, and h is the function

$$(f \circ g \circ h)(x) = f(g(h(x)))$$

In Problems 87 and 88 find $f \circ g \circ h$.

87. $f(x) = x^2 + 6$, $g(x) = 2x + 1$, $h(x) = 3x - 2$

88. $f(x) = \sqrt{x - 5}$, $g(x) = x^2 + 2$, $h(x) = \sqrt{2x + 1}$

In Problems 89 and 90 find a function g.

89. $f(x) = 2x - 5$, $(f \circ g)(x) = -4x + 13$

90. $f(x) = \sqrt{2x + 6}$, $(f \circ g)(x) = 4x^2$

91. Determine whether $f \circ (g + h) = f \circ g + f \circ h$ is true or false.

92. Suppose $[-1, 1]$ is the domain of $f(x) = x^2$. What is the domain of $y = f(x - 2)$?

Let U be the function defined by

$$U(x - a) = \begin{cases} 0, & 0 \le x < a \\ 1, & x \ge a \end{cases}$$

In Problems 93 and 94 graph the given function.

93. $y = U(x - 1)$

94. $y = U(x - 1) + U(x - 2)$

95. Find an equation in terms of $U(x - a)$ for the function illustrated in Figure 1.22.

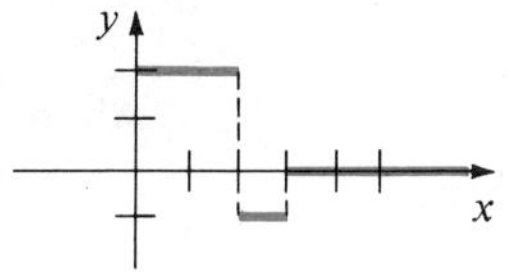

FIGURE 1.22

96. Given that $f(x) = x^2$, compare the graphs of $y = f(x - 3)$ and $y = f(x - 3)U(x - 3)$.

1.2 MORE ON GRAPHS OF FUNCTIONS

In calculus the solution of a problem often depends on one's ability to sketch the graph of a function quickly and accurately. The algebraic techniques considered in this section, when combined with the powerful calculus methods of Chapter 3, provide a means for obtaining the graphs of functions without the drudgery of plotting many points.

Intercepts To graph a straight line we need to plot only two points. If a line with nonzero slope does not pass through the origin, we can find its x- and y-intercepts and draw the line through the corresponding points on the coordinates axes. The y-intercept for a linear function $f(x) = ax + b$, $a \neq 0$ and $b \neq 0$, is $f(0) = b$, and the x-intercept is the solution of $ax + b = 0$; that is, $x = -b/a$. When graphing *any* function, it is recommended that you look first for the intercepts of its graph. The y-coordinate of the point at which the graph of a function $y = f(x)$ crosses the y-axis is its **y-intercept**. This y-coordinate is the value $f(0)$. Note that the graph of a function can have at most one y-intercept. Moreover, if 0 is not in the domain of f, then $f(0)$ is not defined and the graph does not cross the y-axis. The **x-intercepts** of the graph of f are the x-coordinates of the points where the graph crosses the

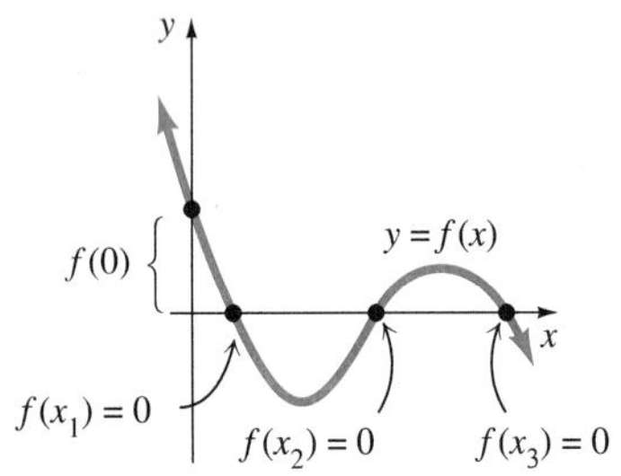

FIGURE 1.23

x-axis. Since $y = 0$ along the x-axis, the x-intercepts of the graph of f are the real solutions of the equation $f(x) = 0$. The values for which $f(x) = 0$ are the **zeros** of the function f; if f has no real zeros, then its graph has no x-intercepts. In Figure 1.23 we have illustrated a function f whose graph has a y-intercept and three x-intercepts, x_1, x_2, and x_3.

EXAMPLE 1

(*a*) The graph of the polynomial function $f(x) = x^2 - x - 6$ has y-intercept $f(0) = -6$. Also, $f(x) = 0$ when $x^2 - x - 6 = 0$ or $(x - 3)(x + 2) = 0$. Thus, 3 and -2 are x-intercepts.

(*b*) The graph of the rational function $f(x) = (3x - 2)/x$ has no y-intercept, since $f(0)$ is not defined (that is, 0 is not in the domain of f). Now, the only way a rational function $f(x) = P(x)/Q(x)$ can be zero is when $P(x) = 0$ and $Q(x) \neq 0$. Accordingly, for the given function, $3x - 2 = 0$ implies that the x-intercept is $\frac{2}{3}$. □

Symmetry Of three symmetries of graphs (symmetry with respect to the y-axis, symmetry with respect to the x-axis, and symmetry with respect to the origin [see page A-16, Appendix I]) we note that the graph of a nonzero function cannot be symmetric with respect to the x-axis. This is because, in view of Definition 1.1, both points (x, y) and $(x, -y)$ cannot be on the graph of a function. A function whose graph is symmetric with respect to the y-axis is

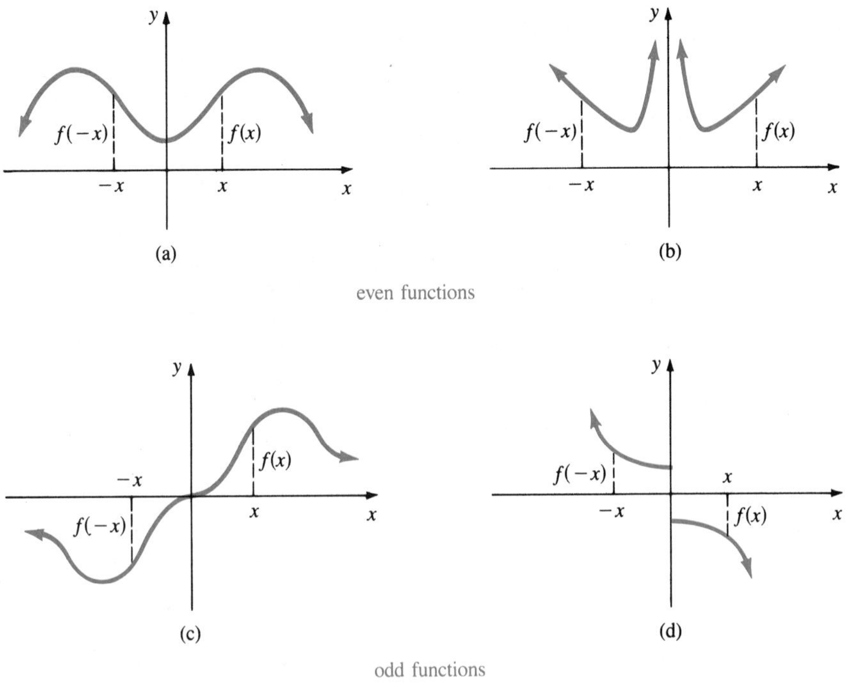

FIGURE 1.24

called an **even function**, whereas a function whose graph possesses symmetry with respect to the origin is called an **odd function**.

Tests for Symmetry

The graph of $y = f(x)$ is **symmetric with respect to the y-axis** if $f(-x) = f(x)$. (1)

The graph of $y = f(x)$ is **symmetric with respect to the origin** if $f(-x) = -f(x)$. (2)

The graphs in Figure 1.24 illustrate each concept.

EXAMPLE 2 Determine whether the following functions are even or odd:

(*a*) $f(x) = \dfrac{3x}{x^2 + 1}$

(*b*) $f(x) = \dfrac{x^2}{x^3 + 1}$

Solution

(*a*) $f(-x) = \dfrac{3(-x)}{(-x)^2 + 1} = -\dfrac{3x}{x^2 + 1} = -f(x)$

The function is odd.

(*b*) $f(-x) = \dfrac{(-x)^2}{(-x)^3 + 1} = \dfrac{x^2}{-x^3 + 1} \neq \pm f(x)$

The function is neither even nor odd. □

EXAMPLE 3 Graph $f(x) = x^3 - 4x$.

Solution The y-intercept is $f(0) = 0$. By writing $f(x) = x(x^2 - 4) = x(x + 2)(x - 2)$, we observe that the x-intercepts are -2, 0, and 2. Furthermore,

$$f(-x) = (-x)^3 - 4(-x) = -x^3 + 4x = -f(x)$$

shows that the graph of f is symmetric with respect to the origin; that is, f is an odd function. Now the three intercepts alone do not suggest the flow of the graph. But by plotting, say, $(1, f(1))$ or $(1, -3)$, we know from symmetry that $(-1, -f(1))$ or $(-1, 3)$ is also on the graph. After connecting the five points shown in Figure 1.25 (a) by a smooth curve, it seems reasonable that the graph of f will be similar to that given in Figure 1.25 (b).

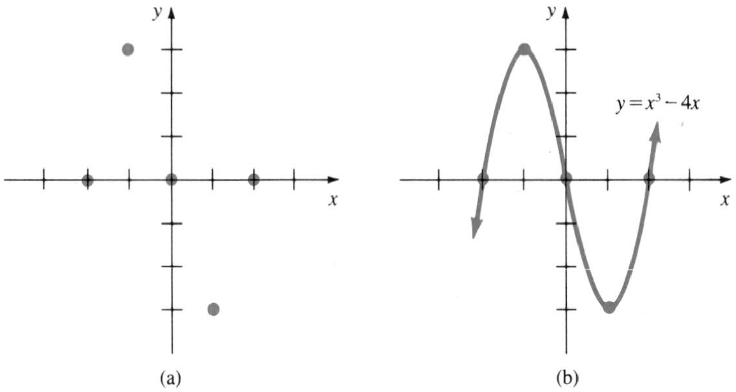

FIGURE 1.25 □

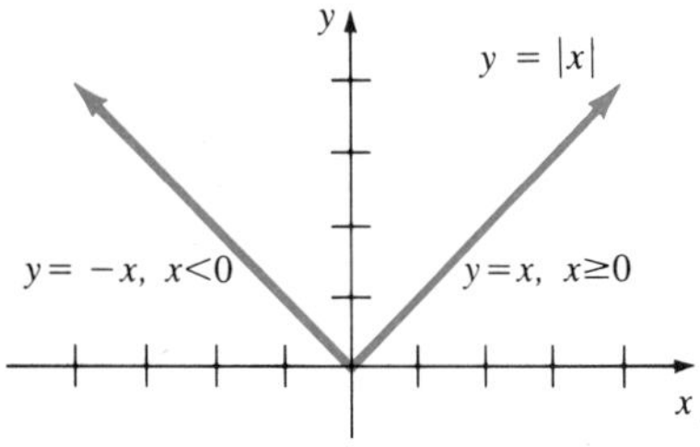

FIGURE 1.26

We note that *polynomial functions* that contain only *even powers* of x are necessarily *even functions*; polynomial functions that consist entirely of *odd powers* of x are *odd functions*. Hence, we can tell by inspection that the graphs of $f(x) = x^2$, $f(x) = x^4$, and $f(x) = x^6 - x^4 + x^2 + 1$ are symmetric with respect to the y-axis, whereas the graphs of $f(x) = x^3$, $f(x) = x^5$, and $f(x) = x^7 - x^3$ are symmetric with respect to the origin. The graph of a polynomial function such as $f(x) = x^3 + 5x^2 - x + 6$ that contains both even and odd powers has neither symmetry. You are cautioned not to generalize to other types of functions; for example, it could be argued that the **absolute value function** $f(x) = |x|$ contains only an odd power of x, but its graph (shown in Figure 1.26) is symmetric with respect to the y-axis. Of course, $f(x) = |x|$ is not a polynomial function.

Shifted Graphs Often we can sketch the graph of a function by **shifting** or **translating** the graph of a simpler function. If c is a constant, the graphs of the sum $y = f(x) + c$ and the difference $y = f(x) - c$ can be obtained from the graph of $y = f(x)$ by a **vertical shift**. The graphs of the compositions $y = f(x + c)$ and $y = f(x - c)$ correspond to **horizontal shifts** in the graph of $y = f(x)$. We summarize the results in the following table for $c > 0$.

Function	Graph
$y = f(x) + c$	Graph of $y = f(x)$ shifted **up** c units
$y = f(x) - c$	Graph of $y = f(x)$ shifted **down** c units
$y = f(x + c)$	Graph of $y = f(x)$ shifted to the **left** c units
$y = f(x - c)$	Graph of $y = f(x)$ shifted to the **right** c units

---

EXAMPLE 4 The graphs of $y = x^2 + 1$, $y = x^2 - 1$, $y = (x + 1)^2$, and $y = (x - 1)^2$ given in Figure 1.27 are obtained from the graph of $f(x) = x^2$ by shifting the graph, in turn, 1 unit up, 1 unit down, 1 unit to the left, and 1 unit to the right.

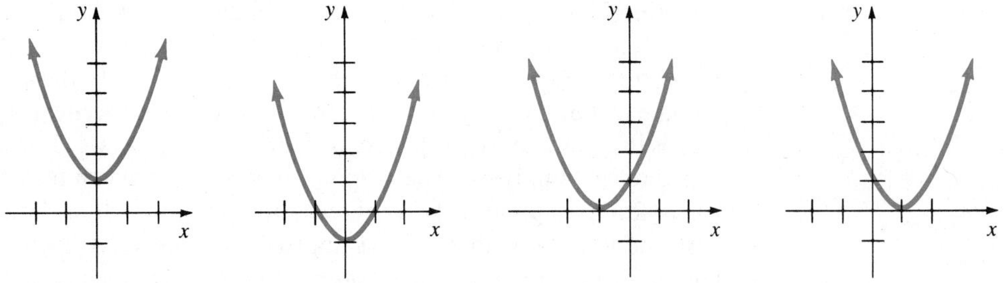

(a) $y = x^2 + 1$ (b) $y = x^2 - 1$ (c) $y = (x + 1)^2$ (d) $y = (x - 1)^2$

FIGURE 1.27 □

The graph of a function $y = f(x \pm c_1) \pm c_2$, $c_1 > 0$ and $c_2 > 0$, combines a horizontal shift (left or right) along with a vertical shift (up or down). For example, the graph of $y = f(x - c_1) + c_2$ is the graph of $y = f(x)$ shifted c_1 units to the right and then c_2 units up.

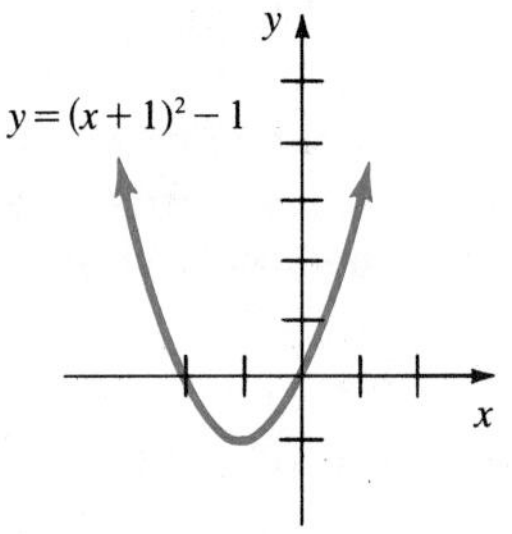

FIGURE 1.28

EXAMPLE 5 Graph $y = (x + 1)^2 - 1$.

Solution From the foregoing discussion we identify the form $y = f(x + c_1) - c_2$, where $c_1 = 1$ and $c_2 = 1$. Thus, the graph of $y = (x + 1)^2 - 1$ is the graph of $f(x) = x^2$ shifted 1 unit to the left and then down 1 unit. The graph is given in Figure 1.28. □

Reflections For $c > 0$ the graph of the product $y = cf(x)$ retains, roughly, the shape of the graph of $y = f(x)$. However, the graph of $y = -cf(x)$, $c > 0$, is a **reflection** of the graph of $y = cf(x)$ in the x-axis; that is, the graph is *turned upside down.*

EXAMPLE 6 In Figure 1.29(a) we have compared the graphs of $y = x^2$ and $y = \frac{1}{4}x^2$. The graph of $y = -\frac{1}{4}x^2$ shown in Figure 1.29(b) is the graph of $y = \frac{1}{4}x^2$ reflected in the x-axis.

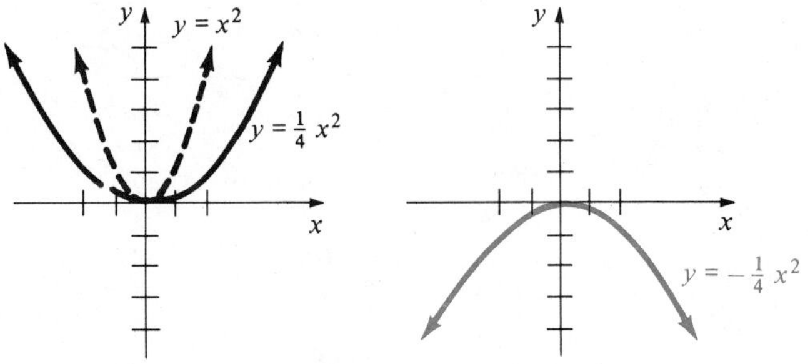

(a) (b)

FIGURE 1.29 □

Asymptotes In your study of precalculus mathematics you may have encountered functions whose graphs possess **asymptotes**. If $f(x) = P(x)/Q(x)$ is a rational function such that $P(a) \neq 0$ and $Q(a) = 0$, then a is not in the domain of f and we say that the vertical line $x = a$ is a **vertical asymptote**. This means that the values $|f(x)|$ increase without bound as we choose x closer and closer to the number a. The graph of a rational function may also possess a **horizontal asymptote**. If $|x|$ is allowed to increase without bound and if the corresponding functional values approach a constant k, then the horizontal line $y = k$ is a horizontal asymptote. Roughly, this means that if we are far from the origin, then the graph of f is close to the line $y = k$. Both of these concepts will be examined again in Chapter 2. You are encouraged to re-examine the graph of the function $y = 1/x$ given in Figure 1.9(a) and verify that $x = 0$ is a vertical asymptote and $y = 0$ is a horizontal asymptote.

EXAMPLE 7 Graph the rational function $f(x) = \dfrac{3 - x}{x + 2}$.

Solution First, we see that the y-intercept is $f(0) = \frac{3}{2}$. Moreover, $f(x) = 0$ when $3 - x = 0$, and so the only x-intercept is 3. Next, since $f(-x)$ is equal to neither $f(x)$ nor $-f(x)$, the graph of f is not symmetric with respect to the y-axis or the origin. Now it is apparent that -2 is not in the domain of f and, if we write $P(x) = 3 - x$ and $Q(x) = x + 2$, we see that $P(-2) \neq 0$ and $Q(-2) = 0$. Thus, $x = -2$ is a vertical asymptote. As shown in the accompanying table, when x is near -2 ($x = -2.1$ and $x = -1.9$), the values of the denominator $Q(x) = x + 2$ are close to zero, and consequently the corresponding functional values ($f(-2.1) = -51$ and $f(-1.9) = 49$) are large in absolute value. Finally, observe from the table that when $|x|$ is large ($x = -100$ and $x = 100$), the function values ($f(-100) \approx -1.05$ and $f(100) \approx -0.95$) are close to -1. The line $y = -1$ is a horizontal asymptote. The graph of f is given in Figure 1.30.

x	$f(x)$
-100	-1.05
-5	$-\frac{8}{3}$
-3	-6
-2.1	-51
-1.9	49
-1	4
0	$\frac{3}{2}$
1	$\frac{2}{3}$
3	0
100	-0.95

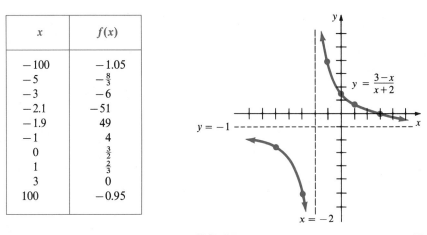

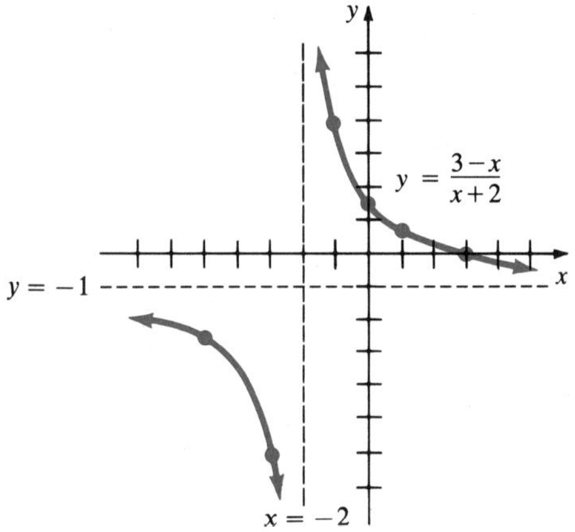

FIGURE 1.30

□

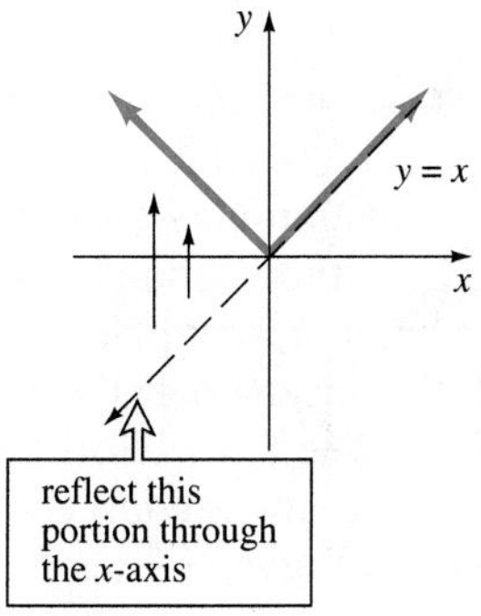

FIGURE 1.31

Absolute Value of a Function In some applications we are interested in the graph of the absolute value of a function $y = f(x)$. Since, by definition, $|f(x)|$ is nonnegative, the graph of the function $y = |f(x)|$ is never below the x-axis. If $f(x) < 0$ on an interval (a, b), then $|f(x)| = -f(x) > 0$ on (a, b), which means that the graph of $y = |f(x)|$ on that interval is above the x-axis and is a reflection of the graph of $y = f(x)$ in the x-axis. For example, to obtain the graph $y = |x|$ we can either graph the piecewise-defined function

$$y = \begin{cases} -x, & x < 0 \\ x, & x \geq 0 \end{cases}$$

as indicated in Figure 1.26 or simply draw the graph of $f(x) = x$ and then reflect in the x-axis that portion of the graph of f that is below the x-axis. See Figure 1.31.

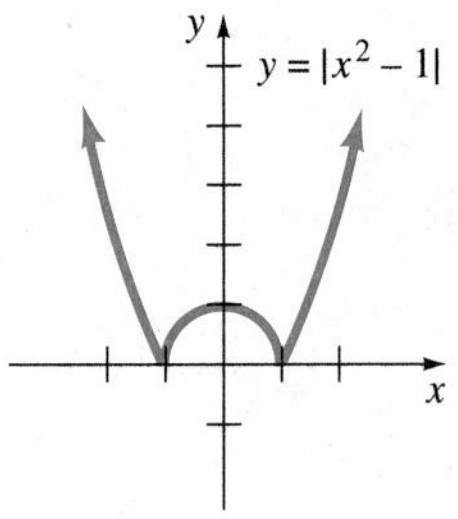

FIGURE 1.32

EXAMPLE 8 Graph $y = |x^2 - 1|$.

Solution We first graph $f(x) = x^2 - 1$ in the usual manner. In this case inspection of Figure 1.27(b) shows that the graph of f is below the x-axis on the interval $(-1, 1)$. By reflecting only that portion of the graph of f on $(-1, 1)$ in the x-axis, we obtain the graph of $y = |f(x)|$, that is, $y = |x^2 - 1|$, as shown in Figure 1.32. □

Use of Calculators/Computers In calculus we deal principally with functions; in solving problems it is often important to have a "feel" or insight for the behavior and the graph of a function by simply inspecting the function itself. Yet many functions encountered in the "real life" of science and engineering are so complicated that asking the questions about intercepts, symmetry, and so on, posed in this section produces little information. Consider the polynomial function of degree 8:

$$f(x) = x^8 - 2x^5 - 8x^4 - x^3 + 7x^2 + 9x + 1$$

Suppose we wish to find the real zeros of f. Since there is no formula for solving the equation $x^8 - 2x^5 - 8x^4 - x^3 + 7x^2 + 9x + 1 = 0$, we would have to either try to factor the expression or try a method for approximating its roots. (See Sections 1.8 and 2.11.) But since the real zeros of f are the x-intercepts of its graph, it makes sense to examine the graph of f. Although we can say that the y-intercept is $f(0) = 1$ and that the graph of f is not symmetric with respect to either the y-axis or the origin (note the mixed powers of x), this is admittedly very meager information about the function. Since plotting points would lead to instant insanity, we naturally turn to the computer or the graphic calculator. With the aid of graphing software we obtain the graph of f shown in Figure 1.33(a) and see immediately that f has one zero in the interval $[-2, -1]$, a zero close to the origin, and two zeros in the interval $[1, 2]$. Although the graph appears to pass through the origin, bear in mind that this cannot happen, since we already know that the y-intercept is $f(0) = 1$ and so the graph actually passes through the point $(0, 1)$. We can also investigate questions such as: "Is there a zero at a certain point?" and "Are there

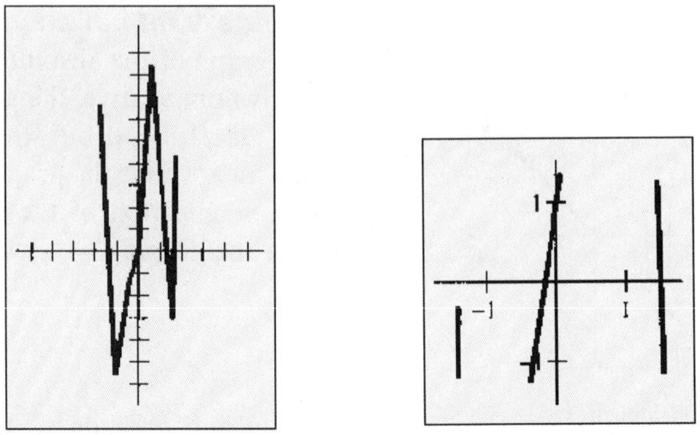

(a) (b)

FIGURE 1.33

any more zeros?" by utilizing the zoom-in and zoom-out capability of a computer or graphic calculator. For example, in Figure 1.33(b) we see clearly from a zoom-in of the graph of the polynomial function f in the neighborhood of the origin that there is definitely no zero there; the zero lies in the interval $[-1, 0)$.

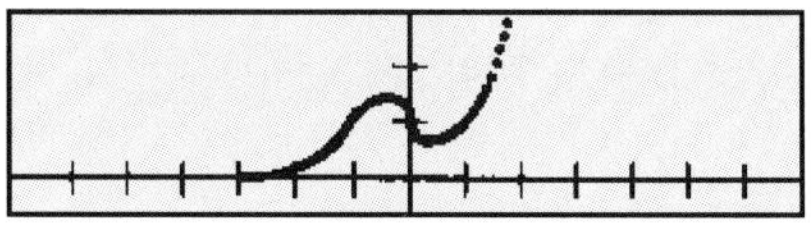

FIGURE 1.34

As another example, consider the function $f(x) = |x|^x$. With the definition that $f(0) = 1$, the domain of f is the set of real numbers. Since $f(-x)$ does not equal either $f(x)$ or $-f(x)$, the graph of f is not symmetric. But that is about all that is apparent. Plotting points with the aid of a calculator would be tedious but not difficult. However, with the aid of a graphic calculator (in this case the HP-28S) we obtain the graph shown in Figure 1.34.

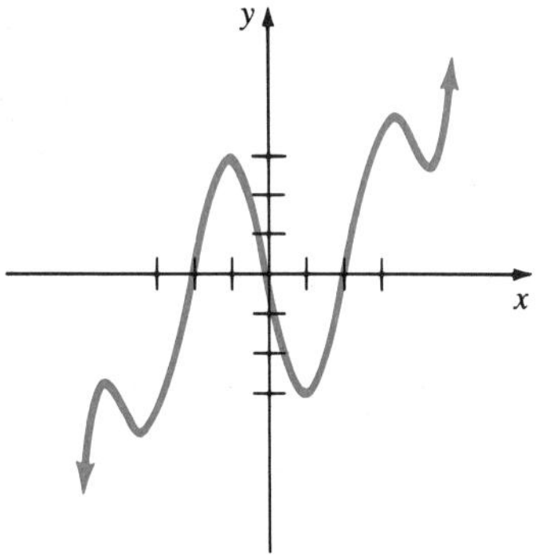

FIGURE 1.35

▲ ***Remark*** Throughout this text we shall use arrowheads on a drawn graph to suggest that there are no surprises; that is, nothing new happens to the graph when the curve is extended in the indicated direction. But how do we know, without plotting more and more points or zooming out with a computer or calculator, that the graph given in Figure 1.25 isn't really as shown in Figure 1.35? With calculus it is a simple matter to show that the graph of $f(x) = x^3 - 4x$ cannot be the graph given in Figure 1.35. Furthermore, Figure 1.25(b) is slightly inaccurate, since the highest point on the graph on the interval $[-2, 0]$ can, with the aid of calculus techniques, be shown to occur at $x = -2/\sqrt{3} \approx -1.15$ rather than at $x = -1$. Similarly, on the interval $[0, 2]$ the lowest point on the graph occurs not at $x = 1$ but at $x = 2/\sqrt{3} \approx 1.15$.

EXERCISES 1.2

Answers to odd-numbered problems begin on page A-64.

In Problems 1–30 graph the given function. Before graphing determine, when possible, any intercepts, symmetry, and asymptotes. Use shifted graphs and reflections where appropriate.

1. $y = -x^2 + 4$
2. $y = -x^2 - 4$
3. $y = -(x - 2)^2$
4. $y = -(x + 3)^2$
5. $y = (x - 3)^2 - 2$
6. $y = (x + 1)^2 + 5$
7. $y = -x^3 + 4x$
8. $y = (x - 1)^3 - 4(x - 1)$

9. $y = x^3 - x$

10. $y = x^3 - x + 2$

11. $y = x^2(x + 3)$

12. $y = (x - 3)^2 x$

13. $y = x^3 - x^2 - 2x$

14. $y = (x - 1)(x - 2)(x - 3)$

15. $y = \dfrac{x}{x - 1}$

16. $y = \dfrac{x}{1 - x}$

17. $y = \dfrac{x}{x^2 - 1}$

18. $y = \dfrac{1 - x}{x}$

19. $y = \dfrac{x(x - 2)}{x - 3}$

20. $y = \dfrac{2x + 4}{x + 1}$

21. $y = |2x + 1|$

22. $y = |x + 2| + 3$

23. $y = |-x^2 + 4|$

24. $y = |-x^2 - 1|$

25. $y = |x^2 - 4x|$

26. $y = |1 - (x - 2)^2|$

27. $y = |x^3 - 4x|$

28. $y = |x^2(x + 3)|$

29. $y = \dfrac{1}{|x|}$

30. $y = \dfrac{2}{|1 - x^2|}$

In Problems 31–34 graph each function using the graph of $f(x) = \sqrt{x}$ given in Figure 1.36.

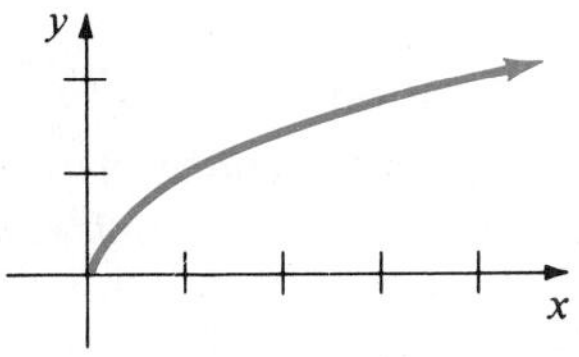

FIGURE 1.36

31. $y = \sqrt{x - 1}$

32. $y = \sqrt{x} + 3$

33. $y = -\sqrt{x}$

34. $y = 1 + \sqrt{x - 3}$

In Problems 35–40 graph each function using the graph of $f(x) = x^3$ given in Figure 1.37.

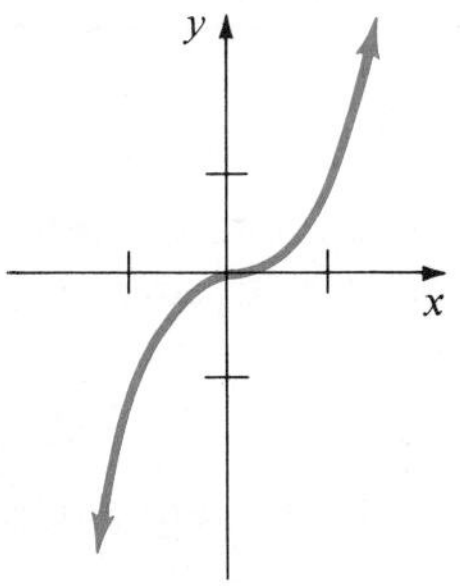

FIGURE 1.37

35. $y = (x - 2)^3$

36. $y = -(x + 1)^3$

37. $y = 1 - x^3$

38. $y = (x + 1)^3 - 1$

39. $y = |x^3|$

40. $y = |x^3 - 1|$

In Problems 41–44 each graph is a shifted graph of the indicated function. Find an equation of the graph.

41. $y = |x|$

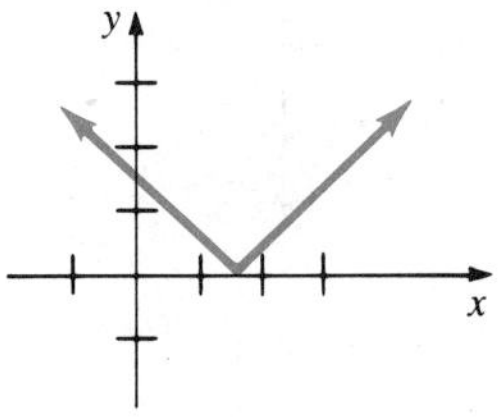

FIGURE 1.38

42. $y = \dfrac{2}{x^2 + 1}$

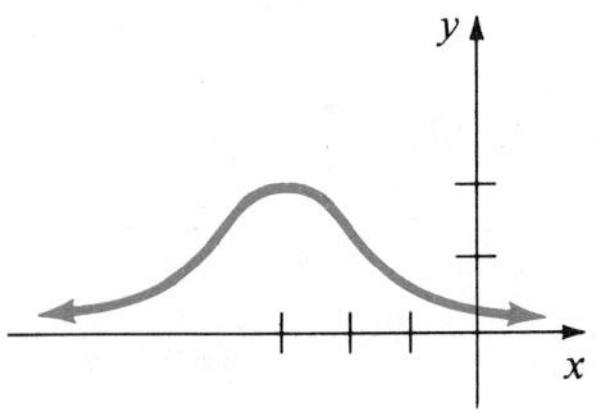

FIGURE 1.39

43. $y = -x^2$

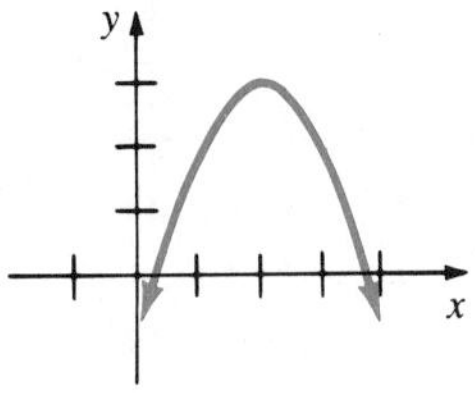

FIGURE 1.40

44. $y = -x^4$

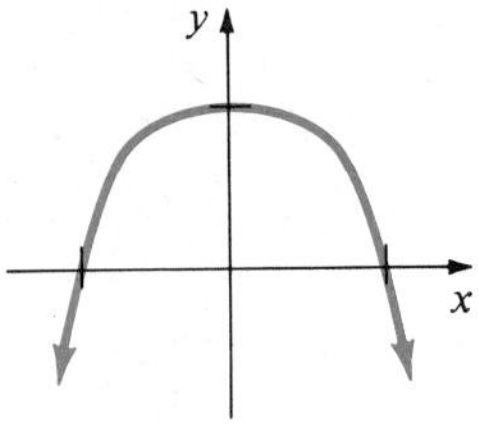

FIGURE 1.41

Complete the square in Problems 45 and 46, and express the given function in the form $y = f(x \pm c_1) \pm c_2$, where $f(x) = x^2$. Graph.

45. $y = x^2 + 8x + 21$

46. $y = x^2 - 6x + 10$

In Problems 47–52 the graph of $y = f(x)$ shown in Figure 1.42 is shifted/reflected. Match the given graph with one of the functions listed here.

(a) $y = f(x - 1)$
(b) $y = f(x + 1)$
(c) $y = 1 - f(x)$
(d) $y = f(x) + 1$
(e) $y = f(x) - 1$
(f) $y = -f(x + 1)$

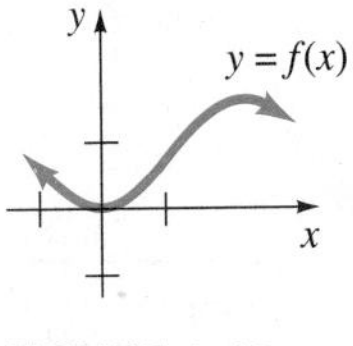

FIGURE 1.42

47.

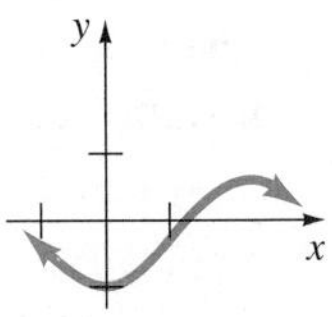

FIGURE 1.43

48.

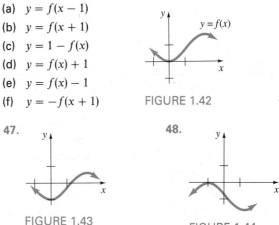

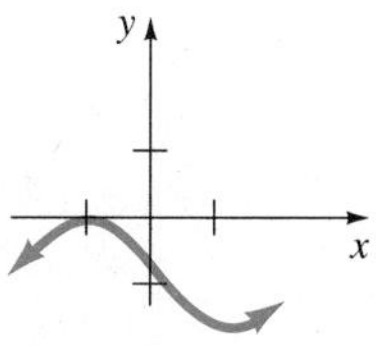

FIGURE 1.44

49.

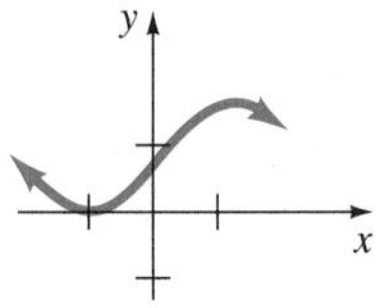

FIGURE 1.45

50.

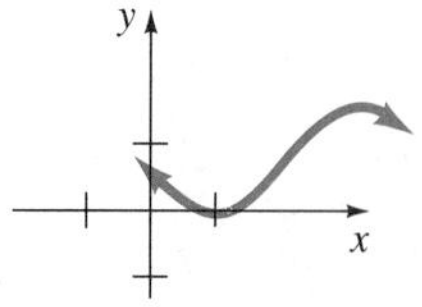

FIGURE 1.46

51.

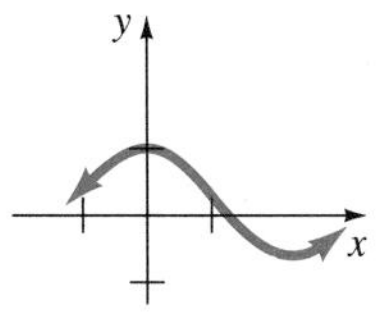

FIGURE 1.47

52.

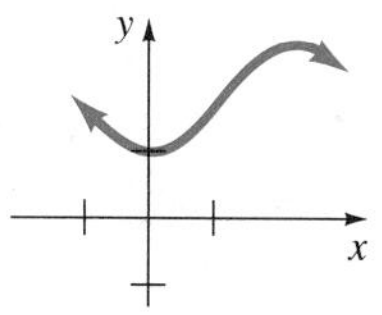

FIGURE 1.48

In Problems 53 and 54 use the given graph of $y = f(x)$ to sketch a possible graph of $y = 1/f(x)$.

53.

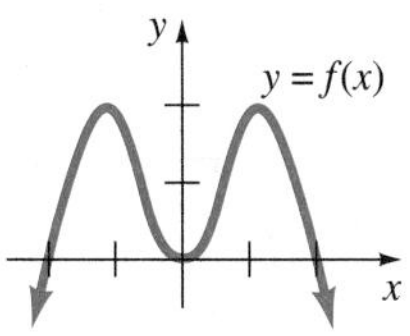

FIGURE 1.49

54.

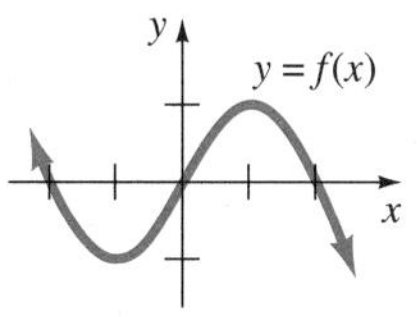

FIGURE 1.50

Calculator/Computer Problems

In Problems 55–62 use a calculator or computer to obtain the graph of the given function.

55. $f(x) = x|x|^x$

56. $f(x) = |x^2 - 4|^x$

57. $f(x) = \left(1 + \dfrac{3}{2x}\right)^x$

58. $f(x) = 2^x|x|^{-x}$

59. $f(x) = x^5 - x^3 - x$

60. $f(x) = x^7 - x^4 + 5x^2 + x + 4$

61. $f(x) = 2x^8 - x^6 - x^4 - x^2 - 3$

62. $f(x) = 2x^9 - x^8 + 6x^3 + 2$

63. Graphs of quadratic functions, when we ignore differences such as "skinniness," "fatness," and upward or downward orientation, have *one* basic shape described by the word *parabolic*. Use a calculator or computer to obtain the graph of each cubic polynomial function. Conjecture the number of basic shapes that graphs of cubic polynomials can have. Sketch these basic shapes.

(a) $f(x) = x^3 + x$
(b) $f(x) = x^3 - 3x$
(c) $f(x) = -2x^3 + 4x - 5$
(d) $f(x) = -x^3 + 2x^2 + 2$
(e) $f(x) = x^3 - x^2 + x - 2$
(f) $f(x) = x^3 - 4x^2 + 6x$

64. Repeat Problem 63 and conjecture the number of basic shapes that graphs of quartic polynomials can have. Sketch these basic shapes.

(a) $f(x) = x^4 - 3x^2$
(b) $f(x) = -x^4 - 3x^3$
(c) $f(x) = -x^4 - 3x^2 + x$
(d) $f(x) = x^4 + 2x^3 - 2x + 3$
(e) $f(x) = 2x^4 + x^3 + 4x^2 + 3$
(f) $f(x) = x^4 - x^3 - 4x^2 + x - 1$

Miscellaneous Problems

65. Given that f is an odd function and g is an odd function, determine whether the functions $f + g$, $f - g$, fg, and f/g are even, odd, or neither even nor odd.

66. Given that f is an odd function and g is an even function, determine whether the functions $f + g$, $f - g$, fg, and f/g, are even, odd, or neither even nor odd.

67. Any function f can be written as $f = f_e + f_o$, where $f_e = \frac{1}{2}[f(x) + f(-x)]$ and $f_o = \frac{1}{2}[f(x) - f(-x)]$. Show that f_e is an even function and f_o is an odd function.

1.3 TRIGONOMETRIC FUNCTIONS

Cosine and Sine In trigonometry recall that the **cosine function** and **sine function** of an angle t, denoted by $\cos t$ and $\sin t$, respectively, can be interpreted in two ways:

(*i*) as the x- and y-coordinates of a point on a unit circle, as shown in Figure 1.51(a), or (*ii*) as the quotient of lengths of sides of a right triangle, as shown in Figure 1.51(b):

$$\cos t = \frac{\text{side adjacent}}{\text{hypotenuse}} \qquad \sin t = \frac{\text{side opposite}}{\text{hypotenuse}}$$

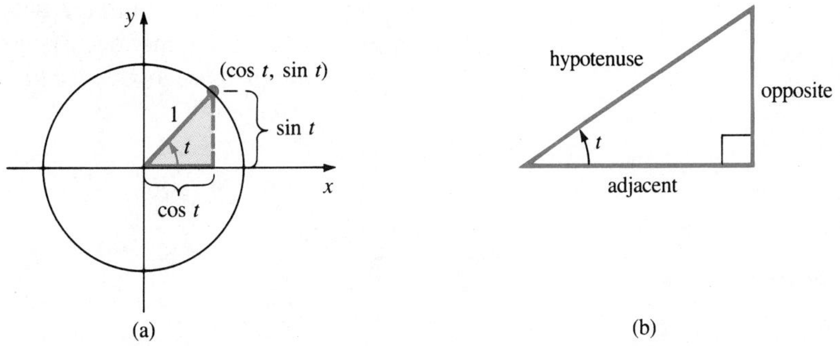

FIGURE 1.51

In the following discussion, we shall focus on the former interpretation.

Measure of Angles An angle is measured in either **degrees** or **radians.** As shown in Figure 1.52(a), an angle of **one radian** subtends an arc of length 1 unit on the circumference of a unit circle. The semicircumference is equivalent to an angle of π radians, where π denotes the irrational number 3.1415926 . . . . The equivalence

$$\pi \text{ radians} = 180 \text{ degrees } (180^\circ)$$

can be used in the form

$$1 \text{ radian} = \left(\frac{180}{\pi}\right)^\circ \quad \text{or} \quad 1^\circ = \frac{\pi}{180} \text{ radians}$$

to convert from one measure to the other. By division we obtain the approximation 1 radian $\approx$ 57.296°.

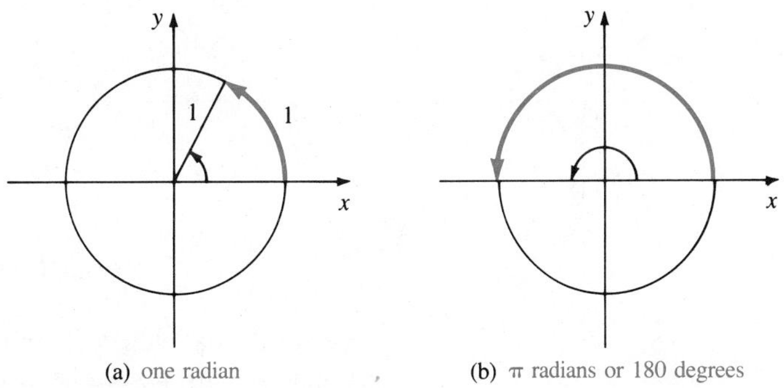

(a) one radian (b) π radians or 180 degrees

FIGURE 1.52

EXAMPLE 1

(*a*) $\pi/9$ radians $= (\pi/9)(180/\pi) = 20°$

(*b*) $15° = 15(\pi/180) = \pi/12$ radians □

Since radian measure is used almost exclusively throughout calculus, you are urged to become familiar with, and to be able to obtain, the radian equivalents of frequently occurring angles such as 30°, 45°, 60°, 90°, 120°, and so on. Some of these angles are given in the following table.

Degrees	0°	30°	45°	60°	90°	120°	135°	150°	180°	270°	360°
Radians	0	$\frac{\pi}{6}$	$\frac{\pi}{4}$	$\frac{\pi}{3}$	$\frac{\pi}{2}$	$\frac{2\pi}{3}$	$\frac{3\pi}{4}$	$\frac{5\pi}{6}$	π	$\frac{3\pi}{2}$	2π

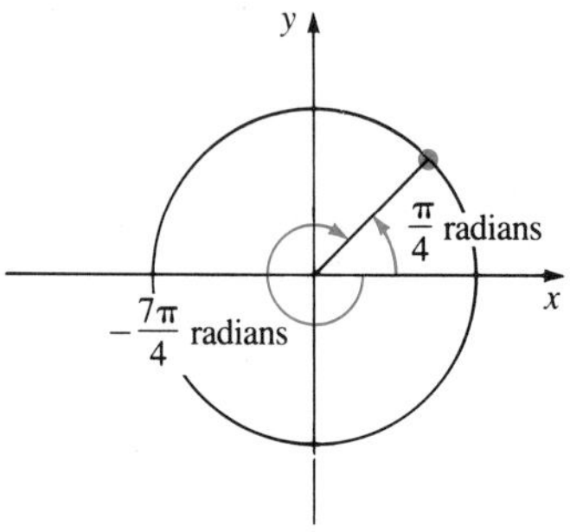

FIGURE 1.53

On a unit circle, a central angle t is said to be in **standard position** if its vertex is at the origin and its initial side coincides with the positive x-axis. Angles measured counterclockwise are given a **positive** measure, whereas angles measured clockwise are **negative**. Two angles in standard position are **coterminal** if they have the same terminal side. In Figure 1.53 the coterminal angles $\pi/4$ and $-7\pi/4$ determine the same point on the unit circle.

Additional Trigonometric Functions Four additional trigonometric functions are defined in terms of $\sin t$ and $\cos t$:

tangent: $\tan t = \dfrac{\sin t}{\cos t}$ **secant:** $\sec t = \dfrac{1}{\cos t}$

cotangent: $\cot t = \dfrac{\cos t}{\sin t} = \dfrac{1}{\tan t}$ **cosecant:** $\csc t = \dfrac{1}{\sin t}$

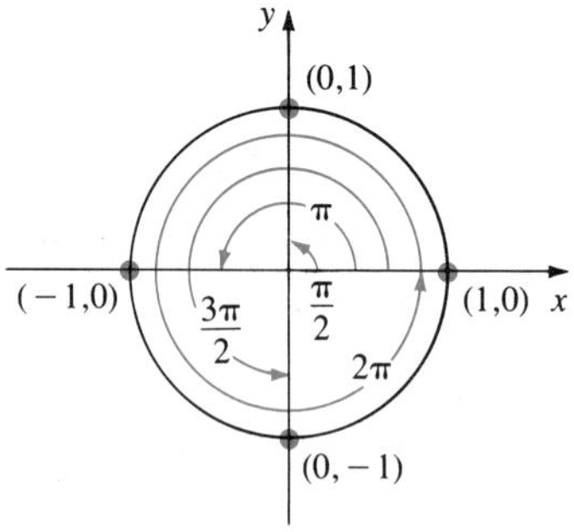

FIGURE 1.54

Numerical Values From Figure 1.54 the following **numerical values** of $\sin t$ and $\cos t$ are evident.

$$\sin 0 = 0 \qquad \cos 0 = 1$$

$$\sin\frac{\pi}{2} = 1 \qquad \cos\frac{\pi}{2} = 0$$

$$\sin \pi = 0 \qquad \cos \pi = -1$$

$$\sin\frac{3\pi}{2} = -1 \qquad \cos\frac{3\pi}{2} = 0$$

$$\sin 2\pi = 0 \qquad \cos 2\pi = 1$$

Some additional values of $\sin t$ and $\cos t$, given in the next table, will be used throughout this text. These values occur often enough to warrant their memorization.

t (radians)	0	$\frac{\pi}{6}$	$\frac{\pi}{4}$	$\frac{\pi}{3}$	$\frac{\pi}{2}$	$\frac{2\pi}{3}$	$\frac{3\pi}{4}$	$\frac{5\pi}{6}$	π	$\frac{3\pi}{2}$	2π
$\sin t$	0	$\frac{1}{2}$	$\frac{\sqrt{2}}{2}$	$\frac{\sqrt{3}}{2}$	1	$\frac{\sqrt{3}}{2}$	$\frac{\sqrt{2}}{2}$	$\frac{1}{2}$	0	-1	0
$\cos t$	1	$\frac{\sqrt{3}}{2}$	$\frac{\sqrt{2}}{2}$	$\frac{1}{2}$	0	$-\frac{1}{2}$	$-\frac{\sqrt{2}}{2}$	$-\frac{\sqrt{3}}{2}$	-1	0	1

The values of the other trigonometric functions can be obtained from the values of the sine and cosine.

EXAMPLE 2

$$(a)\quad \tan\frac{\pi}{3} = \frac{\sin\frac{\pi}{3}}{\cos\frac{\pi}{3}} = \frac{\sqrt{3}/2}{1/2} = \sqrt{3} \qquad (b)\quad \sec\pi = \frac{1}{\cos\pi} = -1$$

$$(c)\quad \cot 2\pi = \frac{\cos 2\pi}{\sin 2\pi} \text{ is not defined, since } \sin 2\pi = 0$$

□

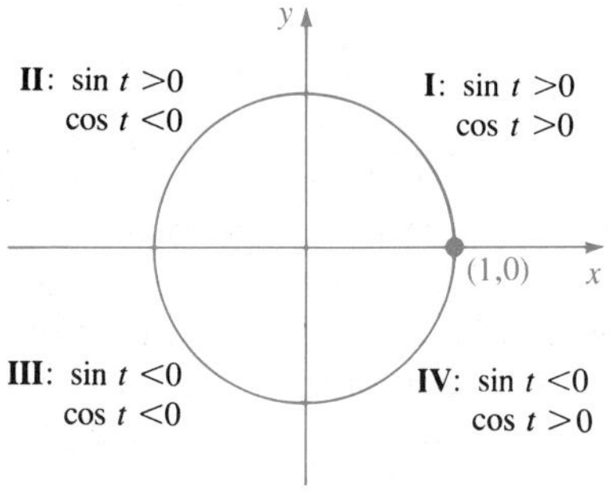

FIGURE 1.55

Figure 1.55 shows the algebraic signs of $\sin t$ and $\cos t$ in the four quadrants of the Cartesian plane. With the aid of this figure you should be able to extend the preceding table to angles such as $7\pi/6$ radians, $5\pi/3$ radians, $7\pi/4$ radians, and so on.

Some Basic Identities Since the angles t and $t + 2\pi$ are coterminal, the values of sine and cosine functions repeat every 2π radians:

$$\sin t = \sin(t + 2\pi) \qquad \cos t = \cos(t + 2\pi) \tag{1}$$

Because of the identities in (1), $\sin t$ and $\cos t$ are said to be **periodic** with **period** $\mathbf{2\pi}$. It can be proved that $\tan t$ is periodic with period π:

$$\tan t = \tan(t + \pi) \tag{2}$$

Moreover, the cosine and sine functions are related by the fundamental identity:

$$\cos^2 t + \sin^2 t = 1 \tag{3}$$

where $\cos^2 t = (\cos t)^2$ and $\sin^2 t = (\sin t)^2$. This result follows from the fact that the coordinates of the point $(\cos t, \sin t)$ must satisfy an equation of a unit circle, $x^2 + y^2 = 1$. Dividing (3) in turn by $\cos^2 t$ and $\sin^2 t$ yields

$$1 + \tan^2 t = \sec^2 t \tag{4}$$

$$1 + \cot^2 t = \csc^2 t \tag{5}$$

There are many other identities that involve the trigonometric functions. For future reference, some of the more important ones follow.

Odd/Even Formulas

$$\sin(-t) = -\sin t \tag{6}$$

$$\cos(-t) = \cos t \tag{7}$$

Addition Formulas

$$\sin(t_1 \pm t_2) = \sin t_1 \cos t_2 \pm \cos t_1 \sin t_2 \tag{8}$$

$$\cos(t_1 \pm t_2) = \cos t_1 \cos t_2 \mp \sin t_1 \sin t_2 \tag{9}$$

Double-Angle Formulas

$$\sin 2t = 2 \sin t \cos t \tag{10}$$

$$\cos 2t = \cos^2 t - \sin^2 t \tag{11}$$

Half-Angle Formulas

$$\sin^2\frac{t}{2} = \frac{1}{2}(1 - \cos t) \tag{12}$$

$$\cos^2\frac{t}{2} = \frac{1}{2}(1 + \cos t) \tag{13}$$

Formulas (12) and (13) will be particularly useful later on in their equivalent forms:

$$\sin^2 t = \frac{1}{2}(1 - \cos 2t) \quad \text{and} \quad \cos^2 t = \frac{1}{2}(1 + \cos 2t) \tag{14}$$

Trigonometric Equations There are times when, in the course of solving a problem, it is to your benefit quickly to recall the numerical values of some trigonometric functions. For example, in this text you will eventually run into a trigonometric equation such as $\sin t = 0$. You should know that $\sin t = 0$ when $t = 0, -\pi, \pi, -2\pi, 2\pi$, and so on. In other words, $\sin t = 0$ when $t = n\pi$, where n is an integer. What are the solutions of $\cos t = 0$?

EXAMPLE 3 Find all angles t such that $\sin t = -\frac{1}{2}$.

Solution From the table on page 27, we know that $\sin(\pi/6) = \frac{1}{2}$. In addition, Figure 1.55 indicates that the sine is negative in quadrants III and IV. Thus, if we use $\pi/6$ as a reference angle, the values of t in these quadrants for which $\sin t = -\frac{1}{2}$ are

$$t = \pi + \frac{\pi}{6} = \frac{7\pi}{6} \quad \text{and} \quad t = 2\pi - \frac{\pi}{6} = \frac{11\pi}{6}$$

But since the sine function has period 2π, the angles $t = 7\pi/6 + 2\pi$, $7\pi/6 + 4\pi, \ldots$, and $t = 11\pi/6 + 2\pi$, $11\pi/6 + 4\pi, \ldots$ are also solutions to the original equation. In other words, the solutions of $\sin t = -\frac{1}{2}$ are

$$t = \frac{7\pi}{6} + 2n\pi \quad \text{or} \quad t = \frac{11\pi}{6} + 2n\pi, \qquad n \text{ an integer}$$

□

The sine and cosine functions are functions in the strictest sense of Definition 1.1; for each angle t there is only one value of $\sin t$ and one value of $\cos t$. Since arc length s on the circumference of a circle of radius r is related to its central angle t by $s = rt$, where t is measured in radians, it follows that, for a unit circle, $s = t$. In other words, for any real number s there is an angle whose measure is s radians. Thus, the sine and cosine functions have as their common domain the set R of real numbers. For example, when we write $\sin 2$, it is understood that we mean $\sin(2 \text{ radians})$ and *not* $\sin 2°$. By introducing the usual symbols x and y to denote the independent and dependent variables, we write

$$y = \sin x \quad \text{and} \quad y = \cos x$$

Pertinent information about the domain and range of each trigonometric function is summarized in the graphs in Figure 1.56.

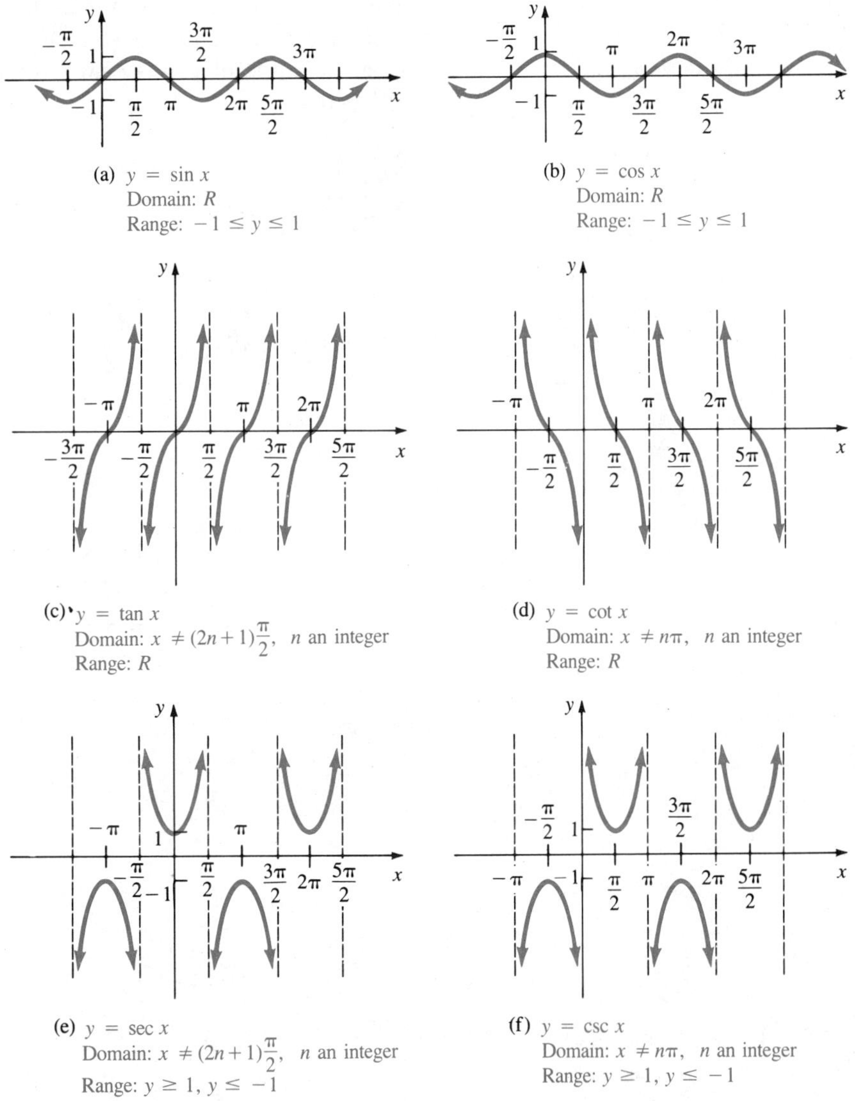

(a) $y = \sin x$
Domain: R
Range: $-1 \le y \le 1$

(b) $y = \cos x$
Domain: R
Range: $-1 \le y \le 1$

(c) $y = \tan x$
Domain: $x \ne (2n+1)\frac{\pi}{2}$, n an integer
Range: R

(d) $y = \cot x$
Domain: $x \ne n\pi$, n an integer
Range: R

(e) $y = \sec x$
Domain: $x \ne (2n+1)\frac{\pi}{2}$, n an integer
Range: $y \ge 1$, $y \le -1$

(f) $y = \csc x$
Domain: $x \ne n\pi$, n an integer
Range: $y \ge 1$, $y \le -1$

FIGURE 1.56

Examination of the graphs in Figure 1.56 clearly reveals the periodic nature of the trigonometric functions; for example, the portion of the graph of $y = \sin x$ on the interval $[0, 2\pi]$ repeats every 2π units. Also, the sine and cosine functions have **amplitude** 1, since the maximum distance a point on the graphs of $y = \sin x$ and $y = \cos x$ can be from the x-axis is 1 unit. In general, the functions

$$y = A \sin kx \quad \text{and} \quad y = A \cos kx, \qquad k > 0$$

have amplitude $|A|$ and period $2\pi/k$. Note that the remaining four trigonometric functions do not have amplitudes.

EXAMPLE 4 Compare the graphs of $y = 2 \sin x$ and $y = -2 \sin x$.

Solution In each case the function has amplitude 2 and period 2π. For comparison, the dashed curve in Figure 1.57 is the graph of $y = \sin x$. Of course, the graph of $y = -2 \sin x$ is a reflection of the graph of $y = 2 \sin x$ in the x-axis.

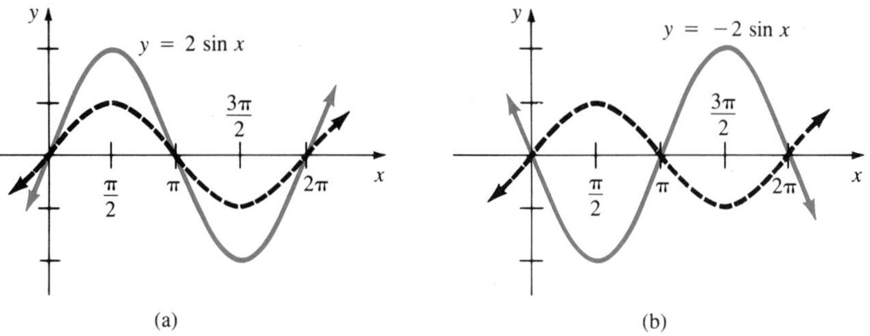

FIGURE 1.57 □

EXAMPLE 5 Graph $y = \cos 4x$.

Solution The amplitude of the function is 1 and its period is $2\pi/4 = \pi/2$. For comparison, the dashed curve in Figure 1.58 is the graph of $y = \cos x$.

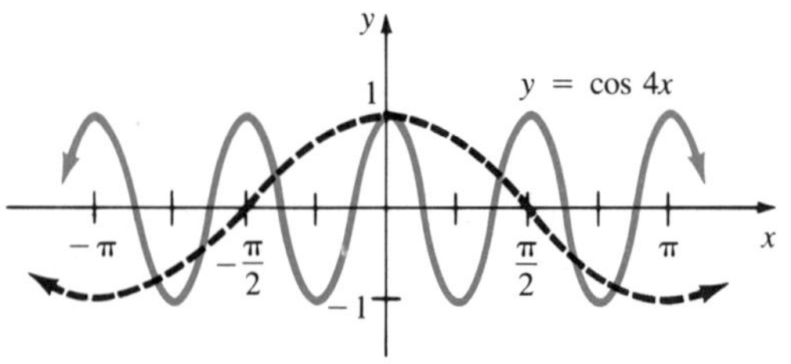

FIGURE 1.58 □

EXAMPLE 6 Graph $y = \sin(x - \pi/2)$.

Solution From Section 1.2 we know that the graph of $y = \sin(x - \pi/2)$ is the graph of $y = \sin x$ shifted $\pi/2$ units to the right. Indeed, the graph of $y = \sin(x - \pi/2)$ given in Figure 1.59 is also the graph of $y = -\cos x$, since (8) implies $\sin(x - \pi/2) = -\cos x$.

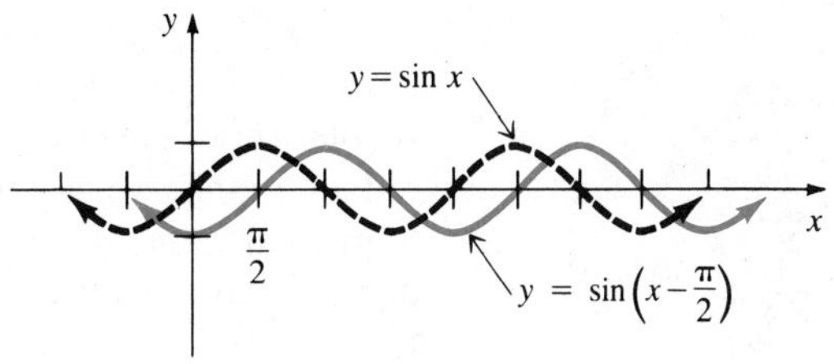

FIGURE 1.59 □

EXAMPLE 7 If $f(x) = \sin x$ and $g(x) = x^2$, graph the composite function $y = (f \circ g)(x)$.

Solution By the definition of $f \circ g$ we have

$$(f \circ g)(x) = f(g(x)) = \sin(g(x)) = \sin x^2$$

Although the sine function $f(x) = \sin x$ is an odd function, when we compose it with the even function $g(x) = x^2$, the resulting function is even. To see this note that $\sin(-x)^2 = \sin x^2$ and so the graph of $y = \sin x^2$ is symmetric with respect to the y-axis. However, because of the x^2 term we would not expect this composite function to be periodic. With the aid of a computer we obtain the graph shown in Figure 1.60. As predicted, the function is nonperiodic but its graph is clearly symmetric with respect to the y-axis. The graph in Figure 1.60 can also be obtained from a graphic calculator.

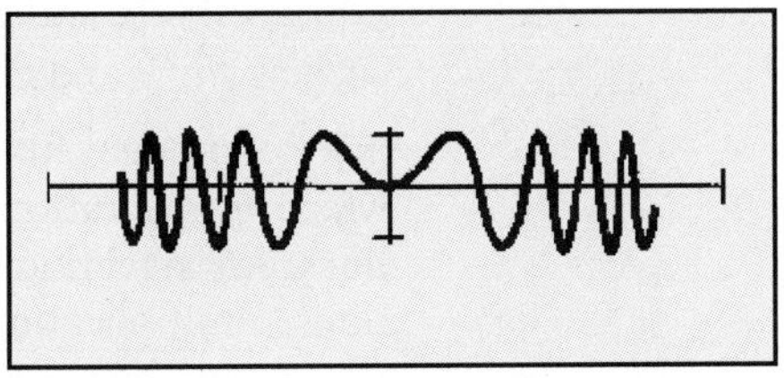

FIGURE 1.60 □

A Word on Calculators A calculator is an invaluable tool in trigonometry. What are the values of cos 2.5 and cos 2.5°? First, with a calculator set in radian mode, we get $\cos 2.5 \approx -0.8011$. Switching the calculator to degree mode, we obtain $\cos 2.5° \approx 0.9990$. A calculator is also helpful in solving equations such as $\sin t = -\frac{1}{2}$ by means of the **inverse trigonometric functions**. You have seen these functions in precalculus mathematics and we will review them in greater detail in Chapter 7. It suffices to recall at this point that inverse trigonometric functions are angles. Most scientific calculators have keys $\boxed{\sin^{-1}}$, $\boxed{\cos^{-1}}$, and $\boxed{\tan^{-1}}$ that refer to the inverse sine, inverse cosine, and inverse tangent, respectively. If your calculator is set in radian mode, then for a number x in the domain of these functions a calculator will give a *unique* answer such that

$$-\frac{\pi}{2} \le \sin^{-1}x \le \frac{\pi}{2}, \quad 0 \le \cos^{-1}x \le \pi, \quad \text{and} \quad -\frac{\pi}{2} < \tan^{-1}x < \frac{\pi}{2}$$

A solution of $\sin t = -\frac{1}{2}$ is written $t = \sin^{-1}(-\frac{1}{2})$. With the calculator set in radian mode, we obtain $t \approx -0.5236$. You should verify that this is the numerical value of $-\pi/6$ (see the last solution in Example 3 with $n = -1$). In degree mode a calculator gives $\sin^{-1}(-\frac{1}{2}) = -30$.

EXAMPLE 8 Use a calculator to solve $\sin^2\theta - 3\sin\theta + 1 = 0$.

Solution The given equation is a quadratic equation in $\sin\theta$. Since the equation does not factor, we use the quadratic formula to solve for $\sin\theta$:

$$\sin\theta = \frac{-(-3) \pm \sqrt{(-3)^2 - 4(1)(1)}}{2}$$

Hence, we have $\sin\theta = (3 + \sqrt{5})/2$ and $\sin\theta = (3 - \sqrt{5})/2$. Now the equation $\sin\theta = (3 + \sqrt{5})/2$ has no solution, since $(3 + \sqrt{5})/2 > 1$. To solve the second equation we write $\theta = \sin^{-1}[(3 - \sqrt{5})/2]$. A calculator set in radian mode then gives $\theta \approx 0.3919$. □

Bear in mind that the equation in Example 8, like the equation in Example 3, has many solutions.

▲ ***Remarks*** (*i*) As a rule, positive integer powers of trigonometric functions are written without parentheses; for example, $(\cos x)^3$ and $(\tan x)^5$ are the same as $\cos^3 x$ and $\tan^5 x$, respectively. Be careful not to confuse expressions such as $\sin^2 x$ and $\cot^6 x$ with $\sin x^2$ and $\cot x^6$.

(*ii*) Polynomial functions, rational functions, and power functions $y = kx^n$, where n is a rational number, belong to a class known as **algebraic functions**. An algebraic function involves a finite number of additions, subtractions, multiplications, divisions, and roots of polynomials; for example, $y = \sqrt{x + \sqrt{x^2 + 5}}$ is an algebraic function. The six trigonometric functions introduced in this section belong to a different class known as **transcendental functions**. A transcendental function is one that is not algebraic. We shall encounter other transcendental functions in Chapters 6 and 7.

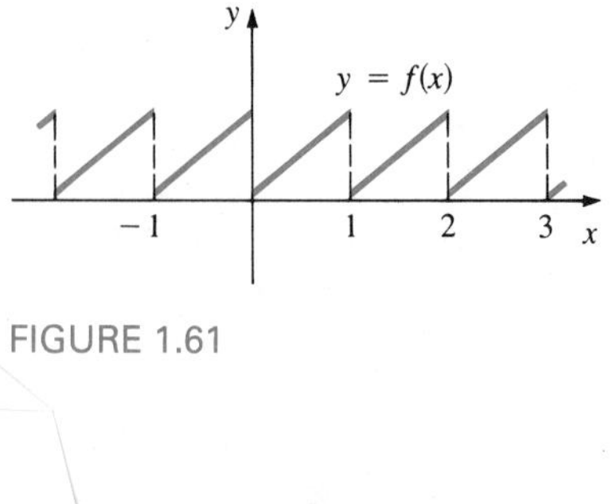

FIGURE 1.61

(*iii*) We have seen that each of the trigonometric functions is periodic. In a more general context, a function f is **periodic** if there exists a positive number p such that $f(x + p) = f(x)$ for every number x in the domain of f. If p is the smallest positive number for which $f(x + p) = f(x)$, then p is the **period** of the function f. The function illustrated in Figure 1.61 has period 1.

EXERCISES 1.3 *Answers to odd-numbered problems begin on page A-66.*

In Problems 1–6 convert from radian measure to degree measure.

1. $\pi/20$ **2.** $2\pi/9$ **3.** $11\pi/6$

4. $5\pi/18$ **5.** $-4\pi/3$ **6.** $-9\pi/4$

In Problems 7–12 convert from degree measure to radian measure.

7. $210°$ **8.** $225°$ **9.** $300°$

10. $315°$ **11.** $-150°$ **12.** $-420°$

In Problems 13–24 find the value of the given quantity.

13. $\sin(-\pi/6)$ **14.** $\cos(9\pi/4)$

15. $\sin(4\pi/3)$ **16.** $\tan(5\pi/4)$

17. $\sec(\pi/6)$ **18.** $\csc(-3\pi/4)$

19. $\cot(2\pi/3)$ **20.** $\sin(11\pi/6)$

21. $\tan(7\pi/6)$ **22.** $\csc(5\pi/6)$

23. $\cos(5\pi/2)$ **24.** $\cot(-\pi/3)$

25. Given that $\sin t = -2/\sqrt{5}$ and $\cos t = 1/\sqrt{5}$, find the values of the remaining four trigonometric functions.

26. Given that $\cos t = \frac{1}{10}$, find all possible values of $\sin t$.

27. Given that $\sin t = -\frac{1}{4}$ and the terminal side of the angle t is in the fourth quadrant, find the values of the remaining five trigonometric functions.

28. Given that $\tan t = -3$ and the terminal side of the angle t is in the second quadrant, find the values of the remaining five trigonometric functions.

29. Find all solutions of the equation $\sin t = -\sqrt{3}/2$.

30. Find all solutions of the equation $\tan t = 1$.

31. Find all solutions of the equation $\sqrt{2}\cos^2 t - \cos t = 0$ in the interval $[0, 2\pi]$.

32. Find all solutions of the equation $\cos 2t + 3\sin t - 2 = 0$ in the interval $[0, 2\pi]$.

In Problems 33–42 use identities (6)–(13) to find the values of the given quantity or to prove the given identity.

33. $\sin(\pi/12)$ **34.** $\cos(5\pi/12)$

35. $\cos(5\pi/8)$ **36.** $\sin(3\pi/8)$

37. $\sin\left(t + \frac{\pi}{2}\right) = \cos t$ **38.** $\cos\left(t + \frac{3\pi}{2}\right) = \sin t$

39. $\cos(t + \pi) = -\cos t$ **40.** $\sin(\pi - t) = \sin t$

41. $\tan(t + \pi) = \tan t$ **42.** $\cot(t - \pi) = \cot t$

In Problems 43 and 44 explain why there is no angle t that satisfies the given equation.

43. $\sin t = \frac{4}{3}$ **44.** $\sec t = \frac{1}{2}$

In Problems 45–60 graph the given function.

45. $y = -\cos x$ **46.** $y = 2\cos x$

47. $y = 1 + \sin x$ **48.** $y = 1 + \cos x$

49. $y = 2 - \sin x$ **50.** $y = -1 + \sin x$

51. $y = -\tan x$ **52.** $y = 2 + \sec x$

53. $y = 3\cos 2x$ **54.** $y = \frac{5}{2}\sin 2\pi x$

55. $y = 2\sin(-\pi x)$ **56.** $y = \cos\frac{1}{2}x$

57. $y = \cos(x - \pi/2)$ **58.** $y = \sin(x + \pi)$

59. $y = |\sin x|$ **60.** $y = |2\cos x|$

In Problems 61–64 the graph is a shifted or reflected graph of the given function. Find an equation of the graph.

61. $y = \sin 4x$ **62.** $y = 3\sin x$

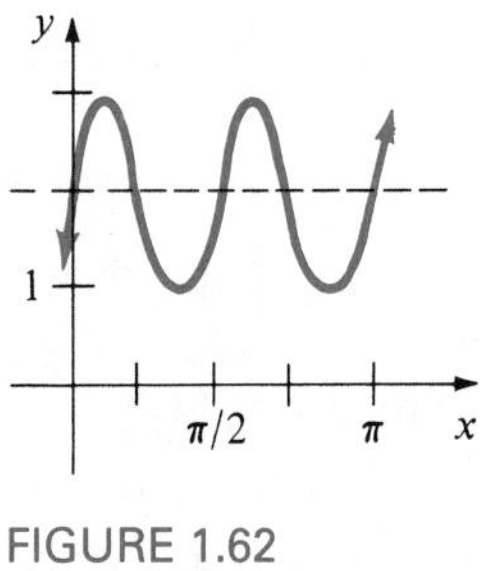

FIGURE 1.62

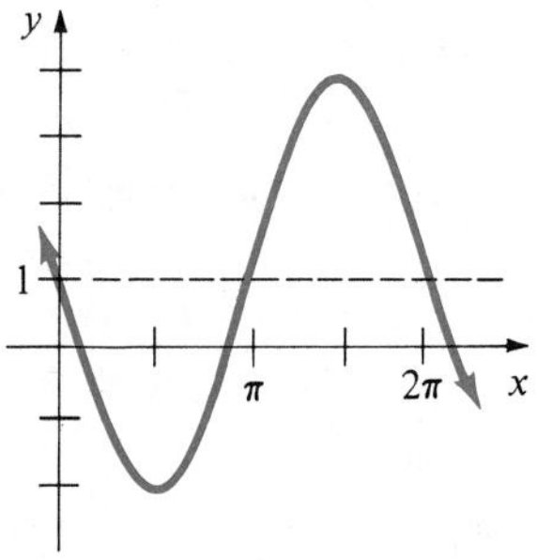

FIGURE 1.63

63. $y = \sin x$

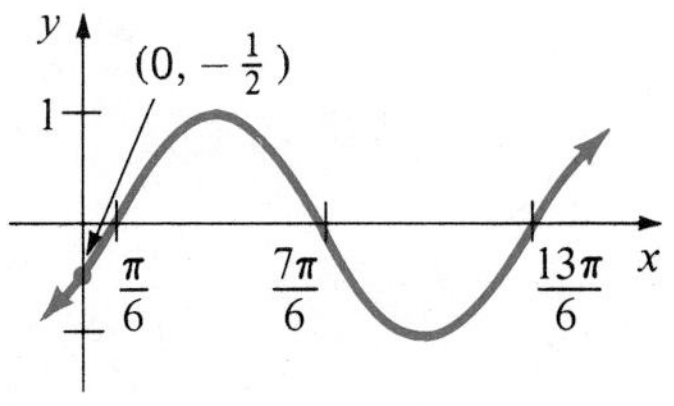

FIGURE 1.64

64. $y = \cos x$

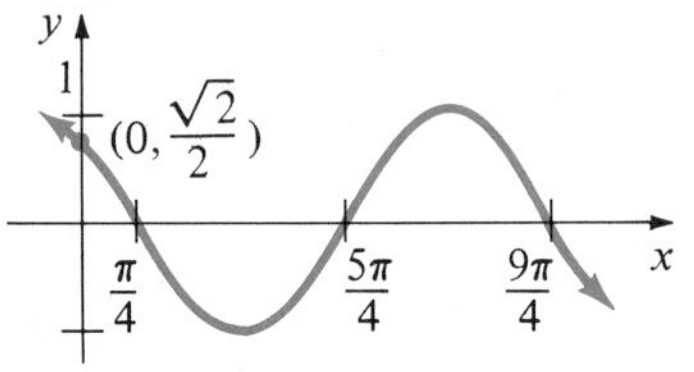

FIGURE 1.65

In Problems 65–70 find fg, f/g, $f \circ g$, and $g \circ f$.

65. $f(x) = 1 + 4x$
$g(x) = \cos x$

66. $f(x) = x^2$
$g(x) = \sin x^2$

67. $f(x) = \sin x$
$g(x) = \cos x$

68. $f(x) = 1 + x^2$
$g(x) = \tan x$

69. $f(x) = 1 + \cos 2x$
$g(x) = \sqrt{x}$

70. $f(x) = \tan 2x$
$g(x) = \cot 2x$

71. Trigonometric functions of the form

$$y = a + b \sin \omega(t - t_0)$$

where a, b, ω, and t_0 are constants, are often used in ecological modeling to simulate periodic phenomena such as temperature variation, height of tides, and day length. Suppose

$$T(t) = 70 + 10 \sin \frac{\pi}{12}(t - 9)$$

represents temperature.

(a) Show that $T(t + 24) = T(t)$.

(b) At what time in [0, 24] does $T(t) = 70$?

(c) What is the maximum temperature? At what time in [0, 24] does the maximum occur?

(d) What is the minimum temperature? At what time in [0, 24] does the minimum occur?

72. In the walk of a human the vertical component of the force of a foot on the ground can be approximated by a trigonometric function of the form $F(t) = K[\cos(\pi t/T) - q \cos(3\pi t/T)]$, where $K > 0$ and $0 < q < 1$. The constants K and q are dependent on the particular nature of the gait. The foot strikes the ground at time $t = -T/2$ and is lifted from the ground at time $t = T/2$.

(a) Show that $F(-T/2) = F(T/2) = 0$.

(b) Show that $F(t + 2T) = F(t)$.

(c) For $T = 1$, $K = 2$, and $q = \frac{1}{2}$, graph the function F on the interval $[-\frac{1}{2}, \frac{1}{2}]$.

In Problems 73–78 determine whether the given function is even, odd, or neither even nor odd.

73. $f(x) = \dfrac{\sin x}{x}$

74. $f(x) = x \cos x$

75. $f(x) = x + \tan x$

76. $f(x) = x^2 \sec x$

77. $f(x) = \cos(x + \pi)$

78. $f(x) = 2 + 5 \csc x$

Calculator/Computer Problems

In Problems 79–84 use a calculator or computer to obtain the graph of the given function. Before graphing, decide whether the graph will possess any symmetry. In Problems 81 and 82 use zoom-in to examine the behavior of the graph near $x = 0$.

79. $f(x) = x + \sin x$

80. $f(x) = \sin x + \cos x$

81. $f(x) = \sin(1/x)$

82. $f(x) = x \sin(1/x)$

83. $f(x) = (\cos x)^2$

84. $f(x) = \sin x^3$

In Problems 85–88 use a calculator to solve the given trigonometric equation. Use radians.

85. $12 \cos^2\theta - 5 \cos \theta - 2 = 0$

86. $\sin^2 2t + 5 \sin 2t - 1 = 0$

87. $\tan^2 t - 3 \tan t + 1 = 0$

88. $\cot^2 x - 9 = 0$

Miscellaneous Problems

89. If $f(x) = A \sin kx$, $k > 0$, show that $f\left(x + \dfrac{2\pi}{k}\right) = f(x)$.

90. If $f(x) = A \tan kx$, $k > 0$, show that $f\left(x + \dfrac{\pi}{k}\right) = f(x)$.

1.4 SETTING UP FUNCTIONS

In calculus it is often important to construct or set up a function from the words that describe a problem. For problems of this sort, it always seems to help to draw a picture when it is possible. Do not hesitate to label *all* the unknowns even though some may not even be mentioned in the narrative of the problem.

EXAMPLE 1 Express the volume of a cube as a function of its surface area.

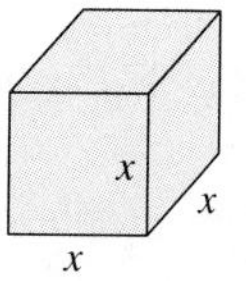

FIGURE 1.66

Solution Let the volume be denoted by V and the surface area by S. If we let x represent the length of one edge of the cube, then the volume is related to x by $V = x^3$. Now, as seen in Figure 1.66, a cube has six sides, and each side has area x^2. Thus, the total surface area of the cube is $S = 6x^2$. We now have the volume and the surface area as functions of x. From the last function we can solve for x in terms of S: $x = \sqrt{S/6}$. Substituting this result into $V = x^3$ gives the volume as a function of the surface area:

$$V = \left(\sqrt{\frac{S}{6}}\right)^3 \quad \text{or} \quad V(S) = \frac{\sqrt{6}}{36}S^{3/2} \qquad \square$$

EXAMPLE 2 A company wants to construct an open box with a volume of 450 in^3 so that the length of its base is 3 times its width. Express the surface area as a function of the width.

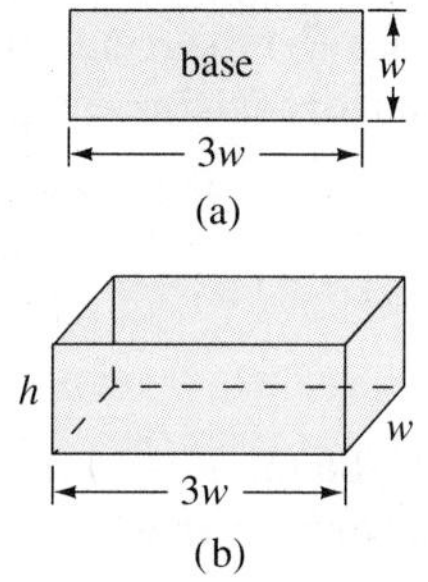

FIGURE 1.67

Solution As in the previous example, let the surface area be denoted by the letter S. Let the width of the base be denoted by w. In Figure 1.67(a) the base of the box is illustrated, whereas Figure 1.67(b) shows the open box; its height, which is not mentioned in the problem, is labeled by the letter h. Now, the surface area of the box is the sum of the areas of the five sides. The area of one side is $3w \times h$, the area of one end is $w \times h$, and the area of the base is $3w \times w$. Therefore, the surface area is

$$S = \underbrace{2(3wh)}_{\text{two sides}} + \underbrace{2(wh)}_{\text{two ends}} + \underbrace{3w^2}_{\text{base}} \tag{1}$$

So far we have not made use of the fact the volume of the box is specified as 450 in^3. In symbols, this means that

$$450 = \text{length} \times \text{width} \times \text{height} \quad \text{or} \quad 450 = 3w^2h$$

This last equation provides a relationship between w and h that allows us to eliminate h from the expression in (1). With $h = 150/w^2$, (1) becomes

$$S = 6w\left(\frac{150}{w^2}\right) + 2w\left(\frac{150}{w^2}\right) + 3w^2 \quad \text{or} \quad S(w) = \frac{1200}{w} + 3w^2 \qquad \square$$

EXAMPLE 3 A right circular cone is inscribed in a sphere of radius 5. Express the volume of the cone as a function of its height.

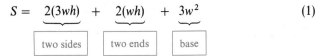

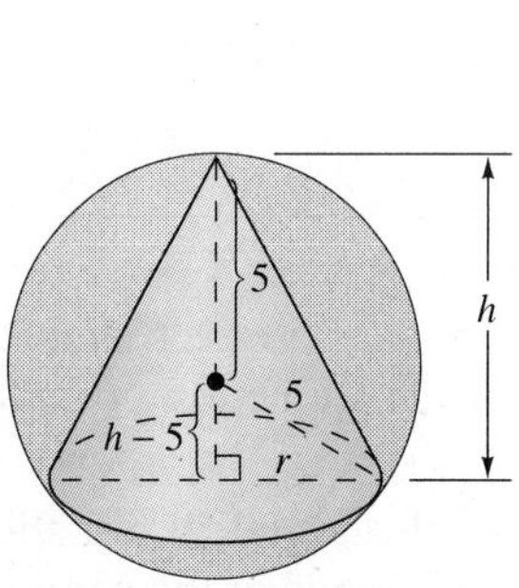

FIGURE 1.68

Solution In Figure 1.68 we have labeled the radius of the cone as r and its height as h. The volume V of a cone depends on both its radius and height:

$$V = \frac{\pi}{3}r^2h \tag{2}$$

(See Appendix V for a collection of geometric formulas.) An inspection of Figure 1.68 shows a right triangle with a base of length r. The Pythagorean Theorem yields

$$(h-5)^2 + r^2 = 5^2 \quad \text{or after simplifying} \quad r^2 = 10h - h^2$$

Hence from (2), the volume of the inscribed cone is

$$V = \frac{\pi}{3}(10h - h^2)h \quad \text{or} \quad V(h) = \frac{\pi}{3}(10h^2 - h^3)$$

□

EXAMPLE 4 At $t = 0$, two planes with a vertical separation of 6000 ft, pass each other going in opposite directions. If the planes are flying horizontally at rates of 500 mi/h and 550 mi/h, express the distance between them as a function of time t.

Solution Using distance = rate × time, we find the horizontal distance traversed by the planes after t hours is $500t$ and $550t$. See Figure 1.69(a) and (b). The distance between the planes after t hours is then the diagonal distance d indicated in Figure 1.69(c). By the Pythagorean Theorem, this diagonal distance as a function of time t is then

$$d(t) = \sqrt{\left(\frac{75}{66}\right)^2 + (1050t)^2}$$

where we have expressed 6000 ft as 6000/5280 = 75/66 mi. For example, after 1 min—that is, $t = 1/60$ h—the distance between the planes is

$$d\left(\frac{1}{60}\right) = \sqrt{\frac{5625}{4356} + \frac{1{,}102{,}500}{3600}} \approx 17.54 \text{ mi}$$

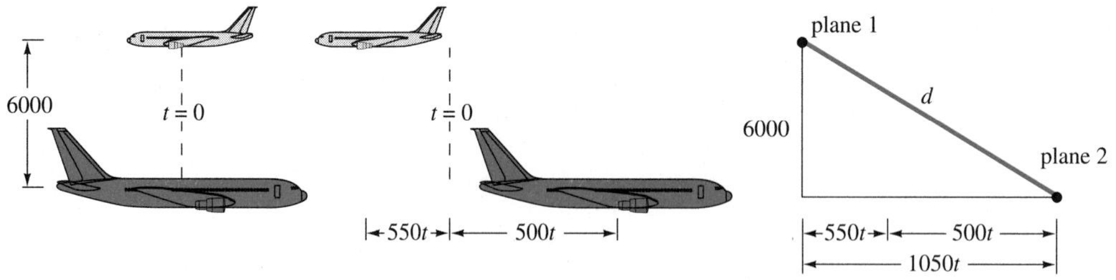

(a) $t = 0$ (b) $t > 0$ (c)

FIGURE 1.69

□

EXAMPLE 5 A tree is planted 30 ft from the base of a street lamp that is 25 ft tall. Express the length of the tree's shadow as a function of its height.

Solution As shown in Figure 1.70(a), the height of the tree at any time is h and the length of its shadow is s. Because the triangles shown Figure 1.70(b)

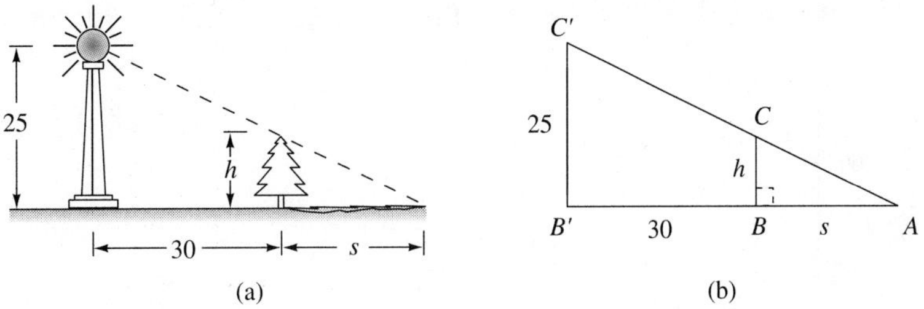

FIGURE 1.70

are right triangles, we might think, as we did in the previous two examples, of using the Pythagorean Theorem. For this problem, however, the Pythagorean Theorem would lead us astray. The important thing to notice here is that triangles ABC and $AB'C'$ are similar. Since the lengths of corresponding sides in similar triangles are proportional, we can write

$$\frac{h}{s} = \frac{25}{s + 30} \quad \text{or} \quad (s + 30)h = 25s$$

By solving the last equation for s in terms of h, we obtain the rational function

$$s(h) = \frac{30h}{25 - h}$$

□

Note in Example 5 that it makes sense to take the domain of the function $s(h) = 30h/(25 - h)$ to be $0 \leq h < 25$. What happens to $s(h)$ as the height of the tree approaches 25 ft?

EXAMPLE 6 Some banks stipulate that a borrower must pay a 1% "insurance" fee on a mortgage loan if he or she is unable to make a down payment of at least 10% on the price of a home. Suppose Mr. and Mrs. Penniless have on hand \$6000 for a down payment plus any insurance fee. Express the fee they must pay as a function of the price of the house.

Solution Let x be the price of a house and F the fee. If $x \leq 60{,}000$, then $F = 0$, since Mr. and Mrs. Penniless can make a 10% down payment. If $x > 60{,}000$, then the F is given by $F = 0.01M$, where M is the mortgage. If D denotes the down payment on a house, then we must have

$$D + F = 6000 \quad \text{and} \quad M = x - D$$

These equations imply $M = x - (6000 - F)$. Substituting this expression into $F = 0.01M$ and solving for F yield $F = (x - 6000)/99$. In other words, the fee is given by the piecewise-defined function

$$F(x) = \begin{cases} 0, & 0 \leq x \leq 60{,}000 \\ \dfrac{x - 6000}{99}, & x > 60{,}000 \end{cases}$$

For example, if a house costs \$62,000, the fee is

$$F(62{,}000) = \frac{62{,}000 - 6000}{99} = \$565.66$$

The graph of F is given in Figure 1.71. □

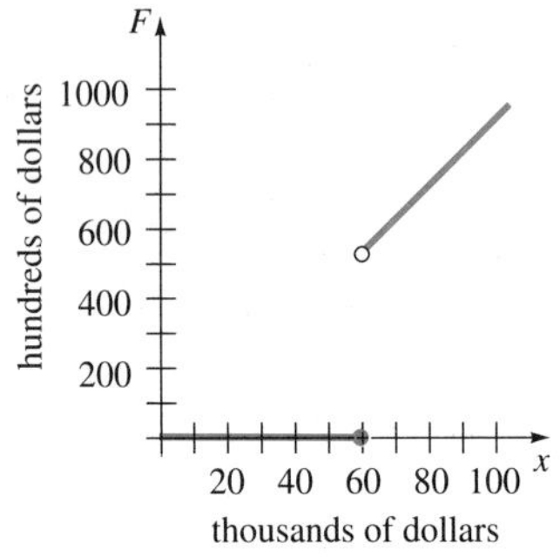

FIGURE 1.71

EXAMPLE 7 A plane flies at a constant height of 3000 ft over level ground away from an observer on the ground. Express the horizontal distance between the plane and the observer as a function of the angle of elevation of the plane measured by the observer.

Solution As shown in Figure 1.72, let P denote the position of the observer on the ground, x the horizontal distance between the plane and the observer, and θ the angle of elevation. By definition,

$$\frac{3000}{x} = \tan\theta \quad \text{and so} \quad x(\theta) = 3000 \cot\theta$$ □

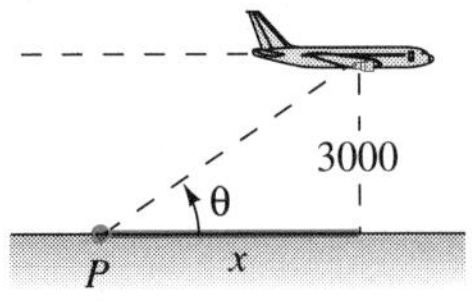

FIGURE 1.72

EXERCISES 1.4 *Answers to odd-numbered problems begin on page A-67.*

1. Express the perimeter of a square as a function of its area.

2. Express the area of a circle as a function of its circumference.

3. Express the area of an equilateral triangle as a function of the length of one side.

4. Express the area of an equilateral triangle as a function of its height.

5. Express the volume of a cube as a function of the area of its base.

6. Express the volume of a right circular cone with a vertex angle of 90° as a function of its height.

7. Express the distance from any point (x, y) on the graph of $y = x^2$ to the point $(3, 0)$ as a function of x.

8. Express the slope of the line from any point (x, y) on the graph of $y = x/(x - 1)$ to the point $(2, -3)$ as a function of x.

9. A rectangle is inscribed in a circle of radius r. Express the area of the rectangle as a function of the length of one of its sides.

10. An isosceles triangle is inscribed in a circle of radius 4. Express the area of the triangle as a function of the length of its base.

11. Express as a function of x the area of a rectangle whose lengths of sides satisfy the **golden ratio**:

$$\frac{y}{x} = \frac{x}{x + y}$$

See Figure 1.73.

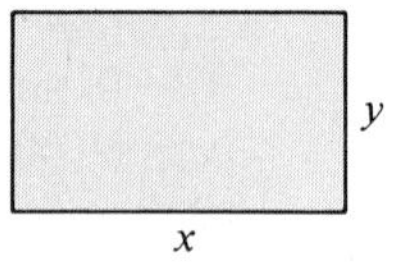

FIGURE 1.73

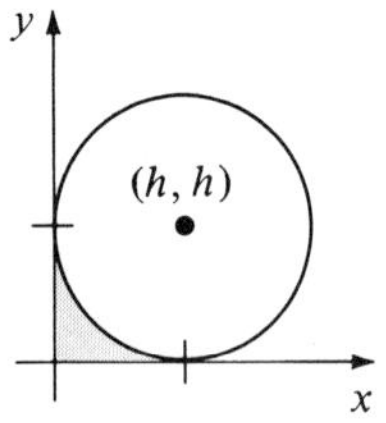

FIGURE 1.74

12. Consider the circle of radius h with center (h, h) shown in Figure 1.74. Express the area of the shaded region as a function of h.

13. (*This problem could present a challenge.*) In an engineering text the area A of the octagon shown in Figure 1.75 is given as $A = 3.314r^2$. Show that this formula is an approximation to

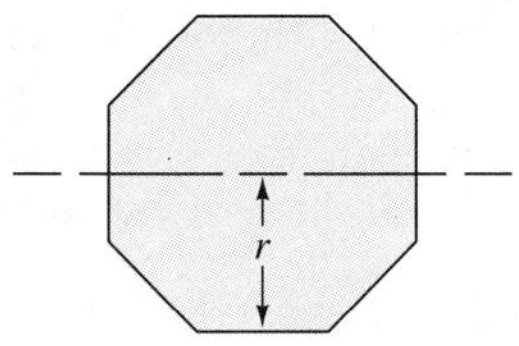

FIGURE 1.75

the area; that is, find an expression that gives the exact area A as a function of r.

14. A wire of length L is cut x units from one end. One piece of the wire is bent into a square and the other piece is bent into a circle. Express the sum of the areas as a function of x.

15. An open box is made from a cardboard square that is 40 cm on a side by cutting out a square with length of side x from each corner and then turning up the remaining sides. Express the area of the base of the box as a function of x.

16. Express the volume of the box in Problem 15 as a function of x.

17. A liquid hydrogen fuel tank for a rocket consists of a right circular cylinder 40 ft long with hemispherical caps at each end of the cylindrical portion of the tank. Express the volume of the tank as a function of the radius r of each cap.

18. A 6-in line is drawn across the corner of a rectangular piece of paper, cutting off a triangular region. Express the area of the triangle as a function of the length of the base of the triangle.

19. Car A passes point O heading east at a constant rate of 40 mi/h; car B passes the same point 1 h later heading north at a constant rate of 60 mi/h. Express the distance d between the cars as a function of time t, where t is measured starting when car B passes point O. See Figure 1.76.

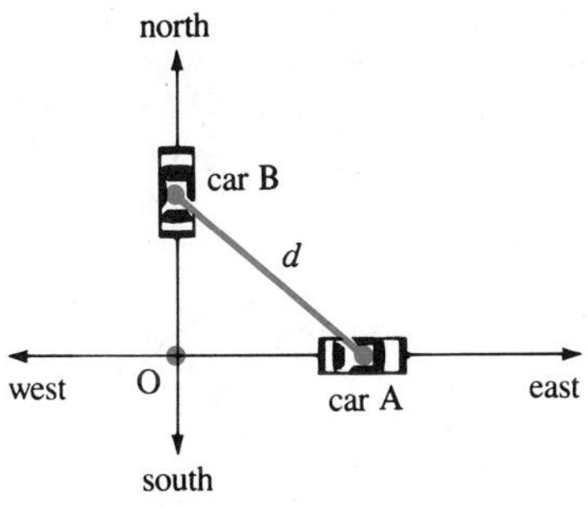

FIGURE 1.76

20. A baseball diamond is a square 90 ft on a side. After a player hits a home run, he jogs around the bases at a rate of 6 ft/s.

(a) As the player runs between home base and first base, express his distance from home base as a function of time t, where $t = 0$ corresponds to the time he left home base—that is, $0 \le t \le 15$.

(b) As he runs between home base and first base, express his distance from second base as a function of time t, where $0 \le t \le 15$.

21. The swimming pool shown in Figure 1.77 is 3 ft deep at the shallow end, 8 ft deep at the deepest end, 40 ft long, 30 ft wide, and the bottom is an inclined plane. Express the volume of the water in the pool as a function of the height h of the water above the deep end. [*Hint:* The volume will be a piecewise-defined function.]

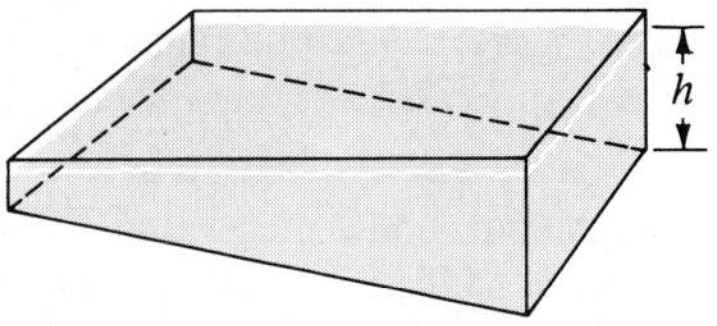

FIGURE 1.77

22. A part of Schedule X of the 1988 Federal Income Tax form 1040 reads:

Schedule X—Use if your filing status is Single

If the amount on Form 1040, line 37, is: Over—	But not over—	Enter on Form 1040, line 38	of the amount over—
\$0	\$17,850	-------- 15%	\$0
17,850	43,150	\$2,677.50 + 28%	17,850
43,150	89,560	9,761.50 + 33%	43,150

If x is the amount on line 37 (taxable income), express the amount to be entered on line 38 (tax owed) as a piecewise-defined function.

23. Water is being pumped into a conical tank whose dimensions are shown in Figure 1.78. Express the volume of the water at any time as a function of its depth h.

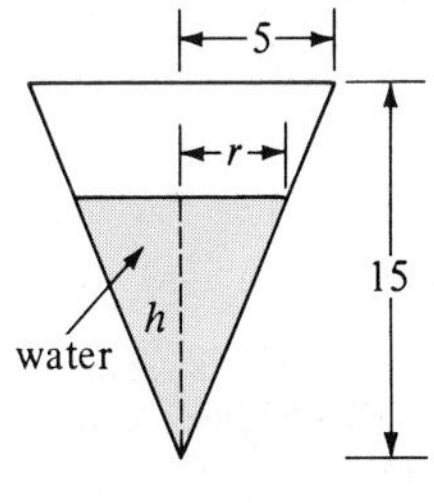

FIGURE 1.78

24. A woman 5 ft tall is standing next to a street lamp that is 25 ft tall. Express the length of the woman's shadow as a function of her distance from the street lamp.

25. A metal soup can is constructed in the shape of a right circular cylinder. If the can is to have a specified volume V_0, express the amount of metal needed to manufacture the can as a function of its radius.

26. A rancher wishes to enclose a 1000-ft^2 rectangular corral using two different construction materials. Along two parallel sides, the material costs \$4 per foot; for the other two parallel sides, the material costs \$1.60 per foot. Express the total cost of construction of the corral as a function of the length of the sides with material that costs \$4 per foot.

27. As shown in Figure 1.79, an oil pipeline is to connect a drilling site and a seaport. The cost per mile of laying the pipe under a $\frac{1}{2}$-mi-wide river is determined to be $\sqrt{5}$ times the cost per mile of laying the pipe under the ground. Express the total cost of construction as a function of the variable x indicated in the figure.

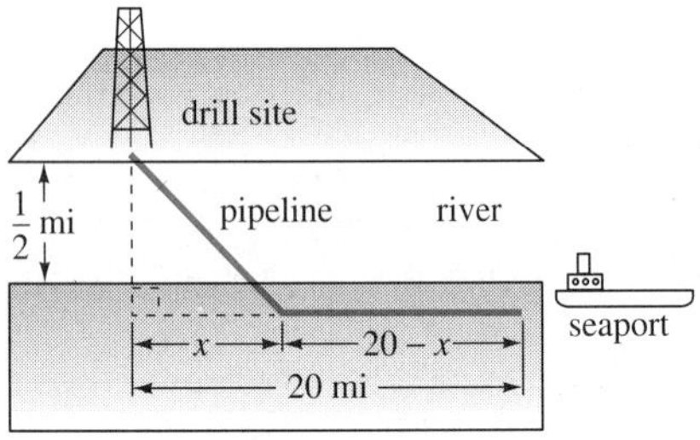

FIGURE 1.79

28. An airline company has a fleet of planes, with a total of 1500 seats, that fly daily between Los Angeles and New York. When the fares are set so that the company makes a profit of \$100 on each fare, the planes fly completely full. Over the years the company observes that for each \$5 increase in profit from increasing the fare rate, it loses 50 passengers each day. Express the company's daily profit as a function of the number of fare increases.

29. Express the height of the balloon shown in Figure 1.80 as a function of its angle of elevation.

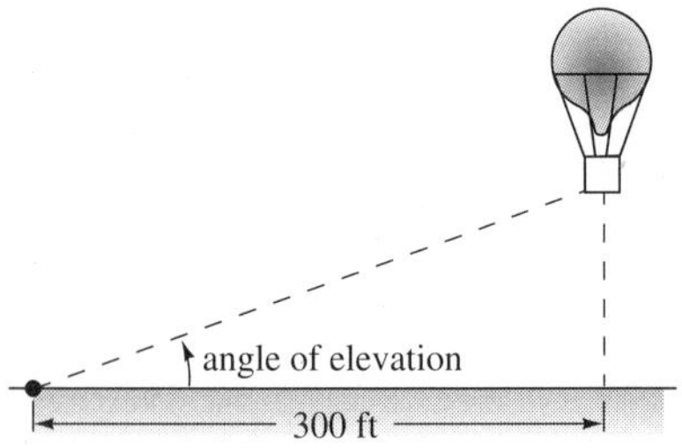

FIGURE 1.80

30. A long sheet of metal 40 in wide is made into a V-shaped trough by bending it in the middle along its length. Express the area of the triangular cross-section of the trough as a function of the angle θ at the vertex of the V.

31. As shown in Figure 1.81, a plank is supported by a sawhorse so that one end rests on the ground and the other end rests against a building. Express the length of the plank as a function of the indicated angle θ.

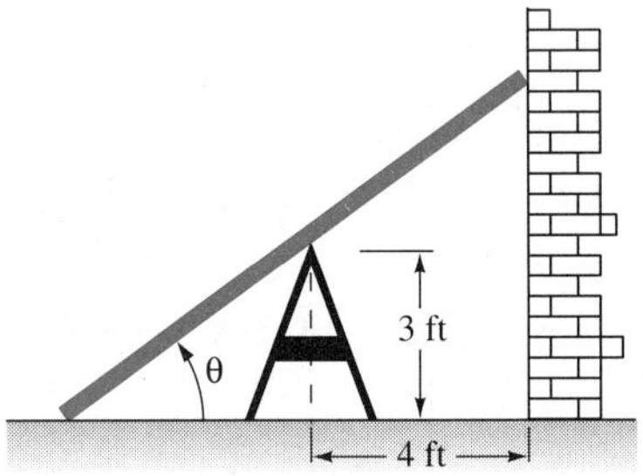

FIGURE 1.81

32. The rectangle shown in color in Figure 1.82 is circumscribed about a rectangle of width w and length l. Express the area of the circumscribed rectangle as a function of the indicated angle θ.

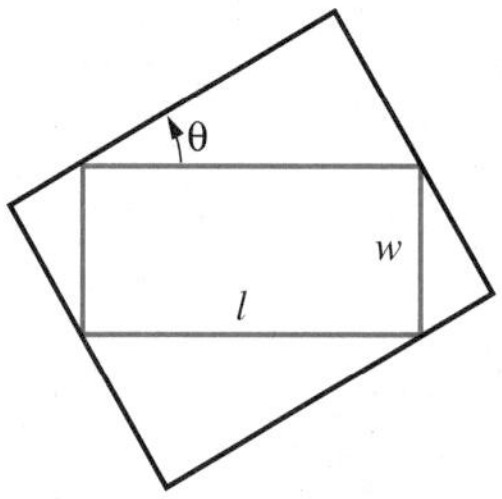

FIGURE 1.82

33. Consider the right circular cylinder inscribed in a sphere of radius R as shown in Figure 1.83. Express the volume and the total surface area of the cylinder as a function of the indicated angle θ.

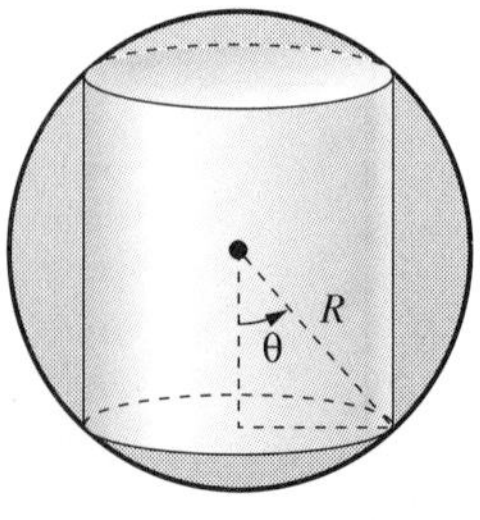

FIGURE 1.83

34. A farmer wishes to enclose a pasture in the form of a right triangle using 2000 ft of fencing on hand. See Figure 1.84. Show that the area of the pasture as a function of the indicated angle θ is

$$A(\theta) = \frac{1}{2}\cot\theta\left(\frac{2000}{1 + \cot\theta + \csc\theta}\right)^2$$

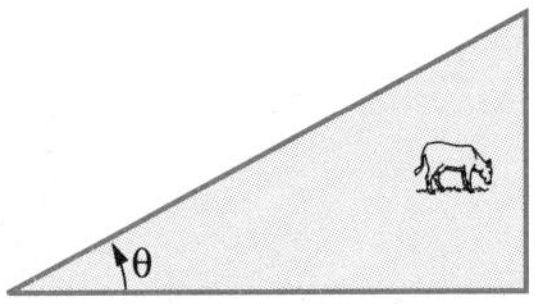

FIGURE 1.84

35. A person on an island wishes to reach a point R on a straight shore on the mainland starting from a point P on the island. The point P is 9 mi from the shore and 15 mi from point R. See Figure 1.85. If the person rows a boat at a rate of 3 mi/h to a point Q on land, and then walks the rest of the way at a rate of 5 mi/h, express the total time it takes the person to reach point R as a function of the indicated angle θ. [*Hint:* Distance = rate × time.]

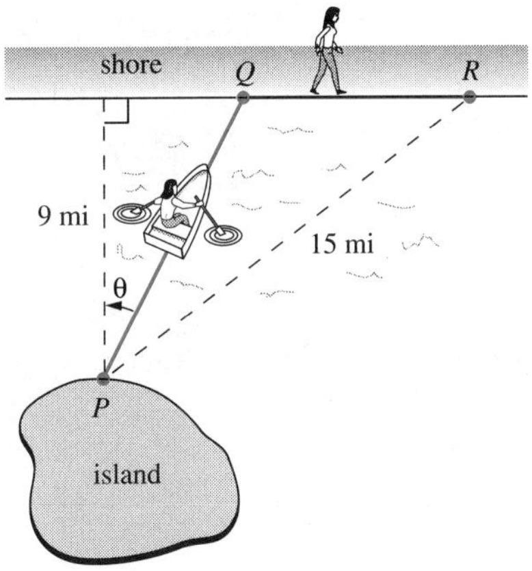

FIGURE 1.85

36. Express the volume of the box shown in Figure 1.86 as a function of the indicated angle θ.

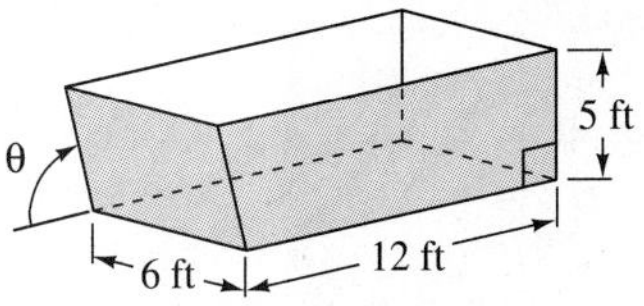

FIGURE 1.86

37. The container shown in Figure 1.87 consists of an inverted cone (open at its top) attached to the bottom of a right circular cylinder (open at its top and bottom) of fixed radius R. The container has a constant volume V. Express the total surface area of the container as a function of the indicated angle θ.

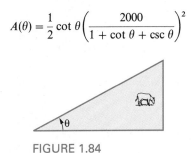

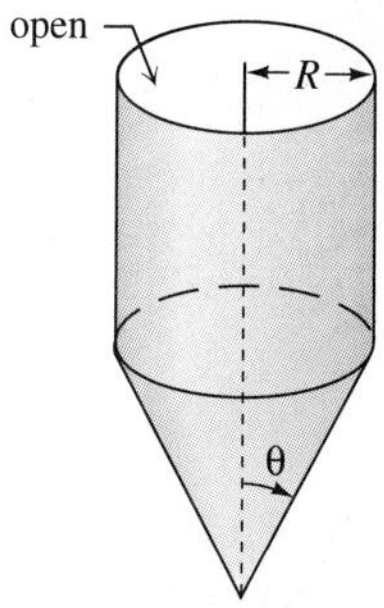

FIGURE 1.87

1.5 LIMIT OF A FUNCTION AT A POINT

1.5.1 Intuitive Interpretation

Limit of a Function As x Approaches a Number Consider the function

$$f(x) = \frac{16 - x^2}{4 + x}$$

whose domain is the set of all real numbers except -4. Although $f(-4)$ is not defined, $f(x)$ can be calculated for any value of x near -4. The tables accompanying Figure 1.88 show that, as x approaches -4 from either the

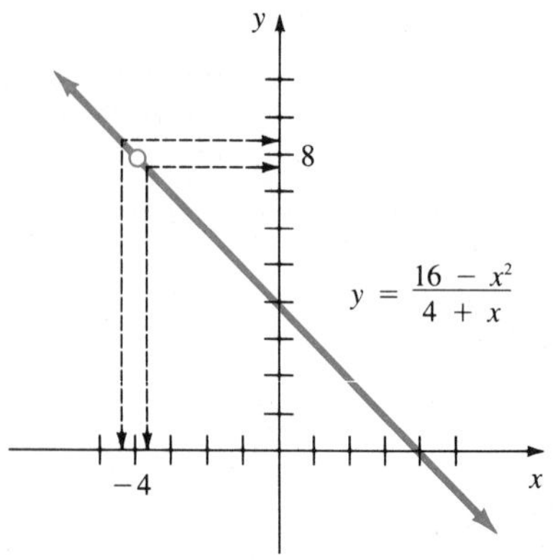

FIGURE 1.88

left or right, the functional values $f(x)$ are approaching 8; that is, when x is near -4, $f(x)$ is near 8. We say 8 is the **limit** of $f(x)$ as x approaches -4 and write

$$f(x) \to 8 \quad \text{as} \quad x \to -4 \quad \text{or} \quad \lim_{x \to -4} \frac{16 - x^2}{4 + x} = 8$$

x	$f(x)$
-4.1	8.1
-4.01	8.01
-4.001	8.001

x	$f(x)$
-3.9	7.9
-3.99	7.99
-3.999	7.999

For $x \neq -4$, f can be simplified by cancellation:

$$f(x) = \frac{16 - x^2}{4 + x} = \frac{(4 + x)(4 - x)}{4 + x} = 4 - x$$

As seen in Figure 1.88, the graph of f is essentially the graph of $y = 4 - x$ with the exception that the graph of f has a hole at the point that corresponds to $x = -4$. As x gets closer and closer to -4, represented by the two arrowheads on the x-axis, the two arrowheads on the y-axis simultaneously get closer and closer to the number 8.

Intuitive Definition Suppose L denotes a finite number. The notion of **$f(x)$ approaching L as x approaches a point a** can be defined informally in the following manner.

If $f(x)$ can be made arbitrarily close to L by taking x sufficiently close to but different from a, from both the left and right sides of a, then $\lim_{x \to a} f(x) = L$.

Notation We shall use the notation $x \to a^-$ to denote that x approaches a from the *left* and $x \to a^+$ to mean that x approaches a from the *right*. Thus, if the one-sided limits $\lim_{x \to a^-} f(x)$ and $\lim_{x \to a^+} f(x)$ have a common value L,

$$\lim_{x \to a^-} f(x) = \lim_{x \to a^+} f(x) = L$$

we then say $\lim_{x \to a} f(x)$ *exists* and write

$$\lim_{x \to a} f(x) = L$$

It is common practice to refer to the number L as the **limit of f at the point a**. However, we should observe:

The existence of a limit of a function f at a does not depend on whether f is actually defined at a but only on whether f is defined for x near a.

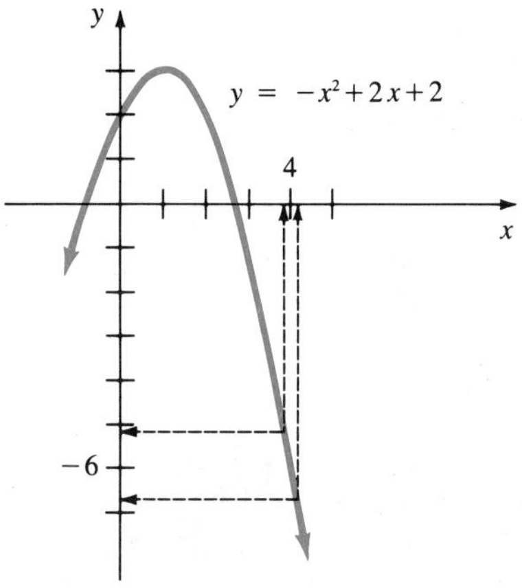

FIGURE 1.89

EXAMPLE 1 The graph of the function $f(x) = -x^2 + 2x + 2$ is shown in Figure 1.89. As seen from the graph and the accompanying tables, it seems plausible that

$$\lim_{x\to 4^-} f(x) = -6 \quad \text{and} \quad \lim_{x\to 4^+} f(x) = -6$$

and consequently

$$\lim_{x\to 4} f(x) = -6$$

$x \to 4^-$	$f(x)$
3.9	−5.41000
3.99	−5.94010
3.999	−5.99400

$x \to 4^+$	$f(x)$
4.1	−6.61000
4.01	−6.06010
4.001	−6.00600

□

Note that in Example 1 the given function is certainly defined at $x = 4$, but at no time did we substitute $x = 4$ into the function to find the value of $\lim_{x\to 4} f(x)$. In the next example we see that $\lim_{x\to 2} f(x)$ exists but $f(2)$ is not defined.

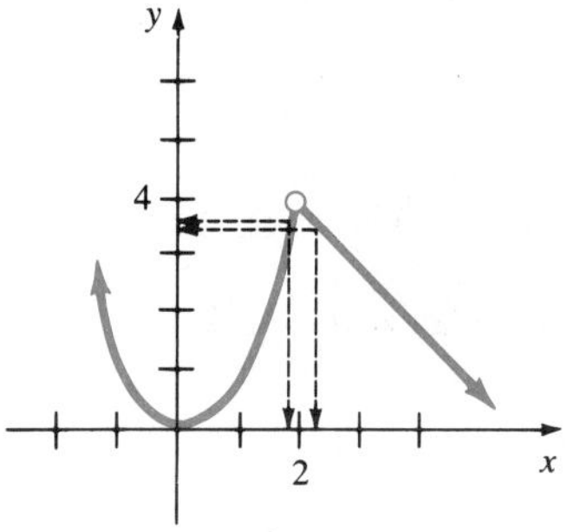

FIGURE 1.90

EXAMPLE 2 The graph of the piecewise-defined function

$$f(x) = \begin{cases} x^2, & x < 2 \\ -x + 6, & x > 2 \end{cases}$$

is given in Figure 1.90. From the graph and the accompanying tables, we see that when x is close to 2, $f(x)$ is close to 4; that is, $\lim_{x\to 2} f(x) = 4$.

$x \to 2^-$	$f(x)$
1.9	3.61000
1.99	3.96010
1.999	3.99600

$x \to 2^+$	$f(x)$
2.1	3.90000
2.01	3.99000
2.001	3.99900

□

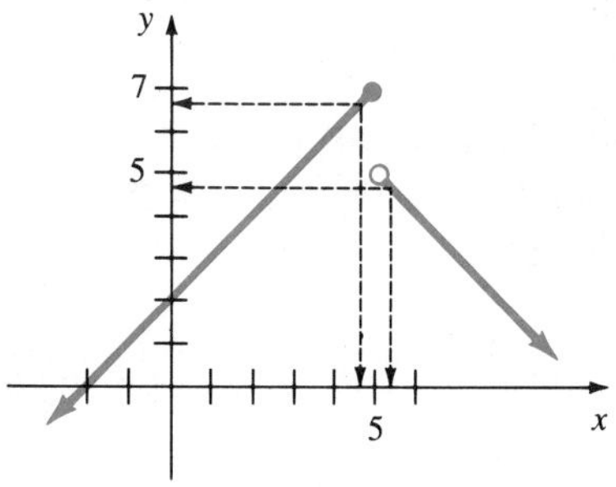

FIGURE 1.91

EXAMPLE 3 The graph of the piecewise-defined function

$$f(x) = \begin{cases} x + 2, & x \le 5 \\ -x + 10, & x > 5 \end{cases}$$

is given in Figure 1.91. From the graph and the accompanying tables, we see that

$$\lim_{x\to 5^-} f(x) = 7 \quad \text{and} \quad \lim_{x\to 5^+} f(x) = 5$$

Since $\lim_{x\to 5^-} f(x) \neq \lim_{x\to 5^+} f(x)$, we conclude that $\lim_{x\to 5} f(x)$ does not exist.

$x \to 5^-$	$f(x)$
4.9	6.9
4.99	6.99
4.999	6.999

$x \to 5^+$	$f(x)$
5.1	4.9
5.01	4.99
5.001	4.999

□

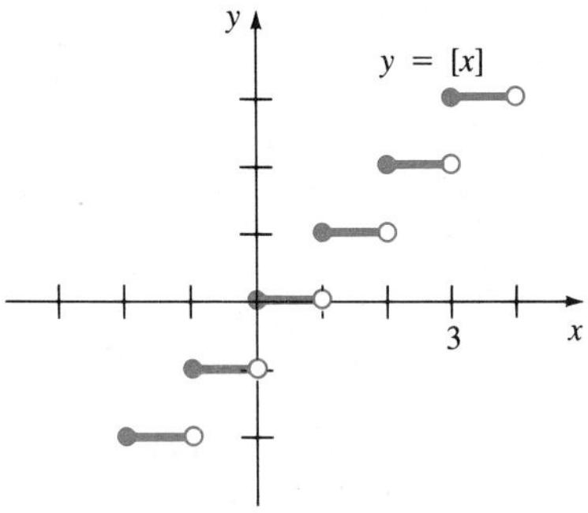

FIGURE 1.92

EXAMPLE 4 If x is any real number, the "output" $f(x)$ of the **greatest integer function** $f(x) = [x]$ is defined to be the greatest integer not exceeding x; for example,

$$f\left(\frac{1}{2}\right) = 0, \quad f\left(-\frac{1}{4}\right) = -1, \quad f(4.8) = 4, \quad \text{and} \quad f(3) = 3$$

The graph of f is given in Figure 1.92. From this graph we see that $f(n)$ is defined for every integer n; nonetheless, $\lim_{x\to n} f(x)$ does not exist. For example, as x approaches the number 3, the two one-sided limits exist but have different values:

$$\lim_{x\to 3^-} f(x) = 2 \quad \text{whereas} \quad \lim_{x\to 3^+} f(x) = 3$$ □

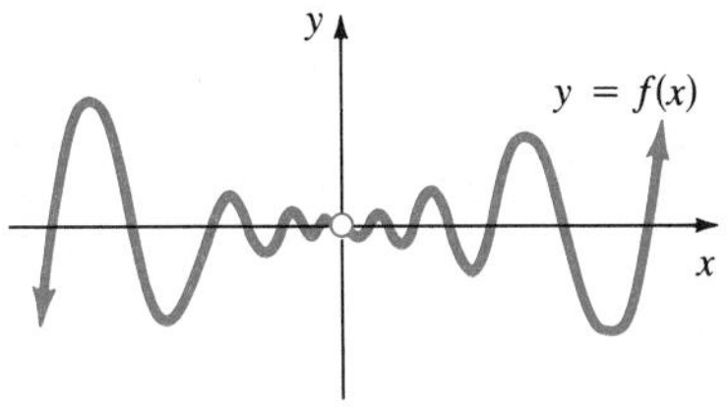

FIGURE 1.93

EXAMPLE 5 In Figure 1.93 the graph of $y = f(x)$ shows that

$$\lim_{x\to 0^-} f(x) = 0 \quad \text{and} \quad \lim_{x\to 0^+} f(x) = 0$$

Hence,

$$\lim_{x\to 0} f(x) = 0$$ □

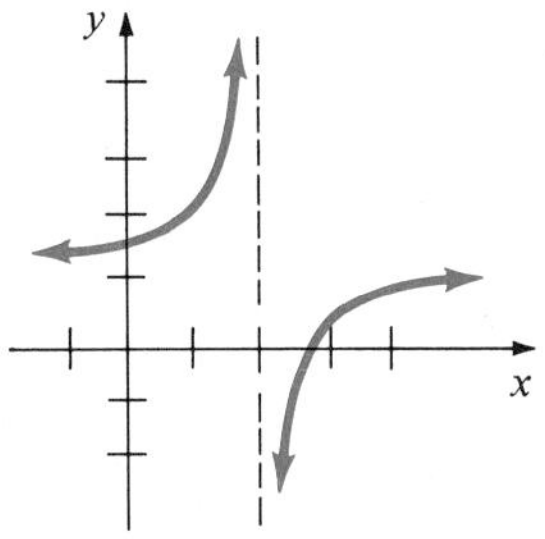

FIGURE 1.94

EXAMPLE 6 In Figure 1.94 we see that there is a break in the graph of $y = f(x)$ at 2. If we let x approach 2 from, say, the left, the functional values $f(x)$ become larger and larger positive numbers. In other words, $f(x)$ is not approaching a finite number as $x \to 2^-$. Since $\lim_{x\to 2^-} f(x)$ does not exist, we conclude that $\lim_{x\to 2} f(x)$ does not exist. □

To determine whether $\lim_{x\to a} f(x)$ exists by graphing a function f is not always an easy task. In some instances numerical calculations can provide a convincing argument for the existence of a limit.

EXAMPLE 7 Based solely on the numerical data in the accompanying tables, we naturally conclude that

$$\lim_{x\to 0} \frac{\sin x}{x} = 1$$

$x \to 0^-$	$\frac{\sin x}{x}$
-0.1	0.9983341
-0.01	0.9999833
-0.001	0.9999998

$x \to 0^+$	$\frac{\sin x}{x}$
0.1	0.9983341
0.01	0.9999833
0.001	0.9999998

□

EXAMPLE 8 Careful inspection of the two tables suggests that

$$\lim_{x \to 0} \frac{1 - \cos x}{x} = 0$$

$x \to 0^-$	$\frac{1 - \cos x}{x}$
-0.1	-0.0499583
-0.01	-0.0049999
-0.001	-0.0005001
-0.0001	-0.0000510

$x \to 0^+$	$\frac{1 - \cos x}{x}$
0.1	0.0499583
0.01	0.0049999
0.001	0.0005001
0.0001	0.0000510

□

The computer-generated graphs of $f(x) = (\sin x)/x$ and $g(x) = (1 - \cos x)/x$ in Figures 1.95 and 1.96, respectively, show clearly that when x is close to 0, the values of $f(x)$ are close to 1 and the values of $g(x)$ are close to 0. Note too that the graphs in Figures 1.95 and 1.96 give the false impression that the functions f and g are defined at $x = 0$; they are not.

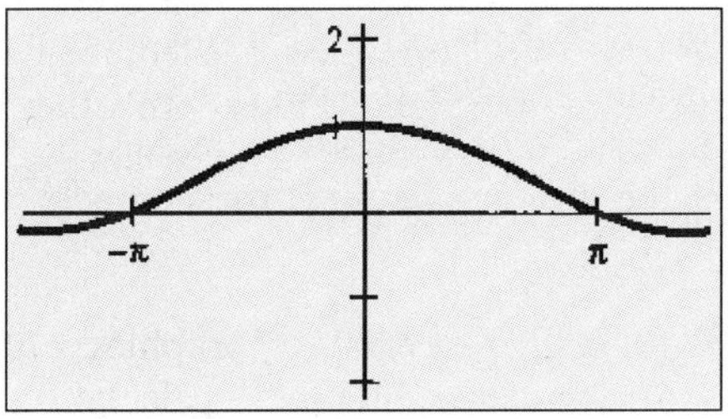

FIGURE 1.95

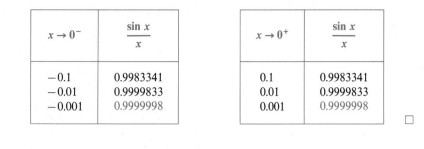

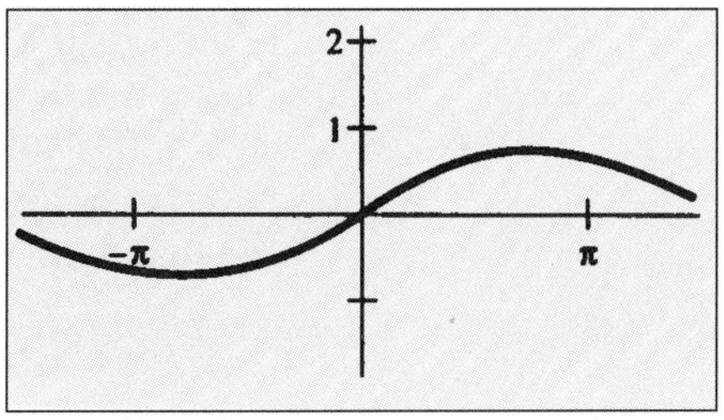

FIGURE 1.96

BASIC Program The tables in Examples 7 and 8 can be verified by means of a calculator or a computer. For those of you who have access to a computer, the following BASIC program may be useful in investigating the existence of $\lim_{x \to a} f(x)$.

```
10  REM LIMIT OF A FUNCTION
20  DEF FNY(X) = . . .
30  INPUT "THE VALUE OF A IS"; A
40  INPUT "THE NUMBER OF TERMS IS"; K
50  PRINT "A – H", "FNY(A – H)", "A + H", "FNY(A + H)"
60  FOR N = 1 TO K
70  LET H = 1/10^N
80  PRINT A – H, FNY(A – H), A + H, FNY(A + H)
90  NEXT N
100  END
```

To use this program you must supply A, define the function FNY(X), and specify K. In Example 8, A = 0, FNY(X) = (SIN (X))/X, and K = 3. A word of caution is in order here: For limits of the form $\lim_{x\to a} f(x)/g(x)$, where $f(x) \to 0$ as $x \to a$ *and* $g(x) \to 0$ as $x \to a$, the results from this program are heavily dependent on the computer used. In calculating numbers that are nearly 0/0, a computer may give a *division by zero* error message.*

[O] 1.5.2 ε–δ Definition of a Limit

In the discussion that follows we will consider an alternative approach to the notion of a limit that is based on analytical concepts rather than on intuitive concepts. While graphs and tables of functional values may be convincing for determining whether a limit does or does not exist, you are certainly aware that all calculators and computers work only with approximations and that graphs can be drawn inaccurately. A **proof** of the existence of a limit can never be based on one's ability to draw pictures. Although a good intuitive understanding of $\lim_{x\to a} f(x)$ is sufficient for proceeding with the study of the calculus in this text, an intuitive understanding is admittedly too vague to be of any use in proving theorems. To give a rigorous demonstration of the existence of a limit, or to prove the theorems of Section 1.6, we must start with the precise definition of a limit.

Let us try to prove that $\lim_{x\to 2}(2x + 6) = 10$ by elaborating on the following idea, "If $f(x) = 2x + 6$ can be made arbitrarily close to 10 by taking x sufficiently close to 2, from either side but different from 2, then $\lim_{x\to 2} f(x) = 10$." We need to make the concepts of "arbitrarily close" and "sufficiently close" precise. In order to set a standard of arbitrary closeness, let us demand that the distance between the numbers $f(x)$ and 10 be less than 0.1; that is,

$$|f(x) - 10| < 0.1 \quad \text{or} \quad 9.9 < f(x) < 10.1 \tag{1}$$

Then, how close must x be to 2 to accomplish (1)? To find out, we can use ordinary algebra to solve the inequality

$$9.9 < 2x + 6 < 10.1$$

*You might also try in line 70 of the program, H = 1/N^2 ($1/n^2$) or H = 1/2^N ($1/2^n$).

and discover that

$$1.95 < x < 2.05 \quad \text{or} \quad -0.05 < x - 2 < 0.05$$

Using absolute values and considering that $x \neq 2$, we can write the last inequality as $0 < |x - 2| < 0.05$. Thus, for an "arbitrary closeness to 10" of 0.1, "sufficiently close to 2" means within 0.05. In other words, if x is a number different from 2 such that its distance from 2 satisfies $|x - 2| < 0.05$, then the distance of $f(x)$ from 10 is guaranteed to satisfy $|f(x) - 10| < 0.1$. Expressed in yet another way, when x is a number different from 2 in the open interval (1.95, 2.05), then $f(x)$ is in the interval (9.9, 10.1).

Using the same example, let us try to generalize. Suppose ε (epsilon) denotes any *small positive number* that is our measure of arbitrary closeness to the number 10. If we demand that

$$|f(x) - 10| < \varepsilon \quad \text{or} \quad 10 - \varepsilon < f(x) < 10 + \varepsilon \tag{2}$$

then from $f(x) = 2x + 6$ and algebra, we find

$$2 - \frac{\varepsilon}{2} < x < 2 + \frac{\varepsilon}{2} \quad \text{or} \quad -\frac{\varepsilon}{2} < x - 2 < \frac{\varepsilon}{2} \tag{3}$$

Again using absolute values and remembering that $x \neq 2$, we can write the last inequality as

$$0 < |x - 2| < \frac{\varepsilon}{2} \tag{4}$$

If we denote $\varepsilon/2$ by the new symbol δ (delta), (2) and (4) can be written as

$$|f(x) - 10| < \varepsilon \quad \text{whenever} \quad 0 < |x - 2| < \delta$$

Thus, for a new value for ε, say $\varepsilon = 0.001$, $\delta = \varepsilon/2 = 0.0005$ tells us the corresponding closeness to 2. For any number x different from 2 in (1.9995, 2.0005),* we can be sure $f(x)$ is in (9.999, 10.001). See Figure 1.97.

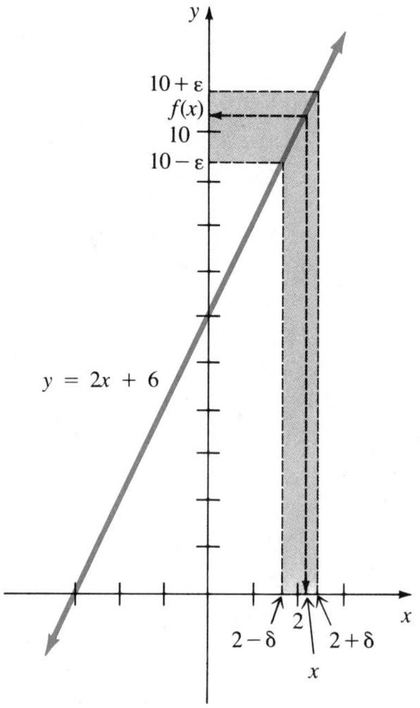

FIGURE 1.97

Definition of a Limit The foregoing discussion leads us to the so-called ε–δ definition of a limit.

> **DEFINITION 1.2 ε–δ Definition of a Limit**
>
> Suppose a function f is defined everywhere on an open interval, except possibly at a point a in the interval. Then
>
> $$\lim_{x \to a} f(x) = L$$
>
> means for every $\varepsilon > 0$ there exists a $\delta > 0$ such that
>
> $$|f(x) - L| < \varepsilon \quad \text{whenever} \quad 0 < |x - a| < \delta$$

*For this reason, we use $0 < |x - 2| < \delta$ rather than $|x - 2| < \delta$. Keep in mind when considering $\lim_{x \to 2} f(x)$, we do not care about f *at* 2.

Let $\lim_{x \to a} f(x) = L$ and suppose $\delta > 0$ is the number that "works" in the sense of Definition 1.2 for a given $\varepsilon > 0$. As shown in Figure 1.98(a), every x in $(a - \delta, a + \delta)$, with the possible exception of a itself, will then have an image $f(x)$ in $(L - \varepsilon, L + \varepsilon)$. Furthermore, as in Figure 1.98(b), a choice $\delta_1 < \delta$ for the same ε also "works" in that every x not equal to a in $(a - \delta_1, a + \delta_1)$ gives $f(x)$ in $(L - \varepsilon, L + \varepsilon)$. However, Figure 1.98(c) shows that choosing a smaller ε_1, $0 < \varepsilon_1 < \varepsilon$, will demand finding a new value of δ.

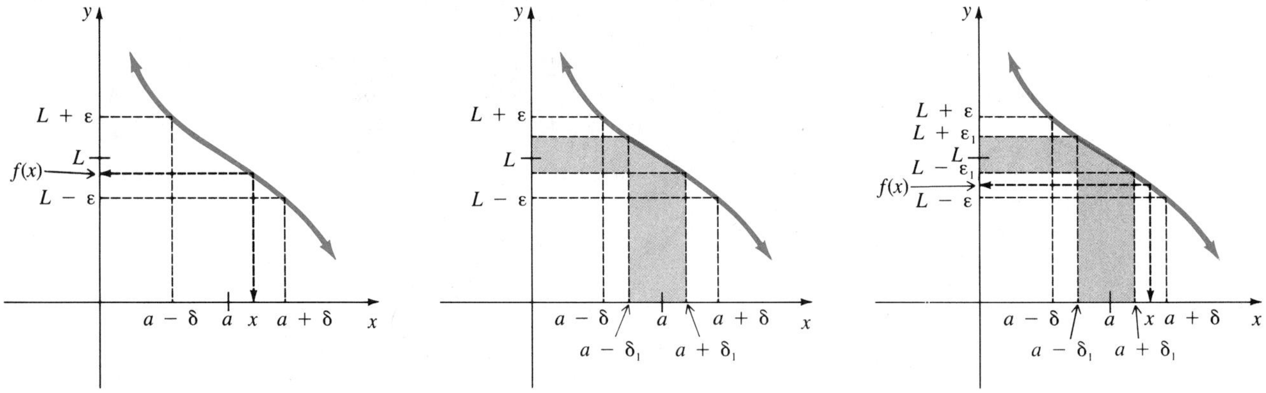

(a) A δ that works for a given ε.

(b) A smaller δ_1 will also work for the same ε.

(c) A smaller ε_1 will require a $\delta_1 < \delta$. For x in $(a - \delta, a + \delta)$, $f(x)$ is not necessarily in $(L - \varepsilon_1, L + \varepsilon_1)$.

FIGURE 1.98

EXAMPLE 9 Prove that $\lim_{x \to 3} (5x + 2) = 17$.

Solution For any given $\varepsilon > 0$, regardless how small, we wish to find a δ so that

$$|(5x + 2) - 17| < \varepsilon \quad \text{whenever} \quad 0 < |x - 3| < \delta$$

To do this consider

$$|(5x + 2) - 17| = |5x - 15| = 5|x - 3|$$

Thus, to make $|(5x + 2) - 17| = 5|x - 3| < \varepsilon$, we need only make $0 < |x - 3| < \varepsilon/5$; that is, choose $\delta = \varepsilon/5$.

Verification If $0 < |x - 3| < \varepsilon/5$, then $5|x - 3| < \varepsilon$ implies

$$|5x - 15| < \varepsilon \quad \text{or} \quad |(5x + 2) - 17| < \varepsilon$$ □

EXAMPLE 10 Prove that $\lim_{x \to -4} \dfrac{16 - x^2}{4 + x} = 8$.

Solution

$$\left|\frac{16 - x^2}{4 + x} - 8\right| = |4 - x - 8| = |-x - 4| = |x + 4| = |x - (-4)|$$

Thus,
$$\left|\frac{16 - x^2}{4 + x} - 8\right| = |x - (-4)| < \varepsilon$$

whenever we have $0 < |x - (-4)| < \varepsilon$; that is, choose $\delta = \varepsilon$. □

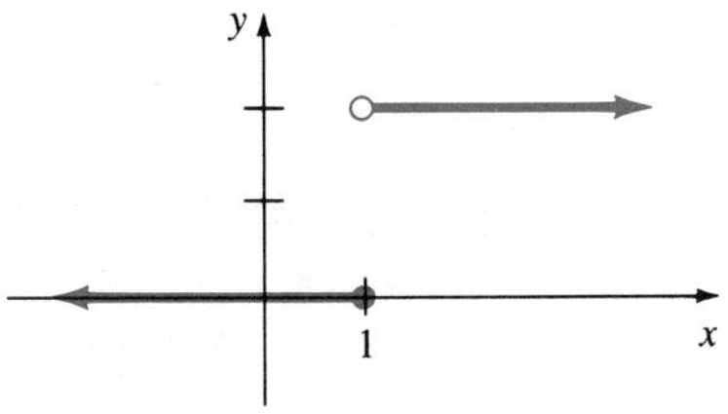

FIGURE 1.99

EXAMPLE 11 Consider the function

$$f(x) = \begin{cases} 0, & x \leq 1 \\ 2, & x > 1 \end{cases}$$

Intuitively, we can see from the graph of f in Figure 1.99 that $\lim_{x \to 1} f(x)$ does not exist. However, to *prove* this, we shall proceed indirectly.

Assume $\lim_{x \to 1} f(x) = L$. Then from Definition 1.2 we know that for the choice $\varepsilon = \frac{1}{2}$ there must exist a $\delta > 0$ so that

$$|f(x) - L| < \frac{1}{2} \quad \text{whenever} \quad 0 < |x - 1| < \delta$$

Now to the right of 1, let us choose $x = 1 + \delta/2$. Since

$$0 < \left|1 + \frac{\delta}{2} - 1\right| = \left|\frac{\delta}{2}\right| < \delta$$

we must have
$$\left|f\left(1 + \frac{\delta}{2}\right) - L\right| = |2 - L| < \frac{1}{2} \tag{5}$$

To the left of 1, choose $x = 1 - \delta/2$. But

$$0 < \left|1 - \frac{\delta}{2} - 1\right| = \left|-\frac{\delta}{2}\right| < \delta$$

implies
$$\left|f\left(1 - \frac{\delta}{2}\right) - L\right| = |0 - L| = |L| < \frac{1}{2} \tag{6}$$

From (5) and (6) we find, respectively,

$$\frac{3}{2} < L < \frac{5}{2} \quad \text{and} \quad -\frac{1}{2} < L < \frac{1}{2}$$

Since no such L can satisfy both of these inequalities, we conclude that $\lim_{x \to 1} f(x)$ does not exist. □

In the next example we reconsider the limit in Example 1. We shall see that finding the δ in this case requires a bit more ingenuity.

EXAMPLE 12 Prove that $\lim_{x \to 4}(-x^2 + 2x + 2) = -6$.

Solution For $\varepsilon > 0$ we must find a $\delta > 0$ so that

$$|-x^2 + 2x + 2 - (-6)| < \varepsilon \quad \text{whenever} \quad 0 < |x - 4| < \delta$$

Now

$$\begin{aligned} |-x^2 + 2x + 2 - (-6)| &= |(-1)(x^2 - 2x - 8)| \\ &= |(x + 2)(x - 4)| \\ &= |x + 2||x - 4| \end{aligned} \tag{7}$$

In other words, we want to make $|x + 2||x - 4| < \varepsilon$. But since we have agreed to examine values of x *near* 4, let us consider only those values for which $|x - 4| < 1$. This last inequality gives $3 < x < 5$ and $5 < x + 2 < 7$. Consequently we can write $|x + 2| < 7$. Hence from (7),

$$0 < |x - 4| < 1 \quad \text{implies} \quad |-x^2 + 2x + 2 - (-6)| < 7|x - 4|$$

If we now choose δ to be the minimum of the two numbers, 1 and $\varepsilon/7$, written $\delta = \min\{1, \varepsilon/7\}$, we have

$$0 < |x - 4| < \delta \quad \text{implies} \quad |-x^2 + 2x + 2 - (-6)| < 7|x - 4| < 7 \cdot \frac{\varepsilon}{7} = \varepsilon$$

□

The reasoning in Example 12 is subtle. It is worth a few minutes of your time to reread the discussion immediately following Definition 1.2, reexamine Figure 1.98(b), and then think again about why $\delta = \min\{1, \varepsilon/7\}$ is the δ that "works" in the example. Remember, you can pick the ε arbitrarily; think about δ for, say, $\varepsilon = 8$, $\varepsilon = 6$, and $\varepsilon = 0.01$.

One-Sided Limits In conclusion, we state the definitions of the **one-sided limits,** $\lim_{x \to a^-} f(x)$ and $\lim_{x \to a^+} f(x)$.

DEFINITION 1.3 Left-Hand Limit

Suppose a function f is defined on an open interval (b, a). Then

$$\lim_{x \to a^-} f(x) = L$$

means for every $\varepsilon > 0$ there exists a $\delta > 0$ such that

$$|f(x) - L| < \varepsilon \quad \text{whenever} \quad a - \delta < x < a$$

DEFINITION 1.4 Right-Hand Limit

Suppose a function f is defined on an open interval (a, c). Then

$$\lim_{x \to a^+} f(x) = L$$

means for every $\varepsilon > 0$ there exists a $\delta > 0$ such that

$$|f(x) - L| < \varepsilon \quad \text{whenever} \quad a < x < a + \delta$$

EXAMPLE 13 Prove $\lim_{x \to 0^+} \sqrt{x} = 0$.

Solution

$$|\sqrt{x} - 0| = |\sqrt{x}| = \sqrt{x}$$

Thus, $|\sqrt{x} - 0| < \varepsilon$ whenever $0 < x < 0 + \varepsilon^2$; that is, choose $\delta = \varepsilon^2$.

Verification If $0 < x < \varepsilon^2$, then $0 < \sqrt{x} < \varepsilon$ implies

$$|\sqrt{x}| < \varepsilon \quad \text{or} \quad |\sqrt{x} - 0| < \varepsilon$$

□

▲ ***Remark*** After this section you may agree with W. Whewell, who wrote in 1858 that "A limit is a peculiar . . . conception." For many years after the invention of calculus in the seventeenth century, mathematicians argued and debated the nature of a limit. There was an awareness that intuition, graphs, and numerical examples of ratios of vanishing quantities provide at best a shaky foundation for such a fundamental concept. As you will see beginning in the next chapter, the limit concept plays a central role in calculus. The study of calculus went through several periods of increased mathematical rigor beginning with the French mathematician Augustin-Louis Cauchy and continuing later with the German mathematician Karl Wilhelm Weierstrass.

Augustin-Louis Cauchy (1789–1857) Born during an era of upheaval in French history, Augustin-Louis Cauchy was destined to initiate a revolution in mathematics. For many contributions, but especially for his efforts in clarifying mathematical obscurities, his incessant demand for satisfactory definitions and rigorous proofs of theorems, Cauchy is often called "the father of modern analysis." A prolific writer whose output has been surpassed by only a few, Cauchy produced nearly 800 papers in astronomy, physics, and mathematics. But the same mind that was always open and inquiring in science and mathematics was also narrow and unquestioning in many other areas. Outspoken and arrogant, Cauchy's passionate stands on political and religious issues often alienated him from his colleagues.

Karl Wilhelm Weierstrass (1815–1897) One of the foremost mathematical analysts of the nineteenth century never earned an academic degree! After majoring in law at the University of Bonn, but concentrating in fencing and beer drinking for four years, Karl Wilhelm Weierstrass "graduated" to real life with no degree. In need of a job, Weierstrass passed a state examination and received a teaching certificate in 1841. During a period of 15 years as a secondary school teacher, his dormant mathematical genius blossomed. Although the quantity of his research publications was modest, especially when compared with that of Cauchy, the quality of these works so impressed the German mathematical community that he was awarded a doctorate, *honoris causa*, from the University of Königsberg and eventually was appointed a professor at the University of Berlin. While there, he achieved worldwide recognition both as a mathematician and as a teacher of mathematics. One of his students was Sonja Kowalewski, the greatest female mathematician of the nineteenth century. It was Karl Wilhelm Weierstrass who was responsible for putting the concept of a limit on a firm foundation with the ε–δ definition.

EXERCISES 1.5 *Answers to odd-numbered problems begin on page A-67.*

[1.5.1]

In Problems 1–14 use a graph to find the given limit, if it exists.

1. $\lim_{x\to 2}(3x+2)$

2. $\lim_{x\to -2}(x^2-1)$

3. $\lim_{x\to 0}\left(1+\frac{1}{x}\right)$

4. $\lim_{x\to 5}\sqrt{x-1}$

5. $\lim_{x\to 1}\frac{x^2-1}{x-1}$

6. $\lim_{x\to 0}\frac{x^2-3x}{x}$

7. $\lim_{x\to 0}\frac{|x|}{x}$

8. $\lim_{x\to 0}\frac{|x|-x}{x}$

9. $\lim_{x\to 0}\frac{x^3}{x}$

10. $\lim_{x\to 1}\frac{x^4-1}{x^2-1}$

11. $\lim_{x\to 0} f(x)$ where $f(x)=\begin{cases} x+3, & x<0 \\ -x+3, & x\ge 0\end{cases}$

12. $\lim_{x\to 2} f(x)$ where $f(x)=\begin{cases} x, & x<2 \\ x+1, & x\ge 2\end{cases}$

13. $\lim_{x\to 2} f(x)$ where $f(x)=\begin{cases} x^2-2x, & x<2 \\ 1, & x=2 \\ x^2-6x+8, & x>2\end{cases}$

14. $\lim_{x\to 0} f(x)$ where $f(x)=\begin{cases} x^2, & x<0 \\ 2, & x=0 \\ \sqrt{x}-1, & x>0\end{cases}$

In Problems 15–18 use the given graph to find each limit, if it exists.

(a) $\lim_{x\to 1^+} f(x)$ (b) $\lim_{x\to 1^-} f(x)$ (c) $\lim_{x\to 1} f(x)$

15.

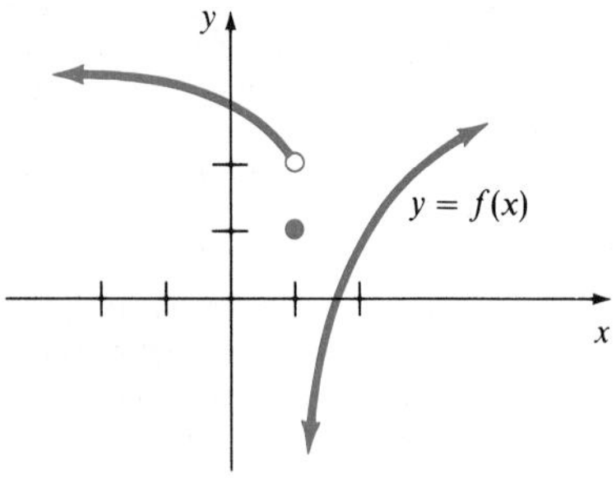

FIGURE 1.100

16.

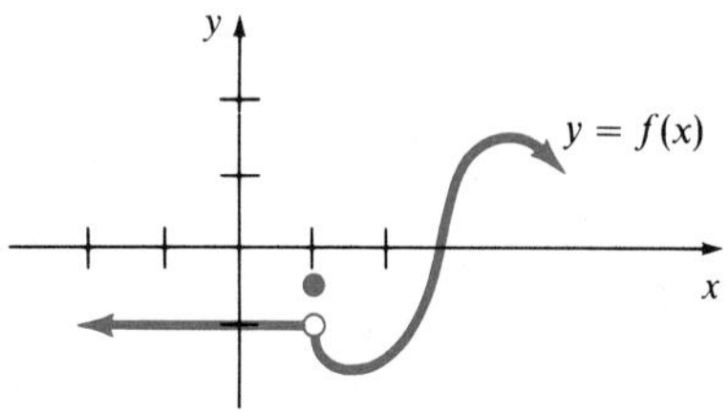

FIGURE 1.101

17.

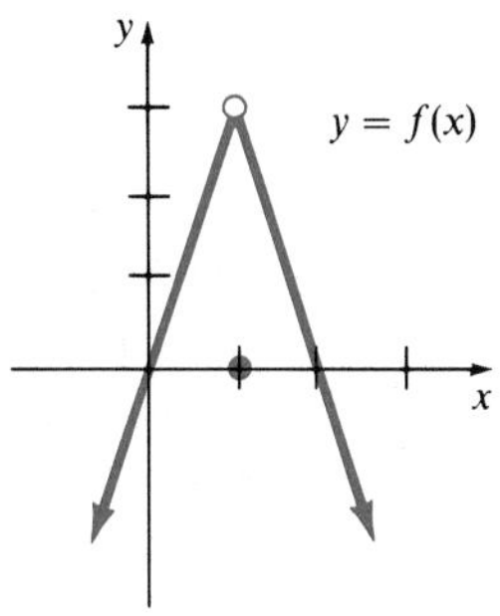

FIGURE 1.102

18.

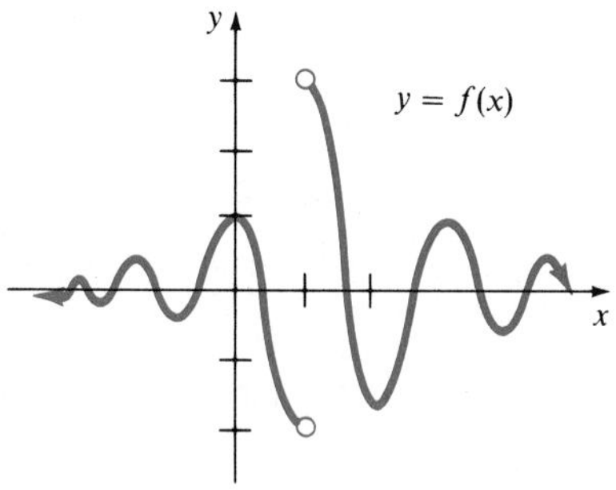

FIGURE 1.103

In Problems 19–22 use the given graph to find each limit, if it exists.

(a) $\lim_{x\to -2} f(x)$ (b) $\lim_{x\to 0} f(x)$

(c) $\lim_{x\to 1} f(x)$ (d) $\lim_{x\to 2} f(x)$

19.

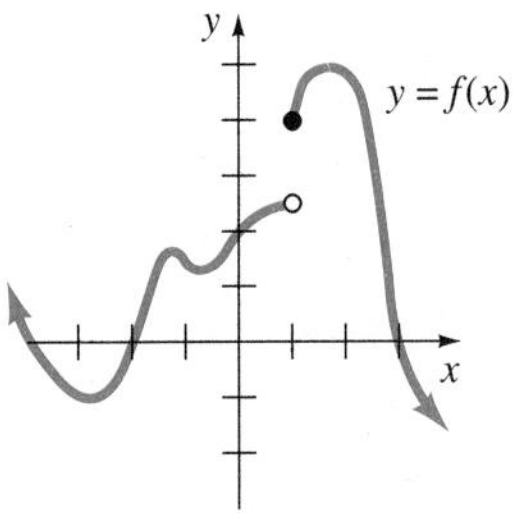

FIGURE 1.104

20.

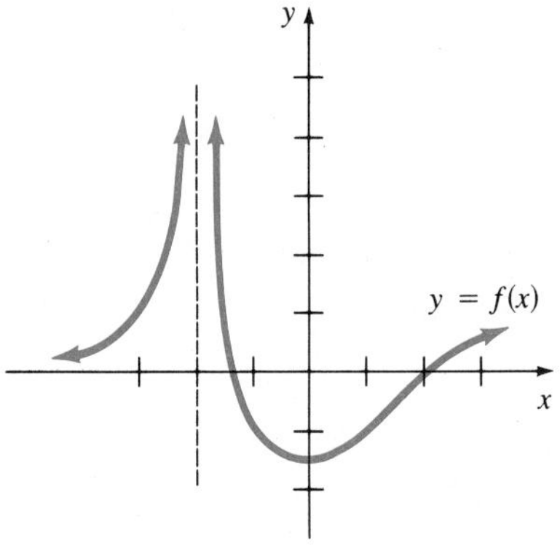

FIGURE 1.105

21.

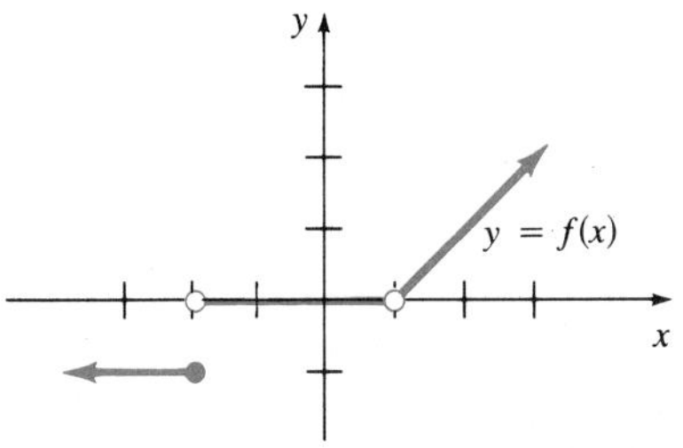

FIGURE 1.106

22.

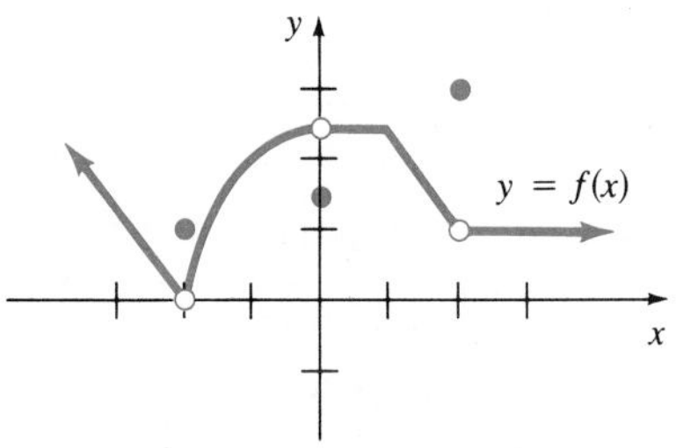

FIGURE 1.107

In Problems 23 and 24 use the graph of the given function to determine whether $\lim_{x\to 0} f(x)$ exists.

23. $f(x) = \sin \dfrac{1}{x}$

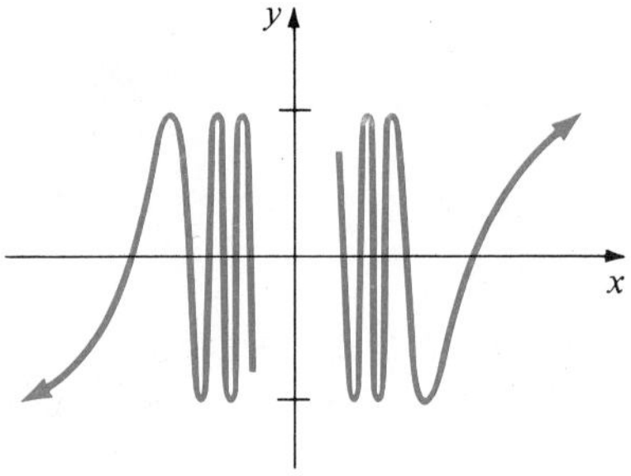

FIGURE 1.108

24. $f(x) = x \sin \dfrac{1}{x}$

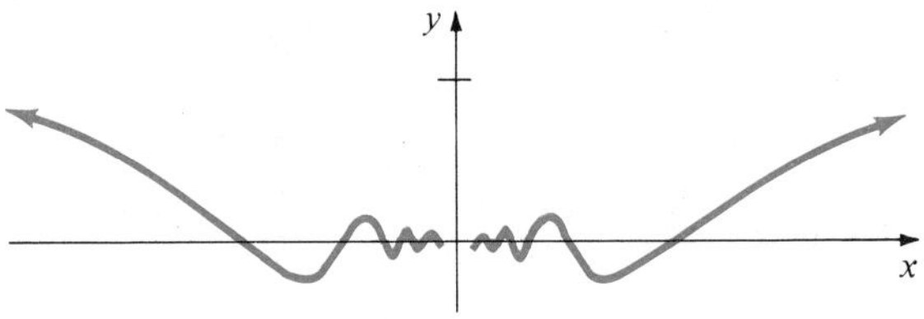

FIGURE 1.109

In Problems 25–28 use a graph to determine whether the given limit statement is correct. If incorrect, give a correct statement.

25. $\lim_{x\to 0} \sqrt{x} = 0$

26. $\lim_{x\to 1} \sqrt{1-x} = 0$

27. $\lim_{x\to 0} \sqrt[3]{x} = 0$

28. $\lim_{x\to 2} \sqrt{4-x^2} = 0$

Calculator/Computer Problems

In Problems 29–38 use a calculator or computer to investigate the given limit. Conjecture its value.

29. $\lim_{x\to 0} \dfrac{2-\sqrt{x+4}}{x}$

30. $\lim_{x\to 0} \dfrac{1-\cos x}{x^2}$

31. $\lim_{x\to 0} \dfrac{x}{\sin 3x}$

32. $\lim_{x\to 0} \dfrac{\tan x}{x}$

33. $\lim_{x\to 4} \dfrac{\sqrt{x}-2}{x-4}$

34. $\lim_{x\to 3}\left[\frac{6}{x^2-9}-\frac{6\sqrt{x-2}}{x^2-9}\right]$

35. $\lim_{x\to 1}\frac{x^4+x-2}{x-1}$

36. $\lim_{x\to -2}\frac{x^3+8}{x+2}$

37. $\lim_{x\to 0}\frac{|x|}{4-\sqrt{|x|+16}}$

38. $\lim_{x\to 2} f(x)$ where

$$f(x)=\begin{cases}\sqrt[4]{4x^7+5x^5-47}, & x<2\\ \dfrac{x^3+x^2-12}{x^4-2x^3+x^2-9x+14}, & x>2\end{cases}$$

[1.5.2]

In Problems 39–62 use Definition 1.2, 1.3, or 1.4 to prove the given limit result.

39. $\lim_{x\to 5} 10=10$

40. $\lim_{x\to -2} \pi=\pi$

41. $\lim_{x\to 3} x=3$

42. $\lim_{x\to 4} 2x=8$

43. $\lim_{x\to -1}(x+6)=5$

44. $\lim_{x\to 0}(x-4)=-4$

45. $\lim_{x\to 0}(3x+7)=7$

46. $\lim_{x\to 1}(9-6x)=3$

47. $\lim_{x\to 2}\frac{2x-3}{4}=\frac{1}{4}$

48. $\lim_{x\to 1/2} 8(2x+5)=48$

49. $\lim_{x\to -5}\frac{x^2-25}{x+5}=-10$

50. $\lim_{x\to 3}\frac{x^2-7x+12}{2x-6}=-\frac{1}{2}$

51. $\lim_{x\to 0}\frac{8x^5+12x^4}{x^4}=12$

52. $\lim_{x\to 1}\frac{2x^3+5x^2-2x-5}{x^2-1}=7$

53. $\lim_{x\to 0} x^2=0$

54. $\lim_{x\to 0} 8x^3=0$

55. $\lim_{x\to 0^+}\sqrt{5x}=0$

56. $\lim_{x\to (1/2)^+}\sqrt{2x-1}=0$

57. $\lim_{x\to 0^-} f(x)=-1,\ f(x)=\begin{cases}2x-1, & x<0\\ 2x+1, & x>0\end{cases}$

58. $\lim_{x\to 1^+} f(x)=3,\ f(x)=\begin{cases}0, & x\le 1\\ 3, & x>1\end{cases}$

59. $\lim_{x\to 3} x^2=9$

60. $\lim_{x\to 2}(2x^2+4)=12$

61. $\lim_{x\to 1}(x^2-2x+4)=3$

62. $\lim_{x\to 5}(x^2+2x)=35$

63. Prove that $\lim_{x\to a}\sqrt{x}=\sqrt{a},\ a>0$. [*Hint:* Use the identity

$$|\sqrt{x}-\sqrt{a}|=|\sqrt{x}-\sqrt{a}|\cdot\frac{\sqrt{x}+\sqrt{a}}{\sqrt{x}+\sqrt{a}}=\frac{|x-a|}{\sqrt{x}+\sqrt{a}}$$

and the fact that $\sqrt{x}\ge 0$.]

64. Prove that $\lim_{x\to 2} 1/x=\frac{1}{2}$. [*Hint:* Consider only those numbers x for which $1<x<3$.]

In Problems 65–68 prove that $\lim_{x\to a} f(x)$ does not exist.

65. $f(x)=\begin{cases}2, & x<1\\ 0, & x\ge 1\end{cases};\quad a=1$

66. $f(x)=\begin{cases}1, & x\le 3\\ -1, & x>3\end{cases};\quad a=3$

67. $f(x)=\begin{cases}x, & x\le 0\\ 2-x, & x>0\end{cases};\quad a=0$

68. $f(x)=\frac{1}{x};\quad a=0$

69. Prove that $\lim_{x\to 0} f(x)=0,\ f(x)=\begin{cases}x, & x \text{ rational}\\ 0, & x \text{ irrational}.\end{cases}$

1.6 THEOREMS ON LIMITS

The intention of the informal discussion in the first subsection of Section 1.5 was to give you an intuitive grasp of when a limit does or does not exist. However, it is neither desirable nor practical, in every instance, to reach a conclusion about the existence of a limit based on a graph or table of functional values. We must be able to evaluate a limit, or discern its nonexistence, in a somewhat mechanical fashion. The theorems that we shall consider in this section establish such a means. The proofs, using Definition 1.2, of some of these results are given in Appendix II.

THEOREM 1.1

(*i*) $\lim_{x \to a} c = c,$ where c is a constant

(*ii*) $\lim_{x \to a} x = a$

EXAMPLE 1 From Theorem 1.1(*i*),

(*a*) $\lim_{x \to 2} 10 = 10$ and (*b*) $\lim_{x \to 0} \pi = \pi$

From Theorem 1.1(*ii*),

(*c*) $\lim_{x \to 2} x = 2$ and (*d*) $\lim_{x \to 0} x = 0$ □

THEOREM 1.2

If c is a constant, then

$$\lim_{x \to a} cf(x) = c \lim_{x \to a} f(x)$$

EXAMPLE 2 From Theorems 1.1(*ii*) and 1.2,

(*a*) $\lim_{x \to 8} 5x = 5 \lim_{x \to 8} x = 5 \cdot 8 = 40$

(*b*) $\lim_{x \to -2} (-\tfrac{3}{2}x) = -\tfrac{3}{2} \lim_{x \to -2} x = (-\tfrac{3}{2}) \cdot (-2) = 3$ □

THEOREM 1.3 Existence Implies Uniqueness

If $\lim_{x \to a} f(x)$ exists, then it is unique.

THEOREM 1.4 Limit of a Sum, Product, and Quotient

If $\lim_{x \to a} f(x) = L_1$ and $\lim_{x \to a} g(x) = L_2$, then

(*i*) $\lim_{x \to a} [f(x) + g(x)] = \lim_{x \to a} f(x) + \lim_{x \to a} g(x) = L_1 + L_2,$

(*ii*) $\lim_{x \to a} f(x) \cdot g(x) = \lim_{x \to a} f(x) \cdot \lim_{x \to a} g(x) = L_1 L_2,$ and

(*iii*) $\lim_{x \to a} \dfrac{f(x)}{g(x)} = \dfrac{\lim_{x \to a} f(x)}{\lim_{x \to a} g(x)} = \dfrac{L_1}{L_2}, \quad L_2 \neq 0.$

In other words, Theorem 1.4 can be stated as: When the limits exist,

(*i*) *the limit of a sum is the sum of the limits,*

(*ii*) *the limit of a product is the product of the limits, and*

(*iii*) *the limit of a quotient is the quotient of the limits provided the limit of the denominator is not zero.*

Note: If the limits exist, then Theorem 1.4 is also applicable to one-sided limits. Moreover, Theorem 1.4 extends to differences as well as sums, products, and quotients that involve more than two functions.

EXAMPLE 3 Evaluate $\lim_{x \to 5}(10x + 7)$.

Solution From Theorems 1.1 and 1.2, we know $\lim_{x \to 5} 7$ and $\lim_{x \to 5} 10x$ exist. Hence, from Theorem 1.4(*i*),

$$\lim_{x \to 5}(10x + 7) = \lim_{x \to 5} 10x + \lim_{x \to 5} 7$$

$$= 10 \lim_{x \to 5} x + \lim_{x \to 5} 7 = 10 \cdot 5 + 7 = 57$$

□

Limit of a Power Theorem 1.4(*ii*) can be used to calculate the limit of a positive integral power of a function. For example, if $\lim_{x \to a} f(x) = L$, then

$$\lim_{x \to a}[f(x)]^2 = \lim_{x \to a} f(x) \cdot f(x) = \lim_{x \to a} f(x) \cdot \lim_{x \to a} f(x) = L^2$$

The next theorem states the general result.

> **THEOREM 1.5 Limit of a Power**
>
> Let $\lim_{x \to a} f(x) = L$ and n be a positive integer. Then
>
> $$\lim_{x \to a}[f(x)]^n = \left[\lim_{x \to a} f(x)\right]^n = L^n$$

For the special case $f(x) = x$, the result given in Theorem 1.5 yields

$$\lim_{x \to a} x^n = a^n \tag{1}$$

EXAMPLE 4 Evaluate $\lim_{x \to 10} x^3$.

Solution From (1), $\lim_{x \to 10} x^3 = 10^3 = 1000$ □

EXAMPLE 5 Evaluate $\lim_{x\to 3}(x^2 - 5x + 6)$.

Solution Since all limits exist,

$$\lim_{x\to 3}(x^2 - 5x + 6) = \lim_{x\to 3} x^2 - \lim_{x\to 3} 5x + \lim_{x\to 3} 6 = 3^2 - 5\cdot 3 + 6 = 0 \qquad \square$$

EXAMPLE 6 Evaluate $\lim_{x\to 1}(4x - 1)^5$.

Solution First, we see that

$$\lim_{x\to 1}(4x - 1) = \lim_{x\to 1} 4x - \lim_{x\to 1} 1 = 3$$

It then follows from Theorem 1.5 that

$$\lim_{x\to 1}(4x - 1)^5 = 3^5 = 243 \qquad \square$$

Limit of a Polynomial Function We can use (1) and Theorem 1.4(*i*) to compute the limit of a general polynomial function. If

$$f(x) = c_n x^n + c_{n-1}x^{n-1} + \cdots + c_1 x + c_0$$

is a polynomial function, then

$$\begin{aligned}\lim_{x\to a} f(x) &= \lim_{x\to a}(c_n x^n + c_{n-1}x^{n-1} + \cdots + c_1 x + c_0)\\ &= \lim_{x\to a} c_n x^n + \lim_{x\to a} c_{n-1}x^{n-1} + \cdots + \lim_{x\to a} c_1 x + \lim_{x\to a} c_0\\ &= c_n a^n + c_{n-1}a^{n-1} + \cdots + c_1 a + c_0\end{aligned}$$

In other words, to compute a limit of a polynomial function f as x approaches a real number a, we need only evaluate the function at a:

$$\lim_{x\to a} f(x) = f(a) \tag{2}$$

A reexamination of Example 5 shows that $\lim_{x\to 3} f(x)$, where $f(x) = x^2 - 5x + 6$, is given by $f(3) = 0$.

We can often use (2) in conjunction with Theorem 1.4(*iii*) to find a limit of a rational function.

EXAMPLE 7 Evaluate $\lim_{x\to -1}\dfrac{3x - 4}{6x + 2}$.

Solution From (2), $\lim_{x\to -1}(3x - 4) = -7$ and $\lim_{x\to -1}(6x + 2) = -4$. Since the limit of the denominator is not zero, we see from Theorem 1.4(*iii*)

that

$$\lim_{x\to-1}\frac{3x-4}{6x+2}=\frac{\lim_{x\to-1}(3x-4)}{\lim_{x\to-1}(6x+2)}=\frac{-7}{-4}=\frac{7}{4}$$ □

You should not get the impression that we can *always* find a limit of a function by substituting, or "plugging in," the real number a.

EXAMPLE 8 Evaluate $\lim_{x\to1}\dfrac{x-1}{x^2+x-2}$.

Solution The limit of this quotient cannot be written immediately as the quotient of limits because $\lim_{x\to1}(x^2+x-2)=0$. However, by simplifying *first*, we can then apply Theorem 1.4(*iii*):

$$\lim_{x\to1}\frac{x-1}{x^2+x-2}=\lim_{x\to1}\frac{x-1}{(x-1)(x+2)}$$

$$=\lim_{x\to1}\frac{1}{x+2}$$

Cancellation is valid for $x \neq 1$.

$$=\frac{\lim_{x\to1}1}{\lim_{x\to1}(x+2)}=\frac{1}{3}$$ □

Limit of a Root The limit of the nth root of a function is the nth root of the limit whenever the limit exists and has a real nth root.

THEOREM 1.6 Limit of a Root

Let $\lim_{x\to a}f(x)=L$ and n be a positive integer. If $L\geq0$ when n is any positive integer or if $L<0$ when n is an odd positive integer, then

$$\lim_{x\to a}\sqrt[n]{f(x)}=\sqrt[n]{\lim_{x\to a}f(x)}=\sqrt[n]{L}$$

EXAMPLE 9 Evaluate $\lim_{x\to9}\sqrt{x}$.

Solution Since $\lim_{x\to9}x=9>0$, we know from Theorem 1.6 that

$$\lim_{x\to9}\sqrt{x}=[\lim_{x\to9}x]^{1/2}=9^{1/2}=3$$ □

EXAMPLE 10 Evaluate $\lim_{x \to -8} \dfrac{x - \sqrt[3]{x}}{2x + 10}$.

Solution Since $\lim_{x \to -8}(2x + 10) = -6 \neq 0$, we see from Theorems 1.4(*iii*) and 1.6 that

$$\lim_{x \to -8} \frac{x - \sqrt[3]{x}}{2x + 10} = \frac{\lim_{x \to -8} x - [\lim_{x \to -8} x]^{1/3}}{\lim_{x \to -8} (2x + 10)} = \frac{-8 - (-8)^{1/3}}{-6} = \frac{-6}{-6} = 1 \quad \square$$

Sometimes you can tell at a glance when *a limit does not exist.*

THEOREM 1.7 A Limit That Doesn't Exist

If $\lim_{x \to a} f(x) = L_1 \neq 0$ and $\lim_{x \to a} g(x) = 0$, then $\lim_{x \to a} f(x)/g(x)$ does not exist.

Proof We shall give an indirect proof of this result based on Theorem 1.4. Suppose $\lim_{x \to a} f(x) = L_1 \neq 0$, $\lim_{x \to a} g(x) = 0$, and $\lim_{x \to a} f(x)/g(x)$ exists and equals L_2. Then

$$\begin{aligned} L_1 = \lim_{x \to a} f(x) &= \lim_{x \to a} g(x) \cdot \frac{f(x)}{g(x)} \qquad [g(x) \neq 0] \\ &= \lim_{x \to a} g(x) \cdot \lim_{x \to a} \frac{f(x)}{g(x)} = 0 \cdot L_2 = 0 \end{aligned}$$

By contradicting the assumption that $L_1 \neq 0$, we have proved the theorem. ■

EXAMPLE 11 From Theorem 1.7, we can see that

$$\lim_{x \to 5} \frac{4x + 7}{x^2 - 25}$$

does not exist because $\lim_{x \to 5}(4x + 7) = 27 \neq 0$ and $\lim_{x \to 5}(x^2 - 25) = 0$. □

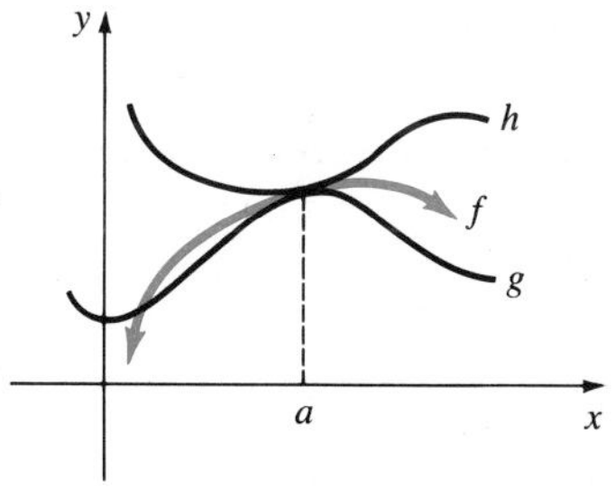

FIGURE 1.110

The last theorem of this section has various names, **Squeeze Theorem, Pinching Theorem, Sandwiching Theorem, Squeeze Play Theorem**, and even **Flyswatter Theorem**. As shown in Figure 1.110, if $f(x)$ is "squeezed" between $g(x)$ and $h(x)$ for all x close to a, and if we know that the functions g and h have a common limit L as $x \to a$, it stands to reason that f also approaches L as $x \to a$.

THEOREM 1.8 Squeeze Theorem

If f, g, and h are functions for which $g(x) \le f(x) \le h(x)$ for all x in an open interval that contains a, except possibly at a itself, and if $\lim_{x\to a} g(x) = \lim_{x\to a} h(x) = L$, then $\lim_{x\to a} f(x) = L$.

EXAMPLE 12 Evaluate $\displaystyle\lim_{x\to 0} x^2 \sin\frac{1}{x}$.

Solution For $x \ne 0$ we have $-1 \le \sin(1/x) \le 1$. Therefore,

$$-x^2 \le x^2 \sin\frac{1}{x} \le x^2$$

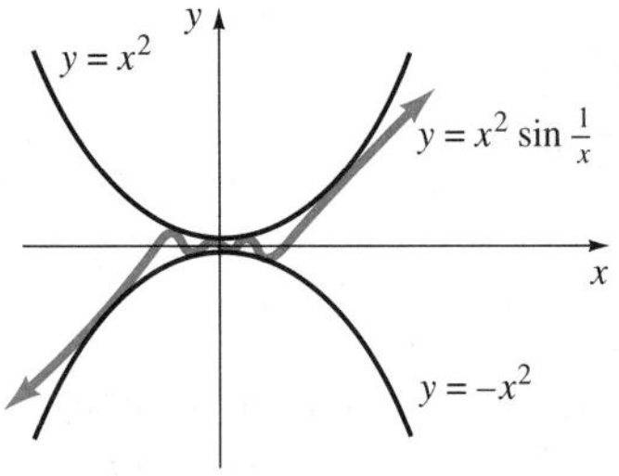

FIGURE 1.111

Now if we make the identifications $g(x) = -x^2$ and $h(x) = x^2$, it follows from (2) that $\lim_{x\to 0} g(x) = \lim_{x\to 0} h(x) = 0$. Thus, from the Squeeze Theorem we conclude that

$$\lim_{x\to 0} x^2 \sin\frac{1}{x} = 0$$

See Figure 1.111. □

▲ ***Remarks*** (*i*) In mathematics a student should be aware of what a definition or theorem does *not* say. Thus, from time to time, we shall conclude the section discussion with a few words of caution on the possibility of giving a definition or theorem an unwarranted extension or limitation.

(*ii*) Theorem 1.4(*i*) does not say that a limit of a sum is *always* the sum of the limits. The graph of $f(x) = 1/x$ in Figure 1.9(a) shows that $\lim_{x\to 0} 1/x$ does not exist. Therefore,

$$\lim_{x\to 0}\left[\frac{1}{x} - \frac{1}{x}\right] \ne \lim_{x\to 0}\frac{1}{x} - \lim_{x\to 0}\frac{1}{x}$$

Nonetheless, the limit of the difference exists:

$$\lim_{x\to 0}\left[\frac{1}{x} - \frac{1}{x}\right] = \lim_{x\to 0} 0 = 0$$

(*iii*) The limit of a product may exist and yet not be the product of the limits. For example,

$$\lim_{x\to 0} x \cdot \frac{1}{x} \ne \lim_{x\to 0} x \cdot \lim_{x\to 0}\frac{1}{x}$$

although

$$\lim_{x\to 0} x \cdot \frac{1}{x} = \lim_{x\to 0} 1 = 1$$

(*iv*) Theorem 1.7 does not say that the limit of a quotient fails to exist whenever the limit of the denominator is zero. Example 8 provides a counterexample to that interpretation. However, Theorem 1.7 states that a limit of a quotient does not exist whenever the limit of the denominator is zero *and* the limit of the numerator is not zero. Furthermore, to conclude that Theorem 1.7 provides the only way a limit can fail to exist would be a sweeping generalization.

EXERCISES 1.6 *Answers to odd-numbered problems begin on page A-67.*

In Problems 1–56 find the given limit, if it exists.

1. $\lim_{x\to -4} 15$
2. $\lim_{x\to 0} \cos\pi$
3. $\lim_{x\to 3}(-4)x$
4. $\lim_{x\to 2}(3x-9)$
5. $\lim_{x\to -2} x^2$
6. $\lim_{x\to 5}(-x^3)$
7. $\lim_{x\to -1}(x^3-4x+1)$
8. $\lim_{x\to 6}(-5x^2+6x+8)$
9. $\lim_{x\to 2}\dfrac{2x+4}{x-7}$
10. $\lim_{x\to 0}\dfrac{x+5}{3x}$
11. $\lim_{t\to 1}(3t-1)(5t^2+2)$
12. $\lim_{t\to -2}(t+4)^2$
13. $\lim_{s\to 7}\dfrac{s^2-21}{s+2}$
14. $\lim_{x\to 6}\dfrac{x^2-6x}{x^2-7x+6}$
15. $\lim_{x\to 6}\sqrt{2x-5}$
16. $\lim_{s\to 8}(1+\sqrt[3]{s})$
17. $\lim_{t\to 1}\dfrac{\sqrt{t}}{t^2+t-2}$
18. $\lim_{x\to 2}x^2\sqrt{x^2+5x+2}$
19. $\lim_{x\to 0}\left(x-\dfrac{1}{x-1}\right)$
20. $\lim_{t\to 0}\left(\dfrac{4}{t}-1\right)t$
21. $\lim_{y\to -5}\dfrac{y^2-25}{y+5}$
22. $\lim_{u\to 8}\dfrac{u^2-5u-24}{u-8}$
23. $\lim_{x\to 1}\dfrac{x^3-1}{x-1}$
24. $\lim_{t\to -1}\dfrac{t^3+1}{t^2-1}$
25. $\lim_{x\to 10}\dfrac{(x-2)(x+5)}{(x-8)}$
26. $\lim_{x\to -3}\dfrac{2x+6}{4x^2-36}$
27. $\lim_{x\to 2}\dfrac{x^3+3x^2-10x}{x-2}$
28. $\lim_{x\to 1.5}\dfrac{2x^2+3x-9}{x-1.5}$
29. $\lim_{t\to 1}\dfrac{t^3-2t+1}{t^3+t^2-2}$
30. $\lim_{x\to 0}x^3(x^4+2x^3)^{-1}$
31. $\lim_{x\to 0^+}\dfrac{(x+2)(x^5-1)^3}{(\sqrt{x}+4)^2}$
32. $\lim_{x\to -2}x\sqrt{x+4}\,\sqrt[3]{x-6}$
33. $\lim_{x\to 0}\left[\dfrac{x^2+3x-1}{x}+\dfrac{1}{x}\right]$
34. $\lim_{x\to 2}\left[\dfrac{1}{x-2}-\dfrac{6}{x^2+2x-8}\right]$
35. $\lim_{x\to 3^+}\dfrac{(x+3)^2}{\sqrt{x-3}}$
36. $\lim_{x\to 3}(x-4)^{99}(x^2-7)^{10}$
37. $\lim_{x\to 10}\sqrt{\dfrac{10x}{2x+5}}$
38. $\lim_{r\to 1}\dfrac{\sqrt{(r^2+3r-2)^3}}{\sqrt[3]{(5r-3)^2}}$
39. $\lim_{h\to 4}\sqrt{\dfrac{h}{h+5}}\left(\dfrac{h^2-16}{h-4}\right)^2$
40. $\lim_{t\to 2}(t+2)^{3/2}(2t+4)^{1/3}$
41. $\lim_{t\to 1^+}\dfrac{3t}{-1+\sqrt{t}}$
42. $\lim_{s\to 8^-}\dfrac{16-s^{4/3}}{4-s^{2/3}}$
43. $\lim_{x\to 0^-}\sqrt[5]{\dfrac{x^3-64x}{x^2+2x}}$
44. $\lim_{x\to -1^+}\left(8x+\dfrac{2}{x}\right)^5$
45. $\lim_{t\to 1}(at^2-bt)^2$
46. $\lim_{x\to -1}\sqrt{u^2x^2+2xu+1}$
47. $\lim_{h\to 0}\dfrac{(x+h)^2-x^2}{h}$
48. $\lim_{h\to 0}\dfrac{1}{h}[(x+h)^3-x^3]$
49. $\lim_{h\to 0}\dfrac{1}{h}\left(\dfrac{1}{x+h}-\dfrac{1}{x}\right)$
50. $\lim_{h\to 0}\dfrac{\sqrt{x+h}-\sqrt{x}}{h}\quad(x>0)$
51. $\lim_{t\to 1}\dfrac{\sqrt{t}-1}{t-1}$
52. $\lim_{u\to 5}\dfrac{\sqrt{u+4}-3}{u-5}$
53. $\lim_{v\to 0}\dfrac{\sqrt{25+v}-5}{\sqrt{1+v}-1}$
54. $\lim_{x\to 1}\dfrac{4-\sqrt{x+15}}{x^2-1}$
55. $\lim_{x\to 4}f(x)$ where $f(x)=\begin{cases}\sqrt{3x^2+16}, & x<4\\ \sqrt[3]{4x^2-9x^3}, & x>4\end{cases}$

56. $\lim_{x \to 3} f(x)$ where $f(x) = \begin{cases} (x^2 - 4x + 1)^5, & x < 3 \\ -x^3 + x^2 - 14, & x > 3 \end{cases}$

In Problems 57 and 58 find $\lim_{h \to 0} \dfrac{f(2+h) - f(2)}{h}$ for the given function.

57. $f(x) = 3x^2 + 1$

58. $f(x) = x^2 - 2x + 7$

In Problems 59 and 60 use the Squeeze Theorem to establish the given limit.

59. $\lim_{x \to 0} x^2 \sin^2 \dfrac{1}{x} = 0$

60. $\lim_{x \to 0} x \sin \dfrac{1}{x} = 0$

In Problems 61 and 62 use the Squeeze Theorem to evaluate the given limit.

61. $\lim_{x \to 2} f(x)$ where $2x - 1 \le f(x) \le x^2 - 2x + 3, \quad x \ne 2$

62. $\lim_{x \to 0} f(x)$ where $|f(x) - 1| \le x^2, \quad x \ne 0$

63. If $|f(x)| \le B$ for all x, show that $\lim_{x \to 0} x^2 f(x) = 0$.

1.7 LIMITS THAT INVOLVE INFINITY

1.7.1 Intuitive Interpretation

In Sections 1.2 and 1.3 we considered some functions whose graphs possessed asymptotes. As we shall see, the concepts of vertical and horizontal asymptotes of a graph are defined in terms of limits.

Infinite Limits The limit of a function f *will fail to exist* as x approaches a point a whenever the functional values increase or decrease without bound. In Figure 1.112(a) the fact that the functional values $f(x)$ increase without bound as x approaches a is denoted by

$$f(x) \to \infty \quad \text{as} \quad x \to a \quad \text{or} \quad \lim_{x \to a} f(x) = \infty$$

In Figure 1.112(b) the functional values decrease without bound as x approaches a, and so we write

$$f(x) \to -\infty \quad \text{as} \quad x \to a \quad \text{or} \quad \lim_{x \to a} f(x) = -\infty$$

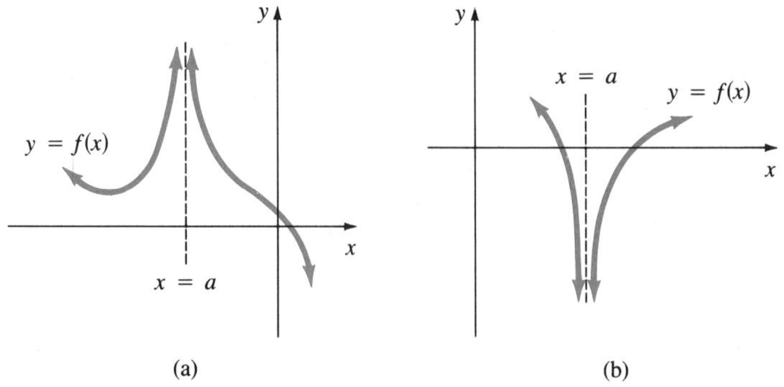

FIGURE 1.112

Similarly, Figure 1.113 shows the unbounded behavior of a function as x approaches a from one side.

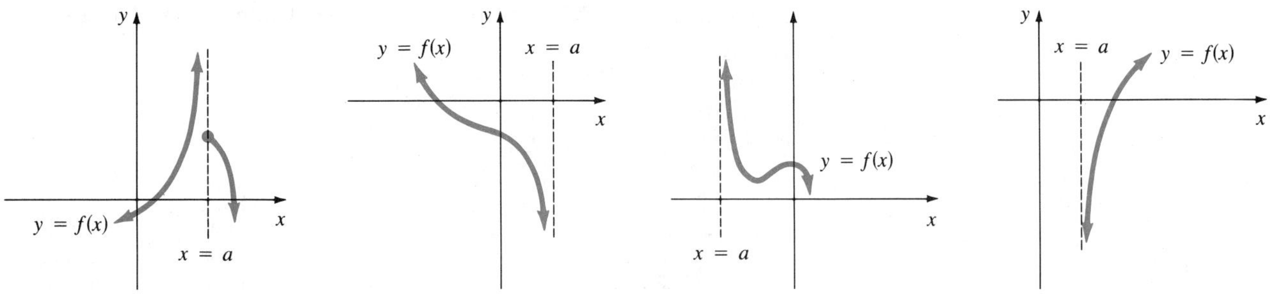

(a) $\lim_{x \to a^-} f(x) = \infty$ (b) $\lim_{x \to a^-} f(x) = -\infty$ (c) $\lim_{x \to a^+} f(x) = \infty$ (d) $\lim_{x \to a^+} f(x) = -\infty$

FIGURE 1.113

In general, any limit of the type

$$\lim_{x\to a} f(x) = \infty, \quad \lim_{x\to a^+} f(x) = \infty, \quad \lim_{x\to a^-} f(x) = \infty, \tag{1}$$
$$\lim_{x\to a} f(x) = -\infty, \quad \lim_{x\to a^+} f(x) = -\infty, \quad \lim_{x\to a^-} f(x) = -\infty,$$

is called an **infinite limit**.

Vertical Asymptotes If any *one* of the conditions in (1) hold, then the line $x = a$ is a **vertical asymptote** for the graph of f.

EXAMPLE 1 From the accompanying tables and the graph in Figure 1.114 we see that the values of the function $f(x) = 2/(x - 1)$ decrease without bound as $x \to 1^-$ and increase without bound as $x \to 1^+$; that is,

$$\lim_{x\to 1^-} f(x) = -\infty \quad \text{and} \quad \lim_{x\to 1^+} f(x) = \infty$$

In the figure the dashed line $x = 1$ is a vertical asymptote.

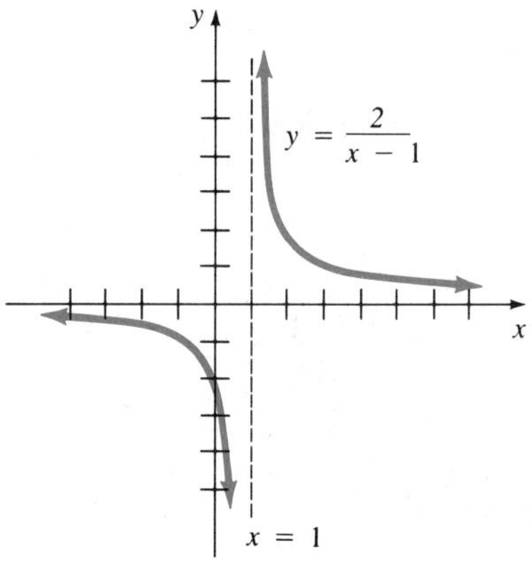

FIGURE 1.114

$x \to 1^-$	$f(x)$
0.9	−20
0.99	−200
0.999	−2000

$x \to 1^+$	$f(x)$
1.1	20
1.01	200
1.001	2000

□

EXAMPLE 2 Graph the function $f(x) = \dfrac{x}{\sqrt{x+2}}$.

Solution Inspection of f reveals that its domain is $(-2, \infty)$ and the y-intercept is 0. From the accompanying table we conclude that f decreases without bound as x approaches -2 from the right:

$$\lim_{x\to -2^+} f(x) = -\infty$$

Hence, the line $x = -2$ is a vertical asymptote. The graph of f is given in Figure 1.115.

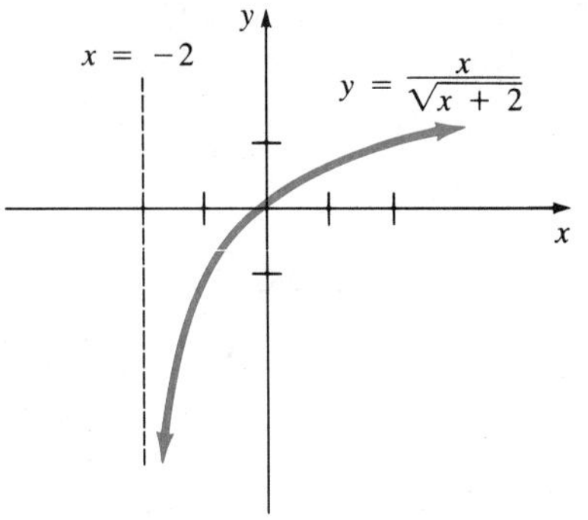

$x \to -2^+$	$f(x)$
−1.9	−6.01
−1.99	−19.90
−1.999	−63.21
−1.9999	−199.90

FIGURE 1.115

□

EXAMPLE 3 Since the domain of $f(x) = 1/x^2$ is the set of all real numbers except zero, the graph of f has no y-intercept. Also, $f(-x) = 1/(-x)^2 = 1/x^2 = f(x)$ implies that f is an even function, and so its graph is symmetric with respect to the y-axis. Finally, when $x \to 0^+$, the functional values $f(x)$ become larger and larger without bound. This fact, along with symmetry, implies

$$\lim_{x \to 0} \frac{1}{x^2} = \infty$$

Therefore the y-axis, that is, the line $x = 0$, is a vertical asymptote. The graph of f is given in Figure 1.116. □

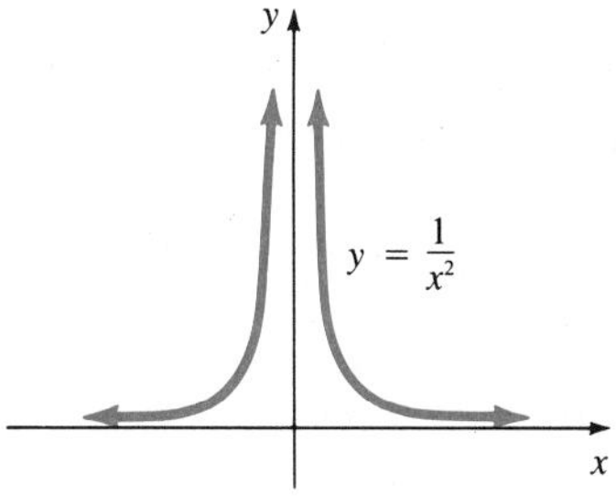

FIGURE 1.116

EXAMPLE 4 The graph of $f(x) = 1/(x-2)^3$ given in Figure 1.117 shows that

$$\lim_{x \to 2^-} \frac{1}{(x-2)^3} = -\infty \quad \text{and} \quad \lim_{x \to 2^+} \frac{1}{(x-2)^3} = \infty$$

The line $x = 2$ is a vertical asymptote.

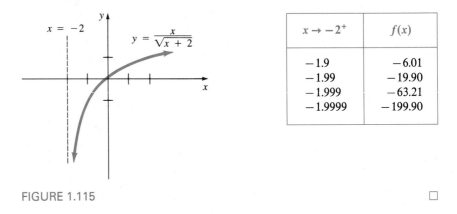

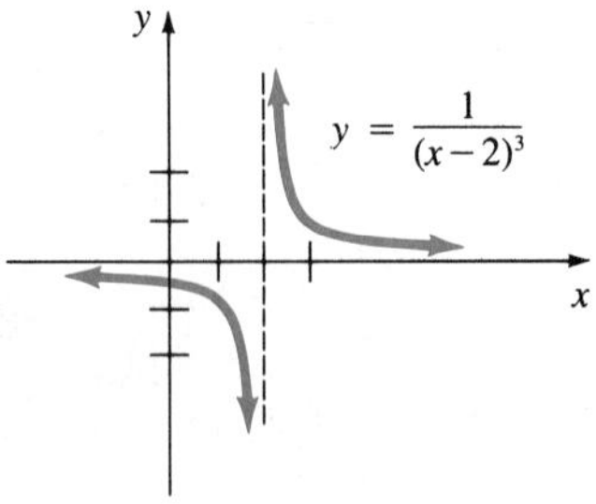

FIGURE 1.117

□

The preceding two examples illustrate the following theorem.

THEOREM 1.9

(*i*) If n is an even positive integer, then

$$\lim_{x\to a} \frac{1}{(x-a)^n} = \infty$$

(*ii*) If n is an odd positive integer, then

$$\lim_{x\to a^-} \frac{1}{(x-a)^n} = -\infty \quad \text{and} \quad \lim_{x\to a^+} \frac{1}{(x-a)^n} = \infty$$

Recall from Section 1.2 that we can find vertical asymptotes for a rational function by inspection. Suppose $f(x) = P(x)/Q(x)$, where P and Q are polynomial functions. If $Q(a) = 0$ and $P(a) \neq 0$, then $x = a$ is a vertical asymptote for the graph of f.

EXAMPLE 5 Determine vertical asymptotes for

$$f(x) = \frac{x^2 + x + 1}{(x^2 + 1)(x^2 - 6x + 8)}$$

Solution f is a rational function with $P(x) = x^2 + x + 1$ and

$$Q(x) = (x^2 + 1)(x^2 - 6x + 8) = (x^2 + 1)(x - 2)(x - 4)$$

Since the denominator $Q = 0$ for only 2 and 4 and since $P(2) \neq 0$ and $P(4) \neq 0$, it follows that the lines $x = 2$ and $x = 4$ are vertical asymptotes. □

EXAMPLE 6 The rational function $f(x) = (x^2 - 9)/(x - 3)$ does *not* have a vertical asymptote at $x = 3$. Why not? □

Limits at Infinity A function f might approach a constant value L as the independent variable x increases or decreases without bound. We write

$$\lim_{x\to\infty} f(x) = L \quad \text{or} \quad \lim_{x\to-\infty} f(x) = L$$

to denote a **limit at infinity**. Figure 1.118 shows four possibilities for the behavior of a function f as x becomes large in absolute value.

Horizontal Asymptotes If $f(x) \to L$ as either $x \to \infty$ or $x \to -\infty$, then we say that the line $y = L$ is a **horizontal asymptote** for the graph of f. In Figure 1.118(c) the function has two horizontal asymptotes, $y = L_1$ and $y = L_2$.

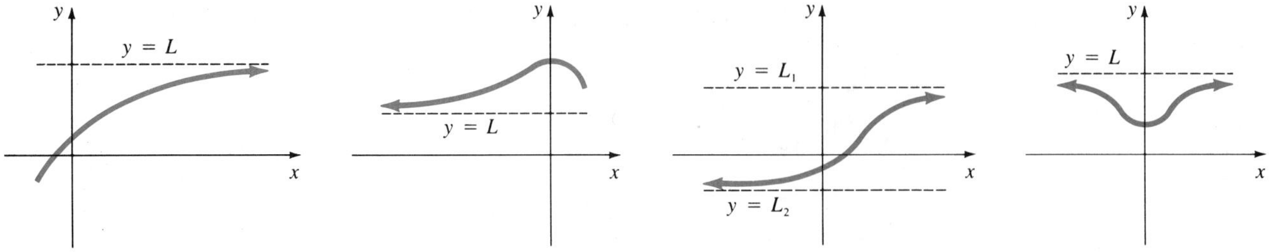

(a) $\lim_{x \to \infty} f(x) = L$

(b) $\lim_{x \to -\infty} f(x) = L$

(c) $\lim_{x \to \infty} f(x) = L_1$, $\lim_{x \to -\infty} f(x) = L_2$

(d) $\lim_{x \to \infty} f(x) = L$, $\lim_{x \to -\infty} f(x) = L$

FIGURE 1.118

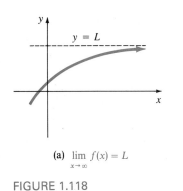

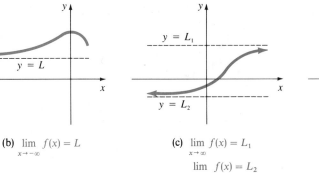

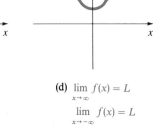

EXAMPLE 7 Reinspection of Figures 1.114 and 1.116 shows that

$$\lim_{x \to \pm\infty} \frac{2}{x-1} = 0 \quad \text{and} \quad \lim_{x \to \pm\infty} \frac{1}{x^2} = 0$$

Thus, $y = 0$ is a horizontal asymptote for the graphs of both $f(x) = 2/(x-1)$ and $f(x) = 1/x^2$. □

The next theorem is useful for evaluating limits at infinity.

> **THEOREM 1.10**
>
> Let t be a positive rational number. If x^t is defined, then
>
> $$\lim_{x \to \infty} \frac{1}{x^t} = 0 \quad \text{and} \quad \lim_{x \to -\infty} \frac{1}{x^t} = 0$$

EXAMPLE 8

(*a*) From Theorem 1.10 we have

$$\lim_{x \to \infty} \frac{1}{\sqrt{x}} = \lim_{x \to \infty} \frac{1}{x^{1/2}} = 0$$

Note that since $x^{1/2}$ is not defined for $x < 0$, it does not make sense to consider $\lim_{x \to -\infty} 1/x^{1/2}$.

(*b*) Since $\sqrt[3]{x} = x^{1/3}$ is defined for $x < 0$, it follows from Theorems 1.2 and 1.10 that

$$\lim_{x \to -\infty} \frac{6}{\sqrt[3]{x}} = 0$$

□

In general, if $F(x) = f(x)/g(x)$, then the following table summarizes the limit results for the forms $\lim_{x\to a} F(x)$, $\lim_{x\to\infty} F(x)$, and $\lim_{x\to-\infty} F(x)$. The symbol L denotes a finite number.

limit form: $x \to a, \infty$, *or* $-\infty$	$\dfrac{L}{\pm\infty}$	$\dfrac{\pm\infty}{L}$	$\dfrac{L}{0}, L \neq 0$
limit is	0	infinite	infinite

Furthermore, the limit results of Theorem 1.4 hold by replacing the symbol a by ∞ or $-\infty$ provided the limits exist. For example,

$$\lim_{x\to\infty} \frac{f(x)}{g(x)} = \frac{\lim_{x\to\infty} f(x)}{\lim_{x\to\infty} g(x)} \tag{2}$$

whenever $\lim_{x\to\infty} f(x)$ and $\lim_{x\to\infty} g(x)$ exist and $\lim_{x\to\infty} g(x) \neq 0$.

EXAMPLE 9 Evaluate $\displaystyle\lim_{x\to\infty} \frac{-6x^4 + x^2 + 1}{2x^4 - x}$.

Solution We cannot apply (2) to the function as given, since $\lim_{x\to\infty}(-6x^4 + x^2 + 1) = -\infty$ and $\lim_{x\to\infty}(2x^4 - x) = \infty$. However, by dividing the numerator and the denominator by x^4, we can write

$$\lim_{x\to\infty} \frac{-6x^4 + x^2 + 1}{2x^4 - x} = \lim_{x\to\infty} \frac{-6 + (1/x^2) + (1/x^4)}{2 - (1/x^3)}$$

$$= \frac{\lim_{x\to\infty} [-6 + (1/x^2) + (1/x^4)]}{\lim_{x\to\infty} [2 - (1/x^3)]} = \frac{-6 + 0 + 0}{2 - 0} = -3$$

This means the line $y = -3$ is a horizontal asymptote for the graph of the function. □

Example 9 illustrates a general procedure for determining the behavior of a rational function $f(x) = P(x)/Q(x)$ as $x \to \infty$ or $x \to -\infty$:

Divide the numerator and denominator by the highest power of x in the denominator.

EXAMPLE 10 Evaluate $\displaystyle\lim_{x\to\infty} \frac{1 - x^3}{3x + 2}$.

Solution By dividing the numerator and the denominator by x:

$$\frac{1 - x^3}{3x + 2} = \frac{(1/x) - x^2}{3 + (2/x)}$$

we see that the function has the limit form $-\infty/3$ as $x \to \infty$. Thus,

$$\lim_{x\to\infty} \frac{1 - x^3}{3x + 2} = -\infty$$

In other words, the limit does not exist. □

EXAMPLE 11 Graph the function $f(x) = \dfrac{x^2}{1 - x^2}$.

Solution Inspection of the function f reveals that its graph is symmetric with respect to the y-axis, the y-intercept is 0, and vertical asymptotes are $x = -1$ and $x = 1$. Now, by dividing numerator and denominator by x^2, we get

$$f(x) = \frac{x^2}{1 - x^2} = \frac{1}{(1/x^2) - 1}$$

and consequently

$$\lim_{x\to\infty} f(x) = \frac{\lim\limits_{x\to\infty} 1}{\lim\limits_{x\to\infty} [(1/x^2) - 1]} = \frac{1}{0 - 1} = -1$$

Hence, the line $y = -1$ is a horizontal asymptote. The graph of f is given in Figure 1.119. □

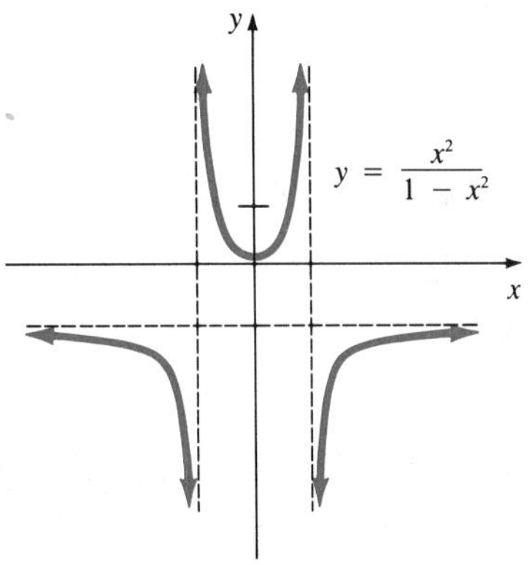

FIGURE 1.119

EXAMPLE 12 Determine whether the graph of $f(x) = \dfrac{5x}{\sqrt{x^2 + 4}}$ has any horizontal asymptotes.

Solution To investigate the limit of f as $x \to \infty$, we proceed as in Examples 9, 10, and 11. First, for $x \geq 0$, $\sqrt{x^2} = x$; therefore, the denominator of f can be written as

$$\sqrt{x^2 + 4} = \sqrt{x^2}\sqrt{1 + 4/x^2} = x\sqrt{1 + 4/x^2}$$

By dividing the numerator and denominator by x, we can apply Theorems 1.4(iii), 1.6, and 1.10:

$$\lim_{x\to\infty} \frac{5x}{\sqrt{x^2 + 4}} = \lim_{x\to\infty} \frac{5}{\sqrt{1 + 4/x^2}} = \frac{\lim\limits_{x\to\infty} 5}{\sqrt{\lim\limits_{x\to\infty} 1 + \lim\limits_{x\to\infty} (4/x^2)}} = \frac{5}{1} = 5$$

Now since $\sqrt{x^2} = |x|$, we have for $x < 0$, $\sqrt{x^2} = -x$ and so

$$\sqrt{x^2 + 4} = \sqrt{x^2}\sqrt{1 + 4/x^2} = -x\sqrt{1 + 4/x^2}$$

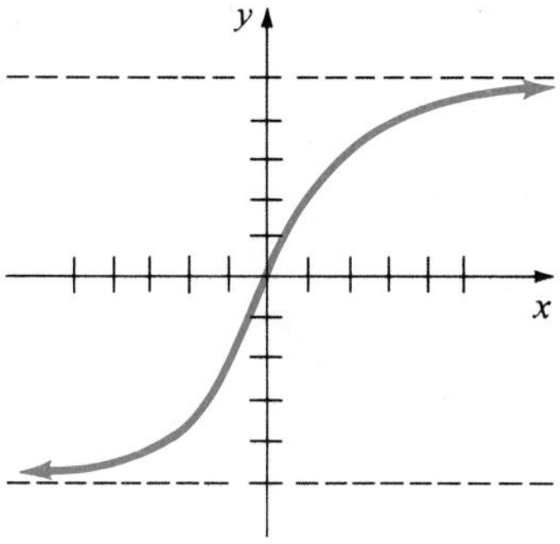

FIGURE 1.120

In this case, dividing the numerator and the denominator by $-x$ yields

$$\lim_{x\to-\infty}\frac{5x}{\sqrt{x^2+4}} = \lim_{x\to-\infty}\frac{-5}{\sqrt{1+4/x^2}} = \frac{-5}{1} = -5$$

Thus, the graph of f has the horizontal asymptotes $y = 5$ and $y = -5$. You should verify that the graph of f is that given in Figure 1.120. □

EXAMPLE 13 Alcohol is removed from the body by the lungs, by the kidneys, and by chemical processes in the liver. With alcohol at moderate concentration levels, the majority of the work of removing the alcohol is done by the liver; less than 5% of the alcohol is eliminated by the lungs and kidneys. The rate r at which the liver processes alcohol from the bloodstream is related to the blood alcohol concentration x by a rational function of the form

$$r(x) = \frac{\alpha x}{x+\beta}$$

for some positive constants α and β. This is a special case of the so-called **Michaelis–Menten Law**. Note that $r(0) = 0$ and

$$\lim_{x\to\infty}\frac{\alpha x}{x+\beta} = \alpha$$

Since the values of r increase as x increases, we can interpret α as the maximum possible rate of removal. A typical value of α for humans is 0.22 gram/liter/minute. □

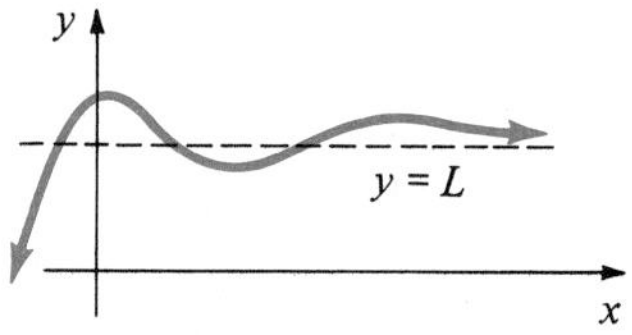

FIGURE 1.121

▲ ***Remarks*** *(i)* You may have heard an asymptote described as "a line that the graph of a function approaches but never crosses." If a function possesses a vertical asymptote, then its graph can never cross it. (Why?) However, Figure 1.121 shows that a graph may cross a horizontal asymptote many times.

(ii) The graph of a function can have many vertical asymptotes. You should think about and supply an answer to the question: How many horizontal asymptotes can the graph of a function have?

(iii) On one hand, statements such as $\lim_{x\to a} f(x) = \infty$ and $\lim_{x\to\infty} f(x) = \infty$ are unfortunate, since the use of the equality sign carries with it the connotation of existence. We reemphasize that in each case *the limits do not exist.* The symbol ∞ does not represent a number and should not be treated as a number. On the other hand, we retain this notation because it is a nice concise way to describe the *manner* in which the limit fails to exist. For example, from $\lim_{x\to a} f(x) = \infty$, one can form a mental picture of the functional values $f(x)$ increasing without bound whenever x is close to a.

[O] 1.7.2 Definitions of $\lim_{x\to a} f(x) = \infty$ and $\lim_{x\to\infty} f(x) = L$

The two concepts

$$f(x)\to\infty \text{ (or } -\infty) \text{ as } x\to a \quad\text{and}\quad f(x)\to L \text{ as } x\to\infty \text{ (or } -\infty)$$

are formalized in the next two definitions.

DEFINITION 1.5 $\lim_{x \to a} f(x) = \infty$ or $-\infty$

(*i*) $\lim_{x \to a} f(x) = \infty$ means for each $M > 0$, there exists a $\delta > 0$ such that $f(x) > M$ whenever $0 < |x - a| < \delta$.

(*ii*) $\lim_{x \to a} f(x) = -\infty$ means for each $M < 0$, there exists a $\delta > 0$ such that $f(x) < M$ whenever $0 < |x - a| < \delta$.

Parts (*i*) and (*ii*) of Definition 1.5 are illustrated in Figure 1.122(a) and 1.122(b), respectively.

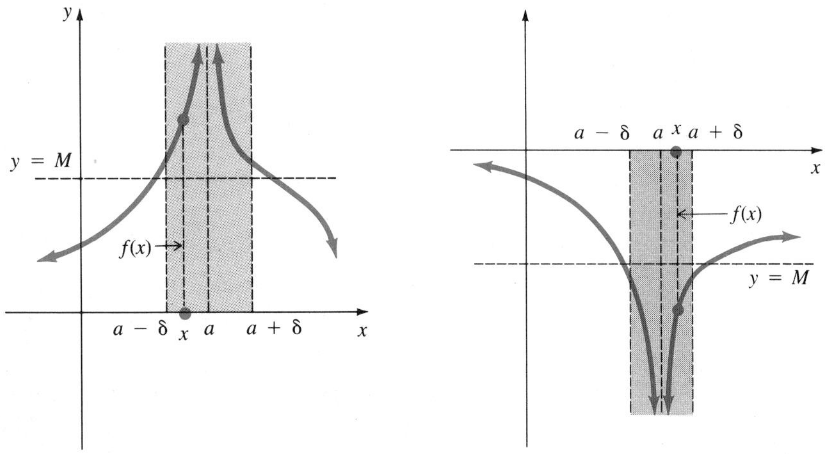

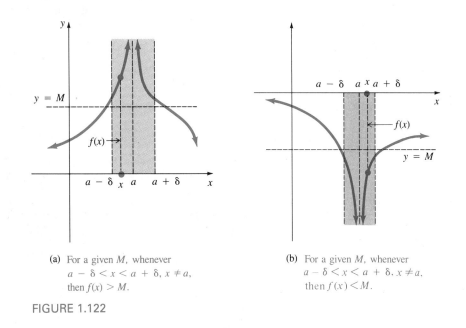

(a) For a given M, whenever $a - \delta < x < a + \delta$, $x \neq a$, then $f(x) > M$.

(b) For a given M, whenever $a - \delta < x < a + \delta$, $x \neq a$, then $f(x) < M$.

FIGURE 1.122

DEFINITION 1.6 $\lim_{x \to \infty \text{ or } x \to -\infty} f(x) = L$

(*i*) $\lim_{x \to \infty} f(x) = L$ if for each $\varepsilon > 0$ there exists an $N > 0$ such that $|f(x) - L| < \varepsilon$ whenever $x > N$.

(*ii*) $\lim_{x \to -\infty} f(x) = L$ if for each $\varepsilon > 0$ there exists an $N < 0$ such that $|f(x) - L| < \varepsilon$ whenever $x < N$.

Parts (*i*) and (*ii*) of Definition 1.6 are illustrated in Figure 1.123(a) and 1.123(b), respectively.

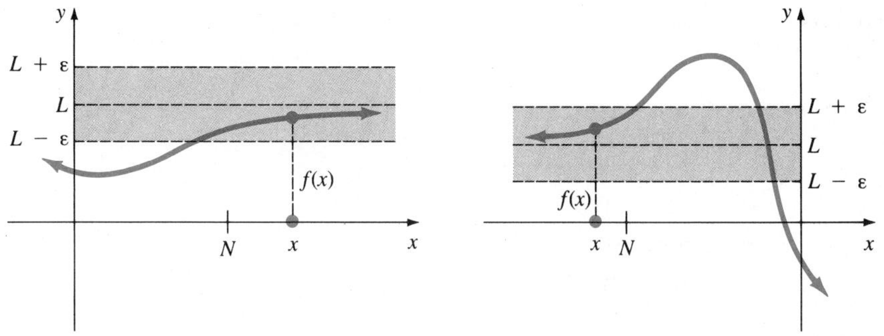

(a) For a given ε, $x > N$ implies $L - \varepsilon < f(x) < L + \varepsilon$.

(b) For a given ε, $x < N$ implies $L - \varepsilon < f(x) < L + \varepsilon$.

FIGURE 1.123

EXAMPLE 14 Prove that $\lim_{x \to \infty} \dfrac{3x}{x+1} = 3$.

Solution By Definition 1.6(i), for any $\varepsilon > 0$, we must find an $N > 0$ such that

$$\left| \frac{3x}{x+1} - 3 \right| < \varepsilon \quad \text{whenever} \quad x > N$$

Now, by considering $x > 0$, we have

$$\left| \frac{3x}{x+1} - 3 \right| = \left| \frac{-3}{x+1} \right| = \frac{3}{x+1} < \frac{3}{x} < \varepsilon$$

whenever $x > 3/\varepsilon$. Hence, choose $N = 3/\varepsilon$. For example, if $\varepsilon = 0.01$, then $N = 3/(0.01) = 300$ will guarantee that $|f(x) - 3| < 0.01$ whenever $x > 300$. □

EXERCISES 1.7

Answers to odd-numbered problems begin on page A-67.

[1.7.1]

In Problems 1–18 sketch the graph of the given function. Identify any vertical and horizontal asymptotes.

1. $f(x) = \dfrac{1}{x+3}$
2. $f(x) = \dfrac{3}{5-x}$
3. $f(x) = \dfrac{x}{x-2}$
4. $f(x) = \dfrac{-2x+1}{x+1}$
5. $f(x) = \dfrac{1}{(x-2)^2}$
6. $f(x) = \dfrac{1}{x^3}$
7. $f(x) = \dfrac{1}{x^2-4}$
8. $f(x) = \dfrac{1}{x^2-x-6}$
9. $f(x) = \dfrac{1}{x^2+1}$
10. $f(x) = \dfrac{x}{x^2+1}$
11. $f(x) = \dfrac{x^2}{x+1}$
12. $f(x) = \dfrac{x^2-x}{x^2-1}$
13. $f(x) = \dfrac{1}{x^2(x-2)}$
14. $f(x) = \dfrac{4x^2}{x^2+4}$
15. $f(x) = \sqrt{\dfrac{x}{x-1}}$
16. $f(x) = \dfrac{1-\sqrt{x}}{\sqrt{x}}$
17. $f(x) = \dfrac{x-2}{\sqrt{x^2+1}}$
18. $f(x) = \dfrac{x+3}{\sqrt{x^2-1}}$

In Problems 19 and 20 use the given graph to find the indicated quantities.

19.

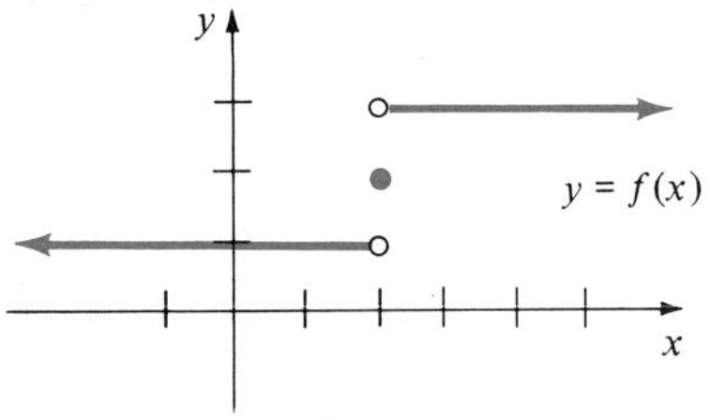

FIGURE 1.124

(a) $\lim_{x\to 2^-} f(x)$ (b) $\lim_{x\to 2^+} f(x)$ (c) $\lim_{x\to 2} f(x)$
(d) $f(2)$ (e) $\lim_{x\to -\infty} f(x)$ (f) $\lim_{x\to \infty} f(x)$

20.

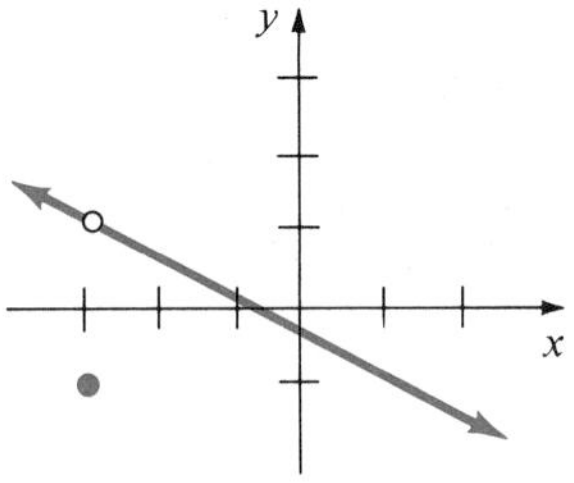

FIGURE 1.125

(a) $\lim_{x\to -3^-} f(x)$ (b) $\lim_{x\to -3^+} f(x)$ (c) $\lim_{x\to -3} f(x)$
(d) $f(-3)$ (e) $\lim_{x\to -\infty} f(x)$ (f) $\lim_{x\to \infty} f(x)$

In Problems 21–24 use the given graph to find:

(a) $\lim_{x\to 2^-} f(x)$ (b) $\lim_{x\to 2^+} f(x)$
(c) $\lim_{x\to -\infty} f(x)$ (d) $\lim_{x\to \infty} f(x)$

21.

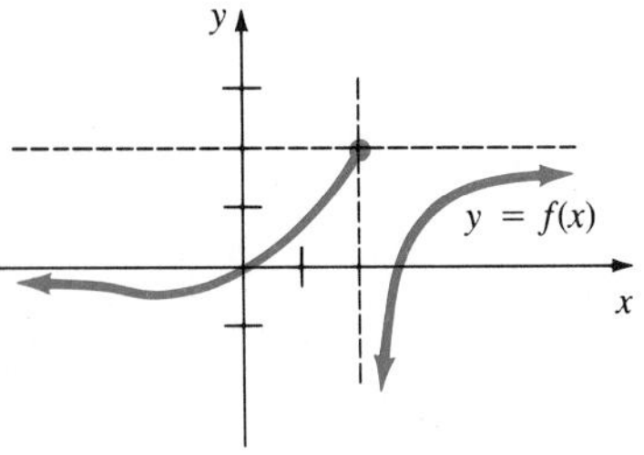

FIGURE 1.126

22.

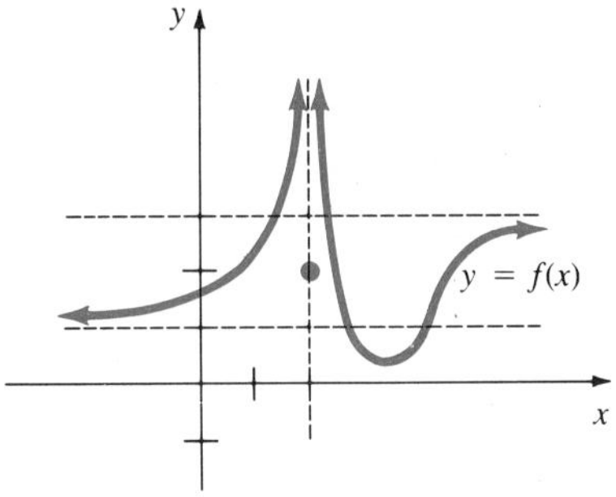

FIGURE 1.127

23.

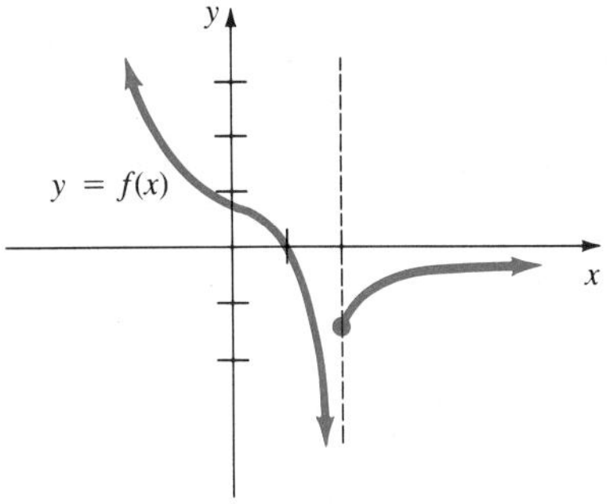

FIGURE 1.128

24.

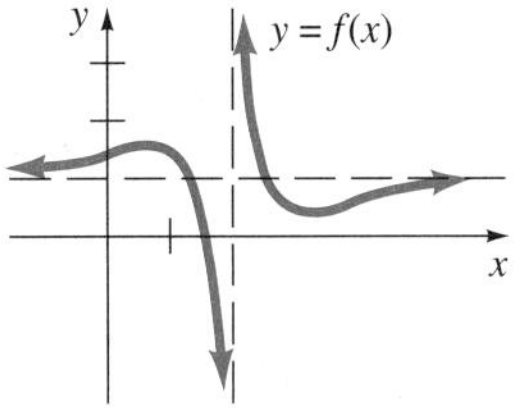

FIGURE 1.129

In Problems 25–52 find the given limit.

25. $\lim_{x\to 5^-} \dfrac{1}{x-5}$

26. $\lim_{x\to 6} \dfrac{4}{(x-6)^2}$

27. $\lim_{x\to -4^+} \dfrac{2}{(x+4)^3}$

28. $\lim_{x\to 2^-} \dfrac{10}{x^2-4}$

29. $\lim_{x\to 1} \dfrac{1}{(x-1)^4}$

30. $\lim_{x\to 0^+} \dfrac{-1}{\sqrt{x}}$

31. $\lim_{x\to 0^+} \dfrac{2+\sin x}{x}$

32. $\lim_{x\to \pi^+} \csc x$

33. $\lim_{x\to \infty} \dfrac{x^2-3x}{4x^2+5}$

34. $\lim_{x\to \infty} \dfrac{x^2}{1+x^{-2}}$

35. $\lim_{x\to\infty}\left(5-\dfrac{2}{x^4}\right)$

36. $\lim_{x\to-\infty}\left(\dfrac{6}{\sqrt[3]{x}}+\dfrac{1}{\sqrt[5]{x}}\right)$

37. $\lim_{x\to-\infty}\dfrac{2-1/x}{x^2+1}$

38. $\lim_{x\to-\infty}\dfrac{x-x^{-3}}{3x+x^{-2}}$

39. $\lim_{x\to\infty}\dfrac{8-\sqrt{x}}{1+4\sqrt{x}}$

40. $\lim_{x\to-\infty}\dfrac{1+7\sqrt[3]{x}}{2\sqrt[3]{x}}$

41. $\lim_{x\to\infty}\left(\dfrac{3x}{x+2}-\dfrac{x-1}{2x+6}\right)$

42. $\lim_{x\to\infty}\left(\dfrac{x}{3x+1}\right)\left(\dfrac{4x^2+1}{2x^2+x}\right)^3$

43. $\lim_{x\to\infty}\dfrac{4x+1}{\sqrt{x^2+1}}$

44. $\lim_{x\to\infty}\dfrac{\sqrt{9x^2+6}}{5x-1}$

45. $\lim_{x\to-\infty}\dfrac{2x+1}{\sqrt{3x^2+1}}$

46. $\lim_{x\to-\infty}\dfrac{-5x^2+6x+3}{\sqrt{x^4+x^2+1}}$

47. $\lim_{x\to\infty}\sqrt{\dfrac{3x+2}{6x-8}}$

48. $\lim_{x\to-\infty}\sqrt[3]{\dfrac{2x-1}{7-16x}}$

49. $\lim_{x\to\infty}(x-\sqrt{x^2+1})$

50. $\lim_{x\to\infty}(\sqrt{x^2+5x}-x)$

[*Hint:* Rationalize the numerator.]

51. $\lim_{x\to-\infty}\dfrac{|x-5|}{x-5}$

52. $\lim_{x\to\infty}\dfrac{|4x|+|x-1|}{x}$

In Problems 53–56 sketch a graph of a function f that satisfies the given conditions.

53. $\lim_{x\to1^+}f(x)=-\infty,\quad \lim_{x\to1^-}f(x)=-\infty,$
$f(2)=0,\quad \lim_{x\to\infty}f(x)=0$

54. $f(0)=1,\quad \lim_{x\to-\infty}f(x)=3,\quad \lim_{x\to\infty}f(x)=-2$

55. $\lim_{x\to2}f(x)=\infty,\quad \lim_{x\to-\infty}f(x)=\infty,\quad \lim_{x\to\infty}f(x)=1$

56. $\lim_{x\to1^-}f(x)=2,\quad \lim_{x\to1^+}f(x)=-\infty,\quad f(\tfrac{3}{2})=0,$
$f(3)=0,\quad \lim_{x\to-\infty}f(x)=0,\quad \lim_{x\to\infty}f(x)=0$

57. According to Einstein's theory of relativity, the mass m of a body moving with velocity v is $m=m_0/\sqrt{1-v^2/c^2}$, where m_0 is the initial mass and c is the speed of light. What happens to m as $v\to c^-$?

58. An important problem in fishery science is to estimate the number of fish presently spawning in streams and use this information to predict the number of mature fish or "recruits" that will return to the rivers during the next reproductive period. If S is the number of spawners and R the number of recruits, the **Beverton–Holt spawner recruit function** is $R(S)=S/(\alpha S+\beta)$, where α and β are positive constants. Show that this function predicts approximately constant recruitment when the number of spawners is sufficiently large.

In Problems 59–64 let $f(x)=P(x)/Q(x)$ be a rational function,

where $\quad P(x)=a_nx^n+a_{n-1}x^{n-1}+\cdots+a_0$

and $\quad Q(x)=b_mx^m+b_{m-1}x^{m-1}+\cdots+b_0$

Find $\lim_{x\to\infty}f(x)$.

59. Degree of $P<$ degree of Q

60. Degree of $P=$ degree of Q

61. Degree of $P>$ degree of Q

62. $P(x)=1,\quad$ degree of $Q\geq1$

63. $Q(x)=1,\quad$ degree of $P\geq1$

64. $Q(x)=xP(x)$

Calculator/Computer Problems

In Problems 65–68 use a calculator or computer to investigate the given limit. Conjecture its value.

65. $\lim_{x\to\infty}x^2\sin\dfrac{2}{x^2}$

66. $\lim_{x\to\infty}x\sin\dfrac{3}{x}$

67. $\lim_{x\to\infty}\left(\cos\dfrac{1}{x}\right)^x$

68. $\lim_{x\to(\pi/2)^-}\dfrac{\tan x}{\tan 5x}$

69. Use a calculator or computer to obtain the graph of $f(x)=(1+x)^{1/x}$. Use the graph to conjecture the values of $f(x)$ as (a) $x\to-1^+$, (b) $x\to0$, and (c) $x\to\infty$.

70. As shown in Figure 1.130 a regular n-sided polygon is inscribed in a circle of radius r.

(a) Using $2n$ congruent right triangles, show that the area of the polygon is

$$A_n=\frac{n}{2}r^2\sin\left(\frac{2\pi}{n}\right)$$

(b) Compute A_{100} and A_{1000}.

(c) Use the substitution $x=2\pi/n$ and the result given in Example 7 of Section 1.5 to find $\lim_{n\to\infty}A_n$.

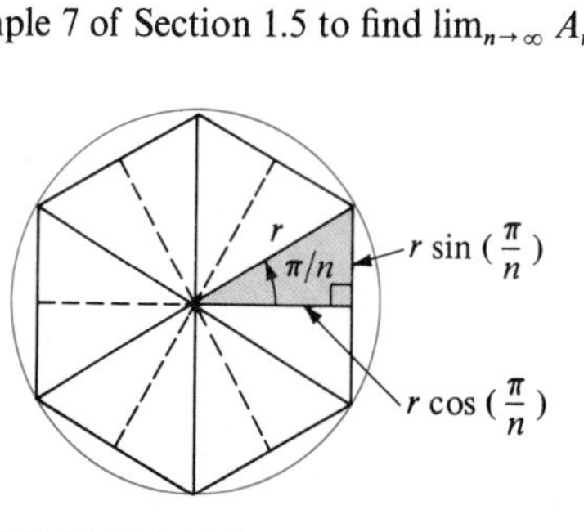

FIGURE 1.130

[1.7.2]

In Problems 71–74 use Definition 1.6 to prove the given limit result.

71. $\lim_{x\to\infty} \dfrac{5x-1}{2x+1} = \dfrac{5}{2}$

72. $\lim_{x\to\infty} \dfrac{2x}{3x+8} = \dfrac{2}{3}$

73. $\lim_{x\to-\infty} \dfrac{10x}{x-3} = 10$

74. $\lim_{x\to-\infty} \dfrac{x^2}{x^2+1} = 1$

1.8 CONTINUITY

In previous discussions about graphing, we used the phrase "connect the points with a smooth curve." This phrase invokes an image of a graph that is a nice *continuous* curve—that is, a curve with no gaps or breaks. Indeed, a **continuous function** is often described as one whose graph can be drawn without lifting pencil from paper.

Before stating the precise definition of continuity, we illustrate in Figure 1.131 some intuitive examples of graphs of functions that are not continuous, or are **discontinuous**, at a point a.

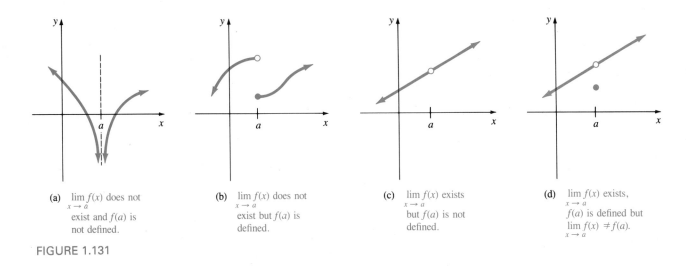

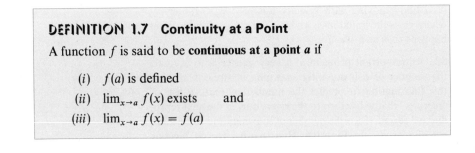

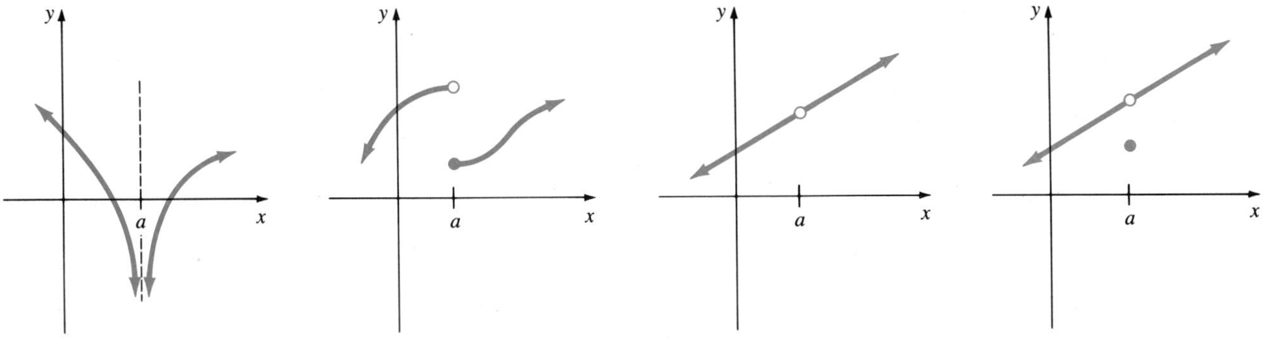

(a) $\lim_{x\to a} f(x)$ does not exist and $f(a)$ is not defined.

(b) $\lim_{x\to a} f(x)$ does not exist but $f(a)$ is defined.

(c) $\lim_{x\to a} f(x)$ exists but $f(a)$ is not defined.

(d) $\lim_{x\to a} f(x)$ exists, $f(a)$ is defined but $\lim_{x\to a} f(x) \neq f(a)$.

FIGURE 1.131

Continuity at a Point Figure 1.131 suggests the following threefold condition of continuity of a function f at a point a.

DEFINITION 1.7 Continuity at a Point

A function f is said to be **continuous at a point *a*** if

(*i*) $f(a)$ is defined

(*ii*) $\lim_{x\to a} f(x)$ exists and

(*iii*) $\lim_{x\to a} f(x) = f(a)$

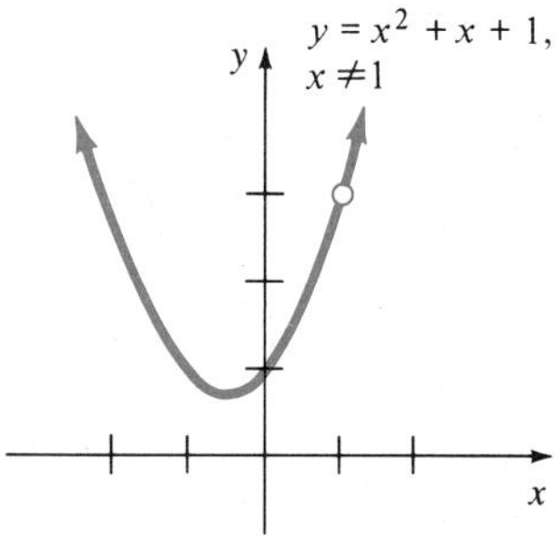

FIGURE 1.132

EXAMPLE 1 The rational function

$$f(x) = \frac{x^3 - 1}{x - 1} = x^2 + x + 1, \qquad x \neq 1$$

is discontinuous at 1, since $f(1)$ is not defined. This is analogous to the case illustrated in Figure 1.131(c), since we have $\lim_{x\to 1} f(x) = 3$. As can be seen in Figure 1.132, f is continuous at all points different from 1. □

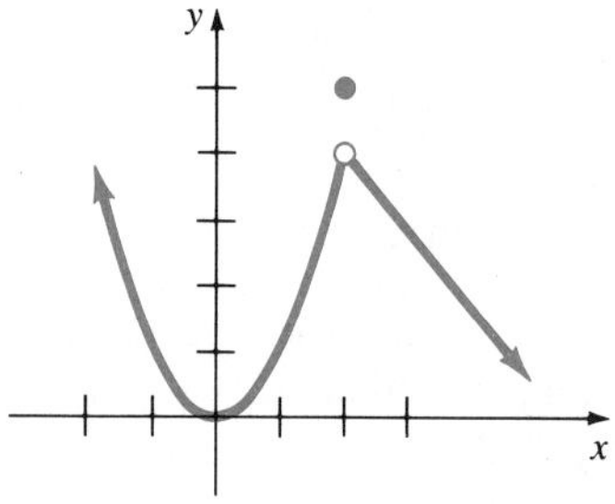

FIGURE 1.133

EXAMPLE 2 Figure 1.133 shows the graph of the piecewise-defined function

$$f(x) = \begin{cases} x^2, & x < 2 \\ 5, & x = 2 \\ -x + 6, & x > 2 \end{cases}$$

Now, $f(2)$ is defined and equals 5. Next

$$\left.\begin{aligned} \lim_{x\to 2^-} f(x) &= \lim_{x\to 2^-} x^2 = 4 \\ \lim_{x\to 2^+} f(x) &= \lim_{x\to 2^+} (-x + 6) = 4 \end{aligned}\right\} \text{ implies } \lim_{x\to 2} f(x) = 4$$

Since $\lim_{x\to 2} f(x) \neq f(2) = 5$, we see from (*iii*) of Definition 1.7 that f is discontinuous at 2. □

Continuity on an Interval We will now extend the notion of continuity at a point to **continuity on an interval**.

DEFINITION 1.8 Continuity on an Interval

A function f is continuous:

(*i*) on an **open interval** (a, b) if it is continuous at every point in the interval; and

(*ii*) on a **closed interval** $[a, b]$ if it is continuous on the open interval (a, b) and, in addition,

$$\lim_{x\to a^+} f(x) = f(a) \quad \text{and} \quad \lim_{x\to b^-} f(x) = f(b)$$

If the right-hand limit condition $\lim_{x\to a^+} f(x) = f(a)$ given in (*ii*) of Definition 1.8 is satisfied, we say that ***f* is continuous from the right at *a***; if $\lim_{x\to b^-} f(x) = f(b)$, then ***f* is continuous from the left at *b***.

Extensions of these concepts to intervals such as $[a, b)$, $(a, b]$, (a, ∞), $(-\infty, b)$, $(-\infty, \infty)$, $[a, \infty)$, and $(-\infty, b]$ are made in the expected manner. For example, f is continuous on $[a, b)$ if it is continuous on (a, b) and continuous from the right at a.

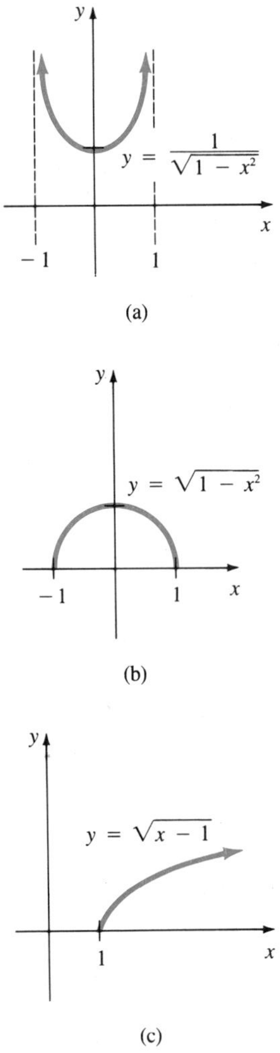

FIGURE 1.134

EXAMPLE 3

(*a*) As we see from Figure 1.134(a), $f(x) = 1/\sqrt{1-x^2}$ is continuous on the open interval $(-1, 1)$ but is not continuous on the closed interval $[-1, 1]$, since neither $f(-1)$ nor $f(1)$ is defined.

(*b*) $f(x) = \sqrt{1-x^2}$ is continuous on $[-1, 1]$. Observe from Figure 1.134(b) that

$$\lim_{x \to -1^+} f(x) = f(-1) = 0 \quad \text{and} \quad \lim_{x \to 1^-} f(x) = f(1) = 0$$

(*c*) $f(x) = \sqrt{x-1}$ is continuous on $[1, \infty)$, since

$$\lim_{x \to a} f(x) = \lim_{x \to a} \sqrt{x-1} = \sqrt{a-1} = f(a), \qquad a > 1$$

and

$$\lim_{x \to 1^+} \sqrt{x-1} = f(1) = 0$$

See Figure 1.134(c). □

A review of the graphs in Figure 1.56 indicates that the sine and cosine functions are continuous on $(-\infty, \infty)$. The tangent and secant functions are discontinuous at $x = (2n+1)\pi/2$, $n = 0, \pm 1, \pm 2, \ldots$. The cotangent and cosecant functions are discontinuous at $x = n\pi$, $n = 0, \pm 1, \pm 2, \ldots$.

Continuity of a Sum, Product, and Quotient

THEOREM 1.11 Continuity of a Sum, Product, and Quotient

If f and g are functions continuous at a, then cf (c a constant), $f + g$, fg, and f/g ($g(a) \neq 0$) are also continuous at a.

Proof: Continuity of the Product fg Since f and g are continuous at a number a, then

$$\lim_{x \to a} f(x) = f(a) \quad \text{and} \quad \lim_{x \to a} g(x) = g(a)$$

Hence, from Theorem 1.4(*ii*), we have

$$\lim_{x \to a} f(x)g(x) = (\lim_{x \to a} f(x)) \cdot (\lim_{x \to a} g(x)) = f(a)g(a)$$ ■

The proofs of the remaining parts of Theorem 1.11 are obtained in a similar manner.

Since Definition 1.7 implies that $f(x) = x$ is continuous at any x, we see from successive applications of Theorem 1.11 that the functions

$$x, x^2, x^3, \ldots, x^n$$

are also continuous $(-\infty, \infty)$. Thus, another application of Theorem 1.11 shows that

a polynomial function is continuous on $(-\infty, \infty)$.

Functions such as polynomials that are continuous on $(-\infty, \infty)$ are sometimes said to be **continuous everywhere** or simply **continuous**. Now, if P and Q are polynomial functions, it also follows directly from Theorem 1.11 that

a rational function $f(x) = P(x)/Q(x)$ is continuous except at points at which the denominator $Q(x)$ is zero.

Limits of a Composite Function The next theorem tells us that if a function f is continuous, then the limit of the function is the function of the limit.

THEOREM 1.12

If $\lim_{x\to a} g(x) = L$ and f is continuous at L, then

$$\lim_{x\to a} f(g(x)) = f(\lim_{x\to a} g(x)) = f(L)$$

Theorem 1.12 is useful in proving other theorems. If the function g is continuous at a and f is continuous at $g(a)$, then we see that

$$\lim_{x\to a} f(g(x)) = f(\lim_{x\to a} g(x)) = f(g(a))$$

In other words,

the composite function $f \circ g$ of two continuous functions f and g is continuous.

We leave it as an exercise for you to prove Theorem 1.5 from Theorem 1.12.

EXAMPLE 4 $f(x) = \sqrt{x}$ is continuous on $[0, \infty)$ and $g(x) = 2 + \sin x$ is continuous on $(-\infty, \infty)$. But, since $g(x) \geq 1 > 0$ for all x, the composite function

$$(f \circ g)(x) = f(g(x)) = \sqrt{2 + \sin x}$$

is continuous everywhere. □

Figure 1.135 illustrates the plausibility of the next result about continuous functions.

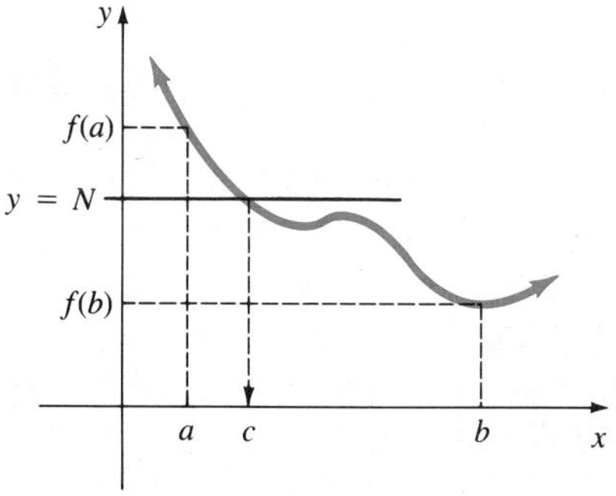

FIGURE 1.135

THEOREM 1.13 Intermediate Value Theorem

If f denotes a function continuous on a closed interval $[a, b]$ for which $f(a) \neq f(b)$, and if N is any number between $f(a)$ and $f(b)$, then there exists at least one number c between a and b such that $f(c) = N$.

The Intermediate Value Theorem states that a function f continuous on a closed interval $[a, b]$ takes on all values between $f(a)$ and $f(b)$. Put another way, f does not "skip" any values.

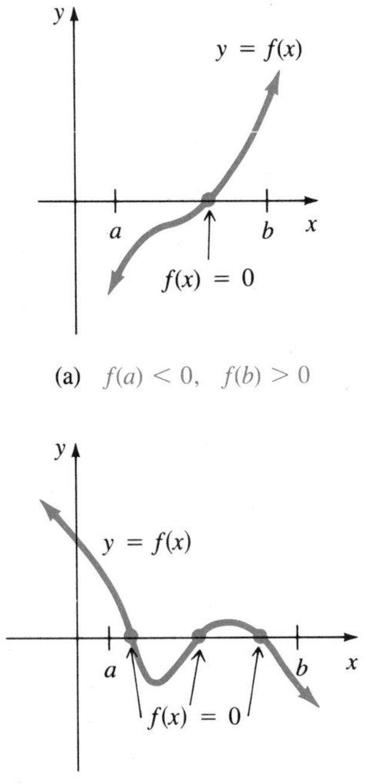

(a) $f(a) < 0,\ f(b) > 0$

(b) $f(a) > 0,\ f(b) < 0$

FIGURE 1.136

EXAMPLE 5 The polynomial function $f(x) = x^2 - x - 5$ is continuous on the interval $[-1, 4]$ and $f(-1) = -3$, $f(4) = 7$. For any number N for which $-3 \le N \le 7$, Theorem 1.13 guarantees that there is a solution to $c^2 - c - 5 = N$ in $[-1, 4]$. Specifically, if we choose $N = 1$, then $c^2 - c - 5 = 1$ is equivalent to

$$c^2 - c - 6 = 0 \quad \text{or} \quad (c - 3)(c + 2) = 0$$

Although the latter equation has two solutions, only the value $c = 3$ is between -1 and 4. □

The foregoing example suggests a corollary to the Intermediate Value Theorem.

If f satisfies the hypotheses of Theorem 1.13 and $f(a)$ and $f(b)$ have opposite algebraic signs, then there exists some value of x between a and b for which $f(x) = 0$.

This fact is often used in locating real zeros of a continuous function f. Figure 1.136 shows several possibilities. In Example 5, by knowing that $f(-1) < 0$ and $f(4) > 0$, we are led to the fact that $x^2 - x - 5 = 0$ for at least one value of x between -1 and 4.

Bisection Method As a direct consequence of the Intermediate Value Theorem, we can devise a means of approximating the zeros of a continuous function to any degree of accuracy. Suppose $y = f(x)$ is continuous on the closed interval $[a, b]$ such that $f(a)$ and $f(b)$ have opposite algebraic signs. Then, as we have just seen, f has a zero in $[a, b]$. Suppose we bisect the interval $[a, b]$ by finding its midpoint $m_1 = (a + b)/2$. If $f(m_1) = 0$, then m_1 is a zero of f and we proceed no further, but if $f(m_1) \ne 0$, then we can say that:

If $f(a)$ and $f(m_1)$ have opposite algebraic signs, than f has a zero in $[a, m_1]$.

If $f(m_1)$ and $f(b)$ have opposite algebraic signs, then f has a zero in $[m_1, b]$.

That is, if $f(m_1) \ne 0$, then f has a zero in an interval that is one-half the length of the original interval. See Figure 1.137. We now repeat the process by bisecting this new interval by finding its midpoint m_2. If m_2 is a zero of f, we stop, but if $f(m_2) \ne 0$, we have located a zero in an interval that is one-fourth the length of $[a, b]$. We continue this process of locating a zero of f in shorter and shorter intervals indefinitely. This method of approximating a zero of a continuous function by a sequence of midpoints is called the **bisection method**. Reinspection of Figure 1.137 shows that the error in an approximation to a zero in an interval is less than one-half the length of the interval.

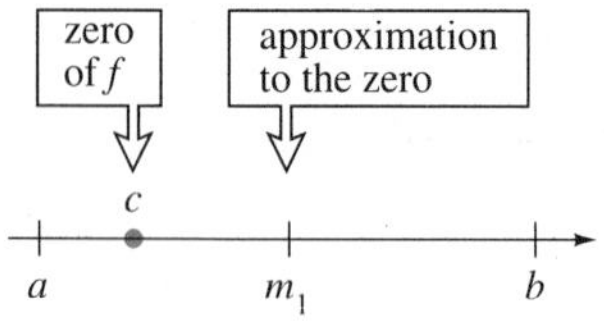

FIGURE 1.137

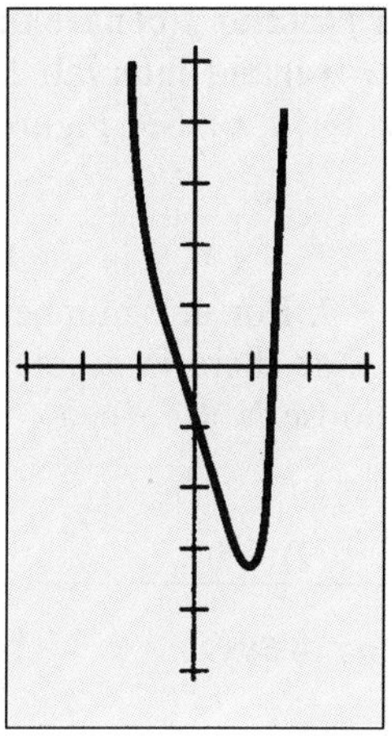

FIGURE 1.138

EXAMPLE 6 The computer-generated graph of $f(x) = x^6 - 3x - 1$ in Figure 1.138 shows that f has a real zero in $[-1, 0]$ and in $[1, 2]$. Approximate the zero in $[1, 2]$ to two decimal places.

Solution To begin, observe that $f(1) = -3 < 0$ and $f(2) = 57 > 0$. Now the first approximation to the zero is the midpoint of $[1, 2]$:

$$m_1 = \frac{1+2}{2} = \frac{3}{2} = 1.5, \qquad \text{error} < \frac{1}{2}(2-1) = 0.5$$

Now since $f(m_1) = f(\frac{3}{2}) > 0$ and $f(1) < 0$, we know that the zero lies in the interval $[1, \frac{3}{2}]$.

The second approximation is the midpoint of $[1, \frac{3}{2}]$:

$$m_2 = \frac{1+\frac{3}{2}}{2} = \frac{5}{4} = 1.25, \qquad \text{error} < \frac{1}{2}\left(\frac{3}{2} - 1\right) = 0.25$$

Since $f(m_2) = f(\frac{5}{4}) < 0$, the zero lies in the interval $[\frac{5}{4}, \frac{3}{2}]$.

The third approximation is the midpoint of $[\frac{5}{4}, \frac{3}{2}]$:

$$m_3 = \frac{\frac{5}{4}+\frac{3}{2}}{2} = \frac{11}{8} = 1.375, \qquad \text{error} < \frac{1}{2}\left(\frac{3}{2} - \frac{5}{4}\right) = 0.125$$

After eight iterations, we find that $m_8 = 1.300781$ with error less than 0.005.* Hence, 1.30 is an approximation to the zero of f in $[1, 2]$ that is accurate to two decimal places. □

▲ ***Remarks*** (*i*) We often give a discontinuity of a function a special name.

- If $x = a$ is a vertical asymptote for the graph of $y = f(x)$, then f is said to have an **infinite discontinuity** at a.
- If $\lim_{x \to a^-} f(x) = L_1$ and $\lim_{x \to a^-} f(x) = L_2$ and $L_1 \neq L_2$, then f is said to have a **finite discontinuity** or a **jump discontinuity** at a. The function $y = f(x)$ given in Figure 1.139 has a jump discontinuity at 0, since $\lim_{x \to 0^-} f(x) = -1$ and $\lim_{x \to 0^+} f(x) = 1$. The greatest integer function $f(x) = [x]$ has a jump discontinuity at every integer value of x.†
- If $\lim_{x \to a} f(x)$ exists but either f is not defined at a or $f(a) \neq \lim_{x \to a} f(x)$, then f is said to have a **removable discontinuity** at a. For example, the function $f(x) = (x^2 - 1)/(x - 1)$ is not defined at 1 but $\lim_{x \to 1} f(x) = 2$. By *defining* $f(1) = 2$, the new function

$$f(x) = \begin{cases} \dfrac{x^2 - 1}{x - 1}, & x \neq 1 \\ 2, & x = 1 \end{cases}$$

is continuous everywhere.

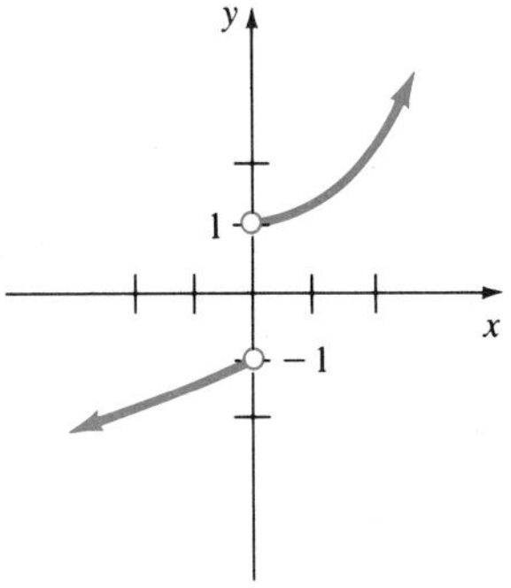

FIGURE 1.139

*If we wish the approximation to be accurate to three decimal places, we continue until the error becomes less than 0.0005, and so on.

† See Example 4 of Section 1.5.

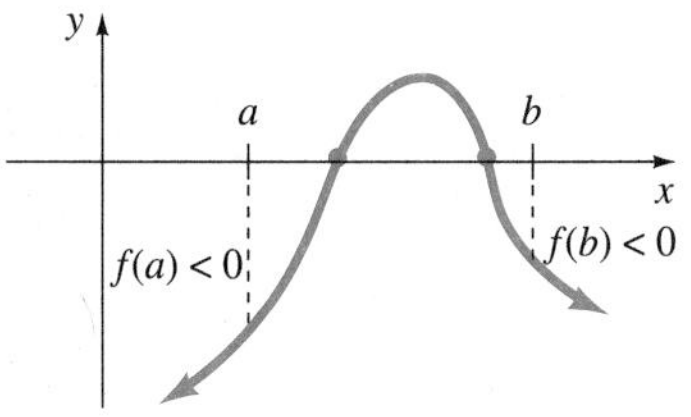

FIGURE 1.140

(*ii*) Suppose $y = f(x)$ is continuous on $[a, b]$ and $f(a)$ and $f(b)$ have the *same* algebraic sign. Does this imply that f has no zeros in the interval? No; the function f could have one or more (or no) zeros in $[a, b]$. See Figure 1.140.

EXERCISES 1.8

Answers to odd-numbered problems begin on page A-68.

In Problems 1–10 determine the points, if any, at which the given function is discontinuous.

1. $f(x) = x^3 - 4x^2 + 7$
2. $f(x) = \dfrac{x}{x^2 + 4}$
3. $f(x) = (x^2 - 9x + 18)^{-1}$
4. $f(x) = \dfrac{x^2 - 1}{x^4 - 1}$
5. $f(x) = \dfrac{x - 1}{\sin 2x}$
6. $f(x) = \dfrac{\tan x}{x + 3}$
7. $f(x) = \begin{cases} x, & x < 0 \\ x^2, & 0 \le x < 2 \\ x, & x > 2 \end{cases}$
8. $f(x) = \begin{cases} \dfrac{|x|}{x}, & x \ne 0 \\ 1, & x = 0 \end{cases}$
9. $f(x) = \begin{cases} \dfrac{\sin x}{x}, & x \ne 0 \\ \dfrac{1}{2}, & x = 0 \end{cases}$
10. $f(x) = \begin{cases} \dfrac{x^2 - 25}{x - 5}, & x \ne 5 \\ 10, & x = 5 \end{cases}$

In Problems 11–22 determine whether the given function is continuous on the indicated intervals.

11. $f(x) = x^2 + 1$ (a) $[-1, 4]$ (b) $[5, \infty)$
12. $f(x) = \dfrac{1}{x}$ (a) $(-\infty, \infty)$ (b) $(0, \infty)$
13. $f(x) = \dfrac{1}{\sqrt{x}}$ (a) $(0, 4]$ (b) $[1, 9]$
14. $f(x) = \sqrt{x^2 - 9}$ (a) $[-3, 3]$ (b) $[3, \infty)$
15. $f(x) = \tan x$ (a) $[0, \pi]$ (b) $\left[-\dfrac{\pi}{2}, \dfrac{\pi}{2}\right]$
16. $f(x) = \csc x$ (a) $(0, \pi)$ (b) $(2\pi, 3\pi)$
17. $f(x) = \dfrac{x}{x^3 + 8}$ (a) $[-4, -3]$ (b) $(-\infty, \infty)$
18. $f(x) = \dfrac{1}{|x| - 4}$ (a) $(-\infty, -1]$ (b) $[1, 6]$
19. $f(x) = \dfrac{x}{2 + \sec x}$ (a) $(-\infty, \infty)$ (b) $\left[\dfrac{\pi}{2}, \dfrac{3\pi}{2}\right]$
20. $f(x) = \sin \dfrac{1}{x}$ (a) $\left[\dfrac{1}{\pi}, \infty\right)$ (b) $\left[-\dfrac{2}{\pi}, \dfrac{2}{\pi}\right]$
21.

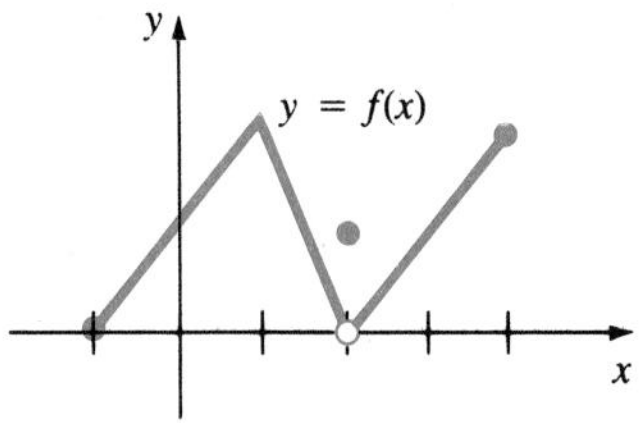

FIGURE 1.141

(a) $[-1, 3]$ (b) $(2, 4]$

22.

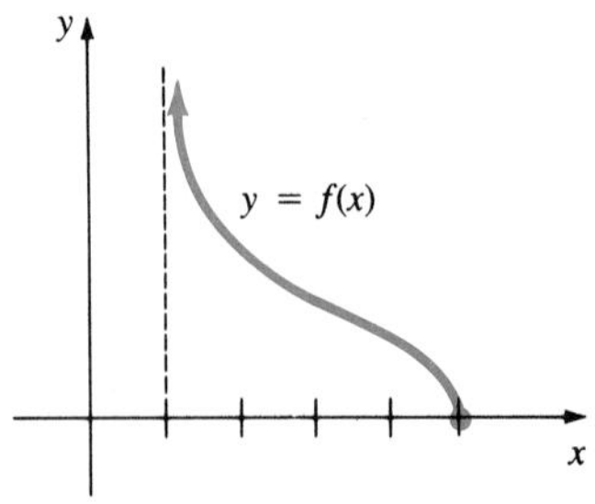

FIGURE 1.142

(a) $[2, 4]$ (b) $[1, 5]$

In Problems 23–26 find values of m and n so that the given function is continuous.

23. $f(x) = \begin{cases} mx, & x < 4 \\ x^2, & x \geq 4 \end{cases}$

24. $f(x) = \begin{cases} \dfrac{x^2 - 4}{x - 2}, & x \neq 2 \\ m, & x = 2 \end{cases}$

25. $f(x) = \begin{cases} mx, & x < 3 \\ n, & x = 3 \\ -2x + 9, & x > 3 \end{cases}$

26. $f(x) = \begin{cases} mx - n, & x < 1 \\ 5, & x = 1 \\ 2mx + n, & x > 1 \end{cases}$

In Problems 27 and 28 $[x]$ denotes the greatest integer not exceeding x. Sketch a graph to determine the points at which the given function is discontinuous.

27. $f(x) = [2x - 1]$

28. $f(x) = [x] - x$

In Problems 29–36 use Theorem 1.12 to evaluate the given limit.

29. $\lim_{x \to \pi/6} \sin(2x + \pi/3)$

30. $\lim_{x \to \pi^2} \cos \sqrt{x}$

31. $\lim_{x \to \pi/2} \sin(\cos x)$

32. $\lim_{x \to \pi/2} (1 + \cos(\cos x))$

33. $\lim_{t \to \pi} \cos\left(\dfrac{t^2 - \pi^2}{t - \pi}\right)$

34. $\lim_{t \to 0} \tan\left(\dfrac{\pi t}{t^2 + 3t}\right)$

35. $\lim_{t \to \pi} \sqrt{t - \pi + \cos^2 t}$

36. $\lim_{t \to 1} (4t + \sin 2\pi t)^3$

In Problems 37 and 38 determine the interval(s) where $f \circ g$ is continuous.

37. $f(x) = \dfrac{1}{\sqrt{x - 1}}, \quad g(x) = x + 4$

38. $f(x) = \dfrac{5x}{x - 1}, \quad g(x) = (x - 2)^2$

In Problems 39–42 verify the Intermediate Value Theorem for f on the given interval. Find a c in the interval for the indicated value of N.

39. $f(x) = x^2 - 2x$, $[1, 5]$; $N = 8$

40. $f(x) = x^2 + x + 1$, $[-2, 3]$; $N = 6$

41. $f(x) = x^3 - 2x + 1$, $[-2, 2]$; $N = 1$

42. $f(x) = \dfrac{10}{x^2 + 1}$, $[0, 1]$; $N = 8$

43. Given that f is continuous on $[a, b]$ and $f(a) = 5$ and $f(b) = 20$, prove that there is a c in (a, b) such that $f(c) = 10$.

44. Given that f and g are continuous on $[a, b]$ such that $f(a) > g(a)$ and $f(b) < g(b)$, prove that there is a c in (a, b) such that $f(c) = g(c)$.

45. Let f be continuous and $f(x) > 0$ for all x in $(1, 2]$. Given that $f(1) = -3$, prove that f is discontinuous on $[1, 2]$.

46. Prove that the **Dirichlet function**

$$f(x) = \begin{cases} 1, & x \text{ rational} \\ 0, & x \text{ irrational} \end{cases}$$

is discontinuous at every real a. What does the graph of f look like?

47. Let $f(x) = [x]$ be the greatest integer function and $g(x) = \cos x$. Determine the points at which $f \circ g$ is discontinuous.

48. Prove that $(\sin x)/x = \frac{1}{2}$ for some value of x between 0 and π.

49. How should $f(x) = (x - 9)/(\sqrt{x} - 3)$ be defined at 9 so that the resulting function is continuous at this point?

50. Can $f(x) = 1/(x - 1)$ be defined at 1 so that the resulting function is continuous at this point?

51. Suppose f is continuous and $f(x) \neq 0$ for all x in $[a, b]$. Prove that $f(x) < 0$ or $f(x) > 0$ for all x in the interval.

52. Consider the functions

$$f(x) = |x| \quad \text{and} \quad g(x) = \begin{cases} x + 1, & x < 0 \\ x - 1, & x \geq 0 \end{cases}$$

Sketch the graphs of $f \circ g$ and $g \circ f$. Determine whether $f \circ g$ and $g \circ f$ are continuous at 0.

53. Given that $f(x) = x^5 + 2x - 7$, prove that there is a number c such that $f(c) = 50$.

54. Prove that the equation

$$\frac{x^2 + 1}{x + 3} + \frac{x^4 + 1}{x - 4} = 0$$

has a solution in the interval $(-3, 4)$.

55. Prove that the equation $2x^7 = 1 - x$ has a solution in $[0, 1]$.

Calculator/Computer Problems

56. Write a BASIC computer program for the bisection method. [*Hint:* If $f(x_1)f(x_2) < 0$, then $f(x_1)$ and $f(x_2)$ have opposite algebraic signs.]

In Problems 57 and 58 use a calculator or computer to obtain the graph of the given function. Use the bisection method to approximate, to an accuracy of two decimal places, the real zeros of f that you discover from the graph.

57. $f(x) = 3x^5 - 5x^3 - 1$

58. $f(x) = x^5 + x - 1$

59. Use the bisection method to approximate the value of c in Problem 53 to an accuracy of two decimal places.

60. Use the bisection method to approximate the solution in Problem 54 to an accuracy of two decimal places.

61. Use the bisection method to approximate the solution in Problem 55 to an accuracy of two decimal places.

62. Suppose a right circular cylinder has a given volume V and surface area S.

(a) Show that the radius r of the cylinder must satisfy the equation $2\pi r^3 - Sr + 2V = 0$.

(b) Suppose $V = 3000$ ft^3 and $S = 1800$ ft^2. Use a calculator or computer to obtain the graph of $f(r) = 2\pi r^3 - 1800r + 6000$.

(c) Use the graph in part **(b)** and the bisection method to find the dimensions of the cylinder corresponding to the volume and surface area given in part **(b)**. Use an accuracy of two decimal places.

Miscellaneous Problems

63. Prove Theorem 1.5.

64. Given that f and g are continuous at a, prove that $f + g$ is continuous at a.

65. Given that f and g are continuous at a and $g(a) \neq 0$, prove that f/g is continuous at a.

CHAPTER 1 REVIEW EXERCISES *Answers begin on page A-68.*

In Problems 1–30 answer true or false.

1. No vertical line intersects the graph of a function more than once. ________

2. If f is a function and $f(a) = f(b)$, then $a = b$. ________

3. The domain of the composition $f \circ g$ is the intersection of the domain of f and the domain of g. ________

4. The function $f(x) = x^5 - 4x^3 + 2$ is an odd function. ________

5. The function $f(x) = 5x^2 \cos x$ is an even function. ________

6. The graph of $y = f(x + 3)$ is the graph of $y = f(x)$ shifted 3 units to the right. ________

7. The function $y = -10 \sec x$ has amplitude 10. ________

8. The range of the function $f(x) = 2 + \cos x$ is $[1, 3]$. ________

9. The graph of the function $f(x) = \dfrac{1}{x-1} + \dfrac{1}{x-2}$ has no x-intercept. ________

10. The number 5 is not in the range of the function

$$f(x) = \begin{cases} 2x, & -2 \le x < 2 \\ 3, & x = 2 \\ x + 4, & x > 2 \end{cases}$$ ________

11. $\displaystyle\lim_{x \to 2} \frac{x^3 - 8}{x - 2} = 12$ ________

12. $\displaystyle\lim_{x \to 5} \sqrt{x - 5} = 0$ ________

13. $\displaystyle\lim_{x \to 0} \frac{|x|}{x} = 1$ ________

14. $\displaystyle\lim_{z \to 1} \frac{z^3 + 8z - 2}{z^2 + 9z - 10}$ does not exist. ________

15. If $\lim_{x \to a} f(x) = 3$ and $\lim_{x \to a} g(x) = 0$, then $\lim_{x \to a} f(x)/g(x)$ does not exist. ________

16. If $\lim_{x \to a} f(x)$ does not exist and $\lim_{x \to a} g(x)$ exists, then $\lim_{x \to a} f(x)g(x)$ does not exist. ________

17. An asymptote is a line that the graph of a function approaches but never crosses. ________

18. If $f(x) = P(x)/Q(x)$ is a rational function and $Q(a) = 0$, then the line $x = a$ is a vertical asymptote. ________

19. The graph of a rational function $f(x) = P(x)/Q(x)$ has a horizontal asymptote only if the degree of $P(x)$ equals the degree of $Q(x)$. ________

20. If $\lim_{x \to a} f(x) = \infty$ and $\lim_{x \to a} g(x) = \infty$, then $\lim_{x \to a} f(x)/g(x) = 1$. ________

21. Any polynomial function is continuous on $(-\infty, \infty)$. ________

22. If $f(x) = x^5 + 3x - 1$, then there exists a number c in $[-2, 2]$ such that $f(c) = 0$. ________

23. If f and g are continuous at 2, then f/g is continuous at 2. ________

24. Suppose $f(x) = [x]$ is the greatest integer function. f is not continuous on the interval $[0, 1]$. ________

25. If $\lim_{x\to a^-} f(x)$ and $\lim_{x\to a^+} f(x)$ exist, then $\lim_{x\to a} f(x)$ exists. ________

26. If $f(x) = \begin{cases} 0, & x \text{ irrational} \\ \dfrac{1}{q}, & x \text{ rational, } p/q \text{ in lowest terms, } q > 0 \end{cases}$

then f is continuous at all irrational points and discontinuous at all rational points. ________

27. The graph of a function can have at most two horizontal asymptotes. ________

28. The function $f(x) = \begin{cases} \dfrac{x^2 - 6x + 5}{x - 5}, & x \neq 5 \\ 4, & x = 5 \end{cases}$

is discontinuous at 5. ________

29. The function $f(x) = \begin{cases} \dfrac{\sin x}{x}, & x \neq 0 \\ 1, & x = 0 \end{cases}$

is continuous at 0. ________

30. If a function f is discontinuous at 3, then $f(3)$ is not defined.

In Problems 31–50 fill in the blanks.

31. If a function f takes a number x, subtracts 2 from it, squares the result, and divides x by the result, then $f(\frac{2}{5}) =$ ________.

32. The domain of the function $f(x) = \sqrt{x + 2}/x$ is ________.

33. The range of the function $f(x) = 10/(x^2 + 1)$ is ________.

34. The y-intercept of the graph of $f(x) = (2x - 4)/(5 - x)$ is ________.

35. The x-intercepts of the graph of $f(x) = x^2 + 2x - 35$ are ________.

36. If $f(x) = 4x^2 + 7$ and $g(x) = 2x + 3$, then $(f \circ g)(1) =$ ________, $(g \circ f)(1) =$ ________, and $(f \circ f)(1) =$ ________.

37. $\lim_{x\to 2} (3x^2 - 4x) =$ ________

38. $\lim_{x\to 3} (5x^2)^0 =$ ________

39. $\lim_{t\to\infty} \dfrac{2t - 1}{3 - 10t} =$ ________

40. $\lim_{x\to -\infty} \dfrac{\sqrt{x^2 + 1}}{2x + 1} =$ ________

41. $\lim_{x\to \_\_\_} \dfrac{1}{x - 3} = -\infty$

42. $\lim_{x\to \_\_\_} (5x + 2) = 22$

43. $\lim_{x\to \_\_\_} x^3 = -\infty$

44. $\lim_{x\to \_\_\_} \dfrac{1}{\sqrt{x}} = \infty$

45. If $f(x) = 2(x - 4)/|x - 4|$, $x \neq 4$, and $f(4) = 9$, then $\lim_{x\to 4^-} f(x) =$ ________.

46. Suppose $x^2 - x^4/3 \le f(x) \le x^2$ for all x. Then $\lim_{x\to 0} f(x)/x^2 =$ ________.

47. If f is continuous at a number a and $\lim_{x\to a} f(x) = 10$, then $f(a) =$ ________.

48. $f(x) = \dfrac{2x - 1}{4x^2 - 1}$ is discontinuous at $x = \frac{1}{2}$ because ________ ________.

49. $f(x) = \begin{cases} kx + 1, & x \le 3 \\ 2 - kx, & x > 3 \end{cases}$ is continuous at 3 if $k =$ ________.

50. If $\lim_{x\to -5} g(x) = -9$ and $f(x) = x^2$, then $\lim_{x\to -5} f(g(x)) =$ ________.

51. If $f(x) = \cos x$, show that for any real number h,

$$\frac{f(x + h) - f(x)}{h} = \cos x\left(\frac{\cos h - 1}{h}\right) - \sin x\left(\frac{\sin h}{h}\right)$$

52. Suppose $f(x) = \sqrt{x + 4}$ and $g(x) = \sqrt{2 - x}$.

(a) What is the domain of $f(-2x)$?

(b) What is the domain of $f(x^2)$?

(c) What is the domain of $g(x^2)$?

(d) What is the domain of $(f + g)(x)$?

(e) What is the domain of $(f/g)(x)$?

53. Consider the interval $[x_1, x_2]$ and the linear function $f(x) = ax + b$, $a \neq 0$. Show that

$$f\left(\frac{x_1 + x_2}{2}\right) = \frac{f(x_1) + f(x_2)}{2}$$

Interpret the result geometrically for $a > 0$.

54. Find all numbers in the domain of $f(x) = 2x^2 - 7x$ that correspond to the number 4 in its range.

55. The width of a rectangular box is 3 times its length and its height is 2 times its length.

(a) Express the volume V of the box as a function of its length l.

(b) As a function of its width w.

(c) As a function of its height h.

56. A closed box in the form of a cube is to be constructed from two different materials. The material for the sides costs 1 cent per square centimeter and the material for the top and bottom costs 2.5 cents per square centimeter. Express the total cost C of construction as a function of the length x of a side.

In Problems 57 and 58 complete the graph.

57. f is an even function.

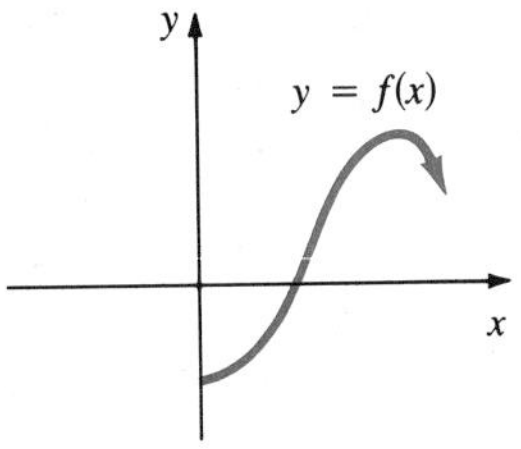

FIGURE 1.143

58. f is an odd function.

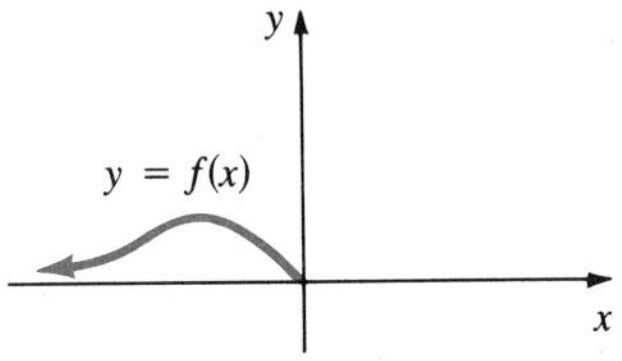

FIGURE 1.144

59. Fill in the blanks by referring to the graph of the function $y = f(x)$ given in Figure 1.145.

$f(-4) =$ ________ $f(-3) =$ ________
$f(-2) =$ ________ $f(-1) =$ ________
$f(0) =$ ________ $f(1) =$ ________
$f(1.5) =$ ________ $f(2) =$ ________
$f(3.5) =$ ________ $f(4) =$ ________

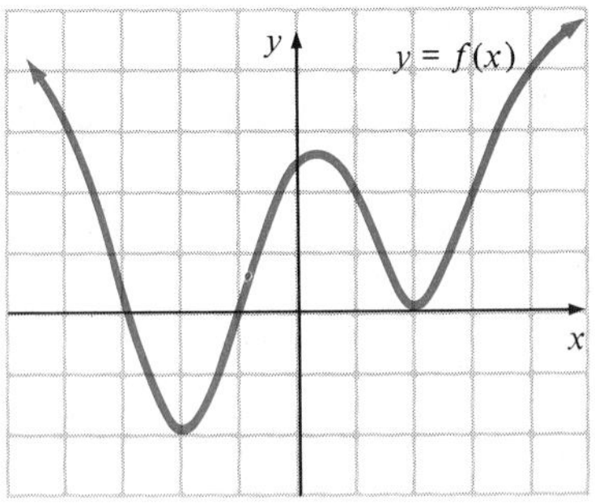

FIGURE 1.145

60. Using the graph of $y = f(x)$ given in Figure 1.146, sketch the graphs:

(a) $y = f(x + 1)$ (b) $y = f(x) + 1$
(c) $y = f(x) - 2$ (d) $y = f(x - 2)$
(e) $y = -2f(x)$ (f) $y = f(x - \frac{1}{2}) + 1$

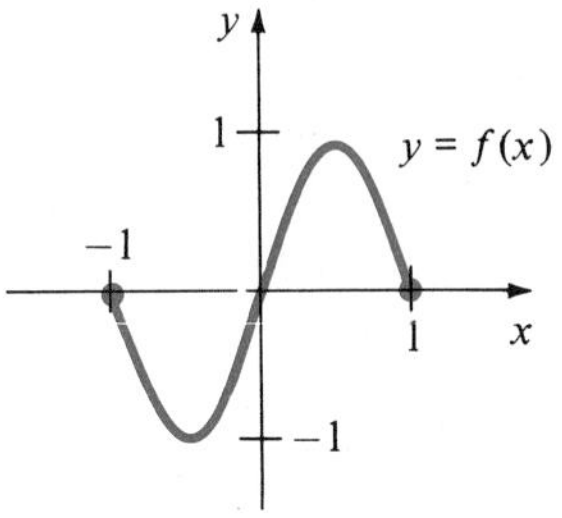

FIGURE 1.146

In Problems 61–70 assume $\lim_{x \to a} f(x) = 4$ and $\lim_{x \to a} g(x) = 2$. Find the given limit, if it exists.

61. $\lim\limits_{x \to a} [5f(x) + 6]$

62. $\lim\limits_{x \to a} [f(x)]^3$

63. $\lim\limits_{x \to a} \dfrac{1}{g(x)}$

64. $\lim\limits_{x \to a} \sqrt{\dfrac{f(x)}{g(x)}}$

65. $\lim\limits_{x \to a} \dfrac{f(x)}{g(x) - 2}$

66. $\lim\limits_{x \to a} \dfrac{[f(x)]^2 - 4[g(x)]^2}{f(x) - 2g(x)}$

67. $\lim\limits_{x \to a^+} f(x)$

68. $\lim\limits_{x \to a^-} f(x)g(x)$

69. $\lim\limits_{x \to a} xg(x)$

70. $\lim\limits_{x \to a} \dfrac{6x + 3}{xf(x) + g(x)}, \quad a \neq -\dfrac{1}{2}$

In Problems 71 and 72 sketch the graph of the given function. Determine the numbers, if any, at which f is discontinuous.

71. $f(x) = |x| + x$

72. $f(x) = \begin{cases} x + 1, & x < 2 \\ 3, & 2 < x < 4 \\ -x + 7, & x > 4 \end{cases}$

In Problems 73 and 74 determine intervals on which the given function is continuous.

73. $f(x) = \dfrac{\sqrt{4 - x^2}}{x^2 - 4x + 3}$

74. $f(x) = \dfrac{x + 1}{\sqrt{x} \sin x}$

75. In Example 12 of Section 1.6 we proved that

$$\lim_{x \to 0} x^2 \sin \frac{1}{x}$$

exists. Does it follow from this that

$$\lim_{x \to 0} x^3 \sin \frac{1}{x}$$

exists?

76. Sketch a graph of a function f that satisfies the following conditions:

$$f(0) = 1, \quad f(4) = 0, \quad f(6) = 0, \quad \lim_{x \to 3^-} f(x) = 2,$$

$$\lim_{x \to 3^+} f(x) = \infty, \quad \lim_{x \to -\infty} f(x) = 0, \quad \lim_{x \to \infty} f(x) = 2$$

In Problems 77 and 78 find the vertical and horizontal asymptotes for the graph of f.

77. $f(x) = \dfrac{2x^3 - 7}{(4x^2 - 25)(x + 9)}$

78. $f(x) = \sqrt{\dfrac{2x + 4}{x - 3}}$

In Problems 79–86 match the given function with one of the graphs (a)–(h).

79. $f(x) = \dfrac{2 - 2x^2}{x^2 + 1}$

80. $f(x) = 2 - \dfrac{2}{x^2}$

81. $f(x) = \dfrac{2x^2}{(x^2 - 4)^2}$

82. $f(x) = \dfrac{x^3}{x^3 + 8}$

83. $f(x) = \dfrac{x^2 - 8}{2x - 4}$

84. $f(x) = \dfrac{x^2}{\sqrt{x^2 - 4}}$

85. $f(x) = \dfrac{2x}{\sqrt{x^2 - 4}}$

86. $f(x) = \dfrac{2x}{\sqrt{x^2 + 4}}$

(a)

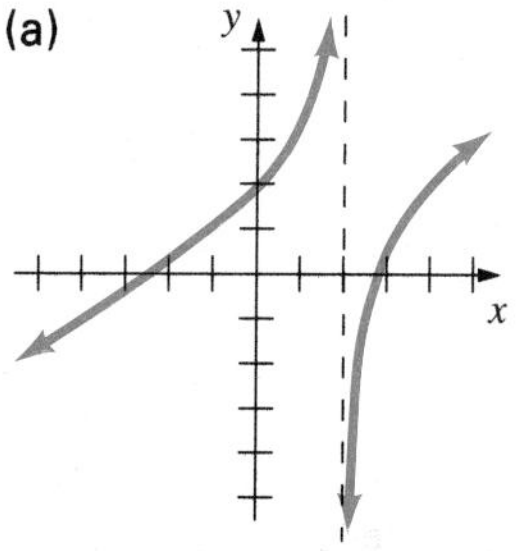

FIGURE 1.147

(b)

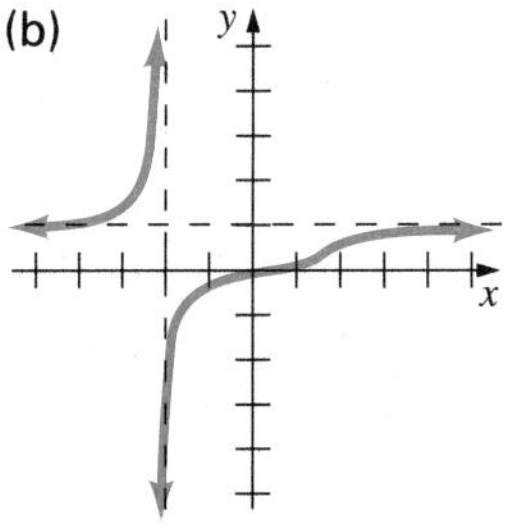

FIGURE 1.148

(c)

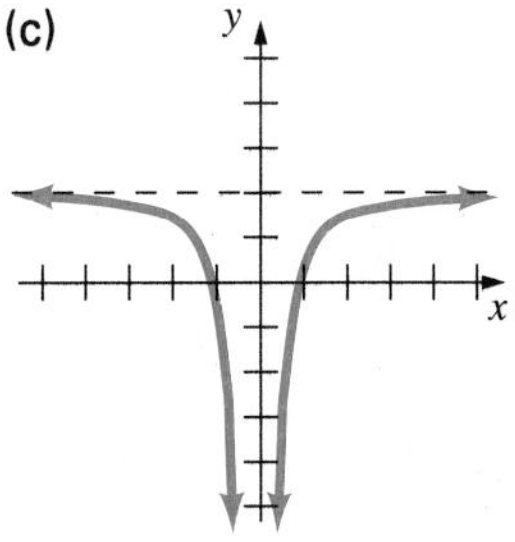

FIGURE 1.149

(d)

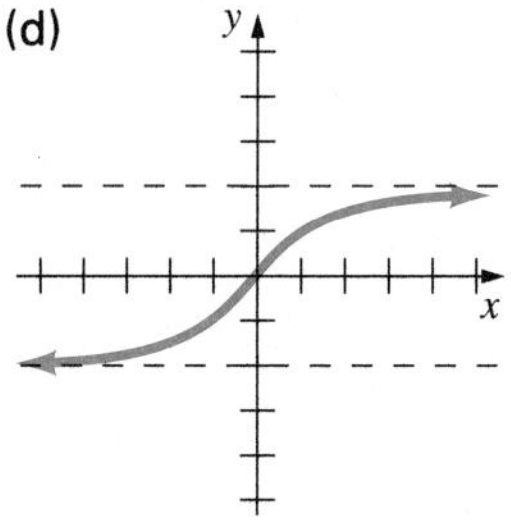

FIGURE 1.150

(e)

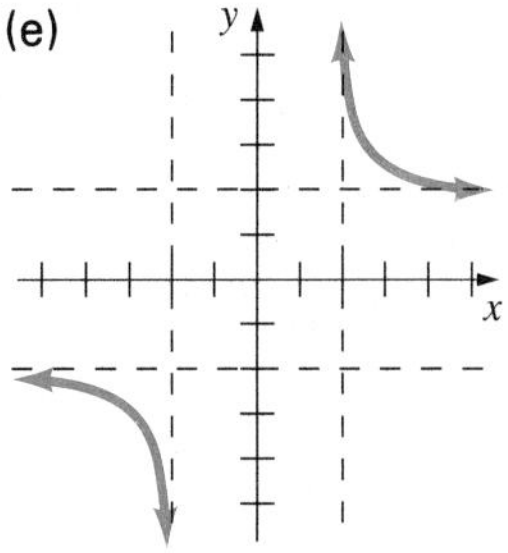

FIGURE 1.151

(f)

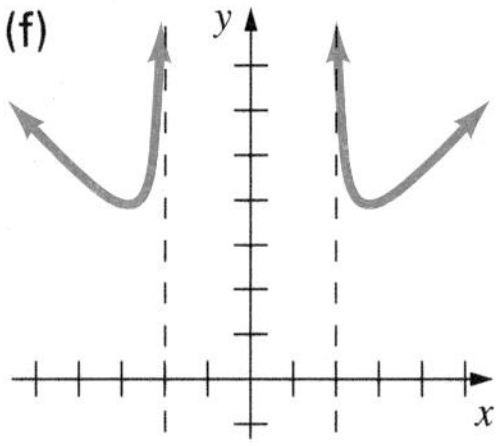

FIGURE 1.152

(g)

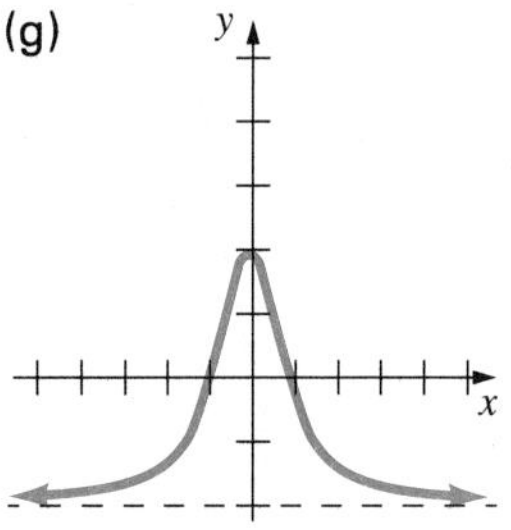

FIGURE 1.153

(h)

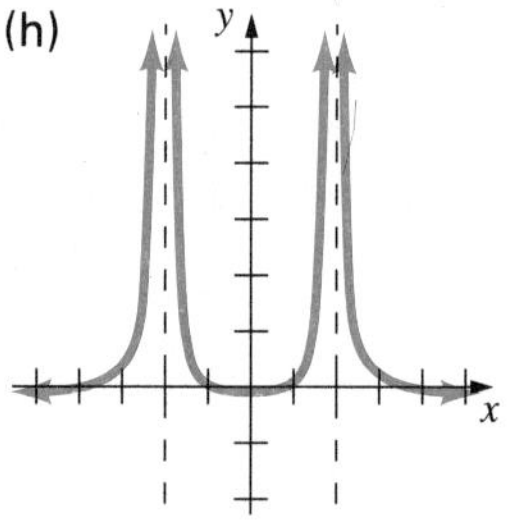

FIGURE 1.154

87. Determine which of the graphs of $y = f(x)$ given in Figure 1.155 seem to match the given condition.

(i) $f(a)$ is not defined.

(ii) $f(a) = L$ ________

(iii) f is continuous at $x = a$ ________

(iv) f is continuous on $[0, a]$ ________

(v) $\lim_{x \to a^+} f(x) = L$ ________

(vi) $\lim_{x \to a} f(x) = L$ ________

(vii) $\lim_{x \to a} |f(x)| = \infty$ ________

(viii) $\lim_{x \to \infty} f(x) = L$ ________

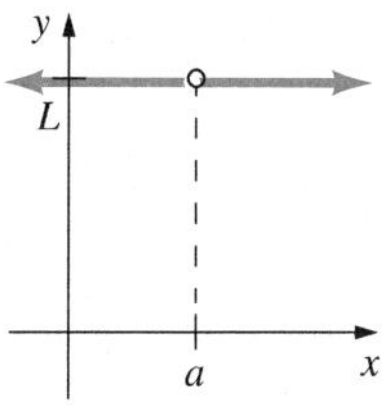

(a)

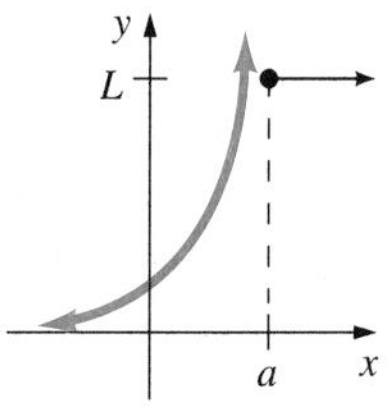

(b)

FIGURE 1.155 *(continues)*

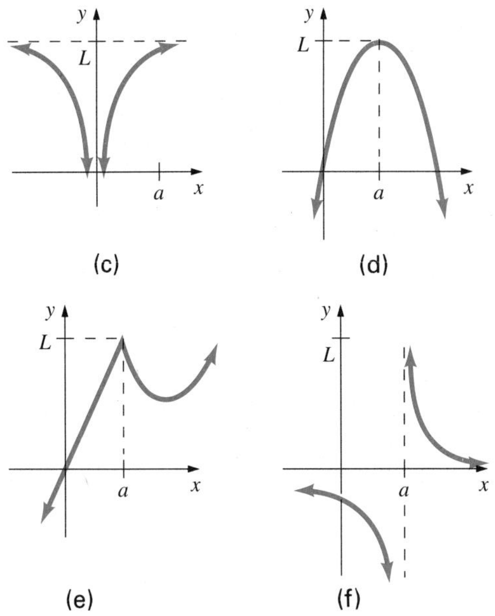

FIGURE 1.155 (*continued*)

88. (a) Find the polynomial equation for the length of an edge of a cube if its volume is tripled when the length of each edge is increased by 1 ft.

(b) Use a computer or calculator to graph the polynomial obtained in part (**a**).

(c) Give the interval(s) where the polynomial in part (**a**) crosses the x-axis.

(d) Use the bisection method to find the length of an edge. Use an accuracy of two decimal places.

2

THE DERIVATIVE

INTRODUCTION

The word *calculus* is a diminutive form of the Latin word *calx*, which means "stone." In ancient civilizations small stones or pebbles were often used as a means of reckoning. Consequently, the word *calculus* can refer to any systematic method of computation. However, over the last several hundred years *calculus* has evolved to mean that branch of mathematics concerned with the calculation and application of entities known as derivatives and integrals. Thus, the subject known as **calculus** has been divided into two rather broad but related areas: **differential calculus** and **integral calculus**. In this chapter we shall begin our study of differential calculus.

The word *tangent* stems from the Latin word *tangere*, meaning "to touch." Recall from plane geometry that a tangent to a circle at a point P is a line that intersects, or touches, the circle at the point P. It is not so easy to define a tangent to a graph of a function f. The line L' shown in the second figure intersects the graph of f at only one point P, yet it violates our intuitive notion of just touching. On the other hand, the line L intersects the graph in two points but seems to be tangent to the graph at P. As we will see in Section 2.1, the notion of touching is important in defining a tangent to a graph, but it is not the whole story.

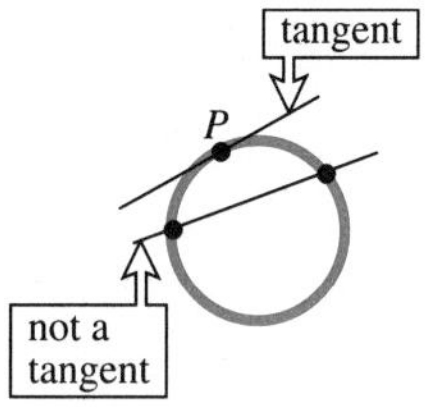

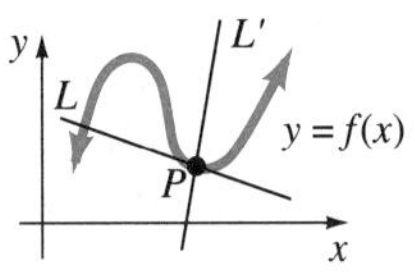

Imagine dropping a ball from a raised platform. The ball is moving; it travels a certain vertical distance in a certain time. In other words, the ball has velocity. What is the velocity? It might surprise you to learn that this and the tangent problem are the same. Read on.

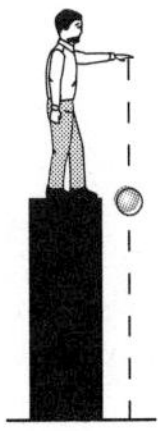

IMPORTANT CONCEPTS

tangent to a graph
average rate of change
instantaneous rate of change
average velocity
rectilinear motion
instantaneous velocity
derivative of a function
differentiation
differentiable function
horizontal tangent
vertical tangent
differentiability on an interval
power rule
sum rule
product rule
quotient rule
chain rule
higher-order derivatives
explicit function
implicit function
implicit differentiation
differentials
Newton's method

2.1 RATE OF CHANGE OF A FUNCTION

REVIEW

For two points $P(x_1, y_1)$ and $P(x_2, y_2)$, the **x-increment** and **y-increment** are the differences in the x- and y-coordinates, respectively:

$$\Delta x = x_2 - x_1 \quad \text{and} \quad \Delta y = y_2 - y_1$$

The symbols Δx and Δy are read, in turn, "delta x" and "delta y."

2.1.1 Tangent to a Graph

Suppose $y = f(x)$ is a continuous function. If, as illustrated in Figure 2.1, the graph of f possesses a tangent line L at a point P, then we would like to find its equation. To do so we need: (1) the coordinates of P and (2) the slope $m_{\tan}$ of L. The coordinates of P pose no difficulty, since a point on a graph is obtained by specifying a value of x, say $x = a$, in the domain of f. The coordinates of the point of tangency are $(a, f(a))$.

As a means of *approximating* the slope $m_{\tan}$, we can readily find the slopes of *secant lines* that pass through the fixed point P and any other point Q on the graph. If P has coordinates $(a, f(a))$ and if we let Q have coordinates

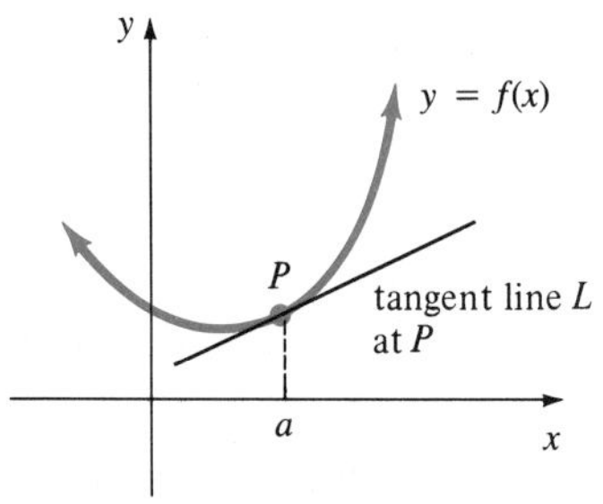

FIGURE 2.1

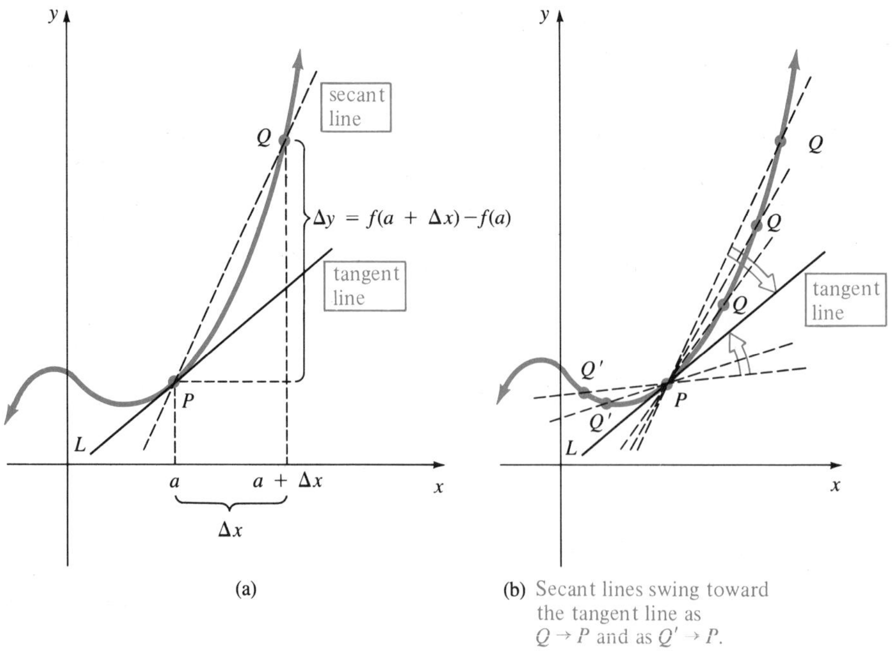

(b) Secant lines swing toward the tangent line as $Q \to P$ and as $Q' \to P$.

FIGURE 2.2

$(a + \Delta x, f(a + \Delta x))$, then, as shown in Figure 2.2(a), the slope of the secant line through P and Q is

$$m_{\text{sec}} = \frac{\text{change in } y\text{-coordinate}}{\text{change in } x\text{-coordinate}} = \frac{f(a + \Delta x) - f(a)}{(a + \Delta x) - a}$$

If
$$\Delta y = f(a + \Delta x) - f(a)$$

then
$$m_{\text{sec}} = \frac{\Delta y}{\Delta x}$$

When the value of Δx is close to zero, either positive or negative, we get points Q and Q' on the graph of f on each side of, but close to, the point P. In turn, we expect that the slopes m_{PQ} and $m_{PQ'}$ are very close to the slope of the tangent line L. See Figure 2.2(b).

EXAMPLE 1 Find the slope of the tangent line to the graph of $f(x) = x^2$ at (1, 1).

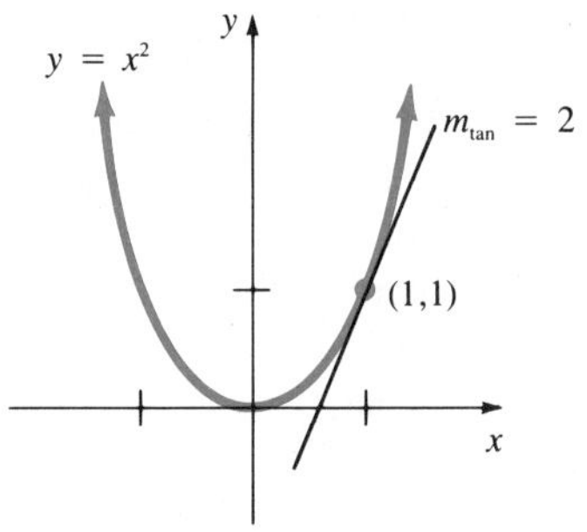

FIGURE 2.3

Solution As a start, let us choose $\Delta x = 0.1$ and find the slope of the secant line through (1, 1) and $(1.1, (1.1)^2)$:

(i) $f(1.1) = (1.1)^2 = 1.21$

(ii) $\Delta y = f(1.1) - f(1) = 1.21 - 1 = 0.21$

(iii) $\dfrac{\Delta y}{\Delta x} = \dfrac{0.21}{0.1} = 2.1$

The rightmost column of the table should convince you that the slope of the tangent line shown in Figure 2.3 is $m_{\tan} = 2$.

Δx	$1 + \Delta x$	$f(1)$	$f(1 + \Delta x)$	Δy	$\Delta y/\Delta x$
0.1	1.1	1	1.21	0.21	2.1
0.01	1.01	1	1.0201	0.0201	2.01
0.001	1.001	1	1.002001	0.002001	2.001
−0.1	0.9	1	0.81	−0.19	1.9
−0.01	0.99	1	0.9801	−0.0199	1.99
−0.001	0.999	1	0.998001	−0.001999	1.999

□

On the basis of Figure 2.2(b), Example 1, and our intuition, we are prompted to say that if a graph of a function $y = f(x)$ has a tangent line L at a point P, then L must be the line that is the *limit* of the secants through P and Q as $Q \to P$ and of the secants through P and Q' as $Q' \to P$. Moreover, the slope $m_{\tan}$ of L should be the *limiting value* of the values m_{sec} as $\Delta x \to 0$. This is summarized as follows:

DEFINITION 2.1 Tangent Line with Slope

Let $y = f(x)$ be continuous at a. If

$$m_{\tan} = \lim_{\Delta x \to 0} \frac{f(a + \Delta x) - f(a)}{\Delta x} = \lim_{\Delta x \to 0} \frac{\Delta y}{\Delta x} \tag{1}$$

exists, then the tangent line to the graph at $(a, f(a))$ is that line passing through the point with slope $m_{\tan}$.

The slope of the tangent line at $(a, f(a))$ is also called the **slope of the curve** at the point. Definition 2.1 implies that a tangent at $(a, f(a))$ is *unique*, since a point and a slope determine a single line.

We synthesize the application of Definition 2.1 to four steps.

(*i*) Evaluate f at a and $a + \Delta x$.

$$f(a) \quad \text{and} \quad f(a + \Delta x)$$

(*ii*) Compute Δy:

$$\Delta y = f(a + \Delta x) - f(a)$$

(*iii*) Divide Δy by $\Delta x \neq 0$ and simplify:

$$\frac{\Delta y}{\Delta x} = \frac{f(a + \Delta x) - f(a)}{\Delta x}$$

(*iv*) Compute the limit as $\Delta x \to 0$:

$$m_{\tan} = \lim_{\Delta x \to 0} \frac{\Delta y}{\Delta x}$$

EXAMPLE 2 Use Definition 2.1 to find the slope of the tangent line to the graph of $f(x) = x^2$ at $(1, f(1))$.

Solution

(*i*) $f(1) = 1^2 = 1$. For any $\Delta x \neq 0$,

$$f(1 + \Delta x) = (1 + \Delta x)^2 = 1 + 2\,\Delta x + (\Delta x)^2$$

(*ii*) $\Delta y = f(1 + \Delta x) - f(1) = [1 + 2\,\Delta x + (\Delta x)^2] - 1$
$= 2\,\Delta x + (\Delta x)^2 = \Delta x(2 + \Delta x)$

$$(iii)\ \frac{\Delta y}{\Delta x} = \frac{\Delta x(2 + \Delta x)}{\Delta x} = 2 + \Delta x$$

Thus, the slope of the tangent at $(1, f(1))$ is given by

$$(iv)\ m_{\tan} = \lim_{\Delta x \to 0} \frac{\Delta y}{\Delta x} = \lim_{\Delta x \to 0} (2 + \Delta x) = 2 \qquad \square$$

EXAMPLE 3 Find the slope of the tangent line to the graph of $f(x) = 5x + 6$ at any point $(a, f(a))$.

Solution

(i) $f(a) = 5a + 6$. For any $\Delta x \neq 0$,

$$f(a + \Delta x) = 5(a + \Delta x) + 6 = 5a + 5\,\Delta x + 6$$

$$\begin{aligned}(ii)\ \Delta y &= f(a + \Delta x) - f(a)\\ &= [5a + 5\,\Delta x + 6] - [5a + 6] = 5\,\Delta x\end{aligned}$$

$$(iii)\ \frac{\Delta y}{\Delta x} = \frac{5\,\Delta x}{\Delta x} = 5$$

Thus, at any point on the graph of $f(x) = 5x + 6$, we have

$$(iv)\ m_{\tan} = \lim_{\Delta x \to 0} \frac{\Delta y}{\Delta x} = \lim_{\Delta x \to 0} 5 = 5 \qquad \square$$

You should be able to explain why the answer in Example 3 is no surprise.

EXAMPLE 4 Find an equation of the tangent line to the graph of $f(x) = x^2$ at $(1, 1)$.

Solution From Example 2 the slope of the tangent at $(1, 1)$ is $m_{\tan} = 2$. The point–slope form of a line then gives

$$y - 1 = 2(x - 1) \quad \text{or} \quad y = 2x - 1 \qquad \square$$

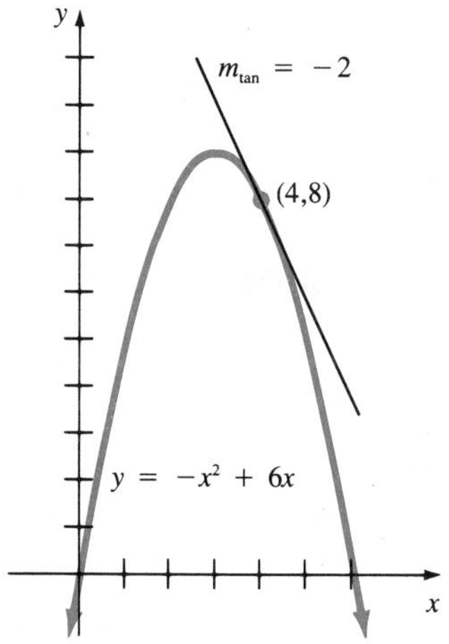

FIGURE 2.4

EXAMPLE 5 Find the slope of the tangent line to the graph of $f(x) = -x^2 + 6x$ at $(4, f(4))$.

Solution

(i) $f(4) = -4^2 + 6(4) = 8$. For any $\Delta x \neq 0$,

$$\begin{aligned}f(4 + \Delta x) &= -(4 + \Delta x)^2 + 6(4 + \Delta x)\\ &= -16 - 8\,\Delta x - (\Delta x)^2 + 24 + 6\,\Delta x\\ &= 8 - 2\,\Delta x - (\Delta x)^2\end{aligned}$$

$$\begin{aligned}(ii)\ \Delta y &= f(4 + \Delta x) - f(4) = [8 - 2\,\Delta x - (\Delta x)^2] - 8\\ &= -2\,\Delta x - (\Delta x)^2 = \Delta x(-2 - \Delta x)\end{aligned}$$

(*iii*) $\dfrac{\Delta y}{\Delta x} = \dfrac{\Delta x(-2 - \Delta x)}{\Delta x} = -2 - \Delta x$

(*iv*) $m_{\tan} = \lim_{\Delta x \to 0} (-2 - \Delta x) = -2$

Figure 2.4 shows the graph of f and the tangent line with slope -2 at (4, 8). □

Vertical Tangents The limit (1) can fail to exist for a function f at a and yet there may be a tangent at $(a, f(a))$. The tangent line to a graph at a point can be **vertical**, in which case its slope is undefined. We will consider the concept of a vertical tangent in Section 2.2.

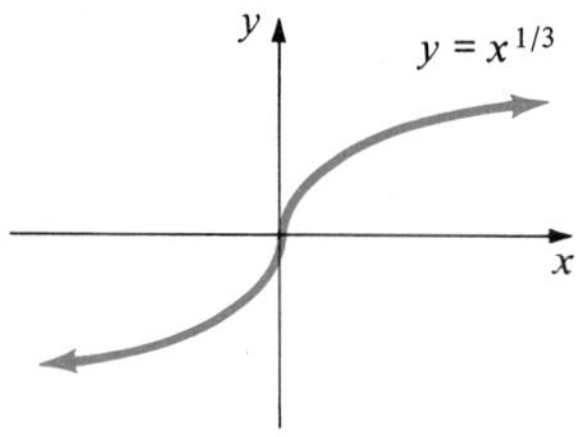

FIGURE 2.5

EXAMPLE 6 Although we shall not pursue the details at this time, it can be shown that the graph of $y = x^{1/3}$ possesses a vertical tangent line at the origin. In Figure 2.5 we see that the y-axis, that is, the line $x = 0$, is tangent to the graph at the point (0, 0). □

A Tangent May Not Exist The graph of a function f *will not* have a tangent line at a point whenever:

(*i*) f is discontinuous at $x = a$, or

(*ii*) the graph of f has a corner at $(a, f(a))$.

Figure 2.6(a) and (b) shows the graphs of two functions that are discontinuous at $x = a$. Neither graph has a tangent at $x = a$. Figure 2.6(c) indicates what goes wrong when the graph of a function f has a "corner." In this case f is continuous at a, but the secant lines that pass through P and Q approach L_2 as $Q \to P$, and the secant lines through P and Q' approach a different line L_1 as $Q' \to P$. Keep in mind: *when we say a tangent line exists at a point P on the graph of a function, we mean that there is only one tangent line.*

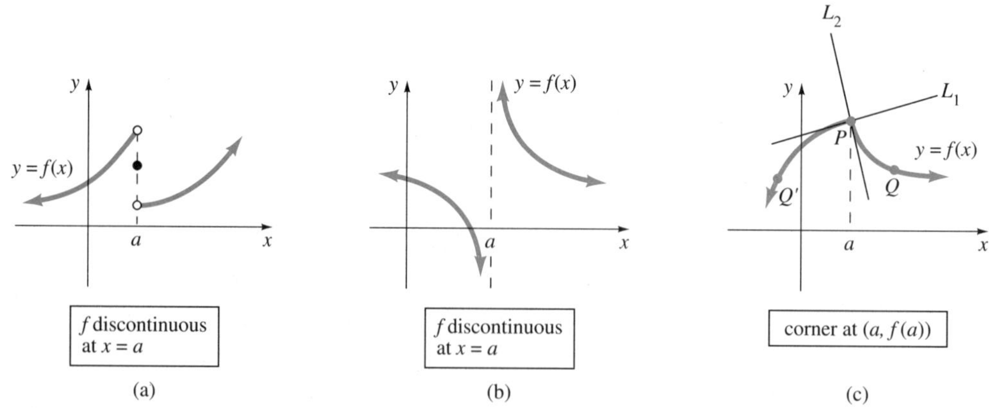

FIGURE 2.6

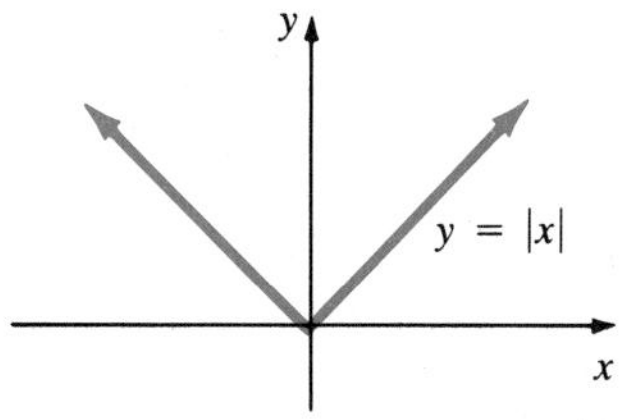

FIGURE 2.7

EXAMPLE 7 Show that the graph of $f(x) = |x|$ does not have a tangent at (0, 0).

Solution An inspection of the graph of f in Figure 2.7 reveals a corner at the origin. To prove that the tangent does not exist at (0, 0), we must examine

$$\lim_{\Delta x \to 0} \frac{f(0 + \Delta x) - f(0)}{\Delta x} = \lim_{\Delta x \to 0} \frac{|0 + \Delta x| - |0|}{\Delta x} = \lim_{\Delta x \to 0} \frac{|\Delta x|}{\Delta x} \tag{2}$$

Now, for $\Delta x > 0$, $$\frac{|\Delta x|}{\Delta x} = \frac{\Delta x}{\Delta x} = 1$$

whereas for $\Delta x < 0$, $$\frac{|\Delta x|}{\Delta x} = \frac{-\Delta x}{\Delta x} = -1$$

Since $\lim_{\Delta x \to 0^+} |\Delta x|/\Delta x = 1$ and $\lim_{\Delta x \to 0^-} |\Delta x|/\Delta x = -1$, we conclude that the limit (2) does not exist. Thus, the graph possesses no tangent at (0, 0). □

Rate of Change The slope $\Delta y/\Delta x$ of a secant line through $(a, f(a))$ is also called the **average rate of change** of f at a. The slope $m_{\tan} = \lim_{\Delta x \to 0} \Delta y/\Delta x$ is said to be the **instantaneous rate of change** of the function at a. For example, if $m_{\tan} = \frac{1}{10}$ at a point $(a, f(a))$, we would not expect the values of f to change drastically for x-values near a.

2.1.2 Instantaneous Velocity

Almost everyone has an intuitive notion of speed or velocity as a rate at which a distance is covered in a certain length of time. When, say, a bus travels 60 miles in one hour, the *average velocity* of the bus must have been 60 mi/h. Of course, it is difficult to maintain the rate of 60 mi/h for the entire trip because the bus slows down for towns and speeds up when it passes cars. In other words, the velocity changes with time. If a bus company's schedule demands that the bus travel the 60 mi from one town to another in 1 h, the driver knows instinctively that he must compensate for velocities or speeds less than 60 mi/h by traveling at speeds greater than this at other points in the journey. Knowing that the average velocity is 60 mi/h does not, however, answer the question: What is the velocity of the bus at a particular instant?*

Average Velocity In general, the **average velocity** or **average speed** of a moving object is the time rate of change of position defined by

$$v_{\text{ave}} = \frac{\text{distance traveled}}{\text{time of travel}} \tag{3}$$

*The bus driver need only look at the speedometer and observe, for example, "I am now going 45 mi/h." This may not be obvious to an observer who watches the bus move down the road.

Consider a runner who finishes a 10-km race in an elapsed time of 1 h 15 min (1.25 h). The runner's average velocity or average speed for the race was

$$v_{\text{ave}} = \frac{10}{1.25} = 8 \text{ km/h}$$

But suppose we now wish to determine the runner's *exact* velocity v at the instant the runner is one-half hour into the race. If the distance run in the time interval from 0 h to 0.5 h is measured to be 5 km, then

$$v_{\text{ave}} = \frac{5}{0.5} = 10 \text{ km/h}$$

Again, this number is not a measure, or necessarily even a good indicator, of the instantaneous rate v at which the runner is moving 0.5 h into the race. If we determine that at 0.6 h the runner is 5.7 km from the starting line, then the average velocity from 0 h to 0.6 h is $v_{\text{ave}} = 5.7/0.6 = 9.5$ km/h. However, during the time interval from 0.5 h to 0.6 h,

$$v_{\text{ave}} = \frac{5.7 - 5}{0.6 - 0.5} = 7 \text{ km/h}$$

The latter number is a more realistic measure of the rate v. See Figure 2.8. By "shrinking" the time interval between 0.5 h and the time that corresponds to a measured position close to 5 km, we expect to obtain even better approximations to the runner's velocity at time 0.5 h.

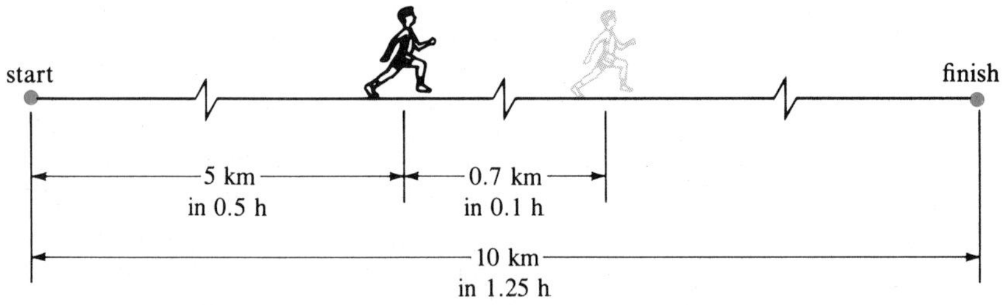

FIGURE 2.8

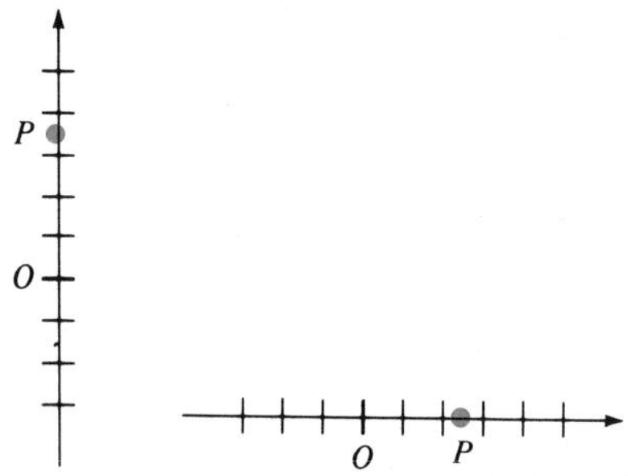

FIGURE 2.9

Rectilinear Motion To generalize the preceding concepts, let us suppose an object, or particle, at point P moves along either a vertical or horizontal coordinate line as shown in Figure 2.9. Furthermore, let the particle move in such a manner that its position, or coordinate, on the line is given by a function $s = f(t)$, where t represents time. The values of s are directed distances measured from O in units such as centimeters, meters, feet, or miles. When P is to the right or above O, we take $s > 0$, whereas $s < 0$ when P is to the left of or below O. Motion in a straight line is called **rectilinear motion.**

If a particle is at point P at time t_1 and at P' at time $t_1 + \Delta t$, then the coordinates of the points, shown in Figure 2.10, are $f(t_1)$ and $f(t_1 + \Delta t)$. By

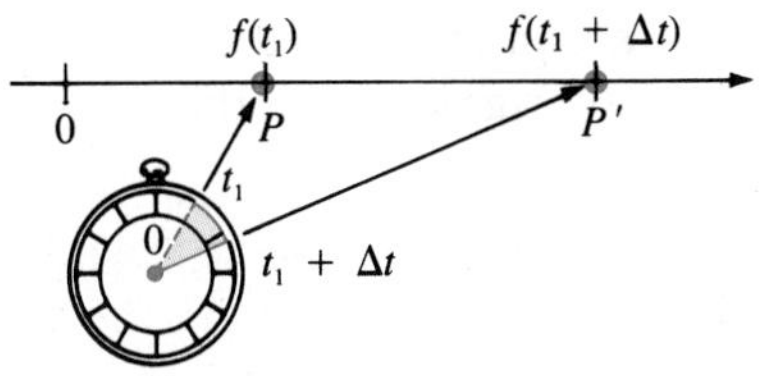

FIGURE 2.10

(3) the average velocity of the particle in the time interval $[t_1, t_1 + \Delta t]$ is

$$v_{\text{ave}} = \frac{\text{change in position}}{\text{change in time}} = \frac{f(t_1 + \Delta t) - f(t_1)}{(t_1 + \Delta t) - t_1}$$

or

$$v_{\text{ave}} = \frac{\Delta s}{\Delta t} \tag{4}$$

This suggests that the limit of (4) as $\Delta t \to 0$ gives the **instantaneous rate of change** of $f(t)$ at t_1, or the **instantaneous velocity**.

DEFINITION 2.2 Instantaneous Velocity

Let $s = f(t)$ be a function that gives the position of an object moving in a straight line. The **instantaneous velocity** at time t_1 is

$$v(t_1) = \lim_{\Delta t \to 0} \frac{f(t_1 + \Delta t) - f(t_1)}{\Delta t} = \lim_{\Delta t \to 0} \frac{\Delta s}{\Delta t} \tag{5}$$

whenever the limit exists.

Note: Except for notation and interpretation, there is no mathematical difference between (1) and (5). Also, the word *instantaneous* is often dropped, and so one often speaks of the "rate of change" of a function or the "velocity" of a moving particle.

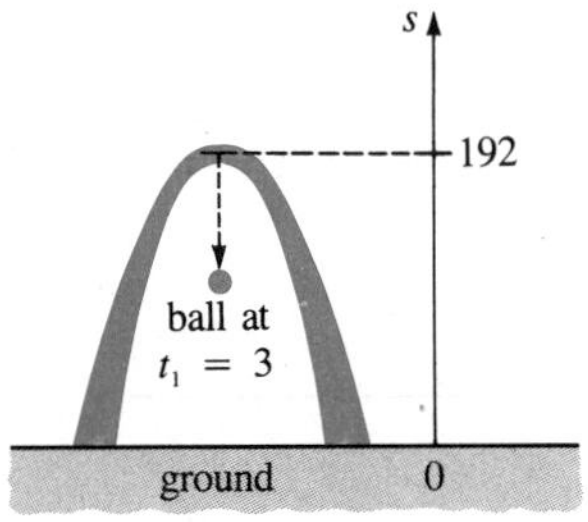

FIGURE 2.11

EXAMPLE 8 The height s above ground of a ball dropped from the top of the St. Louis Gateway Arch is given by $s = -4.9t^2 + 192$, where s is measured in meters and t in seconds.* See Figure 2.11. Find the instantaneous velocity of the falling ball at $t_1 = 3$ s.

Solution We use the same four-step procedure as in the earlier examples.

(*i*) $f(3) = -4.9(9) + 192 = 147.9$. For any $\Delta t \neq 0$,

$$f(3 + \Delta t) = -4.9(3 + \Delta t)^2 + 192 = -4.9(\Delta t)^2 - 29.4\,\Delta t + 147.9$$

(*ii*)
$$\begin{aligned} \Delta s &= f(3 + \Delta t) - f(3) \\ &= [-4.9(\Delta t)^2 - 29.4\,\Delta t + 147.9] - 147.9 \\ &= -4.9(\Delta t)^2 - 29.4\,\Delta t = \Delta t(-4.9\,\Delta t - 29.4) \end{aligned}$$

(*iii*) $$\frac{\Delta s}{\Delta t} = \frac{\Delta t(-4.9\,\Delta t - 29.4)}{\Delta t} = -4.9\,\Delta t - 29.4$$

*This is not a made-up formula; the height of the arch is 630 ft, or 192 m. We shall see how the formula is derived in Section 5.7.

$$(iv)\ v(3) = \lim_{\Delta t \to 0} \frac{\Delta s}{\Delta t} = \lim_{\Delta t \to 0} (-4.9\,\Delta t - 29.4) = -29.4 \text{ m/s}$$

The minus sign is significant because the ball is moving opposite to the positive or upward direction. The number $f(3) = 147.9$ m is the height of the ball above the ground at 3 s. □

We shall study rectilinear motion in greater detail in Section 3.1.

▲ ***Remark*** You have probably noticed that the goal in utilizing (1) and (5) is to divide out Δx and Δt in the quotients $\Delta y/\Delta x$ and $\Delta s/\Delta t$, respectively, *before* passing to the limit. If you are not able to do this in any of the following problems, you should double check your algebra.

EXERCISES 2.1

Answers to odd-numbered problems begin on page A-69.

[2.1.1]

In Problems 1–6 sketch the graph of the function and the tangent line at the given point. Find the slope of the secant line through the points that correspond to the indicated values of x.

1. $f(x) = -x^2 + 9$, $(2, 5)$; $x = 2$, $x = 2.5$
2. $f(x) = x^2 + 4x$, $(0, 0)$; $x = -\frac{1}{4}$, $x = 0$
3. $f(x) = x^3$, $(-2, -8)$; $x = -2$, $x = -1$
4. $f(x) = 1/x$, $(1, 1)$; $x = 0.9$, $x = 1$
5. $f(x) = \sin x$, $(\pi/2, 1)$; $x = \pi/2$, $x = 2\pi/3$
6. $f(x) = \cos x$, $(-\pi/3, 1/2)$; $x = -\pi/2$, $x = -\pi/3$

In Problems 7–16 find the slope of the tangent line to the graph of the given function at the indicated point.

7. $f(x) = 2x - 1$; $(4, 7)$
8. $f(x) = -\frac{1}{2}x + 3$; $(a, f(a))$
9. $f(x) = x^2$; $(3, 9)$
10. $f(x) = x^2 + 4$; $(-1, 5)$
11. $f(x) = 2x^2 + 8x$; $(0, 0)$
12. $f(x) = x^2 - 5x + 4$; $(2, -2)$
13. $f(x) = x^3$; $(1, f(1))$
14. $f(x) = -x^3 + x^2$; $(2, f(2))$
15. $f(x) = 1/x$; $(1/3, f(1/3))$
16. $f(x) = 1/(x - 1)^2$; $(0, f(0))$
17. Find the slope of the tangent line to the graph of $f(x) = 1/x^2$ at the point $(x_0, f(x_0))$. What is the slope of the tangent line at the point where $x_0 = -2$? where $x_0 = \frac{1}{3}$?
18. Find the slope of the tangent line to the graph of $f(x) = 5x^2 - x + 2$ at the point $(x, f(x))$. What is the slope of the tangent line at the point where $x = \frac{3}{5}$? where $x = 4$?

In Problems 19–24 find an equation of the tangent line to the graph of the given function at the indicated value of x.

19. $f(x) = 1 - x^2$; $x = 5$
20. $f(x) = (x + 2)^2$; $x = -3$
21. $f(x) = (x - 1)^4 + 6x$; $x = 1$
22. $f(x) = 4x + 10$; $x = 7$
23. $f(x) = \dfrac{1}{1 + 2x}$; $x = 0$
24. $f(x) = 4 - \dfrac{8}{x}$; $x = -1$

In Problems 25 and 26 find the average rate of change of the given function on the indicated interval.

25. $f(x) = x^3 + 2x^2 - 4x$; $[-1, 2]$
26. $f(x) = \cos x$; $[-\pi, \pi]$

In Problems 27–30 specify the values of x for which the tangent line to the graph is possibly horizontal, vertical, or does not exist.

27.

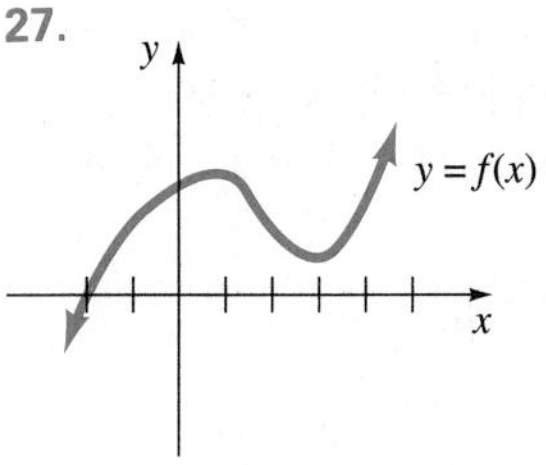

FIGURE 2.12

28.

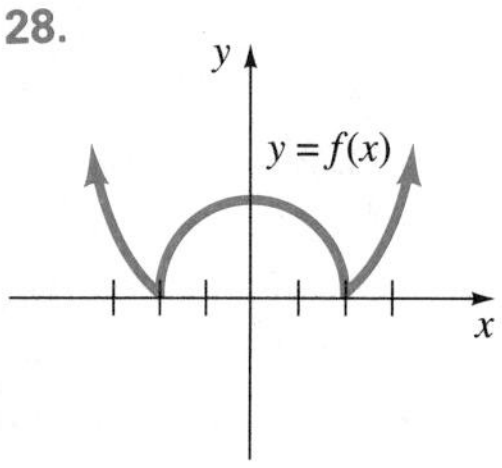

FIGURE 2.13

29.

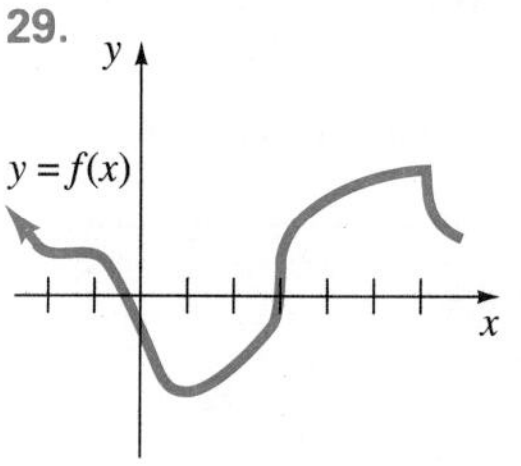

FIGURE 2.14

30.

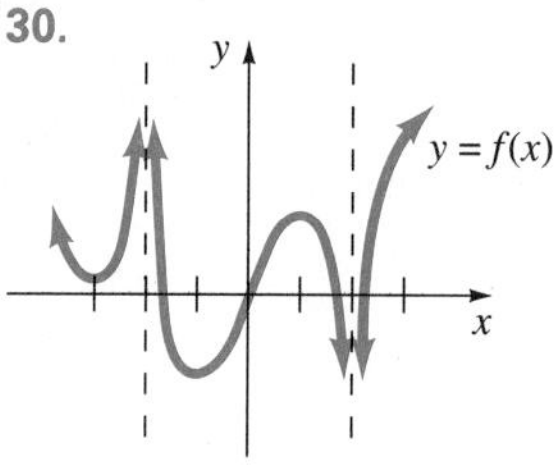

FIGURE 2.15

In Problems 31 and 32 determine whether the line(s) through the given point(s) is (are) tangent to the graph of $f(x) = x^2$ at the point(s) indicated in color.

31.

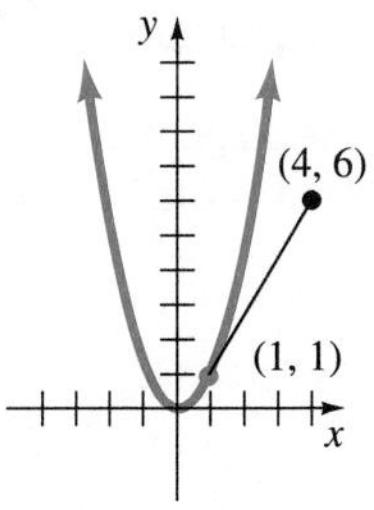

FIGURE 2.16

32.

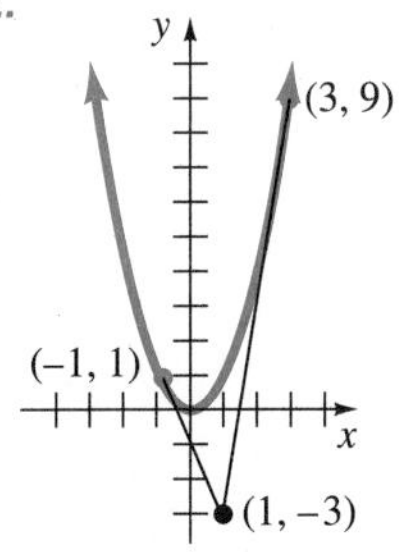

FIGURE 2.17

Calculator/Computer Problems

33. Consider the function $f(x) = \sqrt{3x + 1}$.

(a) What is $\Delta y/\Delta x$ at $x = 0$?

(b) Complete the table using values with four decimal places. Conjecture the slope of the tangent to the graph of f at $x = 0$.

Δx	Δy	$\Delta y/\Delta x$
0.2		
0.1		
0.01		
−0.2		
−0.1		
−0.01		

34. Consider the function $f(x) = \tan x$.

(a) Compute the quotient $(\tan 0.1)/0.1$.

(b) At what point on the graph of f is the number in part (**a**) an approximation to the slope of the tangent line?

Miscellaneous Problems

35. Let $y = f(x)$ be an even function whose graph possesses a tangent line with slope m at (x_0, y_0). Show that the slope of the tangent at $(-x_0, y_0)$ is $-m$. [*Hint:* $f(-x_0 + \Delta x) = f(x_0 - \Delta x)$.]

36. Let $y = f(x)$ be an odd function whose graph possesses a tangent line with slope m at (x_0, y_0). Show that the slope of the tangent at $(-x_0, y_0)$ is m.

[2.1.2]

37. A car travels the 290 mi between Los Angeles and Las Vegas in 5 h. What is its average speed?

38. Two marks on a highway are $\frac{1}{2}$ mi apart. A Highway Patrol plane observes that a car traverses the distance between the marks in 40 s. Will the car be stopped for speeding? (Assume the speed limit is 55 mi/h.)

39. A jet airplane averages 920 km/h to fly the 3500 km between Hawaii and San Francisco. How many hours does the flight take?

40. A marathon race is run over a straight 26-mi course. The race begins at noon. At 1:30 P.M. a contestant passes the 10-mi mark and at 3:10 P.M. the contestant passes the 20-mi mark. What is the contestant's average running speed between 1:30 P.M. and 3:10 P.M.?

In Problems 41 and 42 the position of a particle on a horizontal coordinate line is given by the function. Find the instantaneous velocity of the particle at the indicated time.

41. $f(t) = -4t^2 + 10t + 6;\ t = 3$

42. $f(t) = t^2 + \dfrac{1}{5t + 1};\ t = 0$

43. The height above ground of a ball dropped from an initial altitude of 122.5 m is given by $s = 122.5 - 4.9t^2$, where s is measured in meters and t in seconds.

(a) What is the instantaneous velocity at $t = \frac{1}{2}$?

(b) At what time does the ball hit the ground?

(c) What is the impact velocity?

44. Ignoring air resistance, if an object is dropped from an initial height h, then its height above ground at any time $t > 0$ is given by $s = -\frac{1}{2}gt^2 + h$, where g is the acceleration of gravity.

(a) At what time does the object hit the ground?

(b) If $h = 100$ ft, compare the impact times for Earth ($g = 32$ ft/s^2), for Mars ($g = 12$ ft/s^2), and for the moon ($g = 5.5$ ft/s^2).

(c) Using the times found in part **(b)**, find the corresponding impact velocities for Earth, Mars, and the moon.

45. The height of a projectile shot from ground level is given by $s = -16t^2 + 256t$, where s is measured in feet and t in seconds.

(a) Determine the height of the projectile at $t = 2$, $t = 6$, $t = 9$, and $t = 10$.

(b) What is the average velocity of the projectile between $t = 2$ and $t = 5$.

(c) Show that the average velocity between $t = 7$ and $t = 9$ is zero. Interpret physically.

(d) At what time does the projectile hit the ground?

(e) Determine the instantaneous velocity at time t_1.

(f) What is the impact velocity?

(g) What is the maximum height the projectile attains?

46. Suppose the function $f(t) = t^3 - 3t^2 + 10$ gives the position at any time t of a particle moving in a straight line.

(a) Find the velocity of the particle at any time t.

(b) Find the instantaneous rate of change of the velocity at any time t.

47. Suppose the graph shown in Figure 2.18 is that of position function $s = f(t)$ of a particle moving in a straight line, where s is measured in meters and t in seconds.

(a) Estimate the position of the particle at $t = 4$ and at $t = 6$.

(b) Estimate the average velocity of the particle between $t = 4$ and $t = 6$.

(c) Estimate the initial velocity of the particle—that is, its velocity at $t = 0$.

(d) Estimate a time at which the velocity of the particle is zero.

(e) Determine an interval on which the velocity of the particle is decreasing.

(f) Determine an interval on which the velocity of the particle is increasing.

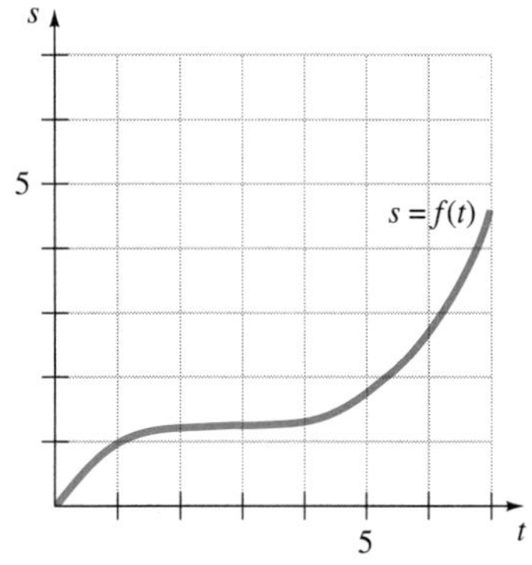

FIGURE 2.18

2.2 THE DERIVATIVE

In the last section we saw that if a graph of a function $y = f(x)$ possesses a tangent at a point $(a, f(a))$, then the slope of this tangent is

$$m_{\tan} = \lim_{\Delta x \to 0} \frac{f(a + \Delta x) - f(a)}{\Delta x}$$

For a given function it is usually possible to obtain a general formula, or rule, that gives the value of the slope of a tangent line. This is accomplished by computing

$$\lim_{\Delta x \to 0} \frac{f(x + \Delta x) - f(x)}{\Delta x} \tag{1}$$

for *any* x (for which the limit exists). We then substitute a value of x *after* the limit has been found. The limit (1) is said to be the **derivative** of f and is denoted by f'.

DEFINITION 2.3 Derivative of a Function

The **derivative** of a function $y = f(x)$ with respect to x is

$$f'(x) = \lim_{\Delta x \to 0} \frac{f(x + \Delta x) - f(x)}{\Delta x} = \lim_{\Delta x \to 0} \frac{\Delta y}{\Delta x} \tag{2}$$

whenever this limit exists.

The derivative $f'(x)$ is also said to be the **instantaneous rate of change** of the function $y = f(x)$ with respect to the variable x.

Let us now reconsider Examples 1 and 2 of Section 2.1.

EXAMPLE 1 Find the derivative of $f(x) = x^2$.

Solution As before, the process consists of four steps:

(*i*) $f(x + \Delta x) = (x + \Delta x)^2 = x^2 + 2x\,\Delta x + (\Delta x)^2$

(*ii*) $\Delta y = f(x + \Delta x) - f(x)$

$\quad = [x^2 + 2x\,\Delta x + (\Delta x)^2] - x^2 = \Delta x[2x + \Delta x]$

(*iii*) $\dfrac{\Delta y}{\Delta x} = \dfrac{\Delta x[2x + \Delta x]}{\Delta x} = 2x + \Delta x$

(*iv*) $f'(x) = \lim_{\Delta x \to 0} \dfrac{\Delta y}{\Delta x} = \lim_{\Delta x \to 0} [2x + \Delta x] = 2x$ □

In Example 1, $f'(x) = 2x$ is another function of x that is *derived* (whence the name derivative) from the original function. Also, observe that the result in Example 2 of Section 2.1 is obtained by evaluating $f'(x)$ *at* $x = 1$.

EXAMPLE 2 For $f(x) = x^2$, find $f'(-2)$, $f'(0)$, $f'(\frac{1}{2})$, and $f'(3)$. Interpret.

Solution From Example 1 we know that $f'(x) = 2x$. Hence,

$$f'(-2) = -4, \quad f'(0) = 0, \quad f'\left(\frac{1}{2}\right) = 1, \quad \text{and} \quad f'(3) = 6$$

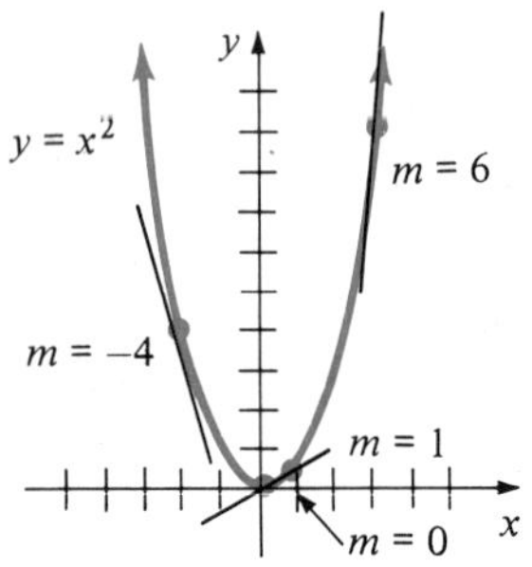

FIGURE 2.19

As shown in Figure 2.19, these values represent, in turn, the slopes of the tangent lines to the graph of $y = x^2$ at the points

$$(-2, 4),\quad (0, 0),\quad \left(\frac{1}{2}, \frac{1}{4}\right),\quad \text{and}\quad (3, 9)$$ □

EXAMPLE 3 Find the derivative of $f(x) = x^3$.

Solution To calculate $f(x + \Delta x)$, we use the Binomial Theorem.

(*i*) $f(x + \Delta x) = (x + \Delta x)^3 = x^3 + 3x^2\,\Delta x + 3x(\Delta x)^2 + (\Delta x)^3$

(*ii*) $\Delta y = f(x + \Delta x) - f(x) = [x^3 + 3x^2\,\Delta x + 3x(\Delta x)^2 + (\Delta x)^3] - x^3$
$= \Delta x[3x^2 + 3x\,\Delta x + (\Delta x)^2]$

(*iii*) $\dfrac{\Delta y}{\Delta x} = \dfrac{\Delta x[3x^2 + 3x\,\Delta x + (\Delta x)^2]}{\Delta x} = 3x^2 + 3x\,\Delta x + (\Delta x)^2$

(*iv*) $f'(x) = \lim\limits_{\Delta x \to 0} \dfrac{\Delta y}{\Delta x} = \lim\limits_{\Delta x \to 0} [3x^2 + 3x\,\Delta x + (\Delta x)^2] = 3x^2$ □

EXAMPLE 4 Find an equation of the tangent line to the graph of $f(x) = x^3$ at $x = \frac{1}{2}$.

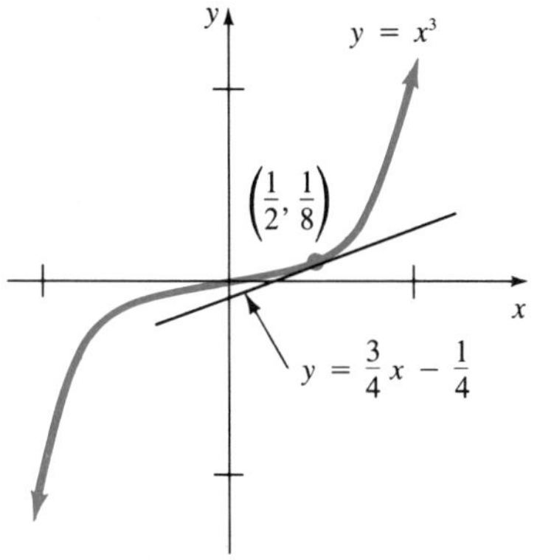

FIGURE 2.20

Solution Since we have just seen that $f'(x) = 3x^2$, it follows that the slope of the tangent at $x = \frac{1}{2}$ is

$$f'\left(\frac{1}{2}\right) = 3\left(\frac{1}{2}\right)^2 = \frac{3}{4}$$

The y-coordinate of the point of tangency is $f(\frac{1}{2}) = (\frac{1}{2})^3 = \frac{1}{8}$. Thus, at $(\frac{1}{2}, \frac{1}{8})$ an equation of the tangent line is given by

$$y - \frac{1}{8} = \frac{3}{4}\left(x - \frac{1}{2}\right) \quad \text{or} \quad y = \frac{3}{4}x - \frac{1}{4}$$

The graph of the function and the tangent line are given in Figure 2.20. □

EXAMPLE 5 Find the derivative of $f(x) = \dfrac{1}{x - 2}$.

Solution In this case you should be able to show that

$$\Delta y = f(x + \Delta x) - f(x) = \frac{1}{x + \Delta x - 2} - \frac{1}{x - 2} = \frac{-\Delta x}{(x + \Delta x - 2)(x - 2)}$$

Therefore,

$$f'(x) = \lim_{\Delta x \to 0} \frac{\Delta y}{\Delta x} = \lim_{\Delta x \to 0} \frac{-\Delta x}{\Delta x(x + \Delta x - 2)(x - 2)}$$

$$= \lim_{\Delta x \to 0} \frac{-1}{(x + \Delta x - 2)(x - 2)} = \frac{-1}{(x - 2)^2}$$ □

Symbols The following is a list of some of the **symbols** used throughout mathematical literature to denote the derivative of a function:

$$f'(x), \quad \frac{dy}{dx}, \quad y', \quad Dy, \quad D_xy$$

The second symbol obviously has its origins in the Δy and Δx notation: $dy/dx = \lim_{\Delta x \to 0} \Delta y/\Delta x$. The dy/dx notation is due to Leibniz. For a function such as $f(x) = x^2$, we write $f'(x) = 2x$; if the same function is written $y = x^2$, we then utilize $dy/dx = 2x$, or $y' = 2x$, or $D_xy = 2x$.

Value of a Derivative The **value of the derivative at *a*** is denoted by

$$f'(a), \quad \left.\frac{dy}{dx}\right|_{x=a}, \quad y'(a), \quad \left.D_xy\right|_{x=a}$$

EXAMPLE 6 The value of the derivative of $y = x^2$ at 5 is

$$\left.\frac{dy}{dx}\right|_{x=5} = \left.2x\right|_{x=5} = 10$$ □

Operation The process of finding a derivative is called **differentiation.** If $f'(a)$ exists, the function f is said to be **differentiable** at a. The **operation** of differentiation with respect to the variable x is represented by the symbols d/dx and D_x. For instance, the results in Examples 1, 3, and 5 can be expressed, in turn, as

$$\frac{d}{dx}x^2 = 2x, \quad \frac{d}{dx}x^3 = 3x^2, \quad D_x\frac{1}{x-2} = \frac{-1}{(x-2)^2}$$

Isaac Newton (1642–1727) Newton used the notation y to represent a *fluxion*, or derivative of a function. This symbol never achieved overwhelming popularity among mathematicians and is used today primarily by physicists. For typographical reasons, the so-called fly-speck notation has been superseded by the prime notation.

It is acknowledged that Newton, an English mathematician and physicist, was the first to set forth many of the basic principles of calculus in unpublished manuscripts on the *method of fluxions*, dated 1665. The word *fluxion* originated from the concept of quantities that "flow"—that is, quantities that change at a certain rate. Newton attained everlasting fame with the publication of his law of universal gravitation in his monumental treatise *Philosophiae Naturalis Principia Mathematica* in 1687. Newton was also the first to prove, using the calculus and his law of gravitation, Kepler's three empirical laws of planetary motion and was the first to prove that white light was composed of all colors. He was elected to Parliament, was appointed Warden of the Mint, and was knighted in 1705. Sir Isaac Newton said about his many accomplishments: "If I have seen farther than others, it is by standing on shoulders of giants."

Gottfried Wilhelm Leibniz (1646–1716) The dy/dx notation for a derivative of a function is due to Leibniz. In fact, it was Leibniz who introduced the word *function* into mathematical literature. A German mathematician, lawyer, and philosopher, Leibniz published a short version of his calculus in an article in a periodical journal in 1684. But, since it was well known that Newton's manuscripts on the *method of fluxions* dated from 1665. Leibniz was accused of appropriating his ideas from these unpublished works. Fueled by nationalistic prides, a controversy about who was the first to "invent" calculus raged for many years. Historians now agree that both Leibniz and Newton arrived at many of the major premises of calculus independent of each other. Leibniz and Newton are considered the "co-inventors" of the subject.

Horizontal Tangents If $y = f(x)$ is continuous at a and $f'(a) = 0$, then the tangent line at $(a, f(a))$ is said to be **horizontal**. In Example 2 we saw that for the continuous function $f(x) = x^2$, $f'(x) = 2x$, so that $f'(0) = 0$. Thus, the tangent line to the graph is horizontal at $(0, f(0))$ or $(0, 0)$. You should verify by Definition 2.3 that the derivative of the continuous function $f(x) = -x^2 + 4x + 1$ is $f'(x) = -2x + 4$. Observe in this latter case that $f'(x) = 0$ when $-2x + 4 = 0$ or $x = 2$. There is a horizontal tangent at $(2, f(2))$ or $(2, 5)$.

Domain of f′ The domain of f', defined by (2), is the set of numbers x for which the limit exists. A derivative fails to exist at a for the same reasons a tangent to its graph fails to exist:

(*i*) the function f is discontinuous at a, or

(*ii*) the graph of f has a corner at $(a, f(a))$.

In addition, since the derivative gives slope, f' will fail to exist

(*iii*) at a point $(a, f(a))$ at which the tangent line to the graph is vertical.

The domain of f' is necessarily a subset of the domain of f.

EXAMPLE 7

(*a*) The function $f(x) = 1/(2x - 1)$ is discontinuous at $\frac{1}{2}$. Therefore, f is not differentiable at this point.

(*b*) The graph of $f(x) = |x|$ possesses a corner at the origin. In Example 7 of Section 2.1, we saw that $\lim_{\Delta x \to 0} \Delta y/\Delta x$ fails to exist at 0 and so f is not differentiable there. □

Vertical Tangents Let $y = f(x)$ be continuous at a point a. If $\lim_{x \to a} |f'(x)| = \infty$, then the graph of f is said to have a **vertical tangent** at $(a, f(a))$. In Example 6 of Section 2.1 we mentioned that the graph of $y = x^{1/3}$ possesses a vertical tangent line at $(0, 0)$. We prove this assertion in the next example.

EXAMPLE 8 It is left as an exercise to use (2) to prove that the derivative of $f(x) = x^{1/3}$ is given by

$$f'(x) = \frac{1}{3x^{2/3}}$$

(See Problem 28 in Exercises 2.2.) Although f is continuous at 0, it is clear that f' is not defined at that value. In other words, f is not differentiable at 0. However, since

$$\lim_{x \to 0^+} f'(x) = \infty \quad \text{and} \quad \lim_{x \to 0^-} f'(x) = \infty$$

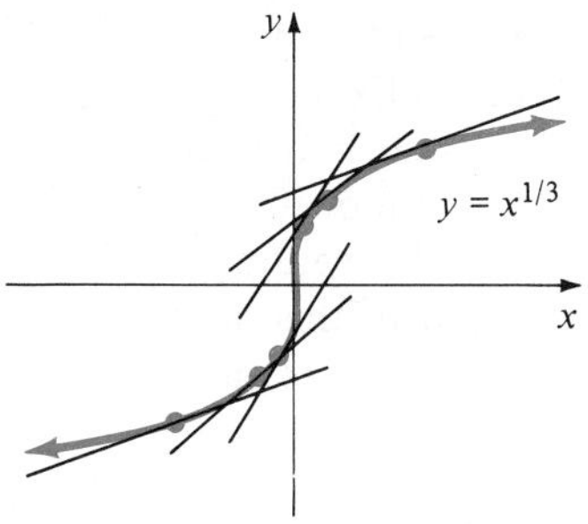

FIGURE 2.21

we have $|f'(x)| \to \infty$ as $x \to 0$. This is sufficient to say that there is a tangent line at $(0, f(0))$ or $(0, 0)$ and that it is vertical. Figure 2.21 shows that the tangent lines to the graph on either side of the origin become steeper and steeper as $x \to 0$. □

Differentiability on an Interval A function f is said to be **differentiable**

(*i*) **on an open interval** (a, b) when f' exists at every point in the interval, and

(*ii*) **on a closed interval** $[a, b]$ when f is differentiable on (a, b) and

$$f'_+(a) = \lim_{\Delta x \to 0^+} \frac{f(a + \Delta x) - f(a)}{\Delta x}$$
$$f'_-(b) = \lim_{\Delta x \to 0^-} \frac{f(b + \Delta x) - f(b)}{\Delta x} \tag{3}$$

both exist.

The limits in (3) are called **right-hand** and **left-hand derivatives**, respectively. A function is differentiable on $[a, \infty)$ when it is differentiable on (a, ∞) and has a right-hand derivative at a. A similar definition in terms of a left-hand derivative holds for differentiability on $(-\infty, a]$. Moreover, it can be shown that

f is differentiable at a point c in an interval (a, b) if and only if $f'_+(c) = f'_-(c)$.

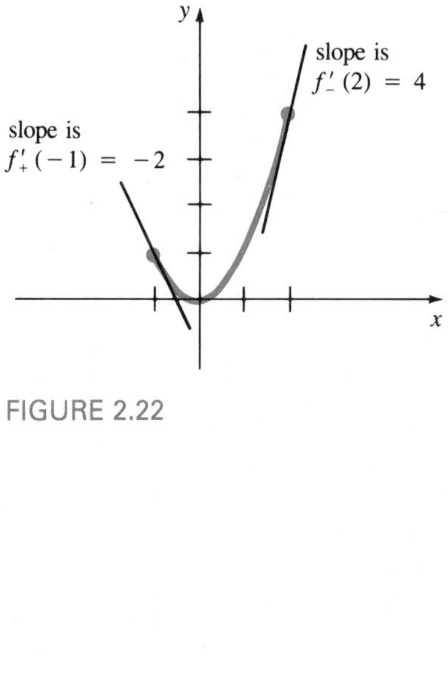

FIGURE 2.22

EXAMPLE 9 The function

$$f(x) = x^2, \qquad -1 \le x \le 2$$

is differentiable on $[-1, 2]$, since $f'(x) = 2x$ for every x in $(-1, 2)$ and

$$f'_-(2) = 4 \quad \text{and} \quad f'_+(-1) = -2$$

See Figure 2.22. □

EXAMPLE 10 The function $f(x) = x^2$ is differentiable on $(-\infty, \infty)$. □

EXAMPLE 11 Since $f(x) = 1/x$ is discontinuous at $x = 0$, f is not differentiable on any interval containing 0. □

EXAMPLE 12 Show that the function

$$f(x) = \begin{cases} 2, & x < 1 \\ x + 1, & x \ge 1 \end{cases}$$

is not differentiable at $x = 1$.

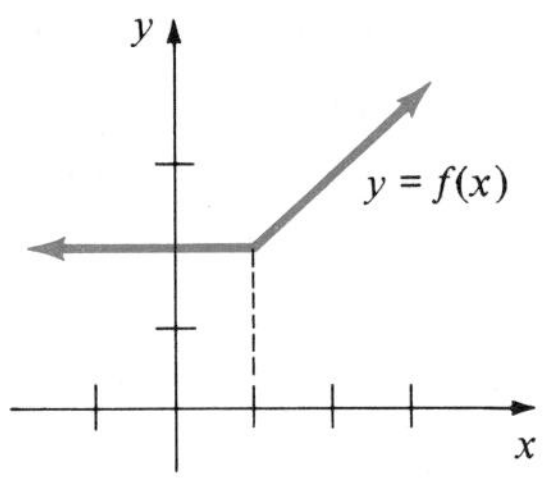

FIGURE 2.23

Solution Figure 2.23 shows that the graph of f has a corner at (1, 2). Now,

$$f'_+(1) = \lim_{\Delta x \to 0^+} \frac{f(1 + \Delta x) - f(1)}{\Delta x} = \lim_{\Delta x \to 0^+} \frac{[(1 + \Delta x) + 1] - 2}{\Delta x} = 1$$

whereas

$$f'_-(1) = \lim_{\Delta x \to 0^-} \frac{f(1 + \Delta x) - f(1)}{\Delta x} = \lim_{\Delta x \to 0^-} \frac{2 - 2}{\Delta x} = 0$$

Since $f'_+(1) \neq f'_-(1)$, f is not differentiable at 1. Consequently, f is not differentiable on any interval that contains 1. □

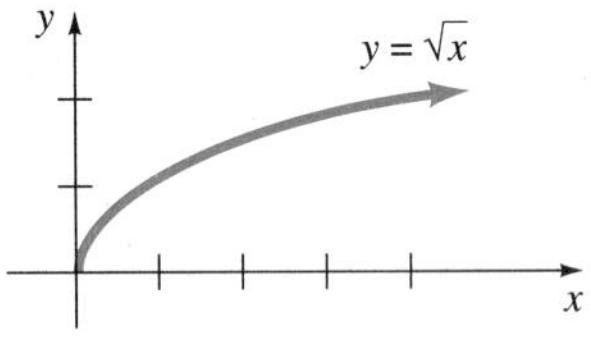

FIGURE 2.24

EXAMPLE 13 It is left as an exercise to show that the function $f(x) = \sqrt{x}$ is not differentiable on the interval $[0, \infty)$. (See Problem 27 in Exercises 2.2.) The graph of f given in Figure 2.24 should provide you with a clue as to why $f'_+(0)$ does not exist. The function $f(x) = \sqrt{x}$ is, however, differentiable on $(0, \infty)$ and on any subinterval of $(0, \infty)$. □

We have seen that $f(x) = |x|$ and $f(x) = x^{1/3}$ are continuous at 0 but not differentiable at that point. In Example 12 the function f is continuous at 1 but $f'(1)$ does not exist. In other words, if f is continuous at a point a, it is not necessarily differentiable at that point. However, if the graph of a function f possesses a tangent line at $(a, f(a))$—that is, if f is differentiable at a—then f must be continuous at that point. We summarize this last fact in the next theorem.

THEOREM 2.1 Differentiability Implies Continuity

If f is differentiable at a point a, then f is continuous at a.

This result is not difficult to prove and is left as an exercise. (See Problem 54 in Exercise 2.2.)

▲ ***Remark*** (*i*) In the preceding discussion, we saw that the derivative of a function is itself a function that gives the slope of a tangent line. The derivative is, however, *not* an equation of a tangent line. Also, to say that $y - y_0 = f'(x) \cdot (x - x_0)$ is an equation of the tangent at (x_0, y_0) is incorrect. Remember that $f'(x)$ must be evaluated at x_0 *before* it is used in the point–slope form. If f is differentiable at x_0, then an equation of the tangent line at (x_0, y_0) is $y - y_0 = f'(x_0) \cdot (x - x_0)$.

(*ii*) Mathematicians from the seventeenth to the nineteenth centuries believed that a continuous function *usually* possessed a derivative. (We have noted exceptions in this section.) In 1872 the German mathematician Karl Weierstrass conclusively destroyed this tenet by publishing an example of a function that was continuous everywhere but nowhere differentiable.

EXERCISES 2.2 *Answers to odd-numbered problems begin on page A-69.*

In Problems 1–20 use Definition 2.3 to find the derivative of the given function.

1. $f(x) = 10$
2. $f(x) = x - 1$
3. $f(x) = -3x + 5$
4. $f(x) = \pi x$
5. $f(x) = 3x^2$
6. $f(x) = -x^2 + 1$
7. $f(x) = 4x^2 - x + 6$
8. $f(x) = 3x^2 + 6x - 7$
9. $f(x) = (x + 1)^2$
10. $f(x) = (2x - 5)^2$
11. $f(x) = x^3 + x$
12. $f(x) = 2x^3 + x^2$
13. $y = -x^3 + 15x^2 - x$
14. $y = x^4$
15. $y = \dfrac{1}{x}$
16. $y = \dfrac{2}{x + 1}$
17. $y = \dfrac{x}{x - 1}$
18. $y = \dfrac{2x + 3}{x + 4}$
19. $f(x) = \dfrac{1}{x} + \dfrac{1}{x^2}$
20. $f(x) = \dfrac{4}{x^3}$

In Problems 21–24 find the derivative of the given function. Find an equation of the tangent line to the graph of the function at the indicated value of x.

21. $f(x) = 4x^2 + 7x;\ x = -1$
22. $f(x) = \frac{1}{3}x^3 + 2x - 4;\ x = 0$
23. $y = x - \dfrac{1}{x};\ x = 1$
24. $y = 2x + 1 + \dfrac{6}{x};\ x = 2$

In Problems 25 and 26 find the point(s) on the graph of the given function where the tangent line is horizontal.

25. $f(x) = x^2 + 8x + 10$
26. $f(x) = x^3 - x^2 + 1$
27. (a) Find the derivative of $f(x) = \sqrt{x}$. [*Hint:* Multiply the numerator and denominator of $\Delta y/\Delta x$ by $(x + \Delta x)^{1/2} + x^{1/2}$.]
    (b) What is the domain of f'?
    (c) Show that $f'_+(0)$ does not exist.
28. (a) Find the derivative of $f(x) = x^{1/3}$. [*Hint:* Multiply the numerator and denominator of $\Delta y/\Delta x$ by $(x + \Delta x)^{2/3} + (x + \Delta x)^{1/3}x^{1/3} + x^{2/3}$.]
    (b) What is the domain of f'?

In Problems 29–32 use the procedures outlined in Problems 27 and 28 to find the derivative of the given function. State the domain of f'.

29. $f(x) = \sqrt{2x + 1}$
30. $f(x) = 1/\sqrt{x}$
31. $f(x) = (x - 4)^{1/3}$
32. $f(x) = (4x)^{1/3} + 7$
33. Use the derivative obtained in Problem 31 to show that the graph of $f(x) = (x - 4)^{1/3}$ has a vertical tangent at (4, 0).
34. Use the derivative obtained in Problem 32 to show that the graph of $f(x) = (4x)^{1/3} + 7$ has a vertical tangent at (0, 7).

In Problems 35 and 36 show that the given function is not differentiable at the indicated value of x.

35. $f(x) = \begin{cases} -x + 2, & x \le 2 \\ 2x - 4, & x > 2 \end{cases};\quad x = 2$
36. $f(x) = \begin{cases} 3x, & x < 0 \\ -4x, & x \ge 0 \end{cases};\quad x = 0$

In Problems 37 and 38 show that the given function is differentiable at the indicated value of x.

37. $f(x) = \begin{cases} 4x, & x \le 1 \\ 2x^2 + 2, & x > 1 \end{cases};\quad x = 1$
38. $f(x) = \begin{cases} 1, & x < 0 \\ x^2 + 1, & x \ge 0 \end{cases};\quad x = 0$

In Problems 39 and 40 determine whether the given function is differentiable at the indicated value of x.

39. $f(x) = x|x|;\ x = 0$
40. $f(x) = |x^2 - 1|;\ x = 1$
41. Show that if $f(x) = k$ is a constant function, then $f'(x) = 0$. Interpret geometrically.
42. Show that if $f(x) = ax + b$, $a \ne 0$, is a linear function, then $f'(x) = a$. Interpret geometrically.

In Problems 43–48 sketch the graph of f' from the graph of f.

43.

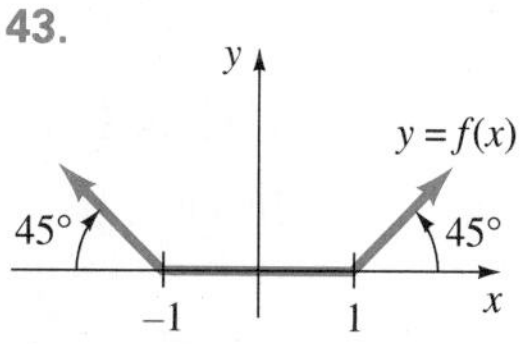

FIGURE 2.25

44.

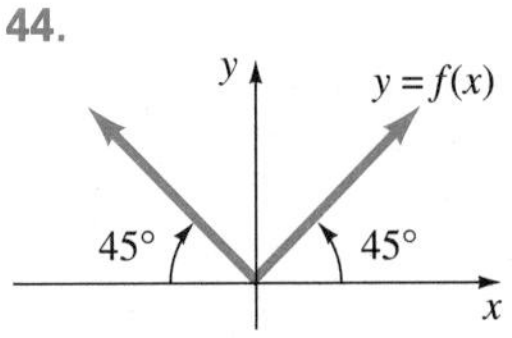

FIGURE 2.26

45.

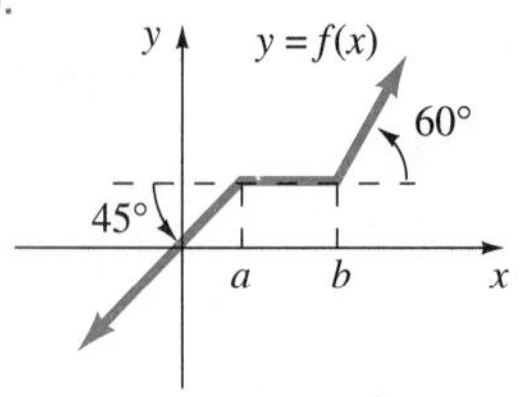

FIGURE 2.27

46.

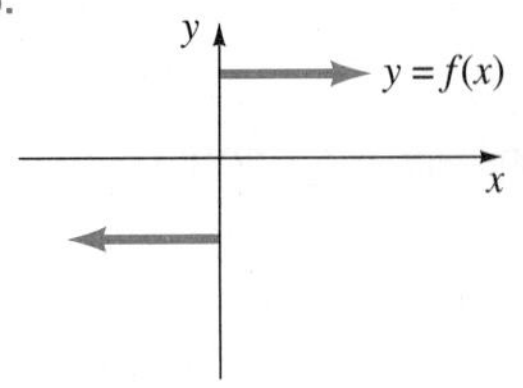

FIGURE 2.28

47.

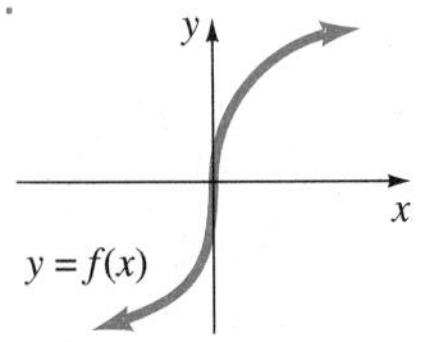

FIGURE 2.29

48.

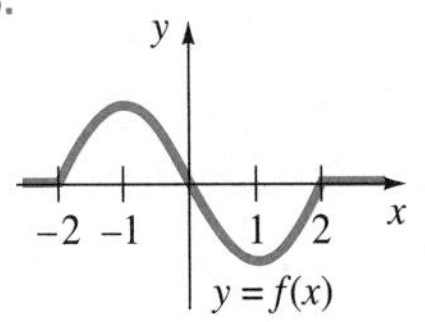

FIGURE 2.30

49. Let c_1 and c_2 be arbitrary constants. Show that $D_x(c_1x^2 + c_2x^3) = c_1D_xx^2 + c_2D_xx^3$.

50. Suppose f is an even differentiable function on $(-\infty, \infty)$. On geometric grounds, explain why $f'(-x) = -f'(x)$; that is, f' is an odd function.

51. An alternative formulation of the derivative of a function f at a is

$$f'(a) = \lim_{x \to a} \frac{f(x) - f(a)}{x - a} \tag{4}$$

whenever the limit exists. Show that (4) is equivalent to (2) when $x = a$. [*Hint:* Let $\Delta x = x - a$.]

52. Use (4) to find the derivative of $f(x) = 6x^2$ at any number a.

53. Use (4) to find the derivative of $f(x) = x^{3/2}$ at any number a.

54. Use (4) to prove Theorem 2.1. [*Hint:* Take the limit of

$$f(x) - f(a) = \frac{f(x) - f(a)}{x - a} \cdot (x - a) \quad \text{as } x \to a$$

and use the fact that a limit of a difference is the difference of limits and the limit of a product is the product of the limits whenever all limits exist.]

2.3 RULES OF DIFFERENTIATION I: POWER AND SUM RULES

The definition of a derivative has the obvious drawback of being rather clumsy and tiresome to apply. We shall now see that the derivative of a function such as $f(x) = 6x^{100} + x^{35}$ can be obtained, so to speak, with a flick of a pencil.

Power Rule The same procedure outlined in Examples 1 and 3 of Section 2.2 to differentiate $f(x) = x^2$ and $f(x) = x^3$ can be employed to find the derivative of $f(x) = x^n$ for any positive integer n. The starting point is again the Binomial Theorem. Recall, for real numbers a and b, and n a positive integer,

$$(a + b)^n = a^n + \frac{n}{1!}a^{n-1}b + \frac{n(n-1)}{2!}a^{n-2}b^2 + \cdots + \frac{n(n-1)\cdots(n-r+1)}{r!}a^{n-r}b^r + \cdots + b^n$$

where $r! = 1 \cdot 2 \cdot 3 \cdots (r-1)r$.

THEOREM 2.2 Power Rule (Positive Integer Exponents)

If n is a positive integer, then

$$\frac{d}{dx}x^n = nx^{n-1} \tag{1}$$

Proof Let $f(x) = x^n$, n a positive integer. By the Binomial Theorem we can write

$$f(x + \Delta x) = (x + \Delta x)^n = x^n + nx^{n-1}\,\Delta x + \frac{n(n-1)}{2}x^{n-2}(\Delta x)^2 + \cdots + (\Delta x)^n$$

Thus,

$$\Delta y = f(x + \Delta x) - f(x) = \left[x^n + nx^{n-1}\,\Delta x + \frac{n(n-1)}{2}x^{n-2}(\Delta x)^2 + \cdots + (\Delta x)^n\right] - x^n$$

$$= \Delta x\left[nx^{n-1} + \frac{n(n-1)}{2}x^{n-2}\,\Delta x + \cdots + (\Delta x)^{n-1}\right]$$

and
$$\frac{\Delta y}{\Delta x} = \frac{\Delta x\left[nx^{n-1} + \frac{n(n-1)}{2}x^{n-2}\,\Delta x + \cdots + (\Delta x)^{n-1}\right]}{\Delta x}$$

$$= nx^{n-1} + \frac{n(n-1)}{2}x^{n-2}\,\Delta x + \cdots + (\Delta x)^{n-1} \qquad (2)$$

Since each term in (2) after the first contains a factor of Δx, it follows that $f'(x) = \lim_{\Delta x \to 0} \Delta y/\Delta x = nx^{n-1}$. ■

The Power Rule simply states that to differentiate x^n:

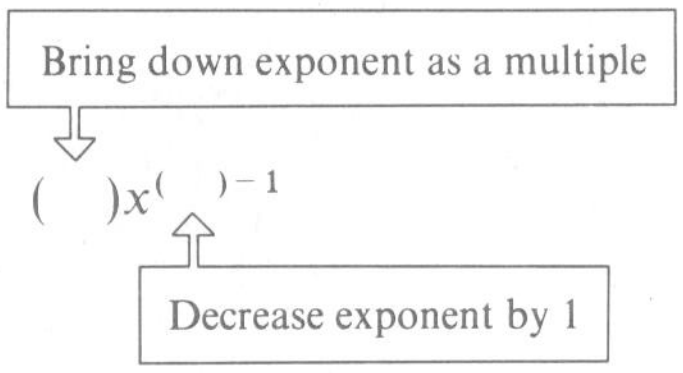

Observe that the derivatives of $y = x^2$ and $y = x^3$ follow from (1):

$$y = x^2, \qquad \frac{dy}{dx} = 2x^{2-1} = 2x$$

$$y = x^3, \qquad \frac{dy}{dx} = 3x^{3-1} = 3x^2$$

EXAMPLE 1 Differentiate $y = x^6$.

Solution From the Power Rule (1),

$$\frac{dy}{dx} = 6x^{6-1} = 6x^5$$ □

EXAMPLE 2 Differentiate $y = x$.

Solution Identifying $n = 1$, we have from (1)

$$\frac{dy}{dx} = 1x^{1-1} = 1x^0 = 1 \qquad \square$$

The next result tells us that the derivative of a constant function is zero. The proof is immediate.

THEOREM 2.3 Constant Function

If $f(x) = k$ is a constant function, then $f'(x) = 0$. (3)

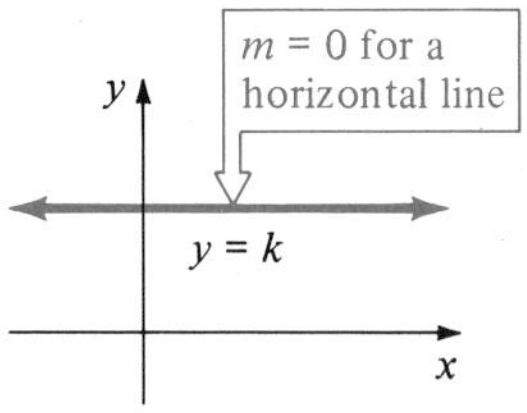

FIGURE 2.31

Proof $\Delta y = f(x + \Delta x) - f(x) = k - k = 0$

Hence, $\lim_{\Delta x \to 0} \Delta y/\Delta x = 0$. ■

Theorem 2.3 has an obvious geometric interpretation. As shown in Figure 2.31, the slope m of the horizontal line $y = k$ is, of course, zero.

EXAMPLE 3 From (3) we have (*a*) $\dfrac{d}{dx} 5 = 0$ and (*b*) $\dfrac{d}{dx} \pi^3 = 0$. □

Note the result in part (*b*) of Example 3. A *common mistake* is to apply the Power Rule and write $d/dx(\pi^3) = 3 \cdot \pi^3$. Bear in mind that (1) applies only to a *variable base* x. Furthermore, (3) implies that (1) holds for the case $n = 0$: If $x \neq 0$, $f(x) = x^0 = 1$ and $f'(x) = 0 = 0 \cdot x^{0-1}$.

THEOREM 2.4 Constant Multiple of a Function

If c is any constant and f is differentiable, then

$$\frac{d}{dx}[cf(x)] = cf'(x) \qquad (4)$$

Proof Let $G(x) = cf(x)$. Then

$$\begin{aligned} G'(x) = \lim_{\Delta x \to 0} \frac{G(x+\Delta x) - G(x)}{\Delta x} &= \lim_{\Delta x \to 0} \frac{cf(x+\Delta x) - cf(x)}{\Delta x} \\ &= \lim_{\Delta x \to 0} c\left[\frac{f(x+\Delta x) - f(x)}{\Delta x}\right] \\ &= c \lim_{\Delta x \to 0} \frac{f(x+\Delta x) - f(x)}{\Delta x} = cf'(x) \end{aligned}$$ ■

EXAMPLE 4 Differentiate $y = 5x^3$.

Solution From (1) and (4),

$$\frac{dy}{dx} = 5 \cdot \frac{d}{dx} x^3 = 5(3x^2) = 15x^2 \qquad \square$$

THEOREM 2.5 Sum Rule

If f and g are differentiable functions, then

$$\frac{d}{dx}[f(x) + g(x)] = f'(x) + g'(x) \tag{5}$$

Proof Let $G(x) = f(x) + g(x)$. Then

$$\begin{aligned} G'(x) = \lim_{\Delta x \to 0} \frac{G(x + \Delta x) - G(x)}{\Delta x} &= \lim_{\Delta x \to 0} \frac{[f(x + \Delta x) + g(x + \Delta x)] - [f(x) + g(x)]}{\Delta x} \\ &= \lim_{\Delta x \to 0} \frac{f(x + \Delta x) - f(x) + g(x + \Delta x) - g(x)}{\Delta x} \\ &= \lim_{\Delta x \to 0} \frac{f(x + \Delta x) - f(x)}{\Delta x} + \lim_{\Delta x \to 0} \frac{g(x + \Delta x) - g(x)}{\Delta x} \\ &= f'(x) + g'(x) \qquad \blacksquare \end{aligned}$$

EXAMPLE 5 Differentiate $y = x^5 + x^2$.

Solution From (1) and (5) we have

$$\frac{dy}{dx} = \frac{d}{dx} x^5 + \frac{d}{dx} x^2 = 5x^4 + 2x \qquad \square$$

The derivative of the function $f(x) = 6x^{100} + x^{35}$, mentioned in the introductory remark, is now readily seen to be $f'(x) = 600x^{99} + 35x^{34}$.

Since $f(x) - g(x) = f(x) + [-g(x)]$, the Sum Rule also holds for the *difference* of two functions.

EXAMPLE 6 Differentiate $y = 9x^7 - 3x^4$.

Solution In view of (1), (4), and (5), we can write

$$\frac{dy}{dx} = 9 \frac{d}{dx} x^7 - 3 \frac{d}{dx} x^4 = 63x^6 - 12x^3 \qquad \square$$

The Sum Rule (5) also extends to any finite sum of differentiable functions:

$$\frac{d}{dx}[f_1(x) + f_2(x) + \cdots + f_n(x)] = f'_1(x) + f'_2(x) + \cdots + f'_n(x)$$

From this latter result, (1) and (4), we see that any polynomial function is differentiable.

EXAMPLE 7 Differentiate $y = 4x^5 - \frac{1}{2}x^4 + 9x^3 + 10x^2 - 13x + 6$.

Solution

$$\frac{dy}{dx} = 4\frac{d}{dx}x^5 - \frac{1}{2}\frac{d}{dx}x^4 + 9\frac{d}{dx}x^3 + 10\frac{d}{dx}x^2 - 13\frac{d}{dx}x + \frac{d}{dx}6$$

Since $\frac{d}{dx}6 = 0$, we obtain

$$\frac{dy}{dx} = 20x^4 - 2x^3 + 27x^2 + 20x - 13$$ □

EXAMPLE 8 Find an equation of the tangent line to the graph of $f(x) = 3x^4 + 2x^3 - 7x$ at $x = -1$.

Solution The y-coordinate of the point that corresponds to $x = -1$ is $f(-1) = 8$. Now,

$$f'(x) = 12x^3 + 6x^2 - 7 \quad \text{and so} \quad f'(-1) = -13$$

At $(-1, 8)$ the point–slope form gives an equation of the tangent line:

$$y - 8 = -13(x + 1) \quad \text{or} \quad y = -13x - 5$$ □

Normal Line A **normal line** to a graph at a point P is one that is perpendicular to the tangent line at P.

EXAMPLE 9 Find an equation of the normal line to the graph of $y = x^2$ at $x = 1$.

Solution Since $dy/dx = 2x$, we know that $m_{\tan} = 2$ at (1, 1). Thus, the slope of the normal line shown in Figure 2.32 is $m = -\frac{1}{2}$. Its equation is

$$y - 1 = -\frac{1}{2}(x - 1) \quad \text{or} \quad y = -\frac{1}{2}x + \frac{3}{2}$$ □

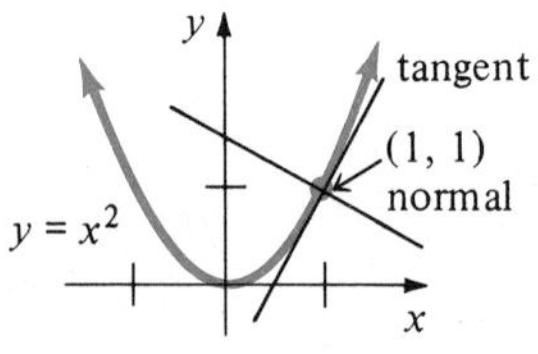

FIGURE 2.32

Calculator/Computers Although at this writing there are several hand calculators with graphing capabilities, only the HP-28S has the additional capability of being able to do some symbolic calculus. There are, however, currently several sophisticated computer programs, called **computer algebra systems**, that are able to perform a wide range of mathematical operations such as graphing, numerical approximations, and symbolic computations. The first such programs developed were MACSYMA and muMath. MACSYMA, the more powerful of the two programs, was originally developed for mainframe computers but is now available for microcomputers. Mathematica, Maple, and DERIVE are three other popular and impressive systems with the ability to do symbolic computations on a personal computer. But the drawback of some of these software packages, besides the expense, is that they require a computer with at least several megabytes of memory and a hard disk drive.

Differentiating a polynomial such as $f(x) = 5x^4 - 6x^2 + 4x - 3$ on a calculator or a computer is like using a machine gun to kill a fly. Nevertheless, for the sake of illustration we will use the HP-28S. We first depress, in turn, the ['] and [d/dx] keys, key in the symbols X(5*X^4 – 6*X^2 + 4*X – 3)′, and then [ENTER]

'∂X(5*X^4 – 6*X^2 + 4*X – 3)′

The symbol ∂X indicates we are performing differentiation with respect to x. We then use the [EVAL] key six times in succession to obtain '5*(4*X^3) – 6*(2*X) + 4′, in other words, $20x^3 - 12x + 4$. In Mathematica and Maple we type the respective commands

```
D[5 x^4 – 6 x^2 + 4 x – 3, x]
diff(5*x^4 – 6*x^2 + 4*x – 3, x);
```

and then [ENTER]. The result is given almost immediately in standard notation: $20x^3 - 12x + 4$.

▲ ***Remark*** In the different contexts of science, engineering, and business, functions are often expressed in variables other than x and y. Correspondingly, we must adapt the derivative notation to new symbols; for example,

Function	*Derivative*
$v(t) = 32t$	$v'(t) = \frac{dv}{dt} = 32$
$A(r) = \pi r^2$	$A'(r) = \frac{dA}{dr} = 2\pi r$
$H(z) = \frac{1}{4}z^6$	$H'(z) = \frac{dH}{dz} = \frac{3}{2}z^5$
$D(p) = 800 - 120p + 5p^2$	$D'(p) = \frac{dD}{dp} = -120 + 10p$
$r(\theta) = 4\theta^2 - 3\theta$	$r'(\theta) = \frac{dr}{d\theta} = 8\theta - 3$

EXERCISES 2.3

Answers to odd-numbered problems begin on page A-69.

In Problems 1–20 find the derivative of the given function.

1. $y = x^9$
2. $y = 4x^{12}$
3. $y = \pi^6/12$
4. $y = -18$
5. $y = 7x^2 - 4x$
6. $y = 6x^3 + 3x^2 - 10$
7. $f(x) = \frac{1}{5}x^5 - 3x^4 + 9x^2 + 1$
8. $f(x) = -\frac{2}{3}x^6 + 4x^5 - 13x^2 + 8x + 2$
9. $f(x) = x^3(4x^2 - 5x - 6)$
10. $f(x) = \dfrac{2x^5 + 3x^4 - x^3}{x^2}$
11. $f(x) = (x + 1)^2$
12. $f(x) = (9 + x)(9 - x)$
13. $h(u) = (4u)^3$
14. $P(t) = (2t)^4 - (2t)^2$
15. $F(z) = 6z^3 + az^2 + a^3$, a a constant.
16. $g(w) = \dfrac{w^n - 5^n}{n}$, n a positive integer
17. $G(\beta) = -3\beta^4 + 7\beta^3 - 5\beta^2 + 2$
18. $Q(u) = \dfrac{u^5 + 4u^2 - 3}{6}$
19. $f(x) = \dfrac{x^2}{(x^2 + 5)^{-2}}$
20. $f(x) = (x^3 + x^2)^3$

In Problems 21–24 find an equation of the tangent line to the graph of the given function at the indicated value of x.

21. $y = 2x^3 - 1$; $x = -1$
22. $y = \frac{1}{2}x^2 + 3$; $x = 2$
23. $y = 4x^2 - 4x - 20$; $x = 3$
24. $y = -x^3 + 6x^2$; $x = 1$

In Problems 25–28 find the point(s) on the graph of the given function at which the tangent is horizontal.

25. $y = x^2 - 8x + 5$
26. $y = \frac{1}{3}x^3 - \frac{1}{2}x^2$
27. $y = x^3 - 3x^2 - 9x + 2$
28. $y = x^4 - 4x^3$

In Problems 29–32 find an equation of the normal line to the graph of the given function at the indicated value of x.

29. $y = -x^2 + 1$; $x = 2$
30. $y = x^3$; $x = 1$
31. $y = \frac{1}{3}x^3 - 2x^2$; $x = 4$
32. $y = x^4 - x$; $x = -1$

In Problems 33 and 34 determine intervals for which $f'(x) > 0$ and intervals for which $f'(x) < 0$.

33. $f(x) = x^2 + 8x - 4$
34. $f(x) = x^3 - 3x^2 - 9x$

35. Let f be a differentiable function. If $f'(x) > 0$ for all x in the interval (a, b), sketch a possible graph of f on the interval. Describe in words the behavior of the graph of f on the interval.

36. Repeat Problem 35 if $f'(x) < 0$ for all x in the interval (a, b).

37. Find the point on the graph of $f(x) = 2x^2 - 3x + 6$ at which the slope of the tangent line is 5.

38. Find the point on the graph of $f(x) = x^2 - x$ at which the tangent line is $3x - 9y - 4 = 0$.

39. Find the point on the graph of $f(x) = x^2 - x$ at which the slope of the normal line is 2.

40. Find the point(s) on the graph of $f(x) = x^2$ such that the tangent line at the point(s) has y-intercept -2.

41. Find the point(s) on the graph of $f(x) = x^2 - 5$ such that the tangent line at the point(s) has x-intercept -3.

42. Find the point on the graph of $f(x) = \frac{1}{4}x^2 - 2x$ at which the tangent line is parallel to the line $3x - 2y + 1 = 0$.

43. Find an equation of a tangent line to the graph of $f(x) = x^3$ that is perpendicular to the line $y = -3x$.

44. Find equations of the tangent lines that pass through $(0, -1)$ tangent to the graph of $f(x) = x^2 + 2x$.

45. Find a point on the graph of $f(x) = x^2 + x$ and a point on the graph of $g(x) = 2x^2 + 4x + 1$ at which the tangent lines are parallel.

46. Find values of a and b such that the slope of the tangent to the graph of $f(x) = ax^2 + bx$ at $(1, 4)$ is -5.

47. Find the values of b and c so that the graph of $f(x) = x^2 + bx$ possesses the tangent line $y = 2x + c$ at $x = -3$.

48. Find an equation of the line(s) that passes through $(\frac{3}{2}, 1)$ and is tangent to the graph of $f(x) = x^2 + 2x + 2$.

49. Find the slopes of all the normal lines to the graph of $f(x) = x^2$ that pass through the point $(2, 4)$. [*Hint:* Draw a figure and note that *at* $(2, 4)$ there is only one normal line.]

50. The graphs of $y = f(x)$ and $y = g(x)$ are said to be **orthogonal** if the tangent lines to each graph are perpendicular at each point of intersection. Show that the graphs of $y = \frac{1}{8}x^2$ and $y = -\frac{1}{4}x^2 + 3$ are orthogonal.

51. Determine the length of the chord of the graph of $f(x) = x^2 - 2$ that is normal to the graph at $(1, -1)$.

52. The height s above ground of a projectile at time t is given by

$$s(t) = \frac{1}{2}gt^2 + v_0t + s_0$$

where g, v_0, and s_0 are constants. Find the instantaneous rate of change of s with respect to t at $t = 4$.

53. The volume V of a sphere of radius r is $V = (4\pi/3)r^3$. Find the surface area S of the sphere if S is the instantaneous rate of change of the volume with respect to the radius.

54. According to Poiseuille,* the velocity v of blood in an artery with a circular cross-section radius R is $v(r) = (P/4vl)(R^2 - r^2)$, where P, v, and l are constants. What is the velocity of blood at the value of r for which $v'(r) = 0$? (See Figure 5.120 for an interpretation of the variable r.)

55. The potential energy of a spring-mass system when the spring is stretched a distance of x units is $U(x) = (k/2)x^2$, where k is the spring constant. The force exerted on the mass is $F = -dU/dx$. Find the force if the spring constant is 30 N/m and the amount of stretch is $\frac{1}{2}$ m.

2.4 RULES OF DIFFERENTIATION II: PRODUCT AND QUOTIENT RULES

In the preceding section we saw that the Sum Rule follows from the limit property that the limit of a sum is the sum of limits whenever the limits exist. We also know that when the limits exist, the limit of a product is the product of the limits. Therefore, a natural conjecture would be that the derivative of a product is the product of the derivatives. However, the rule for differentiating the product of two functions is *not* that simple.

THEOREM 2.6 Product Rule

If f and g are differentiable functions, then

$$\frac{d}{dx}[f(x)g(x)] = f(x)g'(x) + g(x)f'(x) \tag{1}$$

Proof Let $G(x) = f(x)g(x)$. Then

$$G'(x) = \lim_{\Delta x \to 0} \frac{G(x + \Delta x) - G(x)}{\Delta x} = \lim_{\Delta x \to 0} \frac{f(x + \Delta x)g(x + \Delta x) - f(x)g(x)}{\Delta x}$$

$$= \lim_{\Delta x \to 0} \frac{f(x + \Delta x)g(x + \Delta x) \overbrace{- f(x + \Delta x)g(x) + f(x + \Delta x)g(x)}^{\text{zero}} - f(x)g(x)}{\Delta x}$$

$$= \lim_{\Delta x \to 0} \left[f(x + \Delta x)\frac{g(x + \Delta x) - g(x)}{\Delta x} + g(x)\frac{f(x + \Delta x) - f(x)}{\Delta x} \right]$$

$$= \lim_{\Delta x \to 0} f(x + \Delta x) \cdot \lim_{\Delta x \to 0} \frac{g(x + \Delta x) - g(x)}{\Delta x} + \lim_{\Delta x \to 0} g(x) \cdot \lim_{\Delta x \to 0} \frac{f(x + \Delta x) - f(x)}{\Delta x} \tag{2}$$

*Jean Louis Poiseuille (1799–1869), a French physician.

Since f is differentiable, it is continuous and therefore $\lim_{\Delta x \to 0} f(x + \Delta x) = f(x)$. Furthermore, $\lim_{\Delta x \to 0} g(x) = g(x)$. Hence, (2) becomes

$$G'(x) = f(x)g'(x) + g(x)f'(x). \qquad \blacksquare$$

The Product Rule is usually memorized in words:

the first function times the derivative of the second plus the second function times the derivative of the first.

EXAMPLE 1 Differentiate $y = (x^3 - 2x^2 + 4)(8x^2 + 5x)$.

Solution From the Product Rule (1),

first — derivative of second — second — derivative of first

$$\frac{dy}{dx} = (x^3 - 2x^2 + 4) \cdot \frac{d}{dx}(8x^2 + 5x) + (8x^2 + 5x) \cdot \frac{d}{dx}(x^3 - 2x^2 + 4)$$

$$= (x^3 - 2x^2 + 4)(16x + 5) + (8x^2 + 5x)(3x^2 - 4x) \qquad \square$$

Although (1) is stated for only the product of two functions, it can also be applied to functions with a greater number of factors.

EXAMPLE 2 Differentiate $y = (4x + 1)(2x^2 - x)(x^3 - 8x)$.

Solution We identify the first two factors as "the first function."

first — derivative of second — second — derivative of first

$$\frac{dy}{dx} = \overbrace{(4x + 1)(2x^2 - x)}\frac{d}{dx}(x^3 - 8x) + (x^3 - 8x)\frac{d}{dx}(4x + 1)(2x^2 - x)$$

To find the derivative of the first function, we must apply the Product Rule a second time. Thus,

Product Rule again

$$\frac{dy}{dx} = (4x + 1)(2x^2 - x) \cdot (3x^2 - 8) + (x^3 - 8x) \cdot [(4x + 1)(4x - 1) + (2x^2 - x) \cdot 4]$$

$$= (4x + 1)(2x^2 - x)(3x^2 - 8) + (16x^2 - 1)(x^3 - 8x) + 4(2x^2 - x)(x^3 - 8x) \qquad \square$$

The derivative of the quotient of two functions is given by the following:

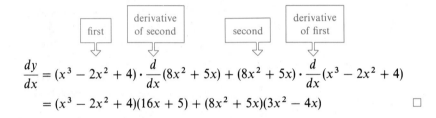

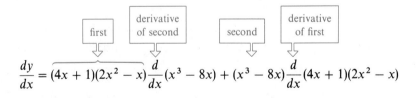

THEOREM 2.7 Quotient Rule

If f and g are differentiable functions and $g(x) \neq 0$, then

$$\frac{d}{dx}\left[\frac{f(x)}{g(x)}\right] = \frac{g(x)f'(x) - f(x)g'(x)}{[g(x)]^2} \tag{3}$$

Proof Let $G(x) = f(x)/g(x)$. Then

$$G'(x) = \lim_{\Delta x \to 0} \frac{G(x + \Delta x) - G(x)}{\Delta x} = \lim_{\Delta x \to 0} \frac{\dfrac{f(x + \Delta x)}{g(x + \Delta x)} - \dfrac{f(x)}{g(x)}}{\Delta x}$$

$$= \lim_{\Delta x \to 0} \frac{g(x)f(x + \Delta x) - f(x)g(x + \Delta x)}{\Delta x\, g(x + \Delta x)g(x)}$$

$$= \lim_{\Delta x \to 0} \frac{g(x)f(x + \Delta x) \overbrace{- g(x)f(x) + g(x)f(x)}^{\text{zero}} - f(x)g(x + \Delta x)}{\Delta x\, g(x + \Delta x)g(x)}$$

$$= \lim_{\Delta x \to 0} \frac{g(x)\left[\dfrac{f(x + \Delta x) - f(x)}{\Delta x}\right] - f(x)\left[\dfrac{g(x + \Delta x) - g(x)}{\Delta x}\right]}{g(x + \Delta x)g(x)}$$

$$= \frac{g(x)f'(x) - f(x)g'(x)}{[g(x)]^2}$$

■

In words, the Quotient Rule is:

the denominator times the derivative of the numerator minus the numerator times the derivative of the denominator all divided by the denominator squared.

EXAMPLE 3 Differentiate $y = \dfrac{3x^2 - 1}{2x^3 + 5x^2 + 7}$.

Solution From the Quotient Rule (3),

denominator · derivative of numerator · numerator · derivative of denominator

$$\frac{dy}{dx} = \frac{\underbrace{(2x^3 + 5x^2 + 7)}_{\text{denominator}} \cdot \underbrace{\frac{d}{dx}(3x^2 - 1)}_{\text{derivative of numerator}} - \underbrace{(3x^2 - 1)}_{\text{numerator}} \cdot \underbrace{\frac{d}{dx}(2x^3 + 5x^2 + 7)}_{\text{derivative of denominator}}}{\underbrace{(2x^3 + 5x^2 + 7)^2}_{\text{denominator squared}}}$$

$$= \frac{(2x^3 + 5x^2 + 7) \cdot 6x - (3x^2 - 1) \cdot (6x^2 + 10x)}{(2x^3 + 5x^2 + 7)^2}$$

$$= \frac{-6x^4 + 6x^2 + 52x}{(2x^3 + 5x^2 + 7)^2}$$

□

EXAMPLE 4 Find an equation of the tangent line to the graph of $f(x) = \dfrac{6x^3}{x^3+1}$ at $x = 1$.

Solution We use the Quotient Rule to find the derivative;

$$f'(x) = \frac{(x^3+1)\cdot \dfrac{d}{dx}6x^3 - 6x^3 \cdot \dfrac{d}{dx}(x^3+1)}{(x^3+1)^2}$$

$$= \frac{(x^3+1)\cdot 18x^2 - 6x^3\cdot(3x^2)}{(x^3+1)^2} = \frac{18x^2}{(x^3+1)^2}$$

When $x = 1$, the slope of the tangent line is

$$f'(1) = \frac{18}{4} = \frac{9}{2}$$

The point of tangency is $(1, f(1))$ or $(1, 3)$. Hence, an equation of the tangent line is

$$y - 3 = \frac{9}{2}(x-1) \quad \text{or} \quad y = \frac{9}{2}x - \frac{3}{2}$$ □

The next example shows that the derivative of a function may utilize a combination of rules.

EXAMPLE 5 Differentiate $y = \dfrac{(x^2+1)(2x^2+1)}{(3x^2+1)}$.

Solution We begin with the Quotient Rule and then use the Product Rule when differentiating the numerator:

Product Rule here

$$\frac{dy}{dx} = \frac{(3x^2+1)\cdot \dfrac{d}{dx}[(x^2+1)(2x^2+1)] - (x^2+1)(2x^2+1)\cdot \dfrac{d}{dx}(3x^2+1)}{(3x^2+1)^2}$$

$$= \frac{(3x^2+1)[(x^2+1)4x + (2x^2+1)2x] - (x^2+1)(2x^2+1)6x}{(3x^2+1)^2}$$

$$= \frac{12x^5 + 8x^3}{(3x^2+1)^2}$$ □

So far the Power Rule, (1) of Section 2.3, is limited to the case where the exponent is a positive integer or zero. We shall now see that this rule is valid even when the exponent is a negative integer.

THEOREM 2.8 Power Rule (Negative Integer Exponents)

If n is a positive integer, then

$$\frac{d}{dx}x^{-n} = -nx^{-n-1} \tag{4}$$

Proof If n denotes a positive integer, then $-n$ is a negative integer. Since $x^{-n} = 1/x^n$, it follows that we can obtain the derivative of x^{-n} by the Quotient Rule and the laws of exponents:

$$\frac{d}{dx}\left[\frac{1}{x^n}\right] = \frac{x^n \cdot \overbrace{\frac{d}{dx}1}^{\boxed{\text{zero}}} - 1 \cdot \frac{d}{dx}x^n}{(x^n)^2} = -\frac{nx^{n-1}}{x^{2n}} = -nx^{-n-1} \qquad \blacksquare$$

EXAMPLE 6 Differentiate $y = x^{-2}$.

Solution As (4) shows, the procedure for differentiating a power with a negative integer exponent is the same as before: bring down the exponent as a multiple and decrease the exponent by 1. We have

$$\frac{dy}{dx} = -2x^{-2-1} = -2x^{-3} = -\frac{2}{x^3} \qquad \square$$

EXAMPLE 7 Differentiate $y = 5x^3 - \dfrac{1}{x^4}$.

Solution We first write the given function as $y = 5x^3 - x^{-4}$. Thus,

$$\frac{dy}{dx} = 5 \cdot 3x^2 - (-4)x^{-5} = 15x^2 + \frac{4}{x^5} \qquad \square$$

▲ ***Remarks*** (*i*) The Product and Quotient Rules will usually lead to expressions that demand simplification. If your answer to a problem does not look like the one in the text answer section, you may not have performed sufficient simplifications. Do not be content simply to carry through the mechanics of the various rules; it is always a good idea to practice your algebraic skills.

(*ii*) You should also note that the Quotient Rule is often used when it is not required. For example, although the Quotient Rule can be utilized in differentiating

$$y = \frac{x^5}{6} \quad \text{and} \quad y = \frac{10}{x^3}$$

it is simpler to write

$$y = \frac{1}{6}x^5 \quad \text{and} \quad y = 10x^{-3}$$

and then use the Power Rule:

$$\frac{dy}{dx} = \frac{5}{6}x^4 \quad \text{and} \quad \frac{dy}{dx} = -30x^{-4}$$

EXERCISES 2.4

Answers to odd-numbered problems begin on page A-70.

In Problems 1–30 find the derivative of the given function.

1. $y = 1/x$
2. $y = (2/x^3)^2$
3. $y = (5x)^{-2}$
4. $y = \pi^4 x^{-4}$
5. $y = 6x^2 + x^{-2}$
6. $y = 5x^4 - 1/2x^5$
7. $y = (x^2 - 7)(x^3 + 4x + 2)$
8. $y = (7x + 1)(x^4 - x^3 - 9x)$
9. $f(x) = \left(4 + \frac{1}{x}\right)\left(2x - \frac{1}{x^2}\right)$
10. $f(x) = \left(x^2 - \frac{1}{x^2}\right)\left(x^2 + \frac{1}{x^2}\right)$
11. $f(x) = \frac{10}{x^2 + 1}$
12. $f(x) = 5(4x - 3)^{-1}$
13. $G(x) = \frac{3x + 1}{2x - 5}$
14. $F(x) = \frac{2 - 3x}{7 - x}$
15. $y = (6x - 1)^2$
16. $y = (x^4 + 5x)^2$
17. $g(t) = \frac{t^2}{2t^2 + t + 1}$
18. $p(y) = \frac{y^2 - 10y + 2}{y(y^2 - 1)}$
19. $H(z) = (z + 1)(2z + 1)(3z + 1)$
20. $Q(r) = (r^2 + 1)(r^3 - r)(3r^4 + 2r - 1)$
21. $y = \frac{(2x + 1)(x - 5)}{3x + 2}$
22. $y = \frac{x^5}{(x^2 + 1)(x^3 + 4)}$
23. $y = \frac{2 - 1/x^3}{3 + 1/x^2}$
24. $y = \frac{x^{-2}}{x^{-3} + x^{-2} + 1}$
25. $f(u) = \frac{1}{u} + \frac{1}{u^2} + \frac{1}{u^3} + \frac{1}{u^4}$
26. $h(v) = \frac{1}{v + v^2 + v^3 + v^4}$
27. $y = \left(\frac{x + 1}{x + 3}\right)(x^2 - 2x - 1)$
28. $y = (x + 1)\left(x + 1 - \frac{1}{x + 2}\right)$
29. $f(x) = (3x + 1)^{-2}$
30. $g(x) = (x + 1)^3$

In Problems 31 and 32 find dy/dx without the aid of the Quotient Rule.

31. $y = \frac{6x^2 - 5x}{x}$
32. $y = \frac{x^4 + 2x^3 - 1}{x^2}$

In Problems 33–36 find an equation of the tangent line to the graph of the given function at the indicated value of x.

33. $y = 1/x^2;\ x = \frac{1}{2}$
34. $y = 4x - 1/x;\ x = -1$
35. $y = (2x^2 - 4)(x^3 + 5x + 3);\ x = 0$
36. $y = \frac{5x}{x^2 + 1};\ x = 2$

In Problems 37–40 find the point(s) on the graph of the given function at which the tangent is horizontal.

37. $y = (x^2 - 4)(x^2 - 6)$
38. $y = x(x - 1)^2$
39. $y = \frac{x^2}{x^4 + 1}$
40. $y = \frac{1}{x^2 - 6x}$

In Problems 41–44 find the point(s) on the graph of the given function at which the tangent has the indicated slope.

41. $y = 4x^{-1};\ m_{\tan} = -64$
42. $y = x - 1/x^2;\ m_{\tan} = 55$
43. $y = \frac{x + 3}{x + 1};\ m_{\tan} = -\frac{1}{8}$
44. $y = (x + 1)(2x + 5);\ m_{\tan} = -3$

In Problems 45–50 use the information $f(1) = 2$, $f'(1) = -3$, and $g(1) = g'(1) = 6$ to evaluate the given derivative.

45. $\left.\dfrac{d}{dx}(f(x)g(x))\right|_{x=1}$

46. $\left.\dfrac{d}{dx}\left(\dfrac{g(x)}{f(x)}\right)\right|_{x=1}$

47. $\left.\dfrac{d}{dx}\left(\dfrac{1+2f(x)}{x-g(x)}\right)\right|_{x=1}$

48. $\left.\dfrac{d}{dx}\left(\dfrac{4}{x}+f(x)\right)g(x)\right|_{x=1}$

49. $\left.D_x(x^2f(x)g(x))\right|_{x=1}$

50. $\left.D_x\left(\dfrac{xf(x)}{g(x)}\right)\right|_{x=1}$

51. Find an equation of the line(s) with slope -8 and tangent to the graph of $f(x) = 2/x$.

52. Find the value of k such that the tangent line to the graph of $f(x) = (k + x)/x^2$ has slope 5 at $x = 2$.

53. (a) Graph the function $f(x) = 2/(x^2 + 1)$.
(b) Find all the points on the graph of f such that the normal lines pass through the origin.

54. Show that at $x = 1$ the tangent to the graph of $f(x) = (x^2 + 14)/(x^2 + 9)$ is perpendicular to the tangent to the graph of $g(x) = (1 + x^2)(1 + 2x)$.

In Problems 55–58 find the values of x for which $f'(x) > 0$.

55. $f(x) = \dfrac{5}{x^2}$

56. $f(x) = \dfrac{x^2+3}{x+1}$

57. $f(x) = (2x + 1)(x + 4)$

58. $f(x) = x + 4/x$

59. The Universal Law of Gravitation states that the force F between two bodies of masses m_1 and m_2 separated by a distance r is $F = km_1m_2/r^2$, where k is constant. What is the instantaneous rate of change of F with respect to r when $r = \frac{1}{2}$ km?

60. The potential energy U between two atoms in a diatomic molecule is given by $U(x) = q_1/x^{12} - q_2/x^6$, where q_1 and q_2 are positive constants and x is the distance between the atoms. The force between the atoms is defined as $F(x) = -U'(x)$. Show that $F(\sqrt[6]{2q_1/q_2}) = 0$.

61. The **van der Waals equation of state** for an ideal gas is

$$\left(P + \frac{a}{V^2}\right)(V - b) = RT$$

where P is pressure, V is volume per mole, R is the universal gas constant, T is temperature, and a and b are constants depending on the gas. Find dP/dV in the case where T is constant.

62. Verify that the function $y = 5x + 1 + 2/x$, $x \neq 0$, satisfies the equation $xy' + y = 10x + 1$.

63. Let $y = f(x)g(x)h(x)$, where f, g, and h are differentiable functions. Use the Product Rule to show that

$$\frac{dy}{dx} = f(x)g(x)h'(x) + f(x)h(x)g'(x) + g(x)h(x)f'(x)$$

64. Let $y = f(x)$ be a differentiable function.
(a) Find dy/dx for $y = [f(x)]^2$.
(b) Find dy/dx for $y = [f(x)]^3$.
(c) Conjecture a rule for finding the derivative of $y = [f(x)]^n$, where n is a positive integer.

65. Use the definition of the derivative to prove that when g is differentiable and $g(x) \neq 0$, then

$$\frac{d}{dx}\left[\frac{1}{g(x)}\right] = -\frac{g'(x)}{[g(x)]^2}$$

66. Use the result of Problem 65, the fact that $f(x)/g(x) = f(x) \cdot [1/g(x)]$, and the Product Rule to derive the Quotient Rule (3).

2.5 DERIVATIVES OF THE TRIGONOMETRIC FUNCTIONS

2.5.1 Some Preliminary Limit Results

Recall from Sections 1.3 and 1.8 that the sine and cosine functions are continuous on $(-\infty, \infty)$. Hence, the following two limit results follow immediately from Definition 1.7:

$$\lim_{t\to 0} \sin t = 0 \tag{1}$$

$$\lim_{t\to 0} \cos t = 1 \tag{2}$$

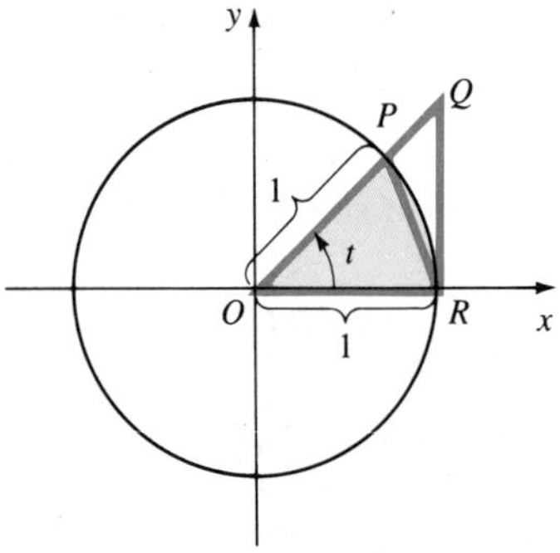

FIGURE 2.33

Also, in Example 7 of Section 1.5, we illustrated the fact that

$$\lim_{t\to 0}\frac{\sin t}{t} = 1 \tag{3}$$

We now give a partial proof this result.

Consider a circle centered at the origin with radius 1. As shown in Figure 2.33, let the shaded region OPR be a sector with central angle t such that $0 < t < \pi/2$. We see from the same figure that the areas of triangles OPR and OQR give lower and upper bounds, respectively, for the area of the sector OPR:

$$\text{area of } \triangle OPR < \text{area of sector } OPR < \text{area of } \triangle OQR \tag{4}$$

The height of $\triangle OPR$ is $\overline{OP}\sin t = 1\cdot\sin t = \sin t$, and so

$$\text{area of } \triangle OPR = \frac{1}{2}\,\overline{OR}\cdot(\text{height}) = \frac{1}{2}\cdot 1\cdot\sin t = \frac{1}{2}\sin t \tag{5}$$

In addition, $\overline{QR}/\overline{OR} = \tan t$ or $\overline{QR} = \tan t$, so that

$$\text{area of } \triangle OQR = \frac{1}{2}\,\overline{OR}\cdot\overline{QR} = \frac{1}{2}\cdot 1\cdot\tan t = \frac{1}{2}\tan t \tag{6}$$

Finally, we know that the area of a sector of a circle is $\frac{1}{2}r^2\theta$, where r is the radius of the circle and θ is the central angle measured in radians. Thus,

$$\text{area of sector } OPR = \frac{1}{2}\cdot 1\cdot t = \frac{1}{2}t \tag{7}$$

Using (5), (6), and (7) in (4) gives

$$\frac{1}{2}\sin t < \frac{1}{2}t < \frac{1}{2}\tan t \quad\text{or}\quad 1 < \frac{t}{\sin t} < \frac{1}{\cos t}$$

From the properties of inequalities, the last inequality is equivalent to

$$\cos t < \frac{\sin t}{t} < 1$$

We now let $t \to 0^+$ in the last result. Since $(\sin t)/t$ is "squeezed" between 1 and $\cos t$ (which is approaching 1), it follows from the Squeeze Theorem that $\lim_{t\to 0^+}(\sin t)/t = 1$. While we have assumed $0 < t < \pi/2$, the same result holds for $t \to 0^-$ when $-\pi/2 < t < 0$. (See Problem 31 in Exercises 2.5.)

EXAMPLE 1 Evaluate $\displaystyle\lim_{t\to 0}\frac{\sin 4t}{t}$.

Solution Let $u = 4t$ so that $t = u/4$. We see that as $t \to 0$, necessarily $u \to 0$, and hence,

$$\lim_{t\to 0}\frac{\sin 4t}{t} = \lim_{u\to 0}\frac{\sin u}{u/4} = 4\lim_{u\to 0}\frac{\sin u}{u} = 4\cdot 1 = 4 \qquad \square$$

Using an argument similar to that illustrated in Example 1, we find

$$\lim_{t\to 0}\frac{\sin kt}{t}=k \tag{8}$$

Also,

$$\lim_{t\to 0}\frac{1}{(\sin t)/t}=\frac{\lim\limits_{t\to 0}1}{\lim\limits_{t\to 0}(\sin t)/t}=1$$

implies

$$\lim_{t\to 0}\frac{t}{\sin t}=1 \tag{9}$$

EXAMPLE 2 Evaluate $\lim\limits_{t\to 0}\dfrac{\sin^2 5t}{t^2}$.

Solution We rewrite the function as

$$\frac{\sin 5t}{t}\cdot\frac{\sin 5t}{t}$$

and use (8):

$$\lim_{t\to 0}\frac{\sin^2 5t}{t^2}=\lim_{t\to 0}\frac{\sin 5t}{t}\cdot\frac{\sin 5t}{t}=\lim_{t\to 0}\frac{\sin 5t}{t}\cdot\lim_{t\to 0}\frac{\sin 5t}{t}=5\cdot 5=25 \quad \square$$

Another limit result that we will use immediately is

$$\lim_{t\to 0}\frac{1-\cos t}{t}=0 \tag{10}$$

To see this, we observe that

$$\begin{aligned}\frac{1-\cos t}{t}&=\frac{(1-\cos t)(1+\cos t)}{t(1+\cos t)}=\frac{1-\cos^2 t}{t(1+\cos t)}\\&=\frac{\sin^2 t}{t(1+\cos t)}=\frac{\sin t}{t}\cdot\frac{\sin t}{1+\cos t}\end{aligned}$$

and so

$$\lim_{t\to 0}\frac{1-\cos t}{t}=\lim_{t\to 0}\frac{\sin t}{t}\cdot\lim_{t\to 0}\frac{\sin t}{1+\cos t}=1\cdot 0=0$$

2.5.2 Derivatives

sin x and cos x We find the derivative of $f(x)=\sin x$ by resorting to the definition $f'(x)=\lim_{\Delta x\to 0}\Delta y/\Delta x$. In this case,

$$\begin{aligned}\Delta y&=f(x+\Delta x)-f(x)\\&=\sin(x+\Delta x)-\sin x\\&=\sin x\cos\Delta x+\cos x\sin\Delta x-\sin x \qquad \boxed{\text{Addition Formula}}\\&=\sin x(\cos\Delta x-1)+\cos x\sin\Delta x\end{aligned}$$

and $$\frac{\Delta y}{\Delta x} = \sin x \frac{\cos \Delta x - 1}{\Delta x} + \cos x \frac{\sin \Delta x}{\Delta x}$$

Employing Theorem 1.4(*ii*) and results (3) and (10) enables us to write

$$\begin{aligned} f'(x) &= \lim_{\Delta x \to 0} \frac{\Delta y}{\Delta x} = \sin x \cdot \lim_{\Delta x \to 0} \frac{\cos \Delta x - 1}{\Delta x} + \cos x \cdot \lim_{\Delta x \to 0} \frac{\sin \Delta x}{\Delta x} \\ &= \sin x \cdot 0 + \cos x \cdot 1 \end{aligned}$$

We conclude that

$$\frac{d}{dx} \sin x = \cos x \tag{11}$$

In a similar manner it can be shown that

$$\frac{d}{dx} \cos x = -\sin x \tag{12}$$

(See Problem 74 in Exercises 2.5.)

EXAMPLE 3 Find the slope of the tangent line to the graph of $f(x) = \sin x$ at $x = \pi/2$ and $x = 4\pi/3$.

Solution We know from (11) that $f'(x) = \cos x$, and so

$$f'\left(\frac{\pi}{2}\right) = \cos\left(\frac{\pi}{2}\right) = 0$$

$$f'\left(\frac{4\pi}{3}\right) = \cos\left(\frac{4\pi}{3}\right) = -\frac{1}{2}$$

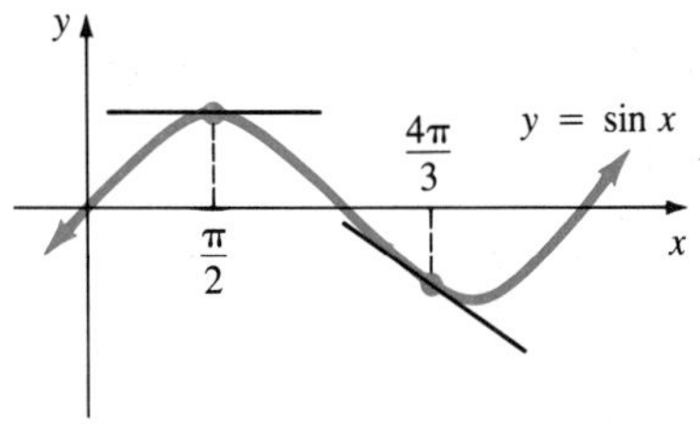

FIGURE 2.34

In Figure 2.34 we see that the tangent line is horizontal at $(\pi/2, 1)$. □

The Other Trigonometric Functions The results in (11) and (12) can be used in conjunction with the rules of differentiation to find the derivatives of the tangent, cotangent, secant, and cosecant functions.

To differentiate $\tan x = (\sin x)/\cos x$, we can use the Quotient Rule:

$$\frac{d}{dx} \frac{\sin x}{\cos x} = \frac{\cos x \dfrac{d}{dx} \sin x - \sin x \dfrac{d}{dx} \cos x}{(\cos x)^2}$$

$$= \frac{\cos x(\cos x) - \sin x(-\sin x)}{\cos^2 x} = \frac{\overbrace{\cos^2 x + \sin^2 x}^{\boxed{1}}}{\cos^2 x}$$

Since $1/\cos^2 x = (1/\cos x)^2 = \sec^2 x$, we have the following result:

$$\frac{d}{dx} \tan x = \sec^2 x \tag{13}$$

The derivative formula for the cotangent is obtained in an analogous fashion and is left as an exercise. (See Problem 75 in Exercises 2.5.)

$$\frac{d}{dx}\cot x = -\csc^2 x \tag{14}$$

Now, $\sec x = 1/\cos x$. Therefore, we can use the Quotient Rule, again, to find the derivative of the secant function:

$$\frac{d}{dx}\sec x = \frac{d}{dx}\frac{1}{\cos x} = \frac{\cos x \dfrac{d}{dx} 1 - 1 \cdot \dfrac{d}{dx}\cos x}{\cos^2 x} = \frac{\sin x}{\cos^2 x} \tag{15}$$

By writing $$\frac{\sin x}{\cos^2 x} = \frac{1}{\cos x} \cdot \frac{\sin x}{\cos x} = \sec x \tan x$$

we can express (15) as

$$\frac{d}{dx}\sec x = \sec x \tan x \tag{16}$$

The last result also follows immediately from the Quotient Rule:

$$\frac{d}{dx}\csc x = -\csc x \cot x \tag{17}$$

(See Problem 76 in Exercises 2.5.)

EXAMPLE 4 Differentiate $y = x^2 \sin x$.

Solution The Product Rule along with (11) yields

$$\begin{aligned}\frac{dy}{dx} &= x^2 \frac{d}{dx}\sin x + \sin x \frac{d}{dx} x^2 \\ &= x^2 \cos x + 2x \sin x\end{aligned}$$

□

EXAMPLE 5 Differentiate $y = (\cos x)(x - \cot x)$.

Solution From the Product Rule, (12), and (14),

$$\begin{aligned}\frac{dy}{dx} &= (\cos x)\frac{d}{dx}(x - \cot x) + (x - \cot x)\frac{d}{dx}\cos x \\ &= (\cos x)(1 + \csc^2 x) + (x - \cot x)(-\sin x) \\ &= 2\cos x - x \sin x + \cos x \csc^2 x\end{aligned}$$

□

EXAMPLE 6 Differentiate $y = \dfrac{\sin x}{2 + \sec x}$.

Solution From the Quotient Rule, (11), and (16),

$$\frac{dy}{dx} = \frac{(2 + \sec x)\dfrac{d}{dx}\sin x - \sin x\dfrac{d}{dx}(2 + \sec x)}{(2 + \sec x)^2}$$

$$= \frac{(2 + \sec x)\cos x - \sin x(\sec x \tan x)}{(2 + \sec x)^2}$$

$$= \frac{1 + 2\cos x - \tan^2 x}{(2 + \sec x)^2}$$ □

For future reference we summarize the derivative formulas introduced in this section.

Derivatives of the Trigonometric Functions

I	$\dfrac{d}{dx}\sin x = \cos x$	II	$\dfrac{d}{dx}\cos x = -\sin x$
III	$\dfrac{d}{dx}\tan x = \sec^2 x$	IV	$\dfrac{d}{dx}\cot x = -\csc^2 x$
V	$\dfrac{d}{dx}\sec x = \sec x \tan x$	VI	$\dfrac{d}{dx}\csc x = -\csc x \cot x$

▲ ***Remarks*** (*i*) Throughout this section and, for that matter, throughout calculus, angles are measured in radians. If t is measured in degrees, the fundamental limit result

$$\lim_{t \to 0} \frac{\sin t}{t} = 1$$

is not true. For this reason, if x is measured in degrees, derivative formulas such $d/dx(\sin x) = \cos x$ are not valid. See Problems 29 and 77 in Exercises 2.5.

(*ii*) It is sometimes possible to obtain a correct answer by incorrect reasoning. An examination question based on the result in (8) often yields the right answer but no points. The following is a very typical mistake:

$$\lim_{t \to 0} \frac{\sin 6t}{t} = \lim_{t \to 0} \frac{6 \sin t}{t}$$

$$= 6 \lim_{t \to 0} \frac{\sin t}{t}$$

$$= 6 \cdot 1 = 6$$

While 6 is the value of the limit, the assumption that $\sin 6t = 6 \sin t$ is false. The correct procedure is illustrated in Example 1.

EXERCISES 2.5 *Answers to odd-numbered problems begin on page A-70.*

[2.5.1]

In Problems 1–28 find the value of each limit, if it exists.

1. $\lim_{t\to 0} \dfrac{\sin 3t}{2t}$
2. $\lim_{t\to 0} \dfrac{\sin(-4t)}{t}$
3. $\lim_{x\to 0} \dfrac{\sin x}{4+\cos x}$
4. $\lim_{x\to 0} \dfrac{1+\sin x}{1+\cos x}$
5. $\lim_{x\to 0} \dfrac{\cos 2x}{\cos 3x}$
6. $\lim_{x\to 0} \dfrac{\tan x}{3x}$
7. $\lim_{t\to 0} \dfrac{1}{t \sec t \csc 4t}$
8. $\lim_{t\to 0} 5t \cot 2t$
9. $\lim_{t\to 0} \dfrac{2\sin^2 t}{t\cos^2 t}$
10. $\lim_{t\to 0} \dfrac{\sin^2(t/2)}{\sin t}$
11. $\lim_{t\to 0} \dfrac{\sin^2 6t}{t^2}$
12. $\lim_{t\to 0} \dfrac{t^3}{\sin^2 3t}$
13. $\lim_{x\to 1} \dfrac{\sin(x-1)}{2x-2}$
14. $\lim_{x\to 2\pi} \dfrac{x-2\pi}{\sin x}$
15. $\lim_{x\to 0} \dfrac{\cos x}{x}$
16. $\lim_{t\to \pi/2} \dfrac{1+\sin t}{\cos t}$
17. $\lim_{x\to 0} \dfrac{\cos(3x-\pi/2)}{x}$
18. $\lim_{x\to -2} \dfrac{\sin(5x+10)}{4x+8}$
19. $\lim_{t\to 0} \dfrac{\sin 3t}{\sin 7t}$
20. $\lim_{t\to 0} \sin 2t \csc 3t$
21. $\lim_{t\to 0^+} \dfrac{\sin t}{\sqrt{t}}$
22. $\lim_{t\to 0^+} \dfrac{1-\cos\sqrt{t}}{\sqrt{t}}$
23. $\lim_{t\to 0} \dfrac{t^2-5t\sin t}{t^2}$
24. $\lim_{t\to 0} \dfrac{\cos 4t}{\cos 8t}$
25. $\lim_{x\to 0^+} \dfrac{(x+2\sqrt{\sin x})^2}{x}$
26. $\lim_{x\to 0} \dfrac{(1-\cos x)^2}{x}$
27. $\lim_{x\to 0} \dfrac{\cos x - 1}{\cos^2 x - 1}$
28. $\lim_{x\to 0} \dfrac{\sin x + \tan x}{x}$

Calculator/Computer Problems

29. (a) Let t be measured in degrees. Complete the following table using values with four decimal places.

t	45	30	15	5	1	0.1
$\dfrac{\sin t}{t}$						

(b) Compute $\pi/180$ to four decimal places.

(c) Given that t is measured in degrees, make a conjecture about the value of $\lim_{t\to 0}(\sin t)/t$.

30. Let t be measured in radians. Use a calculator to investigate whether $\lim_{t\to 0}(1-\cos t)/t^2$ exists.

Miscellaneous Problems

31. Prove $\lim_{t\to 0^-} \dfrac{\sin t}{t} = 1$ when $-\pi/2 < t < 0$.
32. Show that $\lim_{x\to\infty} x\sin(1/x) = 1$.

[2.5.2]

In Problems 33–56 find the derivative of the given function.

33. $y = x^2 - \cos x$
34. $y = 4x^3 + x + \sin x$
35. $y = 1 + 7\sin x - \tan x$
36. $y = 3\cos x - 5\cot x$
37. $y = x\sin x$
38. $y = (x^3-2)\tan x$
39. $y = \sin x \cos x$
40. $y = \cos x \cot x$
41. $y = (x^2+\sin x)\sec x$
42. $y = \csc x \tan x$
43. $f(x) = (\csc x)^{-1}$
44. $f(x) = \dfrac{2}{\cos x \cot x}$
45. $f(x) = \dfrac{\cot x}{x+1}$
46. $f(x) = \dfrac{x^2-6x}{1+\cos x}$
47. $y = \dfrac{x^2}{1+2\tan x}$
48. $y = \dfrac{2+\sin x}{x}$
49. $F(\theta) = \dfrac{\sin\theta}{1+\cos\theta}$
50. $g(z) = \dfrac{1+\csc z}{1+\sec z}$
51. $G(u) = \sin^2 u$
52. $H(v) = (1+\cos v)(v-\sin v)$
53. $y = \cos^2 x + \sin^2 x$
54. $y = x^3\cos x - x^3\sin x$
55. $y = x^2\sin x\tan x$
56. $y = \dfrac{1+\sin x}{x\cos x}$

In Problems 57 and 58 consider the graph of the given function on the interval $[0, 2\pi]$. Find the point(s) at which the tangent is horizontal.

57. $f(x) = x + \cos x$
58. $f(x) = \sin x + \cos x$

In Problems 59–62 find an equation of the tangent line to the graph of the given function at the indicated value of x.

59. $f(x) = \cos x;\ x = \pi/3$

60. $f(x) = \tan x;\ x = \pi$

61. $f(x) = \sec x;\ x = \pi/6$

62. $f(x) = \csc x;\ x = \pi/2$

In Problems 63–66 find an equation of the normal line to the graph of the given function at the indicated value of x.

63. $f(x) = \sin x;\ x = 4\pi/3$

64. $f(x) = \tan^2 x;\ x = \pi/4$

65. $f(x) = x \cos x;\ x = \pi$

66. $f(x) = \dfrac{x}{1 + \sin x};\ x = \pi/2$

In Problems 67 and 68 find the derivative of the given function by first employing a trigonometric identity.

67. $f(x) = \sin 2x$

68. $f(x) = \cos^2(x/2)$

69. When the angle of elevation of the sun is θ, a telephone pole 40 ft high casts a shadow of length s as shown in Figure 2.35. Find the rate of change of s with respect to θ when $\theta = \pi/3$ radians. Explain the significance of the minus sign in the answer.

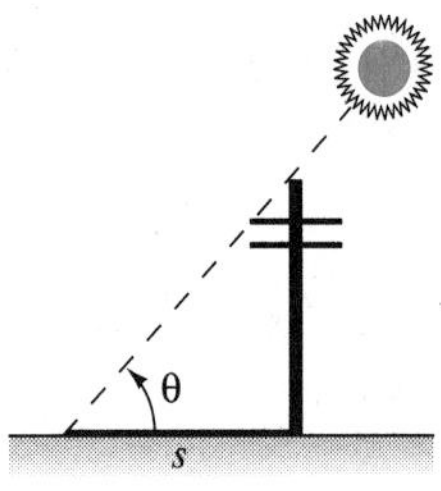

FIGURE 2.35

70. The two ends of a 10-ft board are attached to perpendicular rails, as shown in Figure 2.36, so that point P is free to move vertically and point R is free to move horizontally.

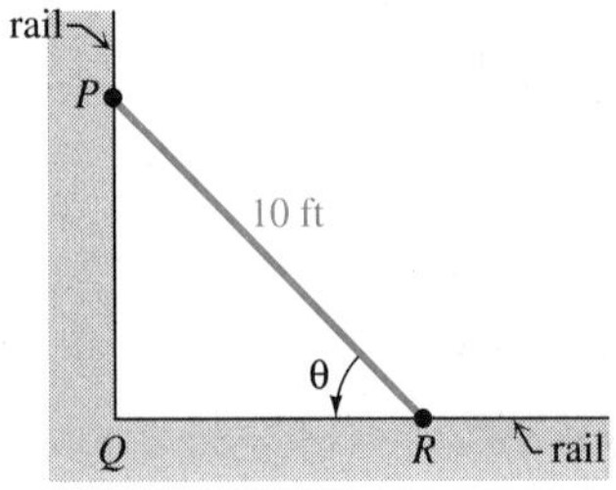

FIGURE 2.36

(a) Express the area A of triangle PQR as a function of the indicated angle θ.

(b) Find the rate of change of A with respect to θ.

(c) Initially the board rests flat on the horizontal rail. Suppose point R is then moved in the direction of point Q, thereby forcing point P to move up the vertical rail. Initially the area of the triangle is 0 ($\theta = 0$), but then it increases for a while as θ increases and then decreases as R approaches Q. When the board is vertical, the area of the triangle is again 0 ($\theta = \pi/2$). Graph the derivative $dA/d\theta$. Interpret this graph to find values of θ for which A is increasing and values of θ for which A is decreasing. Now verify your interpretation of the graph of the derivative by graphing $A(\theta)$.

(d) Use the graphs in part (c) to find the value of θ for which the area of the triangle is the greatest.

Calculator/Computer Problems

In Problems 71 and 72 use a calculator or computer to obtain the graph of the given function. By inspection of the graph indicate where the function may not be differentiable.

71. $f(x) = 0.5(\sin x + |\sin x|)$

72. $f(x) = |x + \sin x|$

73. As shown in Figure 2.37, a boy pulls a sled on which his little sister is seated. If the sled and girl weigh a total of 70 lb, and if the coefficient of sliding friction of snow-covered ground is 0.2, then the magnitude F of the force (measured in pounds) required to move the sled is

$$F = \frac{70(0.2)}{0.2 \sin \theta + \cos \theta}$$

where θ is the angle the tow rope makes with the horizontal.

(a) Use a calculator or computer to obtain the graph of F on the interval $[-1, 1]$.

(b) Find the derivative $dF/d\theta$.

(c) Find the angle (in radians) for which $dF/d\theta = 0$.

(d) Find the value of F corresponding to the angle found in part (**c**).

(e) Use the graph in part (**a**) as an aid in interpreting the numbers found in parts (**c**) and (**d**).

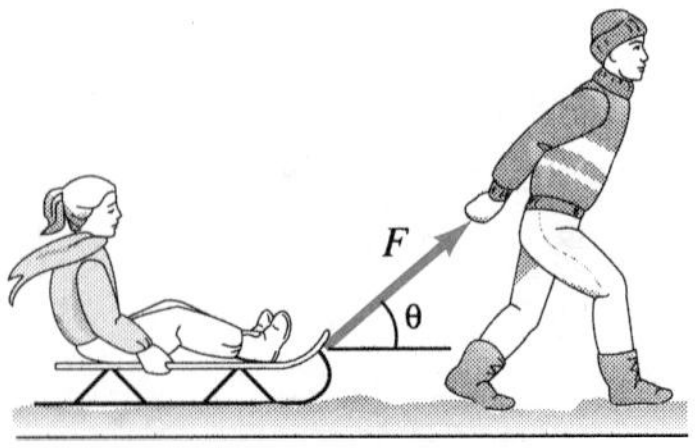

FIGURE 2.37

Miscellaneous Problems

74. Prove (12).

75. Prove (14).

76. Prove (17).

77. If t is measured in degrees, then $\lim_{t\to 0}(\sin t)/t = \pi/180$ and $\lim_{t\to 0}(1 - \cos t)/t = 0$. Show that if x is measured in degrees, then

$$\frac{d}{dx}\sin x = \frac{\pi}{180}\cos x \quad \text{and} \quad \frac{d}{dx}\cos x = -\frac{\pi}{180}\sin x$$

2.6 RULES OF DIFFERENTIATION III: CHAIN RULE

Suppose we wish to differentiate

$$y = (x^5 + 1)^2 \tag{1}$$

By writing (1) as $y = (x^5 + 1) \cdot (x^5 + 1)$, we can find the derivative using the Product Rule:

$$\begin{aligned}\frac{dy}{dx} &= (x^5 + 1)\frac{d}{dx}(x^5 + 1) + (x^5 + 1)\frac{d}{dx}(x^5 + 1)\\ &= (x^5 + 1) \cdot 5x^4 + (x^5 + 1) \cdot 5x^4 = 2(x^5 + 1) \cdot 5x^4 \end{aligned} \tag{2}$$

Similarly, to differentiate $y = (x^5 + 1)^3$, we can write $y = (x^5 + 1)^2 \cdot (x^5 + 1)$ and use both the Product Rule and the result given in (2). It is readily shown that

$$\frac{d}{dx}(x^5 + 1)^3 = 3(x^5 + 1)^2 \cdot 5x^4 \tag{3}$$

Power Rule for Functions Inspection of (2) and (3) reveals a pattern for differentiating a power of a function. For example, in (3) we see

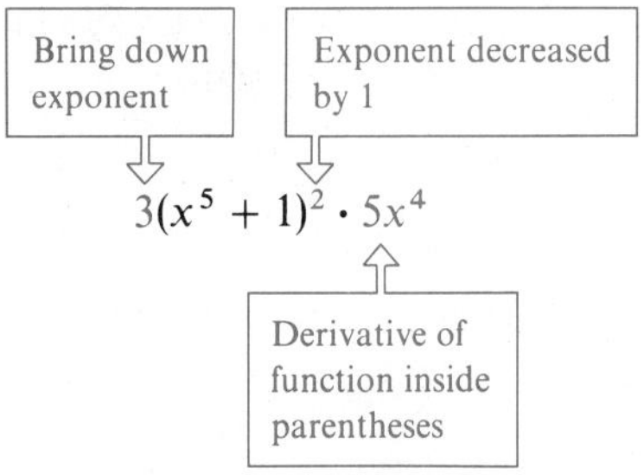

For emphasis, if we denote a differentiable function by [ ], it appears that

$$\frac{d}{dx}[\quad]^n = n[\quad]^{n-1}\frac{d}{dx}[\quad] \tag{4}$$

Later in this section we will prove that (4) holds for *any* integer n. The general result is stated in the next theorem.

THEOREM 2.9 Power Rule for Functions (Integer Exponents)

If n is an integer and g is a differentiable function, then

$$\frac{d}{dx}[g(x)]^n = n[g(x)]^{n-1}g'(x) \tag{5}$$

EXAMPLE 1 Differentiate $y = (4x)^{100}$.

Solution With the identification that $g(x) = 4x$, we see from (5) that

$$\frac{dy}{dx} = \underbrace{100(4x)^{99}}_{n\,[g(x)]^{n-1}} \cdot \underbrace{\frac{d}{dx}4x}_{g'(x)} = 100(4x)^{99} \cdot 4 = 400(4x)^{99}$$ □

EXAMPLE 2 Differentiate $y = (2x^3 + 4x + 1)^4$.

Solution In this case we identify $g(x) = 2x^3 + 4x + 1$ and $n = 4$. From (5) it then follows that

$$\frac{dy}{dx} = 4(2x^3 + 4x + 1)^3 \frac{d}{dx}(2x^3 + 4x + 1) = 4(2x^3 + 4x + 1)^3(6x^2 + 4)$$ □

EXAMPLE 3 Differentiate $y = \dfrac{(x^2 - 1)^3}{(5x + 1)^8}$.

Solution We first apply the Quotient Rule followed by the Power Rule for Functions:

Power Rule

$$\frac{dy}{dx} = \frac{(5x + 1)^8 \frac{d}{dx}(x^2 - 1)^3 - (x^2 - 1)^3 \frac{d}{dx}(5x + 1)^8}{(5x + 1)^{16}}$$

$$= \frac{(5x + 1)^8 \cdot 3(x^2 - 1)^2 \cdot 2x - (x^2 - 1)^3 \cdot 8(5x + 1)^7 \cdot 5}{(5x + 1)^{16}}$$

$$= \frac{6x(5x + 1)^8(x^2 - 1)^2 - 40(5x + 1)^7(x^2 - 1)^3}{(5x + 1)^{16}}$$

$$= \frac{(x^2 - 1)^2(-10x^2 + 6x + 40)}{(5x + 1)^9}$$ □

EXAMPLE 4 To differentiate $y = 1/(x^2 + 1)$, we could, of course, use the Quotient Rule. However, by writing $y = (x^2 + 1)^{-1}$, it is also possible to use the Power Rule for Functions with $n = -1$:

$$\frac{dy}{dx} = (-1)(x^2 + 1)^{-2}\frac{d}{dx}(x^2 + 1) = (-1)(x^2 + 1)^{-2}2x = \frac{-2x}{(x^2 + 1)^2} \qquad \square$$

EXAMPLE 5 Differentiate $y = \dfrac{1}{(7x^5 - x^4 + 2)^{10}}$.

Solution Write the given function as $y = (7x^5 - x^4 + 2)^{-10}$. Identify $n = -10$ and use the Power Rule (5):

$$\frac{dy}{dx} = -10(7x^5 - x^4 + 2)^{-11}\frac{d}{dx}(7x^5 - x^4 + 2) = \frac{-10(35x^4 - 4x^3)}{(7x^5 - x^4 + 2)^{11}} \qquad \square$$

EXAMPLE 6 Differentiate $y = \tan^2 x$.

Solution For emphasis, we first write $y = (\tan x)^2$ and use (5):

$$\frac{dy}{dx} = 2(\tan x)\frac{d}{dx}\tan x$$

Employing (13) of Section 2.5 then yields

$$\frac{dy}{dx} = 2\tan x\sec^2 x \qquad \square$$

Chain Rule A power of a function can be written as a composite function. If $f(x) = x^n$ and $u = g(x)$, then $f(u) = f(g(x)) = [g(x)]^n$. The Power Rule (5) is a special case of the **Chain Rule** for differentiating composite functions.

THEOREM 2.10 Chain Rule

If $y = f(u)$ is a differentiable function of u and $u = g(x)$ is a differentiable function of x, then

$$\frac{dy}{dx} = \frac{dy}{du}\cdot\frac{du}{dx} = f'(g(x))\cdot g'(x) \qquad (6)$$

Proof for $\Delta u \neq 0$ For $\Delta x \neq 0$,

$$\Delta u = g(x + \Delta x) - g(x) \qquad (7)$$

or $$g(x + \Delta x) = g(x) + \Delta u = u + \Delta u$$

In addition,

$$\Delta y = f(u + \Delta u) - f(u) = f(g(x + \Delta x)) - f(g(x))$$

When x and $x + \Delta x$ are in some open interval for which $\Delta u \neq 0$, we can write

$$\frac{\Delta y}{\Delta x} = \frac{\Delta y}{\Delta u} \cdot \frac{\Delta u}{\Delta x}$$

Since g is assumed to be differentiable, it is continuous. Consequently, as $\Delta x \to 0$, $g(x + \Delta x) \to g(x)$, and so from (7) we see that $\Delta u \to 0$. Thus,

$$\begin{aligned}\lim_{\Delta x \to 0} \frac{\Delta y}{\Delta x} &= \left(\lim_{\Delta x \to 0} \frac{\Delta y}{\Delta u}\right)\left(\lim_{\Delta x \to 0} \frac{\Delta u}{\Delta x}\right) \\ &= \left(\lim_{\Delta u \to 0} \frac{\Delta y}{\Delta u}\right)\left(\lim_{\Delta x \to 0} \frac{\Delta u}{\Delta x}\right)\end{aligned}$$

From the definition of the derivative, it follows that

$$\frac{dy}{dx} = \frac{dy}{du} \cdot \frac{du}{dx}$$

■

The assumption that $\Delta u \neq 0$ on some interval does not hold true for every differentiable function g. Although the result given in (6) remains valid when $\Delta u = 0$, the preceding proof does not.

Proof of the Power Rule for Functions As noted previously, a power of a function can be written as $y = u^n$, where n is an integer and $u = g(x)$. Since $dy/du = nu^{n-1}$ and $du/dx = g'(x)$, we see from the Chain Rule that

$$\frac{dy}{dx} = \frac{dy}{du} \cdot \frac{du}{dx} = nu^{n-1} \frac{du}{dx} = n[g(x)]^{n-1} g'(x)$$

This is the Power Rule (5).

Trigonometric Functions We obtain the derivatives of the trigonometric functions composed with a differentiable function g as another direct consequence of the Chain Rule. For example, if $y = \sin u$, where $u = g(x)$, then $dy/du = \cos u$. Hence, (6) implies

$$\frac{dy}{dx} = \frac{dy}{du} \cdot \frac{du}{dx} = \cos u \frac{du}{dx}$$

or equivalently, $$\frac{d}{dx} \sin[\ \ ] = \cos[\ \ ] \frac{d}{dx}[\ \ ]$$

We summarize the six results.

Derivatives of the Trigonometric Functions

I	$\frac{d}{dx}\sin u = \cos u \frac{du}{dx}$	II	$\frac{d}{dx}\cos u = -\sin u \frac{du}{dx}$
III	$\frac{d}{dx}\tan u = \sec^2 u \frac{du}{dx}$	IV	$\frac{d}{dx}\cot u = -\csc^2 u \frac{du}{dx}$
V	$\frac{d}{dx}\sec u = \sec u \tan u \frac{du}{dx}$	VI	$\frac{d}{dx}\csc u = -\csc u \cot u \frac{du}{dx}$

EXAMPLE 7 Differentiate $y = \tan(6x^2 + 1)$.

Solution With $u = 6x^2 + 1$, we see from III that

$$\frac{dy}{dx} = \overbrace{\sec^2(6x^2 + 1)}^{\boxed{\sec^2 u}} \cdot \overbrace{\frac{d}{dx}(6x^2 + 1)}^{\boxed{\frac{du}{dx}}} = 12x \sec^2(6x^2 + 1)$$ □

EXAMPLE 8 Differentiate $y = (9x^3 + 1)^2 \sin 5x$.

Solution We first use the Product Rule:

$$\frac{dy}{dx} = (9x^3 + 1)^2 \frac{d}{dx} \sin 5x + \sin 5x \frac{d}{dx}(9x^3 + 1)^2$$

followed by the Power Rule (5) and I:

$$\begin{aligned}\frac{dy}{dx} &= (9x^3 + 1)^2 \cdot 5 \cos 5x + \sin 5x \cdot 2(9x^3 + 1) \cdot 27x^2 \\ &= (9x^3 + 1)(45x^3 \cos 5x + 54x^2 \sin 5x + 5 \cos 5x)\end{aligned}$$ □

EXAMPLE 9 Differentiate $y = \cos^4(7x^3 + 6x - 1)$.

Solution We first use the Power Rule (5) followed by II:

$$\begin{aligned}\frac{dy}{dx} &= 4 \cos^3(7x^3 + 6x - 1) \cdot \frac{d}{dx} \cos(7x^3 + 6x - 1) \\ &= 4 \cos^3(7x^3 + 6x - 1)\left[-\sin(7x^3 + 6x - 1) \cdot \frac{d}{dx}(7x^3 + 6x - 1)\right] \\ &= -4(21x^2 + 6)\cos^3(7x^3 + 6x - 1)\sin(7x^3 + 6x - 1)\end{aligned}$$ □

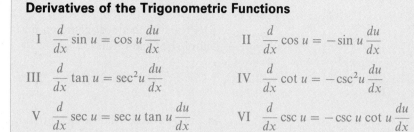

▲ ***Remarks*** It is appropriate to mention two often costly mistakes (on examinations) at this time.

(*i*) Probably the most common mistake is to forget to carry out the second half of the Chain Rule

$$\text{that is, the } \frac{du}{dx} \text{ part in } \frac{dy}{dx} = \frac{dy}{du}\frac{du}{dx}$$

In other words, when differentiating a function, we fail to carry out the multiplication by the derivative of the function *inside* the given function. For instance, the derivative of $y = (1 - x)^{57}$ is not $dy/dx = 57(1 - x)^{56}$, since we did not multiply by (-1), which is the derivative of the function inside the parentheses. Also, the derivative of $y = \sin 9x$ is not $dy/dx = \cos 9x$, since the derivative of the function inside the sine function is 9, and so the correct result is $dy/dx = (\cos 9x)(9) = 9 \cos 9x$.

(*ii*) A less common but probably a worse mistake than the first is to differentiate inside the given function. A student recently wrote on an examination paper that the derivative of $y = \cos(x^2 + 1)$ was $dy/dx = -\sin(2x)$; the reasoning being that the derivative of the cosine is -1 times the sine and the derivative of $x^2 + 1$ is $2x$. Both of these observations are correct, but how they are put together is incorrect. Bear in mind that the derivative of the function inside goes *outside* as a multiple. The derivative of $y = \cos(x^2 + 1)$ is

$$\frac{dy}{dx} = -\sin(x^2 + 1)\frac{d}{dx}(x^2 + 1) = -2x \sin(x^2 + 1)$$

EXERCISES 2.6 *Answers to odd-numbered problems begin on page A-71.*

In Problems 1–38 find the derivative of the given function.

1. $y = (-5x)^{30}$
2. $y = (3/x)^{14}$
3. $y = (2x^2 + x)^{200}$
4. $y = \left(x - \dfrac{1}{x^2}\right)^5$
5. $y = \dfrac{1}{(x^3 - 2x^2 + 7)^4}$
6. $y = \dfrac{1}{x^4 + x^2 + 1}$
7. $y = (3x - 1)^4(-2x + 9)^5$
8. $y = x^4(x^2 + 1)^6$
9. $y = \sin^3 x$
10. $y = \sec^2 x$
11. $f(x) = \left(\dfrac{x^2 - 1}{x^2 + 1}\right)^2$
12. $f(x) = \dfrac{3x - 4}{(5x + 2)^3}$
13. $f(x) = [x + (x^2 - 4)^3]^{10}$
14. $f(x) = \left[\dfrac{1}{(x^3 - x + 1)^2}\right]^4$
15. $g(t) = (t^{-1} + t^{-2} + t^{-3})^{-4}$
16. $F(\theta) = (2\theta + 1)^3 \tan^2\theta$
17. $H(u) = (2 + u \sin u)^{-3}$
18. $q(t) = \dfrac{(1 + \cos t)^2}{(1 + \sin t)^3}$
19. $P(v) = \dfrac{v(2v - 5)^4}{(v + 1)^8}$
20. $R(s) = (s + 1)^2(s + 2)^3(s + 3)^4$
21. $y = \sin(\pi x + 1)$
22. $y = -2\cos(-3x + 7)$
23. $y = \sin^2 4x$
24. $y = \sin(4x)^2$
25. $y = x^3 \cos x^3$
26. $y = \sin 2x \cos 3x$
27. $y = \tan(1/x)$
28. $y = \cot^3 8x$
29. $F(\theta) = \dfrac{\sin 5\theta}{\cos 6\theta}$
30. $h(t) = \dfrac{t + \sin 4t}{10 + \cos 3t}$
31. $f(x) = (\sec 4x + \tan 2x)^5$
32. $f(x) = \csc^2 x - \csc x^2$
33. $f(x) = \sin(\sin x)$
34. $f(x) = \tan\left(\cos \dfrac{x}{2}\right)$

35. $f(x) = \sin^3(4x^2 - 1)$

36. $f(x) = [\cos(x^3 + x^2)]^{-4}$

37. $g(t) = (1 + (1 + (1 + t^3)^4)^5)^6$

38. $r(v) = \left[v^2 - \left(1 + \frac{1}{v}\right)^{-3}\right]^2$

In Problems 39–42 find the slope of the tangent line to the graph of the given function at the indicated value of x.

39. $y = (x^2 + 2)^3$; $x = -1$

40. $y = \frac{1}{(3x + 1)^2}$; $x = 0$

41. $y = \sin 3x + 4x \cos 5x$; $x = \pi$

42. $y = 50x - \tan^3 2x$; $x = \pi/6$

In Problems 43–46 find an equation of the tangent line to the graph of the given function at the indicated value of x.

43. $y = \left(\frac{x}{x + 1}\right)^2$; $x = -\frac{1}{2}$

44. $y = x^2(x - 1)^3$; $x = 2$

45. $y = \tan 3x$; $x = \pi/4$

46. $y = (-1 + \cos 4x)^3$; $x = \pi/8$

In Problems 47 and 48 find an equation of the normal line to the graph of the given function at the indicated value of x.

47. $y = \sin \frac{\pi}{6x} \cos \pi x^2$; $x = \frac{1}{2}$

48. $y = \sin^3 \frac{x}{3}$; $x = \pi$

49. Find the point(s) on the graph of $f(x) = x/(x^2 + 1)^2$ where the tangent line is horizontal. Does the graph of f have any vertical tangents?

50. The function $H = (k/mg)\sin^2\theta$, where k, m, and g are constants, represents the height attained by a grasshopper whose takeoff angle is θ. Determine the values of θ at which $dH/d\theta = 0$.

51. The function $R = (v_0^2/g)\sin 2\theta$ gives the range of a projectile fired at an angle θ from the horizontal with an initial velocity v_0. If v_0 and g are constants, find those values of θ at which $dR/d\theta = 0$.

52. The volume of a spherical balloon of radius r is $V = (4\pi/3)r^3$. The radius is a function of time t and increases at a rate of 5 in/min. What is the instantaneous rate of change of V with respect to t?

53. Suppose a spherical balloon is being filled at a constant rate $dV/dt = 10$ in^3/min. At what rate is its radius increasing when $r = 2$ in?

54. Determine the values of t at which the instantaneous rate of change of $g(t) = \sin t + \frac{1}{2}\cos 2t$ is zero.

In Problems 55 and 56 let $y = f(u)$ and $u = g(x)$ be differentiable functions.

(a) Compute dy/dx without forming $f(g(x))$.

(b) Verify the answer in part **(a)** by finding $f(g(x))$ and $\frac{d}{dx} f(g(x))$.

55. $y = u^3 + 2u$, $u = x^9 + 4x^2$

56. $y = u^4 + 5u^3 - 7u^2 + 8u - 1$, $u = x^3$

57. Let F be a differentiable function. What is $\frac{d}{dx} F(3x)$?

58. Let G be a differentiable function. What is $\frac{d}{dx}[G(-x^2)]^2$?

59. Suppose $\frac{d}{du} f(u) = \frac{1}{u}$. What is $\frac{d}{dx} f(-10x + 7)$?

60. Suppose $\frac{d}{dx} f(x) = \frac{1}{1 + x^2}$. What is $\frac{d}{dx} f(x^3)$?

61. Given that f is an odd differentiable function, use the Chain Rule to show that f' is an even function.

62. Given that f is an even differentiable function, use the Chain Rule to show that f' is an odd function.

63. Assuming differentiability of all functions, find dy/dx for $y = f(g(h(x)))$.

64. Suppose $g(t) = h(f(t))$, where $f(1) = 3$, $f'(1) = 6$, and $h'(3) = -2$. What is $g'(1)$?

2.7 HIGHER-ORDER DERIVATIVES

The Second Derivative The derivative $f'(x)$ is a function derived from a function $y = f(x)$. By differentiating the first derivative $f'(x)$, we obtain yet another function called the **second derivative**, which is denoted by $f''(x)$.

In terms of the operation symbol d/dx, we define the second derivative with respect to x as the function obtained by differentiating $y = f(x)$ twice in succession:

$$\frac{d}{dx}\left(\frac{dy}{dx}\right)$$

The second derivative is commonly denoted by the symbols

$$f''(x), \quad y'', \quad \frac{d^2y}{dx^2}, \quad D_x^2 y$$

EXAMPLE 1 Find the second derivative of $y = x^3 - 2x^2$.

Solution The first derivative is

$$\frac{dy}{dx} = 3x^2 - 4x$$

The second derivative follows from differentiating the first derivative:

$$\frac{d^2y}{dx^2} = \frac{d}{dx}(3x^2 - 4x) = 6x - 4$$

□

EXAMPLE 2 The first derivative of $f(x) = \sin x$ is

$$f'(x) = \cos x$$

Then the second derivative is

$$f''(x) = -\sin x$$

□

EXAMPLE 3 Find the second derivative of $y = (x^3 + 1)^4$.

Solution We obtain the first derivative from the Power Rule for Functions:

$$\frac{dy}{dx} = 4(x^3 + 1)^3 \frac{d}{dx}(x^3 + 1) = 12x^2(x^3 + 1)^3$$

To find the second derivative, we now use the Product and Power Rules:

$$\begin{aligned} \frac{d^2y}{dx^2} &= 12x^2 \frac{d}{dx}(x^3 + 1)^3 + (x^3 + 1)^3 \frac{d}{dx} 12x^2 \\ &= 108x^4(x^3 + 1)^2 + 24x(x^3 + 1)^3 \\ &= (x^3 + 1)^2(132x^4 + 24x) \end{aligned}$$

Factor and simplify

□

Assuming all derivatives exist, we can differentiate a function $y = f(x)$ as many times as we want. The **third derivative** is the derivative of the second derivative; the **fourth derivative** is the derivative of the third derivative; and so on. We denote the third and fourth derivatives by d^3y/dx^3 and d^4y/dx^4, respectively, and define them by

$$\frac{d^3y}{dx^3} = \frac{d}{dx}\left(\frac{d^2y}{dx^2}\right) \quad \text{and} \quad \frac{d^4y}{dx^4} = \frac{d}{dx}\left(\frac{d^3y}{dx^3}\right)$$

In general, if n is a positive integer, then the nth derivative is defined by

$$\frac{d^ny}{dx^n} = \frac{d}{dx}\left(\frac{d^{n-1}y}{dx^{n-1}}\right)$$

Other notations for the first n derivatives are

$$f'(x), f''(x), f'''(x), f^{(4)}(x), \ldots, f^{(n)}(x)$$
$$y', y'', y''', y^{(4)}, \ldots, y^{(n)}$$
$$D_xy, D_x^2y, D_x^3y, D_x^4y, \ldots, D_x^ny$$

EXAMPLE 4 Find the first five derivatives of

$$f(x) = 2x^4 - 6x^3 + 7x^2 + 5x - 10.$$

Solution We have

$$\begin{aligned} f'(x) &= 8x^3 - 18x^2 + 14x + 5 \\ f''(x) &= 24x^2 - 36x + 14 \\ f'''(x) &= 48x - 36 \\ f^{(4)}(x) &= 48 \\ f^{(5)}(x) &= 0 \end{aligned}$$

□

After reflecting a moment, you should be convinced that the $(n + 1)$st derivative of an nth-degree polynomial function is zero.

EXAMPLE 5 Find the third derivative of $y = \dfrac{1}{x^3}$.

Solution We first simplify the function by writing $y = x^{-3}$. By the Power Rule (Theorem 2.8), we have

$$\frac{dy}{dx} = -3x^{-4}$$

Hence,

$$\frac{d^2y}{dx^2} = (-3)(-4)x^{-5} = 12x^{-5}$$

and

$$\frac{d^3y}{dx^3} = (12)(-5)x^{-6} = -\frac{60}{x^6}$$

□

▲ ***Remarks*** You may be wondering what interpretation can be given to higher-order derivatives. In Section 2.1 we saw that if f is differentiable, then the first derivative f' gives the rate of change of the function. Similarly, if f' is differentiable, then the second derivative f'' gives the rate of change of the first derivative f'. We shall see the physical and geometric significance of this last statement in Sections 3.1, 3.5, and 3.6 (see also Problem 45 in Exercises 2.7). If we think in terms of graphs, then the second derived function f'' gives the slope of tangent lines to the graph of f'; f''' gives the slope of tangent lines to graph of f'', and so on.

EXERCISES 2.7

Answers to odd-numbered problems begin on page A-71.

In Problems 1–20 find the second derivative of the given function.

1. $y = -x^2 + 3x - 7$
2. $y = 15x^2 - \pi^2$
3. $y = (-4x + 9)^2$
4. $y = 2x^5 + 4x^3 - 6x^2$
5. $y = 10x^{-2}$
6. $y = (2/x^2)^3$
7. $y = x^3 + 8x^2 - \dfrac{2}{x^4}$
8. $y = \dfrac{x^6 - 7x^3 + 1}{x^2}$
9. $f(x) = x^2(3x - 4)^3$
10. $f(x) = (x^2 + 5x - 1)^4$
11. $g(t) = \dfrac{2t - 3}{4t + 2}$
12. $h(z) = \dfrac{z^2}{z + 1}$
13. $f(x) = \cos 10x$
14. $f(x) = \tan(x/2)$
15. $f(x) = x \sin x$
16. $f(x) = \sin^2 5x$
17. $r(\theta) = \dfrac{1}{3 + 2\cos\theta}$
18. $H(\theta) = \dfrac{\cos\theta}{\theta}$
19. $f(x) = \sec x$
20. $f(x) = \cos x^2$

In Problems 21–24 find the indicated derivative.

21. $y = 4x^6 + x^5 - x^3$; d^4y/dx^4
22. $y = \dfrac{2}{x}$; $D_x^5 y$
23. $f(x) = \sin \pi x$; $f'''(x)$
24. $f(x) = \dfrac{1}{\sec(2x + 1)}$; $f^{(5)}(x)$
25. Show that **(a)** $D_x^4 \sin x = \sin x$ and **(b)** $D_x^4 \cos x = \cos x$.
26. Use the results from Problem 25 to find **(a)** $D_x^{30} \sin x$ and **(b)** $D_x^{67} \cos x$.

Given that n is a positive integer in Problems 27 and 28, find a formula for the given derivative.

27. $\dfrac{d^n}{dx^n} x^n$
28. $\dfrac{d^n}{dx^n}\left(\dfrac{1}{1 - 2x}\right)$

In Problems 29 and 30 find the point(s) on the graph of f at which $f''(x) = 0$.

29. $f(x) = x^3 + 12x^2 + 20x$
30. $f(x) = x^4 - 2x^3$

In Problems 31 and 32 determine intervals for which $f''(x) > 0$ and intervals for which $f''(x) < 0$.

31. $f(x) = (x - 1)^3$
32. $f(x) = x^3 + x^2$
33. Find an equation of the tangent line to the graph of $y = x^3 + 3x^2 - 4x + 1$ at the point where the value of the second derivative is zero.
34. Find an equation of the tangent line to the graph of $y = x^4$ at the point where the value of the third derivative is 12.
35. If $f(x) = \cos(x/3)$, what is the slope of the tangent line to the graph of f' at $x = 2\pi$?
36. If $f(x) = (1 + x)/x$, what is the slope of the tangent line to the graph of f'' at $x = 2$?
37. Find a quadratic function f such that $f(-1) = -11$, $f'(-1) = 7$, and $f''(-1) = -4$.
38. Find the point(s) on the graph of $f(x) = \frac{1}{2}x^2 - 5x + 1$ at which **(a)** $f''(x) = f(x)$ and **(b)** $f''(x) = f'(x)$.
39. If $f(3) = -4$, $f'(3) = 2$, and $f''(3) = 5$, what is

$$\left.\frac{d^2}{dx^2} f^2(x)\right|_{x=3} ?$$

40. If $f'(0) = -1$ and $g'(0) = 6$, what is

$$\left.\frac{d^2}{dx^2}[xf(x) + xg(x)]\right|_{x=0}?$$

41. If $g(1) = 2$, $g'(1) = 3$, $g''(1) = -1$, $f'(2) = 4$, and $f''(2) = 3$, what is

$$\left.\frac{d^2}{dx^2} f(g(x))\right|_{x=1}?$$

42. In Problem 61 in Exercises 2.4 we saw that the van der Waals equation of state for an ideal gas is

$$\left(P + \frac{a}{V^2}\right)(V - b) = RT$$

where a and b are empirical constants, R is the universal gas constant, and T is temperature.

(a) Isotherms for an ideal van der Waals gas are graphs of the equation of state for various constant values of the temperature T. Assuming a constant temperature, the critical molar volume satisfies the equations

$$\frac{dP}{dV} = 0 \quad \text{and} \quad \frac{d^2P}{dV^2} = 0$$

Show that the critical molar volume is $V_{cr} = 3b$ and that the corresponding critical temperature must be $T_{cr} = 8a/27bR$.

(b) Show that the critical pressure is $P_{cr} = a/27b^2$.

An equation containing one or more derivatives of a function and possibly the function itself is called a **differential equation.** Problems 43 and 44 deal with differential equations.

43. In a series circuit containing an inductor and a capacitor (called an L-C circuit), the charge q on the capacitor at any time $t > 0$ satisfies the differential equation

$$L\frac{d^2q}{dt^2} + \frac{1}{C}q = 0$$

where L and C are constants known as the inductance and capacitance, respectively. See Figure 2.38.

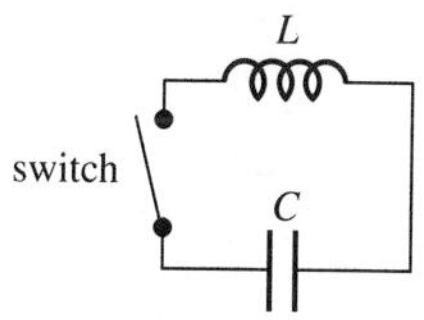

FIGURE 2.38

(a) If C_1 and C_2 are arbitrary constants, verify that

$$q(t) = C_1\cos\frac{1}{\sqrt{LC}}t + C_2\sin\frac{1}{\sqrt{LC}}t$$

satisfies the differential equation.

(b) Use $q(t)$ given in part (**b**) to find a solution of the differential equation that satisfies the conditions $q(0) = q_0$ and $q'(0) = 0$.

44. When a voltage $E(t) = E_0\cos\gamma t$ is impressed on an L-C series circuit, the charge q on the capacitor at any time $t > 0$ satisfies the differential equation

$$L\frac{d^2q}{dt^2} + \frac{1}{C}q = E_0\cos\gamma t$$

(a) Verify that

$$q(t) = \left(q_0 - \frac{E_0C}{1 - \gamma^2LC}\right)\cos\frac{1}{\sqrt{LC}}t + i_0\sqrt{LC}\sin\frac{1}{\sqrt{LC}}t + \frac{E_0C}{1 - \gamma^2LC}\cos\gamma t$$

satisfies the differential equation and the side conditions $q(0) = q_0$ and $q'(0) = i_0$.

(b) The current $i(t)$ in the circuit is defined by $i(t) = dq/dt$. Find $i(t)$.

Calculator/Computer Problems

45. (a) Use a calculator or computer to obtain the graph of $f(x) = x^4 - 4x^3 - 2x^2 + 12x - 2$.

(b) Evaluate $f''(x)$ at $x = -2$, $x = -1$, $x = 0$, $x = 1$, $x = 2$, $x = 3$, and $x = 4$.

(c) From the data in part (**b**), do you see any relationship between the shape of the graph of f and the algebraic signs of f''?

Miscellaneous Problems

46. Let $f(x) = x^3 + 2x$.

(a) Find $f'(x)$ and $f''(x)$.

(b) In general,

$$f''(x) = \lim_{\Delta x \to 0} \frac{f'(x + \Delta x) - f'(x)}{\Delta x}$$

provided this limit exists. Use $f'(x)$ obtained in part (**a**) and the foregoing definition to find $f''(x)$.

Suppose $y = f(x)$ is a twice differentiable function. The **curvature** of the graph of f at a point (x, y) is defined to be

$$\kappa = \frac{|f''(x)|}{[1 + (f'(x))^2]^{3/2}}$$

A small value of κ at a point indicates that the graph is nearly straight near the point.

47. Show that the curvature of a graph of a linear function is zero at every point.

48. Show that $\kappa = 1$ at each point of the semicircle defined by $y = \sqrt{1 - x^2}$.

49. Calculate the curvature of the graph of $f(x) = x^2$ at $x = 0$. As $x \to \infty$, what is the limiting value of κ? Interpret geometrically.

50. (a) Show that

$$\frac{d^2}{dx^2}(fg) = f''g + 2f'g' + fg''$$

$$\frac{d^3}{dx^3}(fg) = f'''g + 3f''g' + 3f'g'' + fg'''$$

(b) Discern the pattern of the derivatives in part **(a)** and then give

$$\frac{d^4}{dx^4}(fg)$$

2.8 IMPLICIT DIFFERENTIATION

Explicit and Implicit Functions A function in which the dependent variable is expressed solely in terms of the independent variable x, namely, $y = f(x)$, is said to be an **explicit** function; for example, $y = \frac{1}{2}x^3 - 1$ is an explicit function. On the other hand, an equivalent equation $2y - x^3 + 2 = 0$ is said to define the function **implicitly**, or y is an **implicit function** of x.

Now, as we know, the equation

$$x^2 + y^2 = 4 \tag{1}$$

describes a circle of radius 2 centered at the origin. Equation (1) is not a function, since for any choice of x in the interval $-2 < x < 2$ there correspond two values of y. However, as shown in Figure 2.39, by considering either the top half or the bottom half of the circle, we obtain a function. We say that (1) defines *at least* two implicit functions of x on the interval $-2 \le x \le 2$. In this case, we observe that the top half of the circle is described by the function

$$f(x) = \sqrt{4 - x^2}, \qquad -2 \le x \le 2 \tag{2}$$

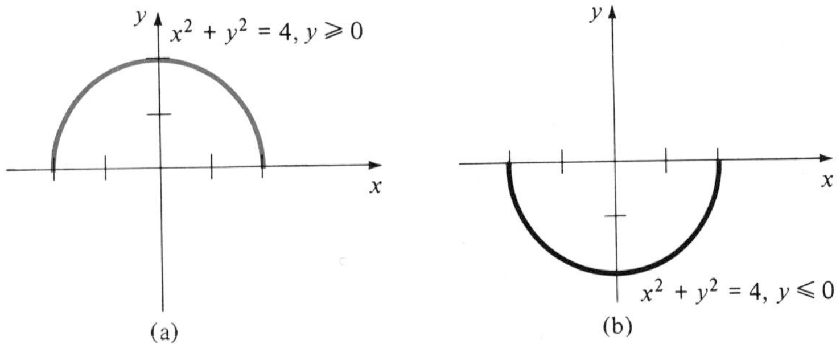

FIGURE 2.39

whereas the bottom half is given by

$$g(x) = -\sqrt{4 - x^2}, \qquad -2 \le x \le 2 \tag{3}$$

Note that both equations

$$x^2 + [f(x)]^2 = 4 \quad \text{and} \quad x^2 + [g(x)]^2 = 4$$

are identities on the interval $-2 \le x \le 2$.

In general, if an equation $F(x, y) = 0$ defines a function f implicitly on some interval, then $F(x, f(x)) = 0$ is an identity on the interval. The graph of f is a portion (or all) of the graph of the equation $F(x, y) = 0$.

A more complicated equation such as

$$x^4 + x^2y^3 - y^5 = 2x + 1$$

may determine several implicit functions on a suitably restricted interval of the x-axis and yet it may not be possible to solve for y in terms of x. However, in some cases we can determine the derivative dy/dx by a process known as **implicit differentiation.** This process consists of differentiating both sides of an equation with respect to x, using the rules of differentiation, and then solving for dy/dx. Since we think of y as being determined by the given equation as a differentiable function, the Chain Rule, in the form of the Power Rule for Functions, gives the useful result

$$\frac{d}{dx}\, y^n = ny^{n-1}\frac{dy}{dx} \tag{4}$$

where n is an integer.

In the following examples we shall assume that the given equation determines at least one differentiable implicit function.

EXAMPLE 1 Find dy/dx if $x^2 + y^2 = 4$.

Solution We differentiate both sides of the equation and then utilize (4):

Power Rule

$$\frac{d}{dx}\, x^2 + \frac{d}{dx}\, y^2 = \frac{d}{dx}\, 4$$

$$2x + 2y\frac{dy}{dx} = 0$$

Solving for the derivative yields

$$\frac{dy}{dx} = -\frac{x}{y} \tag{5}$$ □

As illustrated in (5) of Example 1, implicit differentiation usually yields a derivative that depends on both variables x and y. In our introductory discussion we saw that the equation $x^2 + y^2 = 4$ defines two differentiable implicit functions on the open interval $-2 < x < 2$. The symbolism $dy/dx = -x/y$ represents the derivative of either function on the interval. Note that this derivative clearly indicates that (2) and (3) are not differentiable at $(-2, 0)$ and $(2, 0)$. In general, implicit differentiation yields the derivative of any differentiable implicit function defined by an equation $F(x, y) = 0$.

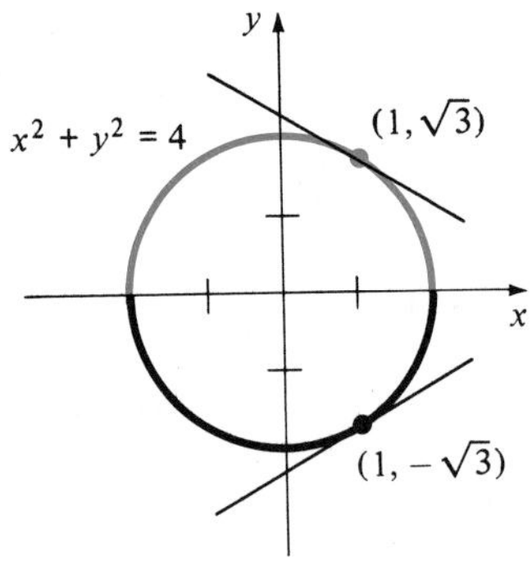

FIGURE 2.40

EXAMPLE 2 Find the slopes of the tangent lines to the graph of $x^2 + y^2 = 4$ at the points corresponding to $x = 1$.

Solution Substituting $x = 1$ into the given equation gives $y^2 = 3$ or $y = \pm\sqrt{3}$. Hence, there are tangent lines at $(1, \sqrt{3})$ and $(1, -\sqrt{3})$. Although $(1, \sqrt{3})$ and $(1, -\sqrt{3})$ are points on the graphs of two different implicit functions, indicated by the different colors in Figure 2.40, (5) of Example 1 gives the correct slope at each point. We have

$$\left.\frac{dy}{dx}\right|_{(1,\sqrt{3})} = -\frac{1}{\sqrt{3}} \quad \text{and} \quad \left.\frac{dy}{dx}\right|_{(1,-\sqrt{3})} = -\frac{1}{-\sqrt{3}} = \frac{1}{\sqrt{3}}$$ □

EXAMPLE 3 Find dy/dx if $x^4 + x^2y^3 - y^5 = 2x + 1$.

Solution In this case, we use (4) and the Product Rule:

$$\frac{d}{dx}x^4 + \frac{d}{dx}x^2y^3 - \frac{d}{dx}y^5 = \frac{d}{dx}2x + \frac{d}{dx}1$$

Product Rule here (under $\frac{d}{dx}x^2y^3$) Power Rule here (under $\frac{d}{dx}y^5$)

$$4x^3 + x^2 \cdot 3y^2\frac{dy}{dx} + 2xy^3 - 5y^4\frac{dy}{dx} = 2$$

$$(3x^2y^2 - 5y^4)\frac{dy}{dx} = 2 - 4x^3 - 2xy^3$$

$$\frac{dy}{dx} = \frac{2 - 4x^3 - 2xy^3}{3x^2y^2 - 5y^4}$$ □

EXAMPLE 4 Find d^2y/dx^2 if $x^2 + y^2 = 4$.

Solution From Example 1, we already know that the first derivative is $dy/dx = -x/y$. Hence, by the Quotient Rule

Substituting for dy/dx

$$\frac{d^2y}{dx^2} = -\frac{d}{dx}\left(\frac{x}{y}\right) = -\frac{y \cdot 1 - x \cdot \frac{dy}{dx}}{y^2} = -\frac{y - x\left(-\frac{x}{y}\right)}{y^2} = -\frac{y^2 + x^2}{y^3}$$

Noting that $x^2 + y^2 = 4$ permits us to rewrite the second derivative as

$$\frac{d^2y}{dx^2} = -\frac{4}{y^3}$$ □

EXAMPLE 5 Find dy/dx if $\sin y = y \cos 2x$.

Solution From the Chain Rule and Product Rule we obtain

$$\frac{d}{dx}\sin y = \frac{d}{dx}\, y \cos 2x$$

$$\cos y \cdot \frac{dy}{dx} = y(-\sin 2x \cdot 2) + \cos 2x \cdot \frac{dy}{dx}$$

$$(\cos y - \cos 2x)\frac{dy}{dx} = -2y \sin 2x$$

$$\frac{dy}{dx} = -\frac{2y \sin 2x}{\cos y - \cos 2x}$$ □

▲ ***Remark*** To determine when an equation defines a function implicitly is not an easy matter. Thus, in the absence of any stated criteria, it is understood that finding dy/dx by implicit differentiation could, in some cases, be nothing more than formal symbol manipulation. For example, you should verify that $x^2 + y^2 = c$ will give $dy/dx = -x/y$ for any choice of the constant c. But for $c < 0$ the equation yields no real function and so dy/dx is meaningless for these values. (See Problems 35 and 36 in Exercises 2.8)

EXERCISES 2.8

Answers to odd-numbered problems begin on page A-71.

In Problems 1–20 assume that the given equation defines at least one differentiable implicit function. Use implicit differentiation to find dy/dx.

1. $y^2 - 2y = x$
2. $4x^2 + y^2 = 8$
3. $xy^2 - x^2 + 4 = 0$
4. $(y - 1)^2 = 4(x + 2)$
5. $x + xy - y^2 - 20 = 0$
6. $y^3 - 2y + 3x^3 = 4x + 1$
7. $x^3y^2 = 2x^2 + y^2$
8. $x^5 - 6xy^3 + y^4 = 1$
9. $(x^2 + y^2)^6 = x^3 - y^3$
10. $y = (x - y)^2$
11. $y^{-3}x^6 + y^6x^{-3} = 2x + 1$
12. $y^4 - y^2 = 10x - 3$
13. $(x - 1)^2 + (y + 4)^2 = 25$
14. $\dfrac{x + y}{x - y} = x$
15. $y^2 = \dfrac{x - 1}{x + 2}$
16. $\dfrac{x}{y^2} + \dfrac{y^2}{x} = 5$
17. $xy = \sin(x + y)$
18. $x + y = \cos xy$
19. $x = \sec y$
20. $x \sin y - y \cos x = 1$

In Problems 21 and 22 find the indicated derivative.

21. $r^2 = \sin 2\theta$; $dr/d\theta$
22. $\pi r^2 h = 100$; dh/dr

In Problems 23 and 24 find dy/dx at the indicated point.

23. $xy^2 + 4y^3 + 3x = 0$; $(1, -1)$
24. $y = \sin xy$; $(\pi/2, 1)$

In Problems 25 and 26 find dy/dx at the points that correspond to the indicated value.

25. $2y^2 + 2xy - 1 = 0$; $x = \frac{1}{2}$
26. $y^3 + 2x^2 = 11y$; $y = 1$

In Problems 27–30 find an equation of the tangent line at the indicated point or value.

27. $x^4 + y^3 = 24$; $(-2, 2)$

28. $\dfrac{1}{x} + \dfrac{1}{y} = 1$; $x = 3$

29. $\tan y = x$; $y = \pi/4$

30. $3y + \cos y = x^2$; $(1, 0)$

In Problems 31 and 32 determine the point(s) on the graph of the given equation where the tangent is horizontal.

31. $x^2 - xy + y^2 = 3$

32. $y^2 = x^2 - 4x + 7$

33. Find the point(s) on the graph of $x^2 + y^2 = 25$ at which the slope of the tangent is $\frac{1}{2}$.

34. Find the point of intersection of the tangents to the graph of $x^2 + y^2 = 25$ at $(-3, 4)$ and $(-3, -4)$.

In Problems 35 and 36 find dy/dx but show that the given equation does not define any real function.

35. $x^2 - 6x + y^2 + 8y + 27 = 0$ [*Hint:* Complete the square.]

36. $x^4 + 3x^2y^2 + 5 = 0$

In Problems 37–44 find d^2y/dx^2.

37. $4y^3 = 6x^2 + 1$

38. $xy^4 = 5$

39. $x^2 - y^2 = 25$

40. $x^2 + 4y^2 = 16$

41. $x + y = \sin y$

42. $y^2 - x^2 = \tan 2x$

43. $x^2 + 2xy - y^2 = 1$

44. $x^3 + y^3 = 27$

In Problems 45 and 46 first use implicit differentiation to find dy/dx. Then solve for y explicitly in terms of x and differentiate. Show that the two answers are equivalent.

45. $x^3y = x + 1$

46. $y \sin x = x - 2y$

In Problems 47–50 determine an implicit function from the given equation such that its graph is the colored curve in the figure.

47. $(y - 1)^2 = x - 2$

48. $x^2 + xy + y^2 = 4$

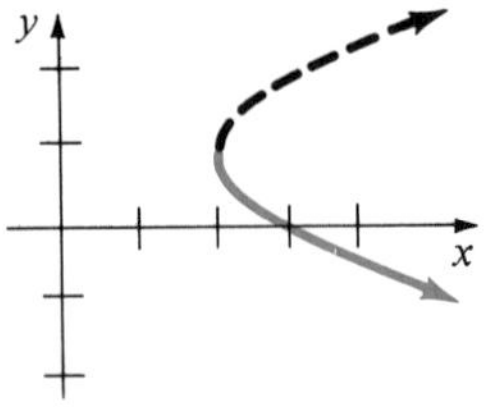

FIGURE 2.41

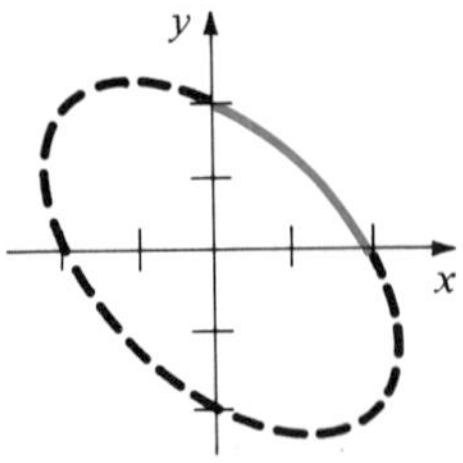

FIGURE 2.42

49. $x^2 + y^2 = 4$

50. $y^2 = x^2(2 - x)$

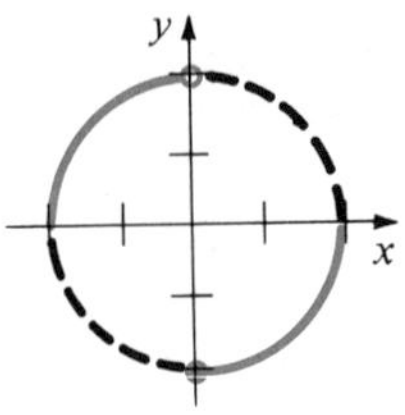

FIGURE 2.43

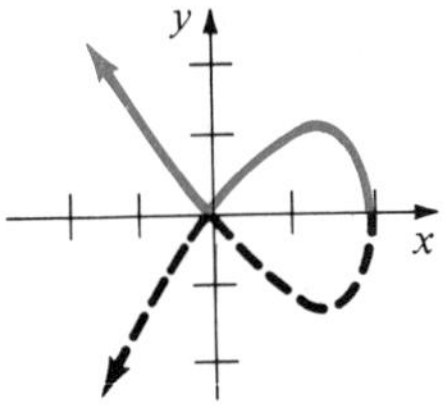

FIGURE 2.44

In Problems 51 and 52 show that the graphs of the given equations are orthogonal at the indicated point of intersection.*

51. $y^2 = x^3$, $2x^2 + 3y^2 = 5$; $(1, 1)$

52. $y^3 + 3x^2y = 13$, $2x^2 - 2y^2 = 3x$; $(2, 1)$

In Problems 53 and 54 assume that both x and y are differentiable functions of a variable t. Find dy/dt in terms of x, y, and dx/dt.

53. $x^2 + y^2 = 25$

54. $x^2 + xy + y^2 - y = 9$

55. The graph of the equation $x^3 + y^3 = 3xy$ shown in Figure 2.45 is called the **folium of Descartes**.

(a) Find an equation of the tangent line at the point in the first quadrant where the folium intersects the graph of $y = x$.

(b) Find the point in the first quadrant at which the tangent line is horizontal.

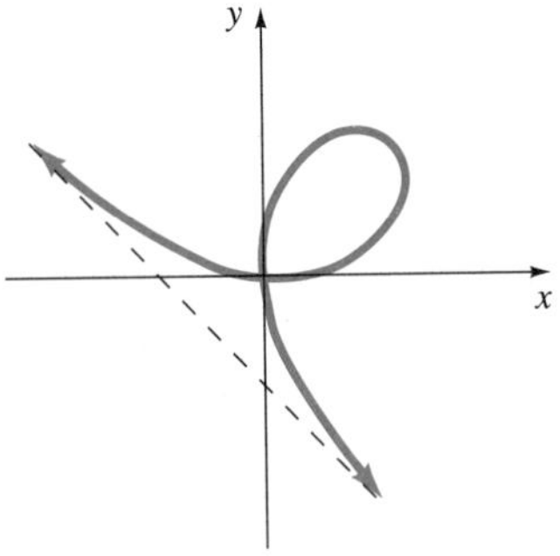

FIGURE 2.45

56. A woman drives toward a freeway sign as shown in Figure 2.46. Let θ be her viewing angle of the sign and let x be her distance (measured in feet) to that sign.

*See Problem 50 in Exercises 2.3.

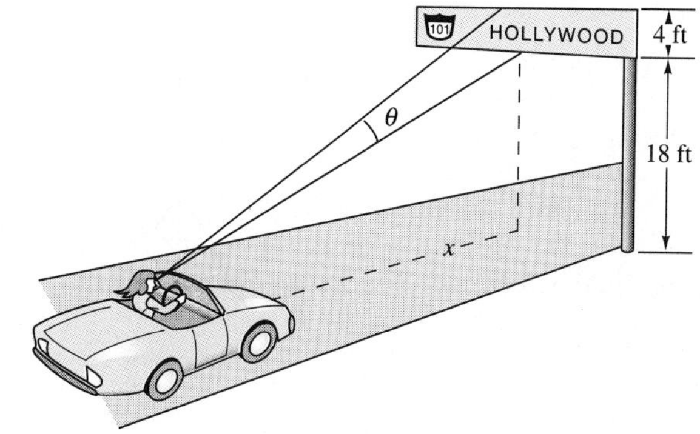

FIGURE 2.46

(a) If her eye level is 4 ft from the surface of the road, show that

$$\tan \theta = \frac{4x}{x^2 + 252}$$

(b) Find the rate at which θ changes with respect to x.

(c) At what distance is the rate in part (**b**) equal to zero?

57. The graph of $(x^2 + y^2)^2 = 4(x^2 - y^2)$ shown in Figure 2.47 is called a **lemniscate**.

(a) Find the points on the graph that correspond to $x = 1$.

(b) Find an equation of the tangent line to the graph at each point found in part (**a**).

(c) Find the points on the graph at which the tangent is horizontal.

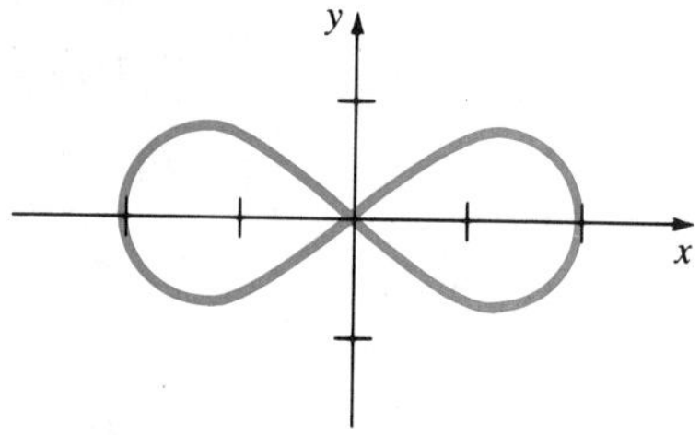

FIGURE 2.47

58. The angle θ $(0 < \theta < \pi)$ between two curves is defined to be the angle between their tangent lines at the point P of intersection. If m_1 and m_2 are the slopes of the tangent lines at P, it can be shown that $\tan \theta = (m_1 - m_2)/(1 + m_1 m_2)$. Determine the angle between the graphs of $x^2 + y^2 + 4y = 6$ and $x^2 + 2x + y^2 = 4$ at $(1, 1)$.

2.9 RULES OF DIFFERENTIATION IV: EXTENDED POWER RULES

In Sections 2.3 and 2.6 we restricted the Power Rule and the Power Rule for Functions to *integer* exponents:

$$\frac{d}{dx} x^n = nx^{n-1} \tag{1}$$

$$\frac{d}{dx} u^n = nu^{n-1} \frac{du}{dx}, \qquad u = g(x) \tag{2}$$

Implicit differentiation provides a means of extending both (1) and (2) to *rational* exponents. If p and q are integers, $q \neq 0$, then for values of x for which $x^{p/q}$ is a real number, the function

$$y = x^{p/q} \quad \text{gives} \quad y^q = x^p \tag{3}$$

Now assuming y' exists and $y \neq 0$, implicit differentiation

$$\frac{d}{dx} y^q = \frac{d}{dx} x^p \quad \text{yields} \quad qy^{q-1} \frac{dy}{dx} = px^{p-1}$$

Thus, $$\frac{dy}{dx} = \frac{p}{q}\frac{x^{p-1}}{y^{q-1}} = \frac{p}{q}\frac{x^{p-1}}{(x^{p/q})^{q-1}} = \frac{p}{q}x^{(p/q)-1}$$

This last result leads to the following extension of (1):

THEOREM 2.11 Power Rule (Rational Exponents)

If p/q is a rational number, then

$$\frac{d}{dx}x^{p/q} = \frac{p}{q}x^{(p/q)-1} \tag{4}$$

We note that (4) reduces to (1) when $p = n$ and $q = 1$. Furthermore, there is nothing new to memorize in (4); it is the same rule as before: *Bring down the exponent as a multiple and decrease the exponent by* 1.

EXAMPLE 1 Differentiate $y = \sqrt{x}$.

Solution First write the given function as $y = x^{1/2}$ and then use (4):

$$\frac{dy}{dx} = \frac{1}{2}x^{(1/2)-1} = \frac{1}{2}x^{-1/2} = \frac{1}{2\sqrt{x}}$$ □

EXAMPLE 2 Find an equation of the tangent line to the graph of $y = \sqrt{x}$ at $x = 4$.

Solution When $x = 4$, $y = \sqrt{4} = 2$. It follows from Example 1 that the slope of the tangent line at (4, 2) is

$$\left.\frac{dy}{dx}\right|_{x=4} = \frac{1}{2\sqrt{4}} = \frac{1}{4}$$

Then an equation of the tangent line is

$$y - 2 = \frac{1}{4}(x - 4) \quad \text{or} \quad y = \frac{1}{4}x + 1$$

Figure 2.48 shows the graphs of the function and the tangent line. □

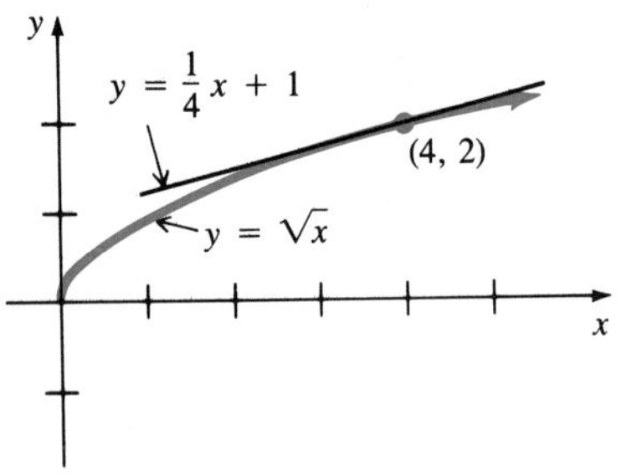

FIGURE 2.48

EXAMPLE 3 Differentiate $y = \dfrac{1}{\sqrt{x}}$.

Solution Since $y = x^{-1/2}$, it follows from (4) that

$$\frac{dy}{dx} = -\frac{1}{2}x^{(-1/2)-1} = -\frac{1}{2}x^{-3/2}$$ □

EXAMPLE 4 Differentiate $y = 9\sqrt[3]{x} + 4\sqrt{x^3}$.

Solution Using rational exponents, we can express the given function as $y = 9x^{1/3} + 4x^{3/2}$. Thus,

$$\frac{dy}{dx} = 9\left(\frac{1}{3}\right)x^{(1/3)-1} + 4\left(\frac{3}{2}\right)x^{(3/2)-1} = 3x^{-2/3} + 6x^{1/2}$$ □

We can show, in a manner similar to that leading to (4), that the Power Rule for Functions (2) is also true for rational exponents. This result is summarized in the next theorem.

THEOREM 2.12 Power Rule for Functions (Rational Exponents)

If p/q is a rational number and g is a differentiable function, then

$$\frac{d}{dx}[g(x)]^{p/q} = \frac{p}{q}[g(x)]^{(p/q)-1} \cdot g'(x) \tag{5}$$

EXAMPLE 5 Differentiate $y = (4x + 1)^{1/3}$.

Solution From (5),

$$\frac{dy}{dx} = \frac{1}{3}(4x + 1)^{(1/3)-1}\frac{d}{dx}(4x + 1) = \frac{4}{3}(4x + 1)^{-2/3}$$ □

EXAMPLE 6 Differentiate $y = \sqrt{\dfrac{1+x}{1-x}}$.

Solution We use (5) followed by the Quotient Rule:

$$\begin{aligned}\frac{dy}{dx} &= \frac{1}{2}\left[\frac{1+x}{1-x}\right]^{-1/2} \cdot \frac{d}{dx}\left(\frac{1+x}{1-x}\right)\\ &= \frac{1}{2}\left[\frac{1+x}{1-x}\right]^{-1/2} \cdot \frac{(1-x)\cdot 1 - (1+x)(-1)}{(1-x)^2}\\ &= \frac{1}{(1-x)^2}\left[\frac{1+x}{1-x}\right]^{-1/2} = \frac{1}{(1+x)^{1/2}(1-x)^{3/2}}\end{aligned}$$ □

EXAMPLE 7 Differentiate $y = \cos \sqrt[5]{x^2 + 1}$.

Solution We first use the Chain Rule followed by the Power Rule for Functions (5):

$$\frac{dy}{dx} = -\sin \sqrt[5]{x^2 + 1} \cdot \frac{d}{dx}(\sqrt[5]{x^2 + 1})$$

$$= -\sin \sqrt[5]{x^2 + 1} \cdot \left[\frac{1}{5}(x^2 + 1)^{-4/5} \cdot 2x\right]$$

$$= -\sin \sqrt[5]{x^2 + 1} \left[\frac{2x}{5(x^2 + 1)^{4/5}}\right]$$ □

Recall from Section 2.2 that the graph of a function f continuous at a number a has a vertical tangent at $(a, f(a))$ if $\lim_{x\to a}|f'(x)| = \infty$. The graphs of many functions with rational exponents possess vertical tangents.

EXAMPLE 8 Find $f(x) = x^{2/3}$, we have

$$f'(x) = \frac{2}{3}x^{-1/3} = \frac{2}{3x^{1/3}}$$

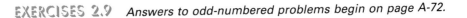

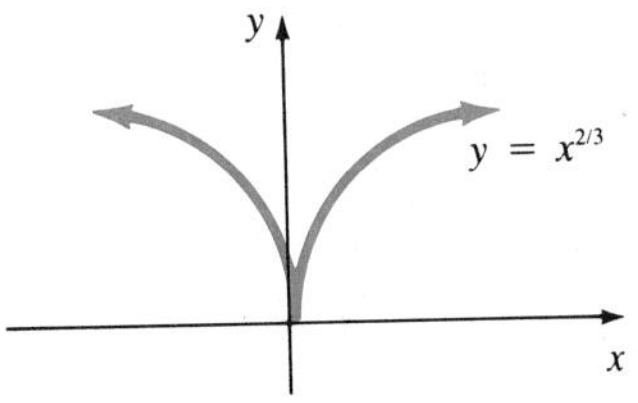

FIGURE 2.49

Notice that $\lim_{x\to 0^+} f'(x) = \infty$, whereas $\lim_{x\to 0^-} f'(x) = -\infty$. Since f is continuous at $x = 0$ and $|f'(x)| \to \infty$ as $x \to 0$, we conclude that the y-axis is a vertical tangent at (0, 0). This fact is apparent from the graph in Figure 2.49. □

The graph of $f(x) = x^{2/3}$ in Example 8 is said to have a **cusp** at the origin. In general, the graph of the function $y = f(x)$ has a cusp at $(a, f(a))$ if f is continuous at a, f' has opposite signs on either side of a, and $|f'(x)| \to \infty$ as $x \to a$.

▲ ***Remark*** In Chapter 7 the results in (1) and (2) will be extended even further to any real exponent (that is, rational *or* irrational).

EXERCISES 2.9 *Answers to odd-numbered problems begin on page A-72.*

In Problems 1–20 find the derivative of the given function.

1. $y = 10x^{3/2}$
2. $y = 8x^{-1.2}$
3. $y = \dfrac{1}{\sqrt[3]{x^4}}$
4. $y = \sqrt[5]{8x^3}$
5. $y = \sqrt{x} + \dfrac{1}{\sqrt{x}}$
6. $y = \sqrt{4x^2 + 9}$
7. $y = (x^2 + 1)(x^2 - 4)^{2/3}$
8. $y = \dfrac{\sqrt{x}}{x^2 + 1}$
9. $y = \left(\dfrac{x - 9}{x + 2}\right)^{3/2}$
10. $y = \sqrt{2x + 1}\sqrt[3]{3x - 1}$
11. $y = \sin \sqrt{x}$
12. $y = x \cot \dfrac{1}{\sqrt{x}}$

13. $f(x) = x + \sqrt{x^2 + 1}$

14. $f(x) = \sqrt{x + \sqrt{x}}$

15. $g(t) = [(t^2 - 1)(t^3 + 4t)]^{1/3}$

16. $H(z) = \dfrac{\sqrt{z} + 1}{\sqrt{z} + 3}$

17. $q(\theta) = \dfrac{1}{\sqrt{\theta + \sin\theta}}$

18. $r(\theta) = \sqrt{\cos 4\theta}$

19. $F(s) = \sqrt[5]{(s^4 + 1)^2}$

20. $g(u) = u^{2/3}(u^{1/3} + 1)^3$

In Problems 21–26 find the second derivative of the given function.

21. $y = x + \sqrt{x}$

22. $y = 18x^{4/3}$

23. $y = (\sqrt{x} + 1)^4$

24. $f(t) = \left(\dfrac{t}{t+6}\right)^{0.3}$

25. $F(\theta) = (\sin\theta)^{2.4}$

26. $g(t) = \tan t^{1.6}$

In Problems 27–30 find an equation of the tangent line to the graph of the given function at the indicated value of x.

27. $y = x^{1/3}$; $x = 8$

28. $y = \sqrt{2x + 1}$; $x = 4$

29. $y = \dfrac{x^2}{\sqrt{x^2 + 3}}$; $x = 1$

30. $y = (\tan 2x)^{1/3}$; $x = \pi/8$

In Problems 31–34 use implicit differentiation to find dy/dx.

31. $x + \sqrt{xy} + y = 1$

32. $xy^2 = \sqrt{x + 1}$

33. $2y^{3/2} = 3x(x^2 - 1)^{5/2}$

34. $(6x)^{1/2} + (8y)^{3/2} = 2$

35. The graph of $x^{2/3} + y^{2/3} = 1$, shown in Figure 2.50, is called a **hypocycloid**. Find equations of the tangent lines to the graph at the points corresponding to $x = \frac{1}{8}$.

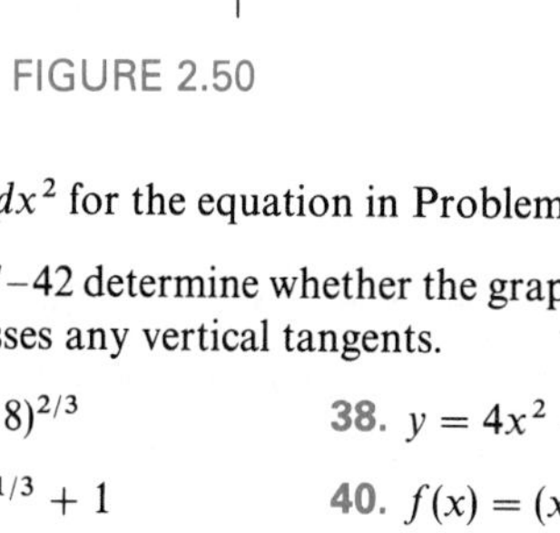

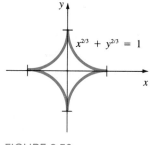

FIGURE 2.50

36. Find d^2y/dx^2 for the equation in Problem 35.

In Problems 37–42 determine whether the graph of the given function possesses any vertical tangents.

37. $y = (2x - 8)^{2/3}$

38. $y = 4x^2 + 6x^{1/3}$

39. $f(x) = x^{-1/3} + 1$

40. $f(x) = (x^2 + 9)^{1/3}$

41. $f(x) = 1/(x^{1/3} + 1)$

42. $f(x) = (x + 1)^{1/3}(x - 5)^{2/3}$

43. The function $T = 2\pi\sqrt{L/g}$, g a constant, gives the period of a simple pendulum of length L. Given that $g = 32$ ft/s^2, find dT/dL when $L = 2$ ft.

44. According to G. K. Zipf, the number N of cities in the United States that have a population over q million is estimated by $N = 24q^{-1.3}$. Find the instantaneous rate of change of N with respect to q.

45. The velocity v of a rocket y kilometers above the center of the earth is given by $v = \sqrt{2k/y - 2k/R + v_0^2}$, where k, R, and v_0 are constants. Find dv/dy.

46. According to the theory of relativity, the mass m of a body moving with velocity v is $m = m_0/\sqrt{1 - v^2/c^2}$, where m_0 is the initial mass and c is the speed of light. Find dm/dv.

47. In special circumstances the function

$$g(\gamma) = \frac{F_0}{\sqrt{(\omega^2 - \gamma^2)^2 + 4\lambda^2\gamma^2}}$$

where F_0, ω, and λ are constants, gives the amplitude of motion of a mass on a vibrating spring. Verify that $g'(\sqrt{\omega^2 - 2\lambda^2}) = 0$.

48. The amount of substance X present at time t during a third-order chemical reaction is given by

$$X(t) = \left(\frac{X_0^2}{2kX_0^2 t + 1}\right)^{1/2}$$

where X_0 and k are constants. Find $X'(1)$.

49. In a binary ionic crystalline compound such as Na^+Cl^-, the atomic fraction of cationic vacancies N_c and anionic vacancies N_a must satisfy the relation

$$N_a N_c = k^2$$

where k is a constant depending only on temperature. If there are doubling charged impurities such as Ca^{2+} at an atomic fraction C, then electrical neutrality requires that

$$N_a + C = N_c$$

Solve these two equations for N_a in terms of C and k. Find a formula for the rate of change of N_a with respect to C.

50. The surface area S of a human who has weight W is estimated by $S = 0.11W^{2/3}$. Find dS/dW.

51. If C_1 and C_2 are arbitrary constants, verify that

$$y(x) = C_1\frac{\sin x}{\sqrt{x}} + C_2\frac{\cos x}{\sqrt{x}}$$

satisfies the differential equation

$$x^2y'' + xy' + \left(x^2 - \frac{1}{4}\right)y = 0$$

52. Find $(d^2/dx^2)\sqrt{f(x^2+1)}$.

Calculator/Computer Problems

53. An estimation of the percent saturation of hemoglobin in a human is given by

$$f(P) = \frac{0.013P^{2.7}}{1 + 0.00013P^{2.7}}$$

where P represents the partial pressure of oxygen in plasma.

(a) Find the instantaneous rate of change of f when $P = 40$.

(b) Find the values of $P > 0$ for which $f''(P) = 0$.

(c) Use a calculator or computer to obtain the graph of f on the interval [0, 110].

2.10 DIFFERENTIALS AND LINEAR APPROXIMATION

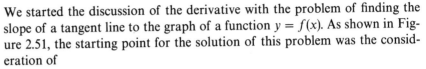

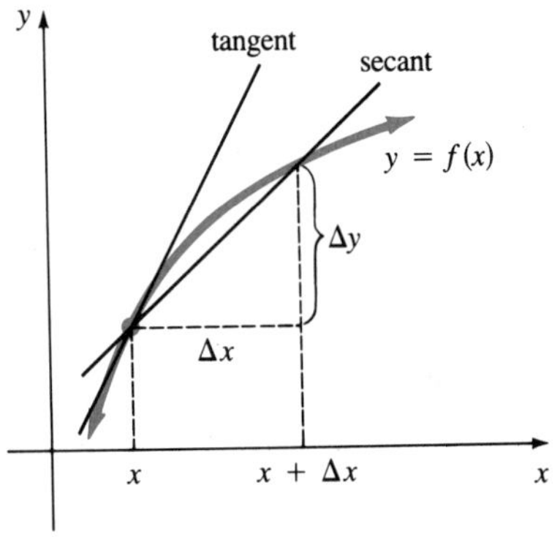

FIGURE 2.51

We started the discussion of the derivative with the problem of finding the slope of a tangent line to the graph of a function $y = f(x)$. As shown in Figure 2.51, the starting point for the solution of this problem was the consideration of

$$m_{\text{sec}} = \frac{f(x + \Delta x) - f(x)}{\Delta x} = \frac{\Delta y}{\Delta x}$$

For small values of Δx, $m_{\text{sec}} \approx m_{\text{tan}}$ or $\Delta y/\Delta x \approx m_{\text{tan}}$. But knowing that $m_{\text{tan}} = f'(x)$ enables us to write

$$\frac{\Delta y}{\Delta x} \approx f'(x) \quad \text{or} \quad \Delta y \approx f'(x)\,\Delta x$$

For convenience we rename the number Δx as follows:

DEFINITION 2.4 Differential of Independent Variable

The increment Δx is called the **differential of the independent variable** x and is denoted by dx; that is, $dx = \Delta x$.

We also rename the quantity $f'(x)\,\Delta x$.

DEFINITION 2.5 Differential of Dependent Variable

The function $f'(x)\,\Delta x$ is called the **differential of the dependent variable** y and is denoted by dy; that is, $dy = f'(x)\,\Delta x = f'(x)\,dx$.

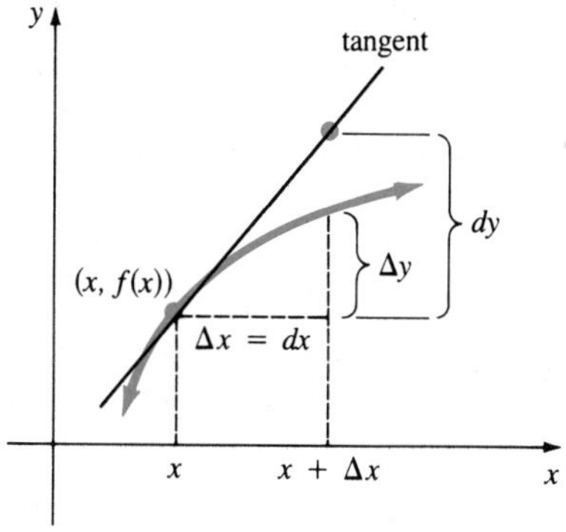

FIGURE 2.52

Since the slope of a tangent to a graph is

$$m_{\tan} = \frac{\text{rise}}{\text{run}} = f'(x) = \frac{f'(x)\,\Delta x}{\Delta x}, \qquad \Delta x \neq 0$$

it follows that the rise of the tangent line can be interpreted as dy.* From Figure 2.52 we see that when Δx is very small ($\Delta x \approx 0$),

$$\Delta y \approx dy \tag{1}$$

EXAMPLE 1

(*a*) Find Δy and dy for $y = 5x^2 + 4x + 1$.

(*b*) Compare the values of Δy and dy for $x = 6$, $\Delta x = dx = 0.02$.

Solution (*a*)

$$\begin{aligned} \Delta y &= f(x + \Delta x) - f(x) \\ &= [5(x + \Delta x)^2 + 4(x + \Delta x) + 1] - [5x^2 + 4x + 1] \\ &= 10x\,\Delta x + 4\,\Delta x + 5(\Delta x)^2 \end{aligned}$$

Now, by Definition 2.5,

$$dy = (10x + 4)\,dx \tag{2}$$

Since $dx = \Delta x$, observe that $\Delta y = (10x + 4)\,\Delta x + 5(\Delta x)^2$ and $dy = (10x + 4)\,\Delta x$ differ by the amount $5(\Delta x)^2$.

(*b*) When $x = 6$, $\Delta x = 0.02$:

$$\Delta y = 10(6)(0.02) + 4(0.02) + 5(0.02)^2 = 1.282$$

whereas

$$dy = (10(6) + 4)(0.02) = 1.28$$

The difference in answers is, of course, $5(0.02)^2 = 0.002$. □

In Example 1 the value $\Delta y = 1.282$ is the *exact* amount by which the function $f(x) = 5x^2 + 4x + 1$ changes as x changes from 6 to 6.02. The differential $dy = 1.28$ represents the *approximate* amount by which the function changes. In other words, for a small change or increment Δx in the independent variable, the corresponding change Δy in the dependent variable can be approximated by the differential dy. Now suppose x changes from 10 to 10.1. It follows from (2) with $x = 10$ and $\Delta x = 0.1$ that the approximate amount of change in the functional values is $dy = 10.4$. Moreover, since $f(10) = 541$, we must have $f(10.1) \approx 541 + 10.4 = 551.4$. Note too that if x changes from, say, 10 to 9.9, then we take $\Delta x = -0.1$.

---

*For this reason, the derivative symbol dy/dx has the appearance of a quotient. To calculate dy, it looks as if both sides of the equality $dy/dx = f'(x)$ are multiplied by the denominator of the left member. While this is, strictly speaking, *not* the case, one can proceed *formally* in this manner.

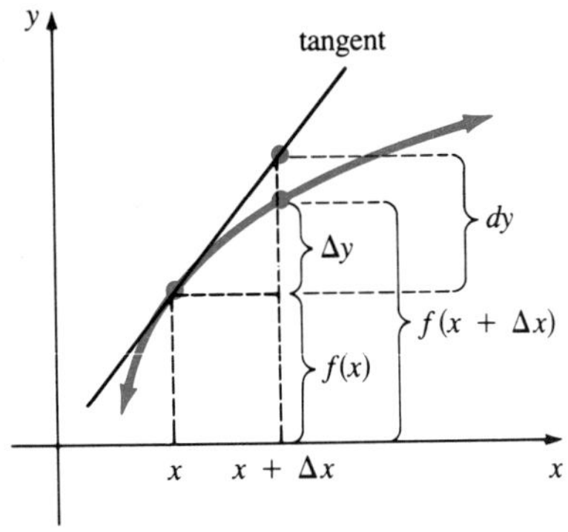

FIGURE 2.53

Linear Approximation As we have just seen, when $\Delta x \approx 0$, differentials give a way of "predicting" the value of $f(x + \Delta x)$ by knowing the value of the function f and its derivative at x. As we see in Figure 2.53, if x changes by an amount Δx, then the corresponding change in the function is $\Delta y = f(x + \Delta x) - f(x)$, and so

$$f(x + \Delta x) = f(x) + \Delta y$$

But in view of (1), for a small change in x we can write

$$f(x + \Delta x) \approx f(x) + dy$$

that is,

$$f(x + \Delta x) \approx f(x) + f'(x)\,dx \tag{3}$$

The result in (3) is called a **linear approximation formula**, since we are using the y-coordinate of a point on a tangent line to approximate the y-coordinate of a point on the graph.

EXAMPLE 2 Use (3) to find an approximation to $\sqrt{25.4}$.

Solution First identify the function $f(x) = \sqrt{x}$. We wish to calculate the approximate value of $f(x + \Delta x) = \sqrt{x + \Delta x}$ when $x = 25$ and $\Delta x = 0.4$. Now

$$dy = \frac{1}{2}x^{-1/2}\,dx = \frac{1}{2\sqrt{x}}\Delta x$$

so that (3) yields

$$\sqrt{x + \Delta x} \approx \sqrt{x} + \frac{1}{2\sqrt{x}}\Delta x$$

Thus, with $x = 25$ and $\Delta x = 0.4$, the preceding formula gives

$$\sqrt{25.4} \approx \sqrt{25} + \frac{1}{2\sqrt{25}}(0.4) = 5.04$$ □

Error The **error** in a calculation is defined to be

$$\textbf{error} = \textbf{true value} - \textbf{approximate value} \tag{4}$$

However, in practice the

$$\textbf{relative error} = \frac{\textbf{error}}{\textbf{true value}} \tag{5}$$

is usually more important than the error. Moreover, (relative error) · 100 is called the **percentage error**. With the aid of a hand calculator, $\sqrt{25.4} = 5.03984$ is correct to five decimal places. Thus, in Example 2 the error is -0.00016, the relative error is -0.00003, and the percentage error is -0.003%.

EXAMPLE 3 A side of a cube is measured to be 30 cm with a possible error of ± 0.02 cm. What is the approximate maximum possible error in the volume of the cube?

Solution The volume of a cube is $V = x^3$, where x is the length of one side. If Δx represents the error in the length of one side, then the corresponding error in the volume is

$$\Delta V = (x + \Delta x)^3 - x^3$$

To simplify matters, we utilize the differential $dV = 3x^2\, dx = 3x^2\, \Delta x$ as an approximation to ΔV. Thus, for $x = 30$ and $\Delta x = \pm 0.02$, the approximate maximum error is

$$dV = 3(30)^2(\pm 0.02) = \pm 54 \text{ cm}^3$$

□

In Example 3, an error of about 54 cm^3 in the volume for an error of 0.02 cm in the length of a side seems considerable. But, observe, if the relative error is $\Delta V/V$, then the approximate relative error is dV/V. When $x = 30$ and $V = 27{,}000$, the approximate maximum relative error is $\pm 54/27{,}000 = \pm 1/500$, and the maximum percentage error is approximately $\pm 0.2\%$.

Rules for Differentials The rules for differentiation considered in this chapter can be rephrased in terms of differentials; for example, if $u = f(x)$ and $v = g(x)$ and $y = f(x) + g(x)$, then $dy/dx = f'(x) + g'(x)$. Hence, $dv = [f'(x) + g'(x)]\, dx = f'(x)\, dx + g'(x)\, dx = du + dy$. We summarize the equivalents of the sum, product, and quotient rules:

$$d(u + v) = du + dv \tag{6}$$

$$d(uv) = u\, dv + v\, du \tag{7}$$

$$d(u/v) = \frac{v\, du - u\, dv}{v^2} \tag{8}$$

As the next example shows, there is little need for memorizing (6), (7), and (8).

EXAMPLE 4 Find dy for $y = x^2 \cos 3x$.

Solution To find the differential of a function, we can simply multiply its derivative by dx. Thus, by the Product Rule,

$$\frac{dy}{dx} = x^2(-\sin 3x \cdot 3) + \cos 3x(2x)$$

$$dy = \left(\frac{dy}{dx}\right) dx = (-3x^2 \sin 3x + 2x \cos 3x)\, dx \tag{9}$$

Alternative Solution Applying (7) gives

$$\begin{aligned} dy &= x^2\, d(\cos 3x) + \cos 3x\, d(x^2) \\ &= x^2(-\sin 3x \cdot 3\, dx) + \cos 3x(2x\, dx) \end{aligned} \tag{10}$$

Factoring dx from (10) yields (9). □

EXERCISES 2.10 *Answers to odd-numbered problems begin on page A-72.*

In Problems 1–10 find the differential dy.

1. $y = 10$
2. $y = x^4 + 3x^2$
3. $y = 1/\sqrt{2x}$
4. $y = \sqrt[4]{x^7}$
5. $y = 12(x^4 - 1)^{1/3}$
6. $y = x^2(1 - x)^5$
7. $y = \dfrac{x^2 - 1}{x^2 + 1}$
8. $y = \dfrac{x}{(3x - 1)^4}$
9. $y = x \cos x - \sin x$
10. $y = (2x + 1)\csc 2x$

In Problems 11–18 find Δy and dy.

11. $y = x^2 + 1$
12. $y = 3x^2 - 5x + 6$
13. $y = (x + 1)^2$
14. $y = x^3$
15. $y = \dfrac{3x + 1}{x}$
16. $y = \dfrac{1}{x^2}$
17. $y = \sin x$
18. $y = -4 \cos 2x$

In Problems 19 and 20 complete the following table for each function.

x	Δx	Δy	dy	$\Delta y - dy$
2	1			
2	0.5			
2	0.1			
2	0.01			

19. $y = 5x^2$
20. $y = 1/x$

In Problems 21–30 use the concept of the differential to find an approximation to the given expression.

21. $\sqrt{37}$
22. $1/\sqrt{96}$
23. $(1.8)^5$
24. $9^{2/3}$
25. $\dfrac{(0.9)^4}{(0.9) + 1}$
26. $(1.1)^3 + 6(1.1)^2$
27. $\cos\left(\dfrac{\pi}{2} - 0.4\right)$
28. $\sin 1°$
29. $\sin 33°$
30. $\tan\left(\dfrac{\pi}{4} + 0.1\right)$

31. Compute the approximate amount by which the function $f(x) = 4x^2 + 5x + 8$ changes as x changes from:
    (a) 4 to 4.03 (b) 3 to 2.9

32. (a) Find an equation of the tangent line to the graph of $f(x) = x^3 + 3x^2$ at $x = 1$.
    (b) Find the y-coordinate of the point of the tangent line in part **(a)** that corresponds to $x = 1.02$.
    (c) Use (3) to find an approximation to $f(1.02)$. Compare your answer with that of part **(b)**.

33. The area of a circle with radius r is $A = \pi r^2$.
    (a) Given that the radius of a circle changes from 4 cm to 5 cm, find the exact change in the area.
    (b) What is the approximate change in the area?

34. According to Poiseuille, the resistance R of a blood vessel of length l and radius r is $R = kl/r^4$, where k is a constant. Given that l is constant, find the approximate change in R when r changes from 0.2 mm to 0.3 mm.

35. Many golf balls consist of a spherical cover over a solid core. Find the exact volume of the cover if its thickness is t and the radius of the core is r. [*Hint:* The volume of a sphere is $V = \frac{4}{3}\pi r^3$. Consider concentric spheres having radii r and $r + \Delta r$.] Use differentials to find an approximation to the volume of the cover. See Figure 2.54. Find an approximation to the volume of the cover if $r = 0.8$ in and $t = 0.04$ in.

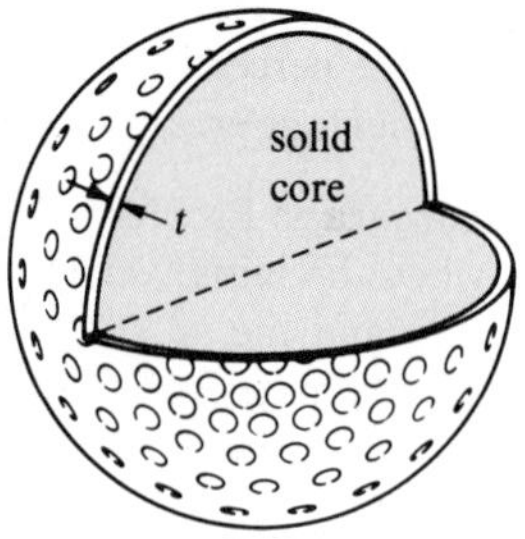

FIGURE 2.54

36. A hollow metal pipe is 1.5 m long. Find an approximation to the volume of the metal if the inner radius of the pipe is 2 cm and the thickness of the metal is 0.25 cm. See Figure 2.55.

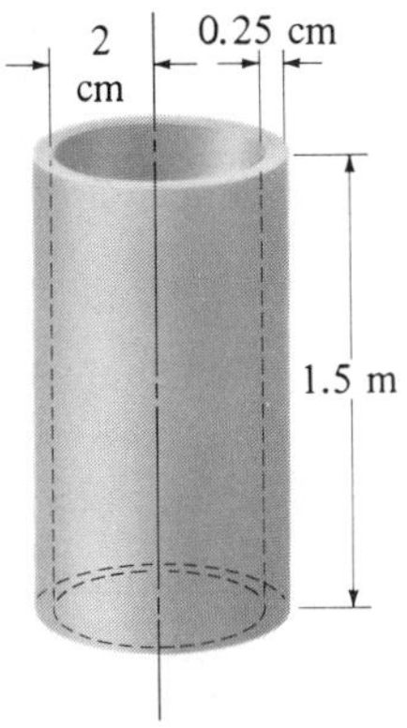

FIGURE 2.55

37. The side of a square is measured to be 10 cm with a possible error of ± 0.3 cm. Use differentials to find an approximation to the maximum error in the area. Find the approximate relative error and the approximate percentage error.

38. An oil storage tank in the form of a circular cylinder has a height of 5 m. The radius is measured to be 8 m with a possible error of ± 0.25 m. Use differentials to estimate the maximum error in the volume. Find the approximate relative error and the approximate percentage error.

39. In the study of some adiabatic processes, the pressure P of a gas is related to the volume V that it occupies by $P = c/V^{\gamma}$, where c and γ are constants. Show that the approximate relative error in P is proportional to the approximate relative error in V.

40. The range R of a projectile with an initial velocity v_0 and angle of elevation θ is given by $R = (v_0^2/g)\sin 2\theta$, where g is the acceleration of gravity. If v_0 and θ are held constant, then show that the percentage error in the range is proportional to the percentage error in g.

41. Use the formula in Problem 40 to determine the range of a projectile when the initial velocity is 256 ft/s, the angle of elevation is 45°, and the acceleration of gravity is 32 ft/s^2. What is the approximate change in the range of the projectile if the initial velocity is increased to 266 ft/s?

42. The acceleration of gravity g is not constant but changes with altitude. For practical purposes, at the surface of the earth g is taken to be 32 ft/s^2, 980 cm/s^2, or 9.8 m/s^2.

(a) From the Law of Universal Gravitation, the force F between a body of mass m_1 and the earth of mass m_2 is $F = km_1m_2/r^2$, where k is constant and r is the distance to the center of the earth. Alternatively, Newton's second law of motion implies $F = m_1 g$. Show that $g = km_2/r^2$.

(b) Use part (**a**) to show $dg/g = -2dr/r$.

(c) Let $r = 6400$ km at the surface of the earth. Use part (**b**) to show that the approximate value of g at an altitude of 16 km is 9.75 m/s^2.

43. The acceleration of gravity g also changes with latitude. The International Geodesy Association has defined g (at sea level) as a function of latitude θ as follows:

$$g = 978.0318(1 + 53.024 \times 10^{-4} \sin^2\theta - 5.9 \times 10^{-6} \sin^2 2\theta)$$

where g is measured in cm/s^2.

(a) According to this function, where is g a minimum? Where is g a maximum?

(b) What is the value of g at latitude 60°N?

(c) What is the approximate change in g as θ changes from 60°N to 61°N? [*Hint:* Remember to use radian measure.]

44. The period (in seconds) of a simple pendulum of length L is $T = 2\pi\sqrt{L/g}$, where g is the acceleration of gravity. Compute the exact change in the period if L is increased from 4 to 5 m. Then use differentials to find an approximation to the change in the period. Assume $g = 9.8$ m/s^2.

45. In Problem 44, given that L is fixed at 4 m, find an approximation to the change in the period if the pendulum is moved to an altitude where $g = 9.75$ m/s^2.

46. Since license plates are all approximately the same size (12 in across), a computerized optical sensor mounted at the front of car A could register the distance D to car B directly in front of car A by measuring the angle θ subtended by car B's license plate. See Figure 2.56.

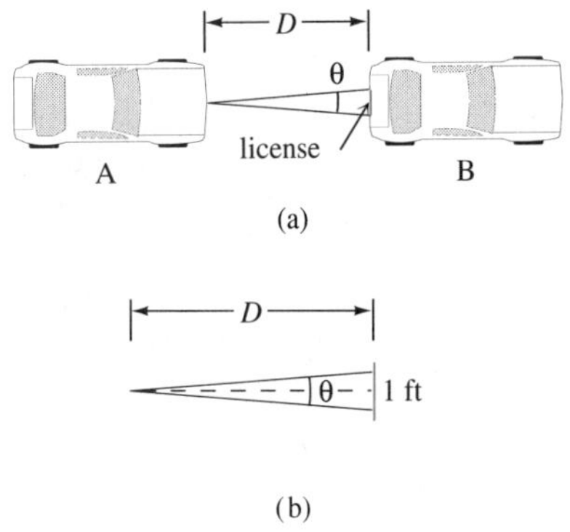

FIGURE 2.56

(a) Express D as a function of the subtended angle θ.

(b) Find the distance to the front car if the subtended angle θ is 30 minutes of an arc (that is, $\frac{1}{2}$°).

(c) Suppose in part (**b**) that θ is decreasing at the rate of 2 minutes of arc per second, and that car A is traveling at a rate of 30 mi/h. At what rate is car B moving?

(d) Show that the approximate relative error in measuring D is given by

$$\frac{dD}{D} = -\frac{d\theta}{\sin\theta}$$

where $d\theta$ is the approximate error (in radians) in measuring θ. What is the approximate relative error in D in part (**b**) if the subtended angle θ is measured with a possible error of ± 1 minute of arc?

2.11 NEWTON'S METHOD

Finding the roots of certain kinds of equations was a problem that captivated mathematicians for centuries. The zeros of a *polynomial* function f of degree 4 or less—that is, the roots of the equation $f(x) = 0$—can always be found by means of an algebraic formula that expresses the unknown x in terms of the coefficients of f. For example, the polynomial equation of degree 2, $ax^2 + bx + c$, $a \neq 0$, can be solved by the quadratic formula. One of the major achievements in the history of mathematics was the proof in the nineteenth century that polynomial equations of degree greater than 4 cannot be solved by means of an algebraic formula—in other words, in terms of radicals.* Thus, solving an algebraic equation such as

$$x^5 - 3x^2 + 4x - 6 = 0 \tag{1}$$

poses a quandary unless the fifth-degree polynomial $x^5 - 3x^2 + 4x - 6$ factors. Furthermore, in scientific analyses, one is often asked to find roots of transcendental equations such as

$$2x = \tan x \tag{2}$$

In the case of problems such as (1) and (2) it is common practice to employ some method that yields an approximation or estimation of the roots. We have already examined one such technique in Section 1.8; if the function f is continuous on an interval $[a, b]$ and $f(a)$ and $f(b)$ have different algebraic signs, then the zeros of f in $[a, b]$ can be approximated through interval bisection. In this section we consider another approximation technique. This new method, known as **Newton's Method** or the **Newton–Raphson Method**, makes use of the derivative of a function.

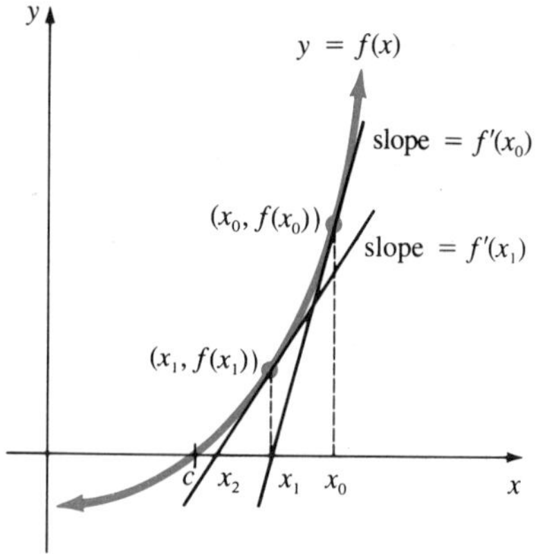

FIGURE 2.57

An Iterative Technique Suppose f is differentiable and suppose c represents an unknown root of $f(x) = 0$; that is, $f(c) = 0$. Let x_0 denote a number that is chosen arbitrarily as a first guess to c. If $f(x_0) \neq 0$, compute $f'(x_0)$ and, as shown in Figure 2.57, construct a tangent to the graph of f at

*Around 1540 the Italian mathematician **Lodovico Farrari** (1522–1565) discovered an algebraic formula for the roots of the general fourth-degree polynomial equation. For nearly the next 300 years mathematicians labored to find such formulas for polynomial equations of degree 5 or greater. In 1824 the Norwegian mathematician **Neils Henrik Abel** (1802–1829), was the first to prove that it is impossible to solve the general fifth-degree polynomial equation $ax^5 + bx^4 + cx^3 + dx^2 + ex + f = 0$ in terms of radicals.

$(x_0, f(x_0))$. If we now let x_1 denote the x-intercept of this line, we must have

$$\text{slope of line} = f'(x_0) = \frac{f(x_0)}{x_0 - x_1}$$

Solving for x_1 then gives

$$x_1 = x_0 - \frac{f(x_0)}{f'(x_0)}$$

Repeat the procedure at $(x_1, f(x_1))$ and let x_2 be the x-intercept of the second tangent line. From

$$f'(x_1) = \frac{f(x_1)}{x_1 - x_2} \quad \text{we find} \quad x_2 = x_1 - \frac{f(x_1)}{f'(x_1)}$$

Continuing in this fashion, we determine x_{n+1} from

$$x_{n+1} = x_n - \frac{f(x_n)}{f'(x_n)} \tag{3}$$

The repetitive use, or **iteration**, of (3) yields a sequence $x_1, x_2, x_3, \ldots$ of approximations that we expect *converges* to the root c; that is, $x_n \to c$ as n increases.

Graphical Analysis Before applying (3), let's try to determine the existence and number of real roots of $f(x) = 0$ through graphical means. For example, the irrational number $\sqrt{3}$ can be interpreted as either

(*i*) a root of the quadratic equation $x^2 - 3 = 0$ and hence, a zero of the continuous function $f(x) = x^2 - 3$, or

(*ii*) the x-coordinate of a point of intersection of the graphs of $y = x^2$ and $y = 3$.

Both interpretations are illustrated in Figure 2.58. Of course, another reason for a graph is to enable us to choose the initial guess x_0 so that it is close to the root c.

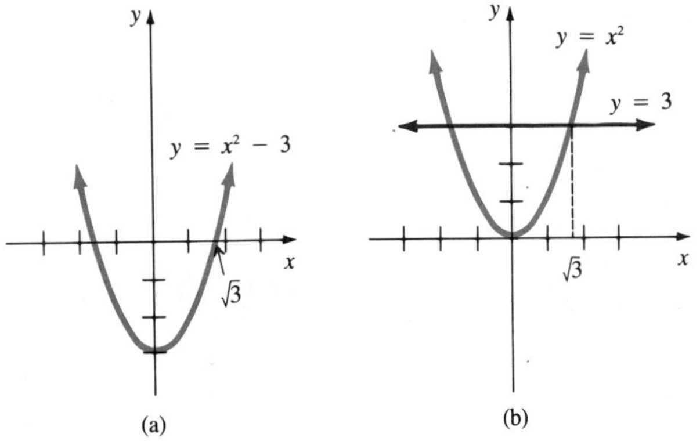

FIGURE 2.58

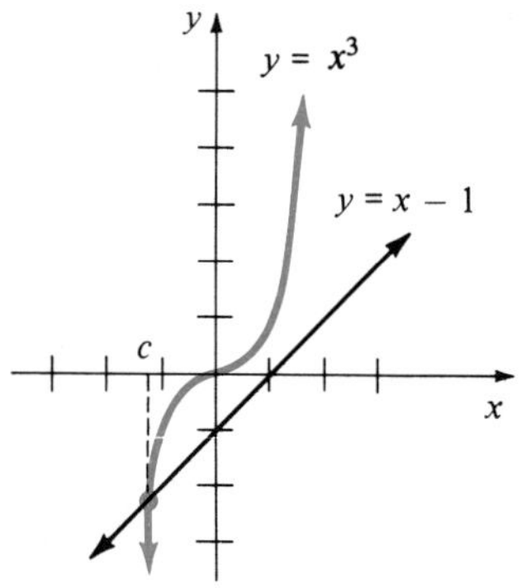

FIGURE 2.59

EXAMPLE 1 Determine the number of real roots of $x^3 - x + 1 = 0$.

Solution From Figure 2.59 we see that the graphs of the functions

$$y = x^3 \quad \text{and} \quad y = x - 1$$

intersect at one point. Hence, we conclude that the equation

$$x^3 = x - 1 \quad \text{or} \quad x^3 - x + 1 = 0$$

possesses only one real root. □

Although the actual calculation of the number $\sqrt{3}$ is trivial on a calculator, its calculation serves nicely as an introduction to the use of Newton's Method.

EXAMPLE 2 Approximate $\sqrt{3}$ by Newton's Method.

Solution If we define $f(x) = x^2 - 3$, then $f'(x) = 2x$ and (3) becomes

$$x_{n+1} = x_n - \frac{x_n^2 - 3}{2x_n} \quad \text{or} \quad x_{n+1} = \frac{1}{2}\left(x_n + \frac{3}{x_n}\right)$$

Since $1 < \sqrt{3} < 2$, it seems reasonable to choose $x_0 = 1$. Thus,

$$x_1 = \frac{1}{2}\left(x_0 + \frac{3}{x_0}\right) = \frac{1}{2}(1 + 3) = 2$$

$$x_2 = \frac{1}{2}\left(x_1 + \frac{3}{x_1}\right) = \frac{1}{2}\left(2 + \frac{3}{2}\right) = 1.75$$

$$x_3 = \frac{1}{2}\left(x_2 + \frac{3}{x_2}\right) = \frac{1}{2}\left(\frac{7}{4} + \frac{12}{7}\right) \approx 1.7321$$

$$x_4 = \frac{1}{2}\left(x_3 + \frac{3}{x_3}\right) \approx 1.7321$$

Since there is no significant difference in x_3 and x_4, it makes sense to stop the iteration. Indeed $\sqrt{3} = 1.73205$ is accurate to five decimal places. □

EXAMPLE 3 Use Newton's Method to find an approximation to the real root of the equation $x^3 - x + 1 = 0$.

Solution Let $f(x) = x^3 - x + 1$, so that $f'(x) = 3x^2 - 1$. Hence, (3) is

$$x_{n+1} = x_n - \frac{x_n^3 - x_n + 1}{3x_n^2 - 1} \quad \text{or} \quad x_{n+1} = \frac{2x_n^3 - 1}{3x_n^2 - 1}$$

If we are interested in three and possibly four decimal place accuracy, we carry out the iteration until two successive iterants agree to four decimal places. Also, Figure 2.59 prompts us to make $x_0 = -1.5$ the initial guess. Consequently,

$$x_1 = \frac{2x_0^3 - 1}{3x_0^2 - 1} = \frac{2(-1.5)^3 - 1}{3(-1.5)^2 - 1} \approx -1.3478$$

$$x_2 = \frac{2x_1^3 - 1}{3x_1^2 - 1} \approx -1.3252$$

$$x_3 = \frac{2x_2^3 - 1}{3x_2^2 - 1} \approx -1.3247$$

$$x_4 = \frac{2x_3^3 - 1}{3x_3^2 - 1} \approx -1.3247$$

Hence, the root of the given equation is approximately -1.3247. □

Basic Program A BASIC program for Newton's Method is given here. In the program, FND(X) denotes the derivative of the function $y = f(x)$.

```
10  REM NEWTON'S METHOD FOR FINDING ROOTS
20  INPUT "THE INITIAL VALUE IS"; X0
30  DEF FNY(X) = . . .
40  DEF FND(X) = . . .
50  X1 = X0 − FNY(X0)/FND(X0)                    (4)
60  PRINT X1
70  X0 = X1
80  GO TO 50
90  END
```

Those familiar with BASIC should provide an exit from the loop in the program.

Polynomial Equations In general, an nth-degree polynomial equation with integer coefficients

$$a_n x^n + a_{n-1} x^{n-1} + \cdots + a_1 x + a_0 = 0 \tag{5}$$

has at most n real roots. An odd-degree polynomial equation, such as the equation in Example 3, always has at least one real root because complex roots must always appear in conjugate pairs $a + bi$ and $a - bi$, where $i^2 = -1$, $b \neq 0$. Real roots can be rational or irrational numbers. Recall from algebra that if (5) has a rational root p/q (p and q integers, $q \neq 0$), then p is a factor of a_0 and q is a factor of the lead coefficient a_n.

The next example requires either a calculator with graphing capabilities or a computer with graphing software. In addition, we shall make use of the BASIC program in (4) to carry out the iteration of (3).

EXAMPLE 4 Use Newton's Method to approximate the real roots of

$$3x^{11} + 5x^9 - 15x^5 - 8x^4 + 5x^2 - x + 6 = 0 \tag{6}$$

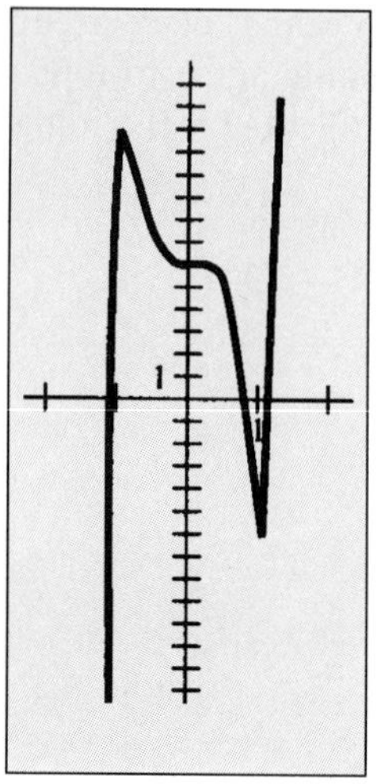

FIGURE 2.60

Solution Using a computer, we obtain the graph of the polynomial function $f(x) = 3x^{11} + 5x^9 - 15x^5 - 8x^4 + 5x^2 - x + 6$ shown in Figure 2.60. The figure suggests that (6) has three real roots in the neighborhood of the origin. Now to check whether these roots are rational, we list all the factors of a_0 and a_n, respectively:

$$p: \quad \pm 1, \pm 2, \pm 3, \pm 6$$
$$q: \quad \pm 1, \pm 3$$

Thus, the possible rational roots of (6) are

$$\frac{p}{q}: \quad -1, 1, -2, 2, -3, 3, -6, 6, -\frac{1}{3}, \frac{1}{3}, -\frac{2}{3}, \frac{2}{3}$$

Since Figure 2.60 clearly indicates that the roots are in the intervals $(-2, -1)$, $(0, 1)$, and $(1, 2)$, the only candidates for rational roots in those intervals are $\frac{1}{3}$ and $\frac{2}{3}$. Although the graph indicates that these are unlikely roots, we can test these numbers either by direct substitution in (6) or, better, by synthetic division. In this manner we find that neither $\frac{1}{3}$ nor $\frac{2}{3}$ is a root. We conclude that the roots of (6) in the three intervals are irrational.

Now, $f'(x) = 33x^{10} + 45x^8 - 75x^4 - 32x^3 + 10x - 1$, so that (3) becomes, after simplifying,

$$x_{n+1} = x_n - \frac{3x_n^{11} + 5x_n^9 - 15x_n^5 - 8x_n^4 + 5x_n^2 - x_n + 6}{33x_n^{10} + 45x_n^8 - 75x_n^4 - 32x_n^3 + 10x_n - 1} \tag{7}$$

The last expression will not be simplified, as in Example 3, in view of line 50 in the BASIC program (4).

We begin by approximating the root in the interval (0, 1). Figure 2.60 suggests that $x_0 = 0.9$ is a reasonable initial guess. Iteration of (7) then gives

$$x_1 \approx 0.8382729$$
$$x_2 \approx 0.8372948$$
$$x_3 \approx 0.8372939$$
$$x_4 \approx 0.8372939$$

We conclude that the root in (0, 1) is approximately 0.8372939. Using the initial values $x_0 = 1.1$ and $x_0 = -1.1$, we find in the same manner that approximations to the roots in (1, 2) and $(-2, -1)$ are 1.160470 and -1.140433, respectively. □

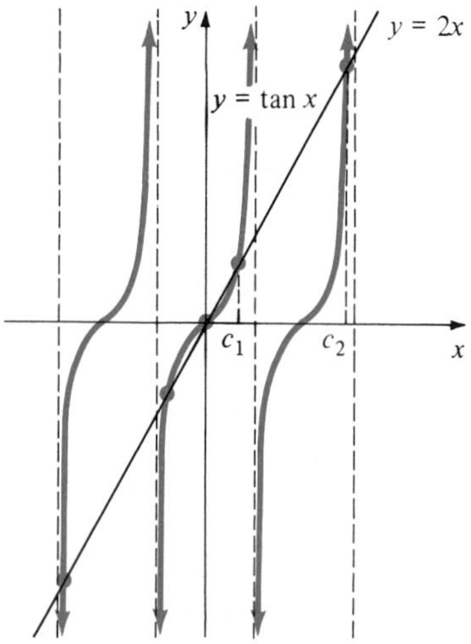

FIGURE 2.61

EXAMPLE 5 Find the smallest positive root of $2x = \tan x$.

Solution Figure 2.61 shows that the equation has an infinite number of roots. With $f(x) = 2x - \tan x$ and $f'(x) = 2 - \sec^2 x$, (3) becomes

$$x_{n+1} = x_n - \frac{2x_n - \tan x_n}{2 - \sec^2 x_n}$$

Since calculators and computers do not possess a secant routine, we express the last equation in terms of sin x and cos x:

$$x_{n+1} = x_n - \frac{2x_n\cos^2 x_n - \sin x_n \cos x_n}{2\cos^2 x_n - 1} \tag{8}$$

It appears from Figure 2.61 that the first positive root is near $x_0 = 1$. Using the program (4), iteration of (8) then yields

$$\begin{aligned} x_1 &\approx 1.310478 \\ x_2 &\approx 1.223929 \\ x_3 &\approx 1.176051 \\ x_4 &\approx 1.165927 \\ x_5 &\approx 1.165562 \\ x_6 &\approx 1.165561 \\ x_7 &\approx 1.165561 \end{aligned}$$

We conclude that the first positive root is approximately 1.165561. □

Example 5 illustrates the importance of the selection of the initial value x_0. You should verify that the choice $x_0 = \frac{1}{2}$ in (8) leads to a sequence of values $x_1, x_2, x_3, \ldots$ that converge to the one obvious root $c = 0$.

▲ ***Remarks*** There are problems with Newton's Method.

(*i*) We must compute $f'(x)$. Needless to say, the form of $f'(x)$ could be formidable when the equation $f(x) = 0$ is complicated.

(*ii*) If the root c of $f(x) = 0$ is near a value for which $f'(x) = 0$, then the denominator in (3) is approaching zero. This necessitates a computation of $f(x_n)$ and $f'(x_n)$ to a high degree of accuracy. A calculation of this kind usually requires a computer with a double precision routine.

(*iii*) It is necessary to find an approximate location of a root of $f(x) = 0$ before x_0 is chosen. Attendant to this are the usual difficulties in graphing. But, worse, the iteration of (3) *may not converge* for an imprudent or perhaps blindly chosen x_0. In Figure 2.62(a) we see that x_2 is undefined because $f'(x_1) = 0$. In

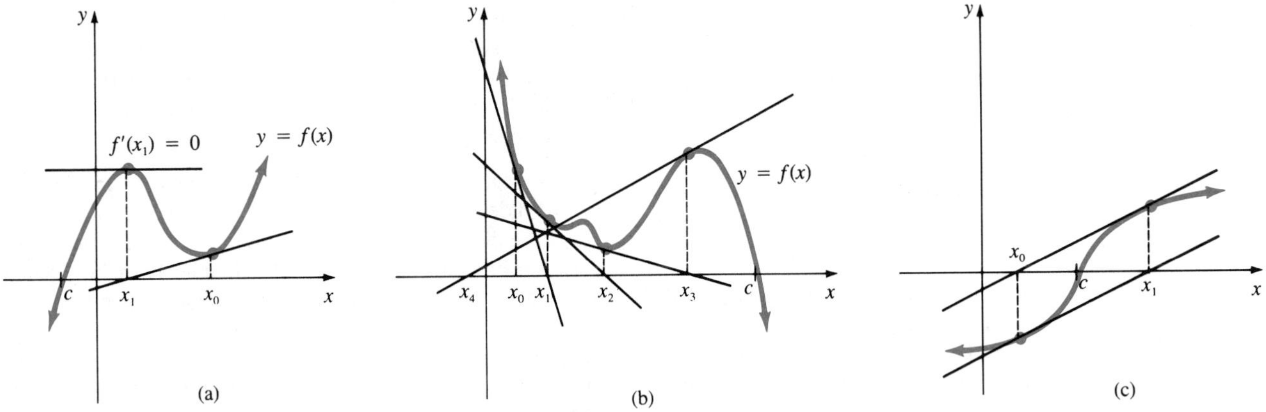

FIGURE 2.62

Figure 2.62(b) we see what can happen to the tangent lines when x_0 is not close to c. In Figure 2.62(c) observe that when $f(x_0) = -f(x_1)$ and $f'(x_0) = f'(x_1)$, the tangent lines "bounce" back and forth between two points $(x_0, f(x_0))$ and $(x_1, f(x_1))$. (See Problems 27 and 28 in Exercises 2.11.)

These three problems notwithstanding, the major advantage of Newton's Method is that when it converges to a root, it usually does so rather rapidly. It can be shown that under certain conditions Newton's Method converges *quadratically*. Very roughly, this means that the number of places of accuracy can, but do not necessarily, double with each iteration.

EXERCISES 2.11

Answers to odd-numbered problems begin on page A-72.

Where appropriate carry out the iteration of (3) *until the successive iterants agree to four decimal places.*

In Problems 1–4 determine graphically whether the given equation possesses any real roots.

1. $x^3 = -2 + \sin x$
2. $x^3 - 3x = x^2 - 1$
3. $x^4 + x^2 - 2x + 3 = 0$
4. $\tan x = \cos x$

In Problems 5–8 use Newton's Method to find an approximation for the given number.

5. $\sqrt{10}$
6. $1 + \sqrt{5}$
7. $\sqrt[3]{4}$
8. $\sqrt[5]{2}$

In Problems 9–14 use Newton's Method, if necessary, to find approximations to all real roots of the given equation.

9. $x^3 = -x + 1$
10. $x^3 - x^2 + 1 = 0$
11. $x^4 + x^2 - 3 = 0$
12. $x^4 = 2x + 1$
13. $x^2 = \sin x$
14. $x + \cos x = 0$

15. Find the smallest positive x-intercept of the graph of $f(x) = 3 \cos x + 4 \sin x$.

16. Consider the function $f(x) = x^5 + x^2$. Use Newton's Method to approximate the smallest positive number for which $f(x) = 4$.

17. A cantilever beam 20 ft long with a load of 600 lb at its end is deflected by an amount $d = (60x^2 - x^3)/16{,}000$, where d is measured in inches and x in feet. See Figure 2.63. Use Newton's Method to approximate the value of x that corresponds to a deflection of 0.01 in.

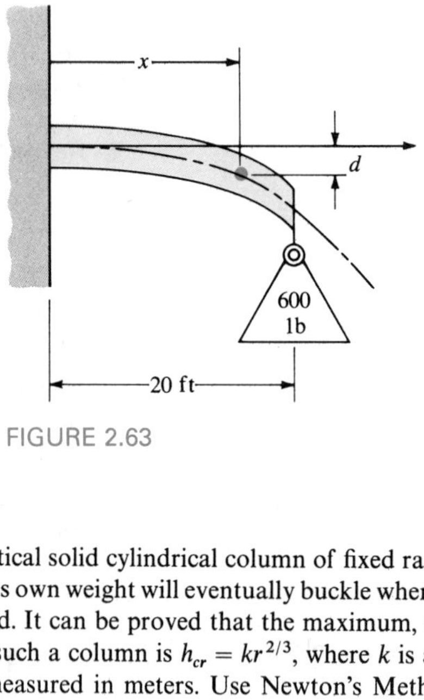

FIGURE 2.63

18. A vertical solid cylindrical column of fixed radius r that supports its own weight will eventually buckle when its height is increased. It can be proved that the maximum, or critical, height of such a column is $h_{cr} = kr^{2/3}$, where k is a constant and r is measured in meters. Use Newton's Method to approximate the diameter of a column for which $h_{cr} = 10$ m and $k = 35$.

19. A beam of light originating at point P in medium A, whose index of refraction is n_1, strikes the surface of medium B, whose index of refraction is n_2. We can prove from Snell's Law (see Problem 57 in Exercises 3.7) that the beam is refracted tangent to the surface for the critical angle determined from $\sin \theta_c = n_2/n_1$, $0 < \theta_c < 90°$. For angles of incidence greater than the critical angle, all light is reflected internally to medium A. See Figure 2.64. If $n_2 = 1$ for air and $n_1 = 1.5$ for glass, use Newton's Method to approximate θ_c in radians.

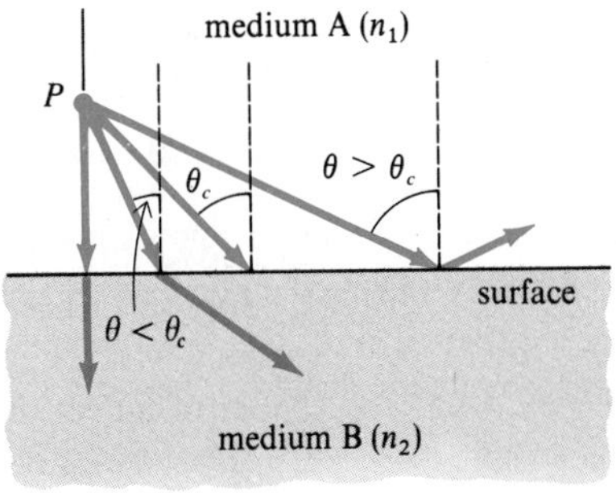

FIGURE 2.64

20. For a suspension bridge, the length s of a cable between two vertical supports whose span is l (horizontal distance) is related to the sag d of the cable by

$$s = l + \frac{8d^2}{3l} - \frac{32d^4}{5l^3}$$

See Figure 2.65. If $s = 404$ ft and $l = 400$ ft, use Newton's Method to approximate the sag. Round your answer to one decimal place.* [*Hint:* The root c satisfies $20 < c < 30$.]

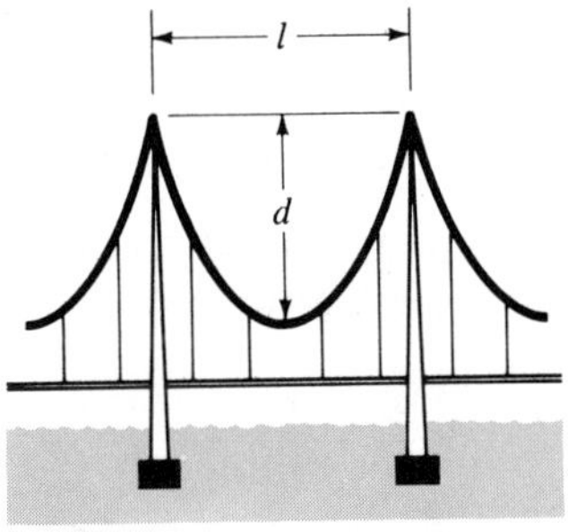

FIGURE 2.65

21. A rectangular block of steel is hollowed out, making a tub with a uniform thickness t. The dimensions of the tub are shown in Figure 2.66(a). For the tub to float in water, as shown in Figure 2.66(b), the weight of the water displaced must equal the weight of the tub (Archimedes' principle). If the weight density of water is 62.4 lb/ft^3 and the weight density of the steel is 490 lb/ft^3, then

$$\text{weight of water displaced} = 62.4 \times \begin{pmatrix}\text{volume of water}\\ \text{displaced}\end{pmatrix}$$

$$\text{weight of tub} = 490 \times (\text{volume of steel in tub})$$

*The formula for s is itself only an approximation.

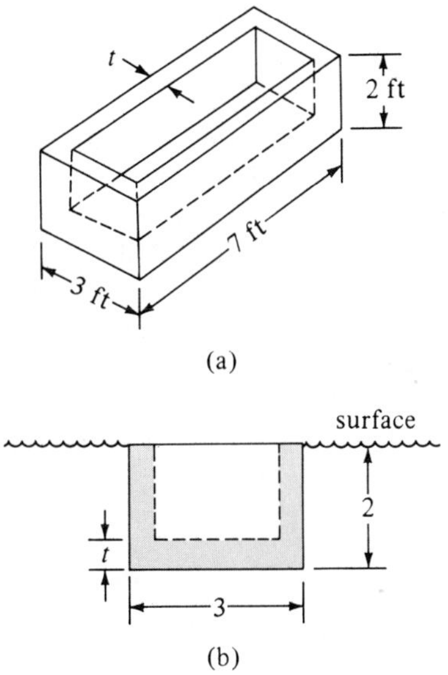

FIGURE 2.66

(a) Show that t satisfies the equation

$$t^3 - 7t^2 + \frac{61}{4}t - \frac{1638}{1225} = 0$$

(b) Use Newton's Method to approximate the largest positive root of the equation in part (**a**).

22. A flexible strip of metal 10 ft long is bent into the shape of a circular arc by securing the ends together by means of a cable that is 8 ft long. See Figure 2.67. Use Newton's Method to approximate the radius r of the circular arc.

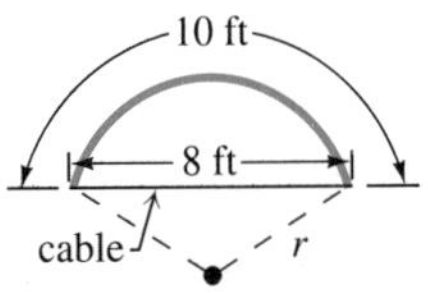

FIGURE 2.67

23. Two ends of a railroad track L feet long are pushed ℓ feet closer together so that the track bows upward in the arc of a circle of radius R. See Figure 2.68. The question is, what is the height h above ground of the highest point on the track?

(a) Use Figure 2.68 to show that

$$h = \frac{L(1 - \ell/L)^2\theta}{2(1 + \sqrt{1 - (1 - \ell/L)^2\theta^2})}$$

where $\theta > 0$ satisfies $\sin\theta = (1 - \ell/L)\theta$. [*Hint:* In a circular sector, how are the arc length, the radius, and the central angle related?]

(b) If $L = 5280$ ft and $\ell = 1$ ft, use Newton's Method to approximate θ and then solve for the corresponding value of h.

(c) If ℓ/L and θ are very small, then $h \approx L\theta/4$ and $\sin\theta \approx \theta - \theta^3/6$. Use these two approximations to show that $h \approx \sqrt{3\ell L/8}$. Use this formula with $L = 5280$ ft and $\ell = 1$ ft, and compare with the result in part (**b**).

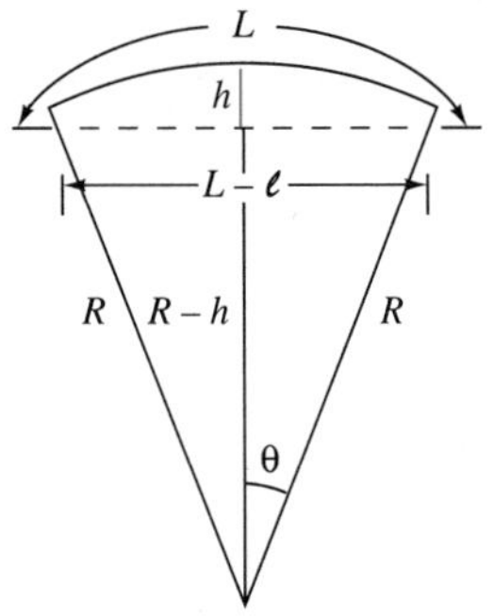

FIGURE 2.68

24. At a foundry a metal sphere of radius 2 ft is recast in the form of a rod that is a right circular cylinder 15 ft long surmounted by a hemisphere at one end. The radius r of the hemisphere is the same as the base radius of the cylinder. Use Newton's Method to approximate r.

25. A round but unbalanced wheel of mass M and radius r is connected by a rope and frictionless pulleys to a mass m as shown in Figure 2.69. O is the center of the wheel and P is its center of mass. If it is released from rest, it can be shown that the angle θ at which the wheel first stops satisfies the equation

$$Mg\frac{r}{2}\sin\theta - mgr\theta = 0$$

where g is the acceleration of gravity. Use Newton's Method to approximate θ if the mass of the wheel is four times the mass m.

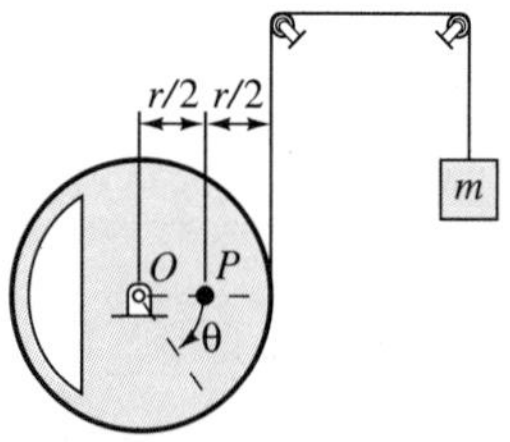

FIGURE 2.69

26. Two ladders of lengths $L_1 = 40$ ft and $L_2 = 30$ ft are placed against two vertical walls as shown in Figure 2.70. The height of the point where the ladders cross is $h = 10$ ft.

(a) Show that the indicated height x in the figure can be determined from the equation

$$x^4 - 2hx^3 + (L_1^2 - L_2^2)x^2 - 2h(L_1^2 - L_2^2)x + h^2(L_1^2 - L_2^2) = 0$$

(b) Use Newton's Method to approximate the solution of the equation in part (**a**). Why does it make sense to choose $x_0 \geq 10$?

(c) Approximate the distance z between the two walls.

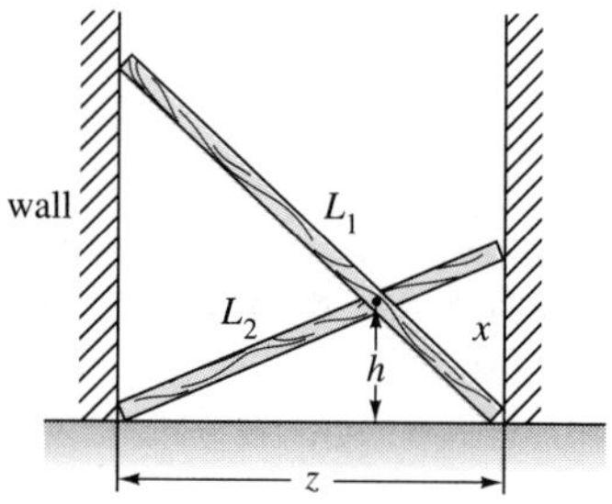

FIGURE 2.70

27. Let f be a differentiable function. Show that if $f(x_0) = -f(x_1)$ and $f'(x_0) = f'(x_1)$, then (3) implies $x_2 = x_0$.

28. Given

$$f(x) = \begin{cases} -\sqrt{4-x}, & x < 4 \\ \sqrt{x-4}, & x \geq 4 \end{cases}$$

observe $f(4) = 0$. Show that for any choice of x_0, Newton's Method will fail to converge to the root. [*Hint:* See Problem 27.]

Calculator/Computer Problems

In Problems 29 and 30 use a calculator or computer to obtain the graph of the given function. Use Newton's Method to approximate the roots of $f(x) = 0$ that you discover from the graph.

29. $f(x) = 2x^5 + 3x^4 - 7x^3 + 2x^2 + 8x - 8$

30. $f(x) = 4x^{12} + x^{11} - 4x^8 + 3x^3 + 2x^2 + x - 10$

31. (a) Use a calculator or computer to obtain the graphs of $f(x) = 0.5x^3 - x$ and $g(x) = \cos x$ on the same coordinate axes.

(b) Use a calculator or computer to obtain the graph of $y = f(x) - g(x)$, where f and g are as given in part (**a**).

(c) Use the graphs in part (**a**) or the graph in part (**b**) to determine the number of roots of the equation $0.5x^3 - x = \cos x$.

(d) Use Newton's Method to approximate the roots of the equation in part (**c**).

CHAPTER 2 REVIEW EXERCISES

Answers begin on page A-73.

In Problems 1–16 answer true or false.

1. The instantaneous rate of change of $y = f(x)$ with respect to x at x_0 is the slope of the tangent line to the graph at $(x_0, f(x_0))$. ________

2. If f is differentiable for every value of x, then f is continuous for every value of x. ________

3. If f is not differentiable at $x = a$, then there is no tangent to the graph at $(a, f(a))$. ________

4. For $f(x) = -x^2 + 5x + 1$ an equation of the tangent line is $f'(x) = -2x + 5$. ________

5. The function $f(x) = x/(x^2 + 9)$ is differentiable on the interval $[-3, 3]$. ________

6. The function $f(x) = |x - 2|$ is not differentiable on $(-\infty, \infty)$. ________

7. The derivative of a product is the product of the derivatives. ________

8. A polynomial function has a tangent line at every point on its graph. ________

9. If $f'(x) = g'(x)$, then $f(x) = g(x)$. ________

10. The Power Rules considered in this chapter for differentiating $f(x) = x^n$ imply $(d/dx)x^{\sqrt{2}} = \sqrt{2}x^{\sqrt{2}-1}$. ________

11. If $f(x) = \sqrt{2kx}$, k is a constant, then $f'(x) = k/f(x)$. ________

12. At $x = -1$, the tangent line to the graph of $f(x) = x^3 - 3x^2 - 9x$ is parallel to the line $y = 2$. ________

13. If f is continuous and $f(a)f(b) < 0$, there is a root of $f(x) = 0$ in the interval $[a, b]$. ________

14. The equation $x^4 = -x + 3$ possesses two real roots. ________

15. Newton's Method always converges when the initial guess x_0 is chosen very close to the root c. ________

16. Newton's Method will fail to converge to the real root of $x^3 - 2x^2 + x - 7 = 0$ for $x_0 = 1$. ________

In Problems 17–30 fill in the blank.

17. If k is a constant and n a positive integer, then $(d/dx)k^n$ = ________.

18. If $y = f(x)$ is a continuous function, then the average rate of change of f on the interval $[a, a + \Delta x]$ is ________.

19. For $f(x) = x^3$ the slope of the tangent at $(3, f(3))$ is

$$m_{\tan} = \lim_{\Delta x \to 0} \frac{(\_\_ + \Delta x)^3 - (\_\_)}{\Delta x}$$

20. The function $f(x) = \cot x$ is not differentiable on $[0, \pi]$ because ________.

21. If $f'(x) = x^2$, then $(d/dx)f(x^3) =$ ________.

22. If $f(2) = 1$, $f'(2) = 5$, $g(2) = 2$, and $g'(2) = -3$, then

$$\left.\frac{d}{dx}\frac{x^2 f(x)}{g(x)}\right|_{x=2} = \_\_\_\_\_\_\_\_$$

23. If $y = \sin x$, then $\dfrac{d^4y}{dx^4} =$ ________.

24. If $y = f(x)$ is a polynomial function of degree 3, then $\dfrac{d^4}{dx^4} f(x) =$ ________.

25. If $f'(4) = 6$ and $g'(4) = 3$, then the slope of the tangent to the graph of $y = 2f(x) - 5g(x)$ at $x = 4$ is ________.

26. $\displaystyle\lim_{x \to 0} \frac{\sin 3x}{5x} =$ ________

27. $\displaystyle\lim_{t \to 1} \frac{1 - \cos^2(t - 1)}{t - 1} =$ ________

28. The slope of the line perpendicular to the tangent line to the graph of $f(x) = \tan x$ at $x = \pi/3$ is ________.

29. For $f(x) = 1/(1 - 3x)$ the instantaneous rate of change of f' at $x = 0$ is ________.

30. The domain of f' for $f(x) = \sqrt{x} + \sqrt{x - 1}$ is ________.

31. Given that $y = \cos x^2$, find all values of x for which $dy/dx = 0$.

32. Find an equation of the tangent line to the graph of $y = (x + 3)/(x - 2)$ at $x = 0$.

33. Find equations for the lines through $(0, -9)$ tangent to the graph of $y = x^2$.

34. (a) Find the x-intercept of the tangent line to the graph of $y = x^2$ at $x = 1$.

(b) Find an equation of the line with the same x-intercept that is perpendicular to the tangent line in part **(a)**.

(c) Find the point(s) where the line in part **(b)** intersects the graph of $y = x^2$.

35. Find the point on the graph of $f(x) = \sqrt{x}$ at which the tangent line is parallel to the secant line through $(1, f(1))$ and $(9, f(9))$.

36. Find two distinct points P_1 and P_2 on the graph of $y = \sin x$ so that the tangent line at P_1 is parallel to the tangent line at P_2.

37. Find two distinct points P_1 and P_2 on the graph of $y = \cos x$ so that the tangent line at P_1 is perpendicular to the tangent line at P_2.

38. Determine the points on the graph of $f(x) = \sqrt[3]{(x-2)/(x-5)}$ at which the tangent is vertical.

39. Find all points in the interval $[0, 2\pi]$ at which the tangent to the graph of $f(x) = 5 - 2\cos x$ is parallel to the line $y = \sqrt{3}x + 1$.

40. Find all points in the interval $[0, 2\pi]$ at which the tangent to the graph of $f(x) = 2\cos x + \cos 2x$ is horizontal.

41. If F is a differentiable function, find

$$\frac{d^2}{dx^2} F(\sin 4x)$$

42. Find values of a and b such that the function

$$f(x) = \begin{cases} ax + b, & x \le 3 \\ x^2, & x > 3 \end{cases}$$

is differentiable at $x = 3$.

43. Sketch the graph of f' from the graph of f given in Figure 2.71.

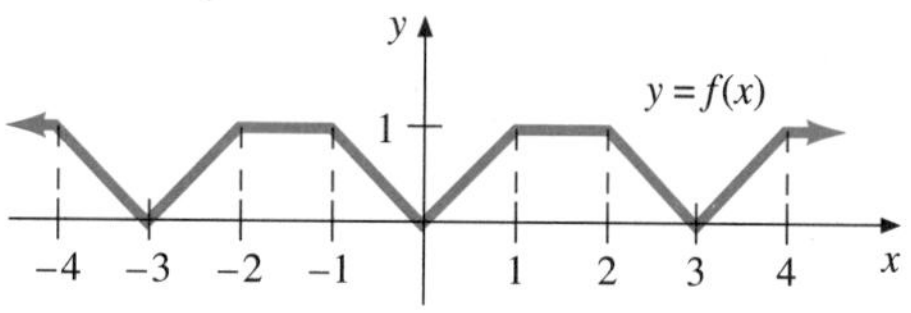

FIGURE 2.71

44. Find Δy and dy for the function $y = x + 1/x$.

45. Use a differential to find an approximation for $1/\sqrt[3]{68}$.

46. (a) Find the points on the graph of $y^3 + x^2y + 2x^2 - 6y^2 = 0$ corresponding to $x = 2$.

(b) Find the slopes of the tangent lines at the points found in part (a).

47. Given $x^{1/3} + y^{1/3} = 1$, find d^2y/dx^2.

48. A jet aircraft "loops the loop" in a circle of radius 1 km as shown in Figure 2.72. Suppose a Cartesian coordinate system is chosen so that the origin is at the center of the circular loop. The aircraft releases a missile that flies on a straight-line path

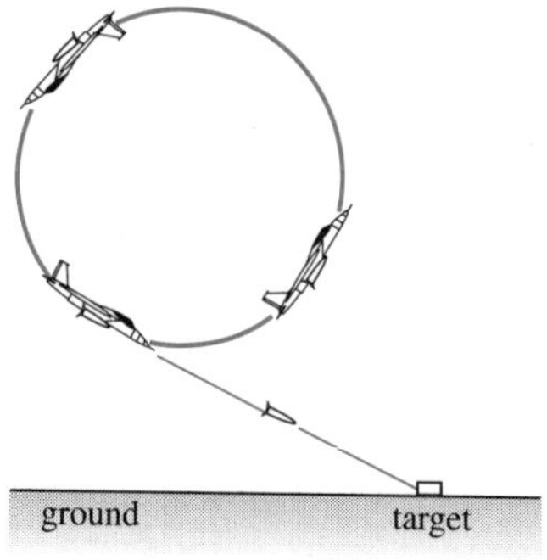

FIGURE 2.72

that is tangent to the circle and hits a target on the ground whose coordinates are $(2, -2)$.

(a) Determine the point on the circle where the missile was released.

(b) If a missile is released at the point $(-1/2, -\sqrt{3}/2)$ on the circle, at what point does it hit the ground?

In Problems 49–60 find the derivative of the given function.

49. $y = \dfrac{\cos 4x}{4x + 1}$

50. $y = 10 \cot 8x$

51. $f(x) = 2 + 2x + x^{-2} + x^2$

52. $f(x) = x^3 \sin^2 5x$

53. $F(t) = (t + \sqrt{t^2 + 1})^{10}$

54. $g(u) = \sqrt{\dfrac{6u - 1}{u + 7}}$

55. $G(x) = \dfrac{4x^{0.3}}{5x^{0.2}}$

56. $h(\theta) = \theta^{1.5}(\theta^2 + 1)^{0.5}$

57. $y = \sqrt[4]{x^4 + 16\sqrt[3]{x^3 + 8}}$

58. $y = \tan^2(\cos 2x)$

59. $y = \dfrac{\sqrt[3]{x^2}}{1 + \sqrt[3]{x}}$

60. $y = \dfrac{1}{x^3 + 4x^2 - 6x + 11}$

In Problems 61–64 find the indicated derivative.

61. $y = (3x)^{5/2}; \dfrac{d^3y}{dx^3}$

62. $y = \sin(x^3 - 2x); \dfrac{d^2y}{dx^2}$

63. $s = t^2 + \dfrac{1}{t^2}; \dfrac{d^2s}{dt^2}$

64. $W = \dfrac{v - 1}{v + 1}; \dfrac{d^3W}{dv^3}$

In Problems 65 and 66 use Newton's Method to find the indicated root. Carry out the method until two successive iterants agree to four decimal places.

65. $x^3 - 4x + 2 = 0$, the largest positive root

66. $\left(\dfrac{\sin x}{x}\right)^2 = \dfrac{1}{2}$, the smallest positive root

3

APPLICATIONS OF THE DERIVATIVE

INTRODUCTION

The derivative gives a rate of change. Geometrically, this rate of change is the slope of a tangent line to a graph. In Section 3.1 we shall elaborate on a concept that was considered briefly in Section 2.1; namely, the rate of change with respect to time of a function that gives the position of a moving object is the velocity of the object. However, in Section 3.2 we shall see that a time rate of change has other interpretations. Rates of change along with the problem of finding the maximum and minimum values of a function are the central topics of study in this chapter.

Suppose it is night and you are walking away from a street lamp at a constant rate. The length of your shadow is changing and the tip of your shadow is also moving away from the street lamp. Think about this. The rate at which your shadow changes is related to the rate at which you are walking. You will be asked to find these rates in Exercises 3.2.

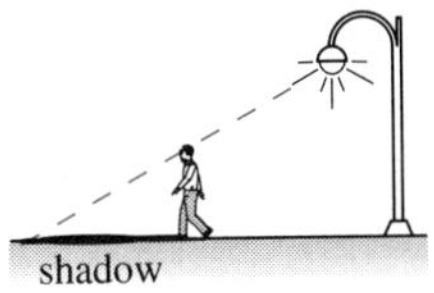

Imagine that you have designed a structure that calls for a beam to be embedded at both ends in concrete walls. The beam must support a constant load uniformly distributed along its length. After the load has been applied, the beam will be distorted. The shape of the beam is approximated by a fourth-degree polynomial function whose graph is called a deflection curve. What will the distorted beam look like? What is the maximum deflection of the beam? You will be asked to solve this problem in Exercises 3.7.

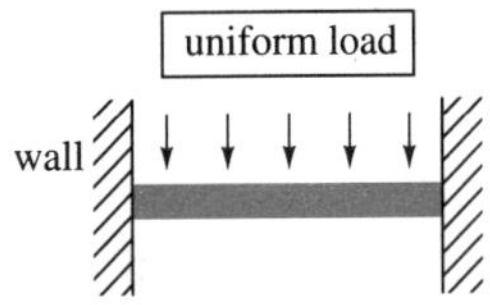

IMPORTANT CONCEPTS

- rectilinear motion
- average velocity
- velocity function
- acceleration function
- related rates
- absolute extrema
- extreme value theorem
- relative extrema
- critical point
- Rolle's theorem
- mean value theorem for derivatives
- increasing and decreasing functions
- first derivative test
- concavity
- point of inflection
- second derivative test
- constraint
- marginal functions

3.1 RECTILINEAR MOTION AND THE DERIVATIVE

In Section 2.1 motion of an object in a straight line, either horizontally or vertically, was said to be **rectilinear motion**. A function s that gives the coordinate of the object on a horizontal or vertical line is called a **position function**. The variable t represents time and $s(t)$ is a directed distance, which is measured in centimeters, meters, feet, miles, and so on, from a reference point $s = 0$. Recall that on a horizontal scale, we take the positive s-direction to be to the right of $s = 0$, and on a vertical scale we take the positive s-direction to be upward.

EXAMPLE 1 A particle moves on a horizontal line according to the position function $s(t) = -t^2 + 4t + 3$, where s is measured in meters and t in seconds. What is the position of the particle at 0, 2, and 6 seconds?

Solution Substitution into the position function gives

$$s(0) = 3, \quad s(2) = 7, \quad \text{and} \quad s(6) = -9$$

As shown in Figure 3.1, $s(6) = -9 < 0$ means that the position of the particle is to the left of the reference point $s = 0$.

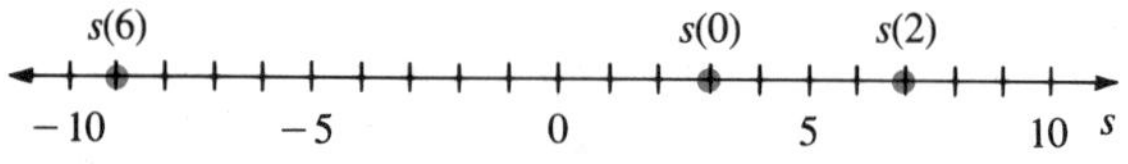

FIGURE 3.1

□

Velocity and Acceleration If the **average velocity** of a body in motion over a time interval of length Δt is

$$\frac{\text{change in position}}{\text{change in time}} = \frac{s(t + \Delta t) - s(t)}{\Delta t}$$

then the instantaneous rate of change is velocity given by

$$v(t) = \lim_{\Delta t \to 0} \frac{s(t + \Delta t) - s(t)}{\Delta t}$$

Thus, we have the following:

> **DEFINITION 3.1 Velocity Function**
>
> If $s(t)$ is a position function of an object that moves rectilinearly, then its **velocity function** at time t is
>
> $$v(t) = \frac{ds}{dt}$$
>
> The **speed** of the object time t is $|v(t)|$.

Velocity is measured in centimeters per second (cm/s), meters per second (m/s), feet per second (ft/s), kilometers per hour (km/h), miles per hour (mi/h), and so on.

We can also compute the rate of change of velocity.

> **DEFINITION 3.2 Acceleration Function**
>
> If $v(t)$ is the velocity of an object that moves rectilinearly, then its **acceleration function** at time t is
>
> $$a(t) = \frac{dv}{dt} = \frac{d^2s}{dt^2}$$

Typical units for measuring acceleration are meters per second per second (m/s^2), feet per second per second (ft/s^2), miles per hour per hour (mi/h^2), and so on. Often we read units of acceleration literally as "meters per second squared."

Significance of Algebraic Signs In Section 3.4 we shall see that whenever the derivative of a function f is *positive* on an interval I, then f is *increasing* on I. Geometrically, the graph of an increasing function rises as x increases. Similarly, if the derivative of a function f is *negative* on I, then f is *decreasing*, which means its graph goes down as x increases. Thus, when $v(t) = s'(t) > 0$, we can say $s(t)$ is increasing and the object is moving to the *right*. A negative velocity indicates motion to the *left*. See Figure 3.2. Similarly, when $a(t) > 0$, the velocity is *increasing*, whereas when $a(t) < 0$, the velocity is *decreasing*. For example, an acceleration of -25 m/s^2 means that the velocity is decreasing by 25 m/s every second. Do not confuse "velocity decreasing" with the concept of "slowing down"; for example, consider a stone that is dropped from the top of a tall building. The acceleration of gravity is a negative constant, -9.8 m/s^2. The negative sign means that the velocity of the

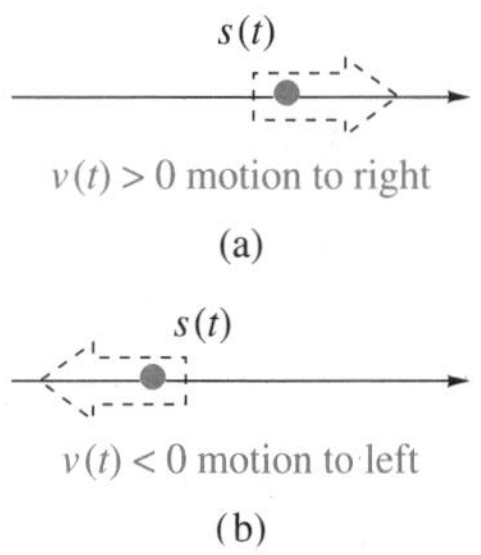

FIGURE 3.2

stone decreases starting from zero. When the stone hits the ground, its speed $|v(t)|$ is fairly large, but $v(t) < 0$. Note that an object that moves rectilinearly on a horizontal line will slow down when $v(t) > 0$ (motion to right) and $a(t) < 0$ (velocity decreasing) or when $v(t) < 0$ (motion to left) and $a(t) > 0$ (velocity increasing). In other words, an object is slowing down when its speed $|v(t)|$ is decreasing. In physics when a moving body is slowing down, the term *deceleration* is used.

EXAMPLE 2 In Example 1 the velocity and acceleration functions for the particle are, respectively,

$$v(t) = \frac{ds}{dt} = -2t + 4 \quad \text{and} \quad a(t) = \frac{dv}{dt} = -2$$

At times 0, 2, and 6 s, the velocities are $v(0) = 4$ cm/s, $v(2) = 0$ cm/s, and $v(6) = -8$ cm/s, respectively. Since the acceleration is always negative, the velocity is always decreasing. Notice that $v(t) = 2(-t + 2) > 0$ for $t < 2$ and $v(t) = 2(-t + 2) < 0$ for $t > 2$. If the time t is allowed to be negative as well as positive, then the particle is moving to the right for the time interval $(-\infty, 2)$ and moving to the left for the time interval $(2, \infty)$. The motion can be represented by the graph given in Figure 3.3(a). Since the motion actually takes place *on* the horizontal line, you should envision the movement of a point P that corresponds to the projection of a point on the graph onto the horizontal line. See Figure 3.3(b).

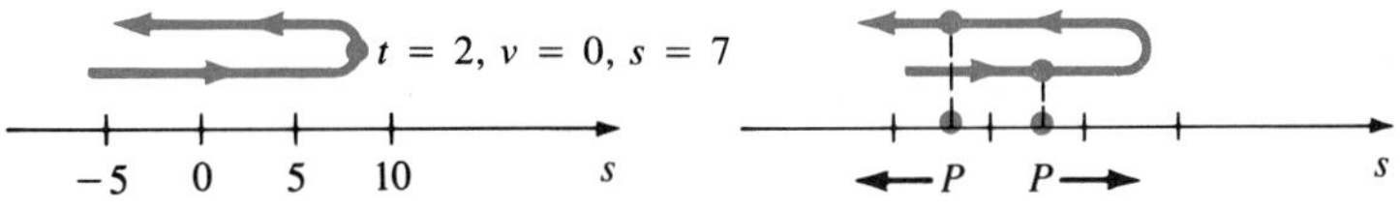

(a) $s(t) = -t^2 + 4t + 3$ (b)

FIGURE 3.3

□

EXAMPLE 3 A particle moves on a horizontal line according to the position function $s(t) = \frac{1}{3}t^3 - t$. Determine the time intervals on which the particle is slowing down.

Solution The algebraic signs of

$$v(t) = t^2 - 1 = (t + 1)(t - 1) \quad \text{and} \quad a(t) = 2t$$

are shown on the time scale in Figure 3.4. Since $v(t)$ and $a(t)$ have opposite signs on $(-\infty, -1)$ and $(0, 1)$, the particle is slowing down on these time intervals. □

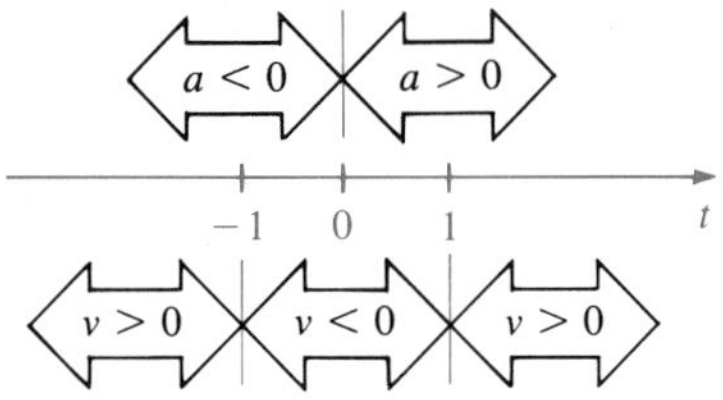

FIGURE 3.4

EXAMPLE 4 An object moves on a horizontal line according to the position function $s(t) = t^4 - 18t^2 + 25$, where s is measured in centimeters and t in seconds. Use a graph to represent the motion during the time interval $[-4, 4]$.

Solution The velocity function is

$$v(t) = \frac{ds}{dt} = 4t^3 - 36t = 4t(t + 3)(t - 3)$$

and the acceleration function is

$$a(t) = \frac{d^2s}{dt^2} = 12t^2 - 36 = 12(t + \sqrt{3})(t - \sqrt{3})$$

Now, from the solutions of $v(t) = 0$, we can determine the time intervals for which $s(t)$ is increasing or decreasing. From the information given in the accompanying tables, we construct the graph shown in Figure 3.5. Inspection of the figure shows that the particle slows down on the time intervals $(-4, -3)$, $(-\sqrt{3}, 0)$, and $(\sqrt{3}, 3)$.

Time Interval	Sign of $v(t)$	Direction of Motion
$(-4, -3)$	$-$	left
$(-3, 0)$	$+$	right
$(0, 3)$	$-$	left
$(3, 4)$	$+$	right

Time	Position	Velocity	Acceleration
-4	-7	-112	156
-3	-56	0	72
0	25	0	-36
3	-56	0	72
4	-7	112	156

Time Interval	Sign of $a(t)$	Velocity
$(-4, -\sqrt{3})$	$+$	increasing
$(-\sqrt{3}, \sqrt{3})$	$-$	decreasing
$(\sqrt{3}, 4)$	$+$	increasing

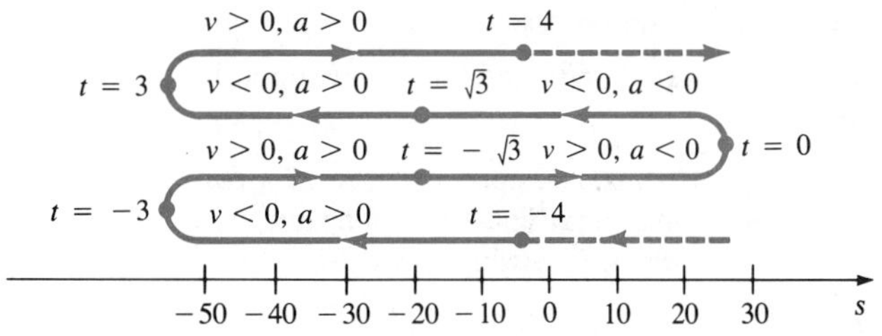

FIGURE 3.5

□

EXERCISES 3.1

Answers to odd-numbered problems begin on page A-73.

In Problems 1–8 $s(t)$ is a position function of a particle that moves on a horizontal line. Find the position, velocity, speed, and acceleration of the particle at the indicated times.

1. $s(t) = 4t^2 - 6t + 1;\ t = \frac{1}{2},\ t = 3$
2. $s(t) = (2t - 6)^2;\ t = 1,\ t = 4$
3. $s(t) = -t^3 + 3t^2 + t;\ t = -2,\ t = 2$
4. $s(t) = t^4 - t^3 + t,\ t = -1,\ t = 3$
5. $s(t) = t - \dfrac{1}{t};\ t = \frac{1}{4},\ t = 1$
6. $s(t) = \dfrac{t}{t + 2};\ t = -1,\ t = 0$
7. $s(t) = t + \sin \pi t;\ t = 1,\ t = \frac{3}{2}$
8. $s(t) = t \cos \pi t;\ t = \frac{1}{2},\ t = 1$

In Problems 9–12 $s(t)$ is a position function of a particle that moves on a horizontal line.

9. $s(t) = t^2 - 4t - 5$
   (a) What is the velocity of the particle when $s(t) = 0$?
   (b) What is the velocity of the particle when $s(t) = 7$?
10. $s(t) = t^2 + 6t + 10$
    (a) What is the position of the particle when $s(t) = v(t)$?
    (b) What is the velocity of the particle when $v(t) = -a(t)$?
11. $s(t) = t^3 - 4t$
    (a) What is the acceleration of the particle when $v(t) = 2$?
    (b) What is the position of the particle when $a(t) = 18$?
    (c) What is the velocity of the particle when $s(t) = 0$?
12. $s(t) = t^3 - 3t^2 + 8$
    (a) What is the position of the particle when $v(t) = 0$?
    (b) What is the position of the particle when $a(t) = 0$?
    (c) At what times do $v(t)$ and $a(t)$ have opposite algebraic signs?

In Problems 13 and 14 $s(t)$ is a position function of a particle that moves on a horizontal line. Determine the time intervals for which the particle is slowing down.

13. $s(t) = t^3 - 27t$
14. $s(t) = t^4 - t^3$

In Problems 15–26 $s(t)$ is a position function of a particle that moves on a horizontal line. Find the velocity and acceleration functions. Represent the motion during the indicated time interval with a graph.

15. $s(t) = t^2;\ [-1, 3]$
16. $s(t) = t^3;\ [-2, 2]$
17. $s(t) = t^2 - 4t - 2;\ [-1, 5]$
18. $s(t) = (t + 3)(t - 1);\ [-3, 1]$
19. $s(t) = 2t^3 - 6t^2;\ [-2, 3]$
20. $s(t) = (t - 1)^2(t - 2);\ [-2, 3]$
21. $s(t) = 3t^4 - 8t^3;\ [-1, 3]$
22. $st = t^4 - 4t^3 - 8t^2 + 60;\ [-2, 5]$
23. $s(t) = t - 4\sqrt{t};\ [1, 9]$
24. $s(t) = 1 + \cos \pi t;\ [-\frac{1}{2}, \frac{5}{2}]$
25. $s(t) = \sin \dfrac{\pi}{2} t;\ [0, 4]$
26. $s(t) = \sin \pi t - \cos \pi t;\ [0, 2]$
27. The graph of a position function in the st-plane is given in Figure 3.6. Complete the accompanying table by stating whether $v(t)$ and $a(t)$ are positive, negative, or zero.

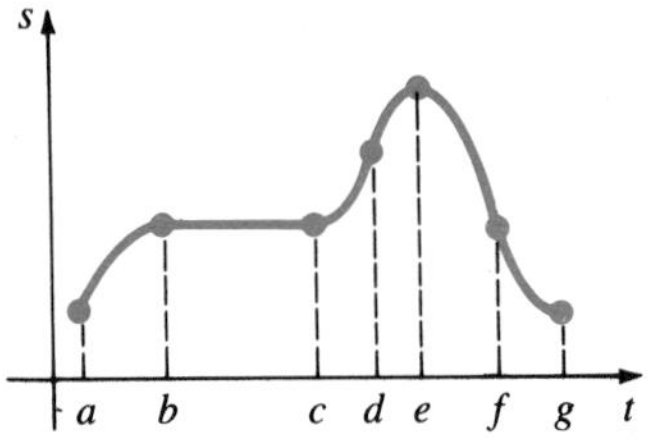

FIGURE 3.6

Interval	$v(t)$	$a(t)$
(a, b)		
(b, c)		
(c, d)		
(d, e)		
(e, f)		
(f, g)		

28. The graph of the velocity function v for a particle that moves on a horizontal line is given in Figure 3.7. Make a graph of a position function s with this velocity function.

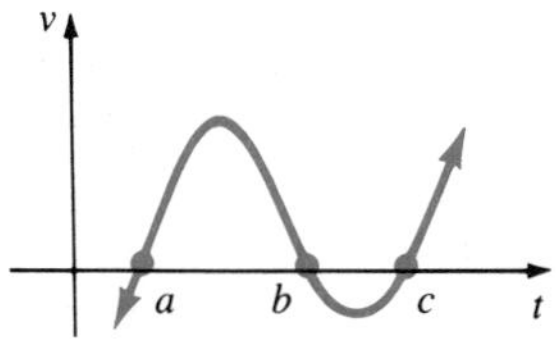

FIGURE 3.7

29. The height (in feet) of a projectile shot vertically upward from a point 6 ft above ground level is given by $s(t) = -16t^2 + 48t + 6$. See Figure 3.8.

(a) Determine the time interval for which $v > 0$ and the time interval for which $v < 0$.

(b) Find the maximum height attained by the projectile.

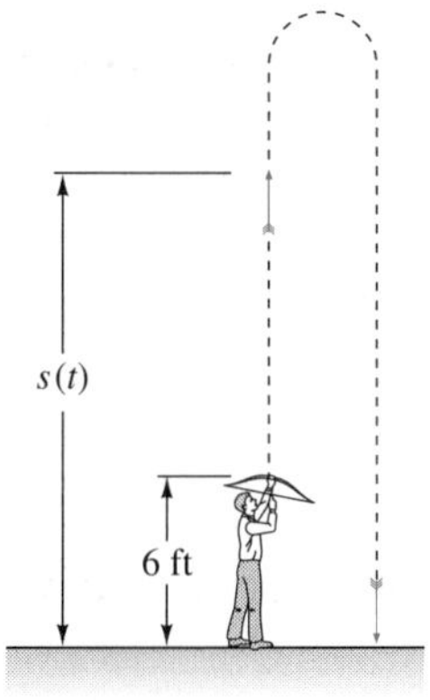

FIGURE 3.8

30. A particle moves on a horizontal line according to the position function $s(t) = -t^2 + 10t - 20$, where s is measured in centimeters and t in seconds. Determine the total distance traveled by the particle during the time interval $[-1, 6]$.

In Problems 31 and 32 use the following information. When friction is ignored, the distance s (in feet) that a body moves down an inclined plane of inclination θ is given by $s(t) = 16t^2 \sin\theta$, $[0, t_1]$, where $s(0) = 0$, $s(t_1) = L$, and t is measured in seconds. See Figure 3.9.

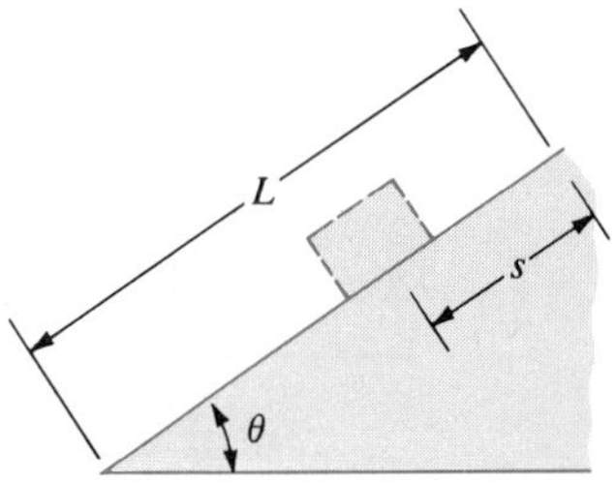

FIGURE 3.9

31. An object is sliding down a 256-ft-long hill with an inclination of 30°. What are the velocity and acceleration of the object at the bottom of the hill?

32. An entry in a soap box derby rolls down the hill shown in Figure 3.10. What are its velocity and acceleration at the bottom of the hill?

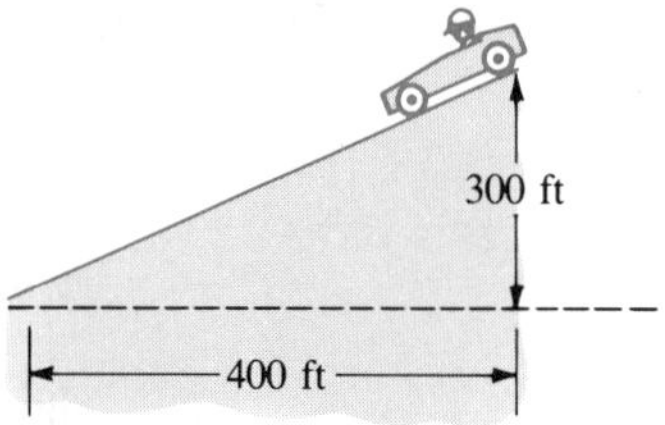

FIGURE 3.10

33. A bucket, attached to a circular windlass by a rope, is permitted to fall in a straight line under the influence of gravity. If the rotational inertia of the windlass is ignored, then the distance the bucket falls is equal to the radian measure of the angle indicated in Figure 3.11—that is, $\theta = \frac{1}{2}gt^2$, where $g = 32$ ft/s^2 is the acceleration of gravity. Find the rate at which the y-coordinate of a point P on the circumference of the windlass changes at $t = \sqrt{\pi}/4$ s. Interpret the result.

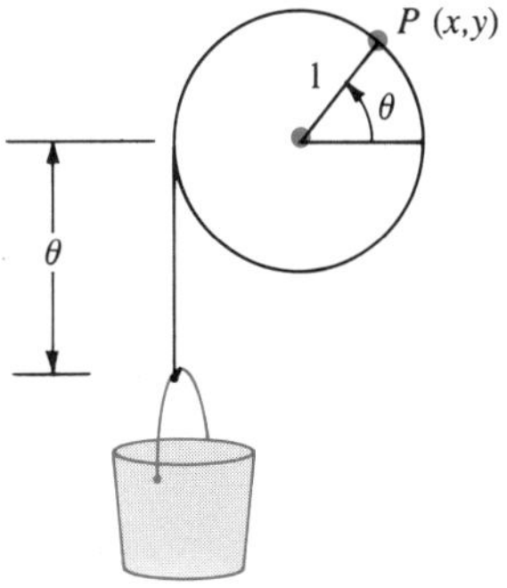

FIGURE 3.11

34. In mechanics the force F acting on a body is defined as the rate of change of its momentum: $F = (d/dt)(mv)$. When m is constant, we obtain from this definition the familiar formula known as Newton's Second Law $F = ma$, where the acceleration is $a = dv/dt$. According to Einstein's theory of relativity,

when a particle of rest mass m_0 moves rectilinearly at a great velocity (such as in a linear accelerator), its mass varies with the velocity v according to the formula $m = m_0/\sqrt{1 - v^2/c^2}$, where c is the constant speed of light. Show that in the theory of relativity the force F acting on a particle is

$$F = \frac{m_0 a}{\sqrt{(1 - v^2/c^2)^3}}$$

where a is acceleration.

3.2 RELATED RATES

In this section we are concerned with **related rates**. Since the problems will be stated in words, you must interpret these words in terms of mathematical symbols.

The derivative dy/dx of a function $y = f(x)$ is its instantaneous rate of change with respect to the variable x. When a function describes either position or distance, its time rate of change is interpreted as velocity. In general, a time rate of change is the answer to the question: How *fast* is a quantity changing? For example, if V stands for volume that is changing in time, then dV/dt is the rate, or how fast, the volume is changing with respect to time t. A rate of, say, $dV/dt = 10$ cm^3/s means that the volume is increasing 10 cubic centimeters each second. Similarly, if a person is walking *toward* the street lamp shown in Figure 3.12(a) at a constant rate of 3 ft/s, then we know that $dx/dt = -3$ ft/s. On the other hand, if the person is walking *away* from the street lamp, then $dx/dt = 3$ ft/s. The negative and positive rates mean, of course, that the distance x is decreasing and increasing, respectively.

Recall that if y denotes a function of x, then the Power Rule for Functions gives

$$\frac{d}{dx} y^n = n y^{n-1} \frac{dy}{dx} \tag{1}$$

where n is a rational number. Of course (1) is applicable to any function, say r, z, or x, that depends on t:

$$\frac{d}{dt} r^n = n r^{n-1} \frac{dr}{dt} \qquad \frac{d}{dt} x^n = n x^{n-1} \frac{dx}{dt} \qquad \frac{d}{dt} z^n = n z^{n-1} \frac{dz}{dt} \tag{2}$$

and so on.

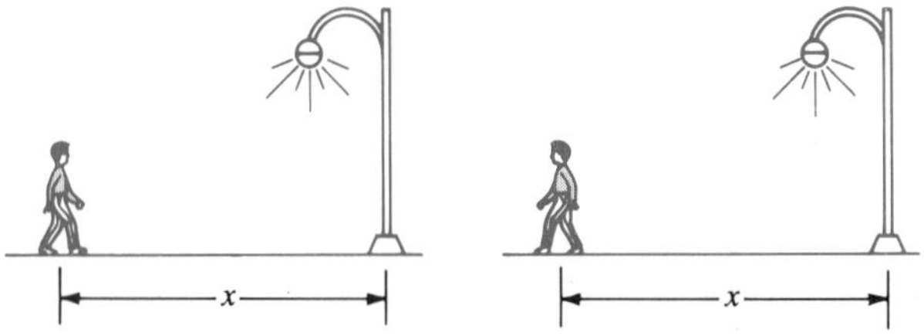

(a) $dx/dt < 0$ (b) $dx/dt > 0$

FIGURE 3.12

EXAMPLE 1 A square is expanding with time. How is the rate at which the area increases related to the rate at which the length of a side increases?

Solution At any time the area A of a square is a function of the length of one side x:

$$A = x^2 \tag{3}$$

Thus, the related rates are obtained from the time derivative of (3). With the help of the second result in (2), we see that

$$\frac{dA}{dt} = \frac{d}{dt}x^2$$

is the same as

$$\frac{dA}{dt} = 2x\frac{dx}{dt}$$

related rates

□

The key to solving word problems is organization. It is suggested that you follow these steps:

(*i*) Draw a picture if possible.

(*ii*) Label with symbols all quantities that change in time.

(*iii*) Carefully read the words of the problem and distinguish which rates are given and which rate is to be found.

(*iv*) Set up an equation that relates all the variables you have introduced.

(*v*) Differentiate the equation found in step (*iv*) with respect to time t. This step may require the use of implicit differentiation. The equation that results after differentiation relates the rates at which the variables change.

In the remainder of the examples, we shall emphasize steps (*iii*)–(*v*) by organizing the solutions in four parts:

Given: Using derivatives, list all the rates that are specified.

Want: Using derivative notation, list the rate that is required.

Know: Find an equation or function that involves all the variables introduced.

Analysis: This is the differentiation process of step (*v*).

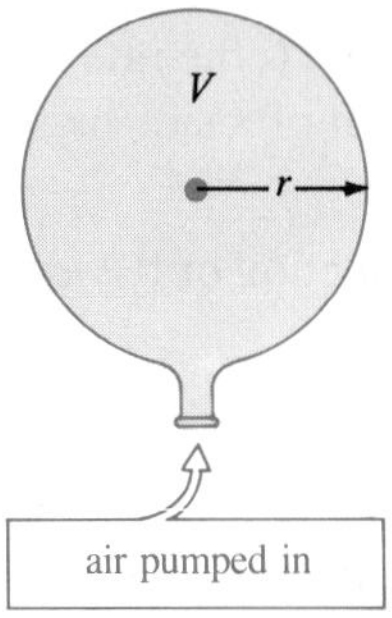

FIGURE 3.13

EXAMPLE 2 Air is being pumped into a spherical balloon at a rate of 20 ft^3/min. At what rate is the radius changing when the radius is 3 ft?

Solution As shown in Figure 3.13, we denote the radius of the balloon by r and its volume by V. Now, the interpretation of "air is being pumped . . . at a rate of 20 ft^3/min" means that we are

Given: $$\frac{dV}{dt} = 20 \text{ ft}^3/\text{min}$$

Want: $$\left.\frac{dr}{dt}\right|_{r=3}$$

Know: A relationship between V and r is given by the formula for the volume of a sphere:

$$V = \frac{4}{3}\pi r^3 \tag{4}$$

Analysis: Differentiating (4) with respect to t and using the first result in (2) give

$$\frac{dV}{dt} = \frac{4}{3}\pi \frac{d}{dt} r^3 = \frac{4}{3}\pi\left(3r^2 \frac{dr}{dt}\right)$$

$$= 4\pi r^2 \frac{dr}{dt}$$

But $dV/dt = 20$; therefore, $20 = 4\pi r^2 (dr/dt)$ yields

$$\frac{dr}{dt} = \frac{20}{4\pi r^2} = \frac{5}{\pi r^2}$$

Thus, $$\left.\frac{dr}{dt}\right|_{r=3} = \frac{5}{9\pi} \text{ ft/min} \approx 0.18 \text{ ft/min}$$ □

EXAMPLE 3 A woman jogging at a constant rate of 10 km/h crosses a point P heading north. Ten minutes later a man jogging at a constant rate of 9 km/h crosses the same point heading east. How fast is the distance between the joggers changing 20 min after the man crosses P?

Solution Let time be measured in hours from the instant the man crosses point P. As shown in Figure 3.14, at $t > 0$ let the man M and woman W be located x and y kilometers, respectively, from point P. Let z be the corresponding distance between the two joggers. Now,

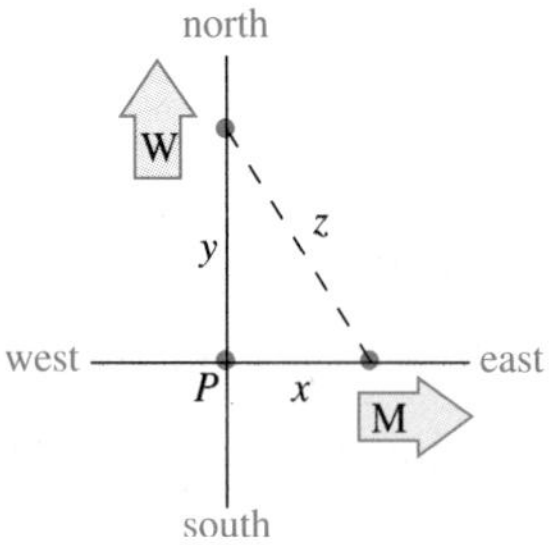

FIGURE 3.14

Given: $$\frac{dx}{dt} = 9 \text{ km/h} \qquad \frac{dy}{dt} = 10 \text{ km/h}$$

Want: $$\left.\frac{dz}{dt}\right|_{t=1/3} \qquad [20 \text{ min} = \tfrac{1}{3} \text{ h}]$$

Know: From the Pythagorean Theorem, the variables x, y, and z are related by

$$z^2 = x^2 + y^2 \tag{5}$$

Analysis: Differentiating (5) with respect to t gives

$$2z\frac{dz}{dt} = 2x\frac{dx}{dt} + 2y\frac{dy}{dt} \tag{6}$$

Using the given rates in (6) then yields

$$z\frac{dz}{dt} = 9x + 10y$$

When $t = \frac{1}{3}$ h we use distance = (rate) $\cdot$ (time) to obtain $x = 9 \cdot (\frac{1}{3}) = 3$ km. Since the woman has run $\frac{1}{6}$ h (10 min) longer, we find $y = 10 \cdot (\frac{1}{3} + \frac{1}{6}) = 5$ km. At $t = \frac{1}{3}$ h, it follows that $z = \sqrt{3^2 + 5^2} = \sqrt{34}$ km. Finally,

$$\sqrt{34}\left.\frac{dz}{dt}\right|_{t=1/3} = 9 \cdot 3 + 10 \cdot 5 \quad \text{or} \quad \left.\frac{dz}{dt}\right|_{t=1/3} = \frac{77}{\sqrt{34}} \approx 13.21 \text{ km/h} \qquad \square$$

EXAMPLE 4 A lighthouse is located on a small island 2 mi off a straight shore. The beacon of the lighthouse revolves at a constant rate of 6 deg/s. How fast is the light beam moving along the shore at a point 3 mi from a point on the shore closest to the lighthouse?

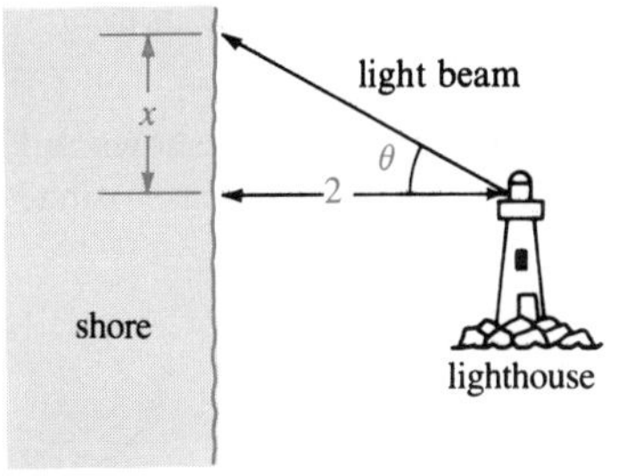

FIGURE 3.15

Solution We first introduce the variables θ and x as shown in Figure 3.15. In addition we change the information of θ to radian measure by recalling that $1°$ is equivalent to $\pi/180$ radians.

Given: $$\frac{d\theta}{dt} = 6 \cdot \frac{\pi}{180} = \frac{\pi}{30} \text{ rad/s}$$

Want: $$\left.\frac{dx}{dt}\right|_{x=3}$$

Know: From right triangle trigonometry we see that

$$\frac{x}{2} = \tan\theta \quad \text{or} \quad x = 2\tan\theta$$

Analysis: Differentiating the last equation with respect to t and using the given rate yield

$$\frac{dx}{dt} = 2\sec^2\theta\frac{d\theta}{dt} = \frac{\pi}{15}\sec^2\theta$$

At the instant $x = 3$, $\tan\theta = \frac{3}{2}$, so that from the trigonometric identity $1 + \tan^2\theta = \sec^2\theta$ we get $\sec^2\theta = \frac{13}{4}$. Hence,

$$\left.\frac{dx}{dt}\right|_{x=3} = \frac{\pi}{15} \cdot \frac{13}{4} = \frac{13\pi}{60} \text{ mi/s} \qquad \square$$

EXERCISES 3.2 *Answers to odd-numbered problems begin on page A-73.*

In the following problems, a solution may require a special formula with which you are not familiar. If necessary, consult the list of formulas given in Appendix V.

1. A cube is expanding with time. How is the rate at which the volume increases related to the rate at which the length of a side increases?

2. The volume of a rectangular box is $V = xyz$. Given that each side expands at a constant rate of 10 cm/min, find the rate at which the volume is expanding when $x = 1$ cm, $y = 2$ cm, and $z = 3$ cm.

3. A plate in the shape of an equilateral triangle expands with time. The length of a side increases at a constant rate of 2 cm/h. At what rate is the area increasing when a side is 8 cm?

4. In Problem 3 at what rate is the area increasing at the instant when the area is $\sqrt{75}$ cm^2?

5. A rectangle expands with time. The diagonal of the rectangle increases at a rate of 1 in/h and the length increases at a rate of $\frac{1}{4}$ in/h. How fast is its width increasing when the width is 6 in and the length is 8 in?

6. The lengths of the sides of a cube increase at a rate of 5 cm/h. At what rate does the length of the diagonal of the cube increase?

7. A boat is sailing toward the vertical cliff shown in Figure 3.16. How are the rates at which x, s, and θ change related?

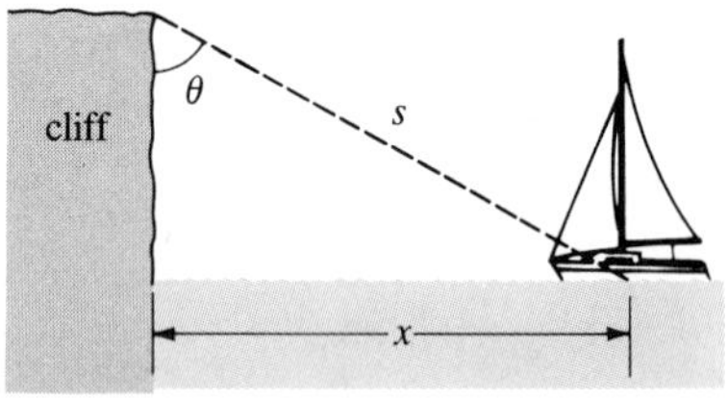

FIGURE 3.16

8. The total resistance R in a parallel circuit that contains two resistors of resistances R_1 and R_2 is given by $1/R = 1/R_1 + 1/R_2$. Each resistance changes with time. How are dR/dt, dR_1/dt, and dR_2/dt related?

9. A bug crawls along the graph of $y = x^2 + 4x + 1$, where x and y are measured in centimeters. If the abscissa x changes at a constant rate of 3 cm/min, how fast is the ordinate changing at the point (2, 13)?

10. In Problem 9 how fast is the ordinate changing when the bug is 6 cm above the x-axis?

11. A particle moves on the graph of $y^2 = x + 1$ so that $dx/dt = 4x + 4$. What is dy/dt when $x = 8$?

12. A particle in continuous motion moves on the graph of $4y = x^2 + x$. Find the point on the graph at which the rate of change of the abscissa and the rate of change of the ordinate are the same.

13. The x-coordinate of the point P shown in Figure 3.17 increases at a rate of $\frac{1}{3}$ cm/h. How fast is the area of the right triangle OPA increasing when P has coordinates (8, 2)?

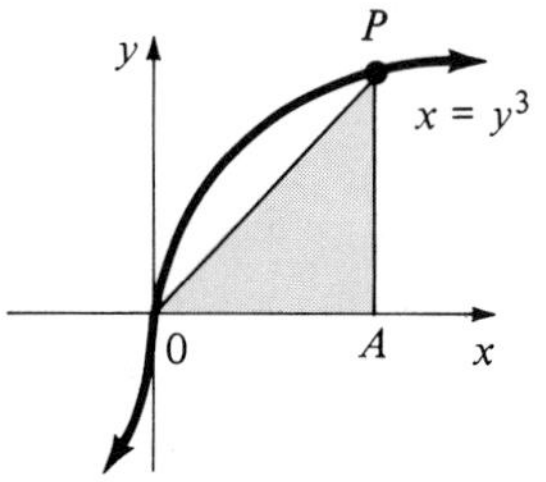

FIGURE 3.17

14. A suitcase is carried up the conveyor belt shown in Figure 3.18 at a rate of 2 ft/s. How fast is the suitcase rising?

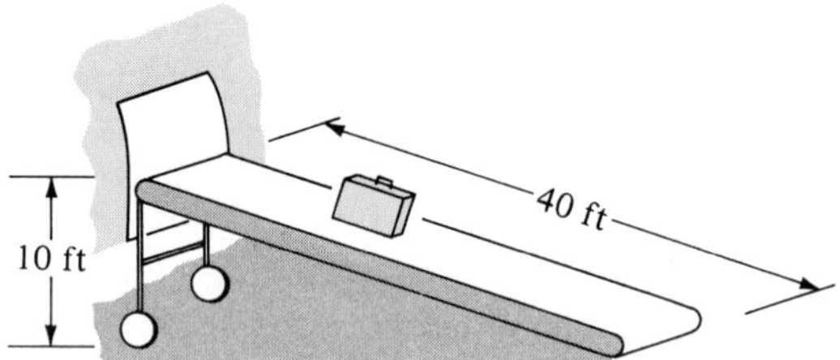

FIGURE 3.18

15. In the adiabatic expansion of air, pressure P and volume V are related by $PV^{1.4} = k$, where k is a constant. At a certain instant the pressure is 100 lb/in^2 and the volume is 32 in^3. At what rate is the pressure changing at that instant if the volume is decreasing at a rate of 2 in^3/s?

16. A stone dropped into a still pond causes a circular wave. Assume the radius of the wave expands at a constant rate of 2 ft/s.

(a) How fast does the diameter increase?

(b) How fast does the circumference increase?

(c) How fast does the area expand when the radius is 3 ft?

(d) How fast does the area expand when the area is 8π ft^2?

17. An oil tank in the shape of a right circular cylinder of radius 8 m is being filled at a constant rate of 10 m^3/min. How fast is the level of the oil rising?

18. A water tank in the shape of a right circular cylinder of diameter 40 ft is being drained so that the level of the water decreases at a constant rate of $\frac{3}{2}$ ft/min. How fast is the volume of the water decreasing?

19. Assume that a cube of ice melts in such a manner that it always retains its cubical shape. If the volume of the cube decreases at a rate of $\frac{1}{4}$ in^3/min, how fast is the surface area of the cube changing when the surface area is 54 in^2?

20. As shown in Figure 3.19, a 5-ft-wide rectangular water tank is divided into two tanks by a partition that moves in the direction indicated at a rate of 1 in/min as water is pumped into the front tank at a rate of 1 ft^3/min. At what rate is the level of the water changing when the volume of the water in the front tank is 40 ft^3 and $x = 4$ ft? Is the level of the water rising or falling at that instant?

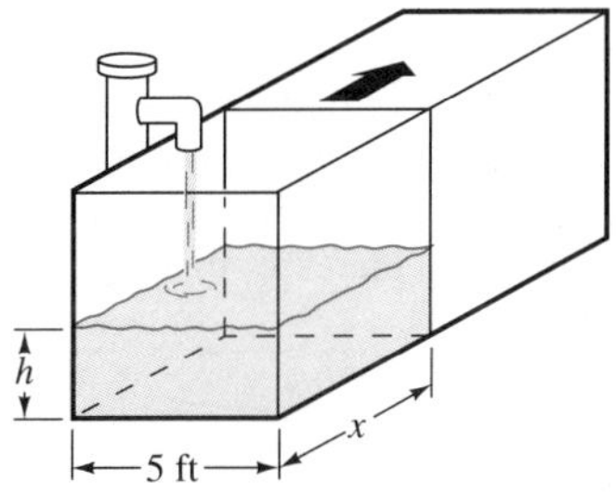

FIGURE 3.19

21. Each vertical end of a 20-ft-long water trough is an equilateral triangle with vertex down. Water is being pumped in at a constant rate of 4 ft^3/min.

(a) How fast is the level h of the water rising when the water is 1 ft deep?

(b) If h_0 is the initial depth of water in the trough, show that

$$\frac{dh}{dt} = \frac{\sqrt{3}}{10}\left(h_0^2 + \frac{\sqrt{3}}{5}t\right)^{-1/2}$$

[*Hint:* Consider the difference in volumes after t minutes.]

(c) If $h_0 = \frac{1}{2}$ ft and the height of the triangular end is 5 ft, determine the time when the trough is full. How fast is the level of the water rising when the trough is full?

22. A water trough with vertical ends in the form of isosceles trapezoids has dimensions as shown in Figure 3.20. If water is pumped in at a constant rate of $\frac{1}{2}$ m^3/s, how fast is the level of the water rising when the water is $\frac{1}{4}$ m deep?

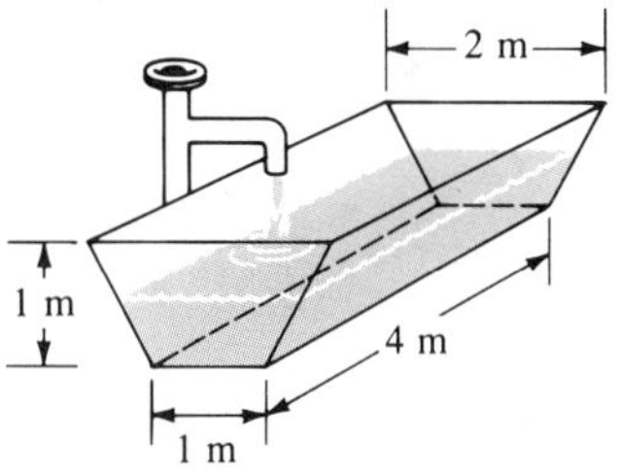

FIGURE 3.20

23. Water leaks out the bottom of the conical tank shown in Figure 3.21 at a constant rate of 1 ft^3/min.

(a) At what rate is the level of the water changing when the water is 6 ft deep?

(b) At what rate is the radius of the water changing when the water is 6 ft deep?

(c) Assume the tank was full at $t = 0$. At what rate is the radius of the water changing at $t = 6$ min?

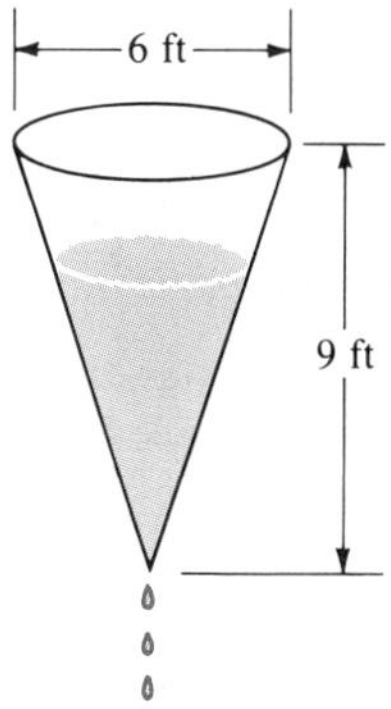

FIGURE 3.21

24. A 5-ft-tall person walks away from a 20-ft-tall street lamp at a constant rate of 3 ft/s. See Figure 3.12(b).

(a) At what rate is the length of the person's shadow increasing?

(b) At what rate is the tip of the shadow moving away from the base of the street lamp?

25. A 15-ft ladder is leaning against a wall of a house. The bottom of the ladder is pulled away from the base of the wall at a constant rate of 2 ft/min. At what rate is the top of the ladder sliding down the wall when the bottom of the ladder is 5 ft from the wall?

26. A kite string is paid out at a constant rate of 3 ft/s. If the wind carries the kite horizontally at an altitude of 200 ft,

how fast is the kite moving when 400 ft of string have been paid out?

27. A plane flying parallel to level ground at a constant rate of 600 mi/h approaches a radar station. If the altitude of the plane is 2 mi, how fast is the distance between the plane and the radar station decreasing when the horizontal distance between them is 1.5 mi? See Figure 3.22.

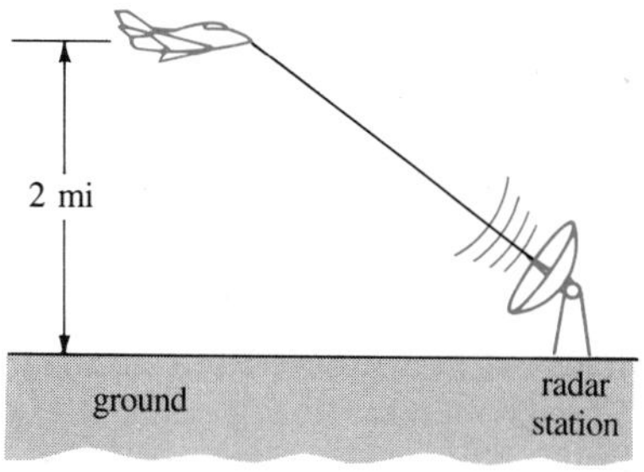

FIGURE 3.22

28. In Problem 27, at the point directly above the radar station, the plane goes into a 30° climb while retaining the same speed. How fast is the distance between the plane and radar station increasing 1 min later? [*Hint:* Review the law of cosines.]

29. A plane at an altitude of 4 km passes directly over a tracking telescope on the ground. When the angle of elevation is 60°, it is observed that this angle is decreasing at a rate of 30 deg/min. How fast is the plane traveling?

30. A rocket is traveling at a constant rate of 1000 mi/h at an angle of 60° to the horizontal. See Figure 3.23.

(a) At what rate is its altitude increasing?

(b) What is the ground speed of the rocket?

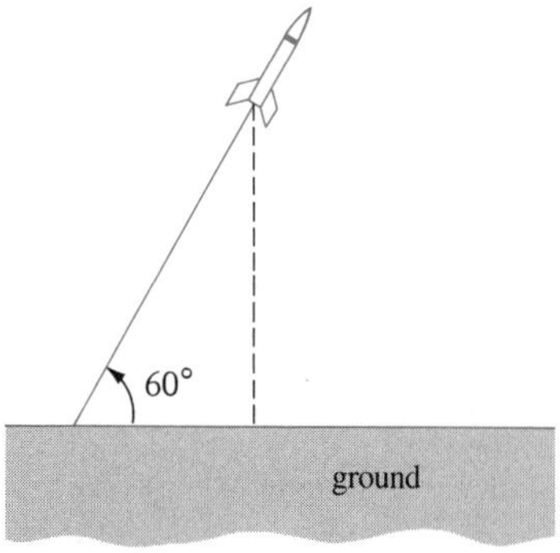

FIGURE 3.23

31. A tracking telescope, located 1.25 km from the point of launching, follows a vertically ascending rocket. When the angle of elevation is 60°, the rate at which the angle is increasing is 3 deg/s. At what rate is the rocket moving at that instant?

32. The volume V between two concentric spheres is expanding. The radius of the outer sphere increases at a constant rate of 2 m/h, whereas the radius of the inner sphere decreases at a constant rate of $\frac{1}{2}$ m/h. At what rate is V changing when the outer radius is 3 m and the inner radius is 1 m?

33. Many spherical objects such as raindrops, snowballs, and mothballs evaporate at a rate proportional to their surface areas. In this case show that the radius of the object decreases at a constant rate.

34. If the rate at which the volume of a sphere changes is constant, show that the rate at which its surface area changes is inversely proportional to the radius.

35. Two tankers depart from the same floating oil terminal. One tanker sails east at noon at a rate of 10 knots. (1 knot = 1 nautical mile/hour. A nautical mile is 6080 ft or 1.15 statute miles.) The other tanker sails north at 1:00 P.M. at a rate of 15 knots. At 2:00 P.M. at what rate is the distance between the two ships changing?

36. At 8:00 A.M. ship S_1 is 20 km due north of ship S_2. Ship S_1 sails south at a rate of 9 km/h and ship S_2 sails west at a rate of 12 km/h. At 9:20 A.M. at what rate is the distance between the two ships changing?

37. Sand flows from the top half of the conical hourglass shown in Figure 3.24 to the bottom half at a constant rate of 4 cm^3/s. At any time, assume that the height of the sand is h and that the sand has the shape of a frustum of a cone. Express the rate at which h increases in terms of h.

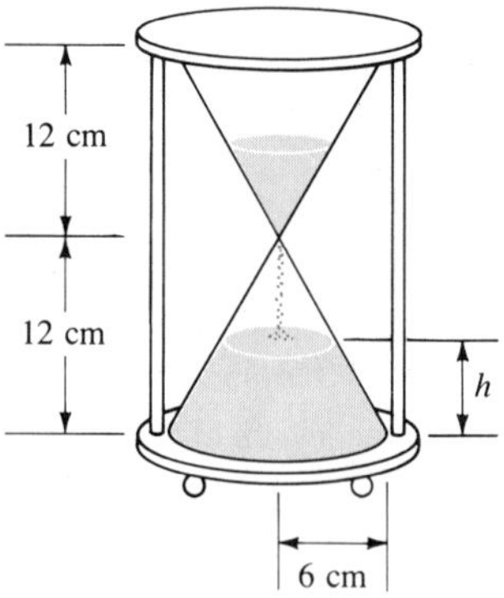

FIGURE 3.24

38. The Ferris wheel shown in Figure 3.25 revolves counterclockwise once every 2 min. How fast is a passenger rising at the instant when she is 64 ft above the ground? How fast is she moving horizontally at the same instant?

39. Suppose the Ferris wheel in Problem 38 is equipped with bidirectional colored spotlights fixed at various points on its

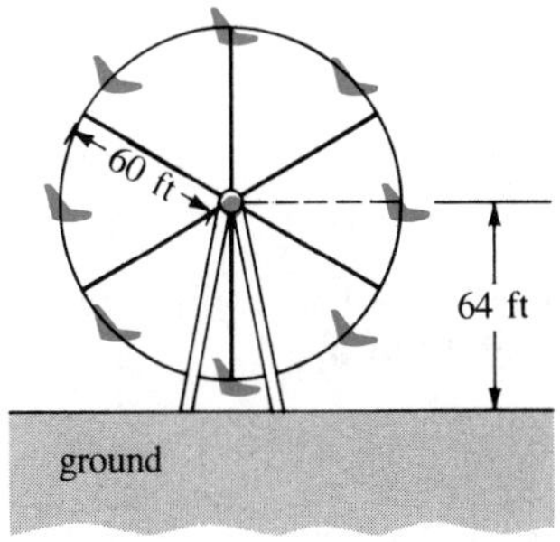

FIGURE 3.25

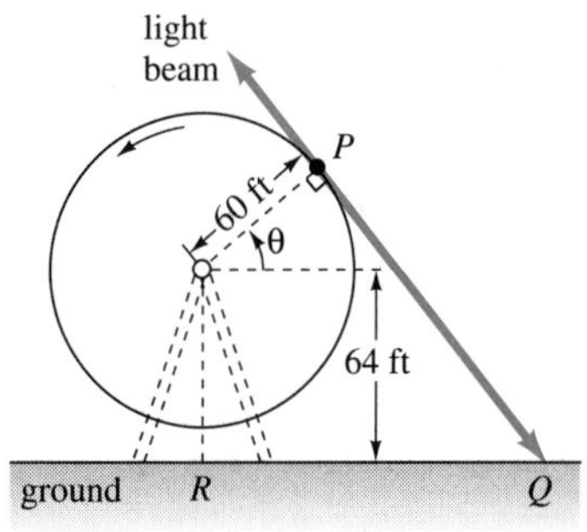

FIGURE 3.26

circumference. Consider the spotlight located at point P in Figure 3.26. If the light beams are tangent to the wheel at point P, at what rate is the spot Q on the ground moving away from point R when $\theta = \pi/4$?

40. In physics the momentum p of a body of mass m that moves in a straight line with velocity v is given by $p = mv$. An airplane of mass 10^5 kg flies in a straight line while ice builds up on the leading edges of its wings at a constant rate of 30 kg/h. See Figure 3.27.

(a) At what rate is the momentum of the plane changing if it is flying at a constant rate of 800 km/h?

(b) At what rate is the momentum of the plane changing at $t = 1$ h if at that instant its velocity is 750 km/h and is increasing at a rate of 20 km/h^2?

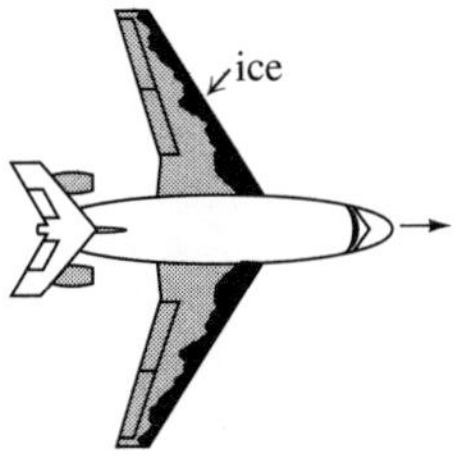

FIGURE 3.27

41. A study of crayfish (*Orconectes virilis*) indicates that the carapace length C is related to the total length T according to the formula $C = 0.493T - 0.913$, where C and T are measured in millimeters. See Figure 3.28.

(a) As the crayfish grows, does the ratio R of the carapace length to the total length increase or decrease?

(b) If the crayfish grows in length at the rate of 1 mm per day, at what rate is the ratio of the carapace to the total length changing when the carapace is one-third of the total length?

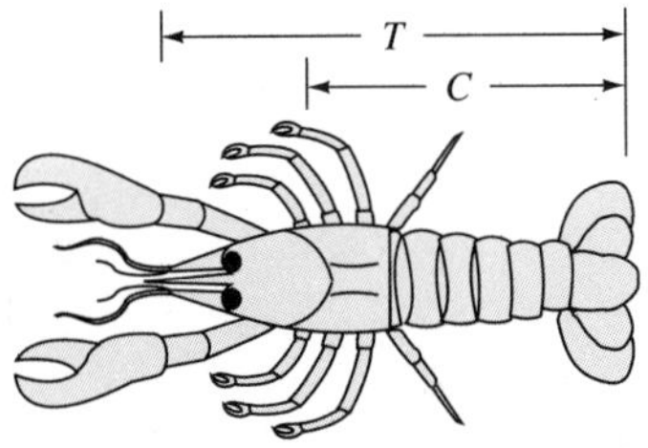

FIGURE 3.28

42. According to allometric studies, brain weight E in fish is related to body weight P by $E = 0.007P^{2/3}$, and body weight is related to body length L by $P = 0.12L^{2.53}$, where E and P are measured in grams and L is measured in centimeters. Suppose that the length of a certain species of fish evolved at a constant rate from 10 cm to 18 cm over the course of 20 million years. At what rate, in grams per million years, was this species's brain growing when the fish was half its final body weight?

3.3 EXTREMA OF FUNCTIONS

Absolute Extrema Suppose a function f is defined on an interval I. The **maximum** and **minimum** values of f on I (if there are any) are said to be **extrema** of the function. In the next two definitions, we distinguish two kinds of extrema.

DEFINITION 3.3 Absolute Extrema

(*i*) A number $f(c_1)$ is an **absolute maximum** of a function f if $f(x) \le f(c_1)$ for every x in the domain of f.

(*ii*) A number $f(c_1)$ is an **absolute minimum** of a function f if $f(x) \ge f(c_1)$ for every x in the domain of f.

Absolute extrema are also called **global extrema**. Figure 3.29 shows several possibilities.

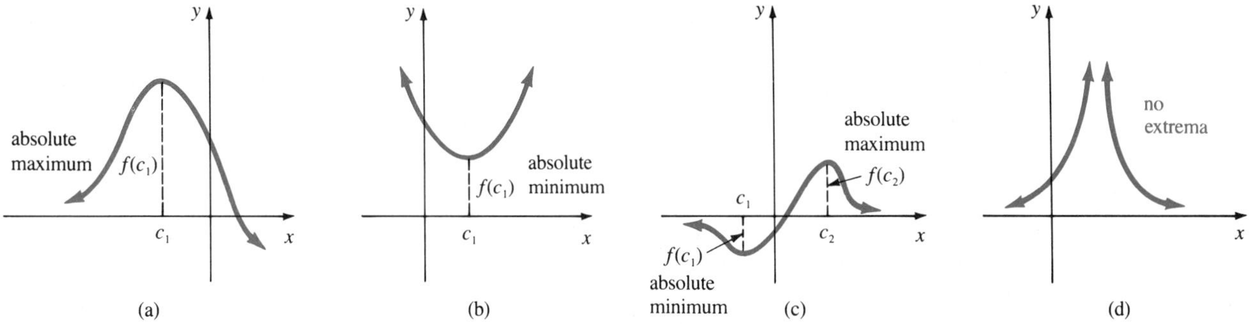

FIGURE 3.29

EXAMPLE 1

(*a*) For $f(x) = \sin x$, $f(\pi/2) = 1$ is its absolute maximum and $f(3\pi/2) = -1$ is its absolute minimum. By periodicity, the maximum and minimum values also occur at $x = \pi/2 + 2n\pi$ and $x = 3\pi/2 + 2n\pi$, $n = \pm 1, \pm 2, \ldots$, respectively.

(*b*) The function $f(x) = x^2$ has the absolute minimum $f(0) = 0$ but has no absolute maximum.

(*c*) $f(x) = 1/x$ has neither an absolute maximum nor an absolute minimum. □

The interval on which a function is defined is very important in the consideration of extrema.

EXAMPLE 2

(*a*) $f(x) = x^2$, defined only on the *closed* interval $[1, 2]$, has the absolute maximum $f(2) = 4$ and the absolute minimum $f(1) = 1$. See Figure 3.30(a).

(*b*) On the other hand, if $f(x) = x^2$ is defined on the *open* interval $(1, 2)$, then f has no absolute extrema. In this case, $f(1)$ and $f(2)$ are not defined.

(*c*) $f(x) = x^2$, defined on $[-1, 2]$, has the absolute maximum $f(2) = 4$, but now the absolute minimum is $f(0) = 0$. See Figure 3.30(b).

(*d*) $f(x) = x^2$, defined on $[-1, 2)$, has an absolute minimum $f(0) = 0$ but no absolute maximum.

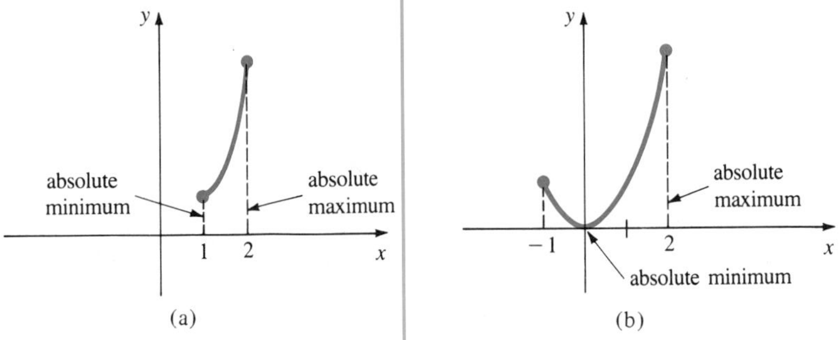

FIGURE 3.30

□

Parts (*a*) and (*c*) of Example 2 illustrate the following general result.

> **THEOREM 3.1 Extreme Value Theorem**
>
> A function f continuous on a closed interval $[a, b]$ always has an absolute maximum and an absolute minimum on the interval.

In other words, when f is continuous on $[a, b]$, there are numbers $f(c_1)$ and $f(c_2)$ such that $f(c_1) \le f(x) \le f(c_2)$ for all x in $[a, b]$. See Figure 3.31.

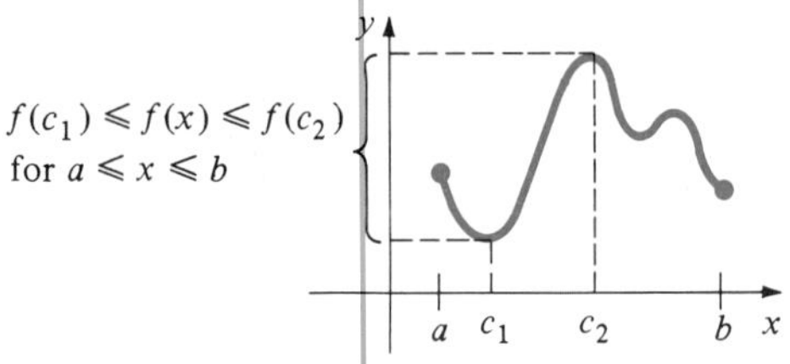

FIGURE 3.31

Endpoint Extrema When an absolute extremum of a function occurs at an endpoint of an interval I, as in parts (*a*) and (*c*) of Example 2, we say it is an **endpoint extremum**. When I is not a closed interval, that is, when I is an interval such as $(a, b]$, $(-\infty, b]$, or $[a, \infty)$, then even when f is continuous there is no guarantee that an absolute extremum exists. See Figure 3.32.

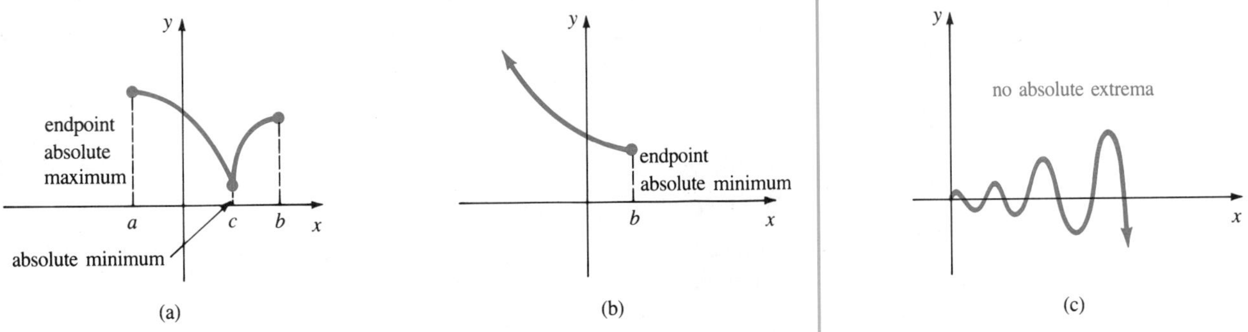

FIGURE 3.32

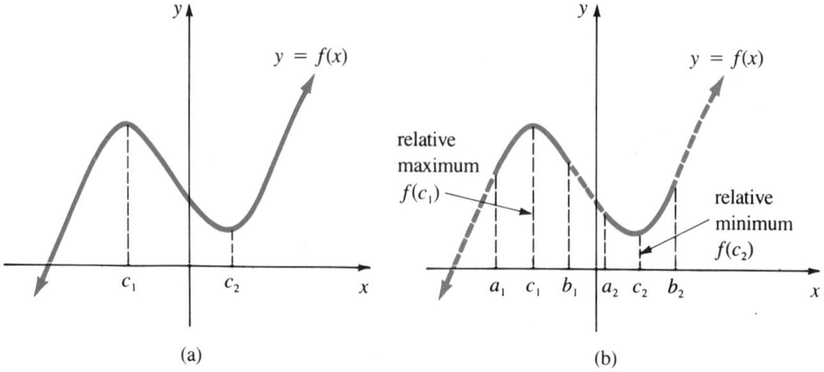

FIGURE 3.33

Relative Extrema The function pictured in Figure 3.33(a) has no absolute extrema. However, suppose we focus our attention on values of x that are close to, or in a *neighborhood* of, the numbers c_1 and c_2. As shown in Figure 3.33(b), $f(c_1)$ is the maximum value of the function in the interval (a_1, b_1), and $f(c_2)$ is a minimum value in the interval (a_2, b_2). These **local** or **relative extrema** are defined as follows.

DEFINITION 3.4 Relative Extrema

(*i*) A number $f(c_1)$ is a **relative maximum** of a function f if $f(x) \leq f(c_1)$ for every x in some open interval that contains c_1.

(*ii*) A number $f(c_1)$ is a **relative minimum** of a function f if $f(x) \geq f(c_1)$ for every x in some open interval that contains c_1.

As a consequence of Definition 3.4, we can conclude that every absolute extremum, with the *exception* of an endpoint extremum, is also a relative extremum. An endpoint absolute extremum is precluded from being a relative extremum on the technicality that an open interval contained in the domain of the function cannot be found around an endpoint of the interval.

An examination of Figures 3.33 and 3.34 suggests that if c is a point at which a function f has a relative extremum, then either $f'(c) = 0$ or $f'(c)$ does not exist.

DEFINITION 3.5 Critical Point

A **critical point** of a function f is a number c in its domain for which $f'(c) = 0$ or $f'(c)$ does not exist.

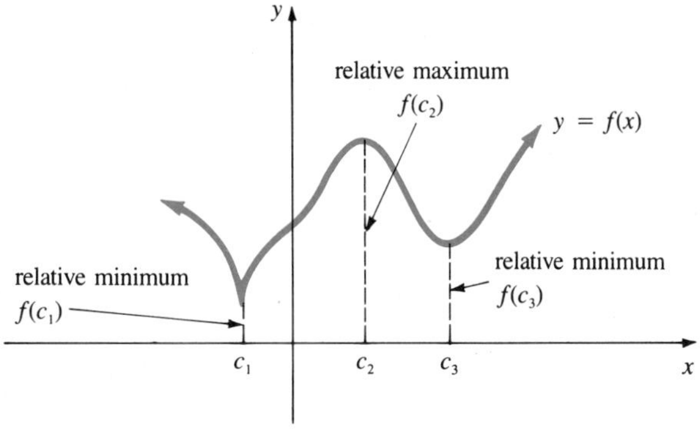

FIGURE 3.34

In some texts a critical point is referred to as a **critical number**.

EXAMPLE 3 Find the critical points of $f(x) = x^3 - 15x + 6$.

Solution
$$f'(x) = 3x^2 - 15 = 3(x + \sqrt{5})(x - \sqrt{5})$$

The critical points are those numbers for which $f'(x) = 0$, namely, $-\sqrt{5}$ and $\sqrt{5}$. □

EXAMPLE 4 Find the critical points of $f(x) = (x + 4)^{2/3}$.

Solution By the Power Rule for Functions,

$$f'(x) = \frac{2}{3}(x + 4)^{-1/3} = \frac{2}{3(x + 4)^{1/3}}$$

In this instance we see that $f'(x)$ does not exist when $x = -4$. Since -4 is in the domain of f, we conclude that it is a critical point. □

EXAMPLE 5 Find the critical points of $f(x) = \dfrac{x^2}{x - 1}$.

Solution By the Quotient Rule, we find after simplifying

$$f'(x) = \frac{x(x - 2)}{(x - 1)^2}$$

Now $f'(x) = 0$ when $x = 0$ and $x = 2$, whereas $f'(x)$ does not exist when $x = 1$. However, inspection of f reveals $x = 1$ is not in its domain, and so the only critical points are 0 and 2. □

THEOREM 3.2 Relative Extrema Occur at Critical Points

If a function f has a relative extremum at $x = c$, then c is a critical point.

Proof Assume $f(c)$ is a relative extremum.

(*i*) If $f'(c)$ does not exist, then c is a critical point by Definition 3.5.

(*ii*) If $f'(c)$ exists, there are three possibilities: $f'(c) > 0$, $f'(c) < 0$, or $f'(c) = 0$. For the sake of argument, let us further assume that $f(c)$ is a relative maximum. Hence, by Definition 3.4 there is some open interval that contains c in which

$$f(c + \Delta x) \leq f(c) \tag{1}$$

where the number Δx is sufficiently small in absolute value. The inequality in (1) then implies that

$$\frac{f(c + \Delta x) - f(c)}{\Delta x} \leq 0 \quad \text{for } \Delta x > 0 \quad \text{and} \quad \frac{f(c + \Delta x) - f(c)}{\Delta x} \geq 0 \quad \text{for} \quad \Delta x < 0 \tag{2}$$

But since $\lim_{\Delta x \to 0}[f(c + \Delta x) - f(c)]/\Delta x$ exists and equals $f'(c)$, the inequalities in (2) show that $f'(c) \leq 0$ *and* $f'(c) \geq 0$, respectively. The only way this can happen is to have $f'(c) = 0$. We leave the proof of the case when $f(c)$ is a relative minimum as an exercise. ■

Extrema of Functions Defined on a Closed Interval We have seen that a function continuous on a *closed* interval has both an absolute maximum and an absolute minimum. The next theorem tells us where these extrema can occur.

THEOREM 3.3 Finding Absolute Extrema

If f is continuous on a closed interval $[a, b]$, then an absolute extremum occurs either at an endpoint of the interval or at a critical point in the open interval (a, b).

We summarize Theorem 3.3 in the following manner. To find an absolute extremum of a function f continuous on $[a, b]$:

(*i*) Evaluate f at a and b.

(*ii*) Find all critical points $c_1, c_2, \ldots, c_n$ in (a, b).

(*iii*) Evaluate f at all critical points.

(*iv*) The largest and smallest values in the list

$$f(a), f(b), f(c_1), \ldots, f(c_n)$$

are the absolute maximum and the absolute minimum, respectively, of f on the interval $[a, b]$.

EXAMPLE 6 Find the absolute extrema of $f(x) = x^3 - 3x^2 - 24x + 2$ on the intervals $[-3, 1]$ and $[-3, 8]$.

Solution We need only evaluate f at the endpoints of each interval and at critical points within each open interval. From the derivative

$$f'(x) = 3x^2 - 6x - 24 = 3(x + 2)(x - 4)$$

we see that the critical points of the function f are -2 and 4.

From the data in the accompanying tables it is evident that the absolute maximum of f on the interval $[-3, 1]$ is $f(-2) = 30$, and the absolute minimum is the endpoint extremum $f(1) = -24$. On the interval $[-3, 8]$ we see that $f(4) = -78$ is an absolute minimum and $f(8) = 130$ is an endpoint absolute maximum.

On $[-3, 1]$			
x	-3	-2	1
$f(x)$	20	30	-24

On $[-3, 8]$				
x	-3	-2	4	8
$f(x)$	20	30	-78	130

□

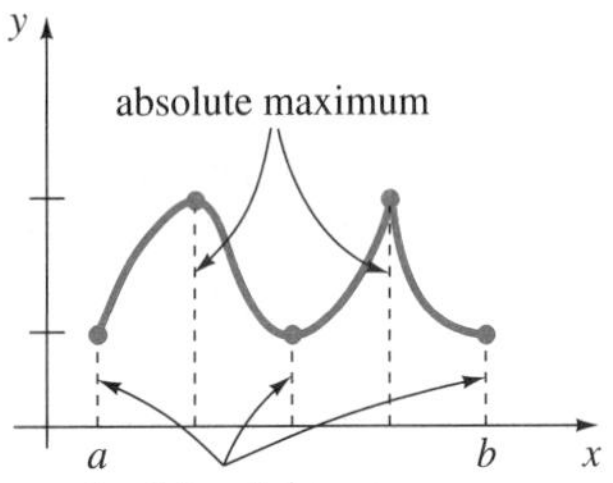

FIGURE 3.35

▲ **Remarks** (*i*) A function may, of course, assume its maximum and minimum values more than once on an interval. In Figure 3.35, the function illustrated attains its absolute maximum two times and attains its absolute minimum three times.

(*ii*) The converse of Theorem 3.2 is not necessarily true; that is, a critical point of a function need not correspond to a relative extremum. Consider $f(x) = x^3$ and $g(x) = x^{1/3}$. The derivatives $f'(x) = 3x^2$ and $g'(x) = (1/3)x^{-2/3}$ show that 0 is a critical point of both functions. But from the graphs of f and g in Figure 3.36 we see that neither function possesses any extrema.

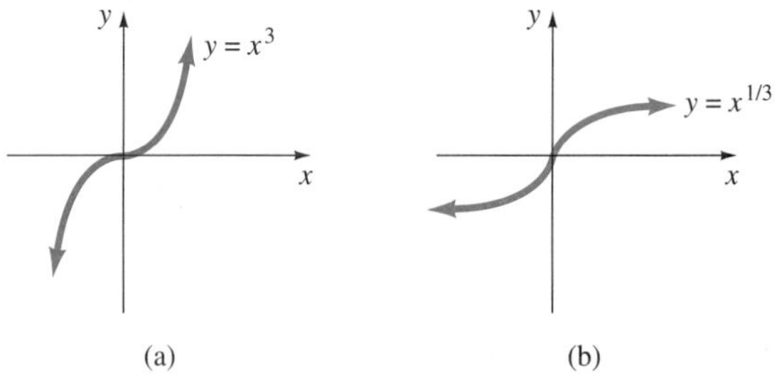

FIGURE 3.36

EXERCISES 3.3 *Answers to odd-numbered problems begin on page A-74.*

In Problems 1–6 use the graph of the given function as an aid in determining any absolute extrema on the indicated intervals.

1. $f(x) = x - 4$
   (a) $[-1, 2]$ (b) $[3, 7]$ (c) $(2, 5)$
   (d) $[1, 4]$
2. $f(x) = |x - 4|$
   (a) $[-1, 2]$ (b) $[3, 7]$ (c) $(2, 5)$
   (d) $[1, 4]$
3. $f(x) = x^2 - 4x$
   (a) $[1, 4]$ (b) $[1, 3]$ (c) $(-1, 3)$
   (d) $(4, 5]$
4. $f(x) = \sqrt{9 - x^2}$
   (a) $[-3, 3]$ (b) $(-3, 3)$ (c) $[0, 3)$
   (d) $[-1, 1]$
5. $f(x) = \tan x$
   (a) $[-\pi/2, \pi/2]$ (b) $[-\pi/4, \pi/4]$
   (c) $[0, \pi/3]$ (d) $[0, \pi]$
6. $f(x) = 2 \cos x$
   (a) $[-\pi, \pi]$ (b) $[-\pi/2, \pi/2]$
   (c) $[\pi/3, 2\pi/3]$ (d) $[-\pi/2, 3\pi/2]$

In Problems 7–20 find the critical points of the given function.

7. $f(x) = 2x^2 - 6x + 8$
8. $f(x) = x^3 + x - 2$
9. $f(x) = 2x^3 - 15x^2 - 36x$
10. $f(x) = x^4 - 4x^3 + 7$
11. $f(x) = (x - 2)^2(x - 1)$
12. $f(x) = x^2(x + 1)^3$
13. $f(x) = \dfrac{1 + x}{\sqrt{x}}$
14. $f(x) = \dfrac{x}{x^2 + 2}$
15. $f(x) = (4x - 3)^{1/3}$
16. $f(x) = x^{2/3} + x$
17. $f(x) = (x - 1)^2\sqrt[3]{x + 2}$
18. $f(x) = \dfrac{x + 4}{\sqrt[3]{x + 1}}$
19. $f(x) = -x + \sin x$
20. $f(x) = \cos 4x$

In Problems 21–34 find the absolute extrema of the given function on the indicated interval.

21. $f(x) = -x^2 + 6x; [1, 4]$
22. $f(x) = (x - 1)^2; [2, 5]$
23. $f(x) = x^{2/3}; [-1, 8]$
24. $f(x) = x^{2/3}(x^2 - 1); [-1, 1]$
25. $f(x) = x^3 - 6x^2 + 2; [-3, 2]$
26. $f(x) = -x^3 - x^2 + 5x; [-2, 2]$
27. $f(x) = x^3 - 3x^2 + 3x - 1; [-4, 3]$
28. $f(x) = x^4 + 4x^3 - 10; [0, 4]$
29. $f(x) = x^4(x - 1)^2; [-1, 2]$
30. $f(x) = \dfrac{\sqrt{x}}{x^2 + 1}; [\frac{1}{4}, \frac{1}{2}]$
31. $f(x) = 2 \cos 2x - \cos 4x; [0, 2\pi]$
32. $f(x) = 1 + 5 \sin 3x; [0, \pi/2]$
33. $f(x) = 3 + 2 \sin^2 4x; [0, \pi]$
34. $f(x) = 2x - \tan x; [-1, 1.5]$

In Problems 35 and 36 find all critical points. Distinguish between absolute, endpoint absolute, and relative extrema.

35. $f(x) = x^2 - 2|x|; [-2, 3]$
36. $f(x) = \begin{cases} 4x + 12, & -5 \le x \le -2 \\ x^2, & -2 < x \le 1 \end{cases}$

37. Consider the continuous function f defined on $[a, b]$ shown in Figure 3.37. Given that c_1 through c_{10} are critical

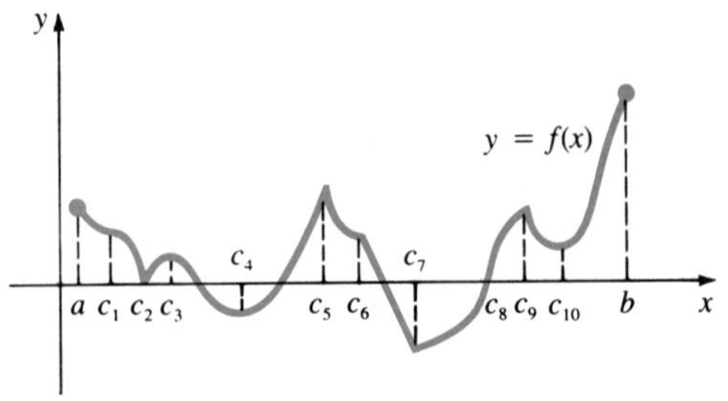

FIGURE 3.37

points:

(a) List critical points at which $f'(x) = 0$.

(b) List critical points at which $f'(x)$ is not defined.

(c) Distinguish between the absolute and endpoint absolute extrema.

(d) Distinguish between the relative maxima and the relative minima.

38. Consider the function $f(x) = x + 1/x$. Show that the relative minimum is greater than the relative maximum.

39. Draw a graph of a continuous function that possesses no absolute extrema but has a relative maximum and a relative minimum that are the same value.

40. Give an example of a continuous function, defined on a closed interval $[a, b]$, for which the absolute maximum is the same as the absolute minimum.

41. Let $f(x) = [x]$ be the greatest integer function. Show that every value of x is a critical value.

42. Show that $f(x) = (ax + b)/(cx + d)$ has no critical points when $ad - bc \neq 0$. What happens when $ad - bc = 0$?

43. The height of a projectile launched from ground level is given by $s(t) = -16t^2 + 320t$, where t is measured in seconds and s in feet.

(a) $s(t)$ is defined only on the time interval $[0, 20]$. Why?

(b) Use the results of Theorem 3.3 to determine the maximum height attained by the projectile.

44. The French physician Jean Louis Poiseuille discovered that the velocity v (in cm/s) of blood flowing through an artery with circular cross-section of radius R is given by $v(r) = (P/4vl)(R^2 - r^2)$, where P, v, and l are positive constants. See Figure 3.38.

(a) Determine a closed interval on which v is defined.

(b) Determine the maximum and minimum velocities of the blood.

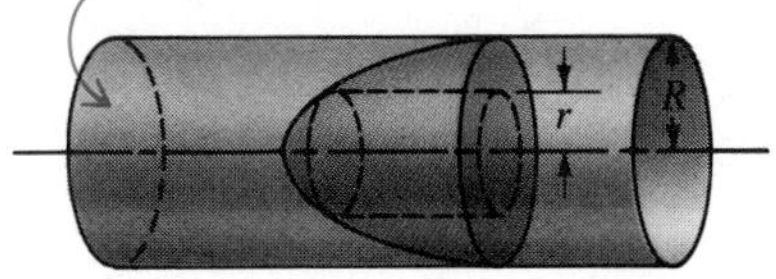

FIGURE 3.38

45. Let $f(x) = x^n$, where n is a positive integer. Determine the values of n for which f has a relative extremum.

46. Prove that a polynomial function of degree n can have at most $n - 1$ critical points.

47. Suppose f is a continuous even function such that $f(a)$ is a relative minimum. What can be said about $f(-a)$?

48. Suppose f is a continuous odd function such that $f(a)$ is a relative maximum. What can be said about $f(-a)$?

49. Suppose f is a differentiable function that possesses a single critical point c. If $k \neq 0$, find the critical points of:

(a) $k + f(x)$ (b) $kf(x)$

(c) $f(x + k)$ (d) $f(kx)$

50. Does the function f whose graph is shown in Figure 3.39 possess any absolute extrema?

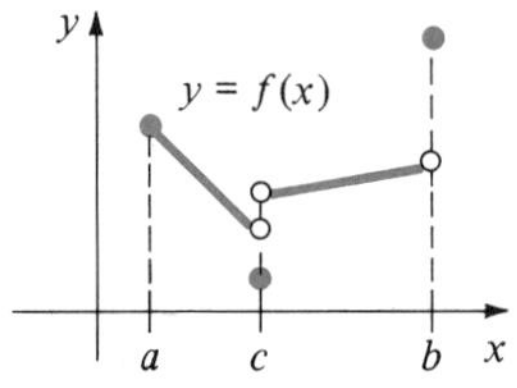

FIGURE 3.39

Calculator/Computer Problems

51. (a) Use a calculator or computer to obtain the graph of $f(x) = -2 \cos x + \cos 2x$.

(b) Find the critical points of f in the interval $[0, 2\pi]$.

(c) Find the absolute extrema of f on $[0, 2\pi]$.

52. In his 1989 study of snow-crystal growth, Colbeck uses the formula

$$I(t) = \frac{b}{\pi} + \frac{b}{2} \sin \omega t - \frac{2b}{3\pi} \cos 2\omega t$$

to model the daily variation in the intensity of solar radiation penetrating the surface of snow. Here t represents time measured in hours after sunrise ($t = 0$) and $\omega = 2\pi/24$.

(a) Use a calculator or computer to obtain the graph of I on the interval $[0, 24]$. Use $b = 1$.

(b) Find the critical points of I in the interval $[0, 24]$.*

(c) Find the absolute extrema of I on $[0, 24]$.

(d) Use the bisection method or Newton's Method to approximate the first time for which $I(t) = 0$.

53. (a) Use a calculator or computer to obtain the graph of $f(x) = x^{10} - x^6 - 2x^3 + x - 3$.

*Remember, the $\boxed{\sin^{-1}}$ key on a calculator gives only angles that satisfy $-\pi/2 \leq \sin^{-1} x \leq \pi/2$.

(b) Use a calculator or computer to obtain the graph of $f'(x) = 10x^9 - 6x^5 - 6x^2 + 1$.

(c) Use the bisection method or Newton's Method to approximate the critical points of f. Use an accuracy of two decimal places.

(d) Compute the approximate values of the relative extrema.

Miscellaneous Problems

54. Prove Theorem 3.2 in the case when $f(c)$ is a relative minimum.

3.4 ROLLE'S THEOREM AND THE MEAN VALUE THEOREM

When $y = f(x)$ is continuous and differentiable on an interval $[a, b]$, it seems plausible that if $f(a) = f(b) = 0$, then its graph must be as indicated in Figure 3.40. The graphs in turn suggest that there must be at least one point on the graph that corresponds to a number c in (a, b) at which the tangent is horizontal. See Figure 3.40(c) and (d). This result is formalized as Rolle's Theorem.*

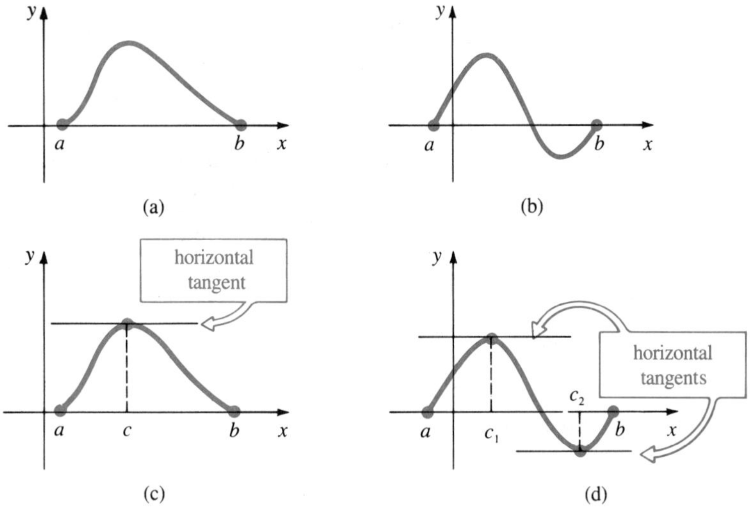

FIGURE 3.40

THEOREM 3.4 Rolle's Theorem

Let f be a function that is continuous on $[a, b]$ and differentiable on (a, b). If $f(a) = f(b) = 0$, then there exists a number c in (a, b) such that $f'(c) = 0$.

***Michel Rolle**, a French mathematician (1652–1719), was one of the early and vocal critics of the, then new, calculus. Rolle did not prove the theorem that bears his name.

Proof Either f is a constant function on the interval $[a, b]$ or it is not.

If f is a constant function on $[a, b]$, then we must have $f'(c) = 0$ for every number c in (a, b).

Now, if f is not a constant function on $[a, b]$, there must be some number x in (a, b) at which either $f(x) > 0$ or $f(x) < 0$. Suppose $f(x) > 0$. Since f is continuous on $[a, b]$, we know from the Extreme Value Theorem that f attains an absolute maximum at some number c in $[a, b]$. But from $f(a) = f(b) = 0$ and $f(x) > 0$ for some x in (a, b), we conclude that the number c cannot be an endpoint of $[a, b]$. Consequently, c is in (a, b). Since f is differentiable on (a, b), it is differentiable at c. Hence, from Theorem 3.2, we have $f'(c) = 0$. The proof of the case when $f(x) < 0$ follows in a similar manner. ■

EXAMPLE 1 Consider the function $f(x) = -x^3 + x$ defined on $[-1, 1]$. Since f is a polynomial function, it is continuous on $[-1, 1]$ and differentiable on $(-1, 1)$. Also, $f(-1) = f(1) = 0$. Thus, the hypotheses of Rolle's Theorem are satisfied. We conclude that there must be at least one number in $(-1, 1)$ for which $f'(x) = -3x^2 + 1$ is zero. To find this number, we solve $f'(c) = 0$ or $-3c^2 + 1 = 0$. The latter leads to *two* solutions in the interval, $c_1 = -\sqrt{3}/3$ and $c_2 = \sqrt{3}/3$. □

Note in the preceding example that the given function f satisfies the hypotheses of Rolle's Theorem on $[0, 1]$ as well as on $[-1, 1]$. In the case of the interval $[0, 1]$, $f'(c) = -3c^2 + 1 = 0$ yields the single solution $c = \sqrt{3}/3$.

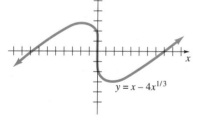

FIGURE 3.41

EXAMPLE 2 The function $f(x) = x - 4x^{1/3}$, shown in Figure 3.41, is continuous on $[-8, 8]$ and satisfies $f(-8) = f(8) = 0$. But f is not differentiable on $(-8, 8)$, since there is a vertical tangent at the origin. Nevertheless, as the figure suggests, there are two numbers c_1 and c_2 in $(-8, 8)$ at which $f'(x) = 0$. You should verify that $f'(-8\sqrt{3/9}) = 0$ and $f'(8\sqrt{3/9}) = 0$. Bear in mind that the hypotheses of Rolle's Theorem are sufficient but not necessary conditions. In other words, if one or more of the three hypotheses: continuity on $[a, b]$, differentiability on (a, b), and $f(a) = f(b) = 0$, do not hold, the conclusion that there exists a c in (a, b) such that $f'(c) = 0$ may or may not hold.

Consider another function $g(x) = 1 - x^{2/3}$. This function is continuous on $[-1, 1]$ and $f(-1) = f(1) = 0$. But like the foregoing function f, g is not differentiable at $x = 0$ and so is not differentiable on the open interval $(-1, 1)$. In this case, however, there is no c in $(-1, 1)$ for which $f'(c) = 0$. You are encouraged to supply the graph of g. □

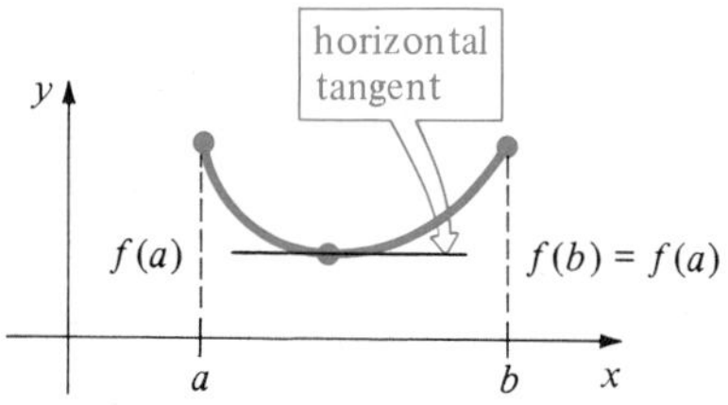

FIGURE 3.42

The conclusion of Rolle's Theorem also holds when the condition $f(a) = f(b) = 0$ is replaced with $f(a) = f(b)$. The plausibility of this fact is illustrated in Figure 3.42.

Rolle's Theorem is helpful in proving the next important result.

THEOREM 3.5 The Mean Value Theorem for Derivatives

Let f be a function that is continuous on $[a, b]$ and differentiable on (a, b). Then there exists a number c in (a, b) such that

$$f'(c) = \frac{f(b) - f(a)}{b - a}$$

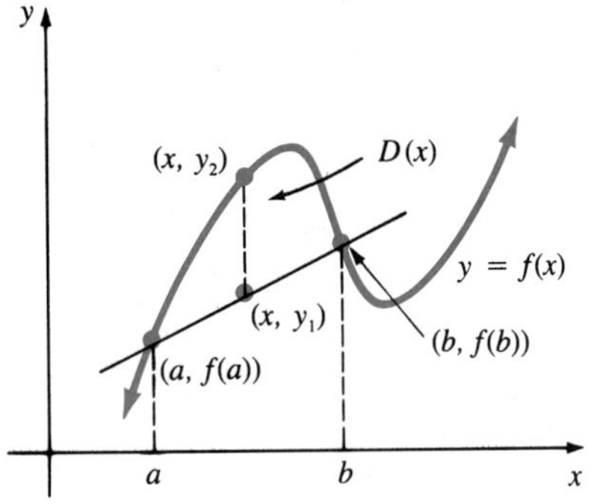

FIGURE 3.43

Proof As shown in Figure 3.43, let $D(x)$ denote the vertical distance between a point on the graph of $y = f(x)$ and the secant line through $(a, f(a))$ and $(b, f(b))$. Since the equation of the secant line is

$$y - f(b) = \frac{f(b) - f(a)}{b - a}(x - b)$$

we have, as shown in the figure,

$$D(x) = f(x) - \left[\frac{f(b) - f(a)}{b - a}(x - b) + f(b)\right]$$

Since $D(a) = D(b) = 0$ and D is continuous on $[a, b]$ and differentiable on (a, b), Rolle's Theorem implies there is some number c in (a, b) for which $D'(c) = 0$. Now

$$D'(x) = f'(x) - \frac{f(b) - f(a)}{b - a}$$

and so $D'(c) = 0$ is the same as

$$f'(c) = \frac{f(b) - f(a)}{b - a}$$

■

Theorem 3.5 is also called the **Theorem of the Mean.**

Geometrically, the Mean Value Theorem asserts that the slope of the tangent line at $(c, f(c))$ is the same as the slope of the secant through $(a, f(a))$,

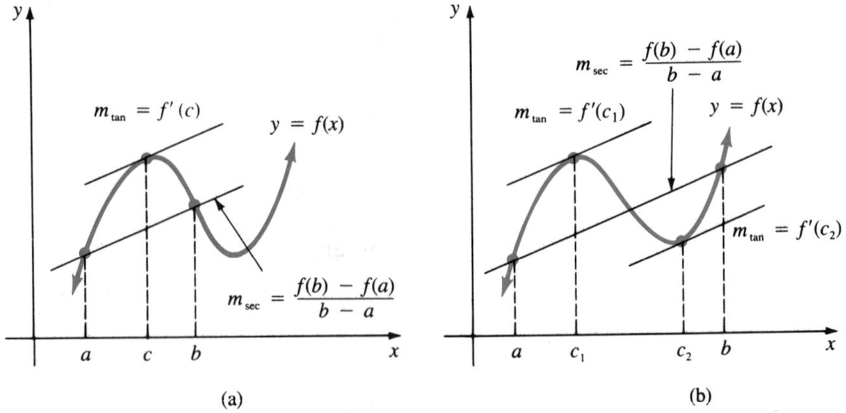

FIGURE 3.44

$(b, f(b))$. See Figure 3.44(a). Also, as indicated in Figure 3.44(b), there may be more than one number c in (a, b) for which the tangent and secant lines are parallel.

EXAMPLE 3 Given the function $f(x) = x^3 - 12x$ defined on $[-1, 3]$, does there exist a number c in $(-1, 3)$ that satisfies the conclusion of the Mean Value Theorem?

Solution Since f is a polynomial function, it is continuous on $[-1, 3]$ and differentiable on $(-1, 3)$. Now, $f(3) = -9$, $f(-1) = 11$,

$$f'(x) = 3x^2 - 12 \quad \text{and} \quad f'(c) = 3c^2 - 12$$

Hence, we must have

$$\frac{f(3) - f(-1)}{3 - (-1)} = \frac{-20}{4} = 3c^2 - 12$$

Thus, $3c^2 = 7$. Although the last equation has two solutions, the only solution in $(-1, 3)$ is $c = \sqrt{7/3} \approx 1.53$. □

The Mean Value Theorem is very useful in proving other theorems. Recall from Section 2.3 that if $f(x) = k$ is a constant function, then $f'(x) = 0$. The converse of this result is given by the following:

THEOREM 3.6

If $f'(x) = 0$ for all x in an interval $[a, b]$, then $f(x)$ is a constant on the interval.

Proof Let x_1 and x_2 be any numbers in $[a, b]$ such that $x_1 < x_2$. By the Mean Value Theorem, there is a number c in (x_1, x_2) such that

$$\frac{f(x_2) - f(x_1)}{x_2 - x_1} = f'(c)$$

But $f'(c) = 0$ by hypothesis. Hence, $f(x_2) - f(x_1) = 0$ or $f(x_1) = f(x_2)$. Since x_1 and x_2 are arbitrarily chosen, the function f has the same value at all points in the interval. Thus, f is constant. ■

Increasing and Decreasing Functions We shall use the Mean Value Theorem to relate the concepts of increasing and decreasing functions with the notion of a derivative.

DEFINITION 3.6 Increasing and Decreasing Functions

Let f be a function defined on an interval I, and let x_1, and x_2 denote any numbers in I.

(*i*) f is **increasing** on I if $f(x_1) < f(x_2)$ whenever $x_1 < x_2$.
(*ii*) f is **decreasing** on I if $f(x_1) > f(x_2)$ whenever $x_1 < x_2$.

In other words, the graph of an increasing function rises as x increases, whereas the graph of a decreasing function falls as x increases. The graph in Figure 3.45 illustrates a function f that is increasing on $[b, c]$ and $[d, e]$ and decreasing on $[a, b]$, $[c, d]$, and $[e, h]$.

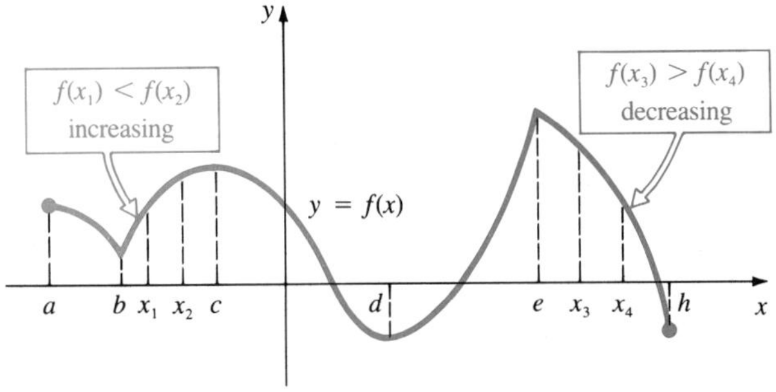

FIGURE 3.45

THEOREM 3.7 Test for Increasing/Decreasing

Let f be a function that is continuous on $[a, b]$ and differentiable on (a, b).

(*i*) If $f'(x) > 0$ for all x in (a, b), then f is increasing on $[a, b]$.
(*ii*) If $f'(x) < 0$ for all x in (a, b), then f is decreasing on $[a, b]$.

Proof of (i) Let x_1 and x_2 be any numbers in $[a, b]$ such that $x_1 < x_2$. By the Mean Value Theorem, there is a number c in (x_1, x_2) such that $[f(x_2) - f(x_1)]/(x_2 - x_1) = f'(c)$. But $f'(c) > 0$ by hypothesis. Hence, $f(x_2) - f(x_1) > 0$ or $f(x_1) < f(x_2)$. Since x_1 and x_2 are arbitrarily chosen, it follows from Definition 3.6 that f is increasing on $[a, b]$. ■

EXAMPLE 4 Determine the intervals on which $f(x) = x^3 - 3x^2 - 24x$ is increasing and the intervals on which f is decreasing.

Solution The derivative is

$$f'(x) = 3x^2 - 6x - 24 = 3(x + 2)(x - 4)$$

To determine when $f'(x) > 0$ and $f'(x) < 0$, we must solve

$$(x + 2)(x - 4) > 0 \quad \text{and} \quad (x + 2)(x - 4) < 0$$

respectively. One way of solving these inequalities is to examine the algebraic signs of the factors $(x + 2)$ and $(x - 4)$ in the intervals of the number line determined by the critical points -2 and 4: $(-\infty, -2]$, $[-2, 4]$, $[4, \infty)$. See Figure 3.46.

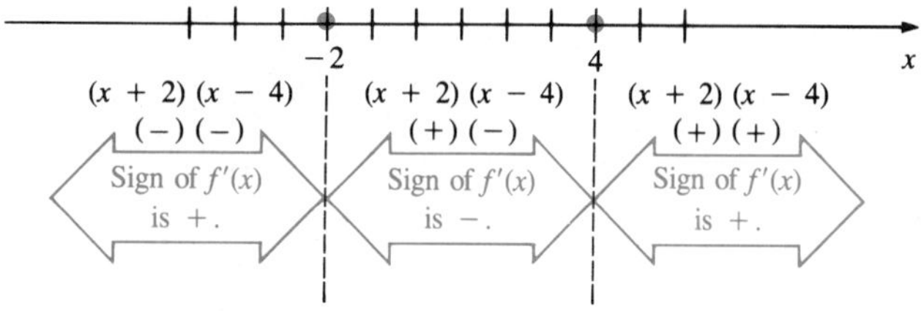

FIGURE 3.46

The information is summarized in the accompanying table.

Interval	Sign of $f'(x)$	$y = f(x)$
$(-\infty, -2)$	+	increasing on $(-\infty, -2]$
$(-2, 4)$	−	decreasing on $[-2, 4]$
$(4, \infty)$	+	increasing on $[4, \infty)$

□

EXAMPLE 5 Determine the intervals on which $f(x) = x^{2/3}$ is increasing and the intervals on which f is decreasing.

Solution Observe that the derivative

$$f'(x) = \frac{2}{3}x^{-1/3} = \frac{2}{3\sqrt[3]{x}}$$

is undefined at 0. Since 0 is in the domain of f, we conclude it is a critical point. Using the facts that $\sqrt[3]{x} < 0$ for $x < 0$ and $\sqrt[3]{x} > 0$ for $x > 0$, we are led to the information given in the accompanying table.

Interval	Sign of $f'(x)$	$y = f(x)$
$(-\infty, 0)$	−	decreasing on $(-\infty, 0]$
$(0, \infty)$	+	increasing on $[0, \infty)$

□

If a function f is discontinuous at one or both endpoints of $[a, b]$, then $f'(x) > 0$ (or $f'(x) < 0$) on (a, b) implies f is increasing (or decreasing) on the open interval (a, b).

▲ ***Remark*** The converses of parts (*i*) and (*ii*) of Theorem 3.7 are not necessarily true. In other words, when f is an increasing (or decreasing) function on an interval, it does not follow that $f'(x) > 0$ (or $f'(x) < 0$). A function could be, say, increasing yet not differentiable.

EXERCISES 3.4 *Answers to odd-numbered problems begin on page A-74.*

In Problems 1–10 determine whether the given function satisfies the hypotheses of Rolle's Theorem on the indicated interval. If so, find all values of c that satisfy the conclusion of the theorem.

1. $f(x) = x^2 - 4; [-2, 2]$
2. $f(x) = x^2 - 6x + 5; [1, 5]$
3. $f(x) = x^3 + 27; [-3, -2]$
4. $f(x) = x^3 - 5x^2 + 4x; [0, 4]$
5. $f(x) = x^3 + x^2; [-1, 0]$
6. $f(x) = x(x - 1)^2; [0, 1]$
7. $f(x) = \sin x; [-\pi, 2\pi]$
8. $f(x) = \tan x; [0, \pi]$
9. $f(x) = x^{2/3} - 1; [-1, 1]$
10. $f(x) = x^{2/3} - 3x^{1/3} + 2; [1, 8]$

In Problems 11 and 12 state why the function f whose graph is given does not satisfy the hypotheses of Rolle's Theorem on $[a, b]$.

11. 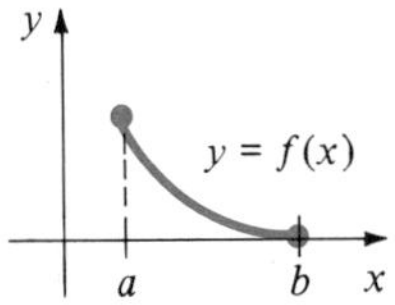

FIGURE 3.47

12. 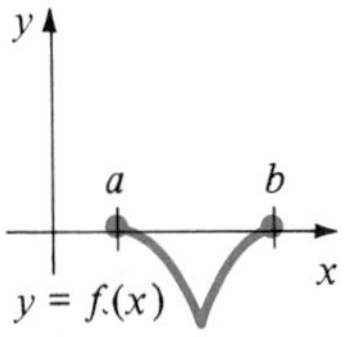

FIGURE 3.48

In Problems 13–22 determine whether the given function satisfies the hypotheses of the Mean Value Theorem on the indicated interval. If so, find all values of c that satisfy the conclusion of the theorem.

13. $f(x) = x^2; [-1, 7]$
14. $f(x) = -x^2 + 8x - 6; [2, 3]$
15. $f(x) = x^3 + x + 2; [2, 5]$
16. $f(x) = x^4 - 2x^2; [-3, 3]$
17. $f(x) = 1/x; [-10, 10]$
18. $f(x) = x + \dfrac{1}{x}; [1, 5]$
19. $f(x) = 1 + \sqrt{x}; [0, 9]$
20. $f(x) = \sqrt{4x + 1}; [2, 6]$
21. $f(x) = \dfrac{x + 1}{x - 1}; [-2, -1]$
22. $f(x) = x^{1/3} - x; [-8, 1]$

In Problems 23 and 24 state why the function f whose graph is given does not satisfy the hypotheses of the Mean Value Theorem on $[a, b]$.

23. 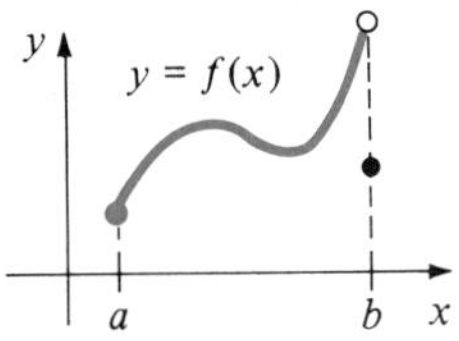

FIGURE 3.49

24. 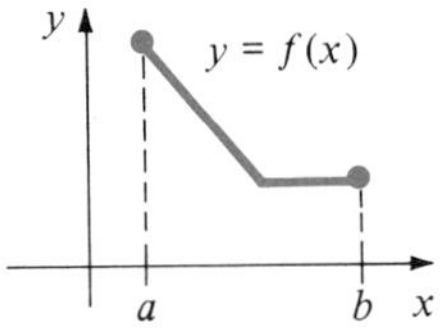

FIGURE 3.50

In Problems 25–44 determine the intervals on which the given function f is increasing and the intervals on which f is decreasing.

25. $f(x) = x^2 + 5$
26. $f(x) = x^3$
27. $f(x) = x^2 + 6x - 1$
28. $f(x) = -x^2 + 10x + 3$
29. $f(x) = x^3 - 3x^2$
30. $f(x) = \frac{1}{3}x^3 - x^2 - 8x + 1$
31. $f(x) = x^4 - 4x^3 + 9$
32. $f(x) = 4x^5 - 10x^4 + 2$
33. $f(x) = 1 - x^{1/3}$
34. $f(x) = x^{2/3} - 2x^{1/3}$
35. $f(x) = x + \dfrac{1}{x}$
36. $f(x) = \dfrac{1}{x} + \dfrac{1}{x^2}$
37. $f(x) = x\sqrt{8 - x^2}$
38. $f(x) = \dfrac{x + 1}{\sqrt{x^2 + 1}}$

39. $f(x) = \dfrac{5}{x^2 + 1}$

40. $f(x) = \dfrac{x^2}{x + 1}$

41. $f(x) = x(x - 3)^2$

42. $f(x) = (x^2 - 1)^3$

43. $f(x) = \sin x$

44. $f(x) = -x + \tan x$

In Problems 45 and 46 show, without graphing, that the given function has no relative extrema.

45. $f(x) = 4x^3 + x$

46. $f(x) = -x + \sqrt{2 - x}$

47. A motorist enters a tollway and is given a stub stamped 1:15 P.M. Sixty miles down the road, when the motorist pays the toll at 2:15 P.M., he is also given a traffic ticket. Explain this by the Mean Value Theorem. Assume the speed limit is 55 mi/h.

48. In the analysis of the human cough, the trachea, or windpipe, is considered to be a cylindrical tube. The volume flow of air (in cm^3/s) through the trachea during its contraction is given by

$$V(r) = kr^4(r_0 - r), \qquad r_0/2 \le r \le r_0$$

where k is a positive constant and r_0 is its radius when there is no pressure difference at the ends of the tracheal tube. Determine an interval for which V is increasing and an interval for which V is decreasing. What radius will give the maximum volume flow of air?

49. Consider the function $f(x) = x^4 + x^3 - x - 1$. Use this function and Rolle's Theorem to show that the equation $4x^3 + 3x^2 - 1 = 0$ has at least one root in $[-1, 1]$.

50. Suppose the functions f and g are increasing functions on an interval I. Use Definition 3.6 to show that $f + g$ is an increasing function on I.

51. Suppose f and g are continuous on $[a, b]$ and differentiable on (a, b) such that $f'(x) > 0$ and $g'(x) > 0$ for all x in (a, b). Give a condition on $f(x)$ and $g(x)$ that will guarantee that the product fg is increasing on $[a, b]$.

52. Show that the equation $ax^3 + bx + c = 0$, $a > 0$, $b > 0$, cannot have two real roots. [*Hint:* Consider the function $f(x) = ax^3 + bx + c$. Suppose there are two numbers r_1 and r_2 such that $f(r_1) = f(r_2) = 0$.]

53. Show that the equation $ax^2 + bx + c = 0$ has at most two real roots. [*Hint:* Consider the function $f(x) = ax^2 + bx + c$. Suppose there are three distinct numbers r_1, r_2, and r_3 such that $f(r_1) = f(r_2) = f(r_3) = 0$.]

54. For a quadratic polynomial function $f(x) = ax^2 + bx + c$ show that the value of x_3 that satisfies the conclusion of the Mean Value Theorem on any interval $[x_1, x_2]$ is $x_3 = (x_1 + x_2)/2$.

In Problems 55 and 56 use a calculator to find a value of c that satisfies the conclusion of the Mean Value Theorem.

55. $f(x) = \cos 2x$; $[0, \pi/4]$

56. $f(x) = 1 + \sin x$; $[\pi/4, \pi/2]$

3.5 GRAPHING AND THE FIRST DERIVATIVE

Knowing that a function does, or does not, possess relative extrema is a great aid in drawing its graph. Recall that when a function has a relative extremum it must occur at a critical point. By finding the critical points of a function, we have a *list of abscissas that possibly correspond to relative extrema.* We shall now combine the ideas of the two preceding sections to devise two tests for determining when a critical point actually is the x-coordinate of a relative extremum.

First Derivative Test Suppose f is differentiable on (a, b) and c is a critical point in the interval. If $f'(x) > 0$ for all x in (a, c) and $f'(x) < 0$ for all x in (c, b), then on the interval (a, b) the graph of f must be as indicated in Figure 3.51(a); that is, $f(c)$ is a relative maximum. On the other hand, when $f'(x) < 0$ for all x in (a, c) and $f'(x) > 0$ for all x in (c, b), then, as shown in Figure 3.51(b), $f(c)$ is a relative minimum. We have demonstrated a special case of the next theorem.

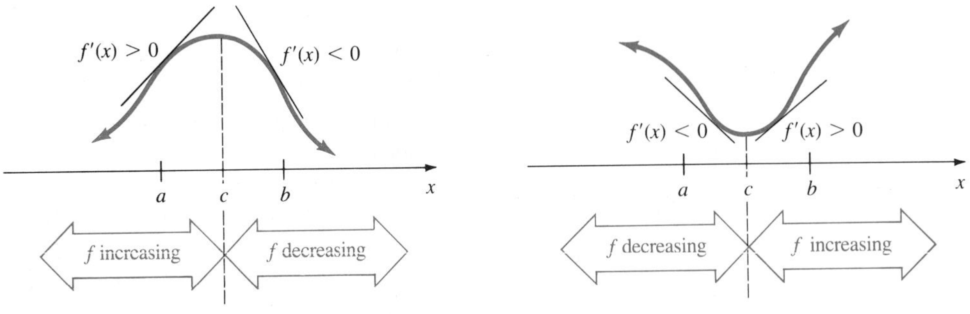

(a) relative maximum (b) relative minimum

FIGURE 3.51

> **THEOREM 3.8 First Derivative Test for Relative Extrema**
>
> Let f be continuous on $[a, b]$ and differentiable on (a, b) except possibly at the critical point c.
>
> (*i*) If $f'(x) > 0$ for $a < x < c$ and $f'(x) < 0$ for $c < x < b$, then $f(c)$ is a relative maximum.
>
> (*ii*) If $f'(x) < 0'$ for $a < x < c$ and $f'(x) > 0$ for $c < x < b$, then $f(c)$ is a relative minimum.
>
> (*iii*) If $f'(x)$ has the same algebraic sign on $a < x < c$ and $c < x < b$, then $f(c)$ is not an extremum.

EXAMPLE 1 Graph $f(x) = x^3 - 3x^2 - 9x + 2$.

Solution The first derivative

$$f'(x) = 3x^2 - 6x - 9 = 3(x + 1)(x - 3) \tag{1}$$

yields the critical points -1 and 3. Now the First Derivative Test is essentially the procedure used in finding the intervals on which f is either increasing or decreasing. Using (1), we see in Figure 3.52(a) that $f'(x) > 0$ for $-\infty < x < -1$ and $f'(x) < 0$ for $-1 < x < 3$. It follows from part (*i*) of Theorem 3.8 that $f(-1) = 7$ is a relative maximum. Similarly, $f'(x) < 0$ for $-1 < x < 3$ and $f'(x) > 0$ for $3 < x < \infty$. Thus, from part (*ii*) of Theorem 3.8, $f(3) = -25$ is a relative minimum.

Now, the graph of the function has the y-intercept $f(0) = 2$. Furthermore, testing the equation $x^3 - 3x^2 - 9x + 2 = 0$ for rational roots (see Example 4 in Section 2.11) reveals that $x = -2$ is a real root and, hence, an x-intercept. Division by the factor $x + 2$ then gives $(x + 2) \cdot (x^2 - 5x + 1) = 0$. The quadratic formula reveals two additional x-intercepts:

$$\frac{5 - \sqrt{21}}{2} \approx 0.21 \quad \text{and} \quad \frac{5 + \sqrt{21}}{2} \approx 4.79$$

Putting all this information together leads to the graph given in Figure 3.52(b).

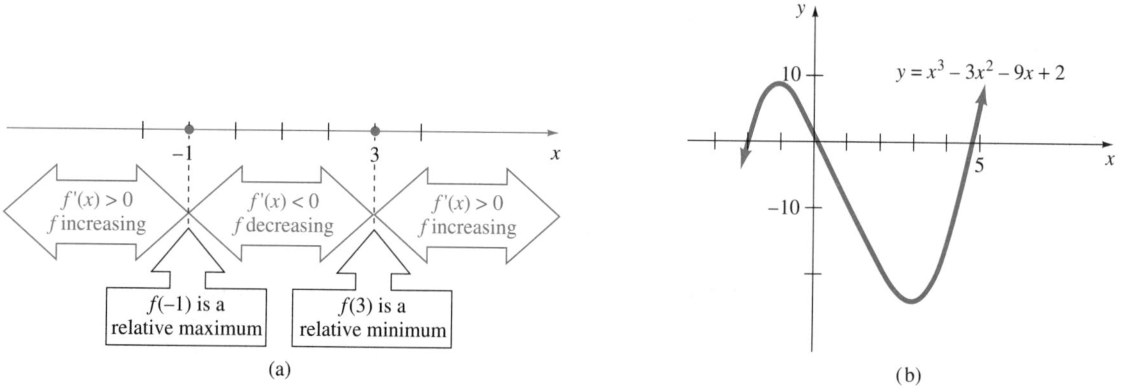

FIGURE 3.52

□

EXAMPLE 2 Graph $f(x) = x + \dfrac{1}{x}$.

Solution Inspection of the function in the form $f(x) = (x^2 + 1)/x$ reveals that the graph of f has no intercepts, $x = 0$ is a vertical asymptote, and the graph of f is symmetric with respect to the origin, since $f(-x) = -f(x)$. Now,

$$f'(x) = 1 - \frac{1}{x^2} = \frac{x^2 - 1}{x^2} = \frac{(x+1)(x-1)}{x^2}$$

shows that -1 and 1 are critical points. In Figure 3.53(a) we see $f'(x) > 0$ for $-\infty < x < -1$ and $f'(x) < 0$ for $-1 < x < 0$. Thus, from the First Derivative Test, we conclude that $f(-1) = -2$ is a relative maximum. By symmetry it follows that $f(1) = 2$ is a relative minimum. The graph of f is given in Figure 3.53(b).

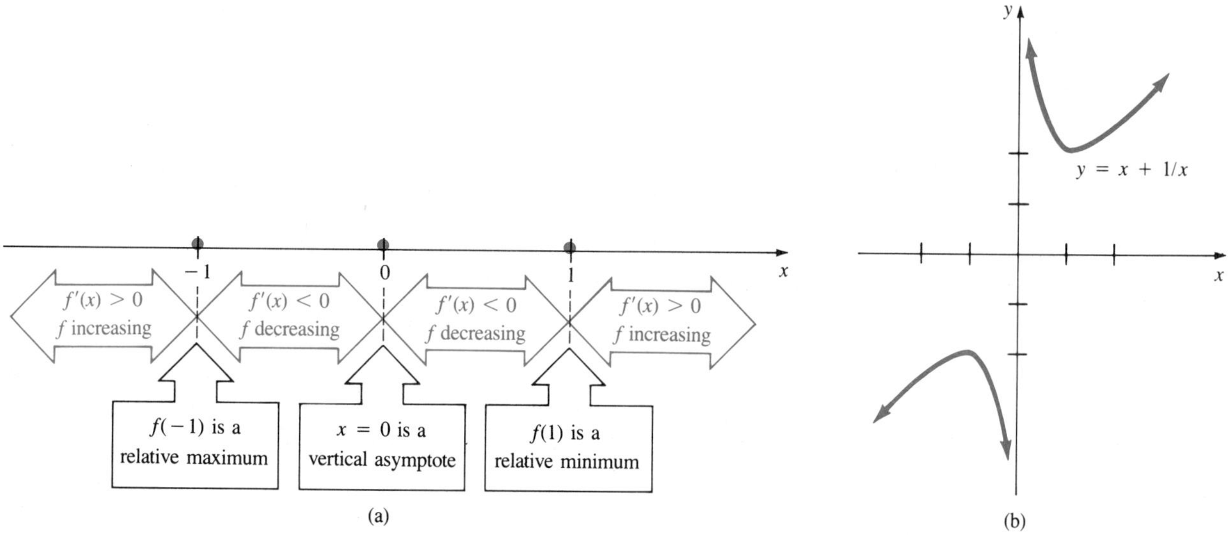

FIGURE 3.53

□

EXAMPLE 3 Graph $f(x) = -x^{5/3} + 5x^{2/3}$.

Solution The derivative is

$$f'(x) = -\frac{5}{3}x^{2/3} + \frac{10}{3}x^{-1/3} = \frac{5}{3}x^{-1/3}(-x+2)$$

Notice that f' does not exist at 0 but 0 is in the domain of the function, since $f(0) = 0$. The critical points are 0 and 2. The First Derivative Test, illustrated in Figure 3.54(a), shows that $f(0) = 0$ is a relative minimum and that $f(2) = -(2)^{5/3} + 5(2)^{2/3} \approx 4.76$ is a relative maximum. Moreover, since $|f'(x)| \to \infty$ as $x \to 0$, there is a vertical tangent at (0, 0). Finally, by writing $f(x) = x^{2/3}(-x+5)$, we see that the x-intercepts are 0 and 5. The graph of f is given in Figure 3.54(b).

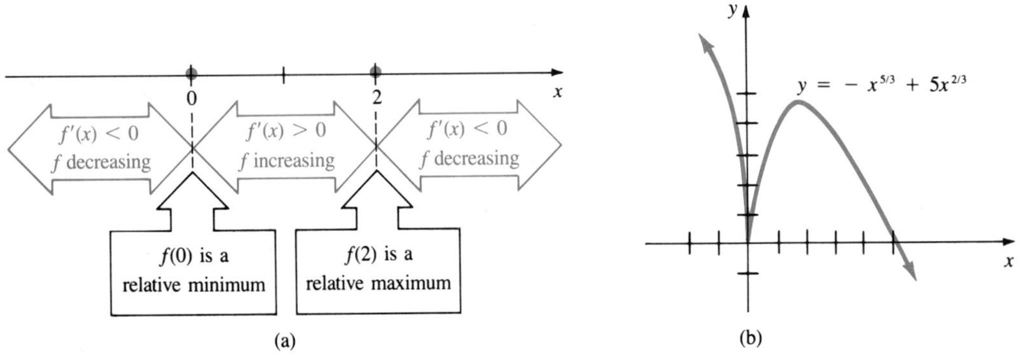

FIGURE 3.54

□

EXAMPLE 4 Graph $f(x) = x^4 - 4x^3 + 10$.

Solution The derivative

$$f'(x) = 4x^3 - 12x^2 = 4x^2(x-3)$$

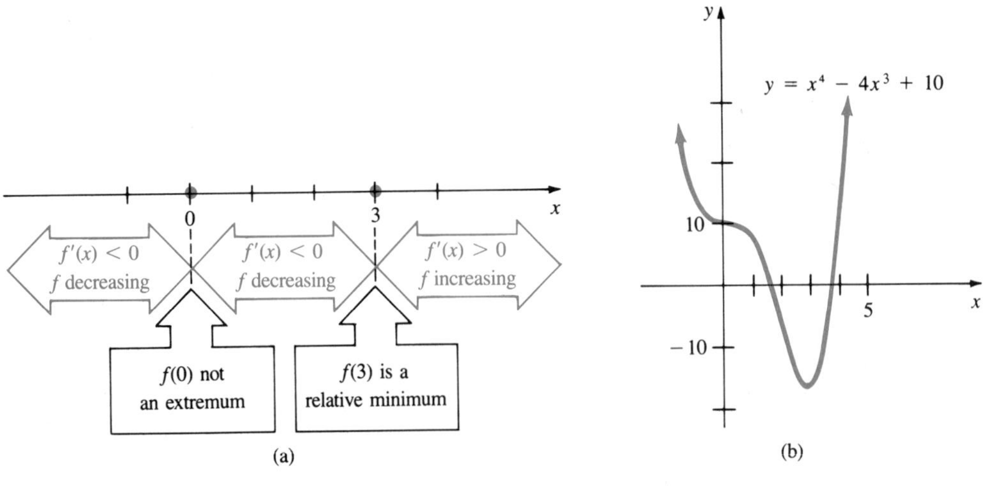

FIGURE 3.55

shows that 0 and 3 are critical points. Now, as seen in Figure 3.55(a), f' has the same algebraic sign in both $(-\infty, 0)$ and $(0, 3)$. Hence, $f(0) = 10$ is not an extremum. In this case $f'(0) = 0$ means there is only a horizontal tangent at $(0, 10)$. However, it is evident from the First Derivative Test that $f(3) = -17$ is a relative minimum. Indeed, the graph of f given in Figure 3.55(b) shows that $f(3)$ is also an absolute minimum. In conclusion, we see that the graph of f has two x-intercepts. By the interval bisection method or by Newton's Method we find that the x-intercepts are approximately 1.61 and 3.82. □

EXAMPLE 5 Graph $f(x) = \dfrac{x^2 - 3}{x^2 + 1}$.

Solution

y-intercept: $f(0) = -3$

x-intercepts: $f(x) = 0$ when $x^2 - 3 = 0$. Thus, $-\sqrt{3}$ and $\sqrt{3}$ are x-intercepts.

Symmetry: y-axis, since $f(-x) = f(x)$.

Vertical asymptotes: None, since $x^2 + 1 \neq 0$ for all real numbers.

Horizontal asymptotes: $\displaystyle\lim_{x\to\infty} \frac{x^2 - 3}{x^2 + 1} = 1$. Symmetry implies

$\displaystyle\lim_{x\to-\infty} \frac{x^2 - 3}{x^2 + 1} = 1$. $y = 1$ is a horizontal asymptote.

Derivative: $f'(x) = \dfrac{8x}{(x^2 + 1)^2}$

Critical points: $f'(x) = 0$ when $8x = 0$. Therefore, 0 is the only critical point.

First Derivative Test: See Figure 3.56(a). $f(0) = -3$ is a relative minimum.

Graph: See Figure 3.56(b).

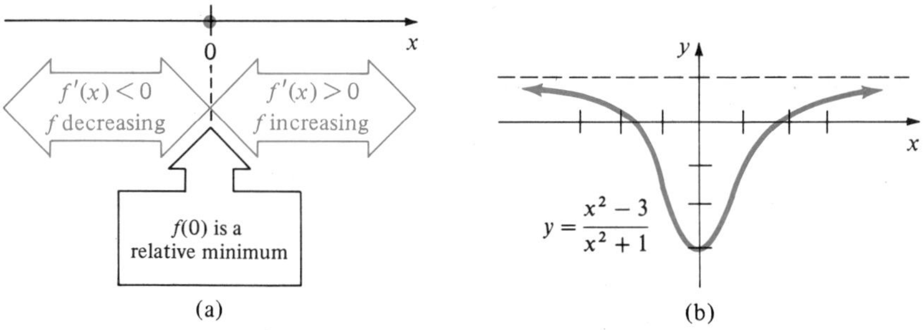

FIGURE 3.56 □

▲ ***Remark*** We conclude with a tabular summary of Theorem 3.8 you can use for easy reference while working the problems in Exercises 3.5.

First Derivative Test			
Critical Value	**Interval**	**Sign of $f'(x)$ on the Interval**	**Conclusion**
c	(a, c) (c, b)	$+$ $-$	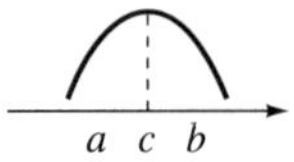 $f(c)$ is a relative maximum
c	(a, c) (c, b)	$-$ $+$	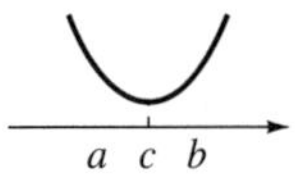 $f(c)$ is a relative minimum
c	(a, c) (c, b)	$+$ $+$	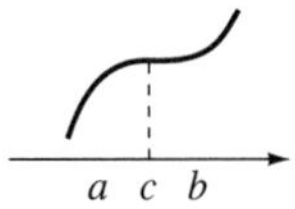 no extremum
c	(a, c) (c, b)	$-$ $-$	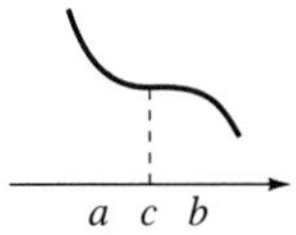 no extremum

EXERCISES 3.5 *Answers to odd-numbered problems begin on page A-75.*

In Problems 1–30 use the First Derivative Test to find the relative extrema of the given function. Graph. Find intercepts when possible.

1. $f(x) = -x^2 + 2x + 1$

2. $f(x) = (x - 1)(x + 3)$

3. $f(x) = x^3 - 3x$

4. $f(x) = \frac{1}{3}x^3 - \frac{1}{2}x^2 + 1$

5. $f(x) = x(x - 2)^2$

6. $f(x) = -x^3 + 3x^2 + 9x - 1$

7. $f(x) = x^3 + x - 3$

8. $f(x) = x^3 + 3x^2 + 3x - 3$

9. $f(x) = x^4 + 4x$

10. $f(x) = (x^2 - 1)^2$

11. $f(x) = \frac{1}{4}x^4 + \frac{4}{3}x^3 + 2x^2$

12. $f(x) = 2x^4 - 16x^2 + 3$

13. $f(x) = -x^2(x - 3)^2$

14. $f(x) = -3x^4 + 8x^3 - 6x^2 - 2$

15. $f(x) = 4x^5 - 5x^4$

16. $f(x) = (x - 2)^2(x + 3)^3$

17. $f(x) = \dfrac{x^2 + 3}{x + 1}$

18. $f(x) = x + \dfrac{25}{x}$

19. $f(x) = \dfrac{1}{x} - \dfrac{1}{x^3}$

20. $f(x) = \dfrac{x^2}{x^2 - 4}$

21. $f(x) = \dfrac{10}{x^2 + 1}$

22. $f(x) = \dfrac{x^2}{x^4 + 1}$

23. $f(x) = (x^2 - 4)^{2/3}$

24. $f(x) = (x^2 - 1)^{1/3}$

25. $f(x) = x\sqrt{1 - x^2}$

26. $f(x) = x(x^2 - 5)^{1/3}$

27. $f(x) = x - 12x^{1/3}$

28. $f(x) = x^{4/3} + 32x^{1/3}$

29. $f(x) = x^{2/3}(x^2 - 16)$

30. $f(x) = (x - 1)^{2/3}(x - 11)$

In Problems 31–34 sketch a graph of a function f whose derivative f' has the given graph.

31.

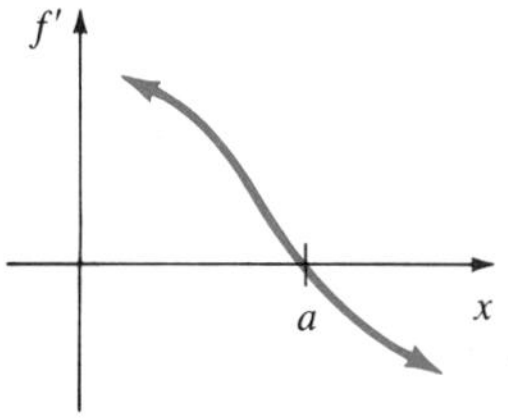

FIGURE 3.57

32.

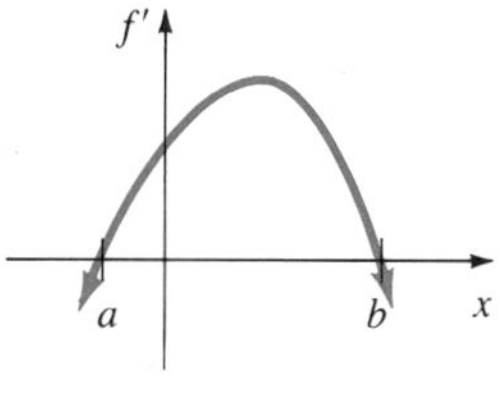

FIGURE 3.58

33.

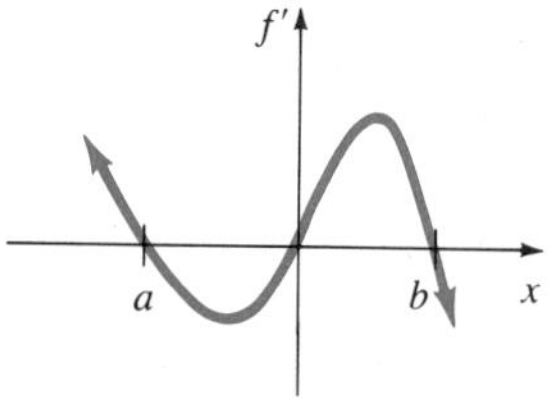

FIGURE 3.59

34.

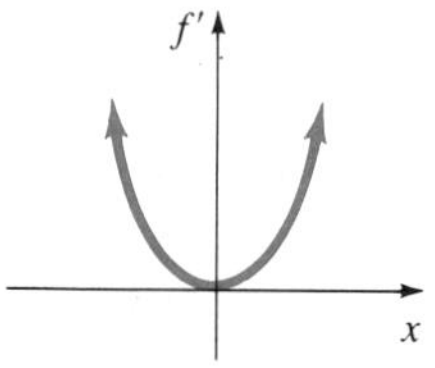

FIGURE 3.60

In Problems 35 and 36 sketch the graph of f' from the graph of f.

35.

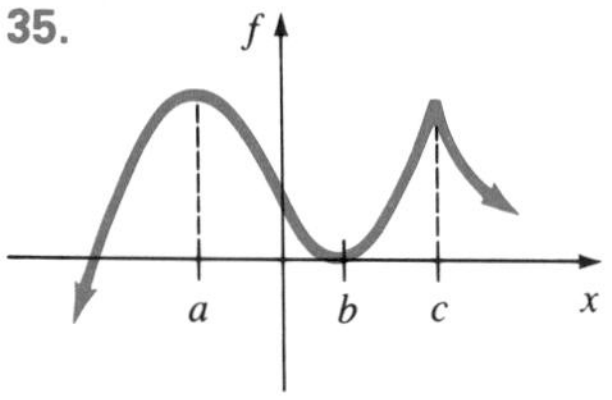

FIGURE 3.61

36.

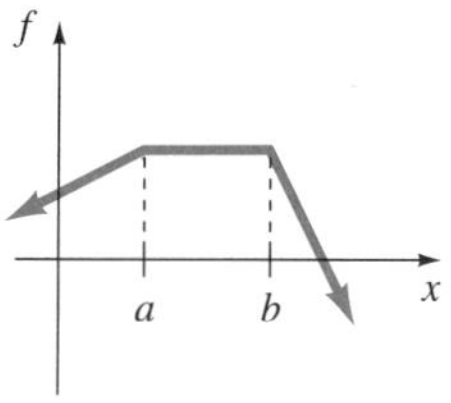

FIGURE 3.62

In Problems 37–40 sketch a graph of a function f that has the given properties.

37. $f(-1) = 0, f(0) = 1$
$f'(3)$ does not exist, $f'(5) = 0$
$f'(x) > 0, x < 3$ and $x > 5$
$f'(x) < 0, 3 < x < 5$

38. $f(0) = 0$
$f'(-1) = 0, f'(0) = 0, f'(1) = 0$
$f'(x) < 0, x < -1, -1 < x < 0$
$f'(x) > 0, 0 < x < 1, x > 1$

39. $f(-x) = f(x)$
$f(2) = 3$
$f'(x) < 0, 0 < x < 2$
$f'(x) > 0, x > 2$

40. $f(1) = -2, f(0) = -1$
$\lim_{x \to 3} f(x) = \infty$
$f'(4) = 0$
$f'(x) < 0, x < 1$
$f'(x) < 0, x > 4$

In Problems 41 and 42 determine where the slope of the tangent to the graph of the given function has a relative maximum or a relative minimum.

41. $f(x) = x^3 + 6x^2 - x$ **42.** $f(x) = x^4 - 6x^2$

43. (a) From the graph of $g(x) = \sin 2x$ determine the intervals for which $g(x) > 0$ and the intervals for which $g(x) < 0$.

(b) Find the critical values of $f(x) = \sin^2 x$. Use the First Derivative Test and the information in part **(a)** to find the relative extrema of f.

(c) Sketch the graph of the function f in part **(b)**.

44. (a) Find the critical values of $f(x) = x - \sin x$.

(b) Show that f has no relative extrema.

(c) Sketch the graph of f.

45. Find values of a, b, and c such that $f(x) = ax^2 + bx + c$ has a relative maximum 6 at $x = 2$ and the graph of f has y-intercept 4.

46. Find values of a, b, c, and d such that $f(x) = ax^3 + bx^2 + cx + d$ has a relative minimum -3 at $x = 0$ and a relative maximum 4 at $x = 1$.

47. Suppose f is a differentiable function whose graph is symmetric about the y-axis. Prove that $f'(0) = 0$. Does f necessarily have a relative extremum at $x = 0$?

48. Let m and n denote positive integers. Show that $f(x) = x^m(x-1)^n$ always has a relative minimum.

49. Suppose f and g are differentiable functions and have relative maxima at the same critical point c.

(a) Show that c is a critical point for the functions $f + g$, $f - g$, and fg.

(b) Does it follow that the functions $f + g$, $f - g$, and fg have relative maxima at c? Prove your assertions.

50. The **arithmetic mean**, or **average**, of the n numbers $a_1, a_2, \ldots, a_n$ is given by

$$\bar{x} = \frac{a_1 + a_2 + \cdots + a_n}{n}$$

(a) Show that $\bar{x}$ is a critical point of the function

$$f(x) = (x - a_1)^2 + (x - a_2)^2 + \cdots + (x - a_n)^2$$

(b) Show that $f(\bar{x})$ is a relative minimum.

51. When sound passes from one medium to another, some of its energy can be lost because of a difference in the acoustic resistances of the two media. (Acoustic resistance is the product of density and elasticity.) The fraction of energy transmitted is given by

$$T(r) = \frac{4r}{(r+1)^2}$$

where r is the ratio of the acoustic resistances of the two media.

(a) Show that $T(r) = T(1/r)$. Explain what this means physically.

(b) Use the First Derivative Test to find the relative extrema of T.

(c) Sketch the graph of the function T for $r \geq 0$.

3.6 GRAPHING AND THE SECOND DERIVATIVE

In the discussion that follows, our goal is to relate the concept of the concavity of a graph with the second derivative of a function.

Concavity You probably have an intuitive idea of what is meant by concavity. Figure 3.63(a) and (b) illustrates geometric shapes that are **concave upward** and **concave downward**, respectively. For example, the Gateway Arch in St. Louis is concave downward; the cables between that supports of the Golden Gate Bridge are concave upward. Often a shape that is concave upward is said to "hold water," whereas a shape that is concave downward "spills water." The graph given in Figure 3.64 is concave upward on the interval (b, c) and concave downward on (a, b) and (c, d).

We state the definition of concavity in terms of the derivative.

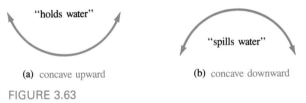

(a) concave upward (b) concave downward

FIGURE 3.63

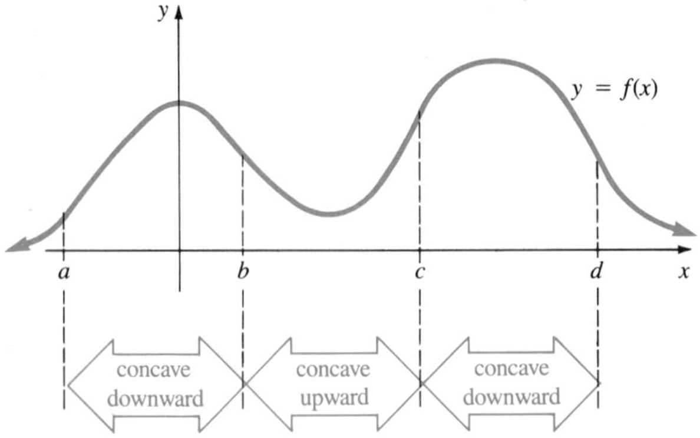

FIGURE 3.64

DEFINITION 3.7 Concavity

Let f be a differentiable function on (a, b).

(*i*) If f' is an increasing function on (a, b), then the graph of f is **concave upward** on the interval.

(*ii*) If f' is a decreasing function on (a, b), then the graph of f is **concave downward** on the interval.

In other words, if the slope of the tangent line increases (decreases) on (a, b), then the graph of the function is concave upward (downward) on the interval. The plausibility of Definition 3.7 is illustrated in Figure 3.65. Equivalently, the graph of a function is concave upward on an interval if the graph at any point lies *above* the tangent at the point. A graph that is concave downward on an interval then lies *below* the tangent lines.

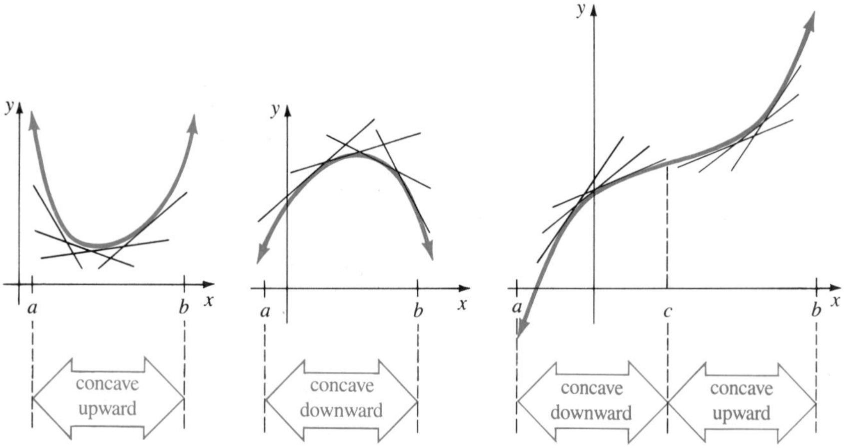

(a) f' increasing from $-$ to $+$

(b) f' decreasing from $+$ to $-$

(c) f' decreasing on (a,c); f' increasing on (c,b)

FIGURE 3.65

Concavity and the Second Derivative From Section 3.4 remember that the algebraic sign of the derivative of a function indicates when the function is increasing or decreasing on an interval. Specifically, if the function referred to in the preceding sentence is the derivative f', then we can conclude that the algebraic sign of the derivative f'' indicates when f' is either increasing or decreasing on an interval. For example, if $f''(x) > 0$ on (a, b), then f' increases on (a, b). In view of Definition 3.7, if f' increases on (a, b), then the graph of f is concave upward on the interval. Therefore, we are led to the following test for concavity.

THEOREM 3.9 Test for Concavity

Let f be a function for which f'' exists on (a, b).

(*i*) If $f''(x) > 0$ for all x in (a, b), then the graph of f is concave upward on (a, b).

(*ii*) If $f''(x) < 0$ for all x in (a, b), then the graph of f is concave downward on (a, b).

EXAMPLE 1 Determine the intervals on which the graph of $f(x) = -x^3 + \frac{9}{2}x^2$ is concave upward and the intervals on which the graph is concave downward.

Solution From

$$f'(x) = -3x^2 + 9x$$
$$f''(x) = -6x + 9 = 6(-x + \tfrac{3}{2})$$

we see that $f''(x) > 0$ when $6(-x + \frac{3}{2}) > 0$ or $x < \frac{3}{2}$ and that $f''(x) < 0$ when $6(-x + \frac{3}{2}) < 0$ or $x > \frac{3}{2}$. It follows from Theorem 3.9 that the graph of f is concave upward on $(-\infty, \frac{3}{2})$ and concave downward on $(\frac{3}{2}, \infty)$. □

Point of Inflection The graph of the function in Example 1 changes concavity at the point that corresponds to $x = \frac{3}{2}$. As x increases through $\frac{3}{2}$, the graph of f changes from concave upward to concave downward *at* the point $(\frac{3}{2}, \frac{27}{4})$. A point on the graph of a function where the concavity changes from upward to downward, or vice versa, is given a special name.

DEFINITION 3.8 Point of Inflection

Let f be continuous at c. A point $(c, f(c))$ is a **point of inflection** if there exists an open interval (a, b) that contains c such that the graph of f is either

(*i*) concave upward on (a, c) and concave downward on (c, b), or

(*ii*) concave downward on (a, c) and concave upward on (c, b).

Figure 3.66 shows a graph of a function $y = f(x)$ that has three points of inflection: $(a, f(a))$, $(b, f(b))$, and $(c, f(c))$. Notice $(d, f(d))$ is not a point of inflection, since the graph is concave upward on both intervals (c, d) and (d, e).

As a consequence of Definitions 3.7 and 3.8, we observe:

A point of inflection $(c, f(c))$ occurs at a number c for which $f''(c) = 0$ or $f''(c)$ does not exist.

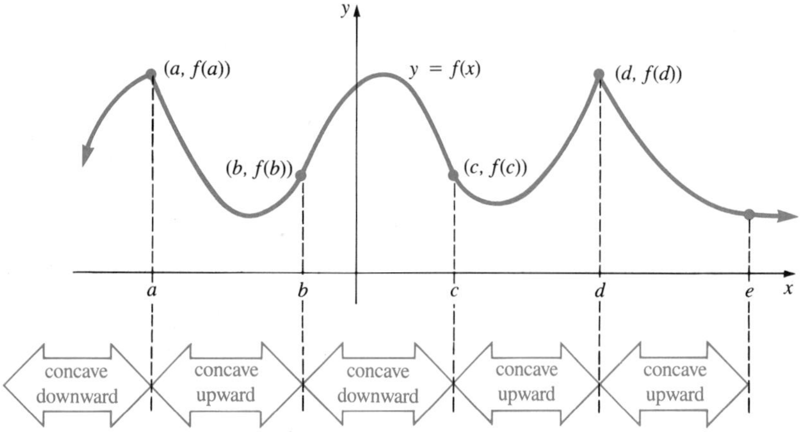

FIGURE 3.66

EXAMPLE 2 Find any points of inflection of $f(x) = -x^3 + x^2$.

Solution The first and second derivatives of f are, respectively,

$$f'(x) = -3x^2 + 2x \quad \text{and} \quad f''(x) = -6x + 2$$

Since $f''(x) = 0$ at $\frac{1}{3}$, the point $(\frac{1}{3}, \frac{2}{27})$ is the only *possible* point of inflection. Now,

$$f''(x) = 6(-x + \tfrac{1}{3}) > 0 \quad \text{for } x < \tfrac{1}{3}$$
$$f''(x) = 6(-x + \tfrac{1}{3}) < 0 \quad \text{for } x > \tfrac{1}{3}$$

implies that the graph of f is concave upward on $(-\infty, \frac{1}{3})$ and concave downward on $(\frac{1}{3}, \infty)$. Thus, $(\frac{1}{3}, f(\frac{1}{3}))$ or $(\frac{1}{3}, \frac{2}{27})$ is a point of inflection. □

EXAMPLE 3 Find any points of inflection of $f(x) = 5x - (x - 4)^{1/3}$.

Solution From $$f'(x) = 5 - \frac{1}{3}(x - 4)^{-2/3} \quad \text{and} \quad f''(x) = \frac{2}{9}(x - 4)^{-5/3}$$

we see that f'' does not exist at 4. Since $f''(x) < 0$ for $x < 4$ and $f''(x) > 0$ for $x > 4$, the graph of f is concave downward on $(-\infty, 4)$ and concave upward on $(4, \infty)$. Now, 4 is in the domain of f and so $(4, f(4))$ or $(4, 20)$ is a point of inflection. □

EXAMPLE 4 An inspection of the graphs of $y = \sin x$, $y = \cos x$, and $y = \tan x$ in Figure 3.67 seems to indicate that the x-intercepts of the graph of each function are the abscissas of points of inflection. You are asked to prove that this is indeed the case. See Problems 19, 20, and 22 in Exercises 3.6.

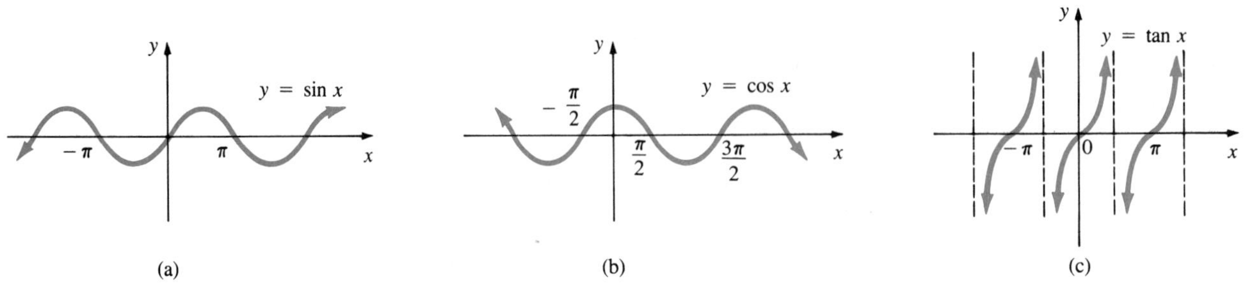

FIGURE 3.67

□

The concavity of a graph can be related to the notion of relative extrema.

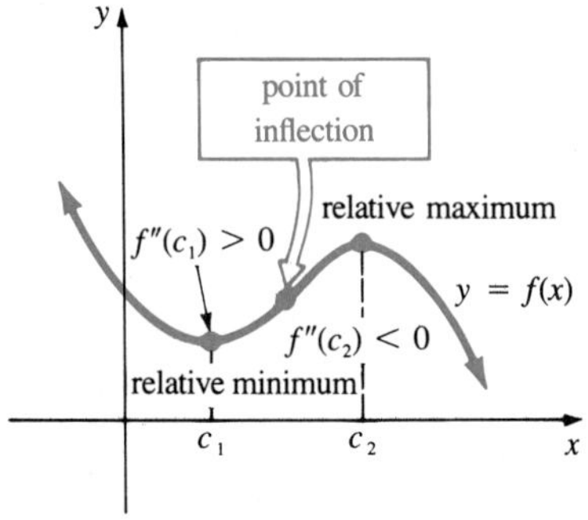

FIGURE 3.68

Second Derivative Test If c is a critical value of $y = f(x)$ and, say, $f''(c) > 0$, then the graph of f is concave upward on some interval (a, b) that contains c. Necessarily then, $f(c)$ is a relative minimum. Similarly, $f''(c) < 0$ at a critical value c implies $f(c)$ is a relative maximum. This so-called **Second Derivative Test** is illustrated in Figure 3.68.

THEOREM 3.10 Second Derivative Test for Relative Extrema

Let f be a function for which f'' exists on an interval (a, b) that contains the critical number c.

(*i*) If $f''(c) > 0$, then $f(c)$ is a relative minimum.
(*ii*) If $f''(c) < 0$, then $f(c)$ is a relative maximum.

At this point one might ask, Why do we need another test for relative extrema when we already have the First Derivative Test? If the function under examination is a polynomial, it is very easy to compute the second derivative. In using Theorem 3.10 we need only determine the algebraic sign of f'' at the critical value. Contrast this with determining the sign of f' at numbers to the right and left of the critical value. If f' is not readily factored, the latter procedure may be somewhat difficult. On the other hand, it may be equally tedious to use Theorem 3.10 in the case of some functions that involve products, quotients, powers, and so on. Thus, Theorems 3.8 and 3.10 both have advantages and disadvantages.

EXAMPLE 5 Graph $f(x) = x^4 - x^2$.

Solution $f'(x) = 4x^3 - 2x = 2x(2x^2 - 1)$ and $f''(x) = 12x^2 - 2$

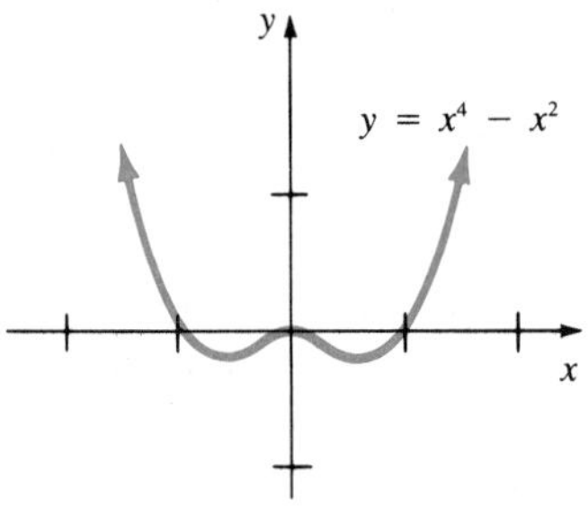

FIGURE 3.69

Hence, the critical points of f are 0, $-\sqrt{2}/2$, and $\sqrt{2}/2$. The Second Derivative Test is summarized in the accompanying table.

x	Sign of $f''(x)$	$f(x)$	Conclusion
0	−	0	rel. max.
$\sqrt{2}/2$	+	$-1/4$	rel. min.
$-\sqrt{2}/2$	+	$-1/4$	rel. min.

Now from $f(x) = x^2(x^2 - 1) = x^2(x + 1)(x - 1)$, we see that the graph of f passes through (0, 0), (−1, 0), and (1, 0). Furthermore, since f is a polynomial with only even powers, we conclude that its graph is symmetric with respect to the y-axis (even function). See Figure 3.69. You should also verify that the graph possesses two points of inflection: $(-\sqrt{6}/6, -\frac{5}{36})$ and $(\sqrt{6}/6, -\frac{5}{36})$. □

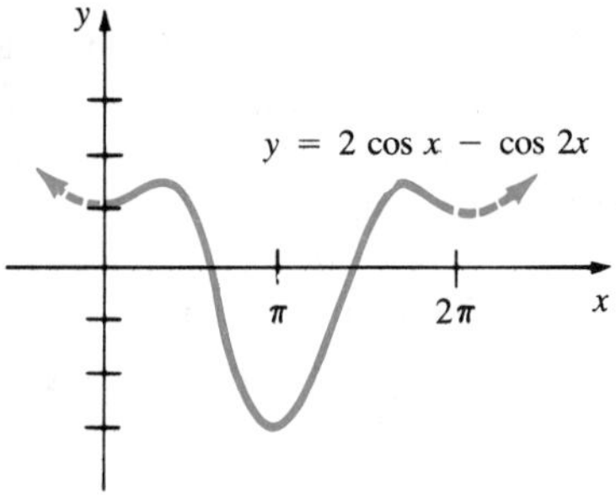

FIGURE 3.70

EXAMPLE 6 Graph $f(x) = 2 \cos x - \cos 2x$.

Solution $f'(x) = -2 \sin x + 2 \sin 2x$ and $f''(x) = -2 \cos x + 4 \cos 2x$

Using the trigonometric identity $\sin 2x = 2 \sin x \cos x$, we can simplify the equation $f'(x) = 0$ to $\sin x(1 - 2 \cos x) = 0$. The solutions of $\sin x = 0$ are $0, \pm\pi, \pm 2\pi, \ldots$ and the solutions of $\cos x = \frac{1}{2}$ are $\pm\pi/3, \pm 5\pi/3, \ldots$. But since f is 2π periodic (show this!), it suffices to consider only those critical values in $[0, 2\pi]$, namely, 0, $\pi/3$, π, $5\pi/3$, and 2π. The Second Derivative Test applied to these values is summarized in the accompanying table. The graph of f is given in Figure 3.70.

x	Sign of $f''(x)$	$f(x)$	Conclusion
0	+	1	rel. min.
$\pi/3$	−	$\frac{3}{2}$	rel. max.
π	+	-3	rel. min.
$5\pi/3$	−	$\frac{3}{2}$	rel. max.
2π	+	1	rel. min.

□

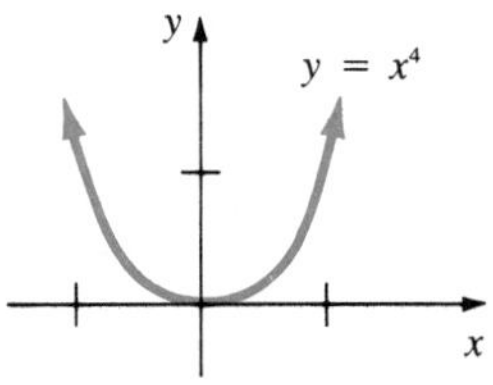

FIGURE 3.71

▲ ***Remarks*** (*i*) One should not get the impression from Theorem 3.9 that when a graph is concave upward (or downward) on an interval (a, b), $f''(x) > 0$ (or $f''(x) < 0$) for *all* x in the interval. The conditions stated in parts (*i*) and (*ii*) of Theorem 3.9 are sufficient but are not necessary. For example, it is clear from Figure 3.71 and Definition 3.7 that the twice-differentiable function $f(x) = x^4$ is concave upward on any interval that contains the origin. But from $f''(x) = 12x^2$ we see $f''(0) = 0$.

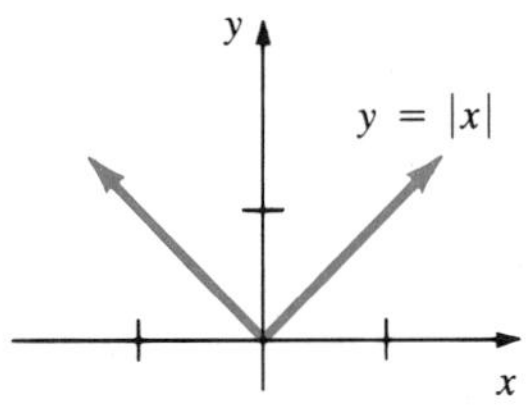

FIGURE 3.72

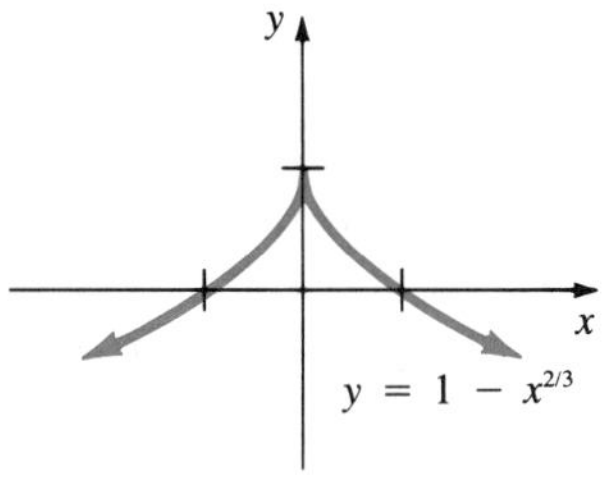

FIGURE 3.73

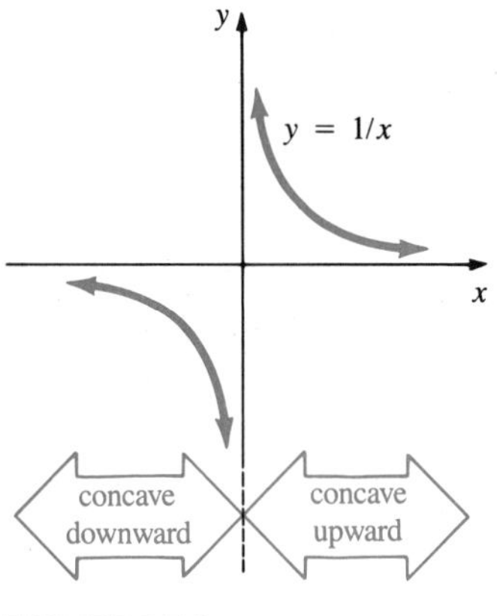

FIGURE 3.74

(*ii*) The intuitive notions of "holding water" and "spilling water" are, of course, not synonymous with the concepts of concave upward and concave downward, respectively. We could argue from Figures 3.72 and 3.73 that the graph of $f(x) = |x|$ "holds water" on $(-1, 1)$ and the graph of $g(x) = 1 - x^{2/3}$ "spills water" on $(-1, 1)$. But we cannot assign any concavity to either graph on $(-1, 1)$, since both f and g are not differentiable on the interval. Furthermore, there are differentiable functions whose graphs possess no concavity. You are asked to supply an example of such a function in Problem 54 of Exercises 3.6. (This is not very difficult.)

(*iii*) Recall, if $(c, f(c))$ is a point of inflection, then $f''(c) = 0$ or $f''(c)$ does not exist. The converse of this statement is not necessarily true. We cannot conclude, simply from the fact that $f''(c) = 0$ or $f''(c)$ does not exist, that $(c, f(c))$ is a point of inflection. For example, the second derivative of the function $f(x) = x^4$ is zero at $x = 0$. We see in Figure 3.71 that $(0, f(0))$ is not a point of inflection, since the graph is concave upward on $(-\infty, 0)$ and on $(0, \infty)$. Also, for $f(x) = 1/x$, we see that $f''(x) = 2/x^3$ is undefined at $x = 0$ and $f''(x) < 0$ for $x < 0$ and $f''(x) > 0$ for $x > 0$. However, 0 is not the x-coordinate of a point of inflection, since f is not continuous at this value. See Figure 3.74.

(*iv*) It is important to note that the Second Derivative Test does *not* state that if $f(c)$ is a relative extremum, then either $f''(c) > 0$ or $f''(c) < 0$. In Figure 3.73 we see that $f(x) = 1 - x^{2/3}$ has an absolute maximum at 0 but $f''(0)$ does not exist. Similarly, $f(x) = x^4$ has an absolute minimum at 0 but $f''(0) = 0$. Thus *the Second Derivative Test can lead to no conclusion.* Whenever the Second Derivative Test fails, the First Derivative Test should be used.

(*v*) In conclusion, we summarize Theorem 3.10 and the preceding remark.

Second Derivative Test

Critical Value	**$f''(x)$ at Critical Value**	**Conclusion**
c	$+$	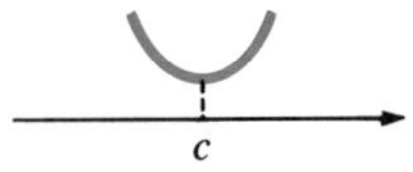 $f(c)$ is a relative minimum
c	$-$	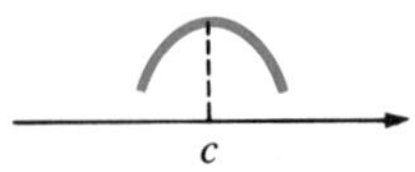 $f(c)$ is a relative maximum
c	0	no conclusion

EXERCISES 3.6 *Answers to odd-numbered problems begin on page A-77.*

In Problems 1–12 use the second derivative to determine the intervals on which the given function is concave upward and the intervals on which it is concave downward.

1. $f(x) = -x^2 + 7x$
2. $f(x) = -(x + 2)^2 + 8$
3. $f(x) = -x^3 + 6x^2 + x - 1$
4. $f(x) = (x + 5)^3$
5. $f(x) = x(x - 4)^3$
6. $f(x) = 6x^4 + 2x^3 - 12x^2 + 3$
7. $f(x) = x^{1/3} + 2x$
8. $f(x) = x^{8/3} - 20x^{2/3}$
9. $f(x) = x + \dfrac{9}{x}$
10. $f(x) = \sqrt{x^2 + 10}$
11. $f(x) = \dfrac{1}{x^2 + 3}$
12. $f(x) = \dfrac{x - 1}{x + 2}$

In Problems 13 and 14 estimate from the graph of the given function f the intervals on which f' is increasing and the intervals on which f' is decreasing.

13.

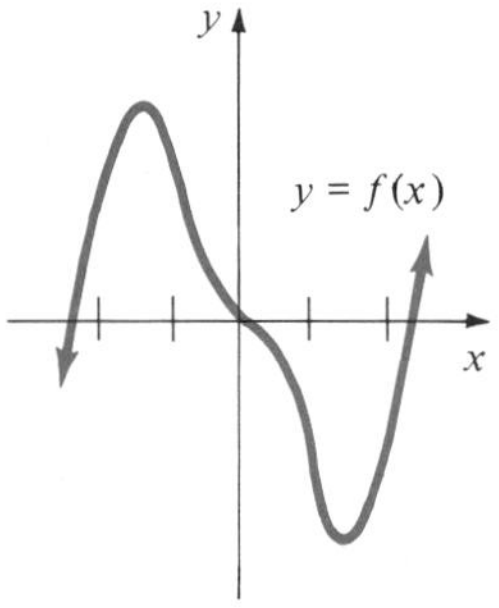

FIGURE 3.75

14.

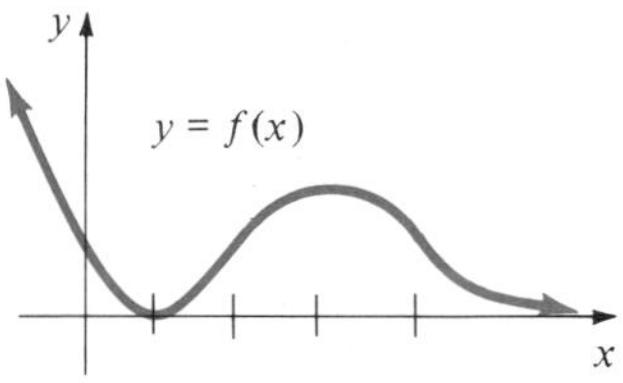

FIGURE 3.76

15. Show that the graph of $f(x) = \sec x$ is concave upward on those intervals on which $\cos x > 0$, and concave downward on those intervals on which $\cos x < 0$.

16. Show that the graph of $f(x) = \csc x$ is concave upward on those intervals on which $\sin x > 0$, and concave downward on those intervals on which $\sin x < 0$.

In Problems 17–22 use the second derivative to locate all points of inflection.

17. $f(x) = x^4 - 12x^2 + x - 1$
18. $f(x) = x^{5/3} + 4x$
19. $f(x) = \sin x$
20. $f(x) = \cos x$
21. $f(x) = x - \sin x$
22. $f(x) = \tan x$

In Problems 23–38 use the Second Derivative Test, when applicable, to find the relative extrema of the given function. Graph. Find points of inflection and intercepts when possible.

23. $f(x) = -(2x - 5)^2$
24. $f(x) = \frac{1}{3}x^3 - 2x^2 - 12x$
25. $f(x) = x^3 + 3x^2 + 3x + 1$
26. $f(x) = \frac{1}{4}x^4 - 2x^2$
27. $f(x) = 6x^5 - 10x^3$
28. $f(x) = x^3(x + 1)^2$
29. $f(x) = \dfrac{x}{x^2 + 2}$
30. $f(x) = x^2 + \dfrac{1}{x^2}$
31. $f(x) = \sqrt{9 - x^2}$
32. $f(x) = x\sqrt{x - 6}$
33. $f(x) = x^{1/3}(x + 1)$
34. $f(x) = x^{1/2} - \frac{1}{4}x$
35. $f(x) = \cos 3x$, $[0, 2\pi]$
36. $f(x) = 2 + \sin 2x$, $[0, 2\pi]$
37. $f(x) = \cos x + \sin x$, $[0, 2\pi]$
38. $f(x) = 2 \sin x + \sin 2x$, $[0, 2\pi]$

In Problems 39–42 determine whether the given function has a relative extremum at the indicated critical value.

39. $f(x) = \sin x \cos x$; $\pi/4$
40. $f(x) = x \sin x$; 0
41. $f(x) = \tan^2 x$; π
42. $f(x) = (1 + \sin 4x)^3$; $\pi/8$

In Problems 43–46 sketch a graph of a function f that has the given properties.

43. $f(-2) = 0$, $f(4) = 0$
$f'(3) = 0$, $f''(1) = 0$, $f''(2) = 0$
$f''(x) < 0$, $x < 1$, $x > 2$
$f''(x) > 0$, $1 < x < 2$

44. $f(0) = 5, f(2) = 0$
$f'(2) = 0, f''(3)$ does not exist
$f''(x) > 0, x < 3$
$f''(x) < 0, x > 3$

45. $f(0) = -1, f(\pi/2) > 0$
$f'(x) \geq 0$ for all x
$f''(x) > 0, (2n-1)\frac{\pi}{2} < x < (2n+1)\frac{\pi}{2}$, n even
$f''(x) < 0, (2n-1)\frac{\pi}{2} < x < (2n+1)\frac{\pi}{2}$, n odd

46. $f(-x) = -f(x)$
vertical asymptote $x = 2$, $\lim_{x\to\infty} f(x) = 0$
$f''(x) < 0, 0 < x < 2$
$f''(x) > 0, x > 2$

47. Find values of a, b, and c such that the graph of $f(x) = ax^3 + bx^2 + cx$ passes through $(-1, 0)$ and has a point of inflection at $(1, 1)$.

48. Find values of a, b, and c such that the graph of $f(x) = ax^3 + bx^2 + cx$ has a horizontal tangent at the point of inflection $(1, 1)$.

49. Use the Second Derivative Test as an aid in graphing $f(x) = \sin(1/x)$. Observe that f is discontinuous at $x = 0$.

50. Show that the graph of a general polynomial function

$$f(x) = a_n x^n + a_{n-1}x^{n-1} + \cdots + a_1 x + a_0, \qquad a_n \neq 0$$

can have at most $n - 2$ points of inflection.

51. Let $f(x) = (x - x_0)^n$ where n is a positive integer.

(a) Show that $(x_0, 0)$ is a point of inflection of the graph of f if n is an odd integer.

(b) Show that $(x_0, 0)$ is not a point of inflection of the graph of f but corresponds to a relative minimum when n is an even integer.

52. Prove that the graph of a quadratic polynomial function $f(x) = ax^2 + bx + c$, $a \neq 0$, is concave upward on the x-axis when $a > 0$ and concave downward on the x-axis when $a < 0$.

53. Let f be a function for which f''' exists on an interval (a, b) that contains the number c. If $f''(c) = 0$ and $f'''(c) \neq 0$, what can be said about $(c, f(c))$?

54. Give an example of a differentiable function whose graph possesses no concavity.

55. Prove or disprove: A point of inflection for a function f must occur at a critical value of f'.

56. Prove or disprove: The function

$$f(x) = \begin{cases} 1/x, & x \neq 0 \\ 0, & x = 0 \end{cases}$$

has a point of inflection at $(0, 0)$.

3.7 FURTHER APPLICATIONS OF EXTREMA

In science, engineering, and business one is often interested in the maximum and minimum values of functions; for example, a company is naturally interested in maximizing revenue while minimizing cost. The next time you go to a supermarket try this experiment: Take along a small ruler and measure the height and diameter of *all* cans that contain, say, 16 ounces of food (28.9 in^3). The fact that all cans of this specified volume have the same shape is no coincidence, since there are specific dimensions that minimize the amount of metal used and, hence, minimize the cost of construction to a company. In the same vein, many of the so-called economy cars have appearances that are remarkably the same. This is not just a simple matter of one company copying the success of another company, but, rather, for a given volume engineers strive for a design that minimizes the amount of material used.

Helpful Hints In the examples and problems that follow either we will be *given* a function or we will have to interpret the words to *set up* a function for which we seek a maximum or a minimum value. These are the kinds of word problems that show off the power of the calculus and provide one of many possible answers to the age-old question: What's it good for? Here are seven important steps in solving an applied max-min problem.

(*i*) Develop a positive and analytical attitude. Read the problem slowly. Do not merely strive for an answer.

(*ii*) Draw a picture when necessary.

(*iii*) Introduce variables and note any relationship among the variables.

(*iv*) Using all necessary variables, set up a function to be maximized or minimized. If more than one variable is used, then employ a relationship between the variables to reduce the function to one variable.

(*v*) Make note of the interval on which the function is defined. Determine all critical points.

(*vi*) If the function to be maximized or minimized is continuous and defined on a closed interval $[a, b]$, then test for endpoint extrema. If the desired extremum does not occur at an endpoint, it must occur at a critical point within the open interval (a, b).

(*vii*) If the function to be maximized or minimized is defined on an interval that is not closed, then a derivative test should be used at each critical point.

In the examples that follow we shall emphasize steps (iv)–(vii) by outlining the solutions in three parts: **function**, **interval**, and **analysis**. Under the label "Analysis" we shall take derivatives, find critical points, evaluate functions at endpoints and critical points, and perform derivative tests.

EXAMPLE 1 In physics it is shown that when air resistance is ignored, the horizontal range R of a projectile is given by

$$R = \frac{v_0^2}{g} \sin 2\theta \tag{1}$$

where v_0 is the constant initial velocity, g is the acceleration of gravity, and θ is the angle of elevation or departure. Find the maximum range of the projectile.

Solution As a physical model of the problem, let us imagine that the projectile is a cannonball. See Figure 3.77.

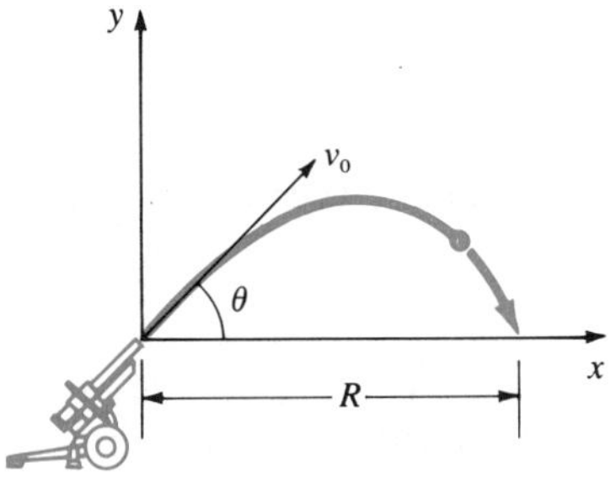

FIGURE 3.77

Function: The function is given in (1).

Interval: For angles greater than $\pi/2$, the cannon shown in Figure 3.77 would shoot backward. Thus, it makes physical sense to restrict the function in (1) to the closed interval $[0, \pi/2]$.

Analysis: From

$$\frac{dR}{d\theta} = \frac{v_0^2}{g} 2 \cos 2\theta$$

we see that $dR/d\theta = 0$ when $\cos 2\theta = 0$ or $2\theta = \pi/2$, and so the only critical point in the open interval $(0, \pi/2)$ is $\pi/4$. Evaluating the function at the endpoints and the critical points gives

$$R(0) = 0 \qquad R\left(\frac{\pi}{4}\right) = \frac{v_0^2}{g} \qquad R\left(\frac{\pi}{2}\right) = 0$$

These values indicate that the minimum range is $R(0) = R(\pi/2) = 0$ and the maximum range is $R(\pi/4) = v_0^2/g$.* In other words, to achieve maximum distance, the projectile should be launched at an angle of 45° to the horizontal. □

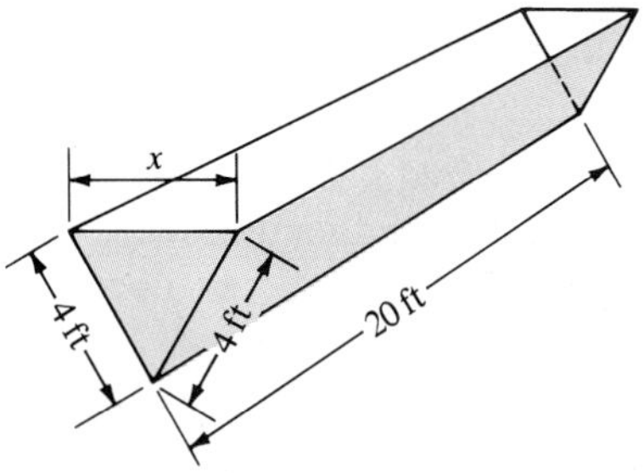

FIGURE 3.78

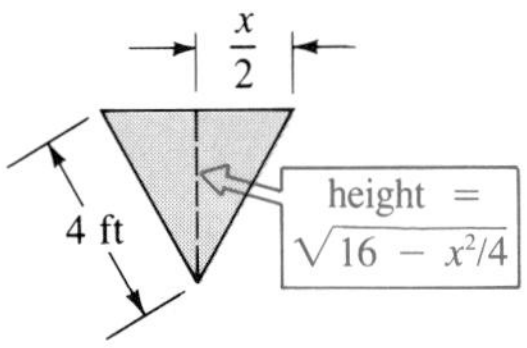

FIGURE 3.79

EXAMPLE 2 A 20-ft-long water trough has ends in the form of isosceles triangles with sides that are 4 ft long. Determine the dimension across the top of the triangular end so that the volume of the trough is a maximum. Find the maximum volume.

Solution The trough with the unknown dimension x is shown in Figure 3.78.

Function: The volume V of the trough is

$$V = (\text{area of triangular end}) \cdot (\text{length})$$

From Figure 3.79 and the Pythagorean Theorem, the area of the triangular end as a function of x is $\frac{1}{2}x\sqrt{16 - x^2/4}$. Consequently, the volume of the trough as a function of x is

$$V(x) = \left(\frac{1}{2}x\sqrt{16 - \frac{x^2}{4}}\right) \cdot 20 = 5x\sqrt{64 - x^2}$$

Interval: The function $V(x)$ makes sense only on the closed interval $[0, 8]$. (Why?)

Analysis: Taking the derivative and simplifying yield

$$V'(x) = -10\frac{x^2 - 32}{\sqrt{64 - x^2}}$$

Although $V'(x) = 0$ for $x = \pm 4\sqrt{2}$, the only critical point in $(0, 8)$ is $4\sqrt{2}$. Since $V(0) = V(8) = 0$, we conclude that the maximum volume occurs when the width across the top of the trough is $4\sqrt{2}$ ft. The maximum volume is $V(4\sqrt{2}) = 160 \text{ ft}^3$. □

Note: Often a problem can be solved in more than one way. In hindsight, you should verify that the solution of Example 2 is slightly "cleaner" if the dimension across the top of the end of the trough is labeled $2x$ rather than x. Indeed, as the next example shows, Example 2 can be solved using an entirely different variable.

*With air resistance, a projectile will fall short of this value. Remember to allow for this the next time you fire a cannon. Also, take a minute and compare the maximum distances a golf ball can be hit on Earth when $v_0 = 160$ ft/s and $g = 32 \text{ ft/s}^2$ and on the moon where $g = 5.4 \text{ ft/s}^2$.

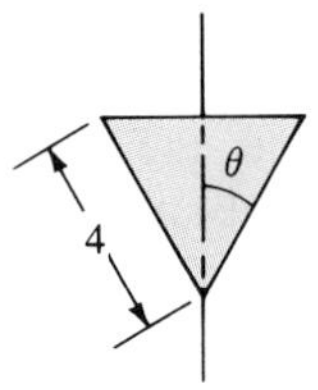

FIGURE 3.80

EXAMPLE 3 *Alternative Solution to Example 2* As shown in Figure 3.80, we let θ denote the angle between the vertical and one of the sides. From trigonometry the height and base of the triangular end are $4 \cos \theta$ and $8 \sin \theta$, respectively. Expressed as a function of θ, V becomes

$$\begin{aligned} V(\theta) &= \frac{1}{2}(4 \cos \theta)(8 \sin \theta) \cdot 20 \\ &= 320 \sin \theta \cos \theta \\ &= 160(2 \sin \theta \cos \theta) \\ &= 160 \sin 2\theta \qquad \boxed{\text{Double Angle Formula}} \end{aligned}$$

where $0 \le \theta \le \pi/2$. Proceeding as in Example 1, we find that the maximum value $V = 160 \text{ ft}^3$ occurs at $\theta = \pi/4$. The dimension across the top of the trough, or the base of the isosceles triangle, is $8 \sin(\pi/4) = 4\sqrt{2}$ ft. □

Problems with Constraints It is often more convenient to set up a function in terms of two variables instead of one. In this case we need to find some kind of subsidiary equation, or **constraint**, involving these variables that can be used to eliminate one of the variables from the function under consideration. The next three examples illustrate this concept.

EXAMPLE 4 Find two nonnegative numbers whose sum is 15 such that the product of one and the square of the other is a maximum.

Solution Suppose x and y denote the two numbers; that is, $x \ge 0$ and $y \ge 0$.

Function: If we let P denote the product, then

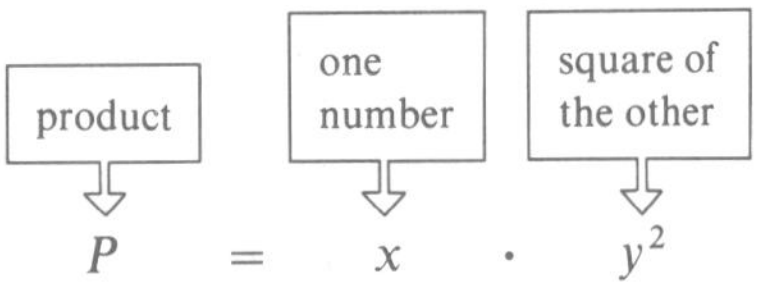

The information that the sum of the two numbers is 15 leads to the additional equation $x + y = 15$. This is the constraint for the problem. We use this last equation in the form $y = 15 - x$ to express P in terms of x alone:

$$P(x) = x(15 - x)^2$$

Interval: The function $P(x)$ is defined only on the closed interval $[0, 15]$, since, if $x > 15$, then, contrary to the stated conditions, $y = 15 - x$ would be negative.

Analysis: By the Product Rule and Power Rule for Functions,

$$P'(x) = x \cdot 2(15 - x)(-1) + (15 - x)^2 = (15 - x)(15 - 3x)$$

Inspection of this derivative reveals that 5 is the only critical point in the open interval (0, 15). Testing the endpoints of the interval shows that $P(0) = P(15) = 0$ is obviously the minimum product. Hence, $P(5) = 5(10)^2 = 500$ must be the maximum value. The two nonnegative numbers are 5 and 10. □

In Example 4 we could just as well express P as a function of y. Substituting $x = 15 - y$ in $P = xy^2$, we then get

$$P(y) = (15 - y)y^2 = 15y^2 - y^3$$

So what? you ask. It is always nice to make life just a little simpler. The derivative P' can now be found without the use of the Product and Power Rules.

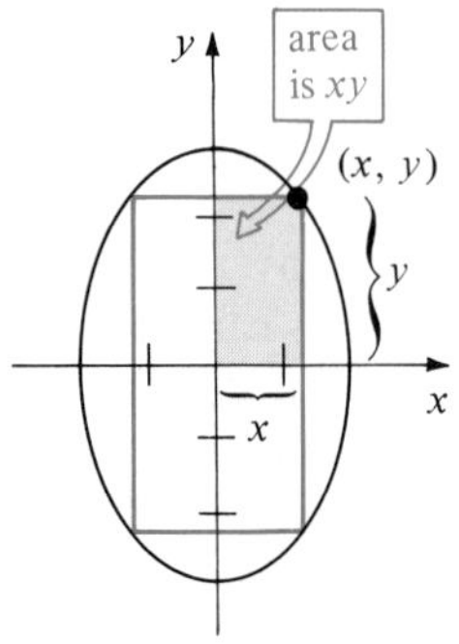

FIGURE 3.81

EXAMPLE 5 Find the dimensions of a rectangle with greatest area that can be inscribed in the ellipse $x^2/4 + y^2/9 = 1$.

Solution Let the corner of the rectangle indicated by the black dot in Figure 3.81 have coordinates (x, y). The symbols x and y will represent dimensions of the shaded rectangle in the figure and so must satisfy $x \geq 0$ and $y \geq 0$.

Function: It is apparent from symmetry that the area of the inscribed rectangle is 4 times the area of the shaded rectangle:

$$A = 4xy$$

The constraint for this problem is the fact that x and y must satisfy the equation of the ellipse. Solving $x^2/4 + y^2/9 = 1$ for y yields $y = \frac{3}{2}\sqrt{4 - x^2}$. Hence, the area as a function of x is given by

$$A(x) = 4x(\tfrac{3}{2}\sqrt{4 - x^2}) = 6x\sqrt{4 - x^2}$$

Interval: Considered out of the context of this problem, the domain of the function $A(x)$ is $[-2, 2]$. Because x is assumed to be nonnegative, $A(x)$ is actually defined on $[0, 2]$.

Analysis: Since the function $A(x)$ is very similar to $V(x)$ in Example 2 we leave it to you to show that $\sqrt{2}$ is the only critical point and that $A(\sqrt{2}) = 12$ square units is the maximum area. Substituting $x = \sqrt{2}$ in the constraint gives $y = 3\sqrt{2}/2$. The dimensions of the inscribed rectangle are then $2\sqrt{2} \times 3\sqrt{2}$. □

EXAMPLE 6 Find the point in the first quadrant on the circle $x^2 + y^2 = 1$ that is closest to (2, 4).

Solution Let (x, y), $x > 0$, $y > 0$, denote the point on the circle closest to the point (2, 4). See Figure 3.82.

Function: As shown in the figure, the distance d between (x, y) and (2, 4) is

$$d = \sqrt{(x - 2)^2 + (y - 4)^2} \quad \text{or} \quad d^2 = (x - 2)^2 + (y - 4)^2$$

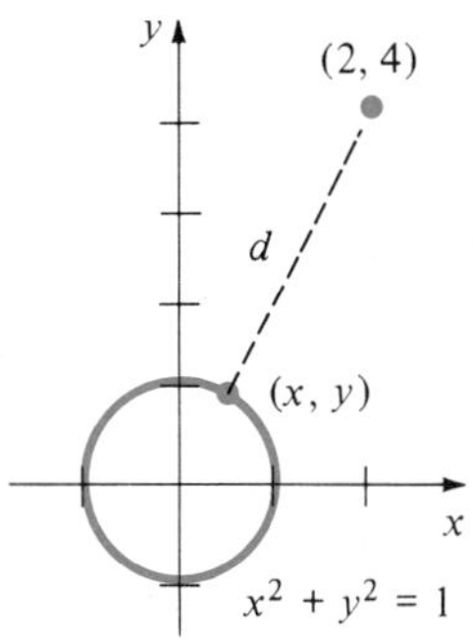

FIGURE 3.82

Now the point that minimizes the square of the distance d^2 also minimizes the distance d. Let us write $D = d^2$. By expanding $(x - 2)^2$ and $(y - 4)^2$ and using the constraint $x^2 + y^2 = 1$ in the form $y = \sqrt{1 - x^2}$, we find

$$D(x) = x^2 - 4x + 4 + \overbrace{(1 - x^2)}^{\boxed{y^2}} - 8\overbrace{\sqrt{1 - x^2}}^{\boxed{y}} + 16$$
$$= -4x - 8\sqrt{1 - x^2} + 21$$

Interval: Because we have assumed x and y to be positive, the domain of the foregoing function is the open interval $(0, 1)$. However, the solution of the problem will not be affected in any way by assuming that the domain is the closed interval $[0, 1]$.

Analysis: Differentiation gives

$$D'(x) = -4 - 4(1 - x^2)^{-1/2}(-2x) = \frac{-4\sqrt{1 - x^2} + 8x}{\sqrt{1 - x^2}}$$

Now $D'(x) = 0$ only if $-4\sqrt{1 - x^2} + 8x = 0$ or $2x = \sqrt{1 - x^2}$. After squaring both sides and simplifying, we find that $\sqrt{5}/5$ is a critical point in the interval $(0, 1)$. From the functional values $D(0) = 13$, $D(\sqrt{5}/5) = 21 - 4\sqrt{5} \approx 12.06$, and $D(1) = 17$, we conclude that D and, hence, the distance d are minimum when $x = \sqrt{5}/5$. Using the constraint, we find correspondingly that $y = 2\sqrt{5}/5$. This means $(\sqrt{5}/5, 2\sqrt{5}/5)$ is the point on the circle closest to $(2, 4)$. □

EXAMPLE 7 A rancher intends to mark off a rectangular plot of land that will have an area of 1500 m². The plot will be fenced and divided into two equal portions by an additional fence parallel to two sides. Find the dimensions of the land that require the least amount of fencing.

Solution As shown in Figure 3.83, we let x and y denote the dimensions of the fenced-in land.

Function: The function we wish to minimize is the sum of the lengths of the five portions of fence. If we denote this sum by the symbol L, we have

$$L = 2x + 3y$$

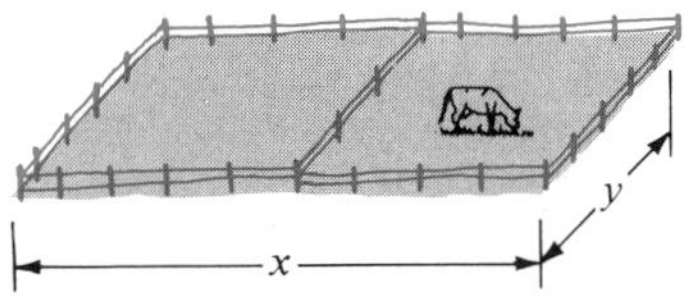

FIGURE 3.83

But the fenced-in land is to have an area of 1500 m², and so x and y must be related by $xy = 1500$ or $y = 1500/x$. With this last equation we can eliminate y and write L as a function of x:

$$L(x) = 2x + \frac{4500}{x}$$

Interval: Since x represents a physical dimension that satisfies $xy = 1500$, we conclude that it is positive.* But other than that, there is no restriction on x. Thus, unlike the prior examples, the function we are considering is not defined on a closed interval; $L(x)$ is defined on the infinite interval $(0, \infty)$.

Analysis: Setting the derivative

$$L'(x) = 2 - \frac{4500}{x^2}$$

equal to zero and solving for x, we find that the only critical point is $15\sqrt{10}$. Since the second derivative is easy to compute, we shall use the Second Derivative Test. From

$$L''(x) = \frac{9000}{x^3}$$

we observe that $L''(15\sqrt{10}) > 0$. It follows from Theorem 3.10 that $L(15\sqrt{10}) = 2(15\sqrt{10}) + 4500/15\sqrt{10} = 60\sqrt{10}$ m is the required minimum amount of fencing. Returning to the relationship $xy = 1500$, we find the corresponding value of y is $10\sqrt{10}$. Therefore, the dimensions of the land should be $15\sqrt{10}$ m × $10\sqrt{10}$ m. □

▲ ***Remark*** As an observant reader, you may question at least two aspects of Example 7. Where did the assumption that the land be divided into two equal portions enter into the solution? In point of fact, it did not. What is important is that the dividing fence be parallel to the two ends. Ask yourself what $L(x)$ would be if this were *not* the case. However, the actual positioning of the dividing fence between the ends is irrelevant as long as it is parallel to them.

In an applied problem we are naturally interested in only absolute extrema. Therefore, another question might be: Since the function L is not defined on a closed interval and since the Second Derivative Test does not guarantee absolute extrema, how can we be certain that $L(15\sqrt{10})$ is an absolute minimum? Sometimes we can argue the existence of an absolute extremum from the physical context of the problem, but perhaps a better procedure is, when in doubt, draw a graph. Figure 3.84 answers the question for $L(x)$.

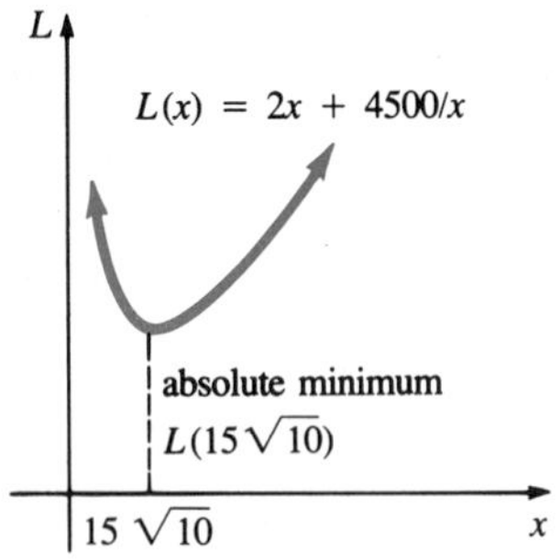

FIGURE 3.84

You should also be able to convince yourself of the validity of the following result:

Suppose a differentiable function f has only one critical point c in an open interval (a, b).† *If a derivative test indicates that $f(c)$ is a relative maximum or a relative minimum, then $f(c)$ is an absolute maximum or an absolute minimum.*

*In other words, we rule out the possibility that $x = 0$. This is reinforced by the fact that $L(x)$ is not defined at $x = 0$.

†This includes intervals such as (a, ∞), $(-\infty, b)$, and $(-\infty, \infty)$.

EXERCISES 3.7 *Answers to odd-numbered problems begin on page A-78.*

In the following problems, a solution may require a special formula with which you are not familiar. If necessary, consult the list of formulas given in Appendix V.

1. Find two nonnegative numbers whose sum is 60 and whose product is a maximum.

2. Find two nonnegative numbers whose product is 50 and whose sum is a minimum.

3. Find a number that exceeds its square by the greatest amount.

4. Let m and n be positive integers. Find two nonnegative numbers whose sum is S such that the product of the mth power of one and the nth power of the other is a maximum.

5. Find two nonnegative numbers whose sum is 1 such that the sum of the square of one and twice the square of the other is a minimum.

6. Consider the graphs of $y = x^2 - 1$ and $y = 1 - x$ given in Figure 3.85. Find the maximum vertical distance between the graphs on the interval $-2 \le x \le 1$.

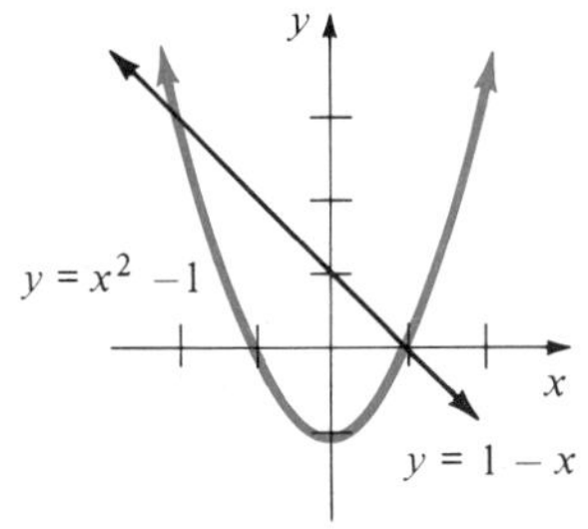

FIGURE 3.85

7. Find the point(s) on the graph of $y^2 = 6x$ closest to

(a) (5, 0) (b) (3, 0)

8. Find the point on the graph of $x + y = 1$ closest to (2, 3).

9. Determine the point on the graph of $y = x^3 - 4x^2$ at which the tangent line has minimum slope.

10. Determine the point on the graph of $y = 8x^2 + 1/x$ at which the tangent line has maximum slope.

In Problems 11–14 find the dimensions of the shaded region such that its area is a maximum.

11.

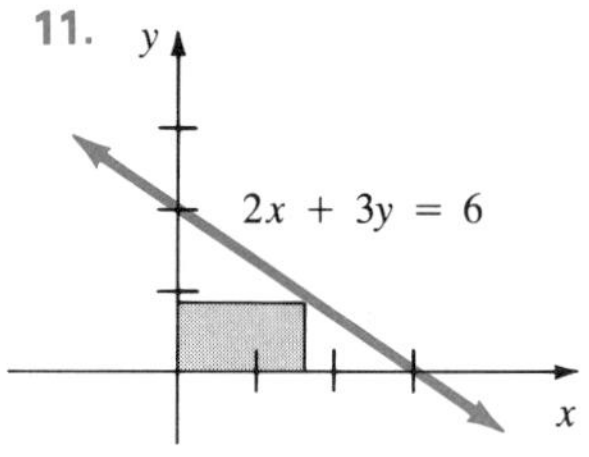

rectangle

FIGURE 3.86

12.

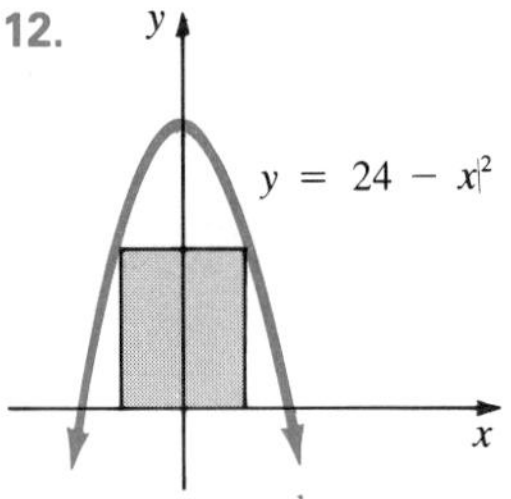

rectangle

FIGURE 3.87

13.

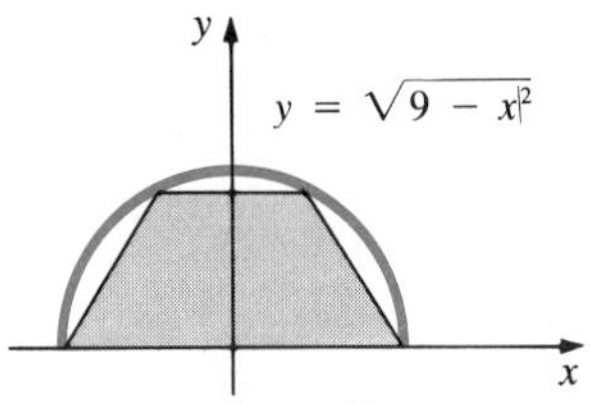

trapezoid

FIGURE 3.88

14.

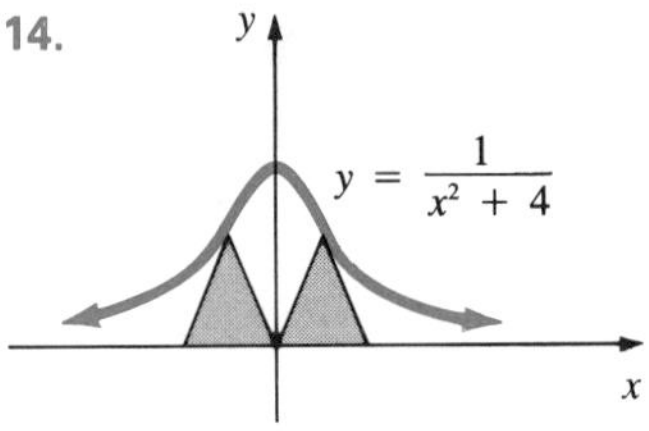

two isosceles triangles with a common vertex at the origin

FIGURE 3.89

15. A rancher has 3000 ft of fencing on hand. Determine the dimensions of a rectangular corral that encloses a maximum area.

16. A rectangular plot of land will be fenced into three equal portions by two dividing fences parallel to two sides. See Figure 3.90. If the area to be enclosed is 4000 m^2, find the dimensions of the land that require the least amount of fence.

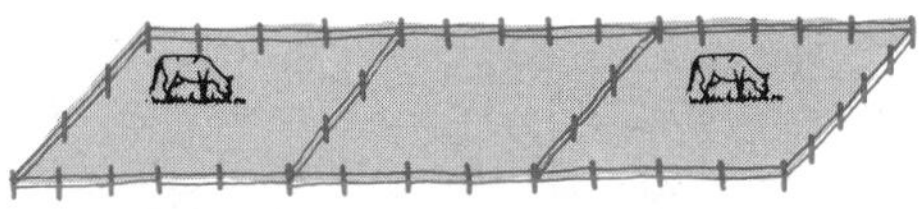

FIGURE 3.90

17. If the total fence to be used is 8000 m, find the dimensions of the enclosed land in Figure 3.90 that has the greatest area.

18. A rancher wishes to build a rectangular corral of 128,000 ft^2 with one side along a vertical cliff. The fencing along the cliff costs \$1.50 per foot, whereas along the other three sides

the fencing costs \$2.50 per foot. Find the dimensions of the corral so that the cost of construction is a minimum.

19. An open rectangular box is to be constructed with a square base and a volume of 32,000 cm^3. Find the dimensions of the box that require the least amount of material.

20. In Problem 19 find the dimensions of a closed box that require the least amount of material.

21. A box, open at the top, is to be made from a square piece of cardboard by cutting a square out of each corner and turning up the sides. Given that the cardboard measures 40 cm on a side, find the dimensions of the box that will give the maximum volume. What is the maximum volume? See Figure 3.91.

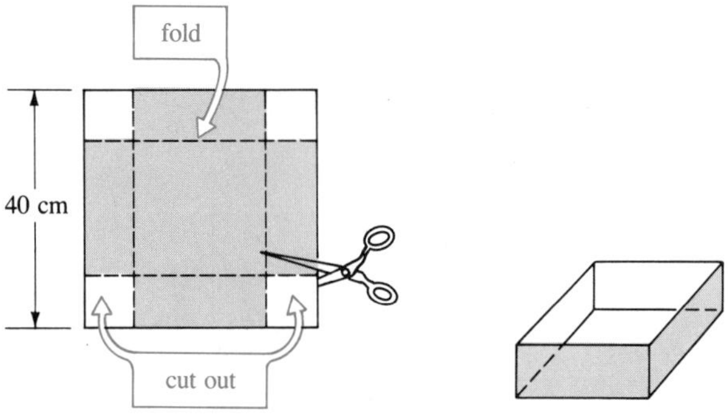

(a) (b)

FIGURE 3.91

22. A box, open at the top, is to be made from a rectangular piece of cardboard that is 30 in long and 20 in wide. The box can hold itself together by cutting a square out of each corner, cutting on the interior solid lines, and then folding the cardboard on the dashed lines. See Figure 3.92(a) and (b). Express the volume of the box as a function of the indicated variable x. Find the dimensions of the box that give the maximum volume. What is the maximum volume?

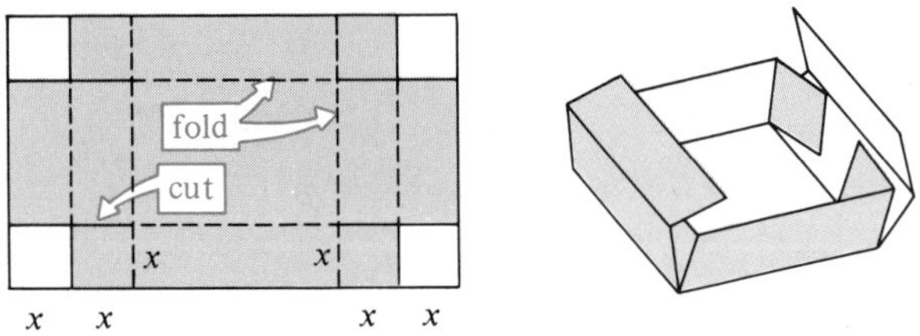

(a) (b)

FIGURE 3.92

23. A rectangular sheet of metal with perimeter 4 m will be rolled and formed into the lateral side of a cylindrical container. Find the dimensions of the container with the largest volume.

24. A printed page will have 2-in margins of white space on the sides and 1-in margins of white space on the top and bottom. The area of the printed portion is 32 in^2. Determine the dimensions of the page so that the least amount of paper is used.

25. Find the dimensions of the right circular cylinder with greatest volume that can be inscribed in a right circular cone of radius R and height H. See Figure 3.93.

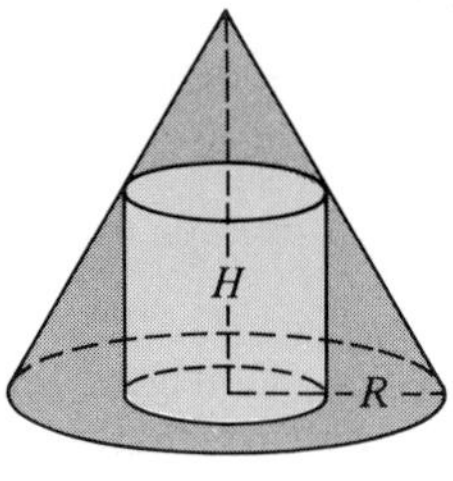

FIGURE 3.93

26. Find the dimensions of the right circular cone with the greatest volume that can be inscribed in a sphere of radius R.

27. Find the dimensions of the rectangle of greatest area that can be circumscribed about a rectangle of length a and width b. See Figure 3.94.

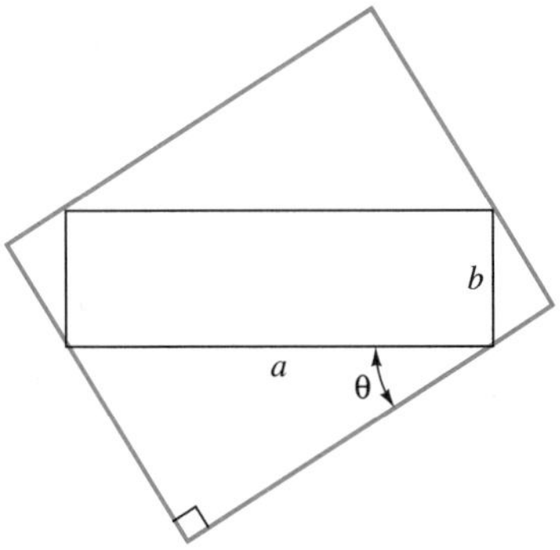

FIGURE 3.94

28. (*This problem could present a challenge.*) Show that the isosceles triangle circumscribed about the ellipse shown in Figure 3.95 has the minimum area when its altitude is $3b$.

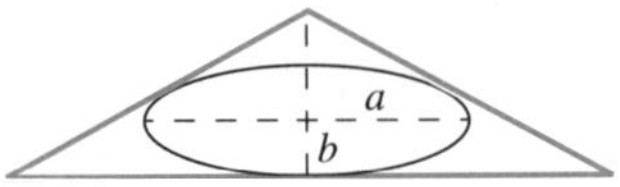

FIGURE 3.95

29. A gutter with a rectangular cross-section is made by bending up equal amounts from the ends of a 30-cm-wide piece of tin. What are the dimensions of the cross-section so that the volume is a maximum?*

30. A gutter will be made so that its cross-section is an isosceles trapezoid with dimensions as indicated in Figure 3.96. Determine the value of θ so that the volume is a maximum.

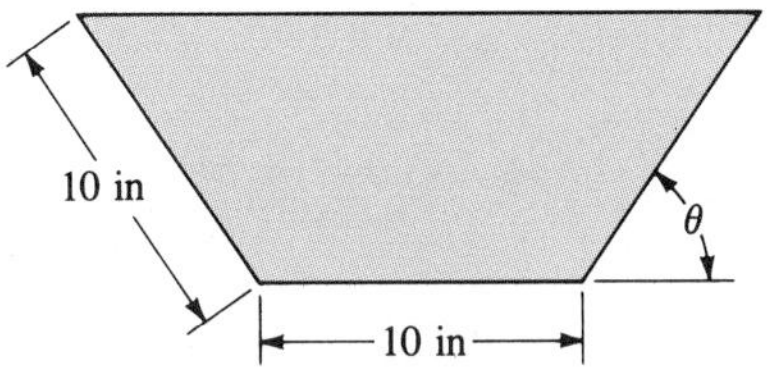

FIGURE 3.96

31. Find the dimensions of a cylindrical juice can that utilize the least amount of material when the volume of the can is 32 in^3.

32. A plastic drinking cup in the shape of a right circular cone will have a volume of 24π cm^3. Find the dimensions that will minimize the amount of material used. [*Hint:* The lateral surface area of a cone is $A = \pi rL$, where r is its radius and L is its slant height.]

33. A conical cup is made from a circular piece of paper of radius R by cutting out a circular sector and then joining the dashed edges shown in Figure 3.97(a).

(a) Determine the value of r indicated in Figure 3.97(b) so that the volume of the cup is a maximum.

(b) What is the maximum volume of the cup?

(c) Find the central angle θ of the circular sector so that the volume of the conical cup is a maximum.

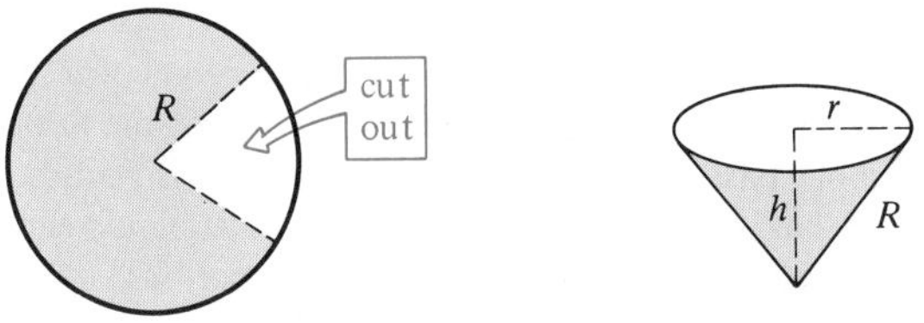

(a) (b)

FIGURE 3.97

34. The container shown in Figure 3.98 is to be constructed by attaching an inverted cone (open at its top) to the bottom

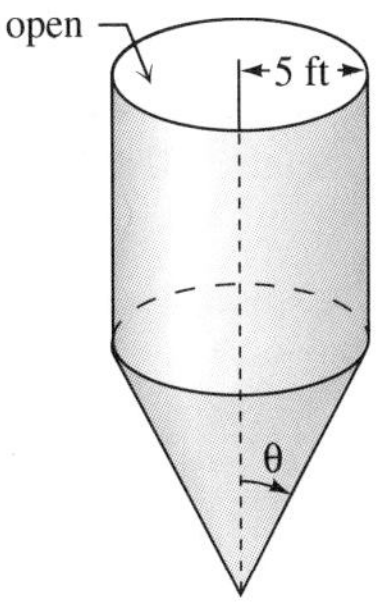

FIGURE 3.98

of a right circular cylinder (open at its top and bottom) of radius 5 ft. The container is to have a volume of 100 ft^3. Find the value of the indicated angle so that the total surface area of the container is a minimum. What is the minimum surface area?

35. A rectangle is to be cut from the circular sector as shown in Figure 3.99. The central angle is $\pi/3$. Find the value of the indicated angle so that the area of the rectangle is a maximum. What is the maximum area?

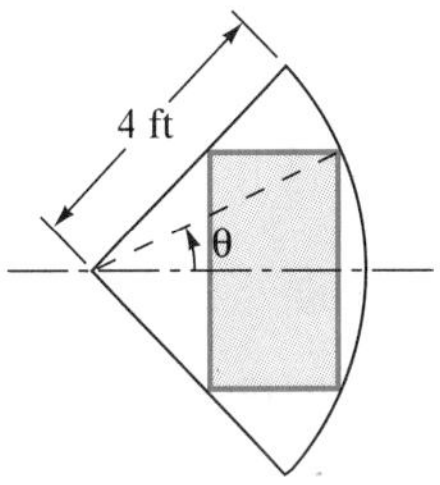

FIGURE 3.99

36. The frame of a kite consists of six pieces of lightweight plastic. As shown in Figure 3.100, the outer frame of the kite consists of four pre-cut pieces, two pieces of length 2 ft and two pieces of length 3 ft. The two remaining crossbar pieces, shown in color in the figure, are to be cut to lengths so that the kite is as large as possible. Find these lengths.

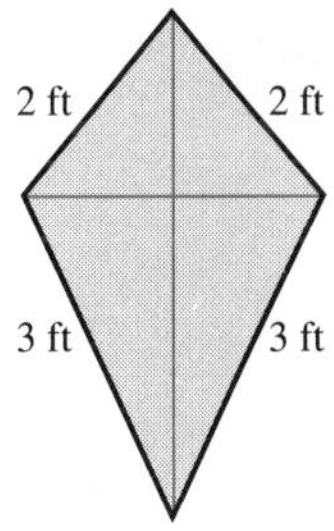

FIGURE 3.100

*Reread Example 2 and ponder why the length of the piece of metal does not have to be specified.

37. A Norman window consists of a rectangle surmounted by a semicircle. Find the dimensions of the window with largest area if its perimeter is 10 m. See Figure 3.101.

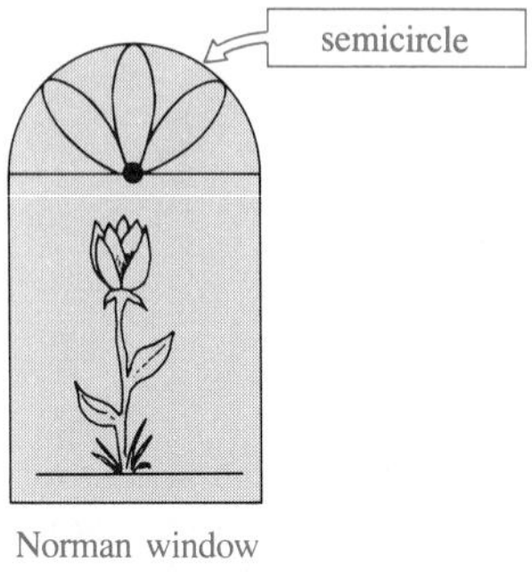

FIGURE 3.101

38. Rework Problem 37 given that the rectangle is surmounted by an equilateral triangle.

39. A person would like to cut a 1-m-long piece of wire into two pieces. One piece will be bent into the shape of a circle and the other into the shape of a square. How should the wire be cut so that the sum of the areas is a maximum?

40. In Problem 39, suppose one piece of wire will be bent into the shape of a circle and the other into the shape of an equilateral triangle. How should the wire be cut so that the sum of the areas is a minimum? a maximum?

41. A cross-section of a rectangular wooden beam cut from a circular log of diameter d has width x and depth y. See Figure 3.102. The strength of the beam varies directly as the product of the width and the square of the depth. Find the dimensions of the cross-section of the beam of greatest strength.

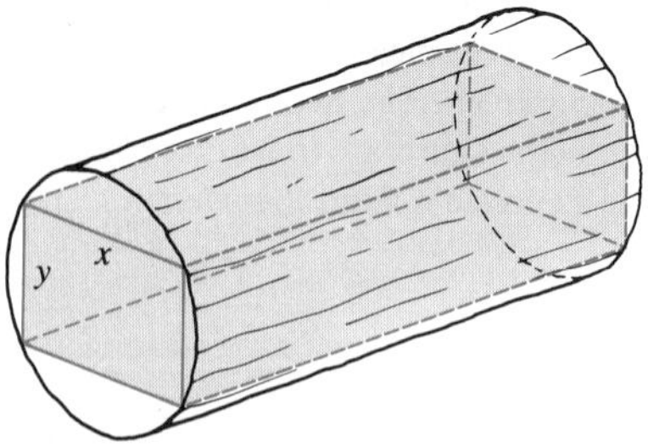

FIGURE 3.102

42. U.S. Postal Service regulations state that a rectangular box sent by fourth-class mail must satisfy the requirement that the length plus the perimeter of one end must not exceed 100 in. Given that a box is to be constructed so that its height is one-half its width, find the dimensions of the box that has a maximum volume.

43. A metal container for transporting nuclear waste consists of a right circular cylinder with hemispherical ends. See Figure 3.103. The container is to have a volume of 30π ft^3. The cost of the metal per square foot for the ends is one and a half times the cost per square foot of the metal used in the cylindrical part. Find the dimensions of the container so that its cost of construction is a minimum.

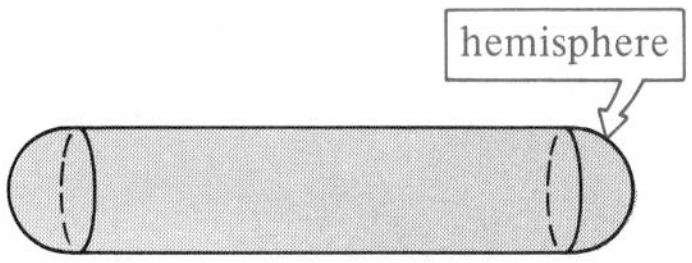

FIGURE 3.103

44. A 10-ft wall stands 5 ft away from a building, as shown in Figure 3.104. Find the length of the shortest ladder, supported by the wall, that reaches from the ground to the building.

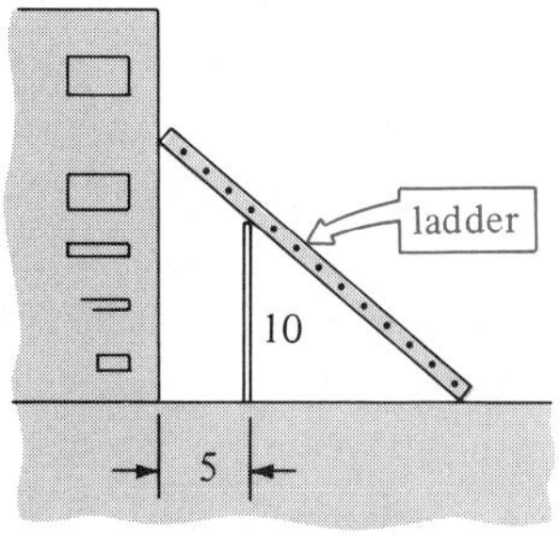

FIGURE 3.104

45. If the number of passengers on a bus tour of a city is exactly 30, the bus company charges \$20 per person. For each additional passenger, the charge per person is reduced by \$0.50. What is the number of passengers a bus should carry in order to maximize the company's revenue per bus?

46. If the U-Drive Truck Rental Company rents 50 trucks per day, it makes a profit of \$84 per truck. Because of increased maintenance and employee costs, for each additional truck rented per day the profit per truck is reduced by \$1.00. Determine the number of trucks that should be rented so that the profit is a maximum. What is the maximum profit?

47. The potential energy between two atoms in a diatomic molecule is given by $U(x) = 2/x^{12} - 1/x^6$. Find the minimum potential energy between the two atoms.

48. The height of a projectile launched with a constant initial

velocity v_0 at an angle of elevation θ_0 is given by

$$y = (\tan\theta_0)x - (g/2v_0^2\cos^2\theta_0)x^2$$

where x is its horizontal displacement measured from the point of launch. Show that the maximum height attained by the projectile is $h = (v_0^2/2g)\sin^2\theta_0$.

49. When a hole is punched into the lateral side of a cylindrical tank full of water, the resulting stream hits the ground at a distance x ft from the base where $x = 2\sqrt{y(h-y)}$. See Figure 3.105. At what point should the hole be punched in the side so that the stream attains a maximum distance from the base? What is the maximum distance?

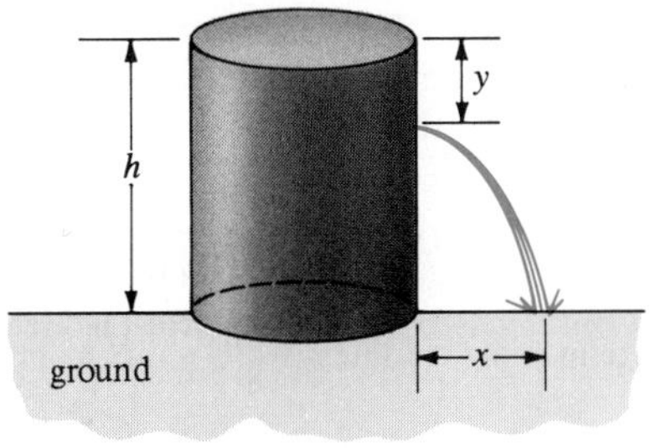

FIGURE 3.105

50. A company's total yearly cost for resupplying depleted inventories of a single item is sometimes given by $C(x) = (a/2)x + (b + cx)q/x$, where a, b, and c are positive constants and x represents the size of the reorder lot. Determine the minimum yearly cost.

51. The illuminance E due to a light source of intensity I at a distance r from the source is given by $E = I/r^2$. The total illuminance from two light bulbs of intensities $I_1 = 125$ and $I_2 = 216$ is the sum of the illuminances. Find the point P between the two light bulbs 10 m apart at which the total illuminance is a minimum. See Figure 3.106.

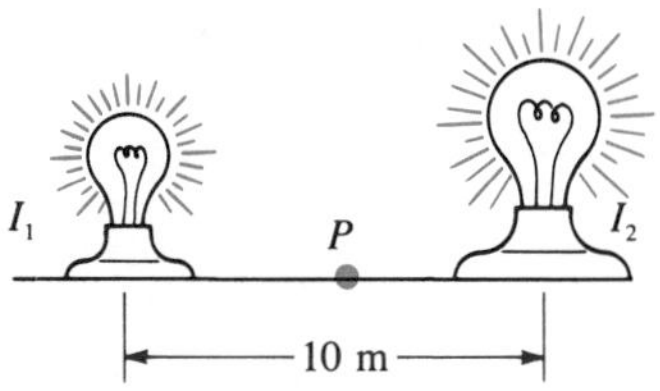

FIGURE 3.106

52. The illuminance E at any point P on the edge of a circular table caused by a light placed directly above its center is given by $E = (I\cos\theta)/r^2$. See Figure 3.107. Given that the radius of the table is 1 m and $I = 100$, find the height at which the light should be placed so that E is a maximum.

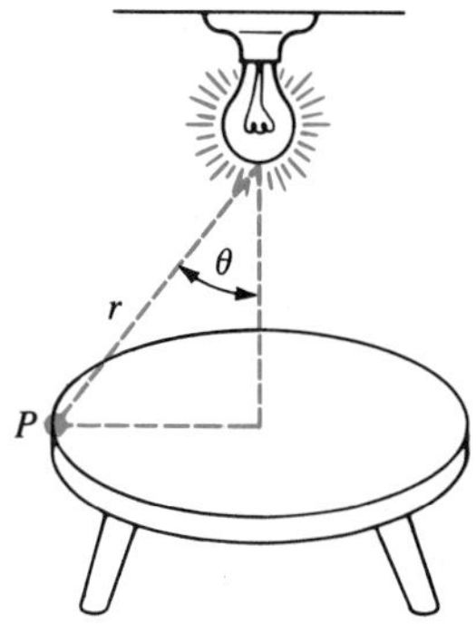

FIGURE 3.107

53. Determine the maximum length of a thin board that can be carried horizontally around the right-angle corner shown in Figure 3.108. [*Hint:* Use similar triangles.]

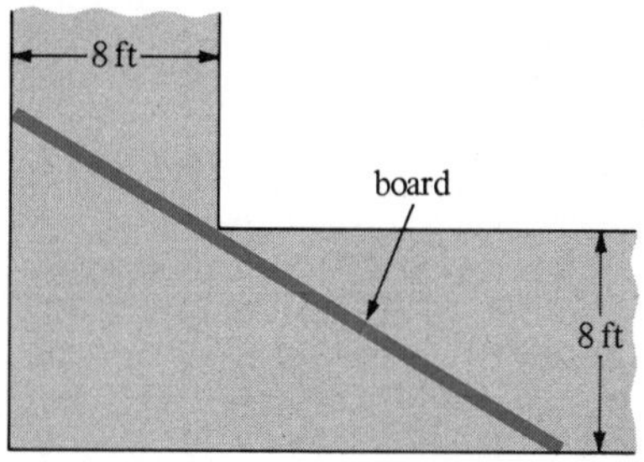

FIGURE 3.108

54. At midnight ship A is 50 km north of ship B. Ship A is sailing south at 20 km/h and ship B is sailing west at 10 km/h. At what time will the distance between the ships be a minimum? [*Hint:* Use distance = rate × time.]

55. A pipeline is to be constructed from a refinery across a swamp to storage tanks. See Figure 3.109. The cost of construction is $25,000 per mile over the swamp and $20,000

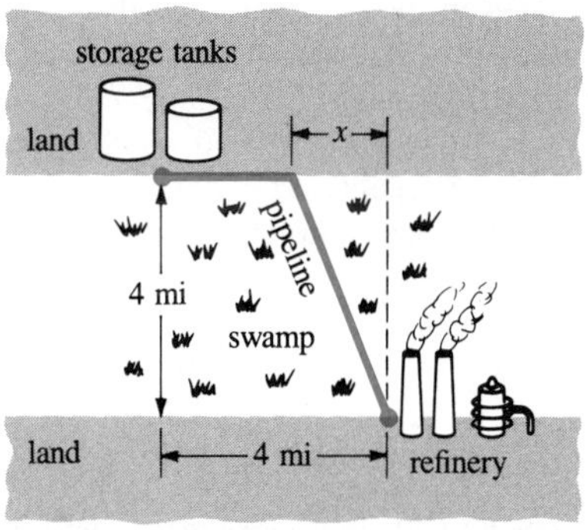

FIGURE 3.109

per mile over land. How should the pipeline be made so that the cost of construction is a minimum?

56. Rework Problem 55 given that the cost per mile across the swamp is twice the cost per mile over land.

57. **Fermat's principle** in optics states that light travels from point A (in the xy-plane) in one medium to point B in another medium on a path that requires minimum time. Denote the speed of light in the medium that contains point A by c_1 and the speed of light in the medium that contains point B by c_2. Show that the time of travel from A to B is a minimum when the angles θ_1 and θ_2, shown in Figure 3.110, satisfy **Snell's law**:

$$\frac{\sin\theta_1}{c_1} = \frac{\sin\theta_2}{c_2}$$

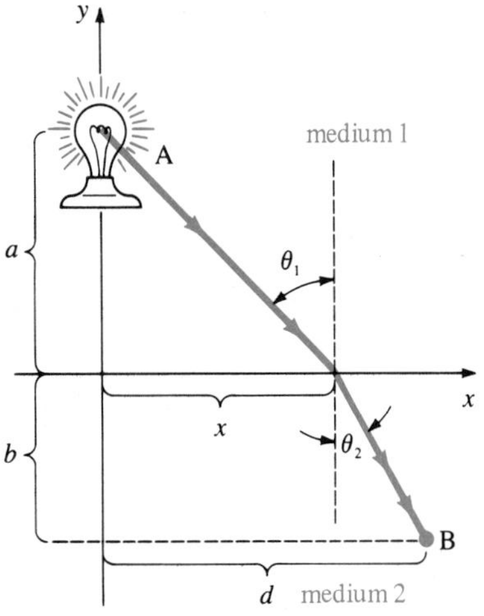

FIGURE 3.110

58. Blood is carried throughout the body by the vascular system, which consists of capillaries, veins, arterioles, and arteries. One consideration of the problem of minimizing the energy expended in moving the blood through the various organs is to find an optimum angle θ for *vascular branching* such that the total resistance to the blood along a path from a larger blood vessel to a smaller blood vessel is a minimum. See Figure 3.111. Use Poiseuille's law, which is that the resistance R of a blood vessel of length l and radius r is $R = kl/r^4$ (see Problem 34 in Exercises 2.10), where k is a constant, to show that the total resistance

$$R = k\left(\frac{x}{r_1^4}\right) + k\left(\frac{y}{r_2^4}\right)$$

along the path $P_1P_2P_3$ is a minimum when $\cos\theta = r_2^4/r_1^4$. [*Hint:* Express x and y in terms of θ and a.]

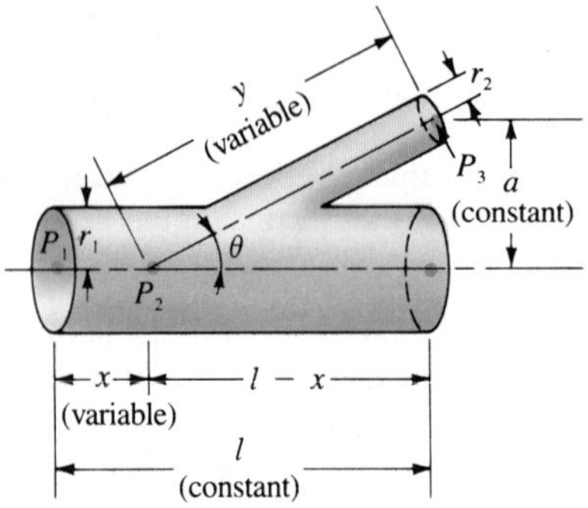

FIGURE 3.111

59. Several rules of thumb are used by physicians to calibrate a child's dose D_c of a particular drug in terms of an adult's dose D_a. Young's rule states that

$$D_c = \frac{t}{t+12} D_a$$

where t is age in years, while Cowling's rule asserts that

$$D_c = \frac{t+1}{24} D_a$$

At what age is the difference between the two rules a maximum? What is the maximum difference?

60. In his study of the collapse of bubbles in a liquid, Lord Rayleigh obtained the following formula for the pressure p in the liquid at a distance r from the center of a spherical bubble that has collapsed from an initial radius R_0 to a radius R:

$$\frac{p(r)}{P} = 1 + \frac{R}{3r}(z-4) - \frac{R^4}{3r^4}(z-1)$$

where $P > 0$ is the ambient pressure of the liquid and $z = (R_0/R)^3$. See Figure 3.112. The formula of course holds only for $r > R$. Note also that $z > 1$.

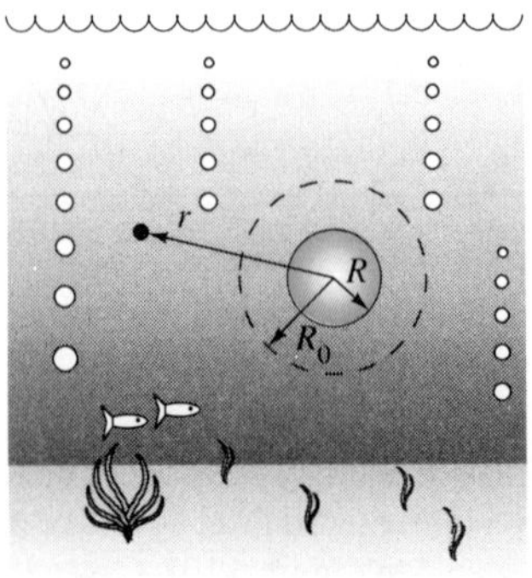

FIGURE 3.112

(a) Show that $\lim_{r\to\infty} p(r) = P$ and $\lim_{r\to R} p(r) = 0$.

(b) Show that $\lim_{r\to R} p'(r) = Pz/R$.

(c) Find the distance r at which $p(r)$ is a maximum. Use the Second Derivative Test.

61. The long bones in mammals may be represented as hollow cylindrical tubes, filled with marrow, of outer radius R and inner radius r. Bones should be constructed to be lightweight yet capable of withstanding certain bending moments. In order to withstand a bending moment M, it can be shown that the mass m per unit length of the bone and marrow is given by

$$m = \pi\rho\left[\frac{M}{K(1 - x^4)}\right]^{2/3}\left(1 - \frac{1}{2}x^2\right)$$

where ρ is the density of the bone and K is a positive constant. If $x = r/R$, show that m is a minimum when $r = 0.63R$ (approximately).

62. The rate P (in mg carbon/m^3/h) at which photosynthesis takes place for a certain species of phytoplankton is related to the light intensity I (in 10^3 ft-candles) by the function

$$P = \frac{100I}{I^2 + I + 4}$$

At what light intensity is P the largest?

63. A beam of length L is embedded in concrete walls as shown in Figure 3.113. When a constant load w_0 is uniformly distributed along its length, the deflection curve $y(x)$ for the beam is given by

$$y(x) = \frac{w_0L^2}{24EI}x^2 - \frac{w_0L}{12EI}x^3 + \frac{w_0}{24EI}x^4$$

where E and I are constants. (E is Young's modulus of elasticity and I is a moment of inertia of a cross-section of the beam.) The deflection curve approximates the shape of the beam.

(a) Determine the maximum deflection of the beam.

(b) Sketch a graph of $y(x)$.

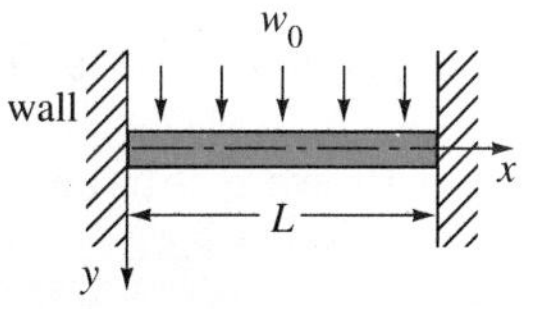

FIGURE 3.113

Calculator/Computer Problems

64. A two-story house under construction consists of two structures A and B with rectangular cross-sections and dimensions as indicated in Figure 3.114. The framing for structure B requires temporary wooden reinforcing buttresses from ground level that rest against structure A as shown.

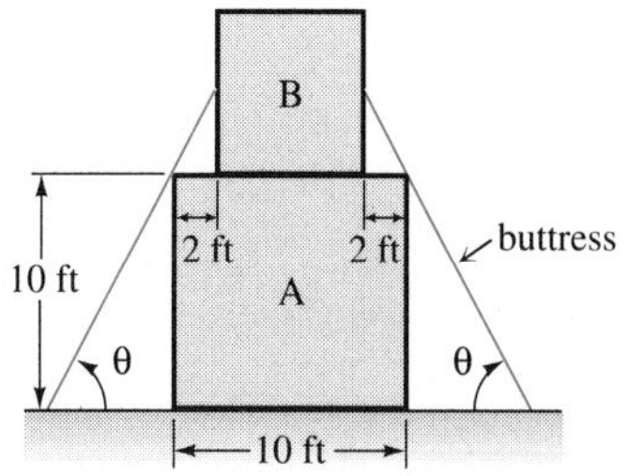

FIGURE 3.114

(a) Express the length L of a buttress as a function of the indicated angle θ.

(b) Find $L'(\theta)$.

(c) Use a computer to graph $L'(\theta)$ on the interval $(0, \pi/2)$. Use this graph to show that L has only one critical point θ_c in $(0, \pi/2)$. Use this graph to determine the algebraic sign of $L'(\theta)$ for $0 < \theta < \theta_c$ and the algebraic sign of $L'(\theta)$ for $\theta_c < \theta < \pi/2$. What is your conclusion?

(d) Find the minimum length of a buttress.

65. Consider the three cables shown in Figure 3.115.

(a) Express the total length L of the three cables shown in Figure 3.115(a) as a function of the length of the cable AB.

(b) Use a calculator or computer to verify that the graph of L has a minimum.

(c) Find the length of the cable AB so that the total length L of the lengths of the three cables is a minimum.

(d) Express the total length L of the three cables shown in Figure 3.115(b) as a function of the length of the cable AB.

(e) Use a calculator or computer to verify that the graph of L has a minimum.

(f) Use the graph obtained in part **(e)** as an aid in estimating the length of the cable AB that minimizes the function L obtained in part **(d)**.

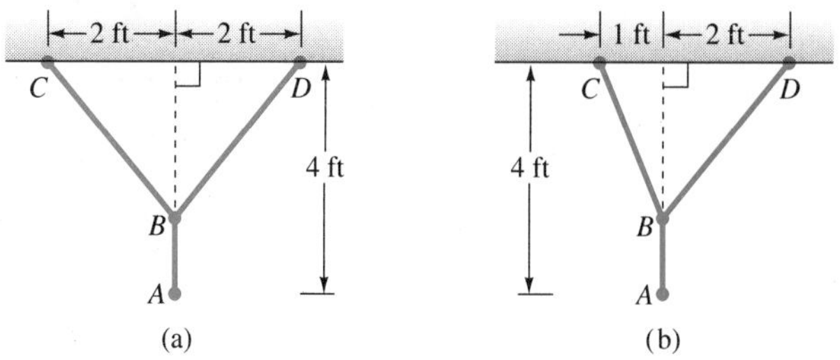

FIGURE 3.115

[O] 3.8 APPLICATIONS OF THE DERIVATIVE IN ECONOMICS

Revenue, Cost, and Profit When a company markets a product at p dollars per unit, the revenue realized in the production of x units is

$$R = px \tag{1}$$

Often the price itself depends on the number of units produced in a linear manner $p = ax + b$ so that (1) becomes

$$R(x) = x(ax + b) = ax^2 + bx \tag{2}$$

The function in (2) is an example of a **revenue function**. Furthermore, if $C(x)$ denotes the **cost** for producing x units, then the company's **profit** and **average cost** (cost per unit) are, respectively,

$$P(x) = R(x) - C(x) \quad \text{and} \quad Q(x) = \frac{C(x)}{x}$$

Naturally a company is interested in maximizing profit, $P(x)$, and minimizing the cost per unit, $Q(x)$. A typical cost function consists of

$$C = \text{variable costs} + \text{fixed costs}$$

Thus, in

$$C(x) = 200x + 600 \quad \text{and} \quad C(x) = x^2 + 640x + 950 \tag{3}$$

the constants 600 and 950 are the fixed costs and could represent rent, insurance premiums, and so on. Since it is assumed that $x \geq 0$, note that the minimum value of each function in (3) is $C(0)$.

EXAMPLE 1 A company determines that in the production of x units of a commodity its revenue and cost functions are, respectively, $R(x) = -3x^2 + 970x$ and $C(x) = 2x^2 + 500$. Find the maximum profit and minimum average cost.

Solution The profit for $x \geq 0$ is

$$P(x) = (-3x^2 + 970x) - (2x^2 + 500) = -5x^2 + 970x - 500$$

From $P'(x) = -10x + 970$ and $P''(x) = -10$, we see that $x = 97$ is a critical point and that $P''(97) < 0$. Thus, the Second Derivative Test implies that $P(97) = 46{,}545$ units of dollars is a maximum.

Now for $x > 0$ the average cost is

$$Q(x) = \frac{2x^2 + 500}{x} = 2x + \frac{500}{x}$$

so that $Q'(x) = 2 - 500/x^2$ and $Q''(x) = 1000/x^3$. Solving $Q'(x) = 0$ gives $x = \sqrt{250}$. Hence, $Q''(\sqrt{250}) > 0$ implies that $Q(\sqrt{250})$ is a minimum. Because the company can produce only an integral number of units, we use $\sqrt{250} \approx 16$ to find the *approximate* minimum average cost, or $Q(16) = 43.25$ units of dollars. □

You should use Example 1 to verify that the maximum profit does not necessarily occur at the same production level that corresponds to the maximum revenue.

Marginal Functions In economics the term **marginal function** usually refers to the derivative of that function.

DEFINITION 3.9 Marginal Functions

Marginal revenue:	$MR = R'(x)$
Marginal cost:	$MC = C'(x)$
Marginal profit:	$MP = P'(x)$

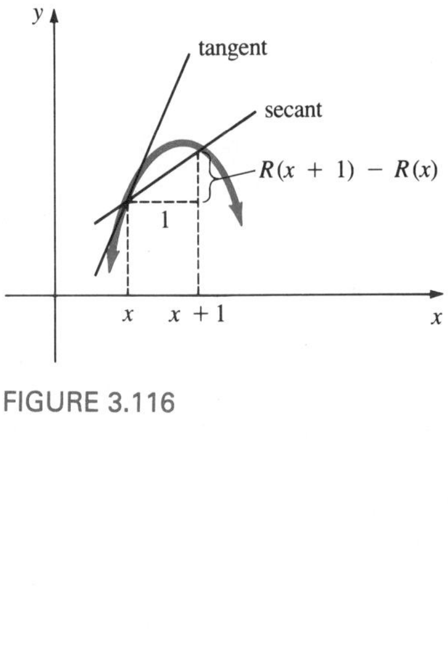

FIGURE 3.116

To get a feeling for the use of these functions, consider the case of the revenue function R. When $\Delta x = 1$, the quotient

$$\frac{\Delta R}{\Delta x} = \frac{R(x + \Delta x) - R(x)}{\Delta x} = R(x + 1) - R(x)$$

gives the slope of the secant line through the points $(x, R(x))$ and $(x + 1, R(x + 1))$ on the graph of R. See Figure 3.116. Since the slope of this secant is an approximation to the slope of the tangent at $(x, R(x))$, the derivative $R'(x)$ gives the *approximate value of the change in revenue for a unit increase in production.*

EXAMPLE 2 For the revenue function given in Example 1, find the revenue gained from the production of the forty-first unit. Approximate this value by means of the marginal revenue.

Solution From $R(x) = -3x^2 + 970x$, we see that the revenue from producing 40 units is $R(40) = 34{,}000$ and the revenue from producing 41 units is $R(41) = 34{,}727$. Hence, the revenue from the production of the forty-first, or one more unit, is

$$R(41) - R(40) = 727 \text{ units of dollars}$$

By way of comparison, $MR = R'(x) = -6x + 970$ and $MR(40) = 730$ units of dollars. □

You should realize that all we are doing in the preceding example is approximating the change in a function by means of a differential; namely, $\Delta R \approx dR$ when $\Delta x = 1$.

EXAMPLE 3 The cost for producing x units of a product is given by $C(x) = x^2 + 560x + 1000$. Find the approximate cost for producing the fiftieth unit.

Solution Rather than computing the exact cost $C(50) - C(49)$, we utilize the concept of the marginal cost. The derivative $MC = C'(x) = 2x + 560$ evaluated at $x = 49$ gives an approximation to the cost of producing one more unit (the fiftieth): $MC(49) = 658$ units of dollars. □

Demand The number D of units of a product **demanded** by consumers is a function of the price p of each unit. Intuitively, we would expect $D(p)$ to be small when p is large.

For a change Δp in price, the quotient

$$\frac{\dfrac{D(p + \Delta p) - D(p)}{D(p)}}{\dfrac{\Delta p}{p}} \tag{4}$$

is the proportionate change in demand divided by the proportionate change in price. Simplifying (4) and taking the limit as $\Delta p \to 0$ give

$$\lim_{\Delta p \to 0} \frac{p}{D(p)} \cdot \frac{D(p + \Delta p) - D(p)}{\Delta p} = \frac{p}{D(p)} \cdot D'(p)$$

Since $D(p)$ is a decreasing function, we expect $D'(p) < 0$. For this reason it is common practice to introduce a minus sign and to refer to the resulting quantity as the **elasticity of demand**:

$$\eta = -\frac{p}{D(p)} \cdot D'(p)$$

Cases When $\eta < 1$, economists say that the demand is **inelastic**; in this case the percentage change in demand is less than the percentage change in price. If $\eta > 1$, the demand is said to be **elastic** and the percentage change in demand is greater than the percentage change in price. When $\eta = 1$, the percentage change in demand equals the percentage change in price.

EXAMPLE 4 If $D(p) = -p^2 + 400$, $0 \le p \le 20$, determine whether the demand is elastic or inelastic at $p = 6$.

Solution $D'(p) = -2p$, $D'(6) = -12$, and $D(6) = 364$. Thus,

$$\eta = -\frac{6}{D(6)} \cdot D'(6) = \frac{72}{364} \approx 0.2 < 1$$

implies the demand is inelastic. □

The result of Example 4 can be interpreted to mean that at $p = 6$ there is approximately a 0.2% decrease in demand for a 1% increase in price. If we suppose the price increases, say, 10%, then the demand would decrease by approximately $10(0.2) = 2\%$.

In general, when a demand is either elastic or inelastic at a price level p, then a percentage increase in price brings with it a decrease in demand. This decrease is greater for an elastic demand than it is for an inelastic demand. In either case the revenue must decrease. However, when $\eta = 1$, an increase in price results in no change in revenue.

EXERCISES 3.8

Answers to odd-numbered problems begin on page A-78.

In Problems 1 and 2 find the maximum revenue, maximum profit, and minimum average cost.

1. $R(x) = -x^2 + 400x$, $C(x) = x^2 + 40x + 100$
2. $R(x) = x(-2x + 60)$, $C(x) = 2x^2 + 12x + 18$
3. Given the revenue function $R(x) = -x^2 + 80x$:
   (a) Find the marginal revenue at $x = 10$.
   (b) Compare the result of part (a) with $R(11) - R(10)$.
4. A company finds that its cost for producing x units of a commodity is $C(x) = 3x^2 + 5x + 10$. Find the approximate cost for making the twenty-first unit.
5. Let $C(x) = 3x^2 + 100$ denote the cost for making x units of a product. Compare the exact cost for producing the thirty-first unit with the marginal costs at $x = 30$ and $x = 31$.
6. Suppose $R(x) = -x^2 + 1000x$ and $C(x) = 20x + 600$ are revenue and cost functions, respectively, for producing x units of a commodity. What profit is realized from the sale of 50 items? What is the approximate amount the profit changes from the sale of one more item?
7. Given that $R(x) = x(-x + 300)$ is a revenue function, show that the marginal revenue is always decreasing.
8. Given that $C(x) = 2x^3 - 21x^2 + 36x + 1000$ is a cost function, determine the interval(s) for which the cost is increasing. Determine whether there are any intervals on which the marginal cost increases.
9. Show that the maximum profit occurs when the marginal revenue equals the marginal cost.
10. Show that the minimum average cost occurs when the average cost equals the marginal cost.

In Problems 11–16 compute the elasticity of demand for the given demand function at the indicated price. State whether the demand is elastic or inelastic.

11. $D(p) = -4p + 500$, $0 \le p \le 125$; $p = 50$
12. $D(p) = -10p + 850$, $0 \le p \le 85$; $p = 40$
13. $D(p) = -2p^2 + 200$, $0 \le p \le 10$; $p = 6$
14. $D(p) = (20 - p)^2$, $0 \le p \le 20$; $p = 10$
15. $D(p) = 800\sqrt{30 - p}$, $0 \le p \le 30$; $p = 14$
16. $D(p) = 1000 + \dfrac{200}{\sqrt{p + 4}}$, $p \ge 0$; $p = 21$
17. Compute the elasticity of demand for $D(p) = -2p^2 + 800$, $0 \le p \le 20$, at $p = 15$. If the price increases by 6%, determine the approximate change in demand.
18. Compute the elasticity of demand for $D(p) = \sqrt{25 - p^2}$, $0 \le p \le 5$, at $p = 2$. If the price decreases by 21%, determine the approximate change in demand.
19. For the demand function in Problem 18 find the price level for which $\eta = 4$.
20. For the demand function $D(p) = -4p + 1000$, $0 \le p \le 250$, determine the prices for which the demand is elastic.

CHAPTER 3 REVIEW EXERCISES

Answers begin on page A-78.

In Problems 1–10 answer true or false.

1. If f is increasing on an interval, then $f'(x) > 0$ on the interval. ________

2. A function f has an extremum at c when $f'(c) = 0$. ________

3. A particle moving rectilinearly slows down when the velocity $v(t)$ decreases. ________

4. For a particle moving rectilinearly, acceleration is the first derivative of the velocity. ________

5. If $f''(x) < 0$ for all x in interval (a, b), then the graph of f is concave downward on the interval. ________

6. If $f''(c) = 0$, then $(c, f(c))$ is a point of inflection. ________

7. If f is continuous on $[a, b]$ and $f(a) = f(b) = 0$, then there exists some c in (a, b) such that $f'(c) = 0$. ________

8. The graph of a cubic polynomial can have at most one point of inflection. ________

9. A function continuous on a closed interval $[a, b]$ has both an absolute maximum and an absolute minimum. ________

10. Every absolute extremum is also a relative extremum. ________

In Problems 11 and 12 supply the reason(s) why the given statement is false.

11. If $f(c)$ is a relative maximum, then $f'(c) = 0$ and $f'(x) > 0$ for $x < c$ and $f'(x) < 0$ for $x > c$.

12. If $f(c)$ is a relative minimum, then $f''(c) > 0$.

In Problems 13–16 find the absolute extrema of the given function on the indicated interval.

13. $f(x) = x^3 - 75x + 150$; $[-3, 4]$

14. $f(x) = 4x^2 - \dfrac{1}{x}$; $[\frac{1}{4}, 1]$

15. $f(x) = \dfrac{x^2}{x+4}$; $[-1, 3]$

16. $f(x) = (x^2 - 3x + 5)^{1/2}$; $[1, 3]$

17. Sketch a graph of a continuous function that has the properties:

$$f(0) = 1, \quad f(2) = 3$$
$$f'(0) = 0, \quad f'(2) \text{ does not exist}$$
$$f'(x) > 0, \quad x < 0$$
$$f'(x) > 0, \quad 0 < x < 2$$
$$f'(x) < 0, \quad x > 2$$

18. Use the first and second derivatives as an aid in comparing the graphs of

$$y = x + \sin x \quad \text{and} \quad y = x + \sin 2x$$

19. The position of a particle moving on a horizontal line is given by $s(t) = -t^3 + 6t^2$. Graph the motion on the time interval $[-1, 5]$. At what point is the velocity function a maximum? Does this point correspond to the maximum speed?

20. The height above ground of a projectile fired vertically is $s(t) = -4.9t^2 + 14.7t + 49$, where s is measured in meters and t in seconds. What is the maximum height attained by the projectile? At what speed does the projectile strike the ground?

21. Consider the function $f(x) = x \sin x$. Use f and Rolle's Theorem to show that the equation $\cot x = -1/x$ has a solution on the interval $(0, \pi)$.

22. Show that the function $f(x) = x^{1/3}$ does not satisfy the hypothesis of the Mean Value Theorem on the interval $[-1, 8]$ yet a number c can be found in $(-1, 8)$ such that $f'(c) = [f(b) - f(a)]/(b - a)$. Explain.

23. Suppose f is a polynomial function with zeros of multiplicity 2 at $x = a$ and $x = b$; that is,

$$f(x) = (x - a)^2(x - b)^2 g(x)$$

where g is a polynomial function.

(a) Show that f' has at least three zeros in the closed interval $[a, b]$.

(b) If $g(x)$ is a constant, find the zeros of f' in $[a, b]$.

24. Give an example of a function $y = f(x)$ that is concave upward on $(-\infty, 0)$, concave downward on $(0, \infty)$, and increasing on $(-\infty, \infty)$.

In Problems 25–28 find the relative extrema of the given function. Graph.

25. $f(x) = 2x^3 + 3x^2 - 36x$

26. $f(x) = x^5 - \frac{5}{3}x^3 + 2$

27. $f(x) = 4x - 6x^{2/3} + 2$

28. $f(x) = \dfrac{x^2 - 2x + 2}{x - 1}$

In Problems 29–32 find the relative extrema and the points of inflection of the given function. Do not graph.

29. $f(x) = x^4 + 8x^3 + 18x^2$

30. $f(x) = x^6 - 3x^4 + 5$

31. $f(x) = 10 - (x - 3)^{1/3}$

32. $f(x) = x(x - 1)^{5/2}$

33. Let a, b, and c be real numbers. Find the x-coordinate of the point of inflection for the graph of

$$f(x) = (x - a)(x - b)(x - c)$$

34. A triangle is expanding with time. The area of the triangle is increasing at a rate of 15 in^2/min, whereas the length of its base is decreasing at a rate of $\frac{1}{2}$ in/min. At what rate is the altitude of the triangle changing when the altitude is 8 in and the base is 6 in?

35. A pulley is secured to the edge of a dock that is 15 ft above the surface of the water. A small boat is being pulled toward the dock by means of a rope on the pulley. The rope is attached to the bow of the boat 3 ft above the water line. If the rope is pulled in at a constant rate of 1 ft/s, how fast does the boat approach the dock when it is 16 ft from the dock?

36. Water drips into a hemispherical tank of radius 10 m at a rate of $\frac{1}{10}$ m^3/min and drips out a hole in the bottom of the tank at a rate of $\frac{1}{5}$ m^3/min. It can be shown that the volume of the water in the tank at any time is

$$V = 10\pi h^2 - \frac{\pi}{3}h^3$$

See Figure 3.117. Is the depth of the water increasing or decreasing? At what rate is the depth of the water changing when the depth is 5 m?

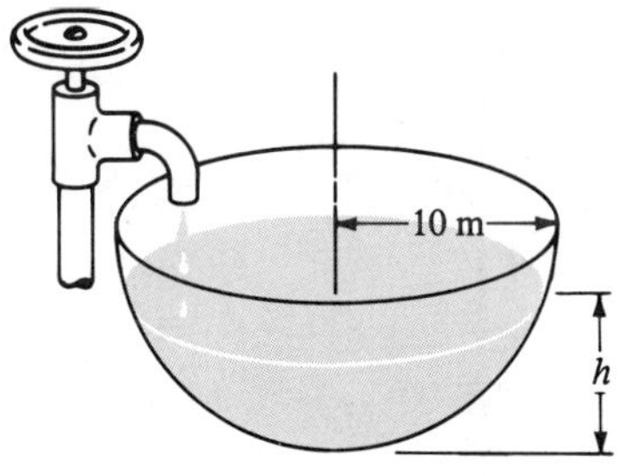

FIGURE 3.117

37. Consider the ladder whose bottom is sliding away from the base of the vertical wall shown in Figure 3.118. Show that the rate at which θ_1 is increasing is the same as the rate at which θ_2 is decreasing.

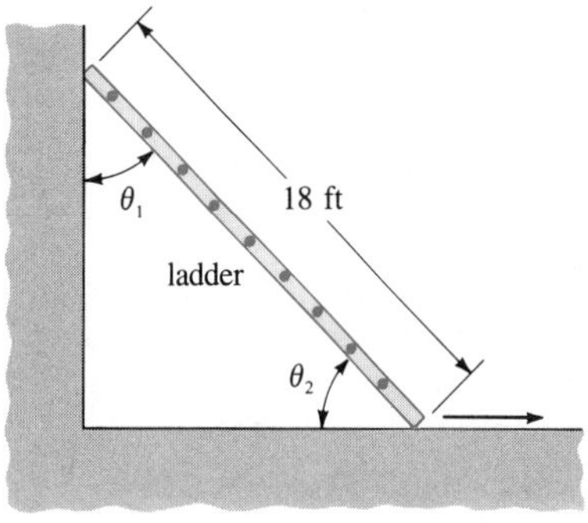

FIGURE 3.118

38. Find two nonnegative numbers whose sum is 8 such that the sum of their squares is a maximum.

39. Find the minimum value of the sum of a nonnegative number and its reciprocal.

40. A battery with constant emf E and constant internal resistance r is wired in series with a resistor that has resistance R. The current in the circuit is then $I = E/(r + R)$. Find the value of R for which the power $P = RI^2$ dissipated in the external load is a maximum. This is called *impedance matching.*

41. Two coils that carry the same current produce a magnetic field at point Q on the x-axis of strength

$$B = \frac{1}{2}\mu_0 r_0^2 I\left\{\left[r_0^2 + \left(x + \frac{r_0}{2}\right)^2\right]^{-3/2} + \left[r_0^2 + \left(x - \frac{r_0}{2}\right)^2\right]^{-3/2}\right\}$$

where μ_0, r_0, and I are constants. See Figure 3.119. Show that the maximum value of B occurs at $x = 0$.

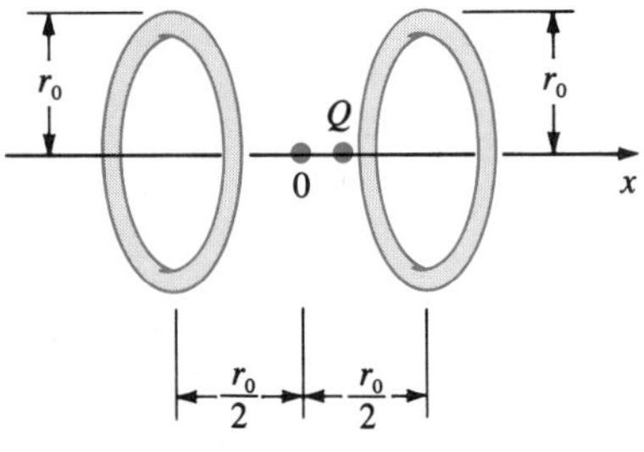

FIGURE 3.119

42. The velocity of air through the trachea (in cm/s) of radius r is

$$v = \frac{V(r)}{\pi r^2}, \qquad \frac{r_0}{2} \le r \le r_0$$

where

$$V(r) = kr^4(r_0 - r), \qquad k > 0, \quad r_0 > 0$$

is the volume flow of air. What radius will give the maximum velocity of air?

43. Some birds fly more slowly over water than over land. A bird flies at constant rates of 6 km/h over water and 10 km/h over land. Use the information in Figure 3.120 to find the path the bird should take to minimize the total flying time between the shore of one island and its nest on the shore of another island.

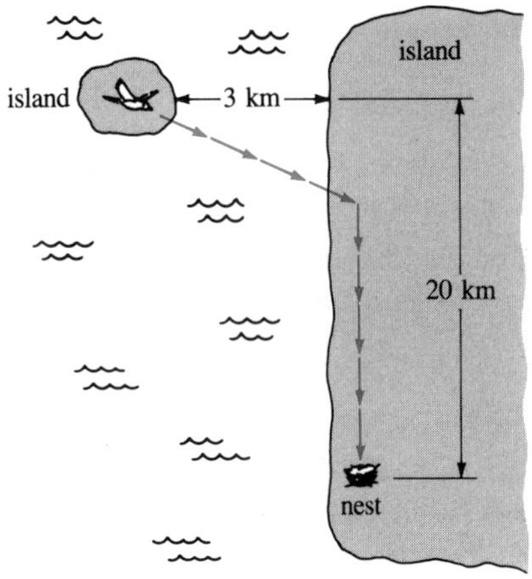

FIGURE 3.120

44. The area of a circular sector of radius r and arc length s is $A = \frac{1}{2}rs$. See Figure 3.121. Find the maximum area of a sector enclosed by a perimeter of 60 cm.

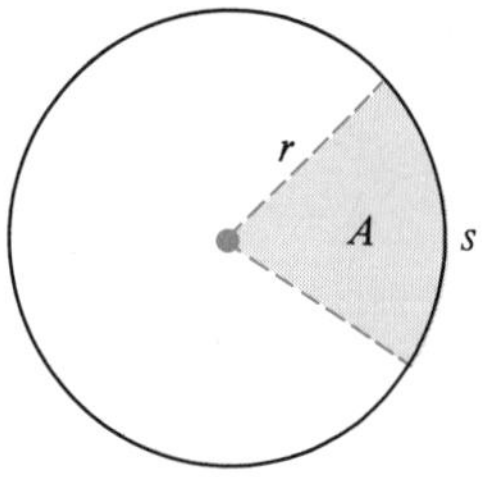

FIGURE 3.121

45. A rectangular yard is to be enclosed with a fence by attaching it to a house whose length is 40 ft. See Figure 3.122. The amount of fence to be used is 160 ft. Describe how the fence should be used so that the greatest area is enclosed.

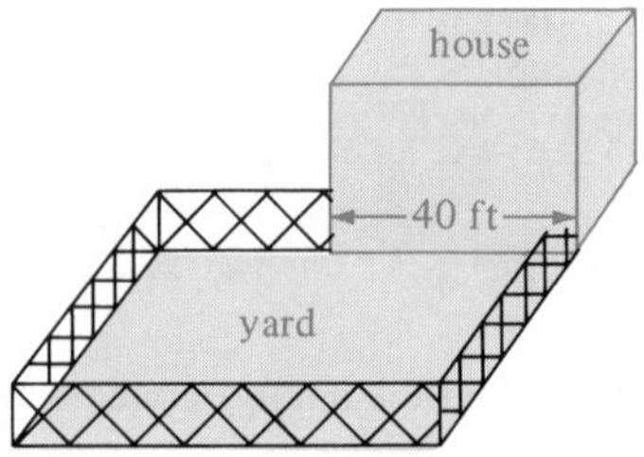

FIGURE 3.122

46. Solve Problem 45 if the amount of fence to be used is 80 ft.

47. A pigpen, attached to a barn, is enclosed using fence on two sides, as shown in Figure 3.123. The amount of fence to be used is 585 ft. Find the values of x and y indicated in the figure so that the greatest area is enclosed. What is the greatest area?

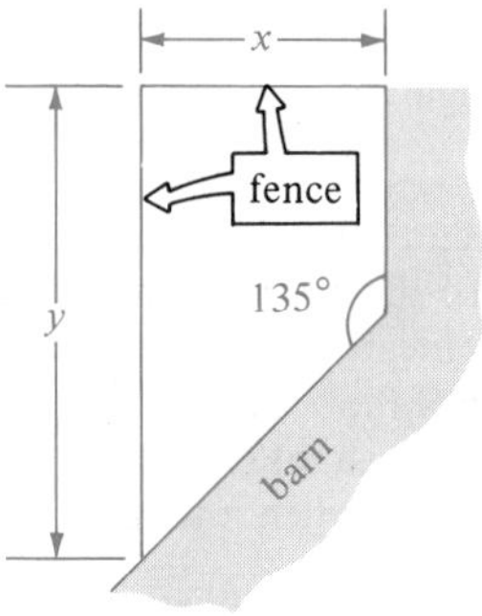

FIGURE 3.123

48. A rancher wants to use 100 m of fence to construct a diagonal fence connecting two existing walls that meet at a right angle. How should this be done so that the area enclosed by the walls and the fence is a maximum?

49. Two television masts on a roof are secured by wires that are attached at a single point between the masts. See Figure 3.124. Where should the point be located to minimize the amount of wire used?

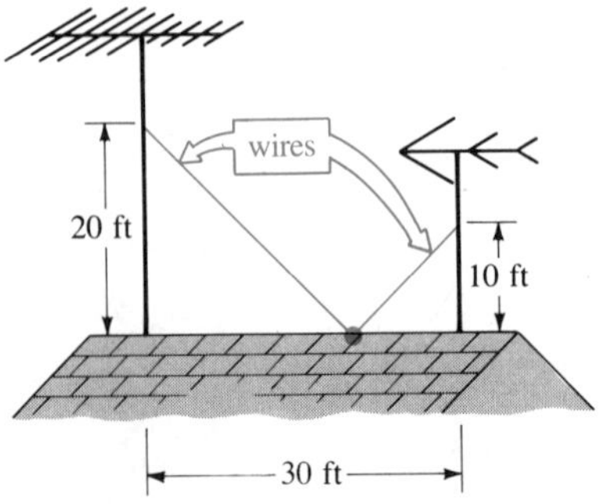

FIGURE 3.124

50. A statue is placed on a pedestal as shown in Figure 3.125. How far should a person stand from the pedestal to maximize the viewing angle θ? [*Hint:* Review the trigonometric identity for $\tan(\theta_2 - \theta_1)$. Also, it suffices to maximize $\tan\theta$ rather than θ. Why?]

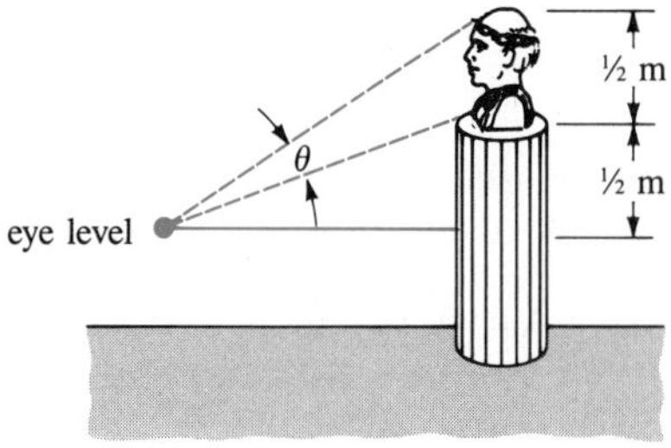

FIGURE 3.125

51. According to Fermat's principle, a ray of light originating at point A and reflected from a plane surface to point B travels on a path requiring the least time. See Figure 3.126. Assume that the speed of light c as well as h_1, h_2, and d are constants. Show that the time is a minimum when $\tan\theta_1 = \tan\theta_2$. Since $0 < \theta_1 < \pi/2$ and $0 < \theta_2 < \pi/2$, it follows that $\theta_1 = \theta_2$. In other words, the angle of incidence equals the angle of reflection. [*Note:* Figure 3.126 is inaccurate on purpose.]

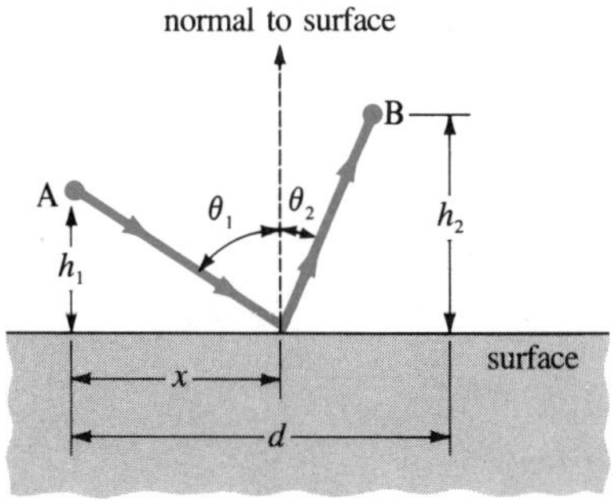

FIGURE 3.126

52. Determine the dimensions of a right circular cone having minimum volume V that circumscribes a sphere of radius r. See Figure 3.127. [*Hint:* Use similar triangles.]

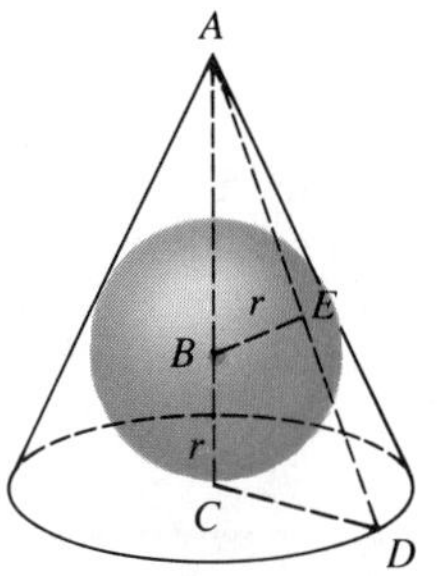

FIGURE 3.127

53. A container in the form of a right circular cylinder has a volume of $100\ \text{in}^3$. The top of the container costs 3 times as much per unit area as the bottom and the sides. Show that the dimension that gives the least cost of construction is a height that is 4 times the radius.

54. A piece of paper is 8 in wide. One corner is folded over to the other edge of the paper as shown in Figure 3.128. Find the width x of the fold so that the length L of the crease is a minimum.

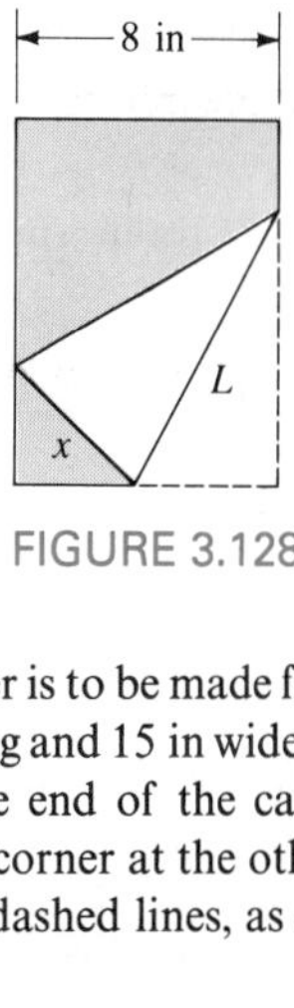

FIGURE 3.128

55. A box with a cover is to be made from a rectangular piece of cardboard 30 in long and 15 in wide by cutting a square out of each corner at one end of the cardboard and cutting a rectangle out of each corner at the other end. The cardboard is then folded on the dashed lines, as shown in Figure 3.129.

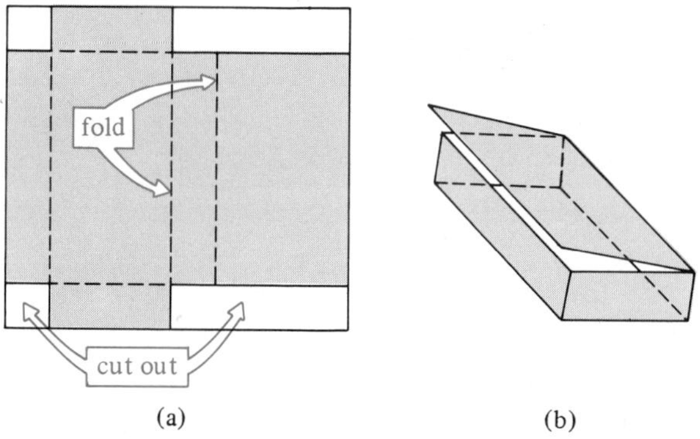

FIGURE 3.129

Find the dimensions of the box that will give the maximum volume. What is the maximum volume?

56. The running track shown in Figure 3.130 is to consist of two parallel straight parts and two semicircular parts. The length of the track is to be 2 km. Find the design of the track so that the rectangular plot of land enclosed by the track is a maximum.

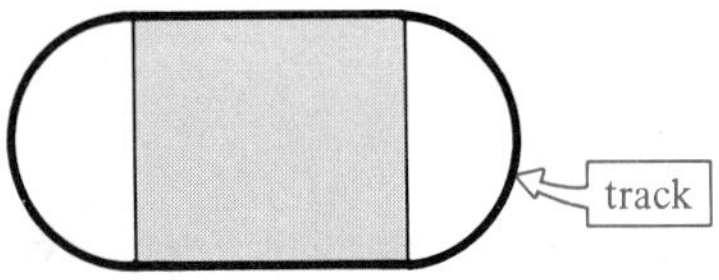

FIGURE 3.130

57. A company determines that in the production and sale of x units of a commodity its revenue and cost functions are $R(x) = x(-3x + 660)$ and $C(x) = x^2 + 196x + 400$, respectively. Find:

(a) The maximum revenue

(b) The maximum profit

(c) The minimum average cost

(d) The approximate revenue from the sale of the eleventh item

(e) The approximate profit from the sale of the thirty-first item

(f) How much does the company lose if no units are sold?

58. Suppose the function $D(p) = -2p + 50$, $0 \leq p \leq 25$, represents the demand for a product. Approximately how much does the demand change when the price of each unit of the product changes from \$10.00 to \$11.80?

59. The relationship between the height h and the diameter d of a tree can be approximated by the quadratic expression $h = 137 + ad - bd^2$, where h and d are measured in centimeters, and a and b are positive parameters that depend on the type of tree. See Figure 3.131.

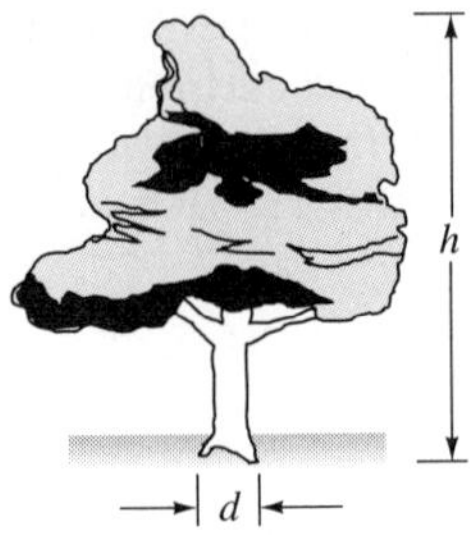

FIGURE 3.131

(a) Suppose a tree attains its maximum height of H centimeters at a diameter of D centimeters. Show that

$$h = 137 + 2\frac{H - 137}{D}d - \frac{H - 137}{D^2}d^2$$

(b) Suppose a certain tree reaches its maximum possible height (according to the formula) of 15 m at a diameter of 0.8 m. What was the diameter of the tree when the tree was 10 m tall?

60. The annual survival rate in a species of mammal can be related to the skull length of the species by a function of the form

$$S = \frac{100x^2}{x^2 + C}$$

where S is the annual survival rate (in percent), x is the skull length, and C is a constant, approximately equal to 200 if x is measured in centimeters.

(a) Show that increased skull length corresponds to an increased rate of survival.

(b) At what skull length is the survival rate increasing most rapidly with respect to skull length? Use $C = 200$.

4

THE INTEGRAL

INTRODUCTION

In the last two chapters we have been concerned with the definition, properties, and applications of the derivative. We turn now from differential to integral calculus. Leibniz originally called this second of the two major divisions of calculus, *calculus summatorius.* In 1696, at the persuasion of the Swiss mathematician Johann Bernoulli, Leibniz changed its name to *calculus integralis.* As the original Latin words suggest, the notion of a sum plays an important role in the full development of the integral.

In Chapter 2 we saw that the tangent problem leads naturally to the derivative of a function. In the area problem, the motivational problem for integral calculus, we want to find the area bounded by the graph of a function and the x-axis. This problem leads to the concept of the integral.

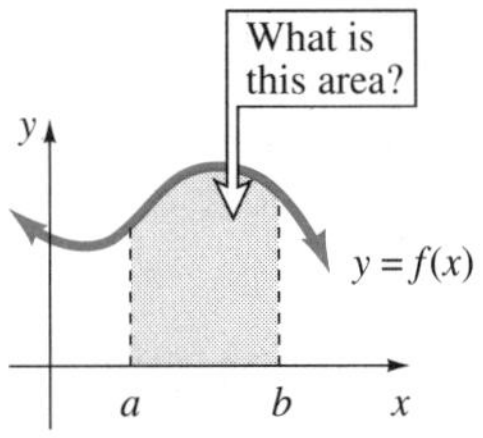

The process of integration that enables us, in many cases, to find areas easily also enables us to solve a problem such as finding the escape velocity of rocket shot vertically upward from the ground. See Exercises 4.1.

IMPORTANT CONCEPTS

antiderivative
indefinite integral
integrand
differential equation
solution of a differential equation
separable differential equation
summation notation
index of summation
summation formulas
area under a graph
definite integral
partition
norm of a partition
Riemann sums
integrable function
limits of integration
regular partition
dummy variable of integration
fundamental theorem of calculus
derivative form
antiderivative form
even function rule
odd function rule
piecewise continuous function
midpoint rule
trapezoidal rule
Simpson's rule
error bounds

4.1 ANTIDERIVATIVES

In Chapters 2 and 3 we were concerned only with the basic problem:

Given a function f, find its derivative f'.

In this chapter and in subsequent chapters of this text we shall see that an equally important problem is:

Given a function f, find a function whose derivative is the same as f.

That is, for a given function f, we wish to find another function F for which $F'(x) = f(x)$ for all x on some interval.

DEFINITION 4.1 Antiderivative of a Function

A function F is said to be an **antiderivative** of a function f on some interval if $F'(x) = f(x)$.

EXAMPLE 1 An antiderivative of $f(x) = 2x$ is $F(x) = x^2$, since $F'(x) = 2x$. □

There is always more than one antiderivative of a function. For instance, in the foregoing example, $F_1(x) = x^2 - 1$ and $F_2(x) = x^2 + 10$ are also antiderivatives of $f(x) = 2x$, since $F_1'(x) = F_2'(x) = f(x)$. Indeed, if F is an antiderivative of a function f, then so is $G(x) = F(x) + C$ for any constant C. This is a consequence of the fact that

$$G'(x) = \frac{d}{dx}(F(x) + C) = F'(x) + 0 = F'(x) = f(x)$$

Thus, $F(x) + C$ stands for a *set* or *family of functions* each member of which has a derivative equal to $f(x)$. We shall now prove that any antiderivative of f must be of the form $G(x) = F(x) + C$; that is, *two antiderivatives of the same function can differ by at most a constant.* Hence, $F(x) + C$ is *the most general antiderivative* of $f(x)$.

THEOREM 4.1 Antiderivatives Differ by a Constant

If $G'(x) = F'(x)$ for all x in some interval $[a, b]$, then

$$G(x) = F(x) + C$$

for all x in the interval.

Proof Suppose we define $g(x) = G(x) - F(x)$. Then, since $G'(x) = F'(x)$, it follows that $g'(x) = G'(x) - F'(x) = 0$ for all x in $[a, b]$. If x_1 and x_2 are any two numbers that satisfy $a \le x_1 < x_2 \le b$, it follows from the Mean Value Theorem (Theorem 3.5) that a number k exists in the open interval (x_1, x_2) for which

$$g'(k) = \frac{g(x_2) - g(x_1)}{x_2 - x_1} \quad \text{or} \quad g(x_2) - g(x_1) = g'(k)(x_2 - x_1)$$

But $g'(x) = 0$ for all x in $[a, b]$; in particular, $g'(k) = 0$. Hence, $g(x_2) - g(x_1) = 0$ or $g(x_2) = g(x_1)$. Now, by assumption, x_1 and x_2 are any two, but different, numbers in the interval. Since the function values $g(x_1)$ and $g(x_2)$ are the same, we must conclude that the function $g(x)$ is a constant C. Thus, $g(x) = C$ implies

$$G(x) - F(x) = C \quad \text{or} \quad G(x) = F(x) + C$$

■

EXAMPLE 2

(*a*) The most general antiderivative of $f(x) = 2x$ is $G(x) = x^2 + C$.

(*b*) The most general antiderivative of $f(x) = 2x + 5$ is $G(x) = x^2 + 5x + C$, since $G'(x) = 2x + 5$. □

Indefinite Integral Notation For convenience, let us introduce a notation for an antiderivative of a function. If $F'(x) = f(x)$, we shall represent the most general antiderivative of f by

$$\int f(x)\,dx = F(x) + C$$

The symbol $\int$ was introduced by Leibniz and is called an **integral sign**. The notation $\int f(x)\,dx$ is called the **indefinite integral** of $f(x)$ with respect to x. The function $f(x)$ is called the **integrand**. The process of finding an antiderivative is called **antidifferentiation** or **integration**. The number C is called a **constant of integration**. Just as d/dx ( ) denotes differentiation *with respect to* x, the symbolism $\int(\ )\,dx$ denotes integration *with respect to* x.

The Indefinite Integral of a Power When differentiating the power x^n, we multiply by the exponent n and decrease the exponent by 1. To find an antiderivative of x^n, the reverse of the differentiation rule is: *Increase the exponent by* 1 *and divide by the new exponent* $n + 1$. The indefinite integral analogue of the Power Rule of differentiation is given by the following:

THEOREM 4.2 Indefinite Integral of a Power of x

If n is a rational number, then for $n \neq -1$,

$$\int x^n\,dx = \frac{x^{n+1}}{n+1} + C \tag{1}$$

Proof   From Theorem 2.11,

$$\frac{d}{dx}\left(\frac{x^{n+1}}{n+1}+C\right)=(n+1)\frac{x^{(n+1)-1}}{n+1}+0=x^n$$ ■

Note: The result given in (1) does not include the case when $n=-1$. We have not as yet encountered any function whose derivative is $x^{-1}=1/x$. The evaluation of

$$\int\frac{1}{x}\,dx$$

will be considered in Chapter 6.

EXAMPLE 3   Evaluate $\displaystyle\int x^6\,dx$.

Solution   With $n=6$ it follows from (1) that

$$\int x^6\,dx=\frac{x^7}{7}+C$$ □

It is often necessary to rewrite an integrand $f(x)$ before carrying out the integration.

EXAMPLE 4   Evaluate (*a*) $\displaystyle\int\frac{1}{x^5}\,dx$ and   (*b*) $\displaystyle\int\sqrt{x}\,dx$.

Solution

(*a*)   By rewriting $1/x^5$ as x^{-5} and identifying $n=-5$, we have from (1)

$$\int x^{-5}\,dx=\frac{x^{-4}}{-4}+C=-\frac{1}{4x^4}+C$$

(*b*)   We first rewrite the radical $\sqrt{x}$ as $x^{1/2}$ and then use (1) with $n=\frac{1}{2}$:

$$\int x^{1/2}\,dx=\frac{x^{3/2}}{3/2}+C=\frac{2}{3}x^{3/2}+C$$ □

It should be kept in mind that the *results of integration can always be checked by differentiation*; for example,

integration

$$\int x^2\,dx=\frac{x^3}{3}+C$$

check by
differentiation

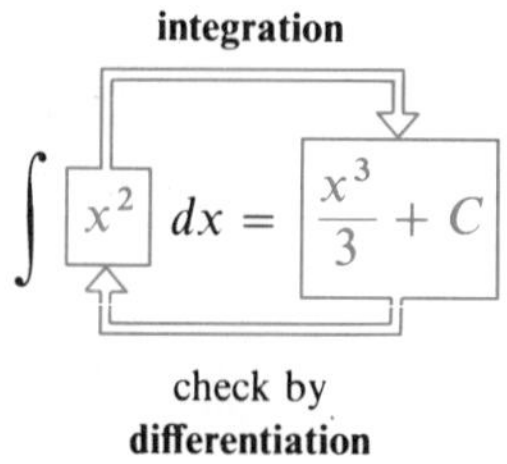

EXAMPLE 5 Evaluate $\int dx$.

Solution Since $\int dx = \int 1 \cdot dx$, and since $d/dx\,(x + C) = 1 + 0 = 1$, it follows from the definition of an antiderivative that

$$\int dx = x + C \qquad \square$$

The result in Example 5 can also be obtained from (1) with $n = 0$.

Some properties of the indefinite integral are given in the next theorem.

THEOREM 4.3 Properties

Let $F'(x) = f(x)$ and $G'(x) = g(x)$. Then

(*i*) $\int kf(x)\,dx = k\int f(x)\,dx = kF(x) + C$, where k is any constant,

(*ii*) $\int [f(x) \pm g(x)]\,dx = \int f(x)\,dx \pm \int g(x)\,dx = F(x) \pm G(x) + C$.

These properties follow immediately from the properties of the derivative. For example, (*ii*) is a consequence of the fact that the derivative of a sum is the sum of the derivatives.

Observe in Theorem 4.3(*ii*) that there is no reason to use two constants of integration, since

$$\begin{aligned}\int [f(x) \pm g(x)]\,dx &= (F(x) + C_1) \pm (G(x) + C_2)\\ &= F(x) \pm G(x) + (C_1 \pm C_2) = F(x) \pm G(x) + C\end{aligned}$$

where we have replaced $C_1 \pm C_2$ by the single constant C.

The antiderivative, or indefinite integral, of any finite sum can be obtained by integrating each term.

EXAMPLE 6 Evaluate $\int \left(4x - \dfrac{2}{\sqrt[3]{x}} + \dfrac{5}{x^2}\right) dx$.

Solution From Theorems 4.2 and 4.3 and the preceding discussion, it follows that

$$\begin{aligned}\int (4x - 2x^{-1/3} + 5x^{-2})\,dx &= 4\int x\,dx - 2\int x^{-1/3} + 5\int x^{-2}\,dx\\ &= 4\cdot\frac{x^2}{2} - 2\cdot\frac{x^{2/3}}{2/3} + 5\cdot\frac{x^{-1}}{-1} + C\\ &= 2x^2 - 3x^{2/3} - 5x^{-1} + C\end{aligned} \qquad \square$$

Differentiation and integration are fundamentally inverse operations. If $\int f(x)\,dx = F(x) + C$, then $F'(x) = f(x)$ so that

$$\int F'(x)\,dx = F(x) + C$$

Moreover, $$\frac{d}{dx}\int f(x)\,dx = \frac{d}{dx}(F(x) + C) = F'(x) = f(x)$$

In other words,

An antiderivative of the derivative of a function is that function plus a constant.

A derivative of an antiderivative of a function is that function.

In this manner, whenever we obtain the derivative of a function, we get at the same time an integration formula. For example, since

$$\frac{d}{dx}\sin x = \cos x \quad \text{then} \quad \int \cos x\,dx = \int \left(\frac{d}{dx}\sin x\right) dx = \sin x + C$$

For convenience we summarize some of our known differentiation results along with their equivalent integration formulas.

Differentiation	**Integration**
$\frac{d}{dx}x^n = nx^{n-1}$	$\int x^n\,dx = \frac{x^{n+1}}{n+1} + C, \quad n \neq -1$
$\frac{d}{dx}\sin x = \cos x$	$\int \cos x\,dx = \sin x + C$
$\frac{d}{dx}\cos x = -\sin x$	$\int \sin x\,dx = -\cos x + C$
$\frac{d}{dx}\tan x = \sec^2 x$	$\int \sec^2 x\,dx = \tan x + C$
$\frac{d}{dx}\cot x = -\csc^2 x$	$\int \csc^2 x\,dx = -\cot x + C$
$\frac{d}{dx}\sec x = \sec x \tan x$	$\int \sec x \tan x\,dx = \sec x + C$
$\frac{d}{dx}\csc x = -\csc x \cot x$	$\int \csc x \cot x\,dx = -\csc x + C$

Calculator/Computers The HP-28S calculator is capable of giving an exact symbolic result for $\int f(x)\,dx$ only when $f(x)$ is a polynomial. In Maple an antiderivative with respect to x is found by typing the command int($f(x)$, x);. In Mathematica an antiderivative is obtained by typing Integrate[$f(x)$, x].

Differential Equations In several exercise sets you were asked to verify that a given function satisfies a **differential equation**. Roughly, a differential equation is an equation that involves derivatives or the differential of an unknown function. For example,

highest derivative

$$\frac{d^2y}{dx^2} + 4\frac{dy}{dx} + 8y = 0 \tag{2}$$

is a differential equation. Differential equations are classified by the **order** of the highest derivative appearing in the equation. Thus, we say that (2) is a second-order differential equation.

The goal is to *solve* differential equations. A **solution** is any differentiable function, defined explicitly or implicitly, that, when substituted into the differential equation, reduces it to an identity. A first-order differential equation of the form

$$\frac{dy}{dx} = g(x)$$

can be solved by finding the most general antiderivative of g; that is, $y = \int g(x)\, dx$.

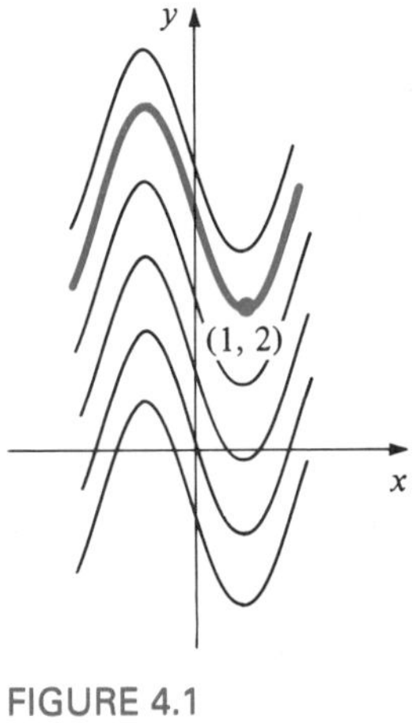

FIGURE 4.1

EXAMPLE 7 Find a function $y = f(x)$ whose graph passes through the point (1, 2) that also satisfies $dy/dx = 3x^2 - 3$.

Solution From the definition of an antiderivative, if

$$\frac{dy}{dx} = 3x^2 - 3 \quad \text{then} \quad y = \int (3x^2 - 3)\, dx$$

That is, $y = x^3 - 3x + C$. Now when $x = 1$, $y = 2$, so that $2 = 1 - 3 + C$ or $C = 4$. Hence, $y = x^3 - 3x + 4$. Thus, out of the family of antiderivatives of $3x^2 - 3$, we see in Figure 4.1 that there is only one whose graph passes through (1, 2). □

When solving a differential equation such as $dy/dx = 3x^2 - 3$ from Example 7, the specified side condition that the graph pass through (1, 2), that is, $f(1) = 2$, is called an **initial condition**. It is common practice to write an initial condition such as this as $y(1) = 2$. The solution $y = x^3 - 3x + 4$ that was determined from the family of solutions $y = x^3 - 3x + C$ by the initial condition is called a **particular solution**.

We note that an nth-order differential equation of the form $d^ny/dx^n = g(x)$ can be solved by integrating g n times. In this case the family of solutions will contain n constants of integration.

Another special kind of first-order differential equation can be solved by integration.

DEFINITION 4.2 Separable Differential Equation

A **separable first-order differential equation** is an equation $F(x, y, y') = 0$ that can be put into the form

$$h(y)\frac{dy}{dx} = g(x) \tag{3}$$

For example,

$$\frac{dy}{dx} = -\frac{x}{y}$$

is separable, since it can be rewritten as

$$y\frac{dy}{dx} = -x \tag{4}$$

It is usual practice to write a separable equation in terms of differentials. The equation in (4) can be written alternatively as

$$y\,dy = -x\,dx$$

In general, (3) is written $h(y)\,dy = g(x)\,dx$.

Method of Solution *If* $y = f(x)$ denotes a solution of (3), we must have

$$h(f(x))f'(x) = g(x)$$

and therefore, by integration,

$$\int h(f(x))f'(x)\,dx = \int g(x)\,dx \tag{5}$$

But $dy = f'(x)\,dx$, so (5) is the same as

$$\int h(y)\,dy = \int g(x)\,dx \tag{6}$$

Equation (6) indicates the procedure for solving separable differential equations: Integrate both sides of

$$h(y)\,dy = g(x)\,dx$$

EXAMPLE 8 Solve $\dfrac{dy}{dx} = -\dfrac{x}{y}$.

Solution Rewriting the given equation in differential form

$$y\,dy = -x\,dx$$

and integrating both sides give

$$\int y\,dy = -\int x\,dx \quad \text{or} \quad \frac{y^2}{2} = -\frac{x^2}{2} + C$$

Thus, a family of solutions is defined by $x^2 + y^2 = C_1$, where we have replaced the arbitrary constant $2C$ by C_1. □

▲ ***Remarks*** (*i*) It was mentioned previously that it is often necessary to rewrite an integrand in a more tractable form before carrying out the integration. This rewriting may sometimes entail division. For example, to evaluate $\int \left(\frac{6x^3 - 1}{x^2}\right) dx$ we first divide termwise:

$$\int \left(6x - \frac{1}{x^2}\right) dx = \int (6x - x^{-2})\,dx$$

In the last form we are now in position to use (1).

(*ii*) Of course, not every first-order differential equation is separable. You should verify that the first-order equation $dy/dx + 2y = x$ is not separable. We shall study first-order differential equations in more detail in Chapter 6.

(*iii*) Even though we are carrying out two integrations in the solution of a separable differential equation, there is no need to introduce two constants of integration. The justification for this statement is the same as that in the note following Theorem 4.2. As a rule, a family of solutions for a first-order differential equation contains one arbitrary constant, for a second-order differential equation a family of solutions contains two constants, and so on.

(*iv*) Constants of integration can be relabeled (as was done in Example 8) in a manner that may prove more convenient in a given problem.

EXERCISES 4.1

Answers to odd-numbered problems begin on page A-79.

In Problems 1–30 evaluate the given indefinite integral.

1. $\int 3\,dx$
2. $\int (\pi^2 - 1)\,dx$
3. $\int x^5\,dx$
4. $\int 5x^{1/4}\,dx$
5. $\int \frac{dx}{\sqrt[3]{x}}$
6. $\int \sqrt[3]{x^2}\,dx$
7. $\int (1 - t^{-0.52})\,dt$
8. $\int 10w\sqrt{w}\,dw$
9. $\int (3x^2 + 2x - 1)\,dw$
10. $\int \left(2\sqrt{t} - t - \frac{9}{t^2}\right) dt$
11. $\int \sqrt{x}(x^2 - 2)\,dx$
12. $\int \left(\frac{5}{\sqrt[3]{s^2}} + \frac{2}{\sqrt{s^3}}\right) ds$
13. $\int (4x + 1)^2\,dx$
14. $\int (\sqrt{x} - 1)^2\,dx$
15. $\int (x + 2)(x - 2)\,dx$
16. $\int \frac{x^3 + 8}{x + 2}\,dx$
17. $\int \frac{r - 10}{r^3}\,dr$
18. $\int \frac{(x + 1)^2}{\sqrt{x}}\,dx$
19. $\int \frac{x^{-1} - x^{-2} + x^{-3}}{x^2}\,dx$
20. $\int \frac{t^2 - 8t + 1}{(2t)^4}\,dt$

21. $\int (4w - 1)^3\, dw$

22. $\int (5u - 1)(3u^3 + 2)\, du$

23. $\int (4 \sin x - x^{-5})\, dx$

24. $\int (-3 \cos x + 4 \sec^2 x)\, dx$

25. $\int \csc x(\csc x - \cot x)\, dx$

26. $\int \frac{\sin t}{\cos^2 t}\, dt$

27. $\int \frac{2 + 3 \sin^2 x}{\sin^2 x}\, dx$

28. $\int \left(4\theta - \frac{2}{\sec \theta}\right) d\theta$

29. $\int (x \cos^2 x + x \sin^2 x)\, dx$

30. $\int \tan^2 x\, dx$

In Problems 31–36 verify the given result by differentiation.

31. $\int \frac{1}{\sqrt{2x+1}}\, dx = \sqrt{2x+1} + C$

32. $\int (2x^2 - 4x)^9(x - 1)\, dx = \frac{1}{40}(2x^2 - 4x)^{10} + C$

33. $\int \cos 4x\, dx = \frac{1}{4} \sin 4x + C$

34. $\int \sin x \cos x\, dx = \frac{1}{2} \sin^2 x + C$

35. $\int x \sin x^2\, dx = -\frac{1}{2} \cos x^2 + C$

36. $\int \frac{\cos x}{\sin^3 x}\, dx = -\frac{1}{2 \sin^2 x} + C$

In Problems 37 and 38 perform the indicated operations.

37. $\frac{d}{dx} \int (x^2 - 4x + 5)\, dx$

38. $\int \frac{d}{dx}(x^2 - 4x + 5)\, dx$

In Problems 39–44 solve the given differential equation.

39. $\frac{dy}{dx} = 6x^2 + 9$

40. $\frac{dy}{dx} = 10x + 3\sqrt{x}$

41. $\frac{dy}{dx} = \frac{1}{x^2}$

42. $\frac{dy}{dx} = \frac{(2 + x)^2}{x^5}$

43. $\frac{dy}{dx} = 1 - 2x + \sin x$

44. $\frac{dy}{dx} = \frac{1}{\cos^2 x}$

45. Find a function $y = f(x)$ whose graph passes through the point (2, 3) that also satisfies $dy/dx = 2x - 1$.

46. Find a function $y = f(x)$ so that $dy/dx = 1/\sqrt{x}$ and $f(9) = 1$.

47. If $f''(x) = 2x$, find $f'(x)$ and $f(x)$.

48. Find a function f such that $f''(x) = 1$, $f'(-1) = 2$, and $f(-1) = 0$.

49. Find a function f such that $f''(x) = 12x^2 + 2$ for which the slope of the tangent line to its graph at (1, 1) is 3.

50. If $f^{(n)}(x) = 0$, what is f?

In Problems 51–56 solve the given differential equation by separation of variables.

51. $\frac{dy}{dx} = \frac{y^3}{x^2}$

52. $\frac{dy}{dx} = \frac{1}{5y^4}$

53. $\frac{dy}{dx} = \left(\frac{1 + x}{1 + y}\right)^2$

54. $\frac{dy}{dx} = \sqrt{xy}$

55. $\frac{dy}{dx} = \frac{1 + 5x^2}{x^2 \sin y}$

56. $\frac{dy}{dx} = y^3 \cos x$

In Problems 57 and 58 solve the given differential equation subject to the indicated initial condition.

57. $\frac{dy}{dx} = \frac{1}{(xy)^2}$, $y(1) = 3$

58. $\frac{dy}{dx} = \frac{2x + \sec^2 x}{2y}$, $y(0) = -2$

59. A bucket that contains liquid is rotating about a vertical axis at a constant angular velocity ω. The shape of the cross-section of the rotating liquid in the xy-plane is determined from

$$\frac{dy}{dx} = \frac{\omega^2}{g} x$$

With coordinate axes as shown in Figure 4.2, find $y = f(x)$.

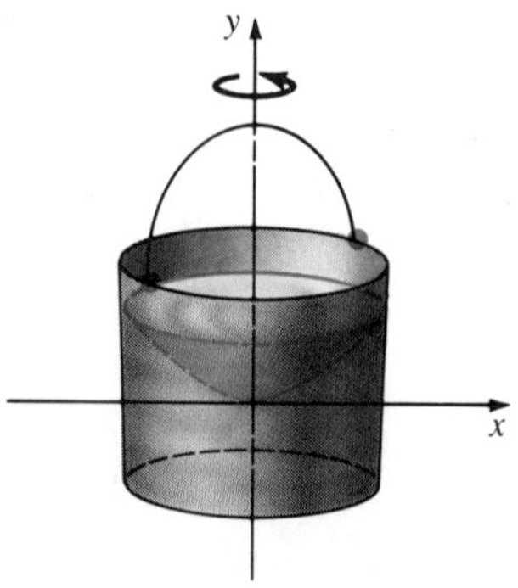

FIGURE 4.2

60. The ends of a beam of length L rest on two supports as shown in Figure 4.3. With a uniform load on the beam its shape (or elastic curve) is determined from

$$EIy'' = \frac{qL}{2}x - \frac{q}{2}x^2$$

where E, I, and q are constants. Find $y = f(x)$ if $f(0) = 0$ and $f'(L/2) = 0$.

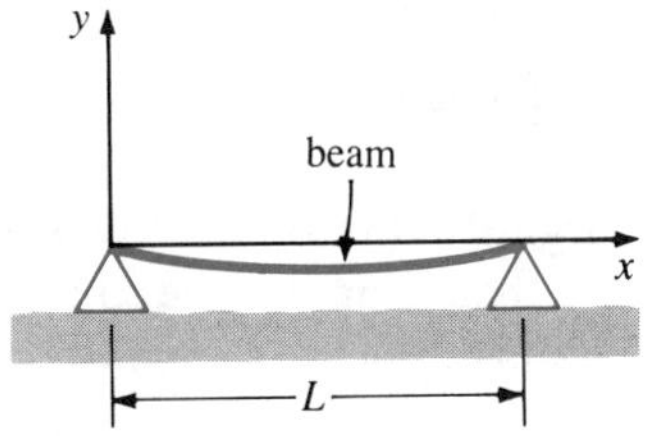

FIGURE 4.3

61. The height h of water that is flowing through an orifice at the bottom of a tank in the form of right circular cylinder (standing on end) is given by

$$\frac{dh}{dt} = -\frac{A_o}{A_w}\sqrt{2gh}$$

where $g = 32$ ft/s^2, and A_o and A_w are the cross-sectional areas of the tank and orifice, respectively. See Figure 4.4.

(a) Solve the differential equation if the initial height of the water is 20 ft and $A_o = 1/4$ ft^2 and $A_w = 50$ ft^2.

(b) At what time is the tank empty?

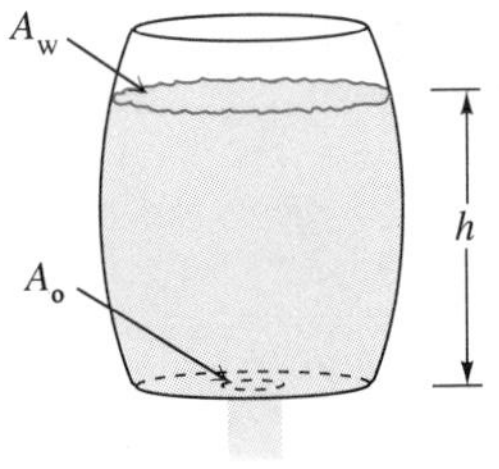

FIGURE 4.4

62. A rocket is shot vertically upward from the ground with an initial velocity v_0. See Figure 4.5. If the positive direction is taken to be upward, then in the absence of air resistance, the differential equation for the velocity v after fuel burnout is

$$v\frac{dv}{dy} = -\frac{k}{y^2}$$

where y is a positive constant.

(a) Solve the differential equation.

(b) If $k = gR^2$ and $R = 4000$ m, use a calculator to show that the "escape velocity" of a rocket is approximately $v_0 = 25{,}000$ mi/h.

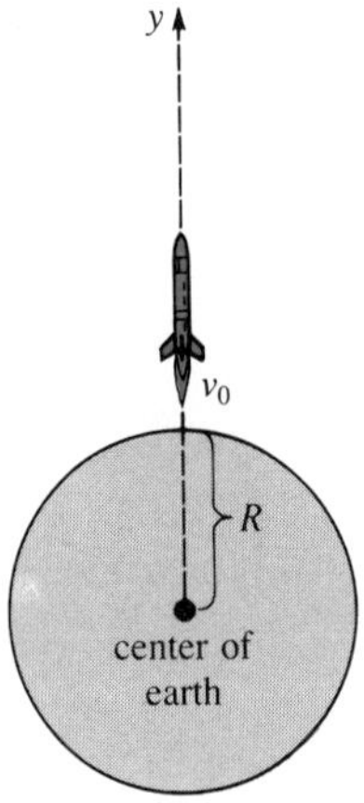

FIGURE 4.5

63. Suppose a sphere of ice melts at a rate proportional to its surface area. Determine the volume V of the sphere at any time t.

4.2 INDEFINITE INTEGRALS AND THE *u*-SUBSTITUTION

So far we have considered only antiderivatives of rational powers of x:

$$\int x^n\,dx = \frac{x^{n+1}}{n+1} + C, \qquad n \neq -1 \tag{1}$$

and of the six trigonometric functions $\cos x$, $\sin x$, $\sec^2 x$, $\csc^2 x$, $\sec x \tan x$, and $\csc x \cot x$. In the present exposition, we shall examine the indefinite analogues of the Power Rule for Functions (Theorem 2.12) and the indefinite integral analogues of the derivative formulas I–VI of Section 2.6.

The Indefinite Integral of a Power of a Function If we wish to find a function F such that

$$\int (5x+1)^{1/2}\,dx = F(x) + C$$

we must have

$$F'(x) = (5x+1)^{1/2}$$

By reasoning "backward," we could argue that to obtain $(5x+1)^{1/2}$ we must have differentiated $(5x+1)^{3/2}$. It would then seem that we could proceed as in (1)—namely, increase the power by 1 and divide by the new power:

$$\int (5x+1)^{1/2}\,dx = \frac{(5x+1)^{3/2}}{3/2} + C = \frac{2}{3}(5x+1)^{3/2} + C \tag{2}$$

Regrettably the "answer" in (2) does not check, since the Power Rule for Functions gives

$$\begin{aligned}\frac{d}{dx}\left[\frac{2}{3}(5x+1)^{3/2} + C\right] &= \frac{2}{3}\frac{3}{2}(5x+1)^{1/2}\cdot 5\\ &= 5(5x+1)^{1/2} \neq (5x+1)^{1/2}\end{aligned}$$

To account for the missing factor of 5 in (2) we use Theorem 4.3 and a little bit of cleverness:

$$\begin{aligned}\int (5x+1)^{1/2}\,dx &= \int (5x+1)^{1/2}\boxed{\frac{1}{5}\cdot 5}\,dx \qquad \leftarrow \text{equal to 1}\\ &= \frac{1}{5}\int \boxed{(5x+1)^{1/2}5}\,dx \qquad \leftarrow \text{derivative of } \frac{(5x+1)^{3/2}}{3/2}\\ &= \frac{1}{5}\frac{2}{3}(5x+1)^{3/2} + C\\ &= \frac{2}{15}(5x+1)^{3/2} + C\end{aligned}$$

You should now verify by differentiation that the last function is indeed an antiderivative of $(5x+1)^{1/2}$.

The key to evaluating indefinite integrals such as

$$\int \frac{x}{(4x^2+3)^6}\,dx \quad \text{and} \quad \int \sin 10x\,dx$$

lies in the next result, which is the indefinite integration form of the Chain Rule.

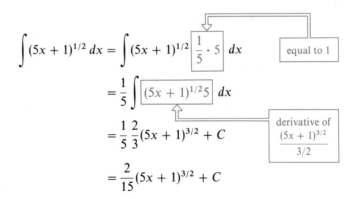

THEOREM 4.4 Indefinite Integral Form of Chain Rule

If F is an antiderivative of f, then

$$\int f(g(x))g'(x)\,dx = F(g(x)) + C \tag{3}$$

Proof By the Chain Rule,

$$\frac{d}{dx}F(g(x)) = F'(g(x))g'(x) = f(g(x))g'(x)$$

Hence, from the definition of an antiderivative,

$$\int f(g(x))g'(x)\,dx = F(g(x)) + C$$

■

To apply (3) we must be certain that we have the exact form

derivative of function inside f

$$\int f(\boxed{g(x)})g'(x)\,dx$$

In particular, if $F(x) = x^{n+1}/(n+1)$, n a rational number, $n \neq -1$, and if $u = g(x)$ is a differentiable function, then

$$F(g(x)) = \frac{[g(x)]^{n+1}}{n+1} \quad \text{and} \quad \frac{d}{dx}F(g(x)) = [g(x)]^n g'(x)$$

Hence, Theorem 4.4 immediately implies

$$\int [g(x)]^n g'(x)\,dx = \frac{[g(x)]^{n+1}}{n+1} + C \tag{4}$$

On a practical level it is often helpful to **change the variable** in an integration problem by the **substitutions**

$$u = g(x) \quad \text{and} \quad du = g'(x)\,dx$$

in (3). Thus, (4) can be summarized in the following manner.

THEOREM 4.5 Indefinite Integral of a Power of a Function

If n is a rational number and $u = g(x)$ is a differentiable function, then for $n \neq -1$,

$$\int u^n\,du = \frac{u^{n+1}}{n+1} + C \tag{5}$$

EXAMPLE 1 Evaluate $\int \frac{x}{(4x^2+3)^6}\,dx$.

Solution Let us rewrite the integral as

$$\int (4x^2+3)^{-6}x\,dx$$

and make the identifications

$$u = 4x^2 + 3 \quad \text{and} \quad du = 8x\,dx$$

Now, to get the precise form $\int u^{-6}\,du$ we must adjust the integrand by multiplying and dividing by 8. By (5) we obtain

$$\begin{aligned}\int (4x^2+3)^{-6}x\,dx &= \frac{1}{8}\int \overbrace{(4x^2+3)^{-6}}^{u^{-6}}\overbrace{(8x\,dx)}^{du} \qquad \boxed{\text{substitution}}\\ &= \frac{1}{8}\int u^{-6}\,du\\ &= \frac{1}{8}\cdot\frac{u^{-5}}{-5} + C \qquad \boxed{\text{resubstitution}}\\ &= -\frac{1}{40}(4x^2+3)^{-5} + C\end{aligned}$$

Check By the Power Rule for Functions,

$$\frac{d}{dx}\left[-\frac{1}{40}(4x^2+3)^{-5} + C\right] = \left(-\frac{1}{40}\right)(-5)(4x^2+3)^{-6}(8x) = \frac{x}{(4x^2+3)^6}$$

□

EXAMPLE 2 Evaluate $\int (x^2+2)^3 x\,dx$.

Solution If $u = x^2 + 2$ then $du = 2x\,dx$

Thus from (5),

$$\begin{aligned}\int (x^2+2)^3x\,dx = \frac{1}{2}\int \overbrace{(x^2+2)^3}^{u^3}\overbrace{(2x\,dx)}^{du} &= \frac{1}{2}\int u^3\,du\\ &= \frac{1}{2}\cdot\frac{u^4}{4} + C_1\\ &= \frac{1}{8}(x^2+2)^4 + C_1 \qquad (6)\end{aligned}$$

Alternative Solution If we use the Binomial Theorem before integrating, we have

$$\begin{aligned}\int (x^2+2)^3 x\,dx &= \int (x^6+6x^4+12x^2+8)x\,dx\\ &= \int (x^7+6x^5+12x^3+8x)\,dx\\ &= \frac{x^8}{8}+6\cdot\frac{x^6}{6}+12\cdot\frac{x^4}{4}+8\cdot\frac{x^2}{2}+C_2\\ &= \frac{1}{8}x^8+x^6+3x^4+4x^2+C_2 \qquad (7)\end{aligned}$$

Note that (6) can be written as

$$\frac{1}{8}(x^2+2)^4+C_1=\frac{1}{8}x^8+x^6+3x^4+4x^2+2+C_1$$

Although (6) and (7) are not exactly the same, the two results differ only by a constant. □

EXAMPLE 3 Evaluate $\int \sqrt[3]{(7-2x^3)^4}\,x^2\,dx$.

Solution We first write the integral as

$$\int (7-2x^3)^{4/3}x^2\,dx$$

and then make the identifications

$$u=7-2x^3 \quad\text{and}\quad du=-6x^2\,dx$$

Hence,

$$\begin{aligned}\int (7-2x^3)^{4/3}x^2\,dx &= -\frac{1}{6}\int (7-2x^3)^{4/3}(-6x^2\,dx)\\ &= -\frac{1}{6}\int u^{4/3}\,du\\ &= -\frac{1}{6}\frac{u^{7/3}}{7/3}+C\\ &= -\frac{1}{14}(7-2x^3)^{7/3}+C\end{aligned}$$

□

Indefinite Integrals of Trigonometric Functions If $u=g(x)$ is a differentiable function, then the differentiation formulas

$$\frac{d}{dx}\sin u=\cos u\frac{du}{dx} \quad\text{and}\quad \frac{d}{dx}(-\cos u)=\sin u\frac{du}{dx}$$

yield, in turn, the integration formulas

$$\int \cos u \frac{du}{dx}\, dx = \sin u + C \tag{8}$$

$$\int \sin u \frac{du}{dx}\, dx = -\cos u + C \tag{9}$$

Since $du = g'(x)\, dx = \dfrac{du}{dx}\, dx$, (8) and (9) are equivalent to

$$\int \cos u\, du = \sin u + C \quad \text{and} \quad \int \sin u\, du = -\cos u + C$$

In general, I–VI of Section 2.6 give the following integration results.

Indefinite Integrals of Trigonometric Functions

I′ $\int \cos u\, du = \sin u + C$	II′ $\int \sin u\, du = -\cos u + C$
III′ $\int \sec^2 u\, du = \tan u + C$	IV′ $\int \csc^2 u\, du = -\cot u + C$
V′ $\int \sec u \tan u\, du = \sec u + C$	VI′ $\int \csc u \cot u\, du = -\csc u + C$

EXAMPLE 4 Evaluate $\int 3 \cos 3x\, dx$.

Solution If $\quad u = 3x \quad$ then $\quad du = 3\, dx$

Hence, we recognize the given problem as being exactly of form I′:

$$\int \cos \underbrace{3x}_{u}\underbrace{(3\, dx)}_{du} = \int \cos u\, du = \sin u + C = \sin 3x + C$$ □

As in the discussion of (5), the differential du is probably the most important part in each of I′–VI′. Before applying one of these results, we may need to "fix up" or adjust an integrand by multiplying and dividing by a constant in order to obtain the appropriate du.

EXAMPLE 5 Evaluate $\int \sin 10x\, dx$.

Solution If $u = 10x$ then we need $du = 10\, dx$

Accordingly, we write

$$\int \sin 10x\, dx = \frac{1}{10}\int \sin \overbrace{10x}^{u}\overbrace{(10\, dx)}^{du} = \frac{1}{10}\int \sin u\, du$$

$$= \frac{1}{10}(-\cos u) + C \qquad \boxed{\text{from II}'}$$

$$= -\frac{1}{10}\cos 10x + C$$

□

After a while, try to perform basic integrations doing the *u*-substitution mentally.

EXAMPLE 6 Evaluate $\int \sec^2(1 - 4x)\, dx$.

Solution
$$\int \sec^2(1 - 4x)\, dx = -\frac{1}{4}\int \sec^2(1 - 4x)(-4\, dx)$$

$$= -\frac{1}{4}\tan(1 - 4x) + C \qquad \boxed{\text{from III}'}$$

Check
$$\frac{d}{dx}\left[-\frac{1}{4}\tan(1 - 4x) + C\right] = -\frac{1}{4}\sec^2(1 - 4x)\frac{d}{dx}(1 - 4x)$$

$$= -\frac{1}{4}\sec^2(1 - 4x)(-4) = \sec^2(1 - 4x)$$

□

The next example shows that not every indefinite integral of a trigonometric function is one of the types I′–VI′.

EXAMPLE 7 Evaluate $\int \cos^4 x \sin x\, dx$.

Solution For emphasis we rewrite the problem as

$$\int (\cos x)^4 \sin x\, dx$$

With the identifications

$$u = \cos x \quad \text{and} \quad du = -\sin x\, dx \cdot$$

we recognize

$$\int (\cos x)^4 \sin x\, dx = -\int \overbrace{(\cos x)^4}^{\boxed{u^4}}\overbrace{(-\sin x\, dx)}^{\boxed{du}} = -\int u^4\, du$$

Hence, from (5) we obtain

$$\int (\cos x)^4 \sin x\, dx = -\frac{u^5}{5} + C = -\frac{1}{5}\cos^5 x + C \qquad \square$$

Useful Identities Sometimes it may be necessary to use a trigonometric identity to solve a problem. The half-angle formulas

$$\cos^2 x = \frac{1 + \cos 2x}{2} \quad \text{and} \quad \sin^2 x = \frac{1 - \cos 2x}{2}$$

are particularly useful in problems that require the antiderivatives of $\cos^2 x$ and $\sin^2 x$.

EXAMPLE 8 Evaluate $\int \cos^2 x\, dx$.

Solution It should be verified that the integral is *not* of the form $\int u^2\, du$. Now using the half-angle formula $\cos^2 x = (1 + \cos 2x)/2$, we obtain

$$\begin{aligned}
\int \cos^2 x\, dx &= \int \frac{1 + \cos 2x}{2}\, dx \\
&= \frac{1}{2}\int [1 + \cos 2x]\, dx \\
&= \frac{1}{2}\left[\int dx + \frac{1}{2}\int \cos 2x(2\, dx)\right] \\
&= \frac{1}{2}\left[x + \frac{1}{2}\sin 2x\right] + C \qquad \boxed{\text{from (1) and I}'} \\
&= \frac{1}{2}x + \frac{1}{4}\sin 2x + C
\end{aligned} \qquad \square$$

We shall consider substitutions and integrals of powers of trigonometric functions in greater detail in Chapter 8.

▲ ***Remark*** The following example illustrates a common, but *totally incorrect*, procedure for evaluating an indefinite integral. Since $2x/2x = 1$,

$$\int (4 + x^2)^{1/2}\,dx = \int (4 + x^2)^{1/2}\frac{2x}{2x}\,dx = \frac{1}{2x}\int (4 + x^2)^{1/2}2x\,dx$$
$$= \frac{1}{2x}\int u^{1/2}\,du$$
$$= \frac{1}{2x}\cdot\frac{2}{3}(4 + x^2)^{3/2} + C$$

You should verify that differentiation of the latter function does *not* yield $(4 + x^2)^{1/2}$. The mistake is in the first line of the "solution." *Variables*, in this case $2x$, *cannot be brought outside an integral symbol.* If $u = x^2 + 4$, then the integrand lacks the function $du = 2x\,dx$; in fact, there is no way of adjusting the problem to fit the form given in (5). At this juncture the integral $\int (4 + x^2)^{1/2}\,dx$ simply cannot be evaluated.

EXERCISES 4.2 *Answers to odd-numbered problems begin on page A-79.*

In Problems 1–48 evaluate the given indefinite integral.

1. $\int \sqrt{1 - 4x}\,dx$
2. $\int (8x + 2)^{1/3}\,dx$
3. $\int \frac{dx}{(5x + 1)^3}$
4. $\int (7 - x)^{49}\,dx$
5. $\int \sqrt[5]{(3 - 4x)^3}\,dx$
6. $\int \frac{dx}{\sqrt[3]{(2x + 1)^2}}$
7. $\int 2x\sqrt{x^2 + 4}\,dx$
8. $\int x\sqrt[3]{7x^2 + 1}\,dx$
9. $\int \frac{z}{\sqrt[3]{z^2 + 9}}\,dz$
10. $\int \frac{x}{\sqrt[3]{(1 - x^2)^4}}\,dx$
11. $\int (4x^2 - 16x + 7)^4(x - 2)\,dx$
12. $\int (x^2 + 2x - 10)^{2/3}(5x + 5)\,dx$
13. $\int \frac{x^2 + 1}{\sqrt[3]{x^3 + 3x - 16}}\,dx$
14. $\int \frac{s(s^3 - 4)}{\sqrt{s^5 - 10s^2 + 6}}\,ds$
15. $\int \sqrt{3 - \frac{2}{v}}\,\frac{dv}{v^2}$
16. $\int \sqrt{\frac{x^3 + 1}{x^3}}\,\frac{dx}{x^4}$
17. $\int \sqrt[3]{\frac{1 - \sqrt[3]{x}}{x^2}}\,dx$
18. $\int \sqrt{\frac{2 + 3\sqrt{x}}{x}}\,dx$
19. $\int \frac{dt}{\sqrt{t}(4 + \sqrt{t})^5}$
20. $\int \frac{\sqrt{z}}{(9 + 6z^{3/2})^3}\,dz$
21. $\int (x^2 - 2x + 1)^3\,dx$
22. $\int (4y^2 + 4y + 1)^{2/3}\,dy$
23. $\int \sin 4x\,dx$
24. $\int 5\cos\frac{x}{2}\,dx$
25. $\int (\sqrt{2t} - \cos 6t)\,dt$
26. $\int \csc^2(0.1x)\,dx$
27. $\int x\cos x^2\,dx$
28. $\int x^2\sec^2 x^3\,dx$
29. $\int \frac{1}{\sec(5x + 1)}\,dx$
30. $\int \sin(2 - 3x)\,dx$
31. $\int \frac{\csc\sqrt{x}\cot\sqrt{x}}{\sqrt{x}}\,dx$
32. $\int \frac{\sin(1/x)}{x^2}\,dx$
33. $\int (z + 1)\csc^2(z^2 + 2z)\,dz$
34. $\int \frac{\cos\sqrt[3]{1 - 3x}}{\sqrt[3]{(1 - 3x)^2}}\,dx$
35. $\int \sin^5 3x\cos 3x\,dx$
36. $\int \frac{\sin t}{\sqrt{4 + \cos t}}\,dt$

37. $\int \tan^2 2x \sec^2 2x \, dx$

38. $\int \tan x \sec^2 x \, dx$

39. $\int \tan^2 7x \, dx$

40. $\int \tan 5v \sec 5v \, dv$

41. $\int \frac{2 + \cos x}{\sin^2 x} \, dx$

42. $\int \frac{(1 + \sin x)^4}{\sec x + \tan x} \, dx$

43. $\int \sin^2 x \, dx$

44. $\int \cos^2 \pi x \, dx$

45. $\int \cos^2 4x \, dx$

46. $\int \sin^2 \frac{x}{5} \, dx$

47. $\int (1 + \cos 2x)^2 \, dx$

48. $\int \cos 2x \sin^2 x \, dx$

In Problems 49–52 evaluate the given indefinite integral.

49. $\int \cos^3 x \, dx$ [*Hint:* $\cos^3 x = \cos^2 x \cdot \cos x$]

50. $\int \sin^3 2x \, dx$

51. $\int \frac{t}{\sqrt{t + 2}} \, dt$ [*Hint:* $t = t + 2 - 2$]

52. $\int \frac{4z + 3}{(4z + 5)^3} \, dz$

In Problems 53–56 solve the given differential equation.

53. $\frac{dy}{dx} = \sqrt[3]{1 - x}$

54. $\frac{dy}{dx} = \frac{(1 - \tan x)^5}{\cos^2 x}$

55. $\frac{dy}{dx} = (\sin x \cos y)^2$

56. $\frac{dy}{dx} = \frac{(1 - 2x)^3 \sqrt{1 + 2y^2}}{y}$

57. Find a function $y = f(x)$ whose graph passes through the point $(\pi, -1)$ that also satisfies $dy/dx = 1 - \sin x$.

58. Find a function f such that $f''(x) = (1 + 2x)^5$, $f(0) = 0$, and $f'(0) = 0$.

59. Show that

(a) $\int \sin x \cos x \, dx = \frac{1}{2} \sin^2 x + C_1$

(b) $\int \sin x \cos x \, dx = -\frac{1}{2} \cos^2 x + C_2$

(c) $\int \sin x \cos x \, dx = -\frac{1}{4} \cos 2x + C_3$

60. In Problem 59:

(a) Verify that the derivative of each answer in parts (**a**), (**b**), and (**c**) is $\sin x \cos x$.

(b) By a trigonometric identity, show how the result in part (**b**) can be obtained from the answer in part (**a**).

(c) By adding the results in parts (**a**) and (**b**), obtain the result in part (**c**).

4.3 SIGMA NOTATION

In Section 4.1 we defined the *indefinite integral.* Later we shall define a related but different concept, the *definite integral.* We shall see that the definite integral is defined as the limit of a certain kind of *sum.* Therefore, it is helpful to introduce a special notation that enables us to write an indicated sum of constants such as

$$1 + 2 + 3 + \cdots + n$$

$$2^2 + 4^2 + 6^2 + \cdots + (2n)^2$$

and

$$\frac{1}{3} + \frac{1}{5} + \frac{1}{7} + \cdots + \frac{1}{2n + 1}$$

in a concise manner.

Let a_k be a real number that depends on an integer k. We denote the sum

$a_1 + a_2 + a_3 + \cdots + a_n$ by the symbol $\Sigma_{k=1}^{n} a_k$; that is,

$$\sum_{k=1}^{n} a_k = a_1 + a_2 + a_3 + \cdots + a_n \tag{1}$$

Since Σ is the capital Greek letter sigma, (1) is called **sigma notation** or **summation notation**. The variable k is called the **index of summation**. Thus, $\Sigma_{k=1}^{n} a_k$ is the sum of all numbers of the form a_k as k takes on the successive values $k = 1$, $k = 2, \ldots$, and concludes with $k = n$.

EXAMPLE 1

(*a*) $$\sum_{k=1}^{4} (3k - 1) = [3(1) - 1] + [3(2) - 1] + [3(3) - 1] + [3(4) - 1]$$
$$= 2 + 5 + 8 + 11$$

(*b*) $$\sum_{k=1}^{5} \frac{1}{(k+1)^2} = \frac{1}{2^2} + \frac{1}{3^2} + \frac{1}{4^2} + \frac{1}{5^2} + \frac{1}{6^2}$$

(*c*) $$\sum_{k=1}^{100} k^3 = 1^3 + 2^3 + 3^3 + \cdots + 98^3 + 99^3 + 100^3$$ □

EXAMPLE 2

(*a*) The sum of the first ten positive odd integers

$$1 + 3 + 5 + 7 + \cdots + 19$$

can be written succinctly as $\sum_{k=1}^{10} (2k - 1)$.

(*b*) It is also easily verified that the sum of the first ten positive even integers

$$2 + 4 + 6 + 8 + \cdots + 20$$

is $\sum_{k=1}^{10} 2k$. □

The index of summation need not start at the value $k = 1$; for example,

$$\sum_{k=3}^{5} 2^k = 2^3 + 2^4 + 2^5 \quad \text{and} \quad \sum_{k=0}^{5} 2^k = 2^0 + 2^1 + 2^2 + 2^3 + 2^4 + 2^5$$

Note that the sum in part (*a*) of Example 2 can also be written as $\Sigma_{k=0}^{9} (2k + 1)$. However, in a general discussion we shall always assume that the summation index starts at $k = 1$. This assumption is for convenience rather than necessity.

The index of summation is often called a **dummy variable**, since the symbol itself is not important; it is the successive integer values of the index and the corresponding sum that are important. In general,

$$\sum_{k=1}^{n} a_k = \sum_{i=1}^{n} a_i = \sum_{j=1}^{n} a_j = \sum_{m=1}^{n} a_m$$

and so on.

EXAMPLE 3

$$\sum_{k=1}^{10} 4^k = \sum_{i=1}^{10} 4^i = \sum_{j=1}^{10} 4^j = 4^1 + 4^2 + 4^3 + \cdots + 4^{10} \qquad \square$$

Properties The following is a list of some important properties of the sigma notation.

THEOREM 4.6 Properties

For positive integers m and n,

(*i*) $\sum_{k=1}^{n} ca_k = c \sum_{k=1}^{n} a_k$, where c is any constant

(*ii*) $\sum_{k=1}^{n} (a_k \pm b_k) = \sum_{k=1}^{n} a_k \pm \sum_{k=1}^{n} b_k$

(*iii*) $\sum_{k=1}^{n} a_k = \sum_{k=1}^{m} a_k + \sum_{k=m+1}^{n} a_k, \quad m < n$

The proof of formula (*i*) is an immediate consequence of the distributive law. The proofs of (*ii*) and (*iii*) are left as exercises.

EXAMPLE 4

(*a*) From Theorem 4.6(*i*) and (*ii*), we find

$$\sum_{k=1}^{20} (3k^2 + 4k) = 3 \sum_{k=1}^{20} k^2 + 4 \sum_{k=1}^{20} k$$

(*b*) From Theorem 4.6(*iii*), we can write

$$\sum_{k=1}^{50} k^2 = \sum_{k=1}^{3} k^2 + \sum_{k=4}^{50} k^2 = (1^2 + 2^2 + 3^2) + (4^2 + 5^2 + 6^2 + \cdots + 50^2) \qquad \square$$

If c is a constant—that is, independent of a summation index k—then $\sum_{k=1}^{n} c$ means

$$c + c + c + \cdots + c$$

Since there are n c's in this sum, we have

$$\sum_{k=1}^{n} c = n \cdot c \tag{2}$$

EXAMPLE 5 From (2),

$$\sum_{k=1}^{75} 6 = 75 \cdot 6 = 450 \qquad \square$$

The sum of the first n positive integers can be written $\Sigma_{k=1}^{n} k$. If this sum is denoted by S, then

$$S = 1 + 2 + 3 + \cdots + (n-1) + n \tag{3}$$

can also be written as

$$S = n + (n-1) + (n-2) + \cdots + 1 \tag{4}$$

If we add (3) and (4), then

$$2S = \underbrace{(n+1) + (n+1) + (n+1) + \cdots + (n+1)}_{\boxed{n \text{ terms of } n+1}} = n(n+1)$$

Solving for S gives $S = n(n+1)/2$ or

$$\sum_{k=1}^{n} k = \frac{n(n+1)}{2} \tag{5}$$

EXAMPLE 6 Find the sum of the first 100 consecutive positive integers.

Solution The required sum is

$$1 + 2 + 3 + \cdots + 99 + 100 = \sum_{k=1}^{100} k$$

With $n = 100$, it follows from (5) that

$$\sum_{k=1}^{100} k = \frac{n(n+1)}{2} = \frac{100(101)}{2} = 5050$$

□

Summation Formulas Formulas (2) and (5) are two of several **summation formulas** that will be of use in the succeeding sections. For completeness we include them in the following list. The number n is a positive integer.

I $\displaystyle\sum_{k=1}^{n} c = nc$ II $\displaystyle\sum_{k=1}^{n} k = \frac{n(n+1)}{2}$

III $\displaystyle\sum_{k=1}^{n} k^2 = \frac{n(n+1)(2n+1)}{6}$ IV $\displaystyle\sum_{k=1}^{n} k^3 = \frac{n^2(n+1)^2}{4}$

V $\displaystyle\sum_{k=1}^{n} k^4 = \frac{n(n+1)(6n^3+9n^2+n-1)}{30}$

As we have already seen, I and II can be derived readily; derivations of the remaining formulas are not quite so simple. You should be able to derive III with the aid of hints supplied in Problems 46, 47, and 48.

EXAMPLE 7 Evaluate $\displaystyle\sum_{k=1}^{10} k^2$.

Solution With $n = 10$, from formula III we have

$$\sum_{k=1}^{10} k^2 = \frac{10(11)(21)}{6} = 385 \qquad \square$$

EXAMPLE 8 Evaluate $\sum_{k=1}^{10} (k+2)^3$.

Solution By the Binomial Theorem and Theorem 4.6(*ii*), we can write

$$\sum_{k=1}^{10} (k+2)^3 = \sum_{k=1}^{10} (k^3 + 6k^2 + 12k + 8)$$
$$= \sum_{k=1}^{10} k^3 + 6\sum_{k=1}^{10} k^2 + 12\sum_{k=1}^{10} k + \sum_{k=1}^{10} 8$$

With $n = 10$, it follows from summation formulas IV, III, II, and I, respectively, that

$$\sum_{k=1}^{10} (k+2)^3 = \frac{10^2 11^2}{4} + 6\frac{10(11)(21)}{6} + 12\frac{10(11)}{2} + 10 \cdot 8$$
$$= 3025 + 2310 + 660 + 80 = 6075 \qquad \square$$

EXERCISES 4.3

Answers to odd-numbered problems begin on page A-79.

In Problems 1–10 expand the indicated sum.

1. $\sum_{k=1}^{5} 3k$
2. $\sum_{k=1}^{5} (2k - 3)$
3. $\sum_{k=1}^{4} \frac{2^k}{k}$
4. $\sum_{k=1}^{4} \frac{3^k}{k}$
5. $\sum_{k=1}^{10} \frac{(-1)^k}{2k+5}$
6. $\sum_{k=1}^{10} \frac{(-1)^{k-1}}{k^2}$
7. $\sum_{j=2}^{5} (j^2 - 2j)$
8. $\sum_{m=0}^{4} (m+1)^2$
9. $\sum_{k=1}^{5} \cos k\pi$
10. $\sum_{k=1}^{5} \frac{\sin(k\pi/2)}{k}$

In Problems 11–22 write the given sum using sigma notation.

11. $2^2 + 4^2 + 6^2 + 8^2 + 10^2 + 12^2$
12. $1 + 2^2 + 3^2 + 4^2 + 5^2$
13. $3 + 5 + 7 + 9 + 11 + 13 + 15$
14. $2 + 4 + 8 + 16 + 32 + 64$
15. $1 + 4 + 7 + 10 + \cdots + 37$
16. $2 + 6 + 10 + 14 + \cdots + 38$
17. $1 - \frac{1}{2} + \frac{1}{3} - \frac{1}{4} + \frac{1}{5}$
18. $-\frac{1}{2} + \frac{2}{3} - \frac{3}{4} + \frac{4}{5} - \frac{5}{6}$
19. $6 + 6 + 6 + 6 + 6 + 6 + 6 + 6$
20. $1 + \sqrt{2} + \sqrt{3} + 2 + \sqrt{5} + \cdots + 3$
21. $\cos\frac{\pi}{p}x - \frac{1}{4}\cos\frac{2\pi}{p}x + \frac{1}{9}\cos\frac{3\pi}{p}x - \frac{1}{16}\cos\frac{4\pi}{p}x$
22. $-f'(1)(x-1) + \frac{f''(1)}{3}(x-1)^2 - \frac{f'''(1)}{5}(x-1)^3 + \frac{f^{(4)}(1)}{7}(x-1)^4 - \frac{f^{(5)}(1)}{9}(x-1)^5$

In Problems 23–32 find the value of the given sum.

23. $\sum_{k=1}^{20} 2k$
24. $\sum_{k=0}^{50} (-3k)$
25. $\sum_{k=1}^{10} (k+1)$
26. $\sum_{k=1}^{1000} (2k-1)$

27. $\sum_{k=1}^{6} (k^2 + 3)$

28. $\sum_{k=1}^{5} (6k^2 - k)$

29. $\sum_{p=0}^{10} (p^3 + 4)$

30. $\sum_{i=-1}^{10} (2i^3 - 5i + 3)$

31. $\sum_{j=1}^{10} (4j^4 - 7)$

32. $\sum_{k=1}^{5} (k^2 + 1)^2$

33. Find the value of $\sum_{k=21}^{60} k^2$.

$\left[\textit{Hint:} \text{ Examine } \sum_{k=1}^{60} k^2 - \sum_{k=1}^{20} k^2.\right]$

34. Find the value of $\sum_{k=-30}^{30} k^2$.

35. Determine whether $\sum_{k=1}^{n} (k + 1)(k + 2)$ and $\sum_{j=4}^{n+3} (j - 1)(j - 2)$ are equal.

36. Show that $\sum_{k=3}^{10} (2k - 5)$ and $\sum_{j=0}^{7} (2j + 1)$ are equal.

In Problems 37 and 38 write the given sum using sigma notation so that the index of summation starts with **(a)** $k = 0$, **(b)** $k = 1$, and **(c)** $k = 2$.

37. $6 + 7 + 8 + 9 + 10 + 11 + 12$

38. $7 + 10 + 13 + 16 + 19$

In Problems 39 and 40 write the given decimal number using sigma notation.

39. 0.1111111

40. 0.232323

In Problems 41 and 42 write the given sum of n terms using sigma notation. Use summation formulas I–III to evaluate each sum.

41. $\left(3 + \frac{2}{n}\right)\frac{2}{n} + \left(3 + \frac{4}{n}\right)\frac{2}{n} + \left(3 + \frac{6}{n}\right)\frac{2}{n} + \cdots + \left(3 + \frac{2n}{n}\right)\frac{2}{n}$

42. $\left(\frac{9}{n^2}\right)\frac{3}{n} + \left(\frac{9 \cdot 4}{n^2}\right)\frac{3}{n} + \left(\frac{9 \cdot 9}{n^2}\right)\frac{3}{n} + \cdots + \left(\frac{9 \cdot n^2}{n^2}\right)\frac{3}{n}$

43. Find the value of $\sum_{k=1}^{400} (\sqrt{k} - \sqrt{k - 1})$.

44. Find the value of $\sum_{k=1}^{100} \frac{1}{k(k + 1)}$.

$\left[\textit{Hint:} \frac{1}{k(k + 1)} = \frac{1}{k} - \frac{1}{k + 1}\right]$

45. (a) Evaluate $\sum_{k=1}^{n} [f(k) - f(k - 1)]$.

(b) In Problems 43 and 44 identify a function f so that the given sums are of the form given in part **(a)**. A sum of the form given in part **(a)** is said to **telescope**.

46. (a) Use Problem 45 part **(a)** to show that

$$\sum_{k=1}^{n} [(k + 1)^2 - k^2] = -1 + (n + 1)^2 = n^2 + 2n$$

(b) Use the fact that $(k + 1)^2 - k^2 = 2k + 1$ to show that

$$\sum_{k=1}^{n} [(k + 1)^2 - k^2] = n + 2 \sum_{k=1}^{n} k$$

(c) Compare the results in parts **(a)** and **(b)** to derive summation formula II.

47. Apply the procedure outlined in Problem 46 to the sum $\Sigma_{k=1}^{n} [(k + 1)^3 - k^3]$ to derive summation formula III.

48. Consider the ratio

$$\frac{1^2 + 2^2 + 3^2 + \cdots + n^2}{1 + 2 + 3 + \cdots + n}$$

for $n = 1, 2, 3, 4$, and 5. Discern the value of the ratio for any positive integer value of n. Use this result to obtain summation formula III.

49. Derive a formula for the sum of the first n positive even integers.

50. In his experiments on gravity, Galileo found that the distance a mass moves down an inclined plane in consecutive time intervals is proportional to a positive odd integer. Hence, the total distance s a mass moves in n seconds, with n a positive integer, is proportional to $1 + 3 + 5 + \cdots + 2n - 1$. Show that the total distance a mass moves down an inclined plane is proportional to the square of the elapsed time n.

51. In a study of pregnancy and lactation in free-ranging little brown bats (*Myotis lucifugus*), Kurtz and colleagues determined that the daily milk-energy production is predicted by the formula

$$E_d = 10.34 + 0.50d$$

where E_d is milk-energy in kilojoules (kJ) on day d. The formula applies to only the first 17 days of lactation, during which milk is the baby bat's only food source. Find the total milk-energy production for the 17-day period to which the formula applies; that is, find the value of $\Sigma_{d=1}^{17} E_d$.

52. The Tower of Hanoi is a stack of circular disks, each larger than the one above it, set on a pole through holes in the disks' centers. See Figure 4.6. An ancient king once commanded that such a tower be built of gold disks to the following specifications: Each disk was to be one finger width thick, and the diameter of each disk was to be one finger width larger than the disk above it. The hole through the centers of the disks was to be one finger width in diameter, and the top disk was to be two finger widths in diameter. Assume that a

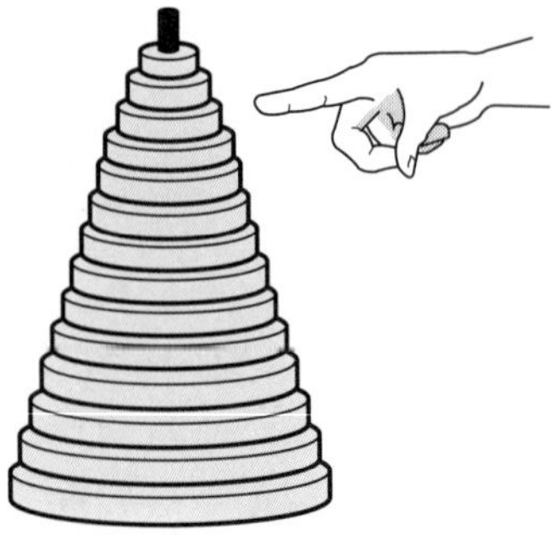

FIGURE 4.6

finger width is 1.5 cm and gold weighs 19.3 g/cm^3 and is valued at $14 per gram. Find a formula for the value of gold in the king's Tower of Hanoi if the tower has n disks.

In Problems 53 and 54 solve for $\bar{x}$.

53. $\sum_{k=1}^{n} (x_k - \bar{x}) = 0$

54. $\sum_{k=1}^{n} (x_k - \bar{x})^2 = 0$

Miscellaneous Problems

55. Prove Theorem 4.6(*ii*).

56. Prove Theorem 4.6(*iii*).

4.4 AREA UNDER A GRAPH

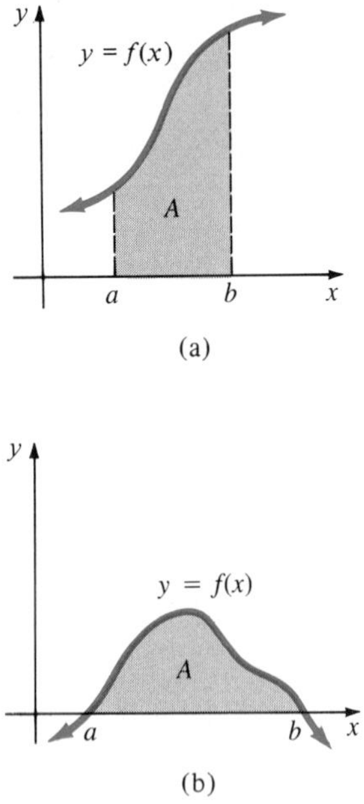

FIGURE 4.7

As the derivative is motivated by the geometric problem of constructing a tangent to a curve, the historical problem leading to the definition of a definite integral is the problem of finding area. Specifically, we are interested in finding the area A of a region bounded by the x-axis, the graph of a *nonnegative* function $y = f(x)$ defined on some interval $[a, b]$,* and

(*i*) the vertical lines $x = a$ and $x = b$, as shown in Figure 4.7(a), or

(*ii*) the x-intercepts of the graph shown in Figure 4.7(b).

We shall call this area the **area under the graph** of f on the interval $[a, b]$.

Assume for the moment that we do not know a formula for calculating the area A of the right triangle given in Figure 4.8(a). By superimposing a

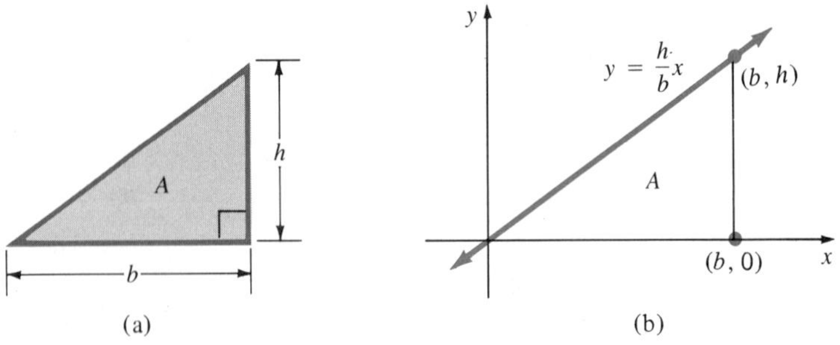

FIGURE 4.8

*The requirement that f be nonnegative on $[a, b]$ means that no portion of its graph on the interval is below the x-axis.

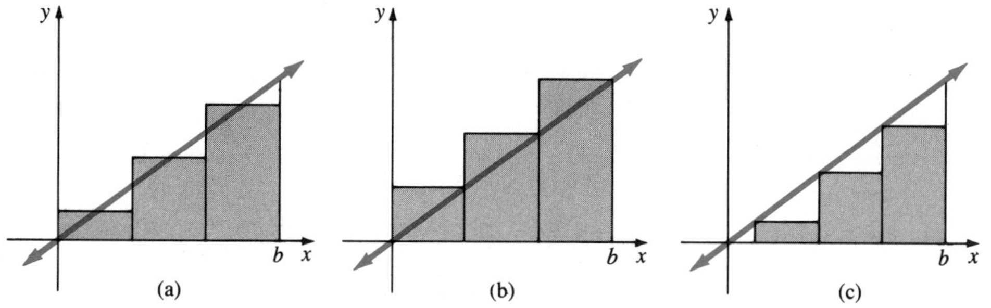

FIGURE 4.9

Cartesian coordinate system on the triangle, as shown in Figure 4.8(b), we see that the problem is the same as finding the area in the first quadrant bounded by the straight lines $y = (h/b)x$, $y = 0$ (the x-axis), and $x = b$. In other words, we wish to find the area under the graph of $y = (h/b)x$ on the interval $[0, b]$.

Using rectangles, Figure 4.9 indicates three different ways of *approximating* the area A. For convenience, let us pursue the procedure hinted at in Figure 4.9(b) in greater detail. We begin by dividing the interval $[0, b]$ into n subintervals of equal length $\Delta x = b/n$. If the right endpoint of each of these intervals is denoted by x_k^*, then

$$x_1^* = \Delta x = \frac{b}{n}$$

$$x_2^* = 2\,\Delta x = 2\left(\frac{b}{n}\right)$$

$$\vdots$$

$$x_k^* = k\,\Delta x = k\left(\frac{b}{n}\right)$$

$$\vdots$$

$$x_n^* = n\,\Delta x = n\left(\frac{b}{n}\right) = b$$

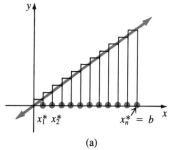

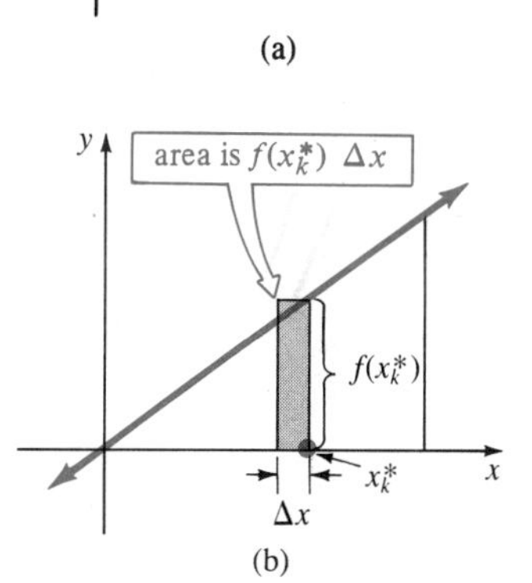

FIGURE 4.10

As shown in Figure 4.10(a), we now construct a rectangle of length $f(x_k^*)$ and width Δx on each of the n subintervals. Since the area of a rectangle is *length* × *width*, the area of each rectangle is $f(x_k^*)\,\Delta x$. See Figure 4.10(b). The sum of the areas of the n rectangles is an approximation to the number A. We write

$$A \approx f(x_1^*)\,\Delta x + f(x_2^*)\,\Delta x + \cdots + f(x_n^*)\,\Delta x$$

or, in sigma notation,

$$A \approx \sum_{k=1}^{n} f(x_k^*)\,\Delta x \tag{1}$$

It seems plausible that we can reduce the error introduced by this method of approximation (the area of each rectangle is larger than the area under the graph on a subinterval $[x_{k-1}, x_k]$) by **partitioning** $[0, b]$ into finer subdivisions. In other words, we expect that a better approximation to A can be obtained by using more and more rectangles ($n \to \infty$) of decreasing widths ($\Delta x \to 0$).

Now,

$$f(x) = \frac{h}{b}x, \quad x_k^* = k\left(\frac{b}{n}\right), \quad f(x_k^*) = \frac{h}{n} \cdot k, \quad \text{and} \quad \Delta x = \frac{b}{n}$$

so that with the aid of summation formula II of Section 4.3, (1) becomes

$$A \approx \sum_{k=1}^{n} \left(\frac{h}{n} \cdot k\right)\frac{b}{n} = \frac{bh}{n^2}\sum_{k=1}^{n} k = \frac{bh}{n^2} \cdot \frac{n(n+1)}{2} = \frac{bh}{2}\left(1 + \frac{1}{n}\right)$$

Finally, as $n \to \infty$, we obtain the familiar formula

$$A = \frac{1}{2}bh \lim_{n\to\infty}\left(1 + \frac{1}{n}\right) = \frac{1}{2}bh$$

The General Problem Now, let us turn from the preceding specific example to the general problem of finding the area A under the graph of a continuous function $y = f(x)$ on an interval $[a, b]$. As shown in Figure 4.11(a), we shall also assume that $f(x) \geq 0$ for all x in the interval. As suggested in Figure 4.11(b), the area A can be approximated by adding the areas of n rectangles that are constructed on the interval.

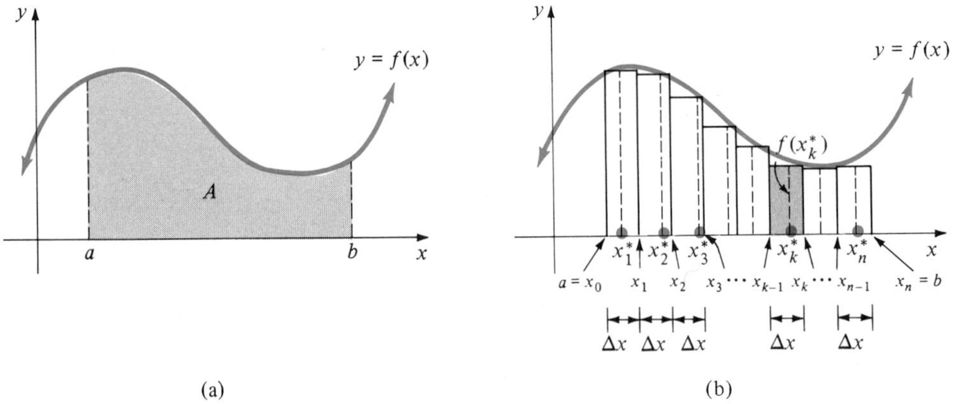

FIGURE 4.11

One possible procedure for determining A is summarized as follows:

(*1*) Partition the interval $[a, b]$ into n subintervals $[x_{k-1}, x_k]$ of equal length, $k = 1, 2, \ldots, n$, where $x_0 = a$, $x_n = b$, and

$$a = x_0 < x_1 < x_2 < \cdots < x_{n-1} < x_n = b$$

(*2*) Denote the length of each subinterval by Δx, where $\Delta x = x_k - x_{k-1}$.

(3) Choose any number x_k^* in each subinterval $[x_{k-1}, x_k]$ and form the product $f(x_k^*)\,\Delta x$. This represents the area of a rectangle on the kth subinterval.

(4) Form the sum $\Sigma_{k=1}^{n} f(x_k^*)\,\Delta x$. This is the sum of the areas of the n rectangles and represents an approximation to the value of A.

With these preliminaries, we are now in a position to define the concept of area under a graph.

DEFINITION 4.3 Area

Let f be continuous on $[a, b]$ and $f(x) \geq 0$ for all x in the interval. We define the area A under the graph on the interval to be

$$A = \lim_{n \to \infty} \sum_{k=1}^{n} f(x_k^*)\,\Delta x \tag{2}$$

It is usually proved in advanced calculus that when f is continuous, the limit in (2) always exists regardless of the manner used to partition $[a, b]$; that is, the subintervals may or may not be taken of equal length, and the points x_k^* can be chosen quite arbitrarily in the subintervals $[x_{k-1}, x_k]$. However, if the subintervals are not of equal length, then a different kind of limiting process is necessary in (2). We must replace $n \to \infty$ with the requirement that the length of the longest subinterval approach zero.

A Practical Form of (2) To use (2), suppose we choose x_k^* as we did in the discussion of Figure 4.9; namely, let x_k^* be the right endpoint of each subinterval. Since the length of each of the n equal subintervals is $\Delta x = (b - a)/n$, we have

$$x_1^* = x_0 + \Delta x = a + \frac{b-a}{n}$$

$$x_2^* = x_0 + 2\,\Delta x = a + 2\left(\frac{b-a}{n}\right)$$

$$\vdots$$

$$x_k^* = x_0 + k\,\Delta x = a + k\left(\frac{b-a}{n}\right)$$

$$\vdots$$

$$x_n^* = x_0 + n\,\Delta x = a + n\left(\frac{b-a}{n}\right) = b$$

It follows that by substituting $a + k\left(\dfrac{b-a}{n}\right)$ for x_k^* and $(b - a)/n$ for Δx

in (2), the area A is also given by

$$A = \lim_{n\to\infty} \sum_{k=1}^{n} f\left(a + k\frac{b-a}{n}\right)\cdot\frac{b-a}{n} \tag{3}$$

We note that since $\Delta x = (b-a)/n$, $n \to \infty$ implies $\Delta x \to 0$.

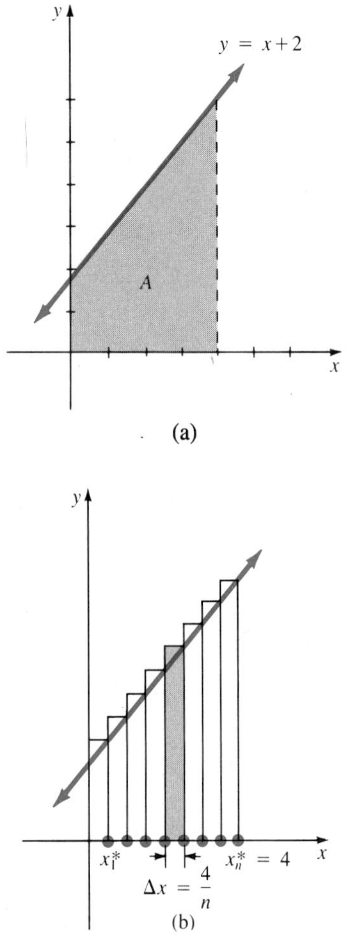

FIGURE 4.12

EXAMPLE 1 Find the area A under the graph of $f(x) = x + 2$ on the interval $[0, 4]$.

Solution The area is bounded by the trapezoid indicated in Figure 4.12(a). By identifying $a = 0$ and $b = 4$, we find

$$\Delta x = \frac{4-0}{n} = \frac{4}{n}$$

Thus, (3) becomes

$$\begin{aligned} A &= \lim_{n\to\infty} \sum_{k=1}^{n} f\left(0 + k\frac{4}{n}\right)\frac{4}{n} = \lim_{n\to\infty} \frac{4}{n}\sum_{k=1}^{n} f\left(\frac{4k}{n}\right) \\ &= \lim_{n\to\infty} \frac{4}{n}\sum_{k=1}^{n}\left(\frac{4k}{n} + 2\right) \\ &= \lim_{n\to\infty} \frac{4}{n}\left[\frac{4}{n}\sum_{k=1}^{n} k + 2\sum_{k=1}^{n} 1\right] \end{aligned}$$

Now, by summation formulas I and II of Section 4.3, we can write

$$\begin{aligned} A = \lim_{n\to\infty} \frac{4}{n}\left[\frac{4}{n}\cdot\frac{n(n+1)}{2} + 2n\right] &= \lim_{n\to\infty}\left[\frac{16}{2}\frac{n(n+1)}{n^2} + 8\right] \\ &= \lim_{n\to\infty}\left[8\left(1 + \frac{1}{n}\right) + 8\right] \\ &= 8\lim_{n\to\infty}\left(1 + \frac{1}{n}\right) + 8\lim_{n\to\infty} 1 \\ &= 8 + 8 = 16 \text{ square units} \end{aligned}$$

□

EXAMPLE 2 Find the area A under the graph of $f(x) = 4 - x^2$ on the interval $[-1, 2]$.

Solution The area is indicated in Figure 4.13(a). Since $a = -1$ and $b = 2$, it follows that

$$\Delta x = \frac{2-(-1)}{n} = \frac{3}{n}$$

Let us review the steps leading up to (3). The width of each rectangle is given by $\Delta x = 3/n$. Now, starting at $x = -1$, the right endpoint of each sub-

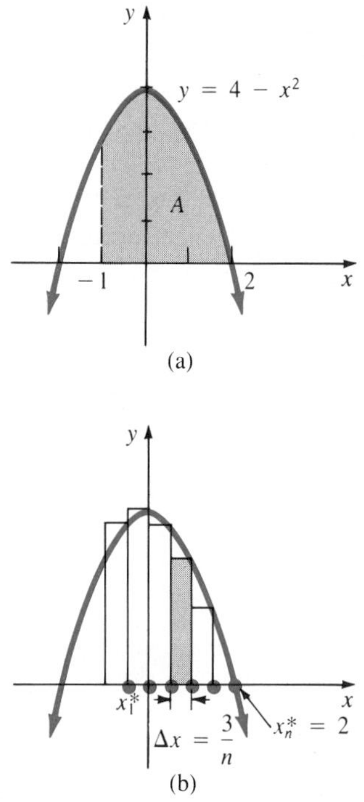

FIGURE 4.13

interval is

$$x_1^* = -1 + \frac{3}{n}$$

$$x_2^* = -1 + 2\left(\frac{3}{n}\right)$$

$$\vdots$$

$$x_k^* = -1 + k\left(\frac{3}{n}\right)$$

$$\vdots$$

$$x_n^* = -1 + n\left(\frac{3}{n}\right) = 2$$

The length of each rectangle is then

$$f(x_1^*) = f\left(-1 + \frac{3}{n}\right) = 4 - \left[-1 + \frac{3}{n}\right]^2$$

$$f(x_2^*) = f\left(-1 + 2\left(\frac{3}{n}\right)\right) = 4 - \left[-1 + 2\left(\frac{3}{n}\right)\right]^2$$

$$\vdots$$

$$f(x_k^*) = f\left(-1 + k\left(\frac{3}{n}\right)\right) = 4 - \left[-1 + k\left(\frac{3}{n}\right)\right]^2$$

$$\vdots$$

$$f(x_n^*) = f\left(-1 + n\left(\frac{3}{n}\right)\right) = f(2) = 4 - (2)^2 = 0$$

The area of the kth rectangle is length × width:

$$f(x_k^*)\frac{3}{n} = \left(4 - \left[-1 + k\frac{3}{n}\right]^2\right)\frac{3}{n} = \left(3 + 6\frac{k}{n} - 9\frac{k^2}{n^2}\right)\frac{3}{n}$$

Adding the areas of the n rectangles gives an approximation to the area under the graph on the interval: $A \approx \Sigma_{k=1}^{n} f(x_k^*)(3/n)$. As the number n of rectangles increases without bound, we obtain

$$A = \lim_{n\to\infty} \sum_{k=1}^{n} \left(3 + 6\frac{k}{n} - 9\frac{k^2}{n^2}\right)\frac{3}{n} = \lim_{n\to\infty} \frac{3}{n} \sum_{k=1}^{n} \left(3 + 6\frac{k}{n} - 9\frac{k^2}{n^2}\right)$$

$$= \lim_{n\to\infty} \frac{3}{n}\left[3\sum_{k=1}^{n} 1 + \frac{6}{n}\sum_{k=1}^{n} k - \frac{9}{n^2}\sum_{k=1}^{n} k^2\right]$$

Using summation formulas I, II, and III of the last section, we get

$$A = \lim_{n\to\infty} \frac{3}{n}\left[3n + \frac{6}{n}\cdot\frac{n(n+1)}{2} - \frac{9}{n^2}\cdot\frac{n(n+1)(2n+1)}{6}\right]$$

$$= \lim_{n\to\infty}\left[9 + 9\left(1 + \frac{1}{n}\right) - \frac{9}{2}\left(1 + \frac{1}{n}\right)\left(2 + \frac{1}{n}\right)\right]$$

$$= 9 + 9 - 9 = 9 \text{ square units}$$

□

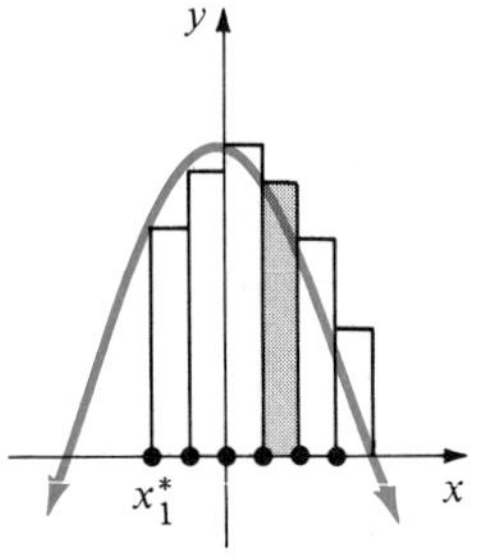

FIGURE 4.14

▲ ***Remark*** There is nothing special about choosing x_k^* to be the right endpoint of each subinterval. We reemphasize that x_k^* can be taken to be any convenient number in $[x_{k-1}, x_k]$. Had we chosen x_k^* to be the left endpoint of each subinterval in Example 2, then the corresponding rectangles would be as shown in Figure 4.14. In this case, we would have $x_k^* = -1 + (k-1)(3/n)$, $k = 1, 2, 3, \ldots, n$. In Problems 26 and 27 of Exercises 4.4 you will be asked to solve the area problem in Example 2 by choosing x_k^* to be first the left endpoint and then the midpoint of each subinterval $[x_{k-1}, x_k]$.

EXERCISES 4.4

Answers to odd-numbered problems begin on page A-80.

In Problems 1–22 find the area under the graph of the given function on the indicated interval.

1. $f(x) = 4$, $[2, 5]$
2. $f(x) = 7$, $[-4, 8]$
3. $f(x) = x$, $[0, 6]$
4. $f(x) = 2x$, $[1, 3]$
5. $f(x) = 2x + 1$, $[1, 5]$
6. $f(x) = 3x - 6$, $[2, 4]$
7. $f(x) = x^2$, $[0, 2]$
8. $f(x) = x^2$, $[-2, 1]$
9. $f(x) = 1 - x^2$, $[-1, 1]$
10. $f(x) = 2x^2 + 3$, $[-3, -1]$
11. $f(x) = x^2 + 2x$, $[1, 2]$
12. $f(x) = (x - 1)^2$, $[0, 2]$
13. $f(x) = x^3$, $[0, 1]$
14. $f(x) = -x^3$, $[-3, 0]$
15. $f(x) = x^3 - 3x^2 + 4$, $[0, 2]$
16. $f(x) = (x + 1)^3$, $[-1, 1]$
17. $f(x) = x^4$, $[0, 2]$
18. $f(x) = 16 - x^4$, $[-1, 0]$
19. $f(x) = |x|$, $[-1, 3]$
20. $f(x) = |x^2 - 1|$, $[0, 2]$
21. $f(x) = \begin{cases} 2, & 0 \le x < 1 \\ x + 1, & 1 \le x \le 4 \end{cases}$
22. $f(x) = \begin{cases} -x + 1, & 0 \le x < 1 \\ x + 2, & 1 \le x \le 3 \end{cases}$
23. Sketch the graph of $y = 1/x$ on $\frac{1}{2} \le x \le \frac{5}{2}$. By dividing the interval into four subintervals of equal lengths, construct rectangles that approximate the area A under the graph. First use the right endpoint of each subinterval, and then use the left endpoint.
24. Repeat Problem 23 for $y = \cos x$ on $-\pi/2 \le x \le \pi/2$.
25. Derive a formula analogous to (3) in which x_k^* is chosen as **(a)** the left endpoint of each subinterval and **(b)** the midpoint of each subinterval.
26. Rework Example 2 by choosing x_k^* to be the left endpoint of each subinterval.
27. Rework Example 2 by choosing x_k^* to be the midpoint of each subinterval.
28. Find the area of the region in the first quadrant bounded by the graphs of $x = 0$, $y = 9$, and $y = x^2$.
29. Find the area of the region in the first quadrant bounded by the graphs of $y = x$ and $y = x^2$.
30. The area of the trapezoid given in Figure 4.15 is

$$A = \left(\frac{h_1 + h_2}{2}\right) b$$

Derive this formula.

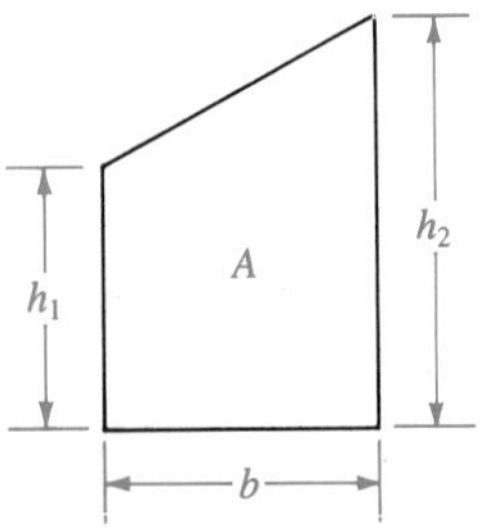

FIGURE 4.15

In Problems 31 and 32 determine a region whose area A is given by the formula. Do not try to evaluate.

31. $A = \lim_{n\to\infty} \sum_{k=1}^{n} \sqrt{4 - \frac{4k^2}{n^2}}\,\frac{2}{n}$

32. $A = \lim_{n\to\infty} \sum_{k=1}^{n} \left(\sin \frac{k\pi}{n}\right)\frac{\pi}{n}$

33. Find the area under the graph of $y = \sqrt{x}$ on $0 \le x \le 1$ by considering the area under the graph of $y = x^2$ on $0 \le x \le 1$.

34. Find the area under the graph of $y = \sqrt[3]{x}$ on $0 \le x \le 8$ by considering the area under the graph of $y = x^3$ on $0 \le x \le 2$.

35. (a) Suppose $y = ax^2 + bx + c \ge 0$ on the interval $[0, x_0]$. Show that the area under the graph on $[0, x_0]$ is given by

$$A = a\frac{x_0^3}{3} + b\frac{x_0^2}{2} + cx_0$$

(b) Use the result in part (a) to find the area under the graph of $y = 6x^2 + 2x + 1$ on the interval $[2, 5]$.

4.5 DEFINITE INTEGRAL

4.5.1 Definition of a Definite Integral

We are now in a position to define the concept of a **definite integral**. To do this, consider the following five steps.

$y = f(x)$

(*1*) Let f be defined on a closed interval $[a, b]$.

(*2*) Partition the interval $[a, b]$ into n subintervals $[x_{k-1}, x_k]$ of length $\Delta x_k = x_k - x_{k-1}$. Let P denote the partition

$$a = x_0 < x_1 < x_2 < \cdots < x_{n-1} < x_n = b$$

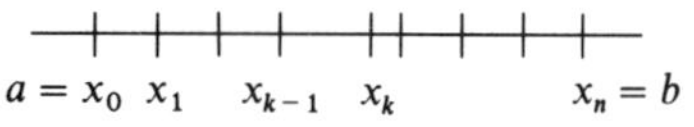

(*3*) Let $\|P\|$ be the length of the longest subinterval. The number $\|P\|$ is called the **norm** of the partition P.

(*4*) Choose a number x_k^* in each subinterval.

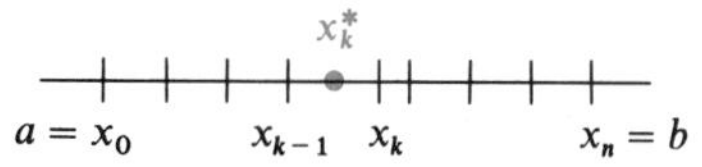

(*5*) Form the sum

$$\sum_{k=1}^{n} f(x_k^*)\,\Delta x_k \qquad (1)$$

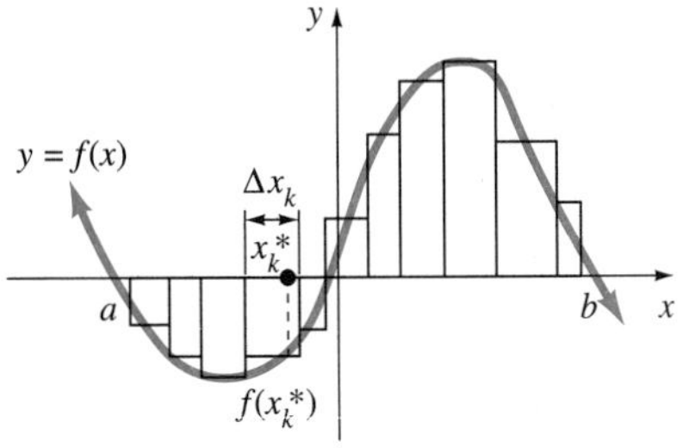

FIGURE 4.16

Sums such as (1) for the various partitions of $[a, b]$ are known as **Riemann sums** and are named for the famous German mathematician, Georg Friedrich Bernhard Riemann.

Although the procedure looks very similar to the five steps leading up to the definition of area under a graph, there are some important differences. Observe that a Riemann sum does not require that f be either continuous or nonnegative on the interval $[a, b]$. Thus, (1) does not necessarily represent an approximation to the area under a graph. Keep in mind that "area under a graph" refers to *the area bounded between the graph of a nonnegative function and the x-axis.* As shown in Figure 4.16, if $f(x) < 0$ for some x in $[a, b]$, a Riemann sum could contain terms $f(x_k^*)\,\Delta x_k$, where $f(x_k^*) < 0$. In this case the products $f(x_k^*)\,\Delta x_k$ are numbers that are the negatives of the areas of rectangles drawn below the x-axis.

EXAMPLE 1 Compute the Riemann sum for $f(x) = x^2 - 4$ on $[-2, 3]$ with five subintervals determined by

$$x_0 = -2, \quad x_1 = -\frac{1}{2}, \quad x_2 = 0, \quad x_3 = 1, \quad x_4 = \frac{7}{4}, \quad x_5 = 3$$

and

$$x_1^* = -1, \quad x_2^* = -\frac{1}{4}, \quad x_3^* = \frac{1}{2}, \quad x_4^* = \frac{3}{2}, \quad x_5^* = \frac{5}{2}$$

Solution Figure 4.17 shows the various points x_k and x_k^* on the interval.

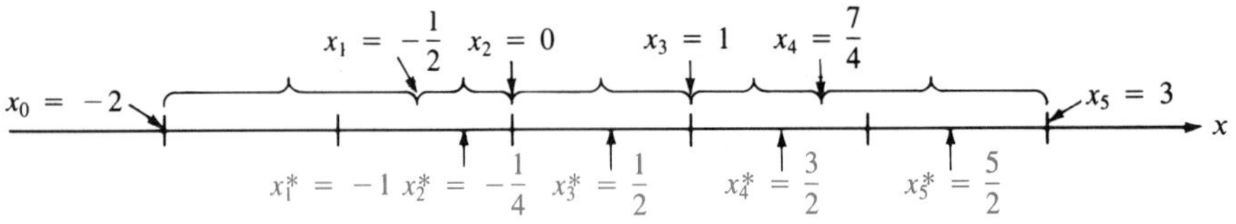

FIGURE 4.17

Georg Friedrich Bernhard Riemann (1826–1866)
Bernhard Riemann, born in Hanover, Germany, in 1826, was the son of a Lutheran minister. Although a devout Christian, Riemann was disinclined to follow his father's vocation and abandoned the study of theology at the University of Göttingen in favor of a course of studies in which his genius was obvious: mathematics. It is likely that the concept of Riemann sums grew out of a course on the definite integral that he had at the university; this concept reflects his attempt to give a precise mathematical meaning to the definite integral of Newton and Leibniz. After submitting his doctoral dissertation on the foundations of functions of a complex variable to the examining committee at the University of Göttingen, Carl Friedrich Gauss, the "prince of mathematicians," paid Riemann a very rare compliment: "The dissertation offers convincing evidence ... of a creative, active, truly mathematical mind ... of glorious fertile originality." Riemann, like so many other promising scholars of that time, possessed a fragile constitution. He died at age 39 of pleurisy. His original contributions to differential geometry, topology, non-Euclidean geometry, and his bold investigations into the nature of space, electricity, and magnetism, foreshadowed the work of Einstein in the next century.

Now,

$$\Delta x_1 = x_1 - x_0 = -\frac{1}{2} - (-2) = \frac{3}{2} \qquad f(x_1^*) = f(-1) = -3$$

$$\Delta x_2 = x_2 - x_1 = 0 - \left(-\frac{1}{2}\right) = \frac{1}{2} \qquad f(x_2^*) = f\left(-\frac{1}{4}\right) = -\frac{63}{16}$$

$$\Delta x_3 = x_3 - x_2 = 1 - 0 = 1 \qquad f(x_3^*) = f\left(\frac{1}{2}\right) = -\frac{15}{4}$$

$$\Delta x_4 = x_4 - x_3 = \frac{7}{4} - 1 = \frac{3}{4} \qquad f(x_4^*) = f\left(\frac{3}{2}\right) = -\frac{7}{4}$$

$$\Delta x_5 = x_5 - x_4 = 3 - \frac{7}{4} = \frac{5}{4} \qquad f(x_5^*) = f\left(\frac{5}{2}\right) = \frac{9}{4}$$

and so the Riemann sum is

$$f(x_1^*)\,\Delta x_1 + f(x_2^*)\,\Delta x_2 + f(x_3^*)\,\Delta x_3 + f(x_4^*)\,\Delta x_4 + f(x_5^*)\,\Delta x_5$$

$$= (-3)\left(\frac{3}{2}\right) + \left(-\frac{63}{16}\right)\left(\frac{1}{2}\right) + \left(-\frac{15}{4}\right)(1) + \left(-\frac{7}{4}\right)\left(\frac{3}{4}\right) + \left(\frac{9}{4}\right)\left(\frac{5}{4}\right) = -\frac{279}{32} \approx -8.72 \qquad \square$$

For a function f defined on an interval $[a, b]$, there are an infinite number of possible Riemann sums for a given partition P of the interval, since the numbers x_k^* can be chosen arbitrarily in each subinterval $[x_{k-1}, x_k]$.

EXAMPLE 2 Compute the Riemann sum for the function and partition of $[-2, 3]$ in Example 1 if $x_1^* = -\frac{3}{2}$, $x_2^* = -\frac{1}{8}$, $x_3^* = \frac{3}{4}$, $x_4^* = \frac{3}{2}$, and $x_5^* = 2.1$.

Solution

$$f(x_1^*) = f\left(-\frac{3}{2}\right) = -\frac{7}{4}$$

$$f(x_2^*) = f\left(-\frac{1}{8}\right) = -\frac{255}{64}$$

$$f(x_3^*) = f\left(\frac{3}{4}\right) = -\frac{55}{16}$$

$$f(x_4^*) = f\left(\frac{3}{2}\right) = -\frac{7}{4}$$

$$f(x_5^*) = f(2.1) = 0.41$$

Since the numbers Δx_k are the same as before, we have

$$f(x_1^*)\,\Delta x_1 + f(x_2^*)\,\Delta x_2 + f(x_3^*)\,\Delta x_3 + f(x_4^*)\,\Delta x_4 + f(x_5^*)\,\Delta x_5$$

$$= \left(-\frac{7}{4}\right)\left(\frac{3}{2}\right) + \left(-\frac{255}{64}\right)\left(\frac{1}{2}\right) + \left(-\frac{55}{16}\right)(1) + \left(-\frac{7}{4}\right)\left(\frac{3}{4}\right) + (0.41)\left(\frac{5}{4}\right) \approx -8.85 \qquad \square$$

We are interested in a special kind of limit of (1). If the Riemann sums $\Sigma_{k=1}^{n} f(x_k^*)\,\Delta x_k$ are close to a number L for *every* partition P of $[a, b]$ for which the norm $\|P\|$ is close to zero, we then write

$$\lim_{\|P\| \to 0} \sum_{k=1}^{n} f(x_k^*)\,\Delta x_k = L \tag{2}$$

and say that L is the **definite integral** of f on the interval $[a, b]$. If the limit in (2) exists, the function f is said to be **integrable** on the interval. In the following definition we introduce a new symbol for the number L.

DEFINITION 4.4 The Definite Integral

Let f be a function defined on a closed interval $[a, b]$. Then the **definite integral of f from a to b**, denoted by $\int_a^b f(x)\,dx$, is defined to be

$$\int_a^b f(x)\,dx = \lim_{\|P\| \to 0} \sum_{k=1}^{n} f(x_k^*)\,\Delta x_k \tag{3}$$

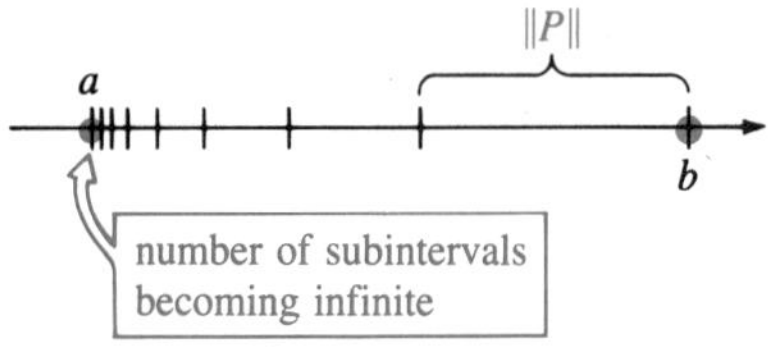

FIGURE 4.18

The numbers a and b in the preceding definition are called the **lower** and **upper limits of integration**, respectively. The integral symbol $\int$, as used by Leibniz, is an elongated S for the word "sum." Also, note that $\|P\| \to 0$ always implies that the number of subintervals n becomes infinite in number ($n \to \infty$). However, as shown in Figure 4.18, the fact that $n \to \infty$ does not necessarily imply $\|P\| \to 0$.

Integrability In the next two theorems we state conditions that are sufficient for a function f to be integrable on an interval $[a, b]$. The proofs of the theorems will not be given.

THEOREM 4.7 Continuity Implies Integrability

If f is continuous on $[a, b]$, then $\int_a^b f(x)\,dx$ exists; that is, f is integrable on the interval.

There are functions that are defined for every value of x in $[a, b]$ for which the limit in (3) does not exist. Also, if the function f is not defined for all values of x in the interval, the definite integral *may* not exist; for example, later on we shall see why an integral such as $\int_{-3}^{2} (1/x)\,dx$ does not exist. Notice that $y = 1/x$ is discontinuous at $x = 0$ and is unbounded on the interval. However, one should not conclude from this one example that when a function f has a discontinuity in $[a, b]$, $\int_a^b f(x)\,dx$ necessarily does not exist. Continuity of a function f on $[a, b]$ is sufficient but not necessary to guarantee the existence of $\int_a^b f(x)\,dx$. The set of functions continuous on $[a, b]$ is a subset of the set of functions that are integrable on the interval.

THEOREM 4.8 Other Sufficient Conditions for Integrability

If a function f is bounded on $[a, b]$, that is, if there exists a positive constant B such that $-B \le f(x) \le B$ for all x in the interval, and has a finite number of discontinuities in $[a, b]$, then f is integrable on the interval.

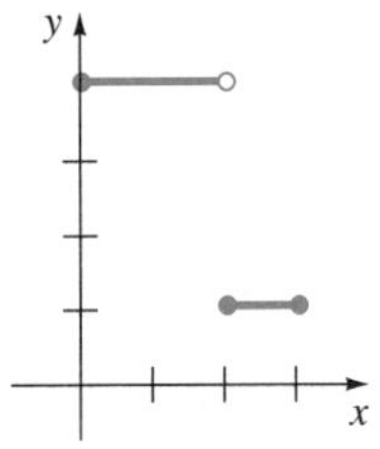

FIGURE 4.19

When a function f is bounded, its complete graph must lie between two horizontal lines, $y = B$ and $y = -B$. In other words, $|f(x)| \le B$ for all x in $[a, b]$. The function

$$f(x) = \begin{cases} 4, & 0 \le x < 2 \\ 1, & 2 \le x \le 3 \end{cases}$$

shown in Figure 4.19 is discontinuous at $x = 2$ but is bounded on $[0, 3]$, since $|f(x)| \le 4$ for all x in $[0, 3]$. (For that matter, $1 \le f(x) \le 4$ for all x in $[0, 3]$ shows that f is bounded on the interval.) Hence, $\int_0^3 f(x)\,dx$ exists.

Regular Partition When a definite integral exists,

the limit in (3) exists for every possible way of partitioning $[a, b]$ and for every way of choosing x_k^* in the subintervals $[x_{k-1}, x_k]$.

In particular, by choosing the subintervals of equal length

$$\Delta x = \frac{b-a}{n} \quad \text{and} \quad x_k^* = a + k\frac{b-a}{n}, \qquad k = 1, 2, \ldots, n$$

we can write

$$\int_a^b f(x)\,dx = \lim_{n\to\infty} \sum_{k=1}^{n} f\left(a + k\frac{b-a}{n}\right)\frac{b-a}{n} \tag{4}$$

A partition P of $[a, b]$ in which the subintervals are of the same length is called a **regular partition**.

Area You might conclude that the formulations of $\int_a^b f(x)\,dx$ given in (3) and (4) are exactly the same as (2) and (3) of Section 4.4 for the general case of finding the area under the curve $y = f(x)$ on $[a, b]$. In a way this is correct; however, Definition 4.4 is a more general concept, since, as noted before, we are not requiring that f be continuous on $[a, b]$ or that $f(x) \ge 0$ on the interval. Thus, a *definite integral need not be area.* What then is a definite integral? For now, accept the fact that a definite integral is simply a real number. Contrast this with the indefinite integral, which is a function (or a set of functions). Is the area under the graph of a continuous nonnegative function a definite integral? The answer is yes.

THEOREM 4.9 The Definite Integral As Area

If f is continuous on $[a, b]$ and $f(x) \geq 0$ for all x in the interval, then the area A under the graph of f on $[a, b]$ is

$$A = \int_a^b f(x)\,dx$$

We shall return to the question of finding areas after we have studied some properties of the definite integral in the next section.

EXAMPLE 3 Evaluate $\int_{-2}^{1} x^3\,dx$.

Solution Since $f(x) = x^3$ is continuous on $[-2, 1]$, we know from Theorem 4.7 that the definite integral exists. We use a regular partition and the result given in (4). Choosing

$$\Delta x = \frac{1-(-2)}{n} = \frac{3}{n} \quad \text{and} \quad x_k^* = -2 + k \cdot \frac{3}{n}$$

we have

$$f\left(-2+\frac{3k}{n}\right) = \left(-2+\frac{3k}{n}\right)^3 = -8 + 36\left(\frac{k}{n}\right) - 54\left(\frac{k^2}{n^2}\right) + 27\left(\frac{k^3}{n^3}\right)$$

It then follows from (4) and summation formulas I–IV that

$$\begin{aligned}
\int_{-2}^{1} x^3\,dx &= \lim_{n\to\infty} \sum_{k=1}^{n} f\left(-2+\frac{3k}{n}\right)\frac{3}{n} \\
&= \lim_{n\to\infty} \frac{3}{n}\sum_{k=1}^{n}\left[-8 + 36\left(\frac{k}{n}\right) - 54\left(\frac{k^2}{n^2}\right) + 27\left(\frac{k^3}{n^3}\right)\right] \\
&= \lim_{n\to\infty} \frac{3}{n}\left[-8n + \frac{36}{n}\cdot\frac{n(n+1)}{2} - \frac{54}{n^2}\cdot\frac{n(n+1)(2n+1)}{6} \right.\\
&\qquad \left. + \frac{27}{n^3}\cdot\frac{n^2(n+1)^2}{4}\right] \\
&= \lim_{n\to\infty}\left[-24 + 54\left(1+\frac{1}{n}\right) - 27\left(1+\frac{1}{n}\right)\left(2+\frac{1}{n}\right) \right.\\
&\qquad \left. + \frac{81}{4}\left(1+\frac{1}{n}\right)\left(1+\frac{1}{n}\right)\right] \\
&= -24 + 54 - 27(2) + \frac{81}{4} = -\frac{15}{4}
\end{aligned}$$

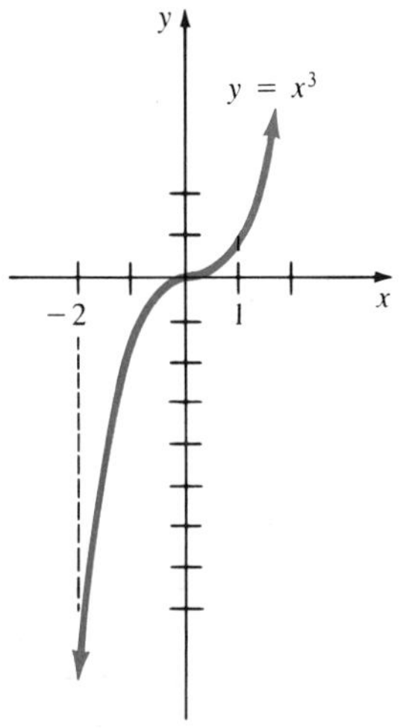

FIGURE 4.20

Figure 4.20 shows that we are not considering area. □

EXAMPLE 4 The values of the Riemann sums in Examples 1 and 2 are approximations to the value of the definite integral $\int_{-2}^{3} (x^2 - 4)\,dx$. It is left as an exercise to show that

$$\int_{-2}^{3} (x^2 - 4)\,dx = -\frac{25}{3} \approx -8.33$$

See Problem 18 in Exercises 4.5. □

BASIC Program The following listing is a BASIC program for the approximation

$$\int_a^b f(x)\,dx \approx \sum_{k=1}^{n} f\left(a + k\frac{b-a}{n}\right)\frac{b-a}{n}$$

```
10  REM EVALUATION OF DEFINITE INTEGRALS
    USING RIEMANN SUMS
20  DEF FNY(X) = . . .
30  INPUT "WHAT IS THE INTERVAL?";A,B
40  INPUT "HOW MANY SUBDIVISIONS?";N
50  LET H = (B − A)/N
60  FOR X = A + H TO B STEP H
70  LET R = R + FNY(X)
80  NEXT X
90  LET R = R * H
100  PRINT "USING";N; "SUBDIVISIONS, THE
     RIEMANN SUM YIELDS";R
110  END
```

[O] 4.5.2 ε–δ Definition

Let f be a function defined on $[a, b]$ and let L denote a real number. The intuitive concept that Riemann sums are close to L whenever the norm $\|P\|$ of a partition P is close to zero can be expressed in a precise manner using the ε–δ symbols introduced in Section 1.5. To say that f is integrable on $[a, b]$, we mean that for every real number $\varepsilon > 0$ there exists a real number $\delta > 0$ such that

$$\left| \sum_{k=1}^{n} f(x_k^*)\,\Delta x_k - L \right| < \varepsilon \tag{5}$$

whenever P is a partition of $[a, b]$ for which $\|P\| < \delta$ and the x_k^* are numbers in $[x_{k-1}, x_k]$. In other words,

$$\lim_{\|P\| \to 0} \sum_{k=1}^{n} f(x_k^*)\,\Delta x_k$$

exists and is equal to the number L.

▲ ***Remark*** The procedure outlined in (4) has limited utility as a practical means of computing a definite integral. We shall see in the next several sections that sometimes there is an easier way of finding the number $\int_a^b f(x)\,dx$.

EXERCISES 4.5 *Answers to odd-numbered problems begin on page A-80.*

[4.5.1]

In Problems 1–8 compute the Riemann sum for the given function on the indicated interval. Specify $\|P\|$.

1. $f(x) = x$, $[0, 4]$, two subintervals;
$x_0 = 0, x_1 = \frac{5}{2}, x_2 = 4$;
$x_1^* = 2, x_2^* = 3$

2. $f(x) = -2x$, $[-1, 2]$, three subintervals;
$x_0 = -1, x_1 = -\frac{1}{2}, x_2 = 1, x_3 = 2$;
$x_1^* = -\frac{1}{2}, x_2^* = 0, x_3^* = \frac{3}{2}$

3. $f(x) = 3x + 1$, $[0, 3]$, four subintervals;
$x_0 = 0, x_1 = 1, x_2 = \frac{5}{3}, x_3 = \frac{7}{3}, x_4 = 3$;
$x_1^* = \frac{1}{2}, x_2^* = \frac{4}{3}, x_3^* = 2, x_4^* = \frac{8}{3}$

4. $f(x) = x - 4$, $[-2, 5]$, five subintervals;
$x_0 = -2, x_1 = -1, x_2 = -\frac{1}{2}, x_3 = \frac{1}{2}, x_4 = 3, x_5 = 5$;
$x_1^* = -\frac{3}{2}, x_2^* = -\frac{1}{2}, x_3^* = 0, x_4^* = 2, x_5^* = 4$

5. $f(x) = x^2$, $[-1, 1]$, four subintervals;
$x_0 = -1, x_1 = -\frac{1}{4}, x_2 = \frac{1}{4}, x_3 = \frac{3}{4}, x_4 = 1$;
$x_1^* = -\frac{3}{4}, x_2^* = 0, x_3^* = \frac{1}{2}, x_4^* = \frac{7}{8}$

6. $f(x) = x^2 + 1$, $[1, 3]$, three subintervals;
$x_0 = 1, x_1 = \frac{3}{2}, x_2 = \frac{5}{2}, x_3 = 3$;
$x_1^* = \frac{5}{4}, x_2^* = \frac{7}{4}, x_3^* = 3$

7. $f(x) = \sin x$, $[0, 2\pi]$, three subintervals;
$$x_0 = 0, x_1 = \pi, x_2 = \frac{3\pi}{2}, x_3 = 2\pi,$$
$$x_1^* = \frac{\pi}{2}, x_2^* = \frac{7\pi}{6}, x_3^* = \frac{7\pi}{4}$$

8. $f(x) = \cos x$, $\left[-\frac{\pi}{2}, \frac{\pi}{2}\right]$, four subintervals;
$$x_0 = -\frac{\pi}{2}, x_1 = -\frac{\pi}{4}, x_2 = 0, x_3 = \frac{\pi}{3}, x_4 = \frac{\pi}{2};$$
$$x_1^* = -\frac{\pi}{3}, x_2^* = -\frac{\pi}{6}, x_3^* = \frac{\pi}{4}, x_4^* = \frac{\pi}{3}$$

9. Given $f(x) = x - 2$ on $[0, 5]$, compute the Riemann sum using a partition with five subintervals of equal length. Let x_k^*, $k = 1, 2, \ldots, 5$, be the right endpoint of each subinterval.

10. Given $f(x) = x^2 - x + 1$ on $[0, 1]$, compute the Riemann sum using a partition with three subintervals of equal length. Let x_k^*, $k = 1, 2, 3$, be the left endpoint of each subinterval.

In Problems 11–20 use (4) to evaluate the given definite integral.

11. $\int_1^9 3\,dx$

12. $\int_{-1}^4 (-2)\,dx$

13. $\int_{-3}^1 x\,dx$

14. $\int_0^3 x\,dx$

15. $\int_0^2 (x^2 - 1)\,dx$

16. $\int_0^3 (x^2 - 2x)\,dx$

17. $\int_1^2 (x^2 - x)\,dx$

18. $\int_{-2}^3 (x^2 - 4)\,dx$

19. $\int_0^1 (x^3 - 1)\,dx$

20. $\int_0^2 (3 - x^3)\,dx$

21. Let $f(x) = k$ be a constant function on $[a, b]$. Use (4) to show that $\int_a^b k\,dx = k(b - a)$.

22. Let P be any partition of $[a, b]$. Use Problem 21 to show that
$$\lim_{\|P\| \to 0} \sum_{k=1}^{n} \Delta x_k = b - a$$
Interpret this result geometrically.

In Problems 23 and 24 use (4) to obtain the given result.

23. $\int_a^b x\,dx = \frac{1}{2}(b^2 - a^2)$

24. $\int_a^b x^2\,dx = \frac{1}{3}(b^3 - a^3)$

25. Evaluate the definite integral $\int_0^1 \sqrt{x}\,dx$ by using a partition of $[0, 1]$ in which the subintervals $[x_{k-1}, x_k]$ are defined by $[(k - 1)^2/n^2, k^2/n^2]$ and choosing x_k^* to be the right endpoint of each subinterval.

26. Evaluate the definite integral $\int_0^{\pi/2} \cos x\,dx$ by using a regular partition of $[0, \frac{\pi}{2}]$ and choosing x_k^* to be the midpoint of each subinterval $[x_{k-1}, x_k]$. Use the results

(*i*) $\cos\theta + \cos 3\theta + \cdots + \cos(2n - 1)\theta = \dfrac{\sin 2n\theta}{2\sin\theta}$

(*ii*) $\displaystyle\lim_{n\to\infty} \frac{1}{n\sin(\pi/4n)} = \frac{4}{\pi}$

In Problems 27 and 28 let P be a partition of the indicated interval and x_k^* a number in the kth subinterval. Write the given sum as a definite integral.

27. $\lim_{\|P\| \to 0} \sum_{k=1}^{n} \sqrt{9 + (x_k^*)^2}\, \Delta x_k;\ [-2, 4]$

28. $\lim_{\|P\| \to 0} \sum_{k=1}^{n} (\tan x_k^*)\, \Delta x_k;\ \left[0, \frac{\pi}{4}\right]$

In Problems 29 and 30 let P be a regular partition of the indicated interval and x_k^* the right endpoint of each subinterval. Write the given sum as a definite integral.

29. $\lim_{n \to \infty} \sum_{k=1}^{n} \frac{4k + 2n}{n^2};\ [0, 2]$

30. $\lim_{n \to \infty} \sum_{k=1}^{n} \frac{3}{n + 3k};\ [1, 4]$

31. Consider the function defined for all x in the interval $[-1, 1]$:

$$f(x) = \begin{cases} 0, & x \text{ rational} \\ 1, & x \text{ irrational} \end{cases}$$

Show that $\int_{-1}^{1} f(x)\, dx$ does not exist. [*Hint:* The result in Problem 22 may be useful.]

Calculator/Computer Problems

32. Find an approximate value for $\int_0^1 (x^3 + 1)^{1/2}\, dx$ by using a Riemann sum, a regular partition of [0, 1] into five subintervals, and x_k^* the right endpoint of each subinterval.

33. Use the BASIC program of this section to obtain approximate values for $\int_0^1 (x^3 + 1)^{1/2}\, dx$ by choosing $n = 500$ and $n = 1000$.

[4.5.2]

34. Consider the discontinuous function defined for all x in the interval [0, 2]:

$$f(x) = \begin{cases} 0, & x \neq 1 \\ 1, & x = 1 \end{cases}$$

Show that f is integrable on the interval and $\int_0^2 f(x)\, dx = 0$. [*Hint:* Use (5).]

4.6 PROPERTIES OF THE DEFINITE INTEGRAL

The following two definitions are useful when working with definite integrals.

DEFINITION 4.5

If $f(a)$ exists, then

$$\int_a^a f(x)\, dx = 0$$

Definition 4.5 can be motivated by thinking that the area under the graph of f and above a single point a on the x-axis is zero.

DEFINITION 4.6

If f is integrable on $[a, b]$, then

$$\int_b^a f(x)\, dx = -\int_a^b f(x)\, dx$$

In the definition of $\int_a^b f(x)\,dx$ it was assumed that $a < b$, and so the usual "direction" of definite integration is left to right. Definition 4.6 states that reversing this direction of integration results in the negative of the integral.

EXAMPLE 1 By Definition 4.5,

$$\int_1^1 (x^3 + 3x)\,dx = 0 \qquad \square$$

EXAMPLE 2 In Example 3 of Section 4.5 we saw that $\int_{-2}^1 x^3\,dx = -\frac{15}{4}$. It follows from Definition 4.6 that

$$\int_1^{-2} x^3\,dx = -\int_{-2}^1 x^3\,dx = -\left(-\frac{15}{4}\right) = \frac{15}{4} \qquad \square$$

The next theorem gives some of the basic properties of the definite integral. These properties are analogous to the properties of the sigma notation given in Theorem 4.6 as well as the properties of the indefinite integral, or antiderivative, discussed in Section 4.1.

THEOREM 4.10 Properties

Let f and g be integrable functions on $[a, b]$. Then

(*i*) $\int_a^b kf(x)\,dx = k\int_a^b f(x)\,dx$, where k is any constant

(*ii*) $\int_a^b [f(x) \pm g(x)]\,dx = \int_a^b f(x)\,dx \pm \int_a^b g(x)\,dx$

Theorem 4.10(*ii*) extends to any finite sum of integrable functions on the interval:

$$\int_a^b [f_1(x) + f_2(x) + \cdots + f_n(x)]\,dx$$
$$= \int_a^b f_1(x)\,dx + \int_a^b f_2(x)\,dx + \cdots + \int_a^b f_n(x)\,dx$$

The independent variable x in a definite integral is called a **dummy variable** of integration. The value of the integral does not depend on the symbol used. In other words,

$$\int_a^b f(x)\,dx = \int_a^b f(r)\,dr = \int_a^b f(s)\,ds = \int_a^b f(t)\,dt$$

and so on.

EXAMPLE 3 $\int_{-2}^{1} x^3\,dx = \int_{-2}^{1} r^3\,dr = \int_{-2}^{1} t^3\,dt = -\frac{15}{4}$ □

THEOREM 4.11 Additive Interval Property

Let f be an integrable function on $[a, b]$. If c is any number in $[a, b]$, then

$$\int_a^b f(x)\,dx = \int_a^c f(x)\,dx + \int_c^b f(x)\,dx$$

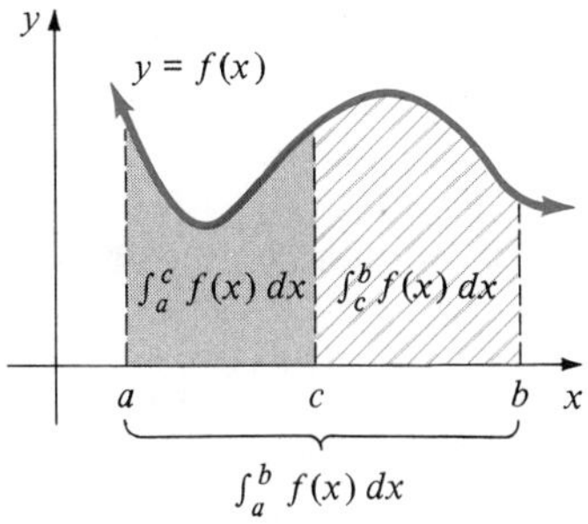

FIGURE 4.21

It is easy to interpret the additive property given in Theorem 4.11 in the special case when f is continuous on $[a, b]$ and $f(x) \geq 0$ for all x in the interval. As seen in Figure 4.21, the area under the graph of f on $[a, c]$ plus the area under the graph on $[c, b]$ is the same as the area under the graph on the entire interval $[a, b]$.

Note: The conclusion of Theorem 4.11 holds when a, b, and c are *any* three numbers in some closed interval. In other words, it is not necessary to have $a < c < b$.

For a given partition P of an interval $[a, b]$, intuitively, it seems that the sum

$$\lim_{\|P\| \to 0} \sum_{k=1}^{n} \Delta x_k$$

is simply the length of the interval, $b - a$. Thus, we have the following. See Problem 22 in Exercises 4.5.

THEOREM 4.12 Definite Integral of a Constant

For any constant k,

$$\int_a^b k\,dx = k\int_a^b dx = k(b - a)$$

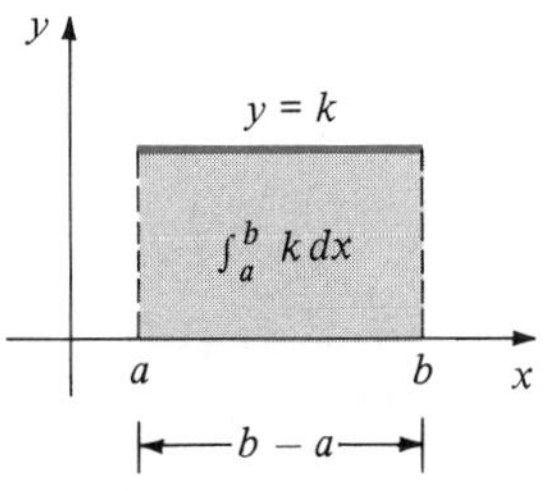

FIGURE 4.22

If $k > 0$, then Theorem 4.12 implies that $\int_a^b k\,dx$ is simply the area of a rectangle of width $b - a$ and height k. See Figure 4.22.

EXAMPLE 4 From Theorem 4.12,

$$\int_2^8 5\,dx = 5\int_2^8 dx = 5(8 - 2) = 30$$ □

EXAMPLE 5 Evaluate $\int_{-2}^{1} (x^3 + 4)\, dx$.

Solution From Theorem 4.10(*ii*) we can write

$$\int_{-2}^{1} (x^3 + 4)\, dx = \int_{-2}^{1} x^3\, dx + \int_{-2}^{1} 4\, dx$$

Now, from Example 3 of Section 4.5 we know that $\int_{-2}^{1} x^3\, dx = -15/4$ and so it follows from Theorem 4.12 that

$$\int_{-2}^{1} (x^3 + 4)\, dx = \left(-\frac{15}{4}\right) + 4[1 - (-2)] = \frac{33}{4}$$ □

Finally, when $f(x) \geq 0$ on $[a, b]$, the following result is not surprising.

THEOREM 4.13 Definite Integral of a Nonnegative Function Is Nonnegative

Let f be integrable on $[a, b]$ and $f(x) \geq 0$ for all x in $[a, b]$. Then

$$\int_{a}^{b} f(x)\, dx \geq 0$$

EXERCISES 4.6

Answers to odd-numbered problems begin on page A-80.

In Problems 1–40 use the definitions and theorems of this section to evaluate the given definite integral. Where appropriate, use the known results

$$\int_{-1}^{3} x^3\, dx = 20 \quad \int_{-1}^{3} x^2\, dx = \frac{28}{3}, \quad \int_{-1}^{3} x\, dx = 4,$$

$$\int_{0}^{\pi/3} \sin x\, dx = \frac{1}{2}, \quad \text{and} \quad \int_{0}^{\pi/2} \cos x\, dx = 1$$

1. $\int_{3}^{6} 4\, dx$
2. $\int_{-2}^{5} (-2)\, dx$
3. $\int_{-1}^{1} 5\, dx$
4. $\int_{-3}^{-3} 6\, dx$
5. $\int_{1}^{-2} \left(\frac{1}{2}\right) dx$
6. $\int_{4}^{-4} (-7)\, dx$
7. $\int_{3}^{-1} x^3\, dx$
8. $-\int_{3}^{-1} 10x\, dx$
9. $\int_{-1}^{3} 6x^2\, dx$
10. $\int_{-1}^{3} \frac{1}{5}x^3\, dx$
11. $\int_{-1}^{3} (-x)^3\, dx$
12. $\int_{-1}^{3} (2x)^3\, dx$
13. $\int_{-1}^{3} (x^3 + 1)\, dx$
14. $\int_{-1}^{3} (x^3 - 2)\, dx$
15. $\int_{-1}^{3} (2x^3 - 10)\, dx$
16. $\int_{-1}^{3} \left(5 - \frac{1}{4}x^3\right) dx$
17. $\int_{-1}^{3} t^3\, dt$
18. $\int_{-1}^{3} (u^3 + u)\, du$
19. $\int_{-1}^{3} (-x^3 + 2x - 11)\, dx$
20. $\int_{-1}^{3} x(x - 1)(x + 2)\, dx$

21. $\int_{-1}^{3} (x+1)^3\,dx$

22. $\int_{-1}^{3} (t+2)(t^2-2t+4)\,dt$

23. $\int_{-1}^{0} x^3\,dx + \int_{0}^{3} x^3\,dx$

24. $\int_{-1}^{2} x^3\,dx + \int_{2}^{3} x^3\,dx$

25. $\int_{-1}^{1/2} 2t\,dt - \int_{3}^{1/2} 2t\,dt$

26. $\int_{-1}^{1} (5u^3+1)\,du - \int_{3}^{1} (5u^3+1)\,du$

27. $\int_{0}^{3} x^3\,dx + \int_{3}^{0} x^3\,dx$

28. $\int_{-1}^{-1} x^3\,dx + \int_{-1}^{3} (x^3-4)\,dx$

29. $\int_{0}^{\pi/3} 6\sin\theta\,d\theta$

30. $\int_{0}^{\pi/3} (9-2\sin x)\,dx$

31. $\int_{0}^{\pi/4} \sin x\,dx + \int_{\pi/4}^{\pi/3} \sin x\,dx$

32. $\int_{1}^{5} \cos^2 x\,dx + \int_{1}^{5} \sin^2 x\,dx$

33. $\int_{0}^{\pi/2} \cos^2 \frac{x}{2}\,dx$

34. $\int_{0}^{\pi/3} \frac{\sin 2x}{\cos x}\,dx$

35. $\int_{\pi/4}^{\pi/4} \frac{dt}{\sin t}$

36. $\int_{0}^{\pi/3} (1+\cos x)\,dx + \int_{\pi/3}^{\pi/2} \cos x\,dx$

37. $\int_{0}^{4} x\,dx + \int_{0}^{4} (9-x)\,dx$

38. $\int_{-1}^{0} t^2\,dt + \int_{0}^{2} x^2\,dx + \int_{2}^{3} u^2\,du$

39. $\int_{\pi/2}^{0} 10\cos(t+\pi)\,dt$

40. $\int_{0}^{\pi/3} (1+\cos x\tan x)\,dx$

41. If f is integrable on $[a, b]$, then so is f^2. Show that $\int_a^b [f(x)]^2\,dx \geq 0$.

42. If f and g are integrable on $[a, b]$ and $f(x) \geq g(x)$ for all x in $[a, b]$, show that $\int_a^b f(x)\,dx \geq \int_a^b g(x)\,dx$. [*Hint:* Consider $f(x) - g(x)$ and Theorem 4.13.]

In Problems 43 and 44, use Problem 42 to establish the given result.

43. $\int_{0}^{1} x^3\,dx \leq \int_{0}^{1} x^2\,dx$

44. $\int_{0}^{\pi/4} (\cos x - \sin x)\,dx \geq 0$

45. By using areas, Figure 4.23 illustrates the plausibility of the following **comparison property**: If f is continuous on $[a, b]$ and $m \leq f(x) \leq M$ for all x in the interval, then

$$m(b-a) \leq \int_a^b f(x)\,dx \leq M(b-a)$$

Use this property to show that

$$1 \leq \int_0^1 (x^3+1)^{1/2}\,dx \leq 1.42$$

(See Problem 32 in Exercises 4.5.)

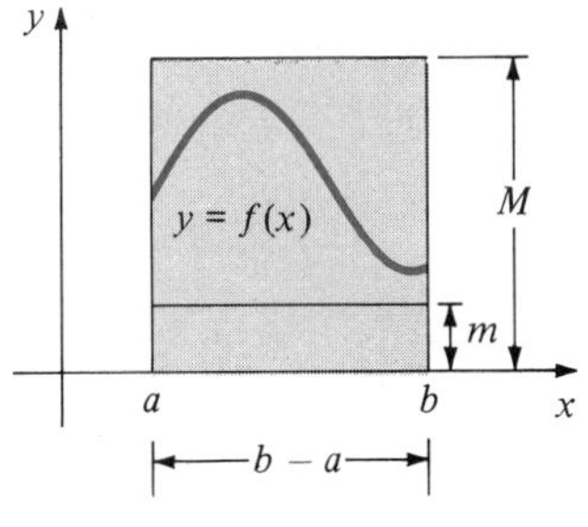

FIGURE 4.23

4.7 FUNDAMENTAL THEOREM OF CALCULUS

At the end of Section 4.5, we indicated that there is an easier way of evaluating a definite integral than by computing a limit of a sum. This "easier way" is by means of the so-called *Fundamental Theorem of Calculus*. In this theorem we shall see that the concept of an antiderivative of a continuous function provides the bridge between the differential calculus and the integral calculus.

Fundamental Theorem of Calculus—First Form Suppose f is continuous and $f(t) \geq 0$ for all t on some interval $[a, b]$. See Figure 4.24(a). Thus, the integral $\int_a^b f(t)\,dt$ exists and represents the area under the graph of f on the interval. Now, if x is any number in $[a, b]$, then, as shown in Figure 4.24(b), the function

$$A(x) = \int_a^x f(t)\,dt \tag{1}$$

gives the area under the graph on the interval $[a, x]$. If $\Delta x > 0$, then

$$A(x + \Delta x) = \int_a^{x+\Delta x} f(t)\,dt \tag{2}$$

is the area indicated in Figure 4.24(c), whereas the difference

$$A(x + \Delta x) - A(x) \tag{3}$$

is the area shown in Figure 4.24(d).

Since f is continuous on the closed interval $[x, x + \Delta x]$, we know from the Extreme Value Theorem, Theorem 3.1, that it attains a minimum value m and a maximum value M on the interval. As suggested in Figure 4.25, the the difference (3) is bracketed between the numbers $m\,\Delta x$ and $M\,\Delta x$:

$$m\,\Delta x \leq A(x + \Delta x) - A(x) \leq M\,\Delta x$$

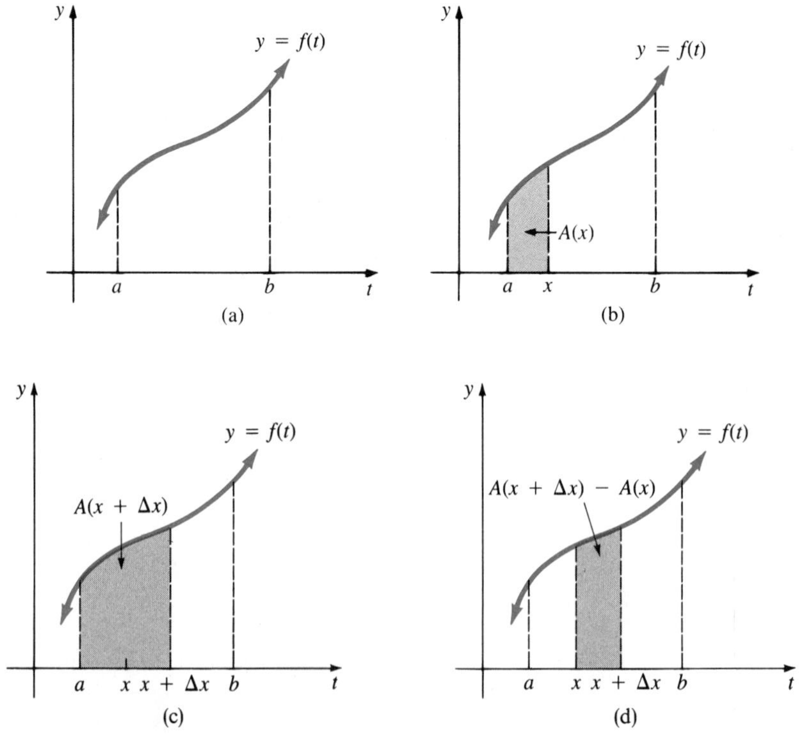

FIGURE 4.24

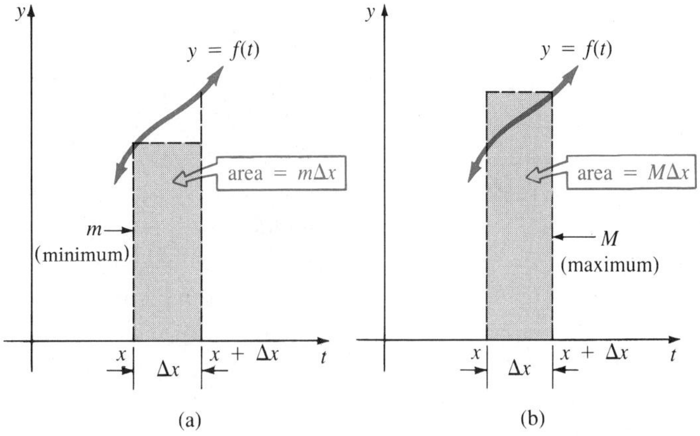

FIGURE 4.25

Although, for the sake of illustration, we have shown f to be an increasing function in Figure 4.25, the foregoing inequalities hold for any function continuous on $[x, x + \Delta x]$. Dividing by Δx then gives

$$m \le \frac{A(x + \Delta x) - A(x)}{\Delta x} \le M$$

Using the fact that $\lim_{\Delta x \to 0^+} m = \lim_{\Delta x \to 0^+} M = f(x)$, we get

$$\lim_{\Delta x \to 0^+} \frac{A(x + \Delta x) - A(x)}{\Delta x} = f(x) \tag{4}$$

From the definition of a derivative, we can write (4) as $A'(x) = f(x)$. A similar argument holds for $\Delta x < 0$.

The preceding discussion suggests the following theorem, known as the first form of the **Fundamental Theorem of Calculus.**

THEOREM 4.14 Fundamental Theorem of Calculus—Derivative Form

Let f be continuous on $[a, b]$ and let x be any number in the interval. If $G(x)$ is a function defined by

$$G(x) = \int_a^x f(t)\,dt \quad \text{then} \quad G'(x) = f(x)$$

Here we have chosen to denote the integral $\int_a^x f(t)\,dt$ by $G(x)$, since in the context of Theorem 4.14, the function f need not be nonnegative on $[a, b]$ and so the integral need not represent a measure of area as did $A(x)$ in (1).

EXAMPLE 1

(a) $\dfrac{d}{dx}\displaystyle\int_{-2}^{x} t^3\,dt = x^3$ (b) $\dfrac{d}{dx}\displaystyle\int_{1}^{x} \sqrt{t^2+1}\,dt = \sqrt{x^2+1}$ □

Fundamental Theorem of Calculus—Second Form For a continuous function f, the statement $G'(x) = f(x)$ for $G(x) = \int_a^x f(t)\,dt$ means that $G(x)$ is an antiderivative of the integrand. If F is any antiderivative of f, we know from Theorem 4.1 and that $G(x) - F(x) = C$ or $G(x) = F(x) + C$, where C is any arbitrary constant. It follows for any x in $[a, b]$ that

$$F(x) + C = \int_a^x f(t)\,dt \tag{5}$$

If we substitute $x = a$ in (5), then

$$F(a) + C = \int_a^a f(t)\,dt$$

implies $C = -F(a)$, since $\int_a^a f(t)\,dt = 0$. Thus, (5) becomes

$$F(x) - F(a) = \int_a^x f(t)\,dt$$

Since the latter equation is valid at $x = b$, we find

$$F(b) - F(a) = \int_a^b f(t)\,dt$$

We have proved an exceedingly useful second form of the **Fundamental Theorem of Calculus.**

THEOREM 4.15 Fundamental Theorem of Calculus—Antiderivative Form

Let f be continuous on $[a, b]$ and let F be any function for which $F'(x) = f(x)$. Then

$$\int_a^b f(x)\,dx = F(b) - F(a) \tag{6}$$

The difference $F(b) - F(a)$ in (6) is usually denoted by

$$F(x)\Big]_a^b$$

that is,

$$\underbrace{\int_a^b f(x)\,dx}_{\boxed{\text{definite integral}}} = \underbrace{\int f(x)\,dx}_{\boxed{\text{indefinite integral or antiderivative}}}\Bigg]_a^b = F(x)\Bigg]_a^b$$

Since Theorem 4.15 indicates that we may use *any* antiderivative, we may always choose the constant of integration C to be zero. Observe that if $C \neq 0$, then

$$(F(x) + C)\Big]_a^b = (F(b) + C) - (F(a) + C) = F(b) - F(a) = F(x)\Big]_a^b$$

Alternative Proof of Theorem 4.15 It is worthwhile to examine yet another proof of Theorem 4.15 using the basic premise that a definite integral is a limit of a sum. If F is an antiderivative of f, then $F'(x) = f(x)$. Since F is differentiable on (a, b), the Mean Value Theorem for derivatives guarantees that there exists an x_k^* in each subinterval (x_{k-1}, x_k) of the partition $P\colon a = x_0 < x_1 < x_2 < \cdots < x_{n-1} < x_n = b$ such that

$$F(x_k) - F(x_{k-1}) = F'(x_k^*)(x_k - x_{k-1}) = F'(x_k^*)\,\Delta x_k = f(x_k^*)\,\Delta x_k$$

Now,

$$\begin{aligned} F(b) - F(a) &= F(x_n) - F(x_0) \\ &= F(x_n) + \underbrace{F(x_1) - F(x_1) + F(x_2) - F(x_2) + \cdots + F(x_{n-1}) - F(x_{n-1})}_{\boxed{\text{zero}}} - F(x_0) \\ &= [F(x_1) - F(x_0)] + [F(x_2) - F(x_1)] + \cdots + [F(x_n) - F(x_{n-1})] = \sum_{k=1}^{n} [F(x_k) - F(x_{k-1})] \\ &= \sum_{k=1}^{n} F'(x_k^*)\,\Delta x_k = \sum_{k=1}^{n} f(x_k^*)\,\Delta x_k \end{aligned} \tag{7}$$

But $\lim_{\|P\| \to 0}[F(b) - F(a)] = F(b) - F(a)$, and so the limit of (7) as $\|P\| \to 0$ is

$$F(b) - F(a) = \lim_{\|P\| \to 0} \sum_{k=1}^{n} f(x_k^*)\,\Delta x_k \tag{8}$$

From Definition 4.4, the right-hand member of (8) is $\int_a^b f(x)\,dx$. ■

EXAMPLE 2 In Example 3 of Section 4.5 we resorted to the rather lengthy definition of the definite integral to show that

$$\int_{-2}^{1} x^3\,dx = -\frac{15}{4}$$

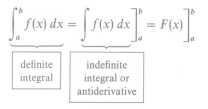

Since $F(x) = x^4/4$ is an antiderivative of $f(x) = x^3$, we now obtain immediately

$$\int_{-2}^{1} x^3\,dx = \frac{x^4}{4}\Big]_{-2}^{1} = \frac{1}{4} - \frac{(-2)^4}{4} = \frac{1}{4} - \frac{16}{4} = -\frac{15}{4} \qquad \square$$

EXAMPLE 3 Evaluate $\int_1^3 x\,dx$.

Solution An antiderivative of $f(x) = x$ is $F(x) = x^2/2$. Consequently, Theorem 4.15 gives

$$\int_1^3 x\,dx = \frac{x^2}{2}\Big]_1^3 = \frac{9}{2} - \frac{1}{2} = 4 \qquad \square$$

EXAMPLE 4 Evaluate $\int_{-2}^{2} (3x^2 - x + 1)\,dx$.

Solution We apply (1) of Section 4.1 to each term of the integrand and then use the Fundamental Theorem:

$$\begin{aligned}\int_{-2}^{2} (3x^2 - x + 1)\,dx &= \left(x^3 - \frac{x^2}{2} + x\right)\Big]_{-2}^{2} \\ &= (8 - 2 + 2) - (-8 - 2 - 2) = 20 \qquad \square\end{aligned}$$

EXAMPLE 5 Evaluate $\int_{\pi/6}^{\pi} \cos x\,dx$.

Solution An antiderivative of $f(x) = \cos x$ is $F(x) = \sin x$. Therefore,

$$\int_{\pi/6}^{\pi} \cos x\,dx = \sin x\Big]_{\pi/6}^{\pi} = \sin \pi - \sin\frac{\pi}{6} = 0 - \frac{1}{2} = -\frac{1}{2} \qquad \square$$

EXAMPLE 6 Evaluate $\int_0^3 |x - 2|\,dx$.

Solution From the definition of absolute value, we know

$$|x-2| = \begin{cases} x - 2 & \text{if } x - 2 \geq 0 \\ -(x-2) & \text{if } x - 2 < 0 \end{cases} \quad \text{or} \quad |x-2| = \begin{cases} x - 2 & \text{if } x \geq 2 \\ -x + 2 & \text{if } x < 2 \end{cases}$$

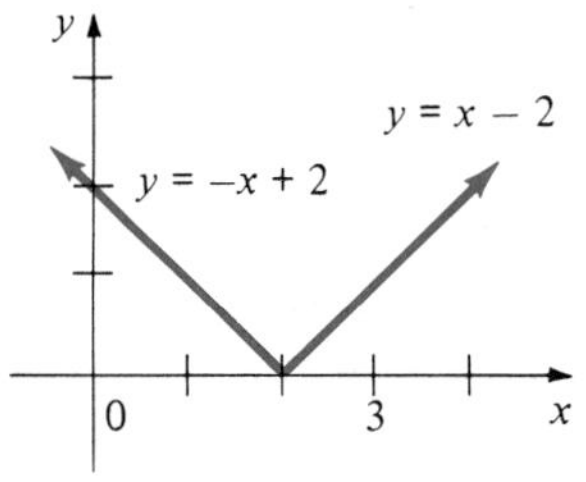

FIGURE 4.26

The graph of $f(x) = |x - 2|$ is given in Figure 4.26. Now in view of Theorem 4.11 we can write

$$\begin{aligned}\int_0^3 |x-2|\,dx &= \int_0^2 |x-2|\,dx + \int_2^3 |x-2|\,dx \\ &= \int_0^2 (-x+2)\,dx + \int_2^3 (x-2)\,dx \\ &= \left(-\frac{x^2}{2} + 2x\right)\Bigg]_0^2 + \left(\frac{x^2}{2} - 2x\right)\Bigg]_2^3 \\ &= (-2+4) + \left(\frac{9}{2} - 6\right) - (2-4) = \frac{5}{2}\end{aligned}$$

□

Substitution in a Definite Integral Recall from Section 4.2 that we sometimes used a substitution as an aid in evaluating an indefinite integral of the form $\int f(g(x))g'(x)\,dx$. Care should be exercised when using a substitution in a definite integral $\int_a^b f(g(x))g'(x)\,dx$, since we can proceed in *two ways*:

(*i*) Evaluate the definite integral by means of the substitution $u = g(x)$. Resubstitute $u = g(x)$ in the antiderivative and then apply the Fundamental Theorem of Calculus by using the original limits of integration $x = a$ and $x = b$.

(*ii*) Alternatively, the resubstitution can be avoided by changing the limits of integration to correspond to the value of u at $x = a$ and u at $x = b$.

The latter method, which is usually quicker, is summarized in the next theorem.

THEOREM 4.16 Substitution in a Definite Integral

Let $u = g(x)$ be a function that has a continuous derivative on the interval $[a, b]$, and let f be a function that is continuous on the range of g. If $F'(u) = f(u)$ and $c = g(a)$, $d = g(b)$, then

$$\int_a^b f(g(x))g'(x)\,dx = F(d) - F(c)$$

Proof

$$\int_a^b f(g(x))g'(x)\,dx = \int_c^d f(u)\frac{du}{dx}\,dx = \int_c^d f(u)\,du = F(u)\Big]_c^d = F(d) - F(c)$$

■

EXAMPLE 7 Evaluate $\int_0^2 \sqrt{2x^2+1}\,x\,dx$.

Solution We shall first illustrate the procedure outlined in method (*i*) above. To evaluate the indefinite integral $\int \sqrt{2x^2+1}\,x\,dx$, we use

$$u = 2x^2 + 1 \quad \text{and} \quad du = 4x\,dx$$

Thus,

$$\begin{aligned}\int \sqrt{2x^2+1}\,x\,dx &= \frac{1}{4}\int \sqrt{2x^2+1}(4x\,dx) && \boxed{\text{substitution}}\\ &= \frac{1}{4}\int u^{1/2}\,du\\ &= \frac{1}{4}\frac{u^{3/2}}{3/2} + C && \boxed{\text{resubstitution}}\\ &= \frac{1}{6}(2x^2+1)^{3/2} + C\end{aligned}$$

Therefore, by Theorem 4.15,

$$\begin{aligned}\int_0^2 \sqrt{2x^2+1}\,x\,dx &= \frac{1}{6}(2x^2+1)^{3/2}\Big]_0^2\\ &= \frac{1}{6}[9^{3/2} - 1^{3/2}]\\ &= \frac{1}{6}[27-1] = \frac{13}{3}\end{aligned}$$

□

The second procedure given in (*ii*) is illustrated in the following example.

EXAMPLE 8 In Example 7, if $u = 2x^2 + 1$, notice that $x = 0$ implies $u = 1$ and that $x = 2$ gives $u = 9$. Thus, by Theorem 4.16,

$$\begin{aligned}\int_0^2 \sqrt{2x^2+1}\,x\,dx &= \frac{1}{4}\int_1^9 u^{1/2}\,du\\ &= \frac{1}{4}\frac{u^{3/2}}{3/2}\Big]_1^9\\ &= \frac{1}{6}[9^{3/2} - 1^{3/2}] = \frac{13}{3}\end{aligned}$$

u limits: integration with respect to *u*

□

When the graph of a function $y = f(x)$ is symmetric with respect to either the y-axis (even function) or the origin (odd function) on a symmetric interval $[-a, a]$, then the definite integral $\int_{-a}^{a} f(x)\,dx$ can be evaluated by means of a "shortcut."

THEOREM 4.17 Even Function Rule

Let f be an even integrable function on $[-a, a]$. Then

$$\int_{-a}^{a} f(x)\,dx = 2\int_{0}^{a} f(x)\,dx$$

THEOREM 4.18 Odd Function Rule

Let f be an odd integrable function on $[-a, a]$. Then

$$\int_{-a}^{a} f(x)\,dx = 0$$

Although we shall leave the proofs of these two theorems as exercises, the geometric motivations for the results are given in Figure 4.27. The point in Theorem 4.18 is this: When we integrate an odd integrable function f on a symmetric interval $[-a, a]$, there is no need to find an antiderivative of f in order to utilize Theorem 4.15; the value of the integral is always zero.

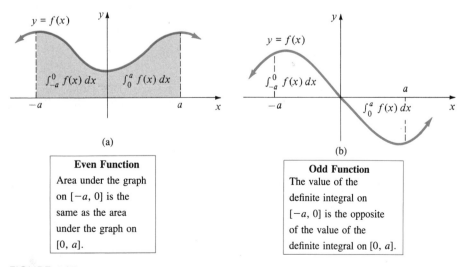

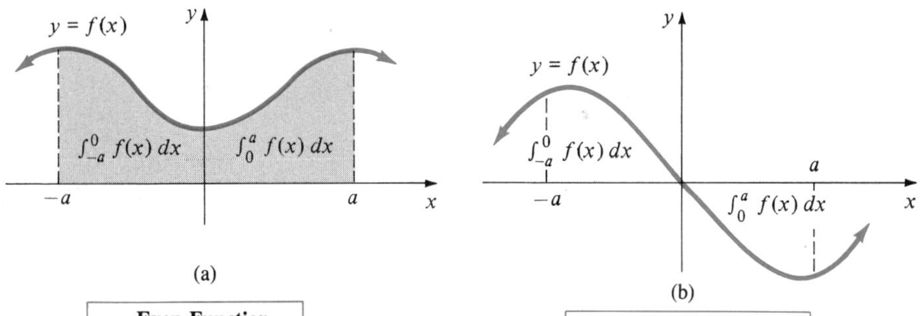

Even Function
Area under the graph on $[-a, 0]$ is the same as the area under the graph on $[0, a]$.

Odd Function
The value of the definite integral on $[-a, 0]$ is the opposite of the value of the definite integral on $[0, a]$.

FIGURE 4.27

EXAMPLE 9 Evaluate $\int_{-1}^{1} (x^4 + x^2)\, dx$.

Solution Since $f(-x) = (-x)^4 + (-x)^2 = x^4 + x^2 = f(x)$, the integrand is an even function on the symmetric interval $[-1, 1]$. It follows from Theorem 4.17 that

$$\begin{aligned}\int_{-1}^{1} (x^4 + x^2)\, dx &= 2\int_{0}^{1} (x^4 + x^2)\, dx \\ &= 2\left(\frac{x^5}{5} + \frac{x^3}{3}\right)\Bigg]_0^1 \\ &= 2\left(\frac{1}{5} + \frac{1}{3}\right) = \frac{16}{15}\end{aligned}$$

□

EXAMPLE 10 Evaluate $\int_{-\pi/2}^{\pi/2} \sin x\, dx$.

Solution In this case $f(x) = \sin x$ is an odd function on the symmetric interval $[-\pi/2, \pi/2]$. Thus, by Theorem 4.18 we have immediately

$$\int_{-\pi/2}^{\pi/2} \sin x\, dx = 0$$

□

Piecewise Continuous Functions A function f is said to be **piecewise continuous** on an interval $[a, b]$ if there are at most a finite number of points c_k, $k = 1, 2, \ldots, n$, $(c_{k-1} < c_k)$ at which f has a finite, or jump, discontinuity and f is continuous on each open interval (c_{k-1}, c_k). See Figure 4.28.

By Theorem 4.8, a piecewise continuous function is integrable. A definite integral of a piecewise continuous function on $[a, b]$ can be evaluated with

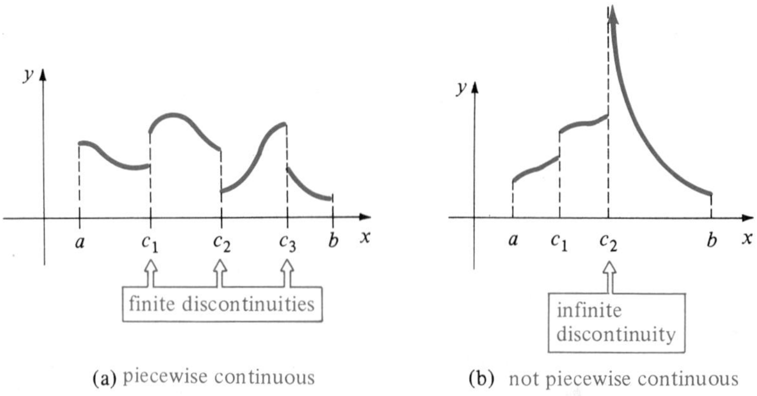

FIGURE 4.28

the help of Theorem 4.10:

$$\int_a^b f(x)\,dx = \int_a^{c_1} f(x)\,dx + \int_{c_1}^{c_2} f(x)\,dx + \cdots + \int_{c_n}^{b} f(x)\,dx$$

and by simply treating the integrands of the definite integrals on the right side of the above equation as if they were continuous on the closed intervals $[a, c_1], [c_1, c_2], \ldots, [c_n, b]$.

EXAMPLE 11 Evaluate

$$\int_{-1}^{4} f(x)\,dx, \quad \text{where } f(x) = \begin{cases} x + 1, & -1 \le x < 0 \\ x, & 0 \le x < 2 \\ 3, & 2 \le x \le 4 \end{cases}$$

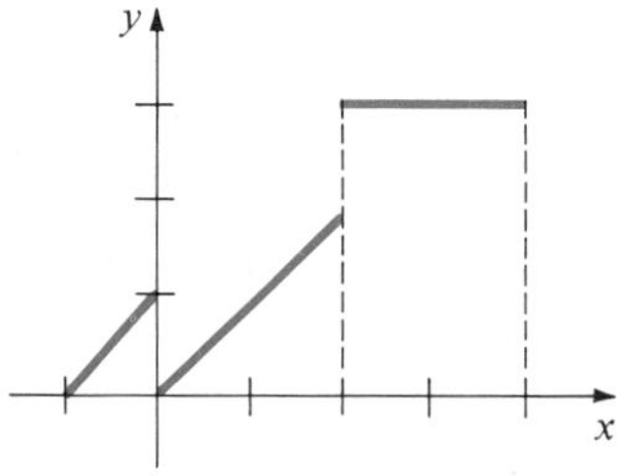

FIGURE 4.29

Solution The graph of the piecewise continuous function f is given in Figure 4.29. Now, from the preceding discussion,

$$\begin{aligned}\int_{-1}^{4} f(x)\,dx &= \int_{-1}^{0} f(x)\,dx + \int_{0}^{2} f(x)\,dx + \int_{2}^{4} f(x)\,dx \\ &= \int_{-1}^{0} (x+1)\,dx + \int_{0}^{2} x\,dx + \int_{2}^{4} 3\,dx \\ &= \left(\frac{x^2}{2} + x\right)\Big]_{-1}^{0} + \frac{x^2}{2}\Big]_{0}^{2} + 3x\Big]_{2}^{4} = \frac{17}{2}\end{aligned}$$

□

Calculator/Computers The HP-28S calculator does not give exact results for definite integrals. It utilizes a method to approximate the value of $\int_a^b f(x)\,dx$. In Maple a definite integral such as $\int_{-2}^{1} x^3\,dx$ is written int(x^3, x = −2..1);. In Mathematica the definite integral is written Integrate[x^3, {x, −2, 1}]. With either program we obtain the exact result $-\frac{15}{4}$.

▲ ***Remark*** The antiderivative form of the Fundamental Theorem of Calculus is an extremely important and powerful tool for evaluating definite integrals. Why should we bother with a clumsy limit of a sum when the value of $\int_a^b f(x)\,dx$ can be found by computing $\int f(x)\,dx$ at the two numbers a and b? This is true up to a point—however, it is time to learn another fact of mathematical life: There are functions for which the antiderivative $\int f(x)\,dx$ *cannot* be expressed in terms of *elementary functions.** The simple continuous function $f(x) = \sqrt{x^3 + 1}$ possesses no antiderivative that is an elementary function. Although, in view of Theorem 4.7, we can say that the definite integral $\int_0^1 \sqrt{x^3 + 1}\,dx$ exists, Theorem 4.15 provides no help in finding its value. We will consider problems of this sort in the next section.

---

*The *elementary functions* that we have seen so far are sums, products, quotients, and powers of polynomial and trigonometric functions. Later on we shall add to this list the logarithmic, exponential, and inverse trigonometric functions.

EXERCISES 4.7

Answers to odd-numbered problems begin on page A-80.

In Problems 1 and 2 evaluate the given expression.

1. $\dfrac{d}{dx}\displaystyle\int_0^x (3t^2 - 2t)^6\,dt$

2. $\dfrac{d}{dx}\displaystyle\int_x^9 \sqrt[3]{u^2+2}\,du$

In Problems 3 and 4 verify the given result by first evaluating the definite integral and then differentiating.

3. $\dfrac{d}{dx}\displaystyle\int_a^x (6t^2 - 8t + 5)\,dt = 6x^2 - 8x + 5$

4. $\dfrac{d}{dt}\displaystyle\int_\pi^t \sin\frac{x}{3}\,dx = \sin\frac{t}{3}$

In Problems 5–48 evaluate the given definite integral.

5. $\displaystyle\int_3^7 dx$

6. $\displaystyle\int_2^{10} (-4)\,dx$

7. $\displaystyle\int_{-1}^2 (2x+3)\,dx$

8. $\displaystyle\int_{-5}^4 t^2\,dt$

9. $\displaystyle\int_1^3 (6x^2 - 4x + 5)\,dx$

10. $\displaystyle\int_{-2}^1 (12x^5 - 36)\,dx$

11. $\displaystyle\int_0^{\pi/2} \sin x\,dx$

12. $\displaystyle\int_{-\pi/3}^{\pi/4} \cos\theta\,d\theta$

13. $\displaystyle\int_{\pi/4}^{\pi/2} \cos 3t\,dt$

14. $\displaystyle\int_{1/2}^1 \sin 2\pi x\,dx$

15. $\displaystyle\int_{1/2}^{3/4} \frac{1}{u^2}\,du$

16. $\displaystyle\int_{1/4}^1 \frac{2}{\sqrt{x}}\,dx$

17. $\displaystyle\int_{-1}^1 \sqrt[3]{x}\,dx$

18. $\displaystyle\int_0^8 \sqrt[3]{t^2}\,dt$

19. $\displaystyle\int_0^2 x(1-x)\,dx$

20. $\displaystyle\int_3^2 x(x-2)(x+2)\,dx$

21. $\displaystyle\int_{-1}^1 (7x^3 - 2x^2 + 5x - 4)\,dx$

22. $\displaystyle\int_{-3}^{-1} (x^2 - 4x + 8)\,dx$

23. $\displaystyle\int_1^4 \frac{x-1}{\sqrt{x}}\,dx$

24. $\displaystyle\int_2^4 \frac{x^2+8}{x^2}\,dx$

25. $\displaystyle\int_0^5 (2-\sqrt{x})^2\,dx + \int_5^9 (2-\sqrt{x})^2\,dx$

26. $\displaystyle\int_{-2}^0 (x+1)^2\,dx + \int_0^3 (x+1)^2\,dx$

27. $\displaystyle\int_{-4}^2 \left(\frac{x}{2}+1\right)^5 dx$

28. $\displaystyle\int_1^3 \frac{1}{x^2+10x+25}\,dx$

29. $\displaystyle\int_{-4}^{12} \sqrt{z+4}\,dz$

30. $\displaystyle\int_0^{7/2} (2x+1)^{-1/3}\,dx$

31. $\displaystyle\int_0^3 \frac{x}{\sqrt{x^2+16}}\,dx$

32. $\displaystyle\int_{-2}^1 \frac{t}{(t^2+1)^2}\,dt$

33. $\displaystyle\int_{1/2}^1 \left(1+\frac{1}{x}\right)^3 \frac{1}{x^2}\,dx$

34. $\displaystyle\int_1^4 \frac{\sqrt[3]{1+4\sqrt{x}}}{\sqrt{x}}\,dx$

35. $\displaystyle\int_0^1 \frac{x+1}{\sqrt{x^2+2x+3}}\,dx$

36. $\displaystyle\int_{-1}^1 \frac{u^3+u}{(u^4+2u^2+1)^5}\,du$

37. $\displaystyle\int_0^{\pi/8} \sec^2 2x\,dx$

38. $\displaystyle\int_{\sqrt{\pi}/2}^{\sqrt{\pi/2}} x\csc x^2 \cot x^2\,dx$

39. $\displaystyle\int_{-1/2}^{3/2} (x - \cos\pi x)\,dx$

40. $\displaystyle\int_1^4 \frac{\cos\sqrt{x}}{2\sqrt{x}}\,dx$

41. $\displaystyle\int_0^{\pi/2} \sqrt{\cos x}\,\sin x\,dx$

42. $\displaystyle\int_{\pi/6}^{\pi/3} \sin x\cos x\,dx$

43. $\displaystyle\int_{\pi/6}^{\pi/2} \frac{1+\cos\theta}{(\theta+\sin\theta)^2}\,d\theta$

44. $\displaystyle\int_{-\pi/4}^{\pi/4} (\sec x + \tan x)^2\,dx$

45. $\displaystyle\int_{-2}^2 \sin 3x\,dx$

46. $\displaystyle\int_{-\pi/2}^{\pi/2} \cos^2 x\,dx$

47. $\displaystyle\int_{-5}^5 (x^3+x)^9\,dx$

48. $\displaystyle\int_{-1}^1 \tan x\,dx$

In Problems 49–54 evaluate the given definite integral.

49. $\displaystyle\int_{-3}^1 |x|\,dx$

50. $\displaystyle\int_0^4 |2x-6|\,dx$

51. $\displaystyle\int_{-8}^3 \sqrt{|x|+1}\,dx$

52. $\displaystyle\int_0^2 |x^2-1|\,dx$

53. $\displaystyle\int_{-\pi}^{\pi} |\sin x|\,dx$

54. $\displaystyle\int_0^{\pi} |\cos x|\,dx$

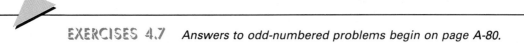

In Problems 55 and 56 evaluate $\int_{-1}^{2} f(x)\,dx$ for the given function.

55. $f(x) = \begin{cases} -x, & x < 0 \\ x^2, & x \geq 0 \end{cases}$

56. $f(x) = \begin{cases} 2x + 3, & x \leq 0 \\ 3, & x > 0 \end{cases}$

In Problems 57–60 evaluate the definite integral of the given piecewise continuous function.

57. $\int_0^3 f(x)\,dx$, where $f(x) = \begin{cases} 4, & 0 \leq x < 2 \\ 1, & 2 \leq x \leq 3 \end{cases}$

58. $\int_0^{\pi} f(x)\,dx$, where $f(x) = \begin{cases} \sin x, & 0 \leq x < \pi/2 \\ \cos x, & \pi/2 < x \leq \pi \end{cases}$

59. $\int_{-2}^{2} f(x)$, where $f(x) = \begin{cases} x^2, & -2 \leq x < -1 \\ 4, & -1 \leq x < 1 \\ x^2, & 1 \leq x \leq 2 \end{cases}$

60. $\int_0^4 f(x)\,dx$, where $f(x) = [x]$ is the greatest integer function.

61. Use the substitution $u = t + 1$ to evaluate

$$\int_0^3 t\sqrt{t + 1}\,dt$$

62. Use the substitution $u = 2x + 1$ to evaluate

$$\int_0^4 \frac{x^2}{\sqrt{2x + 1}}\,dx$$

In Problems 63 and 64 evaluate the given definite integral.

63. $\int_{-1}^{2} \left\{ \int_1^x 12t^2\,dt \right\} dx$

64. $\int_0^{\pi/2} \left\{ \int_0^t \sin x\,dx \right\} dt$

In Problems 65 and 66 let $G(x) = \int_a^x f(t)\,dt$ and $G'(x) = f(x)$.

65. (a) What is $G(x^2)$? (b) What is $[G(x^2)]'$?

66. (a) What is $G(x^3 + 2x)$?
(b) What is $[G(x^3 + 2x)]'$?

In Problems 67 and 68 determine $F'(x)$.

67. $F(x) = \int_{6x-1}^{0} \sqrt{4t + 9}\,dt$

68. $F(x) = \int_{3x}^{x^2} \frac{1}{t^2 + 1}\,dt$ [*Hint:* Use two integrals.]

In Problems 69 and 70 let P be a partition of the indicated interval and x_k^* a number in the kth subinterval. Determine the value of the given limit.

69. $\lim_{\|P\| \to 0} \sum_{k=1}^{n} (2x_k^* + 5)\,\Delta x_k$; $[-1, 3]$

70. $\lim_{\|P\| \to 0} \sum_{k=1}^{n} \cos \frac{x_k^*}{4}\,\Delta x_k$; $[0, 2\pi]$

In Problems 71 and 72 let P be a regular partition of the indicated interval and x_k^* a number in the kth subinterval. Establish the given result.

71. $\lim_{n \to \infty} \frac{\pi}{n} \sum_{k=1}^{n} \sin x_k^* = 2$; $[0, \pi]$

72. $\lim_{n \to \infty} \frac{2}{n} \sum_{k=1}^{n} x_k^* = 0$; $[-1, 1]$

73. Reread Theorem 4.15 and give a reason why the following procedure is *incorrect*:

$$\int_{-1}^{1} x^{-2}\,dx = -x^{-1}\Big]_{-1}^{1} = -1 + (-1) = -2$$

74. Suppose f is an odd function that is defined on the interval $[-4, 4]$. Suppose further that f is differentiable on the interval, has zeros at -3 and 3, has critical points at only -2 and 2, and $f(-2) = 3.5$.

(a) What is $f(0)$?

(b) Sketch a rough graph of f.

(c) Suppose F is a function defined on $[-4, 4]$ by $F(x) = \int_{-3}^{x} f(t)\,dt$. Find $F(-3)$ and $F(3)$.

(d) Sketch a rough graph of F.

(e) Find the critical points and points of inflection of F.

75. Prove the even function rule, Theorem 4.17. [*Hint:* $\int_{-a}^{a} f(x)\,dx = \int_{-a}^{0} f(x)\,dx + \int_0^a f(x)\,dx$. Let $t = -x$ in the integral $\int_{-a}^{0} f(x)\,dx$ and use $f(-x) = f(x)$.]

76. Prove the odd function rule, Theorem 4.18. [*Hint:* Proceed as in Problem 75.]

Calculator/Computer Problems

77. (a) Use a calculator or computer to obtain the graphs of $f(x) = \cos^3 x$ and $f(x) = \sin^3 x$.

(b) Based on your interpretation of the graphs in part **(a)**, conjecture the values of $\int_0^{2\pi} \cos^3 x\,dx$ and $\int_0^{2\pi} \sin^3 x\,dx$.

(c) Use Theorem 4.15 to evaluate the integrals in part **(b)**. [*Hint:* See Problem 49 in Exercises 4.2.]

78. The time t required for an expanding oil spill to evaporate is determined by the formula

$$\frac{RT}{Pv} = \int_0^t \frac{KA(u)}{V_0}\,du$$

where $A(u)$ is the area of the spill at any time u, RT/Pv is a dimensionless thermodynamic term, K is a "mass transfer" coefficient, and V_0 is the initial volume of the spill.

(a) Suppose the oil spill is expanding in the form of a circle whose initial radius is r_0. See Figure 4.30. If the radius r of the spill is increasing at a rate $dr/dt = C$ (in meters per second), solve for t in terms of the other symbols.

(b) Typical values for RT/Pv and K are 1.9×10^6 (for tridecane) and 0.01 mm/s, respectively. If $c = 0.01$ m/s, $r_0 = 100$ m, and $V_0 = 10{,}000$ m^3, determine how long it will take for the oil to evaporate.

(c) Using the result in part (**b**), determine the final area of the oil spill.

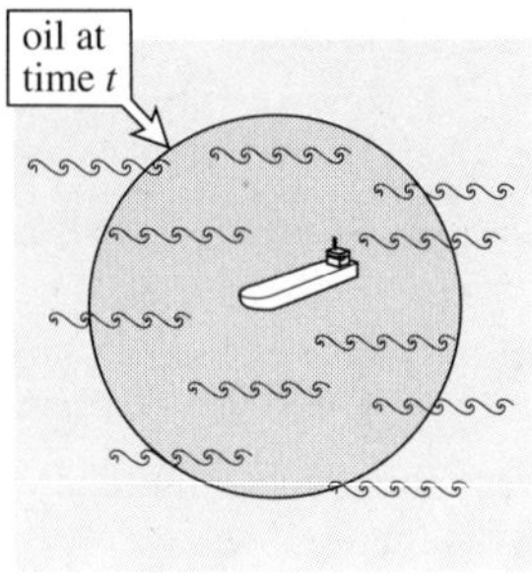

FIGURE 4.30

4.8 APPROXIMATE INTEGRATION

Life in mathematics would be extremely pleasant if the antiderivative of every continuous function could be expressed in terms of elementary functions such as polynomial, rational, or trigonometric functions. This is not the case. Hence, Theorem 4.15 cannot be used to evaluate every definite integral. Sometimes the best we can do is to *approximate* the value of $\int_a^b f(x)\,dx$. In this concluding section, we shall consider three such numerical procedures.

In the following discussion it will again be useful to interpret the definite integral $\int_a^b f(x)\,dx$ as the area under the graph of f on $[a, b]$. Although continuity of f is essential, there is no actual requirement that $f(x) \geq 0$ on the interval.

Midpoint Rule One way of approximating a definite integral is to proceed in the same manner as we did in the initial discussion about finding the area under a graph—namely, construct rectangular elements and add their areas. In particular, let us suppose that $y = f(x)$ is continuous on $[a, b]$ and that this interval is partitioned into n subintervals of equal length $\Delta x = (b - a)/n$. (Recall this is called a regular partition.) A simple, but fairly accurate, approximation rule consists of adding the areas of n rectangular elements whose lengths are calculated at the midpoint of each subinterval. See Figure 4.31(a). Now, if $m_k = (x_{k-1} + x_k)/2$ is the midpoint of $[x_{k-1}, x_k]$, then the

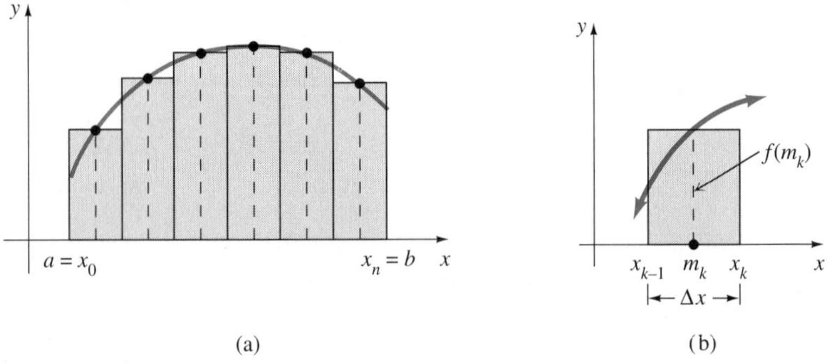

FIGURE 4.31

area of the rectangular element shown in Figure 4.31(b) is

$$A_k = f(m_k)\,\Delta x = f\left(\frac{x_{k-1} + x_k}{2}\right)\Delta x$$

Identifying $a = x_0$ and $b = x_n$ and summing the n areas, we obtain

$$\int_a^b f(x)\,dx \approx f\left(\frac{x_0 + x_1}{2}\right)\Delta x + f\left(\frac{x_1 + x_2}{2}\right)\Delta x + \cdots + f\left(\frac{x_{n-1} + x_n}{2}\right)\Delta x$$

If we replace Δx by $(b - a)/n$, this midpoint approximation rule can be summarized as follows:

DEFINITION 4.7 Midpoint Rule

The **Midpoint Rule** is the approximation $\int_a^b f(x)\,dx \approx M_n$, where

$$M_n = \frac{b-a}{n}\left[f\left(\frac{x_0 + x_1}{2}\right) + f\left(\frac{x_1 + x_2}{2}\right) + \cdots + f\left(\frac{x_{n-1} + x_n}{2}\right)\right] \quad (1)$$

Since the function $f(x) = 1/x$ is continuous on any interval $[a, b]$ that does not include the origin, we know that $\int_a^b dx/x$ exists. However, we have not seen, as yet, any function F such that $F'(x) = 1/x$; that is, we do not know an antiderivative of f.

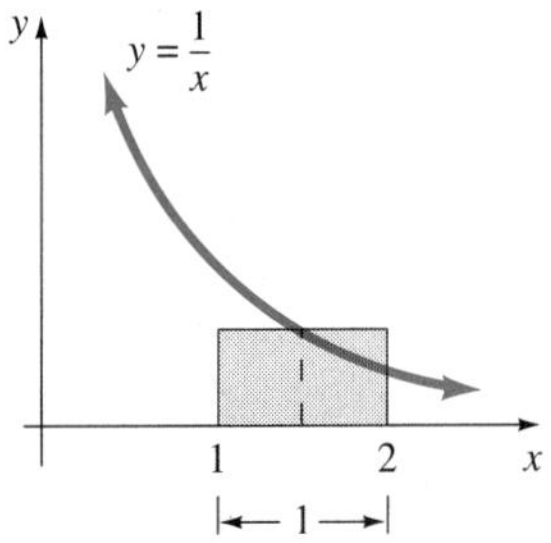

FIGURE 4.32

EXAMPLE 1 Approximate $\int_1^2 dx/x$ by the Midpoint Rule for $n = 1$, $n = 2$, and $n = 5$.

Solution As shown in Figure 4.32, the case $n = 1$ is one rectangle in which $\Delta x = 1$. The midpoint of the interval is $m_1 = \frac{3}{2}$ and $f(\frac{3}{2}) = \frac{2}{3}$. Therefore, from (1),

$$M_1 = 1 \cdot \frac{2}{3} \approx 0.6666$$

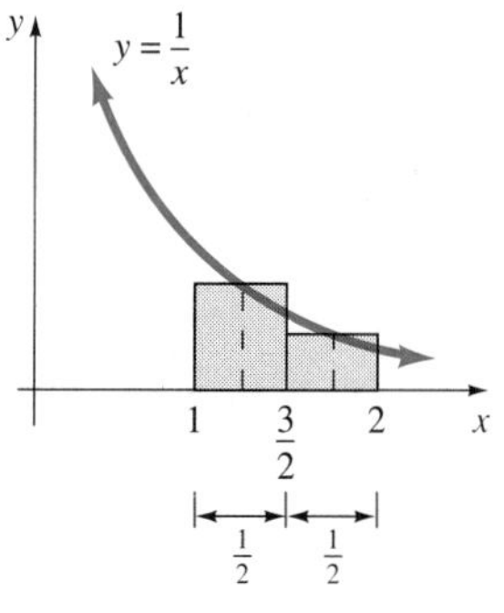

FIGURE 4.33

When $n = 2$, Figure 4.33 shows $\Delta x = \frac{1}{2}$, $x_0 = 1$, $x_1 = 1 + \Delta x = \frac{3}{2}$, and $x_2 = 1 + 2\,\Delta x = 2$. The midpoints of $[1, \frac{3}{2}]$, and $[\frac{3}{2}, 2]$ are, respectively, $m_1 = \frac{5}{4}$ and $m_2 = \frac{7}{4}$ and so $f(\frac{5}{4}) = \frac{4}{5}$ and $f(\frac{7}{4}) = \frac{4}{7}$. Hence, (1) gives

$$M_2 = \frac{1}{2}\left[\frac{4}{5} + \frac{4}{7}\right] \approx 0.6857$$

Finally, for $n = 5$, $\Delta x = \frac{1}{5}$, $x_0 = 1$, $x_1 = 1 + \Delta x = \frac{6}{5}$, $x_2 = 1 + 2\,\Delta x = \frac{7}{5}, \ldots, x_5 = 1 + 5\,\Delta x = 2$. The midpoints of the five subintervals $[1, \frac{6}{5}]$, $[\frac{6}{5}, \frac{7}{5}]$, $[\frac{7}{5}, \frac{8}{5}]$, $[\frac{8}{5}, \frac{9}{5}]$, $[\frac{9}{5}, 2]$ and the corresponding functional values are shown in the table.

k	1	2	3	4	5
m_k	$\frac{11}{10}$	$\frac{13}{10}$	$\frac{15}{10}$	$\frac{17}{10}$	$\frac{19}{10}$
$f(m_k)$	$\frac{10}{11}$	$\frac{10}{13}$	$\frac{10}{15}$	$\frac{10}{17}$	$\frac{10}{19}$

The information in the table then gives

$$M_5 = \frac{1}{5}\left[\frac{10}{11} + \frac{10}{13} + \frac{10}{15} + \frac{10}{17} + \frac{10}{19}\right] \approx 0.6919$$

In other words, $\int_1^2 dx/x \approx M_5$ or $\int_1^2 dx/x \approx 0.6919$. □

Error in the Midpoint Rule Suppose $I = \int_a^b f(x)\,dx$ and M_n is an approximation to I using n rectangles. We define the error in the method to be

$$E_n = |I - M_n|$$

An upper bound for the error can be obtained by means of the next result. The proof is omitted.

THEOREM 4.19 Error Bound for Midpoint Rule

If there exists a number $M > 0$ such that $|f''(x)| \leq M$ for all x in $[a, b]$, then

$$E_n \leq \frac{M(b-a)^3}{24n^2} \tag{2}$$

Observe that this upper bound for the error E_n is inversely proportional to n^2. Hence, the accuracy in the method improves as we take more and more rectangles. For example, if the number of rectangles is doubled, the error E_{2n} is less than one-fourth the error bound for E_n. Thus, we see that $\lim_{n\to\infty} M_n = I$.

The next example illustrates how the error bound (2) can be utilized to determine the number of rectangles that will yield a prescribed accuracy.

EXAMPLE 2 Determine a value of n so that (1) will give an approximation to $\int_1^2 dx/x$ that is accurate to two decimal places.

Solution The Midpoint Rule will be accurate to two decimal places for those values of n for which the upper bound $M(b-a)^3/24n^2$ for the error is strictly less than 0.005.*

*If we want accuracy to three decimal places, we use 0.0005, and so on.

For $f(x) = 1/x$, we have $f''(x) = 2/x^3$. Since f'' decreases on $[1, 2]$, it follows that $f''(x) \leq f''(1) = 2$ for all x in the interval. Thus, with $M = 2$, $b - a = 1$, we want

$$\frac{2(1)^3}{24n^2} < 0.005 \quad \text{or} \quad n^2 > \frac{50}{3} \approx 16.67$$

By taking $n \geq 5$ we obtain the desired accuracy. □

Example 2 indicates that the third approximation $\int_1^2 dx/x = 0.6919$ obtained in Example 1 is accurate to two decimal places. By way of comparison, it is known that the estimate $\int_1^2 dx/x \approx 0.6931$ is correct to four decimal places. Thus, for $n = 5$ the error in the method E_n is approximately 0.0012.

Trapezoidal Rule A more popular method for approximating an integral is based on the plausibility that a better estimate of $\int_a^b f(x)\, dx$ can be obtained by adding the areas of trapezoids instead of the areas of rectangles. See Figure 4.34(a). The area of the trapezoid shown in Figure 4.34(b) is $h(l_1 + l_2)/2$. Thus, the area A_k of the trapezoidal element shown in Figure 4.34(c) is

$$A_k = \Delta x \frac{f(x_{k-1}) + f(x_k)}{2}$$

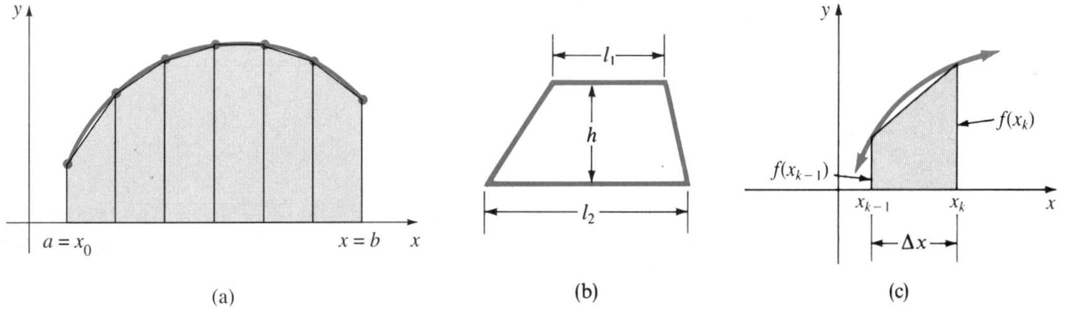

FIGURE 4.34

Thus, for a regular partition of the interval $[a, b]$ on which f is continuous, we obtain

$$\int_a^b f(x)\, dx \approx \Delta x \frac{f(x_0) + f(x_1)}{2} + \Delta x \frac{f(x_1) + f(x_2)}{2} + \cdots + \Delta x \frac{f(x_{n-1}) + f(x_n)}{2}$$

We summarize this new approximation rule in the next definition after we combine like terms and substitute $\Delta x = (b - a)/n$.

DEFINITION 4.8 Trapezoidal Rule

The **Trapezoidal Rule** is the approximation $\int_a^b f(x)\,dx \approx T_n$, where

$$T_n = \frac{b-a}{2n}[f(x_0) + 2f(x_1) + 2f(x_2) + \cdots + 2f(x_{n-1}) + f(x_n)] \quad (3)$$

Error in the Trapezoidal Rule The error in the method for the Trapezoidal Rule is given by $E_n = |I - T_n|$, where $I = \int_a^b f(x)\,dx$. As the next theorem shows, the error bound for the Trapezoidal Rule is almost the same as that for the Midpoint Rule.

THEOREM 4.20 Error Bound for Trapezoidal Rule

If there exists a number $M > 0$ such that $|f''(x)| \le M$ for all x in $[a, b]$, then

$$E_n \le \frac{M(b-a)^3}{12n^2} \quad (4)$$

EXAMPLE 3 Determine a value of n so that the Trapezoidal Rule will give an approximation to $\int_1^2 dx/x$ that is accurate to two decimal places. Approximate the integral.

Solution Using the information in Example 2, we have immediately:

$$\frac{2(1)^3}{12n^2} < 0.005 \quad \text{or} \quad n^2 > \frac{100}{3} \approx 33.33$$

In this case we must take $n \ge 6$ to obtain the desired accuracy. Hence, $\Delta x = \frac{1}{6}$, $x_0 = 1, x_1 = 1 + \Delta x = \frac{7}{6}$, and so on. With the information in the accompanying table,

k	0	1	2	3	4	5	6
x_k	1	$\frac{7}{6}$	$\frac{4}{3}$	$\frac{3}{2}$	$\frac{5}{3}$	$\frac{11}{6}$	2
$f(x_k)$	1	$\frac{6}{7}$	$\frac{3}{4}$	$\frac{2}{3}$	$\frac{3}{5}$	$\frac{6}{11}$	$\frac{1}{2}$

(3) gives

$$T_6 = \frac{1}{12}\left[1 + 2\left(\frac{6}{7}\right) + 2\left(\frac{3}{4}\right) + 2\left(\frac{2}{3}\right) + 2\left(\frac{3}{5}\right) + 2\left(\frac{6}{11}\right) + \frac{1}{2}\right] \approx 0.6949 \quad \square$$

EXAMPLE 4 Approximate $\int_{1/2}^{1} \cos\sqrt{x}\,dx$ by the Trapezoidal Rule so that the error is less than 0.001.

Solution The second derivative of $f(x) = \cos\sqrt{x}$ is

$$f''(x) = \frac{1}{4x}\left(\frac{\sin\sqrt{x}}{\sqrt{x}} - \cos\sqrt{x}\right)$$

For x in the interval $[\frac{1}{2}, 1]$ we have $0 < (\sin\sqrt{x})/\sqrt{x} \le 1$ and $0 < \cos\sqrt{x} \le 1$ and consequently $|f''(x)| \le \frac{1}{4}x$. Therefore, on the interval, $|f''(x)| \le \frac{1}{2}$. Thus, with $M = \frac{1}{2}$ and $b - a = \frac{1}{2}$, it follows from (4) that we want

$$\frac{\frac{1}{2}(\frac{1}{2})^3}{12n^2} < 0.001 \quad \text{or} \quad n^2 > \frac{125}{24} \approx 5.21$$

k	x_k	$f(x_k)$
0	$\frac{1}{2}$	0.7602
1	$\frac{2}{3}$	0.6848
2	$\frac{5}{6}$	0.6115
3	1	0.5403

Hence, to obtain the desired accuracy it suffices to choose $n = 3$ and $\Delta x = \frac{1}{6}$. With the aid of a calculator to obtain the information in the accompanying table, we find the following approximation for $\int_{1/2}^{1} \cos\sqrt{x}\,dx$ from (3):

$$T_3 = \frac{1}{12}\left[\cos\sqrt{\frac{1}{2}} + 2\cos\sqrt{\frac{2}{3}} + 2\cos\sqrt{\frac{5}{6}} + \cos 1\right] \approx 0.3244 \qquad \square$$

Although not obvious from a figure, an improved method of approximating a definite integral $\int_a^b f(x)\,dx$ can be obtained by considering a series of parabolic arcs instead of a series of chords used in the Trapezoidal Rule. It can be proved, under certain conditions, that a parabolic arc passing through *three* specified points will "fit" the graph of f better than a single straight line. See Figure 4.35. By adding the areas under the parabolic arcs, we obtain an approximation to the integral.

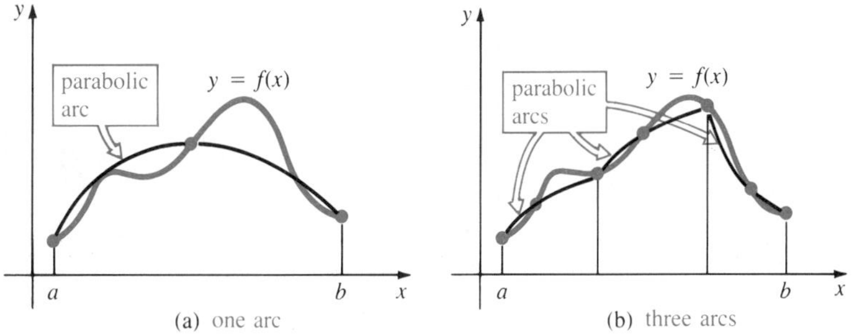

FIGURE 4.35

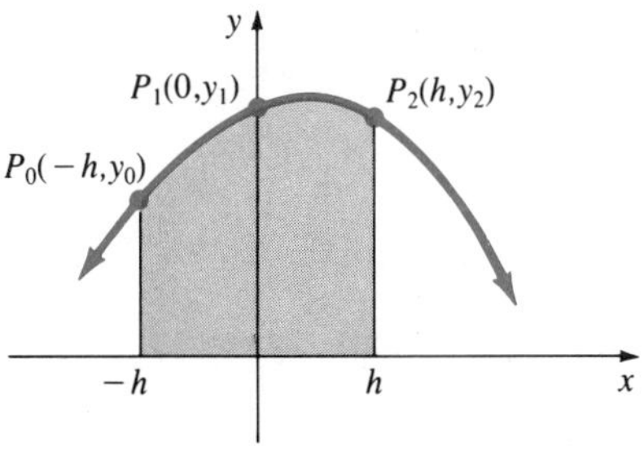

FIGURE 4.36

To begin, let us find the area under an arc of a parabola that passes through three points $P_0(x_0, y_0)$, $P_1(x_1, y_1)$, and $P_2(x_2, y_2)$, where $x_0 < x_1 < x_2$ and $x_1 - x_0 = x_2 - x_1 = h$. As shown in Figure 4.36, this can be done by finding the area under the graph of $y = Ax^2 + Bx + C$ on the interval $[-h, h]$ so that P_0, P_1, and P_2 have coordinates $(-h, y_0)$ $(0, y_1)$, and (h, y_2), respectively. The interval $[-h, h]$ is chosen for simplicity; the area in question does not depend on the location of the y-axis. The area is

$$\int_{-h}^{h} (Ax^2 + Bx + C)\, dx = \frac{h}{3}(2Ah^2 + 6C) \tag{5}$$

But, since the graph is to pass through $(-h, y_0)$, $(0, y_1)$, and (h, y_2), we must have

$$y_0 = Ah^2 - Bh + C \tag{6}$$
$$y_1 = C \tag{7}$$
$$y_2 = Ah^2 + Bh + C \tag{8}$$

By adding (6) and (8) and using (7), we find $2Ah^2 = y_0 + y_2 - 2y_1$. Thus, (5) can be expressed as

$$\text{area} = \frac{h}{3}(y_0 + 4y_1 + y_2) \tag{9}$$

Simpson's Rule Now suppose that the interval $[a, b]$ is partitioned into n subintervals of equal width $\Delta x = (b - a)/n$, where *n is an even integer*. As shown in Figure 4.37, on each subinterval $[x_{k-2}, x_k]$ of width $2\,\Delta x$ we approximate the graph of f by an arc of a parabola through points P_{k-2}, P_{k-1}, and P_k on the graph that corresponds to the endpoints and midpoint of the subinterval. If A_k denotes the area under the parabola on $[x_{k-2}, x_k]$, it follows from (9) that

$$A_k = \frac{\Delta x}{3}[f(x_{k-2}) + 4f(x_{k-1}) + f(x_k)]$$

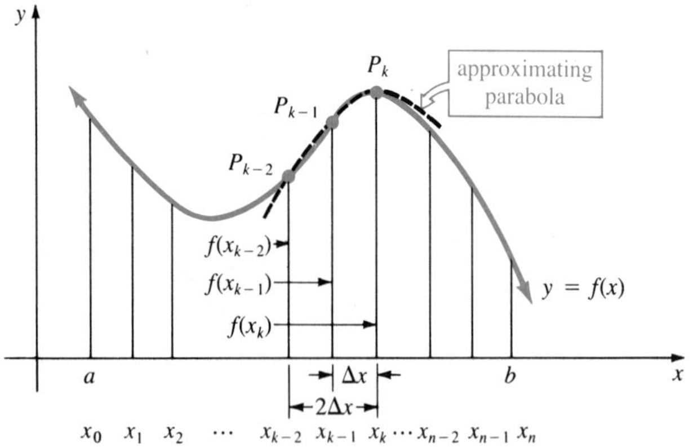

FIGURE 4.37

Thus, summing all the A_k gives

$$\int_a^b f(x)\, dx \approx \frac{\Delta x}{3}[f(x_0) + 4f(x_1) + f(x_2)] + \frac{\Delta x}{3}[f(x_2) + 4f(x_3) + f(x_4)]$$
$$+ \cdots + \frac{\Delta x}{3}[f(x_{n-2}) + 4f(x_{n-1}) + f(x_n)]$$

This approximation rule, named after the English mathematician Thomas Simpson (1710–1761), is summarized in the following definition.

DEFINITION 4.9 Simpson's Rule

Simpson's Rule is the approximation $\int_a^b f(x)\,dx \approx S_n$, where

$$S_n = \frac{b-a}{3n}[f(x_0) + 4f(x_1) + 2f(x_2) + 4f(x_3) + \cdots + 2f(x_{n-2}) + 4f(x_{n-1}) + f(x_n)] \quad (10)$$

We note again that the integer n in (10) must be even, since each A_k represents the area under a parabolic arc on a subinterval of width $2\,\Delta x$.

Error in Simpson's Rule If $I = \int_a^b f(x)\,dx$, the next theorem establishes an upper bound for the error in the method $E_n = |I - S_n|$ using an upper bound on the fourth derivative.

THEOREM 4.21 Error Bound for Simpson's Rule

If there exists a number $M > 0$ such that $|f^{(4)}(x)| \leq M$ for all x in $[a, b]$, then

$$E_n \leq \frac{M(b-a)^5}{180n^4} \quad (11)$$

EXAMPLE 5 Determine a value of n so that (10) will give an approximation to $\int_1^2 dx/x$ that is accurate to two decimal places.

Solution For $f(x) = 1/x$, $f^{(4)}(x) = 24/x^5$ and on $[1, 2]$, $f^{(4)}(x) \leq f^{(4)}(1) = 24$. Thus, with $M = 24$ it follows from (11) that

$$\frac{24(1)^6}{180n^4} < 0.005 \quad \text{or} \quad n^4 > \frac{80}{3} \approx 26.67$$

and so $n > 2.27$. Since n must be an even integer, it suffices to take $n \geq 4$. □

k	x_k	$f(x_k)$
0	1	1
1	$\frac{5}{4}$	$\frac{4}{5}$
2	$\frac{3}{2}$	$\frac{2}{3}$
3	$\frac{7}{4}$	$\frac{4}{7}$
4	2	$\frac{1}{2}$

EXAMPLE 6 Approximate $\int_1^2 dx/x$ by Simpson's Rule for $n = 4$.

Solution When $n = 4$, we have $\Delta x = \frac{1}{4}$. From (10) and the accompanying table we obtain

$$S_4 = \frac{1}{12}\left[1 + 4\left(\frac{4}{5}\right) + 2\left(\frac{2}{3}\right) + 4\left(\frac{4}{7}\right) + \frac{1}{2}\right] \approx 0.6933$$ □

In Example 6, keep in mind that even though we are using $n = 4$, the definite integral $\int_1^2 dx/x$ is being approximated by the area under only two parabolic arcs. Recall that the Midpoint Rule gave $\int_1^2 dx/x \approx 0.6919$ with $n = 5$, the Trapezoidal Rule gave $\int_1^2 dx/x \approx 0.6949$ with $n = 6$, and 0.6931 is an estimation of the integral correct to four decimal places.

In some applications it may only be possible to obtain numerical values of a quantity $Q(x)$—say, by measurements or by experiment—at specific points in some interval $[a, b]$, and yet it may be necessary to have some idea of the value of the definite integral $\int_a^b Q(x)\,dx$. Even though Q is not defined by means of an explicit formula, we may still be able to apply the Trapezoidal Rule or Simpson's Rule to approximate the integral.

EXAMPLE 7 Suppose we wish to find the area of an irregularly shaped piece of land that is bounded between a straight road and the shore of a lake. The boundaries of the land are indicated by the dashed lines in Figure 4.38(a). Suppose we divide the indicated 1-mi boundary along the road into, say, $n = 8$ subintervals and then, as shown in Figure 4.38(b), measure the perpendicular distances from the road to the shore of the lake. We are now in position to approximate the area of the land $A = \int_a^b f(x)\,dx$ by Simpson's Rule. With $b - a = 1$ mi $= 5280$ ft, $\Delta x = 5280/8 = 660$, and the identifications $f(x_0) = 83, \ldots, f(x_8) = 28$, (10) gives the following approximation for A:

$$S_8 = \frac{660}{3}[83 + 4(82) + 2(96) + 4(100) + 2(82) + 4(55) + 2(63) + 4(54) + 28] = 386{,}540 \text{ ft}^2$$

Using the fact that 1 acre $= 43{,}560$ ft^2, we see that the land is approximately 8.9 acres.

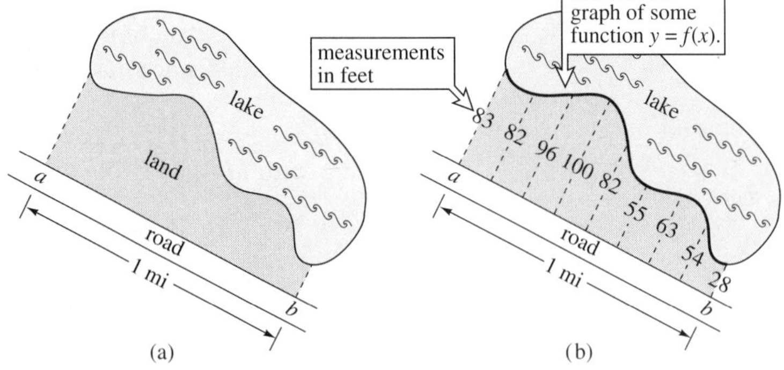

FIGURE 4.38 □

BASIC Programs The following listings are BASIC programs for the Trapezoidal Rule and Simpson's Rule:

```
10 REM EVALUATION OF DEFINITE INTEGRALS
       USING THE TRAPEZOIDAL RULE
20 DEF FNY(X) = . . .
```

```
30  INPUT "WHAT IS THE INTERVAL?"; A, B
40  INPUT "HOW MANY SUBDIVISIONS?"; N
50  LET H = (B - A)/N
60  LET T = FNY(A) + FNY(B)
70  FOR X = A + H TO B - H/2 STEP H
80  LET T = T + 2 * FNY(X)
90  NEXT X
100  LET T = T * H/2
110  PRINT "USING"; N; "SUBDIVISIONS, THE
     TRAPEZOID RULE YIELDS"; T
120  END
```

```
10  REM EVALUATION OF DEFINITE INTEGRALS
    USING SIMPSON'S RULE
20  DEF FNY(X) = . . .
30  INPUT "WHAT IS THE INTERVAL?"; A, B
40  INPUT "HOW MANY SUBDIVISIONS?"; N
50  LET H = (B - A)/N
60  LET S = FNY(A) + FNY(B)
70  FOR X = A + H TO B - H/2 STEP 2 * H
80  LET S = S + 4 * FNY(X)
90  NEXT X
100  FOR X = A + 2 * H TO B -- 3 * H/2 STEP 2 * H
110  LET S = S + 2 * FNY(X)
120  NEXT X
130  LET S = H * S/3
140  PRINT "USING"; N; "SUBDIVISIONS,
     SIMPSON'S RULE YIELDS"; S
150  END
```

It is left as an exercise to write a BASIC program for the Midpoint Rule.

▲ ***Remarks*** (*i*) The popularity of the Trapezoidal Rule notwithstanding, a direct comparison of the error bounds (2) and (4) shows that the Midpoint Rule is actually more accurate than the Trapezoidal Rule. Specifically, (2) suggests that in some cases the error in the Midpoint Rule can be one-half the error in the Trapezoidal Rule. See Problem 33.

(*ii*) Under some circumstances the rules considered in the foregoing discussion will give the exact value of an integral $\int_a^b f(x)\,dx$. The error bounds (2) and (4) indicate that M_n and T_n will yield the precise value whenever f is a linear function. See Problems 31, 32, and 35. Simpson's Rule will give the exact value of $\int_a^b f(x)\,dx$ whenever f is a linear, quadratic, or cubic polynomial function. See Problems 34 and 36.

(*iii*) In general, Simpson's Rule will give greater accuracy than either the Midpoint or the Trapezoidal Rule. So why should we even bother with these other rules? In some instances, the slightly simpler Midpoint and Trapezoidal Rules will yield accuracy that is sufficient for the purpose at hand. Furthermore, the requirement that n must be an even integer in Simpson's Rule may prevent its application to a given problem. Also, to find an error bound for Simpson's Rule, we must compute and then find an upper bound for the fourth derivative. The expression for $f^{(4)}(x)$ can, of course, be very complicated. The error bounds for the other two rules depend on the second derivative.

(*iv*) Since the Trapezoidal Rule and Simpson's Rule are based upon fitting linear and quadratic functions to the graph of a given function f, you might question whether the next step is to try to fit the graph of f with arcs of cubic or even quartic functions. Indeed, you can. A cubic approximation would utilize four points over three intervals of equal width Δx; a quartic arc would use five points; and so on. In this manner, you generate a sequence of approximation formulas known as **Newton–Cotes formulas**. Because of increasing complexity and other inherent problems, Newton–Cotes formulas of order higher than 2 are seldom used.

EXERCISES 4.8 *Answers to odd-numbered problems begin on page A-80.*

In Problems 1 and 2 compare the exact value of the integral with the approximation obtained from the Midpoint Rule for the indicated value of n.

1. $\int_1^4 (3x^2 + 2x)\,dx;\ n = 3$
2. $\int_0^{\pi/6} \cos x\,dx;\ n = 4$

In Problems 3 and 4 compare the exact value of the integral with the approximation obtained from the Trapezoidal Rule for the indicated value of n.

3. $\int_1^3 (x^3 + 1)\,dx;\ n = 4$
4. $\int_0^2 \sqrt{x+1}\,dx;\ n = 6$

In Problems 5–12 use the Midpoint Rule and the Trapezoidal Rule to obtain an approximation to the given integral for the indicated value of n.

5. $\int_1^6 \frac{dx}{x};\ n = 5$
6. $\int_0^2 \frac{dx}{3x+1};\ n = 4$
7. $\int_0^1 \sqrt{x^2+1}\,dx;\ n = 10$
8. $\int_1^2 \frac{dx}{\sqrt{x^3+1}};\ n = 5$
9. $\int_0^\pi \frac{\sin x}{x+\pi}\,dx;\ n = 6$
10. $\int_0^{\pi/4} \tan x\,dx;\ n = 3$
11. $\int_0^2 \cos x^2\,dx;\ n = 6$
12. $\int_0^1 \frac{\sin x}{x}\,dx;\ n = 5$ [*Hint:* Define $f(0) = 1$.]

In Problems 13 and 14 compare the exact value of the integral with the approximation obtained from Simpson's Rule for the indicated value of n.

13. $\int_0^4 \sqrt{2x+1}\,dx;\ n = 4$
14. $\int_0^{\pi/2} \sin^2 x\,dx;\ n = 2$

In Problems 15–22 use Simpson's Rule to obtain an approximation to the given integral for the indicated value of n.

15. $\int_{1/2}^{5/2} \frac{dx}{x};\ n = 4$
16. $\int_0^5 \frac{dx}{x+2};\ n = 6$
17. $\int_0^1 \frac{dx}{1+x^2};\ n = 4$
18. $\int_{-1}^1 \sqrt{x^2+1}\,dx;\ n = 2$
19. $\int_0^\pi \frac{\sin x}{x+\pi}\,dx;\ n = 6$
20. $\int_0^1 \cos\sqrt{x}\,dx;\ n = 4$
21. $\int_2^4 \sqrt{x^3+x}\,dx;\ n = 4$
22. $\int_{\pi/4}^{\pi/2} \frac{dx}{2+\sin x};\ n = 2$

23. Determine the number of rectangles needed so that an approximation to $\int_{-1}^2 dx/(x+3)$ is accurate to two decimal places.

24. Determine the number of trapezoids needed so that the error in an approximation to $\int_0^{1.5} \sin^2 x\,dx$ is less than 0.0001.

25. Use the Trapezoidal Rule so that an approximation to the area under the graph of $f(x) = 1/(1+x^2)$ on $[0, 2]$ is accurate to two decimal places. [*Hint:* Examine $f'''(x)$.]

26. The domain of $f(x) = 10^x$ is the set of real numbers and $f(x) > 0$ for all x. Use the Trapezoidal Rule to approximate the area under the graph of f on $[-2, 2]$ with $n = 4$.

27. Using Simpson's Rule, determine n so that the error in approximating $\int_1^3 dx/x$ is less than 10^{-5}. Compare with the n needed in the Trapezoidal Rule to give the same accuracy.

28. Find an upper bound for the error in approximating $\int_0^3 dx/(2x+1)$ by Simpson's Rule with $n = 6$.

In Problems 29 and 30 use the data given in the table and an appropriate rule to approximate the indicated definite integral.

29.

x	$f(x)$
2.05	4.91
2.10	4.80
2.15	4.66
2.20	4.41
2.25	3.93
2.30	3.58

; $\int_{2.05}^{2.30} f(x)\,dx$

30.

x	$f(x)$
0.0	−0.72
0.1	−0.55
0.2	−0.16
0.4	0.62
0.6	0.78
0.8	1.34
0.9	1.47
1.00	1.61
1.20	1.51

; $\int_{0}^{1.20} f(x)\,dx$

31. Compare the exact value of the integral $\int_0^4 (2x+5)\,dx$ with the approximation obtained from the Midpoint Rule with $n = 2$ and $n = 4$.

32. Repeat Problem 31 using the Trapezoidal Rule.

33. (a) Find the exact value of the integral $I = \int_{-1}^{1} (x^3 + x^2)\,dx$.

(b) Use the Midpoint Rule with $n = 8$ to find an approximation to I.

(c) Use the Trapezoidal Rule with $n = 8$ to find an approximation to I.

(d) Compare the errors $E_8 = |I - M_8|$ and $E_8 = |I - T_8|$.

34. Compare the exact value of the integral $\int_{-1}^{3} (x^3 - x^2)\,dx$ with the approximations obtained from Simpson's Rule with $n = 2$ and $n = 4$.

35. Prove that the Trapezoidal Rule will give the exact value of $\int_a^b f(x)\,dx$ when $f(x) = c_1x + c_0$, with c_0 and c_1 constants. Geometrically, why does this make sense?

36. Prove that Simpson's Rule will give the exact value of $\int_a^b f(x)\,dx$ where $f(x) = c_3x^3 + c_2x^2 + c_1x + c_0$, with c_0, c_1, c_2, and c_3 constants.

37. Use the data given in Figure 4.39 and Simpson's Rule to find an approximation to the area under the graph of the continuous function f on the interval [1, 4].

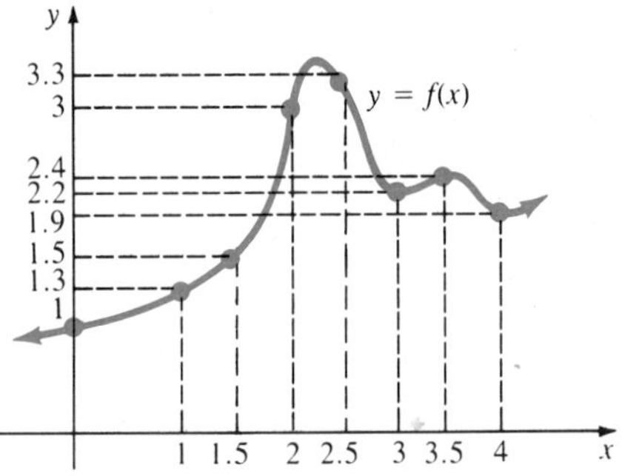

FIGURE 4.39

38. Use the Trapezoidal Rule with $n = 9$ to find an approximation to the area of the shaded region in Figure 4.40. Does the Trapezoidal Rule give the exact value of the area?

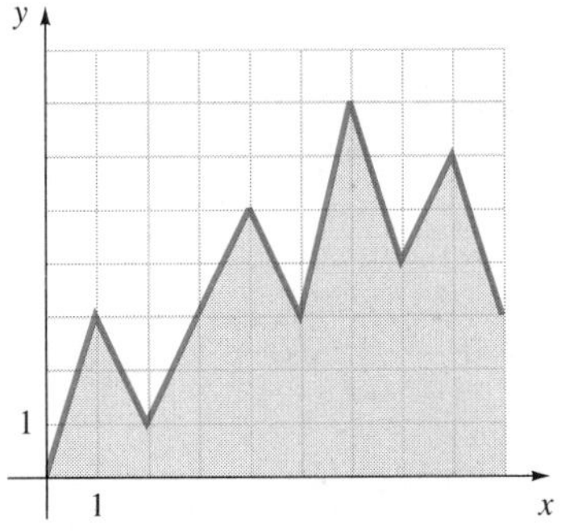

FIGURE 4.40

39. The large irregularly shaped fish pond shown in Figure 4.41 is filled with water to a uniform depth of 4 ft. Use Simpson's Rule to find an approximation to the number of gallons of water in the tank. Measurements are in feet. There are 7.48 gallons in 1 cubic foot of water.

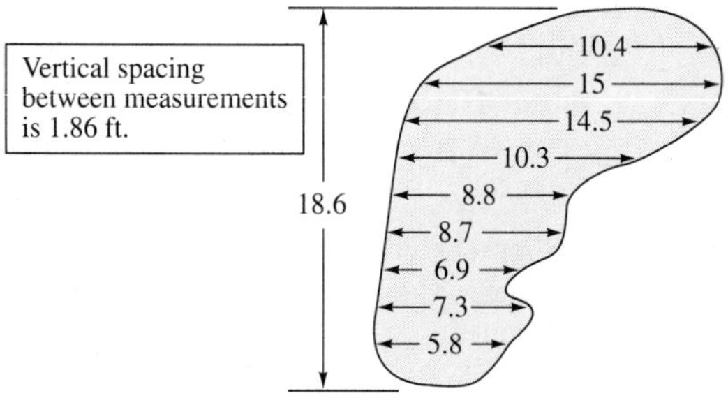

FIGURE 4.41

40. The so-called moment of inertia I of a three-bladed ship's propeller whose dimensions are shown in Figure 4.42(a) is given by

$$I = \frac{3\rho\pi}{2g} + \frac{3\rho}{g}\int_1^{4.5} r^2 A\, dr$$

where ρ is the density of the metal, g is the acceleration of gravity, and A is the area of a cross-section of the propeller at a distance r feet from the center of the hub. If $\rho = 570$ lb/ft^3 for bronze, use the data in Figure 4.42(b) and the Trapezoidal Rule to find an approximation to I.

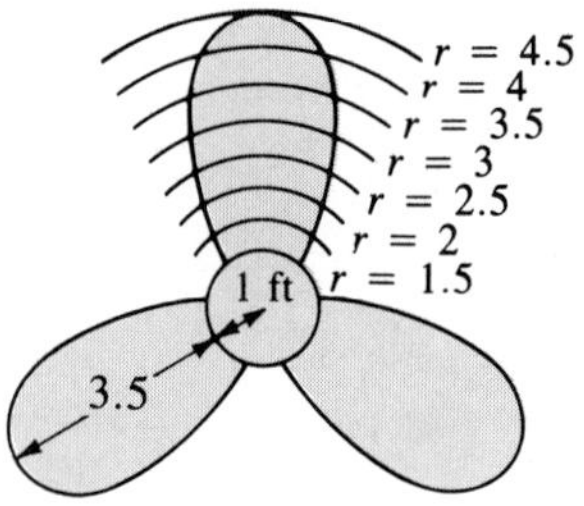

(a)

r (ft)	1	1.5	2	2.5	3	3.5	4	4.5
A (ft)	0.3	0.50	0.62	0.70	0.60	0.50	0.27	0

(b)

FIGURE 4.42

Calculator/Computer Problems

In Problems 41 and 42 use a calculator or computer to obtain the graph of the given function. Use Simpson's Rule to approximate the area bounded by the graph of f and the x-axis on the indicated interval. Use $n = 10$.

41. $f(x) = \sqrt[5]{(5^{2.5} - |x|^{2.5})^2}$; $[-5, 5]$

42. $f(x) = 1 + |\sin x|^x$; $[0, 2\pi]$ [*Hint:* Use the graph to discern $f(0)$.]

43. Write a BASIC program for the Midpoint Rule.

CHAPTER 4 REVIEW EXERCISES

Answers begin on page A-81.

In Problems 1–20 answer true or false.

1. If $f'(x) = 3x^2 + 2x$, then, necessarily, $f(x) = x^3 + x^2$. ________

2. $\sum_{k=2}^{6} (2k - 3) = \sum_{j=0}^{4} (2j + 1)$ ________

3. $\sum_{k=1}^{40} 5 = \sum_{k=1}^{20} 10$ ________

4. $\int_1^3 \sqrt{t^2 + 7}\, dt = -\int_3^1 \sqrt{t^2 + 7}\, dt$ ________

5. $\int_0^1 t^2\, dt + \int_1^0 x^2\, dx = 0$ ________

6. If f is integrable, then f is continuous. ________

7. A definite integral is always an area under a graph. ________

8. $\int \sin x\, dx = \cos x + C$ ________

9. If P is a partition of $[a, b]$ into n subintervals, then $n \to \infty$ implies $\|P\| \to 0$. ________

10. If $\int_a^b f(x)\, dx > 0$, then $f(x) > 0$ for all x in $[a, b]$. ________

11. If $F'(x) = 0$ for all x, then $F(x) = C$ for all x. ________

12. If f is an odd integrable function on $[-\pi, \pi]$, then $\int_{-\pi}^{\pi} f(x)\, dx = 0$. ________

13. $\int_{-1}^{1} |x|\,dx = 2\int_{0}^{1} x\,dx$ ________

14. $\int x\cos x\,dx = x\sin x + \cos x + C$ ________

15. Simpson's Rule generally gives a better approximation to a definite integral than the Trapezoidal Rule. ________

16. The Trapezoidal Rule can be used only for integrals $\int_a^b f(x)\,dx$ for which $f(x) \geq 0$ on $[a, b]$. ________

17. To use Simpson's Rule the interval $[a, b]$ must be partitioned into an even number of subintervals. ________

18. Simpson's Rule will give the exact value of

$$\int_0^5 (x^3 - 4x^2 + 8x + 10)\,dx$$

with $n = 100$. ________

19. If $|f^{(4)}(x)| \leq M$ for all x in $[a, b]$, then the error E_{2n} in Simpson's Rule is less than or equal to $\frac{1}{16}$ the error bound for E_n. ________

20. To approximate

$$\int_{-1}^{2} f(x)\,dx$$

using a regular partition for which $\Delta x = \frac{3}{10}$, Simpson's Rule will utilize five parabolic arcs. ________

In Problems 21–30 fill in the blanks.

21. If G is an antiderivative of f, then $G'(x) =$ ________.

22. $\int \frac{d}{dx} x^2\,dx =$ ________

23. The value of $\frac{d}{dx}\int_3^x \sqrt{t^2 + 5}\,dt$ at $x = 1$ is ________.

24. Using sigma notation, the sum $\frac{1}{3} + \frac{2}{5} + \frac{3}{7} + \frac{4}{9} + \frac{5}{11}$ can be expressed as ________.

25. The value of $\sum_{k=1}^{15} 3k^2$ is ________.

26. If $u = t^2 + 1$, then the definite integral

$$\int_2^4 t(t^2 + 1)^{1/3}\,dt \quad \text{becomes} \quad \frac{1}{2}\int_{\_\_}^{\_\_} u^{1/3}\,du$$

27. The area under the graph of $f(x) = 5x$ on the interval $[0, 10]$ is ________.

28. If the interval $[1, 6]$ is partitioned into four subintervals determined by $x_0 = 1, x_1 = 2, x_2 = \frac{5}{2}, x_3 = 5$, and $x_4 = 6$, the norm of the partition is ________.

29. A partition of an interval in which the subintervals are ________ is said to be a regular partition.

30. If P is a partition of $[0, 4]$ and x_k^* is a number in the kth subinterval, then

$$\lim_{\|P\|\to 0} \sum_{k=1}^{n} \sqrt{x_k^*}\,\Delta x_k$$

is the definition of the definite integral ________. By the Fundamental Theorem of Calculus, the value of this definite integral is ________.

In Problems 31–40 evaluate the given integral.

31. $\int_{-1}^{1} (4x^3 - 6x^2 + 2x - 1)\,dx$

32. $\int_1^9 \frac{6}{\sqrt{x}}\,dx$

33. $\int (5t + 1)^{100}\,dt$

34. $\int \frac{w}{\sqrt{3w^2 - 1}}\,dw$

35. $\int_0^{\pi/4} (\sin 2x - 5\cos 4x)\,dx$

36. $\int_{\pi^2/9}^{\pi^2} \frac{\sin\sqrt{z}}{\sqrt{z}}\,dz$

37. $\int_4^4 (-2x^2 + x^{1/2})\,dx$

38. $\int_{-\pi/4}^{\pi/4} dx + \int_{-\pi/4}^{\pi/4} \tan^2 x\,dx$

39. $\int \cot^6 8x \csc^2 8x\,dx$

40. $\int \csc 3x \cot 3x\,dx$

41. If $\int_1^4 f(x)\,dx = 2$ and $\int_4^9 f(x)\,dx = -8$, evaluate $\int_1^9 f(x)\,dx$.

42. If $\int_0^5 f(x)\,dx = -3$ and $\int_0^7 f(x)\,dx = 2$, evaluate $\int_5^7 f(x)\,dx$.

In Problems 43–48 evaluate the given integral.

43. $\int_0^3 (1 + |x - 1|)\,dx$

44. $\int_0^1 \frac{d}{dt}\left[\frac{10t^4}{(2t^3 + 6t + 1)^2}\right] dt$

45. $\int_{\pi/2}^{\pi/2} \frac{\sin^{10} t}{16t^7 + 1}\,dt$

46. $\int_{-1}^{1} t^5 \sin t^2\,dt$

47. $\int_{-1}^{1} \frac{2\,dx}{1 + x^2}$, where $\int_0^1 \frac{dx}{1 + x^2} = \frac{\pi}{4}$

48. $\int_{-2}^{2} f(x)\,dx$, where $f(x) = \begin{cases} x^3, & x \le 0 \\ x^2, & 0 < x \le 1 \\ x, & x > 1 \end{cases}$

49. Suppose the Trapezoidal Rule is used to approximate the integral

$$\int_0^2 x^2\,dx$$

with $n = 4$. What is the error in the method?

50. Suppose it is desired to approximate

$$\int_{1/2}^{3} \frac{dx}{x}$$

to an accuracy of three decimal places. Compare the values of n needed for the Midpoint Rule, the Trapezoidal Rule, and Simpson's Rule.

In Problems 51 and 52 use **(a)** the Midpoint Rule, **(b)** the Trapezoidal Rule, and **(c)** Simpson's Rule to obtain an approximation to the given integral for the indicated value of n.

51. $\int_0^1 \sin x^2\,dx$; $n = 4$

52. $\int_1^7 x^2\sqrt{2x - 1}\,dx$; $n = 6$

53. Use Simpson's Rule to find an approximation to the area of the archway shown in Figure 4.43. [*Hint:* The shape of the archway above the 9-ft level is not parabolic.]

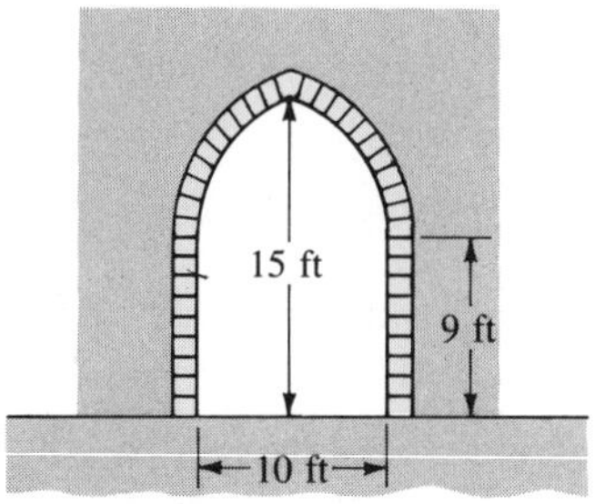

FIGURE 4.43

54. A bucket with dimensions (in feet) shown in Figure 4.44 is filled at a constant rate of $dV/dt = \frac{1}{4}$ ft^3/min. At $t = 0$ the scale reads 31.2 lb. If water weighs 62.4 lb/ft^3, what does the scale read at the end of 8 min? When is the bucket full? [*Hint:* See Appendix V for the formula for the volume of a frustum of a cone. Also, ignore the weight of the bucket.]

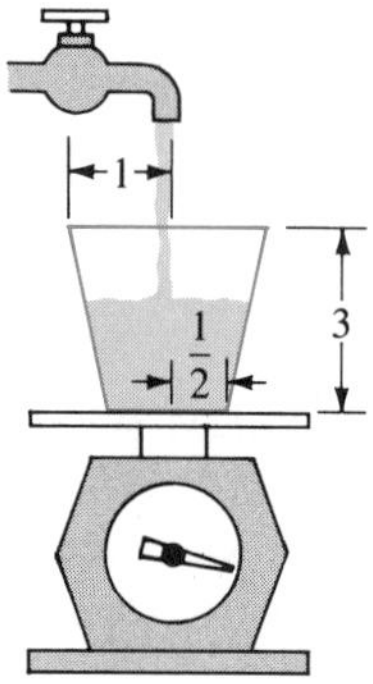

FIGURE 4.44

5

APPLICATIONS OF THE INTEGRAL

INTRODUCTION

Although we return to the problem of finding areas by using the definite integral in Section 5.1, you will see in subsequent sections of this chapter that integrals, definite and indefinite, have many other interpretations besides areas.

Imagine that you are in a spacecraft orbiting the earth. What portion—that is, percentage of the earth's surface—can you see from the spacecraft? Intuitively, we expect the answer to this question to depend on the distance h of the spacecraft from the earth's surface. Certainly one would expect to see more of the surface from the distance of the moon than, say, from an altitude of 2000 km. Think about this: What is the maximum percentage of the earth's surface that can be seen from an orbit? You will be asked to solve this problem in Exercises 5.5.

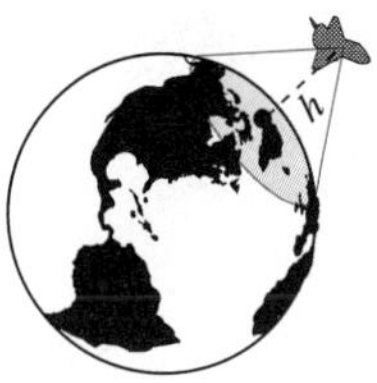

IMPORTANT CONCEPTS

area as a definite integral
area bounded between two graphs
volume of a solid by slicing
volume of a solid of revolution
- **disk method**
- **shell method**

smooth function
arc length
rectifiable graph
differential of arc length
area of a surface of revolution
average value of a function
mean value theorem for definite integrals
work done by a variable force
pressure exerted by a liquid
force exerted by a liquid
centers of mass
centroid of a lamina

5.1 AREA AND AREA BOUNDED BY TWO GRAPHS

REVIEW

Suppose the function $y = f(x)$ is defined on the interval $[a, b]$ and that $f(x) < 0$ on $[a, c)$ and $f(x) \geq 0$ on $[c, b]$. Then the absolute value of the function, $y = |f(x)|$, is nonnegative on the interval and is defined in a piecewise manner

$$|f(x)| = \begin{cases} -f(x), & a \leq x < c \\ f(x), & c \leq x \leq b \end{cases}$$

As shown in the accompanying figures, the graph of $y = |f(x)|$ on the interval $[a, c)$ is obtained by reflecting that portion of the graph of $y = f(x)$ through the x-axis. On the interval $[c, b]$, the graphs of $y = f(x)$ and $y = |f(x)|$ are the same.

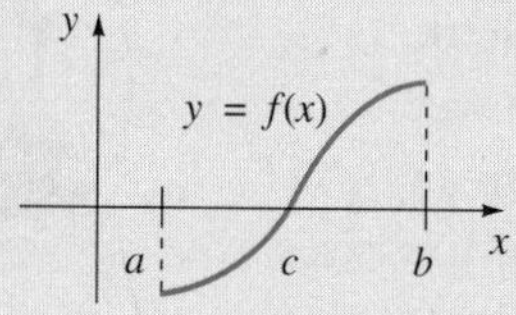

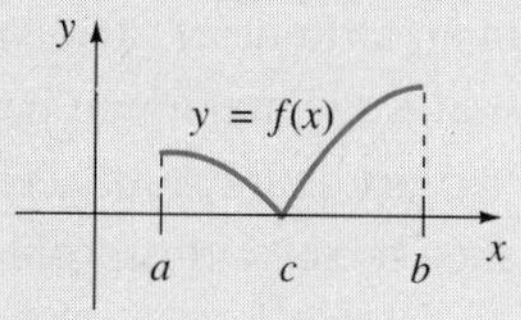

Area If f is a nonnegative continuous function on $[a, b]$, then, as we have already seen, the area under the graph of f on the interval is

$$A = \int_a^b f(x)\,dx$$

Suppose now $f(x) \leq 0$ for all x in $[a, b]$, as shown in Figure 5.1(a). Since $-f(x) \geq 0$, we define the area bounded by the graph of $y = f(x)$ and the x-axis from $x = a$ to $x = b$ to be the area A under the graph of $y = -f(x)$ on $[a, b]$.

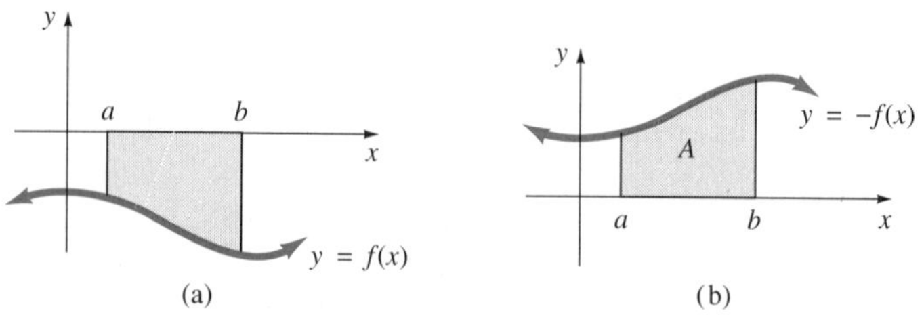

FIGURE 5.1

The area A is shown in Figure 5.1(b). Written as a definite integral, it is

$$A = \int_a^b -f(x)\,dx = -\int_a^b f(x)\,dx$$

Since $-f(x) = |f(x)|$ on $[a, b]$, we are prompted to give the following definition:

DEFINITION 5.1 Area

If $y = f(x)$ is continuous on $[a, b]$, then the **area** A bounded by its graph on the interval and the x-axis is given by

$$A = \int_a^b |f(x)|\,dx \tag{1}$$

EXAMPLE 1 Find the area bounded by the graph of $y = x^3$ and the x-axis on $[-2, 1]$.

Solution From (1) we have

$$A = \int_{-2}^{1} |x^3|\,dx$$

In Figure 5.2 we have compared the graph of $y = x^3$ with the graph of

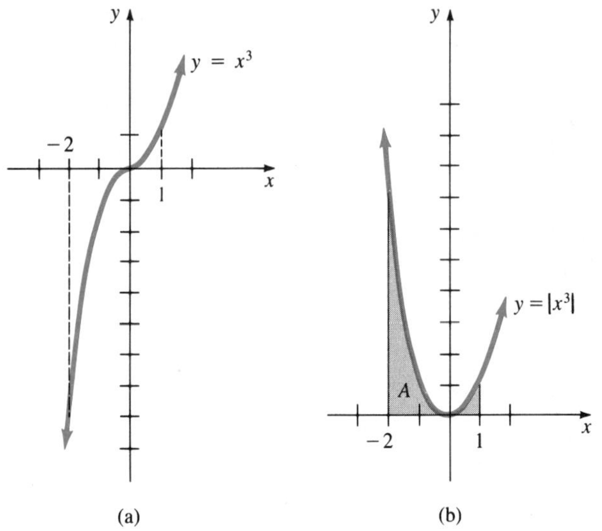

FIGURE 5.2

$y = |x^3|$. Since $x^3 < 0$ for $x < 0$, we have

$$|f(x)| = \begin{cases} -x^3, & x < 0 \\ x^3, & x \geq 0 \end{cases}$$

Thus, by Definition 5.1 and Theorem 4.8 the desired area is

$$\begin{aligned} A = \int_{-2}^{1} |x^3|\, dx &= \int_{-2}^{0} |x^3|\, dx + \int_{0}^{1} |x^3|\, dx \\ &= \int_{-2}^{0} -x^3\, dx + \int_{0}^{1} x^3\, dx \\ &= -\frac{x^4}{4}\bigg]_{-2}^{0} + \frac{x^4}{4}\bigg]_{0}^{1} \\ &= 0 - \left(-\frac{16}{4}\right) + \frac{1}{4} - 0 = \frac{17}{4} \text{ square units} \end{aligned}$$

□

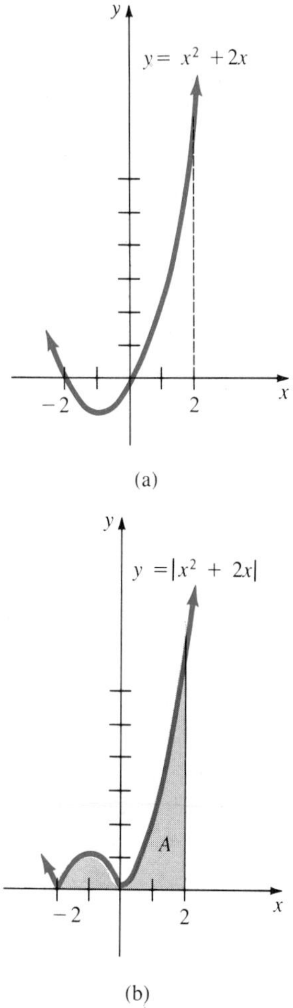

FIGURE 5.3

EXAMPLE 2 Find the area bounded by the graph of $y = x^2 + 2x$ and the x-axis on $[-2, 2]$.

Solution The graphs of $y = f(x)$ and $y = |f(x)|$ are given in Figure 5.3. Now, from Figure 5.3(a) we see that

$$|f(x)| = \begin{cases} -(x^2 + 2x), & -2 < x < 0 \\ x^2 + 2x, & x \leq -2 \text{ or } x \geq 0 \end{cases}$$

Therefore, on the interval $[-2, 2]$ we have

$$\begin{aligned} A = \int_{-2}^{2} |x^2 + 2x|\, dx &= \int_{-2}^{0} |x^2 + 2x|\, dx + \int_{0}^{2} |x^2 + 2x|\, dx \\ &= \int_{-2}^{0} -(x^2 + 2x)\, dx + \int_{0}^{2} (x^2 + 2x)\, dx \\ &= \left(-\frac{x^3}{3} - x^2\right)\bigg]_{-2}^{0} + \left(\frac{x^3}{3} + x^2\right)\bigg]_{0}^{2} \\ &= 0 - \left(\frac{8}{3} - 4\right) + \left(\frac{8}{3} + 4\right) - 0 \\ &= 8 \text{ square units} \end{aligned}$$

□

EXAMPLE 3 Find the area bounded by the graph of $y = \sin x$ and the x-axis on $[0, 2\pi]$.

Solution From (1) we see that

$$A = \int_{0}^{2\pi} |\sin x|\, dx$$

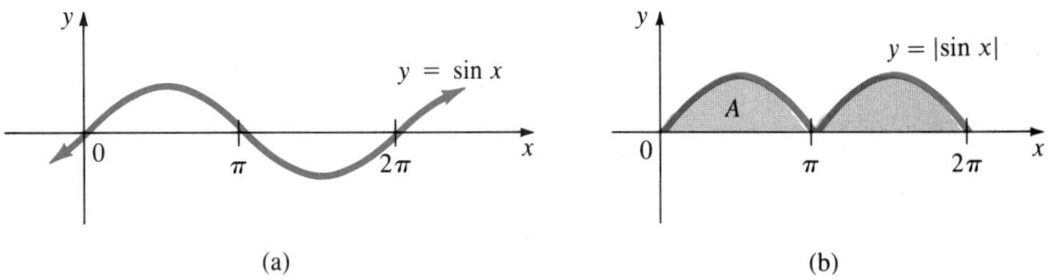

FIGURE 5.4

As indicated in Figure 5.4(a), $\sin x < 0$ on the interval $(\pi, 2\pi)$ and so

$$|f(x)| = \begin{cases} \sin x, & 0 \le x \le \pi \\ -\sin x, & \pi < x < 2\pi \end{cases}$$

Therefore,

$$\begin{aligned} A &= \int_0^{\pi} |\sin x|\, dx + \int_{\pi}^{2\pi} |\sin x|\, dx \\ &= \int_0^{\pi} \sin x\, dx + \int_{\pi}^{2\pi} (-\sin x)\, dx \\ &= -\cos x\Big]_0^{\pi} + \cos x\Big]_{\pi}^{2\pi} \\ &= -\cos \pi + \cos 0 + (\cos 2\pi - \cos \pi) \\ &= -(-1) + 1 + (1 - (-1)) = 4 \text{ square units} \end{aligned}$$

□

Area Bounded by Two Graphs The foregoing discussion is a special case of the more general problem of finding the area of a region bounded by two graphs. The area *under* the graph of a continuous nonnegative function $y = f(x)$ on $[a, b]$ is the area of the region bounded by its graph and the graph of the function $y = 0$ (the x-axis) from $x = a$ to $x = b$.

Suppose $y = f(x)$ and $y = g(x)$ are continuous on $[a, b]$ and $f(x) \ge g(x)$ for all x in the interval. See Figure 5.5. Let P be a partition of the interval $[a, b]$ into n subintervals $[x_{k-1}, x_k]$. If we choose an x_k^* in each subinterval, we can then construct n corresponding rectangles that have area

$$\Delta A_k = [f(x_k^*) - g(x_k^*)]\, \Delta x_k$$

The area of the region bounded by the two graphs is approximately

$$\sum_{k=1}^{n} \Delta A_k = \sum_{k=1}^{n} [f(x_k^*) - g(x_k^*)]\, \Delta x_k$$

and this in turn suggests that the exact area is

$$A = \lim_{\|P\| \to 0} \sum_{k=1}^{n} [f(x_k^*) - g(x_k^*)]\, \Delta x_k$$

Since f and g are continuous, so is $f - g$. Hence, the above limit exists and is, by definition, the definite integral $\int_a^b [f(x) - g(x)]\, dx$.

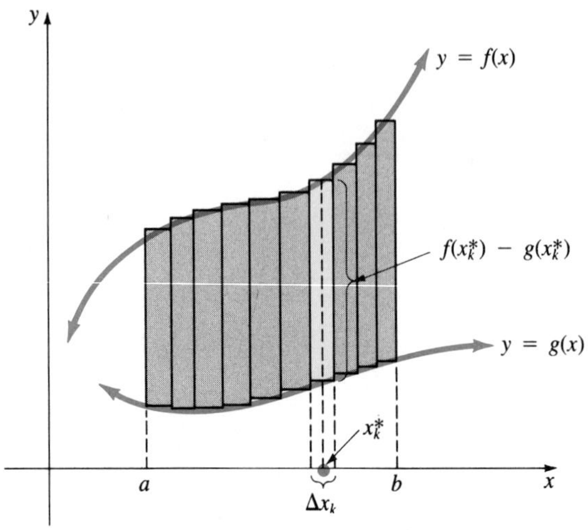

FIGURE 5.5

In general, we have the following definition:

DEFINITION 5.2 Area Bounded by Two Graphs

If f and g are continuous functions on an interval $[a, b]$ and $f(x) \geq g(x)$ for all x in the interval, then the **area** A of the region bounded by their graphs on the interval is given by

$$A = \int_a^b [f(x) - g(x)]\,dx \tag{2}$$

Note that (2) reduces to (1) when $g(x) = 0$ for all x in $[a, b]$. Also, (2) applies to regions for which one or both of the functions f and g have negative values. See Figure 5.6.

You are urged *not to memorize* a formula such as (2) but to sketch the necessary graphs. If the curves intersect on the interval, then the relative

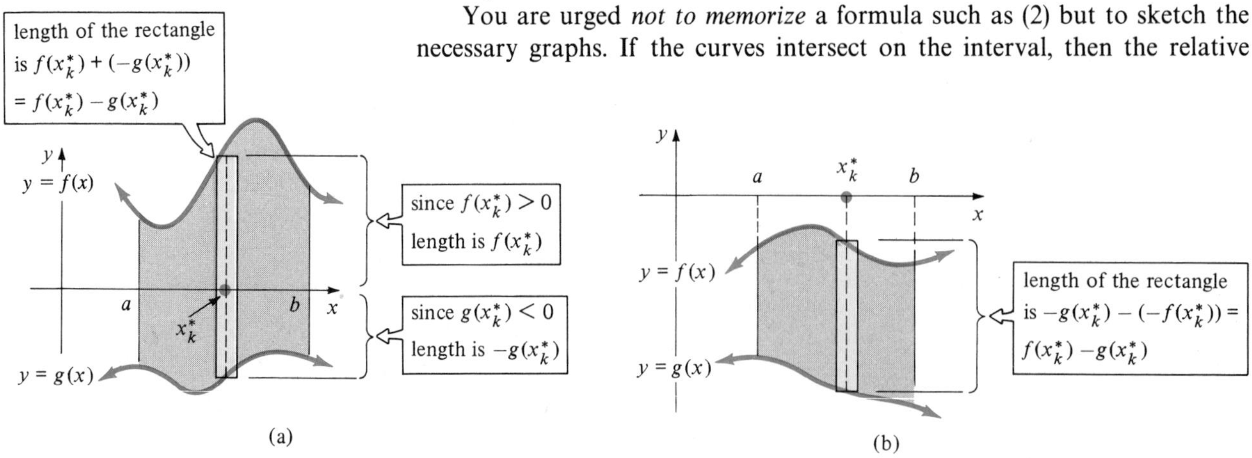

FIGURE 5.6

position of the curves may change. In any event, on any subinterval of $[a, b]$ the appropriate integrand is always

(upper graph ordinate) – *(lower graph ordinate)*

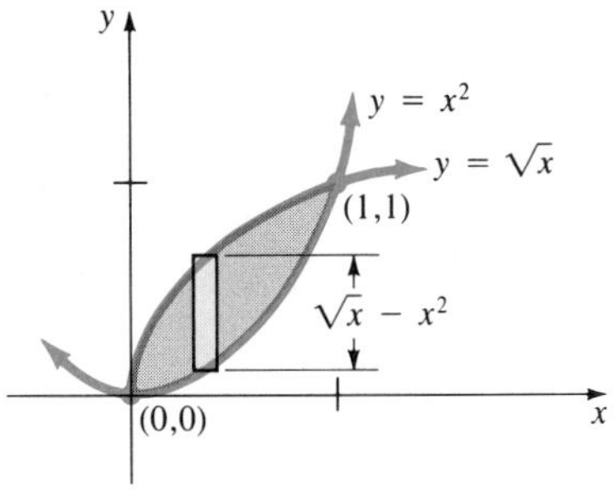

FIGURE 5.7

EXAMPLE 4 Find the area of the region bounded by the graphs of $y = \sqrt{x}$ and $y = x^2$.

Solution As shown in Figure 5.7, the area in question is located in the first quadrant, and the graphs intersect at the points (0, 0) and (1, 1); that is, 0 and 1 are the solutions of $x^2 = \sqrt{x}$. Since $y = \sqrt{x}$ is the upper graph on the interval (0, 1), it follows from (2) that

$$A = \int_0^1 (\sqrt{x} - x^2)\, dx = \left(\frac{2}{3}x^{3/2} - \frac{x^3}{3}\right)\Big]_0^1$$
$$= \frac{2}{3} - \frac{1}{3} - 0 = \frac{1}{3} \text{ square unit}$$ □

EXAMPLE 5 Find the area of the region bounded by the graphs of $y = x^2 + 2x$ and $y = -x + 4$ on $[-4, 2]$.

Solution Let us denote the given functions by

$$y_1 = x^2 + 2x \quad \text{and} \quad y_2 = -x + 4$$

It is easily verified that the graphs intersect at the points $(-4, 8)$ and $(1, 3)$. In addition, inspection of Figure 5.8 shows that on the interval $(-4, 1)$, $y_2 = -x + 4$ is the upper graph, whereas on the interval $(1, 2)$, $y_1 = x^2 + 2x$ is the upper graph. Hence, the total area is the sum of

$$A_1 = \int_{-4}^{1} (y_2 - y_1)\, dx \quad \text{and} \quad A_2 = \int_1^2 (y_1 - y_2)\, dx$$

Thus, $A = A_1 + A_2$ is given by

$$A = \int_{-4}^{1} (y_2 - y_1)\, dx + \int_1^2 (y_1 - y_2)\, dx$$
$$= \int_{-4}^{1} [(-x + 4) - (x^2 + 2x)]\, dx + \int_1^2 [(x^2 + 2x) - (-x + 4)]\, dx$$
$$= \int_{-4}^{1} (-x^2 - 3x + 4)\, dx + \int_1^2 (x^2 + 3x - 4)\, dx$$
$$= \left(-\frac{x^3}{3} - \frac{3}{2}x^2 + 4x\right)\Big]_{-4}^{1} + \left(\frac{x^3}{3} + \frac{3}{2}x^2 - 4x\right)\Big]_1^2$$
$$= \left(-\frac{1}{3} - \frac{3}{2} + 4\right) - \left(\frac{64}{3} - 24 - 16\right) + \left(\frac{8}{3} + 6 - 8\right) - \left(\frac{1}{3} + \frac{3}{2} - 4\right)$$
$$= \frac{71}{3} \text{ square units}$$ □

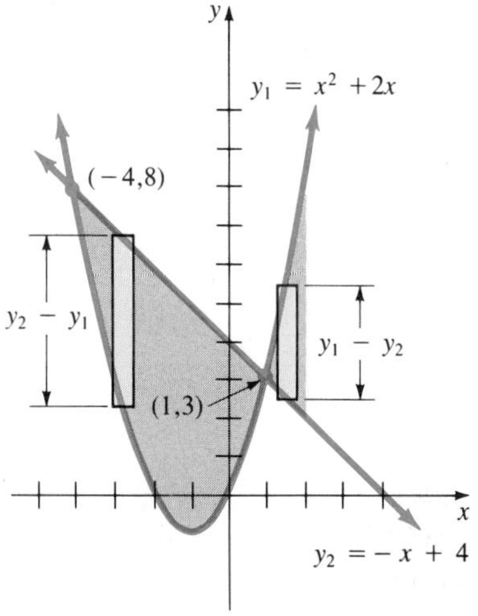

FIGURE 5.8

In finding the area bounded by two graphs, it is sometimes inconvenient to integrate with respect to the variable x.

EXAMPLE 6 Find the area of the region bounded by the graphs of $y^2 = 1 - x$ and $2y = x + 2$.

Solution We note that the equation $y^2 = 1 - x$ is equivalent to two functions, $y_2 = \sqrt{1-x}$ and $y_1 = -\sqrt{1-x}$ for $x \le 1$. If we define $y_3 = \frac{1}{2}x + 1$, we see from Figure 5.9 that the height of an element of area on the interval $(-8, 0)$ is $y_3 - y_1$, whereas the height of an element on the interval $(0, 1)$ is $y_2 - y_1$. Thus, if we integrate with respect to x, the area is the sum of

$$A_1 = \int_{-8}^{0} (y_3 - y_1)\,dx \quad \text{and} \quad A_2 = \int_{0}^{1} (y_2 - y_1)\,dx$$

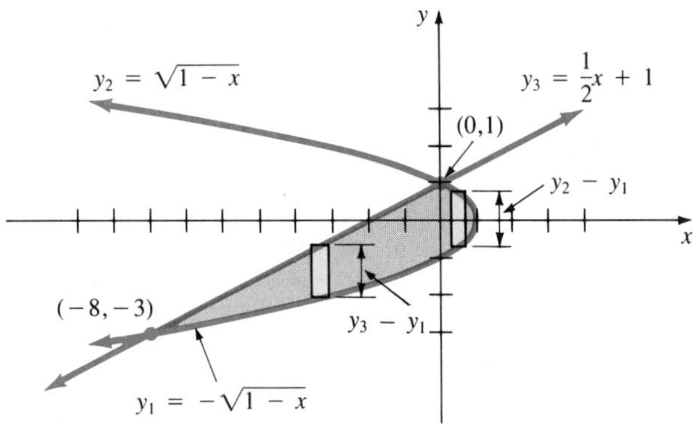

FIGURE 5.9

That is,

$$A = \int_{-8}^{0} \left[\left(\frac{1}{2}x + 1\right) - (-\sqrt{1-x})\right] dx + \int_{0}^{1} [\sqrt{1-x} - (-\sqrt{1-x})]\,dx$$

$$= \int_{-8}^{0} \left(\frac{1}{2}x + 1 + \sqrt{1-x}\right) dx + 2\int_{0}^{1} \sqrt{1-x}\,dx$$

$$= \left(\frac{1}{4}x^2 + x - \frac{2}{3}(1-x)^{3/2}\right)\Bigg]_{-8}^{0} - \frac{4}{3}(1-x)^{3/2}\Bigg]_{0}^{1}$$

$$= -\frac{2}{3}\cdot 1^{3/2} - \left(16 - 8 - \frac{2}{3}\cdot 9^{3/2}\right) - \frac{4}{3}\cdot 0 + \frac{4}{3}\cdot 1^{3/2} = \frac{32}{3} \text{ square units} \quad \square$$

EXAMPLE 7 ***Alternative Solution to Example 6*** The necessity of using two integrals in Example 6 to find the area is avoided by constructing horizontal rectangles and using y as the independent variable. If we define $x_2 = 1 - y^2$

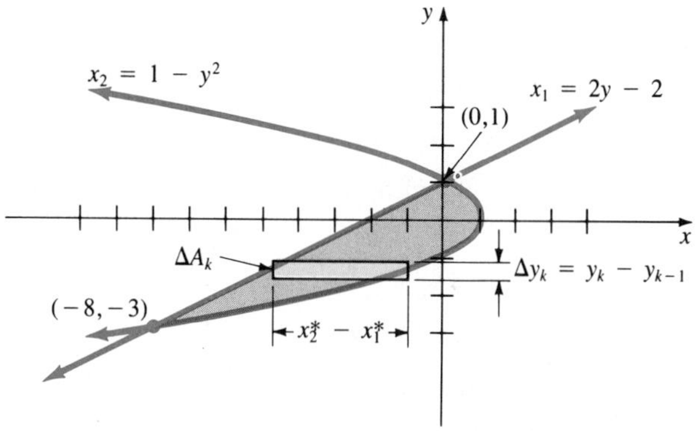

FIGURE 5.10

and $x_1 = 2y - 2$, then, as shown in Figure 5.10, the area of a horizontal element is

$$\Delta A_k = [(\text{right graph abscissa}) - (\text{left graph abscissa})] \cdot \text{width}$$

That is, $$\Delta A_k = [x_2^* - x_1^*]\,\Delta y_k$$

where

$$x_2^* = 1 - (y_k^*)^2, \quad x_1^* = 2y_k^* - 2, \quad \text{and} \quad \Delta y_k = y_k - y_{k-1}$$

Summing the rectangles in the positive y-direction leads to

$$A = \lim_{\|P\| \to 0} \sum_{k=1}^{n} [x_2^*(y) - x_1^*(y)]\,\Delta y_k$$

where $\|P\|$ is the norm of a partition P of the interval $-3 \le y \le 1$. In other words,

$$A = \int_{-3}^{1} (x_2 - x_1)\,dy$$

where the lower limit -3 and the upper limit 1 are the y-coordinates of the points of intersection $(-8, -3)$ and $(0, 1)$, respectively. Substituting the indicated values of x_2 and x_1 then gives

$$\begin{aligned} A &= \int_{-3}^{1} [(1 - y^2) - (2y - 2)]\,dy \\ &= \int_{-3}^{1} (-y^2 - 2y + 3)\,dy \\ &= \left(-\frac{y^3}{3} - y^2 + 3y\right)\Bigg]_{-3}^{1} \\ &= \left(-\frac{1}{3} - 1 + 3\right) - \left(\frac{27}{3} - 9 - 9\right) = \frac{32}{3} \text{ square units} \end{aligned}$$

□

EXERCISES 5.1

Answers to odd-numbered problems begin on page A-81.

In Problems 1–22 find the area bounded by the graph of the given function and the x-axis on the indicated interval.

1. $y = x^2 - 1$; $[-1, 1]$
2. $y = x^2 - 1$; $[0, 2]$
3. $y = x^3$; $[-3, 0]$
4. $y = 1 - x^3$; $[0, 2]$
5. $y = x^2 - 3x$; $[0, 3]$
6. $y = -(x + 1)^2$; $[-1, 0]$
7. $y = x^3 - 6x$; $[-1, 1]$
8. $y = x^3 - 3x^2 + 2$; $[0, 2]$
9. $y = (x - 1)(x - 2)(x - 3)$; $[0, 3]$
10. $y = x(x + 1)(x - 1)$; $[-1, 1]$
11. $y = \dfrac{x^2 - 1}{x^2}$; $[\frac{1}{2}, 3]$
12. $y = \dfrac{x^2 - 1}{x^2}$; $[1, 2]$
13. $y = \sqrt{x} - 1$; $[0, 4]$
14. $y = 2 - \sqrt{x}$; $[0, 9]$
15. $y = \sqrt[3]{x}$; $[-2, 3]$
16. $y = 2 - \sqrt[3]{x}$; $[-1, 8]$
17. $y = \sin x$; $[-\pi, \pi]$
18. $y = 1 + \cos x$; $[0, 3\pi]$
19. $y = -1 + \sin x$; $[-3\pi/2, \pi/2]$
20. $y = \sec^2 x$; $[0, \pi/3]$
21. $y = \begin{cases} x, & -2 \le x < 0 \\ x^2, & 0 \le x \le 1 \end{cases}$; $[-2, 1]$
22. $y = \begin{cases} x + 2, & -3 \le x \le 0 \\ 2 - x^2, & 0 < x \le 2 \end{cases}$; $[-3, 2]$

In Problems 23–50 find the area of the region bounded by the graphs of the given equations.

23. $y = x$, $y = -2x$, $x = 3$
24. $y = x$, $y = 4x$, $x = 2$
25. $y = x^2$, $y = 4$
26. $y = x^2$, $y = x$
27. $y = x^3$, $y = 8$, $x = -1$
28. $y = x^3$, $y = \sqrt[3]{x}$, first quadrant
29. $y = 4(1 - x^2)$, $y = 1 - x^2$
30. $y = 2(1 - x^2)$, $y = x^2 - 1$
31. $y = x$, $y = 1/x^2$, $x = 3$
32. $y = x^2$, $y = 1/x^2$, $y = 9$, first quadrant
33. $y = -x^2 + 6$, $y = x^2 + 4x$
34. $y = x^2$, $y = -x^2 + 3x$
35. $y = x^{2/3}$, $y = 4$
36. $y = 1 - x^{2/3}$, $y = x^{2/3} - 1$
37. $y = x^2 - 2x - 3$, $y = 2x + 2$, on $[-1, 6]$
38. $y = -x^2 + 4x$, $y = \frac{3}{2}x$
39. $y = x^3$, $y = x + 6$, $y = -\frac{1}{2}x$
40. $x = y^2$, $x = 0$, $y = 1$
41. $x = -y$, $x = 2 - y^2$
42. $x = y^2$, $x = 6 - y^2$
43. $x = y^2 + 2y + 2$, $x = -y^2 - 2y + 2$
44. $x = y^2 - 6y + 1$, $x = -y^2 + 2y + 1$
45. $y = x^3 - x$, $y = x + 4$, $x = -1$, $x = 1$
46. $x = y^3 - y$, $x = 0$
47. $y = \cos x$, $y = \sin x$, $x = 0$, $x = \pi/2$
48. $y = 2 \sin x$, $y = -x$, $x = \pi/2$
49. $y = 4 \sin x$, $y = 2$, on $[\pi/6, 5\pi/6]$
50. $y = 2 \cos x$, $y = -\cos x$, on $[-\pi/2, \pi/2]$

In Problems 51 and 52 find the area of the shaded region.

51.

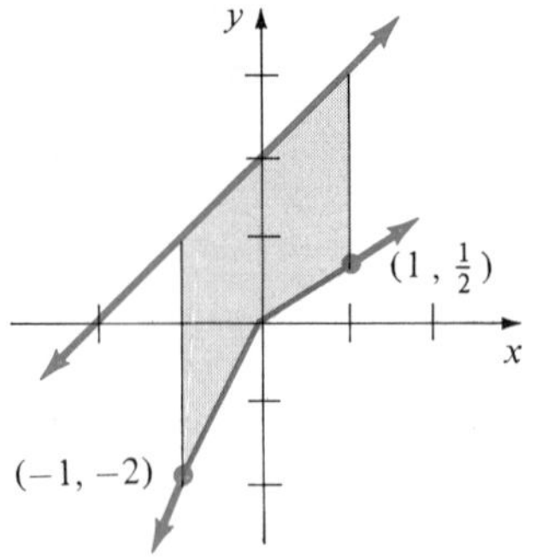

FIGURE 5.11

52.

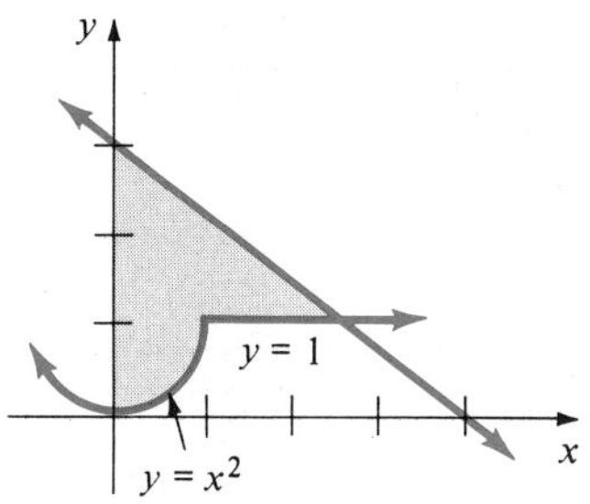

FIGURE 5.12

53. Find the area of the triangle with vertices at (1, 1), (2, 4), and (3, 2).

54. Set up a definite integral that represents the area of an ellipse $x^2/a^2 + y^2/b^2 = 1$. Do not try to evaluate.

In Problems 55 and 56 use the fact that the area of a circle of radius a is πa^2 to evaluate the given definite integral.

55. $\displaystyle\int_0^a \sqrt{a^2 - x^2}\,dx$ 56. $\displaystyle\int_{-a}^a \sqrt{a^2 - x^2}\,dx$

57. Consider the region bounded by the graphs of $y^2 = -x - 2$, $y = 2$, $y = -2$, and $y = 2(x - 1)$. Compute the area of the region by integrating with respect to x.

58. Compute the area of the region given in Problem 57 by integrating with respect to y.

59. A trapezoid is bounded by the graphs of $f(x) = Ax + B$, $x = a$, $x = b$, and $x = 0$. Show that the area of the trapezoid is $\dfrac{f(a) + f(b)}{2}(b - a)$.

60. Given that a function f is continuous on an interval $[a, b]$, it can be proved that there exists a number c in the open interval (a, b) for which the number $f(c)(b - a)$ is the same as $\int_a^b f(x)\,dx$. Interpret this result geometrically in terms of area under the graph of f on the interval.

61. The line segment between Q and R shown in Figure 5.13 is tangent to the graph of $y = 1/x$ at point P. Show that the area of triangle QOR is independent of the coordinates of P.

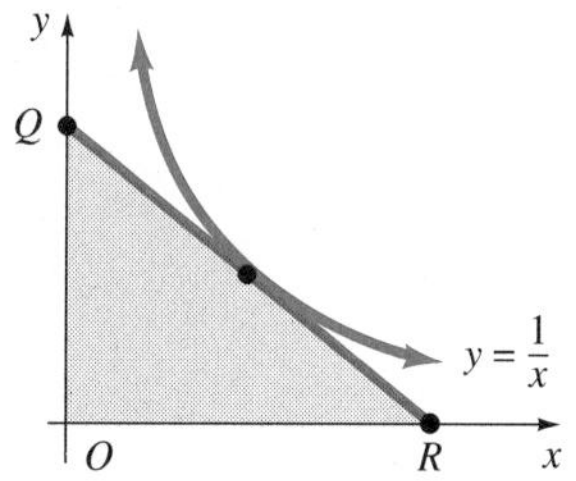

FIGURE 5.13

62. Use the Midpoint Rule to find an approximation to the area bounded by the graph given in Figure 5.14 and the x-axis on the interval [0, 7]. Use $n = 7$.

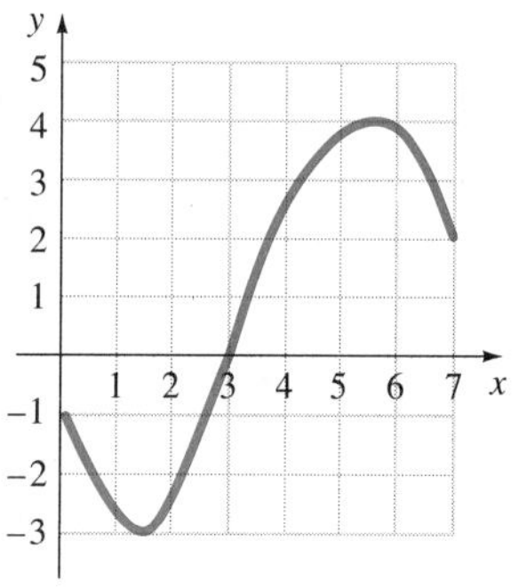

FIGURE 5.14

5.2 VOLUMES BY SLICING

REVIEW

The classic example of a **right cylinder** is the right *circular* cylinder—that is, the shape of a tin can. But a right cylinder need not be circular. From geometry, a **right cylinder** is defined as a solid bounded by two congruent plane regions, in parallel planes, and a lateral surface that is generated by a moving line segment that is perpendicular to both planes and whose ends are on the boundaries of the plane regions.

When the regions are circles, we obtain the right circular cylinder. If the regions are rectangles, the cylinder is a rectangular parallelepiped. The volume V of any right cylinder is given by

$$V = B \cdot h$$

where B denotes the area of a base (that is, the area of one of the plane regions) and h denotes the height of the cylinder (that is, the perpendicular distance between the plane regions).

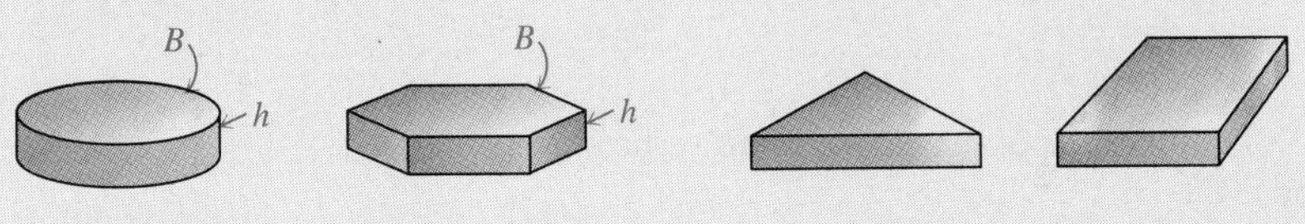

In this section and the following two, we shall show how the definite integral can be used to compute the volumes of certain solids.

Volume of a Solid: Slicing Method Suppose V is the volume of the solid shown in Figure 5.15 bounded by planes that are perpendicular to the x-axis at $x = a$ and $x = b$. Furthermore, suppose we know a continuous function $A(x)$ that gives the area of a plane cross-sectional region, or **slice**, which is formed by cutting the solid by a plane perpendicular to the x-axis. For example, for $a < x_1 < x_2 < b$ the areas of the cross-sections shown in Figure 5.15 are $A(x_1)$ and $A(x_2)$.

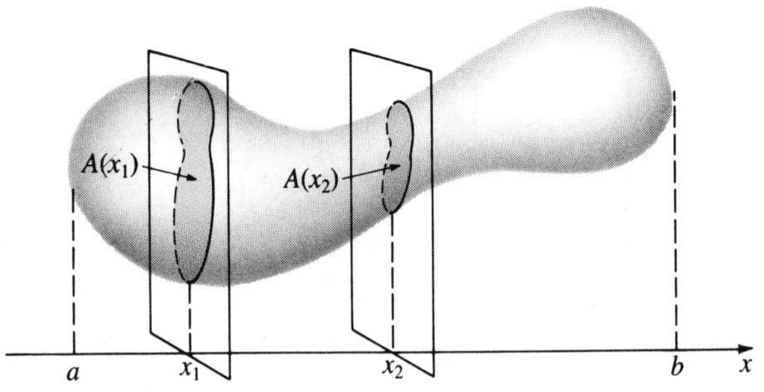

FIGURE 5.15

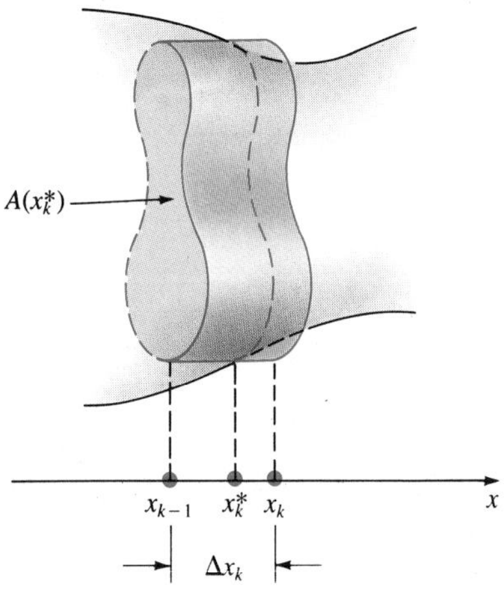

FIGURE 5.16

Now, let P be the partition

$$a = x_0 < x_1 < x_2 < \cdots < x_n = b$$

and x_k^* any number in $[x_{k-1}, x_k]$. Then an approximation to the volume of the solid on this subinterval is the volume of the right cylinder shown in the enlargement in Figure 5.16. Since the volume of a right cylinder is

$$\Delta V_k = (\text{area of base}) \cdot (\text{height}) = A(x_k^*)(x_k - x_{k-1}) = A(x_k^*)\,\Delta x_k$$

it follows that an approximation to the volume of the solid on $[a, b]$ is

$$\sum_{k=1}^{n} \Delta V_k = \sum_{k=1}^{n} A(x_k^*)\,\Delta x_k$$

Thus, we conclude that the exact volume is given by the definite integral

$$V = \lim_{\|P\| \to 0} \sum_{k=1}^{n} A(x_k^*)\,\Delta x_k = \int_a^b A(x)\,dx$$

We state the formal result as a definition.

DEFINITION 5.3 Volume by Slicing

Let V be the volume of a solid bounded by planes that are perpendicular to the x-axis at $x = a$ and $x = b$. If $A(x)$ is a continuous function that gives the area of a cross-section of the solid formed by a plane perpendicular to the x-axis at any point in $[a, b]$, then the **volume** of the solid is

$$V = \int_a^b A(x)\,dx \qquad (1)$$

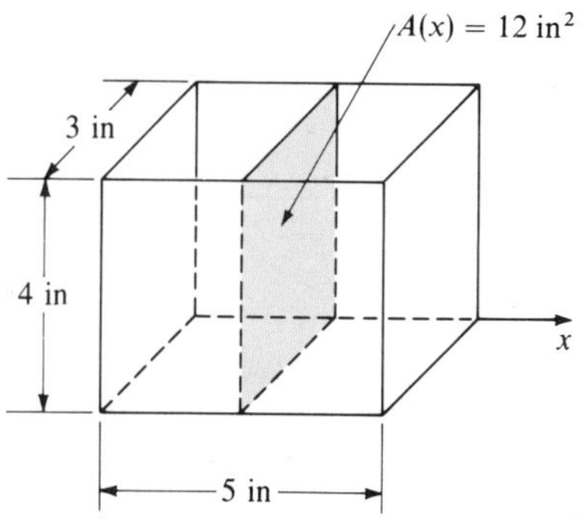

FIGURE 5.17

EXAMPLE 1 Consider the rather simple example of finding the volume of the rectangular box shown in Figure 5.17. The cross-sectional area is the constant function

$$A(x) = 12 \text{ in}^2$$

Although we know the volume is $V = 3 \times 4 \times 5 = 60$ in^3, it also follows from Definition 5.3 that

$$V = \int_0^5 12\,dx = 12x\Big]_0^5 = 12[5 - 0] = 60 \text{ in}^3 \qquad \square$$

EXAMPLE 2 Find the volume V of the right circular cone shown in Figure 5.18(a).

Solution The area of a cross-section, obtained by a horizontal slice x units from the bottom of the cone, is πR^2. Inspection of Figure 5.18(b) shows that R is related to x by similar triangles. We have

$$\frac{h}{r} = \frac{h - x}{R} \quad \text{so that} \quad R = \frac{r}{h}(h - x)$$

Therefore,

$$A(x) = \pi R^2 = \pi \frac{r^2}{h^2}(h - x)^2$$

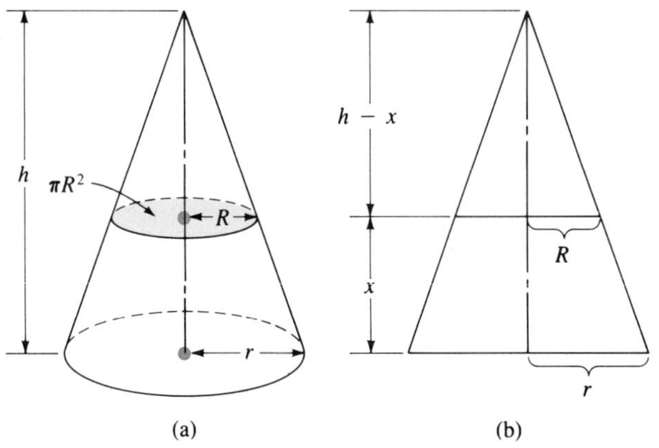

FIGURE 5.18

By Definition 5.3, the volume of the cone is given by

$$V = \int_0^h \pi \frac{r^2}{h^2}(h - x)^2\, dx = -\pi \frac{r^2}{h^2} \int_0^h (h - x)^2(-dx)$$

$$= -\pi \frac{r^2}{h^2} \frac{(h - x)^3}{3}\Big]_0^h$$

$$= -\pi \frac{r^2}{h^2}\left[0 - \frac{h^3}{3}\right]$$

$$= \frac{\pi}{3} r^2 h$$

□

EXERCISES 5.2 *Answers to odd-numbered problems begin on page A-81.*

1. A 75-ft-high power-line pole has a cross-section in the form of an equilateral triangle. Given that the length of a side is $(75 - x)/10$, where x is the distance in feet from the ground, find the volume of the pole.

2. The cross-section of a pyramid is a square x feet on a side, x feet from its top. Given that the pyramid is 100 ft high, find its volume.

3. The cross-sections perpendicular to a diameter of a circular base are squares. Given that the radius of the base is 4 ft, find the volume of the solid. See Figure 5.19.

4. The cross-sections perpendicular to a diameter of a circular base are equilateral triangles. Given that the radius of the base is 4 ft, find the volume of the solid. See Figure 5.20.

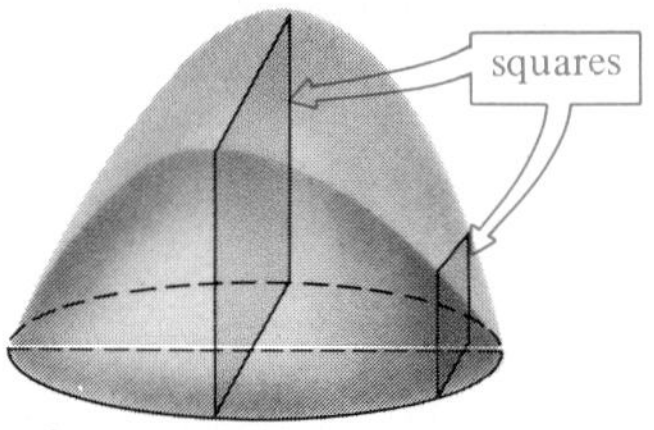

FIGURE 5.19

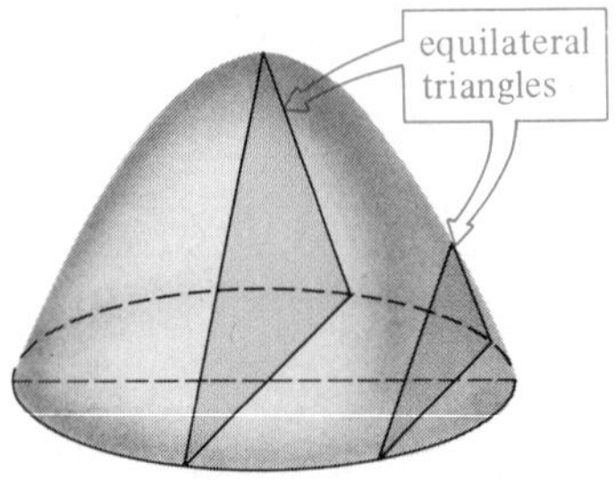

FIGURE 5.20

5. The base of a solid is bounded by the curves $x = y^2$ and $x = 4$ in the xy-plane. The cross-sections perpendicular to the x-axis are rectangles for which the height is four times the base. Find the volume of the solid.

6. The base of a solid is bounded by the curve $y = 4 - x^2$ and the x-axis. The cross-sections perpendicular to the x-axis are equilateral triangles. Find the volume of the solid.

7. The base of a solid is an isosceles triangle whose base is 4 ft and height is 5 ft. The cross-sections perpendicular to the altitude are semicircles. Find the volume of the solid.

8. The axes of two right circular cylinders, each having radius $r = 3$ ft, intersect at right angles. Find the value of the resulting volume.

9. The base of a solid is a right isosceles triangle that is formed by the coordinate axes and the line $x + y = 3$. The cross-sections perpendicular to the y-axis are squares. Find the volume of the solid.

10. A hole of radius 1 ft is drilled through the middle of the solid sphere of radius $r = 2$ ft. Find the volume of the remaining solid.

11. Consider the right circular cylinder of radius a shown in Figure 5.21. A plane inclined at an angle of 45° to the base of the cylinder passes through a diameter of the base. Find the volume of the resulting wedge cut from the cylinder.

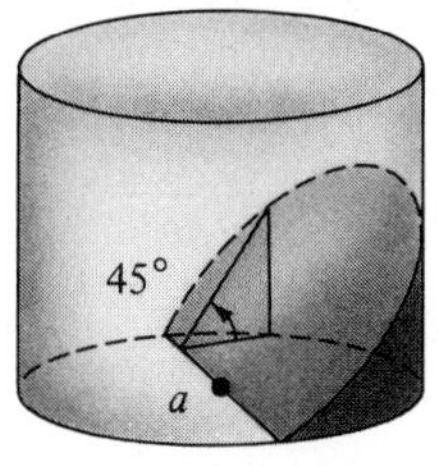

FIGURE 5.21

12. Rework Problem 11 if the plane passing through the diameter of the base is inclined at an angle of 60°.

13. We have seen that a right cylinder need not be a right circular cylinder. Similarly, a cone need not be a right circular cone. If the x-axis, shown in Figure 5.22(a), is perpendicular to the plane containing the base of the cone, the cross-sectional regions formed by slices perpendicular to the axis need not be circles.

(a) Now it is known that the area of a cross-sectional region is directly proportional to the square of the distance x from the origin 0. Use this information to show that the volume of a cone is given by $V = Bh/3$, where B is the area of the base, and h is its height.

(b) What is the volume V when the cross-sectional slices are circles?

(c) Use the information in part **(a)** to show that the volume of a frustum of the cone, shown in Figure 5.22(b), is given by $V = (h/3)[B_1 + \sqrt{B_1 B_2} + B_2]$, where h is the height of the frustum and B_1 and B_2 are the areas of the top and base, respectively.

(d) Use the volume formula in part **(c)** to find the volume of a frustum of a right circular cone. (See Appendix V.)

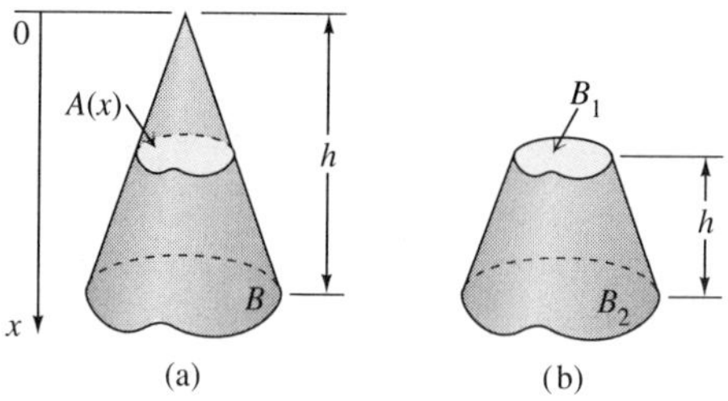

FIGURE 5.22

14. Use Simpson's Rule to find an approximation to the volume of the rocket whose cross-sectional areas, shown in Figure 5.23, are spaced at intervals of width 4 m.

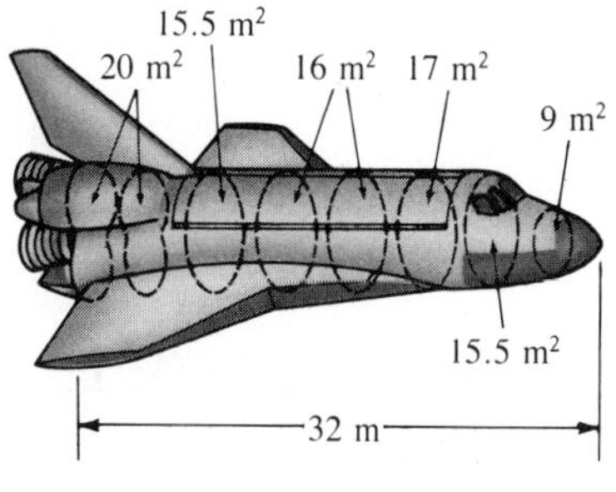

FIGURE 5.23

5.3 SOLIDS OF REVOLUTION

If a region R in the xy-plane is revolved about an axis L, it will generate a solid called a **solid of revolution**. See Figure 5.24. As shown in Section 5.2, we can find the volume V of a solid by means of a definite integral provided that we

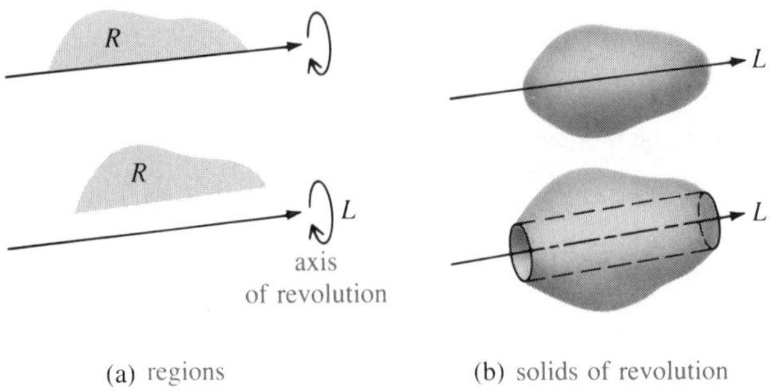

FIGURE 5.24

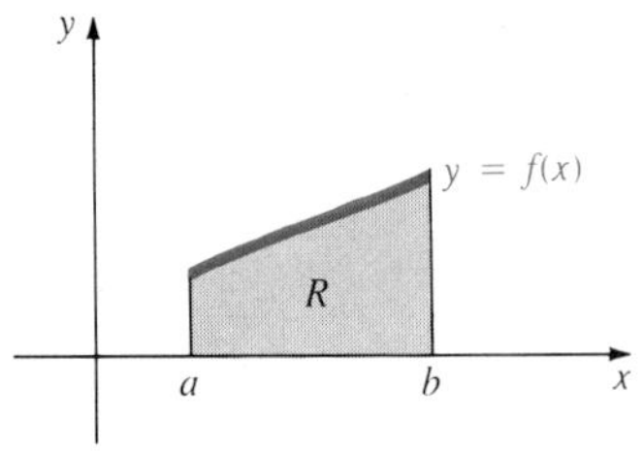

FIGURE 5.25

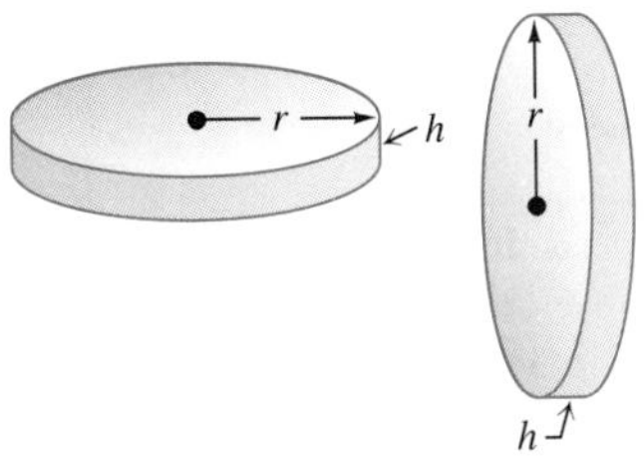

FIGURE 5.26

know the function $A(x)$ that gives the cross-sectional area of a slice formed by passing a plane through the solid perpendicular to an axis. In the case of finding the volume of a solid of revolution, it is always possible to find $A(x)$; the axis in question is the axis of revolution. The first two methods considered in this section are just special cases of Definition 5.3.

Disk Method Let R be the region bounded by the graph of a nonnegative continuous function $y = f(x)$, the x-axis, and the vertical lines $x = a$ and $x = b$, as shown in Figure 5.25. If this region is revolved about the x-axis, let us find the volume V of the resulting solid of revolution.

Let P be a partition of $[a, b]$, and let x_k^* be any number in the kth subinterval $[x_{k-1}, x_k]$ as shown in Figure 5.27(a). As the rectangular element of width $\Delta x_k = x_k - x_{k-1}$ and height $f(x_k^*)$ is revolved about the x-axis, it generates a **disk**. Now the volume of a right circular cylinder, or disk, of radius r and height h is

$$\text{(area of base)} \cdot \text{(height)} = \pi r^2 h$$

See Figure 5.26. Thus, if we make the identification $r = f(x_k^*)$ and $h = \Delta x_k$, the volume of the representative disk in Figure 5.27(b) is

$$\Delta V_k = \pi[f(x_k^*)]^2 \, \Delta x_k$$

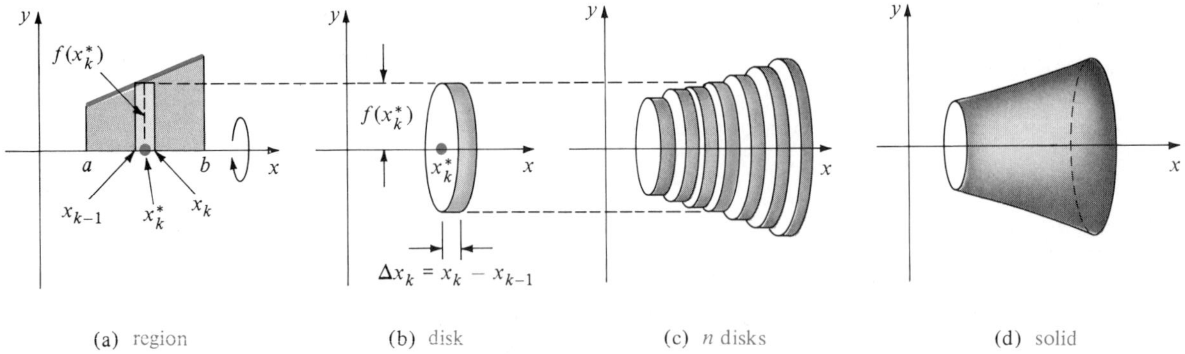

FIGURE 5.27

As seen in Figure 5.27(c) a partition with n subintervals yields n such disks. Hence, the sum

$$\sum_{k=1}^{n} \Delta V_k = \sum_{k=1}^{n} \pi[f(x_k^*)]^2 \Delta x_k$$

represents an approximation to the volume of the solid shown in Figure 5.27(d). This suggests that the exact volume is

$$V = \lim_{\|P\| \to 0} \sum_{k=1}^{n} \pi[f(x_k^*)]^2 \Delta x_k$$

or

$$V = \pi \int_a^b [f(x)]^2 \, dx \tag{1}$$

Note that (1) is a special case of Definition 5.3 with $A(x) = \pi[f(x)]^2$.

If a region R is revolved about some other axis, then (1) may simply not be applicable to the problem of finding the volume of the resulting solid. Rather than apply a formula blindly, you should set up an appropriate integral by carefully analyzing the geometry of each problem.

EXAMPLE 1 Find the volume V of the solid formed by revolving the region bounded by the graphs of $y = \sqrt{x}$, $y = 0$, and $x = 4$ about the x-axis.

Solution Figure 5.28(a) shows the region in question. Now, the volume of the disk shown in Figure 5.28(b) is

$$\Delta V_k = (\text{area of base}) \cdot (\text{height}) = \pi[f(x_k^*)]^2 \Delta x_k = \pi[(x_k^*)^{1/2}]^2 \Delta x_k = \pi x_k^* \Delta x_k$$

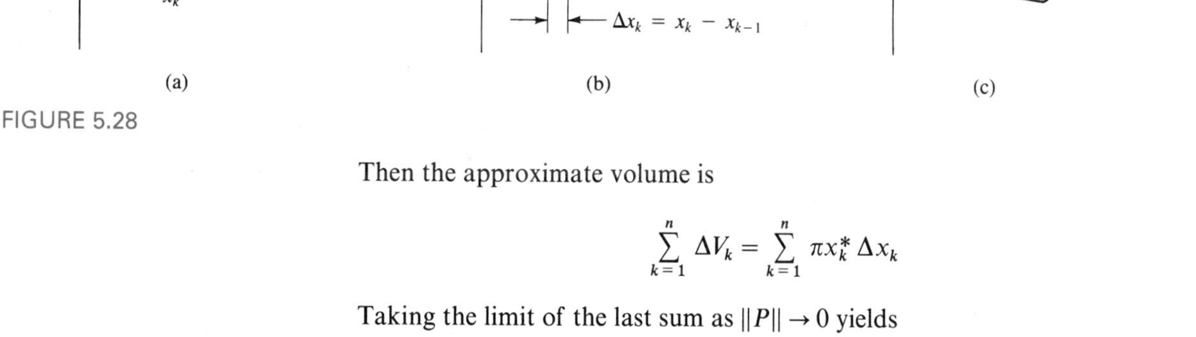

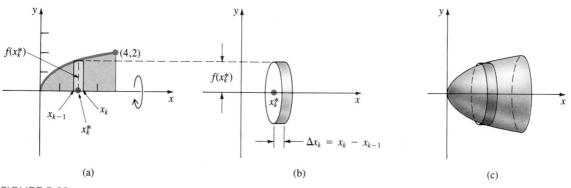

FIGURE 5.28

Then the approximate volume is

$$\sum_{k=1}^{n} \Delta V_k = \sum_{k=1}^{n} \pi x_k^* \Delta x_k$$

Taking the limit of the last sum as $\|P\| \to 0$ yields

$$V = \pi \int_0^4 x \, dx = \pi \frac{x^2}{2}\Big]_0^4 = 8\pi \text{ cubic units} \qquad \square$$

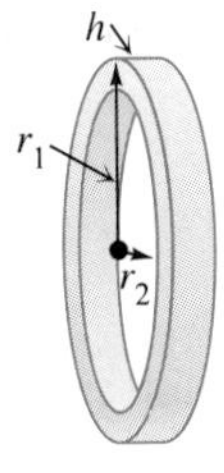

FIGURE 5.29

Washer Method Let the region R bounded by the graphs of the continuous functions $y = f(x)$, $y = g(x)$, and the lines $x = a$ and $x = b$, as shown in Figure 5.30(a), be revolved about the x-axis. Then the rectangular element between the two graphs on $[x_{k-1}, x_k]$ will generate a circular ring or **washer**. The volume of a washer with outer radius r_1, inner radius r_2, and thickness h is

$$(\text{volume of disk}) - (\text{volume of hole}) = \pi r_1^2 h - \pi r_2^2 h = \pi(r_1^2 - r_2^2)h$$

See Figure 5.29. With the identifications $r_1 = f(x_k^*)$, $r_2 = g(x_k^*)$, and $h = \Delta x_k$, the volume of the representative washer in Figure 5.30(b) is

$$\Delta V_k = \pi([f(x_k^*)]^2 - [g(x_k^*)]^2)\,\Delta x_k$$

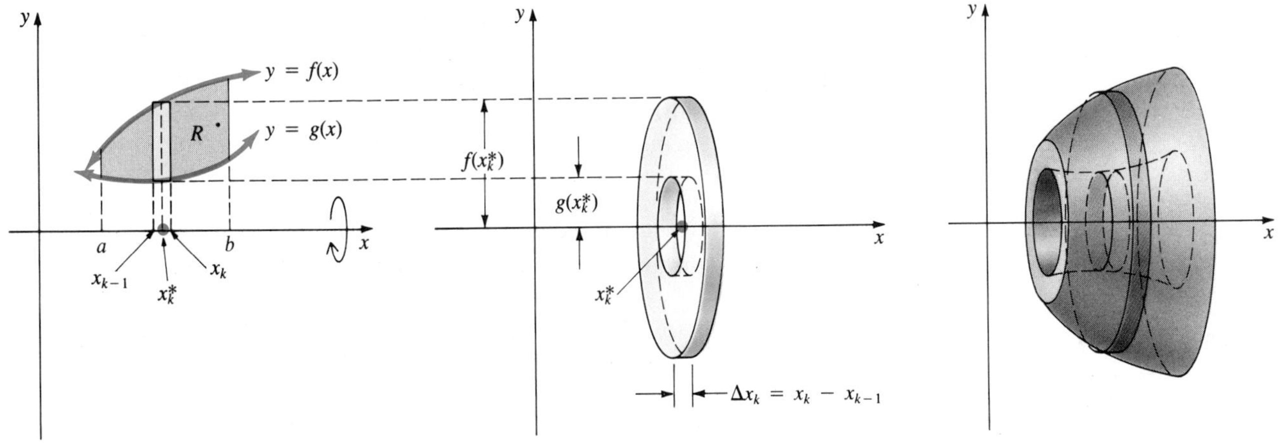

(a) region (b) washer (c) solid

FIGURE 5.30

An approximation to the volume of the solid in Figure 5.30(c) is then

$$\sum_{k=1}^{n} \Delta V_k = \sum_{k=1}^{n} \pi([f(x_k^*)]^2 - [g(x_k^*)]^2)\,\Delta x_k$$

This suggests that as $\|P\| \to 0$, the exact volume of the solid is given by

$$V = \pi \int_a^b ([f(x)]^2 - [g(x)]^2)\,dx \qquad (2)$$

Formula (2) is again a special case of Definition 5.3 with $A(x) = \pi([f(x)]^2 - [g(x)]^2)$. In addition, (2) reduces to (1) when $g(x) = 0$.

EXAMPLE 2 Find the volume V of the solid formed by revolving the region bounded by the graphs of $y = \sqrt{x}$, $y = 0$, and $x = 4$ about the y-axis.

Solution The region is shown in Figure 5.31(a). As the horizontal rectangular element of height $\Delta y_k = y_k - y_{k-1}$ is revolved about the y-axis, we obtain the

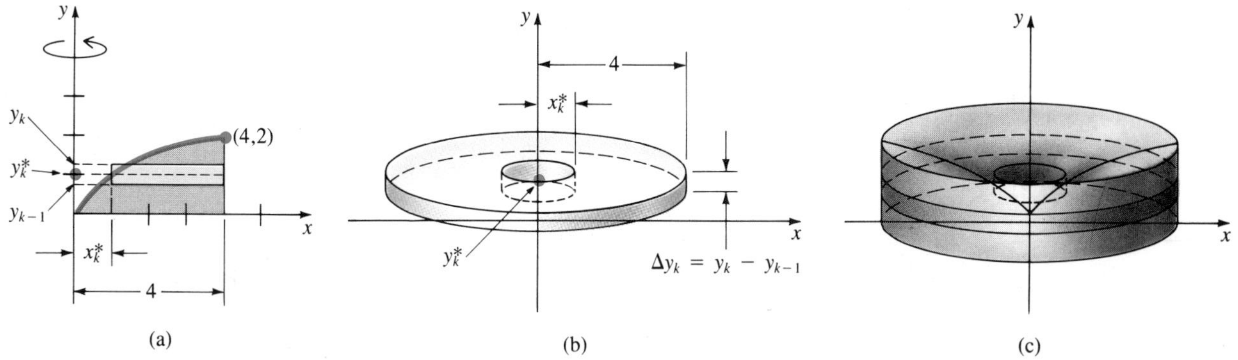

FIGURE 5.31

washer illustrated in Figure 5.31(b). As in the general discussion above, the volume of the washer is

$$\Delta V_k = (\text{volume of disk}) - (\text{volume of hole}) = \pi 4^2\,\Delta y_k - \pi(x_k^*)^2\,\Delta y_k = \pi[16 - (x_k^*)^2]\,\Delta y_k$$

In this case the given function implies $x_k^* = (y_k^*)^2$. Thus, the approximate volume of the solid is

$$\sum_{k=1}^{n} \Delta V_k = \sum_{k=1}^{n} \pi[16 - (y_k^*)^4]\,\Delta y_k$$

The use of a horizontal rectangle of height Δy_k corresponds to a partition of the interval [0, 2] on the y-axis. Hence, we conclude that the exact volume is given by the definite integral

$$V = \pi \int_0^2 (16 - y^4)\,dy = \pi\left(16y - \frac{y^5}{5}\right)\Bigg]_0^2 = \frac{128\pi}{5} \text{ cubic units} \qquad \square$$

EXAMPLE 3 Find the volume V of the solid formed by revolving the region bounded by the graphs of $y = x + 2$, $y = x$, $x = 0$, and $x = 3$ about the x-axis.

Solution As seen in Figure 5.32 on page 332, a vertical rectangular element of width Δx_k, when revolved about the x-axis, yields a washer having volume

$$\Delta V_k = \pi(x_k^* + 2)^2\,\Delta x_k - \pi(x_k^*)^2\,\Delta x_k = \pi(4x_k^* + 4)\,\Delta x_k$$

The usual summing and limiting process yields

$$V = \pi \int_0^3 (4x + 4)\,dx = \pi(2x^2 + 4x)\Big]_0^3 = 30\pi \text{ cubic units} \qquad \square$$

Revolution About a Line The next example shows how to find the volume of a solid of revolution when a region is revolved about an axis that is not a coordinate axis.

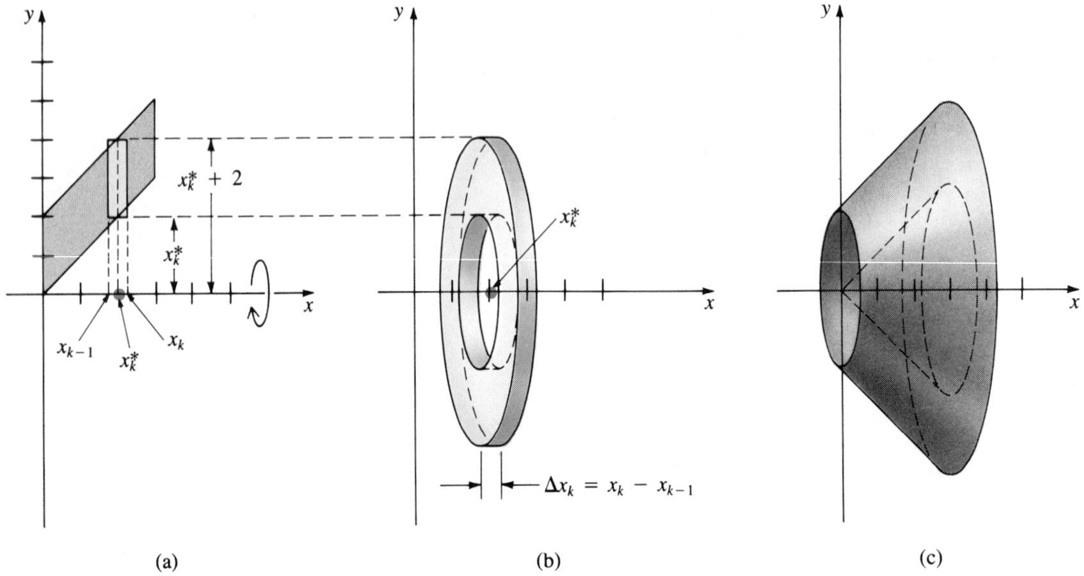

FIGURE 5.32

EXAMPLE 4 Find the volume V of the solid that is formed by revolving the region given in Example 2 about the line $x = 4$.

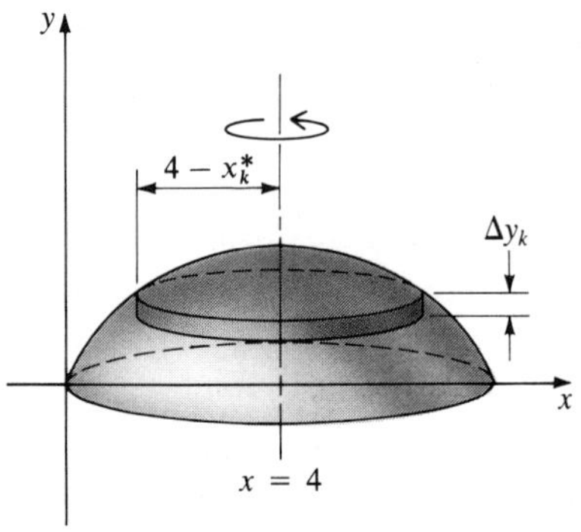

FIGURE 5.33

Solution From inspection of Figure 5.33, we see that a horizontal rectangular element of width Δy_k generates a solid disk of volume

$$\Delta V_k = \pi(4 - x_k^*)^2\, \Delta y_k$$

where, as before, $x_k^* = (y_k^*)^2$. This leads to the integral

$$V = \pi \int_0^2 (4 - y^2)^2\, dy = \pi \int_0^2 (16 - 8y^2 + y^4)\, dy$$

$$= \pi\left(16y - \frac{8}{3}y^3 + \frac{y^5}{5}\right)\Bigg]_0^2 = \frac{256\pi}{15} \text{ cubic units} \qquad \square$$

Shell Method We have just seen that a rectangular element of area that is perpendicular to an axis of revolution will generate a disk or circular ring (washer). Note, however, that if we were to revolve the rectangular element shown in Figure 5.34(a) about a line parallel to the element, in this case the y-axis, we generate a hollow **shell** as shown in Figure 5.34(b).

The volume of a shell with outer radius r_1, inner radius r_2, and height h is the difference

$$\begin{aligned}&(\text{volume of outer cylinder}) - (\text{volume of inner cylinder})\\ &= \pi r_1^2 h - \pi r_2^2 h = \pi(r_1^2 - r_2^2)h = \pi(r_1 + r_2)(r_1 - r_2)h \qquad (3)\end{aligned}$$

See Figure 5.35.

To find the volume of the solid shown in Figure 5.34(c), we form a partition P of $[a, b]$ into n subintervals $[x_{k-1}, x_k]$. If we identify $r_1 = x_k$

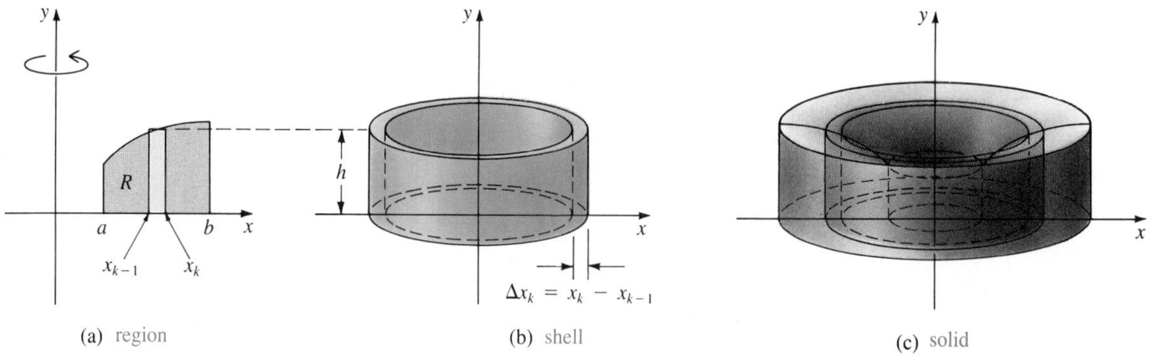

FIGURE 5.34

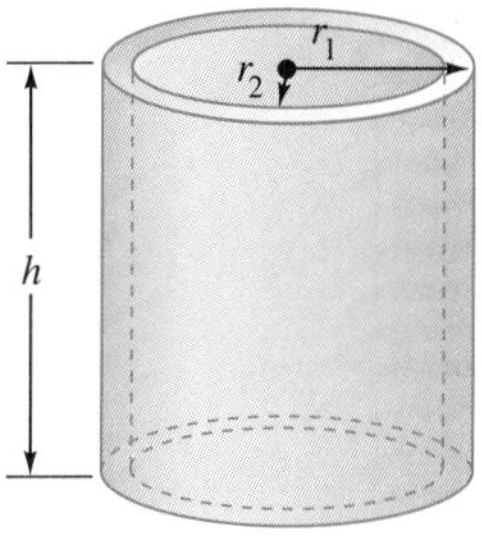

FIGURE 5.35

and $r_2 = x_{k-1}$ and define $x_k^* = \frac{1}{2}(x_k + x_{k-1})$, then x_k^* is the midpoint of the subinterval of length $\Delta x_k = x_k - x_{k-1}$. In addition, $x_k + x_{k-1} = 2x_k^*$. Therefore, with the further identification $h = f(x_k^*)$, it follows from (3) that the volume of the representative shell in Figure 5.34(b) can be written as

$$\Delta V_k = 2\pi x_k^* f(x_k^*)\, \Delta x_k$$

An approximation to the volume of the solid is

$$\sum_{k=1}^{n} \Delta V_k = \sum_{k=1}^{n} 2\pi x_k^* f(x_k^*)\, \Delta x_k \tag{4}$$

As the norm $\|P\|$ of the partition approaches zero, we expect that the limit of (4) is the definite integral

$$V = 2\pi \int_a^b x f(x)\, dx \tag{5}$$

Since it is impossible to analyze every possible case, we urge you, again, not to memorize a particular formula such as (5). On the other hand, it is important to be able to set up the integral for a given problem without going through a lengthy analysis. To accomplish this, imagine that a shell is cut down its side and flattened out to form a thin rectangular solid as in Figure 5.36(b). The volume of the shell is then

$$\begin{aligned} \text{volume} &= (\text{length}) \cdot (\text{width}) \cdot (\text{thickness}) \\ &= (\text{circumference of the cylinder}) \cdot (\text{height}) \cdot (\text{thickness}) \\ &= 2\pi r h t \end{aligned} \tag{6}$$

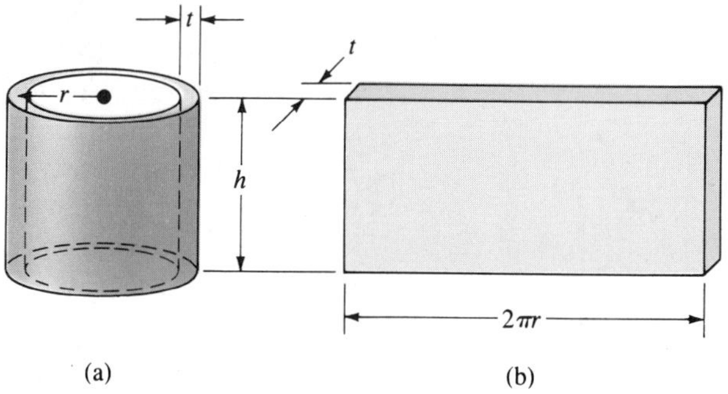

FIGURE 5.36

EXAMPLE 5 Use the shell method to find the volume V of the solid of revolution that is given in Example 2.

Solution From Figure 5.37 we can make the identification $r = x_k^*$, $h = y_k^*$, $t = \Delta x_k$. Thus, from (6),

$$\Delta V_k = 2\pi x_k^* y_k^* \, \Delta x_k = 2\pi x_k^* \sqrt{x_k^*} \, \Delta x_k$$

and so $V = 2\pi \int_0^4 x \cdot x^{1/2} \, dx$. That is,

$$V = 2\pi \int_0^4 x^{3/2} \, dx = 2\pi \frac{2}{5} x^{5/2} \Big]_0^4 = \frac{128\pi}{5} \text{ cubic units}$$

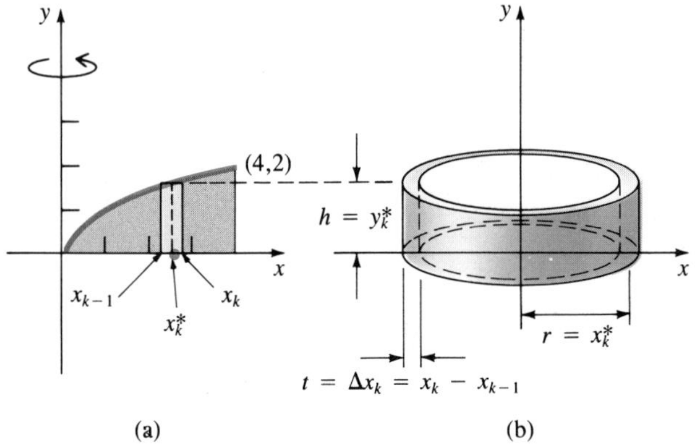

FIGURE 5.37 □

It is not always convenient or even possible to use the disk or washer method to find the volume of a solid of revolution.

EXAMPLE 6 Find the volume V of the solid that is formed by revolving the region bounded by the graphs of $x = y^2 - 2y$ and $x = 3$ about the line $y = 1$.

Solution In this case a rectangular element of area that is perpendicular to a horizontal line and revolved about the line $y = 1$ would generate a disk. Since the radius of the disk is not measured from the x-axis but from the line $y = 1$, it would be necessary to solve $x = y^2 - 2y$ for y in terms of x. We can avoid this inconvenience by using horizontal elements of area, which then generate the shell indicated in Figure 5.38(b). Note that when $x = 3$, the equation $3 = y^2 - 2y$, or equivalently $(y + 1)(y - 3) = 0$, has solutions -1 and 3. Thus, we need only partition the interval $[1, 3]$ on the y-axis. After making the identifications $r = y_k^* - 1$, $h = 3 - x_k^*$, and $t = \Delta y_k$, it follows from (6) that the volume of a shell is

$$\Delta V_k = 2\pi(y_k^* - 1)(3 - x_k^*) \, \Delta y_k = 2\pi(y_k^* - 1)(3 - [(y_k^*)^2 - 2y_k^*]) \, \Delta y_k$$

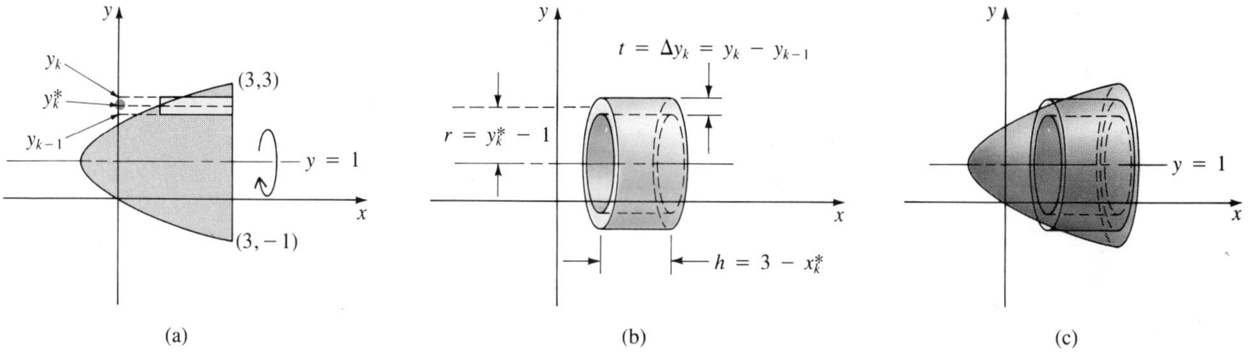

FIGURE 5.38

The volume is

$$V = 2\pi \int_1^3 (y - 1)(3 - y^2 + 2y)\, dy$$

$$= 2\pi \int_1^3 (-y^3 + 3y^2 + y - 3)\, dy$$

$$= 2\pi\left(-\frac{y^4}{4} + y^3 + \frac{y^2}{2} - 3y\right)\Bigg]_1^3$$

$$= 2\pi\left[\left(-\frac{81}{4} + 27 + \frac{9}{2} - 9\right) - \left(-\frac{1}{4} + 1 + \frac{1}{2} - 3\right)\right] = 8\pi \text{ cubic units} \quad \square$$

EXERCISES 5.3

Answers to odd-numbered problems begin on page A-81.

In Problems 1–6 refer to Figure 5.39. Use the disk or washer method to find the volume of the solid of revolution that is formed by revolving the given region about the indicated line.

1. R_1 about OC
2. R_1 about OA
3. R_2 about OA
4. R_2 about OC
5. R_1 about AB
6. R_2 about AB

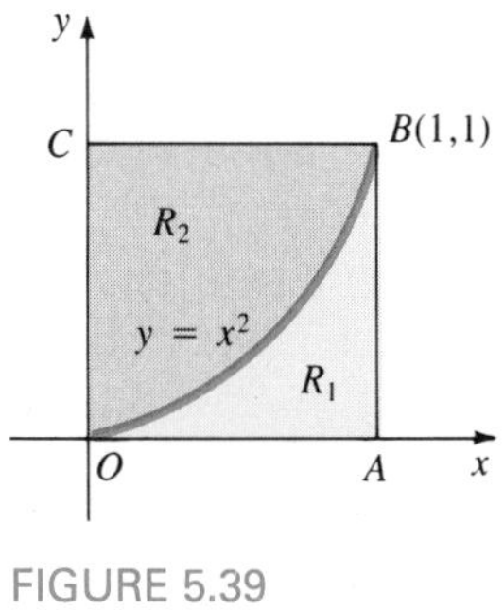

FIGURE 5.39

In Problems 7–30 use the disk or washer method to find the volume of the solid of revolution that is formed by revolving the region bounded by the graphs of the given equations about the indicated line or axis.

7. $y = 9 - x^2$, $y = 0$; x-axis
8. $y = x^2 + 1$, $x = 0$, $y = 5$; y-axis
9. $y = \dfrac{1}{x}$, $x = 1$, $y = \dfrac{1}{2}$; y-axis
10. $y = \dfrac{1}{x}$, $x = \dfrac{1}{2}$, $x = 3$, $y = 0$; x-axis
11. $y = (x - 2)^2$, $x = 0$, $y = 0$; x-axis
12. $y = (x + 1)^2$, $x = 0$, $y = 0$; y-axis
13. $y = 4 - x^2$, $y = 1 - \frac{1}{4}x^2$; x-axis
14. $y = 1 - x^2$, $y = x^2 - 1$, $x = 0$, first quadrant; y-axis
15. $y = x$, $y = x + 1$, $x = 0$, $y = 2$; y-axis

16. $x + y = 2, x = 0, y = 0, y = 1$; x-axis

17. $y = \sqrt{x-1}, x = 5, y = 0$; $x = 5$

18. $x = y^2, x = 1$; $x = 1$

19. $y = x^{1/3}, x = 0, y = 1$; $y = 2$

20. $x = -y^2 + 2y, x = 0$; $x = 2$

21. $x^2 - y^2 = 16, x = 5$; y-axis

22. $y = x^2 - 6x + 9, y = 9 - \frac{1}{2}x^2$; x-axis

23 $x = y^2, y = x - 6$; y-axis

24. $y = x^3 + 1, x = 0, y = 9$; y-axis

25. $y = x^3 - x, y = 0$; x-axis

26. $y = \sin x, y = 0, 0 \le x \le \pi$; x-axis

27. $y = |\cos x|, y = 0, 0 \le x \le 2\pi$; x-axis

28. $y = \sec x, x = -\frac{\pi}{4}, x = \frac{\pi}{4}, y = 0$; x-axis

29. $y = \tan x, y = 0, x = \frac{\pi}{4}$; x-axis

30. $y = \sin x, y = \cos x, x = 0$, first quadrant; x-axis

In Problems 31–36 refer to Figure 5.40. Use the shell method to find the volume of the solid of revolution that is formed by revolving the given region about the indicated line.

31. R_1 about OC

32. R_1 about OA

33. R_2 about BC

34. R_2 about OA

35. R_1 about AB

36. R_2 about AB

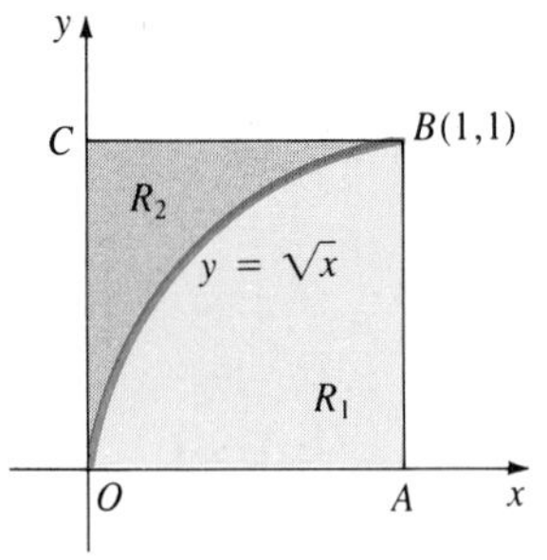

FIGURE 5.40

In Problems 37–60 use the shell method to find the volume of the solid of revolution that is formed by revolving the region bounded by the graphs of the given equations about the indicated line or axis.

37. $y = x, x = 0, y = 5$; x-axis

38. $y = 1 - x, x = 0, y = 0$; $y = -2$

39. $y = x^2, x = 0, y = 3$, first quadrant; x-axis

40. $y = x^2, x = 2, y = 0$; y-axis

41. $y = x^2, x = 1, y = 0$; $x = 3$

42. $y = x^2, y = 9$; x-axis

43. $y = x^2 + 4, x = 0, x = 2, y = 2$; y-axis

44. $y = x^2 - 5x + 4, y = 0$; y-axis

45. $y = (x-1)^2, y = 1$; x-axis

46. $y = (x-2)^2, y = 4$; $x = 4$

47. $y = x^{1/3}, x = 1, y = 0$; $y = -1$

48. $y = x^{1/3} + 1, y = -x + 1, x = 1$; $x = 1$

49. $y = x^2, y = x$; y-axis

50. $y = x^2, y = x$; $x = 2$

51. $y = x^3, y = x + 6, x = 0$; y-axis

52. $y = x^3 - x, y = 0$, second quadrant; y-axis

53. $y = x^2 - 2, y = -x^2 + 2, x = 0$, second and third quadrants; y-axis

54. $y = x^2 - 4x, y = -x^2 + 4x$; $x = -1$

55. $x = y^2 - 5y, x = 0$; x-axis

56. $x = y^2 + 2, y = x - 4, y = 1$; x-axis

57. $y = \sqrt{x-1}, x = 0, y = 0, y = 3$; $y = 3$

58. $y = \sqrt{x}, y = \sqrt{1-x}, y = 0$; x-axis

59. $y = \sin x^2, x = 0, y = 1$; y-axis

60. $x^2 - y^2 = 1, x = \sqrt{5}, y = 0$, first quadrant; y-axis

Choose an appropriate method to solve Problems 61–66.

In Problems 61–66 use the information in part **(a)** of the figure to find the volume of the solid given in part **(b)**.

61.

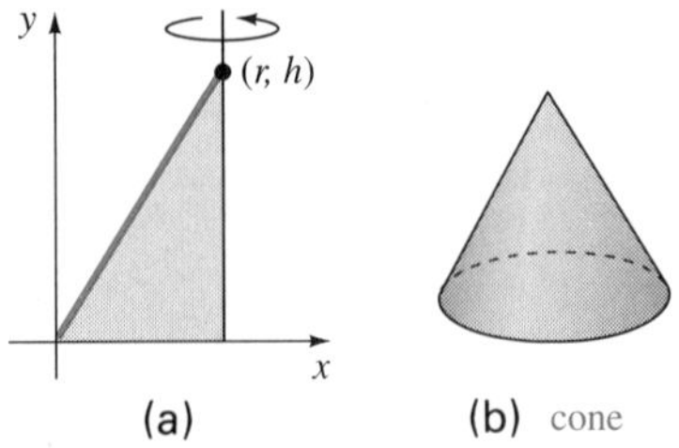

FIGURE 5.41

62.

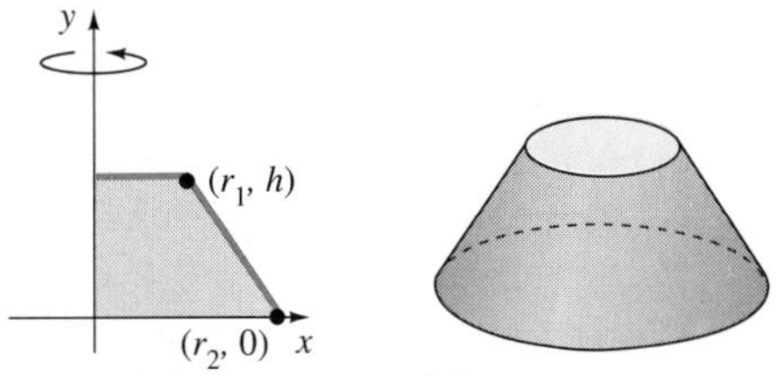

(a) (b) frustum of a cone

FIGURE 5.42

63.

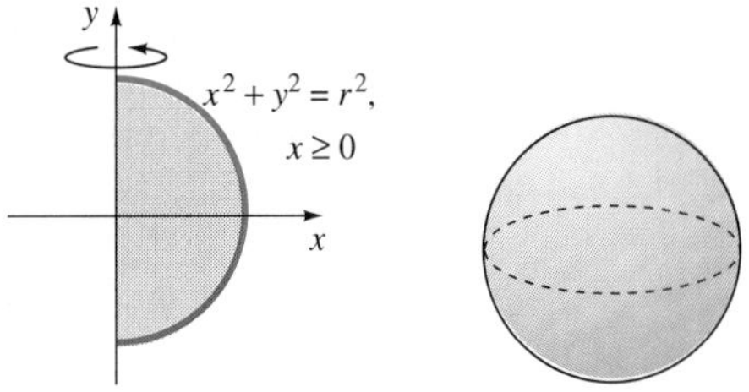

(a) (b) sphere

FIGURE 5.43

64.

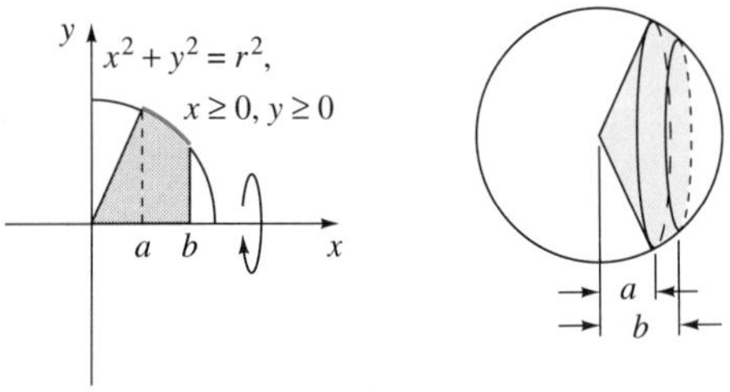

(a) (b) spherical sector

FIGURE 5.44

65.

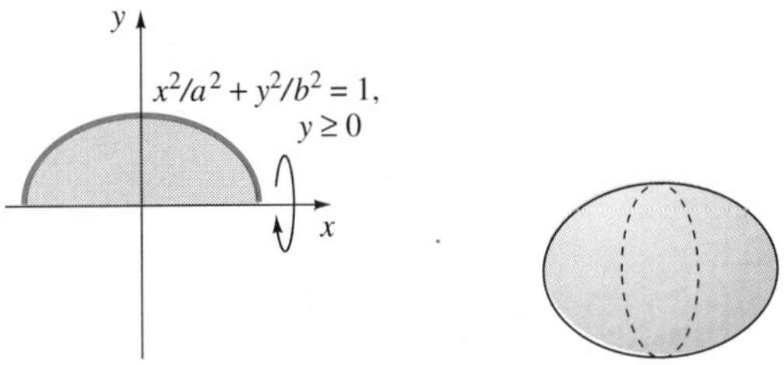

(a) (b) prolate spheroid

FIGURE 5.45

66.

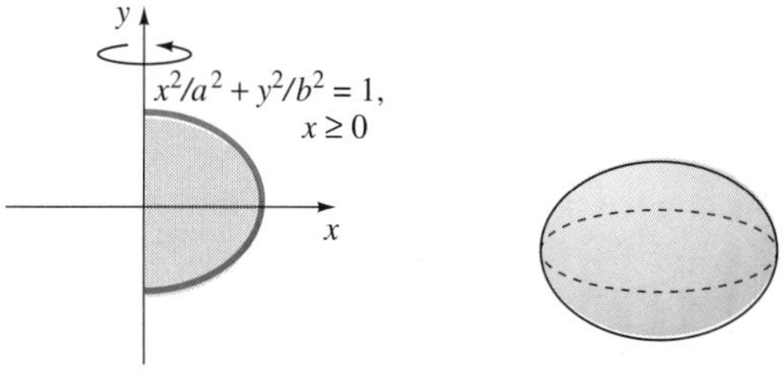

(a) (b) oblate spheroid

FIGURE 5.46

67. The shaded regions in Figure 5.47 are revolved about the y-axis. Use the shell method to find the volume of the solid of revolution.

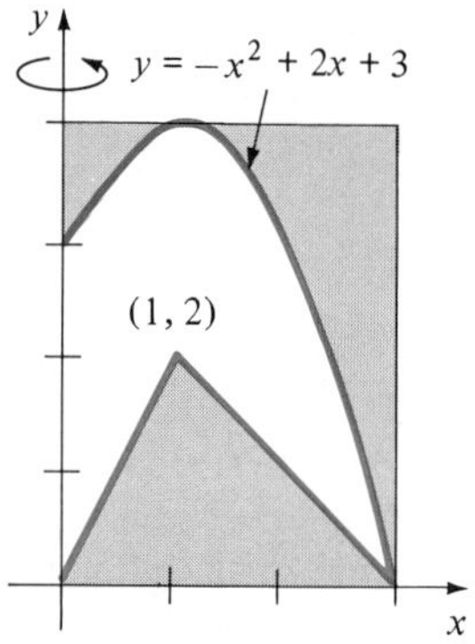

FIGURE 5.47

68. A cylindrical bucket of radius 2 ft that contains a liquid is rotating about the y-axis with a constant angular velocity ω. The surface of the liquid has a parabolic cross-section given by $y = \omega^2 x^2/2g$, $-2 \leq x \leq 2$. Use the shell method to find the volume of the liquid in the rotating bucket given that the height of the bucket is 3 ft. See Figure 5.48.

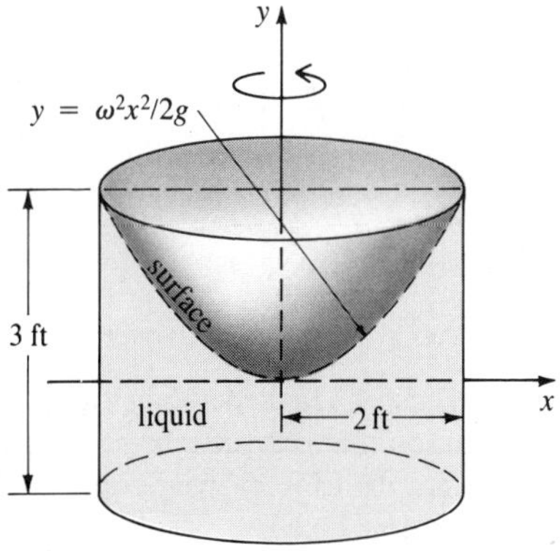

FIGURE 5.48

Calculator/Computer Problems

69. A model for the shape of an egg can be obtained by revolving the region bounded by the graphs of $y = 0$ and the function $f(x) = P(x)\sqrt{1 - x^2}$, where $P(x) = ax^3 + bx^2 + cx + d$ is a cubic polynomial, about the x-axis. For example, an egg of the Common Murre corresponds to $P(x) = -0.07x^3 - 0.02x^2 + 0.2x + 0.56$. Figure 5.49(a) shows the graph of f obtained with the aid of a computer.

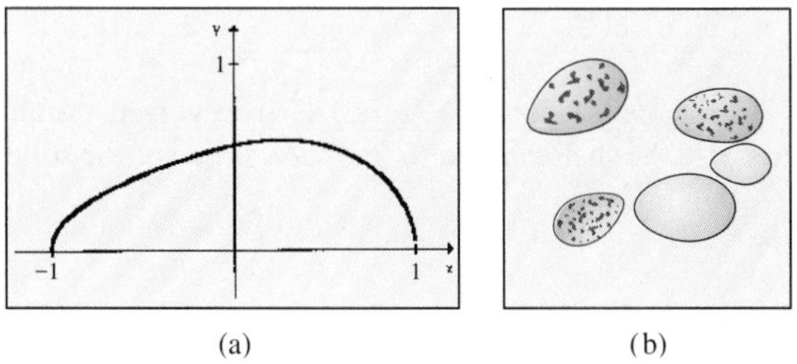

FIGURE 5.49 (a) A model for the shape of the egg of the Common Murre. (b) Bird eggs of various shapes.

(a) Find a general formula for the volume V of an egg based on the mathematical model $f(x) = P(x)\sqrt{1 - x^2}$, where $P(x) = ax^3 + bx^2 + cx + d$. [*Hint:* This problem can be done by hand calculation but it is long and "messy." Use a computer algebra system to carry out the integration.]

(b) Use the formula obtained in part (**a**) to estimate the volume of an egg of the Common Murre.

(c) An egg of a Red-throated Loon corresponds to $P(x) = -0.06x^3 + 0.04x^2 + 0.1x + 0.54$. Use a calculator or computer to obtain the graph of f.

(d) Use part (**a**) to estimate the volume of an egg of a Red-throated Loon.

70. A wooden spherical ball of radius r is floating on a pond of still water. Let h denote the depth that the ball will sink into the water. See Figure 5.50.

(a) Show that the volume of the submerged portion of the ball is given by $V = \pi r^2 h - (\pi/3)h^3$.

(b) Suppose that the weight density of the ball is denoted by ρ_{ball} and the weight density of the water is ρ_{water} (measured in lb/ft^3). If $r = 3$ in and $\rho_{\text{ball}} = 0.4\rho_{\text{water}}$, use Newton's Method and Archimedes' principle, the weight of the ball equals the weight of the water displaced, to determine the approximate depth h that the ball will sink.

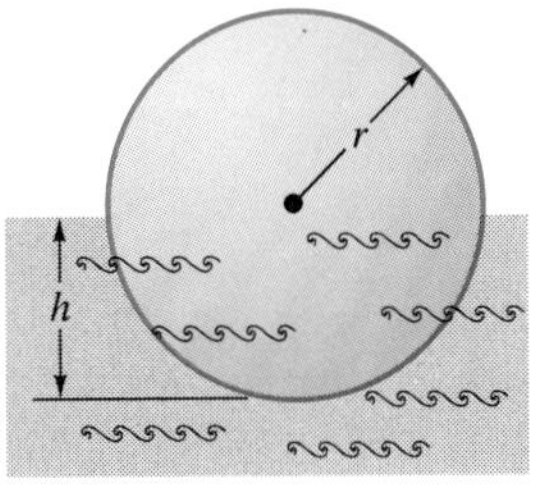

FIGURE 5.50

5.4 ARC LENGTH

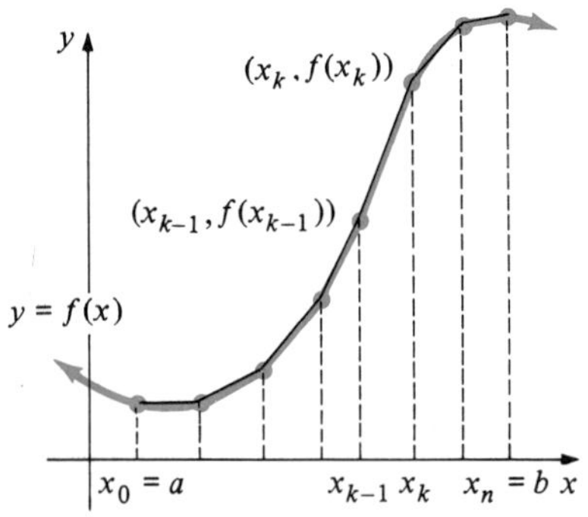

FIGURE 5.51

Smooth Functions If $y = f(x)$ has a continuous first derivative on an interval $[a, b]$, then its graph is said to be **smooth** and f is called a **smooth function**. As the name implies, a smooth graph has no sharp points. In this section we will find the length of a smooth graph.

Let f have a smooth graph on $[a, b]$ and let P denote the arbitrary partition

$$a = x_0 < x_1 < x_2 < \cdots < x_{n-1} < x_n = b$$

As usual, let the length of each subinterval be given by $\Delta x_k = x_k - x_{k-1}$ and let $\|P\|$ be the length of the longest subinterval. As shown in Figure 5.51, the length of the chord between $(x_{k-1}, f(x_{k-1}))$ and $(x_k, f(x_k))$ is an approximation to the length of the graph between these points. The length of the chord is

$$\Delta s_k = \sqrt{(x_k - x_{k-1})^2 + (f(x_k) - f(x_{k-1}))^2} \tag{1}$$

By the Mean Value Theorem, we know there exists an x_k^* in each open subinterval (x_{k-1}, x_k) such that

$$\frac{f(x_k) - f(x_{k-1})}{x_k - x_{k-1}} = f'(x_k^*) \quad \text{or} \quad f(x_k) - f(x_{k-1}) = f'(x_k^*)(x_k - x_{k-1})$$

Substituting from this last equation into (1) gives

$$\begin{aligned}\Delta s_k &= \sqrt{(x_k - x_{k-1})^2 + [f'(x_k^*)]^2(x_k - x_{k-1})^2}\\ &= \sqrt{1 + [f'(x_k^*)]^2}(x_k - x_{k-1})\\ &= \sqrt{1 + [f'(x_k^*)]^2}\,\Delta x_k\end{aligned}$$

The sum

$$\sum_{k=1}^{n} \Delta s_k = \sum_{k=1}^{n} \sqrt{1 + [f'(x_k^*)]^2}\,\Delta x_k$$

gives an approximation to the total length of the graph on $[a, b]$. As $\|P\| \to 0$, we obtain

$$\lim_{\|P\| \to 0} \sum_{k=1}^{n} \sqrt{1 + [f'(x_k^*)]^2}\,\Delta x_k = \int_a^b \sqrt{1 + [f'(x)]^2}\,dx \tag{2}$$

The foregoing discussion prompts us to use equation (2) as the definition of the length of the graph on the interval.

DEFINITION 5.4 Arc Length

Let f be a function for which f' is continuous on an interval $[a, b]$. Then the **length** s of the graph of $y = f(x)$ on the interval is given by

$$s = \int_a^b \sqrt{1 + [f'(x)]^2}\,dx \tag{3}$$

A graph that has arc length is said to be **rectifiable**.

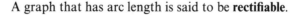

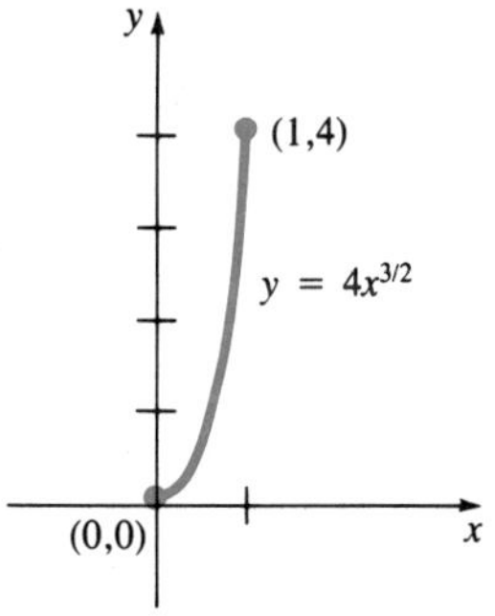

FIGURE 5.52

EXAMPLE 1 Find the length of the graph of $y = 4x^{3/2}$ from the origin (0, 0) to the point (1, 4).

Solution The graph of the function on [0, 1] is given in Figure 5.52. Now,

$$\frac{dy}{dx} = f'(x) = 6x^{1/2}$$

is continuous on the interval. Therefore, it follows from (3) that

$$s = \int_0^1 \sqrt{1 + [6x^{1/2}]^2}\, dx = \int_0^1 (1 + 36x)^{1/2}\, dx$$
$$= \frac{1}{36}\int_0^1 (1 + 36x)^{1/2}(36\, dx)$$
$$= \frac{1}{54}(1 + 36x)^{3/2}\Big]_0^1 = \frac{1}{54}[37^{3/2} - 1] \approx 4.1493 \quad \square$$

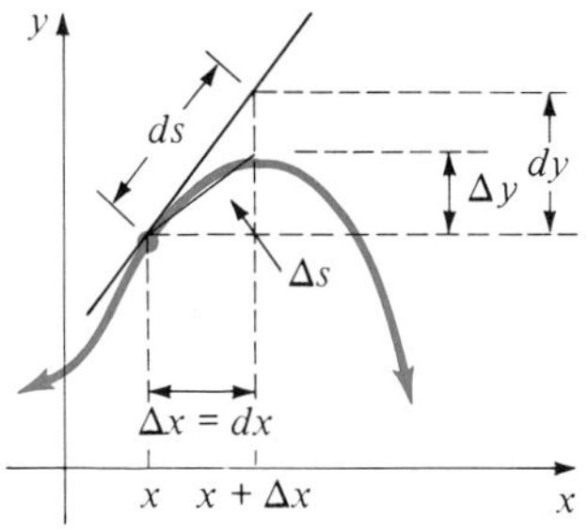

FIGURE 5.53

Differential of Arc Length If $s = \int_a^x \sqrt{1 + [f'(t)]^2}\, dt$, then by Theorem 4.13, $ds/dx = \sqrt{1 + [f'(x)]^2}$ and, consequently,

$$ds = \sqrt{1 + [f'(x)]^2}\, dx \tag{4}$$

The latter function is called the **differential of the arc length** and can be used to approximate lengths of curves. With $dy = f'(x)\, dx$, (4) is often written as

$$ds = \sqrt{(dx)^2 + (dy)^2} \quad \text{or} \quad (ds)^2 = (dx)^2 + (dy)^2 \tag{5}$$

Figure 5.53 shows that the differential ds can be interpreted as the hypotenuse of a right triangle with sides dx and dy.

▲ ***Remark*** The integral in (3) often leads to problems in which special techniques of integration are necessary. See Chapter 8. But even with these subsequent procedures, it is not *always* possible to evaluate the indefinite integral $\int \sqrt{1 + [f'(x)]^2}\, dx$ in terms of the familiar elementary functions even for some of the simplest functions. See Problem 25.

EXERCISES 5.4

Answers to odd-numbered problems begin on page A-81.

In Problems 1–12 find the length of the graph of the given equation on the indicated interval. If available, use a calculator or computer to obtain the graph.

1. $y = x$; $[-1, 1]$
2. $y = 2x + 1$; $[0, 3]$
3. $y = x^{3/2} + 4$; $(0, 4)$ to $(1, 5)$
4. $y = 3x^{2/3}$; $[1, 8]$
5. $y = \frac{2}{3}(x^2 + 1)^{3/2}$; $[1, 4]$
6. $(y + 1)^2 = 4(x + 1)^3$; $(-1, -1)$ to $(0, 1)$
7. $y = \frac{1}{3}x^{3/2} - x^{1/2}$; $[1, 4]$
8. $y = \dfrac{1}{6}x^3 + \dfrac{1}{2x}$; $[2, 4]$
9. $y = \dfrac{1}{4}x^4 + \dfrac{1}{8x^2}$; $[2, 3]$
10. $y = \dfrac{1}{5}x^5 + \dfrac{1}{12x^3}$; $[1, 2]$
11. $y = (4 - x^{2/3})^{3/2}$; $[1, 8]$
12. $y = \begin{cases} x - 2, & 2 \le x < 3 \\ (x - 2)^{2/3}, & 3 \le x < 10; \\ \frac{1}{2}(x - 6)^{3/2}, & 10 \le x \le 15 \end{cases}$ $[2, 15]$

In Problems 13–16 set up, but do not evaluate, an integral for the length of the graph of the given function on the indicated interval.

13. $y = x^2$; $[-1, 3]$
14. $y = 2\sqrt{x + 1}$; $[-1, 3]$
15. $y = \sin x$; $[0, \pi]$
16. $y = \tan x$; $[-\pi/4, \pi/4]$

Let $x = g(y)$, where g' is continuous on an interval $[c, d]$ on the y-axis. The length of the graph on the interval is then

$$s = \int_c^d \sqrt{1 + [g'(y)]^2}\, dy \qquad (6)$$

In Problems 17 and 18 use (6) to find the length of the graph of the given equation on the indicated interval.

17. $x = 4 - y^{2/3}$; $[0, 8]$

18. $5x = y^{5/2} + 5y^{-1/2}$; $[4, 9]$

19. Consider the length of the graph of $x^{2/3} + y^{2/3} = 1$ in the first quadrant.

(a) Show that the use of (3) leads to a discontinuous integrand.

(b) By assuming that the Fundamental Theorem of Calculus can be used to evaluate the integral obtained in part (**a**), find the total length of the graph.

20. Set up, but make no attempt to evaluate, an integral that gives the total length of the ellipse $x^2/a^2 + y^2/b^2 = 1$, $a > b > 0$.

21. Given that the circumference of a circle of radius r is $2\pi r$, find the value of the integral

$$\int_0^1 \frac{dx}{\sqrt{1 - x^2}}$$

22. Use (4) to approximate the length of the graph of $y = x^4/4$ from $(2, 4)$ to $(2.1, 4.862025)$.

23. (a) Use the distance formula to find the length of the graph of $y = mx + b$ on the interval $[x_1, x_2]$.

(b) Use (3) to verify your answer to part (**a**).

24. (a) Estimate the length of the graph given in Figure 5.54 on the interval $[1, 8]$.

(b) Explain why using the Trapezoidal Rule with $n = 7$ is not particularly a good idea.

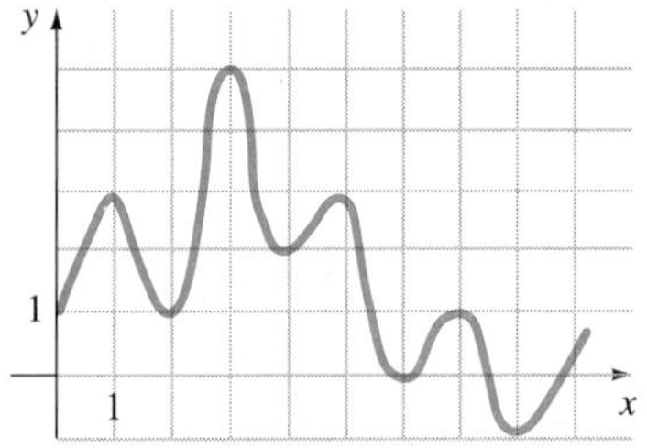

FIGURE 5.54

Calculator/Computer Problems

25. Use the Trapezoidal Rule with $n = 10$ to find an approximation to the length of the graph of $y = x^2$ on the interval $[0, 1]$.

26. Use Simpson's Rule with $n = 4$ to find an approximation to the length of the graph of $y = \frac{1}{3}x^3 + 1$ from $(0, 1)$ to $(2, \frac{11}{3})$.

27. Use Simpson's Rule with $n = 6$ to find an approximation to the length of the graph of $f(x) = \int_1^x (1/t)\, dt$ on the interval $[1, 2]$.

5.5 SURFACES OF REVOLUTION

Surface of Revolution—x-Axis As we have seen in Sections 5.3 and 5.4, when the graph of a continuous function $y = f(x)$ on an interval $[a, b]$ is revolved about the x-axis, it generates a solid of revolution. In this section, we are interested in finding the area of the corresponding surface—that is, a **surface of revolution** as shown in Figure 5.55(b).

To derive the formula for the area of a surface of revolution, we first note that the lateral area (top and bottom excluded) of a *frustum* of a right circular cone shown in Figure 5.56 is given by

$$\pi[r_1 + r_2]L \qquad (1)$$

where r_1 and r_2 are the radii of the top and bottom and L is the slant height. See Problem 18 in Exercises 5.5.

Now, suppose $y = f(x)$ is a smooth function and $f(x) \geq 0$ on the interval $[a, b]$. Let P be a partition of the interval:

$$a = x_0 < x_1 < x_2 < \cdots < x_{n-1} < x_n = b$$

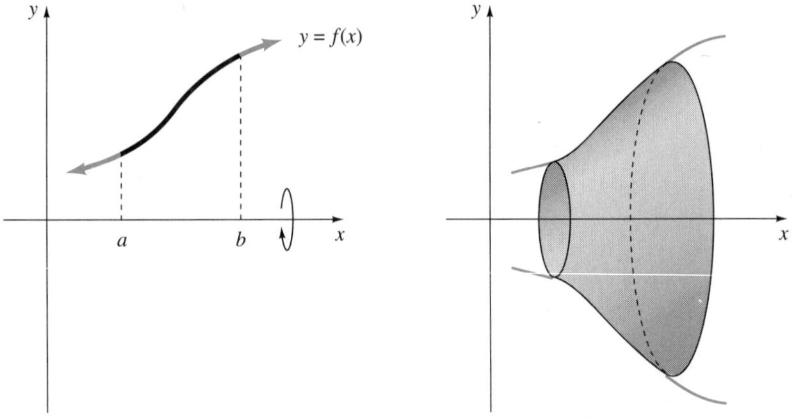

(a) graph (b) surface

FIGURE 5.55

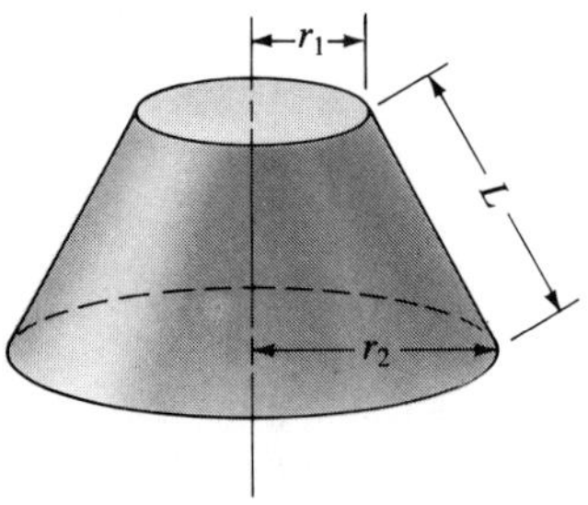

FIGURE 5.56

Now, if we connect the points $(x_{k-1}, f(x_{k-1}))$ and $(x_k, f(x_k))$ shown in Figure 5.57(a) by a chord, we form a trapezoid. When revolved about the x-axis, this trapezoid generates a frustum of a cone with radii $f(x_{k-1})$ and $f(x_k)$. See Figure 5.57(b).

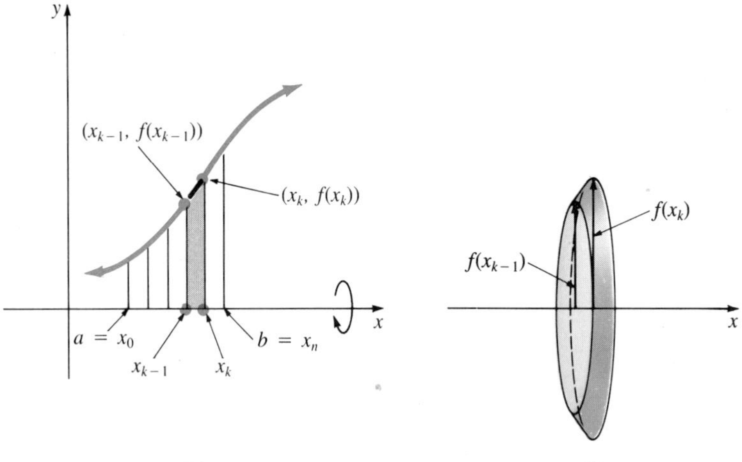

(a) (b)

FIGURE 5.57

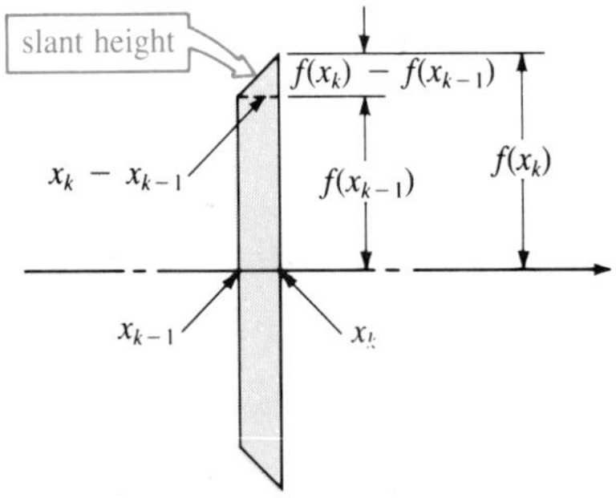

side view of the frustum

FIGURE 5.58

As shown in cross-section in Figure 5.58, the slant height can be obtained from the Pythagorean Theorem.

$$\sqrt{(x_k - x_{k-1})^2 + (f(x_k) - f(x_{k-1}))^2}$$

Thus, from (1) the surface area of this element is

$$\begin{aligned}
\Delta S_k &= \pi[f(x_k) + f(x_{k-1})]\sqrt{(x_k - x_{k-1})^2 + (f(x_k) - f(x_{k-1}))^2} \\
&= \pi[f(x_k) + f(x_{k-1})]\sqrt{1 + \left(\frac{f(x_k) - f(x_{k-1})}{x_k - x_{k-1}}\right)^2}(x_k - x_{k-1}) \\
&= \pi[f(x_k) + f(x_{k-1})]\sqrt{1 + \left(\frac{f(x_k) - f(x_{k-1})}{x_k - x_{k-1}}\right)^2}\,\Delta x_k
\end{aligned}$$

where $\Delta x_k = x_k - x_{k-1}$. This last quantity is an approximation to the actual area of the surface of revolution on the subinterval $[x_{k-1}, x_k]$.

Now, as in the discussion of arc length, we invoke the Mean Value Theorem for derivatives to assert that there exists an x_k^* in (x_{k-1}, x_k) such that

$$f'(x_k^*) = \frac{f(x_k) - f(x_{k-1})}{x_k - x_{k-1}}$$

This suggests that the surface area is given by

$$\lim_{\|P\| \to 0} \pi \sum_{k=1}^{n} [f(x_k) + f(x_{k-1})]\sqrt{1 + [f'(x_k^*)]^2}\, \Delta x_k$$

Since we also expect $f(x_{k-1})$ and $f(x_k)$ to approach the common limit $f(x)$ as $\|P\| \to 0$, the last equation becomes

$$2\pi \int_a^b f(x)\sqrt{1 + [f'(x)]^2}\, dx$$

The foregoing is a plausibility argument for the following definition.

DEFINITION 5.5 Area of a Surface of Revolution

Let f be a function for which f' is continuous and $f(x) \geq 0$ for all x in the interval $[a, b]$. The **area** S of the surface that is obtained by revolving the graph of f on the interval about the x-axis is given by

$$S = 2\pi \int_a^b f(x)\sqrt{1 + [f'(x)]^2}\, dx \tag{2}$$

EXAMPLE 1 Find the area S of the surface that is formed by revolving the graph of $y = \sqrt{x}$ on the interval $[1, 4]$ about the x-axis.

Solution We have $f(x) = x^{1/2}$, $f'(x) = \frac{1}{2}x^{-1/2} = 1/(2\sqrt{x})$, and from (2) of Definition 5.5,

$$S = 2\pi \int_1^4 \sqrt{x}\sqrt{1 + \left(\frac{1}{2\sqrt{x}}\right)^2}\, dx$$

$$= 2\pi \int_1^4 \sqrt{x}\sqrt{1 + \frac{1}{4x}}\, dx$$

$$= 2\pi \int_1^4 \sqrt{x}\sqrt{\frac{4x+1}{4x}}\, dx = \pi \int_1^4 \sqrt{4x+1}\, dx$$

$$= \frac{\pi}{4}\int_1^4 (4x+1)^{1/2}(4\, dx) = \frac{\pi}{6}(4x+1)^{3/2}\Big]_1^4$$

$$= \frac{\pi}{6}[17^{3/2} - 5^{3/2}] \approx 30.85 \text{ square units}$$

See Figure 5.59. □

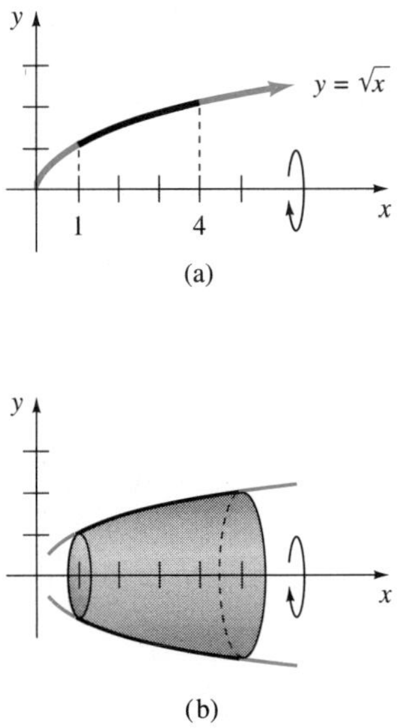

FIGURE 5.59

Surface of Revolution—y-Axis It can be shown that if the graph of a continuous function $y = f(x)$ on $[a, b]$, $0 \le a < b$, is revolved about the y-axis, then the area S of the resulting surface of revolution is given by

$$S = 2\pi \int_a^b x\sqrt{1 + [f'(x)]^2}\, dx \tag{3}$$

As in (2), we assume in (3) that $f'(x)$ is continuous on the interval $[a, b]$.

EXAMPLE 2 Find the area S of the surface formed by revolving the graph of $y = 3x^{1/3}$ on the interval $[1, 8]$ about the y-axis.

Solution We have $f'(x) = x^{-2/3}$, so that from (3) it follows that

$$\begin{aligned} S &= 2\pi \int_1^8 x\sqrt{1 + x^{-4/3}}\, dx \\ &= 2\pi \int_1^8 x^{1/3}\sqrt{x^{4/3} + 1}\, dx \\ &= 2\pi\left(\frac{3}{4}\right)\int_1^8 (x^{4/3} + 1)^{1/2}\left(\frac{4}{3}x^{1/3}\, dx\right) = \pi(x^{4/3} + 1)^{3/2}\Big]_1^8 \\ &= \pi[17^{3/2} - 2^{3/2}] \approx 211.32 \text{ square units} \end{aligned}$$

See Figure 5.60.

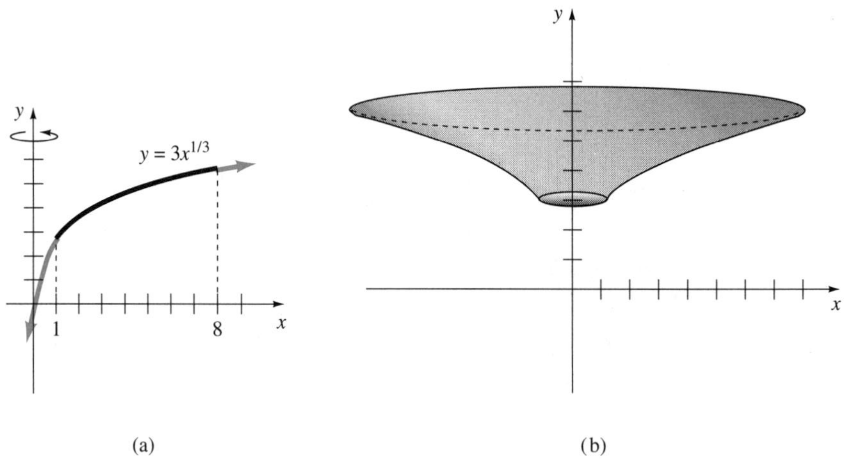

FIGURE 5.60

□

EXERCISES 5.5 *Answers to odd-numbered problems begin on page A-81.*

In Problems 1–10 find the area of the surface that is formed by revolving each graph on the given interval about the indicated axis.

1. $y = 2\sqrt{x}$, $[0, 8]$; x-axis
2. $y = \sqrt{x+1}$, $[1, 5]$; x-axis
3. $y = x^3$, $[0, 1]$; x-axis
4. $y = x^{1/3}$, $[1, 8]$; y-axis
5. $y = x^2 + 1$, $[0, 3]$; y-axis
6. $y = 4 - x^2$, $[0, 2]$; y-axis
7. $y = 2x + 1$, $[2, 7]$; x-axis
8. $y = \sqrt{16 - x^2}$, $[0, \sqrt{7}]$; y-axis
9. $y = \dfrac{1}{4}x^4 + \dfrac{1}{8x^2}$, $[1, 2]$; y-axis
10. $y = \dfrac{1}{3}x^3 + \dfrac{1}{4x}$, $[1, 2]$; x-axis

11. (a) The shape of a dish antenna is a parabola revolved about its axis called a **paraboloid of revolution**. Find the surface area of an antenna of radius r and depth h obtained by revolving the graph of $f(x) = r\sqrt{1 - x/h}$ about the x-axis. See Figure 5.61.

(b) The depth of a dish antenna ranges from 10% to 20% of its radius. If the depth h of the antenna in part (**a**) is 10% of the radius, show that the surface area of the antenna is approximately the same as the area of a circle of radius r. What is the percentage error in this case?

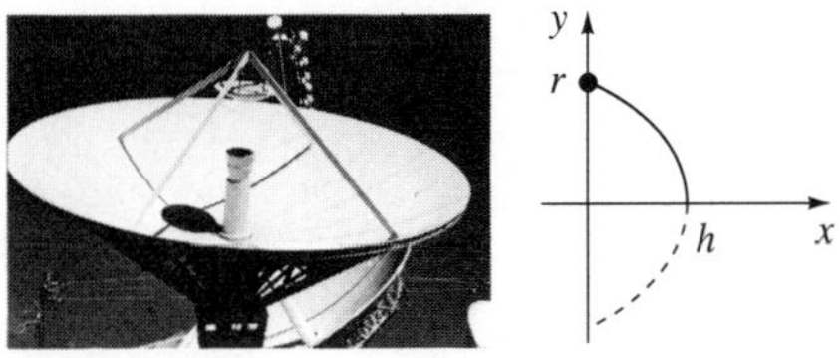

FIGURE 5.61 Dish antennas are paraboloids of revolution.

12. The surface formed by two parallel planes cutting a sphere of radius r is called a **spherical zone**. Find the area of the spherical zone shown in Figure 5.62.

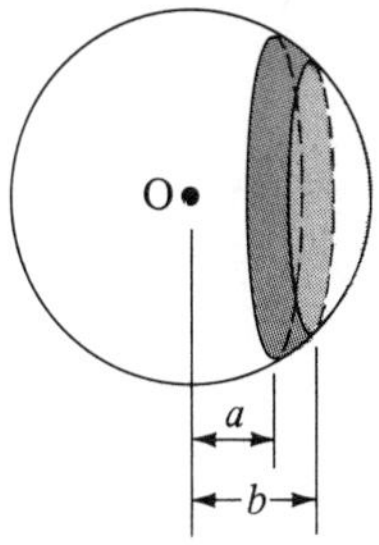

FIGURE 5.62

13. (*This problem could present a challenge.*) (a) From a spacecraft orbiting the earth at a distance h from the surface, an astronaut can observe only a portion A_s of the earth's total surface area A_e. See Figure 5.63(a). Find a formula for the fractional expression A_s/A_e as a function of h. In Figure 5.63(b) we have shown the earth in cross-section as a circle with center C and radius R. Let the x- and y-axes be as shown and let the y-coordinates of the points B and E be y_B and $y_E = R$, respectively.

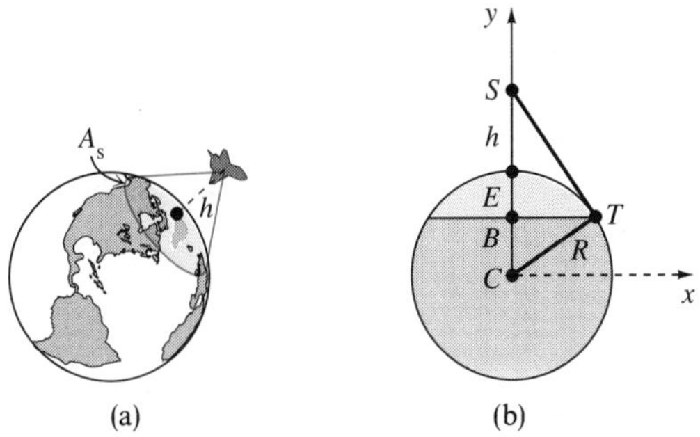

FIGURE 5.63

(b) What percentage of the earth's surface will an astronaut see from a height of 2000 km? Take the radius of the earth to be $R = 6380$ km.

(c) At what height h will an astronaut see one-fourth of the earth's surface?

(d) What is the limit of A_s/A_e as the height h increases without bound ($h \to \infty$)? Why does the answer make intuitive sense?

(e) What percentage of the earth's surface will an astronaut see from the moon if $h = 3.76 \times 10^5$ km?

14. The graph of $y = |x + 2|$ on $[-4, 2]$, shown in Figure 5.64, is revolved about the x-axis. Find the area S of the surface of revolution.

15. Find the area of the surface that is formed by revolving $x^{2/3} + y^{2/3} = a^{2/3}$, $[-a, a]$, about the x-axis.

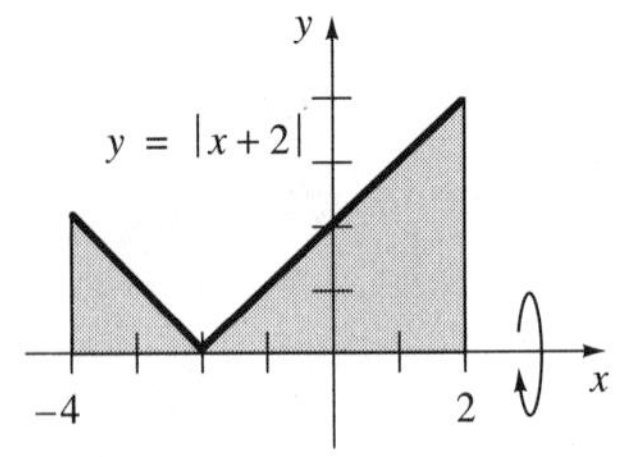

FIGURE 5.64

16. Show that the lateral surface area of a right circular cone of radius r and slant height L is πrL. [*Hint:* A cone cut down its side and flattened forms a circular sector with area $\frac{1}{2}L^2\theta$.]

17. Use Problem 16 to show that the lateral surface area of a right circular cone of radius r and height h is given by $\pi r\sqrt{r^2 + h^2}$. Derive the same result using (1) or (2).

18. Use the result of Problem 16 to derive formula (1). [*Hint:* Consider a complete cone of radius r_2 and slant height L_2. Cut the conical top off. Similar triangles might help.]

19. Show that the surface area of a frustum of a right circular cone of radii r_1 and r_2 and height h is given by $\pi(r_1 + r_2)\sqrt{h^2 + (r_2 - r_1)^2}$.

20. Let $y = f(x)$ be a continuous nonnegative function $[a, b]$ that has a continuous first derivative on the interval. Prove that if the graph of f is revolved around a horizontal line $y = L$, then the area S of the resulting surface of revolution is given by

$$S = 2\pi \int_a^b |f(x) - L|\sqrt{1 + [f'(x)]^2}\,dx$$

21. Use the result of Problem 20 to find a definite integral that gives the area of the surface that is formed by revolving $y = x^{2/3}$, $[1, 8]$, about the line $y = 4$. Do not evaluate.

Calculator/Computer Problems

22. Use the Midpoint Rule with $n = 5$ to find an approximation to the area of the surface that is formed by revolving the graph of $y = \frac{1}{2}x^2$ on the interval $[0, 2]$ about the x-axis.

23. Use Simpson's Rule with $n = 6$ to find an approximation to the area of the surface that is formed by revolving the graph of $x = y^2 + 1$ for $-1 \le y \le 1$ about the y-axis.

5.6 AVERAGE VALUE OF A FUNCTION AND THE MEAN VALUE THEOREM

Averages Every student is aware of averages. If a student takes four examinations in a semester and scores 80%, 75%, 85%, and 92% on them, then the student's average score is

$$\frac{80 + 75 + 85 + 92}{4}$$

or 83%. In general, given n numbers $a_1, a_2, \ldots, a_n$, we say that their **arithmetic mean**, or **average**, is

$$\frac{a_1 + a_2 + \cdots + a_n}{n} = \frac{1}{n}\sum_{k=1}^{n} a_k$$

Suppose now that we have a continuous function f defined on an interval $[a, b]$. For the arbitrarily chosen numbers x_i, $i = 1, 2, \ldots, n$ such that $a < x_1 < x_2 < \cdots < x_n < b$, the average of the set of corresponding functional values is

$$\frac{f(x_1) + f(x_2) + \cdots + f(x_n)}{n} = \frac{1}{n}\sum_{k=1}^{n} f(x_k) \tag{1}$$

If we now consider the set of values $f(x)$ that corresponds to *all* points x in an interval, it should be clear that we cannot use a discrete sum as in (1). For

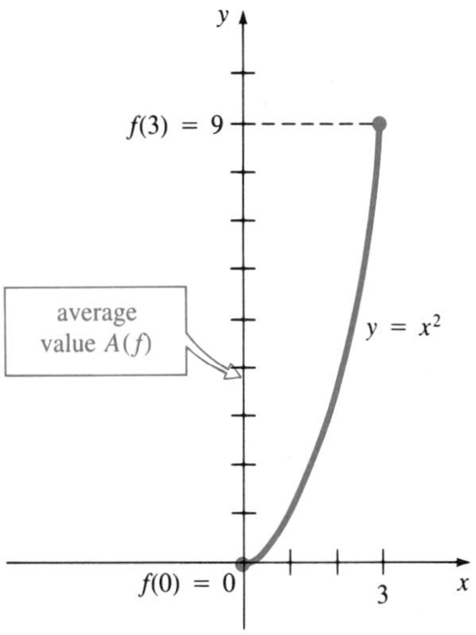

FIGURE 5.65

example, for $f(x) = x^2$ on $[0, 3]$, the values of the function range from a minimum of $f(0) = 0$ to a maximum of $f(3) = 9$. As indicated in Figure 5.65, we intuitively expect that there exists an average value $A(f)$ such that $f(0) \le A(f) \le f(3)$.

Returning to the general case of a continuous function defined on a closed interval $[a, b]$, we let P be a partition of the interval into n subintervals of equal length $\Delta x = (b - a)/n$. If x_k^* is a number chosen in each subinterval, then the average

$$\frac{f(x_1^*) + f(x_2^*) + \cdots + f(x_n^*)}{n}$$

can be written as

$$\frac{f(x_1^*) + f(x_2^*) + \cdots + f(x_n^*)}{\dfrac{b-a}{\Delta x}} \tag{2}$$

since $n = (b - a)/\Delta x$. Rewriting (2) as

$$\frac{1}{b-a}\sum_{k=1}^{n} f(x_k^*)\,\Delta x$$

and taking the limit of this last expression as $\|P\| = \Delta x \to 0$, we obtain the definite integral

$$\frac{1}{b-a}\int_a^b f(x)\,dx \tag{3}$$

Since we have assumed that f is continuous on $[a, b]$, let us denote its minimum and maximum on the interval by m and M, respectively. If we multiply

$$m \le f(x_k^*) \le M$$

by $\Delta x > 0$ and sum, we obtain

$$\sum_{k=1}^{n} m\,\Delta x \le \sum_{k=1}^{n} f(x_k^*)\,\Delta x \le \sum_{k=1}^{n} M\,\Delta x$$

Since $\Sigma_{k=1}^{n}\,\Delta x = b - a$, this last expression can be written as

$$(b-a)m \le \sum_{k=1}^{n} f(x_k^*)\,\Delta x \le (b-a)M$$

And so as $\Delta x \to 0$, it follows that

$$(b-a)m \le \int_a^b f(x)\,dx \le (b-a)M$$

We conclude that the number from (3) satisfies

$$m \le \frac{1}{b-a}\int_a^b f(x) \le M$$

By the Intermediate Value Theorem, f takes on all values between m and M. Hence, the number given by (3) actually corresponds to a value of the function on the interval. This prompts us to state the following definition:

DEFINITION 5.6 Average Value of a Function

Let f be continuous on $[a, b]$. The **average value** of f on the interval is the number

$$A(f) = \frac{1}{b-a}\int_a^b f(x)\,dx \tag{4}$$

Although we are interested primarily in continuous functions, Definition 5.6 is valid for any integrable function on the interval.

EXAMPLE 1 Find the average value of $f(x) = x^2$ on $[0, 3]$.

Solution From (4) of Definition 5.6, we obtain

$$A(f) = \frac{1}{3-0}\int_0^3 x^2\,dx = \frac{1}{3}\left(\frac{x^3}{3}\right)\Bigg]_0^3 = 3$$ □

It is sometimes possible to determine the value of x in the interval that corresponds to the average value of a function.

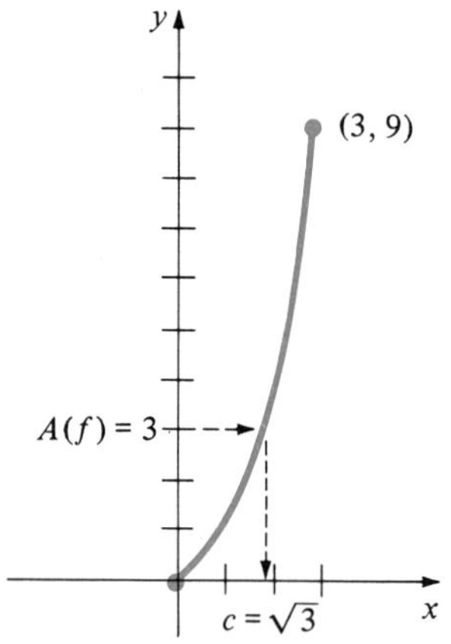

FIGURE 5.66

EXAMPLE 2 Determine the value of x in the interval $[0, 3]$ that corresponds to the average value of the function $f(x) = x^2$.

Solution Since the function $f(x) = x^2$ is continuous on the closed interval $[0, 3]$, we know from Theorem 1.13 that there exists a number c between 0 and 3 so that

$$f(c) = c^2 = A(f)$$

But, from Example 1, we know $A(f) = 3$. Thus, the equation

$$c^2 = 3 \quad \text{has solutions} \quad c = \pm\sqrt{3}$$

As shown in Figure 5.66, the only solution of this equation in $[0, 3]$ is $c = \sqrt{3}$. □

Mean Value Theorem for Definite Integrals The following is an immediate consequence of the foregoing discussion.

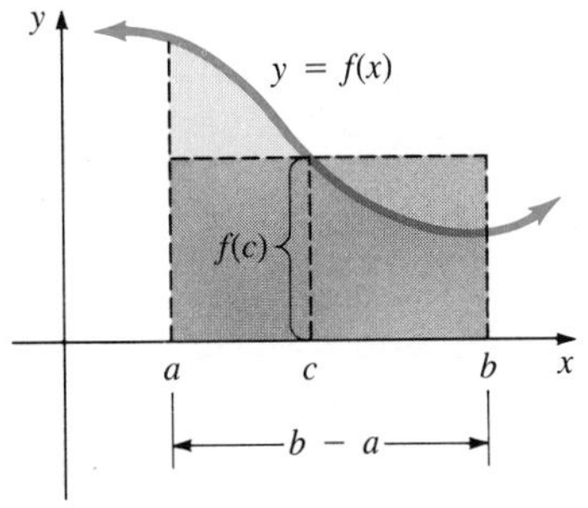

FIGURE 5.67

THEOREM 5.1 The Mean Value Theorem for Definite Integrals

Let f be continuous on $[a, b]$. Then there exists a number c in the open interval (a, b) such that

$$f(c)(b-a) = \int_a^b f(x)\,dx$$

If $f(x) \geq 0$ for all x in $[a, b]$, then Theorem 5.1 simply states that there is some value c in (a, b) for which the area of a rectangle of height $f(c)$ and width $b - a$ is the same as the area under the graph indicated in Figure 5.67.

EXAMPLE 3 Find the height $f(c)$ of a rectangle so that the area A under the graph of $y = x^2 + 1$ on $[-2, 2]$ is the same as $f(c)[2 - (-2)] = 4f(c)$.

Solution This is basically the same type of problem as illustrated in Example 2. Now, the area under the graph is

$$A = \int_{-2}^{2} (x^2 + 1)\,dx = \left(\frac{x^3}{3} + x\right)\Bigg]_{-2}^{2} = \frac{28}{3}$$

Also, $4f(c) = 4(c^2 + 1)$, so that $4(c^2 + 1) = \frac{28}{3}$ implies $c^2 = \frac{4}{3}$. The solutions $c = 2/\sqrt{3}$ and $c = -2/\sqrt{3}$ are both in the interval $(-2, 2)$. The height of the rectangle is $f(c) = (\pm 2/\sqrt{3})^2 + 1 = \frac{7}{3}$. See Figure 5.68.

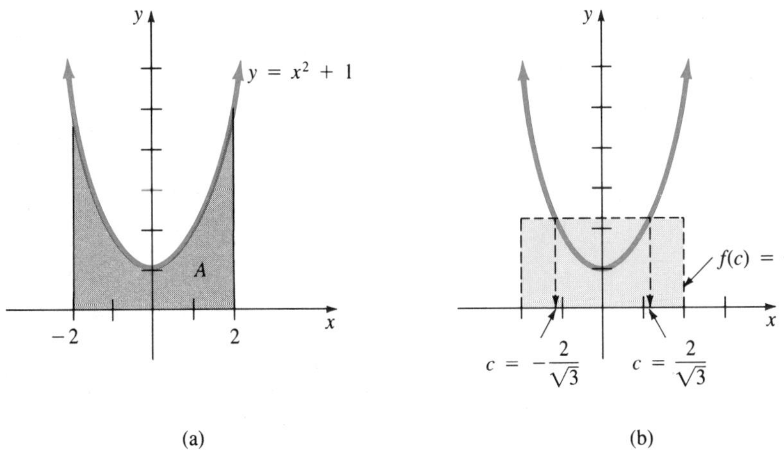

FIGURE 5.68 □

EXERCISES 5.6

Answers to odd-numbered problems begin on page A-82.

In Problems 1–20 find the average value $A(f)$ of the given function on the indicated interval.

1. $f(x) = 4x$; $[-3, 1]$
2. $f(x) = 2x + 3$; $[-2, 5]$
3. $f(x) = x^2 + 10$; $[0, 2]$
4. $f(x) = 2x^3 - 3x^2 + 4x - 1$; $[-1, 1]$
5. $f(x) = 3x^2 - 4x$; $[-1, 3]$
6. $f(x) = (x + 1)^2$; $[0, 2]$
7. $f(x) = x^3$; $[-2, 2]$
8. $f(x) = x(3x - 1)^2$; $[0, 1]$
9. $f(x) = \sqrt{x}$; $[0, 9]$
10. $f(x) = \sqrt{5x + 1}$; $[0, 3]$
11. $f(x) = x\sqrt{x^2 + 16}$; $[0, 3]$
12. $f(x) = \left(1 + \dfrac{1}{x}\right)^{1/3} \dfrac{1}{x^2}$; $\left[\dfrac{1}{2}, 1\right]$
13. $f(x) = \dfrac{1}{x^3}$; $\left[\dfrac{1}{4}, \dfrac{1}{2}\right]$
14. $f(x) = x^{2/3} - x^{-2/3}$; $[1, 4]$
15. $f(x) = \dfrac{2}{(x + 1)^2}$; $[3, 5]$
16. $f(x) = \dfrac{(\sqrt{x} - 1)^3}{\sqrt{x}}$; $[4, 9]$
17. $f(x) = \sin x$; $[-\pi, \pi]$
18. $f(x) = \cos 2x$; $\left[0, \dfrac{\pi}{4}\right]$
19. $f(x) = \csc^2 x$; $\left[\dfrac{\pi}{6}, \dfrac{\pi}{2}\right]$
20. $f(x) = \dfrac{\sin \pi x}{\cos^2 \pi x}$; $\left[-\dfrac{1}{3}, \dfrac{1}{3}\right]$

21. For $f(x) = x^2 + 2x$, find a value of c in the interval $[-1, 1]$ for which $f(c) = A(f)$.

22. For $f(x) = \sqrt{x + 3}$, find a value of c in the interval $[1, 6]$ for which $f(c) = A(f)$.

23. The average value of a continuous nonnegative function $y = f(x)$ on the interval $[1, 5]$ is $A(f) = 3$. What is the area under the graph on the interval?

24. For $f(x) = 1 - \sqrt{x}$, find a value of b so that $A(f) = 0$ on $[0, b]$. Interpret geometrically.

25. The function $T(t) = 100 + 3t - \frac{1}{2}t^2$ approximates the temperature at t hours past noon on a typical August day in Las Vegas. Find the average temperature between noon and 6 P.M.

26. A company determines that the revenue obtained after the sale of x units of a product is given by $R(x) = 50 + 4x + 3x^2$. Find the average revenue for sales $x = 1$ to $x = 5$. Compare the result with the average $\frac{1}{5}\sum_{k=1}^{5} R(k)$.

27. Let $s(t)$ denote the position of a particle on a horizontal axis as a function of time t. The average velocity v_{ave} during the time interval $[t_1, t_2]$ is $v_{\text{ave}} = [s(t_2) - s(t_1)]/(t_2 - t_1)$. Show that $v_{\text{ave}} = A(v)$.

28. In the absence of damping, the position of a mass m on a freely vibrating spring is given by the function $x(t) = A\cos(\omega t + \phi)$, where A, ω, and ϕ are constants. The period of oscillation is $2\pi/\omega$. The potential energy of the system is $U(x) = \frac{1}{2}kx^2$, where k is the so-called spring constant. The kinetic energy of the system is $K = \frac{1}{2}mv^2$, where $v = dx/dt$. If $\omega^2 = k/m$, show that the average potential energy and the average kinetic energy over one period are the same and that each equals $\frac{1}{4}kA^2$.

29. The *impulse–momentum theorem* states that the change in momentum of a body in a time interval $[t_0, t_1]$ is $mv_1 - mv_0 = (t_1 - t_0)\bar{F}$, where mv_0 is the initial momentum, mv_1 is the final momentum, and $\bar{F}$ is the average force acting on the body during the interval. Find the change in momentum of a pile driver dropped on a piling between times $t = 0$ and $t = t_1$ if

$$F(t) = k\left[1 - \left(\frac{2t}{t_1} - 1\right)^2\right]$$

where k is a constant.

30. In a small artery the velocity of blood (in cm/s) is given by $v(r) = (P/4vl)(R^2 - r^2)$, $0 \le r \le R$, where P is blood pressure, v the viscosity of the blood, l the length of the artery, and R the radius of the artery. Find the average of $v(r)$ on the interval $[0, R]$.

31. Figure 5.69 shows a car's velocity in miles per hour as a function of miles traveled. Interpret the graph. Find the

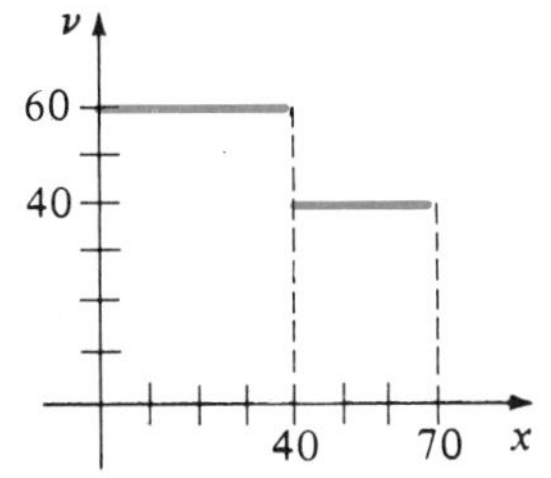

FIGURE 5.69

average velocity of the car with respect to distance. What is the average velocity of the car with respect to time?

32. Show that the average value of the greatest integer function $f(x) = [x]$ on the interval $[0, n]$, where n is a positive integer, is $\frac{1}{2}(n-1)$.

33. If $y = f(x)$ is a differentiable function, find the average value of f' on the interval $[x, x+h]$, where $h > 0$.

34. For a linear function $f(x) = ax + b$, $a > 0$, $b > 0$, show that the average value of the function on $[x_1, x_2]$ is $A(f) = aX + b$, where X is the x-coordinate of the midpoint of the interval.

35. Given that n is a positive integer and $a > 1$, show that the average value of $f(x) = (n+1)x^n$ on the interval $[1, a]$ is $A(f) = a^n + a^{n-1} + \cdots + a + 1$.

36. The **mean square** of a continuous function on an interval $[a, b]$ is defined as

$$\frac{1}{b-a}\int_a^b [f(x)]^2\,dx$$

Compute the average value and the mean square of $f(x) = (x+1)^{-1/3}$ on $[0, 7]$.

37. (*This problem could present a challenge.*) The following formula is often used to approximate the surface area S of a human limb:

$$S \approx \text{average circumference} \times \text{length of limb}$$

(a) As shown in Figure 5.70, a limb can be considered to be a solid of revolution. For many limbs, $f'(x)$ is small. If $|f'(x)| \le \varepsilon$ for $a \le x \le b$, show that

$$\int_a^b 2\pi f(x)\,dx \le S \le \sqrt{1+\varepsilon^2}\int_a^b 2\pi f(x)\,dx$$

(b) Show that $\bar{C}L \le S \le \sqrt{1+\varepsilon^2}\,\bar{C}L$, where $\bar{C}$ is the average circumference of the limb over the interval $[a, b]$. Thus, the approximation formula stated above always underestimates S but does well when ε is small (such as for the forearm to wrist).

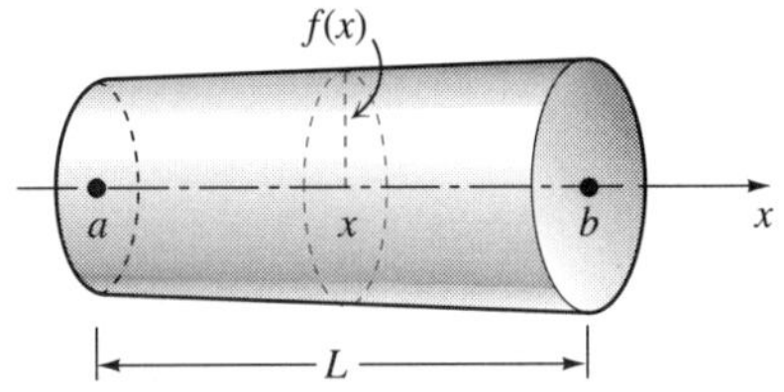

FIGURE 5.70

5.7 RECTILINEAR MOTION AND THE INTEGRAL

If $s = f(t)$ is the position function of an object moving rectilinearly—that is, in a straight line, then we know

$$\text{velocity} = v(t) = \frac{ds}{dt} \quad \text{and} \quad \text{acceleration} = a(t) = \frac{dv}{dt}$$

As an immediate consequence of the definition of an antiderivative, the quantities s and v can be written as indefinite integrals

$$s(t) = \int v(t)\,dt \quad \text{and} \quad v(t) = \int a(t)\,dt \tag{1}$$

By knowing the **initial position** $s(0)$ and the **initial velocity** $v(0)$, we can find specific values of the constants of integration used in (1).

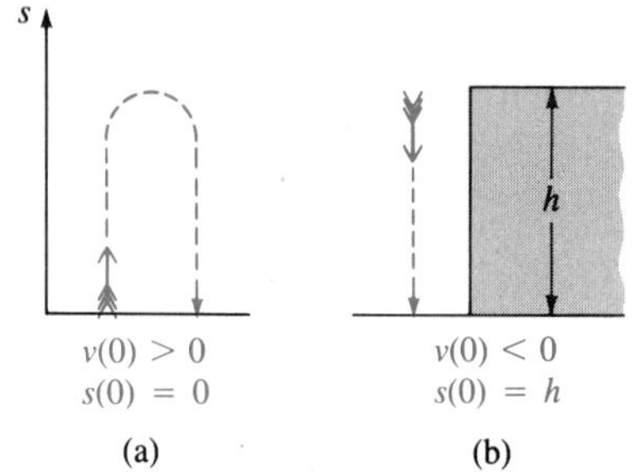

FIGURE 5.71

Recall that when a body moves horizontally on a line, the positive direction is to the right. For motion in a vertical line, the positive direction is upward. As shown in Figure 5.71, if an arrow is shot upward from ground level, then $s(0) = 0$, $v(0) > 0$, whereas if the arrow is shot downward from some initial height, say h meters off the ground, then $s(0) = h$, $v(0) < 0$. A body that moves in a vertical line close to the surface of the earth, such as the arrow shot upward, is acted upon by the force of gravity. This force causes a body to accelerate. Near the surface of the earth the acceleration due to gravity, $a(t) = -g$, is assumed to be a constant. The magnitude g of this acceleration is approximately

$$32 \text{ ft/s}^2, \quad 9.8 \text{ m/s}^2, \quad \text{or} \quad 980 \text{ cm/s}^2$$

EXAMPLE 1 A projectile is shot vertically upward from ground level with an initial velocity of 49 m/s. What is its velocity at $t = 2$ s? What is the maximum height attained by the projectile? How long is the projectile in the air? What is its impact velocity?

Solution We use $a(t) = -9.8$ and

$$v(t) = \int (-9.8)\, dt = -9.8t + C_1 \tag{2}$$

From the given initial condition $v(0) = 49$, we see that (2) implies $C_1 = 49$. Hence,

$$v(t) = -9.8t + 49$$

and so $v(2) = -9.8(2) + 49 = 29.4$ m/s. Notice that $v(2) > 0$ implies the projectile is traveling upward.

Now, the height of the projectile, measured from ground level, is

$$s(t) = \int v(t)\, dt = \int (-9.8t + 49)\, dt = -4.9t^2 + 49t + C_2 \tag{3}$$

Since the projectile starts from ground level, $s(0) = 0$ and (3) gives $C_2 = 0$. Hence,

$$s(t) = -4.9t^2 + 49t \tag{4}$$

When the projectile attains its maximum height, $v(t) = 0$. Solving $-9.8t + 49 = 0$ then gives $t = 5$. From (4) we find the corresponding height to be $s(5) = 122.5$ m.

Finally, to find the time that the projectile hits the ground, we solve $s(t) = 0$ or $-4.9t^2 + 49t = 0$. Writing the latter equation as $-4.9t(t - 10) = 0$, we see the projectile is in the air for 10 s. The impact velocity is $v(10) = -49$ m/s*. □

---

*When air resistance is ignored, the magnitude of the impact velocity (speed) is the same as the initial upward velocity from ground level. See Problem 26. This is not true when air resistance is taken into consideration.

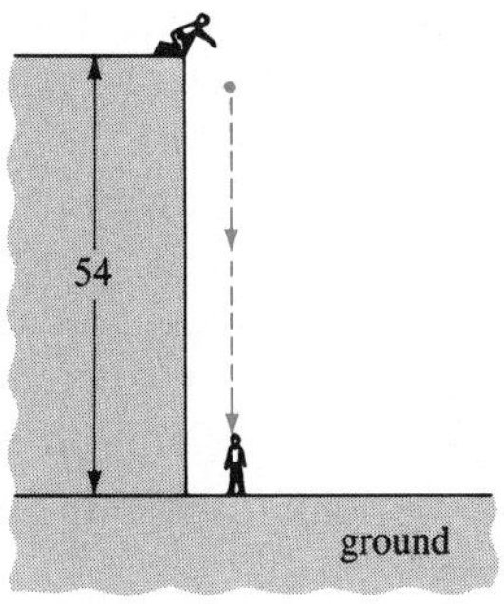

FIGURE 5.72

EXAMPLE 2 A tennis ball is thrown vertically downward from a height of 54 ft with an initial velocity of 8 ft/s. What is its impact velocity if it hits a 6-ft-tall person on the head? See Figure 5.72.

Solution In this case $a(t) = -32$, $s(0) = 54$, and, since the ball is thrown downward, $v(0) = -8$. Now

$$v(t) = \int (-32)\, dt = -32t + C_1$$

Using the initial velocity $v(0) = -8$, we find $C_1 = -8$. Therefore,

$$v(t) = -32t - 8$$

Continuing, we find

$$s(t) = \int (-32t - 8)\, dt = -16t^2 - 8t + C_2$$

When $t = 0$, we know $s = 54$ and so the last equation implies $C_2 = 54$. Hence,

$$s(t) = -16t^2 - 8t + 54$$

To determine the time that corresponds to $s = 6$ we solve

$$-16t^2 - 8t + 54 = 6$$

Simplifying gives $-8(2t - 3)(t + 2) = 0$ and $t = \frac{3}{2}$. The velocity of the ball when it hits the person is then $v(\frac{3}{2}) = -56$ ft/s. □

Distance The total **distance** an object travels in a straight line in a time interval $[t_1, t_2]$ is given by the definite integral

$$\int_{t_1}^{t_2} |v(t)|\, dt \tag{5}$$

The absolute value is necessary in (5), since the object may be moving to the left and hence has negative velocity for some part of the time.

EXAMPLE 3 The position function of an object that moves on a coordinate line is $s(t) = t^2 - 6t$, where s is measured in centimeters and t in seconds. Find the distance traveled in the time interval [0, 9].

Solution The velocity function $v(t) = ds/dt = 2t - 6 = 2(t - 3)$ shows that the motion is as indicated in Figure 5.73; namely, $v < 0$ for $0 \le t < 3$ and $v \ge 0$ for $3 \le t \le 9$. Hence, from (5) the distance traveled is

$$\begin{aligned}\int_0^9 |2t - 6|\, dt &= \int_0^3 |2t - 6|\, dt + \int_3^9 |2t - 6|\, dt \\ &= \int_0^3 -(2t - 6)\, dt + \int_3^9 (2t - 6)\, dt \\ &= (-t^2 + 6t)\Big]_0^3 + (t^2 - 6t)\Big]_3^9 = 45 \text{ cm}\end{aligned}$$

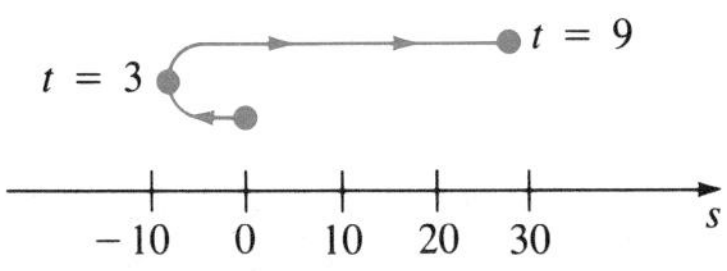

FIGURE 5.73

Of course, the last result must check with the number obtained by simply counting units in Figure 5.73 between $s(0)$ and $s(3)$, and between $s(3)$ and $s(9)$. □

EXERCISES 5.7

Answers to odd-numbered problems begin on page A-82.

In Problems 1–6 a body moves in a straight line with velocity $v(t)$. Find the position function $s(t)$.

1. $v(t) = 6$; $s = 5$ when $t = 2$
2. $v(t) = 2t + 1$; $s = 0$ when $t = 1$
3. $v(t) = t^2 - 4t$; $s = 6$ when $t = 3$
4. $v(t) = \sqrt{4t + 5}$; $s = 2$ when $t = 1$
5. $v(t) = -10\cos(4t + \pi/6)$; $s = \frac{5}{4}$ when $t = 0$
6. $v(t) = 2\sin 3t$; $s = 0$ when $t = \pi$

In Problems 7–12 a body moves in a straight line with acceleration $a(t)$. Find $v(t)$ and $s(t)$.

7. $a(t) = -5$; $v = 4$ and $s = 2$ when $t = 1$
8. $a(t) = 6t$; $v = 0$ and $s = -5$ when $t = 2$
9. $a(t) = 3t^2 - 4t + 5$; $v = -3$ and $s = 10$ when $t = 0$
10. $a(t) = (t - 1)^2$; $v = 4$ and $s = 6$ when $t = 1$
11. $a(t) = 7t^{1/3} - 1$; $v = 50$ and $s = 0$ when $t = 8$
12. $a(t) = 100\cos 5t$; $v = -20$ and $s = 15$ when $t = \pi/2$

13. A driver of a car that is traveling at a constant 88 km/h takes his eyes off the road for 2 s. How far does the car move in this time?

14. A ball is dropped (released from rest) from a height of 144 ft. How long does it take for the ball to hit the ground? At what speed does it hit the ground?

15. An egg is dropped from the top of a building and hits the ground 4 s from release. How tall is the building?

16. A stone is dropped into a well and the splash is heard 2 s later. If the speed of sound in air is 1080 ft/s, find the depth of the well.

17. An arrow is projected vertically upward from ground level with an initial velocity of 24.5 m/s. How high does it rise?

18. How high would the arrow in Problem 17 rise on the planet Mars where $g = 3.6$ m/s^2?

19. A golf ball is thrown vertically upward from the roof of a 384-ft-high building with an initial velocity of 32 ft/s. What is the maximum height attained by the ball? At what time does the ball hit the ground?

20. In Problem 19 what is the velocity of the golf ball as it passes an observer in a window that is 256 ft off the ground?

21. A person throws a marshmallow vertically downward with an initial velocity of 16 ft/s from a window that is 102 ft off the ground. If the marshmallow hits a 6-ft-tall person on the head, what is the impact velocity?

22. The person hit on the head in Problem 21 climbs to the top of a 22-ft-high ladder and throws a stone vertically upward with an initial velocity of 96 ft/s. If the stone hits the culprit at the 102-ft level, what is the impact velocity?

23. In March 1979, the Voyager 1 space probe photographed an active volcanic eruption on Io, one of the moons of Jupiter. See Figure 5.74. Find the ejection velocity of a rock from the volcano Loki if the rock attains a height of 200 km above the summit of the volcano. On Io $g = 1.8$ m/s^2.

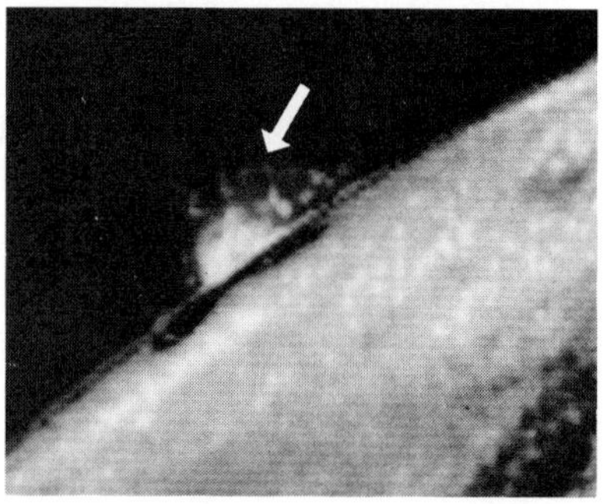

FIGURE 5.74 Eruption of the volcano Loki on Io, a moon of Jupiter, photographed from space. Photo courtesy of NASA.

24. As shown in Figure 5.75, at a point 30 ft from a 25-ft-tall street lamp a ball is thrown vertically downward from a height of 25 ft with an initial velocity of 2 ft/s.

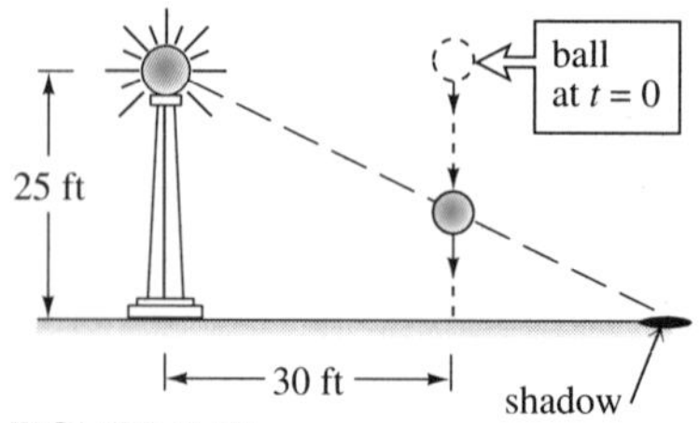

FIGURE 5.75

(a) Find the rate at which the shadow of the ball is moving toward the base of the street lamp.

(b) Find the rate at which the shadow of the ball is moving toward the base of the street lamp at $t = \frac{1}{2}$.

25. If a body is moving rectilinearly with a constant acceleration a and $v = v_0$ when $s = 0$, show that

$$v^2 = v_0^2 + 2as \qquad \left[\textit{Hint: } \frac{dv}{dt} = \frac{dv}{ds}\frac{ds}{dt} = \frac{dv}{ds}v\right]$$

26. Show that, when air resistance is ignored, a projectile shot vertically upward from ground level hits the ground again with a speed equal to the initial velocity v_0.

27. Suppose the acceleration of gravity on a planet is one-half that on the earth. Prove that a ball tossed vertically upward from the surface of the planet would attain a maximum height twice that on the earth when the same initial velocity is used.

28. In Problem 27 suppose the initial velocity of the ball on the planet is v_0 and the initial velocity of the ball on the earth is $2v_0$. Compare the maximum heights attained. Determine the initial velocity of the ball on the earth (in terms of v_0) so that the maximum height attained is the same as on the planet.

In Problems 29–34 an object moves in a straight line according to the given position function. If s is measured in centimeters, find the total distance traveled by the object in the indicated time interval.

29. $s(t) = t^2 - 2t$; $[0, 5]$

30. $s(t) = -t^2 + 4t + 7$; $[0, 6]$

31. $s(t) = t^3 - 3t^2 - 9t$; $[0, 4]$

32. $s(t) = t^4 - 32t^2$; $[1, 5]$

33. $s(t) = 6 \sin \pi t$; $[1, 3]$

34. $s(t) = (t - 3)^2$; $[2, 7]$

5.8 WORK

In physics, when a *constant* force F moves an object a distance d in the same direction of the force, the **work** done is defined to be the product

$$W = Fd \tag{1}$$

For example, if a 10-lb force moves an object 7 ft in the same direction as the force, then the work done is 70 ft-lb.

Units Commonly used **units** are listed in the following table.

Quantity	Engineering system	SI	cgs
Force	pound (lb)	newton (N)	dyne
Distance	foot (ft)	meter (m)	centimeter (cm)
Work	foot-pound (ft-lb)	newton-meter (joule)	dyne-centimeter (erg)

Thus, if a force of 300 N moves an object 15 m, the work done is 4500 N-m or 4500 joules. For comparison, we note that

$$1 \text{ N} = 10^5 \text{ dynes} = 0.2247 \text{ lb}$$
$$1 \text{ ft-lb} = 1.356 \text{ joules} = 1.356 \times 10^7 \text{ ergs}$$

Hence, 70 ft-lb is equivalent to 94.92 joules, and 4500 joules is equivalent to 3318.58 ft-lb.

Now, if $F(x)$ is a continuous *variable* force acting across an interval $[a, b]$, then the work is not simply a product as in (1). Suppose P is the partition

$$a = x_0 < x_1 < x_2 < \cdots < x_n = b$$

and Δx_k is the length of the kth subinterval $[x_{k-1}, x_k]$. Let x_k^* be chosen arbitrarily in each subinterval. If the numbers Δx_k are small, we can consider the force that acts over each subinterval as constant. Hence, the work done from x_{k-1} to x_k is given by the approximation

$$\Delta W_k = F(x_k^*)\,\Delta x_k$$

Thus, an approximation to the work done from a to b is

$$F(x_1^*)\,\Delta x_1 + F(x_2^*)\,\Delta x_2 + \cdots + F(x_n^*)\,\Delta x_n = \sum_{k=1}^{n} F(x_k^*)\,\Delta x_k = \sum_{k=1}^{n} \Delta W_k$$

It is natural to assume that the exact work done by F over the interval is

$$W = \lim_{\|P\| \to 0} \sum_{k=1}^{n} F(x_k^*)\,\Delta x_k$$

We summarize the foregoing discussion with the following definition:

DEFINITION 5.7 Work

Let F be continuous on the interval $[a, b]$ and let $F(x)$ represent force at a value x in the interval. Then the **work** W done by the force in moving an object from a to b is

$$W = \int_a^b F(x)\,dx \tag{2}$$

Note: If F is constant, $F(x) = k$ for all x in the interval, then (2) becomes $W = \int_a^b k\,dx = kx]_a^b = k(b - a)$, which is consistent with (1).

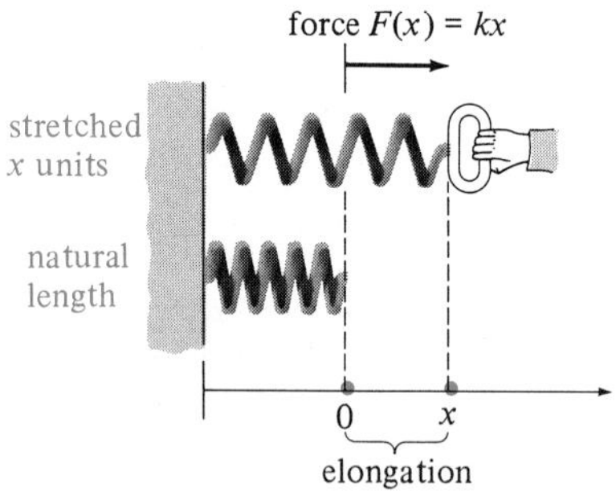

FIGURE 5.76

Spring Problems Hooke's Law* states that, when a spring is stretched (or compressed) beyond its natural length, the restoring force exerted by the spring is directly proportional to the amount of elongation (or compression) x. Thus, in order to stretch a spring x units beyond its natural length, we need to apply the force

$$F(x) = kx \tag{3}$$

where k is a constant of proportionality called the **spring constant**. See Figure 5.76.

EXAMPLE 1 It takes a force of 130 N to stretch a spring 50 cm. Find the work done in stretching the spring 20 cm beyond its natural (unstretched) length.

*Published by the English physicist Robert Hooke (1635–1703) in 1678.

Solution When a force is measured in newtons, distances are commonly expressed in meters. Since $x = 50$ cm $= \frac{1}{2}$ m when $F = 130$ N, (3) becomes $130 = k(\frac{1}{2})$, which implies the spring constant is $k = 260$ N/m. Thus, $F = 260x$. Now, 20 cm $= \frac{1}{5}$ m, so that the work done in stretching the spring by this amount is

$$W = \int_0^{1/5} 260x\, dx = 130x^2 \Big]_0^{1/5} = \frac{26}{5} = 5.2 \text{ joules} \qquad \square$$

Note: Suppose the natural length of the spring in Example 1 is 40 cm. An equivalent way of stating the problem is: Find the work done in stretching the spring to a length of 60 cm. Since the elongation is $60 - 40 = 20$ cm $= \frac{1}{5}$ m, we still integrate $F = 260x$ on the interval $[0, \frac{1}{5}]$. However, if the problem were to find the work done in stretching the same spring from 50 cm to 60 cm, we would then integrate on the interval $[\frac{1}{10}, \frac{1}{5}]$. In this situation we are starting from a position where the spring is already stretched 10 cm ($\frac{1}{10}$ m).

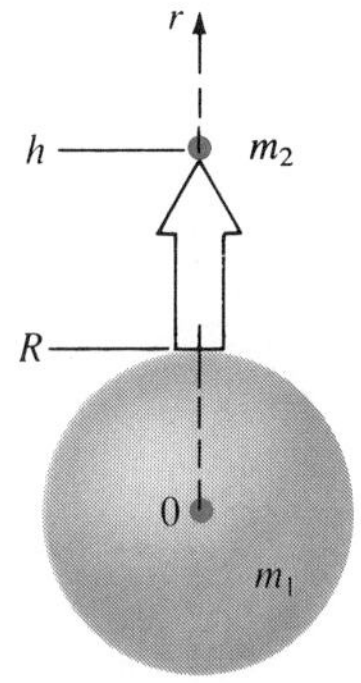

FIGURE 5.77

Work Done Against Gravity From the Universal Law of Gravitation, the force between a planet (or moon) of mass m_1 and a body of mass m_2 is given by

$$F = k\frac{m_1 m_2}{r^2} \tag{4}$$

where k is a constant, called the **gravitational constant**, and r is the distance from the center of the planet to mass m_2. See Figure 5.77. In lifting the mass m_2 off the surface of a planet of radius R to a height h, the work done can be obtained by using (4) in (2):

$$W = \int_R^{R+h} \frac{km_1m_2}{r^2}\, dr = km_1m_2\left(-\frac{1}{r}\right)\Big]_R^{R+h} = km_1m_2\left(\frac{1}{R} - \frac{1}{R+h}\right) \tag{5}$$

In SI units $k = 6.67 \times 10^{-11}$ N $\cdot$ m^2/kg^2. Some masses and values of R are given in the accompanying table.

	m_1 (in kg)	R (in m)
Venus	4.9×10^{24}	6.2×10^6
Earth	6.0×10^{24}	6.4×10^6
Moon	7.3×10^{22}	1.7×10^6
Mars	6.4×10^{23}	3.3×10^6

EXAMPLE 2 The work done in lifting a 5000-kg payload from the surface of the earth to a height of 30,000 m (0.03×10^6 m) follows from (5) and the preceding table:

$$W = (6.67 \times 10^{-11})(6.0 \times 10^{24})(5000)\left(\frac{1}{6.4 \times 10^6} - \frac{1}{6.43 \times 10^6}\right)$$

$$\approx 1.46 \times 10^9 \text{ joules} \qquad \square$$

Pump Problems When a liquid that weighs ρ lb/ft^3 is pumped from a tank, the work done in moving a fixed volume or layer of liquid d ft in a vertical direction is

$$W = (\text{force}) \cdot (\text{distance}) = (\text{weight per unit volume}) \cdot (\text{volume}) \cdot (\text{distance})$$
$$= \rho \cdot (\text{volume}) \cdot d \qquad (6)$$

In physics the quantity ρ is called the **weight density** of the fluid. For water, $\rho = 62.4$ lb/ft^3 or 9800 N/m^3.

EXAMPLE 3 A hemispherical tank of radius 20 ft is filled with water to a 15-ft depth. Find the work done in pumping all the water to the top of the tank.

Solution As shown in Figure 5.78, we let the positive x-axis be directed *downward* and let the origin be at the center-top of the tank. Since the cross-section of the tank is a semicircle, x and y are related by $x^2 + y^2 = (20)^2$, $0 \le x \le 20$. Now suppose the subinterval $[5, 20]$ is partitioned into n subintervals $[x_{k-1}, x_k]$ of length Δx_k. Let x_k^* be any point in the kth subinterval and let ΔW_k denote an approximation to the work done by the pump in lifting a layer of water of thickness Δx_k to the top of the tank. It follows from (6) that

$$\Delta W_k = \underbrace{(62.4\pi(y_k^*)^2\,\Delta x_k)}_{\text{force}} \cdot \underbrace{x_k^*}_{\text{distance}}$$

where $(y_k^*)^2 = 400 - (x_k^*)^2$. Hence, the work done by the pump is approximately

$$\sum_{k=1}^{n} \Delta W_k = \sum_{k=1}^{n} 62.4\pi[400 - (x_k^*)^2]x_k^*\,\Delta x_k$$

The work done in pumping all the water to the top of the tank is the limit of this last expression as $\|P\| \to 0$; that is,

$$W = \int_5^{20} 62.4\pi(400 - x^2)x\,dx = 62.4\pi\left(200x^2 - \frac{x^4}{4}\right)\Big]_5^{20} \approx 6{,}891{,}869 \text{ ft-lb}$$

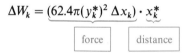

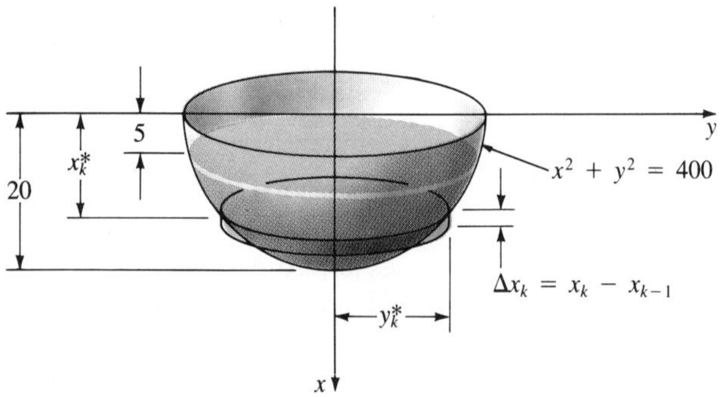

FIGURE 5.78

□

It is worth pursuing the analysis of Example 3 for the case where the positive x-axis is taken in the *upward* direction and the origin is at the center-bottom of the tank.

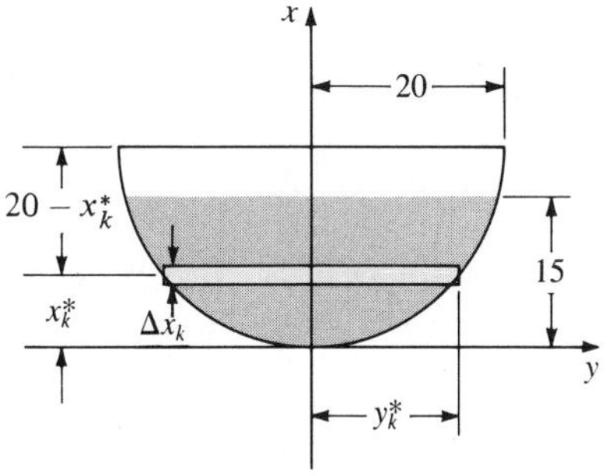

FIGURE 5.79

EXAMPLE 4 ***Alternative Solution to Example 3*** With the axes as shown in Figure 5.79, we see that a layer of water must be lifted a distance of $20 - x_k^*$ ft. But, since the center of the semicircle is at (20, 0), x and y are now related by $(x - 20)^2 + y^2 = 400$. Hence,

$$\Delta W_k = \underbrace{(62.4\pi(y_k^*)^2\,\Delta x_k)}_{\text{force}} \cdot \underbrace{(20 - x_k^*)}_{\text{distance}}$$

$$= 62.4\pi[400 - (x - 20)^2](20 - x_k^*)\,\Delta x_k$$

$$W = 62.4\pi \int_0^{15} [400 - (x - 20)^2](20 - x)\,dx$$

$$= 62.4\pi \int_0^{15} (x^3 - 60x^2 + 800x)\,dx$$

Note the new limits of integration. You should verify that the value of W in this case is the same as in Example 3. □

EXAMPLE 5 In Example 3, find the work done in pumping all the water to a point 10 ft above the hemispherical tank.

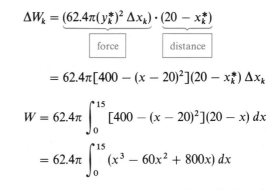

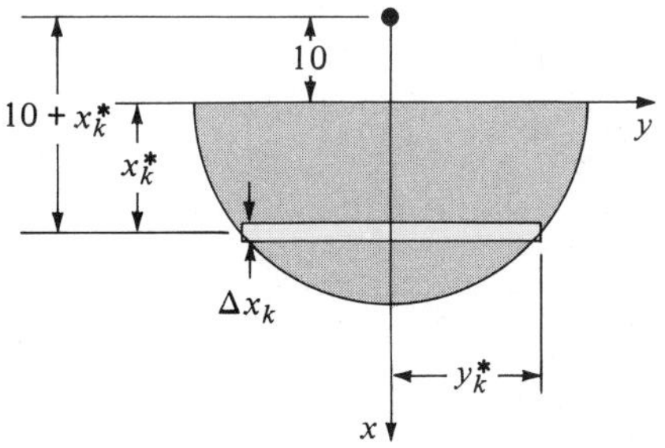

FIGURE 5.80

Solution As in Figure 5.78, we position the positive x-axis downward. Now, from Figure 5.80 we see

$$\Delta W_k = (62.4\pi(y_k^*)^2\,\Delta x_k) \cdot (10 + x_k^*)$$
$$= 62.4\pi[400 - (x_k^*)^2](10 + x_k^*)\,\Delta x_k$$

Hence,

$$W = 62.4\pi \int_5^{20} (400 - x^2)(10 + x)\,dx$$

$$= 62.4\pi \int_5^{20} (-x^3 - 10x^2 + 400x + 4000)\,dx$$

$$= 62.4\pi\left(-\frac{x^4}{4} - \frac{10}{3}x^3 + 200x^2 + 4000x\right)\Bigg]_5^{20}$$

$$= 13{,}508{,}063 \text{ ft-lb}$$ □

Cable Problems The next example illustrates the fact that when you are calculating the work done in lifting an object by means of a cable (heavy rope or chain), the weight of the cable must be taken into consideration.

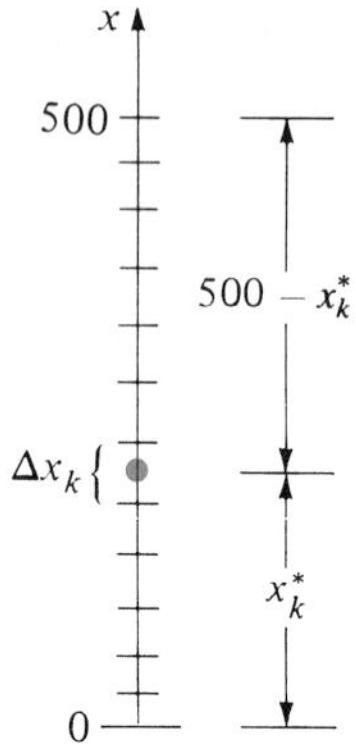

FIGURE 5.81

EXAMPLE 6 A cable weighing 6 lb/ft is connected to a construction elevator weighing 1500 lb. Find the work done in lifting the elevator to a height of 500 ft.

Solution The work done in lifting the elevator a distance of 500 ft is simply

$$W_e = (1500) \cdot (500) = 750{,}000 \text{ ft-lb}$$

Now let W_c denote the work done in lifting the cable. As shown in Figure 5.81, suppose the positive x-axis is directed upward and the interval $[0, 500]$ is partitioned into n subintervals with lengths Δx_k. At a height of x_k^* ft off the ground, a segment of cable corresponding to the subinterval $[x_{k-1}, x_k]$ weighs $6\,\Delta x_k$ and must be pulled up an additional $500 - x_k^*$ ft. Hence, we can write

$$(\Delta W_c)_k = \underbrace{(6\,\Delta x_k)}_{\text{force}} \cdot \underbrace{(500 - x_k^*)}_{\text{distance}} = (3000 - 6x_k^*)\,\Delta x_k$$

and so

$$W_c = \int_0^{500} (3000 - 6x)\,dx = (3000x - 3x^2)\Big]_0^{500} = 750{,}000 \text{ ft-lb}$$

Thus, the total work done in lifting the elevator is

$$W = W_e + W_c = 1{,}500{,}000 \text{ ft-lb}$$

□

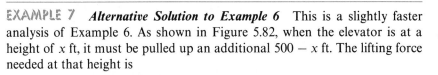

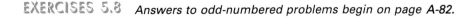

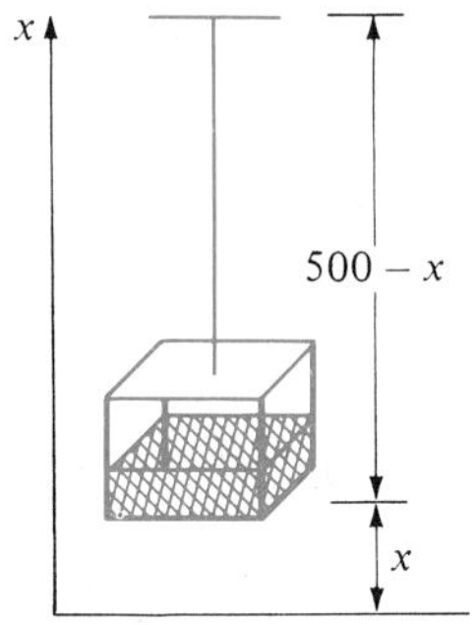

FIGURE 5.82

EXAMPLE 7 ***Alternative Solution to Example 6*** This is a slightly faster analysis of Example 6. As shown in Figure 5.82, when the elevator is at a height of x ft, it must be pulled up an additional $500 - x$ ft. The lifting force needed at that height is

$$\underbrace{1500}_{\text{weight of elevator}} + \underbrace{6(500 - x)}_{\text{weight of cable}} = 4500 - 6x$$

Thus, by (2) the work done is

$$W = \int_0^{500} (4500 - 6x)\,dx = 1{,}500{,}000 \text{ ft-lb}$$

□

EXERCISES 5.8 *Answers to odd-numbered problems begin on page A-82.*

1. Find the work done when a 55-lb force moves an object 20 yd in the same direction of the force.

2. A force of 100 N is applied to an object at an angle of 30° measured from the horizontal. If the object moves 8 m horizontally, find the work done by the force.

3. In 1977 George Willig, who weighed 165 lb, scaled the

outside of the World Trade Center building in New York City to a height of 1350 ft in 3.5 h at a rate of 6.4 ft/min. How much work did George do?

4. A person pushes against an immovable wall with a horizontal force of 75 lb. How much work is done?

5. A mass weighing 10 lb stretches a spring $\frac{1}{2}$ ft. How much will a mass weighing 8 lb stretch the same spring?

6. A spring has a natural length of 0.5 m. A force of 50 N stretches the spring to a length of 0.6 m.

(a) What force is needed to stretch the spring x meters?

(b) What force is required to stretch the spring to a length of 1 m?

(c) How long is the spring when stretched by a force of 200 N?

7. In Problem 6:

(a) Find the work done in stretching the spring 0.2 m.

(b) Find the work done in stretching the spring from a length of 1 m to a length of 1.1 m.

8. A force of $F = \frac{3}{2}x$ lb is needed to stretch a 10-in spring an additional x in.

(a) Find the work done in stretching the spring to a length of 16 in.

(b) Find the work done in stretching the spring 16 in.

9. A mass weighing 10 lb is suspended from a 2-ft spring. The spring is stretched 8 in and then the mass is removed.

(a) Find the work done in stretching the spring to a length of 3 ft.

(b) Find the work done in stretching the spring from a length of 4 ft to a length of 5 ft.

10. A 50-lb force compresses a 15-in-long spring by 3 in. Find the work done in compressing the spring to a final length of 5 in.

11. Find the work done in lifting a mass of 10,000 kg from the surface of the earth to a height of 500 km.

12. Find the work done in lifting a mass of 50,000 kg from the surface of the moon to a height of 200 km.

13. A right cylindrical tank of height 12 ft and radius 3 ft is filled with water. Find the work done in pumping all the water to the top of the tank.

14. A tank in the form of a right circular cone, vertex down, is filled with water to a depth of one-half its height. If the height of the tank is 20 ft and its diameter is 8 ft, find the work done in pumping all the water to the top of the tank. [*Hint:* Assume that the origin is at the vertex of the cone.]

15. For the tank in Problem 14, find the work done in pumping all the water to a point 5 ft above the top of the tank.

16. A horizontal trough with semicircular cross-sections contains oil whose weight density is 80 lb/ft^3. The dimensions of the tank (in feet) are shown in Figure 5.83. If the depth of the oil is 3 ft, find the work done in pumping all the oil to the top of the tank.

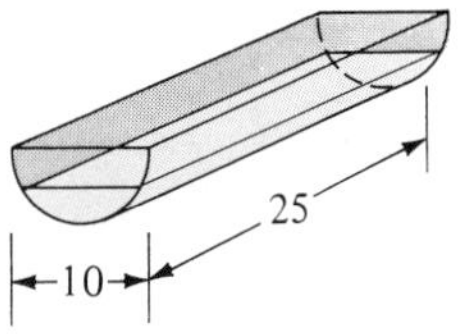

FIGURE 5.83

17. A tank has cross-sections in the form of isosceles triangles, vertex down. The top of the tank is 6 ft wide, its height is 4 ft, and its length is 10 ft. Find the work done in filling the tank with water through a hole in its bottom by a pump located 5 ft below its vertex.

18. A 100-ft anchor chain, weighing 20 lb/ft, is hanging vertically over the side of a boat. How much work is performed by pulling in 40 ft of the chain?

19. A ship is anchored in 200 ft of water. In water the ship's anchor weighs 3000 lb and its anchor chain weighs 40 lb/ft. If the anchor chain hangs vertically, how much work is done in pulling in 100 ft of the chain?

20. A bucket of sand weighing 80 lb is lifted vertically by means of a rope and pulley to a height of 65 ft. Find the work done if **(a)** the weight of the rope is negligible and **(b)** the rope weighs $\frac{1}{2}$ lb/ft.

21. A bucket, initially containing 20 ft^3 of water, is lifted vertically from ground level. If the water leaks out at a rate of $\frac{1}{2}$ ft^3 per vertical foot, find the work done in lifting the bucket to a height at which it is empty.

22. The force of attraction between an electron and the nucleus of an atom is inversely proportional to the square of the distance separating them. If the initial distance between nucleus and electron is 1 unit, find the work done by an external force that moves the electron out to a distance four times the initial distance.

23. A rocket weighing 2,500,000 lb when fueled carries a 200,000-lb shuttle orbiter. Assume, in the early stages of the launch, that the rocket burns fuel at a rate of 100 lb/ft.

(a) Express the total weight of the system in terms of its altitude above the surface of the earth. See Figure 5.84.

(b) Find the work done in lifting the system to an altitude of 1000 ft.

24. In thermodynamics, if a gas enclosed in a cylinder expands against a piston so that the volume of the gas changes

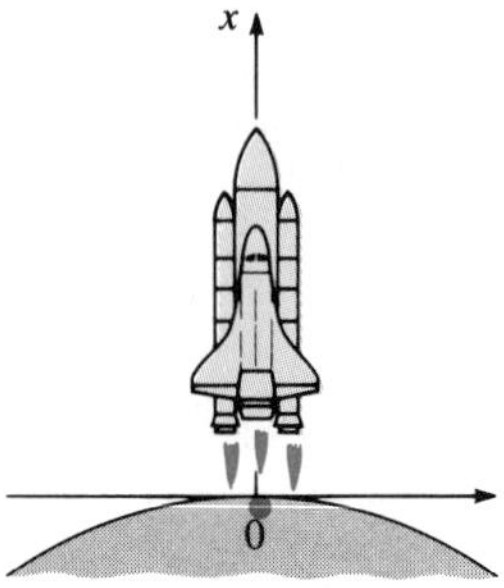

FIGURE 5.84

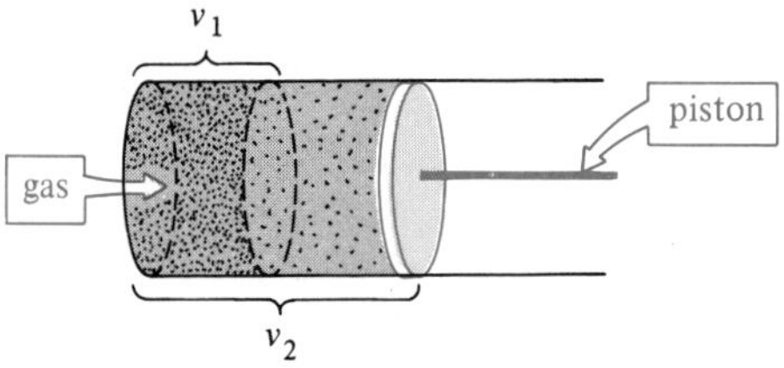

FIGURE 5.85

from v_1 to v_2, then the work done on the piston is given by $W = \int_{v_1}^{v_2} p\, dv$, where p is pressure (force per unit area). See Figure 5.85. In an adiabatic expansion of an ideal gas,* pressure and volume are related by $pv^{\gamma} = k$, where γ and k are constants. Show that if $\gamma \neq 1$, then

$$W = \frac{p_2 v_2 - p_1 v_1}{1 - \gamma}$$

25. As shown in Figure 5.86, a body of mass m is moved by a horizontal force F on a frictionless surface from a position at x_1 to a position at x_2. At these respective points, the body is moving at velocities v_1 and v_2, where $v_2 > v_1$. Show that the work done by the force is the increase in kinetic energy

$$W = \frac{1}{2}mv_2^2 - \frac{1}{2}mv_1^2$$

[*Hint:* Use Newton's second law $F = ma$, and express the acceleration a in terms of velocity v.]

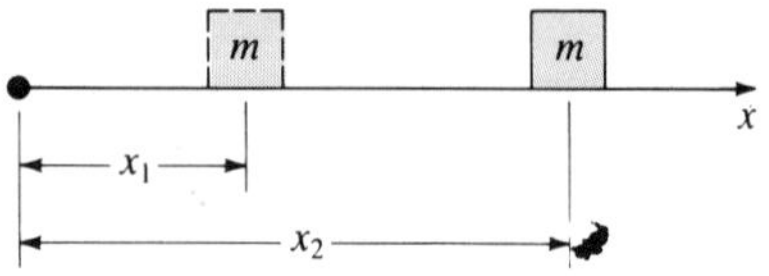

FIGURE 5.86

*This means that no heat is entering or leaving the system.

26. Prove that when a body of weight mg is lifted vertically from a point y_1 to a point y_2, $y_2 > y_1$, the work done is the change in potential energy $W = mgy_2 - mgy_1$.

27. A continuous variable force $F(x)$ acts over the interval $[0, 1]$, where F is measured in newtons and x in meters. It is determined that

x (m)	0	0.2	0.4	0.6	0.8	1
$F(x)$ (N)	0	50	90	150	210	260

Use an appropriate numerical technique to approximate the work done over the interval.

28. The graph of a variable force F is given in Figure 5.87. Using rectangular elements of area, find an approximation to the work done by the force in moving a particle from $x = 1$ to $x = 5$. Then use the Trapezoidal Rule.

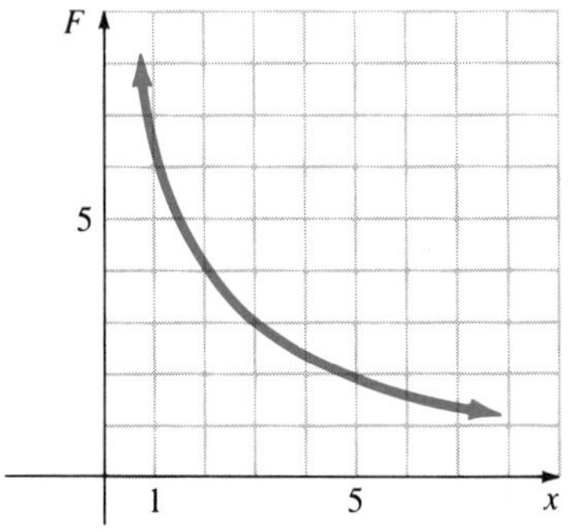

FIGURE 5.87

29. The graph of a variable force F is given in Figure 5.88. Find the work done by the force in moving a particle from $x = 0$ to $x = 6$.

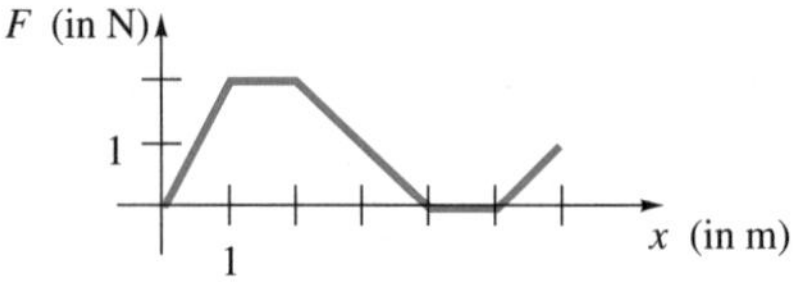

FIGURE 5.88

30. As shown in Figure 5.89, a bucket containing concrete that is suspended from a cable is pushed horizontally from the vertical by a construction worker. The length of the cable is 30 m and the combined mass m of the bucket and concrete is 550 kg. From the principles of physics it can be shown that the force required to move the bucket x meters is given by $F = mg \tan\theta$, where g is the acceleration of gravity. Find

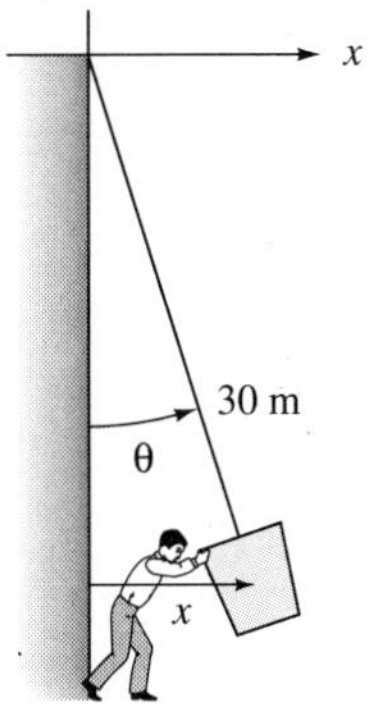

FIGURE 5.89

the work done by the construction worker in pushing the bucket a horizontal distance of 3 m. [*Hint:* Use (2) and a substitution.]

31. (a) Use (5) to show that the work required to lift a mass m_2 to a point an "infinite distance" from the surface of the planet is $W = km_1m_2/R$.

(b) If the mass m_2 is imparted a velocity v_0 at the surface of the planet to enable it to attain an "infinite distance" from its surface, then we must have

$$\frac{1}{2}m_2v_0^2 = k\frac{m_1m_2}{R}$$

Use this relation to find the "escape velocity" v_0 of any planet. Use the data in the table on page 357 to find v_0 for Earth and for Mars.

5.9 LIQUID PRESSURE AND FORCE

When a *horizontal* flat plate is submerged below the surface of a liquid, the **force** exerted on the plate by the liquid above it is given by

$$\begin{aligned} F &= (\text{pressure of liquid}) \cdot (\text{area of surface}) \\ &= (\text{force per unit area}) \cdot (\text{area of surface}) \end{aligned} \qquad (1)$$

If ρ is the weight density of the liquid (weight per unit volume) and A is the area of the horizontal plate submerged to a depth h, shown in Figure 5.90(a), then (1) is the same as

$$F = \underbrace{(\text{weight per unit volume}) \cdot (\text{depth})}_{\text{pressure}} \cdot (\text{area of surface}) = \rho h A \qquad (2)$$

However, when a *vertical* flat plate is submerged, the pressure of the liquid varies with the depth. See Figure 5.90(b). For example, the liquid pressure on a vertical dam is less at the top than at its base.

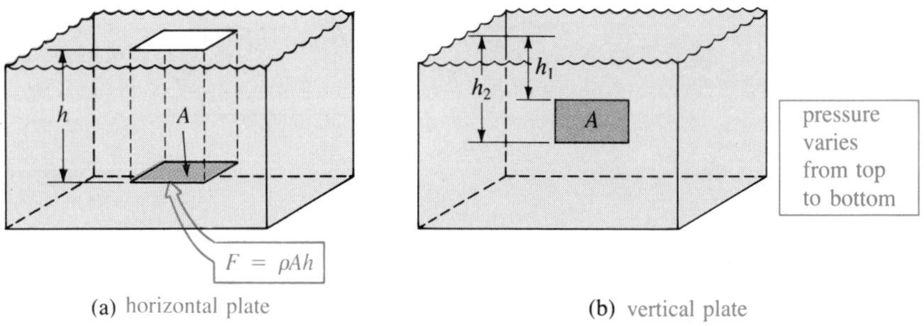

FIGURE 5.90

EXAMPLE 1 A flat rectangular plate with dimensions 5 ft × 6 ft is submerged horizontally in water at a depth of 10 ft. Determine the pressure and the force exerted on the plate by the water above it.

Solution Recall that the weight density for water is 62.4 lb/ft³. Hence,

$$\text{pressure} = \rho h = (62.4) \cdot 10 = 624 \text{ lb/ft}^2$$

Since the surface area of the plate is 30 ft², it follows from (2) that the force on the plate is

$$F = 624 \cdot 30 = 18{,}720 \text{ lb}$$ □

To determine the total force F exerted by a liquid on one side of a vertically submerged flat surface, we employ one form of Pascal's principle:*

The pressure exerted by a liquid at a depth h is the same in all directions.

Thus, if a large container with a flat bottom and vertical sidewalls is filled with water to a depth of 10 ft, the pressure of 624 lb/ft² at its bottom applies equally to the sidewalls.

Now, suppose a vertical flat plate, bounded by the horizontal lines $x = a$ and $x = b$ and the graphs of $y = f(x)$ and $y = g(x)$, is submerged in a liquid as shown in Figure 5.91. Let $f(x) - g(x)$ denote the width of the plate at any point x in $[a, b]$ and let P be any partition of the interval. If x_k^* is any point in the kth subinterval $[x_{k-1}, x_k]$, we conclude from (2) that the force exerted by the liquid on the corresponding rectangular element is approximated by

$$\Delta F_k = \rho \cdot x_k^* \cdot [f(x_k^*) - g(x_k^*)]\,\Delta x_k$$

where, as before, ρ denotes the weight density of the liquid. Thus, an approximation to the force on one side of the plate is

$$\sum_{k=1}^{n} \Delta F_k = \sum_{k=1}^{n} \rho x_k^*[f(x_k^*) - g(x_k^*)]\,\Delta x_k$$

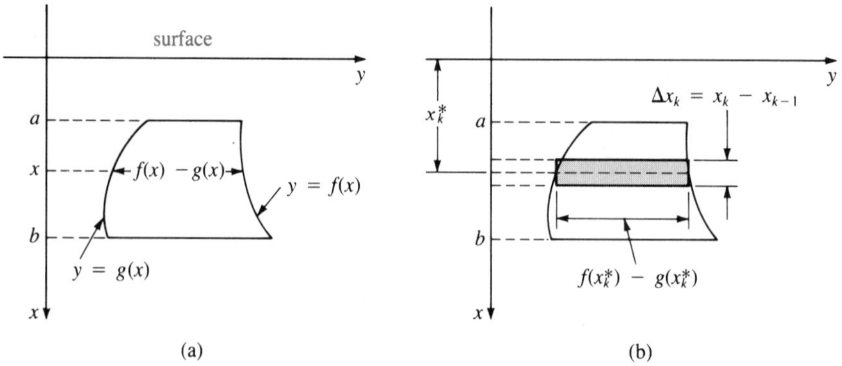

FIGURE 5.91

*Blaise Pascal (1623–1662), a French mathematician, scientist, and philosopher.

This suggests that the total force is

$$F = \lim_{\|P\| \to 0} \sum_{k=1}^{n} \rho x_k^*[f(x_k^*) - g(x_k^*)]\, \Delta x_k$$

DEFINITION 5.8 Force Exerted by a Liquid

Let ρ be the weight density of a liquid and let f and g be functions continuous on $[a, b]$. The **force** F exerted by the liquid on one side of the submerged plate shown in Figure 5.91 is

$$F = \int_a^b \rho x[f(x) - g(x)]\, dx$$

EXAMPLE 2 A plate in the shape of an isosceles triangle 3 ft high and 4 ft wide is submerged vertically, base downward, with the base 5 ft below the surface of the water. Find the force exerted by the water on one side of the plate.

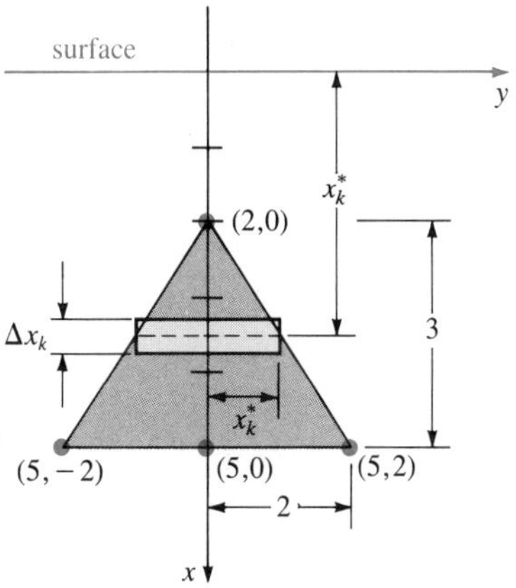

FIGURE 5.92

Solution For convenience, we place the positive x-axis along the axis of symmetry of the triangular plate with the origin at the surface of the water. As indicated in Figure 5.92, we partition the interval [2, 5] into n subintervals $[x_{k-1}, x_k]$ and choose a point x_k^* in each subinterval. Since the equation of the straight line that contains points (2, 0) and (5, 2) is

$$y = \frac{2}{3}x - \frac{4}{3}$$

we conclude by symmetry that the width of the rectangular element shown in Figure 5.92 is

$$2y_k^* = 2\left(\frac{2}{3}x_k^* - \frac{4}{3}\right)$$

Now $\rho = 62.4\ \text{lb/ft}^3$ so that the force on that portion of the plate that corresponds to the kth subinterval is approximated by

$$\Delta F_k = (62.4) \cdot x_k^* \cdot 2\left(\frac{2}{3}x_k^* - \frac{4}{3}\right) \Delta x_k$$

Forming the sum $\Sigma_{k=1}^{n} \Delta F_k$ and passing to the limit as $\|P\| \to 0$ give

$$\begin{aligned} F &= \int_2^5 (62.4)2x\left(\frac{2}{3}x - \frac{4}{3}\right) dx \\ &= 124.8 \int_2^5 \left(\frac{2}{3}x^2 - \frac{4}{3}x\right) dx = 124.8\left(\frac{2}{9}x^3 - \frac{2}{3}x^2\right)\Bigg]_2^5 \\ &= 124.8\left(\frac{108}{9}\right) \approx 1497.6 \text{ lb} \end{aligned}$$

□

EXERCISES 5.9 *Answers to odd-numbered problems begin on page A-82.*

1. Consider the tanks with flat circular bottoms shown in Figure 5.93. Each tank is full of water whose weight density is 9800 N/m^3. Find the pressure and force exerted by the water on the bottom of each tank.

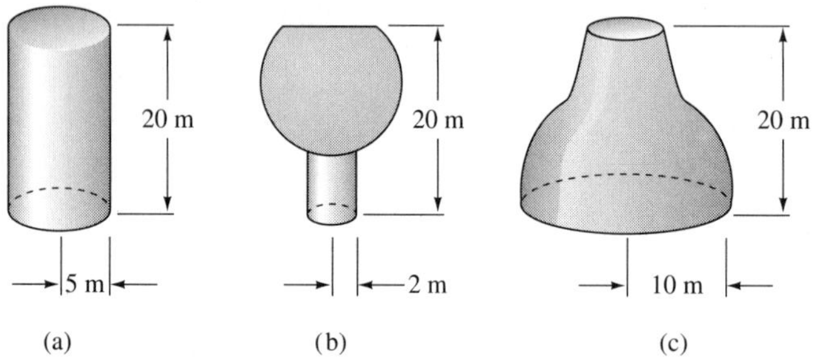

FIGURE 5.93

2. The tanker shown in Figure 5.94 has a flat bottom and is filled with oil whose weight density is 55 lb/ft^3. The tanker is 350 ft long.

(a) What is the pressure exerted on the bottom of the tanker by the oil?

(b) What is the pressure exerted on the bottom of the tanker by the water?

(c) What is the force exerted on the bottom of the tanker by the oil?

(d) What is the force exerted on the bottom of the tanker by the water?

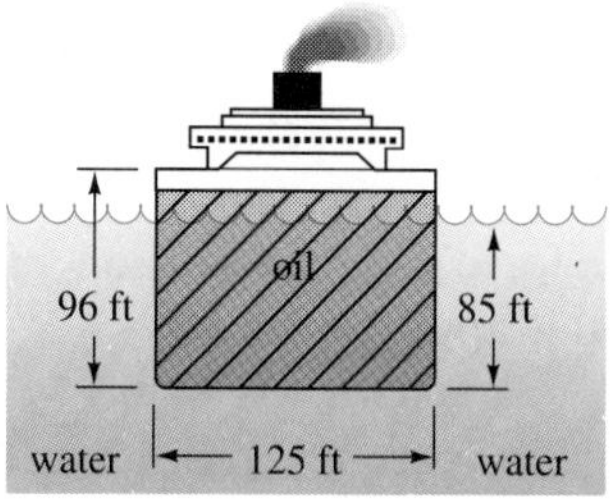

FIGURE 5.94

3. A rectangular swimming pool in the form of a rectangular parallelepiped has dimensions of 30 ft × 15 ft × 9 ft.

(a) Find the pressure and force exerted on the flat bottom if the pool is filled with water to a depth of 8 ft. See Figure 5.95.

(b) Find the force exerted by the water on a vertical sidewall and on a vertical end.

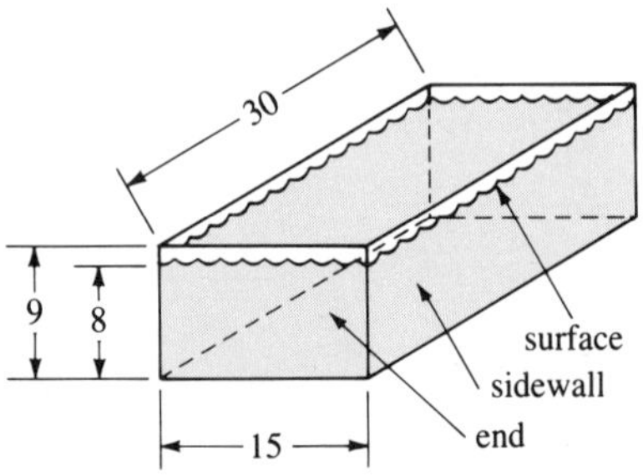

FIGURE 5.95

4. A plate in the shape of an equilateral triangle $\sqrt{3}$ ft on a side is submerged vertically, base downward, with vertex 1 ft below the surface of the water. Find the force exerted by the water on one side of the plate.

5. Find the force on one side of the plate in Problem 4 if it is suspended with base upward 1 ft below the surface of the water.

6. A triangular plate is submerged vertically in water as shown in Figure 5.96. Find the force exerted by the water on one side of the plate.

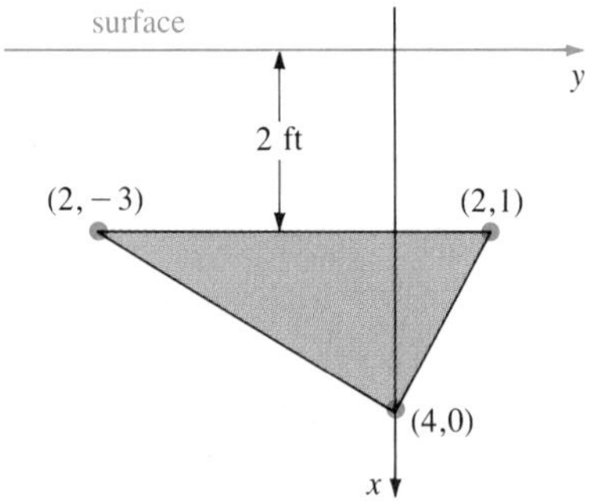

FIGURE 5.96

7. Assuming the positive x-axis is downward, a plate bounded by the parabola $x = y^2$ and the line $x = 4$ is submerged vertically in oil that has weight density 50 lb/ft^3. If the vertex of the parabola is at the surface, find the force exerted by the oil on one side of the plate.

8. Assuming the positive x-axis is downward, a plate bounded by the parabola $x = y^2$ and the line $y = -x + 2$ is submerged vertically in water. If the vertex of the parabola is at the surface, find the force exerted by the water on one side of the plate.

9. A full water trough has vertical ends in the form of trapezoids as shown in Figure 5.97. Find the force exerted by the water on one end of the trough.

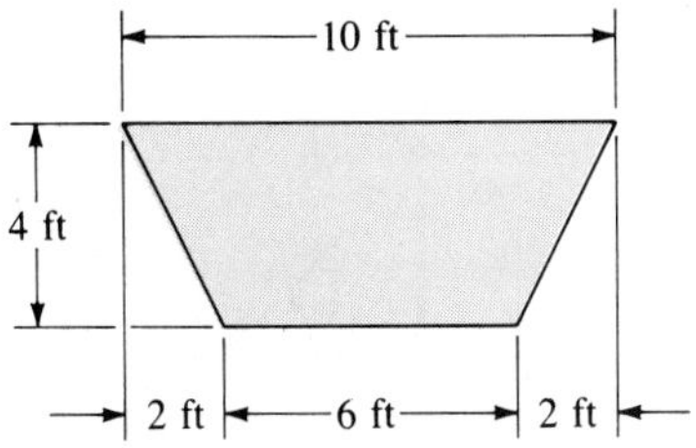

FIGURE 5.97

10. A full water trough has vertical ends in the form shown in Figure 5.98. Find the force exerted by the water on one end of the trough.

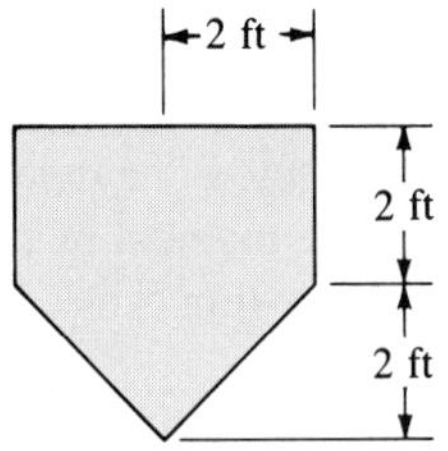

FIGURE 5.98

11. A vertical end of a full swimming pool has the shape given in Figure 5.99. Find the force exerted by the water on this end of the pool.

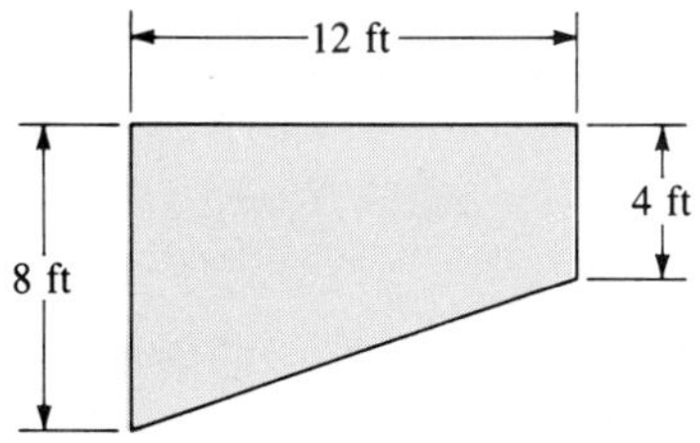

FIGURE 5.99

12. A tank in the shape of right circular cylinder of diameter 10 ft is lying on its side. The tank is half full of oil that has weight density 60 lb/ft^3. Find the force exerted by the oil on one end of the tank.

13. A circular plate of radius 4 ft is submerged vertically so that the center of the plate is 10 ft below the surface of the water. Find the force exerted by the water on one side of the plate. [*Hint:* For simplicity, take the origin to be the center of the plate, positive x-axis downward. See Problem 56 in Exercises 5.1.]

14. A tank whose ends are in the form of an ellipse $x^2/4 + y^2/9 = 1$ is submerged in a liquid that has weight density ρ so that the end plates are vertical. Find the force exerted by the liquid on one end if its center is 10 ft below the surface of the liquid. [*Hint:* Proceed as in Problem 13 and use the fact that the area of an ellipse $x^2/a^2 + y^2/b^2 = 1$ is πab.]

15. A solid block in the shape of a cube 2 ft on a side is submerged in a large tank of water. The top of the block is horizontal and is 3 ft below the surface of the water. Find the total force on the block (six sides) that is caused by liquid pressure. See Figure 5.100.

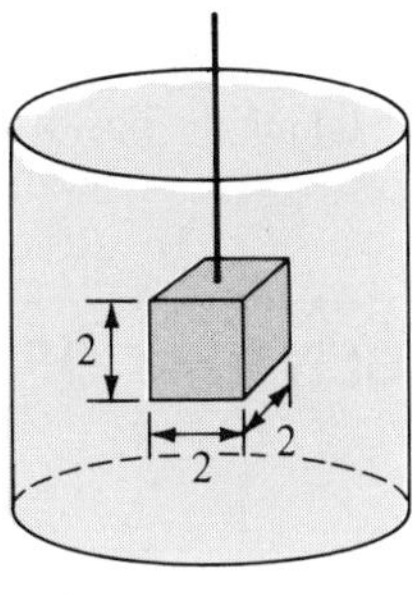

FIGURE 5.100

16. In Problem 15 what is the difference between the force on the bottom of the block and the force on the top of the block? This difference is the buoyant force of the water and, by Archimedes' principle, is equal to the weight of the water displaced. What is the weight of the water displaced by the block?

17. Consider the rectangular swimming pool shown in Figure 5.101(a) whose ends are trapezoids. The pool is full of water. By taking the positive x-axis, as shown in Figure 5.101(b), find the force exerted by the water on the bottom of the pool. [*Hint:* Express the depth d in terms of x.]

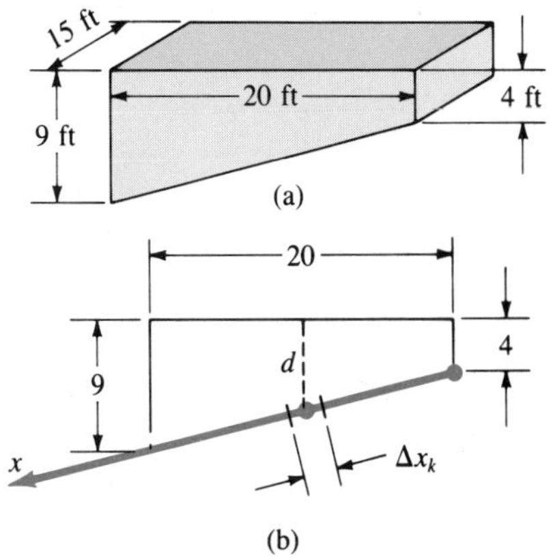

FIGURE 5.101

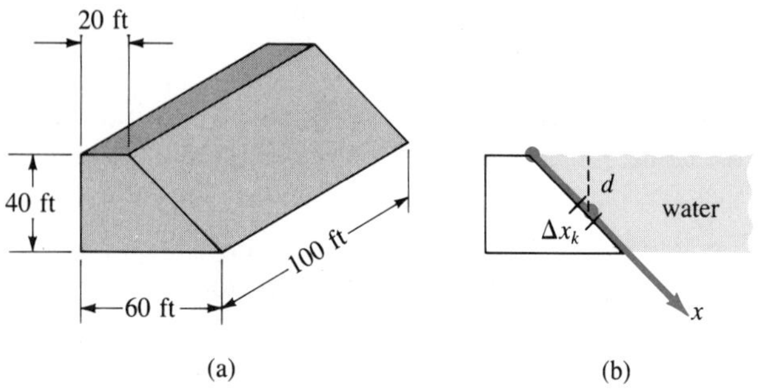

FIGURE 5.102

18. An earth dam is constructed with dimensions as shown in Figure 5.102(a). By taking the positive x-axis, as shown in Figure 5.102(b), find the force exerted by the water on the slanted wall of the dam.

19. Analyze Problem 18 with the positive x-axis shown in Figure 5.103.

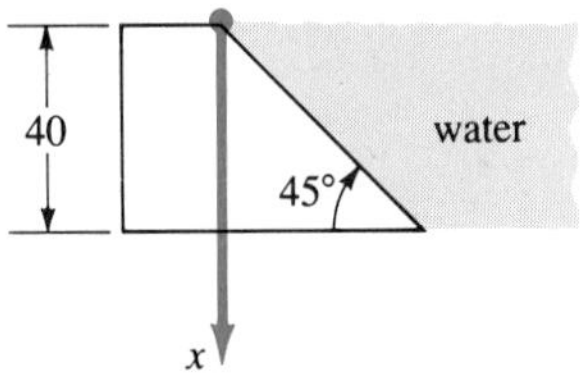

FIGURE 5.103

5.10 CENTERS OF MASS

Moment and Mass If x denotes the directed distance from the origin O to a mass m, we say that the product mx is the **moment of the mass** about the origin. Some units are summarized in the following table.

Quantity	Engineering system	SI	cgs
Mass	slug	kilogram (kg)	gram (g)
Moment of mass	slug-feet	kilogram-meter	gram-centimeter

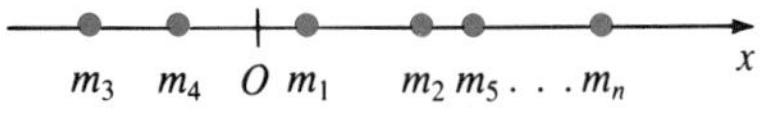

FIGURE 5.104

Now, for n masses $m_1, m_2, \ldots, m_n$ at directed distances $x_1, x_2, \ldots, x_n$, respectively, from O, as in Figure 5.104, we say that

$$m = m_1 + m_2 + \cdots + m_n = \sum_{k=1}^{n} m_k$$

is the **total mass of the system**, and

$$M_O = m_1x_1 + m_2x_2 + \cdots + m_nx_n = \sum_{k=1}^{n} m_kx_k$$

is the **moment of the system about the origin**. If $\Sigma_{k=1}^{n} m_kx_k = 0$, the system is said to be in **equilibrium**. See Figure 5.105. If the system of masses in Figure 5.104 is not in equilibrium, there is a point P with coordinate $\bar{x}$ such that

$$\sum_{k=1}^{n} m_k(x_k - \bar{x}) = 0 \quad \text{or} \quad \sum_{k=1}^{n} m_kx_k - \bar{x}\sum_{k=1}^{n} m_k = 0$$

Solving for $\bar{x}$ gives

$$\bar{x} = \frac{M_O}{m} = \frac{\sum_{k=1}^{n} m_kx_k}{\sum_{k=1}^{n} m_k} \tag{1}$$

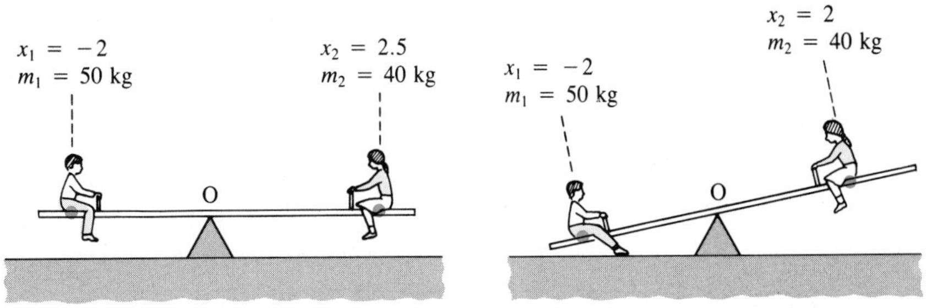

(a) seesaw in equilibrium since $m_1x_1 + m_2x_2 = 0$

(b) seesaw not in equilibrium since $m_1x_1 + m_2x_2 \neq 0$

FIGURE 5.105

The point with coordinate $\bar{x}$ is called the **center of mass** or the **center of gravity** of the system.*

Since (1) implies $\bar{x}(\Sigma_{k=1}^{n} m_k) = \Sigma_{k=1}^{n} m_k x_k$, it follows that $\bar{x}$ is the directed distance from the origin to a point at which the total mass of the system can be considered to be concentrated.

EXAMPLE 1 Three bodies of masses 4 kg, 6 kg, and 10 kg are located at $x_1 = -2$, $x_2 = 4$, and $x_3 = 9$, respectively. Distances are measured in meters. Find the center of mass.

Solution From (1),

$$\bar{x} = \frac{4 \cdot (-2) + 6 \cdot 4 + 10 \cdot 9}{4 + 6 + 10} = \frac{106}{20} = 5.3$$

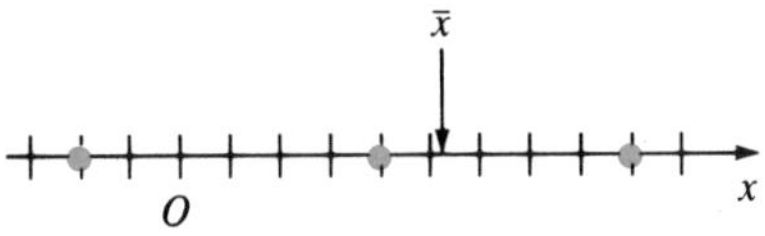

FIGURE 5.106

Figure 5.106 shows that the center of mass is 5.3 m to the right of the origin. □

Rod with Variable Density Now, let us consider the problem of finding the center of mass of a rod of length L that has a **variable linear density** ρ. We assume that the rod coincides with the x-axis on the interval $[0, L]$, as shown in Figure 5.107, and that the density is a continuous function $\rho(x)$ measured in slugs/ft, kg/m, or g/cm. After forming a partition P of the interval, we choose a point x_k^* in $[x_{k-1}, x_k]$. The number

$$\Delta m_k = \rho(x_k^*)\,\Delta x_k$$

is an approximation to the mass of that portion of the rod on the subinterval. Also, the moment of this element of mass about the origin is approximated by

$$(\Delta M_O)_k = x_k^* \rho(x_k^*)\,\Delta x_k$$

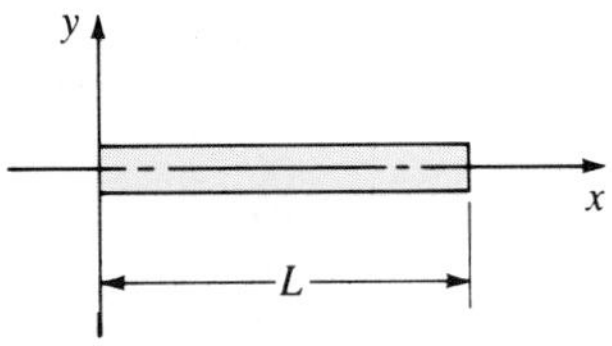

FIGURE 5.107

* In a system in which the acceleration of gravity varies from mass to mass, the center of gravity is not the same as the center of mass.

Thus, we conclude that

$$m = \lim_{\|P\| \to 0} \sum_{k=1}^{n} \rho(x_k^*)\,\Delta x_k = \int_0^L \rho(x)\,dx$$

and

$$M_O = \lim_{\|P\| \to 0} \sum_{k=1}^{n} x_k^* \rho(x_k^*)\,\Delta x_k = \int_0^L x\rho(x)\,dx$$

are the **mass of the rod** and its **moment about the origin**, respectively.

It then follows from $\bar{x} = M_O/m$ that the center of mass of the rod is given by

$$\bar{x} = \frac{\displaystyle\int_0^L x\rho(x)\,dx}{\displaystyle\int_0^L \rho(x)\,dx} \tag{2}$$

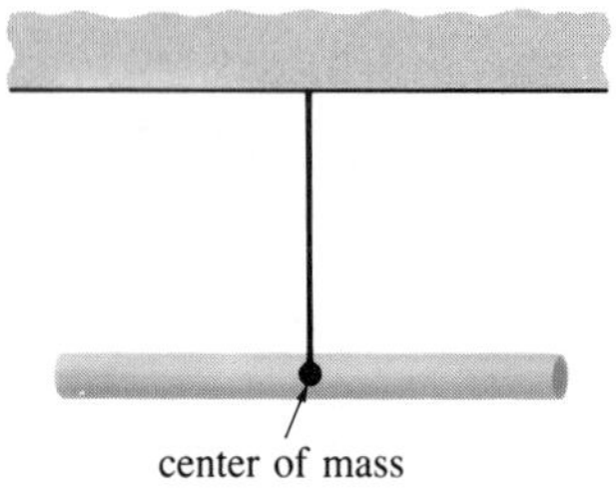

FIGURE 5.108

As shown in Figure 5.108, a rod suspended by a string attached to its center of mass would hang in perfect balance.

EXAMPLE 2 A 16-cm-long rod has a linear density, measured in g/cm, given by $\rho(x) = \sqrt{x}$, $0 \le x \le 16$. Find its center of mass.

Solution The mass of the rod in grams is

$$m = \int_0^{16} x^{1/2}\,dx = \frac{2}{3}x^{3/2}\Big]_0^{16} = \frac{128}{3}$$

The moment about the origin (in g-cm) is

$$M_O = \int_0^{16} x \cdot x^{1/2}\,dx = \frac{2}{5}x^{5/2}\Big]_0^{16} = \frac{2048}{5}$$

From (2) we find

$$\bar{x} = \frac{2048/5}{128/3} = 9.6$$

That is, the center of mass of the rod is 9.6 cm from the left end of the rod that coincides with the origin. □

For n masses located in the xy-plane, as indicated in Figure 5.109, the **center of mass of the system** is defined to be the point $(\bar{x}, \bar{y})$, where

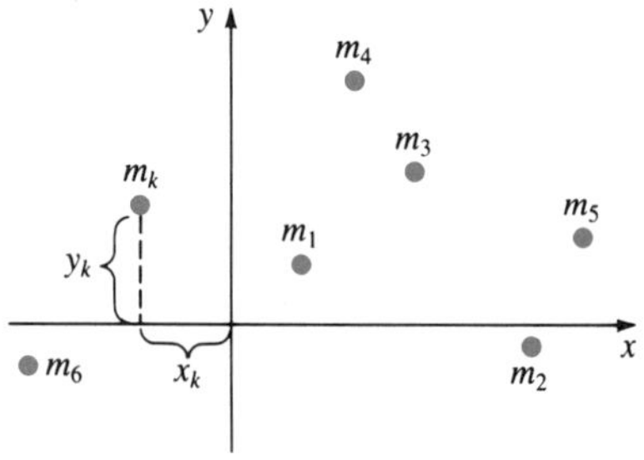

FIGURE 5.109

$$\bar{x} = \frac{M_y}{m} = \frac{\displaystyle\sum_{k=1}^{n} m_k x_k}{\displaystyle\sum_{k=1}^{n} m_k} = \frac{\text{moment of system about } y\text{-axis}}{\text{total mass}}$$

$$\bar{y} = \frac{M_x}{m} = \frac{\displaystyle\sum_{k=1}^{n} m_k y_k}{\displaystyle\sum_{k=1}^{n} m_k} = \frac{\text{moment of system about } x\text{-axis}}{\text{total mass}}$$

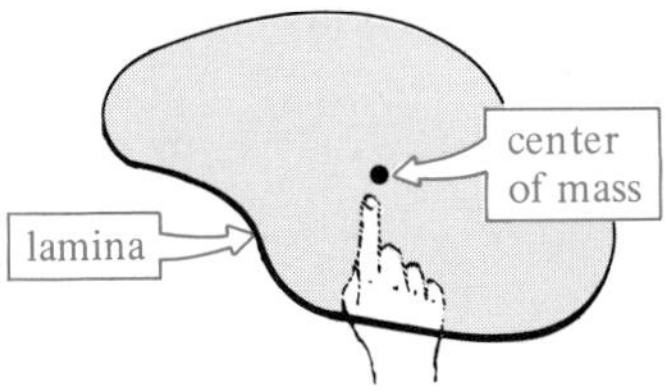

FIGURE 5.110

Lamina Now let's turn to the problem of finding the center of mass, or balancing point, of a thin two-dimensional smear of matter, or **lamina**, that has a constant density ρ (mass per unit area). See Figure 5.110. When ρ is constant, the lamina is said to be **homogeneous**. As shown in Figure 5.111(a), let us suppose that the lamina coincides with a region R in the xy-plane bounded by the graph of a continuous nonnegative function $y = f(x)$, the x-axis, and

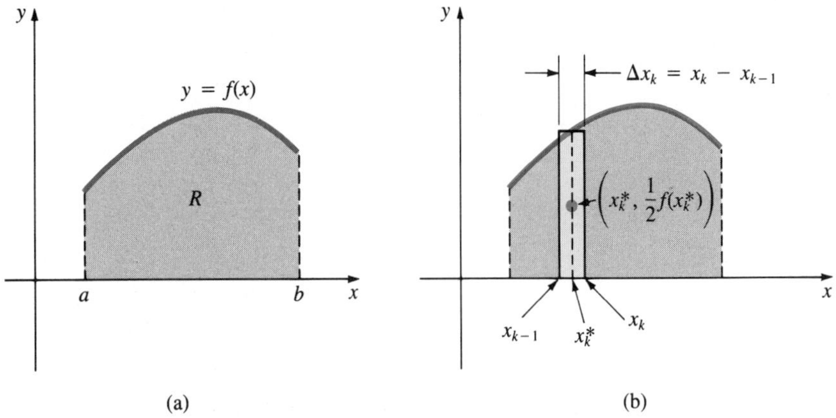

FIGURE 5.111

the vertical lines $x = a$ and $x = b$. If P is a partition of the interval $[a, b]$, then the mass of the rectangular element shown in Figure 5.111(b) is

$$\Delta m_k = \rho\, \Delta A_k = \rho f(x_k^*)\, \Delta x_k$$

where, in this case, we take x_k^* to be the midpoint of the subinterval $[x_{k-1}, x_k]$ and ρ is the constant density. The moment of this element about the y-axis is

$$(\Delta M_y)_k = x_k^*\, \Delta m_k = x_k^*(\rho\, \Delta A_k) = \rho x_k^* f(x_k^*)\, \Delta x_k$$

Since the density is constant, the center of mass of the element is necessarily at its geometric center $(x_k^*, \frac{1}{2}f(x_k^*))$. Hence, the moment of the element about the x-axis is

$$(\Delta M_x)_k = \frac{1}{2} f(x_k^*)[\rho\, \Delta A_k] = \frac{1}{2}\rho[f(x_k^*)]^2\, \Delta x_k$$

We conclude that

$$m = \lim_{\|P\| \to 0} \sum_{k=1}^{n} \rho f(x_k^*)\, \Delta x_k = \int_a^b \rho f(x)\, dx$$

$$M_y = \lim_{\|P\| \to 0} \sum_{k=1}^{n} \rho x_k^* f(x_k^*)\, \Delta x_k = \int_a^b \rho x f(x)\, dx$$

and

$$M_x = \lim_{\|P\| \to 0} \frac{1}{2} \sum_{k=1}^{n} \rho[f(x_k^*)]^2\, \Delta x_k = \frac{1}{2}\int_a^b \rho[f(x)]^2\, dx$$

Thus, the coordinates of the center of mass of the lamina are defined to be

$$\bar{x} = \frac{M_y}{m} = \frac{\displaystyle\int_a^b \rho x f(x)\,dx}{\displaystyle\int_a^b \rho f(x)\,dx} \qquad \bar{y} = \frac{M_x}{m} = \frac{\displaystyle\frac{1}{2}\int_a^b \rho[f(x)]^2\,dx}{\displaystyle\int_a^b \rho f(x)\,dx} \tag{3}$$

Centroid We note that the constant density ρ will cancel in equations (3) for $\bar{x}$ and $\bar{y}$, and that the denominator $\int_a^b f(x)\,dx$ is then the area A of the region R. In other words, the center of mass depends only on the shape of R:

$$\bar{x} = \frac{M_y}{A} = \frac{\displaystyle\int_a^b x f(x)\,dx}{\displaystyle\int_a^b f(x)\,dx} \qquad \bar{y} = \frac{M_x}{A} = \frac{\displaystyle\frac{1}{2}\int_a^b [f(x)]^2\,dx}{\displaystyle\int_a^b f(x)\,dx} \tag{4}$$

To emphasize the distinction, albeit minor, between the physical object, which is the homogeneous lamina, and the geometric object, which is the plane region R, we say the equations in (4) define the coordinates of the **centroid** of the region.

EXAMPLE 3 Find the centroid of the region in the first quadrant bounded by the graph of $y = 9 - x^2$, the x-axis, and the y-axis.

Solution The region is shown in Figure 5.112. Now, if $f(x) = 9 - x^2$, then

$$\Delta A_k = f(x_k^*)\,\Delta x_k$$
$$(\Delta M_y)_k = x f(x_k^*)\,\Delta x_k$$
$$(\Delta M_x)_k = \frac{1}{2} f(x_k^*)[f(x_k^*)\,\Delta x_k] = \frac{1}{2}[f(x_k^*)]^2\,\Delta x_k$$

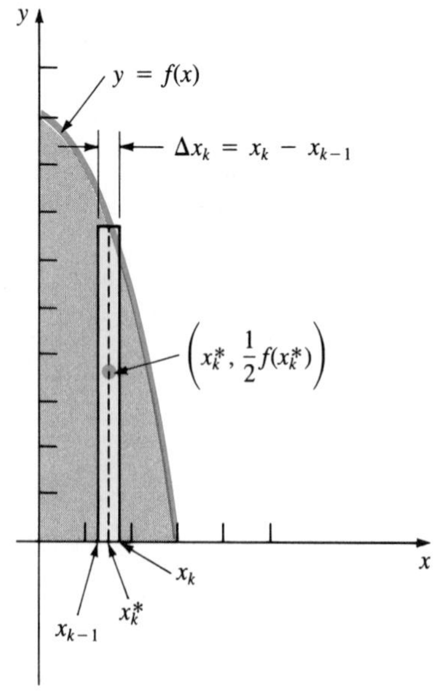

FIGURE 5.112

Hence,

$$A = \int_0^3 (9 - x^2)\,dx = \left(9x - \frac{x^3}{3}\right)\Big]_0^3 = 18$$

$$M_y = \int_0^3 x(9 - x^2)\,dx = \left(\frac{9}{2}x^2 - \frac{x^4}{4}\right)\Big]_0^3 = \frac{81}{4}$$

$$M_x = \frac{1}{2}\int_0^3 (9 - x^2)^2\,dx$$
$$= \frac{1}{2}\int_0^3 (81 - 18x^2 + x^4)\,dx$$
$$= \frac{1}{2}\left(81x - 6x^3 + \frac{x^5}{5}\right)\Big]_0^3 = \frac{324}{5}$$

It follows from (4) that the coordinates of the centroid are

$$\bar{x} = \frac{M_y}{A} = \frac{81/4}{18} = \frac{9}{8} \qquad \bar{y} = \frac{M_x}{A} = \frac{324/5}{18} = \frac{54}{15}$$

□

EXAMPLE 4 Find the centroid of the region bounded by the graphs of $x = y^2 + 1$, $x = 0$, $y = 2$, and $y = -2$.

Solution The region is shown in Figure 5.113. Inspection of the figure suggests that we use horizontal rectangular elements. If $f(y) = y^2 + 1$, then

$$\Delta A_k = f(y_k^*)\,\Delta y_k$$
$$(\Delta M_x)_k = y_k^* f(y_k^*)\,\Delta y_k$$
$$(\Delta M_y)_k = \frac{1}{2} f(y_k^*)[f(y_k^*)\,\Delta y_k] = \frac{1}{2}[f(y_k^*)]^2\,\Delta y_k$$

and so

$$A = \int_{-2}^{2} (y^2 + 1)\,dy = \left(\frac{y^3}{3} + y\right)\Bigg]_{-2}^{2} = \frac{28}{3}$$
$$M_x = \int_{-2}^{2} y(y^2 + 1)\,dy = \left(\frac{y^4}{4} + \frac{y^2}{2}\right)\Bigg]_{-2}^{2} = 0$$
$$M_y = \frac{1}{2}\int_{-2}^{2} (y^2 + 1)^2\,dy = \frac{1}{2}\int_{-2}^{2} (y^4 + 2y^2 + 1)\,dy$$
$$= \frac{1}{2}\left(\frac{y^5}{5} + \frac{2}{3}y^3 + y\right)\Bigg]_{-2}^{2} = \frac{206}{15}$$

Thus, we have

$$\bar{x} = \frac{M_y}{A} = \frac{206/15}{28/3} = \frac{103}{60} \qquad \bar{y} = \frac{M_x}{A} = \frac{0}{28/3} = 0$$

As we would expect, since the lamina is symmetric with respect to the x-axis, the centroid is on the axis of symmetry. We also note that the centroid is outside the region.

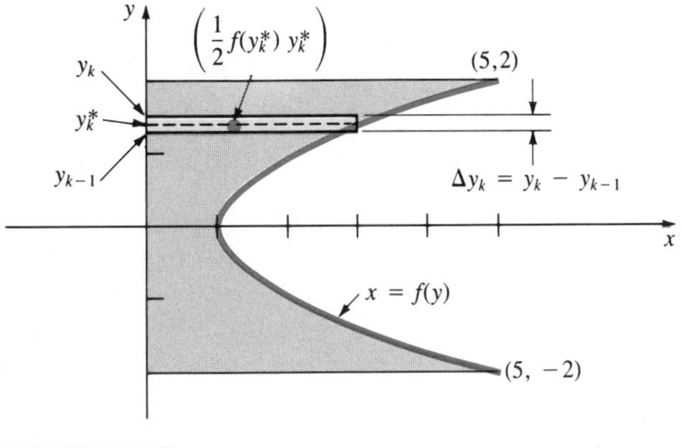

FIGURE 5.113 □

EXAMPLE 5 Find the centroid of the region bounded by the graphs of $y = -x^2 + 3$ and $y = x^2 - 2x - 1$.

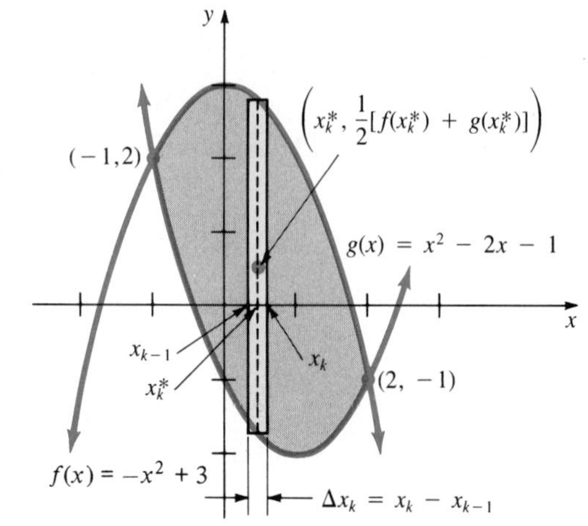

FIGURE 5.114

Solution Figure 5.114 shows the region in question. We note that the points of intersection of the graphs are $(-1, 2)$ and $(2, -1)$. Now, if $f(x) = -x^2 + 3$ and $g(x) = x^2 - 2x - 1$, then the area of the region is

$$A = \int_{-1}^{2} [f(x) - g(x)]\, dx = \int_{-1}^{2} (-2x^2 + 2x + 4)\, dx$$

$$= \left(-\frac{2}{3}x^3 + x^2 + 4x\right)\Bigg]_{-1}^{2} = 9$$

Since the coordinates of the midpoint of the indicated element are $(x_k^*, \frac{1}{2}[f(x_k^*) + g(x_k^*)])$, it follows that

$$M_y = \int_{-1}^{2} x[f(x) - g(x)]\, dx$$

$$= \int_{-1}^{2} (-2x^3 + 2x^2 + 4x)\, dx$$

$$= \left(-\frac{1}{2}x^4 + \frac{2}{3}x^3 + 2x^2\right)\Bigg]_{-1}^{2} = \frac{9}{2}$$

$$M_x = \frac{1}{2}\int_{-1}^{2} [f(x) + g(x)][f(x) - g(x)]\, dx$$

$$= \frac{1}{2}\int_{-1}^{2} ([f(x)]^2 - [g(x)]^2)\, dx$$

$$= \frac{1}{2}\int_{-1}^{2} [(-x^2 + 3)^2 - (x^2 - 2x - 1)^2]\, dx$$

$$= \frac{1}{2}\int_{-1}^{2} (4x^3 - 8x^2 - 4x + 8)\, dx$$

$$= \frac{1}{2}\left(x^4 - \frac{8}{3}x^3 - 2x^2 + 8x\right)\Bigg]_{-1}^{2} = \frac{9}{2}$$

Thus, the coordinates of the centroid are

$$\bar{x} = \frac{M_y}{A} = \frac{9/2}{9} = \frac{1}{2} \qquad \bar{y} = \frac{M_x}{A} = \frac{9/2}{9} = \frac{1}{2}$$ □

EXERCISES 5.10

Answers to odd-numbered problems begin on page A-82.

In Problems 1–4 find the center of mass of the given system of masses. The mass m_k is located on the x-axis at a point whose directed distance from the origin is x_k. Assume that mass is measured in grams and that distance is measured in centimeters.

1. $m_1 = 2, m_2 = 5; x_1 = 4, x_2 = -2$

2. $m_1 = 6, m_2 = 1, m_3 = 3; x_1 = -\frac{1}{2}, x_2 = -3, x_3 = 8$

3. $m_1 = 10, m_2 = 5, m_3 = 8, m_4 = 7; x_1 = -5, x_2 = 2, x_3 = 6, x_4 = -3$

4. $m_1 = 2, m_2 = \frac{3}{2}, m_3 = \frac{7}{2}, m_4 = \frac{1}{2}; x_1 = 9, x_2 = -4, x_3 = -6, x_4 = -10$

5. Two masses are placed at the ends of a uniform board of negligible mass, as shown in Figure 5.115. Where should a fulcrum be placed so that the system is in balance? [*Hint:* Although the origin can be placed anywhere, let us agree to choose it halfway between the masses.]

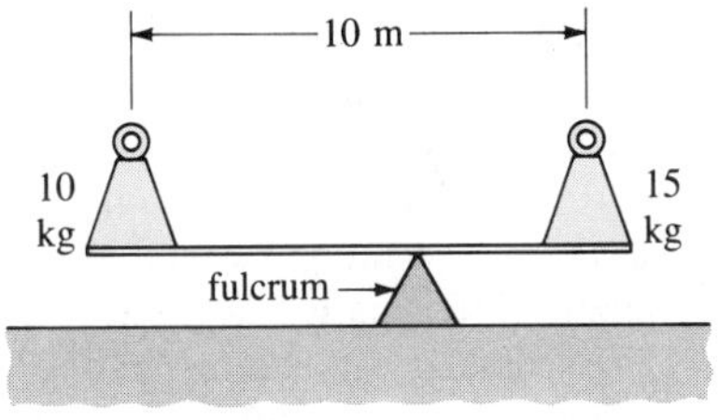

FIGURE 5.115

6. Find the center of mass of the three masses m_1, m_2, and m_3 located at the vertices of the equilateral triangle shown in Figure 5.116. [*Hint:* First find the center of mass of m_1 and m_2.]

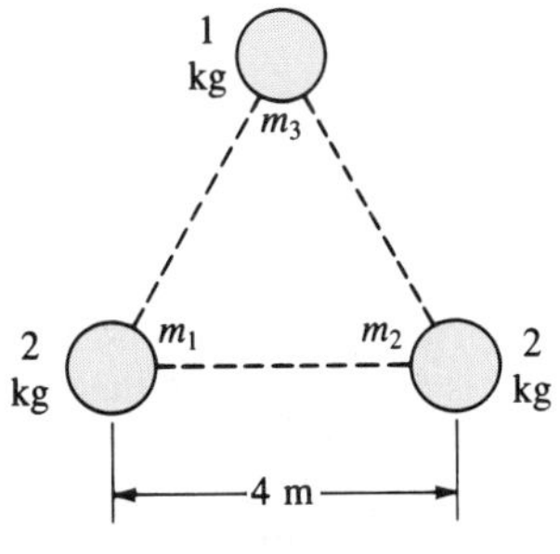

FIGURE 5.116

In Problems 7–14 a rod of linear density $\rho(x)$ kg/m coincides with the x-axis on the interval indicated. Find its center of mass.

7. $\rho(x) = 2x + 1$; $[0, 5]$

8. $\rho(x) = -x^2 + 2x$; $[0, 2]$

9. $\rho(x) = x^{1/3}$; $[0, 1]$

10. $\rho(x) = -x^2 + 1$; $[0, 1]$

11. $\rho(x) = |x - 3|$; $[0, 4]$

12. $\rho(x) = 1 + |x - 1|$; $[0, 3]$

13. $\rho(x) = \begin{cases} x^2, & 0 \le x \le 1 \\ 2 - x, & 1 \le x \le 2 \end{cases}$; $[0, 2]$

14. $\rho(x) = \begin{cases} x, & 0 \le x \le 2 \\ 2 & 2 \le x \le 3 \end{cases}$; $[0, 3]$

15. The density of a 10-ft rod varies as the square of the distance from the left end. Find its center of mass if the density at its center is 12.5 slug/ft.

16. The linear density of a 3-m-long rod varies as the distance from the right end. Find the linear density at the center of the rod if its total mass is 6 kg.

17. A rod of linear density $\rho(x)$ kg/m coincides with the x-axis on the interval $[0, 6]$. If $\rho(x) = x(6 - x) + 1$, where would one intuitively expect the center of mass to be? [*Hint:* Graph $\rho(x)$.] Prove your assertion.

18. The linear density of a rod (in kg/m) is given by $\rho(x) = \sqrt{2x + 1}$. Find the average density on the interval $[0, 4]$.

In Problems 19–22 find the center of mass of the given system of masses. The mass m_k is located at the point P_k. Assume that mass is measured in grams and that distance is measured in centimeters.

19. $m_1 = 3, m_2 = 4; P_1 = (-2, 3), P_2 = (1, 2)$

20. $m_1 = 1, m_2 = 3, m_3 = 2; P_1 = (-4, 1), P_2 = (2, 2), P_3 = (5, -2)$

21. $m_1 = 4, m_2 = 8, m_3 = 10; P_1 = (1, 1), P_2 = (-5, 2), P_3 = (7, -6)$

22. $m_1 = 1, m_2 = \frac{1}{2}, m_3 = 4, m_4 = \frac{5}{2}; P_1 = (9, 3), P_2 = (-4, -6), P_3 = (\frac{3}{2}, -1), P_4 = (-2, 10)$

In Problems 23–40 find the centroid of the region bounded by the graphs of the given equations.

23. $y = 2x + 4, y = 0, x = 0, x = 2$

24. $y = x + 1, y = 0, x = 3$

25. $y = x^2, y = 0, x = 1$

26. $y = x^2 + 2, y = 0, x = -1, x = 2$

27. $y = x^3, y = 0, x = 3$

28. $y = x^3, y = 8, x = 0$

29. $y - \sqrt{x}, y = 0, x = 1, x = 4$

30. $x = y^2, x = 1$

31. $y = x^2, y - x = 2$

32. $y = x^2, y = \sqrt{x}$

33. $y = x^3, y = x^{1/3}$, first quadrant

34. $y = 4 - x^2, y = 0, x = 0$, second quadrant

35. $x^3y = 1, y = 0, x = 1, x = 3$

36. $y = x^2 - 2x + 1, y = -4x + 9$

37. $x = y^2 - 1, y = -1, y = 2, x = -2$

38. $y = x^2 - 4x + 6, y = 0, x = 0, x = 4$

39. $y = 4 - 4x^2, y = 1 - x^2$

40. $y^2 + x = 1, y + x = -1$

In Problems 41 and 42 use symmetry to locate $\bar{x}$ and integration to find $\bar{y}$ of the region bounded by the graphs of the given functions.

41. $y = 1 + \cos x, y = 1, -\pi/2 \le x \le \pi/2$

42. $y = 4 \sin x, y = -\sin x, 0 \le x \le \pi$

43. Let L be an axis in a plane and R a region in the same plane that does not intersect L. The **First Theorem of Pappus*** states that when R is revolved about L, the volume V of the resulting solid of revolution is equal to the area A of R times the length of path traversed by the centroid of R. As shown in Figure 5.117, let the region R be bounded by the graphs of $y = f_1(x)$ and $y = f_2(x)$. Show that if R is revolved about the x-axis, then $V = (2\pi\bar{y})A$, where A is the area of the region.

44. Verify the First Theorem of Pappus by revolving the region bounded by $y = x^2 + 1$, $y = 1$, $x = 2$ about the x-axis.

*Pappus of Alexandria (c. A.D. 350).

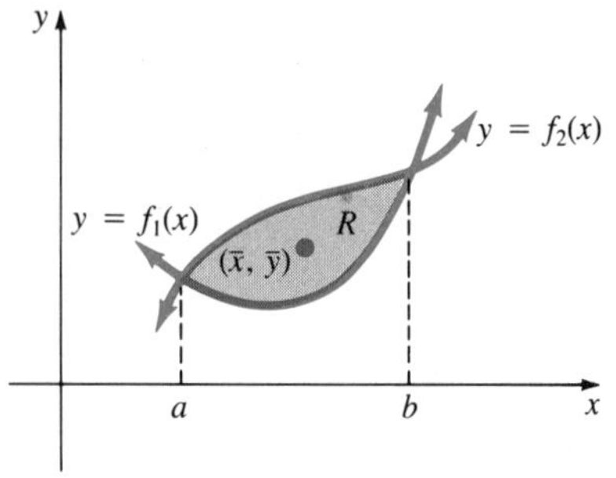

FIGURE 5.117

45. Use the First Theorem of Pappus to find the volume of the torus shown in Figure 5.118.

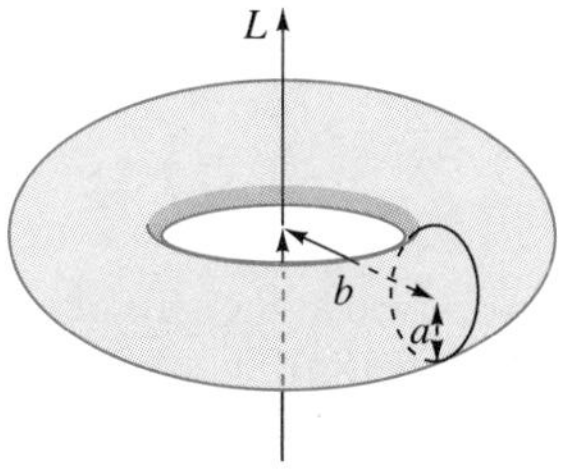

FIGURE 5.118

46. Let L be an axis in a plane and C a curve in the same plane that does not intersect L. The **Second Theorem of Pappus** states that when C is revolved about L, the area S of the resulting surface of revolution is equal to the length s of C times the length of the path traversed by the centroid of C. Now, the coordinates of the centroid of a plane curve C defined by the graph of a continuous function $y = f(x)$, $a \le x \le b$, are defined by

$$\bar{x} = \frac{\int_a^b x\sqrt{1 + [f'(x)]^2}\,dx}{\int_a^b \sqrt{1 + [f'(x)]^2}\,dx}$$

$$\bar{y} = \frac{\int_a^b f(x)\sqrt{1 + [f'(x)]^2}\,dx}{\int_a^b \sqrt{1 + [f'(x)]^2}\,dx}$$

Use this information to prove the Second Theorem of Pappus in the case when C is revolved about the x-axis.

47. Use the Second Theorem of Pappus to find the surface area of the torus shown in Figure 5.118.

[O] 5.11 APPLICATIONS TO BIOLOGY AND BUSINESS

Cardiac Output In physiology, the cardiac output R of the heart is defined as the volume of blood that the heart pumps per unit time. An abnormal cardiac output is an indicator of disease.

One way of measuring this output is called the **dye dilution method**. As shown in Figure 5.119, an amount D, measured in milligrams, of dye is injected into the pulmonary artery near the heart. The dye flows through the lungs and the pulmonary veins into the left atrium of the heart and eventually through the aorta. A probe inserted into the aorta monitors the amount of dye leaving the heart at equally spaced values of time over an interval $[0, T]$ (say, every second up to 30 seconds).

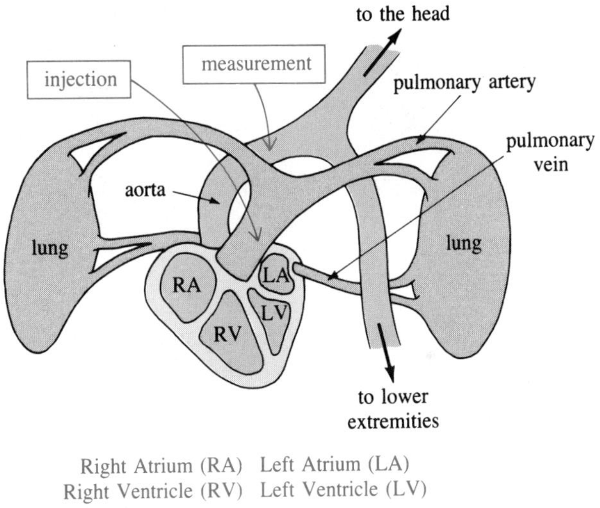

FIGURE 5.119

In determining R, let us assume that the concentration of the dye at any time is given by a continuous function $c(t)$. The interval $[0, T]$ is then subdivided into n subintervals of equal length $\Delta t = t_k - t_{k-1} = T/n$. The amount of dye that flows past the measuring point over a time interval $[t_{k-1}, t_k]$ is given by the approximation

$$\begin{aligned}\Delta D_k &= (\text{concentration}) \cdot (\text{volume}) = (\text{concentration}) \cdot (\text{volume}/\text{time}) \cdot (\text{time}) \\ &= c(t_k^*) \cdot R \cdot \Delta t\end{aligned}$$

where t_k^* is some value in $[t_{k-1}, t_k]$. The approximate amount of dye that flows past the monitoring point in the aorta over the time interval $[0, T]$ is

$$\sum_{k=1}^{n} \Delta D_k = \sum_{k=1}^{n} c(t_k^*) R \, \Delta t = R \sum_{k=1}^{n} c(t_k^*) \, \Delta t \tag{1}$$

Letting $n \to \infty$ in (1) suggests that the total amount of dye that flows past the monitoring point is

$$D = R\int_0^T c(t)\,dt \tag{2}$$

Solving (2) for the flow rate R yields

$$R = \frac{D}{\displaystyle\int_0^T c(t)\,dt}\,* \tag{3}$$

EXAMPLE 1 Five milligrams of dye are injected into the pulmonary artery. Find the cardiac output of the heart over a period of 30 s if the concentration of the dye is $c(t) = -(1/100)t(t - 30)$ milligrams/liter (mg/L).

Solution From (3), we have

$$R = \frac{5}{\displaystyle\int_0^{30} -\frac{1}{100}t(t-30)\,dt} = \frac{5}{\left(-\dfrac{t^3}{300} + \dfrac{3}{20}t^2\right)\Big]_0^{30}}$$

$$= \frac{1}{9}\text{ L/s} = \frac{1}{9}(60) \approx 6.67\text{ L/min}$$ □

Flow of Blood in an Artery The velocity of blood (in cm/s) in an arteriole (small artery) that has a circular cross-section of radius R was given by Poiseuille to be

$$v(r) = \frac{P}{4vl}(R^2 - r^2), \qquad 0 \le r \le R \tag{4}$$

where P is the blood pressure, v the viscosity of the blood, and l the length of the arteriole. Flow characterized by (4) is called *laminar flow*, since, as shown in Figure 5.120, blood is thought to flow in cylindrical shells or *laminae* parallel to the walls of the arteriole at a distance r from the center.

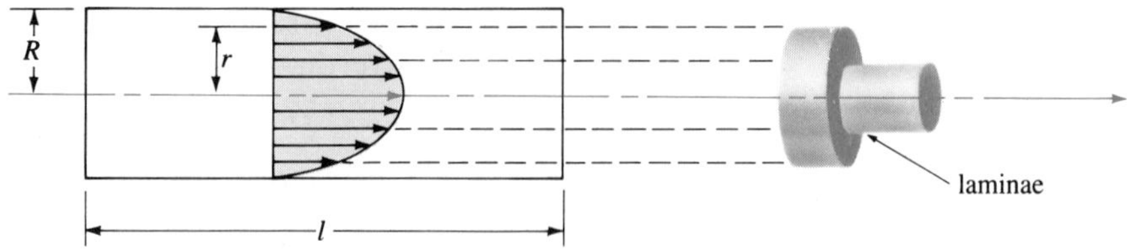

FIGURE 5.120

*The value of R given in (3) is actually an approximation to the cardiac output. In practice, the graph of $c(t)$ is one that is fitted to approximate the set of data points $(0, c(0))$, $(1, c(1))$, ...

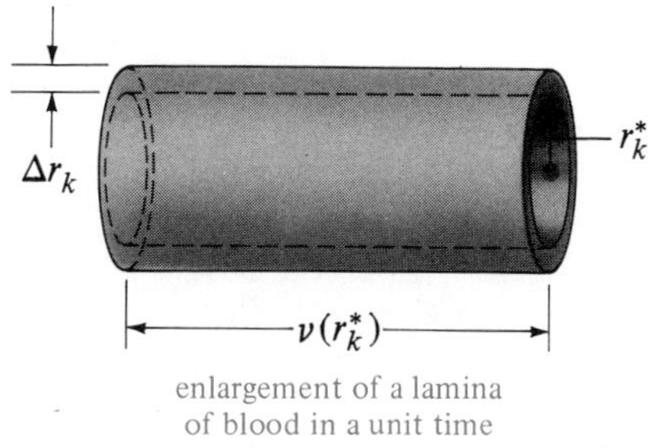

FIGURE 5.121

We can find the volume of blood F, or total flow, inside an artery in a unit time by definite integration. To see this, subdivide $[0, R]$ into n subintervals $[r_{k-1}, r_k]$. From Figure 5.121, we see that the volume of blood per unit time inside a lamina of radius r_k^* and thickness Δr_k is

$$\Delta F_k = 2\pi r_k^* v(r_k^*)\,\Delta r_k$$

Taking the limit as $n \to \infty$ of the sum

$$\sum_{k=1}^{n} \Delta F_k = \sum_{k=1}^{n} 2\pi r_k^* v(r_k^*)\,\Delta r_k$$

suggests

$$F = \int_0^R 2\pi r v(r)\,dr = \frac{\pi P}{2vl}\int_0^R r(R^2 - r^2)\,dr = \frac{\pi P}{2vl}\left(R^2\frac{r^2}{2} - \frac{r^4}{4}\right)\Bigg]_0^R$$

or

$$F = \frac{\pi P R^4}{8vl} \tag{5}$$

In other words, Poiseuille's law (5) indicates that the volume of blood in an arteriole per unit time is proportional to the fourth power of the radius. Formula (5) is used for laminar flow; this usually means that R ranges from $R = 2.5$ cm for the aorta down to $R = 1/1000$ cm for a capillary.

Total Revenue If a landlord receives income at a rate of \$18,000 a year from an apartment house, the revenue that is collected in, say, 15 years is

$$\text{rate} \times \text{time} = \$18{,}000 \times 15 = \$270{,}000$$

Observe that the total revenue, in this case, can also be written as the definite integral

$$\int_0^{15} 18{,}000\,dt = 18{,}000t\Big]_0^{15} = \$270{,}000$$

Suppose now that revenue flows continuously from some source at a rate of $f(t)$ dollars per year over a period of T years. What is the **total revenue** obtained over the interval $[0, T]$? As suggested in the preceding simple example, the answer can be expressed as a definite integral. The analysis is very similar to that of the cardiac output model.

Partition the interval $[0, T]$ into n subintervals of equal length $\Delta t = t_k - t_{k-1} = T/n$. The approximate revenue obtained over a period $[t_{k-1}, t_k]$ is

$$\Delta r_k = f(t_k^*)\,\Delta t$$

where t_k^* is some value in $[t_{k-1}, t_k]$. The sum

$$\sum_{k=1}^{n} \Delta r_k = \sum_{k=1}^{n} f(t_k^*)\,\Delta t$$

is approximately the revenue r obtained in T years. The total revenue r is obtained from this last expression by letting $n \to \infty$:

$$r = \int_0^T f(t)\,dt$$

Consumer and Producer Surpluses Economists have used the definite integral to define the concepts of **consumer** and **producer surpluses**.

The demand for a commodity by consumers, as well as the amount supplied to the market by manufacturers, can often be expressed as a function of the per unit price. Let $D(x)$ and $S(x)$ be the number of units demanded and the number of units supplied, respectively, when the commodity sells at a price x per unit. If the demand equals the supply,

$$D(x) = S(x)$$

the market is said to be in **equilibrium** and the corresponding price of the commodity is called the **equilibrium price**. If p is the equilibrium price and b is the price at which the demand for the commodity is zero ($D(b) = 0$), the integral

$$CS = \int_p^b D(x)\,dx \tag{6}$$

is called the **consumer surplus**. When the market is in equilibrium, the fact that there is a demand for the commodity at even higher prices means that those consumers who would have been willing to pay the higher price will benefit. The CS represents, in units of money, the combined savings realized by these consumers.

The integral

$$PS = \int_c^p S(x)\,dx \tag{7}$$

where $S(c) = 0$, is called the **producer surplus**. With the price of the commodity set at the equilibrium price p, the PS is the combined savings realized by those producers who would have been willing to supply the commodity at even lower prices.

EXAMPLE 2 Suppose the demand and supply of a commodity selling for x dollars a unit are $D(x) = 1000 - 20x$ and $S(x) = x^2 + 10x$, respectively. Find the consumer and producer surpluses.

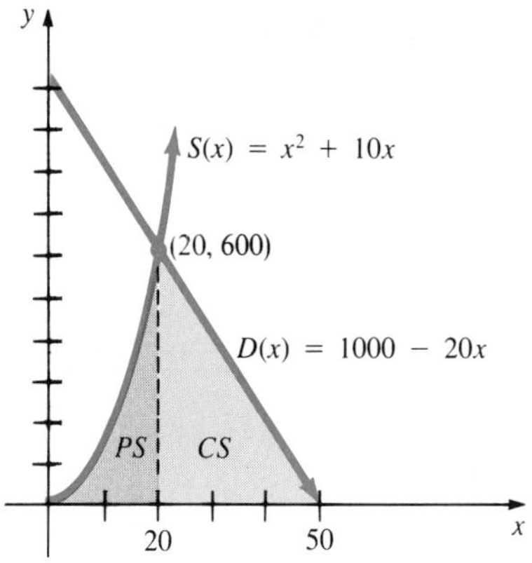

FIGURE 5.122

Solution It is apparent that $D(x) = 0$ when $b = 50$, $S(x) = 0$ when $c = 0$, and $D(x) = S(x)$ for $p = 20$. As shown in Figure 5.122, CS represents the area under the graph of $D(x)$ on the interval $[20, 50]$ and PS is the area under the graph of $S(x)$ on $[0, 20]$. From (6), we obtain

$$CS = \int_{20}^{50} (1000 - 20x)\,dx = (1000x - 10x^2)\Big]_{20}^{50} = \$9000$$

and from (7)

$$PS = \int_{0}^{20} (x^2 + 10x)\,dx = \left(\frac{x^3}{3} + 5x^2\right)\Big]_{0}^{20} \approx \$4666.67$$ □

EXERCISES 5.11

Answers to odd-numbered problems begin on page A-82.

1. Find the cardiac output of a heart over 24 s if 5 mg of dye are used and

$$c(t) = \begin{cases} 0, & 0 \le t < 2 \\ -\dfrac{1}{40}(t^2 - 26t + 48), & 2 \le t \le 24 \end{cases}$$

where $c(t)$ is measured in milligrams/liter.

2. Find the cardiac output of a heart over 30 s if 5 mg of dye are used and

$$c(t) = \begin{cases} \dfrac{4}{21}t, & 0 \le t < 15 \\ -\dfrac{4}{21}t + \dfrac{40}{7}, & 15 \le t \le 30 \end{cases}$$

where $c(t)$ is measured in milligrams/liter.

3. Find the average concentration in Problem 1 on $[0, 24]$.

4. Find the average concentration in Problem 2 on $[0, 30]$.

5. Assuming blood pressure is constant, determine the percentage decrease in the volume of blood that flows through an arteriole per unit time when the radius of the arteriole is constricted from 1 cm to 0.5 cm.

6. The average velocity of blood through an arteriole at a circular cross-section of radius R and area A is defined to be

$$\bar{v} = \frac{1}{A}\int_0^R 2\pi r v(r)\,dr$$

where $v(r)$ is given by (4).

(a) Show that $\bar{v} = PR^2/8vl$.

(b) Show that $\bar{v}$ is one-half the maximum velocity of the blood.

7. Find the total revenue obtained in 4 years if the rate of income in dollars/year is $f(t) = 1200(t - 5)^2$.

8. Find the total revenue obtained in 8 years if the rate of income in dollars/year is $f(t) = 600\sqrt{1 + 3t}$.

9. Find the average rate of flow of revenue over 10 years if the rate of income in dollars/year is given by $f(t) = 3t^2 + 4t + 200$.

10. Find the average rate of flow of revenue over 8 years if the rate of income is as given in Problem 8.

In Problems 11–16 find the consumer and producer surpluses.

11. $S(x) = 2x$, $D(x) = 100 - 2x$

12. $S(x) = \frac{3}{2}x$, $D(x) = 3000 - 6x$

13. $S(x) = x^2 - 4$, $D(x) = -x + 8$

14. $S(x) = 4x^2 + 4x$, $D(x) = -12x + 48$

15. $S(x) = 2x^2 + 3x$, $D(x) = 36 - x^2$

16. $S(x) = x^2 + 16x$, $D(x) = x^2 - 16x + 64$, $0 \le x \le 8$

17. Show that if the marginal revenue is $MR = f(x)$ and the marginal cost is $MC = g(x)$, then the change in the profit P between a production level of $x = a$ units and a production level of $x = b$ units is $\int_a^b [f(x) - g(x)]\,dx$.

18. Use Problem 17 to find the change in the profit when $f(x) = -6x + 60$, $g(x) = 15$, and the production level changes from 3 to 7 units.

19. **Pareto's law of income** states that the number of people with incomes between $x = a$ dollars and $x = b$ dollars is $N = \int_a^b Ax^{-k}\,dx$, where A is a constant and k is an empirical rational constant greater than 1. The average income of all these people is defined to be

$$\bar{x} = \frac{1}{N}\int_a^b Ax^{1-k}\,dx$$

Evaluate $\bar{x}$.

CHAPTER 5 REVIEW EXERCISES

Answers begin on page A-82.

In Problems 1–12 answer true or false.

1. When $\int_a^b f(x)\,dx > 0$, the integral gives the area under the graph of $y = f(x)$ on the interval $[a, b]$. ________

2. $\int_0^3 (x - 1)\,dx$ is the area under the graph of $y = x - 1$ on $[0, 3]$. ________

3. The integral $\int_a^b [f(x) - g(x)]\,dx$ gives the area between the graphs of the continuous functions f and g whenever $f(x) \geq g(x)$ for every x in $[a, b]$. ________

4. The disk and washer methods for finding volumes of solids of revolution are special cases of the slicing method. ________

5. The average value $A(f)$ of a continuous function on an interval $[a, b]$ is necessarily a number that satisfies $m \leq A(f) \leq M$, where m and M are the maximum and minimum values of f on the interval, respectively. ________

6. If f and g are continuous on $[a, b]$, then the average value of $f + g$ is $A(f + g) = A(f) + A(g)$. ________

7. The center of mass of a pencil with a constant linear density ρ is at its geometric center. ________

8. The center of mass of a lamina that coincides with a plane region R is a point in R at which the lamina would hang in balance. ________

9. The pressure on the flat bottom of a swimming pool is the same as the horizontal pressure on the vertical sidewalls at the same depth. ________

10. Consider a circular tin can with radius 6 in and a circular reservoir with radius 50 ft. If each has a flat bottom and is filled with water to a depth of 1 ft, then the liquid pressure on the bottom of the reservoir is greater than the pressure on the bottom of the tin can. ________

11. If $s(t)$ is the position function of a body that moves in a straight line, then $\int_{t_1}^{t_2} v(t)\,dt$ is the distance the body moves in the interval $[t_1, t_2]$. ________

12. In the absence of air resistance, when dropped simultaneously from the same height, a cannonball will hit the ground before a marshmallow. ________

In Problems 13–18 fill in the blanks.

13. The unit of work in the SI system of units is ________.

14. To warm up, a 200-lb jogger pushes against a tree for 5 min with a constant force of 60 lb and then runs 2 mi in 10 min. The total work done is ________.

15. The work done by a 100-lb constant force applied at an angle of 60° to the horizontal over a distance of 50 ft is ________.

16. If 80 N of force stretches a spring that is initially 1 m long into a spring that is 1.5 m long, then the spring will measure ________ m long when 100 N of force is applied.

17. The coordinates of the centroid of a region R are (2, 5) and the moment of the region about the x-axis is 30. Hence, the area of R is ________ square units.

18. The graph of a function with a continuous first derivative is said to be ________.

In Problems 19–28 set up, but do not evaluate, the integral(s) that give the area of the shaded region in each figure.

19.

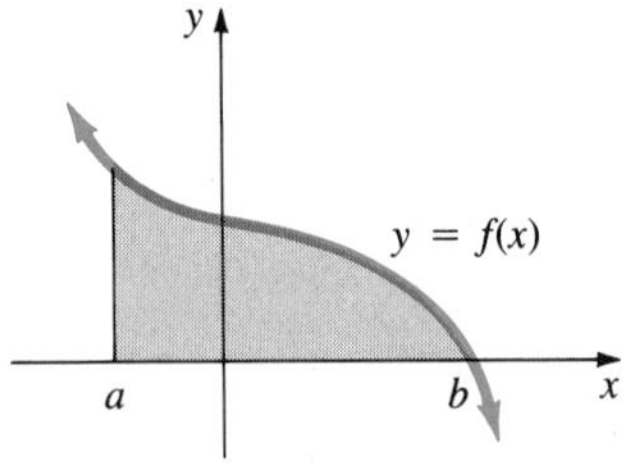

FIGURE 5.123

20.

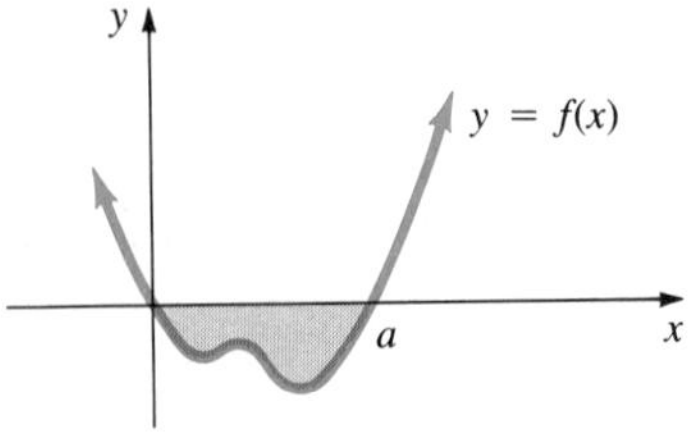

FIGURE 5.124

21.

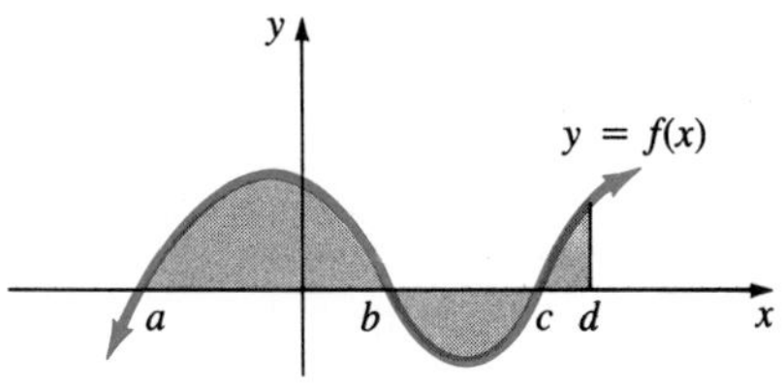

FIGURE 5.125

22.

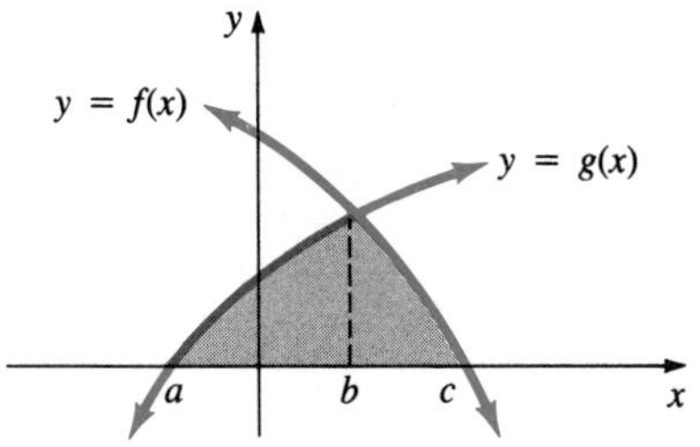

FIGURE 5.126

23.

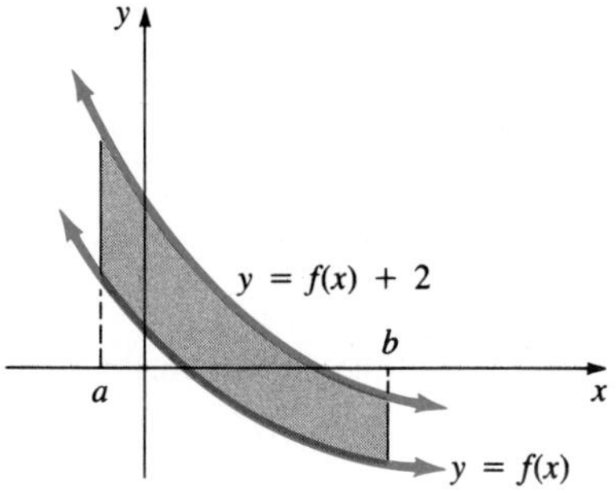

FIGURE 5.127

24.

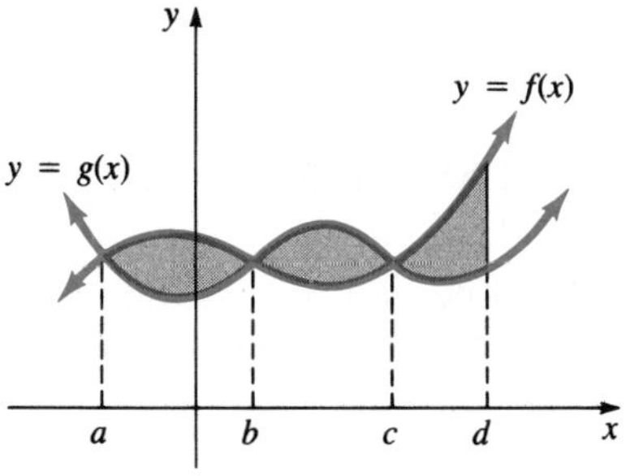

FIGURE 5.128

25.

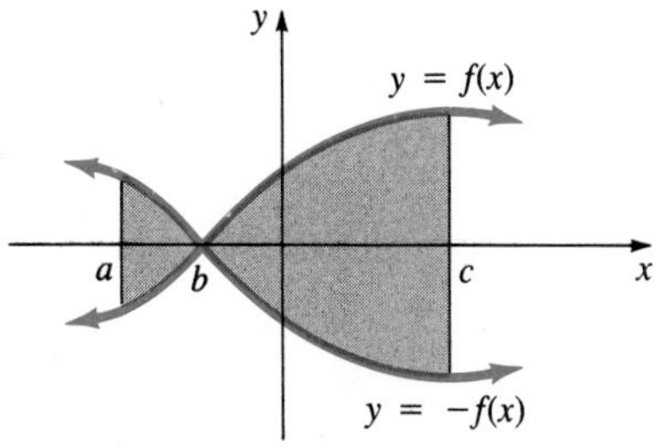

FIGURE 5.129

26.

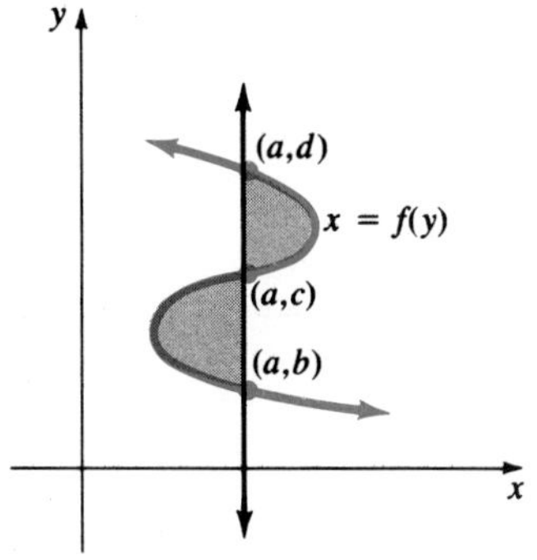

FIGURE 5.130

27.

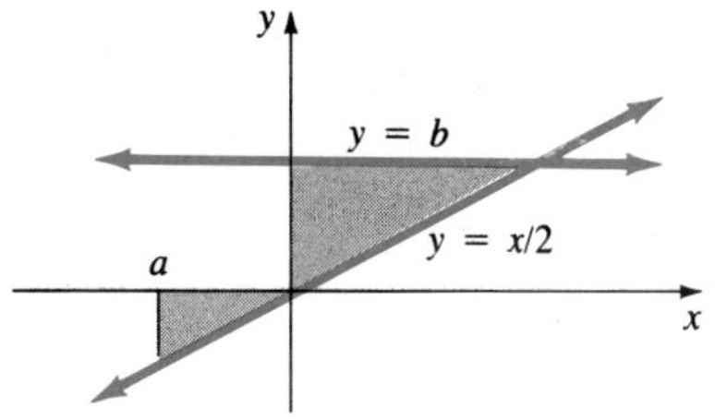

FIGURE 5.131

28.

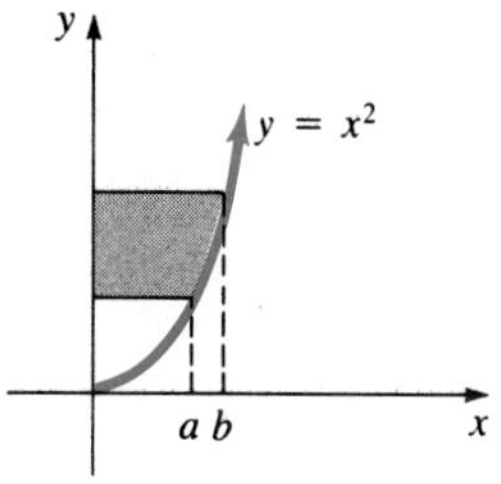

FIGURE 5.132

In Problems 29–33 consider the shaded region R in Figure 5.133. Set up, but do not evaluate, the integral(s) that give the indicated quantity.

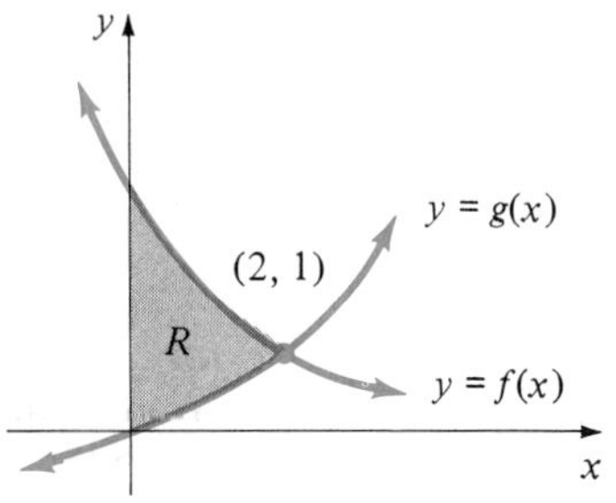

FIGURE 5.133

29. The centroid of the region

30. The volume of the solid of revolution that is formed by revolving R about the x-axis

31. The volume of the solid of revolution that is formed by revolving R about the y-axis

32. The volume of the solid of revolution that is formed by revolving R about the line $x = 2$

33. The volume of the solid with R as its base such that the cross-sections of the solid parallel to the y-axis are squares

34. A solid has as its base the region bounded by the graph of $y = \sin x$ and the x-axis on the interval $[0, \pi]$. As shown in Figure 5.134, cross-sections of the solid perpendicular to the x-axis are right isosceles triangles. Find the volume of the solid.

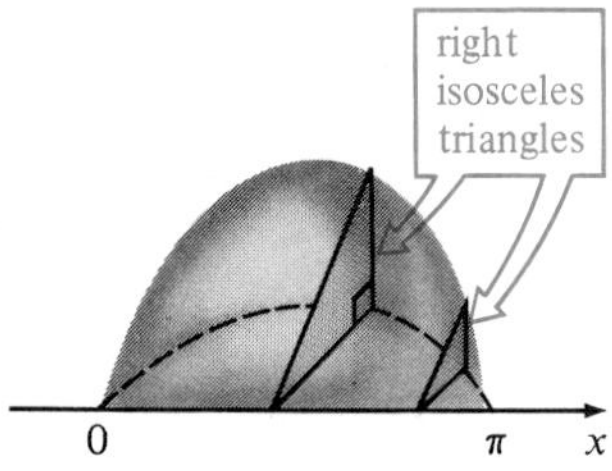

FIGURE 5.134

35. A nose cone of a rocket is to be covered with canvas except for a section at its apex. Find the area of the canvas needed for the nose cone whose radius and altitude are shown in Figure 5.135.

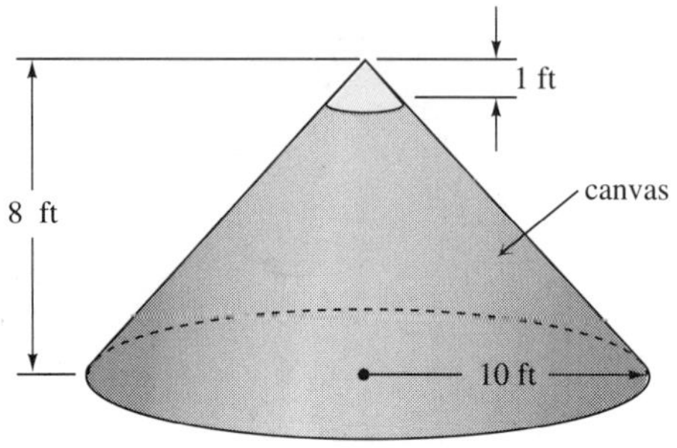

FIGURE 5.135

36. The area under the graph of a continuous nonnegative function $y = f(x)$ on the interval $[-3, 4]$ is 21 square units. What is the average value of the function on the interval?

37. Find the average value of $f(x) = x^{3/2} + x^{1/2}$ on $[1, 4]$.

38. Find a value of x in the interval $[0, 3]$ that corresponds to the average value of the function $f(x) = 2x - 1$.

39. A spring whose unstretched length is $\frac{1}{2}$ m is stretched to a length of 1 m by a force of 50 N. Find the work done in stretching the spring from a length of 1 m to a length of 1.5 m.

40. The work done in stretching a spring 6 in beyond its natural length is 10 ft-lb. Find the spring constant.

41. A water tank, in the form of a cube that is 10 ft on a side, is filled with water. Find the work done in pumping all the water to a point 5 ft above the tank.

42. A bucket weighing 2 lb contains 30 lb of liquid. As the bucket is raised vertically at a rate of 1 ft/s, the liquid leaks out at a rate of $\frac{1}{4}$ lb/s. Find the work done in lifting the bucket a distance of 5 ft.

43. In Problem 42 find the work done in lifting the bucket to a point where it is empty.

44. In Problem 42 find the work done in lifting the leaking bucket a distance of 5 ft if the rope attached to the bucket weighs $\frac{1}{8}$ lb/ft.

45. A tank on top of a tower 15 ft high consists of a frustum of a cone surmounted by a right circular cylinder. The dimensions (in feet) are given in Figure 5.136. Find the work done in filling the tank with water from ground level.

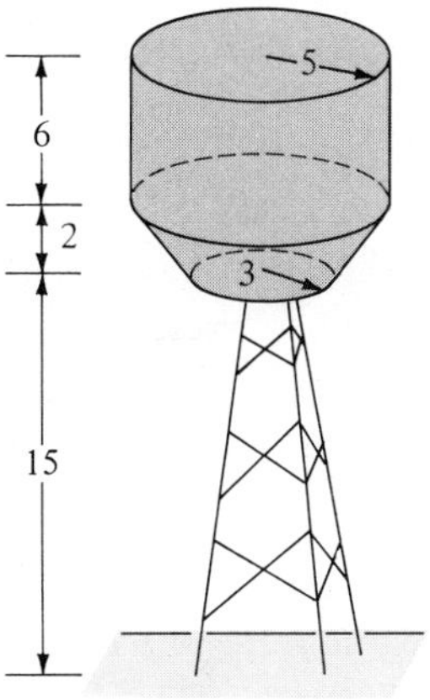

FIGURE 5.136

46. A rock is thrown vertically upward from the surface of the moon with an initial velocity of 44 ft/s.

(a) If the acceleration of gravity on the moon is 5.5 ft/s^2, find the maximum height attained. Compare with the earth.

(b) On the way down, the rock hits a 6-ft-tall astronaut on the head. What is its impact velocity?

47. Find the length of the graph of $y = (x - 1)^{3/2}$ from (1, 0) to (5, 8).

48. The linear density of a 6-m-long rod is a linear function of the distance from its left end. The density in the middle of the rod is 11 kg/m and at the right end the density is 17 kg/m. Find the center of mass of the rod.

49. A flat plate, in the form of a quarter-circle, is submerged vertically in oil as shown in Figure 5.137. If the weight density of the oil is 800 kg/m^3, find the force exerted by the oil on one side of the plate.

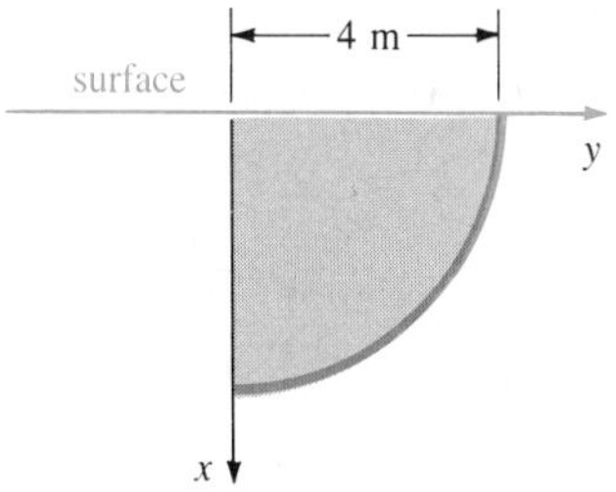

FIGURE 5.137

50. A uniform metal bar of mass 4 kg and length 2 m supports two masses, as shown in Figure 5.138. Where should the wire be attached to the bar so that the system hangs in balance?

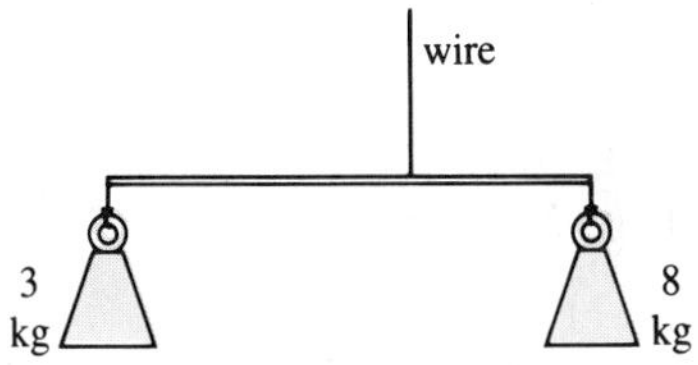

FIGURE 5.138

51. Three masses are suspended from uniform rigid bars of negligible mass as shown in Figure 5.139. Determine where the indicated wires should be attached so that the entire system hangs in balance.

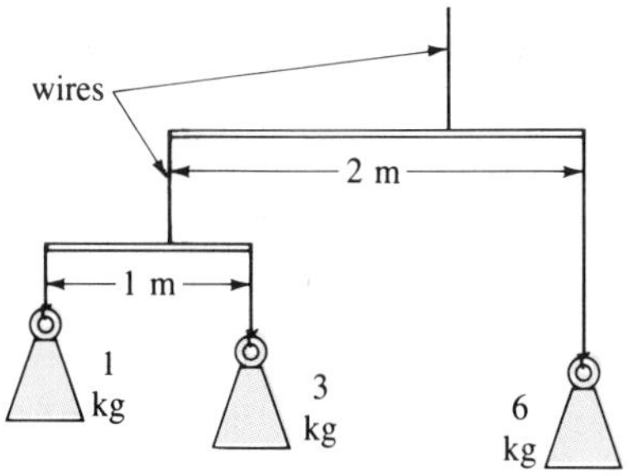

FIGURE 5.139

6

LOGARITHMIC AND EXPONENTIAL FUNCTIONS

INTRODUCTION

There are still a few gaps in our knowledge of the derivative. Up to now, we have been able to differentiate $y = x^n$ only when the exponent n is a rational number. Moreover, we have not encountered any function whose derivative is $1/x = x^{-1}$. This means we can evaluate $\int x^n\,dx$ only when n is rational, with the exception of the rational number $n = -1$.

In this chapter, we shall study two more transcendental functions. These enable us to answer the following questions. Can meaning be given to expressions such as x^π? Is $y = x^\pi$ a function? If so, what is its derivative? What is $\int x^{-1}\,dx$? What is $\int x^n\,dx$ when n is irrational?

In Section 5.5, we found the area of a surface obtained by revolving the graph of a function about an axis. Of all the surfaces generated on the interval $[a, b]$ by revolving graphs about the x-axis, there is only one surface of revolution that has the least area. When two circular rings are immersed together in soapy water, held vertically, and then pulled slowly apart, the soap film stretched between the rings will assume the same shape as this surface of revolution with minimum area. We shall see what graph generates this surface in Section 6.7.

soap film has minimum area

IMPORTANT CONCEPTS

- **one-to-one function**
- **inverse function**
- **natural logarithmic function**
- **derivative of the natural logarithmic function**
- **laws of logarithms**
- **natural exponential function**
- **derivative of the natural exponential function**
- **laws of exponents**
- **power rule for real exponents**
- **logarithmic differentiation**
- **hyperbolic functions**
- **catenary**
- **growth and decay**
- **Newton's law of cooling**
- **linear first-order differential equations**
- **integrating factor**

6.1 INVERSE FUNCTIONS

One-to-One Functions Recall that a function is a rule of correspondence that assigns to each value x, in its domain X, a single, or unique, value y in its range. This rule does not preclude having the same number y associated with *two different* values of x. For example, for $f(x) = x^2$, the value $y = 9$ occurs at either $x = -3$ or $x = 3$; that is, $f(-3) = f(3) = 9$. On the other hand, for the function $f(x) = x^3$, the value $y = 64$ occurs *only* at $x = 4$. Functions of the latter kind are given the special name **one-to-one**.

DEFINITION 6.1 One-to-One Function

A function f is said to be **one-to-one** if every element in its range corresponds to exactly one element in its domain X.

Horizontal Line Test Interpreted geometrically, this means that a horizontal line (y = constant) can intersect the graph of a one-to-one function in at most one point. Furthermore, if every horizontal line that intersects the graph of a function does so in at most one point, then the function is necessarily one-to-one. A function is *not* one-to-one if *some* horizontal line intersects its graph more than once.

EXAMPLE 1 The graphs of $g(x) = -x^2 + 2x + 4$ and $f(x) = -4x + 3$, shown in Figure 6.1, indicate that there are two elements x_1 and x_2 in the domain of g for which $g(x_1) = g(x_2) = c$ but only one element x_1 in the domain of f for which $f(x_1) = c$. Hence, g is not one-to-one but f is one-to-one.

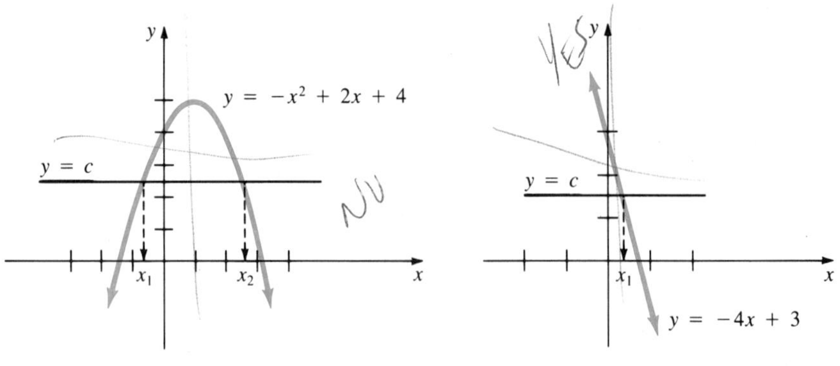

(a) not one-to-one (b) one-to-one

FIGURE 6.1 □

Inverse of a One-to-One Function Suppose f is a one-to-one function that has domain X and range Y. Since every element y of Y corresponds

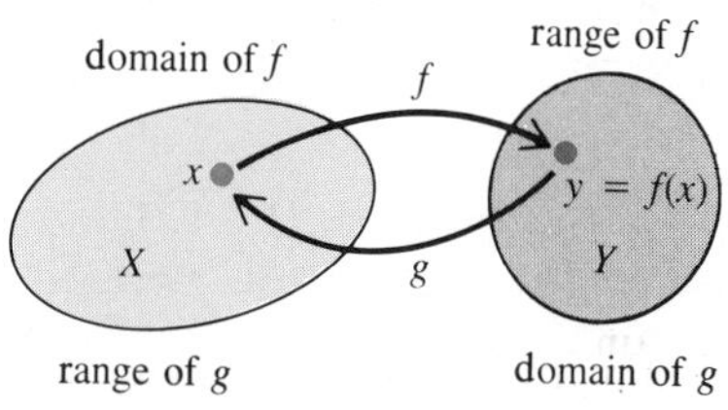

FIGURE 6.2

to precisely one element x of X, the function f must actually determine a "reverse" function g whose domain is Y and range is X. As shown in Figure 6.2, f and g must satisfy

$$f(x) = y \quad \text{and} \quad g(y) = x$$

or

$$f(g(y)) = y \quad \text{and} \quad g(f(x)) = x \tag{1}$$

The function g is given the formal name **inverse of f**. Denoting each independent variable as x, we summarize the results in (1).

Ex 2) $1/(2x-3)$

$x = 1/(2y-3)$

$2y-3 = \frac{1}{x}$

$2y = \frac{1}{x} + 3$

$y = \frac{1}{2x} + \frac{3x}{x}$

$y = \frac{1+3x}{2x}$

DEFINITION 6.2 Inverse of a Function

Let f be a one-to-one function with domain X and range Y. The **inverse** of f is a function g with domain Y and range X for which

$$f(g(x)) = x \quad \text{for every } x \text{ in } Y$$

and

$$g(f(x)) = x \quad \text{for every } x \text{ in } X \tag{2}$$

The inverse of a function f is usually written f^{-1} and is read "f inverse." This latter notation, though standard, is somewhat unfortunate. We hasten to point out that $f^{-1}(x)$ is *not* the same as $[f(x)]^{-1} = 1/f(x)$. In terms of this new notation, (2) becomes

$$f(f^{-1}(x)) = x \quad \text{and} \quad f^{-1}(f(x)) = x \tag{3}$$

Finding f^{-1} The first equation in (3) can be used explicitly to find the inverse of a one-to-one function.

EXAMPLE 2 Find the inverse of $f(x) = \dfrac{1}{2x-3}$, $x \neq \dfrac{3}{2}$.

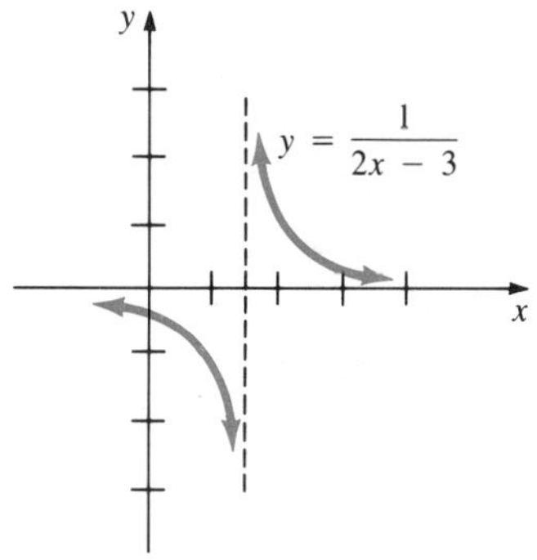

FIGURE 6.3

Solution Inspection of Figure 6.3 shows that f is one-to-one. Hence, from (3) we have

$$f(f^{-1}(x)) = \frac{1}{2f^{-1}(x) - 3} = x$$

Solving the equation $2f^{-1}(x) - 3 = 1/x$ for $f^{-1}(x)$ gives

$$f^{-1}(x) = \frac{3x+1}{2x}$$

□

Observe in Example 2 that, as discussed, the domain of f (all real numbers except $\frac{3}{2}$) and the range of f (all real numbers except 0) are the range and domain of f^{-1}, respectively.

Ex 3 | $-4x+3$

$x = -4y + 3$

$-4y = x - 3$

$y = -\frac{1}{4}x + \frac{3}{4}$

An Alternative Way of Finding f^{-1} The inverse of a function f can be found in a different manner. If g is the inverse of f, then $x = g(y)$. Thus, we need only do the following:

(*i*) Solve $y = f(x)$ for the symbol x and, following convention,

(*ii*) relabel the dependent variable x as y and the independent variable y as x.

The next example illustrates this procedure.

EXAMPLE 3 Find the inverse of $f(x) = -4x + 3$.

Solution From Example 1 we know that f is one-to-one and hence has an inverse. Now, write f as $y = -4x + 3$ and solve for x in terms of y:

$$x = \frac{-y+3}{4}$$

Relabeling variables then yields

$$y = \frac{-x+3}{4} \quad \text{or} \quad f^{-1}(x) = -\frac{1}{4}x + \frac{3}{4}$$ □

Graphs of f and f^{-1} Let (a, b) be any point on the graph of a one-to-one function f. Then $f(a) = b$ and

$$f^{-1}(b) = f^{-1}(f(a)) = a$$

implies that (b, a) is on the graph of f^{-1}. Since the points (a, b) and (b, a) are symmetric with respect to the line $y = x$, we see in Figure 6.4(b) that the graph of f^{-1} is a reflection of the graph of f in the line $y = x$.

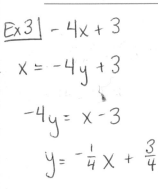

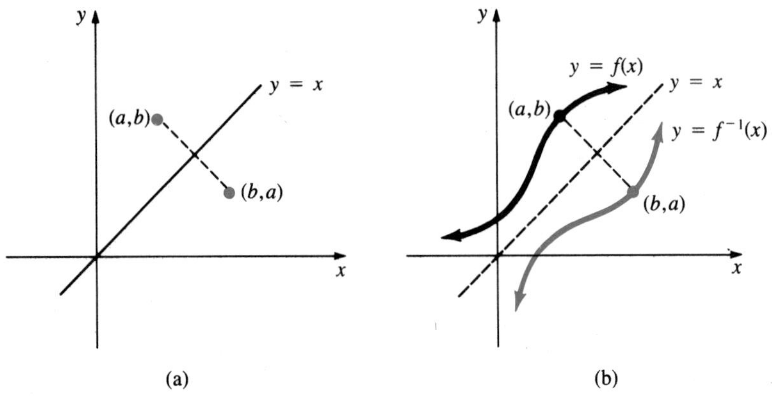

FIGURE 6.4

EXAMPLE 4 Compare the graphs of f and f^{-1} from Example 3.

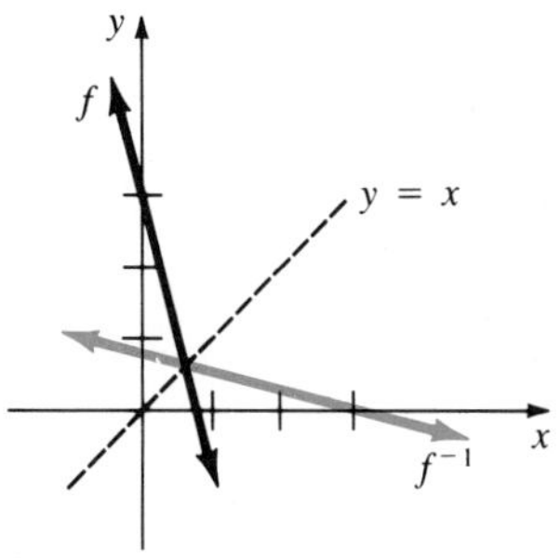

FIGURE 6.5

Solution The graphs of $f(x) = -4x + 3$ and $f^{-1}(x) = -\frac{1}{4}x + \frac{3}{4}$ are straight lines. The x-intercept and y-intercept of the graph of f are $\frac{3}{4}$ and 3, respectively. On the other hand, the x- and y-intercepts of the graph of f^{-1} are, in turn, 3 and $\frac{3}{4}$. Drawing the lines through the corresponding points yields the graphs shown in Figure 6.5. □

EXAMPLE 5 Find the inverse of $f(x) = x^3$ and compare the graphs of f and f^{-1}.

Solution By solving the equation $y = x^3$ for x, we obtain $x = y^{1/3}$. With the usual relabeling of variables, we see that $f^{-1}(x) = x^{1/3}$. The fact that the graphs of f and f^{-1} are reflections of one another in the line $y = x$ should be apparent in Figure 6.6. □

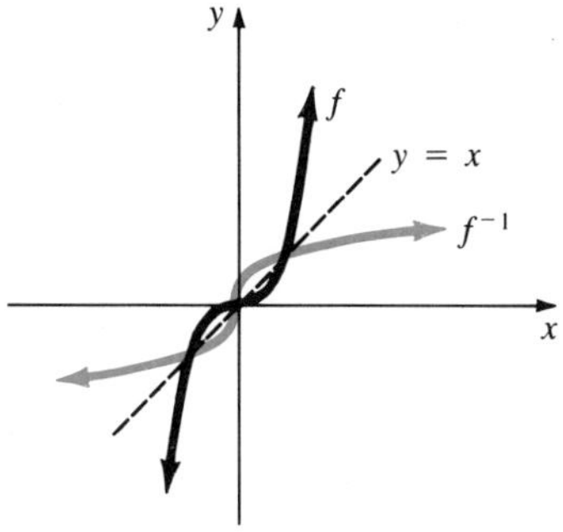

FIGURE 6.6

Continuity of f^{-1} Although we state the next two theorems without proof, their plausibility follows from the fact that the graph of f^{-1} is a reflection of the graph of f in the line $y = x$.

THEOREM 6.1

Let f be a continuous one-to-one function on an interval $[a, b]$. Then f^{-1} is continuous on the interval $[f(a), f(b)]$.

THEOREM 6.2

Let f be continuous and increasing on an interval $[a, b]$. Then f^{-1} exists and is continuous and increasing on $[f(a), f(b)]$.

EXAMPLE 6 Prove that $f(x) = 5x^3 + 8x - 9$ has an inverse.

Solution Since f is a polynomial function, it is continuous. Also,

$$f'(x) = 15x^2 + 8 > 0 \qquad \text{for all } x$$

implies that f is increasing on $(-\infty, \infty)$. It follows from Theorem 6.2 that f^{-1} exists. □

Note: Theorem 6.2 also holds when the word *increasing* is replaced with the word *decreasing*.

Derivative of f^{-1} In the next result it is convenient to denote, as we originally did, the inverse of a function $y = f(x)$ by $x = g(y)$.

POINT (1,4)

Ex. 7) $5x^3 + 8x - 9$

$f'(x) = 15x^2 + 8$

$f'(1) = 15 + 8 = 23$

$(f^{-1}(4))' = 1/23$

Ex. 6

$f'(x) = 15x^2 + 8$ is always positive ∴ for all x

$f(x)$ is always increasing for all x

$f^{-1}(x)$ exists [THEOREM 6.2]

THEOREM 6.3 Derivative of an Inverse Function

Let f be a differentiable function that has an inverse g. Then g is differentiable when $dy/dx = f'(x) \neq 0$ and

$$\frac{dx}{dy} = \frac{1}{dy/dx} \tag{4}$$

Proof Since f is differentiable, it is continuous so that

$$\Delta y = f(x + \Delta x) - f(x) \to 0 \quad \text{when} \quad \Delta x \to 0$$

By Theorem 6.1, g is continuous; therefore, $\Delta x \to 0$ whenever $\Delta y \to 0$. Thus,

$$\frac{dx}{dy} = \lim_{\Delta y \to 0} \frac{\Delta x}{\Delta y} = \lim_{\Delta x \to 0} \frac{1}{\Delta y/\Delta x} = \frac{1}{\displaystyle\lim_{\Delta x \to 0} \frac{\Delta y}{\Delta x}} = \frac{1}{dy/dx}$$ ■

Although the notation in (4) facilitated its proof and has obvious mnemonic significance, it is desirable to write the result as

$$g'(y) = \frac{1}{f'(x)} = \frac{1}{f'(g(y))} \tag{5}$$

With relabeling of variables, (5) becomes

$$g'(x) = \frac{1}{f'(g(x))} \tag{6}$$

Now if f^{-1} denotes the inverse instead of g, then (6) finally becomes

$$(f^{-1})'(x) = \frac{1}{f'(f^{-1}(x))} \tag{7}$$

Equation (7) clearly shows that to compute the derivative of f^{-1} we must know $f^{-1}(x)$ explicitly. However, if (a, b) is a known point on the graph of f, (7) does enable us to evaluate the derivatives of f^{-1} at (b, a) without an equation that defines $f^{-1}(x)$.

EXAMPLE 7 For the function f in Example 6, find the slope of the tangent to the graph of f^{-1} at $(f(1), 1)$.

Solution Since $f(x) = 5x^3 + 8x - 9$, we have $f(1) = 4$,

$$f'(x) = 15x^2 + 8 \quad \text{and} \quad f'(1) = 23$$

But $f(1) = 4$ implies that $f^{-1}(4) = 1$. Thus, (7) gives

$$(f^{-1})'(4) = \frac{1}{f'(f^{-1}(4))} = \frac{1}{f'(1)} = \frac{1}{23}$$

In other words, the slope of the tangent to the graph of f at (1, 4) is 23; the slope of the tangent to the graph of f^{-1} at (4, 1) is the reciprocal, $\frac{1}{23}$. □

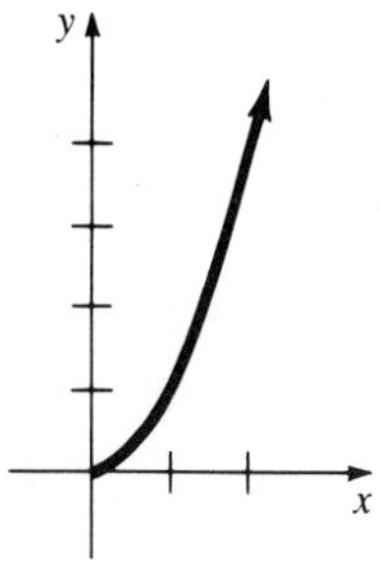

(a) $F(x) = x^2$, $x \geq 0$

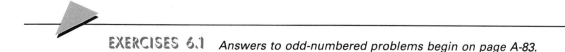

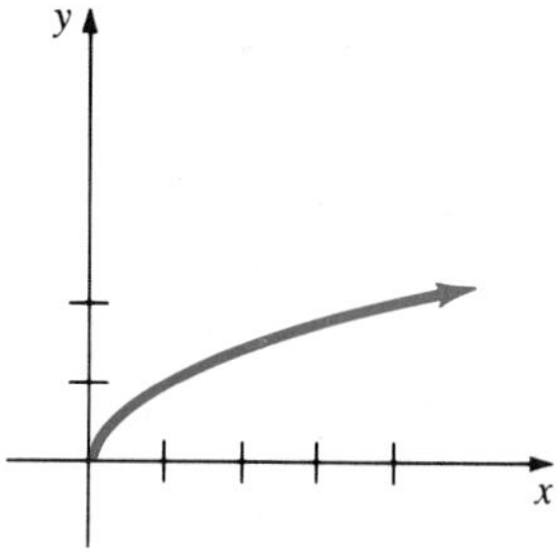

(b) $F^{-1}(x) = \sqrt{x}$, $x \geq 0$

FIGURE 6.7

In the subsequent sections, we shall see that it is often possible to find the derivative of an inverse function by resorting to implicit differentiation.

If a function f is one-to-one, then it has an inverse. Conversely, if f has an inverse, then it is one-to-one. Thus, if a function f is *not* one-to-one, it does *not* possess an inverse. Nonetheless, as the next example shows, it may be possible to restrict the domain of a function that is not one-to-one in such a manner that the newly defined function does, in fact, have an inverse. This notion will be important in Chapter 7.

EXAMPLE 8 In our initial discussion, we saw that $f(x) = x^2$ is not one-to-one. However, by simply requiring x to be nonnegative, we see from Figure 6.7(a) and Theorem 6.2 that the new function

$$F(x) = x^2, x \geq 0 \quad \text{has the inverse} \quad F^{-1}(x) = \sqrt{x}, x \geq 0$$

which is shown in Figure 6.7(b). □

EXERCISES 6.1

Answers to odd-numbered problems begin on page A-83.

In Problems 1–12 determine whether the given function is one-to-one by examining its graph. If the function is one-to-one, find its inverse.

1. $f(x) = 5$

2. $f(x) = 6x - 9$

3. $f(x) = \frac{1}{3}x + 3$

4. $f(x) = |x + 1|$

5. $f(x) = x(x - 5)$

6. $f(x) = (x + 1)^2$

7. $f(x) = x^3 - 8$

8. $f(x) = x^3 - 3x$

9. $f(x) = \dfrac{4}{x}$

10. $f(x) = \dfrac{1}{3x + 6}$

11. $f(x) = x^4 + 2$

12. $f(x) = x^5$

In Problems 13–22 each function is one-to-one. Find its inverse.

13. $f(x) = x + 5$

14. $f(x) = -2x + 8$

15. $f(x) = 3x^3 + 7$

16. $f(x) = (-x + 9)^3$

17. $f(x) = \sqrt[3]{2x - 4}$

18. $f(x) = 6 - (10x - 2)^{1/2}$

19. $f(x) = \dfrac{1}{1 + 8x^3}$

20. $f(x) = 5 - \dfrac{2}{x}$

21. $f(x) = \dfrac{2 - x}{1 - x}$

22. $f(x) = 10 + x^{3/5}$

In Problems 23–26 determine, without graphing, whether the given function has an inverse.

23. $f(x) = 10x^3 + 8x + 12$

24. $f(x) = -7x^5 - 6x^3 - 2x + 17$

25. $f(x) = 2x^3 + 3x^2 - 12x$

26. $f(x) = x^4 - 2x^2$

In Problems 27–32, without finding the inverse, state the domain and range of f^{-1}.

27. $f(x) = \sqrt{x + 2}$

28. $f(x) = 3 + \sqrt{2x - 1}$

29. $f(x) = \dfrac{1}{x + 3}$

30. $f(x) = \dfrac{x - 1}{x - 4}$

31. $f(x) = (x - 5)^2, x \geq 5$

32. $f(x) = x^2 + 2x + 6, x \geq -1$

In Problems 33 and 34 sketch the graph of f^{-1} from the graph of f.

33.

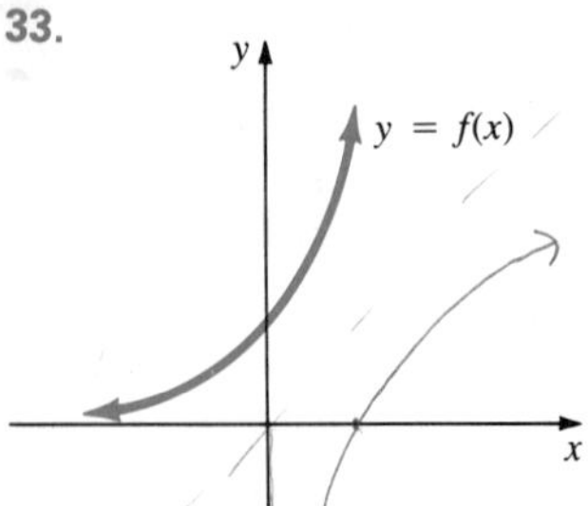

FIGURE 6.8

34.

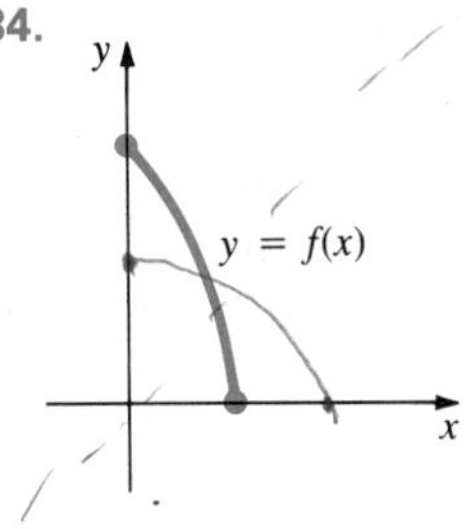

FIGURE 6.9

In Problems 35 and 36 sketch the graph of f from the graph of f^{-1}.

35.

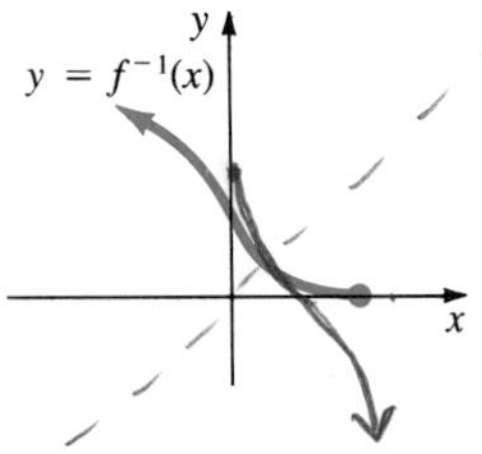

FIGURE 6.10

36.

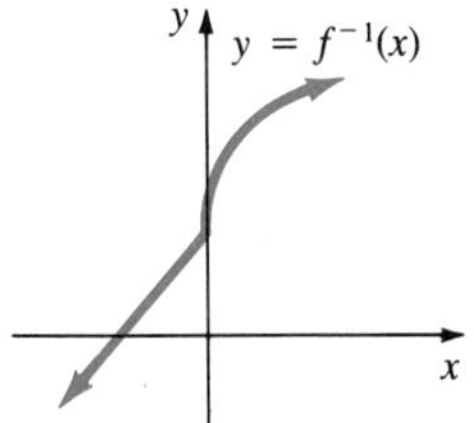

FIGURE 6.11

In Problems 37 and 38 use (7) to find the derivative of the inverse function at the indicated point.

37. $f(x) = 2x^3 + 8;\ (f(\frac{1}{2}), \frac{1}{2})$

38. $f(x) = -x^3 - 3x + 7;\ (f(-1), -1)$

In Problems 39–42, without finding the inverse, find, at the indicated value of x, the corresponding point on the graph of f^{-1} and an equation of the tangent line at this point.

39. $f(x) = \frac{1}{3}x^3 + x - 7;\ x = 3$

40. $f(x) = \dfrac{2x+1}{4x-1};\ x = 0$

41. $f(x) = (x^5 + 1)^3;\ x = 1$

42. $f(x) = 8 - 6\sqrt[3]{x+2};\ x = -3$

In Problems 43 and 44 show, without integrating or graphing, that the given function is one-to-one on the indicated interval. Without finding f^{-1}, find $(f^{-1})'(0)$ in each case.

43. $f(x) = \displaystyle\int_1^x \frac{dt}{t+1},\ x > -1$

44. $f(x) = \displaystyle\int_1^x \sqrt{t^2+4}\,dt,\ (-\infty, \infty)$

In Problems 45 and 46 find f^{-1}. Find $(f^{-1})'$ using (7). Verify the result by direct differentiation of f^{-1}.

45. $f(x) = \dfrac{2x+1}{x}$

46. $f(x) = (5x + 7)^3$

In Problems 47–50 define a new function F, by restricting the domain of f, that is one-to-one and has the same range as f. Find F^{-1}.

47. $f(x) = (5 - 2x)^2$

48. $f(x) = 3x^2 + 9$

49. $f(x) = x^2 + 2x + 4$

50. $f(x) = -x^2 + 8x$

51. Consider the one-to-one function $f(x) = x^3 + x$ on the interval $[1, 2]$. See Figure 6.12. Without finding f^{-1}, determine the value of $\int_{f(1)}^{f(2)} f^{-1}(x)\,dx$.

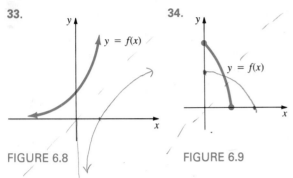

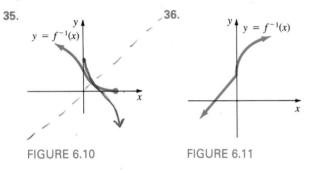

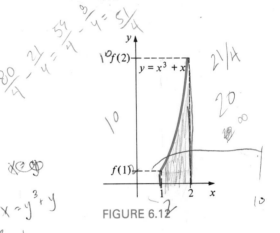

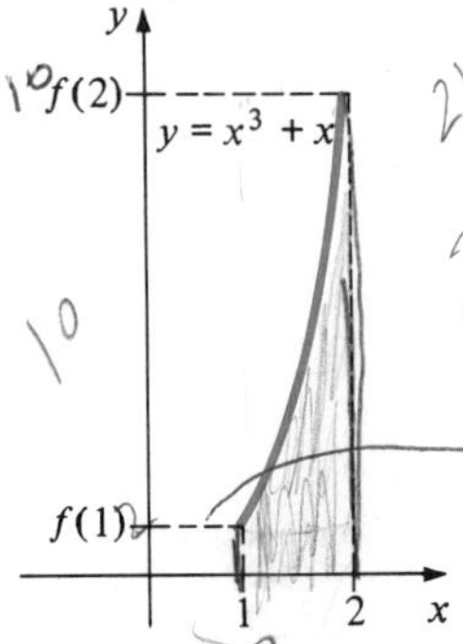

FIGURE 6.12

52. Suppose a function f is one-to-one on $(-\infty, \infty)$ and has the properties $f(0) = 1$; $f(-2) = 0$; $f''(x) > 0$, $x < 0$, $f''(x) < 0$, $x > 0$; $\lim_{x\to-\infty} f(x) = -1$; and $\lim_{x\to\infty} f(x) = 2$. Sketch a graph of f^{-1}.

53. An equivalent definition of a one-to-one function is: A function f is one-to-one if and only if $x_1 \neq x_2$ implies that $f(x_1) \neq f(x_2)$ whenever x_1 and x_2 are in the domain of x. Use this definition to prove that $f(x) = 3x + 7$ is one-to-one.

54. Use the definition of a one-to-one function given in Problem 53 to prove that a function that is either increasing or decreasing is one-to-one. [*Hint:* $x_1 \neq x_2$ implies $x_1 < x_2$ or $x_1 > x_2$.]

55. Suppose f is a periodic function with period p, $f(x + p) = f(x)$. Can f have an inverse function?

56. Let f^{-1} be the inverse of a function f. If f and f^{-1} are differentiable, use implicit differentiation and $f(f^{-1}(x)) = x$ to derive the result given in (7).

57. If f and $(f^{-1})'$ are differentiable, use (7) to show that $(f^{-1})''(x) = -f''(f^{-1}(x))/[f'(f^{-1}(x))]^3$.

58. If the functions f and g have inverses, then it can be proved that

$$(f \circ g)^{-1} = g^{-1} \circ f^{-1}$$

Verify this result for $f(x) = x^3$ and $g(x) = 4x + 5$.

59. The equation $y = \sqrt[3]{x} - \sqrt[3]{y}$ defines a one-to-one function $y = f(x)$. Find $f^{-1}(x)$.

Calculator/Computer Problem

60. (a) Use a calculator or computer to obtain the graph of f^{-1} found in Problem 59.

(b) Sketch the graph of f.

6.2 THE NATURAL LOGARITHMIC FUNCTION

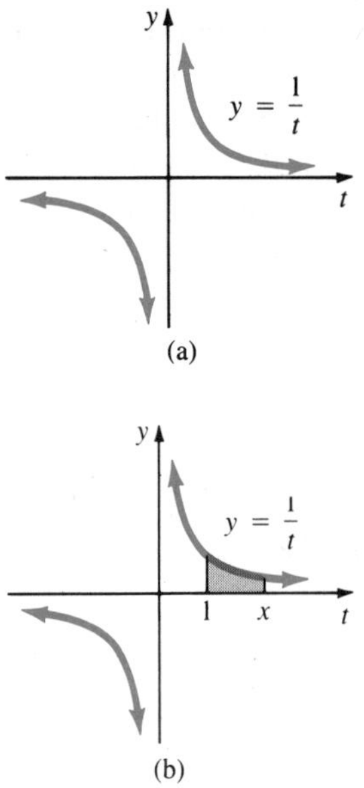

FIGURE 6.13

The function $y = 1/t$, which is discontinuous at $t = 0$, is shown in Figure 6.13(a). As indicated in Figure 6.13(b), $1/t > 0$ for $t > 0$ so that the area under the graph on the interval $[1, x]$ is given by $\int_1^x dt/t$. This seemingly innocent looking integral defines a function of x of such great importance that it is given a special name.

DEFINITION 6.3 Natural Logarithmic Function

The **natural logarithmic function**, denoted by $\ln x$, is defined by

$$\ln x = \int_1^x \frac{dt}{t} \tag{1}$$

for all $x > 0$.

Usually the symbol $\ln x$ is pronounced "ell-en of x."

EXAMPLE 1 Estimate the value of ln 2.

Solution In Section 4.8 we used the Midpoint Rule to estimate $\int_1^2 dt/t$. Example 1 of that section reveals that

$$\ln 2 = \int_1^2 \frac{dt}{t} \approx 0.6919$$ □

Consideration of the areas shown in Figure 6.14 establishes an inequality for the value of ln 2:

$$\frac{1}{2} < \ln 2 < 1 \tag{2}$$

The result in (2) will prove to be useful in graphing the natural logarithmic function.

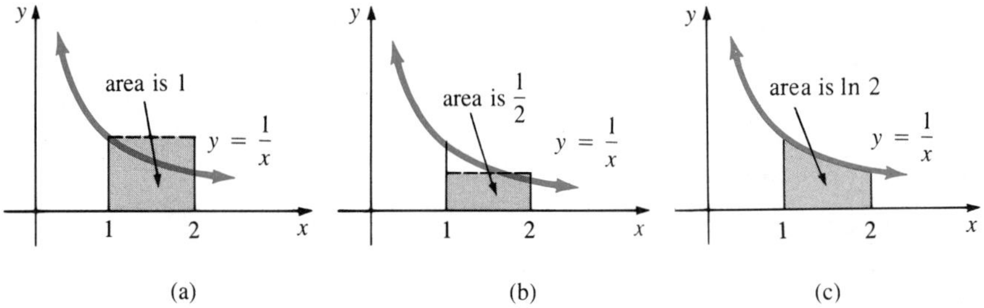

FIGURE 6.14

The word *logarithm* undoubtedly brings back memories from precalculus mathematics of formidable computations with **common**, or **base 10**, logarithms. It is important to realize that (1) is *not* the base 10 logarithm. Although it may seem like strange advice, it is best to forget about the notion of the base of a logarithm for the discussion of this section. Our immediate goal is to show that ln x "behaves like," or more precisely, has the same properties as, the common logarithm. However, you will not be told at this point why (1) is called "a natural" function. By the end of this chapter the answer should be obvious.

Derivative of* ln *x The derivative of the natural logarithmic function can be found readily from the derivative form of the Fundamental Theorem of Calculus. Recall that $\frac{d}{dx}\int_a^x f(t)\,dt = f(x)$. Hence, $\frac{d}{dx}\int_1^x (1/t)\,dt = 1/x$. In other words,

$$\frac{d}{dx}\ln x = \frac{1}{x}, \qquad x > 0 \tag{3}$$

Note: Inherent in (1) and (3) is the answer to one of the questions posed in the introduction to this chapter—namely,

the natural logarithmic function is an antiderivative of x^{-1}.

The result given in (3) generalizes by the Chain Rule to the composition of the natural logarithmic function and a positive differentiable function $u = g(x)$:

$$\frac{d}{dx}\ln u = \frac{1}{u}\frac{du}{dx}, \qquad u > 0 \tag{4}$$

Laws of the Natural Logarithm We put the derivative result in (4) to immediate use in proving the **laws of the natural logarithm**, which are summarized in the next theorem.

THEOREM 6.4 Laws of the Natural Logarithm

Let a and b be positive real numbers and let t be a rational number. Then

(*i*) $\ln ab = \ln a + \ln b$

(*ii*) $\ln \frac{a}{b} = \ln a - \ln b$

(*iii*) $\ln a^t = t \ln a$

Proof of (i) Define a function $F(x) = \ln ax$ and let $f(x) = \ln x$. From (3) and (4),

$$F'(x) = \frac{1}{ax}\frac{d}{dx}ax = \frac{a}{ax} = \frac{1}{x} = f'(x)$$

Thus, from the definition of an antiderivative, $F(x) = f(x) + C$. But, since $\ln 1 = \int_1^1 dt/t = 0$, we have

$$F(1) = f(1) + C$$
$$\ln a = \ln 1 + C$$

and so $C = \ln a$. Hence, $F(x) = \ln x + \ln a$. Substituting $x = b$ gives

$$\ln ab = \ln b + \ln a = \ln a + \ln b$$

Proof of (iii) Define two functions $F(x) = \ln x^t$ and $G(x) = t \ln x$, where t is a rational number. Now from (4),

$$F'(x) = \frac{1}{x^t} \cdot tx^{t-1} = \frac{t}{x}$$

and from (3), $G'(x) = t/x$. Since $F'(x) = G'(x)$, we can write

$$F(x) = G(x) + C$$
$$F(1) = G(1) + C$$
$$\ln 1 = t \ln 1 + C$$

In this case $C = 0$. Substituting $x = a$ in $\ln x^t = t \ln x$ gives

$$\ln a^t = t \ln a$$ ■

The proof of property (*ii*) can be obtained from (*i*) and (*iii*) and is left as an exercise.

Graph of* ln *x First, we observe that

$$\ln 1 = 0 \tag{5}$$

means that the graph of $y = \ln x$ crosses the x-axis at (1, 0). Next,

$$\frac{d}{dx}\ln x = \frac{1}{x} > 0 \qquad \text{for } x > 0 \tag{6}$$

implies that $y = \ln x$ is an increasing function on the interval $(0, \infty)$. Furthermore,

$$\frac{d^2}{dx^2}\ln x = \frac{d}{dx}\frac{1}{x} = -\frac{1}{x^2} < 0 \qquad \text{for } x > 0 \tag{7}$$

shows that the graph of $y = \ln x$ must be concave downward on the interval $(0, \infty)$. We now show that

$$\lim_{x \to \infty} \ln x = \infty \quad \text{and} \quad \lim_{x \to 0^+} \ln x = -\infty \tag{8}$$

These limit results taken together with the Intermediate Value Theorem imply that the range of the natural logarithmic function is $(-\infty, \infty)$, that is, the set R of real numbers. Note also that the second limit in (8) indicates that the y-axis is a vertical asymptote for the graph of $y = \ln x$. To establish the first limit, we make use of the earlier observation that $\frac{1}{2} < \ln 2 < 1$. From this inequality and (*iii*) of the laws of logarithms, it follows that for a positive integer n,

$$\ln 2^n = n \ln 2 > \frac{n}{2}$$

This last inequality shows that $\lim_{n \to \infty} \ln 2^n = \infty$. Now if $x > 2^n$, we can write $\ln x > \ln 2^n > n/2$, since the natural logarithmic function is an increasing function. From this last inequality we can conclude that $\lim_{x \to \infty} \ln x = \infty$. Finally, to establish the second limit in (8) we rewrite $\ln x$ as $\ln x = \ln (x^{-1})^{-1} = -\ln(1/x)$ and make the substitution $t = 1/x$. Since $x \to 0^+$ implies $t \to \infty$, we obtain the desired result:

$$\lim_{x \to 0^+} \ln x = \lim_{t \to \infty}(-\ln t) = -\lim_{t \to \infty} \ln t = -\infty$$

Putting the information in (5), (6), (7), and (8) together yields the graph of $y = \ln x$ shown in Figure 6.15(a). Of course, one can obtain (approximate) points on the graph using the information in Example 1 and the property $\ln 2^n = n \ln 2$. Some numerical values are given in the table in Figure 6.15(b). Numerical values of $\ln x$ can also be obtained from scientific calculators.

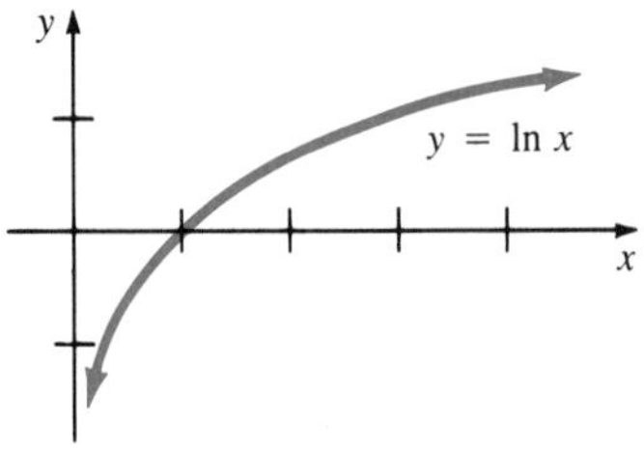

(a)

x	$\ln x$
$\frac{1}{4}$	$-2\ln 2 \approx -1.39$
$\frac{1}{2}$	$-\ln 2 \approx -0.69$
4	$2\ln 2 \approx 1.39$
8	$3\ln 2 \approx 2.08$
64	$6\ln 2 \approx 4.17$

(b)

FIGURE 6.15

y* = ln *x* *Has an Inverse A closer inspection of its graph should convince you that the natural logarithmic function is one-to-one. For any real number r, there is only one value of x_0 for which $r = \ln x_0$. See Figure 6.16. Also, in view of the fact that it is an increasing function for $x > 0$, it follows

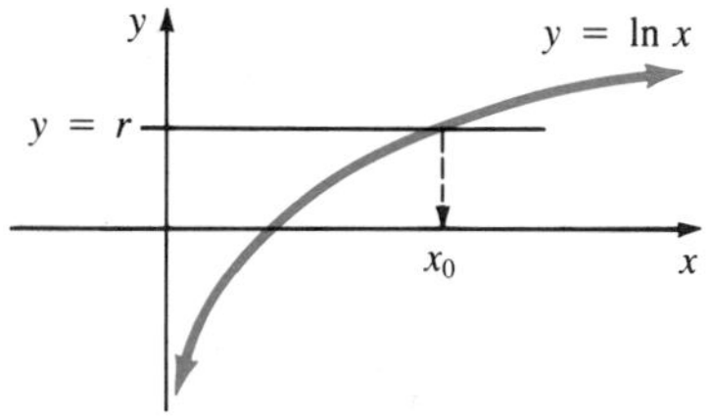

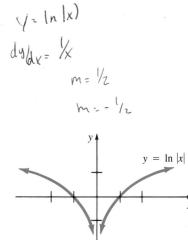

FIGURE 6.16

from Theorem 6.2 that the natural logarithmic function has an inverse. This inverse function will be studied in the next section.

Although the domain of the natural logarithmic function $y = \ln x$ is the set of positive real numbers, the domain of $y = \ln|x|$ extends to the set of real numbers except $x = 0$. Furthermore,

$$\text{for } x > 0, \quad \frac{d}{dx}\ln x = \frac{1}{x}$$

$$\text{for } x < 0, \quad \frac{d}{dx}\ln(-x) = \frac{1}{(-x)}\frac{d}{dx}(-x) = \frac{-1}{-x} = \frac{1}{x}$$

We have shown that

$$\frac{d}{dx}\ln|x| = \frac{1}{x}, \qquad x \neq 0 \tag{9}$$

y = ln |x)
dy/dx = 1/x
m = 1/2
m = -1/2

EXAMPLE 2 Find the slope of the tangent line to the graph of $y = \ln|x|$ at $x = 2$ and at $x = -2$.

Solution Since (9) gives $dy/dx = 1/x$, we have

$$\left.\frac{dy}{dx}\right|_{x=2} = \frac{1}{2} \quad \text{and} \quad \left.\frac{dy}{dx}\right|_{x=-2} = -\frac{1}{2}$$

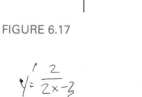

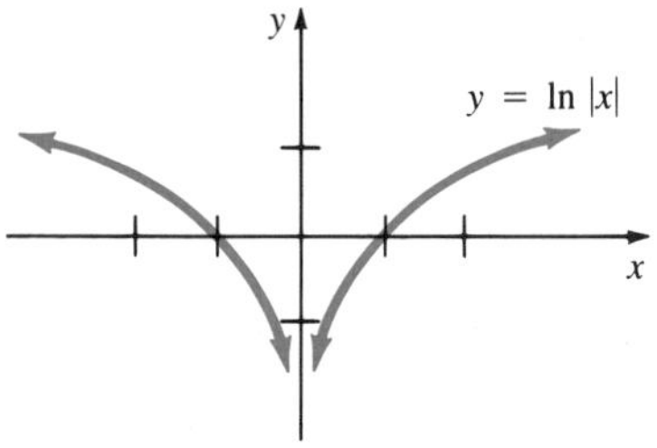

FIGURE 6.17

Observe in Figure 6.17 that the graph of $y = \ln|x|$ is symmetric with respect to the y-axis. □

When $u = g(x)$ is a differentiable function, the Chain Rule gives, additionally,

$$\frac{d}{dx}\ln|u| = \frac{1}{u}\frac{du}{dx}, \qquad u \neq 0 \tag{10}$$

y' = 2/(2x-3)
y' = 2/(2x-3)

EXAMPLE 3 Differentiate (*a*) $y = \ln(2x - 3)$ and (*b*) $y = \ln|2x - 3|$.

Solution

(*a*) For $2x - 3 > 0$, we have from (4),

$$\frac{dy}{dx} = \frac{1}{2x-3}\frac{d}{dx}(2x-3) = \frac{2}{2x-3} \tag{11}$$

(*b*) For $2x - 3 \neq 0$, we have from (10),

$$\frac{dy}{dx} = \frac{1}{2x-3}\frac{d}{dx}(2x-3) = \frac{2}{2x-3} \tag{12}$$

Although (11) and (12) *appear* to be equal, they are definitely not the same function. The difference is simply that the domain of (11) is $(\frac{3}{2}, \infty)$, whereas the domain of (12) is the set of real numbers except $x = \frac{3}{2}$. □

EXAMPLE 4 Differentiate (*a*) $y = \ln|\sin x|$ and (*b*) $y = \ln x^3$.

Solution

(*a*) For $\sin x \neq 0$, it follows from (12) that

$$\frac{dy}{dx} = \frac{1}{\sin x}\frac{d}{dx}\sin x = \frac{\cos x}{\sin x} = \cot x$$

(*b*) Because x^3 must be positive, it is understood that $x > 0$. Hence, from (4),

$$\frac{dy}{dx} = \frac{1}{x^3}\frac{d}{dx}x^3 = \frac{3x^2}{x^3} = \frac{3}{x}$$

Alternative Solution From property (*iii*) we can first write $y = 3\ln x$ and then differentiate to obtain the same result as above. □

EXAMPLE 5 The functions $f(x) = \ln x^4$ and $g(x) = 4\ln x$ are not the same. Since $x^4 > 0$ for all $x \neq 0$, the domain of f is the set of real numbers except $x = 0$. The domain of g is $(0, \infty)$. Thus,

$$f'(x) = \frac{4}{x}, \quad x \neq 0 \quad \text{whereas} \quad g'(x) = \frac{4}{x}, \quad x > 0$$ □

EXAMPLE 6 Differentiate $y = \ln \dfrac{x^{1/2}(2x+7)^4}{(3x^2+1)^2}$.

Solution Using the properties of logarithms, we can write for $x > 0$,

$$\begin{aligned} y &= \ln x^{1/2}(2x+7)^4 - \ln(3x^2+1)^2 \\ &= \ln x^{1/2} + \ln(2x+7)^4 - \ln(3x^2+1)^2 \\ &= \frac{1}{2}\ln x + 4\ln(2x+7) - 2\ln(3x^2+1) \end{aligned}$$

so that

$$\begin{aligned} \frac{dy}{dx} &= \frac{1}{2}\cdot\frac{1}{x} + 4\cdot\frac{1}{2x+7}\cdot 2 - 2\cdot\frac{1}{3x^2+1}\cdot 6x \\ &= \frac{1}{2x} + \frac{8}{2x+7} - \frac{12x}{3x^2+1} \end{aligned}$$ □

EXAMPLE 7 Differentiate $y = \ln(\ln x)$.

Solution From (4),

$$\frac{dy}{dx} = \frac{1}{\ln x}\frac{d}{dx}\ln x = \frac{1}{\ln x}\cdot\frac{1}{x} = \frac{1}{x\ln x}$$ □

▲ ***Remark*** You should exercise care when working with logarithms. Note that

$$\ln x^2 \quad \text{is } not \text{ the same as} \quad (\ln x)^2$$
$$\ln(x^2+4) \quad \text{is } not \text{ the same as} \quad \ln x^2 + \ln 4$$
$$\frac{\ln(x+1)}{\ln(3x+2)} \quad \text{is } not \text{ the same as} \quad \ln(x+1) - \ln(3x+2)$$

and

$$\ln\frac{x}{x+1} \quad \text{is } not \text{ the same as} \quad \frac{\ln x}{\ln(x+1)}$$

EXERCISES 6.2

Answers to odd-numbered problems begin on page A-83.

In Problems 1–4 state the domain of the given function.

1. $f(x) = \ln(x+1)$
2. $f(x) = \ln(3-x)$
3. $f(x) = \ln|x^2-1|$
4. $f(x) = \ln(x^2-1)$

In Problems 5–10 use Theorem 6.4 to determine whether f and g are the same functions.

5. $f(x) = \ln x^6$
   $g(x) = 6\ln x$
6. $f(x) = \ln x^{1/3}$
   $g(x) = \frac{1}{3}\ln x$
7. $f(x) = \ln x(x^4+3)$
   $g(x) = \ln x + \ln(x^4+3)$
8. $f(x) = \ln(2x+7)$
   $g(x) = \ln 2x + \ln 7$
9. $f(x) = \dfrac{\ln(x^2+9)}{\ln(x^2+1)}$
   $g(x) = \ln(x^2+9) - \ln(x^2+1)$
10. $f(x) = \ln(1/x)$
    $g(x) = -\ln x$

In Problems 11–34 find the derivative of the given function.

11. $y = 10\ln x$
12. $y = \ln 10x$
13. $y = \ln x^{1/2}$
14. $y = (\ln x)^{1/2}$
15. $y = \ln(x^4+3x^2+1)$
16. $y = \ln(x^2+1)^{20}$
17. $y = x^2 \ln x^3$
18. $y = x - \ln|5x+1|$
19. $y = \dfrac{\ln x}{x}$
20. $y = x(\ln x)^2$
21. $y = \ln\dfrac{x}{x+1}$
22. $y = \dfrac{\ln 4x}{\ln 2x}$
23. $y = -\ln|\cos x|$
24. $y = \ln(x + \sqrt{x^2-1})$
25. $y = \dfrac{1}{\ln x}$
26. $y = \ln\dfrac{1}{x}$
27. $f(x) = \ln(x\ln x)$
28. $g(x) = \sqrt{\ln\sqrt{x}}$
29. $f(x) = \ln(\ln(\ln x))$
30. $w(\theta) = \theta\sin(\ln 5\theta)$
31. $H(t) = \ln t^2(3t^2+6)$
32. $G(t) = \ln\sqrt{5t+1}(t^3+4)^6$
33. $f(x) = \ln\dfrac{(x+1)(x+2)}{x+3}$
34. $f(x) = \ln\sqrt{\dfrac{(3x+2)^5}{x^4+7}}$

In Problems 35–40 use implicit differentiation to find dy/dx.

35. $y^2 = \ln xy$
36. $y = \ln(x+y)$
37. $x + y^2 = \ln\dfrac{x}{y}$
38. $y = \ln xy^2$
39. $xy = \ln(x^2+y^2)$
40. $x^2+y^2 = \ln(x+y)^2$
41. Find an equation of the tangent line to the graph of $y = \ln x$ at $x = 1$.
42. Find the slope of the tangent to the graph of $y = (\ln|x|)^2$ at $x = 1$.
43. Find an equation of the tangent line to the graph of $y = \ln(x^2-3)$ at $x = 2$.
44. Find the slope of the tangent to the graph of y' at the point where the slope of the tangent to the graph of $y = \ln x^2$ is 4.
45. Determine the point on the graph of $y = \ln 2x$ at which the tangent line is perpendicular to $x + 4y = 1$.
46. If $y = \ln x$, find $d^n y/dx^n$.

In Problems 47–50 sketch the graph of the given function.

47. $y = -\ln x$
48. $y = 2 + \ln x$
49. $y = \ln(x-2)$
50. $y = \ln|x+1|$
51. Answer the following questions about the graph of $f(x) = \ln(x^2+1)$: intercepts? symmetry? asymptotes? relative extrema? concavity? Sketch the graph of f.

52. Compare the graphs of $y = \ln x^2$ and $y = 2 \ln x$.

53. For $x > 0$ verify that both $y = x^{-1/2}$ and $y = x^{-1/2}\ln x$ satisfy the differential equation $4x^2 d^2y/dx^2 + 8x\, dy/dx + y = 0$.

54. For $x > 0$ verify that

$$y = C_1 x^{-1}\cos(\sqrt{2}\ln x) + C_2 x^{-1}\sin(\sqrt{2}\ln x),$$

where C_1 and C_2 are constants, satisfies the differential equation $x^2y'' + 3xy' + 3y = 0$.

Calculator/Computer Problems

In Problems 55 and 56 show graphically that the given equation possesses only one real root. Use Newton's Method to approximate the root to the three decimal places.

55. $\ln x = 2$

56. $x + \ln x - 3 = 0$

57. (a) Use a calculator or computer to obtain the graph of $f(x) = x^3 - \ln|x|$.

(b) Use Newton's Method to approximate any x-intercepts that you discover from the graph in part (**a**).

(c) Find any relative extrema that you discover from the graph in part (**a**).

Miscellaneous Problems

58. Use (*i*) and (*iii*) of Theorem 6.4 to prove (*ii*) of Theorem 6.4. [*Hint:* $\ln(a/b) = \ln(a \cdot 1/b)$.]

6.3 THE EXPONENTIAL FUNCTION

In Section 6.2 we observed that $y = \ln x$ is a one-to-one function and, consequently, possesses an inverse. The inverse of the natural logarithmic function is denoted by $y = \exp x$ and is called the **exponential function** or, sometimes, the **natural exponential function**.

DEFINITION 6.4 Natural Exponential Function

$$y = \exp x \quad \text{if and only if} \quad x = \ln y$$

Range and Domain Since the domain of the natural logarithmic function is the set of positive real numbers, it follows that the range of the exponential function is the same set. In other words, $\exp x > 0$ for all x. Similarly, since the value of $y = \ln x$ can be any real number, the domain of $y = \exp x$ is the set of real numbers.

From (2) of Section 6.1, we have

$$\ln(\exp x) = x \quad \text{for all } x \quad \text{and} \quad \exp(\ln x) = x, \quad x > 0 \tag{1}$$

The Number e The number exp 1 is important in mathematics and is denoted by the special symbol e in honor of the great Swiss mathematician Leonhard Euler; that is,

$$e = \exp 1$$

It can be shown that to 12 decimal places

$$e = 2.718281828459\ldots \tag{2}$$

The number e, like the number π, is an irrational number. You will be asked to prove in Problem 68 in Exercises 6.3 that $2.7 < e < 2.8$.

Now, from (1) we know that $\ln(\exp x) = x$ and therefore $\ln(\exp 1) = 1$. The latter result is equivalent to

$$\ln e = 1 \tag{3}$$

If t is a rational number, we have from (3) and (*iii*) of the laws of the natural logarithm that

$$\ln e^t = t \ln e = t \cdot 1 = t \tag{4}$$

But, in view of Definition 6.4, (4) is the same as

$$e^t = \exp t \tag{5}$$

The result given in (5) suggests that e^x be defined *for any real number* x in the following manner:

DEFINITION 6.5

For any real number x,

$$e^x = \exp x$$

Subsequently, we shall drop the notation $\exp x$ and use e^x exclusively. From now on, the function $y = e^x$ will be called the **exponential function with base *e***. The number x is also called the **exponent** of the base e.

At this point, let us summarize the preceding discussion using the symbol e.

- $y = e^x$ if and only if $x = \ln y$. (6)
- $y = e^x$ is the inverse of $y = \ln x$. (7)
- The domain of $y = e^x$ is $(-\infty, \infty)$. (8)
- The range of $y = e^x$ is $(0, \infty)$. This means $e^x > 0$ for all x. (9)
- $\ln e^x = x$ for all x (10)
- $e^{\ln x} = x,\ x > 0$ (11)

Leonhard Euler (1707–1783) was a man with a prodigious memory and phenomenal powers of concentration. Euler's interests, like those of Descartes, were almost universal: He was a theologian, physicist, astronomer, linguist, physiologist, classical scholar, and foremost a mathematician. Considered to be a true genius in mathematics, he made lasting contributions to algebra, trigonometry, analytic geometry, calculus, calculus of variations, differential equations, complex variables, number theory, and topology. The volume of his mathematical output did not seem to be affected by the distractions of 13 children or the fact that he was totally blind for the last 17 years of his life. Euler wrote over 700 papers and 32 books on mathematics and was responsible for introducing many of the symbols (such as e, π, $i = \sqrt{-1}$) and notations that are still used (such as $f(x)$, Σ, $\sin x$, $\cos x$). Euler was born in Basle, Switzerland, on April 15, 1707, and died of a stroke in St. Petersburg on September 18, 1783, while serving in the court of the Russian empress Catherine the Great.

Numerical values of e^x have been extensively tabulated and most scientific hand calculators have a key labeled $\boxed{e^x}$.

EXAMPLE 1 From a calculator we find $e^2 \approx 7.3891$. □

Laws of Exponents The following laws for e^x are the familiar laws of exponents.

> **THEOREM 6.5 Laws of Exponents**
>
> Let r and s be any real number and let t be a rational number. Then
>
> (*i*) $e^0 = 1$ (*ii*) $e^1 = e$
>
> (*iii*) $e^r e^s = e^{r+s}$ (*iv*) $\dfrac{e^r}{e^s} = e^{r-s}$
>
> (*v*) $(e^r)^t = e^{rt}$ (*vi*) $e^{-r} = \dfrac{1}{e^r}$

Proof of (i) $e^0 = 1$ since $\ln 1 = 0$.

Proof of (iv) Let $M = e^r$ and $N = e^s$ so that $r = \ln M$ and $s = \ln N$, respectively. From (*ii*) of the laws of the natural logarithm,

$$\ln \frac{M}{N} = \ln M - \ln N = r - s$$

Thus, from (6)

$$\frac{M}{N} = e^{r-s} \quad \text{or} \quad \frac{e^r}{e^s} = e^{r-s}$$

Proof of (vi) Let $M = e^{-r}$. Hence, (6) gives

$$\ln M = -r$$
$$-\ln M = r$$
$$\ln M^{-1} = r$$
$$\ln \frac{1}{M} = r$$
$$\frac{1}{M} = e^r$$
$$M = \frac{1}{e^r} \quad \text{or} \quad e^{-r} = \frac{1}{e^r}$$ ■

The proofs of (*iii*) and (*v*) are left as exercises.

EXAMPLE 2 If $y > 0$ is a number such that $\ln y = -1$, then (6) yields $y = e^{-1}$. From (*vi*) of Theorem 6.5,

$$y = \frac{1}{e}$$

and from (2) it follows that $e^{-1} \approx 0.3679$. □

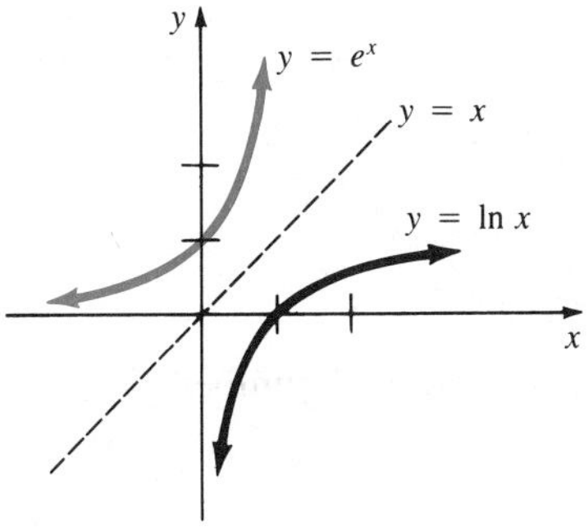

FIGURE 6.18

Graph of $y = e^x$ Since $y = e^x$ is the inverse of the natural logarithmic function, its graph can be obtained by reflection of the graph of $y = \ln x$ in the line $y = x$. Specifically we note that

$$\lim_{x \to 0^+} \ln x = -\infty \quad \text{implies} \quad \lim_{x \to -\infty} e^x = 0$$

and

$$\lim_{x \to \infty} \ln x = \infty \quad \text{implies} \quad \lim_{x \to \infty} e^x = \infty$$

The graphs of $y = \ln x$ and $y = e^x$ are compared in Figure 6.18.

Derivative of e^x We know that if

$$y = e^x \quad \text{then} \quad \ln y = x$$

Differentiating the last equation implicitly with respect to x gives

$$\frac{1}{y}\frac{dy}{dx} = 1 \quad \text{or} \quad \frac{dy}{dx} = y$$

Since $y = e^x$, we obtain the remarkable result*

$$\frac{d}{dx} e^x = e^x \tag{12}$$

Using the Chain Rule, we immediately generalize (12) to

$$\frac{d}{dx} e^u = e^u \frac{du}{dx} \tag{13}$$

where $u = g(x)$ is a differentiable function.

EXAMPLE 3 Differentiate (*a*) $y = e^{4x}$ and (*b*) $y = e^{1/x^3}$.

Solution Both results follow from (13).

(*a*)

$$\frac{dy}{dx} = e^{4x} \cdot \frac{d}{dx}(4x)$$
$$= e^{4x}(4) = 4e^{4x}$$

---

*It can be shown that $y = e^x$ is the *only* function whose graph passes through (0, 1) for which the derivative is the function itself. Any function in the family $y = Ce^x$, C any constant, also satisfies $y' = y$.

(b)
$$\frac{dy}{dx} = e^{1/x^3} \cdot \frac{d}{dx}(x^{-3})$$
$$= e^{1/x^3}(-3x^{-4})$$
$$= -\frac{3e^{1/x^3}}{x^4}$$ □

EXAMPLE 4 Differentiate $y = \ln(e^{4x} + e^{-4x})$.

Solution From (4) of Section 6.2 and (13), we obtain

$$\frac{dy}{dx} = \frac{1}{e^{4x} + e^{-4x}} \cdot \frac{d}{dx}(e^{4x} + e^{-4x})$$
$$= \frac{1}{e^{4x} + e^{-4x}} \cdot (4e^{4x} - 4e^{-4x})$$
$$= \frac{4(e^{4x} - e^{-4x})}{e^{4x} + e^{-4x}}$$ □

$y = e^{2x^{1/2}} \cdot \ln 3x$

$y' = e^{2x^{1/2}}(x^{-1/2}) \ln 3x +$

$e^{2x^{1/2}} \cdot 1/3x \cdot (1/x)$

$y' = e^{2\sqrt{x}}(1/\sqrt{x}) \ln 3x + e^{2\sqrt{x}}(1/x)$

$y'(1) = e^2 \ln 3 + e^2 = e^2(\ln 3 + 1)$

EXAMPLE 5 Find the slope of the tangent line to the graph of $y = e^{2\sqrt{x}}\ln 3x$ at $x = 1$.

Solution By the Product Rule, we have

$$\frac{dy}{dx} = e^{2\sqrt{x}} \cdot \frac{1}{3x} \cdot 3 + \ln 3x \cdot e^{2\sqrt{x}} \cdot 2 \cdot \frac{1}{2}x^{-1/2}$$

Thus,
$$\left.\frac{dy}{dx}\right|_{x=1} = e^2 + e^2 \ln 3 = e^2(1 + \ln 3)$$ □

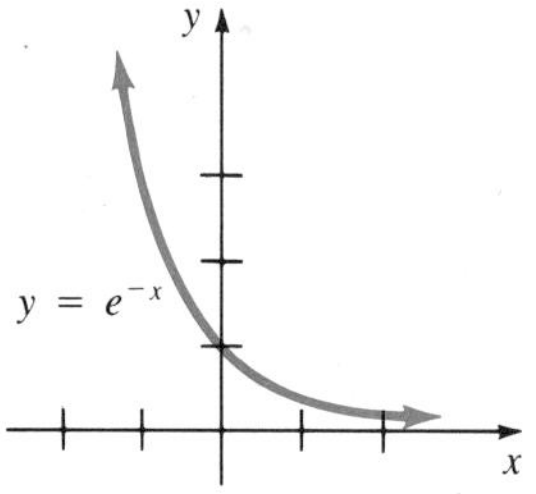

FIGURE 6.19

EXAMPLE 6 The derivative of $f(x) = e^{kx}$ is $f'(x) = ke^{kx}$. Since $e^{kx} > 0$, we see that

$$f'(x) < 0 \quad \text{for } k < 0 \quad \text{and} \quad f'(x) > 0 \quad \text{for } k > 0$$

This shows that f is an increasing function on $(-\infty, \infty)$ when $k > 0$ and a decreasing function on the interval when $k < 0$. The graph of f for the case $k = -1$ is given in Figure 6.19. □

EXAMPLE 7 Graph $y = xe^{x/2}$.

Solution We note that since $f(0) = 0$, the graph passes through the origin. There are no other x-intercepts. Also, when $x < 0$, $y < 0$, and when $x > 0$,

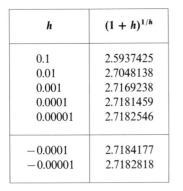

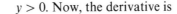

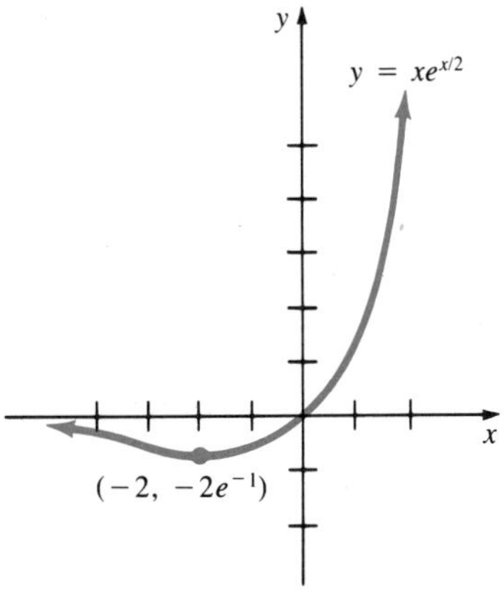

FIGURE 6.20

$y > 0$. Now, the derivative is

$$\frac{dy}{dx} = xe^{x/2} \cdot \frac{1}{2} + e^{x/2} = \frac{1}{2} e^{x/2}(x + 2)$$

Setting this last result equal to zero yields the critical point -2. Since

$$\frac{dy}{dx} < 0 \quad \text{for } x < -2 \quad \text{and} \quad \frac{dy}{dx} > 0 \quad \text{for } x > -2$$

we conclude from the First Derivative Test that $f(-2) = -2e^{-1}$ is a relative minimum. Inspection of Figure 6.20 also indicates that this value is an absolute minimum of the function. □

***e* As a Limit** The number e can be expressed as a limit. In Chapter 9 we shall prove that

$$\lim_{h \to 0} (1 + h)^{1/h} = e \tag{14}$$

h	$(1 + h)^{1/h}$
0.1	2.5937425
0.01	2.7048138
0.001	2.7169238
0.0001	2.7181459
0.00001	2.7182546
−0.0001	2.7184177
−0.00001	2.7182818

The accompanying table merely suggests the foregoing result. If we let $n = 1/h$, $h > 0$, then as $h \to 0$, we have $n \to \infty$; therefore, (14) can be written in the alternative form

$$\lim_{n \to \infty} \left(1 + \frac{1}{n}\right)^n = e \tag{15}$$

These limit results are important and should be remembered. Indeed, (15) is often taken as the *definition* of the number e.

▲ ***Remark*** The numbers e and π are **transcendental** as well as irrational numbers. A transcendental number is one that is *not* a root of a polynomial equation with integer coefficients. For example, $\sqrt{2}$ is irrational but is not transcendental, since it is a root of the polynomial equation $x^2 - 2 = 0$. The number e was proved to be transcendental by the French mathematician Charles Hermite (1822–1901) in 1873, whereas π was proved to be transcendental nine years later by the German mathematician Ferdinand Lindemann. The latter proof showed conclusively that "squaring a circle" with a rule and a compass was impossible.

EXERCISES 6.3

Answers to odd-numbered problems begin on page A-84.

In Problems 1–30 find the derivative of the given function.

1. $y = e^{-x}$
2. $y = e^{2x+3}$
3. $y = e^{\sqrt{x}}$
4. $y = e^{\cos 10x}$
5. $y = \dfrac{e^{-2x}}{x}$
6. $y = x^3 e^{4x}$
7. $y = \sqrt{1 + e^{-5x}}$
8. $y = \dfrac{1}{(e^{2x} - e^{-2x})^2}$
9. $y = \ln(x^4 + e^{x^2})$
10. $y = \ln\sqrt{e^x + e^{-x}}$
11. $y = \dfrac{e^x + e^{-x}}{e^x - e^{-x}}$
12. $y = \dfrac{e^{7x}}{e^{-x}}$

13. $y = e^{\ln x}$

14. $y = \ln e^x$

15. $y = e^{3x}\ln(x^2 + 1)$

16. $y = \dfrac{\ln x}{e^x}$

17. $y = e^{\frac{x+2}{x-2}}$

18. $y = \ln\left|\dfrac{1 + e^{2x}}{1 - e^{2x}}\right|$

19. $y = e^{e^{x^2}}$

20. $y = (e^3)^{x-1}$

21. $y = e^{2x}e^{3x}e^{4x}$

22. $y = e^x + e^{x+e^x}$

23. $F(t) = e^{t^{1/3}} + (e^t)^{1/3}$

24. $g(t) = e^{-t}\tan e^t$

25. $f(x) = (2x + 1)^3 e^{-(1-x)^4}$

26. $f(x) = e^{x\sqrt{x^2+1}}$

27. $f(x) = \dfrac{xe^x}{x + e^x}$

28. $f(x) = xe^{2x}\ln x$

29. $f(x) = \tan e^{2x}$

30. $f(x) = x \sec e^{-x}$

In Problems 31–36 use implicit differentiation to find dy/dx.

31. $y = e^{x+y}$

32. $\ln y = x + e^y$

33. $y = \cos e^{xy}$

34. $y = e^{(x+y)^2}$

35. $x + y^2 = e^{x/y}$

36. $e^x + e^y = y$

In Problems 37–40 find the indicated derivative.

37. $y = e^{-4x}; \dfrac{d^3y}{dx^3}$

38. $y = \sin e^{2x}; \dfrac{d^2y}{dx^2}$

39. $y = \ln(e^x + 1); \dfrac{d^2y}{dx^2}$

40. $y = xe^x; \dfrac{d^4y}{dx^4}$

41. Find an equation of the tangent line to the graph of $y = e^x$ at $x = 1$.

42. Find an equation of the tangent line to the graph of $y = \ln(e^x + 1)$ at $x = 0$.

43. Find the slope of the normal line to the graph of $y = (x - 1)e^{-x}$ at $x = 0$.

44. Find the point on the graph of $y = e^x$ at which the tangent line is parallel to $3x - y = 7$.

In Problems 45–48 sketch the graph of the given function.

45. $y = -e^x$

46. $y = 1 + e^{-x}$

47. $y = 2 - e^{-x}$

48. $y = e^x + e^{-x}$

In Problems 49–54 find the relative extrema of each function. Sketch the graph.

49. $y = xe^{-x}$

50. $y = \dfrac{e^x}{x}$

51. $y = e^{-x^2}$

52. $y = e^{(x-2)^2}$

53. $y = x \ln x$

54. $y = \dfrac{\ln x}{x}$

55. Consider the function $f(x) = e^{2/x}$, $x > 0$.

(a) Prove that f has an inverse.

(b) Sketch a graph of f.

(c) What is the range of f?

(d) Find f^{-1}.

(e) What is the domain of f^{-1}?

56. Consider the function $f(x) = \sqrt{1 + \ln x}$.

(a) What are the domain and range of f?

(b) Prove that f has an inverse.

(c) Sketch a graph of f.

(d) Find f^{-1}.

57. Sketch the graph of $y = e^{|x|}$. Find dy/dx. Is the function differentiable at 0?

58. Show that the function $f(x) = e^{\sin x}$ is periodic with period 2π. Find the relative extrema and points of inflection of f. Sketch the graph of f.

59. Verify that $y = C_1e^{-3x} + C_2e^{2x}$, where C_1 and C_2 are constants, satisfies the differential equation $y'' + y' - 6y = 0$.

60. In science it is sometimes useful to display data using logarithmic coordinates. Which of the following equations determines the graph in Figure 6.21?

(a) $y = 2x + 1$ (b) $y = e + x^2$

(c) $y = ex^2$ (d) $x^2y = e$

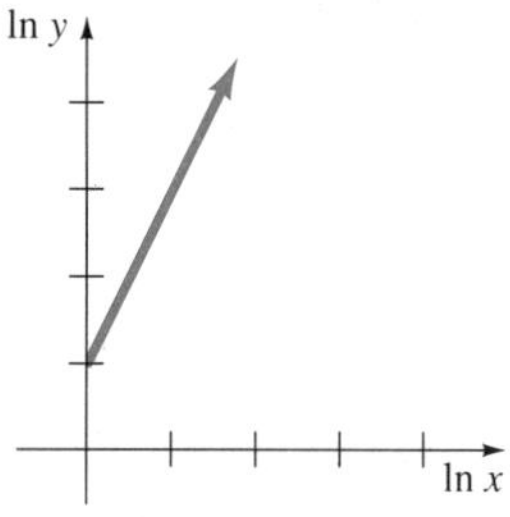

FIGURE 6.21

Calculator/Computer Problems

61. A graph of

$$P(t) = \frac{aP_0}{bP_0 + (a - bP_0)e^{-at}}$$

where a and b are positive constants, is called a **logistic curve** and occurs in a mathematical model of an expanding but limited population. $P(0) = P_0$ is the initial population. Consider the case when $a = 2$, $b = 1$, and $P_0 = 1$.

(a) Find horizontal asymptotes for the graph by determining the limits $\lim_{t\to-\infty} P(t)$ and $\lim_{t\to\infty} P(t)$.

(b) Use a calculator or computer to obtain the graph of $P(t)$.

62. The graph $P(t) = e^{a/b}e^{-ce^{-bt}}$, where a, b, and c are constants, is called a **Gompertz curve** and is named after the English mathematician Benjamin Gompertz (1779–1865).

(a) Find horizontal asymptotes for the graph by determining the limits $\lim_{t\to-\infty} P(t)$ and $\lim_{t\to\infty} P(t)$ and considering the two cases: $b > 0$, $c > 0$ and $b < 0$, $c > 0$.

(b) Use a calculator or computer to obtain the graph of $P(t)$ in the two cases: $a = 1$, $b = 1$, $c = 1$ and $a = -1$, $b = -1$, $c = 1$.

63. The Jenss model (1937) represents the most accurate empirically devised formula for predicting the height h (in centimeters) in terms of age t (in years) for preschool-aged children (3 months to 6 years):

$$h(t) = 79.04 + 6.39t - e^{3.26-0.99t}$$

(a) What height does this model predict for a two-year-old?

(b) How fast is a two-year-old increasing in height?

(c) At what age is the rate of growth most rapid? [*Hint:* $h'(t)$ is a continuous function defined on the closed interval $[\frac{1}{4}, 6]$.]

(d) Use a calculator or computer to obtain the graph of h on the interval $[\frac{1}{4}, 6]$.

(e) Use the graph in part (**d**) to estimate the age of a preschool-aged-child who is 100 cm tall.

64. The weight of many animals can be modeled by a Bertalanffy function $W(t) = a[1 - be^{-ct}]^3$ for some positive constants a, b, and c. For a population of female elephants, the weight (in kilograms) at age t (in years) is given by

$$W(t) = 2600[1 - 0.51e^{-0.075t}]^3$$

(a) Show that $W(t)$ is increasing for $t > 0$.

(b) Compute and interpret $\lim_{t\to\infty} W(t)$.

(c) How fast is a newborn female elephant increasing in weight?

(d) An adult elephant weighs 1600 kg. Determine her age.

(e) At what age is the rate of growth of an elephant a maximum?

65. A thermistor is a semiconductor made of ceramic material. It can act as a thermometer because the resistance of the material varies with temperature according to the formula

$$R(T) = R(T_0)\exp B\left(\frac{1}{T} - \frac{1}{T_0}\right)$$

where $T_0 = 298.16$ K is a reference temperature, $R(T_0)$ is the material's resistance at the reference temperature, and B is a parameter that depends on the material. Typical values are $R(T_0) = 3000$ ohms and $B = 4000$ K.

(a) Show that the resistance R is strictly decreasing with temperature T.

(b) Suppose the resistance is increasing at the rate of 1000 ohms per minute when the temperature is 275 K (just above freezing). At what rate is the temperature dropping?

66. Consider the function $f(n) = (1 + 1/n)^n$. Use a calculator to fill in the following table.

n	$f(n)$
100	
1000	
10,000	
100,000	
1,000,000	

67. (a) Use (14) of this section to show that for any real number r,

$$e^r = \lim_{n\to\infty}\left(1 + \frac{r}{n}\right)^n$$

(b) If P dollars are invested at an annual rate of interest r compounded t times a year, the return S after m years is

$$S = P\left(1 + \frac{r}{t}\right)^{tm}$$

When interest is **compounded continuously**, the return is defined to be

$$S = P\lim_{t\to\infty}\left(1 + \frac{r}{t}\right)^{tm}$$

Use the results of part (**a**) to show that $S = Pe^{rm}$.

(c) Use the result of part (**b**) to compare the return when \$5000 is invested at 10% annual interest compounded quarterly for three years with the return that results when continuous compounding of interest is used.

68. (a) Use the Trapezoidal Rule with $n = 6$ to establish the inequality

$$\int_1^{2.7}\frac{dt}{t} < 1 < \int_1^{2.8}\frac{dt}{t}$$

(b) Use part (**a**) to show that

$$\ln 2.7 < \ln e < \ln 2.8$$

(c) Show that $2.7 < e < 2.8$.

69. Find all intercepts of the graph of the function $f(x) = e^{2x} - e^x - 12$.

70. Show graphically that the equation $e^{-x} = 3x$ possesses only one real root. Use Newton's Method to approximate the root to three decimal places.

Miscellaneous Problems

71. Prove (*iii*) of Theorem 6.5.

72. Prove (*v*) of Theorem 6.5.

73. Prove that the x-intercept of the tangent line to the graph of $y = e^{-x}$ at $x = x_0$ is one unit to the right of x_0.

74. Given $f(x) = \ln x$ and $f^{-1}(x) = e^x$, use (7) of Section 6.1 to show that $(d/dx)e^x = e^x$.

75. Prove that for $x > 0$, $e^x > x + 1$. [*Hint:* Consider $f(x) = e^x - x - 1$.]

6.4 INTEGRALS INVOLVING LOGARITHMIC AND EXPONENTIAL FUNCTIONS

As a consequence of the derivative formulas (9) and (10) of Section 6.2, we obtain the antiderivative or indefinite integral formulas that yield the natural logarithm:

$$\int \frac{dx}{x} = \ln|x| + C \tag{1}$$

$$\int \frac{du}{u} = \ln|u| + C \tag{2}$$

Similarly, (12) and (13) of Section 6.3 translate into

$$\int e^x\,dx = e^x + C \tag{3}$$

$$\int e^u\,du = e^u + C \tag{4}$$

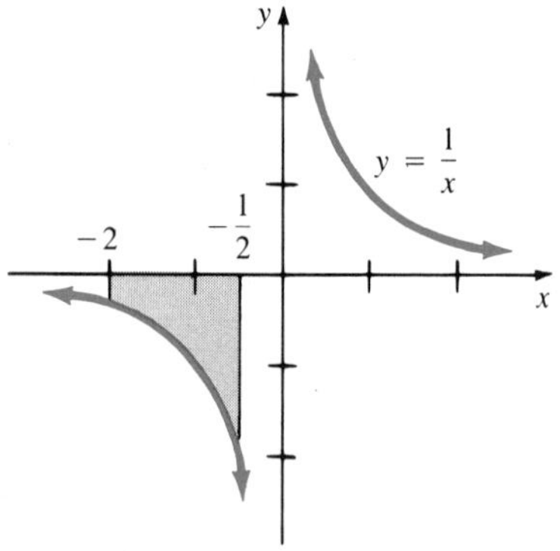

FIGURE 6.22

EXAMPLE 1 Find the area A bounded between the graph of $y = 1/x$ and the x-axis on the interval $[-2, -\frac{1}{2}]$.

Solution The area is indicated in Figure 6.22. Since $f(x) = 1/x < 0$ on the interval, it follows from (1) and Definition 5.1 that

$$\begin{aligned} A &= \int_{-2}^{-1/2} \left|\frac{1}{x}\right| dx = -\int_{-2}^{-1/2} \frac{1}{x}\,dx \\ &= -\ln|x|\Big]_{-2}^{-1/2} \\ &= -\ln\left|-\frac{1}{2}\right| + \ln|-2| \\ &= \ln 2 - (\ln 1 - \ln 2) = 2\ln 2 \approx 1.3863 \text{ square units} \end{aligned}$$ □

EXAMPLE 2 Evaluate $\int \frac{x^2}{x^3 + 5}\, dx$.

Solution If $u = x^3 + 5$, then $du = 3x^2\, dx$. Hence, from (2) we have

$$\begin{aligned}\int \frac{x^2}{x^3+5}\, dx &= \frac{1}{3}\int \frac{3x^2\, dx}{x^3+5}\\ &= \frac{1}{3}\int \frac{du}{u}\\ &= \frac{1}{3}\ln|u| + C\\ &= \frac{1}{3}\ln|x^3+5| + C\end{aligned}$$

□

EXAMPLE 3 Evaluate $\int \frac{dx}{1 + e^{-2x}}$.

Solution The given integral is not of form (2); however, if we multiply the numerator and denominator by e^{2x}, we have

$$\int \frac{dx}{1+e^{-2x}} = \int \frac{e^{2x}}{e^{2x}+1}\, dx$$

Now if $u = e^{2x} + 1$, then $du = 2e^{2x}\, dx$ and, therefore,

$$\begin{aligned}\int \frac{dx}{1+e^{-2x}} &= \int \frac{e^{2x}\, dx}{e^{2x}+1}\\ &= \frac{1}{2}\int \frac{2e^{2x}\, dx}{e^{2x}+1}\\ &= \frac{1}{2}\int \frac{du}{u}\\ &= \frac{1}{2}\ln|u| + C\\ &= \frac{1}{2}\ln(e^{2x}+1) + C\end{aligned}$$

Note that the absolute value can be dropped because $e^{2x} + 1 > 0$. □

EXAMPLE 4 Evaluate $\int e^{5x}\, dx$.

Solution Let $u = 5x$ so that $du = 5\,dx$. Then from (4),

$$\int e^{5x}\,dx = \frac{1}{5}\int e^{5x}(5\,dx)$$
$$= \frac{1}{5}e^{u} + C$$
$$= \frac{1}{5}e^{5x} + C$$

□

EXAMPLE 5 Evaluate $\displaystyle\int \frac{e^{4/x}}{x^2}\,dx$.

Solution Using $u = 4/x$, we have $du = -(4/x^2)\,dx$. From (4) we have

$$\int \frac{e^{4/x}}{x^2}\,dx = -\frac{1}{4}\int e^{4/x}\left(-\frac{4}{x^2}\,dx\right)$$
$$= -\frac{1}{4}\int e^{u}\,du$$
$$= -\frac{1}{4}e^{u} + C$$
$$= -\frac{1}{4}e^{4/x} + C$$

□

EXAMPLE 6 Evaluate $\displaystyle\int \frac{e^x}{\sqrt{1 + e^x}}\,dx$.

Solution Write the integral as

$$\int (1 + e^x)^{-1/2}e^x\,dx$$

and let $u = 1 + e^x$ and $du = e^x\,dx$. We recognize that

$$\int (1 + e^x)^{-1/2}e^x\,dx = \int u^{-1/2}\,du$$
$$= 2u^{1/2} + C$$
$$= 2(1 + e^x)^{1/2} + C$$

□

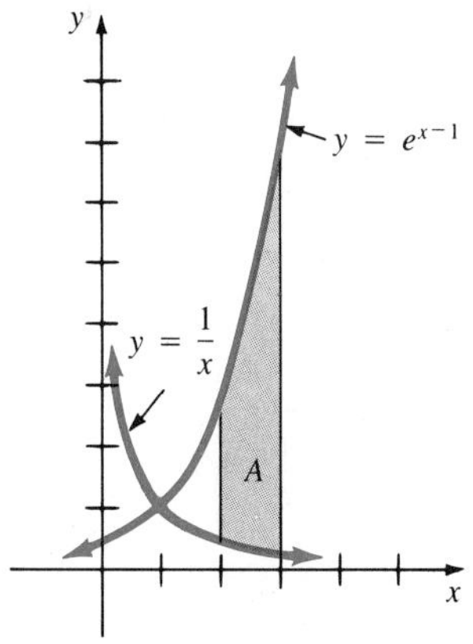

FIGURE 6.23

EXAMPLE 7 Find the area A of the region bounded between the graphs of $y = e^{x-1}$ and $y = 1/x$ on the interval $[2, 3]$.

Solution If we denote the given functions by $f(x) = e^{x-1}$ and $g(x) = 1/x$, then inspection of Figure 6.23 shows that $f(x) - g(x) \geq 0$ on the interval. It follows from Definition 5.2 that

$$\begin{aligned} A &= \int_2^3 \left(e^{x-1} - \frac{1}{x}\right) dx \\ &= (e^{x-1} - \ln x)\Big]_2^3 \\ &= e^2 - e - \ln 3 + \ln 2 \approx 4.2653 \text{ square units} \end{aligned}$$

□

The following integration formulas, which relate some trigonometric functions with the natural logarithmic functions, occur often enough in practice to merit special attention:

$$\int \tan x\, dx = -\ln|\cos x| + C \tag{5}$$

$$\int \cot x\, dx = \ln|\sin x| + C \tag{6}$$

$$\int \sec x\, dx = \ln|\sec x + \tan x| + C \tag{7}$$

$$\int \csc x\, dx = \ln|\csc x - \cot x| + C \tag{8}$$

To obtain (5) we write

$$\begin{aligned} \int \tan x\, dx = \int \frac{\sin x}{\cos x}\, dx &= -\int \frac{(-\sin x)\, dx}{\cos x} \\ &= -\int \frac{du}{u} \\ &= -\ln|u| + C \\ &= -\ln|\cos x| + C \end{aligned}$$

We shall verify (7) by differentiation:

$$\begin{aligned} \frac{d}{dx} \ln|\sec x + \tan x| &= \frac{1}{\sec x + \tan x} \frac{d}{dx}(\sec x + \tan x) \\ &= \frac{\sec x \tan x + \sec^2 x}{\sec x + \tan x} \\ &= \frac{\sec x(\tan x + \sec x)}{\sec x + \tan x} = \sec x \end{aligned}$$

By definition, $\ln|\sec x + \tan x|$ is an antiderivative of $\sec x$. Also, each of the preceding formulas can be written in a general form such as

$$\int \tan u\, du = -\ln|\cos u| + C$$

where $u = g(x)$ and $du = g'(x)\, dx$.

EXAMPLE 8 Evaluate $\int x \sec x^2 \, dx$.

Solution Let $u = x^2$ and $du = 2x\,dx$. Then from (1),

$$\begin{aligned}\int x \sec x^2 \, dx &= \frac{1}{2}\int \sec x^2(2x\,dx)\\ &= \frac{1}{2}\int \sec u \, du\\ &= \frac{1}{2}\ln|\sec u + \tan u| + C\\ &= \frac{1}{2}\ln|\sec x^2 + \tan x^2| + C\end{aligned}$$

□

EXAMPLE 9 Solve $(1 + 3x)\dfrac{dy}{dx} = y$.

Solution The first-order differential equation is separable. Dividing by $(1 + 3x)y$, we can write $dy/y = dx/(1 + 3x)$, from which it follows that

$$\begin{aligned}\int \frac{dy}{y} &= \int \frac{dx}{1 + 3x}\\ \ln|y| &= \frac{1}{3}\ln|1 + 3x| + C\\ \ln|y| - \ln|1 + 3x|^{1/3} &= C\end{aligned}$$

When written as $\ln(|y|/|1+3x|^{1/3}) = C$, the last expression gives $|y|/|1+3x|^{1/3} = e^C$ or $y/(1 + 3x)^{1/3} = \pm e^C$. Relabeling $\pm e^C$ as C_1, we obtain the solution $y = C_1\sqrt[3]{1 + 3x}$. □

EXERCISES 6.4

Answers to odd-numbered problems begin on page A-85.

In Problems 1–50 evaluate the given integral.

1. $\int \dfrac{dx}{3x}$
2. $\int \dfrac{dx}{x + 4}$
3. $\int \dfrac{dx}{2x - 1}$
4. $\int (5x + 6)^{-1}\,dx$
5. $\int \dfrac{x}{x^2 + 1}\,dx$
6. $\int \dfrac{x^2}{5x^3 + 8}\,dx$
7. $\int \dfrac{x + 2}{x^2 + 4x - 3}\,dx$
8. $\int \dfrac{2x^2 + 1}{2x^3 + 3x - 1}\,dx$
9. $\int \dfrac{x}{x + 1}\,dx$
10. $\int \dfrac{x^2 - 2x + 5}{x}\,dx$
11. $\int \dfrac{(x + 3)^2}{x + 2}\,dx$
12. $\int \dfrac{dx}{x^{1/3}(x^{2/3} + 1)}$
13. $\int \dfrac{dx}{x \ln x}$
14. $\int \dfrac{dx}{x \ln \sqrt{x}}$
15. $\int \dfrac{\ln x}{x}\,dx$
16. $\int \dfrac{dx}{x(\ln x)^2}$

17. $\int x^{-1} \sqrt[3]{1 + \ln x}\, dx$

18. $\int \frac{\sin(\ln x)}{x}\, dx$

19. $\int \frac{\cos t}{1 + \sin t}\, dt$

20. $\int \frac{1 - \sin \theta}{\theta + \cos \theta}\, d\theta$

21. $\int e^{10x}\, dx$

22. $\int \frac{dx}{e^{4x}}$

23. $\int x^2 e^{-2x^3}\, dx$

24. $\int \frac{e^{1/x^3}}{x^4}\, dx$

25. $\int \frac{e^{\sqrt{x}}}{\sqrt{x}}\, dx$

26. $\int e^x \tan e^x\, dx$

27. $\int e^{-2x^2+4} x\, dx$

28. $\int (2 - e^{3x})^2\, dx$

29. $\int (1 + e^x)^3\, dx$

30. $\int \frac{e^{2x}}{(1 + e^{2x})^3}\, dx$

31. $\int \frac{1 + e^t}{e^t}\, dt$

32. $\int \frac{e^\theta}{1 + e^\theta}\, d\theta$

33. $\int \frac{(e^{2x} - e^{-2x})^2}{e^x}\, dx$

34. $\int \frac{e^x[\ln(1 + e^x)]^2}{1 + e^x}\, dx$

35. $\int \frac{e^x - e^{-x}}{e^x + e^{-x}}\, dx$

36. $\int \frac{dx}{e^x + 1}$

37. $\int \frac{e^{2x}}{e^x + 1}\, dx$

38. $\int \frac{e^{3x} + e^{2x}}{e^x - 1}\, dx$

39. $\int \sqrt{e^x}\, dx$

40. $\int e^{\cos 2x} \sin 2x\, dx$

41. $\int \tan 5x\, dx$

42. $\int x(1 - \cot x^2)\, dx$

43. $\int (1 + \sec \theta)^2\, d\theta$

44. $\int \frac{dx}{\sin 2x}$

45. $\int_0^4 \frac{dx}{2x + 1}$

46. $\int_{-2}^{-1} \frac{dt}{t - 3}$

47. $\int_{-3}^{3} x e^{-x^2}\, dx$

48. $\int_{-1}^{0} \frac{4 + e^{x+1}}{e^x}\, dx$

49. $\int_1^{e^2} \frac{\ln(2/x)}{x}\, dx$

50. $\int_{\ln 2}^{\ln 3} e^{-x}\, dx$

51. Find the area under the graph of $y = 2x^{-1}$ on the interval $[\frac{1}{2}, 4]$.

52. Find the area of the region bounded by the graphs of $y = x$, $y = 1/x$, and $x = 3$.

53. Find the area of the region bounded by the graphs of $y = 1/x$ and $3x + 3y + 10 = 0$.

54. Find the area bounded by the graph of $y = (x - 1)^{-1}$ and the x-axis on the interval $[-4, -2]$.

55. Find the area under the graph of $y = e^{-2x}$ on the interval $[0, 2]$.

56. Find the area of the region bounded by the graphs of $y = e^x$, $y = -e^{-x}$, $x = 0$, and $x = 1$.

57. Evaluate $\int_{-1}^{1} (e^x - 1)\, dx$ and determine whether the integral represents area. If not, find the area bounded by the graph of the integrand and the x-axis on the given interval.

58. The region in the first quadrant bounded by the graphs of $y = 1/(1 + x^2)$, $x = 0$, $x = \sqrt{3}$, and $y = 0$ is revolved about the y-axis. Find the volume of the solid of revolution.

59. The region in the first quadrant bounded by the graphs of $y = e^x$, $x = 0$, $x = 2$, and $y = 0$ is revolved about the x-axis. Find the volume of the solid of revolution.

60. The region in the first quadrant bounded by the graphs of $y = 1/\sqrt{x}$, $x = 1$, $x = 4$, and $y = 0$ is revolved about the x-axis. Find the volume of the solid of revolution.

61. The region in the first quadrant bounded by the graphs of $y = e^{-x^2}$, $x = 0$, $x = 1$, and $y = 0$ is revolved about the y-axis. Find the volume of the solid of revolution.

62. Find the length of the graph of $y = \ln(\cos x)$ on the interval $[0, \pi/4]$.

63. Find the length of the graph of $y = \frac{1}{2}(e^x + e^{-x})$ on the interval $[0, \ln 2]$.

64. Find the length of the graph of $y = \frac{1}{2}x^2 - \frac{1}{4}\ln x$ on the interval $[1, 4]$.

65. Find the area of the surface that is formed by revolving the graph of the function in Problem 63 on the indicated interval about the x-axis.

66. When gas enclosed in a cylinder expands against a piston so that the volume of the gas changes from v_1 to v_2, the work done on the piston is $W = \int_{v_1}^{v_2} p\, dv$, where p is pressure. If pressure p and volume v are related $pv = k$, show that the work done is $k \ln(v_2/v_1)$.

67. Evaluate:

(a) $e^{4\int dx/x}$ (b) $e^{\int \frac{x-1}{x+1} dx}$

68. If $f(x) = \ln x$, show that the average value of f' on $[a, b]$, $0 < a < b$, is $\frac{1}{b - a} \ln \frac{b}{a}$.

In Problems 69–74 solve the given differential equation by separation of variables.

69. $x^2 y^2 \frac{dy}{dx} = y + 1$

70. $(1 + x)\frac{dy}{dx} - 2y = 0$

71. $\dfrac{dy}{dx} = e^{3x+2y}$

72. $(e^x + 1)^3 e^{-x} \dfrac{dy}{dx} + (e^y + 1)^2 e^{-y} = 0$

73. $\sec y\, dy = 2 \cos x \sin y\, dx$

74. $\sec x^2\, dy = x \cot y\, dx$

In Problems 75 and 76 solve the given differential equation subject to the indicated initial condition.

75. $\dfrac{dy}{dt} + ty = y,\ y(1) = 3$

76. $x^2 y' = y - xy,\ y(-1) = -1$

In Problems 77–80 verify the given result by differentiation.

77. $\displaystyle\int \ln x\, dx = x \ln x - x + C$

78. $\displaystyle\int xe^x\, dx = xe^x - e^x + C$

79. $\displaystyle\int \frac{dx}{a^2 - x^2} = \frac{1}{2a} \ln\left|\frac{a+x}{a-x}\right| + C$

80. $\displaystyle\int \frac{\sqrt{x^2 + a^2}}{x}\, dx = \sqrt{x^2 + a^2} - a \ln\left|\frac{a + \sqrt{x^2 + a^2}}{x}\right| + C$

81. Consider the area under the graph of $y = \ln x$ on the interval $[1, e]$.

(a) Use the result given in Problem 77 to find the area.

(b) Determine an alternative integral method for finding this same area.

82. Verify the formula given in (6).

83. Verify the formula given in (8).

84. Use differentiation to show that

$$\int \csc x\, dx = -\ln|\csc x + \cot x| + C$$

Reconcile this result with (8).

85. If $\dfrac{d}{dx}\left[\dfrac{x^2}{2} \ln x - \dfrac{x^2}{4}\right] = x \ln x$, what is $\displaystyle\int x \ln x\, dx$?

86. If $\dfrac{d}{dx}[x^2 e^x - 2xe^x + 2e^x] = x^2 e^x$, what is $\displaystyle\int x^2 e^x\, dx$?

Calculator/Computer Problems

87. (a) Use the Trapezoidal Rule to obtain an approximation of $\int_0^1 e^{-x^2}\, dx$ that has an error less than 0.01.

(b) How many trapezoids are required to yield an approximation accurate to two decimal places? four decimal places?

88. Repeat Problem 87 using Simpson's Rule.

6.5 EXPONENTIAL AND LOGARITHMIC FUNCTIONS TO OTHER BASES

Inspired by the property

$$a^t = e^{\ln a^t} = e^{t \ln a}$$

where t is a rational number and $a > 0$, we make the following definition:

DEFINITION 6.6

If r is any real number and $a > 0$, then

$$a^r = e^{r \ln a} \tag{1}$$

The explicit use of (1) is to give meaning to a^r when r is an irrational number.

EXAMPLE 1

(*a*) $5^{\pi} = e^{\pi \ln 5}$ (*b*) $10^{\sqrt{3}} = e^{\sqrt{3} \ln 10}$ □

Laws of Exponents The following laws of exponents are immediate results of Definition 6.6. We shall prove the third property and leave some of the others as exercises.

THEOREM 6.6 Laws of Exponents

Let r and s be any real number. For $a > 0$,

(*i*) $a^0 = 1$ (*ii*) $a^1 = a$

(*iii*) $a^r a^s = a^{r+s}$ (*iv*) $(ab)^r = a^r b^r$

(*v*) $\dfrac{a^r}{a^s} = a^{r-s}$ (*vi*) $\left(\dfrac{a}{b}\right)^r = \dfrac{a^r}{b^r}$

(*vii*) $(a^r)^s = a^{rs}$ (*viii*) $a^{-r} = \dfrac{1}{a^r}$

Proof of (iii) From (1) we can write

$$a^r = e^{r \ln a} \quad \text{and} \quad a^s = e^{s \ln a}$$

Therefore, from property (*iii*) of the laws of exponents given in Section 6.3,

$$a^r a^s = e^{r \ln a} e^{s \ln a} = e^{r \ln a + s \ln a} = e^{(r+s) \ln a} = a^{r+s}$$ ■

As a further consequence of Definition 6.6, we can now extend property (*iii*) of the laws of the natural logarithm to include the case when the exponent is an irrational number.

THEOREM 6.7

Let r be any real number. For $a > 0$,

$$\ln a^r = r \ln a$$

Proof Since $y = e^x$ is equivalent to $\ln y = x$, we see that $a^r = e^{r \ln a}$ yields $\ln a^r = r \ln a$. ■

Exponential Function with Base a Recall that $y = e^x$ is a differentiable function whose domain is the set of real numbers. Likewise,

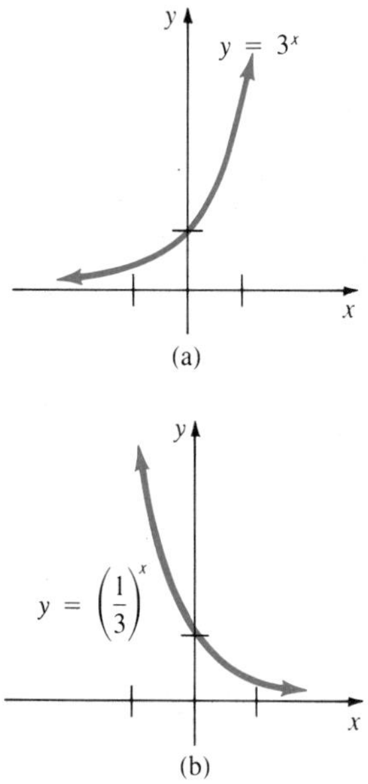

FIGURE 6.24

Definition 6.6 implies that

$$f(x) = a^x, \qquad a > 0 \tag{2}$$

is also a differentiable function that has the same domain. We say that f is an **exponential function with base a**. Furthermore, since $y = e^{kx}$ increases for $k > 0$, it follows from (1) that whenever $a > 1$, $\ln a > 0$, $f(x) = a^x$ is an increasing function on $(-\infty, \infty)$. For $0 < a < 1$, $\ln a < 0$ and so $f(x) = a^x$ is a decreasing function. Typical graphs of exponential functions, which illustrate these two cases, are similar to the graphs of e^x and e^{-x}, respectively. The graphs of $y = 3^x$ and $y = (\frac{1}{3})^x = 3^{-x}$ are shown in Figure 6.24.

Derivative of a^x The derivative of a^x is obtained from the derivative of e^u.

$$\frac{d}{dx}a^x = \frac{d}{dx}e^{x\ln a} = e^{x\ln a}\frac{d}{dx}x\ln a = e^{x\ln a}\ln a$$

Since $e^{x\ln a} = a^x$, this becomes

$$\frac{d}{dx}a^x = a^x\ln a \tag{3}$$

In (3) we can see, again, the importance of the number e; the derivative of an exponential function with a base other than e carries with it the extra factor of $\ln a$.

The Chain Rule gives the general result. If $u = g(x)$ is a differentiable function, then

$$\frac{d}{dx}a^u = a^u\frac{du}{dx}\ln a \tag{4}$$

Note that when $a = e$, $\ln e = 1$ and (3) and (4) reduce, in turn, to (12) and (13) of Section 6.3.

EXAMPLE 2 Differentiate (*a*) $y = 10^x$ and (*b*) $y = 3^{\cos 5x}$.

Solution (*a*) From (3),

$$\frac{dy}{dx} = 10^x\ln 10$$

(*b*) From (4),

$$\frac{dy}{dx} = 3^{\cos 5x}\left(\frac{d}{dx}\cos 5x\right)\ln 3 = 3^{\cos 5x}(-5\sin 5x)\ln 3$$ □

You should verify by differentiation that

$$\int a^x\,dx = \frac{a^x}{\ln a} + C \tag{5}$$

$$\int a^u\,du = \frac{a^u}{\ln a} + C \tag{6}$$

are the integral forms of (3) and (4), respectively.

EXAMPLE 3 Evaluate $\int 8^x\,dx$.

Solution From (5),

$$\int 8^x\,dx = \frac{8^x}{\ln 8} + C \qquad \square$$

EXAMPLE 4 Evaluate $\int \frac{2^{\sqrt{x}}}{\sqrt{x}}\,dx$

Solution Identifying $a = 2$, $u = x^{1/2}$, and $du = \frac{1}{2}x^{-1/2}\,dx$, we see from (6) that

$$\begin{aligned}\int \frac{2^{\sqrt{x}}}{\sqrt{x}}\,dx &= 2\int 2^{\sqrt{x}}\left(\frac{1}{2}x^{-1/2}\,dx\right)\\ &= 2\int 2^u\,du\\ &= 2\left(\frac{2^u}{\ln 2}\right) + C\\ &= 2\left(\frac{2^{\sqrt{x}}}{\ln 2}\right) + C\\ &= \frac{2^{1+\sqrt{x}}}{\ln 2} + C \qquad \square\end{aligned}$$

Power Rule Revisited At last we are able to state and prove the Power Rule of differentiation for any real exponent, *rational or irrational.*

THEOREM 6.8 Power Rule (Real Exponents)

If r is any real number, then

$$\frac{d}{dx}x^r = rx^{r-1} \tag{7}$$

Proof In view of Definition 6.6, we can write

$$x^r = e^{r\ln x}$$

Therefore,

$$\frac{d}{dx}x^r = e^{r\ln x}\frac{d}{dx}(r\ln x) = \frac{r}{x}e^{r\ln x} = \frac{r}{x}x^r = rx^{r-1}$$ ■

Alternatively, (7) can be derived by taking the logarithm of both sides of $y = x^r$ and then using implicit differentiation.

It should be noted that when r is an irrational number, the domain of the function $y = x^r$ is defined to be the set of positive real numbers.

EXAMPLE 5 Differentiate (*a*) $y = \sqrt{3}^x$ and (*b*) $y = x^{\sqrt{3}}$.

Solution (*a*) From (3),

$$\frac{dy}{dx} = \sqrt{3}^x \ln\sqrt{3}$$

(*b*) From (7),

$$\frac{dy}{dx} = \sqrt{3}x^{\sqrt{3}-1}$$ □

The result given in (7) also extends to a power of a function.

EXAMPLE 6 Differentiate $y = (x^4 + e^{-3x})^e$.

Solution

$$\frac{dy}{dx} = e(x^4 + e^{-3x})^{e-1}\frac{d}{dx}(x^4 + e^{-3x})$$
$$= e(x^4 + e^{-3x})^{e-1}(4x^3 - 3e^{-3x})$$ □

The result given in (7) and the fundamental definition of the natural logarithms are the "missing links" that we need to evaluate the indefinite integral $\int x^r\,dx$ for *any* real exponent r:

$$\int x^r\,dx = \begin{cases} \dfrac{x^{r+1}}{r+1} + C, & r \neq -1 \\ \ln|x| + C, & r = -1 \end{cases} \tag{8}$$

EXAMPLE 7 Evaluate $\displaystyle\int x^\pi\,dx$.

Solution From (8),

$$\int x^\pi\,dx = \frac{x^{\pi+1}}{\pi+1} + C$$ □

Logarithmic Function with Base a When the base a is positive and $a \neq 1$, the exponential function $y = a^x$ is one-to-one and, hence, possesses an inverse function denoted by $y = \log_a x$. The latter function is called the **logarithmic function with base *a***. Analogous to (6) of Section 6.3 is

$$y = \log_a x \quad \text{if and only if} \quad x = a^y \tag{9}$$

This is the interpretation of the logarithm that one normally uses in a course in precalculus mathematics. Since the range of the exponential function is the domain of its inverse, the domain of the logarithmic function is the set of positive real numbers.

Because of (6) and (10) of Section 6.3, we are finally able to state that the base of the natural logarithm is the number e:

$$\ln x = \log_e x$$

Laws of Logarithms The laws of logarithms stated for $\ln x$ also hold for $\log_a x$. Of course, as in Theorem 6.7 for $b > 0$, $a > 0$, $a \neq 1$, the property

$$\log_a b^r = r \log_a b$$

is valid for any real number r.

Derivative of $\log_a x$ To find the derivative of the logarithmic function $y = \log_a x$, we take the natural logarithm of both sides of $x = a^y$. Implicit differentiation of

$$\ln x = y \ln a$$

gives

$$\frac{1}{x} = \frac{dy}{dx} \ln a \quad \text{or} \quad \frac{dy}{dx} = \frac{1}{x \ln a}$$

In other words,

$$\frac{d}{dx} \log_a x = \frac{1}{x \ln a} \tag{10}$$

As usual, the general case can be obtained from the Chain Rule. If $u = g(x)$ is a differentiable function, then

$$\frac{d}{dx} \log_a |u| = \frac{1}{u \ln a} \frac{du}{dx} \tag{11}$$

When $a = 10$ in (9), we say that $\log_{10} x$ is the **common logarithm** of the number x.

EXAMPLE 8 Find the slope of the tangent to the graph of $y = \log_{10} x$ at $x = \frac{1}{2}$.

Solution From (10),

$$\frac{dy}{dx} = \frac{1}{x \ln 10}$$

With the aid of a calculator, we find

$$\left.\frac{dy}{dx}\right|_{x=1/2} = \frac{2}{\ln 10} \approx 0.8686 \qquad \square$$

EXAMPLE 9 Differentiate $y = \log_5|x^3 - x|$.

Solution From (11),

$$\frac{dy}{dx} = \frac{1}{(x^3 - x)\ln 5} \cdot \frac{d}{dx}(x^3 - x) = \frac{3x^2 - 1}{(x^3 - x)\ln 5} \qquad \square$$

Logarithmic Differentiation We now know how to differentiate any function of the type

$$y = (\text{constant})^{\text{variable}} \quad \text{and} \quad y = (\text{variable})^{\text{constant}}$$

Using the natural logarithm and its properties, we are also able to differentiate a function $y = f(x)$ where both the base and the exponent are variable—that is,

$$y = (\text{variable})^{\text{variable}} \tag{12}$$

Although we shall not develop a formula for the derivative of (12), dy/dx can be obtained through a technique known as **logarithmic differentiation.*** This procedure consists of three steps:

(*i*) Take the natural logarithm of both sides of $y = f(x)$ and simplify.

(*ii*) Differentiate $\ln y = \ln f(x)$ with respect to x:

$$\frac{1}{y}\frac{dy}{dx} = \frac{d}{dx}\ln f(x)$$

(*iii*) Solve for dy/dx by multiplying both sides of the equation in (*ii*) by $y = f(x)$.

EXAMPLE 10 Differentiate $y = x^{\sqrt{x}}$, $x > 0$.

*This method of differentiation is attributed to the Swiss mathematician Johann Bernoulli (1668–1758).

Solution Taking the logarithm of both sides of the given equation yields

$$\ln y = \ln x^{\sqrt{x}} = \sqrt{x}\ln x$$

Implicit differentiation then yields

$$\frac{1}{y}\frac{dy}{dx} = \sqrt{x}\cdot\frac{1}{x} + \frac{1}{2}x^{-1/2}\cdot\ln x \qquad \boxed{\text{Product Rule}}$$

$$\frac{dy}{dx} = y\left[\frac{1}{\sqrt{x}} + \frac{\ln x}{2\sqrt{x}}\right]$$

$$= x^{\sqrt{x}}\left[\frac{1}{\sqrt{x}} + \frac{\ln x}{2\sqrt{x}}\right]$$

$$= x^{\sqrt{x}-1/2}\left(1 + \frac{1}{2}\ln x\right) \qquad \square$$

EXAMPLE 11 Find the derivative of $y = \dfrac{\sqrt[3]{x^4 + 6x^2}(8x + 3)^5}{(2x^2 + 7)^{2/3}}$.

Solution While we could find dy/dx through application of the Quotient, Product, and Power Rules, this entire procedure can be avoided by first taking the logarithm of the absolute value of both sides of the given equation, simplifying, and *then* differentiating implicitly. We have

$$\ln|y| = \ln\sqrt[3]{x^4 + 6x^2} + \ln|8x + 3|^5 - \ln(2x^2 + 7)^{2/3}$$

$$= \frac{1}{3}\ln(x^4 + 6x^2) + 5\ln|8x + 3| - \frac{2}{3}\ln(2x^2 + 7)$$

Taking the derivative with respect to x gives

$$\frac{1}{y}\frac{dy}{dx} = \frac{1}{3}\cdot\frac{1}{x^4 + 6x^2}\cdot(4x^3 + 12x) + 5\cdot\frac{1}{8x + 3}\cdot 8 - \frac{2}{3}\cdot\frac{1}{2x^2 + 7}\cdot 4x$$

$$\frac{dy}{dx} = y\left[\frac{4x^3 + 12x}{3(x^4 + 6x^2)} + \frac{40}{8x + 3} - \frac{8x}{3(2x^2 + 7)}\right]$$

$$= \frac{\sqrt[3]{x^4 + 6x^2}(8x + 3)^5}{(2x^2 + 7)^{2/3}}\left[\frac{4x^3 + 12x}{3(x^4 + 6x^2)} + \frac{40}{8x + 3} - \frac{8x}{3(2x^2 + 7)}\right] \qquad \square$$

In conclusion, we consider an application of the common logarithm.

Earthquakes The American geologist/seismologist Charles F. Richter (1900–1985) devised a scale for comparing energies of various earthquakes. The so-called **Richter scale**, proposed in 1935, is defined in terms of a common logarithm:

$$R = \log_{10}\frac{A}{A_0} \tag{13}$$

where R represents the magnitude of the earthquake, A is the amplitude of the largest seismic wave that occurs, and A_0 is a reference amplitude corresponding to the magnitude $R = 0$.

FIGURE 6.25 Collapse of the Nimitz Freeway, Oakland, California, October 1989. Los Angeles Times photo by Jayne Kamin-Oncea.

EXAMPLE 12 The Richter magnitude of the 1906 San Francisco earthquake is estimated to have been 8.3. The 1989 "World Series" earthquake in San Francisco had a magnitude of 7.1 on the Richter scale. See Figure 6.25. How much greater was the intensity of the 1906 earthquake than that of the 1989 earthquake?

Solution From (13) we have

$$8.3 = \log_{10}\left(\frac{A}{A_0}\right)_{1906} \quad \text{and} \quad 7.1 = \log_{10}\left(\frac{A}{A_0}\right)_{1989}$$

In view of (9) this means

$$\left(\frac{A}{A_0}\right)_{1906} = 10^{8.3} \quad \text{and} \quad \left(\frac{A}{A_0}\right)_{1989} = 10^{7.1}$$

Using the laws of exponents and a calculator, we find that

$$\left(\frac{A}{A_0}\right)_{1906} = 10^{8.3} = 10^{1.2}10^{7.1} = 10^{1.2}\left(\frac{A}{A_0}\right)_{1989} \approx 15.8\left(\frac{A}{A_0}\right)_{1989}$$

Hence, the 1906 San Francisco earthquake was approximately 16 times stronger than the 1989 earthquake. □

▲ ***Remarks*** (*i*) Note carefully that the Power Rule is not applicable to a function such as $y = x^x$, $x > 0$. In other words, the derivative is *not* xx^{x-1}. On a recent examination several students applied the rule $(d/dx)a^x = a^x \ln a$ with the identification $a = x$. Again this is a misapplication of a rule; the latter rule demands that the base be a constant. The derivative is *not* $x^x \ln x$. It is left as an exercise to show by logarithmic differentiation that the derivative of $y = x^x$ is $dy/dx = x^x(1 + \ln x)$. See Problem 68.

(*ii*) It may be necessary to take the logarithm of both members of a given equation more than once in order to compute the derivative. For example, the logarithm of the function $y = (x^2 + 1)^{(x+1)^x}$, $x > -1$, is

$$\ln y = (x + 1)^x \ln(x^2 + 1)$$

Since this expression still involves a variable base with a variable exponent, we take the logarithm a second time:

$$\ln(\ln y) = x \ln(x + 1) + \ln(\ln(x^2 + 1))$$

We differentiate this last equation and solve for dy/dx. See Problem 76.

EXERCISES 6.5 *Answers to odd-numbered problems begin on page A-85.*

In Problems 1–6 write each number as a power of e.

1. 2^{π}
2. $5^{\sqrt{2}}$
3. 10^{e}
4. $6^{-2.3}$
5. $7^{-\sqrt{5}}$
6. $(\frac{3}{2})^{2\pi}$

In Problems 7–36 find the derivative of the given function.

7. $y = 4^x$
8. $y = 2 - 3^x$
9. $y = 10^{-2x}$
10. $y = 9^{3x+1}$
11. $y = 20^{x^2}$
12. $y = 5^{\sqrt{x}}$
13. $y = \pi^{\sqrt{2}}$
14. $y = 2^{\sin 5x}$
15. $y = x^2 2^x$
16. $y = e^x x^e$
17. $y = x^{\sqrt{5}}$
18. $y = \left(\dfrac{1}{3x}\right)^{\sqrt{3}}$
19. $y = x^{\sin 2}$
20. $y = (\cos x)^{\pi/2}$
21. $f(t) = (t^4 + 3t^2)^{\ln 4}$
22. $g(t) = (t^2)^e$
23. $f(x) = \dfrac{4^x}{1 + e^x}$
24. $f(x) = \dfrac{(x+1)^{\pi}}{\pi^{x+1}}$
25. $f(x) = \sqrt{5^x}$
26. $f(x) = \dfrac{(25)^x}{5^{-x}}$
27. $H(x) = \dfrac{1}{(\ln x)^{1/e}}$
28. $P(v) = \tan(2^{-v^2})$
29. $y = \log_4 x$
30. $y = (\ln 5)^x$
31. $y = (\ln x)(\log_{10} x)$
32. $y = \ln(\log_{10} x)$
33. $y = x^{\pi}\log_3|6x - 4|$
34. $y = \log_{10}\left(\dfrac{x^2+1}{x^4+9}\right)$
35. $f(x) = 3^{x^3}\log_3 x^3$
36. $f(x) = (\log_5 3x)^7$

In Problems 37–40 find the indicated derivative.

37. $y = 9^x;\ \dfrac{d^2y}{dx^2}$
38. $y = x3^x;\ \dfrac{d^2y}{dx^2}$
39. $y = (2x)^e;\ \dfrac{d^3y}{dx^3}$
40. $y = (\log_{10} x)^2;\ \dfrac{d^2y}{dx^2}$

In Problems 41–44 use implicit differentiation to find dy/dx.

41. $2^y = xy$
42. $x^2 + y^2 = 4^x - 4^{-x}$
43. $y + e^y = \log_{10}|x|$
44. $\log_2 xy = \sqrt{2x + 1}$

In Problems 45–60 evaluate the given integral.

45. $\displaystyle\int 7^x\,dx$
46. $\displaystyle\int \frac{dx}{7^{2x}}$
47. $\displaystyle\int (10^x)^{1/3}\,dx$
48. $\displaystyle\int 2^x 4^x\,dx$
49. $\displaystyle\int_2^3 10^{-x}\,dx$
50. $\displaystyle\int_0^1 t^2 3^{-t^3}\,dt$
51. $\displaystyle\int (e^x + x^e + e^e)\,dx$
52. $\displaystyle\int (x+1)^e\,dx$
53. $\displaystyle\int 2^t(1 + 2^t)^{20}\,dt$
54. $\displaystyle\int (1 + 2^t)^2\,dt$
55. $\displaystyle\int_{-1}^1 e^t 2^{e^t}\,dt$
56. $\displaystyle\int_0^{\pi} \frac{\sin 2x}{2^{\cos 2x}}\,dx$
57. $\displaystyle\int (1 + 2^{\tan\theta})^2 \sec^2\theta\,d\theta$
58. $\displaystyle\int_1^{10} \frac{(\log_{10} x)^3}{x}\,dx$
59. $\displaystyle\int \frac{5^x}{1 + 5^x}\,dx$
60. $\displaystyle\int \frac{5^{2x}}{1 + 25^x}\,dx$

61. Find the slope of the tangent line to the graph of $y = 4^x$ at $x = 3$.

62. Find an equation of the tangent line to the graph of $y = x + x^{\pi}$ at $x = 1$.

63. Find the point(s) on the graph of $f(x) = x4^x$ where the tangent line is horizontal.

64. Determine the intervals on which $f(x) = x\log_2|x - 1|$ is concave upward and the intervals on which it is concave downward.

65. Find the area of the region in the first quadrant bounded by the graphs of $y = 2^x$, $y = 2^{-x}$, and $x = 4$.

66. Find the volume of the solid of revolution that is formed by revolving the region bounded by the graphs of $y = 2^{x^2}$, $y = 0$, $x = 0$, and $x = 1$ about the y-axis.

In Problems 67–82 use logarithmic differentiation to find dy/dx.

67. $y = (x^2 + 4)^{2x}$
68. $y = x^x$
69. $y = x^x 2^x$
70. $y = x^{2^x}$
71. $y = x(x - 1)^x$
72. $y = \dfrac{(x^2+1)^x}{x^2}$

73. $y = (\ln|x|)^x$

74. $y = x^{\cos x}$

75. $y = x^{x^x}$

76. $y = (x^2 + 1)^{(x+1)^x}$

77. $y = \tan x^x$

78. $y = x^x e^{x^x}$

79. $y = \sqrt{\dfrac{(2x+1)(3x+2)}{4x+3}}$

80. $y = \dfrac{x^{10}\sqrt{x^2+5}}{\sqrt[3]{8x^2+2}}$

81. $y = \dfrac{(x^3-1)^5(x^4+3x^3)^4}{(7x+5)^9}$

82. $y = x\sqrt{x+1}\,\sqrt[3]{x^2+2}$

83. Find an equation of the tangent line to the graph of $y = x^{x+2}$ at $x = 1$.

84. Find an equation of the tangent line to the graph of $y = x(\ln x)^x$ at $x = e$.

85. Find d^2y/dx^2 for $y = x^{2x}$.

86. By inspection of the graph in Figure 1.34 we see that the function $f(x) = |x|^x$ has two relative extrema. Use logarithmic differentiation to find these extrema.

Calculator/Computer Problems

In Problems 87–90: **(a)** use a calculator or computer to obtain the graph of the given function, and **(b)** find the relative extrema of each function.

87. $f(x) = |x|^{1/x}$

88. $f(x) = |x|^{3x}$

89. $f(x) = 3x|x|^x$

90. $f(x) = |x|^x e^{-x}$

91. In 1979 an earthquake occurred in San Francisco of magnitude 5.9 on the Richter scale. How much greater was the intensity of that city's 1906 earthquake of magnitude 8.3? How does the intensity of the 1979 earthquake compare with the 1971 San Fernando Valley earthquake of magnitude 6.6?

92. An earthquake has a magnitude of 4.7 on the Richter scale. What is the magnitude of an earthquake on the Richter scale if its intensity is 50 times greater?

93. The **pH**, or hydrogen potential, of a solution is defined by $\text{pH} = -\log_{10}[\text{H}^+]$, where $[\text{H}^+]$ is the concentration of hydrogen ions in moles/liter. If $0 < \text{pH} < 7$, the solution is said to be *acid* and if $\text{pH} > 7$, the solution is *base* or *alkaline*. Find the pH of a solution for which $[\text{H}^+] = 3.9 \times 10^{-8}$ moles/liter. Is the solution acid or base?

94. Two liquids have pH values of 4.6 and 6.1. Use Problem 93 to determine how much more acidic the first solution is than the second.

95. The intensity level b of a sound is defined by $b = 10\log_{10}(I/I_0)$, where I is the intensity of the sound (in watts/cm^2) and I_0 is a reference intensity corresponding to $b = 0$. An intensity level is measured in decibels (dB). Now the intensity I of a sound is inversely proportional to the square of its distance d from its source. Use this fact to show that if a sound has an intensity level b_1 at a distance d_1 from its source, then at a distance d_2 the intensity level b_2 is given by $b_2 = b_1 + 20\log_{10}(d_1/d_2)$.

96. The intensity level of a plane at an altitude of 1500 ft and passing a point 5 mi out from an airport is 80 dB. Use Problem 95 to determine the intensity level of another plane passing the same point but at an altitude of 2700 ft.

97. Use (9) to prove the **change of base formula**

$$\log_a x = \frac{\log_b x}{\log_b a}$$

98. Use the result of Problem 97 to show that $\log_{10} e = 1/\ln 10$.

99. The **Naperian logarithm** of a number x is defined as

$$10^7 \log_{1/e}\left(\frac{x}{10^7}\right)$$

Use the change of base formula in Problem 97 to express the Naperian logarithm in terms of the natural logarithm.

John Napier (1550–1617) The first table of logarithms was published in 1614 by the Scotsman John Napier. He used his invention as a means of converting unwieldy calculations involving products and quotients into simpler calculations involving sums and differences. Napier is also credited with popularizing the use of decimal fractions through the notational device known as a decimal point. For the wealthy Napier, mathematics was simply a hobby or diversion in his life devoted to the polemics of politics and religion. His friend and collaborator, the English mathematician Henry Briggs (1561–1639), modified the base of the Naperian logarithm (a number that is close to $1/e$) and published the first tables of the common, or base 10, logarithms. These logarithms are sometimes called Briggsian logarithms.

[O] 6.6 THE NATURAL LOGARITHMIC FUNCTION—A HISTORICAL APPROACH

You may have noticed that the considerations of the logarithmic and exponential functions in the last four sections have been subtle and bear little resemblance to those concepts studied in a course in precalculus mathematics. In an elementary course, first one normally encounters an exponential function

$$y = a^x, \qquad a > 0, \quad a \neq 1, \quad x \text{ any real number}$$

and shows that this function possesses an inverse called the logarithmic function. Recall, *logarithm* is a word meaning the exponent of the base a that gives the number x; that is,

$$y = \log_a x \quad \text{is equivalent to} \quad x = a^y$$

From the laws of exponents, one then proceeds to deduce the properties and the laws of logarithms. Just the reverse of this method was presented in Sections 6.2 and 6.3. Why the difference? The answer is simple. In precalculus mathematics, it is *assumed and never proved* that a^x makes sense, or is well defined, for $a > 0$ and for every real number x. For a rational exponent p/q, where p and q are integers, $a^{p/q} = (a^{1/q})^p$ when $a^{1/q}$ is a real number. But how does one interpret a number with an irrational exponent such as 2^π?

Which interpretation should you use in calculus? To start with the integral definition of the logarithmic function *or* to start with an intuitive notion of a^x and its inverse raises a certain amount of controversy among teachers of mathematics.

You should not get the impression from the foregoing remarks that a^x cannot be defined first in a rigorous manner. Indeed, for any real number x, a^x is defined as follows:

DEFINITION 6.7

Suppose a is any positive real number, x is a fixed real number, and t is a rational number, Then

$$a^x = \lim_{t \to x} a^t$$

EXAMPLE 1 It can be proved that the limit of the sequence of numbers with rational exponents

$$2^3, 2^{3.1}, 2^{3.14}, 2^{3.141}, 2^{3.1416}, \ldots$$

is the number 2^π. □

With Definition 6.7 as a foundation, the many properties of the exponential function can now be proved. However, the proofs—for example, of the laws of exponents—are beyond the level of a first course in calculus.

***Derivative of* $\log_a x$** Although we studied two new functions in Sections 6.2 and 6.3, we obtained the derivatives of these functions without recourse to first principles—namely, $f'(x) = \lim_{\Delta x \to 0} \Delta y/\Delta x$. Independent of prior considerations, let us assume for the remainder of the discussion that $y = a^x$, $a > 0$, $a \neq 1$ is well defined and is a continuous one-to-one function whose inverse is the continuous function $y = \log_a x$, $x > 0$. The following equations show how to find the derivative of the logarithm from the definition of the derivative:

$$\begin{aligned}
\frac{dy}{dx} &= \lim_{\Delta x \to 0} \frac{\log_a(x + \Delta x) - \log_a x}{\Delta x} \\
&= \lim_{\Delta x \to 0} \frac{1}{\Delta x} \log_a \frac{x + \Delta x}{x} && \boxed{\text{algebra and laws of logarithms}} \\
&= \lim_{\Delta x \to 0} \frac{1}{\Delta x} \log_a\left(1 + \frac{\Delta x}{x}\right) && \boxed{\text{algebra}} \\
&= \lim_{\Delta x \to 0} \frac{1}{x} \frac{x}{\Delta x} \log_a\left(1 + \frac{\Delta x}{x}\right) && \boxed{\text{algebra, } x/x = 1} \\
&= \lim_{\Delta x \to 0} \frac{1}{x} \log_a\left(1 + \frac{\Delta x}{x}\right)^{x/\Delta x} && \boxed{\text{laws of logarithms}} \\
&= \frac{1}{x} \log_a\left[\lim_{\Delta x \to 0} \left(1 + \frac{\Delta x}{x}\right)^{x/\Delta x}\right] && (1)
\end{aligned}$$

The last step is justified by invoking the continuity of the function and assuming that the limit inside the brackets exists. Let us make the change of variable $h = \Delta x/x$ in (1). Since x is fixed, $\Delta x \to 0$ implies $h \to 0$. Consequently,

$$\lim_{\Delta x \to 0} \left(1 + \frac{\Delta x}{x}\right)^{x/\Delta x} = \lim_{h \to 0} (1 + h)^{1/h}$$

Now, using the concepts of Sections 9.1 or 10.2, we can prove that

$$\lim_{h \to 0} (1 + h)^{1/h} = e$$

where $e = 2.71828\ldots$. Hence, (1) becomes

$$\frac{d}{dx} \log_a x = \frac{1}{x} \log_a e \tag{2}$$

When the "natural" choice of $a = e$ is made, (2) simplifies to

$$\frac{d}{dx} \log_e x = \frac{1}{x}$$

since $\log_e e = 1$. Accordingly, we say $\log_e x$ is the "natural logarithm" and abbreviate this function by

$$\log_e x = \ln x$$

▲ ***Remark*** Those with sharp eyes and long memories will have noticed that (2) is not the same as (10) of Section 6.5. The results are equivalent, since $\log_a e = 1/\log_e a$. See Problem 97 in Exercises 6.5.

EXERCISES 6.6 *Answers to odd-numbered problems begin on page A-86.*

1. Conjecture the value of

$$\lim_{\Delta x \to 0} \frac{e^{\Delta x} - 1}{\Delta x}$$

by filling in the table.

Δx	$(e^{\Delta x} - 1)/\Delta x$
0.1	
0.01	
0.001	
0.0001	
−0.0001	

2. Use the definition of the derivative along with the result obtained in Problem 1 to find dy/dx for $y = e^x$.

6.7 THE HYPERBOLIC FUNCTIONS

FIGURE 6.26 The shape of an inverted catenary. Gateway Arch photograph by Frank Siteman/Stock, Boston, Inc.

If you have ever toured the 640-ft-high Gateway Arch in St. Louis, Missouri, you may have asked the question: What is the shape of the arch? and received the rather cryptic reply: the shape of an inverted catenary. See Figure 6.26. The word *catenary* stems from the Latin word *catena* and literally means "a hanging chain" (the Romans used the catena as a dog leash). It can be demonstrated that the shape assumed by a long flexible wire, chain, cable, or rope hanging under its own weight between two points is the shape of the graph of the function

$$f(x) = \frac{k}{2}(e^{cx} + e^{-cx})$$

for appropriate choices of the constants c and k. Combinations such as this involving e^x and e^{-x} occur so often in applied mathematics that they warrant special definitions.

DEFINITION 6.8 Hyperbolic Sine and Cosine

For any real number x, the **hyperbolic sine** of x is

$$\sinh x = \frac{e^x - e^{-x}}{2}$$

and the **hyperbolic cosine** of x is

$$\cosh x = \frac{e^x + e^{-x}}{2}$$

Analogous to the trigonometric functions tan x, cot x, sec x, and csc x, which are defined in terms of sin x and cos x, we define four additional hyperbolic functions in terms of sinh x and cosh x.

DEFINITION 6.9 Other Hyperbolic Functions

For a real number x, the

(*i*) **hyperbolic tangent** of x is

$$\tanh x = \frac{\sinh x}{\cosh x} = \frac{e^x - e^{-x}}{e^x + e^{-x}}$$

(*ii*) **hyperbolic cotangent** of x is

$$\coth x = \frac{\cosh x}{\sinh x} = \frac{e^x + e^{-x}}{e^x - e^{-x}}, \qquad x \neq 0$$

(*iii*) **hyperbolic secant** of x is

$$\operatorname{sech} x = \frac{1}{\cosh x} = \frac{2}{e^x + e^{-x}}$$

(*iv*) **hyperbolic cosecant** of x is

$$\operatorname{csch} x = \frac{1}{\sinh x} = \frac{2}{e^x - e^{-x}}, \qquad x \neq 0$$

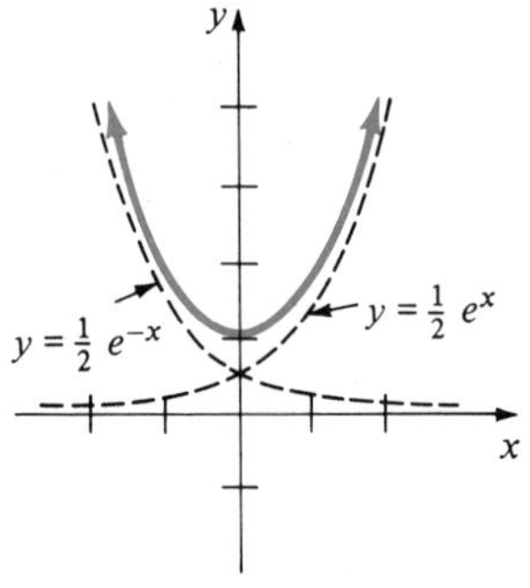

(a) $y = \cosh x$

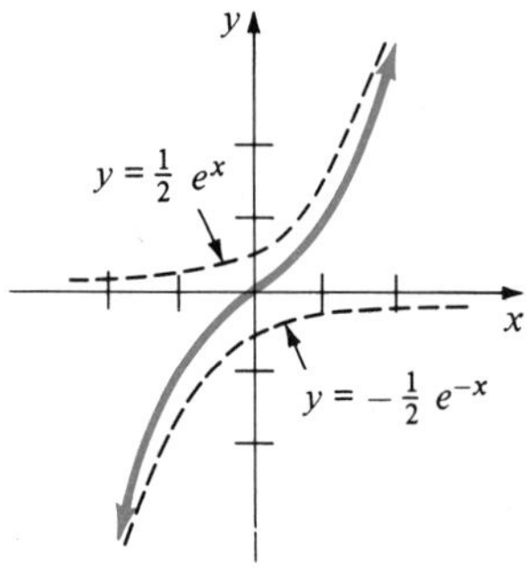

(b) $y = \sinh x$

FIGURE 6.27

Graphs The graphs of the hyperbolic sine and hyperbolic cosine can be obtained by adding ordinates. As illustrated in Figure 6.27(a), the graph of $y = \cosh x$ is gotten by first graphing $\frac{1}{2}e^x$ and $\frac{1}{2}e^{-x}$ and then adding y-coordinates at each point. Similarly, the graph of $y = \sinh x$ shown in Figure 6.27(b) is found by adding the y-coordinates of points on the graphs of $\frac{1}{2}e^x$ and $-\frac{1}{2}e^{-x}$. The graphs of tanh x, coth x, sech x, and csch x are given in Figure 6.28.

Identities Hyperbolic functions possess many identities that are similar to those of the trigonometric functions. Notice that the graphs in Figure 6.28(a) and (b) are symmetric with respect to the y-axis and the origin, respectively. In other words, $y = \cosh x$ is an even function and $y = \sinh x$ is

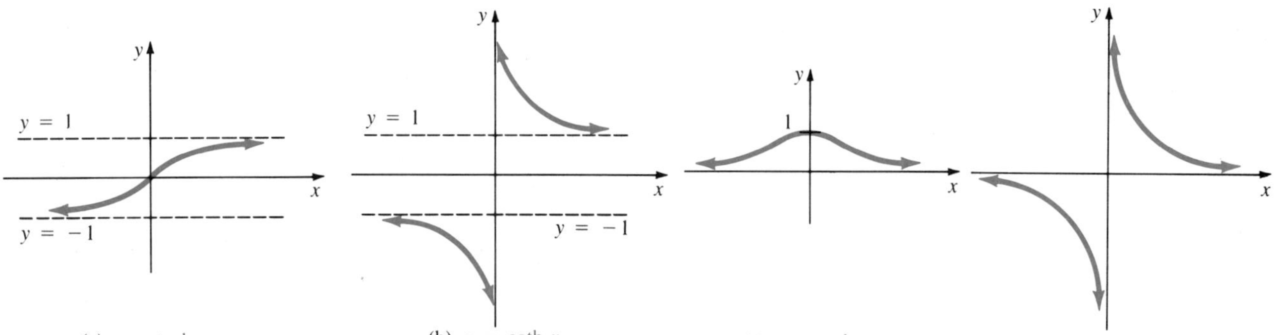

(a) $y = \tanh x$ (b) $y = \coth x$ (c) $y = \operatorname{sech} x$ (d) $y = \operatorname{csch} x$

FIGURE 6.28

an odd function:

$$\cosh(-x) = \cosh x \tag{1}$$
$$\sinh(-x) = -\sinh x \tag{2}$$

In trigonometry a fundamental identity is $\cos^2 x + \sin^2 x = 1$. For hyperbolic functions the analogue of this identity is

$$\cosh^2 x - \sinh^2 x = 1 \tag{3}$$

To prove this we resort to Definition 6.8:

$$\cosh^2 x - \sinh^2 x = \left(\frac{e^x + e^{-x}}{2}\right)^2 - \left(\frac{e^x - e^{-x}}{2}\right)^2$$
$$= \frac{e^{2x} + 2 + e^{-2x}}{4} - \frac{e^{2x} - 2 + e^{-2x}}{4} = 1$$

The proofs of (1), (2), and the following list of identities are left as exercises.

$$1 - \tanh^2 x = \operatorname{sech}^2 x \tag{4}$$
$$\coth^2 x - 1 = \operatorname{csch}^2 x \tag{5}$$
$$\sinh(x + y) = \sinh x \cosh y + \cosh x \sinh y \tag{6}$$
$$\sinh(x - y) = \sinh x \cosh y - \cosh x \sinh y \tag{7}$$
$$\cosh(x + y) = \cosh x \cosh y + \sinh x \sinh y \tag{8}$$
$$\cosh(x - y) = \cosh x \cosh y - \sinh x \sinh y \tag{9}$$
$$\sinh 2x = 2 \sinh x \cosh x \tag{10}$$
$$\cosh 2x = \cosh^2 x + \sinh^2 x \tag{11}$$
$$\cosh^2 x = \frac{1}{2}(1 + \cosh 2x) \tag{12}$$
$$\sinh^2 x = \frac{1}{2}(-1 + \cosh 2x) \tag{13}$$

Derivatives of Hyperbolic Functions The derivatives of the hyperbolic functions follow from (12) of Section 6.3 and the rules of differentiation;

for example,

$$\begin{aligned}\frac{d}{dx}\sinh x &= \frac{d}{dx}\frac{e^x - e^{-x}}{2}\\ &= \frac{1}{2}\left[\frac{d}{dx}e^x - \frac{d}{dx}e^{-x}\right]\\ &= \frac{e^x + e^{-x}}{2}\\ &= \cosh x\end{aligned}$$

Similarly, it should be apparent from the definition of the hyperbolic cosine that

$$\frac{d}{dx}\cosh x = \sinh x$$

To differentiate, say, the hyperbolic tangent, we use the Quotient Rule and the identity given in (3):

$$\begin{aligned}\frac{d}{dx}\tanh x &= \frac{d}{dx}\frac{\sinh x}{\cosh x}\\ &= \frac{\cosh x \cdot \frac{d}{dx}\sinh x - \sinh x \cdot \frac{d}{dx}\cosh x}{\cosh^2 x}\\ &= \frac{\cosh^2 x - \sinh^2 x}{\cosh^2 x}\\ &= \frac{1}{\cosh^2 x}\\ &= \operatorname{sech}^2 x\end{aligned}$$

The derivatives of the six hyperbolic functions in the most general case follow from the Chain Rule. We assume $u = g(x)$ is a differentiable function.

Derivatives of the Hyperbolic Functions

I	$\frac{d}{dx}\sinh u = \cosh u \frac{du}{dx}$	II	$\frac{d}{dx}\cosh u = \sinh u \frac{du}{dx}$
III	$\frac{d}{dx}\tanh u = \operatorname{sech}^2 u \frac{du}{dx}$	IV	$\frac{d}{dx}\coth u = -\operatorname{csch}^2 u \frac{du}{dx}$
V	$\frac{d}{dx}\operatorname{sech} u = -\operatorname{sech} u \tanh u \frac{du}{dx}$	VI	$\frac{d}{dx}\operatorname{csch} u = -\operatorname{csch} u \coth u \frac{du}{dx}$

You should take careful note of the slight difference in the results in II and V and the analogous formulas for the trigonometric functions.

EXAMPLE 1 Differentiate (*a*) $y = \sinh\sqrt{2x+1}$ and (*b*) $y = \coth x^3$.

Solution (*a*) From I,

$$\frac{dy}{dx} = \cosh\sqrt{2x+1}\left(\frac{1}{2}(2x+1)^{-1/2}\cdot 2\right)$$

$$= \frac{\cosh\sqrt{2x+1}}{\sqrt{2x+1}}$$

(*b*) From IV,

$$\frac{dy}{dx} = -\text{csch}^2 x^3 \cdot 3x^2$$

□

EXAMPLE 2 Evaluate the derivative of $y = \dfrac{3x}{4+\cosh 2x}$ at $x = 0$.

Solution From the Quotient Rule,

$$\frac{dy}{dx} = \frac{(4+\cosh 2x)\cdot 3 - 3x(\sinh 2x\cdot 2)}{(4+\cosh 2x)^2}$$

When $x = 0$, it is seen from Definition 6.8 (see also Figure 6.27) that $\sinh 0 = 0$ and $\cosh 0 = 1$. Thus,

$$\left.\frac{dy}{dx}\right|_{x=0} = \frac{15}{25}$$

$$= \frac{3}{5}$$

□

Integrals of Hyperbolic Functions The integral forms of the preceding differentiation formulas are summarized as follows:

Indefinite Integrals of Hyperbolic Functions

I′	$\int \cosh u\, du = \sinh u + C$	II′	$\int \sinh u\, du = \cosh u + C$
III′	$\int \text{sech}^2 u\, du = \tanh u + C$	IV′	$\int \text{csch}^2 u\, du = -\coth u + C$
V′	$\int \text{sech}\, u \tanh u\, du = -\text{sech}\, u + C$	VI′	$\int \text{csch}\, u \coth u\, du = -\text{csch}\, u + C$

EXAMPLE 3 Evaluate $\int \cosh 5x \, dx$.

Solution If $u = 5x$, then $du = 5 \, dx$. Thus, from I′,

$$\begin{aligned}\int \cosh 5x \, dx &= \frac{1}{5}\int \cosh 5x(5 \, dx)\\ &= \frac{1}{5}\int \cosh u \, du\\ &= \frac{1}{5}\sinh u + C\\ &= \frac{1}{5}\sinh 5x + C\end{aligned}$$

□

EXAMPLE 4 Evaluate $\int x \operatorname{csch} x^2 \coth x^2 \, dx$.

Solution If $u = x^2$, then $du = 2x \, dx$. From VI′ we have

$$\begin{aligned}\int x \operatorname{csch} x^2 \coth x^2 \, dx &= \frac{1}{2}\int \operatorname{csch} x^2 \coth x^2(2x \, dx)\\ &= \frac{1}{2}\int \operatorname{csch} u \coth u \, du\\ &= -\frac{1}{2}\operatorname{csch} x^2 + C\end{aligned}$$

□

EXAMPLE 5 Evaluate $\int \tanh x \, dx$.

Solution First, notice that the given integral does not possess any of the six forms given in I′–VI′. However, if we write

$$\int \tanh x \, dx = \int \frac{\sinh x}{\cosh x} dx$$

we can then make the identifications $u = \cosh x$ and $du = \sinh x \, dx$. Therefore,

$$\begin{aligned}\int \tanh x \, dx = \int \frac{\sinh x}{\cosh x} dx &= \int \frac{du}{u}\\ &= \ln|u| + C\\ &= \ln(\cosh x) + C\end{aligned}$$

since $\cosh x > 0$ for all real numbers x. □

▲ ***Remarks*** (*i*) As mentioned in the introduction to this section, the graph of any function of the form $f(x) = (k/2)(e^{cx} + e^{-cx}) = k \cosh cx$, k and c constants, is called a **catenary**. The shape assumed by a wire or heavy rope strung between two posts has the basic shape of a graph of a hyperbolic cosine. Furthermore, if two circular rings are held vertically and are not too far apart, then a soap film stretched between the rings will assume a surface having minimum area. The surface is a portion of a **catenoid**, which is the surface obtained by revolving a catenary about the x-axis. See Figure 6.29 and Problem 54.

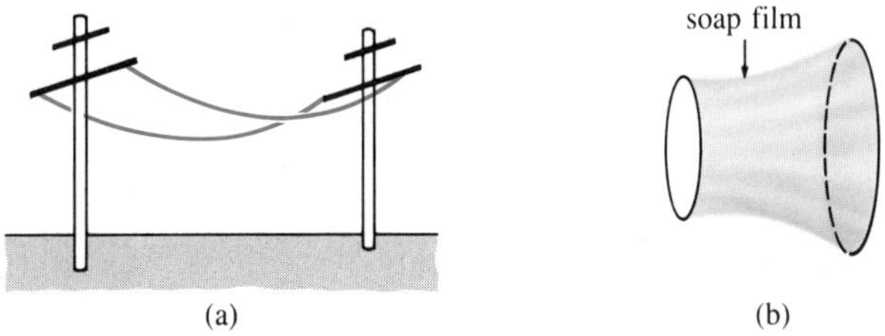

(a) (b)

FIGURE 6.29

(*ii*) The similarity between trigonometric and hyperbolic functions extends beyond the derivative formulas and basic identities. If t is an angle measured in radians whose terminal side is OP, then the coordinates of P on a unit circle $x^2 + y^2 = 1$ are $(\cos t, \sin t)$. Now, the area of the shaded circular sector shown in Figure 6.30 is $A = \frac{1}{2}t$ and so $t = 2A$.* In this manner, the *circular functions* $\cos t$ and $\sin t$ can be considered functions of the area A.

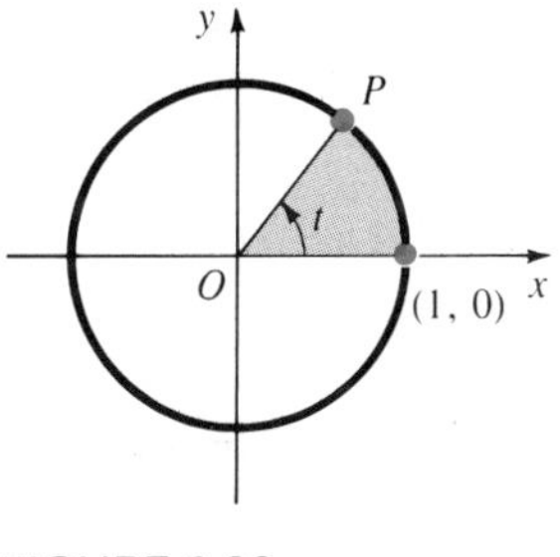

FIGURE 6.30

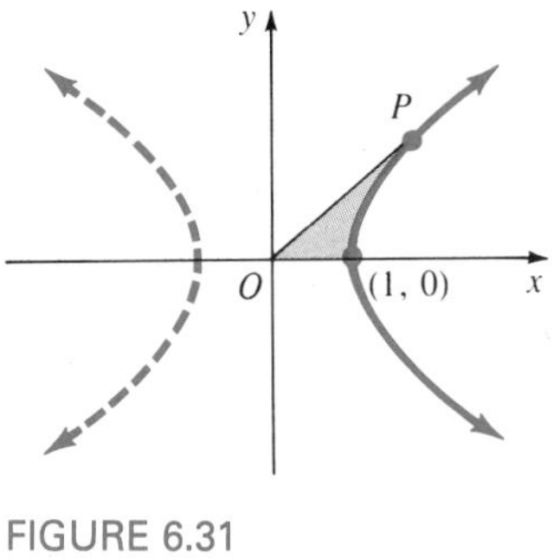

FIGURE 6.31

You might already know that the graph of the equation $x^2 - y^2 = 1$ is called a *hyperbola*. Because $\cosh t \geq 1$ and $\cosh^2 t - \sinh^2 t = 1$, it follows that the coordinates of a point P on the right-hand branch of the hyperbola are $(\cosh t, \sinh t)$. Furthermore, it is not very difficult to show that the area of the hyperbolic sector shown in Figure 6.31 is related to the number t by $t = 2A$. See Problems 65 and 66. Whence we see the origin of the name "hyperbolic function."

(*iii*) Unlike the trigonometric functions, the hyperbolic functions are *not* periodic.

*Recall the area of a circular sector is $A = \frac{1}{2}r^2\theta$, where θ is measured in radians. Set $r = 1$ and $\theta = t$.

EXERCISES 6.7 *Answers to odd-numbered problems begin on page A-86.*

1. If $\sinh x = -\frac{1}{2}$, find the values of the remaining hyperbolic functions.

2. If $\cosh x = 3$, find the values of the remaining hyperbolic functions.

In Problems 3–26 find the derivative of the given function.

3. $y = \cosh 10x$
4. $y = \operatorname{sech} 8x$
5. $y = \tanh \sqrt{x}$
6. $y = \operatorname{csch} \dfrac{1}{x}$
7. $y = \operatorname{sech}(3x - 1)^2$
8. $y = \sinh e^{x^2}$
9. $y = \coth(\cosh 3x)$
10. $y = \tanh(\sinh x^3)$
11. $y = \sinh 2x \cosh 3x$
12. $y = \operatorname{sech} x \coth 4x$
13. $y = x \cosh x^2$
14. $y = \dfrac{\sinh x}{x}$
15. $y = \sinh^3 x$
16. $y = \cosh^4 \sqrt{x}$
17. $f(x) = (x - \cosh x)^{2/3}$
18. $f(x) = \sqrt{4 + \tanh 6x}$
19. $f(x) = \ln(\cosh 4x)$
20. $f(x) = (\ln(\operatorname{sech} x))^2$
21. $f(x) = \dfrac{e^x}{1 + \cosh x}$
22. $f(x) = \dfrac{\ln x}{x^2 + \sinh x}$
23. $F(t) = e^{\sinh t}$
24. $H(t) = e^t e^{\operatorname{csch} t^2}$
25. $g(t) = \dfrac{\sin t}{1 + \sinh 2t}$
26. $w(t) = \dfrac{\tanh t}{(1 + \cosh t)^2}$

27. Find an equation of the tangent line to the graph of $y = \sinh 3x$ at $x = 0$.

28. Find an equation of the tangent line to the graph of $y = \cosh x$ at $x = 1$.

In Problems 29–46 evaluate the given integral.

29. $\displaystyle\int \sinh 8x\, dx$
30. $\displaystyle\int \left(x^2 + \cosh \frac{x}{6}\right) dx$
31. $\displaystyle\int \cosh(5x - 4)\, dx$
32. $\displaystyle\int x \sinh(1 - x^2)\, dx$
33. $\displaystyle\int x^2 \operatorname{sech}^2 x^3\, dx$
34. $\displaystyle\int \frac{\operatorname{csch}^2 \sqrt{x}}{\sqrt{x}}\, dx$
35. $\displaystyle\int \frac{\operatorname{csch} \sqrt[3]{x} \coth \sqrt[3]{x}}{(\sqrt[3]{x})^2}$
36. $\displaystyle\int \operatorname{sech} 2x \tanh 2x\, dx$
37. $\displaystyle\int \sqrt{1 + \sinh 2x} \cosh 2x\, dx$
38. $\displaystyle\int \cosh^2 x \sinh x\, dx$
39. $\displaystyle\int \frac{\sinh 5x}{7 + \cosh 5x}\, dx$
40. $\displaystyle\int x \coth x^2\, dx$
41. $\displaystyle\int e^{-\cosh 3x} \sinh 3x\, dx$
42. $\displaystyle\int \frac{e^{\tanh x}}{\cosh^2 x}\, dx$
43. $\displaystyle\int (\cosh^2 x - 1)^3 \cosh x\, dx$
44. $\displaystyle\int \tanh x \operatorname{sech}^2 x\, dx$
45. $\displaystyle\int e^x \cosh e^x\, dx$
46. $\displaystyle\int e^x \cosh x\, dx$

47. Find the area under the graph of $y = \cosh x$ on the interval $[-1, 1]$.

48. Find the area of the region that is bounded by the graph of $y = \sinh x$ and the x-axis on $[-1, 1]$.

49. Find the area of the region that is bounded by the graphs of $y = \cosh x$, $y = x$, $x = -1$, and $x = 3$.

50. Find the volume of the solid of revolution that is formed by revolving the region bounded by the graphs of $y = \operatorname{sech} x$, $y = 0$, $x = 0$, and $x = 1$ about the x-axis.

51. Find the volume of the solid of revolution that is formed by revolving the region bounded by the graphs of $y = \sinh x^2$, $y = 0$, and $x = \sqrt{3}$ about the y-axis.

52. Find the length of the graph of $y = \cosh x$ on the interval $[0, 2]$.

53. Find the volume of the solid of revolution that is formed by revolving the region bounded by the graphs of $y = \cosh x$, $y = 0$, $x = a$, and $x = b$ about the x-axis. [*Hint:* See (12).]

54. The surface generated by revolving the catenary in Problem 53 about the x-axis is called a **catenoid**. Of all the surfaces generated on $a \le x \le b$ by revolving curves in this manner about the x-axis, the catenoid has the least surface area. Find the surface area of the catenoid.

In Problems 55 and 56 evaluate the given function.

55. $\cosh(\ln x)$
56. $\sinh(\ln x)$

In Problems 57 and 58 find d^2y/dx^2 for the given function.

57. $y = \tanh x$ **58.** $y = \operatorname{sech} x$

59. Verify that $y = C_1\cosh kx + C_2\sinh kx$ satisfies the differential equation $y'' - k^2y = 0$ for any constants C_1 and C_2.

Calculator/Computer Problems

60. (a) Use a calculator or computer to obtain the graph of $f(x) = \tanh(1/x)$.

(b) Use the graph in part (a) to determine whether $x = 0$ is a vertical asymptote for the graph of f. Prove your conjecture.

61. (a) Use a calculator or computer to obtain the graph of $f(x) = x - \tanh x$.

(b) Explain the behavior of the graph as $|x| \to \infty$.

Miscellaneous Problems

62. Prove IV in the form

$$\frac{d}{dx}\coth x = -\operatorname{csch}^2 x$$

63. Prove V in the form

$$\frac{d}{dx}\operatorname{sech} x = -\operatorname{sech} x \tanh x$$

64. Prove VI in the form

$$\frac{d}{dx}\operatorname{csch} x = -\operatorname{csch} x \coth x$$

65. Show that the area of the shaded region in Figure 6.32 is given by

$$A(t) = \frac{1}{2}\cosh t \sinh t - \int_1^{\cosh t} \sqrt{x^2 - 1}\, dx$$

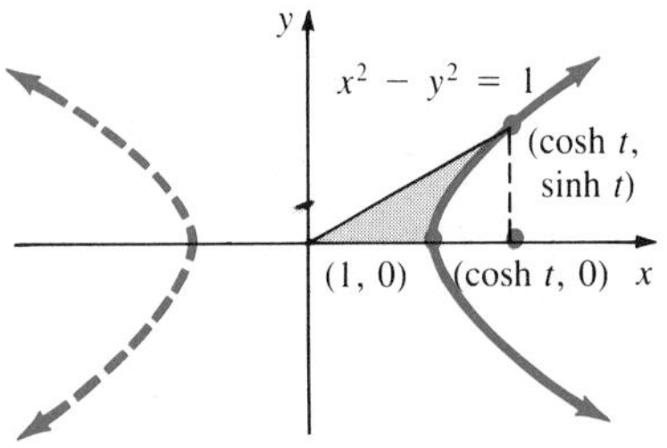

FIGURE 6.32

66. (a) Use the result given in Problem 65 to show that $A(0) = 0$ and $A'(t) = \frac{1}{2}$.

(b) Use part (a) to show that $A(t) = \frac{1}{2}t$.

67. Prove (1). **68.** Prove (2).

69. Prove (4). **70.** Prove (5).

71. Prove (6). **72.** Prove (7).

73. Prove (8). **74.** Prove (9).

75. Prove (10). **76.** Prove (11).

77. Prove (12). **78.** Prove (13).

In Problems 79–82 prove the given identity.

79. $\cosh x + \sinh x = e^x$ **80.** $\cosh x - \sinh x = e^{-x}$

81. $\tanh(x + y) = \dfrac{\tanh x + \tanh y}{1 + \tanh x \tanh y}$

82. $\tanh 2x = \dfrac{2\tanh x}{1 + \tanh^2 x}$

83. Show that for any positive integer n,

$$(\cosh x + \sinh x)^n = \cosh nx + \sinh nx$$

6.8 APPLICATIONS AND FIRST-ORDER DIFFERENTIAL EQUATIONS

6.8.1 Separable First-Order Equations

The simple first-order differential equation

$$\frac{dy}{dt} = ky, \qquad k \text{ a constant} \tag{1}$$

has many applications. The equation can be solved by separation of variables.

EXAMPLE 1 Solve $\dfrac{dy}{dt} = ky$.

Solution We write the differential equation as

$$\frac{dy}{y} = k\,dt \quad \text{and integrate} \quad \int \frac{dy}{y} = k \int dt$$

By assuming that $y > 0$, we have

$$\ln y = kt + C_1 \quad \text{or} \quad y = e^{kt + C_1} = e^{C_1}e^{kt}$$

Relabeling the constant e^{C_1} as C then yields the solution

$$y = Ce^{kt} \tag{2}$$ □

Growth and Decay In biology it is often observed that the rate dN/dt at which a population of certain bacteria grows is proportional to the number N of bacteria present at any time:

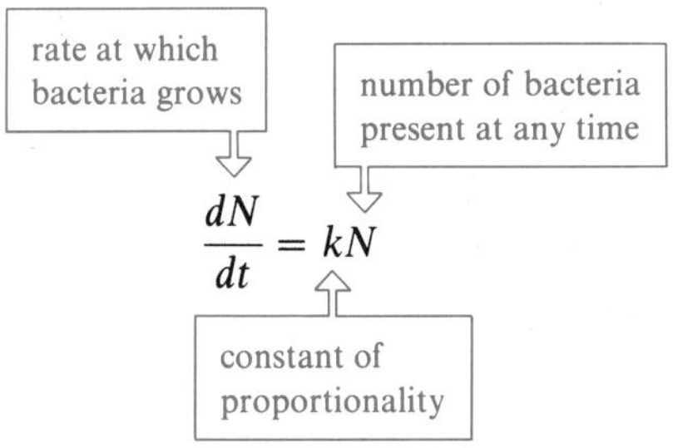

Also, over short periods of time, the population $P(t)$ of small animals such as rodents can be predicted fairly accurately by the solution of the differential equation $dP/dt = kP$. In physics, the solution $A(t)$ of the differential equation $dA/dt = kA$ provides a means for approximating the amount remaining of a substance that is decaying or disintegrating through radioactivity. Thus, one differential equation (1) can serve as the mathematical model for many diverse phenomena. In chemistry, the amount of a substance remaining during a first-order reaction is described by (1).

EXAMPLE 2 Initially, the number of bacteria present in a culture is N_0. At $t = 1$ hour the number of bacteria is measured to be $(\frac{3}{2})N_0$. If the rate of growth is assumed to be proportional to the number of bacteria present, determine the time necessary for the number of bacteria to triple.

Solution We first solve the differential equation

$$\frac{dN}{dt} = kN$$

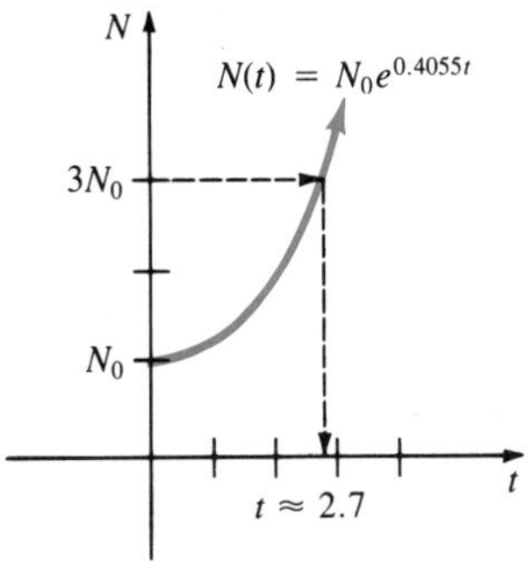

FIGURE 6.33

subject to $N(0) = N_0$. From (2) of Example 1, we can write

$$N(t) = Ce^{kt} \tag{3}$$

At $t = 0$, it follows from (3) that $N_0 = Ce^0 = C$ and so $N(t) = N_0 e^{kt}$. Now at $t = 1$ we have

$$N(1) = \frac{3}{2}N_0 = N_0 e^k \quad \text{or} \quad e^k = \frac{3}{2}$$

To four decimal places, we have

$$k = \ln\frac{3}{2} = 0.4055$$

Thus, for any time $t \geq 0$,

$$N(t) = N_0 e^{0.4055t}$$

To find the time at which the bacteria have tripled we solve

$$3N_0 = N_0 e^{0.4055t}$$

for t:

$$0.4055t = \ln 3$$

$$t = \frac{\ln 3}{0.4055} \approx 2.7 \text{ h}$$

See Figure 6.33. □

Note: We can write the function $N(t)$ obtained in the preceding example in an alternative form. From the laws of exponents,

$$N(t) = N_0 e^{kt} = N_0(e^k)^t = N_0\left(\frac{3}{2}\right)^t$$

since $e^k = \frac{3}{2}$. This latter solution provides a convenient method for computing $N(t)$ for small positive integral values of t; it also clearly shows the influence of the subsequent experimental observation at $t = 1$ on the solution for all time. We notice too that the actual number of bacteria present initially—that is, at time $t = 0$—is quite irrelevant in finding the time required to triple the number in the culture. The necessary time to triple, say, 100 or 10,000 bacteria is still approximately 2.7 h.

Half-life In physics the **half-life** is a measure of the stability of a radioactive substance. The half-life is simply the time it takes for one-half of the atoms in an initial amount A_0 to disintegrate, or transmute, into the atoms of another element. The longer the half-life of a substance, the more stable it is. For example, the half-life of highly radioactive radium, Ra-226, is about 1700 years. In 1700 years one-half of a given quantity of Ra-226 is transmuted into radon, Rn-222. The most commonly occurring uranium isotope, U-238, has a half-life of approximately 4,500,000,000 years. In about 4.5 billion years, one-half of a quantity of U-238 is transmuted into lead, Pb-206.

Carbon Dating In the 1940s the chemist Willard Libby devised a method of using radioactive carbon as a means of determining the approximate ages of fossils. The theory of **carbon dating** is based on the fact that the isotope carbon-14 is produced in the atmosphere by the action of cosmic radiation on nitrogen. The ratio of the amount of C-14 to ordinary carbon in the atmosphere appears to be a constant, and as a consequence the proportionate amount of the isotope present in all living organisms is the same as that in the atmosphere. When an organism dies, the absorption of C-14, by either breathing or eating, ceases. Thus, by comparing the proportionate amount of C-14 present, say, in a fossil with the constant ratio found in the atmosphere, it is possible to obtain a reasonable estimation of its age. The method is based on the knowledge that the half-life of the radioactive C-14 is approximately 5600 years. For his work Libby won the Nobel Prize for chemistry in 1960. Libby's method has been used to date wooden furniture in Egyptian tombs and the woven flax wrappings of the Dead Sea scrolls.

EXAMPLE 3 A fossilized bone is found to contain 1/1000 the original amount of C-14. Determine the age of the fossil.

Solution The starting point is the differential equation $dA/dt = kA$, where A is the amount of C-14 remaining at any time. If A_0 is the initial amount of C-14 in the bone, it follows as in Example 2 that

$$A(t) = A_0 e^{kt}$$

When $t = 5600$ years, $A(t) = A_0/2$, from which we can determine the value of k as follows:

$$\frac{A_0}{2} = A_0 e^{5600k}$$

$$5600k = \ln\left(\frac{1}{2}\right) = -\ln 2$$

$$k = -\frac{\ln 2}{5600} = -0.00012378$$

Therefore, $$A(t) = A_0 e^{-0.00012378t}$$

When $A(t) = A_0/1000$, we have

$$\frac{A_0}{1000} = A_0 e^{-0.00012378t}$$

so that $$-0.00012378t = \ln\left(\frac{1}{1000}\right) = -\ln 1000$$

$$t = \frac{\ln 1000}{0.00012378} \approx 55{,}800 \text{ years}$$ □

The date found in Example 3 is really at the border of accuracy for this method. The usual carbon-14 technique is limited to about 9 half-lives of the isotope or about 50,000 years. One reason is that the chemical analysis needed to obtain an accurate measurement of the remaining C-14 becomes somewhat formidable around the point of $A_0/1000$. Also, this analysis demands the destruction of a rather large sample of the specimen. If this measurement is accomplished indirectly, based on the actual radioactivity of the specimen, then it is very difficult to distinguish between the radiation from the fossil and the normal background radiation.

In recent developments geologists have showed that in some cases, dates determined by carbon dating may be off by as much as 3500 years. One conjecture for this possible error is the fact that carbon-14 levels in the air are known to vary with time. These same scientists have devised another dating technique based on the fact that living organisms ingest traces of uranium. By measuring the relative amounts of uranium and thorium (the isotope into which the uranium decays) and by knowing the half-lives of these elements, scientists can determine the age of a fossil. The advantage of this method is that it can date fossils up to 500,000 years; the disadvantage is that it is effective mostly on marine fossils. Another isotopic technique, using potassium-40 and argon-40, when applicable can give dates of several million years. Nonisotopic methods based on the use of amino acids are also sometimes possible.

Cooling Newton's law of cooling states that the rate at which the temperature $T(t)$ changes in a cooling body is proportional to the difference between the temperature in the body and the constant temperature T_s of the surrounding medium; that is,

$$\frac{dT}{dt} = k(T - T_s) \tag{4}$$

where k is a constant of proportionality.

EXAMPLE 4 When a cake is removed from a baking oven, its temperature is measured at 300°F. Three minutes later its temperature is 200°F. Determine the temperature of the cake at any time after leaving the oven if the room temperature is 70°F.

Solution We identify the temperature of the room (70°F) as T_s. To find the temperature of the cake at any time, we must solve the problem

$$\frac{dT}{dt} = k(T - 70), \qquad T(0) = 300$$

and determine the value of k so that $T(3) = 200$.

Assuming $T > 70$, it follows by separation of variables that

$$\frac{dT}{T - 70} = k\,dt$$

$$\ln(T - 70) = kt + C_1$$

$$T - 70 = C_2 e^{kt} \qquad (C_2 = e^{C_1})$$

$$T = 70 + C_2 e^{kt}$$

When $t = 0$, $T = 300$ so that $300 = 70 + C_2$ gives $C_2 = 230$ and therefore $T = 70 + 230e^{kt}$.

From $T(3) = 200$ we find

$$e^{3k} = \frac{13}{23}$$

and so, to four decimal places, a calculator gives

$$k = \frac{1}{3}\ln\frac{13}{23} = -0.1902$$

Thus,
$$T(t) = 70 + 230e^{-0.1902t}$$

The graph of T along with some calculated values are given in Figure 6.34. □

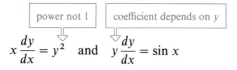

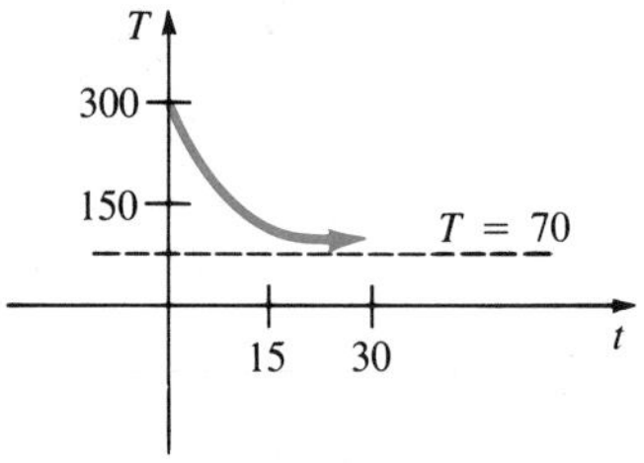

(a)

$T(t)$	t (minutes)
75°	20.1
74°	21.3
73°	22.8
72°	24.9
71°	28.6
70.5°	32.3

(b)

FIGURE 6.34

[O] 6.8.2 Linear First-Order Equations

A **linear** first-order differential equation is any differential equation of the form

$$a_1(x)\frac{dy}{dx} + a_2(x)y = g(x) \tag{5}$$

The separable differential equations (1) and (4), when put in the form

$$\frac{dy}{dt} - ky = 0 \quad \text{and} \quad \frac{dT}{dt} - kT = -kT_s$$

are also recognized as linear equations. In the first equation we identify $a_1(t) = 1$, $a_2(t) = -k$, and $g(t) = 0$; in the second we identify $a_1(t) = 1$, $a_2(t) = -k$, and $g(t) = -kT_s$. The essential properties of a linear equation are that the dependent variable and its derivative are of the first degree, that is, the power of each term involving the dependent variable is 1, and each coefficient depends only on the independent variable. The following separable equations are not linear:

power not 1 — coefficient depends on y

$$x\frac{dy}{dx} = y^2 \quad \text{and} \quad y\frac{dy}{dx} = \sin x$$

Equations that are not linear are called **nonlinear**.

It is important to note that not every first-order linear differential equation can be solved by the method of separation of variables. The linear equation

$$\frac{dy}{dx} + 2y = x$$

is not separable. Hence, we need a new procedure for solving linear equations.

By dividing (1) by $a_1(x)$, we obtain the more useful form of a linear equation

$$\frac{dy}{dx} + P(x)y = f(x) \tag{6}$$

We seek solutions of (6) on an interval I for which P and f are continuous. The left side of (6) has the pleasant property that, when multiplied by the function $e^{\int P(x)\,dx}$, it becomes the derivative of a product. To see this, observe that the Product Rule gives

$$\frac{d}{dx}[e^{\int P(x)\,dx}y] = e^{\int P(x)\,dx}\frac{dy}{dx} + e^{\int P(x)\,dx}P(x)y \tag{7}$$

Thus, if we multiply both sides of (6) by $e^{\int P(x)\,dx}$, we get

$$e^{\int P(x)\,dx}\frac{dy}{dx} + e^{\int P(x)\,dx}P(x)y = e^{\int P(x)\,dx}f(x)$$

it follows from (7) that the last equation is the same as

$$\frac{d}{dx}[e^{\int P(x)\,dx}y] = e^{\int P(x)\,dx}f(x)$$

This last equality enables us to solve linear equations by integration. The function $e^{\int P(x)\,dx}$ that makes this possible is called an **integrating factor**. The procedure is outlined in step-by-step fashion next.

Method of Solution

(*i*) To solve a linear first-order differential equation (5), first put it into form (6); that is, make the coefficient of dy/dx unity.

(*ii*) Identify $P(x)$ and find the integrating factor $e^{\int P(x)\,dx}$

(*iii*) Multiply both sides of the differential equation obtained in (*i*) by the integrating factor.

(*iv*) The left side of the equation in step (*iii*) is the derivative of the integrating factor and the dependent variable; that is,

$$\frac{d}{dx}[e^{\int P(x)\,dx}y] = e^{\int P(x)\,dx}f(x)$$

(*v*) Integrate both sides of the equation found in step (*iv*).

We have already solved the equation $dy/dt - ky = 0$ by separation of variables, but since it is linear, we can also solve it by the foregoing procedure. In this case the integrating factor is $e^{\int(-k)\,dt} = e^{-kt}$ (we need not use a constant in computing $\int P(x)\,dx$) and, after multiplying the equation by this factor, we obtain $d/dt[e^{-kt}y] = 0$. Integrating both sides of this last equation with respect to t gives $e^{-kt}y = C$. From this last expression we get the same solution $y = Ce^{kt}$ as in Example 1.

EXAMPLE 5 Solve $x\dfrac{dy}{dx} + 2y = 6x$.

Solution We first divide by x and write the given equation as

$$\frac{dy}{dx} + \frac{2}{x}y = 6 \tag{8}$$

From the last equation we identify $P(x) = 2/x$. Now the integrating factor is

$$e^{\int 2\,dx/x} = e^{2\ln|x|} = e^{\ln x^2} = x^2$$

Multiplying both sides of (8) by x^2 then gives

$$x^2\frac{dy}{dx} + 2xy = 6x^2 \quad \text{or} \quad \frac{d}{dx}[x^2y] = 6x^2$$

Integrating both sides of the last equation gives

$$x^2y = 2x^3 + C \quad \text{or} \quad y = 2x + Cx^{-2}$$ □

Mixture Problem The mixing of two fluids sometimes gives rise to a linear first-order differential equation. In the next example we consider the mixture of two salt solutions of different concentrations.

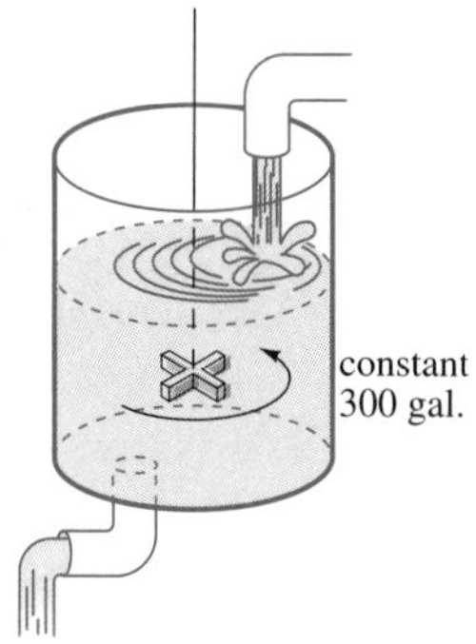

FIGURE 6.35

EXAMPLE 6 Initially 50 lb of salt is dissolved in a large tank holding a 300 gal of water. A brine solution is pumped into the tank at a rate of 3 gal/min, and the well-stirred solution is then pumped out at the same rate. See Figure 6.35. If the concentration of the solution entering is 2 lb/gal, determine the amount of salt in the tank at any time. How much salt is present after 50 min? after a long time?

Solution Let $A(t)$ be the amount of salt (in pounds) in the tank at any time. For problems of this sort, the net rate at which $A(t)$ changes is given by

$$\frac{dA}{dt} = \begin{pmatrix}\text{rate of} \\ \text{substance entering}\end{pmatrix} - \begin{pmatrix}\text{rate of} \\ \text{substance leaving}\end{pmatrix} = R_1 - R_2 \tag{9}$$

Now the rate at which the salt enters the tank is, in pounds per minute,

$$R_1 = (3\text{ gal/min}) \cdot (2\text{ lb/gal}) = 6\text{ lb/min}$$

whereas the rate at which salt is leaving is

$$R_2 = (3\text{ gal/min}) \cdot \left(\frac{A}{300}\text{ lb/gal}\right) = \frac{A}{100}\text{ lb/min}$$

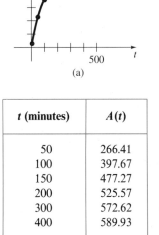

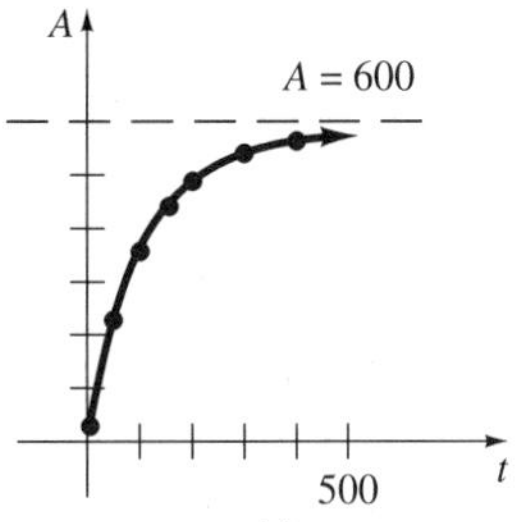

(a)

t (minutes)	A(t)
50	266.41
100	397.67
150	477.27
200	525.57
300	572.62
400	589.93

(b)

FIGURE 6.36

Thus, equation (9) becomes

$$\frac{dA}{dt} = 6 - \frac{A}{100} \quad \text{or} \quad \frac{dA}{dt} + \frac{1}{100}A = 6 \tag{10}$$

which we solve subject to the initial condition $A(0) = 50$.

Since the integrating factor is $e^{t/100}$, we can write (10) as

$$\frac{d}{dt}[e^{t/100}A] = 6e^{t/100}$$

and therefore

$$e^{t/100}A = 600e^{t/100} + C$$

$$A = 600 + Ce^{-t/100} \tag{11}$$

When $t = 0$, $A = 50$, so we find that $C = -550$. Finally, we obtain

$$A(t) = 600 - 550e^{-t/100} \tag{12}$$

At $t = 50$ we find $A(50) = 266.41$ lb. Also, as $t \to \infty$ it is seen from (11) and Figure 6.36 that $A \to 600$. Of course, this is what we would expect; over a long period of time the number of pounds of salt in the solution must be

$$(300 \text{ gal})(2 \text{ lb/gal}) = 600 \text{ lb}$$ □

In Example 6 we assumed that the rate at which the solution was pumped in was the same as the rate at which the solution was pumped out. However, this need not be the case; the mixed brine solution could be pumped out at a rate faster or slower than the rate at which the other solution is pumped in. For example, if the well-stirred solution is pumped out at the slower rate of 2 gal/min, then the solution is accumulating at a rate of $(3 - 2)$ gal/min = 1 gal/min. After t minutes there are $300 + t$ gallons of brine in the tank. The rate at which the salt is leaving is then

$$R_2 = 2 \text{ gal/min} \cdot \frac{A}{300 + t} \text{ lb/gal}$$

$$= \frac{2A}{300 + t} \text{ lb/min}$$

See Problem 40.

▲ ***Remark*** If we solve (6) on an interval on which P and f are continuous, then it can be proved that a one-parameter family of solutions of the equation yields *all* solutions of the equation on the interval. In Example 6, the functions $P(x) = 2/x$ and $f(x) = 6$ are continuous on, say, the interval $(0, \infty)$. In this case, *every* solution of $dy/dx + (2/x)y = 6$ on $(0, \infty)$ can be obtained from $y = 2x + Cx^{-2}$ for appropriate choices of the constant C. For this reason, the family of solutions $y = 2x + Cx^{-2}$ is called the **general solution** of the differential equation.

EXERCISES 6.8 *Answers to odd-numbered problems begin on page A-86.*

[6.8.1]

1. The population of a certain community is known to increase at a rate proportional to the number of people present at any time. Find the population $P(t)$ at any time. If the population has doubled in 5 years, how long will it take to triple? to quadruple?

2. Suppose we know that the population of the community in Problem 1 is 10,000 after 3 years. What was the initial population? What will the population be in 10 years?

3. Initially there were 100 mg of a radioactive substance present. After 6 h the mass decreased by 3%. If the rate of decay is proportional to the amount of the substance present at any time, find the amount $A(t)$ remaining at any time. What amount remains after 24 h? Determine the half-life of the substance.

4. In a piece of burned wood, or charcoal, it was found that 85.5% of the C-14 had decayed. Use the information in Example 3 to determine the approximate age of the wood. (It is precisely these data that archaeologists used to date prehistoric paintings in a cave in Lascaux, France.)

5. A breeder reactor converts the relatively stable uranium-238 into the isotope plutonium-239. After 15 years, it is determined that 0.043% of the initial amount A_0 of the plutonium has disintegrated. If the rate of decay is proportional to the amount of the substance present, determine the half-life of the isotope.

6. When a vertical beam of light passes through a transparent substance, the rate at which its intensity I decreases is proportional to $I(t)$, where t represents the thickness of the medium in feet. In clear seawater the intensity 3 ft below the surface is 25% of the initial intensity I_0 of the incident beam. What is the intensity of the beam 15 ft below the surface?

7. When interest is compounded *continuously*, the amount of money S increases at a rate proportional to the amount present at any time: $dS/dt = rS$, where r is the annual rate of interest.

(a) Find the amount of money accrued at the end of 5 years when \$5000 is deposited in a savings account drawing $5\frac{3}{4}$% annual interest compounded continuously.

(b) In how many years will the initial deposit be doubled?

(c) Use a hand calculator to compare the number obtained in part (a) with

$$S = 5000\left(1 + \frac{0.0575}{4}\right)^{5(4)}$$

This value represents the amount that would be accrued when interest is compounded quarterly.

8. A thermometer is removed from a room where the air temperature is 70°F and placed outside where the temperature is 10°F. After $\frac{1}{2}$ min the thermometer reads 50°F. What is the reading at $t = 1$ min? How long will it take for the thermometer to reach 15°F?

9. A thermometer is taken from an inside room and placed outside where the air temperature is 5°F. After 1 min the thermometer reads 55°F and after 5 min the reading is 30°F. What is the initial temperature of the room?

10. The differential equation (4) also holds for an object that absorbs heat from the surrounding medium. If a small metal bar, whose initial temperature is 20°C, is dropped into a container of boiling water, how long will it take for the bar to reach 90°C if it is known that its temperature increases 2° in 1 s? How long will it take the bar to reach 98°C?

11. The rate at which a drug disseminates into the bloodstream is governed by the differential equation

$$\frac{dX}{dt} = A - BX$$

where A and B are positive constants. The function $X(t)$ describes the concentration of the drug in the bloodstream at any time t. Find X. What is the limiting value of X as $t \to \infty$? At what time is the concentration one-half this limiting value? Assume that $X(0) = 0$.

12. In mathematical models of defense spending and economic growth, gross domestic product Y at time t is divided between military production M and nonmilitary production N. Part of the capital stock K at time t consists of military stock K_m. One model describes the relations among these variables by the equations

$$Y = M + N$$
$$M = mY$$
$$K_m = kM$$
$$(K - K_m)e^{pt} = kN$$
$$\frac{dK}{dt} = cN$$

where m, k, p, and c are constants.

(a) Show that $K = (km + k(1 - m)e^{-pt})Y$.

(b) Show that

$$\frac{dK}{K} = \frac{c(1-m)}{km + k(1-m)e^{-pt}}\,dt$$

and

$$\frac{dY}{Y} = \frac{c(1-m) + pk(1-m)e^{-pt}}{km + k(1-m)e^{-pt}}\,dt$$

(c) Solve the differential equations in part (**b**).

13. Suppose a cell is suspended in a solution containing a solute of constant concentration C_s. Suppose further that the cell has constant volume V and that the area of its permeable membrane is the constant A. By Fick's law,* the rate of change of its mass m is directly proportional to the area A and the difference $C_s - C(t)$, where $C(t)$ is the concentration of the solute inside the cell at any time t. Find $C(t)$ if $m = VC(t)$ and $C(0) = C_0$. See Figure 6.37.

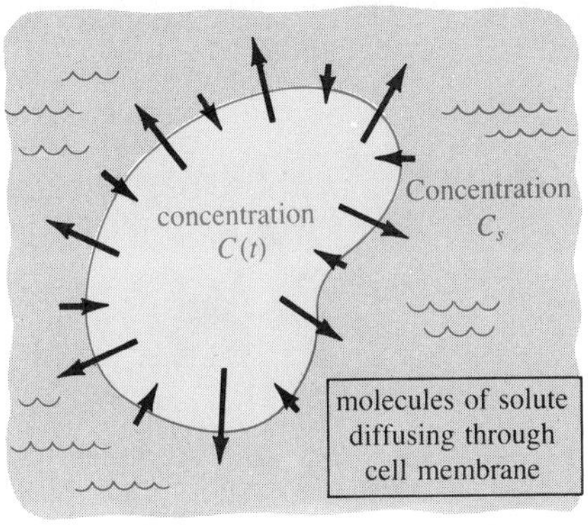

FIGURE 6.37

14. A heart pacemaker, shown in Figure 6.38, consists of a battery, a capacitor, and the heart as a resistor. When the switch S is at P, the capacitor charges; when S is at Q, the capacitor discharges, sending an electrical stimulus to the heart. During this time, the voltage E applied to the heart is given by the linear differential equation

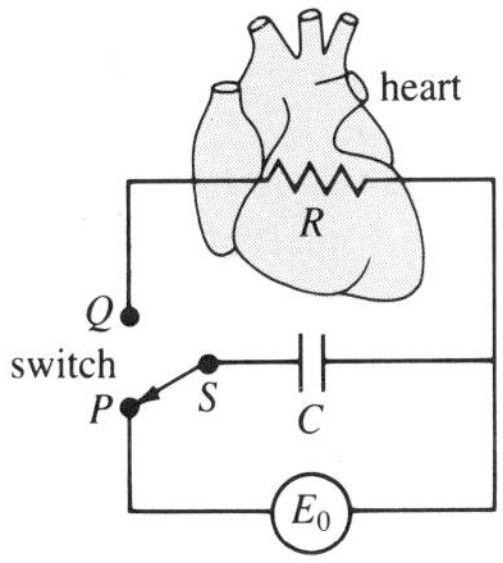

FIGURE 6.38

$$\frac{dE}{dt} = -\frac{1}{RC}E, \qquad t_1 < t < t_2$$

where R and C are constants. Determine $E(t)$ if $E(t_1) = E_0$. (Of course, the opening and closing of the switch are periodic in time to simulate the natural heartbeat.)

15. In a series circuit that contains only a resistor and an inductor, Kirchhoff's second law states that the sum of the voltage drop across the inductor $(L(di/dt))$ and the voltage drop across the resistor (iR) is the same as the impressed voltage (E) on the circuit. See Figure 6.39. Thus, we obtain the differential equation for the current $i(t)$:

$$L\frac{di}{dt} + Ri = E$$

where L and R are constants known as the inductance and the resistance, respectively. Determine the current i if E is 12 volts, the inductance is $\frac{1}{2}$ henry, the resistance is 10 ohms, and $i(0) = 0$.

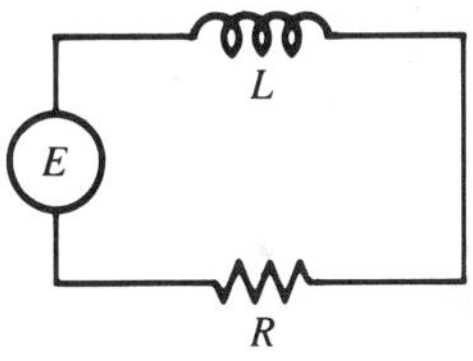

FIGURE 6.39

16. A 30-volt battery is connected to a series circuit in which the inductance is 0.1 henry and the resistance is 50 ohms. Find the current $i(t)$ if $i(0) = 0$. Determine the behavior of the current for large values of time. (See Problem 15.)

17. Under some circumstances, a falling body B of mass m (such as a person hanging from a parachute) encounters air resistance proportional to its instantaneous velocity $v(t)$. See Figure 6.40. Equating the sum of the forces acting on the body, $mg - kv$, with Newton's second law results in the differential equation

$$m\frac{dv}{dt} = mg - kv, \qquad k > 0$$

(a) Solve the equation subject to the initial condition $v(0) = v_0$.

(b) Determine the limiting or terminal velocity $v_t = \lim_{t\to\infty} v(t)$.

*Adolph Fick (1829–1901), a German physiologist.

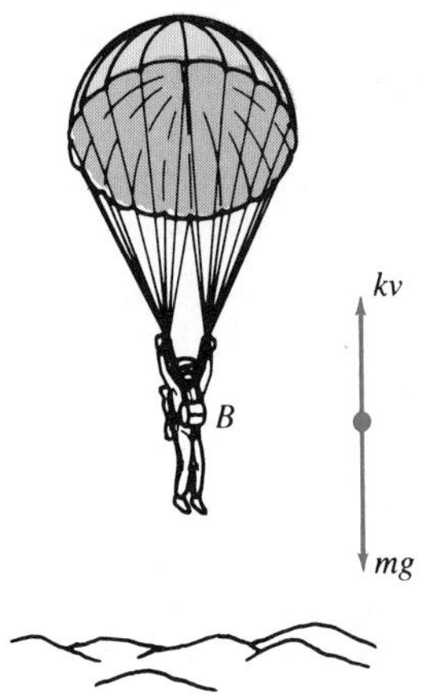

FIGURE 6.40

(c) If distance is related to velocity by $ds/dt = v$, find an explicit expression for s if it is further known that $s(0) = s_0$.

Problem 18 is based on the article "Lagrange Multiplers and the Design of Multistage Rockets" by Anthony L. Peressini in *UMAP Journal* (Fall 1986, Vol. 7, No. 3, pp. 249–262). The first part of the article on **single-stage rockets** is summarized here. See Figure 6.41. The rest of the article is presented as a problem in Exercises 15.11.

18. (*This problem could present a challenge.*) It can be shown that the velocity needed to insert a single-stage rocket into a circular orbit 100 mi above the surface of the earth is approximately 17,500 mi/h. The point of this problem is to show that a typical single-stage rocket is not capable of such an orbital insertion. We make the simplifying assumption that all outside forces, such as aerodynamic drag, are negligible compared to the thrust of the rocket. Let the total mass of a single-stage rocket at any time t be given by $M(t) = P + M_v + M_f(t)$, where P is the constant mass of the payload, M_v is the constant mass of the vehicle, and $M_f(t)$ is the mass of the fuel at any time. The initial mass of the vehicle and fuel is $M_0 = M_v + M_f(0)$.

(a) If the rocket consumes its fuel at a constant rate k, then

$$\frac{dM_f}{dt} = -k$$

Show that $M_f(t) = -kt + M_f(0)$. As a consequence of this result, show that $M(t) = P + M_0 - kt$.

(b) Show that the *burnout time* t_b of the rocket, or the time at which all the fuel is consumed, is $t_b = M_f(0)/k$.

(c) The expressions $R = P/M_0$ and $S = M_v/M_0$ are called, respectively, the *mass ratio* and *structural factor* of the rocket. Show that the burnout time in part (**b**) can be written as $t_b = (1 - S)M_0/k$.

(d) Now, the acceleration of the rocket is given by the *rocket equation*,

$$\frac{dv}{dt} = -\frac{\gamma}{M}\frac{dM}{dt}$$

where γ is the constant speed of the exhaust gases straight out of the back of the rocket. If $v(0) = v_0$, show that the *velocity* of the rocket at any time t is

$$v(t) = v_0 - \gamma \ln\left(1 - \frac{kt}{P + M_0}\right)$$

(e) Show that at burnout, the *velocity increment* $\Delta v = v(t_b) - v_0$ is given by

$$\Delta v = -\gamma \ln\left(\frac{R + S}{R + 1}\right)$$

(f) Using the typical values $R = 0.01$, $S = 0.2$, and $\gamma = 6000$ mi/h, show that the velocity increment in part (**e**) is not sufficient for the single-stage rocket to be inserted in a circular orbit 100 mi above the surface of the earth.

FIGURE 6.41 Single-stage rockets generally are not capable of inserting a payload into an earth orbit. Photo courtesy of NASA.

[6.8.2]

In Problems 19–32 use the method illustrated in Example 5 to solve the given linear differential equation.

19. $\dfrac{dy}{dx} = 4y$

20. $\dfrac{dy}{dx} + 2y = 0$

21. $2\dfrac{dy}{dx} + 10y = 1$

22. $x\dfrac{dy}{dx} + 2y = 3$

23. $\dfrac{dy}{dx} + y = e^{3x}$

24. $\dfrac{dy}{dx} = y + e^x$

25. $x^2y' + xy = 1$

26. $y' + 3x^2y = x^2$

27. $(1 + e^x)\dfrac{dy}{dx} + e^xy = 0$

28. $(1 - x^3)\dfrac{dy}{dx} = 3x^2y$

29. $\cos x \dfrac{dy}{dx} + (\sin x)y = 1$

30. $\dfrac{dy}{dx} + (\cot x)y = 2 \cos x$

31. $\dfrac{dr}{d\theta} + (\sec \theta)r = \cos \theta$

32. $\dfrac{dP}{dt} + 2tP = P + 4t - 2$

In Problems 33 and 34 solve the given linear differential equation subject to the indicated initial condition.

33. $(x + 1)\dfrac{dy}{dx} + y = x^{-1}$, $y(1) = 10$

34. $t\dfrac{dw}{dt} + w = e^t$, $w(1) = 2$

35. A tank contains 200 L of fluid in which 30 g of salt is dissolved. Brine containing 1 g of salt per liter is then pumped into the tank at a rate of 4 L/min. The well-stirred solution is pumped out at the same rate. Find the number of grams of salt $A(t)$ in the tank at any time.

36. Solve Problem 35 assuming pure water is pumped into the tank.

37. A large tank is filled with 500 gal of pure water. Brine containing 2 lb of salt per gallon is pumped into the tank at a rate of 5 gal/min. The well-stirred solution is pumped out at the same rate. Find the number of pounds of salt $A(t)$ in the tank at any time.

38. Solve Problem 37 under the assumption that the solution is pumped out at a faster rate of 10 gal/min. When is the tank empty?

39. A large tank is partially filled with 100 gal of fluid in which 10 lb of salt is dissolved. Brine containing $\frac{1}{2}$ lb of salt per gallon is pumped into the tank at a rate of 6 gal/min. The well-stirred solution is then pumped out at a slower rate of 4 gal/min. Find the number of pounds of salt in the tank after 30 min.

40. Solve the problem in Example 6 when the well-stirred solution is pumped out at a rate of 2 gal/min.

41. Find a continuous solution satisfying the linear differential equation

$$\frac{dy}{dx} + y = f(x) \quad \text{where} \quad f(x) = \begin{cases} 1, & 0 \le x \le 1 \\ 0, & x > 1 \end{cases}$$

and the initial condition $y(0) = 0$. [*Hint:* Solve the problem in two parts.]

CHAPTER 6 REVIEW EXERCISES

Answers begin on page A-86.

In Problems 1–20 answer true or false.

1. If f is one-to-one, then $x_1 \neq x_2$ implies $f(x_1) \neq f(x_2)$. ________

2. The function $f(x) = x^5 + x^3 + x$ does not have an inverse. ________

3. The functions $f(x) = \dfrac{x+1}{x-1}$ and $g(x) = \dfrac{x+1}{x-1}$ are inverses of each other. ________

4. $\dfrac{d}{dx}\log_{10}x = \dfrac{1}{x}$ ________

5. $\dfrac{d}{dx}2^x = x2^{x-1}$ ________

6. $\displaystyle\int_1^3 \frac{dx}{x} = \ln 3$ ________

7. $\displaystyle\int 10^\pi\, dx = \frac{10^{\pi+1}}{\pi+1} + C$ ________

8. The inverse of $y = e^x$ is $y = \ln x$. ________

9. If $\ln x = 1$, then $x = e$. ________

10. If $0 < a < b$, then $\ln a < \ln b$. ________

11. For $a > 0$ and $b > 0$, $\ln(a + b) = \ln a + \ln b$. ________

12. The area under the graph of $y = x^{-1}$ on the interval $[a, 1]$, $a > 0$, is $-\ln a$. ________

13. $\ln x^2 = 2 \ln x$ for all x. ________

14. $e^{\sqrt{x}} = \sqrt{e^x}$ for all x. ________

15. $\exp(x + y) = (\exp x)\cdot(\exp y)$ ________

16. If $f(x) = \ln|x + 1|$ and $g(x) = \ln(x + 1)$, then $f'(x) = g'(x) = 1/(x + 1)$ for all $x \neq -1$. ________

17. $\dfrac{d}{dx}\cosh x = -\sinh x$ ________

18. The hyperbolic cosine is an even function. ________

19. The hyperbolic cosine is never negative. ________

20. $\cosh^2 x + \sinh^2 x = 1$ ________

In Problems 21–30 fill in the blank.

21. $\ln e^3 =$ ________

22. $\ln(\ln e) =$ ________

23. $\ln \dfrac{e^a}{e^b} =$ ________

24. $e^{-2\ln 3} =$ ________

25. The slope of the tangent line to the graph of $y = \ln x$ at $x = \frac{1}{2}$ is ________.

26. For $f(x) = \ln|2x - 4|$, the domain of f' is ________.

27. To differentiate $f(x) = x^x$, we use a process called ________.

28. If $e^{-x} = 5$, then $x =$ ________.

29. The graph of the hyperbolic cosine is called a ________.

30. If f is a one-to-one function such that $f^{-1}(3) = 1$, then $f(1) =$ ________.

In Problems 31–46 find dy/dx.

31. $y = \ln(x\sqrt{4x - 1})$

32. $y = (\ln \cos^2 x)^2$

33. $y = \sin(e^x \ln x)$

34. $y = x^3 \log_3 \sqrt{x}$

35. $y = e^x + e^{2x}$

36. $y = (e + e^2)^x$

37. $y = x^7 + 7^x + 7^\pi$

38. $y = (e^x + 1)^{-e}$

39. $y = (x^2 + e^{2x})^{\ln x}$

40. $y = 5^{x^2} x^{\sin x}$

41. $xy^2 = e^x - e^y$

42. $y = \ln xy$

43. $y = \sinh e^{x^3}$

44. $y = (\tanh 5x)^{-1}$

45. $y = \cosh(\cosh x)$

46. $y = (\operatorname{sech} \sqrt{2x})^{\sqrt{2}}$

In Problems 47–62 evaluate the given integral.

47. $\displaystyle\int \frac{x}{1 - 5x^2}\, dx$

48. $\displaystyle\int_1^2 \left[\frac{1}{x} - \frac{1}{x + 1}\right] dx$

49. $\displaystyle\int \frac{x + 1}{x(x + 2)}\, dx$

50. $\displaystyle\int \frac{3x^2 - 4x + 2}{x}\, dx$

51. $\displaystyle\int_0^1 (x^7 + 7^x + 7^\pi)\, dx$

52. $\displaystyle\int 3^{\cos x} \sin x\, dx$

53. $\displaystyle\int \cot 4x\, dx$

54. $\displaystyle\int x^2 \sec x^3\, dx$

55. $\displaystyle\int_2^5 \frac{dx}{e^{3x}}$

56. $\displaystyle\int \frac{dx}{e^{-x}(4 + e^x)}$

57. $\displaystyle\int 2^t 3^t 4^t\, dt$

58. $\displaystyle\int 4^{e^\theta} e^\theta\, d\theta$

59. $\displaystyle\int \frac{\sinh(1/x)}{x^2}\, dx$

60. $\displaystyle\int \frac{\cosh(\ln 2x)}{x}\, dx$

61. $\displaystyle\int \frac{\cosh 3x}{\sinh^4 3x}\, dx$

62. $\displaystyle\int \frac{\operatorname{sech}^2 x}{1 + \tanh x}\, dx$

63. Given that $f(x) = 8/(1 - x^3)$ is a one-to-one function, find f^{-1} and $(f^{-1})'$.

64. Given that $f(x) = 10 - \sqrt{x + 3}$ is a one-to-one function, without finding an explicit equation for f^{-1} find **(a)** its domain and range, and **(b)** an equation of the tangent line to its graph at $(f(6), 6)$.

65. The region in the first quadrant bounded by the graphs of $y = e^x$, $y = e^{-x}$, and $x = \ln 6$ is revolved about the x-axis. Find the volume of the solid of revolution.

66. Find the point of the graph of $y = \ln 2x$ such that the tangent line passes through the origin.

In Problems 67 and 68 graph the given function.

67. $y = \ln(x^2 - 1)$

68. $y = e^{-1/x}$

In Problems 69–74 use separation of variables.

69. If P_0 is the initial population of a community, show that if P is governed by $dP/dt = kP$, then

$$\left(\frac{P_1}{P_0}\right)^{t_2} = \left(\frac{P_2}{P_0}\right)^{t_1}$$

where $P_1 = P(t_1)$ and $P_2 = P(t_2)$, $t_1 < t_2$.

70. A metal bar is taken out of a furnace whose temperature is 150°C and put into a tank of water whose temperature is maintained at a constant 30°C. After $\frac{1}{4}$ h in the tank, the temperature of the bar is 90°C. What is the temperature of the bar in $\frac{1}{2}$ h? in 1 h?

71. When forgetfulness is taken into account, the rate at which a person can memorize a subject is given by the differential equation

$$\frac{dA}{dt} = k_1(M - A) - k_2 A$$

where k_1 and k_2 are positive constants, $A(t)$ is the amount of material memorized in time t, M is the total amount to be memorized, and $M - A$ is the amount remaining to be memorized.

(a) Solve for $A(t)$ if $A(0) = 0$.

(b) Find the limiting value of A as $t \to \infty$ and interpret the result.

(c) Graph the solution.

72. A projectile is shot vertically into the air with an initial velocity of v_0 ft/s. Assuming that air resistance is proportional to the square of the instantaneous velocity, the motion is

described by the pair of differential equations:

$$m\frac{dv}{dt} = -mg - kv^2, \qquad k > 0$$

positive y-axis up and origin at ground level so that $v = v_0$ at $y = 0$, and

$$m\frac{dv}{dt} = mg - kv^2, \qquad k > 0$$

positive y-axis down and origin at the maximum height so that $v = 0$ at $y = h$. See Figure 6.42. The first and second equations describe the motion of the projectile when rising and falling, respectively. Prove that the impact velocity v_i is less than the initial velocity v_0. [*Hint:* By the Chain Rule, $dv/dt = v\, dv/dy$.]

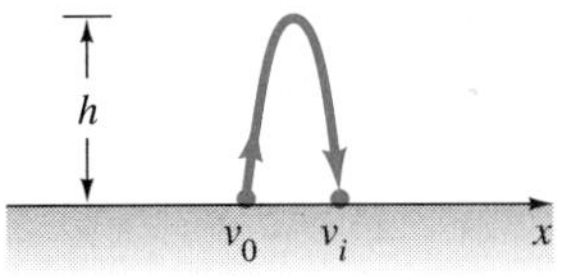

FIGURE 6.42

73. The differential equation $dP/dt = (k \cos t)P$, where k is a positive constant, is often used as a model of a population that undergoes yearly seasonal fluctuations.

(a) Solve for $P(t)$ if $P(0) = P_0$.

(b) Use a calculator or computer to obtain the graph of the function found in part (**a**).

74. Suppose a series circuit contains a capacitor and a variable resistor. If the resistance at any time t is given by $R = k_1 + k_2 t$, where k_1 and k_2 are positive known constants, then the charge q on the capacitor is described by the first-order differential equation

$$(k_1 + k_2 t)\frac{dq}{dt} + \frac{1}{C}q = E(t)$$

where C is a constant called the capacitance and $E(t)$ is the impressed voltage. Show that if $E(t) = E_0$ and $q(0) = q_0$, then

$$q(t) = E_0 C + (q_0 - E_0 C)\left(\frac{k_1}{k_1 + k_2 t}\right)^{1/Ck_2}$$

In Problems 75–80 solve the given differential equation.

75. $\sin x \dfrac{dy}{dx} + (\cos x)y = 0$

76. $\dfrac{dx}{dt} + x = e^{-t}\cos 2t$

77. $t\dfrac{dy}{dt} - 5y = t$

78. $\dfrac{y}{x^2}\dfrac{dy}{dx} + e^{2x^3 + y^2} = 0$

79. $(x^2 + 4)\dfrac{dy}{dx} = 2x - 8xy$

80. $(y^2 + 1)\, dx = y \sec^2 x\, dy$

81. Agricultural studies show that the relationship between the weight of an animal and the amount of feed it has consumed can be modeled by the equation

$$W = A - (A - W_0)e^{-kF/A}$$

where W is the weight of the animal after consuming the total feed quantity F, A is the maximum attainable weight, W_0 is the initial weight when feeding begins, and k is a positive constant depending on the breed of animal. In one study of pigs, researchers found $A = 380$ lb, $W_0 = 26.6$ lb, and $k = 0.409$.

(a) Verify that dW/dF is always less than one as long as $k < 1$. Interpret this statement physically.

(b) Suppose that a pig can be sold for 50 cents per pound and that feed costs 5 cents per pound. At what weight is it most profitable to sell the pig? [*Hint:* profit = revenue − cost.]

82. Seafloor crust, which is produced by upward-welling magma (hot rock) at mid-ocean ridges, subsides (sinks) as it spreads and cools. In a 1982 analysis of the ocean floor, Barry Parsons and John G. Sclater gave the following empirical age–depth relationship for seafloor subsidence:

$$d(t) = \begin{cases} 2500 + 350t^{1/2}, & 0 \le t \le 70 \\ 6400 - 3200e^{-t/62.8}, & t > 70 \end{cases}$$

where $d(t)$ is the seafloor depth in meters and t is the crustal age in millions of years.

(a) Show that $d(t)$ is a discontinuous function.

(b) Suppose we require $d(t)$ to be continuous and differentiable at $t = 70$. Find constants A and B so that

$$d(t) = \begin{cases} 2500 + At^{1/2}, & 0 \le t \le 70 \\ 6400 - Be^{-t/62.8}, & t > 70 \end{cases}$$

satisfies these conditions.

CALCULATOR/COMPUTER ACTIVITIES

CHAPTER 1 Functions and Limits *Answers to odd-numbered problems begin on page A-115.*

1.1 The purpose of a *calculator* or *computer program* is to store a sequence of steps or actions that are to be used repeatedly. For example, to evaluate the function $f(x) = \sqrt{x^2 + 5}$ on a typical calculator (1) enter the desired value of x, (2) square the entry, (3) add 5 to the result, and (4) take the square root. Since these same four steps will be used whenever the function is evaluated, it might save time to store these steps in a program. Consult the operating manual for the programmable calculator, computer language, or mathematical software available to learn how to program the evaluation of a function. On a CASIO fx-7000G calculator, the program steps "X" ? → X: $\sqrt{}(X^2+5) \to$ Y will prompt for input of a value for x, calculate $f(x)$, and store the result in memory Y (displaying the final calculation as well).

In Problems 1–4 write a program to evaluate the given function at the indicated values of x.

1. $f(x) = x^5 - 3x^2 + 10x - 7,\ x = -7, -3.5, 1.2, 8, 25$
2. $f(x) = \sqrt{1 + \sqrt{x}},\ x = 1, 1.4, 2.5, 8, 15, 53, 99$
3. $f(x) = |x^2 - 5x + 1|,\ x = -1, -0.8, 0.5, 2, 2.6, 3, 5.7$
4. $f(x) = \dfrac{7x^3 + 2x + 10}{x^2 + 1},\ x = -25, -10, 5, 0, 15, 35, 52$

1.2 The procedures for obtaining the graph of a function on a graphic calculator and on a computer graphing application are all very similar. First, select the range of values along the x- and y-axes for the desired graph. These two ranges determine the *viewing window* or *rectangle*. There may be a default window if you make no selection. Then key in the formula of the given function. A common difficulty is that the graph of the function may not appear in the selected viewing window. One way to avoid this is to evaluate the function a few times, and then select a range along the y-axis that includes the calculated values. There may be an option to have the y-range of the plot automatically adjusted to show the graph over the selected values for the x-range. There should also be an option to have two or more functions graphed in the same viewing window for comparison.

In Problems 5–8 graph the given function on the indicated viewing window.

5. $f(x) = \sqrt{x^3 - 4},\ 2 \le x \le 10,\ 0 \le y \le 35$
6. $f(x) = \dfrac{4x - 7}{x^2 + 3},\ -10 \le x \le 15,\ -3 \le y \le 2$
7. $f(x) = x^2 - 5,\ g(x) = x^2 + x - 5,\ -4 \le x \le 3,$ $-6 \le y \le 10$
8. $f(x) = 4x^3 - 3x,\ g(x) = 8x^4 - 8x^2 + 1,$ $-1 \le x \le 1,\ -1 \le y \le 1$

Most of the *time* that a calculator or computer takes to generate the graph of a function is spent in repeatedly evaluating the function. There is a way of expressing a function that may speed up this process. Evaluation of a polynomial function in *standard form*, such as $f(x) = 5x^5 - 11x^4 + 7x^3 + 6x^2 - 24x + 50$, requires the use of the complicated power operation. However, if we rewrite the function in *nested form*,

$$f(x) = ((((5x - 11)x + 7)x + 6)x - 24)x + 50$$

then the evaluation requires only the arithmetic operations, which can be executed more quickly and more accurately. On a SHARP EL-5200 calculator, it takes about 49 seconds to generate the graph with the standard form of f in the viewing window $-2 \le x \le 3$, $-270 \le y \le 545$. With the nested form, it takes only 24 seconds to generate the same graph.

In Problems 9 and 10 compare the time it takes for your calculator or graphing application to generate the graph of the given function (**a**) using the standard form and (**b**) using the nested form.

9. $f(x) = 2x^5 + 4x^4 - 15x^3 - 3x^2 + 32x + 7$ $= ((((2x + 4)x - 15)x - 3)x + 32)x + 7$
10. $f(x) = x^5 - 4x^4 - 16x^3 + 46x^2 + 63x - 90$ $= ((((x - 4)x - 16)x + 46)x + 63)x - 90$ $= (x + 3)(x + 2)(x - 1)(x - 3)(x - 5)$

1.3 Some trigonometric and inverse trigonometric functions are available on all scientific calculators and within all computer mathematics applications.

Remember to always use the *radian mode*, rather than the *degree mode*, when using trigonometric functions in calculus.

In Problems 11 and 12 use a graphing utility to examine the graphs of the given class of functions. Find the amplitude and period for each function from its graph.

11. $f(x) = a\sin(bx + c)$ for a, b, and c from the set $\{-2, -1, 0, \frac{1}{2}, 1, 3\}$

12. $f(x) = a\sin bx + c\sin kx$ for a, b, c, and k from the set $\{-10, -4, -\frac{3}{2}, \frac{1}{2}, 1, 5, 8\}$

1.5 One of the most convenient features of calculator/computer graphing is the capability of quickly obtaining different viewing windows for the same function. Usually there is a *zoom-in* command to look at some section of the viewing window in more detail, and a *zoom-out* command to look at more of the graph than is being shown in the current window. Many graphic calculators and computer graphing applications also have a *trace mode* that displays the actual coordinates for calculated points on the graph in the viewing window. The zoom-in feature is especially useful in the grahical examination of limits.

In Problems 13–16 graph the given expression in a viewing window that contains the indicated center. Use the trace mode (if available) to examine the graph near the limiting x-value. Repeatedly zoom-in and trace to see the expression closer and closer to the limiting x-value.

13. $\displaystyle\lim_{x\to 3} \frac{x^3 + 2x^2 - 9x + 18}{x - 3}$ centered on the point (3, 30)

14. $\displaystyle\lim_{x\to 0} \frac{\sin x}{x}$ centered on the point (0, 1)

Note: All calculators and computers use some approximate calculation to represent the sine function. You may be able to zoom-in enough to see where these approximations for the sine function begin to make this limiting expression look a little "rough" even though the true expression is "smooth."

15. $\displaystyle\lim_{x\to 1} (x - 1)\cos\left(\frac{\pi}{x - 1}\right)$ centered on the point (1, 0)

16. $\displaystyle\lim_{x\to \pi/2} \frac{x}{\tan x}$ centered on the point $(\pi/2, 1)$

1.7 Limits that involve infinity describe how the graph of a function will "go off the page" when restricted to a finite viewing window. Vertical asymptotes specify how the graph will leave the viewing window at the top or the bottom, and horizontal asymptotes describe the exit of the graph on the left or right. A vertical asymptote is often represented poorly in many calculator/computer-generated graphs. It is important to understand when and why this happens. A sampling of values from the x-range of the viewing window is selected (usually equally spaced), and the function is evaluated only on this finite set. This gives a finite set of coordinate points that are plotted on the screen in the viewing window. Graphic calculators use about 100 points, and some computer graphing applications allow the user to select the number of points to be used along the curve. It may be possible to have only these calculated points displayed, but usually these points are connected by short line segments. The screen representation of the graph of the function will "miss" a vertical asymptote when the true graph both *leaves* the viewing window at the top or bottom and *returns* between two calculated coordinate points. Frequently, one of these close calculated points is near the top of the viewing window, while the nearest calculated point on the other side of the vertical asymptote is near the bottom. The line segment joining such a pair of calculated points will then be nearly vertical, giving the *false impression* that the graph is continuous. The best way to get a good look at a suspected vertical asymptote on the screen is to narrow the x-range of the viewing window (while expanding the y-range). To see a horizontal asymptote, widen the x-range of the window (while narrowing the y-range).

In Problems 17–20 graph the given function. Find appropriate viewing windows to display vertical and horizontal asymptotes.

17. $f(x) = \dfrac{\sqrt{x + 3}}{x^2 + x - 1}$

18. $f(x) = \dfrac{5x^3 - 20x^2 - 35x + 50}{x^3 - 5x^2 + 2x + 8}$

19. $f(x) = \dfrac{x^3 - 8x^2 + 20x + 2}{|x^2 - 2x - 30|}$

20. $f(x) = \dfrac{x + 5}{|2\sin x - x|}$

1.8 The bisection method for approximating a zero of a function is a direct consequence of the Intermediate Value Theorem: If f is a *continuous* function and $f(a)$ and $f(b)$ have opposite algebraic signs, then there is some number r within the interval $[a, b]$ for which $f(r) = 0$. This same property of continuous functions enables us to approximate a zero graphically using the zoom-in command. First, find a viewing window in which the graph of the function appears to cross the x-axis at $x = r$. If a trace mode is available, move the cursor along the graph to display the coordinate for adjacent calculated points $(a, f(a))$ and $(b, f(b))$, where $a < r < b$ and $f(a)f(b) < 0$. Now make $(a, f(a))$ and $(b, f(b))$ opposite corners for a new viewing window. If no trace is available, zoom-in while retaining the point where the graph crosses the x-axis within the new viewing window. This graphical zooming-in on the zero can be repeated to more accurately determine the location of r. As in the bisection method, the midpoint $(a + b)/2$ of the *final* x-range $[a, b]$ will be taken as the final estimate for the root r with an error no greater than half the length of this interval.

In Problems 21–24 use zoom-in and trace to approximate the zeros of the given function. Continue zooming-in until at least *four correct significant digits* can be justified.

21. $f(x) = x^5 - x + 3$
22. $f(x) = x^3 - \sin x + 0.5$
23. $f(x) = 8x^4 - 14x^3 - 9x^2 + 12x - 2$
24. $f(x) = 3x^5 - 10x^4 + 10x^3 + 3x + 7$

In Problems 25 and 26 the given function has a single root. However, since the graph of each function is very "flat" near this root, it is difficult to approximate this number accurately from the graph. Use zoom-in to approximate this zero as much as is possible.

25. $f(x) = x^3 + 12x^2 + 48x + 64$
26. $f(x) = x^5 - 5x^4 + 10x^3 - 10x^2 + 5x - 1$

CHAPTER 2 The Derivative *Answers to odd-numbered problems begin on page A-116.*

2.1 All calculators and most computer programs can store only a finite number of significant digits. For example, the CASIO fx-7000G displays and stores *ten* significant digits. The TI-81 displays only *ten* digits but stores and does arithmetic with a total of *thirteen*. Single precision calculations on a VAX computer are equivalent to using about *six* significant digits. This means that if Δx is chosen to be too small, the machine will round both a and $a + \Delta x$ to the same stored value. Even assuming that a and $a + \Delta x$ store as nearby but different values in the machine, to take smaller and smaller values for Δx begins to make it likely that $f(a + \Delta x)$ and $f(a)$ will have many digits that are the same. If so, the subtraction in the rate of change or slope calculation Δy will cause this computed value to have far fewer significant digits, a phenomenon called *loss of significance*. This will show up in the slope calculations in the following way. For moderately sized Δx, the computed ratio $\Delta y/\Delta x$ will get closer and closer to the desired limit as Δx decreases. Then when Δx gets smaller relative to the precision of the machine, rounding and loss of significance will cause this computed ratio to go away from the limit. Eventually an extremely small Δx will yield $f(a + \Delta x)$ and $f(a)$ that both store to exactly the same machine number, giving a $\Delta y/\Delta x$ that is computed to be zero.

In Problems 1–4 calculate the rate of change or slope $\Delta y/\Delta x$ for the given function. Take Δx to be smaller and smaller in the pattern given until loss of significance becomes evident. How small you need to go will depend on the precision of the calculator or computer you are using.

1. $f(x) = x^5 - 3x^2 + 10x - 7$, at $a = 2.5$, for $\Delta x = 0.5$, -0.05, 0.005, $-0.0005, \ldots$
2. $f(x) = \sqrt{1 + \sqrt{x}}$, at $a = 1.4$, for $\Delta x = 0.1, 0.001, 0.00001, 0.0000001, \ldots$
3. $f(x) = \sin \pi x$, at $a = 0.37$, for $\Delta x = -0.2, -0.06, -0.002, -0.0006, \ldots$
4. $f(x) = \dfrac{x^3 + 7x - 5}{x - 2}$, at $a = 2.01$, for $\Delta x = 0.5, 0.025, 0.00125, 0.0000625, \ldots$

2.2 Let Δx be chosen from a decreasing positive sequence—for example, 0.1, 0.05, 0.01, 0.005, 0.001, . . . Then we get closer and closer to the *right-hand derivative* $f'_+(c)$ by calculating $[f(c + \Delta x) - f(c)]/\Delta x$. Similarly, we get closer and closer to the *left-hand derivative* $f'_-(c)$ by calculating $[f(c - \Delta x) - f(c)]/(-\Delta x)$. In general, when the derivative $f'(c)$ exists, the average of these two *one-sided* slope calculations will converge to the desired derivative limit at a rate much faster than either of the one-sided slope ratios:

$$m_c = \frac{1}{2}\left\{\frac{f(c + \Delta x) - f(c)}{\Delta x} + \frac{f(c - \Delta x) - f(c)}{-\Delta x}\right\}$$

$$= \frac{f(c + \Delta x) - f(c - \Delta x)}{2\,\Delta x}$$

This is called a **centered derivative approximation** because the point c is in the center of the interval $[c - \Delta x, c + \Delta x]$ over which the slope is calculated. *Note:* Users of the TI-81 calculator can get exactly this centered derivative approximation by using the operation NDeriv.

In Problems 5–8 calculate the centered derivative approximation m_c for the function and point given. Take Δx to be smaller and smaller in the pattern given and compare with the slope calculations in Problems 1–4. Eventually, for small enough Δx, these approximations also suffer from loss of significance.

5. $f(x) = x^5 - 3x^2 + 10x - 7$, at $c = 2.5$, for $\Delta x = 0.5$, $0.05, 0.005, 0.0005, \ldots$

6. $f(x) = \sqrt{1 + \sqrt{x}}$, at $c = 1.4$, for $\Delta x = 0.1, 0.001, 0.00001, 0.0000001, \ldots$

7. $f(x) = \sin \pi x$, at $c = 0.37$, for $\Delta x = 0.2, 0.06, 0.002, 0.0006, \ldots$

8. $f(x) = \dfrac{x^3 + 7x - 5}{x - 2}$, at $c = 2.01$, for $\Delta x = 0.5, 0.025, 0.00125, 0.0000625, \ldots$

2.3 Some calculators and some microcomputer applications have the capability to do *symbolic calculus operations.* For example, you can enter the formula for a function and have the machine find the formula for the derivative. Early in your study of calculus, this capability can allow you to find derivatives in situations where you have not yet learned the appropriate derivative rules. As you learn more and more rules for differentiation, you can use this as a check of your work. Finally, working mathematicians, engineers, and scientists use the computer to do symbolic calculus operations for problems that are so long and involved that people are very likely to make an error when they try to work these same problems by hand. Either explore your campus to find a microcomputer lab with such calculus-performing applications, or borrow one of the Hewlett-Packard graphic calculators that have these capabilities to do the following exercises.

In Problems 9–16 use a symbolic calculus-performing calculator or computer application to find the derivative of the given function. Where possible, identify the derivative rules being applied and check the operations by hand. Some calculators or applications may not be able to handle all of these functions.

9. $f(x) = 5x^4 - 6x^2 + 14x + 1$

10. $f(x) = \sqrt{1 + \sqrt{1 + \sqrt{x}}}$

11. $f(x) = \sin(7x + 2)$

12. $f(x) = \dfrac{x^3 - 9x + 7}{x + 3}$

13. $f(x) = x^x$

14. $f(x) = 2^{\sin x}$

15. $f(x) = |x^2 + x - 3|$

16. $f(x) = x + \cfrac{1}{x + \cfrac{1}{x + \cfrac{1}{x + 1}}}$

2.6 It is very important to begin to develop a graphical understanding of the derivative of a function. This is particularly the case with the trigonometric functions. To aid in this process, it is very helpful to pick range settings on the calculator or computer graph so that the two axes will be scaled equally. On many graphic calculators, the default range settings may be equally scaled. On the TI-81, where the default standard setting is not equally scaled, select ZOOM Square to change any viewing window into an equally scaled window. If in doubt, plot the function $y = x$ to see if this makes the appropriate 45° angle with the x-axis. Some computer graphing applications do not let you specify an equally scaled viewing window. Try leaving the graph of $y = x$ on the screen to set a "square" window and then add the graphs of trigonometric functions and their derivatives.

In Problems 17–24 plot the given function and its derivative on the same equally scaled viewing window. Use the trace to compare four values of the derivative to the slope of the graph of the function. Either print out this graph or sketch what you can see on the screen.

17. $f(x) = \sin(0.5x + 1)$

18. $f(x) = \cos(0.8x - 0.2)$

19. $f(x) = \tan(x - 0.6)$

20. $f(x) = x \sin x$

21. $f(x) = \sin x^2$

22. $f(x) = \cot(2x + 1)$

23. $f(x) = \sin x \cos 2x$

24. $f(x) = \dfrac{1}{\cos x + 1.5}$

2.11 There are many variations on Newton's Method that help with some of the problems mentioned in this section. Here we consider some alternatives that eliminate the need to provide the formula for the derivative (which can be a formidable task to perform by hand). These alternative iterations generally do not coverage as rapidly to the desired root, but this is small price to pay for avoiding the task of differentiating the function. Modify the BASIC program in the text or modify the calculator program you used to perform Newton's Method to demonstrate these variations.

In Problems 25–27 write a short program to implement each iteration. Rework Examples 4 and 5 from Section 2.11 and compare to the iterations for Newton's Method.

25. Give an initial guess x_0 and a fixed Δx. Then use a centered derivative approximation.

$$x_{n+1} = x_n - \frac{2\,\Delta x\, f(x_n)}{f(x_n + \Delta x) - f(x_n - \Delta x)}$$

26. Give two different initial guesses x_0 and x_1. Use a centered derivative approximation with $\Delta x = |x_n - x_{n-1}|$. Thus, $f(x_{n-1})$ equals either $f(x_n + \Delta x)$ or $f(x_n - \Delta x)$.

$$x_{n+1} = x_n - \frac{2(x_n - x_{n-1})f(x_n)}{f(2x_n - x_{n-1}) - f(x_{n-1})}$$

27. Give two different initial guesses x_0 and x_1. Use a one-sided derivative approximation with $\Delta x = x_{n-1} - x_n$. This method is called the **secant method**. A careful implementation will save the value $f(x_n)$ in one step for use in the next step as $f(x_{n-1})$.

$$x_{n+1} = x_n - \frac{(x_n - x_{n-1})f(x_n)}{f(x_n) - f(x_{n-1})}$$

CHAPTER 3 Applications of the Derivative

Answers to odd-numbered problems begin on page A-116.

3.3 Approximations for the extrema of a function can be read off the screen showing the graph of the function. These approximations can be quite accurate from a trace feature, and you can zoom-in for more accuracy.

In Problems 1–4 generate the graph of the given function. Find the critical points and the extrema for the function on the indicated interval. Zoom-in enough to give answers accurate to at least *four* significant digits.

1. $f(x) = |x^3 - 5x^2 + 8x - 105| + 2x$, on $[-4, 10]$
2. $f(x) = \sin(x^2 - 7x + 3)$, on $[0, 6]$
3. $f(x) = \sqrt{x} - 2 \sin x + 1$, on $[-2, 15]$
4. $f(x) = \dfrac{x^3 + 7x - 5}{x - 2}$, on $[-8, 10]$

3.4 Newton's Method (or one of its variations) may be needed to actually find the number c guaranteed by the Mean Value Theorem for derivatives. Use a calculator/computer graph to find an initial guess for the number c and then carry out the following Newton iteration:

$$x_{n+1} = x_n - \frac{f'(x_n) - m}{f''(x_n)}, \qquad \text{where } m = \frac{f(b) - f(a)}{b - a}$$

In Problems 5–8 determine whether the given funcions satisfies the hypotheses of the Mean Value Theorem for derivatives on the indicated interval. If so, find all values of c that satisfy the conclusion of the theorem using Newton's Method.

5. $f(x) = \sin(x^2 - 7x + 3)$, on $[1.5, 3]$
6. $f(x) = x^4 - x^3 - 13x^2 + x + 12$, on $[-4, 4]$
7. $f(x) = |x^4 - 4x^2 - x - 8|$, on $[-5, 1]$
8. $f(x) = |x^4 - 4x^2 - x - 8|$, on $[-2, 2]$

3.5 The efficient method for evaluating polynomials using *nested multiplication* can be related to *synthetic division.* For example, if

$$p(x) = a_n x^n + a_{n-1}x^{n-1} + \cdots + a_1 x + a_0,$$

then the calculation of $p(w)$ via

$$(\cdots(((a_n)w + a_{n-1})w + a_{n-2})\cdots + a_1)w + a_0$$

can be specified as finding $b_n = a_n$, $b_{n-1} = b_n w + a_{n-1}$, $b_{n-2} = b_{n-1}w + a_{n-2}, \ldots, b_1 = b_2 w + a_1$, and $b_0 = b_1 w + a_0$. These are the calculations starting with the innermost parentheses, and these are also the coefficients for synthetic division of the polynomial $p(x)$ by $(x - w)$ with remainder b_0. Thus, $p(x) = \{b_n x^{n-1} + b_{n-1}x^{n-2} + \cdots + b_2 x + b_1\}(x - w) + b_0$. If we repeat this process on the resulting $(n - 1)$-degree polynomial, that is, $c_n = b_n$, $c_{n-1} = c_n w + b_{n-1}$, $c_{n-2} = c_{n-1}w + b_{n-2}, \ldots, c_1 = c_2 w + b_1$, then $p(x) = \{\{c_n x^{n-2} + c_{n-1}x^{n-3} + \cdots + c_2\}(x - w) + c_1\}(x - w) + b_0$. It is easy to perform these calculations in a program loop. Finally, $p(w) = b_0$ and $p'(w) = c_1$.

In Problems 9–12 write a program to evaluate the given polynomial and its first derivative using the repeated nested multiplication. If possible, use this to generate a rapid plot of the polynomial and its derivative point-by-point. Then use Newton's Method and this program to accurately find the roots of the given polynomial on the given interval.

9. $f(x) = 64x^7 + 32x^6 - 112x^5 - 48x^4 + 56x^3 + 18x^2 - 7x - 1$, on $[-1, 1]$
10. $f(x) = 64x^7 - 32x^6 - 112x^5 + 48x^4 + 56x^3 - 18x^2 - 7x + 1$, on $[-1, 1]$
11. $f(x) = 32x^6 - 16x^5 - 48x^4 + 20x^3 + 18x^2 - 5x - 1$, on $[-1.5, 1]$
12. $f(x) = 0.000032x^6 + 0.00016x^5 - 0.0048x^4 - 0.02x^3 + 0.18x^2 + 0.5x - 1$, on $[-10, 10]$

3.6 The second derivative can also be approximated by a **centered approximation**:

$$s_c = \frac{m_c(c + \frac{1}{2}\Delta x) - m_c(c - \frac{1}{2}\Delta x)}{\Delta x}$$

$$= \frac{f(c + \Delta x) - 2f(c) + f(c - \Delta x)}{(\Delta x)^2}$$

Users of the TI-81 graphic calculator effectively get this operation when they store the formula for f in the slot Y_1 and then calculate NDeriv(NDeriv(Y_1, (1/2) Δx), (1/2) Δx) with c stored in X.

In Problems 13–16 graph the given function and its second derivative on the same viewing window over the indicated interval. This may require using the centered second derivative approximation s_c, or it may be built into the computer application available. From the graph, determine the intervals on which the function is concave upward and the intervals on which it is concave downward.

13. $f(x) = \sin\left(\dfrac{1}{x^2 - 7x + 3}\right)$, on $[-\infty, 0.4]$, $[0.5, 6.5]$, and $[6.7, \infty]$ (Also explain why the intervals $[0.4, 0.5]$ and $[6.5, 6.7]$ were left out.)
14. $f(x) = \dfrac{x^4 - x^3 - 13x^2 + x + 5}{x^4 + 1}$, on $[-5, 5]$
15. $f(x) = x^{(2 - x^2)}$, on $[0, 5]$
16. $f(x) = \sin(\sin x)$, on $[-\pi, 2\pi]$

4.3 A calculator or computer program can easily add the terms in a summation. The technique of repeatedly going back to execute a certain section of code (such as here the calculation of one term in the summation) is called looping. In a definite loop there is a loop index to keep track of how many times the loop code has been repeated and make it possible to stop the process after the desired repetitions. Here the loop index will be closely related to the index of the summation. In BASIC or on the HP-28S, the needed structure is a FOR–NEXT loop. The CASIO fx-7000G has the commands Isz and Dsz that can be used for this purpose, and the TI-81 has similar commands Is>(and Ds<(.

In Problems 1–8 write a short program to find the value of the given sum.

1. $\sum_{k=1}^{25} \frac{2^k}{k}$
2. $\sum_{k=1}^{15} \frac{3^k}{k}$
3. $\sum_{k=1}^{100} \frac{1}{k}$
4. $\sum_{k=1}^{50} \frac{(-1)^k}{k}$
5. $\sum_{k=1}^{25} \sqrt{k}$
6. $\sum_{k=2}^{60} \frac{k-1}{k+1}$
7. $\sum_{k=5}^{35} \cos k$
8. $\sum_{i=5}^{35} i^5$

4.4 The summations $\Sigma_{k=1}^{n} f(x_k^*)\,\Delta x$ used in the limiting definition for the area A under the graph of the function f can be computed in a short calculator/computer program. It is much easier to use partitions of the interval $[a, b]$ into equal subintervals with $\Delta x = (b-a)/n$ and to use some regular pattern for choosing the point x_k^* on each subinterval (for example, use right endpoints). Many calculus microcomputer applications provide a graphic illustration of the rectangles in the summation compared to the area under the graph. With a little effort, this picture can be drawn on a graphic calculator screen (with shading on the TI-81). Even though you will soon learn more efficient ways to find areas, both symbolically and numerically, this graphical representation of thin rectangles will be extremely important in the next chapter when new applications for this technique are presented.

In Problems 9–14 compute $\Sigma_{k=1}^{n} f(x_k^*)\,\Delta x$ using equal subintervals and right endpoints for the given function over the indicated interval. If possible, produce a graphical representation of these summations on the screen. Based on this evidence, estimate the area under the graph.

9. $f(x) = \sin(\sin x)$, on $[0, \pi]$ for $n = 20, 30, 40$
10. $f(x) = 5\tan x$, on $\left[\frac{\pi}{6}, \frac{\pi}{3}\right]$ for $n = 15, 30, 45$
11. $f(x) = 1 + \cos x^2$, on $[-1, 2]$ for $n = 12, 24, 48$
12. $f(x) = \sqrt{x^2 - x + 0.3}$, on $[-1, 2]$ for $n = 15, 25, 35$
13. $f(x) = \sqrt[4]{1 - x^4}$, on $[-1, 1]$ for $n = 10, 20, 30$
14. $f(x) = \cos\sqrt{x}$, on $[0, 1]$ for $n = 20, 25, 30$

4.8 Simpson's Rule can also be used to approximate the antiderivative $G(x) = \int_a^x f(t)\,dt$ for x in $[a, b]$. Choose $\Delta x = (b-a)/n$, n even. First, $G(a) = 0$. Next,

$$G(a + 2\,\Delta x) \approx \frac{\Delta x}{3}[f(a) + 4f(a + \Delta x) + f(a + 2\,\Delta x)]$$

$$\begin{aligned} G(a + 4\,\Delta x) &\approx \frac{\Delta x}{3}[f(a) + 4f(a + \Delta x) \\ &\quad + 2f(a + 2\,\Delta x) + 4f(a + 3\,\Delta x) \\ &\quad + f(a + 4\,\Delta x)] \\ &= G(a + 2\,\Delta x) + \frac{\Delta x}{3}[f(a + 2\,\Delta x) \\ &\quad + 4f(a + 3\,\Delta x) + f(a + 4\,\Delta x)] \\ G(a + 6\,\Delta x) &\approx G(a + 4\,\Delta x) + \frac{\Delta x}{3}[f(a + 4\,\Delta x) \\ &\quad + 4f(a + 5\,\Delta x) + f(a + 6\,\Delta x)] \end{aligned}$$

Continuing in this fashion, we eventually reach an approximation for $G(a + n\,\Delta x) = G(b)$ that is the usual Simpson's Rule on $[a, b]$. Recording these intermediate calculations for $G(a + k\,\Delta x)$, $k = 2, 4, 6, 8, \ldots, n$, gives a table of approximate values for the antiderivative G.

In Problems 15–20 apply Simpson's Rule as explained above to construct a table of values approximating the antiderivative $G(x) = \int_a^x f(t)\,dt$ for the given function over the indicated interval. If possible, graph this table to see the approximate graph for this antiderivative on the screen.

15. $f(x) = \cos\sqrt{x}$, on $[0, 4]$ for $n = 30$
16. $f(x) = \tan(0.25x^2)$, on $[-1, 2]$ for $n = 20$
17. $f(x) = \sqrt{1 + x^3}$, on $[-1, 4]$ for $n = 26$
18. $f(x) = \sin(\sin x)$, on $[-\pi, 2\pi]$ for $n = 60$
19. $f(x) = |\cos x^2|$, on $[-2, 2]$ for $n = 30$
20. $f(x) = \frac{x-1}{1+x^3}$, on $[0, 5]$ for $n = 26$

CHAPTER 5 Applications of the Integral

Answers to odd-numbered problems begin on page A-117.

5.1 To find the area of the region bounded by the graphs of two functions, $y = f(x)$ and $y = g(x)$, the locations for the points of intersection of the graphs are often needed. A convenient graphing utility with some zoom-in capability makes it easy to find approximations for these intersection points. For more accuracy, numerical root-finding techniques can be applied to the function $s(x) = f(x) - g(x)$.

In Problems 1–8 approximate the area of the region bounded by the graphs of the given equations. Use a graphing utility to locate the points of intersection to an accuracy of at least two decimal places.

1. $y = x^3 - 2x^2 - x + 1$, $y = \sqrt{20x}$
2. $y = 50\cos(0.5x)$, $y = x^2 - 20$
3. $y = -0.4x^3 + 2x$, $y = \sin x$
4. $y = x\sqrt{x^2 + 15}$, $y^2 = x + 30$
5. $y = 0.01x^5 - 0.1x^4 + 0.8x^3 + 1.4x^2 - 10x + 50$, $y = 18.05x + 114.62$
6. $y = 8x^4 - 8x^2 + x - 1$, $y = -2x^2 - x + 2$
7. $y = \dfrac{2x}{\sqrt{x^2 + 4}}$, $y = |x| - 8$
8. $y = 5\cos(0.4x - 1)$, $y = 2|x|^{2/3}$

5.4 While the integrals in any section of Chapter 5 may require numerical approximation, the integrals representing arc length frequently require approximation because of the square root in the integrand. If you reexamine the functions in this section where the arc length can be evaluated exactly, you will see that these functions must have a very special form. Numerical integration rules, such as the Trapezoidal Rule and Simpson's Rule, must be used for arc lengths involving elementary trigonometric functions and polynomials.

In Problems 9–14 approximate the arc length of the graph of the given equation on the indicated interval. Use a program to implement Simpson's Rule for various numbers of subintervals or use the numerical integration procedure built into the calculator or computer application available. Indicate the expected accuracy achieved.

9. $y = x^2 - x + 1$; $[-2, 3]$
10. $y = x^3 - 3x^2 + 2$; $[-1, 4]$
11. $y = \sin^3 x$; $[0, \pi]$
12. $y = \sqrt[3]{5x + 2}$; $[0, 16]$
13. $y = \dfrac{x}{x^2 + 1}$; $[0, 20]$
14. $y = \sin(\sin x)$; $[0, \pi]$

The TI-81 Graphics Calculator has several "default" viewing windows. For graphing trigonometric functions, the recommended TRIG window in the ZOOM menu has range settings corresponding to $-2\pi \le x \le 2\pi$ and $-3 \le y \le 3$. No matter what graphing utility is used, this is a nice viewing window for the elementary trigonometric functions. The next problems explore the arc lengths of various trigonometric functions as seen in this special window. TI-81 users might also experiment with using a numerical derivative estimate via NDeriv in place of the derivative in the integrand representing arc length. Others not using a TI-81 can try the same thing using the centered derivative approximation discussed in the Calculator/Computer Activities for Section 2.2. Still another variation on these problems might be to have the arc length integrand automatically constructed from the formula for the functions using some symbolic differentiation process within a computer algebra system or symbolic calculator.

In Problems 15–18 approximate the total arc length of the parts of the graph of the given trigonometric function visible in the special viewing window with boundaries $-2\pi \le x \le 2\pi$ and $-3 \le y \le 3$. Use a program to implement Simpson's Rule for various numbers of subintervals or use the numerical integration procedure built into the calculator or computer application available. Indicate the expected accuracy achieved.

15. $y = \sin x$
16. $y = \tan x$
17. $y = \sec x$
18. $y = \sin 2x$

5.5 The definite integrals representing areas of surfaces of revolution also tend to require numerical integration techniques.

In Problems 19–22 approximate the area of the surface that is formed by revolving each graph on the given interval about the indicated axis. Use a program to implement Simpson's Rule for various numbers of subintervals or use the numerical integration procedure built into the calculator or computer application available. Indicate the expected accuracy achieved.

19. $y = \sin x$; $[0, \pi]$, x-axis
20. $y = 3.4 - 0.095x^2$; $[-5.5, 5.5]$, x-axis
21. $y = \cos x$; $[0, \pi/2]$, y-axis
22. $y = x^3$; $[0, 1]$, y-axis

5.6 Finding the number c that demonstrates the Mean Value Theorem for definite integrals, or equivalently where the function equals an average value over an interval, may involve two numerical operations. First, the definite integral may need to be approximated numerically (for example, by using the intrinsic numerical integration on an HP-28S or a computer application). Then the task of finding c where the function equals the desired numerical value can be viewed as a root-finding problem (for example, use the solve operations on an HP-28S or a computer algebra system).

In Problems 23–26 approximate the value of c in the given interval for which $f(c) = A(f)$, where $A(f)$ is the average value of the given function f on the indicated interval. Use programs to implement Simpson's Rule and Newton's Method, or use the numerical integration procedure and solving routine built into the calculator or computer application available.

23.
24. $y = \sqrt{x^3 + x + 1}$; $[0, 5]$
25. $y = x \sin x$; $[0, 2\pi]$
26. $y = \cos(1.5 \sin x)$; $[0, \pi]$
$y = \sqrt{1 - 0.8 \sin^2 x}$; $[0, \pi/2]$

CHAPTER 6 Logarithmic and Exponential Functions

Answers to odd-numbered problems begin on page A-117.

6.2 Most scientific calculators provide both the natural logarithmic function, generally labeled **ln**, and the common logarithm using base 10, generally labeled **log**, which agrees with the notation of this text. Computer programming languages (and the packages and application programs written in them) generally use the notation LOG(X) to represent the natural logarithm of the variable X while using LOG10(X) to indicate a common logarithm. The confusion of different notations continues in engineering and scientific literature, where "log" sometimes indicates the natural logarithm (usually in electronics and physics) and sometimes indicates the common logarithm (usually in chemistry, geophysics, and biology). Find out the syntax needed for the calculator or computer application available for graphing.

In Problems 1–6 graph the given function on the indicated interval. Use the trace and zoom-in features of the graphing utility to locate any x-intercepts, y-intercepts, and relative extrema on the displayed graph to an accuracy of at least two decimal places.

1. $y = \ln(\sin x)$; $[0.1, 3.14]$
2. $y = \sin(\ln x)$; $[0.1, 800]$
3. $y = \dfrac{\ln(4(x+1))}{\ln(2(x+1))}$; $(-\frac{1}{2}, 3]$
4. $y = x^2 \ln x^3$; $(0, 3]$
5. $y = \ln(x(1.2 + \sin x))$; $[0.1, 20]$
6. $y = \ln(x^4 - 4x^2 - x + 5.6)$; $[-3, 4]$

6.3 Most scientific calculators provide both the natural exponential function, generally labeled $\boldsymbol{e^x}$, and the power of 10 function, generally labeled $\mathbf{10^x}$, which agrees with the notation of this text. Computer programming languages (and the packages and application programs written in them) generally use the notation EXP(X) to represent the natural exponential of the variable X, which is also found in this text. Using the intrinsic exponential function is much easier (and generally more accurate) than typing the digits 2.71828182846 to approximate e and using the arithmetic operation for raising one number to the power of another number.

In Problems 7–12 graph the given function on the indicated interval. Use the trace and zoom-in features of the graphing utility to locate any x-intercepts, y-intercepts, and relative extrema on the displayed graph to an accuracy of at least two decimal places.

7. $y = \dfrac{6}{2 + e^{1/x}}$; $[-4, 4]$
8. $y = 2.56e^{-0.22x} \cos 4.9x$; $[0, 12]$
9. $y = \dfrac{2}{\sqrt{\pi}} e^{-x^2}$; $[-3, 3]$
10. $y = \ln(5 + e^x)$; $[-8, 12]$
11. $y = 5 \sin 0.42x + 2.5e^{-0.47x} \cos 4x$; $[0, 20]$
12. $y = e^{\frac{x^3 - 17x + 8}{50}}$; $[-7, 7]$

6.4 Integrals involving logarithmic and exponential functions may require numerical integration. For any programmable calculator or computer, either a standard program should be written to implement Simpson's Rule or some more accurate numerical operation provided should be used.

In Problems 13–18 numerically evaluate the given definite integral. Run the program for Simpson's Rule for several different numbers of subintervals to get some indication of the accuracy, or report the error estimate provided by the numerical integration operation used.

13. $\dfrac{2}{\sqrt{\pi}} \displaystyle\int_0^8 e^{-x^2}\, dx$
14. $\displaystyle\int_5^{4500} \ln(\ln x)\, dx$
15. $\displaystyle\int_{0.5}^{6.5} \frac{e^x}{x}\, dx$
16. $\displaystyle\int_{1.8}^{14} \frac{1}{\ln x}\, dx$

17. $\int_{12}^{24} \ln(2 + e^x)\, dx$ 18. $\int_{0}^{2} e^{-1.7x} \cos x^2\, dx$

6.5 There is an easy way to evaluate logarithmic functions to a base other than e or 10. The **change of base formula** (Problem 97 in Exercises 6.5) implies that

$$\log_a x = \left(\frac{1}{\ln a}\right) \ln x = \left(\frac{1}{\log_{10} a}\right) \log_{10} x$$

Because any logarithmic function can be written as $k \ln x$ for some constant k, little is lost by using only natural logarithms.

In Problems 19–22 graph the given function on the indicated interval. Use the trace and zoom-in features of the graphing utility to locate any x-intercepts, y-intercepts, and relative extrema on the displayed graph to an accuracy of at least two decimal places.

19. $y = \log_a x$, for $a = 0.3$, 1.3, and 3.3; $(0, 15]$
20. $y = (x - 2)^2 \log_2 x$; $(0, 5]$
21. $y = \dfrac{10 \log_4 x}{4^x}$; $(0, 4]$
22. $y = \log_{0.7}(x^4 - 4x^2 + x + 6)$; $[-10, 10]$

6.7 Hyperbolic functions can be evaluated using Definition 6.8 in terms of the exponential function. However, some scientific calculators provide both the hyperbolic functions and their inverses directly. Implementing a hyperbolic function may require pressing a key labeled **hyp** followed by pressing a trigonometric function key (for example **hyp sin** for the hyperbolic sine). On some calculators, the hyperbolic functions can be implemented by means of a menu. The HP-28S has the hyperbolic sine, the hyperbolic cosine, the hyperbolic tangent, and the inverses of these functions in the LOGS menu. Although it is not as common to find hyperbolic functions in standard computer languages, nearly all computer algebra systems and mathematical application programs provide an implementation for the basic hyperbolic functions.

In Problems 23–26 solve the given equation for x using the hyperbolic functions provided on the calculator or computer application available.

23. $\sinh 3.741 = x$
24. $\operatorname{sech} 5.9 + \operatorname{csch} 5.9 = x$
25. $(\sinh^2 1.5)x^2 + (2 \cosh^2(-1.27))x + \tanh 0.18 = 0$
26. $x \ln(\cosh 7.915) + \ln(\sinh 7.915) = 5$

In Problems 27–30 graph the given function on the indicated interval. Use the trace and zoom-in features of the graphing utility to locate any x-intercepts, y-intercepts, and relative extrema on the displayed graph to an accuracy of at least two decimal places.

27. $y = x^2 + 2 - \cosh x$; $[-5, 5]$
28. $y = \sinh^3 x + 6 \sinh x + 15$; $[-4, 6]$
29. $y = \coth x - \operatorname{csch} x$; $[-10, 10]$
30. $y = \ln(\cosh x) - x$; $[-12, 12]$

7

INVERSE TRIGONOMETRIC AND HYPERBOLIC FUNCTIONS

INTRODUCTION

For each of the six trigonometric functions and the six hyperbolic functions we can define an inverse function. Our goal in this chapter is to study the derivatives and integrals that involve these twelve new functions.

Since the hyperbolic functions are combinations of e^x and e^{-x}, we shall see that the inverse hyperbolic functions can be expressed in terms of the natural logarithm.

If we lived in a vacuum (such as on the moon) then the velocity of a falling body would increase without bound until it struck the surface. On earth, due to air resistance, the velocity of a body of mass m falling from a great height (such as a sky diver) increases up to a point and then becomes constant. In Problem 17 of Exercises 6.8 we saw that when air resistance is assumed to be proportional to the instantaneous velocity of the body, this constant, or *terminal* velocity is given by $y_t = mg/k$, where k is the constant of proportionality.

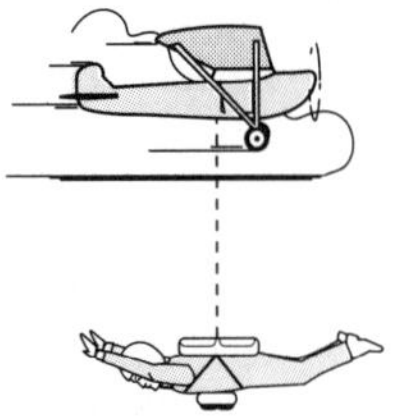

But for high-speed motion through air, it is more realistic to assume that the resistance is proportional to the *square* of the instantaneous velocity. In this case, you are asked to show in Exercises 7.3 that the terminal velocity of a falling body is $v_t = \sqrt{mg/k}$.

IMPORTANT CONCEPTS

inverse trigonometric functions
range of inverse sine
range of inverse cosine
range of inverse tangent
derivatives of inverse trigonometric functions
integrals yielding inverse trigonometric functions
inverse hyperbolic functions
derivatives of inverse hyperbolic functions
integrals yielding inverse hyperbolic functions
logarithmic identities

7.1 INVERSE TRIGONOMETRIC FUNCTIONS

Inspection of their graphs and observations such as

$$\sin(x + 2\pi) = \sin x$$
$$\cos(x + 2\pi) = \cos x$$
$$\tan(x + \pi) = \tan x$$
$$\sin\frac{\pi}{6} = \sin\frac{5\pi}{6} = \frac{1}{2}$$
$$\cos\frac{\pi}{4} = \cos\frac{7\pi}{4} = \frac{\sqrt{2}}{2}$$

and
$$\cot\frac{\pi}{2} = \cot\frac{3\pi}{2} = 0$$

should convince you that none of the trigonometric functions are one-to-one. Nevertheless, by proceeding as in Example 8 of Section 6.1, we can find the inverse of a trigonometric function that is defined *on a restricted domain.*

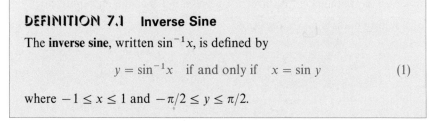

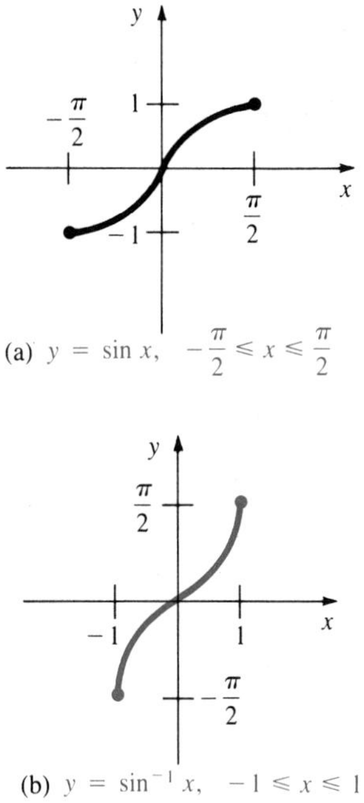

(a) $y = \sin x, \quad -\frac{\pi}{2} \leq x \leq \frac{\pi}{2}$

(b) $y = \sin^{-1} x, \quad -1 \leq x \leq 1$

FIGURE 7.1

Inverse Sine By considering $y = \sin x$ only on the closed interval $-\pi/2 \leq x \leq \pi/2$, it is apparent from Figure 7.1(a) that for each value of y in $[-1, 1]$ there is only one corresponding x in $[-\pi/2, \pi/2]$. In addition, we notice that this new function fulfills the criteria of Theorem 6.2 and thus has an inverse. Recall that the domain and range of the original function become, respectively, the range and domain of the inverse. We summarize these observations in the form of a definition.

DEFINITION 7.1 Inverse Sine

The **inverse sine**, written $\sin^{-1}x$, is defined by

$$y = \sin^{-1}x \quad \text{if and only if} \quad x = \sin y \tag{1}$$

where $-1 \leq x \leq 1$ and $-\pi/2 \leq y \leq \pi/2$.

In other words,

the inverse sine of a number is the angle (in radians) whose sine is x

provided that this angle satisfies $-\pi/2 \leq \sin^{-1}x \leq \pi/2$. The graph of $y = \sin^{-1}x$ is given in Figure 7.1(b).

In view of (3) of Section 6.1 we have

$$\begin{aligned} \sin(\sin^{-1}x) &= x \quad \text{when} \quad -1 \leq x \leq 1 \\ \sin^{-1}(\sin x) &= x \quad \text{when} \quad -\frac{\pi}{2} \leq x \leq \frac{\pi}{2} \end{aligned} \tag{2}$$

EXAMPLE 1 Evaluate $\sin^{-1}\left(\frac{\sqrt{2}}{2}\right)$.

Solution Let $y = \sin^{-1}(\sqrt{2}/2)$. One angle y for which $\sin y = \sqrt{2}/2$ is $y = \pi/4$. Since this is the only number in $[-\pi/2, \pi/2]$ for which this is true, we can write

$$\sin^{-1}\left(\frac{\sqrt{2}}{2}\right) = \frac{\pi}{4} \qquad \square$$

Notation In order not to confuse $\sin^{-1}x$ with the reciprocal $(\sin x)^{-1} = 1/\sin x$, an alternative notation is often used:

$$\arcsin x = \sin^{-1}x$$

It is read as it appears, "arc sine of x."

EXAMPLE 2 Evaluate $\arcsin\left(\frac{1}{2}\right)$.

Solution If $y = \arcsin(\frac{1}{2})$, then $\frac{1}{2} = \sin y$. The only number y in $[-\pi/2, \pi/2]$ for which this is true is $\pi/6$; that is,

$$\arcsin\left(\frac{1}{2}\right) = \frac{\pi}{6} \qquad \square$$

Note: Although some calculators give

$$\arcsin\left(\frac{1}{2}\right) = 30°$$

this is, strictly speaking, *incorrect.* The output of an inverse trigonometric function is an angle in *radian measure.*

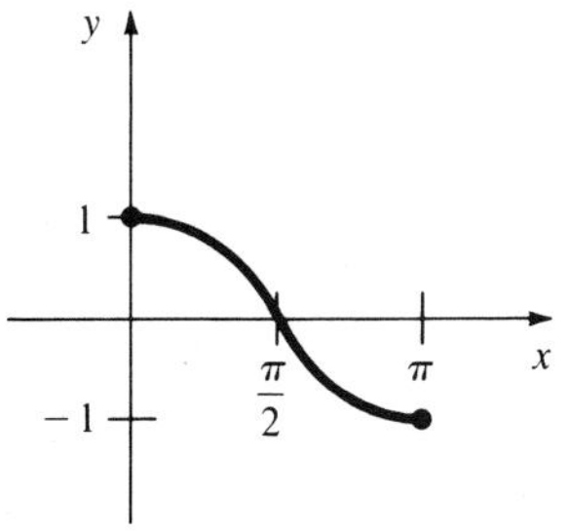

(a) $y = \cos x,\ 0 \leq x \leq \pi$

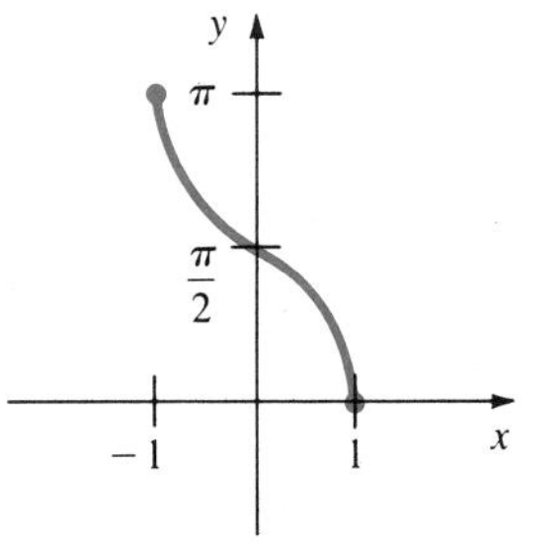

(b) $y = \cos^{-1} x,\ -1 \leq x \leq 1$

FIGURE 7.2

Inverse Cosine Since the function $y = \cos x$ is continuous and decreasing on $[0, \pi]$, we define its inverse on that interval.

DEFINITION 7.2 Inverse Cosine

The **inverse cosine** is defined by

$$y = \cos^{-1}x \quad \text{if and only if} \quad x = \cos y \tag{3}$$

where $-1 \leq x \leq 1$ and $0 \leq y \leq \pi$.

The graphs of $y = \cos x$ and $y = \cos^{-1}x$ are shown in Figure 7.2.

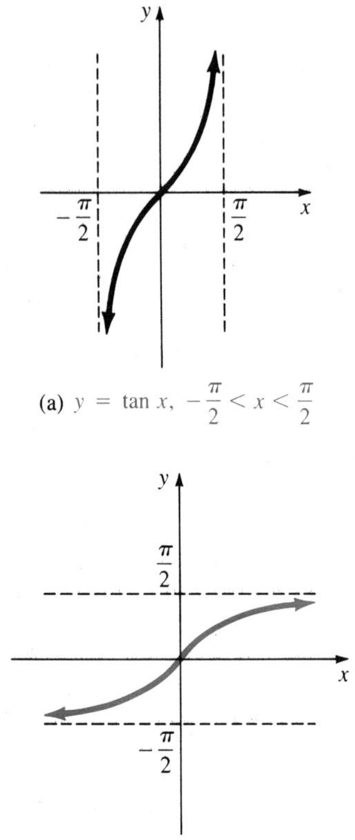

(a) $y = \tan x, \ -\frac{\pi}{2} < x < \frac{\pi}{2}$

(b) $y = \tan^{-1} x, \quad -\infty < x < \infty$

FIGURE 7.3

EXAMPLE 3 Evaluate $\cos^{-1}(-1)$.

Solution Let $y = \cos^{-1}(-1)$. From (3) we see that the only number y in the closed interval $[0, \pi]$ for which $\cos y = -1$ is $y = \pi$. Thus,

$$\cos^{-1}(-1) = \pi$$ □

Analogous to (2) we have

$$\cos(\cos^{-1} x) = x \quad \text{when} \quad -1 \leq x \leq 1$$
$$\cos^{-1}(\cos x) = x \quad \text{when} \quad 0 \leq x \leq \pi$$

Inverse Tangent On $(-\pi/2, \pi/2)$, $y = \tan x$ is continuous and increasing and so has an inverse on that open interval.

> **DEFINITION 7.3 Inverse Tangent**
>
> The **inverse tangent** is defined by
>
> $$y = \tan^{-1} x \quad \text{if and only if} \quad x = \tan y \tag{4}$$
>
> where $-\infty < x < \infty$ and $-\pi/2 < y < \pi/2$.

See Figure 7.3.

With their domains suitably restricted, the remaining trigonometric functions also have inverses. Definition 7.4 summarizes the definitions of the inverses of the cotangent, secant, and cosecant. We urge you to compare the graphs of $y = \cot^{-1} x$, $y = \sec^{-1} x$, and $y = \csc^{-1} x$ given in Figure 7.4 with the pertinent portions of the graphs of $y = \cot x$, $y = \sec x$, and $y = \csc x$ (see page 29).

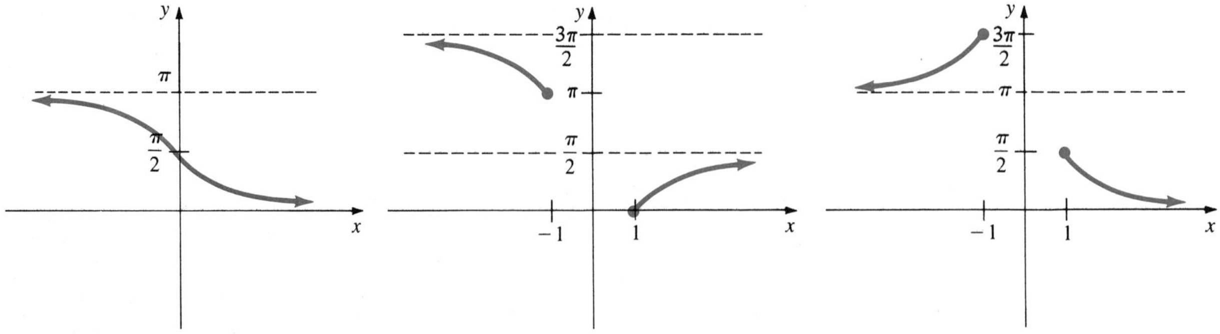

(a) $y = \cot^{-1} x, \quad -\infty < x < \infty$

(b) $y = \sec^{-1} x, \quad |x| \geqslant 1$

(c) $y = \csc^{-1} x, \quad |x| \geqslant 1$

FIGURE 7.4

DEFINITION 7.4 Inverses of Cotangent, Secant, Cosecant

(*i*) $y = \cot^{-1}x$ if and only if $x = \cot y$,
$-\infty < x < \infty$ and $0 < y < \pi$

(*ii*) $y = \sec^{-1}x$ if and only if $x = \sec y$,
$|x| \geq 1$ and y is a number in $[0, \pi/2)$ or $[\pi, 3\pi/2)$

(*iii*) $y = \csc^{-1}x$ if and only if $x = \csc y$,
$|x| \geq 1$ and y is a number in $(0, \pi/2]$ or $(\pi, 3\pi/2]$

As it did for the sine function, the prefix "arc" denotes the inverse of any of the other five trigonometric functions. We shall use both notations interchangeably.

EXAMPLE 4 Evaluate $\text{arcsec}(-\sqrt{2})$.

Solution Let $y = \text{arcsec}(-\sqrt{2})$ so that $-\sqrt{2} = \sec y$ or $\cos y = -1/\sqrt{2}$. The only number y in $[0, \pi/2)$ or $[\pi, 3\pi/2)$ that satisfies these conditions is $y = 5\pi/4$. Since $\sec 5\pi/4 = -\sqrt{2}$, we have

$$\text{arcsec}(-\sqrt{2}) = \frac{5\pi}{4}$$ □

EXAMPLE 5 Evaluate $\cos\left(\tan^{-1}\frac{2}{3}\right)$.

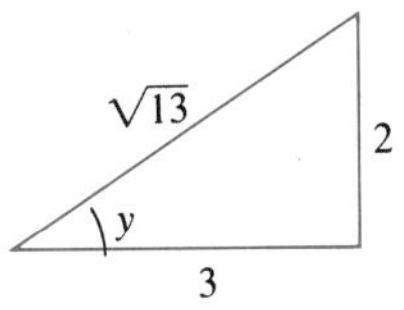

FIGURE 7.5

Solution Let $y = \tan^{-1}\frac{2}{3}$ so that $\frac{2}{3} = \tan y$. Using the triangle in Figure 7.5 as an aid, we see that

$$\cos\left(\tan^{-1}\frac{2}{3}\right) = \cos y = \frac{3}{\sqrt{13}}$$ □

EXAMPLE 6 Evaluate $\cos(\sin^{-1}1)$.

Solution Since $\sin^{-1}1 = \pi/2$, we have

$$\cos(\sin^{-1}1) = \cos\frac{\pi}{2} = 0$$ □

The inverse tangent is especially important in applications.

EXAMPLE 7 On a microcomputer the arctangent function is intrinsic, or built into, the BASIC programming language, whereas the arcsine and arccosine functions are not. In order to write a program that involves, say, the

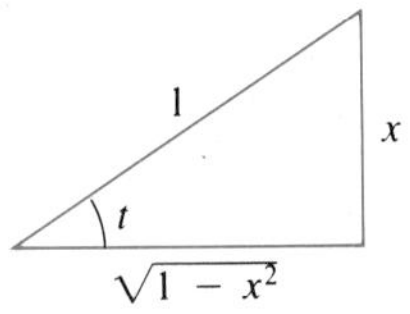

FIGURE 7.6

arcsine, the writer must utilize what is known as a "user-defined function" that expresses the arcsine in terms of the arctangent.

If $t = \sin^{-1}x$, we can interpret x as the side of a right triangle that is opposite t and 1 as the hypotenuse of the triangle. In Figure 7.6, we see that the side adjacent to the angle t is $\sqrt{1-x^2}$. Hence,

$$\tan t = \frac{x}{\sqrt{1-x^2}} \quad \text{implies} \quad t = \tan^{-1}\frac{x}{\sqrt{1-x^2}}$$

Thus,

$$\sin^{-1}x = \tan^{-1}\frac{x}{\sqrt{1-x^2}} \tag{5}$$

A similar result can be derived for the inverse cosine. See Problem 44 in Exercises 7.1. □

Although we derived (5) from a triangle in which $0 < t < \pi/2$, the result is valid as well for $-\pi/2 < t < \pi/2$.

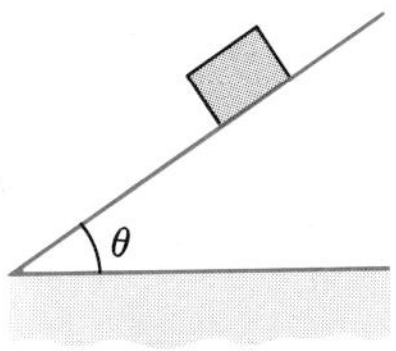

FIGURE 7.7

EXAMPLE 8 The so-called *angle of repose* of a block at the verge of slipping down an inclined plane is defined by

$$\theta = \tan^{-1}f \tag{6}$$

where f is the coefficient of friction. (For wood on wood, $0.3 \le f \le 0.6$.) See Figure 7.7. For an angle greater than $\tan^{-1}f$, the block will slide down the inclined plane. □

▲ ***Remark*** The range of an inverse trigonometric function is arbitrary up to a point, since a trigonometric function has an inverse on *any* interval of the x-axis for which the function is one-to-one. For example, we could just as well limit the domain of $y = \sin x$ to the interval $[\pi/2, 3\pi/2]$, thereby yielding $y = \sin^{-1}x$, $|x| \le 1$, whose range is given by $\pi/2 \le y \le 3\pi/2$. The ranges specified in Definitions 7.1–7.3 and 7.4(*i*) are universally agreed upon and grew out of the most logical and most convenient limitation of the original function. Thus, when we see arccos x or $\tan^{-1}x$ in any context, we know

$$0 \le \arccos x \le \pi \quad \text{and} \quad -\frac{\pi}{2} < \tan^{-1}x < \frac{\pi}{2}$$

These conventions are the same as those used in hand calculators when the $\boxed{\sin^{-1}}$, $\boxed{\cos^{-1}}$, and $\boxed{\tan^{-1}}$ keys are employed. You may be interested to know that there has been no agreement on the ranges of either $y = \sec^{-1}x$ or $y = \csc^{-1}x$. For example, in contrast to (*ii*) and (*iii*) of Definition 7.4, some calculus texts define

$$y = \sec^{-1}x \quad \text{if and only if} \quad x = \sec y$$
$$|x| \ge 1 \quad \text{and} \quad y \text{ is a number in } [0, \pi/2) \quad \text{or} \quad (\pi/2, \pi]$$

and

$$y = \csc^{-1} x \quad \text{if and only if} \quad x = \csc y$$
$$|x| \geq 1 \quad \text{and} \quad y \text{ is a number in } [-\pi/2, 0) \quad \text{or} \quad (0, \pi/2]$$

EXERCISES 7.1

Answers to odd-numbered problems begin on page A-87.

In Problems 1–30 obtain the exact value of the given expression. Do not use a calculator or tables.

1. $\sin^{-1}(0)$

2. $\cos^{-1}\left(\frac{1}{2}\right)$

3. $\arccos\left(-\frac{\sqrt{2}}{2}\right)$

4. $\sin^{-1}(-1)$

5. $\arctan(1)$

6. $\tan^{-1}(\sqrt{3})$

7. $\cot^{-1}(-1)$

8. $\sec^{-1}(-1)$

9. $\arcsin\left(-\frac{\sqrt{3}}{2}\right)$

10. $\operatorname{arccot}(-\sqrt{3})$

11. $\operatorname{arcsec}\left(-\frac{2}{\sqrt{3}}\right)$

12. $\csc^{-1}(-\sqrt{2})$

13. $\cos\left(\sin^{-1}\frac{1}{2}\right)$

14. $\sin(\cos^{-1}0)$

15. $\tan^{-1}(\cos \pi)$

16. $\cos^{-1}\left(\sin \frac{5\pi}{4}\right)$

17. $\sin\left(\arctan \frac{4}{3}\right)$

18. $\cos\left(\sin^{-1}\frac{2}{5}\right)$

19. $\tan\left(\cot^{-1}\frac{1}{2}\right)$

20. $\csc\left(\tan^{-1}\frac{2}{3}\right)$

21. $\sec(\tan^{-1}1)$

22. $\arcsin\left(\cot\left(-\frac{\pi}{4}\right)\right)$

23. $\sin^{-1}\left(\sin \frac{2\pi}{3}\right)$

24. $\cos^{-1}\left(\cos \frac{4\pi}{3}\right)$

25. $\cos^{-1}(\cos 3\pi)$

26. $\sin^{-1}(\sin \pi)$

27. $\tan\left(\sin^{-1}\left(-\frac{1}{2}\right)\right)$

28. $\sec\left(\tan^{-1}\left(-\frac{1}{\sqrt{3}}\right)\right)$

29. $\cos(2 \sin^{-1}(-1))$

30. $\sin\left(2 \cos^{-1}\frac{1}{2}\right)$

In Problems 31–36 evaluate the given expression by means of an appropriate trigonometric identity.

31. $\sin\left(2 \sin^{-1}\frac{1}{3}\right)$

32. $\cos\left(2 \cos^{-1}\frac{3}{4}\right)$

33. $\sin\left(\arcsin \frac{1}{\sqrt{3}} + \arccos \frac{2}{3}\right)$

34. $\sin(\arctan 2 - \arctan 1)$

35. $\cos(\tan^{-1}4 - \tan^{-1}3)$

36. $\cos\left(\sin^{-1}\frac{1}{2} + \cos^{-1}\frac{1}{2}\right)$

In Problems 37–42 write the given expression as an algebraic quantity in x.

37. $\sin(\arccos x)$

38. $\cos\left(\tan^{-1}\frac{x}{2}\right)$

39. $\tan(\sec^{-1}x)$

40. $\sec(\cot^{-1}x)$

41. $\sin(2 \tan^{-1}x)$

42. $\tan(\arccos x)$

In Problems 43 and 44 verify the identities.

43. $\sin^{-1}x + \cos^{-1}x = \frac{\pi}{2}$

44. $\cos^{-1}x = \frac{\pi}{2} - \tan^{-1}\frac{x}{\sqrt{1 - x^2}}$

In Problems 45 and 46 solve for x.

45. $2 \arcsin(2x - 5) = \pi$

46. $3x + \frac{\pi}{3} = \cos^{-1}\left(-\frac{\sqrt{3}}{2}\right)$

47. If $t = \sin^{-1}(-2/\sqrt{5})$, find $\cos t$, $\tan t$, $\cot t$, $\sec t$, and $\csc t$.

48. If $\theta = \arctan(0.6)$, find $\sin \theta$, $\cos \theta$, $\cot \theta$, $\sec \theta$, and $\csc \theta$.

Calculator/Computer Problems

49. Use a calculator to verify:

(a) $\tan(\tan^{-1}1.3) = 1.3$ and $\tan^{-1}(\tan 1.3) = 1.3$

(b) $\tan(\tan^{-1}5) = 5$ and $\tan^{-1}(\tan 5) = -1.2832$
Explain why $\tan^{-1}(\tan 5) \neq 5$.

50. Let $x = 1.7$ radians. Compare, if possible, the values of $\sin^{-1}(\sin x)$ and $\sin(\sin^{-1}x)$. Explain any differences.

In Problems 51 and 52 use a calculator or computer to obtain the graph of the given function. What is $\lim_{x\to\infty} f(x)$ in each case?

51. $f(x) = \tan^{-1}x^2$ 52. $f(x) = (\tan^{-1}x)^2$

In Problems 53 and 54 use a calculator or computer to obtain the graph of the given function where x is any real number. Explain why the graphs are not simply the graphs of $f(x) = x$, $[-\pi/2, \pi/2]$, and $f(x) = x$, $[0, \pi]$, respectively.

53. $f(x) = \sin^{-1}(\sin x)$ 54. $f(x) = \cos^{-1}(\cos x)$

55. (a) For a car moving at velocity v, the banking angle of a curve at which there is no side thrust on the wheels is given by

$$\phi = \tan^{-1}\frac{v^2}{Rg}$$

where g is the acceleration of gravity and R is the radius of the curve. See Figure 7.8. At what angle should a road be banked for a car moving at 55 mph around a curve of radius 700 ft? [*Hint:* Use consistent units.]

(b) If f is the coefficient of friction between the car and the road and $\theta = \tan^{-1}f$ is defined by (6), then the maximum velocity that a car can travel around the curve without slipping is $v_m^2 = gR\tan(\theta + \phi)$, where ϕ is a given banking angle. If $f = 0.25$, find v_m for the road banked at the angle determined in part (**a**).

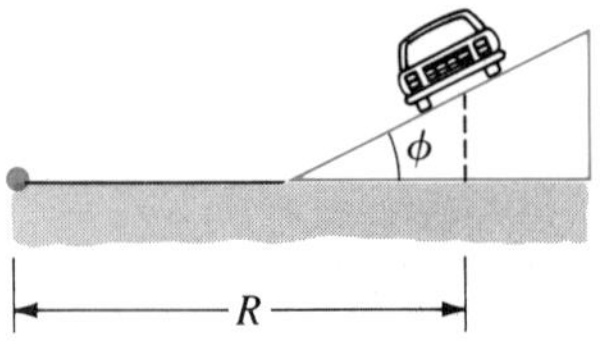

FIGURE 7.8

56. Consider a ladder of length L leaning against a house with a load at point P shown in Figure 7.9. The angle β at which the ladder is at the verge of slipping is defined by

$$\frac{x}{L} = \frac{f}{1+f^2}(f + \tan\beta)$$

where f is the coefficient of friction between the ladder and the ground.

(a) Find β when $f = 1$ and the load is at the top of the ladder.

(b) Find β when $f = 0.5$ and the load is $\frac{3}{4}$ of the way up the ladder.

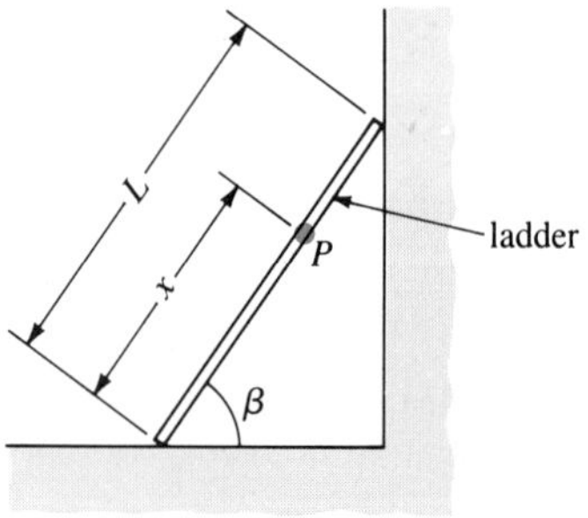

FIGURE 7.9

57. An airplane flies west at a constant speed v_1 and a wind blows from the north at a constant speed v_2. The plane's course south of west is given by $\theta = \tan^{-1}(v_2/v_1)$. See Figure 7.10. Find the course of a plane flying west at 300 km/h if a wind from the north blows at 60 km/h.

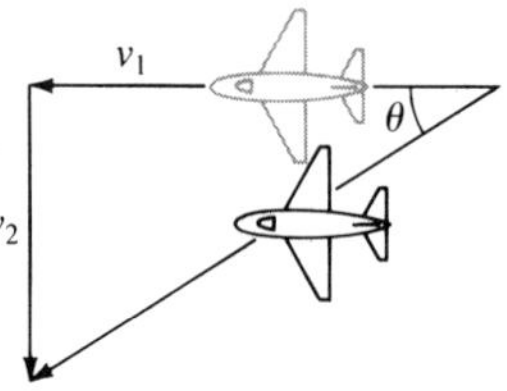

FIGURE 7.10

58. (a) Consider the graph of $f(x) = \cos x$ and the sequence of x-coordinates defined by $x_{n+1} = x_n + \pi$, where $n = 0, 1, 2, \ldots$ and $0 < x_0 < \pi/2$. If the slope of the tangent line to the graph of f at x_0 is $-m$, show that the slopes at $x_1, x_2, x_3, \ldots$ are, respectively, $m, -m, m, \ldots$. See Figure 7.11.

(b) Use part (**a**) to find a value of x_0 so that Newton's Method fails to converge to a positive root of $f(x) = 0$. Do not use a calculator to evaluate x_0.

(c) Now use a calculator to evaluate x_0 obtained in part (**b**) to nine decimal places. With x_0 as the initial value, explain why Newton's Method actually converges to a positive root of $f(x) = 0$.

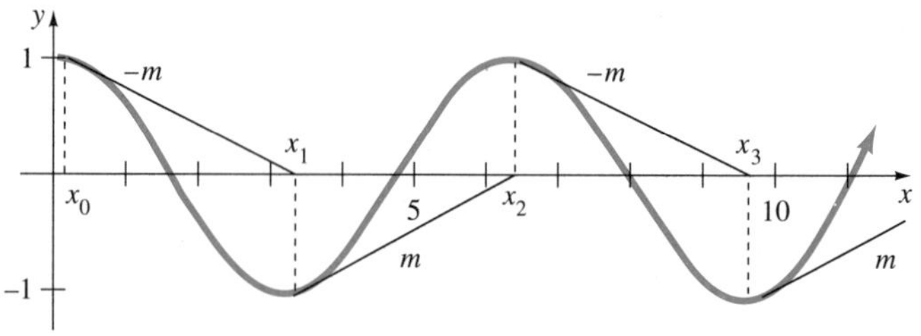

FIGURE 7.11

7.2 DERIVATIVES AND INTEGRALS THAT INVOLVE INVERSE TRIGONOMETRIC FUNCTIONS

Derivatives The derivative of an inverse trigonometric function can be obtained, with minimum effort, through the use of implicit differentiation. A review of Figures 7.3(b) and 7.4(a) reveals that the inverse tangent and inverse cotangent are differentiable for all x. However, the remaining four inverse trigonometric functions are not differentiable at either $x = -1$ or $x = 1$. We shall confine our attention to the derivations of the derivative formulas for the inverse sine, inverse tangent, and inverse secant and leave the others as exercises.

Inverse Sine For $-1 < x < 1$ and $-\pi/2 < y < \pi/2$,

$$y = \sin^{-1}x \quad \text{if and only if} \quad x = \sin y$$

Therefore, implicit differentiation

$$\frac{d}{dx}x = \frac{d}{dx}\sin y \quad \text{gives} \quad 1 = \cos y\frac{dy}{dx}$$

and so

$$\frac{dy}{dx} = \frac{1}{\cos y} \tag{1}$$

For the given restriction on the variable, y, $\cos y > 0$ and $\cos y = \sqrt{1 - \sin^2 y} = \sqrt{1 - x^2}$. By substituting this quantity in (1), we have shown that

$$\frac{d}{dx}\sin^{-1}x = \frac{1}{\sqrt{1 - x^2}} \tag{2}$$

As predicted, note that (2) is not defined at $x = -1$ and $x = 1$.

Inverse Tangent For $-\infty < x < \infty$ and $-\pi/2 < y < \pi/2$,

$$y = \tan^{-1}x \quad \text{if and only if} \quad x = \tan y$$

Thus,

$$\frac{d}{dx}x = \frac{d}{dx}\tan y$$

$$1 = \sec^2 y\frac{dy}{dx}$$

$$\frac{dy}{dx} = \frac{1}{\sec^2 y} \tag{3}$$

Since $\sec^2 y = 1 + \tan^2 y = 1 + x^2$, (3) becomes

$$\frac{d}{dx}\tan^{-1}x = \frac{1}{1 + x^2} \tag{4}$$

Inverse Secant For $|x| > 1$ and $0 < y < \pi/2$ or $\pi < y < 3\pi/2$,

$$y = \sec^{-1}x \quad \text{if and only if} \quad x = \sec y$$

Differentiating the last equation implicitly gives

$$\frac{dy}{dx} = \frac{1}{\sec y \tan y} \tag{5}$$

In view of the restrictions on y, we have

$$\tan y = \sqrt{\sec^2 y - 1} = \sqrt{x^2 - 1}, \qquad |x| > 1$$

Hence, (5) becomes

$$\frac{d}{dx}\sec^{-1}x = \frac{1}{x\sqrt{x^2 - 1}} \tag{6}$$

Of course, (6) is consistent with Figure 7.4(b); the slope of the tangent line to the graph is negative for $x < -1$ and positive for $x > 1$.

The derivative of the composition of an inverse trigonometric function with a differentiable function $u = g(x)$ is obtained from the Chain Rule.

Derivatives of the Inverse Trigonometric Functions

I $\quad \dfrac{d}{dx}\sin^{-1}u = \dfrac{1}{\sqrt{1-u^2}}\dfrac{du}{dx}$ $\qquad$ II $\quad \dfrac{d}{dx}\cos^{-1}u = \dfrac{-1}{\sqrt{1-u^2}}\dfrac{du}{dx}$

III $\quad \dfrac{d}{dx}\tan^{-1}u = \dfrac{1}{1+u^2}\dfrac{du}{dx}$ $\qquad$ IV $\quad \dfrac{d}{dx}\cot^{-1}u = \dfrac{-1}{1+u^2}\dfrac{du}{dx}$

V $\quad \dfrac{d}{dx}\sec^{-1}u = \dfrac{1}{u\sqrt{u^2-1}}\dfrac{du}{dx}$ $\qquad$ VI $\quad \dfrac{d}{dx}\csc^{-1}u = \dfrac{-1}{u\sqrt{u^2-1}}\dfrac{du}{dx}$

EXAMPLE 1 Differentiate $y = \sin^{-1}5x$.

Solution From I,

$$\frac{dy}{dx} = \frac{1}{\sqrt{1-(5x)^2}}\frac{d}{dx}5x = \frac{5}{\sqrt{1-25x^2}}$$ □

EXAMPLE 2 Differentiate $y = \tan^{-1}\sqrt{2x+1}$.

Solution From III,

$$\frac{dy}{dx} = \frac{1}{1 + (\sqrt{2x+1})^2} \frac{d}{dx}(2x+1)^{1/2}$$

$$= \frac{1}{1 + (2x+1)} \cdot \frac{1}{2}(2x+1)^{-1/2} \cdot 2$$

$$= \frac{1}{(2x+2)\sqrt{2x+1}}$$ □

EXAMPLE 3 Differentiate $y = \sec^{-1}x^2$.

Solution For $x^2 > 1$, we have from V,

$$\frac{dy}{dx} = \frac{1}{x^2\sqrt{(x^2)^2 - 1}} \frac{d}{dx} x^2$$

$$= \frac{2x}{x^2\sqrt{x^4 - 1}} = \frac{2}{x\sqrt{x^4 - 1}}$$ □

Integrals From the differentiation formulas I–VI, we obtain equivalent integration formulas. Note, for example, that we can write

$$\int \frac{du}{\sqrt{1-u^2}} = \sin^{-1}u + C \quad \text{or} \quad \int \frac{du}{\sqrt{1-u^2}} = -\cos^{-1}u + C$$

As a consequence we need only summarize the integral analogues of I, III, and V.

Indefinite Integrals Yielding Inverse Trigonometric Functions

I′ $$\int \frac{du}{\sqrt{1-u^2}} = \sin^{-1}u + C$$

III′ $$\int \frac{du}{1+u^2} = \tan^{-1}u + C$$

V′ $$\int \frac{du}{u\sqrt{u^2-1}} = \sec^{-1}u + C$$

It is understood that the variable u is appropriately restricted in I′ and V′. Bear in mind that in I′ we have $|u| < 1$ and in V′, $|u| > 1$.

EXAMPLE 4 Evaluate $\int \frac{dx}{\sqrt{100 - x^2}}$.

Solution By factoring 100 from the radical and identifying

$$u = \frac{x}{10} \quad \text{and} \quad du = \frac{dx}{10}$$

the result is obtained from I′:

$$\int \frac{dx}{\sqrt{100 - x^2}} = \int \frac{\frac{dx}{10}}{\sqrt{1 - \left(\frac{x}{10}\right)^2}}$$
$$= \int \frac{du}{\sqrt{1 - u^2}}$$
$$= \sin^{-1} u + C$$
$$= \sin^{-1} \frac{x}{10} + C$$

□

EXAMPLE 5 Find the area under the graph of $f(x) = \frac{1}{1 + x^2}$ on the interval $[-1, 1]$.

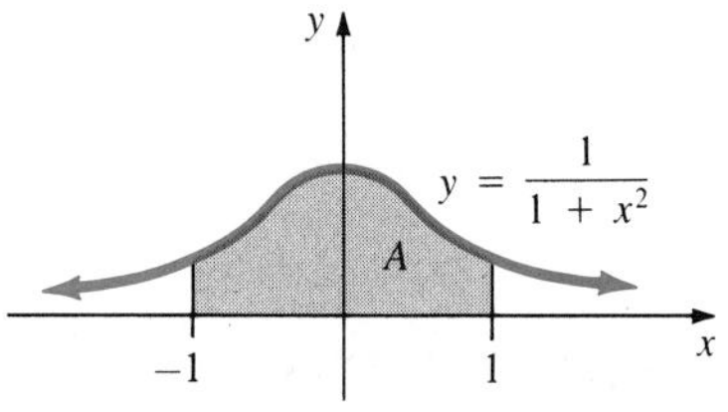

FIGURE 7.12

Solution From Figure 7.12 we see f is a nonnegative function, and so

$$A = \int_{-1}^{1} \frac{dx}{1 + x^2} = \tan^{-1} x \Big]_{-1}^{1}$$
$$= \tan^{-1} 1 - \tan^{-1}(-1)$$
$$= \frac{\pi}{4} - \left(-\frac{\pi}{4}\right) = \frac{\pi}{2} \text{ square units}$$

□

In Example 5 note again the importance of using radian measure when working with inverse trigonometric functions. Had we used a calculator set in degree mode, we would have obtained $\tan^{-1} 1 = 45°$ and $\tan^{-1}(-1) = -45°$. The required area would then have been 90°, which, of course, is a meaningless answer. Bear in mind that $\pi/2 = 1.57079\ldots$ so that the area is approximately 1.57 square units.

EXAMPLE 6 Evaluate $\int \frac{(\tan^{-1} x)^2}{1 + x^2} \, dx$.

Solution It should be apparent that the integral does not fall into any of the three forms I′, III′, or V′. But if we identify

$$u = \tan^{-1}x \quad \text{and} \quad du = \frac{dx}{1+x^2}$$

then

$$\int \frac{(\tan^{-1}x)^2}{1+x^2}\,dx = \int u^2\,du$$

$$= \frac{u^3}{3} + C$$

$$= \frac{(\tan^{-1}x)^3}{3} + C \qquad \square$$

Use of Long Division For the problem of evaluating $\int f(x)\,dx$, where $f(x) = P(x)/Q(x)$ is a rational function, a working rule is:

If the degree of $P(x)$ is greater than or equal to the degree of $Q(x)$, use long division before integration; that is, write

$$\frac{P(x)}{Q(x)} = a \text{ polynomial} + \frac{R(x)}{Q(x)}$$

where the degree of $R(x)$ is less than the degree of $Q(x)$.

EXAMPLE 7 Evaluate $\displaystyle\int \frac{x^2}{1+x^2}\,dx$.

Solution The integrand calls for long division:

$$\frac{x^2}{1+x^2} = 1 - \frac{1}{1+x^2}$$

In view of this result, we have immediately

$$\int \frac{x^2}{1+x^2}\,dx = x - \tan^{-1}x + C \qquad \square$$

For convenience, we will extend I′, III′, and V′ in the following manner. For $a > 0$,

$$\int \frac{du}{\sqrt{a^2 - u^2}} = \sin^{-1}\frac{u}{a} + C \tag{7}$$

$$\int \frac{du}{a^2 + u^2} = \frac{1}{a}\tan^{-1}\frac{u}{a} + C \tag{8}$$

$$\int \frac{du}{u\sqrt{u^2 - a^2}} = \frac{1}{a}\sec^{-1}\frac{u}{a} + C \tag{9}$$

Notice that the result of Example 4 is obtained by identifying $a = 10$ and $u = x$ in (7).

EXAMPLE 8 Evaluate $\displaystyle\int \frac{dx}{x\sqrt{x^4 - 16}}$.

Solution On first inspection, the integral does not appear to be any one of the forms (7)–(9). But, by multiplying the numerator and the denominator of the integrand by x, we recognize $u = x^2$, $du = 2x\,dx$, and $a = 4$. Hence, from (9),

$$\begin{aligned}\int \frac{dx}{x\sqrt{x^4 - 16}} &= \frac{1}{2}\int \frac{2x\,dx}{x^2\sqrt{x^4 - 16}}\\ &= \frac{1}{2}\int \frac{du}{u\sqrt{u^2 - 4^2}}\\ &= \frac{1}{2}\cdot\frac{1}{4}\sec^{-1}\frac{u}{4} + C\\ &= \frac{1}{8}\sec^{-1}\frac{x^2}{4} + C\end{aligned}$$

□

Completing the Square To utilize the integration formulas (7)–(9) it may be necessary to **complete the square**. Recall from algebra that to complete the square for a quadratic expression $x^2 + bx + c$, $b \neq 0$, we add and then subtract the square of one-half the coefficient of x:

$$\begin{aligned}x^2 + bx + c &= \left(x^2 + bx + \left(\frac{b}{2}\right)^2\right) + c - \left(\frac{b}{2}\right)^2\\ &= \left(x^2 + bx + \frac{b^2}{4}\right) + c - \frac{b^2}{4}\\ &= \left(x + \frac{b}{2}\right)^2 + c - \frac{b^2}{4}\end{aligned}$$

$$\boxed{\left(\frac{b}{2}\right)^2 - \left(\frac{b}{2}\right)^2 = 0}$$

EXAMPLE 9 Evaluate $\displaystyle\int \frac{dx}{x^2 + 12x + 37}$.

Solution First complete the square:

$$\begin{aligned}x^2 + 12x + 37 &= (x^2 + 12x \qquad) + 37\\ &= (x^2 + 12x + 36) + 37 - 36\\ &= (x^2 + 12x + 36) + 1\\ &= (x + 6)^2 + 1\end{aligned}$$

It follows from (8) that

$$\int \frac{dx}{x^2 + 12x + 37} = \int \frac{dx}{(x+6)^2 + 1} \qquad \boxed{\begin{array}{l} u = x + 6 \\ du = dx \end{array}}$$
$$= \int \frac{du}{u^2 + 1}$$
$$= \tan^{-1} u + C$$
$$= \tan^{-1}(x+6) + C \qquad \square$$

To complete the square for a quadratic expression $ax^2 + bx + c$, $a \neq 1$, $b \neq 0$, we begin by factoring the lead coefficient a from the x terms:

$$ax^2 + bx + c = a\left(x^2 + \frac{b}{a}x \qquad \right) + c$$

Inside the parentheses we add the square of one-half the coefficient of x. But since we are really adding $a(b/2a)^2 = b^2/4a$ to the expression, we must then subtract this same term to maintain the equality:

$$ax^2 + bx + c = a\left(x^2 + \frac{b}{a}x + \frac{b^2}{4a^2}\right) + c - \frac{b^2}{4a} \qquad \boxed{\frac{b^2}{4a} - \frac{b^2}{4a} = 0}$$
$$= a\left(x + \frac{b}{2a}\right)^2 + c - \frac{b^2}{4a}$$

EXAMPLE 10 Evaluate $\displaystyle\int \frac{dx}{\sqrt{4 - 6x - 9x^2}}$.

Solution Complete the square:

$$4 - 6x - 9x^2 = 4 - (9x^2 + 6x \qquad)$$
$$= 4 - 9\left(x^2 + \frac{2}{3}x + \frac{1}{9}\right) + \frac{9}{9}$$
$$= 5 - 9\left(x + \frac{1}{3}\right)^2 = 5 - (3x+1)^2$$

From (7) we obtain

$$\int \frac{dx}{\sqrt{4 - 6x - 9x^2}} = \frac{1}{3}\int \frac{3\,dx}{\sqrt{5 - (3x+1)^2}} \qquad \boxed{\begin{array}{l} u = 3x + 1 \\ du = 3\,dx \end{array}}$$
$$= \frac{1}{3}\int \frac{du}{\sqrt{(\sqrt{5})^2 - u^2}}$$
$$= \frac{1}{3}\sin^{-1}\frac{u}{\sqrt{5}} + C$$
$$= \frac{1}{3}\sin^{-1}\frac{3x+1}{\sqrt{5}} + C \qquad \square$$

EXERCISES 7.2 *Answers to odd-numbered problems begin on page A-87.*

In Problems 1–20 find the derivative of the given function.

1. $y = \sin^{-1}(5x - 1)$
2. $y = \cos^{-1}\dfrac{x+1}{3}$
3. $y = 4\cot^{-1}\dfrac{x}{2}$
4. $y = 2x - 10\sec^{-1}5x$
5. $y = 2\sqrt{x}\tan^{-1}\sqrt{x}$
6. $y = (\tan^{-1}x)(\cot^{-1}x)$
7. $y = \dfrac{\sin^{-1}2x}{\cos^{-1}2x}$
8. $y = \dfrac{\sin^{-1}x}{\sin x}$
9. $y = \dfrac{1}{\tan^{-1}x^2}$
10. $y = \dfrac{\sec^{-1}x}{x}$
11. $y = 2\sin^{-1}x + x\cos^{-1}x$
12. $y = \cot^{-1}x - \tan^{-1}\dfrac{x}{\sqrt{1-x^2}}$
13. $y = \left(x^2 - 9\tan^{-1}\dfrac{x}{3}\right)^3$
14. $y = \sqrt{x - \cos^{-1}(x+1)}$
15. $F(t) = \arctan\left(\dfrac{t-1}{t+1}\right)$
16. $g(t) = \arccos\sqrt{3t+1}$
17. $f(x) = \arcsin(\cos 4x)$
18. $f(x) = \tan^{-1}\left(\dfrac{\sin x}{2}\right)$
19. $f(x) = \tan(\sin^{-1}x^2)$
20. $f(x) = \cos(x\sin^{-1}x)$

In Problems 21 and 22 use implicit differentiation to find dy/dx.

21. $\tan^{-1}y = x^2 + y^2$
22. $\sin^{-1}y - \cos^{-1}x = 1$

23. Show that if $f(x) = \sin^{-1}x + \cos^{-1}x$, then $f'(x) = 0$. Interpret the result.

24. Repeat Problem 23 for $f(x) = \tan^{-1}x + \tan^{-1}(1/x)$.

In Problems 25–48 evaluate the given integral.

25. $\displaystyle\int \frac{dx}{1+25x^2}$
26. $\displaystyle\int \frac{dx}{\sqrt{9-x^2}}$
27. $\displaystyle\int_0^{1/4}(1-4x^2)^{-1/2}\,dx$
28. $\displaystyle\int \frac{dx}{x\sqrt{x^6-4}}$
29. $\displaystyle\int \frac{2x-3}{\sqrt{1-x^2}}\,dx$
30. $\displaystyle\int_{-1}^{1}\frac{dt}{4+(t-1)^2}$
31. $\displaystyle\int \frac{1-x^2}{1+x^2}\,dx$
32. $\displaystyle\int \frac{x^4 - x^3 + 3x^2 - 3x + 4}{x^2+3}\,dx$
33. $\displaystyle\int \frac{4t}{\sqrt{2-3t^2}}\,dt$
34. $\displaystyle\int \frac{\theta}{\sqrt{1-\theta^4}}\,d\theta$
35. $\displaystyle\int \frac{dx}{5+2x^2}$
36. $\displaystyle\int \frac{(3x^2-7)^{-1/2}}{x}\,dx$
37. $\displaystyle\int_0^{\sqrt{2}/2}\sqrt{\frac{\arcsin x}{1-x^2}}\,dx$
38. $\displaystyle\int_0^{\sqrt{3}}\frac{\tan^{-1}x}{1+x^2}\,dx$
39. $\displaystyle\int \frac{dx}{(x+1)\sqrt{x^2+2x}}$
40. $\displaystyle\int \frac{dx}{16x^2-8x+26}$
41. $\displaystyle\int \frac{x^3+4x^2+8x+1}{x^2+4x+8}\,dx$
42. $\displaystyle\int \frac{dx}{\sqrt{1-6x-x^2}}$
43. $\displaystyle\int \frac{t}{4t^4+4t^2+2}\,dt$
44. $\displaystyle\int \frac{2t+1}{\sqrt{9-(2t+1)^2}}\,dt$
45. $\displaystyle\int_{\pi/12}^{\pi/6}\frac{\cos 2x}{1+4\sin^2 2x}\,dx$
46. $\displaystyle\int \frac{\sin x}{\sqrt{49-\cos^2 x}}\,dx$
47. $\displaystyle\int \frac{e^x}{1+e^{2x}}\,dx$
48. $\displaystyle\int_1^2 \frac{dx}{\sqrt{e^{2x}-1}}$

In Problems 49 and 50 find the slope of the tangent line to the graph of the given function at the indicated value of x.

49. $y = \sin^{-1}\dfrac{x}{2}$; $x = 1$

50. $y = (\cos^{-1}x)^2$; $x = -1/\sqrt{2}$

In Problems 51 and 52 find an equation of the tangent line to the graph of the given function at the indicated value of x.

51. $f(x) = x\tan^{-1}x$; $x = 1$

52. $f(x) = \sin^{-1}(x-1)$; $x = \frac{1}{2}$

53. Find the area under the graph of

$$y = (24 + 2x - x^2)^{-1/2}$$

on the interval [2, 4].

54. The region bounded by the graphs of $y = (4+x^2)^{-1/2}$, $x = 1$, $x = 3$, and $y = 0$ is revolved about the x-axis. Find the volume of the solid of revolution.

55. The region bounded by the graphs of

$$y = \frac{1}{x^2\sqrt{x^4-16}}, \quad x = \frac{5}{2}, \quad x = 4, \quad \text{and} \quad y = 0$$

is revolved around the y-axis. Find the volume of the solid of revolution.

56. Find the length of the graph of $y = \sqrt{2 - x^2}$ on $[0, 1]$.

57. A boat is being pulled toward a dock by means of a winch. The winch is located at the end of the dock and is 10 ft above the level at which the tow rope is attached to the bow of the boat. The rope is pulled in at a constant rate of 1 ft/s. Use an inverse trigonometric function to determine the rate at which the angle of elevation between the bow of the boat and the end of the dock is changing when 30 ft of tow rope is out.

58. A searchlight on a patrol boat that is situated $\frac{1}{2}$ km offshore follows a dune buggy that moves parallel to the water along a straight beach. The dune buggy is traveling at a constant rate of 15 km/h. Use an inverse trigonometric function to determine the rate at which the searchlight is rotating when the dune buggy is $\frac{1}{2}$ km from the point on the shore nearest the boat.

59. A stained glass window measuring 20 ft high is 10 ft above the eye level of an observer. Use inverse trigonometric functions to determine how far the observer should stand from the wall so that the viewing angle at eye level between the bottom and the top of the window is a maximum.

60. Consider a large circular hoop made of a translucent material shown in Figure 7.13. A large bead moves on a thin wire which coincides with a diameter DOB that is parallel to the ground. The bead moves to the right away from point O at a rate of $\frac{1}{2}$ ft/min. A spotlight located at point A is trained on the bead as it moves, causing a shadow S_1 to appear on the hoop and a fainter shadow S_2 to appear on the ground.

(a) Let s be the distance measured along the hoop from point C to S_1. Find a relationship between s and the indicated angle θ.

(b) Express s as a function of x.

(c) Find the rate at which the shadow S_1 moves along the circular hoop away from point C when $x = 5$ ft.

(d) Find the rate at which the shadow S_2 moves along the ground away from point C.

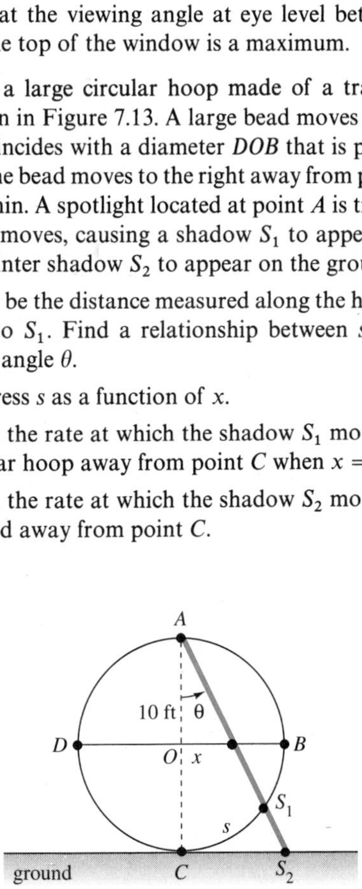

FIGURE 7.13

61. A diver jumps from a high platform with an initial downward velocity of 1 ft/s toward the center of a large circular tank of water. See Figure 7.14.

(a) Find the rate at which the angle θ subtended by the circular tank at the diver's eye is increasing at $t = 3$ s into the dive.

(b) What is the value of θ when the diver hits the water?

(c) What is the rate of change of θ when the diver hits the water?

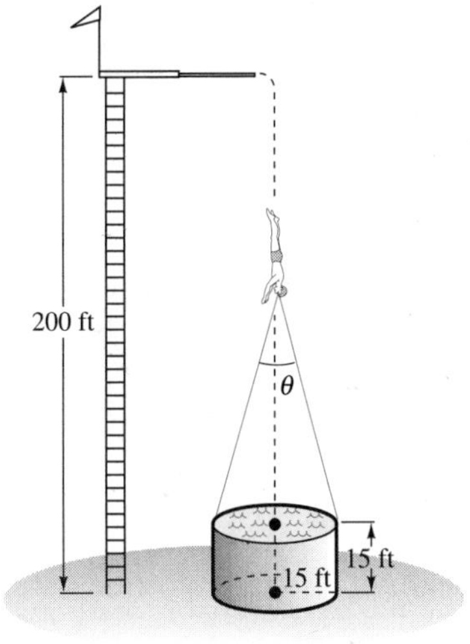

FIGURE 7.14

62. The energy of the spring–mass system shown in Figure 7.15 is the sum of potential and kinetic energies:

$$E = \frac{1}{2}ky^2 + \frac{1}{2}m\left(\frac{dy}{dt}\right)^2$$

where k is the spring constant, m is mass, and y is the displacement of the mass from an equilibrium position. If E is constant, use separation of variables to show that

$$y = \sqrt{\frac{2E}{k}}\sin\left(\pm\sqrt{\frac{k}{m}}\,t + C\sqrt{\frac{k}{2E}}\right)$$

where C is an arbitrary constant.

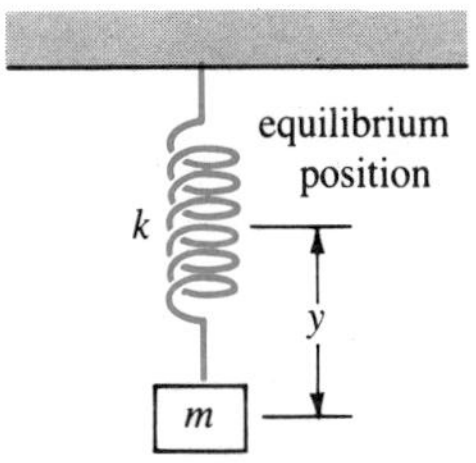

FIGURE 7.15

Calculator/Computer Problems

63. Use Newton's Method to find an approximation to the positive root of $\tan^{-1}x = x/4$.

64. Use Newton's Method to approximate the x-coordinate of the point of intersection of the graphs of $y = \tan^{-1}x$ and $y = \cos^{-1}x$.

65. Use Simpson's Rule with $n = 4$ to find an approximation to $\int_0^1 \tan^{-1}x\, dx$.

66. Consider the region bounded by the graphs of

$$y = \frac{18}{9 + x^2}, \quad x = 0, \quad x = 3, \quad \text{and} \quad y = 0$$

Approximate, to one decimal place, the coordinates of the centroid of the region. [*Hint:* At this time, we do not know the value of $\int_0^3 dx/(9 + x^2)^2$. Use Simpson's Rule with $n = 4$.]

In Problems 67 and 68 use a calculator or a computer to obtain the graph of the given function. Use differentiation to find the relative extrema that you discover from the graph.

67. $f(x) = -2x + 3\tan^{-1}x$

68. $f(x) = 2x - \sin^{-1}x$

Miscellaneous Problems

69. For $|x| < 1$ and $0 < \cos^{-1}x < \pi$, prove that

$$\frac{d}{dx}\cos^{-1}x = \frac{-1}{\sqrt{1 - x^2}}$$

70. For $-\infty < x < \infty$ and $0 < \cot^{-1}x < \pi$, prove that

$$\frac{d}{dx}\cot^{-1}x = \frac{-1}{1 + x^2}$$

71. For $|x| > 1$ and $0 < \csc^{-1}x < \pi/2$ or $\pi < \csc^{-1}x < 3\pi/2$, prove that

$$\frac{d}{dx}\csc^{-1}x = \frac{-1}{x\sqrt{x^2 - 1}}$$

7.3 INVERSE HYPERBOLIC FUNCTIONS

For any real number y in the range of the hyperbolic sine, there corresponds only one real number x in its domain. In other words, $y = \sinh x$ is a one-to-one function and, hence, has an inverse function that is written $y = \sinh^{-1}x$. As in our earlier discussion of the inverse trigonometric functions in Section 7.1, this later notation is equivalent to $x = \sinh y$.

The graphs of $y = \sinh x$ and $y = \sinh^{-1}x$ are compared in Figure 7.16. From Figure 6.27(a) it is apparent that the hyperbolic cosine is not a one-to-one function and so does not possess an inverse function unless its domain is restricted. Inspection of Figure 7.17(a) shows that when the domain of $y = \cosh x$ is restricted to the interval $0 \le x < \infty$, we have correspondingly $1 \le y < \infty$. The inverse function $y = \cosh^{-1}x$ is then defined for $1 \le x < \infty$ $(0 \le y < \infty)$. See Figure 7.17(b).

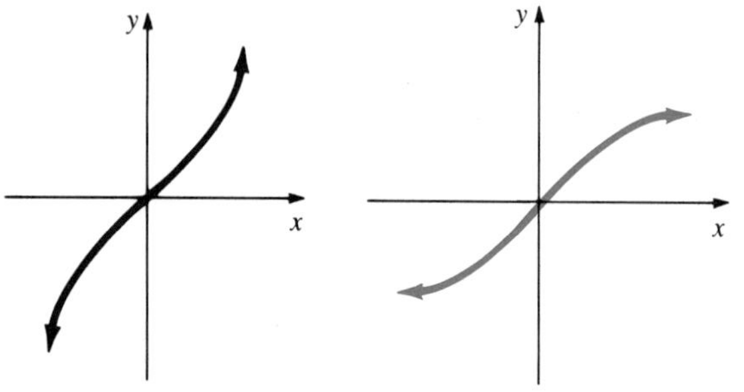

(a) $y = \sinh x$ (b) $y = \sinh^{-1} x$

FIGURE 7.16

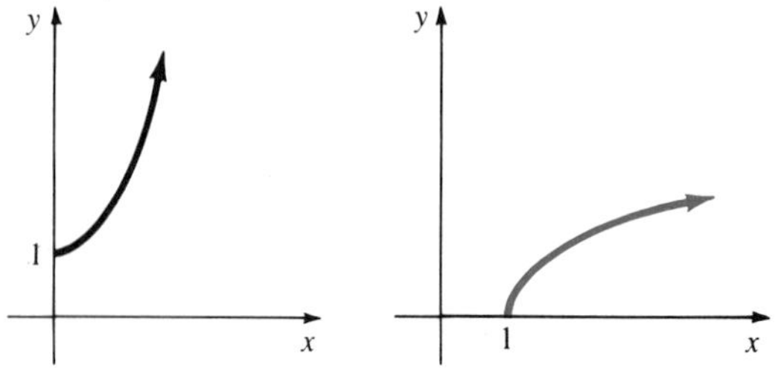

(a) $y = \cosh x,\ x \ge 0$ (b) $y = \cosh^{-1} x,\ x \ge 1$

FIGURE 7.17

A reexamination of Figure 6.28 shows that $y = \tanh x$, $y = \coth x$, and $y = \operatorname{csch} x$ are one-to-one functions. When the domain of $y = \operatorname{sech} x$ is restricted to $0 \leq x < \infty$ $(0 < y \leq 1)$, its inverse function $y = \operatorname{sech}^{-1} x$ is defined for $0 < x \leq 1$ $(0 \leq y < \infty)$.

Definitions of the six inverse hyperbolic functions are summarized as follows:

DEFINITION 7.5 Inverse Hyperbolic Functions

(*i*)	$y = \sinh^{-1} x$	if and only if	$x = \sinh y$,	$-\infty < x < \infty$, $-\infty < y < \infty$
(*ii*)	$y = \cosh^{-1} x$	if and only if	$x = \cosh y$,	$x \geq 1$, $y \geq 0$
(*iii*)	$y = \tanh^{-1} x$	if and only if	$x = \tanh y$,	$\lvert x \rvert < 1$, $-\infty < y < \infty$
(*iv*)	$y = \coth^{-1} x$	if and only if	$x = \coth y$,	$\lvert x \rvert > 1$, $y \neq 0$
(*v*)	$y = \operatorname{sech}^{-1} x$	if and only if	$x = \operatorname{sech} y$,	$0 < x \leq 1$, $y \geq 0$
(*vi*)	$y = \operatorname{csch}^{-1} x$	if and only if	$x = \operatorname{csch} y$,	$x \neq 0$, $y \neq 0$

The graphs of the last four functions in Definition 7.5 are given in Figure 7.18.

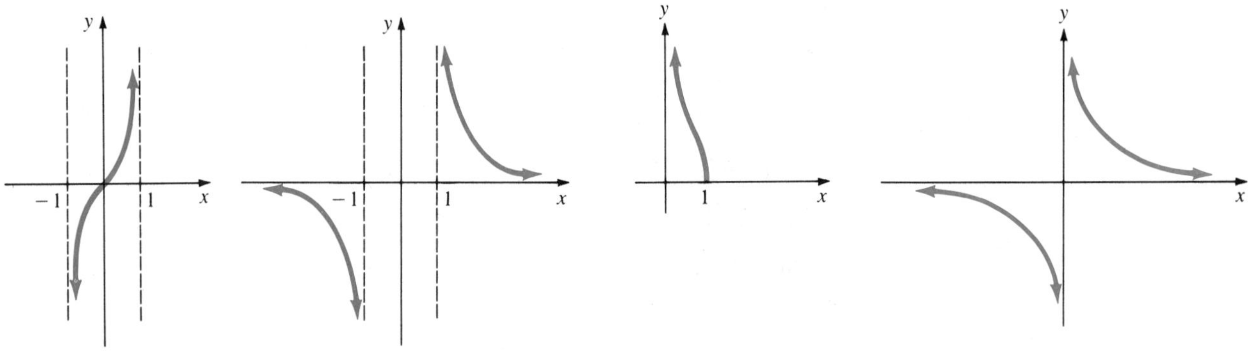

(a) $y = \tanh^{-1} x$, $\lvert x \rvert < 1$ (b) $y = \coth^{-1} x$, $\lvert x \rvert > 1$ (c) $y = \operatorname{sech}^{-1} x$, $0 < x \leq 1$ (d) $y = \operatorname{csch}^{-1} x$, $x \neq 0$

FIGURE 7.18

Inverse Hyperbolic Functions As Logarithms Since all the hyperbolic functions are defined in terms of combinations of e^x, it should not come as any surprise to find that the inverse hyperbolic functions can be expressed in terms of the natural logarithm. For example, $y = \sinh^{-1} x$ is equivalent to $x = \sinh y$, so that

$$x = \frac{e^y - e^{-y}}{2}$$

$$2x = \frac{e^{2y} - 1}{e^y} \quad \text{or} \quad e^{2y} - 2xe^y - 1 = 0$$

Since this last equation is quadratic in e^y, the quadratic formula gives

$$e^y = \frac{2x \pm \sqrt{4x^2 + 4}}{2} = x \pm \sqrt{x^2 + 1}$$

Now the solution corresponding to the minus sign must be rejected since $e^y > 0$ but $x - \sqrt{x^2 + 1} < 0$. Thus, we have

$$e^y = x + \sqrt{x^2 + 1} \quad \text{or} \quad y = \sinh^{-1}x = \ln(x + \sqrt{x^2 + 1})$$

Similarly, for $y = \tanh^{-1}x$, $|x| < 1$,

$$x = \tanh y = \frac{e^y - e^{-y}}{e^y + e^{-y}}$$

gives

$$e^y(1 - x) = (1 + x)e^{-y}$$

$$e^{2y} = \frac{1 + x}{1 - x}$$

$$2y = \ln\left(\frac{1 + x}{1 - x}\right)$$

or

$$y = \tanh^{-1}x = \frac{1}{2}\ln\left(\frac{1 + x}{1 - x}\right)$$

We have proved two of the results in the next theorem.

THEOREM 7.1 Logarithmic Identities

(*i*) $\sinh^{-1}x = \ln(x + \sqrt{x^2 + 1})$

(*ii*) $\cosh^{-1}x = \ln(x + \sqrt{x^2 - 1}), \quad x \geq 1$

(*iii*) $\tanh^{-1}x = \frac{1}{2}\ln\left(\frac{1 + x}{1 - x}\right), \quad |x| < 1$

(*iv*) $\coth^{-1}x = \frac{1}{2}\ln\left(\frac{x + 1}{x - 1}\right), \quad |x| > 1$

(*v*) $\operatorname{sech}^{-1}x = \ln\left(\frac{1}{x} + \frac{\sqrt{1 - x^2}}{x}\right), \quad 0 < x \leq 1$

(*vi*) $\operatorname{csch}^{-1}x = \ln\left(\frac{1}{x} + \frac{\sqrt{1 + x^2}}{|x|}\right), \quad x \neq 0$

The foregoing identities are a convenient means for obtaining the numerical values of an inverse hyperbolic function. For example, with the aid of a calculator we see that $\sinh^{-1}4 = \ln(4 + \sqrt{17}) \approx 2.0947$.

Derivatives To find the derivative of an inverse hyperbolic function, we can proceed in two different ways. For example, if

$$y = \sinh^{-1}x \quad \text{then} \quad x = \sinh y$$

Using implicit differentiation, we can write

$$\frac{d}{dx}(x) = \frac{d}{dx}(\sinh y)$$

$$1 = \cosh y \frac{dy}{dx}$$

$$\frac{dy}{dx} = \frac{1}{\cosh y}$$

$$= \frac{1}{\sqrt{\sinh^2 y + 1}} = \frac{1}{\sqrt{x^2 + 1}}$$

On the other hand, we know from Theorem 7.1(i) that

$$y = \ln(x + \sqrt{x^2 + 1})$$

Therefore, from the derivative of the logarithm, we obtain

$$\frac{dy}{dx} = \frac{1}{x + \sqrt{x^2 + 1}}\left(1 + \frac{1}{2}(x^2 + 1)^{-1/2} \cdot 2x\right)$$

$$= \frac{1}{x + \sqrt{x^2 + 1}} \frac{\sqrt{x^2 + 1} + x}{\sqrt{x^2 + 1}} = \frac{1}{\sqrt{x^2 + 1}}$$

We have proved a special case of part I in the next result. The proofs of the remaining parts are left as exercises. The function $u = g(x)$ is differentiable.

Derivatives of the Inverse Hyperbolic Functions

I $\quad \dfrac{d}{dx}\sinh^{-1}u = \dfrac{1}{\sqrt{u^2 + 1}}\dfrac{du}{dx}$

II $\quad \dfrac{d}{dx}\cosh^{-1}u = \dfrac{1}{\sqrt{u^2 - 1}}\dfrac{du}{dx}, \quad u > 1$

III $\quad \dfrac{d}{dx}\tanh^{-1}u = \dfrac{1}{1 - u^2}\dfrac{du}{dx}, \quad |u| < 1$

IV $\quad \dfrac{d}{dx}\coth^{-1}u = \dfrac{1}{1 - u^2}\dfrac{du}{dx}, \quad |u| > 1$

V $\quad \dfrac{d}{dx}\operatorname{sech}^{-1}u = \dfrac{-1}{u\sqrt{1 - u^2}}\dfrac{du}{dx}, \quad 0 < u < 1$

VI $\quad \dfrac{d}{dx}\operatorname{csch}^{-1}u = \dfrac{-1}{|u|\sqrt{1 + u^2}}\dfrac{du}{dx}, \quad u \neq 0$

EXAMPLE 1 Differentiate $y = \cosh^{-1}(x^2 + 5)$.

Solution From II, $$\frac{dy}{dx} = \frac{1}{\sqrt{(x^2+5)^2 - 1}} \cdot 2x = \frac{2x}{\sqrt{x^4 + 10x^2 + 24}}$$ □

EXAMPLE 2 Differentiate $y = \tanh^{-1}4x$.

Solution From III, $$\frac{dy}{dx} = \frac{1}{1 - (4x)^2} \cdot 4 = \frac{4}{1 - 16x^2}$$ □

EXAMPLE 3 Differentiate $y = e^{x^2}\operatorname{sech}^{-1}x$.

Solution By the Product Rule and V, we have

$$\begin{aligned}\frac{dy}{dx} &= e^{x^2}\left(\frac{-1}{x\sqrt{1-x^2}}\right) + 2xe^{x^2}\operatorname{sech}^{-1}x \\ &= -\frac{e^{x^2}}{x\sqrt{1-x^2}} + 2xe^{x^2}\operatorname{sech}^{-1}x\end{aligned}$$ □

Integrals As in Section 7.2, we limit ourselves to three integral formulas. From I–IV it is readily shown that when a is a positive constant:

Indefinite Integrals Yielding Inverse Hyperbolic Functions

I′ $$\int \frac{du}{\sqrt{u^2 + a^2}} = \sinh^{-1}\frac{u}{a} + C$$

II′ $$\int \frac{du}{\sqrt{u^2 - a^2}} = \cosh^{-1}\frac{u}{a} + C, \quad u > a > 0$$

III′ $$\int \frac{du}{a^2 - u^2} = \begin{cases} \dfrac{1}{a}\tanh^{-1}\dfrac{u}{a} + C, & |u| < a \\ \dfrac{1}{a}\coth^{-1}\dfrac{u}{a} + C, & |u| > a \end{cases}$$

EXAMPLE 4 Evaluate $\displaystyle\int \frac{dx}{\sqrt{3x^2 + 1}}$.

Solution It follows from I′ that

$$\int \frac{dx}{\sqrt{3x^2+1}} = \frac{1}{\sqrt{3}} \int \frac{\sqrt{3}\,dx}{\sqrt{(\sqrt{3}x)^2+1}}$$

$u = \sqrt{3}x$
$du = \sqrt{3}\,dx$

$$= \frac{1}{\sqrt{3}} \int \frac{du}{\sqrt{u^2+1}}$$

$$= \frac{1}{\sqrt{3}} \sinh^{-1} u + C$$

$$= \frac{1}{\sqrt{3}} \sinh^{-1} \sqrt{3}x + C$$

□

EXAMPLE 5 Evaluate $\int \frac{dx}{9-x^2}$.

Solution With $a = 3$ and $u = x$, we obtain from III′

$$\int \frac{dx}{9-x^2} = \begin{cases} \frac{1}{3}\tanh^{-1}\frac{x}{3} + C, & |x| < 3 \\ \frac{1}{3}\coth^{-1}\frac{x}{3} + C, & |x| > 3 \end{cases}$$

□

Sometimes we would like to write the integral formulas I′–III′ in terms of logarithms rather than inverse hyperbolic functions. With the aid of parts (*i*), (*ii*), (*iii*), and (*iv*) of Theorem 7.1, it follows that when a is a positive constant:

$$\int \frac{du}{\sqrt{u^2+a^2}} = \ln(u + \sqrt{u^2+a^2}) + C \tag{1}$$

$$\int \frac{du}{\sqrt{u^2-a^2}} = \ln(u + \sqrt{u^2-a^2}) + C, \quad u > a > 0 \tag{2}$$

$$\int \frac{du}{a^2-u^2} = \begin{cases} \frac{1}{2a}\ln\left(\frac{a+u}{a-u}\right) + C, & |u| < a \\ \frac{1}{2a}\ln\left(\frac{u+a}{u-a}\right) + C, & |u| > a \end{cases} \tag{3}$$

Note that the last integral can be written compactly as

$$\int \frac{du}{a^2-u^2} = \frac{1}{2a}\ln\left|\frac{a+u}{a-u}\right| + C, \quad |u| \neq a \tag{4}$$

EXAMPLE 6 In view of (4), the result in Example 5 can be written as

$$\int \frac{dx}{9 - x^2} = \frac{1}{6} \ln \left| \frac{3 + x}{3 - x} \right| + C$$

□

EXAMPLE 7 Find the area A under the graph of $y = \dfrac{1}{\sqrt{x^2 + 1}}$ on the interval [0, 2].

Solution Figure 7.19 shows the area in question. Since the function is nonnegative, it follows immediately from (1) that

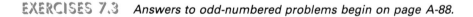

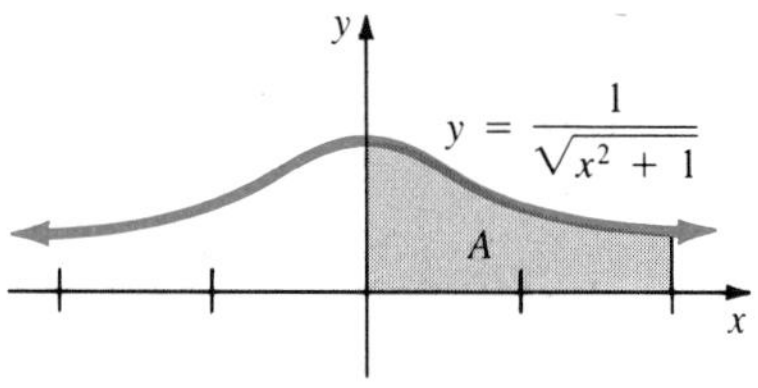

FIGURE 7.19

$$A = \int_0^2 \frac{dx}{\sqrt{x^2 + 1}} = \ln(x + \sqrt{x^2 + 1})\Big]_0^2$$

$$= \ln(2 + \sqrt{5}) \approx 1.4436 \text{ square units}$$

□

▲ ***Remark*** In Section 1.3, we pointed out that functions can be categorized as either algebraic or transcendental. The functions studied in Chapters 6 and 7:

trigonometric, inverse trigonometric,

logarithmic, exponential, hyperbolic, inverse hyperbolic, and

the power function $f(x) = x^r$, r irrational,

are transcendental, since they cannot be constructed out of real numbers and a variable x by a finite number of additions, subtractions, multiplications, divisions, and extraction of roots. Note that $y = x^{1/2}$ is an algebraic function, but $y = x^\pi$ is transcendental.

EXERCISES 7.3 *Answers to odd-numbered problems begin on page A-88.*

In Problems 1–20 find the derivative of the given function.

1. $y = \sinh^{-1} 3x$
2. $y = \cosh^{-1} \dfrac{x}{2}$
3. $y = \tanh^{-1}(1 - x^2)$
4. $y = \coth^{-1} \dfrac{1}{x}$
5. $y = \coth^{-1}(\csc x)$
6. $y = \sinh^{-1}(\sin x)$
7. $y = x \sinh^{-1} x^3$
8. $y = x^2 \operatorname{csch}^{-1} x$
9. $y = \dfrac{\operatorname{sech}^{-1} x}{x}$
10. $y = \dfrac{\coth^{-1} e^{2x}}{e^{2x}}$
11. $y = \ln(\operatorname{sech}^{-1} x)$
12. $y = \ln|\operatorname{csch}^{-1} x|$
13. $y = (\cosh^{-1} 6x)^{1/2}$
14. $y = \dfrac{1}{(\tanh^{-1} 2x)^3}$
15. $y = \sqrt{x^2 - 1} + \cosh^{-1} x$
16. $y = \dfrac{1}{1 - x^2} + \coth^{-1} 4x$
17. $y = \sinh^{-1}(\cosh x)$
18. $y = \sinh(\tanh^{-1} x)$
19. $y = \ln \sqrt{1 - x^2} + x \coth^{-1} x$
20. $y = -\frac{1}{3}\sqrt{9x^2 - 1} + x \cosh^{-1} 3x$

In Problems 21–40 evaluate the given integral.

21. $\int \frac{dx}{\sqrt{4x^2+1}}$

22. $\int \frac{dx}{\sqrt{x^2+4}}$

23. $\int \frac{dx}{1-4x^2}$

24. $\int \frac{dx}{4-x^2}$

25. $\int \frac{dx}{\sqrt{9x^2-16}}$

26. $\int \frac{dx}{\sqrt{25x^2+9}}$

27. $\int \frac{x\,dx}{\sqrt{x^4+1}}$

28. $\int \frac{x\,dx}{1-x^4}$

29. $\int \frac{dx}{x^2+2x}$

30. $\int \frac{dx}{\sqrt{x^2+2x}}$

31. $\int \frac{dx}{\sqrt{x^2+4x+5}}$

32. $\int \frac{dx}{-x^2+6x-8}$

33. $\int \frac{e^x}{\sqrt{e^{2x}-1}}\,dx$

34. $\int \frac{dx}{e^{-x}-e^x}$

35. $\int \frac{\cos\theta}{\sqrt{1+\sin^2\theta}}\,d\theta$

36. $\int \frac{\sin t}{\sqrt{2-\sin^2 t}}\,dt$

37. $\int \frac{t}{\sqrt{t^2+16}}\,dt$

38. $\int \frac{r}{1-r^2}\,dr$

39. $\int \frac{x^2+1}{x^2-1}\,dx$

40. $\int \frac{2+x}{\sqrt{5x^2+1}}\,dx$

In Problems 41 and 42 express the antiderivative of the integral as a natural logarithm. Find the value of the integral.

41. $\int_{-1}^{3} \frac{dx}{\sqrt{x^2+1}}$

42. $\int_{0}^{4} \frac{dt}{\sqrt{t^2+9}}$

43. Find the area under the graph of $y = 1/(1-x^2)$ on the interval $[-\frac{3}{4}, 0]$.

44. Find the area bounded by the graph of $y = 1/(1-x^2)$ and the x-axis on $[2, 5]$.

45. Consider the graph of the function $y = 1/\sqrt{1-x^2}$ shown in Figure 7.20.

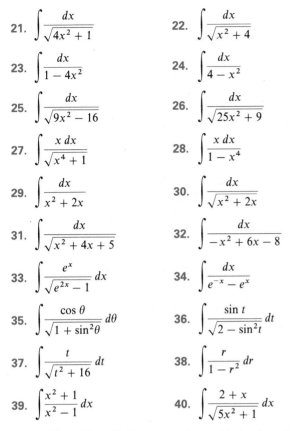

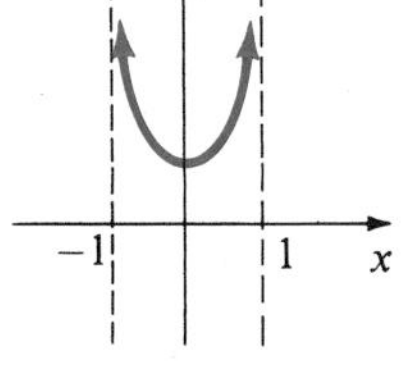

FIGURE 7.20

(a) Find the area under the graph on the interval $[\frac{1}{2}, \sqrt{3}/2]$.

(b) The region bounded by the graphs of $y = 1/\sqrt{1-x^2}$, $y = 0$, $x = \frac{1}{2}$, and $x = \sqrt{3}/2$ is revolved about the y-axis. Find the volume of the solid of revolution.

(c) The region in part **(b)** is revolved about the x-axis. Find the volume of the solid of revolution.

46. Use the shell method to find the volume of the solid of revolution that is formed by revolving the region bounded by the graphs of $y = 1/(1-x^4)$, $x = 0$, $x = \frac{1}{2}$, and $y = 0$ about the y-axis.

47. The differential equation that describes the shape of a wire of constant density w hanging under its own weight is

$$\frac{d^2y}{dx^2} = \frac{w}{T}\sqrt{1 + \left(\frac{dy}{dx}\right)^2}$$

where T is a constant. Use separation of variables and the substitution $p = dy/dx$ to solve this equation subject to $y(0) = 1$ and $y'(0) = 0$.

48. The velocity $v(t)$ of a body of mass m, which is falling through a medium that imparts a resisting force proportional to the square of the velocity, satisfies the differential equation

$$m\frac{dv}{dt} = mg - kv^2$$

where g is the acceleration of gravity and k is a positive constant.

(a) Use separation of variables to show that a solution of this equation that satisfies $v(0) = 0$ is

$$v(t) = \sqrt{\frac{mg}{k}}\tanh\left(\sqrt{\frac{gk}{m}}\,t\right)$$

(b) Determine the terminal velocity of the mass; that is, $v_t = \lim_{t\to\infty} v(t)$. [*Hint:* See Figure 6.28(a).]

(c) Suppose an 80-kg sky diver delays opening the parachute until terminal velocity is attained. Determine v_t if it is known that $k = 0.25$ kg/m.

Miscellaneous Problems

49. Prove II in the form

$$\frac{d}{dx}\cosh^{-1}x = \frac{1}{\sqrt{x^2-1}}$$

50. Prove III in the form

$$\frac{d}{dx}\tanh^{-1}x = \frac{1}{1-x^2}$$

51. Prove IV in the form

$$\frac{d}{dx}\coth^{-1}x = \frac{1}{1-x^2}$$

52. Prove V in the form

$$\frac{d}{dx}\operatorname{sech}^{-1}x = \frac{-1}{x\sqrt{1-x^2}}$$

53. Prove VI in the form

$$\frac{d}{dx}\operatorname{csch}^{-1}x = \frac{-1}{|x|\sqrt{1+x^2}}$$

54. Use the form $\int du/\sqrt{u^2+1}$ with $u = \tan\theta$ to evaluate $\int \sec\theta\, d\theta$ in terms of a natural logarithm.

55. Use the identity $\dfrac{1}{1-x^2} = \dfrac{\frac{1}{2}}{1+x} + \dfrac{\frac{1}{2}}{1-x}$ to evaluate $\int dx/(1-x^2)$.

CHAPTER 7 REVIEW EXERCISES

Answers begin on page A-88.

In Problems 1–10 answer true or false.

1. $f(x) = \sin^{-1}x$ is not differentiable at $x = 1$. ________

2. If $t = \tan^{-1}(-\frac{1}{2})$, then $\cos t = -2/\sqrt{5}$. ________

3. $\arcsin(1) = 90°$ ________

4. $\tan^{-1}x = 1/\tan x$ ________

5. $\sin^{-1}\left(\sin\dfrac{5\pi}{6}\right) = \dfrac{5\pi}{6}$ ________

6. $\sec(\arctan(-1)) = -\sqrt{2}$ ________

7. For $y = \tan^{-1}x$, $dy/dx > 0$ for all x. ________

8. The hyperbolic sine is one-to-one. ________

9. Every inverse hyperbolic function can be expressed as a logarithm. ________

10. $\cosh^{-1}1 = 0$ ________

In Problems 11–24 find dy/dx.

11. $y = \sin^{-1}\dfrac{3}{x}$

12. $y = \cos x \cos^{-1}x$

13. $y = (\cot^{-1}x)^{-1}$

14. $y = x \operatorname{arcsec}(2x-1)$

15. $y = 2\cos^{-1}x + 2x\sqrt{1-x^2}$

16. $y = x^2\tan^{-1}\sqrt{x^2-1}$

17. $y = (\tan^{-1}(\sin x^2))^2$

18. $y = \displaystyle\int_0^x \frac{\sin^{-1}t}{t+1}\,dt, \quad x > 0$

19. $y - x = \sin^{-1}y$

20. $(1+9x^2)y^2 = \cot^{-1}3x$

21. $y = \sinh^{-1}(\sin^{-1}x)$

22. $y = (\tan^{-1}x)(\tanh^{-1}x)$

23. $y = xe^{x\cosh^{-1}x}$

24. $y = \sinh^{-1}\sqrt{x^2-1}$

In Problems 25–36 evaluate the given integral.

25. $\displaystyle\int \frac{dx}{\sqrt{9-4x^2}}$

26. $\displaystyle\int \frac{x^{-1/2}}{\sqrt{1-x}}\,dx$

27. $\displaystyle\int_1^3 \frac{dx}{\sqrt{x}(1+x)}$

28. $\displaystyle\int_0^2 \frac{x^4-15}{x^2+4}\,dx$

29. $\displaystyle\int \frac{dx}{\sqrt{7-6x-x^2}}$

30. $\displaystyle\int \frac{dx}{(x+2)\sqrt{x(x+4)}}$

31. $\displaystyle\int_0^{1/2} \frac{dx}{(\cos^{-1}x)^2\sqrt{1-x^2}}$

32. $\displaystyle\int \frac{\cos x}{\sqrt{25-\sin^2 x}}\,dx$

33. $\displaystyle\int_0^{1/\sqrt[3]{2}} \frac{x^2}{\sqrt{1-x^6}}\,dx$

34. $\displaystyle\int \frac{x}{\sqrt{1-x^2}}\,dx$

35. $\displaystyle\int_0^1 \frac{dx}{\sqrt{9x^2+1}}$

36. $\displaystyle\int_{2e}^{3e} \frac{dx}{x\sqrt{(\ln x)^2-1}}$

In Problems 37 and 38 find the exact value of the given expression.

37. $\cos\left(\sin^{-1}\left(-\dfrac{12}{13}\right)\right)$

38. $\sin\left(2\cos^{-1}\left(-\dfrac{4}{5}\right)\right)$

39. Find the area of the region in the first quadrant bounded between the graphs of $y = 1/(1+x^2)$, $y = 1/\sqrt{1-x^2}$, and $x = \frac{1}{2}$.

40. The region bounded by the graphs of $y = 32/(16+x^4)$, $y = 0$, $x = 0$, and $x = 2$ is revolved around the y-axis. Find the volume of the solid of revolution.

41. Show that the area of the shaded region in Figure 7.21 is $2\sin a - a^2$.

42. Consider the plane pendulum, shown in Figure 7.22, that swings between points A and C. If B is midway between A and

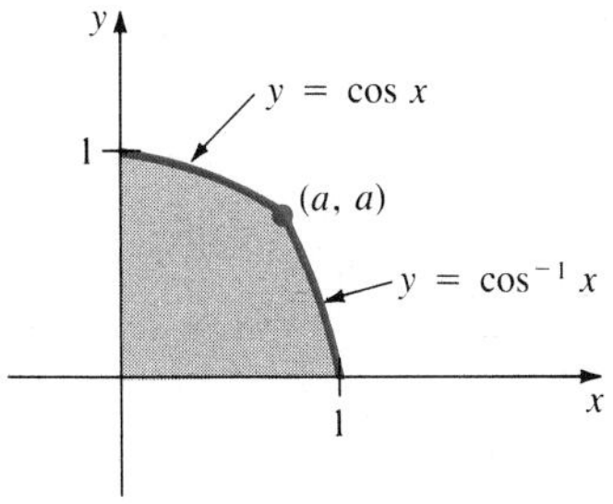

FIGURE 7.21

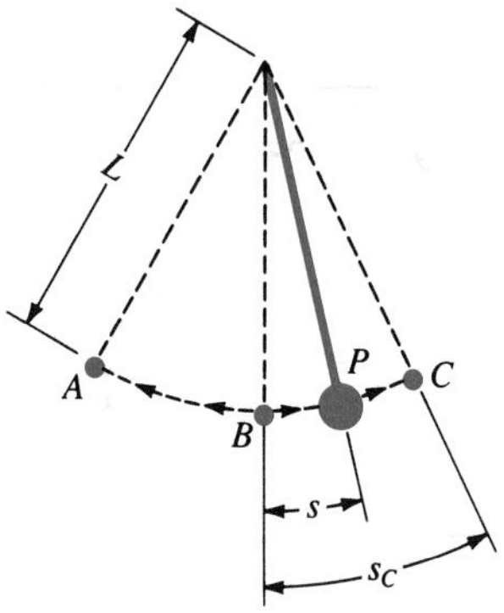

FIGURE 7.22

C, it can be shown that the time the pendulum takes to travel between points B and P is

$$t = \int_0^s \sqrt{\frac{L}{g(s_C^2 - x^2)}}\,dx$$

(a) Show that $t = \sqrt{\dfrac{L}{g}}\sin^{-1}\dfrac{s}{s_C}$.

(b) Use the result in part (a) to determine the time of travel from B to C.

(c) Use (b) to determine the period T of the pendulum—that is, the time of an oscillation from A to C and back to A.

43. Studies indicate that the peak particle velocity from an underground explosion of a vertically aligned, cylindrical charge depends on the radial distance from the charge according to a formula of the form

$$V = k\left(L\int_0^H \frac{dx}{R^2 + (R\tan\theta - x)^2}\right)^a$$

Here V is the peak particle velocity in mm/s, L is the charge density in kg/m, $k \approx 700$ and $a \approx 0.8$ are constants, and H, R, and θ are as shown in Figure 7.23. Evaluate the integral and show that $V = k(L\phi/R)^a$, where ϕ, as shown in the figure, is the angle subtended by the charge as seen from the point at radial distance R.

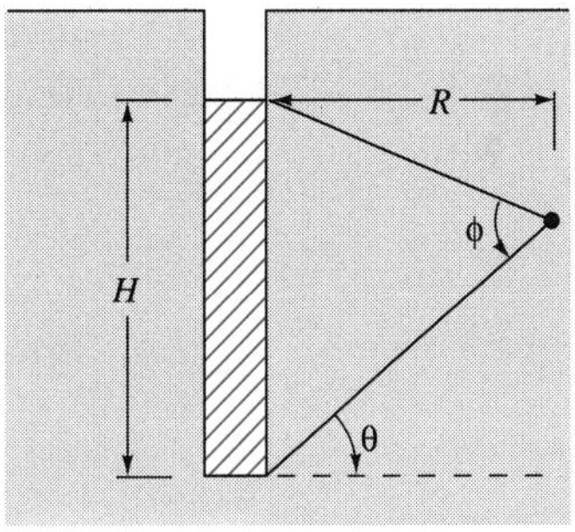

FIGURE 7.23

44. (a) Show that $\tan^{-1}\dfrac{x+1}{x-1} + \tan^{-1}x = C$, where C is a constant.

(b) Find the value of C in part (a).

45. Find the length of the graph of $y = \sin^{-1}2e^x$ from $y = \pi/6$ to $y = \pi/3$.

8

TECHNIQUES OF INTEGRATION

INTRODUCTION

One often encounters an integral that cannot be categorized as a general form such as $\int u^n\,du$ or $\int e^u\,du$. For example, it is not possible to evaluate $\int x^2\sqrt{x+1}\,dx$ by an immediate application of any one of the formulas listed on page 492. However, by applying a **technique of integration**, an integral such as this can sometimes be reduced to one or more of these familiar forms.

Suppose you attach a rope to a rowboat located on the y-axis and then pull it, keeping the rope taut, while walking down the x-axis. Suppose further that the rowboat is initially located at $(0, s)$ and that the rope is of length s. The path that the boat follows is called a **tractrix**. In Exercises 8.4 you will be asked to derive a differential equation that describes this path. In solving this equation by separation of variables we encounter the integral

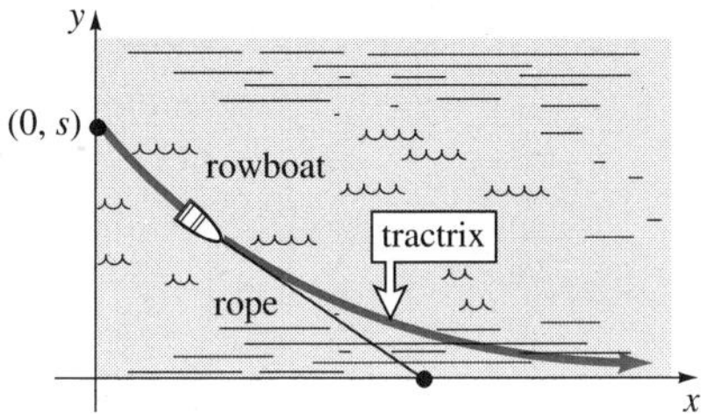

$$\int \frac{\sqrt{s^2 - y^2}}{y}\,dy.$$ Since this integral is not a general form, we need a special method or trick to evaluate it.

IMPORTANT CONCEPTS

algebraic substitutions

integration by parts

tabular integration

trigonometric substitutions

partial fractions

A Review of Integration Formulas

$$\int u^n\,du = \frac{u^{n+1}}{n+1} + C, \quad n \neq -1 \qquad \int \frac{du}{u} = \ln|u| + C$$

$$\int e^u\,du = e^u + C \qquad \int a^u\,du = \frac{a^u}{\ln a} + C$$

$$\int \cos u\,du = \sin u + C \qquad \int \sin u\,du = -\cos u + C$$

$$\int \sec^2 u\,du = \tan u + C \qquad \int \csc^2 u\,du = -\cot u + C$$

$$\int \sec u \tan u\,du = \sec u + C \qquad \int \csc u \cot u\,du = -\csc u + C$$

$$\int \cosh u\,du = \sinh u + C \qquad \int \sinh u\,du = \cosh u + C$$

$$\int \tan u\,du = -\ln|\cos u| + C \qquad \int \cot u\,du = \ln|\sin u| + C$$

$$\int \sec u\,du = \ln|\sec u + \tan u| + C \qquad \int \csc u\,du = \ln|\csc u - \cot u| + C$$

$$\int \frac{du}{\sqrt{a^2 - u^2}} = \sin^{-1}\frac{u}{a} + C \qquad \int \frac{du}{a^2 + u^2} = \frac{1}{a}\tan^{-1}\frac{u}{a} + C$$

$$\int \frac{du}{u\sqrt{u^2 - a^2}} = \frac{1}{a}\sec^{-1}\frac{u}{a} + C \qquad \int \frac{du}{\sqrt{u^2 + a^2}} = \sinh^{-1}\frac{u}{a} + C$$

$$\int \frac{du}{\sqrt{u^2 - a^2}} = \cosh^{-1}\frac{u}{a} + C \qquad \int \frac{du}{u^2 - a^2} = \begin{cases} \dfrac{1}{a}\tanh^{-1}\dfrac{u}{a} + C, & |u| < a \\ \dfrac{1}{a}\coth^{-1}\dfrac{u}{a} + C, & |u| > a \end{cases}$$

8.1 ALGEBRAIC SUBSTITUTIONS

Throughout Chapters 4, 5, 6, and 7, we used a substitution $u = g(x)$ on many occasions to evaluate an integral. For example, $\int e^{x^2} x\,dx$ is recognized as $(\frac{1}{2})\int e^u\,du$ with $u = x^2$ and $du = 2x\,dx$. We now extend the idea of the u-substitution to integrals that are not of the precise form $\int f(g(x))g'(x)\,dx$.

EXAMPLE 1 Evaluate $\displaystyle\int x^2\sqrt{2x+1}\,dx$.

Solution If we let $u = 2x + 1$, then $x = \frac{1}{2}(u - 1)$, $dx = \frac{1}{2}\,du$,

$$x^2 = \frac{1}{4}(u-1)^2 = \frac{1}{4}(u^2 - 2u + 1) \quad \text{and} \quad \sqrt{2x+1} = u^{1/2}$$

Thus,

$$\int x^2\sqrt{2x+1}\,dx = \int \frac{1}{4}(u^2 - 2u + 1)u^{1/2}\frac{1}{2}\,du$$

don't forget to substitute for dx

$$= \frac{1}{8}\int (u^{5/2} - 2u^{3/2} + u^{1/2})\,du$$

$$= \frac{1}{8}\left(\frac{2}{7}u^{7/2} - \frac{4}{5}u^{5/2} + \frac{2}{3}u^{3/2}\right) + C$$

$$= \frac{1}{28}(2x+1)^{7/2} - \frac{1}{10}(2x+1)^{5/2} + \frac{1}{12}(2x+1)^{3/2} + C$$

You should verify that the derivative of the last line actually is $x^2\sqrt{2x+1}$. □

The choice of which, if any, substitution to use is not always obvious. Generally, if the integrand contains a power of a function, then it is a good idea to try to let u be that function *or* the power of the function itself. In Example 1, the alternative substitution $u = \sqrt{2x+1}$ or $u^2 = 2x + 1$ leads to the different integral $\frac{1}{4}\int (1 - u^2)^2 u^2\,du$. The latter can be evaluated by expanding the integrand and integrating each term.

EXAMPLE 2 Evaluate $\displaystyle\int \frac{dx}{1 + \sqrt{x}}$.

Solution Let $u = \sqrt{x}$ so that $x = u^2$ and $dx = 2u\,du$. Then

$$\int \frac{dx}{1+\sqrt{x}} = \int \frac{2u\,du}{1+u}$$

long division

$$= \int \left(2 - \frac{2}{1+u}\right) du$$

$$= 2u - 2\ln|1 + u| + C$$

$$= 2\sqrt{x} - 2\ln(1 + \sqrt{x}) + C$$

□

Integrands Containing a Quadratic Expression As we saw in Chapter 7, if an integrand contains a quadratic expression, $ax^2 + bx + c$, completion of the square may lead to an integral that can be expressed as an inverse trigonometric function or an inverse hyperbolic function. Of course, more complicated integrals can yield other functions as well.

EXAMPLE 3 Evaluate $\displaystyle\int \frac{x+4}{x^2+6x+18}\,dx.$

Solution After completing the square, the given integral can be written as

$$\int \frac{x+4}{x^2+6x+18}\,dx = \int \frac{x+4}{(x+3)^2+9}\,dx$$

Now, if $u = x + 3$, then $x = u - 3$ and $dx = du$. Therefore,

$$\begin{aligned}
\int \frac{x+4}{x^2+6x+18}\,dx &= \int \frac{u+1}{u^2+9}\,du \\
&= \int \frac{u}{u^2+9}\,du + \int \frac{du}{u^2+9} \\
&= \frac{1}{2}\int \frac{2u}{u^2+9}\,du + \int \frac{du}{u^2+9} \\
&= \frac{1}{2}\ln(u^2+9) + \frac{1}{3}\tan^{-1}\frac{u}{3} + C \\
&= \frac{1}{2}\ln[(x+3)^2+9] + \frac{1}{3}\tan^{-1}\frac{x+3}{3} + C \\
&= \frac{1}{2}\ln(x^2+6x+18) + \frac{1}{3}\tan^{-1}\frac{x+3}{3} + C
\end{aligned}$$

termwise division

□

EXAMPLE 4 Evaluate $\displaystyle\int_0^2 \frac{6x+1}{\sqrt[3]{3x+2}}\,dx.$

Solution If $u = 3x + 2$, then $x = \frac{1}{3}(u-2)$, $dx = \frac{1}{3}\,du$,

$$6x + 1 = 2(u-2) + 1 = 2u - 3 \quad \text{and} \quad \sqrt[3]{3x+2} = u^{1/3}$$

Now, observe that when $x = 0$, $u = 2$, and when $x = 2$, $u = 8$. Therefore, integrating on the variable u, we obtain

$$\begin{aligned}
\int_0^2 \frac{6x+1}{\sqrt[3]{3x+2}}\,dx &= \int_2^8 \frac{2u-3}{u^{1/3}}\,\frac{1}{3}\,du \\
&= \int_2^8 \left(\frac{2}{3}u^{2/3} - u^{-1/3}\right) du = \left(\frac{2}{5}u^{5/3} - \frac{3}{2}u^{2/3}\right)\Bigg]_2^8 \\
&= \left(\frac{2}{5}\cdot 2^5 - \frac{3}{2}\cdot 2^2\right) - \left(\frac{2}{5}\cdot 2^{5/3} - \frac{3}{2}\cdot 2^{2/3}\right) \\
&= \frac{34}{5} - \frac{2}{5}\cdot 2^{5/3} + \frac{3}{2}\cdot 2^{2/3} \approx 7.9112
\end{aligned}$$

□

You are encouraged to rework Example 4, this time using the substitution $u = \sqrt[3]{3x+2}$.

▲ ***Remarks*** (*i*) When working the exercises throughout this chapter, do not be overly disturbed if you do not always obtain the same answer as given in the text. Different techniques applied to the same problem can lead to answers that look different. Remember, two antiderivatives of the same function can differ at most by a constant. Try to resolve any conflicts.

(*ii*) It might also prove helpful at this point to recall that integration of the quotient of two polynomial functions, $P(x)/Q(x)$, usually begins with long division if the degree of P is greater than or equal to the degree of Q. See Example 2.

(*iii*) Look for problems that can be solved by previous methods.

EXERCISES 8.1

Answers to odd-numbered problems begin on page A-88.

In Problems 1–40 evaluate the given integral.

1. $\int x(x+1)^3\,dx$
2. $\int \frac{x^2-3}{(x+1)^3}\,dx$
3. $(2x+1)\sqrt{x-5}\,dx$
4. $\int (x^2-1)\sqrt{2x+1}\,dx$
5. $\int \frac{x}{\sqrt{x-1}}\,dx$
6. $\int \frac{x^2}{\sqrt{x+2}}\,dx$
7. $\int \frac{x+3}{(3x-4)^{3/2}}\,dx$
8. $\int (x^2+x)\sqrt[3]{x+7}\,dx$
9. $\int \frac{\sqrt{x}}{x+1}\,dx$
10. $\int \frac{t}{\sqrt{t}+1}\,dt$
11. $\int \frac{\sqrt{t}-3}{\sqrt{t}+1}\,dt$
12. $\int \frac{\sqrt{r}+3}{r+3}\,dr$
13. $\int \frac{x^3}{\sqrt[3]{x^2+1}}\,dx$
14. $\int \frac{x^5}{\sqrt[5]{x^2+4}}\,dx$
15. $\int \frac{x^2}{(x-1)^4}\,dx$
16. $\int \frac{2x+1}{(x+7)^2}\,dx$
17. $\int \frac{dx}{\sqrt{x}-\sqrt[3]{x}}$ [*Hint:* Let $u = x^{1/6}$.]
18. $\int \frac{\sqrt[6]{t}}{\sqrt[3]{t}+1}\,dt$
19. $\int \sqrt{e^x-1}\,dx$
20. $\int \frac{dx}{\sqrt{e^x-1}}$
21. $\int \sqrt{1-\sqrt{v}}\,dv$
22. $\int \frac{\sqrt{w}}{\sqrt{1-\sqrt{w}}}\,dw$
23. $\int \frac{\sqrt{1+\sqrt{t}}}{\sqrt{t}}\,dt$
24. $\int \sqrt{t}\sqrt{1+t\sqrt{t}}\,dt$
25. $\int \frac{2x+7}{x^2+2x+5}\,dx$
26. $\int \frac{6x-1}{4x^2+4x+10}\,dx$
27. $\int \frac{2x+5}{\sqrt{16-6x-x^2}}\,dx$
28. $\int \frac{4x-3}{\sqrt{11+10x-x^2}}\,dx$
29. $\int_0^1 x\sqrt{5x+4}\,dx$
30. $\int_{-1}^0 x\sqrt[3]{x+1}\,dx$
31. $\int_1^{16} \frac{dx}{10+\sqrt{x}}$
32. $\int_4^9 \frac{\sqrt{x}-1}{\sqrt{x}+1}\,dx$
33. $\int_2^9 \frac{5x-6}{\sqrt[3]{x-1}}\,dx$
34. $\int_{-\sqrt{3}}^0 \frac{2x^3}{\sqrt{x^2+1}}\,dx$
35. $\int_0^1 (1-\sqrt{x})^{50}\,dx$
36. $\int_0^4 \frac{dx}{(1+\sqrt{x})^3}$
37. $\int_1^8 \frac{dx}{x^{1/3}+x^{2/3}}$
38. $\int_1^{64} \frac{x^{1/3}}{x^{2/3}+2}\,dx$
39. $\int_0^1 x^2(1-x)^5\,dx$
40. $\int_0^6 \frac{2x+5}{\sqrt{2x+4}}\,dx$

In Problems 41 and 42 use a substitution to establish the given result. Assume $x > 0$.

41. $\int_1^{x^2} \frac{1}{t}\,dt = 2\int_1^x \frac{1}{t}\,dt$ **42.** $\int_1^{\sqrt{x}} \frac{1}{t}\,dt = \frac{1}{2}\int_1^x \frac{1}{t}\,dt$

43. Find the area under the graph of $y = 1/(x^{1/3} + 1)$ on the interval $[0, 1]$.

44. Find the area bounded by the graph of $y = x^3\sqrt{x+1}$ and the x-axis on the interval $[-1, 1]$.

45. Find the volume of the solid of revolution that is formed by revolving the region bounded by the graphs of $y = 1/(\sqrt{x}+1)$, $x = 0$, $x = 4$, and $y = 0$ about the y-axis.

46. Find the volume of the solid of revolution that is formed by revolving the region in Problem 45 about the x-axis.

47. Find the length of the graph of $y = \frac{4}{5}x^{5/4}$ on the interval $[0, 9]$.

48. **Bertalanffy's equation** for the growth of an organism assumes that constructive metabolism (anabolism) proceeds at a rate proportional to the surface area, while destructive metabolism (catabolism) proceeds at a rate proportional to the volume. Assuming that surface area is proportional to the two-thirds power of volume and that the organism's weight w is directly proportional to the volume, we can write Bertalanffy's equation as $dw/dt = Aw^{2/3} - Bw$, where A and B are positive parameters. One can conclude from this that the time it takes such an organism to grow from weight w_1 to w_2 is given by

$$T = \int_{w_1}^{w_2} \frac{dw}{Aw^{2/3} - Bw}$$

Evaluate this integral. Find an upper limit on how large the organism can grow according to Bertalanffy's equation.

8.2 INTEGRATION BY PARTS

In this section we are going to develop a formula that can often be used to integrate the product of two functions. To apply this formula we have to identify one of the functions in the product as a differential. Recall that if $v = g(x)$, then its differential is the function $dv = g'(x)\,dx$.

To derive this integration formula for a product we begin with the Product Rule of differentiation. Suppose $u = f(x)$ and $v = g(x)$ are differentiable functions. Then by the Product Rule we have

$$\frac{d}{dx}[f(x)g(x)] = f(x)g'(x) + g(x)f'(x) \tag{1}$$

In turn, integration of (1)

$$f(x)g(x) = \int f(x)g'(x)\,dx + \int g(x)f'(x)\,dx$$

produces a formula

$$\int f(x)g'(x)\,dx = f(x)g(x) - \int g(x)f'(x)\,dx \tag{2}$$

The basic idea behind (2) is to evaluate the integral $\int f(x)g'(x)\,dx$ by means of evaluating another, and it is hoped simpler, integral $\int g(x)f'(x)\,dx$.

The formula in (2) is usually written in terms of the differentials $du = f'(x)\,dx$ and $dv = g'(x)\,dx$:

$$\int u\,dv = uv - \int v\,du \tag{3}$$

The procedure, known as **integration by parts**, starts with an integration followed by a differentiation:

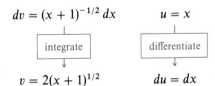

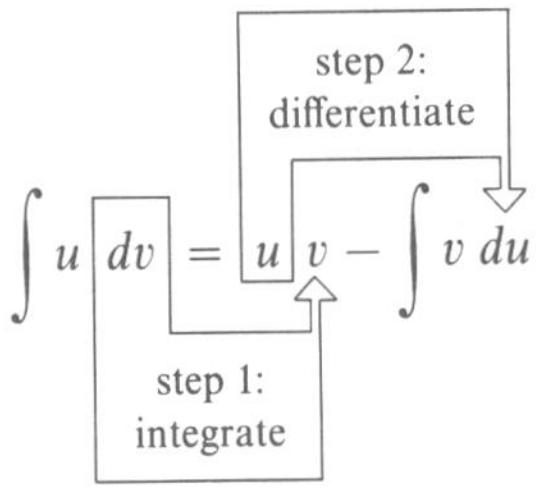

The last step is, of course, the evaluation of $\int v\,du$.

Integration problems can sometimes be done by several methods. In the first example, the integral can be evaluated by means of an algebraic substitution and integration by parts.

EXAMPLE 1 Evaluate $\displaystyle\int \frac{x}{\sqrt{x+1}}\,dx$.

Solution First, we write the integral as

$$\int x(x+1)^{-1/2}\,dx$$

In this latter form, there are several possible choices for the function dv. We could have $dv = (x+1)^{-1/2}\,dx$, $dv = x\,dx$, or simply $dv = dx$. As a practical matter, the function dv is *usually* the most complicated factor in the product that can be integrated using the formulas on page 492. Moreover, the choice of dv is dictated by what happens in the second integral in (3). With these thoughts in mind, if we choose

$$dv = (x+1)^{-1/2}\,dx \qquad\qquad u = x$$

$$\downarrow \text{integrate} \qquad\qquad \downarrow \text{differentiate}$$

then

$$v = 2(x+1)^{1/2} \qquad\qquad du = dx$$

Substituting these functions into (3) yields

$$\int \overbrace{x}^{u}\overbrace{(x+1)^{-1/2}\,dx}^{dv} = \overbrace{x}^{u}\overbrace{[2(x+1)^{1/2}]}^{v} - \int \overbrace{2(x+1)^{1/2}}^{v}\,\overbrace{dx}^{du}$$

$$= 2x(x+1)^{1/2} - 2\cdot\frac{2}{3}(x+1)^{3/2} + C$$

$$= 2x(x+1)^{1/2} - \frac{4}{3}(x+1)^{3/2} + C \qquad \square$$

The key to making integration by parts work is to make the "right" choice for the function dv. It was pointed out in Example 1 that dv is usually the most complicated factor in the product that can be immediately integrated by a previously known formula. Yet this cannot be given as a firm rule. Knowledge that the "right" choice for dv has been made is often based on pragmatic hindsight: Is the second integral $\int v\,du$ less complicated than the first integral $\int u\,dv$? Can we evaluate this second integral? To see what happens when the "wrong" choice is made, let us consider Example 1 again, but this time we select

$$dv = x\,dx \qquad u = (x+1)^{-1/2}$$
$$v = \frac{1}{2}x^2 \qquad du = -\frac{1}{2}(x+1)^{-3/2}\,dx$$

Applying (3) in this instance gives

$$\int x(x+1)^{-1/2}\,dx = \frac{1}{2}x^2(x+1)^{-1/2} + \frac{1}{4}\int x^2(x+1)^{-3/2}\,dx$$

The problem here is apparent; the second integral $\int v\,du$ is more complicated than the original $\int u\,dv$. The alternative selection $dv = dx$ also leads to an impasse.

In the next example we shall evaluate $\int x\tan^{-1}x\,dx$. Obviously the most complicated factor in the product $x\tan^{-1}x\,dx$ is $\tan^{-1}x\,dx$. But the choice $dv = \tan^{-1}x\,dx$ is not a judicious one, since we cannot immediately integrate this function based on previously known results. As the example shows, we choose in this case $dv = x\,dx$.

EXAMPLE 2 Evaluate $\displaystyle\int x\tan^{-1}x\,dx$.

Solution By choosing

$$dv = x\,dx \qquad u = \tan^{-1}x$$
$$v = \frac{x^2}{2} \qquad du = \frac{dx}{1+x^2}$$

we see that (3) gives

$$\int \underbrace{(\tan^{-1}x)}_{u}\underbrace{(x\,dx)}_{dv} = \underbrace{(\tan^{-1}x)}_{u}\underbrace{\frac{x^2}{2}}_{v} - \int \underbrace{\frac{x^2}{2}}_{v}\cdot\underbrace{\frac{dx}{1+x^2}}_{du}$$

To evaluate $\int x^2\,dx/(1+x^2)$, we use long division (see Example 7 of Section 7.2). Hence,

$$\int x\tan^{-1}x\,dx = \frac{x^2}{2}\tan^{-1}x - \frac{1}{2}\int\left(1 - \frac{1}{1+x^2}\right)dx$$
$$= \frac{x^2}{2}\tan^{-1}x - \frac{1}{2}x + \frac{1}{2}\tan^{-1}x + C \qquad \square$$

EXAMPLE 3 Evaluate $\int x^3 \ln x \, dx$.

Solution Let

$$dv = x^3\,dx \qquad u = \ln x$$
$$v = \frac{x^4}{4} \qquad du = \frac{1}{x}\,dx$$

Integrating by parts then gives

$$\begin{aligned}\int x^3 \ln x\,dx &= \frac{x^4}{4}\ln x - \frac{1}{4}\int x^4 \cdot \frac{1}{x}\,dx \qquad \boxed{\text{simplify}}\\ &= \frac{x^4}{4}\ln x - \frac{1}{4}\int x^3\,dx\\ &= \frac{x^4}{4}\ln x - \frac{x^4}{16} + C\end{aligned}$$

□

Successive Integrations A problem may require integration by parts several times in succession. As a rule, integrals of the type

$$\int x^n \sin kx\,dx, \quad \int x^n \cos kx\,dx, \quad \int x^n e^{kx}\,dx, \quad \text{and} \quad \int x^k (\ln x)^n\,dx$$

where n is a positive integer and k a constant, will require integration by parts n times. To illustrate let us consider the integral $\int f(x)g'(x)\,dx$. As we have seen in the previous examples, with differential notation, the starting point in integration by parts is to identify a function as $g'(x)$ and then integrate. For example, in the integral $\int x^2 \cos x\,dx$ let us identify $f(x) = x^2$ and $g'(x) = \cos x$. After integrating $\cos x$, we see from (2) that the second integral also requires integration by parts. In this example, we shall emphasize all steps and certain important algebraic signs:

$$\int x^2 \cos x\,dx = x^2 \sin x - \int 2x \sin x\,dx \qquad \boxed{\text{to use (2) again now integrate } \sin x}$$

$$= x^2 \sin x - \left[2x(-\cos x) - \int 2(-\cos x)\,dx\right]$$

$$= x^2 \sin x - 2x(-\cos x) + \int 2(-\cos x)\,dx$$

$\boxed{f \text{ and successive derivatives}}$

$$= +\ x^2 \sin x \quad 2x(-\cos x) + 2(-\sin x) + C \tag{4}$$

$\boxed{\text{successive integrals of } g'(x) = \cos x}$

$$= x^2 \sin x + 2x \cos x - 2 \sin x + C \tag{5}$$

The result in (5) can be obtained by a systematic shortcut. We display the derivatives and integrals in an array:

$f(x)$ and its derivatives		$g'(x)$ and its integrals
x^2	+	$\cos x$
$2x$	−	$\sin x$
2	+	$-\cos x$
0		$-\sin x$

and then form the products of the functions joined by the arrows and either add or subtract a product according to the algebraic sign indicated in color (as in (4)). The last zero in the derivative column indicates that we need not integrate $g'(x)$ any further; the products from that point on are zero.

EXAMPLE 4 Evaluate $\int x^4 e^{-2x}\,dx$.

Solution We use the array method just discussed to integrate by parts four times. In this case we choose $f(x) = x^4$ and $g'(x) = e^{-2x}$.

$f(x)$ and its derivatives		$g'(x)$ and its integrals
x^4	+	e^{-2x}
$4x^3$	−	$-\frac{1}{2}e^{-2x}$
$12x^2$	+	$\frac{1}{4}e^{-2x}$
$24x$	−	$-\frac{1}{8}e^{-2x}$
24	+	$\frac{1}{16}e^{-2x}$
0		$-\frac{1}{32}e^{-2x}$

Forming the products, we obtain

$$\int x^4e^{-2x}\,dx = +x^4\left(-\frac{1}{2}e^{-2x}\right) - 4x^3\left(\frac{1}{4}e^{-2x}\right) + 12x^2\left(-\frac{1}{8}e^{-2x}\right) - 24x\left(\frac{1}{16}e^{-2x}\right) + 24\left(-\frac{1}{32}e^{-2x}\right) + C$$

$$= -\frac{1}{2}x^4e^{-2x} - x^3e^{-2x} - \frac{3}{2}x^2e^{-2x} - \frac{3}{2}xe^{-2x} - \frac{3}{4}e^{-2x} + C$$

□

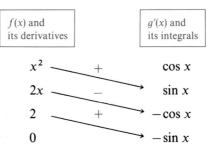

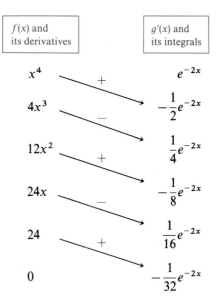

The technique illustrated above for successive integrations is sometimes referred to as **tabular integration** or the **ladder method.***

Solving for Integrals For certain integrals, one or more applications of integration by parts may result in a situation where the original integral occurs on the right-hand side. In this case the problem of evaluating that integral is completed by *solving* for the original integral. The next two examples illustrate the technique.

EXAMPLE 5 Evaluate $\int \sec^3 x\, dx$.

Solution Inspection of the integral reveals no obvious choice for dv. However, by writing $\sec^3 x = \sec x \cdot \sec^2 x$, we can identify

$$dv = \sec^2 x\, dx \qquad u = \sec x$$
$$v = \tan x \qquad du = \sec x \tan x\, dx$$

It follows from (3) that

$$\int \sec^3 x\, dx = \sec x \tan x - \int \tan^2 x \sec x\, dx$$

trig identity

$$= \sec x \tan x - \int (\sec^2 x - 1)\sec x\, dx$$
$$= \sec x \tan x + \int \sec x\, dx - \int \sec^3 x\, dx$$
$$= \sec x \tan x + \ln|\sec x + \tan x| - \int \sec^3 x\, dx$$

We solve the last equation for $\int \sec^3 x\, dx$ and add a constant of integration:

$$2\int \sec^3 x\, dx = \sec x \tan x + \ln|\sec x + \tan x|$$
$$\int \sec^3 x\, dx = \frac{1}{2}\sec x \tan x + \frac{1}{2}\ln|\sec x + \tan x| + C$$ □

Integrals of the type $\int e^{ax}\sin bx\, dx$ and $\int e^{ax}\cos bx\, dx$ require two applications of integration by parts before recovering the original integral on the right-hand side. Moreover, tabular integration discussed in Example 4 can be used to evaluate each integral.

---

*See K. W. Folley, Integration by Parts, *The American Mathematical Monthly*, Vol. 54 (1947), pp. 542–543.

EXAMPLE 6 Evaluate $\int e^{2x}\cos 3x\, dx$.

Solution If we choose $f(x) = \cos 3x$ and $g'(x) = e^{2x}$, we have:

$f(x)$ and its derivatives		$g'(x)$ and its integrals
$\cos 3x$		e^{2x}
$-3\sin 3x$	$+$	$\frac{1}{2}e^{2x}$
$-9\cos 3x$	$-$	$\frac{1}{4}e^{2x}$

We know that we can stop at this point, since the last cosine term in the derivative column means that we have recovered the original integral. The first and last lines of the array are to be interpreted as the integrals $\int e^{2x}\cos 3x\, dx$ and $\int \frac{1}{4}e^{2x}(-9\cos 3x)\, dx$, respectively; that is,

$$\int e^{2x}\cos 3x\, dx = \frac{1}{2}e^{2x}\cos 3x - \frac{1}{4}e^{2x}(-3\sin 3x) + \int \frac{1}{4}e^{2x}(-9\cos 3x)\, dx$$

Solving the last equation for $\int e^{2x}\cos 3x\, dx$ gives

$$\frac{13}{4}\int e^{2x}\cos 3x\, dx = \frac{1}{2}e^{2x}\cos 3x + \frac{3}{4}e^{2x}\sin 3x$$

and so $$\int e^{2x}\cos 3x\, dx = \frac{2}{13}e^{2x}\cos 3x + \frac{3}{13}e^{2x}\sin 3x + C$$ □

In the integrals $\int e^{ax}\sin bx\, dx$ and $\int e^{ax}\cos bx\, dx$ it does not matter which functions are chosen as $f(x)$ and $g'(x)$. You are urged to rework Example 6 using $f(x) = e^{2x}$ and $g'(x) = \cos 3x$.

Definite Integrals A definite integral can be evaluated using integration by parts in the following manner:

$$\int_a^b f(x)g'(x)\, dx = f(x)g(x)\Big]_a^b - \int_a^b g(x)f'(x)\, dx$$

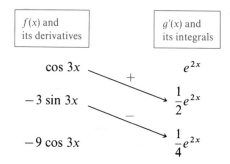

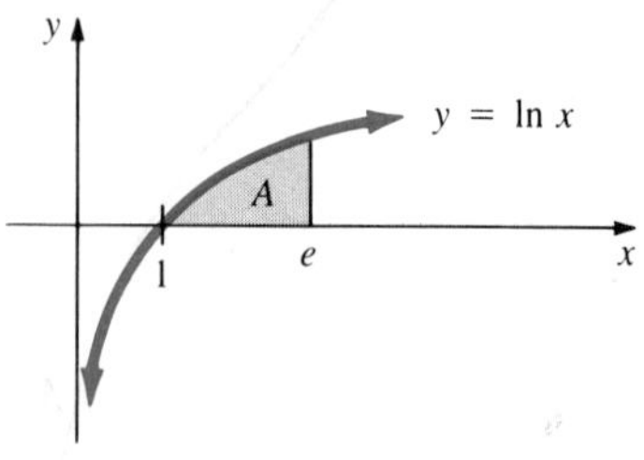

FIGURE 8.1

EXAMPLE 7 Find the area under the graph of $y = \ln x$ on the interval $[1, e]$.

Solution From Figure 8.1 we see that the area A is given by

$$A = \int_1^e \ln x\, dx$$

Choosing $\qquad dv = dx \qquad u = \ln x$

$$v = x \qquad du = \frac{1}{x}\,dx$$

we have

$$A = x \ln x \Big]_1^e - \int_1^e x \cdot \frac{1}{x}\,dx$$

$$= x \ln x \Big]_1^e - \int_1^e dx = x \ln x \Big]_1^e - x \Big]_1^e$$

$$= e \ln e - \ln 1 - e + 1 = 1 \text{ square unit}$$

since $\ln e = 1$ and $\ln 1 = 0$. □

▲ ***Remark*** Whenever we apply integration by parts we must carry out at least two integrations: the initial integration of dv and then the evaluation of $\int v\,du$. Note that throughout the examples in this section we did not introduce a constant in the integration of dv. The single constant of integration at the finish of the problem is a "collective" constant.

EXERCISES 8.2

Answers to odd-numbered problems begin on page A-89.

In Problems 1–48 evaluate the given integral using integration by parts.

1. $\int x\sqrt{x+3}\,dx$

2. $\int \frac{x}{\sqrt{2x-5}}\,dx$

3. $\int \ln 4x\,dx$

4. $\int \ln(x+1)\,dx$

5. $\int x \ln 2x\,dx$

6. $\int x^{1/2}\ln x\,dx$

7. $\int \frac{\ln x}{x^2}\,dx$

8. $\int \frac{\ln x}{\sqrt{x^3}}\,dx$

9. $\int (\ln t)^2\,dt$

10. $\int (t \ln t)^2\,dt$

11. $\int_0^2 x\ln(x+1)\,dx$

12. $\int_1^e x^2 \ln x^2\,dx$

13. $\int_0^1 \tan^{-1}x\,dx$

14. $\int_1^4 \frac{\tan^{-1}\sqrt{x}}{\sqrt{x}}\,dx$

15. $\int \sin^{-1}x\,dx$

16. $\int x^2\tan^{-1}x\,dx$

17. $\int xe^{3x}\,dx$

18. $\int x^2e^{5x}\,dx$

19. $\int x^3e^{-4x}\,dx$

20. $\int x^5e^x\,dx$

21. $\int x^3e^{x^2}\,dx$

22. $\int x^5e^{2x^3}\,dx$

23. $\int_2^4 xe^{-x/2}\,dx$

24. $\int_0^1 (x^2-x)e^{-x}\,dx$

25. $\int t\cos 8t\,dt$

26. $\int x\sinh x\,dx$

27. $\int x^2\sin dx$

28. $\int_0^\pi x^2\cos\frac{x}{2}\,dx$

29. $\int x^3\cos 3x\,dx$

30. $\int x^4\sin 2x\,dx$

31. $\int e^x\sin 4x\,dx$

32. $\int_{-\pi}^{\pi} e^x\cos x\,dx$

33. $\int \theta\sec\theta\tan\theta\,d\theta$

34. $\int e^{2t}\cos e^t\,dt$

35. $\int \sin x\cos 2x\,dx$

36. $\int \cosh x\cosh 2x\,dx$

37. $\int x^3\sqrt{x^2+4}\,dx$

38. $\int \frac{t^5}{(t^3+1)^2}\,dt$

39. $\int \sin(\ln x)\,dx$

40. $\int \cos x \ln(\sin x)\,dx$

41. $\int \csc^3 x\,dx$

42. $\int x \sec^2 x\,dx$

43. $\int x \sin^2 x\,dx$

44. $\int x \tan^2 x\,dx$

45. $\int (\sin^{-1}x)^2\,dx$

46. $\int \ln(x + \sqrt{x^2+1})\,dx$

47. $\int \frac{x^2 e^x}{(x+2)^2}\,dx$

48. $\int e^{2x}\tan^{-1}e^x\,dx$

In Problems 49 and 50 evaluate the given integral using integration by parts.

49. $\int xe^x \sin x\,dx$

50. $\int xe^{-x}\cos 2x\,dx$

In Problems 51–54 use a substitution first, followed by integration by parts.

51. $\int_0^4 \tan^{-1}\sqrt{x}\,dx$

52. $\int xe^{\sqrt{x}}\,dx$

53. $\int \sin\sqrt{x+2}\,dx$

54. $\int_0^{\pi^2} \cos\sqrt{t}\,dt$

55. Evaluate $\int \frac{\sin^{-1}}{\sqrt{1-x^2}}\,dx$ using two different methods.

56. Evaluate $\int \sin mx \cos nx\,dx$ using integration by parts.

57. Find the area under the graph of $y = 1 + \ln x$ on the interval $[e^{-1}, 3]$.

58. Find the area bounded by the graph of $y = \tan^{-1}x$ and the x-axis on the interval $[-1, 1]$.

59. The region in the first quadrant bounded by the graphs of $y = \ln x$, $x = 5$, and $y = 0$ is revolved about the x-axis. Find the volume of the solid of revolution.

60. The region in the first quadrant bounded by the graphs of $y = e^x$, $x = 0$, and $y = 3$ is revolved about the y-axis. Find the volume of the solid of revolution.

61. The region in the first quadrant bounded by the graphs of $y = \sin x$ and $y = 0$, $0 \le x \le \pi$, is revolved about the y-axis. Find the volume of the solid of revolution.

62. Find the average value of $f(x) = \tan^{-1}(x/2)$ on the interval $[0, 2]$.

63. A body moves in a straight line with velocity $v(t) = e^{-t}\sin t$, where v is measured in cm/s. Find the position function $s(t)$ if it is known that $s = 0$ when $t = 0$.

64. A body moves in a straight line with acceleration $a(t) = te^{-t}$, where a is measured in cm/s^2. Find the velocity function $v(t)$ and the position function $s(t)$ if $v(0) = 1$ and $s(0) = -1$.

65. A water tank is formed by revolving the region bounded by the graphs of $y = \sin \pi x$ and $y = 0$, $0 \le x \le 1$, about the x-axis, which is taken in the downward direction. The tank is filled to a depth of $\frac{1}{2}$ ft. Determine the work done in pumping all the water to the top of the tank.

66. Find the force caused by liquid pressure on one side of the vertical plate shown in Figure 8.2. Assume that the plate is submerged in water and that dimensions are in feet.

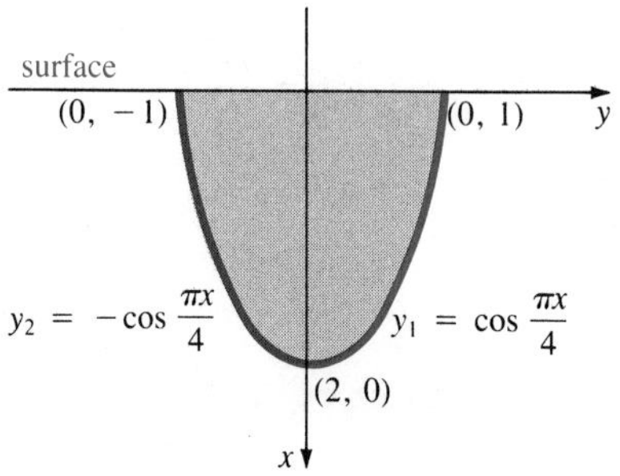

FIGURE 8.2

67. Find the centroid of the region bounded by the graphs of $y = \sin x$, $y = 0$, and $x = \pi/2$.

68. (a) Show that integration by parts applied to $\int dx/(x \ln x)$ with $dv = dx/x$, $u = 1/\ln x$, $v = \ln x$, and $du = -dx/x(\ln x)^2$ leads to $0 = 1$. Explain.

(b) Evaluate the integral in part (a) by an alternative method.

In Problems 69–72 use integration by parts to establish the given **reduction formula.**

69. $\int (\ln x)^n\,dx = x(\ln x)^n - n\int (\ln x)^{n-1}\,dx$

70. $\int \sin^n x\,dx = -\frac{\sin^{n-1}x \cos x}{n} + \frac{n-1}{n}\int \sin^{n-2}x\,dx$

71. $\int \cos^n x\,dx = \frac{\cos^{n-1}x \sin x}{n} + \frac{n-1}{n}\int \cos^{n-2}x\,dx$

72. $\int \sec^n x\,dx = \frac{\sec^{n-2}x \tan x}{n-1} + \frac{n-2}{n-1}\int \sec^{n-2}x\,dx,$

$n \ne 1$

In Problems 73–76 use the reduction formulas in Problems 70–72 to evaluate the given integral.

73. $\int \sin^3 x \, dx$

74. $\int \sec^4 x \, dx$

75. $\int \cos^3 10x \, dx$

76. $\int x^2 \sin^4 x^3 \, dx$

Calculator/Computer Problems

77. **(a)** Use a calculator or computer to obtain the graph of $f(x) = 3 + 2\sin^2 x - 5\sin^4 x$.

(b) Find the area bounded by the graph of the function given in part (**a**) and the x-axis on the interval $[0, 2\pi]$.

8.3 INTEGRATION OF POWERS OF TRIGONOMETRIC FUNCTIONS

***Integrals of the Form* $\sin^m x \cos^n x \, dx$** With the aid of trigonometric identities, it is possible to evaluate integrals of the type

$$\int \sin^m x \cos^n x \, dx \tag{1}$$

We distinguish two cases.

CASE I ***m* or *n* is an odd positive integer.**

Let us first assume that m is an odd positive integer. By writing

$$\sin^m x = \sin^{m-1} x \sin x$$

where $m - 1$ is even, and using $\sin^2 x = 1 - \cos^2 x$, we can express the integrand in (1) as a *sum* of powers of $\cos x$ times $\sin x$. The original integral can then be expressed as a sum of integrals, each having the recognizable form

$$\int \cos^k x \sin x \, dx = -\int \overbrace{\cos^k x}^{u^k}\overbrace{(-\sin x)\, dx}^{du} = -\int u^k \, du$$

We note that the exponent k need not be an integer.

EXAMPLE 1 Evaluate $\int \sin^5 x \cos^2 x \, dx$.

Solution

$$\begin{aligned}\int \sin^5 x \cos^2 x \, dx &= \int \cos^2 x \sin^4 x \sin x \, dx \\ &= \int \cos^2 x (\sin^2 x)^2 \sin x \, dx \\ &= \int \cos^2 x (1 - \cos^2 x)^2 \sin x \, dx\end{aligned}$$

$$= \int \cos^2 x(1 - 2\cos^2 x + \cos^4 x)\sin x\, dx$$

$$= -\int \overbrace{\cos^2 x}^{\boxed{u^2}}\overbrace{(-\sin x)\, dx}^{\boxed{du}} + 2\int \overbrace{\cos^4 x}^{\boxed{u^4}}\overbrace{(-\sin x)\, dx}^{\boxed{du}}$$

$$- \int \overbrace{\cos^6 x}^{\boxed{u^6}}\overbrace{(-\sin x)\, dx}^{\boxed{du}}$$

$$= -\frac{1}{3}\cos^3 x + \frac{2}{5}\cos^5 x - \frac{1}{7}\cos^7 x + C \qquad \square$$

EXAMPLE 2 Evaluate $\int \sin^3 x\, dx$.

Solution

$$\int \sin^3 x\, dx = \int \sin^2 x \sin x\, dx$$

$$= \int (1 - \cos^2 x)\sin x\, dx$$

$$= \int \sin x\, dx + \int \cos^2 x(-\sin x)\, dx$$

$$= -\cos x + \frac{1}{3}\cos^3 x + C \qquad \square$$

If n is an odd positive integer, the procedure for evaluation is the same except that we seek an integrand that is a sum of powers of $\sin x$ times $\cos x$.

EXAMPLE 3 Evaluate $\int \sin^4 x \cos^3 x\, dx$.

Solution

$$\int \sin^4 x \cos^3 x\, dx = \int \sin^4 x \cos^2 x \cos x\, dx$$

$$= \int \sin^4 x(1 - \sin^2 x)\cos x\, dx$$

$$= \int \overbrace{\sin^4 x}^{\boxed{u^4}}\overbrace{(\cos x)\, dx}^{\boxed{du}} - \int \overbrace{\sin^6 x}^{\boxed{u^6}}\overbrace{(\cos x)\, dx}^{\boxed{du}}$$

$$= \frac{1}{5}\sin^5 x - \frac{1}{7}\sin^7 x + C \qquad \square$$

CASE II m and n are both even nonnegative integers.

When both m and n are even nonnegative integers, the evaluation of (1) relies heavily on the identities

$$\sin x \cos x = \frac{1}{2}\sin 2x \qquad \sin^2 x = \frac{1 - \cos 2x}{2} \qquad \cos^2 x = \frac{1 + \cos 2x}{2}$$

We have already seen the following special cases several times:

$$\int \sin^2 x\, dx \quad \text{and} \quad \int \cos^2 x\, dx$$

EXAMPLE 4 Evaluate $\int \sin^2 x \cos^2 x\, dx$.

Solution

$$\begin{aligned}
\int \sin^2 x \cos^2 x\, dx &= \int \frac{1 - \cos 2x}{2} \cdot \frac{1 + \cos 2x}{2}\, dx \\
&= \frac{1}{4}\int (1 - \cos^2 2x)\, dx \\
&= \frac{1}{4}\int \left(1 - \frac{1 + \cos 4x}{2}\right) dx \\
&= \frac{1}{4}\int \left(\frac{1}{2} - \frac{1}{2}\cos 4x\right) dx \\
&= \frac{1}{8}x - \frac{1}{32}\sin 4x + C
\end{aligned}$$

Alternative Solution

$$\begin{aligned}
\int \sin^2 x \cos^2 x\, dx &= \int (\sin x \cos x)^2\, dx \\
&= \int \left(\frac{\sin 2x}{2}\right)^2 dx \\
&= \frac{1}{4}\int \frac{1 - \cos 4x}{2}\, dx
\end{aligned}$$

The remainder of the solution is the same as before. □

EXAMPLE 5 Evaluate $\int \cos^4 x\, dx$.

Solution

$$\begin{aligned}
\int \cos^4 x\, dx &= \int (\cos^2 x)^2\, dx \\
&= \int \left(\frac{1 + \cos 2x}{2}\right)^2 dx
\end{aligned}$$

$$= \frac{1}{4}\int (1 + 2\cos 2x + \cos^2 2x)\,dx$$

$$= \frac{1}{4}\int \left(1 + 2\cos 2x + \frac{1 + \cos 4x}{2}\right) dx$$

$$= \frac{1}{4}\int \left(\frac{3}{2} + 2\cos 2x + \frac{1}{2}\cos 4x\right) dx$$

$$= \frac{3}{8}x + \frac{1}{4}\sin 2x + \frac{1}{32}\sin 4x + C$$ □

The foregoing procedures are summarized in the following table.

Evaluation of $\int \sin^m x \cos^n x\,dx$		
Case	**Procedure**	**Identities used**
I m odd	$u = \cos x$	$\sin^2 x = 1 - \cos^2 x$
n odd	$u = \sin x$	$\cos^2 x = 1 - \sin^2 x$
II m and n even	Reduce the powers of $\sin x$ and $\cos x$ in the integrand by using the identities	$\sin x \cos x = \frac{1}{2}\sin 2x$ $\sin^2 x = \frac{1 - \cos 2x}{2}$ $\cos^2 x = \frac{1 + \cos 2x}{2}$

Integrals of the Form $\int \tan^m x \sec^n x\,dx$ To evaluate an integral of the type

$$\int \tan^m x \sec^n x\,dx \tag{2}$$

we shall consider three cases.

CASE I *m is an odd positive integer.*

When m is an odd positive integer, then $m - 1$ is even. Using

$$\tan^m x \sec^n x = \tan^{m-1} x \sec^{n-1} x \sec x \tan x$$

and $\tan^2 x = \sec^2 x - 1$, we can write the given integral as a sum of integrals each having the form

$$\int \overbrace{\sec^k x}^{u^k}\, \overbrace{\sec x \tan x\,dx}^{du} = \int u^k\,du$$

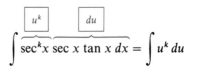

EXAMPLE 6 Evaluate $\int \tan^3 x \sec^7 x\, dx$.

Solution

$$\int \tan^3 x \sec^7 x\, dx = \int \tan^2 x \sec^6 x \sec x \tan x\, dx$$

$$= \int (\sec^2 x - 1)\sec^6 x \sec x \tan x\, dx$$

$$= \int \overbrace{\sec^8 x}^{u^8}\overbrace{(\sec x \tan x)\, dx}^{du} - \int \overbrace{\sec^6 x}^{u^6}\overbrace{(\sec x \tan x)\, dx}^{du}$$

$$= \frac{1}{9}\sec^9 x - \frac{1}{7}\sec^7 x + C$$ □

CASE II ***n* is an even positive integer.**

If n is an even positive integer, the evaluation procedure is similar to case I for integrals of the type given in (1). Employing

$$\sec^n x = \sec^{n-2} x \sec^2 x$$

and the identity $1 + \tan^2 x = \sec^2 x$, we can write the given integral as a sum of integrals of the form

$$\int \overbrace{\tan^k x}^{u^k}\, \overbrace{\sec^2 x\, dx}^{du} = \int u^k\, du$$

EXAMPLE 7 Evaluate $\int \sqrt{\tan x}\, \sec^4 x\, dx$.

Solution $$\int \sqrt{\tan x}\, \sec^4 x\, dx = \int (\tan x)^{1/2} \sec^2 x \sec^2 x\, dx$$

$$= \int (\tan x)^{1/2}(1 + \tan^2 x)\sec^2 x\, dx$$

$$= \int \overbrace{(\tan x)^{1/2}}^{u^{1/2}}\overbrace{\sec^2 x\, dx}^{du} + \int \overbrace{(\tan x)^{5/2}}^{u^{5/2}}\overbrace{\sec^2 x\, dx}^{du}$$

$$= \frac{2}{3}(\tan x)^{3/2} + \frac{2}{7}(\tan x)^{7/2} + C$$ □

CASE III *m* is even and *n* is odd.

Finally, if m is an even positive integer and n is an odd positive integer, we write the integrand in terms of sec x and use integration by parts.

EXAMPLE 8 Evaluate $\int \tan^2 x \sec x \, dx$.

Solution By writing

$$\int \tan^2 x \sec x \, dx = \int (\sec^2 x - 1)\sec x \, dx$$
$$= \int \sec^3 x \, dx - \int \sec x \, dx$$

we have two integrals previously evaluated. Integration by parts gives (see Example 5 in Section 8.2)

$$\int \sec^3 x \, dx = \frac{1}{2} \sec x \tan x + \frac{1}{2} \ln|\sec x + \tan x| + C_1 \tag{3}$$

Also,

$$\int \sec x \, dx = \ln|\sec x + \tan x| + C_2 \tag{4}$$

Subtracting the results in (3) and (4) finally yields

$$\int \tan^2 x \sec x \, dx = \frac{1}{2} \sec x \tan x - \frac{1}{2} \ln|\sec x + \tan x| + C$$ □

The three cases considered in evaluating the integral $\int \tan^m x \sec^n x \, dx$ are summarized in the following table.

Evaluation of $\int \tan^m x \sec^n x \, dx$

Case		Procedure	Identities used
I	m odd	$u = \sec x$	$\tan^2 x = \sec^2 x - 1$
II	n even	$u = \tan x$	$\sec^2 x = 1 + \tan^2 x$
III	m even and n odd	Change integrand to powers of sec x alone. Integration by parts may be required.	$\tan^2 x = \sec^2 x - 1$

▲ ***Remarks*** (*i*) Integrals of the type $\int \cot^m x \csc^n x \, dx$ are handled in a manner analogous to (2). In this case the identity $\csc^2 x = 1 + \cot^2 x$ will be useful.

(*ii*) Some of the following problems can be worked in several different ways. So if your answer looks different than that given in the text, check your algebra, try to find a different way to work the problem, or try a trigonometric identity.

EXERCISES 8.3

Answers to odd-numbered problems begin on page A-89.

In Problems 1–36 evaluate the given integral. Note that some integrals do not, strictly speaking, fall into any of the cases considered in this section.

1. $\int (\sin x)^{1/2} \cos x \, dx$
2. $\int \cos^4 5x \sin 5x \, dx$
3. $\int \cos^3 x \, dx$
4. $\int \sin^5 t \, dt$
5. $\int \sin^3 x \cos^3 x \, dx$
6. $\int \sin^5 2x \cos^2 2x \, dx$
7. $\int_{\pi/3}^{\pi/2} \sin^3\theta \sqrt{\cos\theta} \, d\theta$
8. $\int \frac{\cos^3 x}{\sin^2 x} \, dx$
9. $\int_0^{\pi/2} \sin^5 x \cos^5 x \, dx$
10. $\int_0^{\pi} \sin^3 2t \, dt$
11. $\int \sin^4 t \, dt$
12. $\int \cos^6\theta \, d\theta$
13. $\int \sin^2 x \cos^4 x \, dx$
14. $\int_{-\pi}^{\pi} \sin^4 x \cos^2 x \, dx$
15. $\int \sin^4 x \cos^4 x \, dx$
16. $\int \sin^2 3x \cos^2 3x \, dx$
17. $\int \tan^3 2t \sec^4 2t \, dt$
18. $\int (2 - \sqrt{\tan x})^2 \sec^2 x \, dx$
19. $\int \frac{dx}{\cos^4 x}$
20. $\int_{-\pi/4}^{\pi/4} \tan y \sec^4 y \, dy$
21. $\int \cot^{10} x \csc^4 x \, dx$
22. $\int (1 + \csc^2 t)^2 \, dt$
23. $\int \tan^3 x (\sec x)^{-1/2} \, dx$
24. $\int \left(\tan\frac{x}{2} \sec\frac{x}{2}\right)^3 dx$
25. $\int \tan^2 x \sec^3 x \, dx$
26. $\int (1 + \tan x)^2 \sec x \, dx$
27. $\int \cos^2 x \cot x \, dx$
28. $\int \sin x \sec^7 x \, dx$
29. $\int \frac{\sec^4(1-t)}{\tan^8(1-t)} \, dt$
30. $\int \frac{\sin^3\sqrt{t} \cos^2\sqrt{t}}{\sqrt{t}} \, dt$
31. $\int_0^{\pi/3} \tan^2 x \, dx$
32. $\int (\tan x + \cot x)^2 \, dx$
33. $\int \tan^4 x \, dx$
34. $\int \tan^5 x \, dx$
35. $\int \cot^3 t \, dt$
36. $\int \csc^5 t \, dt$

In Problems 37 and 38 find the volume of the solid of revolution that is formed by revolving the region bounded by the graphs of the given equations about the x-axis.

37. $y = \cos 2x$, $y = 0$, $0 \le x \le \pi/6$
38. $y = \tan^2 x$, $y = 0$, $0 \le x \le \pi/4$

In Problems 39–44 use the trigonometric identities.

$$\sin mx \cos nx = \frac{1}{2}[\sin(m+n)x + \sin(m-n)x]$$

$$\sin mx \sin nx = \frac{1}{2}[\cos(m-n)x - \cos(m+n)x]$$

$$\cos mx \cos nx = \frac{1}{2}[\cos(m-n)x + \cos(m+n)x]$$

to evaluate the given integrals.

39. $\int \sin x \cos 2x \, dx$
40. $\int \cos 3x \cos 5x \, dx$
41. $\int \sin 2x \sin 4x \, dx$
42. $\int \frac{5 - 3\sin 2x}{\sec 6x} \, dx$
43. $\int_0^{\pi/6} \cos 2x \cos x \, dx$
44. $\int_0^{\pi/2} \sin\frac{3}{2}x \sin\frac{1}{2}x \, dx$
45. Show that

$$\int_{-\pi}^{\pi} \sin mx \sin nx \, dx = \begin{cases} 0, & m \ne n \\ \pi, & m = n \end{cases}$$

46. Evaluate $\int_{-\pi}^{\pi} \sin mx \cos nx \, dx$.

47. The equation $r = \left|\sin 4\theta \sin \dfrac{\theta}{2}\right|$, $0 \le \theta \le 2\pi$, describes a shape that looks very much like a horse chestnut leaf. See Figure 8.3. As we shall see in Chapter 12, the area A bounded by this graph is given by $A = \frac{1}{2}\int_0^{2\pi} r^2\, d\theta$. Find this area. [*Hint:* Use the identity given before Problems 39–44 for $\cos mx \cos nx$.]

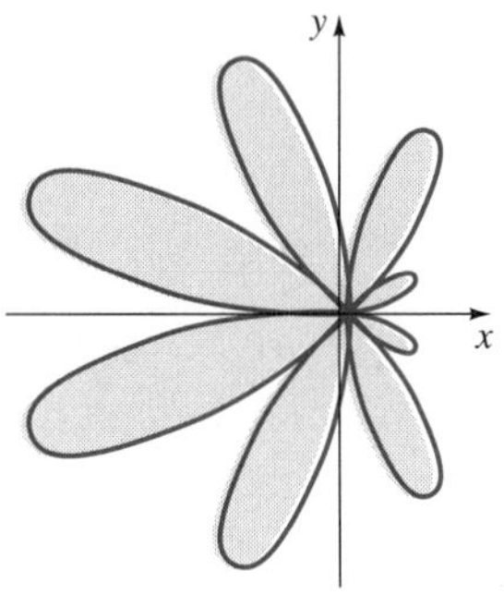

FIGURE 8.3

8.4 TRIGONOMETRIC SUBSTITUTIONS

When an integrand contains integer powers of x and integer powers of

$$\sqrt{a^2 - x^2}, \quad \sqrt{a^2 + x^2}, \quad \text{or} \quad \sqrt{x^2 - a^2}, \qquad a > 0$$

we may be able to evaluate the integral by means of a trigonometric substitution. The three cases we shall now consider depend, in turn, on the fundamental identities:

$$1 - \sin^2\theta = \cos^2\theta \qquad 1 + \tan^2\theta = \sec^2\theta \qquad \sec^2\theta - 1 = \tan^2\theta$$

CASE I Integrands Containing $\sqrt{a^2 - x^2}$, $a > 0$

If we let $x = a \sin \theta$, $-\pi/2 \le \theta \le \pi/2$, then

$$\sqrt{a^2 - x^2} = \sqrt{a^2 - a^2\sin^2\theta} = \sqrt{a^2(1 - \sin^2\theta)} = \sqrt{a^2\cos^2\theta} = a \cos \theta$$

When $\sqrt{a^2 - x^2}$ appears in the denominator of an integrand, there is the further restriction $-\pi/2 < \theta < \pi/2$.

EXAMPLE 1 Evaluate $\displaystyle\int \frac{x^2}{\sqrt{9 - x^2}}\, dx$.

Solution Identifying $a = 3$ leads to the substitutions

$$x = 3 \sin \theta \qquad dx = 3 \cos \theta \, d\theta$$

where $-\pi/2 < \theta < \pi/2$. The integral becomes

$$\int \frac{x^2}{\sqrt{9 - x^2}}\, dx = \int \frac{9 \sin^2\theta}{\sqrt{9 - 9\sin^2\theta}}(3 \cos \theta \, d\theta)$$

$$= 9 \int \sin^2\theta \, d\theta$$

Recall, to evaluate this last trigonometric integral, we make use of the identity $\sin^2\theta = (1 - \cos 2\theta)/2$:

$$\int \frac{x^2}{\sqrt{9 - x^2}}\,dx = \frac{9}{2}\int (1 - \cos 2\theta)\,d\theta$$
$$= \frac{9}{2}\theta - \frac{9}{4}\sin 2\theta + C$$

In order to express this result back in terms of the variable x, we note that $\sin\theta = x/3$, $\cos\theta = \sqrt{1 - \sin^2\theta} = \sqrt{9 - x^2}/3$, and $\theta = \sin^{-1}(x/3)$. Since $\sin 2\theta = 2\sin\theta\cos\theta$, it follows that

$$\int \frac{x^2}{\sqrt{9 - x^2}}\,dx = \frac{9}{2}\sin^{-1}\frac{x}{3} - \frac{1}{2}x\sqrt{9 - x^2} + C \qquad \square$$

EXAMPLE 2 Evaluate $\displaystyle\int \frac{\sqrt{1 - x^2}}{x}\,dx$.

Solution Let $x = \sin\theta$ and $dx = \cos\theta\,d\theta$. Then

$$\int \frac{\sqrt{1 - x^2}}{x}\,dx = \int \frac{\sqrt{1 - \sin^2\theta}}{\sin\theta}(\cos\theta\,d\theta)$$
$$= \int \frac{\cos^2\theta}{\sin\theta}\,d\theta$$
$$= \int \frac{1 - \sin^2\theta}{\sin\theta}\,d\theta$$

termwise division

$$= \int (\csc\theta - \sin\theta)\,d\theta$$
$$= \ln|\csc\theta - \cot\theta| + \cos\theta + C \qquad (1)$$

Since $\cos\theta = \sqrt{1 - \sin^2\theta} = \sqrt{1 - x^2}$, $\csc\theta = 1/\sin\theta = 1/x$, and $\cot\theta = \cos\theta/\sin\theta = \sqrt{1 - x^2}/x$, (1) can be written as

$$\int \frac{\sqrt{1 - x^2}}{x}\,dx = \ln\left|\frac{1 - \sqrt{1 - x^2}}{x}\right| + \sqrt{1 - x^2} + C \qquad \square$$

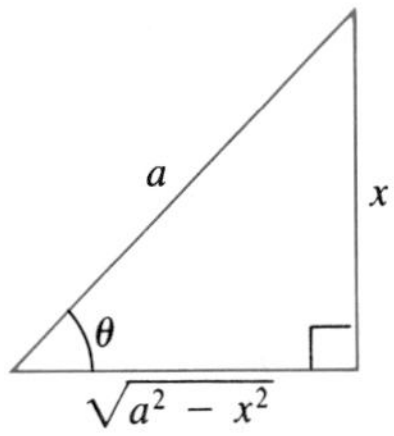

FIGURE 8.4

In Examples 1 and 2 the return to the variable x can be accomplished in an alternative manner. If we construct a right triangle, as shown in Figure 8.4, such that $\sin\theta = x/a$, then the other trigonometric functions can be readily expressed in terms of x. For instance, in Example 2 with $a = 1$, $\sin\theta = x/1$ and so from Figure 8.4 we see that $\cos\theta = \sqrt{1 - x^2}$ and $\cot\theta = \cos\theta/\sin\theta = \sqrt{1 - x^2}/x$.

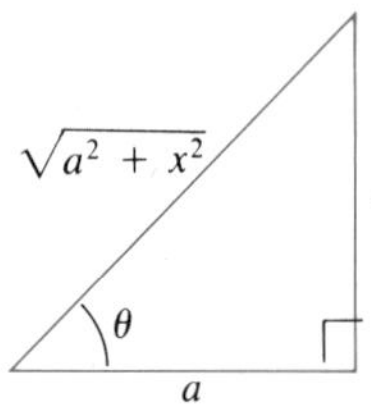

FIGURE 8.5

CASE II Integrands Containing $\sqrt{a^2 + x^2}$, $a > 0$

Suppose $x = a \tan \theta$, where $-\pi/2 < \theta < \pi/2$. Then

$$\sqrt{a^2 + x^2} = \sqrt{a^2 + a^2\tan^2\theta} = \sqrt{a^2(1 + \tan^2\theta)} = \sqrt{a^2\sec^2\theta} = a \sec \theta$$

As in the preceding discussion, an integral that involves an algebraic term $\sqrt{a^2 + x^2}$ is transformed into a trigonometric integral. After integration we can eliminate the variable θ by employing a right triangle where $\tan \theta = x/a$. See Figure 8.5.

EXAMPLE 3 Evaluate $\displaystyle\int \frac{dx}{(4 + x^2)^{3/2}}$.

Solution Observe that the integrand is an integer power of $\sqrt{4 + x^2}$, since $(4 + x^2)^{3/2} = (\sqrt{4 + x^2})^3$. Now, when

$$x = 2 \tan \theta \quad dx = 2 \sec^2\theta \, d\theta$$

we have $\sqrt{4 + x^2} = 2 \sec \theta$ and $(4 + x^2)^{3/2} = 8 \sec^3\theta$. Thus,

$$\begin{aligned} \int \frac{dx}{(4 + x^2)^{3/2}} &= \int \frac{2 \sec^2\theta \, d\theta}{8 \sec^3\theta} \\ &= \frac{1}{4}\int \cos \theta \, d\theta \\ &= \frac{1}{4} \sin \theta + C \end{aligned}$$

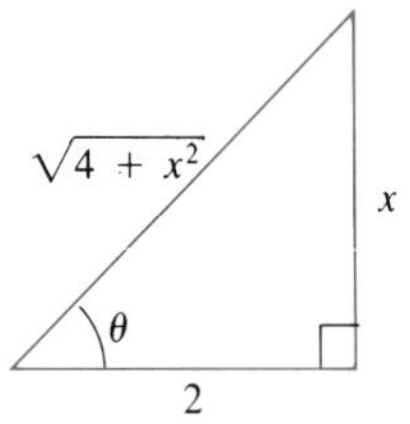

FIGURE 8.6

From the triangle in Figure 8.6, we see that $\sin \theta = x/\sqrt{4 + x^2}$. Hence,

$$\int \frac{dx}{(4 + x^2)^{3/2}} = \frac{1}{4} \frac{x}{\sqrt{4 + x^2}} + C$$

□

CASE III Integrands Containing $\sqrt{x^2 - a^2}$, $a > 0$

In this last case, if we use the substitution $x = a \sec \theta$, where $0 \le \theta < \pi/2$ or $\pi \le \theta < 3\pi/2$, then

$$\begin{aligned} \sqrt{x^2 - a^2} &= \sqrt{a^2\sec^2\theta - a^2} \\ &= \sqrt{a^2(\sec^2\theta - 1)} \\ &= \sqrt{a^2\tan^2\theta} \\ &= a \tan \theta \end{aligned}$$

EXAMPLE 4 Evaluate $\displaystyle\int \frac{\sqrt{x^2 - 16}}{x^4} \, dx$.

Solution Setting $x = 4 \sec \theta$ and $dx = 4 \sec \theta \tan \theta \, d\theta$ yields

$$\int \frac{\sqrt{x^2 - 16}}{x^4} dx = \int \frac{\sqrt{16 \sec^2\theta - 16}}{256 \sec^4\theta} (4 \sec \theta \tan \theta \, d\theta)$$
$$= \frac{1}{16} \int \frac{\tan^2\theta}{\sec^3\theta} d\theta$$
$$= \frac{1}{16} \int \frac{\sin^2\theta}{\cos^2\theta} \cos^3\theta \, d\theta$$
$$= \frac{1}{16} \int \sin^2\theta (\cos \theta \, d\theta)$$
$$= \frac{1}{48} \sin^3\theta + C$$

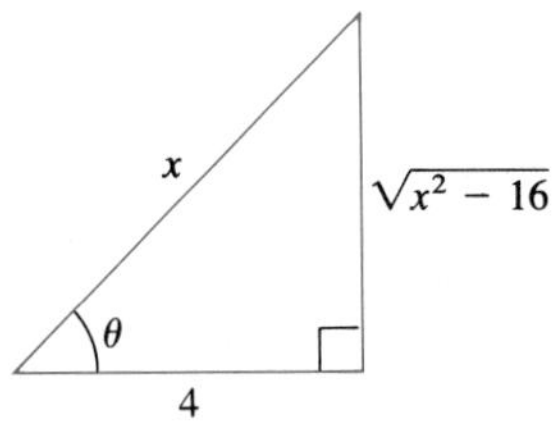

FIGURE 8.7

Referring to the triangle in Figure 8.7, we see that if $\sec \theta = x/4$, then $\cos \theta = 4/x$ and $\sin \theta = \sqrt{x^2 - 16}/x$. It follows that

$$\int \frac{\sqrt{x^2 - 16}}{x^4} dx = \frac{1}{48} \frac{(x^2 - 16)^{3/2}}{x^3} + C$$ □

EXAMPLE 5 Find the length of the graph of $y = \frac{1}{2}x^2 + 3$ on the interval $[0, 1]$.

Solution Recall that the formula for arc length is $s = \int_a^b \sqrt{1 + [f'(x)]^2}\, dx$. Since $dy/dx = x$, we have

$$s = \int_0^1 \sqrt{1 + x^2}\, dx$$

Now, if we substitute $x = \tan \theta$ and $dx = \sec^2\theta \, d\theta$, the limits of integration in the resulting definite integral are $\theta = \tan^{-1}0 = 0$ and $\theta = \tan^{-1}1 = \pi/4$. Therefore,

$$s = \int_0^{\pi/4} \sqrt{1 + \tan^2\theta} \sec^2\theta \, d\theta = \int_0^{\pi/4} \sec^3\theta \, d\theta$$

The antiderivative of $\sec^3\theta$ was found in Example 5 of Section 8.2 using integration by parts:

$$s = \left(\frac{1}{2} \sec \theta \tan \theta + \frac{1}{2} \ln|\sec \theta + \tan \theta|\right)\Big]_0^{\pi/4}$$
$$= \frac{1}{2} \sec \frac{\pi}{4} \tan \frac{\pi}{4} + \frac{1}{2} \ln\left|\sec \frac{\pi}{4} + \tan \frac{\pi}{4}\right|$$
$$= \frac{\sqrt{2}}{2} + \frac{1}{2} \ln(\sqrt{2} + 1) \approx 1.1478$$ □

Integrands Containing a Quadratic Expression By completion of the square, it is possible to express an integrand that contains a quadratic expression in one of the following forms:

$$a^2 - u^2, \quad a^2 + u^2, \quad \text{or} \quad u^2 - a^2$$

The appropriate substitutions are summarized in the accompanying diagrams.

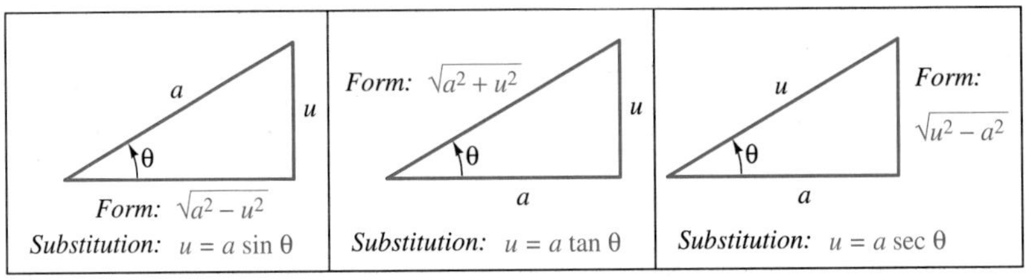

EXAMPLE 6 Evaluate $\displaystyle\int \frac{dx}{(x^2 + 8x + 25)^{3/2}}$.

Solution Since

$$\int \frac{dx}{(x^2 + 8x + 25)^{3/2}} = \int \frac{dx}{[9 + (x + 4)^2]^{3/2}}$$

we identify $a^2 + u^2$ with $a = 3$ and $u = x + 4$. Using

$$x + 4 = 3 \tan \theta \qquad dx = 3 \sec^2\theta \, d\theta$$

we find

$$\begin{aligned}
\int \frac{dx}{(x^2 + 8x + 25)^{3/2}} &= \int \frac{3 \sec^2\theta \, d\theta}{[9 + 9 \tan^2\theta]^{3/2}} \\
&= \frac{1}{9} \int \frac{\sec^2\theta}{\sec^3\theta} \, d\theta \\
&= \frac{1}{9} \int \cos \theta \, d\theta \\
&= \frac{1}{9} \sin \theta + C
\end{aligned}$$

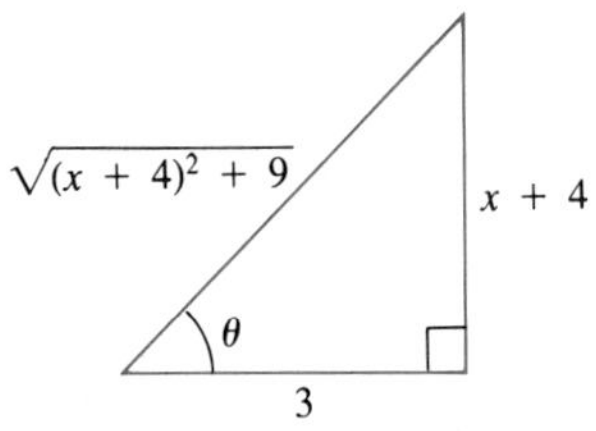

FIGURE 8.8

Inspection of the triangle in Figure 8.8 indicates how to express $\sin \theta$ in terms of x. It follows that

$$\int \frac{dx}{(x^2 + 8x + 25)^{3/2}} = \frac{1}{9} \frac{x + 4}{\sqrt{(x + 4)^2 + 9}} + C = \frac{x + 4}{9\sqrt{x^2 + 8x + 25}} + C \quad \square$$

▲ ***Remarks*** (*i*) A common error when evaluating an integral $\int f(x) \, dx$ by the substitution method is to forget to substitute for the differential dx. If the integral calls for, say, $x = a \sin \theta$, be mindful to replace dx by $a \cos \theta \, d\theta$.

(*ii*) Do you still remember the integral forms

$$\int \frac{du}{\sqrt{a^2 - u^2}} \qquad \int \frac{du}{a^2 + u^2} \qquad \int \frac{du}{u\sqrt{u^2 - a^2}}$$

from Chapter 7? More important, do you remember the value of each integral? If not, and if you do recognize these integrals as candidates for the types of trigonometric substitutions considered in this chapter, then you can easily recover the results on page 475. For example, try substituting $u = a \sin \theta$ in the first integral above and see what happens.

(*iii*) In the three cases just considered in this section, other substitutions are possible though not necessarily desirable. For example, $x = a \sinh t$ can be used to eliminate the radical $\sqrt{a^2 + x^2}$, $a > 0$. See Problem 61. Likewise, we can use $x = a \cos \theta$, $0 \le \theta \le \pi$, for $\sqrt{a^2 - x^2}$, $a > 0$. See Problem 62.

EXERCISES 8.4

Answers to odd-numbered problems begin on page A-89.

In Problems 1–44 evaluate the given integral by a trigonometric substitution when appropriate.*

1. $\displaystyle\int \frac{\sqrt{1 - x^2}}{x^2}\, dx$

2. $\displaystyle\int \frac{x^3}{\sqrt{x^2 - 4}}\, dx$

3. $\displaystyle\int \frac{dx}{\sqrt{x^2 - 36}}$

4. $\displaystyle\int \sqrt{3 - x^2}\, dx$

5. $\displaystyle\int x\sqrt{x^2 + 7}\, dx$

6. $\displaystyle\int (1 - x^2)^{3/2}\, dx$

7. $\displaystyle\int x^3\sqrt{1 - x^2}\, dx$

8. $\displaystyle\int x^3\sqrt{x^2 - 1}\, dx$

9. $\displaystyle\int \frac{dx}{(x^2 - 4)^{3/2}}$

10. $\displaystyle\int (9 - x^2)^{-3/2}\, dx$

11. $\displaystyle\int \sqrt{x^2 + 4}\, dx$

12. $\displaystyle\int \frac{x}{25 + x^2}\, dx$

13. $\displaystyle\int \frac{dx}{\sqrt{25 - x^2}}$

14. $\displaystyle\int \frac{dx}{x\sqrt{x^2 - 25}}$

15. $\displaystyle\int \frac{dx}{x\sqrt{16 - x^2}}$

16. $\displaystyle\int \frac{dx}{x^2\sqrt{16 - x^2}}$

17. $\displaystyle\int \frac{dx}{x\sqrt{1 + x^2}}$

18. $\displaystyle\int \frac{dx}{x^2\sqrt{1 + x^2}}$

19. $\displaystyle\int \frac{\sqrt{1 - x^2}}{x^4}\, dx$

20. $\displaystyle\int \frac{\sqrt{x^2 - 1}}{x^4}\, dx$

21. $\displaystyle\int \frac{x^2}{(9 - x^2)^{3/2}}\, dx$

22. $\displaystyle\int \frac{x^2}{(4 + x^2)^{3/2}}\, dx$

23. $\displaystyle\int \frac{dx}{(1 + x^2)^2}$

24. $\displaystyle\int \frac{x^2}{(x^2 - 1)^2}\, dx$

25. $\displaystyle\int \frac{dx}{(4 + x^2)^{5/2}}$

26. $\displaystyle\int \frac{x^3}{(1 - x^2)^{5/2}}\, dx$

27. $\displaystyle\int \frac{dx}{\sqrt{x^2 + 2x + 10}}$

28. $\displaystyle\int \frac{x}{\sqrt{4x - x^2}}\, dx$

29. $\displaystyle\int \frac{dx}{(x^2 + 6x + 13)^2}$

30. $\displaystyle\int \frac{dx}{(11 - 10x - x^2)^2}$

31. $\displaystyle\int \frac{x - 3}{(5 - 4x - x^2)^{3/2}}\, dx$

32. $\displaystyle\int \frac{dx}{(x^2 + 2x)^{3/2}}$

33. $\displaystyle\int \frac{2x + 4}{x^2 + 4x + 13}\, dx$

34. $\displaystyle\int \frac{dx}{4 + (x - 3)^2}$

35. $\displaystyle\int \frac{x^2}{x^2 + 16}\, dx$

36. $\displaystyle\int \frac{\sqrt{4 - 9x^2}}{x}\, dx$

37. $\displaystyle\int_{-1}^{1} \sqrt{4 - x^2}\, dx$

38. $\displaystyle\int_{-1}^{\sqrt{3}} \frac{x^2}{\sqrt{4 - x^2}}\, dx$

39. $\displaystyle\int_0^5 \frac{dx}{(x^2 + 25)^{3/2}}$

40. $\displaystyle\int_{\sqrt{2}}^{2} \frac{dx}{x^3\sqrt{x^2 - 1}}$

41. $\displaystyle\int_1^{6/5} \frac{16\, dx}{x^4\sqrt{4 - x^2}}$

42. $\displaystyle\int_0^{1/2} x^3(1 + x^2)^{-1/2}\, dx$

43. $\displaystyle\int \frac{dx}{\sqrt{e^{2x} - 1}}$

44. $\displaystyle\int \sqrt{e^{2x} - 1}\, dx$

*Look before you leap.

45. Find the area under the graph of $y = 1/(x\sqrt{3 + x^2})$ on the interval $[1, \sqrt{3}]$.

46. Find the area under the graph of $y = x^5\sqrt{1 - x^2}$ on the interval $[0, 1]$.

47. Show that the area of a circle given by $x^2 + y^2 = a^2$ is πa^2.

48. Show that the area of an ellipse given by $a^2x^2 + b^2y^2 = a^2b^2$ is πab.

49. The region described in Problem 45 is revolved about the x-axis. Find the volume of the solid of revolution.

50. The region in the first quadrant bounded by the graphs of $y = 4/(4 + x^2)$, $x = 2$, and $y = 0$ is revolved about the x-axis. Find the volume of the solid of revolution.

51. The region in the first quadrant bounded by the graphs of $y = x\sqrt{4 + x^2}$, $x = 2$, and $y = 0$ is revolved about the y-axis. Find the volume of the solid of revolution.

52. The region in the first quadrant bounded by the graphs of $y = x/\sqrt{4 - x^2}$, $x = 1$, and $y = 0$ is revolved about the y-axis. Find the volume of the solid of revolution.

53. Find the length of the graph of $y = \ln x$ on the interval $[1, \sqrt{3}]$.

54. Find the length of the graph of $y = -\frac{1}{2}x^2 + 2x$ on the interval $[1, 2]$.

55. A woman, W, starting at the origin, moves in the direction of the positive x-axis pulling a weight along the curve C, called a **tractrix**, indicated in Figure 8.9. The weight, initially located on the y-axis at $(0, s)$, is pulled by a rope of constant length s that is kept taut throughout the motion.

(a) Show that the differential equation of the tractrix is

$$\frac{dy}{dx} = -\frac{y}{\sqrt{s^2 - y^2}}$$

(b) Solve the equation in part (**a**). Assume that the initial point on the y-axis is $(0, 10)$ and the length of the rope is $s = 10$ ft.

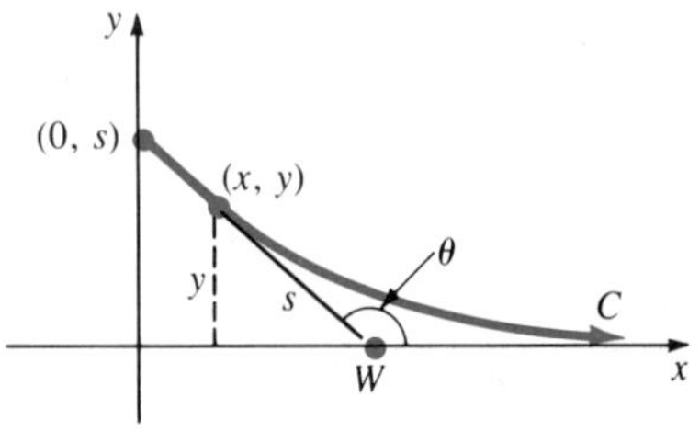

FIGURE 8.9

56. The region bounded by the graph of $(x - a)^2 + y^2 = r^2$, $r < a$, is revolved about the y-axis. Find the volume of the solid of revolution or **torus**. See Figure 8.10.

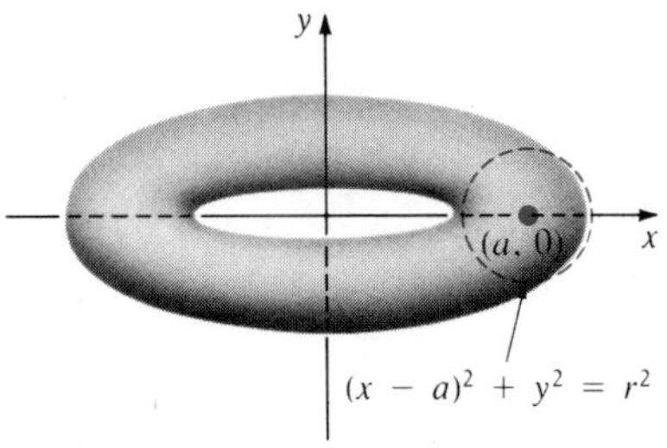

FIGURE 8.10

57. Find the force caused by liquid pressure on one side of the vertical plate shown in Figure 8.11. Assume that the plate is submerged in water and that dimensions are in feet.

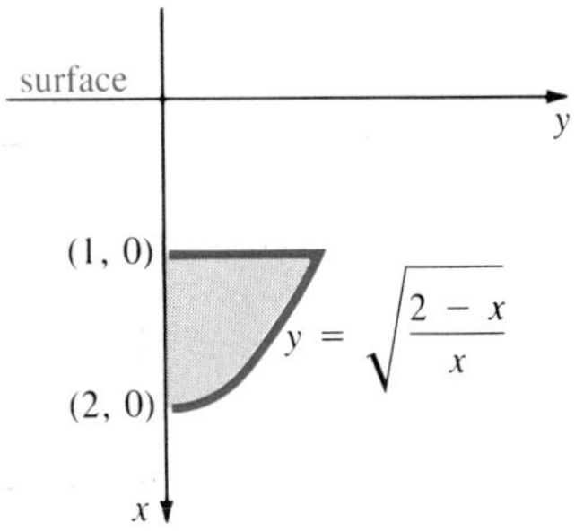

FIGURE 8.11

58. Find the centroid of the region bounded by the graphs of $y = 1/\sqrt{1 + x^2}$, $y = 0$, $x = 0$, and $x = \sqrt{3}$.

In Problems 59 and 60 use integration by parts followed by a trigonometric substitution.

59. $\int x^2 \sin^{-1}x\, dx$

60. $\int x \cos^{-1}x\, dx$

In Problems 61 and 62 use the indicated substitution to evaluate the given integral.

61. $\int \frac{dx}{x^2\sqrt{9 + x^2}}$; $x = 3\sinh t$

62. $\int \frac{(1 + x)^2}{\sqrt{1 - x^2}}\, dx$; $x = \cos\theta$

63. Establish the formula

$$\int \sqrt{u^2 \pm a^2}\, du = \frac{1}{2}u\sqrt{u^2 \pm a^2} \pm \frac{a^2}{2}\ln|u + \sqrt{u^2 \pm a^2}| + C, \quad a > 0$$

8.5 PARTIAL FRACTIONS

8.5.1 Denominators Containing Linear Factors

When the terms in the sum

$$\frac{2}{x+5} + \frac{1}{x+1} \tag{1}$$

are combined by means of a common denominator, we obtain the single rational expression

$$\frac{3x+7}{(x+5)(x+1)} \tag{2}$$

Now suppose that we are faced with the problem of evaluating the integral $\int (3x+7)\,dx/[(x+5)(x+1)]$. Of course, the solution is obvious: We use the equality of (1) and (2) to write

$$\begin{aligned}\int \frac{3x+7}{(x+5)(x+1)}\,dx &= \int \left[\frac{2}{x+5} + \frac{1}{x+1}\right] dx\\ &= 2\ln|x+5| + \ln|x+1| + C\end{aligned}$$

This example illustrates a procedure for integrating certain rational functions $P(x)/Q(x)$, where the degree of $P(x)$ is less than the degree of $Q(x)$. This method, known as **partial fractions**, consists of decomposing such a rational function into simpler component fractions, and then evaluating the integral term-by-term. In this section, we shall study four cases of partial fraction decomposition.

CASE I Nonrepeated Linear Factors

We state the following fact from algebra without proof. If

$$\frac{P(x)}{Q(x)} = \frac{P(x)}{(a_1x+b_1)(a_2x+b_2)\cdots(a_nx+b_n)}$$

where all the factors $a_ix + b_i$, $i = 2, \ldots, n$, are distinct and the degree of $P(x)$ is less than n, then unique real constants $C_1, C_2, \ldots, C_n$ exist such that

$$\frac{P(x)}{Q(x)} = \frac{C_1}{a_1x+b_1} + \frac{C_2}{a_2x+b_2} + \cdots + \frac{C_n}{a_nx+b_n}$$

EXAMPLE 1 Evaluate $\displaystyle\int \frac{2x+1}{(x-1)(x+3)}\,dx$.

Solution We make the assumption that the integrand can be written as

$$\frac{2x+1}{(x-1)(x+3)} = \frac{A}{x-1} + \frac{B}{x+3}$$

Combining the terms of the right-hand member of the equation over a common denominator gives

$$\frac{2x+1}{(x-1)(x+3)} = \frac{A(x+3)+B(x-1)}{(x-1)(x+3)}$$

Since the denominators are identical, the numerators are identical

$$\begin{aligned} 2x+1 &= A(x+3)+B(x-1) \\ &= (A+B)x+(3A-B) \end{aligned} \tag{3}$$

and the coefficients of the powers of x are the same:

$$\begin{aligned} 2 &= A+B \\ 1 &= 3A-B \end{aligned}$$

These simultaneous equations can now be solved for A and B. The results are $A=\frac{3}{4}$ and $B=\frac{5}{4}$. Therefore,

$$\begin{aligned} \int \frac{2x+1}{(x-1)(x+3)}\,dx &= \int \left[\frac{3/4}{x-1}+\frac{5/4}{x+3}\right] dx \\ &= \frac{3}{4}\ln|x-1| + \frac{5}{4}\ln|x+3| + C \end{aligned}$$ □

Note: In the preceding example, the numbers A and B can be determined in an alternative manner. Since (3) is an identity, that is, the equality is true for every value of x, it holds for $x=1$ and $x=-3$ (the zeros of the denominator). Setting $x=1$ in (3) gives $3=4A$, from which it follows that $A=\frac{3}{4}$. Similarly, by setting $x=-3$ in (3), we obtain $-5=(-4)B$ or $B=\frac{5}{4}$.

See the second remark at the end of this section for another quick method for determining the constants.

EXAMPLE 2 Evaluate $\displaystyle\int \frac{x^3-2x}{x^2+3x+2}\,dx$.

Solution We first observe that the degree of the numerator is greater than the degree of the denominator. Hence, long division is called for:

$$\int \frac{x^3-2x}{x^2+3x+2}\,dx = \int \left[x-3+\frac{5x+6}{x^2+3x+2}\right] dx$$

Since $x^2+3x+2=(x+1)(x+2)$, we write

$$\frac{5x+6}{(x+1)(x+2)} = \frac{A}{x+1}+\frac{B}{x+2}$$

and

$$5x+6 = A(x+2)+B(x+1) \tag{4}$$

If we set $x=-2$ and $x=-1$ in (4), we see immediately that $B=4$ and $A=1$, respectively. Thus,

$$\int \frac{x^3 - 2x}{x^2 + 3x + 2}\,dx = \int \left[x - 3 + \frac{1}{x+1} + \frac{4}{x+2}\right] dx$$
$$= \frac{x^2}{2} - 3x + \ln|x+1| + 4\ln|x+2| + C \quad \square$$

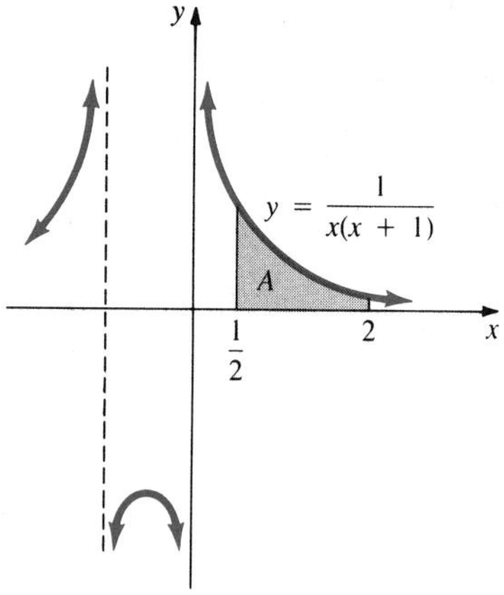

FIGURE 8.12

EXAMPLE 3 Find the area A under the graph of $y = 1/x(x+1)$ on the interval $[\frac{1}{2}, 2]$.

Solution The area in question is shown in Figure 8.12. We have

$$A = \int_{1/2}^{2} \frac{dx}{x(x+1)}$$

Using partial fractions

$$\frac{1}{x(x+1)} = \frac{A}{x} + \frac{B}{x+1} = \frac{A(x+1) + Bx}{x(x+1)}$$

it follows that
$$1 = A(x+1) + Bx$$
$$= (A+B)x + A$$

The solution of the system

$$0 = A + B$$
$$1 = A$$

is immediate: $A = 1$, $B = -1$. Therefore,

$$A = \int_{1/2}^{2} \left[\frac{1}{x} - \frac{1}{x+1}\right] dx = (\ln|x| - \ln|x+1|)\Big]_{1/2}^{2}$$
$$= \ln\left|\frac{x}{x+1}\right|\Bigg]_{1/2}^{2} = \ln 2 \approx 0.6931 \text{ square unit} \quad \square$$

Logistic Equation We saw in Section 6.8 that if a population P is described by $dP/dt = kP$, $k > 0$, then $P(t)$ exhibits unbounded exponential growth. In many instances this differential equation provides an unrealistic model of the growth of a population; that is, what is actually observed differs substantially from what is predicted. Around 1840 the Belgian mathematician-biologist P. F. Verhulst was concerned with mathematical formulations for predicting populations of various countries. One of the equations he studied was

$$\frac{dP}{dt} = P(a - bP)$$

where a and b are positive constants. This differential equation came to be known as the **logistic equation** and its solution is called the **logistic function** (the graph of which is naturally called a logistic curve). Logistic curves have proved to be quite accurate in predicting the growth patterns, in a limited space, of certain types of bacteria, protozoa, water fleas (*Daphnia*), and fruit flies (*Drosophilia*).

EXAMPLE 4 Solve the logistic equation.

Solution By separating variables and using partial fractions, we can write

$$\frac{dP}{P(a - bP)} = dt$$

$$\left[\frac{1/a}{P} + \frac{b/a}{a - bP}\right] dP - dt$$

$$\frac{1}{a}\ln|P| - \frac{1}{a}\ln|a - bP| = t + C$$

$$\ln\left|\frac{P}{a - bP}\right| = at + aC$$

$$\frac{P}{a - bP} = C_1 e^{at}$$

It follows from the last equation that

$$P(t) = \frac{aC_1 e^{at}}{1 + bC_1 e^{at}} = \frac{aC_1}{bC_1 + e^{-at}}$$ □

CASE II Repeated Linear Factors

If
$$\frac{P(x)}{Q(x)} = \frac{P(x)}{(ax + b)^n}$$

where $n > 1$ and the degree of $P(x)$ is less than n, then unique real constants $C_1, C_2, \ldots, C_n$ can be found such that

$$\frac{P(x)}{(ax + b)^n} = \frac{C_1}{ax + b} + \frac{C_2}{(ax + b)^2} + \cdots + \frac{C_n}{(ax + b)^n}$$

EXAMPLE 5 Evaluate $\displaystyle\int \frac{x^2 + 2x + 4}{(x + 1)^3}\,dx.$

Solution The decomposition of the integrand is

$$\frac{x^2 + 2x + 4}{(x + 1)^3} = \frac{A}{x + 1} + \frac{B}{(x + 1)^2} + \frac{C}{(x + 1)^3}$$

By equating numerators,

$$\begin{aligned} x^2 + 2x + 4 &= A(x + 1)^2 + B(x + 1) + C \\ &= Ax^2 + (2A + B)x + (A + B + C) \end{aligned} \tag{5}$$

we obtain the system of equations

$$\begin{aligned} 1 &= A \\ 2 &= 2A + B \\ 4 &= A + B + C \end{aligned}$$

Solving the equations yields $A = 1$, $B = 0$, and $C = 3$. Therefore,

$$\begin{aligned} \int \frac{x^2 + 2x + 4}{(x+1)^3}\,dx &= \int \left[\frac{1}{x+1} + \frac{3}{(x+1)^3}\right] dx \\ &= \int \left[\frac{1}{x+1} + 3(x+1)^{-3}\right] dx \\ &= \ln|x+1| - \frac{3}{2}(x+1)^{-2} + D \end{aligned}$$

Note that setting $x = -1$ (the single zero of the denominator) in (5) yields only $C = 3$. □

Combining the Cases When the denominator $Q(x)$ contains distinct as well as repeated linear factors, we combine the two cases.

EXAMPLE 6 Evaluate $\displaystyle\int \frac{6x - 1}{x^3(2x - 1)}\,dx.$

Solution We write

$$\frac{6x-1}{x^3(2x-1)} = \frac{A}{x} + \frac{B}{x^2} + \frac{C}{x^3} + \frac{D}{2x-1}$$

from which it follows that

$$\begin{aligned} 6x - 1 &= Ax^2(2x-1) + Bx(2x-1) + C(2x-1) + Dx^3 && (6) \\ &= (2A + D)x^3 + (-A + 2B)x^2 + (-B + 2C)x - C && (7) \end{aligned}$$

If we set $x = 0$ and $x = \frac{1}{2}$ in (6), we find $C = 1$ and $D = 16$, respectively. Now, by equating the coefficients of x^3 and x^2 in (7), we get

$$\begin{aligned} 0 &= 2A + D \\ 0 &= -A + 2B \end{aligned}$$

Since we know the value of D, the first equation yields $A = -D/2 = -8$. The second then gives $B = A/2 = -4$. Therefore,

$$\begin{aligned} \int \frac{6x-1}{x^3(2x-1)}\,dx &= \int \left[-\frac{8}{x} - \frac{4}{x^2} + \frac{1}{x^3} + \frac{16}{2x-1}\right] dx \\ &= -8\ln|x| + 4x^{-1} - \frac{1}{2}x^{-2} + 8\ln|2x-1| + E \\ &= 8\ln\left|\frac{2x-1}{x}\right| + 4x^{-1} - \frac{1}{2}x^{-2} + E \end{aligned}$$

□

8.5.2 Denominators Containing Irreducible Quadratic Factors*

CASE III Nonrepeated Quadratic Factors

Suppose the denominator of the rational function $P(x)/Q(x)$ can be expressed as a product of distinct irreducible* quadratic factors $a_i x^2 + b_i x + c_i$, $i = 1, 2, \ldots, n$. If the degree of $P(x)$ is less than $2n$, we can find unique real constants $A_1, A_2, \ldots, A_n, B_1, B_2, \ldots, B_n$ such that

$$\frac{P(x)}{(a_1x^2 + b_1x + c_1)(a_2x^2 + b_2x + c_2)\cdots(a_nx^2 + b_nx + c_n)}$$
$$= \frac{A_1x + B_1}{a_1x^2 + b_1x + c_1} + \frac{A_2x + B_2}{a_2x^2 + b_2x + c_2} + \cdots + \frac{A_nx + B_n}{a_nx^2 + b_nx + c_n}$$

EXAMPLE 7 Evaluate $\displaystyle\int \frac{4x}{(x^2+1)(x^2+2x+3)}\,dx.$

Solution We write

$$\frac{4x}{(x^2+1)(x^2+2x+3)} = \frac{Ax+B}{x^2+1} + \frac{Cx+D}{x^2+2x+3}$$

from which we find

$$\begin{aligned} 4x &= (Ax+B)(x^2+2x+3) + (Cx+D)(x^2+1) \\ &= (A+C)x^3 + (2A+B+D)x^2 + (3A+2B+C)x + (3B+D) \end{aligned}$$

Since the denominator of the integrand has no real roots, we compare coefficients of powers of x:

$$\begin{aligned} 0 &= A + C \\ 0 &= 2A + B + D \\ 4 &= 3A + 2B + C \\ 0 &= 3B + D \end{aligned}$$

Solving the equations yields $A = 1$, $B = 1$, $C = -1$, and $D = -3$. Therefore,

$$\int \frac{4x}{(x^2+1)(x^2+2x+3)}\,dx = \int \left[\frac{x+1}{x^2+1} - \frac{x+3}{x^2+2x+3}\right] dx$$

Now, the integral of each term still presents a slight problem. First, we write

$$\frac{x+1}{x^2+1} = \frac{1}{2}\frac{2x}{x^2+1} + \frac{1}{x^2+1} \qquad (8)$$

*The word *irreducible* means that the quadratic expression $ax^2 + bx + c$ does not factor over the set of real numbers. This situation occurs when $b^2 - 4ac < 0$.

and then, after completing the square,

$$\frac{x+3}{x^2+2x+3} = \frac{x+1+2}{(x+1)^2+2} = \frac{1}{2}\frac{2(x+1)}{(x+1)^2+2} + \frac{2}{(x+1)^2+2} \tag{9}$$

In the right-hand members of (8) and (9), we recognize that the integrals of the first and second terms are, respectively, of the forms $\int du/u$ and $\int du/(u^2+a^2)$. Finally, we obtain

$$\int \frac{4x}{(x^2+1)(x^2+2x+3)}\,dx = \int \left[\frac{1}{2}\frac{2x}{x^2+1} + \frac{1}{x^2+1} - \frac{1}{2}\frac{2(x+1)}{(x+1)^2+2} - \frac{2}{(x+1)^2+(\sqrt{2})^2}\right] dx$$

$$= \frac{1}{2}\ln(x^2+1) + \tan^{-1}x - \frac{1}{2}\ln[(x+1)^2+2] - \sqrt{2}\tan^{-1}\frac{x+1}{\sqrt{2}} + E$$

$$= \frac{1}{2}\ln\left(\frac{x^2+1}{x^2+2x+3}\right) + \tan^{-1}x - \sqrt{2}\tan^{-1}\frac{x+1}{\sqrt{2}} + E \qquad \square$$

CASE IV Repeated Quadratic Factors

We now consider the case when the integrand is $P(x)/(ax^2+bx+c)^n$, where ax^2+bx+c is irreducible and $n > 1$. If the degree of $P(x)$ is less than $2n$, we can find the unique real constants $A_1, A_2, \ldots, A_n, B_1, B_2, \ldots, B_n$ such that

$$\frac{P(x)}{(ax^2+bx+c)^n} = \frac{A_1x+B_1}{ax^2+bx+c} + \frac{A_2x+B_2}{(ax^2+bx+c)^2} + \cdots + \frac{A_nx+B_n}{(ax^2+bx+c)^n}$$

EXAMPLE 8 Evaluate $\displaystyle\int \frac{x^2}{(x^2+4)^2}\,dx$.

Solution The partial fraction decomposition of the integrand

$$\frac{x^2}{(x^2+4)^2} = \frac{Ax+B}{x^2+4} + \frac{Cx+D}{(x^2+4)^2}$$

leads to

$$\begin{aligned} x^2 &= (Ax+B)(x^2+4) + Cx + D \\ &= Ax^3 + Bx^2 + (4A+C)x + (4B+D) \end{aligned}$$

and

$$\begin{aligned} 0 &= A \\ 1 &= B \\ 0 &= 4A + C \\ 0 &= 4B + D \end{aligned}$$

We find that $A = 0$, $B = 1$, $C = 0$, and $D = -4$. Consequently,

$$\int \frac{x^2}{(x^2+4)^2}\,dx = \int \left[\frac{1}{x^2+4} - \frac{4}{(x^2+4)^2}\right] dx$$

The integral of the first term is an inverse tangent. However, to evaluate the integral of the second term, we employ the trigonometric substitution $x = 2\tan\theta$:

$$\begin{aligned}
\int \frac{dx}{(x^2+4)^2} &= \int \frac{2\sec^2\theta\, d\theta}{(4\tan^2\theta + 4)^2} \\
&= \frac{1}{8}\int \frac{\sec^2\theta}{\sec^4\theta}\, d\theta = \frac{1}{8}\int \cos^2\theta\, d\theta \\
&= \frac{1}{16}\int (1+\cos 2\theta)\, d\theta = \frac{1}{16}\left(\theta + \frac{1}{2}\sin 2\theta\right) \\
&= \frac{1}{16}(\theta + \sin\theta\cos\theta) \\
&= \frac{1}{16}\left[\tan^{-1}\frac{x}{2} + \frac{x}{\sqrt{x^2+4}}\cdot\frac{2}{\sqrt{x^2+4}}\right] \\
&= \frac{1}{16}\left[\tan^{-1}\frac{x}{2} + \frac{2x}{x^2+4}\right]
\end{aligned}$$

Therefore, the original integral is

$$\begin{aligned}
\int \frac{x^2}{(x^2+4)^2}\, dx &= \frac{1}{2}\tan^{-1}\frac{x}{2} - 4\left[\frac{1}{16}\tan^{-1}\frac{x}{2} + \frac{1}{8}\frac{x}{x^2+4}\right] + E \\
&= \frac{1}{4}\tan^{-1}\frac{x}{2} - \frac{x}{2(x^2+4)} + E
\end{aligned}$$

□

EXAMPLE 9 Evaluate $\displaystyle\int \frac{x+3}{x^4+9x^2}\, dx.$

Solution From $x^4 + 9x^2 = x^2(x^2+9)$, we see that the problem combines the quadratic factor $x^2 + 9$ with the repeated linear factor x. Accordingly, the partial fraction decomposition is

$$\frac{x+3}{x^2(x^2+9)} = \frac{A}{x} + \frac{B}{x^2} + \frac{Cx+D}{x^2+9}$$

Proceeding as usual, we find

$$x + 3 = (A+C)x^3 + (B+D)x^2 + 9Ax + 9B$$

and

$$\begin{aligned}
0 &= A + C \\
0 &= B + D \\
1 &= 9A \\
3 &= 9B
\end{aligned}$$

Hence, $A = \frac{1}{9}$, $B = \frac{1}{3}$, $C = -\frac{1}{9}$, and $D = -\frac{1}{3}$. This gives

$$\begin{aligned}\int \frac{x+3}{x^2(x^2+9)}\,dx &= \int \left[\frac{1/9}{x} + \frac{1/3}{x^2} - \frac{x/9 + 1/3}{x^2+9}\right] dx \\ &= \int \left[\frac{1/9}{x} + \frac{1/3}{x^2} - \frac{1}{18}\frac{2x}{x^2+9} - \frac{1}{3}\frac{1}{x^2+9}\right] dx \\ &= \frac{1}{9}\ln|x| - \frac{1}{3}x^{-1} - \frac{1}{18}\ln(x^2+9) - \frac{1}{9}\tan^{-1}\frac{x}{3} + E \\ &= \frac{1}{18}\ln\left(\frac{x^2}{x^2+9}\right) - \frac{1}{3}x^{-1} - \frac{1}{9}\tan^{-1}\frac{x}{3} + E \qquad \square\end{aligned}$$

▲ ***Remarks*** (*i*) Integrals such as $\int dx/(x+2)^4$ and $\int (2x+1)\,dx/(x^2+1)^2$ *appear* to be candidates for partial fractions. However, this is not the case. Why? You should be able to evaluate these integrals by alternative means.

(*ii*) There is another way, called the **cover-up method**, of determining the coefficients in a partial fraction decomposition in the special case when $F(x)$ is a quotient of polynomials $P(x)/Q(x)$ and $Q(x)$ is a product of *distinct* linear factors:

$$F(x) = \frac{P(x)}{(x-r_1)(x-r_2)\cdots(x-r_n)}$$

Let's illustrate by means of a specific example. From the foregoing discussion we know there exists unique constants A, B, and C such that

$$\frac{x^2+4x-1}{(x-1)(x-2)(x+3)} = \frac{A}{x-1} + \frac{B}{x-2} + \frac{C}{x+3} \tag{10}$$

Suppose we then multiply both sides of this last expression by $x - 1$, simplify, and then set $x = 1$. Since the coefficients of B and C are zero, we get

$$\left.\frac{x^2+4x-1}{(x-2)(x+3)}\right|_{x=1} = A \quad \text{or} \quad A = -1$$

Written another way,

$$\left.\frac{x^2+4x-1}{(x-1)\,(x-2)(x+3)}\right|_{x=1} = A$$

where we have shaded or *covered-up* the factor that canceled when the left side of (10) was multiplied by $x - 1$. We *do not evaluate this covered-up factor* at $x = 1$. Now to obtain B and C we simply evaluate the left member of (10) while covering, in turn, $x - 2$ and $x + 3$:

$$\left.\frac{x^2+4x-1}{(x-1)\,(x-2)\,(x+3)}\right|_{x=2} = B \quad \text{or} \quad B = \frac{11}{5}$$

$$\left.\frac{x^2+4x-1}{(x-1)(x-2)\,(x+3)}\right|_{x=-3} = C \quad \text{or} \quad C = -\frac{1}{5}$$

Thus, we have

$$\frac{x^2+4x-1}{(x-1)(x-2)(x+3)} = \frac{-1}{x-1} + \frac{11/15}{x-2} + \frac{-1/5}{x+3}$$

EXERCISES 8.5

Answers to odd-numbered problems begin on page A-89.

[8.5.1]

In Problems 1–40 use partial fractions when appropriate to evaluate the given integral.

1. $\int \frac{dx}{x(x-2)}$
2. $\int \frac{dx}{x(2x+3)}$
3. $\int \frac{x+2}{2x^2-x}\,dx$
4. $\int \frac{3x+10}{x^2+2x}\,dx$
5. $\int \frac{dx}{x^2-9}$
6. $\int \frac{dx}{4x^2-25}$
7. $\int \frac{x+1}{x^2-16}\,dx$
8. $\int \frac{x+5}{(x+4)(x^2-1)}\,dx$
9. $\int \frac{dx}{x^2+4x+3}$
10. $\int \frac{dx}{x^2+x-2}$
11. $\int \frac{x}{2x^2+5x+2}\,dx$
12. $\int \frac{x+7}{x^2-3x-10}\,dx$
13. $\int \frac{x^2+2x-6}{x^3-x}\,dx$
14. $\int \frac{5x^2-x+1}{x^3-4x}\,dx$
15. $\int \frac{dx}{(x+1)(x+2)(x+3)}$
16. $\int \frac{dx}{(4x^2-1)(x+7)}$
17. $\int \frac{4t^2+3t-1}{t^3-t^2}\,dt$
18. $\int \frac{2x-11}{x^3+2x^2}\,dx$
19. $\int \frac{dx}{x^3+2x^2+x}$
20. $\int \frac{t-1}{t^4+6t^3+9t^2}\,dt$
21. $\int \frac{dx}{(x-3)^4}$
22. $\int \frac{4x^2-5x+7}{x^3}\,dx$
23. $\int \frac{2x-1}{(x+1)^3}\,dx$
24. $\int \frac{x^2+2x-6}{(x-1)^3}\,dx$
25. $\int \frac{x}{(x^2-1)^2}\,dx$
26. $\int \frac{dx}{x^2(x^2-4)^2}$
27. $\int \frac{dx}{(x^2+6x+5)^2}$
28. $\int \frac{dx}{(x^2-x-6)(x^2-2x-8)}$
29. $\int \frac{x^4+2x^2-x+9}{x^5+2x^4}\,dx$
30. $\int \frac{5x-1}{x(x-3)^2(x+2)^2}\,dx$
31. $\int \frac{x^4+3x^2+4}{(x+1)^2}\,dx$
32. $\int \frac{x^5-10x^3}{x^4-10x^2+9}\,dx$
33. $\int_2^4 \frac{dx}{x^2-6x+5}$
34. $\int_0^1 \frac{dx}{x^2-4}$
35. $\int_0^2 \frac{2x-1}{(x+3)^2}\,dx$
36. $\int_1^5 \frac{2x+6}{x(x+1)^2}\,dx$
37. $\int \frac{\cos x}{\sin^2 x+3\sin x+2}\,dx$
38. $\int \frac{\sin x}{\cos^2 x-\cos^3 x}\,dx$
39. $\int \frac{e^t}{(e^t+1)^2(e^t-2)}\,dt$
40. $\int \frac{e^{2t}}{(e^t+1)^3}\,dt$

In Problems 41 and 42 use the indicated substitution to evaluate the given integral.

41. $\int \frac{\sqrt{1-x^2}}{x^3}\,dx;\ u^2=1-x^2$
42. $\int \sqrt{\frac{x-1}{x+1}}\,dx;\ u^2=\frac{x-1}{x+1}$

43. Find the area under the graph of $y=1/(x^2+2x-3)$ on the interval $[2, 4]$.

44. Find the area bounded by the graph of $y=x/(x+2)(x+3)$ and the x-axis on the interval $[-1, 1]$.

45. The region in the first quadrant bounded by the graphs of $y=2/x(x+1)$, $x=1$, $x=3$, and $y=0$ is revolved about the x-axis. Find the volume of the solid of revolution.

46. The region in the first quadrant bounded by the graphs of $y=1/\sqrt{(x+1)(x+4)}$, $x=0$, $x=2$, and $y=0$ is revolved about the x-axis. Find the volume of the solid of revolution.

47. The region in the first quadrant bounded by the graphs of $y=4/(x+1)^2$, $x=0$, $x=1$, and $y=0$ is revolved about the y-axis. Find the volume of the solid of revolution.

48. Find the length of the graph of $y=e^x$ on the interval $[0, \ln 2]$. [*Hint:* Let $u^2=1+e^{2x}$.]

49. Suppose a student carrying a flu virus returns to an isolated college campus of 1000 students. If it is assumed that the rate at which the virus spreads is proportional not only to the number N of infected students but also to the number $1000 - N$ of students not infected, and if it is further assumed that no one leaves the campus throughout the duration of the disease, then we must solve the logistic differential equation $dN/dt = kN(1000 - N)$ subject to $N(0) = 1$. Determine the number of infected students after 6 days if it is observed that after 4 days $N(4) = 50$.

50. The rate at which a chemical is formed during a second-order chemical reaction is given by $dX/dt = k(a - X)(b - X)$, where k, a, and b are constants. Use separation of variables to solve the differential equation in the case $a \neq b$.

[8.5.2]

In Problems 51–80 use partial fractions when appropriate to evaluate the given integral.

51. $\int \frac{dx}{x^4 + 5x^2 + 4}$

52. $\int \frac{dx}{x^4 + 13x^2 + 36}$

53. $\int \frac{x - 15}{(x^2 + 2x + 5)(x^2 + 6x + 10)} dx$

54. $\int \frac{x^2}{(x^2 + 8x + 20)(x^2 + 4x + 6)} dx$

55. $\int \frac{x - 1}{x(x^2 + 1)} dx$

56. $\int \frac{dx}{(x - 1)(x^2 + 3)}$

57. $\int \frac{2x - 3}{x^3 - 3x^2 + 9x - 27} dx$

58. $\int \frac{x + 4}{x^4 + 9x^2} dx$

59. $\int \frac{x}{(x + 1)^2(x^2 + 1)} dx$

60. $\int \frac{x^2}{(x - 1)^3(x^2 + 4)} dx$

61. $\int \frac{dt}{t^4 - 1}$

62. $\int \frac{t^3}{t^4 - 16} dt$

63. $\int \frac{2x + 1}{(x^2 + 4)^2} dx$

64. $\int \frac{dx}{(x^4 - 16)^2}$

65. $\int \frac{3x^2 - x + 1}{(x + 1)(x^2 + 2x + 2)} dx$

66. $\int \frac{4x - 5}{(x - 2)(x^2 + 4x + 8)} dx$

67. $\int \frac{dx}{x^3 - 1}$

68. $\int \frac{dx}{x^4 + 27x}$

69. $\int \frac{dx}{(x^3 + x)^2}$

70. $\int \frac{dx}{x^3(x^2 + 1)^2}$

71. $\int \frac{x^3 - 2x^2 + x - 3}{x^4 + 8x^2 + 16} dx$

72. $\int \frac{3x^2 + 2x - 4}{x^4 + 6x^2 + 9} dx$

73. $\int \frac{2x}{(4x^2 + 5)^2} dx$

74. $\int \frac{x^2}{(x^2 + 1)^3} dx$

75. $\int \frac{x^2 - 2x + 3}{x(x^2 + 2x + 2)^2} dx$

76. $\int \frac{x}{(x - 1)(x^2 + 4x + 5)^2} dx$

77. $\int_0^1 \frac{dx}{x^3 + x^2 + 2x + 2}$

78. $\int_0^1 \frac{x^2}{x^4 + 8x^2 + 16} dx$

79. $\int_{-1}^1 \frac{2x^3 + 5x}{x^4 + 5x^2 + 6} dx$

80. $\int_1^2 \frac{1}{x^5 + 4x^4 + 5x^3} dx$

In Problems 81 and 82 use the indicated substitution to evaluate the given integral.

81. $\int \frac{\sqrt[3]{x + 1}}{x} dx; u^3 = x + 1$

82. $\int \frac{dx}{\sqrt{x}(1 + \sqrt[3]{x})^2}; u^6 = x$

83. Find the area under the graph of $y = \frac{x^3}{(x^2 + 1)(x^2 + 2)}$ on the interval $[0, 4]$.

84. Find the area bounded by the graph of $y = 3x^2/(x^3 - 1)$ and the x-axis on the interval $[-1, \frac{1}{2}]$.

85. The region in the first quadrant bounded by the graphs of $y = 2x/(x^2 + 1)$, $x = 1$, and $y = 0$ is revolved about the x-axis. Find the volume of the solid of revolution.

86. The region in the first quadrant bounded by the graphs of

$$y = \frac{8}{(x^2 + 1)(x^2 + 4)}$$

$x = 0$, $x = 1$, and $y = 0$ is revolved about the y-axis. Find the volume of the solid of revolution.

87. According to **Stefan's law of radiation**, the rate of change of temperature from a body at absolute temperature T is $dT/dt = k(T^4 - T_s^4)$, where T_s is the constant absolute temperature of the surrounding medium. Find a solution of this differential equation.

[O] 8.6 INTEGRAL TABLES, CALCULATORS, AND COMPUTERS

Scientific calculators have been around for more than 20 years, so it is rare that a student does not possess one. Still, the almost unused tables of numerical values of the trigonometric, logarithmic, and exponential functions are slow to disappear from mathematics texts. With the evolution of calculators to hand-held computers and the development of even more sophisticated numerical, graphical, and symbolic computer software, it is undoubtedly only a matter of time before tables, including a list of integration formulas such as that given in Appendix IV, will be as extinct as slide rules. Indeed, there are those who maintain that many of the mechanical or routine aspects of mathematics such as graphing, solving equations, symbolic differentiation, and integration, will be relegated to the computer or calculator, thereby freeing students to better understand and explore concepts. Perhaps. But the advancement of computer technology over the next several years will surely bring about parallel changes in mathematics curricula and in mathematics texts. Currently in mathematical circles there are a lot of ideas, panel discussions, articles, and experimental calculus programs, but not a lot of agreement.

Tables Some functions defy the techniques of integration presented in this chapter, and some only appear to. Thus, in the absence of an appropriately equipped computer, a table of integrals, if used judiciously, can sometimes speed up the solution of a problem. In the next three examples we shall illustrate the use of a table and point out alternative solutions.

EXAMPLE 1 Evaluate $\int \frac{x^3}{\sqrt{3+2x}}\,dx$ from tables.

Solution With $u = x$, $a = 3$, $b = 2$, and $n = 3$ we see from formula 61 of the Table of Integrals (Appendix IV) that

$$\int \frac{x^3}{\sqrt{3+2x}}\,dx = \frac{2x^3\sqrt{3+2x}}{2\cdot 7} - \frac{2\cdot 3\cdot 3}{2\cdot 7}\int \frac{x^2}{\sqrt{3+2x}}\,dx$$

Continuing, we apply formula 56 to the second integral:

$$\begin{aligned}\int \frac{x^3}{\sqrt{3+2x}}\,dx &= \frac{1}{7}x^3\sqrt{3+2x} - \frac{9}{7}\left[\frac{2}{15\cdot 8}(8\cdot 9 + 3\cdot 4x^2 - 4\cdot 6x)\sqrt{3+2x}\right] + C\\ &= \frac{1}{7}x^3\sqrt{3+2x} - \frac{54}{35}\sqrt{3+2x} - \frac{9}{35}x^2\sqrt{3+2x} + \frac{18}{35}x\sqrt{3+2x} + C\end{aligned}$$

Alternative Solution I Let $u = 3 + 2x$ and proceed as in Section 8.1.

Alternative Solution II Let $dv = (3+2x)^{-1/2}\,dx$ and $u = x^3$ and use integration by parts. □

EXAMPLE 2 Evaluate $\int \sqrt{4x - x^2}\,dx$ from tables.

Solution From formula 120 with $u = x$ and $a = 2$, we get

$$\int \sqrt{4x - x^2}\,dx = \frac{x-2}{2}\sqrt{4x - x^2} + 2\cos^{-1}\left(\frac{2-x}{2}\right) + C$$

Alternative Solution Write $4x - x^2 = 4 - (2 - x)^2$ and use a trigonometric substitution. □

EXAMPLE 3 Evaluate $\int \frac{dx}{1 + e^x}$ from tables.

Solution From formula 109 with $u = x$, $a = 1$, and $b = 1$,

$$\int \frac{dx}{1 + e^x} = x - \ln|1 + e^x| + C$$

Alternative Solution Write

$$\frac{1}{1 + e^x} = \frac{1 + e^x - e^x}{1 + e^x} = 1 - \frac{e^x}{1 + e^x}$$

and integrate term-by-term. □

Calculators As pointed out in Chapter 4, the HP-28S calculator is capable of yielding exact antiderivative results only for polynomials. For other kinds of integrands and for definite integrals, the calculator uses methods to approximate the integral.

Computers In the next three examples we shall illustrate symbolic integration using the computer algebra systems Maple and Mathematica on a Macintosh computer.

EXAMPLE 4 Use a computer to evaluate $\int xe^x \sin x\,dx$.

Solution In Maple the integrand is written x*exp(x)*sin(x). After typing the command

```
int(x*exp(x)*sin(x), x);
```

and pressing the [enter] key, we obtain

```
(-1/2 x + 1/2)exp(x)cos(x) + 1/2 x exp(x)sin(x)
```

Be careful reading this computer output; $1/2x$ does not mean $\frac{1}{2x}$. Using standard notation, we have

$$\int xe^x \sin x\,dx = \left(-\frac{1}{2}x + \frac{1}{2}\right)e^x \cos x + \frac{1}{2}xe^x \sin x + C \qquad \square$$

The integral in Example 4 presents some challenge but nevertheless can also be evaluated using integration by parts. See Problems 49 and 50 of Exercises 8.2. Note that neither the Maple nor the Mathematica program adds a constant of integration nor does either program supply absolute value signs in antiderivatives that are logarithms.*

EXAMPLE 5 Use a computer to evaluate $\int \frac{1}{1+\sqrt{x}}\,dx$.

Solution Typing: int(1/(1 + sqrt(x)), x); and entering, we get the following result from the Maple program:

$$-\ln(-1+x) + 2x^{1/2} + \ln\left(\frac{-1-x+2x^{1/2}}{-1+x}\right) \qquad (1) \quad \square$$

The result obtained in (1) is interesting, since we evaluated the integral in Example 2 of Section 8.1 by means of the algebraic substitution $u = \sqrt{x}$ and long division. In that example we found the simpler answer

$$\int \frac{1}{1+\sqrt{x}}\,dx = \int \frac{2u}{1+u}\,du = 2\sqrt{x} - 2\ln(1+\sqrt{x}) + C \qquad (2)$$

In this case we can obtain (2) by making the substitution *before* using the computer. In other words, int(2*u/(1 + u), u); gives $2u - 2\ln(1+u)$. You are encouraged to verify that (1) and (2) are equivalent.

Integral tables and programs such as Maple and Mathematica are not a cure-all for all integration problems. You will not find integrals such as $\int \frac{(4-e^{-x})^{5/3}}{e^x}\,dx$ and $\int e^{\sin\theta}\sin 2\theta\,d\theta$ in tables. Both these problems can be done the old-fashioned way: a little thought followed by "conventional methods." Also, as of this writing, the Maple program failed to evaluate $\int \frac{1}{\sqrt{1+\sqrt[3]{x}}}\,dx$. This integral possesses an elementary antiderivative, but at times even a computer needs some help. See Problems 27 and 28.

---

*This sometimes leads to a strange-looking answer. For example, for the integral $\int_{-2}^{1} dx/x$ Maple gives $-\ln(-2) + \ln(-1)$.

EXAMPLE 6 Use a computer to evaluate $\int \frac{1}{2 + 2\sin x + \cos x}\,dx$.

Solution In Mathematica we type the command:

$$\text{Integrate}[1/(2 + 2\,\text{Sin}[x] + \text{Cos}[x]), x]$$

Entering then yields

$$\text{Log}\left[\frac{1 + \frac{\text{Sin}[x]}{1 + \text{Cos}[x]}}{3 + \frac{\text{Sin}[x]}{1 + \text{Cos}[x]}}\right] \tag{3}$$

In (3) the symbol log means ln. Simplifying and replacing the brackets by absolute values give

$$\int \frac{1}{2 + 2\sin x + \cos x}\,dx = \ln\left|\frac{1 + \cos x + \sin x}{3 + 3\cos x + \sin x}\right| + C \qquad \square$$

The integral in Example 6 can be evaluated by hand, but it requires a rather complicated trigonometric substitution. It is also interesting to note that Maple gives an entirely different looking answer:

$$-\ln(\tan(1/2\,x) + 3) + \ln(\tan(1/2\,x) + 1)$$

▲ ***Remark*** The Maple and Mathematica programs are very large. Since the Maple library consists of approximately 2500 files and folders, integration, at times, is not particularly fast. On a Macintosh SE with a Motorola 68000 microprocessor it took about 5 minutes to evaluate the definite integral $\int_0^{\pi/3} dx/(1 + \sin x)$. See Problem 39. But even 5 minutes is faster than most people who know how to evaluate this particular integral by "conventional methods."

EXERCISES 8.6

Answers to odd-numbered problems begin on page A-90.

In Problems 1–22 evaluate the given integral using the Table of Integrals in Appendix IV.

1. $\int \frac{dx}{x^2\sqrt{9 + x^2}}$
2. $\int x^2\sqrt{25 - x^2}\,dx$
3. $\int \frac{\sqrt{x^2 - 5}}{x}\,dx$
4. $\int (4 - x^2)^{-3/2}\,dx$
5. $\int \frac{dx}{x(4 + 5x)^2}$
6. $\int \frac{x^4}{\sqrt{1 + x}}\,dx$
7. $\int t^2\sqrt{1 + 2t}\,dt$
8. $\int \frac{\sqrt{1 + u}}{u}\,du$
9. $\int \frac{x^2}{(3 - x)^2}\,dx$
10. $\int \frac{dx}{x\sqrt{5 - x}}$
11. $\int \tan^5\theta\,d\theta$
12. $\int \cos 6y \cos 2y\,dy$
13. $\int \ln(x^2 + 16)\,dx$
14. $\int e^x \ln|e^{2x} - 1|\,dx$

15. $\int \frac{dx}{1+\sin 2x}$

16. $\int \frac{x}{1-\sin 4x}\,dx$

17. $\int \frac{x}{\sqrt{2x-x^2}}\,dx$

18. $\int \frac{\sqrt{6x-x^2}}{x^2}\,dx$

19. $\int_0^{\pi/2} \sin^{10}x\,dx$

20. $\int_1^{e} x^9 \ln x\,dx$

21. $\int_0^{\pi} e^{2t}\sin 3t\,dt$

22. $\int_0^{1} \frac{dt}{2+e^{4t}}$

In Problems 23–42 use a computer to evaluate, if possible, the given integral.

23. $\int x^2 e^{-x}\cos x\,dx$

24. $\int x^2\sin^2 x\,dx$

25. $\int \frac{\sqrt{x}}{x+\sqrt{x+4}}\,dx$

26. $\int e^{\sqrt{x}}x^{3/2}\,dx$

27. $\int \frac{1}{\sqrt{1+\sqrt[3]{x}}}\,dx$

28. $\int \sqrt{\tan x}\,dx$

29. $\int \frac{dx}{2+\cos x}$

30. $\int \frac{dx}{4-5\sin x}$

31. $\int \frac{dx}{1+\sec x}$

32. $\int \frac{\sec x}{\sec x+\tan x-1}\,dx$

33. $\int \frac{dx}{\tan x+\sin x}$

34. $\int \frac{dx}{\cot x+\cos x}$

35. $\int \frac{\sec x}{1+\cos x}\,dx$

36. $\int \frac{\csc x}{1+\sin x}\,dx$

37. $\int \frac{dx}{1-2\sin x}$

38. $\int \frac{dx}{2\sec x-1}$

39. $\int_0^{\pi/3} \frac{dx}{1+\sin x}$

40. $\int_0^{2\pi/3} \frac{\cos x}{\cos x+\sin x}\,dx$

41. $\int_0^{\pi/2} \frac{dx}{3+2\cos x+3\sin x}$

42. $\int_0^{\pi/2} \frac{1+\sin x}{1+\cos x}\,dx$

43. The nonlinear differential equation

$$\left(\frac{dr}{dt}\right)^2 = \frac{2\mu}{r} + 2h$$

where μ and h are nonnegative constants, arises in the study of the two-body problem of celestial mechanics. Here the variable r represents the distance between the two masses. Solve the equation in the case $h > 0$.

CHAPTER 8 REVIEW EXERCISES

Answers begin on page A-90.

In Problems 1–10 answer true or false.

1. Under the change of variable $u = 2x + 3$, the integral $\int_1^5 \frac{4x}{\sqrt{2x+3}}\,dx$ becomes $\int_5^{13} (u^{1/2} - 3u^{-1/2})\,du$. ________

2. The trigonometric substitution $u = a\sec\theta$ is appropriate for integrals that contain $\sqrt{a^2+u^2}$. ________

3. The method of integration by parts is derived from the Product Rule for differentiation. ________

4. $\int_1^e 2x\ln x^2\,dx = e^2 + 1$ ________

5. Partial fractions are not applicable to $\int \frac{dx}{(x-1)^3}$. ________

6. A partial fraction decomposition of $x^2/(x+1)^2$ can be found having the form $A/(x+1) + B/(x+1)^2$, where A and B are constants. ________

7. To evaluate $\int \frac{dx}{(x^2-1)^2}$, we assume constants A, B, C, and D can be found such that

$$\frac{1}{(x^2-1)^2} = \frac{Ax+B}{x^2-1} + \frac{Cx+D}{(x^2-1)^2}$$ ________

8. To evaluate $\int x^n e^x\,dx$, n a positive integer, integration by parts is used $n-1$ times. ________

9. To evaluate $\int \frac{x}{\sqrt{9-x^2}}\,dx$, it is necessary to use $x = 3\sin\theta$. ________

10. When evaluated, the integral $\int \sin^3 x\cos^2 x\,dx$ can be expressed as a sum of powers of $\cos x$. ________

In Problems 11–90 use the methods of this chapter, or previous chapters, to evaluate the given integral.

11. $\int \frac{dx}{\sqrt{x+9}}$

12. $\int e^{\sqrt{x+1}}\,dx$

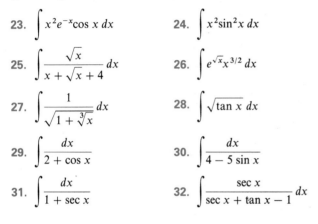

13. $\int \frac{x}{\sqrt{x^2+4}}\,dx$

14. $\int \frac{dx}{\sqrt{x^2+4}}$

15. $\int \frac{dx}{(x^2+4)^3}$

16. $\int \frac{x^2}{x^2+4}\,dx$

17. $\int \frac{x^2+4}{x^2}\,dx$

18. $\int \frac{3x-1}{x(x^2-4)}\,dx$

19. $\int \frac{x-5}{x^2+4}\,dx$

20. $\int \frac{\sqrt[3]{x+27}}{x}\,dx$

21. $\int \frac{(\ln x)^9}{x}\,dx$

22. $\int (\ln 3x)^2\,dx$

23. $\int t \sin^{-1} t\,dt$

24. $\int \frac{\ln x}{(x-1)^2}\,dx$

25. $\int (x+1)^3(x-2)\,dx$

26. $\int \frac{dx}{(x+1)^3(x-2)}$

27. $\int \ln(x^2+4)\,dx$

28. $\int 8te^{2t^2}\,dt$

29. $\int \frac{dx}{x^4+10x^3+25x^2}$

30. $\int \frac{dx}{x^2+8x+25}$

31. $\int \frac{x}{x^3+3x^2-9x-27}\,dx$

32. $\int \frac{x+1}{(x^2-x)(x^2+3)}\,dx$

33. $\int \frac{\sin^2 t}{\cos^2 t}\,dt$

34. $\int \frac{\sin^3\theta}{(\cos\theta)^{3/2}}\,d\theta$

35. $\int \tan^{10}x \sec^4 x\,dx$

36. $\int \frac{x \tan x}{\cos x}\,dx$

37. $\int y \cos y\,dy$

38. $\int x^2 \sin x^3\,dx$

39. $\int (1+\sin^2 t)\cos^3 t\,dt$

40. $\int \frac{\sec^3\theta}{\tan\theta}\,d\theta$

41. $\int e^w(1+e^w)^5\,dw$

42. $\int (x-1)e^{-x}\,dx$

43. $\int \cot^3 4x\,dx$

44. $\int (3-\sec x)^2\,dx$

45. $\int_0^{\pi/4} \cos^2 x \tan x\,dx$

46. $\int_0^{\pi/3} \sin^4 x \tan x\,dx$

47. $\int \frac{\sin x}{1+\sin x}\,dx$

48. $\int \frac{\cos x}{1+\sin x}\,dx$

49. $\int_0^1 \frac{dx}{(x+1)(x+2)(x+3)}$

50. $\int_{\ln 3}^{\ln 2} \sqrt{e^x+1}\,dx$

51. $\int e^x \cos 3x\,dx$

52. $\int x(x-5)^9\,dx$

53. $\int \cos(\ln t)\,dt$

54. $\int \sec^2 x \ln(\tan x)\,dx$

55. $\int \cos\sqrt{x}\,dx$

56. $\int \frac{\cos\sqrt{x}}{\sqrt{x}}\,dx$

57. $\int \cos x \sin 2x\,dx$

58. $\int (\cos^2 x - \sin^2 x)\,dx$

59. $\int \sqrt{x^2+2x+5}\,dx$

60. $\int \frac{dx}{(8-2x-x^2)^{3/2}}$

61. $\int \tan^5 x \sec^3 x\,dx$

62. $\int \cos^4 \frac{x}{2}\,dx$

63. $\int \frac{t^5}{1+t^2}\,dt$

64. $\int \frac{dx}{\sqrt{1-x^2}}$

65. $\int \frac{5x^3+x^2+6x+1}{(x^2+1)^2}\,dx$

66. $\int \frac{\sqrt{x^2+9}}{x^2}\,dx$

67. $\int x \sin^2 x\,dx$

68. $\int (t+1)^2 e^{3t}\,dt$

69. $\int \sin x \cos x\, e^{\sin x}\,dx$

70. $\int e^x \tan^2 e^x\,dx$

71. $\int_0^{\pi/6} \frac{\cos x}{\sqrt{1+\sin x}}\,dx$

72. $\int_0^{\pi/2} \frac{dx}{\sin x + \cos x}$

73. $\int \sinh^{-1} t\,dt$

74. $\int x \cot x^2\,dx$

75. $\int_3^8 \frac{dx}{x\sqrt{x+1}}$

76. $\int \frac{t+3}{t^2+2t+1}\,dt$

77. $\int \frac{\sec^4 3u}{\cot^{12} 3u}\,du$

78. $\int_0^2 x^5\sqrt{x^2+4}\,dx$

79. $\int \frac{3+\sin x}{\cos^2 x}\,dx$

80. $\int \frac{\sin 2x}{5+\cos^2 x}\,dx$

81. $\int x(1+\ln x)^2\,dx$

82. $\int x \cos^2 x\,dx$

83. $\int e^x e^{e^x}\,dx$

84. $\int \frac{dx}{\sqrt{x+1}-\sqrt{x}}$

85. $\int \frac{2t}{1+e^{t^2}}\,dt$

86. $\int \cos x \cos 2x\,dx$

87. $\int \frac{dx}{\sqrt{1-(5x+2)^2}}$

88. $\int (\ln 2x)\ln x\,dx$

89. $\int \cos x \ln|\sin x|\,dx$

90. $\int \ln\left(\frac{x+1}{x-1}\right)dx$

91. In quantitative genetics, the rate at which a recessive gene disappears from a population because of a selective disadvantage can be approximated as

$$\frac{dq}{dt} = -kq^2(1-q)$$

where $q(t)$ is the fraction of the genes that are recessive at time t (usually measured in generations) and k is a constant related to the selective disadvantage.

(a) If $T = \int_0^T dt$, show that the approximate time T required for the recessive gene to decline from an initial fraction $q(0) = q_0$ to a fraction $q(T) = q_1$ is

$$T = \frac{1}{k}\int_{q_1}^{q_0} \frac{dq}{q^2(1-q)}$$

(b) One study of peppered moths, which carry a dominant gene for dark coloration and a recessive gene for light coloration, found that the recessive gene had a selective disadvantage (due to the introduction of dark, industrial soot into the moth's environment, light-colored moths were more visible to predators) with $k \approx 0.53$. Find the approximate number of generations it took to reduce the fraction of recessive genes from 95% down to 5%.

9

INDETERMINATE FORMS AND IMPROPER INTEGRALS

INTRODUCTION

The material that follows in Section 9.2 and Chapter 10 demands that we know more about computing limits. Therefore, in Section 9.1 we shall consider a fairly simple but exceedingly useful rule for computing certain limits by taking derivatives.

In preparation for this chapter, you are encouraged to review Sections 1.5–1.7 and 4.5.

If the region bounded by the graphs of $y = 1/x$ and $y = 0$ on the interval $[1, \infty)$ is revolved around the x-axis, we obtain an unbounded

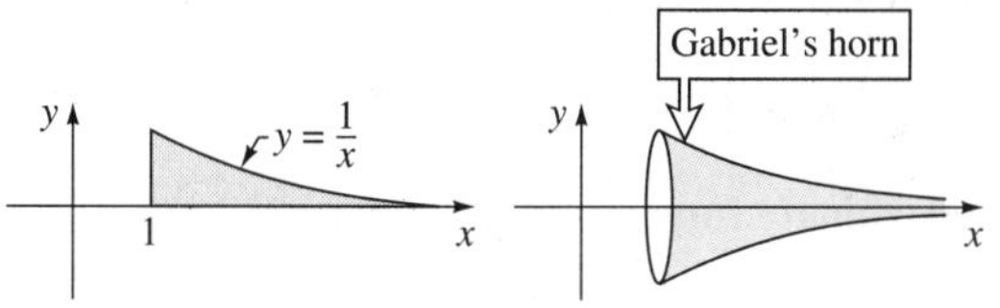

solid of revolution sometimes called "Gabriel's horn." The "horn" has some interesting properties. In Exercises 9.2 you will be asked to show that the area of the original region is infinite but that the resulting "horn" has a finite volume. In addition, you will be asked to show that this "horn" with finite volume has an infinite surface area. Think about this: Could you cover the exterior surface of the "horn" with paint?

IMPORTANT CONCEPTS

indeterminate forms

L'Hôpital's Rule

improper integral

convergent integral

divergent integral

9.1 L'HÔPITAL'S RULE

9.1.1 The Indeterminate Forms 0/0 and ∞/∞

In Chapter 2 we considered limits of quotients such as

$$\lim_{x \to 1} \frac{x^2 + 3x - 4}{x - 1} \quad \text{and} \quad \lim_{x \to \infty} \frac{2x^2 - x}{3x^2 + 1} \tag{1}$$

where in the first limit both the numerator and denominator are approaching zero as $x \to 1$, and in the second both the numerator and denominator are approaching ∞ as $x \to \infty$.

Terminology In general, we say that the limit

$$\lim_{x \to a} \frac{f(x)}{g(x)}$$

has the **indeterminate form 0/0 at *x* = *a*** if

$$f(x) \to 0 \quad \text{and} \quad g(x) \to 0 \quad \text{as} \quad x \to a$$

and has the **indeterminate form ∞/∞ at *x* = *a*** if

$$|f(x)| \to \infty \quad \text{and} \quad |g(x)| \to \infty \quad \text{as} \quad x \to a^*$$

A limit can also have an indeterminate form as

$$x \to a^-, \quad x \to a^+, \quad x \to -\infty, \quad \text{or} \quad x \to \infty$$

EXAMPLE 1

(*a*) $\lim_{x \to 0} \frac{\sin x}{x}$ has the indeterminate form 0/0 at $x = 0$, since

$$\sin x \to 0 \quad \text{and} \quad x \to 0 \quad \text{as} \quad x \to 0$$

(*b*) $\lim_{x \to 3^+} \frac{1/(3 - x)}{1/(3 - x)^2}$ has the indeterminate form ∞/∞ at $x = 3$, since

$$\frac{1}{3 - x} \to -\infty \quad \text{and} \quad \frac{1}{(3 - x)^2} \to \infty \quad \text{as} \quad x \to 3^+$$

---

*The absolute value signs mean that as x approaches a we could have, say, $f(x) \to \infty, g(x) \to \infty$; or $f(x) \to -\infty, g(x) \to \infty$; or $f(x) \to -\infty, g(x) \to -\infty$; and so on.

(c) $\lim_{x\to\infty} \dfrac{\ln x}{e^x}$ has the indeterminate form ∞/∞, since

$$\ln x \to \infty \quad \text{and} \quad e^x \to \infty \quad \text{as} \quad x \to \infty$$ □

Note: Limits of the form

$$\frac{0}{k}, \quad \frac{k}{0}, \quad \frac{\infty}{k}, \quad \text{and} \quad \frac{k}{\infty}$$

where k is a nonzero constant, are *not* indeterminate forms. The value of a limit whose form is $0/k$ or k/∞ is zero, whereas a limit whose form is either $k/0$ or ∞/k does not exist.

In establishing whether limits of quotients such as those given in (1) exist, we resorted to the algebraic manipulations of factoring, canceling, and dividing. However, recall that the proof of $\lim_{x\to 0}(\sin x)/x = 1$ used an elaborate geometric argument. But, algebra and geometric intuition fail miserably when confronted with a problem of the type

$$\lim_{x\to 0} \frac{\sin x}{e^x - e^{-x}}$$

which has the indeterminate form $0/0$. The next theorem will aid us in proving a rule that is extremely helpful in evaluating many limits which have an indeterminate form.

THEOREM 9.1 The Extended Mean Value Theorem

Let f and g be continuous on $[a, b]$ and differentiable on (a, b) and $g'(x) \neq 0$ for all x in (a, b). Then there exists a number c in (a, b) such that

$$\frac{f(b) - f(a)}{g(b) - g(a)} = \frac{f'(c)}{g'(c)}$$

Observe that Theorem 9.1 reduces to the Mean Value Theorem when $g(x) = x$. A proof of this theorem, which is reminiscent of the proof of Theorem 3.5, is outlined in Problems 47 and 48 in Exercises 9.1.

The following rule is named after the French mathematician G. F. A. L'Hôpital.* We assume that f and g are differentiable on the open intervals (r, a) and (a, s) and that $g'(x) \neq 0$ for $x \neq a$.

*It is questionable whether the Marquis Guillaume François Antoine de L'Hôpital (1661–1704) discovered the rule that bears his name. The result is probably due to Johann Bernoulli. However, L'Hôpital was the first to publish the rule in his text *Analyse des Infiniment Petits.* The book was published in 1696 and is considered to be the first textbook on calculus.

THEOREM 9.2 L'Hôpital's Rule

Suppose $\lim_{x\to a} \dfrac{f(x)}{g(x)}$ has an indeterminate form at $x = a$ and $\lim_{x\to a} \dfrac{f'(x)}{g'(x)} = L$ or $\pm\infty$. Then

$$\lim_{x\to a} \frac{f(x)}{g(x)} = \lim_{x\to a} \frac{f'(x)}{g'(x)} \tag{2}$$

Proof of the Case 0/0 Since

$$\lim_{x\to a} f(x) = 0 \quad \text{and} \quad \lim_{x\to a} g(x) = 0$$

it can be assumed that $f(a) = 0$ and $g(a) = 0$. It follows that f and g are continuous at a. Moreover, since f and g are differentiable, these functions are continuous on both (r, a) and (a, s). Consequently, f and g are continuous on the interval (r, s). Now, for any $x \neq a$ in the interval, Theorem 9.1 is applicable to either $[x, a]$ or $[a, x]$. In either case, there exists a number c between x and a such that

$$\frac{f(x) - f(a)}{g(x) - g(a)} = \frac{f(x)}{g(x)} = \frac{f'(c)}{g'(c)}$$

Letting $x \to a$ implies $c \to a$, and so

$$\lim_{x\to a} \frac{f(x)}{g(x)} = \lim_{x\to a} \frac{f'(c)}{g'(c)} = \lim_{c\to a} \frac{f'(c)}{g'(c)} = L$$ ■

EXAMPLE 2 Evaluate $\lim_{x\to 0} \dfrac{\sin x}{x}$ by L'Hôpital's Rule.

Solution In Example 1 we saw that the given limit has the indeterminate form 0/0 at $x = 0$. Thus, from (2) we can write

derivative

$$\lim_{x\to 0} \frac{\sin x}{x} = \lim_{x\to 0} \frac{\cos x}{1} = \frac{1}{1} = 1$$

derivative

□

EXAMPLE 3 Evaluate $\lim_{x\to 0} \dfrac{\sin x}{e^x - e^{-x}}$.

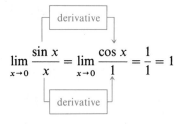

Solution Since the given limit has the indeterminate form 0/0 at $x = 0$, we apply (2):

$$\lim_{x\to 0}\frac{\sin x}{e^x - e^{-x}} = \lim_{x\to 0}\frac{\dfrac{d}{dx}\sin x}{\dfrac{d}{dx}(e^x - e^{-x})}$$

$$= \lim_{x\to 0}\frac{\cos x}{e^x + e^{-x}} = \frac{1}{1+1} = \frac{1}{2}$$ □

The result given in (2) remains valid when $x \to a$ is replaced by one-sided limits or by $x \to \infty$, $x \to -\infty$. The proof of the case $x \to \infty$ can be obtained by using the substitution $x = 1/t$ in $\lim_{x\to\infty} f(x)/g(x)$ and noting that $x \to \infty$ is equivalent to $t \to 0^+$.

EXAMPLE 4 Evaluate $\lim_{x\to\infty}\dfrac{\ln x}{e^x}$.

Solution The limit has the indeterminate form ∞/∞. Thus, from L'Hôpital's Rule we have

derivative

$$\lim_{x\to\infty}\frac{\ln x}{e^x} = \lim_{x\to\infty}\frac{1/x}{e^x} = \lim_{x\to\infty}\frac{1}{xe^x}$$

derivative

In this latter limit, $xe^x \to \infty$ as $x \to \infty$, whereas 1 remains constant. Consequently,

$$\lim_{x\to\infty}\frac{\ln x}{e^x} = \lim_{x\to\infty}\frac{1}{xe^x} = 0$$ □

It may be necessary to apply L'Hôpital's Rule several times in the course of solving a problem.

EXAMPLE 5 Evaluate $\lim_{x\to\infty}\dfrac{6x^2 + 5x + 7}{4x^2 + 2x}$.

Solution The indeterminate form is clearly ∞/∞, and so by (2),

$$\lim_{x\to\infty}\frac{6x^2 + 5x + 7}{4x^2 + 2x} = \lim_{x\to\infty}\frac{12x + 5}{8x + 2}$$

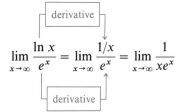

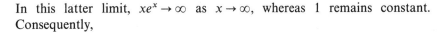

Since the new limit still has the indeterminate form ∞/∞, we apply (2) a second time:

$$\lim_{x\to\infty} \frac{12x+5}{8x+2} = \lim_{x\to\infty} \frac{12}{8} = \frac{3}{2}$$

Thus,
$$\lim_{x\to\infty} \frac{6x^2+5x+7}{4x^2+2x} = \frac{3}{2}$$ □

EXAMPLE 6 Evaluate $\lim_{x\to\infty} \frac{e^{3x}}{x^2}$.

Solution The given limit and the limit obtained after one application of L'Hôpital's Rule have the indeterminate form ∞/∞:

$$\lim_{x\to\infty} \frac{e^{3x}}{x^2} = \lim_{x\to\infty} \frac{3e^{3x}}{2x} = \lim_{x\to\infty} \frac{9e^{3x}}{2}$$

After the second application of (2), we notice $e^{3x} \to \infty$ while the denominator remains constant. From this we conclude that

$$\lim_{x\to\infty} \frac{e^{3x}}{x^2} = \infty$$

In other words, the given limit does not exist. □

EXAMPLE 7 Evaluate $\lim_{x\to\infty} \frac{x^4}{e^{2x}}$.

Solution We apply (2) four times:

$$\begin{aligned}
\lim_{x\to\infty} \frac{x^4}{e^{2x}} &= \lim_{x\to\infty} \frac{4x^3}{2e^{2x}} && (\infty/\infty)\\
&= \lim_{x\to\infty} \frac{12x^2}{4e^{2x}} && (\infty/\infty)\\
&= \lim_{x\to\infty} \frac{6x}{2e^{2x}} && (\infty/\infty)\\
&= \lim_{x\to\infty} \frac{6}{4e^{2x}} = 0
\end{aligned}$$ □

In successive applications of L'Hôpital's Rule, it is sometimes possible to change a limit form from, say, ∞/∞ to $0/0$.

EXAMPLE 8 Evaluate $\lim_{t\to\pi/2^+} \dfrac{\tan t}{\tan 3t}$.

Solution We observe that $\tan t \to -\infty$ and $\tan 3t \to -\infty$ as $t \to \pi/2^+$. Hence, from (2),

$$\lim_{t\to\pi/2^+} \frac{\tan t}{\tan 3t} = \lim_{t\to\pi/2^+} \frac{\sec^2 t}{3\sec^2 3t} \qquad (\infty/\infty)$$
$$= \lim_{t\to\pi/2^+} \frac{\cos^2 3t}{3\cos^2 t}$$

The latter form, which results from using $\sec t = 1/\cos t$, is now $0/0$. Proceeding, we find

$$\lim_{t\to\pi/2^+} \frac{\tan t}{\tan 3t} = \lim_{t\to\pi/2^+} \frac{2\cos 3t(-3\sin 3t)}{6\cos t(-\sin t)}$$
$$= \lim_{t\to\pi/2^+} \frac{2\sin 3t\cos 3t}{2\sin t\cos t}$$

double angle formula

$$= \lim_{t\to\pi/2^+} \frac{\sin 6t}{\sin 2t}$$
$$= \lim_{t\to\pi/2^+} \frac{6\cos 6t}{2\cos 2t} = \frac{-6}{-2} = 3$$ □

EXAMPLE 9 Evaluate $\lim_{x\to 1^+} \dfrac{\ln x}{\sqrt{x-1}}$.

Solution The given limit has the indeterminate form $0/0$ at $x = 1$. Hence, by L'Hôpital's Rule,

$$\lim_{x\to 1^+} \frac{\ln x}{\sqrt{x-1}} = \lim_{x\to 1^+} \frac{1/x}{(1/2)(x-1)^{-1/2}} = \lim_{x\to 1^+} \frac{2\sqrt{x-1}}{x} = \frac{0}{1} = 0$$ □

▲ ***Remarks*** (*i*) In the application of L'Hôpital's Rule, many students will misinterpret

$$\lim_{x\to a} \frac{f'(x)}{g'(x)} \quad \text{as} \quad \lim_{x\to a} \frac{d}{dx}\frac{f(x)}{g(x)}$$

The rule utilizes the *quotient of derivatives* and *not* the *derivative of the quotient.*

(*ii*) Inspect a problem before you leap to its solution. The limit $\lim_{x\to 0}(\cos x)/x$ is the form $1/0$ and, as a consequence, does not exist. Lack of mathematical forethought in writing

$$\lim_{x\to 0} \frac{\cos x}{x} = \lim_{x\to 0} \frac{-\sin x}{1} = 0$$

is an incorrect application of L'Hôpital's Rule. Of course, the "answer" has no significance.

(*iii*) L'Hôpital's rule is not a cure-all for every indeterminate form. For example, $\lim_{x\to\infty} e^x/e^{x^2}$ is certainly of the form ∞/∞, but

$$\lim_{x\to\infty} \frac{e^x}{e^{x^2}} = \lim_{x\to\infty} \frac{e^x}{2xe^{x^2}}$$

is of no immediate help. Nor does L'Hôpital's Rule provide any help in a problem such as $\lim_{x\to 0} e^{1/x}/x$. Why not?

9.1.2 The Indeterminate Forms $\infty - \infty$, $0 \cdot \infty$, 0^0, ∞^0, and 1^∞

There are five additional indeterminate forms:

$$\infty - \infty, \quad 0 \cdot \infty, \quad 0^0, \quad \infty^0, \quad \text{and} \quad 1^\infty$$

By a combination of algebra and a little cleverness we can often convert one of these new limit forms to either $0/0$ or ∞/∞.

The Form $\infty - \infty$ The next example illustrates a limit that has the indeterminate form $\infty - \infty$. This example should destroy any unwarranted convictions that $\infty - \infty = 0$.

EXAMPLE 10 Evaluate $\lim_{x\to 0^+} \left[\frac{1+3x}{\sin x} - \frac{1}{x}\right]$.

Solution We note that $(1+3x)/\sin x \to \infty$ and $1/x \to \infty$ as $x \to 0^+$. However, after writing the difference as a single fraction, we recognize the form $0/0$:

$$\lim_{x\to 0^+} \left[\frac{1+3x}{\sin x} - \frac{1}{x}\right] = \lim_{x\to 0^+} \frac{3x^2 + x - \sin x}{x \sin x}$$

$$= \lim_{x\to 0^+} \frac{6x + 1 - \cos x}{x \cos x + \sin x} \qquad \boxed{\text{applying (2)}}$$

$$= \lim_{x\to 0^+} \frac{6 + \sin x}{-x \sin x + 2\cos x} \qquad \boxed{\text{(2) again}}$$

$$= \frac{6+0}{0+2} = 3$$

□

The Form $0 \cdot \infty$ If

$$f(x) \to 0 \quad \text{and} \quad |g(x)| \to \infty \quad \text{as} \quad x \to a$$

then $\lim_{x\to a} f(x)g(x)$ has the indeterminate form $0 \cdot \infty$. We can change a limit that has this form to one with the form $0/0$ or ∞/∞ by writing, in turn,

$$f(x)g(x) = \frac{f(x)}{1/g(x)} \quad \text{or} \quad f(x)g(x) = \frac{g(x)}{1/f(x)}$$

EXAMPLE 11 Evaluate $\displaystyle\lim_{x\to\infty} x \sin\frac{1}{x}$.

Solution Since $1/x \to 0$, we have $\sin(1/x) \to 0$ as $x \to \infty$. Hence, the limit has the indeterminate form $0 \cdot \infty$. By writing

$$\lim_{x\to\infty} \frac{\sin(1/x)}{1/x}$$

we now have the form $0/0$. Hence,

$$\lim_{x\to\infty} \frac{\sin(1/x)}{1/x} = \lim_{x\to\infty} \frac{(-x^{-2})\cos(1/x)}{(-x^{-2})} = \lim_{x\to\infty} \cos\frac{1}{x} = 1 \qquad \square$$

The Forms 0^0, ∞^0, and 1^∞ Suppose $y = f(x)^{g(x)}$ tends toward 0^0, ∞^0, or 1^∞ as $x \to a$. By taking the natural logarithm of y:

$$\ln y = \ln f(x)^{g(x)} = g(x) \ln f(x)$$

we see

$$\lim_{x\to a} \ln y = \lim_{x\to a} g(x) \ln f(x)$$

has the form $0 \cdot \infty$. If it is assumed that $\lim_{x\to a} \ln y = \ln(\lim_{x\to a} y) = L$, then $\lim_{x\to a} y = e^L$ or

$$\lim_{x\to a} f(x)^{g(x)} = e^L$$

Of course, the procedure just outlined is applicable to limits involving $x \to a^-$, $x \to a^+$, $x \to \infty$, or $x \to -\infty$.

EXAMPLE 12 Evaluate $\displaystyle\lim_{x\to 0^+} x^{1/\ln x}$.

Solution The limit has the indeterminate form 0^0. Now, if we set $y = x^{1/\ln x}$, then

$$\ln y = \frac{1}{\ln x} \ln x = 1$$

Notice that we do not need L'Hôpital's Rule in this case, since

$$\lim_{x\to 0^+} \ln y = 1 \quad \text{or} \quad \ln(\lim_{x\to 0^+} y) = 1$$

Hence

$$\lim_{x\to 0^+} y = e^1 \quad \text{or equivalently} \quad \lim_{x\to 0^+} x^{1/\ln x} = e \qquad \square$$

In the next example we consider an important limit result whose proof was postponed in Section 6.3.

EXAMPLE 13 Evaluate $\lim_{x\to 0}(1+x)^{1/x}$.

Solution The limit has the indeterminate form 1^∞. If $y=(1+x)^{1/x}$, then

$$\ln y = \frac{1}{x}\ln(1+x)$$

Now, $\lim_{x\to 0}\ln(1+x)/x$ has the form $0/0$ and so by (2), $\ln(\lim_{x\to 0} y)$ is

$$\lim_{x\to 0}\frac{\ln(1+x)}{x}=\lim_{x\to 0}\frac{1/(1+x)}{1}=\lim_{x\to 0}\frac{1}{1+x}=1$$

Thus,

$$\lim_{x\to 0}(1+x)^{1/x}=e$$ □

EXAMPLE 14 Evaluate $\lim_{x\to\infty}\left(1-\dfrac{3}{x}\right)^{2x}$.

Solution As in the preceding example, the indeterminate form is 1^∞. If

$$y=\left(1-\frac{3}{x}\right)^{2x}\quad\text{then}\quad \ln y = 2x\ln\left(1-\frac{3}{x}\right)$$

Observe that the form of $\lim_{x\to\infty} 2x\ln(1-3/x)$ is $\infty\cdot 0$, whereas the form of $\lim_{x\to\infty} 2\ln(1-3/x)/(1/x)$ is $0/0$. Applying (2) to the latter limit gives

$$\lim_{x\to\infty} 2\,\frac{\ln(1-3/x)}{1/x}=\lim_{x\to\infty} 2\,\frac{\dfrac{3/x^2}{(1-3/x)}}{-1/x^2}=\lim_{x\to\infty}\frac{-6}{(1-3/x)}=-6$$

Finally, we conclude that

$$\lim_{x\to\infty}\left(1-\frac{3}{x}\right)^{2x}=e^{-6}$$ □

EXERCISES 9.1

Answers to odd-numbered problems begin on page A-91.

[9.1.1]

In Problems 1–42 use L'Hôpital's Rule where appropriate to find the limit if it exists.

1. $\lim_{x\to 0}\dfrac{\cos x - 1}{x}$

2. $\lim_{t\to 3}\dfrac{t^3-27}{t-3}$

3. $\lim_{x\to 1}\dfrac{2x-2}{\ln x}$

4. $\lim_{x\to 0^+}\dfrac{\ln 2x}{\ln 3x}$

5. $\lim_{x\to 0}\dfrac{e^x-1}{3x+x^2}$

6. $\lim_{x\to 0}\dfrac{\tan x}{2x}$

7. $\lim_{t\to\pi} \dfrac{5\sin^2 t}{1+\cos t}$

8. $\lim_{\theta\to 1} \dfrac{\theta^2-1}{e^{\theta^2}-e}$

9. $\lim_{x\to 0} \dfrac{6+6x+3x^2-6e^x}{x-\sin x}$

10. $\lim_{x\to\infty} \dfrac{3x^2-4x^3}{5x+7x^3}$

11. $\lim_{x\to 0^+} \dfrac{\cot 2x}{\cot x}$

12. $\lim_{x\to 0} \dfrac{\arcsin(x/6)}{\arctan(x/2)}$

13. $\lim_{t\to 2} \dfrac{t^2+3t-10}{t^3-2t^2+t-2}$

14. $\lim_{r\to -1} \dfrac{r^3-r^2-5r-3}{(r+1)^2}$

15. $\lim_{x\to 0} \dfrac{x-\sin x}{x^3}$

16. $\lim_{x\to 1} \dfrac{x^2+4}{x^2+1}$

17. $\lim_{x\to 0} \dfrac{\cos 2x}{x^2}$

18. $\lim_{x\to\infty} \dfrac{2e^{4x}+x}{e^{4x}+3x}$

19. $\lim_{x\to 1^+} \dfrac{\ln\sqrt{x}}{x-1}$

20. $\lim_{x\to\infty} \dfrac{\ln(3x^2+5)}{\ln(5x^2+1)}$

21. $\lim_{x\to 2} \dfrac{e^{x^2}-e^{2x}}{x}$

22. $\lim_{x\to 0} \dfrac{4^x-3^x}{x}$

23. $\lim_{x\to\infty} \dfrac{x\ln x}{x^2+1}$

24. $\lim_{t\to 0} \dfrac{1-\cosh t}{t^2}$

25. $\lim_{x\to 0} \dfrac{\sin 5x}{x}$

26. $\lim_{x\to 0} \dfrac{(\sin 2x)^2}{x^2}$

27. $\lim_{x\to\infty} \dfrac{e^x}{x^4}$

28. $\lim_{x\to\infty} \dfrac{e^{1/x}}{\sin(1/x)}$

29. $\lim_{x\to 0} \dfrac{x-\tan^{-1}x}{x-\sin^{-1}x}$

30. $\lim_{t\to 1} \dfrac{t^{1/3}-t^{1/2}}{t-1}$

31. $\lim_{u\to\pi/2} \dfrac{\ln(\sin u)}{(2u-\pi)^2}$

32. $\lim_{\theta\to\pi/2} \dfrac{\tan\theta}{\ln(\cos\theta)}$

33. $\lim_{x\to -\infty} \dfrac{1+e^{-2x}}{1-e^{-2x}}$

34. $\lim_{x\to 0} \dfrac{e^x-x-1}{2x^2}$

35. $\lim_{r\to 0} \dfrac{r-\cos r}{r-\sin r}$

36. $\lim_{t\to\pi} \dfrac{\csc 7t}{\csc 2t}$

37. $\lim_{x\to 0^+} \dfrac{x^2}{\ln^2(1+3x)}$

38. $\lim_{x\to 3} \left(\dfrac{\ln x-\ln 3}{x-3}\right)^2$

39. $\lim_{x\to 0} \dfrac{3x^2+e^x-e^{-x}-2\sin x}{x\sin x}$

40. $\lim_{x\to 8} \dfrac{\sqrt{x+1}-3}{x^2-64}$

41. $\lim_{u\to a} \dfrac{u^q-a^q}{u^p-a^p}$, $a>0$

42. $\lim_{x\to 0} \dfrac{\int_0^{3x} e^{t^2}\,dt}{xe^{9x^2}}$

43. Consider the circle shown in Figure 9.1.

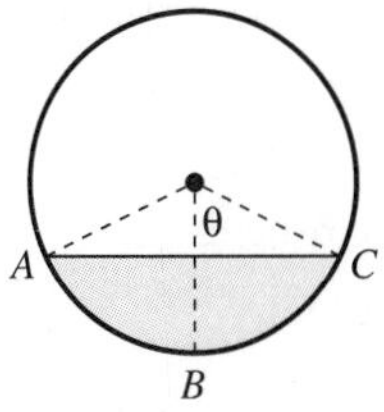

FIGURE 9.1

(a) If the arc ABC is 5 in long, express the area A of the shaded region as a function of the indicated angle θ.

(b) Evaluate $\lim_{\theta\to 0} A$.

(c) Evaluate $\lim_{\theta\to 0} dA/d\theta$.

44. The retina is most sensitive to photons that enter the eye near the center of the pupil, and less sensitive to light that enters near the edge of the pupil. (This phenomenon is known as the **Stiles–Crawford effect** of the first kind.) The percentage σ of photons that reach the photopigments is related to the pupil radius p (measured in mm) by the formula

$$\sigma = \frac{1-10^{-0.05p^2}}{0.115p^2} \times 100$$

See Figure 9.2.

(a) What percentage of photons reach the photopigments when $p = 2$ mm?

(b) What, according to the formula, is the limiting percentage as the pupil radius tends to zero? Can you explain why it seems to be more than 100%?

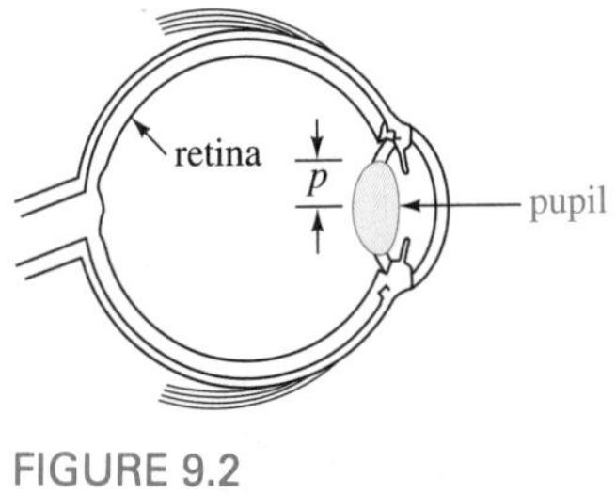

FIGURE 9.2

Calculator/Computer Problems

45. (a) Use a calculator or computer to obtain the graph of

$$f(x) = \frac{x\sin x}{x^2+1}$$

(b) From the graph in part (a), discern the value of $\lim_{x\to\infty} f(x)$.

(c) Explain why L'Hôpital's Rule does not apply to $\lim_{x\to\infty} f(x)$. [*Hint:* Examine the graph of $x \sin x$.]

Miscellaneous Problems

46. Prove the given result.

(a) $\lim_{x\to\infty} \frac{e^x}{x^n} = \infty$, n an integer

(b) $\lim_{x\to\infty} \frac{\ln x}{x^k} = 0$, k a positive constant

47. Review Rolle's Theorem and show that, under the hypotheses of Theorem 9.1, $g(b) - g(a) \neq 0$. [*Hint:* Suppose $g(b) - g(a) = 0$ or $g(b) = g(a)$. What does Rolle's Theorem say?]

48. Define the function

$$\phi(x) = f(x) - f(a) - \frac{f(b) - f(a)}{g(b) - g(a)}[g(x) - g(a)]$$

(a) Show that $\phi(a) = \phi(b) = 0$.

(b) Apply Rolle's Theorem to ϕ to obtain the result in Theorem 9.1.

[9.1.2]

In Problems 49–86 use L'Hôpital's Rule where appropriate to find the limit if it exists.

49. $\lim_{x\to 0} \left(\frac{1}{e^x - 1} - \frac{1}{x}\right)$

50. $\lim_{x\to 0^+} (\cot x - \csc x)$

51. $\lim_{x\to\infty} x(e^{1/x} - 1)$

52. $\lim_{x\to 0^+} x \ln x$

53. $\lim_{x\to 0^+} x^x$

54. $\lim_{x\to 1^-} x^{1/(1-x)}$

55. $\lim_{x\to 0} \left[\frac{1}{x} - \frac{1}{\sin x}\right]$

56. $\lim_{x\to 0} \left[\frac{1}{x} - \sin^{-1}x\right]$

57. $\lim_{x\to 0^+} x\left[\frac{1}{\tan x} - \frac{5}{x}\right]$

58. $\lim_{x\to 0^+} \left[\frac{1}{x} - \frac{1}{\ln(x+1)}\right]$

59. $\lim_{\theta\to 0} \theta \csc 4\theta$

60. $\lim_{x\to\pi/2^+} (\sin^2 x)^{\tan x}$

61. $\lim_{x\to\infty} (2 + e^x)^{e^{-x}}$

62. $\lim_{x\to 0^-} (1 - e^x)^{x^2}$

63. $\lim_{t\to\infty} \left(1 + \frac{3}{t}\right)^t$

64. $\lim_{h\to 0} (1 + 2h)^{4/h}$

65. $\lim_{x\to 0} x^{(1-\cos x)}$

66. $\lim_{\theta\to 0} (\cos 2\theta)^{1/\theta^2}$

67. $\lim_{x\to\infty} \frac{1}{x^2 \sin^2(2/x)}$

68. $\lim_{x\to 1} (x^2 - 1)^{x^2}$

69. $\lim_{x\to 0} \left[\frac{x}{1+x} - \sin x\right]$

70. $\lim_{t\to\pi} (\cos 2t)^{t-\pi}$

71. $\lim_{x\to 1} \left[\frac{1}{x-1} - \frac{5}{x^2 + 3x - 4}\right]$

72. $\lim_{x\to 0} \left[\frac{1}{x^2} - \frac{1}{x}\right]$

73. $\lim_{x\to\infty} x^2 e^{-x}$

74. $\lim_{x\to\infty} (x + e^x)^{2/x}$

75. $\lim_{x\to\infty} x\left(\frac{\pi}{2} - \arctan x\right)$

76. $\lim_{t\to\pi/4} (t - \pi/4)\tan 2t$

77. $\lim_{t\to 3} \left[\frac{\sqrt{t+1}}{t^2 - 9} - \frac{2}{t^2 - 9}\right]$

78. $\lim_{x\to 0^+} x \ln(\sin x)$

79. $\lim_{x\to -\infty} \left[\frac{1}{e^x} - x^2\right]$

80. $\lim_{x\to 0} (1 + 5 \sin x)^{\cot x}$

81. $\lim_{x\to\infty} \left(\frac{3x}{3x+1}\right)^x$

82. $\lim_{\theta\to\pi/2^-} (\sec^3\theta - \tan^3\theta)$

83. $\lim_{x\to 0} (\sinh x)^{\tan x}$

84. $\lim_{x\to 0^+} x^{(\ln x)^2}$

85. $\lim_{x\to 0^+} \frac{1}{x} \ln\left(\frac{e^x - 1}{x}\right)$

86. $\lim_{x\to\infty} \frac{1}{x} \ln\left(\frac{e^x - 1}{x}\right)$

Calculator/Computer Problems

In Problems 87 and 88 use a calculator or computer to fill in the table.

87.

x	$(1+x)^{1/x^2}$
0.1	
0.01	

88.

x	$(1+x)^{1/x^2}$
−0.1	
−0.01	

Miscellaneous Problems

89. Use L'Hôpital's Rule to evaluate

(a) $\lim_{x\to 0^+} (1 + x)^{1/x^2}$ (b) $\lim_{x\to 0^-} (1 + x)^{1/x^2}$

Reexamine Problems 87 and 88.

90. Evaluate $\lim_{x\to\infty}[\sqrt{x^2 + x} - x]$ without the aid of L'Hôpital's Rule.

In Problems 91 and 92 find all positive integers n for which the given limit exists.

91. $\lim_{x\to 0^+} \frac{\ln(x+1)}{x^n}$

92. $\lim_{x\to 1^+} \frac{e^{x-1} - x}{(x-1)^n}$

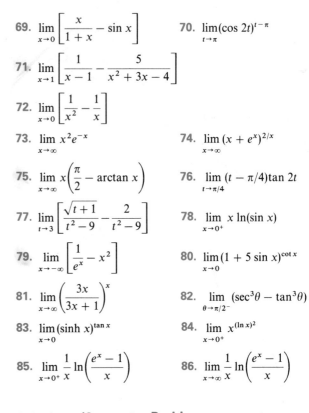

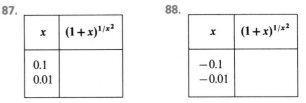

9.2 IMPROPER INTEGRALS

Up to now in our study of the definite integral $\int_a^b f(x)\,dx$, it was understood that

- the limits of integration were finite numbers, and
- the function f either was continuous on $[a, b]$ or, if discontinuous, was bounded on the interval.

When either of these two conditions is dropped, the resulting integral is said to be an **improper integral**. In the first subsection that follows, we shall consider integrals of functions that are defined on unbounded intervals. In the second subsection, we shall examine integrals on bounded intervals of functions that are unbounded. In the latter type of improper integral, an integrand f has an *infinite discontinuity* at some number in the interval of integration.

9.2.1 Infinite Limits of Integration

Improper Integrals—Unbounded Intervals If f is defined on an unbounded interval, then there are three possible **improper** integrals with infinite limits of integration. Their definitions are summarized as follows:

DEFINITION 9.1 Improper Integrals—Unbounded Intervals

(*i*) If f is continuous on $[a, \infty)$, then

$$\int_a^\infty f(x)\,dx = \lim_{t\to\infty} \int_a^t f(x)\,dx \qquad (1)$$

(*ii*) If f is continuous on $(-\infty, a]$, then

$$\int_{-\infty}^a f(x)\,dx = \lim_{s\to-\infty} \int_s^a f(x)\,dx \qquad (2)$$

(*iii*) If f is continuous for all x, and a is any real number, then

$$\int_{-\infty}^\infty f(x)\,dx = \int_{-\infty}^a f(x)\,dx + \int_a^\infty f(x)\,dx \qquad (3)$$

When the limits in (1) and (2) exist, the integrals are said to **converge**. If the limit fails to exist, the integral is said to **diverge**. In (3) the integral $\int_{-\infty}^\infty f(x)\,dx$ converges provided both $\int_{-\infty}^a f(x)\,dx$ and $\int_a^\infty f(x)\,dx$ converge. If either $\int_{-\infty}^a f(x)\,dx$ or $\int_a^\infty f(x)\,dx$ diverges, then $\int_{-\infty}^\infty f(x)\,dx$ diverges.

EXAMPLE 1 Evaluate $\int_1^\infty x^2\,dx$ if possible.

Solution By (1),

$$\int_1^\infty x^2\,dx = \lim_{t\to\infty}\int_1^t x^2\,dx = \lim_{t\to\infty}\frac{x^3}{3}\bigg]_1^t = \lim_{t\to\infty}\left(\frac{t^3}{3}-\frac{1}{3}\right)$$

Since $\lim_{t\to\infty}\left(\frac{t^3}{3}-\frac{1}{3}\right)=\infty$, we conclude that the integral diverges. □

EXAMPLE 2 Evaluate $\int_{-\infty}^\infty x^2\,dx$ if possible.

Solution Since a can be chosen arbitrarily, we pick $a = 1$ and write

$$\int_{-\infty}^\infty x^2\,dx = \int_{-\infty}^1 x^2\,dx + \int_1^\infty x^2\,dx$$

But, in Example 1 we saw that $\int_1^\infty x^2\,dx$ diverges. This is sufficient to conclude that $\int_{-\infty}^\infty x^2\,dx$ also diverges. □

EXAMPLE 3 Evaluate $\int_2^\infty \frac{dx}{x^3}$ if possible.

Solution By (1),

$$\int_2^\infty \frac{dx}{x^3} = \lim_{t\to\infty}\int_2^t x^{-3}\,dx = \lim_{t\to\infty}\frac{x^{-2}}{-2}\bigg]_2^t = \lim_{t\to\infty}\left[\frac{1}{8}-\frac{1}{2t^2}\right]$$

Since $\lim_{t\to\infty}\left[\frac{1}{8}-\frac{1}{2t^2}\right]=\frac{1}{8}$, the integral converges and

$$\int_2^\infty \frac{dx}{x^3} = \frac{1}{8}$$ □

EXAMPLE 4 Evaluate $\int_{-1}^\infty e^{-x}\,dx$ if possible. Interpret geometrically.

Solution By (1),

$$\int_{-1}^\infty e^{-x}\,dx = \lim_{t\to\infty}\int_{-1}^t e^{-x}\,dx = \lim_{t\to\infty}(-e^{-x})\bigg]_{-1}^t = \lim_{t\to\infty}[e-e^{-t}]$$

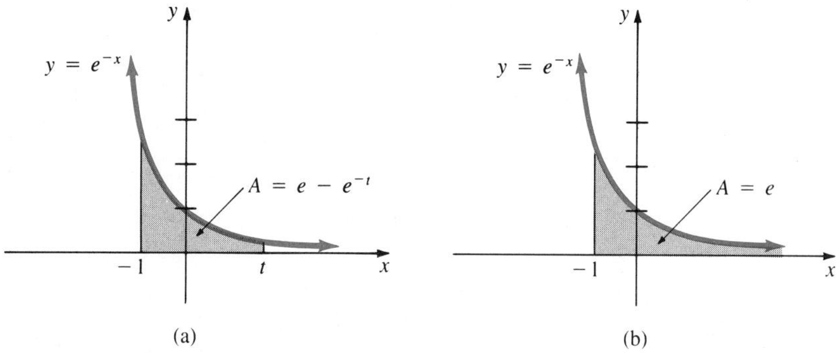

FIGURE 9.3

Since $\lim_{t\to\infty} e^{-t} = 0$, $\lim_{t\to\infty}[e - e^{-t}] = e$; and so the given integral converges to e. In Figure 9.3(a) we see that the area under the graph of the nonnegative function $f(x) = e^{-x}$ on $[-1, t]$ is $e - e^{-t}$. But, by taking $t \to \infty$, $e^{-t} \to 0$, and hence, as shown in Figure 9.3(b), we can interpret $\int_{-1}^{\infty} e^{-x}\,dx = e$ as a measure of the area under the graph of f on $[-1, \infty)$. □

EXAMPLE 5 Evaluate $\displaystyle\int_{-\infty}^{0} \cos x\,dx$ if possible.

Solution By (2),

$$\int_{-\infty}^{0} \cos x\,dx = \lim_{s\to-\infty} \int_{s}^{0} \cos x\,dx = \lim_{s\to-\infty} \sin x\Big]_{s}^{0} = \lim_{s\to-\infty} (-\sin s)$$

Since $\sin s$ oscillates between -1 and 1, we conclude that $\lim_{s\to-\infty}(-\sin s)$ does not exist. Hence, $\int_{-\infty}^{0} \cos x\,dx$ diverges. □

EXAMPLE 6 Evaluate $\displaystyle\int_{-\infty}^{\infty} \frac{e^x}{e^x + 1}\,dx$ if possible.

Solution Choosing $a = 0$, we can write

$$\int_{-\infty}^{\infty} \frac{e^x}{e^x + 1}\,dx = \int_{-\infty}^{0} \frac{e^x}{e^x + 1}\,dx + \int_{0}^{\infty} \frac{e^x}{e^x + 1}\,dx = I_1 + I_2$$

First, let us examine I_1:

$$I_1 = \lim_{s\to-\infty} \int_{s}^{0} \frac{e^x}{e^x + 1}\,dx = \lim_{s\to-\infty} \ln(e^x + 1)\Big]_{s}^{0} = \lim_{s\to-\infty} [\ln 2 - \ln(e^s + 1)]$$

Since $e^s + 1 \to 1$ as $s \to -\infty$, $\ln(e^s + 1) \to 0$. Hence, $I_1 = \ln 2$.

Second, we have

$$I_2 = \lim_{t\to\infty} \int_0^t \frac{e^x}{e^x+1}\,dx = \lim_{t\to\infty} \ln(e^x+1)\Big]_0^t = \lim_{t\to\infty} [\ln(e^t+1) - \ln 2]$$

However, $e^t + 1 \to \infty$ as $t \to \infty$, so $\ln(e^t+1) \to \infty$. Hence, I_2 diverges. It follows that $\int_{-\infty}^{\infty} e^x\,dx/(e^x+1)$ is divergent. □

EXAMPLE 7 It is left as an exercise to show that $\displaystyle\int_{-\infty}^{\infty} \frac{dx}{1+x^2}$ converges and

$$\int_{-\infty}^{\infty} \frac{dx}{1+x^2} = \int_{-\infty}^{0} \frac{dx}{1+x^2} + \int_0^{\infty} \frac{dx}{1+x^2} = -\left(-\frac{\pi}{2}\right) + \frac{\pi}{2} = \pi \qquad \square$$

EXAMPLE 8 In (5) of Section 5.8, we saw that the work done in lifting a mass m_2 off the surface of a planet of mass m_1 to a height h is given by

$$W = \int_R^{R+h} \frac{km_1m_2}{r^2}\,dr$$

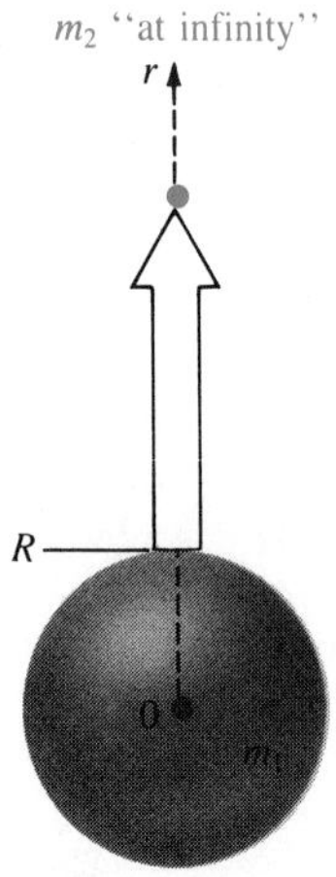

FIGURE 9.4

where R is the radius of the planet. Hence, the amount of work done in lifting m_2 to an unlimited or "infinite distance" from the surface of the planet is

$$W = \int_R^{\infty} \frac{km_1m_2}{r^2}\,dr = \lim_{t\to\infty} \int_R^t \frac{km_1m_2}{r^2}\,dr = \frac{km_1km_2}{R}$$

See Figure 9.4. From the data in Example 2 of Section 5.8, it follows that the work done in lifting a payload of 5000 kg to an "infinite distance" from the surface of the earth is

$$W = \frac{(6.67\times 10^{-11})(6.0\times 10^{24})(5000)}{6.4\times 10^6} \approx 3.13\times 10^{11} \text{ joules} \qquad \square$$

▲ ***Remarks*** (*i*) You should verify that $\int_{-\infty}^{\infty} x\,dx = \int_{-\infty}^{0} x\,dx + \int_0^{\infty} x\,dx$ diverges since both $\int_{-\infty}^{0} x\,dx$ and $\int_0^{\infty} x\,dx$ diverge. A *common mistake* when working with integrals with doubly infinite limits is to use *one* limit:

$$\int_{-\infty}^{\infty} x\,dx = \lim_{t\to\infty} \int_{-t}^{t} x\,dx = \lim_{t\to\infty} \frac{x^2}{2}\Big]_{-t}^{t} = \lim_{t\to\infty} \left[\frac{t^2}{2} - \frac{t^2}{2}\right] = 0$$

Of course, this "answer" is incorrect. Integrals of the type $\int_{-\infty}^{\infty} f(x)\,dx$ require the evaluation of *two independent* limits.

(*ii*) In our previous work we often wrote without thinking

$$\int_a^b [f(x)+g(x)]\,dx = \int_a^b f(x)\,dx + \int_a^b g(x)\,dx \tag{4}$$

For improper integrals one should proceed with a bit more caution. For example, the integral $\int_1^\infty [1/x - 1/(x+1)]\,dx$ converges (see Problem 25 in Exercises 9.2), but

$$\int_1^\infty \left[\frac{1}{x} - \frac{1}{x+1}\right] dx \neq \int_1^\infty \frac{dx}{x} - \int_1^\infty \frac{dx}{x+1}$$

The property in (4) remains valid for improper integrals whenever the integrals on the right side of the equality converge.

9.2.2 Integrals with an Integrand That Becomes Infinite

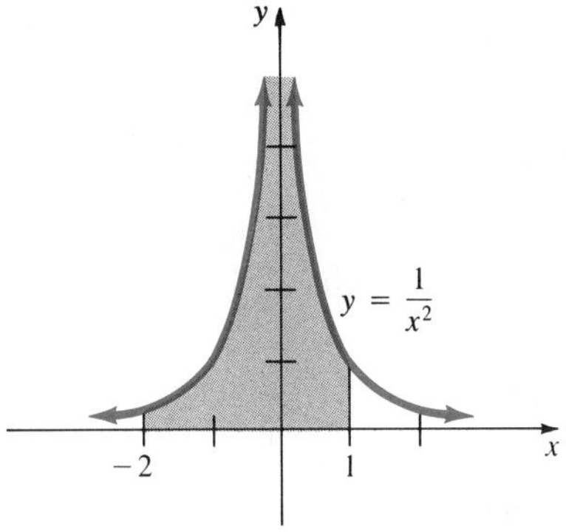

FIGURE 9.5

Recall, if f is continuous on $[a, b]$, then the definite integral $\int_a^b f(x)\,dx$ exists. Moreover, if $F'(x) = f(x)$, then

$$\int_a^b f(x)\,dx = F(b) - F(a) \tag{5}$$

However, we cannot evaluate an integral such as

$$\int_{-2}^1 \frac{1}{x^2}\,dx \tag{6}$$

by the simple and traditional method given in (5), since $f(x) = 1/x^2$ possesses an infinite discontinuity in $[-2, 1]$. See Figure 9.5. In other words, for the integral in (6), the "procedure"

$$-x^{-1}\Big]_{-2}^{1} = (-1) - \left(\frac{1}{2}\right) = -\frac{3}{2}$$

is just meaningless scratchings on paper. Thus, we have another type of integral that demands special handling.

Improper Integrals—Unbounded Functions An integral $\int_a^b f(x)\,dx$ is also said to be **improper** if f is unbounded on $[a, b]$—that is, if f has an infinite discontinuity at some number in the interval of integration. We again distinguish three cases.

DEFINITION 9.2 Improper Integrals—Unbounded Functions

(*i*) If f is continuous on $[a, b)$ and $|f(x)| \to \infty$ as $x \to b^-$, then

$$\int_a^b f(x)\,dx = \lim_{t\to b^-} \int_a^t f(x)\,dx \tag{7}$$

(*ii*) If f is continuous on $(a, b]$ and $|f(x)| \to \infty$ as $x \to a^+$, then

$$\int_a^b f(x)\,dx = \lim_{s\to a^+} \int_s^b f(x)\,dx \tag{8}$$

(*iii*) If $|f(x)| \to \infty$ as $x \to c$ for some c in (a, b) and f is continuous at all other numbers in $[a, b]$, then

$$\int_a^b f(x)\,dx = \int_a^c f(x)\,dx + \int_c^b f(x)\,dx \tag{9}$$

When the limits in (7) and (8) exist, the integrals are said to **converge**. If the limit fails to exist, the integral is said to **diverge**. In (9) the integral $\int_a^b f(x)\,dx$ converges provided both $\int_a^c f(x)\,dx$ and $\int_c^b f(x)\,dx$ converge. If either $\int_a^c f(x)\,dx$ or $\int_c^b f(x)\,dx$ diverges, then $\int_a^b f(x)\,dx$ diverges.

EXAMPLE 9 Evaluate $\displaystyle\int_0^4 \frac{dx}{\sqrt{x}}$ if possible.

Solution Observe that $f(x) = 1/\sqrt{x} \to \infty$ as $s \to 0^+$. Thus, by (8)

$$\int_0^4 \frac{dx}{\sqrt{x}} = \lim_{s\to 0^+}\int_s^4 x^{-1/2}\,dx = \lim_{s\to 0^+} 2x^{1/2}\Big]_s^4 = \lim_{s\to 0^+}[4 - 2s^{1/2}]$$

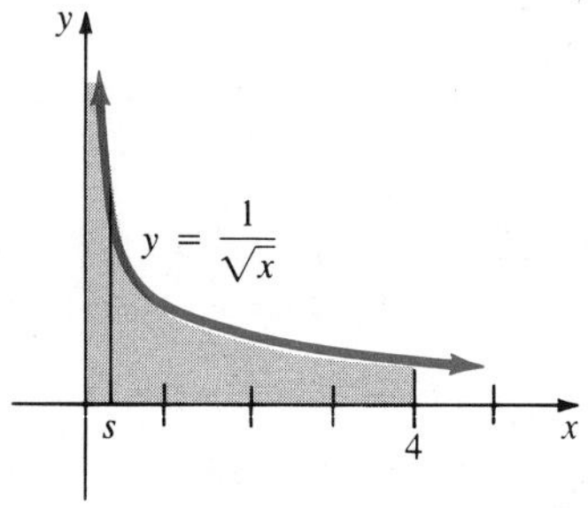

FIGURE 9.6

Since $\lim_{s\to 0^+}[4 - 2s^{1/2}] = 4$, the integral converges and

$$\int_0^4 \frac{dx}{\sqrt{x}} = 4$$

As seen in Figure 9.6, the number 4 can be regarded as a measure of the area under the graph of f on the interval [0, 4]. □

EXAMPLE 10 Evaluate $\displaystyle\int_0^e \ln x\,dx$ if possible.

Solution In this case we know $f(x) = \ln x \to -\infty$ as $x \to 0^+$. Using (8) and integration by parts gives

$$\int_0^e \ln x\,dx = \lim_{s\to 0^+}\int_s^e \ln x\,dx = \lim_{s\to 0^+}(x\ln x - x)\Big]_s^e = \lim_{s\to 0^+} s(1 - \ln s)$$

Now, when the limit is written

$$\lim_{s\to 0^+}\frac{1 - \ln s}{1/s}$$

we recognize the indeterminate form ∞/∞. Thus, by L'Hôpital's Rule,

$$\lim_{s\to 0^+}\frac{1 - \ln s}{1/s} = \lim_{s\to 0^+}\frac{-1/s}{-1/s^2} = \lim_{s\to 0^+} s = 0$$

The integral converges and

$$\int_0^e \ln x \, dx = 0$$ □

EXAMPLE 11 Evaluate $\int_1^5 \frac{dx}{(x-2)^{1/3}}$ if possible.

Solution In the interval [1, 5] the integrand has an infinite discontinuity at 2. Consequently, from (9) we write

$$\int_1^5 \frac{dx}{(x-2)^{1/3}} = \int_1^2 (x-2)^{-1/3}\,dx + \int_2^5 (x-2)^{-1/3}\,dx = I_1 + I_2$$

Now,

$$I_1 = \lim_{t\to 2^-} \int_1^t (x-2)^{-1/3}\,dx = \lim_{t\to 2^-} \frac{3}{2}(x-2)^{2/3}\Big]_1^t$$

$$= \frac{3}{2}\lim_{t\to 2^-} [(t-2)^{2/3} - 1] = -\frac{3}{2}$$

$$I_2 = \lim_{s\to 2^+} \int_s^5 (x-2)^{-1/3}\,dx = \lim_{s\to 2^+} \frac{3}{2}(x-2)^{2/3}\Big]_s^5$$

$$= \frac{3}{2}\lim_{s\to 2^+} [3^{2/3} - (s-2)^{2/3}] = \frac{3^{5/3}}{2}$$

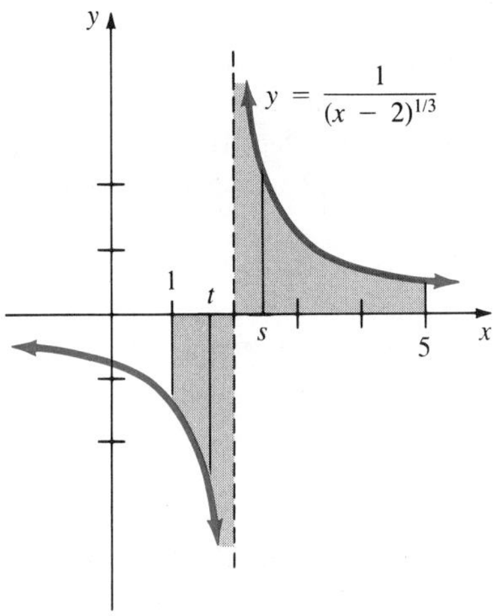

FIGURE 9.7

Since both I_1 and I_2 converge, the given integral converges and

$$\int_1^5 \frac{dx}{(x-2)^{1/3}} = -\frac{3}{2} + \frac{3^{5/3}}{2} \approx 1.62$$

Note from Figure 9.7 that this last number is not a measure of area. Why? □

EXAMPLE 12 Evaluate $\int_{-2}^1 \frac{dx}{x^2}$ if possible.

Solution This is the integral discussed in (6). Since, in the interval [−2, 1], the integrand has an infinite discontinuity at 0, we write

$$\int_{-2}^1 \frac{1}{x^2}\,dx = \int_{-2}^0 \frac{dx}{x^2} + \int_0^1 \frac{dx}{x^2} = I_1 + I_2$$

Now the result

$$I_1 = \int_{-2}^0 \frac{dx}{x^2} = \lim_{t\to 0^-} \int_{-2}^t x^{-2}\,dx = \lim_{t\to 0^-} -x^{-1}\Big]_{-2}^t = \lim_{t\to 0^-} \left[-\frac{1}{t} - \frac{1}{2}\right] = \infty$$

indicates there is no need to evaluate $I_2 = \int_0^1 dx/x^2$. The integral $\int_{-2}^1 dx/x^2$ diverges. □

▲ **Remarks** (*i*) It is possible for an integral to have infinite limits of integration *and* an integrand with an infinite discontinuity. To determine whether an integral such as

$$\int_1^{\infty} \frac{dx}{x\sqrt{x^2-1}}$$

converges, we break up the integration at some convenient point of continuity of the integrand, say, $x = 2$:

$$\int_1^{\infty} \frac{dx}{x\sqrt{x^2-1}} = \int_1^{2} \frac{dx}{x\sqrt{x^2-1}} + \int_2^{\infty} \frac{dx}{x\sqrt{x^2-1}} = I_1 + I_2$$

I_1 and I_2 are improper integrals; I_1 is of the type given in (8) and I_2 is of the type given in (1). If both I_1 *and* I_2 converge, then the original integral is convergent. See Problems 93 and 94 in Exercises 9.2.

(*ii*) The integrand of $\int_a^b f(x)\,dx$ can also have infinite discontinuities at both $x = a$ and $x = b$. In this case the improper integral is defined in a manner similar to (9). Lastly, if an integrand f has an infinite discontinuity at several numbers in (a, b), then the improper integral is defined by a natural extension of (9). See Problems 95 and 96 in Exercises 9.2.

EXERCISES 9.2 *Answers to odd-numbered problems begin on page A-91.*

[9.2.1]

In Problems 1–30 evaluate the given improper integral or show that it diverges.

1. $\int_3^{\infty} \frac{dx}{x^4}$
2. $\int_{-\infty}^{-1} \frac{dx}{\sqrt[3]{x}}$
3. $\int_1^{\infty} \frac{dx}{x^{0.99}}$
4. $\int_1^{\infty} \frac{dx}{x^{1.01}}$
5. $\int_{-\infty}^{3} e^{2x}\,dx$
6. $\int_{-\infty}^{\infty} e^{-x}\,dx$
7. $\int_1^{\infty} \frac{\ln x}{x}\,dx$
8. $\int_1^{\infty} \frac{\ln t}{t^2}\,dt$
9. $\int_e^{\infty} \frac{dx}{x(\ln x)^3}$
10. $\int_e^{\infty} \ln x\,dx$
11. $\int_{-\infty}^{\infty} te^{-t^2}\,dt$
12. $\int_{-\infty}^{\infty} \frac{dx}{1+x^2}$
13. $\int_{-\infty}^{0} \frac{x}{(x^2+9)^2}\,dx$
14. $\int_5^{\infty} \frac{dx}{\sqrt[4]{3x+1}}$
15. $\int_2^{\infty} ue^{-u}\,du$
16. $\int_{-\infty}^{3} \frac{x^3}{x^4+1}\,dx$
17. $\int_{2/\pi}^{\infty} \frac{\sin(1/x)}{x^2}\,dx$
18. $\int_{-\infty}^{\infty} \cos 3\theta\,d\theta$
19. $\int_{-1}^{\infty} \frac{dx}{x^2+2x+2}$
20. $\int_{-\infty}^{0} \frac{dx}{x^2+2x+3}$
21. $\int_0^{\infty} e^{-x}\sin x\,dx$
22. $\int_{-\infty}^{0} e^{x}\cos 2x\,dx$
23. $\int_{1/2}^{\infty} \frac{x+1}{x^3}\,dx$
24. $\int_0^{\infty} (e^{-x}-e^{-2x})^2\,dx$
25. $\int_1^{\infty} \left[\frac{1}{x} - \frac{1}{x+1}\right] dx$
26. $\int_3^{\infty} \left[\frac{1}{x} + \frac{1}{x^2+9}\right] dx$
27. $\int_2^{\infty} \frac{dx}{x^2+6x+5}$
28. $\int_{-\infty}^{0} \frac{dx}{x^2-3x+2}$
29. $\int_{-\infty}^{-2} \frac{x^2}{(x^3+1)^2}\,dx$
30. $\int_0^{\infty} \frac{dx}{e^x+e^{-x}}$

In Problems 31–34 find the area under the graph of the given function on the indicated interval.

31. $f(x) = \frac{1}{(2x+1)^2}$; $[1, \infty)$

32. $f(x) = \dfrac{10}{x^2 + 25}; (-\infty, 5]$

33. $f(x) = e^{-|x|}; (-\infty, \infty)$

34. $f(x) = |x|^3 e^{-x^4}; (-\infty, \infty)$

35. Consider the region that is bounded by the graphs of $y = 1/x$ and $y = 0$ on the interval $[1, \infty)$.

(a) Show that the region does not have finite area.

(b) Show that the solid of revolution that is formed by revolving the region about the x-axis has finite volume. See Figure 9.8.

(c) Show that the surface area of the solid of revolution in part (**b**) is infinite. [*Hint:* Use an algebraic substitution followed by a trigonometric substitution.]

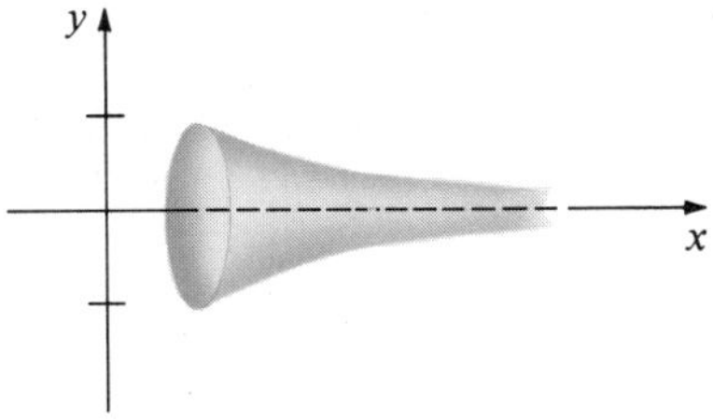

FIGURE 9.8

36. Find the volume of the solid of revolution that is formed by revolving the region bounded by the graphs of $y = xe^{-x}$ and $y = 0$ on $[0, \infty)$ around the x-axis.

37. Find the work done against gravity in lifting a 10,000-kg payload to an infinite distance above the surface of the moon. [*Hint:* Review page 357 of Section 5.8.]

38. The work done by an external force in moving a test charge q_0 radially from point A to point B in the electric field of a charge q is defined to be:

$$W = -\frac{qq_0}{4\pi e_0}\int_{r_A}^{r_B} \frac{dr}{r^2}$$

See Figure 9.9.

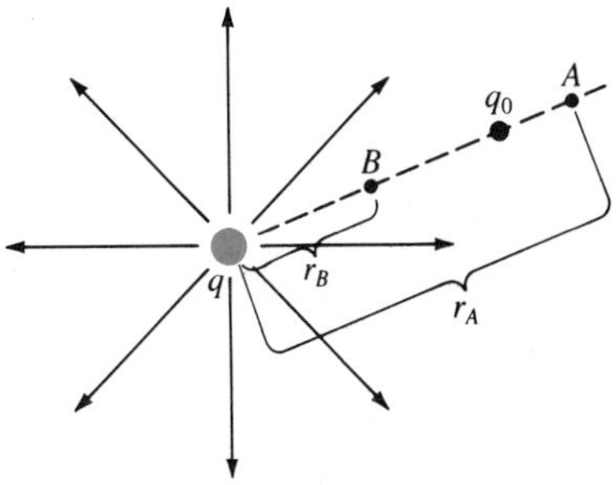

FIGURE 9.9

(a) Show that $W = \dfrac{qq_0}{4\pi e_0}\left(\dfrac{1}{r_B} - \dfrac{1}{r_A}\right)$.

(b) Find the work done in bringing the test charge in from an infinite distance to point B.

39. A **probability density function** is any nonnegative function f defined on an interval $[a, b]$ for which $\int_a^b f(x)\,dx = 1$. Verify that

$$f(x) = \begin{cases} 0, & x < 0 \\ ke^{-kx}, & x \geq 0, \quad k > 0 \end{cases}$$

is a probability density function on $(-\infty, \infty)$.

40. The **capital value** of a perpetual stream of income is defined to be $V = \int_0^\infty f(t)e^{-rt}\,dt$, where f is the rate at which income flows per year and r is the annual rate of interest compounded continuously. Find the capital value of a stream of income whose rate is $f(t) = 500t$ if the interest rate is 10%.

The **Laplace transform** of a function $y = f(x)$ defined by

$$\mathscr{L}\{f(x)\} = \int_0^\infty e^{-st} f(t)\,dt$$

is very useful in some areas of applied mathematics. In Problems 41–48 find the Laplace transform of the given function and state a restriction on s for which the integral converges.

41. $f(x) = 1$

42. $f(x) = x$

Pierre Simon Marquis de Laplace (1749–1827) A noted mathematician, physicist, and astronomer, Laplace was called by some of his more enthusiastic contemporaries the "Newton of France." Although Laplace made use of the transform in his work in probability theory, it is likely that the integral was first discovered by Euler. Laplace's noted treatises were *Mécanique Céleste* and *Théorie Analytique des Probabilités*. Born into a poor farming family, Laplace became a friend of Napoleon but was elevated to the nobility by Louis XVIII after the restoration.

43. $f(x) = e^x$

44. $f(x) = e^{-5x}$

45. $f(x) = \sin x$

46. $f(x) = \cos 2x$

47. $f(x) = \begin{cases} 0, & 0 \le x < 1 \\ 1, & x \ge 1 \end{cases}$

48. $f(x) = \begin{cases} 0, & 0 \le x < 3 \\ e^{-x}, & x \ge 3 \end{cases}$

49. Another integral in applied mathematics is the so-called **gamma function**:

$$\Gamma(\alpha) = \int_0^\infty t^{\alpha-1} e^{-t}\, dt, \qquad \alpha > 0$$

(a) Show that $\Gamma(\alpha + 1) = \alpha\Gamma(\alpha)$.

(b) Use the result in part (**a**) to show that

$$\Gamma(n + 1) = 1 \cdot 2 \cdot 3 \cdots (n - 1) \cdot n = n!*$$

Because of this property, the gamma function is called the **generalized factorial function**.

50. The **gamma distribution** plays an important role in the modeling of arrival times of customers at a check-out counter. In addition, this distribution is used in traffic control studies. In either case a probability density function (see Problem 39) takes the form $f(x) = cx^n e^{-\beta x}$, for $x > 0$, $\beta > 0$, and n a positive integer. Find c so that $\int_0^\infty f(x)\, dx = 1$. [*Hint:* Let $t = \beta x$ and use the result of Problem 49.]

51. Theoretical considerations of rainfall physics suggest that the relative fraction of raindrops having radius less than R is given by an integral of the form $\int_0^R e^{-r/r_0}\, dr/r_0$, where $r_0 > 0$ is a parameter to be experimentally determined. The average radius $\bar{R}$ and average volume $\bar{V}$ of a raindrop are defined as the integrals

$$\bar{R} = \int_0^\infty r\left(\frac{e^{-r/r_0}}{r_0}\right) dr$$

and

$$\bar{V} = \int_0^\infty \frac{4}{3}\pi r^3 \left(\frac{e^{-r/r_0}}{r_0}\right) dr$$

Compute these integrals and show that the "obvious" equation $\bar{V} = \frac{4}{3}\pi \bar{R}^3$ is not correct.

52. A study of the Bombay plague epidemic of 1905–06 found that the death rate for that epidemic could be approximated by the formula

$$R = 890\, \text{sech}^2(0.2t - 3.4)$$

where R is the number of deaths per week and t is the time (in weeks) from the onset of the plague.

(a) What is the peak death rate, and when does it occur?

(b) Estimate the total number of deaths by computing the integral $\int_{-\infty}^\infty R(t)\, dt$.

(c) Show that more than 99% of the deaths occurred in the first 34 weeks of the epidemic; that is, compare $\int_0^{34} R(t)\, dt$ to the result in part (**b**).

(d) Suppose you want to use a "simpler" model for the death rate, of the form

$$R_0 = \frac{a}{t^2 - 2bt + c}$$

where $c > b^2$. You want this model to have the same peak death rate at the same time as the original model and you also want the total number of deaths, $\int_{-\infty}^\infty R_0(t)\, dt$, to be the same. Find coefficients a, b, and c that satisfy these requirements.

(e) For the model in part (**d**), show that less than 89% of the deaths occur in the first 34 weeks of the epidemic.

In Problems 53–56 determine all values of k such that the given integral is convergent.

53. $\displaystyle\int_1^\infty \frac{dx}{x^k}$

54. $\displaystyle\int_{-\infty}^1 x^{2k}\, dx$

55. $\displaystyle\int_0^\infty e^{kx}\, dx$

56. $\displaystyle\int_1^\infty \frac{(\ln x)^k}{x}\, dx$

The following is the **Comparison Test** for convergence. Suppose f and g are continuous and $0 \le f(x) \le g(x)$ for $x \ge a$. If $\int_a^\infty g(x)\, dx$ converges, then $\int_a^\infty f(x)\, dx$ also converges. In Problems 57–60 use this result to show that the given integral converges.

57. $\displaystyle\int_1^\infty \frac{\sin^2 x}{x^2}\, dx$

58. $\displaystyle\int_2^\infty \frac{dx}{x^3 + 4}$

59. $\displaystyle\int_0^\infty \frac{dx}{x + e^x}$

60. $\displaystyle\int_0^\infty e^{-x^2}\, dx$

61. Prove that the surface area of the solid of revolution described in Problem 35(**b**) is not finite by using the Comparison Test and the inequality $\sqrt{x^4 + 1}/x^3 > 1/x$.

Calculator/Computer Problems

62. (a) Show that the convergent integral $\int_1^\infty e^{1/x}\, dx/x^{5/2}$ can be written as $\int_0^1 t^{1/2} e^t\, dt$.

(b) Use the result of part (**a**) and Simpson's Rule with $n = 4$ to find an approximation to the original integral.

*A rigorous demonstration of this fact requires mathematical induction. The symbol $n!$ is read "n factorial."

[9.2.2]

In Problems 63–84 evaluate the given improper integral or show that it diverges.

63. $\int_0^5 \frac{dx}{x}$

64. $\int_0^8 \frac{dx}{x^{2/3}}$

65. $\int_0^1 \frac{dx}{x^{0.99}}$

66. $\int_0^1 \frac{dx}{x^{1.01}}$

67. $\int_0^2 \frac{dx}{\sqrt{2-x}}$

68. $\int_1^3 \frac{dx}{(x-1)^2}$

69. $\int_{-1}^1 \frac{dx}{x^{5/3}}$

70. $\int_0^2 \frac{dx}{\sqrt[3]{x-1}}$

71. $\int_0^2 (x-1)^{-2/3}\,dx$

72. $\int_0^{27} \frac{e^{x^{1/3}}}{x^{2/3}}\,dx$

73. $\int_0^1 x \ln x\,dx$

74. $\int_1^e \frac{dx}{x \ln x}$

75. $\int_0^{\pi/2} \tan t\,dt$

76. $\int_0^{\pi/4} \frac{\sec^2\theta}{\sqrt{\tan\theta}}\,d\theta$

77. $\int_0^{\pi} \frac{\sin x}{1+\cos x}\,dx$

78. $\int_0^{\pi} \frac{\cos x}{\sqrt{1-\sin x}}\,dx$

79. $\int_{-1}^0 \frac{x}{\sqrt{1+x}}\,dx$

80. $\int_0^3 \frac{dx}{x^2-1}$

81. $\int_0^1 \frac{x^2}{\sqrt{1-x^2}}\,dx$

82. $\int_0^2 \frac{e^w}{\sqrt{e^w-1}}\,dw$

83. $\int_1^3 \frac{dx}{\sqrt{3+2x-x^2}}$

84. $\int_0^1 \left[\frac{1}{\sqrt{x}} + \frac{1}{\sqrt{1-x}}\right] dx$

In Problems 85 and 86 determine whether the area under the graph of the given function on the indicated interval is finite.

85. $f(x) = \frac{x}{\sqrt{16-x^2}}$; $[0, 4]$

86. $f(x) = \sec x$; $[0, \pi/2]$

87. Consider the region that is bounded by the graphs of $y = 1/\sqrt{x+2}$ and $y = 0$ on the interval $[-2, 1]$.

(a) Show that the region has finite area.

(b) Show that the solid of revolution that is formed by revolving the region around the x-axis has infinite volume.

88. The region that is bounded by the graphs of $y = 1/x^{3/2}$ and $y = 0$ on the interval $[0, 4]$ is revolved around the y-axis. Determine whether the volume of the solid of revolution is finite.

89. Determine whether the area of the region that is bounded by the graphs of

$$y = \frac{1}{x} \quad \text{and} \quad y = \frac{1}{x(x^2+1)}$$

on the interval $[0, 1]$ is finite.

90. Find the area of the region that is bounded by the graphs of $y = 1/\sqrt{x-1}$ and $y = -1/\sqrt{x-1}$ on the interval $[1, 5]$.

In Problems 91 and 92 determine all values of k such that the given integral is convergent.

91. $\int_0^1 \frac{dx}{x^k}$

92. $\int_0^1 x^k \ln x\,dx$

In Problems 93–96 determine whether the given integral converges or diverges.

93. $\int_1^{\infty} \frac{dx}{x\sqrt{x^2-1}}$

94. $\int_{-\infty}^4 \frac{dx}{(x-1)^{2/3}}$

95. $\int_{-1}^1 \frac{dx}{\sqrt{1-x^2}}$

96. $\int_0^2 \frac{2x-1}{\sqrt[3]{x^2-x}}\,dx$

97. Discuss whether $\int_0^{\pi/2} \frac{\sin x}{x}\,dx$ is an improper integral.

CHAPTER 9 REVIEW EXERCISES

Answers begin on page A-91.

In Problems 1–20 answer true or false.

1. A limit of the form $\infty - \infty$ always has the value 0. ________

2. A limit of the form ∞/∞ is indeterminate. ________

3. A limit of the form $0/\infty$ is indeterminate. ________

4. If $\lim_{x\to a} \frac{f(x)}{g(x)}$ and $\lim_{x\to a} \frac{f'(x)}{g'(x)}$ are both of the form ∞/∞, then the first limit does not exist. ________

5. A limit of the form 1^{∞} is always 1. ________

6. For an indeterminate form, L'Hôpital's Rule states that

the limit of a quotient is the same as the derivative of the quotient. ________

7. $\lim_{x\to\infty} \dfrac{x^n}{e^x} = 0$ for every integer n. ________

8. $\lim_{x\to 0} \dfrac{1-\cos^2 x}{x^2} = \lim_{x\to 0} \dfrac{1-\cos x}{x} \cdot \lim_{x\to 0} \dfrac{1+\cos x}{x}$

____

9. If $\int_a^\infty f(x)\,dx$ and $\int_a^\infty g(x)\,dx$ converge, then $\int_a^\infty [f(x)+g(x)]\,dx$ converges. ________

10. If $\int_a^\infty [f(x)+g(x)]\,dx$ converges, then $\int_a^\infty f(x)\,dx$ converges. ________

11. If f is continuous for all x and $\int_{-\infty}^a f(x)\,dx$ diverges, then $\int_{-\infty}^\infty f(x)\,dx$ diverges. ________

12. The integral $\int_{-\infty}^\infty f(x)\,dx$ is defined by $\lim_{t\to\infty} \int_{-t}^t f(x)\,dx$.

____

13. $\int_0^4 x^{-0.999}\,dx$ converges. ________

14. $\int_1^\infty x^{-0.999}\,dx$ diverges. ________

15. $\int_2^\infty \left[\dfrac{e^x}{e^x+1} - \dfrac{e^x}{e^x-1}\right] dx$ diverges, since $\int_2^\infty \dfrac{e^x}{e^x+1}\,dx$ diverges. ________

16. If $f(x)\to 0$ as $x\to\infty$, then $\int_a^\infty f(x)\,dx$ converges.

____

17. If a positive function f has an infinite discontinuity at a number in $[a, b]$, then the area under the graph on the interval is also infinite. ________

18. A region with infinite area, when revolved about an axis, always generates a solid of revolution of infinite volume.

____

19. $\int_0^\pi \sec^2 x\,dx = \tan x\Big]_0^\pi = 0$ ________

20. $\lim_{x\to\pi/2} \dfrac{\cos x}{(x-\pi/2)^2} = \lim_{x\to\pi/2} \dfrac{-\sin x}{2(x-\pi/2)} = \lim_{x\to\pi/2} \dfrac{-\cos x}{2} = 0$

____

In Problems 21–36 evaluate the given limit if it exists.

21. $\lim_{x\to -2} \dfrac{x^2+4x+4}{x^2+x-2}$

22. $\lim_{x\to 1} \dfrac{x^2-3x+5}{x^2-1}$

23. $\lim_{x\to\sqrt{3}} \dfrac{\sqrt{3}-\tan(\pi/x^2)}{x-\sqrt{3}}$

24. $\lim_{\theta\to -\pi} \dfrac{\sin\theta}{\theta+\pi}$

25. $\lim_{x\to 2} \dfrac{3x^2-4x-2}{x-2}$

26. $\lim_{x\to -\infty} \dfrac{-x^2+3x+1}{4x^2-2x}$

27. $\lim_{\theta\to 0} \dfrac{10\theta - 5\sin 2\theta}{10\theta - 2\sin 5\theta}$

28. $\lim_{x\to\infty} \dfrac{x}{\ln x}$

29. $\lim_{x\to\infty} x\left(\cos\dfrac{1}{x} - e^{2/x}\right)$

30. $\lim_{y\to 0} \left[\dfrac{1}{y} - \dfrac{1}{\ln(y+1)}\right]$

31. $\lim_{t\to 0} \dfrac{(\sin t)^2}{\sin t^2}$

32. $\lim_{x\to 0} \dfrac{\tan 5x}{e^{3x/2} - e^{-x/2}}$

33. $\lim_{x\to 0^+} (3x)^{-1/\ln x}$

34. $\lim_{x\to 0} (2x + e^{3x})^{4/x}$

35. $\lim_{x\to\infty} \ln\left(\dfrac{x+e^{2x}}{1+e^{4x}}\right)$

36. $\lim_{x\to 0^+} x(\ln x)^2$

In Problems 37–48 evaluate the given integral or show that it diverges.

37. $\int_0^3 x(x^2-9)^{-2/3}\,dx$

38. $\int_0^5 x(x^2-9)^{-2/3}\,dx$

39. $\int_{-\infty}^0 (x+1)e^x\,dx$

40. $\int_0^\infty \dfrac{e^{2x}}{e^{4x}+1}\,dx$

41. $\int_3^\infty \dfrac{dx}{1+5x}$

42. $\int_0^\infty \dfrac{x}{(x^2+4)^2}\,dx$

43. $\int_0^e \ln\sqrt{x}\,dx$

44. $\int_0^{\pi/2} \dfrac{\sec^2 t}{\tan^3 t}\,dt$

45. $\int_0^{\pi/2} \dfrac{dx}{1-\cos x}$

46. $\int_0^\infty \dfrac{x}{x+1}\,dx$

47. $\int_0^1 \dfrac{dx}{\sqrt{x}e^{\sqrt{x}}}$

48. $\int_0^\infty \dfrac{dx}{\sqrt{x}e^{\sqrt{x}}}$

49. Find the area of the region that is bounded by the graphs of $y = e^{-x}$ and $y = e^{-3x}$ on $[0, \infty)$.

50. Consider the region that is bounded by the graphs of $y = 1/(1-x)^{1/3}$ and $y = 0$ on the interval $[0, 1]$.

(a) Find the area of the region.

(b) Find the volume of the solid of revolution that is formed by revolving the region about the x-axis.

(c) Find the volume of the solid of revolution that is formed by revolving the region about the line $x = 1$.

51. Consider the graph of $f(x) = (x^2-1)/(x^2+1)$ given in Figure 9.10.

(a) Determine whether the region R_1, which is bounded between the graph of f and its horizontal asymptote, is finite.

(b) Determine whether the regions R_2 and R_3 have finite areas.

52. Prove that $\int_1^\infty \dfrac{\sqrt{x}}{(1+x)^2}\,dx = \dfrac{\pi}{4} + \dfrac{1}{2}$.

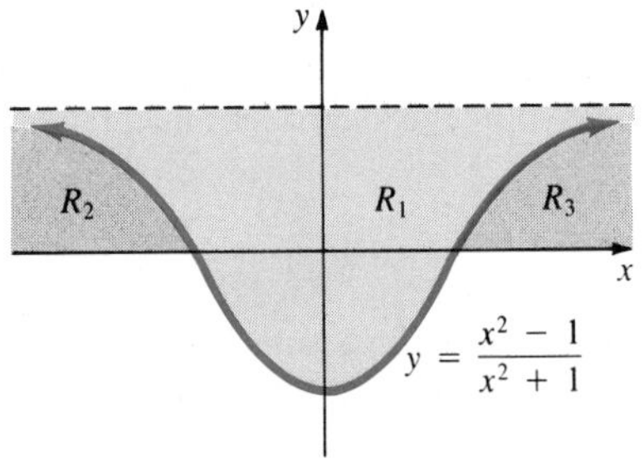

FIGURE 9.10

53. Prove that $\int_0^{\infty} \frac{dx}{\sqrt{x}(x+1)} = \pi$.

54. Prove that if $a > 0$ and $ax^2 + bx + c > 0$ for all x, then

$$\int_{-\infty}^{\infty} \frac{dx}{ax^2 + bx + c} = \frac{2\pi}{\sqrt{4ac - b^2}}$$

In Problems 55 and 56 use the fact that $\lim_{x\to\infty} \int_0^x e^{t^2}\,dt = \infty$ to evaluate the given limit.

55. $\lim_{x\to\infty} \frac{x\int_0^x e^{t^2}\,dt}{e^{x^2}}$ 56. $\lim_{x\to\infty} \frac{\int_0^x e^{t^2}\,dt}{xe^{x^2}}$

57. In Problem 24 of Exercises 5.8 we saw that when a gas expands from pressure p_1 and volume v_1 to pressure p_2 and volume v_2 such that $pv^{\gamma} = k$ (constant) throughout the expansion, if $\gamma \neq 1$, the work done is

$$W = \frac{p_2 v_2 - p_1 v_1}{1 - \gamma}$$

(a) Show that

$$W = p_1 v_1 \left[\frac{(v_2/v_1)^{1-\gamma} - 1}{1 - \gamma}\right]$$

(b) Find the work done in the case when $pv = k$ (constant) throughout the expansion by letting $\gamma \to 1$ in the expression in part (**a**).

58. (a) Show that

$$1 + x + x^2 + x^3 + \cdots + x^n = \frac{x^{n+1} - 1}{x - 1}$$

[*Hint:* Multiply both sides of the equality by $x - 1$.]

(b) Differentiate both sides of the result in part (**a**) to show that

$$1 + 2x + 3x^2 + \cdots + nx^{n-1} = \frac{nx^{n+1} - (n+1)x^n + 1}{(x-1)^2}$$

(c) By taking the limit of both sides of the result in part (**b**) as $x \to 1$ show that

$$1 + 2 + 3 + \cdots + n = \frac{n(n+1)}{2}$$

See (5) of Section 4.3.

59. Suppose f has a second derivative. Evaluate

$$\lim_{h\to 0} \frac{f(x+h) - 2f(x) + f(x-h)}{h^2}$$

60. If $\int_0^{\infty} e^{-x^2}\,dx = \frac{\sqrt{\pi}}{2}$, show that $\int_0^{\infty} \frac{e^{-x}}{\sqrt{x}}\,dx = \sqrt{\pi}$.

61. Find the value of x such that $\int_1^x e^{-2t}\,dt = \int_x^{\infty} e^{-2t}\,dt$.

62. In population biology, the collection of young that are the results of an annual reproductive period is called a *year class*. In many animal population models, it is assumed that the number $N(t)$ of a year class that are still alive after t years is given by $N(t) = N_0 e^{-kt}$, for some $k > 0$. It can be further shown that the **average life length** of an animal is given by $E = (-1/N_0)\int_0^{\infty} tN'(t)\,dt$. Determine the average life length for Pacific halibut if $k = 0.2$.

63. The annual egg production of chickens decreases approximately exponentially with the age of the hen. A poultry farmer is comparing two breeds of chicken, which each begin laying eggs at age six months. Thereafter, the rate of egg production in the first breed is given by the formula $E(t) = 179e^{-0.12t}$, where $t \geq \frac{1}{2}$ is the age of the chicken in years and $E(t)$ is the rate of egg production at age t in eggs per year. In the second breed, the production rate is given by the formula $E(t) = 193e^{-0.14t}$, $t \geq \frac{1}{2}$.

(a) Assuming the birds live forever, which breed has the higher total yield? That is, compare the integrals

$$Y_1 = \int_{1/2}^{\infty} 179e^{-0.12t}\,dt \quad \text{and} \quad Y_2 = \int_{1/2}^{\infty} 193e^{-0.14t}\,dt$$

(b) Making the more reasonable assumption that the birds live for 10 years, which breed has the higher total yield?

(c) By what age does the first breed produce as many eggs as the second breed produces by age 4?

(d) For the first several years, the total yield of the second breed is greater than that of the first breed. At what age is the difference between the yields at a maximum?

(e) To the nearest year, at what age does the total yield of the first breed overtake (that is, equal) that of the second breed? Use the bisection method or Newton's Method.

64. Use Newton's Method to find the point x^* for which the shaded area in Figure 9.11 is 99% of the total area under the graph of $y = xe^{-x}$ on $[0, \infty]$.

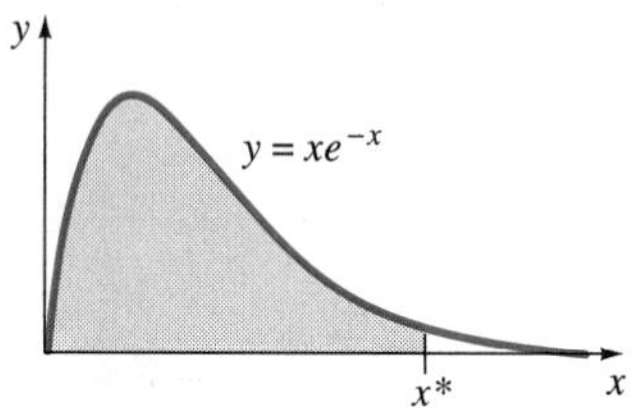

FIGURE 9.11

65. (a) The **logarithmic integral function**, $\mathrm{Li}(x)$, is defined as the integral

$$\mathrm{Li}(x) = \int_2^x \frac{dt}{\ln t}$$

for $x > 2$. In 1896, Jacques Hadamard and Charles-Jean de la Vallée Poussin proved the Prime Number Theorem, which states that the number of prime numbers (2, 3, 5, 7, 11, etc.) less than or equal to x, denoted $\pi(x)$, can be approximated by the logarithmic integral, meaning that

$$\lim_{x\to\infty} \frac{\pi(x)}{\mathrm{Li}(x)} = 1$$

Show that $\pi(x)$ can also be approximated by the function $x/\ln x$ by using L'Hôpital's Rule and the Fundamental Theorem of Calculus to show that

$$\lim_{x\to\infty} \frac{\mathrm{Li}(x)}{x/\ln x} = 1$$

Since there are an infinite number of primes, $\mathrm{Li}(x) \to \infty$ as $x \to \infty$.

(b) Use Simpson's Rule to approximate $\mathrm{Li}(100)$. Compute $x/\ln x$ for $x = 100$. Compare these numbers with the actual number of prime numbers less than 100.

10

SEQUENCES AND SERIES

INTRODUCTION

Everyday experience gives one a feeling for the notion of a **sequence**. For example, the words *sequence of events* or *sequence of numbers* intuitively suggest an arrangement whereby the events E or numbers n are set down in some order: $E_1, E_2, E_3, \ldots$ or $n_1, n_2, n_3, \ldots$.

Every student of mathematics is also familiar with the fact that any real number can be written as a decimal. For example, the rational number $\frac{1}{3}$ has the familiar representation $\frac{1}{3} = 0.333\ldots$, where the mysterious three dots . . . (ellipses) signify that the 3s go on forever. This means $0.333\ldots$ is an infinite sum or the **infinite series**

$$\frac{3}{10} + \frac{3}{100} + \frac{3}{1000} + \frac{3}{10{,}000} + \cdots$$

In this chapter we shall see that the concepts of sequence and infinite series are related.

In 1972 more than 6500 persons were hospitalized and 459 died in the country of Iraq when people ate bread accidentally made from wheat that had been contaminated with highly toxic methylmercury fungicide. Bioscientists often try to learn from a tragedy such as this. The effect on a person obviously depends on the amount of poison ingested over a matter of days. The level of poisoning and the time it takes for certain symptoms to appear were modeled using a special type of series.

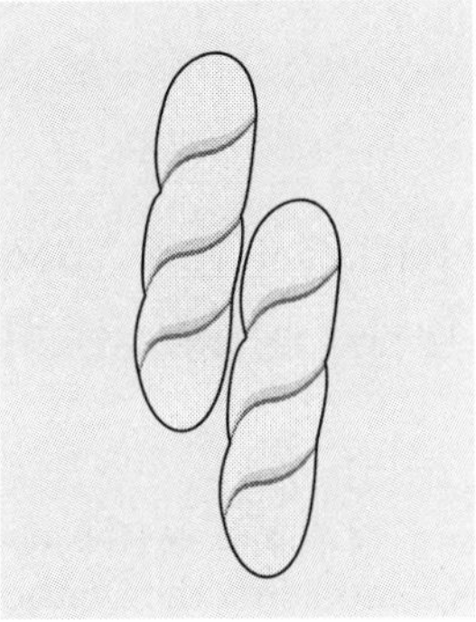

You will be encouraged to work on this mathematical model in Exercises 10.3.

IMPORTANT CONCEPTS

sequence
convergent sequence
divergent sequence
monotonic sequence
infinite series
sequence of partial sums
convergent infinite series
divergent infinite series
geometric series
harmonic series
***n*th term test for divergence**
integral test
comparison test
limit comparison test
ratio test
root test
alternating series
alternating series test
absolute convergence
conditional convergence
power series
interval of convergence
Taylor series
Maclaurin series
binomial series

10.1 SEQUENCES

If the domain of a function f is the set of positive integers, then the elements $f(n)$ in its range can be arranged in order of increasing n:

$$f(1), f(2), f(3), f(4), \ldots, f(n), \ldots$$

EXAMPLE 1 If n is a positive integer, then the first several elements in the range of $f(n) = (1 + 1/n)^n$ are

$$f(1) = 2, \quad f(2) = \frac{9}{4}, \quad f(3) = \frac{64}{27}, \quad \ldots \tag{1}$$ □

Functions whose domains are the entire set of positive integers are given a special name.

> **DEFINITION 10.1 Sequence**
> A **sequence** is a function whose domain is the set of positive integers.*

Terms Instead of the customary function notation $f(n)$, a sequence is usually denoted by the symbol $\{a_n\}$. The **terms** of the sequence are formed by letting n take on the values 1, 2, 3, . . . in the **general term** a_n. Thus, $\{a_n\}$ is equivalent to

$$a_1, a_2, a_3, \ldots, a_n, \ldots \quad \leftarrow \text{elements in range}$$
$$\uparrow \;\; \uparrow \;\; \uparrow$$
$$1 \;\; 2 \;\; 3 \;\; \ldots \;\; n \;\; \ldots \quad \leftarrow \text{elements in domain}$$

For example, the sequence defined in (1) would be written $\{(1 + 1/n)^n\}$.

Note: In some circumstances it is convenient to take the first term of a sequence to be a_0.

EXAMPLE 2 Write out the first four terms of the sequences.

(*a*) $\left\{\dfrac{1}{\sqrt{n}}\right\}$ (*b*) $\{n^2 + n\}$ (*c*) $\{(-1)^n\}$

---

*Some texts use the words *infinite sequence.* When the domain of a function is a finite subset of the set of positive integers, we get a *finite sequence.* All the sequences considered in this discussion will be infinite.

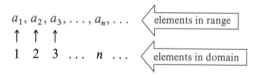

Solution By substituting $n = 1, 2, 3, 4$ in the respective general terms, we obtain

$$(a)\ 1, \frac{1}{\sqrt{2}}, \frac{1}{\sqrt{3}}, \frac{1}{2}, \ldots \qquad (b)\ 2, 6, 12, 20, \ldots \qquad (c)\ -1, 1, -1, 1, \ldots$$ □

Convergent Sequences For the sequence in (*a*) of Example 2, we see that as n becomes progressively larger, the values $a_n = 1/\sqrt{n}$ do not increase without bound. Indeed, as $n \to \infty$, $1/\sqrt{n} \to 0$. We say that the terms $1/\sqrt{n}$ approach the **limit** 0 and that the sequence $\{1/\sqrt{n}\}$ **converges** to 0. In contrast, the terms of the sequences in (*b*) and (*c*) do not approach a limit as $n \to \infty$.

In general we have the following definition:

DEFINITION 10.2 Convergent Sequence

A sequence $\{a_n\}$ is said to **converge** to a number L if for every $\varepsilon > 0$ there exists a positive number N such that

$$|a_n - L| < \varepsilon \quad \text{whenever} \quad n > N \tag{2}$$

The number L is called the **limit** of the sequence.

If a sequence $\{a_n\}$ converges, then its limit L is unique.

Divergent Sequences If $\{a_n\}$ is a convergent sequence, (2) means that the terms a_n can be made arbitrarily close to L for n sufficiently large. We indicate that a sequence converges to a number L by writing

$$\lim_{n\to\infty} a_n = L$$

When $\lim_{n\to\infty} a_n$ does not exist, we say that the sequence **diverges**.

Figure 10.1 illustrates several ways in which a sequence $\{a_n\}$ can converge to a number L. As indicated in Figure 10.1(a), if $\{a_n\}$ converges, then *all but a finite* number of the terms a_n are in the interval $(L - \varepsilon, L + \varepsilon)$.

EXAMPLE 3 Using Definition 10.2, prove that $\{1/\sqrt{n}\}$ converges to 0.

Solution Let $\varepsilon > 0$ be given. Since the terms of the sequence are positive, the inequality $|a_n - 0| < \varepsilon$ is the same as

$$\frac{1}{\sqrt{n}} < \varepsilon$$

This is equivalent to $\sqrt{n} > 1/\varepsilon$ or $n > 1/\varepsilon^2$. Hence, we need only choose N to be the first positive integer greater than or equal to $1/\varepsilon^2$. □

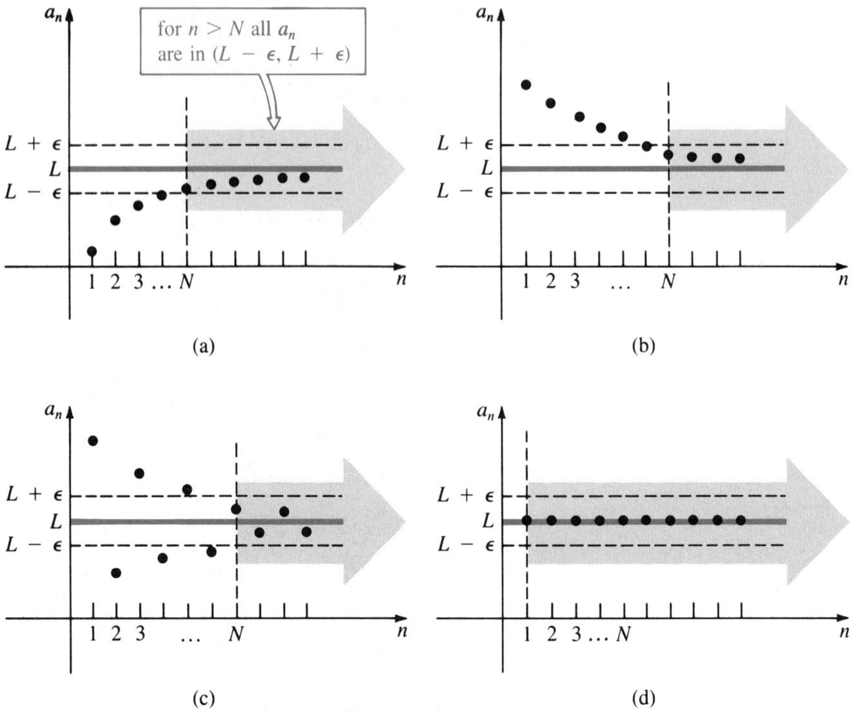

FIGURE 10.1

For instance, had we chosen $\varepsilon = 0.01$ in Example 3, then $|1/\sqrt{n} - 0| = 1/\sqrt{n} < 0.01$ whenever $n > 10{,}000$. That is, we choose $N = 10{,}000$.

In practice, to determine whether a sequence $\{a_n\}$ converges or diverges, we work directly with $\lim_{n\to\infty} a_n$ and proceed as we would in the examination of $\lim_{x\to\infty} f(x)$. If a_n either increases or decreases without bound as $n \to \infty$, then $\{a_n\}$ is necessarily divergent and we write, respectively,

$$\lim_{n\to\infty} a_n = \infty \quad \text{or} \quad \lim_{n\to\infty} a_n = -\infty \tag{3}$$

EXAMPLE 4 The sequence $\{n^2 + n\}$ is divergent, since $\lim_{n\to\infty}(n^2 + n) = \infty$. □

A sequence may diverge in a manner other than that given in (3).

EXAMPLE 5 The sequence $\{(-1)^n\}$ diverges, since $\lim_{n\to\infty}(-1)^n$ does not exist. The general term $(-1)^n$ alternates between 1 and -1 as $n \to \infty$. □

EXAMPLE 6 Determine whether the sequence $\left\{\dfrac{3n(-1)^n}{n+1}\right\}$ converges or diverges.

Solution
$$\lim_{n\to\infty} \frac{3n(-1)^n}{n+1} = \lim_{n\to\infty} \frac{3(-1)^n}{1+1/n}$$

Although $3/(1+1/n) \to 3$ as $n \to \infty$, the foregoing limit still does not exist. Because of the term $(-1)^n$, we see that as $n \to \infty$,

$$a_n \to 3, \quad n \text{ even}, \quad \text{and} \quad a_n \to -3, \quad n \text{ odd}$$

The sequence diverges. □

Constant Sequence A sequence of constants

$$c, c, c, \ldots$$

is written $\{c\}$. Common sense tells us that this sequence converges and that its limit is c. See Figure 10.1(d).

EXAMPLE 7 The sequence $\{\pi\}$ converges to π. □

EXAMPLE 8 Show that the sequence $\{(n+1)^{1/n}\}$ converges to 1.

Solution If $y = (x+1)^{1/x}$, we recognize the indeterminate form ∞^0 as $x \to \infty$. Hence, by L'Hôpital's Rule,

$$\lim_{x\to\infty} \ln y = \lim_{x\to\infty} \frac{\ln(x+1)}{x} = \lim_{x\to\infty} \frac{\frac{1}{x+1}}{1} = 0$$

Consequently,
$$\lim_{n\to\infty} (n+1)^{1/n} = e^0 = 1$$
□

EXAMPLE 9 Determine whether the sequence $\left\{\sqrt{\dfrac{n}{9n+1}}\right\}$ converges or diverges.

Solution It follows either by using L'Hôpital's Rule or by dividing numerator and denominator by n that $n/(9n+1) \to \frac{1}{9}$ as $n \to \infty$. Thus, we may write

$$\lim_{n\to\infty} \sqrt{\frac{n}{9n+1}} = \sqrt{\lim_{n\to\infty} \frac{n}{9n+1}} = \sqrt{\frac{1}{9}} = \frac{1}{3}$$

The sequence converges to $\frac{1}{3}$. □

Properties The following **properties** of sequences are analogous to those given in Theorems 1.2 and 1.4.

THEOREM 10.1 Properties

Let $\{a_n\}$ and $\{b_n\}$ be convergent sequences. If $\lim_{n\to\infty} a_n = L_1$ and $\lim_{n\to\infty} b_n = L_2$, then

(*i*) $\displaystyle\lim_{n\to\infty} ka_n = k\lim_{n\to\infty} a_n = kL_1, \qquad k \text{ a constant}$

(*ii*) $\displaystyle\lim_{n\to\infty} (a_n + b_n) = \lim_{n\to\infty} a_n + \lim_{n\to\infty} b_n = L_1 + L_2$

(*iii*) $\displaystyle\lim_{n\to\infty} a_nb_n = \lim_{n\to\infty} a_n \cdot \lim_{n\to\infty} b_n = L_1 \cdot L_2$

(*iv*) $\displaystyle\lim_{n\to\infty} \frac{a_n}{b_n} = \frac{\lim_{n\to\infty} a_n}{\lim_{n\to\infty} b_n} = \frac{L_1}{L_2}, \qquad L_2 \neq 0$

EXAMPLE 10 In view of Example 3 and Theorem 10.1(*i*), we see that the sequence $\{5/\sqrt{n}\}$ converges to $5 \cdot 0 = 0$. □

EXAMPLE 11 Use Theorem 10.1 to show that $\{1/n\}$ converges to 0.

Solution Using Example 3 and Theorem 10.1(*iii*), we can write

$$\lim_{n\to\infty} \frac{1}{n} = \lim_{n\to\infty} \frac{1}{\sqrt{n}} \cdot \frac{1}{\sqrt{n}} = \lim_{n\to\infty} \frac{1}{\sqrt{n}} \cdot \lim_{n\to\infty} \frac{1}{\sqrt{n}} = 0 \cdot 0 = 0$$ □

The next two theorems should seem plausible.

THEOREM 10.2 Sequences of the Form $\{r^n\}$

(*i*) The sequence $\{r^n\}$ converges to 0 if $|r| < 1$.

(*ii*) The sequence $\{r^n\}$ diverges if $|r| > 1$.

THEOREM 10.3 Sequences of the Form $\{1/n^r\}$

For any positive rational number r, the sequence $\left\{\dfrac{1}{n^r}\right\}$ converges to 0.

The proof of Theorem 10.2(*i*) follows from Definition 10.2 and is left as an exercise.

EXAMPLE 12

(*a*) The sequence $\left\{\left(\frac{3}{2}\right)^n\right\}$, or $\frac{3}{2}, \frac{9}{4}, \frac{27}{8}, \ldots$, diverges by Theorem 10.2(*ii*), since $r = \frac{3}{2} > 1$.

(*b*) The sequence $\{e^{-n}\}$ converges to 0 by Theorem 10.2(*i*), since $\{e^{-n}\} = \left\{\left(\frac{1}{e}\right)^n\right\}$ and $\frac{1}{e} < 1$. □

EXAMPLE 13 Determine whether the sequence $\left\{\dfrac{e^n}{n + 4e^n}\right\}$ converges or diverges.

Solution Observe that $e^n \to \infty$ and $n + 4e^n \to \infty$ as $n \to \infty$. If $f(x) = e^x/(x + 4e^x)$, then $\lim_{x\to\infty} f(x)$ has the indeterminate form ∞/∞. By L'Hôpital's Rule,

$$\lim_{x\to\infty} \frac{e^x}{x + 4e^x} = \lim_{x\to\infty} \frac{e^x}{1 + 4e^x} = \lim_{x\to\infty} \frac{e^x}{4e^x} = \frac{1}{4}$$

Thus,

$$\lim_{n\to\infty} \frac{e^n}{n + 4e^n} = \frac{1}{4}$$

The sequence converges to $\frac{1}{4}$. □

EXAMPLE 14 Determine whether the sequence $\left\{\dfrac{2 - 3e^{-n}}{6 + 4e^{-n}}\right\}$ converges or diverges.

Solution Observe that $2 - 3e^{-n} \to 2$ and $6 + 4e^{-n} \to 6$ as $n \to \infty$. According to Theorem 10.1(*iv*), we have

$$\lim_{n\to\infty} \frac{2 - 3e^{-n}}{6 + 4e^{-n}} = \frac{\lim_{n\to\infty} (2 - 3e^{-n})}{\lim_{n\to\infty} (6 + 4e^{-n})} = \frac{2}{6} = \frac{1}{3}$$

The sequence converges to $\frac{1}{3}$. □

EXAMPLE 15 From Theorems 10.1(*ii*) and 10.3 we see that $\{10 + 4/n^{3/2}\}$ converges to 10. □

Recursive Sequences As the following example indicates, a sequence can be defined by specifying the first term a_1 together with a rule for obtaining the subsequent terms. In this case the sequence is said to be defined **recursively**. Newton's Method, given in (3) in Section 2.11, is an example of a recursively defined sequence.

EXAMPLE 16 Suppose a sequence is defined recursively by $a_{n+1} = 3a_n + 4$, where $a_1 = 2$. Then

$$a_2 = 3a_1 + 4 = 3(2) + 4 = 10 \qquad a_3 = 3a_2 + 4 = 3(10) + 4 = 34$$

and so on. □

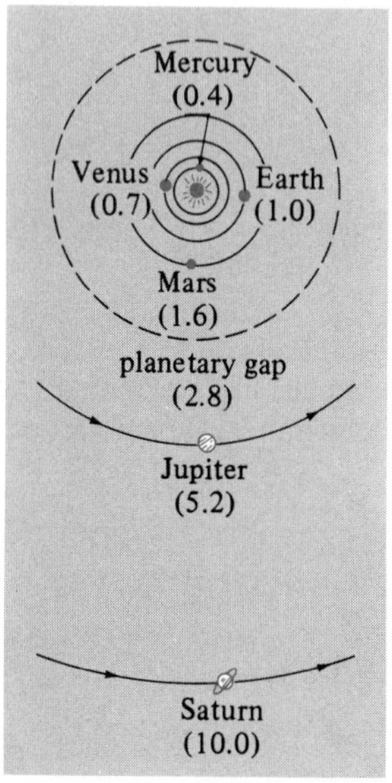

FIGURE 10.2

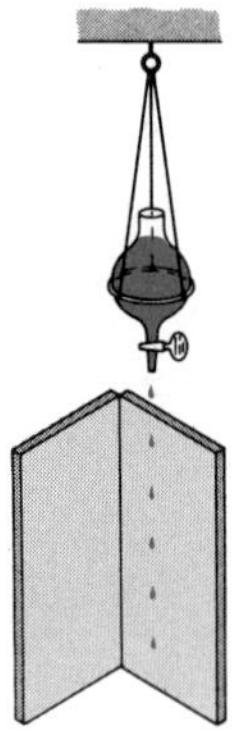
FIGURE 10.3

▲ ***Remarks*** (*i*) In 1772 the German astronomer Johann Elert Bode studied the sequence

$$0, 3, 6, 12, 24, 48, 96, \ldots$$

By adding 4 to each term and dividing the result by 10, he obtained

$$0.4, 0.7, 1.0, 1.6, 2.8, 5.2, 10.0, \ldots$$

If the number 1 represents an astronomical unit (A.U.),* then 0.4, 0.7, 1.0, 1.6, 5.2, and 10 predict fairly accurately the respective distances of the planets Mercury, Venus, Earth, Mars, Jupiter, and Saturn from the sun. However, at that time no planet was known to exist at 2.8 astronomical units from the sun. The discovery of the planet Uranus in 1781, at a distance from the sun consistent with the next term of Bode's sequence (see Problem 66 in Exercises 10.1), brought about a flurry of observational activity by astronomers seeking the missing planet. In 1801 the asteroid Ceres was the first of thousands of asteroids discovered that filled the so-called planetary gap between Mars and Jupiter. Early speculation centered on the belief that the asteroids were remnants of an exploded planet. See Figure 10.2.

Much earlier, Leonardo da Vinci (1452–1519) was able to discern the velocity of a falling body by examining a sequence. Leonardo permitted water drops to fall, at equally spaced intervals of time, between two boards covered with blotting paper. When a spring mechanism was disengaged, the boards were clapped together. See Figure 10.3. By inspecting the sequence of water blots, Leonardo discovered that the distances between consecutive drops increased in "a continuous arithmetic proportion." In this manner he discovered the formula $v = gt$.

(*ii*) The symbol $n!$, read "n factorial," occurs frequently enough in this chapter to warrant a brief review of its definition. If n is a positive integer, then $n!$ is the product of the first n positive integers:

$$n! = 1 \cdot 2 \cdot 3 \cdots (n-1) \cdot n$$

For example, $5! = 1 \cdot 2 \cdot 3 \cdot 4 \cdot 5 = 120$. An important property of the factorial is given by

$$n! = (n-1)!n$$

To see this, consider the case when $n = 6$:

$$6! = 1 \cdot 2 \cdot 3 \cdot 4 \cdot 5 \cdot 6 = (1 \cdot 2 \cdot 3 \cdot 4 \cdot 5)6 = 5!6$$

*Ninety-three million miles, or the mean distance from Earth to the sun.

Stated in a slightly different manner, the property $n! = (n-1)!n$ is equivalent to

$$(n+1)! = n!(n+1)$$

One last point, for purposes of convenience and to ensure that the formula $n! = (n-1)!n$ is valid when $n = 1$, we define $0! = 1$.

EXERCISES 10.1

Answers to odd-numbered problems begin on page A-92.

In Problems 1–12 list the first four terms of the sequence whose general term is a_n.

1. $a_n = \dfrac{1}{2n+1}$
2. $a_n = \dfrac{3}{4n-2}$
3. $a_n = \dfrac{(-1)^n}{n}$
4. $a_n = \dfrac{(-1)^n n^2}{n+1}$
5. $a_n = 3(-1)^{n-1}$
6. $a_n = 10(-1)^n$
7. $a_n = 10^n$
8. $a_n = 10^{-n}$
9. $a_n = \displaystyle\sum_{k=1}^{n} \frac{1}{k}$
10. $a_n = \displaystyle\sum_{k=1}^{n} 2^{-k}$
11. $a_n = 2n!$
12. $a_n = (2n)!$

In Problems 13–18 use Definition 10.2 to show that each sequence converges to the given number L.

13. $\left\{\dfrac{1}{n}\right\}; L = 0$
14. $\left\{\dfrac{1}{n^2}\right\}; L = 0$
15. $\left\{\dfrac{n}{n+1}\right\}; L = 1$
16. $\left\{\dfrac{4n}{2n-1}\right\}; L = 2$
17. $\{10^{-n}\}; L = 0$
18. $\left\{\dfrac{e^n+1}{e^n}\right\}; L = 1$

In Problems 19–52 find $\lim_{n\to\infty} a_n$, if it exists, of each sequence that has the given general term a_n.

19. $a_n = \dfrac{10}{\sqrt{n+1}}$
20. $a_n = \dfrac{1}{n^{3/2}}$
21. $a_n = \dfrac{1}{5n+6}$
22. $a_n = \dfrac{4}{2n+7}$
23. $a_n = \dfrac{3n-2}{6n+1}$
24. $a_n = \dfrac{n}{1-2n}$
25. $a_n = 20(-1)^{n+1}$
26. $a_n = \left(-\dfrac{1}{3}\right)^n$
27. $a_n = \dfrac{n^2-1}{2n}$
28. $a_n = \dfrac{7n}{n^2+1}$
29. $a_n = \dfrac{n^2-3}{4n^2+n}$
30. $a_n = \dfrac{n^2}{1+2n^2}$
31. $a_n = ne^{-n}$
32. $a_n = n^3e^{-n}$
33. $a_n = \dfrac{\sqrt{n+1}}{n}$
34. $a_n = \dfrac{n}{\sqrt{n+1}}$
35. $a_n = \cos n\pi$
36. $a_n = \sin n\pi$
37. $a = \dfrac{\ln n}{n}$
38. $a_n = \dfrac{e^n}{\ln(n+1)}$
39. $a_n = \dfrac{5-2^{-n}}{7+4^{-n}}$
40. $a_n = \dfrac{2^n}{3^n+1}$
41. $a_n = \dfrac{2^n+1}{2^n}$
42. $a_n = 4 + \dfrac{3^n}{2^n}$
43. $a_n = n\sin\dfrac{6}{n}$
44. $a_n = \left(1-\dfrac{5}{n}\right)^n$
45. $a_n = \dfrac{e^n-e^{-n}}{e^n+e^{-n}}$
46. $a_n = \dfrac{\pi}{4} - \arctan(n)$
47. $a_n = n^{2/(n+1)}$
48. $a_n = 10^{(n+1)/n}$
49. $a_n = \ln\left(\dfrac{4n+1}{3n-1}\right)$
50. $a_n = \dfrac{\ln n}{\ln 3n}$
51. $a_n = \sqrt{n+1} - \sqrt{n}$
52. $a_n = \sqrt{n}(\sqrt{n+1} - \sqrt{n})$

In Problems 53–58 write the given sequence in the form $\{a_n\}$.

53. $\dfrac{2}{1}, \dfrac{4}{3}, \dfrac{6}{5}, \dfrac{8}{7}, \dots$
54. $1+\dfrac{1}{2}, \dfrac{1}{2}+\dfrac{1}{3}, \dfrac{1}{3}+\dfrac{1}{4}, \dfrac{1}{4}+\dfrac{1}{5}, \dots$
55. $3, -5, 7, -9, \dots$
56. $-2, 2, -2, 2, \dots$
57. $2, \dfrac{2}{3}, \dfrac{2}{9}, \dfrac{2}{27}, \dots$

58. $\frac{1}{1 \cdot 4}, \frac{1}{2 \cdot 8}, \frac{1}{3 \cdot 16}, \frac{1}{4 \cdot 32}, \ldots$

In Problems 59–62, for each recursively defined sequence, write the next four terms after the indicated initial term(s).

59. $a_{n+1} = \frac{1}{2} a_n$; $a_1 = -1$

60. $a_{n+1} = 2a_n - 1$; $a_1 = 2$

61. $a_{n+1} = \frac{a_n}{a_{n-1}}$; $a_1 = 1$, $a_2 = 3$

62. $a_{n+1} = 2a_n - 3a_{n-1}$; $a_1 = 2$, $a_2 = 4$

63. A ball is dropped from an initial height of 15 ft onto a concrete slab. Each time it bounces, it reaches a height of $\frac{2}{3}$ its preceding height. See Figure 10.4. What height does it reach on its third bounce? on its nth bounce?

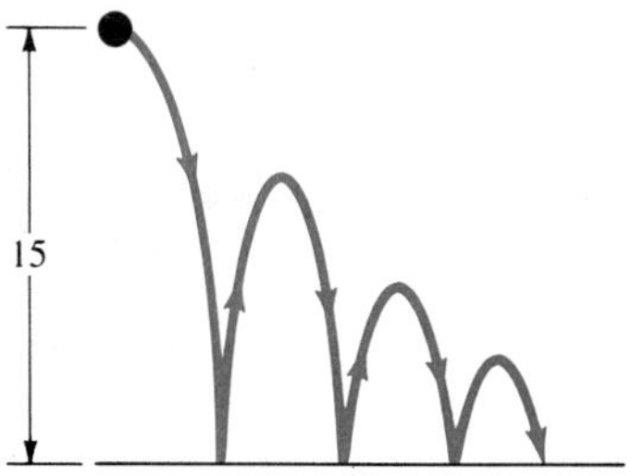

FIGURE 10.4

64. A ball, falling from a great height, travels 16 ft during the first second, 48 ft during the second, 80 ft during the third, and so on. How far does the ball travel during the sixth second?

65. A patient takes 15 mg of a drug each day. Suppose 80% of the drug accumulated is excreted each day by bodily functions. Write out the first six terms of the sequence $\{A_n\}$, where A_n is the amount of the drug present in the patient's body immediately after the nth dose.

66. The mean distances from the sun to the planets Uranus, Neptune, and Pluto are 19.19, 30.07, and 39.46 astronomical units, respectively. Determine how well these numbers agree with the next three terms of Bode's sequence given on page 574.

67. One dollar is deposited in a savings account that pays an annual rate r of interest. If no money is withdrawn, what is the amount accrued in the account after the first, second, and third years?

68. Each person has two parents. Determine how many great-great-great-grandparents each person has.

69. In his work *Liber Abacci*, published in 1202, Leonardo Fibonacci of Pisa speculated on the reproduction of rabbits:

How many pairs of rabbits will be produced in a year beginning with a single pair, if in every month each pair bears a new pair which become productive from the second month on?

Discern the pattern in the following table and complete it.

	Start	After each month											
		1	2	3	4	5	6	7	8	9	10	11	12
Adult pairs	1	1	2	3	5	8	13	21					
Baby pairs	0	1	1	2	3	5	8	13					
Total pairs	1	2	3	5	8	13	21	34					

70. Write out the five terms, after the initial two, of the sequence defined recursively by $F_{n+1} = F_n + F_{n-1}$, $F_1 = 1$, $F_2 = 1$. Reexamine Problem 69.

71. The amount of light that enters a camera depends on the aperture of its lens. This aperture is controlled by an iris diaphragm whose settings are specified in **f-stop numbers**. On a typical camera the first several f-stop numbers are

$$1.4, 2, 2.8, 4, \ldots$$

The lower the f-stop number, the wider the aperture. See Figure 10.5. The f-stop numbers are approximated by the terms of the sequence

$$\sqrt{2}, (\sqrt{2})^2, (\sqrt{2})^3, (\sqrt{2})^4, \ldots$$

(a) Compute the second, third, and fourth terms of this sequence and compare your results with the f-stop numbers given.

(b) Use the sequence to find approximations (rounded to one decimal place) for the next three f-stop numbers.

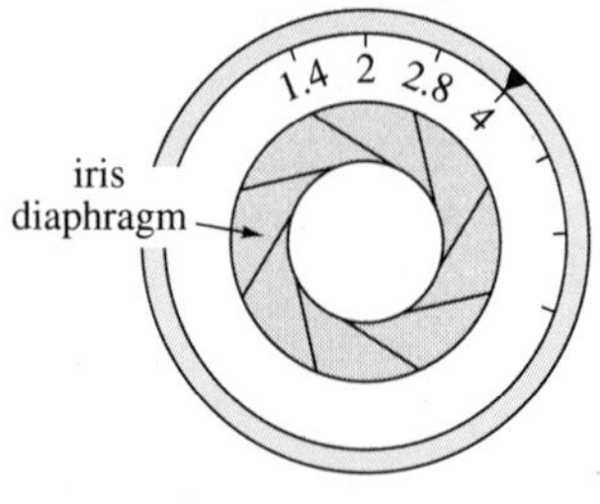

FIGURE 10.5

72. Consider an equilateral triangle with sides of length 1 as shown in Figure 10.6(a). As shown in Figure 10.6(b), on each of the three sides of the first triangle, another equilateral triangle is constructed with sides of length $\frac{1}{3}$. As indicated in Figure 10.6(c) and (d), this construction is continued: Equilateral triangles are constructed on the sides of each previously new triangle such that the length of the sides of the new triangles is $\frac{1}{3}$ the length of the sides of the previous triangle. Let the perimeter of the first figure be P_1, the perimeter of the second figure P_2, and so on.

(a) Find the values of P_1, P_2, P_3, and P_4.

(b) Find a formula for the perimeter P_n of the nth figure.

(c) What is $\lim_{n\to\infty} P_n$? The perimeter of the snowflake-like region obtained by length $n \to \infty$ is called a **Koch curve*** and plays a part in the theory of **fractals**. See Appendix VI.

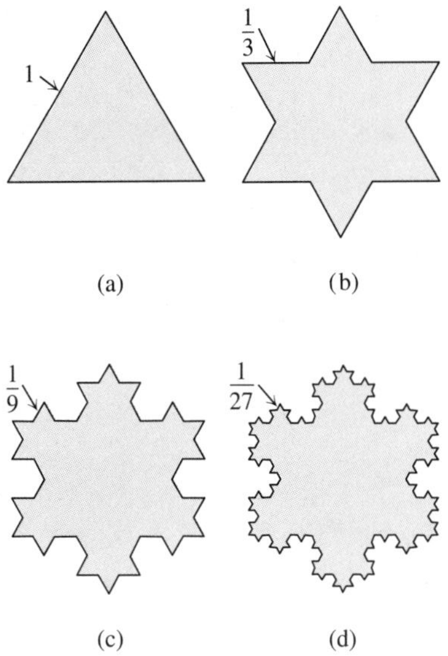

FIGURE 10.6

If $\{a_n\}$, $\{b_n\}$, and $\{c_n\}$ are sequences for which $a_n \le b_n \le c_n$ for all n and $\lim_{n\to\infty} a_n = \lim_{n\to\infty} c_n = L$, then $\lim_{n\to\infty} b_n = L$. In Problems 73–76 use this **Squeeze Theorem** to establish convergence of the given sequence.

73. $\left\{\dfrac{\sin^2 n}{4^n}\right\}$

74. $\left\{\dfrac{|\cos n|}{n^2}\right\}$

75. $\left\{\sqrt{16 + \dfrac{1}{n^2}}\right\}$

76. $\left\{\dfrac{\ln n}{n(n+2)}\right\}$

77. Use Definition 10.2 to prove Theorem 10.2(i) under the assumption $0 < \varepsilon < 1$.

*Invented in 1904 by the Swedish mathematician Helge Von Koch (1870–1924).

78. Show that for any real number x, the sequence $\{(1 + x/n)^n\}$ converges to e^x.

Calculator/Computer Problems

79. Consider the sequence whose first three terms are

$$1 + \frac{1}{2}, \quad 1 + \cfrac{1}{2 + \cfrac{1}{2}}, \quad 1 + \cfrac{1}{2 + \cfrac{1}{2 + \cfrac{1}{2}}}, \quad \ldots$$

(a) What are the fourth and fifth terms?

(b) Calculate the numerical values of the first five terms of the sequence.

(c) Make a conjecture about the convergence or divergence of the sequence.

80. The sequence

$$\left\{1 + \frac{1}{2} + \frac{1}{3} + \cdots + \frac{1}{n} - \ln n\right\}$$

is known to converge to the so-called **Euler–Mascheroni constant** γ. Calculate the first ten terms of the sequence.

81. Conjecture the limit of the convergent sequence $\sqrt{3}$, $\sqrt{3\sqrt{3}}$, $\sqrt{3\sqrt{3\sqrt{3}}}$, ...

82. In Figure 10.7 the square drawn in black is 1 unit on a side. A second square is constructed inside the first square by connecting the midpoints of the sides of the first one. A third square is constructed by connecting the midpoints of the sides of the second square, and so on.

(a) Find a formula for the area A_n of the nth inscribed square.

(b) Consider the sequence $\{S_n\}$, where $S_n = A_1 + A_2 + \cdots + A_n$. Calculate the numerical values of the first ten terms of this sequence.

(c) Make a conjecture about the convergence or divergence of $\{S_n\}$.

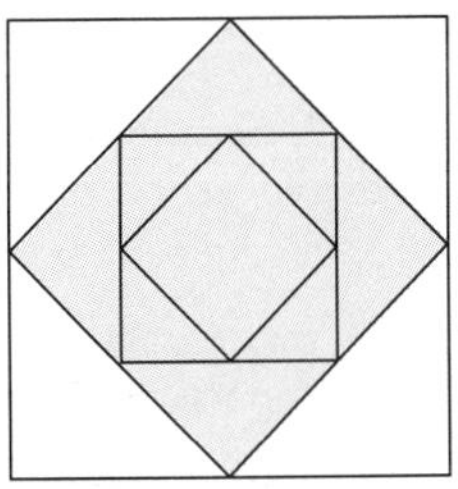

FIGURE 10.7

83. (a) Let AB be a chord of a circle of radius 1, centered at O, let C be the midpoint of the arc AB, let L denote the length of the chord AB, and let l denote the length of the chords AC and BC. See Figure 10.8. Show that $l^2 = 2 - \sqrt{4 - L^2}$. [*Hint*: Use the law of cosines and a half-angle formula.]

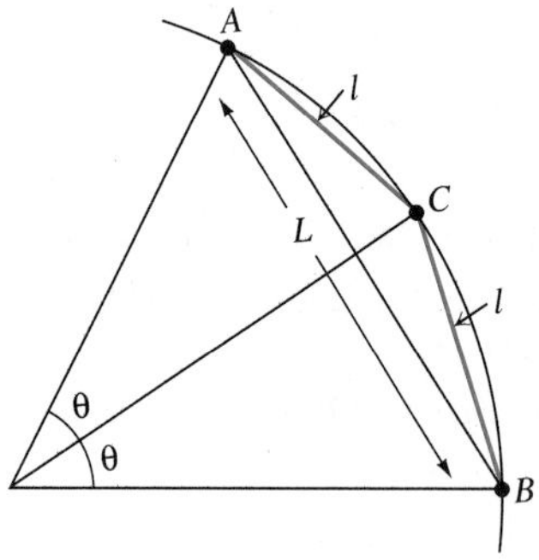

FIGURE 10.8

(b) Consider a semicircle of radius 1 that is divided into three equal arcs. Show that each arc has chord length 1. Suppose each of the three arcs is bisected, then each of the six new arcs is bisected, and so on. After n such bisections, how many chords are there?

(c) Consider the sequence defined recursively by

$$a_0 = 1, \quad a_{n+1} = 2 - \sqrt{4 - a_n}, \qquad n \geq 0$$

Find the next six terms $a_1, \ldots, a_6$. Using parts (**a**) and (**b**), interpret $\sqrt{a_n}$ geometrically.

(d) Let $b_n = 3 \times 2^n \sqrt{a_n}$ for $n \geq 0$. Find $b_0, b_1, \ldots, b_6$. Interpret b_n geometrically. Explain why the sequence $\{b_n\}$ converges to π.

84. Consider the sequence $\{a_n\}$ defined recursively by

$$a_0 = 0.1, \qquad a_{n+1} = (1 + h)a_n - ha_n^2$$

(a) Write a computer program for the iteration of this sequence.

(b) Show that for $h = 1.9$ the sequence appears to converge to 1.

(c) Show that for $h = 2.1$ the terms of the sequence oscillate between two numbers that are (approximately) 1.13 and 0.82.

(d) Show that for $h = 2.5$ the terms of the sequence oscillate between four numbers that are (approximately) 1.22, 0.54, 1.16, and 0.70.

(e) Do the terms of the sequence appear to oscillate between any numbers when $h = 2.55$? when $h = 2.6$?

(f) Show that for $h = 3.0$ the terms of the sequence exhibit no distinguishable pattern.

The behavior illustrated in parts (**c**) and (**d**) is called **period doubling**. As the value of the parameter h is increased, an instability sets in; the sequence transitions between period doubling and the phenomenon in part (**e**), which is called **chaos**.*

*See *Chaos* by James Gleick (New York: Penguin Press, 1987).

10.2 MONOTONIC SEQUENCES

In the preceding section we showed that a sequence $\{a_n\}$ converged by finding $\lim_{n\to\infty} a_n$. However, it is not always easy or even possible to determine whether a sequence $\{a_n\}$ converges by seeking the exact value of $\lim_{n\to\infty} a_n$. For example, does the sequence

$$\left\{1 + \frac{1}{2} + \frac{1}{3} + \cdots + \frac{1}{n} - \ln n\right\}$$

converge? It turns out that this sequence does indeed converge; it is a special type of sequence that can be proved convergent without finding the value of

$$\lim_{n\to\infty}\left(1 + \frac{1}{2} + \frac{1}{3} + \cdots + \frac{1}{n} - \ln n\right)$$

See Problem 28 in Exercises 10.2.

We begin with a definition.

DEFINITION 10.3 Monotonic Sequence

A sequence $\{a_n\}$ is said to be **monotonic** if it is

(*i*) **increasing**: $a_1 < a_2 < a_3 < \cdots < a_n < a_{n+1} < \cdots$,

(*ii*) **decreasing**: $a_1 > a_2 > a_3 > \cdots > a_n > a_{n+1} > \cdots$,

(*iii*) **nondecreasing**: $a_1 \leq a_2 \leq a_3 \leq \cdots \leq a_n \leq a_{n+1} \leq \cdots$, or

(*iv*) **nonincreasing**: $a_1 \geq a_2 \geq a_3 \geq \cdots \geq a_n \geq a_{n+1} \geq \cdots$.

EXAMPLE 1

(*a*) The sequences

$$4, 6, 8, 10 \ldots$$

$$1, \frac{1}{2}, \frac{1}{4}, \frac{1}{8}, \ldots$$

$$5, 5, 4, 4, 4, 3, 3, 3, 3, \ldots$$

are monotonic. They are, respectively, increasing, decreasing, and nonincreasing.

(*b*) The sequence $-1, \frac{1}{2}, -\frac{1}{3}, \frac{1}{4}, -\frac{1}{5}, \ldots$ is not monotonic. □

It is not always evident whether a sequence is increasing, decreasing, and so on. The following table illustrates three ways of demonstrating monotonicity.

Form	If	Then
A **function** $f(x)$ such that $f(n) = a_n$	$f'(x) > 0$ for all x $f'(x) < 0$ for all x	$\{a_n\}$ is increasing $\{a_n\}$ is decreasing
The **ratio** a_{n+1}/a_n where $a_n > 0$ for all n	$a_{n+1}/a_n > 1$ for all n $a_{n+1}/a_n < 1$ for all n	$\{a_n\}$ is increasing $\{a_n\}$ is decreasing
The **difference** $a_{n+1} - a_n$	$a_{n+1} - a_n > 0$ for all n $a_{n+1} - a_n < 0$ for all n	$\{a_n\}$ is increasing $\{a_n\}$ is decreasing

EXAMPLE 2 Show that $\{n/e^n\}$ is a decreasing sequence.

Solution If we define $f(x) = x/e^x$, then $f(n) = a_n$. Now,

$$f'(x) = \frac{1 - x}{e^x} < 0 \qquad \text{for } x > 1$$

implies that f is decreasing on $[1, \infty)$. Thus, the given sequence is decreasing.

Alternative Solution

$$\frac{a_{n+1}}{a_n} = \frac{n+1}{e^{n+1}} \cdot \frac{e^n}{n} = \frac{n+1}{ne} = \frac{1}{e} + \frac{1}{ne} \leq \frac{1}{e} + \frac{1}{e} = \frac{2}{e} < 1$$

implies that $a_{n+1} < a_n$ for $n \geq 1$. □

EXAMPLE 3 The sequence $\left\{\frac{2n+1}{n+1}\right\}$ or $\frac{3}{2}, \frac{5}{3}, \frac{7}{4}, \frac{9}{5}, \ldots$ appears to be increasing. Since

$$a_{n+1} - a_n = \frac{2n+3}{n+2} - \frac{2n+1}{n+1} = \frac{1}{(n+2)(n+1)} > 0 \qquad \text{for all } n$$

we have $a_{n+1} > a_n$ for all n. □

> **DEFINITION 10.4 Bounded Sequence**
>
> A sequence $\{a_n\}$ is said to be **bounded** if there exists a positive number B such that $|a_n| \leq B$ for all n.

EXAMPLE 4 The sequence $\left\{\frac{2n+1}{n+1}\right\}$ is bounded above by 2, since

$$\frac{2n+1}{n+1} \leq \frac{2n+2}{n+1} = \frac{2(n+1)}{n+1} = 2$$

Furthermore, $(2n+1)/(n+1) \geq 0$ shows that the sequence is bounded below by 0.* Thus, $0 \leq (2n+1)/(n+1) \leq 2$ for all n implies that the sequence is bounded. Of course, the terms of the sequence are also bounded below by -2. This enables us to write $|(2n+1)/(n+1)| \leq 2$ for all n. □

The next result will be useful in subsequent sections of this chapter.

> **THEOREM 10.4 Conditions Sufficient for Convergence**
>
> A bounded monotonic sequence converges.

---

*Indeed, from Example 3 we see that the terms are bounded below by the first term of the sequence.

Proof We prove the theorem in the case of a nondecreasing sequence. By assumption, $\{a_n\}$ is bounded and so $|a_n| \le B$ for all n. In turn, this means the infinite set of terms $S = \{a_1, a_2, a_3, \ldots, a_n, \ldots\}$ is bounded above and therefore has a least upper bound L.* The sequence actually converges to L. For $\varepsilon > 0$ we know that $L - \varepsilon < L$, and consequently $L - \varepsilon$ is not an upper bound for S (there are no upper bounds smaller than the least upper bound). Hence, there exists a positive integer N such that $a_N > L - \varepsilon$. But, since $\{a_n\}$ is nondecreasing,

$$L - \varepsilon \le a_N \le a_{N+1} \le a_{N+2} \le a_{N+3} \le \cdots \le L + \varepsilon$$

It follows that for $n > N$, $L - \varepsilon \le a_n \le L + \varepsilon$ or $|a_n - L| < \varepsilon$. From Definition 10.2 we conclude that $\lim_{n\to\infty} a_n = L$. ■

EXAMPLE 5 The sequence $\left\{\dfrac{2n+1}{n+1}\right\}$ was shown to be monotonic (Example 3) and bounded (Example 4). Hence, by Theorem 10.4 the sequence is convergent. □

EXAMPLE 6 Show that the sequence $\left\{\dfrac{1\cdot 3\cdot 5\cdots(2n-1)}{2\cdot 4\cdot 6\cdots(2n)}\right\}$ converges.

Solution First, the ratio

$$\frac{a_{n+1}}{a_n} = \frac{1\cdot 3\cdot 5\cdots(2n-1)(2n+1)}{2\cdot 4\cdot 6\cdots(2n)(2n+2)} \cdot \frac{2\cdot 4\cdot 6\cdots(2n)}{1\cdot 3\cdot 5\cdots(2n-1)} = \frac{2n+1}{2n+2} < 1$$

shows that $a_{n+1} < a_n$ for all n. The sequence is monotonic, since it is decreasing. Next, from

$$0 < \frac{1\cdot 3\cdot 5\cdot 7\cdots(2n-1)}{2\cdot 4\cdot 6\cdot 8\cdots(2n)} = \frac{1}{2}\cdot\frac{3}{4}\cdot\frac{5}{6}\cdot\frac{7}{8}\cdots\frac{2n-1}{2n} < 1 \qquad \text{(Why?)}$$

we see that the sequence is bounded. It follows from Theorem 10.4 that the sequence is convergent. □

▲ ***Remark*** Every convergent sequence $\{a_n\}$ is necessarily bounded. But it does not follow that every bounded sequence is convergent. You will be asked to supply an example that illustrates this last statement in Problem 25 of Exercises 10.2. On the other hand, if a sequence $\{a_n\}$ is unbounded, then it is necessarily divergent.

---

*This is one of the basic axioms in mathematics. It is known as the **completeness property** of the real number system.

EXERCISES 10.2 *Answers to odd-numbered problems begin on page A-92.*

In Problems 1–12 determine whether the given sequence is monotonic. If so, state whether it is increasing, decreasing, nondecreasing, or nonincreasing.

1. $\left\{\dfrac{n}{3n+1}\right\}$

2. $\left\{\dfrac{10+n}{n^2}\right\}$

3. $\{(-1)^n\sqrt{n}\}$

4. $\{(n-1)(n-2)\}$

5. $\left\{\dfrac{e^n}{n}\right\}$

6. $\left\{\dfrac{e^n}{n^5}\right\}$

7. $\left\{\dfrac{2^n}{n!}\right\}$

8. $\left\{\dfrac{2^{2n}(n!)^2}{(2n)!}\right\}$

9. $\left\{n+\dfrac{1}{n}\right\}$

10. $\{n^2+(-1)^n n\}$

11. $\{(\sin 1)(\sin 2)\cdots(\sin n)\}$

12. $\left\{\ln\left(\dfrac{n+2}{n+1}\right)\right\}$

In Problems 13–24 use Theorem 10.4 to show that the given sequence converges.*

13. $\left\{\dfrac{4n-1}{5n+2}\right\}$

14. $\left\{\dfrac{6-4n^2}{1+n^2}\right\}$

15. $\left\{\dfrac{3^n}{1+3^n}\right\}$

16. $\{n5^{-n}\}$

17. $\{e^{1/n}\}$

18. $\left\{\dfrac{n!}{n^n}\right\}$

19. $\left\{\dfrac{n!}{1\cdot 3\cdot 5\cdots(2n-1)}\right\}$

20. $\left\{\dfrac{2\cdot 4\cdot 6\cdots(2n)}{3\cdot 5\cdot 7\cdots(2n+1)}\right\}$

21. $\{\tan^{-1}n\}$

22. $\left\{\dfrac{\ln(n+3)}{n+3}\right\}$

23. $(0.8), (0.8)^2, (0.8)^3, \ldots$

24. $\sqrt{3}, \sqrt{\sqrt{3}}, \sqrt{\sqrt{\sqrt{3}}}, \ldots$

25. Give an example of a bounded sequence that is not convergent.

26. Consider the sequence $\{a_n\}$ defined by

$$a_1 = 2, \quad a_2 = 1, \quad a_{n+1} = \left(1 - \frac{1}{n^2}\right)a_n, \qquad n \geq 2$$

Show that $\{a_n\}$ converges.

27. Show that $\{\int_1^n e^{-t^2}\,dt\}$ converges. [*Hint:* For $x \geq 1$, $e^{-x^2} \leq e^{-x}$.]

28. Prove that the sequence

$$\left\{1 + \frac{1}{2} + \frac{1}{3} + \cdots + \frac{1}{n} - \ln n\right\}$$

is bounded and monotonic and hence convergent.† [*Hint:* First prove the inequality

$$\frac{1}{2} + \frac{1}{3} + \cdots + \frac{1}{n-1} + \frac{1}{n} < \ln n < 1 + \frac{1}{2} + \frac{1}{3} + \cdots + \frac{1}{n-1}$$

by considering the area under the graph of $y = 1/x$ on $[1, n]$.]

29. Certain studies in fishery management hold that the size of an undisturbed fish population changes from one year to the next in accordance with the formula

$$p_{n+1} = \frac{bp_n}{a + p_n}, \qquad n \geq 0$$

where $p_n > 0$ is the population after n years, and a and b are positive parameters that depend on the species and its environment. Suppose that a population size p_0 is introduced in year 0.

(a) Show that the only possible limit values for the sequence $\{p_n\}$ are 0 and $b - a$. [*Hint:* If $\lim_{n\to\infty} p_n = L$, then $\lim_{n\to\infty} p_{n+1} = L$.]

(b) Show that $p_{n+1} < (b/a)p_n$.

(c) Use the result in part (**b**) to show that if $a > b$, then the population dies out: that is, $\lim_{n\to\infty} p_n = 0$.

(d) Now assume $a < b$. Show that if $0 < p_0 < b - a$, then the sequence $\{p_n\}$ is increasing and bounded above by $b - a$. Show that if $0 < b - a < p_0$, then the sequence $\{p_n\}$ is decreasing and bounded below by $b - a$. Conclude that $\lim_{n\to\infty} p_n = b - a$ for any $p_0 > 0$. [*Hint:* Examine $|b - a - p_{n+1}|$, which is the distance between p_{n+1} and $b - a$.]

---

*Of course, in some of these problems the actual limit can be obtained by other methods.

†This limit is denoted by γ. From Problem 80 of Exercises 10.1, $\gamma \approx 0.5772\ldots$.

10.3 INFINITE SERIES

The concept of a *series* is closely related to the concept of a *sequence*. If $\{a_n\}$ is the sequence $a_1, a_2, a_3, \ldots, a_n, \ldots$, then the indicated sum

$$a_1 + a_2 + a_3 + \cdots + a_n + \cdots \tag{1}$$

is called an **infinite series** or simply a **series**. The a_k, $k = 1, 2, 3, \ldots$, are called the **terms** of the series; a_n is called the **general term**. We write (1) compactly as $\Sigma_{k=1}^{\infty} a_k$, or for convenience $\Sigma\, a_k$.

EXAMPLE 1 In the opening remarks to this chapter we noted that the decimal representation for the rational number $\frac{1}{3}$ is, in fact, an infinite series

$$\frac{3}{10} + \frac{3}{10^2} + \frac{3}{10^3} + \cdots = \sum_{k=1}^{\infty} \frac{3}{10^k}$$ □

The question we seek to answer in this and the next several sections is:

When does an infinite series "add up" to a number?

Intuitively, we expect that $\frac{1}{3}$ is the sum of the series $\Sigma_{k=1}^{\infty} 3/10^k$. But, just as intuitively, we expect that an infinite series such as

$$100 + 1000 + 10{,}000 + 100{,}000 + \cdots$$

where the terms are becoming larger and larger, has no sum. In other words, we do not expect the latter series to "add up" or *converge* to any number. The concept of convergence of an infinite series is defined in terms of the convergence of a special kind of sequence.

Sequence of Partial Sums Associated with every infinite series $\Sigma\, a_k$, there is a **sequence of partial sums** $\{S_n\}$ whose terms are defined by

$$\begin{aligned} S_1 &= a_1 \\ S_2 &= a_1 + a_2 \\ S_3 &= a_1 + a_2 + a_3 \\ &\vdots \\ S_n &= a_1 + a_2 + a_3 + \cdots + a_n \\ &\vdots \end{aligned}$$

The general term $S_n = a_1 + a_2 + \cdots + a_n$ of this sequence is called the ***n*th partial sum** of the series.

EXAMPLE 2 The sequence of partial sums for $\sum_{k=1}^{\infty} \frac{3}{10^k}$ is

$$\begin{aligned} S_1 &= \frac{3}{10} \\ S_2 &= \frac{3}{10} + \frac{3}{10^2} \\ S_3 &= \frac{3}{10} + \frac{3}{10^2} + \frac{3}{10^3} \\ &\vdots \\ S_n &= \frac{3}{10} + \frac{3}{10^2} + \frac{3}{10^3} + \cdots + \frac{3}{10^n} \\ &\vdots \end{aligned}$$

□

In Example 2, when n is very large, S_n will give a good approximation to $\frac{1}{3}$, and so it seems reasonable to write

$$\frac{1}{3} = \lim_{n\to\infty} S_n = \lim_{n\to\infty} \sum_{k=1}^{n} \frac{3}{10^k} = \sum_{k=1}^{\infty} \frac{3}{10^k}$$

This leads to the following definition:

DEFINITION 10.5 Convergent Series

An infinite series $\Sigma_{k=1}^{\infty} a_k$ is said to be **convergent** if the sequence of partial sums $\{S_n\}$ converges; that is,

$$\lim_{n\to\infty} S_n = \lim_{n\to\infty} \sum_{k=1}^{n} a_k = S$$

The number S is the **sum** of the series. If $\lim_{n\to\infty} S_n$ does not exist, the series is said to be **divergent**.

EXAMPLE 3 Show that the series $\sum_{k=1}^{\infty} \frac{1}{(k+4)(k+5)}$ is convergent.

Solution By partial fractions the general term of the series can be written as

$$a_n = \frac{1}{n+4} - \frac{1}{n+5}$$

Thus, the nth partial sum of the series is

$$S_n = \left[\frac{1}{5} - \frac{1}{6}\right] + \left[\frac{1}{6} - \frac{1}{7}\right] + \left[\frac{1}{7} - \frac{1}{8}\right] + \cdots + \left[\frac{1}{n+4} - \frac{1}{n+5}\right]$$

$$= \frac{1}{5} - \frac{1}{6} + \frac{1}{6} - \frac{1}{7} + \frac{1}{7} - \frac{1}{8} + \frac{1}{8} - \cdots - \frac{1}{n+4} + \frac{1}{n+4} - \frac{1}{n+5}$$

$$= \frac{1}{5} - \frac{1}{n+5}$$

Since $\lim_{n\to\infty} 1/(n+5) = 0$, we see that $\lim_{n\to\infty} S_n = \frac{1}{5}$. Hence, the series converges and we write

$$\sum_{k=1}^{\infty} \frac{1}{(k+4)(k+5)} = \frac{1}{5}$$ □

Telescoping Series Because of the manner in which the general term of the sequence of partial sums "collapses" to two terms, the series in Example 3 is said to be a **telescoping** series. See Problem 57 in Exercises 10.3.

Geometric Series A series of the form

$$a + ar + ar^2 + \cdots + ar^{n-1} + \cdots = \sum_{k=1}^{\infty} ar^{k-1} \tag{2}$$

is called a **geometric series.**

THEOREM 10.5 Sum of a Geometric Series

A geometric series $\sum_{k=1}^{\infty} ar^{k-1}$, $a \neq 0$, converges to $\dfrac{a}{1-r}$ if $|r| < 1$, and it diverges if $|r| \geq 1$.

Proof Consider the general term of the sequence of partial sums of (2):

$$S_n = a + ar + ar^2 + \cdots + ar^{n-1} \tag{3}$$

Multiplying both sides of (3) by r gives

$$rS_n = ar + ar^2 + ar^3 + \cdots + ar^n \tag{4}$$

We subtract (4) from (3) and solve for S_n:

$$S_n - rS_n = a - ar^n$$

$$(1 - r)S_n = a(1 - r^n)$$

$$S_n = \frac{a(1-r^n)}{1-r}, \qquad r \neq 1 \tag{5}$$

Now, from Theorem 10.2 we know that $\lim_{n\to\infty} r^n = 0$ for $|r| < 1$. Consequently,

$$\lim_{n\to\infty} S_n = \lim_{n\to\infty} \frac{a(1-r^n)}{1-r} = \frac{a}{1-r}, \qquad |r| < 1$$

If $|r| > 1$, $\lim_{n\to\infty} r^n$ does not exist and so the limit of (5) fails to exist. The proof that a geometric series diverges when $r = \pm 1$ is left as an exercise. See Problem 59 in Exercises 10.3. ■

EXAMPLE 4 In the geometric series

$$\sum_{k=1}^{\infty} \left(-\frac{1}{3}\right)^{k-1} = 1 - \frac{1}{3} + \frac{1}{9} - \frac{1}{27} + \cdots$$

we identify $a = 1$ and $r = -\frac{1}{3}$. Since $|-\frac{1}{3}| < 1$, the series converges. From Theorem 10.5 the sum of the series is

$$\sum_{k=1}^{\infty} \left(-\frac{1}{3}\right)^{k-1} = \frac{1}{1-(-1/3)} = \frac{3}{4}$$ □

EXAMPLE 5 The geometric series

$$\sum_{k=1}^{\infty} 5\left(\frac{3}{2}\right)^{k-1} = 5 + \frac{15}{2} + \frac{45}{4} + \frac{135}{8} + \cdots$$

diverges because $|r| = \frac{3}{2} > 1$. □

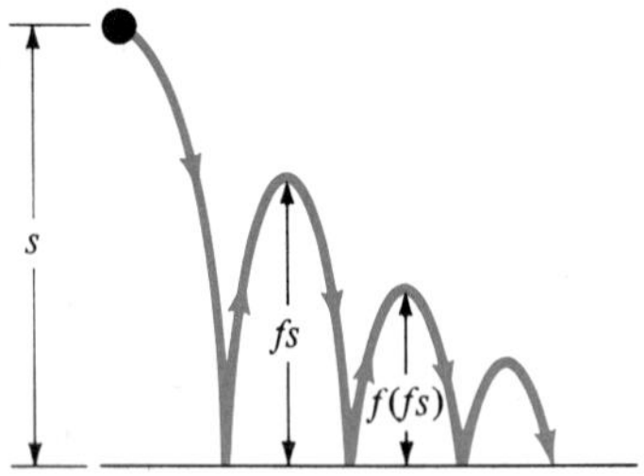

FIGURE 10.9

EXAMPLE 6 When a ball is dropped from a height of s ft off the ground, it takes $t = \sqrt{2s/g}$ s to reach the ground. (See Section 5.7.) Suppose the ball always rebounds to a certain fixed fraction f $(0 < f < 1)$ of its prior height. Find a formula for the time T it takes for the ball to come to rest. See Figure 10.9.

Solution The time to fall from a height of s ft to the ground is: $\sqrt{2s/g}$; the time to rise fs ft and then fall fs ft to the ground is: $2\sqrt{2fs/g}$; the time to rise $f(fs)$ ft and then fall $f(fs)$ ft to the ground is: $2\sqrt{2f^2s/g}$; and so on. Thus, the total time T is given by the infinite series

$$T = \sqrt{2s/g} + 2\sqrt{2fs/g} + 2\sqrt{2f^2s/g} + \cdots + 2\sqrt{2f^ns/g} + \cdots$$
$$= \sqrt{2s/g}\left[1 + 2\sum_{k=1}^{\infty} (\sqrt{f})^k\right]$$

Because $0 < f < 1$, the series $\Sigma_{k=1}^{\infty} (\sqrt{f})^k$ is a convergent geometric series with $a = \sqrt{f}$ and $r = \sqrt{f}$. Consequently,

$$T = \sqrt{2s/g}\left[1 + 2\frac{\sqrt{f}}{1-\sqrt{f}}\right] \quad \text{or} \quad T = \sqrt{2s/g}\left[\frac{1+\sqrt{f}}{1-\sqrt{f}}\right] \qquad \square$$

Harmonic Series Another example of a divergent series is the so-called **harmonic series**:

$$1 + \frac{1}{2} + \frac{1}{3} + \cdots + \frac{1}{n} + \cdots = \sum_{k=1}^{\infty} \frac{1}{k} \tag{6}$$

The general term of the sequence of partial sums for (6) is given by

$$S_n = 1 + \frac{1}{2} + \frac{1}{3} + \cdots + \frac{1}{n}$$

Thus,

$$\begin{aligned} S_{2n} &= 1 + \frac{1}{2} + \frac{1}{3} + \cdots + \frac{1}{n} + \frac{1}{n+1} + \frac{1}{n+2} + \cdots + \frac{1}{2n} \\ &= S_n + \frac{1}{n+1} + \frac{1}{n+2} + \cdots + \frac{1}{2n} \\ &\geq S_n + \underbrace{\frac{1}{2n} + \frac{1}{2n} + \cdots + \frac{1}{2n}}_{n \text{ terms}} = S_n + n \cdot \frac{1}{2n} = S_n + \frac{1}{2} \end{aligned}$$

The inequality $S_{2n} \geq S_n + \frac{1}{2}$ implies that the sequence of partial sums is unbounded. To see this, we observe that

$$\begin{aligned} S_2 &\geq S_1 + \frac{1}{2} = 1 + \frac{1}{2} = \frac{3}{2} \\ S_4 &\geq S_2 + \frac{1}{2} \geq \frac{3}{2} + \frac{1}{2} = 2 \\ S_8 &\geq S_4 + \frac{1}{2} \geq 2 + \frac{1}{2} = \frac{5}{2} \\ S_{16} &\geq S_8 + \frac{1}{2} \geq \frac{5}{2} + \frac{1}{2} = 3 \end{aligned}$$

and so on. Hence, we conclude that the harmonic series is divergent.

If a_n and S_n are the general terms of a series and the corresponding sequence of partial sums, respectively, then $a_n = S_n - S_{n-1}$. Now, if the series converges to a number S, we have $\lim_{n\to\infty} S_n = S$ and $\lim_{n\to\infty} S_{n-1} = S$. This implies that $\lim_{n\to\infty} a_n = \lim_{n\to\infty}(S_n - S_{n-1}) = S - S = 0$. We have established the next theorem.

THEOREM 10.6

If $\Sigma_{k=1}^{\infty} a_k$ converges, then $\lim_{n\to\infty} a_n = 0$.

Test for a Divergent Series Theorem 10.6 simply states that if an infinite series converges, it is necessary that the nth, or general, term of the series approach zero. Equivalently, we conclude:

If the nth term of an infinite series does not approach zero as $n \to \infty$, then the series does not converge.

We formalize this result as a test for divergence.

THEOREM 10.7 The *n*th Term Test for Divergence

If $\lim_{n\to\infty} a_n \neq 0$, then $\Sigma_{k=1}^{\infty} a_k$ diverges.

EXAMPLE 7 Consider the series $\sum_{k=1}^{\infty} \frac{4k-1}{5k+3}$. Since

$$\lim_{n\to\infty} \frac{4n-1}{5n+3} = \lim_{n\to\infty} \frac{4-1/n}{5+3/n} = \frac{4}{5} \neq 0$$

it follows from Theorem 10.7 that the series diverges. □

You are encouraged to read (and remember) Remark (*iii*) at this time. We state the following theorems without proof.

THEOREM 10.8

If c is any nonzero constant, then $\Sigma_{k=1}^{\infty} a_k$ and $\Sigma_{k=1}^{\infty} ca_k$ both converge or both diverge.

THEOREM 10.9

If $\Sigma_{k=1}^{\infty} a_k$ and $\Sigma_{k=1}^{\infty} b_k$ converge to S_1 and S_2, respectively, then $\Sigma_{k=1}^{\infty} (a_k + b_k)$ converges to $S_1 + S_2$.

THEOREM 10.10

If $\Sigma_{k=1}^{\infty} a_k$ is convergent and $\Sigma_{k=1}^{\infty} b_k$ is divergent, then $\Sigma_{k=1}^{\infty} (a_k + b_k)$ is divergent.

EXAMPLE 8 With the aid of Theorem 10.5, we see that the geometric series $\Sigma_{k=1}^{\infty} (\frac{1}{2})^{k-1}$ and $\Sigma_{k=1}^{\infty} (\frac{1}{3})^{k-1}$ converge to 2 and $\frac{3}{2}$, respectively. Hence, from Theorem 10.9, $\Sigma_{k=1}^{\infty} [(\frac{1}{2})^{k-1} + (\frac{1}{3})^{k-1}]$ converges to $2 + \frac{3}{2} = \frac{7}{2}$. □

EXAMPLE 9 From Example 3 we know that $\Sigma_{k=1}^{\infty} 1/[(k + 4)(k + 5)]$ converges. Since $\Sigma_{k=1}^{\infty} 1/k$ is the divergent harmonic series, it follows from Theorem 10.10 that

$$\sum_{k=1}^{\infty} \left[\frac{1}{(k + 4)(k + 5)} + \frac{1}{k}\right]$$

diverges. □

▲ ***Remarks*** (*i*) When written in terms of summation notation, a geometric series may not be immediately recognizable, or if it is, the values of a and r may not be apparent. For example, to see that $\Sigma_{n=3}^{\infty} 4(\frac{1}{2})^{n+2}$ is a geometric series, it is best to write out two or three terms:

$$\sum_{n=3}^{\infty} 4\left(\frac{1}{2}\right)^{n+2} = \overbrace{4\left(\frac{1}{2}\right)^{5}}^{\boxed{a}} + \overbrace{4\left(\frac{1}{2}\right)^{6}}^{\boxed{ar}} + \overbrace{4\left(\frac{1}{2}\right)^{7}}^{\boxed{ar^2}} + \cdots$$

From the right side of the last equality, we can make the identifications $a = 4(\frac{1}{2})^5$ and $r = \frac{1}{2} < 1$. Consequently, the series converges to $\dfrac{4(\frac{1}{2})^5}{1 - \frac{1}{2}} = \dfrac{1}{4}$. If desired, although there is no real need to do this, we can express $\Sigma_{n=3}^{\infty} 4(\frac{1}{2})^{n+2}$ in the more familiar form $\Sigma_{k=1}^{\infty} ar^{k-1}$ by letting $k = n - 2$. The result is

$$\sum_{n=3}^{\infty} 4\left(\frac{1}{2}\right)^{n+2} = \sum_{k=1}^{\infty} 4\left(\frac{1}{2}\right)^{k+4} = \sum_{k=1}^{\infty} \overbrace{4\left(\frac{1}{2}\right)^{5}}^{\boxed{a}}\overbrace{\left(\frac{1}{2}\right)^{k-1}}^{\boxed{r^{k-1}}}$$

(*ii*) Every rational number is either a terminating decimal or a repeating decimal.* Every repeating decimal is a geometric series. Thus, $\Sigma_{k=1}^{\infty} 3/10^k$ converges, since $r = \frac{1}{10} < 1$. With $a = \frac{3}{10}$ we find that

$$\sum_{k=1}^{\infty} \frac{3}{10^k} = \frac{3/10}{1 - 1/10} = \frac{3/10}{9/10} = \frac{1}{3}$$

(*iii*) Note carefully how Theorems 10.6 and 10.7 are stated. Specifically, Theorem 10.6 *does not* say if $\lim_{n\to\infty} a_n = 0$, then $\Sigma\, a_k$ converges. In other

*Even a terminating decimal such as 0.5 is a repeating decimal in the sense that 0.5 = 0.5000 . . . .

words, $\lim_{n\to\infty} a_n = 0$ is not *sufficient* to guarantee that Σa_k converges. In fact, if $\lim_{n\to\infty} a_n = 0$, the series may either converge or diverge. For example, in the harmonic series $\Sigma_{k=1}^{\infty} 1/k$, $a_n = 1/n$ and $\lim_{n\to\infty} 1/n = 0$, but the series diverges.

(*iv*) When determining convergence, it is possible, and sometimes convenient, to **delete** or ignore the first several terms of a series. In other words, infinite series $\Sigma_{k=1}^{\infty} a_k$ and $\Sigma_{k=N}^{\infty} a_k$, $N > 1$, that differ by at most a finite number of terms are either both convergent or both divergent. Of course, deleting the first $N - 1$ terms of a convergent series usually does affect the sum of the series.

EXERCISES 10.3 *Answers to odd-numbered problems begin on page A-92.*

In Problems 1–10 write out the first four terms in each series.

1. $\sum_{k=1}^{\infty} \frac{2k+1}{k}$
2. $\sum_{k=1}^{\infty} \frac{2^k}{k}$
3. $\sum_{k=1}^{\infty} \frac{(-1)^{k-1}}{k(k+1)}$
4. $\sum_{k=1}^{\infty} \frac{(-1)^{k+1}}{k3^k}$
5. $\sum_{n=0}^{\infty} \frac{n+1}{n!}$
6. $\sum_{n=1}^{\infty} \frac{(2n)!}{n^2+1}$
7. $\sum_{m=1}^{\infty} \frac{2 \cdot 4 \cdot 6 \cdots (2m)}{1 \cdot 3 \cdot 5 \cdots (2m-1)}$
8. $\sum_{m=1}^{\infty} \frac{1 \cdot 3 \cdot 5 \cdots (2m-1)}{m!}$
9. $\sum_{j=3}^{\infty} \frac{\cos j\pi}{2j+1}$
10. $\sum_{i=5}^{\infty} i \sin \frac{i\pi}{2}$

In Problems 11–20 determine whether the given geometric series converges or diverges. If convergent, find the sum of the series.

11. $\sum_{k=1}^{\infty} 3\left(\frac{1}{5}\right)^{k-1}$
12. $\sum_{k=1}^{\infty} 10\left(\frac{3}{4}\right)^{k-1}$
13. $\sum_{k=1}^{\infty} \frac{(-1)^{k-1}}{2^{k-1}}$
14. $\sum_{k=1}^{\infty} \left(\frac{\pi}{3}\right)^{k-1}$
15. $\sum_{r=1}^{\infty} 5^r 4^{-r}$
16. $\sum_{s=1}^{\infty} (-3)^s 7^{-s}$
17. $\sum_{n=1}^{\infty} 1000(0.9)^n$
18. $\sum_{n=1}^{\infty} \frac{(1.1)^n}{1000}$
19. $\sum_{k=0}^{\infty} \frac{1}{(\sqrt{3}-\sqrt{2})^k}$
20. $\sum_{k=0}^{\infty} \left(\frac{\sqrt{5}}{1+\sqrt{5}}\right)^k$

In Problems 21–26 write each repeating decimal number as a quotient of integers.

21. 0.222 . . .
22. 0.555 . . .
23. 0.616161 . . .
24. 0.393939 . . .
25. 1.314314 . . .
26. 0.5262626 . . .

In Problems 27–32 find the sum of each convergent series.

27. $\sum_{k=1}^{\infty} \frac{1}{k(k+1)}$
28. $\sum_{k=1}^{\infty} \frac{1}{(k+1)(k+2)}$
29. $\sum_{k=1}^{\infty} \frac{1}{4k^2-1}$
30. $\sum_{k=1}^{\infty} \frac{1}{k^2+7k+12}$
31. $\sum_{k=1}^{\infty} \left[\left(\frac{1}{3}\right)^{k-1} + \left(\frac{1}{4}\right)^{k-1}\right]$
32. $\sum_{k=1}^{\infty} \frac{2^k - 1}{4^k}$

In Problems 33–42 show that each series is divergent.

33. $\sum_{k=1}^{\infty} 10$
34. $\sum_{k=1}^{\infty} (5k+1)$
35. $\sum_{k=1}^{\infty} \frac{k}{2k+1}$
36. $\sum_{k=1}^{\infty} \frac{k^2+1}{k^2+2k+3}$
37. $\sum_{k=1}^{\infty} (-1)^k$
38. $\sum_{k=1}^{\infty} \ln\left(\frac{k}{3k+1}\right)$
39. $\sum_{k=1}^{\infty} \frac{10}{k}$
40. $\sum_{k=1}^{\infty} \frac{1}{6k}$
41. $\sum_{k=1}^{\infty} \left[\frac{1}{2^{k-1}} + \frac{1}{k}\right]$
42. $\sum_{k=1}^{\infty} k \sin \frac{1}{k}$

In Problems 43–46 determine the values of x for which the given series converges.

43. $\sum_{k=1}^{\infty} \left(\frac{x}{2}\right)^{k-1}$
44. $\sum_{k=1}^{\infty} \left(\frac{1}{x}\right)^{k-1}$
45. $\sum_{k=1}^{\infty} (x+1)^k$
46. $\sum_{k=0}^{\infty} 2^k x^{2k}$

47. A ball is dropped from an initial height of 15 ft onto a concrete slab. Each time the ball bounces, it reaches a height

of $\frac{2}{3}$ its preceding height. Use geometric series to determine the distance the ball travels before it comes to rest.

48. In Problem 47 determine the time it takes for the ball to come to rest.

49. To eradicate agricultural pests (such as the Medfly), sterilized male flies are released into the general population at regular time intervals. Let N_0 be the number of flies released each day and let s be the proportion that survive a given day. Of the original N_0 sterilized males, $N_0 s^n$ will survive for n successive weeks. Hence, the total number of such males that survive n weeks after the program has begun is $N_0 + N_0 s + N_0 s^2 + \cdots + N_0 s^n$. What does this sum approach as $n \to \infty$? Suppose $s = 0.9$ and 10,000 sterilized males are needed to control the population in a certain area. Determine the number that should be released each day.

50. In some circumstances the amount of a drug that will accumulate in a patient's body after a long period of time is $A_0 + A_0 e^{-k} + A_0 e^{-2k} + \cdots$, where $k > 0$ is a constant and A_0 is the daily dose of the drug. Find the sum of the series.

51. A patient takes 15 mg of a drug each day. If 80% of the drug accumulated is excreted each day by bodily functions, how much of the drug will accumulate after a long period of time (that is, as $n \to \infty$)? (Assume that the measurement of the accumulation is made immediately after each dose. See Problem 65 in Exercises 10.1.)

52. A force is applied to a particle, which moves in a straight line, in such a fashion that after each second the particle moves only one-half the distance it moved in the preceding second. If the particle moves 20 cm in the first second, how far will it move?

53. In 1972, an outbreak of methylmercury poisoning in Iraq resulted in 459 deaths among 6530 cases of poisoning admitted to hospitals. The outbreak was caused by the consumption of homemade bread prepared from wheat that had been treated with a methylmercury fungicide. The first symptoms of *parasthesia* (loss of sensation at the mouth, hands, and feet) begin to occur when the accumulated level of mercury reaches 25 mg. Symptoms of *ataxia* (loss of coordination in gait) begin to occur at 55 mg, *dysarthia* (slurred speech) at 90 mg, and deafness at 170 mg. Death becomes a possibility when the accumulated mercury level exceeds 200 mg. It was estimated that a typical loaf of bread made from the contaminated wheat contained 1.4 mg of mercury. It is also estimated that the body removes only about 0.9% of the accumulated mercury each day.

(a) Suppose that a person receives a dosage d of mercury each day, and that the body removes a fraction p of the accumulated mercury each day. Find a formula for L_n, the accumulated level after eating on the nth day, and a formula for the limiting level, $\lim_{n\to\infty} L_n$.

(b) Using $d = 1.4$ and $p = 0.009$, find the limiting level of mercury and determine on which days the various symptoms begin to occur.

(c) What would the daily dose have to be in order for death to become possible by the 100th day? (Use $p = 0.009$.)

54. (*This problem could present a challenge.*) In Problem 72 in Exercises 10.1 we considered the perimeters of the regions bounded by the Koch curves shown in Figure 10.6. We saw in part (c) of the problem that the perimeter of the limiting region is infinite. In this problem we consider the *areas* of the sequence of figures. Let the area of the first figure be A_1, the area of the second figure A_2, and so on.

(a) Using the fact that the area of an equilateral triangle with sides of length s is $(\sqrt{3}/4)s^2$, find the values of A_1, A_2, A_3, and A_4.

(b) Show that the area of the nth figure is

$$A_n = \frac{\sqrt{3}}{20}\left[8 - 3\left(\frac{4}{9}\right)^{n-1}\right]$$

(c) What is $\lim_{n\to\infty} A_n$?

55. Prove that $\Sigma_{k=1}^{\infty} 1/\sqrt{k}$ is divergent by showing that $S_n \geq \sqrt{n}$.

56. Use the fact that $k! \geq 2^{k-1}$, $k = 1, 2, 3, \ldots$, to prove that $\Sigma_{k=1}^{\infty} 1/k!$ converges.

57. Prove that if $\lim_{n\to\infty} f(n+1) = L$, where L is a number, then $\Sigma_{k=1}^{\infty} [f(k+1) - f(k)] = L - f(1)$.

58. Find the sum of the series

$$\sum_{k=1}^{\infty}\left(\int_k^{k+1} xe^{-x}\,dx\right)$$

59. Prove that a geometric series $\Sigma_{k=1}^{\infty} ar^{k-1}$ diverges if $r = \pm 1$.

60. Determine whether the sum of two divergent series is necessarily divergent.

61. Suppose the sequence $\{a_n\}$ converges to a number $L \neq 0$. Show that $\Sigma_{k=1}^{\infty} a_k$ diverges.

62. Find all values of x in $(-\pi/2, \pi/2)$ for which

$$\lim_{n\to\infty}\left(\frac{1}{1 - \tan x} - \sum_{k=0}^{n} \tan^k x\right) = 0$$

63. Determine whether the following argument is valid: If $S = 1 + 2 + 4 + 8 + \cdots$, then $2S = 2 + 4 + 8 + \cdots = S - 1$. Solving $2S = S - 1$ gives $S = -1$.

64. Determine whether $\sum_{n=1}^{\infty}\left(\sum_{k=1}^{n}\frac{1}{k}\right)$ converges or diverges.

10.4 INTEGRAL AND COMPARISON TESTS

Unless $\Sigma_{k=1}^{\infty} a_k$ is a telescoping series or a geometric series, it is a difficult, if not futile, task to prove convergence or divergence directly from the sequence of partial sums. However, it is usually possible to determine whether a series converges or diverges by means of a *test* that utilizes only the terms of the series. In this and the next section we shall examine five such tests that are applicable to infinite series of *positive terms.*

Integral Test The first test that we shall consider relates the concepts of convergence and divergence of an improper integral to convergence and divergence of an infinite series.

THEOREM 10.11 Integral Test

Suppose f is a continuous function that is nonnegative and decreasing for $x \geq 1$ such that $f(k) = a_k$ for $k \geq 1$.

(*i*) If $\int_1^{\infty} f(x)\,dx$ converges, then $\Sigma_{k=1}^{\infty} a_k$ converges.
(*ii*) If $\int_1^{\infty} f(x)\,dx$ diverges, then $\Sigma_{k=1}^{\infty} a_k$ diverges.

Proof If the graph of f is given as in Figure 10.10, then by considering the areas of the rectangles shown in the figure, we see that

$$0 \leq a_2 + a_3 + a_4 + \cdots + a_n \leq \int_1^n f(x)\,dx \leq a_1 + a_2 + a_3 + \cdots + a_{n-1}$$

or

$$S_n - a_1 \leq \int_1^n f(x)\,dx \leq S_{n-1}$$

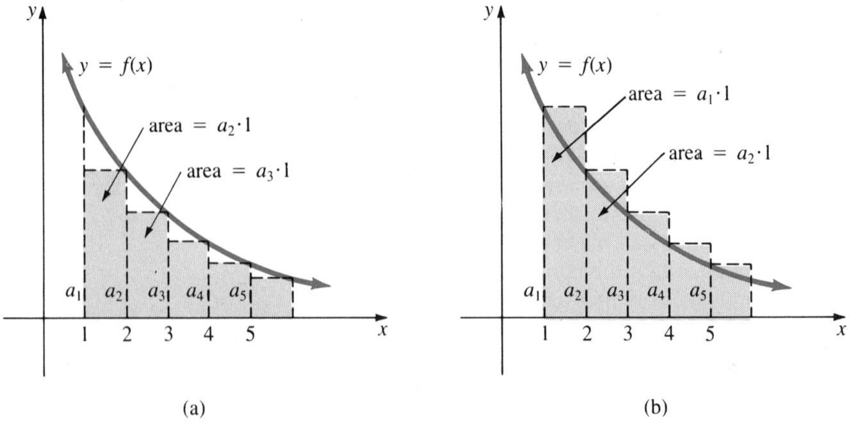

FIGURE 10.10

From the inequality $S_n - a_1 \le \int_1^n f(x)\,dx$, it is apparent that $\lim_{n\to\infty} S_n$ exists whenever $\lim_{n\to\infty} \int_1^n f(x)\,dx$ exists. On the other hand, from $S_{n-1} \ge \int_1^n f(x)\,dx$, we conclude that $\lim_{n\to\infty} S_{n-1}$ fails to exist whenever $\int_1^\infty f(x)\,dx$ diverges. ■

Note: In the Integral Test, if the positive-term series is of the form $\Sigma_{k=N}^{\infty} a_k$, we then use

$$\int_N^\infty f(x)\,dx \quad \text{where} \quad f(k) = a_k$$

EXAMPLE 1 Test for convergence $\displaystyle\sum_{k=3}^{\infty} \frac{\ln k}{k}$.

Solution The function $f(x) = (\ln x)/x$ is continuous, nonnegative, and decreasing* on $[3, \infty)$ and $f(k) = a_k = (\ln k)/k$. Now,

$$\begin{aligned}\int_3^\infty \frac{\ln x}{x}\,dx &= \lim_{t\to\infty} \int_3^t \frac{\ln x}{x}\,dx \\ &= \lim_{t\to\infty} \frac{1}{2}(\ln x)^2\Big]_3^t \\ &= \lim_{t\to\infty} \frac{1}{2}[(\ln t)^2 - (\ln 3)^2] = \infty\end{aligned}$$

shows that the series diverges. □

***p*-Series** The Integral Test is particularly useful on any series of the form

$$\sum_{k=1}^{\infty} \frac{1}{k^p} = 1 + \frac{1}{2^p} + \frac{1}{3^p} + \cdots \tag{1}$$

where p is any fixed real number. The infinite series $\Sigma_{k=1}^{\infty} 1/k^p$ is called the ***p*-series** or **hyperharmonic series**. The next theorem indicates the values of p for which the p-series converges (diverges).

THEOREM 10.12 Convergence of *p*-Series

The p-series $\displaystyle\sum_{k=1}^{\infty} \frac{1}{k^p}$ converges if $p > 1$ and diverges if $p \le 1$.

Proof We distinguish four cases: $p > 1$, $p = 1$, $0 < p < 1$, and $p \le 0$. In the first and third cases we use the Integral Test with $f(x) = 1/x^p = x^{-p}$.

---

*Show this by examining $f'(x)$.

(*i*) If $p > 1$, then $p - 1 > 0$ and so

$$\int_1^{\infty} x^{-p}\,dx = \lim_{t\to\infty} \frac{x^{-p+1}}{-p+1}\bigg|_1^t = \frac{1}{1-p}\lim_{t\to\infty}\left[\frac{1}{t^{p-1}} - 1\right]$$

$$= \frac{1}{1-p}[0 - 1] = \frac{1}{p-1}$$

The p-series is convergent by Theorem 10.11(*i*).

(*ii*) If $p = 1$, then we recognize the p-series as the divergent harmonic series.

(*iii*) If $0 < p < 1$, then $-p + 1 > 0$ and so

$$\int_1^{\infty} x^{-p}\,dx = \lim_{t\to\infty} \frac{x^{-p+1}}{-p+1}\bigg|_1^t = \frac{1}{1-p}\lim_{t\to\infty}[t^{-p+1} - 1] = \infty$$

The p-series is divergent by Theorem 10.11(*ii*)

(*iv*) Last, if $p \le 0$, then $-p \ge 0$ and so $\lim_{n\to\infty} 1/n^p = \lim_{n\to\infty} n^{-p} \ne 0$. The p-series is divergent by Theorem 10.7. ■

EXAMPLE 2

(*a*) The series $\Sigma_{k=1}^{\infty} 1/k^{1/2}$ diverges, since $p = \frac{1}{2} < 1$.

(*b*) The series $\Sigma_{k=1}^{\infty} 1/k^2$ converges, since $p = 2 > 1$. □

Comparison Tests It is often possible to determine convergence or divergence of a series $\Sigma\, a_k$ by *comparing* its terms with the terms of a *test series* $\Sigma\, b_k$ that is known to be either convergent or divergent.

THEOREM 10.13 Comparison Test

Suppose $\Sigma_{k=1}^{\infty} a_k$ and $\Sigma_{k=1}^{\infty} b_k$ are series of positive terms.

(*i*) If $\Sigma_{k=1}^{\infty} b_k$ converges and $a_k \le b_k$ for every positive integer k, then $\Sigma_{k=1}^{\infty} a_k$ converges.

(*ii*) If $\Sigma_{k=1}^{\infty} b_k$ diverges and $a_k \ge b_k$ for every positive integer k, then $\Sigma_{k=1}^{\infty} a_k$ diverges.

Proof Let $S_n = a_1 + a_2 + \cdots + a_n$ and $T_n = b_1 + b_2 + \cdots + b_n$ be the general terms of the sequences of partial sums for $\Sigma\, a_k$ and $\Sigma\, b_k$, respectively.

(*i*) If $\Sigma\, b_k$ is a convergent series for which $a_k \le b_k$, then $S_n \le T_n$. Since $\lim_{n\to\infty} T_n$ exists, $\{S_n\}$ is a bounded increasing sequence and, hence, convergent by Theorem 10.4. Therefore, $\Sigma\, a_k$ is convergent.

(*ii*) If $\Sigma\, b_k$ diverges and $a_k \ge b_k$, then $S_n \ge T_n$. Since T_n increases without bound, so does S_n. Hence, $\Sigma\, a_k$ is divergent. ■

In general, if $\Sigma\, c_k$ and $\Sigma\, d_k$ are two series for which $c_k \le d_k$ for all k, we say that the series $\Sigma\, c_k$ is **dominated** by the series $\Sigma\, d_k$. Thus, for positive-term series, Theorem 10.13(*i*) indicates that a series $\Sigma\, a_k$ is convergent if it is dominated by a convergent series. In Theorem 10.13(*ii*) we see that a series $\Sigma\, a_k$ diverges if it dominates a divergent series.

EXAMPLE 3 Test for convergence $\displaystyle\sum_{k=1}^{\infty} \frac{k}{k^3+4}$.

Solution We observe that

$$\frac{k}{k^3+4} \le \frac{k}{k^3} = \frac{1}{k^2}$$

Because the given series is dominated by the convergent p-series $\Sigma_{k=1}^{\infty} 1/k^2$ (see Example 2), it follows from Theorem 10.13(*i*) that the given series is also convergent. □

EXAMPLE 4 Test for convergence $\displaystyle\sum_{k=1}^{\infty} \frac{\ln(k+2)}{k}$.

Solution Since $\ln(k+2) > 1$ for $k \ge 1$, we have

$$\frac{\ln(k+2)}{k} > \frac{1}{k}$$

In this case the given series has been shown to dominate the divergent harmonic series $\Sigma_{k=1}^{\infty} 1/k$. Hence, by Theorem 10.13(*ii*) the given series diverges. □

Another kind of comparison test involves taking the limit of the ratio of the general term of a series to the general term of a test series that is known to be convergent or divergent.

THEOREM 10.14 Limit Comparison Test

Suppose $\Sigma_{k=1}^{\infty} a_k$ and $\Sigma_{k=1}^{\infty} b_k$ are series of positive terms and that

$$\lim_{n\to\infty} \frac{a_n}{b_n} = L$$

(*i*) If L is a positive constant, then the two series are either both convergent or both divergent.

(*ii*) If $L = 0$ and $\Sigma_{k=1}^{\infty} b_k$ converges, then $\Sigma_{k=1}^{\infty} a_k$ converges.

(*iii*) If $\lim_{n\to\infty} a_n/b_n = \infty$ and $\Sigma_{k=1}^{\infty} b_k$ diverges, then $\Sigma_{k=1}^{\infty} a_k$ diverges.

Proof of (i) Since $\lim_{n\to\infty} a_n/b_n = L > 0$, we can choose n so large, say $n \geq N$ for some positive integer N, that

$$\frac{1}{2}L \leq \frac{a_n}{b_n} \leq \frac{3}{2}L$$

This inequality implies that $a_n \leq \frac{3}{2}Lb_n$ for $n \geq N$. If $\Sigma_{k=1}^{\infty} b_k$ converges, it follows from the Comparison Test that $\Sigma_{k=N}^{\infty} a_k$ and, therefore, $\Sigma_{k=1}^{\infty} a_k$ is convergent. Furthermore, since $\frac{1}{2}Lb_n \leq a_n$ for $n \geq N$, we see that if $\Sigma_{k=1}^{\infty} b_k$ diverges, then $\Sigma_{k=N}^{\infty} a_k$ and $\Sigma_{k=1}^{\infty} a_k$ diverge. ■

The Limit Comparison Test is often applicable to series for which the Comparison Test is inconvenient.

EXAMPLE 5 You should convince yourself that it is difficult to apply the Comparison Test to the series $\Sigma_{k=1}^{\infty} 1/(k^3 - 5k^2 + 1)$. However, we know $\Sigma_{k=1}^{\infty} 1/k^3$ is a convergent p-series ($p = 3 > 1$). Hence, with

$$a_n = \frac{1}{n^3 - 5n^2 + 1} \quad \text{and} \quad b_n = \frac{1}{n^3}$$

we have

$$\lim_{n\to\infty} \frac{a_n}{b_n} = \lim_{n\to\infty} \frac{n^3}{n^3 - 5n^2 + 1} = 1$$

From part (*i*) of Theorem 10.14, it follows that the given series converges. □

If the general term a_n of a series $\Sigma\, a_k$ is a quotient of either rational powers of n or roots of polynomials in n, it is possible to discern the general term b_n of a test series $\Sigma\, b_k$ by examining the "degree behavior" of a_n for large values of n. In other words, to find a candidate for b_n we need only examine the quotient of the *highest powers of* n in the numerator and denominator of a_n.

EXAMPLE 6 Test for convergence $\displaystyle\sum_{k=1}^{\infty} \frac{k}{\sqrt[3]{8k^5 + 7}}$.

Solution For large values of n, $a_n = n/\sqrt[3]{8n^5 + 7}$ "behaves" like a constant multiple of

$$\frac{n}{\sqrt[3]{n^5}} = \frac{n}{n^{5/3}} = \frac{1}{n^{2/3}}$$

Thus, we try the divergent p-series

$$\sum_{k=1}^{\infty} \frac{1}{k^{2/3}} \qquad \boxed{p = \tfrac{2}{3} < 1}$$

as a test series
$$\lim_{n\to\infty} \frac{a_n}{b_n} = \lim_{n\to\infty} \frac{n/\sqrt[3]{8n^5+7}}{1/n^{2/3}}$$
$$= \lim_{n\to\infty} \left(\frac{n^5}{8n^5+7}\right)^{1/3} = \left(\frac{1}{8}\right)^{1/3} = \frac{1}{2}$$

Thus, from part (*i*) of Theorem 10.14 the given series diverges. □

▲ ***Remarks*** (*i*) When applying the Integral Test, you should be aware that the value of the convergent improper integral $\int_1^\infty f(x)\,dx$ is not equal to the sum of the series.

(*ii*) The results of the Integral Test for $\Sigma_{k=n}^\infty a_k$ hold even if the continuous nonnegative function f does not begin to decrease until $x \geq N \geq n$. For the series $\Sigma_{k=1}^\infty (\ln k)/k$, the function $f(x) = (\ln x)/x$ decreases on the interval $[3, \infty)$. Nonetheless, in the Integral Test we may use $\int_1^\infty (\ln x)\,dx/x$.

(*iii*) The hypotheses in the Comparison Test can also be weakened, giving a stronger theorem. It is only required that $a_k \leq b_k$ or $a_k \geq b_k$ for k sufficiently large and not for all positive integers.

(*iv*) In the application of the basic Comparison Test, it is often easy to reach a point where the given series is dominated by a divergent series. For example, $1/(5^k + \sqrt{k}) \leq 1/\sqrt{k}$ is certainly true and $\Sigma_{k=1}^\infty 1/\sqrt{k}$ diverges. This kind of reasoning proves nothing about the series $\Sigma_{k=1}^\infty 1/(5^k + \sqrt{k})$. But this series does converge. Why? Similarly, no conclusion can be reached by showing that a given series dominates a convergent series.

The following table summarizes the Comparison Test. Let $\Sigma\, a_k$ be a series of positive terms.

Comparison of Terms	Test Series $\Sigma\, b_k$	Conclusion about $\Sigma\, a_k$
$a_k \leq b_k$	converges	converges
$a_k \leq b_k$	diverges	none
$a_k \geq b_k$	diverges	diverges
$a_k \geq b_k$	converges	none

EXERCISES 10.4

Answers to odd-numbered problems begin on page A-93.

In Problems 1–40 use the Integral Test, Comparison Test, or Limit Comparison Test to determine whether the given series converges or diverges.

1. $\sum_{k=1}^\infty \frac{1}{k^{1.1}}$

2. $\sum_{k=1}^\infty \frac{1}{k^{0.99}}$

3. $\sum_{k=1}^\infty \frac{1}{2k+7}$

4. $\sum_{k=1}^\infty \frac{1}{10+\sqrt{k}}$

5. $\sum_{k=1}^\infty \frac{k}{3k+1}$

6. $\sum_{k=1}^\infty \frac{1}{k^2+5}$

7. $\sum_{k=1}^\infty ke^{-k^2}$

8. $\sum_{k=1}^\infty \frac{\arctan k}{1+k^2}$

9. $\sum_{k=1}^\infty \frac{1}{(k+1)(k+2)}$

10. $\sum_{k=1}^\infty \frac{1}{k+\sqrt{k}}$

11. $\sum_{k=2}^{\infty} \frac{1}{k \ln k}$

12. $\sum_{k=3}^{\infty} \frac{\ln k}{k^5}$

13. $\sum_{k=2}^{\infty} \frac{(\ln k)^{-2}}{k}$

14. $\sum_{k=2}^{\infty} \frac{1}{k\sqrt{\ln k}}$

15. $\sum_{n=2}^{\infty} \frac{1}{n\sqrt{n^2-1}}$

16. $\sum_{n=1}^{\infty} \frac{1}{\sqrt{(n+1)(n+2)}}$

17. $\sum_{n=1}^{\infty} \frac{n^2-n+2}{3n^5+n^2}$

18. $\sum_{n=2}^{\infty} \frac{n}{(4n+1)^{3/2}}$

19. $\sum_{k=1}^{\infty} \frac{2+\sin k}{\sqrt[3]{k^4+1}}$

20. $\sum_{k=1}^{\infty} \frac{3}{2+\sin k}$

21. $\sum_{k=1}^{\infty} \frac{1+8^k}{3+10^k}$

22. $\sum_{k=1}^{\infty} \frac{(1.1)^k}{4k}$

23. $\sum_{k=1}^{\infty} \frac{1}{3^k+k}$

24. $\sum_{k=1}^{\infty} \frac{1+3^k}{2^k}$

25. $\sum_{k=2}^{\infty} \frac{1}{\ln k}$

26. $\sum_{k=2}^{\infty} \frac{2k+1}{k \ln k}$

27. $\sum_{j=1}^{\infty} \frac{j+e^{-j}}{5^j(j+9)}$

28. $\sum_{n=1}^{\infty} \frac{5n^2+2n}{3n(n^2+1)}$

29. $\sum_{n=1}^{\infty} \frac{1+1/n}{10^n}$

30. $\sum_{i=1}^{\infty} \frac{ie^{-i}}{i+1}$

31. $\sum_{j=1}^{\infty} \ln\left(5+\frac{1}{5^j}\right)$

32. $\sum_{j=1}^{\infty} \ln\left(1+\frac{1}{3^j}\right)$

33. $\sum_{k=2}^{\infty} \frac{k+\ln k}{k^3+2k-1}$

34. $\sum_{k=1}^{\infty} \frac{\sin(1/k)}{k}$

35. $\sum_{k=1}^{\infty} \frac{\sqrt{k+1}}{\sqrt[3]{64k^9+40}}$

36. $\sum_{k=1}^{\infty} \frac{5k^2+k-k^{-1}}{2k^3+2k^2+8}$

37. $\sum_{j=1}^{\infty} \frac{e^{1/j}}{j^2}$

38. $\sum_{k=1}^{\infty} \tan\frac{1}{k}$

39. $\frac{1}{2\cdot 3}+\frac{2}{3\cdot 4}+\frac{3}{4\cdot 5}+\frac{4}{5\cdot 6}+\cdots$

40. $\frac{1}{1\cdot 3}+\frac{1}{2\cdot 9}+\frac{1}{3\cdot 27}+\frac{1}{4\cdot 81}+\cdots$

41. Suppose $a_k > 0$ for $k = 1, 2, 3, \ldots$. Prove that if $\Sigma_{k=1}^{\infty} a_k$ converges, then $\Sigma_{k=1}^{\infty} a_k^2$ converges. Is the converse true?

10.5 RATIO AND ROOT TESTS

In this section, as in the last, the tests that we shall consider are applicable to infinite series of *positive terms.*

Ratio Test The first of these tests employs a limit of the **ratio** of the $(n+1)$st term to the nth term of the series. This test is especially useful when a_k involves factorials, kth powers of a constant, and sometimes, kth powers of k.

THEOREM 10.15 Ratio Test

Suppose $\Sigma_{k=1}^{\infty} a_k$ is a series of positive terms such that

$$\lim_{n\to\infty} \frac{a_{n+1}}{a_n} = L$$

(*i*) If $L < 1$, the series is convergent.

(*ii*) If $L > 1$, or if $\lim_{n\to\infty} a_{n+1}/a_n = \infty$, the series is divergent.

(*iii*) If $L = 1$, the test is inconclusive.

Proof of (i) Let r be a positive number such that $0 \le L \le r < 1$. For n sufficiently large, say $n \ge N$ for some positive integer N, $a_{n+1}/a_n < r$; that is

$$a_{n+1} < ra_n, \qquad n \ge N$$

This inequality implies

$$\begin{aligned} a_{N+1} &< ra_N \\ a_{N+2} &< ra_{N+1} < a_N r^2 \\ a_{N+3} &< ra_{N+2} < a_N r^3 \end{aligned}$$

and so on. Thus, the series $\Sigma_{k=N+1}^{\infty} a_k$ converges by comparison with the convergent geometric series $\Sigma_{k=1}^{\infty} a_N r^k$. Since $\Sigma_{k=1}^{\infty} a_k$ differs from $\Sigma_{k=N+1}^{\infty} a_k$ by at most a finite number of terms, we conclude that the former series also converges.

Proof of (ii) Let r be a finite number such that $1 < r < L$. Then for n sufficiently large, say $n \ge N$ for some positive integer N, $a_{n+1}/a_n > r$ or $a_{n+1} > ra_n$. For $r > 1$ this last inequality implies $a_{n+1} > a_n$ and so $\lim_{n\to\infty} a_n \ne 0$. From Theorem 10.7 we see that $\Sigma_{k=1}^{\infty} a_k$ diverges. ■

In the case when $L = 1$, we must apply another test to the series to determine its convergence or divergence.

EXAMPLE 1 Test for convergence $\displaystyle\sum_{k=1}^{\infty} \frac{5k}{k!}$.

Solution

$$\begin{aligned} \lim_{n\to\infty} \frac{a_{n+1}}{a_n} &= \lim_{n\to\infty} \frac{5^{n+1}}{(n+1)!} \cdot \frac{n!}{5^n} \\ &= \lim_{n\to\infty} 5\frac{n!}{(n+1)!} \\ &= \lim_{n\to\infty} 5\frac{n!}{n!(n+1)} \\ &= \lim_{n\to\infty} \frac{5}{n+1} = 0 \end{aligned}$$

Since $L = 0 < 1$, it follows from Theorem 10.15(i) that the series is convergent. □

EXAMPLE 2 Test for convergence $\displaystyle\sum_{k=1}^{\infty} \frac{k^k}{k!}$.

Solution

$$\lim_{n\to\infty}\frac{a_{n+1}}{a_n} = \lim_{n\to\infty}\frac{(n+1)^{n+1}}{(n+1)!}\cdot\frac{n!}{n^n}$$
$$= \lim_{n\to\infty}\frac{(n+1)^{n+1}}{n+1}\cdot\frac{1}{n^n}$$
$$= \lim_{n\to\infty}\left(\frac{n+1}{n}\right)^n$$
$$= \lim_{n\to\infty}\left(1+\frac{1}{n}\right)^n = e$$

The last limit result follows from (15) of Section 6.3. Since $L = e > 1$, it follows from Theorem 10.15(*ii*) that the series is divergent. □

Root Test If the terms of a series $\Sigma\, a_k$ consist of only kth powers, then the following test, which involves taking the nth **root** of the nth term, may be applicable.

THEOREM 10.16 Root Test

Suppose $\Sigma_{k=1}^{\infty} a_k$ is a series of positive terms such that

$$\lim_{n\to\infty}\sqrt[n]{a_n} = L$$

(*i*) If $L < 1$, the series is convergent.

(*ii*) If $L > 1$, or if $\lim_{n\to\infty}\sqrt[n]{a_n} = \infty$, the series is divergent.

(*iii*) If $L = 1$, the test is inconclusive.

The proof of the Root Test is very similar to the proof of the Ratio Test and will not be given.

EXAMPLE 3 Test for convergence $\displaystyle\sum_{k=1}^{\infty}\left(\frac{5}{k}\right)^k$

Solution By the Root Test we see that

$$\lim_{n\to\infty}\left[\left(\frac{5}{n}\right)^n\right]^{1/n} = \lim_{n\to\infty}\frac{5}{n} = 0$$

Since $L = 0 < 1$, we conclude from Theorem 10.16(*i*) that the series converges. □

▲ ***Remarks*** (*i*) The Ratio Test will always lead to the inconclusive case when applied to a p-series. Try it on $\Sigma_{k=1}^{\infty} 1/k^2$ and see what happens.

(*ii*) The tests examined in this section and the previous section tell us when a series has a sum, but none of these tests gives so much as a clue as to what the actual sum is. Knowing that a series converges, we can now add up five, a hundred, or a thousand terms on a computer to obtain an approximation of the sum.

EXERCISES 10.5

Answers to odd-numbered problems begin on page A-93.

In Problems 1–20 use the Ratio Test or Root Test to determine whether the given series converges or diverges.

1. $\sum_{k=1}^{\infty} \frac{1}{k!}$

2. $\sum_{k=1}^{\infty} \frac{2^k}{k!}$

3. $\sum_{k=1}^{\infty} \frac{k!}{1000^k}$

4. $\sum_{k=1}^{\infty} k\left(\frac{2}{3}\right)^k$

5. $\sum_{j=1}^{\infty} \frac{j^{10}}{(1.1)^j}$

6. $\sum_{j=1}^{\infty} \frac{1}{j^5(0.99)^j}$

7. $\sum_{k=1}^{\infty} \frac{1}{k^k}$

8. $\sum_{k=1}^{\infty} \left(\frac{ke}{k+1}\right)^k$

9. $\sum_{n=1}^{\infty} \frac{4^{n-1}}{n3^{n+2}}$

10. $\sum_{n=1}^{\infty} \frac{n^3 2^{n+3}}{7^{n-1}}$

11. $\sum_{k=1}^{\infty} \frac{k!}{(2k)!}$

12. $\sum_{k=1}^{\infty} \frac{(2k)!}{k!(2k)^k}$

13. $\sum_{k=1}^{\infty} \frac{5^{2k+1}}{k^k}$

14. $\sum_{k=2}^{\infty} \frac{1}{(\ln k)^k}$

15. $\sum_{k=1}^{\infty} \left(\frac{k}{k+1}\right)^{k^2}$

16. $\sum_{k=1}^{\infty} \left(1-\frac{2}{k}\right)^{k^2}$

17. $\sum_{k=1}^{\infty} \frac{1 \cdot 3 \cdot 5 \cdots (2k-1)}{k!}$

18. $\sum_{k=1}^{\infty} \frac{k!}{2 \cdot 4 \cdot 6 \cdots (2k)}$

19. $\sum_{k=1}^{\infty} \frac{k^k}{e^k}$

20. $\sum_{k=1}^{\infty} \frac{k!3^k}{k^k}$

In Problems 21 and 22 determine the nonnegative values of p for which the given series converges.

21. $\sum_{k=1}^{\infty} kp^k$

22. $\sum_{k=1}^{\infty} k^2\left(\frac{2}{p}\right)^k$

In Problems 23 and 24 determine all real values of p for which the given series converges.

23. $\sum_{k=1}^{\infty} \frac{k^p}{k!}$

24. $\sum_{k=2}^{\infty} \frac{\ln k}{k^p}$

25. The Fibonacci sequence* $\{F_n\}$,

$$1, 1, 2, 3, 5, 8, \ldots$$

is defined by the recursion formula $F_{n+1} = F_n + F_{n-1}$, where $F_1 = 1, F_2 = 1$. Verify that the general term of the sequence is

$$F_n = \frac{1}{\sqrt{5}}\left(\frac{1+\sqrt{5}}{2}\right)^n - \frac{1}{\sqrt{5}}\left(\frac{1-\sqrt{5}}{2}\right)^n$$

by showing that this result satisfies the recursion formula.

26. Let F_n be the general term of the Fibonacci sequence given in Problem 25. Prove that

$$\lim_{n\to\infty} \frac{F_{n+1}}{F_n} = \frac{1+\sqrt{5}}{2}$$

27. Let $\{F_n\}$ be the Fibonacci sequence given in Problem 25. Prove that the series

$$\frac{1}{1}+\frac{1}{1}+\frac{1}{2}+\frac{1}{3}+\frac{1}{5}+\frac{1}{8}+\cdots = \sum_{n=1}^{\infty} \frac{1}{F_n}$$

converges.

28. In 1985 William Gosper used the following identity to compute the first 17 million digits of π:

$$\frac{1}{\pi} = \frac{2\sqrt{2}}{9801} \sum_{n=0}^{\infty} (1103 + 26{,}390n)\frac{(4n)!}{(n!)^4(4 \cdot 99)^{4n}}$$

This identity was discovered by Ramanujan† in 1920.

(a) Verify that the infinite series converges.

(b) How many correct decimal places of π does the first term of the series yield?

(c) How many correct decimal places of π do the first two terms of the series yield?

---

*See Problems 69 and 70 of Exercises 10.1.

†The Indian mathematician Srinivasa Ramanujan (1887–1920) was noted for his remarkable insights in handling exceedingly complex algebraic manipulations and calculations.

10.6 ALTERNATING SERIES AND ABSOLUTE CONVERGENCE

A series having either form

$$a_1 - a_2 + a_3 - a_4 + \cdots + (-1)^{n+1}a_n + \cdots = \sum_{k=1}^{\infty} (-1)^{k+1}a_k$$

or

$$-a_1 + a_2 - a_3 + a_4 - \cdots + (-1)^n a_n + \cdots = \sum_{k=1}^{\infty} (-1)^k a_k$$

where $a_k > 0$ for $k = 1, 2, 3, \ldots$, is said to be an **alternating series**. Since $\Sigma_{k=1}^{\infty}(-1)^k a_k$ is just a multiple of $\Sigma_{k=1}^{\infty}(-1)^{k+1}a_k$, we shall limit our discussion to the latter series.

EXAMPLE 1

$$1 - \frac{1}{2} + \frac{1}{3} - \frac{1}{4} + \cdots = \sum_{k=1}^{\infty} \frac{(-1)^{k+1}}{k}$$

and

$$\frac{\ln 2}{4} - \frac{\ln 3}{8} + \frac{\ln 4}{16} - \frac{\ln 5}{32} + \cdots = \sum_{k=2}^{\infty} (-1)^k \frac{\ln k}{2^k}$$

are examples of alternating series. □

Alternating Series Test The first series of Example 1, $1 - \frac{1}{2} + \frac{1}{3} - \frac{1}{4} + \cdots$, is called the **alternating harmonic series**. Although the harmonic series $\Sigma_{k=1}^{\infty} 1/k$ diverges, the introduction of positive and negative terms in the sequence of partial sums for the alternating harmonic series is sufficient to produce a convergent series. We will prove that $\sum_{k=1}^{\infty}(-1)^{k+1}/k$ converges by means of the next test.

THEOREM 10.17 Alternating Series Test

If $\lim_{n\to\infty} a_n = 0$ and $0 < a_{k+1} \le a_k$ for every positive integer k, then $\Sigma_{k=1}^{\infty}(-1)^{k+1}a_k$ converges.

Proof Consider the partial sums that contain $2n$ terms:

$$\begin{aligned} S_{2n} &= a_1 - a_2 + a_3 - a_4 + \cdots + a_{2n-1} - a_{2n} \\ &= (a_1 - a_2) + (a_3 - a_4) + \cdots + (a_{2n-1} - a_{2n}) \end{aligned} \tag{1}$$

Since $a_k - a_{k+1} \ge 0$ for $k = 1, 2, 3, \ldots$, we have

$$S_2 \le S_4 \le S_6 \le \cdots \le S_{2n} \le \cdots$$

Thus, the sequence $\{S_{2n}\}$, whose general term S_{2n} contains an even number of terms of the series, is a monotonic sequence. Rewriting (1) as

$$S_{2n} = a_1 - (a_2 - a_3) - \cdots - a_{2n}$$

shows that $S_{2n} < a_1$ for every positive integer n. Hence, $\{S_{2n}\}$ is bounded. By Theorem 10.4 it follows that $\{S_{2n}\}$ converges to a limit S. Now,

$$S_{2n+1} = S_{2n} + a_{2n+1}$$

implies that

$$\lim_{n\to\infty} S_{2n+1} = \lim_{n\to\infty} S_{2n} + \lim_{n\to\infty} a_{2n+1} = S + 0 = S$$

This shows that the sequence of partial sums $\{S_{2n+1}\}$, whose general term S_{2n+1} contains an odd number of terms, also converges to S. Since both $\{S_{2n}\}$ and $\{S_{2n+1}\}$ converge to S, we conclude that $\{S_n\}$ converges to S. ■

EXAMPLE 2 Show that the alternating harmonic series $\sum_{k=1}^{\infty} \frac{(-1)^{k+1}}{k}$ converges.

Solution With the identification $a_n = 1/n$, we have immediately

$$\lim_{n\to\infty} a_n = \lim_{n\to\infty} \frac{1}{n} = 0 \quad \text{and} \quad a_{k+1} \le a_k$$

since $1/(k+1) \le 1/k$ for $k \ge 1$. It follows from Theorem 10.17 that the alternating series converges. □

EXAMPLE 3 The alternating series $\sum_{k=1}^{\infty} (-1)^{k+1} \frac{2k+1}{3k-1}$ diverges, since

$$\lim_{n\to\infty} a_n = \lim_{n\to\infty} \frac{2n+1}{3n-1} = \frac{2}{3}$$

Recall from Theorem 10.6 that it is necessary that $\lim_{n\to\infty} a_n = 0$ for the convergence of *any* infinite series. □

Although showing that $a_{k+1} \le a_k$ may seem a straightforward task, this is often not the case.

EXAMPLE 4 Test for convergence $\sum_{k=1}^{\infty} (-1)^{k+1} \frac{\sqrt{k}}{k+1}$.

Solution In order to show that the terms of the series satisfy the condition $a_{k+1} \le a_k$, let us consider the function $f(x) = \sqrt{x}/(x+1)$ for which $f(x) = a_k$. From

$$f'(x) = -\frac{x-1}{2\sqrt{x}(x+1)^2} < 0 \qquad \text{for } x > 1$$

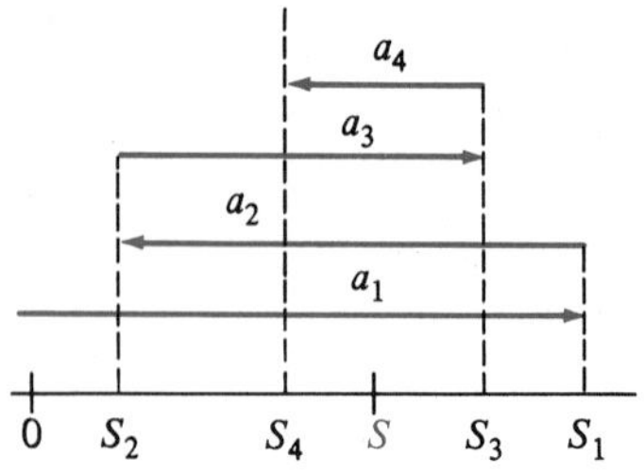

FIGURE 10.11

we see that the function f decreases for $x > 1$. Hence, $a_{k+1} \le a_k$ is true for $k \ge 1$. Now, L'Hôpital's Rule shows that

$$\lim_{x\to\infty} f(x) = 0 \quad \text{and so} \quad \lim_{n\to\infty} f(n) = \lim_{n\to\infty} a_n = 0$$

Hence, the alternating series converges by Theorem 10.17. □

Error in Approximating the Sum of an Alternating Series Suppose the alternating series $\Sigma_{k=1}^{\infty}(-1)^{k+1}a_k$ converges to a number S. The partial sums

$$S_1 = a_1, \quad S_2 = a_1 - a_2, \quad S_3 = a_1 - a_2 + a_3, \quad S_4 = a_1 - a_2 + a_3 - a_4, \quad \ldots$$

can be represented on a number line as shown in Figure 10.11. The sequence $\{S_n\}$ converges in the manner illustrated in Figure 10.1(c): that is, the terms S_n get closer and closer to S as $n \to \infty$ although they oscillate on either side of S. As indicated in Figure 10.11, the even-numbered partial sums are less than S and the odd-numbered partial sums are greater than S. Roughly, the even-numbered partial sums increase to the number S, and, in turn, the odd-numbered partial sums decrease to S. Because of this, *the sum S of the series must lie between consecutive partial sums S_n and S_{n+1}*:

$$S_n \le S \le S_{n+1}, \qquad \text{if } n \text{ is even} \tag{2}$$

or

$$S_{n+1} \le S \le S_n, \qquad \text{if } n \text{ is odd} \tag{3}$$

Now (2) yields $0 \le S - S_n \le S_{n+1} - S_n$ for n even, and (3) implies $0 \le S_n - S \le S_n - S_{n+1}$ for n odd. Thus, in either case $|S_n - S| \le |S_{n+1} - S_n|$. But $S_{n+1} - S_n = a_{n+1}$ for n even and $S_{n+1} - S_n = -a_{n+1}$ for n odd. Thus, $|S_n - S| \le a_{n+1}$ for all n. We state this result as our next theorem.

THEOREM 10.18 Error Bound for Alternating Series

Suppose the alternating series $\Sigma_{k=1}^{\infty}(-1)^{k+1}a_k$, $a_k > 0$, converges to a number S. If S_n is the nth partial sum of the series and $a_{k+1} \le a_k$ for all k, then $|S_n - S| \le a_{n+1}$ for all n.

Theorem 10.18 is useful in approximating the sum of a convergent alternating series. It states that the **error** $|S_n - S|$ between the nth partial sum and the sum of the series is less than the absolute value of the $(n + 1)$st term of the series.

EXAMPLE 5 Approximate the sum of the convergent series $\displaystyle\sum_{k=1}^{\infty} \frac{(-1)^{k+1}}{(2k)!}$ to four decimal places.

Solution Theorem 10.18 indicates that we must have

$$a_{n+1} = \frac{1}{(2n+2)!} < 0.00005$$

From

$$n = 1, \quad a_2 = \frac{1}{4!} \approx 0.041667$$

$$n = 2, \quad a_3 = \frac{1}{6!} \approx 0.001389$$

$$n = 3, \quad a_4 = \frac{1}{8!} \approx 0.000025 < 0.00005$$

we see that

$$S_3 = \frac{1}{2!} - \frac{1}{4!} + \frac{1}{6!} \approx 0.4597$$

has the desired accuracy. □

Absolute and Conditional Convergence An infinite series such as

$$\frac{2}{3} + \left(\frac{2}{3}\right)^2 - \left(\frac{2}{3}\right)^3 - \left(\frac{2}{3}\right)^4 + \left(\frac{2}{3}\right)^5 + \left(\frac{2}{3}\right)^6 - - + + \cdots \tag{4}$$

is not an alternating series and so Theorem 10.17 is not applicable. Nonetheless, the series (4) is convergent because the series of absolute values

$$\frac{2}{3} + \left(\frac{2}{3}\right)^2 + \left(\frac{2}{3}\right)^3 + \left(\frac{2}{3}\right)^4 + \left(\frac{2}{3}\right)^5 + \left(\frac{2}{3}\right)^6 + \cdots$$

is a convergent series (a geometric series with $r = \frac{2}{3} < 1$). The series (4) is an example of a series that **converges absolutely**.

DEFINITION 10.6 Absolute Convergence

A series $\Sigma_{k=1}^{\infty} a_k$ is said to be **absolutely convergent** if $\Sigma_{k=1}^{\infty} |a_k|$ converges.

EXAMPLE 6 The alternating series $\displaystyle\sum_{k=1}^{\infty} \frac{(-1)^{k+1}}{k^2}$ is absolutely convergent, since

$$\sum_{k=1}^{\infty} \left|\frac{(-1)^{k+1}}{k^2}\right| = \sum_{k=1}^{\infty} \frac{1}{k^2}$$

is a convergent p-series with $p = 2 > 1$. □

DEFINITION 10.7 Conditional Convergence

A series $\Sigma_{k=1}^{\infty} a_k$ is said to be **conditionally convergent** if $\Sigma_{k=1}^{\infty} |a_k|$ diverges and $\Sigma_{k=1}^{\infty} a_k$ converges.

EXAMPLE 7 In Example 2 we saw that the alternating harmonic series $\Sigma_{k=1}^{\infty}(-1)^{k+1}/k$ is convergent. But taking the absolute value of each term gives the divergent harmonic series $\Sigma_{k=1}^{\infty} 1/k$. Thus, $\Sigma_{k=1}^{\infty}(-1)^{k+1}/k$ is conditionally convergent. □

The next result shows that every absolute convergent series is also convergent. It is for this reason that the series in (4) converges.

THEOREM 10.19 Absolute Convergence Implies Convergence

If $\Sigma_{k=1}^{\infty} |a_k|$ converges, then $\Sigma_{k=1}^{\infty} a_k$ converges.

Proof If $c_k = a_k + |a_k|$, then $c_k \leq 2|a_k|$. Since $\Sigma |a_k|$ converges, it follows from the Comparison Test that $\Sigma\, c_k$ converges. Furthermore,

$$\sum_{k=1}^{\infty} (c_k - |a_k|)$$

converges, since both $\Sigma\, c_k$ and $\Sigma |a_k|$ converge. But

$$\sum_{k=1}^{\infty} a_k = \sum_{k=1}^{\infty} (c_k - |a_k|)$$

Therefore, $\Sigma\, a_k$ converges. ■

Note that since $\Sigma |a_k|$ is a series of positive terms, the tests of the preceding section can be utilized to determine whether a series converges absolutely.

EXAMPLE 8 Test for convergence $\displaystyle\sum_{k=1}^{\infty} \frac{(-1)^{k+1}}{1+k^2}$.

Solution It follows from the Integral Test that $\Sigma_{k=1}^{\infty} 1/(1+k^2)$ is convergent. (Show this.) Hence, by Definition 10.6, the alternating series is absolutely convergent. From Theorem 10.19 we conclude that the given series is convergent. □

Ratio and Root Tests The following modified forms of the **Ratio Test** and the **Root Test** can be applied directly to an alternating series.

THEOREM 10.20 Ratio Test

Suppose $\Sigma_{k=1}^{\infty} a_k$ is a series of nonzero terms such that

$$\lim_{n\to\infty} \left|\frac{a_{n+1}}{a_n}\right| = L$$

(*i*) If $L < 1$, the series is absolutely convergent.

(*ii*) If $L > 1$, or if $\lim_{n\to\infty} |a_{n+1}/a_n| = \infty$, the series is divergent.

(*iii*) If $L = 1$, the test is inconclusive.

EXAMPLE 9 Test for convergence $\displaystyle\sum_{k=1}^{\infty} \frac{(-1)^{k+1}2^{2k-1}}{k3^k}$.

Solution

$$\lim_{n\to\infty} \left|\frac{a_{n+1}}{a_n}\right| = \lim_{n\to\infty} \left|\frac{(-1)^{n+2}2^{2n+1}}{(n+1)3^{n+1}} \cdot \frac{n3^n}{(-1)^{n+1}2^{2n-1}}\right|$$

$$= \lim_{n\to\infty} \frac{4n}{3(n+1)} = \frac{4}{3}$$

Since $L = \frac{4}{3} > 1$, we see from Theorem 10.20(*ii*) that the alternating series diverges. □

THEOREM 10.21 Root Test

Suppose $\Sigma_{k=1}^{\infty} a_k$ is an infinite series such that

$$\lim_{n\to\infty} \sqrt[n]{|a_n|} = L$$

(*i*) If $L < 1$, the series is absolutely convergent.

(*ii*) If $L > 1$, or if $\lim_{n\to\infty} \sqrt[n]{|a_n|} = \infty$, the series is divergent.

(*iii*) If $L = 1$, the test is inconclusive.

▲ ***Remarks*** (*i*) The conclusion of Theorem 10.17 remains true when the hypothesis "$a_{k+1} \le a_k$ for every positive integer k" is replaced with the statement "$a_{k+1} \le a_k$ for k sufficiently large." For the alternating series $\Sigma_{k=1}^{\infty} (-1)^{k+1}(\ln k)/k^{1/3}$, it is readily shown by the procedure used in Example 4 that $a_{k+1} \le a_k$ for $k \ge 21$. Moreover, $\lim_{n\to\infty} a_n = 0$. Hence, the series converges by the Alternating Series Test.

(*ii*) If the series of absolute values $\Sigma |a_k|$ is found to be divergent, then no conclusion can be drawn concerning the convergence or divergence of the series Σa_k.

(*iii*) If $\Sigma\, a_k$ is absolutely convergent, then the terms of the series can be rearranged or regrouped in any manner and the resulting series will converge to the same number as the original series. In contrast, if the terms of a conditionally convergent series are written in a different order, the new series may diverge or converge to an entirely different number. It is left as an exercise to show that if S is the sum of the convergent alternating harmonic series

$$S = 1 - \frac{1}{2} + \frac{1}{3} - \frac{1}{4} + \frac{1}{5} - \frac{1}{6} + \cdots$$

then the rearranged series

$$1 + \frac{1}{3} - \frac{1}{2} + \frac{1}{5} + \frac{1}{7} - \frac{1}{4} + \cdots$$

converges to $\frac{3}{2}S$. See Problem 52 in Exercises 10.6. You are also encouraged to reflect on the following "reasoning":

$$\begin{aligned} 2S &= 2\left[1 - \frac{1}{2} + \frac{1}{3} - \frac{1}{4} + \frac{1}{5} - \frac{1}{6} + \frac{1}{7} - \frac{1}{8} + \frac{1}{9} - \cdots\right] \\ &= 2 - 1 + \frac{2}{3} - \frac{1}{2} + \frac{2}{5} - \frac{1}{3} + \frac{2}{7} - \frac{1}{4} + \frac{2}{9} - \cdots \\ &= (2-1) - \frac{1}{2} + \left(\frac{2}{3} - \frac{1}{3}\right) - \frac{1}{4} + \left(\frac{2}{5} - \frac{1}{5}\right) - \frac{1}{6} + \cdots = S \end{aligned}$$

By dividing by S, we obtain the interesting result that $2 = 1$.

EXERCISES 10.6

Answers to odd-numbered problems begin on page A-93.

In Problems 1–14 use the Alternating Series Test to determine whether the given series is convergent.

1. $\displaystyle\sum_{k=1}^{\infty} \frac{(-1)^{k+1}}{k+2}$
2. $\displaystyle\sum_{k=1}^{\infty} \frac{(-1)^{k-1}}{\sqrt{k}}$
3. $\displaystyle\sum_{k=1}^{\infty} (-1)^{k-1}\frac{k}{k+1}$
4. $\displaystyle\sum_{k=1}^{\infty} (-1)^{k}\frac{k}{k^2+1}$
5. $\displaystyle\sum_{k=1}^{\infty} (-1)^{k+1}\frac{k^2+2}{k^3}$
6. $\displaystyle\sum_{k=1}^{\infty} (-1)^{k+1}\frac{3k-1}{k+5}$
7. $\displaystyle\sum_{k=1}^{\infty} (-1)^{k-1}\left(\frac{1}{k} + \frac{1}{3^k}\right)$
8. $\displaystyle\sum_{k=1}^{\infty} (-1)^{k+1}\frac{k+1}{4^k}$
9. $\displaystyle\sum_{n=1}^{\infty} (-1)^{n+1}\frac{4\sqrt{n}}{2n+1}$
10. $\displaystyle\sum_{n=1}^{\infty} (-1)^{n-1}\frac{\sqrt[3]{n}}{n+1}$
11. $\displaystyle\sum_{n=2}^{\infty} (\cos n\pi)\frac{\sqrt{n+1}}{n+2}$
12. $\displaystyle\sum_{k=2}^{\infty} (-1)^{k}\frac{\sqrt{k^2+1}}{k^3}$
13. $\displaystyle\sum_{k=2}^{\infty} (-1)^{k}\frac{k}{\ln k}$
14. $\displaystyle\sum_{k=3}^{\infty} (-1)^{k-1}\frac{\ln k^{10}}{k}$

In Problems 15–34 determine whether the given series is absolutely convergent, conditionally convergent, or divergent.

15. $\displaystyle\sum_{k=1}^{\infty} \frac{(-1)^{k+1}}{2k+1}$
16. $\displaystyle\sum_{k=1}^{\infty} \frac{(-1)^{k-1}}{\sqrt{k+5}}$
17. $\displaystyle\sum_{k=1}^{\infty} (-1)^{k+1}\left(\frac{2}{3}\right)^k$
18. $\displaystyle\sum_{k=1}^{\infty} (-1)^{k+1}\frac{2^{2k}}{3^k}$
19. $\displaystyle\sum_{k=1}^{\infty} (-1)^{k}\frac{k}{5^k}$
20. $\displaystyle\sum_{k=1}^{\infty} (-1)^{k}(k2^{-k})^2$
21. $\displaystyle\sum_{k=1}^{\infty} \frac{(-1)^{k}}{k!}$
22. $\displaystyle\sum_{k=1}^{\infty} (-1)^{k}\frac{(k!)^2}{(2k)!}$
23. $\displaystyle\sum_{k=1}^{\infty} (-1)^{k+1}\frac{k!}{100^k}$
24. $\displaystyle\sum_{k=1}^{\infty} (-1)^{k-1}\frac{5^{2k-3}}{10^{k+2}}$

25. $\sum_{k=1}^{\infty} (-1)^{k-1} \frac{k}{1+k^2}$

26. $\sum_{k=1}^{\infty} (-1)^{k+1} \frac{k}{1+k^4}$

27. $\sum_{k=1}^{\infty} \cos k\pi$

28. $\sum_{k=1}^{\infty} \frac{\sin\left(\frac{2k+1}{2}\pi\right)}{\sqrt{k+1}}$

29. $\sum_{k=1}^{\infty} (-1)^{k-1} \sin\left(\frac{1}{k}\right)$

30. $\sum_{k=1}^{\infty} (-1)^{k-1} \frac{\sin\left(\frac{1}{k}\right)}{k^2}$

31. $\sum_{k=1}^{\infty} (-1)^k \left[\frac{1}{k+1} - \frac{1}{k}\right]$

32. $\sum_{k=1}^{\infty} (-1)^k [\sqrt{k+1} - \sqrt{k}]$

33. $\sum_{k=1}^{\infty} (-1)^k \left(\frac{2k}{k+50}\right)^k$

34. $\sum_{k=1}^{\infty} (-1)^{k+1} \frac{6^{3k}}{k^k}$

In Problems 35 and 36 approximate the sum of the convergent series to the indicated number of decimal places.

35. $\sum_{k=1}^{\infty} \frac{(-1)^{k+1}}{(2k-1)!}$; five

36. $\sum_{k=1}^{\infty} \frac{(-1)^{k+1}}{k!}$; three

In Problems 37 and 38 find the smallest positive integer n so that S_n approximates the sum of the convergent series to the indicated number of decimal places.

37. $\sum_{k=1}^{\infty} \frac{(-1)^{k+1}}{k^3}$; two

38. $\sum_{k=1}^{\infty} \frac{(-1)^{k+1}}{\sqrt{k}}$; three

In Problems 39 and 40 approximate the sum of the convergent series so that the error is less than the indicated amount.

39. $1 - \frac{1}{4^2} + \frac{1}{4^3} - \frac{1}{4^4} + \cdots$; 10^{-3}

40. $1 - \frac{2}{5^2} + \frac{3}{5^3} - \frac{4}{5^4} + \cdots$; 10^{-4}

In Problems 41 and 42 estimate the error in using the indicated partial sum as an approximation to the sum of the convergent series.

41. $\sum_{k=1}^{\infty} \frac{(-1)^{k+1}}{k}$; S_{100}

42. $\sum_{k=1}^{\infty} \frac{(-1)^{k+1}}{k\,2^k}$; S_6

In Problems 43–48 state why the Alternating Series Test is not applicable to the given series. Determine whether the given series converges or diverges.

43. $\sum_{k=1}^{\infty} \frac{\sin(k\pi/6)}{\sqrt{k^4+1}}$

44. $\sum_{k=1}^{\infty} \frac{100 + (-1)^k 2^k}{3^k}$

45. $1 - \frac{1}{2} - \frac{1}{4} + \frac{1}{8} + \frac{1}{16} - - + + \cdots$

46. $\frac{1}{1} - \frac{1}{4} - \frac{1}{9} + \frac{1}{16} + \frac{1}{25} + \frac{1}{36} - - - + + + + \cdots$

47. $\frac{2}{1} - \frac{1}{1} + \frac{2}{2} - \frac{1}{2} + \frac{2}{3} - \frac{1}{3} + \frac{2}{4} - \frac{1}{4} + \cdots$ [*Hint:* Consider the partial sums S_{2n} for $n = 1, 2, 3, \ldots$.]

48. $\frac{1}{2} + \frac{1}{2} - \frac{1}{3} - \frac{1}{3} - \frac{1}{3} + \frac{1}{4} + \frac{1}{4} + \frac{1}{4} + \frac{1}{4} - - - - - \cdots$

49. Determine whether each of the following series converges or diverges.

(a) $1 - 1 + 1 - 1 + \cdots$

(b) $(1-1) + (1-1) + (1-1) + \cdots$

(c) $1 + (-1+1) + (-1+1) + \cdots$

(d) $1 + (-1+1) + (-1+1-1) + \cdots$

50. The series $1 - \frac{1}{3} + \frac{1}{9} - \frac{1}{27} + \cdots$ is an absolutely convergent geometric series. Show that its rearrangement

$$-\frac{1}{3} + \frac{1}{1} - \frac{1}{27} + \frac{1}{9} - \cdots$$

is convergent. Try the Ratio Test and Root Test. [*Hint:* Examine $3^{k+(-1)^k}$, $k = 0, 1, 2, \ldots$.]

In Problems 51 and 52 use

$$S = 1 - \frac{1}{2} + \frac{1}{3} - \frac{1}{4} + \frac{1}{5} - \frac{1}{6} + \cdots$$

to prove the given result.

51. $\frac{1}{2}S = 0 + \frac{1}{2} + 0 - \frac{1}{4} + 0 + \frac{1}{6} - \cdots$

52. $\frac{3}{2}S = 1 + \frac{1}{3} - \frac{1}{2} + \frac{1}{5} + \frac{1}{7} - \frac{1}{4} + \cdots$

53. If $\Sigma\, a_k$ is absolutely convergent, prove that $\Sigma\, a_k^2$ converges. [*Hint:* For n sufficiently large, $|a_n| < 1$. Why?]

10.7 POWER SERIES

A series containing nonnegative integral powers of a variable x,

$$c_0 + c_1x + c_2x^2 + \cdots + c_nx^n + \cdots = \sum_{k=0}^{\infty} c_kx^k \tag{1}$$

where the c_k are constants depending on k, is called a **power series in x**. The series in (1) is just a particular case of the more general form

$$c_0 + c_1(x-a) + c_2(x-a)^2 + \cdots + c_n(x-a)^n + \cdots = \sum_{k=0}^{\infty} c_k(x-a)^k \tag{2}$$

which is called a **power series in $x - a$**. The power series (2) is also said to be **centered at a** or have **center a**. The power series in (1) is centered at 0.

The problem we face in this section is:

Find the values of x for which a power series converges.

Observe that (1) and (2) converge to c_0 when $x = 0$ and $x = a$, respectively.*

EXAMPLE 1 The power series

$$1 + x + x^2 + \cdots + x^n + \cdots = \sum_{k=0}^{\infty} x^k$$

is recognized as a geometric series with $r = x$. Hence, the series converges for those values of x that satisfy $|x| < 1$ or $-1 < x < 1$. □

Interval of Convergence The set of all real numbers x for which a power series converges is said to be its **interval of convergence**. For the general case, a power series in $x - a$ may converge

(*i*) on a *finite interval* centered at a: $(a - r, a + r)$, $[a - r, a + r)$, $(a - r, a + r]$, or $[a - r, a + r]$;

(*ii*) on the *infinite interval* $(-\infty, \infty)$; or

(*iii*) at the *single point* $x = a$.

In the respective cases, we say that the **radius of convergence** is r, ∞, or 0. Figure 10.12 illustrates the case $(a - r, a + r)$.

The Ratio Test, as stated in Theorem 10.20, is especially useful in finding an interval of convergence.

*It is convenient to define $x^0 = 1$ and $(x - a)^0 = 1$ even when $x = 0$ and $x = a$, respectively.

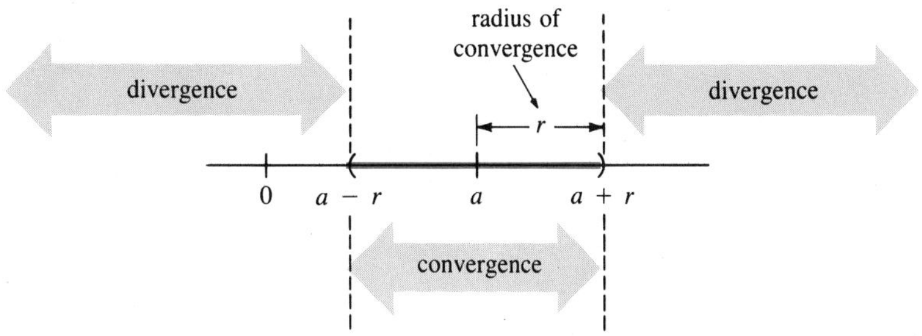

FIGURE 10.12

EXAMPLE 2 Find the interval of convergence for $\sum_{k=0}^{\infty} \frac{x^k}{2^k(k+1)^2}$.

Solution

$$\lim_{n\to\infty}\left|\frac{a_{n+1}}{a_n}\right| = \lim_{n\to\infty}\left|\frac{x^{n+1}}{2^{n+1}(n+2)^2}\cdot\frac{2^n(n+1)^2}{x^n}\right|$$

$$= \lim_{n\to\infty}\left(\frac{n+1}{n+2}\right)^2\frac{|x|}{2} = \frac{|x|}{2}$$

From part (i) of Theorem 10.20, we have absolute convergence whenever this limit is strictly less than 1. Thus, the series is absolutely convergent for those values of x that satisfy $|x|/2 < 1$ or $|x| < 2$; that is, the series will converge for any number x in the open interval $(-2, 2)$. However, if $|x|/2 = 1$, or when $x = 2$ and $x = -2$, the Ratio Test gives no information. We must perform separate checks of the series for convergence at these endpoints. Substituting 2 for x gives $\Sigma_{k=1}^{\infty} 1/(k+1)^2$, which is convergent by comparison with the convergent p-series $\Sigma_{k=1}^{\infty} 1/k^2$. Similarly, substituting -2 for x yields the alternating series $\Sigma_{k=1}^{\infty}(-1)^k/(k+1)^2$, which is evidently convergent. (Why?) We conclude that the interval of convergence is the closed interval $[-2, 2]$. The radius of convergence is 2. The series diverges if $|x| > 2$. □

EXAMPLE 3 Find the interval of convergence for $\sum_{k=0}^{\infty} \frac{x^k}{k!}$.

Solution By Theorem 10.20 we have

$$\lim_{n\to\infty}\left|\frac{a_{n+1}}{a_n}\right| = \lim_{n\to\infty}\left|\frac{x^{n+1}}{(n+1)!}\cdot\frac{n!}{x^n}\right| = \lim_{n\to\infty}\frac{n!}{(n+1)!}|x|$$

$$= \lim_{n\to\infty}\frac{|x|}{n+1}$$

Since $\lim_{n\to\infty}|x|/(n+1) = 0$ for any choice of x, the series converges absolutely for every real number. Thus, the interval of convergence is $(-\infty, \infty)$ and the radius of convergence is ∞. □

EXAMPLE 4 Find the interval of convergence for $\sum_{k=1}^{\infty} \frac{(x-5)^k}{k3^k}$.

Solution

$$\lim_{n\to\infty}\left|\frac{a_{n+1}}{a_n}\right| = \lim_{n\to\infty}\left|\frac{(x-5)^{n+1}}{(n+1)3^{n+1}} \cdot \frac{n3^n}{(x-5)^n}\right|$$

$$= \lim_{n\to\infty}\left(\frac{n}{n+1}\right)\frac{|x-5|}{3} = \frac{|x-5|}{3}$$

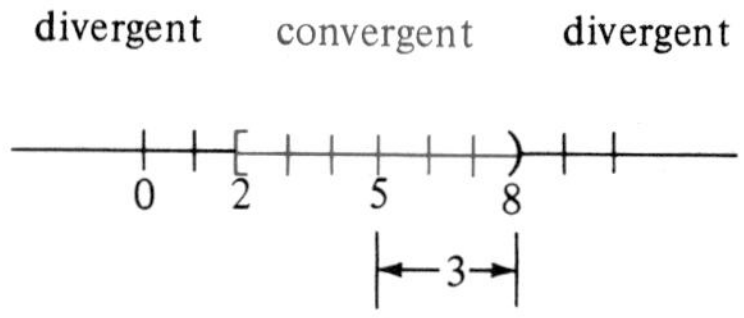

FIGURE 10.13

The series converges absolutely if $|x-5|/3 < 1$ or $|x-5| < 3$. This last inequality yields the open interval $(2, 8)$. At $x = 2$ and $x = 8$, we obtain, in turn, $\Sigma_{k=1}^{\infty}(-1)^k/k$ and $\Sigma_{k=1}^{\infty} 1/k$. The former series converges by the Alternating Series Test; the latter series is the divergent harmonic series. Consequently, the interval of convergence is $[2, 8)$. The radius of convergence is 3. The series diverges if $x < 2$ or $x \geq 8$. See Figure 10.13. □

EXAMPLE 5 Find the interval of convergence for $\sum_{k=1}^{\infty} k!(x+10)^k$.

Solution

$$\lim_{n\to\infty}\left|\frac{a_{n+1}}{a_n}\right| = \lim_{n\to\infty}\left|\frac{(n+1)!(x+10)^{n+1}}{n!(x+10)^n}\right|$$

$$= \lim_{n\to\infty}(n+1)|x+10|$$

$$= \begin{cases}\infty, & x \neq -10 \\ 0, & x = -10\end{cases}$$

The series diverges for every real number x, *except* $x = -10$. At $x = -10$, we obtain a convergent series consisting of all zeros. The radius of convergence is 0. □

EXERCISES 10.7

Answers to odd-numbered problems begin on page A-93.

In Problems 1–24 use the Ratio Test to find the interval of convergence of the given power series.

1. $\sum_{k=1}^{\infty} \frac{(-1)^k}{k} x^k$
2. $\sum_{k=1}^{\infty} \frac{x^k}{k^2}$
3. $\sum_{k=1}^{\infty} \frac{2^k}{k} x^k$
4. $\sum_{k=0}^{\infty} \frac{5^k}{k!} x^k$
5. $\sum_{k=1}^{\infty} \frac{(x-3)^k}{k^3}$
6. $\sum_{k=1}^{\infty} \frac{(x+7)^k}{\sqrt{k}}$
7. $\sum_{k=1}^{\infty} \frac{(-1)^k}{10^k}(x-5)^k$
8. $\sum_{k=1}^{\infty} \frac{k}{(k+2)^2}(x-4)^k$
9. $\sum_{k=0}^{\infty} k!2^k x^k$
10. $\sum_{k=0}^{\infty} \frac{k-1}{k^{2k}} x^k$
11. $\sum_{k=1}^{\infty} \frac{(3x-1)^k}{k^2+k}$
12. $\sum_{k=0}^{\infty} \frac{(4x-5)^k}{3^k}$
13. $\sum_{k=2}^{\infty} \frac{x^k}{\ln k}$
14. $\sum_{k=2}^{\infty} \frac{(-1)^k x^k}{k \ln k}$
15. $\sum_{k=1}^{\infty} \frac{k^2}{3^{2k}}(x+7)^k$
16. $\sum_{k=1}^{\infty} k^3 2^{4k}(x-1)^k$
17. $\sum_{k=1}^{\infty} \frac{2^{5k}}{5^{2k}}\left(\frac{x}{3}\right)^k$
18. $\sum_{k=1}^{\infty} \frac{1000^k}{k^k} x^k$

19. $\sum_{k=0}^{\infty} \frac{(-3)^k}{(k+1)(k+2)}(x-1)^k$

20. $\sum_{k=1}^{\infty} \frac{3^k}{(-2)^k k(k+1)}(x+5)^k$

21. $\sum_{k=1}^{\infty} \frac{(-1)^{k+1}}{(k!)^2}\left(\frac{x-2}{3}\right)^k$

22. $\sum_{k=0}^{\infty} \frac{(6-x)^{k+1}}{\sqrt{2k+1}}$

23. $\sum_{k=0}^{\infty} \frac{(-1)^k}{9^k} x^{2k+1}$

24. $\sum_{k=1}^{\infty} \frac{5^k}{(2k)!} x^{2k}$

In Problems 25–28 use the Root Test, Theorem 10.21, to find the interval of convergence of the given power series.

25. $\sum_{k=2}^{\infty} \frac{x^k}{(\ln k)^k}$

26. $\sum_{k=1}^{\infty} (k+1)^k (x+1)^k$

27. $\sum_{k=1}^{\infty} \left(\frac{4}{3}\right)^k (x+3)^k$

28. $\sum_{k=1}^{\infty} \left(\frac{k}{k+1}\right)^{k^2} (x-e)^k$

In Problems 29 and 30 find the radius of convergence of the given power series.

29. $\sum_{k=1}^{\infty} \frac{k!}{1 \cdot 3 \cdot 5 \cdots (2k-1)}\left(\frac{x}{2}\right)^k$

30. $\sum_{k=2}^{\infty} \frac{1 \cdot 3 \cdot 5 \cdots (2k-3)}{3^k k!}(x-1)^k$

In Problems 31–38 the given series is not a power series. Nonetheless, find all values of x for which the given series converges.

31. $\sum_{k=1}^{\infty} \frac{1}{x^k}$

32. $\sum_{k=1}^{\infty} \frac{7^k}{x^{2k}}$

33. $\sum_{k=1}^{\infty} \left(\frac{x+1}{x}\right)^k$

34. $\sum_{k=1}^{\infty} \frac{1}{2^k}\left(\frac{x}{x+2}\right)^k$

35. $\sum_{k=0}^{\infty} \left(\frac{x^2+2}{6}\right)^{k^2}$

36. $\sum_{k=1}^{\infty} \frac{k!}{(kx)^k}$

37. $\sum_{k=0}^{\infty} e^{kx}$

38. $\sum_{k=0}^{\infty} k! e^{-kx^2}$

39. Find all values of x in $[0, 2\pi]$ for which $\Sigma_{k=0}^{\infty}(2/\sqrt{3})^k \sin^k x$ converges.

40. Show that $\Sigma_{k=1}^{\infty}(\sin kx)/k^2$ converges for all real values of x.

10.8 DIFFERENTIATION AND INTEGRATION OF POWER SERIES

A Power Series Represents a Function For each x in its interval of convergence, a power series* $\Sigma\, c_k x^k$ converges to one number. Thus, a power series defines or *represents* a function f whose domain is the interval of convergence of the power series. For each x in the interval of convergence, we define the functional value $f(x)$ by the sum of the series

$$f(x) = c_0 + c_1 x + c_2 x^2 + \cdots + c_n x^n + \cdots = \sum_{k=0}^{\infty} c_k x^k \tag{1}$$

The next three theorems, which we shall state without proof, answer some fundamental questions about a function represented by a power series.

THEOREM 10.22 A Power Series Is a Continuous Function

If $f(x) = \Sigma_{k=0}^{\infty} c_k x^k$ converges on an interval $(-r, r)$ for which the radius of convergence is either positive or ∞, then f is continuous at each x in $(-r, r)$.

*For convenience, we limit our discussion to power series centered at 0. The results of this section apply equally to power series centered at a.

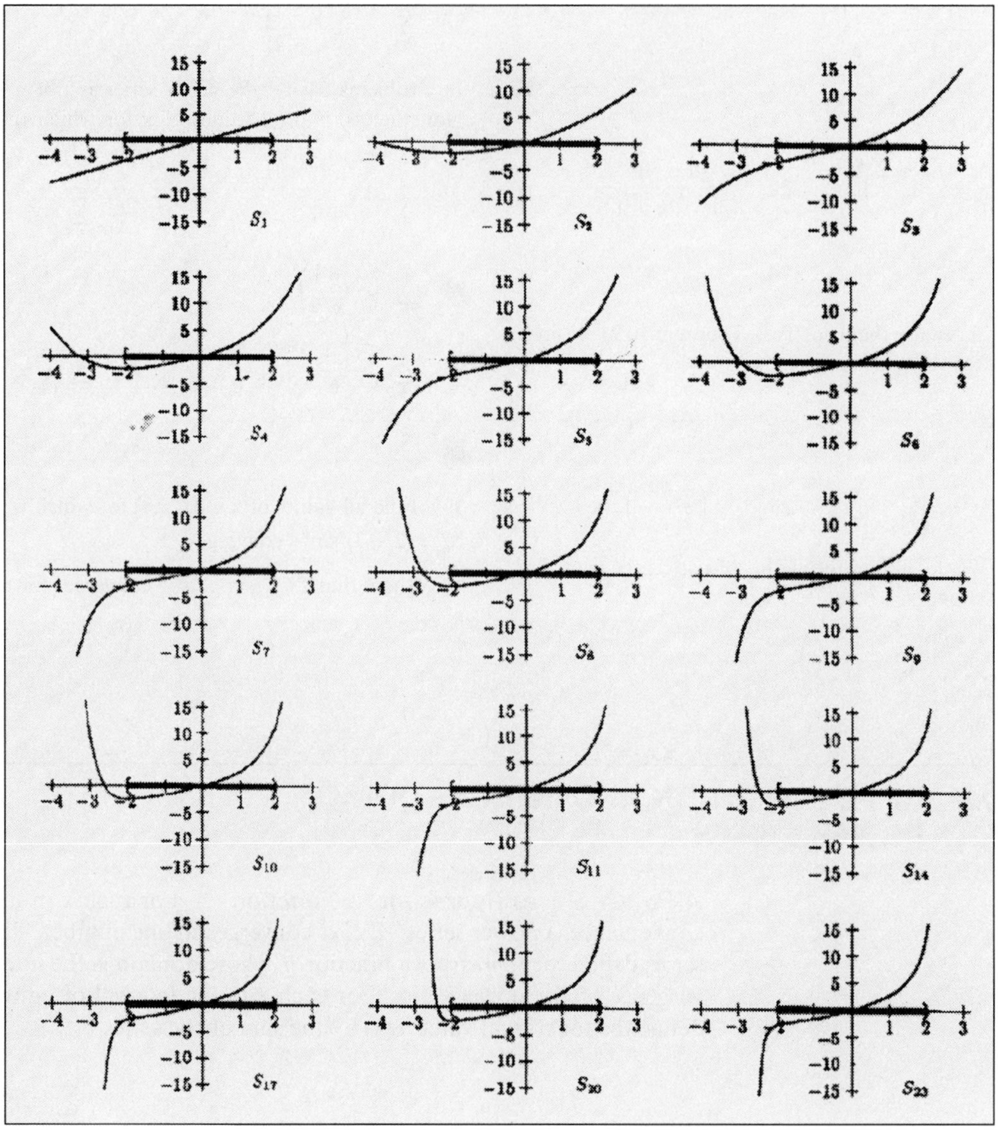

FIGURE 10.14

It is interesting to observe the behavior of the sequence of partial sums of a typical power series on its interval of convergence. Figure 10.14, obtained with the computer algebra system Mathematica, illustrates graphs of some partial sums of the power series $\Sigma_{k=1}^{\infty} x^k/k2^{k-2}$. These partial sums are graphed on the interval $[-4, 3]$, although the interval of convergence, shown in boldface in the figure, is $[-2, 2)$. Observe, for $n \geq 3$, that even though the graphs are fluctuating above and below the x-axis to the left of -2, there is very little change in the behavior of the partial sums *at* $x = -2$. Since the power series converges at $x = -2$, the function $f(x) = \Sigma_{k=1}^{\infty} x^k/k2^{k-2}$ is continuous from the right at -2 and $f(-2)$ is seen to be in the "ballpark" of -3 (in fact, $S_{23}(-2) \approx -2.84766$). Since the series diverges at $x = 2$, f is not continuous from the left. Indeed, it appears that $f(x) \to \infty$ as $x \to 2^-$.

Differentiation of a Power Series

THEOREM 10.23 Differentiation of a Power Series

If $f(x) = \Sigma_{k=0}^{\infty} c_k x^k$ converges on an interval $(-r, r)$ for which the radius of convergence is either positive or ∞, then f is differentiable at each x in $(-r, r)$ and

$$f'(x) = \sum_{k=1}^{\infty} kc_k x^{k-1} \tag{2}$$

The result in (2) simply states that a power series can be differentiated term-by-term as we would a polynomial:

$$\begin{aligned} f'(x) &= \frac{d}{dx}c_0 + \frac{d}{dx}c_1 x + \frac{d}{dx}c_2 x^2 + \frac{d}{dx}c_3 x^3 + \cdots + \frac{d}{dx}c_n x^n + \cdots \\ &= c_1 + 2c_2 x + 3c_3 x^2 + \cdots + nc_n x^{n-1} + \cdots = \sum_{k=1}^{\infty} kc_k x^{k-1} \end{aligned}$$

The radius of convergence of (2) is the same as that of $\Sigma\, c_k x^k$. Thus, by applying Theorem 10.23 to f' defined in (2), we can say f' is differentiable at each x in $(-r, r)$; that is

$$f''(x) = 2c_2 + 3 \cdot 2c_3 x + \cdots + n(n-1)c_x x^{n-2} + \cdots = \sum_{k=2}^{\infty} k(k-1)c_k x^{k-2}$$

Continuing in this manner, it follows that:

A function f represented by a power series on $(-r, r)$, $r > 0$, possesses derivatives of all orders in the interval.

Integration of a Power Series As in (2), the process of integration of a power series may also be carried out term-by-term:

$$\begin{aligned} \int f(x)\,dx &= \int c_0\,dx + \int c_1 x\,dx + \int c_2 x^2\,dx + \cdots + \int c_n x^n\,dx + \cdots \\ &= \left(c_0 x + \frac{c_1}{2}x^2 + \frac{c_2}{3}x^3 + \cdots + \frac{c_n}{n+1}x^{n+1} + \cdots\right) + C \\ &= \sum_{k=0}^{\infty} \frac{c_k}{k+1}x^{k+1} + C \end{aligned}$$

This is summarized in the next theorem.

THEOREM 10.24 Integration of a Power Series

If $f(x) = \Sigma_{k=0}^{\infty} c_k x^k$ converges on an interval $(-r, r)$ for which the radius of convergence is either positive or ∞, then

$$\int f(x)\,dx = \sum_{k=0}^{\infty} \frac{c_k}{k+1}x^{k+1} + C \tag{3}$$

The radius of convergence of (3) is again the same as that of $\Sigma\, c_k x^k$.
For definite integrals, (3) implies that

$$\int_a^b f(x)\,dx = \sum_{k=0}^{\infty} c_k\left(\int_a^b x^k\,dx\right) = \sum_{k=0}^{\infty} c_k \frac{b^{k+1} - a^{k+1}}{k+1}$$

for any numbers a and b in $(-r, r)$.

EXAMPLE 1 Find a power series representation for the function

$$f(x) = \frac{1}{1+x}$$

Solution Recall that a geometric series converges to $a/(1-r)$ if $|r| < 1$. Identifying $a = 1$ and $r = -x$, we see that

$$\frac{1}{1+x} = 1 - x + x^2 - x^3 + \cdots + (-1)^n x^n + \cdots = \sum_{k=0}^{\infty} (-1)^k x^k \tag{4}$$

for any x in $(-1, 1)$. □

EXAMPLE 2 Term-by-term differentiation of (4) yields a power series representation for $1/(1+x)^2$ on $(-1, 1)$:

$$\frac{-1}{(1+x)^2} = -1 + 2x - 3x^2 + \cdots + (-1)^n nx^{n-1} + \cdots$$

or

$$\frac{1}{(1+x)^2} = 1 - 2x + 3x^2 + \cdots + (-1)^{n+1} nx^{n-1} + \cdots$$

$$= \sum_{k=1}^{\infty} (-1)^{k+1} kx^{k-1}$$

□

EXAMPLE 3 Find a power series representation for $\ln(1+x)$ on $(-1, 1)$.

Solution Substituting $x = t$ in (4) gives

$$\frac{1}{1+t} = 1 - t + t^2 - t^3 + \cdots + (-1)^n t^n + \cdots$$

Thus, for any x in $(-1, 1)$,

$$\int_0^x \frac{dt}{1+t} = \int_0^x dt - \int_0^x t\,dt + \int_0^x t^2\,dt - \cdots + (-1)^n \int_0^x t^n\,dt + \cdots$$

$$= x - \frac{x^2}{2} + \frac{x^3}{3} - \cdots + (-1)^n \frac{x^{n+1}}{n+1} + \cdots$$

But, $\int_0^x dt/(1+t) = \ln(1+t)]_0^x = \ln(1+x)$, and so

$$\ln(1+x) = x - \frac{x^2}{2} + \frac{x^3}{3} - \cdots + (-1)^n \frac{x^{n+1}}{n+1} + \cdots = \sum_{k=0}^{\infty} \frac{(-1)^k}{k+1} x^{k+1} \tag{5}$$

□

EXAMPLE 4 Approximate ln(1.2) to four decimal places.

Solution Substituting $x = 0.2$ in (5) gives

$$\ln(1.2) = 0.2 - \frac{(0.2)^2}{2} + \frac{(0.2)^3}{3} - \frac{(0.2)^4}{4} + \frac{(0.2)^5}{5} - \frac{(0.2)^6}{6} + \cdots \tag{6}$$

$$= 0.2 - 0.02 + 0.00267 - 0.0004 + 0.000064 - 0.00001067 + \cdots$$

$$\approx 0.1823 \tag{7}$$

□

If the sum of the series (6) in Example 4 is denoted by S, then we know from Theorem 10.18 that $|S_n - S| \le a_{n+1}$. The number given in (7) is accurate to four decimal places, since, for the fifth partial sum, $|S_5 - S| \le 0.00001067 < 0.00005$.

▲ **Remark** It is interesting to note that if the interval of convergence of a power series representation of a function f is the open interval $(-r, r)$, then the power series representation for $\int_0^x f(t)\,dt$ may converge at one or both endpoints of the interval. You should check that the series in (5) diverges $x = -1$ but converges at $x = 1$. At the latter value we discover that the sum of the alternating harmonic series is given by:

$$\ln 2 = 1 - \frac{1}{2} + \frac{1}{3} - \frac{1}{4} + \cdots$$

EXERCISES 10.8

Answers to odd-numbered problems begin on page A-93.

In Problems 1–14 find a power series representation, centered at 0, for the given function. Give the interval of convergence.

1. $f(x) = \dfrac{1}{1-x}$
2. $f(x) = \dfrac{1}{(1-x)^2}$
3. $f(x) = \dfrac{1}{5+3x}$
4. $f(x) = \dfrac{6}{2-x}$
5. $f(x) = \dfrac{1}{(1-x)^3}$
6. $f(x) = \dfrac{x^2}{(1+x)^3}$
7. $f(x) = \ln(4+x)$
8. $f(x) = \ln(1+2x)$
9. $f(x) = \dfrac{1}{1+x^2}$
10. $f(x) = \dfrac{x}{1+x^2}$
11. $f(x) = \tan^{-1}x$
12. $f(x) = \ln(1+x^2)$
13. $f(x) = \displaystyle\int_0^x \ln(1+t^2)\,dt$
14. $f(x) = \displaystyle\int_0^x \tan^{-1}t\,dt$

In Problems 15 and 16 find the domain of the given function.

15. $f(x) = \dfrac{x}{3} - \dfrac{x^2}{2\cdot 3^2} + \dfrac{x^3}{3\cdot 3^3} - \dfrac{x^4}{4\cdot 3^4} + \cdots$

16. $f(x) = 1 + 2x + \dfrac{4x^2}{1 \cdot 2} + \dfrac{8x^3}{1 \cdot 2 \cdot 3} + \cdots$

In Problems 17–22 use infinite series to approximate the given quantity to four decimal places.

17. $\ln(1.1)$

18. $\tan^{-1}(0.2)$

19. $\displaystyle\int_0^{1/2} \frac{dx}{1 + x^3}$

20. $\displaystyle\int_0^{1/3} \frac{x}{1 + x^4}\, dx$

21. $\displaystyle\int_0^{0.3} x \tan^{-1} x\, dx$

22. $\displaystyle\int_0^{1/2} \tan^{-1} x^2\, dx$

23. If $f(x) = \sum_{k=0}^{\infty} \dfrac{(-1)^k}{(2k + 1)!} x^{2k+1}$ for $-\infty < x < \infty$, show that $f''(x) + f(x) = 0$.

24. (a) Show that if $f(x) = \sum_{k=0}^{\infty} \dfrac{x^k}{k!}$ for $-\infty < x < \infty$, then $f'(x) = f(x)$.

(b) Show that

$$e^x = \sum_{k=0}^{\infty} \frac{x^k}{k!}$$

by solving the differential equation in part (**a**) by separation of variables. [*Hint:* $f(0) = 1$.]

In Problems 25 and 26 use the result of Problem 24(**b**) to find a power series representation of the given function.

25. $f(x) = e^{-x}$

26. $f(x) = e^{x/5}$

In Problems 27–30 use the result of Problem 24(**b**) to approximate the given quantity to four decimal places.

27. $\dfrac{1}{e}$

28. $e^{-1/2}$

29. $\displaystyle\int_0^{0.2} e^{-x^2}\, dx$

30. $\displaystyle\int_0^{1} e^{-x^2/2}\, dx$

31. Use Problem 11 to show that

$$\frac{\pi}{4} = 1 - \frac{1}{3} + \frac{1}{5} - \frac{1}{7} + \cdots$$

32. The series in Problem 31 is known to converge very slowly. Show this by finding the smallest positive integer n so that S_n approximates $\pi/4$ to four decimal places.

10.9 TAYLOR SERIES

Suppose we are given a power series $\Sigma_{k=0}^{\infty} c_k(x - a)^k$ centered at a that is convergent on an interval $(a - r, a + r)$, $r > 0$. Then, as we saw in the preceding section, the power series *defines* or *represents* a function f on the interval $(a, -r, a + r)$. Recall that for each x in the interval, the functional value $f(x)$ is defined by the sum of the series: $f(x) = \Sigma_{k=0}^{\infty} c_k(x - a)^k$. The function f defined in this manner is continuous and possesses derivatives of all orders on the interval $(a - r, a + r)$. But we are really more interested in what could be called the reverse problem:

> *Suppose we are given a function f that possess derivatives of all orders. Can we find a power series that represents it?*

In other words, can we **expand** an infinitely differentiable function, such as $f(x) = \sin x$, $f(x) = \cos x$, or $f(x) = e^x$, into a power series that converges to the correct functional value $f(x)$?

Taylor and Maclaurin Series for a Function f Before answering the last question, let us return to the concept of a power series $\Sigma_{k=0}^{\infty} c_k(x - a)^k$ representing a function f on an interval $(a - r, a + r)$, $r > 0$. Although it certainly is not apparent, there is a relationship between the coefficients c_k of the series and the derivatives of f. Repeated differentiation of

$$f(x) = c_0 + c_1(x - a) + c_2(x - a)^2 + c_3(x - a)^3 + \cdots + c_n(x - a)^n + \cdots \tag{1}$$

yields

$$f'(x) = c_1 + 2c_2(x - a) + 3c_3(x - a)^2 + \cdots \tag{2}$$
$$f''(x) = 2c_2 + 3 \cdot 2c_3(x - a) + \cdots \tag{3}$$
$$f'''(x) = 3 \cdot 2 \cdot 1c_3 + \cdots \tag{4}$$

and so on. By evaluating (1), (2), (3), and (4) at $x = a$, we obtain

$$f(a) = c_0, \quad f'(a) = 1!c_1, \quad f''(a) = 2!c_2, \quad \text{and} \quad f'''(a) = 3!c_3$$

respectively. In general, $f^{(n)}(a) = n!c_n$, or

$$c_n = \frac{f^{(n)}(a)}{n!}, \qquad n \geq 0 \tag{5}$$

When $n = 0$ we interpret the zeroth derivative as $f(a)$ and $0! = 1$. Substituting (5) in (1) yields

$$f(x) = \sum_{k=0}^{\infty} \frac{f^{(k)}(a)}{k!}(x - a)^k \tag{6}$$

which is valid for all values for x in $(a - r, a + r)$, $r > 0$. This series is called the **Taylor series** for f centered at a.* A Taylor series with center $a = 0$,

$$f(x) = \sum_{k=0}^{\infty} \frac{f^{(k)}(0)}{k!} x^k \tag{7}$$

is called the **Maclaurin series** for f.†

The question posed above can now be rephrased as:

Can we expand an infinitely differentiable function f into a Taylor series (6)?

It would appear that the answer is yes—by simply calculating the coefficients as dictated by (5). Unfortunately, the concept of expanding a given infinitely differentiable function f in a Taylor series is not that simple. We must look upon such a power series obtained from (5) and (6) as a formal result or a series that is simply *generated* by the function f. We do not know whether the series generated in this manner converges or, even if it does, whether it converges to $f(x)$.

EXAMPLE 1 Find the Taylor series centered at $a = 1$ generated by

$$f(x) = \ln x$$

Solution We have

$$f(x) = \ln x \qquad\qquad f(1) = 0$$

---

*Named in honor of the English mathematician Brook Taylor (1685–1731), who published this result in 1715. The formula was discovered by Johann Bernoulli about 20 years earlier.

†Named after the Scottish mathematician and former student of Isaac Newton, Colin Maclaurin (1698–1746). It is not clear why Maclaurin's name is associated with this series.

$$f'(x) = \frac{1}{x} \qquad f'(1) = 1$$

$$f''(x) = -\frac{1}{x^2} \qquad f''(1) = -1$$

$$f'''(x) = \frac{1 \cdot 2}{x^3} \qquad f'''(1) = 2!$$

$$\vdots \qquad \vdots$$

$$f^{(n)}(x) = (-1)^{n-1}\frac{(n-1)!}{x^n} \qquad f^{(n)}(1) = (-1)^{n-1}(n-1)!$$

Thus, (5) and (6) yield

$$(x-1) - \frac{1}{2}(x-1)^2 + \frac{1}{3}(x-1)^3 - \cdots = \sum_{k=1}^{\infty} \frac{(-1)^{k-1}}{k}(x-1)^k \tag{8}$$

With the aid of the Ratio Test, it is found that this series converges for all values of x in the interval $(0, 2]$. □

Taylor's Theorem It is apparent from (5) that a function f can have a Taylor series centered at a only if the function possesses finite derivatives of all orders at a. Thus, for example, $f(x) = \ln x$ does not possess a Maclaurin series, since f and all its derivatives are not defined at 0. Moreover, it is important to note that even if a function f possesses derivatives of all orders and generates a Taylor series convergent on some interval, it is possible that the series does not converge to $f(x)$ at every x in the interval. (See Problem 47 in Exercises 10.9.) If the series generated by f actually converges to $f(x)$ on an interval, then we say that the series *represents* the given function on the interval. At this point it has not been established that the series given in (8) represents $\ln x$ on the interval $(0, 2]$. This question can be resolved by **Taylor's Theorem**.

THEOREM 10.25 Taylor's Theorem

Let f be a function such that $f^{(n+1)}(x)$ exists for every x in the interval $(a - r, a + r)$. Then for all x in the interval,

$$f(x) = P_n(x) + R_n(x)$$

where $$P_n(x) = f(a) + \frac{f'(a)}{1!}(x-a) + \cdots + \frac{f^{(n)}(a)}{n!}(x-a)^n \tag{9}$$

is called the nth-degree **Taylor polynomial of f at a**, and

$$R_n(x) = \frac{f^{(n+1)}(c)}{(n+1)!}(x-a)^{n+1} \tag{10}$$

is called the **Lagrange form of the remainder**.* The number c is between a and x.

*There are several forms of the remainder. This form is due to the French mathematician, Joseph Louis Lagrange (1736–1813).

Since the proof of this theorem would deflect us from the main thrust of our discussion, it is given in Appendix III. The importance of Theorem 10.25 lies in the fact that the $P_n(x)$ are the partial sums of the Taylor series and that

$$P_n(x) = f(x) - R_n(x)$$

Hence, from

$$\lim_{n\to\infty} P_n(x) = f(x) - \lim_{n\to\infty} R_n(x)$$

we see that if $R_n(x) \to 0$ as $n \to \infty$, then the sequence of partial sums converges to $f(x)$. We summarize the result.

THEOREM 10.26 Convergence of Taylor Series

If f has derivatives of all orders at every x in the interval $(a - r, a + r)$ and if $\lim_{n\to\infty} R_n(x) = 0$ for every x in the interval, then

$$f(x) = \sum_{k=0}^{\infty} \frac{f^{(k)}(a)}{k!}(x - a)^k$$

In practice, the proof that the remainder R_n approaches zero as $n \to \infty$ often depends on the fact that

$$\lim_{n\to\infty} \frac{|x|^n}{n!} = 0 \tag{11}$$

This latter result follows from applying Theorem 10.6 to the series $\Sigma_{k=0}^{\infty} x^k/k!$, which is known to be absolutely convergent for all real numbers. (See Example 3 in Section 10.7.)

EXAMPLE 2 Represent $f(x) = \cos x$ by Maclaurin series.

Solution We have

$$\begin{aligned} f(x) &= \cos x & f(0) &= 1 \\ f'(x) &= -\sin x & f'(0) &= 0 \\ f''(x) &= -\cos x & f''(0) &= -1 \\ f'''(x) &= \sin x & f'''(0) &= 0 \end{aligned}$$

and so on. From (7) we obtain

$$1 - \frac{x^2}{2!} + \frac{x^4}{4!} - \frac{x^6}{6!} + \cdots = \sum_{k=0}^{\infty} \frac{(-1)^k}{(2k)!} x^{2k} \tag{12}$$

The Ratio Test shows that (12) converges absolutely for all real values of x. To prove that $\cos x$ is indeed represented by the series (12), we must show that $\lim_{n\to\infty} R_n(x) = 0$.

To this end, we note that the derivatives of f satisfy

$$|f^{(n+1)}(x)| = \begin{cases} |\sin x|, & n \text{ even} \\ |\cos x|, & n \text{ odd} \end{cases}$$

In either case, $|f^{(n+1)}(c)| \leq 1$ for any real number c, and so

$$|R_n(x)| = \frac{|f^{(n+1)}(c)|}{(n+1)!}|x|^{n+1} \leq \frac{|x|^{n+1}}{(n+1)!}$$

For any fixed but arbitrary choice of x, $\lim_{n\to\infty}|x|^{n+1}/(n+1)! = 0$ by (11). But $\lim_{n\to\infty}|R_n(x)| = 0$ implies that $\lim_{n\to\infty} R_n(x) = 0$. Therefore,

$$\cos x = 1 - \frac{x^2}{2!} + \frac{x^4}{4!} - \frac{x^6}{6!} + \cdots + (-1)^n \frac{x^{2n}}{(2n)!} + \cdots$$

is a valid representation of $\cos x$ for every real number x. □

EXAMPLE 3 Represent $f(x) = \sin x$ in a Taylor series with center $a = \pi/3$.

Solution We have

$$f(x) = \sin x \qquad f\left(\frac{\pi}{3}\right) = \frac{\sqrt{3}}{2}$$

$$f'(x) = \cos x \qquad f'\left(\frac{\pi}{3}\right) = \frac{1}{2}$$

$$f''(x) = -\sin x \qquad f''\left(\frac{\pi}{3}\right) = -\frac{\sqrt{3}}{2}$$

$$f'''(x) = -\cos x \qquad f'''\left(\frac{\pi}{3}\right) = -\frac{1}{2}$$

and so on. Hence, the Taylor series at $\pi/3$ that corresponds to $\sin x$ is

$$\frac{\sqrt{3}}{2} + \frac{1}{2 \cdot 1!}\left(x - \frac{\pi}{3}\right) - \frac{\sqrt{3}}{2 \cdot 2!}\left(x - \frac{\pi}{3}\right)^2 - \frac{1}{2 \cdot 3!}\left(x - \frac{\pi}{3}\right)^3 + \cdots \qquad (13)$$

Again, from the Ratio Test it follows that (13) converges absolutely for all real values of x. To show that

$$\sin x = \frac{\sqrt{3}}{2} + \frac{1}{2 \cdot 1!}\left(x - \frac{\pi}{3}\right) - \frac{\sqrt{3}}{2 \cdot 2!}\left(x - \frac{\pi}{3}\right)^2 - \frac{1}{2 \cdot 3!}\left(x - \frac{\pi}{3}\right)^3 + \cdots$$

for every real x, we note that, as in the preceding example, $|f^{(n+1)}(c)| \leq 1$. This implies that $|R_n(x)| \leq |x - \pi/3|^{n+1}/(n+1)!$ from which we see, with the help of (11), that $\lim_{n\to\infty} R_n(x) = 0$. □

EXAMPLE 4 Prove that the series (8) represents $f(x) = \ln x$ on the interval $(0, 2]$.

Solution For $f(x) = \ln x$, the nth derivative is given by

$$f^{(n)}(x) = \frac{(-1)^{n-1}(n-1)!}{x^n}$$

Therefore $$f^{(n+1)}(c) = \frac{(-1)^n n!}{c^{n+1}}$$

and $$|R_n(x)| = \left|\frac{(-1)^n n!}{c^{n+1}} \cdot \frac{(x-1)^{n+1}}{(n+1)!}\right| = \frac{1}{n+1}\left|\frac{x-1}{c}\right|^{n+1}$$

where c is some number in $(0, 2]$ between 1 and x.

If $1 \le x \le 2$, then $0 < x - 1 \le 1$. Since $1 < c < x$, we must have $0 < x - 1 \le 1 < c$ and, consequently, $(x-1)/c < 1$. Hence,

$$|R_n(x)| \le \frac{1}{n+1} \quad \text{and} \quad \lim_{n\to\infty} R_n(x) = 0$$

In the case where $0 < x < 1$, we can also show that $\lim_{n\to\infty} R_n(x) = 0$. We omit the proof.* Hence,

$$\ln x = (x-1) - \frac{1}{2}(x-1)^2 + \frac{1}{3}(x-1)^3 - \cdots$$

for all values of x in the interval $(0, 2]$. □

We summarize some important Maclaurin series:

$$e^x = 1 + x + \frac{x^2}{2!} + \frac{x^3}{3!} + \cdots = \sum_{k=0}^{\infty} \frac{x^k}{k!}, \qquad -\infty < x < \infty \tag{14}$$

$$\cos x = 1 - \frac{x^2}{2!} + \frac{x^4}{4!} - \frac{x^6}{6!} + \cdots = \sum_{k=0}^{\infty} \frac{(-1)^k}{(2k)!} x^{2k}, \qquad -\infty < x < \infty \tag{15}$$

$$\sin x = x - \frac{x^3}{3!} + \frac{x^5}{5!} - \frac{x^7}{7!} + \cdots = \sum_{k=0}^{\infty} \frac{(-1)^k}{(2k+1)!} x^{2k+1}, \qquad -\infty < x < \infty \tag{16}$$

$$\cosh x = 1 + \frac{x^2}{2!} + \frac{x^4}{4!} + \frac{x^6}{6!} + \cdots = \sum_{k=0}^{\infty} \frac{x^{2k}}{(2k)!}, \qquad -\infty < x < \infty \tag{17}$$

$$\sinh x = x + \frac{x^3}{3!} + \frac{x^5}{5!} + \frac{x^7}{7!} + \cdots = \sum_{k=0}^{\infty} \frac{x^{2k+1}}{(2k+1)!}, \qquad -\infty < x < \infty \tag{18}$$

$$\ln(1+x) = x - \frac{x^2}{2} + \frac{x^3}{3} - \frac{x^4}{4} + \cdots = \sum_{k=0}^{\infty} \frac{(-1)^k}{(k+1)} x^{k+1}, \qquad -1 < x \le 1 \tag{19}$$

You are asked to demonstrate the validity of the representations (14), (16), (17), and (18) as exercises.

Some Graphs of Taylor Polynomials In Example 2 we saw that the Taylor series of $f(x) = \cos x$ at $a = 0$ represents the function for all x, since $\lim_{n\to\infty} R_n(x) = 0$. It is always of interest to see graphically how partial sums of the Taylor series, which are the Taylor polynomials defined in (9), converge to the function. In Figure 10.15 the graphs of the Taylor polynomials $P_0(x)$, $P_2(x)$,

*This part of the proof is usually based on an integral form of the remainder $R_n(x)$ that we shall not consider.

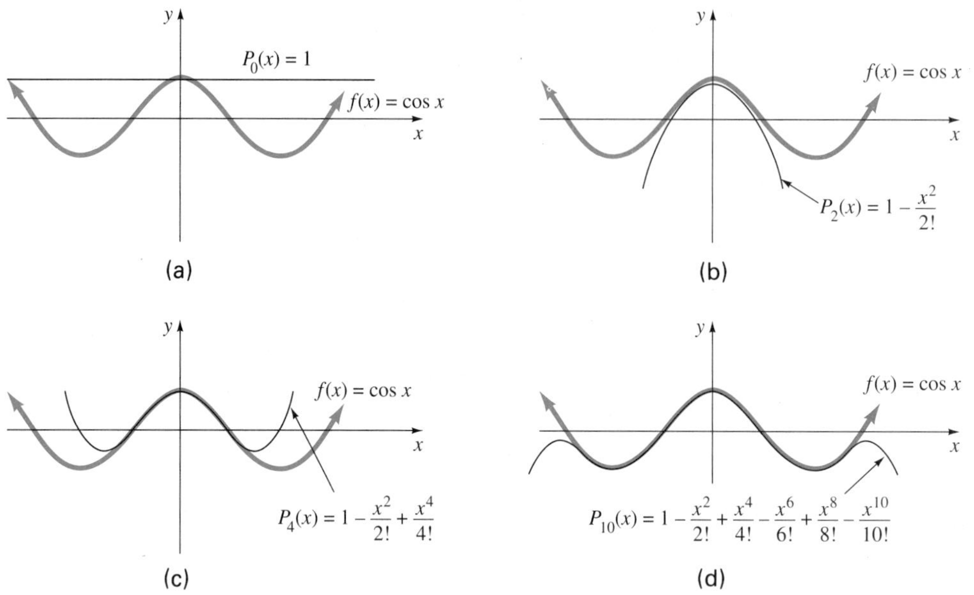

x	$P_2(x)$	$P_4(x)$	$P_{10}(x)$	$\cos x$
$\pi/6$	0.86292	0.86605	0.86603	0.86603
$\pi/4$	0.69157	0.70743	0.70711	0.70711
$\pi/3$	0.45169	0.50180	0.50000	0.5
$\pi/2$	-0.23370	0.01997	0.00000	0

(e)

FIGURE 10.15

$P_4(x)$, and $P_{10}(x)$ at $a = 0$ are compared with the graph $f(x) = \cos x$. A comparison of numerical values is given in Figure 10.15(e).

Approximations with Taylor Polynomials When the value of x is close to the number a ($x \approx a$), the Taylor polynomial $P_n(x)$ of a function f at a can be used to approximate the functional value $f(x)$. The error in this approximation is given by

$$|f(x) - P_n(x)| = |R_n(x)|$$

EXAMPLE 5 Approximate $e^{-0.2}$ by $P_3(x)$. Determine the accuracy of the approximation.

Solution Because the value $x = -0.2$ is close to zero, we use the Taylor polynomial $P_3(x)$ of $f(x) = e^x$ at $a = 0$. It follows from $f(x) = f'(x) = f''(x) = f'''(x) = e^x$ and (9) that

$$P_3(x) = f(0) + \frac{f'(0)}{1!}x + \frac{f''(0)}{2!}x^2 + \frac{f'''(0)}{3!}x^3 = 1 + x + \frac{1}{2}x^2 + \frac{1}{6}x^3$$

and $\quad P_3(-0.2) = 1 + (-0.2) + \frac{1}{2}(-0.2)^2 + \frac{1}{6}(-0.2)^3 \approx 0.81867$

Consequently,

$$e^{-0.2} \approx 0.81867 \tag{20}$$

Now, from (10) we can write

$$|R_3(x)| = \frac{e^c}{4!}|x|^4 < \frac{|x|^4}{4!}$$

since $-0.2 < c < 0$ and $e^c < 1$. Thus,

$$|R_3(-0.2)| < \frac{|-0.2|^4}{24} < 0.0001$$

implies that the result in (20) is accurate to three decimal places. □

▲ ***Remarks*** (*i*) The Taylor series method of finding a power series for a function and then proving that the series represents the function has one big and obvious drawback. Obtaining a general expression for the nth derivative for most functions is nearly impossible. Thus, we are often limited to finding just the first few coefficients c_n. See Problems 25 and 26 in Exercises 10.9.

(*ii*) Taylor's Theorem is also called the **Generalized Mean Value Theorem.*** The case $n = 0$ reduces to the usual Mean Value Theorem given on page 194. Those who read Appendix III will see that the proof of Theorem 10.25, like the proof of Theorem 2.5, relies on the construction of a special function and Rolle's Theorem.

(*iii*) It is easy to pass over the significance of the results in (6) and (7). Suppose we wish to find the Maclaurin series for $f(x) = 1/(2 - x)$. We can, of course, use (7)—and you are asked to do so in Problem 1. On the other hand, you should also recognize, from the discussion of the preceding section, that a power series representation of f can be obtained utilizing geometric series. The point is:

The representation is unique. Thus, on its interval of convergence, a power series representing a function, regardless of how it is obtained, is the Taylor or Maclaurin series of that function.

EXERCISES 10.9

Answers to odd-numbered problems begin on page A-93.

In Problems 1–10 use (7) to find the Maclaurin series for the given function.

1. $f(x) = \dfrac{1}{2 - x}$
2. $f(x) = \dfrac{1}{1 + 5x}$
3. $f(x) = \ln(1 + x)$
4. $f(x) = \ln(1 + 2x)$
5. $f(x) = \sin x$
6. $f(x) = \cos 2x$
7. $f(x) = e^x$
8. $f(x) = e^{-x}$
9. $f(x) = \sinh x$
10. $f(x) = \cosh x$
11. Prove that the series obtained in Problem 5 represents $\sin x$ for every real value of x.

*Do not confuse this with Theorem 9.1.

12. Prove that the series obtained in Problem 7 represents e^x for every real value of x.

13. Prove that the series obtained in Problem 9 represents $\sinh x$ for every real value of x.

14. Prove that the series obtained in Problem 10 represents $\cosh x$ for every real value of x.

In Problems 15–24 use (6) to find the Taylor series for the given function centered at the indicated value of a.

15. $f(x) = \dfrac{1}{1+x}$, $a = 4$ 16. $f(x) = \sqrt{x}$, $a = 1$

17. $f(x) = \sin x$, $a = \pi/4$ 18. $f(x) = \sin x$, $a = \pi/2$

19. $f(x) = \cos x$, $a = \pi/3$ 20. $f(x) = \cos x$, $a = \pi/6$

21. $f(x) = \ln x$, $a = 2$

22. $f(x) = \ln(x + 1)$, $a = 2$

23. $f(x) = e^x$, $a = 1$ 24. $f(x) = e^{-2x}$, $a = 1/2$

In Problems 25 and 26 use (7) to find the first four nonzero terms of the Maclaurin series for the given function.

25. $f(x) = \tan x$ 26. $f(x) = \sin^{-1}x$

In Problems 27–34 use previous results or problems to find the Maclaurin series for the given function.

27. $f(x) = e^{-x^2}$ 28. $f(x) = x^2e^{-3x}$

29. $f(x) = x \cos x$ 30. $f(x) = \sin x^3$

31. $f(x) = \ln(1 - x)$ 32. $f(x) = \ln\left(\dfrac{1+x}{1-x}\right)$

33. $f(x) = \sec^2 x$ 34. $f(x) = \ln(\cos x)$

In Problems 35–38 approximate the given quantity using the Taylor polynomial $P_n(x)$ for the indicated values of n and a. Determine the accuracy of the approximation.

35. $\sin 46°$, $n = 2$, $a = \pi/4$ [*Hint:* Convert 46° to radian measure.]

36. $\cos 29°$, $n = 2$, $a = \pi/6$

37. $e^{0.3}$, $n = 4$, $a = 0$ 38. $\sinh(0.1)$, $n = 3$, $a = 0$

39. In leveling a long roadway of length L, an allowance must be made for the curvature of the earth.

(a) Use (7) to show that the first three nonzero terms of the Maclaurin series for $f(x) = \sec x$ are

$$1 + \frac{1}{2}x^2 + \frac{5}{24}x^4 + \cdots$$

(b) For small values of x, use the approximation

$$\sec x = 1 + \frac{1}{2}x^2$$

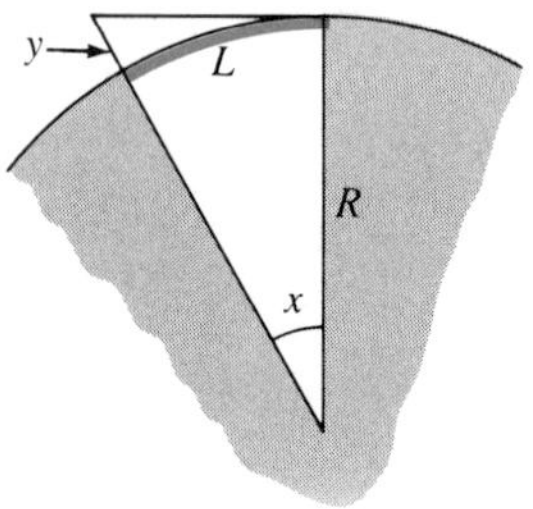

FIGURE 10.16

and Figure 10.16 to show that the leveling correction is $y = L^2/2R$, where R is the radius of the earth.

(c) Find the number of inches of leveling correction needed for 1 mile of roadway. Use $R = 4000$ miles.

40. A wave of length L is traveling left to right across water of depth d (in ft), as illustrated in Figure 10.17. We can show that the speed v of the wave is related to L and d by the function $v = \sqrt{(gL/2\pi)\tanh(2\pi d/L)}$.

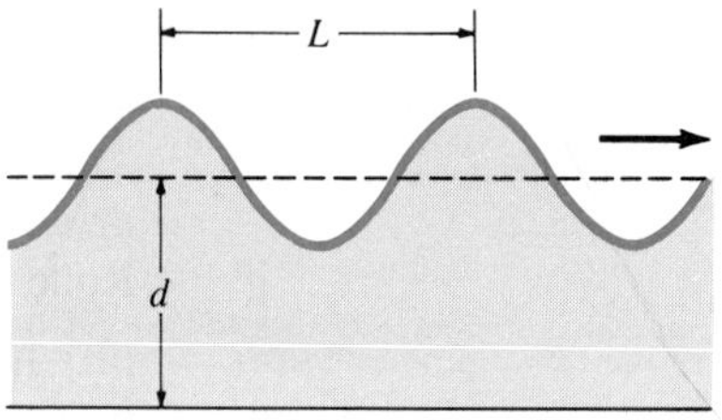

FIGURE 10.17

(a) Show that in deep water $v \approx \sqrt{gL/2\pi}$.

(b) Use (7) to find the first three nonzero terms of the Maclaurin series for $f(x) = \tanh x$. Show that when d/L is small, $v \approx \sqrt{gd}$. In other words, in shallow water the speed of a wave is independent of wave length.

41. Show that

$$\int_0^1 \sin x^2\,dx = \frac{1}{3\cdot 1!} - \frac{1}{7\cdot 3!} + \frac{1}{11\cdot 5!} - \frac{1}{15\cdot 7!} + \cdots$$

42. Show that

$$\int_0^1 \frac{\sin x}{x}\,dx = 1 - \frac{1}{3\cdot 3!} + \frac{1}{5\cdot 5!} - \frac{1}{7\cdot 7!} + \cdots$$

Calculator/Computer Problems

43. (a) The general term of the alternating series in Problem 41 is $a_n = \dfrac{1}{(2n+1)n!}$ for $n = 1, 3, 5, \ldots$. Show

that

$$a_{n+2} = \frac{(2n+1)a_n}{(2n+5)(n+1)(n+2)}$$

(b) Use the following BASIC program to evaluate $\int_0^1 \sin x^2\, dx$ accurate to six decimal places.

```
10  REM SUM OF A SERIES
20  LET N=1
30  LET A=1/3
40  LET S=1/3
50  LET A=-A*(2*N+1)/((2*N+5)*(N+1)*(N+2))
60  LET S=S+A
70  LET N=N+2
80  IF ABS(A)>5*1.0E-07 THEN GOTO 50
90  PRINT "THE SUM IS "; S
100 END
```

44. (a) The general term of the alternating series in Problem 42 is $a_n = \frac{1}{n \cdot n!}$ for $n = 1, 3, 5, \ldots$. Show that

$$a_{n+2} = \frac{na_n}{(n+1)(n+2)^2}$$

(b) Write a BASIC program to evaluate $\int_0^1 \frac{\sin x}{x}\, dx$ accurate to six decimal places.

Miscellaneous Problems

45. Use $i^2 = -1$ and (14) to derive **Euler's formula**

$$e^{i\theta} = \cos\theta + i\sin\theta$$

46. Use Euler's formula given in Problem 45 to show that

(a) $\sin\theta = \dfrac{e^{i\theta} - e^{-i\theta}}{2i}$ (b) $\cos\theta = \dfrac{e^{i\theta} + e^{-i\theta}}{2}$

47. Use (7) to find the Maclaurin series for

$$f(x) = \begin{cases} e^{-1/x^2}, & x \neq 0 \\ 0, & x = 0 \end{cases}$$

[*Hint:* $f'(0) = \lim_{\Delta x \to 0} \dfrac{f(0+\Delta x) - f(0)}{\Delta x}$. Let $t = \Delta x$.]

48. Use the Maclaurin series for $\cos x$ and a trigonometric identity to find the Maclaurin series for $\sin^2 x$.

10.10 BINOMIAL SERIES

Binomial Theorem From basic mathematics we know that

$$(1+x)^2 = 1 + 2x + x^2$$
$$(1+x)^3 = 1 + 3x + 3x^2 + x^3$$

and, in general, if m is a nonnegative integer,

$$(1+x)^m = 1 + mx + \frac{m(m-1)}{2!}x^2 + \cdots + \frac{m(m-1)\cdots(m-n+1)}{n!}x^n + \cdots + x^m \tag{1}$$

The expansion of $(1+x)^m$ in (1) is called the **Binomial Theorem**. The result in (1) inspires the following definition:

DEFINITION 10.8 Binomial Series

For any real number r, the series

$$1 + rx + \frac{r(r-1)}{2!}x^2 + \cdots + \frac{r(r-1)\cdots(r-n+1)}{n!}x^n + \cdots \tag{2}$$

is called the **binomial series**.

Note that (2) terminates only when r is a nonnegative integer. In this case (2) reduces to (1).

The Ratio Test shows that the binomial series converges if $|x| < 1$ and diverges if $|x| > 1$. Thus, on the interval $(-1, 1)$ the binomial series defines an infinitely differentiable function f. It should come as no big surprise to learn that the function represented by (2) is $f(x) = (1 + x)^r$. Since the proof of this leads to a separable differentiable equation, you can practice your skills by following the guided steps given in Problem 18 of Exercises 10.10.

THEOREM 10.27 Sum of a Binomial Series

If $|x| < 1$, then for any real number r,

$$(1 + x)^r = 1 + rx + \frac{r(r-1)}{2!}x^2 + \cdots + \frac{r(r-1)\cdots(r-n+1)}{n!}x^n + \cdots \tag{3}$$

EXAMPLE 1 Find a power series representation for $\sqrt{1 + x}$.

Solution With $r = \frac{1}{2}$ it follows from (3) that for $|x| < 1$,

$$\begin{aligned}\sqrt{1+x} &= 1 + \frac{1}{2}x + \frac{\frac{1}{2}\left(\frac{1}{2}-1\right)}{2!}x^2 + \frac{\frac{1}{2}\left(\frac{1}{2}-1\right)\left(\frac{1}{2}-2\right)}{3!}x^3 + \cdots \\ &\quad + \frac{\frac{1}{2}\left(\frac{1}{2}-1\right)\cdots\left(\frac{1}{2}-n+1\right)}{n!}x^n + \cdots \\ &= 1 + \frac{1}{2}x - \frac{1}{2^2 2!}x^2 + \frac{1\cdot 3}{2^3 3!}x^3 + \cdots \\ &\quad + (-1)^{n+1}\frac{1\cdot 3\cdot 5\cdots(2n-3)}{2^n n!}x^n + \cdots\end{aligned}$$

□

In science a binomial series is often used to find approximations.

EXAMPLE 2 In Einstein's theory of relativity, the mass of a particle moving at a velocity v relative to an observer is

$$m = \frac{m_0}{\sqrt{1 - v^2/c^2}} \tag{4}$$

where m_0 is the rest mass and c is the speed of light.

Many of the results from classical physics do not hold for particles, such as electrons, which may move close to the speed of light. Kinetic energy is no longer $K = \frac{1}{2}m_0v^2$ but is

$$K = mc^2 - m_0c^2 \tag{5}$$

If we identify $r = -\frac{1}{2}$ and $x = -v^2/c^2$ in (4), we have $|x| < 1$, since no particle can surpass the speed of light. Hence, (5) can be written:

$$\begin{aligned} K &= \frac{m_0c^2}{\sqrt{1+x}} - m_0c^2 \\ &= m_0c^2[(1+x)^{-1/2} - 1] \\ &= m_0c^2\left[\left(1 - \frac{1}{2}x + \frac{3}{8}x^2 - \frac{5}{16}x^3 + \cdots\right) - 1\right] \\ &= m_0c^2\left[\frac{1}{2}\left(\frac{v^2}{c^2}\right) + \frac{3}{8}\left(\frac{v^4}{c^4}\right) + \frac{5}{16}\left(\frac{v^6}{c^6}\right) + \cdots\right] \end{aligned} \tag{6}$$

In the case where v is very much smaller than c, terms beyond the first in (6) are negligible. This leads to the well-known result

$$K \approx m_0c^2\left[\frac{1}{2}\left(\frac{v^2}{c^2}\right)\right] = \frac{1}{2}m_0v^2$$

□

▲ ***Remark*** You probably think by now that divergent infinite series are worthless. Not quite so. Mathematicians hate to see anything go to waste. There exists a field of study called **summability** of infinite series. It turns out in this theory that some divergent series are summable. But more important is the use of divergent series in the theory of asymptotic representations of functions. It goes something like this: A divergent series of the form $a_0 + a_1/x + a_2/x^2 + \cdots$ is an **asymptotic representation** of a function $f(x)$ if

$$\lim_{n\to\infty} x^n[f(x) - S_n(x)] = 0$$

where $S_n(x)$ is the $(n + 1)$st partial sum of the divergent series. Some important functions in applied mathematics are defined in this manner.

EXERCISES 10.10

Answers to odd-numbered problems begin on page A-94.

In Problems 1–10 use (3) to find the first four terms of a power series representation of the given function. Give the radius of convergence.

1. $f(x) = \sqrt[3]{1+x}$
2. $f(x) = \sqrt{1-x}$
3. $f(x) = \sqrt{9-x}$
4. $f(x) = \dfrac{1}{\sqrt{1+5x}}$
5. $f(x) = \dfrac{1}{\sqrt{1+x^2}}$
6. $f(x) = \dfrac{x}{\sqrt[3]{1-x^2}}$
7. $f(x) = (4+x)^{3/2}$
8. $f(x) = \dfrac{x}{\sqrt{(1+x)^5}}$
9. $f(x) = \dfrac{x}{(2+x)^2}$
10. $f(x) = x^2(1-x^2)^{-3}$

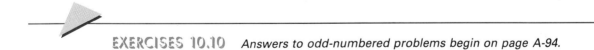

In Problems 11 and 12 explain why the error in the given approximation is less than the indicated amount. [*Hint:* Review Theorem 10.18.]

11. $(1+x)^{1/3} \approx 1 + \frac{x}{3};\ \frac{1}{9}x^2,\ x > 0$

12. $(1+x^2)^{-1/2} \approx 1 - \frac{x^2}{2} + \frac{3}{8}x^4;\ \frac{5}{16}x^6$

13. Find a power series representation for $\sin^{-1}x$ using

$$\sin^{-1}x = \int_0^x \frac{dt}{\sqrt{1-t^2}}$$

14. (a) Show that the length of one-quarter of the ellipse $x^2/a^2 + y^2/b^2 = 1$ is given by

$$a\int_0^{\pi/2} \sqrt{1-k^2\sin^2\theta}\,d\theta$$

where $k^2 = (a^2-b^2)/a^2 < 1$. This integral is called the **complete elliptic integral of the second kind.**

(b) Show that

$$a\int_0^{\pi/2} \sqrt{1-k^2\sin^2\theta}\,d\theta = a\frac{\pi}{2} - \frac{a}{2}\frac{\pi}{4}k^2 - \frac{a}{8}\frac{3\pi}{16}k^4 - \cdots$$

15. In Figure 10.18 a hanging cable is supported at points A and B and carries a uniformly distributed load (such as the floor of a bridge). If $y = (4d/l^2)x^2$ is the equation of the cable, show that its length is given by

$$s = l + \frac{8d^2}{3l} - \frac{32d^4}{5l^3} + \cdots$$

See Problem 20 in Exercises 2.11.

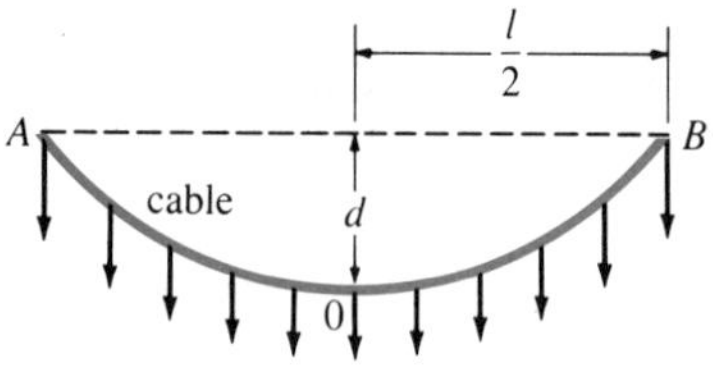

FIGURE 10.18

16. Approximate the following integrals to three decimal places.

(a) $\int_0^{0.2} \sqrt{1+x^3}\,dx$ (b) $\int_0^{1/2} \sqrt[3]{1+x^4}\,dx$

17. By the law of cosines the potential at point A in Figure 10.19 due to a unit charge at point B is $1/R = (1-2xr+r^2)^{-1/2}$, where $x = \cos\theta$. The expression $(1-2xr+r^2)^{-1/2}$ is said to be the **generating function** for the so-called **Legendre polynomials** $P_k(x)$, since

$$(1-2xr+r^2)^{-1/2} = \sum_{k=0}^{\infty} P_k(x)r^k$$

Use (3) to find $P_0(x)$, $P_1(x)$, and $P_2(x)$.

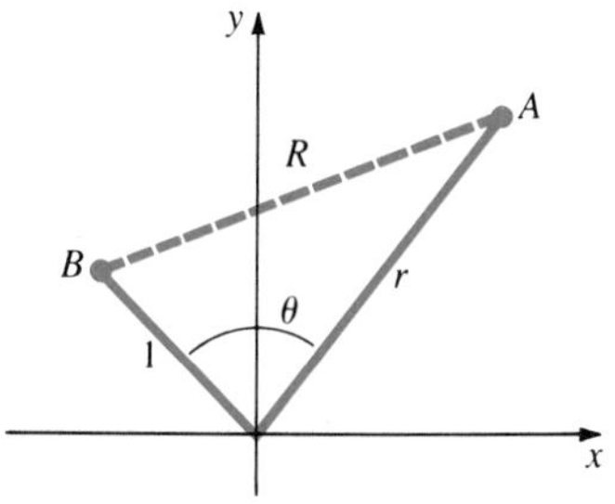

FIGURE 10.19

18. (a) Suppose

$$f(x) = 1 + rx + \frac{r(r-1)}{2!}x^2 + \cdots + \frac{r(r-1)\cdots(r-n+1)}{n!}x^n + \cdots$$

for $|x| < 1$. Find $f'(x)$ and $xf'(x)$.

(b) Show that

$$(n+1)\frac{r(r-1)\cdots(r-n)}{(n+1)!} + n\frac{r(r-1)\cdots(r-n+1)}{n!} = r\frac{r(r-1)\cdots(r-n+1)}{n!}$$

(c) Show that $f'(x) + xf'(x) = rf(x)$.

(d) Solve the differential equation $(1+x)f'(x) = rf(x)$ by separation of variables subject to $f(0) = 1$.

In Problems 19 and 20 use (3) to find a power series representation in $x - 1$ of the given function. [*Hint:* $1 + x = 2 + (x-1)$.]

19. $f(x) = \sqrt{1+x}$

20. $f(x) = (1+x)^{-2}$

CHAPTER 10 REVIEW EXERCISES

Answers begin on page A-94.

In Problems 1–26 answer true or false.

1. Every bounded sequence converges. ________

2. If a sequence is not monotonic, it is not convergent. ________

3. The sequence $\left\{\frac{10^n}{2^{n^2}}\right\}$ is not monotonic. ________

4. If $a_n \le B$ for all n and $a_{n+1}/a_n \ge 1$ for all n, then $\{a_n\}$ converges. ________

5. $\lim_{n\to\infty} \frac{|x|^n}{n!} = 0$ for every value of x. ________

6. If $\{a_n\}$ is a convergent sequence, then $\Sigma_{k=1}^{\infty} a_k$ always converges. ________

7. If $a_n \to 0$ as $n \to \infty$, then $\Sigma\, a_k$ converges. ________

8. If $\Sigma\, a_k^2$ converges, then $\Sigma\, a_k$ converges. ________

9. $\sum_{k=1}^{\infty} \frac{1}{k^p}$ converges for $p = 1.0001$. ________

10. The series $2 + \frac{2}{2} + \frac{2}{3} + \frac{2}{4} + \cdots$ diverges. ________

11. If $\Sigma\, |a_k|$ diverges, then $\Sigma\, a_k$ diverges. ________

12. If $\Sigma_{k=1}^{\infty} a_k$, $a_k > 0$, converges, then $\Sigma_{k=1}^{\infty} (-1)^{k+1} a_k$ converges. ________

13. If $\sum_{k=1}^{\infty} (-1)^{k+1} a_k$ converges absolutely, then $\sum_{k=1}^{\infty} (-1)^{k+1} \frac{a_k}{k}$ converges. ________

14. If $\Sigma\, b_k$ converges and $a_k \ge b_k$ for every positive integer k, then $\Sigma\, a_k$ converges. ________

15. If $\lim_{n\to\infty} \left|\frac{a_{n+1}}{a_n}\right| = 1$, then $\Sigma\, a_k$ converges absolutely. ______

16. Every power series has a nonzero radius of convergence. ________

17. A power series converges absolutely at every value of x in its interval of convergence. ________

18. A power series $\Sigma_{k=0}^{\infty} c_k x^k$ represents an infinitely differentiable function on an interval $(-r, r)$ of convergence. ________

19. A power series $\Sigma_{k=0}^{\infty} c_k x^k$ converges for $-1 < x < 1$ and is convergent at $x = 1$. The series must also converge at $x = -1$. ________

20. If $\Sigma\, a_k x^k$, $a_k > 0$, has the interval of convergence $[-r, r)$, the series converges conditionally, but not absolutely, at $x = -r$. ________

21. Since $\int_0^{\infty} e^{-x}\, dx = 1$, the series $\Sigma_{k=0}^{\infty} e^{-k}$ also converges to 1. ________

22. The sequence $\left\{\frac{(-1)^n n}{n+1}\right\}$ converges. ________

23. The series

$$1 - \frac{1}{2^2} + \frac{1}{3^2} + \frac{1}{4^2} - \frac{1}{5^2} - \frac{1}{6^2} + + + - - - \cdots$$

converges. ________

24. $f(x) = \ln x$ cannot be represented by a Maclaurin series. ________

25. $f(x) = \sum_{k=0}^{\infty} \frac{f^{(k)}(a)}{k!}(x-a)^k$ on an interval if $\lim_{n\to\infty} R_n(x) = 0$. ________

26. The interval of convergence of the power series $x - \frac{x^2}{2} + \frac{x^3}{3} - \frac{x^4}{4} + \cdots$ is $(-1, 1)$. ________

In Problems 27–34 fill in the blanks.

27. If $\{a_n\}$ converges to 4 and $\{b_n\}$ converges to 5, then $\{a_n b_n\}$ converges to ________, $\{a_n + b_n\}$ converges to ________, $\{a_n/b_n\}$ converges to ________, and $\{a_n^2\}$ converges to ________.

28. To approximate the sum of the series $\sum_{k=1}^{\infty} \frac{(-1)^{k+1}}{10^k}$ to four decimal places, we need only use the ________th partial sum.

29. The sum of the series $\sum_{k=0}^{\infty} 4\left(\frac{2}{3}\right)^k$ is ________.

30. If $\Sigma\, a_k$ converges and $\Sigma\, b_k$ diverges, then $\Sigma\,(a_k + b_k)$ ________.

31. The series $\Sigma_{k=1}^{\infty} [\tan^{-1} k - \tan^{-1}(k+1)]$ converges to ________.

32. The power series $\sum_{k=0}^{\infty} \frac{x^k}{k!}$ represents the function $f(x) =$ ________ for all x.

33. The binomial series representation of $f(x) = (4 + x)^{1/2}$ has the radius of convergence ________.

34. The geometric series $\sum_{k=1}^{\infty} \left(\frac{5}{x}\right)^{k-1}$ converges for the following values of x: ________.

In Problems 35–44 determine whether the given series converges or diverges.

35. $\sum_{k=1}^{\infty} \frac{k}{(k^2+1)^2}$

36. $\sum_{k=1}^{\infty} \frac{1}{1+e^{-k}}$

37. $\sum_{k=1}^{\infty} \frac{k!}{e^{k^2}}$

38. $\sum_{k=1}^{\infty} \frac{1}{3k^2+4k+6}$

39. $\sum_{k=1}^{\infty} \frac{\sqrt{k}\ln k}{k^4+4}$

40. $\sum_{k=1}^{\infty} \frac{\sin k}{k^{3/2}}$

41. $\sum_{k=2}^{\infty} \frac{k}{\sqrt[3]{k^6-4k}}$

42. $\sum_{k=2}^{\infty} \frac{1}{k\sqrt{\ln k}}$

43. $\sum_{k=1}^{\infty} \frac{1+(-1)^k}{\sqrt{k}}$

44. $\sum_{k=1}^{\infty} \frac{(k^2)!}{(k!)^2}$

In Problems 45 and 46 find the sum of the given convergent series.

45. $\sum_{k=1}^{\infty} \frac{(-1)^{k-1}+3}{(1.01)^{k-1}}$

46. $\sum_{k=1}^{\infty} \frac{1}{k^2+11k+30}$

In Problems 47–50 find the interval of convergence of the given power series.

47. $\sum_{k=1}^{\infty} \frac{3^k}{k^3} x^k$

48. $\sum_{k=1}^{\infty} \frac{k}{4^k}(2x-1)^k$

49. $\sum_{k=1}^{\infty} k!(x+5)^k$

50. $\sum_{k=2}^{\infty} \frac{(2x)^k}{\ln k}$

51. Find the radius of convergence for

$$\sum_{k=1}^{\infty} \frac{2 \cdot 5 \cdot 8 \cdots (3k-1)}{3 \cdot 7 \cdot 11 \cdots (4k-1)} x^k$$

52. Find the values of x for which $\Sigma_{k=1}^{\infty} (\cos x)^k$ converges.

In Problems 53–56 find, by any method, the first three nonzero terms of the Maclaurin series for the given function.

53. $f(x) = \frac{1}{\sqrt[3]{1+x^5}}$

54. $f(x) = \frac{x}{2-x}$

55. $f(x) = \sin x \cos x$

56. $f(x) = \int_0^x e^{t^2}\,dt$

57. Find the Taylor series for $f(x) = \cos x$ with center $a = \pi/2$.

58. Prove that the series in Problem 57 represents the function by showing that $R_n(x) \to 0$ as $n \to \infty$.

59. A large convention of free-spending mathematicians from out of town contributes \$3 million to the economy of the city of San Francisco. It is estimated that each resident of the city spends $\frac{2}{3}$ of his or her income in the city. Thus, of the amount of the money brought in by the convention, $3(\frac{2}{3}) = \$2$ million, is spent by the people of San Francisco in the city. Of this last amount, $\frac{2}{3}$ is spent in the city, and so on. How much, in the long run, do the residents of San Francisco spend in their city as a result of the convention?

60. If P dollars are invested at an annual rate of interest r compounded annually, the return S after m years is $S = (1+r)^m$. The **Rule of 70**, which is often used by loan officers and stock analysts, says that the time required to double an investment earning an annual rate of interest r is approximately $70/100r$ years. For example, money invested at an annual rate of 5% takes approximately $70/100(0.05) = 14$ years to double.

(a) Show that the true doubling time is $\ln 2/\ln(1+r)$.

(b) Use the Maclaurin series for $\ln(1+r)$ to derive the Rule of 70.

(c) Use the first three terms of the Maclaurin series for $\ln(1+r)$ to approximate that interest rate for which the Rule of 70 gives the true doubling time.

11

ANALYTIC GEOMETRY IN THE PLANE

INTRODUCTION

To the ancient Greek geometers such as Euclid (c. 300 B.C.) and Archimedes (c. 287–212 B.C.), a **conic section** was, as shown in the accompanying figure, a curve of intersection of a plane and a double-napped cone. However, in this chapter we shall see how the **parabola, ellipse**, and **hyperbola** are defined by means of distance. Using a Cartesian coordinate system and the distance formula, we obtain equations for the conics.

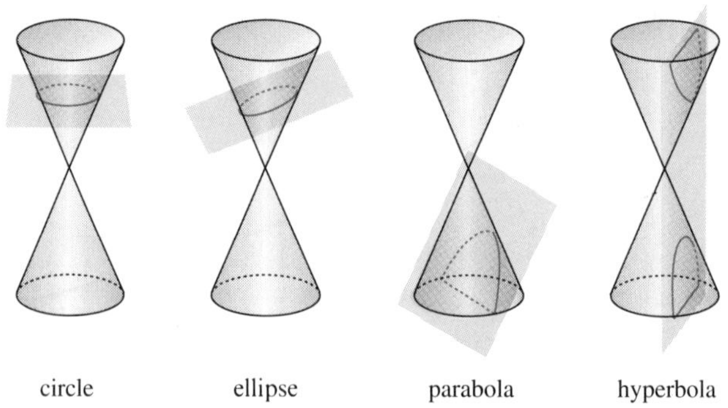

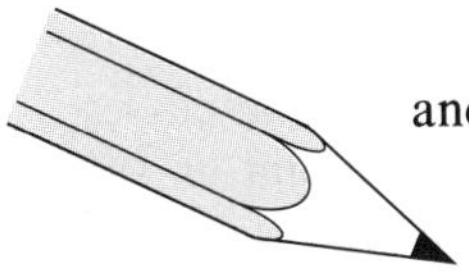

Look around you. It is easy to see the circle used in the construction and manufacture of everyday products. But if you look very carefully, you may be able to spot the other conic sections as well.

IMPORTANT CONCEPTS

vertices
directrix
focus
major axis
minor axis
eccentricity
transverse axis
conjugate axis
asymptote
translation of axes
rotation of axes

11.1 PARABOLA

The graph of a quadratic function $f(x) = ax^2 + bx + c$, $a \neq 0$, is a parabola. However, not every parabola is the graph of a function of x. In general, a parabola is defined in the following manner:

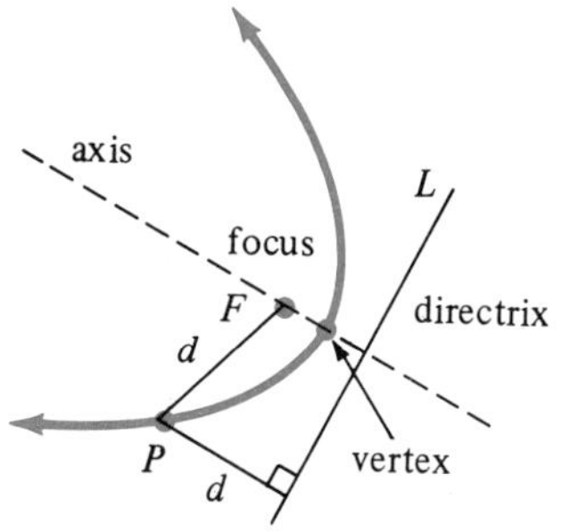

FIGURE 11.1

DEFINITION 11.1 Parabola

A **parabola** is the set of all points P in the plane that are equidistant from a fixed line L, called the **directrix**, and a fixed point F, called the **focus**.

As shown in Figure 11.1, a parabola has an **axis** of symmetry that passes through the focus F and is perpendicular to the directrix. The point on the axis midway between F and L is called the **vertex** of the parabola.

Equation of a Parabola To describe a parabola analytically, let us assume for the sake of discussion that the directrix is the horizontal line $y = -p$ and that the focus is $F(0, p)$. Using Definition 11.1 and Figure 11.2(a), we see that $d(F, P) = d(P, Q)$ is the same as

$$\sqrt{x^2 + (y - p)^2} = y + p$$

Squaring both sides and simplifying lead to

$$x^2 = 4py \tag{1}$$

We say that (1) is the **standard form** for the equation of a parabola with focus $F(0, p)$ and directrix $y = -p$. In like manner, if the directrix and focus are $x = -p$ and $F(p, 0)$, respectively, we find that the standard form for the equation of the parabola is

$$y^2 = 4px \tag{2}$$

See Figure 11.2(b).

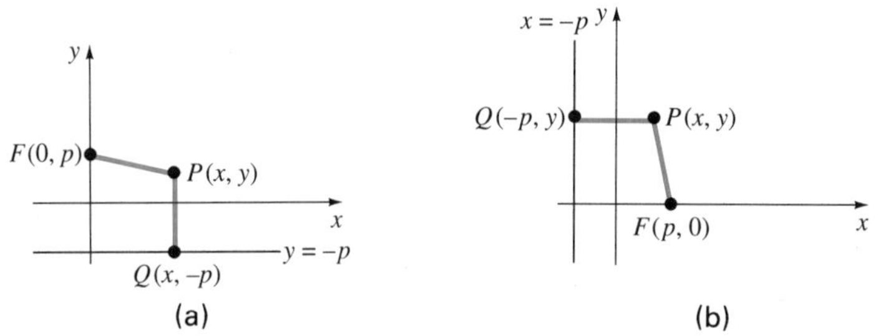

FIGURE 11.2

Although we assume $p > 0$ in Figure 11.2, this need not, of course, be the case. The following table and Figure 11.3 summarize information about equations (1) and (2).

Equation	Vertex	Axis	Focus	Directrix	Graph Opens
$x^2 = 4py$	(0, 0)	$x = 0$	$(0, p)$	$y = -p$	up if $p > 0$; down if $p < 0$
$y^2 = 4px$	(0, 0)	$y = 0$	$(p, 0)$	$x = -p$	right if $p > 0$; left if $p < 0$

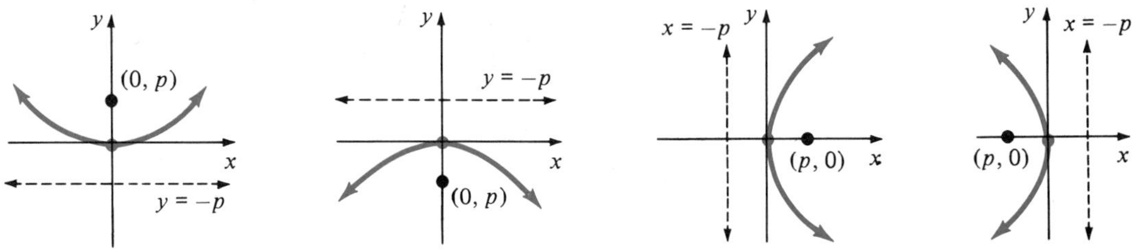

$x^2 = 4py, p > 0$ (a) | $x^2 = 4py, p < 0$ (b) | $y^2 = 4px, p > 0$ (c) | $y^2 = 4px, p < 0$ (d)

FIGURE 11.3

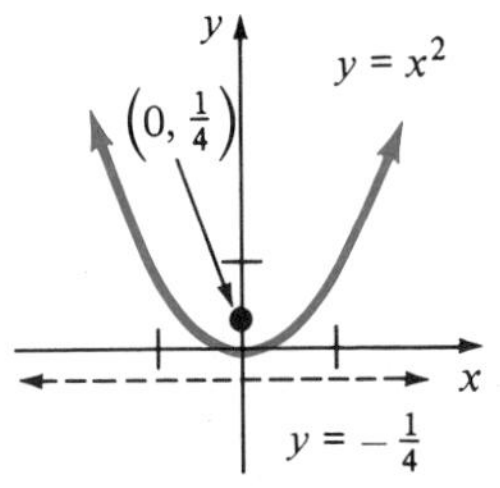

FIGURE 11.4

EXAMPLE 1 Find the focus and directrix of the parabola whose equation is $y = x^2$.

Solution Comparing the equation $y = x^2$ with (1) enables us to identify $4p = 1$ and so $p = \frac{1}{4}$. Hence, the focus of the parabola is $(0, \frac{1}{4})$ and its directrix is the horizontal line $y = -\frac{1}{4}$. The familiar graph, along with the focus and directrix, is given in Figure 11.4. □

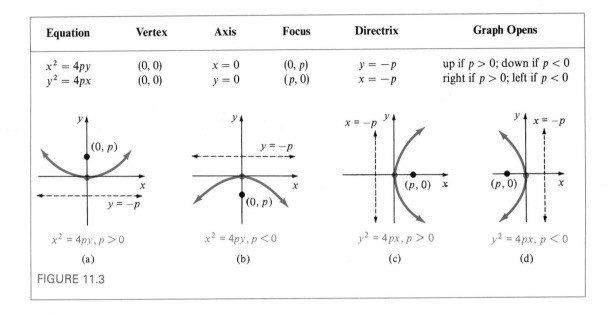

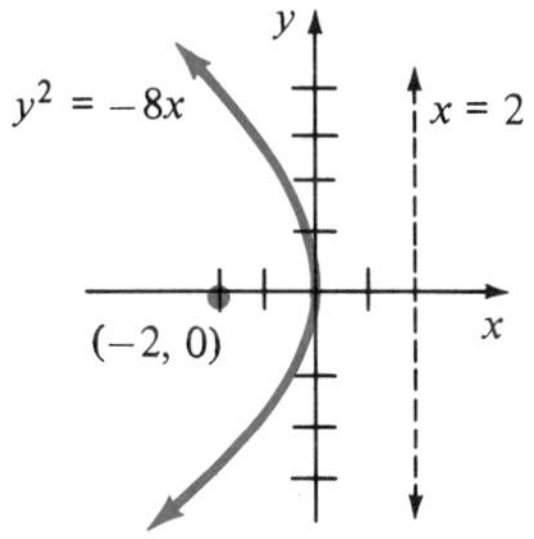

FIGURE 11.5

EXAMPLE 2 Find the focus and directrix of the parabola whose equation is $x = -\frac{1}{8}y^2$. Graph.

Solution By writing the given equation as $y^2 = -8x$, we see from (2) that $4p = -8$ and $p = -2$. Hence, the focus and directrix are $(-2, 0)$ and $x = 2$, respectively. Since $p < 0$, the parabola opens to the left and must have the same basic shape as the graph given in Figure 11.3(d). See Figure 11.5. □

Vertex at (h, k) In general, the standard form for the equation of a parabola with vertex (h, k) is given by either

$$(x - h)^2 = 4p(y - k) \tag{3}$$

or

$$(y - k)^2 = 4p(x - h) \tag{4}$$

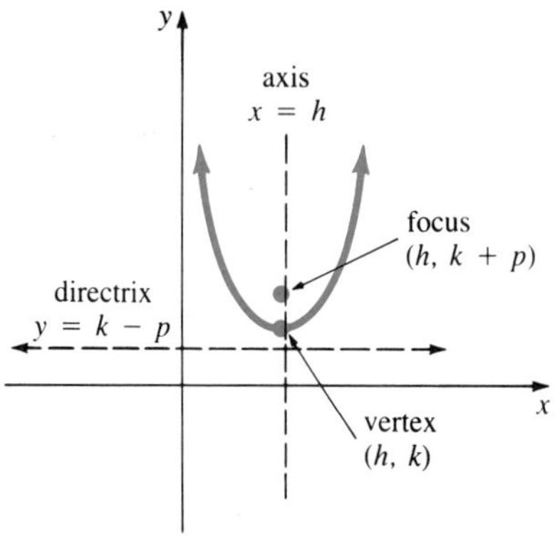

(a) $(x - h)^2 = 4p(y - k), \quad p > 0$

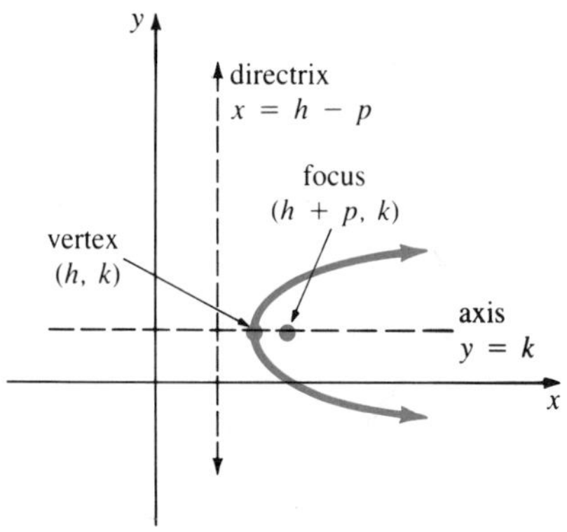

(b) $(y - k)^2 = 4p(x - h), \quad p > 0$

FIGURE 11.6

Figure 11.6 illustrates two possible graphs of (3) and (4). As in (1) and (2), to locate the focus and directrix we need only measure $|p|$ units along the axis of the parabola. We measure either above and below the vertex or to each side of the vertex. The next table summarizes the information about (3) and (4).

Equation	Vertex	Axis	Focus	Directrix	Graph Opens
$(x-h)^2 = 4p(y-k)$	(h, k)	$x = h$	$(h, k+p)$	$y = k - p$	up if $p > 0$; down if $p < 0$
$(y-k)^2 = 4p(x-h)$	(h, k)	$y = k$	$(h+p, k)$	$x = h - p$	right if $p > 0$; left if $p < 0$

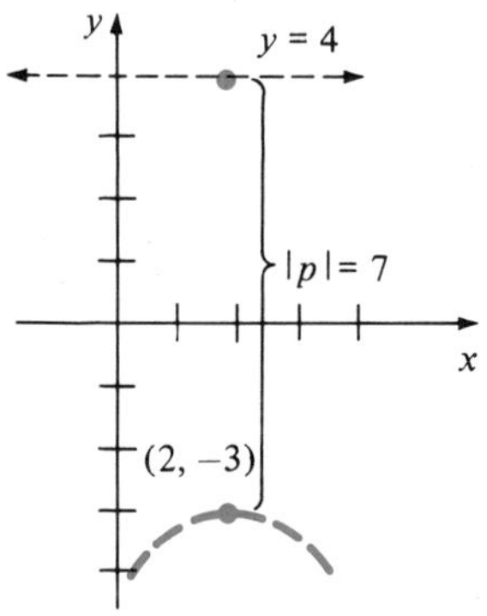

FIGURE 11.7

EXAMPLE 3 Find an equation of the parabola with vertex $(2, -3)$ and directrix $y = 4$. Determine the focus.

Solution With the vertex at $(2, -3)$, it follows that $h = 2$ and $k = -3$. Since the directrix of the parabola is $y =$ constant, the equation we seek must be of form (3). In addition, in Figure 11.7, we see that $|p| = 7$. Furthermore, $p = -7$ because the graph must open down. (We can also find the value of p by simply solving $k - p = 4$.) Thus, the standard form for the equation of the parabola is

$$(x-2)^2 = 4(-7)(y-(-3)) \quad \text{or} \quad (x-2)^2 = -28(y+3)$$

The focus is $(2, -3 + (-7))$ or $(2, -10)$. □

EXAMPLE 4 Find the vertex, axis, focus, directrix, and graph of the parabola whose equation is $y^2 - 8y = 4x - 8$.

Solution Completing the square in y yields

$$y^2 - 8y + 16 = 4x + 8 \quad \text{or} \quad (y-4)^2 = 4(x+2)$$

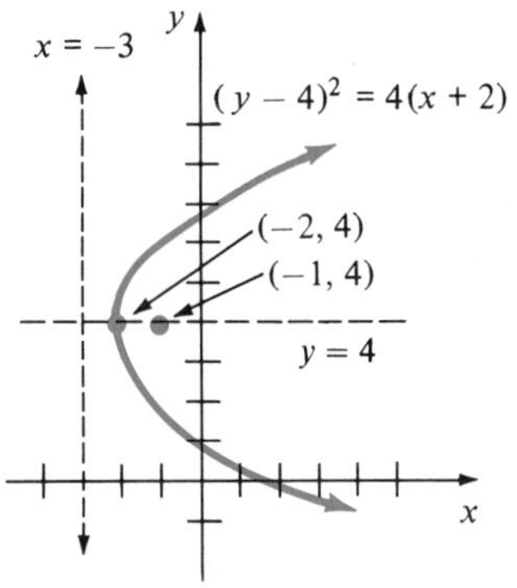

FIGURE 11.8

Identifying the last equation with (4) shows that the vertex is $(-2, 4)$ and that the axis is $y = 4$. Now $4p = 4$ implies that $p = 1 > 0$, and so the parabola opens to the right and its focus and directrix are $(-1, 4)$ and $x = -3$, respectively. You should be able to verify that the graph given in Figure 11.8 has y-intercepts $4 - 2\sqrt{2}$ and $4 + 2\sqrt{2}$, and x-intercept 2. □

▲ ***Remark*** The designs of objects such as searchlights, automobile headlights, reflecting telescopes, and microwave antennas are based on a pleasant property of the parabola. As shown in Figure 11.9(a), if parallel beams of light, say from a distant star, enter a reflecting telescope with a parabolic mirror, then it can be proved that all beams will be reflected to the focus. See Problem 42 in Exercises 11.1. On the other hand, if a light source is placed at the focus of a parabolic reflecting surface, then a beam of light will be reflected from the surface parallel to the axis of the parabola. Other examples of the parabolic form are the cables of a suspension bridge and the trajectory of an obliquely launched projectile. See Figure 11.9(b) and (c). Under some circumstances, the pursuit curve of a shark seeking its prey is also parabolic.

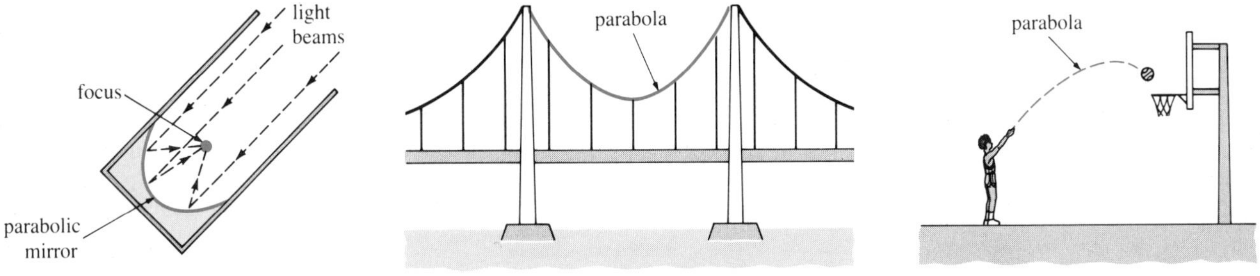

(a) reflecting telescope (b) suspension bridge (c) trajectory

FIGURE 11.9

EXERCISES 11.1 *Answers to odd-numbered problems begin on page A-94.*

just vertex @ graph

In Problems 1–20 find the vertex, ~~focus, directrix~~, and graph of the parabola whose equation is given.

1. $-2x^2 = y$
2. $x^2 = 8y$
3. $y^2 = 12x$
4. $\frac{1}{4}x^2 = -y$
5. $y^2 = -10x$
6. $y^2 = 5x$
7. $(x - 1)^2 = 4y$
8. $(y + 2)^2 = x$
9. $(y - 3)^2 + 8 = 2x$
10. $-(x + 2)^2 = 4y - 16$
11. $(x - 4)^2 = 4(y + 3)$
12. $(2y + 3)^2 = -16(2x + 1)$
13. $y = x^2 + 4x + 6$
14. $3y = -x^2 + 3x + 7$
15. $y^2 - 6y = 4x + 3$
16. $x = y^2 + 10y + 27$
17. $6x^2 + 24x - 8y + 19 = 0$
18. $-x^2 + 6x - 4y - 9 = 0$
19. $2(x + 3) = -y(y + 5)$
20. $2y = x(2 - x)$

In Problems 21–30 find an equation of the parabola that satisfies the given conditions.

21. Vertex (0, 0), axis $x = 0$, through $(-1, 4)$
22. Vertex (1, 2), axis $y = 2$, through (0, 0)
23. Vertex (1, 1), directrix $x = 4$

24. Vertex $(3, -2)$, directrix $y = 2$

25. Focus $(-2, 4)$, vertex $(1, 4)$

26. Focus $(0, 0)$, vertex $(0, \frac{3}{2})$

27. Focus $(0, 4)$, directrix $x = -5$

28. Focus $(2, -\frac{5}{2})$, directrix $2y - 1 = 0$

29. Axis $x = 0$, through $(1, 3)$ and $(2, 6)$

30. Axis $y = 0$, through $(2, 1)$ and $(11, -2)$

In Problems 31 and 32 find an equation of the parabola in part (**a**) of the figure. Use part (**a**) to find an equation of the shifted parabola in part (**b**).

31.

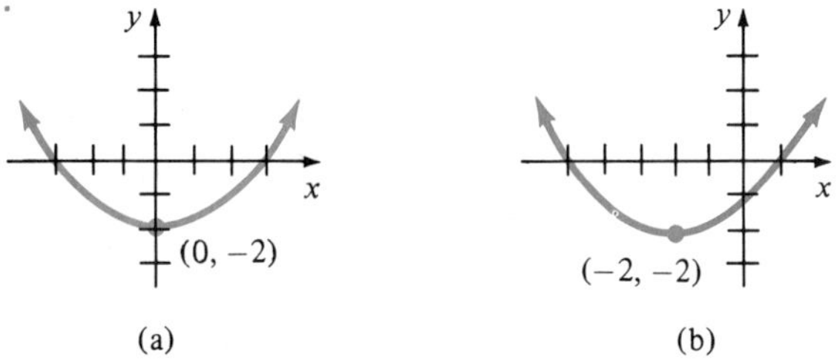

FIGURE 11.10

32.

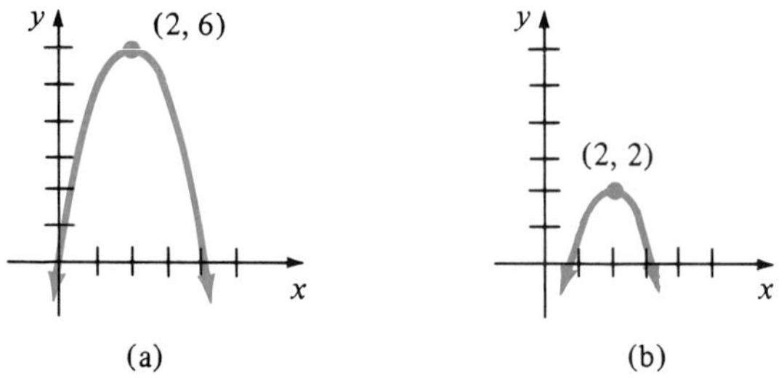

FIGURE 11.11

33. Find an equation of the tangent line to the parabola whose equation is $x = y^2 + 2y + 1$ at $y = 3$.

34. In Problem 33 find the area of the region that is bounded by the parabola and the graph of $x = 4$.

35. Show that the vertex is the point on a parabola closest to the focus. [*Hint:* Consider $x^2 = 4py$.]

36. Oil is gushing from an end of a horizontal pipe 30 m above ground level. Assume the oil stream is an arc of a parabola with vertex at the end of the pipe. Ten meters down from the end of the pipe the horizontal distance to the oil stream is 15 m. See Figure 11.12. Find, relative to the end of the pipe, where the oil hits the ground.

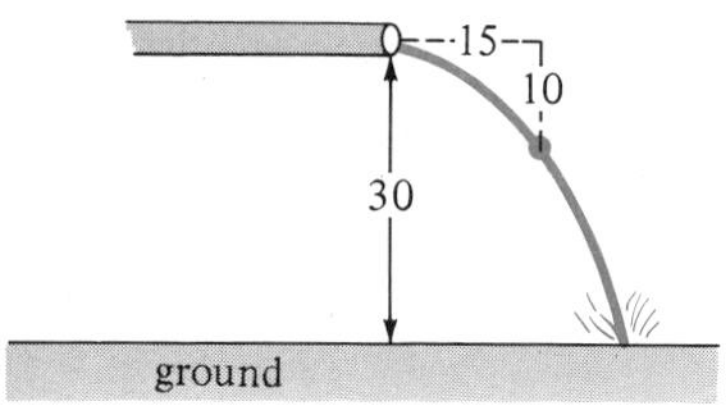

FIGURE 11.12

37. The distance between two vertical supports of a suspension bridge is 100 m and the sag of the cable is 15 m.

(a) Assume that the shape of the cable is a parabola. Find its equation if the vertex is in the center at the lowest point on the cable.

(b) What is the height of the cable 30 m from the center?

38. The reflecting telescope at Mount Palomar uses a circular mirror 200 inches in diameter. A cross-section of the mirror through a diameter is a parabola whose focal length is 55.5 ft. (Focal length is the distance from the focus to the vertex.)

(a) Find an equation of the parabolic cross-section.

(b) What is the maximum depth of the mirror?

39. The **focal chord** is the line segment through the focus, perpendicular to the axis, with endpoints on the parabola. See Figure 11.13. Show that the length of the focal chord for both $x^2 = 4py$ and $y^2 = 4px$ is $4|p|$.

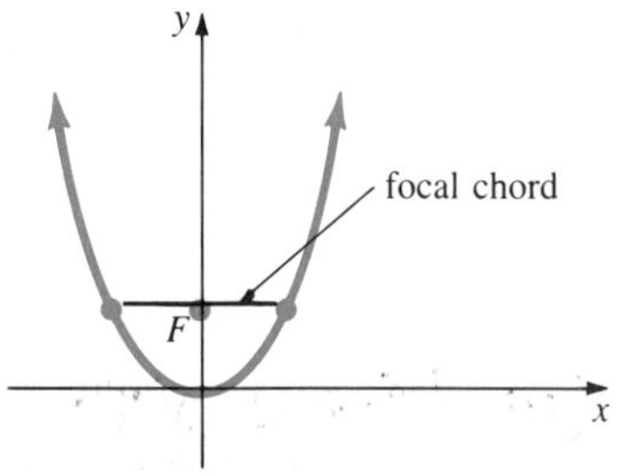

FIGURE 11.13

40. Use Problem 39 to find the length of the focal chord for $y = \frac{1}{3}x^2$.

41. Show that the tangent lines at the endpoints of the focal chord for $x^2 = 4py$ are perpendicular to each other.

42. As shown in Figure 11.14, let L be the tangent line to the graph of $y^2 = 4px$ at $P(x_0, y_0)$.

(a) Find an equation of the tangent line at P.

(b) Show that the x-intercept of L is $(-x_0, 0)$.

(c) Show that the line segments QF and PF have the same length.

(d) Show that $\alpha = \beta = \gamma$.

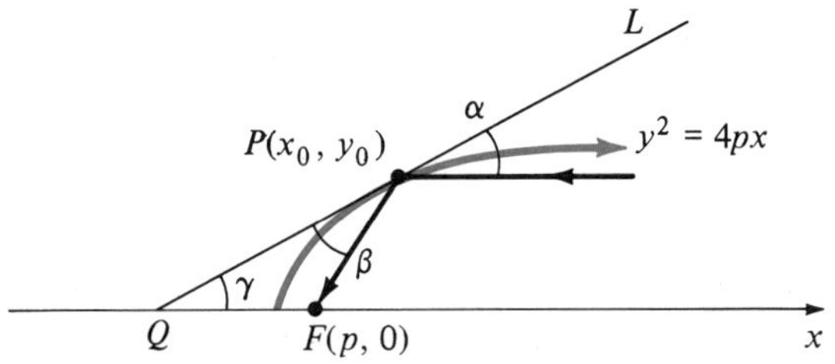

FIGURE 11.14

43. Find the vertex and focus of a parabola whose equation is $y = ax^2 + bx + c$, $a \neq 0$.

44. Using $y = ax^2 + bx + c$, find an equation of the parabola passing through the points (0, 5), (1, 4), and (−2, 13).

45. The distance d from a point (x_0, y_0) to a line $ax + by + c = 0$ is given by $d = |ax_0 + by_0 + c|/\sqrt{a^2 + b^2}$. Use this fact and Definition 11.1 to find an equation of the parabola that has (1, 2) as its focus and the line $x - y - 3 = 0$ as its directrix.

11.2 ELLIPSE

The ellipse is defined as follows:

DEFINITION 11.2 Ellipse

An **ellipse** is the set of points P in the plane such that the sum of the distances between P and two fixed points F_1 and F_2, called **foci**, is a constant.

As shown in Figure 11.15 (a), if P is a point on the ellipse and $d_1 = d(F_1, P)$, $d_2 = d(F_2, P)$, and $k > 0$ is a constant, then Definition 11.2 asserts that

$$d_1 + d_2 = k \tag{1}$$

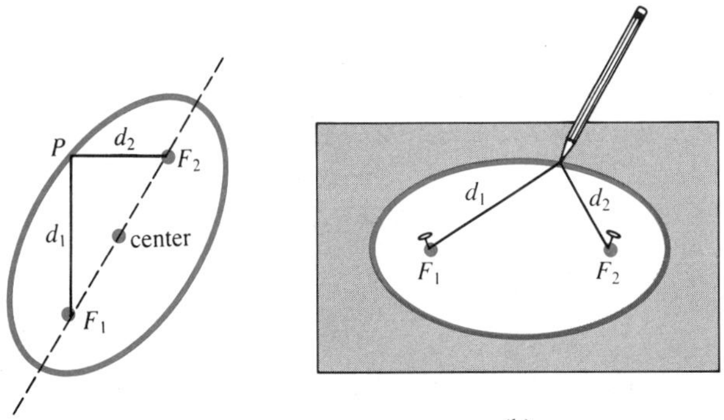

(a) $d_1 + d_2 = k$ (b)

FIGURE 11.15

The midpoint of the line segment F_1F_2 is called the **center** of the ellipse. On a practical level, (1) can be used to sketch an ellipse. Figure 11.15 (b) shows that if a string of length k is attached to a paper by two tacks, then an ellipse can be traced by inserting a pencil against the string and moving it in such a manner that the string remains taut.

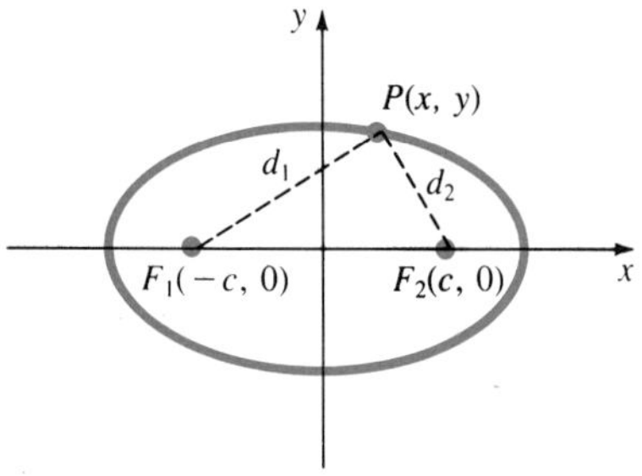

FIGURE 11.16

Equation of an Ellipse For convenience, let us choose $k = 2a$ and put the foci on the x-axis with coordinates $F_1(-c, 0)$ and $F_2(c, 0)$. See Figure 11.16. It follows from (1) that

$$\sqrt{(x+c)^2+y^2}+\sqrt{(x-c)^2+y^2}=2a$$

or

$$\sqrt{(x+c)^2+y^2}=2a-\sqrt{(x-c)^2+y^2} \tag{2}$$

We square both sides of (2) and simplify:

$$(x+c)^2+y^2=4a^2-4a\sqrt{(x-c)^2+y^2}+(x-c)^2+y^2$$
$$a\sqrt{(x-c)^2+y^2}=a^2-cx$$

Squaring again gives

$$a^2[(x-c)^2+y^2]=a^4-2a^2cx+c^2x^2$$
$$(a^2-c)^2x^2+a^2y^2=a^2(a^2-c^2)$$

or

$$\frac{x^2}{a^2}+\frac{y^2}{a^2-c^2}=1 \tag{3}$$

Referring to Figure 11.16, we see that the points F_1, F_2, and P form a triangle. Since the so-called triangle inequality states that the sum of the lengths of any two sides of a triangle is greater than the length of the remaining side, we must have $2a > 2c$ or $a > c$. Hence, $a^2 - c^2 > 0$. When we let

$$b^2=a^2-c^2 \tag{4}$$

the **standard form** for the equation of an ellipse with foci $F_1(-c, 0)$ and $F_2(c, 0)$ becomes

$$\frac{x^2}{a^2}+\frac{y^2}{b^2}=1 \tag{5}$$

The **major axis** of an ellipse is the line segment through its center, containing the foci, and with endpoints on the ellipse. The line segment through the center, perpendicular to the major axis, and with endpoints on the ellipse is called the **minor axis**. Now, setting $y = 0$ in equation (5) reveals that $-a$ and a are the x-intercepts, whereas $x = 0$ leads to the y-intercepts $-b$ and b. The corresponding points $(\pm a, 0)$ and $(0, \pm b)$ are called **vertices**. Since $a > b$, the major axis of an ellipse is necessarily longer than its minor axis.

If the foci are on the y-axis at $F_1(0, -c)$ and $F_2(0, c)$, then a repetition of this analysis will show that an equation of the ellipse is

$$\frac{x^2}{b^2}+\frac{y^2}{a^2}=1 \tag{6}$$

where, as before, $b^2 = a^2 - c^2$.

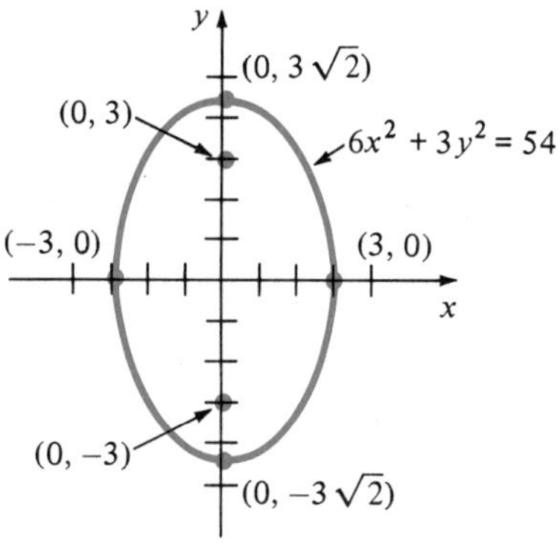

FIGURE 11.17

EXAMPLE 1 Find the vertices and foci of the ellipse whose equation is $6x^2 + 3y^2 = 54$. Graph.

Solution By dividing both sides by 54,

$$\frac{x^2}{9} + \frac{y^2}{18} = 1$$

we see that $18 > 9$ so we identify the equation with (6). From $a^2 = 18$ and $b^2 = 9$, we conclude that the vertices are $(0, \pm 3\sqrt{2})$ and $(\pm 3, 0)$. The major axis is vertical with endpoints at $(0, -3\sqrt{2})$ and $(0, 3\sqrt{2})$. The minor axis is horizontal with endpoints $(-3, 0)$ and $(3, 0)$. Now, $c^2 = a^2 - b^2 = 9$ implies that $c = 3$. Hence, the foci are on the y-axis at $(0, -3)$ and $(0, 3)$. The graph is given in Figure 11.17. □

EXAMPLE 2 Find an equation of an ellipse that has (2, 0) as a focus and x-intercept 5.

Solution Since the given focus is on the x-axis, we can find an equation of form (5). Consequently, $c = 2$, $a = 5$, and from (4), $b^2 = 25 - 4 = 21$. The desired equation is then

$$\frac{x^2}{25} + \frac{y^2}{21} = 1$$ □

Center at (*h*, *k*) When the center is at (h, k), the standard form for the equation of an ellipse is either

$$\frac{(x-h)^2}{a^2} + \frac{(y-k)^2}{b^2} = 1 \qquad (7)$$

or

$$\frac{(x-h)^2}{b^2} + \frac{(y-k)^2}{a^2} = 1 \qquad (8)$$

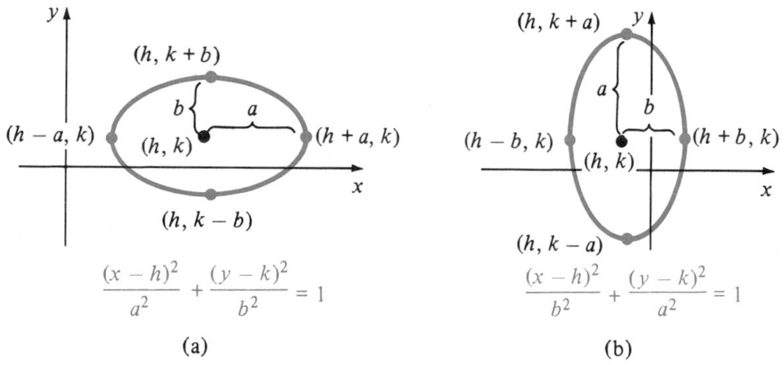

FIGURE 11.18

In each case we still have $a > c$ and $b^2 = a^2 - c^2$. But, for (7) the foci are located on the horizontal line $y = k$ at $(h \pm c, k)$. For (8) the foci are on the vertical line $x = h$ at $(h, k \pm c)$. To find the vertices of an ellipse with center (h, k), we set, in turn, $x = h$ and $y = k$ in its equation and solve for y and x. The coordinates of the vertices are given in Figure 11.18.

EXAMPLE 3 Find the vertices and foci of the ellipse whose equation is $9x^2 - 36x + 25y^2 + 150y + 36 = 0$.

Solution We first write the given equation in one of the standard forms (7) or (8) by completing the square in both x and y:

$$\begin{aligned} 9x^2 - 36x + 25y^2 + 150y &= -36 \\ 9(x^2 - 4x \quad\ ) + 25(y^2 + 6y \quad\ ) &= -36 \\ 9(x^2 - 4x + 4) + 25(y^2 + 6y + 9) &= -36 + 36 + 175 \\ 9(x - 2)^2 + 25(y + 3)^2 &= 175 \end{aligned}$$

Dividing the last equation by 175 leads to

$$\frac{(x-2)^2}{25} + \frac{(y+3)^2}{9} = 1$$

The center of the ellipse is $(2, -3)$ and, because $25 > 9$, we know from (7) that its major and minor axes lie along the lines $y = -3$ and $x = 2$, respectively. Now, solving

$$\frac{(x-2)^2}{25} = 1 \quad \text{and} \quad \frac{(y+3)^2}{9} = 1$$

gives $x = 2 \pm 5$ and $y = -3 \pm 3$. Hence, the vertices are $(7, -3)$, $(-3, -3)$, $(2, 0)$, and $(2, -6)$. Alternatively, the vertices can be found by moving $a = 5$ units to the right and left of the center along $y = -3$ and $b = 3$ units above and below the center along $x = 2$. Finally, $c^2 = a^2 - b^2 = 25 - 9 = 16$ shows that $c = 4$ and that the foci are $(-2, -3)$ and $(6, -3)$. □

Eccentricity Associated with each conic section is a number e called its **eccentricity**.* For an ellipse we have the following definition:

DEFINITION 11.3 Eccentricity

The **eccentricity** of an ellipse is

$$e = \frac{c}{a} = \frac{\sqrt{a^2 - b^2}}{a}$$

Since $0 < \sqrt{a^2 - b^2} < a$, the eccentricity of an ellipse satisfies $0 < e < 1$.

---

*Do not confuse this with the number e of Chapter 6. The eccentricity of a parabola will be discussed in Section 12.6.

EXAMPLE 4 Determine the eccentricity of the ellipse in Example 3.

Solution In the solution of that example we found $a = 5$ and $c = 4$. Hence, the eccentricity of the ellipse is $e = \frac{4}{5}$. □

Eccentricity is an indicator of the shape of an ellipse. When $e \approx 0$ the ellipse is nearly circular, and when $e \approx 1$ the ellipse is flattened or elongated. To see this, observe that if e is close to 0, it then follows from $e = \sqrt{a^2 - b^2}/a$ that $\sqrt{a^2 - b^2} \approx 0$ and consequently $a \approx b$. This means that the shape of the ellipse is close to circular. See Figure 11.19(a). If $e \approx 1$, then $\sqrt{a^2 - b^2} \approx a$ and so $b \approx 0$. Thus, the ellipse is flattened. See Figure 11.19(b).

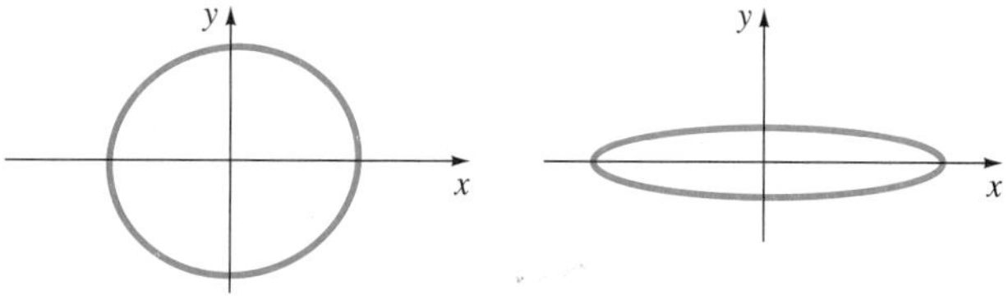

(a) e close to 0 (b) e close to 1

FIGURE 11.19

Using his Law of Universal Gravitation, Isaac Newton was the first to prove Kepler's first law of planetary motion: the orbit of each planet about the sun is an ellipse with the sun at one focus.

EXAMPLE 5 The perihelion distance of the earth (the least distance between the earth and the sun) is approximately 9.16×10^7 miles, and its aphelion distance (the greatest distance between the earth and the sun) is approximately 9.46×10^7 miles. What is the eccentricity of the earth's orbit?

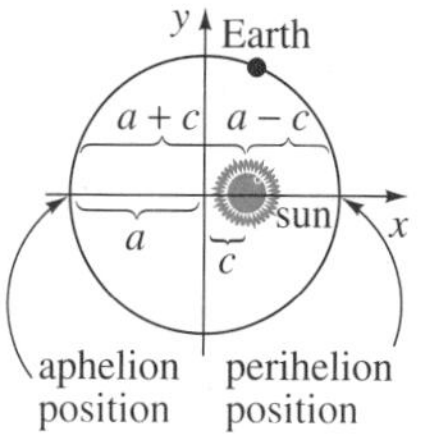

FIGURE 11.20

Solution Let us assume that the orbit of the earth is as shown in Figure 11.20. From the figure we obtain

$$a - c = 9.16 \times 10^7$$
$$a + c = 9.46 \times 10^7$$

Solving this system of equations gives $a = 9.31 \times 10^7$ and $c = 0.15 \times 10^7$. Thus, from Definition 11.3 we obtain

$$e = \frac{0.15 \times 10^7}{9.31 \times 10^7} \approx 0.016$$ □

The orbits of seven of the nine planets have eccentricities less than 0.1 and, hence, are not far from circular. Mercury and Pluto are the exceptions with orbits having eccentricities 0.21 and 0.25, respectively. Many of the asteroids and comets have highly eccentric elliptical orbits. The orbit of the asteroid Hildago is one of the most eccentric, with $e = 0.66$. Another notable case is the orbit of the comet Halley. See Problem 47 in Exercises 11.2.

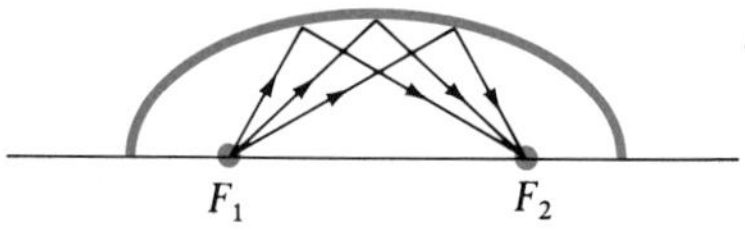

FIGURE 11.21

▲ ***Remark*** The ellipse has a reflecting property similar to that of the parabola. If a light or sound source is placed at one focus of an ellipse, say F_1 in Figure 11.21, then all rays or waves will be reflected to the other focus F_2. For example, if a pool table is constructed in the form of an ellipse with a pocket at one focus, then any shot originating at the other focus will never miss the pocket. Similarly, if a ceiling is elliptical with two foci on the floor, but considerably distant from each other, then anyone whispering at one focus will be heard at the other. Two famous "whispering galleries" are located in Statuary Hall at the Capitol in Washington, D.C., and in the Mormon Tabernacle in Salt Lake City, Utah. See Problem 52 in Exercises 11.2.

EXERCISES 11.2

Answers to odd-numbered problems begin on page A-94.

In Problems 1–20 find the vertices, ~~foci,~~ and graph of the ellipse whose equation is given.

1. $\dfrac{x^2}{16} + \dfrac{y^2}{25} = 1$

2. $\dfrac{x^2}{3^2} + \dfrac{y^2}{2^2} = 1$

3. $x^2 + \dfrac{y^2}{10} = 1$

4. $\dfrac{x^2}{6} + y^2 = 1$

5. $\dfrac{x^2}{8} + \dfrac{y^2}{4} = 2$

6. $9x^2 + y^2 = 9$

7. $4x^2 + 7y^2 = 28$

8. $2x^2 + 4y^2 = 1$

9. $\dfrac{(x-1)^2}{25} + \dfrac{(y-3)^2}{36} = 1$

10. $\dfrac{(x+1)^2}{9} + \dfrac{(y-4)^2}{4} = 1$

11. $\dfrac{(x-2)^2}{64} + \dfrac{(y-2)^2}{36} = 1$

12. $\dfrac{(2x+1)^2}{4} + \dfrac{(3y-2)^2}{16} = 1$

13. $2\left(x + \dfrac{1}{2}\right)^2 + (y-1)^2 = 4$

14. $x^2 + 9(y-2)^2 = 81$

15. $3x^2 + y^2 - 6y = 0$

16. $y^2 = 2x(1-x)$

17. $9x^2 + 25y^2 + 18x + 50y = 191$

18. $x^2 + 2y^2 + x + y = 0$

19. $5x^2 + y^2 - 40x - 4y + 83 = 0$

20. $4x^2 + 12y^2 - 4x - 24y + 1 = 0$

In Problems 21–24 find the eccentricity of the ellipse with the given equation.

21. $\dfrac{(x-3)^2}{169} + \dfrac{(y+4)^2}{144} = 1$

22. $x^2 + \dfrac{(y+5)^2}{50} = 1$

23. $3x^2 + 5y^2 + 12x - 20y + 17 = 0$

24. $x^2 + 4y^2 - 2x - 24y + 21 = 0$

In Problems 25–38 find an equation of the ellipse that satisfies the given conditions.

25. Center $(0, 0)$, vertices $(5, 0)$ and $(0, -2)$

26. Vertices $(\pm 4, 0)$ and $(0, \pm 7)$

27. Vertices $(0, \pm 4)$, foci $(0, \pm 2)$

28. Vertices $(-1, 1)$ and $(5, 1)$, foci $(4, 1)$ and $(0, 1)$

29. Vertices $(-5, -2)$ and $(3, -2)$, foci $(-1, -5)$ and $(-1, 1)$

30. Vertices $(-3, 3)$, $(9, 3)$, $(3, -7)$, and $(3, 13)$

31. Foci $(\pm\sqrt{7}, 0)$, length of minor axis $\sqrt{3}$

32. Foci $(1 \pm \sqrt{3}, 1)$, length of major axis 6

33. Vertices $(\pm 2\sqrt{2}, 0)$, through $(2, -1)$

34. Vertices $(\pm 6, 2)$, through $(4, 4)$

35. Center $(4, -2)$, focus $(4, 4)$, through $(4, 8)$

36. Set of all points in the plane the sum of whose distances from (0, 1) and (0, −1) is 4

37. Eccentricity $\frac{1}{3}$, length of major axis 8

38. Foci (3, −1) and (3, 11), eccentricity $\frac{3}{4}$

39. Find the slope(s) of the tangent line(s) to the ellipse whose equation is $9x^2 + 5y^2 + 18x - 10y - 27 = 0$ at $x = 1$.

40. Find an equation of the normal line to the ellipse whose equation is $x^2 + 4(y - 1)^2 = 16$ at $(2\sqrt{3}, 2)$.

41. Show that the point on the ellipse whose equation is $x^2/a^2 + y^2/b^2 = 1$, which is closest to the focus $(c, 0)$, is the vertex $(a, 0)$.

42. Without solving, determine how many solutions there are to the system of equations

$$25x^2 + y^2 = 25$$
$$9(x - 1)^2 + 4y^2 = 36$$

43. The ellipse whose equation is $x^2/4 + (y - 1)^2/9 = 1$ is shifted 4 units to the right. What are the center, vertices, and foci of the shifted ellipse?

44. The ellipse whose equation is $(x - 1)^2/9 + (y - 4)^2 = 1$ is shifted 5 units to the left and 3 units up. What are the center, vertices, and foci of the shifted ellipse?

45. The planet Mercury has an elliptical orbit with the sun located at one focus. The length of the major axis is 7.2×10^7 miles and the length of the minor axis is 7.04×10^7 miles. What is Mercury's perihelion distance (the least distance between the planet and the sun)? What is Mercury's aphelion distance (the greatest distance between the planet and the sun)?

46. What is the eccentricity of the orbit of Mercury in Problem 45?

47. The orbit of Halley's comet is an ellipse whose major axis is 3.34×10^9 miles long and whose minor axis is 8.5×10^8 miles long. What is the eccentricity of the comet's orbit?

48. A satellite is put into an elliptical orbit around the earth. The radius of the earth is 6000 km (approximately) and its center is located at one focus of the orbit.

(a) Use the information given in Figure 11.22 to find an equation of the orbit.

(b) What is the height of the satellite above the surface of the earth at point P?

49. What is the eccentricity of the orbit of the satellite in Problem 48?

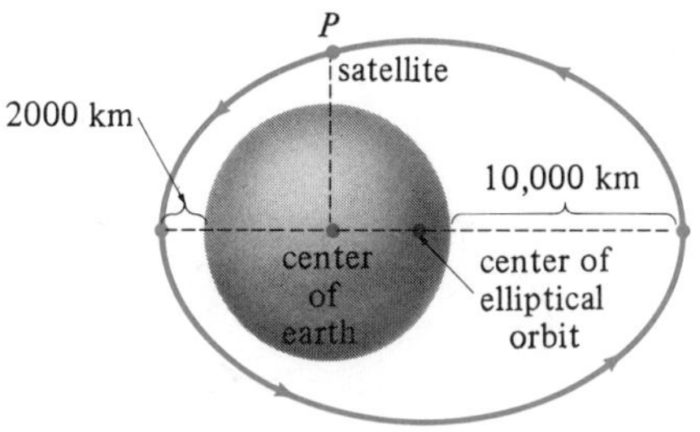

FIGURE 11.22

50. The line segment perpendicular to the major axis through a focus with endpoints on the ellipse is called a **focal chord**. See Figure 11.23. Show that the length of a focal chord of an ellipse with center at the origin is $2b^2/a$.

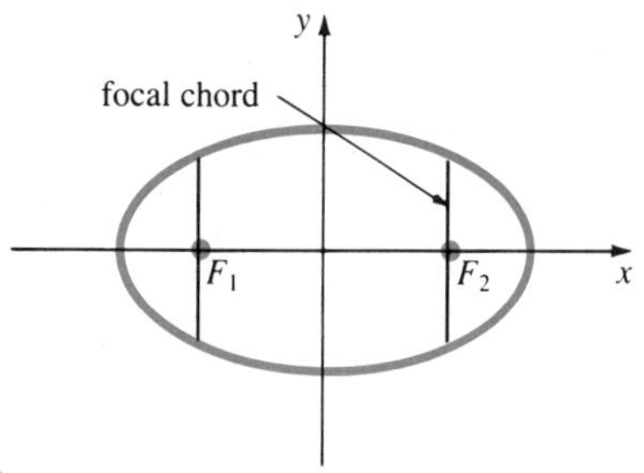

FIGURE 11.23

51. Rework part (**b**) of Problem 48 using the information in Problem 50.

52. As shown in Figure 11.24, let L be the tangent line to the graph of $x^2/a^2 + y^2/b^2 = 1$ at $P(x_0, y_0)$. Show that $\alpha = \beta$.

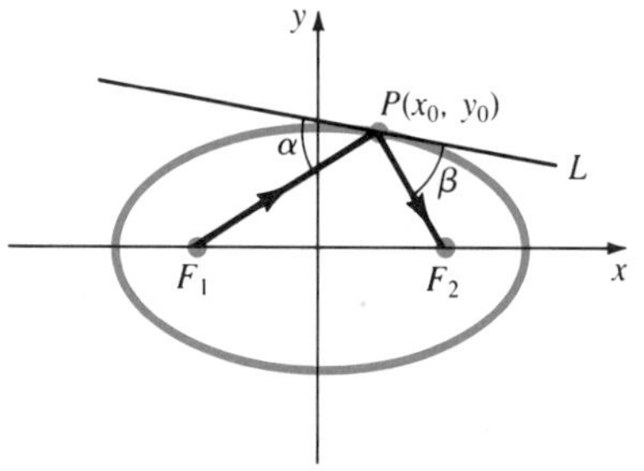

FIGURE 11.24

53. An **oval** is an approximation to an ellipse consisting of arcs from symmetrically placed pairs of circles of different radii, each small circle being tangent to a large circle at two transition points as indicated in Figure 11.25. Architects in the Renaissance and Baroque periods commonly used ovals because they are simpler to construct than ellipses. In this

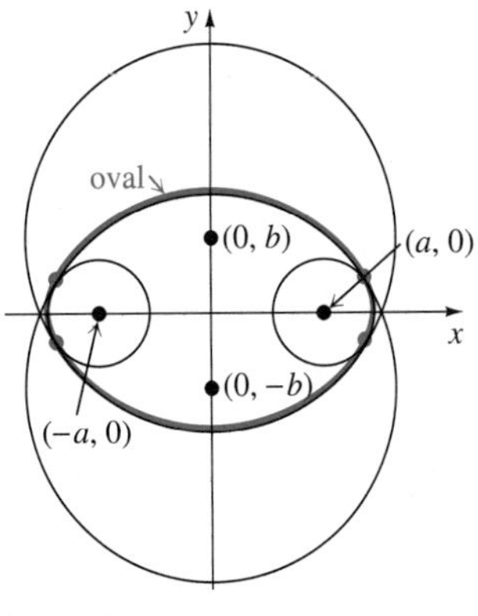

FIGURE 11.25

problem, let the small circles be centered at $(\pm a, 0)$, $a > 0$, with radius r, and let the large circles be centered at $(0, \pm b)$, $b > 0$, with radius R. Also, let $(\pm A, 0)$, $A > 0$, and $(0, \pm B)$, $B > 0$, be the points of intersection of the oval with the x- and y-axes.

(a) Express R in terms of a, b, and r.

(b) Show that $A > B$. This shows that the "major axis" of the oval is always in line with the centers of the small circles, and that the "minor axis" of the oval is always in line with the centers of the large circles. [*Hint:* Show that $A - B = a + b - \sqrt{a^2 + b^2}$.]

54. Show that the circumference of the ellipse $x^2/a^2 + y^2/b^2 = 1$ with eccentricity e is $4a \int_0^{\pi/2} \sqrt{1 - e^2\sin^2\theta}\, d\theta$.

Calculator/Computer Problems

55. Use the result in Problem 54 and Simpson's Rule* with $n = 4$ to find an approximation to the circumference of the ellipse whose equation is $x^2/25 + y^2/9 = 1$.

*The antiderivative $\int \sqrt{1 - e^2\sin^2\theta}\, d\theta$ cannot be expressed as an elementary function.

11.3 HYPERBOLA

The definition of a hyperbola is the same as that of the ellipse with one exception: the word *sum* is replaced by *difference*.

> **DEFINITION 11.4 Hyperbola**
>
> A **hyperbola** is the set of points P in the plane such that the difference of the distances between P and two fixed points F_1 and F_2, called **foci**, is a constant.

As shown in Figure 11.26, a hyperbola consists of two **branches**. The midpoint of the line segment F_1F_2 is the **center** of the hyperbola. If P is a point on the hyperbola, then

$$|d_1 - d_2| = k \tag{1}$$

where $d_1 = d(F_1, P)$ and $d_2 = d(F_2, P)$.

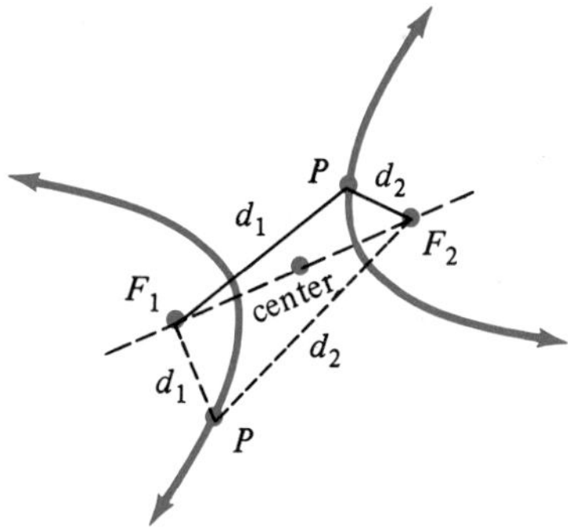

FIGURE 11.26

Equation of a Hyperbola Proceeding as for the ellipse, we place the foci on the x-axis at $F_1(-c, 0)$ and $F_2(c, 0)$ and choose the constant k to be $2a$ for algebraic convenience. Writing (1) as

$$d_1 - d_2 = \pm 2a \tag{2}$$

or

$$\sqrt{(x + c)^2 + y^2} - \sqrt{(x - c)^2 + y^2} = \pm 2a$$

$$\sqrt{(x + c)^2 + y^2} = \pm 2a + \sqrt{(x - c)^2 + y^2}$$

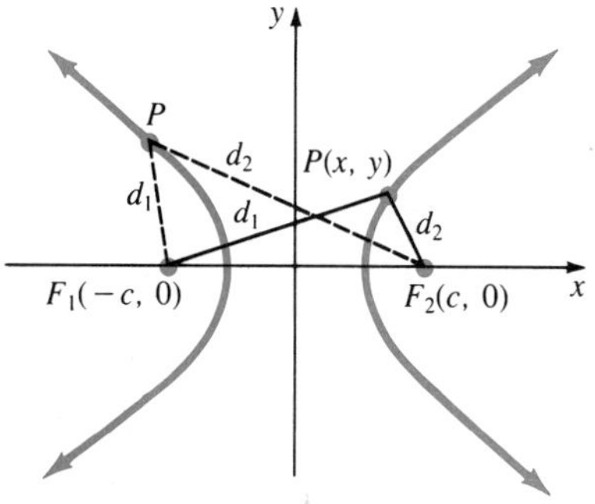

FIGURE 11.27

we square, simplify, and square again:

$$(x + c)^2 + y^2 = 4a^2 \pm 4a\sqrt{(x - c)^2 + y^2} + (x - c)^2 + y^2$$
$$\pm a\sqrt{(x - c)^2 + y^2} = cx - a^2$$
$$a^2[(x - c)^2 + y^2] = c^2x^2 - 2a^2cx + a^4$$
$$x^2(c^2 - a^2) - a^2y^2 = a^2(c^2 - a^2)$$
$$\frac{x^2}{a^2} - \frac{y^2}{c^2 - a^2} = 1 \tag{3}$$

From Figure 11.27, we see that the triangle inequality gives

$$d_1 < d_2 + 2c \quad \text{and} \quad d_2 < d_1 + 2c$$

or

$$d_1 - d_2 < 2c \quad \text{and} \quad d_2 - d_1 < 2c$$

From (2), the foregoing inequalities imply $2a < 2c$ or $a < c$. Thus, if b^2 denotes the positive constant

$$b^2 = c^2 - a^2 \tag{4}$$

in (3), we obtain the **standard form** for the equation of a hyperbola with foci $F_1(-c, 0)$ and $F_2(c, 0)$:

$$\frac{x^2}{a^2} - \frac{y^2}{b^2} = 1 \tag{5}$$

Notice that the graph of (5) has x-intercepts $\pm a$ but has no y-intercepts, since $-y^2/b^2 = 1$ has no real solution. The points $(-a, 0)$ and $(a, 0)$ are **vertices** of the hyperbola. The line segment through the center with the vertices as endpoints is called the **transverse axis**. Also, it is common practice to refer to the line segment that has $(0, -b)$ and $(0, b)$ as endpoints and that passes through the center perpendicular to the transverse axis as the **conjugate axis**. See Figure 11.28(a).

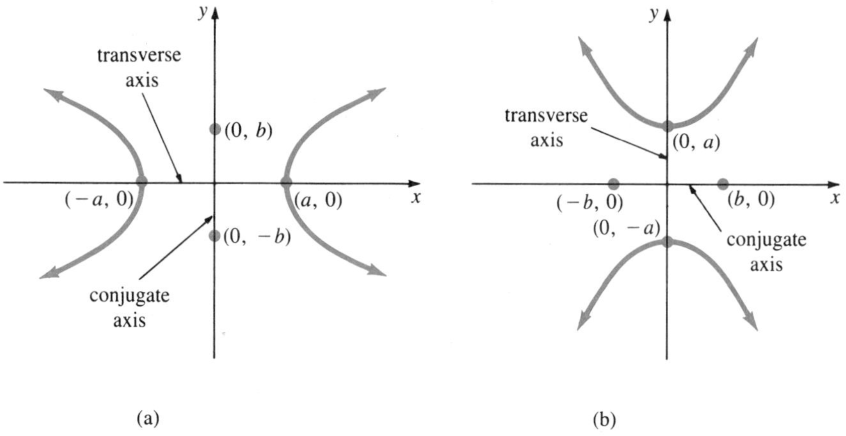

FIGURE 11.28

When the foci are placed on the y-axis at $F_1(0, -c)$ and $F_2(0, c)$, the standard form for the equation of the hyperbola is

$$\frac{y^2}{a^2} - \frac{x^2}{b^2} = 1 \tag{6}$$

where, again, $b^2 = c^2 - a^2$. As indicated in Figure 11.28(b), the transverse axis has endpoints $(0, -a)$ and $(0, a)$ and the conjugate axis has endpoints $(-b, 0)$ and $(b, 0)$.

Asymptotes Solving (5) for y in terms of x gives

$$y = \pm\frac{b}{a}x\sqrt{1 - \frac{a^2}{x^2}}$$

Observe that when the value of $|x|$ becomes very large, the value of the term a^2/x^2 becomes very small. Geometrically, this means that, for large values of $|x|$, the hyperbola is close to the lines

$$y = \frac{b}{a}x \quad \text{and} \quad y = -\frac{b}{a}x$$

These lines, which pass through the center of the hyperbola, are called **asymptotes**. In the case of (6), the asymptotes are $y = (a/b)x$ and $y = -(a/b)x$. Rather than just memorizing formulas, one can find equations for the asymptotes by the following mnemonic: Replace 1 with 0 in the standard equations (5) and (6) and factor the difference of two squares:

$$\frac{x^2}{a^2} - \frac{y^2}{b^2} = 0 \quad \text{or} \quad \frac{y^2}{a^2} - \frac{x^2}{b^2} = 0$$

Then set each factor equal to zero and solve for y.

As shown in Figure 11.29, by drawing the asymptotes first, we have a guideline for drawing a fairly accurate hyperbola.

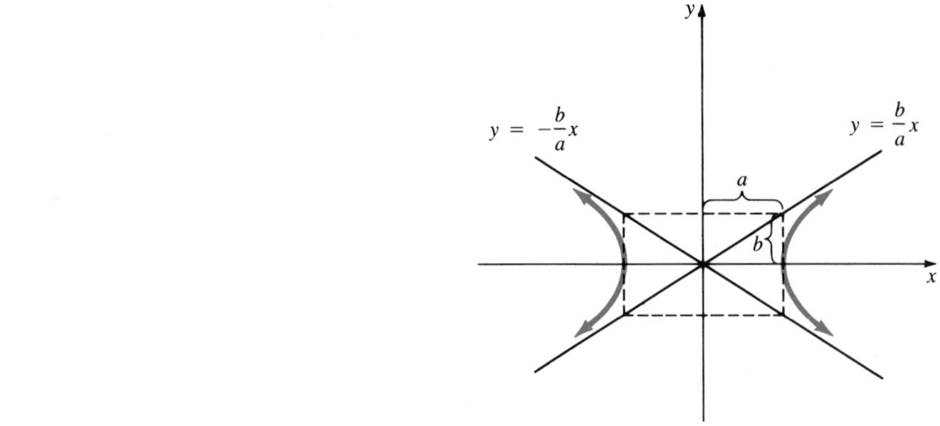

(a) $\frac{x^2}{a^2} - \frac{y^2}{b^2} = 1$

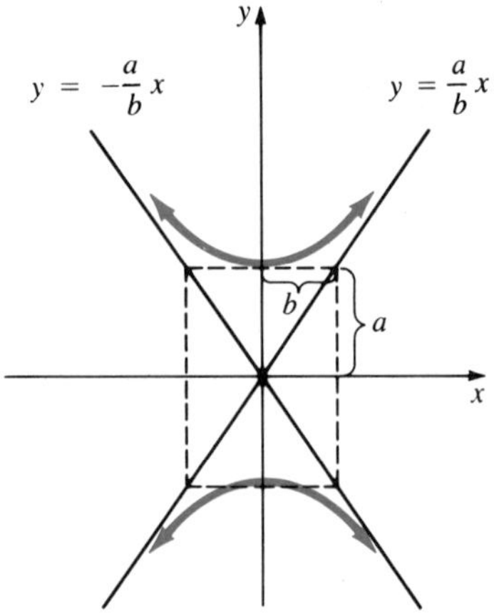

(b) $\frac{y^2}{a^2} - \frac{x^2}{b^2} = 1$

FIGURE 11.29

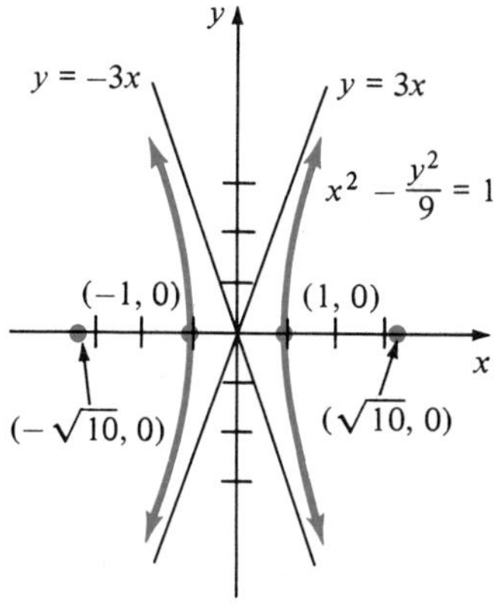

FIGURE 11.30

EXAMPLE 1 Find the vertices, foci, and asymptotes of the hyperbola whose equation is $x^2 - y^2/9 = 1$. Graph.

Solution By writing

$$\frac{x^2}{1^2} - \frac{y^2}{3^2} = 1$$

we identify the equation with (5) and conclude that the transverse axis is horizontal. Setting $y = 0$ gives $x^2 = 1$, and so the vertices are $(-1, 0)$ and $(1, 0)$. From $a^2 = 1$ and $b^2 = 9$, (4) implies that $c^2 = a^2 + b^2 = 10$ and $c = \sqrt{10}$. The coordinates of the foci are $(-\sqrt{10}, 0)$ and $(\sqrt{10}, 0)$. Finally, from

$$x^2 - \frac{y^2}{3^2} = 0 \quad \text{or} \quad \left(x + \frac{y}{3}\right)\left(x - \frac{y}{3}\right) = 0$$

we see that the asymptotes are $y = -3x$ and $y = 3x$. Graphing the asymptotes and vertices leads to the hyperbola in Figure 11.30. □

Center at (h, k) The analogues of (5) and (6) are

$$\frac{(x-h)^2}{a^2} - \frac{(y-k)^2}{b^2} = 1 \tag{7}$$

and

$$\frac{(y-k)^2}{a^2} - \frac{(x-h)^2}{b^2} = 1 \tag{8}$$

when the center of the hyperbola is at (h, k). For each equation, $b^2 = c^2 - a^2$. The asymptotes, which pass through (h, k), can be found from the factors of

$$\frac{(x-h)^2}{a^2} - \frac{(y-k)^2}{b^2} = 0 \quad \text{or} \quad \frac{(y-k)^2}{a^2} - \frac{(x-h)^2}{b^2} = 0$$

To find the two vertices, we set $y = k$ in (7) and solve for x, or set $x = h$ in (8) and solve for y. On a practical level, to find the vertices, we need only measure a units, either horizontally or vertically, from the center. Similarly, the foci are c units from the center: $(h \pm c, k)$ for (7) and $(h, k \pm c)$ for (8). The coordinates of the vertices are given in Figure 11.31.

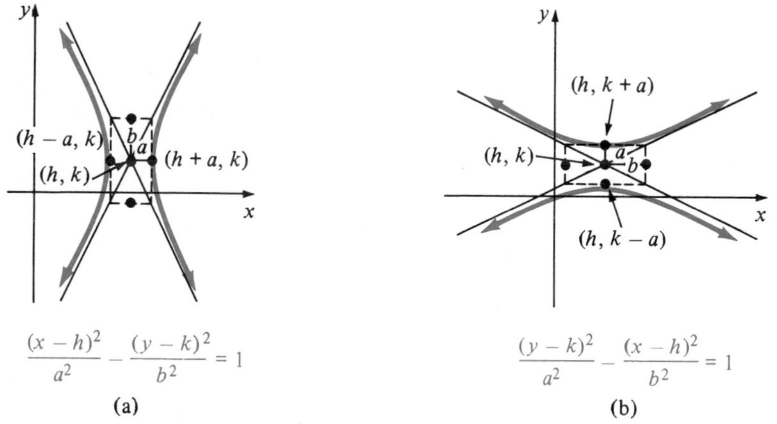

FIGURE 11.31

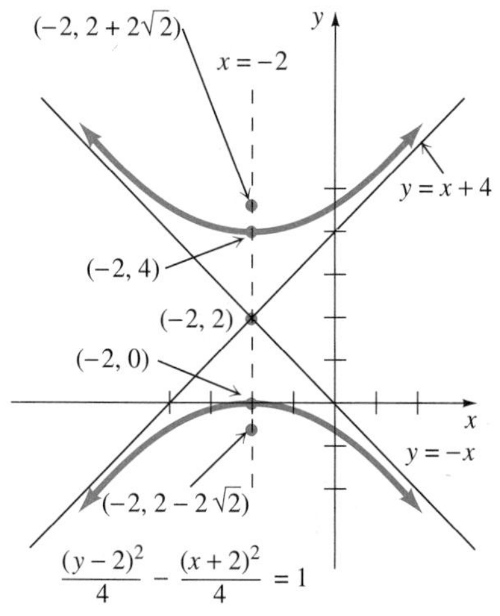

FIGURE 11.32

EXAMPLE 2 Find the vertices, foci, and asymptotes of the hyperbola whose equation is $(y-2)^2/4 - (x+2)^2/4 = 1$. Graph.

Solution Identifying the equation with the standard form (8), we see immediately that the center is $(-2, 2)$, and $a^2 = 4, b^2 = 4, c^2 = 8$. In addition, we conclude the following:

Transverse axis: vertical

Vertices: $a = 2$ units above and below the center along the line $x = -2$; that is $(-2, 4)$ and $(-2, 0)$. (Check this by solving $(y-2)^2/4 = 1$.)

Foci: $c = 2\sqrt{2}$ units above and below the center along the line $x = -2$; that is, $(-2, 2 + 2\sqrt{2})$ and $(-2, 2 - 2\sqrt{2})$.

Asymptotes: $\dfrac{(y-2)^2}{4} - \dfrac{(x+2)^2}{4} = 0$ implies $y - 2 = -(x+2)$ and $y - 2 = x + 2$, or $y = -x$ and $y = x + 4$.

By using the asymptotes and vertices, we find the graph shown in Figure 11.32. □

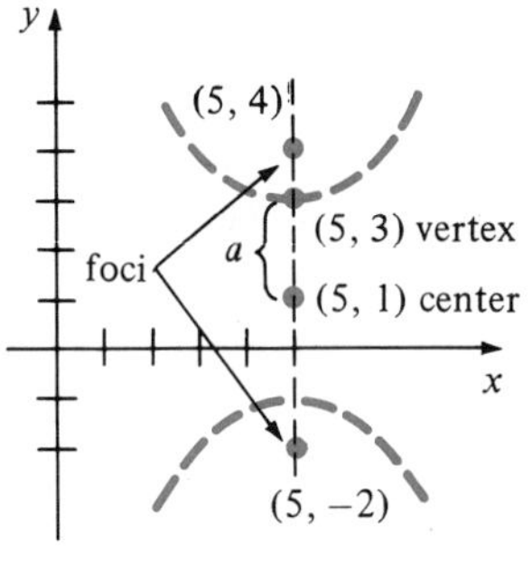

FIGURE 11.33

EXAMPLE 3 Find an equation of the hyperbola with foci $(5, -2)$ and $(5, 4)$ and one vertex $(5, 3)$.

Solution Since the center must be at the midpoint of the line segment between $(5, -2)$ and $(5, 4)$, we have $h = 5$ and $k = (-2 + 4)/2 = 1$. The center is $(5, 1)$. Furthermore, from Figure 11.33 we see that $a = 2$ and $a^2 = 4$. With $c = 3$ and $b^2 = c^2 - a^2 = 9 - 4 = 5$, it follows from (8) that an equation is

$$\frac{(y-1)^2}{4} - \frac{(x-5)^2}{5} = 1$$ □

Eccentricity The eccentricity of a hyperbola is defined in the following manner:

DEFINITION 11.5 Eccentricity

The **eccentricity** of a hyperbola is

$$e = \frac{c}{a} = \frac{\sqrt{a^2 + b^2}}{a}$$

Since $0 < a < \sqrt{a^2 + b^2}$, the eccentricity of a hyperbola satisfies $e > 1$. As with the ellipse, the magnitude of the eccentricity of a hyperbola is an indicator of its shape. Figure 11.34 shows examples of two extreme cases: $e \approx 1$ and e much greater than 1.

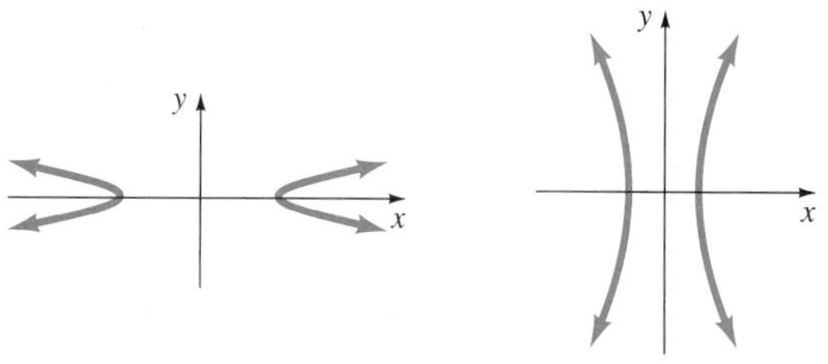

(a) e close to 1 (b) e much greater than 1

FIGURE 11.34

EXAMPLE 4 Find the eccentricity of the hyperbola whose equation is $\frac{y^2}{2} - \frac{(x-1)^2}{36} = 1.$

Solution Identifying $a^2 = 2$ and $b^2 = 36$, we get $c^2 = 2 + 36 = 38$. Thus, from Definition 11.5 the eccentricity is found to be

$$e = \frac{\sqrt{38}}{\sqrt{2}} \approx 4.4$$

Hence, the hyperbola is one whose branches open widely as in Figure 11.34(b). □

▲ ***Remark*** As shown in Figure 11.35(a), a plane flying at a supersonic speed parallel to level ground leaves a hyperbolic sonic "footprint" on the ground. The Cassegranian reflecting telescope shown in Figure 11.35(b) utilizes a convex hyperbolic secondary mirror to reflect a ray of light back through a hole to an eyepiece behind the parabolic primary reflector. This telescope construction utilizes the fact that a beam of light directed along a line through one focus of a hyperbolic mirror will be reflected on a line through the other focus. See Figure 11.38 and Problem 48 in Exercises 11.3.

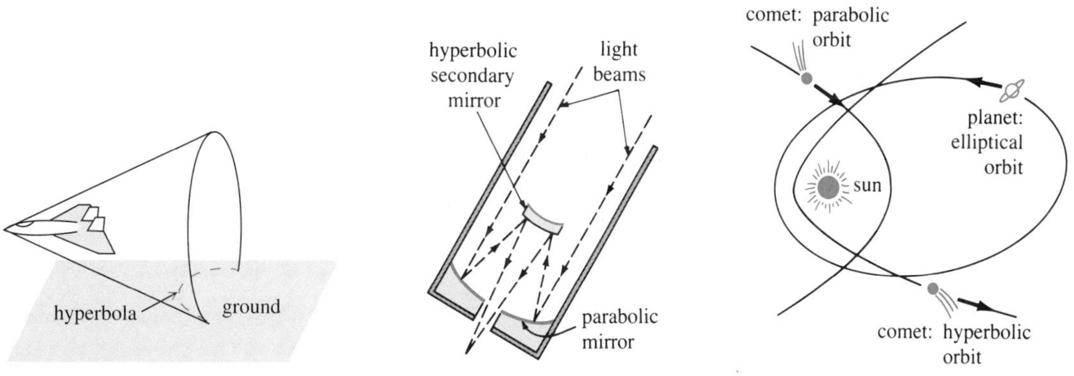

(a) sonic "footprint" (b) Cassegranian telescope (c)

FIGURE 11.35

In intermediate mechanics, it is usually proved that the orbit of a body or atomic particle that moves in a central force field whose magnitude varies as $1/r^2$ must be a conic section. **Kepler's first law of planetary motion** states that the orbit of a planet must be an ellipse with the sun at one focus. The orbit of a "short-period" comet, such as Halley's comet, is also elliptical. The "period" of Halley's comet, or the time it takes the comet to travel once around its orbit, is approximately 76 years. As indicated in Figure 11.35(c) the orbit of a "nonperiodic" comet, or a comet that never returns to the solar system, can be either parabolic or hyperbolic. A modern hypothesis is that the orbits of all comets are elliptical, since they originate in a gigantic spherical cloud of comet nuclei that surrounds the solar system. This cloud, not yet observed, is called the "Oort cloud" after its proposer, the Dutch astronomer Jan Hendrik Oort. The Oort cloud supposedly exists far beyond the orbit of Pluto, extending outward from about 1000 astronomical units to 100,000 astronomical units. (1 A.U. is 93 million miles or the distance from the earth to the sun; the mean distance to the planet Pluto is only 39.5. A.U.) When a star passes close enough to another star to have its orbit disturbed, the orbit will be hyperbolic.

Did you notice the hyperbolic arcs on the pencil in the figure in the chapter opening discussion?

EXERCISES 11.3 *Answers to odd-numbered problems begin on page A-94.*

In Problems 1–20 find the vertices, ~~foci~~, asymptotes, and graph of the hyperbola whose equation is given.

1. $\dfrac{x^2}{4^2} - \dfrac{y^2}{3^2} = 1$
2. $\dfrac{y^2}{16} - \dfrac{x^2}{9} = 1$
3. $y^2 - x^2 = 9$
4. $2x^2 - y^2 = 4$
5. $4x^2 - 25y^2 = 100$
6. $y^2 - x^2 + 10 = 0$
7. $16x^2 - 9y^2 + 144 = 0$
8. $4y^2 - 9x^2 = 1$
9. $(x-1)^2 - (y-2)^2 = 1$
10. $\dfrac{(y-3)^2}{9} - x^2 = 1$
11. $\dfrac{(y+1)^2}{36} - \dfrac{(x-4)^2}{4} = 1$
12. $\dfrac{(x+3/2)^2}{25} - \dfrac{(y-5/2)^2}{9} = 1$

Johannes Kepler (1571–1630) The first law of planetary motion is one of three laws advanced by the German astronomer, and successor to the great astronomical observer Tycho Brahe, Johannes Kepler. The other two laws of planetary motion are:

> As a planet revolves, a line joining it to the sun sweeps out equal areas in equal time intervals.
>
> The square of the period of a planet—that is, the time it takes for a planet to revolve around the sun—is proportional to the cube of the planet's mean distance from the sun.

These laws, based on observations, were subsequently proved by Isaac Newton using his newly formulated law of gravitation. In his writings Kepler hinted at the law of gravitation by arguing that the earth and the moon were held in their orbits by some "vital force." He was also the first to advance the idea that tides were caused by lunar attraction on the oceans. Kepler was also a mathematician, astrologer, and a mystic. His mother was accused of being a witch.

13. $\dfrac{(2x-1)^2}{16} - \dfrac{(3y+4)^2}{36} = 1$

14. $\dfrac{4}{9}(x+2)^2 - \dfrac{16}{9}(y+3)^2 + 1 = 0$

15. $25(y-5)^2 - 4(x+2)^2 = 100$

16. $x^2 + 5x - y^2 + 3y = 1$

17. $x(x-4) = y(y-2)$

18. $(3x - y - 10)(3x + y - 8) = 9$

19. $x^2 - y^2 + 8x - 2y - 10 = 0$

20. $9y^2 - 64x^2 + 90y - 128x = 415$

In Problems 21–24 find the eccentricity of the hyperbola with the given equation.

21. $y^2 - \dfrac{x^2}{8} = 1$

22. $\dfrac{(x+9)^2}{16} - \dfrac{(y-7)^2}{33} = 1$

23. $2x^2 - y^2 + 6y - 11 = 0$

24. $20(x+1)^2 - 16(y-1)^2 = 320$

In Problems 25–38 find an equation of the hyperbola that satisfies the given conditions.

25. Center (0, 0), vertices $(\pm 3, 0)$, one focus (5, 0)

26. Foci $(0, \pm 3)$, one vertex (0, 1)

27. Foci $(2 \pm 5\sqrt{2}, -7)$, length of transverse axis 10

28. Center (0, 0), one focus (5, 0), length of conjugate axis 8

29. Vertices $(1, -1)$ and (1, 5), one focus $(1, -2)$

30. Center $(-2, -5)$, length of horizontal transverse axis 14, length of conjugate axis 12

31. Vertices $(0, \pm 4)$, passing through $(\frac{1}{2}, -3\sqrt{2})$

32. Vertices $(\pm \frac{1}{2}, 0)$, asymptotes $y = \pm \frac{3}{2}x$

33. Foci $(\pm 3, 0)$, asymptotes $y = \pm\sqrt{3}x$

34. Center (2, 4), one vertex (2, 5), one asymptote $2y - x - 6 = 0$

35. Center $(-1, 2)$, focus (5, 2), through (3, 2)

36. Set of all points in the plane the difference of whose distances from (1, 3) and $(1, -1)$ is 2

37. Foci $(0, \pm 10)$, eccentricity $\frac{5}{3}$

38. Endpoints of conjugate axis $(-5, 4)$ and $(-5, 10)$ eccentricity $\sqrt{10}$

39. Determine the point(s) on the hyperbola whose equation is $2x^2 - y^2 = 14$ at which the slope of the tangent is 4.

40. Find an equation of the tangent line to the hyperbola whose equation is $(x-1)^2 - (y+2)^2 = 1$ at $x = 2$. at $x = 3$.

41. Tangent lines to the graph of the hyperbola $x^2 - y^2 = 40$ intersect at (0, 5). Find the points of tangency.

42. For the equation $x^2/a^2 - y^2/b^2 = 1$ show that the area A of the shaded region in Figure 11.36 is

$$A = \frac{1}{2}ab \ln\left(\frac{x_0}{a} + \frac{y_0}{b}\right)$$

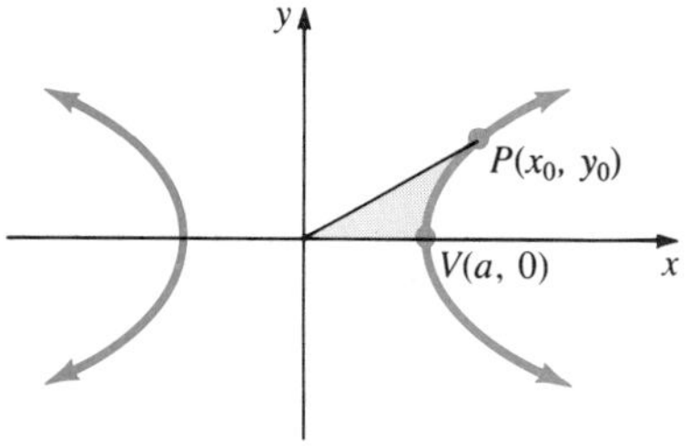

FIGURE 11.36

43. The region in the first quadrant that is bounded by the graphs of $x^2/4 - y^2 = 1$, an asymptote, $y = 3$, and $y = 0$ is revolved about the y-axis. Find the volume of the solid of revolution.

44. A cannon is heard at three points that are given the coordinates $P_1(0, 3)$, $P_2(12, 6)$, and $P_3(0, -3)$, where distance is measured in kilometers. By sound equipment it is determined that the cannon is 2 km closer to P_1 than to P_3.

(a) Show that the position of the cannon must be on a branch of a hyperbola.

(b) Suppose it is determined that the cannon lies on the line through P_2 and P_3. Find its position.*

45. The line segment that is perpendicular to a line containing the transverse axis through a focus with endpoints on the hyperbola is called a **focal chord**. See Figure 11.37. Show that the length of a focal chord of a hyperbola with center at the origin is $2b^2/a$.

*During wartime this method was employed to locate enemy artillery. The same idea is used in LORAN (long-range navigation), where the position of a ship can be determined by calculating the time difference between the radio signals from two known stations.

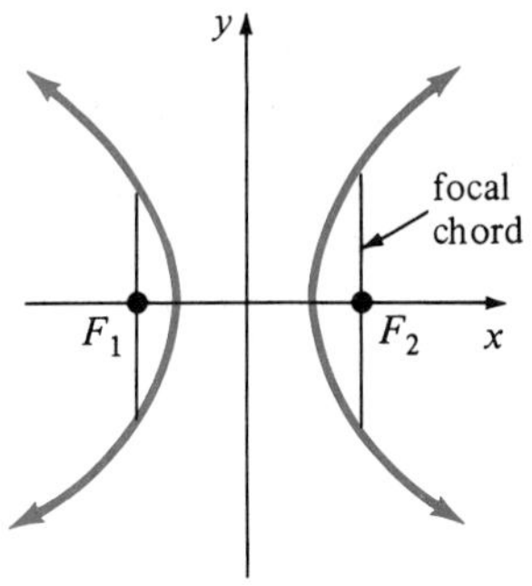

FIGURE 11.37

46. **Conjugate hyperbolas** are two hyperbolas such that the transverse axis of each is the conjugate axis of the other. Using the same coordinate axes, graph $x^2/36 - y^2/4 = 1$ and its conjugate.

47. A **rectangular hyperbola** is one for which the asymptotes are perpendicular. Which of the hyperbolas whose equations are given in Problems 1–20 are rectangular?

48. As shown in Figure 11.38, let L be the tangent line to the graph of $x^2/a^2 - y^2/b^2 = 1$ at $P(x_0, y_0)$. Show that $\alpha = \beta$.

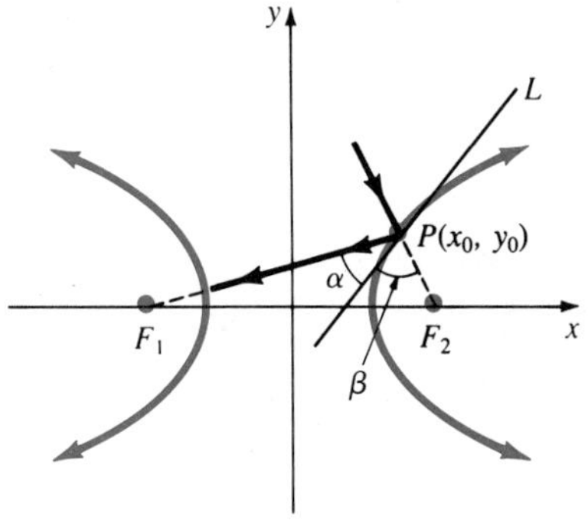

FIGURE 11.38

[O] 11.4 TRANSLATION AND ROTATION OF AXES

Consider the ellipse whose equation is

$$\frac{(x-3)^2}{6^2} + \frac{(y-4)^2}{3^2} = 1$$

If we let $X = x - 3$ and $Y = y - 4$, the equation becomes

$$\frac{X^2}{6^2} + \frac{Y^2}{3^2} = 1$$

In terms of a new XY-coordinate system, the graph of the equation is as shown in Figure 11.39(a). The center of the ellipse corresponds to the simul-

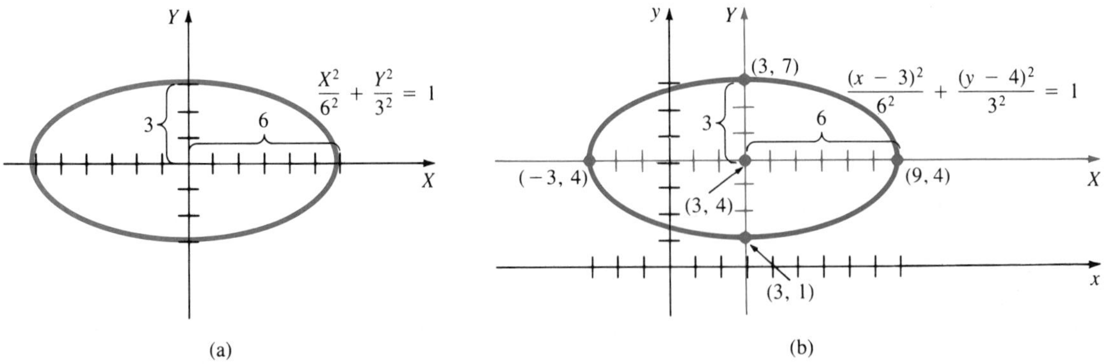

FIGURE 11.39

taneous condition $X = 0$, $Y = 0$, which, in turn, is the point (3, 4) in the xy-plane. By superimposing the xy-plane over the XY-plane, we obtain the points indicated in Figure 11.39(b).

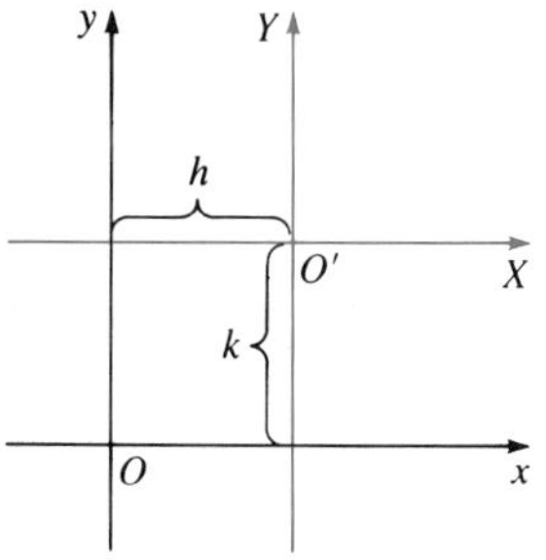

FIGURE 11.40

Translation of Axes Equations such as $(x - h)^2 = 4p(y - k)$ and $(x - h)^2/a^2 + (y - k)^2/b^2 = 1$ can be simplified by means of the substitution $X = x - h$, $Y = y - k$ and readily graphed in terms of X and Y. In general, the equations

$$\begin{aligned} X &= x - h \\ Y &= y - k \end{aligned} \quad \text{or} \quad \begin{aligned} x &= X + h \\ y &= Y + k \end{aligned}$$

define a **translation** of axes or a translation between the xy-plane and the XY-plane. Relative to the xy-plane, the equation of the Y-axis is $x = h$ and the equation of the X-axis is $y = k$. As shown in Figure 11.40, the origin O' in the XY-plane is the point (h, k) in the xy-plane.

Rotation of Axes In contrast to the translation of axes, we obtain yet another coordinate system by a **rotation** of axes. If the positive x-axis is rotated by an amount θ, $0 < \theta < 90°$, about the origin, we obtain the X- and Y-axes shown in Figure 11.41. A straightforward exercise in trigonometry (see Problem 25 in Exercises 11.4) shows that if (X, Y) are the coordinates of a point in the new coordinate system, then its xy-coordinates are

$$\begin{aligned} x &= X \cos\theta - Y \sin\theta \\ y &= X \sin\theta + Y \cos\theta \end{aligned} \tag{1}$$

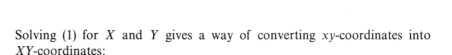

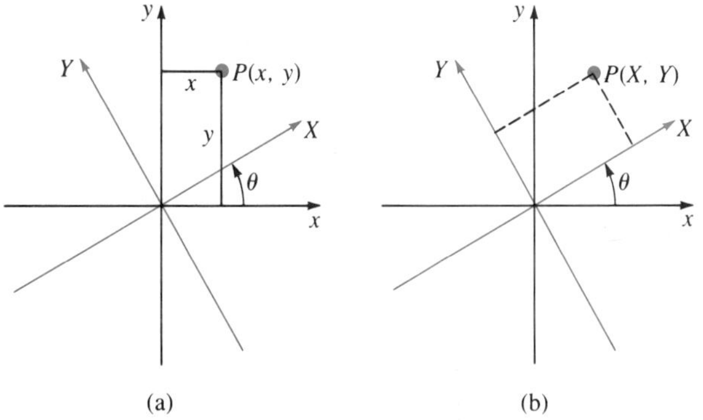

FIGURE 11.41

Solving (1) for X and Y gives a way of converting xy-coordinates into XY-coordinates:

$$\begin{aligned} X &= x \cos\theta + y \sin\theta \\ Y &= -x \sin\theta + y \cos\theta \end{aligned} \tag{2}$$

EXAMPLE 1 The positive x-axis is rotated by 30°. Find the XY-coordinates of the point whose xy-coordinates are (4, 6).

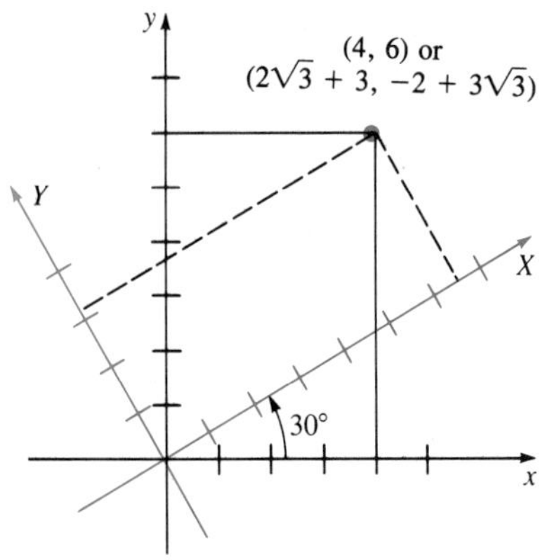

FIGURE 11.42

Solution With $\theta = 30°$, $x = 4$, and $y = 6$, we have, from (2),

$$X = 4\cos 30° + 6\sin 30°$$
$$Y = -4\sin 30° + 6\cos 30°$$

or

$$X = 2\sqrt{3} + 3 \approx 6.5$$
$$Y = -2 + 3\sqrt{3} \approx 3.2$$

The point and the rotated axes are shown in Figure 11.42. □

Elimination of xy-Term An appropriate rotation of axes enables us to eliminate the xy-term in

$$ax^2 + bxy + cy^2 + dx + ey + f = 0 \tag{3}$$

and obtain an equation of the form

$$AX^2 + CY^2 + DX + EY + F = 0 \tag{4}$$

If (4) defines a real locus of points other than a point, a line, or a pair of intersecting lines, its graph will then be a conic section. The cases: a point, a line, and a pair of intersecting lines are said to be **degenerate conics**.

Substituting the equations in (1) into (3) and simplifying reveal that the coefficient of the XY-term is zero provided that

$$2(c - a)\sin\theta\cos\theta + b(\cos^2\theta - \sin^2\theta) = 0$$

or equivalently

$$(a - c)\sin 2\theta = b\cos 2\theta \tag{5}$$

Notice that if $a = c$ in (3), then $a - c = 0$ in (5). The resulting equation, $\cos 2\theta = 0$, implies that a rotation by an amount $\theta = 45°$ will eliminate the xy-term in (3). However, if $a \neq c$, then this elimination can be accomplished by choosing θ to be an angle for which

$$\tan 2\theta = \frac{b}{a - c} \tag{6}$$

EXAMPLE 2 The simple equation $xy = 1$ can be rewritten in terms of X and Y without the product xy. Since $a = c = 0$, we use $\theta = 45°$ in (1):

$$x = \frac{X}{\sqrt{2}} - \frac{Y}{\sqrt{2}}$$
$$y = \frac{X}{\sqrt{2}} + \frac{Y}{\sqrt{2}}$$

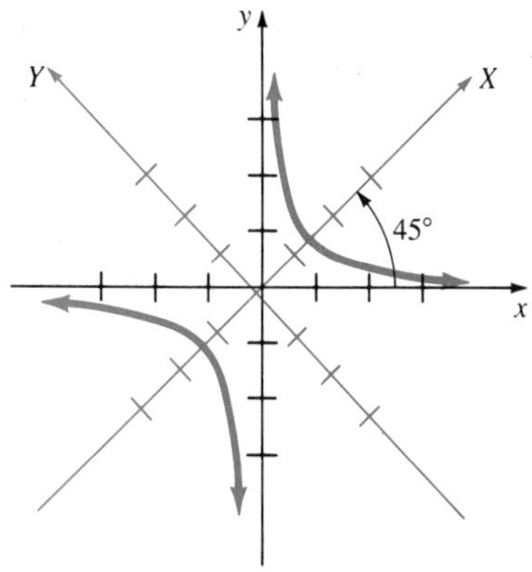

FIGURE 11.43

Thus, $xy = 1$ is the same as

$$\left(\frac{X}{\sqrt{2}} - \frac{Y}{\sqrt{2}}\right)\left(\frac{X}{\sqrt{2}} + \frac{Y}{\sqrt{2}}\right) = 1 \quad \text{or} \quad \frac{X^2}{2} - \frac{Y^2}{2} = 1$$

As illustrated in Figure 11.43, the equation is a standard form for a hyperbola in the XY-coordinate system with vertices $(\pm\sqrt{2}, 0)$. □

In the slightly more complicated case of $a - c \neq 0$, the identities

$$\cos 2\theta = \frac{\pm 1}{\sqrt{1 + \tan^2 2\theta}} \qquad \sin\theta = \sqrt{\frac{1 - \cos 2\theta}{2}} \qquad \cos\theta = \sqrt{\frac{1 + \cos 2\theta}{2}} \tag{7}$$

prove to be very useful.*

EXAMPLE 3 By rotation of axes, eliminate the xy-term in the equation

$$9x^2 + 12xy + 4y^2 + 2x - 3y = 0$$

Identify and graph.

Solution Identifying $a = 9$, $b = 12$, and $c = 4$, we have from (6), $\tan 2\theta = \frac{12}{5}$. Since 2θ is an angle in the first quadrant, (7) yields

$$\cos 2\theta = \frac{5}{13} \qquad \sin\theta = \frac{2}{\sqrt{13}} \qquad \cos\theta = \frac{3}{\sqrt{13}}$$

Hence, the equations in (1) become

$$x = \frac{3}{\sqrt{13}}X - \frac{2}{\sqrt{13}}Y$$

$$y = \frac{2}{\sqrt{13}}X + \frac{3}{\sqrt{13}}Y$$

Substituting these equations into $9x^2 + 12xy + 4y^2 + 2x - 3y = 0$ and simplifying give

$$Y = \sqrt{13}X^2$$

Thus, the graph of the given equation is a parabola. From $\theta = \frac{1}{2}\tan^{-1}(\frac{12}{5})$, we see that the positive x-axis is rotated by $\theta \approx 33.7°$. The graph of the equation is given in Figure 11.44. □

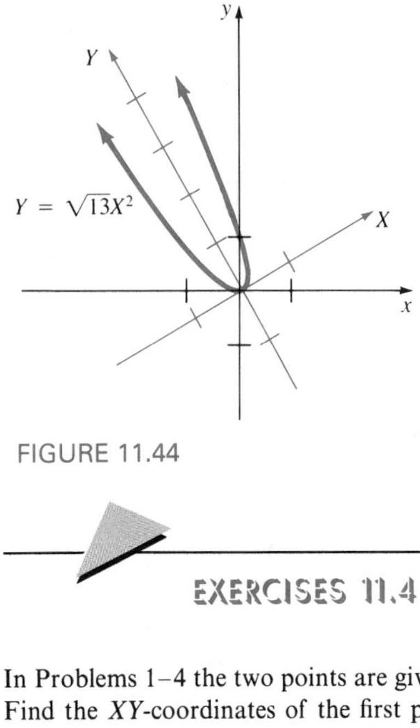

FIGURE 11.44

EXERCISES 11.4 *Answers to odd-numbered problems begin on page A-95.*

In Problems 1–4 the two points are given in xy-coordinates. Find the XY-coordinates of the first point if the translated origin O' is the second point.

1. $(3, 2), (1, 3)$
2. $(-1, 7), (-4, 3)$
3. $(\frac{1}{2}, -\frac{3}{2}), (2, -5)$
4. $(-6, -8), (2, -8)$

In Problems 5–10 use translation of axes to discuss the difference in the graphs of the given pairs of equations.

5. $y = |x|$
   $y = |x - 1| + 4$
6. $y = x^{2/3}$
   $y = (x + 2)^{2/3} - 1$
7. $y = \sin x$
   $y = \sin(x + \pi/2)$
8. $y = x^2$
   $y = x^2 - 6x + 7$

*The first identity comes from $1 + \tan^2 2\theta = \sec^2 2\theta$. The second two are the half-angle formulas given in (12) and (13) of Section 1.3.

9. $x^2 - y^2 = 4$
$(x + 1)^2 - (y - 1)^2 = 4$

10. $4x^2 + y^2 = 16$
$4(x - 5)^2 + y^2 = 16$

In Problems 11–14 the positive x-axis is rotated by the indicated amount. Find the XY-coordinates of the point with the given xy-coordinates.

11. (6, 2); 45°

12. (−2, 8); 30°

13. (−1, −1); 60°

14. (5, 3); 15°

In Problems 15 and 16, without the aid of a calculator or tables, find the XY-coordinates of the point whose xy-coordinates are given for the indicated amount of rotation.

15. $(5, -5)$; $\tan^{-1}\frac{4}{3}$

16. $(20, 10)$; $\tan^{-1}\frac{1}{3}$

In Problems 17–22 use rotation of axes to eliminate the xy-term in the given equation. Identify and graph.

17. $x^2 + xy + y^2 = 4$

18. $2x^2 - 3xy - 2y^2 = 5$

19. $x^2 - 2xy + y^2 = 8x + 8y$

20. $3x^2 + 4xy = 16$

21. $4x^2 - 4xy + 7y^2 + 12x + 6y - 9 = 0$

22. $x^2 - xy + y^2 - 4x - 4y = 20$

23. Given $3x^2 + 2\sqrt{3}xy + y^2 + 2x - 2\sqrt{3}y = 0$.

(a) By rotation of axes show that the graph of the equation is a parabola.

(b) Find the XY-coordinates of the focus. Use this information to find the xy-coordinates of the focus.

(c) Find an equation of the directrix in terms of the XY-coordinates. Use this information to find an equation of the directrix in terms of the xy-coordinates.

24. Given $13x^2 - 8xy + 7y^2 = 30$.

(a) By rotation of axes show that the graph of the equation is an ellipse.

(b) Find the XY-coordinates of the foci. Use this information to find the xy-coordinates of the foci.

(c) Find the xy-coordinates of the vertices.

25. (a) Use Figure 11.45 to show that $X = d\cos\phi$, $Y = d\sin\phi$ and $x = d\cos(\theta + \phi)$, $y = d\sin(\theta + \phi)$.

(b) Use the results in part (**a**) to derive the equations in (1).

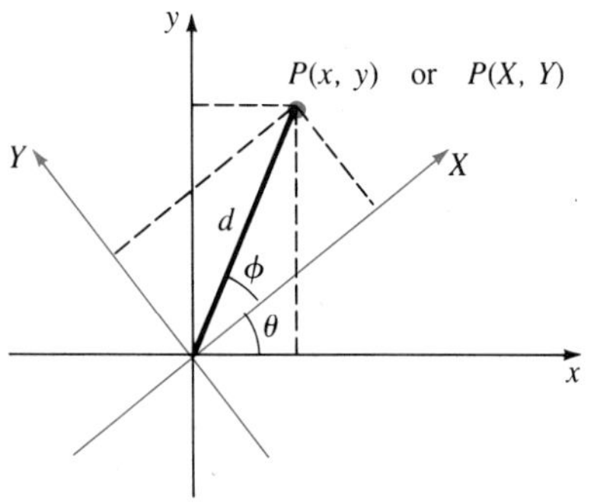

FIGURE 11.45

26. Graph the equation $xy = 3x - 2y + 6$.

Except in degenerate cases, the graph of (3) is

(*i*) a parabola if $b^2 - 4ac = 0$,
(*ii*) an ellipse if $b^2 - 4ac < 0$, or
(*iii*) a hyperbola if $b^2 - 4ac > 0$.

The expression $b^2 - 4ac$ is called the **discriminant** of the quadratic equation. In Problems 27–32 determine the type of graph without actually graphing.

27. $x^2 - 3xy + y^2 = 5$

28. $2x^2 - 2xy + 2y^2 = 1$

29. $4x^2 - 4xy + y^2 - 6 = 0$

30. $x^2 + \sqrt{3}xy - \frac{1}{2}y^2 = 0$

31. $x^2 + xy + y^2 - x + 2y + 1 = 0$

32. $3x^2 + 2\sqrt{3}xy + y^2 - 2x + 2\sqrt{3}y - 4 = 0$

CHAPTER 11 REVIEW EXERCISES

Answers begin on page A-95.

In Problems 1–16 answer true or false.

1. For a parabola, the distance from the vertex to the focus is the same as the distance from the vertex to the directrix. ________

2. The directrix of $y^2 = 4px$, $p < 0$, is perpendicular to the positive y-axis. ________

3. The minor axis of an ellipse bisects the major axis. ________

4. For an ellipse whose equation is $x^2/a^2 + y^2/b^2 = 1$, the numbers a, b, and c are related by $a^2 = b^2 + c^2$. ________

5. For a hyperbola, the numbers c and a satisfy $c > a$. ________

6. The foci of $(x - h)^2/a^2 + (y - k)^2/b^2 = 1$ are $(h, k \pm c)$. ________

7. A beam of light originating at one focus of an ellipse and striking its mirrored surface will necessarily be reflected to the other focus. ________

8. The asymptotes of $x^2/a^2 - y^2/a^2 = 1$ are perpendicular. ________

9. The asymptotes of a hyperbola always pass through the origin. ________

10. The branches of a hyperbola are two parabolas. ________

11. If (0, 4), (3, 6), (6, 4), and (3, 2) are vertices of an ellipse, then its major axis is vertical. ________

12. If P is a point on the left branch of the hyperbola given by $x^2/a^2 - y^2/b^2 = 1$, then $d(P, F_1) - d(P, F_2) < 0$, where $F_1(-c, 0)$ and $F_2(c, 0)$ are the foci. ________

13. The directrix of a parabola must be either vertical or horizontal. ________

14. The y-intercepts of the graph of $x^2/a^2 - y^2/b^2 = 1$ are $\pm b$. ________

15. The point $(-2, 5)$ is on the ellipse $x^2/8 + y^2/50 = 1$. ________

16. The graphs of $y = x^2$ and $y^2 - x^2 = 1$ have at most two points in common. ________

In Problems 17–30 fill in each blank with the answer for the given equation.

17. $y = 2x^2$, focus ________

18. $\dfrac{x^2}{4} - \dfrac{y^2}{12} = 1$, foci ________

19. $4x^2 + 5(y - 2)^2 = 20$, center ________

20. $25y^2 - 4x^2 = 100$, asymptotes ________

21. $8(y + 3) = (x - 1)^2$, directrix ________

22. $\dfrac{(x + 1)^2}{36} + \dfrac{(y + 7)^2}{16} = 1$, vertices ________

23. $x^2 - 2y^2 = 18$, length of conjugate axis ________

24. $y = x^2 + 4x - 6$, vertex ________

25. $(x - 4)^2 - (y + 1)^2 = 4$, endpoints of transverse axis ________

26. $\dfrac{(x - 3)^2}{7} + \dfrac{(y + 3/2)^2}{8} = 1$, equation of line that contains major axis ________

27. $25x^2 + y^2 - 200x + 6y + 384 = 0$, center ________

28. $(x + 1)^2 + (y + 8)^2 = 100$, x-intercepts ________

29. $y^2 - (x - 2)^2 = 1$, y-intercepts ________

30. $y^2 - y + 3x = 3$, slope of tangent line at (1, 1) ________

31. Show that an equation of the tangent line to the graph of an ellipse $x^2/a^2 + y^2/b^2 = 1$ at (x_1, y_1) is given by

$$\frac{xx_1}{a^2} + \frac{yy_1}{b^2} = 1$$

32. Show that an equation of the tangent line to the graph of a hyperbola $x^2/a^2 - y^2/b^2 = 1$ at (x_1, y_1) is given by

$$\frac{xx_1}{a^2} - \frac{yy_1}{b^2} = 1$$

33. A satellite revolves around the planet Jupiter in an elliptical orbit with the center of the planet at one focus. The length of the major axis of the orbit is 10^9 m and the length of the minor axis is 6×10^8 m. Find the minimum distance between the satellite and the center of Jupiter. What is the maximum distance?

34. Find an equation of the hyperbola that has asymptotes $3y = 5x$ and $3y = -5x$ and vertices (0, 10) and (0, −10).

35. The distance between a point (x_0, y_0) and a line $ax + by + c = 0$ is given by $|ax_0 + by_0 + c|/\sqrt{a^2 + b^2}$. Use this result to find an equation of the directrix of a parabola whose vertex is at the origin and whose focus is (2, 2).

36. Use Definition 11.1 to find an equation of the parabola described in Problem 35.

37. The arch of a masonry bridge is a semi-ellipse whose span is 90 ft and whose height in the center is 30 ft. What is the height of the arch 15 ft from the center?

12

PARAMETRIC EQUATIONS AND POLAR COORDINATES

INTRODUCTION

A rectangular or Cartesian equation is not the only, and often not the most convenient, way of describing a curve in the plane. In this chapter we shall consider two additional means by which a curve can be represented. One of the two approaches utilizes an entirely new kind of coordinate system.

Although they are fast disappearing from record stores, most of us still have collections of the large black vinyl 33-rpm records. In these records, sound is encoded on the disk by mechanical means along a continuous groove. When a record is played, a needle starts at a point near the outer edge of the disk and traverses the groove up to a point near its center. Think about this: how long is the groove of a record? If interested, then solve Problem 34 in Exercises 12.5.

IMPORTANT CONCEPTS

- plane curve
- parametric equations
- parameter
- elimination of the parameter
- smooth curve
- piecewise smooth curve
- arc length
- slope of tangent line
- polar coordinate system
- graphs of polar equations
- symmetry of polar graphs
- area in polar coordinates
- arc length in polar coordinates
- conic sections in polar coordinates

12.1 PARAMETRIC EQUATIONS

Curvilinear Motion The motion of a particle along a curve, in contrast to a straight line, is called **curvilinear motion**. In physics it is shown that the motion of a projectile in the xy-plane, such as a struck golf ball,* is governed by the fact that its acceleration in the x- and y-directions satisfies

$$a_x = 0 \qquad a_y = -g \tag{1}$$

where g is the acceleration due to gravity and $a_x = d^2x/dt^2$, $a_y = d^2y/dt^2$. At $t = 0$ we take $x = 0$, $y = 0$, and the x- and y-components of the initial velocity v_0 to be

$$v_0\cos\theta_0 \quad \text{and} \quad v_0\sin\theta_0$$

respectively. See Figure 12.1. Taking two antiderivatives of (1), we see from the initial conditions that the x- and y-coordinates of the ball at time t are

$$x = (v_0\cos\theta_0)t \qquad y = -\frac{1}{2}gt^2 + (v_0\sin\theta_0)t \tag{2}$$

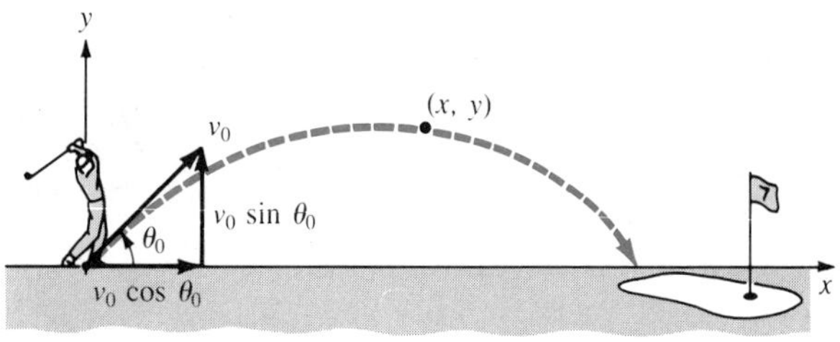

FIGURE 12.1

These equations, which give the ball's position in the xy-plane at a time t, are said to be **parametric equations**. The variable t is called a **parameter** and is restricted to an interval $0 \le t \le T$, where T is the time the ball hits the ground.

In general, a curve can be *defined* in terms of parametric equations.

DEFINITION 12.1 Plane Curve

A **plane curve** is a set C of ordered pairs $(f(t), g(t))$, where f and g are functions continuous on an interval I. The equations $x = f(t)$, $y = g(t)$, for t in I, are called **parametric equations** for C. The variable t is called a **parameter**.

*Assuming no slice or hook. The effects of air resistance are also ignored.

The **graph** of a plane curve C is the set of all points (x, y) in the Cartesian plane corresponding to the ordered pairs $(f(t), g(t))$. Hereafter, we shall refer to a plane curve as a **curve**. Furthermore, for simplicity we shall not belabor the distinction between a *curve* and the *graph of a curve*.

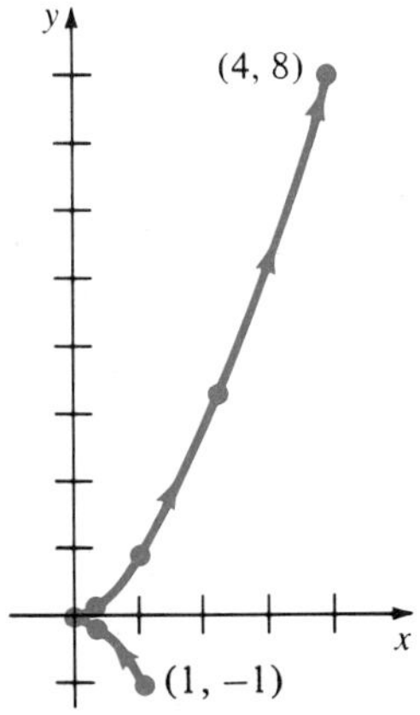

FIGURE 12.2

EXAMPLE 1 Graph the curve that has the parametric equations

$$x = t^2, \qquad y = t^3, \qquad -1 \le t \le 2$$

Solution As shown in the accompanying table, for any choice of t in $[-1, 2]$, we obtain an ordered pair (x, y). By connecting the points with a curve, we obtain the graph in Figure 12.2.

t	-1	$-\frac{1}{2}$	0	$\frac{1}{2}$	1	$\frac{3}{2}$	2
x	1	$\frac{1}{4}$	0	$\frac{1}{4}$	1	$\frac{9}{4}$	4
y	-1	$-\frac{1}{8}$	0	$\frac{1}{8}$	1	$\frac{27}{8}$	8

□

In Example 1, if we think in terms of motion and t as time, then as t increases from -1 to 2, a particle at point P starts from $(1, -1)$, advances up the lower branch, passes to the upper branch, and finally stops at $(4, 8)$. In general, as we plot points corresponding to increasing values of the parameter, the curve C is traced in a certain direction. This direction is called the **orientation** of the curve. The arrowheads on the curve in Figure 12.2 indcate its orientation.

A parameter need have no relation to time. When the interval I is not specified, it is usually understood to be either $-\infty < t < \infty$ or the longest interval on which f and g are both continuous. When I is a closed interval $a \le t \le b$, we say that $(f(a), g(a))$ is the **initial point** of the curve C and $(f(b), g(b))$ is its **terminal point**. In Example 1, the initial point is $(1, -1)$ and the terminal point is $(4, 8)$. If $(f(a), g(a)) = (f(b), g(b))$, then C is a **closed curve**. The next example illustrates a closed curve.

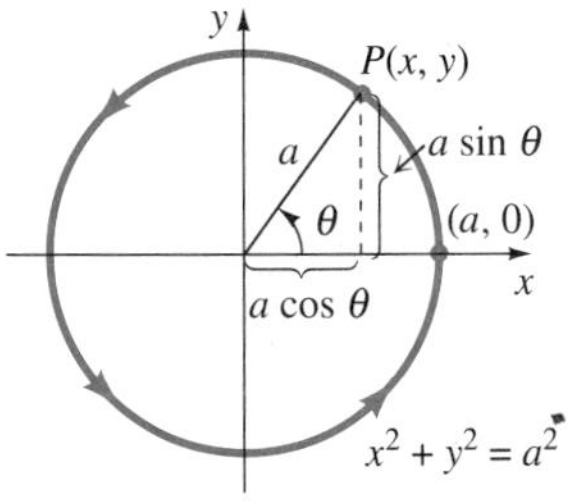

FIGURE 12.3

EXAMPLE 2 The circle $x^2 + y^2 = a^2$ with center at the origin and radius a can be parameterized in terms of a central angle θ. Using Figure 12.3 and trigonometry, we see that the equations

$$x = a\cos\theta, \qquad y = a\sin\theta, \qquad 0 \le \theta \le 2\pi \tag{3}$$

give every point P on the circle. Note in Figure 12.3 that the orientation of the curve is counterclockwise. The initial and terminal point is $(a, 0)$. □

EXAMPLE 3 In Example 2, the semicircle $x^2 + y^2 = a^2, 0 \le y \le a$, is given parametrically by $x = a\cos\theta$, $y = a\sin\theta$, $0 \le \theta \le \pi$. □

Eliminating the Parameter Given a set of parametric equations, we sometimes desire to **eliminate** or **clear the parameter** to obtain a Cartesian equation. To eliminate the parameter in (3), we simply square x and y and add the two equations:

$$x^2 + y^2 = a^2\cos^2\theta + a^2\sin^2\theta \quad \text{implies that} \quad x^2 + y^2 = a^2$$

since $\cos^2\theta + \sin^2\theta = 1$.

EXAMPLE 4

(*a*) From the first equation in (2), $t = x/(v_0\cos\theta_0)$, so that the second equation yields

$$y = -\frac{g}{2(v_0\cos\theta_0)^2}x^2 + (\tan\theta_0)x$$

Thus, the trajectory of any projectile launched at an angle $0 < \theta_0 < \pi/2$ is necessarily a parabola.

(*b*) In Example 1, we can eliminate the parameter by solving the second equation for t in terms of y and substituting in the first equation. We find

$$t = y^{1/3} \quad \text{and so} \quad x = (y^{1/3})^2 = y^{2/3}$$

But for $-1 \le t \le 2$ we have correspondingly $-1 \le y \le 8$. Thus, a Cartesian equation for the curve is given by $x = y^{2/3}$, $-1 \le y \le 8$. □

A curve C can have more than one parameterization. For example, an examination of $x = t$, $y = 2t^2$, $-\infty < t < \infty$, and $x = t^3/4$, $y = t^6/8$, $-\infty < t < \infty$, reveals that both sets of equations represent $y = 2x^2$.* But, one has to be careful when working with parametric equations. Eliminating the parameter in $x = t^2$, $y = 2t^4$, $-\infty < t < \infty$, would seem to yield the same parabola $y = 2x^2$. This is not the case because, for any value of t, $t^2 \ge 0$ and so $x \ge 0$. In other words, the last set is a parametric representation of only the right branch of the parabola: $y = x^2$, $x \ge 0$. You are encouraged to work Problems 17–22 in Exercises 12.1.

EXAMPLE 5 Consider the curve C given parametrically by

$$x = \sin\theta, \qquad y = \cos 2\theta, \qquad 0 \le \theta \le \frac{\pi}{2}$$

Eliminate the parameter and obtain a Cartesian equation that has the same graph.

*Note that a point need not correspond to the same value of the parameter in each set. For example, (1, 2) is obtained from $t = 1$ in the first set but from $t = \sqrt[3]{4}$ in the second set.

Solution Using the double angle formula for cos 2θ, we can write

$$\begin{aligned} y &= \cos^2\theta - \sin^2\theta \\ &= (1 - \sin^2\theta) - \sin^2\theta \\ &= 1 - 2\sin^2\theta \end{aligned}$$

$x = \sin\theta$

or $y = 1 - 2x^2$

Now the curve C described by the parametric equations does not consist of the entire parabola whose equation is $y = 1 - 2x^2$. See Figure 12.4(a). For $0 \le \theta \le \pi/2$ we have $0 \le \sin\theta \le 1$ and $-1 \le \cos 2\theta \le 1$. This means C is only that portion of the parabola for which the coordinates of a point $P(x, y)$ satisfy $0 \le x \le 1$ *and* $-1 \le y \le 1$. Thus, C is the curve shown in color in Figure 12.4(b). A Cartesian equation of C is then $y = 1 - 2x^2$, $0 \le x \le 1$. □

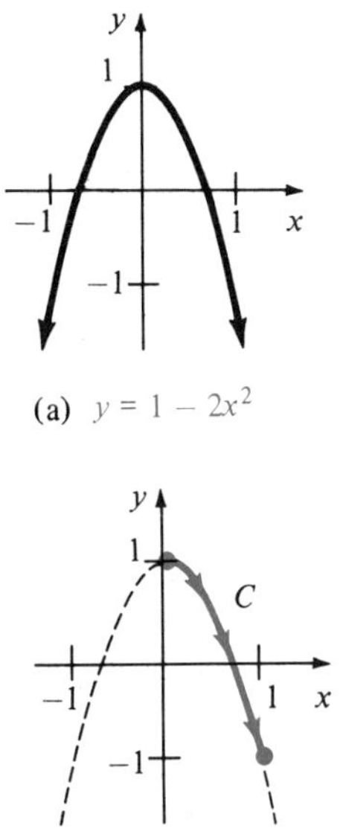

(a) $y = 1 - 2x^2$

(b) $x = \sin\theta$, $y = \cos 2\theta$, $0 \le \theta \le \pi$

FIGURE 12.4

We can get the intercepts of a curve C without finding its Cartesian equation. For instance, in Example 5 we can find the x-intercept by solving $y = 0$. The equation $\cos 2\theta = 0$ yields $2\theta = \pi/2$ or $\theta = \pi/4$. The corresponding point at which C crosses the x-axis is $(1/\sqrt{2}, 0)$. Similarly, the y-intercept of C is found by solving $x = 0$.

Cycloidal Curves **Cycloidal curves** were a popular topic of study by mathematicians in the seventeenth century. Suppose a point $P(x, y)$, marked on a circle of radius a, is at the origin when its diameter lies along the y-axis. As the circle rolls along the x-axis, the point P traces out a curve C that is called a **cycloid**. See Figure 12.5(a). Now suppose a circle of radius a rolls on the *inside* of a larger circle of radius b. If a point $P(x, y)$ on the inside circle

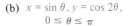

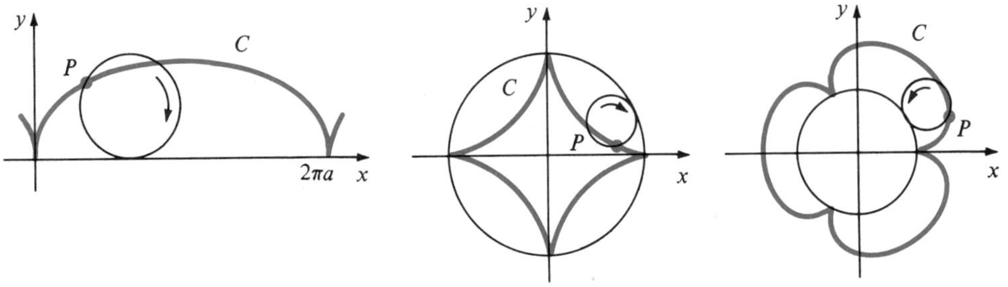

(a) cycloid

(b) hypocycloid of four cusps

(c) epicycloid of three cusps

FIGURE 12.5

starts from $(b, 0)$, the curve C traced out by P is called a **hypocycloid**. Specifically, if $b = 4a$, then C is a **hypocycloid of four cusps**. This is the curve illustrated in Figure 12.5(b). Finally, if a circle of radius a rolls on the *outside* of a circle of radius b, starting from $(b, 0)$, we obtain a curve C known as an **epicycloid**. The **epicycloid of three cusps** illustrated in Figure 12.5(c) results when $b = 3a$. It is left as exercises to obtain the parametric equations of the curves shown in Figure 12.5(b) and (c).

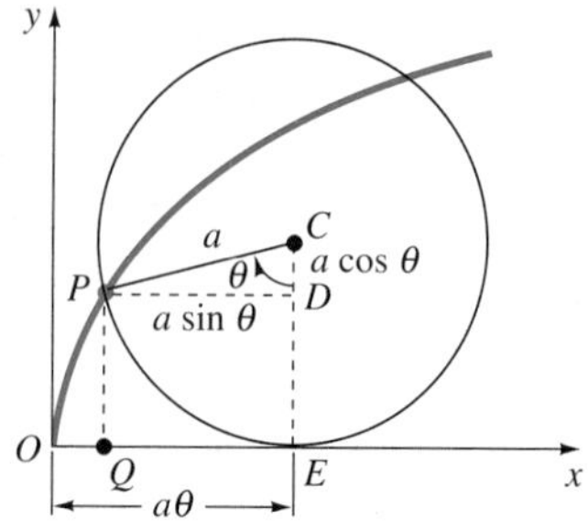

FIGURE 12.6

EXAMPLE 6 Find parametric equations for the cycloid in Figure 12.5(a).

Solution A circle of radius a whose diameter initially lies along the y-axis rolls along the x-axis without slipping. We take as a parameter the angle θ (in radians) through which the circle has rotated. The point $P(x, y)$ starts at the origin, which corresponds to $\theta = 0$. As the circle rolls through an angle θ, its distance from the origin is the arc $PE = \overline{OE} = a\theta$. From Figure 12.6 we then see that the x-coordinate of P is

$$x = \overline{OE} - \overline{QE} = a\theta - a \sin \theta$$

Now the y-coordinate of P is seen to be

$$y = \overline{CE} - \overline{CD} = a - a \cos \theta$$

Hence, parametric equations for the cycloid are

$$x = a(\theta - \sin \theta) \qquad y = a(1 - \cos \theta)$$

As shown in Figure 12.5(a), one arch of a cycloid is generated by one rotation of the circle and corresponds to $0 \le \theta \le 2\pi$. □

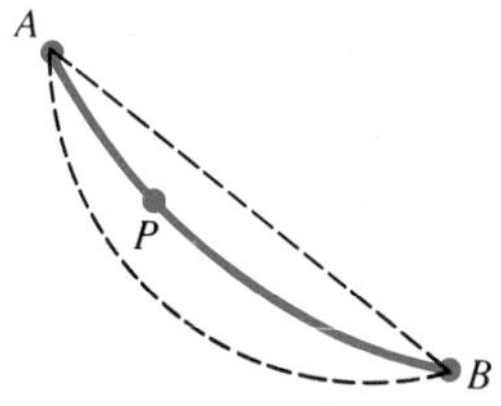

FIGURE 12.7

Two problems were extensively studied in the seventeenth century. Consider a flexible (frictionless) wire fixed at points A and B and a bead free to slide down the wire starting at P. See Figure 12.7. Is there a particular shape of the wire so that, regardless of where the bead starts, the time to slide down the wire to B will be the same? Also, what would the shape of the wire be so that the bead slides from P to B in the shortest time? The so-called **tautochrone** (same time) and **brachistochrone** (least time) were shown to be an inverted half-arch of a cycloid.

Smooth Curves A curve C, given parametrically by

$$x = f(t), \qquad y = g(t), \qquad a \le t \le b$$

Christiaan Huygens (1629–1695) Huygens was a Dutch physicist and astronomer. His lasting contributions to physics were the wave theory of light and the invention of the pendulum clock. It was in his research on the motion of a pendulum that he posed, and then solved, the tautochrone problem. In astronomy he improved the quality of reflecting telescopes and produced the earliest known sketches of the planet Mars. Huygens also discovered the great nebula in the constellation Orion and the rings of Saturn.

Jakob Bernoulli (1654–1705) and Johann Bernoulli (1667–1748) The Bernoullis were a Swiss family of scholars whose contributions to mathematics, physics, astronomy, and history spanned from the sixteenth to the twentieth centuries. Jakob, the elder of the two sons of the patriarch Jaques Bernoulli, made many contributions to the then-new fields of calculus and probability. But it was Johann Bernoulli who challenged mathematicians to solve the brachistochrone problem in 1696. Both he and Jakob, along with Newton and Leibniz, showed that the solution to the problem was a cycloid.

is said to be **smooth** if f' and g' are continuous on $[a, b]$ and not simultaneously zero on (a, b). A curve C is said to be **piecewise smooth** if the interval $[a, b]$ can be partitioned into subintervals such that C is smooth on each subinterval. The curves in Examples 2, 3, and 5 are smooth; the curves in Examples 1 and 6 are piecewise smooth.

▲ ***Remark*** A curve C described by a continuous function $y = f(x)$ can *always* be parameterized by letting $x = t$. A set of parametric equations for C is $x = t$, $y = f(t)$.

EXERCISES 12.1

Answers to odd-numbered problems begin on page A-95.

In Problems 1 and 2 fill in the table for the given set of parametric equations.

1. $x = 2t + 1$, $y = t^2 + t$

t	-3	-2	-1	0	1	2	3
x							
y							

2. $x = \cos t$, $y = \sin^2 t$

t	0	$\pi/6$	$\pi/4$	$\pi/3$	$\pi/2$	$5\pi/6$	$7\pi/4$
x							
y							

In Problems 3–10 graph the curve that has the given set of parametric equations.

3. $x = t - 1$, $y = 2t - 1$; $-1 \le t \le 5$

4. $x = 3t$, $y = t^2 - 1$; $-2 \le t \le 3$

5. $x = \sqrt{t}$, $y = 5 - t$; $t \ge 0$

6. $x = 3 + 2\sin t$, $y = 4 + \sin t$; $-\pi/2 \le t \le \pi/2$

7. $x = 4\cos t$, $y = 4\sin t$; $-\pi/2 \le t \le \pi/2$

8. $x = t^3 + 1$, $y = t^2 - 1$; $-2 \le t \le 2$

9. $x = e^t$, $y = e^{3t}$; $0 \le t \le \ln 2$

10. $x = -e^t$, $y = e^{-t}$; $t \ge 0$

In Problems 11–16 eliminate the parameter from the given set of parametric equations and obtain a Cartesian equation that has the same graph.

11. $x = t^2$, $y = t^4 + 3t^2 - 1$

12. $x = t^3 + t + 4$, $y = -2t^3 - 2t$

13. $x = \cos 2\theta$, $y = \sin\theta$; $-\pi/2 \le \theta \le \pi/2$

14. $x = e^t$, $y = \ln t$; $t > 0$

15. $x = t^3$, $y = 3\ln t$; $t > 0$

16. $x = \tan t$, $y = \sec t$; $-\pi/2 < t < \pi/2$

In Problems 17–22 graphically show the difference between the given curves.

17. $y = x$ and $x = \sin t$, $y = \sin t$

18. $y = x^2$ and $x = -\sqrt{t}$, $y = t$, $t \ge 0$

19. $y = \dfrac{x^2}{4} - 1$ and $x = 2t$, $y = t^2 - 1$, $-1 \le t \le 2$

20. $y = -x^2$ and $x = e^t$, $y = -e^{2t}$, $t \ge 0$

21. $x^2 - y^2 = 1$ and $x = \cosh t$, $y = \sinh t$

22. $y = 2x - 2$ and $x = t^2 - 1$, $y = 2t^2 - 4$

In Problems 23–26 graphically show the difference between the given curves. Assume $a > 0$, $b > 0$.

23. $x = a\cos t$, $y = a\sin t$, $0 \le t \le \pi$
$x = a\sin t$, $y = a\cos t$, $0 \le t \le \pi$

24. $x = a\cos t$, $y = b\sin t$, $a > b$, $\pi \le t \le 2\pi$
$x = a\sin t$, $y = b\cos t$, $a > b$, $\pi \le t \le 2\pi$

25. $x = a\cos t$, $y = a\sin t$, $-\pi/2 \le t \le \pi/2$
$x = a\cos 2t$, $y = a\sin 2t$, $-\pi/2 \le t \le \pi/2$

26. $x = a \cos \frac{t}{2}$, $y = a \sin \frac{t}{2}$, $0 \le t \le \pi$

$x = a \cos\left(-\frac{t}{2}\right)$, $y = a \sin\left(-\frac{t}{2}\right)$, $-\pi \le t \le 0$

In Problems 27 and 28 graph the curve that has the given parametric equations.

27. $x = 1 + 2 \cosh t$, $y = 2 + 3 \sinh t$

28. $x = -3 + 3 \cos t$, $y = 5 + 5 \sin t$

29. Determine which of the following sets of parametric equations have the same graph as the Cartesian equation $xy = 1$.

(a) $x = \dfrac{1}{2t + 1}$, $y = 2t + 1$

(b) $x = t^{1/2}$, $y = t^{-1/2}$

(c) $x = \cos t$, $y = \sec t$

(d) $x = t^2 + 1$, $y = (t^2 + 1)^{-1}$

(e) $x = e^{-2t}$, $y = e^{2t}$

(f) $x = t^3$, $y = t^{-3}$

30. The ends of a rod of length L slide on horizontal and vertical tracks that coincide with the x- and y-axes, respectively.

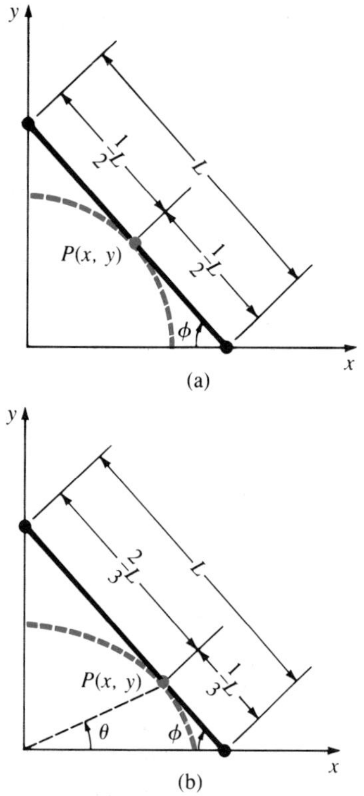

FIGURE 12.8

(a) Parameterize the coordinates of the point P in Figure 12.8(a) in terms of ϕ. Show that P traces a circular path as ϕ varies from 0 to $\pi/2$.

(b) Show that if P is located as shown in Figure 12.8(b), then it traces an elliptical path as ϕ varies from 0 to $\pi/2$.

(c) Parameterize the coordinates of P in Figure 12.8(b) in terms of θ.

31. As shown in Figure 12.9, a piston is attached by means of a rod of length L to a circular crank mechanism of radius r. Parameterize the coordinates of the point P in terms of the angle ϕ.

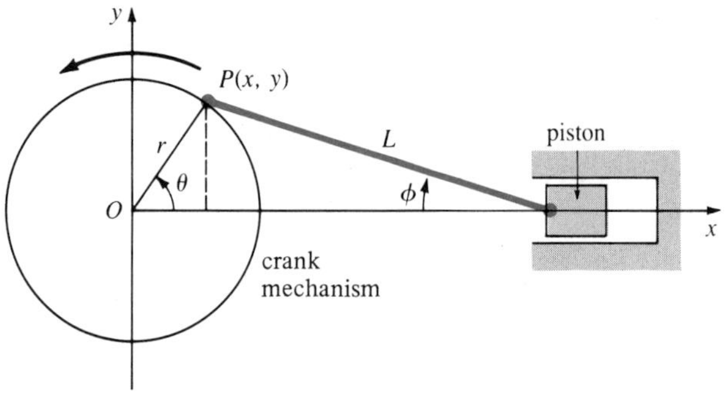

FIGURE 12.9

32. A point Q traces out a circular path of radius r and a point P moves in the manner shown in Figure 12.10. If R is constant, find parametric equations of the path traced by P. This curve is called an **epitrochoid**. (Those knowledgeable about automobiles might recognize the curve traced by P as the shape of the rotor housing in the rotary or Wankel engine.)

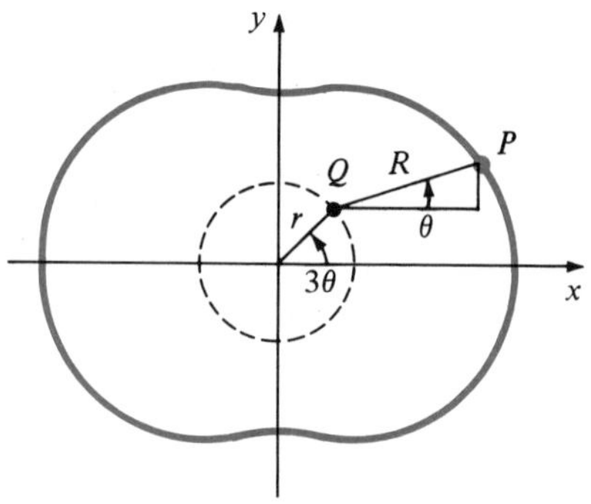

FIGURE 12.10

33. Consider a circle of radius a, which is tangent to the x-axis at the origin O. Let B be a point on a horizontal line through $(O, 2a)$ and let the line segment OB cut the circle at

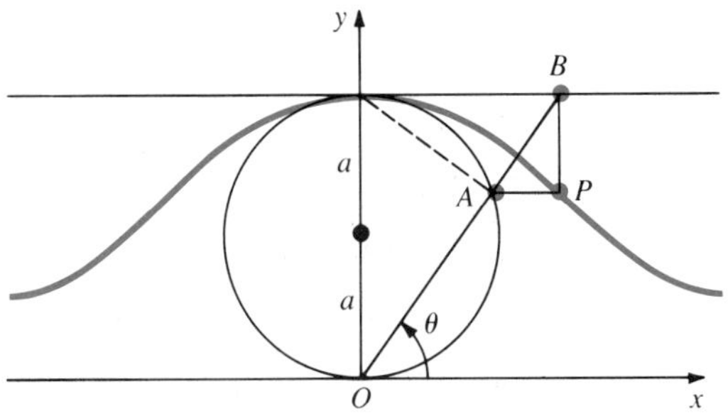

FIGURE 12.11

point A. As shown in Figure 12.11, the projection of AB on the vertical gives the line segment BP. Find parametric equations of the path traced by the point P as A varies around the circle. The curve is called the **Witch of Agnesi**.

34. In Problem 33 eliminate the parameter and show that the curve has the Cartesian equation $y = 8a^3/(x^2 + 4a^2)$.

35. A circular spool wound with thread has its center at the origin. The radius of the spool is a. The end of the thread P, starting from $(a, 0)$, is unwound while the thread is kept taut. See Figure 12.12. Find parametric equations of the path traced by the point P if the thread PR is tangent to the circular spool at R. The curve is called an **involute** of a circle.

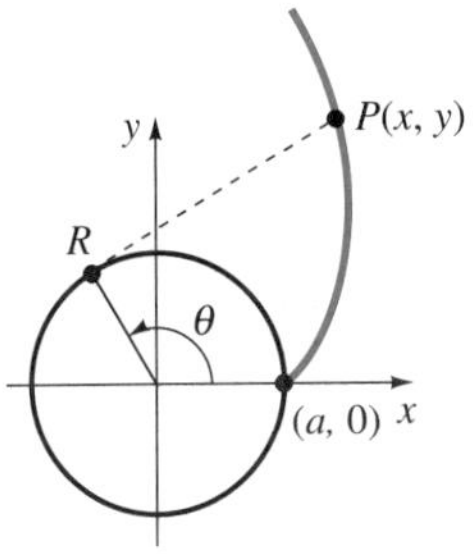

FIGURE 12.12

36. Use Figure 12.13 to show that parametric equations of a **hypocycloid** are

$$x = (b - a)\cos\theta + a\cos\frac{b - a}{a}\theta$$

$$y = (b - a)\sin\theta - a\sin\frac{b - a}{a}\theta$$

Use these equations to show that parametric equations of a **hypocycloid of four cusps** are

$$x = b\cos^3\theta \qquad y = b\sin^3\theta$$

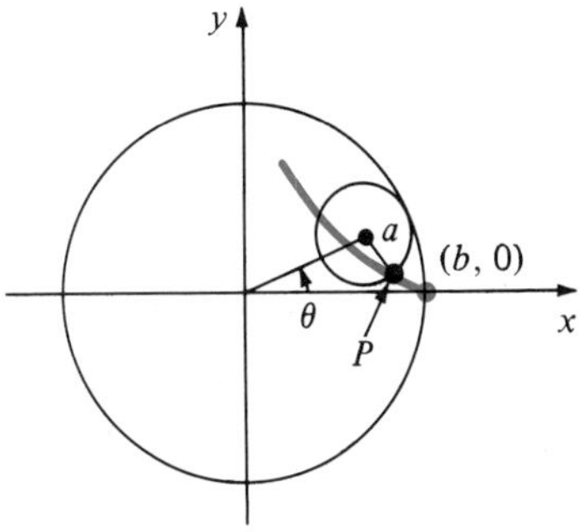

FIGURE 12.13

37. In Problem 36 eliminate the parameter and obtain a Cartesian equation for the hypocycloid of four cusps.

38. Use Figure 12.14 to show that parametric equations of an **epicycloid** are given by

$$x = (a + b)\cos\theta - a\cos\frac{a + b}{a}\theta$$

$$y = (a + b)\sin\theta - a\sin\frac{a + b}{a}\theta$$

Use these equations to show that parametric equations of an **epicycloid of three cusps** are

$$x = 4a\cos\theta - a\cos 4\theta$$

$$y = 4a\sin\theta - a\sin 4\theta$$

Maria Gaetana Agnesi (1718–1799) Agnesi was an Italian mathematician and philosopher. The curve that bears her name was not discovered by her but was first studied by Pierre Fermat and Guido Grandi (1672–1742). Grandi called the curve "versoria," which is Latin for a certain kind of nautical rope. The curve "versoria" appeared in Agnesi's two-volume text *Analytical Institutions,* which was published in 1748. This text on analytical geometry and calculus proved to be so popular that it was soon translated into English. The translator confused the word *versoria* with the Italian word *versiera,* which means "female goblin." In English, "female goblin" became "witch."

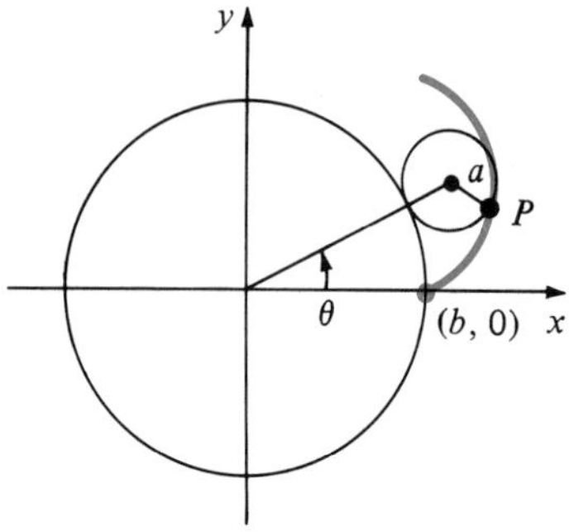

FIGURE 12.14

Calculator/Computer Problems

In Problems 39–44 use a computer to obtain the graph of the given set of parametric equations.

39. $x = 4 \sin 2t$, $y = 2 \sin t$; $0 \le t \le 2\pi$

40. $x = 6 \cos 3t$, $y = 4 \sin 2t$; $0 \le t \le 2\pi$

41. $x = 6 \cos 4t$, $y = 4 \sin t$; $0 \le t \le 2\pi$

42. $x = \cos t + t \sin t$, $y = \sin t - t \cos t$; $0 \le t \le 3\pi$

43. $x = t^3 - 4t + 1$, $y = t^4 - 4t^2$; $-5 \le t \le 5$

44. $x = t^5 - t + 1$, $y = t^3 + 2t - 1$; $-3 \le t \le 6$

12.2 SLOPE OF TANGENT LINES; ARC LENGTH

As with graphs of functions $y = f(x)$, we can often obtain useful information about a curve C defined parametrically by examining the derivative dy/dx.

Slope Let $x = f(t)$ and $y = g(t)$ be parametric equations of a smooth curve C. The **slope** of the tangent line at a point $P(x, y)$ on C is given by dy/dx. To calculate this derivative, we form

$$\Delta x = f(t + \Delta t) - f(t) \quad \text{and} \quad \Delta y = g(t + \Delta t) - g(t)$$

and

$$\frac{\Delta y}{\Delta x} = \frac{\Delta y/\Delta t}{\Delta x/\Delta t}$$

so that

$$\lim_{\Delta t \to 0} \frac{\Delta y}{\Delta x} = \frac{\displaystyle\lim_{\Delta t \to 0} \Delta y/\Delta t}{\displaystyle\lim_{\Delta t \to 0} \Delta x/\Delta t}$$

if the limit of the denominator is not zero. The parametric form of the derivative is summarized in the next theorem.

THEOREM 12.1 Slope of Tangent Line

If $x = f(t)$, $y = g(t)$ define a smooth curve C, then the **slope of the tangent line** at a point $P(x, y)$ on C is

$$\frac{dy}{dx} = \frac{dy/dt}{dx/dt} = \frac{g'(t)}{f'(t)} \tag{1}$$

provided that $f'(t) \ne 0$.

EXAMPLE 1 Find an equation of the tangent line to the curve $x = t^2 - 4t - 2$, $y = t^5 - 4t^3 - 1$ at the point corresponding to $t = 1$.

Solution We first find the slope dy/dx of the tangent line. Since

$$\frac{dx}{dt} = 2t - 4 \quad \text{and} \quad \frac{dy}{dt} = 5t^4 - 12t^2$$

it follows from (1) that

$$\frac{dy}{dx} = \frac{dy/dt}{dx/dt} = \frac{5t^4 - 12t^2}{2t - 4}$$

Thus, at $t = 1$ we have

$$\left.\frac{dy}{dx}\right|_{t=1} = \frac{-7}{-2} = \frac{7}{2}$$

By substituting $t = 1$ back into the original parametric equations, we find the point of tangency to be $(-5, -4)$. Hence, an equation of the tangent line at that point is

$$y - (-4) = \frac{7}{2}(x - (-5)) \quad \text{or} \quad y = \frac{7}{2}x + \frac{27}{2}$$

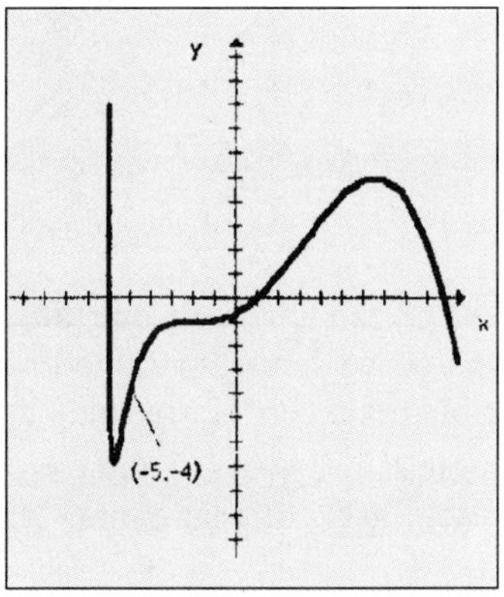

FIGURE 12.15

With the aid of a computer we obtain the curve given in Figure 12.15. □

Horizontal and Vertical Tangents At a point (x, y) on a curve C at which $dy/dt = 0$ and $dx/dt \neq 0$, the tangent line is necessarily **horizontal** because $dy/dx = 0$ at that point. On the other hand, at a point at which $dx/dt = 0$ and $dy/dt \neq 0$, the tangent line is **vertical**. When both dy/dt and dx/dt are zero at a point, we can draw no immediate conclusion about the tangent line.

EXAMPLE 2 Graph the curve that has the parametric equations $x = t^2 - 4$, $y = t^3 - 3t$.

Solution

x-intercepts: $y = 0$ implies $t(t^2 - 3) = 0$ at $t = 0, t = -\sqrt{3}$, and $t = \sqrt{3}$.

y-intercepts: $x = 0$ implies $t^2 - 4 = 0$ at $t = -2$ and $t = 2$.

Horizontal tangents: $\frac{dy}{dt} = 3t^2 - 3$; $\frac{dy}{dt} = 0$ implies $3(t^2 - 1) = 0$ at $t = -1$ and $t = 1$. Note that $dx/dt \neq 0$ at $t = \pm 1$.

Vertical tangents: $\frac{dx}{dt} = 2t$; $\frac{dx}{dt} = 0$ implies $2t = 0$ at $t = 0$. Note that $dy/dt \neq 0$ at $t = 0$.

The points (x, y) on the curve corresponding to these values of the parameter are summarized in the table. A curve plotted through these points, consistent with the orientation and tangent information, is illustrated in Figure 12.16.

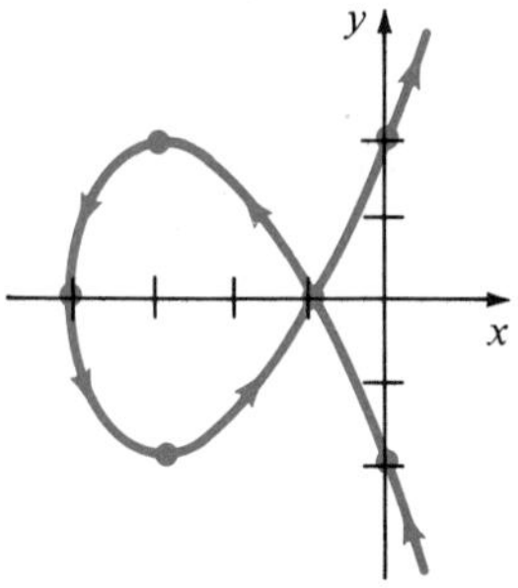

FIGURE 12.16

t	-2	$-\sqrt{3}$	-1	0	1	$\sqrt{3}$	2
x	0	-1	-3	-4	-3	-1	0
y	-2	0	2	0	-2	0	2
	y-int	x-int	hor tan	ver tan, x-int	hor tan	x-int	y-int

□

The graph of a differentiable function $y = f(x)$ can have only one tangent line at a point on its graph. In contrast, since a curve defined parametrically may not be the graph of a function, it is possible that such a curve may have more than one tangent line at a point. In Example 2, for $t = -\sqrt{3}$ and $t = \sqrt{3}$, we get $(-1, 0)$. This means the curve intersects itself at that point. Now, since

$$\left.\frac{dy}{dx}\right|_{t=-\sqrt{3}} = -\sqrt{3} \quad \text{and} \quad \left.\frac{dy}{dx}\right|_{t=\sqrt{3}} = \sqrt{3}$$

we conclude that there are two tangent lines at $(-1, 0)$. This fact should be apparent in Figure 12.16.

Higher-Order Derivatives Higher-order derivatives can be found in exactly the same manner as dy/dx. Suppose (1) is written as

$$\frac{d}{dx}(\quad) = \frac{d(\quad)/dt}{dx/dt} \tag{2}$$

If $y' = dy/dx$ is a differentiable function of t, it follows from (2) by replacing $(\quad)$ by y' that

$$\frac{d^2y}{dx^2} = \frac{d}{dx}y' = \frac{dy'/dt}{dx/dt}$$

Similarly, if $y'' = d^2y/dx^2$ is a differentiable function of t, then the third derivative is

$$\frac{d^3y}{dx^3} = \frac{d}{dx}y'' = \frac{dy''/dt}{dx/dt}$$

and so on.

EXAMPLE 3 Find d^3y/dx^3 for the curve given by $x = 4t + 6$, $y = t^2 + t - 2$.

Solution

$$\frac{dy}{dx} = \frac{dy/dt}{dx/dt} = \frac{2t+1}{4} = y'$$

$$\frac{d^2y}{dx^2} = \frac{dy'/dt}{dx/dt} = \frac{2/4}{4} = \frac{1}{8} = y''$$

$$\frac{d^3y}{dx^3} = \frac{dy''/dt}{dx/dt} = \frac{0}{4} = 0$$

□

Inspection of Example 3 shows that the curve has a horizontal tangent at $t = -\frac{1}{2}$ or $(4, -\frac{9}{4})$. Furthermore, since $d^2y/dx^2 > 0$ for all t, the graph of the curve is concave upward at every point.

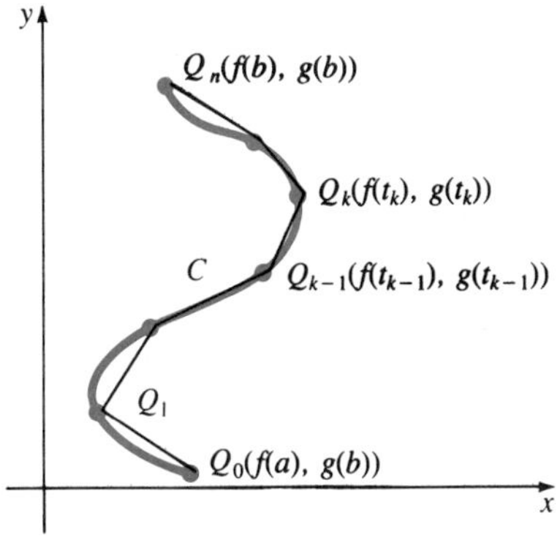

FIGURE 12.17

Length of a Curve In Section 5.4 we were able to find the length s of the graph of a smooth function $y = f(x)$ by means of a definite integral. We can now generalize the result given in (3) of that section to curves defined parametrically. Suppose $x = f(t)$, $y = g(t)$, $a \le t \le b$, are parametric equations of a smooth curve C that does not intersect itself for $a < t < b$. If P is a partition of $[a, b]$ given by

$$a = t_0 < t_1 < t_2 < \cdots < t_{n-1} < t_n = b$$

then, as shown in Figure 12.17, it seems reasonable that C can be approximated by a polygonal path through the points $Q_k(f(t_k), g(t_k))$, $k = 0, 1, \ldots, n$. Denoting the length of the line segment through Q_{k-1} and Q_k by $|Q_{k-1}Q_k|$, we write the approximate length of C as

$$\sum_{k=1}^{n} |Q_{k-1}Q_k| \tag{3}$$

where $\quad |Q_{k-1}Q_k| = \sqrt{[f(t_k) - f(t_{k-1})]^2 + [g(t_k) - g(t_{k-1})]^2}$

Now, since f and g have continuous derivatives, the Mean Value Theorem (see Section 3.4) asserts that there exist numbers u_k^* and v_k^* in (t_{k-1}, t_k) such that

$$f(t_k) - f(t_{k-1}) = f'(u_k^*)(t_k - t_{k-1}) = f'(u_k^*)\,\Delta t_k \tag{4}$$

and

$$g(t_k) - g(t_{k-1}) = g'(v_k^*)(t_k - t_{k-1}) = g'(v_k^*)\,\Delta t_k \tag{5}$$

Using (4) and (5) in (3) and simplifying yield

$$\sum_{k=1}^{n} |Q_{k-1}Q_k| = \sum_{k=1}^{n} \sqrt{[f'(u_k^*)]^2 + [g'(v_k^*)]^2}\,\Delta t_k \tag{6}$$

By taking $\|P\| \to 0$ in (6), we obtain a formula for the length of a smooth curve. Notice that the limit of the sum in (6) is not the usual definition of a definite integral, since we are dealing with two numbers (u_k^* and v_k^*) rather than one in the interval (t_{k-1}, t_k). Nevertheless, it *can* be shown rigorously that the formula given in the next theorem results from (6) by taking $\|P\| \to 0$.

THEOREM 12.2 Arc Length

If $x = f(t)$ and $y = g(t)$, $a \le t \le b$, define a smooth curve C that does not intersect itself for $a < t < b$, then the **length** s of C is

$$s = \int_a^b \sqrt{[f'(t)]^2 + [g'(t)]^2}\, dt = \int_a^b \sqrt{\left(\frac{dx}{dt}\right)^2 + \left(\frac{dy}{dt}\right)^2}\, dt \qquad (7)$$

EXAMPLE 4 Find the length of the curve given by $x = 4t$, $y = t^2$, $0 \le t \le 2$.

Solution Since $f'(t) = 4$ and $g'(t) = 2t$, (7) gives

$$s = \int_0^2 \sqrt{16 + 4t^2}\, dt = 2\int_0^2 \sqrt{4 + t^2}\, dt$$

With the trigonometric substitution $t = 2\tan\theta$, the last integral becomes

$$s = 8\int_0^{\pi/4} \sec^3\theta\, d\theta$$

Integration by parts leads to (see Example 5, Section 8.2)

$$\begin{aligned} s &= [4\sec\theta\tan\theta + 4\ln|\sec\theta + \tan\theta|]_0^{\pi/4} \\ &= 4\sqrt{2} + 4\ln(\sqrt{2} + 1) \approx 9.1823 \end{aligned}$$

□

EXERCISES 12.2

Answers to odd-numbered problems begin on page A-96.

In Problems 1–6 find the slope of the tangent line at the point corresponding to the indicated value of the parameter.

1. $x = t^3 - t^2$, $y = t^2 + 5t$; $t = -1$
2. $x = 4/t$, $y = 2t^3 - t + 1$; $t = 2$
3. $x = \sqrt{t^2 + 1}$, $y = t^4$; $t = \sqrt{3}$
4. $x = e^{2t}$, $y = e^{-4t}$; $t = \ln 2$
5. $x = \cos^2\theta$, $y = \sin\theta$; $\theta = \pi/6$
6. $x = 2\theta - 2\sin\theta$, $y = 2 - 2\cos\theta$; $\theta = \pi/4$

In Problems 7 and 8 find an equation of the tangent line to the given curve at the point corresponding to the indicated value of the parameter.

7. $x = t^3 + 3t$, $y = 6t^2 + 1$; $t = -1$
8. $x = 2t + 4$, $y = t^2 + \ln t$; $t = 1$

In Problems 9 and 10 find an equation of the tangent line to the given curve at the indicated point.

9. $x = t^2 + t$, $y = t^2$; $(2, 4)$
10. $x = t^4 - 9$, $y = t^4 - t^2$; $(0, 6)$
11. What is the slope of the tangent line to the curve given by $x = 4\sin 2t$, $y = 2\cos t$, $0 \le t \le 2\pi$, at the point $(2\sqrt{3}, 1)$?
12. A curve C has parametric equations $x = t^2$, $y = t^3 + 1$. At what point on C is the tangent line given by $y + 3x - 5 = 0$?
13. A curve C has parametric equations $x = 2t - 5$, $y = t^2 - 4t + 3$. Find an equation of the tangent line to C that is parallel to the line $y = 3x + 1$.
14. Verify that the curve given by $x = \cos\theta - 2/\pi$, $y = \sin\theta - 2\theta/\pi$, $-\pi \le \theta \le \pi$, intersects itself. Find equations of tangent lines at the point of intersection.

In Problems 15–18 determine the points on the given curve at which the tangent line is either horizontal or vertical. Graph the curve.

15. $x = t^3 - t,\ y = t^2$

16. $x = \frac{1}{8}t^3 + 1,\ y = t^2 - 2t$

17. $x = t - 1,\ y = t^3 - 3t^2$

18. $x = \sin t,\ y = \cos 3t,\ 0 \le t \le 2\pi$

In Problems 19–22 find dy/dx, d^2y/dx^2, and d^3y/dx^3.

19. $x = 3t^2,\ y = 6t^3$

20. $x = \cos t,\ y = \sin t$

21. $x = e^{-t},\ y = e^{2t} + e^{3t}$

22. $x = \frac{1}{2}t^2 + t,\ y = \frac{1}{2}t^2 - t$

23 Use d^2y/dx^2 to determine the intervals of the parameter for which the curve in Problem 16 is concave upward and the intervals for which it is concave downward.

24. Use d^2y/dx^2 to determine whether the curve given by $x = 2t + 5,\ y = 2t^3 + 6t^2 + 4t$ has any points of inflection.

In Problems 25–30 find the length of the given curve.

25. $x = \frac{5}{3}t^3 + 2,\ y = 4t^3 + 6;\ 0 \le t \le 2$

26. $x = \frac{1}{3}t^3,\ y = \frac{1}{2}t^2;\ 0 \le t \le \sqrt{3}$

27. $x = e^t \sin t,\ y = e^t \cos t;\ 0 \le t \le \pi$

28. One arch of the cycloid:

$$x = a(\theta - \sin\theta),\ y = a(1 - \cos\theta);\ 0 \le \theta \le 2\pi$$

29. One arch of the hypocycloid of four cusps:

$$x = b\cos^3\theta,\ y = b\sin^3\theta;\ 0 \le \theta \le \pi/2$$

30. One arch of the epicycloid of three cusps:

$$x = 4a\cos\theta - a\cos 4\theta,$$

$$y = 4a\sin\theta - a\sin 4\theta,\ 0 \le \theta \le 2\pi/3$$

Calculator/Computer Problems

31. Consider the curve $x = t^2 - 4t - 2,\ y = t^5 - 4t^3 - 1$ from Example 1.

(a) Use a calculator to find the approximate value of the y-intercept shown in Figure 12.15.

(b) Use Newton's Method to find the approximate values of the three x-intercepts shown in Figure 12.15.

Miscellaneous Problems

32. Use $dy/dx = (dy/dt)/(dx/dt)$ and

$$s = \int_{x_1}^{x_2} \sqrt{1 + [f'(x)]^2}\, dx$$

to derive (7).

33. Let C be a curve described by $y = F(x)$, where F is a continuous nonnegative function on $x_1 \le x \le x_2$. Show that if C is given parametrically by $x = f(t),\ y = g(t),\ a \le t \le b,\ f'$ and g continuous, then the **area** under the graph of C is $\int_a^b g(t)f'(t)\,dt$.

34. Use Problem 33 to show that the area under one arch of the cycloid in Figure 12.5(a) is three times the area of the circle.

12.3 POLAR COORDINATE SYSTEM

Up to now we have been using a rectangular coordinate system to specify a point P in the plane. We can regard this system as a grid of horizontal and vertical lines. The coordinates of P are determined by the intersection of two lines, one perpendicular to a horizontal reference line called the x-axis and the other perpendicular to a vertical reference line called the y-axis. As an alternative, in **polar coordinates**, a point P can be described by means of a grid of circles centered at a point O, called the **pole**, and straight lines or rays emanating from O. We take as a reference axis a horizontal half-line directed to the right of O and call it the **polar axis**. By specifying a directed distance r from O and an angle θ, measured in radians, whose initial side is the polar axis and whose terminal side is the ray OP, we can label the coordinates of the point P (r, θ). See Figure 12.18.

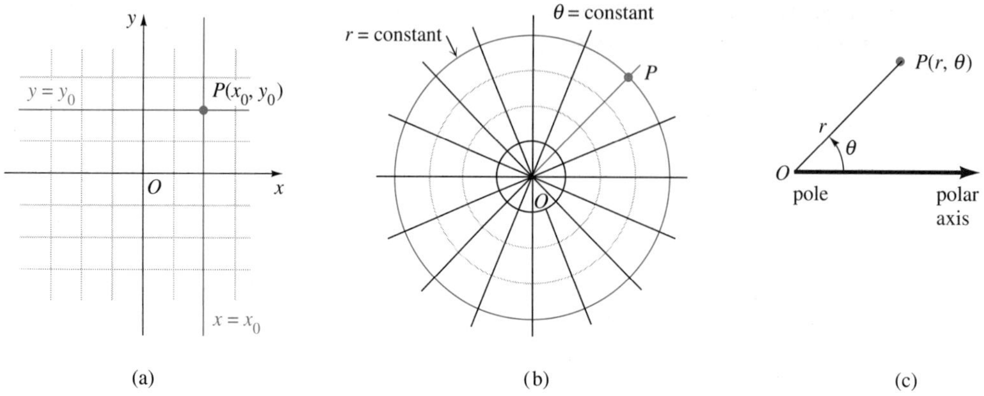

FIGURE 12.18

Conventions In polar coordinates we adopt the following conventions:

(*i*) Angles $\theta > 0$ are measured counterclockwise from the polar axis, whereas angles $\theta < 0$ are measured clockwise.

(*ii*) To graph a point $(-r, \theta)$, $-r < 0$, measure $|r|$ units along the ray $\theta + \pi$.

(*iii*) The coordinates of the pole O are $(0, \theta)$, θ any angle.

EXAMPLE 1 Graph the points whose polar coordinates are given.

(*a*) $\left(4, \dfrac{\pi}{6}\right)$ (*b*) $\left(2, -\dfrac{\pi}{4}\right)$ (*c*) $\left(-3, \dfrac{3\pi}{4}\right)$

Solution (*a*) Measure 4 units along the ray $\pi/6$. See Figure 12.19(a).

(*b*) Measure 2 units along the ray $-\pi/4$. See Figure 12.19(b).

(*c*) Measure 3 units along the ray $3\pi/4 + \pi = 7\pi/4$. Equivalently, we can measure 3 units along the ray $3\pi/4$ extended *backward* through the pole. Note carefully in Figure 12.19(c) that the point $(-3, 3\pi/4)$ is not in the same quadrant as the terminal side of the given angle.

□

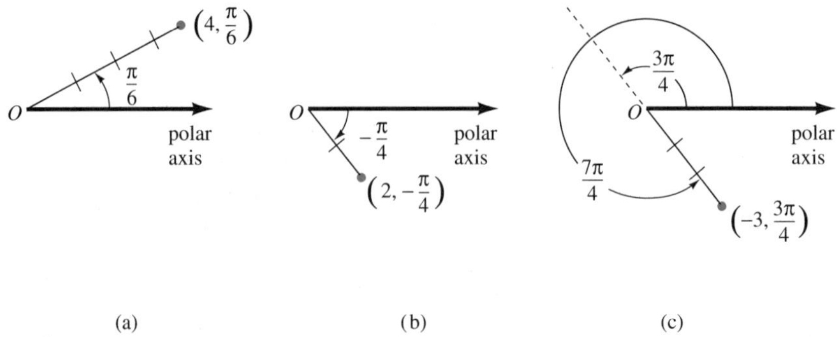

FIGURE 12.19

Unlike in the rectangular coordinate system, the description of a point in polar coordinates is not unique. This is an immediate consequence of the fact that

$$(r, \theta) \quad \text{and} \quad (r, \theta + 2n\pi), \qquad n \text{ an integer}$$

are equivalent. To compound the problem, negative values of r can be used.

EXAMPLE 2 The following are some alternative representations of the point $(2, \pi/6)$:

$$\left(2, \frac{13\pi}{6}\right), \ \left(2, -\frac{11\pi}{6}\right), \ \left(-2, \frac{7\pi}{6}\right), \ \left(-2, -\frac{5\pi}{6}\right) \qquad \square$$

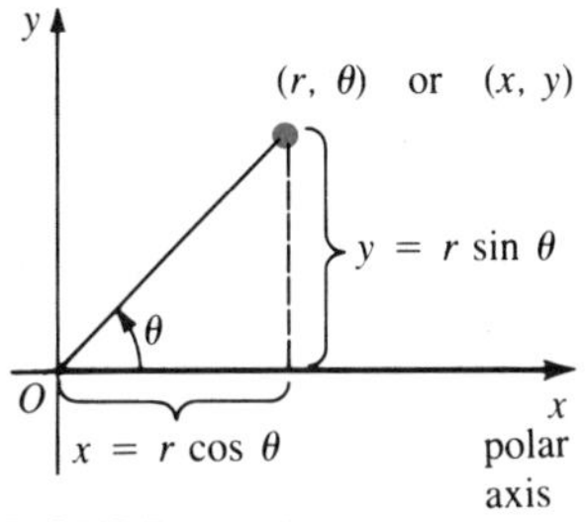

FIGURE 12.20

Conversion of Polar Coordinates to Rectangular Coordinates By superimposing a rectangular coordinate system on a polar coordinate system, as shown in Figure 12.20, we can convert a polar description of a point to rectangular coordinates by using

$$x = r\cos\theta, \qquad y = r\sin\theta \tag{1}$$

These values hold for any value of r.

EXAMPLE 3 Convert $(2, \pi/6)$ in polar coordinates to rectangular coordinates.

Solution With $r = 2$, $\theta = \pi/6$, we have from (1)

$$x = 2\cos\frac{\pi}{6} = 2\left(\frac{\sqrt{3}}{2}\right) = \sqrt{3}$$

$$y = 2\sin\frac{\pi}{6} = 2\left(\frac{1}{2}\right) = 1$$

Thus, $(2, \pi/6)$ is equivalent to $(\sqrt{3}, 1)$ in rectangular coordinates. $\square$

Conversion of Rectangular Coordinates to Polar Coordinates It should be evident from Figure 12.20 that x, y, r, and θ are also related by

$$r^2 = x^2 + y^2, \qquad \tan\theta = \frac{y}{x} \tag{2}$$

These latter equations are used to convert the rectangular coordinates (x, y) to the polar coordinates (r, θ).

EXAMPLE 4 Convert $(-1, 1)$ in rectangular coordinates to polar coordinates.

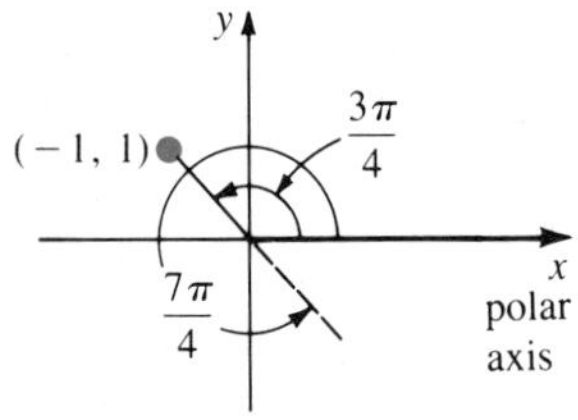

FIGURE 12.21

Solution With $x = -1$, $y = 1$, we have from (2)

$$r^2 = 2 \quad \text{and} \quad \tan\theta = -1$$

Now, $r = \pm\sqrt{2}$, and two, of many possible, angles that satisfy $\tan\theta = -1$ are $3\pi/4$ and $7\pi/4$. From Figure 12.21 we see that two representations for the given point are

$$\left(\sqrt{2}, \frac{3\pi}{4}\right) \quad \text{and} \quad \left(-\sqrt{2}, \frac{7\pi}{4}\right)$$ □

Caution: Note in Example 4 that $(-\sqrt{2}, 3\pi/4)$ and $(\sqrt{2}, 7\pi/4)$ are *not* polar representations of $(-1, 1)$.* In other words, we cannot pair just *any* angle θ and *any* value r that satisfy (2); these solutions must also be consistent with (1).

By virtue of the equations in (1), a Cartesian equation can often be expressed as a polar equation $r = f(\theta)$.

EXAMPLE 5 Find a polar equation that has the same graph as the circle $x^2 + y^2 = 4y$.

Solution Using $x = r\cos\theta$, $y = r\sin\theta$, we write

$$r^2\cos^2\theta + r^2\sin^2\theta = 4r\sin\theta$$
$$r^2(\cos^2\theta + \sin^2\theta) = 4r\sin\theta$$
$$r(r - 4\sin\theta) = 0$$

The latter equation implies that

$$r = 0 \quad \text{or} \quad r = 4\sin\theta$$

Since $r = 0$ determines *only* the pole O, we conclude that a polar equation of the circle is $r = 4\sin\theta$. Observe that relative to this last equation, the coordinates of the pole can be taken to be $(0, \pi)$. □

EXAMPLE 6 Find a polar equation that has the same graph as the parabola $x^2 = 8(2 - y)$.

Solution

$$r^2\cos^2\theta = 8(2 - r\sin\theta)$$
$$r^2(1 - \sin^2\theta) = 16 - 8r\sin\theta$$
$$r^2 = r^2\sin^2\theta - 8r\sin\theta + 16$$
$$r^2 = (r\sin\theta - 4)^2$$
$$r = \pm(r\sin\theta - 4)$$

*$(-\sqrt{2}, 3\pi/4)$ and $(\sqrt{2}, 7\pi/4)$ represent the rectangular coordinates $(1, -1)$.

Solving for r gives

$$r = \frac{4}{1 + \sin\theta} \quad \text{or} \quad r = \frac{-4}{1 - \sin\theta} \tag{3}$$

Since replacement of (r, θ) by $(-r, \theta + \pi)$ in the second equation of (3) yields the first equation,* we may simply take the polar equation of the parabola to be $r = 4/(1 + \sin\theta)$. □

EXAMPLE 7 Find a Cartesian equation that has the same graph as the polar equation $r^2 = 9\cos 2\theta$.

Solution First, use the trigonometric identity for the cosine of a double angle:

$$r^2 = 9(\cos^2\theta - \sin^2\theta)$$

Then, from $r^2 = x^2 + y^2$, $\cos\theta = x/r$, $\sin\theta = y/r$, we have

$$x^2 + y^2 = 9\left(\frac{x^2}{x^2 + y^2} - \frac{y^2}{x^2 + y^2}\right) \quad \text{or} \quad (x^2 + y^2)^2 = 9(x^2 - y^2)$$ □

The next section will be devoted to graphing polar equations.

EXERCISES 12.3 *Answers to odd-numbered problems begin on page A-96.*

In Problems 1–4 graph the point whose polar coordinates are given.

1. $(2, \pi)$
2. $(-4, \pi/3)$
3. $(4, -3\pi/2)$
4. $(-5, -\pi/6)$

In Problems 5–10 find alternative polar coordinate representations of the given point that satisfy

(a) $r > 0, \theta < 0$; (b) $r > 0, \theta > 2\pi$;
(c) $r < 0, \theta > 0$; (d) $r < 0, \theta < 0$.

5. $(6, 3\pi/4)$
6. $(10, \pi/2)$
7. $(2, 2\pi/3)$
8. $(5, \pi/4)$
9. $(1, \pi/6)$
10. $(3, 7\pi/6)$

In Problems 11–16 find the rectangular coordinates of each point whose polar coordinates are given.

11. $(-1, 2\pi/3)$
12. $(1/2, 7\pi/4)$
13. $(-7, -\pi/3)$
14. $(\sqrt{3}, -11\pi/6)$
15. $(4, 5\pi/4)$
16. $(-5, \pi/2)$

In Problems 17–24 find polar coordinates that satisfy **(a)** $r > 0$, $-\pi \le \theta \le \pi$, and **(b)** $r < 0$, $-\pi \le \theta \le \pi$, of each point whose rectangular coordinates are given:

17. $(-3, -3)$
18. $(1, 1)$
19. $(\sqrt{3}, -1)$
20. $(\sqrt{2}, \sqrt{6})$
21. $(0, -5)$
22. $(7, 0)$
23. $(-4, 3)$
24. $(1, 2)$

In Problems 25–34 find a polar equation that has the same graph as the given Cartesian equation.

25. $y = 5$
26. $x + 1 = 0$
27. $y = 7x$
28. $3x + 8y + 6 = 0$
29. $y^2 = -4x + 4$
30. $x^2 - 12y - 36 = 0$

*Remember that by convention (*ii*), (r, θ) and $(-r, \theta + \pi)$ represent the same point.

31. $x^2 + y^2 = 36$

32. $x^2 - y^2 = 25$

33. $x^2 + y^2 + x = \sqrt{x^2 + y^2}$

34. $x^3 + y^3 - xy = 0$

35. Show that a polar equation that has the same graph as a Cartesian equation of a circle through the origin with center at $(a/2, 0)$ is $r = a \cos \theta$.

36. Show that a polar equation that has the same graph as $x^2 + y^2 = ay$ is $r = a \sin \theta$. What is the graph?

In Problems 37–48 find a Cartesian equation that has the same graph as the given polar equation.

37. $r = 2 \sec \theta$

38. $r \cos \theta = -4$

39. $r = 6 \sin 2\theta$

40. $2r = \tan \theta$

41. $r^2 = 4 \sin 2\theta$

42. $r^2\cos 2\theta = 16$

43. $r + 5 \sin \theta = 0$

44. $r = 2 + \cos \theta$

45. $r = \dfrac{2}{1 + 3 \cos \theta}$

46. $r(4 - \sin \theta) = 10$

47. $r = \dfrac{5}{3 \cos \theta + 8 \sin \theta}$

48. $r = 3 + 3 \sec \theta$

49. In Section 11.2, the eccentricity of the ellipse $x^2/a^2 + y^2/b^2 = 1$ was defined to be $e = c/a$, where $c^2 = a^2 - b^2$. Show that a polar equation that has the same graph as the ellipse is $r^2(1 - e^2\cos^2\theta) = b^2$.

50. Show that the distance $d(P_1, P_2)$ between two points whose polar coordinates are $P_1(r_1, \theta_1)$ and $P_2(r_2, \theta_2)$ is given by

$$d(P_1, P_2) = [r_1^2 + r_2^2 - 2r_1r_2\cos(\theta_2 - \theta_1)]^{1/2}$$

51. The two lever arms shown in Figure 12.22 rotate about a common hub so that $\theta_1 = 4t$ and $\theta_2 = 6t$, where t is measured in seconds. At what rate is the distance between the ends of the arms changing when $t = \pi/4$?

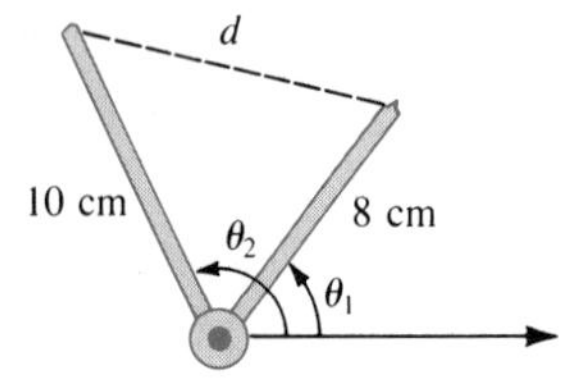

FIGURE 12.22

52. A line L, which is not through the pole, is completely determined by a point and a direction. Suppose (p, α), $p > 0$, are polar coordinates of a point Q. Show that a polar equation of a line through Q and perpendicular to the line segment of length p, given in Figure 12.23, is $r \cos(\theta - \alpha) = p$.

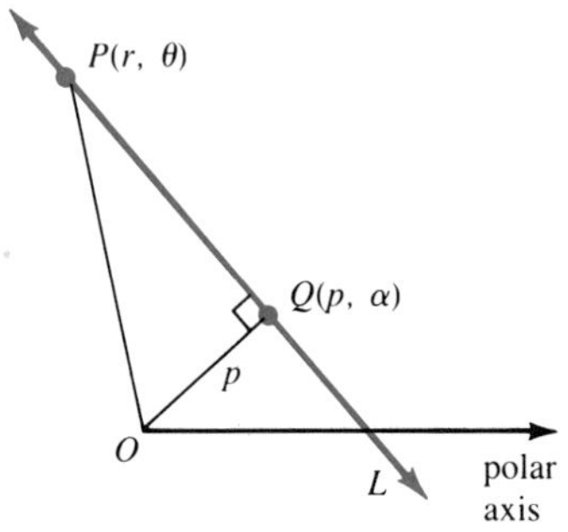

FIGURE 12.23

53. Show that a Cartesian equation for the line L in Problem 52 is $x \cos \alpha + y \sin \alpha = p$.

54. Use the result of Problem 53 to find the distance from the origin to the line $4x - 3y = 9$. [*Hint:* Divide the equation by $\sqrt{4^2 + 3^2} = 5$. Explain this.]

12.4 GRAPHS OF POLAR EQUATIONS

The graph of a polar equation $r = f(\theta)$ is the set of points P with *at least* one set of coordinates that satisfies the equation.

We begin our consideration of graphing polar equations with the polar analogue of the simple Cartesian equation $y = x$.

EXAMPLE 1 Graph $r = \theta$.

Solution As $\theta \geq 0$ increases, r increases and the points (r, θ) wind around the pole in a counterclockwise manner. This is illustrated by the colored portion of the graph in Figure 12.24. The black portion of the graph is obtained by plotting points for $\theta < 0$.

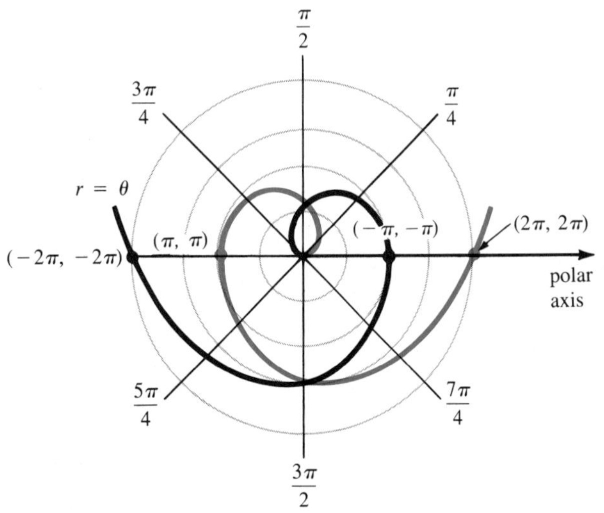

FIGURE 12.24

□

Many graphs in polar coordinates are given special names. The graph given in Example 1 is a special case of $r = a\theta$. A graph of this equation is called a **spiral of Archimedes**.

In addition to basic point plotting, *symmetry* can often be utilized to graph a polar equation.

Symmetry To facilitate graphing and discussion of graphs of polar equations, we shall, as shown in Figure 12.25, superimpose rectangular coordinates over the polar coordinate system. As shown in the figure, a polar

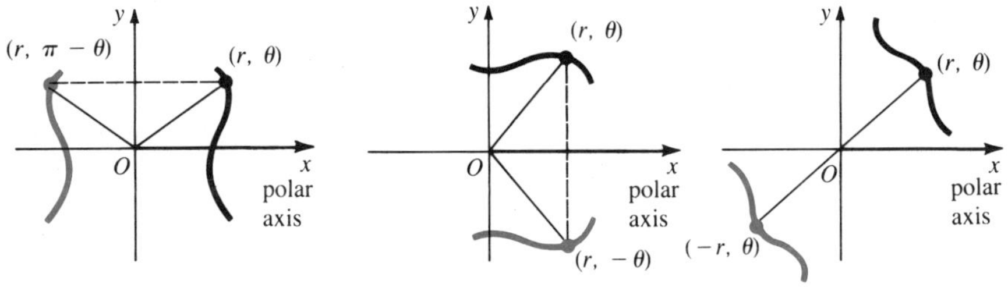

(a) symmetry with respect to y-axis

(b) symmetry with respect to x-axis

(c) symmetry with respect to origin

FIGURE 12.25

graph can have three types of symmetry. A polar graph is:

(i) **symmetric with respect to the y-axis** if, whenever (r, θ) is a point on the graph, $(r, \pi - \theta)$ is also a point on the graph;

(ii) **symmetric with respect to the x-axis** if, whenever (r, θ) is a point on the graph, $(r, -\theta)$ is also a point on the graph; and

(iii) **symmetric with respect to the origin** if, whenever (r, θ) is a point on the graph, $(-r, \theta)$ is also a point on the graph.

Hence, the graph of a polar equation is symmetric with respect to:

(i) the **y-axis** if replacing θ by $\pi - \theta$ results in an equivalent equation;

(ii) the **x-axis** if replacing θ by $-\theta$ results in an equivalent equation; and

(iii) the **origin** if replacing r by $-r$ results in an equivalent equation.

EXAMPLE 2 Since $\sin(\pi - \theta) = \sin \theta$, the graph of $r = 3 - 3 \sin \theta$ is symmetric with respect to the y-axis. Replacing, in turn, θ by $-\theta$ and r by $-r$ fails to give the original equation. Hence, no conclusion can be drawn regarding additional symmetries of the graph. □

EXAMPLE 3 Graph $r = 3 - 3 \sin \theta$.

Solution By using the symmetry information from Example 2 and plotting the points that correspond to the data in the following table, we obtain the graph given in Figure 12.26.

θ	0	$\frac{\pi}{6}$	$\frac{\pi}{3}$	$\frac{\pi}{2}$	$\frac{2\pi}{3}$	$\frac{5\pi}{6}$	π	$\frac{7\pi}{6}$	$\frac{4\pi}{3}$	$\frac{3\pi}{2}$	$\frac{5\pi}{3}$	$\frac{11\pi}{6}$	2π
r	3	$\frac{3}{2}$	0.4	0	0.4	$\frac{3}{2}$	3	$\frac{9}{2}$	5.6	6	5.6	$\frac{9}{2}$	3

□

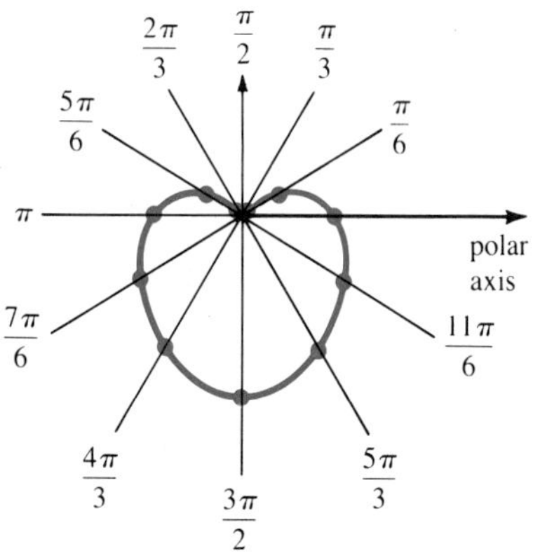

$r = 3 - 3 \sin \theta$

FIGURE 12.26

Cardioids The equation in Example 3 is a member of a family of polar equations that all have a “heart-shaped” graph that passes through the origin. A graph of any polar equation of the form

$$r = a \pm a \sin \theta \quad \text{or} \quad r = a \pm a \cos \theta$$

is called a **cardioid**. The only difference in the graphs of these four equations is their symmetry with respect to the y-axis ($r = a \pm a \sin \theta$) or the x-axis ($r = a \pm a \cos \theta$). See Figure 12.27.

$$r = a - a \sin \theta$$
(a)

$$r = a + a \sin \theta$$
(b)

$$r = a - a \cos \theta$$
(c)

$$r = a + a \cos \theta$$
(d)

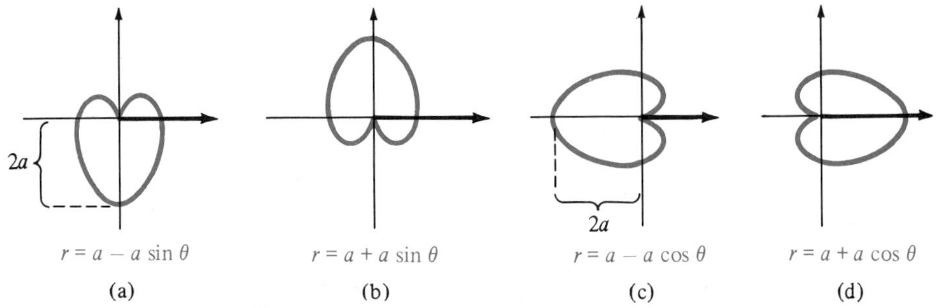

FIGURE 12.27

Limaçons   Cardioids are special cases of polar curves known as **limaçons**:

$$r = a \pm b \sin \theta \quad \text{or} \quad r = a \pm b \cos \theta$$

The shape of a limaçon depends on the relative magnitudes of a and b. Let us assume that $a > 0$ and $b > 0$. When $a = b$, the curve is a cardioid. As the following diagram shows, there are three other shapes determined by whether a is less than b or greater than b:

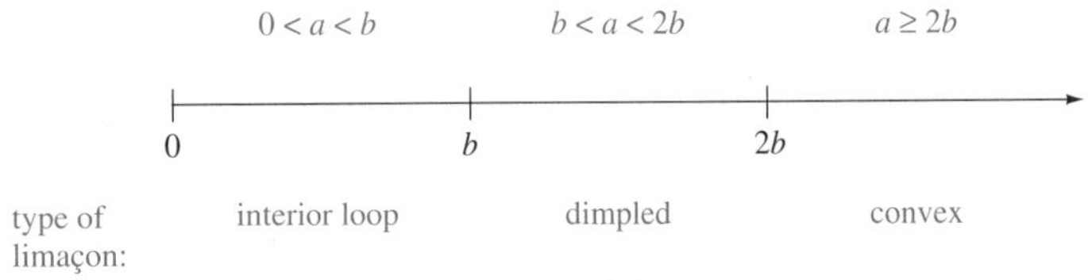

Each type of curve for $a > 0$ and $b > 0$ is illustrated in Figure 12.28. Note that the dimpled limaçon and the convex limaçon do not pass through the origin.

(a) $r = a + b \sin \theta$,
    $0 < a < b$,
    limaçon with
    interior loop

(b) $r = a + b \cos \theta$,
    $b < a < 2b$,
    dimpled limaçon

(c) $r = a - b \cos \theta$,
    $a \geq 2b$,
    convex limaçon

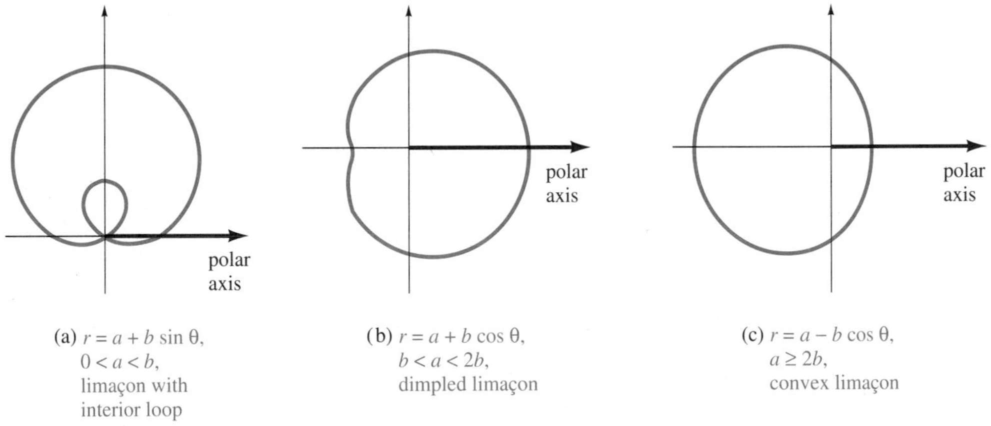

FIGURE 12.28

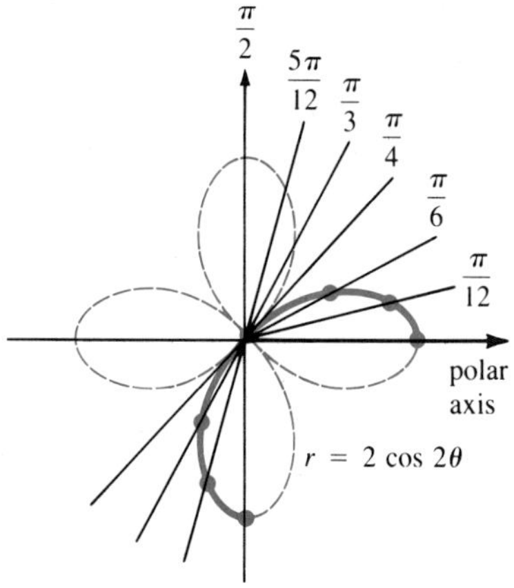

FIGURE 12.29

EXAMPLE 4 The graph of $r = 3 - \sin\theta$ is a convex limaçon, since $a = 3$, $b = 1$, and $a > 2b$. □

EXAMPLE 5 Graph $r = 2\cos 2\theta$.

Solution Since $\cos(-2\theta) = \cos 2\theta$ and $\cos 2(\pi - \theta) = \cos 2\theta$

we conclude that the graph is symmetric with respect to both the x- and y-axes. Although the given equation has period π,* a moment of reflection should convince you that we need only consider $0 \le \theta \le \pi/2$. Using the data in the following table, we see that the dashed portion of the graph given in Figure 12.29 is that completed by symmetry. The graph is called a **rose curve with four petals**.

θ	0	$\frac{\pi}{12}$	$\frac{\pi}{6}$	$\frac{\pi}{4}$	$\frac{\pi}{3}$	$\frac{5\pi}{12}$	$\frac{\pi}{2}$
r	2	1.7	1	0	-1	-1.7	-2

□

Rose Curves In general, if n is a positive integer, the graphs of

$$r = a\sin n\theta \quad \text{or} \quad r = a\cos n\theta, \qquad n \ge 2$$

are called **rose curves**. If n is odd, the number of loops or petals is n; if n is even, there are $2n$ petals. To graph a rose curve we can start by graphing one petal. To begin, we find an angle θ for which r is an extremum. This gives the center line of the petal. We then find corresponding values of θ for which the rose

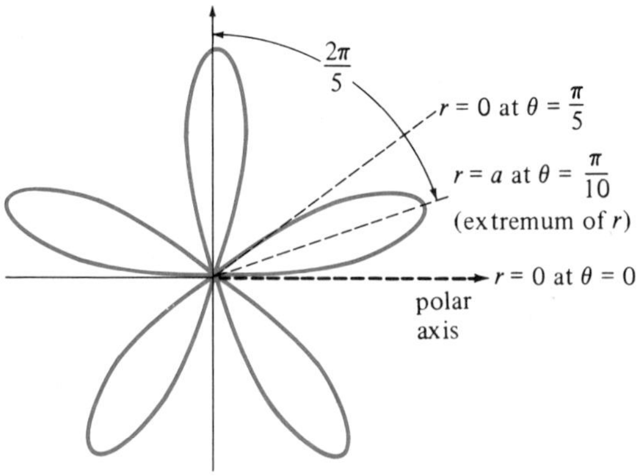

FIGURE 12.30

*This does not mean that the graph is necessarily *complete* after we graph points for $0 \le \theta \le \pi$. Rather, it means the values of r repeat for $\pi \le \theta \le 2\pi$. In this case the graph is complete on $[0, 2\pi]$.

curve enters the pole ($r = 0$). To complete the graph we use the fact that the center lines of the petals are spaced $2\pi/n$ radians ($360/n$ degrees) apart if n is odd, and $2\pi/2n$ radians ($180/n$ degrees) apart if n is even. In Figure 12.30 we have drawn the graph of $r = a \sin 5\theta$, $a > 0$. The spacing between the center lines of the five petals is $2\pi/5$ radians (72°).

Circles The graphs of $r = a \sin n\theta$ or $r = a \cos n\theta$ in the special case $n = 1$:

$$r = a \sin \theta \quad \text{or} \quad r = a \cos \theta$$

are **circles** passing through the origin with diameter $|a|$ and centers on the y-axis and x-axis, respectively. Figure 12.31 illustrates the graphs of $r = a \sin \theta$ and $r = a \cos \theta$ in the case $a > 0$.

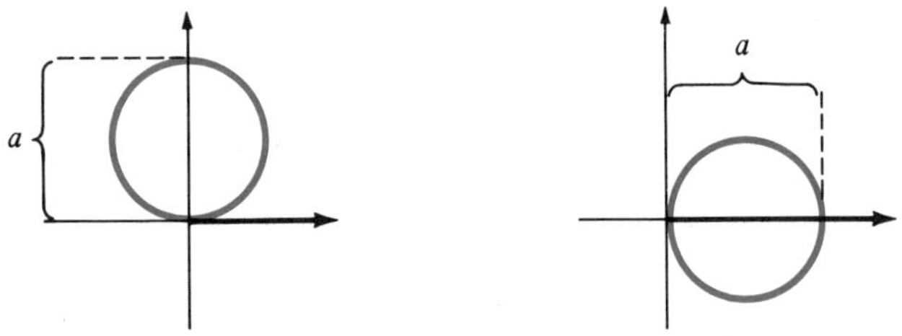

(a) $r = a \sin \theta$, $a > 0$ (b) $r = a \cos \theta$, $a > 0$

FIGURE 12.31

Slope of a Tangent to a Polar Graph Somewhat surprisingly, the slope of a tangent line to the graph of a polar equation $r = f(\theta)$ is *not* the derivative $dr/d\theta$. The slope of a tangent is still dy/dx. To find this latter derivative, we use $r = f(\theta)$ and $x = r\cos\theta$, $y = r\sin\theta$ to write parametric equations of the curve:

$$x = f(\theta)\cos\theta \qquad y = f(\theta)\sin\theta$$

Then from (1) of Section 12.1,

$$\frac{dy}{dx} = \frac{dy/d\theta}{dx/d\theta} = \frac{f(\theta)\cos\theta + f'(\theta)\sin\theta}{-f(\theta)\sin\theta + f'(\theta)\cos\theta}$$

This result is summarized in the next theorem.

THEOREM 12.3 Slope of Tangent Line

If f is a differentiable function of θ, then the **slope of the tangent line** to the graph of $r = f(\theta)$ at a point (r, θ) on the graph is

$$\frac{dy}{dx} = \frac{dy/d\theta}{dx/d\theta} = \frac{f(\theta)\cos\theta + f'(\theta)\sin\theta}{-f(\theta)\sin\theta + f'(\theta)\cos\theta}$$

provided that $dx/d\theta \neq 0$.

The formula in Theorem 12.3 is presented "for the record"; do not memorize it. To find dy/dx in polar coordinates simply form the parametric equations $x = f(\theta)\cos\theta$, $y = f(\theta)\sin\theta$ and then use the parametric form of the derivative.

EXAMPLE 6 Find the slope of the tangent to the graph of $r = 4\sin 3\theta$ at $\theta = \dfrac{\pi}{6}$.

Solution From the parametric equations

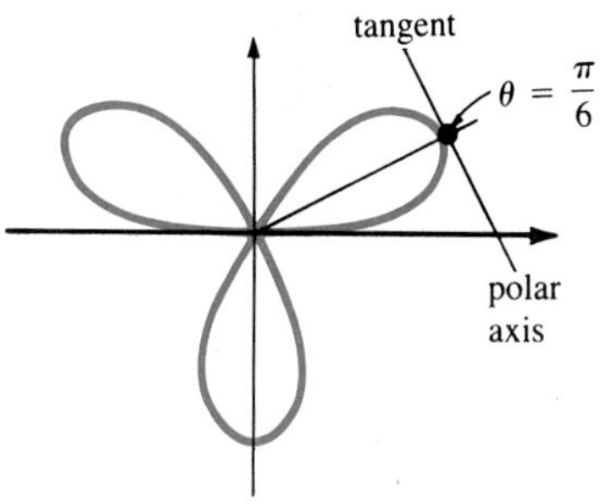

FIGURE 12.32

$$x = 4\sin 3\theta\cos\theta \qquad y = 4\sin 3\theta\sin\theta$$

we find

$$\frac{dy}{dx} = \frac{dy/d\theta}{dx/d\theta} = \frac{4\sin 3\theta\cos\theta + 12\cos 3\theta\sin\theta}{-4\sin 3\theta\sin\theta + 12\cos 3\theta\cos\theta}$$

$$\left.\frac{dy}{dx}\right|_{\theta=\pi/6} = -\sqrt{3}$$

The graph of the equation, which we recognize as a rose curve with three petals, and the tangent line are illustrated in Figure 12.32. □

EXAMPLE 7 Find the points on the graph of $r = 3 - 3\sin\theta$ at which the tangent line is horizontal and the points at which the tangent line is vertical.

Solution Recall from Section 12.2 that a horizontal tangent occurs at a point for which $dy/d\theta = 0$ and $dx/d\theta \neq 0$, whereas a vertical tangent occurs at a point for which $dx/d\theta = 0$ and $dy/d\theta \neq 0$. Now, from the parametric equations

$$x = (3 - 3\sin\theta)\cos\theta \qquad y = (3 - 3\sin\theta)\sin\theta$$

we get

$$\begin{aligned}\frac{dx}{d\theta} &= (3 - 3\sin\theta)(-\sin\theta) + \cos\theta(-3\cos\theta)\\ &= -3\sin\theta + 3\sin^2\theta - 3\cos^2\theta\\ &= -3 - 3\sin\theta + 6\sin^2\theta\\ &= 3(2\sin\theta + 1)(\sin\theta - 1)\\ \frac{dy}{d\theta} &= (3 - 3\sin\theta)\cos\theta + \sin\theta(-3\cos\theta)\\ &= 3\cos\theta(1 - 2\sin\theta)\end{aligned}$$

From these derivatives we see that:

$$\frac{dy}{d\theta} = 0\left(\frac{dx}{d\theta} \neq 0\right) \text{ at } \theta = \frac{\pi}{6}, \quad \theta = \frac{5\pi}{6}, \quad \text{and } \theta = \frac{3\pi}{2}$$

$$\frac{dx}{d\theta} = 0\left(\frac{dy}{d\theta} \neq 0\right) \text{ at } \theta = \frac{7\pi}{6} \quad \text{and} \quad \theta = \frac{11\pi}{6}$$

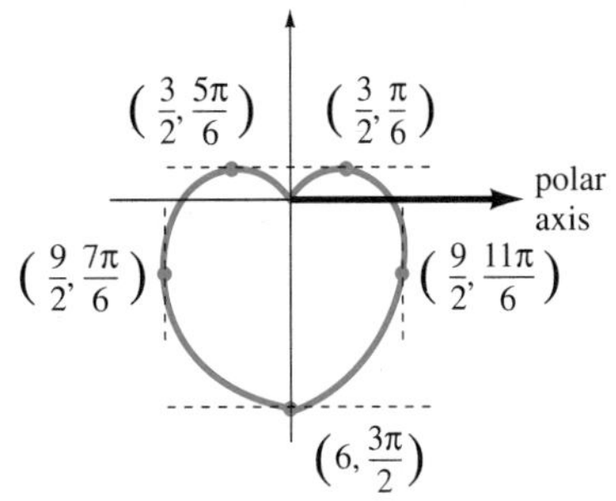

FIGURE 12.33

Thus, there are

$$\textit{horizontal tangents at:} \quad \left(\frac{3}{2}, \frac{\pi}{6}\right), \left(\frac{3}{2}, \frac{5\pi}{6}\right), \left(6, \frac{3\pi}{2}\right)$$

$$\textit{vertical tangents at:} \quad \left(\frac{9}{2}, \frac{7\pi}{6}\right), \left(\frac{9}{2}, \frac{11\pi}{6}\right)$$

These points, along with the tangent lines, are shown in Figure 12.33. □

Note in Example 7 that $dy/d\theta = 0$ and $dx/d\theta = 0$ at $\theta = \pi/2$. We shall not draw any conclusion about the tangent line at the pole $(0, \pi/2)$.

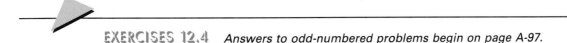

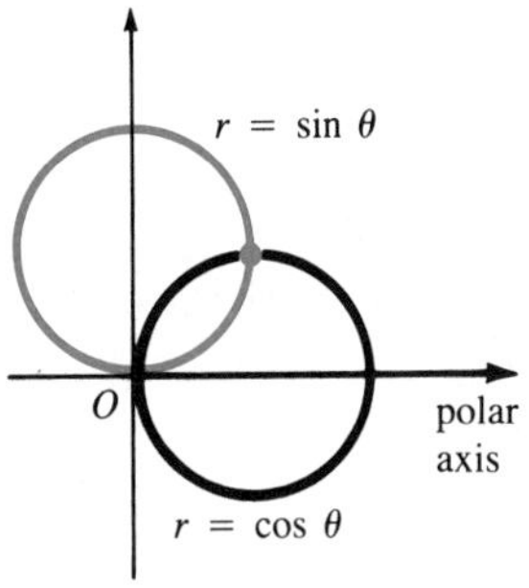

FIGURE 12.34

▲ ***Remarks*** (*i*) This is a good opportunity to review the difference between necessary and sufficient conditions. The tests for symmetry of the graph of a polar equation are sufficient but not necessary. Recall, when a test for symmetry of the graph of a Cartesian equation fails, we can say definitely that the graph does not possess that particular symmetry. In rectangular coordinates, the tests for symmetry are necessary and sufficient. By way of contrast, if, say, the test for symmetry about the x-axis fails for a polar equation, the graph may still have that symmetry. Notice in Example 2 we stated "no conclusion" rather than "not symmetric" when a test failed. Because a point has many polar coordinates, it is possible to devise alternative tests. See Problems 51–54 in Exercises 12.4.

(*ii*) Problems also arise when determining where two polar curves intersect. For example, Figure 12.34 shows that the circles $r = \cos\theta$ and $r = \sin\theta$ have two points of intersection. But solving the two equations simultaneously: $\cos\theta = \sin\theta$ leads to only $(\sqrt{2}/2, \pi/4)$. The problem here is that the pole is $(0, \pi/2)$ on the first curve but is $(0, 0)$ on the second. This is analogous to the curves reaching the same point at different times. Furthermore, *a point can be on the graph of a polar equation even though its given coordinates do not satisfy the equation.* You should verify that $(2, \pi/2)$ is an alternative description of the point $(-2, 3\pi/2)$ on the graph of $r = 1 + 3\sin\theta$. But, note that the coordinates of $(2, \pi/2)$ do not satisfy the equation.

EXERCISES 12.4 *Answers to odd-numbered problems begin on page A-97.*

In Problems 1–24 graph the given polar equation.

1. $r = 6$
2. $r = -1$
3. $\theta = \pi/3$
4. $\theta = 5\pi/6$
5. $r = 2\theta, \theta \le 0$
6. $r = 3\theta, \theta \ge 0$
7. $r = 1 + \cos\theta$
8. $r = 5 - 5\sin\theta$
9. $r = 2(1 + \sin\theta)$
10. $2r = 1 - \cos\theta$
11. $r = 3 + 2\cos\theta$
12. $r = -1 + 2\sin\theta$
13. $r = 2 + 4\sin\theta$
14. $r = 4 - \cos\theta$
15. $r = 3\cos 3\theta$
16. $r = 3\sin 4\theta$
17. $r = \cos 5\theta$
18. $r = 2\sin 9\theta$
19. $r = 6\cos\theta$
20. $r = -2\cos\theta$
21. $r = -3\sin\theta$
22. $r = 5\sin\theta$
23. $r = \tan\theta$
24. $r = \sec\theta$

In Problems 25–30 graph the **lemniscates**.

25. $r^2 = 4\sin 2\theta$ **26.** $r^2 = 4\cos 2\theta$

27. $r^2 = -25\cos 2\theta$ **28.** $r^2 = -9\sin 2\theta$

29. $r^2 = \sin\theta$ **30.** $r^2 = \cos\theta$

In Problems 31 and 32 graph the **spirals**.

31. $r = 2^\theta, \theta \geq 0$ (logarithmic)

32. $r\theta = \pi, \theta > 0$ (hyperbolic)

33. Graph the **cissoid of Diocles**: $r = \sec\theta - \cos\theta$.

34. Graph the **conchoid of Nicomedes**: $r = \csc\theta - 2$.

In Problems 35–38 find the slope of the tangent at the indicated value of θ.

35. $r = \theta;\ \theta = \pi/2$

36. $r = \sin\theta;\ \theta = \pi/6$

37. $r = 2 + 3\cos\theta;\ \theta = \pi/3$

38. $r = 10\sin 5\theta;\ \theta = \pi/4$

39. Find a Cartesian equation of the tangent line to the graph of $r = 1/(1+\cos\theta)$ at $\theta = \pi/2$.

40. Find a polar equation of the tangent line to the graph of $r = 2\cos 3\theta$ at $\theta = 2\pi/3$. [*Hint:* First find a Cartesian equation.]

In Problems 41 and 42 find the points on the graph of the given equation at which the tangent line is horizontal and the points at which the tangent line is vertical.

41. $r = 2 + 2\cos\theta$ **42.** $r = \sin\theta$

In Problems 43–46 find the points of intersection of the graphs of the given pair of polar equations.

43. $r = 2$
$r = 4\sin\theta$

44. $r = \sin\theta$
$r = \sin 2\theta$

45. $r = 1 - \cos\theta$
$r = 1 + \cos\theta$

46. $r = -\sin\theta$
$r = \cos\theta$

In Problems 47 and 48 use the fact that

$$r = f(\theta) \quad \text{and} \quad -r = f(\theta + \pi)$$

describe the same curve as an aid in finding the points of intersection of the given pair of polar equations.

47. $r = 3$
$r = 6\sin 2\theta$

48. $r = \cos 2\theta$
$r = 1 + \cos\theta$

49. Graph the **bifolium** $r = 4\sin\theta\cos^2\theta$ and the circle $r = \sin\theta$. Find all points of intersection of the graphs.

50. By means of carefully drawn graphs, verify that the cardioid $r = 1 + \cos\theta$ and the lemniscate $r^2 = 4\cos\theta$ intersect at four points. Determine whether these points of intersection can be found by solving the equations simultaneously.

In Problems 51–54 identify the symmetries if the given pair of points are on the graph of $r = f(\theta)$.

51. $(r, \theta), (-r, \pi - \theta)$ **52.** $(r, \theta), (r, \theta + \pi)$

53. $(r, \theta), (-r, \theta + 2\pi)$ **54.** $(r, \theta), (-r, -\theta)$

In Problems 55 and 56 let $r = f(\theta)$ be a polar equation. Interpret the given result geometrically.

55. $f(-\theta) = f(\theta)$ (even function)

56. $f(-\theta) = -f(\theta)$ (odd function)

57. (a) Show that the tangent of the angle ψ, measured counterclockwise from the line OP to the tangent to the graph of $r = f(\theta)$ at $P(r, \theta)$, is given by

$$\tan\psi = \frac{r}{dr/d\theta}$$

Assume, as shown in Figure 12.35, that $\phi > \theta$. [*Hint:* How are $\tan\psi$ and $\tan(\phi - \theta)$ related? What is $\tan\phi$?]

(b) Use part (a) to find the angle ψ for $r = 1/(1+\cos\theta)$ at $\theta = \pi/3$.

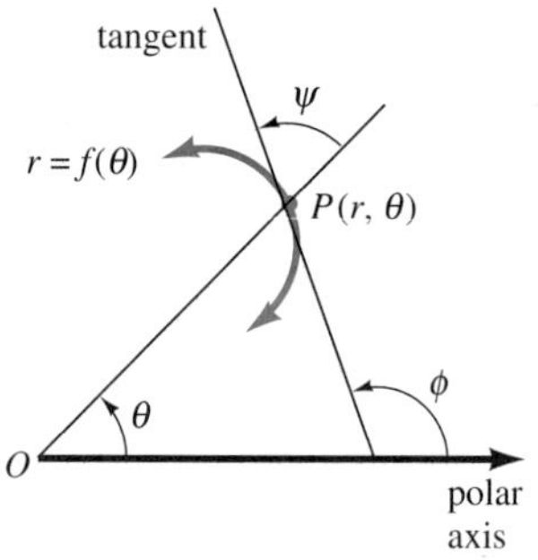

FIGURE 12.35

58. Several objects are released simultaneously from a common point and allowed to slide without friction down ramps at various angles, accelerating due to gravity. See Figure 12.36. Show that at any instant, all the objects lie on a

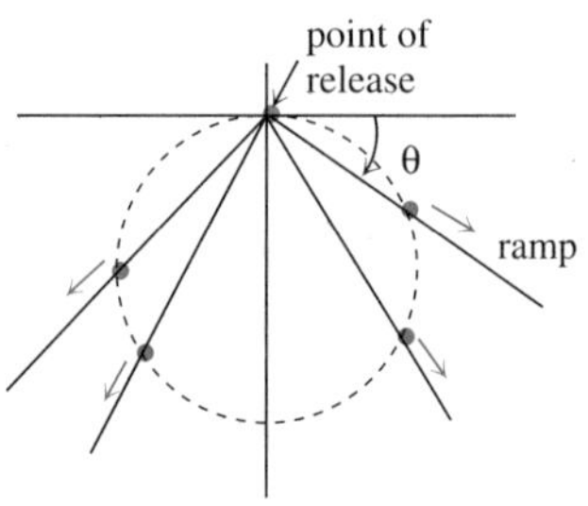

FIGURE 12.36

common circle whose topmost point is the point of release. (This was first proved by Galileo without the benefit of Cartesian or polar coordinates.)

Calculator/Computer Problems

In Problems 59–64 use a computer, if necessary, to match the given graph with the appropriate polar equation.

(a) $r = 2\sin\frac{\theta}{2}$ (b) $r = 2\sin\frac{\theta}{3}$

(c) $r = 2\sin\frac{\theta}{4}$ (d) $r = 2\sin\frac{3\theta}{2}$

(e) $r = 2\cos\frac{\theta}{5}$ (f) $r = 2\cos(2.1)\theta$

59.

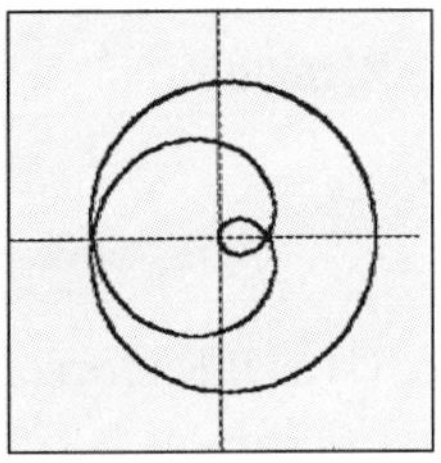

FIGURE 12.37

60.

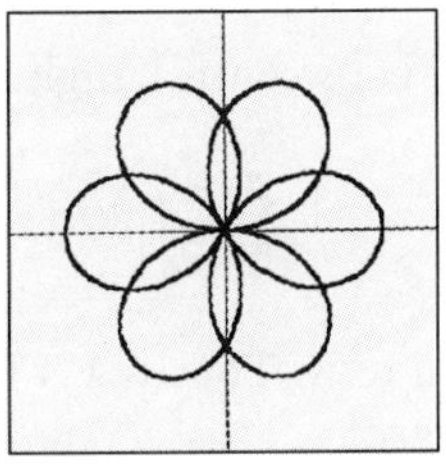

FIGURE 12.38

61.

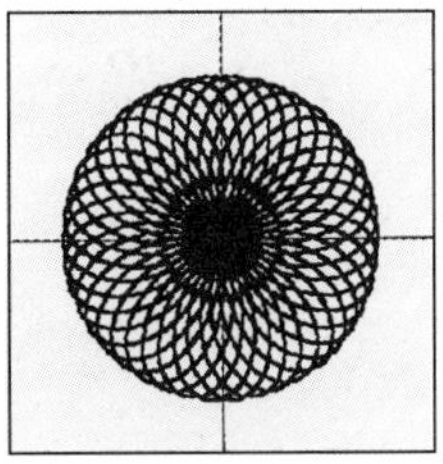

FIGURE 12.39

62.

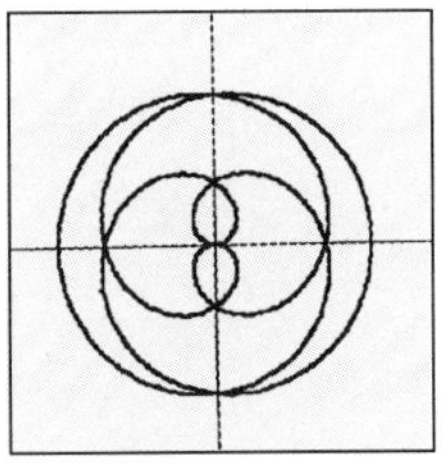

FIGURE 12.40

63.

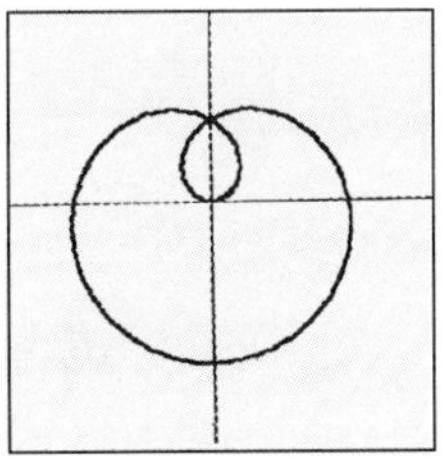

FIGURE 12.41

64.

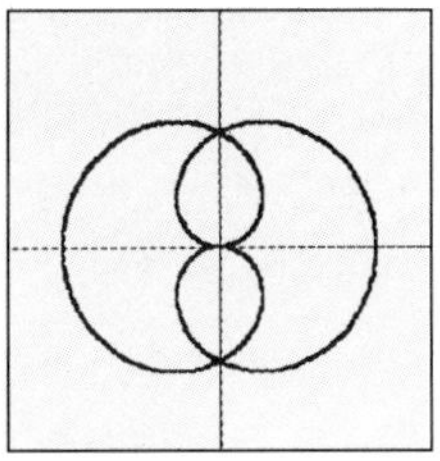

FIGURE 12.42

12.5 AREA IN POLAR COORDINATES; ARC LENGTH

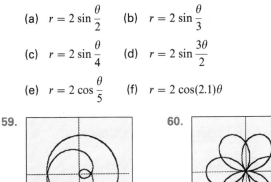

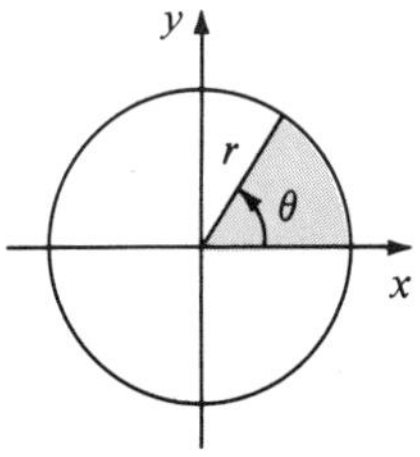

FIGURE 12.43

In the discussion that follows we shall use the fact that the area of a circular sector, shown in Figure 12.43, is given by

$$\text{area} = \tfrac{1}{2}r^2\theta$$

where θ is measured in radians.*

Area Suppose $r = f(\theta)$ is a nonnegative continuous function on $[\alpha, \beta]$, where $0 \le \alpha < \beta \le 2\pi$. To find the area A of the region shown in Figure 12.44(a) that is bounded by the graph of f and the rays $\theta = \alpha$ and $\theta = \beta$, we start by forming a partition P of $[\alpha, \beta]$:

$$\alpha = \theta_0 < \theta_1 < \theta_2 < \cdots < \theta_n = \beta$$

If θ_k^* is any number in the kth subinterval $[\theta_{k-1}, \theta_k]$, then the area of the circular sector of radius $r_k = f(\theta_k^*)$ indicated in Figure 12.44(b) is

$$\Delta A_k = \frac{1}{2}[f(\theta_k^*)]^2\,\Delta\theta_k$$

*Since the area of the circle is πr^2, the area of a sector determined by a central angle θ measured in radians is $(\theta/2\pi)(\pi r^2) = \frac{1}{2}r^2\theta$.

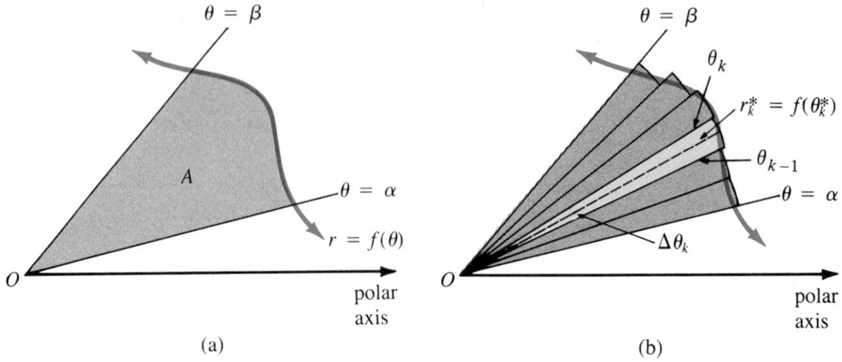

FIGURE 12.44

where $\Delta\theta_k = \theta_k - \theta_{k-1}$ is its central angle. In turn, the Riemann sum

$$\sum_{k=1}^{n} \frac{1}{2}[f(\theta_k^*)]^2 \, \Delta\theta_k \tag{1}$$

gives an approximation to A. The area A is the limit of (1) as the norm of the partition approaches zero:

$$A = \lim_{\|P\| \to 0} \sum_{k=1}^{n} \frac{1}{2}[f(\theta_k^*)]^2 \, \Delta\theta_k$$

THEOREM 12.4 Area in Polar Coordinates

If $r = f(\theta)$ is a nonnegative continuous function on $[\alpha, \beta]$, then the **area** bounded by its graph and the rays $\theta = \alpha$ and $\theta = \beta$ is given by

$$A = \int_{\alpha}^{\beta} \frac{1}{2}[f(\theta)]^2 \, d\theta = \frac{1}{2}\int_{\alpha}^{\beta} r^2 \, d\theta \tag{2}$$

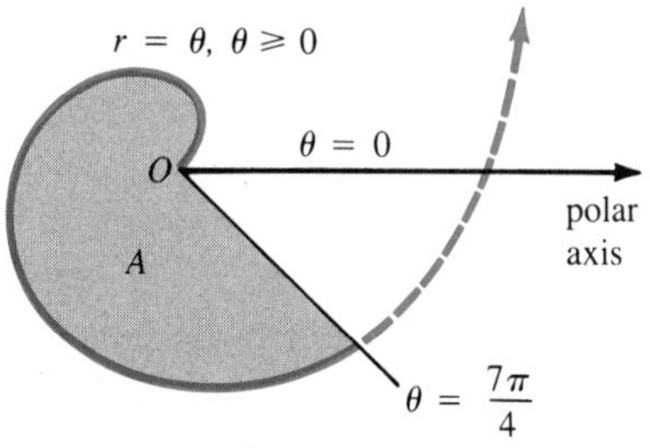

FIGURE 12.45

EXAMPLE 1 Find the area of the region that is bounded by the spiral $r = \theta$, $\theta \geq 0$, between the rays $\theta = 0$ and $\theta = 7\pi/4$.

Solution From (2), the area of the shaded region shown in Figure 12.45 is

$$A = \frac{1}{2}\int_{0}^{7\pi/4} \theta^2 \, d\theta = \frac{1}{2}\frac{\theta^3}{3}\bigg]_0^{7\pi/4} = \frac{343}{384}\pi^3 \approx 27.70 \text{ square units}$$ □

EXAMPLE 2 Find the area of the region that is common to the interiors of the cardioid $r = 2 - 2\cos\theta$ and the limaçon $r = 2 + \cos\theta$.

Solution Inspection of Figure 12.46 shows that we need two integrals. Solving the given equations simultaneously:

$$2 - 2\cos\theta = 2 + \cos\theta \quad \text{or} \quad \cos\theta = 0$$

yields $\theta = \pi/2$ so that a point of intersection is $(2, \pi/2)$. By symmetry, it follows that

$$\begin{aligned}
A &= 2\left\{\frac{1}{2}\int_0^{\pi/2} (2 - 2\cos\theta)^2\,d\theta + \frac{1}{2}\int_{\pi/2}^{\pi} (2 + \cos\theta)^2\,d\theta\right\} \\
&= 4\int_0^{\pi/2} (1 - 2\cos\theta + \cos^2\theta)\,d\theta + \int_{\pi/2}^{\pi} (4 + 4\cos\theta + \cos^2\theta)\,d\theta \\
&= 4\int_0^{\pi/2} \left(1 - 2\cos\theta + \frac{1 + \cos 2\theta}{2}\right) d\theta + \int_{\pi/2}^{\pi} \left(4 + 4\cos\theta + \frac{1 + \cos 2\theta}{2}\right) d\theta \\
&= 4\left[\frac{3}{2}\theta - 2\sin\theta + \frac{1}{4}\sin 2\theta\right]_0^{\pi/2} + \left[\frac{9}{2}\theta + 4\sin\theta + \frac{1}{4}\sin 2\theta\right]_{\pi/2}^{\pi} \\
&= \frac{21}{4}\pi - 12 \approx 4.49 \text{ square units}
\end{aligned}$$

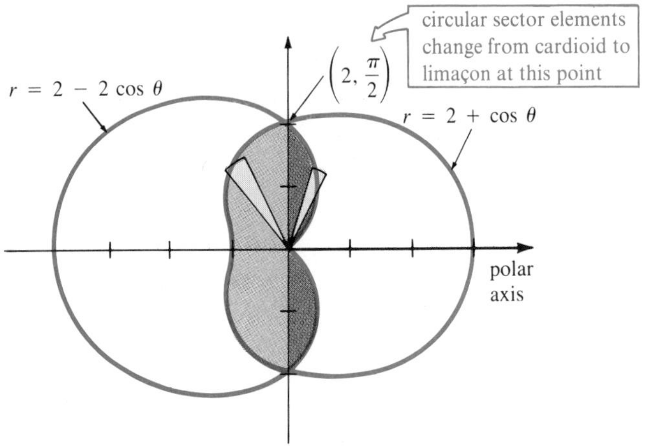

FIGURE 12.46

□

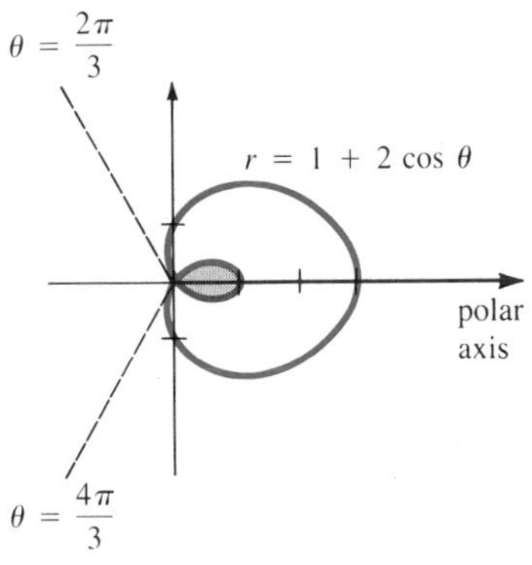

FIGURE 12.47

EXAMPLE 3 Find the area of the interior loop of the limaçon $r = 1 + 2\cos\theta$.

Solution As seen in Figure 12.47, the interior loop corresponds to $2\pi/3 \le \theta \le 4\pi/3$. Although $r \le 0$ for these values of θ, the fact that (2) utilizes $r^2 \ge 0$ enables us to write

$$A = \frac{1}{2}\int_{2\pi/3}^{4\pi/3} (1 + 2\cos\theta)^2\,d\theta = \pi - \frac{3\sqrt{3}}{2} \approx 0.54 \text{ square unit}$$

□

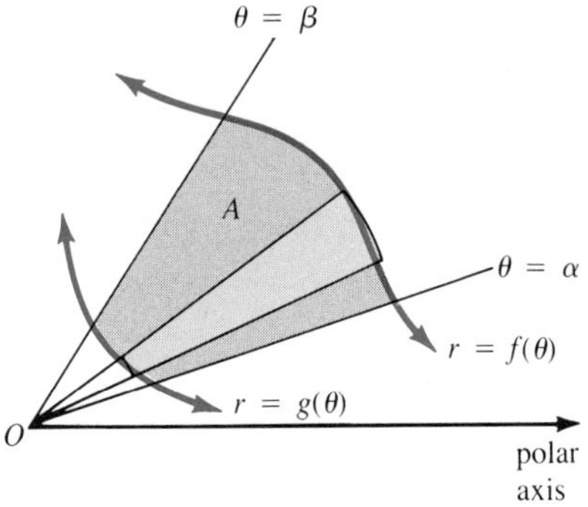

FIGURE 12.48

Area Bounded by Two Graphs The area A of the shaded region shown in Figure 12.48 can be found by subtracting areas. If f and g are continuous on $[\alpha, \beta]$ and $f(\theta) \geq g(\theta)$ on the interval, then the area bounded by the graphs of $r = f(\theta)$, $r = g(\theta)$, $\theta = \alpha$, and $\theta = \beta$ is

$$A = \frac{1}{2}\int_{\alpha}^{\beta} [f(\theta)]^2\, d\theta - \frac{1}{2}\int_{\alpha}^{\beta} [g(\theta)]^2\, d\theta$$

Written as a single integral, the area is given by

$$A = \frac{1}{2}\int_{\alpha}^{\beta} ([f(\theta)]^2 - [g(\theta)]^2)\, d\theta \tag{3}$$

EXAMPLE 4 Find the area of the region in the first quadrant that is outside the circle $r = 1$ and inside the rose curve $r = 2\sin 2\theta$.

Solution Solving the two equations simultaneously:

$$1 = 2\sin 2\theta \quad \text{or} \quad \sin 2\theta = \frac{1}{2}$$

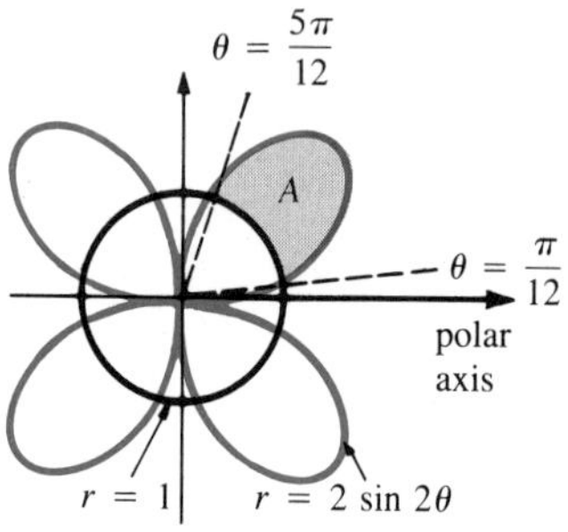

FIGURE 12.49

implies that $2\theta = \pi/6$ and $2\theta = 5\pi/6$. Thus, two points of intersection in the first quadrant are $(1, \pi/12)$ and $(1, 5\pi/12)$. The area in question is shaded in Figure 12.49. From (3),

$$\begin{aligned} A &= \frac{1}{2}\int_{\pi/12}^{5\pi/12} [(2\sin 2\theta)^2 - 1^2]\, d\theta \\ &= \frac{1}{2}\int_{\pi/12}^{5\pi/12} [4\sin^2 2\theta - 1]\, d\theta \\ &= \frac{1}{2}\int_{\pi/12}^{5\pi/12} \left[4\left(\frac{1-\cos 4\theta}{2}\right) - 1\right] d\theta \\ &= \frac{1}{2}\left[\theta - \frac{1}{2}\sin 4\theta\right]_{\pi/12}^{5\pi/12} = \frac{\pi}{6} + \frac{\sqrt{3}}{4} \approx 0.96 \text{ square unit} \end{aligned}$$ □

Arc Length for Polar Graphs We have seen that if $r = f(\theta)$ is the equation of a curve C in polar coordinates, then parametric equations of C are

$$x = f(\theta)\cos\theta, \qquad y = f(\theta)\sin\theta, \qquad \alpha \leq \theta \leq \beta$$

If f has a continuous derivative, then it is a straightforward matter to derive a formula for arc length in polar coordinates. Since

$$\frac{dx}{d\theta} = f'(\theta)\cos\theta - f(\theta)\sin\theta, \qquad \frac{dy}{d\theta} = f'(\theta)\sin\theta + f(\theta)\cos\theta$$

$$\left(\frac{dx}{d\theta}\right)^2 + \left(\frac{dy}{d\theta}\right)^2 = [f'(\theta)]^2 + [f(\theta)]^2 = \left(\frac{dr}{d\theta}\right)^2 + r^2$$

the following results from (7) of Section 12.2:

THEOREM 12.5 Arc Length in Polar Coordinates

Let f be a function for which f' is continuous on an interval $[\alpha, \beta]$. Then the **length** s of the graph of $r = f(\theta)$ on the interval is

$$s = \int_\alpha^\beta \sqrt{\left(\frac{dr}{d\theta}\right)^2 + r^2}\, d\theta \tag{4}$$

EXAMPLE 5 Find the length of the cardioid $r = 1 + \cos\theta$ for $0 \le \theta \le \pi$.

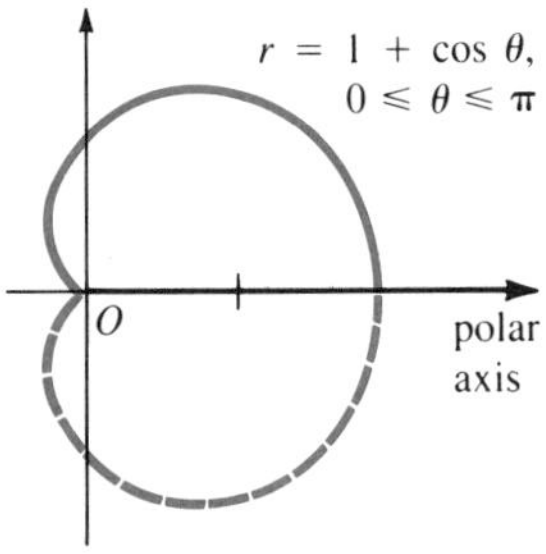

FIGURE 12.50

Solution The portion of the complete graph is shown in Figure 12.50. Now, $dr/d\theta = -\sin\theta$ so that

$$\left(\frac{dr}{d\theta}\right)^2 + r^2 = \sin^2\theta + (1 + 2\cos\theta + \cos^2\theta) = 2 + 2\cos\theta$$

and

$$s = \sqrt{2}\int_0^\pi \sqrt{1 + \cos\theta}\, d\theta$$

To evaluate this integral, we employ the trigonometric identity $\cos^2(\theta/2) = (1 + \cos\theta)/2$:

$$s = 2\int_0^\pi \cos\frac{\theta}{2}\, d\theta = 4\sin\frac{\theta}{2}\bigg]_0^\pi = 4 \qquad \square$$

It is left as an exercise for you to ponder why the length of the complete cardioid $(0 \le \theta \le 2\pi)$ in Example 5 is *not* $s = 2\int_0^{2\pi} \cos(\theta/2)\, d\theta$.

EXERCISES 12.5

Answers to odd-numbered problems begin on page A-98.

In Problems 1–10 find the area of the region that is bounded by the graph of the given polar equation.

1. $r = 2\sin\theta$
2. $r = 10\cos\theta$
3. $r = 4 + 4\cos\theta$
4. $r = 1 - \sin\theta$
5. $r = 3 + 2\sin\theta$
6. $r = 2 + \cos\theta$
7. $r = 3\sin 2\theta$
8. $r = \cos 4\theta$
9. $r = 2\sin 3\theta$
10. $r = 4\cos 5\theta$
11. Consider the lemniscate $r^2 = 9\cos 2\theta$.

    (a) Explain why $\frac{1}{2}\int_0^{2\pi} 9\cos 2\theta\, d\theta$ is not the area of the region bounded by the graph.

    (b) Using an appropriate integral, find the area of the region bounded by the graph.
12. Repeat Problem 11 for $r^2 = \sin 2\theta$.

In Problems 13–18 find the area of the region that is bounded by the graph of the given polar equation and the indicated rays.

13. $r = 2\theta, \theta \ge 0, \theta = 0, \theta = 3\pi/2$
14. $r\theta = \pi, \theta > 0, \theta = \pi/2, \theta = \pi$
15. $r = e^\theta, \theta = 0, \theta = \pi$
16. $r = 10e^{-\theta}, \theta = 1, \theta = 2$

17. $r = \tan\theta$, $\theta = 0$, $\theta = \pi/4$

18. $r\sin\theta = 5$, $\theta = \pi/6$, $\theta = \pi/3$

19. Find the area of the region that is outside the circle $r = 1$ and inside the rose curve $r = 2\cos 3\theta$.

20. Find the area of the region that is common to the interiors of the circles $r = \cos\theta$ and $r = \sin\theta$.

21. Find the area of the region that is inside the circle $r = 5\sin\theta$ and outside the limaçon $r = 3 - \sin\theta$.

22. Find the area of the region that is common to the interiors of the graphs of the equations in Problem 21.

23. Find the area of the region that is inside the cardioid $r = 4 - 4\cos\theta$ and outside the circle $r = 6$.

24. Find the area of the region that is common to the interiors of the graphs of the equations in Problem 23.

25. Find the area of the region that is inside the limaçon $r = 1 + 2\sin\theta$. [*Hint:* The area is *not*

$$\frac{1}{2}\int_0^{2\pi}(1 + 2\sin\theta)^2\,d\theta]$$

26. For the polar equation in Problem 25, find the area that is outside the interior loop but inside the limaçon.

27. Find the area of the region that is outside the circle $r = 1$ and inside the lemniscate $r^2 = 4\cos 2\theta$.

28. Find the area of the region that is common to the interiors of the circle $r = 4\sin\theta$ and the cardioid $r = 1 + \sin\theta$.

In Problems 29–32 find the length of the curve for the indicated values of θ.

29. $r = e^{\theta/2}$, $0 \le \theta \le 4$

30. $r = 2e^{-\theta}$, $0 \le \theta \le \pi$

31. $r = \theta$, $0 \le \theta \le 1$

32. $r = 3 - 3\cos\theta$, $0 \le \theta \le \pi$

33. In polar coordinates the **angular momentum** of a moving particle of mass m is defined to be $L = mr^2\,d\theta/dt$. Assume that the polar coordinates of a planet of mass m are (r_1, θ_1) and (r_2, θ_2) at times $t = a$ and $t = b$, $a < b$, respectively. Since the gravitational force acting on the planet is a central force, the angular momentum L of the planet is a constant. (See Problem 27 in Exercises 14.2.) Use this fact to show that the area A swept out by r is $A = L(b - a)/2m$. When the sun is taken to be at the origin, this equation proves **Kepler's second law of planetary motion**: a line joining a planet with the sun sweeps out equal areas in equal time intervals. See Figure 12.51.

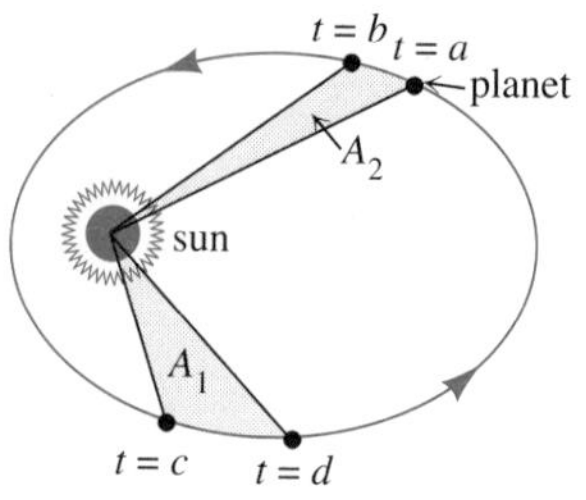

FIGURE 12.51

34. How long is the groove in a record? Suppose a record plays for 20 min at 33 revolutions per minute. As the record plays, the needle goes from an outer radius R_o to an inner radius R_i. See Figure 12.52. Assume that the groove of the record is a spiral that can be described by a polar equation of the form $r = R_o - k\theta$, where k is a constant and θ is measured in radians.

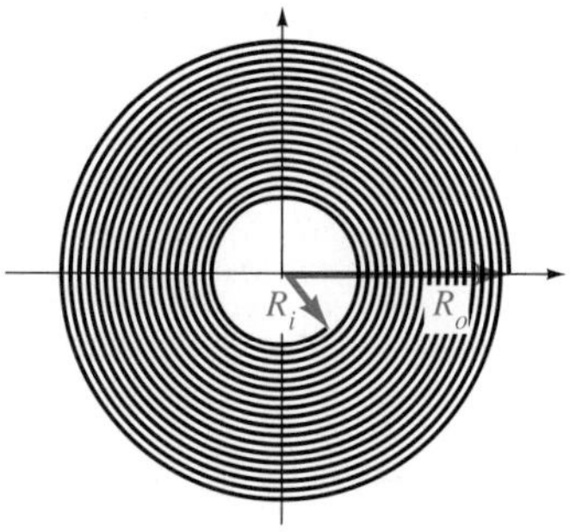

FIGURE 12.52

(a) Express k in terms of R_o, R_i, and N, where N is the number of revolutions completed by the record.

(b) Show that the length L of the record groove is given by

$$L = \frac{1}{k}\int_{R_i}^{R_o}\sqrt{k^2 + u^2}\,du$$

(c) Use a binomial series to establish the approximation

$$\sqrt{k^2 + u^2} \approx u\left[1 + \frac{1}{2}\left(\frac{k}{u}\right)^2\right]$$

(d) Use the result in part (**c**) in (**b**) to show that

$$L \approx \pi N(R_i + R_o) + \frac{R_o - R_i}{4\pi N}\ln\frac{R_o}{R_i}$$

(e) Use the result in part (**d**) to approximate the length L if $R_o = 6$ in and $R_i = 2.5$ in.

(f) Use an appropriate substitution to evaluate the integral in part (**b**) using the specified values of R_o and R_i given in part (**e**). Compare this answer with that obtained in part (**e**).

12.6 CONIC SECTIONS REVISITED

In Chapter 11 we saw that the parabola, ellipse, and hyperbola are defined by three different geometric properties. However, by using the concept of eccentricity it is possible to unify the three definitions.

THEOREM 12.6 Conic Section

The set of all points P in the plane for which the ratio of the distance $d(P, F)$ from a fixed point F to the distance $d(P, Q)$ from a fixed line L is a constant e is a **conic section**. The conic is a parabola if $e = 1$, an ellipse if $0 < e < 1$, and a hyperbola if $e > 1$.

The fixed line L is a **directrix** and the constant ratio $d(P, F)/d(P, Q) = e$ is the **eccentricity** of the conic. In each of the three cases in Theorem 12.6 the fixed point F is a **focus** of the conic section.

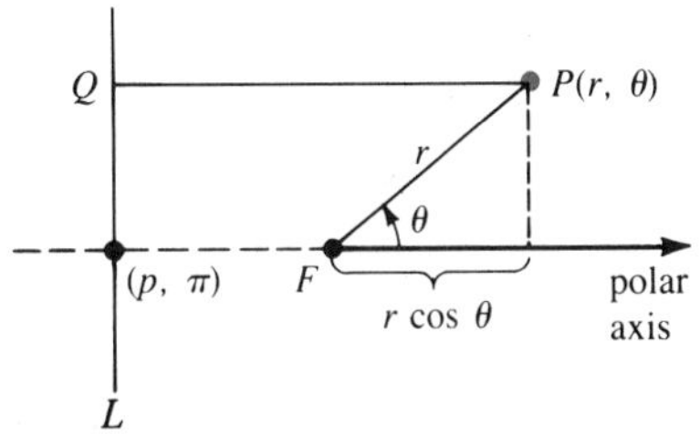

FIGURE 12.53

Polar Equations of Conics The equation

$$\frac{d(P, F)}{d(P, Q)} = e \quad \text{or} \quad d(P, F) = ed(P, Q) \tag{1}$$

is readily interpreted using polar coordinates. Suppose F is placed at the pole and L is p units ($p > 0$) to the left of F perpendicular to the extended polar axis. We see from Figure 12.53 that (1) is the same as

$$r = e(p + r\cos\theta) \quad \text{or} \quad r - er\cos\theta = ep \tag{2}$$

Solving for r yields

$$r = \frac{ep}{1 - e\cos\theta} \tag{3}$$

If the directrix is chosen to the right of F through $(p, 0)$, then the only change in (3) is that the negative sign in the denominator becomes a positive sign.

To see that Theorem 12.6 yields the familiar equations of the conics, let us superimpose a rectangular coordinate system on the polar coordinate system with the origin at the pole and the positive x-axis coinciding with the polar axis. We then express (3) in rectangular coordinates and simplify:

$$\pm\sqrt{x^2 + y^2} - ex = ep$$
$$x^2 + y^2 = e^2x^2 + 2e^2px + e^2p^2$$
$$(1 - e^2)x^2 - 2e^2px + y^2 = e^2p^2 \tag{4}$$

When $e = 1$, (4) becomes

$$-2px + y^2 = p^2 \quad \text{or} \quad y^2 = 2p\left(x + \frac{p}{2}\right)$$

which is an equation of a parabola whose axis is the x-axis, whose vertex is at $(-p/2, 0)$ and, consistent with the placement of F, whose focus is at the origin.

Now suppose that $0 < e < 1$. Completing the square in (3) and simplifying then yields

$$\left(x - \frac{pe^2}{1-e^2}\right)^2 + \frac{y^2}{1-e^2} = \frac{p^2e^2}{(1-e^2)^2} \tag{5}$$

If we let $h = pe^2/(1-e^2)$, $a^2 = p^2e^2/(1-e^2)^2$, and $b^2 = p^2e^2/(1-e^2)$, then (5) is recognized as an equation of an ellipse:

$$\frac{(x-h)^2}{a^2} + \frac{y^2}{b^2} = 1$$

From $c^2 = a^2 - b^2 = p^2e^4/(1-e^2)^2$ we get $c = pe^2/(1-e^2)$, and so we see that the eccentricity is $c/a = e$. Moreover, since the foci are located at $(h \pm c, 0)$ and since $h = c$, we also see that one focus is at the origin.

It is left as an exercise to show that (4) describes a hyperbola with one focus at the origin when $e > 1$.

When the directrix is chosen parallel to the polar axis, then the equation of the conic is found to be either

$$r = \frac{ep}{1 + e\sin\theta} \quad \text{or} \quad r = \frac{ep}{1 - e\sin\theta}$$

Specifically, $r = ep/(1 + e\sin\theta)$ describes a conic whose directrix is parallel to the polar axis and passes through $(p, \pi/2)$, whereas the directrix for $r = ep/(1 - e\sin\theta)$ passes through $(p, 3\pi/2)$.

We summarize the preceding discussion as a theorem.

THEOREM 12.7 Polar Equations of Conics

Any polar equation of the form

$$r = \frac{ep}{1 \pm e\cos\theta} \tag{6}$$

or

$$r = \frac{ep}{1 \pm e\sin\theta} \tag{7}$$

is a conic section with focus at the origin and axis along a coordinate axis. The axis of the conic is along the x-axis for equations of the form given in (6) and along the y-axis for equations of the form given in (7).

EXAMPLE 1 Identify the conics (*a*) $\dfrac{6}{1 + 2\sin\theta}$ and (*b*) $r = \dfrac{4}{3 - \cos\theta}$.

Solution

(*a*) Comparison with (7) enables us to make the identification $e = 2$. The conic section is a hyperbola.

(*b*) After numerator and denominator are divided by 3, the given equation becomes

$$r = \frac{\frac{4}{3}}{1 - \frac{1}{3}\cos\theta}$$

Thus, from (6) we see that $e = \frac{1}{3}$ and so the conic section is an ellipse. □

Graphs A rough **graph** of a conic defined by (6), or (7) can be obtained by knowing the orientation of its axis and by finding x- and y-intercepts and vertices. In the case of (6), the two vertices on the axis of an ellipse or a hyperbola occur at $\theta = 0$ and $\theta = \pi$. When $e = 1$ the denominator of (6) contains $1 - \cos\theta$ or $1 + \cos\theta$ and, consequently, the vertex of a parabola can occur at only one of the values: $\theta = 0$ or $\theta = \pi$. For (7) the vertices of an ellipse or a hyperbola occur at $\theta = \pi/2$ and $\theta = 3\pi/2$. One of the values $\theta = \pi/2$ or $\theta = 3\pi/2$ will yield the vertex of a parabola.

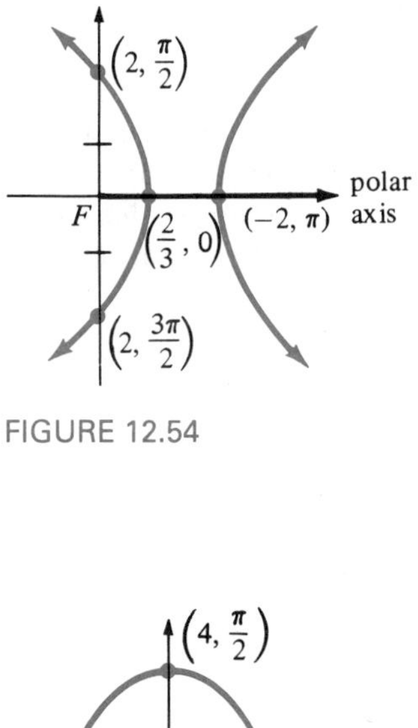

FIGURE 12.54

EXAMPLE 2 Graph $r = \dfrac{2}{1 + 2\cos\theta}$.

Solution From (6) we see that $e = 2$ and so the equation describes a hyperbola whose transverse axis is horizontal (because of $\cos\theta$). In view of the foregoing discussion, we obtain:

vertices: $(\frac{2}{3}, 0), (-2, \pi)$

y-intercepts: $(2, \pi/2), (2, 3\pi/2)$

The graph of the equation is given in Figure 12.54. □

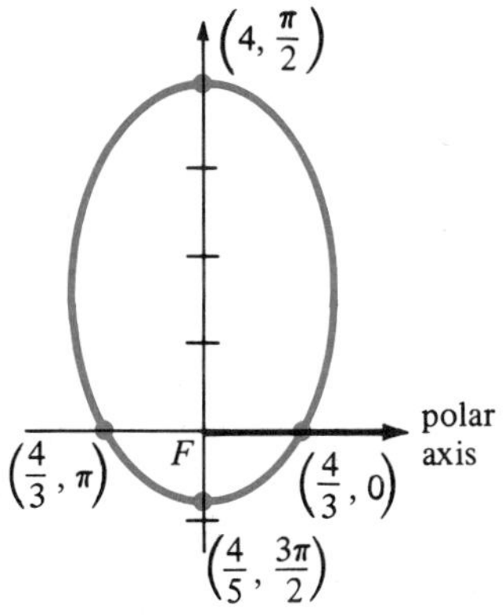

FIGURE 12.55

EXAMPLE 3 Graph $r = \dfrac{4}{3 - 2\sin\theta}$.

Solution By writing

$$r = \frac{\frac{4}{3}}{1 - \frac{2}{3}\sin\theta}$$

we see that $e = \frac{2}{3}$ and so the conic section is an ellipse whose major axis is vertical (because of $\sin\theta$). It follows that:

vertices: $(4, \pi/2), (\frac{4}{5}, 3\pi/2)$

x-intercepts: $(\frac{4}{3}, 0), (\frac{4}{3}, \pi)$

The graph of the equation is given in Figure 12.55. □

In Example 3, the rectangular coordinates of the vertices on the major axis are $(0, 4)$ and $(0, -\frac{4}{5})$. By averaging the y-coordinates of these points, we find that the center of the ellipse is $(0, \frac{8}{5})$. Now, since the distance from the center to the focus at the origin is $\frac{8}{5}$, we conclude that $c = \frac{8}{5}$. Hence, the other focus of the ellipse is at $(0, 4 - \frac{8}{5})$ or $(0, \frac{12}{5})$. Also, one-half of the length of the major axis yields the number $a = \frac{12}{5}$. Using $b^2 = a^2 - c^2$, we get $b = 4\sqrt{5}/5$ and so the vertices on the minor axis are $(-4\sqrt{5}/5, \frac{8}{5})$ and $(4\sqrt{5}/5, \frac{8}{5})$.

EXAMPLE 4 Graph $r = \dfrac{1}{1 - \cos\theta}$.

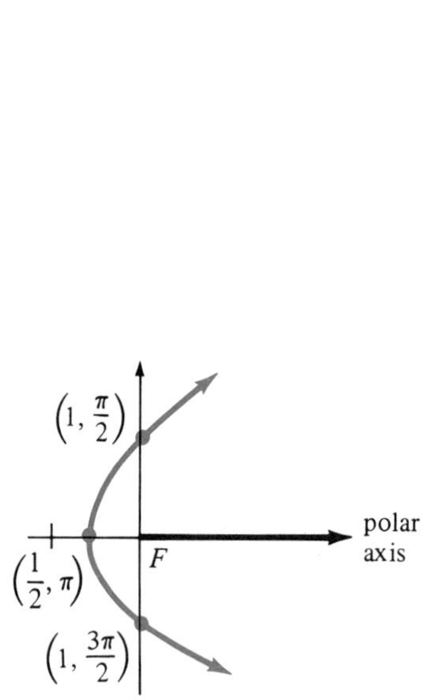

FIGURE 12.56

Solution Inspection of the equation reveals that $e = 1$ and so the conic section is a parabola whose axis is horizontal (because of $\cos\theta$). Since r is undefined at $\theta = 0$, the vertex of the parabola occurs at $\theta = \pi$:

$$\textit{vertex:}\ \left(\tfrac{1}{2}, \pi\right)$$
$$\textit{y-intercepts:}\ (1, \pi/2), (1, 3\pi/2)$$

The graph of the equation is given in Figure 12.56. □

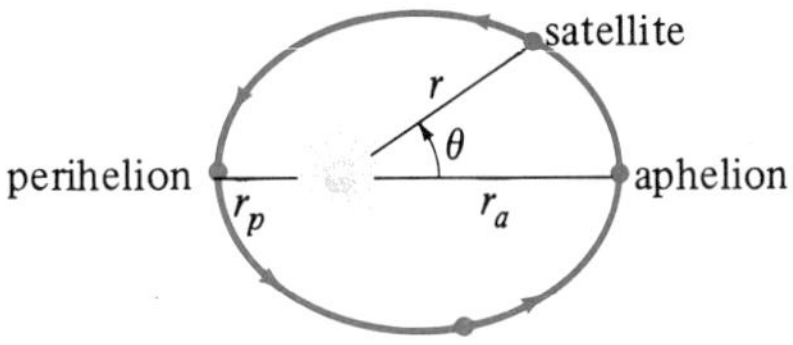

FIGURE 12.57

Orbits The orbit of a satellite around the sun (earth or moon) is an ellipse with the sun (earth or moon) at one focus. Suppose that an equation of the orbit is given by $r = ep/(1 - e\cos\theta)$, $0 < e < 1$, and that r_p is the value of r at **perihelion** (perigee or perilune) and r_a is the value of r at **aphelion** (apogee or apolune). These are the points in the orbit, occurring on the x-axis, at which the satellite is closest and farthest, respectively, from the sun (earth or moon).* See Figure 12.57. It is left as an exercise to show that the eccentricity of the orbit is related to r_p and r_a by

$$e = \frac{r_a - r_p}{r_a + r_p} \tag{8}$$

EXAMPLE 5 Find a polar equation of the orbit of the planet Mercury around the sun if $r_p = 2.85 \times 10^7$ mi and $r_a = 4.36 \times 10^7$ mi.

Solution From (8) the eccentricity of Mercury's orbit

$$e = \frac{(4.36 - 2.85) \times 10^7}{(4.36 + 2.85) \times 10^7} = 0.21$$

Hence,

$$r = \frac{0.21p}{1 - 0.21\cos\theta}$$

*A body such as a comet may have a parabolic or hyperbolic orbit. In this case there is a perihelion but no aphelion.

To find p we note that aphelion occurs at $\theta = 0$:

$$4.36 \times 10^7 = \frac{0.21p}{1 - 0.21}$$

The last equation yields $0.21p = 3.44 \times 10^7$ or $p = 16.38 \times 10^7$. Hence, a polar equation of the orbit is

$$r = \frac{3.44 \times 10^7}{1 - 0.21 \cos \theta}$$ □

EXERCISES 12.6

Answers to odd-numbered problems begin on page A-98.

In Problems 1–10 determine the eccentricity, identify the conic section, and sketch its graph.

1. $r = \dfrac{2}{1 + \cos \theta}$
2. $r = \dfrac{2}{2 - \sin \theta}$
3. $r = \dfrac{15}{4 - \cos \theta}$
4. $r = \dfrac{5}{2 - 2 \sin \theta}$
5. $r = \dfrac{4}{1 + 2 \sin \theta}$
6. $r = \dfrac{12}{6 + 2 \sin \theta}$
7. $r = \dfrac{18}{3 + 6 \cos \theta}$
8. $r = \dfrac{6 \sec \theta}{\sec \theta - 1}$
9. $r = \dfrac{10}{5 + 4 \sin \theta}$
10. $r = \dfrac{2}{2 + 5 \cos \theta}$

11. Consider the ellipse whose equation is

$$r = \frac{6}{2 + \cos \theta}$$

Find the rectangular coordinates of the center, foci, and four vertices.

12. Consider the hyperbola whose equation is

$$r = \frac{8}{1 - 3 \sin \theta}$$

Find the rectangular coordinates of the center and foci. What is the length of the conjugate axis?

In Problems 13–16 find a polar equation of the conic section with a focus at the pole and with the indicated eccentricity and directrix.

13. $e = 1, r = -3 \sec \theta$
14. $e = \frac{2}{3}, r = -2 \csc \theta$
15. $e = \frac{5}{2}, r = 6 \csc \theta$
16. $e = \frac{1}{2}, r = 4 \sec \theta$

In Problems 17 and 18 find a polar equation of the parabola with focus at the pole and vertex at the given point.

17. $(\frac{5}{2}, \pi/2)$
18. $(3, \pi)$

19. Use the equation $r = ep/(1 - e \cos \theta)$ to derive the result given in (8).

20. A communications satellite is 12,000 km above the earth at its apogee. The eccentricity of its orbit is 0.2. Use (8) to find the perigee distance.

21. Determine a polar equation of the orbit of the satellite in Problem 20.

22. Find a polar equation of the orbit of the earth around the sun if $r_p = 1.47 \times 10^8$ km and $r_a = 1.52 \times 10^8$ km.

23. The eccentricity of the orbit of Halley's comet is 0.97 and the length of the major axis of its orbit is 3.34×10^9 mi. Find a polar equation of its orbit of the form $r = ep/(1 - e \cos \theta)$.

24. Use the equation obtained in Problem 23 to obtain r_p and r_a for the orbit of Halley's comet.

25. (a) Derive $r = ep/(1 + e \cos \theta)$.
    (b) Derive $r = ep/(1 \pm e \sin \theta)$.

26. For the orientation of the directrix L shown in Figure 12.58, show that the polar equation of a conic section is $r = ep/[1 + e \cos(\theta + \phi)]$.

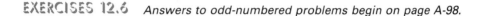

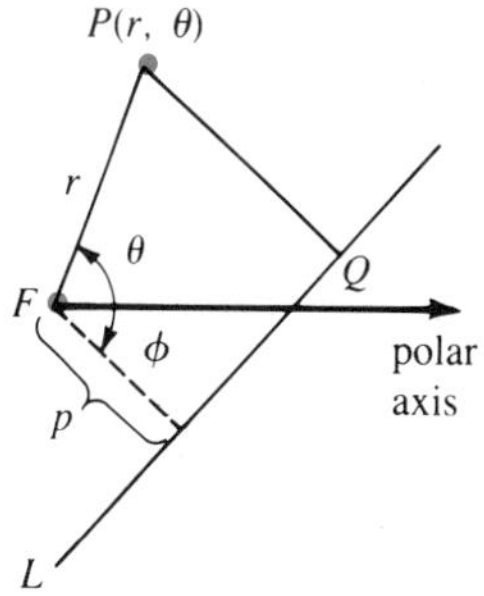

FIGURE 12.58

27. Consider the polar equation $r = ep/(1 - e\cos\theta)$, $e > 1$.

(a) By completing the square, show that the graph of (4) is a hyperbola and that F is a focus at the origin.

(b) Show that $e = c/a$.

Calculator/Computer Problems

28. From (2) of Section 12.5 the area A of the region bounded by the graph of the equation $r = ep/(1 + e\cos\theta)$, $0 < e < 1$, is

$$A = e^2p^2\int_0^{\pi} \frac{1}{(1 + e\cos\theta)^2}\,d\theta$$

Use a computer algebra system to evaluate the integral to show that $A = \pi e^2p^2/(1 - e^2)^{3/2}$.

CHAPTER 12 REVIEW EXERCISES

Answers begin on page A-98.

In Problems 1–20 answer true or false.

1. If for all values of θ the points $(-r, \theta)$ and $(r, \theta + \pi)$ are on the graph of the polar equation $r = f(\theta)$, then the graph is symmetric with respect to the origin. ________

2. The graph of the curve $x = t^2$, $y = t^4 + 1$ is the same as the graph of $y = x^2 + 1$. ________

3. The graph of the curve $x = t^2 + t - 12$, $y = t^3 - 7t$ crosses the y-axis at (0, 6). ________

4. $(3, \pi/6)$ and $(-3, -5\pi/6)$ are polar coordinates of the same point. ________

5. Rectangular coordinates of a point in the plane are unique. ________

6. The graph of the "rose" $r = 5\sin 6\theta$ has six "petals." ________

7. The point $(4, 3\pi/2)$ is not on the graph of $r = 4\cos 2\theta$, since its coordinates do not satisfy the equation. ________

8. The eccentricity of a parabola is $e = 1$. ________

9. The transverse axis of the hyperbola $r = 5/(2 + 3\cos\theta)$ lies along the x-axis. ________

10. The graph of the ellipse $r = 90/(15 - \sin\theta)$ is nearly circular. ________

11. The rectangular coordinates of the point $(-\sqrt{2}, 5\pi/4)$ in polar coordinates are (1, 1). ________

12. The graph of $r = -5\sec\theta$ is a line. ________

13. The terminal side of the angle θ is always in the same quadrant as the point (r, θ). ________

14. The slope of the tangent to the graph of $r = e^{\theta}$ at $\theta = \pi/2$ is -1. ________

15. The graphs of the cardioids $r = 3 + 3\cos\theta$ and $r = -3 + 3\cos\theta$ are the same. ________

16. The area bounded by $r = \cos 2\theta$ is $2\int_{-\pi/4}^{\pi/4}\cos^2 2\theta\,d\theta$. ________

17. The area bounded by $r = 2\sin 3\theta$ is $6\int_0^{\pi/3}\sin^2 3\theta\,d\theta$. ________

18. The area bounded by $r = 1 - 2\cos\theta$ is $\frac{1}{2}\int_0^{2\pi}(1 - 2\cos\theta)^2\,d\theta$. ________

19. The area bounded by $r^2 = 36\cos 2\theta$ is $18\int_0^{2\pi}\cos 2\theta\,d\theta$. ________

20. The θ-coordinate of a point of intersection of the graphs of $r = f(\theta)$ and $r = g(\theta)$ must satisfy the equation $f(\theta) = g(\theta)$. ________

21. Find an equation of the line that is normal to the graph of the curve $x = t - \sin t$, $y = 1 - \cos t$, $0 \le t \le 2\pi$, at $t = \pi/3$.

22. Find the length of the curve given in Problem 21.

23. Find the points on the graph of the curve $x = t^2 + 4$, $y = t^3 - 9t^2 + 2$ at which the tangent line is parallel to $6x + y = 8$.

24. Find the points on the graph of the curve $x = t^2 + 1$, $y = 2t$ at which the tangent line passes through (1, 5).

25. Consider the Cartesian equation $y^2 = 4x^2(1 - x^2)$.

(a) Explain why it is necessary that $|x| \le 1$.

(b) If $x = \sin t$, then $|x| \le 1$. Find parametric equations that have the same graph as the given equation.

(c) Using parametric equations, find the points on the graph of the curve at which the tangent is horizontal.

(d) Graph the curve.

26. Find the area of the region that is outside the circle $r = 4\cos\theta$ and inside the limaçon $r = 3 + \cos\theta$.

27. Find the area of the region that is common to the interiors of the circle $r = 3\sin\theta$ and the cardioid $r = 1 + \sin\theta$.

28. In polar coordinates, sketch the region whose area A is described by $A = \int_0^{\pi/2} (25 - 25 \sin^2\theta)\, d\theta$.

29. Find (**a**) a Cartesian equation and (**b**) a polar equation of the tangent line to the graph of $r = 2 \sin 2\theta$ at $\theta = \pi/4$.

30. Find a Cartesian equation that has the same graph as $r = \cos\theta + \sin\theta$.

31. Find a polar equation that has the same graph as $(x^2 + y^2 - 2x)^2 = 9(x^2 + y^2)$.

32. Determine the rectangular coordinates of the four vertices of the ellipse whose polar equation is $r = 2/(2 - \sin\theta)$.

33. Find a polar equation for the set of points that are equidistant from the pole and the line $r = -\sec\theta$.

34. Find a polar equation of the hyperbola with focus at the origin, vertices (in rectangular coordinates) $(0, -\frac{4}{3})$ and $(0, -4)$, and eccentricity 2.

In Problems 35 and 36 find the area of the shaded region. Each circle has radius 1.

35.

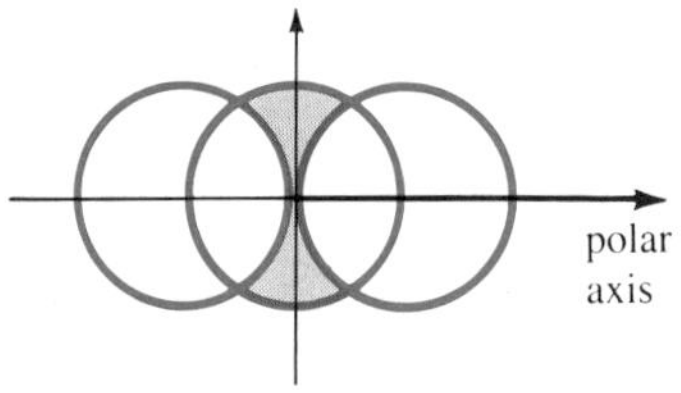

FIGURE 12.59

36.

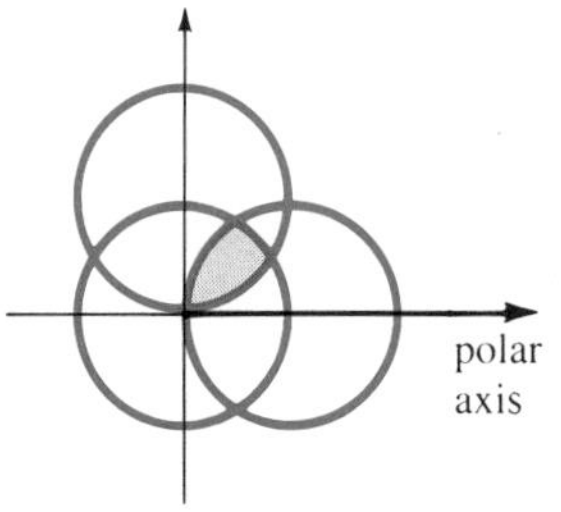

FIGURE 12.60

37. The **folium of Descartes**, shown in Figure 12.61, has the Cartesian equation $x^3 + y^3 = 3axy$. Use the substitution $y = tx$ to find parametric equations for the curve.

38. Use the parametric equations found in Problem 37 to find the points on the folium of Descartes where the tangent line is horizontal.

39. (a) Find a polar equation for the folium of Descartes in Problem 37.

(b) Use the polar equation to find the area bounded by the loop in the first quadrant. [*Hint:* Let $u = \tan\theta$.]

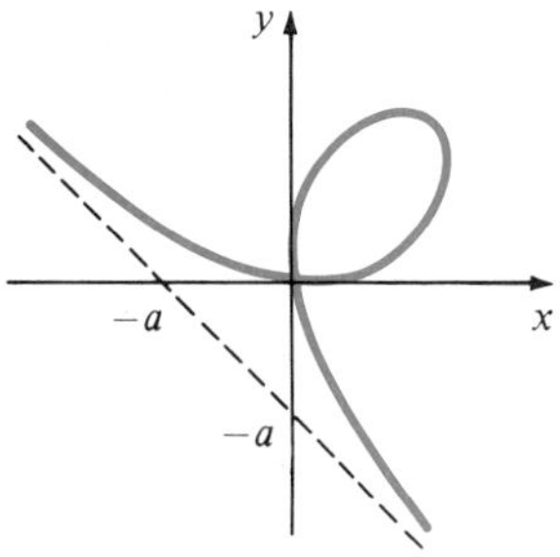

FIGURE 12.61

40. Use the parametric equations found in Problem 37 to show that the folium of Descartes has the asymptote $x + y + a = 0$. This is the dashed line in Figure 12.61. [*Hint:* Consider what happens to x, y, and $x + y$ as $t \to -1$.]

41. The graph of $r = 2 \sin \dfrac{\theta}{3}$ is given in Figure 12.41. Find the area of the interior loop.

42. **Drumlins** are streamlined landforms that are created by glaciers. In a 1959 analysis, Richard J. Chorley concluded that the shape of drumlins can be described as one petal of a rose curve, as shown in Figure 12.62(a), whose equation in polar coordinates is $r = l \cos k\theta$ for $-\pi/2k \le \theta \le \pi/2k$, where l is the length of the drumlin and $k > 1$ is a parameter relating the length to the width.

(a) Show that the area covered by a drumlin of length l and parameter k is $A = l^2\pi/4k$.

(b) Show that the widest point of the drumlin occurs at an angle θ between 0 and $\pi/2k$ that satisfies the equation

$$(k + 1)\cos(k + 1)\theta = (k - 1)\cos(k - 1)\theta$$

See Figure 12.62(b).

(c) Find the approximate width of a drumlin that is 600 yd long and covers 10,000 yd^2. [*Hint:* First find k; then use Newton's Method with $\theta_0 = \pi/4k$ to get an approximate solution to the equation in part (**b**).]

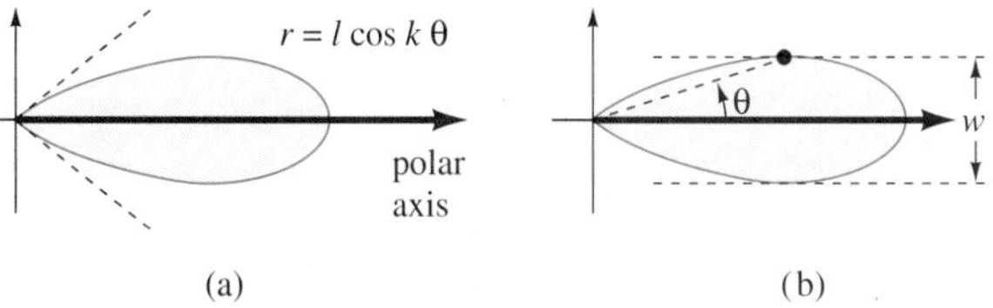

FIGURE 12.62

CALCULATOR/COMPUTER ACTIVITIES

CHAPTER 7 Inverse Trigonometric and Hyperbolic Functions

Answers to odd-numbered problems begin on page A-117.

7.1 All scientific calculators provide the inverse sine, the inverse cosine, and the inverse tangent with keys generally labeled $\sin^{-1}$, $\cos^{-1}$, and $\tan^{-1}$, respectively. Computer programming languages (and the packages and application programs written in them) generally do not use this notation because the superscript is hard to type on a standard keyboard and would lead to confusion with the exponent -1 that indicates a multiplicative inverse. Thus, for example, a computer application may require typing ARCTAN or ATAN to indicate the inverse tangent function. Find out the syntax needed for the calculator or computer application you have available for graphing.

In Problems 1–4 graph the given function on the indicated interval. Use the trace and zoom-in features of the graphing utility to locate any x-intercepts, y-intercepts, and relative extrema on the displayed graph to an accuracy of at least two decimal places.

1. $y = \sin^{-1}x - \dfrac{\pi}{2}x$; $[-1, 1]$
2. $y = \cos^{-1}(\sin x)$; $[-8, 8]$
3. $y = \tan^{-1}(\sin(3\cos^{-1}x))$; $[-1, 1]$
4. $y = \dfrac{1}{\tan^{-1}x}$; $[-10, 10]$

The arccotangent, the arcsecant, and the arccosecant are seldom provided as intrinsic functions on a calculator or computer. In a manner similar to Example 7 in Section 7.1 these inverse trigonometric functions can be obtained using the arctangent. The hard part when programming is to ensure that the resulting function program matches the range specified in the definition. [See the remark at the end of this section for a discussion of this arbitrary selection.]

In Problems 5–7 write a calculator program, a computer program, or some procedure in the application or computer algebra system available to implement each function. Verify that the program operates correctly by testing several known values, making sure to test all subintervals of the desired range.

5. $f(x) = \cot^{-1}x$
6. $f(x) = \sec^{-1}x$
7. $f(x) = \csc^{-1}x$

From Section 7.1, $T_1(x) = \cos(\arccos x) = x$, where x is a number in the interval $[-1, 1]$. Using the double angle formula for the cosine, we can show that $T_2(x) = \cos(2\arccos x) = 2x^2 - 1$, again on $[-1, 1]$. In general, $T_n(x) = \cos(n\arccos x)$ is a polynomial of degree n on $[-1, 1]$. These polynomials are called **Chebyshev polynomials** (or Tschebysheff polynomials, using an alternative transliteration of the Russian name that suggests the notation T). A striking feature of a Chebyshev polynomial is that all of its roots and all of the points on the graph corresponding to local extreme values can be seen in the viewing window $-1 \le x \le 1$, $-1 \le y \le 1$. Chebyshev polynomials have important applications in several areas of applied and computational mathematics.

In Problems 8–10 plot the indicated Chebyshev polynomials on the same screen. From the graphs, make some conjectures about how the roots (or the x-coordinates of points giving extreme values) of the Chebyshev polynomial of one degree are related to those of the next higher degree in the problem.

8. $T_1(x)$, $T_2(x)$, $T_3(x)$
9. $T_2(x)$, $T_4(x)$, $T_6(x)$, $T_8(x)$
10. $T_3(x)$, $T_5(x)$, $T_7(x)$, $T_9(x)$

7.3 Some scientific calculators provide the inverse hyperbolic sine, the inverse hyperbolic cosine, and the inverse hyperbolic tangent by the keystroke **hyp** followed by $\sin^{-1}$, $\cos^{-1}$, and $\tan^{-1}$, respectively. Computer programming languages generally do not include the inverse hyperbolic functions as intrinsic functions, but computer algebra systems often have these functions using the alternative notation arcsinh, arccosh, and arctanh. Find out the syntax needed for the calculator or computer application available for graphing or use the equivalent expressions in terms of the natural logarithmic function given in Theorem 7.1.

In Problems 11–14 graph the given function on the indicated interval. Use the trace and zoom-in features of the graphing utility to locate any x-intercepts, y-intercepts, and relative extrema on the displayed graph to an accuracy of at least two decimal places.

11. $y = \sinh^{-1}(\sin x)$; $[-2\pi, 2\pi]$

12. $y = \tanh^{-1}(\sin x)$; $[-2\pi, 2\pi]$

13. $y = \dfrac{1}{\sinh^{-1}x} - \operatorname{csch}^{-1}x$; $[0, 6]$

14. $y = \cosh^{-1}x - \sqrt{x-1}$; $[1, 25]$

CHAPTER 8 Techniques of Integration

Answers to odd-numbered problems begin on page A-118.

8.5 The better computer algebra systems, such as Maple, Mathematica, Derive, and MACSYMA, have algorithms to factor polynomials. Factorization of the denominator is usually the most difficult step in the partial fraction decomposition of a rational function. Once this factorization is accomplished, the computer algebra system can complete the steps given in this section to find the partial fraction decomposition and then evaluate the indefinite integral of the rational function. Since there are other uses for a partial fraction decomposition, this process can usually be called separately.

In Problems 1–6 use a computer algebra system to find the partial fraction decomposition for the given rational function.

1. $\dfrac{7.4x^3 - 76.91x^2 + 247.89x - 474.68}{x^4 - 14.2x^3 + 53.17x^2 + 6.504x - 190.43}$

2. $\dfrac{159.7x^2 - 183.16x - 2.421}{42.84x^3 - 82.659x^2 + 6.441x + 3.348}$

3. $\dfrac{24x^4 - 51x^3 + 4x^2 - 476x - 105}{2x^5 - 5x^4 - 7x^3 - 161x^2 - 195x - 594}$

4. $\dfrac{5824x^3 - 1726x^2 - 122x + 2}{3264x^4 - 1304x^3 - 190x^2 + 11x + 1}$

5. $\dfrac{12x^4 - 428x^3 + 8601x^2 + 5954x + 223{,}025}{x^5 - 14x^4 - 959x^3 + 7919x^2 - 7202x + 155{,}295}$

6. $\dfrac{28x^5 - 5x^4 + 285x^3 + 75x^2 + 575x - 46}{6x^6 + 7x^5 + 81x^4 + 86x^3 + 160x^2 + 99x - 55}$

Even when the denominator of a rational function is factored, solving the n linear equations for the n unknown constants can be a formidable task by hand. Some calculators can solve up to six linear equations in six unknowns, and many linear algebra applications are available to solve systems of linear equations numerically. Computer algebra systems can generally find exact rational solutions for a linear system.

In Problems 7–10 set up and solve the system of linear equations for the constants in the partial fraction decomposition for the given rational function.

7. $\dfrac{64x^3 - 1585x^2 + 8219x - 11{,}720}{(2x+1)(x-19)(x-18)(x-7)}$

8. $\dfrac{34x^4 + 52x^3 - 746x^2 - 3408x - 4572}{(x^2+5x+10)(x+3)(x-6)(x-1)}$

9. $\dfrac{1103x^4 - 2421x^3 + 21{,}789x^2 - 124{,}455x - 215{,}021}{(5x^2+2x+106)(2x+53)(3x^2+6x+11)}$

10. $\dfrac{35{,}114.7225x^4 + 2283.2483x^3 - 2004.923x^2 + 10.21x + 16.1}{(3.2x+1)(8.7x+1)(5.8x-1)(6.1x-1)(9.3x-1)}$

CHAPTER 9 Indeterminate Forms and Improper Integrals

Answers to odd-numbered problems begin on page A-118.

9.2 Convergent integrals with upper limit of integration ∞ can be approximated by numerically evaluating integrals with larger and larger upper limits. One way to do this efficiently is with the special form of Simpson's Rule used in the calculator/computer activities for Section 4.8, where the approximation for the value of the integral over the subinterval $[k\Delta x, (k+2)\Delta x]$ is added to the previously computed approximation for the value of the integral over the interval $[a, k\Delta x]$. If an automatic numerical integration algorithm is provided on a calculator or within a computer application, then it is a simple matter to repeatedly invoke numerical integration on an interval $[a, t]$, where the value t is made larger and larger for each repetition.

In Problems 1–6 estimate the value of the given improper integral by numerically approximating integrals on larger and larger intervals of integration.

1. $\displaystyle\int_0^\infty e^{-x^2}\,dx$

2. $\Gamma\left(\dfrac{3}{2}\right) = \displaystyle\int_0^\infty t^{1/2}e^{-t}\,dt$

[See Problem 49 in Exercises 9.2.]

3. $\Gamma\left(\dfrac{7}{2}\right) = \displaystyle\int_0^\infty t^{5/2}e^{-t}\,dt$

[See Problem 49 in Exercises 9.2.]

4. $\displaystyle\int_1^{\infty} \frac{\sin^2 x}{x^2}\,dx$

5. $\displaystyle\int_{10}^{\infty} \frac{1-\cos x}{x^3}\,dx$

6. $\displaystyle\int_0^{\infty} e^{-2x}\tan^{-1}x\,dx$

By means of a substitution, an improper integral can sometimes be transformed into a definite integral or a different improper integral that is more tractable for numerical approximation. For example, it is usually easy to find a substitution that turns an integral with ∞ as an upper limit of integration into a definite integral with finite limits of integration (but the new integrand may now become infinite).

In Problems 7–10 make the indicated substitution to transform the infinite interval of integration into a finite interval. Then use Simpson's Rule or an intrinsic numerical integration on the resulting definite integral, taking care to avoid including points where the integrand becomes infinite.

7. $\displaystyle\int_1^{\infty} x^{-2}\sin x\,dx;\quad x = t^{-1}$

8. $\displaystyle\int_1^{\infty} x^{-1}\sin x\,dx;\quad x = t^{-1}$

9. $\displaystyle\int_0^{\infty} e^{-x^2}\,dx;\quad x = -\ln t$

10. $\displaystyle\int_0^{\infty} \sin x^2\,dx;\quad x = \tan t$

CHAPTER 10 Sequences and Series

Answers to odd-numbered problems begin on page A-118.

10.1 For practical calculations, it is important to get some sense of how rapidly a sequence $\{a_n\}$ converges, assuming it does converge to some number L. In terms of Definition 10.2, this means that given a specific ε, say $\varepsilon = 10^{-4}$, we are interested in how large N needs to be in order to guarantee that $|a_n - L| < 10^{-4}$ whenever $n \geq N$.

In Problems 1–6 experiment with a calculator or computer to find N so that the given inequality holds whenever $n \geq N$.

1. $|\sqrt[n]{0.4} - 1| < 10^{-3}$

2. $|(-0.85)^n - 0| < 10^{-4}$

3. $\left|\dfrac{e^n}{n!} - 0\right| < 10^{-7}$

4. $|\sqrt[n]{n} - 1| < 10^{-3}$

5. $\left|\dfrac{5n^2}{4n^2+n+1} - \dfrac{5}{4}\right| < 10^{-5}$

6. $|ne^{-n} - 0| < 10^{-8}$

A value x such that $g(x) = x$ is called a **fixed point** for the function g. A common method for approximating a fixed point for g is to compute the recursive sequence $x_{n+1} = g(x_n)$. This particular sequence has a nice graphical representation commonly called a "cobweb." First plot the graph of $y = g(x)$ and the graph of $y = x$ on the same viewing window. Then a fixed point is the x-coordinate of an intersection point of the two graphs. Given an initial value x_0, draw a vertical line from the x-axis at $x = x_0$ to the curve $y = g(x)$ at the point $(x_0, g(x_0))$ corresponding to the calculation of $g(x_0)$. Draw a horizontal line from this last point to the point $(g(x_0), g(x_0))$ on the line $y = x$ corresponding to the assignment $x_1 = g(x_0)$. Then draw a vertical line from this point to the point $(x_1, g(x_1))$ for the calculation of $g(x_1)$, and draw a horizontal line from this last point on the function curve to the line at $(g(x_1), g(x_1))$ for the next assignment $x_2 = g(x_1)$. Continue in this fashion. Plotting the example $g(x) = \cos x$ with initial value $x_0 = 1.3$ will make it evident why the resulting diagram is called a "cobweb." Newton's Method is, in fact, a special case of this fixed-point recursive sequence where $g(x) = x - f(x)/f'(x)$. Just as with Newton's Method, there is no guarantee of convergence for any initial value in this fixed-point recursive sequence, and convergence may not be to the fixed point expected.

In Problems 7–12 calculate the next ten terms of the fixed-point recursive sequence $x_{n+1} = g(x_n)$ for the given function and initial value. If possible, obtain a plot of the "cobweb" diagram corresponding to each problem.

7. $g(x) = \cos x,\quad x_0 = 0.3$
8. $g(x) = \sqrt{3x+1},\quad x_0 = 0.5$
9. $g(x) = \sqrt{3x+1},\quad x_0 = 15$
10. $g(x) = |2\sin x| + 0.7x,\quad x_0 = 3.5$
11. $g(x) = |2\sin x| + 0.7x,\quad x_0 = 4.0$
12. $g(x) = 2 + \ln x,\quad x_0 = 30$

10.2 Consider the Wallis sequence of partial products in the pattern

$$\frac{2}{1}, \frac{2\times 2}{1\times 3}, \frac{2\times 2\times 4}{1\times 3\times 3}, \frac{2\times 2\times 4\times 4}{1\times 3\times 3\times 5}, \frac{2\times 2\times 4\times 4\times 6}{1\times 3\times 3\times 5\times 5}, \ldots$$

This sequence can be defined recursively by the formulas

$$a_1 = 2,\ a_{2n} = a_{2n-1}\left(\frac{2n}{2n+1}\right),$$
$$a_{2n+1} = a_{2n}\left(\frac{2n+2}{2n+1}\right) \qquad \text{for } n = 1, 2, \ldots$$

The limit of this sequence is called an **infinite product**, and it can be shown that this sequence converges to $\pi/2$.

In Problems 13–17 explore the properties of the Wallis sequence of partial products.

13. Show that the Wallis sequence $\{a_n\}$ is not monotonic.

14. Show that the sequence of even terms $b_n = a_{2n}$ forms a monotonic sequence, and determine whether it is increasing or decreasing.

15. Show that the sequence of odd terms $c_n = a_{2n+1}$ forms a monotonic sequence, and determine whether it is increasing or decreasing.

16. Experiment with a calculator or computer to find N so that $\left|b_n - \frac{\pi}{2}\right| < 0.5 \times 10^{-1}$ holds whenever $n \geq N$.

17. Consider the sequence consisting of the average of the even and odd terms $\left\{\frac{a_{2n} + a_{2n+1}}{2}\right\}$. Experiment with a calculator or computer to find N so that $\left|\frac{a_{2n} + a_{2n+1}}{2} - \frac{\pi}{2}\right| < 0.5 \times 10^{-3}$ holds whenever $n \geq N$.

In Problems 18–21 explore the sequence $\{a_n\}$ defined recursively by $a_{n+1} = \sqrt{a_n}$ and the sequence $\{b_n\}$ defined recursively by $b_{n+1} = b_n^2$.

18. Prove that $\{a_n\}$ is monotonic and verify experimentally that $\lim_{n\to\infty} a_n = 1$ for any positive starting value a_1.

19. Prove that $\{b_n\}$ is monotonic and verify experimentally that $\lim_{n\to\infty} b_n = 0$ for any starting value $0 < b_1 < 1$ and that $\{b_n\}$ diverges for any starting value $b_1 > 1$.

20. Let $a_1 = 10$. Calculate a_{n+1} by taking a square root n times. Then let $b_1 = a_{n+1}$ and calculate b_{n+1} by squaring n times. Theoretically the final result should be 10, but rounding will cause a slightly different answer. Try this for $n = 15, 20, 25, 30, \ldots$ until the effect of rounding is evident on the calculator or computer used.

21. Let $a_1 = 0.2$. Calculate a_{n+1} by taking a square root n times. Then let $b_1 = a_{n+1}$ and calculate b_{n+1} by squaring n times. Theoretically the final result should be 0.2, but rounding will cause a slightly different answer. Try this for $n = 15, 20, 25, 30, \ldots$ until the effect of rounding is evident on the calculator or computer used.

10.3 Adding the terms of an infinite series using finite precision on a calculator or computer can lead to results that are different from the theoretical limit. If we assume that the terms in the infinite series tend to zero and the addition begins with the term largest in absolute value, the calculations eventually reach the situation where the next term a_n and all later terms are so small that adding one of them to the partial sum calculated before that term makes no change on the finite precision result stored. One way to lessen the effect of this rounding is to add the smaller terms in the series *first* rather than starting with the largest.

In Problems 22–27 calculate the sum of the given finite series using a short program on a calculator or computer.

22. $\displaystyle\sum_{k=1}^{1000} \frac{1}{k}$

23. $\displaystyle\sum_{k=1}^{500} \frac{1}{k^{1.5}}$

24. $\displaystyle\sum_{k=1}^{400} \frac{k+1}{k!}$

25. $\displaystyle\sum_{k=1}^{200} \frac{(-1)^k}{k^2}$

26. $\displaystyle\sum_{k=1}^{500} \frac{\cos k}{k(k+1)}$

27. $\displaystyle\sum_{k=1}^{200} \left(\frac{\pi}{k}\right)^k$

10.6 The nth partial sum S_n for an alternating series $\Sigma_{k=1}^{\infty}(-1)^{k+1}a_k$, $a_k > 0$, converging to S satisfies either

$$S_n \leq S \leq S_{n+1} \quad \text{or} \quad S_{n+1} \leq S \leq S_n$$

depending upon whether n is odd or even. Theorem 10.8 states that the error $|S_n - S|$ is less than the absolute value of the $(n+1)$st term, a_{n+1}. The inequalities above suggest that a better approximation than S_n can be obtained using the *average* of the nth and $(n+1)$st partial sums; that is,

$$A_n = \frac{S_n + S_{n+1}}{2} = S_n + \frac{(-1)^{n+2}a_{n+1}}{2}$$

The error $|A_n - S|$ is less than $a_{n+1}/2$. For a slowly converging series, this average can be a better approximation for S than the next partial sum S_{n+1} with error $|S_{n+1} - S|$ less than a_{n+2}.

In Problems 28–31 calculate the nth partial sum S_n and the average A_n for the given values for n using a short program on a calculator or computer. Find the associated error bounds in terms of a_{n+1}.

28. $\displaystyle\sum_{k=1}^{\infty} \frac{(-1)^{k+1}}{k^2}, \quad n = 100, 101$

29. $\displaystyle\sum_{k=1}^{\infty} \frac{(-1)^{k+1}}{k}, \quad n = 800, 900$

30. $\displaystyle\sum_{k=1}^{\infty} \frac{(-1)^{k+1}}{k!}, \quad n = 50, 51$

31. $\displaystyle\sum_{k=1}^{\infty} \frac{(-2)^{k+1}}{k!}, \quad n = 80, 100$

10.7 A power series is a function, and the interval of convergence is the domain of this function. Unfortunately, it is not easy to generate the graph of a power series as a function on a graphic calculator or a graphing application on a computer. Fixing x from the interval of convergence, we can select a positive integer n so that the partial sum up to and including the nth power is sufficiently close to the limit of the infinite series evaluated at that x for graphical purposes. The difficulty is that n may need to be much larger for a different value of x. In general, the first few terms of a power series will give a nice graphical representation for that power series as a function for x-values near the center $x = a$. A very large number of terms will be needed to accurately represent a power series near an endpoint of a finite interval of convergence. Partial sums of a power series are polynomials, which are defined everywhere and which show no hint of the interval of convergence for the power series. Recall that polynomials can be more quickly and accurately evaluated in the nested form discussed in the calculator/computer activities for Section 1.2.

In Problems 32–37 plot the graphs of the indicated partial sums on the given interval.

32. $\sum_{k=1}^{n} \frac{x^k}{k^2}$, $n = 4, 6, 8$; $-2 \le x \le 2$

33. $\sum_{k=1}^{n} \frac{x^k}{k}$, $n = 5, 6, 7, 8$; $-1.6 \le x \le 1.6$

34. $\sum_{k=0}^{n} \frac{(x-2)^k}{(k+1)3^k}$, $n = 3, 4, 5$; $-2 \le x \le 6$

35. $\sum_{k=0}^{n} \frac{(x+5)^k}{k!}$, $n = 3, 5, 7$; $-10 \le x \le -1$

36. $\sum_{k=0}^{n} x^k$, $n = 8, 11, 14$; $-1.6 \le x \le 1.3$

37. $\sum_{k=0}^{n} \frac{(-1)^k(x-4)^{2k}}{(2k)!}$, $n = 3, 4, 5$; $-1 \le x \le 9$

10.9 For a graphical understanding of Taylor polynomials, it is helpful to plot the function and several of its Taylor polynomials in the same viewing window. The suggested viewing window for each problem in this section is carefully chosen to show both where the indicated Taylor polynomials match the function very closely near $x = a$ and where the polynomials and the function differ greatly. If a symbolic calculator or computer algebra system is being used, check to see if there is a command to produce Taylor polynomials for a given function automatically.

In Problems 38–41 plot the graph of the given function and the indicated Taylor polynomials on the same viewing window. Zoom in or out from the suggested window to investigate the relationships between these graphs. Use the trace feature (such as on the TI-81) to compare, if possible, the values of the different Taylor polynomials with the value of the function.

38. $f(x) = e^{x-1}$; $a = 1$; $P_2(x), P_3(x), P_4(x)$; on $-2 \le x \le 4$, $-2 \le y \le 9$
39. $f(x) = \sin x$; $a = 0$; $P_3(x), P_5(x), P_7(x)$; on $-4 \le x \le 4$, $-2 \le y \le 2$
40. $f(x) = \cos x$; $a = \pi/4$; $P_2(x), P_3(x), P_4(x)$; on $-3 \le x \le 4$, $-2 \le y \le 2$
41. $f(x) = \ln(x + 1)$; $a = 2$; $P_1(x), P_2(x), P_3(x)$; on $-4 \le x \le 15$, $-3 \le y \le 7$

In Problems 42–45 construct a table listing the values of each Taylor polynomial at the indicated values of x. This is most easily accomplished by using a short program that prompts for the desired x-value and then displays the value of the polynomial.

42. $f(x) = \sin x$; $a = 0$; $P_3(x), P_5(x), P_7(x)$; for $x = -1, -0.5, 0.3, 0.7, 1.1, 1.5$
43. $f(x) = \sinh x$; $a = 0$; $P_3(x), P_7(x), P_{11}(x)$; for $x = -2, -0.8, 0.2, 1, 1.6, 3$
44. $f(x) = \ln x$; $a = 2$; $P_2(x), P_4(x), P_6(x)$; for $x = 1, 1.8, 2.06, 2.4, 3.5, 5$
45. $f(x) = \cos x$; $a = \pi/3$; $P_3(x), P_4(x), P_5(x)$; for $x = 0.5, 1, 1.5, 2, 2.5, 3$

CHAPTER 11 Analytic Geometry in the Plane

Answers to odd-numbered problems begin on page A-119.

11.1 For most calculator/computer graphing, the viewing window for a plot will use different scales along each axis. In fact, the y-range on the graphical window is sometimes automatically adjusted so that the graph of a function will fit exactly. However, for plotting a parabola (and later an ellipse or hyperbola), it is desirable to have the x- and y-axes equally scaled so that true distances and angles are more accurately represented. On most graphic calculators the default range setting for the viewing window gives equal scaling. For example, the HP-28S uses the default window $-6.8 \le x \le 6.8$, $-1.5 \le y \le 1.6$, where each square pixel on the screen represents a width and height of 0.1. One simple way to preserve equal scaling is to choose settings that are proportional to the equally scaled default settings. Thus, multiplying the default settings for the HP-28S by 2.5 gives $-17 \le x \le 17$, $-3.75 \le y \le 4$. The **Standard** and **Trig** range settings on the TI-81 do not give equally scaled graphical screens. On that calculator use the command **Square** in the ZOOM menu to transform any viewing window into a larger equally scaled window. On a computer graphing application it may take some experimentation to determine appropriate range settings for equal scaling.

In Problems 1–6 graph the given parabola. Use an equally scaled viewing window containing the vertex.

1. $y = 2(x - 3)^2 - 5$
2. $y = 1 \pm \sqrt{2x - 8}$
3. $y = \pm\frac{1}{5}\sqrt{x + 2}$
4. The parabola with focus (2, 3) and vertex (2, 2)
5. The parabola with focus (1, 1) and vertex (−2, 1)
6. The parabola with focus (4, 1) and directrix $y = -2$

11.2 The length s of the circumference of an ellipse $x^2/a^2 + y^2/b^2 = 1$, $a > b$, with eccentricity $e = \sqrt{a^2 - b^2}/a$ can be written in the form

$$s = 4a \int_0^{\pi/2} \sqrt{1 - e^2\sin^2\theta}\, d\theta$$

See Problem 20 of Exercises 5.4, Problem 14 in Exercises 10.10, and Problem 54 in Exercises 11.2 for more information about this arc length and this definite integral. This formula is convenient for satellite orbit calculations where the eccentricity is small and the orbit is nearly circular ($e = 0$). If we assume the Earth's orbit is circular ($e = 0$), then $a = b = 1$ A.U. (astronomical unit ≈ 93,000,000 miles) and the orbit arc length would be $s = 2\pi$ A.U. One of the problems in this section calculates a more accurate estimate for the Earth's orbit.

In Problems 7–10 find the arc length of the elliptical orbit for the given planet or comet. Use a program to implement Simpson's Rule for various numbers of subintervals, use the numerical integration procedure built into the calculator or computer application available, or use the series approximation for this integral found in Problem 14 in Exercises 10.10. Indicate the expected accuracy achieved.

7. Earth: $e = 0.016722$, $a = 1$ A.U., $b = 0.9998601776$ A.U.
8. Mercury: $e = 0.205628$, $a = 0.387099$ A.U., $b = 0.378827$ A.U.
9. Saturn: $e = 0.05565$, $a = 9.53884$ A.U., $b = 9.52406$ A.U.
10. Halley's comet: $e = 0.97$, $a = 18$ A.U., $b = 4.376$ A.U.

11.3 In Problems 11–16 graph the hyperbola with the given equation. Use an equally scaled viewing window containing the vertices. Plot the asymptotes for the given hyperbola on the same screen.

11. $y = \pm 0.8\sqrt{x^2 + 2}$
12. $y = \pm 1.4\sqrt{x^2 - 8}$
13. $y = 5 \pm \sqrt{x^2 + 20}$
14. $y = 2 \pm 0.4\sqrt{x^2 + 4.3x + 7}$
15. $y = -15 \pm 2\sqrt{x^2 - 7x + 35}$
16. $3x^2 - 7y^2 + 2x - 5y = 12$

CHAPTER 12 Parametric Equations and Polar Coordinates

Answers to odd-numbered problems begin on page A-119.

12.1 Most computer graphing applications plot parametric equations in addition to graphs of functions. All graphic calculators can plot parametric equations by using a short program that repeatedly calculates a point on the curve and then plots this point. The TI-81 graphics calculator has parametric graphing capabilities without programming and even has the tracing feature for parametric curves. Actually watching the plotting of parametric equations is more meaningful than looking at the completed graph because the points are plotted in order as the value of the parameter increases. We consider here some special parametric curves called **Bézier curves**, which are commonly used in computer-aided design (CAD), in microcomputer drawing applications, and in the mathematical representation of different fonts for many laser printers. A cubic Bézier curve is specified by four control points in the plane—for example, $P_0(p_0, q_0)$, $P_1(p_1, q_1)$, $P_2(p_2, q_2)$, and $P_3(p_3, q_3)$. The curve starts at the first point for the parameter $t = 0$, ends at the last point for $t = 1$, and roughly "heads toward" the middle points for parameter values between 0 and 1. Artists and engineering designers can move the control points to adjust the end locations and the shape of the parametric curve. The cubic Bézier curve for these four control points has the following parametric equations:

$$x = p_0(1 - t)^3 + 3p_1(1 - t)^2 t + 3p_2(1 - t)t^2 + p_3 t^3$$
$$y = q_0(1 - t)^3 + 3q_1(1 - t)^2 t + 3q_2(1 - t)t^2 + q_3 t^3$$
$$0 \le t \le 1$$

Several Bézier curves can be pieced together continuously by making the last control point on one curve the first control point on the next curve. Equivalently, piecewise parametric equations can be constructed. For example, the next piece can be represented by

$$x = p_3(2-t)^3 + 3p_4(2-t)^2(t-1) + 3p_5(2-t)(t-1)^2 + p_6(t-1)^3$$

$$y = q_3(2-t)^3 + 3q_4(2-t)^2(t-1) + 3q_5(2-t)(t-1)^2 + q_6(t-1)^3$$

$$1 \le t \le 2$$

In Problems 1–6 use a graphing utility to obtain the graph of the piecewise continuous Bézier curve associated with the given control points. Set the viewing window so that all control points are within the window. If possible, use equally scaled axes. Plot the control points on the same screen.

1. $P_0(5, 1)$, $P_1(1, 30)$, $P_2(50, 28)$, $P_3(55, 5)$
2. $P_0(32, 1)$, $P_1(85, 25)$, $P_2(1, 30)$, $P_3(40, 3)$
3. $P_0(10, 5)$, $P_1(16, 4)$, $P_2(25, 28)$, $P_3(30, 30)$, $P_4(18, 1)$, $P_5(40, 18)$, $P_6(16, 20)$
4. $P_0(55, 50)$, $P_1(45, 40)$, $P_2(38, 20)$, $P_3(50, 20)$, $P_4(60, 20)$, $P_5(63, 35)$, $P_6(45, 32)$
5. $P_0(30, 30)$, $P_1(40, 5)$, $P_2(12, 12)$, $P_3(45, 10)$, $P_4(58, 10)$, $P_5(66, 31)$, $P_6(25, 30)$
6. $P_0(48, 20)$, $P_1(20, 15)$, $P_2(20, 50)$, $P_3(48, 45)$, $P_4(28, 47)$, $P_5(28, 18)$, $P_6(48, 20)$, $P_7(48, 36)$, $P_8(52, 32)$, $P_9(40, 32)$

In Problems 7–9 experiment with the locations for control points to obtain piecewise continuous Bézier curves approximating the given shape or object. Give the final control points chosen, and sketch the resulting parametric curve.

7. Historically the letter "S" has been one of the most difficult to represent mathematically. Use two or three Bézier curve pieces to draw a letter "S" in some simple font style.
8. The long cross-section of an egg is not quite an ellipse because one end is more pointed than the other. Use several Bézier curve pieces to represent an approximation for the shape of an egg.
9. Give a curve approximating the shape of the Greek letter epsilon, ε, using as few pieces as possible.

12.2 The tangent lines for Bézier curves have several interesting properties that are explored in the problems in this section. For smoothly connecting different pieces, the tangent lines at the endpoints of a Bézier curve are of special interest. Since the parametric equations for a Bézier curve are two cubic functions in t, the integrand for the definite integral giving the arc length will be the square root of a nonnegative fourth-degree polynomial in t. Generally numerical integration will be needed to evaluate this integral.

In Problems 10–15 graph the indicated quantities and prove the stated facts about Bézier curves. These problems refer to the calculator/computer activities for Section 12.1.

10. Plot the Bézier curve in Problem 1 again and also draw the line segment between P_0 and P_1. Verify graphically that the tangent line to the curve at $t = 0$ has the same slope as this line segment and passes through P_1. Then prove that the tangent line at the initial control point P_0 on any cubic Bézier curve will pass through the second control point P_1.

11. Plot the Bézier curve in Problem 2 again and also draw the line segment between P_2 and P_3. Verify graphically that the tangent line to the curve at $t = 1$ has the same slope as this line segment and passes through P_2. Then prove that the tangent line at the final control point P_3 on any cubic Bézier curve will pass through the third control point P_2.
12. Redraw the piecewise Bézier curve that you chose to represent the letter "S" in Problem 7. Then calculate the slope of the tangent line to this curve in the middle of the center downward part of the letter. Experiment further drawing the letter "S" using pieces that change at this center point for the letter. Control the slope at this center point using the results proven in Problems 10 and 11. Find the slope for this center point for the letter "S" that you feel looks best.
13. Redraw the piecewise Bézier curve that you chose to represent the shape of an egg in Problem 8. In view of the results of Problems 10 and 11, adjust the control points so that the slope of the tangent line varies continuously between pieces.
14. Find the arc length for the Bézier curve in Problem 1. Calculate the length of the piecewise linear curve made up of the line segments from P_0 to P_1, from P_1 to P_2, and from P_2 to P_3. Compare the two lengths.
15. Find the arc length for the Bézier curve in Problem 2. Calculate the length of the piecewise linear curve made up of the line segments from P_0 to P_1, from P_1 to P_2, and from P_2 to P_3. Compare the two lengths. Make a conjecture about a general result concerning these two lengths.

12.3 Most scientific calculators provide the conversions between polar coordinates and rectangular coordinates. A calculator will give only one answer when converting from rectangular coordinates to polar coordinates, generally with $0 \le r$ and $0 \le \theta < 2\pi$. Make sure that the calculator is in *radian mode* when these conversions are performed.

In Problems 16–20 find the indicated conversions on a scientific calculator.

16. Convert the following rectangular coordinates into polar coordinates: $(1, 5)$, $(-2, 7)$, $(4, -3)$, $(-1, -0.2)$.
17. Convert the following polar coordinates into rectangular coordinates: $(1, 5)$, $(5, \pi/15)$, $(4, 1.2)$, $(25, 4.3)$.
18. Unfortunately some calculators do not handle a

negative value for r. Try to convert the following polar coordinates into rectangular coordinates directly. If your calculator gives an error message, rewrite each point in an alternative form with a positive value for r and then calculate the rectangular coordinates: $(-1, \pi/7)$, $(-9, 0.8)$, $(-0.4, 4.5)$.

19. Perhaps it is not surprising that some calculators cannot handle the ambiguous request to convert the rectangular coordinate $(0, 0)$ into polar coordinates. Check out what your calculator does with this conversion. Any answer with $r = 0$ is correct, but some give an error message. Also check out how your calculator handles the request to convert a polar coordinate with $r = 0$ into rectangular coordinates.

20. Check whether your calculator can handle conversion from polar coordinates to rectangular coordinates when the angle θ is negative or greater than 2π. If not, rewrite the following polar coordinates in an alternative form and then calculate the rectangular coordinates: $(3, -\pi/10)$, $(-25, -5)$, $(6, 10)$, $(8, 5\pi)$.

12.4 Computer applications dedicated to calculus may provide direct ways to plot the graph of a polar equation. If such a feature is not available, then a polar equation can easily be plotted using the procedure for parametric equations in the plane. Solve the polar equation for $r = f(\theta)$ and enter the parametric equations $x = f(\theta)\cos\theta$ and $y = f(\theta)\sin\theta$. If these parametric equations are periodic, then choose one period for the parameter interval. On the TI-81 graphics calculator, the variable t must be used as the parameter for the parametric equations. This calculator can be set to a Polar display mode where the coordinates for plotting and for the trace feature are displayed on the screen in polar coordinates. Note, however, that polar coordinates are not unique and the coordinates (r, θ) displayed may not be the coordinates that satisfy the polar equation.

In Problems 21–29 graph the given polar equation.

21. $r = 5 + 2\sin\theta$

22. $r = 5 + 6\cos\theta$

23. $r = 5\cos 4\theta$

24. $r = 5\sin\theta + 2\cos\theta$

25. $r = 4\sin 3\theta + \cos 2\theta$

26. $r = \sin 5\theta + 5\cos 2\theta$

27. $r = \dfrac{\sin\theta + 2\cos\theta}{3 + 2\sin\theta + \cos\theta}$

28. $r = 5\sin 4\theta + 3\cos 6\theta$

29. $r = 4 + 3\cos 5\theta$

The graphs of polar equations can be a great aid for finding points of intersection. Zoom in to take a closer look at any suspected point of intersection. Sometimes it may be necessary to look closely near the origin when several curve segments appear to intersect exactly at the origin and perhaps at other nearby points. Remember that a point in the plane can have more than one polar coordinate representation. A certain polar coordinate representation may cause a point to lie on one curve, while a different polar coordinate representation may cause this same point to lie on another curve. If very high accuracy is required, numerical root-finding can be used once the differences in coordinate representations have been included in the equation to be solved. For example, to find the unique intersection of the curves $r = 2 + 5\cos\theta$ and $r = 2 + \sin\theta$ in the first quadrant, numerically solve the equation

$$-(2 + 5\cos(\theta + \pi)) = 2 + \sin\theta$$

In Problems 30–32 find the points of intersection of the graphs of the given pair of polar equations.

30. $r = 1 + \sin\theta$
$r = 1.2 + 2\cos\theta$

31. $r = 25 - 25\sin\theta$
$r = 20 + 80\cos\theta$

32. $r = 1 - \sin\dfrac{\theta}{2}$
$r = 1.4 + \cos\dfrac{\theta}{2}$

12.5 The definite integral for the area of a region bounded by a polar curve or the arc length of a polar curve may have to be computed numerically. Use Simpson's Rule or the numerical integration procedure available to do these numerical calculations when needed. In addition, points of intersection may need to be found in order to set up the definite integrals giving areas or arc lengths. Zoom in and use the trace, if available, to get initial estimates for points of intersection. Numerically solve the equations for the coordinates of these points.

In Problems 33–38 evaluate or approximate the definite integral representing the indicated area or arc length.

33. The area of the region that is both inside the graph of $r^2 = 5 + \cos 2\theta$ and inside the graph of $r = 3 + 3\cos\theta$

34. The area of the region bounded by the graph of $r = \dfrac{4 + 2.8\cos\theta}{3 - 2\sin\theta}$

35. The length of the graph of $r = 5\sin 2\theta$ for $0 \le \theta \le \dfrac{\pi}{2}$

36. The length of the graph of $r = 7 - 3\sin\theta$ for $0 \le \theta \le 2\pi$

37. The length of the part of the cardioid $r = 4 + 4\cos\theta$ that lies inside the circle $r = 12\cos\theta$

38. The arc length of one petal of the rose curve $r = 5\cos 4\theta$.

12.6 Since the path of a spacecraft moving near the sun (earth, moon, or planet) is approximately a conic section with the sun (earth, moon, or planet) at one focus, the polar equation for this conic section is particularly convenient. A spacecraft or natural heavenly body is called a **satellite** when it travels around repeatedly in an elliptical path or orbit. The spacecraft Pioneer 10 moved in an approximate hyperbolic orbit when it traveled near the planet Jupiter in 1973. As the spacecraft moved away from Jupiter, the gravitational attraction of the sun became more important and the orbit gradually changed to a different conic with the sun as focus. Pioneer 10 is now headed out of our solar system along a hyperbolic orbit with respect to the sun.

In Problems 39 and 40 investigate possible orbits for a spacecraft about the sun of the form

$$r = \frac{ep}{1 + e\cos\theta}$$

39. If the spacecraft passes through the point with polar coordinates $(r, \theta) = (1, 0)$, then the perihelion distance will be 1 (distances are measured in A.U.) Find a formula for p in terms of the eccentricity e for all such orbits passing through this point.

40. Plot the orbits of this form passing through (1, 0) with eccentricities $e = 0.3$, 0.9, 1, 1.3, and 2 in an equally scaled viewing window containing $-27 \le x \le 4$, $-10 \le y \le 10$. Use the interval $-\pi \le \theta \le \pi$ and, if possible, plot the curves in a "dot mode" rather than with connected line segments to see how the points move along these conic curves when equal steps in the parameter θ are taken.

The orbital characteristics (eccentricity, perigee point, and major axis) of a satellite near the earth gradually degrade over time due to many small forces acting on the satellite other than the gravitational force of the earth. These forces include atmospheric drag, the gravitational attractions of the sun and moon, and magnetic forces. Approximately once a month tiny rockets are activated for a few seconds in order to "boost" the orbital characteristics back into the desired ranges. Rockets are turned on longer to make a major change in the orbit of a satellite. The most fuel-efficient way to move from an inner orbit to an outer orbit, called a **Hohmann transfer**, is to add velocity in the direction of flight at the time the satellite reaches perigee on the inner orbit, follow the Hohmann transfer ellipse halfway around to its apogee, and add velocity again to achieve the outer orbit. A similar process (subtracting velocity at apogee on the outer orbit and subtracting velocity at perigee on the Hohmann transfer orbit) moves a satellite from an outer orbit back to an inner one.

In Problems 41–44 plot the indicated orbits to show a Hohmann transfer.

41. Inner orbit $r = \dfrac{24}{1 + 0.2\cos\theta}$; Hohmann transfer $r = \dfrac{32}{1 + 0.6\cos\theta}$; outer orbit $r = \dfrac{56}{1 + 0.3\cos\theta}$

42. Inner orbit $r = \dfrac{5.5}{1 + 0.1\cos\theta}$; Hohmann transfer $r = \dfrac{7.5}{1 + 0.5\cos\theta}$; outer orbit $r = \dfrac{13.5}{1 + 0.1\cos\theta}$

43. Inner orbit $r = 9$; Hohmann transfer $r = \dfrac{15.3}{1 + 0.7\cos\theta}$; outer orbit $r = 51$

44. Inner orbit $r = \dfrac{73.5}{1 + 0.05\cos\theta}$; Hohmann transfer $r = \dfrac{77}{1 + 0.1\cos\theta}$; outer orbit $r = \dfrac{84.7}{1 + 0.01\cos\theta}$

The axis for a conic section with one focus at the origin need not be along one of the coordinate axes. A simple shift of the angle gives a new axis along some line through the origin.

In Problems 45–50 find the graph of the given conic section and locate its axis.

45. $r = \dfrac{16}{1 - 0.2\cos\left(\theta - \dfrac{\pi}{3}\right)}$

46. $r = \dfrac{6}{2 + 3\sin\left(\theta - \dfrac{\pi}{4}\right)}$

47. $r = \dfrac{4}{3 + 3\cos\left(\theta - \dfrac{5\pi}{6}\right)}$

48. $r = \dfrac{10}{1 - 2\cos(\theta - 1)}$

49. $r = \dfrac{13}{2 - 2\cos(\theta + 5)}$

50. $r = \dfrac{25}{6 + \cos\left(\theta - \dfrac{\pi}{10}\right)}$

13

VECTORS AND 3-SPACE

INTRODUCTION

Until now we have carried out most of our endeavors in calculus in the flatland of the two-dimensional Cartesian plane or 2-space. For the next several chapters, we shall be primarily interested in examining mathematical life in three dimensions or 3-space.

The methane molecule CH_4 consists of four hydrogen atoms surrounding a single carbon atom. As shown in the accompanying figure, the hydrogen atoms are located at the vertices of a regular tetrahedron. You will be asked in Exercises 13.3 to use the vector methods developed in this chapter to find the distance between two of the hydrogen atoms.

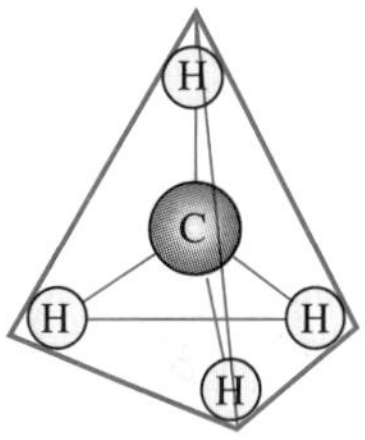

IMPORTANT CONCEPTS

scalar
vector
position vector
addition of vectors
subtraction of vectors
scalar multiplication
components
magnitude or length of a vector
unit vector
rectangular coordinate system in three dimensions
ordered triple
coordinate plane
i, j, k vectors
dot product
orthogonal vectors
direction cosines
component of a vector on another vector
projection of a vector on another vector
cross product
parallel vectors
equations for a line in three dimensions
Cartesian equation of a plane
cylinder
sphere
quadric surfaces
surface of revolution

13.1 VECTORS IN 2-SPACE

Geometric Vectors In science, mathematics, and engineering we distinguish two important quantities: **scalars** and **vectors**. A **scalar** is simply a real number or a quantity that has *magnitude*. For example, length, temperature, and blood pressure are represented by numbers such as 80 m, 20°C, and the systolic–diastolic ratio 120/80. A **vector**, on the other hand, is usually described as a quantity that has both *magnitude* and *direction*. Geometrically, a vector is represented by a directed line segment—that is, an arrow—and is written either as a boldface symbol **v** or as $\vec{v}$ or $\overrightarrow{AB}$. Examples of vector quantities shown in Figure 13.1 are weight **w**, velocity **v**, and the retarding force of friction $\mathbf{F}_f$.

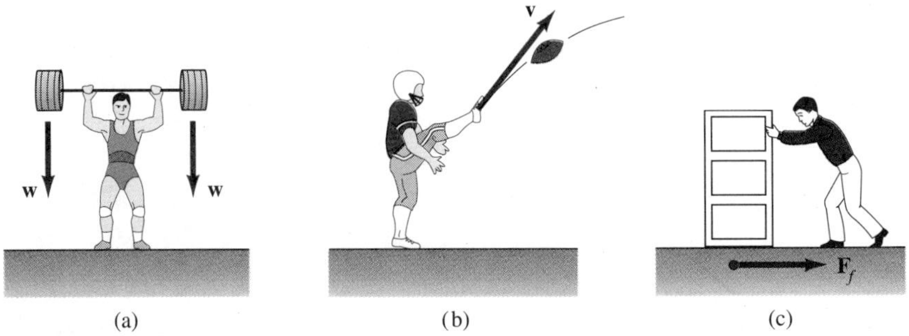

FIGURE 13.1

Notation and Terminology A vector whose initial point (or end) is A and whose terminal point (or tip) is B is written $\overrightarrow{AB}$. The magnitude of a vector is written $|\overrightarrow{AB}|$. Two vectors that have the same magnitude and same direction are said to be **equal**. Thus, in Figure 13.2, we have $\overrightarrow{AB} = \overrightarrow{CD}$. Vectors are said to be **free**, which means that a vector can be moved from one position to another provided its magnitude and direction are not changed. The **negative** of a vector $\overrightarrow{AB}$, written $-\overrightarrow{AB}$, is a vector that has the same magnitude as $\overrightarrow{AB}$ but is opposite in direction. If $k \neq 0$ is a scalar, the **scalar multiple** of a vector, $k\overrightarrow{AB}$, is a vector that is $|k|$ times as long as $\overrightarrow{AB}$. If $k > 0$, then $k\overrightarrow{AB}$ has the same direction as the vector $\overrightarrow{AB}$; if $k < 0$, then $k\overrightarrow{AB}$ has the direction opposite to that of $\overrightarrow{AB}$. When $k = 0$, we say $0\overrightarrow{AB} = \mathbf{0}$ is the **zero vector**.* Two vectors are **parallel** if and only if they are nonzero scalar multiples of each other. See Figure 13.3.

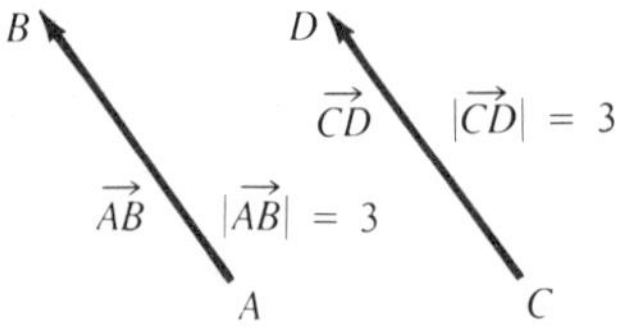

FIGURE 13.2

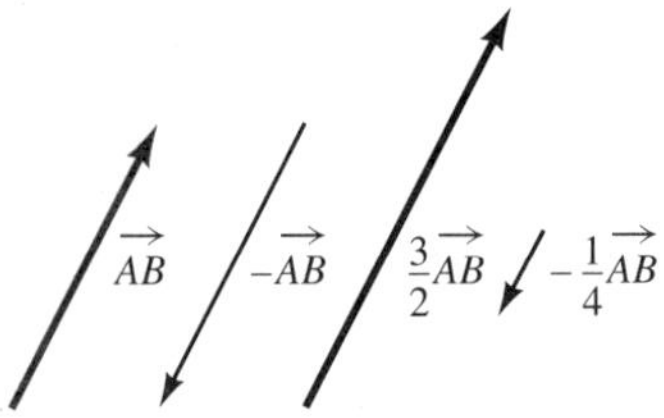

FIGURE 13.3

*The question of what is the direction of **0** is usually answered by saying that the zero vector can be assigned *any* direction. More to the point, **0** is needed in order to have a vector algebra.

Addition and Subtraction Two vectors can be considered as having a common initial point, such as A in Figure 13.4(a). Thus, if nonparallel vectors $\overrightarrow{AB}$ and $\overrightarrow{AC}$ are the sides of a parallelogram in Figure 13.4(b), we say the vector that is the main diagonal, or $\overrightarrow{AD}$, is the **sum** of $\overrightarrow{AB}$ and $\overrightarrow{AC}$. We write

$$\overrightarrow{AD} = \overrightarrow{AB} + \overrightarrow{AC}$$

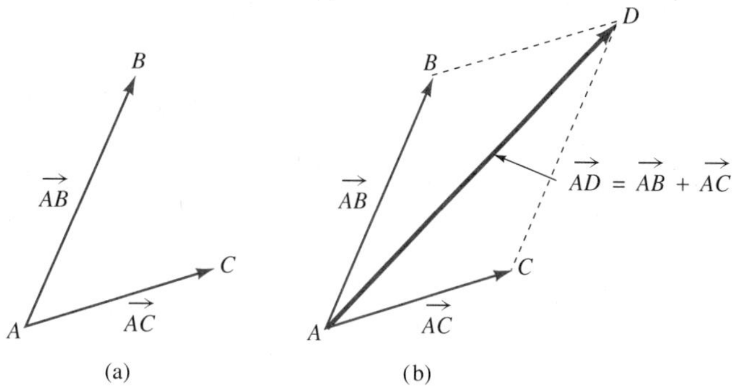

FIGURE 13.4

The **difference** of two vectors $\overrightarrow{AB}$ and $\overrightarrow{AC}$ is defined by

$$\overrightarrow{AB} - \overrightarrow{AC} = \overrightarrow{AB} + (-\overrightarrow{AC})$$

As seen in Figure 13.5(a), the difference $\overrightarrow{AB} - \overrightarrow{AC}$ can be interpreted as the main diagonal of the parallelogram with sides $\overrightarrow{AB}$ and $-\overrightarrow{AC}$. However, as shown in Figure 13.5(b), we can also interpret the same vector difference as the third side of a triangle with sides $\overrightarrow{AB}$ and $\overrightarrow{AC}$. In this second interpretation, observe that the vector difference $\overrightarrow{CB} = \overrightarrow{AB} - \overrightarrow{AC}$ points toward the terminal point of the vector *from* which we are subtracting the second vector. If $\overrightarrow{AB} = \overrightarrow{AC}$, then

$$\overrightarrow{AB} - \overrightarrow{AC} = \mathbf{0}$$

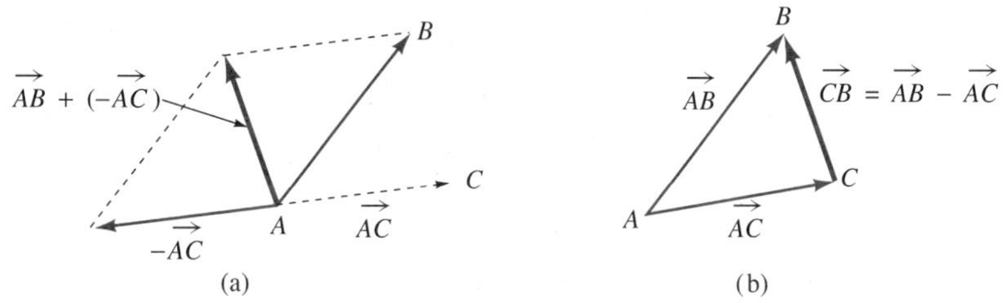

FIGURE 13.5

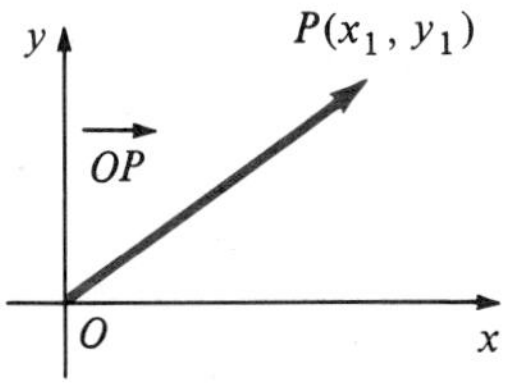

FIGURE 13.6

Vectors in a Coordinate Plane To describe a vector analytically, let us suppose for the remainder of this section that the vectors we are considering lie in a two-dimensional coordinate plane or **2-space**. The vector shown in Figure 13.6, whose initial point is the origin O and whose terminal point is $P(x_1, y_1)$, is called the **position vector** of the point P and is written

$$\overrightarrow{OP} = \langle x_1, y_1 \rangle$$

Components In general, a vector **a** in 2-space is any ordered pair of real numbers,

$$\mathbf{a} = \langle a_1, a_2 \rangle$$

The numbers a_1 and a_2 are said to be the **components** of the vector **a**.

As we see in the first example, the vector **a** is not necessarily a position vector.

EXAMPLE 1 The displacement between the points (x, y) and $(x + 4, y + 3)$ in Figure 13.7(a) is written $\langle 4, 3 \rangle$. As seen in Figure 13.7(b), the position vector of $\langle 4, 3 \rangle$ is the vector emanating from the origin and terminating at the point $P(4, 3)$.

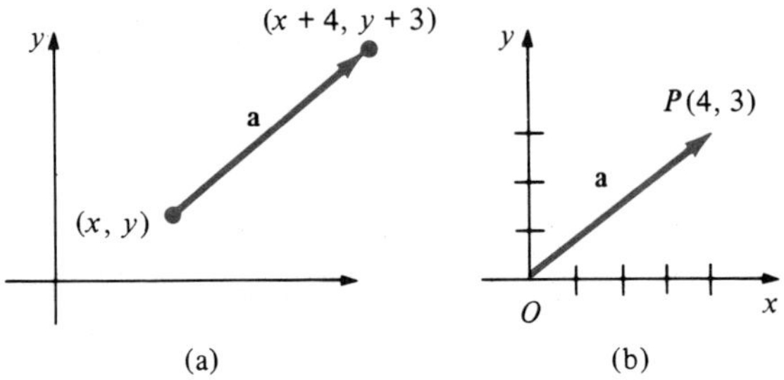

FIGURE 13.7

□

Addition and subtraction of vectors, multiplication of vectors by scalars, and so on, are defined in terms of their components.

DEFINITION 13.1

Let $\mathbf{a} = \langle a_1, a_2 \rangle$ and $\mathbf{b} = \langle b_1, b_2 \rangle$ be vectors in 2-space.

(*i*) Addition: $\mathbf{a} + \mathbf{b} = \langle a_1 + b_1, a_2 + b_2 \rangle$ (1)

(*ii*) Scalar multiplication: $k\mathbf{a} = \langle ka_1, ka_2 \rangle$ (2)

(*iii*) Equality: $\mathbf{a} = \mathbf{b}$ if and only if $a_1 = b_1, a_2 = b_2$ (3)

Subtraction Using (2), we define the **negative** of a vector **b** by

$$-\mathbf{b} = (-1)\mathbf{b} = \langle -b_1, -b_2 \rangle$$

We can then define the **subtraction**, or the difference, of two vectors as

$$\mathbf{a} - \mathbf{b} = \mathbf{a} + (-\mathbf{b}) = \langle a_1 - b_1, a_2 - b_2 \rangle \tag{4}$$

In Figure 13.8(a), we see the sum of two vectors $\overrightarrow{OP_1}$ and $\overrightarrow{OP_2}$ illustrated. In Figure 13.8(b) the vector $\overrightarrow{P_1P_2}$, whose initial point is P_1 and terminal point is

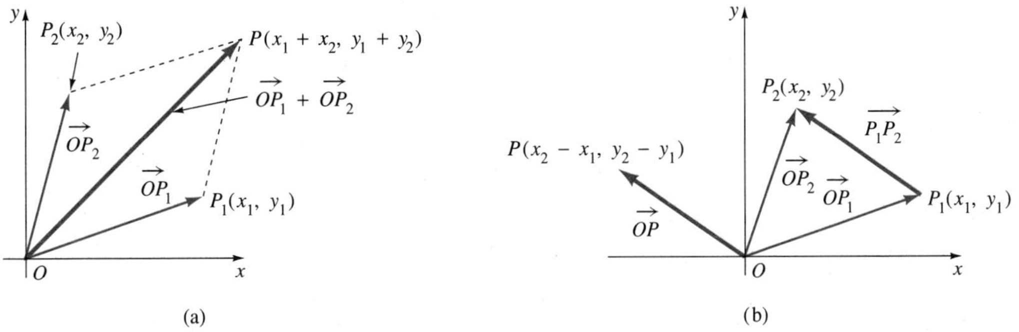

FIGURE 13.8

P_2, is the difference of position vectors

$$\overrightarrow{P_1P_2} = \overrightarrow{OP_2} - \overrightarrow{OP_1} = \langle x_2 - x_1, y_2 - y_1 \rangle$$

As shown in Figure 13.8(b), the vector $\overrightarrow{P_1P_2}$ can be drawn either starting from the terminal point of $\overrightarrow{OP_1}$ and ending at the terminal point of $\overrightarrow{OP_2}$, or as the position vector $\overrightarrow{OP}$ whose terminal point has coordinates $(x_2 - x_1, y_2 - y_1)$. Remember, $\overrightarrow{OP}$ and $\overrightarrow{P_1P_2}$ are considered equal, since they have the same magnitude and direction.

EXAMPLE 2 If $\mathbf{a} = \langle 1, 4 \rangle$ and $\mathbf{b} = \langle -6, 3 \rangle$, find (*a*) $\mathbf{a} + \mathbf{b}$, (*b*) $\mathbf{a} - \mathbf{b}$, and (*c*) $2\mathbf{a} + 3\mathbf{b}$.

Solution We use (1), (2), and (4).

(*a*) $\mathbf{a} + \mathbf{b} = \langle 1 + (-6), 4 + 3 \rangle = \langle -5, 7 \rangle$

(*b*) $\mathbf{a} - \mathbf{b} = \langle 1 - (-6), 4 - 3 \rangle = \langle 7, 1 \rangle$

(*c*) $2\mathbf{a} + 3\mathbf{b} = \langle 2, 8 \rangle + \langle -18, 9 \rangle = \langle -16, 17 \rangle$ □

Properties The component definition of a vector can be used to verify each of the following **properties**:

(*i*) $\mathbf{a} + \mathbf{b} = \mathbf{b} + \mathbf{a}$ — commutative law

(*ii*) $\mathbf{a} + (\mathbf{b} + \mathbf{c}) = (\mathbf{a} + \mathbf{b}) + \mathbf{c}$ — associative law

(*iii*) $\mathbf{a} + \mathbf{0} = \mathbf{a}$ — additive identity

(*iv*) $\mathbf{a} + (-\mathbf{a}) = \mathbf{0}$ — additive inverse

(*v*) $k(\mathbf{a} + \mathbf{b}) = k\mathbf{a} + k\mathbf{b}$, k a scalar

(*vi*) $(k_1 + k_2)\mathbf{a} = k_1\mathbf{a} + k_2\mathbf{a}$, k_1 and k_2 scalars

(*vii*) $(k_1)(k_2\mathbf{a}) = (k_1k_2)\mathbf{a}$, k_1 and k_2 scalars

(*viii*) $1\mathbf{a} = \mathbf{a}$

(*ix*) $0\mathbf{a} = \mathbf{0}$

The **zero vector 0** in properties (*iii*), (*iv*), and (*ix*) is defined as

$$\mathbf{0} = \langle 0, 0 \rangle$$

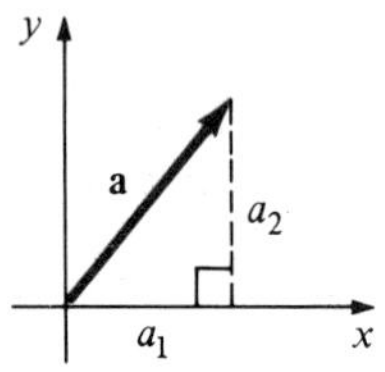

FIGURE 13.9

Magnitude The **magnitude**, **length**, or **norm** of a vector **a** is denoted by $|\mathbf{a}|$. Motivated by the Pythagorean Theorem and Figure 13.9, we define the magnitude of a vector

$$\mathbf{a} = \langle a_1, a_2 \rangle \quad \text{to be} \quad |\mathbf{a}| = \sqrt{a_1^2 + a_2^2}$$

Clearly, $|\mathbf{a}| \geq 0$ for any vector **a**, and $|\mathbf{a}| = 0$ if and only if $\mathbf{a} = \mathbf{0}$. For example, if $\mathbf{a} = \langle 6, -2 \rangle$, then

$$|\mathbf{a}| = \sqrt{6^2 + (-2)^2} = \sqrt{40} = 2\sqrt{10}$$

Unit Vectors A vector that has magnitude 1 is called a **unit vector**. We can obtain a unit vector **u** in the same direction as a nonzero vector **a** by multiplying **a** by the reciprocal of its magnitude. The vector $\mathbf{u} = (1/|\mathbf{a}|)\mathbf{a}$ is a unit vector, since

$$|\mathbf{u}| = \left| \frac{1}{|\mathbf{a}|} \mathbf{a} \right| = \frac{1}{|\mathbf{a}|} |\mathbf{a}| = 1$$

EXAMPLE 3 Given $\mathbf{a} = \langle 2, -1 \rangle$. Form a unit vector (*a*) in the same direction as **a** and (*b*) in the opposite direction of **a**.

Solution First, we find the magnitude of the vector **a**:

$$|\mathbf{a}| = \sqrt{4 + (-1)^2} = \sqrt{5}$$

(*a*) A unit vector in the same direction as **a** is then

$$\mathbf{u} = \frac{1}{\sqrt{5}} \mathbf{a} = \frac{1}{\sqrt{5}} \langle 2, -1 \rangle = \left\langle \frac{2}{\sqrt{5}}, \frac{-1}{\sqrt{5}} \right\rangle$$

(*b*) A unit vector in the opposite direction of **a** is the negative of **u**:

$$-\mathbf{u} = \left\langle -\frac{2}{\sqrt{5}}, \frac{1}{\sqrt{5}} \right\rangle$$

□

If **a** and **b** are vectors and c_1 and c_2 are scalars, then the expression $c_1\mathbf{a} + c_2\mathbf{b}$ is called a **linear combination** of **a** and **b**. As we shall see next, any vector in 2-space can be written as a linear combination of two special vectors.

The i, j Vectors In view of (1) and (2), any vector $\mathbf{a} = \langle a_1, a_2 \rangle$ can be written as a sum:

$$\langle a_1, a_2 \rangle = \langle a_1, 0 \rangle + \langle 0, a_2 \rangle = a_1\langle 1, 0 \rangle + a_2\langle 0, 1 \rangle \tag{5}$$

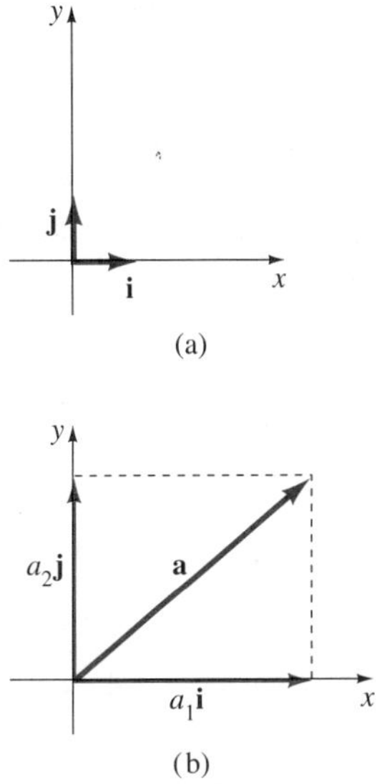

FIGURE 13.10

The unit vectors $\langle 1, 0 \rangle$ and $\langle 0, 1 \rangle$ are usually given the special symbols **i** and **j**. See Figure 13.10(a). Thus, if

$$\mathbf{i} = \langle 1, 0 \rangle \quad \text{and} \quad \mathbf{j} = \langle 0, 1 \rangle$$

then (5) becomes

$$\mathbf{a} = a_1\mathbf{i} + a_2\mathbf{j} \tag{6}$$

The unit vectors **i** and **j** are said to form a **basis** for the system of two-dimensional vectors, since any vector **a** can be written uniquely as a linear combination of **i** and **j**. If $\mathbf{a} = a_1\mathbf{i} + a_2\mathbf{j}$ is a position vector, then Figure 13.10(b) shows that **a** is the sum of the vectors $a_1\mathbf{i}$ and $a_2\mathbf{j}$, which have the origin as a common initial point and which lie on the x- and y-axes, respectively. The scalar a_1 is called the **horizontal component** of **a** and the scalar a_2 is called the **vertical component** of **a**.

EXAMPLE 4

(*a*) $\langle 4, 7 \rangle = 4\mathbf{i} + 7\mathbf{j}$

(*b*) $(2\mathbf{i} - 5\mathbf{j}) + (8\mathbf{i} + 13\mathbf{j}) = 10\mathbf{i} + 8\mathbf{j}$

(*c*) $|\mathbf{i} + \mathbf{j}| = \sqrt{2}$

(*d*) $10(3\mathbf{i} - \mathbf{j}) = 30\mathbf{i} - 10\mathbf{j}$

(*e*) $\mathbf{a} = 6\mathbf{i} + 4\mathbf{j}$ and $\mathbf{b} = 9\mathbf{i} + 6\mathbf{j}$ are parallel, since **b** is a scalar multiple of **a**. We see $\mathbf{b} = \frac{3}{2}\mathbf{a}$. □

EXAMPLE 5 Let $\mathbf{a} = 4\mathbf{i} + 2\mathbf{j}$ and $\mathbf{b} = -2\mathbf{i} + 5\mathbf{j}$. Graph (*a*) $\mathbf{a} + \mathbf{b}$ and (*b*) $\mathbf{a} - \mathbf{b}$.

Solution (*a*) $\mathbf{a} + \mathbf{b} = 2\mathbf{i} + 7\mathbf{j}$ (*b*) $\mathbf{a} - \mathbf{b} = 6\mathbf{i} - 3\mathbf{j}$

The graphs of these two vectors in the xy-plane are given in Figure 13.11.

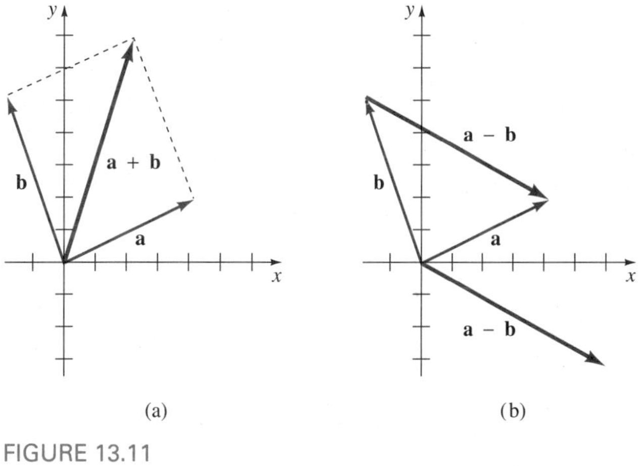

FIGURE 13.11

□

EXERCISES 13.1 *Answers to odd-numbered problems begin on page A-98.*

In Problems 1–8 find **(a)** $3\mathbf{a}$, **(b)** $\mathbf{a}+\mathbf{b}$, **(c)** $\mathbf{a}-\mathbf{b}$, **(d)** $|\mathbf{a}+\mathbf{b}|$, and **(e)** $|\mathbf{a}-\mathbf{b}|$.

1. $\mathbf{a} = 2\mathbf{i} + 4\mathbf{j}$, $\mathbf{b} = -\mathbf{i} + 4\mathbf{j}$

2. $\mathbf{a} = \langle 1, 1\rangle$, $\mathbf{b} = \langle 2, 3\rangle$

3. $\mathbf{a} = \langle 4, 0\rangle$, $\mathbf{b} = \langle 0, -5\rangle$

4. $\mathbf{a} = \frac{1}{6}\mathbf{i} - \frac{1}{6}\mathbf{j}$, $\mathbf{b} = \frac{1}{2}\mathbf{i} + \frac{5}{6}\mathbf{j}$

5. $\mathbf{a} = -3\mathbf{i} + 2\mathbf{j}$, $\mathbf{b} = 7\mathbf{j}$

6. $\mathbf{a} = \langle 1, 3\rangle$, $\mathbf{b} = -5\mathbf{a}$

7. $\mathbf{a} = -\mathbf{b}$, $\mathbf{b} = 2\mathbf{i} - 9\mathbf{j}$

8. $\mathbf{a} = \langle 7, 10\rangle$, $\mathbf{b} = \langle 1, 2\rangle$

In Problems 9–14 find **(a)** $4\mathbf{a} - 2\mathbf{b}$ and **(b)** $-3\mathbf{a} - 5\mathbf{b}$.

9. $\mathbf{a} = \langle 1, -3\rangle$, $\mathbf{b} = \langle -1, 1\rangle$

10. $\mathbf{a} = \mathbf{i} + \mathbf{j}$, $\mathbf{b} = 3\mathbf{i} - 2\mathbf{j}$

11. $\mathbf{a} = \mathbf{i} - \mathbf{j}$, $\mathbf{b} = -3\mathbf{i} + 4\mathbf{j}$

12. $\mathbf{a} = \langle 2, 0\rangle$, $\mathbf{b} = \langle 0, -3\rangle$

13. $\mathbf{a} = \langle 4, 10\rangle$, $\mathbf{b} = -2\langle 1, 3\rangle$

14. $\mathbf{a} = \langle 3, 1\rangle + \langle -1, 2\rangle$, $\mathbf{b} = \langle 6, 5\rangle - \langle 1, 2\rangle$

In Problems 15–18 find the vector $\overrightarrow{P_1P_2}$. Graph $\overrightarrow{P_1P_2}$ and its corresponding position vector.

15. $P_1(3, 2)$, $P_2(5, 7)$

16. $P_1(-2, -1)$, $P_2(4, -5)$

17. $P_1(3, 3)$, $P_2(5, 5)$

18. $P_1(0, 3)$, $P_2(2, 0)$

19. Find the terminal point of the vector $\overrightarrow{P_1P_2} = 4\mathbf{i} + 8\mathbf{j}$ if its initial point is $(-3, 10)$.

20. Find the initial point of the vector $\overrightarrow{P_1P_2} = \langle -5, -1\rangle$ if its terminal point is $(4, 7)$.

21. Determine which of the following vectors are parallel to $\mathbf{a} = 4\mathbf{i} + 6\mathbf{j}$.

(a) $-4\mathbf{i} - 6\mathbf{j}$ (b) $-\mathbf{i} - \frac{3}{2}\mathbf{j}$
(c) $10\mathbf{i} + 15\mathbf{j}$ (d) $2(\mathbf{i} - \mathbf{j}) - 3(\frac{1}{2}\mathbf{i} - \frac{5}{12}\mathbf{j})$
(e) $8\mathbf{i} + 12\mathbf{j}$ (f) $(5\mathbf{i} + \mathbf{j}) - (7\mathbf{i} + 4\mathbf{j})$

22. Determine a scalar c so that $\mathbf{a} = 3\mathbf{i} + c\mathbf{j}$ and $\mathbf{b} = -\mathbf{i} + 9\mathbf{j}$ are parallel.

In Problems 23 and 24 find $\mathbf{a} + (\mathbf{b} + \mathbf{c})$ for the given vectors.

23. $\mathbf{a} = \langle 5, 1\rangle$, $\mathbf{b} = \langle -2, 4\rangle$, $\mathbf{c} = \langle 3, 10\rangle$

24. $\mathbf{a} = \langle 1, 1\rangle$, $\mathbf{b} = \langle 4, 3\rangle$, $\mathbf{c} = \langle 0, -2\rangle$

In Problems 25–28 find a unit vector **(a)** in the same direction as **a**, and **(b)** in the opposite direction of **a**.

25. $\mathbf{a} = \langle 2, 2\rangle$ **26.** $\mathbf{a} = \langle -3, 4\rangle$

27. $\mathbf{a} = \langle 0, -5\rangle$ **28.** $\mathbf{a} = \langle 1, -\sqrt{3}\rangle$

In Problems 29 and 30 $\mathbf{a} = \langle 2, 8\rangle$ and $\mathbf{b} = \langle 3, 4\rangle$. Find a unit vector in the same direction as the given vector.

29. $\mathbf{a} + \mathbf{b}$ **30.** $2\mathbf{a} - 3\mathbf{b}$

In Problems 31 and 32 find a vector **b** that is parallel to the given vector and has the indicated magnitude.

31. $\mathbf{a} = 3\mathbf{i} + 7\mathbf{j}$, $|\mathbf{b}| = 2$ **32.** $\mathbf{a} = \frac{1}{2}\mathbf{i} - \frac{1}{2}\mathbf{j}$, $|\mathbf{b}| = 3$

33. Find a vector in the opposite direction of $\mathbf{a} = \langle 4, 10\rangle$ but $\frac{3}{4}$ as long.

34. Given that $\mathbf{a} = \langle 1, 1\rangle$ and $\mathbf{b} = \langle -1, 0\rangle$, find a vector in the same direction as $\mathbf{a} + \mathbf{b}$ but 5 times as long.

In Problems 35 and 36 use the given figure to draw the indicated vector.

35. $3\mathbf{b} - \mathbf{a}$ **36.** $\mathbf{a} + (\mathbf{b} + \mathbf{c})$

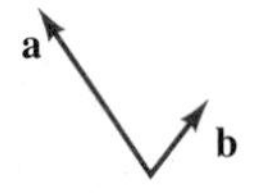

FIGURE 13.12

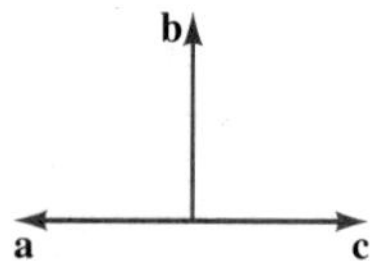

FIGURE 13.13

In Problems 37 and 38 express the vector **x** in terms of vectors **a** and **b**.

37.

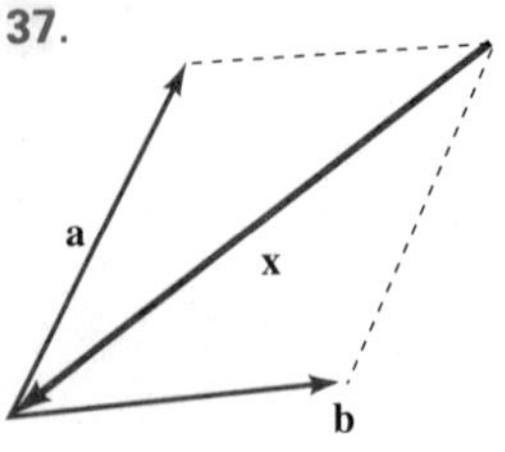

FIGURE 13.14

38.

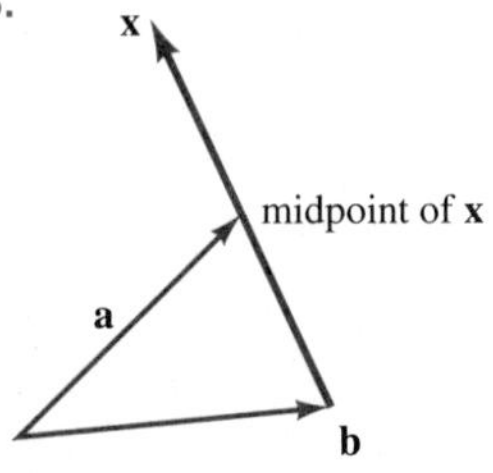

FIGURE 13.15

In Problems 39 and 40 use the given figure to prove the given result.

39. $\mathbf{a} + \mathbf{b} + \mathbf{c} = \mathbf{0}$

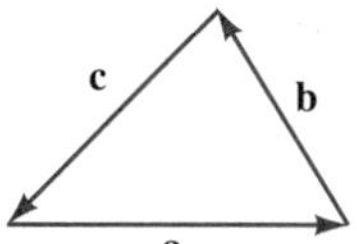

FIGURE 13.16

40. $\mathbf{a} + \mathbf{b} + \mathbf{c} + \mathbf{d} = \mathbf{0}$

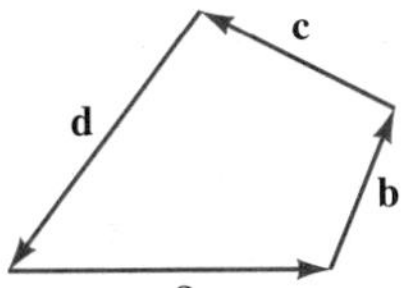

FIGURE 13.17

In Problems 41 and 42 express the vector $\mathbf{a} = 2\mathbf{i} + 3\mathbf{j}$ as a linear combination of the given vectors $\mathbf{b}$ and $\mathbf{c}$.

41. $\mathbf{b} = \mathbf{i} + \mathbf{j}$, $\mathbf{c} = \mathbf{i} - \mathbf{j}$

42. $\mathbf{b} = -2\mathbf{i} + 4\mathbf{j}$, $\mathbf{c} = 5\mathbf{i} + 7\mathbf{j}$

A vector is said to be tangent to a curve at a point if it is parallel to the tangent line at the point. In Problems 43 and 44 find a unit tangent vector to the given curve at the indicated point.

43. $y = \frac{1}{4}x^2 + 1$; $(2, 2)$

44. $y = -x^2 + 3x$; $(0, 0)$

45. Let P_1, P_2, and P_3 be distinct points such that $\mathbf{a} = \overrightarrow{P_1P_2}$, $\mathbf{b} = \overrightarrow{P_2P_3}$, and $\mathbf{a} + \mathbf{b} = \overrightarrow{P_1P_3}$.

(a) What is the relation of $|\mathbf{a} + \mathbf{b}|$ to $|\mathbf{a}| + |\mathbf{b}|$?

(b) Under what condition is $|\mathbf{a} + \mathbf{b}| = |\mathbf{a}| + |\mathbf{b}|$?

46. An electric charge Q is uniformly distributed along the y-axis between $y = -a$ and $y = a$. See Figure 13.18. The total force exerted on the charge q on the x-axis by the charge Q is $\mathbf{F} = F_x\mathbf{i} + F_y\mathbf{j}$, where

$$F_x = \frac{qQ}{4\pi\varepsilon_0}\int_{-a}^{a} \frac{L\,dy}{2a(L^2 + y^2)^{3/2}} \quad \text{and}$$

$$F_y = -\frac{qQ}{4\pi\varepsilon_0}\int_{-a}^{a} \frac{y\,dy}{2a(L^2 + y^2)^{3/2}}$$

Determine $\mathbf{F}$.

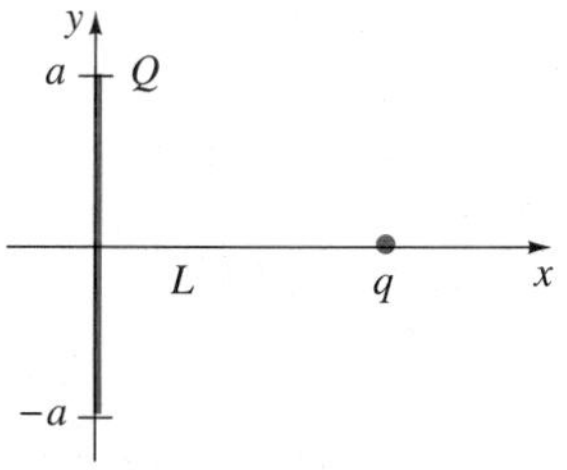

FIGURE 13.18

47. Using vectors, show that the diagonals of a parallelogram bisect each other. [*Hint:* Let M be the midpoint of one diagonal and N the midpoint of the other.]

48. Using vectors, show that the line segment between the midpoints of two sides of a triangle is parallel to the third side and half as long.

49. When walking, a person's foot strikes the ground with a force $\mathbf{F}$ at an angle θ from the vertical. In Figure 13.19, the vector $\mathbf{F}$ is resolved into vector components $\mathbf{F}_g$, which is parallel to the ground, and $\mathbf{F}_n$, which is perpendicular to the ground. In order that the foot does not slip, the force $\mathbf{F}_g$ must be offset by the opposing force $\mathbf{F}_f$ of friction; that is, $\mathbf{F}_f = -\mathbf{F}_g$.

(a) Use the fact that $|\mathbf{F}_f| = \mu|\mathbf{F}_n|$, where μ is the coefficient of friction, to show that $\tan\theta = \mu$. The foot will not slip for angles less than or equal to θ.

(b) Given that $\mu = 0.6$ for a rubber heel striking an asphalt sidewalk, find the "no-slip" angle.

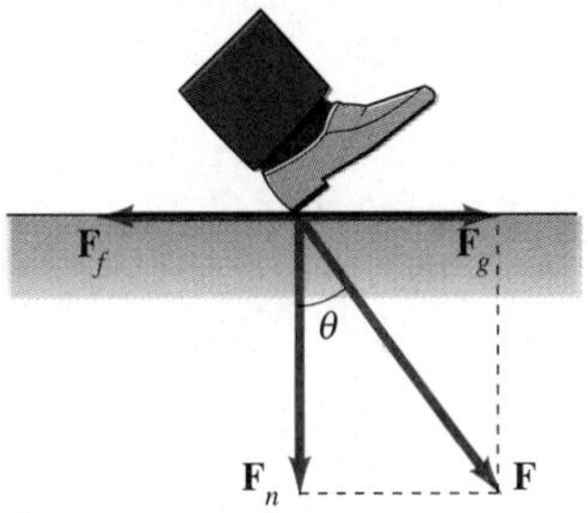

FIGURE 13.19

50. A 200-lb traffic light supported by two cables hangs in equilibrium. As shown in Figure 13.20(b) on page 724, let the weight of the light be represented by $\mathbf{w}$ and the forces in the two cables by $\mathbf{F}_1$ and $\mathbf{F}_2$. From Figure 13.20(c), we see that a condition of equilibrium is

$$\mathbf{w} + \mathbf{F}_1 + \mathbf{F}_2 = \mathbf{0} \tag{7}$$

(See Problem 39.) If

$$\mathbf{w} = -200\mathbf{j}$$
$$\mathbf{F}_1 = (|\mathbf{F}_1|\cos 20^\circ)\mathbf{i} + (|\mathbf{F}_1|\sin 20^\circ)\mathbf{j}$$
$$\mathbf{F}_2 = -(|\mathbf{F}_2|\cos 15^\circ)\mathbf{i} + (|\mathbf{F}_2|\sin 15^\circ)\mathbf{j}$$

use (7) to determine the magnitudes of $\mathbf{F}_1$ and $\mathbf{F}_2$. [*Hint:* Reread (*iii*) of Definition 13.1.]

51. Water rushing from a fire hose exerts a horizontal force $\mathbf{F}_1$ of magnitude 200 lb. See Figure 13.21 on page 724. What is the magnitude of the force $\mathbf{F}_3$ that a firefighter must exert to hold the hose at an angle of 45° from the horizontal?

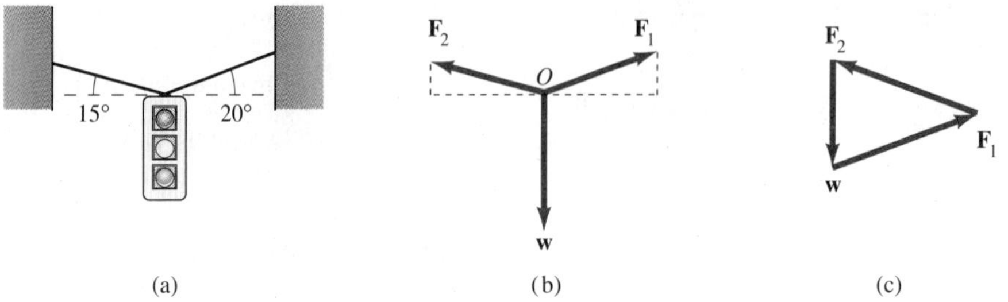

FIGURE 13.20

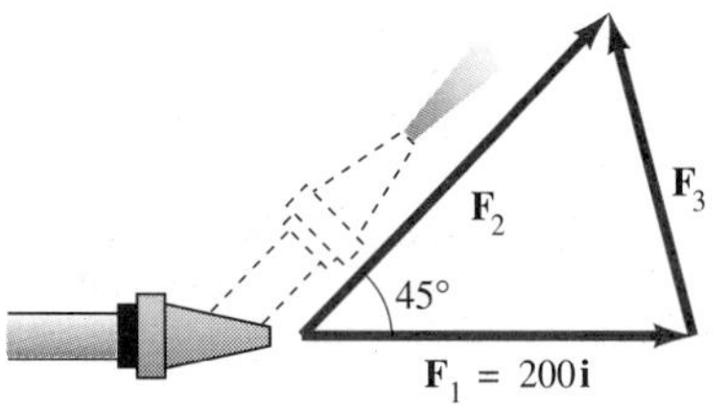

FIGURE 13.21

52. An airplane starts from an airport located at the origin O and flies 150 mi in the direction 20° north of east to city A. From A the airplane then flies 200 mi in the direction 23° west of north to city B. From B the airplane flies 240 mi in the direction 10° south of west to city C. Express the location of city C as a vector $\mathbf{r}$ as shown in Figure 13.22. Find the distance from O to C.

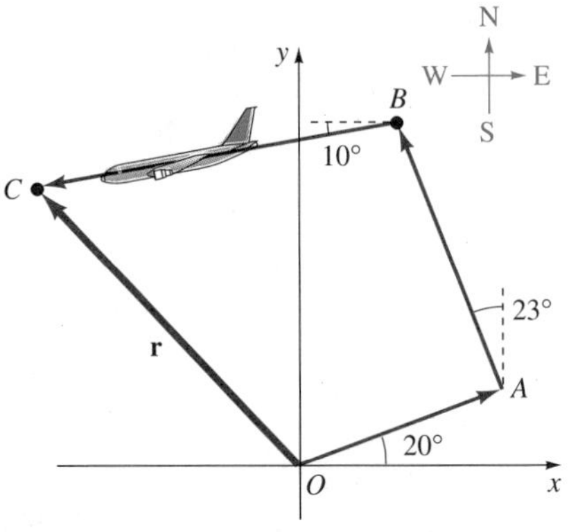

FIGURE 13.22

13.2 3-SPACE

13.2.1 Rectangular Coordinate System in Three Dimensions

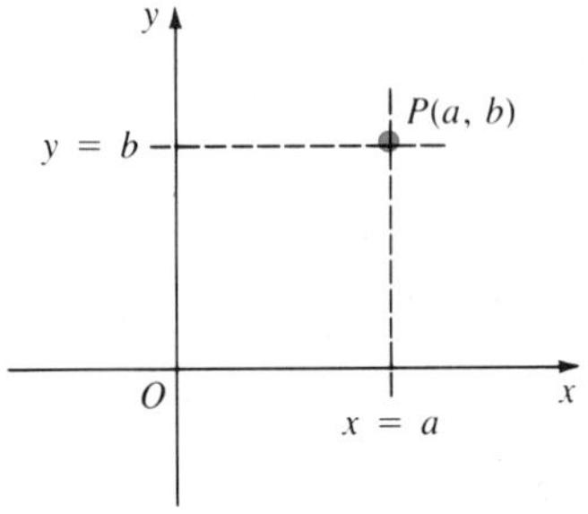

FIGURE 13.23

In the plane, or 2-space, one way of describing the position of a point P is to assign to it coordinates relative to two mutually orthogonal axes called the x- and y-axes. If P is the point of intersection of the line $x = a$ (perpendicular to the x-axis) and the line $y = b$ (perpendicular to the y-axis), then the **ordered pair** (a, b) is said to be the **rectangular** or **Cartesian coordinates** of the point. See Figure 13.23.

In three dimensions, or **3-space**, a rectangular coordinate system is constructed using three mutually orthogonal axes. The point at which these axes intersect is called the **origin** O. These axes, shown in Figure 13.24(a), are labeled in accordance with the so-called **right-hand rule**: If the fingers of the right hand, pointing in the direction of the positive x-axis, are curled toward the positive y-axis, then the thumb will point in the direction of a new axis perpendicular to the plane of the x- and y-axes. This new axis is labeled the z-

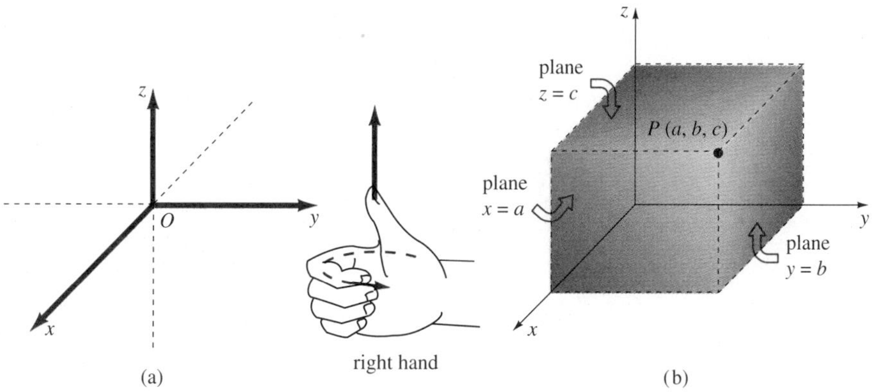

FIGURE 13.24

axis.* The dashed lines in Figure 13.24(a), represent the negative axes. Now, if

$$x = a \qquad y = b \qquad z = c$$

are planes perpendicular to the x-axis, y-axis, and z-axis, respectively, the point P at which these planes intersect can be represented by an **ordered triple** of numbers (a, b, c) said to be the **rectangular** or **Cartesian coordinates** of the point. The numbers a, b, and c are, in turn, called the **x-, y-, and z-coordinates** of $P(a, b, c)$. See Figure 13.24(b).

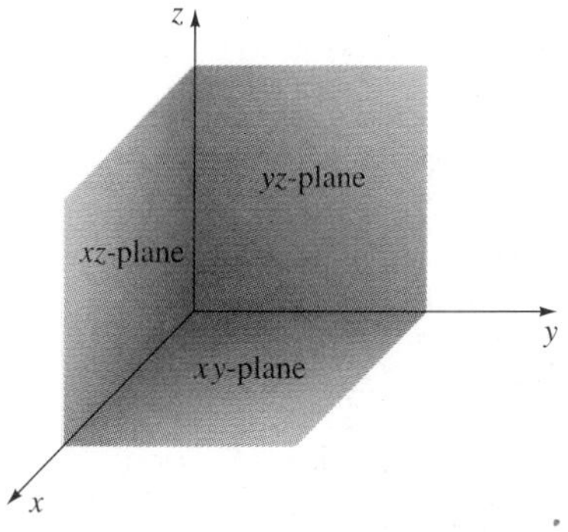

FIGURE 13.25

Octants Each pair of coordinate axes determines a **coordinate plane**. As shown in Figure 13.25, the x- and y-axes determine the xy-plane, the x- and z-axes determine the xz-plane, and so on. The coordinate planes divide 3-space into eight parts known as **octants**. The octant in which all three coordinates of a point are *positive* is called the **first octant**. There is no agreement for naming the other seven octants.

The following table summarizes the coordinates of a point either on a coordinate axis or in a coordinate plane. As seen in the table, we can also describe, say, the xy-plane by the simple equation $z = 0$. Similarly, the xz-plane is $y = 0$ and the yz-plane is $x = 0$.

Axes	Coordinates	Plane	Coordinates
x	$(a, 0, 0)$	xy	$(a, b, 0)$
y	$(0, b, 0)$	xz	$(a, 0, c)$
z	$(0, 0, c)$	yz	$(0, b, c)$

*If the x- and y-axes are interchanged in Figure 13.24(a), the coordinate system is said to be **left-handed**.

EXAMPLE 1 Graph the points (4, 5, 6), (3, −3, −1), and (−2, −2, 0).

Solution Of the three points shown in Figure 13.26 only (4, 5, 6) is in the first octant. The point (−2, −2, 0) is in the xy-plane.

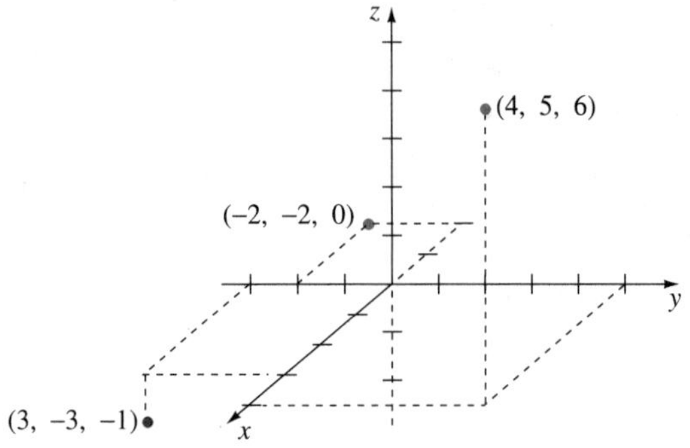

FIGURE 13.26

Distance Formula To find the **distance** between two points $P_1(x_1, y_1, z_1)$ and $P_2(x_2, y_2, z_2)$ in 3-space, let us first consider their projections onto the xy-plane. As seen in Figure 13.27, the distance between $(x_1, y_1, 0)$ and $(x_2, y_2, 0)$ follows from the usual distance formula in the plane and is $\sqrt{(x_2 - x_1)^2 + (y_2 - y_1)^2}$. Hence, from the Pythagorean Theorem applied to the right triangle $P_1P_3P_2$, we have

$$[d(P_1, P_2)]^2 = [\sqrt{(x_2 - x_1)^2 + (y_2 - y_1)^2}]^2 + |z_2 - z_1|^2$$

or

$$d(P_1, P_2) = \sqrt{(x_2 - x_1)^2 + (y_2 - y_1)^2 + (z_2 - z_1)^2} \tag{1}$$

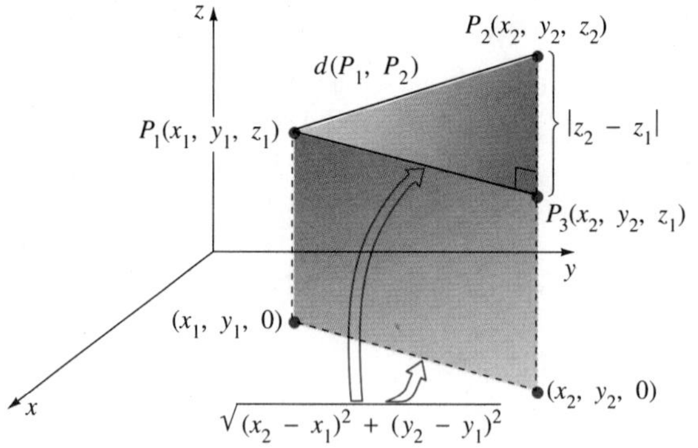

FIGURE 13.27

EXAMPLE 2 Find the distance between (2, −3, 6) and (−1, −7, 4).

Solution From (1),

$$d = \sqrt{(2 - (-1))^2 + (-3 - (-7))^2 + (6 - 4)^2} = \sqrt{29}$$ □

Midpoint Formula The formulas for finding the midpoint of a line segment between two points in 2-space (see Appendix I) carry over in an analogous fashion to 3-space. If $P_1(x_1, y_1, z_1)$ and $P_2(x_2, y_2, z_2)$ are two distinct points, then the **coordinates of the midpoint** of the line segment between them are

$$\left(\frac{x_1 + x_2}{2}, \frac{y_1 + y_2}{2}, \frac{z_1 + z_2}{2}\right) \tag{2}$$

EXAMPLE 3 Find the coordinates of the midpoint of the line segment between the two points in Example 2.

Solution From (2) we obtain

$$\left(\frac{2 + (-1)}{2}, \frac{-3 + (-7)}{2}, \frac{6 + 4}{2}\right) \quad \text{or} \quad \left(\frac{1}{2}, -5, 5\right) \qquad \square$$

13.2.2 Vectors in 3-Space

A **vector a in 3-space** is any ordered triple of real numbers

$$\mathbf{a} = \langle a_1, a_2, a_3 \rangle$$

where a_1, a_2, and a_3 are the **components** of the vector. The **position vector** of a point $P(x_1, y_1, z_1)$ in space is the vector $\overrightarrow{OP} = \langle x_1, y_1, z_1 \rangle$ whose initial point is the origin O and whose terminal point is P. See Figure 13.28.

The component definitions of addition, subtraction, scalar multiplication, and so on, are natural generalizations of those given for vectors in 2-space.

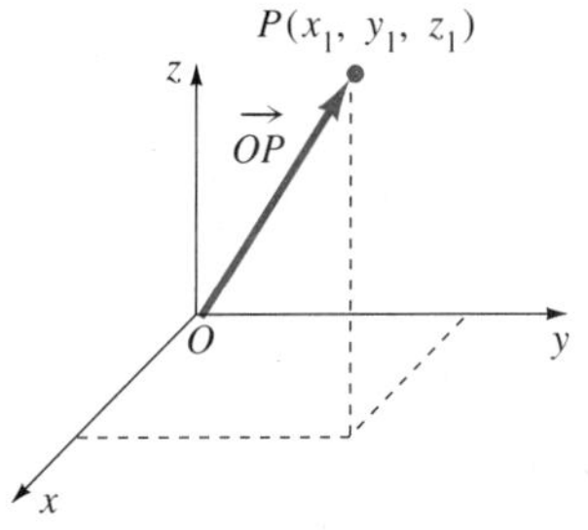

FIGURE 13.28

DEFINITION 13.2

Let $\mathbf{a} = \langle a_1, a_2, a_3 \rangle$ and $\mathbf{b} = \langle b_1, b_2, b_3 \rangle$ be vectors in 3-space.

(*i*) Addition: $\mathbf{a} + \mathbf{b} = \langle a_1 + b_1, a_2 + b_2, a_3 + b_3 \rangle$
(*ii*) Scalar multiplication: $k\mathbf{a} = \langle ka_1, ka_2, ka_3 \rangle$
(*iii*) Equality: $\mathbf{a} = \mathbf{b}$ if and only if $a_1 = b_1, a_2 = b_2, a_3 = b_3$
(*iv*) Negative: $-\mathbf{b} = (-1)\mathbf{b} = \langle -b_1, -b_2, -b_3 \rangle$
(*v*) Subtraction: $\mathbf{a} - \mathbf{b} = \mathbf{a} + (-\mathbf{b}) = \langle a_1 - b_1, a_2 - b_2, a_3 - b_3 \rangle$
(*vi*) Zero vector: $\mathbf{0} = \langle 0, 0, 0 \rangle$
(*vii*) Magnitude: $|\mathbf{a}| = \sqrt{a_1^2 + a_2^2 + a_3^2}$

If $\overrightarrow{OP_1}$ and $\overrightarrow{OP_2}$ are the position vectors of the points $P_1(x_1, y_1, z_1)$ and $P_2(x_2, y_2, z_2)$, then the vector $\overrightarrow{P_1P_2}$ is given by

$$\overrightarrow{P_1P_2} = \overrightarrow{OP_2} - \overrightarrow{OP_1} = \langle x_2 - x_1, y_2 - y_1, z_2 - z_1 \rangle \tag{3}$$

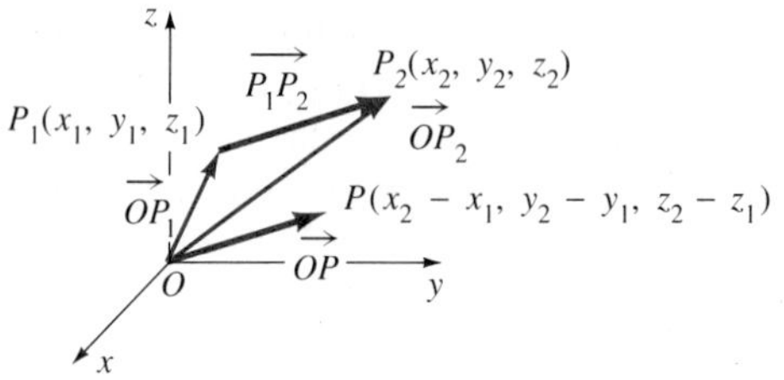

FIGURE 13.29

As in 2-space, $\overrightarrow{P_1P_2}$ can be drawn either as a vector whose initial point is P_1 and whose terminal point is P_2 or as a position vector $\overrightarrow{OP}$ with terminal point $P(x_2 - x_1, y_2 - y_1, z_2 - z_1)$. See Figure 13.29.

EXAMPLE 4 Find the vector $\overrightarrow{P_1P_2}$ if the points P_1 and P_2 are given by P_1 (4, 6, −2) and P_2 (1, 8, 3).

Solution If the position vectors of the points are $\overrightarrow{OP_1} = \langle 4, 6, -2\rangle$ and $\overrightarrow{OP_2} = \langle 1, 8, 3\rangle$, then from (3) we have

$$\overrightarrow{P_1P_2} = \overrightarrow{OP_2} - \overrightarrow{OP_1} = \langle 1-4, 8-6, 3-(-2)\rangle = \langle -3, 2, 5\rangle \qquad \square$$

EXAMPLE 5 From part (*vii*) of Definition 13.2 we see that $\mathbf{a} = \left\langle -\frac{2}{7}, \frac{3}{7}, \frac{6}{7}\right\rangle$ is a unit vector, since

$$|\mathbf{a}| = \sqrt{\left(-\frac{2}{7}\right)^2 + \left(\frac{3}{7}\right)^2 + \left(\frac{6}{7}\right)^2} = \sqrt{\frac{4+9+36}{49}} = 1 \qquad \square$$

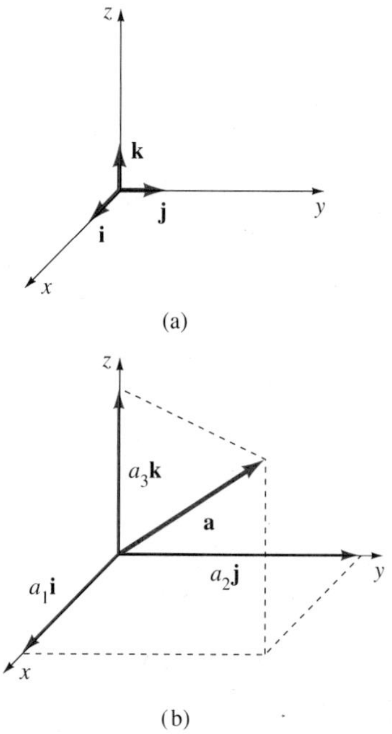

FIGURE 13.30

The i, j, k Vectors We saw in the preceding section that the unit vectors $\mathbf{i} = \langle 1, 0\rangle$ and $\mathbf{j} = \langle 0, 1\rangle$ are a basis for the system of two-dimensional vectors in that any vector **a** in 2-space can be written as a linear combination of **i** and **j**: $\mathbf{a} = a_1\mathbf{i} + a_2\mathbf{j}$. A basis for the system of three-dimensional vectors is given by the set of unit vectors:

$$\mathbf{i} = \langle 1, 0, 0\rangle \qquad \mathbf{j} = \langle 0, 1, 0\rangle \qquad \mathbf{k} = \langle 0, 0, 1\rangle$$

Any vector $\mathbf{a} = \langle a_1, a_2, a_3\rangle$ in 3-space can be expressed as a linear combination of **i**, **j**, and **k**:

$$\begin{aligned}\langle a_1, a_2, a_3\rangle &= \langle a_1, 0, 0\rangle + \langle 0, a_2, 0\rangle + \langle 0, 0, a_3\rangle \\ &= a_1\langle 1, 0, 0\rangle + a_2\langle 0, 1, 0\rangle + a_3\langle 0, 0, 1\rangle\end{aligned}$$

that is,

$$\mathbf{a} = a_1\mathbf{i} + a_2\mathbf{j} + a_3\mathbf{k}$$

The vectors **i**, **j**, and **k** are illustrated in Figure 13.30(a). In Figure 13.30(b) we see that a position vector $\mathbf{a} = a_1\mathbf{i} + a_2\mathbf{j} + a_3\mathbf{k}$ is the sum of the vectors

$a_1\mathbf{i}$, $a_2\mathbf{j}$, and $a_3\mathbf{k}$, which lie along the coordinate axes and have the origin as a common initial point.

EXAMPLE 6 The vector $\mathbf{a} = \langle 7, -5, 13\rangle$ is the same as $\mathbf{a} = 7\mathbf{i} - 5\mathbf{j} + 13\mathbf{k}$. □

When the third dimension is taken into consideration, any vector in the xy-plane is equivalently described as a three-dimensional vector that lies in the coordinate plane $z = 0$. Although the vectors $\langle a_1, a_2\rangle$ and $\langle a_1, a_2, 0\rangle$ are technically not equal, we shall ignore the distinction. That is why, for example, we denoted $\langle 1, 0\rangle$ and $\langle 1, 0, 0\rangle$ by the same symbol **i**. A vector in either the yz-plane or the xz-plane must also have one zero component. In the yz-plane a vector

$$\mathbf{b} = \langle 0, b_2, b_3\rangle \quad \text{is written} \quad \mathbf{b} = b_2\mathbf{j} + b_3\mathbf{k}$$

In the xz-plane a vector

$$\mathbf{c} = \langle c_1, 0, c_3\rangle \quad \text{is the same as} \quad \mathbf{c} = c_1\mathbf{i} + c_3\mathbf{k}$$

EXAMPLE 7

(*a*) The vector $\mathbf{a} = 5\mathbf{i} + 3\mathbf{k}$ is in the xz-coordinate plane.

(*b*) $|5\mathbf{i} + 3\mathbf{k}| = \sqrt{5^2 + 3^2} = \sqrt{34}$ □

EXAMPLE 8 If $\mathbf{a} = 3\mathbf{i} - 4\mathbf{j} + 8\mathbf{k}$ and $\mathbf{b} = \mathbf{i} - 4\mathbf{k}$, find $5\mathbf{a} - 2\mathbf{b}$.

Solution We treat **b** as a three-dimensional vector and write, for emphasis, $\mathbf{b} = \mathbf{i} + 0\mathbf{j} - 4\mathbf{k}$. From

$$5\mathbf{a} = 15\mathbf{i} - 20\mathbf{j} + 40\mathbf{k} \quad \text{and} \quad 2\mathbf{b} = 2\mathbf{i} + 0\mathbf{j} - 8\mathbf{k}$$

we get

$$\begin{aligned} 5\mathbf{a} - 2\mathbf{b} &= (15\mathbf{i} - 20\mathbf{j} + 40\mathbf{k}) - (2\mathbf{i} + 0\mathbf{j} - 8\mathbf{k}) \\ &= 13\mathbf{i} - 20\mathbf{j} + 48\mathbf{k} \end{aligned}$$ □

EXERCISES 13.2

Answers to odd-numbered problems begin on page A-99.

[13.2.1]

In Problems 1–6 graph the given point. Use the same coordinate axes.

1. (1, 1, 5) **2.** (0, 0, 4)

3. (3, 4, 0) **4.** (6, 0, 0)

5. (6, −2, 0) **6.** (5, −4, 3)

In Problems 7–10 describe geometrically all points $P(x, y, z)$ whose coordinates satisfy the given condition.

7. $z = 5$ **8.** $x = 1$

9. $x = 2, y = 3$ **10.** $x = 4, y = -1, z = 7$

11. Give the coordinates of the vertices of the rectangular parallelepiped whose sides are the coordinate planes and the planes $x = 2$, $y = 5$, $z = 8$.

12. In Figure 13.31, two vertices are shown of a rectangular parallelepiped having sides parallel to the coordinate planes. Find the coordinates of the remaining six vertices.

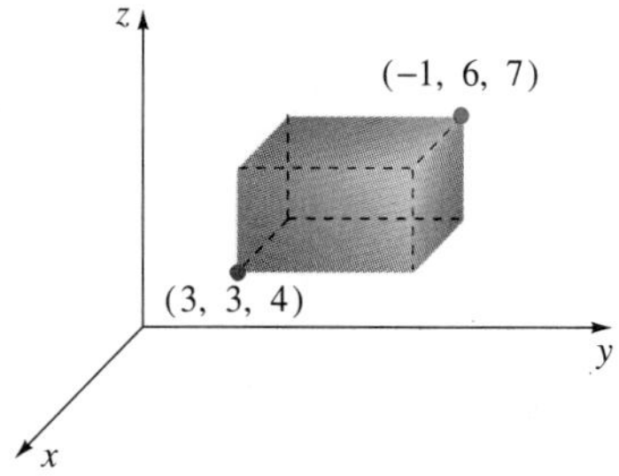

FIGURE 13.31

13. Consider the point $P(-2, 5, 4)$.

(a) If lines are drawn from P perpendicular to the coordinate planes, what are the coordinates of the point at the base of each perpendicular?

(b) If a line is drawn from P to the plane $z = -2$, what are the coordinates of the point at the base of the perpendicular?

(c) Find the point in the plane $x = 3$ that is closest to P.

14. Determine an equation of a plane parallel to a coordinate plane that contains the given pair of points.

(a) $(3, 4, -5), (-2, 8, -5)$

(b) $(1, -1, 1), (1, -1, -1)$

(c) $(-2, 1, 2), (2, 4, 2)$

In Problems 15–20 describe the set of points $P(x, y, z)$ in 3-space whose coordinates satisfy the given equation.

15. $xyz = 0$

16. $x^2 + y^2 + z^2 = 0$

17. $(x + 1)^2 + (y - 2)^2 + (z + 3)^2 = 0$

18. $(x - 2)(z - 8) = 0$

19. $z^2 - 25 = 0$

20. $x = y = z$

In Problems 21 and 22 find the distance between the given points.

21. $(3, -1, 2), (6, 4, 8)$

22. $(-1, -3, 5), (0, 4, 3)$

23. Find the distance from the point $(7, -3, -4)$ to **(a)** the yz-plane and **(b)** the x-axis.

24. Find the distance from the point $(-6, 2, -3)$ to **(a)** the xz-plane and **(b)** the origin.

In Problems 25–28 the given three points form a triangle. Determine which triangles are isosceles and which are right triangles.

25. $(0, 0, 0), (3, 6, -6), (2, 1, 2)$

26. $(0, 0, 0), (1, 2, 4), (3, 2, 2\sqrt{2})$

27. $(1, 2, 3), (4, 1, 3), (4, 6, 4)$

28. $(1, 1, -1), (1, 1, 1), (0, -1, 1)$

In Problems 29 and 30 use the distance formula to prove that the given points are collinear.

29. $P_1(1, 2, 0), P_2(-2, -2, -3), P_3(7, 10, 6)$

30. $P_1(2, 3, 2), P_2(1, 4, 4), P_3(5, 0, -4)$

In Problems 31 and 32 solve for the unknown.

31. $P_1(x, 2, 3), P_2(2, 1, 1); d(P_1, P_2) = \sqrt{21}$

32. $P_1(x, x, 1), P_2(0, 3, 5); d(P_1, P_2) = 5$

In Problems 33 and 34 find the coordinates of the midpoint of the line segment between the given points.

33. $(1, 3, \frac{1}{2}), (7, -2, \frac{5}{2})$

34. $(0, 5, -8), (4, 1, -6)$

35. The coordinates of the midpoint of the line segment between $P_1(x_1, y_1, z_1)$ and $P_2(2, 3, 6)$ are $(-1, -4, 8)$. Find the coordinates of P_1.

36. Let P_3 be the midpoint of the line segment between $P_1(-3, 4, 1)$ and $P_2(-5, 8, 3)$. Find the coordinates of the midpoint of the line segment **(a)** between P_1 and P_3 and **(b)** between P_3 and P_2.

37. As shown in Figure 13.32(a), a spacecraft can perform rotations called **pitch**, **roll**, and **yaw** about three distinct axes. To describe the coordinates of a point P we use two coordinate systems: a fixed three-dimensional Cartesian coordinate system in which the coordinates of P are (x, y, z) and a spacecraft coordinate system that moves with the particular rotation. In Figure 13.32(b) we have illustrated a yaw—that is, a rotation around the z-axis (which is perpendicular to the plane of the page). When the spacecraft performs a pitch, roll, and yaw *in sequence* through the angles α, β, and γ, respectively, the final coordinates of the point P in the spacecraft system (x_S, y_S, z_S) are obtained from the sequence of transformations:

$$\begin{aligned} x_P &= x & x_R &= x_P\cos\beta - z_P\sin\beta & x_S &= x_R\cos\gamma + y_R\sin\gamma \\ y_P &= y\cos\alpha + z\sin\alpha & y_R &= y_P & y_S &= -x_R\sin\gamma + y_R\cos\gamma \\ z_P &= -y\sin\alpha + z\cos\alpha & z_R &= x_P\sin\beta + z_P\cos\beta & z_S &= z_R \end{aligned}$$

Suppose the coordinates of a point are $(1, 1, 1)$ in the fixed coordinate system. Determine the coordinates of the point in the spacecraft system if the spacecraft performs a pitch, roll, and yaw in sequence through the angles $\alpha = 30°$, $\beta = 45°$, $\gamma = 60°$.

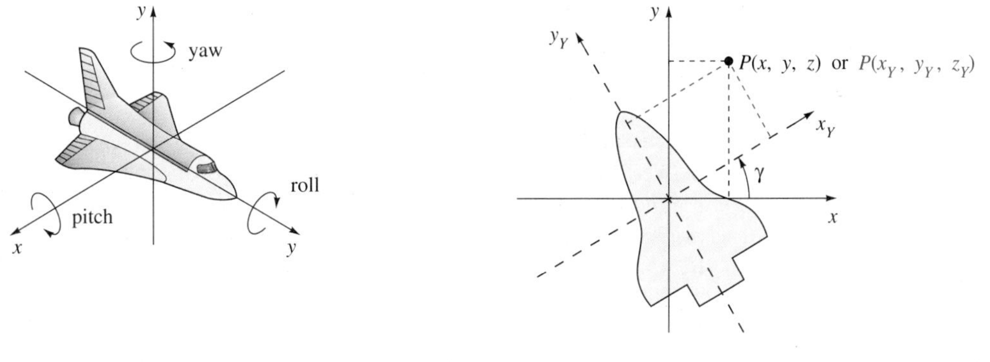

(a) (b)

FIGURE 13.32

38. (*In order to work this problem, you should learn about, or be familiar with, matrix multiplication. See Appendix I.*)

(a) Each system of equations in Problem 37 can be written as a matrix equation. For example, the last system is

$$\begin{bmatrix} x_S \\ y_S \\ z_S \end{bmatrix} = \mathbf{M}_Y \begin{bmatrix} x_R \\ y_R \\ z_R \end{bmatrix}$$

where $\mathbf{M}_Y = \begin{bmatrix} \cos\gamma & \sin\gamma & 0 \\ -\sin\gamma & \cos\gamma & 0 \\ 0 & 0 & 1 \end{bmatrix}$. Identify the matrices $\mathbf{M}_P$ and $\mathbf{M}_R$ and write the first two systems as

$$\begin{bmatrix} x_P \\ y_P \\ z_P \end{bmatrix} = \mathbf{M}_P \begin{bmatrix} x \\ y \\ z \end{bmatrix} \quad \text{and} \quad \begin{bmatrix} x_R \\ y_R \\ z_R \end{bmatrix} = \mathbf{M}_R \begin{bmatrix} x_P \\ y_P \\ z_P \end{bmatrix}$$

(b) Verify that the final coordinates (x_S, y_S, z_S) in the spacecraft system after a pitch, roll, and yaw are obtained from

$$\begin{bmatrix} x_S \\ y_S \\ z_S \end{bmatrix} = \mathbf{M}_Y\mathbf{M}_R\mathbf{M}_P \begin{bmatrix} x \\ y \\ z \end{bmatrix}$$

(c) With $(x, y, z) = (1, 1, 1)$ and $\alpha = 30°$, $\beta = 45°$, $\gamma = 60°$, carry out the indicated matrix multiplication in part (**b**) and verify that your answer is the same as in Problem 37.

[13.2.2]

In Problems 39–42 find the vector $\overrightarrow{P_1P_2}$.

39. $P_1(3, 4, 5)$, $P_2(0, -2, 6)$

40. $P_1(-2, 4, 0)$, $P_2(6, \frac{3}{2}, 8)$

41. $P_1(0, -1, 0)$, $P_2(2, 0, 1)$

42. $P_1(\frac{1}{2}, \frac{3}{4}, 5)$, $P_2(-\frac{5}{2}, -\frac{9}{4}, 12)$

In Problems 43–50 $\mathbf{a} = \langle 1, -3, 2\rangle$, $\mathbf{b} = \langle -1, 1, 1\rangle$, and $\mathbf{c} = \langle 2, 6, 9\rangle$. Find the indicated vector or scalar.

43. $\mathbf{a} + (\mathbf{b} + \mathbf{c})$

44. $2\mathbf{a} - (\mathbf{b} - \mathbf{c})$

45. $\mathbf{b} + 2(\mathbf{a} - 3\mathbf{c})$

46. $4(\mathbf{a} + 2\mathbf{c}) - 6\mathbf{b}$

47. $|\mathbf{a} + \mathbf{c}|$

48. $|\mathbf{c}||2\mathbf{b}|$

49. $\left|\dfrac{\mathbf{a}}{|\mathbf{a}|}\right| + 5\left|\dfrac{\mathbf{b}}{|\mathbf{b}|}\right|$

50. $|\mathbf{b}|\mathbf{a} + |\mathbf{a}|\mathbf{b}$

51. Find a unit vector in the opposite direction of $\mathbf{a} = \langle 10, -5, 10\rangle$.

52. Find a unit vector in the same direction as $\mathbf{a} = \mathbf{i} - 3\mathbf{j} + 2\mathbf{k}$.

53. Find a vector **b** that is four times as long as $\mathbf{a} = \mathbf{i} - \mathbf{j} + \mathbf{k}$ in the same direction as **a**.

54. Find a vector **b** for which $|\mathbf{b}| = \frac{1}{2}$ that is parallel to $\mathbf{a} = \langle -6, 3, -2\rangle$ but has the opposite direction.

55. Using the vectors **a** and **b** shown in Figure 13.33, sketch the "average vector" $\frac{1}{2}(\mathbf{a} + \mathbf{b})$.

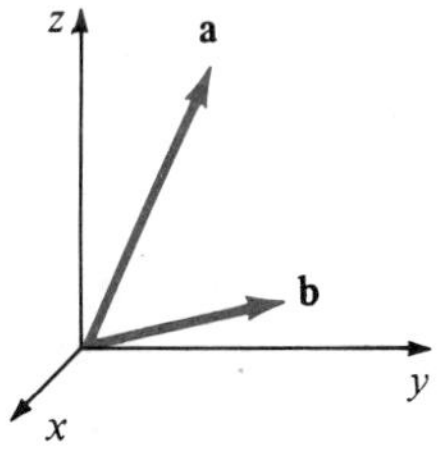

FIGURE 13.33

13.3 DOT PRODUCT

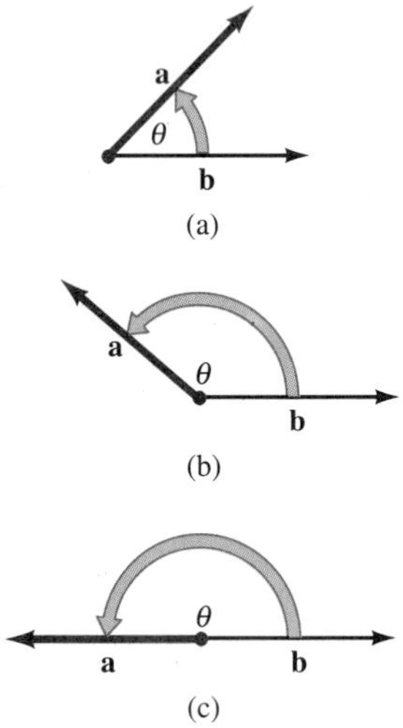

FIGURE 13.34

In this and the following section, we shall consider two kinds of products between vectors that originated in the study of mechanics and electricity and magnetism. The first of these products is known as the **dot product**, **inner product**, or **scalar product**. The dot product of two vectors **a** and **b** is denoted by $\mathbf{a} \cdot \mathbf{b}$ and is a real number or scalar.

DEFINITION 13.3 Dot Product of Two Vectors

The **dot product** of two vectors **a** and **b** is the scalar

$$\mathbf{a} \cdot \mathbf{b} = |\mathbf{a}||\mathbf{b}|\cos\theta \tag{1}$$

where θ is the angle between the vectors such that $0 \le \theta \le \pi$.

Figure 13.34 illustrates the angle θ in three cases. If the vectors **a** and **b** are not parallel, then θ is the *smaller* of the two possible angles between them.

EXAMPLE 1 From (1) we obtain

$$\mathbf{i} \cdot \mathbf{i} = 1 \qquad \mathbf{j} \cdot \mathbf{j} = 1 \qquad \mathbf{k} \cdot \mathbf{k} = 1 \tag{2}$$

since $|\mathbf{i}| = |\mathbf{j}| = |\mathbf{k}| = 1$ and, in each case, $\cos\theta = 1$. □

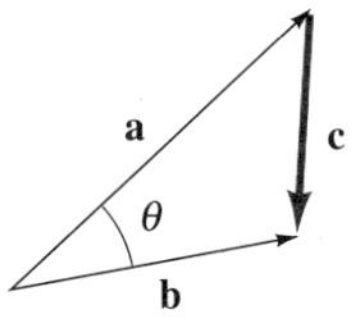

FIGURE 13.35

Component Form of Dot Product The dot product can be expressed in terms of the components of two vectors. Suppose θ is the angle between the vectors $\mathbf{a} = a_1\mathbf{i} + a_2\mathbf{j} + a_3\mathbf{k}$ and $\mathbf{b} = b_1\mathbf{i} + b_2\mathbf{j} + b_3\mathbf{k}$. Then the vector

$$\mathbf{c} = \mathbf{b} - \mathbf{a} = (b_1 - a_1)\mathbf{i} + (b_2 - a_2)\mathbf{j} + (b_3 - a_3)\mathbf{k}$$

is the third side of the triangle indicated in Figure 13.35. By the law of cosines we can write

$$|\mathbf{c}|^2 = |\mathbf{b}|^2 + |\mathbf{a}|^2 - 2|\mathbf{a}||\mathbf{b}|\cos\theta \quad \text{or} \quad |\mathbf{a}||\mathbf{b}|\cos\theta = \tfrac{1}{2}(|\mathbf{b}|^2 + |\mathbf{a}|^2 - |\mathbf{c}|^2) \tag{3}$$

Using $|\mathbf{a}|^2 = a_1^2 + a_2^2 + a_3^2$, $|\mathbf{b}|^2 = b_1^2 + b_2^2 + b_3^2$, and $|\mathbf{b} - \mathbf{a}|^2 = (b_1 - a_1)^2 + (b_2 - a_2)^2 + (b_3 - a_3)^2$, we can simplify the right side of the second equation in (3) to $a_1b_1 + a_2b_2 + a_3b_3$. Since the left side of this equation is the definition of the dot product, we have derived an alternative form of the dot product:

$$\mathbf{a} \cdot \mathbf{b} = a_1b_1 + a_2b_2 + a_3b_3 \tag{4}$$

In other words, the dot product of two vectors is the *sum of the products of their corresponding components.*

EXAMPLE 2 If $\mathbf{a} = 10\mathbf{i} + 2\mathbf{j} - 6\mathbf{k}$ and $\mathbf{b} = -\frac{1}{2}\mathbf{i} + 4\mathbf{j} - 3\mathbf{k}$, then it follows from (4) that $\mathbf{a} \cdot \mathbf{b} = (10)(-\frac{1}{2}) + (2)(4) + (-6)(-3) = 21$. □

Properties The dot product possesses the following properties:

THEOREM 13.1 Properties of the Dot Product

(i) $\mathbf{a} \cdot \mathbf{b} = 0$ if $\mathbf{a} = \mathbf{0}$ or $\mathbf{b} = \mathbf{0}$

(ii) $\mathbf{a} \cdot \mathbf{b} = \mathbf{b} \cdot \mathbf{a}$ — commutative law

(iii) $\mathbf{a} \cdot (\mathbf{b} + \mathbf{c}) = \mathbf{a} \cdot \mathbf{b} + \mathbf{a} \cdot \mathbf{c}$ — distributive law

(iv) $\mathbf{a} \cdot (k\mathbf{b}) = (k\mathbf{a}) \cdot \mathbf{b} = k(\mathbf{a} \cdot \mathbf{b})$, k a scalar

(v) $\mathbf{a} \cdot \mathbf{a} \geq 0$

(vi) $\mathbf{a} \cdot \mathbf{a} = |\mathbf{a}|^2$

Each of these properties, with the possible exception of *(iii)*, should be apparent from (1). Notice that *(vi)* states that the magnitude of a vector

$$\mathbf{a} = a_1\mathbf{i} + a_2\mathbf{j} + a_3\mathbf{k}$$

can be written in terms of the dot product

$$|\mathbf{a}| = \sqrt{\mathbf{a} \cdot \mathbf{a}} = \sqrt{a_1^2 + a_2^2 + a_3^2}$$

We can use (4) to prove *(iii)*. If $\mathbf{a} = a_1\mathbf{i} + a_2\mathbf{j} + a_3\mathbf{k}$, $\mathbf{b} = b_1\mathbf{i} + b_2\mathbf{j} + b_3\mathbf{k}$, and $\mathbf{c} = c_1\mathbf{i} + c_2\mathbf{j} + c_3\mathbf{k}$, then from (4) we have

$$\begin{aligned} \mathbf{a} \cdot (\mathbf{b} + \mathbf{c}) &= a_1(b_1 + c_1) + a_2(b_2 + c_2) + a_3(b_3 + c_3) \\ &= (a_1b_1 + a_2b_2 + a_3b_3) + (a_1c_1 + a_2c_2 + a_3c_3) \\ &= \mathbf{a} \cdot \mathbf{b} + \mathbf{a} \cdot \mathbf{c} \end{aligned}$$

Orthogonal Vectors If **a** and **b** are nonzero vectors, then Definition 13.3 implies that

(i) $\mathbf{a} \cdot \mathbf{b} > 0$ if and only if θ is acute,

(ii) $\mathbf{a} \cdot \mathbf{b} < 0$ if and only if θ is obtuse, and

(iii) $\mathbf{a} \cdot \mathbf{b} = 0$ if and only if $\cos\theta = 0$.

But in the last case the only number in $[0, \pi]$ for which $\cos\theta = 0$ is $\theta = \pi/2$. When $\theta = \pi/2$, we say that the vectors are **perpendicular** or **orthogonal**. Thus, we are led to the following result:

THEOREM 13.2 Criterion for Orthogonal Vectors

Two nonzero vectors **a** and **b** are orthogonal if and only if $\mathbf{a} \cdot \mathbf{b} = 0$.

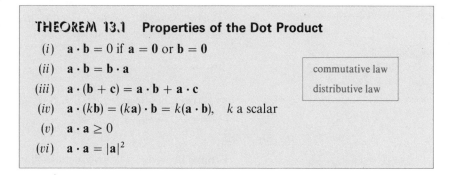

Since $\mathbf{0} \cdot \mathbf{b} = 0$ for every vector $\mathbf{b}$, the zero vector is regarded to be orthogonal to every vector.

EXAMPLE 3   It follows immediately from Theorem 13.1 and the fact that the dot product is commutative that

$$\mathbf{i} \cdot \mathbf{j} = \mathbf{j} \cdot \mathbf{i} = 0$$
$$\mathbf{j} \cdot \mathbf{k} = \mathbf{k} \cdot \mathbf{j} = 0 \qquad (5)$$
$$\mathbf{k} \cdot \mathbf{i} = \mathbf{i} \cdot \mathbf{k} = 0 \qquad \square$$

EXAMPLE 4   If $\mathbf{a} = -3\mathbf{i} - \mathbf{j} + 4\mathbf{k}$ and $\mathbf{b} = 2\mathbf{i} + 14\mathbf{j} + 5\mathbf{k}$, then

$$\mathbf{a} \cdot \mathbf{b} = (-3)(2) + (-1)(14) + (4)(5) = 0$$

From Theorem 13.1, we conclude that $\mathbf{a}$ and $\mathbf{b}$ are orthogonal. $\square$

Angle Between Two Vectors   By equating the two forms of the dot product, (1) and (4), we can determine **the angle between two vectors** from

$$\cos \theta = \frac{a_1 b_1 + a_2 b_2 + a_3 b_3}{|\mathbf{a}||\mathbf{b}|} \qquad (6)$$

EXAMPLE 5   Find the angle between $\mathbf{a} = 2\mathbf{i} + 3\mathbf{j} + \mathbf{k}$ and $\mathbf{b} = -\mathbf{i} + 5\mathbf{j} + \mathbf{k}$.

Solution $|\mathbf{a}| = \sqrt{14} \qquad |\mathbf{b}| = \sqrt{27} \qquad \mathbf{a} \cdot \mathbf{b} = 14$

Hence, (6) gives

$$\cos \theta = \frac{14}{\sqrt{14}\sqrt{27}} = \frac{\sqrt{42}}{9}$$

and so $\theta = \cos^{-1}\left(\frac{\sqrt{42}}{9}\right) \approx 0.77$ radian   or $\theta \approx 44.9°$ $\square$

Direction Cosines   For a nonzero vector $\mathbf{a} = a_1\mathbf{i} + a_2\mathbf{j} + a_3\mathbf{k}$ in 3-space, the angles α, β, and γ between $\mathbf{a}$ and the unit vectors $\mathbf{i}$, $\mathbf{j}$, and $\mathbf{k}$, respectively, are called **direction angles** of $\mathbf{a}$. See Figure 13.36. Now, by (6),

$$\cos \alpha = \frac{\mathbf{a} \cdot \mathbf{i}}{|\mathbf{a}||\mathbf{i}|} \qquad \cos \beta = \frac{\mathbf{a} \cdot \mathbf{j}}{|\mathbf{a}||\mathbf{j}|} \qquad \cos \gamma = \frac{\mathbf{a} \cdot \mathbf{k}}{|\mathbf{a}||\mathbf{k}|}$$

which simplify to

$$\cos \alpha = \frac{a_1}{|\mathbf{a}|} \qquad \cos \beta = \frac{a_2}{|\mathbf{a}|} \qquad \cos \gamma = \frac{a_3}{|\mathbf{a}|}$$

We say that $\cos \alpha$, $\cos \beta$, and $\cos \gamma$ are the **direction cosines** of $\mathbf{a}$. The direction cosines of a nonzero vector $\mathbf{a}$ are simply the components of the unit vector

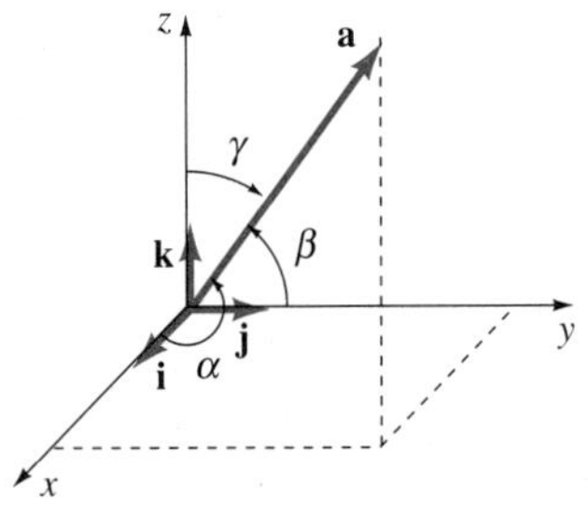

FIGURE 13.36

$(1/|\mathbf{a}|)\mathbf{a}$:

$$\frac{1}{|\mathbf{a}|}\mathbf{a} = \frac{a_1}{|\mathbf{a}|}\mathbf{i} + \frac{a_2}{|\mathbf{a}|}\mathbf{j} + \frac{a_3}{|\mathbf{a}|}\mathbf{k}$$
$$= (\cos\alpha)\mathbf{i} + (\cos\beta)\mathbf{j} + (\cos\gamma)\mathbf{k}$$

Since the magnitude of $(1/|\mathbf{a}|)\mathbf{a}$ is 1, it follows from the last equation that

$$\cos^2\alpha + \cos^2\beta + \cos^2\gamma = 1$$

EXAMPLE 6 Find the direction cosines and direction angles of the vector $\mathbf{a} = 2\mathbf{i} + 5\mathbf{j} + 4\mathbf{k}$.

Solution From $|\mathbf{a}| = \sqrt{2^2 + 5^2 + 4^2} = \sqrt{45} = 3\sqrt{5}$, we see that the direction cosines are

$$\cos\alpha = \frac{2}{3\sqrt{5}} \qquad \cos\beta = \frac{5}{3\sqrt{5}} \qquad \cos\gamma = \frac{4}{3\sqrt{5}}$$

The direction angles are

$$\alpha = \cos^{-1}\left(\frac{2}{3\sqrt{5}}\right) \approx 1.27 \text{ radians} \quad \text{or} \quad \alpha \approx 72.7^\circ$$

$$\beta = \cos^{-1}\left(\frac{5}{3\sqrt{5}}\right) \approx 0.73 \text{ radian} \quad \text{or} \quad \beta \approx 41.8^\circ$$

$$\gamma = \cos^{-1}\left(\frac{4}{3\sqrt{5}}\right) \approx 0.93 \text{ radian} \quad \text{or} \quad \gamma \approx 53.4^\circ$$ □

Observe in Example 6 that

$$\cos^2\alpha + \cos^2\beta + \cos^2\gamma = \frac{4}{45} + \frac{25}{45} + \frac{16}{45} = 1$$

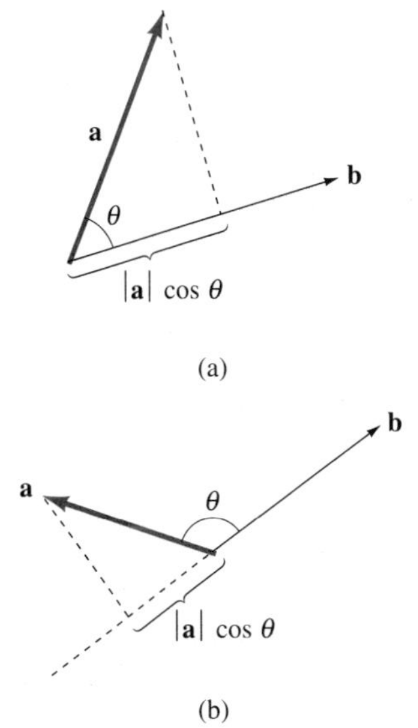

FIGURE 13.37

***Component of* a *on* b** Using the distributive law and (3) enables us to express the components of a vector $\mathbf{a} = a_1\mathbf{i} + a_2\mathbf{j} + a_3\mathbf{k}$ in terms of the dot product:

$$a_1 = \mathbf{a}\cdot\mathbf{i} \qquad a_2 = \mathbf{a}\cdot\mathbf{j} \qquad a_3 = \mathbf{a}\cdot\mathbf{k} \tag{7}$$

Symbolically, we write the components of **a** as

$$\text{comp}_{\mathbf{i}}\mathbf{a} = \mathbf{a}\cdot\mathbf{i} \qquad \text{comp}_{\mathbf{j}}\mathbf{a} = \mathbf{a}\cdot\mathbf{j} \qquad \text{comp}_{\mathbf{k}}\mathbf{a} = \mathbf{a}\cdot\mathbf{k} \tag{8}$$

We shall now see that the procedure indicated in (8) carries over to finding the **component of a on an arbitrary vector b**. Note that in either of the two cases shown in Figure 13.37,

$$\text{comp}_{\mathbf{b}}\mathbf{a} = |\mathbf{a}|\cos\theta \tag{9}$$

In Figure 13.37(b), $\text{comp}_{\mathbf{b}}\mathbf{a} < 0$, since $\pi/2 < \theta \leq \pi$. Now, by writing (9) as

$$\text{comp}_{\mathbf{b}}\mathbf{a} = \frac{|\mathbf{a}||\mathbf{b}|\cos\theta}{|\mathbf{b}|} = \frac{\mathbf{a}\cdot\mathbf{b}}{|\mathbf{b}|}$$

we see that

$$\text{comp}_{\mathbf{b}}\mathbf{a} = \mathbf{a}\cdot\left(\frac{1}{|\mathbf{b}|}\mathbf{b}\right) \tag{10}$$

In other words,

> *to find the component of* **a** *on a vector* **b**, *we dot* **a** *with a unit vector in the direction of* **b**.

EXAMPLE 7 Let $\mathbf{a} = 2\mathbf{i} + 3\mathbf{j} - 4\mathbf{k}$ and $\mathbf{b} = \mathbf{i} + \mathbf{j} + 2\mathbf{k}$. Find (*a*) $\text{comp}_{\mathbf{b}}\mathbf{a}$ and (*b*) $\text{comp}_{\mathbf{a}}\mathbf{b}$.

Solution (*a*) We first form a unit vector in the direction of **b**:

$$|\mathbf{b}| = \sqrt{6} \qquad \frac{1}{|\mathbf{b}|}\mathbf{b} = \frac{1}{\sqrt{6}}(\mathbf{i} + \mathbf{j} + 2\mathbf{k})$$

Then from (10) we have

$$\text{comp}_{\mathbf{b}}\mathbf{a} = (2\mathbf{i} + 3\mathbf{j} - 4\mathbf{k})\cdot\frac{1}{\sqrt{6}}(\mathbf{i} + \mathbf{j} + 2\mathbf{k}) = -\frac{3}{\sqrt{6}}$$

(*b*) By modifying (10) accordingly, we have

$$\text{comp}_{\mathbf{a}}\mathbf{b} = \mathbf{b}\cdot\left(\frac{1}{|\mathbf{a}|}\mathbf{a}\right)$$

Therefore,

$$|\mathbf{a}| = \sqrt{29} \qquad \frac{1}{|\mathbf{a}|}\mathbf{a} = \frac{1}{\sqrt{29}}(2\mathbf{i} + 3\mathbf{j} - 4\mathbf{k})$$

and

$$\text{comp}_{\mathbf{a}}\mathbf{b} = (\mathbf{i} + \mathbf{j} + 2\mathbf{k})\cdot\frac{1}{\sqrt{29}}(2\mathbf{i} + 3\mathbf{j} - 4\mathbf{k}) = -\frac{3}{\sqrt{29}}$$ □

***Projection of* a *onto* b** As illustrated in Figure 13.38(a), the projection of a vector **a** in any of the directions determined by **i**, **j**, **k** is simply the *vector* formed by multiplying the component of **a** in the specified direction with a unit vector in that direction; for example,

$$\text{proj}_{\mathbf{i}}\mathbf{a} = (\text{comp}_{\mathbf{i}}\mathbf{a})\mathbf{i} = (\mathbf{a}\cdot\mathbf{i})\mathbf{i} = a_1\mathbf{i}$$

and so on. Figure 13.38(b) shows the general case of the **projection of a onto b**:

$$\text{proj}_{\mathbf{b}}\mathbf{a} = (\text{comp}_{\mathbf{b}}\mathbf{a})\left(\frac{1}{|\mathbf{b}|}\mathbf{b}\right) \tag{11}$$

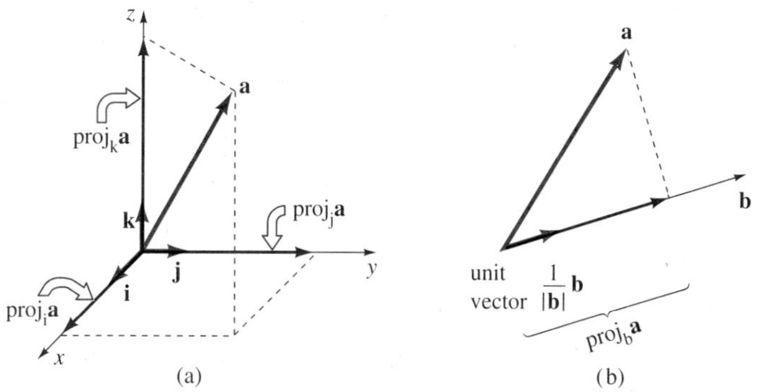

FIGURE 13.38

EXAMPLE 8 Find the projection of $\mathbf{a} = 4\mathbf{i} + \mathbf{j}$ on the vector $\mathbf{b} = 2\mathbf{i} + 3\mathbf{j}$. Graph.

Solution First, we find the component of $\mathbf{a}$ on $\mathbf{b}$. Since $|\mathbf{b}| = \sqrt{13}$, we find from (10),

$$\text{comp}_{\mathbf{b}}\mathbf{a} = (4\mathbf{i} + \mathbf{j}) \cdot \frac{1}{\sqrt{13}}(2\mathbf{i} + 3\mathbf{j}) = \frac{11}{\sqrt{13}}$$

Thus, from (11),

$$\text{proj}_{\mathbf{b}}\mathbf{a} = \left(\frac{11}{\sqrt{13}}\right)\left(\frac{1}{\sqrt{13}}\right)(2\mathbf{i} + 3\mathbf{j}) = \frac{22}{13}\mathbf{i} + \frac{33}{13}\mathbf{j}$$

The graph of this vector is shown in Figure 13.39. □

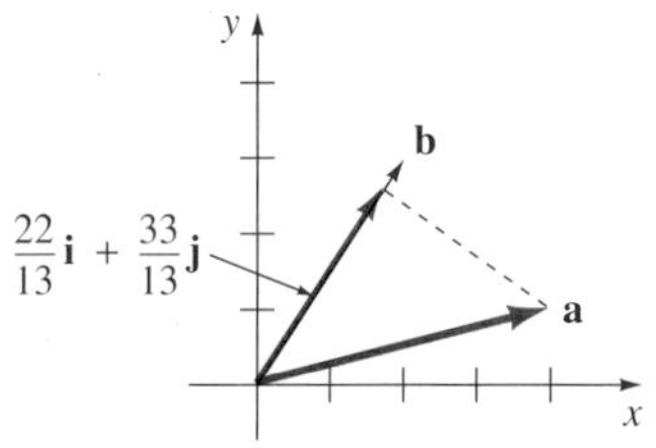

FIGURE 13.39

Projection of a onto $\mathbf{b}^{\perp}$ If $\mathbf{b} \neq \mathbf{0}$, any vector $\mathbf{a}$ can be projected onto $\mathbf{b}$ as well as onto a vector $\mathbf{b}^{\perp}$, of magnitude $|\mathbf{b}|$, that is orthogonal to $\mathbf{b}$. From Figure 13.40, we see that $\mathbf{a}$ can be written as the sum of two projections:

$$\text{proj}_{\mathbf{b}}\mathbf{a} + \text{proj}_{\mathbf{b}^{\perp}}\mathbf{a} = \mathbf{a} \tag{12}$$

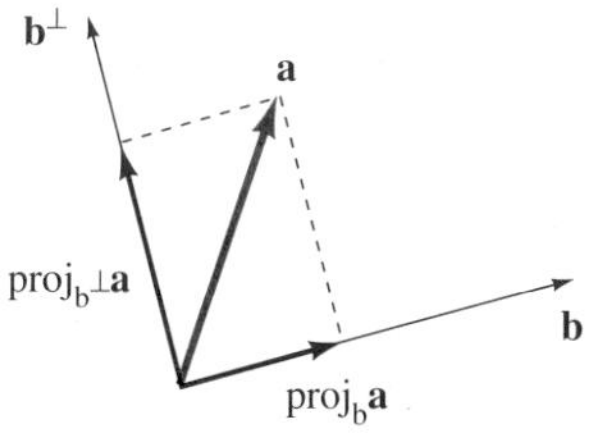

FIGURE 13.40

Equation (12) enables us to define the projection of $\mathbf{a}$ onto $\mathbf{b}^{\perp}$:

$$\text{proj}_{\mathbf{b}^{\perp}}\mathbf{a} = \mathbf{a} - \text{proj}_{\mathbf{b}}\mathbf{a} \tag{13}$$

EXAMPLE 9 Let $\mathbf{a} = 3\mathbf{i} - \mathbf{j} + 5\mathbf{k}$ and $\mathbf{b} = 2\mathbf{i} + \mathbf{j} + 2\mathbf{k}$. Find (*a*) $\text{proj}_{\mathbf{b}}\mathbf{a}$ and (*b*) $\text{proj}_{\mathbf{b}^{\perp}}\mathbf{a}$.

Solution Since $|\mathbf{b}| = 3$, we have

$$\text{comp}_{\mathbf{b}}\mathbf{a} = (3\mathbf{i} - \mathbf{j} + 5\mathbf{k}) \cdot \frac{1}{3}(2\mathbf{i} + \mathbf{j} + 2\mathbf{k}) = \frac{15}{3} = 5$$

(*a*) $\text{proj}_{\mathbf{b}}\mathbf{a} = (5)\left(\frac{1}{3}\right)(2\mathbf{i} + \mathbf{j} + 2\mathbf{k}) = \frac{10}{3}\mathbf{i} + \frac{5}{3}\mathbf{j} + \frac{10}{3}\mathbf{k}$

(*b*)
$$\text{proj}_{\mathbf{b}\perp}\mathbf{a} = \mathbf{a} - \text{proj}_{\mathbf{b}}\mathbf{a}$$
$$= (3\mathbf{i} - \mathbf{j} + 5\mathbf{k}) - \left(\frac{10}{3}\mathbf{i} + \frac{5}{3}\mathbf{j} + \frac{10}{3}\mathbf{k}\right) = -\frac{1}{3}\mathbf{i} - \frac{8}{3}\mathbf{j} + \frac{5}{3}\mathbf{k} \quad \square$$

Physical Interpretation of the Dot Product In Section 5.8 we saw that when a constant force of magnitude F moves an object a distance d in the same direction of the force, the work done is simply

$$W = Fd \tag{14}$$

However, if a constant force $\mathbf{F}$ applied to a body acts at an angle θ to the direction of motion, then the work done by $\mathbf{F}$ is defined to be the product of the component of $\mathbf{F}$ in the direction of the displacement and the distance $|\mathbf{d}|$ that the body moves:

$$W = (|\mathbf{F}|\cos\theta)|\mathbf{d}| = |\mathbf{F}||\mathbf{d}|\cos\theta$$

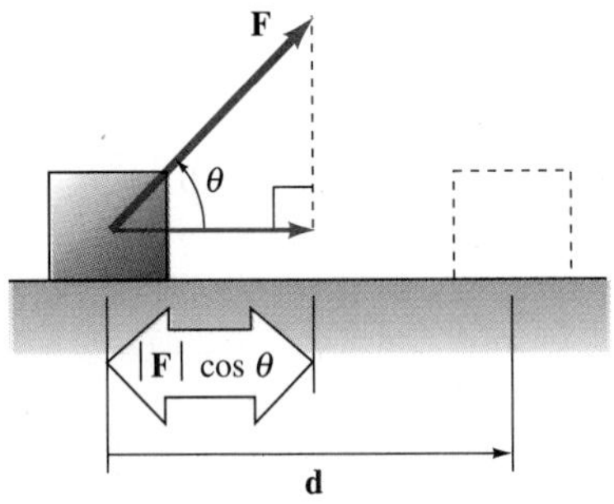

FIGURE 13.41

See Figure 13.41. It follows from Definition 13.3 that if $\mathbf{F}$ causes a displacement $\mathbf{d}$ of a body, then the work done is

$$W = \mathbf{F} \cdot \mathbf{d} \tag{15}$$

Note that (15) reduces to (14) when $\theta = 0$.

EXAMPLE 10 Find the work done by a constant force $\mathbf{F} = 2\mathbf{i} + 4\mathbf{j}$ if its point of application to a block moves from $P_1(1, 1)$ to $P_2(4, 6)$. Assume that $|\mathbf{F}|$ is measured in newtons and $|\mathbf{d}|$ is measured in meters.

Solution The displacement of the block is given by

$$\mathbf{d} = \overrightarrow{P_1P_2} = \overrightarrow{OP_2} - \overrightarrow{OP_1} = 3\mathbf{i} + 5\mathbf{j}$$

It follows from (15) that the work done is

$$W = (2\mathbf{i} + 4\mathbf{j}) \cdot (3\mathbf{i} + 5\mathbf{j}) = 26 \text{ N-m} \quad \square$$

EXERCISES 13.3 *Answers to odd-numbered problems begin on page A-99.*

In Problems 1 and 2 find $\mathbf{a} \cdot \mathbf{b}$ if the smaller angle between $\mathbf{a}$ and $\mathbf{b}$ is as given.

1. $|\mathbf{a}| = 10$, $|\mathbf{b}| = 5$, $\theta = \pi/4$
2. $|\mathbf{a}| = 6$, $|\mathbf{b}| = 12$, $\theta = \pi/6$

In Problems 3–14 $\mathbf{a} = \langle 2, -3, 4\rangle$, $\mathbf{b} = \langle -1, 2, 5\rangle$, and $\mathbf{c} = \langle 3, 6, -1\rangle$. Find the indicated scalar or vector.

3. $\mathbf{a} \cdot \mathbf{b}$
4. $\mathbf{b} \cdot \mathbf{c}$
5. $\mathbf{a} \cdot \mathbf{c}$
6. $\mathbf{a} \cdot (\mathbf{b} + \mathbf{c})$
7. $\mathbf{a} \cdot (4\mathbf{b})$
8. $\mathbf{b} \cdot (\mathbf{a} - \mathbf{c})$
9. $\mathbf{a} \cdot \mathbf{a}$
10. $(2\mathbf{b}) \cdot (3\mathbf{c})$
11. $\mathbf{a} \cdot (\mathbf{a} + \mathbf{b} + \mathbf{c})$
12. $(2\mathbf{a}) \cdot (\mathbf{a} - 2\mathbf{b})$
13. $\left(\dfrac{\mathbf{a} \cdot \mathbf{b}}{\mathbf{b} \cdot \mathbf{b}}\right)\mathbf{b}$
14. $(\mathbf{c} \cdot \mathbf{b})\mathbf{a}$
15. Determine which pairs of the following vectors are orthogonal:

    (a) $\langle 2, 0, 1\rangle$ (b) $3\mathbf{i} + 2\mathbf{j} - \mathbf{k}$

(c) $2\mathbf{i} - \mathbf{j} - \mathbf{k}$ (d) $\mathbf{i} - 4\mathbf{j} + 6\mathbf{k}$

(e) $\langle 1, -1, 1 \rangle$ (f) $\langle -4, 3, 8 \rangle$

16. Determine a scalar c so that the given vectors are orthogonal.

(a) $\mathbf{a} = 2\mathbf{i} - c\mathbf{j} + 3\mathbf{k}$, $\mathbf{b} = 3\mathbf{i} + 2\mathbf{j} + 4\mathbf{k}$

(b) $\mathbf{a} = \langle c, \frac{1}{2}, c \rangle$, $\mathbf{b} = \langle -3, 4, c \rangle$

17. Find a vector $\mathbf{v} = \langle x_1, y_1, 1 \rangle$ that is orthogonal to both $\mathbf{a} = \langle 3, 1, -1 \rangle$ and $\mathbf{b} = \langle -3, 2, 2 \rangle$.

18. A **rhombus** is an oblique-angled parallelogram with all four sides equal. Use the dot product to show that the diagonals of a rhombus are perpendicular.

19. Verify that the vector

$$\mathbf{c} = \mathbf{b} - \frac{\mathbf{a} \cdot \mathbf{b}}{|\mathbf{a}|^2}\mathbf{a}$$

is orthogonal to the vector $\mathbf{a}$.

20. Determine a scalar c so that the angle between $\mathbf{a} = \mathbf{i} + c\mathbf{j}$ and $\mathbf{b} = \mathbf{i} + \mathbf{j}$ is 45°.

In Problems 21–24 find the angle θ between the given vectors.

21. $\mathbf{a} = 3\mathbf{i} - \mathbf{k}$, $\mathbf{b} = 2\mathbf{i} + 2\mathbf{k}$

22. $\mathbf{a} = 2\mathbf{i} + \mathbf{j}$, $\mathbf{b} = -3\mathbf{i} - 4\mathbf{j}$

23. $\mathbf{a} = \langle 2, 4, 0 \rangle$, $\mathbf{b} = \langle -1, -1, 4 \rangle$

24. $\mathbf{a} = \langle \frac{1}{2}, \frac{1}{2}, \frac{3}{2} \rangle$, $\mathbf{b} = \langle 2, -4, 6 \rangle$

In Problems 25–28 find the direction cosines and direction angles of the given vector.

25. $\mathbf{a} = \mathbf{i} + 2\mathbf{j} + 3\mathbf{k}$ **26.** $\mathbf{a} = 6\mathbf{i} + 6\mathbf{j} - 3\mathbf{k}$

27. $\mathbf{a} = \langle 1, 0, -\sqrt{3} \rangle$ **28.** $\mathbf{a} = \langle 5, 7, 2 \rangle$

29. Find the angle between the diagonal $\overrightarrow{AD}$ of the cube shown in Figure 13.42 and the edge $\overrightarrow{AB}$. Find the angle between the diagonal $\overrightarrow{AD}$ and the diagonal $\overrightarrow{AC}$.

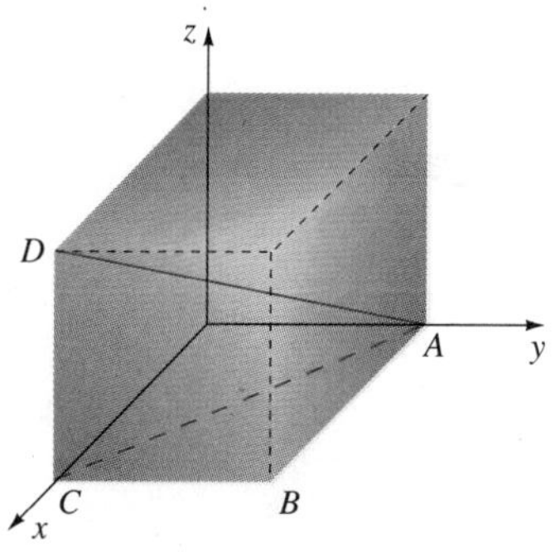

FIGURE 13.42

30. Show that if two nonzero vectors $\mathbf{a}$ and $\mathbf{b}$ are orthogonal, then their direction cosines satisfy

$$\cos\alpha_1\cos\alpha_2 + \cos\beta_1\cos\beta_2 + \cos\gamma_1\cos\gamma_2 = 0$$

31. An airplane is 4 km high, 5 km south, and 7 km east of an airport. See Figure 13.43. Find the direction angles of the plane.

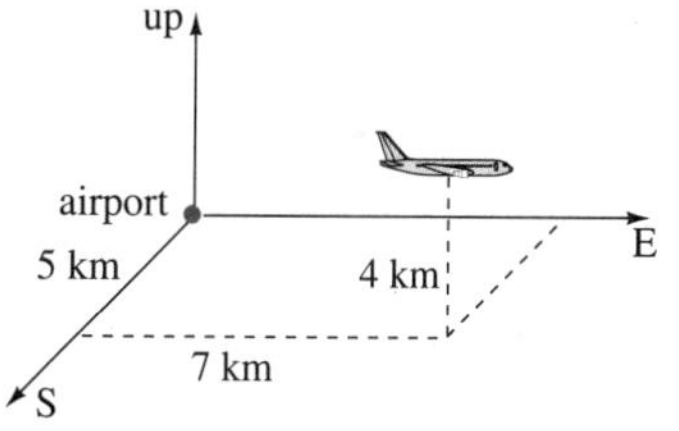

FIGURE 13.43

32. Determine a unit vector whose direction angles, relative to the three coordinate axes, are equal.

In Problems 33–36 $\mathbf{a} = \langle 1, -1, 3 \rangle$ and $\mathbf{b} = \langle 2, 6, 3 \rangle$. Find the indicated number.

33. $\text{comp}_{\mathbf{b}}\mathbf{a}$ **34.** $\text{comp}_{\mathbf{a}}\mathbf{b}$

35. $\text{comp}_{\mathbf{a}}(\mathbf{b} - \mathbf{a})$ **36.** $\text{comp}_{2\mathbf{b}}(\mathbf{a} + \mathbf{b})$

In Problems 37 and 38 find the component of the given vector in the direction from the origin to the indicated point.

37. $\mathbf{a} = 4\mathbf{i} + 6\mathbf{j}$; $P(3, 10)$

38. $\mathbf{a} = \langle 2, 1, -1 \rangle$; $P(1, -1, 1)$

In Problems 39–42 find **(a)** $\text{proj}_{\mathbf{b}}\mathbf{a}$ and **(b)** $\text{proj}_{\mathbf{b}\perp}\mathbf{a}$.

39. $\mathbf{a} = -5\mathbf{i} + 5\mathbf{j}$, $\mathbf{b} = -3\mathbf{i} + 4\mathbf{j}$

40. $\mathbf{a} = 4\mathbf{i} + 2\mathbf{i}$, $\mathbf{b} = -3\mathbf{i} + \mathbf{j}$

41. $\mathbf{a} = -\mathbf{i} - 2\mathbf{j} + 7\mathbf{k}$, $\mathbf{b} = 6\mathbf{i} - 3\mathbf{j} - 2\mathbf{k}$

42. $\mathbf{a} = \langle 1, 1, 1 \rangle$, $\mathbf{b} = \langle -2, 2, -1 \rangle$

In Problems 43 and 44 $\mathbf{a} = 4\mathbf{i} + 3\mathbf{j}$ and $\mathbf{b} = -\mathbf{i} + \mathbf{j}$. Find the indicated vector.

43. $\text{proj}_{(\mathbf{a}+\mathbf{b})}\mathbf{a}$ **44.** $\text{proj}_{(\mathbf{a}-\mathbf{b})\perp}\mathbf{b}$

45. A sled is pulled horizontally over ice by a rope attached to its front. A 20-lb force acting at an angle of 60° with the horizontal moves the sled 100 ft. Find the work done.

46. Find the work done if the point at which the constant force $\mathbf{F} = 4\mathbf{i} + 3\mathbf{j} + 5\mathbf{k}$ is applied to an object moves from $P_1(3, 1, -2)$ to $P_2(2, 4, 6)$. Assume that $|\mathbf{F}|$ is measured in newtons and $|\mathbf{d}|$ is measured in meters.

47. A block with weight $\mathbf{w}$ is pulled along a frictionless horizontal surface by a constant force $\mathbf{F}$ of magnitude 30 newtons in the direction given by a vector **d**. See Figure 13.44. Assume $|\mathbf{d}|$ is measured in meters.

(a) What is the work done by the weight **w**?

(b) What is the work done by the force **F** if $\mathbf{d} = 4\mathbf{i} + 3\mathbf{j}$?

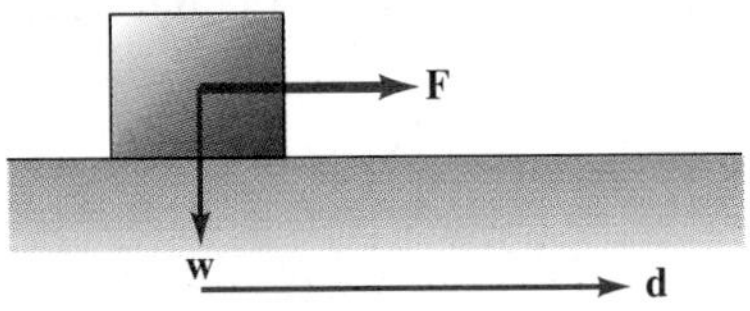

FIGURE 13.44

48. A constant force **F** of magnitude 3 lb is applied to the block shown in Figure 13.45. **F** has the same direction as the vector $\mathbf{a} = 3\mathbf{i} + 4\mathbf{j}$. Find the work done in the direction of motion if the block moves from $P_1(3, 1)$ to $P_2(9, 3)$. Assume distance is measured in feet.

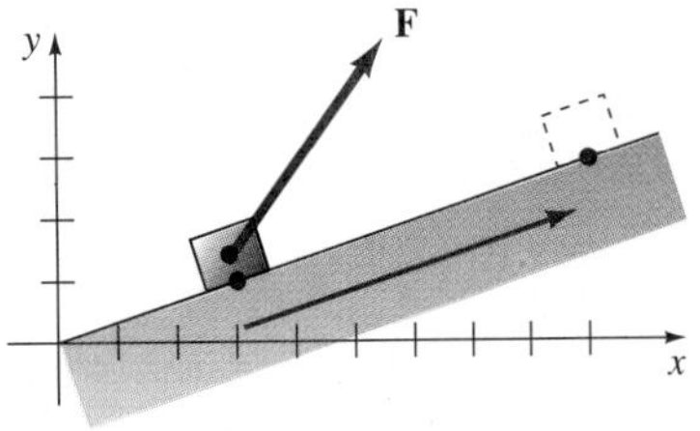

FIGURE 13.45

49. In the methane molecule CH_4, the hydrogen atoms are located at the four vertices of a regular tetrahedron. See Figure 13.46. The distance between the center of a hydrogen atom and the center of a carbon atom is 1.10 angstroms (1 angstrom $= 10^{-10}$ m) and the hydrogen–carbon–hydrogen bond angle is $\theta = 109.5°$. Using vector methods only, find the distance between two hydrogen atoms.

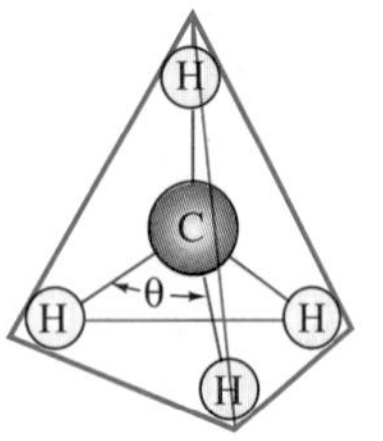

FIGURE 13.46

50. (*This problem could present a challenge.*) Light from a source at point $S(a, b)$ is reflected by a spherical mirror of radius 1, centered at the origin, to an observer located at point $O(c, d)$ as shown in Figure 13.47. The point of reflection $P(x, y)$ from a spherical mirror lies in the plane determined by the source, the observer, and the center of the sphere. (The analysis of spherical mirrors occurs, among other places, in the study of radar design.)

(a) Use Definition 13.3 twice, once with the angle θ and once with the angle ϕ, to show that the coordinates of the point of reflection $P(x, y)$ satisfy the equation

$$\frac{ax + by - 1}{ay - bx} = \frac{cx + dy - 1}{dx - cy}$$

[*Hint:* As shown in the figure, let **N** and **T** denote, respectively, a unit normal vector and a unit tangent to the circle at $P(x, y)$. If $\mathbf{N} = x\mathbf{i} + y\mathbf{j}$, what is **T** in terms of x and y?]

(b) Let $a = 2, b = 0, c = 0$, and $d = 3$. Use the relationship $x^2 + y^2 = 1$ to show that the x-coordinate of the point of reflection is a root of a fourth-degree polynomial equation.

(c) Use Newton's Method to find the point of reflection in part (**b**). You may have to consider all four roots of the equation in part (**b**) to find the one that corresponds to a solution of the equation in part (**a**).

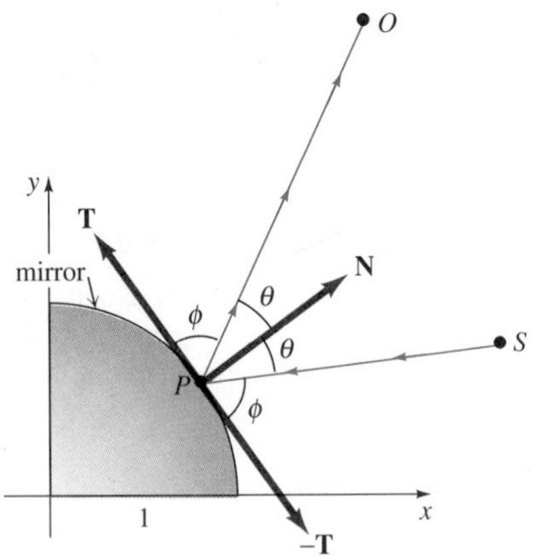

FIGURE 13.47

51. Use the dot product to prove the **Cauchy–Schwarz inequality**: $|\mathbf{a} \cdot \mathbf{b}| \leq |\mathbf{a}||\mathbf{b}|$.

52. Use the dot product to prove the **triangle inequality**. $|\mathbf{a} + \mathbf{b}| \le |\mathbf{a}| + |\mathbf{b}|$. [*Hint:* Consider property (*vi*) in Theorem 13.1]

53. Prove that the vector $\mathbf{n} = a\mathbf{i} + b\mathbf{j}$ is perpendicular to the line whose equation is $ax + by + c = 0$. [*Hint:* Let $P_1(x_1, y_1)$ and $P_2(x_2, y_2)$ be distinct points on the line.]

54. Use the result of Problem 53 and Figure 13.48 to show that the distance d from a point $P_1(x_1, y_1)$ to a line $ax + by + c = 0$ is $d = |ax_1 + by_1 + c|/\sqrt{a^2 + b^2}$.

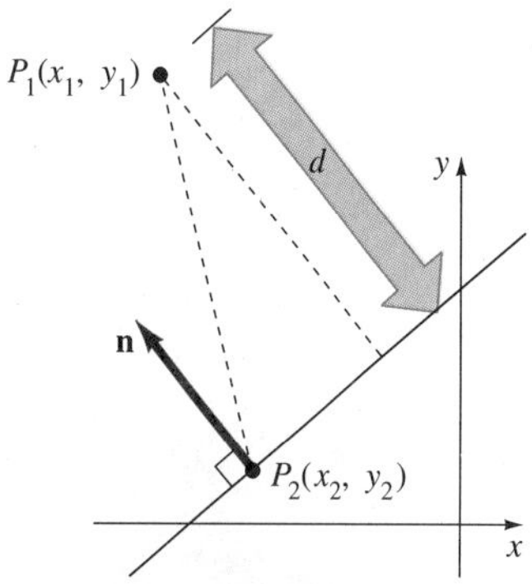

FIGURE 13.48

13.4 CROSS PRODUCT

REVIEW

Since knowledge of determinants of order 2 and order 3 is important to the discussion that follows, we recall the following facts:

(*a*) $\begin{vmatrix} a_1 & a_2 \\ b_1 & b_2 \end{vmatrix} = a_1b_2 - a_2b_1$

(*b*) $\begin{vmatrix} a_1 & a_2 & a_3 \\ b_1 & b_2 & b_3 \\ c_1 & c_2 & c_3 \end{vmatrix} = a_1\begin{vmatrix} b_2 & b_3 \\ c_2 & c_3 \end{vmatrix} - a_2\begin{vmatrix} b_1 & b_3 \\ c_1 & c_3 \end{vmatrix} + a_3\begin{vmatrix} b_1 & b_2 \\ c_1 & c_2 \end{vmatrix}$

This is called **expanding the determinant by cofactors** of the first row.

(*c*) When two rows of a determinant are interchanged, the resulting determinant is the negative of the original.

See Appendix I for a review of the properties of determinants.

In contrast to the dot product, which is a number or a scalar, the next special product of two vectors, called the **cross product**, is a vector.

DEFINITION 13.4 Cross Product of Two Vectors

The **cross product** of two vectors **a** and **b** in 3-space is the vector

$$\mathbf{a} \times \mathbf{b} = (|\mathbf{a}||\mathbf{b}|\sin\theta)\mathbf{n} \tag{1}$$

where θ is the angle between the vectors such that $0 \le \theta \le \pi$ and $\mathbf{n}$ is a unit vector perpendicular to the plane of **a** and **b** with direction given by the right-hand rule.

As seen in Figure 13.49(a), if the fingers of the right hand point along the vector **a** and then curl toward the vector **b**, the thumb will give the direction of **n** and hence **a** × **b**. In Figure 13.49(b), the right-hand rule shows the direction of **b** × **a**.

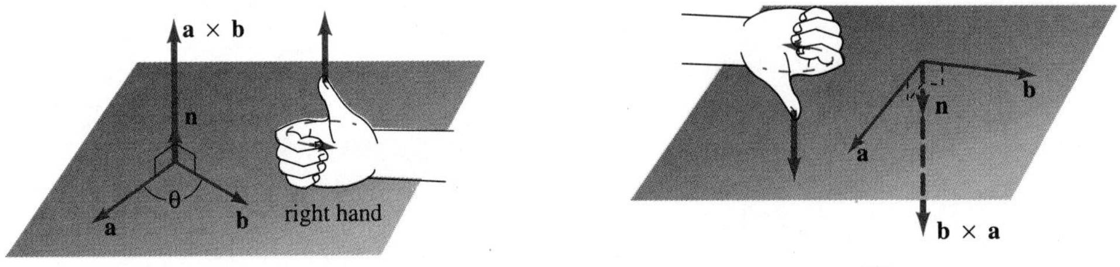

FIGURE 13.49

EXAMPLE 1 In physics a force **F** acting at the end of a position vector **r**, as shown in Figure 13.50, is said to produce a **torque** $\boldsymbol{\tau}$ defined by $\boldsymbol{\tau} = \mathbf{r} \times \mathbf{F}$. For example, if $|\mathbf{F}| = 20$ N, $|\mathbf{r}| = 3.5$ m, and $\theta = 30°$, then from (1), $|\boldsymbol{\tau}| = (3.5)(20)\sin 30° = 35$ N-m. If **F** and **r** are in the plane of the page, the right-hand rule implies that the direction of $\boldsymbol{\tau}$ is outward from, and perpendicular to, the page (toward the reader).

As we see in Figure 13.51, when a force **F** is applied to a wrench, the magnitude of the torque $\boldsymbol{\tau}$ is a measure of the turning effect about the pivot point P and the vector $\boldsymbol{\tau}$ is directed along the axis of the bolt. In this case $\boldsymbol{\tau}$ points inward from the page.

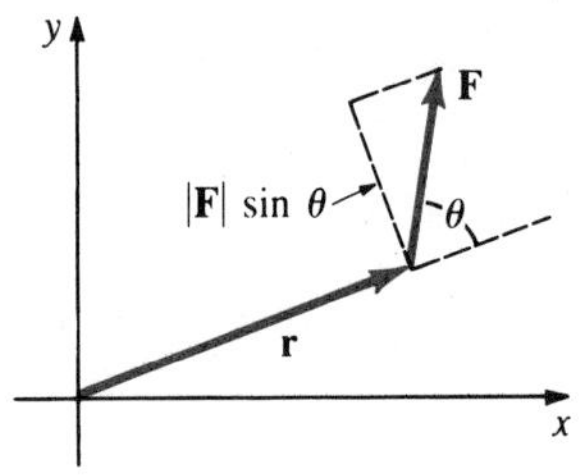

FIGURE 13.50

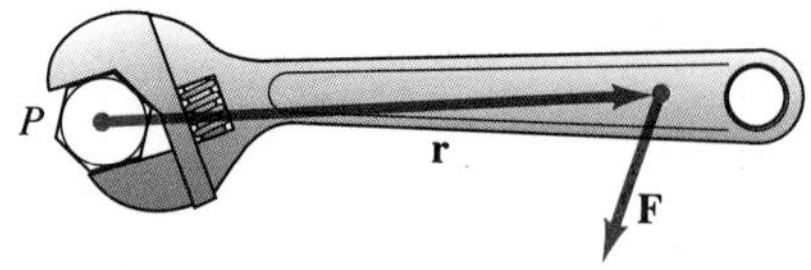

FIGURE 13.51 □

Properties The cross product has the following **properties**.

THEOREM 13.3 Properties of the Cross Product

(*i*) $\mathbf{a} \times \mathbf{b} = \mathbf{0}$ if $\mathbf{a} = \mathbf{0}$ or $\mathbf{b} = \mathbf{0}$

(*ii*) $\mathbf{a} \times \mathbf{b} = -\mathbf{b} \times \mathbf{a}$

(*iii*) $\mathbf{a} \times (\mathbf{b} + \mathbf{c}) = (\mathbf{a} \times \mathbf{b}) + (\mathbf{a} \times \mathbf{c})$

(*iv*) $(\mathbf{a} + \mathbf{b}) \times \mathbf{c} = (\mathbf{a} \times \mathbf{c}) + (\mathbf{b} \times \mathbf{c})$

distributive laws

(*v*) $\mathbf{a} \times (k\mathbf{b}) = (k\mathbf{a}) \times \mathbf{b} = k(\mathbf{a} \times \mathbf{b})$, k a scalar

(*vi*) $\mathbf{a} \times \mathbf{a} = \mathbf{0}$

(*vii*) $\mathbf{a} \cdot (\mathbf{a} \times \mathbf{b}) = 0$

(*viii*) $\mathbf{b} \cdot (\mathbf{a} \times \mathbf{b}) = 0$

Property (*vi*) follows from (1) because $\theta = 0$. Properties (*vii*) and (*viii*) are simply statements of the fact that $\mathbf{a} \times \mathbf{b}$ is perpendicular to the plane containing **a** and **b**. Property (*ii*) should be intuitively clear from Figure 13.49. The proof of the distributive property (*iii*) is too complicated to warrant its inclusion here.

Parallel Vectors When the angle between two nonzero vectors is either $\theta = 0$ or $\theta = \pi$, then $\sin\theta = 0$ and so we must have $\mathbf{a} \times \mathbf{b} = \mathbf{0}$. This is stated formally in the next theorem.

THEOREM 13.4 Criterion for Parallel Vectors

Two nonzero vectors **a** and **b** are **parallel** if and only if $\mathbf{a} \times \mathbf{b} = \mathbf{0}$.

EXAMPLE 2 (*a*) From property (*vi*) we have

$$\mathbf{i} \times \mathbf{i} = \mathbf{0} \qquad \mathbf{j} \times \mathbf{j} = \mathbf{0} \qquad \mathbf{k} \times \mathbf{k} = \mathbf{0} \tag{2}$$

(*b*) If $\mathbf{a} = 2\mathbf{i} + \mathbf{j} - \mathbf{k}$ and $\mathbf{b} = -6\mathbf{i} - 3\mathbf{j} + 3\mathbf{k} = -3\mathbf{a}$, then **a** and **b** are parallel. Hence, from Theorem 13.4, $\mathbf{a} \times \mathbf{b} = \mathbf{0}$. Note that this result also follows by combining properties (*v*) and (*vi*). □

From (1), if $\mathbf{a} = \mathbf{i}$ and $\mathbf{b} = \mathbf{j}$, then

$$\mathbf{i} \times \mathbf{j} = \left(|\mathbf{i}||\mathbf{j}|\sin\frac{\pi}{2}\right)\mathbf{n} = \mathbf{n} \tag{3}$$

But, since a unit vector perpendicular to the plane that contains **i** and **j** with the direction given by the right-hand rule is **k**, it follows from (3) that $\mathbf{n} = \mathbf{k}$. In other words, $\mathbf{i} \times \mathbf{j} = \mathbf{k}$.

EXAMPLE 3 The cross products of any pair of vectors in the set **i**, **j**, **k** can be obtained by the circular mnemonic

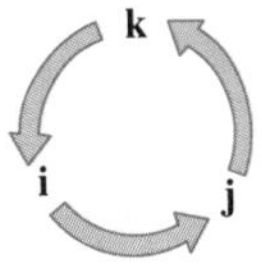

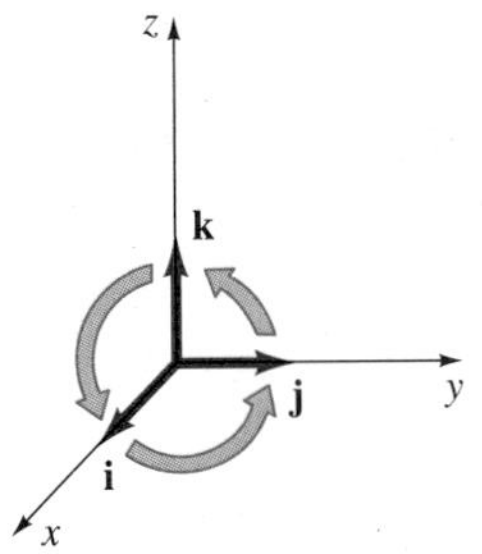

FIGURE 13.52

that is,

$$\left.\begin{array}{l}\mathbf{i}\times\mathbf{j}=\mathbf{k}\\ \mathbf{j}\times\mathbf{k}=\mathbf{i}\\ \mathbf{k}\times\mathbf{i}=\mathbf{j}\end{array}\right\}\quad\text{and from property }(ii)\quad\left\{\begin{array}{l}\mathbf{j}\times\mathbf{i}=-\mathbf{k}\\ \mathbf{k}\times\mathbf{j}=-\mathbf{i}\\ \mathbf{i}\times\mathbf{k}=-\mathbf{j}\end{array}\right.\tag{4}$$

See Figure 13.52. □

Component Form of Cross Product We can use the distributive law (*iii*) to write the cross product $\mathbf{a}\times\mathbf{b}$ in terms of the components of $\mathbf{a}$ and $\mathbf{b}$:

$$\begin{aligned}\mathbf{a}\times\mathbf{b} &= (a_1\mathbf{i}+a_2\mathbf{j}+a_3\mathbf{k})\times(b_1\mathbf{i}+b_2\mathbf{j}+b_3\mathbf{k})\\ &= a_1\mathbf{i}\times(b_1\mathbf{i}+b_2\mathbf{j}+b_3\mathbf{k})+a_2\mathbf{j}\times(b_1\mathbf{i}+b_2\mathbf{j}+b_3\mathbf{k})\\ &\quad + a_3\mathbf{k}\times(b_1\mathbf{i}+b_2\mathbf{j}+b_3\mathbf{k})\\ &= a_1b_1(\mathbf{i}\times\mathbf{i})+a_1b_2(\mathbf{i}\times\mathbf{j})+a_1b_3(\mathbf{i}\times\mathbf{k})\\ &\quad + a_2b_1(\mathbf{j}\times\mathbf{i})+a_2b_2(\mathbf{j}\times\mathbf{j})+a_2b_3(\mathbf{j}\times\mathbf{k})\\ &\quad + a_3b_1(\mathbf{k}\times\mathbf{i})+a_3b_2(\mathbf{k}\times\mathbf{j})+a_3b_3(\mathbf{k}\times\mathbf{k})\end{aligned}\tag{5}$$

With the results in (2) and (4), (5) simplifies to

$$\mathbf{a}\times\mathbf{b}=(a_2b_3-a_3b_2)\mathbf{i}-(a_1b_3-a_3b_1)\mathbf{j}+(a_1b_2-a_2b_1)\mathbf{k}\tag{6}$$

By taking a quick glance at (*a*) of the introductory review, we note that the components of the vector in (6) can be written as determinants of order 2:

$$\mathbf{a}\times\mathbf{b}=\begin{vmatrix}a_2 & a_3\\ b_2 & b_3\end{vmatrix}\mathbf{i}-\begin{vmatrix}a_1 & a_3\\ b_1 & b_3\end{vmatrix}\mathbf{j}+\begin{vmatrix}a_1 & a_2\\ b_1 & b_2\end{vmatrix}\mathbf{k}\tag{7}$$

In turn, inspection of (*b*) of the review suggests that (7) can be written as

$$\mathbf{a}\times\mathbf{b}=\begin{vmatrix}\mathbf{i} & \mathbf{j} & \mathbf{k}\\ a_1 & a_2 & a_3\\ b_1 & b_2 & b_3\end{vmatrix}\tag{8}$$

The expression on the right-hand side of the equality in (8) is not an actual determinant, since its entries are not all scalars; (8) is simply a way of remembering the awkward expression in (6).

EXAMPLE 4 Let $\mathbf{a}=4\mathbf{i}-2\mathbf{j}+5\mathbf{k}$ and $\mathbf{b}=3\mathbf{i}+\mathbf{j}-\mathbf{k}$. Find $\mathbf{a}\times\mathbf{b}$.

Solution From (8) we have

$$\begin{aligned}\mathbf{a}\times\mathbf{b}&=\begin{vmatrix}\mathbf{i} & \mathbf{j} & \mathbf{k}\\ 4 & -2 & 5\\ 3 & 1 & -1\end{vmatrix}=\begin{vmatrix}-2 & 5\\ 1 & -1\end{vmatrix}\mathbf{i}-\begin{vmatrix}4 & 5\\ 3 & -1\end{vmatrix}\mathbf{j}+\begin{vmatrix}4 & -2\\ 3 & 1\end{vmatrix}\mathbf{k}\\ &=-3\mathbf{i}+19\mathbf{j}+10\mathbf{k}\end{aligned}$$

□

Special Products The so-called **triple scalar product** of vectors $\mathbf{a}$, $\mathbf{b}$, and $\mathbf{c}$ is

$$\mathbf{a}\cdot(\mathbf{b}\times\mathbf{c})$$

Now,

$$\mathbf{a}\cdot(\mathbf{b}\times\mathbf{c}) = (a_1\mathbf{i} + a_2\mathbf{j} + a_3\mathbf{k})\cdot\left[\begin{vmatrix} b_2 & b_3 \\ c_2 & c_3 \end{vmatrix}\mathbf{i} - \begin{vmatrix} b_1 & b_3 \\ c_1 & c_3 \end{vmatrix}\mathbf{j} + \begin{vmatrix} b_1 & b_2 \\ c_1 & c_2 \end{vmatrix}\mathbf{k}\right]$$

$$= a_1\begin{vmatrix} b_2 & b_3 \\ c_2 & c_3 \end{vmatrix} - a_2\begin{vmatrix} b_1 & b_3 \\ c_1 & c_3 \end{vmatrix} + a_3\begin{vmatrix} b_1 & b_2 \\ c_1 & c_2 \end{vmatrix}$$

Thus, we see that

$$\mathbf{a}\cdot(\mathbf{b}\times\mathbf{c}) = \begin{vmatrix} a_1 & a_2 & a_3 \\ b_1 & b_2 & b_3 \\ c_1 & c_2 & c_3 \end{vmatrix} \tag{9}$$

Furthermore, from the properties of determinants, we also have

$$\mathbf{a}\cdot(\mathbf{b}\times\mathbf{c}) = (\mathbf{a}\times\mathbf{b})\cdot\mathbf{c}$$

The **triple vector product** of three vectors **a**, **b**, and **c** is

$$\mathbf{a}\times(\mathbf{b}\times\mathbf{c})$$

It is left as an exercise to show that

$$\mathbf{a}\times(\mathbf{b}\times\mathbf{c}) = (\mathbf{a}\cdot\mathbf{c})\mathbf{b} - (\mathbf{a}\cdot\mathbf{b})\mathbf{c} \tag{10}$$

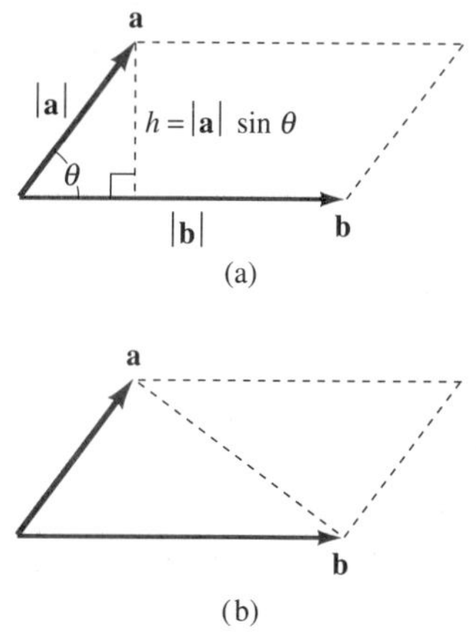

FIGURE 13.53

Areas and Volume Two nonzero and nonparallel vectors **a** and **b** can be considered to be the sides of a parallelogram. The **area A of a parallelogram** is

$$A = (\text{base})(\text{altitude})$$

From Figure 13.53(a), we see that

$$A = |\mathbf{b}|(|\mathbf{a}|\sin\theta) = |\mathbf{a}||\mathbf{b}|\sin\theta$$

or

$$A = |\mathbf{a}\times\mathbf{b}| \tag{11}$$

Likewise from Figure 13.53(b), we see that the **area of a triangle** with sides **a** and **b** is

$$A = \frac{1}{2}|\mathbf{a}\times\mathbf{b}| \tag{12}$$

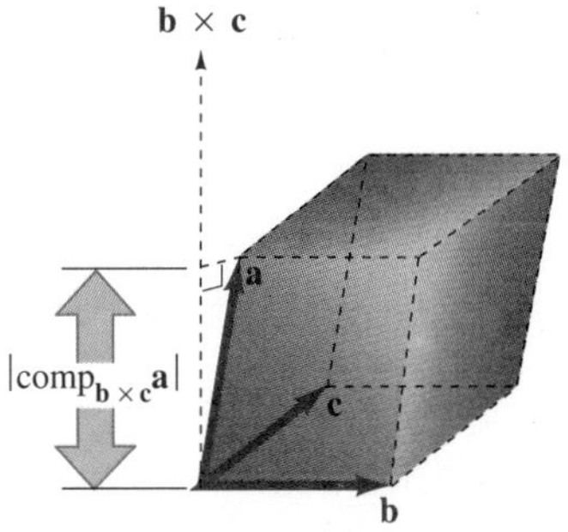

FIGURE 13.54

Similarly, if the vectors **a**, **b**, and **c** do not lie in the same plane, then the **volume of the parallelepiped** with edges **a**, **b**, and **c** shown in Figure 13.54 is

$$V = (\text{area of base})(\text{height})$$

$$= |\mathbf{b}\times\mathbf{c}||\text{comp}_{\mathbf{b}\times\mathbf{c}}\mathbf{a}|$$

$$= |\mathbf{b}\times\mathbf{c}|\left|\mathbf{a}\cdot\left(\frac{1}{|\mathbf{b}\times\mathbf{c}|}\,\mathbf{b}\times\mathbf{c}\right)\right|$$

or

$$V = |\mathbf{a}\cdot(\mathbf{b}\times\mathbf{c})| \tag{13}$$

Because of this last result, the triple scalar product is often referred to as the **box product** of **a**, **b**, and **c**.

EXAMPLE 5 Find the area of the triangle determined by the points $P_1(1, 1, 1)$, $P_2(2, 3, 4)$, and $P_3(3, 0, -1)$.

Solution The vectors $\overrightarrow{P_1P_2}$ and $\overrightarrow{P_2P_3}$ can be taken as two sides of the triangle. Since

$$\overrightarrow{P_1P_2} = \mathbf{i} + 2\mathbf{j} + 3\mathbf{k} \quad \text{and} \quad \overrightarrow{P_2P_3} = \mathbf{i} - 3\mathbf{j} - 5\mathbf{k}$$

we have

$$\overrightarrow{P_1P_2} \times \overrightarrow{P_2P_3} = \begin{vmatrix} \mathbf{i} & \mathbf{j} & \mathbf{k} \\ 1 & 2 & 3 \\ 1 & -3 & -5 \end{vmatrix} = \begin{vmatrix} 2 & 3 \\ -3 & -5 \end{vmatrix}\mathbf{i} - \begin{vmatrix} 1 & 3 \\ 1 & -5 \end{vmatrix}\mathbf{j} + \begin{vmatrix} 1 & 2 \\ 1 & -3 \end{vmatrix}\mathbf{k}$$

$$= -\mathbf{i} + 8\mathbf{j} - 5\mathbf{k}$$

From (12) we see that the area is

$$A = \frac{1}{2}|-\mathbf{i} + 8\mathbf{j} - 5\mathbf{k}| = \frac{3}{2}\sqrt{10} \text{ square units}$$ □

Coplanar Vectors Vectors that lie in the same plane are said to be **coplanar**. We have just seen that if the vectors **a**, **b**, and **c** are not coplanar, then necessarily $\mathbf{a} \cdot (\mathbf{b} \times \mathbf{c}) \neq 0$, since the volume of a parallelepiped with edges **a**, **b**, and **c** has nonzero volume. Equivalently stated, this means that if $\mathbf{a} \cdot (\mathbf{b} \times \mathbf{c}) = 0$, then the vectors **a**, **b**, and **c** are coplanar. Since the converse of this last statement is also true, we have

$$\mathbf{a} \cdot (\mathbf{b} \times \mathbf{c}) = 0 \textit{ if and only if } \mathbf{a}, \mathbf{b}, \textit{and } \mathbf{c} \textit{ are coplanar}$$

▲ ***Remark*** When working with vectors, one should be careful not to mix the symbols · and × with the symbols for ordinary multiplication, and to be especially careful in the use, or lack of use, of parentheses. For example, expressions such as

$$\mathbf{a} \times \mathbf{b} \times \mathbf{c} \qquad \mathbf{a} \cdot \mathbf{b} \times \mathbf{c} \qquad \mathbf{a} \cdot \mathbf{b} \cdot \mathbf{c} \qquad \mathbf{a} \cdot \mathbf{bc}$$

are not meaningful or well-defined.

EXERCISES 13.4

Answers to odd-numbered problems begin on page A-99.

In Problems 1–10 find $\mathbf{a} \times \mathbf{b}$.

1. $\mathbf{a} = \mathbf{i} - \mathbf{j}$, $\mathbf{b} = 3\mathbf{j} + 5\mathbf{k}$
2. $\mathbf{a} = 2\mathbf{i} + \mathbf{j}$, $\mathbf{b} = 4\mathbf{i} - \mathbf{k}$
3. $\mathbf{a} = \langle 1, -3, 1\rangle$, $\mathbf{b} = \langle 2, 0, 4\rangle$
4. $\mathbf{a} = \langle 1, 1, 1\rangle$, $\mathbf{b} = \langle -5, 2, 3\rangle$
5. $\mathbf{a} = 2\mathbf{i} - \mathbf{j} + 2\mathbf{k}$, $\mathbf{b} = -\mathbf{i} + 3\mathbf{j} - \mathbf{k}$
6. $\mathbf{a} = 4\mathbf{i} + \mathbf{j} - 5\mathbf{k}$, $\mathbf{b} = 2\mathbf{i} + 3\mathbf{j} - \mathbf{k}$
7. $\mathbf{a} = \langle \frac{1}{2}, 0, \frac{1}{2}\rangle$, $\mathbf{b} = \langle 4, 6, 0\rangle$
8. $\mathbf{a} = \langle 0, 5, 0\rangle$, $\mathbf{b} = \langle 2, -3, 4\rangle$
9. $\mathbf{a} = \langle 2, 2, -4\rangle$, $\mathbf{b} = \langle -3, -3, 6\rangle$

10. $\mathbf{a} = \langle 8, 1, -6 \rangle$, $\mathbf{b} = \langle 1, -2, 10 \rangle$

In Problems 11 and 12 find $\overrightarrow{P_1P_2} \times \overrightarrow{P_1P_3}$.

11. $P_1(2, 1, 3)$, $P_2(0, 3, -1)$, $P_3(-1, 2, 4)$

12. $P_1(0, 0, 1)$, $P_2(0, 1, 2)$, $P_3(1, 2, 3)$

In Problems 13 and 14 find a nonzero vector that is perpendicular to both **a** and **b**.

13. $\mathbf{a} = 2\mathbf{i} + 7\mathbf{j} - 4\mathbf{k}$, $\mathbf{b} = \mathbf{i} + \mathbf{j} - \mathbf{k}$

14. $\mathbf{a} = \langle -1, -2, 4 \rangle$, $\mathbf{b} = \langle 4, -1, 0 \rangle$

In Problems 15 and 16 verify that $\mathbf{a} \cdot (\mathbf{a} \times \mathbf{b}) = 0$ and $\mathbf{b} \cdot (\mathbf{a} \times \mathbf{b}) = 0$.

15. $\mathbf{a} = \langle 5, -2, 1 \rangle$, $\mathbf{b} = \langle 2, 0, -7 \rangle$

16. $\mathbf{a} = \frac{1}{2}\mathbf{i} - \frac{1}{4}\mathbf{j}$, $\mathbf{b} = 2\mathbf{i} - 2\mathbf{j} + 6\mathbf{k}$

In Problems 17 and 18 (**a**) calculate $\mathbf{b} \times \mathbf{c}$ followed by $\mathbf{a} \times (\mathbf{b} \times \mathbf{c})$. (**b**) Verify the results in part (**a**) by (10) of this section.

17. $\mathbf{a} = \mathbf{i} - \mathbf{j} + 2\mathbf{k}$
$\mathbf{b} = 2\mathbf{i} + \mathbf{j} + \mathbf{k}$
$\mathbf{c} = 3\mathbf{i} + \mathbf{j} + \mathbf{k}$

18. $\mathbf{a} = 3\mathbf{i} - 4\mathbf{k}$
$\mathbf{b} = \mathbf{i} + 2\mathbf{j} - \mathbf{k}$
$\mathbf{c} = -\mathbf{i} + 5\mathbf{j} + 8\mathbf{k}$

In Problems 19–36 find the indicated scalar or vector *without* using (8), (9), or (10).

19. $(2\mathbf{i}) \times \mathbf{j}$

20. $\mathbf{i} \times (-3\mathbf{k})$

21. $\mathbf{k} \times (2\mathbf{i} - \mathbf{j})$

22. $\mathbf{i} \times (\mathbf{j} \times \mathbf{k})$

23. $[(2\mathbf{k}) \times (3\mathbf{j})] \times (4\mathbf{j})$

24. $(2\mathbf{i} - \mathbf{j} + 5\mathbf{k}) \times \mathbf{i}$

25. $(\mathbf{i} + \mathbf{j}) \times (\mathbf{i} + 5\mathbf{k})$

26. $\mathbf{i} \times \mathbf{k} - 2(\mathbf{j} \times \mathbf{i})$

27. $\mathbf{k} \cdot (\mathbf{j} \times \mathbf{k})$

28. $\mathbf{i} \cdot [\mathbf{j} \times (-\mathbf{k})]$

29. $|4\mathbf{j} - 5(\mathbf{i} \times \mathbf{j})|$

30. $(\mathbf{i} \times \mathbf{j}) \cdot (3\mathbf{j} \times \mathbf{i})$

31. $\mathbf{i} \times (\mathbf{i} \times \mathbf{j})$

32. $(\mathbf{i} \times \mathbf{j}) \times \mathbf{i}$

33. $(\mathbf{i} \times \mathbf{i}) \times \mathbf{j}$

34. $(\mathbf{i} \cdot \mathbf{i})(\mathbf{i} \times \mathbf{j})$

35. $2\mathbf{j} \cdot [\mathbf{i} \times (\mathbf{j} - 3\mathbf{k})]$

36. $(\mathbf{i} \times \mathbf{k}) \times (\mathbf{j} \times \mathbf{i})$

In Problems 37–44, $\mathbf{a} \times \mathbf{b} = 4\mathbf{i} - 3\mathbf{j} + 6\mathbf{k}$ and $\mathbf{c} = 2\mathbf{i} + 4\mathbf{j} - \mathbf{k}$. Find the indicated scalar or vector.

37. $\mathbf{a} \times (3\mathbf{b})$

38. $\mathbf{b} \times \mathbf{a}$

39. $(-\mathbf{a}) \times \mathbf{b}$

40. $|\mathbf{a} \times \mathbf{b}|$

41. $(\mathbf{a} \times \mathbf{b}) \times \mathbf{c}$

42. $(\mathbf{a} \times \mathbf{b}) \cdot \mathbf{c}$

43. $\mathbf{a} \cdot (\mathbf{b} \times \mathbf{c})$

44. $(4\mathbf{a}) \cdot (\mathbf{b} \times \mathbf{c})$

In Problems 45 and 46 (**a**) verify that the given quadrilateral is a parallelogram and (**b**) find the area of the parallelogram.

45.

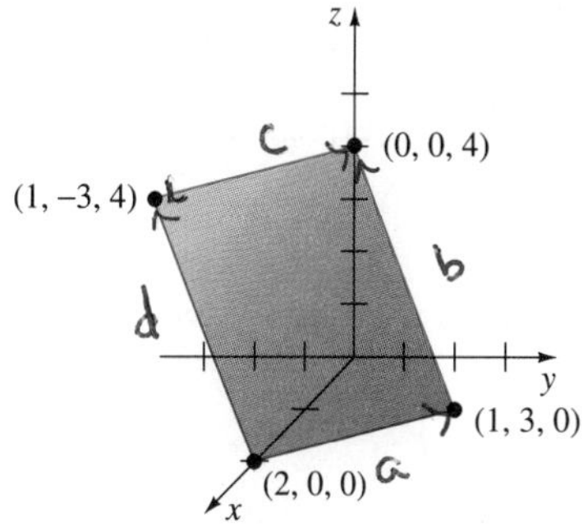

FIGURE 13.55

46.

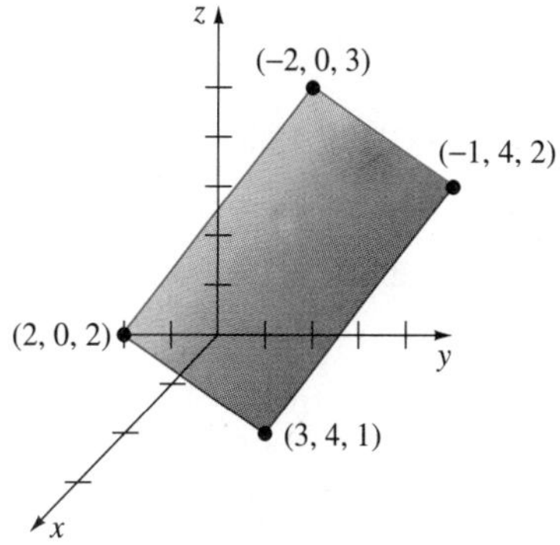

FIGURE 13.56

In Problems 47–50 find the area of the triangle determined by the given points.

47. $P_1(1, 1, 1)$, $P_2(1, 2, 1)$, $P_3(1, 1, 2)$

48. $P_1(0, 0, 0)$, $P_2(0, 1, 2)$, $P_3(2, 2, 0)$

49. $P_1(1, 2, 4)$, $P_2(1, -1, 3)$, $P_3(-1, -1, 2)$

50. $P_1(1, 0, 3)$, $P_2(0, 0, 6)$, $P_3(2, 4, 5)$

In Problems 51 and 52 find the volume of the parallelepiped for which the given vectors are three edges.

51. $\mathbf{a} = \mathbf{i} + \mathbf{j}$, $\mathbf{b} = -\mathbf{i} + 4\mathbf{j}$, $\mathbf{c} = 2\mathbf{i} + 2\mathbf{j} + 2\mathbf{k}$

52. $\mathbf{a} = 3\mathbf{i} + \mathbf{j} + \mathbf{k}$, $\mathbf{b} = \mathbf{i} + 4\mathbf{j} + \mathbf{k}$, $\mathbf{c} = \mathbf{i} + \mathbf{j} + 5\mathbf{k}$

53. Determine whether the vectors $\mathbf{a} = 4\mathbf{i} + 6\mathbf{j}$, $\mathbf{b} = -2\mathbf{i} + 6\mathbf{j} - 6\mathbf{k}$, and $\mathbf{c} = \frac{5}{2}\mathbf{i} + 3\mathbf{j} + \frac{1}{2}\mathbf{k}$ are coplanar.

54. Determine whether the four points $P_1(1, 1, -2)$, $P_2(4, 0, -3)$, $P_3(1, -5, 10)$, and $P_4(-7, 2, 4)$ lie in the same plane.

55. As shown in Figure 13.57, the vector **a** lies in the xy-plane and the vector **b** lies along the positive z-axis. Their magnitudes are $|\mathbf{a}| = 6.4$ and $|\mathbf{b}| = 5$.

(a) Use Definition 13.4 to find $|\mathbf{a} \times \mathbf{b}|$.

(b) Use the right-hand rule to find the direction of $\mathbf{a} \times \mathbf{b}$.

(c) Use part (b) to express $\mathbf{a} \times \mathbf{b}$ in terms of the unit vectors $\mathbf{i}, \mathbf{j}, \mathbf{k}$.

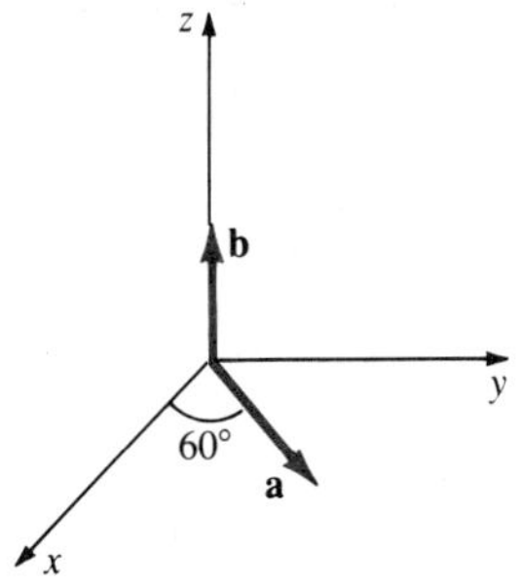

FIGURE 13.57

56. Two vectors $\mathbf{a}$ and $\mathbf{b}$ lie in the xz-plane so that the angle between them is 120°. If $|\mathbf{a}| = \sqrt{27}$ and $|\mathbf{b}| = 8$, find all possible values of $\mathbf{a} \times \mathbf{b}$.

57. Prove or disprove $\mathbf{a} \times (\mathbf{b} \times \mathbf{c}) = (\mathbf{a} \times \mathbf{b}) \times \mathbf{c}$.

58. Prove $\mathbf{a} \times (\mathbf{b} \times \mathbf{c}) = (\mathbf{a} \cdot \mathbf{c})\mathbf{b} - (\mathbf{a} \cdot \mathbf{b})\mathbf{c}$.

59. Prove $\mathbf{a} \cdot (\mathbf{b} \times \mathbf{c}) = (\mathbf{a} \times \mathbf{b}) \cdot \mathbf{c}$.

60. Prove $\mathbf{a} \times (\mathbf{b} \times \mathbf{c}) + \mathbf{b} \times (\mathbf{c} \times \mathbf{a}) + \mathbf{c} \times (\mathbf{a} \times \mathbf{b}) = \mathbf{0}$.

61. Some calculus texts use the expression in (6) as the definition of the cross product $\mathbf{a} \times \mathbf{b}$ and (4) of Section 13.3 as the definition of the dot product $\mathbf{a} \cdot \mathbf{b}$. Use (6) and (4) of Section 13.3 to prove **Lagrange's identity**:

$$|\mathbf{a} \times \mathbf{b}|^2 = |\mathbf{a}|^2|\mathbf{b}|^2 - (\mathbf{a} \cdot \mathbf{b})^2$$

62. Use Lagrange's identity in Problem 61 to prove that

$$|\mathbf{a} \times \mathbf{b}| = |\mathbf{a}||\mathbf{b}|\sin\theta$$

where $0 \le \theta \le \pi$.

63. A three-dimensional lattice is a collection of integer combinations of three noncoplanar basis vectors $\mathbf{a}$, $\mathbf{b}$, and $\mathbf{c}$. In crystallography, a lattice can specify the locations of atoms in a crystal. X-ray diffraction studies of crystals use the "reciprocal lattice," which has basis

$$\mathbf{A} = \frac{\mathbf{b} \times \mathbf{c}}{\mathbf{a} \cdot (\mathbf{b} \times \mathbf{c})} \qquad \mathbf{B} = \frac{\mathbf{c} \times \mathbf{a}}{\mathbf{b} \cdot (\mathbf{c} \times \mathbf{a})} \qquad \mathbf{C} = \frac{\mathbf{a} \times \mathbf{b}}{\mathbf{c} \cdot (\mathbf{a} \times \mathbf{b})}$$

(a) A certain lattice has basis vectors $\mathbf{a} = \mathbf{i}$, $\mathbf{b} = \mathbf{j}$, and $\mathbf{c} = \frac{1}{2}(\mathbf{i} + \mathbf{j} + \mathbf{k})$. Find basis vectors for the reciprocal lattice.

(b) The unit cell of the reciprocal lattice is the parallelepiped with edges $\mathbf{A}$, $\mathbf{B}$, and $\mathbf{C}$, while the unit cell of the original lattice is the parallelepiped with edges $\mathbf{a}$, $\mathbf{b}$, and $\mathbf{c}$. Show that the volume of the unit cell of the reciprocal lattice is the reciprocal of the volume of the unit cell of the original lattice. [*Hint:* Start with $\mathbf{B} \times \mathbf{C}$ and use (10).]

13.5 LINES IN 3-SPACE

Vector Equation As in the plane, any two distinct points in 3-space determine only one line between them. To find an equation of the line through $P_1(x_1, y_1, z_1)$ and $P_2(x_2, y_2, z_2)$ let us assume that $P(x, y, z)$ is *any* point on the line. From Figure 13.58, if $\mathbf{r} = \overrightarrow{OP}$, $\mathbf{r}_1 = \overrightarrow{OP_1}$, and $\mathbf{r}_2 = \overrightarrow{OP_2}$, we see that vector $\mathbf{a} = \mathbf{r}_2 - \mathbf{r}_1$ is parallel to vector $\mathbf{r} - \mathbf{r}_2$. Thus,

$$\mathbf{r} - \mathbf{r}_2 = t(\mathbf{r}_2 - \mathbf{r}_1) \tag{1}$$

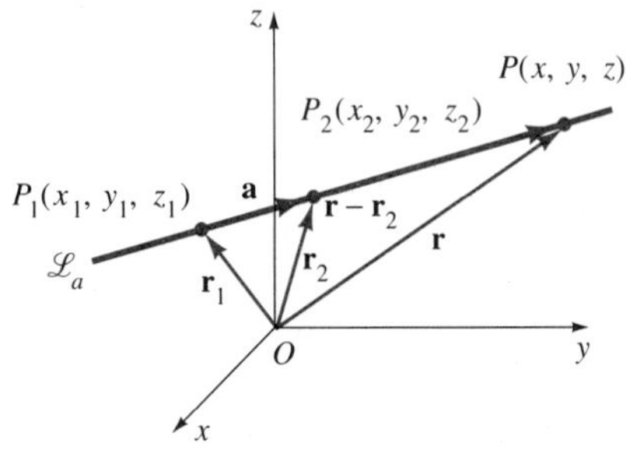

FIGURE 13.58

If we write

$$\mathbf{a} = \mathbf{r}_2 - \mathbf{r}_1 = \langle x_2 - x_1, y_2 - y_1, z_2 - z_1 \rangle = \langle a_1, a_2, a_3 \rangle$$

then (1) implies that a **vector equation** for the line $\mathscr{L}_a$ is

$$\mathbf{r} = \mathbf{r}_2 + t\mathbf{a} \tag{2}$$

The vector $\mathbf{a}$ is called a **direction vector** of the line.

Note: Since $\mathbf{r} - \mathbf{r}_1$ is also parallel to $\mathscr{L}_a$, an alternative vector equation for the line is $\mathbf{r} = \mathbf{r}_1 + t\mathbf{a}$. Indeed, $\mathbf{r} = \mathbf{r}_1 + t(-\mathbf{a})$ and $\mathbf{r} = \mathbf{r}_1 + t(k\mathbf{a})$, k a nonzero scalar, are also equations for $\mathscr{L}_a$.

EXAMPLE 1 Find a vector equation for the line through $(2, -1, 8)$ and $(5, 6, -3)$.

Solution Define

$$\mathbf{a} = \langle 2 - 5, -1 - 6, 8 - (-3)\rangle = \langle -3, -7, 11\rangle$$

The following are three of many possible vector equations for the line:

$$\langle x, y, z\rangle = \langle 2, -1, 8\rangle + t\langle -3, -7, 11\rangle \tag{3}$$
$$\langle x, y, z\rangle = \langle 5, 6, -3\rangle + t\langle -3, -7, 11\rangle \tag{4}$$
$$\langle x, y, z\rangle = \langle 5, 6, -3\rangle + t\langle 3, 7, -11\rangle \tag{5}$$

□

Parametric Equations By writing (2) as

$$\begin{aligned}\langle x, y, z\rangle &= \langle x_2 + t(x_2 - x_1), y_2 + t(y_2 - y_1), z_2 + t(z_2 - z_1)\rangle \\ &= \langle x_2 + a_1t, y_2 + a_2t, z_2 + a_3t\rangle\end{aligned}$$

and equating components, we obtain

$$x = x_2 + a_1t \qquad y = y_2 + a_2t \qquad z = z_2 + a_3t \tag{6}$$

The equations in (6) are called **parametric equations** for the line through P_1 and P_2. As the parameter t increases from $-\infty$ to ∞, we can think of the point $P(x, y, z)$ tracing out the entire line. If the parameter t is restricted to a closed interval $[t_0, t_1]$, then $P(x, y, z)$ traces out a **line segment** starting at the point corresponding to t_0 and ending at the point corresponding to t_1. For example, in Figure 13.58 if $-1 \le t \le 0$, then $P(x, y, z)$ traces out the line segment starting at $P_1(x_1, y_1, z_1)$ and ending at $P_2(x_2, y_2, z_2)$.

EXAMPLE 2 Find parametric equations for the line in Example 1.

Solution From (3) it follows that

$$x = 2 - 3t \qquad y = -1 - 7t \qquad z = 8 + 11t \tag{7}$$

An alternative set of parametric equations can be obtained from (5):

$$x = 5 + 3t \qquad y = 6 + 7t \qquad z = -3 - 11t \tag{8}$$

□

Note in Example 2 that the value $t = 0$ in (7) gives $(2, -1, 8)$, whereas in (8), $t = -1$ must be used to obtain the same point.

EXAMPLE 3 Find a vector **a** that is parallel to the line $\mathscr{L}_a$ whose parametric equations are

$$x = 4 + 9t \qquad y = -14 + 5t \qquad z = 1 - 3t$$

Solution The coefficients (or a nonzero constant multiple of the coefficients) of the parameter in each equation are the components of a vector that is parallel to the line. Thus, $\mathbf{a} = 9\mathbf{i} + 5\mathbf{j} - 3\mathbf{k}$ is parallel to $\mathscr{L}_a$ and, hence, is a direction vector of the line. □

Symmetric Equations From (6) observe that we can eliminate the parameter by writing

$$t = \frac{x - x_2}{a_1} = \frac{y - y_2}{a_2} = \frac{z - z_2}{a_3}$$

provided that the three numbers a_1, a_2, and a_3 are nonzero. The resulting equations

$$\frac{x - x_2}{a_1} = \frac{y - y_2}{a_2} = \frac{z - z_2}{a_3} \tag{9}$$

are said to be **symmetric equations** for the line through P_1 and P_2.

EXAMPLE 4 Find symmetric equations for the line through (4, 10, −6) and (7, 9, 2).

Solution Define $a_1 = 7 - 4 = 3$, $a_2 = 9 - 10 = -1$, and $a_3 = 2 - (-6) = 8$. It follows from (9) that symmetric equations for the line are

$$\frac{x - 7}{3} = \frac{y - 9}{-1} = \frac{z - 2}{8}$$ □

If one of the numbers a_1, a_2, or a_3 is zero in (6), we use the remaining two equations to eliminate the parameter t. For example, if $a_1 = 0, a_2 \neq 0, a_3 \neq 0$, then (6) yields

$$x = x_2 \quad \text{and} \quad t = \frac{y - y_2}{a_2} = \frac{z - z_2}{a_3}$$

In this case,

$$x = x_2, \qquad \frac{y - y_2}{a_2} = \frac{z - z_2}{a_3}$$

are symmetric equations for the line.

EXAMPLE 5 Find symmetric equations for the line through (5, 3, 1) and (2, 1, 1).

Solution Define $a_1 = 5 - 2 = 3$, $a_2 = 3 - 1 = 2$, and $a_3 = 1 - 1 = 0$. From the preceding discussion it follows that symmetric equations for the line are

$$\frac{x-5}{3} = \frac{y-3}{2}, \qquad z = 1$$

In other words, the symmetric equations describe a line in the plane $z = 1$. □

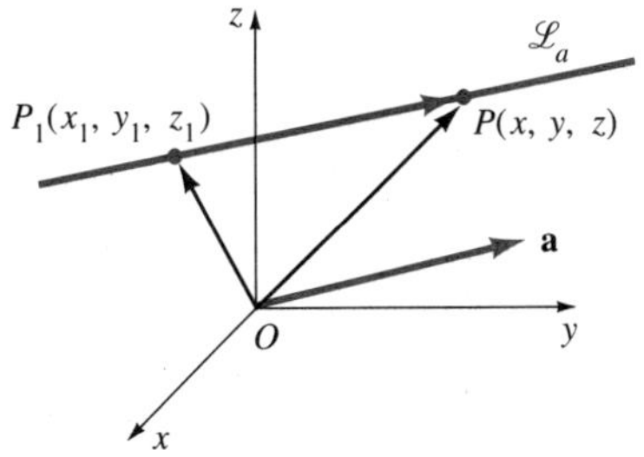

FIGURE 13.59

A line in space is also determined by specifying a point $P_1(x_1, y_1, z_1)$ and a nonzero direction vector **a**. Through the point P_1 there passes only one line $\mathscr{L}_a$ parallel to the given vector. If $P(x, y, z)$ is a point on the line $\mathscr{L}_a$, shown in Figure 13.59, then, as before,

$$\overrightarrow{OP} - \overrightarrow{OP_1} = t\mathbf{a}$$

or

$$\mathbf{r} = \mathbf{r}_1 + t\mathbf{a}$$

EXAMPLE 6 Write vector, parametric, and symmetric equations for the line through $(4, 6, -3)$ and parallel to $\mathbf{a} = 5\mathbf{i} - 10\mathbf{j} + 2\mathbf{k}$.

Solution With $a_1 = 5$, $a_2 = -10$, and $a_3 = 2$ we have immediately

$$\textit{vector:} \quad \langle x, y, z\rangle = \langle 4, 6, -3\rangle + t\langle 5, -10, 2\rangle$$

$$\textit{parametric:} \quad x = 4 + 5t, \qquad y = 6 - 10t, \qquad z = -3 + 2t$$

$$\textit{symmetric:} \quad \frac{x-4}{5} = \frac{y-6}{-10} = \frac{z+3}{2}$$

□

Orthogonal and Parallel Lines Let **a** and **b** be direction vectors for lines $\mathscr{L}_a$ and $\mathscr{L}_b$, respectively.

DEFINITION 13.5 Orthogonal and Parallel Lines

(*i*) $\mathscr{L}_a$ and $\mathscr{L}_b$ are **orthogonal** if $\mathbf{a} \cdot \mathbf{b} = 0$, and

(*ii*) $\mathscr{L}_a$ and $\mathscr{L}_b$ are **parallel** if $\mathbf{a} = k\mathbf{b}$ for some nonzero scalar k.

EXAMPLE 7 The lines

$$\mathscr{L}_a\colon x = 4 - 2t, \qquad y = 1 + 4t, \qquad z = 3 + 10t$$

$$\mathscr{L}_b\colon x = s, \qquad y = 6 - 2s, \qquad z = \frac{1}{2} - 5s$$

are parallel, since $\mathbf{a} = -2\mathbf{b}$ (or $\mathbf{b} = -\frac{1}{2}\mathbf{a}$), where

$$\mathbf{a} = -2\mathbf{i} + 4\mathbf{j} + 10\mathbf{k} \quad \text{and} \quad \mathbf{b} = \mathbf{i} - 2\mathbf{j} - 5\mathbf{k}$$

□

EXAMPLE 8 Determine whether the lines

$$\mathscr{L}_a\colon x = -6 - t, \qquad y = 20 + 3t, \qquad z = 1 + 2t$$
$$\mathscr{L}_b\colon x = 5 + 2s, \qquad y = -9 - 4s, \qquad z = 1 + 7s$$

are orthogonal.

Solution Reading off the coefficients of the parameters, we see that

$$\mathbf{a} = -\mathbf{i} + 3\mathbf{j} + 2\mathbf{k} \quad \text{and} \quad \mathbf{b} = 2\mathbf{i} - 4\mathbf{j} + 7\mathbf{k}$$

are the direction vectors for $\mathscr{L}_a$ and $\mathscr{L}_b$, respectively. Since $\mathbf{a} \cdot \mathbf{b} = -2 - 12 + 14 = 0$, the lines are orthogonal. □

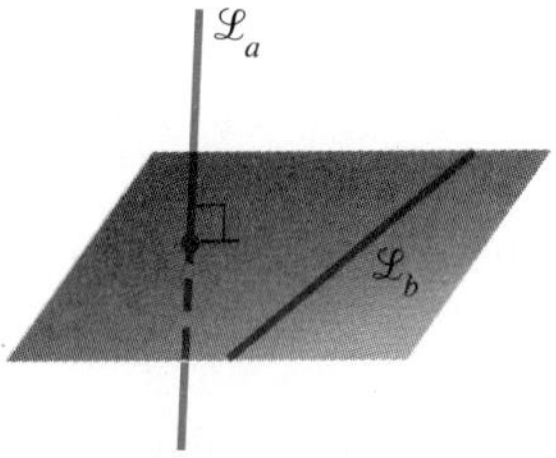

FIGURE 13.60

Notice that (i) of Definition 13.5 does not demand that the two lines intersect in order to be orthogonal. Figure 13.60 shows two orthogonal lines that do not intersect. In other words, $\mathscr{L}_a$ can be perpendicular to the plane containing $\mathscr{L}_b$.

EXAMPLE 9 Determine whether the lines $\mathscr{L}_a$ and $\mathscr{L}_b$ in Example 8 intersect.

Solution Since a point (x, y, z) of intersection is common to both lines, we must have

$$\left.\begin{aligned} -6 - t &= 5 + 2s \\ 20 + 3t &= -9 - 4s \\ 1 + 2t &= 1 + 7s \end{aligned}\right\} \quad \text{or} \quad \left\{\begin{aligned} 2s + t &= -11 \\ 4s + 3t &= -29 \\ -7s + 2t &= 0 \end{aligned}\right. \tag{10}$$

We now solve any *two* of the equations simultaneously and use the remaining equation as a check. Choosing the first and third, we find from

$$\begin{aligned} 2s + t &= -11 \\ -7s + 2t &= 0 \end{aligned}$$

that $s = -2$ and $t = -7$. Substitution of these values in the second equation yields the identity $-8 - 21 = -29$. Thus, $\mathscr{L}_a$ and $\mathscr{L}_b$ intersect. To find the point of intersection, we use, say, $s = -2$:

$$x = 5 + 2(-2) \qquad y = -9 - 4(-2) \qquad z = 1 + 7(-2)$$

or $(1, -1, -13)$. □

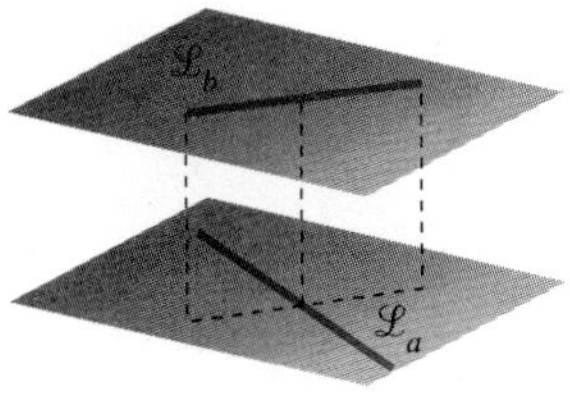

FIGURE 13.61

In Example 9, had the remaining equation not been satisfied when the values $s = -2$ and $t = -7$ were substituted, then the three equations would not be satisfied simultaneously and so the lines would not intersect.

Two lines $\mathscr{L}_a$ and $\mathscr{L}_b$ in 3-space that do not intersect and are not parallel are called **skew lines**. As shown in Figure 13.61, skew lines lie in parallel planes.

EXERCISES 13.5 *Answers to odd-numbered problems begin on page A-99.*

In Problems 1–6 find a vector equation for the line through the given points.

1. $(1, 2, 1), (3, 5, -2)$
2. $(0, 4, 5), (-2, 6, 3)$
3. $(\frac{1}{2}, -\frac{1}{2}, 1), (-\frac{3}{2}, \frac{5}{2}, -\frac{1}{2})$
4. $(10, 2, -10), (5, -3, 5)$
5. $(1, 1, -1), (-4, 1, -1)$
6. $(3, 2, 1), (\frac{5}{2}, 1, -2)$

In Problems 7–12 find parametric equations for the line through the given points.

7. $(2, 3, 5), (6, -1, 8)$
8. $(2, 0, 0), (0, 4, 9)$
9. $(1, 0, 0), (3, -2, -7)$
10. $(0, 0, 5), (-2, 4, 0)$
11. $(4, \frac{1}{2}, \frac{1}{3}), (-6, -\frac{1}{4}, \frac{1}{6})$
12. $(-3, 7, 9), (4, -8, -1)$

In Problems 13–18 find symmetric equations for the line through the given points.

13. $(1, 4, -9), (10, 14, -2)$
14. $(\frac{2}{3}, 0, -\frac{1}{4}), (1, 3, \frac{1}{4})$
15. $(4, 2, 1), (-7, 2, 5)$
16. $(-5, -2, -4), (1, 1, 2)$
17. $(5, 10, -2), (5, 1, -14)$
18. $(\frac{5}{6}, -\frac{1}{4}, \frac{1}{5}), (\frac{1}{3}, \frac{3}{8}, -\frac{1}{10})$

In Problems 19–22 find parametric and symmetric equations for the line through the given point parallel to the given vector.

19. $(4, 6, -7)$, $\mathbf{a} = \langle 3, \frac{1}{2}, -\frac{3}{2} \rangle$
20. $(1, 8, -2)$, $\mathbf{a} = -7\mathbf{i} - 8\mathbf{j}$
21. $(0, 0, 0)$, $\mathbf{a} = 5\mathbf{i} + 9\mathbf{j} + 4\mathbf{k}$
22. $(0, -3, 10)$, $\mathbf{a} = \langle 12, -5, -6 \rangle$

23. Find parametric equations for the line through $(6, 4, -2)$ that is parallel to the line $x/2 = (1 - y)/3 = (z - 5)/6$.

24. Find symmetric equations for the line through $(4, -11, -7)$ that is parallel to the line $x = 2 + 5t$, $y = -1 + t/3$, $z = 9 - 2t$.

25. Find parametric equations for the line through $(2, -2, 15)$ that is parallel to the xz-plane and the xy-plane.

26. Find parametric equations for the line through $(1, 2, 8)$ that is **(a)** parallel to the y-axis and **(b)** perpendicular to the xy-plane.

27. Show that the lines given by $\mathbf{r} = t\langle 1, 1, 1 \rangle$ and $\mathbf{r} = \langle 6, 6, 6 \rangle + t\langle -3, -3, -3 \rangle$ are the same.

28. Determine which of the following lines are orthogonal and which are parallel.

(a) $\mathbf{r} = \langle 1, 0, 2 \rangle + t\langle 9, -12, 6 \rangle$

(b) $x = 1 + 9t$, $y = 12t$, $z = 2 - 6t$

(c) $x = 2t$, $y = -3t$, $z = 4t$

(d) $x = 5 + t$, $y = 4t$, $z = 3 + \frac{5}{2}t$

(e) $x = 1 + t$, $y = \frac{3}{2}t$, $z = 2 - \frac{3}{2}t$

(f) $\dfrac{x + 1}{-3} = \dfrac{y + 6}{4} = \dfrac{z - 3}{-2}$

In Problems 29 and 30 determine the points of intersection of the given line and the three coordinate planes.

29. $x = 4 - 2t$, $y = 1 + 2t$, $z = 9 + 3t$

30. $\dfrac{x - 1}{2} = \dfrac{y + 2}{3} = \dfrac{z - 4}{2}$

In Problems 31–34 determine whether the given lines intersect. If so, find the point of intersection.

31. $x = 4 + t$, $y = 5 + t$, $z = -1 + 2t$
$x = 6 + 2s$, $y = 11 + 4s$, $z = -3 + s$

32. $x = 1 + t$, $y = 2 - t$, $z = 3t$
$x = 2 - s$, $y = 1 + s$, $z = 6s$

33. $x = 2 - t$, $y = 3 + t$, $z = 1 + t$
$x = 4 + s$, $y = 1 + s$, $z = 1 - s$

34. $x = 3 - t$, $y = 2 + t$, $z = 8 + 2t$
$x = 2 + 2s$, $y = -2 + 3s$, $z = -2 + 8s$

In Problems 35 and 36 determine whether the given points lie on the same line.

35. $(4, 3, -5), (10, 15, -11), (-1, -7, 0)$

36. $(1, 6, 6), (-11, 10, -2), (-2, 7, 5)$

37. Find parametric equations for the line segment joining the points $(2, 5, 9)$ and $(6, -1, 3)$.

38. Find parametric equations for the line segment joining the midpoints of the given line segments.

$$x = 1 + 2t,\ y = 2 - t,\ z = 4 - 3t,\ 1 \le t \le 2$$

$$x = -2 + 4t,\ y = 6 + t,\ z = 5 + 6t,\ -1 \le t \le 1$$

The angle between two lines $\mathscr{L}_a$ and $\mathscr{L}_b$ is the angle between their direction vectors **a** and **b**. In Problems 39 and 40 find the angle between the given lines.

39. $x = 4 - t$, $y = 3 + 2t$, $z = -2t$
$x = 5 + 2s$, $y = 1 + 3s$, $z = 5 - 6s$

40. $\dfrac{x - 1}{2} = \dfrac{y + 5}{7} = \dfrac{z - 1}{-1}$; $\dfrac{x + 3}{-2} = y - 9 = \dfrac{z}{4}$

In Problems 41 and 42 the given lines lie in the same plane. Find parametric equations for the line through the indicated point that is perpendicular to this plane:

41. $x = 3 + t,\ y = -2 + t,\ z = 9 + t$
$x = 1 - 2s,\ y = 5 + s,\ z = -2 - 5s$; (4, 1, 6)

42. $\dfrac{x-1}{3} = \dfrac{y+1}{2} = \dfrac{z}{4}$
$\dfrac{x+4}{6} = \dfrac{y-6}{4} = \dfrac{z-10}{8}$; (1, −1, 0)

In Problems 43 and 44 show that the given lines are skew.

43. $x = -3 + t,\ y = 7 + 3t,\ z = 5 + 2t$
$x = 4 + s,\ y = 8 - 2s,\ z = 10 - 4s$

44. $x = 6 + 2t,\ y = 6t,\ z = -8 + 10t$
$x = 7 + 8s,\ y = 4 - 4s,\ z = 3 - 24s$

45. Suppose $\mathscr{L}_a$ and $\mathscr{L}_b$ are skew lines. Let P_1 and P_2 be points on line $\mathscr{L}_a$ and let P_3 and P_4 be points on line $\mathscr{L}_b$. Use the vector $\overrightarrow{P_1P_3}$, shown in Figure 13.62, to show that the shortest distance d between $\mathscr{L}_a$ and $\mathscr{L}_b$ is

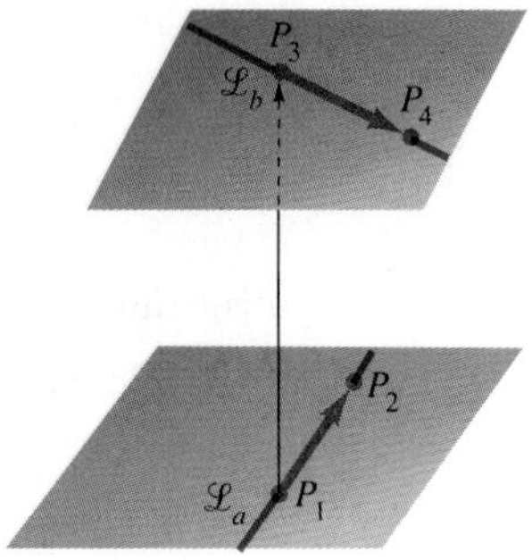

FIGURE 13.62

$$d = \frac{|\overrightarrow{P_1P_3} \cdot (\overrightarrow{P_1P_2} \times \overrightarrow{P_3P_4})|}{|\overrightarrow{P_1P_2} \times \overrightarrow{P_3P_4}|}$$

46. Using the result in Problem 45, find the distance between the skew lines in Problem 43.

13.6 PLANES

Vector Equation Figure 13.63(a) illustrates the fact that through a given point $P_1(x_1, y_1, z_1)$ there pass an infinite number of planes. However, as shown in Figure 13.63(b), if a point P_1 and a vector **n** are specified, there is only *one* plane $\mathscr{P}$ containing P_1 with **n normal**, or perpendicular, to the plane. Moreover, if $P(x, y, z)$ is any point on $\mathscr{P}$, and $\mathbf{r} = OP$, $\mathbf{r} = OP_1$, then, as shown in Figure 13.63(c), $\mathbf{r} - \mathbf{r}_1$ is in the plane. It follows that a **vector equation** of the plane is

$$\mathbf{n} \cdot (\mathbf{r} - \mathbf{r}_1) = 0 \tag{1}$$

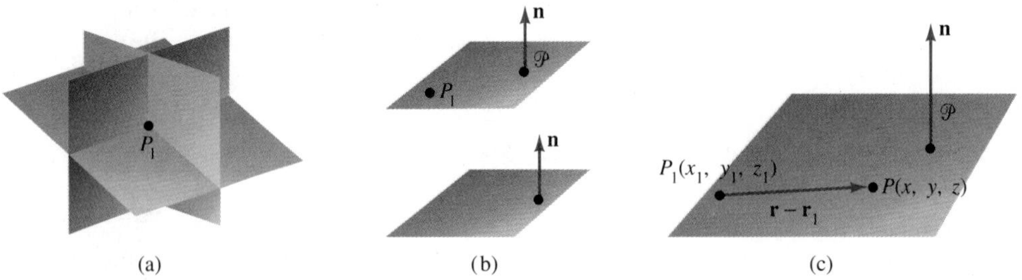

FIGURE 13.63

Cartesian Equation Specifically, if the normal vector is $\mathbf{n} = a\mathbf{i} + b\mathbf{j} + c\mathbf{k}$, then (1) yields a **Cartesian equation** of the plane containing $P_1(x_1, y_1, z_1)$:

$$a(x - x_1) + b(y - y_1) + c(z - z_1) = 0 \tag{2}$$

EXAMPLE 1 Find an equation of the plane that contains the point $(4, -1, 3)$ and is perpendicular to the vector $\mathbf{n} = 2\mathbf{i} + 8\mathbf{j} - 5\mathbf{k}$.

Solution It follows immediately from (2) that

$$2(x - 4) + 8(y + 1) - 5(z - 3) = 0 \quad \text{or} \quad 2x + 8y - 5z + 15 = 0 \qquad \square$$

The equation in (2) can always be written as $ax + by + cz + d = 0$ by identifying $d = -ax_1 - by_1 - cz_1$. Conversely, we shall now prove that any linear equation

$$ax + by + cz + d = 0, \qquad a, b, c \text{ not all zero} \tag{3}$$

is a plane.

> **THEOREM 13.5 Plane and Normal Vector**
>
> The graph of any equation $ax + by + cz + d = 0$, a, b, c not all zero, is a plane with the normal vector $\mathbf{n} = a\mathbf{i} + b\mathbf{j} + c\mathbf{k}$.

Proof Suppose x_0, y_0, and z_0 are numbers that satisfy the given equation. Then, $ax_0 + by_0 + cz_0 + d = 0$ implies that $d = -ax_0 - by_0 - cz_0$. Replacing this latter value of d in the original equation gives, after simplifying,

$$a(x - x_0) + b(y - y_0) + c(z - z_0) = 0$$

or, in terms of vectors,

$$[a\mathbf{i} + b\mathbf{j} + c\mathbf{k}] \cdot [(x - x_0)\mathbf{i} + (y - y_0)\mathbf{j} + (z - z_0)\mathbf{k}] = 0$$

This last equation implies that $a\mathbf{i} + b\mathbf{j} + c\mathbf{k}$ is normal to the plane containing the point (x_0, y_0, z_0) and the vector

$$(x - x_0)\mathbf{i} + (y - y_0)\mathbf{j} + (z - z_0)\mathbf{k} \qquad \blacksquare$$

EXAMPLE 2 A vector normal to the plane $3x - 4y + 10z - 8 = 0$ is $\mathbf{n} = 3\mathbf{i} - 4\mathbf{j} + 10\mathbf{k}$. $\square$

Of course, a nonzero scalar multiple of a normal vector is still perpendicular to the plane.

Three noncollinear points P_1, P_2, and P_3 also determine a plane.* To obtain an equation of the plane, we need only form two vectors between two

---

*If you have ever sat at a four-legged table that rocks, you might consider replacing it with a three-legged table.

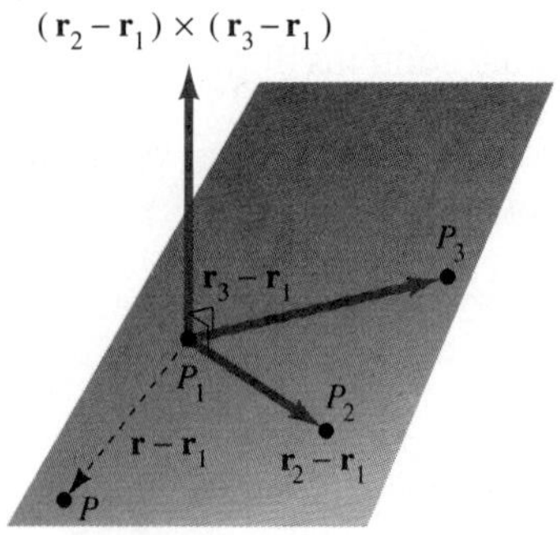

FIGURE 13.64

pairs of points. As shown in Figure 13.64, their cross product is a vector normal to the plane containing these vectors. If $P(x, y, z)$ represents any point on the plane, and $\mathbf{r} = \overrightarrow{OP}$, $\mathbf{r}_1 = \overrightarrow{OP_1}$, $\mathbf{r}_2 = \overrightarrow{OP_2}$, $\mathbf{r}_3 = \overrightarrow{OP_3}$, then $\mathbf{r} - \mathbf{r}_1$ (or, for that matter, $\mathbf{r} - \mathbf{r}_2$ or $\mathbf{r} - \mathbf{r}_3$) is in the plane. Hence,

$$[(\mathbf{r}_2 - \mathbf{r}_1) \times (\mathbf{r}_3 - \mathbf{r}_1)] \cdot (\mathbf{r} - \mathbf{r}_1) = 0 \tag{4}$$

is a vector equation of the plane. You are urged not to memorize the last formula. The procedure is the same as (1) with the exception that the vector normal to the plane is obtained by means of the cross product.

EXAMPLE 3 Find an equation of the plane that contains $(1, 0, -1)$, $(3, 1, 4)$, and $(2, -2, 0)$.

Solution We need three vectors. Pairing the points on the left as shown yields the vectors on the right. The order in which we subtract is irrelevant.

$$\begin{matrix} (1, 0, -1) \\ (3, 1, 4) \\ (2, -2, 0) \\ (x, y, z) \end{matrix} \quad \begin{aligned} \mathbf{u} &= 2\mathbf{i} + \mathbf{j} + 5\mathbf{k} \\ \mathbf{v} &= \mathbf{i} + 3\mathbf{j} + 4\mathbf{k} \\ \mathbf{w} &= (x - 2)\mathbf{i} + (y + 2)\mathbf{j} + z\mathbf{k} \end{aligned}$$

Now,

$$\mathbf{u} \times \mathbf{v} = \begin{vmatrix} \mathbf{i} & \mathbf{j} & \mathbf{k} \\ 2 & 1 & 5 \\ 1 & 3 & 4 \end{vmatrix} = -11\mathbf{i} - 3\mathbf{j} + 5\mathbf{k}$$

is a vector normal to the plane containing the given points. Hence, a vector equation of the plane is $(\mathbf{u} \times \mathbf{v}) \cdot \mathbf{w} = 0$. The latter equation yields

$$-11(x - 2) - 3(y + 2) + 5z = 0 \quad \text{or} \quad -11x - 3y + 5z + 16 = 0 \qquad \square$$

Orthogonal and Parallel Planes Figure 13.65 illustrates the plausibility of the following definition about **orthogonal** and **parallel** planes.

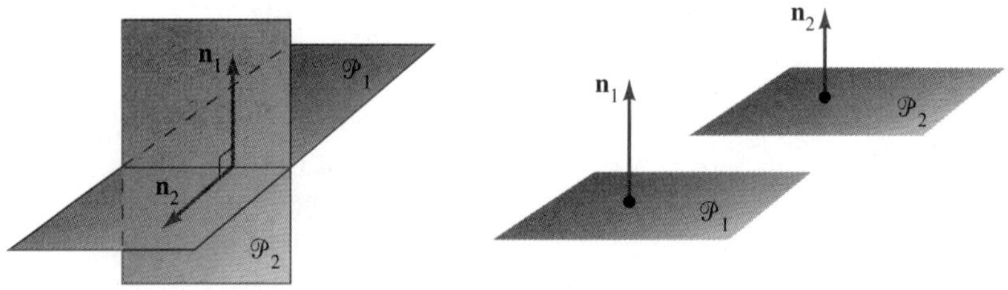

(a) orthogonal planes (b) parallel planes

FIGURE 13.65

DEFINITION 13.6 Orthogonal and Parallel Planes

Let $\mathbf{n}_1$ be a normal vector to plane $\mathscr{P}_1$ and let $\mathbf{n}_2$ be a normal vector to plane $\mathscr{P}_2$. Then

(*i*) $\mathscr{P}_1$ and $\mathscr{P}_2$ are **orthogonal** if $\mathbf{n}_1 \cdot \mathbf{n}_2 = 0$, and

(*ii*) $\mathscr{P}_1$ and $\mathscr{P}_2$ are **parallel** if $\mathbf{n}_1 = k\mathbf{n}_2$ for some nonzero scalar k.

EXAMPLE 4 The planes given by

$$\begin{aligned} \mathscr{P}_1&: \quad 2x - 4y + 8z = 7 \\ \mathscr{P}_2&: \quad x - 2y + 4z = 0 \\ \mathscr{P}_3&: \quad -3x + 6y - 12z = 1 \end{aligned}$$

are parallel, since their respective normal vectors

$$\begin{aligned} \mathbf{n}_1 &= 2\mathbf{i} - 4\mathbf{j} + 8\mathbf{k} \\ \mathbf{n}_2 &= \mathbf{i} - 2\mathbf{j} + 4\mathbf{k} = \tfrac{1}{2}\mathbf{n}_1 \\ \mathbf{n}_3 &= -3\mathbf{i} + 6\mathbf{j} - 12\mathbf{k} = -\tfrac{3}{2}\mathbf{n}_1 \end{aligned}$$

are parallel. □

Graph The graph of (3) with one or even two variables missing is still a plane. For example, we saw in Section 13.2 that the graphs of

$$x = x_0 \qquad y = y_0 \qquad z = z_0$$

where x_0, y_0, and z_0 are constants, are planes perpendicular to the x-, y-, and z-axes, respectively. In general, to graph a plane, we should try to find

(*i*) the x-, y-, and z-intercepts and, if necessary,

(*ii*) the trace of the plane in each coordinate plane.

A **trace** of a plane in a coordinate plane is the line of intersection of the plane with a coordinate plane.

EXAMPLE 5 Graph the equation $2x + 3y + 6z = 18$.

Solution Setting:

$$\begin{aligned} y = 0,\ z = 0 &\quad \text{gives} \quad x = 9 \\ x = 0,\ z = 0 &\quad \text{gives} \quad y = 6 \\ x = 0,\ y = 0 &\quad \text{gives} \quad z = 3 \end{aligned}$$

The x-, y-, and z-intercepts are then 9, 6, and 3, respectively. As shown in Figure 13.66, we use the points (9, 0, 0), (0, 6, 0), and (0, 0, 3) to draw the graph of the plane in the first octant. □

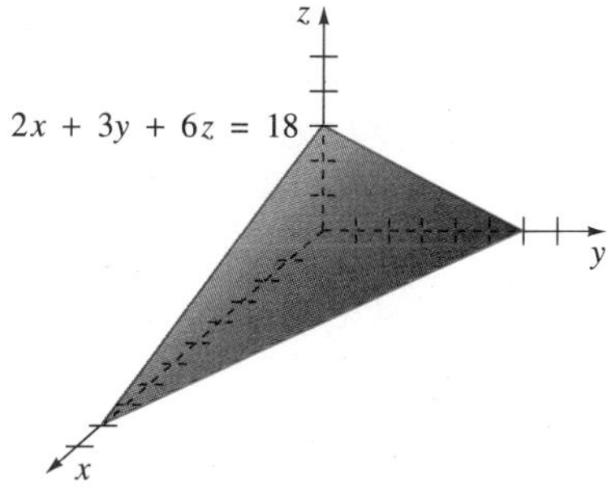

FIGURE 13.66

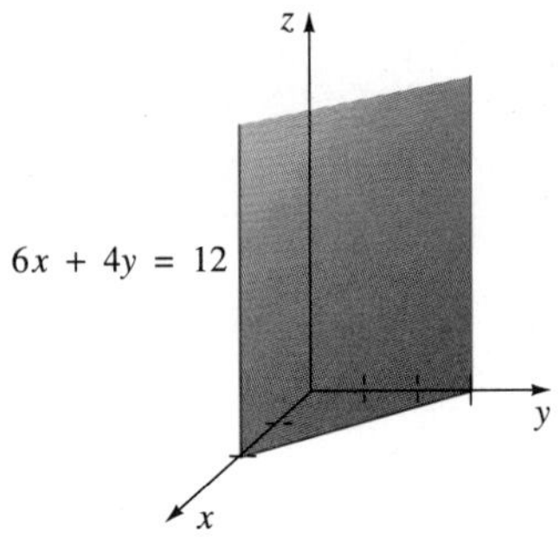

FIGURE 13.67

EXAMPLE 6 Graph the equation $6x + 4y = 12$.

Solution In two dimensions the graph of the equation is a line with x-intercept 2 and y-intercept 3. However, in three dimensions this line is the trace of a plane in the xy-coordinate plane. Since z is not specified, it can be any real number. In other words, (x, y, z) is a point on the plane provided that x and y are related by the given equation. As shown in Figure 13.67, the graph is a plane parallel to the z-axis. □

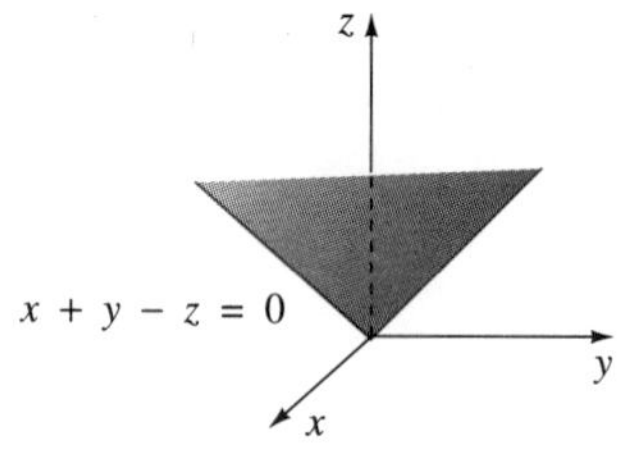

FIGURE 13.68

EXAMPLE 7 Graph the equation $x + y - z = 0$.

Solution First observe that the plane passes through the origin (0, 0, 0). Now, the trace of the plane in the xz-plane ($y = 0$) is $z = x$, whereas its trace in the yz-plane ($x = 0$) is $z = y$. Drawing these two lines leads to the graph given in Figure 13.68. □

Two planes $\mathscr{P}_1$ and $\mathscr{P}_2$ that are not parallel must intersect in a line $\mathscr{L}$. See Figure 13.69. Example 8 illustrates one way of finding parametric equations for the line of intersection. In Example 9 we see how to find a point of intersection (x_0, y_0, z_0) of a plane $\mathscr{P}$ and a line $\mathscr{L}$. See Figure 13.70.

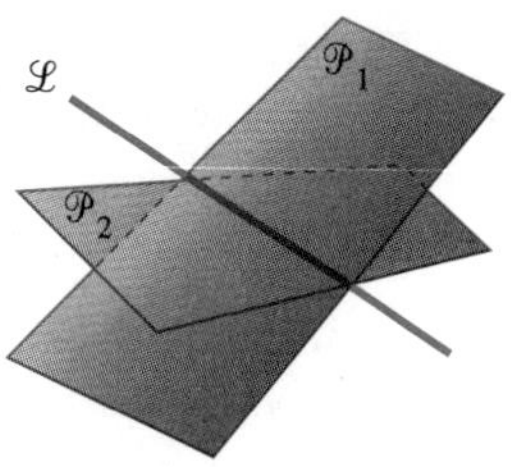

FIGURE 13.69

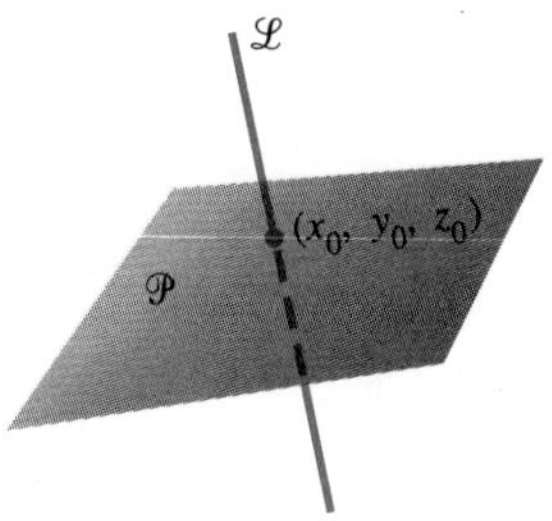

FIGURE 13.70

EXAMPLE 8 Find parametric equations for the line of intersection of

$$\begin{aligned} 2x - 3y + 4z &= 1 \\ x - y - z &= 5 \end{aligned}$$

Solution In a system of two equations and three unknowns, we choose one variable arbitrarily, say $z = t$, and solve for x and y from

$$\begin{aligned} 2x - 3y &= 1 - 4t \\ x - y &= 5 + t \end{aligned}$$

Proceeding, we find

$$x = 14 + 7t \qquad y = 9 + 6t \qquad z = t$$

These are parametric equations for the line of intersection of the given planes. □

EXAMPLE 9 Find the point of intersection of the plane $3x - 2y + z = -5$ and the line $x = 1 + t$, $y = -2 + 2t$, $z = 4t$.

Solution If (x_0, y_0, z_0) denotes the point of intersection, then we must have

$$3x_0 - 2y_0 + z_0 = -5 \quad \text{and} \quad x_0 = 1 + t_0, \quad y = -2 + 2t_0, \quad z = 4t_0$$

for some number t_0. Substituting the latter equations into the equation of the plane gives

$$3(1 + t_0) - 2(-2 + 2t_0) + 4t_0 = -5 \quad \text{or} \quad t_0 = -4$$

From the parametric equations for the line we then obtain $x_0 = -3$, $y_0 = -10$, and $z_0 = -16$. The point of intersection is $(-3, -10, -16)$. □

EXERCISES 13.6

Answers to odd-numbered problems begin on page A-100.

In Problems 1–6 find an equation of the plane that contains the given point and is perpendicular to the indicated vector.

1. $(5, 1, 3)$; $2\mathbf{i} - 3\mathbf{j} + 4\mathbf{k}$
2. $(1, 2, 5)$; $4\mathbf{i} - 2\mathbf{j}$
3. $(6, 10, -7)$; $-5\mathbf{i} + 3\mathbf{k}$
4. $(0, 0, 0)$; $6\mathbf{i} - \mathbf{j} + 3\mathbf{k}$
5. $(\frac{1}{2}, \frac{3}{4}, -\frac{1}{2})$; $6\mathbf{i} + 8\mathbf{j} - 4\mathbf{k}$
6. $(-1, 1, 0)$; $-\mathbf{i} + \mathbf{j} - \mathbf{k}$

In Problems 7–12 find, if possible, an equation of a plane that contains the given points.

7. $(3, 5, 2), (2, 3, 1), (-1, -1, 4)$
8. $(0, 1, 0), (0, 1, 1), (1, 3, -1)$
9. $(0, 0, 0), (1, 1, 1), (3, 2, -1)$
10. $(0, 0, 3), (0, -1, 0), (0, 0, 6)$
11. $(1, 2, -1), (4, 3, 1), (7, 4, 3)$
12. $(2, 1, 2), (4, 1, 0), (5, 0, -5)$

In Problems 13–22 find an equation of the plane that satisfies the given conditions.

13. Contains $(2, 3, -5)$ and is parallel to $x + y - 4z = 1$
14. Contains the origin and is parallel to $5x - y + z = 6$
15. Contains $(3, 6, 12)$ and is parallel to the xy-plane
16. Contains $(-7, -5, 18)$ and is perpendicular to the y-axis
17. Contains the lines $x = 1 + 3t$, $y = 1 - t$, $z = 2 + t$; $x = 4 + 4s$, $y = 2s$, $z = 3 + s$
18. Contains the lines $\dfrac{x-1}{2} = \dfrac{y+1}{-1} = \dfrac{z-5}{6}$; $\mathbf{r} = \langle 1, -1, 5\rangle + t\langle 1, 1, -3\rangle$
19. Contains the parallel lines $x = 1 + t$, $y = 1 + 2t$, $z = 3 + t$; $x = 3 + s$, $y = 2s$, $z = -2 + s$
20. Contains the point $(4, 0, -6)$ and the line $x = 3t$, $y = 2t$, $z = -2t$
21. Contains $(2, 4, 8)$ and is perpendicular to the line $x = 10 - 3t$, $y = 5 + t$, $z = 6 - \frac{1}{2}t$
22. Contains $(1, 1, 1)$ and is perpendicular to the line through $(2, 6, -3)$ and $(1, 0, -2)$
23. Determine which of the following planes are orthogonal and which are parallel.
    (a) $2x - y + 3z = 1$ (b) $x + 2y + 2z = 9$
    (c) $x + y - \frac{3}{2}z = 2$ (d) $-5x + 2y + 4z = 0$
    (e) $-8x - 8y + 12z = 1$ (f) $-2x + y - 3z = 5$
24. Find parametric equations for the line that contains $(-4, 1, 7)$ and is perpendicular to the plane $-7x + 2y + 3z = 1$.
25. Determine which of the following planes are perpendicular to the line $x = 4 - 6t$, $y = 1 + 9t$, $z = 2 + 3t$.
    (a) $4x + y + 2x = 1$ (b) $2x - 3y + z = 4$
    (c) $10x - 15y - 5z = 2$ (d) $-4x + 6y + 2z = 9$
26. Determine which of the following planes are parallel to the line $(1 - x)/2 = (y + 2)/4 = z - 5$.
    (a) $x - y + 3z = 1$ (b) $6x - 3y = 1$
    (c) $x - 2y + 5z = 0$ (d) $-2x + y - 2z = 7$

In Problems 27–30 find parametric equations for the line of intersection of the given planes.

27. $5x - 4y - 9z = 8$
$x + 4y + 3z = 4$

28. $x + 2y - z = 2$
$3x - y + 2z = 1$

29. $4x - 2y - z = 1$
$x + y + 2z = 1$

30. $2x - 5y + z = 0$
$y = 0$

In Problems 31–34 find the point of intersection of the given plane and line.

31. $2x - 3y + 2z = -7; x = 1 + 2t, y = 2 - t, z = -3t$

32. $x + y + 4z = 12; x = 3 - 2t, y = 1 + 6t, z = 2 - \frac{1}{2}t$

33. $x + y - z = 8; x = 1, y = 2, z = 1 + t$

34. $x - 3y + 2z = 0; x = 4 + t, y = 2 + t, z = 1 + 5t$

In Problems 35 and 36 find parametric equations for the line through the indicated point that is parallel to the given planes.

35. $x + y - 4z = 2$
$2x - y + z = 10; (5, 6, -12)$

36. $2x + z = 0$
$-x + 3y + z = 1; (-3, 5, -1)$

In Problems 37 and 38 find an equation of the plane that contains the given line and that is orthogonal to the indicated plane.

37. $x = 4 + 3t, y = -t, z = 1 + 5t; x + y + z = 7$

38. $\dfrac{2-x}{3} = \dfrac{y+2}{5} = \dfrac{z-8}{2}; 2x - 4y - z + 16 = 0$

In Problems 39–44 graph the given equation.

39. $5x + 2y + z = 10$

40. $3x + 2z = 9$

41. $-y - 3z + 6 = 0$

42. $3x + 4y - 2z - 12 = 0$

43. $-x + 2y + z = 4$

44. $3x - y - 6 = 0$

45. Show that the line $x = -2t, y = t, z = -t$:

(a) is parallel to but above the plane $x + y - z = 1$.

(b) is parallel to but below the plane $-3x - 4y + 2z = 8$.

46. Let $P_1(x_1, y_1, z_1)$ be a point on the plane $ax + by + cz + d = 0$ and let **n** be a normal vector to the plane. See Figure 13.71. Show that if $P_2(x_2, y_2, z_2)$ is any point not on the plane, then **the distance D from a point to a plane is given by**

$$D = \frac{|ax_2 + by_2 + cz_2 + d|}{\sqrt{a^2 + b^2 + c^2}}$$

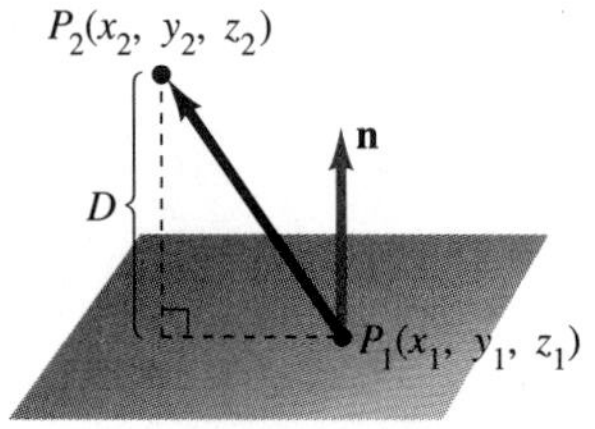

FIGURE 13.71

47. Use the result of Problem 46 to find the distance from the point (2, 1, 4) to the plane $x - 3y + z - 6 = 0$.

48. (a) Show that the planes $x - 2y + 3z = 3$, and $-4x + 8y - 12z = 7$ are parallel.

(b) Find the distance between the planes in part (**a**).

As shown in Figure 13.72, **the angle between two planes** is defined to be the acute angle between their normal vectors. In Problems 49 and 50 find the angles between the given planes.

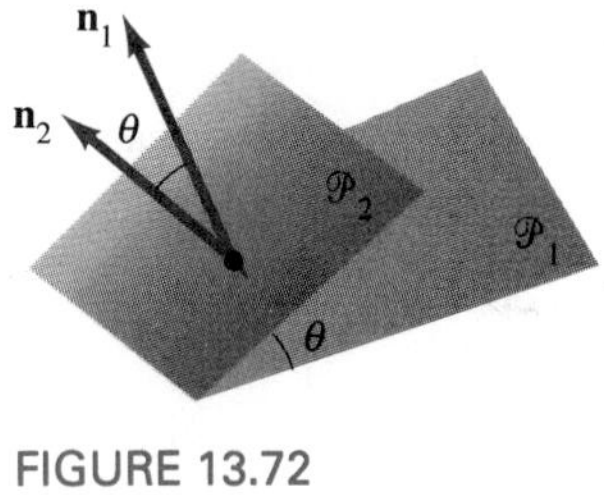

FIGURE 13.72

49. $x - 3y + 2z = 14$
$-x + y + z = 10$

50. $2x + 6y + 3z = 13$
$4x - 2y + 4z = -7$

13.7 CYLINDERS AND SPHERES

Cylinder In 2-space the graph of the equation $x^2 + y^2 = 1$ is a circle centered at the origin. However, in 3-space we can interpret the graph of the set

$$\{(x, y, z) \,|\, x^2 + y^2 = 1, z \text{ arbitrary}\}$$

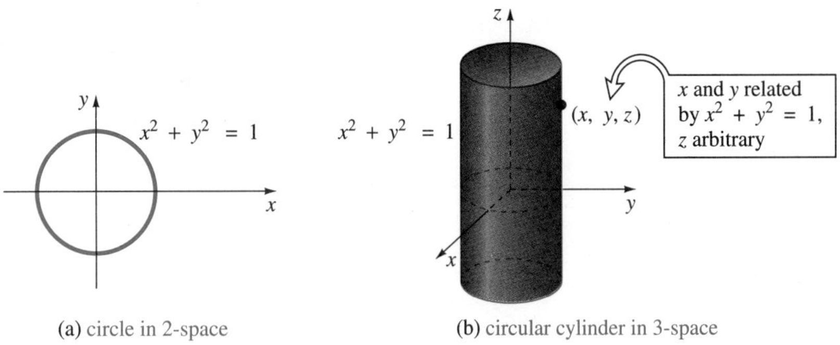

FIGURE 13.73

as a **surface** that is the right circular cylinder shown in Figure 13.73(b). We have already seen in Example 6 of Section 13.6 that the graph of an equation $ax + by + c = 0$ is a line in 2-space but a plane in 3-space.

EXAMPLE 1

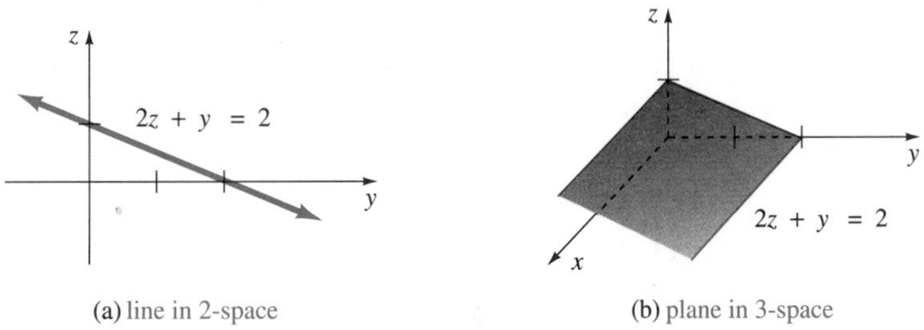

FIGURE 13.74

□

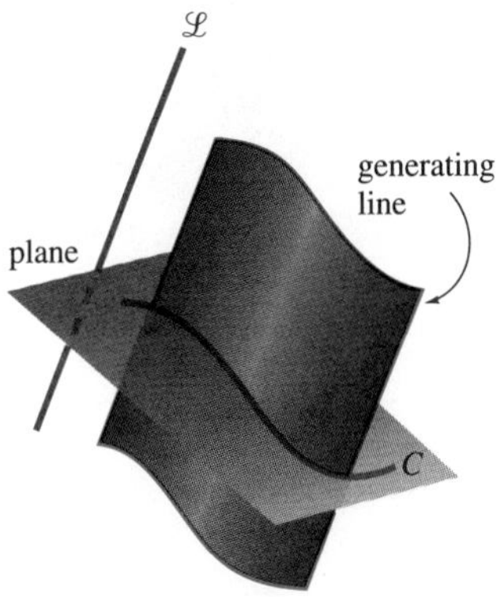

FIGURE 13.75

The surface illustrated in Example 1 is also called a cylinder. We use the term *cylinder* in a more general sense than that of a right circular cylinder. If C is a curve in a plane and $\mathscr{L}$ is a line not parallel to the plane, then the set of all points (x, y, z) generated by a moving line traversing C parallel to $\mathscr{L}$ is called a **cylinder**. The curve C is called the **directrix** of the cylinder. See Figure 13.75.

Thus, an equation of a curve in a coordinate plane, when considered in three dimensions, is an equation of a cylinder perpendicular to that coordinate plane.

If the graphs of $f(x, y) = c_1$, $g(y, z) = c_2$, and $h(x, z) = c_3$ are curves in the 2-space of their respective coordinate planes, then their graphs in 3-space are surfaces called cylinders. A cylinder is generated by a moving line that traverses the curve parallel to the coordinate axis which is represented by the variable missing in its equation.

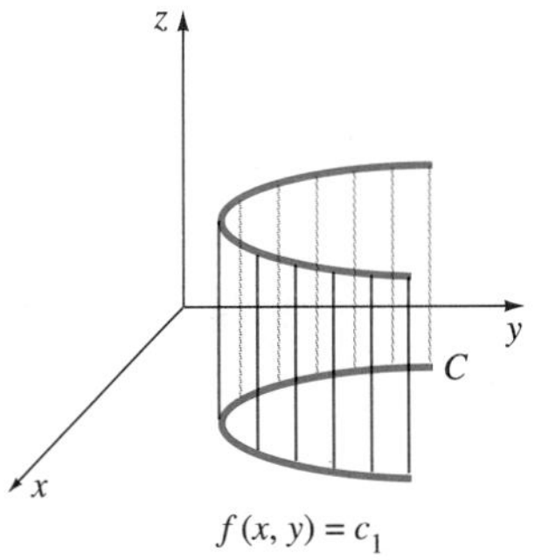

FIGURE 13.76

Figure 13.76 shows a curve C defined by $f(x, y) = c_1$ in the xy-plane and a sequence of lines called **rulings** that represent various positions of a generating line that is traversing C while moving parallel to the z-axis.

In the next example, we compare the graph of an equation in a coordinate plane with its interpretation as a cylinder in 3-space (Figures 13.77–13.80). As in Figure 13.74(b), we shall show only a portion of the cylinder.

EXAMPLE 2

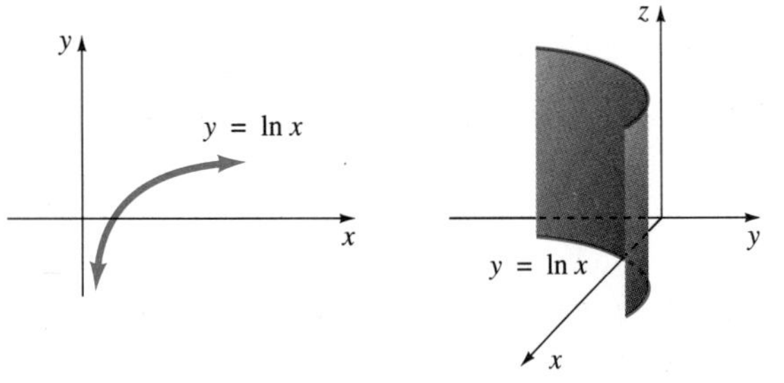

(a) logarithmic curve (b) logarithmic cylinder

FIGURE 13.77

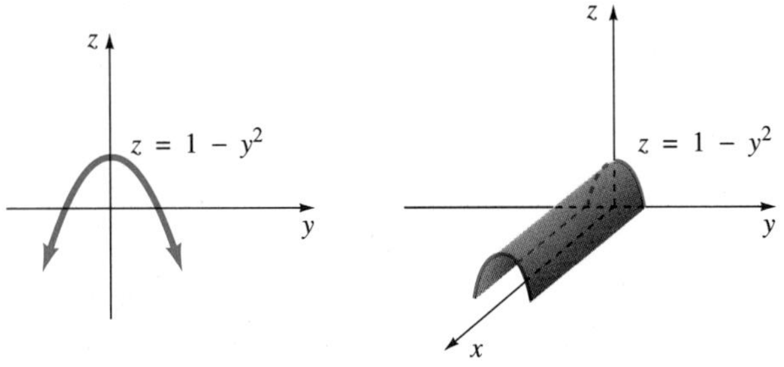

(a) parabola (b) parabolic cylinder

FIGURE 13.78

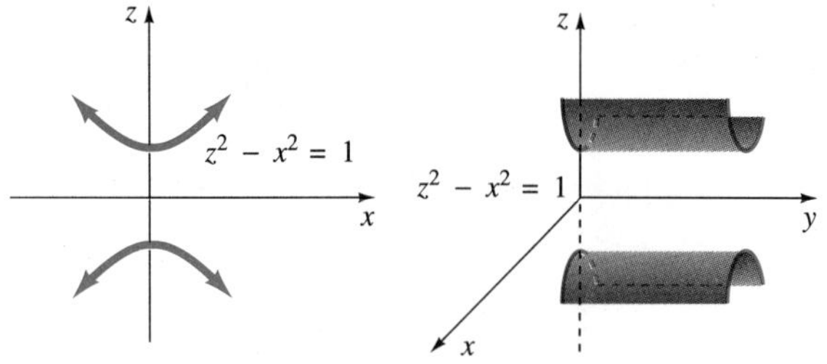

(a) hyperbola (b) hyperbolic cylinder

FIGURE 13.79

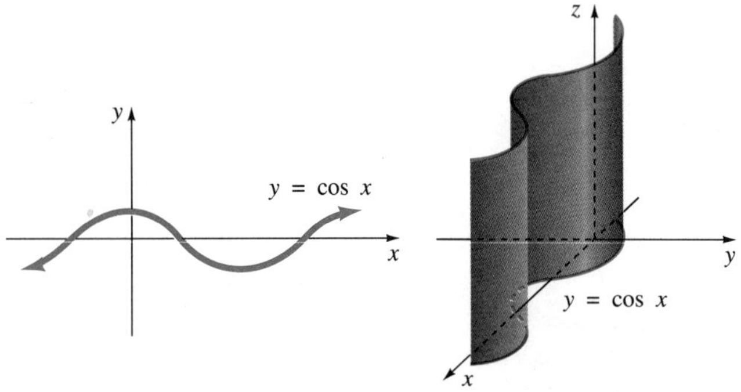

(a) sinusoidal curve* (b) sinusoidal cylinder

FIGURE 13.80

□

Spheres Like a circle, a **sphere** can be defined by means of the distance formula.

> **DEFINITION 13.7 Sphere**
>
> A **sphere** is the set of all points P in 3-space that are equidistant from a fixed point called the **center**.

*The graph of $\cos x$ is the graph of the *sine* function shifted $\pi/2$ radians to the left.

If r denotes the fixed distance, or **radius** of the sphere, and if the center is $P_1(a, b, c)$, then a point $P(x, y, z)$ is on the sphere if and only if $[d(P_1, P)]^2 = r^2$, or

$$(x - a)^2 + (y - b)^2 + (z - c)^2 = r^2 \tag{1}$$

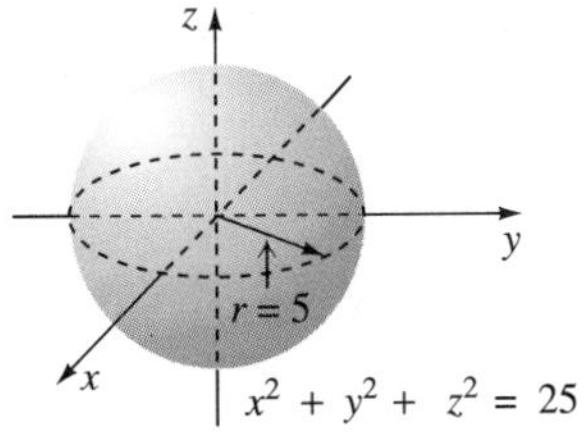

FIGURE 13.81

EXAMPLE 3 Graph $x^2 + y^2 + z^2 = 25$.

Solution We identify $a = 0, b = 0, c = 0$, and $r^2 = 25$ in (1), and so the graph of $x^2 + y^2 + z^2 = 25$ is a sphere of radius 5 whose center is at the origin. The graph of the equation is given in Figure 13.81. □

Trace of a Surface In general, a **trace of a surface** in any plane is the curve formed by the intersection of the surface and the plane. Note that in Figure 13.81 the trace of the sphere in the xy-plane is the dashed circle $x^2 + y^2 = 25$. In the xz- and yz-planes, the traces of the sphere are the circles $x^2 + z^2 = 25$ and $y^2 + z^2 = 25$, respectively.

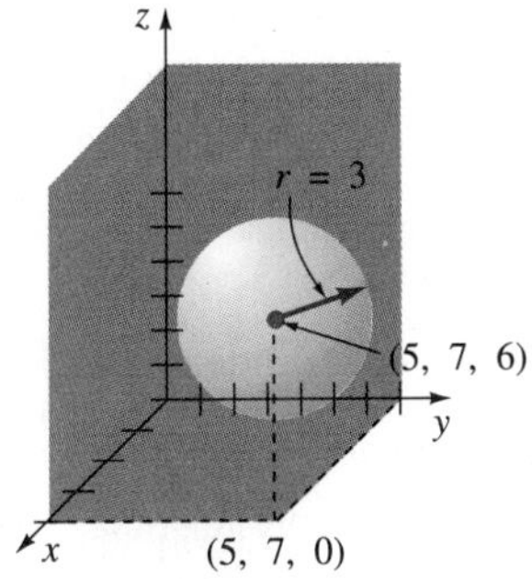

FIGURE 13.82

EXAMPLE 4 Graph $(x - 5)^2 + (y - 7)^2 + (z - 6)^2 = 9$.

Solution In this case we identify $a = 5, b = 7, c = 6$, and $r^2 = 9$. From (1) we see that the graph of $(x - 5)^2 + (y - 7)^2 + (z - 6)^2 = 9$ is a sphere with center $(5, 7, 6)$ and radius 3. Its graph lies entirely in the first octant and is shown in Figure 13.82. □

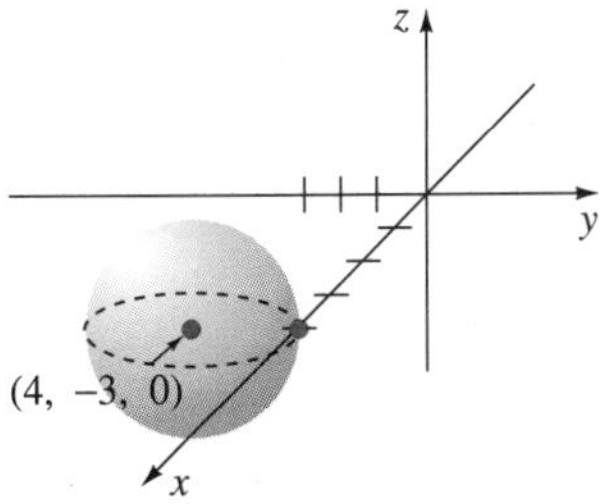

FIGURE 13.83

EXAMPLE 5 Find an equation of the sphere whose center is $(4, -3, 0)$ that is tangent to the xz-plane.

Solution The perpendicular distance from the given point to the xz-plane, and hence the radius of the sphere, is the absolute value of the y-coordinate, $|-3| = 3$. Thus, an equation of the sphere is $(x - 4)^2 + (y + 3)^2 + z^2 = 3^2$. See Figure 13.83. □

EXAMPLE 6 Find the center and radius of the sphere whose equation is

$$16x^2 + 16y^2 + 16z^2 - 16x + 8y - 32z + 16 = 0$$

Solution Dividing by 16 and completing the square in x, y, and z yield

$$\left(x - \frac{1}{2}\right)^2 + \left(y + \frac{1}{4}\right)^2 + (z - 1)^2 = \frac{5}{16}$$

The center and radius of the sphere are $(\frac{1}{2}, -\frac{1}{4}, 1)$ and $\sqrt{5}/4$, respectively. □

EXERCISES 13.7 *Answers to odd-numbered problems begin on page A-100.*

In Problems 1–16 sketch the graph of the given cylinder.

1. $y = x$
2. $z = -y$
3. $y = x^2$
4. $x^2 + z^2 = 25$
5. $y^2 + z^2 = 9$
6. $z = y^2$
7. $z = e^{-x}$
8. $z = 1 - e^y$
9. $y^2 - x^2 = 4$
10. $z = \cosh y$
11. $4x^2 + y^2 = 36$
12. $x = 1 - y^2$
13. $z = \sin x$
14. $y = \dfrac{1}{x^2}$
15. $yz = 1$
16. $z = x^3 - 3x$

In Problems 17–20 sketch the graph of the given equation.

17. $x^2 + y^2 + z^2 = 9$
18. $x^2 + y^2 + (z-3)^2 = 16$
19. $(x-1)^2 + (y-1)^2 + (z-1)^2 = 1$
20. $(x+3)^2 + (y+4)^2 + (z-5)^2 = 4$

In Problems 21–24 find the center and radius of the sphere with the given equation.

21. $x^2 + y^2 + z^2 + 8x - 6y - 4z - 7 = 0$
22. $4x^2 + 4y^2 + 4z^2 + 4x - 12z + 9 = 0$
23. $x^2 + y^2 + z^2 - 16z = 0$
24. $x^2 + y^2 + z^2 - x + y = 0$

In Problems 25–32 find an equation of a sphere that satisfies the given conditions.

25. Center $(-1, 4, 6)$; radius $\sqrt{3}$
26. Center $(0, -3, 0)$; diameter $\frac{5}{2}$
27. Center $(1, 1, 4)$; tangent to the xy-plane
28. Center $(5, 2, -2)$; tangent to the yz-plane
29. Center on the positive y-axis; radius 2; tangent to $x^2 + y^2 + z^2 = 36$
30. Center on the line $x = 2t$, $y = 3t$, $z = 6t$, $t > 0$, at a distance 21 units from the origin; radius 5
31. Diameter has endpoints $(0, -4, 7)$ and $(2, 12, -3)$
32. Center $(-3, 1, 2)$; passing through the origin

In Problems 33–36 describe geometrically all points $P(x, y, z)$ whose coordinates satisfy the given condition(s).

33. $x^2 + y^2 + (z-1)^2 = 4$, $1 \le z \le 3$
34. $x^2 + y^2 + (z-1)^2 = 4$, $z = 2$
35. $x^2 + y^2 + z^2 \ge 1$
36. $0 < (x-1)^2 + (y-2)^2 + (z-3)^2 < 1$

13.8 QUADRIC SURFACES AND SURFACES OF REVOLUTION

The equation of the sphere given in (1) of Section 13.7 is just a particular case of the second-degree equation

$$Ax^2 + By^2 + Cz^2 + Dx + Ey + Fz + G = 0 \tag{1}$$

When A, B, and C are not all zero, the graph of an equation of form (1), describing a real locus, is said to be a **quadric surface**. For example, both the elliptical cylinder $x^2/4 + y^2/9 = 1$ and the parabolic cylinder $z = y^2$ are quadric surfaces. We conclude this section by considering six additional quadric surfaces.

Ellipsoid The graph of any equation of the form

$$\frac{x^2}{a^2} + \frac{y^2}{b^2} + \frac{z^2}{c^2} = 1, \qquad a > 0, b > 0, c > 0 \tag{2}$$

is called an **ellipsoid**. For $|y_0| < b$, the equation

$$\frac{x^2}{a^2} + \frac{z^2}{c^2} = 1 - \frac{y_0^2}{b^2}$$

represents a family of ellipses (or circles if $a = c$) parallel to the xz-plane that are formed by slicing the surface by planes $y = y_0$. By choosing, in turn, $x = x_0$ and $z = z_0$, we would find that slices of the surface are ellipses (or circles) parallel to the yz- and xy-planes, respectively. Figure 13.84 summarizes the traces in the coordinate planes and gives a typical graph.

Coordinate Plane	Trace
$xy\ (z = 0)$	ellipse: $\frac{x^2}{a^2} + \frac{y^2}{b^2} = 1$
$xz\ (y = 0)$	ellipse: $\frac{x^2}{a^2} + \frac{z^2}{c^2} = 1$
$yz\ (x = 0)$	ellipse: $\frac{y^2}{b^2} + \frac{z^2}{c^2} = 1$

(a)

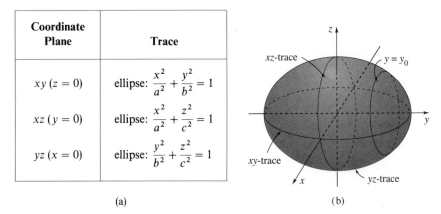

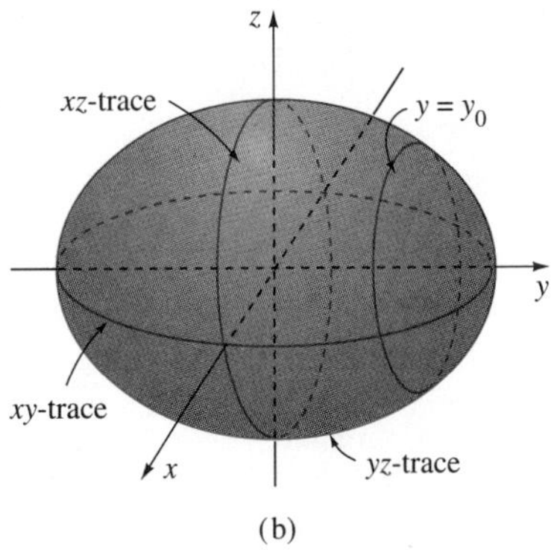

(b)

FIGURE 13.84

Hyperboloid of One Sheet The graph of an equation of the form

$$\frac{x^2}{a^2} + \frac{y^2}{b^2} - \frac{z^2}{c^2} = 1, \qquad a > 0, b > 0, c > 0 \tag{3}$$

is called a **hyperboloid of one sheet**. In this case, a plane $z = z_0$, parallel to the xy-plane, slices the surface into elliptical (or circular if $a = b$) cross-sections. The equations of these ellipses are

$$\frac{x^2}{a^2} + \frac{y^2}{b^2} = 1 + \frac{z_0^2}{c^2}$$

The smallest ellipse, $z_0 = 0$, corresponds to the trace in the xy-plane. A summary of the traces and a typical graph of (3) are given in Figure 13.85.

Coordinate Plane	Trace
$xy\ (z = 0)$	ellipse: $\dfrac{x^2}{a^2} + \dfrac{y^2}{b^2} = 1$
$xz\ (y = 0)$	hyperbola: $\dfrac{x^2}{a^2} - \dfrac{z^2}{c^2} = 1$
$yz\ (x = 0)$	hyperbola: $\dfrac{y^2}{b^2} - \dfrac{z^2}{c^2} = 1$

(a)

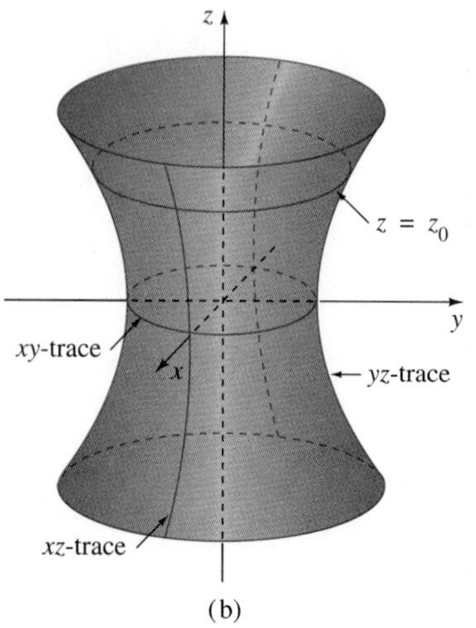

(b)

FIGURE 13.85

Hyperboloid of Two Sheets As seen in Figure 13.86(b), a graph of

$$-\frac{x^2}{a^2} + \frac{y^2}{b^2} - \frac{z^2}{c^2} = 1, \qquad a > 0, b > 0, c > 0 \tag{4}$$

is appropriately called a **hyperboloid of two sheets**. For $|y_0| > b$, the equation $x^2/a^2 + z^2/c^2 = y_0^2/b^2 - 1$ describes the elliptical curve of intersection of the surface with the plane $y = y_0$.

Coordinate Plane	Trace
$xy\ (z = 0)$	hyperbola: $-\dfrac{x^2}{a^2} + \dfrac{y^2}{b^2} = 1$
$xz\ (y = 0)$	no locus
$yz\ (x = 0)$	hyperbola: $\dfrac{y^2}{b^2} - \dfrac{z^2}{c^2} = 1$

(a)

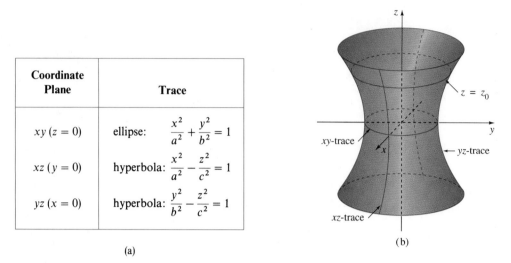

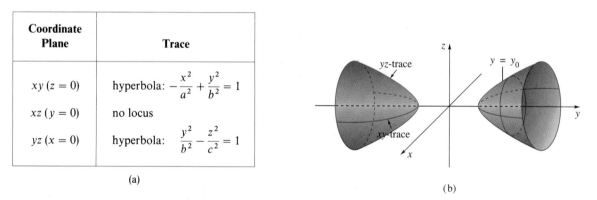

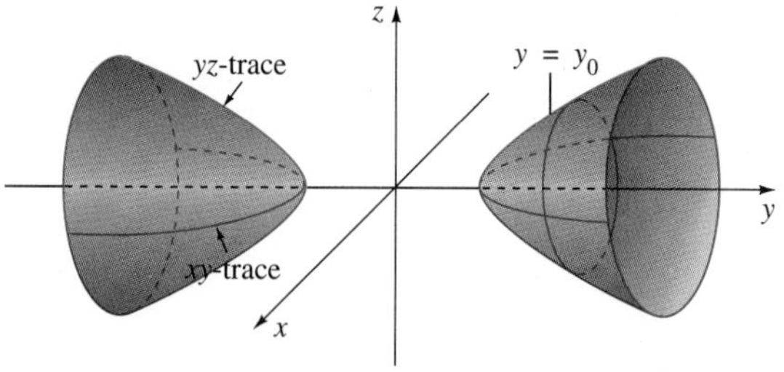

(b)

FIGURE 13.86

Paraboloid The graph of an equation of the form

$$\frac{x^2}{a^2} + \frac{y^2}{b^2} = cz \tag{5}$$

is called a **paraboloid**. In Figure 13.87(b) we see that for $c > 0$, planes $z = z_0 > 0$, parallel to the xy-plane, slice the surface in ellipses whose equations are

$$\frac{x^2}{a^2} + \frac{y^2}{b^2} = cz_0$$

Coordinate Plane	Trace
$xy\ (z = 0)$	point: $(0, 0)$
$xz\ (y = 0)$	parabola: $\dfrac{x^2}{a^2} = cz$
$yz\ (x = 0)$	parabola: $\dfrac{y^2}{b^2} = cz$

(a)

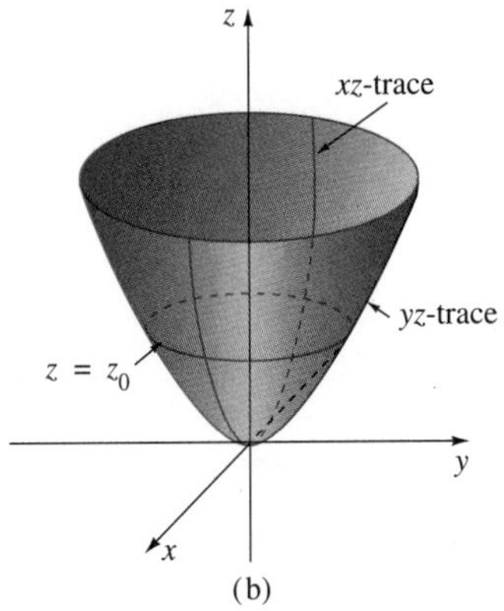

(b)

FIGURE 13.87

Cone The graph of an equation of the form

$$\frac{x^2}{a^2} + \frac{y^2}{b^2} = \frac{z^2}{c^2}, \qquad a > 0, b > 0, c > 0 \tag{6}$$

is called an **elliptical** (or circular if $a = b$) **cone**. For arbitrary z_0, planes parallel to the xy-plane slice the surface in ellipses whose equations are

$$\frac{x^2}{a^2} + \frac{y^2}{b^2} = \frac{z_0^2}{c^2}$$

A typical graph of (6) is shown in Figure 13.88(b).

Coordinate Plane	Trace
$xy\ (z = 0)$	point: $(0, 0)$
$xz\ (y = 0)$	lines: $z = \pm\dfrac{c}{a}x$
$yz\ (x = 0)$	lines: $z = \pm\dfrac{c}{b}y$

(a)

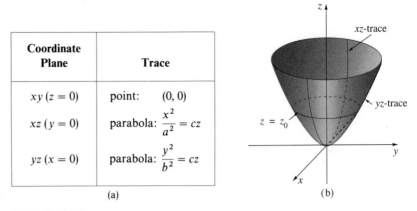

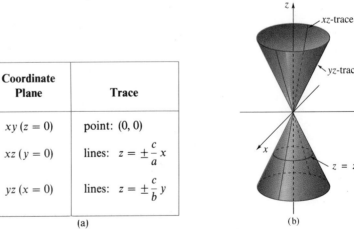

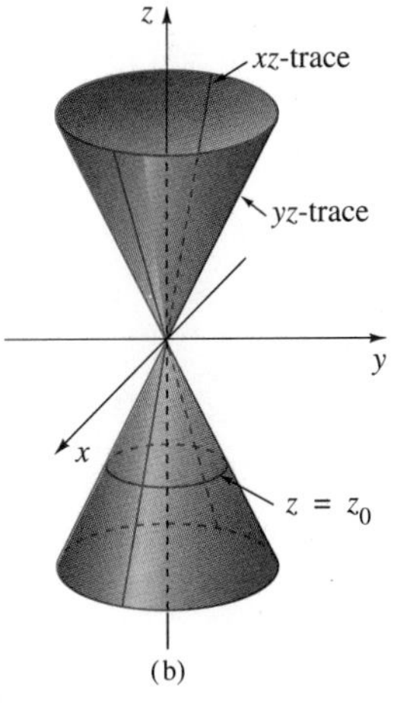

(b)

FIGURE 13.88

Hyperbolic Paraboloid The last quadric surface we shall consider, known as a **hyperbolic paraboloid**, is the graph of any equation of the form

$$\frac{y^2}{a^2} - \frac{x^2}{b^2} = cz, \qquad a > 0, b > 0 \tag{7}$$

Note that for $c > 0$, planes $z = z_0$, parallel to the xy-plane, cut the surface in hyperbolas whose equations are

$$\frac{y^2}{a^2} - \frac{x^2}{b^2} = cz_0$$

The characteristic saddle shape of a hyperbolic paraboloid is shown in Figure 13.89(b).

Coordinate Plane	Trace	
$xy\ (z = 0)$	lines:	$y = \pm\frac{a}{b}x$
$xz\ (y = 0)$	parabola:	$-\frac{x^2}{b^2} = cz$
$yz\ (x = 0)$	parabola:	$\frac{y^2}{a^2} = cz$

(a)

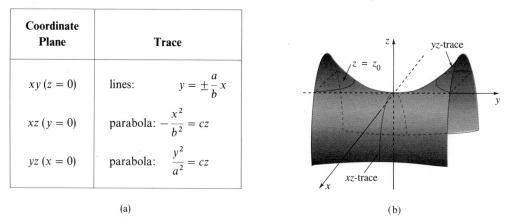

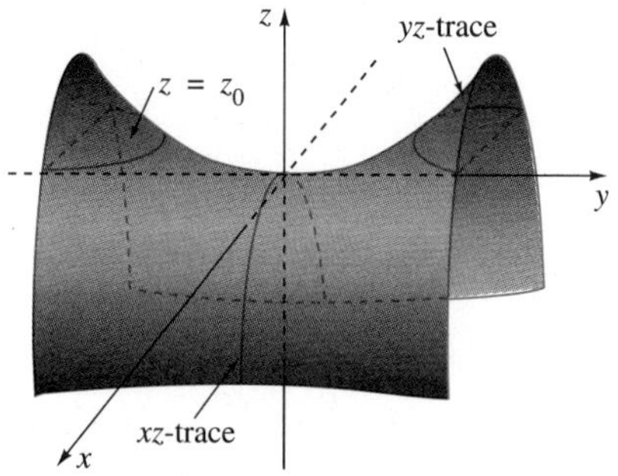

(b)

FIGURE 13.89

Variation of the Equations Interchanging the position of the variables in equations (2)–(7) does not change the basic nature of a surface, but *does* change the surface's orientation in space. For example, graphs of

$$\frac{x^2}{a^2} - \frac{y^2}{b^2} + \frac{z^2}{c^2} = 1 \quad \text{and} \quad -\frac{x^2}{a^2} + \frac{y^2}{b^2} + \frac{z^2}{c^2} = 1 \tag{8}$$

are still hyperboloids of one sheet. Similarly, the two minus signs in (4) that characterize hyperboloids of two sheets can occur anywhere in the equation. Similarly,

$$\frac{x^2}{a^2} + \frac{z^2}{b^2} = cy \quad \text{and} \quad \frac{y^2}{a^2} + \frac{z^2}{b^2} = cx \tag{9}$$

are paraboloids. Graphs of equations of the form

$$\frac{x^2}{a^2} - \frac{z^2}{b^2} = cy \quad \text{and} \quad \frac{y^2}{a^2} - \frac{z^2}{b^2} = cx \tag{10}$$

are hyperbolic paraboloids.

EXAMPLE 1 Identify (*a*) $y = x^2 + z^2$ and (*b*) $y = x^2 - z^2$. Compare the graphs.

Solution From the first equations in (9) and (10) with $a = 1, b = 1$, and $c = 1$, we identify the graph of (*a*) as a paraboloid and the graph of (*b*) as a hyperbolic paraboloid. In the case of equation (a), a plane $y = y_0, y_0 > 0$, slices the surface in circles whose equations are $y_0 = x^2 + z^2$. On the other hand, a plane $y = y_0$ slices the graph of equation (b) in hyperbolas $y_0 = x^2 - z^2$. The graphs are compared in Figure 13.90.

(a) $y = x^2 + z^2$ (b) $y = x^2 - z^2$

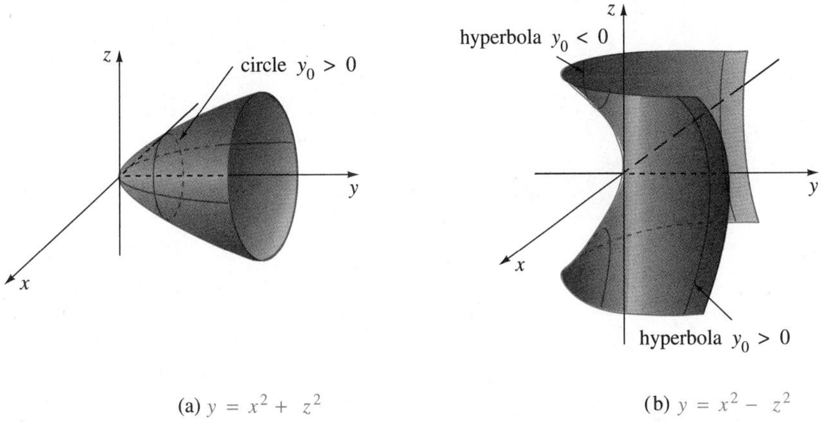

FIGURE 13.90

EXAMPLE 2   Identify (a) $2x^2 - 4y^2 + z^2 = 0$ and
(b) $-2x^2 + 4y^2 + z^2 = -36$.

Solution (a) From $\dfrac{x^2}{2} + \dfrac{z^2}{4} = y^2$, we identify the graph as a cone.

(b) From $\dfrac{x^2}{18} - \dfrac{y^2}{9} - \dfrac{z^2}{36} = 1$, we identify the graph as a hyperboloid of two sheets.

Origin at (h, k, l)   When the origin is translated to (h, k, l), the equations of the quadric surfaces become

$$\frac{(x - h)^2}{a^2} + \frac{(y - k)^2}{b^2} + \frac{(z - l)^2}{c^2} = 1$$

$$\frac{(x - h)^2}{a^2} + \frac{(y - k)^2}{b^2} - \frac{(z - l)^2}{c^2} = 1$$

and so on.

EXAMPLE 3   Graph $z = 4 - x^2 - y^2$.

Solution   By writing the equation as

$$-(z - 4) = x^2 + y^2$$

we recognize the equation of a paraboloid. The minus sign in front of the term on the left side of the equality indicates that the graph of the paraboloid opens downward from $(0, 0, 4)$. See Figure 13.91.

$z = 4 - x^2 - y^2$ $(0, 0, 4)$

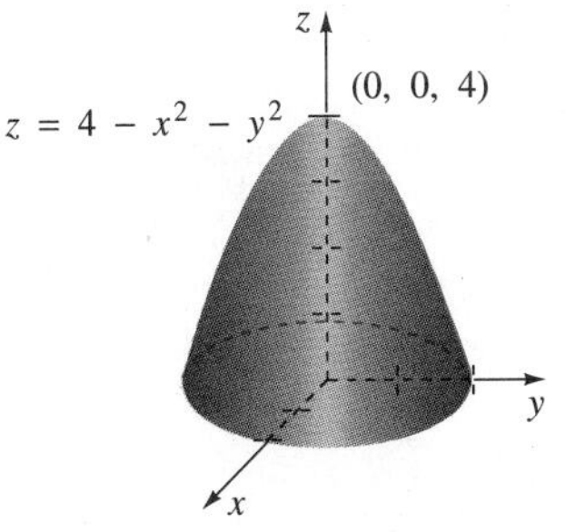

FIGURE 13.91

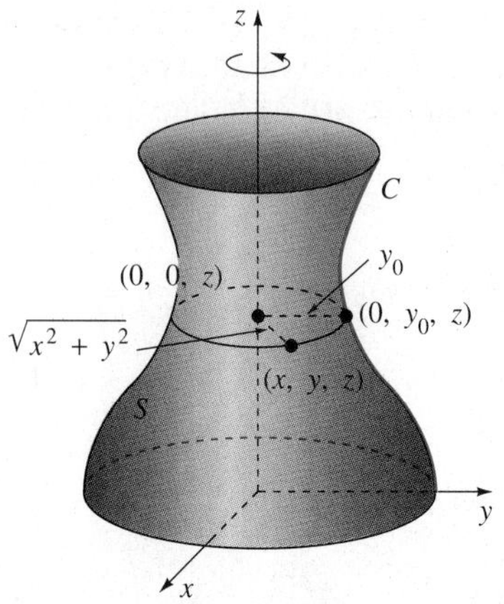

FIGURE 13.92

Surfaces of Revolution In Section 5.5 we saw that a surface S could be generated by revolving a plane curve C about an axis. In the discussion that follows we shall find equations of **surfaces of revolution** when C is a curve in a coordinate plane and the axis of revolution is a coordinate axis.

For the sake of discussion, let us suppose that $f(y, z) = 0$ is an equation of a curve C in the yz-plane and that C is revolved about the z-axis generating a surface S. Let us also suppose for the moment that the y- and z-coordinates of points on C are nonnegative. If (x, y, z) denotes a general point on S that results from revolving the point $(0, y_0, z)$ on C, then we see from Figure 13.92 that the distance from (x, y, z) to $(0, 0, z)$ is the same as the distance from $(0, y_0, z)$ to $(0, 0, z)$; that is, $y_0 = \sqrt{x^2 + y^2}$. From the fact that $f(y_0, z) = 0$ we arrive at an equation for S:

$$f(\sqrt{x^2 + y^2}, z) = 0 \tag{11}$$

A curve in a coordinate plane can, of course, be revolved about each coordinate axis. If the curve C in the yz-plane defined by $f(y, z) = 0$ is now revolved about the y-axis, it can be shown that an equation of the resulting surface of revolution is

$$f(y, \sqrt{x^2 + z^2}) = 0 \tag{12}$$

Finally, we note that if there are points $(0, y, z)$ on C for which the y- or z-coordinates are negative, then we replace $\sqrt{x^2 + y^2}$ in (11) by $\pm\sqrt{x^2 + y^2}$ and $\sqrt{x^2 + z^2}$ in (12) by $\pm\sqrt{x^2 + z^2}$.

Equations of surfaces of revolution generated when a curve in the xy- or xz-plane is revolved about a coordinate axis are analogous to (11) and (12). As the following table shows, an equation of a surface generated by revolving a curve in a coordinate plane about the

$$\left.\begin{matrix} x\text{-axis} \\ y\text{-axis} \\ z\text{-axis} \end{matrix}\right\} \quad \text{involves the term} \quad \left\{\begin{matrix} \sqrt{y^2 + z^2} \\ \sqrt{x^2 + z^2} \\ \sqrt{x^2 + y^2} \end{matrix}\right.$$

Equation of Curve C	**Axis of Revolution**	**Equation of Surface S**
$f(x, y) = 0$	x-axis y-axis	$f(x, \pm\sqrt{y^2 + z^2}) = 0$ $f(\pm\sqrt{x^2 + z^2}, y) = 0$
$f(x, z) = 0$	x-axis z-axis	$f(x, \pm\sqrt{y^2 + z^2}) = 0$ $f(\pm\sqrt{x^2 + y^2}, z) = 0$
$f(y, z) = 0$	y-axis z-axis	$f(y, \pm\sqrt{x^2 + z^2}) = 0$ $f(\pm\sqrt{x^2 + y^2}, z) = 0$

EXAMPLE 4

(*a*) In Example 1, the equation $y = x^2 + z^2$ can be written as

$$y = (\pm\sqrt{x^2 + z^2})^2$$

Hence, from the preceding table we see that the surface is generated by revolving either the parabola $y = z^2$ or the parabola $y = x^2$ about the y-axis. The surface shown in Figure 13.90(a) is called a **paraboloid of revolution.**

(*b*) In Example 3, the equation $-(z - 4) = x^2 + y^2$ can be written as

$$-(z - 4) = (\pm\sqrt{x^2 + y^2})^2$$

The surface is also a paraboloid of revolution. In this case the surface is generated by revolving either the parabola $-(z - 4) = y^2$ or the parabola $-(z - 4) = x^2$ about the z-axis. □

EXAMPLE 5 The graph of $4x^2 + y^2 = 16$ is revolved about the x-axis. Find an equation of the surface of revolution.

Solution The given equation has the form $f(x, y) = 0$. Since the axis of revolution is the x-axis, we see from the table that an equation of the surface of revolution can be found by replacing y by $\pm\sqrt{y^2 + z^2}$. It follows that

$$4x^2 + (\pm\sqrt{y^2 + z^2})^2 = 16$$

or

$$4x^2 + y^2 + z^2 = 16$$

The surface is called an **ellipsoid of revolution.** □

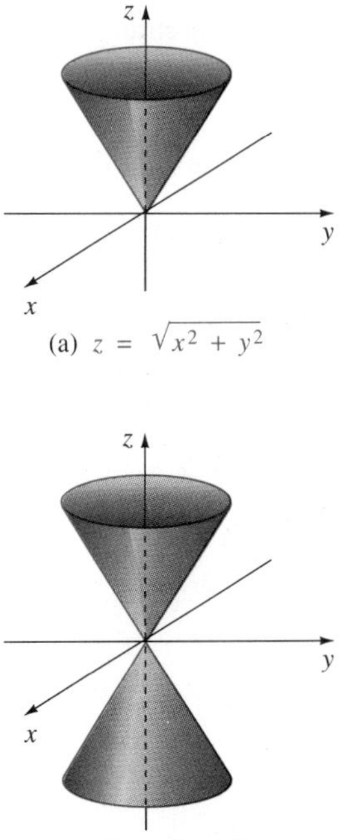

(a) $z = \sqrt{x^2 + y^2}$

(b) $z^2 = x^2 + y^2$

FIGURE 13.93

EXAMPLE 6 The graph of $z = y$, $y \geq 0$, is revolved about the z-axis. Find an equation of the surface of revolution.

Solution Since there are no points on the graph of $z = y$, $y \geq 0$, with a negative y-coordinate, we obtain an equation for the surface of revolution by substituting $\sqrt{x^2 + y^2}$ for y:

$$z = \sqrt{x^2 + y^2} \tag{13}$$

Observe that (13) is not the same as $z^2 = x^2 + y^2$. The latter equation is equivalent to $z = \pm\sqrt{x^2 + y^2}$, which is an equation of the surface obtained by revolving the graph of $z = y$ about the z-axis. In other words, the equation $z = \sqrt{x^2 + y^2}$ describes only the upper nappe of the cone whose equation is $z^2 = x^2 + y^2$. See Figure 13.93. □

EXERCISES 13.8 *Answers to odd-numbered problems begin on page A-101.*

In Problems 1–14 identify and graph the quadric surface.

1. $x^2 + y^2 = z$
2. $-x^2 + y^2 = z^2$
3. $9x^2 + 36y^2 + 4z^2 = 36$
4. $x^2 + y^2 - z^2 = -4$
5. $36x^2 - y^2 + 9z^2 = 144$
6. $4x^2 + 4y^2 + z^2 = 100$
7. $y^2 + 5z^2 = x^2$
8. $-9x^2 + 16y^2 = 144z$
9. $y = 4x^2 - z^2$
10. $9z + x^2 + y^2 = 0$
11. $x^2 - y^2 - z^2 = 4$
12. $-x^2 + 9y^2 + z^2 = 9$
13. $y^2 + 4z^2 = x$
14. $x^2 + y^2 - z^2 = 1$

In Problems 15–18 graph the quadric surface.

15. $z = 3 + x^2 + y^2$
16. $y + x^2 + 4z^2 = 4$
17. $(x - 4)^2 + (y - 6)^2 - z^2 = 1$
18. $5x^2 + (y - 5)^2 + 5z^2 = 25$

In Problems 19–22 the given equation is an equation of a surface of revolution obtained by revolving a curve C in a coordinate plane about a coordinate axis. Find an equation for C and identify the axis of revolution.

19. $x^2 + y^2 + z^2 = 1$
20. $-9x^2 + 4y^2 + 4z^2 = 36$
21. $y = e^{x^2+z^2}$
22. $x^2 + y^2 = \sin^2 z$

In Problems 23–30 the graph of the given equation is revolved about the indicated axis. Find an equation of the surface of revolution.

23. $y = 2x$; y-axis
24. $y = \sqrt{z}$; y-axis
25. $z = 9 - x^2$, $x \geq 0$; x-axis
26. $z = 1 + y^2$, $y \geq 0$; z-axis
27. $x^2 - z^2 = 4$; x-axis
28. $3x^2 + 4z^2 = 12$; z-axis
29. $z = \ln y$; z-axis
30. $xy = 1$; x-axis
31. Which of the surfaces in Problems 1–14 are surfaces of revolution? Identify the axis of revolution.
32. Sketch a graph of the equation in Problem 22 for $0 \leq z \leq 2\pi$.

In Problems 33 and 34 compare the graphs of the given equations.

33. $z + 2 = -\sqrt{x^2 + y^2}$, $(z + 2)^2 = x^2 + y^2$
34. $y - 1 = \sqrt{x^2 + z^2}$, $(y - 1)^2 = x^2 + z^2$
35. Consider the paraboloid

$$z - c = -\left(\frac{x^2}{a^2} + \frac{y^2}{b^2}\right), \qquad c > 0$$

(a) The area of an ellipse $x^2/A^2 + y^2/B^2 = 1$ is πAB. Use this fact to express the area of a cross-section perpendicular to the z-axis as a function of z, $z \leq c$.

(b) Use the slicing method (see Section 5.2) to find the volume of the solid bounded by the paraboloid and the xy-plane.

36. (a) Using the slicing method as in Problem 35, find the volume of the ellipsoid

$$\frac{x^2}{a^2} + \frac{y^2}{b^2} + \frac{z^2}{c^2} = 1$$

(b) What does your answer in part (**a**) become when $a = b = c$?

37. Determine the points where the line

$$\frac{x - 2}{2} = \frac{y + 2}{-3} = \frac{z - 6}{3/2}$$

intersects the ellipsoid $x^2/9 + y^2/36 + z^2/81 = 1$.

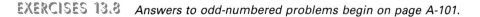

CHAPTER 13 REVIEW EXERCISES

Answers begin on page A-102.

In Problems 1–10 answer true or false.

1. The vectors $\langle -4, -6, 10\rangle$ and $\langle -10, -15, 25\rangle$ are parallel. ________

2. In 3-space any three distinct points determine a plane. ________

3. The line $x = 1 + 5t$, $y = 1 - 2t$, $z = 4 + t$ and the plane $2x + 3y - 4z = 1$ are perpendicular. ________

4. Nonzero vectors $\mathbf{a}$ and $\mathbf{b}$ are parallel if $\mathbf{a} \times \mathbf{b} = \mathbf{0}$. ________

5. If $\mathbf{a} \cdot \mathbf{b} < 0$, the angle between $\mathbf{a}$ and $\mathbf{b}$ is obtuse. ________

6. If $\mathbf{a}$ is a unit vector, then $\mathbf{a} \cdot \mathbf{a} = 1$. ________

7. The cross product of two vectors is not commutative. ________

8. The terminal point of the vector $\mathbf{a} - \mathbf{b}$ is at the terminal point of $\mathbf{a}$. ________

9. $(\mathbf{a} \times \mathbf{b}) \cdot \mathbf{c} = \mathbf{a} \cdot (\mathbf{b} \times \mathbf{c})$ ________

10. If $\mathbf{a}$, $\mathbf{b}$, $\mathbf{c}$, and $\mathbf{d}$ are nonzero coplanar vectors, then $(\mathbf{a} \times \mathbf{b}) \times (\mathbf{c} \times \mathbf{d}) = \mathbf{0}$. ________

In Problems 11–30 fill in the blanks.

11. The sum of $3\mathbf{i} + 4\mathbf{j} + 5\mathbf{k}$ and $6\mathbf{i} - 2\mathbf{j} - 3\mathbf{k}$ is ________.

12. If $\mathbf{a} \cdot \mathbf{b} = 0$, the nonzero vectors $\mathbf{a}$ and $\mathbf{b}$ are ________.

13. $(-\mathbf{k}) \times (5\mathbf{j}) =$ ________

14. $\mathbf{i} \cdot (\mathbf{i} \times \mathbf{j}) =$ ________

15. $|-12\mathbf{i} + 4\mathbf{j} + 6\mathbf{k}| =$ ________

16. $\begin{vmatrix} \mathbf{i} & \mathbf{j} & \mathbf{k} \\ 2 & 1 & 5 \\ 0 & 4 & -1 \end{vmatrix} =$ ________

17. A vector that is normal to the plane $-6x + y - 7z + 10 = 0$ is ________.

18. The trace of the surface $x^2 - y^2 + z^2 = 4$ in the xz-plane is ________.

19. The point of intersection of the line $x - 1 = (y + 2)/3 = (z + 1)/2$ and the plane $x + 2y - z = 13$ is ________.

20. A unit vector that has the opposite direction of $\mathbf{a} = 4\mathbf{i} + 3\mathbf{j} - 5\mathbf{k}$ is ________.

21. If $\overrightarrow{P_1P_2} = \langle 3, 5, -4\rangle$ and P_1 has coordinates $(2, 1, 7)$, then the coordinates of P_2 are ________.

22. The midpoint of the line segment between $P_1(4, 3, 10)$ and $P_2(6, -2, -5)$ has coordinates ________.

23. If $|\mathbf{a}| = 7.2$, $|\mathbf{b}| = 10$, and the angle between $\mathbf{a}$ and $\mathbf{b}$ is $135°$, then $\mathbf{a} \cdot \mathbf{b} =$ ________.

24. If $\mathbf{a} = \langle 3, 1, 0\rangle$, $\mathbf{b} = \langle -1, 2, 1\rangle$, and $\mathbf{c} = \langle 0, -2, 2\rangle$, then $\mathbf{a} \cdot (2\mathbf{b} + 4\mathbf{c}) =$ ________.

25. The x-, y-, and z-intercepts of the plane $2x - 3y + 4z = 24$ are, respectively, ________.

26. The angle θ between the vectors $\mathbf{a} = \mathbf{i} + \mathbf{j}$ and $\mathbf{b} = \mathbf{i} - \mathbf{k}$ is ________.

27. The area of a triangle with two sides given by $\mathbf{a} = \langle 1, 3, -1\rangle$ and $\mathbf{b} = \langle 2, -1, 2\rangle$ is ________.

28. An equation of a sphere with center $(-5, 7, -9)$ and radius $\sqrt{6}$ is ________.

29. The distance from the plane $y = -5$ to the point $(4, -3, 1)$ is ________.

30. The vectors $\langle 1, 3, c\rangle$ and $\langle -2, -6, 5\rangle$ are parallel for $c =$ ________ and orthogonal for $c =$ ________.

31. Find a unit vector that is perpendicular to both $\mathbf{a} = \mathbf{i} + \mathbf{j}$ and $\mathbf{b} = \mathbf{i} - 2\mathbf{i} + \mathbf{k}$.

32. Find the direction cosines and direction angles of the vector $\mathbf{a} = \frac{1}{2}\mathbf{i} + \frac{1}{2}\mathbf{j} - \frac{1}{4}\mathbf{k}$.

In Problems 33–36 let $\mathbf{a} = \langle 1, 2, -2\rangle$ and $\mathbf{b} = \langle 4, 3, 0\rangle$. Find the indicated number or vector.

33. $\text{comp}_{\mathbf{b}}\mathbf{a}$

34. $\text{proj}_{\mathbf{a}}\mathbf{b}$

35. $\text{proj}_{\mathbf{a}\perp}\mathbf{b}$

36. $\text{proj}_{\mathbf{b}\perp}(\mathbf{a} - \mathbf{b})$

In Problems 37–42 identify the surface whose equation is given.

37. $x^2 + 4y^2 = 16$

38. $y + 2x^2 + 4z^2 = 0$

39. $x^2 + 4y^2 - z^2 = -9$

40. $x^2 + y^2 + z^2 = 10z$

41. $9z - x^2 + y^2 = 0$

42. $2x - 3y = 6$

43. Find an equation of the surface of revolution obtained by revolving the graph of $x^2 - y^2 = 1$ about the y-axis. About the x-axis. Identify the surface in each case.

44. A surface of revolution has an equation $y = 1 + \sqrt{x^2 + z^2}$. Find an equation of a curve C in a coordinate plane that, when revolved about a coordinate axis, generates the surface.

45. Let $\mathbf{r}$ be the position vector of a variable point $P(x, y, z)$ in space and let $\mathbf{a}$ be a constant vector. Determine the surface described by the following equations:

(a) $(\mathbf{r} - \mathbf{a}) \cdot \mathbf{r} = 0$ (b) $(\mathbf{r} - \mathbf{a}) \cdot \mathbf{a} = 0$

46. Use the dot product to determine whether the points $(4, 2, -2)$, $(2, 4, -3)$, and $(6, 7, -5)$ are vertices of a right triangle.

47. Find symmetric equations for the line through the point $(7, 3, -5)$ that is parallel to $(x - 3)/4 = (y + 4)/(-2) = (z - 9)/6$.

48. Find parametric equations for the line through the point $(5, -9, 3)$ that is perpendicular to the plane $8x + 3y - 4z = 13$.

49. Show that the lines $x = 1 - 2t$, $y = 3t$, $z = 1 + t$ and $x = 1 + 2s$, $y = -4 + s$, $z = -1 + s$ intersect orthogonally.

50. Find an equation of the plane containing the points $(0, 0, 0)$, $(2, 3, 1)$, $(1, 0, 2)$.

51. Find an equation of the plane containing the lines $x = t$, $y = 4t$, $z = -2t$ and $x = 1 + t$, $y = 1 + 4t$, $z = 3 - 2t$.

52. Find an equation of the plane containing $(1, 7, -1)$ that is perpendicular to the line of intersection of $-x + y - 8z = 4$ and $3x - y + 2z = 0$.

53. Find an equation of the plane containing $(1, -1, 2)$ that is parallel to the vectors $\mathbf{i} - 2\mathbf{j}$ and $2\mathbf{i} + 3\mathbf{k}$.

54. Find an equation of a sphere for which the line segment $x = 4 + 2t$, $y = 7 + 3t$, $z = 8 + 6t$, $-1 \le t \le 0$, is a diameter.

55. Show that the three vectors $\mathbf{a} = 3\mathbf{i} + 5\mathbf{j} + 2\mathbf{k}$, $\mathbf{b} = 3\mathbf{i} + 4\mathbf{j} + \mathbf{k}$, and $\mathbf{c} = 4\mathbf{i} + 5\mathbf{j} + \mathbf{k}$ are coplanar.

56. Consider the right triangle whose sides are the vectors $\mathbf{a}$, $\mathbf{b}$, and $\mathbf{c}$ shown in Figure 13.94. Show that the midpoint M of the hypotenuse is equidistant from all three vertices of the triangle.

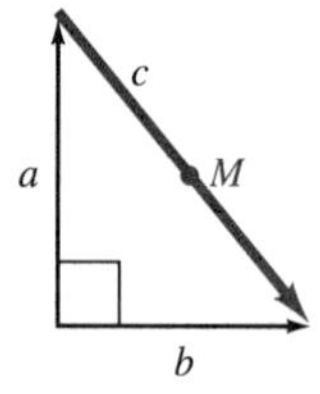

FIGURE 13.94

57. (a) The force $\mathbf{F}$ acting on a particle of charge q moving with velocity $\mathbf{v}$ through a magnetic field $\mathbf{B}$ is given by $\mathbf{F} = q(\mathbf{v} \times \mathbf{B})$. Find $\mathbf{F}$ if $\mathbf{v}$ acts along the positive y-axis and $\mathbf{B}$ acts along the positive x-axis. Assume $|\mathbf{v}| = v$ and $|\mathbf{B}| = B$.

(b) The angular momentum $\mathbf{L}$ of a particle of mass m moving with a linear velocity $\mathbf{v}$ in a circle of radius $\mathbf{r}$ is given by $\mathbf{L} = m(\mathbf{r} \times \mathbf{v})$, where $\mathbf{r}$ is perpendicular to $\mathbf{v}$. Use vector methods to solve for $\mathbf{v}$ in terms of $\mathbf{L}$, $\mathbf{r}$, and m.

58. A constant force of 10 N in the direction of $\mathbf{a} = \mathbf{i} + \mathbf{j}$ moves a block on a frictionless surface from $P_1(4, 1, 0)$ to $P_2(7, 4, 0)$. Suppose distance is measured in meters. Find the work done.

59. In Problem 58 find the work done in moving the block between the same points if another cosntant force of 50 N in the direction of $\mathbf{b} = \mathbf{i}$ acts simultaneously with the original force.

60. A uniform ball of weight 50 lb is supported by two frictionless planes as shown in Figure 13.95. Let the force exerted by the supporting plane $\mathcal{P}_1$ on the ball be $\mathbf{F}_1$ and the force exerted by the plane $\mathcal{P}_2$ be $\mathbf{F}_2$. Since the ball is held in equilibrium, we must have $\mathbf{w} + \mathbf{F}_1 + \mathbf{F}_2 = \mathbf{0}$, where $\mathbf{w} = -50\mathbf{j}$. Find the magnitudes of the forces $\mathbf{F}_1$ and $\mathbf{F}_2$. [*Hint:* Assume the forces $\mathbf{F}_1$ and $\mathbf{F}_2$ are normal to the planes $\mathcal{P}_1$ and $\mathcal{P}_2$, respectively, and act along lines through the center C of the ball. Place the origin of a two-dimensional coordinate system at C.]

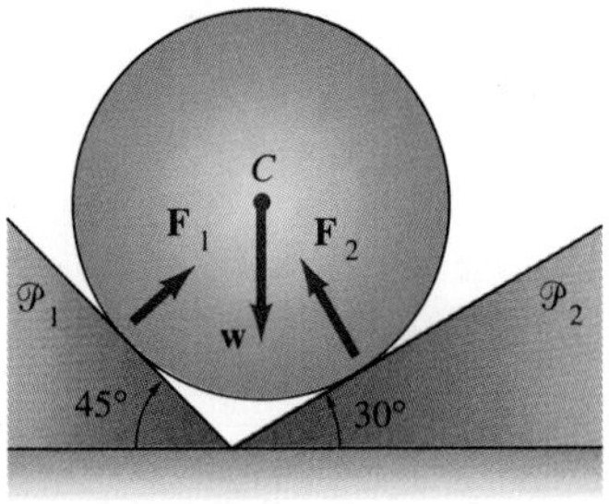

FIGURE 13.95

14

VECTOR-VALUED FUNCTIONS

INTRODUCTION

Curves in the plane as well as in 3-space can be described by means of parametric equations. Using the functions in a set of parametric equations as components, we can construct a vector-valued function whose values are position vectors of points on a curve. In this chapter we shall consider the calculus and applications of these vector functions.

When Isaac Newton proved Johann Kepler's three laws of planetary motion, he did not use vector methods. But, with the modern-day vector interpretations of Newton's own universal law of gravitation and his second law of motion, the proof of Kepler's laws are problems whose solutions are within the reach of a student of calculus. In Exercises 14.2 you will be asked to establish Kepler's first law—that the orbit of a planet around the sun is an ellipse with the sun at one focus. You will see that this is a consequence of the fact that the orbit of a body moving in a central force field is a conic section.

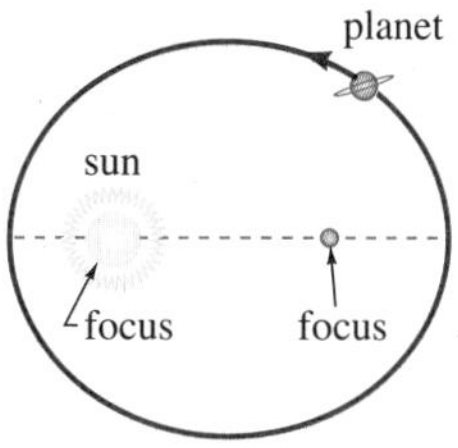

IMPORTANT CONCEPTS

vector-valued function
space curve
smooth function
smooth curve
length of a space curve
velocity
acceleration
curvature
principal normal
tangential component of acceleration
normal component of acceleration
binormal
moving trihedral
radius of curvature
circle of curvature
center of curvature

14.1 VECTOR FUNCTIONS

Vector-Valued Functions Recall that a curve C in the xy-plane can be parameterized by

$$x = f(t) \qquad y = g(t), \qquad a \le t \le b$$

It is often convenient in science and engineering to introduce a vector $\mathbf{r}$ with the functions f and g as components:

$$\mathbf{r}(t) = \langle f(t), g(t) \rangle = f(t)\mathbf{i} + g(t)\mathbf{j}$$

where $\mathbf{i} = \langle 1, 0 \rangle$ and $\mathbf{j} = \langle 0, 1 \rangle$. We say that $\mathbf{r}$ is a **vector-valued function** or simply a **vector function**. Similarly, a **space curve** is parameterized by three equations

$$x = f(t) \qquad y = g(t) \qquad z = h(t), \qquad a \le t \le b \tag{1}$$

Correspondingly, a vector function is given by

$$\mathbf{r}(t) = \langle f(t), g(t), h(t) \rangle = f(t)\mathbf{i} + g(t)\mathbf{j} + h(t)\mathbf{k}$$

where $\mathbf{i} = \langle 1, 0, 0 \rangle$, $\mathbf{j} = \langle 0, 1, 0 \rangle$, and $\mathbf{k} = \langle 0, 0, 1 \rangle$. As shown in Figure 14.1, for a given number t_0, the vector $\mathbf{r}(t_0)$ is the *position vector* of a point P on the curve C. In other words, as t varies, we can envision the curve C being traced out by the moving arrowhead of $\mathbf{r}(t)$.

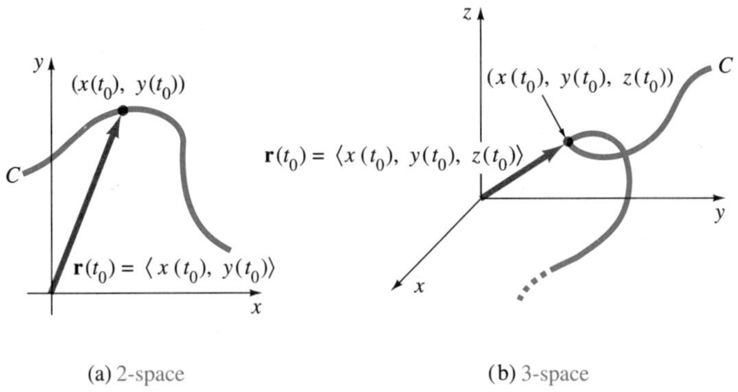

FIGURE 14.1

We have already seen an example of parametric equations as well as the vector function of a space curve in Section 13.5, when we discussed the line in 3-space.

EXAMPLE 1 Graph the curve traced by the vector function

$$\mathbf{r}(t) = 2\cos t\mathbf{i} + 2\sin t\mathbf{j} + t\mathbf{k}, \qquad t \geq 0$$

Solution The parametric equations of the curve are $x = 2\cos t$, $y = 2\sin t$, $z = t$. By eliminating the parameter t from the first two equations,

$$x^2 + y^2 = (2\cos t)^2 + (2\sin t)^2 = 2^2$$

we see that a point on the curve lies on the circular cylinder $x^2 + y^2 = 4$. As seen in Figure 14.2 and the accompanying table, as the value of t increases, the curve winds upward in a spiral or a *circular helix.*

t	0	$\pi/2$	π	$3\pi/2$	2π	$5\pi/2$	3π	$7\pi/2$	4π	$9\pi/2$
x	2	0	-2	0	2	0	-2	0	2	0
y	0	2	0	-2	0	2	0	-2	0	2
z	0	$\pi/2$	π	$3\pi/2$	2π	$5\pi/2$	3π	$7\pi/2$	4π	$9\pi/2$

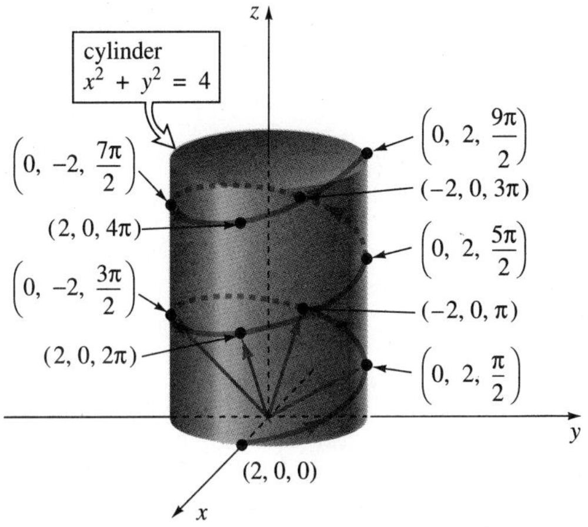

FIGURE 14.2

□

The curve in Example 1 is a special case of the vector function

$$\mathbf{r}(t) = a\cos t\mathbf{i} + b\sin t\mathbf{j} + ct\mathbf{k}, \qquad a > 0, b > 0, c > 0$$

which describes an **elliptical helix**. When $a = b$, the helix is circular. The **pitch** of a helix is defined to be the number $2\pi c$. Problems 9 and 10 in Exercises 14.1 illustrate two other kinds of helixes.

EXAMPLE 2 Graph the curve traced by the vector function

$$\mathbf{r}(t) = 2\cos t\mathbf{i} + 2\sin t\mathbf{j} + 3\mathbf{k}$$

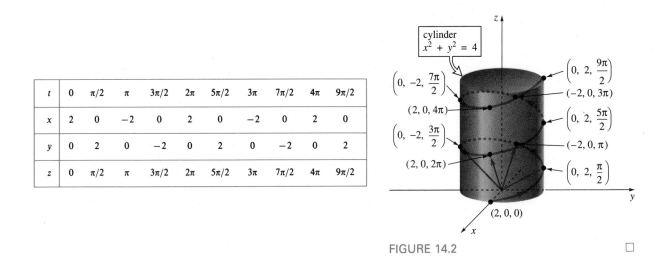

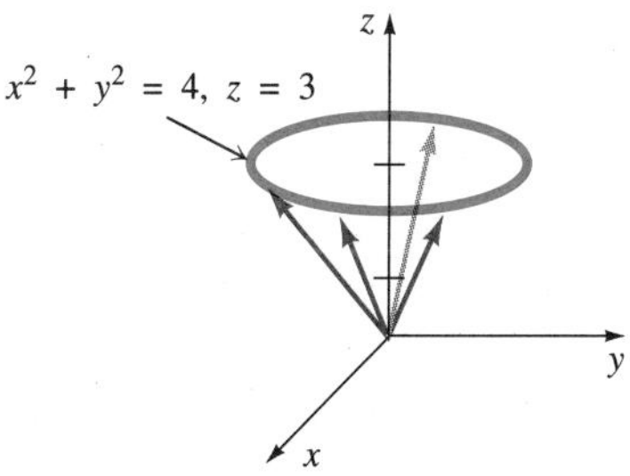

FIGURE 14.3

Solution The parametric equations of this curve are $x = 2\cos t$, $y = 2\sin t$, $z = 3$. As in Example 1, we see that a point on the curve must also lie on the cylinder $x^2 + y^2 = 4$. However, since the z-coordinate of any point has the constant value $z = 3$, the vector function $\mathbf{r}(t)$ traces out a circle 3 units above the xy-plane. See Figure 14.3. □

EXAMPLE 3 Find the vector function that describes the curve C of intersection of the plane $y = 2x$ and the paraboloid $z = 9 - x^2 - y^2$.

Solution We first parameterize the curve C of intersection by letting $x = t$. It follows that $y = 2t$ and $z = 9 - t^2 - (2t)^2 = 9 - 5t^2$. From the parametric equations

$$x = t \qquad y = 2t \qquad z = 9 - 5t^2$$

we see that a vector function describing the trace of the paraboloid in the plane $y = 2x$ is given by

$$\mathbf{r}(t) = t\mathbf{i} + 2t\mathbf{j} + (9 - 5t^2)\mathbf{k}$$

See Figure 14.4.

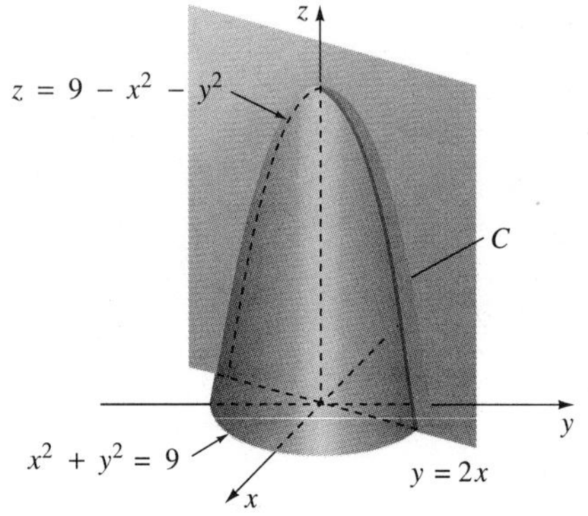

FIGURE 14.4 □

Limits and Continuity The fundamental notion of the **limit** of a vector function $\mathbf{r}(t) = \langle f(t), g(t), h(t)\rangle$ is defined in terms of the limits of the component functions.

DEFINITION 14.1 Limit of a Vector Function

If $\lim_{t\to a} f(t)$, $\lim_{t\to a} g(t)$, and $\lim_{t\to a} h(t)$ exist, then

$$\lim_{t\to a} \mathbf{r}(t) = \left\langle \lim_{t\to a} f(t), \lim_{t\to a} g(t), \lim_{t\to a} h(t)\right\rangle$$

The symbol $t \to a$ in Definition 14.1 can, of course, be replaced by $t \to a^+$, $t \to a^-$, $t \to \infty$, or $t \to -\infty$.

As an immediate consequence of Definition 14.1, we have the following result:

THEOREM 14.1 Properties

If $\lim_{t\to a} \mathbf{r}_1(t) = \mathbf{L}_1$ and $\lim_{t\to a} \mathbf{r}_2(t) = \mathbf{L}_2$, then

(*i*) $\lim_{t\to a} c\mathbf{r}_1(t) = c\mathbf{L}_1$, c a scalar

(*ii*) $\lim_{t\to a}[\mathbf{r}_1(t) + \mathbf{r}_2(t)] = \mathbf{L}_1 + \mathbf{L}_2$

(*iii*) $\lim_{t\to a} \mathbf{r}_1(t) \cdot \mathbf{r}_2(t) = \mathbf{L}_1 \cdot \mathbf{L}_2$

DEFINITION 14.2 Continuity

A vector function $\mathbf{r}$ is said to be **continuous** at $t = a$ if

(*i*) $\mathbf{r}(a)$ is defined, (*ii*) $\lim_{t\to a} \mathbf{r}(t)$ exists, and (*iii*) $\lim_{t\to a} \mathbf{r}(t) = \mathbf{r}(a)$.

Equivalently, $\mathbf{r}(t)$ is continuous at $t = a$ if and only if the component functions f, g, and h are continuous there.

Derivative of a Vector Function

DEFINITION 14.3 Derivative of a Vector Function

The **derivative** of a vector function $\mathbf{r}$ is

$$\mathbf{r}'(t) = \lim_{\Delta t\to 0} \frac{1}{\Delta t}[\mathbf{r}(t + \Delta t) - \mathbf{r}(t)] \tag{2}$$

for all t for which the limit exists.

The derivative of $\mathbf{r}$ is also written $d\mathbf{r}/dt$. The next theorem shows that on a practical level, the derivative of a vector function is obtained by simply differentiating its component functions.

THEOREM 14.2 Differentiation of a Vector Function

If $\mathbf{r}(t) = \langle f(t), g(t), h(t)\rangle$, where f, g, and h are differentiable, then

$$\mathbf{r}'(t) = \langle f'(t), g'(t), h'(t)\rangle$$

Proof From (2) we have

$$\mathbf{r}'(t) = \lim_{\Delta t \to 0} \frac{1}{\Delta t}[\langle f(t + \Delta t), g(t + \Delta t), h(t + \Delta t)\rangle - \langle f(t), g(t), h(t)\rangle]$$

$$= \lim_{\Delta t \to 0} \left\langle \frac{f(t + \Delta t) - f(t)}{\Delta t}, \frac{g(t + \Delta t) - g(t)}{\Delta t}, \frac{h(t + \Delta t) - h(t)}{\Delta t} \right\rangle$$

Taking the limit of each component yields the desired result. ■

Smooth Curves When the component functions of a vector function $\mathbf{r}$ have continuous first derivatives and $\mathbf{r}'(t) \neq \mathbf{0}$ for all t in the open interval (a, b), then $\mathbf{r}$ is said to be a **smooth function** and the curve C traced by $\mathbf{r}$ is called a **smooth curve**.

Geometric Interpretation of $\mathbf{r}'(t)$ If the vector $\mathbf{r}'(t)$ is not $\mathbf{0}$ at a point P, then it may be drawn *tangent to the curve* at P. As seen in Figure 14.5(a) and (b), the vectors

$$\Delta\mathbf{r} = \mathbf{r}(t + \Delta t) - \mathbf{r}(t) \quad \text{and} \quad \frac{\Delta\mathbf{r}}{\Delta t} = \frac{1}{\Delta t}[\mathbf{r}(t + \Delta t) - \mathbf{r}(t)]$$

are parallel. Assuming $\lim_{\Delta t \to 0} \Delta\mathbf{r}/\Delta t$ exists, it seems reasonable to conclude that as $\Delta t \to 0$, $\mathbf{r}(t)$ and $\mathbf{r}(t + \Delta t)$ become close, and, as a consequence, the limiting position of the vector $\Delta\mathbf{r}/\Delta t$ is the tangent line at P. Indeed, the tangent line at P is *defined* as that line through P parallel to $\mathbf{r}'(t)$.

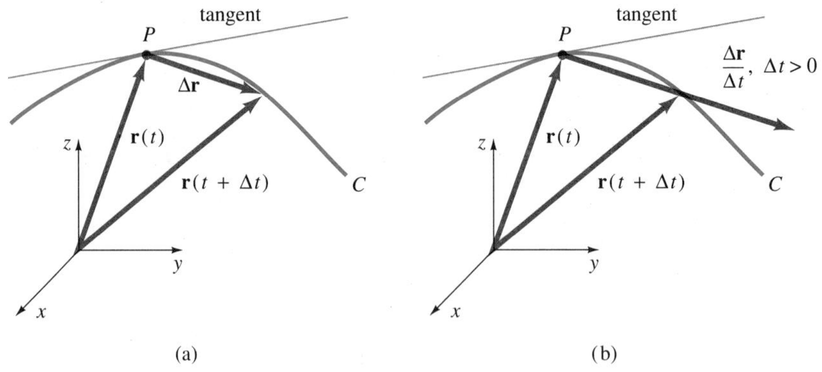

FIGURE 14.5

EXAMPLE 4 Graph the curve C that is traced by a point P whose position is given by $\mathbf{r}(t) = \cos 2t\mathbf{i} + \sin t\mathbf{j}$, $0 \le t \le 2\pi$. Graph $\mathbf{r}'(0)$ and $\mathbf{r}'(\pi/6)$.

Solution We first eliminate the parameter from the parametric equations $x = \cos 2t$, $y = \sin t$:

$$x = \cos 2t = \cos^2 t - \sin^2 t = 1 - 2\sin^2 t = 1 - 2y^2$$

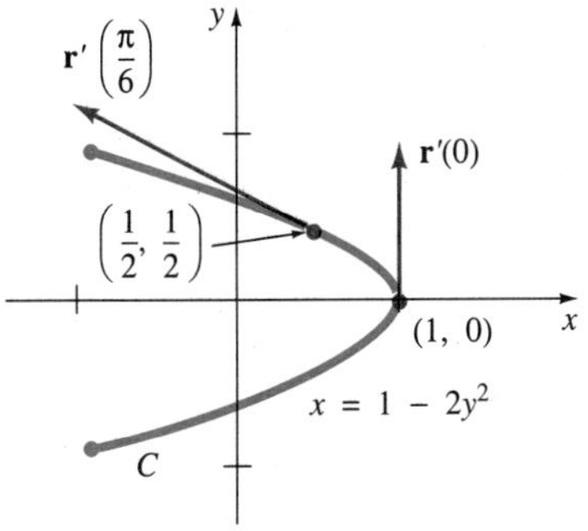

FIGURE 14.6

Since $0 \le t \le 2\pi$, we see that the curve C is the portion of the parabola $x = 1 - 2y^2$ on the interval $-1 \le x \le 1$. Then from

$$\mathbf{r}'(t) = -2 \sin 2t\mathbf{i} + \cos t\mathbf{j}$$

we find
$$\mathbf{r}'(0) = \mathbf{j} \quad \text{and} \quad \mathbf{r}'\left(\frac{\pi}{6}\right) = -\sqrt{3}\mathbf{i} + \frac{\sqrt{3}}{2}\mathbf{j}$$

In Figure 14.6 these vectors are drawn tangent to the curve C at $(1, 0)$ and $(\frac{1}{2}, \frac{1}{2})$, respectively. □

EXAMPLE 5 Find parametric equations of the tangent line to the graph of the curve C whose parametric equations are $x = t^2$, $y = t^2 - t$, $z = -7t$ at $t = 3$.

Solution The vector function that gives the position of a point P on the curve is given by $\mathbf{r}(t) = t^2\mathbf{i} + (t^2 - t)\mathbf{j} - 7t\mathbf{k}$. Now,

$$\mathbf{r}'(t) = 2t\mathbf{i} + (2t - 1)\mathbf{j} - 7\mathbf{k} \quad \text{and so} \quad \mathbf{r}'(3) = 6\mathbf{i} + 5\mathbf{j} - 7\mathbf{k}$$

which is tangent to C at the point whose position vector is

$$\mathbf{r}(3) = 9\mathbf{i} + 6\mathbf{j} - 21\mathbf{k}$$

that is, $P(9, 6, -21)$. Using the components of $\mathbf{r}'(3)$, we see that parametric equations of the tangent line are

$$x = 9 + 6t \qquad y = 6 + 5t \qquad z = -21 - 7t$$ □

Higher-Order Derivatives Higher-order derivatives of a vector function are also obtained by differentiating its components. In the case of the **second derivative**, we have

$$\mathbf{r}''(t) = \langle f''(t), g''(t), h''(t) \rangle = f''(t)\mathbf{i} + g''(t)\mathbf{j} + h''(t)\mathbf{k}$$

EXAMPLE 6 If $\quad \mathbf{r}(t) = (t^3 - 2t^2)\mathbf{i} + 4t\mathbf{j} + e^{-t}\mathbf{k}$

then $\quad \mathbf{r}'(t) = (3t^2 - 4t)\mathbf{i} + 4\mathbf{j} - e^{-t}\mathbf{k}$

and $\quad \mathbf{r}''(t) = (6t - 4)\mathbf{i} + e^{-t}\mathbf{k}$ □

THEOREM 14.3 Chain Rule

If $\mathbf{r}$ is a differentiable vector function and $s = u(t)$ is a differentiable scalar function, then the derivative of $\mathbf{r}(s)$ with respect to t is

$$\frac{d\mathbf{r}}{dt} = \frac{d\mathbf{r}}{ds}\frac{ds}{dt} = \mathbf{r}'(s)u'(t)$$

EXAMPLE 7 If $\mathbf{r}(s) = \cos 2s\mathbf{i} + \sin 2s\mathbf{j} + e^{-3s}\mathbf{k}$, where $s = t^4$, then

$$\frac{d\mathbf{r}}{dt} = [-2\sin 2s\mathbf{i} + 2\cos 2s\mathbf{j} - 3\mathbf{e}^{-3s}\mathbf{k}]4t^3$$
$$= -8t^3\sin(2t^4)\mathbf{i} + 8t^3\cos(2t^4)\mathbf{j} - 12t^3e^{-3t^4}\mathbf{k}$$ □

Details of the proof of the next theorem are left as exercises.

THEOREM 14.4 Rules of Differentiation

Let $\mathbf{r}_1$ and $\mathbf{r}_2$ be differentiable vector functions and $u(t)$ a differentiable scalar function.

(*i*) $\frac{d}{dt}[\mathbf{r}_1(t) + \mathbf{r}_2(t)] = \mathbf{r}_1'(t) + \mathbf{r}_2'(t)$

(*ii*) $\frac{d}{dt}[u(t)\mathbf{r}_1(t)] = u(t)\mathbf{r}_1'(t) + u'(t)\mathbf{r}_1(t)$

(*iii*) $\frac{d}{dt}[\mathbf{r}_1(t)\cdot\mathbf{r}_2(t)] = \mathbf{r}_1(t)\cdot\mathbf{r}_2'(t) + \mathbf{r}_1'(t)\cdot\mathbf{r}_2(t)$

(*iv*) $\frac{d}{dt}[\mathbf{r}_1(t)\times\mathbf{r}_2(t)] = \mathbf{r}_1(t)\times\mathbf{r}_2'(t) + \mathbf{r}_1'(t)\times\mathbf{r}_2(t)$

Note: Since the cross product of two vectors is not commutative, the order in which $\mathbf{r}_1$ and $\mathbf{r}_2$ appear in part (*iv*) must be strictly observed.

Integrals of Vector Functions If f, g, and h are integrable, then the indefinite and definite integrals of a vector function $\mathbf{r}(t) = f(t)\mathbf{i} + g(t)\mathbf{j} + h(t)\mathbf{k}$ are defined, respectively, by

$$\int \mathbf{r}(t)\,dt = \left[\int f(t)\,dt\right]\mathbf{i} + \left[\int g(t)\,dt\right]\mathbf{j} + \left[\int h(t)\,dt\right]\mathbf{k}$$
$$\int_a^b \mathbf{r}(t)\,dt = \left[\int_a^b f(t)\,dt\right]\mathbf{i} + \left[\int_a^b g(t)\,dt\right]\mathbf{j} + \left[\int_a^b h(t)\,dt\right]\mathbf{k}$$

The indefinite integral of $\mathbf{r}$ is another vector $\mathbf{R} + \mathbf{c}$ such that $\mathbf{R}'(t) = \mathbf{r}(t)$.

EXAMPLE 8 If $\mathbf{r}(t) = 6t^2\mathbf{i} + 4e^{-2t}\mathbf{j} + 8\cos 4t\mathbf{k}$

then $$\int \mathbf{r}(t)\,dt = \left[\int 6t^2\,dt\right]\mathbf{i} + \left[\int 4e^{-2t}\,dt\right]\mathbf{j} + \left[\int 8\cos 4t\,dt\right]\mathbf{k}$$
$$= [2t^3 + c_1]\mathbf{i} + [-2e^{-2t} + c_2]\mathbf{j} + [2\sin 4t + c_3]\mathbf{k}$$
$$= 2t^3\mathbf{i} - 2e^{-2t}\mathbf{j} + 2\sin 4t\mathbf{k} + \mathbf{c}$$

where $\mathbf{c} = c_1\mathbf{i} + c_2\mathbf{j} + c_3\mathbf{k}$. □

Length of a Space Curve If $\mathbf{r}(t) = f(t)\mathbf{i} + g(t)\mathbf{j} + h(t)\mathbf{k}$ is a smooth function, then, in a manner similar to that given in Section 12.2, it can be shown that the **length** of the smooth curve traced by $\mathbf{r}$ is given by

$$s = \int_a^b \sqrt{[f'(t)]^2 + [g'(t)]^2 + [h'(t)]^2}\, dt = \int_a^b |\mathbf{r}'(t)|\, dt \tag{3}$$

Arc Length As a Parameter A curve in the plane or in space can be parameterized in terms of the arc length s.

EXAMPLE 9 Consider the helix of Example 1. Since $|\mathbf{r}'(t)| = \sqrt{5}$, it follows from (3) that the length of the curve starting at $\mathbf{r}(0)$ to an arbitrary point $\mathbf{r}(t)$ is

$$s = \int_0^t \sqrt{5}\, du = \sqrt{5}\,t$$

where we have used u as a dummy variable of integration. Using $t = s/\sqrt{5}$, we obtain a vector equation of the helix as a function of arc length:

$$\mathbf{r}(s) = 2\cos\frac{s}{\sqrt{5}}\,\mathbf{i} + 2\sin\frac{s}{\sqrt{5}}\,\mathbf{j} + \frac{s}{\sqrt{5}}\,\mathbf{k} \tag{4}$$

Parametric equations of the helix are then

$$f(s) = 2\cos\frac{s}{\sqrt{5}} \qquad g(s) = 2\sin\frac{s}{\sqrt{5}} \qquad h(s) = \frac{s}{\sqrt{5}}$$ □

The derivative of a vector function $\mathbf{r}(t)$ with respect to the parameter t is a tangent vector to the curve traced by $\mathbf{r}$. However, if the curve is parameterized in terms of arc length s, then

$\mathbf{r}'(s)$ is a unit tangent vector.

To see this, let a curve be described by $\mathbf{r}(s)$, where s is arc length. From (3), the length of the curve from $\mathbf{r}(0)$ to $\mathbf{r}(s)$ is

$$s = \int_0^s |\mathbf{r}'(u)|\, du$$

Differentiation of this last equation with respect to s then yields $|\mathbf{r}'(s)| = 1$.

EXERCISES 14.1

Answers to odd-numbered problems begin on page A-102.

In Problems 1–10 graph the curve traced by the given vector function.

1. $\mathbf{r}(t) = 2\sin t\mathbf{i} + 4\cos t\mathbf{j} + t\mathbf{k},\ t \ge 0$
2. $\mathbf{r}(t) = \cos t\mathbf{i} + t\mathbf{j} + \sin t\mathbf{k},\ t \ge 0$
3. $\mathbf{r}(t) = t\mathbf{i} + 2t\mathbf{j} + \cos t\mathbf{k},\ t \ge 0$
4. $\mathbf{r}(t) = 4\mathbf{i} + 2\cos t\mathbf{j} + 3\sin t\mathbf{k}$
5. $\mathbf{r}(t) = \langle e^t, e^{2t}\rangle$

6. $\mathbf{r}(t) = \cosh t\mathbf{i} + 3\sinh t\mathbf{j}$

7. $\mathbf{r}(t) = \langle\sqrt{2}\sin t, \sqrt{2}\sin t, 2\cos t\rangle, 0 \le t \le \pi/2$

8. $\mathbf{r}(t) = t\mathbf{i} + t^3\mathbf{j} + t\mathbf{k}$

9. $\mathbf{r}(t) = e^t\cos t\mathbf{i} + e^t\sin t\mathbf{j} + e^t\mathbf{k}$

10. $\mathbf{r}(t) = \langle t\cos t, t\sin t, t^2\rangle$

In Problems 11–14 find the vector function that describes the curve C of intersection between the given surfaces. Sketch the curve C. Use the indicated parameter.

11. $z = x^2 + y^2, y = x; x = t$

12. $x^2 + y^2 - z^2 = 1, y = 2x; x = t$

13. $x^2 + y^2 = 9, z = 9 - x^2; x = 3\cos t$

14. $z = x^2 + y^2, z = 1; x = \sin t$

15. Given that $\mathbf{r}(t) = \dfrac{\sin 2t}{t}\mathbf{i} + (t-2)^5\mathbf{j} + t\ln t\mathbf{k}$, find $\lim_{t\to 0^+}\mathbf{r}(t)$.

16. Given that $\lim_{t\to a}\mathbf{r}_1(t) = \mathbf{i} - 2\mathbf{j} + \mathbf{k}$ and $\lim_{t\to a}\mathbf{r}_2(t) = 2\mathbf{i} + 5\mathbf{j} + 7\mathbf{k}$, find
(a) $\lim_{t\to a}[-4\mathbf{r}_1(t) + 3\mathbf{r}_2(t)]$ (b) $\lim_{t\to a}\mathbf{r}_1(t)\cdot\mathbf{r}_2(t)$

In Problems 17–20 find $\mathbf{r}'(t)$ and $\mathbf{r}''(t)$ for the given vector function.

17. $\mathbf{r}(t) = \ln t\mathbf{i} + \dfrac{1}{t}\mathbf{j}, t > 0$

18. $\mathbf{r}(t) = \langle t\cos t - \sin t, t + \cos t\rangle$

19. $\mathbf{r}(t) = \langle te^{2t}, t^3, 4t^2 - t\rangle$

20. $\mathbf{r}(t) = t^2\mathbf{i} + t^3\mathbf{j} + \tan^{-1}t\mathbf{k}$

In Problems 21–24 graph the curve C that is described by $\mathbf{r}$ and graph $\mathbf{r}'$ at the indicated value of t.

21. $\mathbf{r}(t) = 2\cos t\mathbf{i} + 6\sin t\mathbf{j}; t = \pi/6$

22. $\mathbf{r}(t) = t^3\mathbf{i} + t^2\mathbf{j}; t = -1$

23. $\mathbf{r}(t) = 2\mathbf{i} + t\mathbf{j} + \dfrac{4}{1+t^2}\mathbf{k}; t = 1$

24. $\mathbf{r}(t) = 3\cos t\mathbf{i} + 3\sin t\mathbf{j} + 2t\mathbf{k}; t = \pi/4$

In Problems 25 and 26 find parametric equations of the tangent line to the given curve at the indicated value of t.

25. $x = t, y = \dfrac{1}{2}t^2, z = \dfrac{1}{3}t^3; t = 2$

26. $x = t^3 - t, y = \dfrac{6t}{t+1}, z = (2t+1)^2; t = 1$

In Problems 27–32 find the indicated derivative. Assume that all vector functions are differentiable.

27. $\dfrac{d}{dt}[\mathbf{r}(t) \times \mathbf{r}'(t)]$

28. $\dfrac{d}{dt}[\mathbf{r}(t)\cdot(t\mathbf{r}(t))]$

29. $\dfrac{d}{dt}[\mathbf{r}(t)\cdot(\mathbf{r}'(t) \times \mathbf{r}''(t))]$

30. $\dfrac{d}{dt}[\mathbf{r}_1(t) \times (\mathbf{r}_2(t) \times \mathbf{r}_3(t))]$

31. $\dfrac{d}{dt}\left[\mathbf{r}_1(2t) + \mathbf{r}_2\left(\dfrac{1}{t}\right)\right]$

32. $\dfrac{d}{dt}[t^3\mathbf{r}(t^2)]$

In Problems 33–36 evaluate the given integral.

33. $\displaystyle\int_{-1}^{2}(t\mathbf{i} + 3t^2\mathbf{j} + 4t^3\mathbf{k})\,dt$

34. $\displaystyle\int_{0}^{4}(\sqrt{2t+1}\mathbf{i} - \sqrt{t}\mathbf{j} + \sin\pi t\mathbf{k})\,dt$

35. $\displaystyle\int(te^t\mathbf{i} - e^{-2t}\mathbf{j} + te^{t^2}\mathbf{k})\,dt$

36. $\displaystyle\int\frac{1}{1+t^2}(\mathbf{i} + t\mathbf{j} + t^2\mathbf{k})\,dt$

In Problems 37–40 find a vector function $\mathbf{r}$ that satisfies the indicated conditions.

37. $\mathbf{r}'(t) = 6\mathbf{i} + 6t\mathbf{j} + 3t^2\mathbf{k}; \mathbf{r}(0) = \mathbf{i} - 2\mathbf{j} + \mathbf{k}$

38. $\mathbf{r}'(t) = t\sin t^2\mathbf{i} - \cos 2t\mathbf{j}; \mathbf{r}(0) = \frac{3}{2}\mathbf{i}$

39. $\mathbf{r}''(t) = 12t\mathbf{i} - 3t^{-1/2}\mathbf{j} + 2\mathbf{k}; \mathbf{r}'(1) = \mathbf{j}, \mathbf{r}(1) = 2\mathbf{i} - \mathbf{k}$

40. $\mathbf{r}''(t) = \sec^2 t\mathbf{i} + \cos t\mathbf{j} - \sin t\mathbf{k}; \mathbf{r}'(0) = \mathbf{i} + \mathbf{j} + \mathbf{k}$, $\mathbf{r}(0) = -\mathbf{j} + 5\mathbf{k}$

In Problems 41–44 find the length of the curve traced by the given vector function on the indicated interval.

41. $\mathbf{r}(t) = a\cos t\mathbf{j} + a\sin t\mathbf{j} + ct\mathbf{k}; 0 \le t \le 2\pi$

42. $\mathbf{r}(t) = t\mathbf{i} + t\cos t\mathbf{j} + t\sin t\mathbf{k}; 0 \le t \le \pi$

43. $\mathbf{r}(t) = e^t\cos 2t\mathbf{i} + e^t\sin 2t\mathbf{j} + e^t\mathbf{k}; 0 \le t \le 3\pi$

44. $\mathbf{r}(t) = 3t\mathbf{i} + \sqrt{3}t^2\mathbf{j} + \frac{2}{3}t^3\mathbf{k}; 0 \le t \le 1$

45. Express the vector equation of a circle $\mathbf{r}(t) = a\cos t\mathbf{i} + a\sin t\mathbf{j}$ as a function of arc length s. Verify that $\mathbf{r}'(s)$ is a unit vector.

46. If $\mathbf{r}(s)$ is a vector function given in (4), verify that $\mathbf{r}'(s)$ is a unit vector.

47. Suppose $\mathbf{r}$ is a differentiable vector function for which $|\mathbf{r}(t)| = c$ for all t. Show that the tangent vector $\mathbf{r}'(t)$ is perpendicular to the position vector $\mathbf{r}(t)$ for all t.

48. In Problem 47 describe geometrically the kind of curve C for which $|\mathbf{r}(t)| = c$.

Miscellaneous Problems

49. Prove Theorem 14.4(*ii*).

50. Prove Theorem 14.4(*iii*).

51. Prove Theorem 14.4(*iv*).

52. If $\mathbf{v}$ is a constant vector and $\mathbf{r}$ is integrable on $[a, b]$, prove that $\int_a^b \mathbf{v} \cdot \mathbf{r}(t)\, dt = \mathbf{v} \cdot \int_a^b \mathbf{r}(t)\, dt$.

14.2 MOTION ON A CURVE; VELOCITY AND ACCELERATION

Suppose a body or particle moves along a curve C so that its position at time t is given by

$$\mathbf{r}(t) = f(t)\mathbf{i} + g(t)\mathbf{j} + h(t)\mathbf{k}$$

If f, g, and h have second derivatives, then the vectors

$$\mathbf{v}(t) = \mathbf{r}'(t) = f'(t)\mathbf{i} + g'(t)\mathbf{j} + h'(t)\mathbf{k}$$
$$\mathbf{a}(t) = \mathbf{r}''(t) = f''(t)\mathbf{i} + g''(t)\mathbf{j} + h''(t)\mathbf{k}$$

are called the **velocity** and **acceleration** of the particle, respectively. The scalar function $|\mathbf{v}(t)|$ is the **speed** of the particle. Since

$$|\mathbf{v}(t)| = \left|\frac{d\mathbf{r}}{dt}\right| = \sqrt{\left(\frac{dx}{dt}\right)^2 + \left(\frac{dy}{dt}\right)^2 + \left(\frac{dz}{dt}\right)^2}$$

speed is related to arc length s by $s'(t) = |\mathbf{v}(t)|$. In other words, arc length is given by

$$s = \int_{t_0}^{t_1} |\mathbf{v}(t)|\, dt$$

It also follows from the discussion of Section 14.1 that if $P(x_1, y_1, z_1)$ is the position of the particle on C at time t_1, then we may draw

$$\mathbf{v}(t_1) \text{ tangent to } C \text{ at } P$$

Similar remarks hold for curves traced by the vector function

$$\mathbf{r}(t) = f(t)\mathbf{i} + g(t)\mathbf{j}$$

EXAMPLE 1 The position of a moving particle is given by

$$\mathbf{r}(t) = t^2\mathbf{i} + t\mathbf{j} + \frac{5}{2}t\mathbf{k}$$

Graph the curve defined by $\mathbf{r}(t)$ and the vectors $\mathbf{v}(2)$ and $\mathbf{a}(2)$.

Solution Since $x = t^2$, $y = t$, the path of the particle is above the parabola $x = y^2$. When $t = 2$, the position vector $\mathbf{r}(2) = 4\mathbf{i} + 2\mathbf{j} + 5\mathbf{k}$ indicates that the particle is at the point $P(4, 2, 5)$. Now,

$$\mathbf{v}(t) = \mathbf{r}'(t) = 2t\mathbf{i} + \mathbf{j} + \frac{5}{2}\mathbf{k} \quad \text{and} \quad \mathbf{a}(t) = \mathbf{r}''(t) = 2\mathbf{i}$$

so that

$$\mathbf{v}(2) = 4\mathbf{i} + \mathbf{j} + \frac{5}{2}\mathbf{k} \quad \text{and} \quad \mathbf{a}(2) = 2\mathbf{i}$$

These vectors are shown in Figure 14.7. □

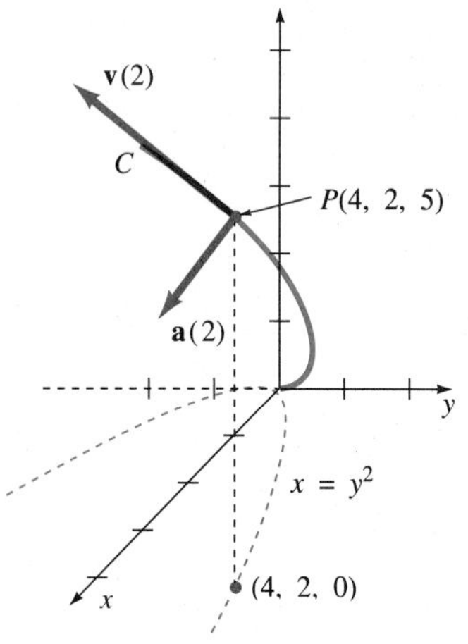

FIGURE 14.7

If a particle moves with a constant speed c, then its acceleration vector is perpendicular to the velocity vector $\mathbf{v}$. To see this, note that

$$|\mathbf{v}|^2 = c^2 \quad \text{or} \quad \mathbf{v} \cdot \mathbf{v} = c^2$$

We differentiate both sides with respect to t, and obtain, with the aid of Theorem 14.4(*iii*),

$$\frac{d}{dt}(\mathbf{v} \cdot \mathbf{v}) = \mathbf{v} \cdot \frac{d\mathbf{v}}{dt} + \frac{d\mathbf{v}}{dt} \cdot \mathbf{v} = 2\mathbf{v} \cdot \frac{d\mathbf{v}}{dt} = 0$$

Thus,

$$\frac{d\mathbf{v}}{dt} \cdot \mathbf{v} = 0 \quad \text{or} \quad \mathbf{a}(t) \cdot \mathbf{v}(t) = 0 \qquad \text{for all } t$$

EXAMPLE 2 Suppose the vector function in Example 2 of Section 14.1 represents the position of a particle moving in a circular orbit. Graph the velocity and acceleration vectors at $t = \pi/4$.

Solution Recall that

$$\mathbf{r}(t) = 2\cos t\mathbf{i} + 2\sin t\mathbf{j} + 3\mathbf{k}$$

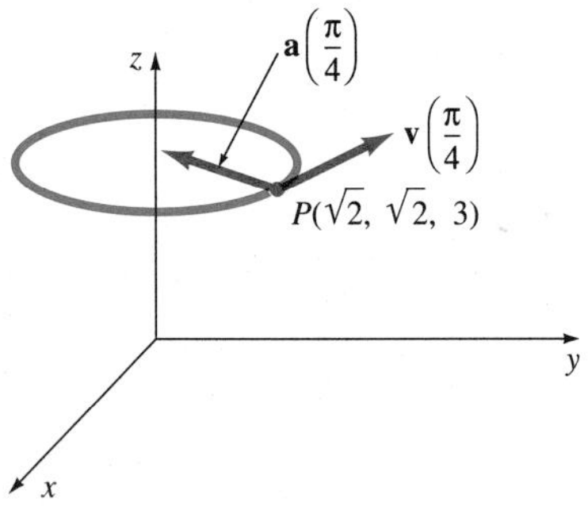

FIGURE 14.8

is the position vector of a particle moving in a circular orbit of radius 2 in the plane $z = 3$. When $t = \pi/4$, the particle is at the point $P(\sqrt{2}, \sqrt{2}, 3)$. Now,

$$\mathbf{v}(t) = \mathbf{r}'(t) = -2\sin t\mathbf{i} + 2\cos t\mathbf{j}$$
$$\mathbf{a}(t) = \mathbf{r}''(t) = -2\cos t\mathbf{i} - 2\sin t\mathbf{j}$$

Since the speed is $|\mathbf{v}(t)| = 2$ for all time t, it follows from the discussion preceding this example that $\mathbf{a}(t)$ is perpendicular to $\mathbf{v}(t)$. (Verify this.) As shown in Figure 14.8, the vectors

$$\mathbf{v}\left(\frac{\pi}{4}\right) = -2\sin\frac{\pi}{4}\mathbf{i} + 2\cos\frac{\pi}{4}\mathbf{j} = -\sqrt{2}\mathbf{i} + \sqrt{2}\mathbf{j}$$

$$\mathbf{a}\left(\frac{\pi}{4}\right) = -2\cos\frac{\pi}{4}\mathbf{i} - 2\sin\frac{\pi}{4}\mathbf{j} = -\sqrt{2}\mathbf{i} - \sqrt{2}\mathbf{j}$$

are drawn at the point P. The vector $\mathbf{v}(\pi/4)$ is tangent to the circular path and is $\mathbf{a}(\pi/4)$ points along a radius toward the center of the circle. □

Centripetal Acceleration For circular motion in the plane, described by $\mathbf{r}(t) = r_0\cos\omega t\mathbf{i} + r_0\sin\omega t\mathbf{j}$, r_0 and ω constants, it is evident that $\mathbf{r}'' =$

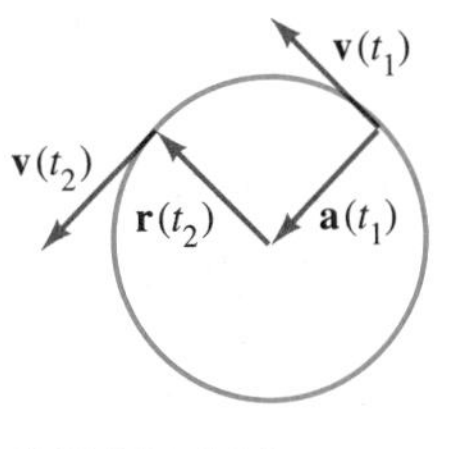

FIGURE 14.9

$-\omega^2\mathbf{r}$. This means that the acceleration vector $\mathbf{a}(t) = \mathbf{r}''(t)$ points in the direction opposite to that of the position vector $\mathbf{r}(t)$. We then say $\mathbf{a}(t)$ is **centripetal acceleration**. See Figure 14.9. If $v = |\mathbf{v}(t)|$ and $a = |\mathbf{a}(t)|$, we leave it as an exercise to show that $a = v^2/r_0$.

Curvilinear Motion in the Plane Many important applications of vector functions occur in describing curvilinear motion in a plane. For example, planetary and projectile motions take place in a plane.

In analyzing the motion of short-range ballistic projectiles,* we begin with the acceleration of gravity written in vector form

$$\mathbf{a}(t) = -g\mathbf{j}$$

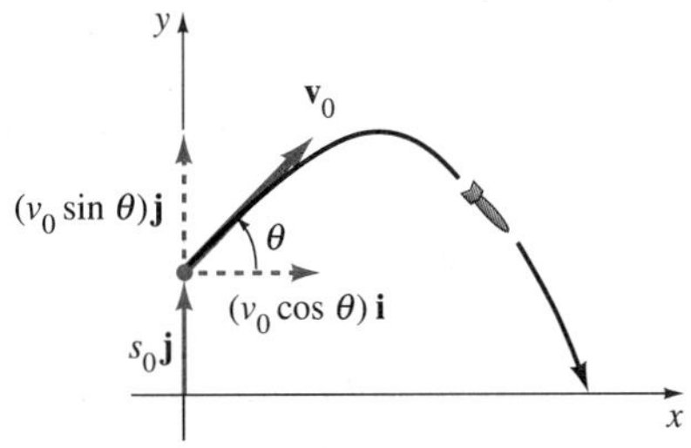

FIGURE 14.10

If, as shown in Figure 14.10, a projectile is launched with an initial velocity $\mathbf{v}_0 = v_0\cos\theta\mathbf{i} + v_0\sin\theta\mathbf{j}$ from an initial height $\mathbf{s}_0 = s_0\mathbf{j}$, then

$$\mathbf{v}(t) = \int(-g\mathbf{j})\,dt = -gt\mathbf{j} + \mathbf{c}_1$$

where $\mathbf{v}(0) = \mathbf{v}_0$ implies that $\mathbf{c}_1 = \mathbf{v}_0$. Therefore,

$$\mathbf{v}(t) = (v_0\cos\theta)\mathbf{i} + (-gt + v_0\sin\theta)\mathbf{j}$$

Integrating again and using $\mathbf{r}(0) = \mathbf{s}_0$ yield

$$\mathbf{r}(t) = (v_0\cos\theta)t\mathbf{i} + \left[-\frac{1}{2}gt^2 + (v_0\sin\theta)t + s_0\right]\mathbf{j}$$

Hence, parametric equations for the trajectory of the projectile are

$$x(t) = (v_0\cos\theta)t \qquad y(t) = -\frac{1}{2}gt^2 + (v_0\sin\theta)t + s_0 \tag{1}$$

We are naturally interested in finding the maximum height H and the range R attained by a projectile. As shown in Figure 14.11, these quantities are the maximum values of $y(t)$ and $x(t)$, respectively.

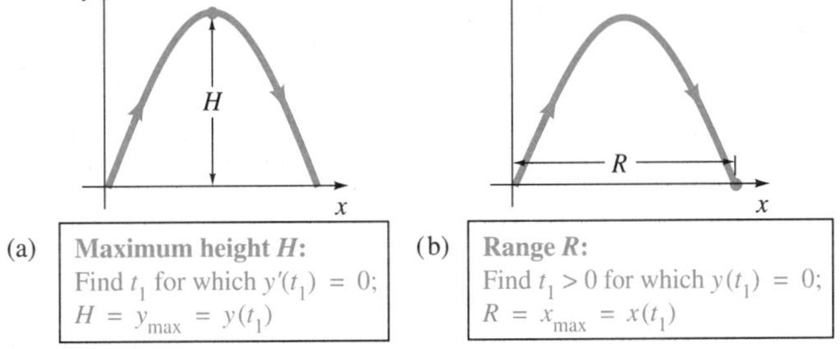

FIGURE 14.11

*The projectile is shot or hurled rather than self-propelled. In the analysis of *long-range* ballistic motion, the curvature of the earth must be taken into consideration.

EXAMPLE 3 A shell is fired from ground level with an initial speed of 768 ft/s at an angle of elevation of 30°. Find (*a*) the vector function and parametric equations of the shell's trajectory, (*b*) the maximum altitude attained, (*c*) the range of the shell, and (*d*) the speed at impact.

Solution (*a*) Initially, we have $\mathbf{s}_0 = \mathbf{0}$ and

$$\begin{aligned}\mathbf{v}_0 &= (768 \cos 30°)\mathbf{i} + (768 \sin 30°)\mathbf{j} \\ &= 384\sqrt{3}\mathbf{i} + 384\mathbf{j}\end{aligned} \tag{2}$$

Integrating $\mathbf{a}(t) = -32\mathbf{j}$ and using (2) give

$$\mathbf{v}(t) = (384\sqrt{3})\mathbf{i} + (-32t + 384)\mathbf{j} \tag{3}$$

Integrating again gives

$$\mathbf{r}(t) = (384\sqrt{3}t)\mathbf{i} + (-16t^2 + 384t)\mathbf{j}$$

Hence, the parametric equations of the shell's trajectory are

$$x(t) = 384\sqrt{3}t \qquad y(t) = -16t^2 + 384t \tag{4}$$

(*b*) From (4) we see that $dy/dt = 0$ when

$$-32t + 384 = 0 \quad \text{or} \quad t = 12$$

Thus, the maximum height H attained by the shell is

$$\begin{aligned}H = y(12) &= -16(12)^2 + 384(12) \\ &= 2304 \text{ ft}\end{aligned}$$

(*c*) From (4) we see that $y(t) = 0$ when

$$-16t(t - 24) = 0 \quad \text{or} \quad t = 0, t = 24$$

The range R is then

$$R = x(24) = 384\sqrt{3}(24) \approx 15{,}963 \text{ ft}$$

(*d*) From (3) we obtain the impact speed of the shell:

$$\begin{aligned}|\mathbf{v}(24)| &= \sqrt{(-384)^2 + (384\sqrt{3})^2} \\ &= 768 \text{ ft/s}\end{aligned}$$

□

▲ **Remark** We have seen that the rate of change of arc length ds/dt is the same as the speed $|\mathbf{v}(t)| = |\mathbf{r}'(t)|$. However, as we shall see in the next section, it does *not* follow that the *scalar acceleration* d^2s/dt^2 is the same as $|\mathbf{a}(t)| = |\mathbf{r}''(t)|$. See Problem 20 in Exercises 14.2.

EXERCISES 14.2 *Answers to odd-numbered problems begin on page A-102.*

In Problems 1–8 $\mathbf{r}(t)$ is the position vector of a moving particle. Graph the curve and the velocity and acceleration vectors at the indicated time. Find the speed at that time.

1. $\mathbf{r}(t) = t^2\mathbf{i} + \frac{1}{4}t^4\mathbf{j}$; $t = 1$

2. $\mathbf{r}(t) = t^2\mathbf{i} + \frac{1}{t^2}\mathbf{j}$; $t = 1$

3. $\mathbf{r}(t) = -\cosh 2t\mathbf{i} + \sinh 2t\mathbf{j}$; $t = 0$

4. $\mathbf{r}(t) = 2\cos t\mathbf{i} + (1 + \sin t)\mathbf{j}$; $t = \pi/3$

5. $\mathbf{r}(t) = 2\mathbf{i} + (t-1)^2\mathbf{j} + t\mathbf{k}$; $t = 2$

6. $\mathbf{r}(t) = t\mathbf{i} + t\mathbf{j} + t^3\mathbf{k}$; $t = 2$

7. $\mathbf{r}(t) = t\mathbf{i} + t^2\mathbf{j} + t^3\mathbf{k}$; $t = 1$

8. $\mathbf{r}(t) = t\mathbf{i} + t^3\mathbf{j} + t\mathbf{k}$; $t = 1$

9. Suppose $\mathbf{r}(t) = t^2\mathbf{i} + (t^3 - 2t)\mathbf{j} + (t^2 - 5t)\mathbf{k}$ is the position vector of a moving particle. At what points does the particle pass through the xy-plane? What are its velocity and acceleration at these points?

10. Suppose a particle moves in space so that $\mathbf{a}(t) = \mathbf{0}$ for all time t. Describe its path.

11. A shell is fired from ground level with an initial speed of 480 ft/s at an angle of elevation of 30°. Find:

 (a) a vector function and parametric equations of the shell's trajectory,

 (b) the maximum altitude attained,

 (c) the range of the shell, and

 (d) the speed at impact.

12. Rework Problem 11 if the shell is fired with the same initial speed and the same angle of elevation but from a cliff 1600 ft high.

13. A used car is pushed off an 81-ft-high sheer seaside cliff with a speed of 4 ft/s. Find the speed at which the car hits the water.

14. A small projectile is launched from ground level with an initial speed of 98 m/s. Find the possible angles of elevation so that its range is 490 m.

15. A football quarterback throws a 100-yd "bomb" at an angle of 45° from the horizontal. What is the initial speed of the football at the point of release?

16. A quarterback throws a football with the same initial speed at an angle of 60° from the horizontal and then at an angle of 30° from the horizontal. Show that the range of the football is the same in each case. Generalize this result for any release angle $0 < \theta < \pi/2$.

17. A projectile is fired from a cannon directly at a target that is dropped from rest simultaneously as the cannon is fired. Show that the projectile will strike the target in midair. See Figure 14.12. [*Hint:* Assume that the origin is at the muzzle of the cannon and that the angle of elevation is θ. If $\mathbf{r}_p$ and $\mathbf{r}_t$ are position vectors of the projectile and target, respectively, is there a time at which $\mathbf{r}_p = \mathbf{r}_t$?]

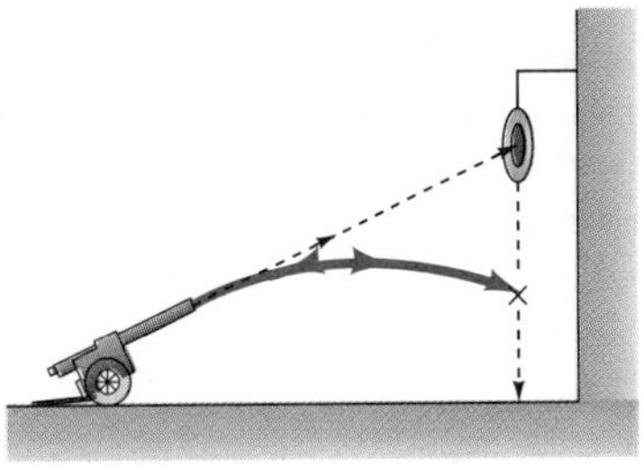

FIGURE 14.12

18. In army field maneuvers sturdy equipment and supply packs are simply dropped from planes that fly horizontally at a slow speed and a low altitude. A supply plane flies horizontally over a target at an altitude of 1024 ft at a constant speed of 180 mph. Use (1) to determine the horizontal distance a supply pack travels relative to the point from which it was dropped. At what line-of-sight angle α should the supply pack be released in order to hit the target indicated in Figure 14.13?

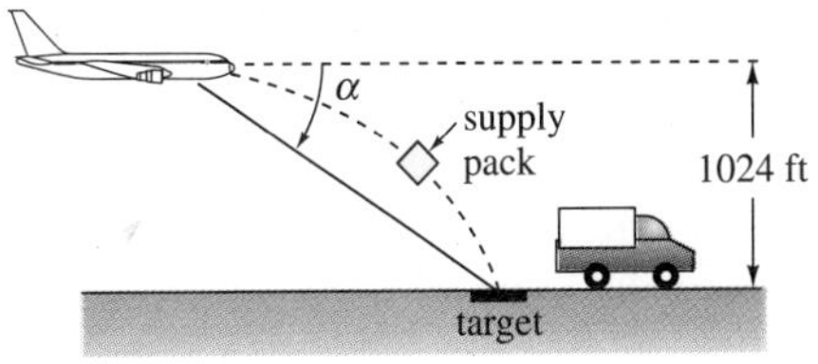

FIGURE 14.13

19. Suppose that $\mathbf{r}(t) = r_0\cos\omega t\mathbf{i} + r_0\sin\omega t\mathbf{j}$ is the position vector of an object that is moving in a circle of radius r_0 in the xy-plane. If $|\mathbf{v}(t)| = v$, show that the magnitude of the centripetal acceleration is $a = |\mathbf{a}(t)| = v^2/r_0$.

20. The motion of a particle in space is described by

$$\mathbf{r}(t) = b\cos t\mathbf{i} + b\sin t\mathbf{j} + ct\mathbf{k}, \qquad t \geq 0$$

(a) Compute $|\mathbf{v}(t)|$.

(b) Compute $s = \int_0^t |\mathbf{v}(t)|\,dt$ and verify that ds/dt is the same as the result of part (a).

(c) Verify that $d^2s/dt^2 \neq |\mathbf{a}(t)|$.

21. The **effective weight** w_e of a body of mass m at the equator of the earth is defined by $w_e = mg - ma$, where a is the magnitude of the centripetal acceleration given in Problem 19. Determine the effective weight of a 192-lb person if the radius of the earth is 4000 mi, $g = 32$ ft/s^2, and $v = 1530$ ft/s.

22. Consider a bicyclist riding on a flat circular track of radius r_0. If m is the combined mass of the rider and bicycle, fill in the blanks in Figure 14.14. [*Hint:* Use Problem 19 and force = mass × acceleration. Assume that the positive directions are upward and to the left.] The **resultant** vector **U** gives the direction the bicyclist must be tipped to avoid falling. Find the angle ϕ from the vertical at which the bicyclist must be tipped if her speed is 44 ft/s and the radius of the track is 60 ft.

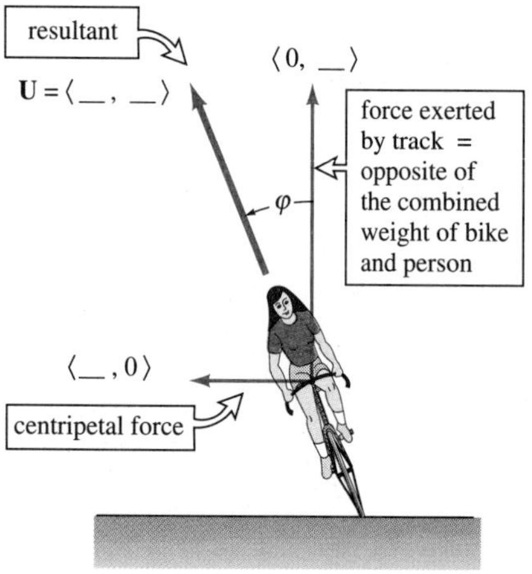

FIGURE 14.14

23. Use the results given in (1) to show that the trajectory of a ballistic projecile is parabolic.

24. A projectile is launched with an initial speed v_0 from ground level at an angle of elevation θ. Use (1) to show that the maximum height and range of the projectile are

$$H = \frac{v_0^2 \sin^2\theta}{2g} \quad \text{and} \quad R = \frac{v_0^2 \sin 2\theta}{g}$$

respectively.

25. The velocity of a particle moving in a fluid is described by means of a **velocity field** $\mathbf{v} = v_1\mathbf{i} + v_2\mathbf{j} + v_3\mathbf{k}$, where the components v_1, v_2, and v_3 are functions of x, y, z, and time t. If the velocity of the particle is $\mathbf{v}(t) = 6t^2x\mathbf{i} - 4ty^2\mathbf{j} + 2t(z+1)\mathbf{k}$, find $\mathbf{r}(t)$. [*Hint:* Use separation of variables.]

26. Suppose m is the mass of a moving particle. Newton's second law of motion can be written in vector form as

$$\mathbf{F} = m\mathbf{a} = \frac{d}{dt}(m\mathbf{v}) = \frac{d\mathbf{p}}{dt}$$

where $\mathbf{p} = m\mathbf{v}$ is called **linear momentum**. The **angular momentum** of the particle with respect to the origin is defined to be $\mathbf{L} = \mathbf{r} \times \mathbf{p}$, where $\mathbf{r}$ is its position vector. If the torque of the particle about the origin is $\boldsymbol{\tau} = \mathbf{r} \times \mathbf{F} = \mathbf{r} \times d\mathbf{p}/dt$, show that $\boldsymbol{\tau}$ is the time rate of change of angular momentum.

27. Suppose the sun is located at the origin. The gravitational force **F** exerted on a planet of mass m by the sun of mass M is

$$\mathbf{F} = -k\frac{Mm}{r^2}\mathbf{u}$$

F is a **central force**—that is, a force directed along the position vector **r** of the planet. Here k is the gravitational constant (see page 357), $r = |\mathbf{r}|$, $\mathbf{u} = \mathbf{r}/r$ is a unit vector in the direction of **r**, and the minus sign indicates that **F** is an attractive force—that is, a force directed toward the sun. See Figure 14.15.

(a) Use Problem 26 to show that the torque acting on the planet due to this central force is **0**.

(b) Explain why the angular momentum **L** of a planet is constant.

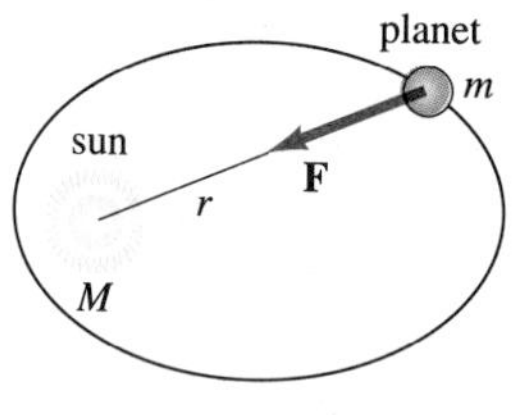

FIGURE 14.15

28. (*This problem could present a challenge.*) In this problem you are going to use the properties in Sections 13.4 and 14.1 to prove **Kepler's first law of planetary motion**: the orbit of a planet is an ellipse with the sun at one focus. We assume that the sun is of mass M and is located at the origin, **r** is the position vector of a body of mass m moving under the gravitational attraction of the sun, and $\mathbf{u} = \mathbf{r}/r$ is a unit vector in the direction of **r**.

(a) Use Problem 27 and Newton's second law of motion $\mathbf{F} = m\mathbf{a}$ to show that

$$\frac{d^2\mathbf{r}}{dt^2} = -kM\frac{\mathbf{u}}{r^2}$$

(b) Use part (a) to show that $\mathbf{r} \times \mathbf{r}'' = \mathbf{0}$.

(c) Use part (b) to show that $\dfrac{d}{dt}(\mathbf{r} \times \mathbf{v}) = \mathbf{0}$.

(d) It follows from part (c) that $\mathbf{r} \times \mathbf{v} = \mathbf{c}$, where $\mathbf{c}$ is a constant vector. Show that $\mathbf{c} = r^2(\mathbf{u} \times \mathbf{u}')$.

(e) Show that $\dfrac{d}{dt}(\mathbf{u} \cdot \mathbf{u}) = 0$ and consequently $\mathbf{u} \cdot \mathbf{u}' = 0$.

(f) Use parts (a), (e), and (d) to show that

$$\frac{d}{dt}(\mathbf{v} \times \mathbf{c}) = kM\frac{d\mathbf{u}}{dt}$$

(g) After integrating the result in part (f) with respect to t, it follows that $\mathbf{v} \times \mathbf{c} = kM\mathbf{u} + \mathbf{d}$, where $\mathbf{d}$ is another constant vector. Dot both sides of this last expression by the vector $\mathbf{r} = r\mathbf{u}$ and use Problem 59 in Exercises 13.4 to show that

$$r = \frac{c^2/kM}{1 + (d/kM)\cos\theta}$$

where $c = |\mathbf{c}|$, $d = |\mathbf{d}|$, and θ is the angle between $\mathbf{d}$ and $\mathbf{r}$.

(h) Explain why the result in part (g) proves Kepler's first law.

(i) At perihelion (see page 700) the vectors $\mathbf{r}$ and $\mathbf{v}$ are perpendicular and have magnitudes r_0 and v_0, respectively. Use this information and parts (d) and (g) to show that $c = r_0 v_0$ and $d = r_0 v_0^2 - kM$.

14.3 CURVATURE; COMPONENTS OF ACCELERATION

Curvature Let C be a smooth curve in 2- or 3-space traced out by a vector function $\mathbf{r}(t)$. In this section we shall consider the acceleration vector $\mathbf{a}(t) = \mathbf{r}''(t)$ introduced in the last section in greater detail. But before doing this, we need to examine a scalar quantity called the **curvature** of a curve. We know that $\mathbf{r}'(t)$ is a tangent vector to the curve C and consequently

$$\mathbf{T} = \frac{\mathbf{r}'(t)}{|\mathbf{r}'(t)|} \tag{1}$$

is a **unit tangent**. But recall from the end of Section 14.1 that if C is parameterized by arc length s, then a unit tangent to the curve is also given by $d\mathbf{r}/ds$. The quantity $|\mathbf{r}'(t)|$ in (1) is related to arc length s by $ds/dt = |\mathbf{r}'(t)|$. Since the curve C is smooth, we know from page 782 that $ds/dt > 0$. Hence, by the Chain Rule,

$$\frac{d\mathbf{r}}{dt} = \frac{d\mathbf{r}}{ds}\frac{ds}{dt}$$

and so

$$\frac{d\mathbf{r}}{ds} = \frac{d\mathbf{r}/dt}{ds/dt} = \frac{\mathbf{r}'(t)}{|\mathbf{r}'(t)|} = \mathbf{T} \tag{2}$$

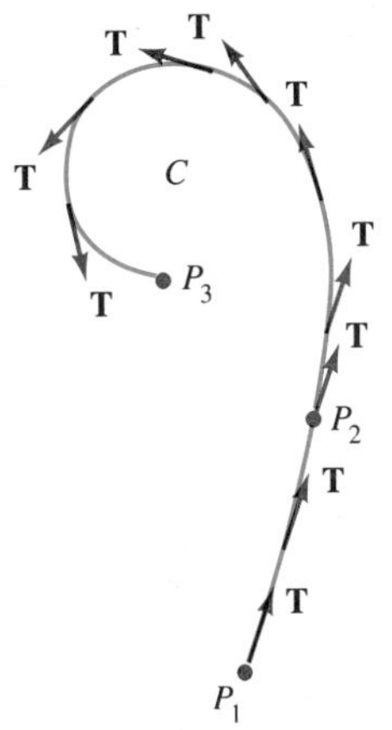

FIGURE 14.16

Now suppose C is as shown in Figure 14.16. As s increases, $\mathbf{T}$ moves along C changing direction but not length (it is always of unit length). Along the portion of the curve between P_1 and P_2 the vector $\mathbf{T}$ varies little in direction; along the curve between P_2 and P_3, where C obviously bends more sharply, the change in the direction of the tangent $\mathbf{T}$ is more pronounced. We use the *rate* at which the unit vector $\mathbf{T}$ changes direction with respect to arc length as an indicator of the curvature of a smooth curve C.

DEFINITION 14.4 Curvature

Let $\mathbf{r}(t)$ be a vector function defining a smooth curve C. If s is the arc length parameter and $\mathbf{T} = d\mathbf{r}/ds$ is the unit tangent vector, then the **curvature** of C at a point is

$$\kappa = \left|\frac{d\mathbf{T}}{ds}\right| \tag{3}$$

The symbol κ in (3) is the Greek letter kappa. Now, since curves are often not parameterized by arc length, it is convenient to express (3) in terms of a general parameter t. Using the Chain Rule again, we can write

$$\frac{d\mathbf{T}}{dt} = \frac{d\mathbf{T}}{ds}\frac{ds}{dt} \quad \text{and consequently} \quad \frac{d\mathbf{T}}{ds} = \frac{d\mathbf{T}/dt}{ds/dt}$$

In other words, curvature is given by

$$\kappa = \frac{|\mathbf{T}'(t)|}{|\mathbf{r}'(t)|} \tag{4}$$

EXAMPLE 1 Find the curvature of a circle of radius a.

Solution A circle can be described by the vector function $\mathbf{r}(t) = a \cos t\mathbf{i} + a \sin t\mathbf{j}$. Now from $\mathbf{r}'(t) = -a \sin t\mathbf{i} + a \cos t\mathbf{j}$ and $|\mathbf{r}'(t)| = a$, we get

$$\mathbf{T}(t) = \frac{\mathbf{r}'(t)}{|\mathbf{r}'(t)|} = -\sin t\mathbf{i} + \cos t\mathbf{j} \quad \text{and} \quad \mathbf{T}'(t) = -\cos t\mathbf{i} - \sin t\mathbf{j}$$

Hence, from (4) the curvature is

$$\kappa = \frac{|\mathbf{T}'(t)|}{|\mathbf{r}'(t)|} = \frac{\sqrt{\cos^2 t + \sin^2 t}}{a} = \frac{1}{a} \tag{5}$$ □

FIGURE 14.17

The result in (5) shows that the curvature at a point on a circle is the reciprocal of the radius of the circle and indicates a fact that is in keeping with our intuition: a circle with a small radius curves more than one with a large radius. See Figure 14.17.

Tangential and Normal Components of Acceleration Suppose a particle moves in 2- or 3-space on a smooth curve C described by the vector function $\mathbf{r}(t)$. Then the velocity of the particle on C is $\mathbf{v}(t) = \mathbf{r}'(t)$, whereas its speed is $ds/dt = v = |\mathbf{v}(t)|$. Thus, (1) implies $\mathbf{v}(t) = v\mathbf{T}$. Differentiating this last expression with respect to t gives acceleration:

$$\mathbf{a}(t) = v\frac{d\mathbf{T}}{dt} + \frac{dv}{dt}\mathbf{T} \tag{6}$$

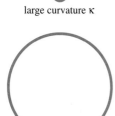

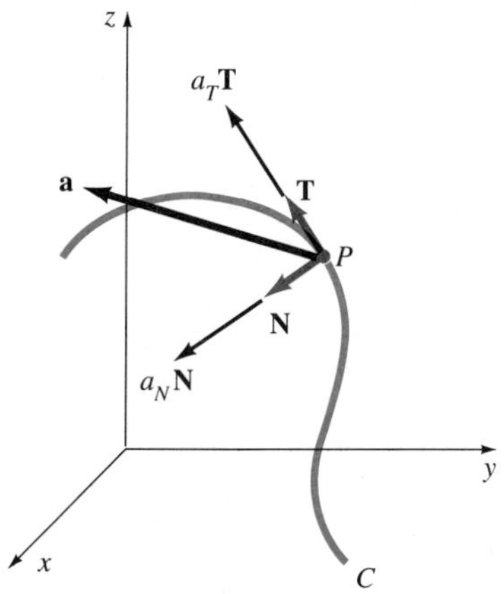

FIGURE 14.18

Furthermore, with the help of Theorem 14.4(*iii*), it follows from differentiation of $\mathbf{T}\cdot\mathbf{T} = 1$ that $\mathbf{T}\cdot d\mathbf{T}/dt = 0$. Hence, at a point P on C the vectors $\mathbf{T}$ and $d\mathbf{T}/dt$ are orthogonal. If $|d\mathbf{T}/dt| \neq 0$, the vector

$$\mathbf{N} = \frac{d\mathbf{T}/dt}{|d\mathbf{T}/dt|} \tag{7}$$

is a unit normal to the curve C at P with direction given by $d\mathbf{T}/dt$. The vector $\mathbf{N}$ is also called the **principal normal**. But, since curvature is $\kappa = |d\mathbf{T}/dt|/v$, it follows from (7) that $d\mathbf{T}/dt = \kappa v\mathbf{N}$. Thus, (6) becomes

$$\mathbf{a}(t) = \kappa v^2\mathbf{N} + \frac{dv}{dt}\mathbf{T} \tag{8}$$

By writing (8) as

$$\mathbf{a}(t) = a_N\mathbf{N} + a_T\mathbf{T} \tag{9}$$

we see that the acceleration vector $\mathbf{a}$ of the moving particle is the sum of two orthogonal vectors $a_N\mathbf{N}$ and $a_T\mathbf{T}$. See Figure 14.18. The scalar functions $a_T = dv/dt$ and $a_N = \kappa v^2$ are called the **tangential** and **normal components of the acceleration**, respectively. Note that the tangential component of the acceleration results from a change in the *magnitude* of the velocity $\mathbf{v}$, whereas the normal component of the acceleration results from a change in the *direction* of $\mathbf{v}$.

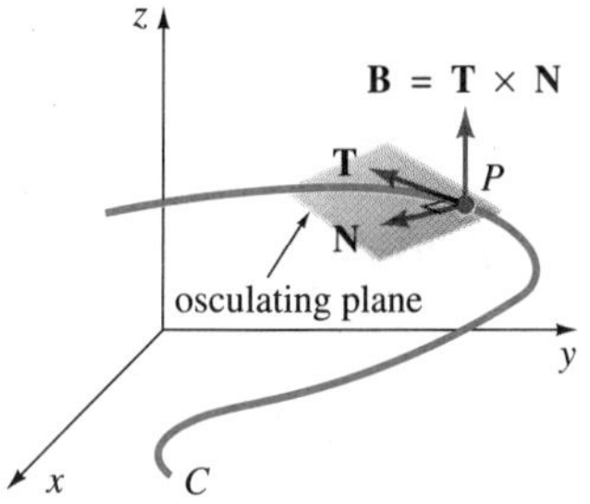

FIGURE 14.19

The Binormal A third unit vector defined by

$$\mathbf{B} = \mathbf{T} \times \mathbf{N}$$

is called the **binormal**. The three unit vectors $\mathbf{T}$, $\mathbf{N}$, and $\mathbf{B}$ form a right-handed set of mutually orthogonal vectors called the **moving trihedral**. The plane of $\mathbf{T}$ and $\mathbf{N}$ is called the **osculating plane**,* the plane of $\mathbf{N}$ and $\mathbf{B}$ is said to be the **normal plane**, and the plane of $\mathbf{T}$ and $\mathbf{B}$ is the **rectifying plane**. See Figure 14.19.

EXAMPLE 2 The position of a moving particle is given by $\mathbf{r}(t) = 2\cos t\mathbf{i} + 2\sin t\mathbf{j} + 3t\mathbf{k}$. Find the vectors $\mathbf{T}$, $\mathbf{N}$, and $\mathbf{B}$. Find the curvature.

Solution Since $\mathbf{r}'(t) = -2\sin t\mathbf{i} + 2\cos t\mathbf{j} + 3\mathbf{k}$, $|\mathbf{r}'(t)| = \sqrt{13}$, and so from (1) we see that a unit tangent is

$$\mathbf{T} = -\frac{2}{\sqrt{13}}\sin t\mathbf{i} + \frac{2}{\sqrt{13}}\cos t\mathbf{j} + \frac{3}{\sqrt{13}}\mathbf{k}$$

Next, we have

$$\frac{d\mathbf{T}}{dt} = -\frac{2}{\sqrt{13}}\cos t\mathbf{i} - \frac{2}{\sqrt{13}}\sin t\mathbf{j} \quad \text{and} \quad \left|\frac{d\mathbf{T}}{dt}\right| = \frac{2}{\sqrt{13}}$$

*Literally, this means the "kissing" plane.

Hence, (7) gives the principal normal

$$\mathbf{N} = -\cos t\mathbf{i} - \sin t\mathbf{j}$$

Now, the binormal is

$$\mathbf{B} = \mathbf{T} \times \mathbf{N} = \begin{vmatrix} \mathbf{i} & \mathbf{j} & \mathbf{k} \\ -\frac{2}{\sqrt{13}}\sin t & \frac{2}{\sqrt{13}}\cos t & \frac{3}{\sqrt{13}} \\ -\cos t & -\sin t & 0 \end{vmatrix}$$

$$= \frac{3}{\sqrt{13}}\sin t\mathbf{i} - \frac{3}{\sqrt{13}}\cos t\mathbf{j} + \frac{2}{\sqrt{13}}\mathbf{k}$$

Finally, using $|d\mathbf{T}/dt| = 2/\sqrt{13}$ and $|\mathbf{r}'(t)| = \sqrt{13}$, we find from (4) that the curvature at any point is the constant

$$\kappa = \frac{2/\sqrt{13}}{\sqrt{13}} = \frac{2}{13}$$ □

The fact that the curvature in Example 2 is constant is not surprising, since the curve defined by $\mathbf{r}(t)$ is a circular helix.

Formulas for a_T, a_N, and Curvature By dotting, and in turn crossing, the vector $\mathbf{v} = v\mathbf{T}$ with (9), it is possible to obtain explicit formulas involving $\mathbf{r}$, $\mathbf{r}'$, and $\mathbf{r}''$ for the tangential and normal components of the acceleration and the curvature. Observe that

$$\mathbf{v} \cdot \mathbf{a} = a_N(\underbrace{v\mathbf{T} \cdot \mathbf{N}}_{0}) + a_T(\underbrace{v\mathbf{T} \cdot \mathbf{T}}_{1}) = a_T v$$

yields the tangential component of acceleration:

$$a_T = \frac{dv}{dt} = \frac{\mathbf{v} \cdot \mathbf{a}}{|\mathbf{v}|} = \frac{\mathbf{r}'(t) \cdot \mathbf{r}''(t)}{|\mathbf{r}'(t)|} \tag{10}$$

On the other hand,

$$\mathbf{v} \times \mathbf{a} = a_N(\underbrace{v\mathbf{T} \times \mathbf{N}}_{\mathbf{B}}) + a_T(\underbrace{v\mathbf{T} \times \mathbf{T}}_{\mathbf{0}}) = a_N v\mathbf{B}$$

Since $|\mathbf{B}| = 1$, it follows that the normal component of acceleration is

$$a_N = \kappa v^2 = \frac{|\mathbf{v} \times \mathbf{a}|}{|\mathbf{v}|} = \frac{|\mathbf{r}'(t) \times \mathbf{r}''(t)|}{|\mathbf{r}'(t)|} \tag{11}$$

Solving (11) for the curvature gives

$$\kappa = \frac{|\mathbf{v} \times \mathbf{a}|}{|\mathbf{v}|^3} = \frac{|\mathbf{r}'(t) \times \mathbf{r}''(t)|}{|\mathbf{r}'(t)|^3} \tag{12}$$

EXAMPLE 3 The curve traced by $\mathbf{r}(t) = t\mathbf{i} + \frac{1}{2}t^2\mathbf{j} + \frac{1}{3}t^3\mathbf{k}$ is said to be a "twisted cubic." If $\mathbf{r}(t)$ is the position vector of a moving particle, find the tangential and normal components of the acceleration at any t. Find the curvature.

Solution

$$\mathbf{v}(t) = \mathbf{r}'(t) = \mathbf{i} + t\mathbf{j} + t^2\mathbf{k}$$
$$\mathbf{a}(t) = \mathbf{r}''(t) = \mathbf{j} + 2t\mathbf{k}$$

Since $\mathbf{v} \cdot \mathbf{a} = t + 2t^3$ and $|\mathbf{v}| = \sqrt{1 + t^2 + t^4}$, it follows from (10) that

$$a_T = \frac{dv}{dt} = \frac{t + 2t^3}{\sqrt{1 + t^2 + t^4}}$$

Now,

$$\mathbf{v} \times \mathbf{a} = \begin{vmatrix} \mathbf{i} & \mathbf{j} & \mathbf{k} \\ 1 & t & t^2 \\ 0 & 1 & 2t \end{vmatrix} = t^2\mathbf{i} - 2t\mathbf{j} + \mathbf{k}$$

and $|\mathbf{v} \times \mathbf{a}| = \sqrt{t^4 + 4t^2 + 1}$. Thus, (11) gives

$$a_N = \kappa v^2 = \frac{\sqrt{t^4 + 4t^2 + 1}}{\sqrt{1 + t^2 + t^4}} = \sqrt{\frac{t^4 + 4t^2 + 1}{t^4 + t^2 + 1}}$$

From (12) we find that the curvature of the twisted cubic is given by

$$\kappa = \frac{(t^4 + 4t^2 + 1)^{1/2}}{(1 + t^2 + t^4)^{3/2}}$$

□

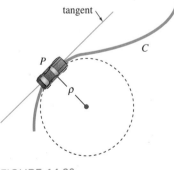

FIGURE 14.20

Radius of Curvature The reciprocal of the curvature, $\rho = 1/\kappa$, is called the **radius of curvature**. The radius of curvature at a point P on a curve C is the radius of a circle that "fits" the curve there better than any other circle. The circle at P is called the **circle of curvature** and its center is the **center of curvature**. The circle of curvature has the same tangent line at P as the curve C, and its center lies on the concave side of C. For example, a car moving on a curved track, as shown in Figure 14.20, can, at any instant, be thought to be moving on a circle of radius ρ. Hence, the normal component of its acceleration $a_N = \kappa v^2$ must be the same as the magnitude of its centripetal acceleration $a = v^2/\rho$. Therefore, $\kappa = 1/\rho$ and $\rho = 1/\kappa$. By knowing the radius of curvature, it is possible to determine the speed v at which a car can negotiate a banked curve without skidding. (This is essentially the idea in Problem 22 of Exercises 14.2.)

▲ **Remark** By writing (6) as

$$\mathbf{a}(t) = \frac{ds}{dt}\frac{d\mathbf{T}}{dt} + \frac{d^2s}{dt^2}\mathbf{T}$$

we note that the so-called scalar acceleration d^2s/dt^2, referred to in the last remark, is now seen to be the tangential component of the acceleration a_T.

EXERCISES 14.3 *Answers to odd-numbered problems begin on page A-103.*

In Problems 1 and 2, for the given position function, find the unit tangent.

1. $\mathbf{r}(t) = (t\cos t - \sin t)\mathbf{i} + (t\sin t + \cos t)\mathbf{j} + t^2\mathbf{k}, t > 0$

2. $\mathbf{r}(t) = e^t\cos t\mathbf{i} + e^t\sin t\mathbf{j} + \sqrt{2}e^t\mathbf{k}$

3. Use the procedure outlined in Example 2 to find **T**, **N**, **B**, and κ for motion on a general circular helix that is described by $\mathbf{r}(t) = a\cos t\mathbf{i} + a\sin t\mathbf{j} + ct\mathbf{k}$.

4. Use the procedure outlined in Example 2 to show on the twisted cubic of Example 3 that at $t = 1$:

$$\mathbf{T} = \frac{1}{\sqrt{3}}(\mathbf{i} + \mathbf{j} + \mathbf{k}) \qquad \mathbf{N} = -\frac{1}{\sqrt{2}}(\mathbf{i} - \mathbf{k})$$

$$\mathbf{B} = -\frac{1}{\sqrt{6}}(-\mathbf{i} + 2\mathbf{j} - \mathbf{k}) \qquad \kappa = \frac{\sqrt{2}}{3}$$

In Problems 5 and 6 find an equation of the osculating plane to the given space curve at the point that corresponds to the indicated value of t.

5. The circular helix of Example 2; $t = \pi/4$

6. The twisted cubic of Example 3; $t = 1$

In Problems 7–16 $\mathbf{r}(t)$ is the position vector of a moving particle. Find the tangential and normal components of the acceleration at any t.

7. $\mathbf{r}(t) = \mathbf{i} + t\mathbf{j} + t^2\mathbf{k}$

8. $\mathbf{r}(t) = 3\cos t\mathbf{i} + 2\sin t\mathbf{j} + t\mathbf{k}$

9. $\mathbf{r}(t) = t^2\mathbf{i} + (t^2 - 1)\mathbf{j} + 2t^2\mathbf{k}$

10. $\mathbf{r}(t) = t^2\mathbf{i} - t^3\mathbf{j} + t^4\mathbf{k}$

11. $\mathbf{r}(t) = 2t\mathbf{i} + t^2\mathbf{j}$

12. $\mathbf{r}(t) = \tan^{-1}t\mathbf{i} + \frac{1}{2}\ln(1 + t^2)\mathbf{j}$

13. $\mathbf{r}(t) = 5\cos t\mathbf{i} + 5\sin t\mathbf{j}$

14. $\mathbf{r}(t) = \cosh t\mathbf{i} + \sinh t\mathbf{j}$

15. $\mathbf{r}(t) = e^{-t}(\mathbf{i} + \mathbf{j} + \mathbf{k})$

16. $\mathbf{r}(t) = t\mathbf{i} + (2t - 1)\mathbf{j} + (4t + 2)\mathbf{k}$

17. Find the curvature of an elliptical helix that is described by $\mathbf{r}(t) = a\cos t\mathbf{i} + b\sin t\mathbf{j} + ct\mathbf{k}, a > 0, b > 0, c > 0$.

18. (a) Find the curvature of an elliptical orbit that is described by $\mathbf{r}(t) = a\cos t\mathbf{i} + b\sin t\mathbf{j} + c\mathbf{k}, a > 0, b > 0, c > 0$.

(b) Show that when $a = b$, the curvature of a circular orbit is the constant $\kappa = 1/a$.

19. Show that the curvature of a straight line is the constant $\kappa = 0$. [*Hint:* Use (2) of Section 13.5.]

20. Find the curvature of the cycloid that is described by

$$\mathbf{r}(t) = a(t - \sin t)\mathbf{i} + a(1 - \cos t)\mathbf{j}, \qquad a > 0$$

at $t = \pi$.

21. Let C be a plane curve traced by $\mathbf{r}(t) = f(t)\mathbf{i} + g(t)\mathbf{j}$, where f and g have second derivatives. Show that the curvature at a point is given by

$$\kappa = \frac{|f'(t)g''(t) - g'(t)f''(t)|}{([f'(t)]^2 + [g'(t)]^2)^{3/2}}$$

22. Show that if $y = F(x)$, the formula for κ in Problem 21 reduces to

$$\kappa = \frac{|F''(x)|}{[1 + (F'(x))^2]^{3/2}}$$

In Problems 23 and 24 use the result of Problem 22 to find the curvature and radius of curvature of the curve at the indicated points. Decide at which point the curve is "sharper."

23. $y = x^2$; $(0, 0), (1, 1)$

24. $y = x^3$; $(-1, -1), (\frac{1}{2}, \frac{1}{8})$

25. Discuss the curvature near a point of inflection of $y = F(x)$.

26. Show that $|\mathbf{a}(t)|^2 = a_N^2 + a_T^2$.

CHAPTER 14 REVIEW EXERCISES *Answers begin on page A-103.*

In Problems 1–10 answer true or false.

1. A particle whose position vector is $\mathbf{r}(t) = \cos t\mathbf{i} + \cos t\mathbf{j} + \sqrt{2}\sin t\mathbf{k}$ moves with constant speed. ________

2. The path of a moving particle whose position vector is $\mathbf{r}(t) = (t^2 + 1)\mathbf{i} + 4\mathbf{j} + t^4\mathbf{k}$ lies in a plane. ________

3. The binormal vector is perpendicular to the osculating plane. ________

4. If $\mathbf{r}(t)$ is the position vector of a moving particle, then the velocity vector $\mathbf{v}(t) = \mathbf{r}'(t)$ and the acceleration vector $\mathbf{a}(t) = \mathbf{r}''(t)$ are orthogonal. ________

5. If s is the arc length of a curve C, then the magnitude of the velocity of a particle moving on C is ds/dt. ________

6. If s is the arc length of a curve C, then the magnitude of the acceleration of a particle on C is d^2s/dt^2. ________

7. If the binormal is defined by $\mathbf{B} = \mathbf{T} \times \mathbf{N}$, then the principal normal is $\mathbf{N} = \mathbf{B} \times \mathbf{T}$. ________

8. If $\lim_{t\to a} \mathbf{r}_1(t) = 2\mathbf{i} + \mathbf{j}$ and $\lim_{t\to a} \mathbf{r}_2(t) = -\mathbf{i} + 2\mathbf{j}$, then $\lim_{t\to a} \mathbf{r}_1(t) \cdot \mathbf{r}_2(t) = 0$. ________

9. $\displaystyle\int_a^b [\mathbf{r}_1(t) \cdot \mathbf{r}_2(t)]\,dt = \left[\int_a^b \mathbf{r}_1(t)\,dt\right] \cdot \left[\int_a^b \mathbf{r}_2(t)\,dt\right]$

____

10. If $\mathbf{r}(t)$ is differentiable, then $\dfrac{d}{dt}|\mathbf{r}(t)|^2 = 2\mathbf{r}(t) \cdot \dfrac{d\mathbf{r}}{dt}$.

____

11. Find the length of the curve that is traced by the vector function $\mathbf{r}(t) = \sin t\mathbf{i} + (1 - \cos t)\mathbf{j} + t\mathbf{k}$ on the interval $0 \le t \le \pi$.

12. The position vector of a moving particle is given by $\mathbf{r}(t) = 5t\mathbf{i} + (1 + t)\mathbf{j} + 7t\mathbf{k}$. Given that the particle starts at the point corresponding to $t = 0$, find the distance the particle travels to the point corresponding to $t = 3$. At what point will the particle have traveled $80\sqrt{3}$ units along the curve?

13. Find parametric equations for the tangent line to the curve that is traced by

$$\mathbf{r}(t) = -3t^2\mathbf{i} + 4\sqrt{t+1}\,\mathbf{j} + (t - 2)\mathbf{k}$$

at $t = 3$.

14. Show that the curve traced by $\mathbf{r}(t) = t\cos t\mathbf{i} + t\sin t\mathbf{j} + t\mathbf{k}$ lies on the surface of a cone. Sketch the curve.

15. Sketch the curve traced by $\mathbf{r}(t) = \cosh t\mathbf{i} + \sinh t\mathbf{j} + t\mathbf{k}$.

16. Given that $\mathbf{r}_1(t) = t^2\mathbf{i} + 2t\mathbf{j} + t^3\mathbf{k}$ and $\mathbf{r}_2(t) = -t\mathbf{i} + t^2\mathbf{j} + (t^2 + 1)\mathbf{k}$, calculate $(d/dt)[\mathbf{r}_1(t) \times \mathbf{r}_2(t)]$ in two different ways.

17. Given that $\mathbf{r}_1(t) = \cos t\mathbf{i} - \sin t\mathbf{j} + 4t^3\mathbf{k}$ and $\mathbf{r}_2(t) = t^2\mathbf{i} + \sin t\mathbf{j} + e^{2t}\mathbf{k}$, calculate $(d/dt)[\mathbf{r}_1(t) \cdot \mathbf{r}_2(t)]$ in two different ways.

18. Given that $\mathbf{r}_1$, $\mathbf{r}_2$, and $\mathbf{r}_3$ are differentiable, find $(d/dt)[\mathbf{r}_1(t) \cdot (\mathbf{r}_2(t) \times \mathbf{r}_3(t))]$.

19. A particle of mass m is acted on by a continuous force of magnitude 2, which is directed parallel to the positive y-axis. If the particle starts with an initial velocity $\mathbf{v}(0) = \mathbf{i} + \mathbf{j} + \mathbf{k}$ from $(1, 1, 0)$, find the position vector of the particle and the parametric equations of its path. [*Hint:* $\mathbf{F} = m\mathbf{a}$.]

20. The position vector of a moving particle is $\mathbf{r}(t) = t\mathbf{i} + (1 - t^3)\mathbf{j}$.

(a) Sketch the path of the particle.

(b) Sketch the velocity and acceleration vectors at $t = 1$.

(c) Find the speed at $t = 1$.

21. Find the velocity and acceleration of a particle whose position vector is $\mathbf{r}(t) = 6t\mathbf{i} + t\mathbf{j} + t^2\mathbf{k}$ as it passes through the plane $-x + y + z = -4$.

22. The velocity of a moving particle is $\mathbf{v}(t) = -10t\mathbf{i} + (3t^2 - 4t)\mathbf{j} + \mathbf{k}$. If the particle starts at $t = 0$ at $(1, 2, 3)$, what is its position at $t = 2$?

23. The acceleration of a moving particle is $\mathbf{a}(t) = \sqrt{2}\sin t\mathbf{i} + \sqrt{2}\cos t\mathbf{j}$. Given that the velocity and position of the particle at $t = \pi/4$ are $\mathbf{v}(\pi/4) = -\mathbf{i} + \mathbf{j} + \mathbf{k}$ and $\mathbf{r}(\pi/4) = \mathbf{i} + 2\mathbf{j} + (\pi/4)\mathbf{k}$, respectively, what was the position of the particle at $t = 3\pi/4$?

24. Given that

$$\mathbf{r}(t) = \frac{1}{2}t^2\mathbf{i} + \frac{1}{3}t^3\mathbf{j} - \frac{1}{2}t^2\mathbf{k}$$

is the position vector of a moving particle, find the tangential and normal components of the acceleration at any t. Find the curvature.

25. Suppose that the vector function of Problem 15 is the position vector of a moving particle. Find the vectors $\mathbf{T}$, $\mathbf{N}$, and $\mathbf{B}$ at $t = 1$. Find the curvature at that point.

15

DIFFERENTIAL CALCULUS OF FUNCTIONS OF SEVERAL VARIABLES

INTRODUCTION

Up to this point in our study of calculus, we have considered only functions of a single variable. Previously considered concepts for functions of a single variable, such as limits, continuity, derivatives, tangents, maxima and minima, integrals, and so on, extend to functions of two or more variables as well. This chapter is devoted primarily to the differential calculus of multivariable functions.

In the Apollo program, astronauts were launched to the moon in huge three-stage Saturn V rockets. Even today, when a satellite is sent into orbit by means other than the space shuttle, a three-stage rocket is used. Obviously there is something special about using three stages. Why not one, two, or four stages? It turns out that this is an optimization problem. If you take on the challenge of working Problem 35 in Exercises 15.11, you should gain some appreciation for the importance of three stages for rockets designed to deliver a payload into orbit.

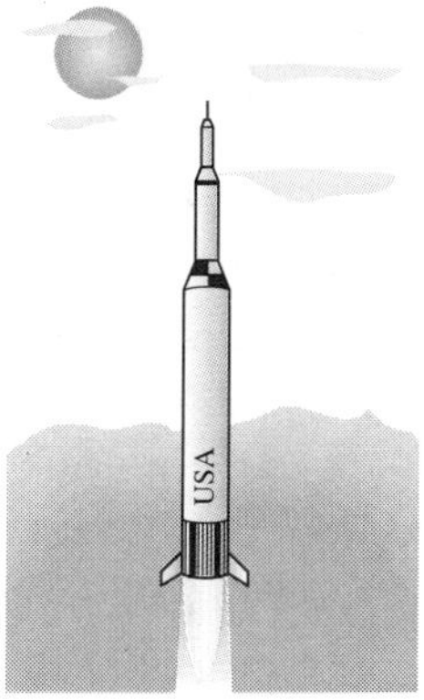

IMPORTANT CONCEPTS

functions of two variables
independent variables
dependent variable
domain
range
level curves
functions of three variables
level surfaces
limits
continuity
closed region
bounded region
unbounded region
partial differentiation
total differential
differentiability
exact differentials
exact differential equations
chain rule for functions of two or more variables
gradient
directional derivative
tangent plane
normal line
critical points
second partials test
boundary extrema
method of least squares
least-squares line
method of Lagrange multipliers

15.1 FUNCTIONS OF TWO OR MORE VARIABLES

Recall that a function of one variable $y = f(x)$ is a *rule of correspondence* that assigns to an element x of a subset of the real numbers R, called the *domain* of f, one and only one real number y. The set $\{y \mid y = f(x)\}$ is called the *range* of f. You are probably already aware of the existence of functions of two or more variables.

EXAMPLE 1

(*a*) $A = xy$, area of a rectangle

(*b*) $V = \pi r^2 h$, volume of a circular cylinder

(*c*) $V = \dfrac{\pi}{3} r^2 h$, volume of a cone

(*d*) $P = 2x + 2y$, perimeter of a rectangle □

EXAMPLE 2

(*a*) The pressure P exerted by an enclosed ideal gas is a function of its temperature T and volume V:

$$P = k\left(\frac{T}{V}\right), \qquad k \text{ a constant}$$

(*b*) The area S of the surface of a human body is a function of its weight w and its height h:

$$S = 0.1091 w^{0.425} h^{0.725}$$ □

Functions of Two Variables The formal definition of a function of two variables follows.

DEFINITION 15.1 Function of Two Variables

A **function of two variables** is a rule of correspondence that assigns to each ordered pair of real numbers (x, y) of a subset of the xy-plane one and only one number z in the set R of real numbers.

The set of ordered pairs (x, y) is called the **domain** of the function and the set of corresponding values of z is called the **range**. A function of two variables is usually written $z = f(x, y)$ and read "f of x, y." The variables x and y are called the **independent variables** of the function and z is called the **dependent variable**.

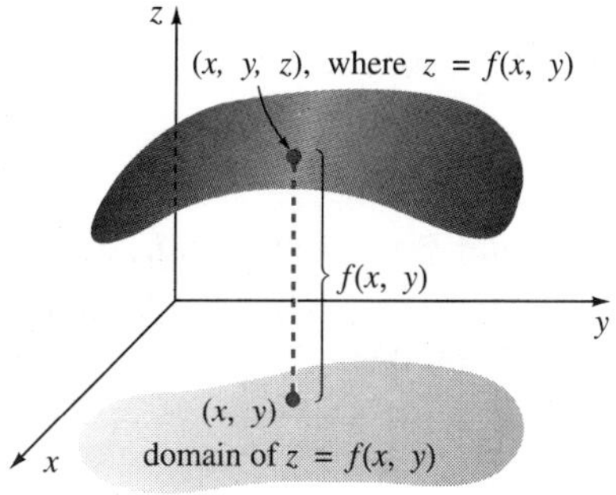

FIGURE 15.1

Graphs The **graph** of a function $z = f(x, y)$ is a *surface* in 3-space. See Figure 15.1.

EXAMPLE 3 An equation of a plane $ax + by + cz = d$, $c \neq 0$, describes a function when written as

$$z = -\frac{a}{c}x - \frac{b}{c}y + \frac{d}{c} \quad \text{or} \quad f(x, y) = -\frac{a}{c}x - \frac{b}{c}y + \frac{d}{c}$$

Since z is defined for any choice of x and y, the domain of the function consists of the entire xy-plane. □

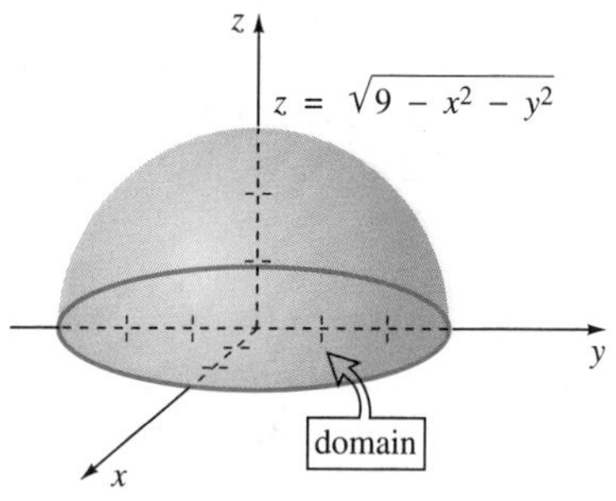

FIGURE 15.2

EXAMPLE 4 The graph of the function

$$f(x, y) = \sqrt{9 - x^2 - y^2}$$

is the hemisphere shown in Figure 15.2.* The domain of the function is the set of ordered pairs (x, y) that satisfies

$$9 - x^2 - y^2 \geq 0 \quad \text{or} \quad x^2 + y^2 \leq 9$$

That is, the domain of f consists of the interior and boundary of the circle $x^2 + y^2 = 9$. Inspection of Figure 15.2 shows that the range of the function is defined by $0 \leq z \leq 3$. □

EXAMPLE 5 From the discussion of quadric surfaces in Section 13.7 you should recognize the graph of the function $f(x, y) = x^2 + 9y^2$ as a paraboloid. Since f is defined for every ordered pair of real numbers, its domain is the entire xy-plane. From the fact that $x^2 \geq 0$ and $y^2 \geq 0$, we see that the range of f is given by $z \geq 0$. □

EXAMPLE 6

(*a*) Given that $f(x, y) = 4 + \sqrt{x^2 - y^2}$, find $f(1, 0)$, $f(5, 3)$, and $f(4, -2)$.

(*b*) Sketch the domain of the function.

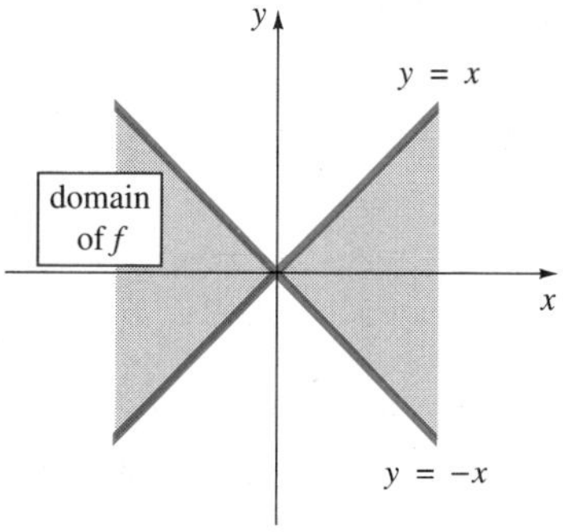

FIGURE 15.3

Solution (*a*)

$$f(1, 0) = 4 + \sqrt{1 - 0} = 5$$
$$f(5, 3) = 4 + \sqrt{25 - 9} = 4 + \sqrt{16} = 8$$
$$f(4, -2) = 4 + \sqrt{16 - (-2)^2} = 4 + \sqrt{12} = 4 + 2\sqrt{3}$$

(*b*) The domain of f consists of all ordered pairs (x, y) for which $x^2 - y^2 \geq 0$. As shown in Figure 15.3, the domain consists of all points on the lines $y = x$ and $y = -x$ and in the shaded regions between them. □

*Verify this by replacing the symbol $f(x, y)$ by z and squaring both sides of the equation.

In science, one often encounters the words **isothermal, equipotential**, and **isobaric**. These terms apply to lines or curves on which either temperature, potential, or barometric pressure is *constant*.

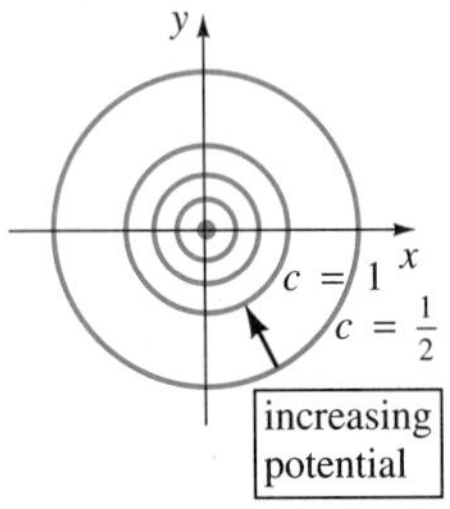

FIGURE 15.4

EXAMPLE 7 The electrostatic potential at a point $P(x, y)$ in the plane due to a unit point charge at the origin is given by $U = 1/\sqrt{x^2 + y^2}$. If the potential is a constant, say $U = c$, where c is a positive constant, then

$$\frac{1}{\sqrt{x^2 + y^2}} = c \quad \text{or} \quad x^2 + y^2 = \frac{1}{c^2}$$

Thus, as shown in Figure 15.4, the curves of equipotential are concentric circles surrounding the charge. □

Note that in Figure 15.4 we can get a feeling for the behavior of the function U, specifically where it is increasing (or decreasing), by observing the direction of increasing c.

Level Curves In general, if a function of two variables is given by $z = f(x, y)$, then the curves defined by $f(x, y) = c$, for suitable c, are called the **level curves** of f. The word *level* arises from the fact that we can interpret $f(x, y) = c$ as the projection onto the xy-plane of the curve of intersection, or trace, of $z = f(x, y)$ and the (horizontal or level) plane $z = c$. See Figure 15.5.

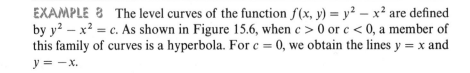

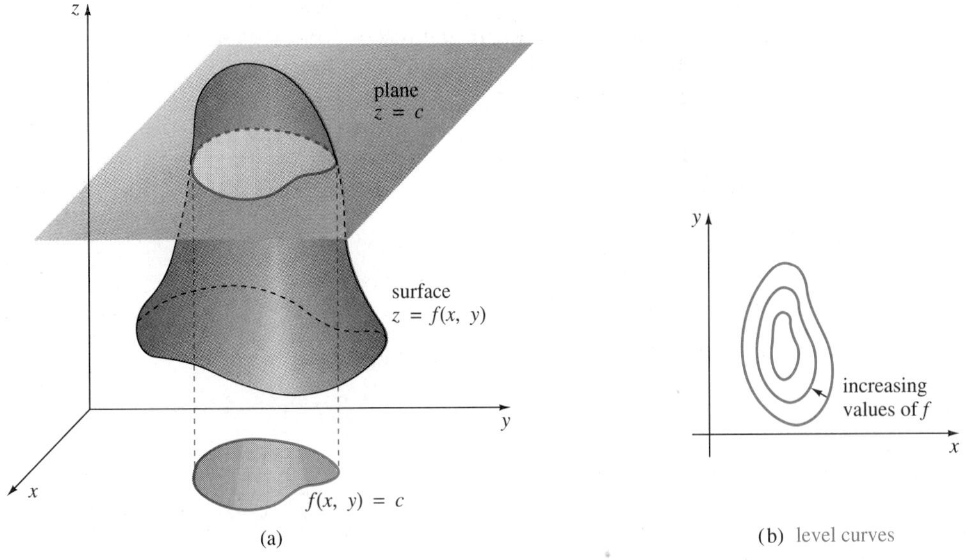

FIGURE 15.5

EXAMPLE 8 The level curves of the function $f(x, y) = y^2 - x^2$ are defined by $y^2 - x^2 = c$. As shown in Figure 15.6, when $c > 0$ or $c < 0$, a member of this family of curves is a hyperbola. For $c = 0$, we obtain the lines $y = x$ and $y = -x$.

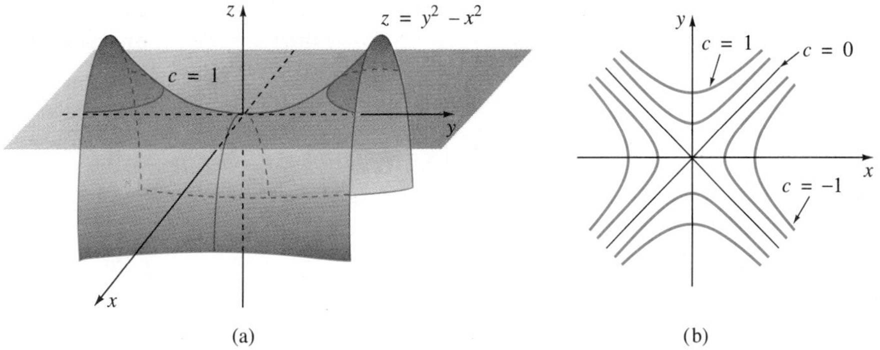

FIGURE 15.6

In the plane, the curves $f(x, y) = c$ are also called **contour lines** of a surface. In Figure 15.7, we see that a contour map illustrates the various segments of a hill that have a given altitude. This is the idea of the contours in Figure 15.8, which show the thickness of volcanic ash surrounding the volcano El Chichon. El Chichon, in the state of Chiapas, Mexico, erupted on March 28 and April 4, 1982.

Computer Graphics In many instances the task of graphing a function of two variables is formidable. The use of **computer graphics** has become widespread in analyzing complicated surfaces in 3-space. See Figures 15.9–15.13. Note the computer-generated level curves in Figures 15.9 and 15.10.

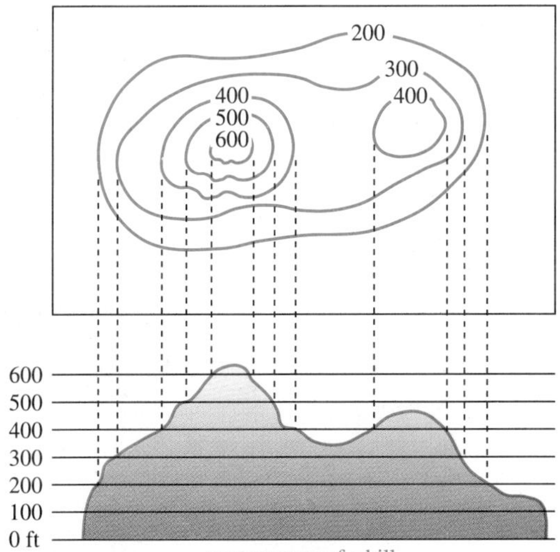

contour map of a hill

FIGURE 15.7

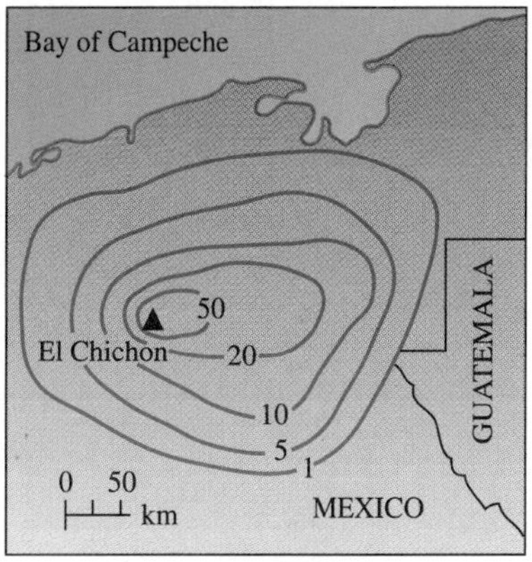

thickness (in mm) of rain-compacted volcanic ash surrounding El Chichon*

FIGURE 15.8

*Adapted, with permission, from *National Geographic* magazine.

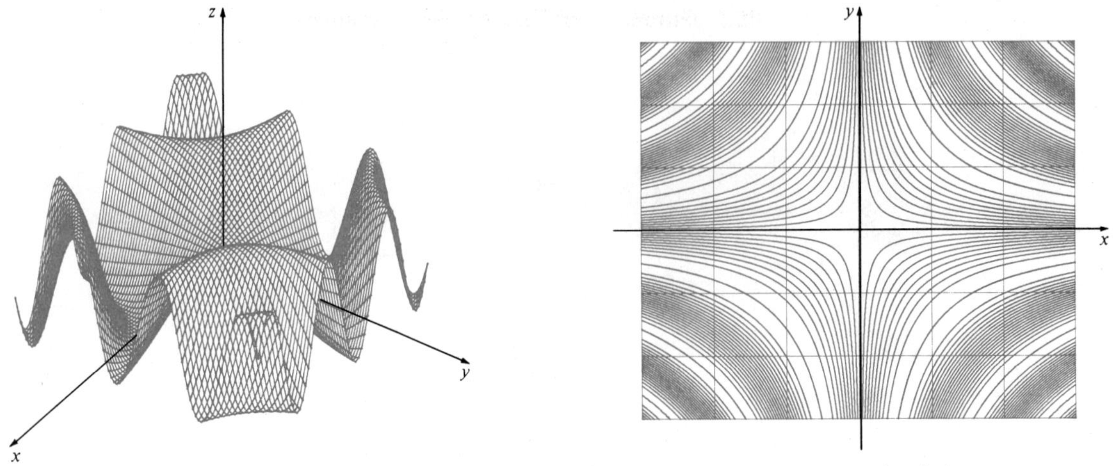

(a) $z = 2 \sin xy$ (b) level curves of $z = 2 \sin xy$

FIGURE 15.9

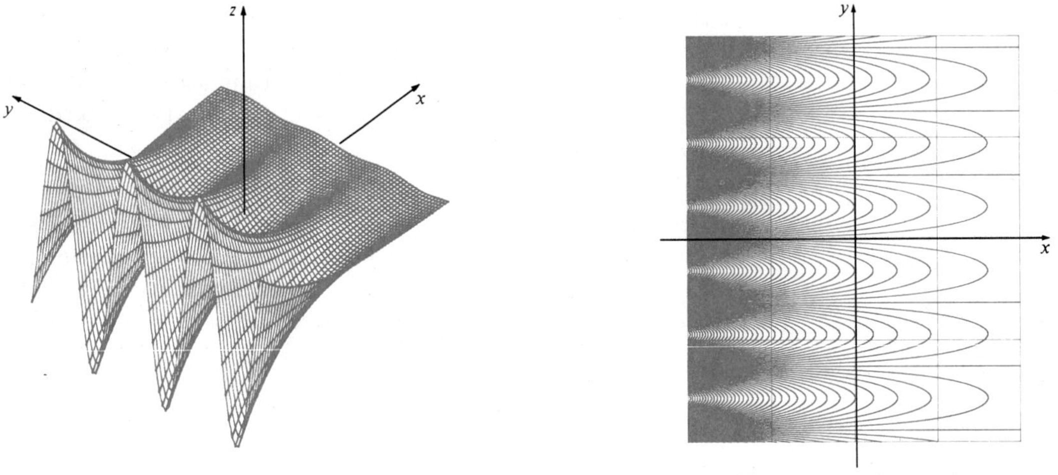

(a) $z = e^{-x} \sin y$ (b) level curves of $z = e^{-x} \sin y$

FIGURE 15.10

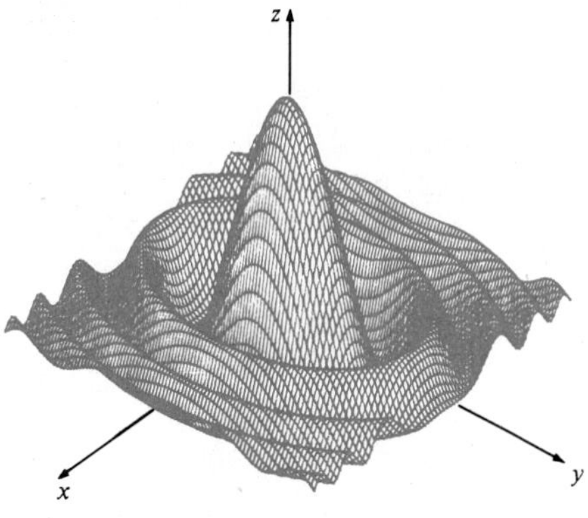

$$z = \frac{\frac{1}{2}\cos(2x^2 + y^2)}{1 + 2x^2 + y^2}$$

FIGURE 15.11

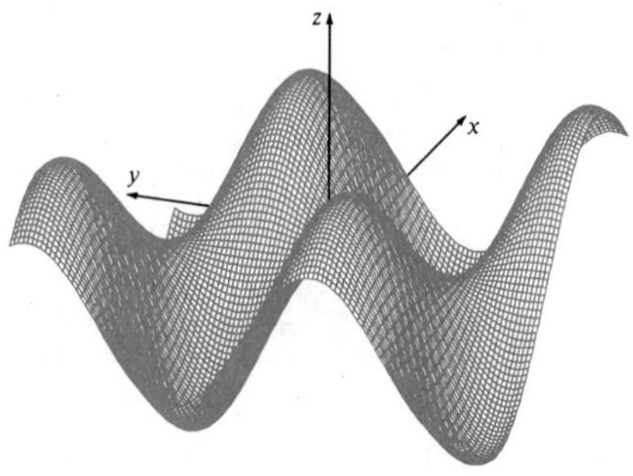

$z = \sin \frac{x}{2} \sin y$

FIGURE 15.12

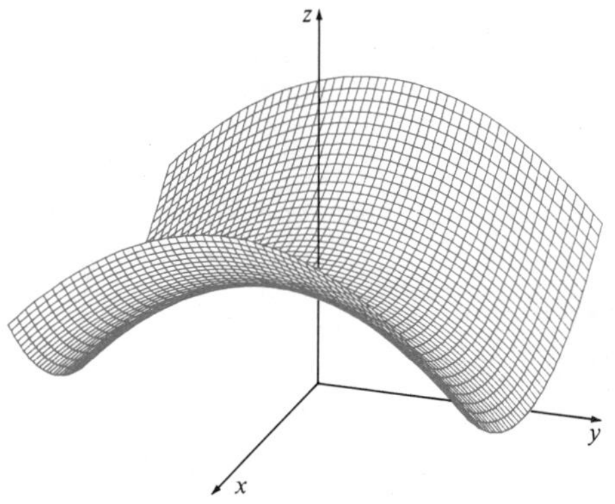

FIGURE 15.13

Functions of Three or More Variables The definitions of functions of three or more variables are simply generalizations of Definition 15.1. For example, a **function of three variables** is a rule of correspondence that assigns to each ordered triple of real numbers (x, y, z) in a subset of 3-space, one and only one number w in the set R of real numbers.

EXAMPLE 9

$$F(x, y, z) = \frac{2x + 3y + z}{4 - x^2 - y^2 - z^2}$$

is an example of a function of three variables. Its domain is the set of points (x, y, z) that satisfy $x^2 + y^2 + z^2 \neq 4$; that is, the domain of F is all of 3-space *except* the points on the surface of a sphere of radius 2 centered at the origin. □

EXAMPLE 10

(*a*) $V = xyz$, volume of a rectangular box

(*b*) Poiseuille's law states that the discharge rate, or rate of flow, of a viscous fluid (such as blood) through a tube (such as an artery) is

$$Q = k\frac{R^4}{L}(p_1 - p_2), \qquad k \text{ a constant}$$

where R is the radius of the tube, L is its length, and p_1 and p_2 are the pressures at the ends of the tube. This is an example of a function of *four* variables. □

Note: Since it would take four dimensions, we cannot graph a function of three variables.

Level Surfaces For a function of three variables, $w = F(x, y, z)$, the surfaces defined by $F(x, y, z) = c$, for suitable values of c, are called **level surfaces.***

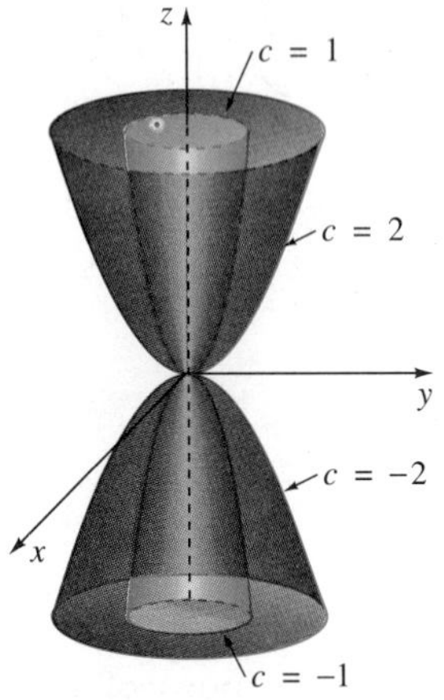

FIGURE 15.14

EXAMPLE 11 The level surfaces of $F(x, y, z) = 2x - 3y + 4z$ are parallel planes defined by $2x - 3y + 4z = c$. □

EXAMPLE 12 Describe the level surfaces of the function $F(x, y, z) = (x^2 + y^2)/z$.

Solution For $c \neq 0$ the level surfaces are given by

$$\frac{x^2 + y^2}{z} = c$$

or

$$x^2 + y^2 = cz$$

A few members of this family of paraboloids are shown in Figure 15.14. □

EXERCISES 15.1 *Answers to odd-numbered problems begin on page A-103.*

In Problems 1–10 find the domain of the given function.

1. $f(x, y) = \dfrac{xy}{x^2 + y^2}$
2. $f(x, y) = (x^2 - 9y^2)^{-2}$
3. $f(x, y) = \dfrac{y^2}{y + x^2}$
4. $f(x, y) = x^2 - y^2\sqrt{4 + y}$
5. $f(s, t) = s^3 - 2t^2 + 8st$
6. $f(u, v) = \dfrac{u}{\ln(u^2 + v^2)}$
7. $g(r, s) = e^{2r}\sqrt{s^2 - 1}$
8. $g(\theta, \phi) = \dfrac{\tan\theta + \tan\phi}{1 - \tan\theta\tan\phi}$
9. $H(u, v, w) = \sqrt{u^2 + v^2 + w^2 - 16}$
10. $F(x, y, z) = \dfrac{\sqrt{25 - x^2 - y^2}}{z - 5}$

In Problems 11–18 match the given function with the figure that most closely resembles its domain.

11. $f(x, y) = \sqrt{y - x^2}$
12. $f(x, y) = \ln(x - y^2)$
13. $f(x, y) = \sqrt{x} + \sqrt{y - x}$
14. $f(x, y) = \sqrt{\dfrac{x}{y} - 1}$
15. $f(x, y) = \sqrt{xy}$
16. $f(x, y) = \sin^{-1}(xy)$
17. $f(x, y) = \dfrac{x^4 + y^4}{xy}$
18. $f(x, y) = \dfrac{\sqrt{x^2 + y^2 - 1}}{y - x}$

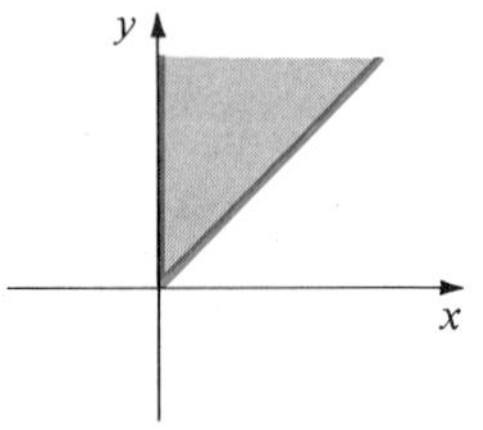

FIGURE 15.15

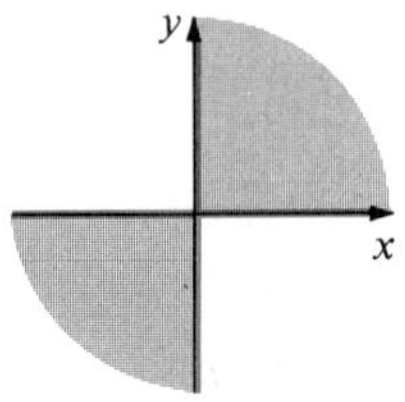

FIGURE 15.16

*An unfortunate, but standard, choice of words, since level surfaces are not usually level.

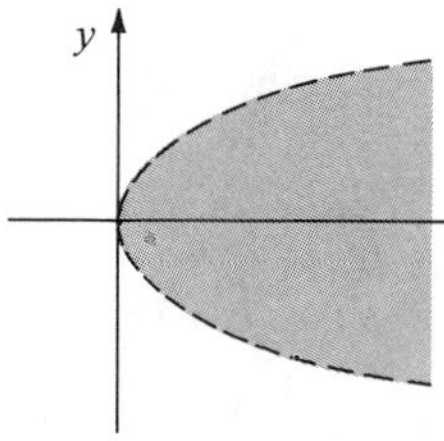

FIGURE 15.17

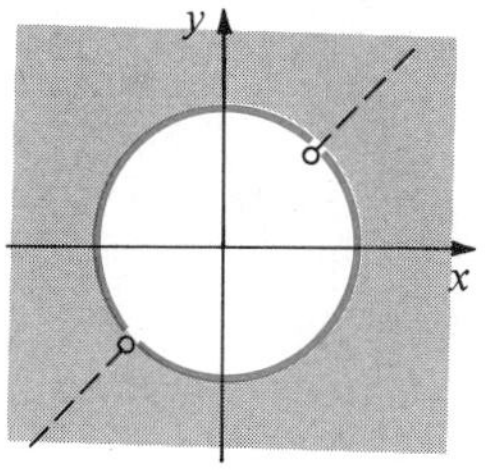

FIGURE 15.18

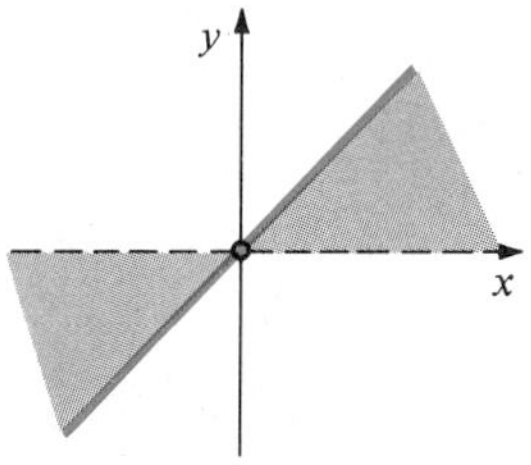

FIGURE 15.19

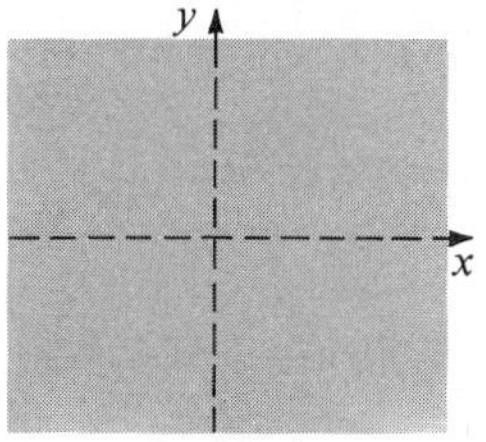

FIGURE 15.20

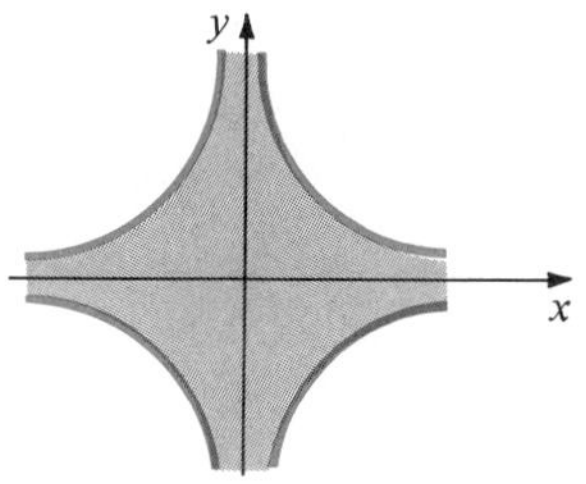

FIGURE 15.21

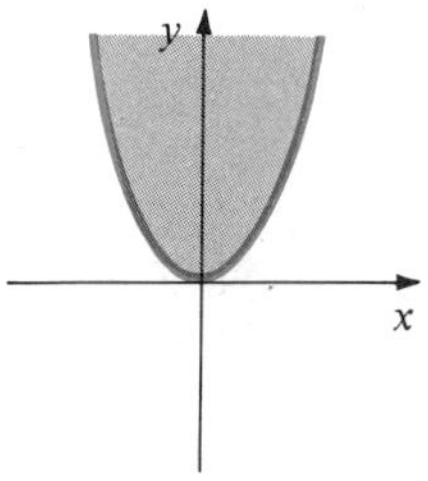

FIGURE 15.22

In Problems 19–22 sketch the domain of the given function.

19. $f(x, y) = \sqrt{x} - \sqrt{y}$

20. $f(x, y) = \sqrt{(1 - x^2)(y^2 - 4)}$

21. $f(x, y) = \sqrt{\ln(y - x + 1)}$

22. $f(x, y) = e^{\sqrt{xy+1}}$

In Problems 23–26 find the range of the given function.

23. $f(x, y) = 10 + x^2 + 2y^2$

24. $f(x, y) = x + y$

25. $F(x, y, z) = \sin(x + 2y + 3z)$

26. $F(x, y, z) = 7 - e^{xyz}$

In Problems 27–30 evaluate the given function at the indicated points.

27. $f(x, y) = \int_x^y (2t - 1)\,dt$; $(2, 4), (-1, 1)$

28. $f(x, y) = \ln \dfrac{x^2}{x^2 + y^2}$; $(3, 0), (5, -5)$

29. $F(x, y, z) = (x + 2y + 3z)^2$; $(-1, 1, -1), (2, 3, -2)$

30. $F(x, y, z) = \dfrac{1}{x^2} + \dfrac{1}{y^2} + \dfrac{1}{z^2}$; $(\sqrt{3}, \sqrt{2}, \sqrt{6}), (\frac{1}{4}, \frac{1}{5}, \frac{1}{3})$

In Problems 31–36 discuss the graph of the given function.

31. $z = x$

32. $z = y^2$

33. $z = \sqrt{x^2 + y^2}$

34. $z = \sqrt{1 + x^2 + y^2}$

35. $z = \sqrt{9 - x^2 - 3y^2}$

36. $z = -\sqrt{16 - x^2 - y^2}$

In Problems 37–42 sketch some of the level curves associated with the given function.

37. $f(x, y) = x + 2y$

38. $f(x, y) = y^2 - x$

39. $f(x, y) = \sqrt{x^2 - y^2 - 1}$

40. $f(x, y) = \sqrt{36 - 4x^2 - 9y^2}$

41. $f(x, y) = e^{y - x^2}$

42. $f(x, y) = \tan^{-1}(y - x)$

In Problems 43–46 describe the level surfaces but do not graph.

43. $F(x, y, z) = \dfrac{x^2}{9} + \dfrac{z^2}{4}$

44. $F(x, y, z) = x^2 + y^2 + z^2$

45. $F(x, y, z) = x^2 + 3y^2 + 6z^2$

46. $F(x, y, z) = 4y - 2z + 1$

47. Graph some of the level surfaces associated with $F(x, y, z) = x^2 + y^2 - z^2$ for $c = 0$, $c > 0$, and $c < 0$.

48. Given that

$$F(x, y, z) = \frac{x^2}{16} + \frac{y^2}{4} + \frac{z^2}{9}$$

find the x-, y-, and z-intercepts of the level surface that passes through $(-4, 2, -3)$.

49. The temperature, pressure, and volume of an enclosed ideal gas are related by $T = 0.01PV$, where T, P, and V are measured in degrees Kelvin, atmospheres, and liters, respectively. Sketch the isotherms $T = 300°\text{K}$, $400°\text{K}$, and $600°\text{K}$.

50. Express the height of a rectangular box with a square bottom as a function of the volume and the length of one side of the box.

51. A soda pop can is constructed with a tin lateral side and an aluminium top and bottom. Given that the cost is 1.8 cents per square unit for the top, 1 cent per square unit for the bottom, and 2.3 cents per square unit for the side, determine the cost function $C(r, h)$, where r is the radius of the can and h is its height.

52. A closed rectangular box is to be constructed from 500 cm^2 of cardboard. Express the volume V as a function of the length x and width y.

53. As shown in Figure 15.23, a conical cap rests on top of a circular cylinder. If the height of the cap is two-thirds the height of the cylinder, express the volume of the solid as a function of the indicated variables.

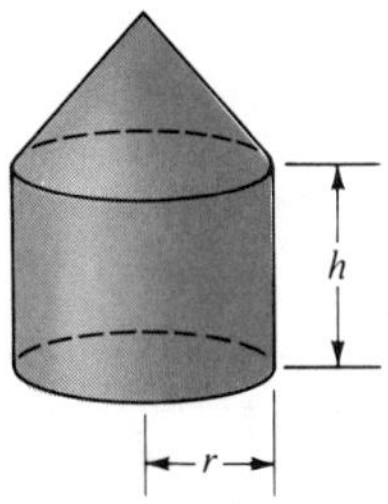

FIGURE 15.23

54. An obliquely cut cylinder (such as used in tissue samples) is shown in Figure 15.24. Express the thickness t of the cut as a function of x, y, and z.

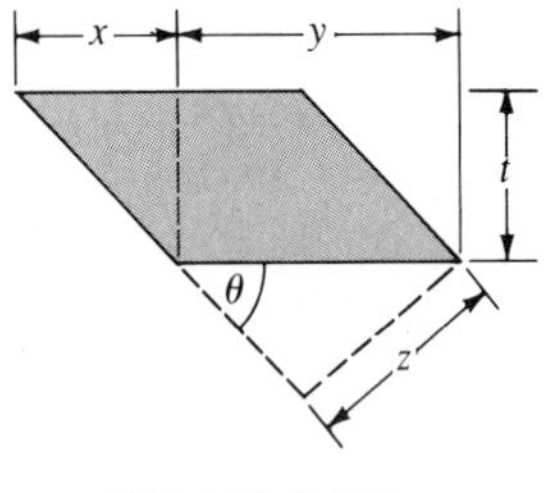

FIGURE 15.24

55. In medicine, formulas for surface area (see Example 2) are often used to calibrate drug doses, since it is assumed that a drug dose D and surface area S are directly proportional. The following simple function can be used to obtain a quick estimate of the body surface area of a human: $S = 2ht$, where h is the height (in cm) and t is the maximum thigh circumference (in cm). Estimate the surface area of a 156-cm-tall person with a maximum thigh circumference of 50 cm. Estimate your own surface area.

56. When water flows from a spigot, as shown in Figure 15.25(a), it contracts as it accelerates downward. It does this

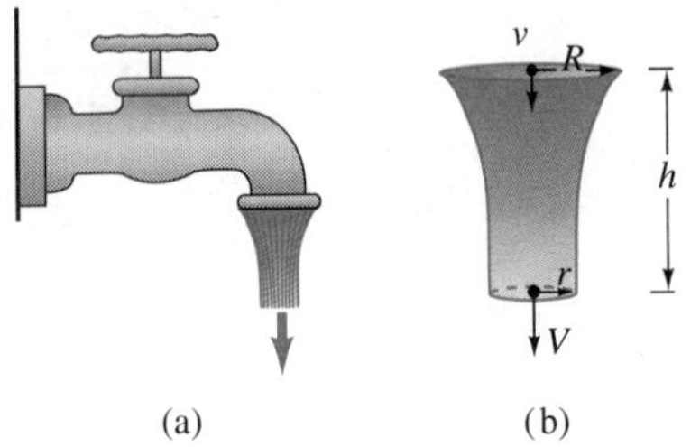

FIGURE 15.25

because the **flow rate** Q, which is defined as velocity times the cross-sectional area of the water column, must be constant at each level. In this problem assume that the cross-sections of the fluid column are circular.

(a) Consider the column of water shown in Figure 15.25(b). Suppose v is the velocity of the water at the top level, V is the velocity of the water at the bottom level a distance h units below the top level, R is the radius of the cross-section at the top level, and r is the radius of the cross-section at the bottom level. Show that the flow rate Q as a function of r and R is

$$Q = \frac{\pi r^2 R^2 \sqrt{2gh}}{\sqrt{R^4 - r^4}}$$

where g is the acceleration due to gravity. [*Hint:* Start by expressing the time t it takes a cross-section of water to fall a distance h in terms of v and V. For convenience take the positive direction to be downward.]

(b) Find the flow rate Q (in cm^3/s) if $g = 980$ cm/s^2, $h = 10$ cm, $R = 1$ cm, and $r = 0.2$ cm.

57. During his investigation of the winter of 1941 in the Antarctic, Dr. Paul A. Siple devised the following function for defining the **wind chill factor**:

$$H(v, T) = (10\sqrt{v} - v + 10.5)(33 - T)$$

where H is measured in kcal/m^{2}h, v is wind velocity in m/s, and T is temperature in °C. An example of this index is: 1000 = very cold, 1200 = bitterly cold, and 1400 = exposed flesh freezes. Determine the wind chill factor at -6.67°C (20°F) with a wind velocity of 20 m/s (45 mi/h).

58. In the clean and jerk competition, a weightlifter in the heavyweight class who weighs 110 kg lifts 210 kg. In the flyweight class, a person who weighs 50 kg lifts 130 kg. How can these feats of strength be compared? In an overall competition, who would be judged the superior lifter? Several different formulas for handicapping lifts have been proposed. Let w_l denote the weight lifted (in kg) and w_b be the body weight of the lifter (in kg).

(a) Used in ABC's Superstars competition: $h = w_l - w_b$
(b) Austin formula: $h = w_l/w_b^{3/4}$
(c) Classical formula: $h = w_l/w_b^{2/3}$
(d) O'Carroll formula: $h = w_l/(w_b - 35)^{1/3}$

Use these formulas to determine whether the heavyweight or the flyweight is the superior lifter.

59. The total energy consumption C in calories per hour of a person with a metabolic rate r, weight w, and height h is given by $C = 0.2rw^{0.425}h^{0.725}$. Find the energy consumption of a jogger who weighs 80 kg, is 1.6 m tall, and whose metabolic rate is 600 cal/m²h.

15.2 LIMITS AND CONTINUITY

15.2.1 An Informal Discussion

For functions of one variable, in many instances we were able to make a judgment about the existence of $\lim_{x\to a} f(x)$ from the graph of $y = f(x)$. Also, we utilized the fact that $\lim_{x\to a} f(x)$ exists if and only if $\lim_{x\to a^-} f(x)$ and $\lim_{x\to a^+} f(x)$ exist and are equal to the same number L, in which case, $\lim_{x\to a} f(x) = L$.

The situation is more demanding when we consider limits of functions of two variables. To analyze a limit by sketching a graph of $z = f(x, y)$ is not convenient or even routinely possible for most functions. Intuitively, f has a limit at a point (a, b) if the functional values $f(x, y)$ are approaching a number L as (x, y) approaches (a, b). We write $f(x, y) \to L$ as $(x, y) \to (a, b)$, or

$$\lim_{(x,y)\to(a,b)} f(x, y) = L$$

To be more precise, f has a limit L at a point (a, b) if the points in space $(x, y, f(x, y))$ can be made arbitrarily close to (a, b, L) whenever (x, y) is close enough to (a, b).

The notation of (x, y) "approaching" a point (a, b) is not as simple as in functions of one variable where $x \to a$ means that x can approach a only from the left and from the right. In the xy-plane, there are an infinite number of ways of approaching a point (a, b). As shown in Figure 15.26, in order that $\lim_{(x,y)\to(a,b)} f(x, y)$ exist, we now require that f approach the same number L along *every* possible curve or **path** through (a, b). Put in a negative way:

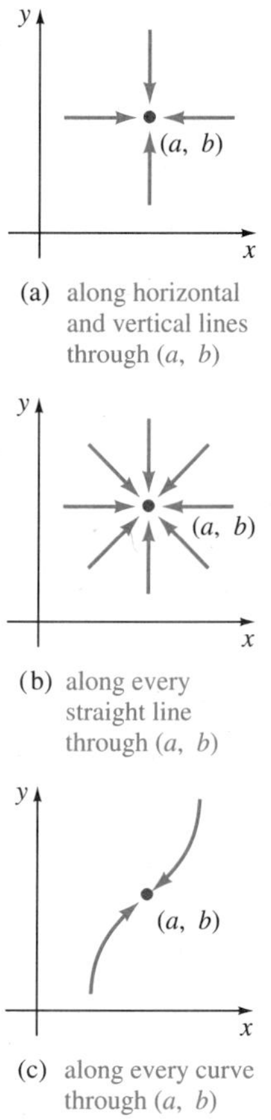

(a) along horizontal and vertical lines through (a, b)

(b) along every straight line through (a, b)

(c) along every curve through (a, b)

FIGURE 15.26

If $f(x, y)$ does not approach the same number L for two different paths to (a, b), then $\lim_{(x,y)\to(a,b)} f(x, y)$ does not exist. (1)

In the discussion of $\lim_{(x,y)\to(a,b)} f(x, y)$ that follows we shall assume that the function f is defined at every point (x, y) in the interior of a circle centered at (a, b) but not necessarily *at* (a, b) itself.

EXAMPLE 1 Show that $\displaystyle\lim_{(x,y)\to(0,0)} \frac{x^2 - 3y^2}{x^2 + 2y^2}$ does not exist.

Solution The function $f(x, y) = (x^2 - 3y^2)/(x^2 + 2y^2)$ is defined everywhere except at (0, 0). Two ways of approaching (0, 0) are along the x-axis ($y = 0$)

and along the y-axis ($x = 0$). We have

$$\text{on } y = 0, \qquad \lim_{(x,0)\to(0,0)} f(x, 0) = \lim_{(x,0)\to(0,0)} \frac{x^2 - 0}{x^2 + 0} = 1$$

$$\text{on } x = 0, \qquad \lim_{(0,y)\to(0,0)} f(0, y) = \lim_{(0,y)\to(0,0)} \frac{0 - 3y^2}{0 + 2y^2} = -\frac{3}{2}$$

In view of (1), we conclude that the limit does not exist. □

EXAMPLE 2 Show that $\displaystyle\lim_{(x,y)\to(0,0)} \frac{xy}{x^2 + y^2}$ does not exist.

Solution In this case the limits along the x- and y-axes are the same:

$$\lim_{(x,0)\to(0,0)} f(x, 0) = \lim_{(0,y)\to(0,0)} f(0, y) = 0$$

However, this does *not* mean $\lim_{(x,y)\to(0,0)} f(x, y)$ exists, since we have not examined every path to (0, 0). We now try any line through the origin given by $y = mx$:

$$\lim_{(x,y)\to(0,0)} f(x, y) = \lim_{(x,y)\to(0,0)} \frac{mx^2}{x^2 + m^2x^2} = \frac{m}{1 + m^2}$$

Since $\lim_{(x,y)\to(0,0)} f(x, y)$ depends on the slope of the line on which we approach the origin, we conclude that the limit does not exist. For example,

$$\text{on } y = x, \qquad \lim_{(x,y)\to(0,0)} f(x, y) = \frac{1}{2}$$

$$\text{on } y = 2x, \qquad \lim_{(x,y)\to(0,0)} f(x, y) = \frac{2}{5}$$ □

EXAMPLE 3 Show that $\displaystyle\lim_{(x,y)\to(0,0)} \frac{x^3y}{x^6 + y^2}$ does not exist.

Solution Let $f(x, y) = x^3y/(x^6 + y^2)$. You are encouraged to show that along the x- and y-axes, along any line $y = mx$, $m \neq 0$ through (0, 0), and along any parabola $y = kx^2$, $k \neq 0$ through (0, 0), $\lim_{(x,y)\to(0,0)} f(x, y) = 0$. Although this is an infinite number of paths to the origin, the limit *still* does not exist, since on $y = x^3$:

$$\lim_{(x,y)\to(0,0)} f(x, y) = \lim_{(x,y)\to(0,0)} \frac{x^6}{x^6 + x^6} = \frac{1}{2}$$ □

We state the next two results without proof. Theorem 15.2 is the analogue, for functions of two variables, of Theorem 1.4.

THEOREM 15.1 Two Simple Limits

(*i*) $\lim_{(x, y) \to (a, b)} x = a$ (*ii*) $\lim_{(x, y) \to (a, b)} y = b$

THEOREM 15.2 Limit of a Sum, Product, Quotient

If $\lim_{(x, y) \to (a, b)} f(x, y) = L_1$ and $\lim_{(x, y) \to (a, b)} g(x, y) = L_2$, then

(*i*)
$$\lim_{(x, y) \to (a, b)} [f(x, y) + g(x, y)] = \lim_{(x, y) \to (a, b)} f(x, y) + \lim_{(x, y) \to (a, b)} g(x, y) = L_1 + L_2$$

(*ii*)
$$\lim_{(x, y) \to (a, b)} f(x, y) \cdot g(x, y) = \lim_{(x, y) \to (a, b)} f(x, y) \cdot \lim_{(x, y) \to (a, b)} g(x, y) = L_1 L_2$$

(*iii*)
$$\lim_{(x, y) \to (a, b)} \frac{f(x, y)}{g(x, y)} = \frac{\lim_{(x, y) \to (a, b)} f(x, y)}{\lim_{(x, y) \to (a, b)} g(x, y)} = \frac{L_1}{L_2}, \quad L_2 \neq 0$$

EXAMPLE 4 Evaluate $\lim_{(x, y) \to (2, 3)} (x + y^2)$.

Solution From Theorem 15.1 we first note that

$$\lim_{(x, y) \to (2, 3)} x = 2 \quad \text{and} \quad \lim_{(x, y) \to (2, 3)} y = 3$$

Then from parts (*i*) and (*ii*) of Theorem 15.2 it follows that

$$\begin{aligned} \lim_{(x, y) \to (2, 3)} (x + y^2) &= \lim_{(x, y) \to (2, 3)} x + \lim_{(x, y) \to (2, 3)} y^2 \\ &= \lim_{(x, y) \to (2, 3)} x + \left(\lim_{(x, y) \to (2, 3)} y \right)\left(\lim_{(x, y) \to (2, 3)} y \right) \\ &= 2 + 3 \cdot 3 = 11 \end{aligned}$$

□

Use of Polar Coordinates In some cases polar coordinates can be used to evaluate a limit of the form $\lim_{(x, y) \to (0, 0)} f(x, y)$. If $x = r \cos \theta$, $y = r \sin \theta$, $r^2 = x^2 + y^2$, then $(x, y) \to (0, 0)$ if and only if $r \to 0$.

EXAMPLE 5 Evaluate $\lim_{(x,y)\to(0,0)} \dfrac{xy^2}{x^2+y^2}$.

Solution If $x = r\cos\theta$, $y = r\sin\theta$, then

$$\frac{xy^2}{x^2+y^2} = \frac{r^3\cos\theta\,\sin^2\theta}{r^2} = r\cos\theta\,\sin^2\theta$$

Since $\lim_{r\to 0} r\cos\theta\,\sin^2\theta = 0$, we conclude that

$$\lim_{(x,y)\to(0,0)} \frac{xy^2}{x^2+y^2} = 0$$

□

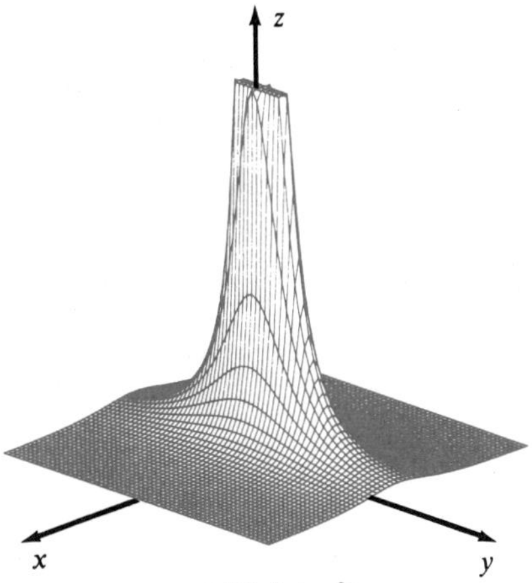

$z = 1/(9x^2 + y^2)$

FIGURE 15.27

Continuity A function $z = f(x, y)$ is **continuous** at (a, b) if $f(a, b)$ is defined, $\lim_{(x,y)\to(a,b)} f(x, y)$ exists, and the limit is the same as $f(a, b)$; that is, f is continuous at (a, b) if

$$\lim_{(x,y)\to(a,b)} f(x, y) = f(a, b)$$

If f is not continuous at (a, b), it is said to be **discontinuous**. A function $z = f(x, y)$ is **continuous on a region *R*** of the xy-plane if f is continuous at every point in R. The **sum** and **product** of two continuous functions are continuous. The **quotient** of two continuous functions is continuous, except at points where the denominator is zero. Also, if g is a function of two variables continuous at (a, b) and if F is a function of one variable continuous at $g(a, b)$, then the **composition** $f(x, y) = F(g(x, y))$ is continuous at (a, b). Last, the **graph** of a continuous function is a surface with no breaks. In Figure 15.27 we see a computer-generated graph of $f(x, y) = 1/(9x^2 + y^2)$. As the figure shows, the function f is discontinuous at (0, 0).

Polynomial and Rational Functions A **polynomial function** of two variables consists of the sum of powers $x^m y^n$, where m and n are nonnegative integers. For example, $f(x, y) = 3xy^2 - 5x^2y + x$ is a polynomial function in two variables. The quotient of two polynomial functions is called a **rational function**. Polynomial functions are continuous throughout the entire xy-plane and rational functions are continuous except at points where the denominator is zero.

EXAMPLE 6 The rational function $f(x, y) = xy/(y - x)$ is continuous except at points on the line $y = x$. □

EXAMPLE 7 Evaluate $\lim_{(x,y)\to(1,4)} \dfrac{x+2y}{x^2+y}$.

Solution Since the rational function $f(x, y) = (x + 2y)/(x^2 + y)$ is continuous at (1, 4), the limit is $f(1, 4) = \frac{9}{5}$. □

EXAMPLE 8

(*a*) The function $f(x, y) = \dfrac{x^4 - y^4}{x^2 + y^2}$ is discontinuous at (0, 0), since $f(0, 0)$ is not defined.

(*b*) The function f defined by

$$f(x, y) = \begin{cases} \dfrac{x^4 - y^4}{x^2 + y^2}, & (x, y) \neq (0, 0) \\ 0, & (x, y) = (0, 0) \end{cases}$$

is continuous at (0, 0), since $f(0, 0) = 0$ and

$$\lim_{(x,y)\to(0,0)} \frac{x^4 - y^4}{x^2 + y^2} = \lim_{(x,y)\to(0,0)} \frac{(x^2 + y^2)(x^2 - y^2)}{x^2 + y^2} = \lim_{(x,y)\to(0,0)} (x^2 - y^2) = 0^2 - 0^2 = 0$$

We see that $\lim_{(x,y)\to(0,0)} f(x, y) = f(0, 0)$. □

EXAMPLE 9 Since $F(x) = e^x$ is continuous for all real numbers, the composite function $f(x, y) = e^{x/(y+2x)}$ is continuous except at points on the line $y = -2x$. □

In Figures 15.28–15.31 we have used a computer to illustrate four functions that are discontinuous at points on a curve.

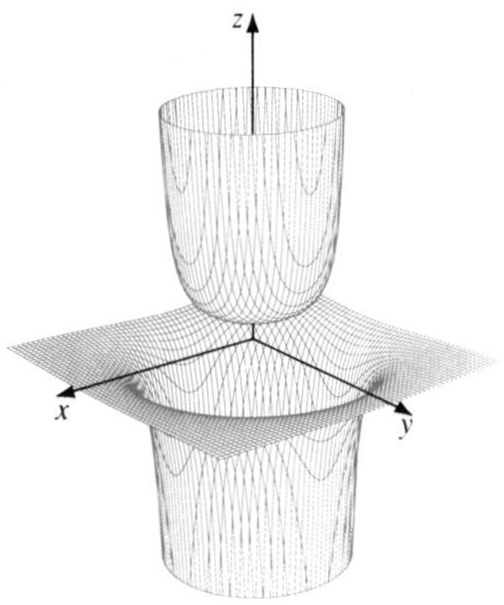

$f(x, y) = \dfrac{4}{6 - x^2 - y^2}$
discontinuous on $x^2 + y^2 = 6$

FIGURE 15.28

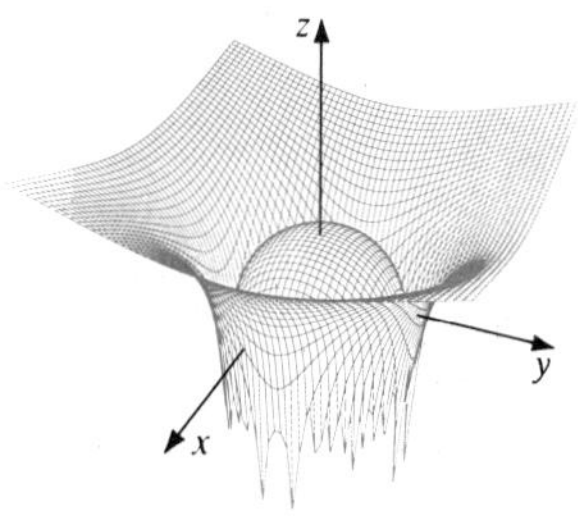

$f(x, y) = \ln|x^2 + y^2 - 4|$
discontinuous on $x^2 + y^2 = 4$

FIGURE 15.29

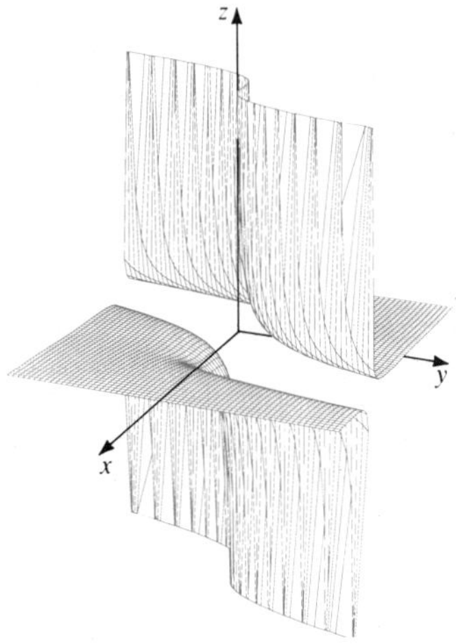

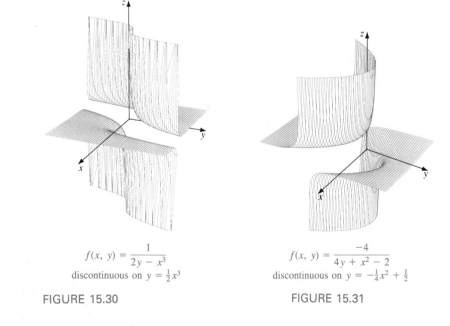

$f(x, y) = \dfrac{1}{2y - x^3}$
discontinuous on $y = \frac{1}{2}x^3$

FIGURE 15.30

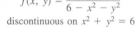

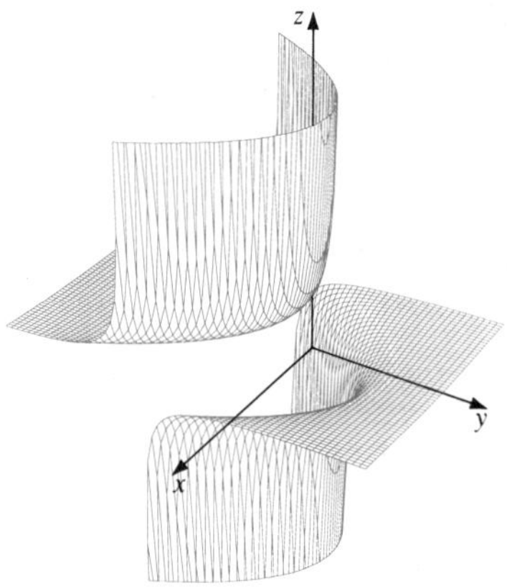

$f(x, y) = \dfrac{-4}{4y + x^2 - 2}$
discontinuous on $y = -\frac{1}{4}x^2 + \frac{1}{2}$

FIGURE 15.31

Terminology At this point we need to introduce some terminology that will be used in the following definition as well as in the next chapter. If R is some region of the xy-plane, then a point (a, b) is said to be an **interior point** of R if there is *some* circle centered at (a, b) whose interior contains only points of R. In contrast, we say that (a, b) is a **boundary point** of R if the interior of every circle centered at (a, b) contains both points in R and points not in R. The region R is said to be **open** if it contains no boundary points and **closed** if it contains all boundary points. See Figure 15.32(a). A region R is said to be **bounded** if it can be contained in a sufficiently large rectangle in the plane. Figure 15.32(b) illustrates a bounded region; the first quadrant illustrated in Figure 15.32(c) is an example of an unbounded region.

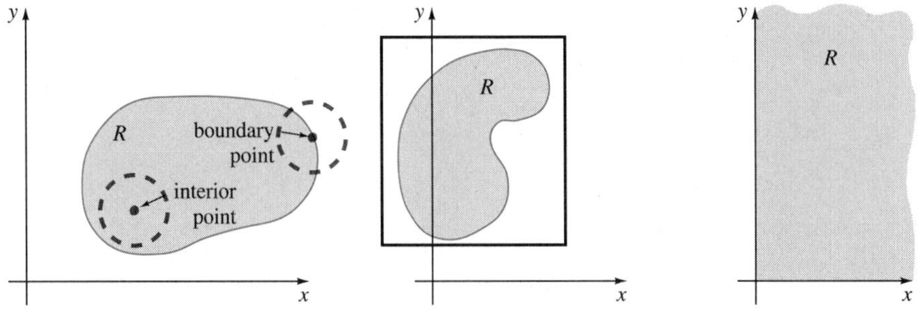

(a) closed region (b) bounded region (c) unbounded region

FIGURE 15.32

[O] 15.2.2 The ε–δ Definition of a Limit

As in Section 2.5, the formal definition of a limit of a function $z = f(x, y)$ at a point (a, b) is given in terms of ε–δ.

DEFINITION 15.2 ϵ–δ Definition of a Limit

Let f be a function of two variables that is defined at every point (x, y) in the interior of a circle centered at (a, b), except possibly at (a, b). Then

$$\lim_{(x,y)\to(a,b)} f(x, y) = L$$

means that for every $\varepsilon > 0$, there exists a number $\delta > 0$ such that

$$|f(x, y) - L| < \varepsilon \quad \text{whenever} \quad 0 < \sqrt{(x-a)^2 + (y-b)^2} < \delta$$

As illustrated in Figure 15.33, when f has a limit at (a, b), for a given $\varepsilon > 0$, regardless how small, we can find a circle of radius δ centered at (a, b)

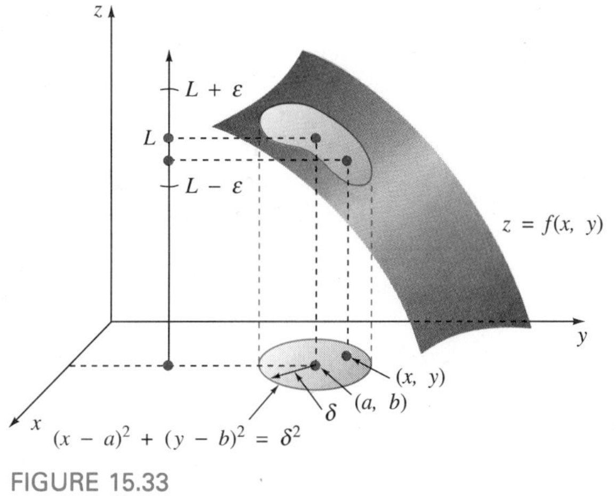

FIGURE 15.33

so that $L - \varepsilon < f(x, y) < L + \varepsilon$ for every interior point $(x, y) \neq (a, b)$ of the circle. As mentioned previously, the values of f are close to L whenever (x, y) is close enough to (a, b). The concept of "close enough" is defined by the number δ.

▲ ***Remark*** The definitions of limit and continuity for functions of three or more variables are natural extensions of those just considered. For example, $w = F(x, y, z)$ is continuous at (a, b, c) if

$$\lim_{(x,y,z)\to(a,b,c)} F(x, y, z) = F(a, b, c)$$

The polynomial function in three variables $F(x, y, z) = xy^2z^3$ is continuous throughout 3-space. The rational function

$$F(x, y, z) = \frac{xy^2}{x^2 + y^2 + (z-1)^2}$$

is continuous except at the point (0, 0, 1). The function

$$F(x, y, z) = \frac{x + 3y}{2x + 5y + z}$$

is continuous except at points on the plane $2x + 5y + z = 0$.

EXERCISES 15.2 *Answers to odd-numbered problems begin on page A-104.*

[15.2.1]

If DNE, Show it.

In Problems 1–30 evaluate the given limit, if it exists.

1. $\lim_{(x,y)\to(5,-1)} (x^2 + y^2)$

2. $\lim_{(x,y)\to(2,1)} \frac{x^2 - y}{x - y}$

3. $\lim_{(x,y)\to(0,0)} \frac{5x^2 + y^2}{x^2 + y^2}$

4. $\lim_{(x,y)\to(1,2)} \frac{4x^2 + y^2}{16x^4 + y^4}$

5. $\lim_{(x,y)\to(1,1)} \frac{4 - x^2 - y^2}{x^2 + y^2}$

6. $\lim_{(x,y)\to(0,0)} \frac{2x^2 - y}{x^2 + 2y^2}$

7. $\lim_{(x,y)\to(0,0)} \frac{x^2y}{x^4 + y^2}$

8. $\lim_{(x,y)\to(0,0)} \frac{6xy^2}{x^2 + y^4}$

9. $\lim_{(x,y)\to(1,2)} x^3y^2(x + y)^3$

10. $\lim_{(x,y)\to(2,3)} \frac{xy}{x^2 - y^2}$

11. $\lim_{(x,y)\to(0,0)} \dfrac{e^{xy}}{x+y+1}$

12. $\lim_{(x,y)\to(0,0)} \dfrac{\sin xy}{x^2+y^2}$

13. $\lim_{(x,y)\to(2,2)} \dfrac{xy}{x^3+y^2}$

14. $\lim_{(x,y)\to(\pi,\pi/4)} \cos(3x+y)$

15. $\lim_{(x,y)\to(0,0)} \dfrac{x^2-3y+1}{x+5y-3}$

16. $\lim_{(x,y)\to(0,0)} \dfrac{x^2y^2}{x^4+5y^4}$

17. $\lim_{(x,y)\to(4,3)} xy^2\left(\dfrac{x+2y}{x-y}\right)$

18. $\lim_{(x,y)\to(1,0)} \dfrac{x^2y}{x^3+y^3}$

19. $\lim_{(x,y)\to(1,1)} \dfrac{xy-x-y+1}{x^2+y^2-2x-2y+2}$

20. $\lim_{(x,y)\to(0,3)} \dfrac{xy-3y}{x^2+y^2-6y+9}$

21. $\lim_{(x,y)\to(0,0)} \dfrac{x^3y+xy^3-3x^2-3y^2}{x^2+y^2}$

22. $\lim_{(x,y)\to(-2,2)} \dfrac{y^3+2x^3}{x+5xy^2}$

23. $\lim_{(x,y)\to(1,1)} \ln(2x^2-y^2)$

24. $\lim_{(x,y)\to(1,2)} \dfrac{\sin^{-1}(x/y)}{\cos^{-1}(x-y)}$

25. $\lim_{(x,y)\to(0,0)} \dfrac{(x^2-y^2)^2}{x^2+y^2}$

26. $\lim_{(x,y)\to(0,0)} \dfrac{\sin(3x^2+3y^2)}{x^2+y^2}$

27. $\lim_{(x,y)\to(0,0)} \dfrac{6xy}{\sqrt{x^2+y^2}}$

28. $\lim_{(x,y)\to(0,0)} \dfrac{x^2-y^2}{\sqrt{x^2+y^2}}$

29. $\lim_{(x,y)\to(0,0)} \dfrac{x^3}{x^2+y^2}$

30. $\lim_{(x,y)\to(0,0)} \dfrac{x^3+y^3}{x^2+y^2}$

In Problems 31–34 determine where the given function is continuous.

31. $f(x, y) = \sqrt{x}\cos\sqrt{x+y}$

32. $f(x, y) = y^2e^{1/xy}$

33. $f(x, y) = \tan\dfrac{x}{y}$

34. $f(x, y) = \ln(4x^2+9y^2+36)$

In Problems 35 and 36 determine whether the given function is continuous on the indicated regions.

35. $f(x, y) = \begin{cases} x+y, & x \geq 2 \\ 0, & x < 2 \end{cases}$

(a) $x^2+y^2<1$

(b) $x \geq 0$

(c) $y > x$

36. $f(x, y) = \dfrac{xy}{\sqrt{x^2+y^2-25}}$

(a) $y \geq 3$

(b) $|x|+|y|<1$

(c) $(x-2)^2+y^2<1$

37. Determine whether the function f defined by

$$f(x, y) = \begin{cases} \dfrac{6x^2y^3}{(x^2+y^2)^2}, & (x, y) \neq (0, 0) \\ 0, & (x, y) = (0, 0) \end{cases}$$

is continuous at (0, 0).

38. Show that

$$f(x, y) = \begin{cases} \dfrac{xy}{2x^2+2y^2}, & (x, y) \neq (0, 0) \\ 0, & (x, y) = (0, 0) \end{cases}$$

is continuous in each variable separately at (0, 0); that is, that $f(x, 0)$ and $f(0, y)$ are continuous at $x = 0$ and $y = 0$, respectively. Show, however, that f is not continuous at (0, 0).

[15.2.2]

In Problems 39 and 40 use Definition 15.2 to prove the given result; that is, find δ for any arbitrary $\varepsilon > 0$. [*Hint:* Use polar coordinates.]

39. $\lim_{(x,y)\to(0,0)} \dfrac{3xy^2}{x^2+y^2} = 0$

40. $\lim_{(x,y)\to(0,0)} \dfrac{x^2y^2}{x^2+y^2} = 0$

41. Use Definition 15.2 to prove that

$$\lim_{(x,y)\to(a,b)} y = b$$

15.3 PARTIAL DIFFERENTIATION

The derivative of a function of **one variable** $y = f(x)$ is given by

$$\frac{dy}{dx} = \lim_{\Delta x \to 0} \frac{f(x+\Delta x)-f(x)}{\Delta x}$$

In exactly the same manner, we can define a derivative of a function of **two variables** $z = f(x, y)$ with respect to *each* variable.

DEFINITION 15.3 Partial Derivatives

If $z = f(x, y)$, then the **partial derivative with respect to x** is

$$\frac{\partial z}{\partial x} = \lim_{\Delta x \to 0} \frac{f(x + \Delta x, y) - f(x, y)}{\Delta x} \tag{1}$$

and the **partial derivative with respect to y** is

$$\frac{\partial z}{\partial y} = \lim_{\Delta y \to 0} \frac{f(x, y + \Delta y) - f(x, y)}{\Delta y} \tag{2}$$

provided each limit exists.

In (1) the variable y does not change in the limiting process; that is, y is held fixed. Similarly, in (2) the variable x is held fixed. The two partial derivatives (1) and (2) then represent the *rates of change* of f with respect to x and y, respectively. On a practical level:

To compute $\partial z/\partial x$, use the laws of ordinary differentiation while treating y as a constant.

To compute $\partial z/\partial y$, use the laws of ordinary differentiation while treating x as a constant.

EXAMPLE 1 If $z = 4x^3y^2 - 4x^2 + y^6 + 1$, find (*a*) $\dfrac{\partial z}{\partial x}$ and (*b*) $\dfrac{\partial z}{\partial y}$.

Solution (*a*) We hold y fixed and treat constants in the usual manner. Thus,

$$\frac{\partial z}{\partial x} = 12x^2y^2 - 8x$$

(*b*) By treating x as a constant, we obtain

$$\frac{\partial z}{\partial y} = 8x^3y + 6y^5$$

□

Alternative Symbols The partial derivatives $\partial z/\partial x$ and $\partial z/\partial y$ are often represented by alternative symbols. If $z = f(x, y)$, then

$$\frac{\partial z}{\partial x} = \frac{\partial f}{\partial x} = z_x = f_x \quad \text{and} \quad \frac{\partial z}{\partial y} = \frac{\partial f}{\partial y} = z_y = f_y$$

A symbol such as $\partial/\partial x$ denotes the *operation* of taking a partial derivative, in this case with respect to x; for example,

$$\frac{\partial}{\partial x}(x^2 - y^2) = 2x$$

EXAMPLE 2 If $f(x, y) = x^5y^{10}\cos(xy^2)$, find f_y.

Solution When x is held fixed, observe that

$$f(x, y) = x^5 \underbrace{y^{10}\cos(xy^2)}_{\text{product of two functions of } y}$$

Hence, by the Product Rule,

$$f_y(x, y) = x^5y^{10}[-2xy\sin(xy^2)] + 10x^5y^9\cos(xy^2)$$ □

EXAMPLE 3 The function $S = 0.1091w^{0.425}h^{0.725}$ relates the surface area (in square feet) of a person's body as a function of weight w (in pounds) and height h (in inches). Find $\partial S/\partial w$ when $w = 150$ and $h = 72$. Interpret.

Solution

$$\frac{\partial S}{\partial w} = (0.1091)(0.425)w^{-0.575}h^{0.725}$$

$$\left.\frac{\partial S}{\partial w}\right|_{(150,\,72)} = (0.1091)(0.425)(150)^{-0.575}(72)^{0.725} \approx 0.058$$

The partial derivative $\partial S/\partial w$ is the rate at which the surface area of a person of fixed height, such as an adult, changes with respect to weight. Since the units for the derivative are ft^2/lb, we see that a gain of 1 lb, while h is fixed at 72, results in an increase in the area of the skin of approximately $0.058 \approx \frac{1}{17}\ \text{ft}^2$. □

Geometric Interpretation As seen in Figure 15.34(a), when y is constant, say $y = b$, the trace of the surface $z = f(x, y)$ in the plane $y = b$ is a curve C. If we define the slope of the secant through the indicated points P and R as

$$\frac{f(a + \Delta x, b) - f(a, b)}{\Delta x}$$

we have

$$\left.\frac{\partial z}{\partial x}\right|_{(a,\,b)} = \lim_{\Delta x \to 0} \frac{f(a + \Delta x, b) - f(a, b)}{\Delta x}$$

In other words, we can interpret $\partial z/\partial x$ as the slope of the tangent at any point (for which the limit exists) on a curve C of intersection between the surface $z = f(x, y)$ and a plane $y =$ constant. In turn, an inspection of Figure 15.34(b)

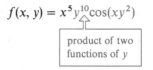

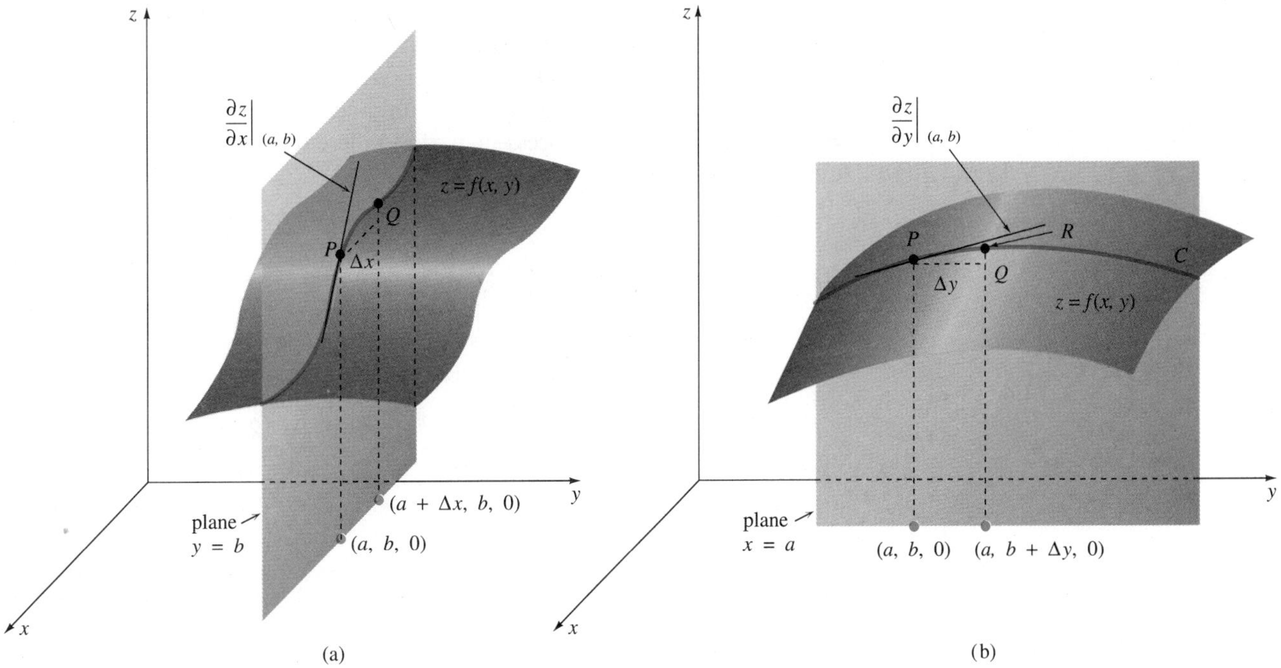

FIGURE 15.34

reveals that $\partial z/\partial y$ is the slope of the tangent at a point on a curve of intersection between the surface $z = f(x, y)$ and a plane $x =$ constant.

EXAMPLE 4 For $z = 9 - x^2 - y^2$, find the slope of the tangent line at (2, 1, 4) in (*a*) the plane $x = 2$ and (*b*) the plane $y = 1$.

Solution

(*a*) By specifying the plane $x = 2$, we are holding all values of x constant. Hence, we compute the partial derivative with respect to y:

$$\frac{\partial z}{\partial y} = -2y$$

At (2, 1, 4) the slope is $$\left.\frac{\partial z}{\partial y}\right|_{(2,1)} = -2.$$

(*b*) In the plane $y = 1$, y is constant and so we find the partial derivative with respect to x:

$$\frac{\partial z}{\partial x} = -2x$$

At (2, 1, 4) the slope is $$\left.\frac{\partial z}{\partial x}\right|_{(2,1)} = -4.$$

See Figure 15.35. □

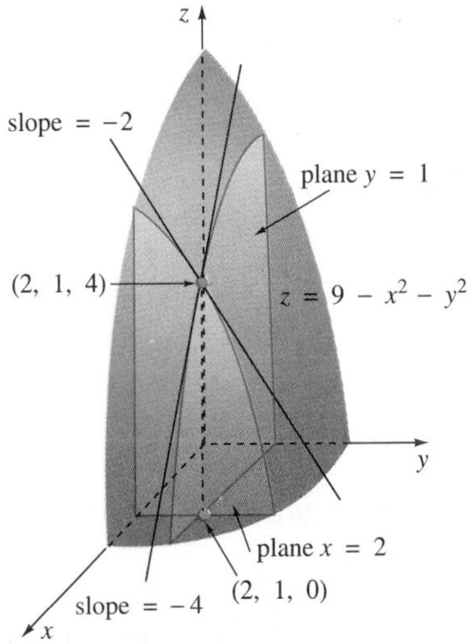

FIGURE 15.35

Functions of Three or More Variables The rates of change of a function of three variables $w = F(x, y, z)$ in the x, y, and z directions are $\partial w/\partial x$, $\partial w/\partial y$, and $\partial w/\partial z$, respectively. To compute, say, $\partial w/\partial x$, we differentiate with respect to x in the usual manner while holding *both* y and z constant. In this manner we extend the process of partial differentiation to functions of any number of variables.

EXAMPLE 5 If $w = \dfrac{x^2 - z^2}{y^2 + z^2}$, find $\dfrac{\partial w}{\partial z}$.

Solution We use the Quotient Rule while holding x and y constant:

$$\frac{\partial w}{\partial z} = \frac{(y^2 + z^2)(-2z) - (x^2 - z^2)2z}{(y^2 + z^2)^2} = -\frac{2z(x^2 + y^2)}{(y^2 + z^2)^2}$$ □

EXAMPLE 6 If $F(x, y, t) = e^{-3\pi t}\cos 4x \sin 6y$, then the partial derivatives with respect to x, y, and t are, in turn,

$$F_x(x, y, t) = -4e^{-3\pi t}\sin 4x \sin 6y$$
$$F_y(x, y, t) = 6e^{-3\pi t}\cos 4x \cos 6y$$
$$F_t(x, y, t) = -3\pi e^{-3\pi t}\cos 4x \sin 6y$$ □

Higher-Order and Mixed Derivatives For a function of two variables $z = f(x, y)$, the partial derivatives $\partial z/\partial x$ and $\partial z/\partial y$ are themselves functions of x and y. Consequently, we can compute **second**, and higher, **partial derivatives**. Indeed, we can find the partial derivative of $\partial z/\partial x$ with respect to y, and the partial derivative of $\partial z/\partial y$ with respect to x. The latter types of partial derivatives are called **mixed partial derivatives**. In summary, for $z = f(x, y)$:

Second-order partial derivatives:

$$\frac{\partial^2 z}{\partial x^2} = \frac{\partial}{\partial x}\left(\frac{\partial z}{\partial x}\right) \quad \text{and} \quad \frac{\partial^2 z}{\partial y^2} = \frac{\partial}{\partial y}\left(\frac{\partial z}{\partial y}\right)$$

Third-order partial derivatives:

$$\frac{\partial^3 z}{\partial x^3} = \frac{\partial}{\partial x}\left(\frac{\partial^2 z}{\partial x^2}\right) \quad \text{and} \quad \frac{\partial^3 z}{\partial y^3} = \frac{\partial}{\partial y}\left(\frac{\partial^2 z}{\partial y^2}\right)$$

Mixed second-order partial derivatives:

$$\frac{\partial^2 z}{\partial x\,\partial y} = \frac{\partial}{\partial x}\left(\frac{\partial z}{\partial y}\right) \quad \text{and} \quad \frac{\partial^2 z}{\partial y\,\partial x} = \frac{\partial}{\partial y}\left(\frac{\partial z}{\partial x}\right)$$

Higher-order partial derivatives for f and for functions of three or more variables are defined in a similar manner.

Alternative Symbols The second- and third-order partial derivatives are denoted by f_{xx}, f_{yy}, f_{xxx}, and so on. The subscript notation for mixed second partial derivatives is f_{xy} or f_{yx}. Note that

$$f_{xy} = (f_x)_y = \frac{\partial}{\partial y}\left(\frac{\partial z}{\partial x}\right) = \frac{\partial^2 z}{\partial y\,\partial x} \quad \text{and} \quad f_{yx} = \frac{\partial^2 z}{\partial x\,\partial y}$$

Equality of Mixed Partials Although we shall not prove it, the next theorem states that under certain conditions the order in which a mixed second partial derivative is done is irrelevant; that is, the mixed partial derivatives f_{xy} and f_{yx} are equal.

THEOREM 15.3 Equality of Mixed Partials

Let f be a function of two variables. If f_x, f_y, f_{xy}, and f_{yx} are continuous on an open region R, then $f_{xy} = f_{yx}$ at each point of R.

See Problem 64 in Exercises 15.3.

EXAMPLE 7 If

$$z = x^2y^2 - y^3 + 3x^4 + 5$$

find

$$\frac{\partial^2 z}{\partial x^2}, \quad \frac{\partial^3 z}{\partial x^3}, \quad \frac{\partial^2 z}{\partial y^2}, \quad \frac{\partial^3 z}{\partial y^3}, \quad \text{and} \quad \frac{\partial^2 z}{\partial x\,\partial y}$$

Solution

$$\frac{\partial z}{\partial x} = 2xy^2 + 12x^3 \qquad \frac{\partial z}{\partial y} = 2x^2y - 3y^2$$

$$\frac{\partial^2 z}{\partial x^2} = 2y^2 + 36x^2 \qquad \frac{\partial^2 z}{\partial y^2} = 2x^2 - 6y$$

$$\frac{\partial^3 z}{\partial x^3} = 72x \qquad \frac{\partial^3 z}{\partial y^3} = -6$$

$$\frac{\partial^2 z}{\partial x\,\partial y} = \frac{\partial}{\partial x}(2x^2y - 3y^2) = 4xy$$

You should also verify that $\dfrac{\partial^2 z}{\partial y\,\partial x} = 4xy$. □

If f is a function of two variables and has continuous first-, second-, and third-order partial derivatives on an open region R, then the mixed third-order derivatives are equal; that is,

$$f_{xyy} = f_{yxy} = f_{yyx} \quad \text{and} \quad f_{yxx} = f_{xyx} = f_{xxy}$$

Similar remarks hold for functions of three or more variables. For example, if F is a function of three variables with continuous partial derivatives of any order, then the mixed partials such as $F_{xyz} = F_{zyx} = F_{yxz}$ are equal.

EXAMPLE 8 If $F(x, y, z) = \sqrt{x^2 + y^4 + z^6}$, find F_{yzz}.

Solution F_{yzz} is a mixed third-order partial derivative. First we find the partial derivative with respect to y by the Power Rule for Functions:

$$F_y = \frac{1}{2}(x^2 + y^4 + z^6)^{-1/2}4y^3 = 2y^3(x^2 + y^4 + z^6)^{-1/2}$$

Then,

$$F_{yz} = (F_y)_z = (2y^3)\left(-\frac{1}{2}\right)(x^2 + y^4 + z^6)^{-3/2}6z^5$$

$$= -6y^3z^5(x^2 + y^4 + z^6)^{-3/2}$$

Finally, by the Product Rule,

$$F_{yzz} = (F_{yz})_z = -6y^3z^5\left(-\frac{3}{2}\right)(x^2 + y^4 + z^6)^{-5/2}(6z^5) - 30y^3z^4(x^2 + y^4 + z^6)^{-3/2}$$

$$= y^3z^4(x^2 + y^4 + z^6)^{-5/2}(24z^6 - 30x^2 - 30y^4)$$

You are encouraged to compute F_{zzy} and F_{zyz} and verify in any open region not containing the origin that $F_{yzz} = F_{zzy} = F_{zyz}$. □

EXERCISES 15.3

Answers to odd-numbered problems begin on page A-104.

In Problems 1–4 use Definition 15.3 to compute $\partial z/\partial x$ and $\partial z/\partial y$ for the given function.

1. $z = 7x + 8y^2$
2. $z = xy$
3. $z = 3x^2y + 4xy^2$
4. $z = \dfrac{x}{x + y}$

In Problems 5–24 find the first partial derivatives of the given function.

5. $z = x^2 - xy^2 + 4y^5$
6. $z = -x^3 + 6x^2y^3 + 5y^2$
7. $z = 5x^4y^3 - x^2y^6 + 6x^5 - 4y$
8. $z = \tan(x^3y^2)$
9. $z = \dfrac{4\sqrt{x}}{3y^2 + 1}$
10. $z = 4x^3 - 5x^2 + 8x$
11. $z = (x^3 - y^2)^{-1}$
12. $z = (-x^4 + 7y^2 + 3y)^6$
13. $z = \cos^2 5x + \sin^2 5y$
14. $z = e^{x^2 \tan^{-1} y^2}$
15. $f(x, y) = xe^{x^3y}$
16. $f(\theta, \phi) = \phi^2 \sin\dfrac{\theta}{\phi}$
17. $f(x, y) = \dfrac{3x - y}{x + 2y}$
18. $f(x, y) = \dfrac{xy}{(x^2 - y^2)^2}$
19. $g(u, v) = \ln(4u^2 + 5v^3)$
20. $h(r, s) = \dfrac{\sqrt{r}}{s} - \dfrac{\sqrt{s}}{r}$
21. $w = 2\sqrt{x}y - ye^{y/z}$
22. $w = xy \ln xz$
23. $F(u, v, x, t) = u^2w^2 - uv^3 + vw\cos(ut^2) + (2x^2t)^4$
24. $G(p, q, r, s) = (p^2q^3)^{r^4s^5}$

In Problems 25 and 26 suppose $z = 4x^3y^4$.

25. Find the slope of the tangent line at $(1, -1, 4)$ in the plane $x = 1$.
26. Find the slope of the tangent line at $(1, -1, 4)$ in the plane $y = -1$.

In Problems 27 and 28 suppose $f(x, y) = \dfrac{18xy}{x + y}$.

27. Find parametric equations for the tangent line at $(-1, 4, -24)$ in the plane $x = -1$.
28. Find symmetric equations for the tangent line at $(-1, 4, -24)$ in the plane $y = 4$.

In Problems 29 and 30 suppose $z = \sqrt{9 - x^2 - y^2}$.

29. At what rate is z changing with respect to x in the plane $y = 2$ at the point (2, 2, 1)?

30. At what rate is z changing with respect to y in the plane $x = \sqrt{2}$ at the point $(\sqrt{2}, \sqrt{3}, 2)$?

In Problems 31–38 find the indicated partial derivative.

31. $z = e^{xy}; \dfrac{\partial^2 z}{\partial x^2}$

32. $z = x^4 y^{-2}; \dfrac{\partial^3 z}{\partial y^3}$

33. $f(x, y) = 5x^2y^2 - 2xy^3; f_{xy}$

34. $f(p, q) = \ln \dfrac{p+q}{q^2}; f_{qp}$

35. $w = u^2v^3t^3; w_{tuv}$

36. $w = \dfrac{\cos(u^2 v)}{t^3}; w_{vvt}$

37. $F(r, \theta) = e^{r^2}\cos\theta; F_{r\theta r}$

38. $H(s, t) = \dfrac{s+t}{s-t}; H_{tts}$

In Problems 39 and 40 verify that $\dfrac{\partial^2 z}{\partial x\, \partial y} = \dfrac{\partial^2 z}{\partial y\, \partial x}$.

39. $z = x^6 - 5x^4y^3 + 4xy^2$

40. $z = \tan^{-1}(2xy)$

In Problems 41 and 42 verify that the indicated partial derivatives are equal.

41. $w = u^3v^4 - 4u^2v^2t^3 + v^2t; w_{uvt}, w_{tvu}, w_{vut}$

42. $F(\eta, \xi, \tau) = (\eta^3 + \xi^2 + \tau)^2; F_{\eta\xi\eta}, F_{\xi\eta\eta}, F_{\eta\eta\xi}$

In Problems 43–46 suppose the given equation defines z as a function of the remaining two variables. Use implicit differentiation to find the first partial derivatives.

43. $x^2 + y^2 + z^2 = 25$

44. $z^2 = x^2 + y^2 z$

45. $z^2 + u^2v^3 - uvz = 0$

46. $se^z - e^{st} + 4s^3t = z$

47. The area A of a parallelogram with base x and height $y \sin\theta$ is $A = xy\sin\theta$. Find all first partial derivatives.

48. The volume of the frustum of a cone shown in Figure 15.36 is $V = (\pi/3)h(r^2 + rR + R^2)$. Find all first partial derivatives.

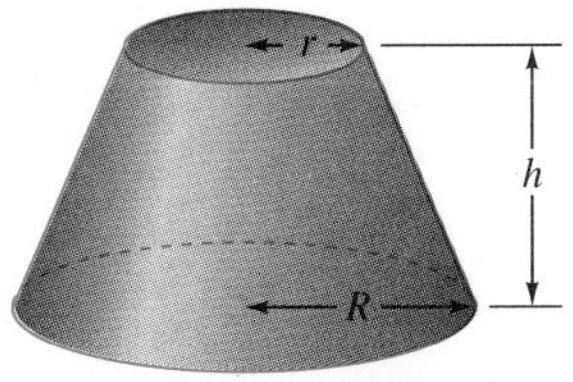

FIGURE 15.36

A solution $u(x, y)$ of **Laplace's equation** in two dimensions

$$\frac{\partial^2 u}{\partial x^2} + \frac{\partial^2 u}{\partial y^2} = 0$$

can be interpreted as the time-independent temperature distribution throughout a thin two-dimensional plate. See Figure 15.37. In Problems 49 and 50 verify that the given temperature distribution satisfies Laplace's equation.

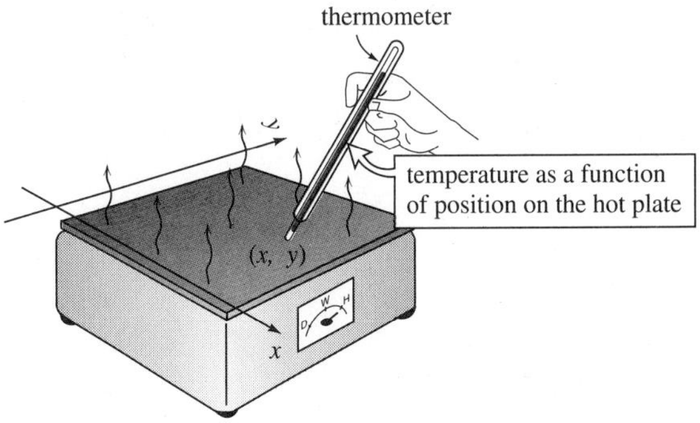

FIGURE 15.37

49. $u(x, y) = (\cosh 2\pi y + \sinh 2\pi y)\sin 2\pi x$

50. $u(x, y) = e^{-(n\pi x/L)}\sin(n\pi y/L)$, n and L constants

In Problems 51 and 52 verify that the given function satisfies Laplace's equation.

51. $u(x, y) = \ln(x^2 + y^2)$

52. $u(x, y) = \tan^{-1}\dfrac{y}{x}$

In Problems 53 and 54 verify that the given function satisfies Laplace's equation in three dimensions $\partial^2 u/\partial x^2 + \partial^2 u/\partial y^2 + \partial^2 u/\partial z^2 = 0$.

53. $u(x, y, z) = \dfrac{1}{\sqrt{x^2 + y^2 + z^2}}$

54. $u(x, y, z) = e^{\sqrt{m^2+n^2}x}\cos my \sin nz$

The one-dimensional **wave equation**

$$a^2 \frac{\partial^2 u}{\partial x^2} = \frac{\partial^2 u}{\partial t^2}$$

occurs in problems involving vibrational phenomena. In Problems 55 and 56 verify that the given function satisfies the wave equation.

55. $u(x, t) = \cos at \sin x$

56. $u(x, t) = \cos(x + at) + \sin(x - at)$

57. The molecular concentration $C(x, t)$ of a liquid is given by $C(x, t) = t^{-1/2}e^{-x^2/kt}$. Verify that this function satisfies the

one-dimensional **diffusion equation**

$$\frac{k}{4}\frac{\partial^2 C}{\partial x^2} = \frac{\partial C}{\partial t}$$

58. The pressure P exerted by an enclosed ideal gas is given by $P = k(T/V)$, where k is a constant, T is temperature, and V is volume. Find:

(a) the rate of change of P with respect to V,

(b) the rate of change of V with respect to T, and

(c) the rate of change of T with respect to P.

59. The vertical displacement of a long string fastened at the origin but falling under its own weight is given by

$$u(x, t) = \begin{cases} -\dfrac{g}{2a^2}(2axt - x^2), & 0 \le x \le at \\ -\dfrac{1}{2}gt^2, & x > at \end{cases}$$

See Figure 15.38.

(a) Find $\partial u/\partial t$. Interpret.

(b) Find $\partial u/\partial x$ for $x > at$. Interpret.

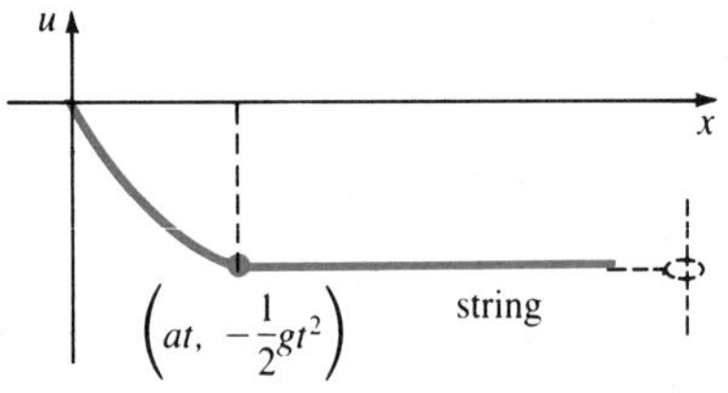

FIGURE 15.38

60. For the skin-area function

$$S = 0.1091w^{0.425}h^{0.725}$$

discussed in Example 3, find $\partial S/\partial h$ at $w = 60$, $h = 36$. If a girl grows in height from 36 to 37 in, while her weight is fixed at 60 lb, what is the approximate increase in the area of skin?

61. State a limit definition that is analogous to Definition 15.3 for

(a) $\dfrac{\partial^2 z}{\partial x^2}$ (b) $\dfrac{\partial^2 z}{\partial y^2}$ (c) $\dfrac{\partial^2 z}{\partial x\,\partial y}$

62. Find a function $z = f(x, y)$ such that

$$\frac{\partial z}{\partial x} = 2xy^3 + 2y + \frac{1}{x} \quad \text{and} \quad \frac{\partial z}{\partial y} = 3x^2y^2 + 2x + 1$$

63. Suppose the function $w = F(x, y, z)$ has continuous third-order partial derivatives. How many different third-order partial derivatives are there?

64. Consider the function $z = f(x, y)$ defined by

$$f(x, y) = \begin{cases} \dfrac{xy(y^2 - x^2)}{x^2 + y^2}, & (x, y) \ne (0, 0) \\ 0, & (x, y) = (0, 0) \end{cases}$$

(a) Compute $\left.\dfrac{\partial z}{\partial x}\right|_{(0,y)}$ and $\left.\dfrac{\partial z}{\partial y}\right|_{(x,0)}$.

(b) Show that $\left.\dfrac{\partial^2 z}{\partial y\,\partial x}\right|_{(0,0)} \ne \left.\dfrac{\partial^2 z}{\partial x\,\partial y}\right|_{(0,0)}$.

65. A function $z = f(x, y)$ may not be continuous at a point but still may have partial derivatives at that point. The function

$$f(x, y) = \begin{cases} \dfrac{xy}{2x^2 + 2y^2}, & (x, y) \ne (0, 0) \\ 0, & (x, y) = (0, 0) \end{cases}$$

is not continuous at (0, 0). (See Problem 38 in Exercises 15.2.) Use (1) and (2) to show that $\left.\dfrac{\partial z}{\partial x}\right|_{(0,0)} = 0$ and $\left.\dfrac{\partial z}{\partial y}\right|_{(0,0)} = 0$.

15.4 THE DIFFERENTIAL AND DIFFERENTIABILITY

Increment of the Dependent Variable The notion of the differentiability of a function of any number of independent variables depends on the **increment** of the dependent variable. Recall that for a function of *one variable* $y = f(x)$,

$$\Delta y = f(x + \Delta x) - f(x)$$

Analogously, for a function of *two variables* $z = f(x, y)$, we define

$$\Delta z = f(x + \Delta x, y + \Delta y) - f(x, y) \tag{1}$$

Figure 15.39 shows that Δz gives the amount of **change** in the function as (x, y) changes to $(x + \Delta x, y + \Delta y)$.

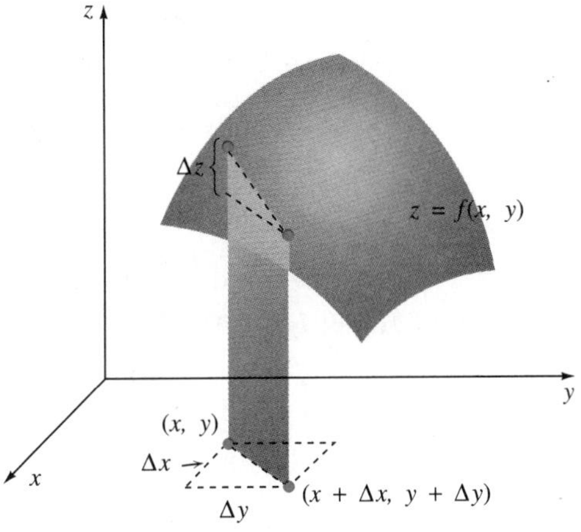

FIGURE 15.39

EXAMPLE 1 Find Δz for $z = x^2 - xy$. What is the change in the function from (1, 1) to (1.2, 0.7)?

Solution From (1),

$$\begin{aligned}\Delta z &= [(x + \Delta x)^2 - (x + \Delta x)(y + \Delta y)] - (x^2 - xy)\\ &= (2x - y)\,\Delta x - x\,\Delta y + (\Delta x)^2 - \Delta x\,\Delta y \qquad (2)\end{aligned}$$

With $x = 1$, $y = 1$, $\Delta x = 0.2$, and $\Delta y = -0.3$,

$$\Delta z = (1)(0.2) - (1)(-0.3) + (0.2)^2 - (0.2)(-0.3) = 0.6$$

□

A Fundamental Increment Formula A brief reinspection of the increment Δz in (2) shows that in the first two terms the coefficients of Δx and Δy are $\partial z/\partial x$ and $\partial z/\partial y$, respectively. The following important theorem shows that this is no accident:

THEOREM 15.4 An Increment Formula

Let $z = f(x, y)$ have continuous partial derivatives $f_x(x, y)$ and $f_y(x, y)$ in a rectangular region that is defined by $a < x < b$, $c < y < d$. If (x, y) is any point in this region, then there exist ε_1 and ε_2, which are functions of Δx and Δy, such that

$$\Delta z = f_x(x, y)\,\Delta x + f_y(x, y)\,\Delta y + \varepsilon_1\,\Delta x + \varepsilon_2\,\Delta y \qquad (3)$$

where $\varepsilon_1 \to 0$ and $\varepsilon_2 \to 0$ when $\Delta x \to 0$ and $\Delta y \to 0$.

Proof By adding and subtracting $f(x, y + \Delta y)$ in (1), we have

$$\Delta z = [f(x + \Delta x, y + \Delta y) - f(x, y + \Delta y)] + [f(x, y + \Delta y) - f(x, y)]$$

Applying the Mean Value Theorem to each set of brackets then gives

$$\Delta z = f_x(x_0, y + \Delta y)\,\Delta x + f_y(x, y_0)\,\Delta y \tag{4}$$

where, as shown in Figure 15.40, $x < x_0 < x + \Delta x$ and $y < y_0 < y + \Delta y$. Now, define

$$\varepsilon_1 = f_x(x_0, y + \Delta y) - f_x(x, y) \qquad \varepsilon_2 = f_y(x, y_0) - f_y(x, y) \tag{5}$$

As $\Delta x \to 0$ and $\Delta y \to 0$, then, as shown in the figure, $P_2 \to P_1$ and $P_3 \to P_1$. Since f_x and f_y are assumed continuous in the region, we have

$$\lim_{(\Delta x, \Delta y)\to(0,0)} \varepsilon_1 = 0 \quad \text{and} \quad \lim_{(\Delta x, \Delta y)\to(0,0)} \varepsilon_2 = 0$$

Solving (5) for $f_x(x_0, y + \Delta y)$ and $f_y(x, y_0)$ and substituting in (4) give (3).

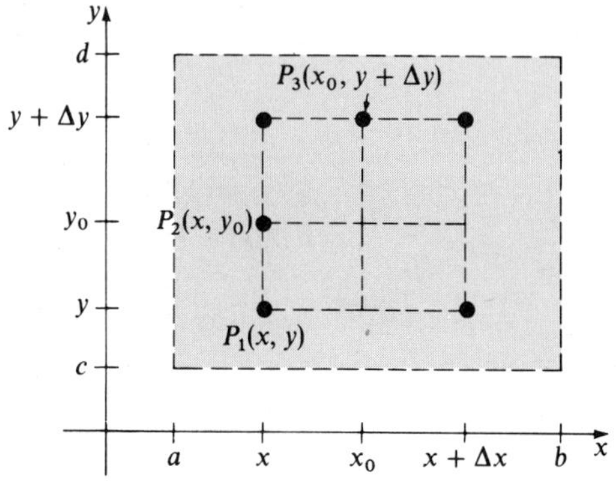

FIGURE 15.40

■

EXAMPLE 2 In Example 1 we can take $\varepsilon_1 = \Delta x$ and $\varepsilon_2 = -\Delta x$. □

Total Differential for $z = f(x, y)$ Inspired by (3), we now define the **total differential**, or simply **differential**, of a function $z = f(x, y)$.

DEFINITION 15.4 Differentials

Let $z = f(x, y)$ be a function for which the first partials f_x and f_y exist.

(*i*) The **differentials** of the independent variables are

$$dx = \Delta x \qquad dy = \Delta y$$

(*ii*) The **total differential** of the function is

$$dz = f_x(x, y)\,dx + f_y(x, y)\,dy = \frac{\partial z}{\partial x}\,dx + \frac{\partial z}{\partial y}\,dy$$

EXAMPLE 3 If $z = x^2 - xy$, then $\dfrac{\partial z}{\partial x} = 2x - y$ and $\dfrac{\partial z}{\partial y} = -x$, so that

$$dz = (2x - y)\,dx - x\,dy$$ □

It follows immediately from Theorem 15.4 that when f_x and f_y are continuous and when Δx and Δy are small, then dz is an approximation for Δz:

$$dz \approx \Delta z$$

EXAMPLE 4 The change in the function in Example 1 can be approximated using the differential in Example 3:

$$dz = (1)(0.2) - (1)(-0.3) = 0.5$$ □

EXAMPLE 5 The human cardiovascular system is similar to electrical series and parallel circuits. For example, when blood flows through two resistances in parallel, as shown in Figure 15.41, then the equivalent resistance R of the network is

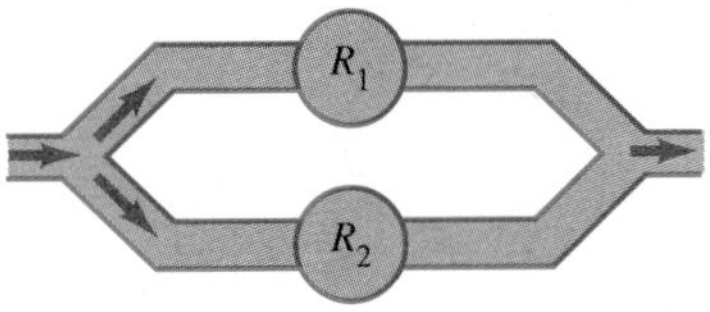

FIGURE 15.41

$$\frac{1}{R} = \frac{1}{R_1} + \frac{1}{R_2} \quad \text{or} \quad R = \frac{R_1R_2}{R_1 + R_2}$$

If the percentage errors in measuring R_1 and R_2 are $\pm 0.2\%$ and $\pm 0.6\%$, respectively, find the approximate maximum percentage error in R.

Solution We have $\Delta R_1 = \pm 0.002R_1$ and $\Delta R_2 = \pm 0.006R_2$. Now,

$$dR = \frac{R_2^2}{(R_1 + R_2)^2}\,dR_1 + \frac{R_1^2}{(R_1 + R_2)^2}\,dR_2$$

and so

$$\begin{aligned}
|\Delta R| \approx |dR| &\le \left|\frac{R_2^2}{(R_1 + R_2)^2}(\pm 0.002R_1)\right| + \left|\frac{R_1^2}{(R_1 + R_2)^2}(\pm 0.006R_2)\right| \\
&= R\left[\frac{0.002R_2}{R_1 + R_2} + \frac{0.006R_1}{R_1 + R_2}\right] \\
&\le R\left[\frac{0.006R_2}{R_1 + R_2} + \frac{0.006R_1}{R_1 + R_2}\right] = (0.006)R
\end{aligned}$$

Thus, the maximum relative error is given by the approximation $|dR|/R \approx 0.006$; therefore, the maximum percentage error is approximately 0.6%. □

Differentiability of $z = f(x, y)$ Although we have considered partial derivatives and the differential of a function $z = f(x, y)$, we have not yet defined **differentiability** of f. The latter concept is defined in terms of the fundamental increment formula (3).

DEFINITION 15.5 Differentiable Function

The function $z = f(x, y)$ is said to be **differentiable** at (x, y) if Δz can be written

$$\Delta z = f_x(x, y)\,\Delta x + f_y(x, y)\,\Delta y + \varepsilon_1\,\Delta x + \varepsilon_2\,\Delta y$$

where
$$\lim_{(\Delta x, \Delta y)\to(0,0)} \varepsilon_1 = \lim_{(\Delta x, \Delta y)\to(0,0)} \varepsilon_2 = 0$$

It is interesting to note that the partial derivatives f_x and f_y may exist at a point (x, y) and yet f may not be differentiable at the point. The next theorem gives us sufficient conditions under which the existence of the partial derivatives implies differentiability.

THEOREM 15.5 Differentiability Over a Rectangular Region

If $f_x(x, y)$ and $f_y(x, y)$ are continuous at every point (x, y) in a rectangular region that is defined by $a < x < b$, $c < y < d$, then $z = f(x, y)$ is differentiable over the region.

The next theorem is the analogue of Theorem 2.4; it states that if $z = f(x, y)$ is differentiable at a point, then it is continuous at that point.

THEOREM 15.6 Differentiability Implies Continuity

If $z = f(x, y)$ is differentiable at (x, y), then f is continuous at (x, y).

Total Differential for $w = F(x, y, z)$ Definition 15.4 generalizes to functions of three or more variables. Specifically, if $w = F(x, y, z)$, then its total differential is given by

$$dw = \frac{\partial w}{\partial x}\,dx + \frac{\partial w}{\partial y}\,dy + \frac{\partial w}{\partial z}\,dz$$

EXAMPLE 6 If $w = x^2 + 2y^3 + 3z^4$, then $\dfrac{\partial w}{\partial x} = 2x$, $\dfrac{\partial w}{\partial y} = 6y^2$, $\dfrac{\partial w}{\partial z} = 12z^3$,

and so
$$dw = 2x\,dx + 6y^2\,dy + 12z^3\,dz$$ □

▲ ***Remarks*** (*i*) Since $dy \approx \Delta y$ when $f'(x)$ exists and Δx is small, it seems reasonable to expect $dz = f_x(x, y)\,\Delta x + f_y(x, y)\,\Delta y$ to give a good approximation for Δz for small Δx and Δy. But life is not so simple for functions of several variables; there are functions for which dz and Δz are not close even when Δx and Δy are small. The guarantee $dz \approx \Delta z$ for small increments in the independent variables comes from the continuity and not simply from the existence of $f_x(x, y)$ and $f_y(x, y)$.

(*ii*) When you work Problems 5–8 in Exercises 15.4 you will discover that the functions ε_1 and ε_2 introduced in (3) of Theorem 15.4 are not unique.

EXERCISES 15.4

Answers to odd-numbered problems begin on page A-104.

In Problems 1–4 compare the values of Δz and dz for the given function as (x, y) varies from the first to the second point.

1. $z = 3x + 4y + 8$; $(2, 4)$, $(2.2, 3.9)$
2. $z = 2x^2y + 5y$; $(0, 0)$, $(0.2, -0.1)$
3. $z = (x + y)^2$; $(3, 1)$, $(3.1, 0.8)$
4. $z = x^2 + x^2y^2 + 2$; $(1, 1)$, $(0.9, 1.1)$

In Problems 5–8 find functions ε_1 and ε_2 from Δz as defined in (3) of Theorem 15.4.

5. $z = 5x^2 + 3y - xy$
6. $z = 10y^2 + 3x - x^2$
7. $z = x^2y^2$
8. $z = x^3 - y^3$

In Problems 9–20 find the total differential of the given function.

9. $z = x^2 \sin 4y$
10. $z = xe^{x^2 - y^2}$
11. $z = \sqrt{2x^2 - 4y^3}$
12. $z = (5x^3y + 4y^5)^3$
13. $f(s, t) = \dfrac{2s - t}{s + 3t}$
14. $g(r, \theta) = r^2 \cos 3\theta$
15. $w = x^2y^4z^{-5}$
16. $w = e^{-z^2}\cos(x^2 + y^4)$
17. $F(r, s, t) = r^3 + s^{-2} - 4t^{1/2}$
18. $G(\rho, \theta, \phi) = \rho \sin \phi \cos \theta$
19. $w = \ln\left(\dfrac{uv}{st}\right)$
20. $w = \sqrt{u^2 + s^2t^2 - v^2}$

21. When blood flows through three resistances R_1, R_2, R_3 in parallel, the equivalent resistance R of the network is

$$\frac{1}{R} = \frac{1}{R_1} + \frac{1}{R_2} + \frac{1}{R_3}$$

Given that the percentage error in measuring each resistance is $\pm 0.9\%$, find the approximate maximum percentage error in R.

22. The pressure P of an enclosed ideal gas is given by $P = k(T/V)$, where V is volume, T is temperature, and k is a constant. Given that the percentage errors in measuring T and V are at most 0.6% and 0.8%, respectively, find the approximate maximum percentage error in P.

23. The tension T in the string of the yo-yo shown in Figure 15.42 is

$$T = mg\frac{R}{2r^2 + R^2}$$

where mg is its constant weight. Find the approximate change in the tension if R and r are increased from 4 cm and 0.8 cm to 4.1 cm and 0.9 cm, respectively. Does the tension increase or decrease?

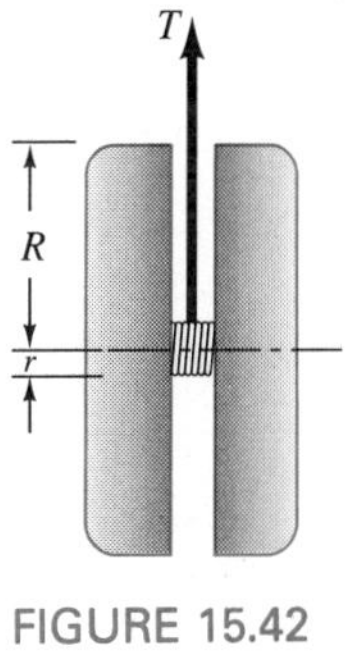

FIGURE 15.42

24. Find the approximate increase in the volume of a right circular cylinder if its height is increased from 10 to 10.5 cm and its radius is increased from 5 to 5.3 cm. What is the approximate new volume?

25. If the length, width, and height of a closed rectangular box are increased by 2%, 5%, and 8%, respectively, what is the approximate percentage increase in volume?

26. In Problem 25 if the original length, width, and height are 3 ft, 1 ft, and 2 ft, respectively, what is the approximate increase in surface area? What is the approximate new surface area?

27. The function $S = 0.1091w^{0.425}h^{0.725}$ gives the surface area of a person's body in terms of weight w and height h. If the error in the measurement of w is at most 3% and the error in the measurement of h is at most 5%, what is the approximate maximum percentage error in the measurement of S?

28. The impedance Z of the series circuit shown in Figure 15.43 is $Z = \sqrt{R^2 + X^2}$, where R is resistance, $X = 1000L - 1/(1000C)$ is net reactance, L is inductance, and C is capacitance. If the values of R, L, and C given in the figure are increased to 425 ohms, 0.45 henry, and 11.1×10^{-5} farad, respectively, what is the approximate change in the impedance of the circuit? What is the approximate new impedance?

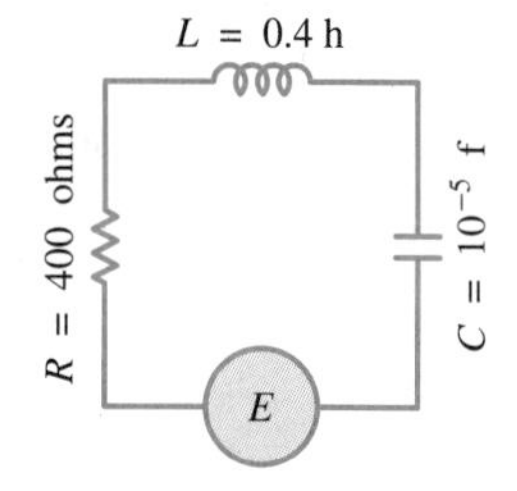

FIGURE 15.43

29. A two-dimensional robot arm whose shoulder is fixed at the origin keeps track of its position by means of a shoulder angle θ and an elbow angle ϕ as shown in Figure 15.44. The shoulder angle is measured counterclockwise from the x-axis, and the elbow angle is measured counterclockwise from the upper arm to the lower arm, which are of length L and l, respectively.

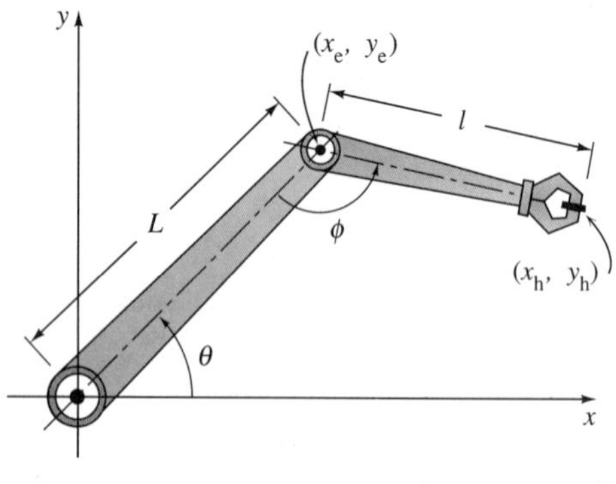

FIGURE 15.44

(a) The location of the elbow joint is given by

$$x_e = L\cos\theta \qquad y_e = L\sin\theta$$

Find corresponding formulas for the location (x_h, y_h) of the hand.

(b) Show that the total differentials of x_h and y_h can be written as

$$dx_h = -y_h\,d\theta + (y_e - y_h)\,d\phi$$

$$dy_h = x_h\,d\theta + (x_e - x_h)\,d\phi$$

(c) Suppose that $L = l$ and that the arm is to be positioned so as to reach the point (L, L). Suppose also that the error in measuring each of the angles θ and ϕ is at most $\pm 1°$. Find the approximate maximum error in the x-coordinate of the hand's location for each of the two possible positionings.

30. A projectile is fired at an angle θ with velocity v across a chasm of width D toward a vertical cliff wall that is essentially infinite in both height and depth. See Figure 15.45.

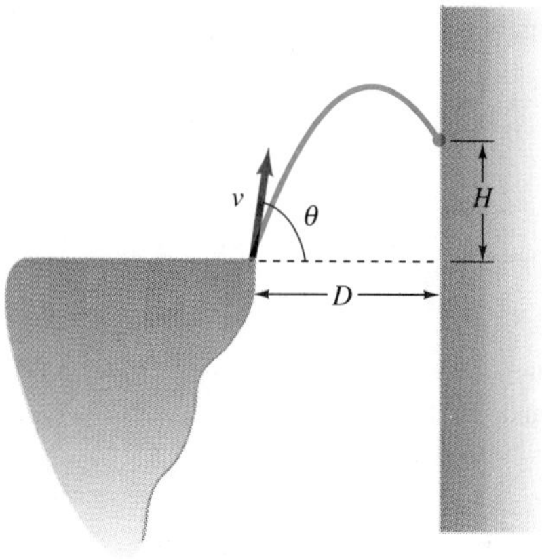

FIGURE 15.45

(a) If the projectile is subject only to the force of gravity, show that the height H at which the projectile strikes the cliff wall as a function of the variables v and θ is given by

$$H = D\tan\theta - \frac{1}{2}g\frac{D^2}{v^2}\sec^2\theta$$

[*Hint:* See Section 12.1]

(b) Find the total differential of H.

(c) Suppose that $D = 100$ ft, $g = 32$ ft/s², $v = 100$ ft/s, and $\theta = 45°$. Find H.

(d) Suppose, for the data in part (c), that the error in measuring v is at most ± 1 ft/s and that the error in measuring θ is at most $\pm 1°$. Find the approximate maximum error in H.

(e) By allowing D to vary, H can also be considered a function of three variables. Find the total differential of H. Using the data from parts (**c**) and (**d**) and assuming that the error in measuring D is at most ± 2 ft, find the approximate maximum error in H.

In Problems 31 and 32 use the concept of the differential to find an approximation to the given number.

31 $\sqrt{102} + \sqrt[4]{80}$

32. $\sqrt{\dfrac{35}{63}}$

15.5 EXACT DIFFERENTIALS

Two Variables Often it is important to be able to tell at a glance when a differential expression

$$P(x, y)\,dx + Q(x, y)\,dy \tag{1}$$

is the total differential of a function f of two variables.

DEFINITION 15.6 Exact Differential

A differential expression (1) is said to be an **exact differential** if there exists a function f such that

$$df = P(x, y)\,dx + Q(x, y)\,dy$$

EXAMPLE 1 $x^2y^3\,dx + x^3y^2\,dy$ is an exact differential, since it is the total differential of $f(x, y) = \frac{1}{3}x^3y^3$. Verify this. □

The following result provides a test for determining when (1) is exact:

THEOREM 15.7 Test for Exactness

Let P and Q be continuous and have continuous first partial derivatives in a rectangular region of the xy-plane. Then (1) is an exact differential if and only if

$$\frac{\partial P}{\partial y} = \frac{\partial Q}{\partial x} \tag{2}$$

for all (x, y) in the region.

Proof If (1) is an exact differential, there exists a function f such that

$$df = P(x, y)\,dx + Q(x, y)\,dy = \frac{\partial f}{\partial x}\,dx + \frac{\partial f}{\partial y}\,dy$$

Since $\partial f/\partial x = P$ and $\partial f/\partial y = Q$, we have

$$\frac{\partial^2 f}{\partial y\,\partial x} = \frac{\partial P}{\partial y} \quad \text{and} \quad \frac{\partial^2 f}{\partial x\,\partial y} = \frac{\partial Q}{\partial x}$$

By continuity, the mixed partials are equal and thus we obtain the results in (2). Conversely, let us now assume $\partial P/\partial y = \partial Q/\partial x$. We wish to find a function f such that $\partial f/\partial x = P$ and $\partial f/\partial y = Q$. It turns out that

$$f(x, y) = \int_{x_0}^{x} P(x, y_0)\,dx + \int_{y_0}^{y} Q(x, y)\,dy \tag{3}$$

where (x_0, y_0) is a fixed point in the region, is the function that we seek. In the second integral in (3) the variable x is held fixed. Now,

$$\begin{aligned}\frac{\partial f}{\partial x} &= \frac{\partial}{\partial x}\int_{x_0}^{x} P(x, y_0)\,dx + \frac{\partial}{\partial x}\int_{y_0}^{y} Q(x, y)\,dy \\ &= P(x, y_0) + \int_{y_0}^{y} \frac{\partial Q}{\partial x}\,dy \\ &\qquad\qquad\qquad\downarrow \\ &= P(x, y_0) + \int_{y_0}^{y} \frac{\partial P}{\partial y}\,dy \qquad \text{(by assumption)} \\ &= P(x, y_0) + P(x, y) - P(x, y_0) = P(x, y)\end{aligned}$$

Verifying that $\partial f/\partial y = Q$ is left as an exercise. ■

EXAMPLE 2 Determine whether the following differential expressions are exact:

(*a*) $(2y^2 - 2y)\,dx + (2xy - x)\,dy$ (*b*) $2xy\,dx + (x^2 - 1)\,dy$

Solution (*a*)

$$P = 2y^2 - 2y \qquad Q = 2xy - x$$

$$\frac{\partial P}{\partial y} = 4y - 2 \qquad \frac{\partial Q}{\partial x} = 2y - 1$$

Since $\partial P/\partial y \neq \partial Q/\partial x$, the expression is not exact. In other words, there is no function f whose total differential is $(2y^2 - 2y)\,dx + (2xy - x)\,dy$.

(*b*)

$$P = 2xy \qquad Q = x^2 - 1$$

$$\frac{\partial P}{\partial y} = 2x \qquad \frac{\partial Q}{\partial x} = 2x$$

Since $\partial P/\partial y = \partial Q/\partial x$, the expression is exact. □

Exact Differential Equations The **first-order differential equation**

$$P(x, y)\,dx + Q(x, y)\,dy = 0 \tag{4}$$

is also said to be **exact** when $P\,dx + Q\,dy$ is an exact differential. When (4) is exact, it is equivalent to $df(x, y) = 0$. Hence, a family of **solutions** of the equation is given implicitly by $f(x, y) = C$.

EXAMPLE 3 Solve $2xy\,dx + (x^2 - 1)\,dy = 0$.

Solution It was proved in Example 2(*b*) that the equation is exact. Consequently, there exists a function f such that

$$\frac{\partial f}{\partial x} = \overbrace{2xy}^{\boxed{P}} \quad\text{and}\quad \frac{\partial f}{\partial y} = \overbrace{x^2 - 1}^{\boxed{Q}}$$

Integrating the first of these equations* with respect to x gives

$$f = x^2y + g(y)$$

where $g(y)$ is the "constant" of integration. Taking the partial derivative of this last expression with respect to y and setting the result equal to Q give

$$\frac{\partial f}{\partial y} = x^2 + g'(y) = x^2 - 1$$

It follows that $g'(y) = -1$, $g(y) = -y$, and so $f = x^2y - y$. A family of solutions of the equation is then $x^2y - y = C$. □

EXAMPLE 4 Solve $(\cos x \sin x - xy^2)\,dx + (y - x^2y)\,dy = 0$ subject to $y(0) = 2$.

Solution The equation is exact, since

$$\frac{\partial P}{\partial y} = -2xy = \frac{\partial Q}{\partial x}$$

Now, $$\frac{\partial f}{\partial y} = \overbrace{y - x^2y}^{\boxed{Q}} \quad \text{implies that} \quad f = \frac{1}{2}y^2 - \frac{1}{2}x^2y^2 + h(x)$$

We then take the partial derivative of this expression with respect to x and equate the result with P.

$$\frac{\partial f}{\partial x} = -xy^2 + h'(x) = \cos x \sin x - xy^2$$

$$h'(x) = \cos x \sin x \qquad h(x) = \int \sin x(\cos x\, dx) = \frac{1}{2}\sin^2 x$$

From $f = \frac{1}{2}(y^2 - x^2y^2 + \sin^2 x)$, we find a family of solutions $y^2 - x^2y^2 + \sin^2 x = C_1$, where we have relabeled $2C$ as C_1. Finally, substituting the given conditions $x = 0$ and $y = 2$ in the last equation immediately gives $C_1 = 4$. Consequently, a solution to the problem is $y^2 - x^2y^2 + \sin^2 x = 4$. □

Three Variables A differential expression

$$P(x, y, z)\,dx + Q(x, y, z)\,dy + R(x, y, z)\,dz \tag{5}$$

is an exact differential if there exists a function F of three variables such that (5) is its total differential dF. The last theorem of this section is a test for exactness for (5).

*We could integrate either of these equations.

THEOREM 15.8 Test for Exactness

Let P, Q, and R be continuous and have continuous first partial derivatives in a rectangular region of 3-space. Then (5) is an exact differential if and only if

$$\frac{\partial P}{\partial y} = \frac{\partial Q}{\partial x} \qquad \frac{\partial P}{\partial z} = \frac{\partial R}{\partial x} \qquad \frac{\partial Q}{\partial z} = \frac{\partial R}{\partial y} \tag{6}$$

EXAMPLE 5 Determine whether

$$(1 + 2xy \sin z)\,dx + x^2 \sin z\,dy + x^2 y \cos z\,dz$$

is exact.

Solution With $P = 1 + 2xy \sin z$, $Q = x^2 \sin z$, and $R = x^2 y \cos z$, we see from (6) that

$$\frac{\partial P}{\partial y} = \frac{\partial Q}{\partial x} = 2x \sin z \qquad \frac{\partial P}{\partial z} = \frac{\partial R}{\partial x} = 2xy \cos z \qquad \frac{\partial Q}{\partial z} = \frac{\partial R}{\partial y} = x^2 \cos z$$

Hence, the given differential expression is an exact differential. □

EXERCISES 15.5

Answers to odd-numbered problems begin on page A-105.

In Problems 1–24 determine whether the given differential equation is exact. If exact, solve.

1. $(2x + 4)\,dx + (3y - 1)\,dy = 0$
2. $(2x + y)\,dx - (x + 6y)\,dy = 0$
3. $(5x + 4y)\,dx + (4x - 8y^3)\,dy = 0$
4. $(\sin y - y \sin x)\,dx + (\cos x + x \cos y - y)\,dy = 0$
5. $(2y^2 x - 3)\,dx + (2yx^2 + 4)\,dy = 0$
6. $\left(2y - \frac{1}{x} + \cos 3x\right)\frac{dy}{dx} + \frac{y}{x^2} - 4x^3 + 3y \sin 3x = 0$
7. $(x + y)(x - y)\,dx + x(x - 2y)\,dy = 0$
8. $\left(1 + \ln x + \frac{y}{x}\right)dx = (1 - \ln x)\,dy$
9. $(y^3 - y^2 \sin x - x)\,dx + (3xy^2 + 2y \cos x)\,dy = 0$
10. $(x^3 + y^3)\,dx + 3xy^2\,dy = 0$
11. $(y \ln y - e^{-xy})\,dx + \left(\frac{1}{y} + x \ln y\right)dy = 0$
12. $\frac{2x}{y}\,dx - \frac{x^2}{y^2}\,dy = 0$
13. $x\frac{dy}{dx} = 2xe^x - y + 6x^2$
14. $(3x^2 y + e^y)\,dx + (x^3 + xe^y - 2y)\,dy = 0$
15. $\left(1 - \frac{3}{x} + y\right)dx + \left(1 - \frac{3}{y} + x\right)dy = 0$
16. $(e^y + 2xy \cosh x)y' + xy^2 \sinh x + y^2 \cosh x = 0$
17. $\left(x^2 y^3 - \frac{1}{1 + 9x^2}\right)dx + x^3 y^2\,dy = 0$
18. $(5y - 2x)y' - 2y = 0$
19. $(\tan x - \sin x \sin y)\,dx + \cos x \cos y\,dy = 0$
20. $(3x \cos 3x + \sin 3x - 3)\,dx + (2y + 5)\,dy = 0$
21. $(1 - 2x^2 - 2y)\frac{dy}{dx} = 4x^3 + 4xy$
22. $(2y \sin x \cos x - y + 2y^2 e^{xy^2})\,dx = (x - \sin^2 x - 4xye^{xy^2})\,dy$

23. $(4x^3y - 15x^2 - y)\,dx + (x^4 + 3y^2 - x)\,dy = 0$

24. $\left(\frac{1}{x} + \frac{1}{x^2} - \frac{y}{x^2 + y^2}\right)dx + \left(ye^y + \frac{x}{x^2 + y^2}\right)dy = 0$

In Problems 25–28 solve the given differential equation subject to the indicated condition.

25. $(x + y)^2\,dx + (2xy + x^2 - 1)\,dy = 0,\ y(1) = 1$

26. $(e^x + y)\,dx + (2 + x + ye^y)\,dx = 0,\ y(0) = 1$

27. $(4y + 2x - 5)\,dx + (6y + 4x - 1)\,dy = 0,\ y(-1) = 2$

28. $(y^2\cos x - 3x^2y - 2x)\,dx + (2y\sin x - x^3 + \ln y)\,dy = 0,$ $y(0) = e$

29. Determine a function $P(x, y)$ such that

$$P(x, y)\,dx + \left(xe^{xy} + 2xy + \frac{1}{x}\right)dy = 0$$

is an exact differential equation.

30. Determine a function $Q(x, y)$ such that

$$\left(y^{1/2}x^{-1/2} + \frac{x}{x^2 + y}\right)dx + Q(x, y)\,dy = 0$$

is an exact differential equation.

31. Find a constant k so that

$$(y^3 + kxy^4 - 2x)\,dx + (3xy^2 + 20x^2y^3)\,dy = 0$$

is an exact differential equation.

32. If f is the function defined in (3), verify that $\partial f/\partial y = Q$.

In Problems 33–40 determine whether the given differential expression is exact.

33. $y\,dx + x\,dy + 10\,dz$

34. $dx + 2y\,dy + 4z\,dz$

35. $y\sin z\,dx + x\sin z\,dy + (y + xy\cos z)\,dz$

36. $(y^2 + z^2)\,dx + 2xy\,dy + 2xz\,dz$

37. $y\sin y\,dx + (xy\cos y + x\sin y)\,dy + (1 + e^{2z})\,dz$

38. $z\sin x\sin y\,dx - z\cos x\cos y\,dy + \sin x\sin y\,dz$

39. $3x^2\ln z\,dx + 4y^3\,dy + \frac{x^3}{z}\,dz$

40. $\frac{xy^2}{z^2}\,dx + \frac{x^2y}{z^2}\,dy - \frac{x^2y^2}{z^3}\,dz$

15.6 THE CHAIN RULE

The Chain Rule for functions of one variable states that if $y = f(u)$ is a differentiable function of u, and $u = g(x)$ is a differentiable function of x, then the derivative of the composite function is

$$\frac{dy}{dx} = \frac{dy}{du}\frac{du}{dx} \tag{1}$$

For a composite function of two variables $z = f(u, v)$, where $u = g(x, y)$ and $v = h(x, y)$, we would naturally expect *two* formulas analogous to (1), since we can compute both $\partial z/\partial x$ and $\partial z/\partial y$. The **Chain Rule** for functions of two variables is summarized as follows:

THEOREM 15.9 Chain Rule

If $z = f(u, v)$ is differentiable and $u = g(x, y)$ and $v = h(x, y)$ have continuous first partial derivatives, then

$$\frac{\partial z}{\partial x} = \frac{\partial z}{\partial u}\frac{\partial u}{\partial x} + \frac{\partial z}{\partial v}\frac{\partial v}{\partial x} \quad\text{and}\quad \frac{\partial z}{\partial y} = \frac{\partial z}{\partial u}\frac{\partial u}{\partial y} + \frac{\partial z}{\partial v}\frac{\partial v}{\partial y} \tag{2}$$

Proof We prove the second of these results. If $\Delta x = 0$, then

$$\Delta z = f(g(x, y + \Delta y), h(x, y + \Delta y)) - f(g(x, y), h(x, y))$$

Now, if

$$\Delta u = g(x, y + \Delta y) - g(x, y) \qquad \Delta v = h(x, y + \Delta y) - h(x, y)$$

then

$$g(x, y + \Delta y) = u + \Delta u \qquad h(x, y + \Delta y) = v + \Delta v$$

Hence, Δz can be rewritten as

$$\Delta z = f(u + \Delta u, v + \Delta v) - f(u, v)$$

Since f is differentiable, it follows from the increment formula (3) of Section 15.4 that

$$\Delta z = \frac{\partial z}{\partial u}\Delta u + \frac{\partial z}{\partial v}\Delta v + \varepsilon_1\,\Delta u + \varepsilon_2\,\Delta v$$

where, recall, $\varepsilon_1(\Delta u, \Delta v)$ and $\varepsilon_2(\Delta u, \Delta v)$ are functions of Δu and Δv with the property $\lim_{(\Delta u, \Delta v)\to(0,0)} \varepsilon_1 = 0$ and $\lim_{(\Delta u, \Delta v)\to(0,0)} \varepsilon_2 = 0$. Since ε_1 and ε_2 are not uniquely defined functions, a pair of functions can always be found for which $\varepsilon_1(0, 0) = 0$, $\varepsilon_2(0, 0) = 0$. Hence, ε_1 and ε_2 are continuous at (0, 0). Therefore,

$$\frac{\Delta z}{\Delta y} = \frac{\partial z}{\partial u}\frac{\Delta u}{\Delta y} + \frac{\partial z}{\partial v}\frac{\Delta v}{\Delta y} + \varepsilon_1\frac{\Delta u}{\Delta y} + \varepsilon_2\frac{\Delta v}{\Delta y}$$

Now, taking the limit of the last line as $\Delta y \to 0$ gives

$$\frac{\partial z}{\partial y} = \frac{\partial z}{\partial u}\frac{\partial u}{\partial y} + \frac{\partial z}{\partial v}\frac{\partial v}{\partial y} + 0\cdot\frac{\partial u}{\partial y} + 0\cdot\frac{\partial v}{\partial y} = \frac{\partial z}{\partial u}\frac{\partial u}{\partial y} + \frac{\partial z}{\partial v}\frac{\partial v}{\partial y}$$

since Δu and Δv both approach zero as $\Delta y \to 0$. ■

EXAMPLE 1 If $z = u^2 - v^3$ and $u = e^{2x-3y}$, $v = \sin(x^2 - y^2)$, find $\partial z/\partial x$ and $\partial z/\partial y$.

Solution Since $\partial z/\partial u = 2u$ and $\partial z/\partial v = -3v^2$, it follows from (2) that

$$\frac{\partial z}{\partial x} = 2u(2e^{2x-3y}) - 3v^2[2x\cos(x^2 - y^2)] = 4ue^{2x-3y} - 6xv^2\cos(x^2 - y^2) \tag{3}$$

$$\frac{\partial z}{\partial y} = 2u(-3e^{2x-3y}) - 3v^2[(-2y)\cos(x^2 - y^2)] = -6ue^{2x-3y} + 6yv^2\cos(x^2 - y^2) \tag{4}$$

□

Of course, in Example 1, we could substitute the expressions for u and v in the original function and then find the partial derivatives directly. In the same manner, the answers (3) and (4) can be expressed in terms of x and y.

Special Case If $z = f(u, v)$ is differentiable and $u = g(t)$ and $v = h(t)$ are differentiable functions of a single variable t, then Theorem 15.9 implies that the ordinary derivative dz/dt is

$$\frac{dz}{dt} = \frac{\partial z}{\partial u}\frac{du}{dt} + \frac{\partial z}{\partial v}\frac{dv}{dt} \tag{5}$$

Generalizations The results given in (2) and (5) immediately generalize to any number of variables. If $z = f(u_1, u_2, \ldots, u_n)$ and each of the variables $u_1, u_2, u_3, \ldots, u_n$ are functions of $x_1, x_2, \ldots, x_k$, then under the same assumptions as in Theorem 15.9, we have

$$\frac{\partial z}{\partial x_i} = \frac{\partial z}{\partial u_1}\frac{\partial u_1}{\partial x_i} + \frac{\partial z}{\partial u_2}\frac{\partial u_2}{\partial x_i} + \cdots + \frac{\partial z}{\partial u_n}\frac{\partial u_n}{\partial x_i} \tag{6}$$

where $i = 1, 2, \ldots, k$. Similarly, if the u_i, $i = 1, \ldots, n$, are differentiable functions of a single variable t, then

$$\frac{dz}{dt} = \frac{\partial z}{\partial u_1}\frac{du_1}{dt} + \frac{\partial z}{\partial u_2}\frac{du_2}{dt} + \cdots + \frac{\partial z}{\partial u_n}\frac{du_n}{dt} \tag{7}$$

Tree Diagrams The results in (2) can be memorized in terms of a **tree diagram**. The dots in the first diagram that follows indicate that z depends on u and v; u and v depend, in turn, on x and y. To compute $\partial z/\partial x$, for example, we read from left to right and follow the *two* colored polygonal paths leading from z to x, multiply the partial derivatives on each path, and then add the products. The result given in (5) is represented by the second tree diagram.

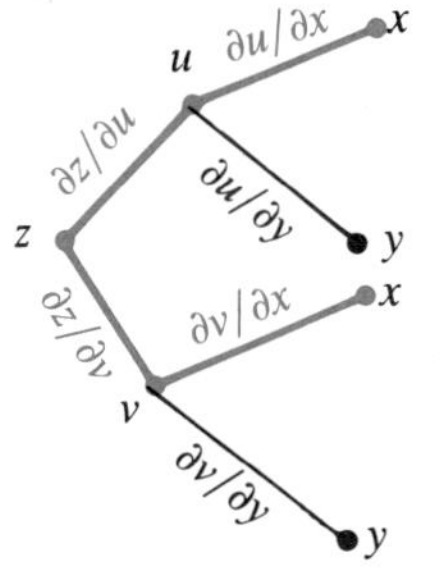

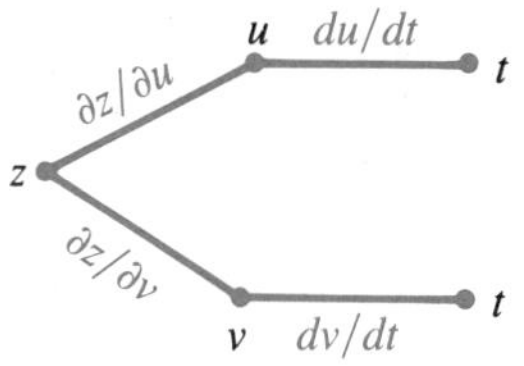

We use tree diagrams in the next two examples to illustrate special cases of (6) and (7).

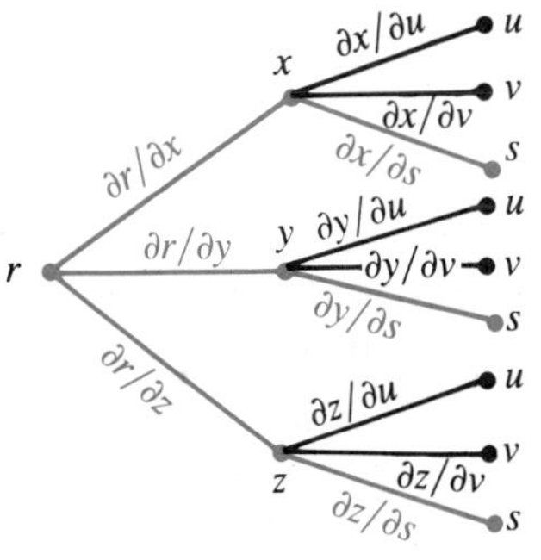

EXAMPLE 2 If $r = x^2 + y^5z^3$ and $x = uve^{2s}$, $y = u^2 - v^2s$, $z = \sin(uvs^2)$, find $\partial r/\partial s$.

Solution From the colored paths of the accompanying tree diagram we obtain

$$\begin{aligned}\frac{\partial r}{\partial s} &= \frac{\partial r}{\partial x}\frac{\partial x}{\partial s} + \frac{\partial r}{\partial y}\frac{\partial y}{\partial s} + \frac{\partial r}{\partial z}\frac{\partial z}{\partial s}\\ &= 2x(2uve^{2s}) + 5y^4z^3(-v^2) + 3y^5z^2(2uvs\cos(uvs^2))\end{aligned}$$

□

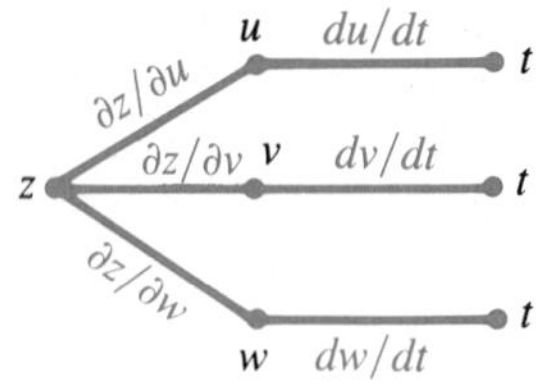

EXAMPLE 3 If $z = u^2v^3w^4$ and $u = t^2$, $v = 5t - 8$, $w = t^3 + t$, find dz/dt.

Solution In this case the tree diagram indicates that

$$\begin{aligned}\frac{dz}{dt} &= \frac{\partial z}{\partial u}\frac{du}{dt} + \frac{\partial z}{\partial v}\frac{dv}{dt} + \frac{\partial z}{\partial w}\frac{dw}{dt}\\ &= 2uv^3w^4(2t) + 3u^2v^2w^4(5) + 4u^2v^3w^3(3t^2 + 1)\end{aligned}$$

Alternative Solution Differentiate $z = t^4(5t - 8)^3(t^3 + t)^4$ by the Product Rule. □

Implicit Differentiation If the equation $F(x, y) = 0$ defines a function $y = f(x)$ implicitly, then $F(x, f(x)) = 0$ for all x in the domain of f. Recall from Section 2.8 that we found the derivative dy/dx by a process called implicit differentiation. The derivative dy/dx can also be determined from the Chain Rule. If we assume $w = F(x, y)$ and $y = f(x)$ are differentiable functions, then from (5) we have

$$\frac{dw}{dx} = F_x(x, y)\frac{dx}{dx} + F_y(x, y)\frac{dy}{dx} \tag{8}$$

Since $w = F(x, y) = 0$ and $dx/dx = 1$, (8) implies

$$F_x(x, y) + F_y(x, y)\frac{dy}{dx} = 0 \quad \text{or} \quad \frac{dy}{dx} = -\frac{F_x(x, y)}{F_y(x, y)}$$

provided $F_y(x, y) \neq 0$.

Moreover, if $F(x, y, z) = 0$ implicitly defines a function $z = f(x, y)$, then $F(x, y, f(x, y)) = 0$ for all (x, y) in the domain of f. If $w = F(x, y, z)$ is a differentiable function and $z = f(x, y)$ is differentiable in x and y, then (6) yields

$$\frac{\partial w}{\partial x} = F_x(x, y, z)\frac{\partial x}{\partial x} + F_y(x, y, z)\frac{\partial y}{\partial x} + F_z(x, y, z)\frac{\partial z}{\partial x} \tag{9}$$

Since $w = F(x, y, z) = 0$, $\partial x/\partial x = 1$, and $\partial y/\partial x = 0$, (9) gives

$$F_x(x, y, z) + F_z(x, y, z)\frac{\partial z}{\partial x} = 0 \quad \text{or} \quad \frac{\partial z}{\partial x} = -\frac{F_x(x, y, z)}{F_z(x, y, z)}$$

provided $F_z(x, y, z) \neq 0$. The partial derivative $\partial z/\partial y$ can be obtained in a similar manner. We summarize these results in the following theorem:

THEOREM 15.10 Implicit Differentiation

(*i*) If $w = F(x, y)$ is differentiable and $y = f(x)$ is a differentiable function of x defined implicitly by $F(x, y) = 0$, then

$$\frac{dy}{dx} = -\frac{F_x(x, y)}{F_y(x, y)}, \qquad F_y(x, y) \neq 0$$

(*ii*) If $w = F(x, y, z)$ is differentiable and $z = f(x, y)$ is a differentiable function in x and y defined implicitly by $F(x, y, z) = 0$, then

$$\frac{\partial z}{\partial x} = -\frac{F_x(x, y, z)}{F_z(x, y, z)} \quad \text{and} \quad \frac{\partial z}{\partial y} = -\frac{F_y(x, y, z)}{F_z(x, y, z)}, \qquad F_z(x, y, z) \neq 0$$

EXAMPLE 4

(*a*) Find dy/dx if $x^2 - 4xy - 3y^2 = 10$.

(*b*) Find $\partial z/\partial y$ if $x^2y - 5xy^2 = 2yz - 4z^3$.

Solution (*a*) Let $F(x, y) = x^2 - 4xy - 3y^2 - 10$. Then y is defined as a function of x by $F(x, y) = 0$. Now $F_x = 2x - 4y$ and $F_y = -4x - 6y$, and so by Theorem 15.10(*i*) we have

$$\frac{dy}{dx} = -\frac{F_x(x, y)}{F_y(x, y)} = -\frac{2x - 4y}{-4x - 6y} = \frac{x - 2y}{2x + 3y}$$

You are encouraged to verify this result by implicit differentiation.

(*b*) Let $F(x, y, z) = x^2y - 5xy^2 - 2yz + 4z^3$. Then z is defined as a function of x and y by $F(x, y, z) = 0$. Since $F_y = x^2 - 10xy - 2z$ and $F_z = -2y + 12z^2$, it follows from the second equation in Theorem 15.10(*ii*) that

$$\frac{\partial z}{\partial y} = -\frac{F_y(x, y, z)}{F_z(x, y, z)} = -\frac{x^2 - 10xy - 2z}{-2y + 12z^2}$$ □

EXERCISES 15.6 *Answers to odd-numbered problems begin on page A-105.*

In Problems 1–10 find the indicated partial derivatives.

1. $z = e^{uv^2}$; $u = x^3$, $v = x - y^2$; $\dfrac{\partial z}{\partial x}, \dfrac{\partial z}{\partial y}$

2. $z = u^2\cos 4v$; $u = x^2y^3$, $v = x^3 + y^3$; $\dfrac{\partial z}{\partial x}, \dfrac{\partial z}{\partial y}$

3. $z = 4x - 5y^2$; $x = u^4 - 8v^3$, $y = (2u - v)^2$; $\dfrac{\partial z}{\partial u}, \dfrac{\partial z}{\partial v}$

4. $z = \dfrac{x - y}{x + y}$; $x = \dfrac{u}{v}$, $y = \dfrac{v^2}{u}$; $\dfrac{\partial z}{\partial u}, \dfrac{\partial z}{\partial v}$

5. $w = (u^2 + v^2)^{3/2}$; $u = e^{-t}\sin\theta$, $v = e^{-t}\cos\theta$; $\dfrac{\partial w}{\partial t}, \dfrac{\partial w}{\partial \theta}$

6. $w = \tan^{-1}\sqrt{uv}$; $u = r^2 - s^2$, $v = r^2s^2$; $\dfrac{\partial w}{\partial r}, \dfrac{\partial w}{\partial s}$

7. $R = rs^2t^4$; $r = ue^{v^2}$, $s = ve^{-u^2}$, $t = e^{u^2v^2}$; $\dfrac{\partial R}{\partial u}, \dfrac{\partial R}{\partial v}$

8. $Q = \ln(pqr)$; $p = t^2\sin^{-1}x$, $q = \dfrac{x}{t^2}$, $r = \tan^{-1}\dfrac{x}{t}$; $\dfrac{\partial Q}{\partial x}, \dfrac{\partial Q}{\partial t}$

9. $w = \sqrt{x^2 + y^2}$; $x = \ln(rs + tu)$, $y = \dfrac{t}{u}\cosh rs$; $\dfrac{\partial w}{\partial t}, \dfrac{\partial w}{\partial r}, \dfrac{\partial w}{\partial u}$

10. $s = p^2 + q^2 - r^2 + 4t$; $p = \phi e^{3\theta}$, $q = \cos(\phi + \theta)$, $r = \phi\theta^2$, $t = 2\phi + 8\theta$; $\dfrac{\partial s}{\partial \phi}, \dfrac{\partial s}{\partial \theta}$

In Problems 11–16 find the indicated derivative.

11. $z = \ln(u^2 + v^2)$; $u = t^2$, $v = t^{-2}$; $\dfrac{dz}{dt}$

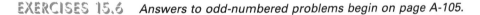

12. $z = u^3v - uv^4;\ u = e^{-5t},\ v = \sec 5t;\ \dfrac{dz}{dt}$

13. $w = \cos(3u + 4v);\ u = 2t + \dfrac{\pi}{2}\ v = -t - \dfrac{\pi}{4};\ \left.\dfrac{dw}{dt}\right|_{t=\pi}$

14. $w = e^{xy};\ x = \dfrac{4}{2t+1},\ y = 3t + 5;\ \left.\dfrac{dw}{dt}\right|_{t=0}$

15. $p = \dfrac{r}{2s+t};\ r = u^2,\ s = \dfrac{1}{u^2},\ t = \sqrt{u};\ \dfrac{dp}{du}$

16. $r = \dfrac{xy^2}{z^3},\ x = \cos s,\ y = \sin s,\ z = \tan s;\ \dfrac{dr}{ds}$

In Problems 17–20 find dy/dx by two methods: (**a**) implicit differentiation and (**b**) Theorem 15.10(*i*).

17. $x^3 - 2x^2y^2 + y = 1$

18. $x + 2y^2 = e^y$

19. $y = \sin xy$

20. $(x + y)^{2/3} = xy$

In Problems 21–24 use Theorem 15.10(*ii*) to find $\partial z/\partial x$ and $\partial z/\partial y$.

21. $x^2 + y^2 - z^2 = 1$

22. $x^{2/3} + y^{2/3} + z^{2/3} = a^{2/3}$

23. $xy^2z^3 + x^2 - y^2 = 5z^2$

24. $z = \ln(xyz)$

25. Suppose $w = F(x, y, z, u)$ is differentiable and $u = f(x, y, z)$ is a differentiable function of x, y, and z defined implicitly by $F(x, y, z, u) = 0$. Find expressions for $\partial u/\partial x$, $\partial u/\partial y$, and $\partial u/\partial z$.

26. Use the results of Problem 25 to find $\partial u/\partial x$, $\partial u/\partial y$, and $\partial u/\partial z$ if u is defined by $-xyz + x^2yu + 2xy^3u - u^4 = 8$.

27. If F and G have second partial derivatives, show that $u(x, t) = F(x + at) + G(x - at)$ satisfies the **wave equation**

$$a^2\frac{\partial^2 u}{\partial x^2} = \frac{\partial^2 u}{\partial t^2}$$

28. Let $\eta = x + at$ and $\xi = x - at$. Show that the wave equation in Problem 27 becomes

$$\frac{\partial^2 u}{\partial \eta\, \partial \xi} = 0$$

where $u = f(\eta, \xi)$.

29. If $u = f(x, y)$ and $x = r\cos\theta$, $y = r\sin\theta$, show that **Laplace's equation** $\partial^2u/\partial x^2 + \partial^2u/\partial y^2 = 0$ becomes

$$\frac{\partial^2 u}{\partial r^2} + \frac{1}{r}\frac{\partial u}{\partial r} + \frac{1}{r^2}\frac{\partial^2 u}{\partial \theta^2} = 0$$

30. If $z = f(u)$ is a differentiable function of one variable and $u = g(x, y)$ possesses first partial derivatives, what are $\partial z/\partial x$ and $\partial z/\partial y$?

31. Use the result of Problem 30 to show that for any differentiable function f, $z = f(y/x)$ satisfies the equation $x\,\partial z/\partial x + y\,\partial z/\partial y = 0$.

32. If $u = f(r)$ and $r = \sqrt{x^2 + y^2}$, show that Laplace's equation $\partial^2u/\partial x^2 + \partial^2u/\partial y^2 = 0$ becomes

$$\frac{d^2u}{dr^2} + \frac{1}{r}\frac{du}{dr} = 0$$

33. The **error function** defined by $erf(x) = \dfrac{2}{\sqrt{\pi}}\displaystyle\int_0^x e^{-v^2}\,dv$ is important in applied mathematics. Show that $u(x, t) = A + B\,erf(x/\sqrt{4kt})$, A and B constants, satisfies the diffusion equation

$$k\frac{\partial^2 u}{\partial x^2} = \frac{\partial u}{\partial t}$$

34. The voltage across a conductor is increasing at a rate of 2 volts/min and the resistance is decreasing at a rate of 1 ohm/min. Use $I = E/R$ and the Chain Rule to find the rate at which the current passing through the conductor is changing when $R = 50$ ohms and $E = 60$ volts.

35. The length of the side labeled x of the triangle in Figure 15.46 increases at a rate of 0.3 cm/s, the side labeled y increases at a rate of 0.5 cm/s, and the included angle θ increases at a rate of 0.1 rad/s. Use the Chain Rule to find the rate at which the area of the triangle is changing at the instant $x = 10$ cm, $y = 8$ cm, and $\theta = \pi/6$.

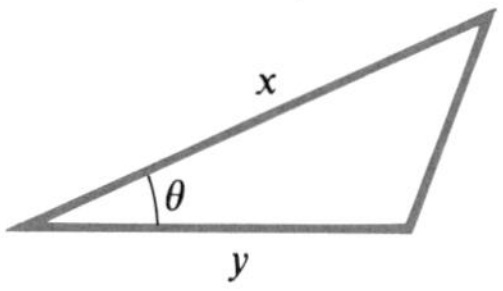

FIGURE 15.46

36. Van der Waals' equation of state for the real gas CO_2 is

$$P = \frac{0.08T}{V - 0.0427} - \frac{3.6}{V^2}$$

If dT/dt and dV/dt are rates at which the temperature and volume change, respectively, use the Chain Rule to find dP/dt.

37. A very young child grows at a rate of 2 in/yr and gains weight at a rate of 4.2 lb/yr. Use $S = 0.1091w^{0.425}h^{0.725}$ and the Chain Rule to find the rate at which the surface area of the child is changing when it weighs 25 lb and is 29 in tall.

38. A particle moves in 3-space so that its coordinates at any time are $x = 4\cos t$, $y = 4\sin t$, $z = 5t$, $t \geq 0$. Use the Chain Rule to find the rate at which its distance

$$w = \sqrt{x^2 + y^2 + z^2}$$

from the origin is changing at $t = 5\pi/2$ seconds.

39. The equation of state for a thermodynamic system is $F(P, V, T) = 0$, where P, V, and T are pressure, volume, and temperature, respectively. If the equation defines V as a function of P and T, and also defines T as a function of V and P, show that

$$\frac{\partial V}{\partial T} = -\frac{\frac{\partial F}{\partial T}}{\frac{\partial F}{\partial V}} = \frac{1}{\frac{\partial T}{\partial V}}$$

40. Two coast guard ships (denoted by A and B in Figure 15.47) located a distance 500 yd apart spot a suspect ship C at relative bearings θ and ϕ as shown in the figure.

(a) Use the law of sines to express the distance r from A to C in terms of θ and ϕ.

(b) How far is C from A when $\theta = 62°$ and $\phi = 75°$?

(c) Suppose that at the moment specified in part (**b**), the angle θ is increasing at the rate of 5° per minute, while ϕ is decreasing at the rate of 10° per minute. Is the distance from C to A increasing or decreasing? at what rate?

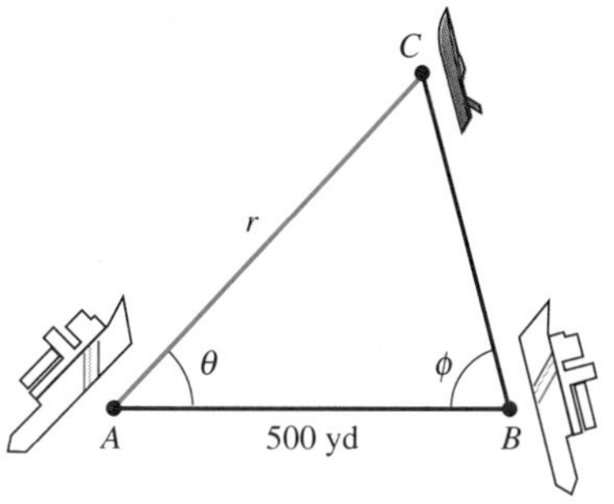

FIGURE 15.47

41. A **Helmholtz resonator** is any container with a neck and an opening (such as a jug or a soda pop bottle). When air is blown across the opening, the resonator produces a characteristic sound whose frequency, in cycles per second, is

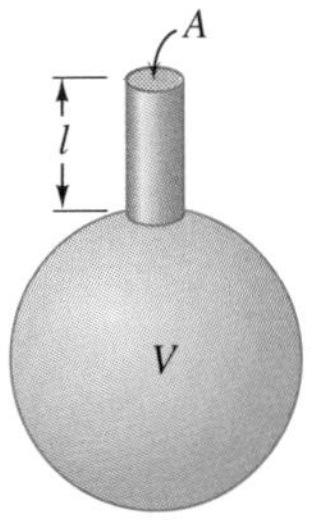

FIGURE 15.48

$$f = \frac{c}{2\pi}\sqrt{\frac{A}{lV}}$$

where A is the cross-sectional area of the opening, l is the length of the neck, V is the volume of the container (not counting the neck), and c is the speed of sound (approximately 330 m/s). See Figure 15.48.

(a) What frequency sound will a bottle make if it has a circular opening 2 cm in diameter, a neck 6 cm long, and a volume of 100 cm^3. [Be sure to convert c to cm/s.]

(b) Suppose the volume of the bottle in part (**a**) is decreasing at a rate of 10 cm^3/s, while its neck is lengthening at the rate of 1 cm/s. At the instant specified in part (**a**) (that is, $V = 100$, $l = 6$) is the frequency increasing or decreasing?

42. (a) A function f is said to be **homogeneous of degree n** if $f(\lambda x, \lambda y) = \lambda^n f(x, y)$. If f has first partial derivatives, show that

$$x\frac{\partial f}{\partial x} + y\frac{\partial f}{\partial y} = nf$$

(b) Verify that $f(x, y) = 4x^2y^3 - 3xy^4 + x^5$ is a homogeneous function of degree 5.

(c) Verify that the function in part (**b**) satisfies the differential equation in part (**a**).

15.7 THE DIRECTIONAL DERIVATIVE

The Gradient of a Function In this and the next section, it is convenient to introduce a new vector based on partial differentation. When the **vector differential operator**

$$\nabla = \mathbf{i}\frac{\partial}{\partial x} + \mathbf{j}\frac{\partial}{\partial y} \quad \text{or} \quad \nabla = \mathbf{i}\frac{\partial}{\partial x} + \mathbf{j}\frac{\partial}{\partial y} + \mathbf{k}\frac{\partial}{\partial z}$$

is applied to a differentiable function $z = f(x, y)$ or $w = F(x, y, z)$, we say that the vectors

$$\nabla f(x, y) = \frac{\partial f}{\partial x}\mathbf{i} + \frac{\partial f}{\partial y}\mathbf{j} \tag{1}$$

$$\nabla F(x, y, z) = = \frac{\partial F}{\partial x}\mathbf{i} + \frac{\partial F}{\partial y}\mathbf{j} + \frac{\partial F}{\partial z}\mathbf{k} \tag{2}$$

are the **gradients** of the respective functions. The symbol ∇, an inverted capital Greek delta, is called "del" or "nabla." The symbol ∇f is usually read "grad f."

EXAMPLE 1 Compute $\nabla f(x, y)$ for $f(x, y) = 5y - x^3y^2$.

Solution From (1),

$$\nabla f(x, y) = \frac{\partial}{\partial x}(5y - x^3y^2)\mathbf{i} + \frac{\partial}{\partial y}(5y - x^3y^2)\mathbf{j}$$
$$= -3x^2y^2\mathbf{i} + (5 - 2x^3y)\mathbf{j}$$ □

EXAMPLE 2 If $F(x, y, z) = xy^2 + 3x^2 - z^3$, find $\nabla F(x, y, z)$ at $(2, -1, 4)$.

Solution From (2),

$$\nabla F(x, y, z) = (y^2 + 6x)\mathbf{i} + 2xy\mathbf{j} - 3z^2\mathbf{k}$$

and so $$\nabla F(2, -1, 4) = 13\mathbf{i} - 4\mathbf{j} - 48\mathbf{k}$$ □

A Generalization of Partial Differentiation The partial derivatives $\partial z/\partial x$ and $\partial z/\partial y$ give the slope of a tangent to the trace, or curve of intersection, of the surface given by $z = f(x, y)$ and vertical planes which are, respectively, parallel to the x- and y-coordinate axes. Equivalently, $\partial z/\partial x$ is the rate of change of the function in the direction given by the vector **i**, and $\partial z/\partial y$ is the rate of change of $z = f(x, y)$ in the **j**-direction. There is no reason to confine our attention to just two directions; we can find the rate of change of a differentiable function in *any* direction. See Figure 15.49.

Suppose $\mathbf{u} = \cos\theta\mathbf{i} + \sin\theta\mathbf{j}$ is a unit vector in the xy-plane that makes an angle θ with the positive x-axis and that is parallel to the vector $\mathbf{v}$ from $(x, y, 0)$ to $(x + \Delta x, y + \Delta y, 0)$. If $h = \sqrt{(\Delta x)^2 + (\Delta y)^2} > 0$, then $\mathbf{v} = h\mathbf{u}$. Furthermore, let the plane perpendicular to the xy-plane that contains these points slice the surface $z = f(x, y)$ in a curve C. We ask: What is the slope of the tangent line to C at a point P with coordinates $(x, y, f(x, y))$ in the direction given by $\mathbf{v}$? See Figure 15.50.

From Figure 15.50, we see that $\Delta x = h\cos\theta$ and $\Delta y = h\sin\theta$, so that the slope of the indicated secant line is

$$\frac{f(x + \Delta x, y + \Delta y) - f(x, y)}{h} = \frac{f(x + h\cos\theta, y + h\sin\theta) - f(x, y)}{h} \tag{3}$$

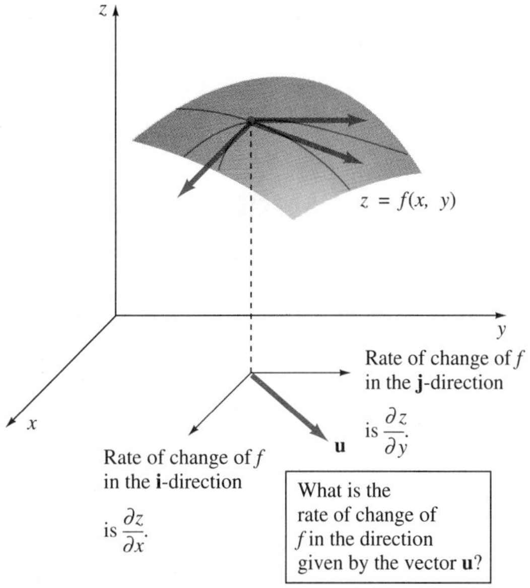
z
y
x
z = f(x, y)
Rate of change of f
in the j-direction
is $\frac{\partial z}{\partial y}$.
u
Rate of change of f
in the i-direction
is $\frac{\partial z}{\partial x}$.
What is the
rate of change of
f in the direction
given by the vector u?

FIGURE 15.49

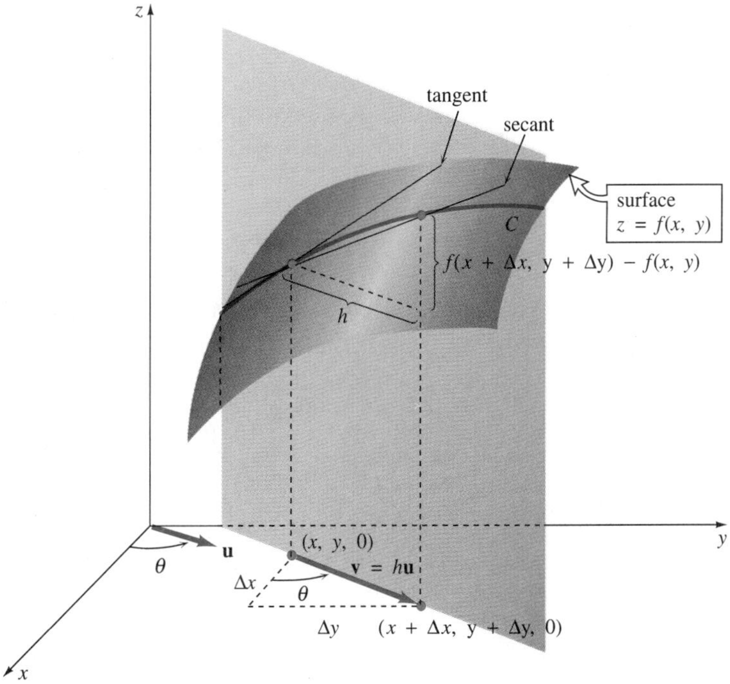
z
tangent
secant
surface
z = f(x, y)
C
f(x + Δx, y + Δy) − f(x, y)
h
y
u
(x, y, 0)
θ
v = hu
Δx
θ
Δy
(x + Δx, y + Δy, 0)
x

FIGURE 15.50

We expect the slope of the tangent at P to be the limit of (3) as $h \to 0$. This slope is the rate of change of f at P in the direction specified by the unit vector $\mathbf{u}$. This leads us to the following definition:

DEFINITION 15.7 Directional Derivative

The **directional derivative** of $z = f(x, y)$ in the direction of a unit vector $\mathbf{u} = \cos\theta\mathbf{i} + \sin\theta\mathbf{j}$ is

$$D_{\mathbf{u}}f(x, y) = \lim_{h\to 0}\frac{f(x + h\cos\theta, y + h\sin\theta) - f(x, y)}{h} \tag{4}$$

provided the limit exists.

Observe that (4) is truly a generalization of (1) and (2) of Section 15.3, since:

$$\theta = 0 \quad \text{implies that} \quad D_{\mathbf{i}}f(x, y) = \lim_{h\to 0}\frac{f(x + h, y) - f(x, y)}{h} = \frac{\partial z}{\partial x}$$

$$\text{and} \quad \theta = \frac{\pi}{2} \quad \text{implies that} \quad D_{\mathbf{j}}f(x, y) = \lim_{h\to 0}\frac{f(x, y + h) - f(x, y)}{h} = \frac{\partial z}{\partial y}$$

Method for Computing the Directional Derivative While (4) could be used to find $D_{\mathbf{u}}f(x, y)$ for a given function, as usual we seek a more efficient procedure. The next theorem shows how the concept of the gradient of a function plays a key role in computing a directional derivative.

THEOREM 15.11 Computing a Directional Derivative

If $z = f(x, y)$ is a differentiable function of x and y and $\mathbf{u} = \cos\theta\mathbf{i} + \sin\theta\mathbf{j}$, then

$$D_{\mathbf{u}}f(x, y) = \nabla f(x, y)\cdot\mathbf{u} \tag{5}$$

Proof Let x, y, and θ be fixed so that

$$g(t) = f(x + t\cos\theta, y + t\sin\theta)$$

is a function of one variable. We wish to compare the value of $g'(0)$, which is found by two different methods. First, by the definition of a derivative,

$$g'(0) = \lim_{h\to 0}\frac{g(0 + h) - g(0)}{h} = \lim_{h\to 0}\frac{f(x + h\cos\theta, y + h\sin\theta) - f(x, y)}{h} \tag{6}$$

Second, by the Chain Rule,

$$\begin{aligned} g'(t) &= f_1(x + t\cos\theta, y + t\sin\theta)\frac{d}{dt}(x + t\cos\theta) \\ &\quad + f_2(x + t\cos\theta, y + t\sin\theta)\frac{d}{dt}(x + t\sin\theta) \\ &= f_1(x + t\cos\theta, y + t\sin\theta)\cos\theta \\ &\quad + f_2(x + t\cos\theta, y + t\sin\theta)\sin\theta \end{aligned} \tag{7}$$

Here the subscripts 1 and 2 refer to the partial derivatives of $f(x + t\cos\theta, y + t\sin\theta)$ with respect to $x + t\cos\theta$ and $y + t\sin\theta$, respectively. When $t = 0$, we note that $x + t\cos\theta$ and $y + t\sin\theta$ are simply x and y, and therefore (7) becomes

$$g'(0) = f_x(x, y)\cos\theta + f_y(x, y)\sin\theta \tag{8}$$

Comparing (4), (6), and (8) then gives

$$\begin{aligned} D_{\mathbf{u}}f(x, y) &= f_x(x, y)\cos\theta + f_y(x, y)\sin\theta \\ &= [f_x(x, y)\mathbf{i} + f_y(x, y)\mathbf{j}] \cdot (\cos\theta\mathbf{i} + \sin\theta\mathbf{j}) \\ &= \nabla f(x, y) \cdot \mathbf{u} \end{aligned}$$

■

EXAMPLE 3 Find the directional derivative of $f(x, y) = 2x^2y^3 + 6xy$ at (1, 1) in the direction of a unit vector whose angle with the positive x-axis is $\pi/6$.

Solution Since $\partial f/\partial x = 4xy^3 + 6y$ and $\partial f/\partial y = 6x^2y^2 + 6x$, we have

$$\nabla f(x, y) = (4xy^3 + 6y)\mathbf{i} + (6x^2y^2 + 6x)\mathbf{j} \quad \text{and} \quad \nabla f(1, 1) = 10\mathbf{i} + 12\mathbf{j}$$

Now, at $\theta = \pi/6$, $\mathbf{u} = \cos\theta\mathbf{i} + \sin\theta\mathbf{j}$ becomes

$$\mathbf{u} = \frac{\sqrt{3}}{2}\mathbf{i} + \frac{1}{2}\mathbf{j}$$

Therefore,

$$D_{\mathbf{u}}f(1, 1) = \nabla f(1, 1) \cdot \mathbf{u} = (10\mathbf{i} + 12\mathbf{j}) \cdot \left(\frac{\sqrt{3}}{2}\mathbf{i} + \frac{1}{2}\mathbf{j}\right) = 5\sqrt{3} + 6$$

□

EXAMPLE 4 Consider the plane that is perpendicular to the xy-plane and passes through the points $P(2, 1)$ and $Q(3, 2)$. What is the slope of the tangent line to the curve of intersection of this plane with the surface $f(x, y) = 4x^2 + y^2$ at (2, 1, 17) in the direction of Q?

Solution We want $D_{\mathbf{u}}f(2, 1)$ in the direction given by the vector $\overrightarrow{PQ} = \mathbf{i} + \mathbf{j}$. But since PQ is not a unit vector, we form $\mathbf{u} = (1/\sqrt{2})\mathbf{i} + (1/\sqrt{2})\mathbf{j}$. Now,

$$\nabla f(x, y) = 8x\mathbf{i} + 2y\mathbf{j} \quad \text{and} \quad \nabla f(2, 1) = 16\mathbf{i} + 2\mathbf{j}$$

Therefore, the required slope is

$$D_{\mathbf{u}}f(2, 1) = (16\mathbf{i} + 2\mathbf{j}) \cdot \left(\frac{1}{\sqrt{2}}\mathbf{i} + \frac{1}{\sqrt{2}}\mathbf{j}\right) = 9\sqrt{2}$$ □

Functions of Three Variables For a function $w = F(x, y, z)$ the directional derivative is defined by

$$D_{\mathbf{u}}f(x, y, z) = \lim_{h \to 0} \frac{F(x + h\cos\alpha, y + h\cos\beta, z + h\cos\gamma) - F(x, y, z)}{h}$$

where α, β, and γ are the direction angles of the vector $\mathbf{u}$ measured relative to the positive x-, y- and z-axes, respectively.* But in the same manner as before, we can show that

$$D_{\mathbf{u}}f(x, y, z) = \nabla F(x, y, z) \cdot \mathbf{u} \tag{9}$$

Notice, since $\mathbf{u}$ is a unit vector, it follows from (10) of Section 13.3 that

$$D_{\mathbf{u}}f(x, y) = \text{comp}_{\mathbf{u}}\nabla f(x, y) \quad \text{and} \quad D_{\mathbf{u}}F(x, y, z) = \text{comp}_{\mathbf{u}}\nabla F(x, y, z)$$

In addition, (9) reveals that

$$D_{\mathbf{k}}F(x, y, z) = \frac{\partial w}{\partial z}$$

EXAMPLE 5 Find the directional derivative of $F(x, y, z) = xy^2 - 4x^2y + z^2$ at $(1, -1, 2)$ in the direction of $6\mathbf{i} + 2\mathbf{j} + 3\mathbf{k}$.

Solution We have $\partial F/\partial x = y^2 - 8xy$, $\partial F/\partial y = 2xy - 4x^2$, and $\partial F/\partial z = 2z$, so that

$$\nabla F(x, y, z) = (y^2 - 8xy)\mathbf{i} + (2xy - 4x^2)\mathbf{j} + 2z\mathbf{k}$$
$$\nabla F(1, -1, 2) = 9\mathbf{i} - 6\mathbf{j} + 4\mathbf{k}$$

Since $$|6\mathbf{i} + 2\mathbf{j} + 3\mathbf{k}| = 7 \quad \text{then} \quad \mathbf{u} = \frac{6}{7}\mathbf{i} + \frac{2}{7}\mathbf{j} + \frac{3}{7}\mathbf{k}$$

---

*Note that the numerator of (4) can be written

$$f(x + h\cos\alpha, y + h\cos\beta) - f(x, y)$$

where $\beta = (\pi/2) - \alpha$.

is a unit vector in the indicated direction. It follows from (9) that

$$D_{\mathbf{u}}F(1, -1, 2) = (9\mathbf{i} - 6\mathbf{j} + 4\mathbf{k}) \cdot \left(\frac{6}{7}\mathbf{i} + \frac{2}{7}\mathbf{j} + \frac{3}{7}\mathbf{k}\right) = \frac{54}{7} \qquad \square$$

Maximum Value of the Directional Derivative Let f represent a function of either two or three variables. Since (5) and (9) express the directional derivative as a dot product, we see from Definition 13.2 that

$$D_{\mathbf{u}}f = |\nabla f||\mathbf{u}|\cos\phi = |\nabla f|\cos\phi \qquad (|\mathbf{u}| = 1)$$

where ϕ is the angle between ∇f and $\mathbf{u}$. Because $0 \le \phi \le \pi$, we have $-1 \le \cos\phi \le 1$ and, consequently,

$$-|\nabla f| \le D_{\mathbf{u}}f \le |\nabla f|$$

In other words,

the maximum value of the directional derivative is $|\nabla f|$ and it occurs when $\mathbf{u}$ has the same direction as ∇f (when $\cos\phi = 1$), (10)

and

the minimum value of the directional derivative is $-|\nabla f|$ and it occurs when $\mathbf{u}$ and ∇f have opposite directions (when $\cos\phi = -1$). (11)

EXAMPLE 6 In Example 5 the maximum value of the directional derivative at F at $(1, -1, 2)$ is $|\nabla F(1, -1, 2)| = \sqrt{133}$. The minimum value of $D_{\mathbf{u}}F(1, -1, 2)$ is then $-\sqrt{133}$. □

Gradient Points in Direction of Most Rapid Increase of f Put yet another way, (10) and (11) state:

The gradient vector ∇f points in the direction in which f increases most rapidly, whereas $-\nabla f$ points in the direction of the most rapid decrease of f.

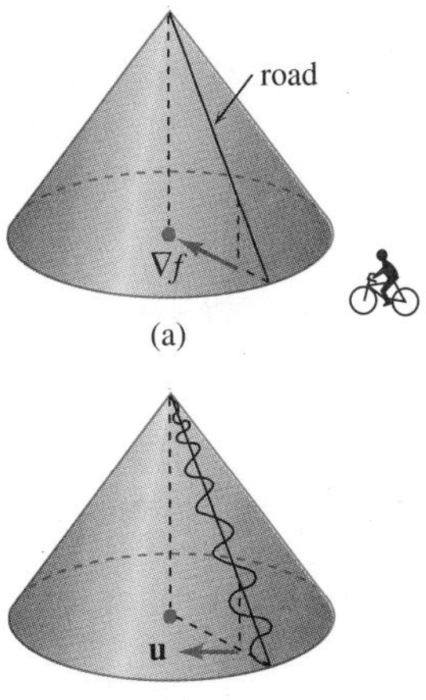

FIGURE 15.51

EXAMPLE 7 Each year in Los Angeles there is a bicycle race up to the top of a hill by a road known to be the steepest in the city. To understand why a bicyclist, with a modicum of sanity, will zig-zag up the road, let us suppose the graph of $f(x, y) = 4 - \frac{2}{3}\sqrt{x^2 + y^2}$, $0 \le z \le 4$, shown in Figure 15.51(a) is a mathematical model of the hill. The gradient of f is

$$\nabla f(x, y) = \frac{2}{3}\left[\frac{-x}{\sqrt{x^2 + y^2}}\mathbf{i} + \frac{-y}{\sqrt{x^2 + y^2}}\mathbf{j}\right] = \frac{2/3}{\sqrt{x^2 + y^2}}\mathbf{r}$$

where $\mathbf{r} = -x\mathbf{i} - y\mathbf{j}$ is a vector pointing to the center of the circular base.

Thus, the steepest ascent up the hill is a straight road whose projection in the xy-plane is a radius of the circular base. Since $D_{\mathbf{u}}f = \text{comp}_{\mathbf{u}}\nabla f$, a bicyclist will zag-zag, or seek a direction $\mathbf{u}$ other than ∇f, in order to reduce this component. See Figure 15.51(b). □

EXAMPLE 8 The temperature in a rectangular box is approximated by

$$T(x, y, z) = xyz(1 - x)(2 - y)(3 - z), \qquad 0 \le x \le 1, 0 \le y \le 2, 0 \le z \le 3$$

If a mosquito is located at $(\frac{1}{2}, 1, 1)$, in which direction should it fly to cool off as rapidly as possible?

Solution The gradient of T is

$$\nabla T(x, y, z) = yz(2 - y)(3 - z)(1 - 2x)\mathbf{i} + xz(1 - x)(3 - z)(2 - 2y)\mathbf{j} + xy(1 - x)(2 - y)(3 - 2z)\mathbf{k}$$

Therefore,
$$\nabla T\left(\frac{1}{2}, 1, 1\right) = \frac{1}{4}\mathbf{k}$$

To cool off most rapidly, the mosquito should fly in the direction of $-\frac{1}{4}\mathbf{k}$; that is, it should dive for the floor of the box, where the temperature is $T(x, y, 0) = 0$. □

EXERCISES 15.7

Answers to odd-numbered problems begin on page A-105.

In Problems 1–4 compute the gradient for the given function.

1. $f(x, y) = x^2 - x^3y^2 + y^4$
2. $f(x, y) = y - e^{-2x^2y}$
3. $F(x, y, z) = \dfrac{xy^2}{z^3}$
4. $F(x, y, z) = xy \cos yz$

In Problems 5–8 find the gradient of the given function at the indicated point.

5. $f(x, y) = x^2 - 4y^2$; $(2, 4)$
6. $f(x, y) = \sqrt{x^3y - y^4}$; $(3, 2)$
7. $F(x, y, z) = x^2z^2\sin 4y$; $(-2, \pi/3, 1)$
8. $F(x, y, z) = \ln(x^2 + y^2 + z^2)$; $(-4, 3, 5)$

In Problems 9 and 10 use Definition 15.7 to find $D_{\mathbf{u}}f(x, y)$ given that $\mathbf{u}$ makes the indicated angle with the positive x-axis.

9. $f(x, y) = x^2 + y^2$; $\theta = 30°$
10. $f(x, y) = 3x - y^2$; $\theta = 45°$

In Problems 11–20 find the directional derivative of the given function at the given point in the indicated direction.

11. $f(x, y) = 5x^3y^6$; $(-1, 1)$, $\theta = \pi/6$
12. $f(x, y) = 4x + xy^2 - 5y$; $(3, -1)$, $\theta = \pi/4$
13. $f(x, y) = \tan^{-1}\dfrac{y}{x}$; $(2, -2)$, $\mathbf{i} - 3\mathbf{j}$
14. $f(x, y) = \dfrac{xy}{x + y}$; $(2, -1)$, $6\mathbf{i} + 8\mathbf{j}$
15. $f(x, y) = (xy + 1)^2$; $(3, 2)$, in the direction of $(5, 3)$
16. $f(x, y) = x^2\tan y$; $\left(\dfrac{1}{2}, \dfrac{\pi}{3}\right)$, in the direction of the negative x-axis
17. $F(x, y, z) = x^2y^2(2z + 1)^2$; $(1, -1, 1)$, $\langle 0, 3, 3\rangle$
18. $F(x, y, z) = \dfrac{x^2 - y^2}{z^2}$; $(2, 4, -1)$, $\mathbf{i} - 2\mathbf{j} + \mathbf{k}$
19. $F(x, y, z) = \sqrt{x^2y + 2y^2z}$; $(-2, 2, 1)$, in the direction of the negative z-axis
20. $F(x, y, z) = 2x - y^2 + z^2$; $(4, -4, 2)$, in the direction of the origin

In Problems 21 and 22 consider the plane through the points P and Q that is perpendicular to the xy-plane. Find the slope of the tangent at the indicated point to the curve of intersection of this plane and the graph of the given function in the direction of Q.

21. $f(x, y) = (x - y)^2$; $P(4, 2)$, $Q(0, 1)$; $(4, 2, 4)$
22. $f(x, y) = x^3 - 5xy + y^2$; $P(1, 1)$, $Q(-1, 6)$; $(1, 1, -3)$

In Problems 23–26 find a vector tha gives the direction in which the given function increases most rapidly at the indicated point. Find the maximum rate.

23. $f(x, y) = e^{2x}\sin y$; $(0, \pi/4)$

24. $f(x, y) = xye^{x-y}$; $(5, 5)$

25. $F(x, y, z) = x^2 + 4xz + 2yz^2$; $(1, 2, -1)$

26. $F(x, y, z) = xyz$; $(3, 1, -5)$

In Problems 27–30 find a vector that gives the direction in which the given function decreases most rapidly at the indicated point. Find the minimum rate.

27. $f(x, y) = \tan(x^2 + y^2)$; $(\sqrt{\pi/6}, \sqrt{\pi/6})$

28. $f(x, y) = x^3 - y^3$; $(2, -2)$

29. $F(x, y, z) = \sqrt{xze^y}$; $(16, 0, 9)$

30. $F(x, y, z) = \ln \dfrac{xy}{z}$; $\left(\dfrac{1}{2}, \dfrac{1}{6}, \dfrac{1}{3}\right)$

31. Find the directional derivative(s) of $f(x, y) = x + y^2$ at $(3, 4)$ in the direction of a tangent vector to the graph of $2x^2 + y^2 = 9$ at $(2, 1)$.

32. If $f(x, y) = x^2 + xy + y^2 - x$, find all points where $D_{\mathbf{u}}f(x, y)$ in the direction of $\mathbf{u} = (1/\sqrt{2})(\mathbf{i} + \mathbf{j})$ is zero.

33. Suppose $\nabla f(a, b) = 4\mathbf{i} + 3\mathbf{j}$. Find a unit vector $\mathbf{u}$ so that

(a) $D_{\mathbf{u}}f(a, b) = 0$

(b) $D_{\mathbf{u}}f(a, b)$ is a maximum

(c) $D_{\mathbf{u}}f(a, b)$ is a minimum

34. Suppose $D_{\mathbf{u}}f(a, b) = 6$. What is the value of $D_{-\mathbf{u}}f(a, b)$?

35. (a) If $f(x, y) = x^3 - 3x^2y^2 + y^3$, find the directional derivative of f at a point (x, y) in the direction of $\mathbf{u} = (1/\sqrt{10})(3\mathbf{i} + \mathbf{j})$.

(b) If $F(x, y) = D_{\mathbf{u}}f(x, y)$ of part (a), find $D_{\mathbf{u}}F(x, y)$.

36. Consider the gravitational potential

$$U(x, y) = \frac{-Gm}{\sqrt{x^2 + y^2}}$$

where G and m are constants. Show that U increases or decreases most rapidly along a line through the origin.

37. If $f(x, y) = x^3 - 12x + y^2 - 10y$, find all points at which $|\nabla f| = 0$.

38. Suppose

$$D_{\mathbf{u}}f(a, b) = 7 \qquad D_{\mathbf{v}}f(a, b) = 3$$

$$\mathbf{u} = \frac{5}{13}\mathbf{i} - \frac{12}{13}\mathbf{j} \qquad \mathbf{v} = \frac{5}{13}\mathbf{i} + \frac{12}{13}\mathbf{j}$$

Find $\nabla f(a, b)$.

39. Consider the rectangular plate shown in Figure 15.52. The temperature at a point (x, y) on the plate is given by $T(x, y) = 5 + 2x^2 + y^2$. Determine the direction an insect should take, starting at $(4, 2)$, in order to cool off as rapidly as possible.

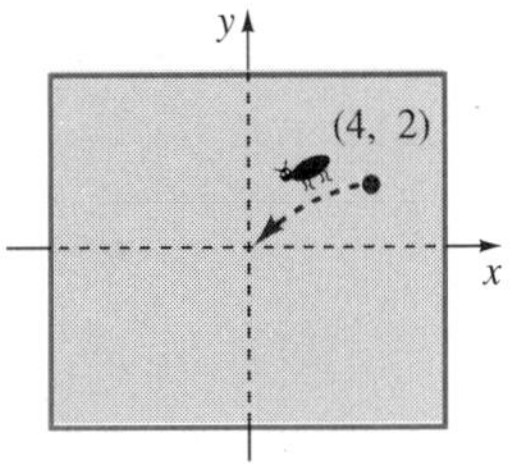

FIGURE 15.52

40. In Problem 39 observe that $(0, 0)$ is the coolest point of the plate. Find the path the cold-seeking insect, starting at $(4, 2)$, will take to the origin. If $\langle x(t), y(t)\rangle$ is the vector equation of the path, then use the fact that $-\nabla T(x, y) = \langle x'(t), y'(t)\rangle$. Why is this? [*Hint:* Remember separation of variables?]

41. The temperature at a point (x, y) on a rectangular metal plate is given by $T(x, y) = 100 - 2x^2 - y^2$. Find the path a heat-seeking particle will take, starting at $(3, 4)$, as it moves in the direction in which the temperature increases most rapidly.

42. The temperature T at a point (x, y, z) in space is inversely proportional to the square of the distance from (x, y, z) to the origin. It is known that $T(0, 0, 1) = 500$. Find the rate of change of T at $(2, 3, 3)$ in the direction of $(3, 1, 1)$. In which direction from $(2, 3, 3)$ does the temperature T increase most rapidly? At $(2, 3, 3)$ what is the maximum rate of change of T?

43. Find a function f such that

$$\nabla f = (3x^2 + y^3 + ye^{xy})\mathbf{i} + (-2y^2 + 3xy^2 + xe^{xy})\mathbf{j}$$

44. Let f_x, f_y, f_{xy}, f_{yx} be continuous and $\mathbf{u}$ and $\mathbf{v}$ be unit vectors. Show that $D_{\mathbf{u}}D_{\mathbf{v}}f = D_{\mathbf{v}}D_{\mathbf{u}}f$.

In Problems 45–48 assume that f and g are differentiable functions of two variables. Prove the given identity.

45. $\nabla(cf) = c\,\nabla f$

46. $\nabla(f + g) = \nabla f + \nabla g$

47. $\nabla(fg) = f\,\nabla g + g\,\nabla f$

48. $\nabla\left(\dfrac{f}{g}\right) = \dfrac{g\,\nabla f - f\,\nabla g}{g^2}$

49. If $F(x, y, z) = f_1(x, y, z)\mathbf{i} + f_2(x, y, z)\mathbf{j} + f_3(x, y, z)\mathbf{k}$ and

$$\nabla = \mathbf{i}\frac{\partial}{\partial x} + \mathbf{j}\frac{\partial}{\partial y} + \mathbf{k}\frac{\partial}{\partial z}$$

show that

$$\nabla \times F = \left(\frac{\partial f_3}{\partial y} - \frac{\partial f_2}{\partial z}\right)\mathbf{i} + \left(\frac{\partial f_1}{\partial z} - \frac{\partial f_3}{\partial x}\right)\mathbf{j} + \left(\frac{\partial f_2}{\partial x} - \frac{\partial f_1}{\partial y}\right)\mathbf{k}$$

15.8 Tangent Planes and Normal Lines

Geometric Interpretation of the Gradient—Functions of Two Variables Suppose $f(x, y) = c$ is the *level curve* of the differentiable function $z = f(x, y)$ that passes through a specified point $P(x_0, y_0)$; that is, $f(x_0, y_0) = c$. If this level curve is parameterized by the differentiable functions

$$x = g(t),\ y = h(t) \quad \text{such that} \quad x_0 = g(t_0),\ y_0 = h(t_0)$$

then the derivative of $f(g(t), h(t)) = c$ with respect to t is

$$\frac{\partial f}{\partial x}\frac{dx}{dt} + \frac{\partial f}{\partial y}\frac{dy}{dt} = 0 \tag{1}$$

By introducing the vectors

$$\nabla f(x, y) = \frac{\partial f}{\partial x}\mathbf{i} + \frac{\partial f}{\partial y}\mathbf{j} \quad \text{and} \quad \mathbf{r}'(t) = \frac{dx}{dt}\mathbf{i} + \frac{dy}{dt}\mathbf{j}$$

(1) becomes $\nabla f \cdot \mathbf{r}' = 0$. Specifically, at $t = t_0$, we have

$$\nabla f(x_0, y_0) \cdot \mathbf{r}'(t_0) = 0 \tag{2}$$

Thus, if $\mathbf{r}'(t_0) \neq 0$, the vector $\nabla f(x_0, y_0)$ is orthogonal to the tangent vector $\mathbf{r}'(t_0)$ at $P(x_0, y_0)$. We interpret this to mean

∇f is perpendicular to the level curve at P.

See Figure 15.53.

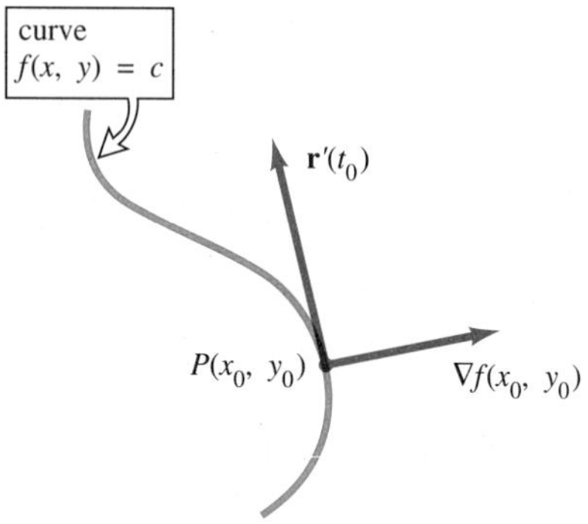

FIGURE 15.53

EXAMPLE 1 Find the level curve of $f(x, y) = -x^2 + y^2$ passing through (2, 3). Graph the gradient at the point.

Solution Since $f(2, 3) = -4 + 9 = 5$, the level curve is the hyperbola $-x^2 + y^2 = 5$. Now,

$$\nabla f(x, y) = -2x\mathbf{i} + 2y\mathbf{j} \quad \text{and} \quad \nabla f(2, 3) = -4\mathbf{i} + 6\mathbf{j}$$

Figure 15.54 shows the level curve and $\nabla f(2, 3)$. □

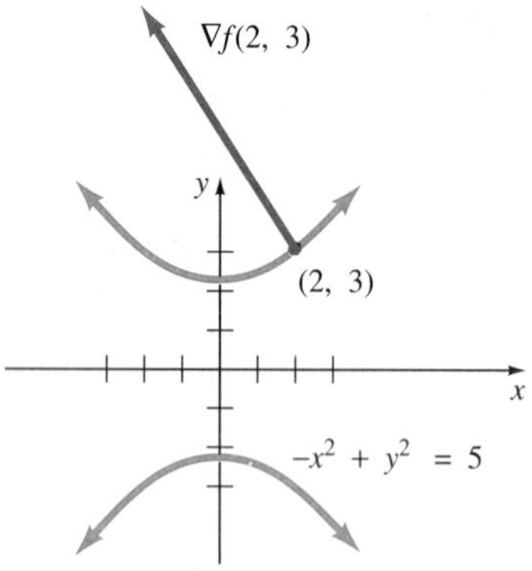

FIGURE 15.54

Geometric Interpretation of the Gradient—Functions of Three Variables Proceeding as before, let $F(x, y, z) = c$ be the *level surface* of a differentiable function $w = F(x, y, z)$ that passes through $P(x_0, y_0, z_0)$. If the differentiable functions

$$x = f(t) \qquad y = g(t) \qquad z = h(t)$$

are the parametric equations of a curve C on the surface for which $x_0 = f(t_0)$, $y_0 = g(t_0)$, $z_0 = h(t_0)$, then the derivative of $F(f(t), g(t), h(t)) = c$

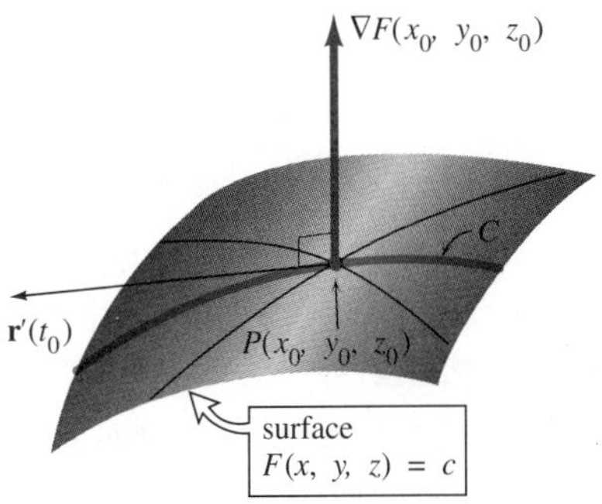

FIGURE 15.55

implies that

$$\frac{\partial F}{\partial x}\frac{dx}{dt} + \frac{\partial F}{\partial y}\frac{dy}{dt} + \frac{\partial F}{\partial z}\frac{dz}{dt} = 0$$

or

$$\left(\frac{\partial F}{\partial x}\mathbf{i} + \frac{\partial F}{\partial y}\mathbf{j} + \frac{\partial F}{\partial z}\mathbf{k}\right)\cdot\left(\frac{dx}{dt}\mathbf{i} + \frac{dy}{dt}\mathbf{j} + \frac{dz}{dt}\mathbf{k}\right) = 0 \tag{3}$$

In particular, at $t = t_0$, (3) is

$$\nabla F(x_0, y_0, z_0) \cdot \mathbf{r}'(t_0) = 0 \tag{4}$$

Thus, when $\mathbf{r}'(t_0) \neq 0$, the vector $\nabla F(x_0, y_0, z_0)$ is orthogonal to the tangent vector $\mathbf{r}'(t_0)$. Since this argument holds for any differentiable curve through $P(x_0, y_0, z_0)$ on the surface, we conclude that

∇F is perpendicular (normal) to the level surface at P.

See Figure 15.55.

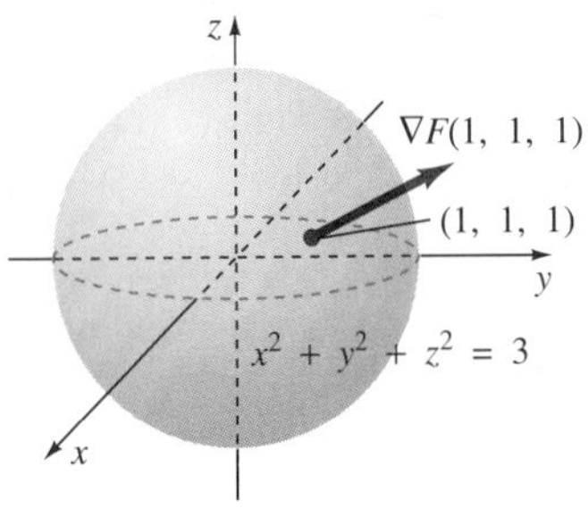

FIGURE 15.56

EXAMPLE 2 Find the level surface of $F(x, y, z) = x^2 + y^2 + z^2$ passing through (1, 1, 1). Graph the gradient at the point.

Solution Since $F(1, 1, 1) = 3$, the level surface passing through (1, 1, 1) is the sphere $x^2 + y^2 + z^2 = 3$. The gradient of the function is

$$\nabla F(x, y, z) = 2x\mathbf{i} + 2y\mathbf{j} + 2z\mathbf{k}$$

and so, at the given point,

$$\nabla F(1, 1, 1) = 2\mathbf{i} + 2\mathbf{j} + 2\mathbf{k}$$

The level surface and $\nabla F(1, 1, 1)$ are illustrated in Figure 15.56. □

Tangent Plane In the earlier chapters we found equations of tangent lines to graphs of functions. In 3-space we can now solve the analogous problem of finding equations of **tangent planes** to surfaces. We assume again that $w = F(x, y, z)$ is a differentiable function and that a surface is given by $F(x, y, z) = c$.

DEFINITION 15.8 Tangent Plane

Let $P(x_0, y_0, z_0)$ be a point on the graph of $F(x, y, z) = c$ where ∇F is not $\mathbf{0}$. The **tangent plane** at P is that plane through P that is perpendicular to ∇F evaluated at P.

Thus, if $P(x, y, z)$ and $P(x_0, y_0, z_0)$ are points on the tangent plane and $\mathbf{r}$ and $\mathbf{r}_0$ are their respective position vectors, a vector equation of the tangent

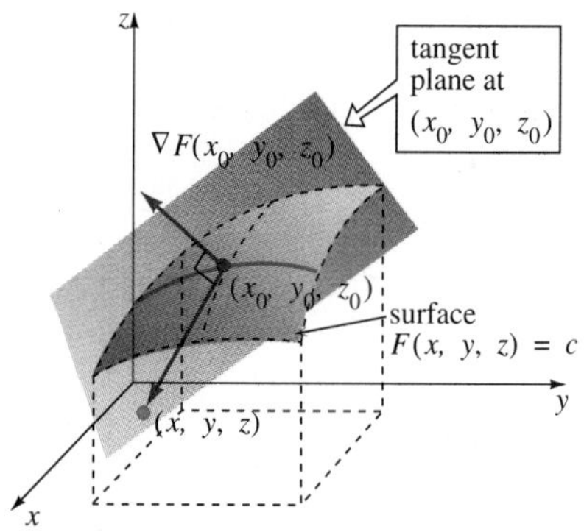

FIGURE 15.57

plane is

$$\nabla F(x_0, y_0, z_0) \cdot (\mathbf{r} - \mathbf{r}_0) = 0$$

See Figure 15.57. We summarize this last result.

THEOREM 15.12 Equation of Tangent Plane

Let $P(x_0, y_0, z_0)$ be a point on the graph of $F(x, y, z) = c$ where ∇F is not $\mathbf{0}$. Then an **equation of the tangent plane** at P is

$$F_x(x_0, y_0, z_0)(x - x_0) + F_y(x_0, y_0, z_0)(y - y_0) + F_z(x_0, y_0, z_0)(z - z_0) = 0 \qquad (5)$$

EXAMPLE 3 Find an equation of the tangent plane to the graph of $x^2 - 4y^2 + z^2 = 16$ at (2, 1, 4).

Solution By defining $F(x, y, z) = x^2 - 4y^2 + z^2$, we find the given surface is the level surface $F(x, y, z) = F(2, 1, 4) = 16$ passing through (2, 1, 4). Now, $F_x(x, y, z) = 2x$, $F_y(x, y, z) = -8y$, and $F_z(x, y, z) = 2z$, so that

$$\nabla F(x, y, z) = 2x\mathbf{i} - 8y\mathbf{j} + 2z\mathbf{k} \quad \text{and} \quad \nabla F(2, 1, 4) = 4\mathbf{i} - 8\mathbf{j} + 8\mathbf{k}$$

It follows from (5) that an equation of the tangent plane is

$$4(x - 2) - 8(y - 1) + 8(z - 4) = 0 \quad \text{or} \quad x - 2y + 2z = 8$$ □

Surfaces Given by z = f(x, y) For a surface given explicitly by a differentiable function $z = f(x, y)$, we define $F(x, y, z) = f(x, y) - z$ or $F(x, y, z) = z - f(x, y)$. Thus a point (x_0, y_0, z_0) is on the graph of $z = f(x, y)$ if and only if it is also on the level surface $F(x, y, z) = 0$. This follows from $F(x_0, y_0, z_0) = f(x_0, y_0) - z_0 = 0$.

EXAMPLE 4 Find an equation of the tangent plane to the graph of $z = \frac{1}{2}x^2 + \frac{1}{2}y^2 + 4$ at $(1, -1, 5)$.

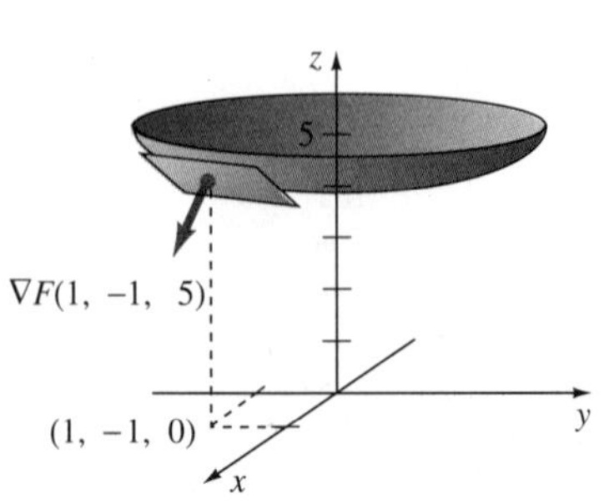

FIGURE 15.58

Solution Define $F(x, y, z) = \frac{1}{2}x^2 + \frac{1}{2}y^2 - z + 4$ so that the level surface of F passing through the given point is $F(x, y, z) = F(1, -1, 5)$ or $F(x, y, z) = 0$. Now, $F_x = x$, $F_y = y$, and $F_z = -1$, so that

$$\nabla F(x, y, z) = x\mathbf{i} + y\mathbf{j} - \mathbf{k} \quad \text{and} \quad \nabla F(1, -1, 5) = \mathbf{i} - \mathbf{j} - \mathbf{k}$$

Thus, from (5) the desired equation is

$$(x + 1) - (y - 1) - (z - 5) = 0 \quad \text{or} \quad -x + y + z = 7$$

See Figure 15.58. □

Normal Line Let $P(x_0, y_0, z_0)$ be a point on the graph of $F(x, y, z) = c$ where ∇F is not $\mathbf{0}$. The line containing $P(x_0, y_0, z_0)$ that is parallel to $\nabla F(x_0, y_0, z_0)$ is called the **normal line** to the surface at P. The normal line is perpendicular to the tangent plane to the surface at P.

EXAMPLE 5 Find parametric equations for the normal line to the surface in Example 4 at $(1, -1, 5)$.

Solution A direction vector for the normal line at $(1, -1, 5)$ is $\nabla F(1, -1, 5) = \mathbf{i} - \mathbf{j} - \mathbf{k}$. It follows that parametric equations for the normal line are $x = 1 + t$, $y = -1 - t$, $z = 5 - t$. □

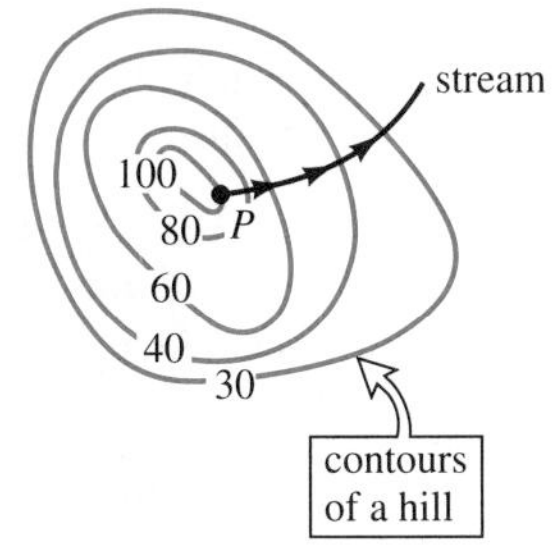

FIGURE 15.59

▲ ***Remark*** Water flowing down a hill chooses a path in the direction of the greatest change in altitude. Figure 15.59 shows the contours, or level curves, of a hill. As shown in the figure, a stream starting at point P will take a path that is perpendicular to the contours. After reading Sections 15.7 and 15.8 you should be able to explain why.

EXERCISES 15.8

Answers to odd-numbered problems begin on page A-105.

In Problems 1–12 sketch the level curve or surface passing through the indicated point. Sketch the gradient at the point.

1. $f(x, y) = x - 2y$; $(6, 1)$
2. $f(x, y) = \dfrac{y + 2x}{x}$; $(1, 3)$
3. $f(x, y) = y - x^2$; $(2, 5)$
4. $f(x, y) = x^2 + y^2$; $(-1, 3)$
5. $f(x, y) = \dfrac{x^2}{4} + \dfrac{y^2}{9}$; $(-2, -3)$
6. $f(x, y) = \dfrac{y^2}{x}$; $(2, 2)$
7. $f(x, y) = (x - 1)^2 - y^2$; $(1, 1)$
8. $f(x, y) = \dfrac{y - 1}{\sin x}$; $\left(\dfrac{\pi}{6}, \dfrac{3}{2}\right)$
9. $F(x, y, z) = y + z$; $(3, 1, 1)$
10. $F(x, y, z) = x^2 + y^2 - z$; $(1, 1, 3)$
11. $F(x, y, z) = \sqrt{x^2 + y^2 + z^2}$; $(3, 4, 0)$
12. $F(x, y, z) = x^2 - y^2 + z$; $(0, -1, 1)$

In Problems 13 and 14 find the points on the given surface at which the gradient is parallel to the indicated vector.

13. $z = x^2 + y^2$; $4\mathbf{i} + \mathbf{j} + \frac{1}{2}\mathbf{k}$
14. $x^3 + y^2 + z = 15$; $27\mathbf{i} + 8\mathbf{j} + \mathbf{k}$

In Problems 15–24 find an equation of the tangent plane to the graph of the given equation at the indicated point.

15. $x^2 + y^2 + z^2 = 9$; $(-2, 2, 1)$
16. $5x^2 - y^2 + 4z^2 = 8$; $(2, 4, 1)$
17. $x^2 - y^2 - 3z^2 = 5$; $(6, 2, 3)$
18. $xy + yz + zx = 7$; $(1, -3, -5)$
19. $z = 25 - x^2 - y^2$; $(3, -4, 0)$
20. $xz = 6$; $(2, 0, 3)$
21. $z = \cos(2x + y)$; $\left(\dfrac{\pi}{2}, \dfrac{\pi}{4}, -\dfrac{1}{\sqrt{2}}\right)$
22. $x^2y^3 + 6z = 10$; $(2, 1, 1)$
23. $z = \ln(x^2 + y^2)$; $\left(\dfrac{1}{\sqrt{2}}, \dfrac{1}{\sqrt{2}}, 0\right)$
24. $z = 8e^{-2y}\sin 4x$; $\left(\dfrac{\pi}{24}, 0, 4\right)$

In Problems 25 and 26 find the points on the given surface at which the tangent plane is parallel to the indicated plane.

25. $x^2 + y^2 + z^2 = 7$; $2x + 4y + 6z = 1$
26. $x^2 - 2y^2 - 3z^2 = 33$; $8x + 4y + 6z = 5$

27. Find points on the surface $x^2 + 4x + y^2 + z^2 - 2z = 11$ at which the tangent plane is horizontal.

28. Find points on the surface $x^2 + 3y^2 + 4z^2 - 2xy = 16$ at which the tangent plane is parallel to **(a)** the xz-plane, **(b)** the yz-plane, and **(c)** the xy-plane.

In Problems 29 and 30 show that the second equation is an equation of the tangent plane to the graph of the first equation at (x_0, y_0, z_0).

29. $\dfrac{x^2}{a^2} + \dfrac{y^2}{b^2} + \dfrac{z^2}{c^2} = 1; \dfrac{xx_0}{a^2} + \dfrac{yy_0}{b^2} + \dfrac{zz_0}{c^2} = 1$

30. $\dfrac{x^2}{a^2} - \dfrac{y^2}{b^2} + \dfrac{z^2}{c^2} = 1; \dfrac{xx_0}{a^2} - \dfrac{yy_0}{b^2} + \dfrac{zz_0}{c^2} = 1$

31. Show that every tangent plane to the graph of $z^2 = x^2 + y^2$ passes through the origin.

32. Show that the sum of the x-, y-, and z-intercepts of every tangent plane to the graph of $\sqrt{x} + \sqrt{y} + \sqrt{z} = \sqrt{a}, a > 0$, is the number a.

In Problems 33 and 34 find parametric equations for the normal line at the indicated point. In Problems 35 and 36 find symmetric equations for the normal line.

33. $x^2 + 2y^2 + z^2 = 4; (1, -1, 1)$

34. $z = 2x^2 - 4y^2; (3, -2, 2)$

35. $z = 4x^2 + 9y^2 + 1; (\frac{1}{2}, \frac{1}{3}, 3)$

36. $x^2 + y^2 - z^2 = 0; (3, 4, 5)$

37. Show that every normal line to the graph of $x^2 + y^2 + z^2 = a^2$ passes through the origin.

38. Two surfaces are said to be **orthogonal** at a point P of intersection if their normal lines at P are orthogonal. Prove that the surfaces given by $F(x, y, z) = 0$ and $G(x, y, z) = 0$ are orthogonal at P if and only if $F_xG_x + F_yG_y + F_zG_z = 0$.

In Problems 39 and 40 use the result of Problem 38 to show that the given surfaces are orthogonal at a point of intersection.

39. $x^2 + y^2 + z^2 = 25; -x^2 + y^2 + z^2 = 0$

40. $x^2 - y^2 + z^2 = 4; z = 1/xy^2$

15.9 EXTREMA FOR FUNCTIONS OF TWO VARIABLES

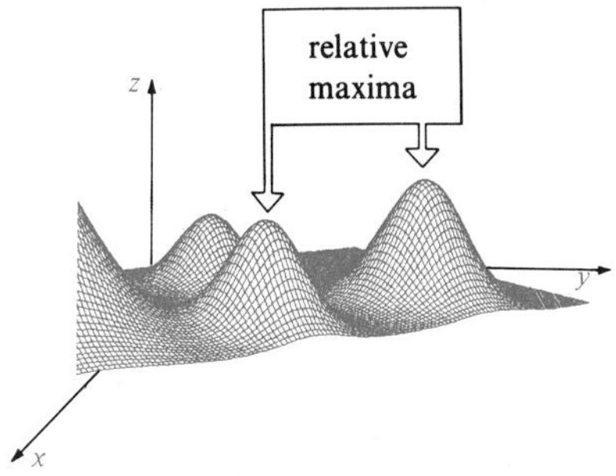

FIGURE 15.60

If $z = f(x, y)$ is continuous on a closed rectangular region R whose sides are parallel to the x- and y-coordinate axes, then, analogous to Theorem 3.1, f has both an **absolute maximum** and an **absolute minimum** on the region. That is, there are points (c, d) and (a, b) in R so that

$$f(c, d) \leq f(x, y) \leq f(a, b)$$

for every point (x, y) in R.

A function f of two variables can, of course, have **relative**, or **local, extrema**. We say that $z = f(x, y)$ has a **relative maximum** at (a, b) if there is some rectangular region containing (a, b) such that $f(x, y) \leq f(a, b)$ for all (x, y) in the region. Similarly, $z = f(x, y)$ has a **relative minimum** at (a, b) if there is some rectangular region containing (a, b) such that $f(x, y) \geq f(a, b)$ for all (x, y) in the region. Figure 15.60 shows a function with several relative maxima.

Suppose for the sake of illustration that (a, b) is an interior point of a rectangular region R at which f has a maximum and, furthermore, suppose that f has first and second partial derivatives. As seen in Figure 15.61, on the curve C_1 of intersection of the surface and the plane $x = a$, we must have

$$f_y(a, b) = 0 \quad \text{and} \quad f_{yy}(a, b) \leq 0^*$$

*Do you remember the relationship between concavity and the second derivative?

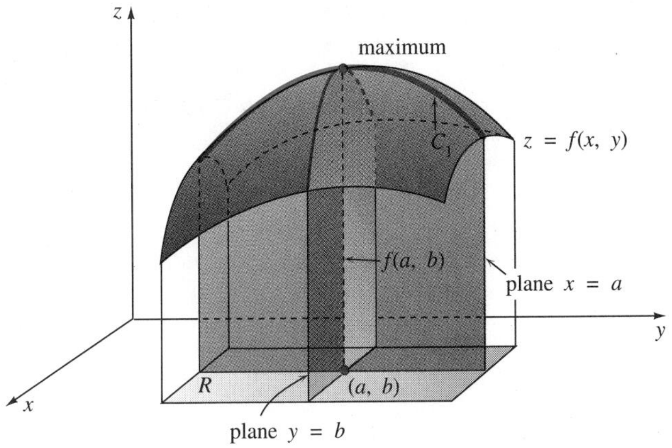

FIGURE 15.61

Similarly, on the curve C_2, which is the trace of the surface in the plane $y = b$, we have

$$f_x(a, b) = 0 \quad \text{and} \quad f_{xx}(a, b) \leq 0$$

In contrast, if $z = f(x, y)$ has a minimum at (a, b) we expect

$$f_x(a, b) = 0, \quad f_y(a, b) = 0 \quad \text{and} \quad f_{xx}(a, b) \geq 0, \quad f_{yy}(a, b) \geq 0$$

Critical Points The foregoing discussion suggests that to find an extremum of $z = f(x, y)$, we must solve $f_x(x, y) = 0$ and $f_y(x, y) = 0$ *simultaneously*.* This leads to the following definition:

DEFINITION 15.9 Critical Points

If $z = f(x, y)$ has first partial derivatives, then the solutions of

$$f_x(x, y) = 0 \quad \text{and} \quad f_y(x, y) = 0$$

are called **critical points.**

The critical points correspond to points where f could *possibly* have a relative extremum. In some texts critical points are also called **stationary points.**

*Alternatively, an equation of the tangent plane to the graph of $z = f(x, y)$ at the point (a, b) is

$$z - f(a, b) = (x - a)f_x(a, b) + (y - b)f_y(a, b)$$

Arguing as we did for tangent lines, we look for horizontal tangent planes. At a point (a, b), where the tangent plane is horizontal, its equation must be $z = f(a, b)$, and so $f_x(a, b) = 0$ and $f_y(a, b) = 0$.

EXAMPLE 1 Find all critical points for $f(x, y) = x^3 + y^3 - 27x - 12y$.

Solution The first partial derivatives are

$$f_x(x, y) = 3x^2 - 27 \quad \text{and} \quad f_y(x, y) = 3y^2 - 12$$

Hence, $f_x(x, y) = 0$ and $f_y(x, y) = 0$ imply that

$$x^2 = 9 \quad \text{and} \quad y^2 = 4$$

and so $x = \pm 3$, $y = \pm 2$. Thus, there are four critical points $(3, 2)$, $(-3, 2)$, $(3, -2)$, and $(-3, -2)$. □

Second Partials Test The following theorem gives sufficient conditions for ascertaining relative extrema. The proof of the theorem will not be given.

THEOREM 15.13 Second Partials Test for Relative Extrema

Let (a, b) be a critical point of $z = f(x, y)$ and suppose f_{xx}, f_{yy}, and f_{xy} are continuous in a rectangular region containing (a, b). Let $D(x, y) = f_{xx}(x, y)f_{yy}(x, y) - [f_{xy}(x, y)]^2$.

(*i*) If $D(a, b) > 0$ and $f_{xx}(a, b) > 0$, then $f(a, b)$ is a relative minimum.

(*ii*) If $D(a, b) > 0$ and $f_{xx}(a, b) < 0$, then $f(a, b)$ is a relative maximum.

(*iii*) If $D(a, b) < 0$, then $f(a, b)$ is not an extremum.

(*iv*) If $D(a, b) = 0$, no conclusion can be drawn concerning a relative extremum.

EXAMPLE 2 Find the extrema for $f(x, y) = 4x^2 + 2y^2 - 2xy - 10y - 2x$.

Solution The first partial derivatives are

$$f_x(x, y) = 8x - 2y - 2 \quad \text{and} \quad f_y(x, y) = 4y - 2x - 10$$

Solving the simultaneous equations

$$8x - 2y = 2 \quad \text{and} \quad -2x + 4y = 10$$

yields the single critical point $(1, 3)$. Now,

$$f_{xx}(x, y) = 8 \qquad f_{yy}(x, y) = 4 \qquad f_{xy}(x, y) = -2$$

and so $D(x, y) = (8)(4) - (-2)^2 = 28$. Since $D(1, 3) > 0$ and $f_{xx}(1, 3) > 0$, it follows from (*i*) of Theorem 15.13 that $f(1, 3) = -16$ is a relative minimum. □

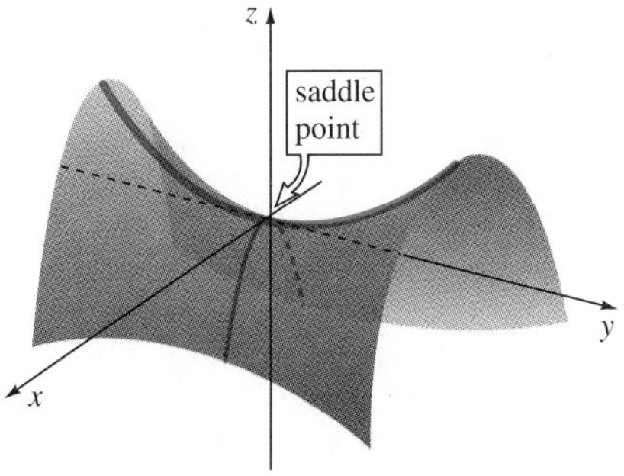

FIGURE 15.62

EXAMPLE 3 The graph of $f(x, y) = y^2 - x^2$ is the hyperbolic paraboloid given in Figure 15.62. From $f_x(x, y) = -2x$ and $f_y(x, y) = 2y$ we see that (0, 0) is a critical point and that $f(0, 0) = 0$ is the only possible extremum of the function. But from

$$f(0, y) = y^2 \geq 0 \quad \text{and} \quad f(x, 0) = -x^2 \leq 0$$

we see that, in a neighborhood of (0, 0), the points along the y-axis correspond to functional values that are greater than or equal to $f(0, 0) = 0$ and the points along the x-axis corresponding to functional values that are less than or equal to $f(0, 0) = 0$. Hence, $f(0, 0) = 0$ is not an extremum. We say that (0, 0) is a **saddle point** of the function.

You should verify that $D(0, 0) < 0$. □

In general, the critical point (a, b) in case (*iii*) of Theorem 15.13 is called a **saddle point**. When $D(a, b) < 0$ for a critical point (a, b), then the graph of the function behaves essentially like the saddle-shaped hyperbolic paraboloid in a neighborhood of (a, b). See Figure 15.63.

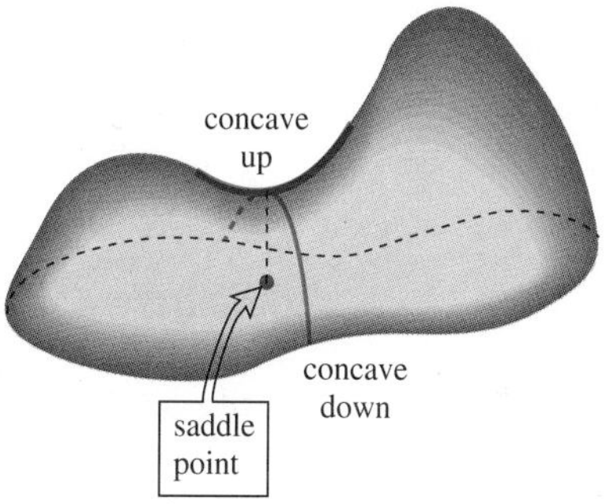

FIGURE 15.63

EXAMPLE 4 Find the extrema for

$$f(x, y) = 4xy - x^2 - y^2 - 14x + 4y + 10$$

Solution From $f_x(x, y) = 4y - 2x - 14$ and $f_y(x, y) = 4x - 2y + 4$, we find the solution of the system

$$4y - 2x - 14 = 0 \quad \text{and} \quad 4x - 2y + 4 = 0$$

to be (1, 4). Now $f_{xx}(x, y) = -2$, $f_{yy}(x, y) = -2$, and $f_{xy}(x, y) = 4$. By themselves the equations

$$f_{xx}(0, 0) = -2 \quad \text{and} \quad f_y(0, 0) = -2$$

would seem to imply that $f(1, 4) = 11$ is a relative maximum. However, $D(1, 4) = (-2)(-2) - (4)^2 < 0$ indicates that $f(1, 4)$ is not *any* extremum and (1, 4) is a saddle point. □

EXAMPLE 5 Find the extrema of $f(x, y) = x^3 + y^3 - 3x^2 - 3y^2 - 9x$.

Solution From $f_x(x, y) = 3x^2 - 6x - 9 = 3(x - 3)(x + 1)$
and $f_y(x, y) = 3y^2 - 6y = 3y(y - 2)$

and the equations

$$(x - 3)(x + 1) = 0 \quad \text{and} \quad y(y - 2) = 0$$

we find our critical points: (3, 0), (3, 2), (−1, 0), (−1, 2). Since

$$f_{xx} = 6x - 6 \qquad f_{yy} = 6y - 6 \qquad f_{xy} = 0$$

we find $D(x, y) = 36(x - 1)(y - 1)$. The second partial derivatives test is summarized in the accompanying table.

Critical Point (a, b)	$D(a, b)$	$f_{xx}(a, b)$	$f(a, b)$	Conclusion
(3, 0)	negative	positive	−27	no extremum
(3, 2)	positive	positive	−31	rel. minimum
(−1, 0)	positive	negative	5	rel. maximum
(−1, 2)	negative	negative	1	no extremum

□

Boundary Extrema Recall that if a function f of a single variable x is continuous on a closed interval $[a, b]$, then f always has an absolute maximum and an absolute minimum on the interval. In this case, an absolute extremum could be an endpoint extremum. For a function f of two variables x and y, the **Extreme Value Theorem** states that if f is continuous on a closed and bounded region R, then f has an absolute maximum and an absolute minimum at points in R. Analogous to endpoint extrema, a function of two variables could have **boundary extrema**.

EXAMPLE 6 The function $f(x, y) = 6x^2 - 8x + 2y^2 - 5$ is continuous on the closed region R defined by $x^2 + y^2 \leq 1$. Find its absolute extrema over R.

Solution We first find any critical points of f in the interior of the region. From $f_x(x, y) = 12x - 8$ and $f_y(x, y) = 4y$, we get the critical point $(\frac{2}{3}, 0)$.

In order to examine f on the boundary of the region, we represent the circle $x^2 + y^2 = 1$ by means of the parametric equations $x = \cos t$, $y = \sin t$, $0 \leq t \leq 2\pi$. Thus, on the boundary we can write f as a function of a single variable t:

$$F(t) = f(\cos t, \sin t) = 6\cos^2 t - 8\cos t + 2\sin^2 t - 5$$

We now proceed as in Section 3.3. Differentiating F with respect to t and simplifying give

$$F'(t) = 8\sin t(-\cos t + 1)$$

Hence, $F'(t) = 0$ implies $\sin t = 0$ or $\cos t = 1$. From these equations we find that the only critical point of F in the open interval $(0, 2\pi)$ is $t = \pi$. The corresponding point in R is then found to be $(-1, 0)$. The endpoints of the interval $[0, 2\pi]$, $t = 0$ and $t = 2\pi$, both correspond to the point $(1, 0)$ in R. From

$$f\left(\frac{2}{3}, 0\right) = -\frac{23}{3} \qquad f(-1, 0) = 9 \qquad f(1, 0) = -7$$

we see that the absolute maximum of f over R is $f(-1, 0) = 9$ and the absolute minimum is $f(\frac{2}{3}, 0) = -\frac{23}{3}$. □

▲ ***Remarks*** (*i*) Recall that a number c in the domain of $y = f(x)$ is a critical point if $f'(c) = 0$ or $f'(c)$ does not exist. Similarly, $z = f(x, y)$ can have an extremum at a point (a, b), where f_x and f_y do not exist, but Theorem 15.13 is certainly not applicable at such a point. See Problems 33 and 36 in Exercises 15.9. For functions of two variables, there is no convenient first partial derivative test to fall back on.

(*ii*) The method of solution for the system $f_x(x, y) = 0$, $f_y(x, y) = 0$ will not always be obvious, especially when f_x and f_y are not linear. Do not be afraid to exercise your algebraic skills in the problems that follow.

EXERCISES 15.9

Answers to odd-numbered problems begin on page A-106.

In Problems 1–20 find any relative extrema of the given function.

1. $f(x, y) = x^2 + y^2 + 5$
2. $f(x, y) = 4x^2 + 8y^2$
3. $f(x, y) = -x^2 - y^2 + 8x + 6y$
4. $f(x, y) = 3x^2 + 2y^2 - 6x + 8y$
5. $f(x, y) = 5x^2 + 5y^2 + 20x - 10y + 40$
6. $f(x, y) = -4x^2 - 2y^2 - 8x + 12y + 5$
7. $f(x, y) = 4x^3 + y^3 - 12x - 3y$
8. $f(x, y) = -x^3 + 2y^3 + 27x - 24y + 3$
9. $f(x, y) = 2x^2 + 4y^2 - 2xy - 10x - 2y + 2$
10. $f(x, y) = 5x^2 + 5y^2 + 5xy - 10x - 5y + 18$
11. $f(x, y) = (2x - 5)(y - 4)$
12. $f(x, y) = (x + 5)(2y + 6)$
13. $f(x, y) = -2x^3 - 2y^3 + 6xy + 10$
14. $f(x, y) = x^3 + y^3 - 6xy + 27$
15. $f(x, y) = xy - \dfrac{2}{x} - \dfrac{4}{y} + 8$
16. $f(x, y) = -3x^2y - 3xy^2 + 36xy$
17. $f(x, y) = xe^x \sin y$
18. $f(x, y) = e^{y^2 - 3y + x^2 + 4x}$
19. $f(x, y) = \sin x + \sin y$
20. $f(x, y) = \sin xy$
21. Find three positive numbers whose sum is 21 such that their product P is a maximum. [*Hint:* Express P as a function of only two variables.]
22. Find the dimensions of a rectangular box with a volume of 1 ft^3 that has a minimal surface area S.
23. Find the point on the plane $x + 2y + z = 1$ closest to the origin. [*Hint:* Consider the square of the distance.]
24. Find the least distance between the point $(2, 3, 1)$ and the plane $x + y + z = 1$.
25. Find all points on the surface $xyz = 8$ that are closest to the origin. Find the least distance.
26. Find the shortest distance between the lines

$$\mathscr{L}_1: x = t,\ y = 4 - 2t,\ z = 1 + t$$

$\mathscr{L}_2$: $x = 3 + 2s$, $y = 6 + 2s$, $z = 8 - 2s$

At what points on the lines does the minimum occur?

27. Find the maximum volume of a rectangular box with sides parallel to the coordinate planes that can be inscribed in the ellipsoid

$$\frac{x^2}{a^2} + \frac{y^2}{b^2} + \frac{z^2}{c^2} = 1, \qquad a > 0, b > 0, c > 0$$

28. The volume of an ellipsoid

$$\frac{x^2}{a^2} + \frac{y^2}{b^2} + \frac{z^2}{c^2} = 1, \qquad a > 0, b > 0, c > 0$$

is $V = 4\pi abc/3$. Show that the ellipsoid of greatest volume that satisfies $a + b + c =$ constant is a sphere.

29. A closed rectangular box is to be made so that its volume is 60 ft^3. The costs of the material for the top and bottom are 10 cents per square foot and 20 cents per square foot, respectively. The cost of the sides is 2 cents per square foot. Determine the cost function $C(x, y)$, where x and y are the length and width of the box, respectively. Find the dimensions of the box that will give a minimum cost.

30. A revenue function is

$$R(x, y) = x(100 - 6x) + y(192 - 4y)$$

where x and y denote the number of items of two commodities sold. Given that the corresponding cost function is

$$C(x, y) = 2x^2 + 2y^2 + 4xy - 8x + 20$$

find the maximum profit. [*Hint:* Profit = revenue − cost.]

31. The pentagon shown in Figure 15.64, formed by an isosceles triangle surmounted on a rectangle, has a fixed perimeter P. Find x, y, and θ so that the area of the pentagon is a maximum.

FIGURE 15.64

32. A 24-in wide piece of tin is bent into a trough whose cross-section is an isosceles trapezoid. See Figure 15.65. Find x and θ so that the cross-sectional area is a maximum. What is the maximum area?

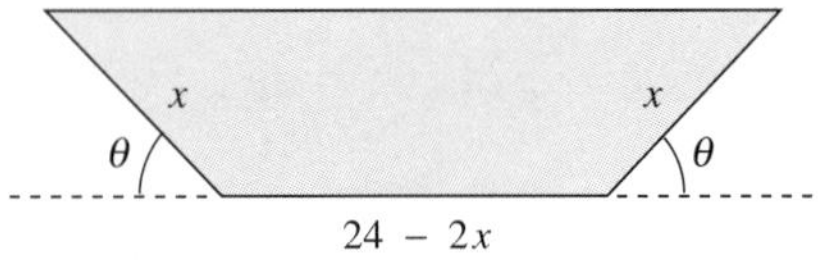

FIGURE 15.65

In Problems 33–36 show that the given function has an absolute extremum but that Theorem 15.13 is not applicable.

33. $f(x, y) = 16 - x^{2/3} - y^{2/3}$

34. $f(x, y) = 1 - x^4y^2$

35. $f(x, y) = 5x^2 + y^4 - 8$

36. $f(x, y) = \sqrt{x^2 + y^2}$

In Problems 37–40 find the absolute extrema of the given continuous function over the closed region R defined by $x^2 + y^2 \le 1$.

37. $f(x, y) = x + \sqrt{3}y$ 38. $f(x, y) = xy$

39. $f(x, y) = x^2 + xy + y^2$

40. $f(x, y) = -x^2 - 3y^2 + 4y + 1$

41. Find the absolute extrema of $f(x, y) = 4x - 6y$ over the closed region R defined by $x^2/4 + y^2 \le 1$.

42. Find the absolute extrema of $f(x, y) = xy - 2x - y + 6$ over the closed triangular region R with vertices (0, 0), (0, 8), and (4, 0).

43. The function $f(x, y) = \sin xy$ is continuous on the closed rectangular region R defined by $0 \le x \le \pi$, $0 \le y \le 1$.

(a) Find the critical points in the region.

(b) Find the points where f has an absolute extremum.

(c) Graph the function on the rectangular region.

[O] 15.10 METHOD OF LEAST SQUARES

When performing experiments, we often tabulate data in the form of ordered pairs (x_1, y_1) $(x_2, y_2), \ldots, (x_n, y_n)$, with each x_i distinct. Given the data, it is then often desirable to be able to extrapolate or predict y from x by finding a mathematical model—that is, a function that approximates or "fits" the data. In other words, we want a function $f(x)$ such that

$$f(x_1) \approx y_1, f(x_2) \approx y_2, \ldots, f(x_n) \approx y_n$$

Naturally, we do not want just any function but a function that fits the data as closely as possible.

In the discussion that follows we shall confine our attention to the problem of finding a linear polynomial $f(x) = mx + b$ or a straight line that "best fits" the dàta $(x_1, y_1), (x_2, y_2), \ldots, (x_n, y_n)$. The procedure for finding this linear function is known as **the method of least squares.**

EXAMPLE 1 Consider the data (1, 1), (2, 3), (3, 4), (4, 6), (5, 5) shown in Figure 15.66(a). Looking at Figure 15.66(b) and seeing that the line $y = x + 1$ passes through two of the data points, we might take this line as the one that best fits the data.

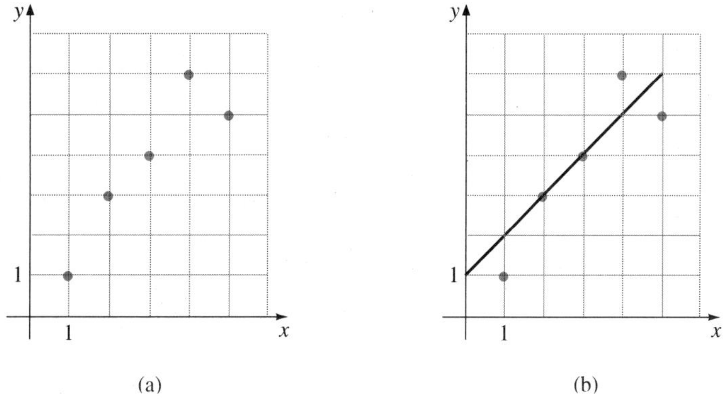

FIGURE 15.66 □

Obviously we need something better than a visual guess to determine the linear function $y = f(x)$ as in Example 1. We need a criterion that defines the concept of "best fit" or, as it is sometimes called, the "goodness of fit."

If we try to match the data points with the function $f(x) = mx + b$, then we wish to find m and b that satisfy the system of equations

$$\begin{aligned} y_1 &= mx_1 + b \\ y_2 &= mx_2 + b \\ &\vdots \\ y_n &= mx_n + b \end{aligned} \tag{1}$$

Unfortunately, (1) is an *overdetermined* system; that is, the number of equations is greater than the number of unknowns. We do not expect such a system to have a solution unless, of course, the data points all lie on the same line.

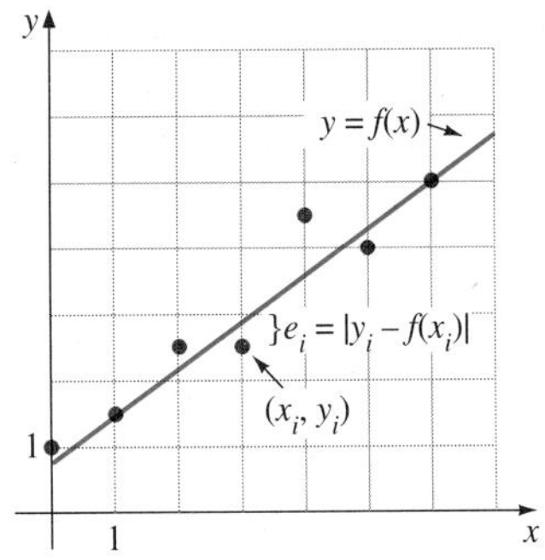

FIGURE 15.67

Least-Squares Line If the data points are $(x_1, y_1), (x_2, y_2), \ldots, (x_n, y_n)$, then one manner of determining how well the linear function $f(x) = mx + b$ fits the data is to measure the vertical distances between the points and the graph of f:

$$e_i = |y_i - f(x_i)|, \qquad i = 1, 2, \ldots, n$$

We can think of each e_i as the error in approximating the data value y_i by the functional value $f(x_i)$. See Figure 15.67. Intuitively, the function f will fit the data well if the sum of all the e_i is a minimum. Actually, a more convenient approach to the problem is to find a linear function f so that the *sum of the squares* of all the e_i is a minimum. We shall define the solution of the system (1) to be those coefficients m and b that minimize the expression

$$\begin{aligned} E &= e_1^2 + e_2^2 + \cdots + e_n^2 \\ &= [y_1 - f(x_1)]^2 + [y_2 - f(x_2)]^2 + \cdots + [y_n - f(x_n)]^2 \\ &= [y_1 - (mx_1 + b)]^2 + [y_2 - (mx_2 + b)]^2 + \cdots + [y_n - (mx_n + b)]^2 \end{aligned}$$

or $$E = \sum_{i=1}^{n} [y_i - mx_i - b]^2 \tag{2}$$

The expression E is called the **sum of the square errors**. The line $y = mx + b$ that minimizes the sum of the square errors (2) is defined to be the **line of best fit** and is called the **least-squares line** or **regression line** for the data $(x_1, y_1), (x_2, y_2), \ldots, (x_n, y_n)$.

The problem remains now, how does one find m and b so that (2) is a minimum? The answer can be found from Theorem 15.13.

If we think of (2) as a function of two variables m and b, then to find the minimum value of E we set the first partial derivatives equal to zero:

$$\frac{\partial E}{\partial m} = 0 \quad \text{and} \quad \frac{\partial E}{\partial b} = 0$$

The last two conditions yield in turn

$$\begin{aligned} -2 \sum_{i=1}^{n} x_i[y_i - mx_i - b] &= 0 \\ -2 \sum_{i=1}^{n} [y_i - mx_i - b] &= 0 \end{aligned} \tag{3}$$

Expanding these sums and using $\Sigma_{i=1}^{n} b = nb$, we find the system (3) is the same as

$$\begin{aligned} \left(\sum_{i=1}^{n} x_i^2\right)m + \left(\sum_{i=1}^{n} x_i\right)b &= \sum_{i=1}^{n} x_i y_i \\ \left(\sum_{i=1}^{n} x_i\right)m + nb &= \sum_{i=1}^{n} y_i \end{aligned} \tag{4}$$

Although we shall not give the details, the values of m and b that satisfy the system (4) yield the minimum value of E. Solving the system (4) gives

$$m = \frac{n \sum_{i=1}^{n} x_i y_i - \sum_{i=1}^{n} x_i \sum_{i=1}^{n} y_i}{n \sum_{i=1}^{n} x_i^2 - \left(\sum_{i=1}^{n} x_i\right)^2}$$

$$b = \frac{\sum_{i=1}^{n} x_i^2 \sum_{i=1}^{n} y_i - \sum_{i=1}^{n} x_i y_i \sum_{i=1}^{n} x_i}{n \sum_{i=1}^{n} x_i^2 - \left(\sum_{i=1}^{n} x_i\right)^2} \qquad (5)$$

EXAMPLE 2 Find the least-squares line for the data in Example 1. Calculate the sum of the square errors E for this line and the line $y = x + 1$.

Solution From the data (1, 1), (2, 3), (3, 4), (4, 6), (5, 5) we identify $x_1 = 1$, $x_2 = 2$, $x_3 = 3$, $x_4 = 4$, $x_5 = 5$, $y_1 = 1$, $y_2 = 3$, $y_3 = 4$, $y_4 = 6$, and $y_5 = 5$. With these values and $n = 5$, we have

$$\sum_{i=1}^{5} x_i y_i = 68 \qquad \sum_{i=1}^{5} x_i = 15 \qquad \sum_{i=1}^{5} y_i = 19 \qquad \sum_{i=1}^{5} x_i^2 = 55$$

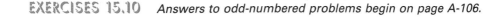

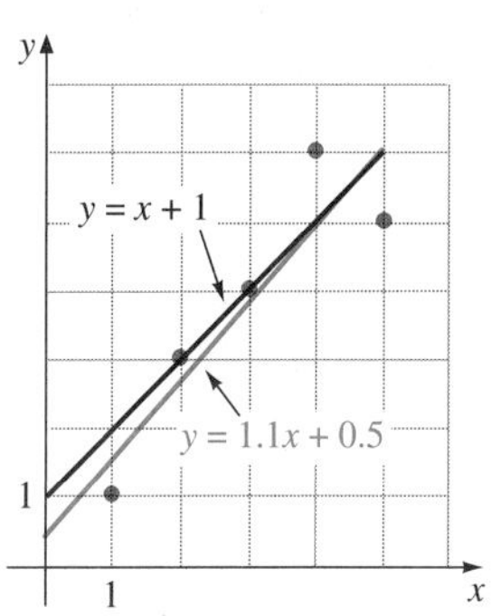

FIGURE 15.68

Substituting these values into the formulas in (5) yields $m = 1.1$ and $b = 0.5$. Thus, the least-squares line is $y = 1.1x + 0.5$. For this line the sum of the square errors is

$$\begin{aligned} E &= [1 - f(1)]^2 + [3 - f(2)]^2 + [4 - f(3)]^2 + [6 - f(4)]^2 + [5 - f(5)]^2 \\ &= [1 - 1.6]^2 + [3 - 2.7]^2 + [4 - 3.8]^2 + [6 - 4.9]^2 + [5 - 6]^2 = 2.7 \end{aligned}$$

For the line $y = x + 1$ that we guessed and that also passed through two of the data points, we find $E = 3.0$.

By way of comparison, Figure 15.68 shows the data, the line $y = x + 1$, and the least-squares line $y = 1.1x + 0.5$. □

▲ ***Remark*** It is possible to generalize the least-squares technique. For example, we might want to fit the given data to a quadratic polynomial $f(x) = ax^2 + bx + c$ instead of a linear polynomial.

EXERCISES 15.10

Answers to odd-numbered problems begin on page A-106.

In Problems 1–6 find the least-squares line for the given data.

1. (2, 1), (3, 2), (4, 3), (5, 2)
2. (0, −1), (1, 3), (2, 5), (3, 7)
3. (1, 1), (2, 1.5), (3, 3), (4, 4.5), (5, 5)
4. (0, 0), (2, 1.5), (3, 3), (4, 4.5), (5, 5)
5. (0, 2), (1, 3), (2, 5), (3, 5), (4, 9), (5, 8), (6, 10)

6. (1, 2), (2, 2.5), (3, 1), (4, 1.5), (5, 2), (6, 3.2), (7, 5)

7. In an experiment the correspondence given in the table was found between temperature T (in °C) and kinematic viscosity v (in Centistokes) of an oil with a certain additive. Find the least-squares line $v = mT + b$. Use this line to estimate the viscosity of the oil at $T = 140$ and $T = 160$.

T	20	40	60	80	100	120
v	220	200	180	170	150	135

8. In an experiment the correspondence given in the table was found between temperature T (in °C) and electrical resistance R (in $M\Omega$). Find the least-squares line $R = mT + b$. Use this line to estimate the resistance at $T = 700$.

T	400	450	500	550	600	650
R	0.47	0.90	2.0	3.7	7.5	15

15.11 LAGRANGE MULTIPLIERS

In Problems 21–28 of Exercises 15.9, you were asked to find the maximum or minimum of a function subject to a given side condition or **constraint**. The side condition was used to eliminate one of the variables in the function so that the second partial derivatives test was applicable. In the present discussion, we examine another procedure for determining the so-called **constrained extrema** of a function.

EXAMPLE 1 Determine geometrically whether $f(x, y) = 9 - x^2 - y^2$ subject to $x + y = 3$ has an extremum.

Solution As seen in Figure 15.69(b), the graph of $x + y = 3$ is a plane that intersects the paraboloid given by $f(x, y) = 9 - x^2 - y^2$. It appears from the

x	y	$f(x, y)$
0.5	2.5	2.5
1	2	4
1.25	1.75	4.375
1.5	1.5	4.5
1.75	1.25	4.375
2	1	4
2.5	0.5	2.5

(a)

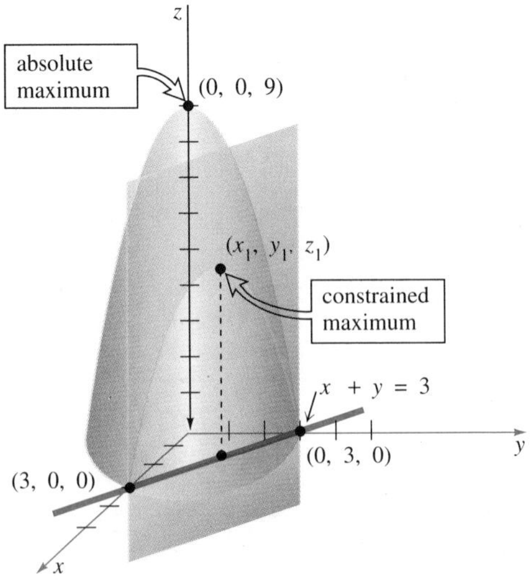

(b)

FIGURE 15.69

figure that the function has a *constrained maximum* for some x_1 and y_1 satisfying $0 < x_1 < 3$, $0 < y_1 < 3$, and $x_1 + y_1 = 3$. The accompanying table would also seem to indicate that this new maximum is $f(1.5, 1.5) = 4.5$. Note that we cannot use numbers such as $x = 1.7$ and $y = 2.4$, since these values do not satisfy the constraint $x + y = 3$. □

Alternatively, we can analyze Example 1 by means of level curves. As shown in Figure 15.70, increasing values of f correspond to increasing c in the level curves $9 - x^2 - y^2 = c$. The maximum value of f (that is, c) subject to the constraint occurs where the level curve $c = \frac{9}{2}$ intersects, or more precisely is tangent to, the line $x + y = 3$. By solving $x^2 + y^2 = \frac{9}{2}$ and $x + y = 3$ simultaneously, we find the point of tangency is $(\frac{3}{2}, \frac{3}{2})$.

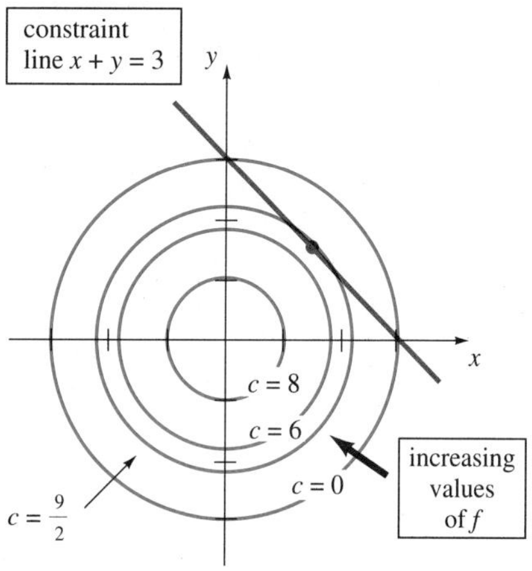

FIGURE 15.70

Functions of Two Variables To generalize the foregoing discussion, suppose we wish to:

find extrema of the function $z = f(x, y)$ subject to a constraint given by $g(x, y) = 0$.

It seems plausible from Figure 15.71 that to find, say, a constrained maximum of f, we need only find the highest level curve $f(x, y) = c$ that is tangent to the

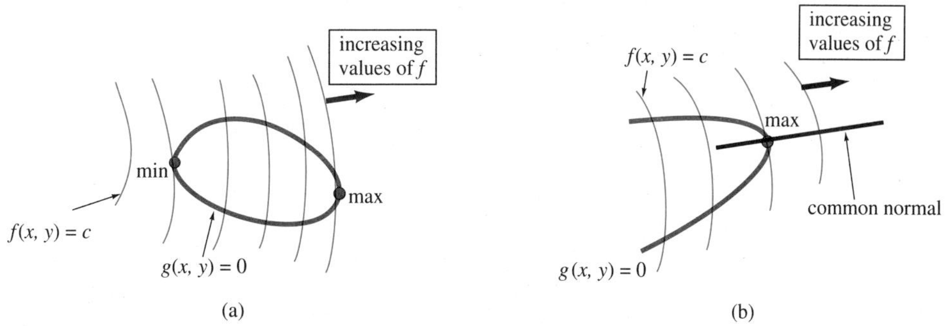

FIGURE 15.71

graph of the constraint equation $g(x, y) = 0$. Now, recall that the gradients ∇f and ∇g are perpendicular to the curves $f(x, y) = c$ and $g(x, y) = 0$, respectively. Hence, if $\nabla g \neq \mathbf{0}$ at a point P of tangency of the curves, then ∇f and ∇g are parallel at P; that is, they lie along a common normal. Therefore, for some nonzero scalar λ (the lowercase Greek letter lambda), we must have $\nabla f = \lambda \nabla g$. We state this result in a formal fashion.

THEOREM 15.14 Lagrange's Theorem

Suppose f has an extremum at a point (x_0, y_0) on the graph of the constraint equation $g(x, y) = 0$. If f and g have continuous first partial derivatives and $\nabla g(x_0, y_0) \neq \mathbf{0}$, then there is a real number λ such that $\nabla f(x_0, y_0) = \lambda \nabla g(x_0, y_0)$.

Method of Lagrange Multipliers The real number λ for which $\nabla f = \lambda \nabla g$, or, after equating components,

$$f_x(x, y) = \lambda g_x(x, y) \qquad f_y(x, y) = \lambda g_y(x, y)$$

is called a **Lagrange multiplier**. If f has a constrained extremum at (x_0, y_0), then we have just seen there is a number λ such that

$$\begin{aligned} f_x(x_0, y_0) &= \lambda g_x(x_0, y_0) \\ f_y(x_0, y_0) &= \lambda g_y(x_0, y_0) \\ g(x_0, y_0) &= 0 \end{aligned} \tag{1}$$

The equations in (1) suggest the following procedure, known as the **method of Lagrange multipliers**, for finding constrained extrema.

To find the extrema of $z = f(x, y)$ subject to the constraint $g(x, y) = 0$, solve the system of equations:

$$\begin{aligned} f_x(x, y) &= \lambda g_x(x, y) \\ f_y(x, y) &= \lambda g_y(x, y) \\ g(x, y) &= 0 \end{aligned} \tag{2}$$

Among the solutions (x, y, λ) of the system will be the points (x_i, y_i), where f has an extremum. When f has a maximum (or minimum), it will be the largest (or smallest) number in the list of functional values $f(x_i, y_i)$.

EXAMPLE 2 Use the method of Lagrange multipliers to find the maximum of $f(x, y) = 9 - x^2 - y^2$ subject to $x + y = 3$.

Solution With $g(x, y) = x + y - 3$ and $f_x = -2x$, $f_y = -2y$, $g_x = 1$, $g_y = 1$, the system in (2) is

$$\begin{aligned} -2x &= \lambda \\ -2y &= \lambda \\ x + y - 3 &= 0 \end{aligned}$$

Equating the first and second equations gives $-2x = -2y$ or $x = y$. Substituting this result into the third equation yields $2y - 3 = 0$ or $y = \frac{3}{2}$. Thus, $x = y = \frac{3}{2}$ and the constrained maximum is $f(\frac{3}{2}, \frac{3}{2}) = \frac{9}{2}$. □

EXAMPLE 3 Find the extrema of $f(x, y) = y^2 - 4x$ subject to $x^2 + y^2 = 9$.

Solution If we define $g(x, y) = x^2 + y^2 - 9$, then $f_x = -4$, $f_y = 2y$, $g_x = 2x$, and $g_y = 2y$. Therefore, (2) becomes

$$\begin{aligned} -4 &= 2x\lambda \\ 2y &= 2y\lambda \\ x^2 + y^2 - 9 &= 0 \end{aligned}$$

From the second of these equations, $y(\lambda - 1) = 0$, we see that either $y = 0$ or $\lambda = 1$. If $y = 0$, the third equation in the system gives $x^2 = 9$ or $x = \pm 3$. Hence, $(-3, 0)$ and $(3, 0)$ are two points at which f might possibly have an extremum. Now if $\lambda = 1$, the first equation yields $x = -2$. Substituting this value into $x^2 + y^2 - 9 = 0$ gives $y^2 = 5$ or $y = \pm\sqrt{5}$. Two more points are $(-2, -\sqrt{5})$ and $(-2, \sqrt{5})$. For the list of functional values

$$f(-3, 0) = 12 \qquad f(3, 0) = -12 \qquad f(-2, -\sqrt{5}) = 13 \qquad f(-2, \sqrt{5}) = 13$$

we conclude that f has a constrained minimum of -12 at $(3, 0)$ and a constrained maximum of 13 at $(-2, -\sqrt{5})$ and at $(-2, \sqrt{5})$.

Figure 15.72 shows the graph of the constraint equation $x^2 + y^2 = 9$ and some of the level curves $y^2 - 4x = c$.

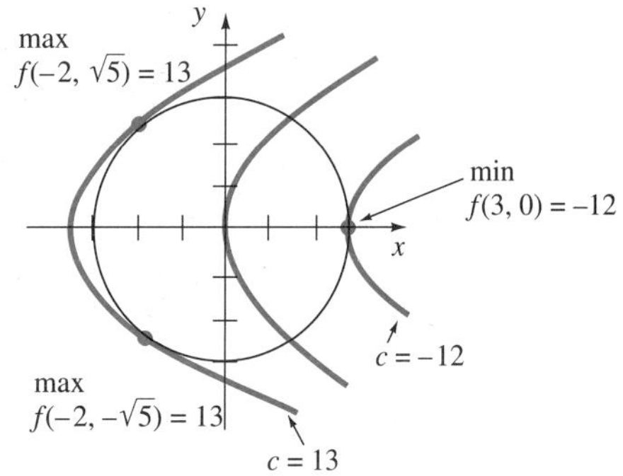

FIGURE 15.72 □

In applying the method of Lagrange multipliers, we really are not very interested in finding the values of λ that satisfy the system (2). Notice in Example 1 that we did not even bother to find λ. In Example 3, we used the value $\lambda = 1$ to help find $x = -2$, but after that we ignored it.

EXAMPLE 4 A closed right circular cylinder will have a volume of 1000 ft^3. The top and bottom of the cylinder are made of metal that costs \$2 per square foot. The lateral side is wrapped in metal costing \$2.50 per square foot. Find the minimum cost of construction.

Solution The cost function is

$$C(r, h) = \underbrace{2(2\pi r^2)}_{\text{cost of bottom and top}} + \underbrace{2.5(2\pi rh)}_{\text{cost of side}}$$

$$= 4\pi r^2 + 5\pi rh$$

Now, from the constraint $1000 = \pi r^2 h$, we can identify $g(r, h) = \pi r^2 h - 1000$, and so $C_r = 8\pi r + 5\pi h$, $C_h = 5\pi r$, $g_r = 2\pi rh$, and $g_h = \pi r^2$. We must then solve the system

$$\begin{aligned} 8\pi r + 5\pi h &= 2\pi rh\lambda \\ 5\pi r &= \pi r^2 \lambda \\ \pi r^2 h - 1000 &= 0 \end{aligned}$$

Multiplying the first equation by r and the second equation by $2h$ and subtracting yield

$$8\pi r^2 - 5\pi rh = 0 \quad \text{or} \quad \pi r(8r - 5h) = 0$$

Since $r = 0$ does not satisfy the constraint equation, we take $r = \frac{5}{8}h$. The constraint then gives

$$h^3 = \frac{1000 \cdot 64}{25\pi} \quad \text{or} \quad h = \frac{40}{\sqrt[3]{25\pi}}$$

Thus, $r = 25/\sqrt[3]{25\pi}$. The constrained minimum cost is

$$C\left(\frac{25}{\sqrt[3]{25\pi}}, \frac{40}{\sqrt[3]{25\pi}}\right) = 4\pi\left(\frac{25}{\sqrt[3]{25\pi}}\right)^2 + 5\pi\left(\frac{25}{\sqrt[3]{25\pi}}\right)\left(\frac{40}{\sqrt[3]{25\pi}}\right)$$

$$= 300\sqrt[3]{25\pi} \approx \$1284.75$$

□

Functions of Three Variables To find the extrema of a function of three variables $w = F(x, y, z)$ subject to the constraint $g(x, y, z) = 0$, we solve a system of four equations:

$$\begin{aligned} F_x(x, y, z) &= \lambda g_x(x, y, z) \\ F_y(x, y, z) &= \lambda g_y(x, y, z) \\ F_z(x, y, z) &= \lambda g_z(x, y, z) \\ g(x, y, z) &= 0 \end{aligned} \tag{3}$$

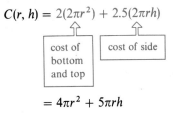

EXAMPLE 5 Find the extrema of $F(x, y, z) = x^2 + y^2 + z^2$ subject to $2x - 2y - z = 5$.

Solution With $g(x, y, z) = 2x - 2y - z - 5$, the system (3) is

$$\begin{aligned} 2x &= 2\lambda \\ 2y &= -2\lambda \\ 2z &= -\lambda \\ 2x - 2y - z - 5 &= 0 \end{aligned}$$

With $\lambda = x = -y = -2z$, the last equation gives $x = \frac{10}{9}$ and so $y = -\frac{10}{9}$, $z = -\frac{5}{9}$. Thus, a constrained extremum is

$$F\left(\frac{10}{9}, -\frac{10}{9}, -\frac{5}{9}\right) = \frac{225}{81}$$ □

Two Constraints In order to optimize a function $w = F(x, y, z)$ subject to *two* constraints, $g(x, y, z) = 0$ and $h(x, y, z) = 0$, we must introduce a second Lagrange multiplier μ (the lowercase Greek letter mu) and solve the system

$$\begin{aligned} F_x(x, y, z) &= \lambda g_x(x, y, z) + \mu h_x(x, y, z) \\ F_y(x, y, z) &= \lambda g_y(x, y, z) + \mu h_y(x, y, z) \\ F_z(x, y, z) &= \lambda g_z(x, y, z) + \mu h_z(x, y, z) \\ g(x, y, z) &= 0 \\ h(x, y, z) &= 0 \end{aligned} \tag{4}$$

EXAMPLE 6 Find the point on the curve C of intersection of the sphere $x^2 + y^2 + z^2 = 9$ and the plane $x - y + 3z = 6$ that is farthest from the xy-plane. Find the point on C that is closest to the xy-plane.

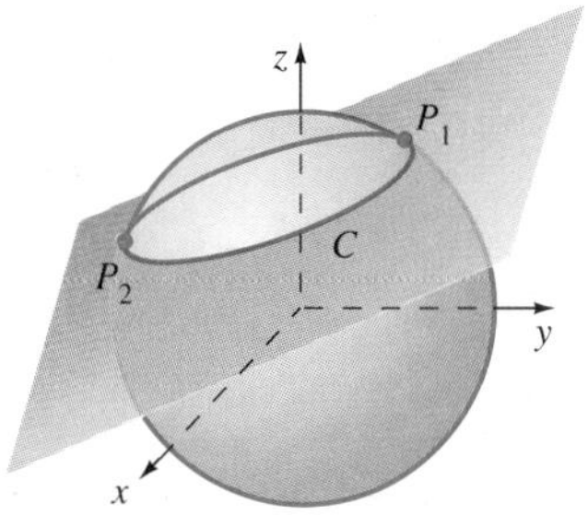

FIGURE 15.73

Solution Figure 15.73 suggests that there does exist two such points P_1 and P_2 with nonnegative z-coordinates. Thus, the distance from these points to the xy-plane is simply $F(x, y, z) = z$. If we take $g(x, y, z) = x^2 + y^2 + z^2 - 9$ and $h(x, y, z) = x - y + 3z - 6$, then the system (4) is

$$\begin{aligned} 0 &= 2x\lambda + \mu \\ 0 &= 2y\lambda - \mu \\ 1 &= 2z\lambda + 3\mu \\ x^2 + y^2 + z^2 - 9 &= 0 \\ x - y + 3z - 6 &= 0 \end{aligned}$$

Adding the first and second equations gives $2\lambda(y + x) = 0$. If $\lambda = 0$, then the first equation implies $\mu = 0$, but the third equation leads to a contradiction. By taking $y = -x$, the last two equations become

$$\begin{aligned} x^2 + x^2 + z^2 - 9 &= 0 \\ x + x + 3z - 6 &= 0 \end{aligned} \quad \text{or} \quad \begin{aligned} 2x^2 + z^2 &= 9 \\ 2x + 3z &= 6 \end{aligned}$$

Solving this system simultaneously leads to

$$x = \frac{6}{11} + \frac{9}{22}\sqrt{14} \qquad z = \frac{18}{11} - \frac{3}{11}\sqrt{14}$$

and

$$x = \frac{6}{11} - \frac{9}{22}\sqrt{14} \qquad z = \frac{18}{11} + \frac{3}{11}\sqrt{14}$$

Thus, the points on C that are farthest and closest to the xy-plane are, respectively,

$$P_1\left(\frac{6}{11} - \frac{9}{22}\sqrt{14}, -\frac{6}{11} + \frac{9}{22}\sqrt{14}, \frac{18}{11} + \frac{3}{11}\sqrt{14}\right) \text{ and } P_2\left(\frac{6}{11} + \frac{9}{22}\sqrt{14}, -\frac{6}{11} - \frac{9}{22}\sqrt{14}, \frac{18}{11} - \frac{3}{11}\sqrt{14}\right)$$

The approximate coordinates of P_1 and P_2 are $(-0.99, 0.99, 2.66)$ and $(2.08, -2.08, 0.62)$. □

▲ ***Remark*** Notice the use of the word *extremum* in Example 5. The method of Lagrange multipliers does not have a built-in indicator that flashes $\boxed{\text{MAX}}$ or $\boxed{\text{MIN}}$ when a single extremum is found. In addition to the graphical procedure discussed at the beginning of this section, another way of convincing oneself as to the nature of the extremum is to compare it with values obtained by calculating the given function at other points that satisfy the constraint equation. Indeed, in this manner we find that $\frac{225}{81}$ of Example 5 is actually a constrained *minimum* of the function.

EXERCISES 15.11 *Answers to odd-numbered problems begin on page A-106.*

In Problems 1 and 2 sketch the graphs of the level curves of the given function f and the indicated constraint equation. Determine whether f has a constrained extremum.

1. $f(x, y) = x + 3y$, subject to $x^2 + y^2 = 1$
2. $f(x, y) = xy$, subject to $\frac{1}{2}x + y = 1$, $x \ge 0$, $y \ge 0$

In Problems 3–20 use the method of Lagrange multipliers to find the constrained extrema of the given function.

3. Problem 1
4. Problem 2
5. $f(x, y) = xy$, subject to $x^2 + y^2 = 2$
6. $f(x, y) = x^2 + y^2$, subject to $2x + y = 5$
7. $f(x, y) = 3x^2 + 3y^2 + 5$, subject to $x - y = 1$
8. $f(x, y) = 4x^2 + 2y^2 + 10$, subject to $4x^2 + y^2 = 4$
9. $f(x, y) = x^2 + y^2$, subject to $x^4 + y^4 = 1$
10. $f(x, y) = 8x^2 - 8xy + 2y^2$, subject to $x^2 + y^2 = 10$
11. $f(x, y) = x^3y$, subject to $\sqrt{x} + \sqrt{y} = 1$
12. $f(x, y) = xy^2$, subject to $x^2 + y^2 = 27$
13. $F(x, y, z) = x + 2y + z$, subject to $x^2 + y^2 + z^2 = 30$
14. $F(x, y, z) = x^2 + y^2 + z^2$, subject to $x + 2y + 3z = 4$
15. $F(x, y, z) = xyz$, subject to $x^2 + y^2/4 + z^2/9 = 1$, $x > 0$, $y > 0$, $z > 0$
16. $F(x, y, z) = xyz + 5$, subject to $x^3 + y^3 + z^3 = 24$
17. $F(x, y, z) = x^3 + y^3 + z^3$, subject to $x + y + z = 1$, $x > 0$, $y > 0$, $z > 0$
18. $F(x, y, z) = 4x^2y^2z^2$, subject to $x^2 + y^2 + z^2 = 9$, $x > 0$, $y > 0$, $z > 0$
19. $F(x, y, z) = x^2 + y^2 + z^2$, subject to $2x + y + z = 1$, $-x + 2y - 3z = 4$
20. $F(x, y, z) = x^2 + y^2 + z^2$, subject to $4x + z = 7$, $z^2 = x^2 + y^2$
21. Give a geometric interpretation of the extrema in Problem 9.
22. Give a geometric interpretation of the extrema in Problem 14.

23. Give a geometric interpretation of the extremum in Problem 19.

24. Give a geometric interpretation of the extremum in Problem 20.

25. Find the maximum area of a right triangle whose perimeter is 4.

26. Find the dimensions of an open rectangular box with maximum volume if its surface area is 75 cm^3. What are the dimensions if the box is closed?

27. A right cylindrical tank is surmounted by a conical cap as shown in Figure 15.74. The radius of the tank is 3 m and its total surface area is $81\pi\ \text{m}^2$. Find heights x and y so that the volume of the tank is a maximum. [*Hint:* The surface area of the cone is $3\pi\sqrt{9+y^2}$.]

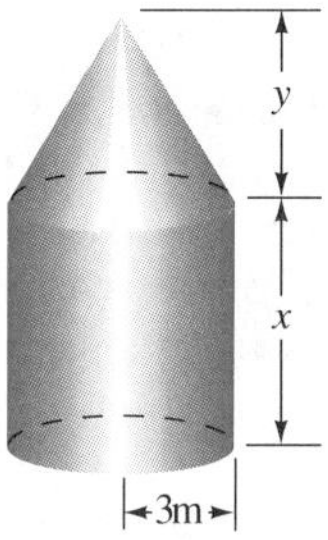

FIGURE 15.74

28. In business a utility index U is a function that gives a measure of satisfaction obtained from the purchasing of variable amounts, x and y, of two commodities that are purchased on a regular basis. If $U(x, y) = x^{1/3}y^{2/3}$ is a utility index, find its extrema subject to $x + 6y = 18$.

29. The Haber–Bosch process* produces ammonia by a direct union of nitrogen and hydrogen under conditions of constant pressure P and constant temperature:

$$N_2 + 3H_2 \overset{\text{catalyst}}{\rightleftharpoons} 2NH_3$$

The partial pressures x, y, and z of hydrogen, nitrogen, and ammonia satisfy $x + y + z = P$ and the equilibrium law $z^2/xy^3 = k$, where k is a constant. The maximum amount of ammonia occurs when the maximum partial pressure of ammonia is obtained. Find the maximum value of z.

30. Find the point $P(x, y)$, $x > 0$, $y > 0$, on the surface $xy^2 = 1$ that is closest to the origin. Show that the line segment from the origin to P is perpendicular to the tangent at P.

31. Find the maximum value of $F(x, y, z) = \sqrt[3]{xyz}$ on the plane $x + y + z = k$.

32. Use the result of Problem 31 to prove the inequality

$$\sqrt[3]{xyz} \le \frac{x+y+z}{3}$$

33. Find the point on the curve C of intersection of the cylinder $x^2 + z^2 = 1$ and the plane $x + y + 2z = 4$ that is farthest from the xz-plane. Find the point on C that is closest to the xz-plane.

34. If a species of animal has n sources of food, the **breadth index** of its ecological niche is defined as

$$\frac{1}{x_1^2 + \cdots + x_n^2}$$

where x_i is the fraction of the animal's diet coming from the ith food source. For example, if a bird's diet consists of 50% insects, 30% worms, and 20% seeds, the breadth index is

$$\frac{1}{(0.50)^2 + (0.30)^2 + (0.20)^2} = \frac{1}{0.25 + 0.09 + 0.04}$$
$$= \frac{1}{0.38} \approx 2.63$$

Note that $x_1 + x_2 + \cdots + x_n = 1$ and $0 \le x_i \le 1$ for all i.

(a) For a species with three food sources, show that the breadth index is maximized if $x_1 = x_2 = x_3 = 1/3$.

(b) Show that the breadth index with n sources is maximized when $x_1 = x_2 = \cdots = x_n = 1/n$.

In order to work the next problem you should work Problem 18 in Exercises 6.8.

35. (*This problem could present a challenge.*) In Problem 18 of Exercises 6.8 we saw that a typical single-stage rocket with mass ratio 0.01 and structural factor 0.2 could not attain a velocity increment great enough to insert the rocket into a circular orbit 100 mi above the earth's surface. To accomplish this insertion, a **multistage rocket** is needed. Suppose $M_0 = M_1 + M_2 + \cdots + M_n$ is the total mass of an n-stage rocket, where each M_i, $i = 1, 2, \ldots, n$, is the total mass of the ith

*Fritz Haber (1868–1934) was a German chemist. For inventing this process, Haber won the Nobel Prize in chemistry in 1918. Carl Bosch (1874–1940) was Haber's brother-in-law and a chemical engineer who made this process practical on a large scale. Bosch won the Nobel Prize in chemistry in 1931. During World War I the German government used the Haber–Bosch process to produce large quantities of fertilizers and explosives. Haber was subsequently expelled from Germany by Adolf Hitler.

stage. We shall assume throughout this problem that, subject to a certain constraint, M_0 has a minimum value for each $n \geq 2$. These minimum values can be found by the method of Lagrange multipliers. We shall also make a simplifying assumption at the outset that each stage has the same structural factor S and the same exhaust speed γ.

(a) If P is the constant mass of the payload of the n-stage rocket, then the velocity increment of the ith stage can be shown to be

$$\Delta v_i = -\gamma \ln\left[1 - \frac{(1-S)M_i}{M_i + M_{i+1} + \cdots + M_n + P}\right]$$

Show that Δv_i can be written as

$$\Delta v_i = \gamma \ln N_i$$

where $$N_i = \frac{M_i + M_{i+1} + \cdots + M_n + P}{SM_i + M_{i+1} + \cdots + M_n + P}$$

(b) It is difficult to minimize M_0 directly. However, since P is constant, minimizing M_0 is equivalent to minimizing $(M_0 + P)/P$ and, in turn, equivalent to minimizing $\ln[(M_0 + P)/P]$. In order to obtain an expression for the latter function, verify that

$$\frac{M_i + M_{i+1} + \cdots + M_n + P}{M_{i+1} + \cdots + M_n + P} = \frac{(1-S)N_i}{1 - SN_i}$$

and that

$$\frac{M_0 + P}{P} = \frac{(1-S)N_1}{1 - SN_1} \cdot \frac{(1-S)N_2}{1 - SN_2} \cdots \frac{(1-S)N_n}{1 - SN_n}$$

(c) Show that

$$\ln\left[\frac{M_0 + P}{P}\right] = \sum_{i=1}^{n} [\ln N_i + \ln(1 - S) - \ln(1 - SN_i)]$$

(d) If the final velocity of the nth stage of the rocket is prescribed to be the constant v_f, then the constraint in this minimization problem is

$$\sum_{i=1}^{n} \Delta v_i = v_f \quad \text{or} \quad \gamma \sum_{i=1}^{n} \ln N_i - v_f = 0$$

Use the method of Lagrange multipliers to show that the values of N_i that minimize the function in part (c) subject to the constraint are

$$N_i = \frac{\lambda\gamma - 1}{\lambda\gamma S}, \qquad i = 1, 2, \ldots, n$$

(e) Show that $v_f = n\gamma \ln\left[\dfrac{\lambda\gamma - 1}{\lambda\gamma S}\right]$.

(f) Use the equation in part (e) to solve for λ. Use this result to show that the common value of N_i is given by $N_i = e^{v_f/n\gamma}$.

(g) Show that the minimum total mass is

$$M_0 = P\left[\frac{(1-S)^n e^{v_f/n\gamma}}{(1 - Se^{v_f/n\gamma})^n} - 1\right]$$

(h) Using part (g) and the values $S = 0.2$, $v_f = 17{,}500$ mi/h, and $\gamma = 6000$ mi/h, compute the minimum total mass M_0 for $n = 2, 3, 4, 5$.

(i) Rockets with four and five stages are rejected due to increased complexity and cost. Determine how much greater the minimum total mass of a two-stage rocket is compared to that of a three-stage rocket. The three-stage rocket is chosen as the best compromise between cost and mass ratio considerations.

FIGURE 15.75 Multistage rockets are needed to insert a payload into an earth orbit. Photo courtesy of NASA.

CHAPTER 15 REVIEW EXERCISES

Answers begin on page A-107.

In Problems 1–10 answer true or false.

1. If $\lim_{(x,y)\to(a,b)} f(x, y)$ has the same value for an infinite number of approaches to (a, b), then the limit exists. ______

2. The domains of the functions

$$f(x, y) = \sqrt{\ln(x^2 + y^2 - 16)}$$

and $$g(x, y) = \ln(x^2 + y^2 - 16)$$

are the same. ______

3. $$f(x, y) = \begin{cases} \dfrac{1 - \cos(x^2 + y^2)}{x^2 + y^2}, & (x, y) \neq (0, 0) \\ 0, & (x, y) = (0, 0) \end{cases}$$

is continuous at $(0, 0)$. ______

4. The function $f(x, y) = x^2 + 2xy + y^3$ is continuous everywhere. ________

5. If $\partial z/\partial x = 0$, then $z =$ constant. ________

6. If $\nabla f = 0$, then $f =$ constant. ________

7. ∇z is perpendicular to the graph of $z = f(x, y)$. ________

8. ∇f points in the direction in which f increases most rapidly. ________

9. If f has continuous second partial derivatives, then $f_{xy} = f_{yx}$. ________

10. If $f_x(x, y) = 0$ and $f_y(x, y) = 0$ at (a, b), then $f(a, b)$ is a relative extremum. ________

In Problems 11–22 fill in the blanks.

11. $\displaystyle\lim_{(x,y)\to(1,1)} \frac{3x^2 + xy^2 - 3xy - 2y^3}{5x^2 - y^2} =$ ________

12. $f(x, y) = \dfrac{xy^2 + 1}{x - y + 1}$ is continuous except for points ________.

13. For $f(x, y) = 3x^2 + y^2$ the level curve that passes through $(2, -4)$ is ________.

14. If $p = g(\eta, \xi)$, $q = h(\eta, \xi)$, then $\dfrac{\partial}{\partial \xi} T(p, q) =$ ________.

15. If $r = g(w)$, $s = h(w)$, then $\dfrac{d}{dw} F(r, s) =$ ________.

16. If s is the distance that a body falls in time t, then the acceleration of gravity g can be obtained from $g = 2s/t^2$. Small errors Δs and Δt in the measurements of s and t, respectively, will result in an approximate error in g of ________.

17. $\dfrac{\partial^4 f}{\partial x\, \partial z\, \partial y^2}$ in subscript notation is ________.

18. f_{xyy} in ∂ notation is ________.

19. If $f(x, y) = \displaystyle\int_x^y F(t)\, dt$, then $\dfrac{\partial f}{\partial x} =$ ________.

20. At (x_0, y_0, z_0) the function $F(x, y, z) = x + y + z$ increases most rapidly in the direction of ________.

21. If $F(x, y, z) = f(x, y)g(y)h(z)$, then $F_{xyz} =$ ________.

22. If $z = f(x, y)$ has continuous partial derivatives of any order, list all possible fourth-order partial derivatives. ________.

In Problems 23–30 compute the indicated derivative.

23. $z = ye^{-x^3y}$; z_y

24. $z = \ln(\cos(uv))$; z_u

25. $f(r, \theta) = \sqrt{r^3 + \theta^2}$; $f_{r\theta}$

26. $f(x, y) = (2x + xy^2)^2$; $\dfrac{\partial^2 f}{\partial x^2}$

27. $z = \cosh(x^2y^3)$; $\dfrac{\partial^2 z}{\partial y^2}$

28. $z = (e^{x^2} + e^{-y^2})^2$; $\dfrac{\partial^3 z}{\partial x^2\, \partial y}$

29. $F(s, t, v) = s^3t^5v^{-4}$; F_{stv}

30. $w = \dfrac{xy}{z} + \dfrac{xz}{y} + \dfrac{yz}{x}$; $\dfrac{\partial^4 w}{\partial x\, \partial y^2\, \partial z}$

In Problems 31 and 32 find the gradient of the given function at the indicated point.

31. $f(x, y) = \tan^{-1}\dfrac{y}{x}$; $(1, -1)$

32. $F(x, y, z) = \dfrac{x^2 - 3y^3}{z^4}$; $(1, 2, 1)$

In Problems 33 and 34 find the directional derivative of the given function in the indicated direction.

33. $f(x, y) = x^2y - y^2x$; $D_{\mathbf{u}}f$ in the direction of $2\mathbf{i} + 6\mathbf{j}$

34. $F(x, y, z) = \ln(x^2 + y^2 + z^2)$; $D_{\mathbf{u}}F$ in the direction of $-2\mathbf{i} + \mathbf{j} + 2\mathbf{k}$.

In Problems 35 and 36 sketch the domain of the given function.

35. $f(x, y) = \sqrt{1 - (x + y)^2}$

36. $f(x, y) = \dfrac{1}{\ln(y - x)}$

In Problems 37 and 38 find Δz for the given function.

37. $z = 2xy - y^2$

38. $z = x^2 - 4y^2 + 7x - 9y + 10$

In Problems 39 and 40 find the total differential of the given function.

39. $z = \dfrac{x - 2y}{4x + 3y}$

40. $A = 2xy + 2yz + 2zx$

In Problems 41 and 42 determine whether the given equation is exact. If exact, solve.

41. $2x \cos y^3\, dx = (3x^2y^2 \sin y^3 + 1)\, dy$

42. $(3x^2 + 2y^3)\, dx + y^2(6x + 1)\, dy = 0$

43. Find symmetric equations of the tangent line at $(-\sqrt{5}, 1, 3)$ to the trace of $z = \sqrt{x^2 + 4y^2}$ in the plane $x = -\sqrt{5}$.

44. Find the slope of the tangent line at (2, 3, 10) to the curve of intersection of the surface $z = xy + x^2$ and the vertical plane that passes through $P(2, 3)$ and $Q(4, 5)$ in the direction of Q.

45. Consider the function $f(x, y) = x^2y^4$. At (1, 1) what is:

(a) the rate of change of f in the direction of $\mathbf{i}$?

(b) the rate of change of f in the direction of $\mathbf{i} - \mathbf{j}$?

(c) the rate of change of f in the direction of $\mathbf{j}$?

46. Let $w = \sqrt{x^2 + y^2 + z^2}$.

(a) If $x = 3 \sin 2t$, $y = 4 \cos 2t$, $z = 5t^3$, find $\dfrac{dw}{dt}$.

(b) If $x = 3 \sin 2\dfrac{t}{r}$, $y = 4 \cos 2\dfrac{r}{t}$, $z = 5t^3r^3$, find $\dfrac{\partial w}{\partial t}$.

47. Find an equation of the tangent plane to the graph of $z = \sin xy$ at $\left(\dfrac{1}{2}, \dfrac{2\pi}{3}, \dfrac{\sqrt{3}}{2}\right)$.

48. Determine whether there are any points on the surface $z^2 + xy - 2x - y^2 = 1$ at which the tangent plane is parallel to $z = 2$.

49. Find an equation of the tangent plane to the cylinder $x^2 + y^2 = 25$ at (3, 4, 6).

50. At what point is the directional derivative of $f(x, y) = x^3 + 3xy + y^3 - 3x^2$ in the direction of $\mathbf{i} + \mathbf{j}$ a minimum?

51. Find the dimensions of the rectangular box of greatest volume that is bounded in the first octant by the coordinate planes and the plane $x + 2y + z = 6$. See Figure 15.76.

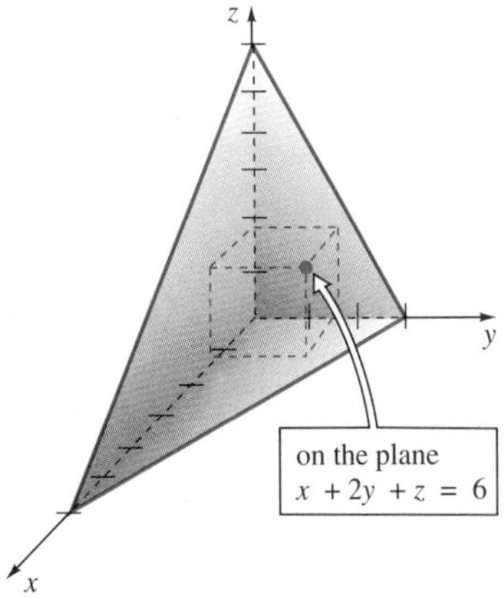

FIGURE 15.76

52. One effect of Einstein's general theory of relativity is that a massive object, such as a galaxy, can act as a "gravitational lens"; that is, if the galaxy is positioned between an observer (on earth) and a light source (such as a quasar), then that light source appears as a ring surrounding the galaxy. If the gravitational lens is much closer to the light source than it is to the observer, then the angular radius θ of the ring (in radians) is related to the mass M of the lens and its distance D from the observer by

$$\theta = \left(\frac{GM}{c^2D}\right)^{1/2}$$

where G is the gravitational constant and c is the speed of light. See Figure 15.77.

(a) Solve for M in terms of θ and D.

(b) Find the total differential of M as a function of θ and D.

(c) If the angular radius θ can be measured with an error no greater than 2% and the distance D to the lens can be estimated with an error no greater than 10%, what is the approximate maximum percentage error in the calculation of the mass M of the lens?

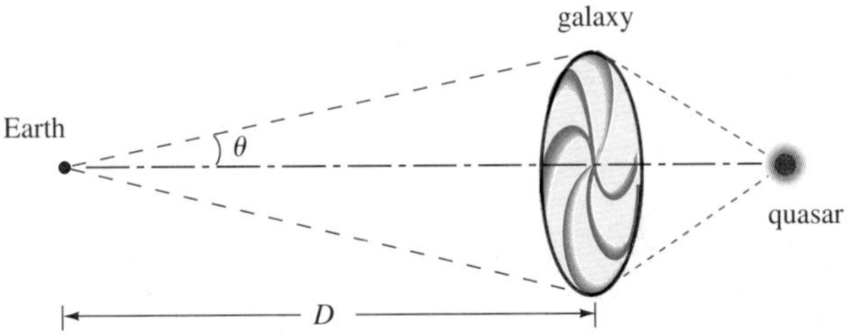

FIGURE 15.77

53. The velocity of the conical pendulum shown in Figure 15.78 is given by $v = r\sqrt{g/y}$, where $g = 980\ \text{cm/s}^2$. If r decreases from 20 cm to 19 cm and y increases from 25 cm to 26 cm, what is the approximate change in the velocity of the pendulum?

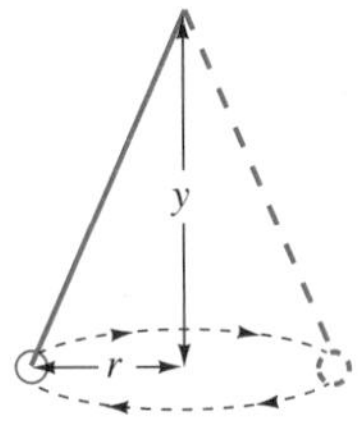

FIGURE 15.78

54. Find the directional derivative of $f(x, y) = x^2 + y^2$ at (3, 4) in the direction of **(a)** $\nabla f(1, -2)$ and **(b)** $\nabla f(3, 4)$.

55. The so-called steady-state temperatures inside a circle of radius R are given by **Poisson's formula***

$$U(r, \theta) = \frac{1}{2\pi}\int_{-\pi}^{\pi} \frac{R^2 - r^2}{R^2 - 2rR\cos(\theta - \phi) + r^2} f(\phi)\, d\phi$$

*Siméon Denis Poisson (1781–1840), a French mathematician and physicist.

By formally differentiating under the integral sign, show that U satisfies

$$r^2U_{rr} + rU_r + U_{\theta\theta} = 0$$

56. The **Cobb–Douglas production function** $z = f(x, y)$ is defined by $z = Ax^{\alpha}y^{\beta}$, where A, α, and β are constants. The value of z is called the efficient output for inputs x and y. Show that

$$f_x = \frac{\alpha z}{x}, \quad f_y = \frac{\beta z}{y}, \quad f_{xx} = \frac{\alpha(\alpha - 1)z}{x^2},$$

$$f_{yy} = \frac{\beta(\beta - 1)z}{y^2}, \quad \text{and} \quad f_{xy} = f_{yx} = \frac{\alpha\beta z}{xy}$$

In Problems 57–60 suppose that $f_x(a, b) = 0$, $f_y(a, b) = 0$. If the given higher-order partial derivatives are evaluated at (a, b), determine, if possible, whether $f(a, b)$ is a relative extremum.

57. $f_{xx} = 4$, $f_{yy} = 6$, $f_{xy} = 5$

58. $f_{xx} = 2$, $f_{yy} = 7$, $f_{xy} = 0$

59. $f_{xx} = -5$, $f_{yy} = -9$, $f_{xy} = 6$

60. $f_{xx} = -2$, $f_{yy} = -8$, $f_{xy} = 4$

61. Express the area A of a right triangle as a function of the length L of its hypotenuse and one of its acute angles θ.

62. In Figure 15.79 express the height h of the mountain as a function of angles θ and ϕ.

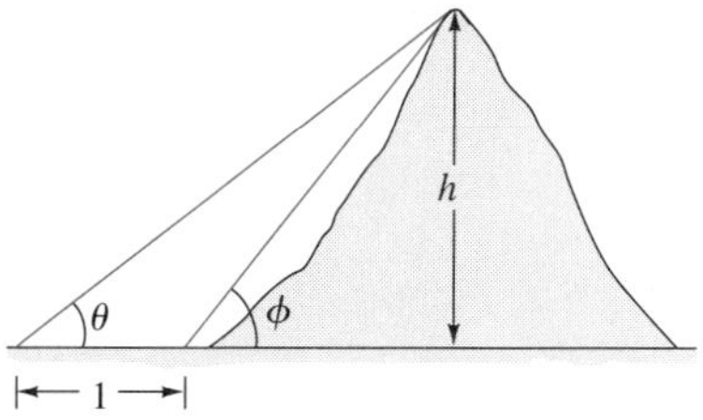

FIGURE 15.79

63. A rectangular walkway shown in Figure 15.80 has a uniform width z. Express the area A of the walkway in terms of x, y, and z.

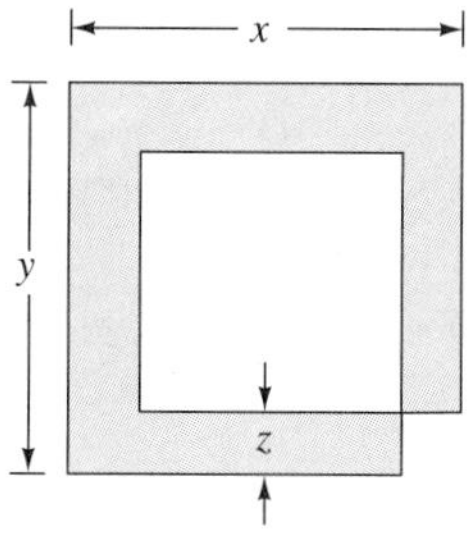

FIGURE 15.80

64. An open box made of plastic has the shape of a rectangular parallelepiped. The outer dimensions of the box are given in Figure 15.81. If the plastic is $\frac{1}{2}$ cm thick, find the approximate volume of the plastic.

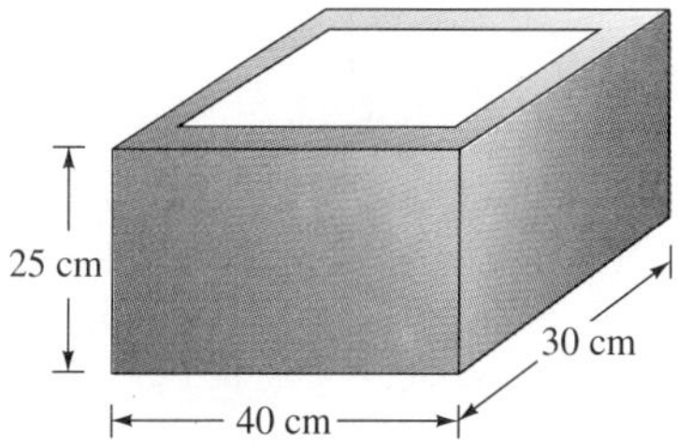

FIGURE 15.81

65. A rectangular box, shown in Figure 15.82, is inscribed in the cone $z = 4 - \sqrt{x^2 + y^2}, 0 \le z \le 4$. Express the volume V of the box in terms of x and y.

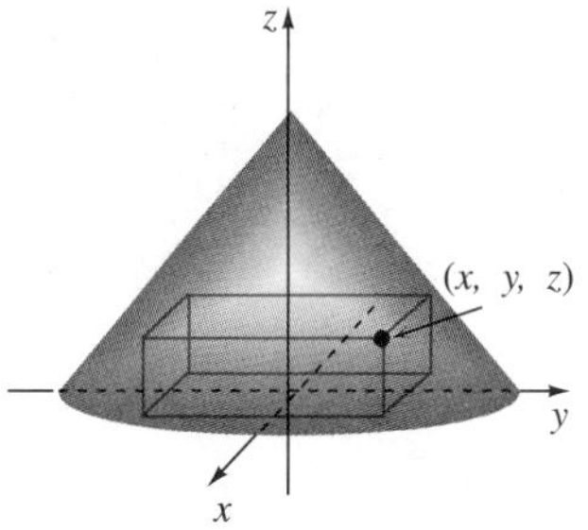

FIGURE 15.82

66. The rectangular box shown in Figure 15.83 has a cover and 12 compartments. The box is made out of heavy plastic that costs 1.5 cents per square inch. Find a function giving the cost C of construction of the box.

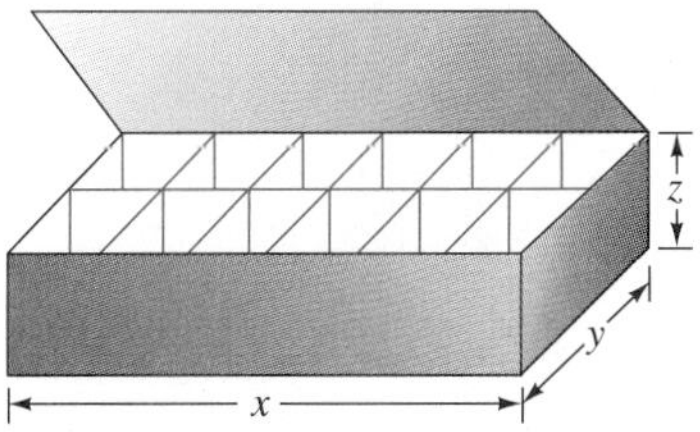

FIGURE 15.83

16

INTEGRAL CALCULUS OF FUNCTIONS OF SEVERAL VARIABLES

INTRODUCTION

We shall conclude our study of the calculus of multivariable functions with the definitions and applications of the two-dimensional and three-dimensional definite integrals. These integrals are more commonly called the **double integral** and the **triple integral**, respectively.

The boundaries of several states, such as North Dakota, South Dakota, and Kansas, are approximately rectangular. Two states, Wyoming and Colorado, have boundaries that are rectangular. In this case, the area of the state is found by multiplying its length times its width. Or is it? Think about this: the boundaries of Wyoming and Colorado lie on the surface of the earth. Hence, the supposedly straight boundaries are actually curves. How does one find the area of a region, such as the state of Colorado, if that region lies on a spherical surface? You will be asked to solve this problem in Exercises 16.6.

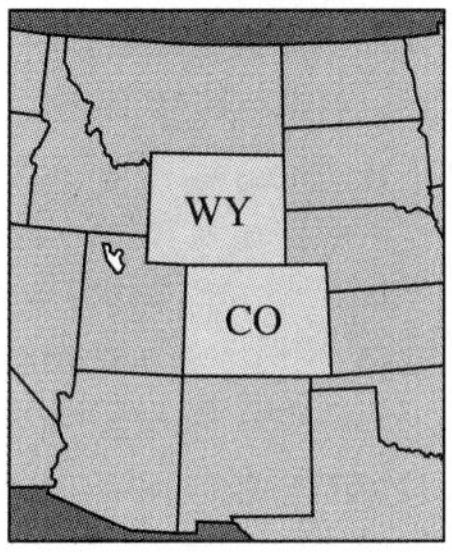

IMPORTANT CONCEPTS

- double integral
- integrability
- Riemann sum
- partial integration
- region of Type I
- region of Type II
- iterated integral
- center of mass
- first moments
- second moments
- double integrals in polar coordinates
- surface area
- triple integral
- cylindrical coordinate system
- triple integrals in cylindrical coordinates
- spherical coordinate system
- triple integrals in spherical coordinates
- one-to-one transformation
- Jacobian
- inverse transformation
- change of variables in multiple integrals

16.1 THE DOUBLE INTEGRAL

Recall from Section 4.5 that the definition of the definite integral of a function of a single variable is given by a limit of a sum:

$$\int_a^b f(x)\,dx = \lim_{\|P\|\to 0} \sum_{k=1}^{n} f(x_k^*)\,\Delta x_k$$

The five steps leading to this definition are outlined in the left-hand column of the following table. The analogous steps, that lead to the concept of a *two-dimensional definite integral* known simply as a **double integral** of a function f of two variables are given in the right-hand column.

$y = f(x)$	$z = f(x, y)$
1. Let f be defined on a closed interval $[a, b]$.	Let f be defined in a closed and bounded region R.
2. Form a partition P of the interval $[a, b]$ into n subintervals of lengths Δx_k.	By means of a grid of vertical and horizontal lines parallel to the coordinate axes, form a partition P of R into n rectangular subregions R_k of areas ΔA_k that lie entirely in R.
3. Let $\|P\|$ be the norm of the partition or the length of the longest subinterval.	Let $\|P\|$ be the norm of the partition or the length of the longest diagonal of the R_k.
4. Choose a number x_k^* in each subinterval $[x_{k-1}, x_k]$.	Choose a point (x_k^*, y_k^*) in each subregion R_k.
5. Form the sum $\sum_{k=1}^{n} f(x_k^*)\,\Delta x_k$.	Form the sum $\sum_{k=1}^{n} f(x_k^*, y_k^*)\,\Delta A_k$.

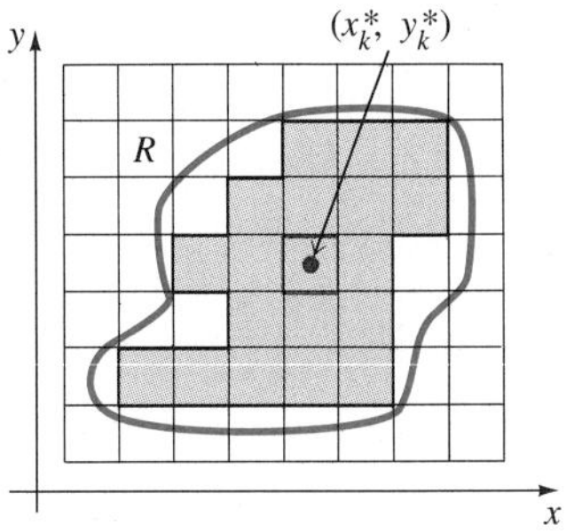

Thus, we have the following definition:

Definition 16.1 The Double Integral

Let f be a function of two variables defined on a closed region R. Then the **double integral of f over R** is given by

$$\iint_R f(x, y)\,dA = \lim_{\|P\|\to 0} \sum_{k=1}^{n} f(x_k^*, y_k^*)\,\Delta A_k \tag{1}$$

Integrability If the limit in (1) exists, we say that f is **integrable** over R and that R is the **region of integration**. When f is continuous on R, f is necessarily integrable over R. For a partition P of R into subregions R_k with (x_k^*, y_k^*) in R_k, a sum of the form $\sum_{k=1}^{n} f(x_k^*, y_k^*)\,\Delta A_k$ is called a **Riemann sum**. The partition described in step 2, where the R_k lie entirely in R, is called an **inner partition** of R. The collection of shaded rectangles in the next two figures illustrate an inner partition.

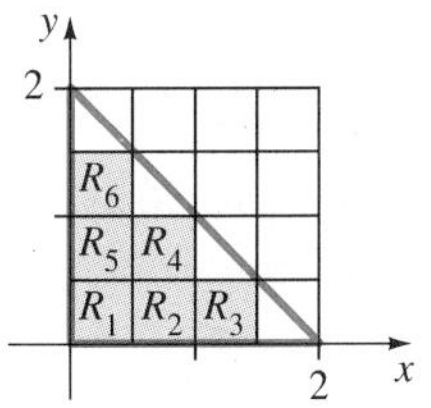

FIGURE 16.1

EXAMPLE 1 Consider the region R in the first quadrant bounded by the graphs of $x + y = 2$, $y = 0$, and $x = 0$. Approximate the double integral $\iint_R (6x + 2y + 3)\,dA$ using a Riemann sum, the R_k shown in Figure 16.1, and (x_k^*, y_k^*), the geometric center of each R_k.

Solution From Figure 16.1 we see that $\Delta A_k = \frac{1}{2}\cdot\frac{1}{2} = \frac{1}{4}$, $k = 1, 2, \ldots, 6$, and the (x_k^*, y_k^*), $k = 1, 2, \ldots, 6$, are in turn $(\frac{1}{4}, \frac{1}{4})$,$(\frac{3}{4}, \frac{1}{4})$, $(\frac{5}{4}, \frac{1}{4})$, $(\frac{3}{4}, \frac{3}{4})$, $(\frac{1}{4}, \frac{3}{4})$, $(\frac{1}{4}, \frac{5}{4})$. Hence, the Riemann sum is

$$\sum_{k=1}^{n} f(x_k^*, y_k^*)\,\Delta A_k = f(\tfrac{1}{4}, \tfrac{1}{4})\tfrac{1}{4} + f(\tfrac{3}{4}, \tfrac{1}{4})\tfrac{1}{4} + f(\tfrac{5}{4}, \tfrac{1}{4})\tfrac{1}{4} + f(\tfrac{3}{4}, \tfrac{3}{4})\tfrac{1}{4} + f(\tfrac{1}{4}, \tfrac{3}{4})\tfrac{1}{4} + f(\tfrac{1}{4}, \tfrac{5}{4})\tfrac{1}{4}$$

$$= \tfrac{5}{4} + \tfrac{8}{4} + \tfrac{11}{4} + \tfrac{9}{4} + \tfrac{6}{4} + \tfrac{7}{4} = 11.5 \qquad \square$$

Area When $f(x, y) = 1$ on R, then $\lim_{\|P\|\to 0} \sum_{k=1}^{n} \Delta A_k$ will simply give the **area** A of the region; that is,

$$A = \iint_R dA \tag{2}$$

Volume If $f(x, y) \ge 0$ on R, then, as shown in Figure 16.2, the product $f(x_k^*, y_k^*)\,\Delta A_k$ can be interpreted as the volume of a rectangular prism of

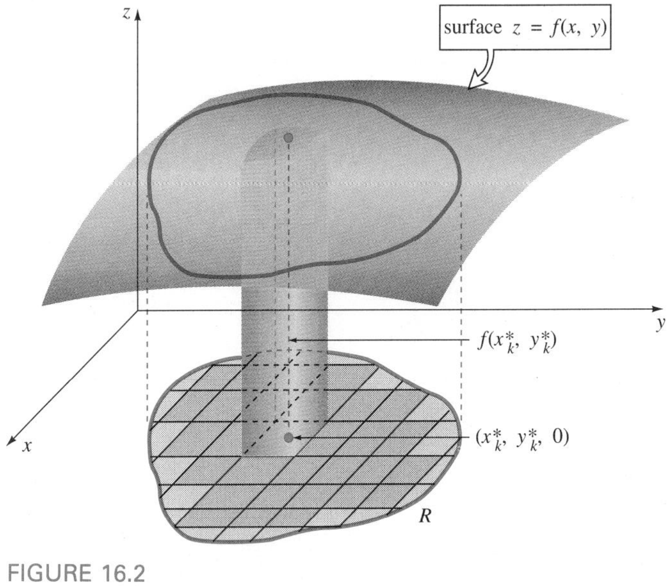

FIGURE 16.2

height $f(x_k^*, y_k^*)$ and base of area ΔA_k. The summation of volumes $\Sigma_{k=1}^n f(x_k^*, y_k^*)\,\Delta A_k$ is an approximation to the **volume** V of the solid *above* the region R and *below* the surface $z = f(x, y)$. The limit of this sum as $\|P\| \to 0$, if it exists, will give the exact volume of this solid; that is, if f is nonnegative on R, then

$$V = \iint_R f(x, y)\, dA \tag{3}$$

Properties The following properties of the double integral are analogous to those of the definite integral given in Theorems 4.10 and 4.11.

THEOREM 16.1 Properties

Let f and g be functions of two variables that are integrable over a region R. Then

(*i*) $\displaystyle\iint_R kf(x, y)\, dA = k\iint_R f(x, y)\, dA,$ where k is any constant

(*ii*) $\displaystyle\iint_R [f(x, y) \pm g(x, y)]\, dA = \iint_R f(x, y)\, dA \pm \iint_R g(x, y)\, dA$

(*iii*) $\displaystyle\iint_R f(x, y)\, dA = \iint_{R_1} f(x, y)\, dA + \iint_{R_2} f(x, y)\, dA,$ where R_1 and R_2 are subregions of R that do not overlap and $R = R_1 \cup R_2$

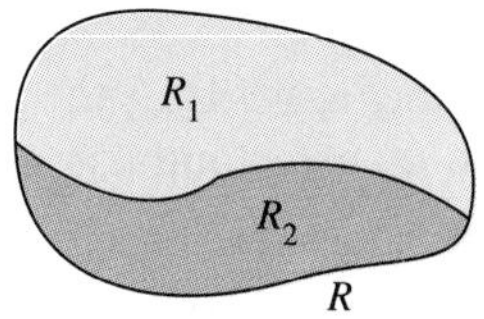

$R = R_1 \cup R_2$

FIGURE 16.3

Part (*iii*) of Theorem 16.1 is the two-dimensional equivalent of $\int_a^b f(x)\,dx = \int_a^c f(x)\,dx + \int_c^b f(x)\,dx$. Figure 16.3 illustrates the division of a region into

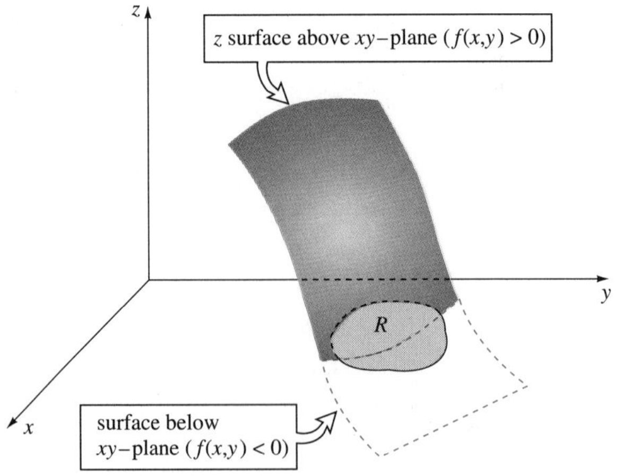

FIGURE 16.4

subregions R_1 and R_2 for which $R = R_1 \cup R_2$. R_1 and R_2 can have no points in common except possibly on their common border. Furthermore, Theorem 16.1(*iii*) extends to any finite number of nonoverlapping subregions whose union is R.

▲ ***Remark*** Of course, not every double integral gives volume. For the surface $z = f(x, y)$ shown in Figure 16.4, $\iint_R f(x, y)\, dA$ is a real number but it is not volume since f is not nonnegative on R.

EXERCISES 16.1

Answers to odd-numbered problems begin on page A-107.

1. Consider the region R in the first quadrant that is bounded by the graphs of $x^2 + y^2 = 16$, $y = 0$, and $x = 0$. Approximate the double integral $\iint_R (x + 3y + 1)\, dA$ using a Riemann sum and the R_k shown in Figure 16.5. Choose (x_k^*, y_k^*) at the geometric center of each R_k.

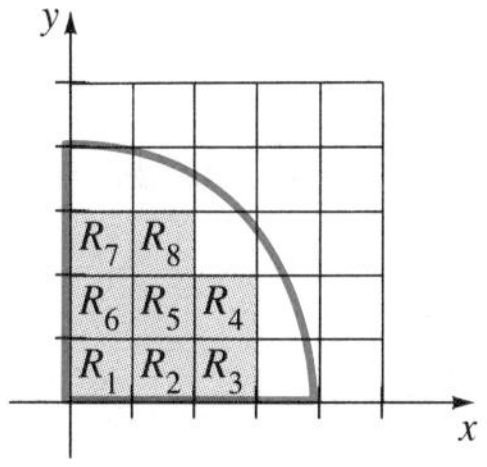

FIGURE 16.5

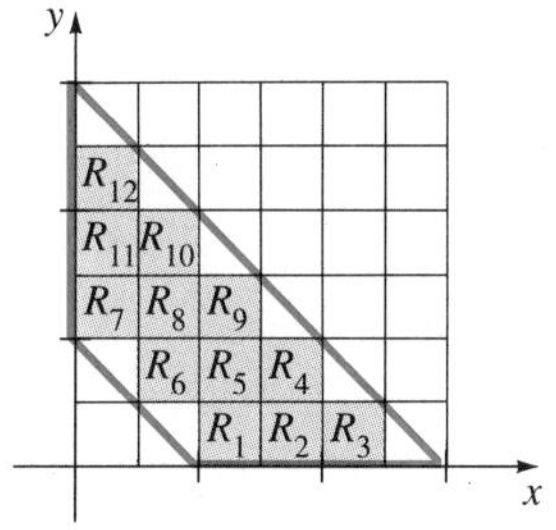

FIGURE 16.6

2. Consider the region R in the first quadrant bounded by the graphs of $x + y = 1$, $x + y = 3$, $y = 0$, $x = 0$. Approximate the double integral $\iint_R (2x + 4y)\, dA$ using a Riemann sum and the R_k shown in Figure 16.6. Choose the (x_k^*, y_k^*) at the upper right-hand corner of each R_k.

3. Consider the rectangular region R shown in Figure 16.7. Approximate the double integral $\iint_R (x + y)\, dA$ using a Riemann sum and the R_k shown in the figure. Choose the (x_k^*, y_k^*) at (**a**) the geometric center of each R_k and (**b**) the upper left-hand corner of each R_k.

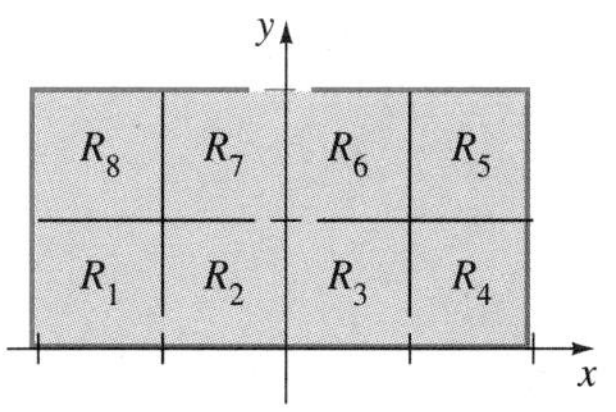

FIGURE 16.7

4. Consider the region R bounded by the graphs of $y = x^2$ and $y = 4$. Place a rectangular grid over R corresponding to the lines $x = -2$, $x = -\frac{3}{2}$, $x = -1, \ldots, x = 2$, and $y = 0$, $y = \frac{1}{2}$, $y = 1, \ldots, y = 4$. Approximate the double integral $\iint_R xy\, dA$ using a Riemann sum, where the (x_k^*, y_k^*) are chosen at the lower right-hand corner of each complete rectangular R_k in R.

In Problems 5–8 evaluate $\iint_R 10\, dA$ over the given region R.

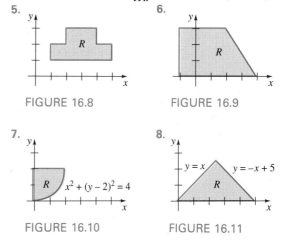

5.

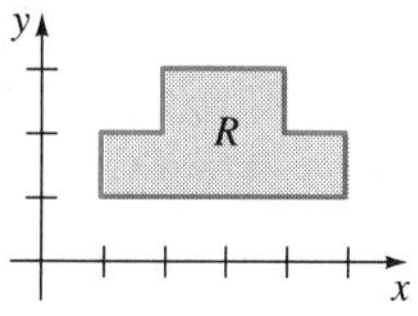

FIGURE 16.8

6.

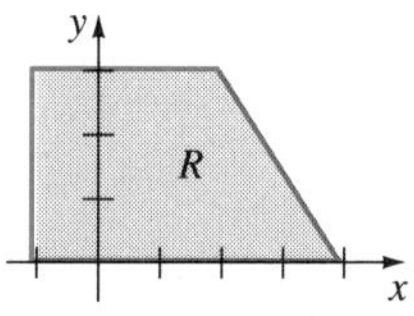

FIGURE 16.9

7.

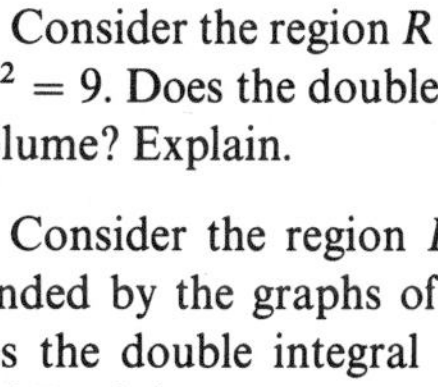

FIGURE 16.10

8.

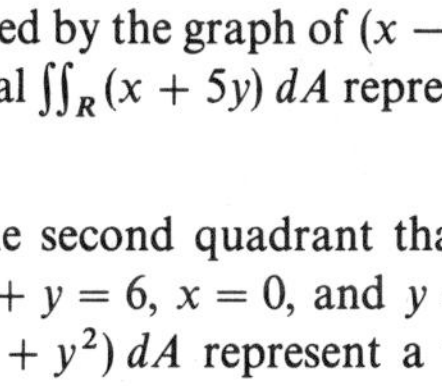

FIGURE 16.11

9. Consider the region R bounded by the graph of $(x - 3)^2 + y^2 = 9$. Does the double integral $\iint_R (x + 5y)\, dA$ represent a volume? Explain.

10. Consider the region R in the second quadrant that is bounded by the graphs of $-2x + y = 6$, $x = 0$, and $y = 0$. Does the double integral $\iint_R (x^2 + y^2)\, dA$ represent a volume? Explain.

In Problems 11–16 suppose that

$$\iint_R x\, dA = 3 \qquad \iint_R y\, dA = 7$$

and the area of R is 8. Evaluate the given double integral.

11. $\iint_R 10\, dA$

12. $\iint_R -5x\, dA$

13. $\iint_R (2x + 4y)\, dA$

14. $\iint_R (x - y)\, dA$

15. $\iint_R (3x + 7y + 1)\, dA$

16. $\iint_R y^2\, dA - \iint_R (2 + y)^2\, dA$

In Problems 17 and 18 let R_1 and R_2 be nonoverlapping regions for which $R = R_1 \cup R_2$.

17. If $\iint_{R_1} f(x, y)\, dA = 4$ and $\iint_{R_2} f(x, y)\, dA = 14$, what is the value of $\iint_R f(x, y)\, dA$?

18. Suppose $\iint_R f(x, y)\, dA = 25$ and $\iint_{R_1} f(x, y)\, dA = 30$. What is the value of $\iint_{R_2} f(x, y)\, dA$?

16.2 ITERATED INTEGRALS

Partial Integration Analogous to the process of partial differentiation we can define **partial integration.**

If $F(x, y)$ is a function such that $F_y(x, y) = f(x, y)$, then the **partial integral of f with respect to y** is

$$\int_{g_1(x)}^{g_2(x)} f(x, y)\, dy = F(x, y)\Big]_{g_1(x)}^{g_2(x)} = F(x, g_2(x)) - F(x, g_1(x))$$

Similarly, if $G(x, y)$ is a function such that $G_x(x, y) = f(x, y)$, then the **partial integral of f with respect to x** is

$$\int_{h_1(y)}^{h_2(y)} f(x, y)\, dx = G(x, y)\Big]_{h_1(y)}^{h_2(y)} = G(h_2(y), y) - G(h_1(y), y)$$

In other words, to evaluate $\int_{g_1(x)}^{g_2(x)} f(x, y)\, dy$, we hold x fixed, whereas in $\int_{h_1(y)}^{h_2(y)} f(x, y)\, dx$ we hold y fixed.

EXAMPLE 1 Evaluate

$$(a)\ \int_1^2 \left(6xy^2 - 4\frac{x}{y}\right) dy \quad \text{and} \quad (b)\ \int_{-1}^3 \left(6xy^2 - 4\frac{x}{y}\right) dx$$

Solution

x fixed

$$(a)\ \int_1^2 \left(6xy^2 - 4\frac{x}{y}\right) dy = \left[2xy^3 - 4x \ln|y|\right]_1^2$$

x fixed

$$= (16x - 4x \ln 2) - (2x - 4x \ln 1)$$
$$= 14x - 4x \ln 2$$

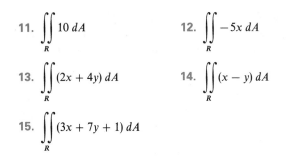

(b) y fixed

$$\int_{-1}^{3}\left(6xy^2 - 4\frac{x}{y}\right)dx = \left[3x^2y^2 - 2\frac{x^2}{y}\right]_{-1}^{3}$$

y fixed

$$= \left(27y^2 - \frac{18}{y}\right) - \left(3y^2 - \frac{2}{y}\right)$$

$$= 24y^2 - \frac{16}{y}$$ □

In Example 1 you should note that

$$\frac{\partial}{\partial y}[2xy^3 - 4x\ln|y|] = 6xy^2 - 4\frac{x}{y} \quad \text{and} \quad \frac{\partial}{\partial x}\left[3x^2y^2 - 2\frac{x^2}{y}\right] = 6xy^2 - 4\frac{x}{y}$$

EXAMPLE 2 Evaluate $\int_{x^2}^{x} \sin xy\, dy$.

Solution By treating x as a constant, we obtain

$$\int_{x^2}^{x} \sin xy\, dy = -\frac{\cos xy}{x}\Bigg]_{x^2}^{x} = -\frac{\cos x^2}{x} + \frac{\cos x^3}{x}$$ □

Regions of Type I and II The region shown in Figure 16.12(a),

$$R: a \le x \le b, \qquad g_1(x) \le y \le g_2(x)$$

where the boundary functions g_1 and g_2 are continuous, is called a **region of Type I**. In Figure 16.12(b), the region

$$R: c \le y \le d, \qquad h_1(y) \le x \le h_2(y)$$

where h_1 and h_2 are continuous, is called a **region of Type II**.

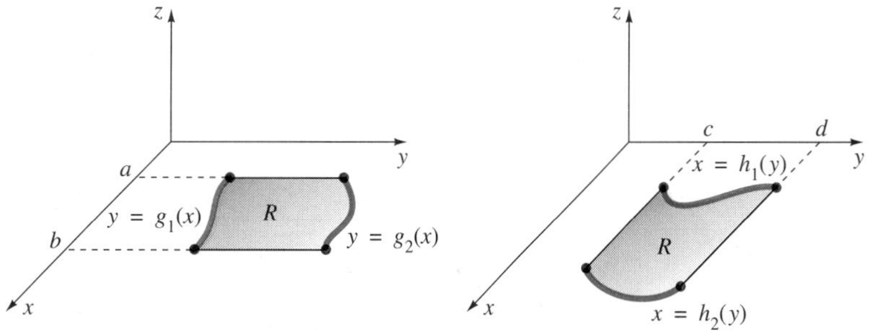

(a) Type I region (b) Type II region

FIGURE 16.12

Iterated Integrals Since the partial integral $\int_{g_1(x)}^{g_2(x)} f(x, y)\, dy$ is a function of x alone, we may in turn integrate the resulting function now with respect to x. If f is continuous on a region of Type I, we define an **iterated integral** of f over the region by

$$\int_a^b \int_{g_1(x)}^{g_2(x)} f(x, y)\, dy\, dx = \int_a^b \left[\int_{g_1(x)}^{g_2(x)} f(x, y)\, dy \right] dx \tag{1}$$

The basic idea in (1) is to carry out *successive integrations*. The partial integral gives a function of x, which is then integrated in the usual manner from $x = a$ to $x = b$. The end result of both integrations will be a real number. In a similar manner, we define an iterated integral of a continuous function f over a region of Type II by

$$\int_c^d \int_{h_1(y)}^{h_2(y)} f(x, y)\, dx\, dy = \int_c^d \left[\int_{h_1(y)}^{h_2(y)} f(x, y)\, dx \right] dy \tag{2}$$

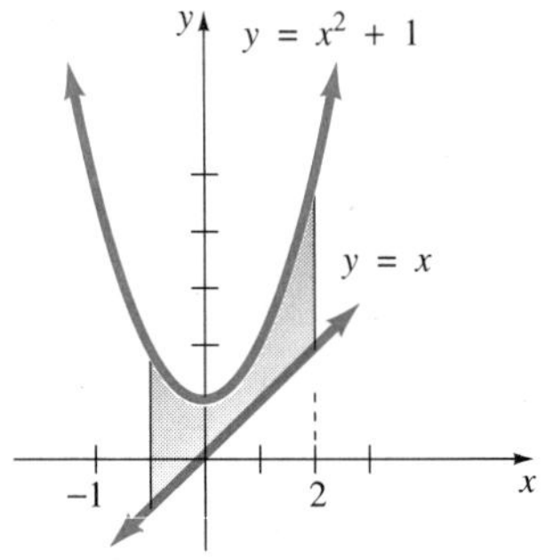

FIGURE 16.13

EXAMPLE 3 Evaluate the iterated integral of $f(x, y) = 2xy$ over the region shown in Figure 16.13.

Solution The region is of Type I and so by (1) we have

$$\int_{-1}^{2} \int_{x}^{x^2+1} 2xy\, dy\, dx = \int_{-1}^{2} \left[\int_{x}^{x^2+1} 2xy\, dy \right] dx = \int_{-1}^{2} \left[xy^2 \right]_x^{x^2+1} dx$$

$$= \int_{-1}^{2} [x(x^2+1)^2 - x^3]\, dx$$

$$= \left[\frac{1}{6}(x^2+1)^3 - \frac{x^4}{4} \right]_{-1}^{2} = \frac{63}{4}$$ □

EXAMPLE 4 Evaluate $\int_0^4 \int_y^{2y} (8x + e^y)\, dx\, dy$.

Solution From (2) we see that

$$\int_0^4 \int_y^{2y} (8x + e^y)\, dx\, dy = \int_0^4 \left[\int_y^{2y} (8x + e^y)\, dx \right] dy = \int_0^4 \left[4x^2 + xe^y \right]_y^{2y} dy$$

$$= \int_0^4 [(16y^2 + 2ye^y) - (4y^2 + ye^y)]\, dy$$

$$= \int_0^4 (12y^2 + ye^y)\, dy \qquad \text{integration by parts}$$

$$= \left[4y^3 + ye^y - e^y \right]_0^4 = 257 + 3e^4 \approx 420.79$$

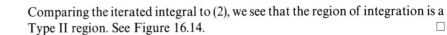

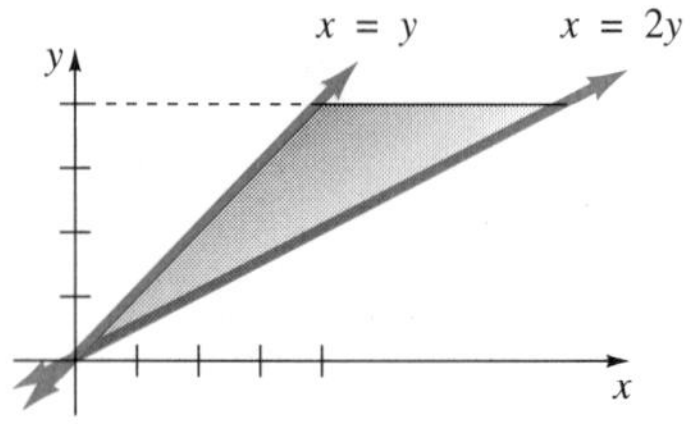

FIGURE 16.14

Comparing the iterated integral to (2), we see that the region of integration is a Type II region. See Figure 16.14. □

EXAMPLE 5 Evaluate $\int_{-1}^{3}\int_{1}^{2}\left(6xy^2 - 4\frac{x}{y}\right) dy\, dx.$

Solution From the result in Example 1(*a*), we have

$$\int_{-1}^{3}\int_{1}^{2}\left(6xy^2 - 4\frac{x}{y}\right) dy\, dx = \int_{-1}^{3}\left[\int_{1}^{2}\left(6xy^2 - 4\frac{x}{y}\right) dy\right] dx$$
$$= \int_{-1}^{3}(14x - 4x\ln 2)\, dx$$
$$= \left[7x^2 - 2x^2\ln 2\right]_{-1}^{3} = 56 - 16\ln 2 \approx 44.91$$

□

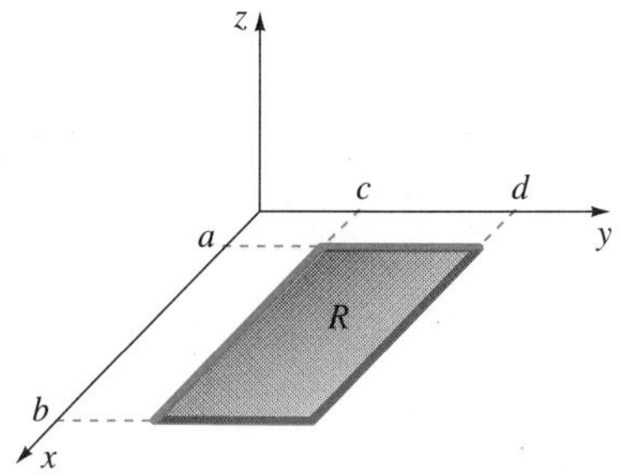

FIGURE 16.15

Inspection of Figure 16.15 should convince you that a rectangular region R defined by $a \le x \le b$, $c \le y \le d$ is simultaneously a Type I and a Type II region. If f is continuous on R, it can be proved that

$$\int_{a}^{b}\int_{c}^{d} f(x, y)\, dy\, dx = \int_{c}^{d}\int_{a}^{b} f(x, y)\, dx\, dy$$

You should verify that

$$\int_{1}^{2}\int_{-1}^{3}\left(6xy^2 - 4\frac{x}{y}\right) dx\, dy$$

gives the same result as the given iterated integral in Example 5.

▲ ***Remark*** A rectangular region is not the only region that can be both Type I and Type II. If f is continuous on a region R that is simultaneously a Type I and a Type II region, the two iterated integrals of f over R are equal. See Problems 39 and 40.

EXERCISES 16.2 *Answers to odd-numbered problems begin on page A-107.*

In Problems 1–8 evaluate the given partial integral.

1. $\int_{-1}^{3}(6xy - 5e^y)\, dx$
2. $\int_{1}^{2}\tan xy\, dy$
3. $\int_{1}^{3x}x^3e^{xy}\, dy$
4. $\int_{\sqrt{y}}^{y^3}(8x^3y - 4xy^2)\, dx$
5. $\int_{0}^{2x}\frac{xy}{x^2+y^2}\, dy$
6. $\int_{x^3}^{x}e^{2y/x}\, dy$
7. $\int_{\tan y}^{\sec y}(2x + \cos y)\, dx$
8. $\int_{\sqrt{y}}^{1}y\ln x\, dx$

In Problems 9–30 evaluate the given iterated integral.

9. $\int_{1}^{2}\int_{-x}^{x^2}(8x - 10y + 2)\, dy\, dx$
10. $\int_{-1}^{1}\int_{0}^{y}(x + y)^2\, dx\, dy$
11. $\int_{0}^{\sqrt{2}}\int_{-\sqrt{2-y^2}}^{\sqrt{2-y^2}}(2x - y)\, dx\, dy$
12. $\int_{0}^{\pi/4}\int_{0}^{\cos x}(1 + 4y\tan^2 x)\, dy\, dx$

13. $\int_0^{\pi}\int_y^{3y} \cos(2x + y)\, dx\, dy$

14. $\int_1^{2}\int_0^{\sqrt{x}} 2y \sin \pi x^2\, dy\, dx$

15. $\int_1^{\ln 3}\int_0^{x} 6e^{x+2y}\, dy\, dx$

16. $\int_0^{1}\int_0^{2y} e^{-y^2}\, dx\, dy$

17. $\int_0^{3}\int_{x+1}^{2x+1} \frac{1}{\sqrt{y-x}}\, dy\, dx$

18. $\int_0^{1}\int_0^{y} x(y^2 - x^2)^{3/2}\, dx\, dy$

19. $\int_1^{9}\int_0^{x} \frac{1}{x^2 + y^2}\, dy\, dx$

20. $\int_0^{1/2}\int_0^{y} \frac{1}{\sqrt{1-x^2}}\, dx\, dy$

21. $\int_1^{e}\int_1^{y} \frac{y}{x}\, dx\, dy$

22. $\int_1^{4}\int_1^{\sqrt{x}} 2ye^{-x}\, dy\, dx$

23. $\int_0^{6}\int_0^{\sqrt{25-y^2}/2} \frac{1}{\sqrt{(25 - y^2) - x^2}}\, dx\, dy$

24. $\int_0^{2}\int_{y^2}^{\sqrt{20-y^2}} y\, dx\, dy$

25. $\int_{\pi/2}^{\pi}\int_{\cos y}^{0} e^x \sin y\, dx\, dy$

26. $\int_0^{1}\int_0^{y^{1/3}} 6x^2 \ln(y + 1)\, dx\, dy$

27. $\int_{\pi}^{2\pi}\int_0^{x} (\cos x - \sin y)\, dy\, dx$

28. $\int_1^{3}\int_0^{1/x} \frac{1}{x+1}\, dy\, dx$

29. $\int_{\pi/12}^{5\pi/12}\int_1^{\sqrt{2 \sin 2\theta}} r\, dr\, d\theta$

30. $\int_0^{\pi/3}\int_{3\cos\theta}^{1+\cos\theta} r\, dr\, d\theta$

In Problems 31–34 sketch the region of integration for the given iterated integral.

31. $\int_0^{2}\int_1^{2x+1} f(x, y)\, dy\, dx$

32. $\int_1^{4}\int_{-\sqrt{y}}^{\sqrt{y}} f(x, y)\, dx\, dy$

33. $\int_{-1}^{3}\int_0^{\sqrt{16-y^2}} f(x, y)\, dx\, dy$

34. $\int_{-1}^{2}\int_{-x^2}^{x^2+1} f(x, y)\, dy\, dx$

In Problems 35–38 verify the given equality.

35. $\int_{-1}^{2}\int_0^{3} x^2\, dy\, dx = \int_0^{3}\int_{-1}^{2} x^2\, dx\, dy$

36. $\int_{-2}^{2}\int_2^{4} (2x + 4y)\, dx\, dy = \int_2^{4}\int_{-2}^{2} (2x + 4y)\, dy\, dx$

37. $\int_1^{3}\int_0^{\pi} (3x^2y - 4 \sin y)\, dy\, dx =$

$$\int_0^{\pi}\int_1^{3} (3x^2y - 4 \sin y)\, dx\, dy$$

38. $\int_0^{1}\int_0^{2} \left(\frac{8y}{x+1} - \frac{2x}{y^2+1}\right) dx\, dy =$

$$\int_0^{2}\int_0^{1} \left(\frac{8y}{x+1} - \frac{2x}{y^2+1}\right) dy\, dx$$

In Problems 39 and 40 the region given in the figure is both Type I and Type II. Verify that the iterated integrals are equal.

39.

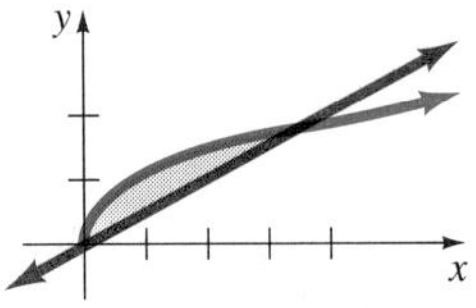

FIGURE 16.16

Type I: $\frac{1}{2}x \le y \le \sqrt{x}, 0 \le x \le 4$

Type II: $y^2 \le x \le 2y, 0 \le y \le 2$

$$\int_0^{4}\int_{x/2}^{\sqrt{x}} x^2y\, dy\, dx = \int_0^{2}\int_{y^2}^{2y} x^2y\, dx\, dy$$

40.

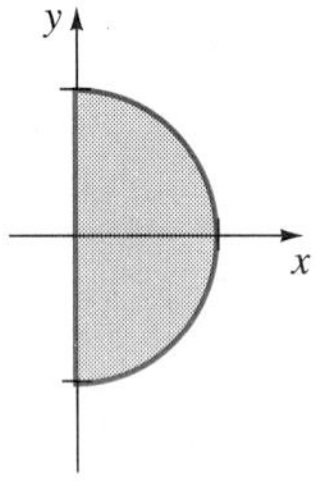

FIGURE 16.17

Type I: $-\sqrt{1-x^2} \le y \le \sqrt{1-x^2}, 0 \le x \le 1$

Type II: $0 \le x \le \sqrt{1-y^2}, -1 \le y \le 1$

$$\int_0^{1}\int_{-\sqrt{1-x^2}}^{\sqrt{1-x^2}} 2x\, dy\, dx = \int_{-1}^{1}\int_0^{\sqrt{1-y^2}} 2x\, dx\, dy$$

41. If f and g are integrable, prove that

$$\int_c^{d}\int_a^{b} f(x)g(y)\, dx\, dy = \left(\int_a^{b} f(x)\, dx\right)\left(\int_c^{d} g(y)\, dy\right)$$

16.3 EVALUATION OF DOUBLE INTEGRALS

The iterated integrals of the preceding section provide the means for evaluating a double integral $\iint_R f(x, y)\, dA$ over a region of Type I or Type II or a region that can be expressed as a union of a finite number of these regions. The following result is due to the Italian mathematician Guido Fubini (1879–1943).

THEOREM 16.2 Fubini's Theorem

Let f be continuous on a region R.

(*i*) If R is of Type I, then

$$\iint_R f(x, y)\, dA = \int_a^b \int_{g_1(x)}^{g_2(x)} f(x, y)\, dy\, dx \tag{1}$$

(*ii*) If R is of Type II, then

$$\iint_R f(x, y)\, dA = \int_c^d \int_{h_1(y)}^{h_2(y)} f(x, y)\, dx\, dy \tag{2}$$

Theorem 16.2 is the double integral analogue of Theorem 4.15, the Fundamental Theorem of Calculus. While Theorem 16.2 is difficult to prove, we can get some intuitive feeling for its significance by considering volumes. Let R be a Type I region and $z = f(x, y)$ be continuous and nonnegative on R. The area A of the vertical plane, as shown in Figure 16.18, is the area under the trace of the surface $z = f(x, y)$ in the plane $x =$ constant and hence is given by the partial integral

$$A(x) = \int_{g_1(x)}^{g_2(x)} f(x, y)\, dy$$

By summing all these areas from $x = a$ to $x = b$, we obtain the volume V of the solid above R and below the surface:

$$V = \int_a^b A(x)\, dx = \int_a^b \int_{g_1(x)}^{g_2(x)} f(x, y)\, dy\, dx$$

But, as we have already seen in (3) of Section 16.1, this volume is also given by the double integral

$$V = \iint_R f(x, y)\, dA$$

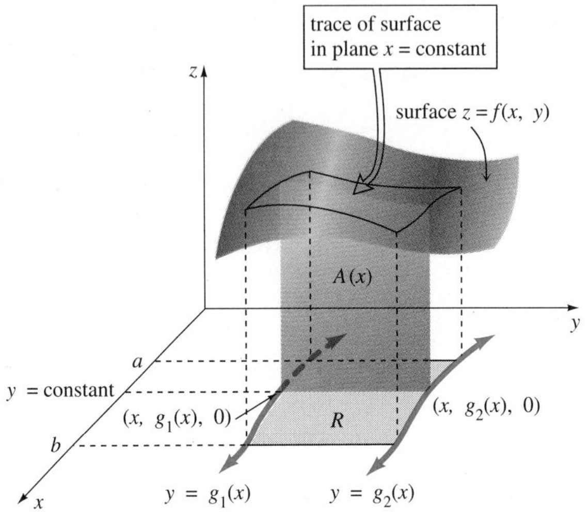

FIGURE 16.18

EXAMPLE 1 Evaluate the double integral $\iint_R e^{x+3y}\,dA$ over the region bounded by the graphs of $y = 1$, $y = 2$, $y = x$, and $y = -x + 5$.

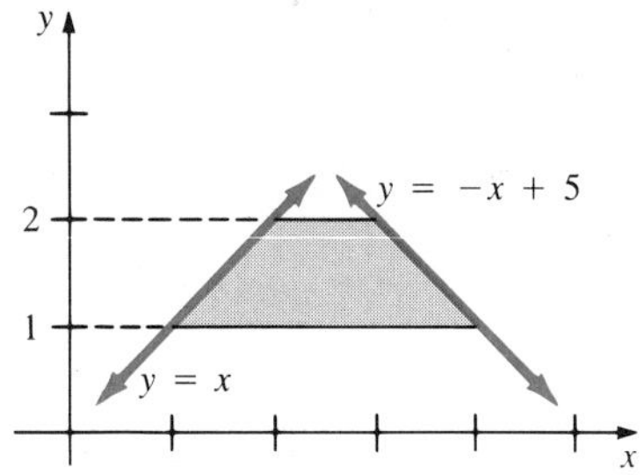

FIGURE 16.19

Solution As seen in Figure 16.19, the region is of Type II; hence, by (2) we integrate first with respect to x from the left boundary $x = y$ to the right boundary $x = 5 - y$:

$$\iint_R e^{x+3y}\,dA = \int_1^2 \int_y^{5-y} e^{x+3y}\,dx\,dy$$

$$= \int_1^2 e^{x+3y}\Big]_y^{5-y}\,dy$$

$$= \int_1^2 (e^{5+2y} - e^{4y})\,dy$$

$$= \left[\frac{1}{2}e^{5+2y} - \frac{1}{4}e^{4y}\right]_1^2$$

$$= \frac{1}{2}e^9 - \frac{1}{4}e^8 - \frac{1}{2}e^7 + \frac{1}{4}e^4 \approx 2771.64 \qquad \square$$

As an aid in reducing a double integral to an iterated integral with correct limits of integration, it is useful to visualize, as suggested in the foregoing discussion, the double integral as a double summation process.* Over a Type I

*Although we shall not pursue the details, the double integral can be defined in terms of a limit of a double sum such as

$$\sum_i \sum_j f(x_i^*, y_j^*)\,\Delta y_j\,\Delta x_i \quad \text{or} \quad \sum_j \sum_i f(x_i^*, y_j^*)\,\Delta x_i\,\Delta y_j$$

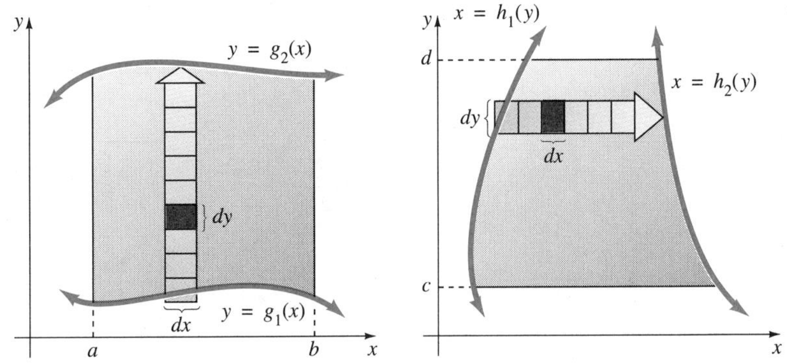

(a) region of Type I (b) region of Type II

FIGURE 16.20

region the iterated integral $\int_a^b \int_{g_1(x)}^{g_2(x)} f(x, y)\, dy\, dx$ is first a summation in the y-direction. Pictorially, this is indicated by the vertical arrow in Figure 16.20(a); the typical rectangle in the arrow has area $dy\, dx$. The dy placed before the dx signifies that the "volumes" $f(x, y)\, dy\, dx$ of prisms built up on the rectangles are summed vertically with respect to y from the lower boundary curve g_1 to the upper boundary curve g_2. The dx following the dy signifies that the result of each vertical summation is then summed horizontally with respect to x from left ($x = a$) to right ($x = b$). Similar remarks hold for double integrals over regions of Type II. See Figure 16.20(b). Recall from (2) of Section 16.1 that when $f(x, y) = 1$, the double integral $A = \iint_R dA$ gives the area of the region. Thus, Figure 16.20(a) shows that $\int_a^b \int_{g_1(x)}^{g_2(x)} dy\, dx$ adds the rectangular *areas* vertically and then horizontally, whereas Figure 16.20(b) shows that $\int_c^d \int_{h_1(y)}^{h_2(y)} dx\, dy$ adds the rectangular areas horizontally and then vertically.

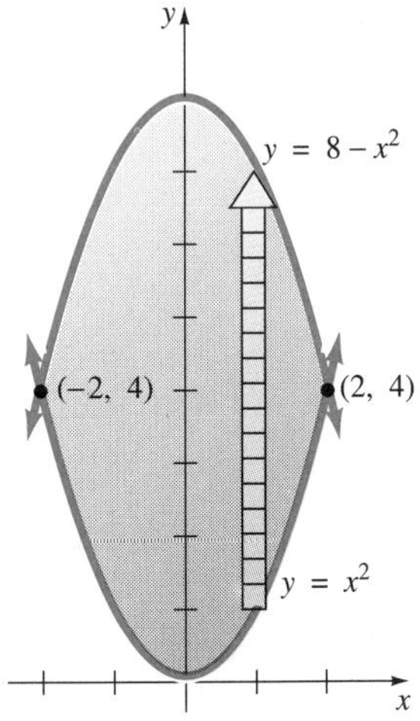

FIGURE 16.21

EXAMPLE 2 Use a double integral to find the area of the region bounded by the graphs of $y = x^2$ and $y = 8 - x^2$.

Solution The graphs and their points of intersection are shown in Figure 16.21. Since R is evidently of Type I, we have from (1)

$$A = \iint_R dA = \int_{2}^{2} \int_{x^2}^{8-x^2} dy\, dx = \int_{-2}^{2} [(8 - x^2) - x^2]\, dx$$

$$= \int_{-2}^{2} (8 - 2x^2)\, dx$$

$$= \left[8x - \frac{2}{3}x^3 \right]_{-2}^{2} = \frac{64}{3} \text{ square units} \qquad \square$$

You should recognize

$$A = \iint_R dA = \int_a^b \int_{g_1(x)}^{g_2(x)} dy\, dx = \int_a^b [g_2(x) - g_1(x)]\, dx$$

as the formula, discussed in Section 5.1, for the area bounded by two graphs.

EXAMPLE 3 Use a double integral to find the volume V of the solid in the first octant that is bounded by the coordinate planes and the graphs of $x^2 + y^2 = 1$ and $z = 3 - x - y$.

Solution From Figure 16.22(a) we see that the volume is given by $V = \iint_R (3 - x - y)\, dA$. Since Figure 16.22(b) shows that R is of Type I, we have from (1),

$$V = \int_0^1 \int_0^{\sqrt{1-x^2}} (3 - x - y)\, dy\, dx = \int_0^1 \left[3y - xy - \frac{y^2}{2}\right]_0^{\sqrt{1-x^2}} dx$$

$$= \int_0^1 \left(3\sqrt{1-x^2} - x\sqrt{1-x^2} - \frac{1}{2} + \frac{1}{2}x^2\right) dx$$

trig substitution

$$= \left[\frac{3}{2}\sin^{-1}x + \frac{3}{2}x\sqrt{1-x^2} + \frac{1}{3}(1-x^2)^{3/2} - \frac{1}{2}x + \frac{1}{6}x^3\right]_0^1$$

$$= \frac{3\pi}{4} - \frac{2}{3} \approx 1.69 \text{ cubic units}$$

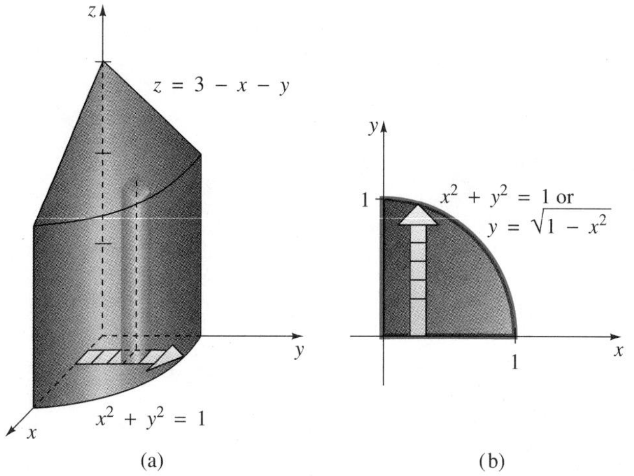

FIGURE 16.22

□

The reduction of a double integral to either of the iterated integrals (1) or (2) depends on (a) the type of region and (b) the function itself. The next two examples illustrate each case.

EXAMPLE 4 Evaluate $\iint_R (x + y)\, dA$ over the region bounded by the graphs of $x = y^2$ and $y = \frac{1}{2}x - \frac{3}{2}$.

Solution The region, which is shown in Figure 16.23(a), can be written as the union $R = R_1 \cup R_2$ of two Type I regions. By solving $y^2 = 2y + 3$ we find that the points of intersection of the two graphs are $(1, -1)$ and $(9, 3)$.

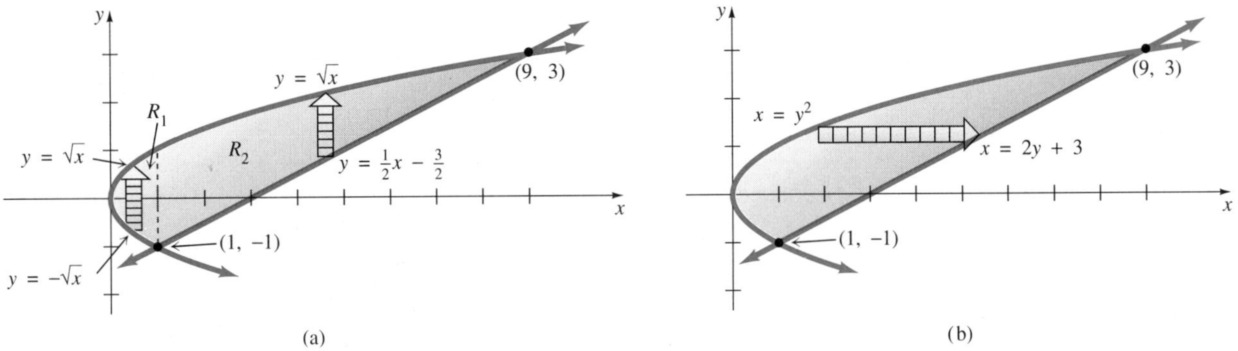

FIGURE 16.23

Thus, from (1) and Theorem 16.1(*iii*), we have

$$\iint_R (x+y)\,dA = \iint_{R_1} (x+y)\,dA + \iint_{R_2} (x+y)\,dA$$

$$= \int_0^1 \int_{-\sqrt{x}}^{\sqrt{x}} (x+y)\,dy\,dx + \int_1^9 \int_{x/2-3/2}^{\sqrt{x}} (x+y)\,dy\,dx$$

$$= \int_0^1 \left[xy + \frac{y^2}{2}\right]_{-\sqrt{x}}^{\sqrt{x}} dx + \int_1^9 \left[xy + \frac{y^2}{2}\right]_{x/2-3/2}^{\sqrt{x}} dx$$

$$= \int_0^1 2x^{3/2}\,dx + \int_1^9 \left(x^{3/2} + \frac{11}{4}x - \frac{5}{8}x^2 - \frac{9}{8}\right) dx$$

$$= \frac{4}{5}x^{5/2}\Big]_0^1 + \left[\frac{2}{5}x^{5/2} + \frac{11}{8}x^2 - \frac{5}{24}x^3 - \frac{9}{8}x\right]_1^9 \approx 46.93$$

Alternative Solution By interpreting the region as a single region of Type II, we see from Figure 16.23(b) that

$$\iint_R (x+y)\,dA = \int_{-1}^3 \int_{y^2}^{2y+3} (x+y)\,dx\,dy$$

$$= \int_{-1}^3 \left[\frac{x^2}{2} + xy\right]_{y^2}^{2y+3} dy$$

$$= \int_{-1}^3 \left(-\frac{y^4}{2} - y^3 + 4y^2 + 9y + \frac{9}{2}\right) dy$$

$$= \left[-\frac{y^5}{10} - \frac{y^4}{4} + \frac{4}{3}y^3 + \frac{9}{2}y^2 + \frac{9}{2}y\right]_{-1}^3 \approx 46.93 \qquad \square$$

Note that the answer in Example 4 does not represent the volume of the solid above R and below the plane $z = x + y$. Why not?

Reversing the Order of Integration As Example 4 illustrates, a problem may become easier when the order of integration is **changed** or **reversed**.

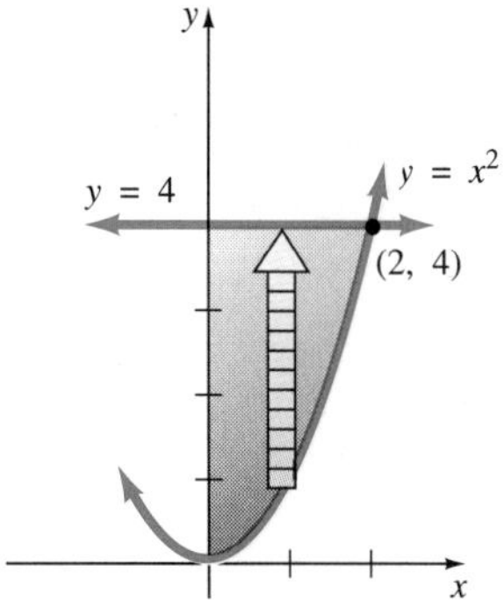

(a) Type I

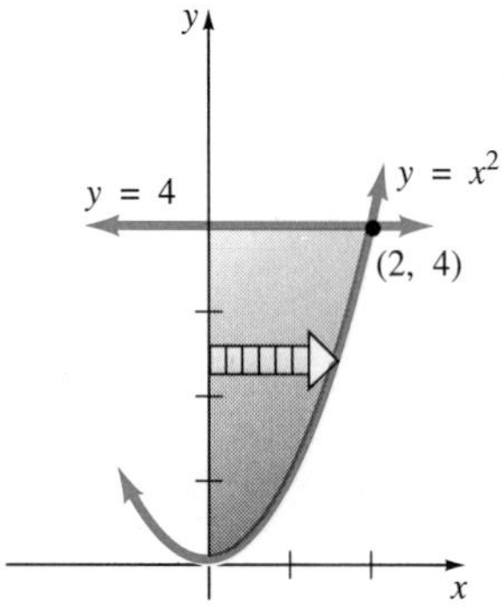

(b) Type II

FIGURE 16.24

Also, some iterated integrals that may be impossible to evaluate using one order of integration can, perhaps, be evaluated using the reverse order of integration.

EXAMPLE 5 Evaluate $\iint_R xe^{y^2}\,dA$ over the region R in the first quadrant bounded by the graphs of $y = x^2$, $x = 0$, $y = 4$.

Solution When viewed as a region of Type I, we have from Figure 16.24(a), $0 \le x \le 2$, $x^2 \le y \le 4$, and so

$$\iint_R xe^{y^2}\,dA = \int_0^2 \int_{x^2}^4 xe^{y^2}\,dy\,dx$$

The difficulty here is that the partial integral $\int_{x^2}^4 xe^{y^2}\,dy$ cannot be evaluated, since e^{y^2} has no elementary antiderivative with respect to y. However, as we see in Figure 16.24(b), we can interpret the same region as a Type II region defined by $0 \le y \le 4$, $0 \le x \le \sqrt{y}$. Hence, from (2),

$$\begin{aligned}\iint_R xe^{y^2}\,dA &= \int_0^4 \int_0^{\sqrt{y}} xe^{y^2}\,dx\,dy \\ &= \int_0^4 \frac{x^2}{2}e^{y^2}\Big]_0^{\sqrt{y}}\,dy \\ &= \int_0^4 \frac{1}{2}ye^{y^2}\,dy \\ &= \frac{1}{4}e^{y^2}\Big]_0^4 = \frac{1}{4}(e^{16} - 1)\end{aligned}$$ □

▲ ***Remarks*** (*i*) You are encouraged to take advantage of symmetries to minimize your work when finding areas and volumes by double integration. In the case of volumes, make sure *both* the region and the surface over the region possess corresponding symmetries. See Problem 19 in Exercises 16.3.

(*ii*) Before attempting to evaluate a double integral, *always* try to sketch an accurate picture of the region R of integration.

EXERCISES 16.3

Answers to odd-numbered problems begin on page A-107.

In Problems 1–10 evaluate the double integral over the region R that is bounded by the graphs of the given equations. Choose the most convenient order of integration.

1. $\iint_R x^3y^2\,dA$, $y = x$, $y = 0$, $x = 1$

2. $\iint_R (x + 1)\,dA$, $y = x$, $x + y = 4$, $x = 0$

3. $\iint_R (2x + 4y + 1)\,dA$, $y = x^2$, $y = x^3$

4. $\iint_R xe^y\, dA$, R the same as in Problem 1

5. $\iint_R 2xy\, dA$, $y = x^3$, $y = 8$, $x = 0$

6. $\iint_R \frac{x}{\sqrt{y}}\, dA$, $y = x^2 + 1$, $y = 3 - x^2$

7. $\iint_R \frac{y}{1 + xy}\, dA$, $y = 0$, $y = 1$, $x = 0$, $x = 1$

8. $\iint_R \sin \frac{\pi x}{y}\, dA$, $x = y^2$, $x = 0$, $y = 1$, $y = 2$

9. $\iint_R \sqrt{x^2 + 1}\, dA$, $x = y$, $x = -y$, $x = \sqrt{3}$

10. $\iint_R x\, dA$, $y = \tan^{-1}x$, $y = 0$, $x = 1$

In Problems 11 and 12 evaluate $\iint_R (x + y)\, dA$, where R is the given region.

11.

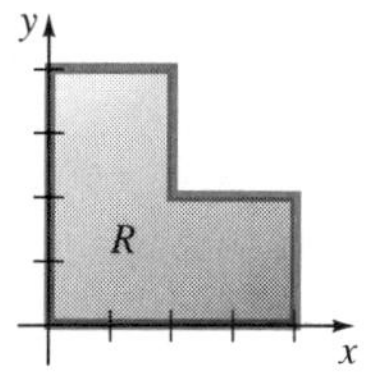

FIGURE 16.25

12.

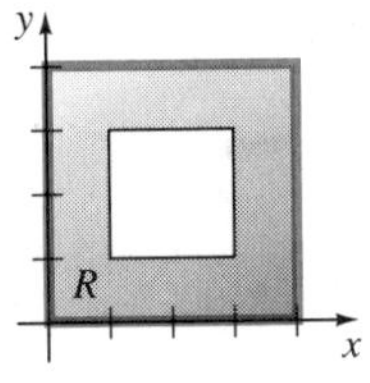

FIGURE 16.26

In Problems 13–18 use a double integral to find the area of the region R that is bounded by the graphs of the given equations.

13. $y = -x$, $y = 2x - x^2$

14. $x = y^2$, $x = 2 - y^2$

15. $y = e^x$, $y = \ln x$, $x = 1$, $x = 4$

16. $\sqrt{x} + \sqrt{y} = 2$, $x + y = 4$

17. $y = -2x + 3$, $y = x^3$, $x = -2$

18. $y = -x^2 + 3x$, $y = -2x + 4$, $y = 0$, $0 \le x \le 2$

19. Consider the solid bounded by the graphs of $x^2 + y^2 = 4$, $z = 4 - y$, and $z = 0$ shown in Figure 16.27. Choose and evaluate the correct integral representing the volume V of the solid.

(a) $4 \int_0^2 \int_0^{\sqrt{4-x^2}} (4 - y)\, dy\, dx$

(b) $2 \int_{-2}^2 \int_0^{\sqrt{4-x^2}} (4 - y)\, dy\, dx$

(c) $2 \int_{-2}^2 \int_0^{\sqrt{4-y^2}} (4 - y)\, dx\, dy$

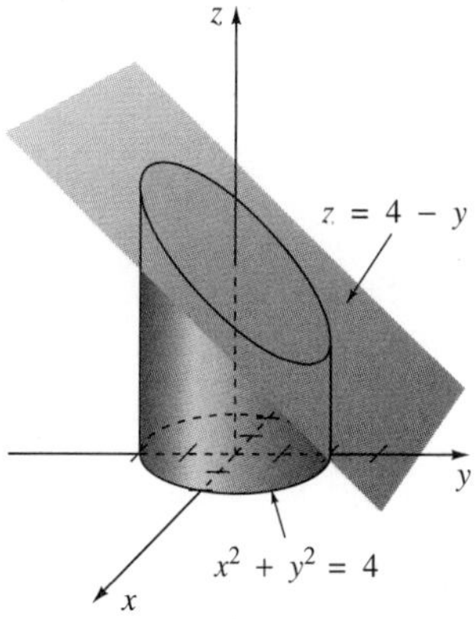

FIGURE 16.27

20. Consider the solid bounded by the graphs of $x^2 + y^2 = 4$ and $y^2 + z^2 = 4$. An eighth of the solid is shown in Figure 16.28. Choose and evaluate the correct integral corresponding the volume V of the solid.

(a) $4 \int_{-2}^2 \int_{-\sqrt{4-x^2}}^{\sqrt{4-x^2}} (4 - y^2)^{1/2}\, dy\, dx$

(b) $8 \int_0^2 \int_0^{\sqrt{4-y^2}} (4 - y^2)^{1/2}\, dx\, dy$

(c) $8 \int_0^2 \int_0^{\sqrt{4-x^2}} (4 - x^2)^{1/2}\, dy\, dx$

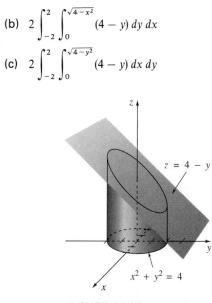

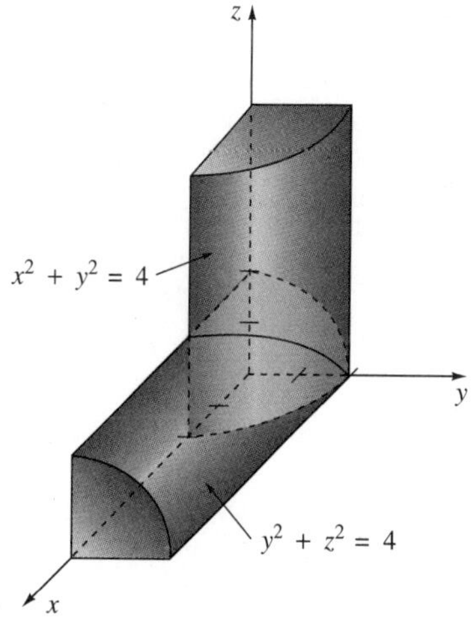

FIGURE 16.28

In Problems 21–30 find the volume of the solid bounded by the graphs of the given equations.

21. $2x + y + z = 6, x = 0, y = 0, z = 0$, first octant

22. $z = 4 - y^2, x = 3, x = 0, y = 0, z = 0$, first octant

23. $x^2 + y^2 = 4, x - y + 2z = 4, x = 0, y = 0, z = 0$, first octant

24. $y = x^2, y + z = 3, z = 0$

25. $z = 1 + x^2 + y^2, 3x + y = 3, x = 0, y = 0, z = 0$, first octant

26. $z = x + y, x^2 + y^2 = 9, x = 0, y = 0, z = 0$, first octant

27. $yz = 6, x = 0, x = 5, y = 1, y = 6, z = 0$

28. $z = 4 - x^2 - \frac{1}{4}y^2, z = 0$

29. $z = 4 - y^2, x^2 + y^2 = 2x, z = 0$

30. $z = 1 - x^2, z = 1 - y^2, x = 0, y = 0, z = 0$, first octant

If $f_2(x, y) \geq f_1(x, y)$ for all (x, y) in a region R, then the volume of the solid bounded by the two surfaces over R is $V = \iint_R [f_2(x, y) - f_1(x, y)]\, dA$. In Problems 31–34 use this result to find the volume bounded by the graphs of the given equations.

31. $x + 2y + z = 4, z = x + y, x = 0, y = 0$, first octant

32. $z = x^2 + y^2, z = 9$

33. $z = x^2, z = -x + 2, x = 0, y = 0, y = 5$, first octant

34. $2z = 4 - x^2 - y^2, z = 2 - y$

In Problems 35–40 reverse the order of integration.

35. $\displaystyle\int_0^2 \int_0^{y^2} f(x, y)\, dx\, dy$

36. $\displaystyle\int_{-5}^5 \int_0^{\sqrt{25-y^2}} f(x, y)\, dx\, dy$

37. $\displaystyle\int_0^3 \int_1^{e^x} f(x, y)\, dy\, dx$

38. $\displaystyle\int_0^2 \int_{y/2}^{3-y} f(x, y)\, dx\, dy$

39. $\displaystyle\int_0^1 \int_0^{\sqrt[3]{x}} f(x, y)\, dy\, dx + \int_1^2 \int_0^{2-x} f(x, y)\, dy\, dx$

40. $\displaystyle\int_0^1 \int_0^{\sqrt{y}} f(x, y)\, dx\, dy + \int_1^2 \int_0^{\sqrt{2-y}} f(x, y)\, dx\, dy$

In Problems 41–46 evaluate the given iterated integral by reversing the order of integration.

41. $\displaystyle\int_0^1 \int_x^1 x^2\sqrt{1 + y^4}\, dy\, dx$

42. $\displaystyle\int_0^1 \int_{2y}^2 e^{-y/x}\, dx\, dy$

43. $\displaystyle\int_0^2 \int_{y^2}^4 \cos\sqrt{x^3}\, dx\, dy$

44. $\displaystyle\int_{-1}^1 \int_{-\sqrt{1-x^2}}^{\sqrt{1-x^2}} x\sqrt{1 - x^2 - y^2}\, dy\, dx$

45. $\displaystyle\int_0^1 \int_x^1 \frac{1}{1 + y^4}\, dy\, dx$

46. $\displaystyle\int_0^4 \int_{\sqrt{y}}^2 \sqrt{x^3 + 1}\, dx\, dy$

47. Let R be a rectangular region bounded by the lines $x = a$, $x = b$, $y = c$, and $y = d$, where $a < b$, $c < d$.

(a) Show that

$$\iint_R \cos 2\pi(x + y)\, dA = \frac{1}{4\pi^2}(S_1S_2 - C_1C_2)$$

$$\iint_R \sin 2\pi(x + y)\, dA = -\frac{1}{4\pi^2}(C_1S_2 + S_1C_2)$$

where

$$S_1 = \sin 2\pi b - \sin 2\pi a \qquad S_2 = \sin 2\pi d - \sin 2\pi c$$

$$C_1 = \cos 2\pi b - \cos 2\pi a \qquad C_2 = \cos 2\pi d - \cos 2\pi c$$

(b) Show that if at least one of the two perpendicular sides of R has an integer length, then

$$\iint_R \cos 2\pi(x + y)\, dA = 0 \quad \text{and}$$

$$\iint_R \sin 2\pi(x + y)\, dA = 0$$

(c) Conversely, show that if

$$\iint_R \cos 2\pi(x + y)\, dA = 0 \quad \text{and}$$

$$\iint_R \sin 2\pi(x + y)\, dA = 0$$

then at least one of the two perpendicular sides of R must have integer length. [*Hint:* Consider $0 = (S_1S_2 - C_1C_2)^2 + (C_1S_2 + S_1C_2)^2$.]

48. Let R be a rectangular region that has been partitioned into n nonoverlapping rectangular subregions $R_1, R_2, \ldots, R_n$ whose sides are all parallel to the horizontal and vertical sides of R. See Figure 16.29. Suppose that each interior rectangle has the property that one of its two perpendicular sides has integer length. Show that R has the same property. [*Hint:* Use Problem 47 and Theorem 16.1(*iii*).]

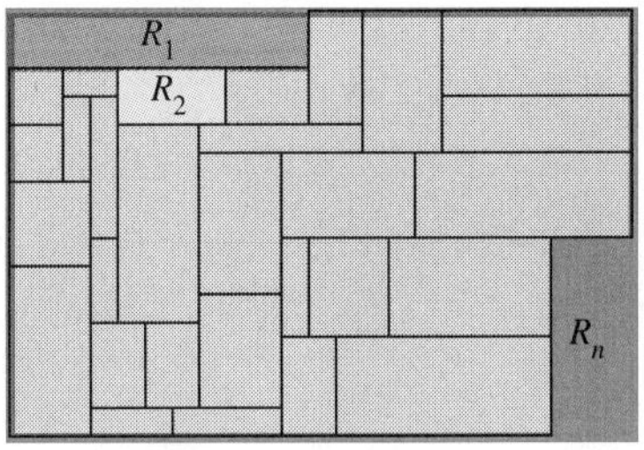

FIGURE 16.29

16.4 CENTER OF MASS AND MOMENTS

Laminas with Variable Density—Center of Mass In Section 5.10 we saw that if ρ is a constant density (mass per unit area), then the mass of the lamina coinciding with a region bounded by the graphs of $y = f(x)$, the x-axis, and the lines $x = a$ and $x = b$ is

$$m = \lim_{\|P\| \to 0} \sum_{k=1}^{n} \rho f(x_k^*)\, \Delta x_k = \int_a^b \rho f(x)\, dx \tag{1}$$

If a lamina corresponding to a region R has a *variable density* $\rho(x, y)$, where ρ is nonnegative and continuous on R, then analogous to (1) we define its mass m by the double integral

$$m = \lim_{\|P\| \to 0} \sum_{k=1}^{n} \rho(x_k^*, y_k^*)\, \Delta A_k = \iint_R \rho(x, y)\, dA \tag{2}$$

As in Section 5.10, we define the coordinates of the **center of mass** of the lamina by

$$\bar{x} = \frac{M_y}{m} \qquad \bar{y} = \frac{M_x}{m} \tag{3}$$

where
$$M_y = \iint_R x\rho(x, y)\, dA \quad \text{and} \quad M_x = \iint_R y\rho(x, y)\, dA \tag{4}$$

are the **moments** of the lamina about the y- and x-axes, respectively. The center of mass is the point where we consider all the mass of the lamina to be concentrated. If $\rho(x, y)$ is a constant, the center of mass is called the **centroid** of the lamina.

EXAMPLE 1 A lamina has the shape of the region in the first quadrant that is bounded by the graphs of $y = \sin x$ and $y = \cos x$ between $x = 0$ and $x = \pi/4$. Find its center of mass if the density is $\rho(x, y) = y$.

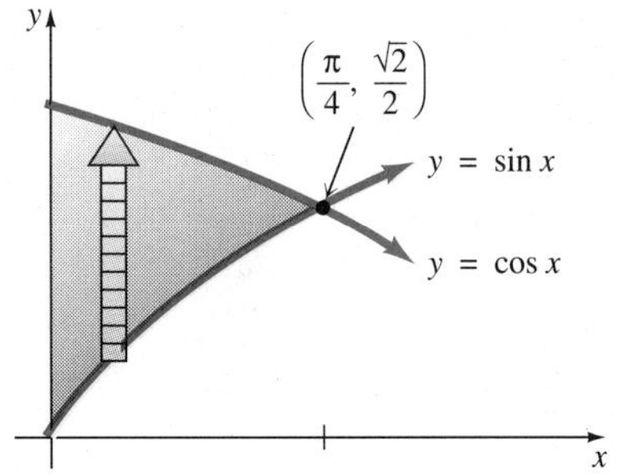

FIGURE 16.30

Solution From Figure 16.30 we see that

$$m = \iint_R y\,dA = \int_0^{\pi/4}\int_{\sin x}^{\cos x} y\,dy\,dx$$
$$= \int_0^{\pi/4} \frac{y^2}{2}\Bigg]_{\sin x}^{\cos x} dx$$
$$= \frac{1}{2}\int_0^{\pi/4} (\cos^2 x - \sin^2 x)\,dx$$

double angle formula

$$= \frac{1}{2}\int_0^{\pi/4} \cos 2x\,dx = \frac{1}{4}\sin 2x\Bigg]_0^{\pi/4} = \frac{1}{4}$$

Now,

$$M_y = \iint_R xy\,dA = \int_0^{\pi/4}\int_{\sin x}^{\cos x} xy\,dy\,dx$$
$$= \int_0^{\pi/4} \frac{1}{2}xy^2\Bigg]_{\sin x}^{\cos x} dx$$
$$= \frac{1}{2}\int_0^{\pi/4} x\cos 2x\,dx$$

integration by parts

$$= \left[\frac{1}{4}x\sin 2x + \frac{1}{8}\cos 2x\right]_0^{\pi/4} = \frac{\pi - 2}{16}$$

Similarly,

$$M_x = \iint_R y^2\,dA = \int_0^{\pi/4}\int_{\sin x}^{\cos x} y^2\,dy\,dx$$
$$= \frac{1}{3}\int_0^{\pi/4} (\cos^3 x - \sin^3 x)\,dx$$
$$= \frac{1}{3}\int_0^{\pi/4} [\cos x(1 - \sin^2 x) - \sin x(1 - \cos^2 x)]\,dx$$
$$= \frac{1}{3}\left[\sin x - \frac{1}{3}\sin^3 x + \cos x - \frac{1}{3}\cos^3 x\right]_0^{\pi/4} = \frac{5\sqrt{2} - 4}{18}$$

Hence, from (3),

$$\bar{x} = \frac{M_y}{m} = \frac{(\pi - 2)/16}{1/4} = \frac{\pi - 2}{4} \approx 0.29$$
$$\bar{y} = \frac{M_x}{m} = \frac{(5\sqrt{2} - 4)/18}{1/4} = \frac{10\sqrt{2} - 8}{9} \approx 0.68$$

Thus, the center of mass has the approximate coordinates (0.29, 0.68). □

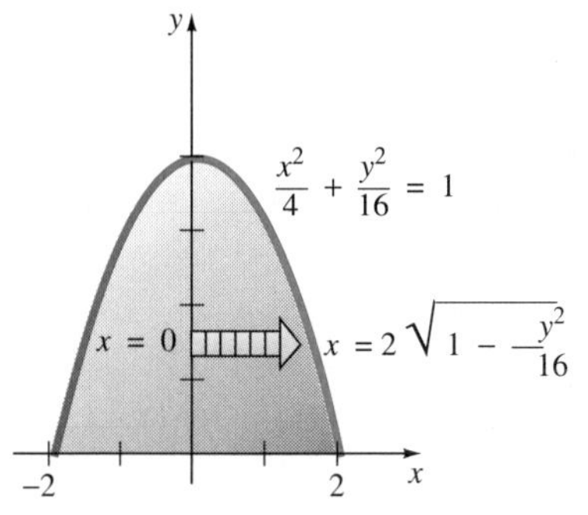

FIGURE 16.31

EXAMPLE 2 A lamina has the shape of the region bounded by the graph of $x^2/4 + y^2/16 = 1, 0 \le y \le 4$, and $y = 0$. Find its center of mass if the density is $\rho(x, y) = |x|y$.

Solution From Figure 16.31 we see that the region is symmetric with respect to the y-axis. Furthermore, since $\rho(-x, y) = \rho(x, y)$, ρ is symmetric about this axis. Since the y-coordinate of the center of mass must lie on the axis of symmetry, we have $\bar{x} = 0$. Utilizing symmetry, we have

$$m = \iint_R |x|y\,dA = 2\int_0^4 \int_0^{2\sqrt{1-y^2/16}} xy\,dx\,dy$$

$$= \int_0^4 x^2y\Big]_0^{2\sqrt{1-y^2/16}} dy$$

$$= 4\int_0^4 \left(y - \frac{1}{16}y^3\right) dy$$

$$= 4\left[\frac{y^2}{2} - \frac{1}{64}y^4\right]_0^4 = 16$$

Similarly,

$$M_x = \iint_R |x|y^2 dA = 2\int_0^4 \int_0^{2\sqrt{1-y^2/16}} xy^2\,dx\,dy = \frac{512}{15}$$

From (3)

$$\bar{y} = \frac{512/15}{16} = \frac{32}{15}$$

The coordinates of the center of mass are then $(0, \frac{32}{15})$. □

Moments of Inertia The integrals M_x and M_y in (4) are also called the **first moments** of a lamina about the x-axis and y-axis, respectively. The so-called **second moments** of a lamina or **moments of inertia** about the x- and y-axes are, in turn, defined by the double integrals

$$I_x = \iint_R y^2\rho(x, y)\,dA \quad \text{and} \quad I_y = \iint_R x^2\rho(x, y)\,dA \tag{5}$$

A moment of inertia is the rotational equivalent of mass. For translational motion, kinetic energy is given by $K = \frac{1}{2}mv^2$, where m is mass and v is linear speed. The kinetic energy of a particle of mass m rotating at a distance r from an axis is $K = \frac{1}{2}mv^2 = \frac{1}{2}m(r\omega)^2 = \frac{1}{2}(mr^2)\omega^2 = \frac{1}{2}I\omega^2$, where $I = mr^2$ is its moment of inertia about the axis of rotation and ω is angular speed.

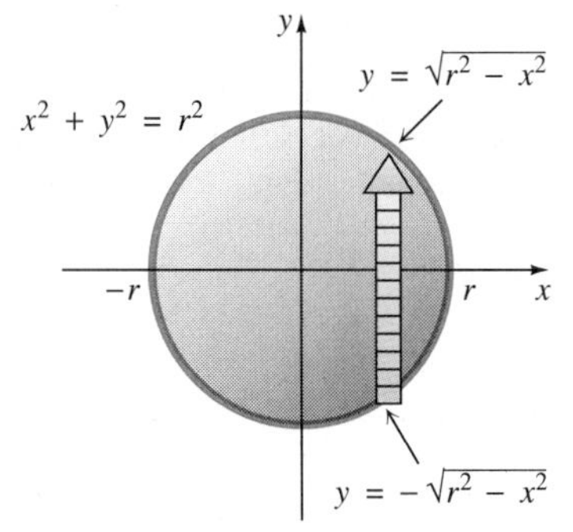

FIGURE 16.32

EXAMPLE 3 Find the moment of inertia about the y-axis of the thin homogeneous disk of mass m shown in Figure 16.32.

Solution Since the disk is homogeneous, its density is the constant $\rho(x, y) = m/\pi r^2$. Hence, from (5),

$$\begin{aligned} I_y = \iint_R x^2\left(\frac{m}{\pi r^2}\right) dA &= \frac{m}{\pi r^2}\int_{-r}^{r}\int_{-\sqrt{r^2-x^2}}^{\sqrt{r^2-x^2}} x^2\, dy\, dx \\ &= \frac{2m}{\pi r^2}\int_{-r}^{r} x^2\sqrt{r^2-x^2}\, dx \qquad \boxed{\text{trig substitution}} \\ &= \frac{2mr^2}{\pi}\int_{-\pi/2}^{\pi/2} \sin^2\theta\cos^2\theta\, d\theta \\ &= \frac{mr^2}{2\pi}\int_{-\pi/2}^{\pi/2} \sin^2 2\theta\, d\theta \\ &= \frac{mr^2}{4\pi}\int_{-\pi/2}^{\pi/2} (1-\cos 4\theta)\, d\theta = \frac{1}{4}mr^2 \end{aligned}$$ □

Radius of Gyration The **radius of gyration** of a lamina of mass m and the moment of inertia I about an axis is defined by

$$R_g = \sqrt{\frac{I}{m}} \tag{6}$$

Since (6) implies that $I = mR_g^2$, the radius of gyration is interpreted as the radial distance the lamina, considered as a point mass, can rotate about the axis without changing the rotational inertia of the body. In Example 3 the radius of gyration is $R_g = \sqrt{I_y/m} = \sqrt{(mr^2/4)/m} = r/2$.

▲ ***Remark*** Do not conclude from Example 2 that the center of mass must always lie on an axis of symmetry of a lamina. Bear in mind that the density function $\rho(x, y)$ must also be symmetric with respect to that axis.

EXERCISES 16.4

Answers to odd-numbered problems begin on page A-107.

In Problems 1–10 find the center of mass of the lamina that has the given shape and density.

1. $x = 0, x = 4, y = 0, y = 3; \rho(x, y) = xy$
2. $x = 0, y = 0, 2x + y = 4; \rho(x, y) = x^2$
3. $y = x, x + y = 6, y = 0; \rho(x, y) = 2y$
4. $y = |x|, y = 3; \rho(x, y) = x^2 + y^2$
5. $y = x^2, x = 1, y = 0; \rho(x, y) = x + y$
6. $x = y^2, x = 4; \rho(x, y) = y + 5$
7. $y = 1 - x^2$, $y = 0$; density at a point P directly proportional to the distance from the x-axis
8. $y = \sin x, 0 \le x \le \pi, y = 0$; density at a point P directly proportional to the distance from the y-axis

9. $y = e^x, x = 0, x = 1, y = 0; \rho(x, y) = y^3$

10. $y = \sqrt{9 - x^2}, y = 0; \rho(x, y) = x^2$

In Problems 11–14 find the moment of inertia about the x-axis of the lamina that has the given shape and density.

11. $x = y - y^2, x = 0; \rho(x, y) = 2x$

12. $y = x^2, y = \sqrt{x}; \rho(x, y) = x^2$

13. $y = \cos x, -\pi/2 \le x \le \pi/2, y = 0; \rho(x, y) = k$ (constant)

14. $y = \sqrt{4 - x^2}, x = 0, y = 0$, first quadrant; $\rho(x, y) = y$

In Problems 15–18 find the moment of inertia about the y-axis of the lamina that has the given shape and density.

15. $y = x^2, x = 0, y = 4$, first quadrant; $\rho(x, y) = y$

16. $y = x^2, y = \sqrt{x}; \rho(x, y) = x^2$

17. $y = x, y = 0, y = 1, x = 3; \rho(x, y) = 4x + 3y$

18. Same R and density as in Problem 7

In Problems 19 and 20 find the radius of gyration about the indicated axis of the lamina that has the given shape and density.

19. $x = \sqrt{a^2 - y^2}, x = 0; \rho(x, y) = x$; y-axis

20. $x + y = a, a > 0, x = 0, y = 0; \rho(x, y) = k$ (constant); x-axis

21. A lamina has the shape of the region bounded by the graph of the ellipse $x^2/a^2 + y^2/b^2 = 1$. If its density is $\rho(x, y) = 1$, find:

(a) the moment of inertia about the x-axis of the lamina,

(b) the moment of inertia about the y-axis of the lamina,

(c) the radius of gyration about the x-axis [*Hint:* The area of the ellipse is πab. See Problem 48 in Exercises 8.4.], and

(d) the radius of gyration about the y-axis.

22. A cross-section of an experimental airfoil is the lamina shown in Figure 16.33. The arc ABC is elliptical, whereas the two arcs AD and CD are parabolic. Find the moment of inertia about the x-axis of the lamina under the assumption that the density is $\rho(x, y) = 1$.

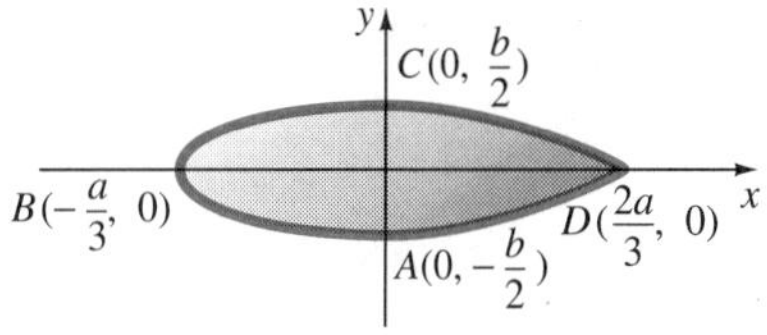

FIGURE 16.33

The **polar moment of inertia** of a lamina with respect to the origin is defined to be

$$I_0 = \iint_R (x^2 + y^2)\rho(x, y)\, dA = I_x + I_y$$

In Problems 23–26 find the polar moment of inertia of the lamina that has the given shape and density.

23. $x + y = a, a > 0, x = 0, y = 0; \rho(x, y) = k$ (constant)

24. $y = x^2, y = \sqrt{x}; \rho(x, y) = x^2$ [*Hint:* See Problems 12 and 16.]

25. $x = y^2 + 2, x = 6 - y^2$; density at a point P inversely proportional to the square of the distance from the origin

26. $y = x, y = 0, y = 3, x = 4; \rho(x, y) = k$ (constant)

27. Find the radius of gyration in Problem 23.

28. Show that the polar moment of inertia about the center of a thin homogeneous rectangular plate of mass m, width w, and length l is $I_0 = m(l^2 + w^2)/12$.

16.5 DOUBLE INTEGRALS IN POLAR COORDINATES

Suppose R is a region bounded by the graphs of the polar equations $r = g_1(\theta)$, $r = g_2(\theta)$, and the rays $\theta = \alpha$, $\theta = \beta$, and f is a function of r and θ that is continuous on R. In order to define the double integral of f over R, we use rays and concentric circles to partition the region into a grid of "polar rectangles" or subregions R_k. See Figure 16.34(a) and (b). The area ΔA_k of typical subregion R_k, shown in Figure 16.34(c), is the difference of areas of two

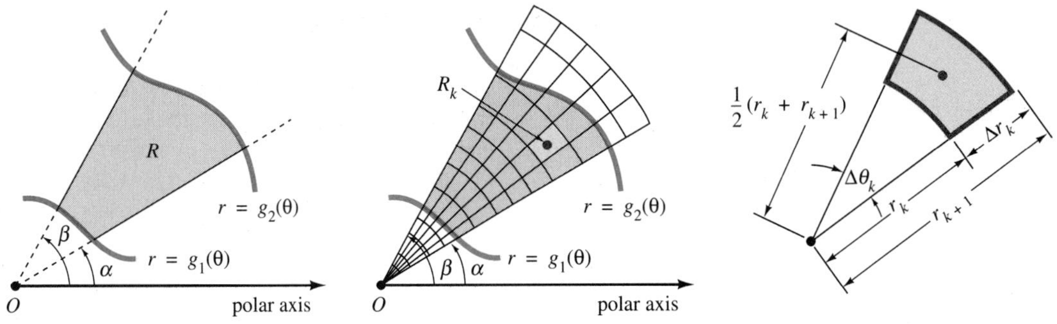

(a) region R (b) subregion R_k (c) enlargement of R_k

FIGURE 16.34

circular sectors:

$$\Delta A_k = \frac{1}{2}r_{k+1}^2\,\Delta\theta_k - \frac{1}{2}r_k^2\,\Delta\theta_k = \frac{1}{2}(r_{k+1}^2 - r_k^2)\,\Delta\theta_k$$

$$= \frac{1}{2}(r_{k+1} + r_k)(r_{k+1} - r_k)\,\Delta\theta_k = r_k^*\,\Delta r_k\,\Delta\theta_k$$

where $\Delta r_k = r_{k+1} - r_k$ and r_k^* denotes the average radius $(r_{k+1} + r_k)/2$. By choosing (r_k^*, θ_k^*) on each R_k, the double integral of f over R is

$$\lim_{\|P\|\to 0} \sum_{k=1}^{n} f(r_k^*, \theta_k^*) r_k^*\,\Delta r_k\,\Delta\theta_k = \iint_R f(r, \theta)\,dA$$

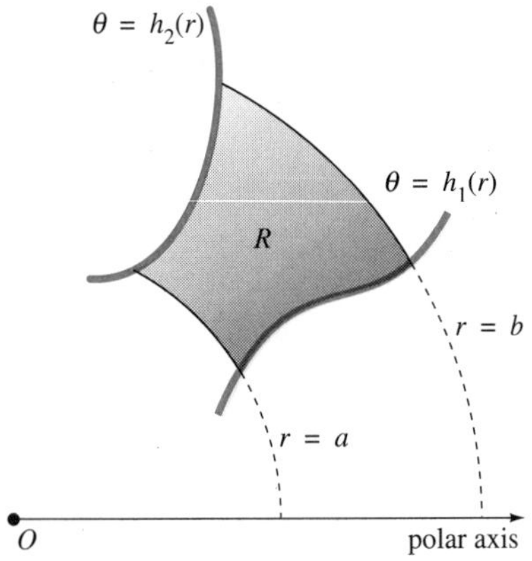

FIGURE 16.35

The double integral is then evaluated by means of the iterated integral

$$\iint_R f(r, \theta)\,dA = \int_\alpha^\beta \int_{g_1(\theta)}^{g_2(\theta)} f(r, \theta) r\,dr\,d\theta \tag{1}$$

On the other hand, if the region R is as given in Figure 16.35, the double integral of f over R is then

$$\iint_R f(r, \theta)\,dA = \int_a^b \int_{h_1(r)}^{h_2(r)} f(r, \theta) r\,d\theta\,dr \tag{2}$$

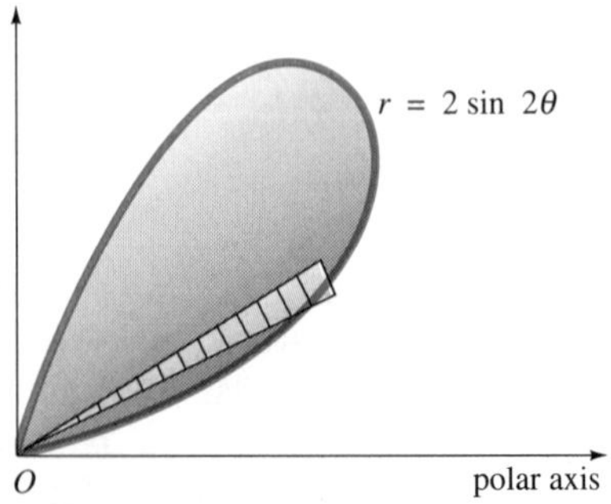

FIGURE 16.36

EXAMPLE 1 Find the center of mass of the lamina that corresponds to the region bounded by one leaf of the rose $r = 2 \sin 2\theta$ in the first quadrant if the density at a point P in the lamina is directly proportional to the distance from the pole.

Solution By varying θ from 0 to $\pi/2$, we obtain the graph in Figure 16.36. Now, $d(0, P) = |r|$. Hence, $\rho(r, \theta) = k|r|$, where k is a constant of proportionality. From (2) of Section 16.4, we have

$$m = \iint_R k|r|\,dA = k\int_0^{\pi/2}\int_0^{2\sin 2\theta}(r)r\,dr\,d\theta$$

$$= k\int_0^{\pi/2}\frac{r^3}{3}\Big]_0^{2\sin 2\theta}d\theta = \frac{8}{3}k\int_0^{\pi/2}\sin^3 2\theta\,d\theta$$

$$= \frac{8}{3}k\int_0^{\pi/2}\sin^2 2\theta\sin 2\theta\,d\theta$$

trig identity

$$= \frac{8}{3}k\int_0^{\pi/2}(1-\cos^2 2\theta)\sin 2\theta\,d\theta$$

$$= \frac{8}{3}k\left[-\frac{1}{2}\cos 2\theta + \frac{1}{6}\cos^3 2\theta\right]_0^{\pi/2} = \frac{16}{9}k$$

Since $x = r\cos\theta$, we can write $M_y = k\iint_R x|r|\,dA$ as

$$M_y = \int_0^{\pi/2}\int_0^{2\sin 2\theta} r^3\cos\theta\,dr\,d\theta$$

$$= k\int_0^{\pi/2}\frac{r^4}{4}\cos\theta\Big]_0^{2\sin 2\theta}d\theta$$

$$= 4k\int_0^{\pi/2}\sin^4 2\theta\cos\theta\,d\theta$$

double angle formula

$$= 4k\int_0^{\pi/2}16\sin^4\theta\cos^4\theta\cos\theta\,d\theta$$

$$= 64k\int_0^{\pi/2}\sin^4\theta\cos^5\theta\,d\theta$$

$$= 64k\int_0^{\pi/2}\sin^4\theta(1-\sin^2\theta)^2\cos\theta\,d\theta$$

$$= 64k\int_0^{\pi/2}(\sin^4\theta - 2\sin^6\theta + \sin^8\theta)\cos\theta\,d\theta$$

$$= 64k\left[\frac{1}{5}\sin^5\theta - \frac{2}{7}\sin^7\theta + \frac{1}{9}\sin^9\theta\right]_0^{\pi/2} = \frac{512}{315}k$$

Similarly, by using $x = r\sin\theta$, we find*

$$M_x = k\int_0^{\pi/2}\int_0^{2\sin 2\theta} r^2\sin\theta\,dr\,d\theta = \frac{512}{315}k$$

Here the rectangular coordinates of the center of mass are

$$\bar{x} = \bar{y} = \frac{512k/315}{16k/9} = \frac{32}{35}$$ □

*We could have argued to the fact that $M_x = M_y$ and hence $\bar{x} = \bar{y}$ from the fact that the lamina and the density function are symmetric about the ray $\theta = \pi/4$.

Change of Variables: Rectangular to Polar Coordinates In some instances a double integral $\iint_R f(x, y)\, dA$ that is difficult or even impossible to evaluate using rectangular coordinates may be readily evaluated when a change of variables is used. If we assume that f is continuous on the region R and if R can be described in polar coordinates as $0 \le g_1(\theta) \le r \le g_2(\theta)$, $\alpha \le \theta \le \beta$, $0 < \beta - \alpha \le 2\pi$, then

$$\iint_R f(x, y)\, dA = \int_\alpha^\beta \int_{g_1(\theta)}^{g_2(\theta)} f(r\cos\theta, r\sin\theta) r\, dr\, d\theta \tag{3}$$

Equation (3) is particularly useful when f contains the expression $x^2 + y^2$, since, in polar coordinates, we can now write

$$x^2 + y^2 = r^2 \quad \text{and} \quad \sqrt{x^2 + y^2} = r$$

EXAMPLE 2 Use polar coordinates to evaluate

$$\int_0^2 \int_x^{\sqrt{8-x^2}} \frac{1}{5 + x^2 + y^2}\, dy\, dx$$

Solution From $x \le y \le \sqrt{8 - x^2}$, $0 \le x \le 2$, we have sketched the region R of integration in Figure 16.37. Since $x^2 + y^2 = r^2$, the polar description of the circle $x^2 + y^2 = 8$ is $r = \sqrt{8}$. Hence, in polar coordinates, the region of R is given by $0 \le r \le \sqrt{8}$, $\pi/4 \le \theta \le \pi/2$. From $1/(5 + x^2 + y^2) = 1/(5 + r^2)$ the original integral becomes

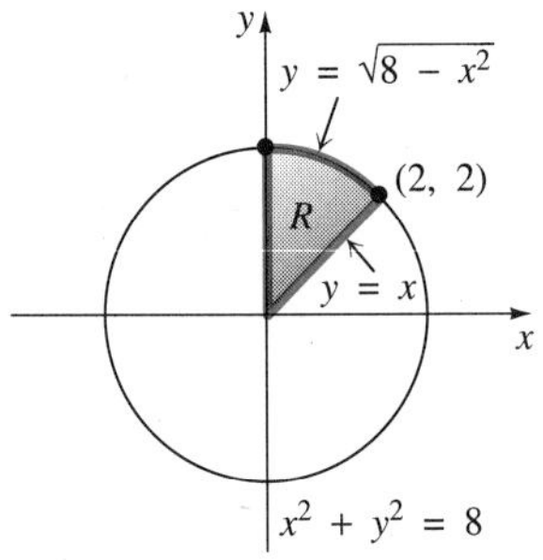

FIGURE 16.37

$$\begin{aligned}\int_0^2 \int_x^{\sqrt{8-x^2}} \frac{1}{5 + x^2 + y^2}\, dy\, dx &= \int_{\pi/4}^{\pi/2} \int_0^{\sqrt{8}} \frac{1}{5 + r^2}\, r\, dr\, d\theta \\ &= \frac{1}{2}\int_{\pi/4}^{\pi/2} \int_0^{\sqrt{8}} \frac{2r\, dr}{5 + r^2}\, d\theta \\ &= \frac{1}{2}\int_{\pi/4}^{\pi/2} \ln(5 + r^2)\Big]_0^{\sqrt{8}}\, d\theta \\ &= \frac{1}{2}(\ln 13 - \ln 5)\int_{\pi/4}^{\pi/2} d\theta \\ &= \frac{1}{2}(\ln 13 - \ln 5)\left(\frac{\pi}{2} - \frac{\pi}{4}\right) = \frac{\pi}{8}\ln\frac{13}{5}\end{aligned}$$ □

EXAMPLE 3 Find the volume of the solid that is under the hemisphere $z = \sqrt{1 - x^2 - y^2}$ and above the region bounded by the graph of the circle $x^2 + y^2 - y = 0$.

Solution From Figure 16.38 we see that

$$V = \iint_R \sqrt{1 - x^2 - y^2}\, dA$$

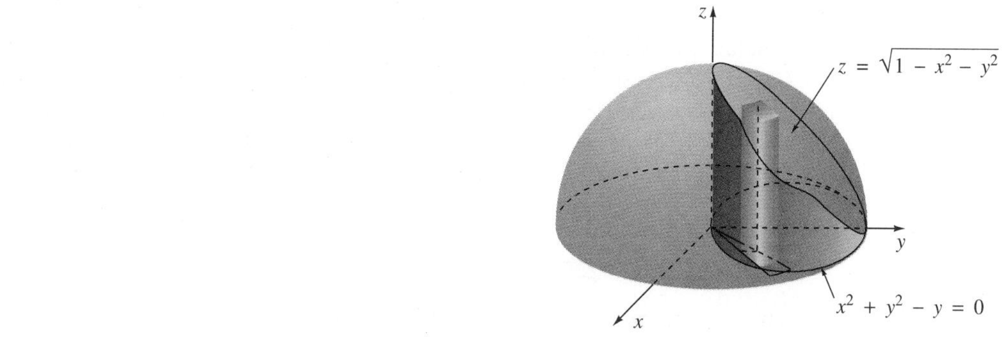

FIGURE 16.38

In polar coordinates the equations of the hemisphere and the circle become, respectively, $z = \sqrt{1 - r^2}$ and $r = \sin\theta$. Now, using symmetry, we have

$$\begin{aligned}
V = \iint_R \sqrt{1 - r^2}\, dA &= 2\int_0^{\pi/2}\int_0^{\sin\theta} (1 - r^2)^{1/2} r\, dr\, d\theta \\
&= 2\int_0^{\pi/2} \left[-\frac{1}{3}(1 - r^2)^{3/2}\right]_0^{\sin\theta} d\theta \\
&= \frac{2}{3}\int_0^{\pi/2} [1 - (1 - \sin^2\theta)^{3/2}]\, d\theta \\
&= \frac{2}{3}\int_0^{\pi/2} [1 - (\cos^2\theta)^{3/2}]\, d\theta \\
&= \frac{2}{3}\int_0^{\pi/2} [1 - \cos^3\theta]\, d\theta \\
&= \frac{2}{3}\int_0^{\pi/2} [1 - (1 - \sin^2\theta)\cos\theta]\, d\theta \\
&= \frac{2}{3}\left[\theta - \sin\theta + \frac{1}{3}\sin^3\theta\right]_0^{\pi/2} = \frac{\pi}{3} - \frac{4}{9} \approx 0.60 \text{ cubic unit}
\end{aligned}$$

□

Area Note that in (1) if $f(r, \theta) = 1$, then the **area** of the region R in Figure 16.34(a) is given by

$$A = \iint_R dA = \int_\alpha^\beta \int_{g_1(\theta)}^{g_2(\theta)} r\, dr\, d\theta$$

The same observation holds for (2) and Figure 16.35 when $f(r, \theta) = 1$.

▲ ***Remark*** You are invited to reexamine Example 3. The graph of the circle $r = \sin\theta$ is obtained by varying θ from 0 to π. However, carry out the iterated integration

$$V = \int_0^{\pi} \int_0^{\sin\theta} (1 - r^2)^{1/2} r\, dr\, d\theta$$

and see if you obtain the *incorrect* answer $\pi/3$. What goes wrong?

EXERCISES 16.5

Answers to odd-numbered problems begin on page A-107.

In Problems 1–4 use a double integral in polar coordinates to find the area of the region bounded by the graphs of the given polar equations.

1. $r = 3 + 3\sin\theta$
2. $r = 2 + \cos\theta$
3. $r = 2\sin\theta$, $r = 1$, common area
4. $r = 8\sin 4\theta$, one petal

In Problems 5–10 find the volume of the solid bounded by the graphs of the given equations.

5. One petal of $r = 5\cos 3\theta$, $z = 0$, $z = 4$
6. $x^2 + y^2 = 4$, $z = \sqrt{9 - x^2 - y^2}$, $z = 0$
7. Between $x^2 + y^2 = 1$ and $x^2 + y^2 = 9$, $z = \sqrt{16 - x^2 - y^2}$, $z = 0$
8. $z = \sqrt{x^2 + y^2}$, $x^2 + y^2 = 25$, $z = 0$
9. $r = 1 + \cos\theta$, $z = y$, $z = 0$, first octant
10. $r = \cos\theta$, $z = 2 + x^2 + y^2$, $z = 0$

In Problems 11–16 find the center of mass of the lamina that has the given shape and density.

11. $r = 1$, $r = 3$, $x = 0$, $y = 0$, first quadrant; $\rho(r, \theta) = k$ (constant)
12. $r = \cos\theta$; density at a point P directly proportional to the distance from the pole
13. $y = \sqrt{3}x$, $y = 0$, $x = 3$; $\rho(r, \theta) = r^2$
14. $r = 4\cos 2\theta$, petal on the polar axis; $\rho(r, \theta) = k$ (constant)
15. Outside $r = 2$ and inside $r = 2 + 2\cos\theta$, $y = 0$, first quadrant; density at a point P inversely proportional to the distance from the pole
16. $r = 2 + 2\cos\theta$, $y = 0$, first and second quadrants; $\rho(r, \theta) = k$ (constant)

In Problems 17–20 find the indicated moment of inertia of the lamina that has the given shape and density.

17. $r = a$; $\rho(r, \theta) = k$ (constant); I_x
18. $r = a$; $\rho(r, \theta) = \dfrac{1}{1 + r^4}$; I_x
19. Outside $r = a$ and inside $r = 2a\cos\theta$; density at a point P inversely proportional to the cube of the distance from the pole; I_y
20. Outside $r = 1$ and inside $r = 2\sin 2\theta$, first quadrant; $\rho(r, \theta) = \sec^2\theta$; I_y

In Problems 21–24 find the **polar moment of inertia** $I_0 = \iint_R r^2 \rho(r, \theta)\, dA = I_x + I_y$ of the lamina that has the given shape and density.

21. $r = a$; $\rho(r, \theta) = k$ (constant) [*Hint:* Use Problem 17 and the fact that $I_x = I_y$.]
22. $r = \theta$, $0 \le \theta \le \pi$, $y = 0$; density at a point P proportional to the distance from the pole
23. $r\theta = 1$, $\frac{1}{3} \le \theta \le 1$, $r = 1$, $r = 3$, $y = 0$; density at a point P inversely proportional to the distance from the pole [*Hint:* Integrate first with respect to θ.]
24. $r = 2a\cos\theta$; $\rho(r, \theta) = k$ (constant)

In Problems 25–32 evaluate the given iterated integral by changing to polar coordinates.

25. $\displaystyle\int_{-3}^{3} \int_0^{\sqrt{9 - x^2}} \sqrt{x^2 + y^2}\, dy\, dx$
26. $\displaystyle\int_0^{\sqrt{2}/2} \int_y^{\sqrt{1 - y^2}} \frac{y^2}{\sqrt{x^2 + y^2}}\, dx\, dy$
27. $\displaystyle\int_0^1 \int_0^{\sqrt{1 - y^2}} e^{x^2 + y^2}\, dx\, dy$
28. $\displaystyle\int_{-\sqrt{\pi}}^{\sqrt{\pi}} \int_0^{\sqrt{\pi - x^2}} \sin(x^2 + y^2)\, dy\, dx$
29. $\displaystyle\int_0^1 \int_{\sqrt{1 - x^2}}^{\sqrt{4 - x^2}} \frac{x^2}{x^2 + y^2}\, dy\, dx + \int_1^2 \int_0^{\sqrt{4 - x^2}} \frac{x^2}{x^2 + y^2}\, dy\, dx$
30. $\displaystyle\int_0^1 \int_0^{\sqrt{2y - y^2}} (1 - x^2 - y^2)\, dx\, dy$

31. $\int_{-5}^{5} \int_{0}^{\sqrt{25-x^2}} (4x + 3y)\, dy\, dx$

32. $\int_{0}^{1} \int_{0}^{\sqrt{1-y^2}} \frac{1}{1 + \sqrt{x^2 + y^2}}\, dx\, dy$

33. The liquid hydrogen tank in the space shuttle has the form of a right circular cylinder with a semiellipsoidal cap at each end. The radius of the cylindrical part of the tank is 4.2 m. Find the volume of the tank shown in Figure 16.39.

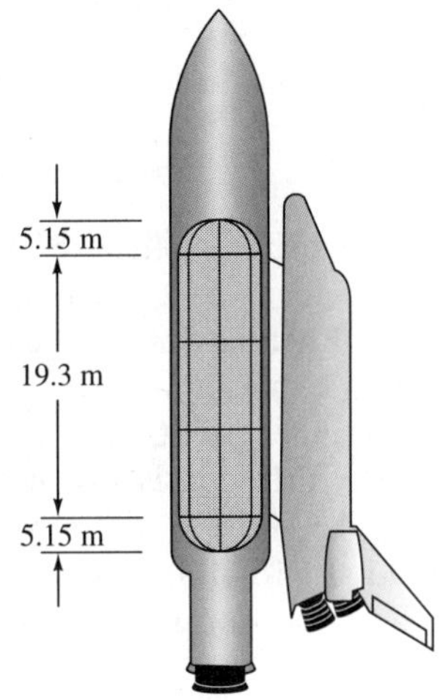

FIGURE 16.39

34. Evaluate $\iint_R (x + y)\, dA$ over the region shown in Figure 16.40.

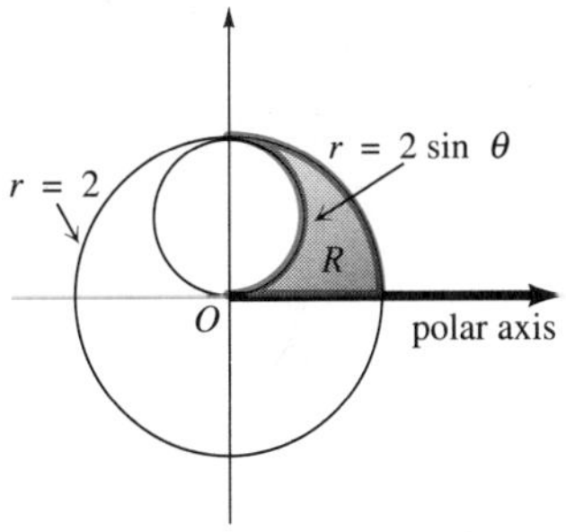

FIGURE 16.40

35. The improper integral $\int_0^\infty e^{-x^2}\, dx$ is important in the theory of probability, statistics, and other areas of applied mathematics. If I denotes the integral, then

$$I = \int_0^\infty e^{-x^2}\, dx \quad \text{and} \quad I = \int_0^\infty e^{-y^2}\, dy$$

In view of Problem 41 of Exercises 16.2 we have

$$I^2 = \left(\int_0^\infty e^{-x^2}\, dx\right)\left(\int_0^\infty e^{-y^2}\, dy\right)$$

$$= \int_0^\infty \int_0^\infty e^{-(x^2+y^2)}\, dx\, dy$$

Use polar coordinates to evaluate the last integral. Find the value of I.

36. In some studies of the spread of plant disease, the number of infections per unit area as a function of the distance from an infected source plant is described by a formula of the form

$$I(r) = a(r + c)^{-b}$$

where $I(r)$ is the number of infections per unit area at a radial distance r from the infected source plant, and a, b, and c are (positive) parameters depending on the disease.

(a) Derive a formula for the total number of infections within a circle of radius R centered at the infected source plant; that is, evaluate $\iint_C I(r)\, dA$, where C is a circular region of radius R centered at the origin. Assume that $b \neq 1$ or 2.

(b) Show that if $b > 2$, then the result in part (**a**) tends to a finite limit as $R \to \infty$.

(c) For common maize rust, the number of infections per square meter is modeled as

$$I(r) = 68.585(r + 0.248)^{-2.351}$$

where r is measured in meters. Find the total number of infections in the plane.

37. Urban population densities fall off exponentially with distance from the central business district (CBD); that is,

$$D(r) = D_0 e^{-r/d}$$

where $D(r)$ is the population density at a radial distance r from the CBD, D_0 is the density at the center, and d is a parameter.

(a) Using the formula $P = \iint_C D(r)\, dA$, find an expression for the total population living within a circular region C of radius R of the CBD.

(b) Using $$\frac{\iint_C rD(r)\, dA}{\iint_C D(r)\, dA}$$

find an expression for the average commute (distance traveled) to the CBD for the people living within the region C.

(c) Using the results in part (**a**) and (**b**), find the total population and average commute as $R \to \infty$.

38. It's arguable that the cost, in terms of time, money, or effort, of collecting or distributing material to or from a single location is proportional to the integral $\iint_R r\,dA$, where R is the region being covered and r denotes distance to the collection/distribution site. Suppose, for example, that a snowplow is sent to clear off a circular parking area of diameter D. Show that plowing all the snow to a single point on the perimeter is approximately 70% more costly than plowing everything to the center of the parking lot. [*Hint:* Set up the integral for each case separately, using a polar coordinate equation for the circle with the collection site at the origin.]

16.6 SURFACE AREA

In the plane we saw that the length of an arc of the graph of $y = f(x)$ from $x = a$ to $x = b$ was given by

$$s = \int_a^b \sqrt{1 + \left(\frac{dy}{dx}\right)^2}\,dx \qquad (1)$$

The problem in three dimensions, which is the counterpart of the arc length problem, is to find the area $A(S)$ of that portion of the surface S given by a function $z = f(x, y)$ having continuous first partial derivatives on a closed region R in the xy-plane. Such a surface S is said to be **smooth**.

Suppose, as shown in Figure 16.41(a), that an inner partition P of R is formed using lines parallel to the x- and y-axes. P then consists of n rectangular elements R_k of area $\Delta A_k = \Delta x_k\,\Delta y_k$ that lie entirely within R. Let $(x_k, y_k, 0)$ denote any point in an R_k. As we see in Figure 16.41(a), by projecting the sides of R_k upward, we determine two quantities: a portion S_k of the surface

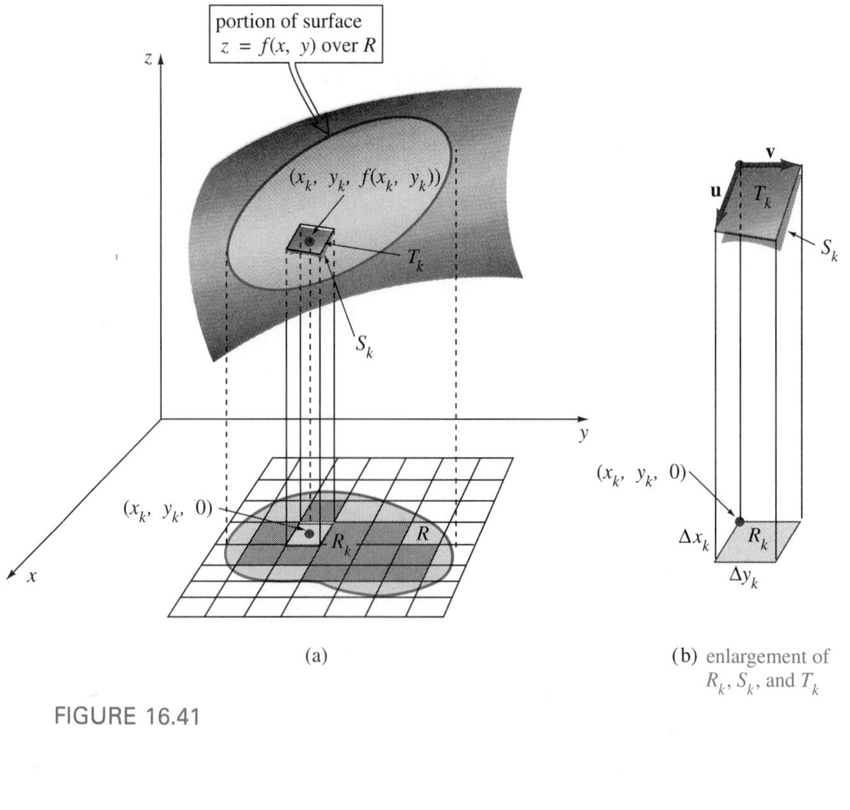

FIGURE 16.41

and a portion of T_k of a tangent plane at $(x_k, y_k, f(x_k, y_k))$. It seems reasonable to assume that when R_k is small, the area ΔT_k of T_k is approximately the same as the area ΔS_k of S_k.

To find the area of T_k let us choose $(x_k, y_k, 0)$ at a corner of R_k as shown in Figure 16.41(b). The indicated vectors **u** and **v**, which form two sides of T_k, are given by

$$\mathbf{u} = \Delta x_k \mathbf{i} + f_x(x_k, y_k)\,\Delta x_k \mathbf{k} \quad \text{and} \quad \mathbf{v} = \Delta y_k \mathbf{j} + f_y(x_k, y_k)\,\Delta y_k \mathbf{k}$$

where $f_x(x_k, y_k)$ and $f_y(x_k, y_k)$ are the slopes of the lines containing **u** and **v**, respectively. Now from (11) of Section 13.4 we know that

$$\Delta T_k = |\mathbf{u} \times \mathbf{v}|$$

where
$$\mathbf{u} \times \mathbf{v} = \begin{vmatrix} \mathbf{i} & \mathbf{j} & \mathbf{k} \\ \Delta x_k & 0 & f_x(x_k, y_k)\,\Delta x_k \\ 0 & \Delta y_k & f_y(x_k, y_k)\,\Delta y_k \end{vmatrix}$$
$$= [-f_x(x_k, y_k)\mathbf{i} - f_y(x_k, y_k)\mathbf{j} + \mathbf{k}]\,\Delta x_k\,\Delta y_k$$

In other words,

$$\Delta T_k = \sqrt{[f_x(x_k, y_k)]^2 + [f_y(x_k, y_k)]^2 + 1}\,\Delta x_k\,\Delta y_k$$

Consequently, the area A is approximately

$$\sum_{k=1}^{n} \sqrt{1 + [f_x(x_k, y_k)]^2 + [f_y(x_k, y_k)]^2}\,\Delta x_k\,\Delta y_k$$

Taking the limit of the foregoing sum as $\|P\| \to 0$ leads us to the next definition.

DEFINITION 16.2 Surface Area

Let f be a function for which the first partial derivatives f_x and f_y are continuous on a closed region R. Then the **area of the surface over R** is given by

$$A(S) = \iint_R \sqrt{1 + [f_x(x, y)]^2 + [f_y(x, y)]^2}\,dA \qquad (2)$$

Note: One could have almost guessed the form of (2) by naturally extending the one-variable structure of (1) to two variables.

EXAMPLE 1 Find the surface area of that portion of the sphere $x^2 + y^2 + z^2 = a^2$ that is above the xy-plane and within the cylinder $x^2 + y^2 = b^2$, $0 < b < a$.

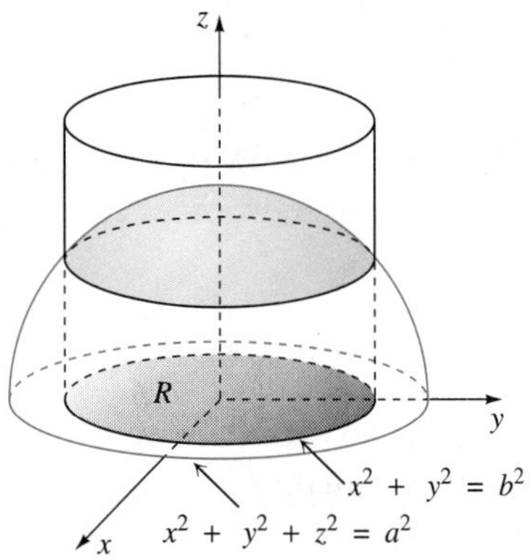

FIGURE 16.42

Solution If we define $z = f(x, y)$ by $f(x, y) = \sqrt{a^2 - x^2 - y^2}$, then

$$f_x(x, y) = \frac{-x}{\sqrt{a^2 - x^2 - y^2}} \quad \text{and} \quad f_y(x, y) = \frac{-y}{\sqrt{a^2 - x^2 - y^2}}$$

and so $$1 + [f_x(x, y)]^2 + [f_y(x, y)]^2 = \frac{a^2}{a^2 - x^2 - y^2}$$

Hence, (2) is $$A(S) = \iint_R \frac{a}{\sqrt{a^2 - x^2 - y^2}}\, dA$$

where R is indicated in Figure 16.42. To evaluate this double integral, we change to polar coordinates:

$$\begin{aligned} A(S) &= a \int_0^{2\pi} \int_0^b (a^2 - r^2)^{-1/2} r\, dr\, d\theta \\ &= a \int_0^{2\pi} \left[-(a^2 - r^2)^{1/2} \right]_0^b d\theta \\ &= a(a - \sqrt{a^2 - b^2}) \int_0^{2\pi} d\theta \\ &= 2\pi a(a - \sqrt{a^2 - b^2}) \text{ square units} \end{aligned}$$

□

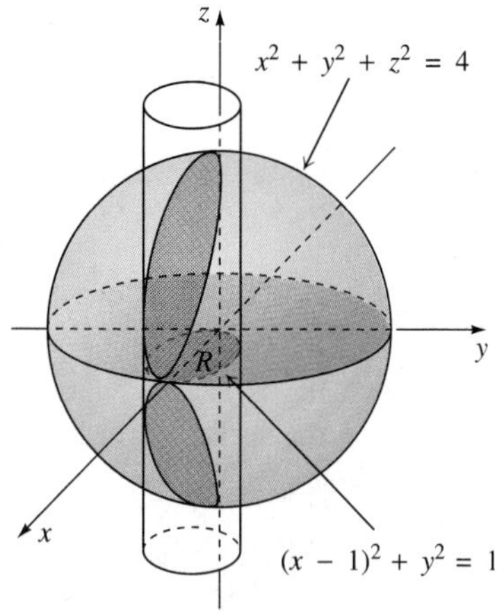

FIGURE 16.43

EXAMPLE 2 Find the surface area of the portions of the sphere $x^2 + y^2 + z^2 = 4$ that are within the cylinder $(x - 1)^2 + y^2 = 1$.

Solution The surface area, which is above and below the xy-plane, is shaded gray in Figure 16.43. As in Example 1, (2) simplifies to

$$A(S) = 2 \iint_R \frac{2}{\sqrt{4 - x^2 - y^2}}\, dA$$

where R is the region bounded by the graph of $(x - 1)^2 + y^2 = 1$. The extra factor of 2 comes from using symmetry. Now, in polar coordinates the boundary of R is simply $r = 2\cos\theta$. Thus,

$$\begin{aligned} A(S) &= 4 \int_0^{\pi} \int_0^{2\cos\theta} (4 - r^2)^{-1/2} r\, dr\, d\theta \\ &= 8(\pi - 2) \approx 9.13 \text{ square units} \end{aligned}$$

□

Differential of Surface Area The function

$$dS = \sqrt{1 + [f_x(x, y)]^2 + [f_y(x, y)]^2}\, dA$$

is called the **differential of the surface area**. We will use this function in Section 17.4.

EXERCISES 16.6

Answers to odd-numbered problems begin on page A-108.

1. Find the surface area of that portion of the plane $2x + 3y + 4z = 12$ that is bounded by the coordinate planes in the first octant.

2. Find the surface area of that portion of the plane $2x + 3y + 4z = 12$ that is above the region in the first quadrant bounded by the graph of $r = \sin 2\theta$.

3. Find the surface area of that portion of the cylinder $x^2 + z^2 = 16$ that is above the region in the first quadrant bounded by the graphs of $x = 0$, $x = 2$, $y = 0$, $y = 5$.

4. Find the surface area of that portion of the paraboloid $z = x^2 + y^2$ that is below the plane $z = 2$.

5. Find the surface area of that portion of the paraboloid $z = 4 - x^2 - y^2$ that is above the xy-plane.

6. Find the surface area of the portions of the sphere $x^2 + y^2 + z^2 = 2$ that are within the cone $z^2 = x^2 + y^2$.

7. Find the surface area of that portion of the sphere $x^2 + y^2 + z^2 = 25$ that is above the region in the first quadrant bounded by the graphs of $x = 0$, $y = 0$, $4x^2 + y^2 = 25$. [*Hint:* Integrate first with respect to x.]

8. Find the surface area of that portion of the graph of $z = x^2 - y^2$ that is in the first octant within the cylinder $x^2 + y^2 = 4$.

9. Find the surface area of the portions of the sphere $x^2 + y^2 + z^2 = a^2$ that are within the cylinder $x^2 + y^2 = ay$.

10. Find the surface area of the portions of the cone $z^2 = \frac{1}{4}(x^2 + y^2)$ that are within the cylinder $(x - 1)^2 + y^2 = 1$.

11. Find the surface area of the portions of the cylinder $y^2 + z^2 = a^2$ that are within the cylinder $x^2 + y^2 = a^2$. [*Hint:* See Figure 16.28.]

12. Use the result given in Example 1 to prove that the surface area of a sphere of radius a is $4\pi a^2$. [*Hint:* Consider a limit as $b \to a$.]

13. Find the surface area of that portion of the sphere $x^2 + y^2 + z^2 = a^2$ that is bounded between $y = c_1$ and $y = c_2$, $0 < c_1 < c_2 < a$. [*Hint:* Use polar coordinates in the xz-plane.]

14. Show that the area found in Problem 13 is the same as the surface area of the cylinder $x^2 + z^2 = a^2$ between $y = c_1$ and $y = c_2$.

15. As shown in Figure 16.44, a sphere of radius 1 has its center on the surface of a sphere of radius $a > 1$. Find the surface area of that portion of the larger sphere cut out by the smaller sphere.

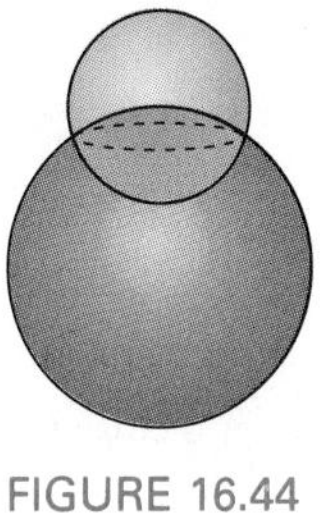

FIGURE 16.44

16. On the surface of a globe or, more precisely, on the surface of the earth, the boundaries of the states of Colorado and Wyoming are both "spherical rectangles." (In this problem we assume that the earth is a perfect sphere.) Colorado is bounded by the lines of longitude 102°W and 109°W and the lines of latitude 37°N and 41°N. Wyoming is bounded by longitudes 104°W and 111°W and latitudes 41°N and 45°N. See Figure 16.45.

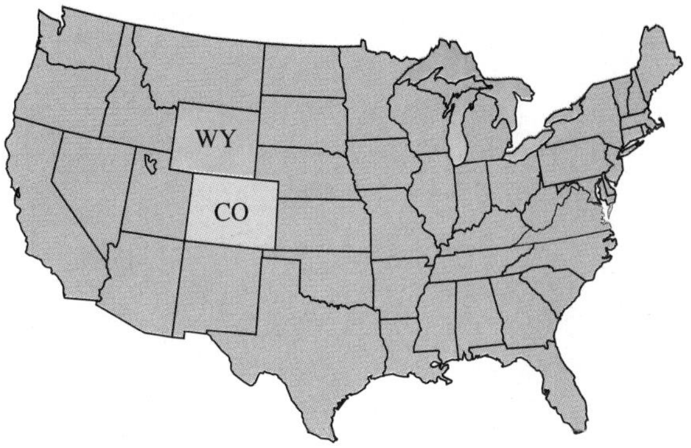

FIGURE 16.45

(a) Without explicitly computing their areas, determine which state is larger and explain why.

(b) By what percentage is Wyoming larger (or smaller) than Colorado? [*Hint:* Suppose the radius of the earth is R. Project a spherical rectangle in the Northern Hemisphere that is determined by latitudes θ_1 and θ_2 and longitudes ϕ_1 and ϕ_2 onto the xy-plane.]

(c) One reference book gives the areas of the two states as 104,247 and 97,914 mi^2. How does this answer compare with the answer in part (**b**)?

16.7 THE TRIPLE INTEGRAL

The steps leading to the definition of the *three-dimensional definite integral*, or **triple integral**, $\iiint_D F(x, y, z)\, dV$ are quite similar to those for the double integral.

$w = F(x, y, z)$

1. Let F be defined over a closed and bounded region D of space.
2. By means of a three-dimensional grid of vertical and horizontal planes parallel to the coordinate planes, form a partition P of D into n subregions (boxes) D_k of volumes ΔV_k that lie entirely in D.
3. Let $\|P\|$ be the norm of the partition or the length of the longest diagonal of the D_k.
4. Choose a point (x_k^*, y_k^*, z_k^*) in each subregion D_k.
5. Form the sum $\sum_{k=1}^{n} F(x_k^*, y_k^*, z_k^*)\, \Delta V_k$.

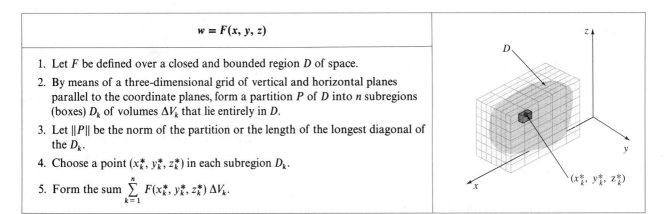

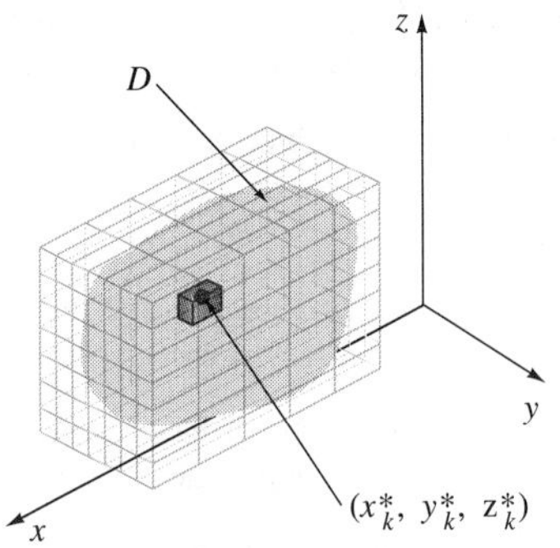

A sum of the form $\Sigma_{k=1}^{n} F(x_k^*, y_k^*, z_k^*)\, \Delta V_k$, where (x_k^*, y_k^*, z_k^*) is an arbitrary point within each D_k and ΔV_k denotes the volume of each D_k, is called a **Riemann sum**. The type of partition used in step 2, where all the D_k lie completely within D, is called an **inner partition** of D.

DEFINITION 16.3 The Triple Integral

Let F be a function of three variables defined over a closed region D of space. Then the **triple integral of F over D** is given by

$$\iiint_D F(x, y, z)\, dV = \lim_{\|P\| \to 0} \sum_{k=1}^{n} F(x_k^*, y_k^*, z_k^*)\, \Delta V_k \tag{1}$$

As in our previous discussions on the integral, when F is continuous over D, then the limit in (1) exists; that is, F is **integrable** over D.

Evaluation by Iterated Integrals If the region D is bounded above by the graph of $z = f_2(x, y)$ and bounded below by the graph of $z = f_1(x, y)$, then it can be shown that the triple integral (1) can be expressed as a double integral of the partial integral $\int_{f_1(x,y)}^{f_2(x,y)} F(x, y, z)\, dz$; that is,

$$\iiint_D F(x, y, z)\, dV = \iint_R \left[\int_{f_1(x,y)}^{f_2(x,y)} F(x, y, z)\, dz \right] dA$$

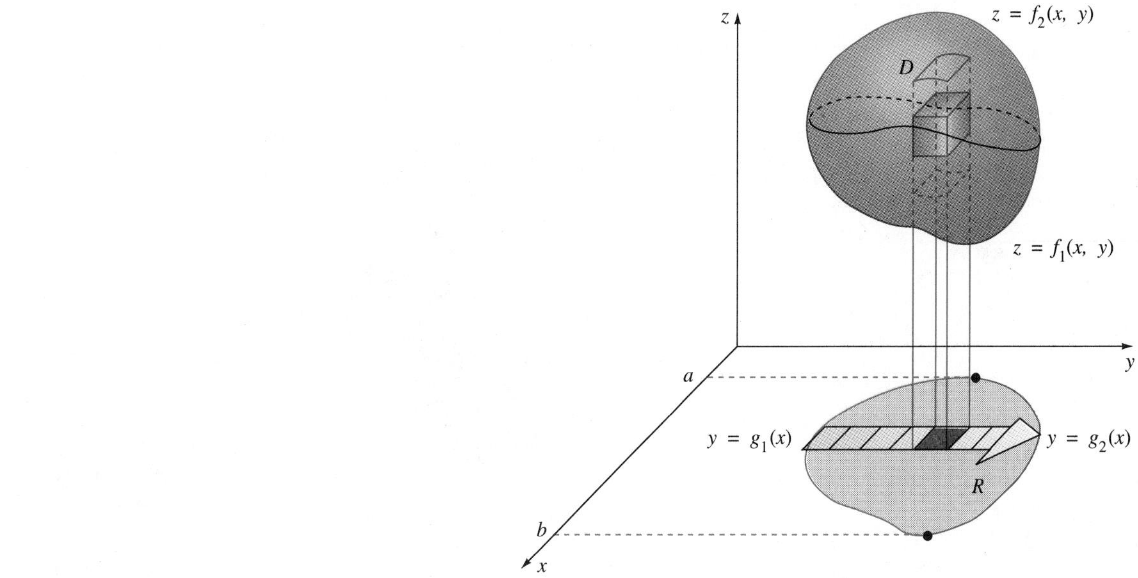

FIGURE 16.46

where R is the orthogonal projection of D onto the xy-plane. In particular, if R is a Type I region, then, as shown in Figure 16.46, the triple integral of F over D can be written as an iterated integral:

$$\iiint_D F(x, y, z)\, dV = \int_a^b \int_{g_1(x)}^{g_2(x)} \int_{f_1(x,y)}^{f_2(x,y)} F(x, y, z)\, dz\, dy\, dx \tag{2}$$

To evaluate the iterated integral in (2) we begin by evaluating the partial integral

$$\int_{f_1(x,y)}^{f_2(x,y)} F(x, y, z)\, dz$$

in which *both* x and y are held fixed.

In a double integral there are only two possible orders of integration, $dy\,dx$ and $dx\,dy$. The triple integral in (2) illustrates one of *six* possible orders of integration:

$$\begin{array}{ccc} dz\,dy\,dx & dz\,dx\,dy & dy\,dx\,dz \\ dx\,dy\,dz & dx\,dz\,dy & dy\,dz\,dx \end{array}$$

The last two differentials tell us the coordinate plane in which the region R is situated. For example, the iterated integral corresponding to the order of integration $dx\,dz\,dy$ must have the form

$$\int_c^d \int_{k_1(y)}^{k_2(y)} \int_{h_1(y,z)}^{h_2(y,z)} F(x, y, z)\, dx\, dz\, dy$$

The geometric interpretation of this integral and the region R of integration in the yz-plane are shown in Figure 16.47.

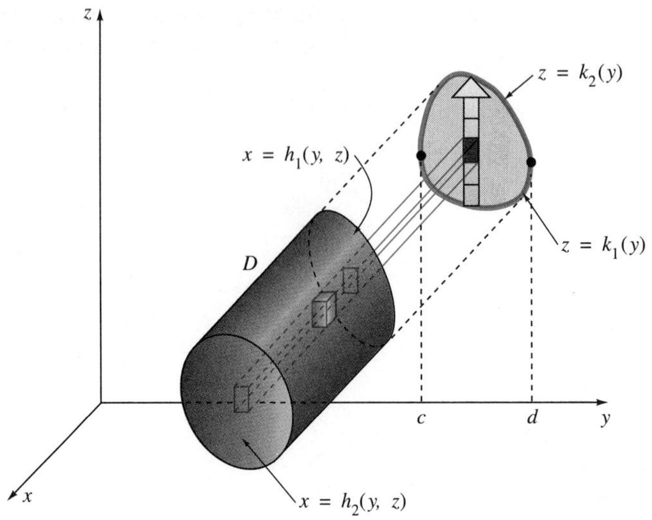

FIGURE 16.47

Applications A list of some of the standard applications of the triple integral follows.

Volume If $F(x, y, z) = 1$, then the **volume** of the solid D is

$$V = \iiint_D dV$$

Mass If $\rho(x, y, z)$ is density, then the **mass** of the solid D is given by

$$m = \iiint_D \rho(x, y, z)\, dV$$

First Moments The **first moments** of the solid about the coordinate planes indicated by the subscripts are given by

$$M_{xy} = \iiint_D z\rho(x, y, z)\, dV \qquad M_{xz} = \iiint_D y\rho(x, y, z)\, dV$$

$$M_{yz} = \iiint_D x\rho(x, y, z)\, dV$$

Center of Mass The coordinates of the **center of mass** of D are given by

$$\bar{x} = \frac{M_{yz}}{m} \qquad \bar{y} = \frac{M_{xz}}{m} \qquad \bar{z} = \frac{M_{xy}}{m}$$

Centroid If $\rho(x, y, z) =$ a constant, the center of mass is called the **centroid** of the solid.

Second Moments The **second moments**, or **moments of inertia** of D about the coordinate axes indicated by the subscripts, are given by

$$I_x = \iiint_D (y^2 + z^2)\rho(x, y, z)\, dV \qquad I_y = \iiint_D (x^2 + z^2)\rho(x, y, z)\, dV$$

$$I_z = \iiint_D (x^2 + y^2)\rho(x, y, z)\, dV$$

Radius of Gyration As in Section 16.4, if I is a moment of inertia of the solid about a given axis, then the **radius of gyration** is

$$R_g = \sqrt{\frac{I}{m}}$$

EXAMPLE 1 Find the volume of the solid in the first octant bounded by the graphs of $z = 1 - y^2$, $y = 2x$, and $x = 3$.

Solution As indicated in Figure 16.48(a), the first integration with respect to z will be from 0 to $1 - y^2$. Furthermore, from Figure 16.48(b) we see that the projection of the solid D in the xy-plane is a region of Type II. Hence, we next integrate, with respect to x, from $y/2$ to 3. The last integration is with respect to y from 0 to 1. Thus,

$$V = \iiint_D dV = \int_0^1 \int_{y/2}^3 \int_0^{1-y^2} dz\, dx\, dy = \int_0^1 \int_{y/2}^3 (1 - y^2)\, dx\, dy$$

$$= \int_0^1 \left[x - xy^2 \right]_{y/2}^3 dy = \int_0^1 \left(3 - 3y^2 - \frac{1}{2}y + \frac{1}{2}y^3 \right) dy$$

$$= \left[3y - y^3 - \frac{1}{4}y^2 + \frac{1}{8}y^4 \right]_0^1 = \frac{15}{8} \text{ cubic units}$$

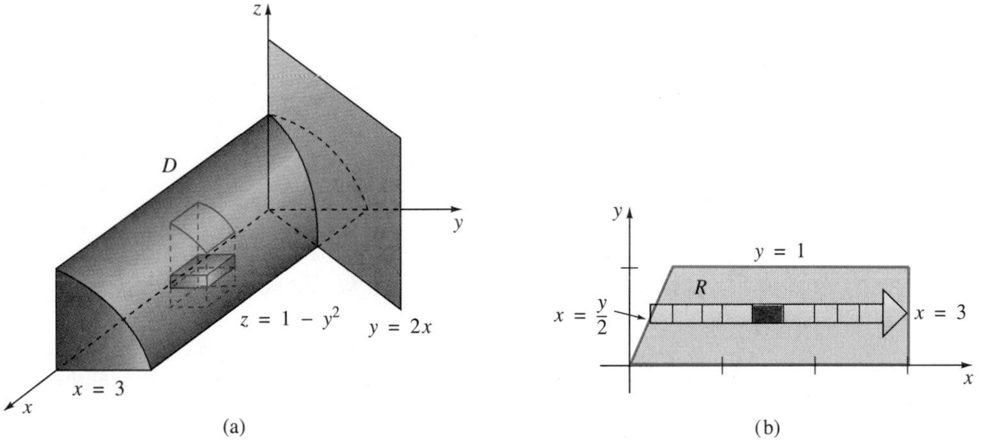

FIGURE 16.48

□

You should observe that the volume in Example 1 could have been obtained just as easily by means of a double integral.

EXAMPLE 2 Find the triple integral that gives the volume of the solid that has the shape determined by the cone $x = \sqrt{y^2 + z^2}$ and the paraboloid $x = 6 - y^2 - z^2$.

Solution By substituting $y^2 + z^2 = x^2$ in $y^2 + z^2 = 6 - x$, we find that $x^2 = 6 - x$ or $(x + 3)(x - 2) = 0$. Thus, the two surfaces intersect in the plane $x = 2$. The projection onto the yz-plane of the curve of intersection is $y^2 + z^2 = 4$. Using symmetry and referring to Figure 16.49(a) and (b), we see that

$$V = \iiint_D dV = 4 \int_0^2 \int_0^{\sqrt{4-y^2}} \int_{\sqrt{y^2+z^2}}^{6-y^2-z^2} dx\,dz\,dy$$

While evaluation of this integral is straightforward, it is admittedly "messy." We shall return to this integral in the next section after we have examined triple integrals in other coordinate systems.

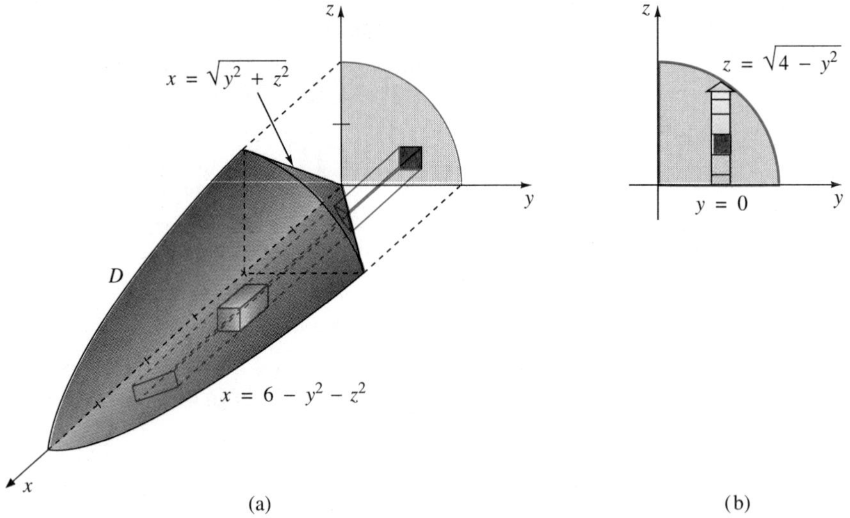

FIGURE 16.49 □

EXAMPLE 3 A solid has the shape determined by the graphs of the cylinder $|x| + |y| = 1$* and the planes $z = 2$ and $z = 4$. Find its center of mass if the density is given by $\rho(x, y, z) = kz$, k a constant.

Solution The solid and its orthogonal projection onto a region R of Type I in the xy-plane are shown in Figure 16.50(a). Since the density function

*This is equivalent to four lines: $x + y = 1$, $x > 0$, $y > 0$; $x - y = 1$, $x > 0$, $y < 0$; and so on.

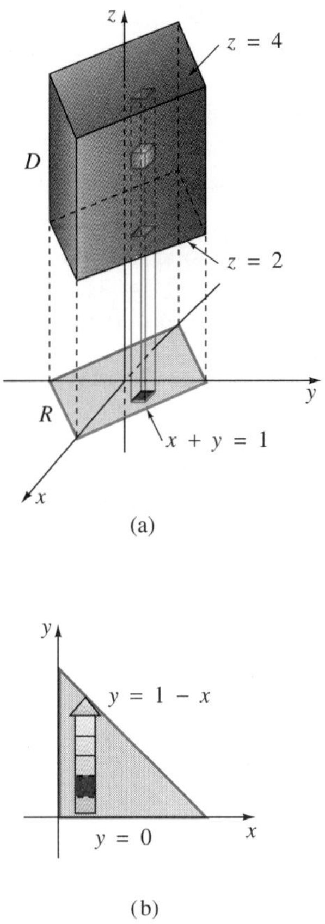

FIGURE 16.50

$\rho(x, y, z) = kz$ is symmetric over R, we conclude that the center of mass is on the z-axis; that is, we need compute only m and M_{xy}. From symmetry and Figure 16.50(b) it follows that

$$m = 4\int_0^1\int_0^{1-x}\int_2^4 kz\,dz\,dy\,dx = 4k\int_0^1\int_0^{1-x}\left.\frac{z^2}{2}\right]_2^4 dy\,dx$$

$$= 24k\int_0^1\int_0^{1-x} dy\,dx$$

$$= 24k\int_0^1 (1-x)\,dx$$

$$= 24k\left[x - \frac{x^2}{2}\right]_0^1 = 12k$$

$$M_{xy} = 4\int_0^1\int_0^{1-x}\int_2^4 kz^2\,dz\,dy\,dx = 4k\int_0^1\int_0^{1-x}\left.\frac{z^3}{3}\right]_2^4 dy\,dx$$

$$= \frac{224}{3}k\int_0^1\int_0^{1-x} dy\,dx = \frac{112}{3}k$$

Hence,

$$\bar{z} = \frac{M_{xy}}{m} = \frac{112k/3}{12k} = \frac{28}{9}$$

The coordinates of the center of mass are then $(0, 0, \frac{28}{9})$. □

EXAMPLE 4 Find the moment of inertia of the solid in Example 3 about the z-axis. Find the radius of gyration.

Solution We know that $I_z = \iiint_D (x^2 + y^2)kz\,dV$. Using symmetry again, we can write this triple integral as

$$I_z = 4k\int_0^1\int_0^{1-x}\int_2^4 (x^2+y^2)z\,dz\,dy\,dx = 4k\int_0^1\int_0^{1-x}\left.(x^2+y^2)\frac{z^2}{2}\right]_2^4 dy\,dx$$

$$= 24k\int_0^1\int_0^{1-x}(x^2+y^2)\,dy\,dx$$

$$= 24k\int_0^1\left[x^2y + \frac{y^3}{3}\right]_0^{1-x} dx$$

$$= 24k\int_0^1\left[x^2 - x^3 + \frac{1}{3}(1-x)^3\right]dx$$

$$= 24k\left[\frac{x^3}{3} - \frac{x^4}{4} - \frac{1}{12}(1-x)^4\right]_0^1 = 4k$$

Since $m = 12k$, it follows that the radius of gyration is

$$R_g = \sqrt{\frac{I_z}{m}} = \sqrt{\frac{4k}{12k}} = \frac{\sqrt{3}}{3}$$ □

EXAMPLE 5 Change the order of integration in

$$\int_0^6 \int_0^{4-2x/3} \int_0^{3-x/2-3y/4} F(x, y, z)\, dz\, dy\, dx$$

to $dy\, dx\, dz$.

Solution As seen in Figure 16.51(a), the region D is the solid in the first octant bounded by the three coordinate planes and the plane $2x + 3y + 4z = 12$. Referring to Figure 16.51(b) and the included table, we conclude that

$$\int_0^6 \int_0^{4-2x/3} \int_0^{3-x/2-3y/4} F(x, y, z)\, dz\, dy\, dx = \int_0^3 \int_0^{6-2z} \int_0^{4-2x/3-4z/3} F(x, y, z)\, dy\, dx\, dz$$

Order of Integration	First Integration	Second Integration	Third Integration
$dz\, dy\, dx$	0 to $3 - x/2 - 3y/4$	0 to $4 - 2x/3$	0 to 6
$dy\, dx\, dz$	0 to $4 - 2x/3 - 4z/3$	0 to $6 - 2z$	0 to 3

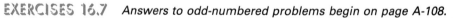

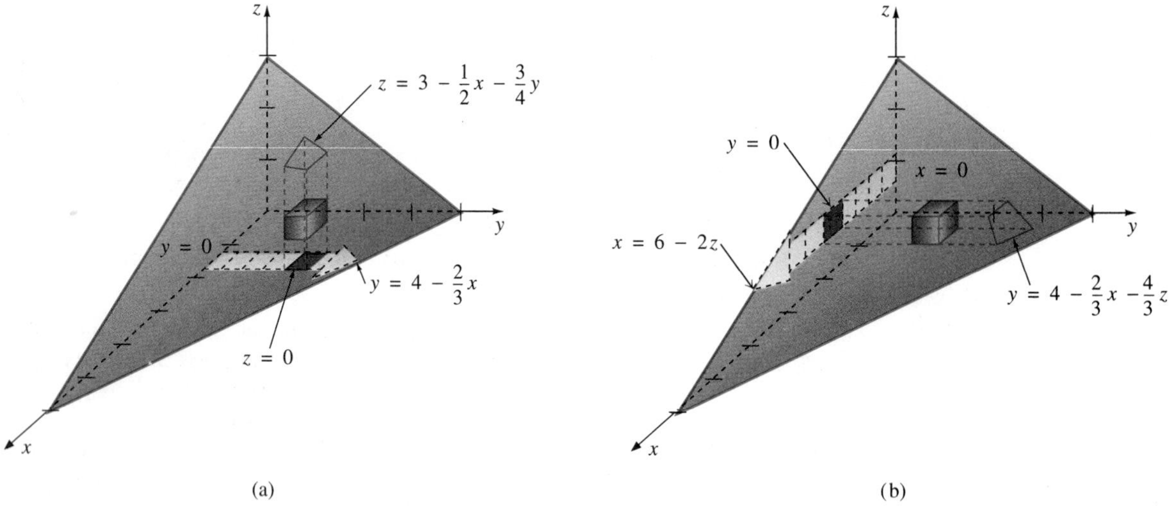

FIGURE 16.51 □

EXERCISES 16.7 *Answers to odd-numbered problems begin on page A-108.*

In Problems 1–8 evaluate the given iterated integral.

1. $\int_2^4 \int_{-2}^2 \int_{-1}^1 (x + y + z)\, dx\, dy\, dz$

2. $\int_1^3 \int_1^x \int_2^{xy} 24xy\, dz\, dy\, dx$

3. $\int_0^6 \int_0^{6-x} \int_0^{6-x-z} dy\, dz\, dx$

4. $\int_0^1 \int_0^{1-x} \int_0^{\sqrt{y}} 4x^2z^3\, dz\, dy\, dx$

5. $\int_0^{\pi/2}\int_0^{y^2}\int_0^{y} \cos\left(\frac{x}{y}\right) dz\, dx\, dy$

6. $\int_0^{\sqrt{2}}\int_{\sqrt{y}}^{2}\int_0^{e^{x^2}} x\, dz\, dx\, dy$

7. $\int_0^{1}\int_0^{1}\int_0^{2-x^2-y^2} xye^z\, dz\, dx\, dy$

8. $\int_0^{4}\int_0^{1/2}\int_0^{x^2} \frac{1}{\sqrt{x^2-y^2}}\, dy\, dx\, dz$

9. Evaluate $\iiint_D z\, dV$, where D is the region in the first octant bounded by the graphs of $y = x$, $y = x - 2$, $y = 1$, $y = 3$, $z = 0$, and $z = 5$.

10. Evaluate $\iiint_D (x^2 + y^2)\, dV$, where D is the region bounded by the graphs of $y = x^2$, $z = 4 - y$, and $z = 0$.

In Problems 11 and 12 change the indicated order of integration to each of the other five orders.

11. $\int_0^{2}\int_0^{4-2y}\int_{x+2y}^{4} F(x, y, z)\, dz\, dx\, dy$

12. $\int_0^{2}\int_0^{\sqrt{36-9x^2}/2}\int_1^{3} F(x.\ y, z)\, dz\, dy\, dx$

In Problems 13 and 14 consider the solid given in the figure. Set up, but do not evaluate, the integrals, giving the volume V of the solid using the indicated orders of integration.

13. (a) $dz\, dy\, dx$ (b) $dx\, dz\, dy$ (c) $dy\, dx\, dz$

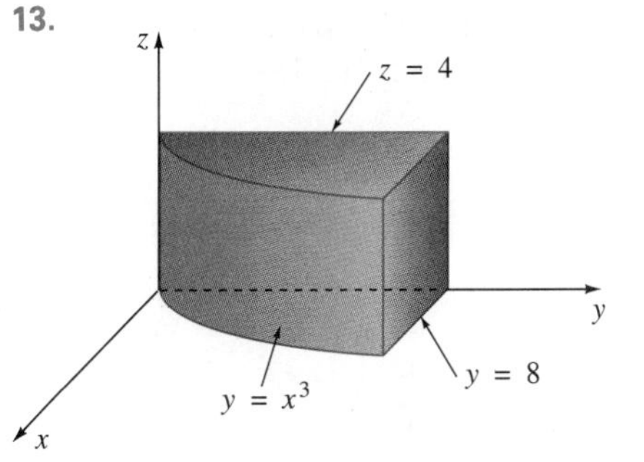

FIGURE 16.52

14. (a) $dx\, dz\, dy$ (b) $dy\, dx\, dz$ (c) $dz\, dx\, dy$ [*Hint:* This will require two integrals.]

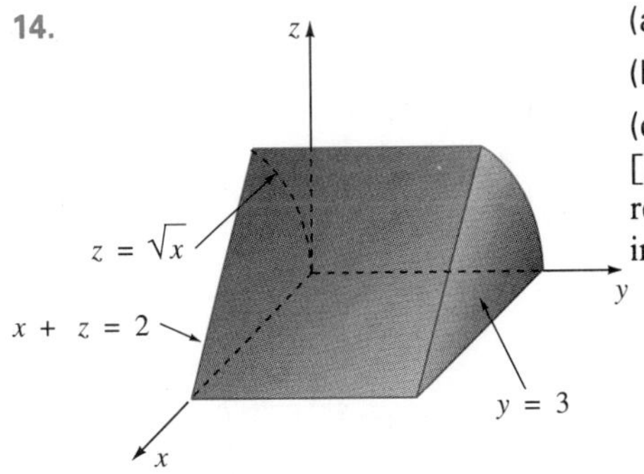

FIGURE 16.53

In Problems 15–20 sketch the region D whose volume V is given by the iterated integral.

15. $\int_0^{4}\int_0^{3}\int_0^{2-2z/3} dx\, dz\, dy$

16. $4\int_0^{3}\int_0^{\sqrt{9-y^2}}\int_4^{\sqrt{25-x^2-y^2}} dz\, dx\, dy$

17. $\int_{-1}^{1}\int_{-\sqrt{1-x^2}}^{\sqrt{1-x^2}}\int_0^{5} dz\, dy\, dx$

18. $\int_0^{2}\int_0^{\sqrt{4-x^2}}\int_{x^2+y^2}^{4} dz\, dy\, dx$

19. $\int_0^{2}\int_0^{2-y}\int_{-\sqrt{y}}^{\sqrt{y}} dx\, dz\, dy$

20. $\int_1^{3}\int_0^{1/x}\int_0^{3} dy\, dz\, dx$

In Problems 21–24 find the volume of the solid bounded by the graphs of the given equations.

21. $x = y^2$, $4 - x = y^2$, $z = 0$, $z = 3$

22. $x^2 + y^2 = 4$, $z = x + y$, the coordinate planes, first octant

23. $y = x^2 + z^2$, $y = 8 - x^2 - z^2$

24. $x = 2$, $y = x$, $y = 0$, $z = x^2 + y^2$, $z = 0$

25. Find the center of mass of the solid given in Figure 16.52 if the density at a point P is directly proportional to the distance from the xy-plane.

26. Find the centroid of the solid in Figure 16.53 if the density is constant.

27. Find the center of mass of the solid bounded by the graphs of $x^2 + z^2 = 4$, $y = 0$, and $y = 3$ if the density at a point P is directly proportional to the distance from the xz-plane.

28. Find the center of mass of the solid bounded by the graphs of $y = x^2$, $y = x$, $z = y + 2$, and $z = 0$ if the density at a point P is directly proportional to the distance from the xy-plane.

In Problems 29 and 30 set up, but do not evaluate, the iterated integrals giving the mass of the solid that has the given shape and density.

29. $x^2 + y^2 = 1$, $z + y = 8$, $z - 2y = 2$; $\rho(x, y, z) = x + y + 4$

30. $x^2 + y^2 - z^2 = 1$, $z = -1$, $z = 2$; $\rho(x, y, z) = z^2$ [*Hint:* Do not use $dz\, dy\, dx$.]

31. Find the moment of inertia of the solid in Figure 16.52 about the y-axis if the density is as given in Problem 25. Find the radius of gyration.

32. Find the moment of inertia of the solid in Figure 16.53 about the x-axis if the density is constant. Find the radius of gyration.

33. Find the moment of inertia about the z-axis of the solid in the first octant that is bounded by the coordinate planes and the graph of $x + y + z = 1$ if the density is constant.

34. Find the moment of inertia about the y-axis of the solid bounded by the graphs of $z = y$, $z = 4 - y$, $z = 1$, $z = 0$, $x = 2$, and $x = 0$ if the density at a point P is directly proportional to the distance from the yz-plane.

In Problems 35 and 36 set up, but do not evaluate, the iterated integral giving the indicated moment of inertia of the solid having the given shape and density.

35. $z = \sqrt{x^2 + y^2}$, $z = 5$; density at a point P directly proportional to the distance from the origin; I_z

36. $x^2 + z^2 = 1$, $y^2 + z^2 = 1$; density at a point P directly proportional to the distance from the yz-plane; I_y

16.8 TRIPLE INTEGRALS IN OTHER COORDINATE SYSTEMS

Depending on the geometry of a region in 3-space, the evaluation of a triple integral over that region may be made easier by utilizing a new coordinate system.

16.8.1 Cylindrical Coordinates

The **cylindrical coordinate system** combines the polar description of a point in the plane with the rectangular description of the z-component of a point in space. As seen in Figure 16.54(a), the cylindrical coordinates of a point P are denoted by the ordered triple (r, θ, z). The word *cylindrical* arises from the fact that a point P in space is determined by the intersection of the planes $z =$ constant, $\theta =$ constant, with a cylinder $r =$ constant. See Figure 16.54(b).

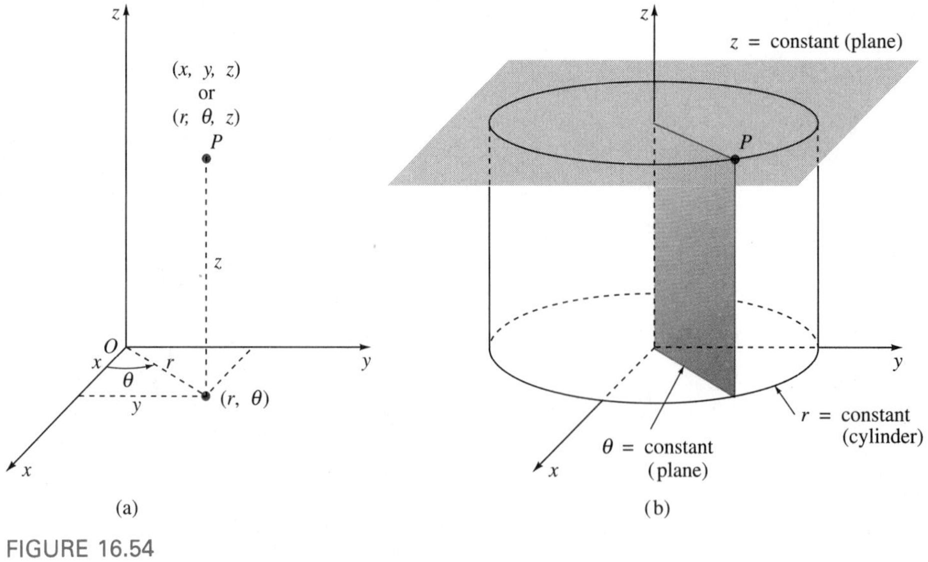

FIGURE 16.54

Conversion of Cylindrical Coordinates to Rectangular Coordinates From Figure 16.54(a) we also see that the rectangular coordinates (x, y, z) of a point can be obtained from the cylindrical coordinates (r, θ, z) by means of

$$x = r\cos\theta \qquad y = r\sin\theta \qquad z = z \tag{1}$$

EXAMPLE 1 Convert $(8, \pi/3, 7)$ in cylindrical coordinates to rectangular coordinates.

Solution From (1),

$$x = 8\cos\frac{\pi}{3} = 8\left(\frac{1}{2}\right) = 4$$

$$y = 8\sin\frac{\pi}{3} = 8\left(\frac{\sqrt{3}}{2}\right) = 4\sqrt{3}$$

$$z = 7$$

Thus, $(8, \pi/3, 7)$ is equivalent to $(4, 4\sqrt{3}, 7)$ in rectangular coordinates. □

Conversion of Rectangular Coordinates to Cylindrical Coordinates To express rectangular coordinates (x, y, z) as cylindrical coordinates, we use

$$r^2 = x^2 + y^2 \qquad \tan\theta = \frac{y}{x} \qquad z = z \tag{2}$$

EXAMPLE 2 Convert $(-\sqrt{2}, \sqrt{2}, 1)$ in rectangular coordinates to cylindrical coordinates.

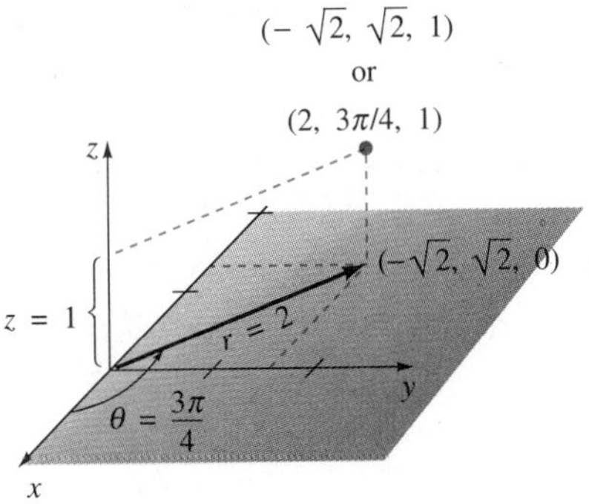

FIGURE 16.55

Solution From (2) we see that

$$r^2 = (-\sqrt{2})^2 + (\sqrt{2})^2 = 4$$

$$\tan\theta = \frac{\sqrt{2}}{-\sqrt{2}} = -1$$

$$z = 1$$

If we take $r = 2$, then, consistent with the fact that $x < 0$ and $y > 0$, we take $\theta = 3\pi/4$.† Consequently, $(-\sqrt{2}, \sqrt{2}, 1)$ is equivalent to $(2, 3\pi/4, 1)$ in cylindrical coordinates. See Figure 16.55. □

Triple Integrals in Cylindrical Coordinates Recall from Section 16.5 that the area of a "polar rectangle" is $\Delta A = r^*\,\Delta r\,\Delta\theta$, where r^* is the average radius. From Figure 16.56(a) we see that the volume of a "cylindrical wedge" is simply

$$\Delta V = (\text{area of base})\cdot(\text{height}) = r^*\,\Delta r\,\Delta\theta\,\Delta z$$

---

†If we use $\theta = \tan^{-1}(-1) = -\pi/4$, then we can use $r = -2$. Notice that the combinations $r = 2$, $\theta = -\pi/4$ and $r = -2$, $\theta = 3\pi/4$ are inconsistent.

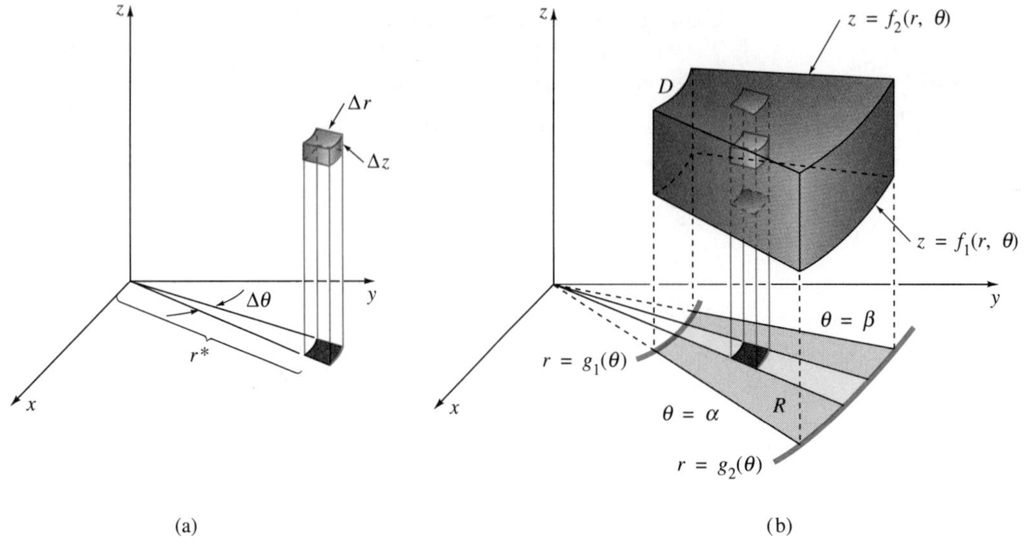

FIGURE 16.56

Thus, if $F(r, \theta, z)$ is a continuous function over the region D, as shown in Figure 16.56(b), then the triple integral of F over D is given by

$$\iiint_D F(r, \theta, z)\, dV = \iint_R \left[\int_{f_1(r,\theta)}^{f_2(r,\theta)} F(r, \theta, z)\, dz\right] dA = \int_\alpha^\beta \int_{g_1(\theta)}^{g_2(\theta)} \int_{f_1(r,\theta)}^{f_2(r,\theta)} F(r, \theta, z) r\, dz\, dr\, d\theta$$

EXAMPLE 3 A solid in the first octant has the shape determined by the graph of the cone $z = \sqrt{x^2 + y^2}$ and the planes $z = 1$, $x = 0$, and $y = 0$. Find the center of mass if the density is given by $\rho(r, \theta, z) = r$.

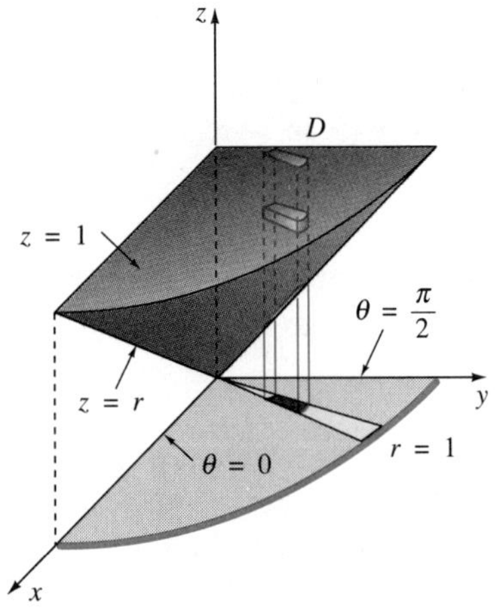

FIGURE 16.57

Solution In view of (2), the equation of the cone is $z = r$. Hence, we see from Figure 16.57 that

$$m = \iiint_D r\, dV = \int_0^{\pi/2} \int_0^1 \int_r^1 r(r\, dz\, dr\, d\theta)$$

$$= \int_0^{\pi/2} \int_0^1 r^2 z \Big]_r^1 dr\, d\theta$$

$$= \int_0^{\pi/2} \int_0^1 (r^2 - r^3)\, dr\, d\theta = \frac{\pi}{24}$$

$$M_{xy} = \iiint_D zr\, dV = \int_0^{\pi/2} \int_0^1 \int_r^1 zr^2\, dz\, dr\, d\theta$$

$$= \int_0^{\pi/2} \int_0^1 \frac{z^2}{2} r^2 \Big]_r^1 dr\, d\theta$$

$$= \frac{1}{2} \int_0^{\pi/2} \int_0^1 (r^2 - r^4)\, dr\, d\theta = \frac{\pi}{30}$$

Using $y = r\sin\theta$ and $x = r\cos\theta$, we also have

$$M_{xz} = \iiint_D r^2\sin\theta\, dV = \int_0^{\pi/2}\int_0^1\int_r^1 r^3\sin\theta\, dz\, dr\, d\theta$$

$$= \int_0^{\pi/2}\int_0^1 r^3 z\sin\theta\Big]_r^1 dr\, d\theta$$

$$= \int_0^{\pi/2}\int_0^1 (r^3 - r^4)\sin\theta\, dr\, d\theta = \frac{1}{20}$$

$$M_{yz} = \iiint_D r^2\cos\theta\, dV = \int_0^{\pi/2}\int_0^1\int_r^1 r^3\cos\theta\, dz\, dr\, d\theta = \frac{1}{20}$$

Hence,

$$\bar{x} = \frac{M_{yz}}{m} = \frac{1/20}{\pi/24} = \frac{6}{5\pi} \approx 0.38$$

$$\bar{y} = \frac{M_{xz}}{m} = \frac{1/20}{\pi/24} = \frac{6}{5\pi} \approx 0.38$$

$$\bar{z} = \frac{M_{xy}}{m} = \frac{\pi/30}{\pi/24} = \frac{4}{5} = 0.8$$

The center of mass has the approximate coordinates (0.38, 0.38, 0.8). □

EXAMPLE 4 Evaluate the volume integral

$$V = 4\int_0^2\int_0^{\sqrt{4-y^2}}\int_{\sqrt{y^2+z^2}}^{6-y^2-z^2} dx\, dz\, dy$$

of Example 2 in Section 16.7.

Solution If we introduce polar coordinates in the yz-plane by $y = r\cos\theta$, $z = r\sin\theta$, then the cylindrical coordinates of a point in 3-space are (r, θ, x). The polar analogue of Figure 16.49(b) is given in Figure 16.58. Now, since $y^2 + z^2 = r^2$, we have

$$x = \sqrt{y^2 + z^2} = r \quad \text{and} \quad x = 6 - y^2 - z^2 = 6 - r^2$$

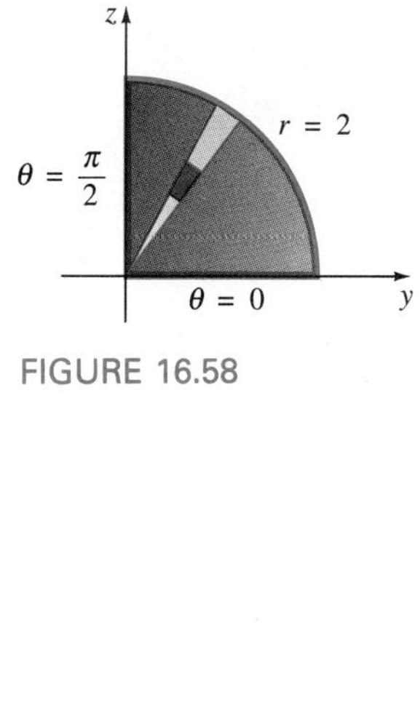

FIGURE 16.58

Hence, the integral becomes

$$V = 4\int_0^{\pi/2}\int_0^2\int_r^{6-r^2} r\, dx\, dr\, d\theta = 4\int_0^{\pi/2}\int_0^2 rx\Big]_r^{6-r^2} dr\, d\theta$$

$$= 4\int_0^{\pi/2}\int_0^2 (6r - r^3 - r^2)\, dr\, d\theta$$

$$= 4\int_0^{\pi/2}\left[3r^2 - \frac{r^4}{4} - \frac{r^3}{3}\right]_0^2 d\theta$$

$$= \frac{64}{3}\int_0^{\pi/2} d\theta = \frac{32\pi}{3} \text{ cubic units}$$ □

16.8.2 Spherical Coordinates

As seen in Figure 16.59(a), the **spherical coordinates** of a point P are given by the ordered triple (ρ, ϕ, θ), where ρ is the distance from the origin to P, ϕ is the angle between the positive z-axis and the vector $\overrightarrow{OP}$, and θ is the angle measured from the positive x-axis to the vector projection $\overrightarrow{OQ}$ of $\overrightarrow{OP}$.* Figure 16.59(b) shows that a point P in space is determined by the intersection of a cone $\phi =$ constant, a plane $\theta =$ constant, and a sphere $\rho =$ constant; whence arises the name "spherical" coordinates.

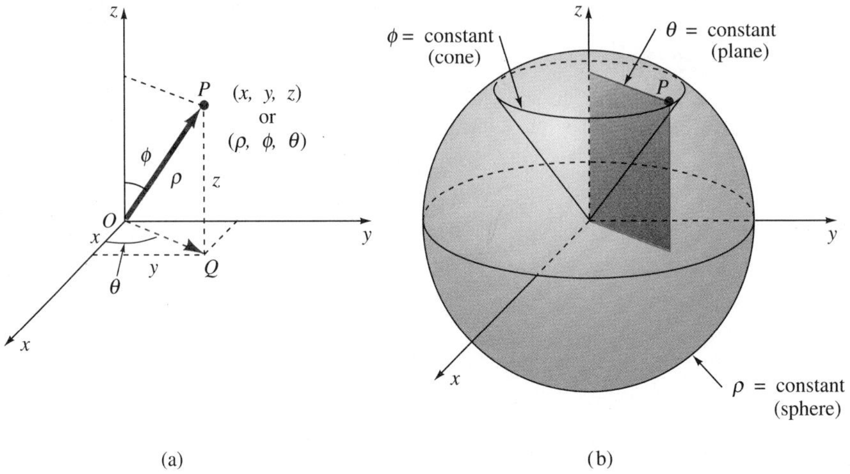

FIGURE 16.59

Conversion of Spherical Coordinates to Rectangular and Cylindrical Coordinates To transform from spherical coordinates (ρ, ϕ, θ) to rectangular coordinates (x, y, z), we observe from Figure 16.59(a) that

$$x = |\overrightarrow{OQ}|\cos\theta \qquad y = |\overrightarrow{OQ}|\sin\theta \qquad z = |\overrightarrow{OP}|\cos\phi$$

Since $|\overrightarrow{OQ}| = \rho\sin\phi$ and $|\overrightarrow{OP}| = \rho$, the foregoing equations become

$$x = \rho\sin\phi\cos\theta \qquad y = \rho\sin\phi\sin\theta \qquad z = \rho\cos\phi \tag{3}$$

It is customary to take $\rho \geq 0$ and $0 \leq \phi \leq \pi$. Also, since $|\overrightarrow{OQ}| = \rho\sin\phi = r$, the formulas

$$r = \rho\sin\phi \qquad \theta = \theta \qquad z = \rho\cos\phi \tag{4}$$

enable us to transform spherical coordinates (ρ, ϕ, θ) into cylindrical coordinates (r, θ, z).

*θ is the same angle as in polar and cylindrical coordinates.

EXAMPLE 5 Convert $(6, \pi/4, \pi/3)$ in spherical coordinates to (*a*) rectangular coordinates and (*b*) cylindrical coordinates.

Solution

(*a*) Identifying $\rho = 6$, $\phi = \pi/4$, and $\theta = \pi/3$, we find from (3) that the rectangular coordinates of the point are given by

$$x = 6 \sin\frac{\pi}{4}\cos\frac{\pi}{3} = 6\left(\frac{\sqrt{2}}{2}\right)\left(\frac{1}{2}\right) = \frac{3\sqrt{2}}{2}$$

$$y = 6 \sin\frac{\pi}{4}\sin\frac{\pi}{3} = 6\left(\frac{\sqrt{2}}{2}\right)\left(\frac{\sqrt{3}}{2}\right) = \frac{3\sqrt{6}}{2}$$

$$z = 6 \cos\frac{\pi}{4} = 6\left(\frac{\sqrt{2}}{2}\right) = 3\sqrt{2}$$

(*b*) From (4) we obtain

$$r = 6 \sin\frac{\pi}{4} = 6\left(\frac{\sqrt{2}}{2}\right) = 3\sqrt{2}$$

$$\theta = \frac{\pi}{3}$$

$$z = 6 \cos\frac{\pi}{4} = 6\left(\frac{\sqrt{2}}{2}\right) = 3\sqrt{2}$$

Thus, the cylindrical coordinates of the point are $(3\sqrt{2}, \pi/3, 3\sqrt{2})$. □

Conversion of Rectangular Coordinates to Spherical Coordinates To transform rectangular coordinates into spherical coordinates, we use

$$\rho^2 = x^2 + y^2 + z^2 \qquad \tan\theta = \frac{y}{x} \qquad \cos\phi = \frac{z}{\sqrt{x^2 + y^2 + z^2}} \tag{5}$$

Triple Integrals in Spherical Coordinates As seen in Figure 16.60, the volume of a "spherical wedge" is given by the approximation

$$\Delta V \approx \rho^2 \sin\phi \,\Delta\rho\,\Delta\phi\,\Delta\theta$$

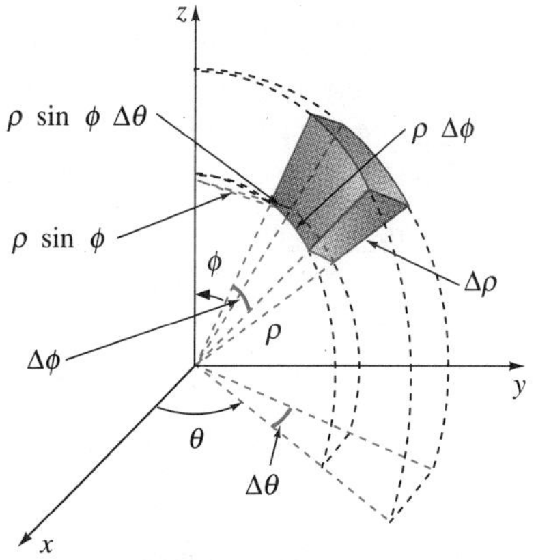

FIGURE 16.60

Thus, in a triple integral of a continuous spherical-coordinate function $F(\rho, \phi, \theta)$, the differential of volume dV is given by

$$dV = \rho^2 \sin\phi \, d\rho \, d\phi \, d\theta$$

A typical triple integral in spherical coordinates has the form

$$\iiint_D F(\rho, \phi, \theta)\, dV = \int_\alpha^\beta \int_{g_1(\theta)}^{g_2(\theta)} \int_{f_1(\phi,\theta)}^{f_2(\phi,\theta)} F(\rho, \phi, \theta)\rho^2 \sin\phi \, d\rho \, d\phi \, d\theta$$

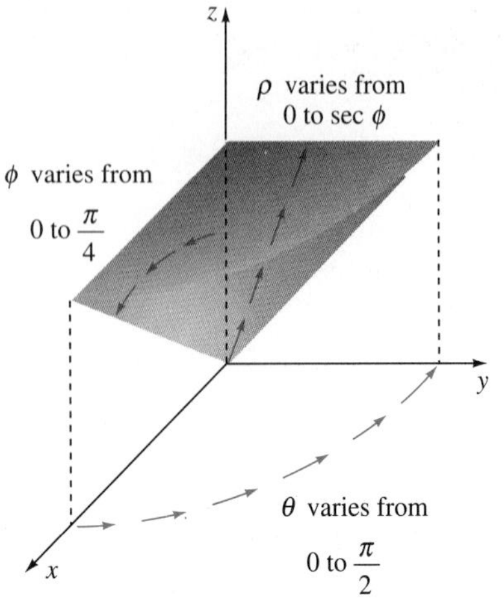

FIGURE 16.61

EXAMPLE 6 Use spherical coordinates to find the volume of the solid in Example 3.

Solution Using (3),

$$z = 1 \quad \text{becomes} \quad \rho = \sec\phi \qquad \text{and} \qquad z = \sqrt{x^2 + y^2} \quad \text{becomes} \quad \phi = \pi/4$$

As indicated in Figure 16.61, $V = \iiint_D dV$ written as an iterated integral is

$$\begin{aligned} V &= \int_0^{\pi/2}\int_0^{\pi/4}\int_0^{\sec\phi} \rho^2 \sin\phi \, d\rho \, d\phi \, d\theta \\ &= \int_0^{\pi/2}\int_0^{\pi/4} \frac{\rho^3}{3}\bigg]_0^{\sec\phi} \sin\phi \, d\phi \, d\theta \\ &= \frac{1}{3}\int_0^{\pi/2}\int_0^{\pi/4} \sec^3\phi \sin\phi \, d\phi \, d\theta \\ &= \frac{1}{3}\int_0^{\pi/2}\int_0^{\pi/4} \tan\phi \sec^2\phi \, d\phi \, d\theta \\ &= \frac{1}{3}\int_0^{\pi/2} \frac{1}{2}\tan^2\phi\bigg]_0^{\pi/4} d\theta \\ &= \frac{1}{6}\int_0^{\pi/2} d\theta = \frac{\pi}{12} \text{ cubic units} \end{aligned}$$

□

EXAMPLE 7 Find the moment of inertia about the z-axis of the homogeneous solid bounded between the spheres $x^2 + y^2 + z^2 = a^2$ and $x^2 + y^2 + z^2 = b^2$, $a < b$.

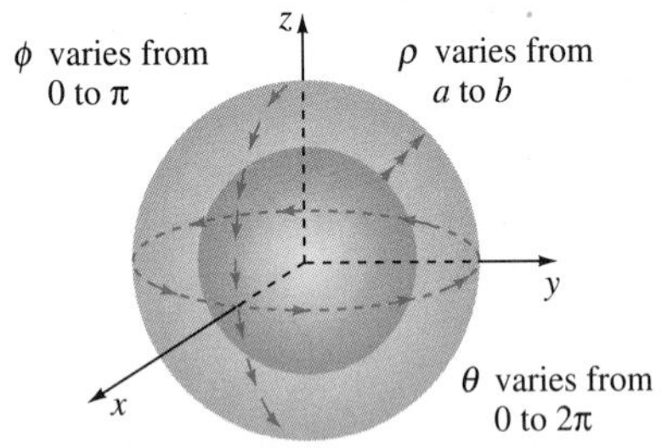

FIGURE 16.62

Solution If $\delta(\rho, \phi, \theta) = k$ is the density,* then

$$I_z = \iiint_D (x^2 + y^2)k \, dV$$

From (3) we find $x^2 + y^2 = \rho^2\sin^2\phi$, and from the first equation in (5) we see that the equations of the spheres are simply $\rho = a$ and $\rho = b$. See Figure 16.62. Consequently, in spherical coordinates the foregoing integral becomes

$$\begin{aligned} I_z &= k\int_0^{2\pi}\int_0^{\pi}\int_a^b \rho^2\sin^2\phi(\rho^2\sin\phi \, d\rho \, d\phi \, d\theta) \\ &= k\int_0^{2\pi}\int_0^{\pi}\int_a^b \rho^4\sin^3\phi \, d\rho \, d\phi \, d\theta \end{aligned}$$

*We must use a different symbol to denote density to avoid confusion with the symbol ρ of spherical coordinates.

$$= k\int_0^{2\pi}\int_0^{\pi}\frac{\rho^5}{5}\sin^3\phi\Big]_a^b\,d\phi\,d\theta$$

$$= \frac{k}{5}(b^5 - a^5)\int_0^{2\pi}\int_0^{\pi}(1 - \cos^2\phi)\sin\phi\,d\phi\,d\theta$$

$$= \frac{k}{5}(b^5 - a^5)\int_0^{2\pi}\left[-\cos\phi + \frac{1}{3}\cos^3\phi\right]_0^{\pi}d\theta$$

$$= \frac{4k}{15}(b^5 - a^5)\int_0^{2\pi}d\theta = \frac{8\pi k}{15}(b^5 - a^5) \qquad \square$$

▲ ***Remark*** Spherical coordinates are used in navigation. If we think of the earth as a sphere of fixed radius centered at the origin, then a point P can be located by specifying two angles θ and ϕ. As shown in Figure 16.63, when ϕ is held constant the resulting curve is called a **parallel**. Fixed values of θ result in curves called **great circles**. Half of one of these great circles joining the north and south poles is called a **meridian**. The intersection of a parallel and a meridian gives the position of a point P. If $0° \le \phi \le 180°$ and $-180° \le \theta \le 180°$, the angles $90° - \phi$ and θ are said to be the **latitude** and **longitude** of P, respectively. The **prime meridian** corresponds to a longitude of $0°$. The latitude of the equator is $0°$; the latitudes of the north and south poles are, in turn, $+90°$ (or $90°$ north) and $-90°$ (or $90°$ south).

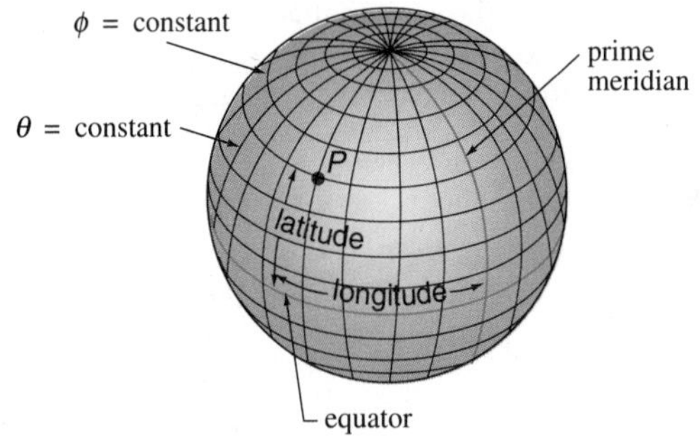

FIGURE 16.63

EXERCISES 16.8 *Answers to odd-numbered problems begin on page A-108.*

[16.8.1]

In Problems 1–6 convert the point given in cylindrical coordinates to rectangular coordinates.

1. $\left(10, \frac{3\pi}{4}, 5\right)$
2. $\left(2, \frac{5\pi}{6}, -3\right)$
3. $\left(\sqrt{3}, \frac{\pi}{3}, -4\right)$
4. $\left(4, \frac{7\pi}{4}, 0\right)$
5. $\left(5, \frac{\pi}{2}, 1\right)$
6. $\left(10, \frac{5\pi}{3}, 2\right)$

In Problems 7–12 convert the point given in rectangular coordinates to cylindrical coordinates.

7. $(1, -1, -9)$
8. $(2\sqrt{3}, 2, 17)$
9. $(-\sqrt{2}, \sqrt{6}, 2)$
10. $(1, 2, 7)$
11. $(0, -4, 0)$
12. $(\sqrt{7}, -\sqrt{7}, 3)$

In Problems 13–16 convert the given equation to cylindrical coordinates.

13. $x^2 + y^2 + z^2 = 25$

14. $x + y - z = 1$

15. $x^2 + y^2 - z^2 = 1$

16. $x^2 + z^2 = 16$

In Problems 17–20 convert the given equation to rectangular coordinates.

17. $z = r^2$

18. $z = 2r \sin \theta$

19. $r = 5 \sec \theta$

20. $\theta = \pi/6$

In the following problems use triple integrals and cylindrical coordinates.

In Problems 21–24 find the volume of the solid that is bounded by the graphs of the given equations.

21. $x^2 + y^2 = 4,\ x^2 + y^2 + z^2 = 16,\ z = 0$

22. $z = 10 - x^2 - y^2,\ z = 1$

23. $z = x^2 + y^2,\ x^2 + y^2 = 25,\ z = 0$

24. $y = x^2 + z^2,\ 2y = x^2 + z^2 + 4$

25. Find the centroid of the homogeneous solid that is bounded by the hemisphere $z = \sqrt{a^2 - x^2 - y^2}$ and the plane $z = 0$.

26. Find the center of mass of the solid that is bounded by the graphs of $y^2 + z^2 = 16$, $x = 0$, and $x = 5$ if the density at a point P is directly proportional to distance from the yz-plane.

27. Find the moment of inertia about the z-axis of the solid that is bounded above by the hemisphere $z = \sqrt{9 - x^2 - y^2}$ and below by the plane $z = 2$ if the density at a point P is inversely proportional to the square of the distance from the z-axis.

28. Find the moment of inertia about the x-axis of the solid that is bounded by the cone $z = \sqrt{x^2 + y^2}$ and the plane $z = 1$ if the density at a point P is directly proportional to the distance from the z-axis.

[16.8.2]

In Problems 29–34 convert the point given in spherical coordinates to **(a)** rectangular coordinates and **(b)** cylindrical coordinates.

29. $\left(\frac{2}{3}, \frac{\pi}{2}, \frac{\pi}{6}\right)$

30. $\left(5, \frac{5\pi}{4}, \frac{2\pi}{3}\right)$

31. $\left(8, \frac{\pi}{4}, \frac{3\pi}{4}\right)$

32. $\left(\frac{1}{3}, \frac{5\pi}{3}, \frac{\pi}{6}\right)$

33. $\left(4, \frac{3\pi}{4}, 0\right)$

34. $\left(1, \frac{11\pi}{6}, \pi\right)$

In Problems 35–40 convert the points given in rectangular coordinates to spherical coordinates.

35. $(-5, -5, 0)$

36. $(1, -\sqrt{3}, 1)$

37. $\left(\frac{\sqrt{3}}{2}, \frac{1}{2}, 1\right)$

38. $\left(-\frac{\sqrt{3}}{2}, 0, -\frac{1}{2}\right)$

39. $(3, -3, 3\sqrt{2})$

40. $(1, 1, -\sqrt{6})$

In Problems 41–44 convert the given equation to spherical coordinates.

41. $x^2 + y^2 + z^2 = 64$

42. $x^2 + y^2 + z^2 = 4z$

43. $z^2 = 3x^2 + 3y^2$

44. $-x^2 - y^2 + z^2 = 1$

In Problems 45–48 convert the given equation to rectangular coordinates.

45. $\rho = 10$

46. $\phi = \pi/3$

47. $\rho = 2 \sec \phi$

48. $\rho \sin^2\phi = \cos \phi$

In the remaining problems use triple integrals and spherical coordinates.

In Problems 49–52 find the volume of the solid that is bounded by the graphs of the given equations.

49. $z = \sqrt{x^2 + y^2},\ x^2 + y^2 + z^2 = 9$

50. $x^2 + y^2 + z^2 = 4,\ y = x,\ y = \sqrt{3}x,\ z = 0$, first octant

51. $z^2 = 3x^2 + 3y^2,\ x = 0,\ y = 0,\ z = 2$, first octant

52. Inside $x^2 + y^2 + z^2 = 1$ and outside $z^2 = x^2 + y^2$

53. Find the centroid of the homogeneous solid that is bounded by the cone $z = \sqrt{x^2 + y^2}$ and the sphere $x^2 + y^2 + z^2 = 2z$.

54. Find the center of mass of the solid that is bounded by the hemisphere $z = \sqrt{1 - x^2 - y^2}$ and the plane $z = 0$ if the density at a point P is directly proportional to distance from the xy-plane.

55. Find the mass of the solid that is bounded above by the hemisphere $z = \sqrt{25 - x^2 - y^2}$ and below by the plane $z = 4$ if the density at a point P is inversely proportional to the distance from the origin. [*Hint:* Express the upper ϕ limit of integration as an inverse cosine.]

56. Find the moment of inertia about the z-axis of the solid that is bounded by the sphere $x^2 + y^2 + z^2 = a^2$ if the density at a point P is directly proportional to the distance from the origin.

16.9 CHANGE OF VARIABLES IN MULTIPLE INTEGRALS

Single Integral In many instances it is convenient to make a substitution, or change of variable, in an integral in order to evaluate it. The idea in Theorem 4.14 can be rephrased as follows: If f is continuous and $x = g(u)$ has a continuous derivative and $dx = g'(u)\,du$, then

$$\int_a^b f(x)\,dx = \int_c^d f(g(u))g'(u)\,du \tag{1}$$

where $c = g(a)$ and $d = g(b)$. There are three things that bear emphasizing in (1). To change the variable in a definite integral we replace x where it appears in the integrand by $g(u)$, we change the interval of integration $[a, b]$ on the x-axis to the corresponding interval $[c, d]$ on the u-axis, and we replace dx by a function multiple (namely, the derivative of g) of du. If we write $J(u) = dx/du$, then (1) has the form

$$\int_a^b f(x)\,dx = \int_c^d f(g(u))J(u)\,du \tag{2}$$

For example, using $x = 2\sin\theta$, $-\pi/2 \le \theta \le \pi/2$, we have

$$\underbrace{\int_0^2}_{x\text{-limits}} \overbrace{\sqrt{4-x^2}}^{f(x)}\,dx = \underbrace{\int_0^{\pi/2}}_{\theta\text{-limits}} \overbrace{2\cos\theta}^{f(2\sin\theta)}\;\overbrace{(2\cos\theta)}^{J(\theta)}\,d\theta = 4\int_0^{\pi/2}\cos^2\theta\,d\theta = \pi$$

Double Integrals Although changing variables in a multiple integral is not as straightforward as the procedure in (1), the basic idea illustrated in (2) will carry over. To change variables in a double integral we need two equations, such as

$$x = f(u, v) \qquad y = g(u, v) \tag{3}$$

To be analogous with (2), we expect that a change of variables in a double integral would take the form

$$\iint_R F(x, y)\,dA = \iint_S F(f(u, v), g(u, v))J(u, v)\,dA' \tag{4}$$

where S is the region in the uv-plane corresponding to the region R in the xy-plane, and $J(u, v)$ is some function that depends on partial derivatives of the equations in (3). The symbol dA' on the right side of (4) represents either $du\,dv$ or $dv\,du$.

In Section 16.5 we briefly discussed how to change a double integral $\iint_R F(x, y)\,dA$ from rectangular coordinates to polar coordinates. Recall, in

Example 2 of that section, the substitutions

$$x = r\cos\theta \quad \text{and} \quad y = r\sin\theta \tag{5}$$

led to

$$\int_0^2 \int_x^{\sqrt{8-x^2}} \frac{1}{5+x^2+y^2}\,dy\,dx = \int_{\pi/4}^{\pi/2} \int_0^{\sqrt{8}} \frac{1}{5+r^2}\,r\,dr\,d\theta \tag{6}$$

As we see in Figure 16.64, the introduction of polar coordinates changes the original region of integration R in the xy-plane to the more convenient *rectangular region* of integration S in the $r\theta$-plane. We note too that by comparing (4) with (6), we can identify $J(r, \theta) = r$ and $dA' = dr\,d\theta$.

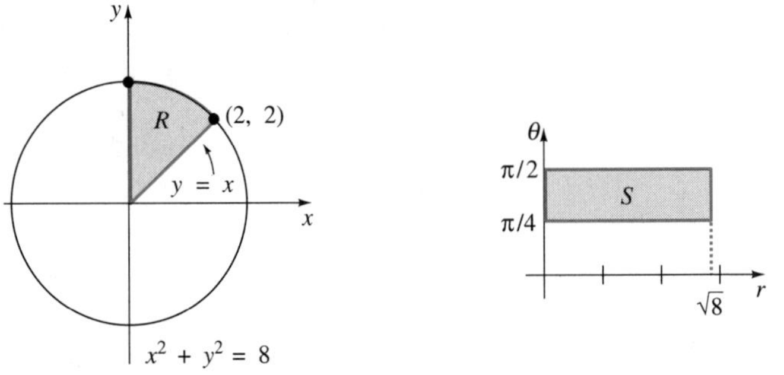

(a) region R in xy-plane (b) region S in $r\theta$-plane

FIGURE 16.64

The change-of-variable equations in (3) define a **transformation** or **mapping** T from the uv-plane to the xy-plane. A point (x_0, y_0) in the xy-plane determined from $x_0 = f(u_0, v_0)$, $y_0 = g(u_0, v_0)$ is said to be an **image** of (u_0, v_0).

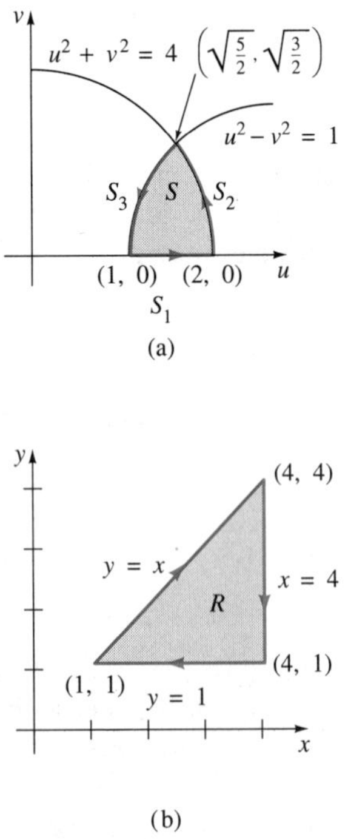

FIGURE 16.65

EXAMPLE 1 Find the image of the region S shown in Figure 16.65(a), under the transformation $x = u^2 + v^2$, $y = u^2 - v^2$.

Solution We begin by finding the images of the sides of S that we have indicated by S_1, S_2, and S_3.

S_1: On this side $v = 0$ so that $x = u^2$, $y = u^2$. Eliminating u then gives $y = x$. Now, imagine moving along the boundary from (1, 0) to (2, 0) (that is, $1 \le u \le 2$). The equations $x = u^2$ and $y = u^2$ then indicate that x ranges from $x = 1$ to $x = 4$ and y ranges simultaneously from $y = 1$ to $y = 4$. In other words, in the xy-plane the image of S_1 is the line segment $y = x$ from (1, 1) to (4, 4).

S_2: On this boundary $u^2 + v^2 = 4$ and so $x = 4$. Now, as we move from the point (2, 0) to $(\sqrt{\tfrac{5}{2}}, \sqrt{\tfrac{3}{2}})$, the remaining equation $y = u^2 - v^2$ indicates that y ranges from $y = 2^2 - 0^2 = 4$ to $y = (\sqrt{\tfrac{5}{2}})^2 - (\sqrt{\tfrac{3}{2}})^2 = 1$. In this case the image of S_2 is the vertical line segment $x = 4$ starting at (4, 4) and going down to (4, 1).

S_3: Since $u^2 - v^2 = 1$, we get $y = 1$. But as we move on this boundary from $(\sqrt{\frac{5}{2}}, \sqrt{\frac{3}{2}})$ to (1, 0), the equation $x = u^2 + v^2$ indicates that x ranges from $x = 4$ to $x = 1$. The image of S_3 is the horizontal line segment $y = 1$ starting at (4, 1) and ending at (1, 1).

The image of S is the region R given in Figure 16.65(b). □

Observe in Example 1 that as we traverse the boundary of S in the counterclockwise direction, the boundary of R is traversed in a clockwise manner. We say that the transformation of the boundary of S has *induced* an orientation on the boundary of R.

Although a proof of the formula for changing variables in a multiple integral is beyond the level of this text, we will give *some* of the underlying assumptions that are made about equations (3) and the regions R and S. We make the following assumptions:

The functions f and g have continuous first partial derivatives on S.

The transformation is one-to-one.

Each of the regions R and S consists of a piecewise smooth simple closed curve and its interior.

The determinant

$$\begin{vmatrix} \dfrac{\partial x}{\partial u} & \dfrac{\partial x}{\partial v} \\ \dfrac{\partial y}{\partial u} & \dfrac{\partial y}{\partial v} \end{vmatrix} = \frac{\partial x}{\partial u}\frac{\partial y}{\partial v} - \frac{\partial x}{\partial v}\frac{\partial y}{\partial u} \tag{7}$$

is not zero on S.

Jacobian A transformation T is said to be **one-to-one** if each point (x_0, y_0) in R is the image under T of a unique point (u_0, v_0) in S. Put another way, no two points in S have the same image in R. With the restrictions that $r \geq 0$ and $0 \leq \theta < 2\pi$, the equations in (5) define a one-to-one transformation from the $r\theta$-plane to the xy-plane. The determinant in (7) is called the **Jacobian determinant**, or simply **Jacobian,*** of the transformation T and is the key to

***Carl Gustav Jacob Jacobi (1804–1851)** Born into a rich German family, the young Carl Gustav excelled in many areas of study, but his ability and love for intricate algebraic calculations led him to the life of a working, and impoverished, mathematician and teacher. His Ph.D. dissertation was related to a topic now known to every student of calculus: partial fractions. But Jacobi's greatest contributions to mathematics were in the field of elliptic functions and number theory. He also made major contributions to the theory of determinants and to the simplification of that theory. Jacobi was principally a "pure" mathematician; nonetheless, every student of dynamics and quantum mechanics will recognize Jacobi's contribution to these areas through the famous Hamilton–Jacobi equations.

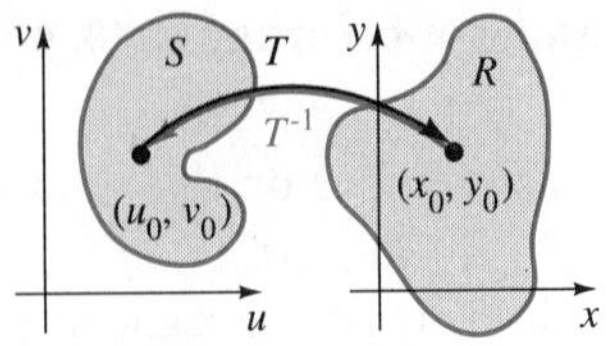

FIGURE 16.66

changing variables in a multiple integral. The Jacobian of the transformation defined by the equations in (3) is denoted by the symbol

$$\frac{\partial(x, y)}{\partial(u, v)}$$

Similar to the notion of a one-to-one function introduced in Section 6.1, a one-to-one transformation T has an **inverse transformation** T^{-1} such that (u_0, v_0) is the image under T^{-1} of (x_0, y_0). See Figure 16.66. If it is possible to solve (3) for u and v in terms of x and y, then the inverse transformation is defined by the pair of equations

$$u = h(x, y) \quad \text{and} \quad v = k(x, y) \tag{8}$$

The Jacobian of the inverse transformation T^{-1} is

$$\frac{\partial(u, v)}{\partial(x, y)} = \begin{vmatrix} \dfrac{\partial u}{\partial x} & \dfrac{\partial u}{\partial y} \\ \dfrac{\partial v}{\partial x} & \dfrac{\partial v}{\partial y} \end{vmatrix} \tag{9}$$

and is related to the Jacobian of the transformation T by

$$\frac{\partial(x, y)}{\partial(u, v)} \frac{\partial(u, v)}{\partial(x, y)} = 1 \tag{10}$$

EXAMPLE 2 The Jacobian of the transformation $x = r \cos \theta$, $y = r \sin \theta$ is

$$\frac{\partial(x, y)}{\partial(r, \theta)} = \begin{vmatrix} \dfrac{\partial x}{\partial r} & \dfrac{\partial x}{\partial \theta} \\ \dfrac{\partial y}{\partial r} & \dfrac{\partial y}{\partial \theta} \end{vmatrix} = \begin{vmatrix} \cos \theta & -r \sin \theta \\ \sin \theta & r \cos \theta \end{vmatrix} = r(\cos^2\theta + \sin^2\theta) = r \qquad \square$$

We shall now turn our attention to the main point of this discussion: how to change variables in a multiple integral. The idea expressed in (4) is valid; the function $J(u, v)$ turns out to be the absolute value of the Jacobian; that is, $J(u, v) = |\partial(x, y)/\partial(u, v)|$. Under the assumptions made above, we have the following result for double integrals.

THEOREM 16.3 Change of Variables in a Double Integral

If F is continuous on a region R and if $x = f(u, v)$, $y = g(u, v)$ is a transformation from the uv-plane to the xy-plane, then

$$\iint_R F(x, y)\, dA = \iint_S F(f(u, v), g(u, v)) \left| \frac{\partial(x, y)}{\partial(u, v)} \right| dA' \tag{11}$$

Formula (3) of Section 16.5 for changing a double integral to polar coordinates is just a special case of (11) with

$$\left|\frac{\partial(x, y)}{\partial(r, \theta)}\right| = |r| = r$$

since $r \geq 0$. In (6), then, we have $J(r, \theta) = |\partial(x, y)/\partial(r, \theta)| = r$.

A change of variables in a multiple integral can be used for either a simplification of the integrand or a simplification of the region of integration. The actual change of variables used is often inspired by the structure of the integrand $F(x, y)$ or by equations that define the region R. As a consequence, the transformation is defined by equations of the form given in (8); that is, we are dealing with the inverse transformation. The next two examples illustrate these ideas.

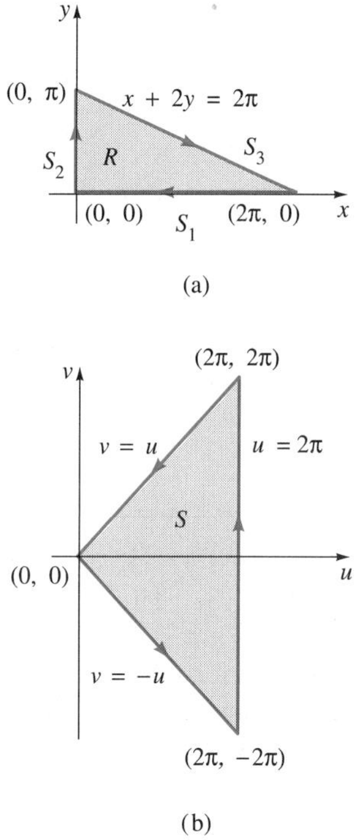

FIGURE 16.67

EXAMPLE 3 Evaluate $\iint_R \sin(x + 2y)\cos(x - 2y)\, dA$ over the region R shown in Figure 16.67(a).

Solution The difficulty in evaluating this double integral is clearly the integrand. The presence of the terms $x + 2y$ and $x - 2y$ prompts us to define the change of variables

$$u = x + 2y \quad \text{and} \quad v = x - 2y$$

These equations will map R onto the region S in the uv-plane. As in Example 1, we transform the sides of the region.

S_1: $y = 0$ implies $u = x$ and $v = x$ or $v = u$. As we move from $(2\pi, 0)$ to $(0, 0)$, we see that the corresponding image points in the uv-plane lie on the line segment $v = u$ from $(2\pi, 2\pi)$ to $(0, 0)$.

S_2: $x = 0$ implies $u = 2y$ and $v = -2y$ or $v = -u$. As we move from $(0, 0)$ to $(0, \pi)$, the corresponding image points in the uv-plane lie on the line segment $v = -u$ from $(0, 0)$ to $(2\pi, -2\pi)$.

S_3: $x + 2y = 2\pi$ implies $u = 2\pi$. As we move from $(0, \pi)$ to $(2\pi, 0)$, the equation $v = x - 2y$ shows that v ranges from $v = -2\pi$ to $v = 2\pi$. Thus, the image of S_3 is the vertical line segment $u = 2\pi$ starting at $(2\pi, -2\pi)$ and extending up to $(2\pi, 2\pi)$. See Figure 16.67(b).

Now, solving for x and y in terms of u and v gives

$$x = \frac{1}{2}(u + v) \qquad y = \frac{1}{4}(u - v)$$

Therefore,

$$\frac{\partial(x, y)}{\partial(u, v)} = \begin{vmatrix} \dfrac{\partial x}{\partial u} & \dfrac{\partial x}{\partial v} \\ \dfrac{\partial y}{\partial u} & \dfrac{\partial y}{\partial v} \end{vmatrix} = \begin{vmatrix} \dfrac{1}{2} & \dfrac{1}{2} \\ \dfrac{1}{4} & -\dfrac{1}{4} \end{vmatrix} = -\frac{1}{4}$$

Hence, from (11) we find that

$$\iint_R \sin(x + 2y)\cos(x - 2y)\, dA = \iint_S \sin u \cos v \left|-\frac{1}{4}\right| dA'$$
$$= \frac{1}{4}\int_0^{2\pi}\int_{-u}^{u} \sin u \cos v\, dv\, du$$
$$= \frac{1}{4}\int_0^{2\pi} \sin u \sin v\Big]_{-u}^{u} du$$
$$= \frac{1}{2}\int_0^{2\pi} \sin^2 u\, du$$
$$= \frac{1}{4}\int_0^{2\pi} (1 - \cos 2u)\, du$$
$$= \frac{1}{4}\left[u - \frac{1}{2}\sin 2u\right]_0^{2\pi} = \frac{\pi}{2}$$ □

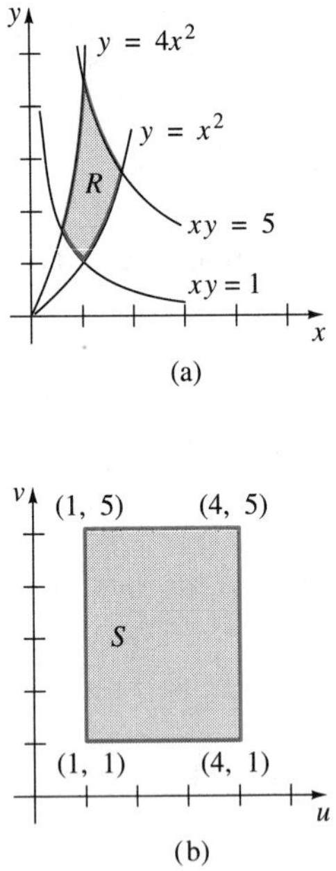

FIGURE 16.68

EXAMPLE 4 Evaluate $\iint_R xy\, dA$ over the region R shown in Figure 16.68(a).

Solution In this case the integrand is fairly simple, but integration over the region R would be tedious, since we would have to express $\iint_R xy\, dA$ as the sum of three integrals. (Verify this.)

The equations of the boundaries of R suggest the change of variables

$$u = \frac{y}{x^2} \quad \text{and} \quad v = xy \tag{12}$$

Obtaining the image of R is a straightforward matter in this case, since the images of the curves that make up the four boundaries are simply $u = 1, u = 4$, $v = 1$, and $v = 5$. In other words, the image of the region R is the rectangular region S: $1 \le u \le 4$, $1 \le v \le 5$. See Figure 16.68(b).

Now instead of trying to solve the equations in (12) for x and y in terms of u and v, we can compute the Jacobian $\partial(x, y)/\partial(u, v)$ by computing $\partial(u, v)/\partial(x, y)$ and using (10). We have

$$\frac{\partial(u, v)}{\partial(x, y)} = \begin{vmatrix} \dfrac{\partial u}{\partial x} & \dfrac{\partial u}{\partial y} \\ \dfrac{\partial v}{\partial x} & \dfrac{\partial v}{\partial y} \end{vmatrix} = \begin{vmatrix} -\dfrac{2y}{x^3} & \dfrac{1}{x^2} \\ y & x \end{vmatrix} = -\frac{3y}{x^2}$$

and so from (10),

$$\frac{\partial(x, y)}{\partial(u, v)} = \frac{1}{\dfrac{\partial(u, v)}{\partial(x, y)}} = -\frac{x^2}{3y} = -\frac{1}{3u}$$

Hence,
$$\iint_R xy\,dA = \iint_S v\left|-\frac{1}{3u}\right| dA'$$
$$= \frac{1}{3}\int_1^4 \int_1^5 \frac{v}{u}\,dv\,du$$
$$= \frac{1}{3}\int_1^4 \frac{v^2}{2u}\bigg]_1^5 du$$
$$= 4\int_1^4 \frac{1}{u}\,du = 4\ln u\bigg]_1^4 = 4\ln 4$$
□

Triple Integrals To change variables in a triple integral, let

$$x = f(u, v, w) \qquad y = g(u, v, w) \qquad z = h(u, v, w)$$

be a one-to-one transformation T from a region E in uvw-space to a region D in xyz-space. If the functions f, g, and h satisfy conditions analogous to those listed on page 931 and the Jacobian determinant

$$\frac{\partial(x, y, z)}{\partial(u, v, w)} = \begin{vmatrix} \dfrac{\partial x}{\partial u} & \dfrac{\partial x}{\partial v} & \dfrac{\partial x}{\partial w} \\ \dfrac{\partial y}{\partial u} & \dfrac{\partial y}{\partial v} & \dfrac{\partial y}{\partial w} \\ \dfrac{\partial z}{\partial u} & \dfrac{\partial z}{\partial v} & \dfrac{\partial z}{\partial w} \end{vmatrix}$$

is not zero on E, then we have the following result.

THEOREM 16.4 Change of Variables in a Triple Integral

If F is continuous on a region D and if $x = f(u, v, w)$, $y = g(u, v, w)$, $z = h(u, v, w)$ is a transformation from the uvw-space to the xyz-space, then

$$\iiint_D F(x, y, z)\,dV = \iiint_E F(f(u, v, w), g(u, v, w), h(u, v, w))\left|\frac{\partial(x, y, z)}{\partial(u, v, w)}\right| dV' \tag{13}$$

We leave it as an exercise to show that if T is the transformation from spherical to rectangular coordinates defined by

$$x = \rho\sin\phi\cos\theta \qquad y = \rho\sin\phi\sin\theta \qquad z = \rho\cos\phi$$

then
$$\frac{\partial(x, y, z)}{\partial(\rho, \phi, \theta)} = \rho^2\sin\phi \tag{14}$$

EXERCISES 16.9 *Answers to odd-numbered problems begin on page A-108.*

1. Consider a transformation T defined by $x = 4u - v$, $y = 5u + 4v$. Find the images of the points (0, 0), (0, 2), (4, 0), and (4, 2) in the uv-plane under T.

2. Consider a transformation T defined by $x = \sqrt{v - u}$, $y = v + u$. Find the images of the points (1, 1), (1, 3), and $(\sqrt{2}, 2)$ in the xy-plane under T^{-1}.

In Problems 3–6 find the image of the set S under the given transformation.

3. $S: 0 \le u \le 2, 0 \le v \le u; x = 2u + v, y = u - 3v$

4. $S: -1 \le u \le 4, 1 \le v \le 5; u = x - y, v = x + 2y$

5. $S: 0 \le u < 1, 0 \le v < 2; x = u^2 - v^2, y = uv$

6. $S: 1 \le u \le 2, 1 \le v \le 2; x = uv, y = v^2$

In Problems 7–10 find the Jacobian of the transformation T from the uv-plane to the xy-plane.

7. $x = ve^{-u}, y = ve^u$

8. $x = e^{3u}\sin v, y = e^{3u}\cos v$

9. $u = \dfrac{y}{x^2}, v = \dfrac{y^2}{x}$

10. $u = \dfrac{2x}{x^2 + y^2}, v = \dfrac{-2y}{x^2 + y^2}$

11. (a) Find the image of the region $S: 0 \le u \le 1, 0 \le v \le 1$ under the transformation $x = u - uv$, $y = uv$.

(b) Explain why the transformation is not one-to-one on the boundary of S.

12. Determine where the Jacobian $\partial(x, y)/\partial(u, v)$ of the transformation in Problem 11 is zero.

In Problems 13–22 evaluate the given integral by means of the indicated change of variables.

13. $\iint_R (x + y)\, dA$, where R is the region bounded by the graphs of $x - 2y = -6$, $x - 2y = 6$, $x + y = -1$, $x + y = 3$; $u = x - 2y, v = x + y$

14. $\displaystyle\iint_R \frac{\cos \frac{1}{2}(x - y)}{3x + y}\, dA$, where R is the region bounded by the graphs of $y = x$, $y = x - \pi$, $y = -3x + 3$, $y = -3x + 6$; $u = x - y, v = 3x + y$

15. $\displaystyle\iint_R \frac{y^2}{x}\, dA$, where R is the region bounded by the graphs of $y = x^2$, $y = \frac{1}{2}x^2$, $x = y^2$, $x = \frac{1}{2}y^2$; $u = \dfrac{x^2}{y}, v = \dfrac{y^2}{x}$

16. $\iint_R (x^2 + y^2)^{-3}\, dA$, where R is the region bounded the circles $x^2 + y^2 = 2x$, $x^2 + y^2 = 4x$, $x^2 + y^2 = 2y$, $x^2 + y^2 = 6y$; $u = \dfrac{2x}{x^2 + y^2}, v = \dfrac{2y}{x^2 + y^2}$ [*Hint:* Form $u^2 + v^2$.]

17. $\iint_R (x^2 + y^2)\, dA$, where R is the region in the first quadrant bounded by the graphs of $x^2 - y^2 = a$, $x^2 - y^2 = b$, $2xy = c$, $2xy = d$, $0 < a < b$, $0 < c < d$; $u = x^2 - y^2, v = 2xy$

18. $\iint_R (x^2 + y^2)\sin xy\, dA$, where R is the region bounded by the graphs of $x^2 - y^2 = 1$, $x^2 - y^2 = 9$, $xy = 2$, $xy = -2$; $u = x^2 - y^2, v = xy$

19. $\displaystyle\iint_R \frac{x}{y + x^2}\, dA$, where R is the region in the first quadrant bounded by the graphs of $x = 1$, $y = x^2$, $y = 4 - x^2$; $x = \sqrt{v - u}, y = v + u$

20. $\iint_R y\, dA$, where R is the triangular region with vertices (0, 0), (2, 3), and (−4, 1); $x = 2u - 4v, y = 3u + v$

21. $\iint_R y^4\, dA$, where R is the region in the first quadrant bounded by the graphs of $xy = 1$, $xy = 4$, $y = x$, $y = 4x$; $u = xy, v = y/x$

22. $\iiint_D (4z + 2x - 2y)\, dV$, where D is the parallelepiped $1 \le y + z \le 3$, $-1 \le -y + z \le 1$, $0 \le x - y \le 3$; $u = y + z$, $v = y + z$, $w = x - y$

In Problems 23–26 evaluate the double integral by means of an appropriate change of variables.

23. $\displaystyle\int_0^1 \int_0^{1-x} e^{(y-x)/(y+x)}\, dy\, dx$

24. $\displaystyle\int_{-2}^0 \int_0^{x+2} e^{y^2 - 2xy + x^2}\, dy\, dx$

25. $\iint_R (6x + 3y)\, dA$, where R is the trapezoidal region in the first quadrant with vertices (1, 0), (4, 0), (2, 4), and $(\frac{1}{2}, 1)$

26. $\iint_R (x + y)^4 e^{x-y}\, dA$, where R is the square region with vertices (1, 0), (0, 1), (1, 2), and (2, 1)

27. A problem in thermodynamics is to find the work done by an ideal Carnot engine. This work is defined to be area of the region R in the first quadrant bounded by the isothermals $xy = a$, $xy = b$, $0 < a < b$, and the adiabatics $xy^{1.4} = c$, $xy^{1.4} = d$, $0 < c < d$. Use $A = \iint_R dA$ and an appropriate substitution to find the area shown in Figure 16.69.

28. Use $V = \iiint_D dV$ and the substitutions $u = x/a$, $v = y/b$, $w = z/c$ to show that the volume of the ellipsoid $x^2/a^2 + y^2/b^2 + z^2/c^2 = 1$ is $V = \frac{4}{3}\pi abc$.

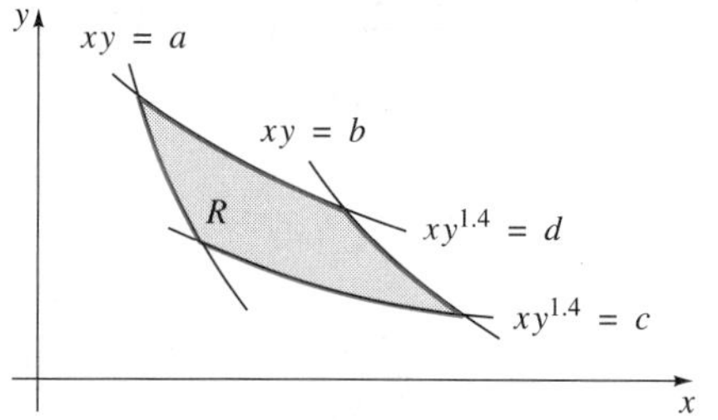

FIGURE 16.69

29. Evaluate the double integral $\iint_R \left(\frac{x^2}{25} + \frac{y^2}{9}\right) dA$, where R is the elliptical region whose boundary is the graph of $x^2/25 + y^2/9 = 1$. Use the substitution $u = x/5$, $v = y/3$ and polar coordinates.

30. Verify that the Jacobian of the transformation given in (14) is $\partial(x, y, z)/\partial(\rho, \phi, \theta) = \rho^2 \sin\phi$.

CHAPTER 16 REVIEW EXERCISES

Answers begin on page A-108.

In Problems 1–4 answer true or false.

1. $\int_{-2}^{3}\int_{1}^{5} e^{x^2-y}\,dx\,dy = \int_{1}^{5}\int_{-2}^{3} e^{x^2-y}\,dy\,dx$ ______

2. If I is the partial integral $\int_{g_1(x)}^{g_2(x)} f(x, y)\,dy$, then $\partial I/\partial y = 0$. ______

3. For every continuous function f, $\int_{-1}^{1}\int_{x^2}^{1} f(x, y)\,dy\,dx = 2\int_{0}^{1}\int_{x^2}^{1} f(x, y)\,dy\,dx$ ______

4. The center of mass of a lamina possessing symmetry always lies on the axis of symmetry. ______

In Problems 5–16 fill in the blanks.

5. $\int_{y^2+1}^{5}\left(8y^3 - \frac{5y}{x}\right) dx =$ ______

6. If R_1 and R_2 are nonoverlapping regions such that $R = R_1 \cup R_2$, $\iint_R f(x, y)\,dA = 10$, and $\iint_{R_2} f(x, y)\,dA = -6$, then $\iint_{R_1} f(x, y)\,dA =$ ______.

7. $\int_{-a}^{a}\int_{-a}^{a} dx\,dy$ gives the area of a ______.

8. The region bounded by the graphs of $9x^2 + y^2 = 36$, $y = -2$, $y = 5$ is of Type ______.

9. $\int_{2}^{4} f_y(x, y)\,dy =$ ______

10. If $\rho(x, y, z)$ is density, then the iterated integral giving the mass of the solid bounded by the ellipsoid $x^2/a^2 + y^2/b^2 + z^2/c^2 = 1$ is ______.

11. $\int_{0}^{2}\int_{y^2}^{2y} f(x, y)\,dx\,dy = \int_{\_\_}^{\_\_}\int_{\_\_}^{\_\_} f(x, y)\,dy\,dx$

12. The rectangular coordinates of the point $\left(6, \frac{5\pi}{3}, \frac{5\pi}{6}\right)$ given in spherical coordinates are ______.

13. The cylindrical coordinates of the point $\left(2, \frac{\pi}{4}, \frac{2\pi}{3}\right)$ given in spherical coordinates are ______.

14. The region R bounded by the graphs of $y = 4 - x^2$ and $y = 0$ is both Type I and Type II. Interpreted as a Type II region, $\iint_R f(x, y)\,dA = \int_{\_\_}^{\_\_}\int_{\_\_}^{\_\_} f(x, y)$ ______ ______.

15. The equation of the paraboloid $z = x^2 + y^2$ in cylindrical coordinates is ______, whereas in spherical coordinates the equation is ______.

16. In cylindrical *and* spherical coordinates the equation of the plane $y = x$ is ______.

In Problems 17–28 evaluate the given integral.

17. $\int_{y^3}^{y} y^2 \sin xy\,dx$

18. $\int_{1/x}^{e^x} \frac{x}{y^2}\,dy$

19. $\int_{0}^{2}\int_{0}^{2x} ye^{y-x}\,dy\,dx$

20. $\int_{0}^{4}\int_{x}^{4} \frac{1}{16 + x^2}\,dy\,dx$

21. $\int_{0}^{1}\int_{x}^{\sqrt{x}} \frac{\sin y}{y}\,dy\,dx$

22. $\int_{e}^{e^2}\int_{0}^{1/x} \ln x\,dy\,dx$

23. $\int_{0}^{5}\int_{0}^{\pi/2}\int_{0}^{\cos\theta} 3r^2\,dr\,d\theta\,dz$

24. $\int_{\pi/4}^{\pi/2}\int_{0}^{\sin z}\int_{0}^{\ln x} e^y\,dy\,dx\,dz$

25. $\iint_R 5\,dA$, where R is bounded by the circle $x^2 + y^2 = 64$

26. $\iint_R dA$, where R is bounded by the cardioid $r = 1 + \cos\theta$

27. $\iint_R (2x + y)\,dA$, where R is bounded by the graphs of $y = \frac{1}{2}x$, $x = y^2 + 1$, $y = 0$

28. $\iiint_D x\,dV$, where D is bounded by the planes $z = x + y$, $z = 6 - x - y$, $x = 0$, $y = 0$

29. Using rectangular coordinates, express as an iterated integral:

$$\iint_R \frac{1}{x^2 + y^2}\,dA$$

where R is the region in the first quadrant that is bounded by the graphs of $x^2 + y^2 = 1$, $x^2 + y^2 = 9$, $x = 0$, and $y = x$. Do not evaluate.

30. Evaluate the double integral in Problem 29 using polar coordinates.

In Problems 31 and 32 sketch the region of integration.

31. $\int_{-2}^{2}\int_{-x^2}^{x^2} f(x, y)\,dy\,dx$

32. $\int_{-1}^{1}\int_{-1}^{1}\int_{0}^{x^2+y^2} F(x, y, z)\,dz\,dx\,dy$

33. Reverse the order of integration and evaluate

$$\int_0^1\int_y^{\sqrt[3]{y}} \cos x^2\,dx\,dy$$

34. Consider $\iiint_D F(x, y, z)\,dV$, where D is the region in the first octant bounded by the planes $z = 8 - 2x - y$, $z = 4$, $x = 0$, $y = 0$. Express the triple integral as six different iterated integrals.

In Problems 35 and 36 use an appropriate coordinate system to evaluate the given integral.

35. $\int_0^2\int_{1/2}^{1}\int_0^{\sqrt{x-x^2}} (4z + 1)\,dy\,dx\,dz$

36. $\int_0^1\int_0^{\sqrt{1-x^2}}\int_{-\sqrt{1-x^2-y^2}}^{\sqrt{1-x^2-y^2}} (x^2 + y^2 + z^2)^4\,dz\,dy\,dx$

37. Find the surface area of that portion of the graph of $z = xy$ within the cylinder $x^2 + y^2 = 1$.

38. Use a double integral to find the volume of the solid shown in Figure 16.70.

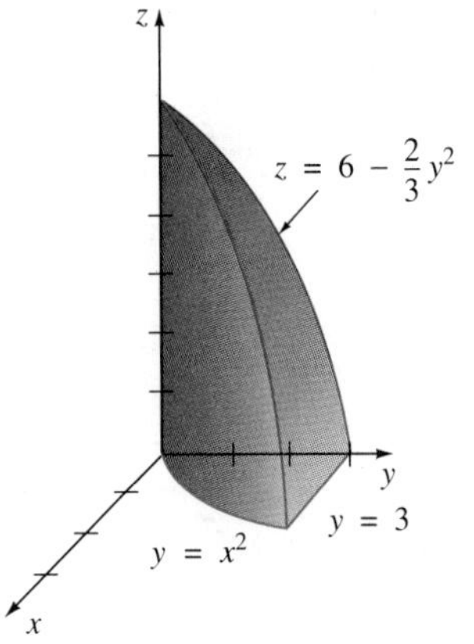

FIGURE 16.70

39. Express the volume of the solid shown in Figure 16.71 as one or more iterated integrals using the order of integration **(a)** $dy\,dx$ and **(b)** $dx\,dy$. Choose either part **(a)** or part **(b)** to find the volume.

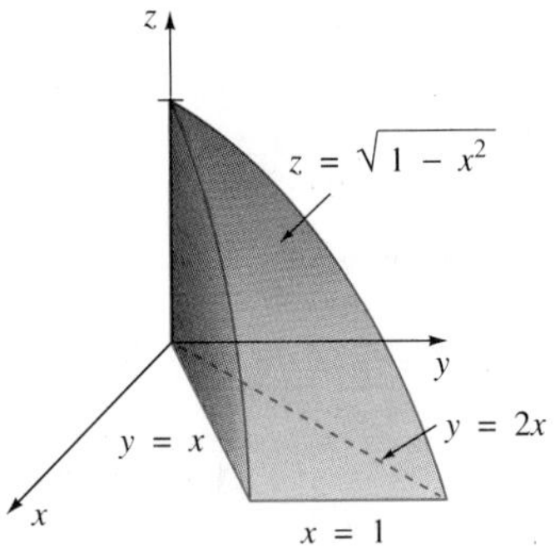

FIGURE 16.71

40. A lamina has the shape of the region in the first quadrant bounded by the graphs of $y = x^2$ and $y = x^3$. Find the center of mass if the density at a point P is directly proportional to the square of the distance from the origin.

41. Find the moment of inertia of the lamina described in Problem 40 about the y-axis.

42. Find the volume of the sphere $x^2 + y^2 + z^2 = a^2$ using a triple integral in **(a)** rectangular coordinates, **(b)** cylindrical coordinates, and **(c)** spherical coordinates.

43. Find the volume of the solid that is bounded between the cones $z = \sqrt{x^2 + y^2}$, $z = \sqrt{9x^2 + 9y^2}$, and the plane $z = 3$.

44. Find the volume of the solid shown in Figure 16.72.

45. Evaluate the integral $\iint_R (x^2 + y^2)\sqrt[3]{x^2 - y^2}\, dA$, where R is the region bounded by the graphs of $x = 0$, $x = 1$, $y = 0$, and $y = 1$ by means of the change of variables $u = 2xy$, $v = x^2 - y^2$.

46. Evaluate the integral $\iint_R (1/\sqrt{(x-y)^2 + 2(x+y) + 1})\, dA$, where R is the region bounded by the graphs of $y = x$, $x = 2$, and $y = 0$ by means of the change of variables $x = u + uv$, $y = v + uv$.

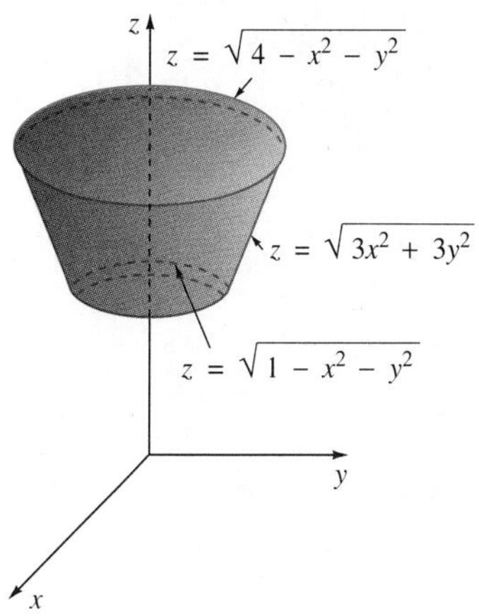

FIGURE 16.72

17

VECTOR INTEGRAL CALCULUS

INTRODUCTION

So far we have studied three kinds of integrals: the definite integral, the double integral, and the triple integral. In this chapter we will introduce two new kinds of integrals: **line** and **surface integrals**. The development of these new concepts depends heavily on vector methods.

Why do some objects float in water and others sink? The answer comes from **Archimedes principle**, which states: When an object is submerged in a fluid, the fluid exerts an upward force on it, called buoyant force, with a magnitude that is equal to the weight of the fluid displaced. Thus, a cork has positive buoyancy or floats since the weight of the cork is less than the magnitude of the buoyant force. A submarine will attain negative buoyancy and sink by filling its ballast tanks with water, thereby making its weight greater than the magnitude of the buoyant force exerted on it. You will be asked to prove this famous principle in Exercises 17.7 using vector methods and a new theorem called the Divergence Theorem.

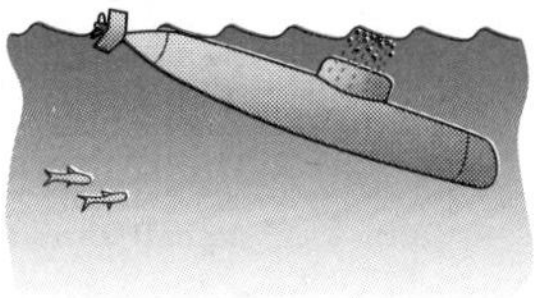

IMPORTANT CONCEPTS

- smooth curve
- piecewise smooth curve
- closed curve
- simple closed curve
- line integral
- vector field
- work as a line integral
- path independence
- fundamental theorem for line integrals
- simply connected region
- multiply connected region
- conservative vector field
- Green's Theorem
- surface integral
- orientable surface
- flux of a vector field
- divergence of a vector field
- curl of a vector field
- Stokes' Theorem
- Divergence Theorem

17.1 LINE INTEGRALS

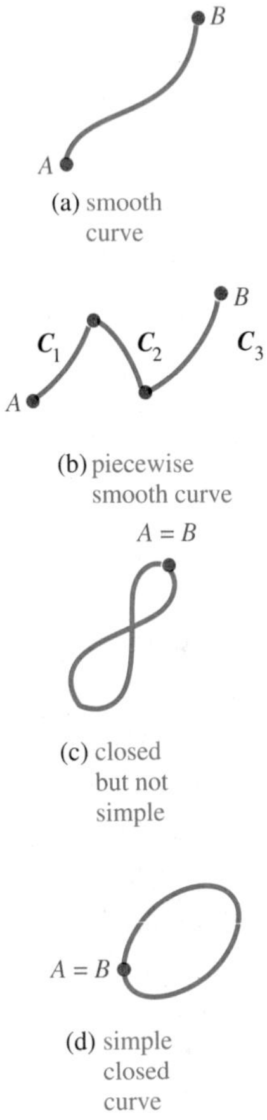

FIGURE 17.1

Terminology The notion of the definite integral $\int_a^b f(x)\,dx$—that is, *integration of a function defined over an interval*—can be generalized to *integration of a function defined along a curve*. To this end we need to introduce some terminology about curves. Suppose C is a curve parameterized by $x = f(t)$, $y = g(t)$, $a \le t \le b$, and A and B are the initial and terminal points $(f(a), g(a))$ and $(f(b), g(b))$, respectively. We say that:

(*i*) C is a **smooth curve** if f' and g' are continuous on the closed interval $[a, b]$ and not simultaneously zero on the open interval (a, b).

(*ii*) C is **piecewise smooth** if it consists of a finite number of smooth curves $C_1, C_2, \ldots, C_n$ joined end to end; that is, $C = C_1 \cup C_2 \cup \cdots \cup C_n$.

(*iii*) C is a **closed curve** if $A = B$.

(*iv*) C is a **simple closed curve** if $A = B$ and the curve does not cross itself.

(*v*) If C is not a closed curve, then the **positive direction** on C is the direction corresponding to the increasing values of t.

Each type of curve defined in (*i*)–(*iv*) is illustrated in Figure 17.1(a)–(d).

This same terminology carries over in a natural manner to curves in 3-space. For example, a space curve C defined by $x = f(t)$, $y = g(t)$, $z = h(t)$, $a \le t \le b$, is smooth if the derivatives f', g', and h' are continuous on $[a, b]$ and not simultaneously zero on (a, b).

Line Integrals in the Plane The following five steps lead to the definitions of three **line integrals*** in the plane:

$z = G(x, y)$

1. Let G be defined in some region that contains the smooth curve C defined by $x = f(t)$, $y = g(t)$, $a \le t \le b$.
2. Divide C into n subarcs of lengths Δs_k according to the partition $a = t_0 < t_1 < t_2 < \cdots < t_n = b$ of $[a, b]$. Let the projection of each subarc onto the x- and y-axes have lengths Δx_k and Δy_k, respectively.
3. Let $\|P\|$ be the **norm** of the partition or the length of the longest subarc.
4. Choose a point (x_k^*, y_k^*) on each subarc.
5. Form the sums $\sum_{k=1}^{n} G(x_k^*, y_k^*)\,\Delta x_k$, $\sum_{k=1}^{n} G(x_k^*, y_k^*)\,\Delta y_k$, and $\sum_{k=1}^{n} G(x_k^*, y_k^*)\,\Delta s_k$.

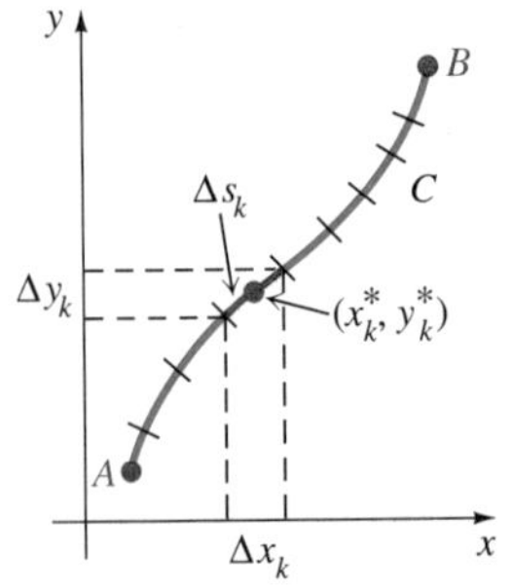

*Another unfortunate choice of names. **Curve integrals** would be more appropriate.

DEFINITION 17.1 Line Integrals in the Plane

Let G be a function of two variables x and y defined on a region of the plane containing a smooth curve C.

(*i*) The **line integral of G along C from A to B with respect to x** is

$$\int_C G(x, y)\,dx = \lim_{\|P\|\to 0} \sum_{k=1}^{n} G(x_k^*, y_k^*)\,\Delta x_k$$

(*ii*) The **line integral of G along C from A to B with respect to y** is

$$\int_C G(x, y)\,dy = \lim_{\|P\|\to 0} \sum_{k=1}^{n} G(x_k^*, y_k^*)\,\Delta y_k$$

(*iii*) The **line integral of G along C from A to B with respect to arc length** is

$$\int_C G(x, y)\,ds = \lim_{\|P\|\to 0} \sum_{k=1}^{n} G(x_k^*, y_k^*)\,\Delta s_k$$

It can be proved that if $G(x, y)$ is continuous on C, then the integrals defined in (*i*), (*ii*), and (*iii*) exist. We shall assume continuity of G as a matter of course.

Method of Evaluation–Curve Defined Parametrically The line integrals in Definition 17.1 can be evaluated in two ways, depending on whether the curve C is defined parametrically or by an explicit function. In either case, the basic idea is to convert the line integral to a definite integral in a single variable. If C is a smooth curve parameterized by $x = f(t)$, $y = g(t)$, $a \le t \le b$, then $dx = f'(t)\,dt$, $dy = g'(t)\,dt$ and so

$$\int_C G(x, y)\,dx = \int_a^b G(f(t), g(t))\,f'(t)\,dt \tag{1}$$

$$\int_C G(x, y)\,dy = \int_a^b G(f(t), g(t))g'(t)\,dt \tag{2}$$

Furthermore, using (14) of Section 5.5 and the given parameterization, we find that $ds = \sqrt{[f'(t)]^2 + [g'(t)]^2}\,dt$. Hence,

$$\int_C G(x, y)\,ds = \int_a^b G(f(t), g(t))\sqrt{[f'(t)]^2 + [g'(t)]^2}\,dt \tag{3}$$

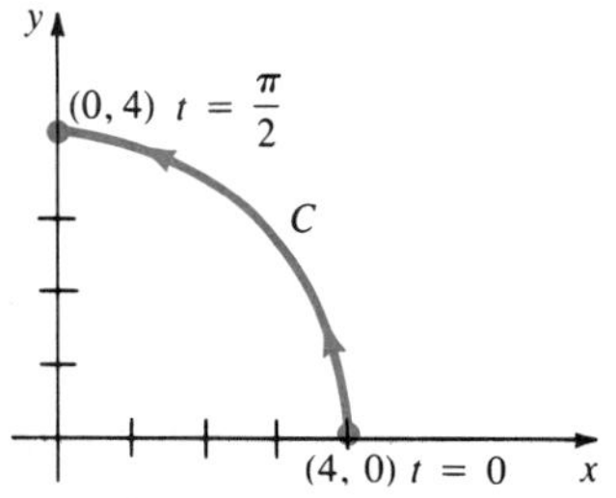

FIGURE 17.2

EXAMPLE 1 Evaluate (*a*) $\int_C xy^2\,dx$, (*b*) $\int_C xy^2\,dy$, and (*c*) $\int_C xy^2\,ds$ on the quarter-circle C defined by $x = 4\cos t$, $y = 4\sin t$, $0 \le t \le \pi/2$. See Figure 17.2.

Solution (*a*) From (1),

$$\int_C xy^2\,dx = \int_0^{\pi/2} \overbrace{(4\cos t)}^{x}\overbrace{(16\sin^2 t)}^{y^2}\overbrace{(-4\sin t\,dt)}^{dx}$$

$$= -256\int_0^{\pi/2} \sin^3 t\cos t\,dt$$

$$= -256\left[\frac{1}{4}\sin^4 t\right]_0^{\pi/2} = -64$$

(*b*) From (2),

$$\int_C xy^2\,dy = \int_0^{\pi/2} \overbrace{(4\cos t)}^{x}\overbrace{(16\sin^2 t)}^{y^2}\overbrace{(4\cos t\,dt)}^{dy}$$

$$= 256\int_0^{\pi/2} \sin^2 t\cos^2 t\,dt$$

$$= 256\int_0^{\pi/2} \frac{1}{4}\sin^2 2t\,dt$$

$$= 64\int_0^{\pi/2} \frac{1}{2}(1-\cos 4t)\,dt$$

$$= 32\left[t - \frac{1}{4}\sin 4t\right]_0^{\pi/2} = 16\pi$$

trig identities

(*c*) From (3),

$$\int_C xy^2\,ds = \int_0^{\pi/2} \overbrace{(4\cos t)}^{x}\overbrace{(16\sin^2 t)}^{y^2}\overbrace{\sqrt{16(\cos^2 t+\sin^2 t)}\,dt}^{ds}$$

$$= 256\int_0^{\pi/2} \sin^2 t\cos t\,dt$$

$$= 256\left[\frac{1}{3}\sin^3 t\right]_0^{\pi/2} = \frac{256}{3}$$ □

Method of Evaluation–Curve Defined by $y = f(x)$ If the curve C is defined by an explicit function $y = f(x)$, $a \le x \le b$, we can use x as a parameter. With $dy = f'(x)\,dx$ and $ds = \sqrt{1 + [f'(x)]^2}\,dx$, the foregoing line integrals become, in turn,

$$\int_C G(x, y)\,dx = \int_a^b G(x, f(x))\,dx \tag{4}$$

$$\int_C G(x, y)\, dy = \int_a^b G(x, f(x))f'(x)\, dx \tag{5}$$

$$\int_C G(x, y)\, ds = \int_a^b G(x, f(x))\sqrt{1 + [f'(x)]^2}\, dx \tag{6}$$

A line integral along a *piecewise smooth* curve C is defined as the *sum* of the integrals over the various smooth curves whose union comprises C. For example, if C is composed of smooth curves C_1 and C_2, then

$$\int_C G(x, y)\, ds = \int_{C_1} G(x, y)\, ds + \int_{C_2} G(x, y)\, ds$$

Notation In many applications, line integrals appear as a sum

$$\int_C P(x, y)\, dx + \int_C Q(x, y)\, dy$$

It is common practice to write this sum as one integral without parantheses as

$$\int_C P(x, y)\, dx + Q(x, y)\, dy \quad \text{or simply} \quad \int_C P\, dx + Q\, dy \tag{7}$$

A line integral along a *closed* curve C is very often denoted by

$$\oint_C P\, dx + Q\, dy$$

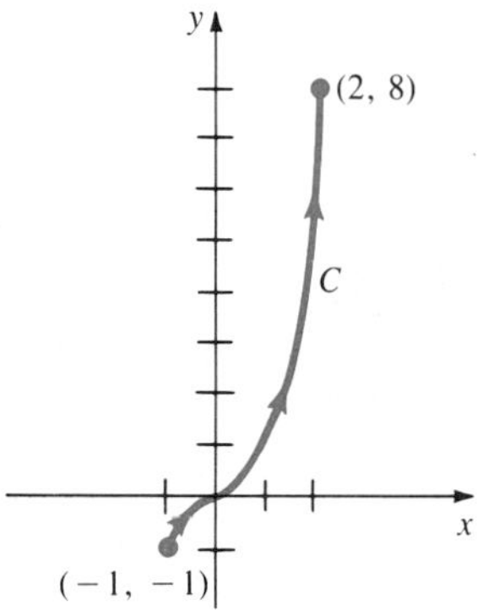

FIGURE 17.3

EXAMPLE 2 Evaluate $\int_C xy\, dx + x^2\, dy$, where C is given by $y = x^3$, $-1 \le x \le 2$.

Solution The curve C is illustrated in Figure 17.3 and is defined by the explicit function $y = x^3$. Hence, we can use x as the parameter. With $dy = 3x^2\, dx$, it follows from (4) and (5) that

$$\begin{aligned}\int_C xy\, dx + x^2\, dy &= \int_{-1}^{2} x\overbrace{(x^3)}^{\boxed{y}}\, dx + x^2\overbrace{(3x^2\, dx)}^{\boxed{dy}}\\ &= \int_{-1}^{2} 4x^4\, dx = \frac{4}{5}x^5\Big]_{-1}^{2} = \frac{132}{5}\end{aligned}$$ □

EXAMPLE 3 Evaluate $\oint_C x\, dx$, where C is the circle $x = \cos t$, $y = \sin t$, $0 \le t \le 2\pi$.

Solution From (1),

$$\oint_C x\, dx = \int_0^{2\pi} \cos t(-\sin t\, dt) = \frac{1}{2}\cos^2 t\Big]_0^{2\pi} = \frac{1}{2}[1 - 1] = 0$$ □

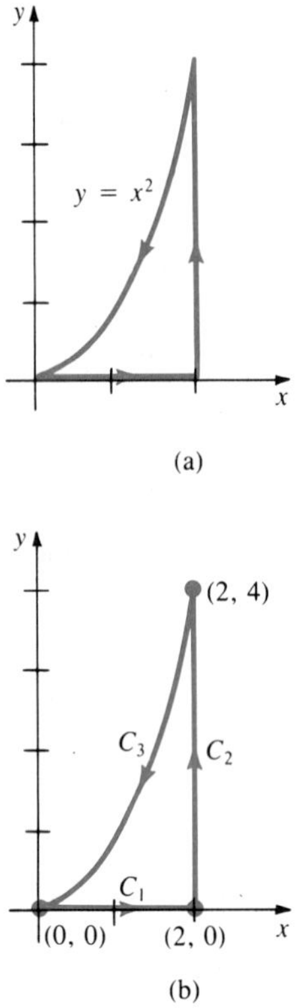

FIGURE 17.4

EXAMPLE 4 Evaluate $\oint_C y^2\,dx - x^2\,dy$ on the closed curve C that is shown in Figure 17.4(a).

Solution Since C is piecewise smooth, we express the integral as a sum of integrals. Symbolically, we write

$$\oint_C = \int_{C_1} + \int_{C_2} + \int_{C_3}$$

where C_1, C_2, and C_3 are the curves shown in Figure 17.4(b). On C_1, we use x as a parameter. Since $y = 0$, $dy = 0$,

$$\int_{C_1} y^2\,dx - x^2\,dy = \int_0^2 0\,dx - x^2(0) = 0$$

On C_2, we use y as a parameter. From $x = 2$, $dx = 0$, we have

$$\int_{C_2} y^2\,dx - x^2\,dy = \int_0^4 y^2(0) - 4\,dy$$
$$= -\int_0^4 4\,dy = -4y\Big]_0^4 = -16$$

Finally, on C_3, we again use x as a parameter. From $y = x^2$, we get $dy = 2x\,dx$ and so

$$\int_{C_3} y^2\,dx - x^2\,dy = \int_2^0 x^4\,dx - x^2(2x\,dx)$$
$$= \int_2^0 (x^4 - 2x^3)\,dx$$
$$= \left(\frac{1}{5}x^5 - \frac{1}{2}x^4\right)\Big]_2^0 = \frac{8}{5}$$

Hence, $$\oint_C y^2\,dx - x^2\,dy = 0 - 16 + \frac{8}{5} = -\frac{72}{5}$$ □

It is important to note that

a line integral is independent of the parameterization of the curve C provided C is given the same orientation by all sets of parametric equations defining the curve.

See Problem 43. Also, recall for definite integrals that

$$\int_b^a f(x)\,dx = -\int_a^b f(x)\,dx$$

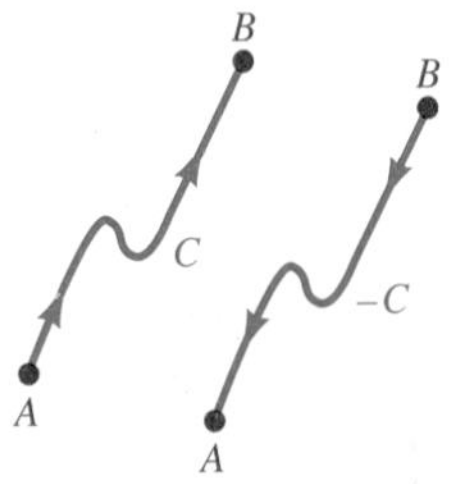

FIGURE 17.5

Line integrals possess a similar property. Suppose, as shown in Figure 17.5, that $-C$ denotes the curve having the opposite orientation of C. Then it can be shown that

$$\int_{-C} P\,dx + Q\,dy = -\int_C P\,dx + Q\,dy$$

or equivalently,

$$\int_{-C} P\,dx + Q\,dy + \int_C P\,dx + Q\,dy = 0 \tag{8}$$

For example, in part (*a*) of Example 1, $\int_{-C} xy^2\,dx = 64$.

Line Integrals in Space Suppose C is a smooth curve in 3-space defined by the parametric equations $x = f(t)$, $y = g(t)$, $z = h(t)$, $a \le t \le b$. If G is a function of three variables defined in some region of 3-space that contains C, we can define *four* line integrals along the curve:

$$\int_C G(x, y, z)\,dx \qquad \int_C G(x, y, z)\,dy \qquad \int_C G(x, y, z)\,dz \qquad \int_C G(x, y, z)\,ds$$

The first, second, and fourth integrals are defined in a manner analogous to Definition 17.1. For example, if C is divided into n subarcs of lengths Δs_k, as shown in Figure 17.6, then

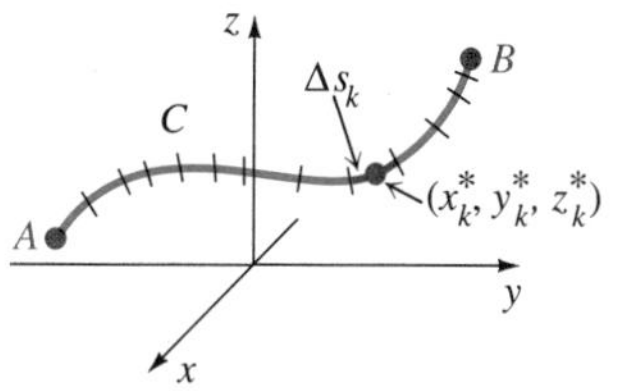

FIGURE 17.6

$$\int_C G(x, y, z)\,ds = \lim_{\|P\| \to 0} \sum_{k=1}^{n} G(x_k^*, y_k^*, z_k^*)\,\Delta s_k$$

The new integral in the list, the **line integral along C with respect to z**, is defined as

$$\int_C G(x, y, z)\,dz = \lim_{\|P\| \to 0} \sum_{k=1}^{n} G(x_k^*, y_k^*, z_k^*)\,\Delta z_k \tag{9}$$

Method of Evaluation Using the parametric equations $x = f(t)$, $y = g(t)$, $z = h(t)$, $a \le t \le b$, we can evaluate the line integrals along the space curve C in the following manner:

$$\int_C G(x, y, z)\,dx = \int_a^b G(f(t), g(t), h(t))f'(t)\,dt$$

$$\int_C G(x, y, z)\,dy = \int_a^b G(f(t), g(t), h(t))g'(t)\,dt$$

$$\int_C G(x, y, z)\,dz = \int_a^b G(f(t), g(t), h(t))h'(t)\,dt$$

$$\int_C G(x, y, z)\,ds = \int_a^b G(f(t), g(t), h(t))\sqrt{[f'(t)]^2 + [g'(t)]^2 + [h'(t)]^2}\,dt$$

As in (7), in 3-space we are often concerned with line integrals in the form of a sum:

$$\int_C P(x, y, z)\,dx + Q(x, y, z)\,dy + R(x, y, z)\,dz$$

EXAMPLE 5 Evaluate $\int_C y\,dx + x\,dy + z\,dz$, where C is the helix $x = 2\cos t$, $y = 2\sin t$, $z = t$, $0 \le t \le 2\pi$.

Solution Substituting the expressions for x, y, and z along with $dx = -2\sin t\,dt$, $dy = 2\cos dt$, $dz = dt$, we get

$$\int_C y\,dx + x\,dy + z\,dz = \int_0^{2\pi} \underbrace{-4\sin^2 t\,dt + 4\cos^2 t\,dt} + t\,dt$$

$$= \int_0^{2\pi} (4\cos 2t + t)\,dt \quad \boxed{\text{double angle formula}}$$

$$= \left(2\sin 2t + \frac{t^2}{2}\right)\Bigg]_0^{2\pi} = 2\pi^2 \qquad \square$$

Vector Fields Vector functions of two and three variables:

$$\mathbf{F}(x, y) = P(x, y)\mathbf{i} + Q(x, y)\mathbf{j}$$

and
$$\mathbf{F}(x, y, z) = P(x, y, z)\mathbf{i} + Q(x, y, z)\mathbf{j} + R(x, y, z)\mathbf{k}$$

are also called **vector fields**. For example, the motion of a wind or a fluid can be described by a *velocity field* in that a vector can be assigned at each point representing the velocity of a particle at the point. See Figure 17.7(a) and (b). The concept of a *force field* plays an important role in mechanics, electricity, and magnetism. See Figure 17.7(c) and (d).

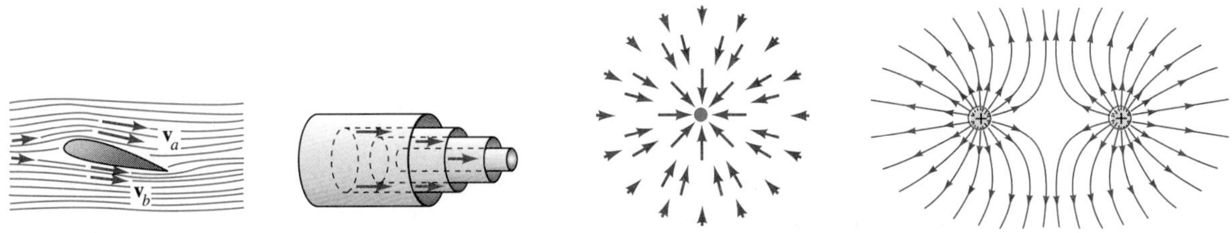

(a) airflow around an airplane wing; $|\mathbf{v}_a| > |\mathbf{v}_b|$

(b) laminar flow of blood in an artery; cylindrical layers of blood flow faster near the center of the artery

(c) inverse square force field; magnitude of the attractive force is large near the particle

(d) lines of force around two equal positive charges

FIGURE 17.7

EXAMPLE 6 Graph the two-dimensional vector field $\mathbf{F}(x, y) = -y\mathbf{i} + x\mathbf{j}$.

Solution One manner of proceeding is simply to choose points in the xy-plane and then graph the vector $\mathbf{F}$ at each point. For example, at (1, 1) we would draw the vector $\mathbf{F}(1, 1) = -\mathbf{i} + \mathbf{j}$.

For the given vector field it is possible to systematically draw vectors of the same length. Observe that $|\mathbf{F}| = \sqrt{x^2 + y^2}$, and so vectors of the same length k must lie along the curve defined by $\sqrt{x^2 + y^2} = k$; that is, at any point on the circle $x^2 + y^2 = k^2$, a vector would have length k. For simplicity let us choose circles that have some points on them with integer coordinates. For example, for $k = 1$, $k = \sqrt{2}$, and $k = 2$ we have:

$x^2 + y^2 = 1$: At the points (1, 0), (0, 1), (−1, 0), (0, −1), the corresponding vectors $\mathbf{j}$, $-\mathbf{i}$, $-\mathbf{j}$, $\mathbf{i}$ have the same length 1.

$x^2 + y^2 = 2$: At the points (1, 1), (−1, 1), (−1, −1), (1, −1), the corresponding vectors $-\mathbf{i} + \mathbf{j}$, $-\mathbf{i} - \mathbf{j}$, $\mathbf{i} - \mathbf{j}$, $\mathbf{i} + \mathbf{j}$ have the same length $\sqrt{2}$.

$x^2 + y^2 = 4$:. At the points (2, 0), (0, 2), (−2, 0), (0, −2), the corresponding vectors $2\mathbf{j}$, $-2\mathbf{i}$, $-2\mathbf{j}$, $2\mathbf{i}$ have the same length 2.

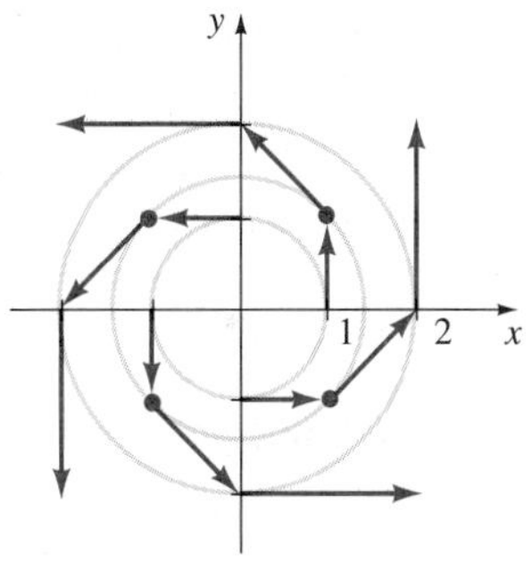

FIGURE 17.8

The vectors at these points are shown in Figure 17.8. □

We can use the concept of a vector function of several variables to write a general line integral in a compact fashion. For example, suppose the vector-valued function $\mathbf{F}(x, y) = P(x, y)\mathbf{i} + Q(x, y)\mathbf{j}$ is defined along a curve C: $x = f(t)$, $y = g(t)$, $a \le t \le b$, and suppose $\mathbf{r}(t) = f(t)\mathbf{i} + g(t)\mathbf{j}$ is the position vector of points on C. Then the derivative of $\mathbf{r}(t)$,

$$\frac{d\mathbf{r}}{dt} = f'(t)\mathbf{i} + g'(t)\mathbf{j} = \frac{dx}{dt}\mathbf{i} + \frac{dy}{dt}\mathbf{j}$$

prompts us to define

$$d\mathbf{r} = \frac{d\mathbf{r}}{dt}\,dt = dx\mathbf{i} + dy\mathbf{j}$$

Since $$\mathbf{F}(x, y) \cdot d\mathbf{r} = P(x, y)\,dx + Q(x, y)\,dy$$

we can then write

$$\int_C P(x, y)\,dx + Q(x, y)\,dy = \int_C \mathbf{F} \cdot d\mathbf{r} \tag{10}$$

Similarly, for a line integral on a space curve,

$$\int_C P(x, y, z)\,dx + Q(x, y, z)\,dy + R(x, y, z)\,dz = \int_C \mathbf{F} \cdot d\mathbf{r} \tag{11}$$

where $\mathbf{F}(x, y, z) = P(x, y, z)\mathbf{i} + Q(x, y, z)\mathbf{j} + R(x, y, z)\mathbf{k}$ and $d\mathbf{r} = dx\mathbf{i} + dy\mathbf{j} + dz\mathbf{k}$.

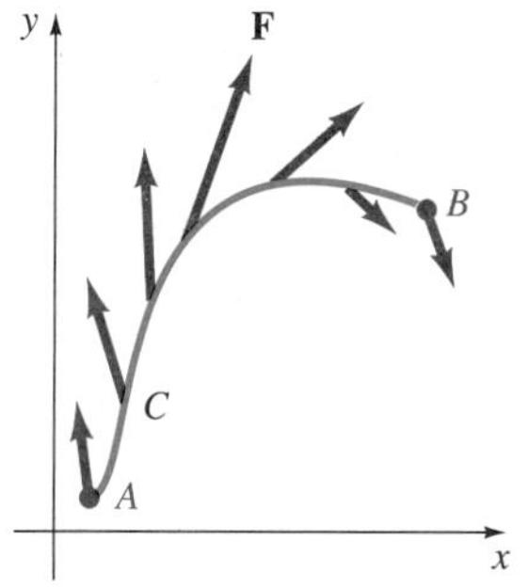

(a)

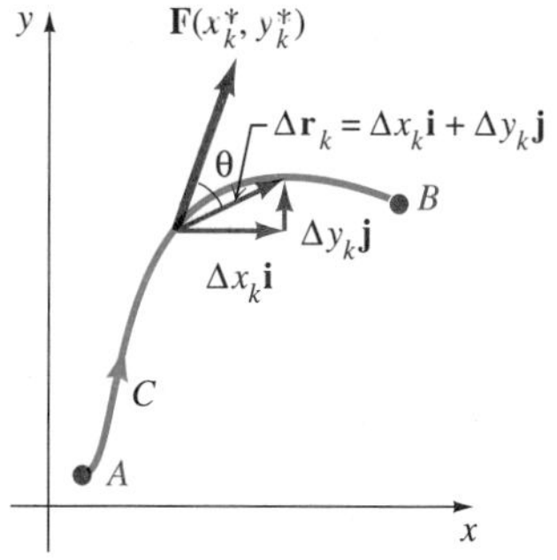

(b)

FIGURE 17.9

Work In Section 13.3 we saw that the work W done by a constant force $\mathbf{F}$ that causes a straight-line displacement $\mathbf{d}$ of an object is $W = \mathbf{F} \cdot \mathbf{d}$. In Section 6.9 it was shown that the work done in moving an object from $x = a$ to $x = b$ by a force $F(x)$ that varies in magnitude but not in direction is given by the definite integral $W = \int_a^b F(x)\,dx$. In general, a force field $\mathbf{F}(x, y) = P(x, y)\mathbf{i} + Q(x, y)\mathbf{j}$ acting at each point on a smooth curve C: $x = f(t)$, $y = g(t)$, $a \le t \le b$, varies in both magnitude and direction. See Figure 17.9(a). If A and B are the points $(f(a), g(a))$ and $(f(b), g(b))$, respectively, we ask: What is the work done by $\mathbf{F}$ as its point of application moves along C from A to B? To answer this question, suppose C is divided into n subarcs of lengths Δs_k. On each subarc $\mathbf{F}(x_k^*, y_k^*)$ is a constant force. If, as shown in Figure 17.9(b), the length of the vector $\Delta\mathbf{r}_k = (x_k - x_{k-1})\mathbf{i} + (y_k - y_{k-1})\mathbf{j} = \Delta x_k\mathbf{i} + \Delta y_k\mathbf{j}$ is an approximation to the length of the kth subarc, then the approximate work done by $\mathbf{F}$ over the subarc is

$$(|\mathbf{F}(x_k^*, y_k^*)|\cos\theta)|\,\Delta\mathbf{r}_k| = \mathbf{F}(x_k^*, y_k^*)\cdot\Delta\mathbf{r}_k$$
$$= P(x_k^*, y_k^*)\,\Delta x_k + Q(x_k^*, y_k^*)\,\Delta y_k$$

By summing these elements of work and passing to the limit, we can naturally define the **work done by F along *C*** as the line integral

$$W = \int_C P(x, y)\,dx + Q(x, y)\,dy \qquad W = \int_C \mathbf{F}\cdot d\mathbf{r} \tag{12}$$

Of course, (12) extends to force fields that act at points on a space curve. In this case, work $\int_C \mathbf{F}\cdot d\mathbf{r}$ is defined as in (11).

Now, since

$$\frac{d\mathbf{r}}{dt} = \frac{d\mathbf{r}}{ds}\frac{ds}{dt}$$

we let $d\mathbf{r} = \mathbf{T}\,ds$, where $\mathbf{T} = d\mathbf{r}/ds$ is a unit tangent* to C. Hence,

$$W = \int_C \mathbf{F}\cdot d\mathbf{r} = \int_C \mathbf{F}\cdot\mathbf{T}\,ds = \int_C \operatorname{comp}_{\mathbf{T}}\mathbf{F}\,ds \tag{13}$$

In other words,

the work done by a force **F** *along a curve C is due entirely to the tangential component of* **F**.

EXAMPLE 7 Find the work done by (*a*) $\mathbf{F} = x\mathbf{i} + y\mathbf{j}$ and (*b*) $\mathbf{F} = \frac{3}{4}\mathbf{i} + \frac{1}{2}\mathbf{j}$ along the curve C traced by $\mathbf{r}(t) = \cos t\mathbf{i} + \sin t\mathbf{j}$ from $t = 0$ to $t = \pi$.

Solution

(*a*) The vector function $\mathbf{r}(t)$ gives the parametric equations $x = \cos t$, $y = \sin t$, $0 \le t \le \pi$, which we recognize as a half-circle. As seen in Figure 17.10, the force field $\mathbf{F}$ is perpendicular to C at every point. (See Problem 29.) Since the tangential components of $\mathbf{F}$ are zero, the work done along C is zero. To see this we use (12):

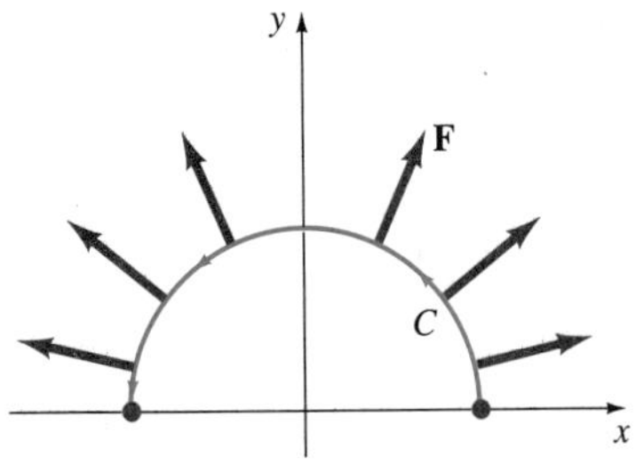

FIGURE 17.10

$$W = \int_C \mathbf{F}\cdot d\mathbf{r} = \int_C (x\mathbf{i} + y\mathbf{j})\cdot d\mathbf{r}$$
$$= \int_0^\pi (\cos t\mathbf{i} + \sin t\mathbf{j})\cdot(-\sin t\mathbf{i} + \cos t\mathbf{j})\,dt$$
$$= \int_0^\pi (-\cos t\sin t + \sin t\cos t)\,dt = 0$$

*See Section 14.1.

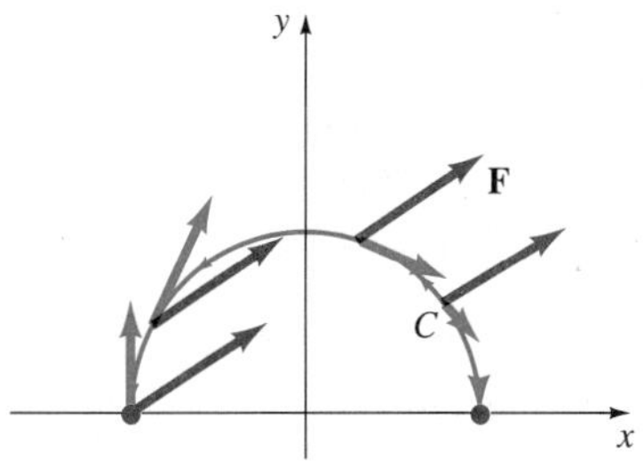

FIGURE 17.11

(*b*) In Figure 17.11 the vectors in green are the projections of **F** on the unit tangent vectors. The work done by **F** is

$$W = \int_C \mathbf{F} \cdot d\mathbf{r} = \int_C \left(\frac{3}{4}\mathbf{i} + \frac{1}{2}\mathbf{j}\right) \cdot d\mathbf{r}$$
$$= \int_0^\pi \left(\frac{3}{4}\mathbf{i} + \frac{1}{2}\mathbf{j}\right) \cdot (-\sin t\mathbf{i} + \cos t\mathbf{j})\, dt$$
$$= \int_0^\pi \left(-\frac{3}{4}\sin t + \frac{1}{2}\cos t\right) dt$$
$$= \left(\frac{3}{4}\cos t + \frac{1}{2}\sin t\right)\Big]_0^\pi = -\frac{3}{2}$$

The units of work depend on the units of $|\mathbf{F}|$ and on the units of distance. □

Circulation A line integral of a vector field **F** around a simple closed curve C is said to be the **circulation** of **F** around C; that is,

$$\text{circulation} = \oint_C \mathbf{F} \cdot d\mathbf{r} = \oint_C \mathbf{F} \cdot \mathbf{T}\, ds$$

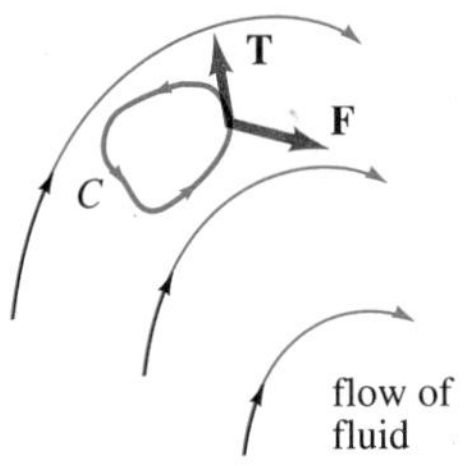

FIGURE 17.12

In particular, if **F** is the velocity field of a fluid, then the circulation is a measure of the amount by which the fluid tends to turn the curve C by rotating, or circulating, around it. For example, if **F** is perpendicular to **T** for every (x, y) on C, then $\int_C \mathbf{F} \cdot \mathbf{T}\, ds = 0$ and the curve does not move at all. On the other hand, $\int_C \mathbf{F} \cdot \mathbf{T}\, ds > 0$ and $\int_C \mathbf{F} \cdot \mathbf{T}\, ds < 0$ mean that the fluid tends to rotate C in the counterclockwise and clockwise directions, respectively. See Figure 17.12.

▲ ***Remark*** In the case of two variables, the line integral with respect to arc length $\int_C G(x, y)\, ds$ can be interpreted in a geometric manner when $G(x, y) \geq 0$ on C. In Definition 17.1 the symbol Δs_k represents the length of the kth subarc on the curve C. But from the figure preceding that definition, we have the approximation $\Delta s_k = \sqrt{(\Delta x_k)^2 + (\Delta y_k)^2}$. With this interpretation of Δs_k, we see from Figure 17.13(a) that the product of $G(x_k^*, y_k^*)\, \Delta s_k$ is the area of a

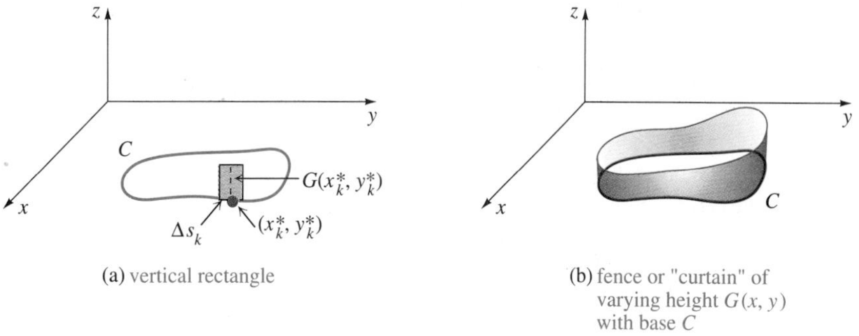

FIGURE 17.13

vertical rectangle of height $G(x_k^*, y_k^*)$ and width Δs_k. The integral $\int_C G(x, y)\, ds$ then represents the area of one side of a "fence" or "curtain" extending from the curve C in the xy-plane up to the graph of $G(x, y)$ that corresponds to points (x, y) on C. See Figure 17.13(b).

EXERCISES 17.1

Answers to odd-numbered problems begin on page A-109.

In Problems 1–4 evaluate $\int_C G(x, y)\, dx$, $\int_C G(x, y)\, dy$, and $\int_C G(x, y)\, ds$ on the indicated curve C.

1. $G(x, y) = 2xy$; $x = 5\cos t$, $y = 5\sin t$, $0 \le t \le \pi/4$
2. $G(x, y) = x^3 + 2xy^2 + 2x$; $x = 2t$, $y = t^2$, $0 \le t \le 1$
3. $G(x, y) = 3x^2 + 6y^2$; $y = 2x + 1$, $-1 \le x \le 0$
4. $G(x, y) = x^2/y^3$; $2y = 3x^{2/3}$, $1 \le x \le 8$

In Problems 5 and 6 evaluate $\int_C G(x, y, z)\, dx$, $\int_C G(x, y, z)\, dy$, $\int_C G(x, y, z)\, dz$, and $\int_C G(x, y, z)\, ds$ on the indicated curve C.

5. $G(x, y, z) = z$; $x = \cos t$, $y = \sin t$, $z = t$, $0 \le t \le \pi/2$
6. $G(x, y, z) = 4xyz$; $x = \frac{1}{3}t^3$, $y = t^2$, $z = 2t$, $0 \le t \le 1$

In Problems 7–10 evaluate $\int_C (2x + y)\, dx + xy\, dy$ on the given curve C between $(-1, 2)$ and $(2, 5)$.

7. $y = x + 3$
8. $y = x^2 + 1$
9.

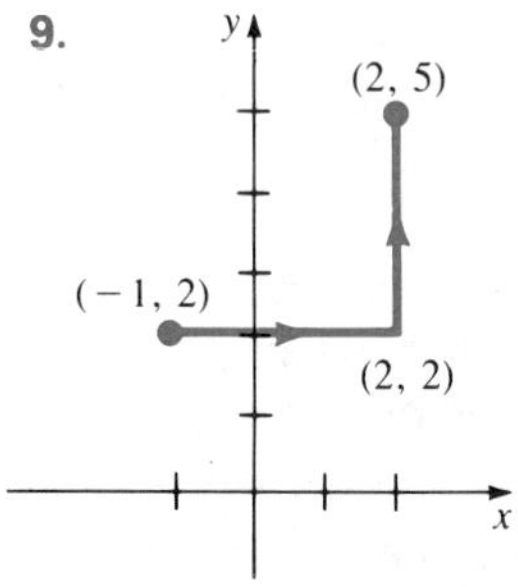

FIGURE 17.14

10.

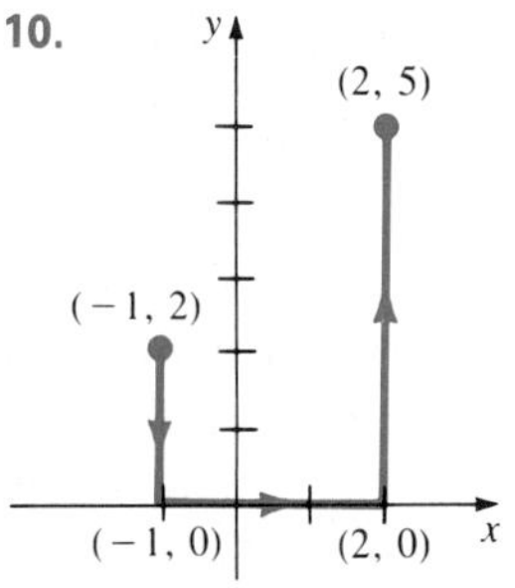

FIGURE 17.15

In Problems 11–14 evaluate $\int_C y\, dx + x\, dy$ on the given curve C between $(0, 0)$ and $(1, 1)$.

11. $y = x^2$
12. $y = x$
13. C consists of the line segments from $(0, 0)$ to $(0, 1)$ and from $(0, 1)$ to $(1, 1)$.
14. C consists of the line segments from $(0, 0)$ to $(1, 0)$ and from $(1, 0)$ to $(1, 1)$.
15. Evaluate $\int_C (6x^2 + 2y^2)\, dx + 4xy\, dy$, where C is given by $x = \sqrt{t}$, $y = t$, $4 \le t \le 9$.
16. Evaluate $\int_C -y^2\, dx + xy\, dy$, where C is given by $x = 2t$, $y = t^3$, $0 \le t \le 2$.
17. Evaluate $\int_C 2x^3 y\, dx + (3x + y)\, dy$, where C is given by $x = y^2$ from $(1, -1)$ to $(1, 1)$.
18. Evaluate $\int_C 4x\, dx + 2y\, dy$, where C is given by $x = y^3 + 1$ from $(0, -1)$, to $(9, 2)$.

In Problems 19 and 20 evaluate $\oint_C (x^2 + y^2)\, dx - 2xy\, dy$ on the given closed curve C.

19.

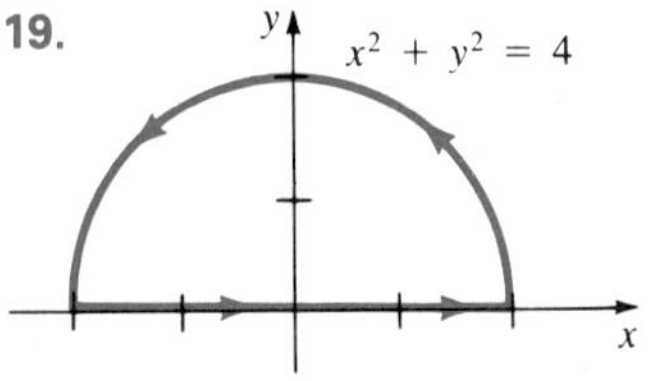

FIGURE 17.16

20.

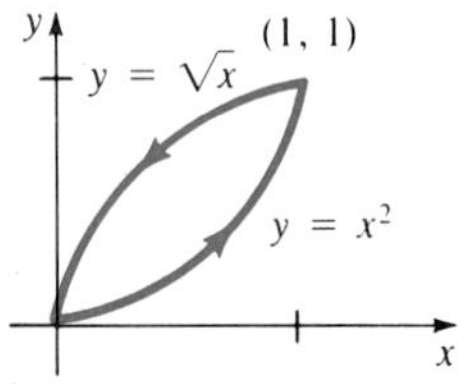

FIGURE 17.17

In Problems 21 and 22 evaluate $\oint_C x^2 y^3\, dx - xy^2\, dy$ on the given closed curve C.

21.

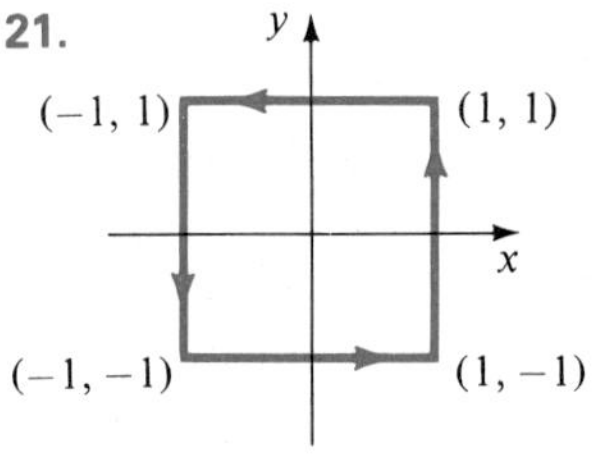

FIGURE 17.18

22.

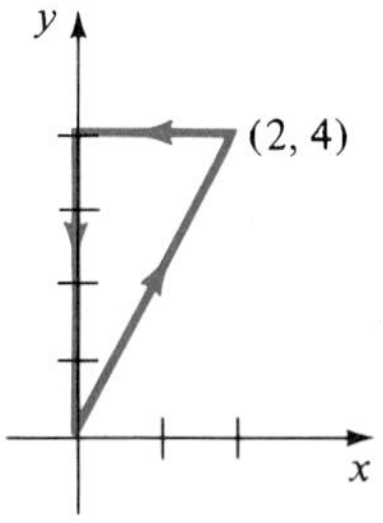

FIGURE 17.19

23. Evaluate $\oint_C (x^2 - y^2)\, ds$, where C is given by

$$x = 5\cos t, \quad y = 5\sin t, \quad 0 \le t \le 2\pi$$

24. Evaluate $\int_{-C} y\, dx - x\, dy$, where C is given by

$$x = 2\cos t, \quad y = 3\sin t, \quad 0 \le t \le \pi$$

In Problems 25–28 evaluate $\int_C y\, dx + z\, dy + x\, dz$ on the given curve C between $(0, 0, 0)$ and $(6, 8, 5)$.

25. C consists of the line segments from $(0, 0, 0)$ to $(2, 3, 4)$ and from $(2, 3, 4)$ to $(6, 8, 5)$.

26. $x = 3t$, $y = t^3$, $z = \frac{5}{4}t^2$, $0 \le t \le 2$

27.

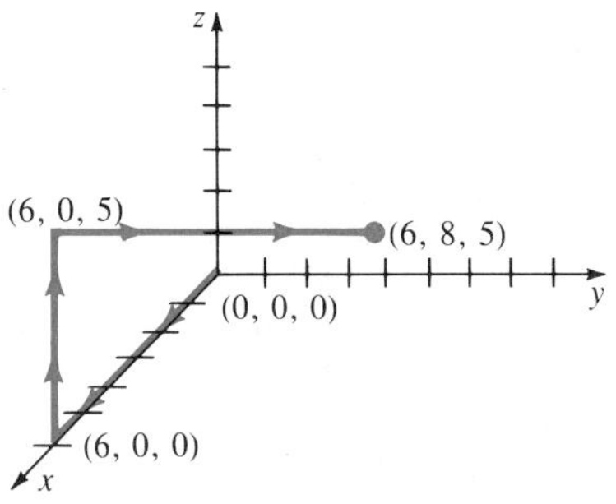

FIGURE 17.20

28.

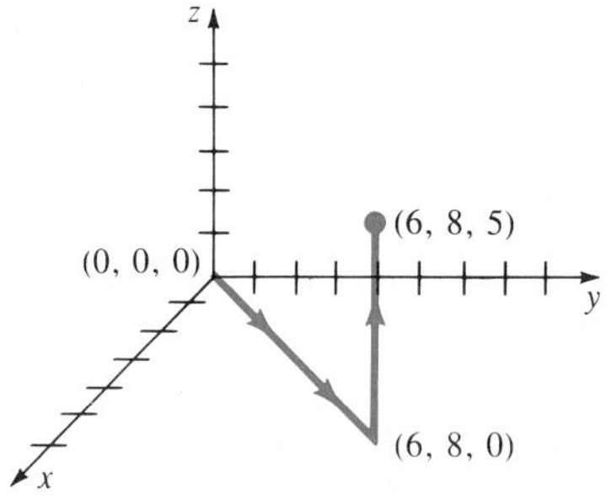

FIGURE 17.21

In Problems 29–34 graph some representative vectors in the given vector field.

29. $\mathbf{F}(x, y) = x\mathbf{i} + y\mathbf{j}$

30. $\mathbf{F}(x, y) = -x\mathbf{i} + y\mathbf{j}$

31. $\mathbf{F}(x, y) = y\mathbf{i} + x\mathbf{j}$

32. $\mathbf{F}(x, y) = x\mathbf{i} + 2y\mathbf{j}$

33. $\mathbf{F}(x, y) = y\mathbf{j}$

34. $\mathbf{F}(x, y) = x\mathbf{j}$

In Problems 35 and 36 evaluate $\int_C \mathbf{F} \cdot d\mathbf{r}$.

35. $\mathbf{F}(x, y) = y^3\mathbf{i} - x^2y\mathbf{j}$; $\mathbf{r}(t) = e^{-2t}\mathbf{i} + e^t\mathbf{j}$, $0 \le t \le \ln 2$

36. $\mathbf{F}(x, y, z) = e^x\mathbf{i} + xe^{xy}\mathbf{j} + xye^{xyz}\mathbf{k}$; $\mathbf{r}(t) = t\mathbf{i} + t^2\mathbf{j} + t^3\mathbf{k}$, $0 \le t \le 1$

37. Find the work done by the force $\mathbf{F}(x, y) = y\mathbf{i} + x\mathbf{j}$ acting along $y = \ln x$ from $(1, 0)$ to $(e, 1)$.

38. Find the work done by the force $\mathbf{F}(x, y) = 2xy\mathbf{i} + 4y^2\mathbf{j}$ acting along the piecewise smooth curve consisting of the line segments from $(-2, 2)$ to $(0, 0)$ and from $(0, 0)$ to $(2, 3)$.

39. Find the work done by the force $\mathbf{F}(x, y) = (x + 2y)\mathbf{i} + (6y - 2x)\mathbf{j}$ acting counterclockwise once around the triangle with vertices $(1, 1)$, $(3, 1)$, and $(3, 2)$.

40. Find the work done by the force $\mathbf{F}(x, y, z) = yz\mathbf{i} + xz\mathbf{j} + xy\mathbf{k}$ acting along the curve given by $\mathbf{r}(t) = t^3\mathbf{i} + t^2\mathbf{j} + t\mathbf{k}$ from $t = 1$ to $t = 3$.

41. Find the work done by a constant force $\mathbf{F}(x, y) = a\mathbf{i} + b\mathbf{j}$ acting counterclockwise once around the circle $x^2 + y^2 = 9$.

42. In an inverse square force field $\mathbf{F} = c\mathbf{r}/|\mathbf{r}|^3$, where c is a constant and $\mathbf{r} = x\mathbf{i} + y\mathbf{j} + z\mathbf{k}$,* find the work done in moving a particle along the line from $(1, 1, 1)$ to $(3, 3, 3)$.

43. Verify that the line integral $\int_C y^2\,dx + xy\,dy$ has the same value on C for each of the following parameterizations:

$$C: \quad x = 2t + 1, \quad y = 4t + 2, \quad 0 \le t \le 1$$
$$C: \quad x = t^2, \quad y = 2t^2, \quad 1 \le t \le \sqrt{3}$$
$$C: \quad x = \ln t, \quad y = 2\ln t, \quad e \le t \le e^3$$

44. Consider the three curves between $(0, 0)$ and $(2, 4)$:

$$C_1: \quad x = t, \quad y = 2t, \quad 0 \le t \le 2$$
$$C_2: \quad x = t, \quad y = t^2, \quad 0 \le t \le 2$$
$$C_3: \quad x = 2t - 4, \quad y = 4t - 8, \quad 2 \le t \le 3$$

Show that $\int_{C_1} xy\,ds = \int_{C_3} xy\,ds$, but $\int_{C_1} xy\,ds \ne \int_{C_2} xy\,ds$. Explain.

45. Assume a smooth curve C is described by the vector function $\mathbf{r}(t)$ for $a \le t \le b$. Let acceleration, velocity, and speed be given by $\mathbf{a} = d\mathbf{v}/dt$, $\mathbf{v} = d\mathbf{r}/dt$, and $v = |\mathbf{v}|$, respectively. Using Newton's second law $\mathbf{F} = m\mathbf{a}$, show that, in the absence of friction, the work done by $\mathbf{F}$ in moving a particle of constant mass m from point A at $t = a$ to point B at $t = b$ is the same as the change in kinetic energy:

$$K(B) - K(A) = \frac{1}{2}m[v(b)]^2 - \frac{1}{2}m[v(a)]^2$$

$$\left[\textit{Hint:} \text{ Consider } \frac{d}{dt}v^2 = \frac{d}{dt}\mathbf{v} \cdot \mathbf{v}.\right]$$

46. If $\rho(x, y)$ is the density of a wire (mass per unit length), then $m = \int_C \rho(x, y)\,ds$ is the mass of the wire. Find the mass of a wire having the shape of the semicircle $x = 1 + \cos t$, $y = \sin t$, $0 \le t \le \pi$, if the density at a point P is directly proportional to distance from the y-axis.

47. The coordinates of the center of mass of a wire with variable density are given by $\bar{x} = M_y/m$, $\bar{y} = M_x/m$, where

$$m = \int_C \rho(x, y)\,ds, \qquad M_x = \int_C y\rho(x, y)\,ds$$

$$M_y = \int_C x\rho(x, y)\,ds$$

Find the center of mass of the wire in Problem 46.

*Note that the magnitude of $\mathbf{F}$ is inversely proportional to $|\mathbf{r}|^2$.

48. A force field $\mathbf{F}(x, y)$ acts at each point on the curve C, which is the union of C_1, C_2, and C_3 shown in Figure 17.22. $|\mathbf{F}|$ is measured in pounds and distance is measured in feet using the scale given in the figure. Use the representative vectors shown to approximate the work done by $\mathbf{F}$ along C. [*Hint:* Use $W = \int_C \mathbf{F} \cdot \mathbf{T}\, ds$.]

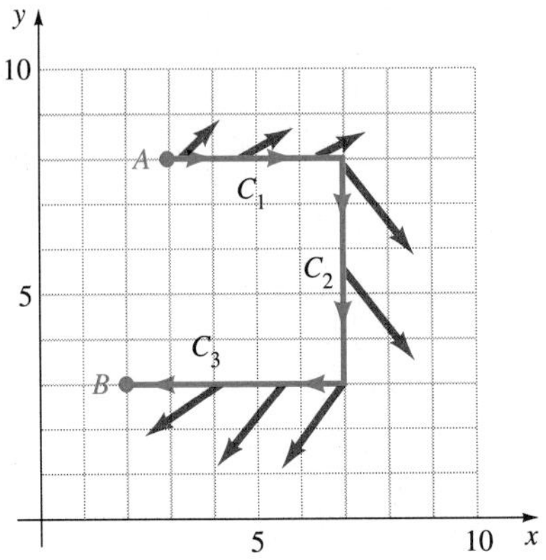

FIGURE 17.22

17.2 LINE INTEGRALS INDEPENDENT OF THE PATH

The concept of an exact differential plays an important role in the discussion that follows. A review of Section 15.5 is recommended.

The value of a line integral $\int_C P\,dx + Q\,dy$ generally depends on the curve or **path** between two points A and B. However, there are exceptions. Let the functions P and Q be continuous in an open region R of the xy-plane that contains A and B. Then a line integral whose value is the same for *every* curve C connecting A and B is said to be **independent of the path.**

EXAMPLE 1 The integral $\int_C y\,dx + x\,dy$ has the same value on each path C between (0, 0) and (1, 1) shown in Figure 17.23. You may recall from Problems 11–14 of Exercises 17.1 that on these paths

$$\int_C y\,dx + x\,dy = 1$$

In Example 2 we shall prove that the given integral is independent of the path.

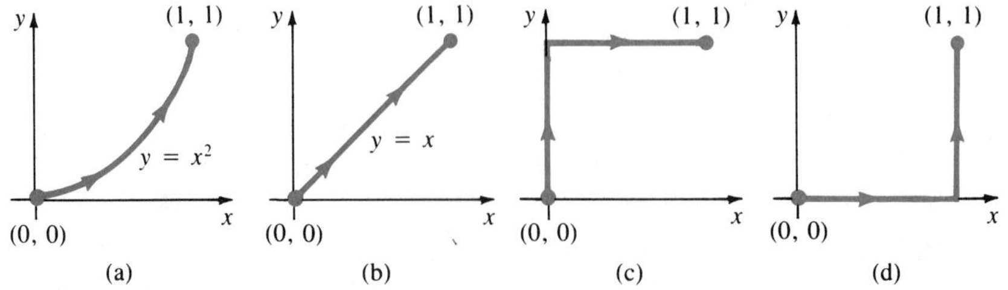

FIGURE 17.23 □

A Fundamental Theorem The next theorem establishes an important relationship between the seemingly disparate notions of path independence and exact differentials. In addition, it provides a means of evaluating path-independent line integrals in a manner analogous to the Fundamental Theorem of Calculus:

$$\int_a^b f'(x)\,dx = f(b) - f(a)$$

Throughout this discussion we assume P and Q are continuous in some open region R of the xy-plane.

THEOREM 17.1 Fundamental Theorem for Line Integrals

Suppose there exists a function $\phi(x, y)$ such that $d\phi = P\,dx + Q\,dy$ in R; that is, $P\,dx + Q\,dy$ is an exact differential. Then

$$\int_C P\,dx + Q\,dy = \phi(B) - \phi(A)$$

Proof Let C be a smooth path defined parametrically by $x = f(t)$, $y = g(t)$, $a \le t \le b$, and let the coordinates of A and B be $(f(a), g(a))$ and $(f(b), g(b))$, respectively. Then by the Chain Rule,

$$\begin{aligned}\int_C P\,dx + Q\,dy &= \int_a^b \left(\frac{\partial \phi}{\partial x}\frac{dx}{dt} + \frac{\partial \phi}{\partial y}\frac{dy}{dt}\right) dt \\ &= \int_a^b \frac{d\phi}{dt}\,dt = \phi(f(t), g(t))\Big|_a^b = \phi(B) - \phi(A)\end{aligned}$$ ■

In Theorem 17.1 we have shown that when $P\,dx + Q\,dy$ is an exact differential, the value of the line integral $\int_C P\,dx + Q\,dy$ depends on only the initial and terminal points A and B of the curve C, and not on C itself. In other words, the line integral is independent of the path. Theorem 17.1 is also valid for piecewise smooth curves, but the foregoing proof must be modified by considering each smooth arc of the curve C.

Notation A line integral $\int_C P\,dx + Q\,dy$, which is independent of the path between the endpoints A and B, is often written

$$\int_A^B P\,dx + Q\,dy$$

EXAMPLE 2 In Example 1 note that $d(xy) = y\,dx + x\,dy$; that is, $y\,dx + x\,dy$ is an exact differential. Hence, $\int_C y\,dx + x\,dy$ is independent of the path between any two points A and B. Specifically, if A and B are, respectively,

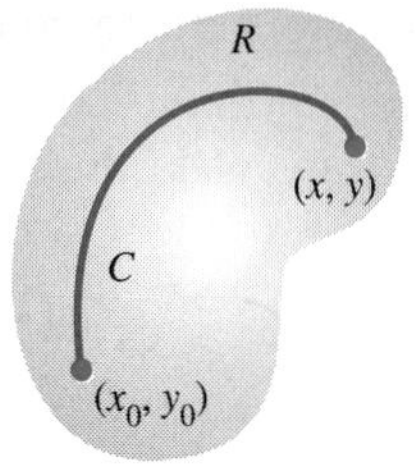

(a)

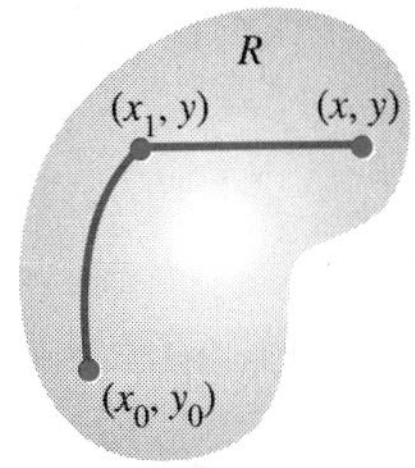

(b)

FIGURE 17.24

(0, 0) and (1, 1), we then have, from Theorem 17.1,

$$\int_{(0,0)}^{(1,1)} y\,dx + x\,dy = \int_{(0,0)}^{(1,1)} d(xy) = xy\Big]_{(0,0)}^{(1,1)} = 1 \qquad \square$$

The converse of Theorem 17.1 is also true.

THEOREM 17.2

If the line integral $\int_C P\,dx + Q\,dy$ is independent of the path in R, then $P\,dx + Q\,dy$ is an exact differential in R.

Proof Let (x_0, y_0) be an arbitrary point in the region R and let the function $\phi(x, y)$ be defined as

$$\phi(x, y) = \int_{(x_0, y_0)}^{(x, y)} P\,dx + Q\,dy$$

where C is any arbitrary path in R from (x_0, y_0) to (x, y). See Figure 17.24(a). Now choose a point (x_1, y), $x_1 \neq x$, so that the line segment from (x_1, y) to (x, y) is in R. See Figure 17.24(b). Then by path independence we can write

$$\phi(x, y) = \int_{(x_0, y_0)}^{(x_1, y)} P\,dx + Q\,dy + \int_{(x_1, y)}^{(x, y)} P\,dx + Q\,dy$$

Now,

$$\frac{\partial \phi}{\partial x} = 0 + \frac{\partial}{\partial x}\int_{(x_1, y)}^{(x, y)} P\,dx + Q\,dy$$

since the first integral does not depend on x. But on the line segment between (x_1, y) and (x, y), y is constant so that $dy = 0$. Hence, $\int_{(x_1, y)}^{(x, y)} P\,dx + Q\,dy = \int_{(x_1, y)}^{(x, y)} P\,dx$. By the Fundamental Theorem of Calculus we then have

$$\frac{\partial \phi}{\partial x} = \frac{\partial}{\partial x}\int_{(x_1, y)}^{(x, y)} P(x, y)\,dx = P(x, y)$$

In the same manner we can also show that $\partial\phi/\partial y = Q(x, y)$. Hence,

$$P\,dx + Q\,dy = \frac{\partial \phi}{\partial x}\,dx + \frac{\partial \phi}{\partial y}\,dy$$

is an exact differential. ■

Theorems 17.1 and 17.2 yield the result:

A line integral $\int_C P\,dx + Q\,dy$ is independent of the path if and only if $P\,dx + Q\,dy$ is an exact differential. (1)

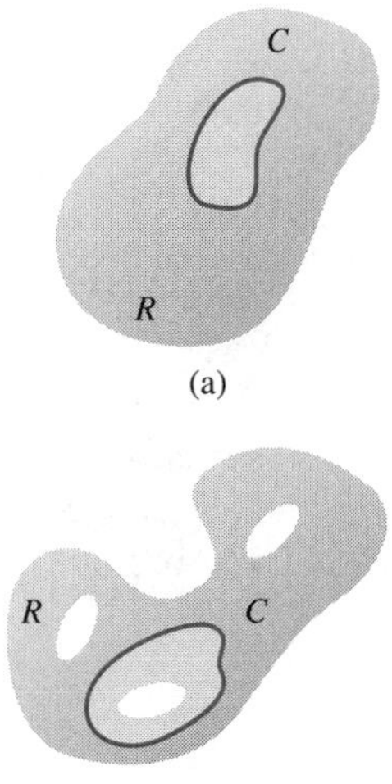

(a)

(b)

FIGURE 17.25

In order to be able to the state the next result we need to consider a particular kind of region in the plane. We say that a region R of the plane is **simply connected** if:

- R is connected; that is, every pair of points in the region can be joined by a piecewise-smooth curve that lies entirely in R, and
- every simple closed curve C lying entirely within R can be shrunk, or contracted, to a point without leaving R.

The last condition means that if C is any simple closed curve lying entirely in R, then the region in the interior of C also lies entirely in R. Roughly put, a simply connected region has no holes in it. In Figure 17.25(a) we have illustrated a simply connected region. The representative simple closed curve C could be shrunk to a point without leaving the region. In Figure 17.25(b) the region shown has three holes in it; since the representative curve C surrounds one of the holes, it could not be shrunk to a point without leaving the region. This last region is said to be **multiply connected**.

Test for Path Independence in the Plane In view of the statement given in (1), the same criteria given in Theorem 15.6 for determining an exact differential become the criteria, or tests, for path independence in the xy-plane.

THEOREM 17.3 Criteria for Path Independence

Let P and Q be continuous and have continuous first partial derivatives in an open simply connected region R. Then $\int_C P\,dx + Q\,dy$ is **independent** of the path C if and only if

$$\frac{\partial P}{\partial y} = \frac{\partial Q}{\partial x}$$

for all (x, y) in R.

EXAMPLE 3 Show that the integral $\int_C (x^2 - 2y^3)\,dx + (x + 5y)\,dy$ is not independent of the path C.

Solution From $P = x^2 - 2y^3$ and $Q = x + 5y$, we find

$$\frac{\partial P}{\partial y} = -6y^2 \quad \text{and} \quad \frac{\partial Q}{\partial x} = 1$$

Since $\partial P/\partial y \neq \partial Q/\partial x$, it follows from Theorem 17.3 that the integral is not independent of the path. □

EXAMPLE 4 Show that $\int_C (y^2 - 6xy + 6)\,dx + (2xy - 3x^2)\,dy$ is independent of any path C between $(-1, 0)$ and $(3, 4)$. Evaluate.

Solution Identifying $P = y^2 - 6xy + 6$ and $Q = 2xy - 3x^2$ yields

$$\frac{\partial P}{\partial y} = 2y - 6x \quad \text{and} \quad \frac{\partial Q}{\partial x} = 2y - 6x$$

Since $\partial P/\partial y = \partial Q/\partial x$, the integral is independent of the path and so there exists a function ϕ such that $\partial \phi/\partial x = y^2 - 6xy + 6$ and $\partial \phi/\partial y = 2xy - 3x^2$. Employing the partial integration method of Section 15.5, we find

$$\phi = xy^2 - 3x^2y + 6x$$

In other words,

$$d(xy^2 - 3x^2y + 6x) = (y^2 - 6xy + 6)\,dx + (2xy - 3x^2)\,dy$$

Hence,

$$\begin{aligned}\int_{(-1,0)}^{(3,4)} (y^2 - 6xy + 6)\,dx + (2xy - 3x^2)\,dy &= \int_{(-1,0)}^{(3,4)} d(xy^2 - 3x^2y + 6x)\\ &= (xy^2 - 3x^2y + 6x)\Big]_{(-1,0)}^{(3,4)}\\ &= (48 - 108 + 18) - (-6)\\ &= -36\end{aligned}$$

Alternative Solution Since the integral is independent of the path, we can integrate along any convenient curve connecting the given points. In particular, $y = x + 1$ is such a curve. Using x as a parameter then gives

$$\begin{aligned}\int_C (y^2 - 6xy + 6)\,dx &+ (2xy - 3x^2)\,dy\\ &= \int_{-1}^{3} [(x + 1)^2 - 6x(x + 1) + 6]\,dx + [2x(x + 1) - 3x^2]\,dx\\ &= \int_{-1}^{3} (-6x^2 - 2x + 7)\,dx = -36\end{aligned}$$

□

Conservative Vector Fields If $\int_C P\,dx + Q\,dy$ is independent of the path C, we know there exists a function ϕ such that

$$\begin{aligned}d\phi &= \frac{\partial \phi}{\partial x}\,dx + \frac{\partial \phi}{\partial y}\,dy = P\,dx + Q\,dy\\ &= (P\mathbf{i} + Q\mathbf{j}) \cdot (dx\mathbf{i} + dy\mathbf{j}) = \mathbf{F} \cdot d\mathbf{r}\end{aligned}$$

where $\mathbf{F} = P\mathbf{i} + Q\mathbf{j}$ is a vector field and $P = \partial \phi/\partial x$, $Q = \partial \phi/\partial y$. In other words, the vector field $\mathbf{F}$ is a gradient of the function ϕ. Since $\mathbf{F} = \nabla \phi$, $\mathbf{F}$ is sometimes said to be a **gradient field** and the function ϕ is then said to be a **potential function** for **F**. In a gradient force field **F**, the work done by the force on a particle moving from position A to position B is the same for all paths between these points. Moreover, the work done by the force along a closed path is *zero*. See Problem 29 in Exercises 17.2. For this reason, such a force field is also said to be **conservative**. In a conservative field **F** the *law of conservation of mechanical energy* holds: For a particle moving along a path in a conservative field,

$$\textit{kinetic energy} + \textit{potential energy} = \textit{constant}$$

See Problem 31. In a simply connected domain the hypotheses of Theorem 17.3 imply that a force field $\mathbf{F}(x, y) = P(x, y)\mathbf{i} + Q(x, y)\mathbf{j}$ is a gradient field (that is, conservative) if and only if

$$\frac{\partial P}{\partial y} = \frac{\partial Q}{\partial x} \tag{2}$$

Thus, (1) can be expressed in two alternative ways:

A line integral $\int_C \mathbf{F} \cdot d\mathbf{r}$ is independent of the path if and only if **F** *is a gradient field.*

A line integral $\int_C \mathbf{F} \cdot d\mathbf{r}$ is independent of the path if and only if **F** *is conservative.*

EXAMPLE 5 Show that the vector field $\mathbf{F} = (y^2 + 5)\mathbf{i} + (2xy - 8)\mathbf{j}$ is a gradient field. Find a potential function for **F**.

Solution By identifying $P = y^2 + 5$ and $Q = 2xy - 8$, we see from (2) that

$$\frac{\partial P}{\partial y} = \frac{\partial Q}{\partial x} = 2y$$

Hence, **F** is a gradient field and so there exists a potential function ϕ satisfying

$$\frac{\partial \phi}{\partial x} = y^2 + 5 \quad \text{and} \quad \frac{\partial \phi}{\partial y} = 2xy - 8$$

Proceeding as in Section 15.5, we find that $\phi = xy^2 - 8y + 5x$.

Check $$\nabla\phi = \frac{\partial \phi}{\partial x}\mathbf{i} + \frac{\partial \phi}{\partial y}\mathbf{j} = (y^2 + 5)\mathbf{i} + (2xy - 8)\mathbf{j}$$ □

Test for Path Independence in Space If C is a space curve, a line integral $\int_C \mathbf{F} \cdot d\mathbf{r}$ is independent of the path whenever the differential expression $P(x, y, z)\,dx + Q(x, y, z)\,dy + R(x, y, z)\,dz$ is an exact differential. The three-dimensional analogue of Theorem 17.3 is given next.

THEOREM 17.4 Criteria for Path Independence

Let P, Q, and R have continuous first partial derivatives in an open simply connected region of space. Then $\int_C P\,dx + Q\,dy + R\,dz$ is independent of the path C if and only if

$$\frac{\partial P}{\partial y} = \frac{\partial Q}{\partial x}, \quad \frac{\partial P}{\partial z} = \frac{\partial R}{\partial x}, \quad \frac{\partial Q}{\partial z} = \frac{\partial R}{\partial y}$$

EXAMPLE 6 Show that

$$\int_C (y + yz)\,dx + (x + 3z^3 + xz)\,dy + (9yz^2 + xy - 1)\,dz$$

is independent of any path C between (1, 1, 1) and (2, 1, 4). Evaluate.

Solution With the identifications

$$P = y + yz, \quad Q = x + 3z^3 + xz, \quad \text{and} \quad R = 9yz^2 + xy - 1$$

we see that

$$\frac{\partial P}{\partial y} = 1 + z = \frac{\partial Q}{\partial x}, \quad \frac{\partial P}{\partial z} = y = \frac{\partial R}{\partial x}, \quad \text{and} \quad \frac{\partial Q}{\partial z} = 9z^2 + x = \frac{\partial R}{\partial y}$$

From Theorem 17.3 we conclude that the integral is independent of the path. Moreover, $(y + yz)\,dx + (x + 3z^3 + xz)\,dy + (9yz^2 + xy - 1)\,dz$ is an exact differential and so there exists a function $\phi(x, y, z)$ such that

$$\frac{\partial \phi}{\partial x} = P, \quad \frac{\partial \phi}{\partial y} = Q, \quad \text{and} \quad \frac{\partial \phi}{\partial z} = R$$

Integrating the first of these three equations with respect to x gives

$$\phi = xy + xyz + g(y, z)$$

The derivative of this last expression with respect to y must then be equal to Q:

$$\frac{\partial \phi}{\partial y} = x + xz + \frac{\partial g}{\partial y} = x + 3z^3 + xz$$

Thus, $$\frac{\partial g}{\partial y} = 3z^3 \quad \text{and so} \quad g = 3yz^3 + h(z)$$

Consequently, $$\phi = xy + xyz + 3yz^3 + h(z)$$

The partial derivative of this last expression with respect to z must now be equal to R:

$$\frac{\partial \phi}{\partial z} = xy + 9yz^2 + h'(z) = 9yz^2 + xy - 1$$

From this we get $h'(z) = -1$ and $h(z) = -z + C$. Disregarding C, we can write

$$\phi = xy + xyz + 3yz^3 - z \tag{3}$$

Finally, we obtain

$$\begin{aligned}\int_{(1,1,1)}^{(2,1,4)} & (y + yz)\,dx + (x + 3z^3 + xz)\,dy + (9yz^2 + xy - 1)\,dz \\ &= \int_{(1,1,1)}^{(2,1,4)} d(xy + xyz + 3yz^3 - z) \\ &= (xy + xyz + 3yz^3 - z)\Big|_{(1,1,1)}^{(2,1,4)} = 198 - 4 = 194\end{aligned}$$

□

In Example 6 the vector field defined by

$$\mathbf{F}(x, y, z) = (y + yz)\mathbf{i} + (x + 3z^3 + xz)\mathbf{j} + (9yz^2 + xy - 1)\mathbf{k}$$

can be interpreted as a gradient field or a conservative force field, since the proof that the differential form $(y + yz)\,dx + (x + 3z^3 + xz)\,dy + (9yz^2 + xy - 1)\,dz$ is exact also proves that $\mathbf{F} = \nabla\phi$, where ϕ is the potential function given in (3).

▲ ***Remark*** A frictional force such as air resistance is *nonconservative*. Nonconservative forces are *dissipative* in that their action reduces kinetic energy without a corresponding increase in potential energy. In other words, if the work done $\int_C \mathbf{F} \cdot d\mathbf{r}$ depends on the path, then $\mathbf{F}$ is nonconservative.

EXERCISE 17.2

Answers to odd-numbered problems begin on page A-109.

In Problems 1–10 show that the given integral is independent of the path. Evaluate in two ways: **(a)** find a function ϕ such that $d\phi = P\,dx + Q\,dy$, and **(b)** integrate along any convenient path between the points.

1. $\displaystyle\int_{(0,0)}^{(2,2)} x^2\,dx + y^2\,dy$
2. $\displaystyle\int_{(1,1)}^{(2,4)} 2xy\,dx + x^2\,dy$
3. $\displaystyle\int_{(1,0)}^{(3,2)} (x + 2y)\,dx + (2x - y)\,dy$
4. $\displaystyle\int_{(0,0)}^{(\pi/2,0)} \cos x \cos y\,dx + (1 - \sin x \sin y)\,dy$
5. $\displaystyle\int_{(4,1)}^{(4,4)} \frac{-y\,dx + x\,dy}{y^2}$ on any path not crossing the x-axis
6. $\displaystyle\int_{(1,0)}^{(3,4)} \frac{x\,dx + y\,dy}{\sqrt{x^2 + y^2}}$ on any path not through the origin
7. $\displaystyle\int_{(1,2)}^{(3,6)} (2y^2x - 3)\,dx + (2yx^2 + 4)\,dy$
8. $\displaystyle\int_{(-1,1)}^{(0,0)} (5x + 4y)\,dx + (4x - 8y^3)\,dy$
9. $\displaystyle\int_{(0,0)}^{(2,8)} (y^3 + 3x^2y)\,dx + (x^3 + 3y^2x + 1)\,dy$
10. $\displaystyle\int_{(-2,0)}^{(1,0)} (2x - y \sin xy - 5y^4)\,dx - (20xy^3 + x \sin xy)\,dy$

In Problems 11–16 determine whether the given vector field is a gradient field. If so, find the potential function ϕ for $\mathbf{F}$.

11. $\mathbf{F}(x, y) = (4x^3y^3 + 3)\mathbf{i} + (3x^4y^2 + 1)\mathbf{j}$
12. $\mathbf{F}(x, y) = 2xy^3\mathbf{i} + 3y^2(x^2 + 1)\mathbf{j}$

13. $\mathbf{F}(x, y) = y^2 \cos xy^2\mathbf{i} - 2xy \sin xy^2\mathbf{j}$

14. $\mathbf{F}(x, y) = (x^2 + y^2 + 1)^{-2}(x\mathbf{i} + y\mathbf{j})$

15. $\mathbf{F}(x, y) = (x^3 + y)\mathbf{i} + (x + y^3)\mathbf{j}$

16. $\mathbf{F}(x, y) = 2e^{2y}\mathbf{i} + xe^{2y}\mathbf{j}$

In Problems 17 and 18 find the work done by the force $\mathbf{F}(x, y) = (2x + e^{-y})\mathbf{i} + (4y - xe^{-y})\mathbf{j}$ along the indicated curve.

17.

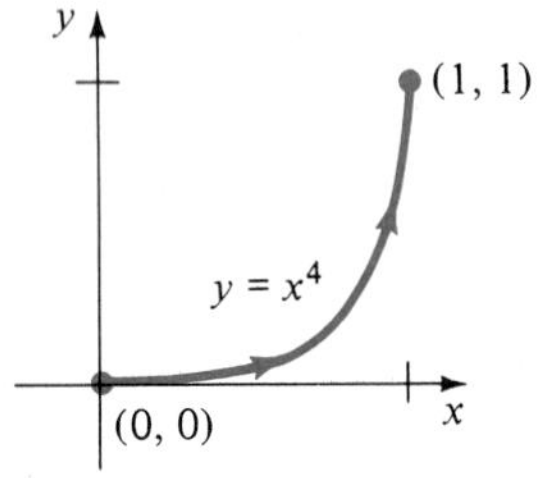

FIGURE 17.26

18.

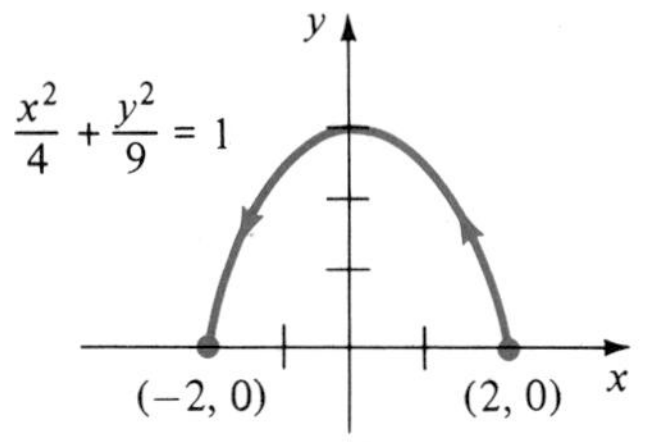

FIGURE 17.27

In Problems 19–24 show that the given integral is independent of the path. Evaluate.

19. $\int_{(1,1,1)}^{(2,4,8)} yz\,dx + xz\,dy + xy\,dz$

20. $\int_{(0,0,0)}^{(1,1,1)} 2x\,dx + 3y^2\,dy + 4z^3\,dz$

21. $\int_{(1,0,0)}^{(2,\pi/2,1)} (2x \sin y + e^{3z})\,dx + x^2 \cos y\,dy + (3xe^{3z} + 5)\,dz$

22. $\int_{(1,2,1)}^{(3,4,1)} (2x + 1)\,dx + 3y^2\,dy + \frac{1}{z}\,dz$

23. $\int_{(1,1,\ln 3)}^{(2,2,\ln 3)} e^{2z}\,dx + 3y^2\,dy + 2xe^{2z}\,dz$

24. $\int_{(-2,3,1)}^{(0,0,0)} 2xz\,dx + 2yz\,dy + (x^2 + y^2)\,dz$

In Problems 25 and 26 evaluate $\int_C \mathbf{F} \cdot d\mathbf{r}$.

25. $\mathbf{F}(x, y, z) = (y - yz \sin x)\mathbf{i} + (x + z \cos x)\mathbf{j} + y \cos x\mathbf{k}$; $\mathbf{r}(t) = 2t\mathbf{i} + (1 + \cos t)^2\mathbf{j} + 4 \sin^3 t\mathbf{k}$, $0 \le t \le \pi/2$

26. $\mathbf{F}(x, y, z) = (2 - e^z)\mathbf{i} + (2y - 1)\mathbf{j} + (2 - xe^z)\mathbf{k}$; $\mathbf{r}(t) = t\mathbf{i} + t^2\mathbf{j} + t^3\mathbf{k}$, $(-1, 1, -1)$ to $(2, 2, 2)$

27. The inverse square law of gravitational attraction between two masses m_1 and m_2 is given by $\mathbf{F} = -Gm_1m_2\mathbf{r}/|\mathbf{r}|^3$, where $\mathbf{r} = x\mathbf{i} + y\mathbf{j} + z\mathbf{k}$. Show that $\mathbf{F}$ is conservative. Find a potential function for $\mathbf{F}$.

28. Find the work done by the force

$$\mathbf{F}(x, y, z) = 8xy^3z\mathbf{i} + 12x^2y^2z\mathbf{j} + 4x^2y^3\mathbf{k}$$

acting along the helix $\mathbf{r}(t) = 2 \cos t\mathbf{i} + 2 \sin t\mathbf{j} + t\mathbf{k}$ from $(2, 0, 0)$ to $(1, \sqrt{3}, \pi/3)$, from $(2, 0, 0)$ to $(0, 2, \pi/2)$. [*Hint:* Show that $\mathbf{F}$ is conservative.]

29. If $\mathbf{F}$ is a conservative force field, show that the work done along any simple closed path is zero.

30. A particle in the plane is attracted to the origin with a force $\mathbf{F} = |\mathbf{r}|^n\mathbf{r}$, where n is a positive integer and $\mathbf{r} = x\mathbf{i} + y\mathbf{j}$ is the position vector of the particle. Show that $\mathbf{F}$ is conservative. Find the work done in moving the particle between (x_1, y_1) and (x_2, y_2).

31. Suppose $\mathbf{F}$ is a conservative force field with potential function ϕ. In physics the function $p = -\phi$ is called **potential energy**. Since $\mathbf{F} = -\nabla p$, Newton's second law becomes

$$m\mathbf{r}'' = -\nabla p \quad \text{or} \quad m\frac{d\mathbf{v}}{dt} + \nabla p = \mathbf{0}$$

By integrating $m\frac{d\mathbf{v}}{dt} \cdot \frac{d\mathbf{r}}{dt} + \nabla p \cdot \frac{d\mathbf{r}}{dt} = 0$ with respect to t, derive the law of conservation of mechanical energy: $\frac{1}{2}mv^2 + p = \text{constant}$. [*Hint:* See Problem 45 in Exercises 17.1.]

32. Suppose C is a smooth curve between points A (at $t = a$) and B (at $t = b$) and that p is potential energy, defined in Problem 31. If $\mathbf{F}$ is a conservative force field and $K = \frac{1}{2}mv^2$ is kinetic energy, show that $p(B) + K(B) = p(A) + K(A)$.

17.3 GREEN'S THEOREM

Line Integrals Along Simple Closed Curves One of the most important theorems in vector integral calculus relates a line integral around a piecewise smooth *simple closed curve* with a double integral over the region bounded by the curve.

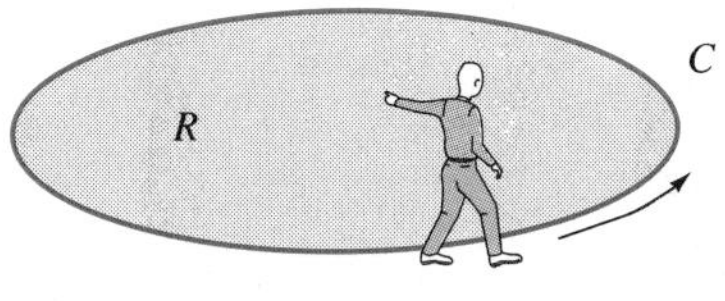

(a) positive direction

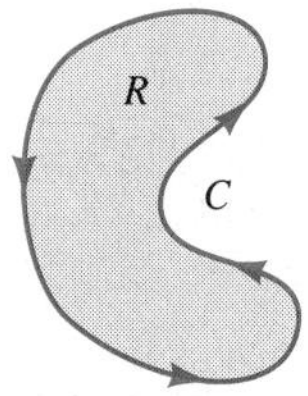

(b) positive direction

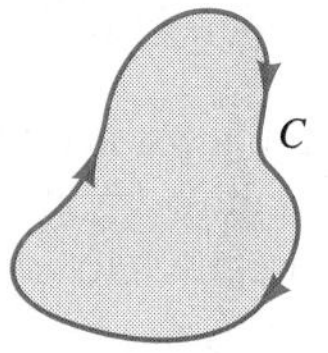

(c) negative direction

FIGURE 17.28

We say the **positive direction** around a simple closed curve C is that direction a point on the curve must move, or the direction a person must walk, on C in order to keep the region R bounded by C to the left. See Figure 17.28(a). Roughly, as shown in Figure 17.28(b) and (c), the *positive* and *negative* directions correspond to the *counterclockwise* and *clockwise directions*, respectively. Line integrals on simple closed curves are written

$$\ointctrclockwise_C P(x, y)\,dx + Q(x, y)\,dy \qquad \ointclockwise_C P(x, y)\,dx + Q(x, y)\,dy \qquad \ointctrclockwise_C F(x, y)\,ds$$

and so on. The symbols $\ointctrclockwise_C$ and $\ointclockwise_C$ refer, in turn, to integrations in the positive and negative directions.

THEOREM 17.5 Green's Theorem in the Plane*

Suppose that C is a piecewise smooth simple closed curve bounding a region R. If P, Q, $\partial P/\partial y$, and $\partial Q/\partial x$ are continuous on R, then

$$\ointctrclockwise_C P\,dx + Q\,dy = \iint_R \left(\frac{\partial Q}{\partial x} - \frac{\partial P}{\partial y}\right) dA \qquad (1)$$

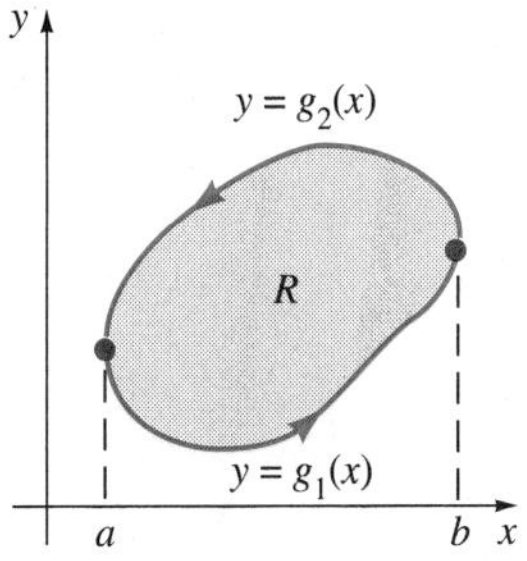

(a) R as a Type I region

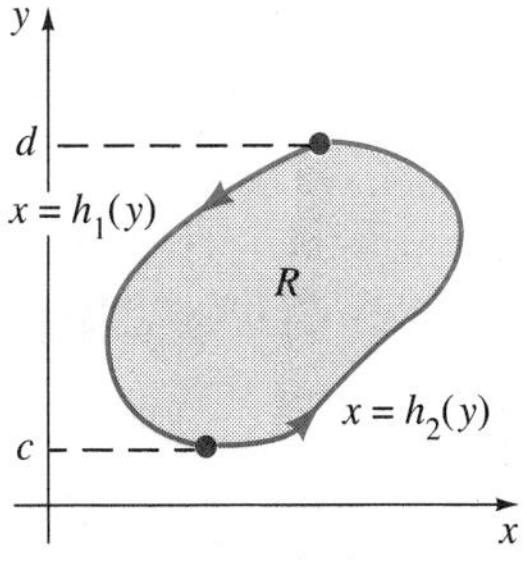

(b) R as a Type II region

FIGURE 17.29

Partial Proof We shall prove (1) only for a region R that is simultaneously of Type I and Type II:

$$R: \quad g_1(x) \le y \le g_2(x), \qquad a \le x \le b$$
$$R: \quad h_1(y) \le x \le h_2(y), \qquad c \le y \le d$$

Using Figure 17.29(a), we have

$$\begin{aligned} -\iint_R \frac{\partial P}{\partial y}\,dA &= -\int_a^b \int_{g_1(x)}^{g_2(x)} \frac{\partial P}{\partial y}\,dy\,dx \\ &= -\int_a^b [P(x, g_2(x)) - P(x, g_1(x))]\,dx \\ &= \int_a^b P(x, g_1(x))\,dx + \int_b^a P(x, g_2(x))\,dx \\ &= \ointctrclockwise_C P(x, y)\,dx \end{aligned} \qquad (2)$$

*Named after George Green (1793–1841), an English mathematician and physicist. The words *in the plane* suggest that the theorem generalizes to 3-space. It does—read on.

Similarly, from Figure 17.29(b),

$$\iint_R \frac{\partial Q}{\partial x}\, dA = \int_c^d \int_{h_1(y)}^{h_2(y)} \frac{\partial Q}{\partial x}\, dx\, dy$$

$$= \int_c^d [Q(h_2(y), y) - Q(h_1(y), y)]\, dy$$

$$= \int_c^d Q(h_2(y), y)\, dy + \int_d^c Q(h_1(y), y)\, dy$$

$$= \oint_C Q(x, y)\, dy \qquad (3)$$

Adding the results in (2) and (3) yields (1). ■

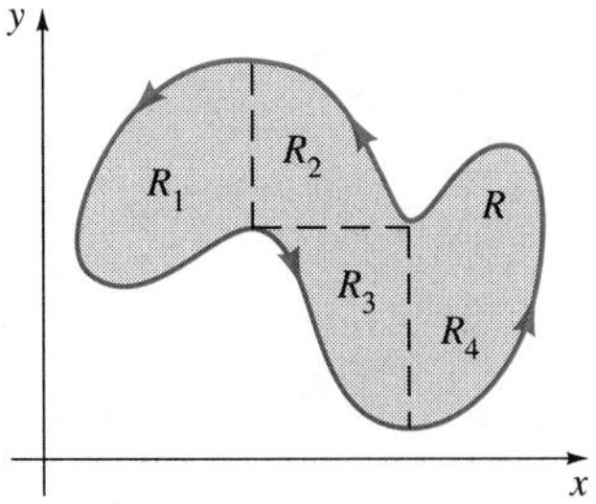

FIGURE 17.30

Although the foregoing proof is not valid, the theorem is applicable to more complicated regions, such as those shown in Figure 17.30. The proof consists of decomposing R into a finite number of subregions to which (1) can be applied and then adding the results.

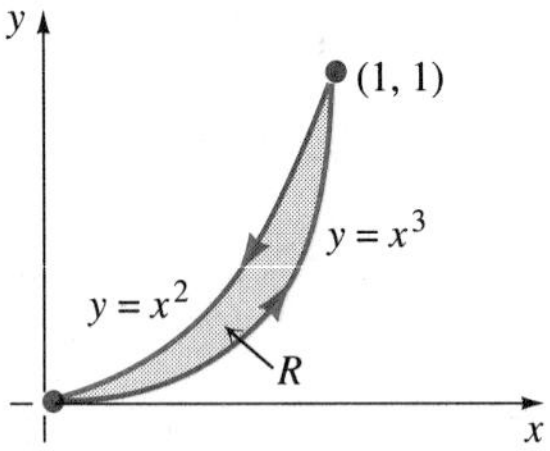

FIGURE 17.31

EXAMPLE 1 Evaluate $\oint_C (x^2 - y^2)\, dx + (2y - x)\, dy$, where C consists of the boundary of the region in the first quadrant that is bounded by the graphs of $y = x^2$ and $y = x^3$.

Solution If $P(x, y) = x^2 - y^2$ and $Q(x, y) = 2y - x$, then $\partial P/\partial y = -2y$ and $\partial Q/\partial x = -1$. From (1) and Figure 17.31 we have

$$\oint_C (x^2 - y^2)\, dx + (2y - x)\, dy = \iint_R (-1 + 2y)\, dA$$

$$= \int_0^1 \int_{x^3}^{x^2} (-1 + 2y) dy\, dx$$

$$= \int_0^1 (-y + y^2) \Big]_{x^3}^{x^2} dx$$

$$= \int_0^1 (-x^6 + x^4 + x^3 - x^2)\, dx = \frac{11}{420} \qquad \square$$

We note that the line integral in Example 1 could have been evaluated in a straightforward manner using the variable x as a parameter. However, in the next example, ponder the problem of evaluating the given line integral in the usual manner.

EXAMPLE 2 Evaluate $\oint_C (x^5 + 3y)\, dx + (2x - e^{y^3})\, dy$, where C is the circle $(x - 1)^2 + (y - 5)^2 = 4$.

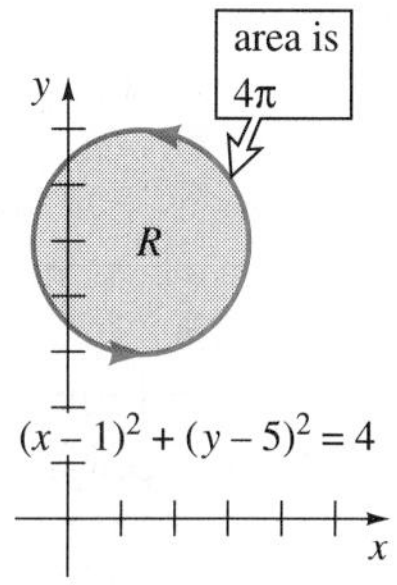

FIGURE 17.32

Solution Identifying $P(x, y) = x^5 + 3y$ and $Q(x, y) = 2x - e^{y^3}$, we have $\partial P/\partial y = 3$ and $\partial Q/\partial x = 2$. Hence, (1) gives

$$\oint_C (x^5 + 3y)\,dx + (2x - e^{y^3})\,dy = \iint_R (2 - 3)\,dA = -\iint_R dA$$

Now the double integral $\iint_R dA$ gives the area of the region R bounded by the circle of radius 2 shown in Figure 17.32. Since the area of the circle is $\pi 2^2 = 4\pi$, it follows that

$$\oint_C (x^5 + 3y)\,dx + (2x - e^{y^3})\,dy = -4\pi$$ □

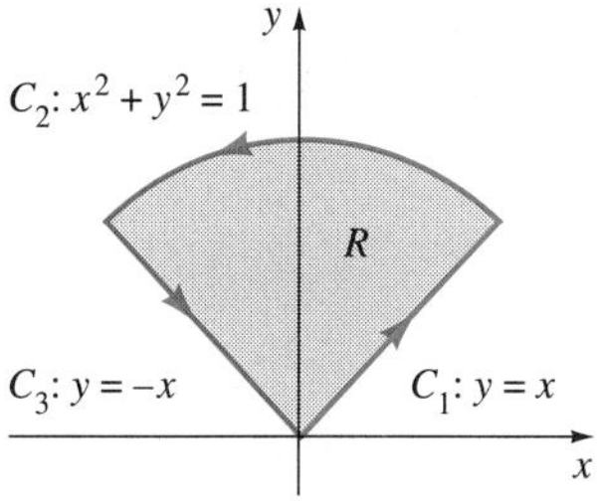

FIGURE 17.33

EXAMPLE 3 Find the work done by the force $\mathbf{F} = (-16y + \sin x^2)\mathbf{i} + (4e^y + 3x^2)\mathbf{j}$ acting along the simple closed curve C shown in Figure 17.33.

Solution From (12) of Section 17.1 the work done by **F** is given by

$$W = \oint_C \mathbf{F}\cdot d\mathbf{r} = \oint_C (-16y + \sin x^2)\,dx + (4e^y + 3x^2)\,dy$$

and so by Green's Theorem,

$$W = \iint_R (6x + 16)\,dA$$

In view of the region R the last integral is best handled in polar coordinates. Since R is defined by $0 \le r \le 1$, $\pi/4 \le \theta \le 3\pi/4$,

$$\begin{aligned} W &= \int_{\pi/4}^{3\pi/4}\int_0^1 (6r\cos\theta + 16)r\,dr\,d\theta \\ &= \int_{\pi/4}^{3\pi/4} (2r^3\cos\theta + 8r^2)\Big]_0^1\,d\theta \\ &= \int_{\pi/4}^{3\pi/4} (2\cos\theta + 8)\,d\theta = 4\pi \end{aligned}$$ □

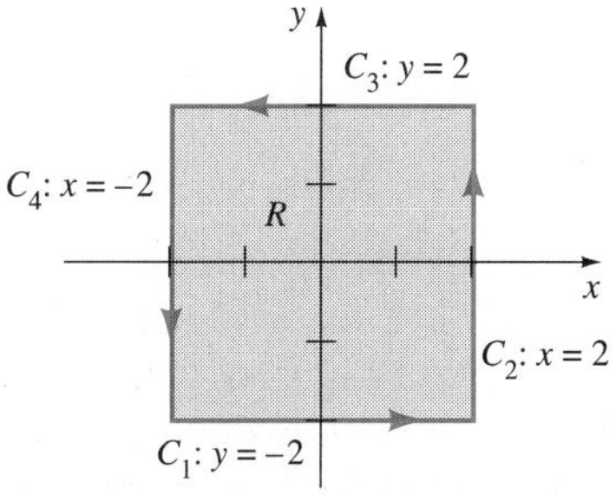

FIGURE 17.34

EXAMPLE 4 Let C be the closed curve consisting of the four straight line segments C_1, C_2, C_3, and C_4 shown in Figure 17.34. Green's Theorem is *not* applicable to the line integral

$$\oint_C \frac{-y}{x^2 + y^2}\,dx + \frac{x}{x^2 + y^2}\,dy$$

since P, Q, $\partial P/\partial y$, and $\partial Q/\partial x$ are not continuous at the origin. □

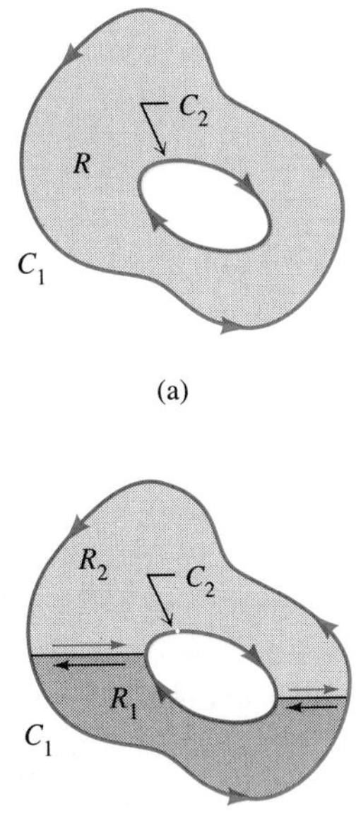

(a)

(b)

FIGURE 17.35

Region with Holes Green's Theorem can also be extended to a region R with "holes"—that is, a region bounded between two or more piecewise smooth simple closed curves. In Figure 17.35(a) we have shown a region R bounded by a curve C that consists of two simple closed curves C_1 and C_2; that is, $C = C_1 \cup C_2$. The curve C is positively oriented, since if we traverse C_1 in a counterclockwise direction and C_2 in a clockwise direction, the region R is always to the left. If we now introduce crosscuts as shown in Figure 17.35(b), the region R is divided into two subregions R_1 and R_2. By applying Green's Theorem to R_1 and R_2, we obtain

$$\begin{aligned}\iint_R \left(\frac{\partial Q}{\partial x} - \frac{\partial P}{\partial y}\right) dA &= \iint_{R_1} \left(\frac{\partial Q}{\partial x} - \frac{\partial P}{\partial y}\right) dA + \iint_{R_2} \left(\frac{\partial Q}{\partial x} - \frac{\partial P}{\partial y}\right) dA \\ &= \oint_{C_1} P\,dx + Q\,dy + \oint_{C_2} P\,dx + Q\,dy \qquad (4) \\ &= \oint_C P\,dx + Q\,dy\end{aligned}$$

The last result follows from the fact that the line integrals on the crosscuts (paths with opposite orientations) will cancel each other. See (8) of Section 17.1.

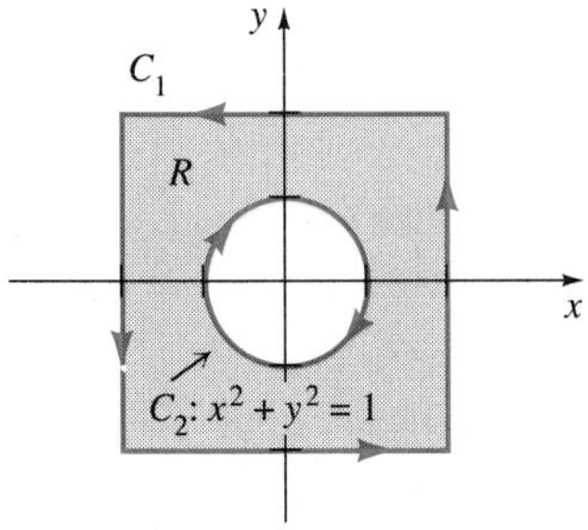

FIGURE 17.36

EXAMPLE 5 Evaluate $\oint_C \frac{-y}{x^2+y^2}\,dx + \frac{x}{x^2+y^2}\,dy$, where $C = C_1 \cup C_2$ is the boundary of the shaded region R shown in Figure 17.36.

Solution Because

$$P(x, y) = \frac{-y}{x^2+y^2} \qquad Q(x, y) = \frac{x}{x^2+y^2}$$

$$\frac{\partial P}{\partial y} = \frac{y^2-x^2}{(x^2+y^2)^2} \qquad \frac{\partial Q}{\partial x} = \frac{y^2-x^2}{(x^2+y^2)^2}$$

are continuous on the region R bounded by C, it follows from the above discussion that

$$\oint_C \frac{-y}{x^2+y^2}\,dx + \frac{x}{x^2+y^2}\,dy = \iint_R \left[\frac{y^2-x^2}{(x^2+y^2)^2} - \frac{y^2-x^2}{(x^2+y^2)^2}\right] dA = 0 \qquad \square$$

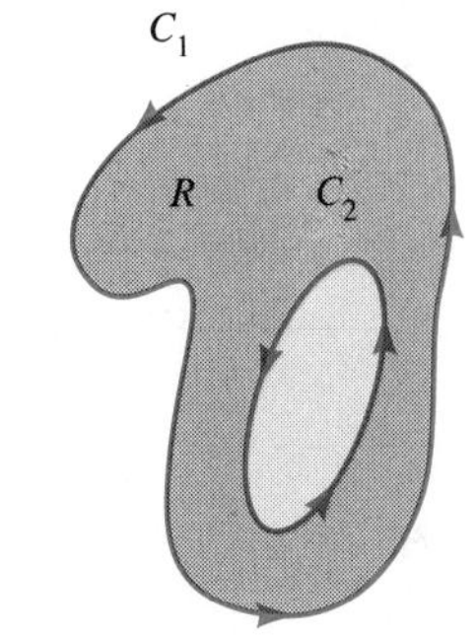

FIGURE 17.37

As a consequence of the discussion preceding Example 5 we can establish a result for line integrals that enables us, under certain circumstances, to replace a complicated closed path with a path that is simpler. Suppose, as shown in Figure 17.37, that C_1 and C_2 are two nonintersecting piecewise smooth simple closed paths that have the same counterclockwise orientation. Suppose further that P and Q have continuous first partial derivatives such

that

$$\frac{\partial P}{\partial y} = \frac{\partial Q}{\partial x}$$

in the region R bounded between C_1 and C_2. Then from (4) above and (8) of Section 17.1 we have

$$\oint_{C_1} P\,dx + Q\,dy + \oint_{-C_2} P\,dx + Q\,dy = 0$$

or

$$\oint_{C_1} P\,dx + Q\,dy = \oint_{C_2} P\,dx + Q\,dy \tag{5}$$

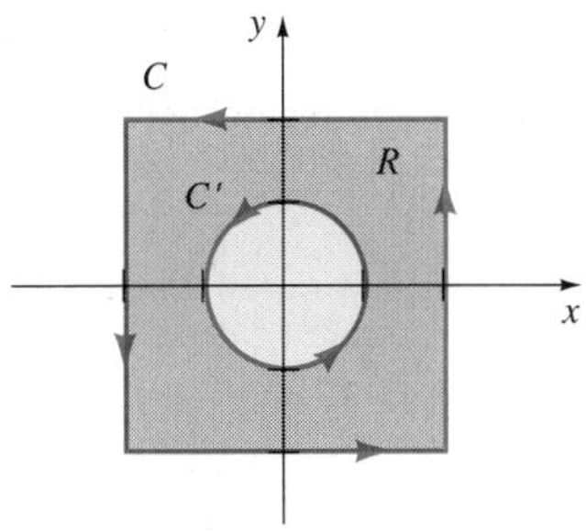

FIGURE 17.38

EXAMPLE 6 Evaluate the line integral in Example 4.

Solution One method of evaluating the line integral is to write

$$\oint_C = \int_{C_1} + \int_{C_2} + \int_{C_3} + \int_{C_4}$$

and then evaluate the four integrals on the line segments C_1, C_2, C_3, and C_4 shown in Figure 17.34. Alternatively, if we note that the circle C': $x^2 + y^2 = 1$ lies entirely within C (see Figure 17.38), then from Example 5 it is apparent that $P = -y/(x^2 + y^2)$ and $Q = x/(x^2 + y^2)$ have continuous first partial derivatives in the region R bounded between C and C'. Moreover,

$$\frac{\partial P}{\partial y} = \frac{y^2 - x^2}{(x^2 + y^2)^2} = \frac{\partial Q}{\partial x}$$

in R. Hence, it follows from (5) that

$$\oint_C \frac{-y}{x^2 + y^2}\,dx + \frac{x}{x^2 + y^2}\,dy = \oint_{C'} \frac{-y}{x^2 + y^2}\,dx + \frac{x}{x^2 + y^2}\,dy$$

Using the parameterization for C', $x = \cos t$, $y = \sin t$, $0 \le t \le 2\pi$, we obtain

$$\begin{aligned}\oint_C \frac{-y}{x^2 + y^2}\,dx + \frac{x}{x^2 + y^2}\,dy &= \int_0^{2\pi} [-\sin t(-\sin t) + \cos t(\cos t)]\,dt \\ &= \int_0^{2\pi} (\sin^2 t + \cos^2 t)\,dt \\ &= \int_0^{2\pi} dt = 2\pi \end{aligned} \tag{6}$$

□

It is interesting to note that the result in (6):

$$\oint_C \frac{-y}{x^2 + y^2}\,dx + \frac{x}{x^2 + y^2}\,dy = 2\pi$$

is true for every piecewise smooth simple closed curve C with the origin in its interior. We need only choose C' to be $x^2 + y^2 = a^2$, where a is small enough so that the circle lies entirely within C.

EXERCISES 17.3

Answers to odd-numbered problems begin on page A-109.

In Problems 1–4 verify Green's Theorem by evaluating both integrals.

1. $\oint_C (x - y)\,dx + xy\,dy = \iint_R (y + 1)\,dA$, where C is the triangle with vertices (0, 0), (1, 0), (1, 3)

2. $\oint_C 3x^2y\,dx + (x^2 - 5y)\,dy = \iint_R (2x - 3x^2)\,dA$, where C is the rectangle with vertices $(-1, 0), (1, 0), (1, 1), (-1, 1)$

3. $\oint_C -y^2\,dx + x^2\,dy = \iint_R (2x + 2y)\,dA$, where C is the circle $x = 3\cos t$, $y = 3\sin t$, $0 \le t \le 2\pi$

4. $\oint_C -2y^2\,dx + 4xy\,dy = \iint_R 8y\,dA$, where C is the boundary of the region in the first quadrant determined by the graphs of $y = 0$, $y = \sqrt{x}$, $y = -x + 2$

In Problems 5–14 use Green's Theorem to evaluate the given line integral.

5. $\oint_C 2y\,dx + 5x\,dy$, where C is the circle $(x - 1)^2 + (y + 3)^2 = 25$

6. $\oint_C (x + y^2)\,dx + (2x^2 - y)\,dy$, where C is the boundary of the region determined by the graphs of $y = x^2$, $y = 4$

7. $\oint_C (x^4 - 2y^3)\,dx + (2x^3 - y^4)\,dy$, where C is the circle $x^2 + y^2 = 4$

8. $\oint_C (x - 3y)\,dx + (4x + y)\,dy$, where C is the rectangle with vertices $(-2, 0), (3, 0), (3, 2), (-2, 2)$

9. $\oint_C 2xy\,dx + 3xy^2\,dy$, where C is the triangle with vertices (1, 2), (2, 2), (2, 4)

10. $\oint_C e^{2x}\sin 2y\,dx + e^{2x}\cos 2y\,dy$, where C is the ellipse $9(x - 1)^2 + 4(y - 3)^2 = 36$

11. $\oint_C xy\,dx + x^2\,dy$, where C is the boundary of the region determined by the graphs of $x = 0$, $x^2 + y^2 = 1$, $x \ge 0$

12. $\oint_C e^{x^2}\,dx + 2\tan^{-1}x\,dy$, where C is the triangle with vertices $(0, 0), (0, 1), (-1, 1)$

13. $\oint_C \frac{1}{3}y^3\,dx + (xy + xy^2)\,dy$, where C is the boundary of the region in the first quadrant determined by the graphs of $y = 0$, $x = y^2$, $x = 1 - y^2$

14. $\oint_C xy^2\,dx + 3\cos y\,dy$, where C is the boundary of the region in the first quadrant determined by the graphs of $y = x^2$, $y = x^3$

In Problems 15 and 16 evaluate the given integral on any piecewise smooth simple closed curve C.

15. $\displaystyle\oint_C ay\,dx + bx\,dy$

16. $\displaystyle\oint_C P(x)\,dx + Q(y)\,dy$

In Problems 17 and 18 let R be the region bounded by a piecewise smooth simple closed curve C. Prove the given result.

17. $\displaystyle\oint_C x\,dy = -\oint_C y\,dx = \text{area of } R$

18. $\displaystyle\frac{1}{2}\oint_C -y\,dx + x\,dy = \text{area of } R$

In Problems 19 and 20 use the results of Problems 17 and 18 to find the area of the region bounded by the given closed curve.

19. The hypocycloid $x = a\cos^3 t$, $y = a\sin^3 t$, $a > 0$, $0 \le t \le 2\pi$

20. The ellipse $x = a\cos t$, $y = b\sin t$, $a > 0$, $b > 0$, $0 \le t \le 2\pi$

21. (a) Show that

$$\int_C -y\,dx + x\,dy = x_1y_2 - x_2y_1$$

where C is the line segment from the point (x_1, y_1) to (x_2, y_2).

(b) Use part (**a**) and Problem 18 to show that the area A of a polygon with vertices $(x_1, y_1), (x_2, y_2), \ldots, (x_n, y_n)$, labeled counterclockwise, is

$$A = \frac{1}{2}(x_1y_2 - x_2y_1) + \frac{1}{2}(x_2y_3 - x_3y_2)$$

$$+ \frac{1}{2}(x_{n-1}y_n - x_ny_{n-1}) + \frac{1}{2}(x_ny_1 - x_1y_n)$$

22. Use part (**b**) of Problem 21 to find the area of the quadrilateral with vertices $(-1, 3)$, (1, 1), (4, 2), and (3, 5).

In Problems 23 and 24 evaluate the given line integral where $C = C_1 \cup C_2$ is the boundary of the shaded region R.

23. $\oint_C (4x^2 - y^3)\,dx + (x^3 + y^2)\,dy$

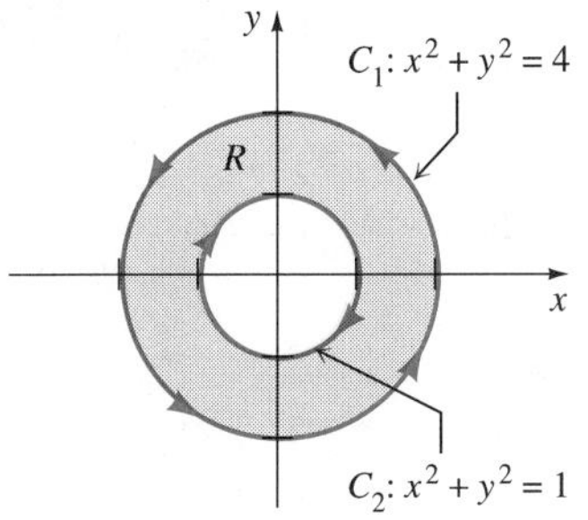

FIGURE 17.39

24. $\oint_C (\cos x^2 - y)\,dx + \sqrt{y^3 + 1}\,dy$

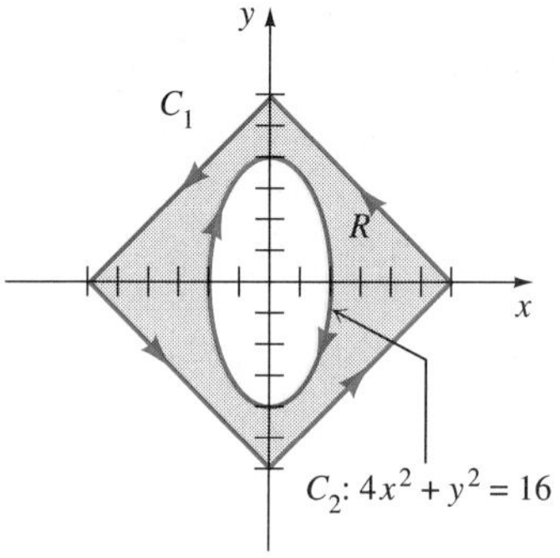

FIGURE 17.40

In Problems 25 and 26 proceed as in Example 6 to evaluate the given line integral.

25. $\oint_C \dfrac{-y^3\,dx + xy^2\,dy}{(x^2 + y^2)^2}$, where C is the ellipse $x^2 + 4y^2 = 4$

26. $\oint_C \dfrac{-y}{(x + 1)^2 + 4y^2}\,dx + \dfrac{x + 1}{(x + 1)^2 + 4y^2}\,dy$, where C is the circle $x^2 + y^2 = 16$

In Problems 27 and 28 use Green's Theorem to evaluate the given double integral by means of a line integral. [*Hint:* Find appropriate functions P and Q.]

27. $\iint_R x^2\,dA$, where R is the region bounded by the ellipse $x^2/9 + y^2/4 = 1$

28. $\iint_R [1 - 2(y - 1)]\,dA$, where R is the region in the first quadrant bounded by the circle $x^2 + (y - 1)^2 = 1$ and $x = 0$

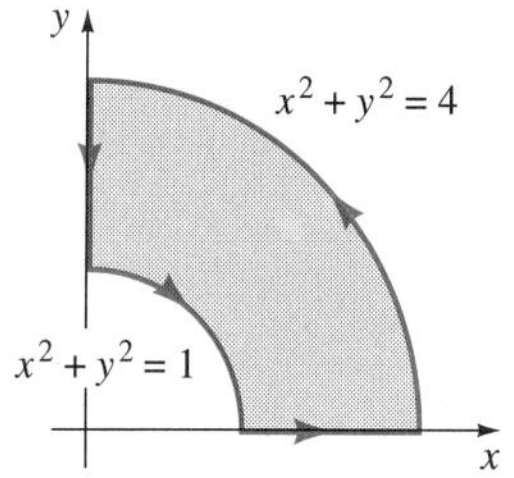

FIGURE 17.41

In Problems 29 and 30 use Green's Theorem to find the work done by the given force **F** around the closed curve in Figure 17.41.

29. $\mathbf{F} = (x - y)\mathbf{i} + (x + y)\mathbf{j}$ **30.** $\mathbf{F} = -xy^2\mathbf{i} + x^2y\mathbf{j}$

31. Let P and Q be continuous and have continuous first partial derivatives in a simply connected region of the xy-plane. If $\int_A^B P\,dx + Q\,dy$ is independent of the path, show that $\oint_C P\,dx + Q\,dy = 0$ on every piecewise smooth simple closed curve C in the region.

32. Let R be a region bounded by a piecewise smooth simple closed curve C. Show that the coordinates of the centroid of the region are given by

$$\bar{x} = \frac{1}{2A}\oint_C x^2\,dy \qquad \bar{y} = -\frac{1}{2A}\oint_C y^2\,dx$$

33. Find the work done by the force $\mathbf{F} = -y\mathbf{i} + x\mathbf{j}$ acting along the cardioid $r = 1 + \cos\theta$.

17.4 SURFACE INTEGRALS

The last kind of integral that we shall consider in this text is called a **surface integral** and involves a function G of three variables defined on a surface S. The five steps preparatory to the definition of this integral are similar to combinations of the steps leading to the line integral, with respect to arc length, and the steps leading to the double integral.

$w = G(x, y, z)$

1. Let G be defined in a region of 3-space that contains a surface S, which is the graph of a function $z = f(x, y)$. Let the projection R of the surface onto the xy-plane be either a Type I or a Type II region.
2. Divide the surface into n pieces of areas ΔS_k corresponding to a partition P of R into n rectangles R_k of areas ΔA_k.
3. Let $\|P\|$ be the **norm** of the partition or the length of the longest diagonal of the R_k.
4. Choose a point (x_k^*, y_k^*, z_k^*) on each element of surface area.
5. Form the sum $\Sigma_{k=1}^{n} G(x_k^*, y_k^*, z_k^*)\,\Delta S_k$.

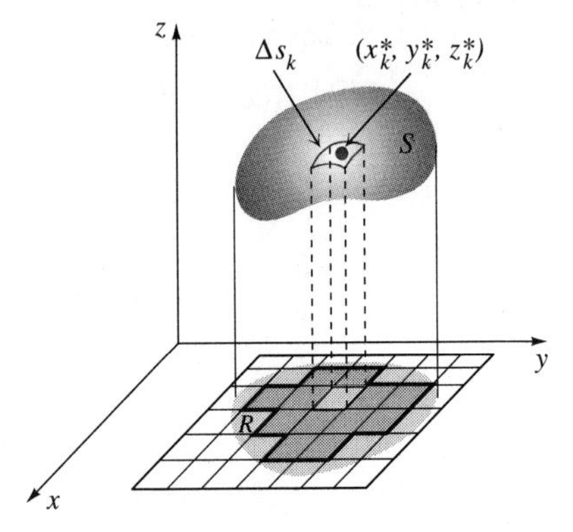

DEFINITION 17.2 Surface Integral

Let G be a function of three variables defined over a region of space containing the surface S. Then the **surface integral of G over S** is given by

$$\iint_S G(x, y, z)\,dS = \lim_{\|P\|\to 0} \sum_{k=1}^{n} G(x_k^*, y_k^*, z_k^*)\,\Delta S_k \tag{1}$$

Method of Evaluation Recall from Section 16.6 that if $z = f(x, y)$ is the equation of a surface S, then the differential of the surface is

$$dS = \sqrt{1 + [f_x(x, y)]^2 + [f_y(x, y)]^2}\,dA$$

Thus, if G, f, f_x, and f_y are continuous throughout a region containing S, we can evaluate (1) by means of a double integral:

$$\iint_S G(x, y, z)\,dS = \iint_R G(x, y, f(x, y))\sqrt{1 + [f_x(x, y)]^2 + [f_y(x, y)]^2}\,dA \tag{2}$$

Note that when $G = 1$, (2) reduces to the formula for surface area.

Projection of S into Other Planes If $y = g(x, z)$ is the equation of a surface S that projects onto a region R of the xz-plane, then

$$\iint_S G(x, y, z)\,dS = \iint_R G(x, g(x, z), z)\sqrt{1 + [g_x(x, z)]^2 + [g_z(x, z)]^2}\,dA \tag{3}$$

Similarly, if $x = h(y, z)$ is the equation of a surface S that projects onto the yz-plane, then the analogue of (2) is

$$\iint_S G(x, y, z)\, dS = \iint_R G(h(y, z), y, z)\sqrt{1 + [h_y(y, z)]^2 + [h_z(y, z)]^2}\, dA \tag{4}$$

Mass of a Surface Suppose $\rho(x, y, z)$ represents the density of a surface at any point, or the mass per unit of surface area. Then the **mass** m of the surface is

$$m = \iint_S \rho(x, y, z)\, dS \tag{5}$$

EXAMPLE 1 Find the mass of the surface of the paraboloid $z = 1 + x^2 + y^2$ in the first octant for $1 \le z \le 5$ if the density at a point P on the surface is directly proportional to distance from the xy-plane.

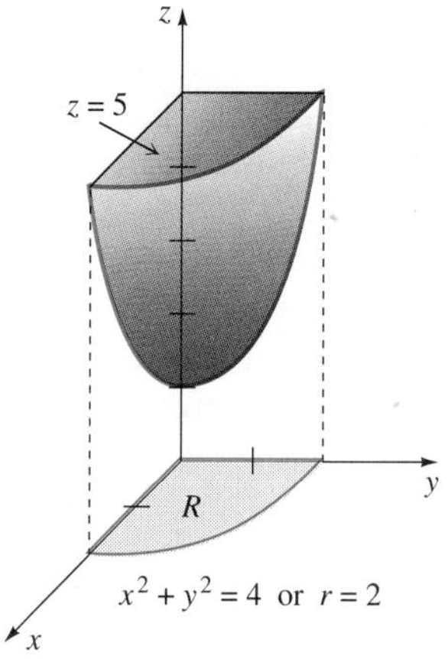

FIGURE 17.42

Solution The surface in question and its projection onto the xy-plane are shown in Figure 17.42. Now, since $\rho(x, y, z) = kz$ and $z = 1 + x^2 + y^2$, (5) and (2) give

$$m = \iint_S kz\, dS = k\iint_R (1 + x^2 + y^2)\sqrt{1 + 4x^2 + 4y^2}\, dA$$

By changing to polar coordinates, we obtain

$$\begin{aligned} m &= k\int_0^{\pi/2}\int_0^2 (1 + r^2)\sqrt{1 + 4r^2}\, r\, dr\, d\theta \\ &= k\int_0^{\pi/2}\int_0^2 [r(1 + 4r^2)^{1/2} + r^3(1 + 4r^2)^{1/2}]\, dr\, d\theta \qquad \boxed{\text{integration by parts}} \\ &= k\int_0^{\pi/2}\left[\frac{1}{12}(1 + 4r^2)^{3/2} + \frac{1}{12}r^2(1 + 4r^2)^{3/2} - \frac{1}{120}(1 + 4r^2)^{5/2}\right]_0^2 d\theta \\ &= \frac{k\pi}{2}\left[\frac{5(17)^{3/2}}{12} - \frac{17^{5/2}}{120} - \frac{3}{40}\right] \approx 19.2k \end{aligned}$$

□

EXAMPLE 2 Evaluate $\iint_S xz^2\, dS$, where S is that portion of the cylinder $y = 2x^2 + 1$ in the first octant bounded by $x = 0$, $x = 2$, $z = 4$, and $z = 8$.

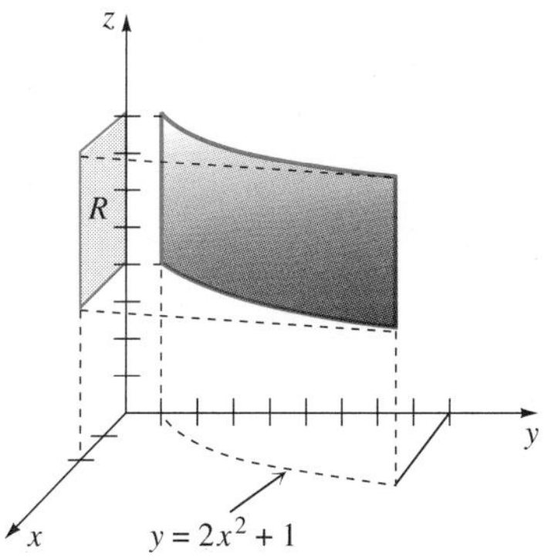

FIGURE 17.43

Solution We shall use (3) with $g(x, z) = 2x^2 + 1$ and R the rectangular region in the xz-plane shown in Figure 17.43. Since $g_x(x, z) = 4x$ and $g_z(x, z) = 0$, it follows that

$$\iint_S xz^2\,dS = \int_0^2 \int_4^8 xz^2\sqrt{1+16x^2}\,dz\,dx$$

$$= \int_0^2 \frac{z^3}{3} x\sqrt{1+16x^2}\Bigg]_4^8 dx$$

$$= \frac{448}{3}\int_0^2 x(1+16x^2)^{1/2}\,dx = \frac{28}{9}(1+16x^2)^{3/2}\Bigg]_0^2$$

$$= \frac{28}{9}[65^{3/2} - 1] \approx 1627.3 \qquad \square$$

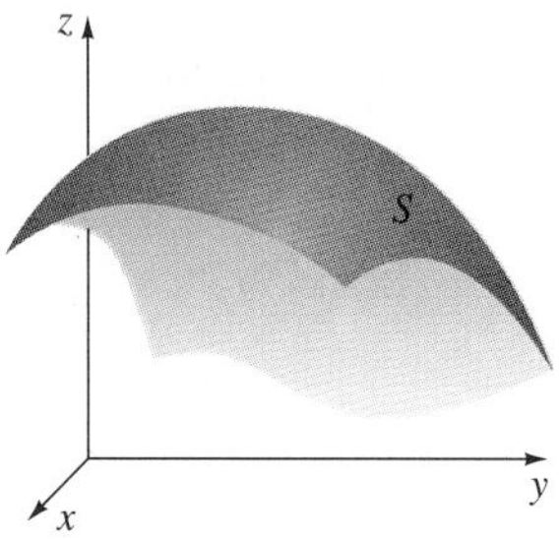

(a) two-sided surface

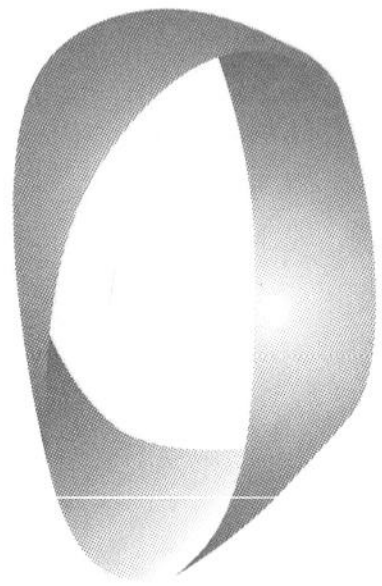

(b) one-sided surface

FIGURE 17.44

Orientable Surfaces In Example 4 we shall evaluate a surface integral of a vector field. In order to do this we need to examine the concept of an **orientable surface**. Roughly, an orientable surface S, such as that given in Figure 17.44(a), has two sides that could be painted different colors. The Möbius strip* shown in Figure 17.44(b) is not an orientable surface and is one-sided. A person who starts to paint the surface of a Möbius strip at a point will paint the entire surface and return to the starting point.

Specifically, we say a smooth surface S is **orientable** or is an **oriented surface** if there exists a continuous unit normal function $\mathbf{n}$ defined at each point (x, y, z) on the surface. The vector field $\mathbf{n}(x, y, z)$ is called the **orientation** of S. But since a unit normal to the surface S at (x, y, z) can be either $\mathbf{n}(x, y, z)$ or $-\mathbf{n}(x, y, z)$, an orientable surface has two orientations. See Figure 17.45(a), (b), and (c). The Möbius strip shown again in Figure 17.45(d) is not an oriented surface, since if a unit normal $\mathbf{n}$ starts at P on the surface and moves *once* around the strip on the curve C, it ends up on the "opposite side" of the strip at P and so points in the opposite direction. A surface S defined by $z = f(x, y)$ has an **upward orientation** (Figure 17.45(b)) when the unit normals are directed upward, that is, have positive $\mathbf{k}$ components, and has a **downward orientation**

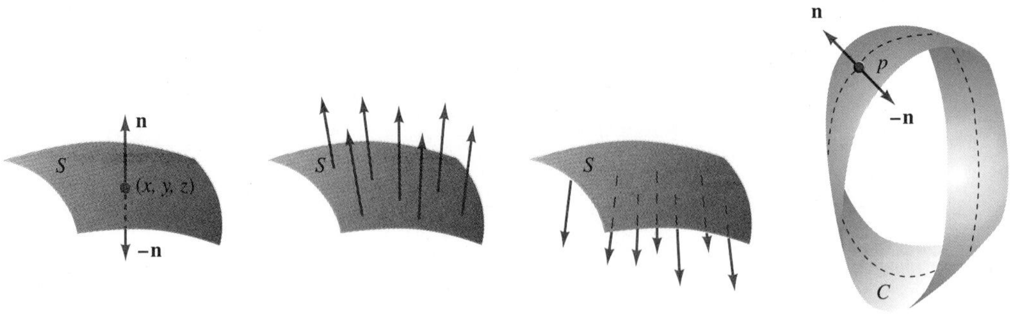

(a) (b) (c) (d)

FIGURE 17.45

*To construct a Möbius strip cut out a long strip of paper, give one end a half-twist, and then attach both ends by tape.

(Figure 17.45(c)) when the unit normals are directed downward, that is, have negative **k** components.

If a smooth surface S is defined by $g(x, y, z) = 0$, then recall that a unit normal is

$$\mathbf{n} = \frac{1}{|\nabla g|} \nabla g$$

where $\nabla g = (\partial g/\partial x)\mathbf{i} + (\partial g/\partial y)\mathbf{j} + (\partial g/\partial z)\mathbf{k}$ is the gradient of g. If S is defined by $z = f(x, y)$, then we can use $g(x, y, z) = z - f(x, y) = 0$ or $g(x, y, z) = f(x, y) - z = 0$ depending on the orientation of S.

As we see in the next example, the two orientations of a orientable closed surface are **outward** and **inward**.

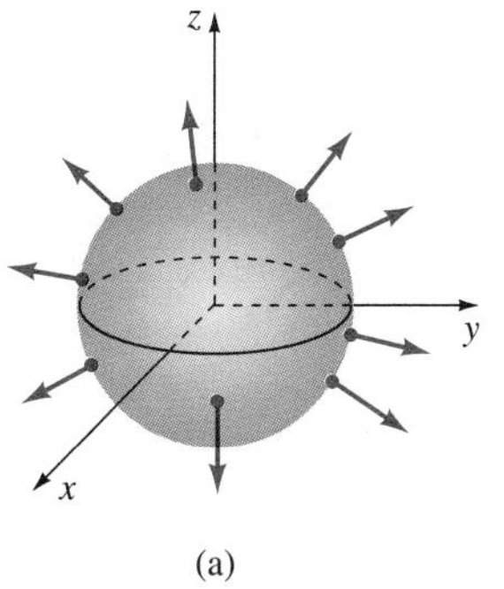

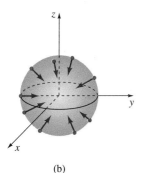

FIGURE 17.46

EXAMPLE 3 Consider the sphere of radius $a > 0$: $x^2 + y^2 + z^2 = a^2$. If we define $g(x, y, z) = x^2 + y^2 + z^2 - a^2$, then

$$\nabla g = 2x\mathbf{i} + 2y\mathbf{j} + 2z\mathbf{k} \quad \text{and} \quad |\nabla g| = \sqrt{4x^2 + 4y^2 + 4z^2} = 2a$$

Then the two orientations of the surface are

$$\mathbf{n} = \frac{x}{a}\mathbf{i} + \frac{y}{a}\mathbf{j} + \frac{z}{a}\mathbf{k} \quad \text{and} \quad \mathbf{n}_1 = -\mathbf{n} = -\frac{x}{a}\mathbf{i} - \frac{y}{a}\mathbf{j} - \frac{z}{a}\mathbf{k}$$

The vector field **n** defines an outward orientation, whereas $\mathbf{n}_1 = -\mathbf{n}$ defines an inward orientation. See Figure 17.46. □

Integrals of Vector Fields If

$$\mathbf{F}(x, y, z) = P(x, y, z)\mathbf{i} + Q(x, y, z)\mathbf{j} + R(x, y, z)\mathbf{k}$$

is the velocity field of fluid, then, as shown in Figure 17.47(b), the volume of the fluid flowing through an element of surface area ΔS per unit time is approximated by

$$(\text{height}) \cdot (\text{area of base}) = (\text{comp}_\mathbf{n}\mathbf{F})\,\Delta S = (\mathbf{F} \cdot \mathbf{n})\,\Delta S$$

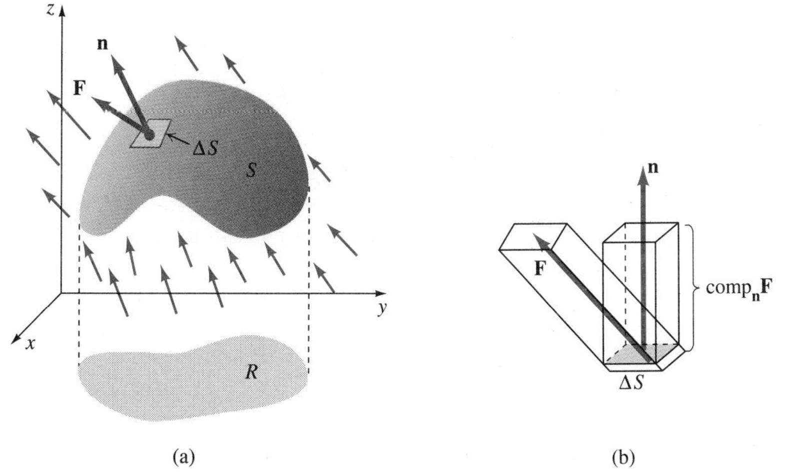

FIGURE 17.47

where **n** is a unit normal to the surface. The total volume of a fluid passing through S per unit time is called the **flux of F through S** and is given by

$$\text{flux} = \iint_S (\mathbf{F} \cdot \mathbf{n})\, dS \tag{6}$$

In the case of a closed surface S, if **n** is the outer (inner) normal, then (6) gives the volume of fluid flowing out (in) through S per unit time.

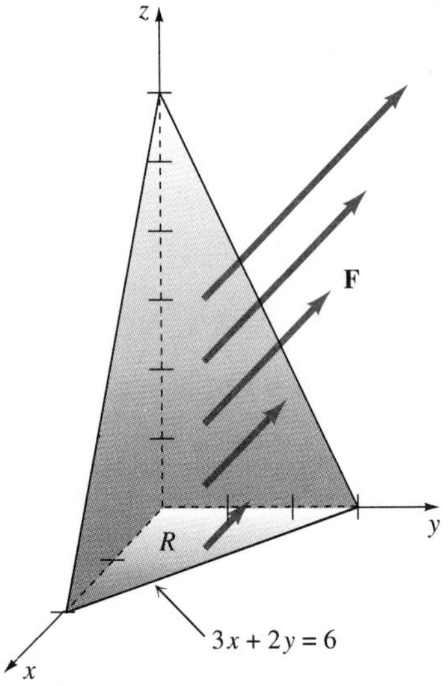

FIGURE 17.48

EXAMPLE 4 Let $\mathbf{F}(x, y, z) = z\mathbf{j} + z\mathbf{k}$ represent the flow of a liquid. Find the flux of **F** through the surface S given by that part of the plane $z = 6 - 3x - 2y$ in the first octant oriented upward.

Solution The vector field and the surface are illustrated in Figure 17.48. By defining the plane by $g(x, y, z) = 3x + 2y + z - 6 = 0$, we see that a unit normal with positive **k** component is

$$\mathbf{n} = \frac{\nabla g}{|\nabla g|} = \frac{3}{\sqrt{14}}\mathbf{i} + \frac{2}{\sqrt{14}}\mathbf{j} + \frac{1}{\sqrt{14}}\mathbf{k}$$

Hence,

$$\text{flux} = \iint_S (\mathbf{F} \cdot \mathbf{n})\, dS = \frac{1}{\sqrt{14}} \iint_S 3z\, dS$$

With R the projection of the surface onto the xy-plane, we find from (6) that

$$\begin{aligned}\text{flux} &= \frac{1}{\sqrt{14}} \iint_R 3(6 - 3x - 2y)(\sqrt{14}\, dA) \\ &= 3\int_0^2 \int_0^{3-3x/2} (6 - 3x - 2y)\, dy\, dx = 18\end{aligned}$$

□

Depending on the nature of the vector field, the integral in (6) can represent other kinds of flux. For example, (6) could also give electric flux, magnetic flux, flux of heat, and so on.

▲ ***Remark*** If the surface S is piecewise-defined, we express a surface integral over S as the sum of the surface integrals over the various pieces of the surface. For example, suppose S is the orientable piecewise smooth closed surface bounded by the paraboloid $z = x^2 + y^2$ (S_1) and the plane $z = 1$ (S_2). Then the flux of a vector field **F** out of the surface S is

$$\iint_S \mathbf{F} \cdot \mathbf{n}\, dS = \iint_{S_1} \mathbf{F} \cdot \mathbf{n}\, dS + \iint_{S_2} \mathbf{F} \cdot \mathbf{n}\, dS$$

where we take S_1 oriented upward and S_2 oriented downward. See Figure 17.49 and Problem 21.

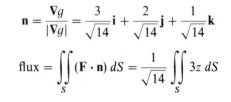

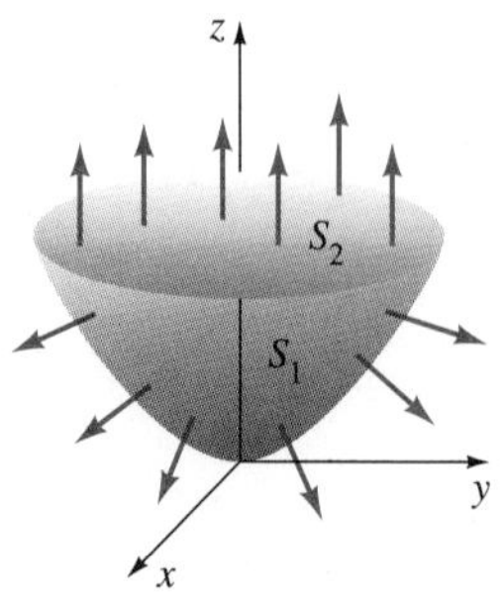

FIGURE 17.49

EXERCISES 17.4 *Answers to odd-numbered problems begin on page A-109.*

In Problems 1–10 evaluate $\iint_S G(x, y, z)\, dS$.

1. $G(x, y, z) = x$; S the portion of the cylinder $z = 2 - x^2$ in the first octant bounded by $x = 0$, $y = 0$, $y = 4$, $z = 0$

2. $G(x, y, z) = xy(9 - 4z)$; same surface as in Problem 1

3. $G(x, y, z) = xz^3$; S the cone $z = \sqrt{x^2 + y^2}$ inside the cylinder $x^2 + y^2 = 1$

4. $G(x, y, z) = x + y + z$; S the cone $z = \sqrt{x^2 + y^2}$ between $z = 1$ and $z = 4$

5. $G(x, y, z) = (x^2 + y^2)z$; S that portion of the sphere $x^2 + y^2 + z^2 = 36$ in the first octant

6. $G(x, y, z) = z^2$; S that portion of the plane $z = x + 1$ within the cylinder $y = 1 - x^2$, $0 \le y \le 1$

7. $G(x, y, z) = xy$; S that portion of the paraboloid $2z = 4 - x^2 - y^2$ within $0 \le x \le 1$, $0 \le y \le 1$

8. $G(x, y, z) = 2z$; S that portion of the paraboloid $2z = 1 + x^2 + y^2$ in the first octant bounded by $x = 0$, $y = \sqrt{3}x$, $z = 1$

9. $G(x, y, z) = 24\sqrt{yz}$; S that portion of the cylinder $y = x^2$ in the first octant bounded by $y = 0$, $y = 4$, $z = 0$, $z = 3$

10. $G(x, y, z) = (1 + 4y^2 + 4z^2)^{1/2}$; S that portion of the paraboloid $x = 4 - y^2 - z^2$ in the first octant outside the cylinder $y^2 + z^2 = 1$

In Problems 11 and 12, evaluate $\iint_S (3z^2 + 4yz)\, dS$, where S is that portion of the plane $x + 2y + 3z = 6$ in the first octant. Use the projection of S onto the coordinate plane indicated in the given figure.

11.

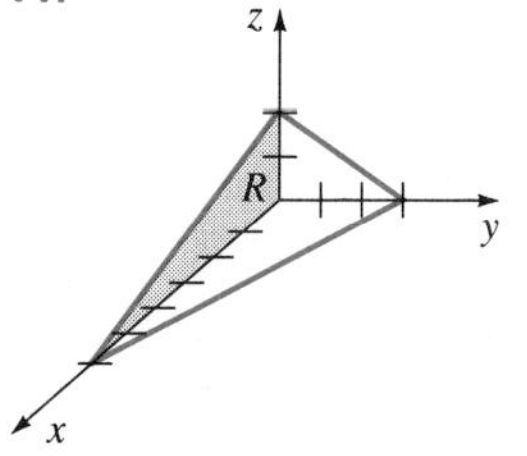

FIGURE 17.50

12.

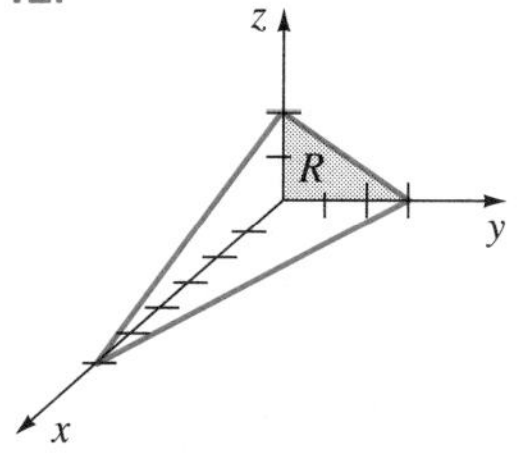

FIGURE 17.51

In Problems 13 and 14 find the mass of the given surface with the indicated density function.

13. S that portion of the plane $x + y + z = 1$ in the first octant; density at a point P directly proportional to the square of the distance from the yz-plane

14. S the hemisphere $z = \sqrt{4 - x^2 - y^2}$; $\rho(x, y, z) = |xy|$

In Problems 15–20 let **F** be a vector field. Find the flux of **F** through the given surface. Assume the surface S is oriented upward.

15. $\mathbf{F} = x\mathbf{i} + 2z\mathbf{j} + y\mathbf{k}$; S that portion of the cylinder $y^2 + z^2 = 4$ in the first octant bounded by $x = 0$, $x = 3$, $y = 0$, $z = 0$

16. $\mathbf{F} = z\mathbf{k}$; S that part of the paraboloid $z = 5 - x^2 - y^2$ inside the cylinder $x^2 + y^2 = 4$

17. $\mathbf{F} = x\mathbf{i} + y\mathbf{j} + z\mathbf{k}$; same surface S as in Problem 16

18. $\mathbf{F} = -x^3y\mathbf{i} + yz^3\mathbf{j} + xy^3\mathbf{k}$; S that portion of the plane $z = x + 3$ in the first octant within the cylinder $x^2 + y^2 = 2x$

19. $\mathbf{F} = \frac{1}{2}x^2\mathbf{i} + \frac{1}{2}y^2\mathbf{j} + z\mathbf{k}$; S that portion of the paraboloid $z = 4 - x^2 - y^2$ for $0 \le z \le 4$

20. $\mathbf{F} = e^y\mathbf{i} + e^x\mathbf{j} + 18y\mathbf{k}$; S that portion of the plane $x + y + z = 6$ in the first octant

21. Find the flux of $\mathbf{F} = y^2\mathbf{i} + x^2\mathbf{j} + 5z\mathbf{k}$ out of the closed surface S given in Figure 17.49.

22. Find the flux of $\mathbf{F} = -y\mathbf{i} + x\mathbf{j} + 6z^2\mathbf{k}$ out of the closed surface S bounded by the paraboloids $z = 4 - x^2 - y^2$ and $z = x^2 + y^2$.

23. Let $T(x, y, z) = x^2 + y^2 + z^2$ represent temperature and let the "flow" of heat be given by the vector field $\mathbf{F} = -\nabla T$. Find the flux of heat out of the sphere $x^2 + y^2 + z^2 = a^2$. [*Hint:* The surface area of a sphere of radius a is $4\pi a^2$.]

24. Find the flux of $\mathbf{F} = x\mathbf{i} + y\mathbf{j} + z\mathbf{k}$ out of the unit cube $0 \le x \le 1$, $0 \le y \le 1$, $0 \le z \le 1$. See Figure 17.52. Use the fact that the flux out of the cube is the sum of the fluxes out of the sides.

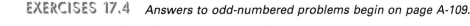

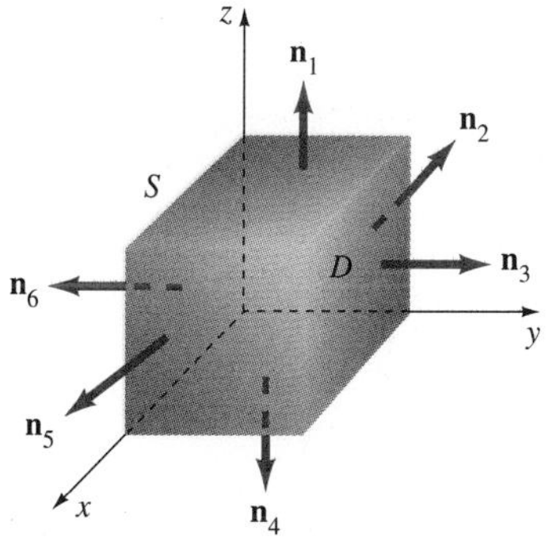

FIGURE 17.52

25. Coulomb's law states that the electric field $\mathbf{E}$ due to a point charge q at the origin is given by $\mathbf{E} = kq\mathbf{r}/|\mathbf{r}|^3$, where k is a constant and $\mathbf{r} = x\mathbf{i} + y\mathbf{j} + z\mathbf{k}$. Determine the flux out of a sphere $x^2 + y^2 + z^2 = a^2$.

26. If $\sigma(x, y, z)$ is charge density in an electrostatic field, then the total charge on a surface S is $Q = \iint_S \sigma(x, y, z)\, dS$. Find the total charge on that part of the hemisphere $z = \sqrt{16 - x^2 - y^2}$ that is inside the cylinder $x^2 + y^2 = 9$ if the charge density at a point P on the surface is directly proportional to distance from the xy-plane.

27. The coordinates of the centroid of a surface are given by

$$\bar{x} = \frac{\iint_S x\, dS}{A(S)} \qquad \bar{y} = \frac{\iint_S y\, dS}{A(S)} \qquad \bar{z} = \frac{\iint_S z\, dS}{A(S)}$$

where $A(S)$ is the area of the surface. Find the centroid of that portion of the plane $2x + 3y + z = 6$ in the first octant.

28. Use the information in Problem 27 to find the centroid of the hemisphere $z = \sqrt{a^2 - x^2 - y^2}$.

29. The moment of inertia of a surface S with density $\rho(x, y, z)$ at a point (x, y, z) about the z-axis is given by

$$I_z = \iint_S (x^2 + y^2)\rho(x, y, z)\, dS$$

Consider the conical surface $z = 4 - \sqrt{x^2 + y^2}$, $0 \le z \le 4$, with constant density k.

(a) Use Problem 27 to find the centroid of the surface.

(b) Find the moment of inertia of the surface about the z-axis.

30. Let $z = f(x, y)$ be the equation of a surface S and let $\mathbf{F}$ be the vector field $\mathbf{F}(x, y, z) = P(x, y, z)\mathbf{i} + Q(x, y, z)\mathbf{j} + R(x, y, z)\mathbf{k}$. Show that

$$\iint_S (\mathbf{F} \cdot \mathbf{n})\, dS = \iint_R \left[-P(x, y, z)\frac{\partial z}{\partial x} - Q(x, y, z)\frac{\partial z}{\partial y} + R(x, y, z) \right] dA$$

17.5 DIVERGENCE AND CURL

We have seen that if a vector force field $\mathbf{F}$ is conservative, then it can be written as the gradient of a potential function ϕ:

$$\mathbf{F} = \nabla\phi = \frac{\partial \phi}{\partial x}\mathbf{i} + \frac{\partial \phi}{\partial y}\mathbf{j} + \frac{\partial \phi}{\partial z}\mathbf{k}$$

The del operator

$$\nabla = \mathbf{i}\frac{\partial}{\partial x} + \mathbf{j}\frac{\partial}{\partial y} + \mathbf{k}\frac{\partial}{\partial z}$$

used in the gradient can also be combined with a vector field $\mathbf{F}(x, y, z) = P(x, y, z)\mathbf{i} + Q(x, y, z)\mathbf{j} + R(x, y, z)\mathbf{k}$ in two different ways: in one case producing another vector field and in the other producing a function. We will assume hereafter that P, Q, and R have continuous partial derivatives.

DEFINITION 17.3 Curl

The **curl** of a vector field $\mathbf{F} = P\mathbf{i} + Q\mathbf{j} + R\mathbf{k}$ is the vector field

$$\text{curl } \mathbf{F} = \left(\frac{\partial R}{\partial y} - \frac{\partial Q}{\partial z}\right)\mathbf{i} + \left(\frac{\partial P}{\partial z} - \frac{\partial R}{\partial x}\right)\mathbf{j} + \left(\frac{\partial Q}{\partial x} - \frac{\partial P}{\partial y}\right)\mathbf{k}$$

In practice, curl **F** can be computed from the cross product of the del operator and the vector **F**:

$$\text{curl }\mathbf{F} = \nabla \times \mathbf{F} = \begin{vmatrix} \mathbf{i} & \mathbf{j} & \mathbf{k} \\ \dfrac{\partial}{\partial x} & \dfrac{\partial}{\partial y} & \dfrac{\partial}{\partial z} \\ P & Q & R \end{vmatrix} \tag{1}$$

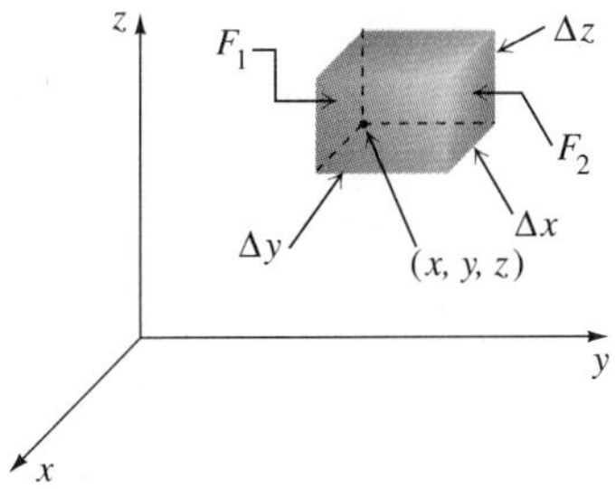

FIGURE 17.53

There is another combination of partial derivatives of the component functions of a vector field that occurs frequently in science and engineering. Before stating the next definition, consider the following motivation.

If $\mathbf{F}(x, y, z) = P(x, y, z)\mathbf{i} + Q(x, y, z)\mathbf{j} + R(x, y, z)\mathbf{k}$ is the velocity field of a fluid, then as we saw in Figure 17.47(b), the volume of the fluid flowing through an element of surface area ΔS per unit time—that is, the **flux** of the vector field **F** through the area ΔS—is approximately

$$(\text{height}) \cdot (\text{area of base}) = (\text{comp}_{\mathbf{n}}\mathbf{F})\,\Delta S = (\mathbf{F} \cdot \mathbf{n})\,\Delta S \tag{2}$$

where **n** is a unit vector normal to the surface. Now consider the rectangular parallelepiped shown in Figure 17.53. To compute the total flux of **F** through its six sides in the outward direction, we first compute the total flux out of parallel faces. The area of face F_1 is $\Delta x\,\Delta z$ and its outward unit normal is $-\mathbf{j}$, and so by (2) the flux of **F** through F_1 is

$$\mathbf{F} \cdot (-\mathbf{j})\,\Delta x\,\Delta z = -Q(x, y, z)\,\Delta x\,\Delta z$$

The flux out of face F_2, whose outward normal is **j**, is given by

$$(\mathbf{F} \cdot \mathbf{j})\,\Delta x\,\Delta z = Q(x, y + \Delta y, z)\,\Delta x\,\Delta z$$

Consequently, the total flux out of these parallel faces is

$$Q(x, y + \Delta y, z)\,\Delta x\,\Delta z + (-Q(x, y, z)\,\Delta x\,\Delta z) = [Q(x, y + \Delta y, z) - Q(x, y, z)]\,\Delta x\,\Delta z \tag{3}$$

By multiplying (3) by $\Delta y/\Delta y$ and using the definition of a partial derivative, we get for small Δy

$$\frac{[Q(x, y + \Delta y, z) - Q(x, y, z)]}{\Delta y}\,\Delta x\,\Delta y\,\Delta z \approx \frac{\partial Q}{\partial y}\,\Delta x\,\Delta y\,\Delta z$$

Arguing in exactly the same manner, we see that the contributions to the total flux out of the parallelepiped from the two faces parallel to the yz-plane and from the two faces parallel to the xy-plane are, in turn,

$$\frac{\partial P}{\partial x}\,\Delta x\,\Delta y\,\Delta z \quad \text{and} \quad \frac{\partial R}{\partial z}\,\Delta x\,\Delta y\,\Delta z$$

Adding the results, we see that the total flux of **F** out of the parallelepiped is approximately

$$\left(\frac{\partial P}{\partial x} + \frac{\partial Q}{\partial y} + \frac{\partial R}{\partial z}\right)\Delta x\,\Delta y\,\Delta z$$

By dividing the last expression by $\Delta x\,\Delta y\,\Delta z$, we get the outward flux of **F** per unit volume:

$$\frac{\partial P}{\partial x} + \frac{\partial Q}{\partial y} + \frac{\partial R}{\partial z}$$

It is this combination of partial derivatives that is given a special name:

DEFINITION 17.4 Divergence

The **divergence** of a vector field $\mathbf{F} = P\mathbf{i} + Q\mathbf{j} + R\mathbf{k}$ is the scalar function

$$\text{div}\,\mathbf{F} = \frac{\partial P}{\partial x} + \frac{\partial Q}{\partial y} + \frac{\partial R}{\partial z}$$

Observe that div **F** can also be written in terms of the del operator as

$$\text{div}\,\mathbf{F} = \nabla \cdot \mathbf{F} = \frac{\partial}{\partial x} P(x, y, z) + \frac{\partial}{\partial y} Q(x, y, z) + \frac{\partial}{\partial z} R(x, y, z) \tag{4}$$

EXAMPLE 1 If $\mathbf{F} = (x^2y^3 - z^4)\mathbf{i} + 4x^5y^2z\mathbf{j} - y^4z^6\mathbf{k}$, find (*a*) curl **F** and (*b*) div **F**.

Solution (*a*) From (1),

$$\text{curl}\,\mathbf{F} = \nabla \times \mathbf{F} = \begin{vmatrix} \mathbf{i} & \mathbf{j} & \mathbf{k} \\ \dfrac{\partial}{\partial x} & \dfrac{\partial}{\partial y} & \dfrac{\partial}{\partial z} \\ x^2y^3 - z^4 & 4x^5y^2z & -y^4z^6 \end{vmatrix}$$

$$= \left[\frac{\partial}{\partial y}(-y^4z^6) - \frac{\partial}{\partial z}(4x^5y^2z)\right]\mathbf{i} - \left[\frac{\partial}{\partial x}(-y^4z^6) - \frac{\partial}{\partial z}(x^2y^3 - z^4)\right]\mathbf{j}$$

$$+ \left[\frac{\partial}{\partial x}(4x^5y^2z) - \frac{\partial}{\partial y}(x^2y^3 - z^4)\right]\mathbf{k}$$

$$= (-4y^3z^6 - 4x^5y^2)\mathbf{i} - 4z^3\mathbf{j} + (20x^4y^2z - 3x^2y^2)\mathbf{k}$$

(*b*) From (4),

$$\text{div}\,\mathbf{F} = \nabla \cdot \mathbf{F} = \frac{\partial}{\partial x}(x^2y^3 - z^4) + \frac{\partial}{\partial y}(4x^5y^2z) + \frac{\partial}{\partial z}(-y^4z^6)$$

$$= 2xy^3 + 8x^5yz - 6y^4z^5$$

□

We ask you to prove the following two important properties. If f is a *scalar function* with continuous second partial derivatives, then

$$\text{curl}(\text{grad } f) = \nabla \times \nabla f = \mathbf{0} \tag{5}$$

Also, if $\mathbf{F}$ is a *vector field* having continuous second partial derivatives, then

$$\text{div}(\text{curl } \mathbf{F}) = \nabla \cdot (\nabla \times \mathbf{F}) = 0 \tag{6}$$

See Problems 23 and 24 in Exercises 17.5.

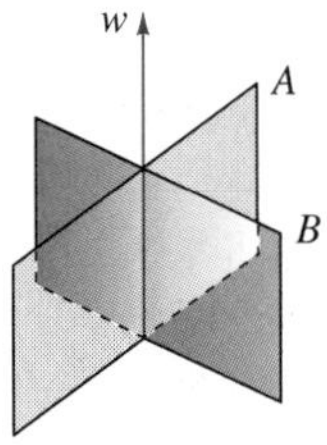

FIGURE 17.54

Physical Interpretations The word *curl* was introduced by Maxwell* in his studies of electromagnetic fields. However, the curl is easily understood in connection with the flow of fluids. If a paddle device, such as shown in Figure 17.54, is inserted in a flowing fluid, then the curl of the velocity field $\mathbf{F}$ is a measure of the tendency of the fluid to turn the device about its vertical axis w. If curl $\mathbf{F} = \mathbf{0}$, then the flow of the fluid is said to be **irrotational**, which means that it is free of vortices or whirlpools that would cause the paddle to rotate.† In Figure 17.55 the axis w of the paddle points straight out of the page. Note from Figure 17.55(b) that "irrotational" does *not* mean that the fluid does not rotate. See Problem 34 for another interpretation of the curl.

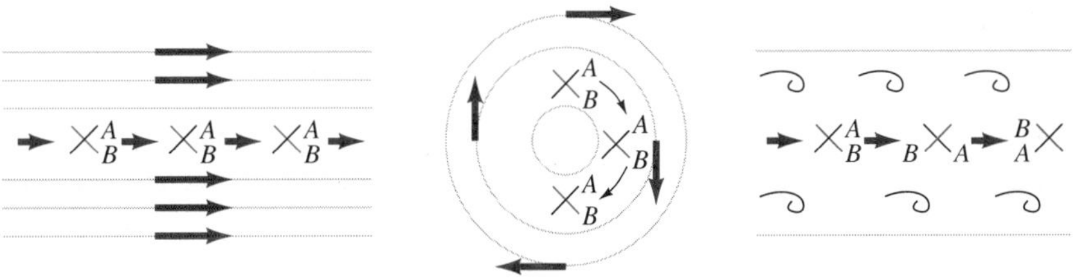

(a) irrotational flow (b) irrotational flow (c) rotational flow

FIGURE 17.55

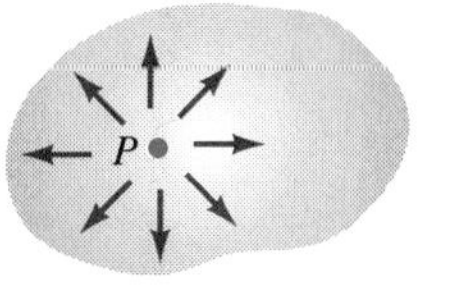

(a) div $\mathbf{F}(P) > 0$; P a source

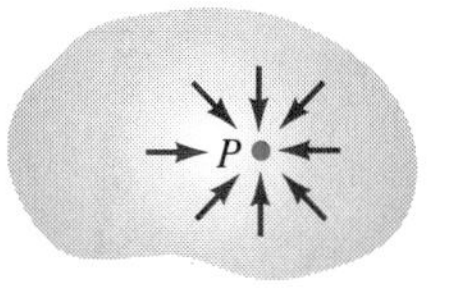

(b) div $\mathbf{F}(P) < 0$; P a sink

FIGURE 17.56

In the motivational discussion leading to Definition 17.4, we saw that the divergence of a velocity field $\mathbf{F}$ near a point $P(x, y, z)$ is the flux per unit volume. If div $\mathbf{F}(P) > 0$, then P is said to be a **source** for $\mathbf{F}$, since there is a net outward flow of fluid near P; if div $\mathbf{F}(P) < 0$, then P is said to be a **sink** for $\mathbf{F}$, since there is a net inward flow of fluid near P; if div $\mathbf{F}(P) = 0$, there are no sources or sinks near P. See Figure 17.56.

The divergence of a vector field has another interpretation in the context of fluid flow. A measure of the rate of change of the density of the fluid at a point is simply div $\mathbf{F}$. In other words, div $\mathbf{F}$ is a measure of the fluid's compressibility. If $\nabla \cdot \mathbf{F} = 0$, the fluid is said to be **incompressible**. In electromagnetic theory, if $\nabla \cdot \mathbf{F} = 0$, the vector field $\mathbf{F}$ is said to be **solenoidal**.

---

*James Clerk Maxwell (1831–1879), a Scottish physicist.

†In science texts the word *rotation* is sometimes used instead of *curl*. The symbol curl $\mathbf{F}$ is then replaced by rot $\mathbf{F}$.

EXERCISES 17.5 *Answers to odd-numbered problems begin on page A-109.*

In Problems 1–10 find the curl and the divergence of the given vector field.

1. $\mathbf{F}(x, y, z) = xz\mathbf{i} + yz\mathbf{j} + xy\mathbf{k}$
2. $\mathbf{F}(x, y, z) = 10yz\mathbf{i} + 2x^2z\mathbf{j} + 6x^3\mathbf{k}$
3. $\mathbf{F}(x, y, z) = 4xy\mathbf{i} + (2x^2 + 2yz)\mathbf{j} + (3z^2 + y^2)\mathbf{k}$
4. $\mathbf{F}(x, y, z) = (x - y)^3\mathbf{i} + e^{-yz}\mathbf{j} + xye^{2y}\mathbf{k}$
5. $\mathbf{F}(x, y, z) = 3x^2y\mathbf{i} + 2xz^3\mathbf{j} + y^4\mathbf{k}$
6. $\mathbf{F}(x, y, z) = 5y^3\mathbf{i} + (\frac{1}{2}x^3y^2 - xy)\mathbf{j} - (x^3yz - xz)\mathbf{k}$
7. $\mathbf{F}(x, y, z) = xe^{-z}\mathbf{i} + 4yz^2\mathbf{j} + 3ye^{-z}\mathbf{k}$
8. $\mathbf{F}(x, y, z) = yz \ln x\mathbf{i} + (2x - 3yz)\mathbf{j} + xy^2z^3\mathbf{k}$
9. $\mathbf{F}(x, y, z) = xye^x\mathbf{i} - x^3yze^z\mathbf{j} + xy^2e^y\mathbf{k}$
10. $\mathbf{F}(x, y, z) = x^2\sin yz\mathbf{i} + z \cos xz^3\mathbf{j} + ye^{5xy}\mathbf{k}$

In Problems 11–18 let $\mathbf{a}$ be a constant vector and $\mathbf{r} = x\mathbf{i} + y\mathbf{j} + z\mathbf{k}$. Verify the given identity.

11. $\text{div}\, \mathbf{r} = 3$
12. $\text{curl}\, \mathbf{r} = \mathbf{0}$
13. $(\mathbf{a} \times \nabla) \times \mathbf{r} = -2\mathbf{a}$
14. $\nabla \times (\mathbf{a} \times \mathbf{r}) = 2\mathbf{a}$
15. $\nabla \cdot (\mathbf{a} \times \mathbf{r}) = 0$
16. $\mathbf{a} \times (\nabla \times \mathbf{r}) = \mathbf{0}$
17. $\nabla \times [(\mathbf{r} \cdot \mathbf{r})\mathbf{a}] = 2(\mathbf{r} \times \mathbf{a})$
18. $\nabla \cdot [(\mathbf{r} \cdot \mathbf{r})\mathbf{a}] = 2(\mathbf{r} \cdot \mathbf{a})$

In Problems 19–26 verify the given identity. Assume continuity of all partial derivatives.

19. $\nabla \cdot (\mathbf{F} + \mathbf{G}) = \nabla \cdot \mathbf{F} + \nabla \cdot \mathbf{G}$
20. $\nabla \times (\mathbf{F} + \mathbf{G}) = \nabla \times \mathbf{F} + \nabla \times \mathbf{G}$
21. $\nabla \cdot (f\mathbf{F}) = f(\nabla \cdot \mathbf{F}) + \mathbf{F} \cdot \nabla f$
22. $\nabla \times (f\mathbf{F}) = f(\nabla \times \mathbf{F}) + (\nabla f) \times \mathbf{F}$
23. $\text{curl}(\text{grad}\, f) = \mathbf{0}$
24. $\text{div}(\text{curl}\, \mathbf{F}) = 0$
25. $\text{div}(\mathbf{F} \times \mathbf{G}) = \mathbf{G} \cdot \text{curl}\, \mathbf{F} - \mathbf{F} \cdot \text{curl}\, \mathbf{G}$
26. $\text{curl}(\text{curl}\, \mathbf{F} + \text{grad}\, f) = \text{curl}(\text{curl}\, \mathbf{F})$
27. Show that

$$\nabla \cdot \nabla f = \frac{\partial^2 f}{\partial x^2} + \frac{\partial^2 f}{\partial y^2} + \frac{\partial^2 f}{\partial z^2}$$

This is known as the **Laplacian** and is also written $\nabla^2 f$.

28. Show that $\nabla \cdot (f\, \nabla f) = f\, \nabla^2 f + |\nabla f|^2$, where $\nabla^2 f$ is the Laplacian defined in Problem 27. [*Hint:* See Problem 21.]

29. Find curl(curl $\mathbf{F}$) for the vector field $\mathbf{F}(x, y, z) = xy\mathbf{i} + 4yz^2\mathbf{j} + 2xz\mathbf{k}$.

30. (a) Assuming continuity of all partial derivatives, show that $\text{curl}(\text{curl}\, \mathbf{F}) = -\nabla^2\mathbf{F} + \text{grad}(\text{div}\, \mathbf{F})$, where

$$\nabla^2\mathbf{F} = \nabla^2(P\mathbf{i} + Q\mathbf{j} + R\mathbf{k}) = \nabla^2P\mathbf{i} + \nabla^2Q\mathbf{j} + \nabla^2R\mathbf{k}$$

(b) Use the identity in part (**a**) to obtain the result in Problem 29.

31. Any scalar function f for which $\nabla^2 f = 0$ is said to be **harmonic**. Verify that $f(x, y, z) = 3x^2 + 5y^2 + 4xy - 9xz - 8z^2$ is harmonic. $\nabla^2 f = 0$ is called **Laplace's equation.**

32. Verify that

$$f(x, y) = \arctan\left(\frac{2y}{x^2 + y^2 - 1}\right), \qquad x^2 + y^2 \neq 1$$

satisfies Laplace's equation in two variables

$$\nabla^2 f = \frac{\partial^2 f}{\partial x^2} + \frac{\partial^2 f}{\partial y^2} = 0$$

33. Let $\mathbf{r} = x\mathbf{i} + y\mathbf{j} + z\mathbf{k}$ be the position vector of a mass m_1 and let the mass m_2 be located at the origin. If the force of gravitational attraction is

$$\mathbf{F} = -\frac{Gm_1m_2}{|\mathbf{r}|^3}\mathbf{r}$$

verify that curl $\mathbf{F} = \mathbf{0}$ and div $\mathbf{F} = 0$, $\mathbf{r} \neq \mathbf{0}$.

34. Suppose a body rotates with a constant angular velocity $\boldsymbol{\omega}$ about an axis. If $\mathbf{r}$ is the position vector of a point P on the body measured from the origin, then the linear velocity vector $\mathbf{v}$ of rotation is $\mathbf{v} = \boldsymbol{\omega} \times \mathbf{r}$. See Figure 17.57. If $\mathbf{r} = x\mathbf{i} + y\mathbf{j} + z\mathbf{k}$ and $\boldsymbol{\omega} = \omega_1\mathbf{i} + \omega_2\mathbf{j} + \omega_3\mathbf{k}$, show that $\boldsymbol{\omega} = \frac{1}{2}\text{curl}\, \mathbf{v}$.

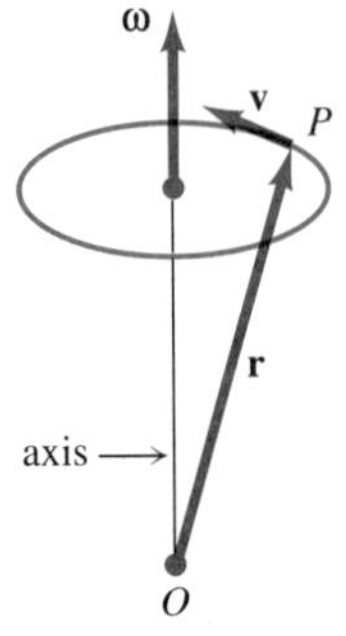

FIGURE 17.57

In Problems 35 and 36 assume that f and g have continuous second partial derivatives. Show that the given vector field is solenoidal. [*Hint:* See Problem 25.]

35. $\mathbf{F} = \nabla f \times \nabla g$

36. $\mathbf{F} = \nabla f \times (f\,\nabla g)$

37. If $\mathbf{F} = y^3\mathbf{i} + x^3\mathbf{j} + z^3\mathbf{k}$, find the flux of $\nabla \times \mathbf{F}$ through that portion of the ellipsoid $x^2 + y^2 + 4z^2 = 4$ in the first octant that is bounded by $y = 0$, $y = x$, $z = 0$. Assume the surface is oriented upward.

38. The velocity vector field for the two-dimensional flow of an ideal fluid around a cylinder is given by

$$\mathbf{F}(x, y) = A\left[\left(1 - \frac{x^2 - y^2}{(x^2 + y^2)^2}\right)\mathbf{i} - \frac{2xy}{(x^2 + y^2)^2}\mathbf{j}\right]$$

for some positive constant A. See Figure 17.58.

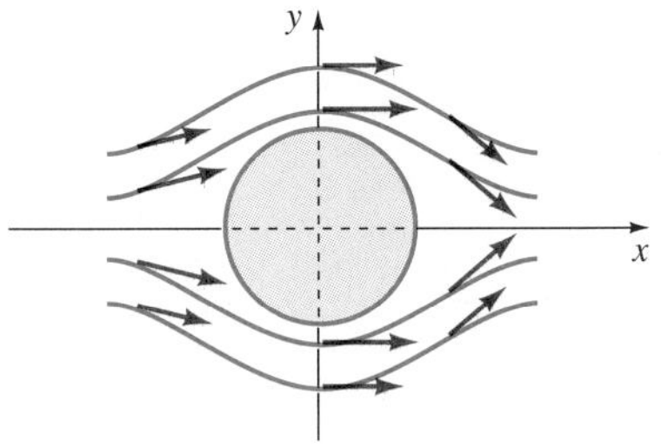

FIGURE 17.58

(a) Show that when the point (x, y) is far from the origin, $\mathbf{F}(x, y) \approx A\mathbf{i}$.

(b) Show that $\mathbf{F}$ is irrotational.

(c) Show that $\mathbf{F}$ is incompressible.

39. If $\mathbf{E} = \mathbf{E}(x, y, z, t)$ and $\mathbf{H} = \mathbf{H}(x, y, z, t)$ represent electric and magnetic fields in empty space, the Maxwell's equations are

$$\text{div } \mathbf{E} = 0 \qquad \text{curl } \mathbf{E} = -\frac{1}{c}\frac{\partial \mathbf{H}}{\partial t}$$

$$\text{div } \mathbf{H} = 0 \qquad \text{curl } \mathbf{H} = \frac{1}{c}\frac{\partial \mathbf{E}}{\partial t}$$

where c is the speed of light. Use the identity in Problem 30(a) to show that $\mathbf{E}$ and $\mathbf{H}$ satisfy

$$\nabla^2\mathbf{E} = \frac{1}{c^2}\frac{\partial^2 \mathbf{E}}{\partial t^2} \qquad \nabla^2\mathbf{H} = \frac{1}{c^2}\frac{\partial^2 \mathbf{H}}{\partial t^2}$$

40. Consider the vector field $\mathbf{F} = x^2yz\mathbf{i} - xy^2z\mathbf{j} + (z + 5x)\mathbf{k}$. Explain why $\mathbf{F}$ is not the curl of another vector field $\mathbf{G}$.

17.6 STOKES' THEOREM

A Vector Form of Green's Theorem Green's Theorem of Section 17.3 has two vector forms. In this and the next section we shall generalize these forms to three dimensions. If $\mathbf{F}(x, y) = P(x, y)\mathbf{i} + Q(x, y)\mathbf{j}$ is a two-dimensional vector field, then

$$\text{curl } \mathbf{F} = \nabla \times \mathbf{F} = \begin{vmatrix} \mathbf{i} & \mathbf{j} & \mathbf{k} \\ \frac{\partial}{\partial x} & \frac{\partial}{\partial y} & \frac{\partial}{\partial z} \\ P & Q & 0 \end{vmatrix} = \left(\frac{\partial Q}{\partial x} - \frac{\partial P}{\partial y}\right)\mathbf{k}$$

From (12) and (13) of Section 17.1, Green's Theorem can be written in vector notation as

$$\oint_C \mathbf{F} \cdot d\mathbf{r} = \oint_C (\mathbf{F} \cdot \mathbf{T})\, ds = \iint_R (\text{curl } \mathbf{F}) \cdot \mathbf{k}\, dA \qquad (1)$$

That is, the line integral of the tangential component of $\mathbf{F}$ is the double integral of the normal component of curl $\mathbf{F}$.

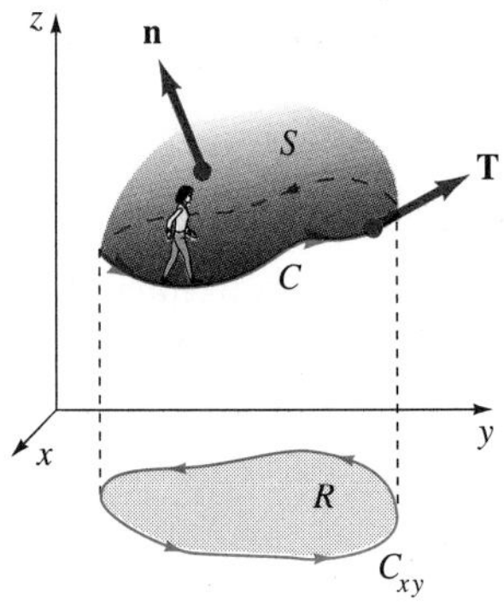

FIGURE 17.59

Green's Theorem in 3-Space The vector form of Green's Theorem given in (1) relates a line integral around a piecewise smooth simple closed curve C forming the boundary of a plane region R to a double integral over R. Green's Theorem in 3-space relates a line integral around a piecewise smooth simple closed space curve C forming the boundary of a surface S with a surface integral over S. Suppose $z = f(x, y)$ is a continuous function whose graph is a piecewise smooth orientable surface over a region R on the xy-plane. Let C form the boundary of S and let the projection of C onto the xy-plane form the boundary of R. The positive direction on C is induced by the orientation of the surface S; the positive direction on C corresponds to the direction a person would have to walk on C to have his or her head point in the direction of the orientation of the surface while keeping the surface to the left. See Figure 17.59. More precisely, the positive orientation of C is in accordance with the right-hand rule: If the thumb of the right hand points in the direction of the orientation of the surface, then roughly the fingers of the right hand wrap around the surface in the positive direction. Finally, let $\mathbf{T}$ be a unit tangent vector to C that points in the positive direction. The three-dimensional form of Green's Theorem, which we shall now give, is called **Stokes' Theorem.***

THEOREM 17.6 Stokes' Theorem

Let S be a piecewise smooth orientable surface bounded by a piecewise smooth simple closed curve C. Let $\mathbf{F}(x, y, z) = P(x, y, z)\mathbf{i} + Q(x, y, z)\mathbf{j} + R(x, y, z)\mathbf{k}$ be a vector field for which P, Q, and R are continuous and have continuous first partial derivatives in a region of 3-space containing S. If C is traversed in the positive direction, then

$$\oint_C \mathbf{F} \cdot d\mathbf{r} = \oint_C (\mathbf{F} \cdot \mathbf{T})\, ds = \iint_S (\text{curl } \mathbf{F}) \cdot \mathbf{n}\, dS \qquad (2)$$

where $\mathbf{n}$ is a unit normal to S in the direction of the orientation of S.

Partial Proof Suppose the surface S is oriented upward and is defined by a function $z = f(x, y)$ that has continuous second partial derivatives. From Definition 17.3 we have

$$\text{curl } \mathbf{F} = \left(\frac{\partial R}{\partial y} - \frac{\partial Q}{\partial z}\right)\mathbf{i} + \left(\frac{\partial P}{\partial z} - \frac{\partial R}{\partial x}\right)\mathbf{j} + \left(\frac{\partial Q}{\partial x} - \frac{\partial P}{\partial y}\right)\mathbf{k}$$

***George G. Stokes (1819–1903)**
Stokes was an Irish mathematician and physicist. Like George Green, Stokes was a don at Cambridge University. In 1854 Stokes posed this theorem as a problem on a prize examination for Cambridge students. It is not known whether anyone solved the problem.

Furthermore, if we write $g(x, y, z) = z - f(x, y) = 0$, then

$$\mathbf{n} = \frac{-\dfrac{\partial f}{\partial x}\mathbf{i} - \dfrac{\partial f}{\partial y}\mathbf{j} + \mathbf{k}}{\sqrt{1 + \left(\dfrac{\partial f}{\partial x}\right)^2 + \left(\dfrac{\partial f}{\partial y}\right)^2}}$$

Hence,

$$\iint_S (\text{curl } \mathbf{F}) \cdot \mathbf{n}\, dS = \iint_R \left[-\left(\frac{\partial R}{\partial y} - \frac{\partial Q}{\partial z}\right)\frac{\partial f}{\partial x} - \left(\frac{\partial P}{\partial z} - \frac{\partial R}{\partial x}\right)\frac{\partial f}{\partial y} + \left(\frac{\partial Q}{\partial x} - \frac{\partial P}{\partial y}\right)\right] dA \tag{3}$$

Our goal is now to show that $\oint_C \mathbf{F} \cdot d\mathbf{r}$ reduces to (3).

If C_{xy} is the projection of C onto the xy-plane and has the parametric equations $x = x(t)$, $y = y(t)$, $a \le t \le b$, then parametric equations for C are $x = x(t)$, $y = y(t)$, $z = f(x(t), y(t))$, $a \le t \le b$. Thus,

Chain Rule

$$\oint_C \mathbf{F} \cdot d\mathbf{r} = \oint_C P\,dx + Q\,dy + R\,dz = \int_a^b \left[P\frac{dx}{dt} + Q\frac{dy}{dt} + R\left(\frac{\partial f}{\partial x}\frac{dx}{dt} + \frac{\partial f}{\partial y}\frac{dy}{dt}\right)\right] dt$$

$$= \oint_{C_{xy}} \left(P + R\frac{\partial f}{\partial x}\right) dx + \left(Q + R\frac{\partial f}{\partial y}\right) dy$$

$$= \iint_R \left[\frac{\partial}{\partial x}\left(Q + R\frac{\partial f}{\partial y}\right) - \frac{\partial}{\partial y}\left(P + R\frac{\partial f}{\partial x}\right)\right] dA \tag{4}$$

Green's Theorem

Now, $\dfrac{\partial}{\partial x}\left(Q + R\dfrac{\partial f}{\partial y}\right) = \dfrac{\partial}{\partial x}\left[Q(x, y, f(x, y)) + R(x, y, f(x, y))\dfrac{\partial f}{\partial y}\right]$

Chain and Product Rules

$$= \frac{\partial Q}{\partial x} + \frac{\partial Q}{\partial z}\frac{\partial f}{\partial x} + R\frac{\partial^2 f}{\partial x\,\partial y} + \frac{\partial f}{\partial y}\left(\frac{\partial R}{\partial x} + \frac{\partial R}{\partial z}\frac{\partial f}{\partial x}\right)$$

$$= \frac{\partial Q}{\partial x} + \frac{\partial Q}{\partial z}\frac{\partial f}{\partial x} + R\frac{\partial^2 f}{\partial x\,\partial y} + \frac{\partial R}{\partial x}\frac{\partial f}{\partial y} + \frac{\partial R}{\partial z}\frac{\partial f}{\partial y}\frac{\partial f}{\partial x} \tag{5}$$

Similarly,

$$\frac{\partial}{\partial y}\left(P + R\frac{\partial f}{\partial x}\right) = \frac{\partial P}{\partial y} + \frac{\partial P}{\partial z}\frac{\partial f}{\partial y} + R\frac{\partial^2 f}{\partial y\,\partial x} + \frac{\partial R}{\partial y}\frac{\partial f}{\partial x} + \frac{\partial R}{\partial z}\frac{\partial f}{\partial x}\frac{\partial f}{\partial y} \tag{6}$$

Subtracting (6) from (5) and using the fact that $\partial^2 f/\partial x\,\partial y = \partial^2 f/\partial y\,\partial x$, we see that (4) becomes, after rearranging,

$$\iint_R \left[-\left(\frac{\partial R}{\partial y} - \frac{\partial Q}{\partial z}\right)\frac{\partial f}{\partial x} - \left(\frac{\partial P}{\partial z} - \frac{\partial R}{\partial x}\right)\frac{\partial f}{\partial y} + \left(\frac{\partial Q}{\partial x} - \frac{\partial P}{\partial y}\right)\right] dA$$

This last expression is the same as the right side of (3), which was to be shown. ■

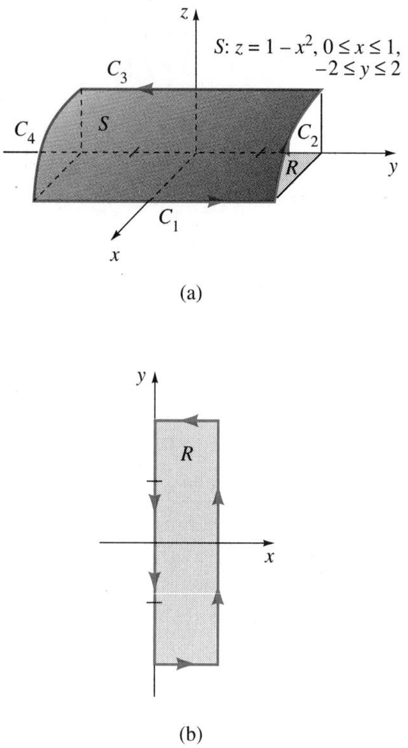

FIGURE 17.60

EXAMPLE 1 Let S be the part of the cylinder $z = 1 - x^2$ for $0 \le x \le 1$, $-2 \le y \le 2$. Verify Stokes' Theorem if $\mathbf{F} = xy\mathbf{i} + yz\mathbf{j} + xz\mathbf{k}$. Assume S is oriented upward.

Solution The surface S, the curve C (which is composed of the union of C_1, C_2, C_3, and C_4), and the region R are shown in Figure 17.60.

The Surface Integral: For $\mathbf{F} = xy\mathbf{i} + yz\mathbf{j} + xz\mathbf{k}$, we find

$$\text{curl}\,\mathbf{F} = \begin{vmatrix} \mathbf{i} & \mathbf{j} & \mathbf{k} \\ \dfrac{\partial}{\partial x} & \dfrac{\partial}{\partial y} & \dfrac{\partial}{\partial z} \\ xy & yz & xz \end{vmatrix} = -y\mathbf{j} - z\mathbf{j} - x\mathbf{k}$$

Now, if $g(x, y, z) = z + x^2 - 1 = 0$ defines the cylinder, the upper normal is

$$\mathbf{n} = \frac{\nabla g}{|\nabla g|} = \frac{2x\mathbf{i} + \mathbf{k}}{\sqrt{4x^2 + 1}}$$

Therefore,
$$\iint_S (\text{curl}\,\mathbf{F} \cdot \mathbf{n})\, dS = \iint_S \frac{-2xy - x}{\sqrt{4x^2 + 1}}\, dS$$

To evaluate the latter surface integral, we use (2) of Section 17.4:

$$\begin{aligned} \iint_S \frac{-2xy - x}{\sqrt{4x^2 + 1}}\, dS &= \iint_R (-2xy - x)\, dA \\ &= \int_0^1 \int_{-2}^2 (-2xy - x)\, dy\, dx \\ &= \int_0^1 \Big[-xy^2 - xy\Big]_{-2}^2\, dx \\ &= \int_0^1 (-4x)\, dx = -2 \end{aligned} \tag{7}$$

The Line Integral: We write $\oint_C = \int_{C_1} + \int_{C_2} + \int_{C_3} + \int_{C_4}$. On C_1: $x = 1$, $z = 0$, $dx = 0$, $dz = 0$, so

$$\int_{C_1} y(0) + y(0)\, dy + 0 = 0$$

On C_2: $y = 2$, $z = 1 - x^2$, $dy = 0$, $dz = -2x\, dx$, so

$$\int_{C_2} 2x\, dx + 2(1 - x^2)0 + x(1 - x^2)(-2x\, dx) = \int_1^0 (2x - 2x^2 + 2x^4)\, dx = -\frac{11}{15}$$

On C_3: $x = 0$, $z = 1$, $dx = 0$, $dz = 0$, so

$$\int_{C_3} 0 + y\, dy + 0 = \int_2^{-2} y\, dy = 0$$

On C_4: $y = -2, z = 1 - x^2, dy = 0, dz = -2x\,dx$, so

$$\int_{C_4} -2x\,dx - 2(1 - x^2)0 + x(1 - x^2)(-2x\,dx) = \int_0^1 (-2x - 2x^2 + 2x^4)\,dx = -\frac{19}{15}$$

Hence,

$$\oint_C xy\,dx + yz\,dy + xz\,dz = 0 - \frac{11}{15} + 0 - \frac{19}{15} = -2$$

which, of course, agrees with (7). □

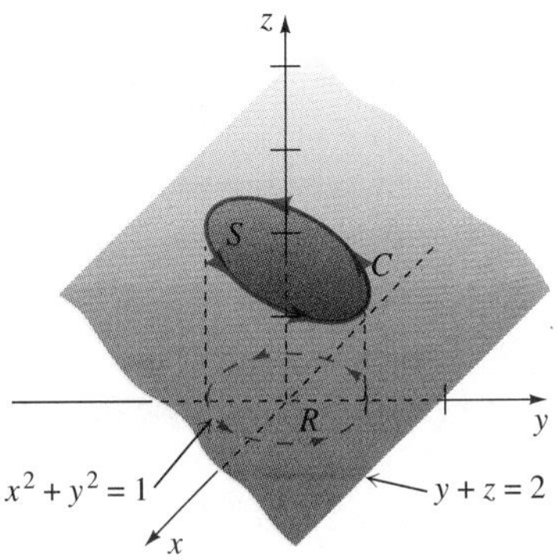

FIGURE 17.61

EXAMPLE 2 Evaluate $\oint_C z\,dx + x\,dy + y\,dz$, where C is the trace of the cylinder $x^2 + y^2 = 1$ in the plane $y + z = 2$. Orient C counterclockwise as viewed from above. See Figure 17.61.

Solution If $\mathbf{F} = z\mathbf{i} + x\mathbf{j} + y\mathbf{k}$, then

$$\text{curl}\,\mathbf{F} = \begin{vmatrix} \mathbf{i} & \mathbf{j} & \mathbf{k} \\ \dfrac{\partial}{\partial x} & \dfrac{\partial}{\partial y} & \dfrac{\partial}{\partial z} \\ z & x & y \end{vmatrix} = \mathbf{i} + \mathbf{j} + \mathbf{k}$$

The given orientation of C corresponds to an upward orientation of the surface S. Thus, if $g(x, y, z) = y + z - 2 = 0$ defines the plane, then the upper normal is

$$\mathbf{n} = \frac{\nabla g}{|\nabla g|} = \frac{1}{\sqrt{2}}\mathbf{j} + \frac{1}{\sqrt{2}}\mathbf{k}$$

Hence, from (2),

$$\oint_C \mathbf{F} \cdot d\mathbf{r} = \iint_S \left[(\mathbf{i} + \mathbf{j} + \mathbf{k}) \cdot \left(\frac{1}{\sqrt{2}}\mathbf{j} + \frac{1}{\sqrt{2}}\mathbf{k}\right)\right] dS$$

$$= \sqrt{2} \iint_S dS = \sqrt{2} \iint_R \sqrt{2}\,dA = 2\pi$$ □

Note that if $\mathbf{F}$ is the gradient of a scalar function, then, in view of (5) of Section 17.5, (2) implies that the circulation $\oint_C \mathbf{F} \cdot d\mathbf{r}$ is zero. Conversely, it can be shown that if the circulation is zero for every simple closed curve, then $\mathbf{F}$ is the gradient of a scalar function. In other words, $\mathbf{F}$ is irrotational if and only if $\mathbf{F} = \nabla\phi$, where ϕ is the potential for $\mathbf{F}$. Equivalently, this gives a test for a conservative vector field.

$\mathbf{F}$ *is a conservative vector field if and only if* curl $\mathbf{F} = \mathbf{0}$.

Physical Interpretation of Curl In Section 17.1 we saw that if $\mathbf{F}$ is a velocity field of a fluid, then the circulation $\oint_C \mathbf{F} \cdot d\mathbf{r}$ of $\mathbf{F}$ around C is a

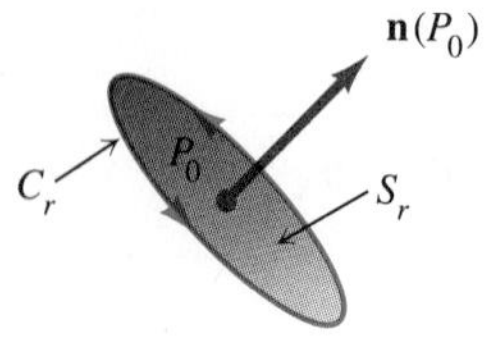

FIGURE 17.62

measure of the amount by which the fluid tends to turn the curve C by circulating around it. The circulation of **F** is closely related to the curl of **F**. To see this, suppose $P_0(x_0, y_0, z_0)$ is any point in the fluid and C_r is a small circle of radius r centered at P_0. See Figure 17.62. Then by Stokes' Theorem,

$$\oint_{C_r} \mathbf{F} \cdot d\mathbf{r} = \iint_{S_r} (\text{curl } \mathbf{F}) \cdot \mathbf{n} \, dS \tag{8}$$

Now at all points $P(x, y, z)$ within the small circle C_r, if we take curl $\mathbf{F}(P) \approx$ curl $\mathbf{F}(P_0)$, then (8) gives the approximation

$$\begin{aligned}\oint_{C_r} \mathbf{F} \cdot d\mathbf{r} &\approx \iint_{S_r} (\text{curl } \mathbf{F}(P_0)) \cdot \mathbf{n}(P_0) \, dS \\ &= (\text{curl } \mathbf{F}(P_0)) \cdot \mathbf{n}(P_0) \iint_{S_r} dS \\ &= (\text{curl } \mathbf{F}(P_0)) \cdot \mathbf{n}(P_0) A_r \end{aligned} \tag{9}$$

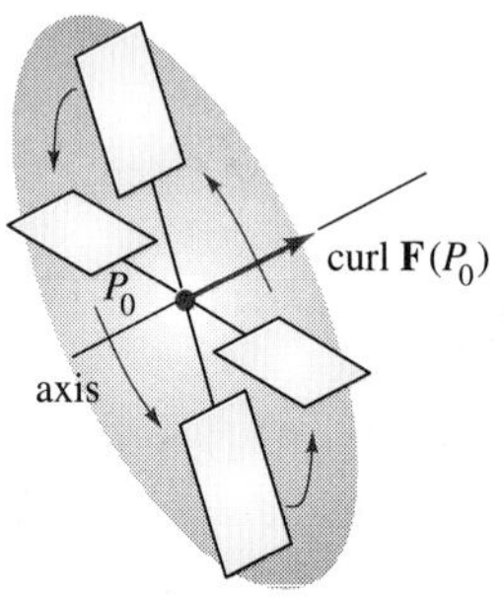

FIGURE 17.63

where A_r is the area (πr^2) of the circular surface S_r. As we let $r \to 0$, the approximation curl $\mathbf{F}(P) \approx$ curl $\mathbf{F}(P_0)$ becomes better and so (9) yields

$$(\text{curl } \mathbf{F}(P_0)) \cdot \mathbf{n}(P_0) = \lim_{r \to 0} \frac{1}{A_r} \oint_{C_r} \mathbf{F} \cdot d\mathbf{r} \tag{10}$$

Thus, we see that the normal component of curl **F** is the limiting value of the ratio of the circulation of **F** to the area of the circular surface. For a small but fixed value of r, we have

$$(\text{curl } \mathbf{F}(P_0)) \cdot \mathbf{n}(P_0) \approx \frac{1}{A_r} \oint_{C_r} \mathbf{F} \cdot d\mathbf{r} \tag{11}$$

Roughly then, curl of **F** is the circulation of **F** per unit area. If curl $\mathbf{F}(P_0) \neq \mathbf{0}$, then the left-hand side of (11) is a maximum when the circle C_r is situated in a manner so that $\mathbf{n}(P_0)$ points in the same direction as curl $\mathbf{F}(P_0)$. In this case, the circulation on the right side of (11) is also a maximum. Thus, a paddle wheel inserted into the fluid at P_0 will rotate fastest when its axis points in the direction of curl $\mathbf{F}(P_0)$. See Figure 17.63. Note, too, that the paddle wheel will not rotate if its axis is perpendicular to curl $\mathbf{F}(P_0)$.

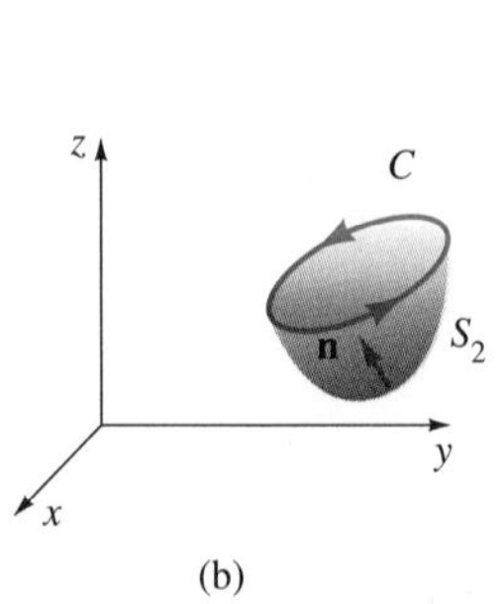

FIGURE 17.64

▲ ***Remark*** The value of the surface integral in (2) is determined solely by the integral around its boundary C. This basically means that the shape of the surface S is irrelevant. Assuming that the hypotheses of Theorem 17.6 are satisfied, then for two different surfaces S_1 and S_2 with the same orientation and with the same boundary C, we have

$$\oint_C \mathbf{F} \cdot d\mathbf{r} = \iint_{S_1} (\text{curl } \mathbf{F}) \cdot \mathbf{n} \, dS = \iint_{S_2} (\text{curl } \mathbf{F}) \cdot \mathbf{n} \, dS$$

See Figure 17.64 and Problems 17 and 18 of Exercises 17.6.

EXERCISES 17.6 *Answers to odd-numbered problems begin on page A-109.*

In Problems 1–4 verify Stokes' Theorem. Assume that the surface S is oriented upward.

1. $\mathbf{F} = 5y\mathbf{i} - 5x\mathbf{j} + 3\mathbf{k}$; S that portion of the plane $z = 1$ within the cylinder $x^2 + y^2 = 4$

2. $\mathbf{F} = 2z\mathbf{i} - 3x\mathbf{j} + 4y\mathbf{k}$; S that portion of the paraboloid $z = 16 - x^2 - y^2$ for $z \geq 0$

3. $\mathbf{F} = z\mathbf{i} + x\mathbf{j} + y\mathbf{k}$; S that portion of the plane $2x + y + 2z = 6$ in the first octant

4. $\mathbf{F} = x\mathbf{i} + y\mathbf{j} + z\mathbf{k}$; S that portion of the sphere $x^2 + y^2 + z^2 = 1$ for $z \geq 0$

In Problems 5–12 use Stokes' Theorem to evaluate $\oint_C \mathbf{F} \cdot d\mathbf{r}$. Assume C is oriented counterclockwise as viewed from above.

5. $\mathbf{F} = (2z + x)\mathbf{i} + (y - z)\mathbf{j} + (x + y)\mathbf{k}$; C the triangle with vertices (1, 0, 0), (0, 1, 0), (0, 0, 1)

6. $\mathbf{F} = z^2 y \cos xy\mathbf{i} + z^2 x(1 + \cos xy)\mathbf{j} + 2z \sin xy\mathbf{k}$; C the boundary of the plane $z = 1 - y$ shown in Figure 17.65

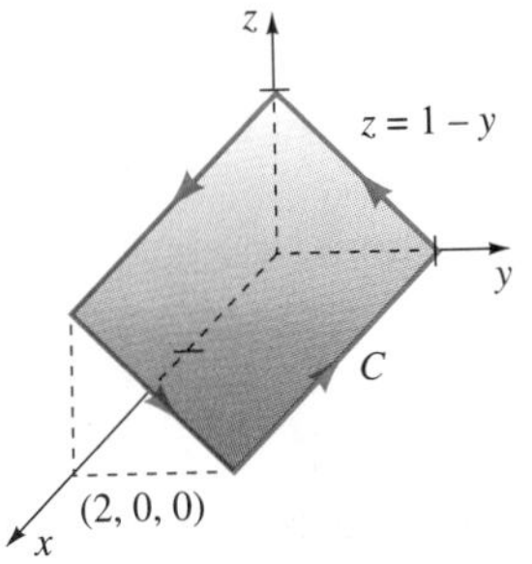

FIGURE 17.65

7. $\mathbf{F} = xy\mathbf{i} + 2yz\mathbf{j} + xz\mathbf{k}$; C the boundary given in Problem 6

8. $\mathbf{F} = (x + 2z)\mathbf{i} + (3x + y)\mathbf{j} + (2y - z)\mathbf{k}$; C the curve of intersection of the plane $x + 2y + z = 4$ with the coordinate planes

9. $\mathbf{F} = y^3\mathbf{i} - x^3\mathbf{j} + z^3\mathbf{k}$; C the trace of the cylinder $x^2 + y^2 = 1$ in the plane $x + y + z = 1$

10. $\mathbf{F} = x^2 y\mathbf{i} + (x + y^2)\mathbf{j} + xy^2 z\mathbf{k}$; C the boundary of the surface shown in Figure 17.66

11. $\mathbf{F} = x\mathbf{i} + x^3 y^2\mathbf{j} + z\mathbf{k}$; C the boundary of the semi-ellipsoid $z = \sqrt{4 - 4x^2 - y^2}$ in the plane $z = 0$

12. $\mathbf{F} = z\mathbf{i} + x\mathbf{j} + y\mathbf{k}$; C the curve of intersection of the plane $x + y + z = 0$ and the sphere $x^2 + y^2 + z^2 = 1$

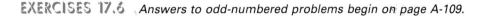

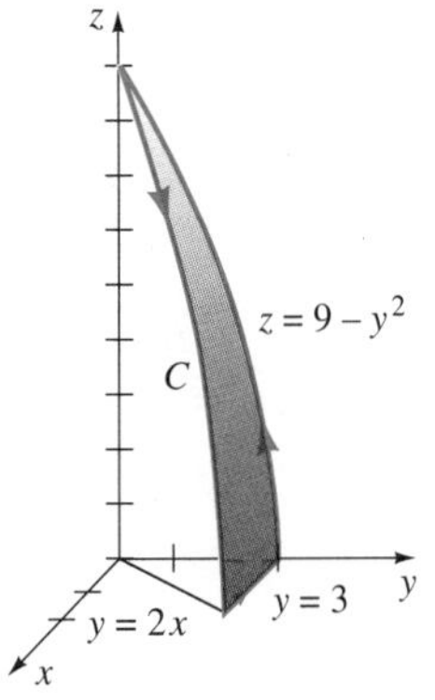

FIGURE 17.66

[*Hint:* See Section 12.4. Recall that the area of an ellipse $x^2/a^2 + y^2/b^2 = 1$ is πab.]

In Problems 13–16 use Stokes' Theorem to evaluate $\iint_S (\text{curl } \mathbf{F}) \cdot \mathbf{n}\, dS$. Assume that the surface S is oriented upward.

13. $\mathbf{F} = 6yz\mathbf{i} + 5x\mathbf{j} + yze^{x^2}\mathbf{k}$; S that portion of the paraboloid $z = \frac{1}{4}x^2 + y^2$ for $0 \leq z \leq 4$

14. $\mathbf{F} = y\mathbf{i} + (y - x)\mathbf{j} + z^2\mathbf{k}$; S that portion of the sphere $x^2 + y^2 + (z - 4)^2 = 25$ for $z \geq 0$

15. $\mathbf{F} = 3x^2\mathbf{i} + 8x^3 y\mathbf{j} + 3x^2 y\mathbf{k}$; S that portion of the plane $z = x$ that lies inside the rectangular cylinder defined by the planes $x = 0$, $y = 0$, $x = 2$, $y = 2$

16. $\mathbf{F} = 2xy^2 z\mathbf{i} + 2x^2 yz\mathbf{j} + (x^2 y^2 - 6x)\mathbf{k}$; S that portion of the plane $z = y$ that lies inside the cylinder $x^2 + y^2 = 1$

17. Use Stokes' Theorem to evaluate

$$\oint_C z^2 e^{x^2}\, dx + xy^2\, dy + \tan^{-1} y\, dz$$

where C is the circle $x^2 + y^2 = 9$ by finding a surface S with C as its boundary and such that the orientation of C is counterclockwise as viewed from above.

18. Consider the surface integral $\iint_S (\text{curl } \mathbf{F}) \cdot \mathbf{n}\, dS$, where $\mathbf{F} = xyz\mathbf{k}$ and S is that portion of the paraboloid $z = 1 - x^2 - y^2$ for $z \geq 0$ oriented upward.

(a) Evaluate the surface integral by the method of Section 17.4; that is, do not use Stokes' Theorem.

(b) Evaluate the surface integral by finding a simpler surface that is oriented upward and has the same boundary as the paraboloid.

(c) Use Stokes' Theorem to verify your result in part (**b**).

17.7 DIVERGENCE THEOREM

Another Vector Form of Green's Theorem Let $\mathbf{F} = P\mathbf{i} + Q\mathbf{j}$ and let $\mathbf{T} = (dx/ds)\mathbf{i} + (dy/ds)\mathbf{j}$ be a *unit tangent* to a simple closed plane curve C. In Section 17.6 we saw that $\oint_C (\mathbf{F} \cdot \mathbf{T})\, ds$ can be evaluated by a double integral involving curl $\mathbf{F}$. Similarly, if $\mathbf{n} = (dy/ds)\mathbf{i} - (dx/ds)\mathbf{j}$ is a *unit normal* to C (check $\mathbf{T} \cdot \mathbf{n}$), then $\oint_C (\mathbf{F} \cdot \mathbf{n})\, ds$ can be expressed in terms of a double integral of div $\mathbf{F}$. From Green's Theorem,

$$\oint_C (\mathbf{F} \cdot \mathbf{n})\, ds = \oint_C P\, dy - Q\, dx = \iint_R \left[\frac{\partial P}{\partial x} - \left(-\frac{\partial Q}{\partial y}\right)\right] dA = \iint_R \left[\frac{\partial P}{\partial x} + \frac{\partial Q}{\partial y}\right] dA$$

That is,

$$\oint_C (\mathbf{F} \cdot \mathbf{n})\, ds = \iint_R \operatorname{div} \mathbf{F}\, dA \qquad (1)$$

The result in (1) is a special case of the **Divergence** or **Gauss' Theorem**.* The following is a generalization of (1) to 3-space:

THEOREM 17.7 Divergence Theorem

Let D be a closed and bounded region in 3-space with a piecewise smooth boundary S that is oriented outward. Let $\mathbf{F}(x, y, z) = P(x, y, z)\mathbf{i} + Q(x, y, z)\mathbf{j} + R(x, y, z)\mathbf{k}$ be a vector field for which P, Q, and R are continuous and have continuous first partial derivatives in a region of 3-space containing D. Then

$$\iint_S (\mathbf{F} \cdot \mathbf{n})\, dS = \iiint_D \operatorname{div} \mathbf{F}\, dV \qquad (2)$$

Partial Proof We shall prove (2) for the special region D shown in Figure 17.67 whose surface S consists of three pieces:

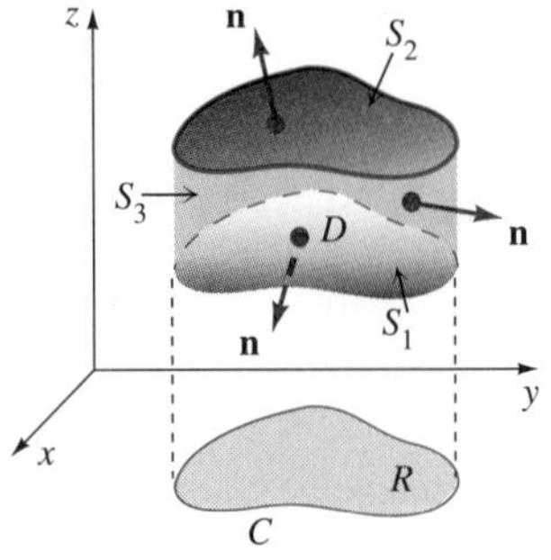

FIGURE 17.67

$$\begin{aligned} &(\text{bottom})\ S_1\colon && z = f_1(x, y), && (x, y) \text{ in } R \\ &(\text{top})\ S_2\colon && z = f_2(x, y), && (x, y) \text{ in } R \\ &(\text{side})\ S_3\colon && f_1(x, y) \le z \le f_2(x, y), && (x, y) \text{ on } C \end{aligned}$$

where R is the projection of D onto the xy-plane and C is the boundary of R. Since

$$\operatorname{div} \mathbf{F} = \frac{\partial P}{\partial x} + \frac{\partial Q}{\partial y} + \frac{\partial R}{\partial z} \quad \text{and} \quad \mathbf{F} \cdot \mathbf{n} = P(\mathbf{i} \cdot \mathbf{n}) + Q(\mathbf{j} \cdot \mathbf{n}) + R(\mathbf{k} \cdot \mathbf{n})$$

***Karl Friedrich Gauss (1777–1855)** Gauss was the first of a new breed of precise and demanding mathematicians—the "rigorists." We have seen in earlier biographical sketches that Augustin Louis Cauchy and Karl Wilhelm Weierstrass were two mathematicians who followed in his footsteps. Gauss was a child prodigy in mathematics. As an adult, he often remarked that he could calculate or "reckon" before he could talk. However, as a college student, Gauss was torn between two loves: philology and mathematics. He was inspired by some original mathematical achievements as a teenager and encouraged by the mathematician Wolfgang Bolyai, so the choice was not too difficult. At the age of 20, Gauss settled on a career in mathematics. At the age of 22, he completed a book on number theory, *Disquisitiones Arithmeticae.* Published in 1801, this text was recognized as a masterpiece and even today remains a classic in its field. Gauss' doctoral dissertation of 1799 is also a memorable document. Using the theory of functions of a complex variable, he was the first to prove the so-called fundamental theorem of algebra: Every polynomial equation has at least one root.

Although Gauss was certainly recognized and respected as an outstanding mathematician during his lifetime, the full extent of his genius was not realized until the publication of his scientific diary in 1898, 44 years after his death. Much to the chagrin of some nineteenth-century mathematicians, the diary revealed that Gauss had foreseen, sometimes by decades, many of their discoveries or, perhaps more accurately, rediscoveries. He was oblivious to fame; his mathematical researches were often pursued, like a child playing on a beach, simply for pleasure and self-satisfaction and not for the instruction that could be given to others through publication.

On any list of "Greatest Mathematicians Who Ever Lived," Karl Friedrich Gauss must surely rank near or at the top. For his profound impact on so many branches of mathematics, Gauss is sometimes referred to as "the prince of mathematicians."

we can write

$$\iint_S (\mathbf{F}\cdot\mathbf{n})\,dS = \iint_S P(\mathbf{i}\cdot\mathbf{n})\,dS + \iint_S Q(\mathbf{j}\cdot\mathbf{n})\,dS + \iint_S R(\mathbf{k}\cdot\mathbf{n})\,dS$$

and $$\iiint_D \operatorname{div}\mathbf{F}\,dV = \iiint_D \frac{\partial P}{\partial x}\,dV + \iiint_D \frac{\partial Q}{\partial y}\,dV + \iiint_D \frac{\partial R}{\partial z}\,dV$$

To prove (2) we need only establish that

$$\iint_S P(\mathbf{i}\cdot\mathbf{n})\,dS = \iiint_D \frac{\partial P}{\partial x}\,dV \tag{3}$$

$$\iint_S Q(\mathbf{j}\cdot\mathbf{n})\,dS = \iiint_D \frac{\partial Q}{\partial y}\,dV \tag{4}$$

$$\iint_S R(\mathbf{k}\cdot\mathbf{n})\,dS = \iiint_D \frac{\partial R}{\partial z}\,dV \tag{5}$$

Indeed, we shall only prove (5) because the proofs of (3) and (4) follow in a similar manner. Now,

$$\iiint_D \frac{\partial R}{\partial z}\,dV = \iint_R \left[\int_{f_1(x,y)}^{f_2(x,y)} \frac{\partial R}{\partial z}\,dz\right] dA = \iint_R [R(x, y, f_2(x, y)) - R(x, y, f_1(x, y))]\,dA \tag{6}$$

Next we write

$$\iint_S R(\mathbf{k}\cdot\mathbf{n})\,dS = \iint_{S_1} R(\mathbf{k}\cdot\mathbf{n})\,dS + \iint_{S_2} R(\mathbf{k}\cdot\mathbf{n})\,dS + \iint_{S_3} R(\mathbf{k}\cdot\mathbf{n})\,dS$$

On S_1: Since the outward normal points downward, we describe the surface as $g(x, y, z) = f_1(x, y) - z = 0$. Thus,

$$\mathbf{n} = \frac{\dfrac{\partial f_1}{\partial x}\mathbf{i} + \dfrac{\partial f_1}{\partial y}\mathbf{j} - \mathbf{k}}{\sqrt{1 + \left(\dfrac{\partial f_1}{\partial x}\right)^2 + \left(\dfrac{\partial f_1}{\partial y}\right)^2}} \quad \text{so that} \quad \mathbf{k}\cdot\mathbf{n} = \frac{-1}{\sqrt{1 + \left(\dfrac{\partial f_1}{\partial x}\right)^2 + \left(\dfrac{\partial f_1}{\partial y}\right)^2}}$$

From the definition of dS we then have

$$\iint_{S_1} R(\mathbf{k}\cdot\mathbf{n})\,dS = -\iint_R R(x, y, f_1(x, y))\,dA \tag{7}$$

On S_2: The outward normal points upward so

$$\mathbf{n} = \frac{-\dfrac{\partial f_2}{\partial x}\mathbf{i} - \dfrac{\partial f_2}{\partial y}\mathbf{j} + \mathbf{k}}{\sqrt{1 + \left(\dfrac{\partial f_2}{\partial x}\right)^2 + \left(\dfrac{\partial f_2}{\partial y}\right)^2}} \quad \text{so that} \quad \mathbf{k}\cdot\mathbf{n} = \frac{1}{\sqrt{1 + \left(\dfrac{\partial f_2}{\partial x}\right)^2 + \left(\dfrac{\partial f_2}{\partial y}\right)^2}}$$

from which we find

$$\iint_{S_2} R(\mathbf{k}\cdot\mathbf{n})\,dS = \iint_R R(x, y, f_2(x, y))\,dA \tag{8}$$

On S_3: Since this side is vertical, $\mathbf{k}$ is perpendicular to $\mathbf{n}$. Consequently, $\mathbf{k}\cdot\mathbf{n} = 0$ and

$$\iint_{S_3} R(\mathbf{k}\cdot\mathbf{n})\,dS = 0 \tag{9}$$

Finally, adding (7), (8), and (9), we get

$$\iint_R [R(x, y, f_2(x, y)) - R(x, y, f_1(x, y))]\,dA$$

which is the same as (6). ■

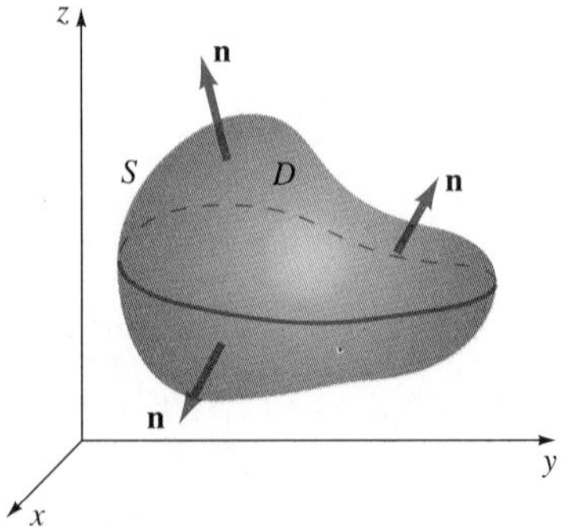

FIGURE 17.68

Although we proved (2) for a special region D that has a vertical side, we note that this type of region is not required in Theorem 17.7. A region D with no vertical side is illustrated in Figure 17.68; a region bounded by a sphere or an ellipsoid also does not have a vertical side. The Divergence Theorem also

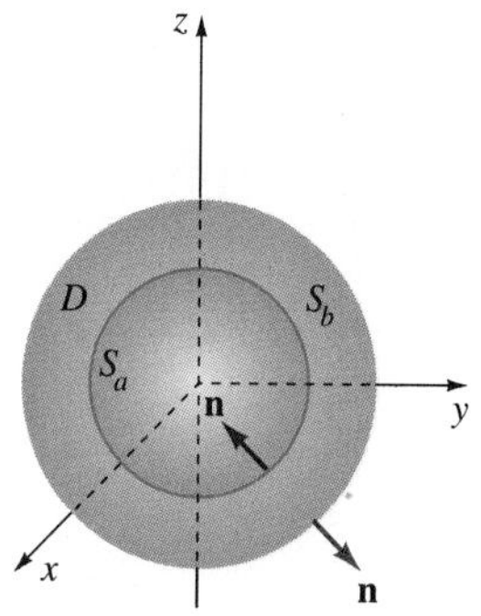

FIGURE 17.69

holds for the region D bounded between two closed surfaces such as the concentric spheres S_a and S_b shown in Figure 17.69; the boundary surface S of D is the union of S_a and S_b. In this case, $\iint_S (\mathbf{F} \cdot \mathbf{n})\, dS = \iiint_D \operatorname{div} \mathbf{F}\, dV$ becomes

$$\iint_{S_b} (\mathbf{F} \cdot \mathbf{n})\, dS + \iint_{S_a} (\mathbf{F} \cdot \mathbf{n})\, dS = \iiint_D \operatorname{div} \mathbf{F}\, dV$$

where $\mathbf{n}$ points outward from D; that is, on S_b, $\mathbf{n}$ points away from the origin and on S_a, $\mathbf{n}$ points toward the origin.

EXAMPLE 1 Let D be the closed region bounded by the hemisphere $x^2 + y^2 + (z - 1)^2 = 9$, $1 \le z \le 4$, and the plane $z = 1$. Verify the Divergence Theorem if $\mathbf{F} = x\mathbf{i} + y\mathbf{j} + (z - 1)\mathbf{k}$.

Solution The closed region is shown in Figure 17.70.

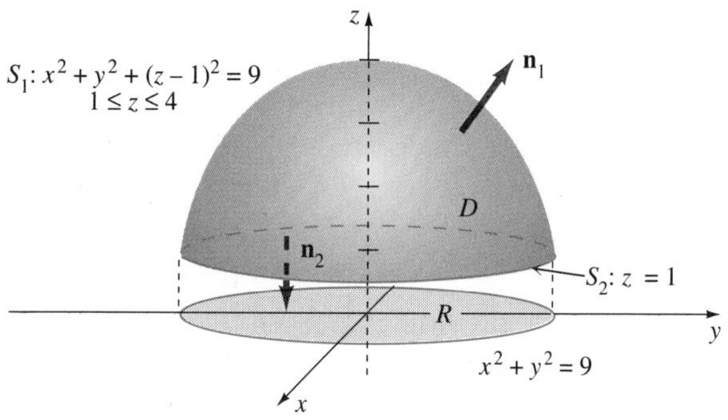

FIGURE 17.70

The Triple Integral: Since $\mathbf{F} = x\mathbf{i} + y\mathbf{j} + (z - 1)\mathbf{k}$, we see $\operatorname{div} \mathbf{F} = 3$. Hence,

$$\iiint_D \operatorname{div} \mathbf{F}\, dV = \iiint_D 3\, dV = 3 \iiint_D dV = 3\left[\frac{2}{3}\pi 3^3\right] = 54\pi \tag{10}$$

In the last calculation, we used the fact that $\iiint_D dV$ gives the volume of the hemisphere.

The Surface Integral: We write $\iint_S = \iint_{S_1} + \iint_{S_2}$, where S_1 is the hemisphere and S_2 is the plane $z = 1$. If S_1 is a level surface of $g(x, y, z) = x^2 + y^2 + (z - 1)^2$, then a unit outer normal is

$$\mathbf{n} = \frac{\nabla g}{|\nabla g|} = \frac{x\mathbf{i} + y\mathbf{j} + (z - 1)\mathbf{k}}{\sqrt{x^2 + y^2 + (z - 1)^2}} = \frac{x}{3}\mathbf{i} + \frac{y}{3}\mathbf{j} + \frac{z - 1}{3}\mathbf{k}$$

Now,
$$\mathbf{F} \cdot \mathbf{n} = \frac{x^2}{3} + \frac{y^2}{3} + \frac{(z - 1)^2}{3} = 3$$

and so
$$\iint_{S_1} (\mathbf{F}\cdot\mathbf{n})\, dS = \iint_R (3)\left(\frac{3}{\sqrt{9-x^2-y^2}}\, dA\right)$$

polar coordinates

$$= 9\int_0^{2\pi}\int_0^3 (9-r^2)^{-1/2} r\, dr\, d\theta = 54\pi$$

On S_2, we take $\mathbf{n} = -\mathbf{k}$ so that $\mathbf{F}\cdot\mathbf{n} = -z + 1$

But, since $z = 1$, $\iint_{S_2} (-z+1)\, dS = 0$. Hence, we see that

$$\iint_S (\mathbf{F}\cdot\mathbf{n})\, dS = 54\pi + 0 = 54\pi$$

agrees with (10). □

EXAMPLE 2 If $\mathbf{F} = xy\mathbf{i} + y^2z\mathbf{j} + z^3\mathbf{k}$, evaluate $\iint_S (\mathbf{F}\cdot\mathbf{n})\, dS$, where S is the unit cube defined by $0 \le x \le 1, 0 \le y \le 1, 0 \le z \le 1$.

Solution See Figure 17.52 and Problem 24 of Exercises 17.4. Rather than evaluate six surface integrals, we apply the Divergence Theorem. Since div $\mathbf{F}$ $= \nabla\cdot\mathbf{F} = y + 2yz + 3z^2$, we have from (2),

$$\iint_S (\mathbf{F}\cdot\mathbf{n})\, dS = \iiint_D (y + 2yz + 3z^2)\, dV$$
$$= \int_0^1\int_0^1\int_0^1 (y + 2yz + 3z^2)\, dx\, dy\, dz$$
$$= \int_0^1\int_0^1 (y + 2yz + 3z^2)\, dy\, dz$$
$$= \int_0^1 \left(\frac{y^2}{2} + y^2z + 3yz^2\right)\Bigg]_0^1 dz$$
$$= \int_0^1 \left(\frac{1}{2} + z + 3z^2\right) dz = \left(\frac{1}{2}z + \frac{z^2}{2} + z^3\right)\Bigg]_0^1 = 2$$ □

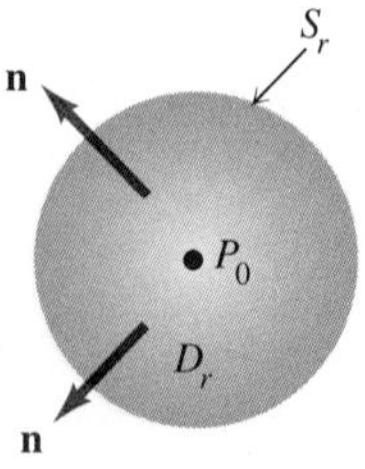

FIGURE 17.71

Physical Interpretation of Divergence In Section 17.6 we saw that we could express the normal component of the curl of a vector field $\mathbf{F}$ at a point as a limit involving the circulation of $\mathbf{F}$. In view of (2) it possible to interpret the divergence of $\mathbf{F}$ at a point as a limit involving the flux of $\mathbf{F}$. Recall from Section 17.4 that the flux of the velocity field $\mathbf{F}$ of a fluid is the rate of fluid flow—that is, the volume of fluid flowing through a surface per unit time. In Section 17.5 we saw that the divergence of $\mathbf{F}$ is the flux per unit volume. To reinforce this last idea, let us suppose $P_0(x_0, y_0, z_0)$ is any point in the fluid and S_r is a small sphere of radius r centered at P_0. See Figure 17.71. If D_r is the sphere S_r and its interior, then the Divergence Theorem gives

$$\iint_{S_r} (\mathbf{F} \cdot \mathbf{n})\, dS = \iiint_{D_r} \operatorname{div} \mathbf{F}\, dV \tag{11}$$

If we take the approximation div $\mathbf{F}(P) \approx$ div $\mathbf{F}(P_0)$ at all points $P(x, y, z)$ within the small sphere, then (11) gives

$$\begin{aligned} \iint_{S_r} (\mathbf{F} \cdot \mathbf{n})\, dS &\approx \iiint_{D_r} \operatorname{div} \mathbf{F}(P_0)\, dV \\ &= \operatorname{div} \mathbf{F}(P_0) \iiint_{D_r} dV \\ &= \operatorname{div} \mathbf{F}(P_0) V_r \end{aligned} \tag{12}$$

where V_r is the volume $(\frac{4}{3}\pi r^3)$ of the spherical region D_r. By letting $r \to 0$, we see from (12) that the divergence of $\mathbf{F}$ is the limiting value of the ratio of the flux of $\mathbf{F}$ to the volume of the spherical region:

$$\operatorname{div} \mathbf{F}(P_0) = \lim_{r \to 0} \frac{1}{V_r} \iint_{S_r} (\mathbf{F} \cdot \mathbf{n})\, dS$$

Hence, the divergence of $\mathbf{F}$ is flux per unit volume.

The Divergence Theorem is extremely useful in the derivation of some of the famous equations in electricity and magnetism and hydrodynamics. In the discussion that follows we shall consider an example from the study of fluids.

Continuity Equation At the end of Section 17.5 we mentioned that one interpretation of div $\mathbf{F}$ is a measure of the rate of change of the density of a fluid at a point. To see why this is so, let us suppose that $\mathbf{F}$ is a velocity field of a fluid and that $\rho(x, y, z, t)$ is the density of the fluid at a point $P(x, y, z)$ at a time t. Let D be the closed region consisting of a sphere S and its interior. We know from Section 16.7 that the total mass m of the fluid in D is given by

$$m = \iiint_D \rho(x, y, z, t)\, dV$$

The rate at which the mass increases in D is given by

$$\frac{dm}{dt} = \frac{d}{dt} \iiint_D \rho(x, y, z, t)\, dV = \iiint_D \frac{\partial \rho}{\partial t}\, dV \tag{13}$$

Now from Figure 17.47(b) we saw that the volume of fluid flowing through an element of surface area ΔS per unit time is approximated by

$$(\mathbf{F} \cdot \mathbf{n})\, \Delta S$$

The mass of the fluid flowing through an element of surface area ΔS per unit time is then

$$(\rho\mathbf{F}\cdot\mathbf{n})\,\Delta S$$

If we assume that the change in mass in D is due only to the flow in and out of D, then the *volume of fluid* flowing out of D per unit time is given by (6) of Section 17.4, $\iint_S (\mathbf{F}\cdot\mathbf{n})\,dS$, whereas the *mass of the fluid* flowing out of D per unit time is $\iint_S (\rho\mathbf{F}\cdot\mathbf{n})\,dS$. Hence, an alternative expression for the rate at which mass increases in D is

$$-\iint_S (\rho\mathbf{F}\cdot\mathbf{n})\,dS \tag{14}$$

By the Divergence Theorem, (14) is the same as

$$-\iiint_D \operatorname{div}(\rho\mathbf{F})\,dV \tag{15}$$

Equating (13) and (15) then yields

$$\iiint_D \frac{\partial\rho}{\partial t}\,dV = -\iiint_D \operatorname{div}(\rho\mathbf{F})\,dV \quad \text{or} \quad \iiint_D \left(\frac{\partial\rho}{\partial t} + \operatorname{div}(\rho\mathbf{F})\right) dV = 0$$

Since this last result is to hold for every sphere, we obtain **the equation of continuity** for fluid flows:

$$\frac{\partial\rho}{\partial t} + \operatorname{div}(\rho\mathbf{F}) = 0 \tag{16}$$

On page 979 we stated that if $\operatorname{div}\mathbf{F} = \nabla\cdot\mathbf{F} = 0$, then a fluid is incompressible. This fact follows immediately from (16). If a fluid is incompressible (such as water), then ρ is constant, so consequently $\nabla\cdot(\rho\mathbf{F}) = \rho\,\nabla\cdot\mathbf{F}$. But in addition $\partial\rho/\partial t = 0$, and so (16) implies $\nabla\cdot\mathbf{F} = 0$.

EXERCISES 17.7

Answers to odd-numbered problems begin on page A-110.

In Problems 1 and 2 verify the Divergence Theorem.

1. $\mathbf{F} = xy\mathbf{i} + yz\mathbf{j} + xz\mathbf{k}$; D the region bounded by the unit cube defined by $0 \le x \le 1$, $0 \le y \le 1$, $0 \le z \le 1$

2. $\mathbf{F} = 6xy\mathbf{i} + 4yz\mathbf{j} + xe^{-y}\mathbf{k}$; D the region bounded by the three coordinate planes and the plane $x + y + z = 1$

In Problems 3–14 use the Divergence Theorem to find outward flux $\iint_S (\mathbf{F}\cdot\mathbf{n})\,dS$ of the given vector field $\mathbf{F}$.

3. $\mathbf{F} = x^3\mathbf{i} + y^3\mathbf{j} + z^3\mathbf{k}$; D the region bounded by the sphere $x^2 + y^2 + z^2 = a^2$

4. $\mathbf{F} = 4x\mathbf{i} + y\mathbf{j} + 4z\mathbf{k}$; D the region bounded by the sphere $x^2 + y^2 + z^2 = 4$

5. $\mathbf{F} = y^2\mathbf{i} + xz^3\mathbf{j} + (z-1)^2\mathbf{k}$; D the region bounded by the cylinder $x^2 + y^2 = 16$ and the planes $z = 1$, $z = 5$

6. $\mathbf{F} = x^2\mathbf{i} + 2yz\mathbf{j} + 4z^3\mathbf{k}$; D the region bounded by the parallelepiped defined by $0 \le x \le 1$, $0 \le y \le 2$, $0 \le z \le 3$

7. $\mathbf{F} = y^3\mathbf{i} + x^3\mathbf{j} + z^3\mathbf{k}$; D the region bounded within by $z = \sqrt{4 - x^2 - y^2}$, $x^2 + y^2 = 3$, $z = 0$

8. $\mathbf{F} = (x^2 + \sin y)\mathbf{i} + z^2\mathbf{j} + xy^3\mathbf{k}$; D the region bounded by $y = x^2$, $z = 9 - y$, $z = 0$

9. $\mathbf{F} = (x\mathbf{i} + y\mathbf{j} + z\mathbf{k})/(x^2 + y^2 + z^2)$; D the region bounded by the concentric spheres $x^2 + y^2 + z^2 = a^2$, $x^2 + y^2 + z^2 = b^2$, $b > a$

10. $\mathbf{F} = 2yz\mathbf{i} + x^3\mathbf{j} + xy^2\mathbf{k}$; D the region bounded by the ellipsoid $x^2/a^2 + y^2/b^2 + z^2/c^2 = 1$

11. $\mathbf{F} = 2xz\mathbf{i} + 5y^2\mathbf{j} - z^2\mathbf{k}$; D the region bounded by $z = y$, $z = 4 - y$, $z = 2 - \frac{1}{2}x^2$, $x = 0$, $z = 0$ (See Figure 17.72.)

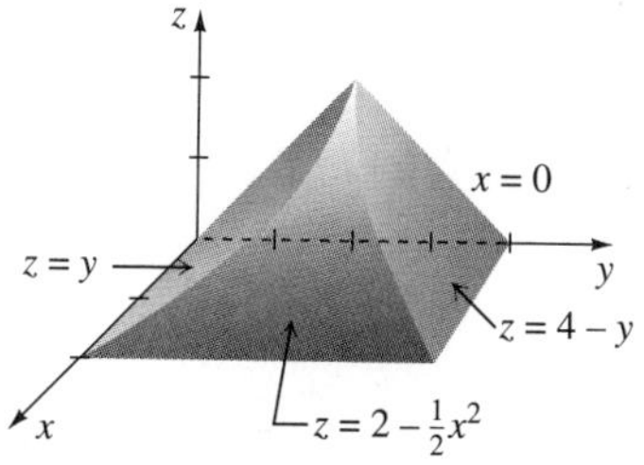

FIGURE 17.72

12. $\mathbf{F} = 15x^2y\mathbf{i} + x^2z\mathbf{j} + y^4\mathbf{k}$; D the region bounded by $x + y = 2$, $z = x + y$, $z = 3$, $x = 0$, $y = 0$

13. $\mathbf{F} = 3x^2y^2\mathbf{i} + y\mathbf{j} - 6zxy^2\mathbf{k}$; D the region bounded by the paraboloid $z = x^2 + y^2$ and the plane $z = 2y$

14. $\mathbf{F} = xy^2\mathbf{i} + x^2y\mathbf{j} + 6\sin xz\mathbf{k}$; D the region bounded by the cone $z = \sqrt{x^2 + y^2}$ and the planes $z = 2$, $z = 4$

15. The electric field at a point $P(x, y, z)$ due to a point charge q located at the origin is given by the inverse square field

$$\mathbf{E} = q\frac{\mathbf{r}}{|\mathbf{r}|^3}$$

where $\mathbf{r} = x\mathbf{i} + y\mathbf{j} + z\mathbf{k}$.

(a) Suppose S is a closed surface, S_a is a sphere $x^2 + y^2 + z^2 = a^2$ lying completely within S, and D is the region bounded between S and S_a. See Figure 17.73.

Show that the outward flux of $\mathbf{E}$ for the region D is zero.

(b) Use the result of part (**a**) to prove **Gauss' Law**:

$$\iint_S (\mathbf{E}\cdot\mathbf{n})\, dS = 4\pi q$$

That is, the outward flux of the electric field $\mathbf{E}$ through *any* closed surface (for which the Divergence Theorem applies) containing the origin is $4\pi q$.

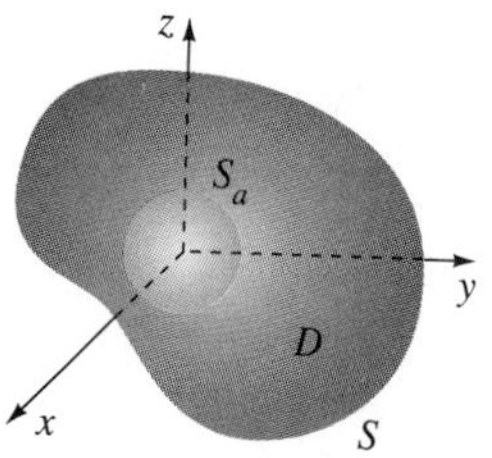

FIGURE 17.73

16. Suppose there is a continuous distribution of charge throughout a closed and bounded region D enclosed by a surface S. Then the natural extension of Gauss' Law is given by

$$\iint_S (\mathbf{E}\cdot\mathbf{n})\, dS = \iiint_D 4\pi q\, dV$$

where $\rho(x, y, z)$ is the charge density or charge per unit volume.

(a) Proceed as in the derivation of the continuity equation (16) to show that div $\mathbf{E} = 4\pi q$.

(b) Given that $\mathbf{E}$ is an irrotational vector field, show that the potential ϕ for $\mathbf{E}$ satisfies Poisson's equation $\nabla^2\phi = 4\pi\rho$.

In Problems 17–21 assume that S forms the boundary of a closed and bounded region D.

17. If $\mathbf{a}$ is a constant vector, show that $\iint_S (\mathbf{a}\cdot\mathbf{n})\, dS = 0$.

18. If $\mathbf{F} = P\mathbf{i} + Q\mathbf{j} + R\mathbf{k}$ and P, Q, and R have continuous second partial derivatives, show that $\iint_S (\text{curl}\,\mathbf{F}\cdot\mathbf{n})\, dS = 0$.

In Problems 19 and 20 assume that f and g are scalar functions with continuous second partial derivatives. Use the Divergence Theorem to establish **Green's identities**.

19. $$\iint_S (f\,\nabla g)\cdot\mathbf{n}\, dS = \iiint_D (f\,\nabla^2 g + \nabla f\cdot\nabla g)\, dV$$

20. $$\iint_S (f\,\nabla g - g\nabla f)\cdot\mathbf{n}\, dS = \iiint_D (f\,\nabla^2 g - g\,\nabla^2 f)\, dV$$

21. If f is a scalar function with continuous first partial derivatives, show that

$$\iint_S f\mathbf{n}\, dS = \iiint_D \nabla f\, dV$$

[*Hint:* Use (2) on $f\mathbf{a}$, where $\mathbf{a}$ is a constant vector, and Problem 21 in Exercises 17.5.]

22. The buoyant force on a floating object is $\mathbf{B} = -\iint_S p\mathbf{n}\, dS$, where p is the fluid pressure. The pressure p is related to the density of the fluid $\rho(x, y, z)$ by a law of hydrostatics: $\nabla p = \rho(x, y, z)\mathbf{g}$, where $\mathbf{g}$ is the constant acceleration of gravity. If the weight of the object is $\mathbf{W} = m\mathbf{g}$, use the result of Problem 21 to prove Archimedes principle, $\mathbf{B} + \mathbf{W} = \mathbf{0}$. See Figure 17.74.

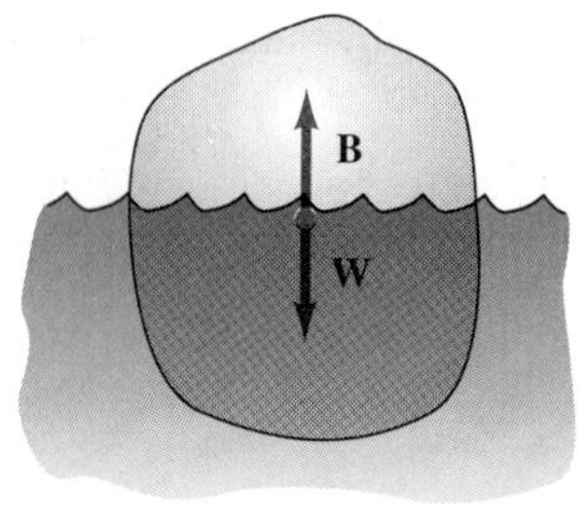

FIGURE 17.74

CHAPTER 17 REVIEW EXERCISES

Answers begin on page A-110.

In Problems 1–12 answer true or false. Where appropriate, assume continuity of P, Q, and their first partial derivatives.

1. The integral $\int_C (x^2 + y^2)\, dx + 2xy\, dy$, where C is given by $y = x^3$ from (0, 0) to (1, 1) has the same value on the curve $y = x^6$ from (0, 0) to (1, 1). ________

2. The value of the integral $\int_C 2xy\, dx - x^2\, dy$ between two points A and B depends on the path C. ________

3. If C_1 and C_2 are two smooth curves such that $\int_{C_1} P\, dx + Q\, dy = \int_{C_2} P\, dx + Q\, dy$, then $\int_C P\, dx + Q\, dy$ is independent of the path. ________

4. If the work $\int_C \mathbf{F} \cdot d\mathbf{r}$ depends on the curve C, then $\mathbf{F}$ is nonconservative. ________

5. If $\partial P/\partial x = \partial Q/\partial y$, then $\int_C P\, dx + Q\, dy$ is independent of the path. ________

6. In a conservative force field $\mathbf{F}$, the work done by $\mathbf{F}$ around a simple closed curve is zero. ________

7. Assuming continuity of all partial derivatives, $\nabla \times \nabla f = \mathbf{0}$. ________

8. The surface integral of the normal component of the curl of a conservative vector field $\mathbf{F}$ over a surface S is equal to zero. ________

9. The work done by a force $\mathbf{F}$ along a curve C is due entirely to the tangential component of $\mathbf{F}$. ________

10. For a two-dimensional vector field $\mathbf{F}$ in the plane $z = 0$, Stokes' Theorem is the same as Green's Theorem. ________

11. If $\mathbf{F}$ is a conservative force field, then the sum of the potential and kinetic energies of an object is constant. ________

12. If $\int_C P\, dx + Q\, dy$ is independent of the path, then $\mathbf{F} = P\mathbf{i} + Q\mathbf{j}$ is the gradient of some function ϕ. ________

In Problems 13–20 fill in the blanks.

13. If $\phi = \dfrac{1}{\sqrt{x^2 + y^2}}$ is a potential function for a conservative force field $\mathbf{F}$, then $\mathbf{F} =$ ________.

14. If $\mathbf{F} = f(x)\mathbf{i} + g(y)\mathbf{j} + h(z)\mathbf{k}$, then curl $\mathbf{F} =$ ________.

In Problems 15–18 $\mathbf{F} = x^2y\mathbf{i} + xy^2\mathbf{j} + 2xyz\mathbf{k}$.

15. $\nabla \cdot \mathbf{F} =$ ________

16. $\nabla \times \mathbf{F} =$ ________

17. $\nabla \cdot (\nabla \times \mathbf{F}) =$ ________

18. $\nabla(\nabla \cdot \mathbf{F}) =$ ________

19. If C is the ellipse $2(x - 10)^2 + 9(y + 13)^2 = 3$, then $\oint_C (y - 7e^{x^3})\, dx + (x + \ln\sqrt{y})\, dy =$ ________.

20. If $\mathbf{F}$ is a velocity field of a fluid for which curl $\mathbf{F} = \mathbf{0}$, then $\mathbf{F}$ is said to be ________.

21. Evaluate $\displaystyle\int_C \frac{z^2}{x^2 + y^2}\, ds$, where C is given by

$$x = \cos 2t, \quad y = \sin 2t, \quad z = 2t, \qquad \pi \le t \le 2\pi$$

22. Evaluate $\int_C (xy + 4x)\, ds$, where C is given by $2x + y = 2$ from (1, 0) to (0, 2).

23. Evaluate $\int_C 3x^2y^2\, dx + (2x^3y - 3y^2)\, dy$, where C is given by $y = 5x^4 + 7x^2 - 14x$ from (0, 0) to (1, −2).

24. Evaluate $\oint_C (x^2 + y^2)\, dx + (x^2 - y^2)\, dy$, where C is the circle $x^2 + y^2 = 9$.

25. Evaluate $\int_C y \sin \pi z\, dx + x^2e^y\, dy + 3xyz\, dz$, where C is given by $x = t$, $y = t^2$, $z = t^3$ from (0, 0, 0) to (1, 1, 1).

26. If $\mathbf{F} = 4y\mathbf{i} + 6x\mathbf{j}$ and C is given by $x^2 + y^2 = 1$, evaluate $\oint_C \mathbf{F} \cdot d\mathbf{r}$ in two different ways.

27. Find the work done by the force $\mathbf{F} = x \sin y\mathbf{i} + y \sin x\mathbf{j}$ acting along the line segments from $(0, 0)$ to $(\pi/2, 0)$ and from $(\pi/2, 0)$ to $(\pi/2, \pi)$.

28. Find the work done by $\mathbf{F} = \dfrac{2}{x^2 + y^2}\mathbf{i} + \dfrac{1}{x^2 + y^2}\mathbf{j}$ from $(-\frac{1}{2}, \frac{1}{2})$ to $(1, \sqrt{3})$ acting on the path shown in Figure 17.75.

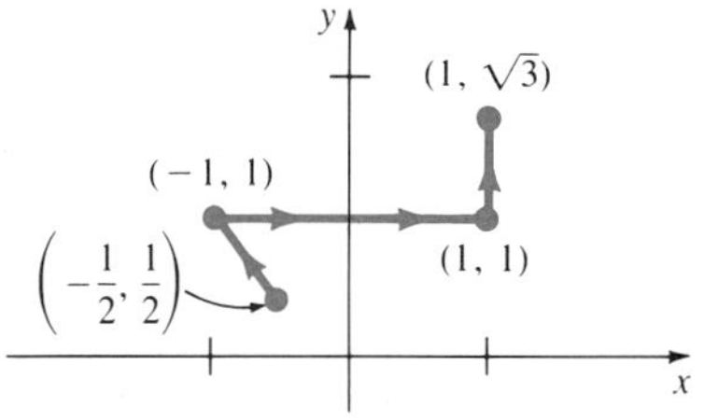

FIGURE 17.75

In Problems 29 and 30 show that the given integral is independent of the path. Evaluate.

29. $\displaystyle\int_{(1,1,0)}^{(1,1,\pi)} 2xy\,dx + (x^2 + 2yz)\,dy + (y^2 + 4)\,dz$

30. $\displaystyle\int_{(0,0,1)}^{(3,2,0)} (2x + 2ze^{2x})\,dx + (2y - 1)\,dy + e^{2x}\,dz$

31. Evaluate $\oint_C -4y\,dx + 8x\,dy$, where $C = C_1 \cup C_2$ is the boundary of the region R shown in Figure 17.76.

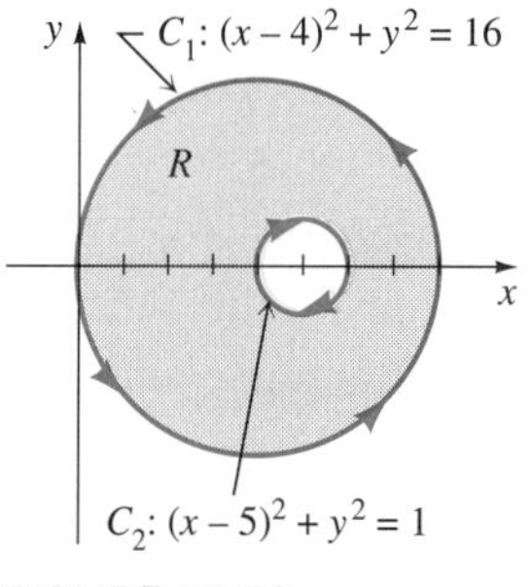

FIGURE 17.76

32. Let C be a piecewise smooth simple closed curve. Show that

$$\oint_C \frac{y-1}{(x-1)^2 + (y-1)^2}\,dx + \frac{1-x}{(x-1)^2 + (y-1)^2}\,dy = \begin{cases} -2\pi, & \text{if } (1, 1) \text{ is inside } C \\ 0, & \text{if } (1, 1) \text{ is outside } C \end{cases}$$

33. Evaluate $\iint_S (z/xy)\,dS$, where S is that portion of the cylinder $z = x^2$ in the first octant that is bounded by $y = 1$, $y = 3$, $z = 1$, $z = 4$.

34. If $\mathbf{F} = \mathbf{i} + 2\mathbf{j} + 3\mathbf{k}$, find the flux of $\mathbf{F}$ through the square $0 \le x \le 1, 0 \le y \le 1, z = 2$.

35. Let the surface S be that portion of the cylinder $y = 2 - e^{-x}$ whose projection onto the xz-plane is the rectangular region R defined by $0 \le x \le 3$, $0 \le z \le 2$. See Figure 17.77(a). Find the flux of $\mathbf{F} = 4\mathbf{i} + (2 - y)\mathbf{j} + 9\mathbf{k}$ through the surface if S is oriented away from the xz-plane.

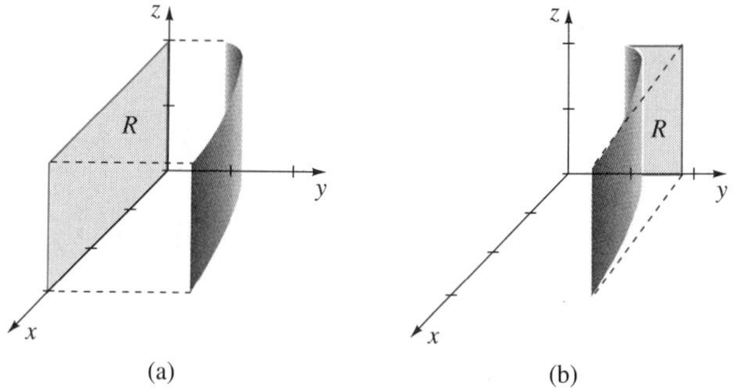

FIGURE 17.77

36. Rework Problem 35 using the region R in the yz-plane that corresponds to $0 \le x \le 3, 0 \le z \le 2$. See Figure 17.77(b).

37. If $\mathbf{F} = c\nabla(1/r)$, where c is constant and $|\mathbf{r}| = r$, $\mathbf{r} = x\mathbf{i} + y\mathbf{j} + z\mathbf{k}$, find the flux of $\mathbf{F}$ through the sphere $x^2 + y^2 + z^2 = a^2$.

38. Explain why the Divergence Theorem is not applicable in Problem 37.

39. Find the flux of $\mathbf{F} = c\nabla(1/r)$ through any surface S that forms the boundary of a closed bounded region of space not containing the origin.

40. If $\mathbf{F} = 6x\mathbf{i} + 7z\mathbf{j} + 8y\mathbf{k}$, use Stokes' Theorem to evaluate $\iint_S (\text{curl }\mathbf{F} \cdot \mathbf{n})\,dS$, where S is that portion of the paraboloid $z = 9 - x^2 - y^2$ within the cylinder $x^2 + y^2 = 4$.

41. Use Stokes' Theorem to evaluate $\oint_C -2y\,dx + 3x\,dy + 10z\,dz$, where C is the circle $(x - 1)^2 + (y - 3)^2 = 25$, $z = 3$.

42. Find the work $\oint_C \mathbf{F} \cdot d\mathbf{r}$ done by the force $\mathbf{F} = x^2\mathbf{i} + y^2\mathbf{j} + z^2\mathbf{k}$ around the curve C that is formed by the intersection of the plane $z = 2 - y$ and the sphere $x^2 + y^2 + z^2 = 4z$.

43. If $\mathbf{F} = x\mathbf{i} + y\mathbf{j} + z\mathbf{k}$, use the Divergence Theorem to evaluate $\iint_S (\mathbf{F} \cdot \mathbf{n})\,dS$, where S is the surface of the region bounded by $x^2 + y^2 = 1$, $z = 0$, $z = 1$.

44. Repeat Problem 43 for $\mathbf{F} = \frac{1}{3}x^3\mathbf{i} + \frac{1}{3}y^3\mathbf{j} + \frac{1}{3}z^3\mathbf{k}$.

45. If $\mathbf{F} = (x^2 - e^y \tan^{-1} z)\mathbf{i} + (x + y)^2\mathbf{j} - (2yz + x^{10})\mathbf{k}$, use the Divergence Theorem to evaluate $\iint_S (\mathbf{F} \cdot \mathbf{n})\,dS$, where S is the surface of the region in the first octant bounded by $z = 1 - x^2$, $z = 0$, $z = 2 - y$, $y = 0$.

46. Suppose $\mathbf{F} = x\mathbf{i} + y\mathbf{j} + (z^2 + 1)\mathbf{k}$ and S is the surface of the region bounded by $x^2 + y^2 = a^2$, $z = 0$, $z = c$. Evaluate $\iint_S (\mathbf{F} \cdot \mathbf{n})\,dS$ without the aid of the Divergence Theorem. [*Hint:* The lateral surface area of the cylinder is $2\pi ac$.]

18

DIFFERENTIAL EQUATIONS

INTRODUCTION

We have examined three types of first-order differential equations $F(x, y, y') = 0$ in earlier sections. In this further, and admittedly brief, discussion of differential equations, we shall focus our attention primarily on an important class of second-order differential equations, $F(x, y, y', y'') = 0$. The basic idea in the study of differential equations is to solve them—that is, find suitable differentiable functions, defined explicitly or implicitly, that, when substituted into the equation, reduce it to an identity.

Consider the following thought experiment. Suppose a shaft is drilled through the center of the earth as shown in the accompanying figure. Suppose, too, that a mass, such as a bowling ball, is dropped into the shaft. Describe the motion of the mass. In other words, does the bowling ball fall straight through the earth? Does it stop at the center? If you are interested in the solution of this problem, see Exercises 18.3.

IMPORTANT CONCEPTS

- ordinary differential equation
- partial differential equation
- order of an equation
- linear equations
- nonlinear equations
- solution of a differential equation
- explicit solution
- implicit solution
- trivial solution
- homogeneous first-order equations
- homogeneous linear second-order equations
- superposition principle
- general solution
- nonhomogeneous linear second-order equations
- method of undetermined coefficients
- power series solution
- simple harmonic motion
- free damped motion
- forced motion

18.1 BASIC DEFINITIONS AND TERMINOLOGY

An equation containing the derivatives or differentials of one or more dependent variables, with respect to one or more independent variables, is said to be a **differential equation**. Differential equations are classified according to *type*, *order*, and *linearity*.

Classification by Type If an equation contains only ordinary derivatives of one or more dependent variables, with respect to a single independent variable, it is then said to be an **ordinary differential equation.**

EXAMPLE 1 The equations

$$\frac{dy}{dx} - 5y = 1, \qquad (x + y)\,dx - 4y\,dy = 0$$

$$\frac{du}{dx} - \frac{dv}{dx} = x, \qquad \frac{d^2y}{dx^2} - 2\frac{dy}{dx} + 6y = 0$$

are ordinary differential equations. □

An equation involving the partial derivatives of one or more dependent variables of two or more independent variables is called a **partial differential equation**.

EXAMPLE 2 The equations

$$\frac{\partial u}{\partial y} = -\frac{\partial v}{\partial x}, \qquad x\frac{\partial u}{\partial x} + y\frac{\partial u}{\partial y} = u$$

$$\frac{\partial^2 u}{\partial x\,\partial y} = x + y, \qquad a^2\frac{\partial^2 u}{\partial x^2} = \frac{\partial^2 u}{\partial t^2} - 2k\frac{\partial u}{\partial t}$$

are partial differential equations. □

Classification by Order The order of the highest derivative in a differential equation is called the **order of the equation.**

EXAMPLE 3

(*a*) The equation $\dfrac{d^2y}{dx^2} + 5\left(\dfrac{dy}{dx}\right)^3 - 4y = x$ is a second-order ordinary differential equation.

(*b*) Since the differential equation $x^2\,dy + y\,dx = 0$ can be put into the form

$$x^2\frac{dy}{dx} + y = 0$$

by dividing by the differential dx, it is an example of a first-order ordinary differential equation.

(c) The equation $c^2 \dfrac{\partial^4 u}{\partial x^4} + \dfrac{\partial^2 u}{\partial t^2} = 0$ is a fourth-order partial differential equation. □

Although partial differential equations are very important, their study demands a good foundation in the theory of ordinary differential equations. Consequently, in this chapter we shall confine our attention to ordinary differential equations.

A general nth-order ordinary differential equation is often represented by the symbolism

$$F\left(x, y, \frac{dy}{dx}, \ldots, \frac{d^n y}{dx^n}\right) = 0 \tag{1}$$

The following is a special case of (1).

Classification As Linear or Nonlinear A differential equation is said to be **linear** if it has the form

$$a_n(x)\frac{d^n y}{dx^n} + a_{n-1}(x)\frac{d^{n-1} y}{dx^{n-1}} + \cdots + a_1(x)\frac{dy}{dx} + a_0(x)y = g(x)$$

It should be observed that linear differential equations are characterized by two properties: (1) the dependent variable y and all its derivatives are of the first degree; that is, the power of each term involving y is 1; and (2) each coefficient depends on only the independent variable x. An equation that is not linear is said to be **nonlinear**.

EXAMPLE 4 The equations

$$x\,dy + y\,dx = 0$$
$$y'' - 2y' + y = 0$$

and

$$x^3 \frac{d^3 y}{dx^3} - x^2 \frac{d^2 y}{dx^2} + 3x \frac{dy}{dx} + 5y = e^x$$

are linear first-, second-, and third-order ordinary differential equations, respectively. On the other hand,

coefficient depends on y ⇩ power not 1 ⇩

$$yy'' - 2y' = x \quad \text{and} \quad \frac{d^3 y}{dx^3} + y^2 = 0$$

are nonlinear second- and third-order ordinary differential equations, respectively. □

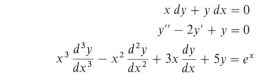

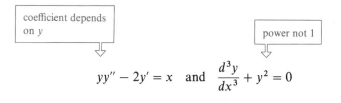

Solutions As mentioned before, our goal is to solve, or find solutions of, differential equations.

DEFINITION 18.1 Solution of a Differential Equation

Any function f defined on some interval that, when substituted into a differential equation, reduces the equation to an identity is said to be a **solution** of the equation.

EXAMPLE 5 The function $y = e^{x^2}$ is a solution of

$$\frac{dy}{dx} - 2xy = 0 \tag{2}$$

Since

$$\frac{dy}{dx} = 2xe^{x^2}$$

we see that

$$\frac{dy}{dx} - 2xy = 2xe^{x^2} - 2x(e^{x^2}) = 0$$ □

EXAMPLE 6 The first-order differential equations

$$\left(\frac{dy}{dx}\right)^2 + 1 = 0 \quad \text{and} \quad (y')^2 + y^2 + 4 = 0$$

possess no real solutions. Why? □

Explicit and Implicit Solutions Solutions of differential equations are further distinguished as either **explicit** or **implicit solutions**. For example, we have already seen that $y = e^{x^2}$ is an explicit solution of (2). A relation $G(x, y) = 0$ is said to define a solution of (1) implicitly on an interval I provided it defines one or more solutions on I.

EXAMPLE 7 $y = xe^x$ is an explicit solution of $y'' - 2y' + y = 0$. To see this, we compute

$$y' = xe^x + e^x \quad \text{and} \quad y'' = xe^x + 2e^x$$

Observe that $y'' - 2y' + y = (xe^x + 2e^x) - 2(xe^x + e^x) + xe^x = 0$. □

EXAMPLE 8 For $-2 < x < 2$, the relation $x^2 + y^2 - 4 = 0$ is an implicit solution of the differential equation

$$\frac{dy}{dx} = -\frac{x}{y}$$

By implicit differentiation it follows that

$$\frac{d}{dx}(x^2) + \frac{d}{dx}(y^2) - \frac{d}{dx}(4) = 0$$

$$2x + 2y\frac{dy}{dx} = 0 \quad \text{or} \quad \frac{dy}{dx} = -\frac{x}{y} \qquad \square$$

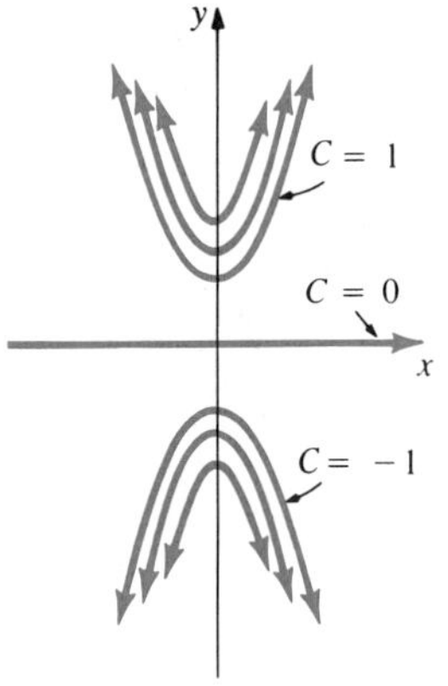

FIGURE 18.1

You should become accustomed to the fact that a given differential equation usually possesses an infinite number of solutions. By direct substitution we can show that any function in the one-parameter family $y = Ce^{x^2}$, where C is any arbitrary constant, also satisfies (2). As indicated in Figure 18.1, $y=0$, obtained by setting $C=0$, is also a solution of the equation. In Example 7 we saw that $y = xe^x$ is a solution of $y'' - 2y' + y = 0$; tracing back through the work reveals that $y = Cxe^x$ represents a family of solutions. By again choosing $C = 0$ in the family of solutions $y = Cxe^x$, we see that $y = 0$ is also a solution of $y'' - 2y' + y = 0$. The constant function $y = 0$, $-\infty < x < \infty$, that satisfies a given differential equation is often referred to as a **trivial solution**.

More Terminology The study of differential equations is similar to integral calculus. When evaluating an antiderivative or indefinite integral, we utilize a single constant of integration. When solving a first-order differential equation, we have already seen that we obtain a one-parameter family of functions $G(x, y, C) = 0$ such that each member of the family is a solution of the differential equation. Naturally, we expect an ***n*-parameter family of solutions** $G(x, y, C_1, \ldots, C_n) = 0$ when solving an nth-order differential equation $F(x, y, y', \ldots, y^{(n)}) = 0$.

A solution of a differential equation that is free of arbitrary parameters is called a **particular solution**. One way of obtaining a particular solution is to choose specific values of the parameters in a family of solutions. For example, it is readily seen that $y = Ce^x$ is a one-parameter family of solutions of the simple first-order equation $y' = y$. For $C = 0, -2$, and 5, we get the particular solutions $y = 0$, $y = -2e^x$, and $y = 5e^x$, respectively.

Sometimes a differential equation possesses a solution that cannot be obtained by specializing the parameters in a family of solutions. Such a solution is called a **singular solution**. The concept of a singular solution is left for an advanced course.

If *every* solution of an nth-order differential equation can be obtained from $G(x, y, C_1, \ldots, C_n) = 0$ by appropriate choices of the C_i, $i = 1, 2, \ldots, n$, we then say that the n-parameter family is the **general** or **complete** solution of the equation.

EXERCISES 18.1

Answers to odd-numbered problems begin on page A-110.

In Problems 1–10 state whether the given differential equation is linear or nonlinear. Give the order of each equation.

1. $(1 - x)y'' - 4xy' + 5y = \cos x$
2. $x\dfrac{d^3y}{dx^3} - 2\left(\dfrac{dy}{dx}\right)^4 + y = 0$
3. $yy' + 2y = 1 + x^2$

4. $x^2\,dy + (y - xy - xe^x)\,dx = 0$

5. $x^3y^{(4)} - x^2y'' + 4xy' - 3y = 0$

6. $\dfrac{d^2y}{dx^2} + 9y = \sin y$

7. $\dfrac{dy}{dx} = \sqrt{1 + \left(\dfrac{d^2y}{dx^2}\right)^2}$

8. $\dfrac{d^2r}{dt^2} = -\dfrac{k}{r^2}$

9. $(\sin x)y''' - (\cos x)y' = 2$

10. $(1 - y^2)\,dx + x\,dy = 0$

In Problems 11–34 verify that the indicated function is a solution of the given differential equation. Where appropriate, C_1 and C_2 denote constants.

11. $2y' + y = 0$; $y = e^{-x/2}$

12. $y' + 4y = 32$; $y = 8$

13. $\dfrac{dy}{dx} - 2y = e^{3x}$; $y = e^{3x} + 10e^{2x}$

14. $\dfrac{dy}{dt} + 20y = 24$; $y = \dfrac{6}{5} - \dfrac{6}{5}e^{-20t}$

15. $y' = 25 + y^2$; $y = 5\tan 5x$

16. $\dfrac{dy}{dx} = \sqrt{\dfrac{y}{x}}$; $y = (\sqrt{x} + C_1)^2$, $x > 0$, $C_1 > 0$

17. $y' + y = \sin x$; $y = \dfrac{1}{2}\sin x - \dfrac{1}{2}\cos x + 10e^{-x}$

18. $2xy\,dx + (x^2 + 2y)\,dy = 0$; $x^2y + y^2 = C_1$

19. $x^2\,dy + 2xy\,dx = 0$; $y = -\dfrac{1}{x^2}$

20. $(y')^3 + xy' = y$; $y = x + 1$

21. $y = 2xy' + y(y')^2$; $y^2 = C_1\left(x + \dfrac{1}{4}C_1\right)$

22. $y' = 2\sqrt{|y|}$; $y = x|x|$

23. $y' - \dfrac{1}{x}y = 1$; $y = x\ln x$, $x > 0$

24. $\dfrac{dP}{dt} = P(a - bP)$; $P = \dfrac{aC_1e^{at}}{1 + bC_1e^{at}}$

25. $\dfrac{dX}{dt} = (2 - X)(1 - X)$; $\ln\dfrac{2 - X}{1 - X} = t$

26. $y' + 2xy = 1$; $y = e^{-x^2}\displaystyle\int_0^x e^{t^2}\,dt + C_1e^{-x^2}$

27. $(x^2 + y^2)\,dx + (x^2 - xy)\,dy = 0$; $C_1(x + y)^2 = xe^{y/x}$

28. $y'' + y' - 12y = 0$; $y = C_1e^{3x} + C_2e^{-4x}$

29. $y'' - 6y' + 13y = 0$; $y = e^{3x}\cos 2x$

30. $\dfrac{d^2y}{dx^2} - 4\dfrac{dy}{dx} + 4y = 0$; $y = e^{2x} + xe^{2x}$

31. $y'' = y$; $y = \cosh x + \sinh x$

32. $y'' + 25y = 0$; $y = C_1\cos 5x$

33. $y'' + (y')^2 = 0$; $y \ln|x + C_1| + C_2$

34. $y'' + y = \tan x$; $y = -\cos x\,\ln(\sec x + \tan x)$

In Problems 35 and 36 find values of m so that $y = e^{mx}$ is a solution of the given differential equation.

35. $y'' - 5y' + 6y = 0$

36. $y'' + 10y' + 25y = 0$

In Problems 37 and 38 find values of m so that $y = x^m$ is a solution of the given differential equation.

37. $xy'' - 3y' = 0$

38. $x^2y'' + 6xy' + 4y = 0$

18.2 HOMOGENEOUS FIRST-ORDER DIFFERENTIAL EQUATIONS

NOTE

We have studied three types of first-order differential equations in earlier chapters: **separable equations** in Section 4.1, **linear equations** in Section 6.8, and **exact equations** in Section 15.5. Before working through this section, you are strongly encouraged to review the material on separable first-order differential equations.

Suppose a first-order differential equation $F(x, y, y') = 0$ has the differential form

$$P(x, y)\,dx + Q(x, y)\,dy = 0$$

and the property that

$$P(tx, ty) = t^n P(x, y) \quad \text{and} \quad Q(tx, ty) = t^n Q(x, y)$$

We then say the equation has **homogeneous coefficients** or is a **homogeneous equation**. The important point in the subsequent discussion is that a homogeneous differential equation *can always be reduced to a separable equation* through an appropriate algebraic substitution. Before pursuing the method of solution for this type of differential equation, let us closely examine the nature of homogeneous functions.

DEFINITION 18.2 Homogeneous Function

If $f(tx, ty) = t^n f(x, y)$ for some real number n, then $f(x, y)$ is said to be a **homogeneous function** of degree n.

EXAMPLE 1

(*a*) $f(x, y) = x - 3\sqrt{xy} + 5y$

$$\begin{aligned} f(tx, ty) &= (tx) - 3\sqrt{(tx)(ty)} + 5(ty) \\ &= tx - 3\sqrt{t^2xy} + 5ty \\ &= t[x - 3\sqrt{xy} + 5y] = tf(x, y) \end{aligned}$$

The function is homogeneous of degree 1.

(*b*) $f(x, y) = \sqrt{x^3 + y^3}$

$f(tx, ty) = \sqrt{t^3x^3 + t^3y^3} = t^{3/2}\sqrt{x^3 + y^3} = t^{3/2}f(x, y)$

The function is homogeneous of degree $\frac{3}{2}$.

(*c*) $f(x, y) = x^2 + y^2 + 1$

$f(tx, ty) = t^2x^2 + t^2y^2 + 1 \neq t^2 f(x, y)$

since $t^2 f(x, y) = t^2x^2 + t^2y^2 + t^2$. The function is not homogeneous.

(*d*) $f(x, y) = \dfrac{x}{2y} + 4$

$$f(tx, ty) = \frac{tx}{2ty} + 4 = \frac{x}{2y} + 4 = t^0 f(x, y)$$

The function is homogeneous of degree 0. □

As parts (*c*) and (*d*) of Example 1 show, a constant added to a function destroys homogeneity, *unless* the function is homogeneous of degree 0. Also, in many instances a homogeneous function can be recognized by examining the total degree of each term.

EXAMPLE 2

degree 1 degree 3 degree 4

(*a*) $$f(x, y) = 6xy^3 - x^2y^2$$

degree 2 degree 4

degree 2

The function is homogeneous of degree 4.

degree 2

(*b*) $$f(x, y) = x^2 - y$$

degree 1

The function is not homogeneous. □

The Method of Solution An equation of the form $P(x, y)\,dx + Q(x, y)\,dy = 0$, where P and Q have the same degree of homogeneity, can be reduced to separable variables by *either* substitution $y = ux$ or $x = vy$, where u and v are new dependent variables. In particular, if we choose $y = ux$, then $dy = u\,dx + x\,du$. Hence, the differential equation becomes

$$P(x, ux)\,dx + Q(x, ux)[u\,dx + x\,du] = 0$$

Now, by homogeneity of P and Q we can write

$$x^nP(1, u)\,dx + x^nQ(1, u)[u\,dx + x\,du] = 0$$

or $$[P(1, u) + uQ(1, u)]\,dx + xQ(1, u)\,du = 0$$

which gives $$\frac{dx}{x} + \frac{Q(1, u)\,du}{P(1, u) + uQ(1, u)} = 0$$

We hasten to point out that the preceding formula should *not* be memorized; rather, *the procedure should be worked through each time.* The proof that the substitution $x = vy$ also leads to a separable equation is left as an exercise.

EXAMPLE 3 Solve $(x^2 + y^2)\,dx + (x^2 - xy)\,dy = 0$.

Solution Both $P(x, y)$ and $Q(x, y)$ are homogeneous of degree 2. If we let $y = ux$, it follows that

$$(x^2 + u^2x^2)\,dx + (x^2 - ux^2)[u\,dx + x\,du] = 0$$

$$x^2(1 + u)\,dx + x^3(1 - u)\,du = 0$$

$$\frac{1 - u}{1 + u}\,du + \frac{dx}{x} = 0$$

$$\left[-1 + \frac{2}{1 + u}\right]du + \frac{dx}{x} = 0$$

Integrating then yields

$$-u + 2\ln|1+u| + \ln|x| + \ln|C| = 0^*$$

$$-\frac{y}{x} + 2\ln\left|1+\frac{y}{x}\right| + \ln|x| + \ln|C| = 0$$

Using the properties of logarithms, we can write this solution in the alternative form

$$C(x+y)^2 = xe^{y/x}$$ □

Although the substitution $y = ux$ can be used for every homogeneous equation, in practice we use the alternative substitution $x = vy$ whenever the function $P(x, y)$ is simpler than $Q(x, y)$. Also, it could happen that after using one substitution we may encounter integrals that are difficult or impossible to evaluate; switching substitutions may result in an easier problem.

EXAMPLE 4 Solve $2x^3y\,dx + (x^4 + y^4)\,dy = 0$.

Solution Each coefficient is a homogeneous function of degree 4. Since the coefficient of dx is slightly simpler than the coefficient of dy, we try $x = vy$. It follows that

$$2v^3y^4[v\,dy + y\,dv] + (v^4y^4 + y^4)\,dy = 0$$
$$2v^3y^5\,dv + y^4(3v^4 + 1)\,dy = 0$$
$$\frac{2v^3\,dv}{3v^4+1} + \frac{dy}{y} = 0$$
$$\frac{1}{6}\ln(3v^4+1) + \ln|y| = \ln|C_1|$$

or

$$3x^4y^2 + y^6 = C$$

Had $y = ux$ been used, then

$$\frac{dx}{x} + \frac{u^4+1}{u^5+3u}\,du = 0$$

It is worth a minute of your time to reflect on how to evaluate the integral of the second term in the last equation. □

EXERCISES 18.2 *Answers to odd-numbered problems begin on page A-110.*

In Problems 1–20 solve the given differential equation by using an appropriate substitution.

1. $(x - y)\,dx + x\,dy = 0$
2. $(x + y)\,dx + x\,dy = 0$
3. $x\,dx + (y - 2x)\,dy = 0$

*If most of the integrations result in logarithms, a judicious choice of the constant of integration is $\ln|C|$ rather than C.

4. $y\,dx = 2(x + y)\,dy$

5. $(y^2 + yx)\,dx - x^2\,dy = 0$

6. $(y^2 + yx)\,dx + x^2\,dy = 0$

7. $\dfrac{dy}{dx} = \dfrac{y - x}{y + x}$

8. $\dfrac{dy}{dx} = \dfrac{x + 3y}{3x + y}$

9. $-y\,dx + (x + \sqrt{xy})\,dy = 0$

10. $x\dfrac{dy}{dx} - y = \sqrt{x^2 + y^2}$

11. $2x^2y\,dx = (3x^3 + y^3)\,dy$

12. $(x^4 + y^4)\,dx - 2x^3y\,dy = 0$

13. $\dfrac{dy}{dx} = \dfrac{y}{x} + \dfrac{x}{y}$

14. $\dfrac{dy}{dx} = \dfrac{y}{x} + \dfrac{x^2}{y^2} + 1$

15. $y\dfrac{dx}{dy} = x + 4ye^{-2x/y}$

16. $(x^2e^{-y/x} + y^2)\,dx = xy\,dy$

17. $\left(y + x\cot\dfrac{y}{x}\right)dx - x\,dy = 0$

18. $\dfrac{dy}{dx} = \dfrac{y}{x}\ln\dfrac{y}{x}$

19. $(x^2 + xy - y^2)\,dx + xy\,dy = 0$

20. $(x^2 + xy + 3y^2)\,dx - (x^2 + 2xy)\,dy = 0$

21. Show that the substitution $x = vy$ reduces a homogeneous equation to one with separate variables.

22. If $f(x, y)$ is a homogeneous function of degree n, show that

$$x\frac{\partial f}{\partial x} + y\frac{\partial f}{\partial y} = nf$$

18.3 LINEAR SECOND-ORDER DIFFERENTIAL EQUATIONS

A linear nth-order differential equation

$$a_n(x)\frac{d^ny}{dx^n} + a_{n-1}(x)\frac{d^{n-1}y}{dx^{n-1}} + \cdots + a_1(x)\frac{dy}{dx} + a_0(x)y = g(x)$$

is said to be **nonhomogeneous** if $g(x) \neq 0$ for some x. If $g(x) = 0$ for every x, then the differential equation is said to be **homogeneous.*** In this and the following section we shall be concerned only with finding solutions of linear *second-order* equations with real *constant* coefficients:

$$ay'' + by' + cy = g(x)$$

We begin by considering the homogeneous equation

$$ay'' + by' + cy = 0 \tag{1}$$

THEOREM 18.1 Superposition Principle

Let y_1 and y_2 be solutions of the homogeneous second-order differential equation (1). Then the linear combination

$$y = C_1y_1(x) + C_2y_2(x)$$

where C_1 and C_2 are arbitrary constants, is also a solution of the equation.

*In this context the word *homogeneous* does not refer to coefficients that are homogeneous functions.

Proof If $y = C_1y_1(x) + C_2y_2(x)$, then

$$a[C_1y_1'' + C_2y_2''] + b[C_1y_1' + C_2y_2'] + c[C_1y_1 + C_2y_2]$$
$$= C_1\underbrace{[ay_1'' + by_1' + cy_1]}_{\boxed{\text{zero}}} + C_2\underbrace{[ay_2'' + by_2' + cy_2]}_{\boxed{\text{zero}}}$$
$$= C_1 \cdot 0 + C_2 \cdot 0 = 0$$

■

Corollaries (*a*) A constant multiple $y = C_1y_1(x)$ of a solution $y_1(x)$ of (1) is also a solution. (*b*) Equation (1) always possesses the trivial solution $y = 0$.

THEOREM 18.2

Let y_1 and y_2 be solutions of the homogeneous second-order differential equation (2) such that neither function is a constant multiple of the other. Then every solution of (1) can be obtained from the linear combination

$$y = C_1y_1(x) + C_2y_2(x) \tag{2}$$

General Solution In Theorem 18.2 we say (2) is the **general solution** of the differential equation. In other words, any solution of (1) can be obtained from (2) by specializing the constants C_1 and C_2. Also, two solutions of (1) such that neither is a constant multiple of the other are said to be **linearly independent**. The general solution of (1) consists of a linear combination of linearly independent solutions.

EXAMPLE 1 You should verify that $y_1 = 0$ and $y_2 = e^{2x}$ are both solutions of the equation $y'' + 2y' - 8y = 0$. Although y_2 is not a constant multiple of y_1, y_1 and y_2 are not linearly independent, since $y_1 = 0 \cdot y_2$. □

The surprising fact about equation (1) is that *all* solutions either are exponential functions or are constructed out of exponential functions.

The Auxiliary Equation If we try a solution of the form $y = e^{mx}$, then $y' = me^{mx}$ and $y'' = m^2e^{mx}$ so that (1) becomes

$$am^2e^{mx} + bme^{mx} + ce^{mx} = 0 \quad \text{or} \quad e^{mx}[am^2 + bm + c] = 0$$

Because e^{mx} is never zero for real values of x, it is apparent that the only way that this exponential function can satisfy the differential equation is to choose m so that it is a root of the quadratic equation

$$am^2 + bm + c = 0$$

This latter equation is called the **auxiliary equation** or **characteristic equation** of the differential equation (1). We shall consider three cases—namely, the solutions corresponding to real distinct roots, real but equal roots, and complex conjugate roots.

Case I Under the assumption that the auxiliary equation of (1) has two unequal real roots m_1 and m_2, we find two solutions

$$y_1 = e^{m_1 x} \quad \text{and} \quad y_2 = e^{m_2 x}$$

Since y_1 and y_2 are linearly independent, it follows that the general solution of the equation is

$$y = C_1 e^{m_1 x} + C_2 e^{m_2 x} \tag{3}$$

EXAMPLE 2 Solve $2y'' - 5y' - 3y = 0$.

Solution We first solve the auxiliary equation:

$$2m^2 - 5m - 3 = 0 \quad \text{or} \quad (2m + 1)(m - 3) = 0$$

$$m_1 = -\frac{1}{2}, \qquad m_2 = 3$$

Hence, by (3) the general solution is

$$y = C_1 e^{-x/2} + C_2 e^{3x}$$ □

Case II When $m_1 = m_2$, we necessarily obtain only one exponential solution $y_1 = e^{m_1 x}$. However, it is a straightforward matter of substitution into (1) to show that $y = u(x)e^{m_1 x}$ is also a solution whenever $u(x) = x$. See Problem 39 in Exercises 18.3. That is, $y_1 = e^{m_1 x}$ and $y_2 = xe^{m_1 x}$ are linearly independent solutions. Hence, the general solution is

$$y = C_1 e^{m_1 x} + C_2 x e^{m_1 x} \tag{4}$$

EXAMPLE 3 Solve $y'' - 10y' + 25y = 0$.

Solution From the auxiliary equation $m^2 - 10m + 25 = 0$ or $(m - 5)^2 = 0$, we see that $m_1 = m_2 = 5$. Thus, by (4) the general solution is

$$y = C_1 e^{5x} + C_2 x e^{5x}$$ □

Complex Numbers The last case deals with complex numbers.* Recall from algebra that a number of the form $z = \alpha + i\beta$, where α and β are real numbers and $i^2 = -1$ (sometimes written $i = \sqrt{-1}$), is called a **complex number**. The complex number $\bar{z} = \alpha - i\beta$ is called the **complex conjugate** of z. Now, from the quadratic formula the roots of $am^2 + bm + c = 0$ can be

---

*See Appendix I.

written

$$m_1 = \frac{-b + \sqrt{b^2 - 4ac}}{2a}, \qquad m_2 = \frac{-b - \sqrt{b^2 - 4ac}}{2a}$$

When $b^2 - 4ac < 0$, m_1 and m_2 are complex conjugates.

Case III If m_1 and m_2 are complex, then we can write

$$m_1 = \alpha + i\beta \quad \text{and} \quad m_2 = \alpha - i\beta$$

where α and β are real and $i^2 = -1$. Formally there is no difference between this case and case I, and hence the general solution is

$$y = c_1 e^{(\alpha + i\beta)x} + c_2 e^{(\alpha - i\beta)x} \tag{5}$$

However, in practice we would prefer to work with real functions instead of complex exponentials. Now, we can rewrite (5) in a more practical form by using **Euler's formula***

$$e^{i\theta} = \cos\theta + i\sin\theta$$

where θ is any real number. From this result we can write

$$e^{i\beta x} = \cos\beta x + i\sin\beta x \quad \text{and} \quad e^{-i\beta x} = \cos\beta x - i\sin\beta x$$

where we have used $\cos(-\beta x) = \cos\beta x$ and $\sin(-\beta x) = -\sin\beta x$. Thus, (5) becomes

$$\begin{aligned} y &= e^{\alpha x}[c_1 e^{i\beta x} + c_2 e^{-i\beta x}] \\ &= e^{\alpha x}[c_1\{\cos\beta x + i\sin\beta x\} + c_2\{\cos\beta x - i\sin\beta x\}] \\ &= e^{\alpha x}[(c_1 + c_2)\cos\beta x + (c_1 i - c_2 i)\sin\beta x] \end{aligned}$$

Since $e^{\alpha x}\cos\beta x$ and $e^{\alpha x}\sin\beta x$ are themselves linearly independent solutions of the given differential equation, we can simply relabel $c_1 + c_2$ as C_1 and $c_1 i - c_2 i$ as C_2 and use the superposition principle to write the general solution

$$\begin{aligned} y &= C_1 e^{\alpha x}\cos\beta x + C_2 e^{\alpha x}\sin\beta x \\ &= e^{\alpha x}(C_1\cos\beta x + C_2\sin\beta x) \end{aligned} \tag{6}$$

EXAMPLE 4 Solve $y'' + y' + y = 0$.

Solution From the quadratic formula we find that the auxiliary equation $m^2 + m + 1 = 0$ has the complex roots

$$m_1 = -\frac{1}{2} + \frac{\sqrt{3}}{2}i \quad \text{and} \quad m_2 = -\frac{1}{2} - \frac{\sqrt{3}}{2}i$$

*You were asked to prove this in Problem 45 of Exercises 10.9.

Identifying $\alpha = -\frac{1}{2}$ and $\beta = \sqrt{3}/2$, we see from (6) that the general solution of the equation is

$$y = e^{-x/2}\left(C_1\cos\frac{\sqrt{3}}{2}x + C_2\sin\frac{\sqrt{3}}{2}x\right) \qquad \square$$

EXAMPLE 5 The two equations

$$y'' + k^2y = 0 \quad \text{and} \quad y'' - k^2y = 0$$

are frequently encountered in the study of applied mathematics. For the former equation the auxiliary equation $m^2 + k^2 = 0$ has complex roots $m_1 = ki$ and $m_2 = -ki$. It follows from (6) that its general solution is

$$y = C_1\cos kx + C_2\sin kx$$

The latter differential equation has the auxiliary equation $m^2 - k^2 = 0$ with real roots $m_1 = k$ and $m_2 = -k$, so that its general solution is

$$y = C_1e^{kx} + C_2e^{-kx} \tag{7}$$

Notice that if we choose $C_1 = C_2 = \frac{1}{2}$ in (7), then

$$y = \frac{e^{kx} + e^{-kx}}{2} = \cosh kx$$

is also a solution. Furthermore, when $C_1 = \frac{1}{2}$, $C_2 = -\frac{1}{2}$, then (7) becomes

$$y = \frac{e^{kx} - e^{-kx}}{2} = \sinh kx$$

Since $\cosh kx$ and $\sinh kx$ are linearly independent, we obtain an alternative form of the general solution for $y'' - k^2y = 0$:

$$y = C_1\cosh kx + C_2\sinh kx \qquad \square$$

Initial-Value Problem The problem

Solve: $\quad ay'' + by' + cy = g(x)$

Subject to: $\quad y(x_0) = y_0, \quad y'(x_0) = y'_0$

where y_0 and y'_0 are arbitrary constants, is called an **initial-value problem**. A solution of the problem is a function whose graph passes through (x_0, y_0) such that the slope of the tangent to the curve at that point is y'_0. The next example illustrates an initial-value problem for a homogeneous equation.

EXAMPLE 6 Solve $y'' - 4y' + 13y = 0$ subject to $y(0) = -1$, $y'(0) = 2$.

Solution The roots of the auxiliary equation

$$m^2 - 4m + 13 = 0$$

are $m_1 = 2 + 3i$ and $m_2 = 2 - 3i$

so that $y = e^{2x}(C_1\cos 3x + C_2\sin 3x)$

The condition $y(0) = -1$ implies that

$$-1 = e^0(C_1\cos 0 + C_2\sin 0) = C_1$$

from which we can write

$$y = e^{2x}(-\cos 3x + C_2\sin 3x)$$

Differentiating this latter expression and using the second condition give

$$y' = e^{2x}(3\sin 3x + 3C_2\cos 3x) + 2e^{2x}(-\cos 3x + C_2\sin 3x)$$
$$2 = 3C_2 - 2$$

so that $C_2 = \frac{4}{3}$. Hence,

$$y = e^{2x}\left(-\cos 3x + \frac{4}{3}\sin 3x\right)$$ □

EXERCISES 18.3

Answers to odd-numbered problems begin on page A-110.

In Problems 1–20 find the general solution of the given differential equation.

1. $3y'' - y' = 0$
2. $2y'' + 5y' = 0$
3. $y'' - 16y = 0$
4. $y'' - 8y = 0$
5. $y'' + 9y = 0$
6. $4y'' + y = 0$
7. $y'' - 3y' + 2y = 0$
8. $y'' - y' - 6y = 0$
9. $\dfrac{d^2y}{dx^2} + 8\dfrac{dy}{dx} + 16y = 0$
10. $\dfrac{d^2y}{dx^2} - 10\dfrac{dy}{dx} + 25y = 0$
11. $y'' + 3y' - 5y = 0$
12. $y'' + 4y' - y = 0$
13. $12y'' - 5y' - 2y = 0$
14. $8y'' + 2y' - y = 0$
15. $y'' - 4y' + 5y = 0$
16. $2y'' - 3y' + 4y = 0$
17. $3y'' + 2y' + y = 0$
18. $2y'' + 2y' + y = 0$
19. $9y'' + 6y' + y = 0$
20. $15y'' - 16y' - 7y = 0$

In Problems 21–30 solve the given differential equation subject to the indicated initial conditions.

21. $y'' + 16y = 0$, $y(0) = 2$, $y'(0) = -2$
22. $y'' - y = 0$, $y(0) = y'(0) = 1$
23. $y'' + 6y' + 5y = 0$, $y(0) = 0$, $y'(0) = 3$
24. $y'' - 8y' + 17y = 0$, $y(0) = 4$, $y'(0) = -1$
25. $2y'' - 2y' + y = 0$, $y(0) = -1$, $y'(0) = 0$
26. $y'' - 2y' + y = 0$, $y(0) = 5$, $y'(0) = 10$
27. $y'' + y' + 2y = 0$, $y(0) = y'(0) = 0$
28. $4y'' - 4y' - 3y = 0$, $y(0) = 1$, $y'(0) = 5$
29. $y'' - 3y' + 2y = 0$, $y(1) = 0$, $y'(1) = 1$
30. $y'' + y = 0$, $y(\pi/3) = 0$, $y'(\pi/3) = 2$
31. The roots of an auxiliary equation are $m_1 = 4$ and $m_2 = -5$. What is the corresponding differential equation?
32. The roots of an auxiliary equation are $m_1 = 3 + i$ and $m_2 = 3 - i$. What is the corresponding differential equation?
33. (a) In the single-loop series circuit shown in Figure 18.2, Kirchhoff's second law states that the impressed voltage $E(t)$ must equal the sum of the voltage drops across the

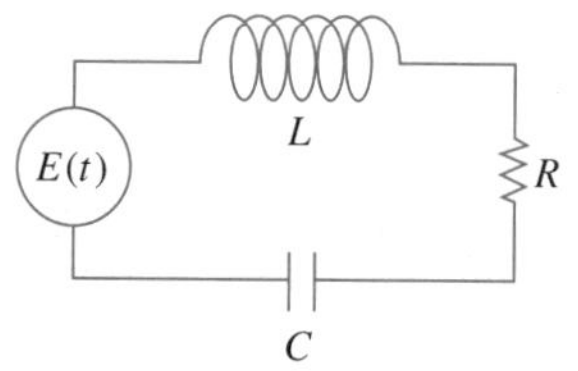

FIGURE 18.2

inductor, resistor, and capacitor. These voltage drops are, respectively, $L\,di/dt$, iR, and q/C, where $i(t)$ is the current in the circuit after a switch is closed, $q(t)$ is the charge on the capacitor at any time t, and L, C, and R are constants known as inductance, capacitance, and resistance. If current i is related to charge q by $i = dq/dt$, find a second-order differential equation for q.

(b) Solve the differential equation in part (**a**) in the case where $L = 0.25$ henry, $R = 10$ ohms, $C = 0.0001$ farad, $E(t) = 0$, $q(0) = q_0$ coulombs, and $i(0) = 0$.

34. Suppose a shaft is drilled through the earth so that it passes through its center. A body with mass m is dropped into the shaft. Let the distance from the center of the earth to the mass at any time t be denoted by r. See Figure 18.3.

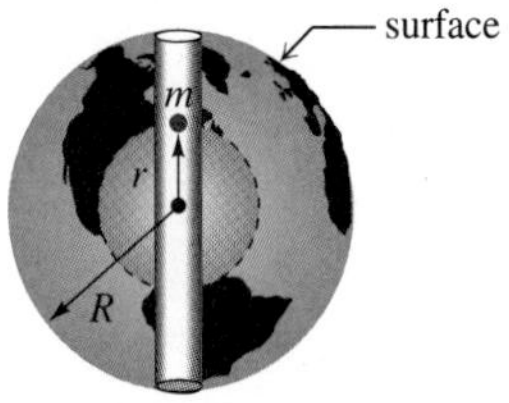

FIGURE 18.3

(a) Let M denote the mass of the earth and M_r denote the mass of that portion of the earth within a sphere of radius r. The gravitational force on m is $F = -kM_rm/r^2$, where the minus sign indicates that the force is one of attraction. Use this fact to show that

$$F = -k\frac{mM}{R^3}r$$

[*Hint:* Assume that the earth is homogeneous—that is, has a constant density ρ. Use mass = density × volume.]

(b) Use Newton's second law $F = ma$ and the result in part (**a**) to derive the differential equation

$$\frac{d^2r}{dt^2} + \omega^2 r = 0$$

where $\omega^2 = kM/R^3 = g/R$.

(c) Solve the differential equation in part (**b**) if the mass m is released from rest at the surface of the earth. Interpret your answer using $R = 3960$ mi.

In Problems 35 and 36 use the assumed solution $y = e^{mx}$ to find the auxiliary equation, roots, and general solution of the given third-order differential equation.

35. $y''' - 4y'' - 5y' = 0$

36. $y''' + 3y'' - 4y' - 12y = 0$

In Problems 37 and 38 find the general solution of the given third-order differential equation if it is known that y_1 is a solution.

37. $y''' - 9y'' + 25y' - 17y = 0$; $y_1 = e^x$

38. $y''' + 6y'' + y' - 34y = 0$; $y_1 = e^{-4x}\cos x$

39. (a) In case II show that if $m_1 = m_2$, the differential equation (1) must be of the form $y'' - 2m_1y' + m_1^2y = 0$.

(b) Show that the substitution $y = u(x)e^{m_1x}$ in the differential equation in part (**a**) leads to $u(x) = x$.

18.4 NONHOMOGENEOUS LINEAR SECOND-ORDER DIFFERENTIAL EQUATIONS

Particular Solutions We now turn our attention to finding the general solution of the nonhomogeneous equation

$$ay'' + by' + cy = g(x) \tag{1}$$

where a, b, and c are constants and g is continuous. Any function y_p free of arbitrary parameters that satisfies (1) is said to be a **particular solution** of the equation.

EXAMPLE 1 $y_p = x^3 - x$ is a particular solution of

$$y'' - y' + 6y = 6x^3 - 3x^2 + 1$$

since $y_p' = 3x^2 - 1$, $y_p'' = 6x$, and

$$\begin{aligned} y'' - y' + 6y &= 6x - (3x^2 - 1) + 6(x^3 - x) \\ &= 6x^3 - 3x^2 + 1 \end{aligned}$$ □

The General Solution The procedure for solving the nonhomogeneous equation (1) consists of two steps:

(*i*) Solve the associated homogeneous equation

$$ay'' + by' + cy = 0$$

(*ii*) and then find *any* particular solution of the nonhomogeneous equation (1).

The sum of the solutions in parts (*i*) and (*ii*) constitutes the **general solution** of (1).

THEOREM 18.3 General Solution

Let y_p be a given solution of the nonhomogeneous equation (1), and let

$$y_c = C_1 y_1(x) + C_2 y_2(x)$$

be the general solution of the associated homogeneous equation. Then the **general solution** of the nonhomogeneous equation is

$$y = C_1 y_1(x) + C_2 y_2(x) + y_p(x) = y_c(x) + y_p(x) \tag{2}$$

The proof that (2) is a solution of (1) is left as an exercise.

Complementary Function In Theorem 18.3 the linear combination $y_c(x) = C_1 y_1(x) + C_2 y_2(x)$ is called the **complementary function** of equation (1). In other words, the general solution of a nonhomogeneous equation is

$$y = \textit{complementary function} + \textit{any particular solution}$$

Variation of Parameters One of the most popular ways of determining a particular solution y_p of the linear second-order differential equation (1) is called the method of **variation of parameters**. To apply this method we put (1) into the form

$$y'' + Py' + Qy = f(x) \tag{3}$$

by dividing through by a.

Suppose y_1 and y_2 are linearly independent solutions of the associated homogeneous form of (3); that is,

$$y_1'' + Py_1' + Qy_1 = 0 \quad \text{and} \quad y_2'' + Py_2' + Qy_2 = 0$$

Now we ask: Can two functions u_1 and u_2 be found so that

$$y_p = u_1(x)y_1(x) + u_2(x)y_2(x) \tag{4}$$

is a particular solution of (1)? Notice that our assumption for y_p is the same as $y_c = C_1y_1 + C_2y_2$, but we have replaced C_1 and C_2 by the "variable parameters" u_1 and u_2. Using the Product Rule to differentiate y_p, we get

$$y_p' = u_1y_1' + y_1u_1' + u_2y_2' + y_2u_2' \tag{5}$$

If we make the further demand that u_1 and u_2 be functions for which

$$y_1u_1' + y_2u_2' = 0 \tag{6}$$

then (5) becomes

$$y_p' = u_1y_1' + u_2y_2'$$

Continuing, we find

$$y_p'' = u_1y_1'' + y_1'u_1' + u_2y_2'' + y_2'u_2'$$

and hence

$$\begin{aligned} y_p'' + Py_p' + Qy_p &= u_1y_1'' + y_1'u_1' + u_2y_2'' + y_2'u_2' \\ &\quad + Pu_1y_1' + Pu_2y_2' + Qu_1y_1 + Qu_2y_2 \\ &= u_1\underbrace{[y_1'' + Py_1' + Qy_1]}_{\text{zero}} + u_2\underbrace{[y_2'' + Py_2' + Qy_2]}_{\text{zero}} \\ &\quad + y_1'u_1' + y_2'u_2' = f(x) \end{aligned}$$

In other words, u_1 and u_2 must be functions that also satisfy the condition

$$y_1'u_1' + y_2'u_2' = f(x) \tag{7}$$

Equations (6) and (7) constitute a linear system of equations for the derivatives u_1' and u_2'. That is, we must solve

$$\begin{aligned} y_1u_1' + y_2u_2' &= 0 \\ y_1'u_1' + y_2'u_2' &= f(x) \end{aligned}$$

By Cramer's rule* we obtain

$$u_1' = \frac{\begin{vmatrix} 0 & y_2 \\ f(x) & y_2' \end{vmatrix}}{\begin{vmatrix} y_1 & y_2 \\ y_1' & y_2' \end{vmatrix}} \quad \text{and} \quad u_2' = \frac{\begin{vmatrix} y_1 & 0 \\ y_1' & f(x) \end{vmatrix}}{\begin{vmatrix} y_1 & y_2 \\ y_1' & y_2' \end{vmatrix}} \tag{8}$$

---

*See Appendix I.

The determinant $\begin{vmatrix} y_1 & y_2 \\ y'_1 & y'_2 \end{vmatrix}$ is called the **Wronskian*** and is usually denoted by W.

EXAMPLE 2 Solve $4y'' + 36y = \csc 3x$.

Solution First write the equation in the form

$$y'' + 9y = \frac{1}{4}\csc 3x$$

Since the roots of the auxiliary equation $m^2 + 9 = 0$ are $m_1 = 3i$ and $m_2 = -3i$,

$$y_c = C_1\cos 3x + C_2\sin 3x$$
$$W = \begin{vmatrix} \cos 3x & \sin 3x \\ -3\sin 3x & 3\cos 3x \end{vmatrix} = 3$$

From (8) we then find

$$u'_1 = -\frac{(\sin 3x)(\frac{1}{4}\csc 3x)}{3} = -\frac{1}{12}$$

and

$$u'_2 = \frac{(\cos 3x)(\frac{1}{4}\csc 3x)}{3} = \frac{1}{12}\frac{\cos 3x}{\sin 3x}$$

Integrating u'_1 and u'_2 yields

$$u_1 = -\frac{1}{12}x \quad \text{and} \quad u_2 = \frac{1}{36}\ln|\sin 3x|$$

Hence, from (4)

$$y_p = -\frac{1}{12}x\cos 3x + \frac{1}{36}(\sin 3x)\ln|\sin 3x|$$

Therefore, the general solution is

$$y = y_c + y_p = C_1\cos 3x + C_2\sin 3x - \frac{1}{12}x\cos 3x + \frac{1}{36}(\sin 3x)\ln|\sin 3x| \quad \square$$

Constants of Integration When computing the indefinite integrals of u'_1 and u'_2, we need not introduce any constants. This is because

$$\begin{aligned} y = y_c + y_p &= C_1y_1 + C_2y_2 + (u_1 + a_1)y_1 + (u_2 + b_1)y_2 \\ &= (C_1 + a_1)y_1 + (C_2 + b_1)y_2 + u_1y_1 + u_2y_2 \\ &= c_1y_1 + c_2y_2 + u_1y_1 + u_2y_2 \end{aligned}$$

where $c_1 = C_1 + a_1$ and $c_2 = C_2 + b_1$.

*Named after the Polish philosopher-mathematician, Joseph M. H. Wronski (1778–1853.)

Undetermined Coefficients When $g(x)$ consists of

(*i*) a constant k,

(*ii*) a polynomial in x,

(*iii*) an exponential function $e^{\alpha x}$,

(*iv*) $\sin \beta x$, $\cos \beta x$,

or finite sums and products of these functions, it is possible to find a particular solution of (1) by the **method of undetermined coefficients**. In the special case* where the n distinct functions $f_k(x)$ appearing in $g(x)$ *or* in its derivatives *do not* appear in the complementary function y_c, a particular solution y_p can be found consisting of a linear combination:

$$y = \sum_{k=1}^{n} A_k f_k(x)$$

We find the coefficients A_k by substituting this expression into the given differential equation.

EXAMPLE 3 Solve $\dfrac{d^2y}{dx^2} + 3\dfrac{dy}{dx} + 2y = 4x^2$.

Solution The complementary function is

$$y_c = C_1 e^{-x} + C_2 e^{-2x}$$

Now, since

$g(x) = 4x^2$	$g'(x) = 8x$	$g''(x) = 8 \cdot 1$
$f_1(x)$	$f_2(x)$	$f_3(x)$

we seek a particular solution having the basic structure

$$y_p = Ax^2 + Bx + C \tag{9}$$

Differentiating (9) and substituting into the given equation give

$$y_p'' + 3y_p' + 2y_p = 2A + 3B + 2C + (6A + 2B)x + 2Ax^2 = 4x^2$$

By equating coefficients in the last identity we obtain the system of equations

$$\begin{aligned} 2A + 3B + 2C &= 0 \\ 6A + 2B &= 0 \\ 2A &= 4 \end{aligned}$$

---

*See the author's *A First Course in Differential Equations* for further details of this method.

Solving gives $A = 2$, $B = -6$, and $C = 7$. Thus, $y_p = 2x^2 - 6x + 7$ and the general solution of the differential equation is $y = y_c + y_p$, or

$$y = C_1e^{-x} + C_2e^{-2x} + 2x^2 - 6x + 7 \qquad \square$$

EXAMPLE 4 Solve $y'' + 2y' + 2y = -10xe^x + 5\sin x$.

Solution The roots of the auxiliary equation $m^2 + 2m + 2 = 0$ are $m_1 = -1 + i$ and $m_2 = -1 - i$, and so

$$y_c = e^{-x}(C_1\cos x + C_2\sin x)$$

In this case,

$$g(x) = \underbrace{-10xe^x}_{f_1(x)} + \underbrace{5\sin x}_{f_2(x)} \qquad g'(x) = 10xe^x - \underbrace{10e^x}_{f_3(x)} + \underbrace{5\cos x}_{f_4(x)}$$

indicates that a particular solution can be found of the form

$$y_p = Axe^x + Be^x + C\sin x + D\cos x$$

Substituting y_p in the differential equation and simplifying yield

$$\begin{aligned} y_p'' + 2y_p' + 2y_p &= 5Axe^x + (4A + 5B)e^x + (C - 2D)\sin x + (2C + D)\cos x \\ &= -10xe^x + 5\sin x \end{aligned}$$

and

$$\begin{aligned} 5A \qquad &= -10 \\ 4A + 5B &= 0 \\ C - 2D &= 5 \\ 2C + \quad D &= 0 \end{aligned}$$

We find $A = -2$, $B = \frac{8}{5}$, $C = 1$, and $D = -2$. Hence,

$$y_p = -2xe^x + \frac{8}{5}e^x + \sin x - 2\cos x$$

The general solution is then

$$y = e^{-x}(C_1\cos x + C_2\sin x) - 2xe^x + \frac{8}{5}e^x + \sin x - 2\cos x \qquad \square$$

EXERCISES 18.4

Answers to odd-numbered problems begin on page A-110.

In Problems 1–20 solve the given differential equation by variation of parameters.

1. $y'' + y = \sec x$
2. $y'' + y = \tan x$
3. $y'' + y = \sin x$
4. $y'' + y = \sec x \tan x$
5. $y'' + y = \cos^2 x$
6. $y'' + y = \sec^2 x$
7. $y'' - y = \cosh x$
8. $y'' - y = \sinh 2x$

9. $y'' - 4y = e^{2x}/x$

10. $y'' - 9y = 9x/e^{3x}$

11. $y'' + 3y' + 2y = 1/(1 + e^x)$

12. $y'' - 3y' + 2y = e^{3x}/(1 + e^x)$

13. $y'' + 3y' + 2y = \sin e^x$

14. $y'' - 2y' + y = e^x \arctan x$

15. $y'' - 2y' + y = e^x/(1 + x^2)$

16. $y'' - 2y' + 2y = e^x \sec x$

17. $y'' + 2y' + y = e^{-x} \ln x$

18. $y'' + 10y' + 25y = e^{-10x}/x^2$

19. $4y'' - 4y' + y = 8e^{-x} + x$

20. $4y'' - 4y' + y = e^{x/2}\sqrt{1 - x^2}$

In Problems 21 and 22 solve the given differential equation by variation of parameters subject to the initial conditions $y(0) = 1$ and $y'(0) = 0$.

21. $y'' - y = xe^x$

22. $2y'' + y' - y = x + 1$

In Problems 23–32 solve the given differential equation by undetermined coefficients.

23. $y'' - 9y = 54$

24. $2y'' - 7y' + 5y = -29$

25. $y'' + 4y' + 4y = 2x + 6$

26. $y'' - 2y' + y = x^3 + 4x$

27. $y'' + 25y = 6 \sin x$

28. $y'' - 4y = 7e^{4x}$

29. $y'' - 2y' - 3y = 4e^{2x} + 2x^3$

30. $y'' + y' + y = x^2e^x + 3$

31. $y'' - 8y' + 25y = e^{3x} - 6 \cos 2x$

32. $y'' - 5y' + 4y = 2 \sinh 3x$

In Problems 33 and 34 solve the given differential equation by undetermined coefficients subject to the initial conditions $y(0) = 1$ and $y'(0) = 0$.

33. $y'' - 64y = 16$

34. $y'' + 5y' - 6y = 10e^{2x}$

35. In Problem 33 in Exercises 18.3 we saw that the differential equation for the charge $q(t)$ on the capacitor in a single-loop series circuit is given by

$$L\frac{d^2q}{dt^2} + R\frac{dq}{dt} + \frac{1}{C}q = E(t)$$

where $E(t)$ is the impressed voltage. In the case when $R \neq 0$, the complementary function $q_c(t)$ of the equation is called a **transient solution**, since $q_c(t) \to 0$ as $t \to \infty$. If $E(t)$ is periodic or a constant, then the particular solution $q_p(t)$ is called a **steady-state solution** and $i_p(t) = q_p'(t)$ is called **steady-state current**. Find the steady-state current in a series circuit when $L = \frac{1}{2}$ henry, $R = 20$ ohms, $C = 0.001$ farad, and $E(t) = 100 \sin 60t + 200 \cos 40t$.

36. Since phosphate is often the limiting nutrient for algae growth in lakes, it is important for the management of water quality to be able to predict phosphate input into lakes. One source is from the sediment in the lakebed. A mathematical model that describes phosphate concentration in lakebed sediment is the

$$\frac{d^2C}{dx^2} = \frac{C(x) - C(\infty)}{\lambda^2}$$

where $C(x)$ is the phosphate concentration at depth x from the surface of the sediment, $C(\infty)$ is the equilibrium concentration at "infinite" depth, that is, $C(\infty) = \lim_{x\to\infty} C(x)$, and $\lambda > 0$ is a "yardstick of thickness" parameter involving the porosity of the sediment, the diffusion coefficient of the phosphate ion, and an adsorption rate constant. Solve this differential equation subject to the initial condition $C(0) = 0$.

37. Given that $y_1 = x$ and $y_2 = x \ln x$ are linearly independent solutions of $x^2y'' - xy' + y = 0$, use variation of parameters to solve $x^2y'' - xy' + y = 4x \ln x$.

38. Given that $y_1 = x^2$ and $y_2 = x^3$ are linearly independent solutions of $x^2y'' - 4xy' + 6y = 0$, use variation of parameters to solve $x^2y'' - 4xy' + 6y = 1/x$.

39. Prove that (2) is a solution of (1).

18.5 POWER SERIES SOLUTIONS

Some homogeneous linear second-order differential equations with *variable coefficients* can be solved by use of power series. The procedure consists of assuming a solution of the form

$$y = c_0 + c_1x + c_2x^2 + c_3x^3 + \cdots = \sum_{n=0}^{\infty} c_nx^n$$

differentiating

$$\frac{dy}{dx} = c_1 + 2c_2x + 3c_3x^2 + \cdots = \sum_{n=1}^{\infty} nc_nx^{n-1}$$

$$\frac{d^2y}{dx^2} = 2c_2 + 6c_3x + \cdots = \sum_{n=2}^{\infty} n(n-1)c_nx^{n-2}$$

and substituting the results into the differential equation with the expectation of determining a **recurrence relation** that will yield the coefficients c_n. The power series will then represent a solution of the differential equation on some interval of convergence.

EXAMPLE 1 Solve $y'' - 2xy = 0$.

Solution If we assume a solution $y = \Sigma_{n=0}^{\infty} c_nx^n$, then, as just shown,

$$y'' = \sum_{n=2}^{\infty} n(n-1)c_nx^{n-2}$$

Hence, we have

$$y'' - 2xy = \sum_{n=2}^{\infty} n(n-1)c_nx^{n-2} - \sum_{n=0}^{\infty} 2c_nx^{n+1}$$

$$= 2\cdot 1c_2x^0 + \underbrace{\sum_{n=3}^{\infty} n(n-1)c_nx^{n-2} - \sum_{n=0}^{\infty} 2c_nx^{n+1}}_{\text{both series start with } x^1}$$

Letting $k = n - 2$ in the first series and $k = n + 1$ in the second yields

$$y'' - 2xy = 2c_2 + \sum_{k=1}^{\infty} (k+2)(k+1)c_{k+2}x^k - \sum_{k=1}^{\infty} 2c_{k-1}x^k$$

$$= 2c_2 + \sum_{k=1}^{\infty} [(k+2)(k+1)c_{k+2} - 2c_{k-1}]x^k = 0$$

We must then have

$$2c_2 = 0 \quad \text{and} \quad (k+2)(k+1)c_{k+2} - 2c_{k-1} = 0$$

The last expression written as

$$c_{k+2} = \frac{2c_{k-1}}{(k+2)(k+1)}, \qquad k = 1, 2, 3, \ldots$$

is called a *recurrence relation* for the c_n. Iteration of this formula gives

$$c_3 = \frac{2c_0}{3\cdot 2}$$

$$c_4 = \frac{2c_1}{4\cdot 3}$$

$$c_5 = \frac{2c_2}{5 \cdot 4} = 0$$

$$c_6 = \frac{2c_3}{6 \cdot 5} = \frac{2^2}{6 \cdot 5 \cdot 3 \cdot 2} c_0$$

$$c_7 = \frac{2c_4}{7 \cdot 6} = \frac{2^2}{7 \cdot 6 \cdot 4 \cdot 3} c_1$$

$$c_8 = \frac{2c_5}{8 \cdot 7} = 0$$

$$c_9 = \frac{2c_6}{9 \cdot 8} = \frac{2^3}{9 \cdot 8 \cdot 6 \cdot 5 \cdot 3 \cdot 2} c_0$$

$$c_{10} = \frac{2c_7}{10 \cdot 9} = \frac{2^3}{10 \cdot 9 \cdot 7 \cdot 6 \cdot 4 \cdot 3} c_1$$

$$c_{11} = \frac{2c_8}{11 \cdot 10} = 0$$

and so on. It should be apparent that both c_0 and c_1 are arbitrary. Now

$$\begin{aligned} y &= c_0 + c_1 x + c_2 x^2 + c_3 x^3 + c_4 x^4 + c_5 x^5 + c_6 x^6 + c_7 x^7 + c_8 x^8 \\ &\quad + c_9 x^9 + c_{10} x^{10} + c_{11} x^{11} + \cdots \\ &= c_0 + c_1 x + 0 + \frac{2}{3 \cdot 2} c_0 x^3 + \frac{2}{4 \cdot 3} c_1 x^4 + 0 + \frac{2^2}{6 \cdot 5 \cdot 3 \cdot 2} c_0 x^6 \\ &\quad + \frac{2^2}{7 \cdot 6 \cdot 4 \cdot 3} c_1 x^7 + 0 + \frac{2^3}{9 \cdot 8 \cdot 6 \cdot 5 \cdot 3 \cdot 2} c_0 x^9 \\ &\quad + \frac{2^3}{10 \cdot 9 \cdot 7 \cdot 6 \cdot 4 \cdot 3} c_1 x^{10} + 0 + \cdots \\ &= c_0 \left[1 + \frac{2}{3 \cdot 2} x^3 + \frac{2^2}{6 \cdot 5 \cdot 3 \cdot 2} x^6 + \frac{2^3}{9 \cdot 8 \cdot 6 \cdot 5 \cdot 3 \cdot 2} x^9 + \cdots \right] \\ &\quad + c_1 \left[x + \frac{2}{4 \cdot 3} x^4 + \frac{2^2}{7 \cdot 6 \cdot 4 \cdot 3} x^7 + \frac{2^3}{10 \cdot 9 \cdot 7 \cdot 6 \cdot 4 \cdot 3} x^{10} + \cdots \right] \quad \square \end{aligned}$$

Although the pattern of the coefficients in Example 1 should be clear, it is sometimes useful to write the solutions in terms of summation notation. By using the properties of the factorial we can write

$$y_1(x) = c_0 \left[1 + \sum_{k=1}^{\infty} \frac{2^k [1 \cdot 4 \cdot 7 \cdots (3k-2)]}{(3k)!} x^{3k} \right]$$

and

$$y_2(x) = c_1 \left[x + \sum_{k=1}^{\infty} \frac{2^k [2 \cdot 5 \cdot 8 \cdots (3k-1)]}{(3k+1)!} x^{3k+1} \right]$$

In this form the Ratio Test can be used to show that each series converges for $|x| < \infty$.

EXAMPLE 2 Solve $(x^2 + 1)y'' + xy' - y = 0$.

Solution The assumption $y = \Sigma_{n=0}^{\infty} c_n x^n$ leads to

$$(x^2 + 1)\sum_{n=2}^{\infty} n(n-1)c_n x^{n-2} + x\sum_{n=1}^{\infty} nc_n x^{n-1} - \sum_{n=0}^{\infty} c_n x^n$$

$$= \sum_{n=2}^{\infty} n(n-1)c_n x^n + \sum_{n=2}^{\infty} n(n-1)c_n x^{n-2} + \sum_{n=1}^{\infty} nc_n x^n - \sum_{n=0}^{\infty} c_n x^n$$

$$= 2c_2x^0 - c_0x^0 + 6c_3x + c_1x - c_1x + \underbrace{\sum_{n=2}^{\infty} n(n-1)c_n x^n}_{\boxed{k=n}}$$

$$+ \underbrace{\sum_{n=4}^{\infty} n(n-1)c_n x^{n-2}}_{\boxed{k=n-2}} + \underbrace{\sum_{n=2}^{\infty} nc_n x^n}_{\boxed{k=n}} - \underbrace{\sum_{n=2}^{\infty} c_n x^n}_{\boxed{k=n}}$$

$$= 2c_2 - c_0 + 6c_3x + \sum_{k=2}^{\infty} [k(k-1)c_k + (k+2)(k+1)c_{k+2} + kc_k - c_k]x^k$$

$$= 2c_2 - c_0 + 6c_3x + \sum_{k=2}^{\infty} [(k+1)(k-1)c_k + (k+2)(k+1)c_{k+2}]x^k = 0$$

Thus, we must have

$$2c_2 - c_0 = 0$$
$$c_3 = 0$$
$$(k+1)(k-1)c_k + (k+2)(k+1)c_{k+2} = 0$$

or, after dividing by $k + 1$,

$$c_2 = \frac{1}{2}c_0$$
$$c_3 = 0$$
$$c_{k+2} = -\frac{k-1}{k+2}c_k, \qquad k = 2, 3, 4 \ldots$$

Iteration of the last formula gives

$$c_4 = -\frac{1}{4}c_2 = -\frac{1}{2 \cdot 4}c_0 = -\frac{1}{2^2 2!}c_0$$

$$c_5 = -\frac{2}{5}c_3 = 0$$

$$c_6 = -\frac{3}{6}c_4 = \frac{3}{2 \cdot 4 \cdot 6}c_0 = \frac{1 \cdot 3}{2^3 3!}c_0$$

$$c_7 = -\frac{4}{7}c_5 = 0$$

$$c_8 = -\frac{5}{8}c_6 = -\frac{3 \cdot 5}{2 \cdot 4 \cdot 6 \cdot 8}c_0 = -\frac{1 \cdot 3 \cdot 5}{2^2 4!}c_0$$

$$c_9 = -\frac{6}{9}c_7 = 0$$

$$c_{10} = -\frac{7}{10}c_8 = \frac{3 \cdot 5 \cdot 7}{2 \cdot 4 \cdot 6 \cdot 8 \cdot 10}c_0 = \frac{1 \cdot 3 \cdot 5 \cdot 7}{2^5 5!}c_0$$

and so on. Therefore,

$$\begin{aligned} y &= c_0 + c_1x + c_2x^2 + c_3x^3 + c_4x^4 + c_5x^5 + c_6x^6 + c_7x^7 + c_8x^8 + \cdots \\ &= c_1x + c_0\left[1 + \frac{1}{2}x^2 - \frac{1}{2^2 2!}x^4 + \frac{1 \cdot 3}{2^3 3!}x^6 - \frac{1 \cdot 3 \cdot 5}{2^4 4!}x^8 + \frac{1 \cdot 3 \cdot 5 \cdot 7}{2^5 5!}x^{10} + \cdots\right] \end{aligned}$$

The solutions are

$$y_1(x) = c_0\left[1 + \frac{1}{2}x^2 + \sum_{n=2}^{\infty}(-1)^{n-1}\frac{1 \cdot 3 \cdot 5 \cdots (2n-3)}{2^n n!}x^{2n}\right]$$

$$y_2(x) = c_1x$$

□

EXERCISES 18.5

Answers to odd-numbered problems begin on page A-110.

In Problems 1–18 find the power series solutions of the given differential equation.

1. $y'' + y = 0$
2. $y'' - y = 0$
3. $y'' = y'$
4. $2y'' + y' = 0$
5. $y'' = xy$
6. $y'' + x^2y = 0$
7. $y'' - 2xy' + y = 0$
8. $y'' - xy' + 2y = 0$
9. $y'' + x^2y' + xy = 0$
10. $y'' + 2xy' + 2y = 0$
11. $(x - 1)y'' + y' = 0$
12. $(x + 2)y'' + xy' - y = 0$
13. $(x^2 - 1)y'' + 4xy' + 2y = 0$
14. $(x^2 + 1)y'' - 6y = 0$
15. $(x^2 + 2)y'' + 3xy' - y = 0$
16. $(x^2 - 1)y'' + xy' - y = 0$
17. $y'' - (x + 1)y' - y = 0$
18. $y'' - xy' - (x + 2)y = 0$

In Problems 19 and 20 use the power series method to solve the given differential equation subject to the indicated initial conditions.

19. $(x - 1)y'' - xy' + y = 0$; $y(0) = -2$, $y'(0) = 6$
20. $y'' - 2xy' + 8y = 0$; $y(0) = 3$, $y'(0) = 0$

18.6 VIBRATIONAL MODELS

Hooke's Law Suppose, as in Figure 18.4(b), a mass m_1 is attached to a flexible spring suspended from a rigid support. When m_1 is replaced with a different mass m_2, the amount of stretch, or elongation of the spring, will of course be different.

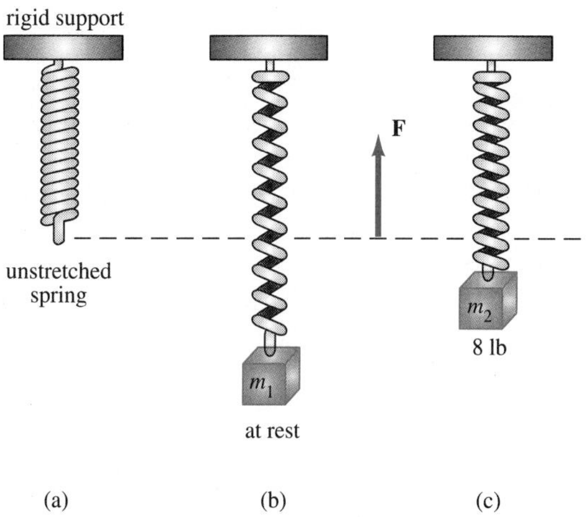

FIGURE 18.4

By Hooke's law, the spring itself exerts a restoring force F opposite to the direction of elongation and proportional to the amount of elongation s. Simply stated, $F = ks$, where k is a constant of proportionality. Although masses with different weights stretch a spring by different amounts, the spring is essentially characterized by the number k. For example, if a mass weighing 10 lb stretches a spring $\frac{1}{2}$ ft, then

$$10 = k\left(\frac{1}{2}\right) \quad \text{implies} \quad k = 20 \text{ lb/ft}$$

Necessarily then, a mass weighing 8 lb stretches the same spring $\frac{2}{5}$ ft.

Newton's Second Law After a mass m is attached to a spring, it will stretch the spring by an amount s and attain a position of equilibrium at which its weight W is balanced by the restoring force ks. Recall that weight is defined by $W = mg$, where the mass m is measured in slugs, kilograms, or grams, and $g = 32 \text{ ft/s}^2$, 9.8 m/s^2, or 980 cm/s^2, respectively. As indicated in Figure 18.5(b), the condition of equilibrium is $mg = ks$ or $mg - ks = 0$.

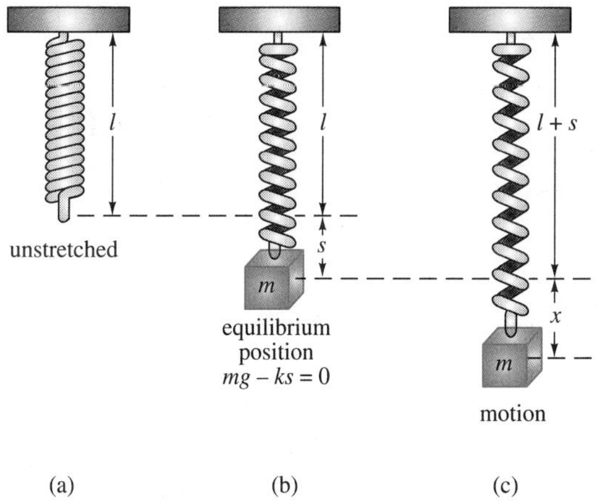

FIGURE 18.5

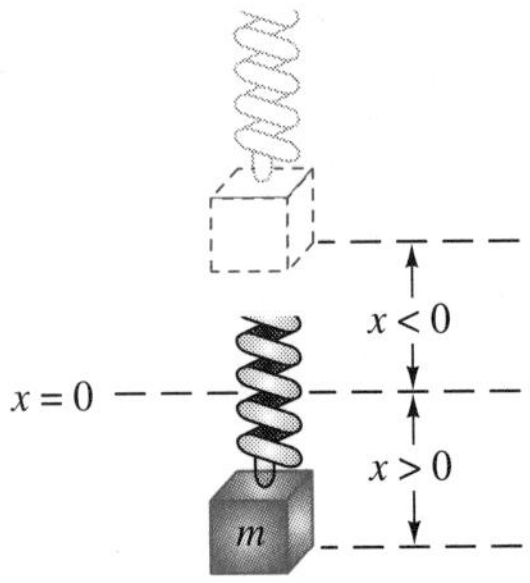

FIGURE 18.6

If the mass is then displaced by an amount x from its equilibrium position and released, the net force F in this dynamic case is given by **Newton's second law of motion**, $F = ma$, where a is the acceleration d^2x/dt^2. Assuming that there are no retarding forces acting on the system and assuming that the mass vibrates free of other external influencing forces—**free motion**—we can equate F to the resultant force of the weight and the restoring force:

$$m\frac{d^2x}{dt^2} = -k(s + x) + mg$$

$$= -kx + \underbrace{mg - ks}_{\text{zero}} = -kx \tag{1}$$

The negative sign in (1) indicates that the restoring force of the spring acts opposite to the direction of motion. Furthermore, we shall adopt the convention that displacements measured *below* the equilibrium position are positive. See Figure 18.6.

Free Undamped Motion By dividing (1) by the mass m we obtain the second-order differential equation

$$\frac{d^2x}{dt^2} + \frac{k}{m}x = 0 \quad \text{or} \quad \frac{d^2x}{dt^2} + \omega^2 x = 0 \tag{2}$$

where $\omega^2 = k/m$. Equation (2) is said to describe **simple harmonic motion**, or **free undamped motion**. There are two obvious initial conditions associated with this differential equation:

$$x(0) = \alpha \qquad \left.\frac{dx}{dt}\right|_{t=0} = \beta \tag{3}$$

representing the amount of initial displacement and the initial velocity, respectively. For example, if $\alpha > 0$, $\beta < 0$, the mass would start from a point *below* the equilibrium position with an imparted *upward* velocity. If $\alpha < 0$, $\beta = 0$, the mass would be released from *rest* from a point $|\alpha|$ units *above* the equilibrium position, and so on.

The Solution and the Equation of Motion To solve (2) we note that the solutions of the auxiliary equation $m^2 + \omega^2 = 0$ are the complex numbers

$$m_1 = \omega i \qquad m_2 = -\omega i$$

Thus, from (6) of Section 18.3, we find the general solution of the equation to be

$$x(t) = C_1 \cos \omega t + C_2 \sin \omega t \tag{4}$$

The **period** of free vibrations described by (4) is $T = 2\pi/\omega$ and the **frequency** is $f = 1/T = \omega/2\pi$. Finally, when the initial conditions (3) are used to determine the constants C_1 and C_2 in (4), we say that the resulting particular solution is the **equation of motion** of the mass.

EXAMPLE 1 Solve and interpret the initial-value problem

$$\frac{d^2x}{dt^2} + 16x = 0$$

$$x(0) = 10 \qquad \left.\frac{dx}{dt}\right|_{t=0} = 0$$

Solution The problem is equivalent to pulling a mass on a spring down 10 units below the equilibrium position, holding it until $t = 0$, and then releasing it from rest. Applying the initial conditions to the solution

$$x(t) = C_1\cos 4t + C_2\sin 4t$$

gives

$$x(0) = 10 = C_1 \cdot 1 + C_2 \cdot 0$$

so that $C_1 = 10$, and hence

$$x(t) = 10\cos 4t + C_2\sin 4t$$

$$\frac{dx}{dt} = -40\sin 4t + 4C_2\cos 4t$$

$$\left.\frac{dx}{dt}\right|_{t=0} = 0 = 4C_2 \cdot 1$$

The latter equation implies that $C_2 = 0$, so the equation of motion is $x(t) = 10\cos 4t$.

The solution clearly shows that once the system is set in motion, it stays in motion with the mass bouncing back and forth 10 units on either side of the equilibrium position $x = 0$. As shown in Figure 18.7(b), the period of oscillation is $2\pi/4 = \pi/2$ seconds.

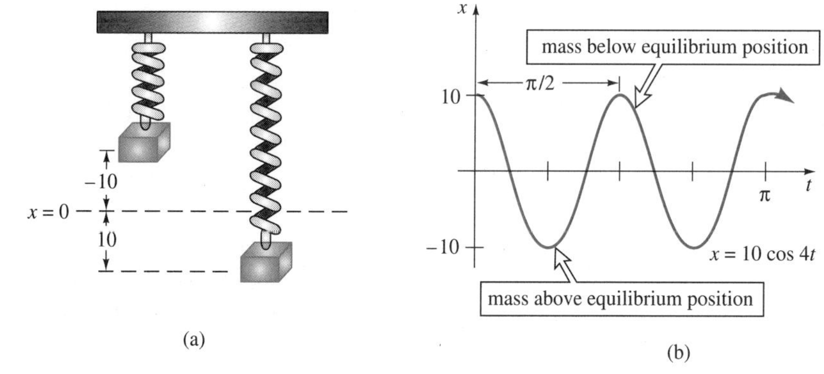

FIGURE 18.7 □

Free Damped Motion The discussion of simple harmonic motion in Example 1 is somewhat unrealistic. Unless the mass is suspended in a perfect vacuum, there will be at least a resisting force due to the surrounding medium. For example, as Figure 18.8 shows, the mass m could be suspended in a viscous medium or connected to a dashpot damping device. In the study of mechanics, damping forces acting on a body are considered to be proportional to a power

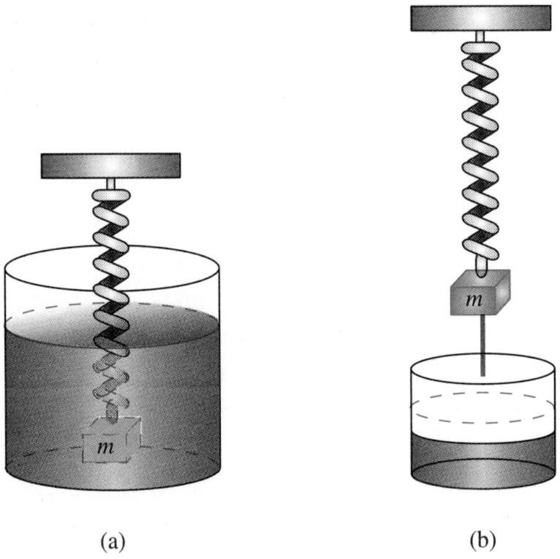

FIGURE 18.8

of the instantaneous velocity. In particular, we shall assume that this force is given by a constant multiple of dx/dt. Thus, when no other external forces are impressed on the system, it follows from Newton's second law that

$$m\frac{d^2x}{dt^2} = -kx - \beta\frac{dx}{dt} \tag{5}$$

where β is a positive *damping constant* and the negative sign is a consequence of the fact that the damping force acts in a direction opposite to the motion. When we divide (5) by the mass m, the differential equation of **free damped motion** is

$$\frac{d^2x}{dt^2} + \frac{\beta}{m}\frac{dx}{dt} + \frac{k}{m}x = 0 \quad \text{or} \quad \frac{d^2x}{dt^2} + 2\lambda\frac{dx}{dt} + \omega^2 x = 0 \tag{6}$$

where $2\lambda = \beta/m$ and $\omega^2 = k/m$. The symbol 2λ is used only for algebraic convenience, since the auxiliary equation is $m^2 + 2\lambda m + \omega^2 = 0$ and the corresponding roots are then

$$m_1 = -\lambda + \sqrt{\lambda^2 - \omega^2} \qquad m_2 = -\lambda - \sqrt{\lambda^2 - \omega^2}$$

We can now distinguish three possible cases depending on the algebraic sign of $\lambda^2 - \omega^2$. Since each solution will contain the *damping factor* $e^{-\lambda t}$, $\lambda > 0$, the displacements of the mass will become negligible for large time.

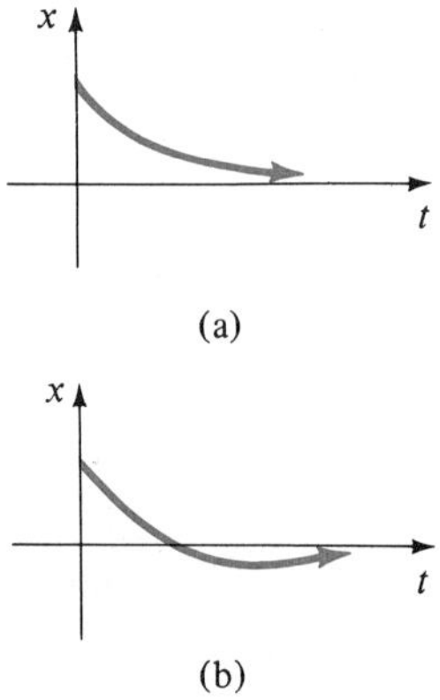

FIGURE 18.9

Case I $\lambda^2 - \omega^2 > 0$. In this situation the system is said to be **overdamped**, since the damping coefficient β is large when compared to the spring constant k. The corresponding solution of (6) is

$$x(t) = C_1 e^{m_1 t} + C_2 e^{m_2 t}$$

or

$$x(t) = e^{-\lambda t}\left(C_1 e^{\sqrt{\lambda^2 - \omega^2}\,t} + C_2 e^{-\sqrt{\lambda^2 - \omega^2}\,t}\right) \tag{7}$$

This equation represents a smooth and nonoscillatory motion. Figure 18.9 shows two possible graphs of $x(t)$.

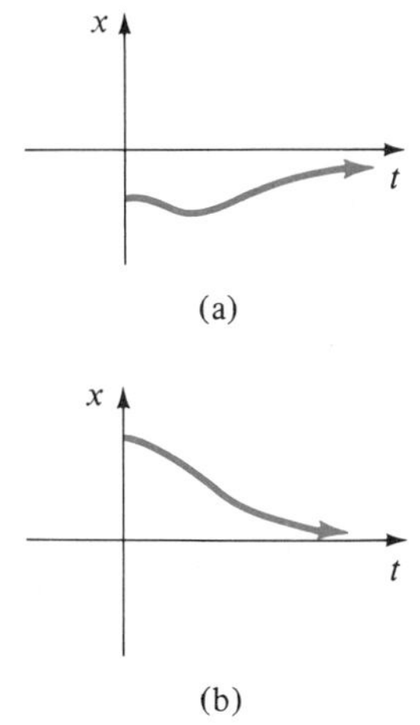

FIGURE 18.10

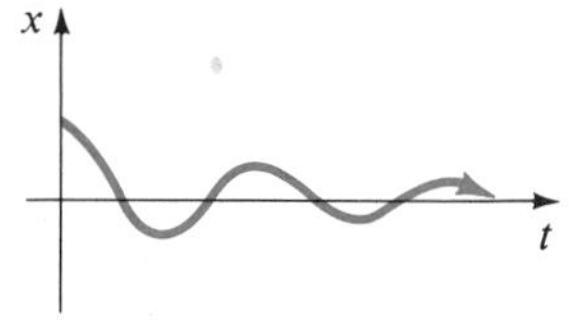

FIGURE 18.11

Case II $\lambda^2 - \omega^2 = 0$. The system is said to be **critically damped**, since any slight decrease in the damping force would result in oscillatory motion. The general solution of (6) is

$$x(t) = C_1 e^{m_1 t} + C_2 t e^{m_1 t}$$

or

$$x(t) = e^{-\lambda t}(C_1 + C_2 t) \tag{8}$$

Some graphs of typical motion are given in Figure 18.10. Notice that the motion is quite similar to that of an overdamped system. It is also apparent from (8) that the mass can pass through the equilibrium position at most one time.

Case III $\lambda^2 - \omega^2 < 0$. In this case the system is said to be **underdamped**, since the damping coefficient is small compared to the spring constant. The roots m_1 and m_2 are now complex:

$$m_1 = -\lambda + \sqrt{\omega^2 - \lambda^2}\, i \qquad m_2 = -\lambda - \sqrt{\omega^2 - \lambda^2}\, i$$

and so the general solution of equation (6) is

$$x(t) = e^{-\lambda t}(C_1 \cos \sqrt{\omega^2 - \lambda^2}\, t + C_2 \sin \sqrt{\omega^2 - \lambda^2}\, t) \tag{9}$$

As indicated in Figure 18.11, the motion described by (9) is oscillatory, but because of the coefficient $e^{-\lambda t}$ we see that the amplitudes of vibration $\to 0$ as $t \to \infty$.

EXAMPLE 2 An 8-lb weight stretches a spring 2 ft. Assuming a damping force numerically equal to two times the instantaneous velocity acts on the system, determine the equation of motion if the weight is released from the equilibrium position with an upward velocity of 3 ft/s.

Solution From Hooke's law we have

$$8 = k(2) \qquad k = 4 \text{ lb/ft}$$

and from $m = W/g$,

$$m = \frac{8}{32} = \frac{1}{4} \text{ slug}$$

Thus, the differential equation of motion is

$$\frac{1}{4}\frac{d^2x}{dt^2} = -4x - 2\frac{dx}{dt} \quad \text{or} \quad \frac{d^2x}{dt^2} + 8\frac{dx}{dt} + 16x = 0$$

The initial conditions are

$$x(0) = 0 \qquad \left.\frac{dx}{dt}\right|_{t=0} = -3$$

Now the auxiliary equation of the differential equation is

$$m^2 + 8m + 16 = (m + 4)^2 = 0$$

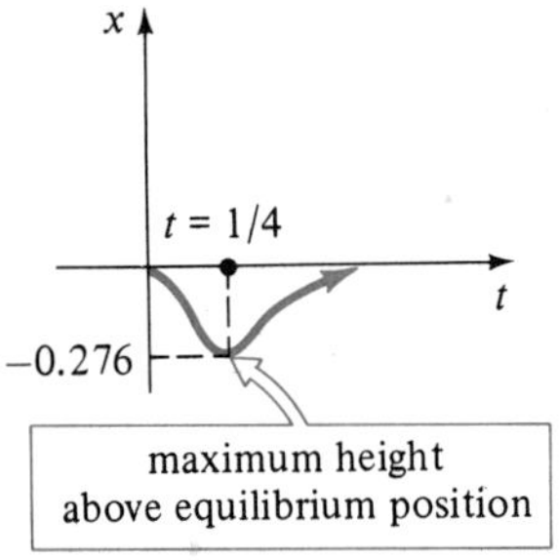

FIGURE 18.12

so that $m_1 = m_2 = -4$. Hence, the system is critically damped and

$$x(t) = C_1e^{-4t} + C_2te^{-4t}$$

The initial condition $x(0) = 0$ immediately demands that $C_1 = 0$, whereas using $x'(0) = -3$ gives $C_2 = -3$. Thus, the equation of motion is

$$x(t) = -3te^{-4t}$$

To graph $x(t)$ we find the time at which $x'(t) = 0$:

$$x'(t) = -3(-4te^{-4t} + e^{-4t}) = -3e^{-4t}(1 - 4t)$$

Clearly, $x'(t) = 0$ when $t = \frac{1}{4}$. The corresponding displacement is

$$x\left(\frac{1}{4}\right) = -3\left(\frac{1}{4}\right)e^{-1} = -0.276 \text{ ft}$$

As shown in Figure 18.12, we interpret this value to mean that the weight reaches a maximum height of 0.276 ft above the equilibrium position. □

EXAMPLE 3 A 16-lb weight is attached to a 5-ft-long spring. At equilibrium the spring measures 8.2 ft. If the weight is pushed up and released from rest at a point 2 ft above the equilibrium position, find the displacements $x(t)$ if it is further known that the surrounding medium offers a resistance numerically equal to the instantaneous velocity.

Solution The elongation of the spring after the weight is attached is $8.2 - 5 = 3.2$ ft, so it follows from Hooke's law that

$$16 = k(3.2) \qquad k = 5 \text{ lb/ft}$$

In addition,

$$m = \frac{16}{32} = \frac{1}{2} \text{ slug}$$

so that the differential equation is given by

$$\frac{1}{2}\frac{d^2x}{dt^2} = -5x - \frac{dx}{dt} \quad \text{or} \quad \frac{d^2x}{dt^2} + 2\frac{dx}{dt} + 10x = 0$$

This latter equation is solved subject to the conditions

$$x(0) = -2 \qquad \left.\frac{dx}{dt}\right|_{t=0} = 0$$

Proceeding, we find that the roots of $m^2 + 2m + 10 = 0$ are $m_1 = -1 + 3i$ and $m_2 = -1 - 3i$, which then implies the system is underdamped and

$$x(t) = e^{-t}(C_1\cos 3t + C_2\sin 3t)$$

Now $x(0) = -2 = C_1$

$$x(t) = e^{-t}(-2\cos 3t + C_2\sin 3t)$$
$$x'(t) = e^{-t}(6\sin 3t + 3C_2\cos 3t) - e^{-t}(-2\cos 3t + C_2\sin 3t)$$
$$x'(0) = 0 = 3C_2 + 2$$

which gives $C_2 = -\frac{2}{3}$. Thus, we finally obtain

$$x(t) = e^{-t}\left(-2\cos 3t - \frac{2}{3}\sin 3t\right)$$ □

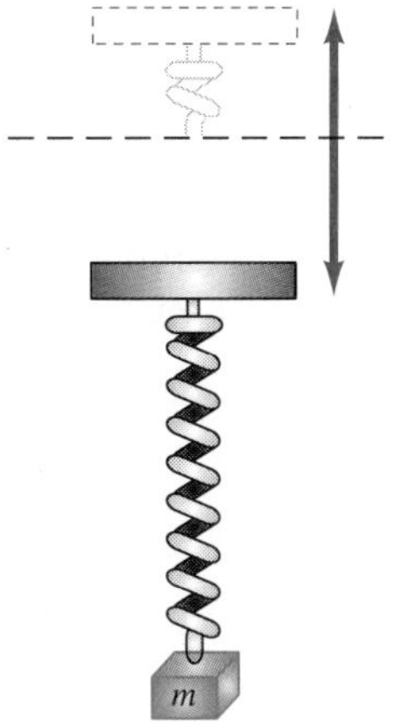

FIGURE 18.13

Forced Motion Suppose we now take into consideration an external force $f(t)$ acting on a vibrating mass on a spring. For example, $f(t)$ could represent a driving force causing an oscillatory vertical motion of the support of the spring (Figure 18.13). The inclusion of $f(t)$ in the formulation of Newton's second law gives

$$m\frac{d^2x}{dt^2} = -kx - \beta\frac{dx}{dt} + f(t)$$

$$\frac{d^2x}{dt^2} + \frac{\beta}{m}\frac{dx}{dt} + \frac{k}{m}x = \frac{f(t)}{m} \quad \text{or} \quad \frac{d^2x}{dt^2} + 2\lambda\frac{dx}{dt} + \omega^2 x = F(t)$$

where $F(t) = f(t)/m$ and, as before, $2\lambda = \beta/m$, $\omega^2 = k/m$. To solve the latter nonhomogeneous equation, we can employ either the method of undetermined coefficients or variation of parameters.

The next example illustrates undamped forced motion.

EXAMPLE 4 Solve the initial-value problem

$$\frac{d^2x}{dt^2} + \omega^2 x = F_0\sin\gamma t, \qquad F_0 = \text{constant}$$

$$x(0) = 0 \qquad \left.\frac{dx}{dt}\right|_{t=0} = 0$$

Solution The complementary function is $x_c(t) = C_1\cos\omega t + C_2\sin\omega t$. To obtain a particular solution we use the method of undetermined coefficients and assume $x_p(t) = A\cos\gamma t + B\sin\gamma t$, so that

$$x_p' = -A\gamma\sin\gamma t + B\gamma\cos\gamma t$$
$$x_p'' = -A\gamma^2\cos\gamma t - B\gamma^2\sin\gamma t$$
$$x_p'' + \omega^2 x_p = A(\omega^2 - \gamma^2)\cos\gamma t + B(\omega^2 - \gamma^2)\sin\gamma t$$
$$= F_0\sin\gamma t$$

It follows that $A = 0$ and $B = \dfrac{F_0}{\omega^2 - \gamma^2}$, $\gamma \neq \omega$

Therefore,

$$x_p(t) = \frac{F_0}{\omega^2 - \gamma^2}\sin\gamma t$$

Applying the given initial conditions to the general solution

$$x(t) = C_1\cos \omega t + C_2\sin \omega t + \frac{F_0}{\omega^2 - \gamma^2}\sin \gamma t$$

yields $C_1 = 0$ and $C_2 = -\gamma F_0/\omega(\omega^2 - \gamma^2)$. Thus, the solution is

$$x(t) = \frac{F_0}{\omega(\omega^2 - \gamma^2)}(-\gamma \sin \omega t + \omega \sin \gamma t), \qquad \gamma \neq \omega \qquad (10)$$ □

Pure Resonance Although (10) is not defined for $\gamma = \omega$, it is interesting to observe that its limiting value as $\gamma \to \omega$ can be obtained by applying L'Hôpital's Rule. This limiting process is analogous to "tuning in" the frequency of the driving force $(\gamma/2\pi)$ to the frequency of free vibrations $(\omega/2\pi)$. Intuitively, we expect that over a length of time we should be able to substantially increase the amplitudes of vibration. For $\gamma = \omega$, we define the solution to be

$$\begin{aligned} x(t) &= \lim_{\gamma \to \omega} F_0 \frac{-\gamma \sin \omega t + \omega \sin \gamma t}{\omega(\omega^2 - \gamma^2)} \\ &= F_0 \lim_{\gamma \to \omega} \frac{\dfrac{d}{d\gamma}[-\gamma \sin \omega t + \omega \sin \gamma t]}{\dfrac{d}{d\gamma}[\omega^3 - \omega\gamma^2]} \\ &= F_0 \lim_{\gamma \to \omega} \frac{-\sin \omega t + \omega t \cos \gamma t}{-2\omega\gamma} \\ &= \frac{F_0}{2\omega^2}[\sin \omega t - \omega t \cos \omega t] \\ &= \frac{F_0}{2\omega^2}\sin \omega t - \frac{F_0}{2\omega} t \cos \omega t \end{aligned} \qquad (11)$$

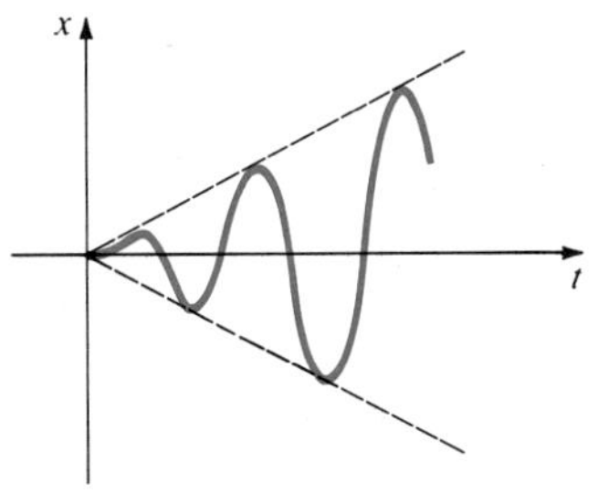

FIGURE 18.14

As suspected, when $t \to \infty$ the displacements become large; in fact, $|x(t)| \to \infty$. The phenomenon we have just described is known as **pure resonance**. The graph in Figure 18.14 displays typical motion in this case.

In conclusion, it should be noted that there is no actual need to use a limiting process on (10) to obtain the solution for $\gamma = \omega$. Alternatively, (11) follows by solving the initial-value problem

$$\frac{d^2x}{dt^2} + \omega^2 x = F_0\sin \omega t$$

$$x(0) = 0 \qquad \left.\frac{dx}{dt}\right|_{t=0} = 0$$

directly by conventional methods.

▲ ***Remark*** If a mechanical system were actually described by a function such as (11), it would necessarily fail; large oscillations of a weight on a spring would eventually force the spring beyond its elastic limit. One might argue too that the resonating model presented (Figure 18.14) is completely unrealistic, since it ignores the retarding effects of ever-present damping forces. While it is true that pure resonance cannot occur when the smallest amount of damping is taken into consideration, nevertheless, large and equally destructive amplitudes of vibration (though bounded as $t \to \infty$) could take place.

If you have ever looked out a window while in flight, you have probably observed that the wings on an airplane are not perfectly rigid. A reasonable amount of flutter is not only tolerated but necessary to prevent the wing from snapping like a stick of peppermint candy. In late 1959 and early 1960 two commercial plane crashes occurred, with a then relatively new model of prop-jet, that illustrate the destructive effects of large mechanical oscillations.

The unusual aspect of these crashes was that they both happened while the planes were in midflight. Barring mid-air collisions, the safest period during any flight is when the plane has attained its cruising altitude. It is well known that a plane is most vulnerable to an accident when it is least maneuverable—namely, during either takeoff or landing. So having two planes simply fall out of the sky was not only a tragedy but an embarrassment to the aircraft industry as well as a thoroughly puzzling problem to aerodynamic engineers. In crashes of this sort, a structural failure of some kind is immediately suspected. After a subsequent and massive technical investigation, the problem was eventually traced in each case to an outboard engine and engine housing. Roughly, it was determined that when each plane surpassed a critical speed of approximately 400 mph, a propeller and engine began to wobble, causing a gyroscopic force, which could not be quelled or damped by the engine housing. This external vibrational force was then transferred to the already oscillating wing. This in itself need not have been destructively dangerous, since aircraft wings are designed to withstand the stress of unusual and excessive forces. (In fact, the particular wing in question was so incredibly strong that test engineers and pilots, who were deliberately trying, failed to snap a wing under every conceivable flight condition.) Unfortunately, after a short period of time during which the engine wobbled rapidly, the frequency of the impressed force actually slowed to a point where it approached and finally coincided with the maximum frequency of wing flutter (around 3 cycles per second). The resulting resonance situation finally accomplished what the test engineers could not do; namely, the amplitudes of wing flutter became large enough to snap the wing. See Figure 18.15. The problem was solved in two steps. All models of this particular plane were required to fly at speeds substantially below 400 mph until each plane could be modified by considerably strengthening (or stiffening) the engine housings. A strengthened engine housing was shown to be able to impart a damping effect capable of preventing the critical resonance phenomenon even in the unlikely event of a subsequent engine wobble.*

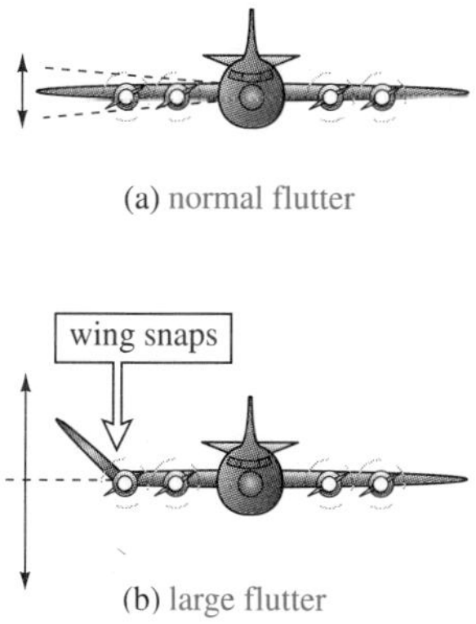

(a) normal flutter

(b) large flutter

FIGURE 18.15

*For a fascinating nontechnical account of the investigation, see Robert J. Serling, *Loud and Clear* (New York: Dell, 1970), Chapter 5.

FIGURE 18.16 Frequencies of the forces applied to the bridge by the wind were close to the natural frequencies of the bridge. Photo courtesy of Wide World Photos/AP.

You may be aware that soldiers usually do not march in step across bridges. The reason for breaking stride is simply to avoid any possibility of resonance occurring between the natural vibrations inherent in the bridge's structure and the frequency of the external force of a multitude of feet stomping in unison on the bridge.

Bridges are good examples of vibrating mechanical systems that are constantly being subjected to external forces, from people walking, cars and trucks driving on them, water pushing against their foundations, or wind blowing against their superstructures. On November 7, 1940, the Tacoma Narrows Bridge at Puget Sound in the state of Washington collapsed. However, the crash came as no surprise, since "Galloping Gertie," as the bridge was called by local residents, was famous for a vertical undulating motion of its roadway, which gave many motorists a very exciting crossing. On November 7, only four months after its grand opening, the amplitudes of these undulations became so large that the bridge failed and a substantial portion was sent splashing into the water below. In the investigation that followed, it was found that a poorly designed superstructure caused the wind blowing across it to vortex in a periodic manner. When the frequencies of these periodic forces approached the natural frequencies of the bridge, large vertical upheavals of the road resulted. In a word, the bridge was another victim of the destructive effect of mechanical resonance. Since this disaster developed over a matter of months, there was sufficient opportunity to record on film the strange and frightening phenomenon of a bucking and heaving bridge and its ultimate collapse.* See Figure 18.16.

Acoustic vibrations can be as destructive as large mechanical vibrations. In television commercials, jazz singers have inflicted destruction on the lowly wine glass. See Figure 18.17. Sounds from organs and piccolos have been known to crack windows.

As the horns blew, the people began to shout. When they heard the signal horn, they raised a tremendous shout. The wall collapsed. . . . (*Joshua 6:20*)

FIGURE 18.17 The shattering effect of acoustic resonance. Photo © 1987 Memtek Products.

*National Committee for Fluid Mechanics Films, Educational Services, Inc., Watertown, Mass. See also *American Society of Civil Engineers: Proceedings*, "Failure of the Tacoma Narrows Bridge," Vol. 69, Dec. 1943, pp. 1555–86.

Did the power of acoustic resonance cause the walls of Jericho to tumble down? This is the conjecture of some contemporary scholars.

The phenomenon of resonance is not always destructive. It is resonance of an electrical circuit that enables a radio to be tuned to a specific station.

EXERCISES 18.6

Answers to odd-numbered problems begin on page A-110.

In Problems 1 and 2 state in words a possible physical interpretation of the given initial-value problem.

1. $\frac{4}{32}x'' + 3x = 0$; $x(0) = -3$, $x'(0) = -2$

2. $\frac{1}{16}x'' + 4x = 0$; $x(0) = 0.7$, $x'(0) = 0$

3. An 8-lb weight attached to a spring exhibits simple harmonic motion. Determine the equation of motion if the spring constant is 1 lb/ft and if the weight is released 6 in below the equilibrium position with a downward velocity of $\frac{3}{2}$ ft/s.

4. A 24-lb weight attached to a spring exhibits simple harmonic motion. When placed on the spring, the weight stretched the spring 4 in. Find the equation of motion if the weight is released from rest from a point 3 in above the equilibrium position.

5. A force of 400 newtons stretches a spring 2 m. A mass of 50 kg attached to the spring exhibits simple harmonic motion. Find the equation of motion if the mass is released from the equilibrium position with an upward velocity of 10 m/s.

6. A mass weighing 2 lb attached to a spring exhibits simple harmonic motion. At $t = 0$ the mass is released from a point 8 in below the equilibrium position with an upward velocity of $\frac{4}{3}$ ft/s. If the spring constant is $k = 4$ lb/ft, find the equation of motion.

In Problems 7 and 8 state in words a possible physical interpretation of the given initial-value problem.

7. $\frac{1}{16}x'' + 2x' + x = 0$; $x(0) = 0$, $\left.\frac{dx}{dt}\right|_{t=0} = -1.5$

8. $\frac{16}{32}x'' + x' + 2x = 0$; $x(0) = -2$, $x'(0) = 1$

9. A 4-lb weight attached to a spring exhibits free damped motion. The spring constant is 2 lb/ft and the medium offers a resistance to the motion of the weight numerically equal to the instantaneous velocity. If the weight is released from a point 1 ft above the equilibrium position with a downward velocity of 8 ft/s, determine the time that the weight passes through the equilibrium position. Find the time for which the weight attains its maximum displacement from the equilibrium position. What is the position of the weight at this instant?

10. A mass of 40 g stretches a spring 10 cm. A damping device imparts a resistance to motion numerically equal to 560 times the instantaneous velocity. Find the equation of free motion if the mass is released from the equilibrium position with a downward velocity of 2 cm/s.

11. After a 10-lb weight is attached to a 5-ft spring, the spring measures 7 ft. The 10-lb weight is removed and replaced with an 8-lb weight and the entire system is placed in a medium offering a resistance numerically equal to the instantaneous velocity. Find the equation of motion if the weight is released $\frac{1}{2}$ ft below the equilibrium position with a downward velocity of 1 ft/s.

12. A 24-lb weight stretches a spring 4 ft. The subsequent motion takes place in a medium offering a resistance numerically equal to β ($\beta > 0$) times the instantaneous velocity. If the weight starts from the equilibrium position with an upward velocity of 2 ft/s, show that if $\beta > 3\sqrt{2}$, the equation of motion is

$$x(t) = \frac{-3}{\sqrt{\beta^2 - 18}} e^{-2\beta t/3} \sinh \frac{2}{3}\sqrt{\beta^2 - 18}\,t$$

13. A 10-lb weight attached to a spring stretches it 2 ft. The system is then set in motion in a medium that offers a resistance numerically equal to β ($\beta > 0$) times the instantaneous velocity. Determine the values of β so that the motion is **(a)** overdamped, **(b)** critically damped, and **(c)** underdamped.

14. A mass of 1 slug when attached to a spring stretches it 2 ft and then comes to rest in the equilibrium position. Starting at $t = 0$, an external force equal to $f(t) = 8 \sin 4t$ is applied to the system. Find the equation of motion if the surrounding medium offers a damping force numerically equal to 8 times the instantaneous velocity.

15. Solve Problem 14 when $f(t) = e^{-t} \sin 4t$. Analyze the displacements for $t \to \infty$.

16. Solve and interpret the initial-value problem:

$$\frac{d^2x}{dt^2} + 9x = 5 \sin 3t; \quad x(0) = 2, \quad \left.\frac{dx}{dt}\right|_{t=0} = 0$$

CHAPTER 18 REVIEW EXERCISES *Answers begin on page A-111.*

In Problems 1–8 answer true or false.

1. The function $f(x, y) = x^2/y^2 + 9$ is homogeneous of degree 0. ________

2. The first-order differential equation $x^2y' + 2xy = y^2$ is homogeneous. ________

3. Every separable first-order differential equation is homogeneous. ________

4. If y_1 is a solution of $ay'' + by' + cy = 0$, a, b, c constants, then C_1y_1 is also a solution for every real number C_1. ________

5. The general solution of $y'' + y = 0$ is $y = C_1\cos x + C_2\sin x$. ________

6. $y_1 = e^{-x}$ and $y_2 = 0$ are linearly independent solutions of the differential equation $y'' + y' = 0$. ________

7. Undamped and unforced motion of a mass on a spring is called simple harmonic motion. ________

8. Pure resonance cannot occur when damping is present. ________

In Problems 9 and 10 solve the given differential equation.

9. $y\,dx + x\,dy = 0$

10. $xy\dfrac{dy}{dx} = 3y^2 + x^2$

In Problems 11 and 12 solve the given initial-value problem.

11. $(x + ye^{y/x})\,dx - xe^{y/x}\,dy = 0$, $y(1) = 0$

12. $y\,dx + \left(y\cos\dfrac{x}{y} - x\right)dy = 0$, $y(0) = 2$

In Problems 13–18 find the general solution of the given differential equation.

13. $y'' - 2y' - 2y = 0$

14. $y'' - 8y = 0$

15. $y'' - 3y' - 10y = 0$

16. $4y'' + 20y' + 25y = 0$

17. $9y'' + y = 0$

18. $2y'' - 5y' = 0$

In Problems 19 and 20 solve the given initial-value problem.

19. $y'' + 36y = 0$, $y(\pi/2) = 24$, $y'(\pi/2) = -18$

20. $y'' + 4y' + 4y = 0$, $y(0) = 0$, $y'(0) = 0$

In Problems 21 and 22 find power series solutions of the given differential equation.

21. $y'' + xy = 0$

22. $(x - 1)y'' + 3y = 0$

In Problems 23 and 24 solve each differential equation by the method of variation of parameters.

23. $y'' - 2y' + 2y = e^x\tan x$

24. $y'' - y = 2e^x/(e^x + e^{-x})$

25. Solve $y'' + y = \sec^3 x$ subject to $y(0) = 1$, $y'(0) = \frac{1}{2}$.

26. Without solving, determine the form of a particular solution of

$$y'' + 6y' + 9y = 5x^2 - 6x + 4x^2e^{2x} + 3e^{5x} - 2e^x\sin 3x$$

27. Use the method of undetermined coefficients to solve $y'' - y' - 12y = (x + 1)e^{2x}$.

28. The vertical motion of a weight attached to a spring is described by the initial-value problem

$$\frac{1}{4}\frac{d^2x}{dt^2} + \frac{dx}{dt} + x = 0; \quad x(0) = 4, \quad x'(0) = 2$$

Determine the maximum vertical displacement.

29. A spring with constant $k = 2$ is suspended in a liquid that offers a damping force numerically equal to 4 times the instantaneous velocity. If a mass m is suspended from the spring, determine the values of m for which the subsequent free motion is nonoscillatory.

30. Find a particular solution for $\dfrac{d^2x}{dt^2} + 2\lambda\dfrac{dx}{dt} + \omega^2x = A$, where A is a constant force.

31. A 4-lb weight is suspended from a spring whose constant is 3 lb/ft. The entire system is immersed in a fluid offering a damping force numerically equal to the instantaneous velocity. Beginning at $t = 0$, an external force equal to $f(t) = e^{-t}$ is impressed on the system. Determine the equation of motion if the weight is released from rest at a point 2 ft below the equilibrium position.

32. A weight of W lb stretches one spring $\frac{1}{2}$ ft and stretches a different spring $\frac{1}{4}$ ft. If the two springs are attached in series, the effective spring constant k is given by $1/k = 1/k_1 + 1/k_2$. The weight W is then attached to the double spring, as shown in Figure 18.18. Assume that the motion is free and that there is no damping force present. Determine the equation of motion if the weight is released at a point 1 ft below the equilibrium position with a downward velocity of $\frac{2}{3}$ ft/s. Show that the maximum speed of the weight is $\frac{2}{3}\sqrt{3g + 1}$.

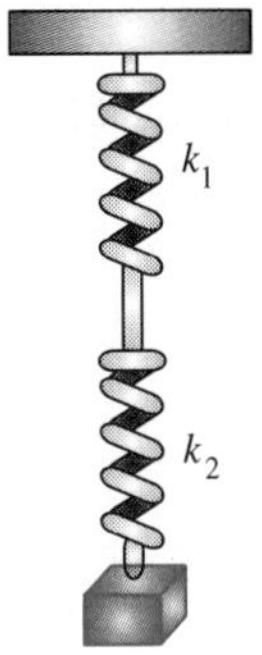

FIGURE 18.18

CALCULATOR/COMPUTER ACTIVITIES

CHAPTER 13 Vectors and 3-Space *Answers to odd-numbered problems begin on page A-120.*

13.1 A vector in 2-space can be represented as a line segment on the graphical screen of a calculator. To do this, an initial point and a terminal point in the plane must be specified. As with a conic section or a polar equation, this will be more meaningful if the scales along the two axes are equal. A short calculator program could even be written to draw a small arrow tip at one end to give the plotted vector a sense of direction.

In Problems 1–6 graph the given vector in 2-space by locating the indicated initial point.

1. $\mathbf{a} = \langle 2.75, 1.29 \rangle$ initially at the origin; $\mathbf{b} = \langle -4.55, 3.8 \rangle$ initially at the terminal point of $\mathbf{a}$; $\mathbf{c} = \mathbf{a} + \mathbf{b}$ initially at the origin

2. $\mathbf{a} = \langle -63, 21 \rangle$ initially at the origin; $\mathbf{b} = \langle 45, -17 \rangle$ initially at the terminal point of $\mathbf{a}$; $\mathbf{c} = \langle 39, 52 \rangle$ initially at the terminal point of $\mathbf{b}$; $\mathbf{w} = \langle -81, -38 \rangle$ initially at the terminal point of $\mathbf{b}$

3. $\mathbf{a} = \left\langle \cos\dfrac{\pi}{15}, \sin\dfrac{\pi}{15} \right\rangle$ initially at the origin;
$\mathbf{b} = \left\langle \cos\dfrac{6\pi}{15}, \sin\dfrac{6\pi}{15} \right\rangle$ initially at the origin;
$\mathbf{c} = \left\langle \cos\dfrac{11\pi}{15}, \sin\dfrac{11\pi}{15} \right\rangle$ initially at the origin;
$\mathbf{w} = \left\langle \cos\dfrac{16\pi}{15}, \sin\dfrac{16\pi}{15} \right\rangle$ initially at the origin

4. $\mathbf{a} = \langle \cos 1, \sin 1 \rangle$ initially at the origin; $\mathbf{b} = \langle \cos 2, \sin 2 \rangle$ initially at the origin; $\mathbf{c} = \langle \cos 3, \sin 3 \rangle$ initially at the origin; $\mathbf{w} = \langle \cos 4, \sin 4 \rangle$ initially at the origin

5. $\mathbf{a} = \langle 35.2, -17.9 \rangle$ initially at the origin; $\mathbf{b} = \langle 25.1, 48.5 \rangle$ once initially at the origin and a second time initially at the terminal point of $\mathbf{b}$; $\mathbf{c} = \mathbf{a} + \mathbf{b}$ initially at the origin; $\mathbf{w} = \mathbf{b} - \mathbf{a}$ initially at the terminal point of $\mathbf{a}$

6. $\mathbf{a} = \langle 1.11, 5.32 \rangle$ initially at the origin; $\mathbf{b} = \langle 2.75, 4.99 \rangle$ initially at the origin and then initially at the terminal point of $\mathbf{b}$; $\mathbf{c} = \mathbf{a} - \mathbf{b}$ initially at the origin; $\mathbf{w} = \mathbf{b} - \mathbf{a}$ initially at the terminal point of $\mathbf{a}$

13.2 A point or a vector in 3-space is harder to represent on the graphical screen of a calculator or computer because the screen really has only two dimensions. Figures in a text use shading and many auxiliary lines to give the illusion of three dimensions. The storage memory of a calculator or computer can be used to retain the three coordinates of a point or the three components of a vector in 3-space. A vector such as $\mathbf{a} = \langle 2, -5, 3 \rangle$ may be stored in the form [2 −5 3] or [2, −5, 3] on a Hewlett-Packard HP-28S calculator or in computer applications such as MATLAB or Maple. In most programming languages, the best storage syntax for a vector is a one-dimensional array **A**, where $A(1) = 2$, $A(2) = -5$, and $A(3) = 3$. The CASIO family of graphic calculators also allows array-type memories, but $A[1]$ is really the standard memory B, $A[2]$ is the same as C, and $A[3]$ is just D. Thus, on such a CASIO, care must be taken *not* to use both A and one of the following two letters in the alphabet at the same time as array-type memories to represent two vectors in 3-space. A two-dimensional array can be considered as a **matrix**. (See Appendix I.) For example, the 2×3 array with $B(1, 1) = 2$, $B(1, 2) = -5$, $B(1, 3) = 3$, $B(2, 1) = 7$, $B(2, 2) = 0$, and $B(2, 3) = -1$ represents the matrix

$$\mathbf{B} = \begin{bmatrix} 2 & -5 & 3 \\ 7 & 0 & -1 \end{bmatrix}$$

This two-dimensional array is said to have 2 rows and 3 columns. The SHARP EL-5200 and the Texas Instruments TI-81 graphic calculators have only two-dimensional arrays. One way to store a three-dimensional vector is to use a 1×3 (or 3×1) matrix. Since the TI-81 is limited to only three matrices, another way to store more vectors is to consider each row of a 6×3 matrix as a different vector. Programming languages and computer algebra systems even support higher-dimensional arrays. Most calculators and computer applications provide for some of the vector algebraic operations such as multiplying a vector by a scalar, adding two vectors, or finding the length of a vector.

In Problems 7–12 store the given vectors in a calculator or computer application and compute the indicated quantities. Try to use as many built-in operations as are provided to carry out these tasks.

7. $\mathbf{a} = \langle 2, -5, 3\rangle, \mathbf{b} = \langle 4, 13, -9\rangle$; $\mathbf{a} + \mathbf{b}, \mathbf{b} - \mathbf{a}, |\mathbf{a}|, |\mathbf{b}|$

8. $\mathbf{a} = \langle 75.214, 15.281, 302.64\rangle$, $\mathbf{b} = \langle -29.417, 1.3656, 964.18\rangle$; $7\mathbf{a} + 3\mathbf{b}, \mathbf{b} - \mathbf{a}, |\mathbf{a}|, |5\mathbf{b}|$

9. $\mathbf{a} = \langle -125, 459, 21\rangle, \mathbf{b} = \langle 572, 337, 803\rangle$, $\mathbf{c} = \langle 232, 928, -640\rangle$; $0.0571\mathbf{c}, \mathbf{b} - \mathbf{c}, |\mathbf{a} + \mathbf{c}|, |\mathbf{a} + \mathbf{b} + \mathbf{c}|$

10. $\mathbf{a} = \langle 4831, -7329, 3602\rangle, \mathbf{b} = \langle 1185, -5492, 999\rangle$, $\mathbf{c} = \langle -8010, -3787, -267\rangle$; $\dfrac{\mathbf{b}}{|\mathbf{b}|}, \mathbf{b} + \mathbf{c}, |47\mathbf{a} + 23\mathbf{b}|$, $|\mathbf{a} - \mathbf{b} + \mathbf{c}|$

11. $\mathbf{a} = 81\mathbf{i} + 68\mathbf{j} - 24\mathbf{k}, \mathbf{b} = -47\mathbf{i} + 35\mathbf{j} - 77\mathbf{k}$, $\mathbf{c} = 53\mathbf{i} + 21\mathbf{j} + 94\mathbf{k}$; $\dfrac{\mathbf{c}}{|\mathbf{c}|}, \dfrac{\mathbf{b} + \mathbf{c}}{|\mathbf{b} + \mathbf{c}|}, |18\mathbf{a} + 67\mathbf{c}|, \mathbf{a} - \mathbf{b} + \mathbf{c}$

12. $\mathbf{a} = 0.062\mathbf{i} - 0.094\mathbf{j} - 0.058\mathbf{k}$, $\mathbf{b} = -0.26\mathbf{i} + 0.72\mathbf{j} + 0.34\mathbf{k}, \mathbf{c} = 7\mathbf{i} - 4\mathbf{j} + 5\mathbf{k}$; $|\mathbf{a}|, |\mathbf{b}|, |\mathbf{c}|, 10\mathbf{a} - \mathbf{c}, \mathbf{a} + \mathbf{b} + \mathbf{c}$

13.3 The **dot product** is closely related to **matrix multiplication**. Although the dot product is implemented only on the most advanced calculators and computer algebra systems, matrix multiplication is much more commonly provided. Consider, for example, the case where a 2×3 matrix is multiplied by a 3×3 matrix:

$$\begin{bmatrix} 2 & -4 & 9 \\ -3 & 1 & 0 \end{bmatrix}\begin{bmatrix} 1 & 1 & 5 \\ 7 & 8 & -3 \\ 2 & 6 & 4 \end{bmatrix} = \begin{bmatrix} -8 & 24 & 22 \\ 4 & 5 & -18 \end{bmatrix}$$

For example, note that

$$\langle 2, -4, 9\rangle \cdot \langle 1, 7, 2\rangle = -8, \quad \langle 2, -4, 9\rangle \cdot \langle 1, 8, 6\rangle = 24,$$

and

$$\langle -3, 1, 0\rangle \cdot \langle 5, -3, 4\rangle = -18$$

In general, the dot product of a row vector from the first matrix (reading left to right) and a column vector from the second matrix will be an entry in the matrix product or third matrix. Another matrix operation useful here is the **transpose**, which consists of interchanging rows and columns. Thus, if a vector **a** is stored in a 1×3 matrix **A** and a vector **b** is stored in a 1×3 matrix **B**, then

$$\mathbf{a} \cdot \mathbf{b} = \mathbf{A}\mathbf{B}^t$$

where $\mathbf{B}^t$ denotes the transpose of the matrix **B**. This formula also provides a method for computing the magnitude of a vector **a**, since $|\mathbf{a}| = \sqrt{\mathbf{a} \cdot \mathbf{a}}$.

In Problems 13–18 store the given vectors in a calculator or computer application and compute the indicated quantities. Try to use as many built-in operations as are provided to carry out these tasks.

13. $\mathbf{a} = \langle 7, -1, 3\rangle, \mathbf{b} = \langle 5, 18, -4\rangle$; $\mathbf{a} \cdot \mathbf{b}, \mathbf{b} \cdot \mathbf{a}, |\mathbf{a}|, |\mathbf{b}|$

14. $\mathbf{a} = \langle 75.214, 15.281, 302.64\rangle$, $\mathbf{b} = \langle -29.417, 1.3656, 964.18\rangle$; $7\mathbf{a} \cdot 3\mathbf{b}, |\mathbf{b} - \mathbf{a}|, |\mathbf{a}|$, $(\mathbf{a} \cdot \mathbf{b})\mathbf{a}$

15. $\mathbf{a} = \langle -125, 459, 21\rangle, \mathbf{b} = \langle 572, 337, 803\rangle$, $\mathbf{c} = \langle 232, 928, -640\rangle$; $\mathbf{a} \cdot \mathbf{c}, \mathbf{b} \cdot \mathbf{c}, |(\mathbf{a} \cdot \mathbf{c})\mathbf{a}|, |(\mathbf{a} + \mathbf{b}) \cdot \mathbf{c}|$

16. $\mathbf{a} = \langle 4831, -7329, 3602\rangle, \mathbf{b} = \langle 1185, -5492, 999\rangle$, $\mathbf{c} = \langle -8010, -3787, -267\rangle$; $\dfrac{\mathbf{b} \cdot \mathbf{c}}{|\mathbf{b}||\mathbf{c}|}, \mathbf{a} \cdot \mathbf{c}, |\mathbf{a} \cdot \mathbf{b}|, |(\mathbf{a} \cdot \mathbf{b})\mathbf{c}|$

17. $\mathbf{a} = 81\mathbf{i} + 68\mathbf{j} - 24\mathbf{k}, \mathbf{b} = -47\mathbf{i} + 35\mathbf{j} - 77\mathbf{k}$, $\mathbf{c} = 53\mathbf{i} + 21\mathbf{j} + 94\mathbf{k}$; $|\mathbf{c}|^2, (\mathbf{a} + \mathbf{c}) \cdot \mathbf{b}, (\mathbf{a} + \mathbf{b}) \cdot \mathbf{c}$, $\mathbf{a} \cdot (\mathbf{b} + \mathbf{c})$

18. $\mathbf{a} = 0.062\mathbf{i} - 0.094\mathbf{j} - 0.058\mathbf{k}$, $\mathbf{b} = -0.26\mathbf{i} + 0.72\mathbf{j} + 0.34\mathbf{k}, \mathbf{c} = 7\mathbf{i} - 4\mathbf{j} + 5\mathbf{k}$; $\dfrac{\mathbf{a} \cdot \mathbf{b}}{|\mathbf{a}||\mathbf{b}|}, \dfrac{\mathbf{a} \cdot \mathbf{c}}{|\mathbf{a}||\mathbf{c}|}, \dfrac{\mathbf{b} \cdot \mathbf{c}}{|\mathbf{b}||\mathbf{c}|}$

13.4 The **cross product** is closely related to the **determinant of a matrix** (see Appendix I). While the cross product is implemented on only the most advanced calculators and computer algebra systems, the determinant of a matrix is commonly provided. It is very unlikely that a calculator will be able to use the form of the cross product given in (8) of this section. Given the difficulty selecting the parts of each vector needed in the determinants of order 2 in (7), it may be easier to simply program formula (6). The **triple scalar product** is a determinant of order 3 that can be computed by storing the three vectors as the three rows of a 3×3 matrix and using the command to compute the determinant of this matrix.

In Problems 19–22 store the given vectors in a calculator or computer application and compute the indicated quantities. Try to use as many built-in operations as are provided to carry out these tasks.

19. $\mathbf{a} = \langle -12, 59, 21\rangle, \mathbf{b} = \langle 52, 33, 87\rangle$, $\mathbf{c} = \langle 32, -92, 64\rangle$; $\mathbf{a} \times \mathbf{c}, \mathbf{b} \times \mathbf{c}, \mathbf{a} \times \mathbf{b}, \mathbf{a} \cdot (\mathbf{b} \times \mathbf{c})$.

20. $\mathbf{a} = \langle 831, -129, 702\rangle, \mathbf{b} = \langle 285, -54, 39\rangle$, $\mathbf{c} = \langle -610, 378, -467\rangle$; $\mathbf{a} \times \mathbf{c}, \mathbf{b} \times \mathbf{c}, \mathbf{a} \times (\mathbf{b} \times \mathbf{c})$, $\mathbf{a} \cdot (\mathbf{b} \times \mathbf{c})$.

21. $\mathbf{a} = 25\mathbf{i} + 17\mathbf{j} - 91\mathbf{k}, \mathbf{b} = 34\mathbf{i} - 67\mathbf{j} - 58\mathbf{k}$, $\mathbf{c} = 14\mathbf{i} + 26\mathbf{j} + 84\mathbf{k}$; $\mathbf{a} \times \mathbf{c}, \mathbf{b} \times \mathbf{c}, \mathbf{a} \times (\mathbf{b} \times \mathbf{c}), \mathbf{a} \cdot (\mathbf{b} \times \mathbf{c})$.

22. $\mathbf{a} = 0.011\mathbf{i} - 0.064\mathbf{j} + 0.039\mathbf{k}$, $\mathbf{b} = -0.93\mathbf{i} + 0.52\mathbf{j} + 0.84\mathbf{k}, \mathbf{c} = 3\mathbf{i} - 4\mathbf{j} - \mathbf{k}$; $\mathbf{a} \times \mathbf{c}$, $\mathbf{b} \times \mathbf{c}, \mathbf{a} \times \mathbf{b}, \mathbf{a} \cdot (\mathbf{b} \times \mathbf{c})$.

13.8 Graphing quadric surfaces or surfaces of revolution requires a special three-dimensional graphing program on a microcomputer or work station with a high-resolution monitor. It is possible to write programs that give 3-dimensional graphs on a graphic calculator, but the low-resolution calculator screen makes this activity not worth the trouble. (As the difference between calculators and small laptop computers narrows, this may change in the future.) Good graphing programs do a number of complicated things to give the illusion of three dimensions. The simplest graph presented is a "wire frame" model of the surface. The intersection points of the "wires" are computed, and line segments or curves are drawn between points. Line segments that would be "hidden" by the surface from the view presented may be erased. During some three-dimensional graphing, you can actually see the back lines being drawn and then erased as the plot moves forward on the surface. Adding coloring or shading makes the result look more like a surface, and the coloring can even imitate the effect of a light source shining on the surface.

In Problems 23–28 use a computer to obtain the graph of the given quadric surface. If necessary, solve the equation for z and graph pieces of the surface separately.

23. $z = 18 + 3x^2 + 5y^2$
24. $10x^2 + 35y^2 + 24z^2 = 560$
25. $28x^2 - 5y^2 + 42z^2 = 300$
26. $425x^2 + 117y^2 - 839z^2 = -2506$
27. $7x^2 + 3y^2 = 11z^2$
28. $10x^2 - 35y^2 + 24z = 0$

CHAPTER 14 Vector-Valued Functions

Answers to odd-numbered problems begin on page A-121.

14.1 The **Bézier curves** considered in the calculator/computer activities for Section 12.1 can be generalized to space curves and represented using vector functions. Let four vectors $\mathbf{c}_0 = \langle u_0, v_0, w_0 \rangle$, $\mathbf{c}_1 = \langle u_1, v_1, w_1 \rangle$, $\mathbf{c}_2 = \langle u_2, v_2, w_2 \rangle$, and $\mathbf{c}_3 = \langle u_3, v_3, w_3 \rangle$ be given. Then a cubic Bézier curve in 3-space is specified by the vector function

$$\mathbf{r}(t) = (1-t)^3\mathbf{c}_0 + 3(1-t)^2 t\mathbf{c}_1 + 3(1-t)t^2\mathbf{c}_2 + t^3\mathbf{c}_3, \qquad 0 \le t \le 1$$

This curve will start at the point $P_0(u_0, v_0, w_0)$ for $t = 0$, end at the point $P_3(u_3, v_3, w_3)$ for $t = 1$, and "head toward" the intermediate control points $P_1(u_1, v_1, w_1)$ and $P_2(u_2, v_2, w_2)$ between the endpoints. Bézier space curves can also be pieced together in a continuous fashion. The next piece can be represented by

$$\mathbf{r}(t) = (2-t)^3\mathbf{c}_3 + 3(2-t)^2(t-1)\mathbf{c}_4 + 3(2-t)(t-1)^2\mathbf{c}_5 + (t-1)^3\mathbf{c}_6, \qquad 1 \le t \le 2$$

In Problems 1–6 use a graphing utility to obtain the graph of the piecewise continuous Bézier space curve associated with the given control points. Set the viewing box so that all control points are within the box in 3-space. If possible, use equally scaled axes. Plot the control points on the screen.

1. $P_0(13, 45, 5)$, $P_1(30, 5, 25)$, $P_2(52, 75, 6)$, $P_3(80, 75, 25)$
2. $P_0(1, 1, 1)$, $P_1(25, 1, 25)$, $P_2(1, 50, 50)$, $P_3(75, 75, 75)$
3. $P_0(40, 40, 20)$, $P_1(0, 50, 50)$, $P_2(50, 50, 0)$, $P_3(60, 15, 25)$
4. $P_0(85, 10, 10)$, $P_1(10, 85, 10)$, $P_2(10, 10, 85)$, $P_3(85, 85, 85)$
5. $P_0(1, 1, 1)$, $P_1(100, 1, 100)$, $P_2(100, 100, 1)$, $P_3(1, 100, 1)$
6. $P_0(60, 1, 1)$, $P_1(60, 1, 100)$, $P_2(0, 40, 100)$, $P_3(50, 50, 50)$, $P_4(100, 60, 0)$, $P_5(100, 1, 40)$, $P_6(1, 1, 40)$

In Problems 7–11 graph the indicated quantities, prove the stated facts about Bézier space curves, and calculate the indicated arc lengths.

7. Plot the Bézier curve in Problem 1 again and also draw the line segment between P_0 and P_1. Verify graphically that the tangent line to the curve at $t = 0$ has the same direction numbers as this line segment and passes through P_1. Then prove that the tangent line at the initial control point P_0 on any cubic Bézier space curve will pass through the second control point P_1.
8. Plot the Bézier curve in Problem 2 again and also draw the line segment between P_2 and P_3. Verify graphically that the tangent line to the curve at $t = 1$ has the same direction numbers as this line segment and passes through P_2. Then prove that the tangent line at the final control point P_3 on any cubic Bézier space curve will pass through the third control point P_2.
9. Find the arc length for the Bézier curve in Problem 1. Calculate the length of the piecewise linear curve made up of the line segments from P_0 to P_1, from P_1 to P_2, and from P_2 to P_3. Compare the two lengths.
10. Find the arc length for the Bézier curve in Problem 2. Calculate the length of the piecewise linear curve made up of the line segments from P_0 to P_1, from P_1 to P_2, and from P_2 to P_3. Compare the two lengths. Make a conjecture about a general result concerning these two lengths.
11. Verify that the four control points in Problem 3 all lie on the plane $x + y + z = 100$. Prove that the resulting cubic Bézier space curve will also lie in this plane.

14.2 Most graphing programs plot a set of parametric equations or a curve traced by a vector function in the following way. Suppose $a = t_0 < t_1 < \cdots < t_n = b$ is a partition of the parameter interval $a \le t \le b$ such that the values t_i are equally spaced; that is, $t_{i+1} - t_i = \Delta t$, for $i = 1, 2, \ldots, n$. Individual points on the curve are calculated and generally these points are connected with short line segments. In some programs it is possible to omit the line segments and plot only the points. On the TI-81 this is called the **dot mode** for plotting a function or parametric curve. Suppose that $\mathbf{r}(t) = f(t)\mathbf{i} + g(t)\mathbf{j}$ gives the position of a moving particle in the plane for $a \le t \le b$. Further suppose that the values t_i of the parameter are equally spaced. Then the spacing of the points $\mathbf{r}(t_i)$ will give a visual indication of speed along the curve since, numerically, the velocity and the acceleration can be approximated by the centered differences

$$\mathbf{v}(t_i) \approx \frac{\mathbf{r}(t_{i+1}) - \mathbf{r}(t_{i-1})}{2\,\Delta t}$$

and
$$\mathbf{a}(t_i) \approx \frac{\mathbf{r}(t_{i+1}) - 2\mathbf{r}(t_i) + \mathbf{r}(t_{i-1})}{(\Delta t)^2}$$

The mental image for the velocity that is given by the "dot" spacing is roughly equivalent to the vector $\mathbf{r}(t_{i+1}) - \mathbf{r}(t_{i-1})$. The mental image for the acceleration vector is less precise, but the plot does give some idea about how the velocity vectors are changing. This roughly corresponds to $[\mathbf{r}(t_{i+1}) - \mathbf{r}(t_i)] - [\mathbf{r}(t_i) - \mathbf{r}(t_{i-1})]$, which can indicate the general direction of the acceleration vector.

In Problems 12–17 $\mathbf{r}(t)$ is the position vector of a particle moving in the plane. Graph the curve, using a dot mode with equal time increments, if possible.

12. $\mathbf{r}(t) = 25 \cos t\mathbf{i} + 12 \sin t\mathbf{j}$; $0 \le t \le 2\pi$
13. $\mathbf{r}(t) = 37 \cos^2 t\mathbf{i} + 19 \sin^2 t\mathbf{j}$; $0 \le t \le 2\pi$
14. $\mathbf{r}(t) = \dfrac{56}{20 - 11 \sin t} \cos t\mathbf{i} + \dfrac{56}{20 - 11 \sin t} \sin t\mathbf{j}$; $0 \le t \le 2\pi$
15. $\mathbf{r}(t) = \dfrac{5}{\sqrt{3} + 2 \cos t} \cos t\mathbf{i} + \dfrac{5}{\sqrt{3} + 2 \cos t} \sin t\mathbf{j}$; $-\dfrac{5\pi}{6} < t < \dfrac{5\pi}{6}$
16. $\mathbf{r}(t) = 10 \sin 3t \cos t\mathbf{i} + 10 \sin 3t \sin t\mathbf{j}$; $0 \le t \le \pi$
17. $\mathbf{r}(t) = \sqrt{35 \cos 2t} \cos t\mathbf{i} + \sqrt{35 \cos 2t} \sin t\mathbf{j}$; $-\dfrac{\pi}{4} \le t \le \dfrac{\pi}{4}$

14.3 It is instructive to plot the **circle of curvature** to illustrate the geometric significance of curvature. This can be done on a graphic calculator or computer for curves in the plane traced by the vector function $\mathbf{r}(t) = f(t)\mathbf{i} + g(t)\mathbf{j}$ with $a \le t \le b$. Then the **curvature** κ is given by

$$\kappa = \frac{|f'(t)g''(t) - g'(t)f''(t)|}{([f'(t)]^2 + [g'(t)]^2)^{3/2}}$$

(See Problem 21 in Exercises 14.3.) The **radius of curvature** is $\rho = 1/\kappa$ and the center of the desired circle lies on the concave side of the curve on the normal line. Once the radius ρ and the center (h, k) are determined, the circle can be plotted as the curve traced by the vector function $\mathbf{r}(t) = (h + \rho \cos t)\mathbf{i} + (k + \rho \sin t)\mathbf{j}$ for $0 \le t \le 2\pi$.

In Problems 18–21 graph the given vector function. Plot the circles of curvatures at the indicated points along the curve.

18. $\mathbf{r}(t) = t\mathbf{i} - \cos(\pi\sqrt{t})\mathbf{j}$; $0 \le t \le 25$ at $(1, 1), (4, -1), (9, 1), (16, -1), (25, 1)$
19. $\mathbf{r}(t) = t\mathbf{i} + \cos t\mathbf{j}$; $-\pi \le t \le \pi$ at $\left(-\dfrac{2\pi}{3}, -\dfrac{1}{2}\right)$, $(0, 1)$, $\left(\dfrac{\pi}{6}, \dfrac{\sqrt{3}}{2}\right)$, $(2, \cos 2)$
20. $\mathbf{r}(t) = 2\sqrt{\sin 2t} \cos t\mathbf{i} + 2\sqrt{\sin 2t} \sin t\mathbf{j}$; $0 \le t \le \dfrac{\pi}{2}$ at $\mathbf{r}\left(\dfrac{\pi}{6}\right), \mathbf{r}\left(\dfrac{\pi}{4}\right), \mathbf{r}\left(\dfrac{5\pi}{12}\right)$
21. $\mathbf{r}(t) = 3 \cos 2t \cos t\mathbf{i} + 3 \cos 2t \sin t\mathbf{j}$; $-\dfrac{\pi}{4} \le t \le \dfrac{3\pi}{4}$ at $\mathbf{r}\left(-\dfrac{\pi}{4}\right), \mathbf{r}(0), \mathbf{r}\left(\dfrac{\pi}{6}\right), \mathbf{r}\left(\dfrac{\pi}{4}\right), \mathbf{r}\left(\dfrac{\pi}{2}\right), \mathbf{r}\left(\dfrac{3\pi}{4}\right)$

In Problems 22–25 plot the indicated circle of curvature and answer the question for the given ellipse.

22. $\mathbf{r}(t) = 10 \cos t\mathbf{i} + 4 \sin t\mathbf{j}$; $-\pi \le t \le \pi$ at the vertex $(10, 0)$. Does this circle of curvature lie within or contain the ellipse?
23. $\mathbf{r}(t) = 7 \cos t\mathbf{i} + 5 \sin t\mathbf{j}$; $-\pi \le t \le \pi$ at the vertex $(0, 5)$. Does this circle of curvature lie within or contain the ellipse?
24. $\mathbf{r}(t) = 12 \cos t\mathbf{i} + 8 \sin t\mathbf{j}$; $-\pi \le t \le \pi$ at the point $(6, 4\sqrt{3})$ where $t = \pi/3$. Does this circle of curvature lie within or contain the ellipse?
25. $\mathbf{r}(t) = 10 \cos t\mathbf{i} + b \sin t\mathbf{j}$; $-\pi \le t \le \pi$ at the vertex $(10, 0)$. Find the value of b so that the circle of curvature at $(10, 0)$ passes through the origin.

The curvature κ is a measure of the **bending** of the curve at the point. This notion is made more precise in elasticity theory. The potential energy due to the bending of a thin flexible elastic rod into a shape in the plane is shown to be proportional to

$$\int_0^L \kappa^2(s)\,ds$$

where s represents arc length along a curve of length L. This integral sums the bending energy of each point along the curve, and it is assumed that the rod does not crack and that the bending is small enough that the elastic rod will return to a straight shape when released. The constant of proportionality contains the units of energy as well as factors related to the cross-sectional area of the rod and the flexibility of the material used to make the elastic rod. If the curve in the plane can be represented as a function $y = F(x)$ for $a \le x \le b$, then this integral becomes

$$\int_0^L \kappa^2(s)\,ds = \int_a^b \left\{\frac{|F''(x)|}{(1+[F'(x)]^2)^{3/2}}\right\}^2 \sqrt{1+[F'(x)]^2}\,dx$$

$$= \int_a^b \frac{|F''(x)|^2}{(1+[F'(x)]^2)^{5/2}}\,dx$$

(See Problem 22 in Exercises 14.3.)

In Problems 26–28 numerically approximate the "bending energy" integral for the given function.

26. $F(x) = \sin x;\ 0 \le x \le 2\pi$
27. $F(x) = x^{-1};\ 3^{-1} \le x \le 3$
28. $F(x) = x^3 + 10x^2 - 25x + 40;\ -15 \le x \le 8$

CHAPTER 15 Differential Calculus of Functions of Several Variables

Answers to odd-numbered problems begin on page A-122.

15.1 A good three-dimensional graphing application on a computer should offer several options for graphing a function of the form $z = f(x, y)$. The default mode usually is some kind of wire frame model of the surface of the graph with "hidden lines" removed to give the illusion of three dimensions. Coloring or shading may be added to the patches outlined by the wire frame for a more solid image. Sometimes different colors or shadings are used to represent different heights along the z-axis. More difficult is a coloring scheme designed to simulate the effect of a light source shining on the surface where patches facing the light are lighter in color and patches hidden from the light are darker. Truly advanced graphical pictures requiring hours of supercomputer time operate on the same principles. The advanced graphs omit the wire frame, use millions of patches, and even calculate the effect of reflections of the light from the source between several parts of the surface.

The graphing tool shows a viewing box in 3-space. In addition, the plot gives the illusion of viewing this box from some vantage point outside of the box. There will be commands to expand or shrink the viewing box and to change the point of view. By changing the point of view several times, you can rotate the box to show the essential features of the surface that may be hidden in any one view.

In Problems 1–8 use a computer to obtain the graph of the given function on the indicated viewing box. Change the point of view, and print one of the views if possible. Also obtain a plot of the level curves for the same function.

1. $f(x, y) = \dfrac{x^3}{3} + x(y^2 - 1) - \dfrac{y}{2} + 1,\ -3 \le x \le 3,$ $-3 \le y \le 3,\ -35 \le z \le 35$

2. $f(x, y) = 2xy - 5x^2 - 2y^2 + 4x + 4y + 20,$ $-2 \le x \le 2,\ -2 \le y \le 2,\ -20 \le z \le 30$

3. $f(x, y) = x^4 - 3xy^2 + y^3 - x - y,\ -3 \le x \le 4,$ $-3 \le y \le 3,\ -60 \le z \le 200$

4. $f(x, y) = \dfrac{-4x}{x^2 + y^2 + 1},\ -4 \le x \le 4,\ -3 \le y \le 3,$ $-1.5 \le z \le 1.5$

5. $f(x, y) = |xy|,\ -3 \le x \le 3,\ -3 \le y \le 3,$ $-1 \le z \le 12$

6. $f(x, y) = e^{1-x^2-y^2},\ -3 \le x \le 3,\ -3 \le y \le 3,$ $0 \le z \le 3$

7. $f(x, y) = \cos\left(\dfrac{x^2 + 2y^2}{4}\right),\ -6 \le x \le 6,\ -3 \le y \le 3,$ $-1 \le z \le 1$

8. $f(x, y) = (\cos x^2)e^{-\sqrt{\frac{x^2+3y^2}{10}}},\ -5 \le x \le 5,$ $-3 \le y \le 3,\ -1 \le z \le 1$

15.2 Even with a computer, the graph of a surface is very difficult to obtain when the function has discontinuities. The wire frame representation may place steep, nearly vertical, segments into the plot where the graph really should be separated into two or more pieces.

In Problems 9–12 use a computer to obtain the graph of the given function over a region containing points of discontinuity. Plot level curves over the same region.

9. $f(x, y) = \dfrac{x - 3y}{2y - x^2}$ **10.** $f(x, y) = \dfrac{4x^2y}{x^4 + y^2}$

11. $f(x, y) = \dfrac{x^2y^4}{x^4 + y^2}$ **12.** $f(x, y) = \dfrac{x^2y^2}{3x^2 + y^2}$

15.3 Taylor polynomials can be generalized to two variables. For a function $f(x, y)$ with f_x, f_y, f_{xy}, and f_{yx} continuous in an open region containing (a, b), the first- and second-degree Taylor polynomials are given by the following:

$$p_1(x, y) = f(a, b) + f_x(a, b)(x - a) + f_y(a, b)(y - b)$$

$$p_2(x, y) = p_1(a, b) + \frac{1}{2}[f_{xx}(a, b)(x - a)^2 + 2f_{xy}(a, b)(x - a)(y - b) + f_{yy}(a, b)(y - b)^2]$$

In Problems 13–17 find the first- and second-degree Taylor polynomials for the given function at the indicated point. Use a computer to obtain the graph of the given function over a region containing the point (a, b). Compare this graph with the graphs of the first- and second-degree Taylor polynomials over the same region. Also plot the level curves for the function and these polynomials.

13. $f(x, y) = \sin 2xy$ at $(a, b) = (0, 0)$
14. $f(x, y) = \sin \pi x \cos \pi y$ at $(a, b) = (0.5, 1)$
15. $f(x, y) = \sqrt{|xy|}$ at $(a, b) = (2, 5)$
16. $f(x, y) = 5xe^{-x^2-y^2}$ at $(a, b) = (1, 0)$
17. $f(x, y) = 5xe^{-x^2-y^2}$ at $(a, b) = (3, 2)$

15.4 There is a close connection between the increment $\Delta z = f(x + \Delta x, y + \Delta y) - f(x, y)$ and the differential $dz = f_x(x, y)\,\Delta x + f_y(x, y)\,\Delta y$ when the first partial derivatives are continuous over some region containing (x, y). When the partial derivatives are easy to find and evaluate, the differential conveniently approximates the increment. Here we explore the connection in the other direction when it might be possible to evaluate changes in the function for various increments but the partial derivatives are not easy to obtain. Selecting $\Delta x = h$ and $\Delta y = 0$ gives

$$f(x + h, y) - f(x, y) \approx f_x(x, y)h$$

or

$$f_x(x, y) \approx \frac{f(x + h, y) - f(x, y)}{h}$$

Selecting $\Delta x = -h$ and $\Delta y = 0$ gives

$$f(x - h, y) - f(x, y) \approx f_x(x, y)(-h)$$

or

$$f_x(x, y) \approx \frac{f(x, y) - f(x - h, y)}{h}$$

Averaging these two partial derivative approximations yields the centered partial derivative approximation

$$f_x(x, y) \approx \frac{f(x + h, y) - f(x - h, y)}{2h} \quad \text{for small } h > 0$$

Similarly, for small $k > 0$,

$$f_y(x, y) \approx \frac{f(x, y + k) - f(x, y - k)}{2k}$$

$$f_{xx}(x, y) \approx \frac{f(x + h, y) - 2f(x, y) + f(x - h, y)}{h^2}$$

and

$$f_{yy}(x, y) \approx \frac{f(x, y + k) - 2f(x, y) + f(x, y - k)}{k^2}$$

Assuming that f_{xy} and f_{yx} are also continuous in the region near (x, y), we have

$$f_{xy}(x, y) = f_{yx}(x, y)$$

$$\approx \frac{\left[\dfrac{f(x + h, y + k) - f(x - h, y + k)}{2h} - \dfrac{f(x + h, y - k) - f(x - h, y - k)}{2h}\right]}{2k}$$

$$= \frac{\left[\begin{array}{l} f(x + h, y + k) - f(x - h, y + k) \\ \quad - f(x + h, y - k) + f(x - h, y - k) \end{array}\right]}{4hk}$$

In Problems 18–21 approximate each partial derivative using $h = 0.2$, $k = 0.5$, and the following data:

$$f(0.8, 1.5) = 6.604,\ f(1.0, 1.5) = 7.189,$$
$$f(1.2, 1.5) = 7.721,\ f(1.4, 1.5) = 8.493$$
$$f(0.8, 2.0) = 6.017,\ f(1.0, 2.0) = 6.316,$$
$$f(1.2, 2.0) = 6.698,\ f(1.4, 2.0) = 7.235$$
$$f(0.8, 2.5) = 5.726,\ f(1.0, 2.5) = 5.912,$$
$$f(1.2, 2.5) = 6.145,\ f(1.4, 2.5) = 6.479$$

18. $f_x(1.0, 1.5)$, $f_{xx}(1.0, 1.5)$, $f_x(1.2, 1.5)$, $f_{xx}(1.2, 1.5)$
19. $f_y(0.8, 2.0)$, $f_{yy}(0.8, 2.0)$, $f_y(1.4, 2.0)$, $f_{yy}(1.4, 2.0)$
20. $f_x(1.0, 2.0)$, $f_y(1.0, 2.0)$, $f_{xx}(1.0, 2.0)$, $f_{xy}(1.0, 2.0)$, $f_{yy}(1.0, 2.0)$
21. $f_x(1.2, 2.0)$, $f_y(1.2, 2.0)$, $f_{xx}(1.2, 2.0)$, $f_{xy}(1.2, 2.0)$, $f_{yy}(1.2, 2.0)$
22. Use the partial derivatives approximations found in Problem 20 to construct the second-degree Taylor polynomial centered at the point (1.0, 2.0). Evaluate this polynomial at the 12 grid points to see how closely it matches the given data.
23. Use the partial derivatives approximations found in Problem 21 to construct the second-degree Taylor polynomial centered at the point (1.2, 2.0). Evaluate this polynomial at the 12 grid points to see how closely it matches the given data.

15.9 It should be possible to approximate extrema of a function of two variables from a computer-generated graph. Look for some kind of capability to "zoom in" on the graph over a smaller region giving a maximum or minimum. Unfortunately, this feature is not often available on three-dimensional graphing applications. If you lack such a special feature, simply redraw the graph for a new smaller viewing box where you estimate that the extrema can be seen more closely. Even less common is to find some kind of tracing capability that allows the user to select a point in the domain of the viewing box and then see the coordinates of the graph over that point. Such graphical tools should become common as more powerful computer workstations become widely available.

In Problems 24–27 use a computer-generated graph to locate the extrema for the given function over the indicated region.

24. $f(x, y) = \dfrac{10x}{x^2 + y^2 + 1}$, $-4 \le x \le 4$, $-3 \le y \le 3$
25. $f(x, y) = 2xy - 5x^2 - 2y^2 + 4x + 4y + 20$, $-3 \le x \le 5$, $-2 \le y \le 4$
26. $f(x, y) = 5e^{1-x^2-y^2} + x + y$, $-2 \le x \le 2$, $-2 \le y \le 2$
27. $f(x, y) = x^3 + 5xy + y^2$, $x^2 + y^2 \le 1$

15.10 A least-squares line to fit a set of data (x_1, y_1), $(x_2, y_2), \ldots, (x_n, y_n)$ should be used only when a **linear polynomial model** $f(x) = mx + b$ is appropriate for the behavior of the data. Unfortunately, the least-squares line is sometimes used when there is no reason to expect that the data lie roughly on a straight line. A common technique for effectively using alternative models is to transform the data into a form where a straight line is then appropriate. For example, if we consider the least-squares line to the data $(\ln x_1, y_1)$, $(\ln x_2, y_2), \ldots, (\ln x_n, y_n)$, then we are assuming a **logarithmic function model** of the form $f(x) = m \ln x + b$. The least-squares line to the data $(x_1, \ln y_1)$, $(x_2, \ln y_2), \ldots, (x_n, \ln y_n)$ assumes an **exponential function model** of the form $\ln f(x) = mx + b$ or $f(x) = ce^{mx}$ where $c = e^b$. Finally, taking logarithms of both coordinate values to find a line of best fit for the data $(\ln x_1, \ln y_1), (\ln x_2, \ln y_2), \ldots, (\ln x_n, \ln y_n)$ assumes a **power model** of the form $\ln f(x) = m \ln x + b$ or $f(x) = cx^m$ where $c = e^b$. A researcher, with no idea about what model to use, might try several models. A key step in exploratory data analysis is to plot both the data and the model on the same coordinate system. The fit can be examined visually and the appropriateness of a model selection can be verified. All statistical packages and most scientific calculators provide the linear regression operation. Consult your manual for exactly how to enter and edit data. The TI-81 graphic calculator provides all four models discussed here, and each can be plotted together with a scatter plot of the original data. If no such automatic process is provided, then take the logarithms of selected coordinates by hand and enter "new data" for a linear regression. Interpret the regression coefficients in terms of the alternative function model.

In Problems 28–32 find the best fit for the given data using the indicated function models. Plot the resulting function and the original data on the same axes and examine the fit.

28. (1900, 75.994575), (1920, 105.710620), (1940, 131.669275), (1960, 179.32175), (1980, 226.545805) [U.S. census data in millions]; linear polynomial model and exponential function model
29. (100 m, 9.93 s), (200 m, 19.72 s), (400 m, 43.86 s), (800 m, 101.73 s), (1000 m, 132.18 s) [men's world running records in seconds over different distances in meters as of 1985]; linear polynomial model, exponential function model, and power model
30. (1940, 12.688), (1950, 51.100), (1960, 113.120), (1970, 232.877), (1980, 574.244) [total yearly taxes paid in the United States in billions of dollars]; exponential function model and logarithmic function model
31. (−81.091, 48), (−81.058, 72), (−80.818, 96), (−80.387, 120), (−79.789, 144) [NASA satellite data giving geodetic latitude in degrees at time in seconds]; linear polynomial model and exponential function model
32. (6, 89), (18, 84), (23, 97), (30, 83), (35, 63), (42, 68), (45, 45), (52, 56), (56, 50); linear polynomial model and logarithmic model

CHAPTER 16 Integral Calculus of Functions of Several Variables

Answers to odd-numbered problems begin on page A-124.

16.1 The definition for a double integral of f over a region R suggests a numerical approximation for the double integral in terms of a Riemann sum. Choosing the point (x_k^*, y_y^*) to be the midpoint in each subregion R_k will lead to more accurate approximations, but it is easier to program the choice of one of the corner points. When the region R has a curved or slanted boundary, it can be challenging to determine in the program exactly which subregions R_k lie entirely in R.

In Problems 1–4 write a calculator or computer program to find the indicated Riemann sums.

1. Consider the region R in the first quadrant that is bounded by the graph of $9x^2 + 25y^2 = 225$, $y = 0$, $x = 0$. Approximate the double integral $\iint_R e^{xy}\,dA$ using Riemann sums constructed in the following way. Given an integer n, let $0 = x_0 < x_1 < \cdots < x_n = 5$ with $x_i - x_{i-1} = \Delta x = 5/n$ and $0 = y_0 < y_1 < \cdots < y_n = 3$ with $y_j - y_{j-1} = \Delta y = 3/n$. For each rectangular subregion R_k with opposite corners (x_{i-1}, y_{j-1}) and (x_i, y_j), select the point (x_i, y_j). Test this point to determine whether it is contained in the region R. If so, add the corresponding term $f(x_i, y_j)\,\Delta A$ in the summation. Calculate the Riemann sum of this form for $n = 5, 10, 15, 20$.

2. Consider the region R in the first quadrant that is bounded by the graph of $81x^4 + 625y^4 = 50{,}625$, $y = 0$, $x = 0$. Approximate the double integral $\iint_R \cos(x + y^2)\,dA$ using Riemann sums constructed in the following way. Given an integer n, let $0 = x_0 < x_1 < \cdots < x_n = 5$ with $x_i - x_{i-1} = \Delta x = 5/n$ and $0 = y_0 < y_1 < \cdots < y_n = 3$ with $y_j - y_{j-1} = \Delta y = 3/n$. For each rectangular subregion R_k with opposite corners (x_{i-1}, y_{j-1}) and (x_i, y_j), select the point (x_i, y_j). Test this point to determine whether it is contained in the region R. If so, add the corresponding term $f(x_i, y_j)\,\Delta A$ in the summation. Calculate the Riemann sum of this form for $n = 5, 10, 15, 20$.

3. Consider the region R in the second quadrant that is bounded by the graph of $9x^2 + 16y^2 = 144$, $y = 0$, $x = 0$. Approximate the double integral $\iint_R \sqrt{144 - 9x^2 - 16y^2}\,dA$ using Riemann sums constructed in the following way. Given an integer n, let $-4 = x_0 < x_1 < \cdots < x_n = 0$ with $x_i - x_{i-1} = \Delta x = 4/n$ and $0 = y_0 < y_1 < \cdots < y_n = 3$ with $y_j - y_{j-1} = \Delta y = 3/n$. For each rectangular subregion R_k with opposite corners (x_{i-1}, y_{j-1}) and (x_i, y_j), select the point (x_{i-1}, y_j). Test this point to determine whether it is contained in the region R. If so, add the corresponding term $f(x_{i-1}, y_j)\,\Delta A$ in the summation. Calculate the Riemann sum of this form for $n = 5, 10, 15, 20$.

4. Consider the region R in the first and second quadrants that is bounded by the graph of $25x^2 + 9y^2 = 225$, $y = 0$. Approximate the double integral $\iint_R \ln(x^2 y + 1)\,dA$ using Riemann sums constructed in the following way. Given an integer n, let $-3 = x_0 < x_1 < \cdots < x_n = 3$ with $x_i - x_{i-1} = \Delta x = 6/n$ and $0 = y_0 < y_1 < \cdots < y_n = 5$ with $y_j - y_{j-1} = \Delta y = 5/n$. For each rectangular subregion R_k with opposite corners (x_{i-1}, y_{j-1}) and (x_i, y_j), test the corners (x_{i-1}, y_j) and (x_i, y_j) to determine whether this subregion is contained in the region R. If so, add the corresponding term $f(x_{i-1} + \Delta x/2, y_{j-1} + \Delta y/2)\,\Delta A$ in the summation. Calculate the Riemann sum of this form for $n = 5, 10, 15, 20$.

16.2 When the limits of integration for an iterated double integral are all constants, then an **iterated Trapezoidal Rule** can easily be applied over this rectangular region R. Let $a = x_0 < x_1 < \cdots < x_n = b$ and $c = y_0 < y_1 < \cdots < y_m = d$, where $(b - a)/n = \Delta x = x_{i+1} - x_i$ and $(d - c)/m = \Delta y = y_{j+1} - y_j$. The iterated Trapezoidal Rule is then

$$\int_a^b \int_c^d f(x, y)\,dy\,dx$$

$$\approx \frac{b-a}{2n}\left\{\int_c^d f(a, y)\,dy + 2\sum_{i=1}^{n-1}\int_c^d f(x_i, y)\,dy + \int_c^d f(b, y)\,dy\right\}$$

$$\approx \frac{(b-a)(d-c)}{4nm}\left\{f(a, c) + f(a, d) + f(b, c) + f(b, d) + 2\sum_{j=1}^{m-1}(f(a, y_j) + f(b, y_j)) + 2\sum_{i=1}^{n-1}(f(x_i, c) + f(x_i, d)) + 4\sum_{i=1}^{n-1}\sum_{j=1}^{m-1} f(x_i, y_j)\right\}$$

In Problems 5–8 use a calculator or computer program implementing the iterated Trapezoidal Rule to obtain an approximation to the given double integral for the indicated values of n and m.

5. $\displaystyle\int_{-1}^{2}\int_0^5 \sin(x^2 + y)\,dy\,dx$; $n = 10$, $m = 15$ and $n = 15$, $m = 30$

6. $\displaystyle\int_0^4\int_0^4 e^{-x^2 - y^2}\,dy\,dx$; $n = 12$, $m = 12$ and $n = 20$, $m = 20$

7. $\displaystyle\int_0^{2\pi}\int_0^{5\pi} \sqrt[3]{\sin(x + 0.4y) + \cos x + 8}\,dy\,dx$; $n = 10$, $m = 25$ and $n = 14$, $m = 30$

8. $\displaystyle\int_{-4}^{3}\int_{-1}^{1} \ln(xy + 5)\,dy\,dx$; $n = 25$, $m = 10$ and $n = 35$, $m = 18$

In a similar fashion, Simpson's Rule can be iterated. Let $a = x_0 < x_1 < \cdots < x_{2n} = b$ and $c = y_0 < y_1 < \cdots < y_{2m} = d$, where $(b - a)/(2n) = \Delta x = x_{i+1} - x_i$ and $(d - c)/(2m) = \Delta y = y_{j+1} - y_j$. Then

$$\int_{x_{2i}}^{x_{2i+2}}\int_{y_{2j}}^{y_{2j+2}} f(x, y)\,dy\,dx$$

$$\approx \frac{\Delta x\,\Delta y}{9}\{f(x_{2i}, y_{2j}) + f(x_{2i+2}, y_{2j}) + f(x_{2i}, y_{2j+2}) + f(x_{2i+2}, y_{2j+2})$$

$$+ 4f(x_{2i}, y_{2j+1}) + 4f(x_{2i+2}, y_{2j+1})$$
$$+ 4f(x_{2i+1}, y_{2j}) + 4f(x_{2i+1}, y_{2j+2})$$
$$+ 16f(x_{2i+1}, y_{2j+1})\}$$

for $i = 0, 1, 2, \ldots, n-1$ and $j = 0, 1, 2, \ldots, m-1$. Summing over all of these subrectangles (and collecting like terms) then gives an iterated rule for approximating $\int_a^b \int_c^d f(x, y)\, dy\, dx$.

In Problems 9–12 use a calculator or computer program implementing the iterated Simpson's Rule to obtain an approximation to the given double integral for the indicated values of n and m.

9. $\int_0^3 \int_2^4 \sqrt{x^4 + y^2}\, dy\, dx$; $n = 9, m = 8$ and $n = 12, m = 16$

10. $\int_{-1}^1 \int_1^2 \sin(x^2y^2)\, dy\, dx$; $n = 8, m = 4$ and $n = 12, m = 10$

11. $\int_0^{10} \int_0^{25} e^{-x^2-y^2}\, dy\, dx$; $n = 10, m = 25$ and $n = 15, m = 50$

12. $\int_{-0.5}^{0.5} \int_{-0.2}^{0.2} \sqrt{9 - 4x^2 - 3y^2}\, dy\, dx$; $n = 5, m = 4$ and $n = 10, m = 8$

16.3 Computer algebra systems can evaluate some iterated double integrals exactly. Since these applications work in a symbolic manner, they can work on the "inner" integration with parameters and can handle a variable upper limit of integration. Consider the double integral

$$\int_0^2 \int_0^{\sqrt{2-x}} x^2(y^2 + 1)\, dy\, dx$$

The Maple session

```
>
  int(int((x^2)*(y^2+1),
                    y=0..sqrt(2-x)),x=0..2);
```

$$2/27\, 2^{9/2} - 2/21\, 2^{7/2} - 16/15\, 2^{5/2} + 8/3\, 2^{3/2}$$

```
>
  evalf('');
                          2.107103382
>
  evalf('''',25);
                 2.107103381821500347844208
```

correctly evaluates the integral, first in algebraic form and then in floating-point form with 10 and 25 significant digits.

If the integrand cannot be completely evaluated by Maple, then numerical evaluation as a floating-point number is still possible.

```
>
  int(int(sin(x^2*y^2),
  y=-1..x^2),x=0..1);
```

$$\int_0^1 \left(\frac{\mathrm{Pi}^{1/2} \mathrm{S}\left(\frac{2^{1/2} x^3}{\mathrm{Pi}^{1/2}}\right)}{2^{1/2} x} - \frac{\mathrm{Pi}^{1/2} \mathrm{S}\left(-\frac{2^{1/2} x}{\mathrm{Pi}^{1/2}}\right)}{2^{1/2} x} \right) dx$$

```
>
  evalf('');
                    .1437036625
```

In Problems 13–16 use a computer algebra system to evaluate the given double integral.

13. $\int_0^3 \int_0^{\sqrt{4-x}} x^2(y^2 + 1)\, dy\, dx$

14. $\int_0^8 \int_0^{y^2+y} \sqrt{5x^4 + 3x^2y^2 + 7y^4}\, dx\, dy$

15. $\int_{-1}^3 \int_{-5}^{5x} e^{5-x^2-6xy-y^2}\, dy\, dx$

16. $\int_2^{20} \int_0^{\ln x} \ln(x^2 + y)\, dy\, dx$

16.6 Computer algebra systems and numerical integration routines can be quite helpful for many of the double integrals representing surface area. If the computer application has a partial differentiation operation, then the process of finding the integrand for the integral representing surface area can be completely automated.

In Problems 17–20 use a calculator or computer to evaluate the resulting double integrals.

17. Find the surface area of that portion of the ellipsoid $12x^2 + 7y^2 + 3z^2 = 25$ that is above the circular region in the xy-plane bounded by $x^2 + y^2 = 1$.

18. Find the surface area of that portion of the graph of $z = \dfrac{-4x}{x^2 + y^2 + 1}$ that lies above the region in the xy-plane that is bounded by the ellipse $x^2 + 4y^2 = 9$.

19. Find the surface area of that portion of the hyperbolic paraboloid $z = \dfrac{y^2}{4} - \dfrac{x^2}{9}$ that lies above the region in the xy-plane that is bounded by the ellipse $\dfrac{x^2}{9} + \dfrac{y^2}{4} = 1$.

20. Find the surface area of that portion of the graph of $z = x^2y^2$ that lies above the rectangle in the xy-plane satisfying $|x - y| \le 3$ and $|x + y| \le 1$.

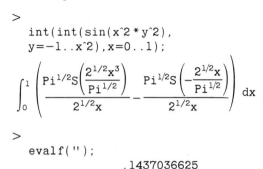

16.7 An iterated triple integral can be approximated using an iterated Trapezoidal Rule or an iterated Simpson's Rule, but these rules are not easy to implement when iterated three times. A computer algebra system provides a very convenient method for the symbolic integration of triple integrals. For example, the Maple session for a simple triple integral would look like this:

```
>
int(int(int(x^3-3*x*y^2+y*z^2,
z=x..y+1),y=5*x..10),x=0..1);
```

$$\frac{134683}{24}$$

In Problems 21–24 use a calculator or computer to evaluate the given triple integrals.

21. $\displaystyle\int_2^5 \int_x^{2x+1} \int_{-\sqrt{25-x^2}}^{y+\sqrt{25-x^2}} \sin(x + y + z^2)\, dz\, dy\, dx$

22. $\displaystyle\int_0^1 \int_0^{1-x^2} \int_0^{4\sqrt{1-x^2-y}} \ln(x + y + z + 1)\, dz\, dy\, dx$

23. $\displaystyle\int_{-1}^1 \int_{-1}^1 \int_{-1}^1 e^{2-x^2-y^2-z^2}\, dx\, dy\, dz$

24. $\displaystyle\int_{-2}^2 \int_1^{4-z^2} \int_0^{\sin y} \frac{x \sin y}{z-3}\, dx\, dy\, dz$

CHAPTER 17 Vector Integral Calculus

Answers to odd-numbered problems begin on page A-124.

17.1 A line integral along a smooth curve C can be expressed as a definite integral in a single variable. Symbolic computer operations and numerical methods such as Simpson's Rule can be applied to the definite integral. Implementing this will usually be much easier than attempting a numerical method more directly based on Definition 17.1 and its extensions.

In Problems 1–4 use a calculator or computer to approximate $\int_C G(x, y)\, dx$, $\int_C G(x, y)\, dy$, and $\int_C G(x, y)\, ds$ on the indicated curve C.

1. $G(x, y) = \ln(1 + x^2 + y^2)$; $x = 3\cos t$, $y = 5\sin t$, $0 \le t \le \pi$

2. $G(x, y) = e^{(1-x^2-y^2)}$; $x = \dfrac{t\cos t}{5}$, $y = \dfrac{t\sin t}{5}$, $\dfrac{\pi}{2} \le t \le \dfrac{9\pi}{2}$

3. $G(x, y) = \dfrac{10}{x^4 + y^4}$; $x = t^3 + 4t - 3$, $y = t^2 - 7t + 1$, $0 \le t \le 1$

4. $G(x, y) = \sqrt{xy + 3x + 9y}$; $x = t^2$, $y = e^t$, $-2 \le t \le 1$

In Problems 5 and 6 use a calculator or computer to approximate $\int_C G(x, y, z)\, dx$, $\int_C G(x, y, z)\, dy$, $\int_C G(x, y, z)\, dz$, and $\int_C G(x, y, z)\, ds$ on the indicated curve C.

5. $G(x, y, z) = \dfrac{6 + 3xy}{2x + z^2}$; $x = \sqrt{t}$, $y = \cosh t$, $z = \sinh t$, $0.5 \le t \le 3$

6. $G(x, y, z) = \cos(x + 2y + z)$; $x = t^{1/3}$, $y = \cos(0.2t) - 1$, $z = t^2$, $1 \le t \le 2$

Suppose that the parametric equations for the curve C are not given but we have a sequence of points $(x_0, y_0), (x_1, y_1), (x_2, y_2), \ldots, (x_n, y_n)$ along the curve (including the endpoints). Further assume that the continuous function $G(x, y)$ is not given by a simple formula but is known only on the sequence of points. Then the various line integrals of G along C from (x_0, y_0) to (x_n, y_n) can be approximated by a generalized Trapezoidal Rule (for non-equally spaced subintervals). For example,

$$\int_C G(x, y)\, ds \approx \sum_{k=1}^{n} \frac{1}{2}\{G(x_{k-1}, y_{k-1}) + G(x_k, y_k)\}\sqrt{(\Delta x_k)^2 + (\Delta y_k)^2}$$

In Problems 7–10 calculate a generalized Trapezoidal Rule approximation for the indicated line integral using the following table:

x	y	$G(x, y)$
30.000	30.000	6.944
43.508	24.714	1.172
51.664	23.112	−3.100
55.416	24.078	−4.661
55.712	26.496	−4.932
53.500	29.250	−4.854
49.728	31.224	−4.285
45.344	31.302	−2.599
41.296	28.368	0.858
38.532	21.306	6.556
38.000	9.000	14.241

7. $\displaystyle\int_C G(x, y)\, ds$

8. $\displaystyle\int_C G(x, y)\, dx$

9. $\int_C \left(\frac{G(x, y) + 5}{4}\right) ds$

10. $\int_C \left(\frac{G(x, y) + 5}{4}\right) dy$

17.2 When a line integral $\int_C P(x, y)\, dx + Q(x, y)\, dy$ is not independent of path, then it is interesting to compare the results of integrating over a family of curves all beginning at the same point A and ending at the same point B. A convenient way to obtain such a family is to use the Bézier curve representation introduced in the calculator/computer activities for Section 12.1 Recall that the cubic Bézier curve for four control points has the following parametric equations:

$$x = p_0(1 - t)^3 + 3p_1(1 - t)^2 t + 3p_2(1 - t)t^2 + p_3 t^3$$
$$y = q_0(1 - t)^3 + 3q_1(1 - t)^2 t + 3q_2(1 - t)t^2 + q_3 t^3$$
$$0 \le t \le 1$$

In Problems 11–14 approximate the line integral $\int_C \sin(x + y)\, dx + \cos(xy)\, dy$ over the cubic Bézier curves with the given control points. Plot the family of curves and discuss how the value of the line integral changes.

11. $A = P_0 = (0, 0)$; $B = P_3 = (1, 1)$; $P_1 = (w, 0)$; $P_2 = (1, 1 - w)$ for $w = 0.2, 0.4, 0.6, 0.8, 1$
12. $A = P_0 = (0, 0)$; $B = P_3 = (1, 1)$; $P_1 = (w, 0)$; $P_2 = (1 - w, 1)$ for $w = 0.2, 0.4, 0.6, 0.8, 1$
13. $A = P_0 = (0, 0)$; $B = P_3 = (1, 1)$; $P_1 = (0.3, w)$; $P_2 = (0.7, 1 - w)$ for $w = 0.3, 0.5, 0.7, 1$
14. $A = P_0 = (0, 0)$; $B = P_3 = (1, 0)$; $P_1 = (0.3, w)$; $P_2 = (0.7, w)$ for $w = 0.2, 0.4, 0.6, 0.8, 1$

17.3 Green's Theorem is useful when one of the integrals is easier to compute. Generally it is much easier to numerically approximate a line integral over a curve C than to approximate a double integral over a region R. However, if the curve bounding the region has several pieces, it may be easier to set up and solve a single iterated integral than to set up several line integrals over the different pieces.

In Problems 15–17 verify Green's Theorem by evaluating both integrals numerically. Discuss the computational effort required to evaluate each integral.

15. $\oint_C e^{-xy}\, dx + \ln(x^3 y + 1)\, dy = \iint_R \left(\frac{3x^2 y}{x^3 y + 1} + xe^{-xy}\right) dA$, where R is the region defined by $0 \le x \le 1, 0 \le y \le 1$

16. $\oint_C \cos(2xy + 5x)\, dx + \sin(x + y + 1)\, dy = \iint_R \{\cos(x + y + 1) + 2x \sin(2xy + 5x)\}\, dA$, where C is the curve given by $x = -1 + 12t - 12t^2$, $y = 12t - 36t^2 + 24t^3, 0 \le t \le 1$

17. $\oint_C \frac{10x}{x^4 + y^4}\, dx + \frac{5y}{x^2 + y^2}\, dy = \iint_R \left(-\frac{10xy}{(x^2 + y^2)^2} + \frac{40xy^3}{(x^4 + y^4)^2}\right) dA$, where R is the region bounded by the ellipse $4(x - 6)^2 + 25y^2 = 100$

17.4 Given an equation describing how a surface S in three-dimensional space projects onto a region R in one of the coordinate planes, a surface integral $\iint_S G(x, y, z)\, dS$ can be represented as a double integral over the region R. Symbolic computer operations and numerical methods can be applied to the double integral. Implementing this will usually be much easier than attempting a numerical method more directly based on Definition 17.2.

In Problems 18–21 use a calculator or computer to approximate $\iint_S G(x, y, z)\, dS$ on the indicated surface S.

18. $G(x, y, z) = \frac{15}{1 + x^2 y^2 + z^4}$; S is the part of the paraboloid $z = 5 - 2x^2 - y^2$ that lies above the xy-plane.
19. $G(x, y, z) = \cos(x + 4y - xz^2)$; S is the part of the sphere $x^2 + y^2 + z^2 = 18$ that lies inside the cylinder $x^2 + z^2 = 12$ and intersects the positive y-axis.
20. $G(x, y, z) = \sqrt{5x^3 + 4yz + 2}$; S is the part of the surface $z = \cos(x - 1) + \cos(y - 2)$ that lies above the rectangle $0 \le x \le 2, 1 \le y \le 3$ in the xy-plane.
21. $G(x, y, z) = \ln(1 + x + 3yz)$; S is the part of the surface $x = 5 - z^2 + y^2$ that lies above the rectangle $0 \le y \le 1, 0 \le z \le 2$ in the yz-plane.

17.5 Symbolic computer algebra systems often provide curl and divergence as operations that can be performed on a vector function or field. The curl (like the cross product) is defined only for a vector field $\mathbf{F}(x, y, z) = P(x, y, z)\mathbf{i} + Q(x, y, z)\mathbf{j} + R(x, y, z)\mathbf{k}$ in three dimensions. For example, in Maple:

```
f:=[x^2*y,5*x+3*y^3*z,x+y+5*z^4];
v:=[x,y,z];
curl(f,v);       yields  [1-3y³, -1, 5-x²]
diverge(f,v);    yields  2xy+9y²z+20z³
```

The divergence (like the gradient) is defined for a vector field over any number of variables:

```
g:=sin(5*x^2*y^5+y);
w:=[x,y];
grad(g,w);
  array (1 .. 2,
        [10 cos(5x²y⁵+y)xy⁵,
        cos(5x²y⁵+y)(25x²y⁴+1)])
```

```
diverge('',w);
  - 100 sin(5x^2y^5 + y)x^2y^10
  + 10 cos(5x^2y^5 + y)y^5
  - sin(5x^2y^5 + y)(25x^2y^4 + 1)
  + 100 cos(5x^2y^5 + y)x^2y^3
```

In Problems 22–29 use a calculator or computer to find the indicated expressions.

22. curl **F** and div **F**, where
$\mathbf{F}(x, y, z) = e^x\sin(yz^2)\mathbf{i} + \ln(x^3 + 2y + z)\mathbf{j} + x^5y^2z\mathbf{k}$

23. curl **F** and div **F**, where
$\mathbf{F}(x, y, z) = \cos(3x + y^2z)\mathbf{i} + \sin(x^3 + 7yz)\mathbf{j} + \cos(x^2y^3z^5)\mathbf{k}$

24. div(grad f), where
$f(x, y, z) = \sqrt{\cos(10x^5z^3 + 25xyz) + x^2 + 2} + \dfrac{x}{y - 3z}$

25. div(grad f), where $f(x, y, z) = \dfrac{2 + 7xy - z^2}{\sin(1 + x + y^2 + z^3)}$

26. curl **F** and curl(curl **F**), where
$$\mathbf{F}(x, y, z) = \frac{x^3 + 2x^2y + 3xy^2 + 4y^3}{z^3 + 2z^2y + 3zy^2 + 4y^3}\mathbf{i} + \frac{x^2 + 4xz + 5z^2}{y^2 + 4yz + 5z^2}\mathbf{j} + \frac{x + y + z}{xyz}\mathbf{k}$$

27. curl **F** and curl(curl **F**), where
$\mathbf{F}(x, y, z) = (5x^3 + 6x^2y + 12xy^2 + 7y^3)\mathbf{i} + x^5y^8z^2\mathbf{j} + (2 + y - z)\mathbf{k}$

28. curl((grad f) × (grad g)), where
$f(x, y, z) = x^3 + 5z^2y - 18xy^2 + 2z^3$ and
$g(x, y, z) = 9z^3 + 4x^2y + 13xz^2 - 6y^3$

29. grad(div **F**), where
$\mathbf{F}(x, y, z) = \sec(xy + z^2)\mathbf{i} + \cot(x^4 + 5yz)\mathbf{j} + \csc(x^3y^5z^2)\mathbf{k}$

CHAPTER 18 Differential Equations *Answers to odd-numbered problems begin on page A-125.*

18.1 A one-parameter family of explicit solutions for a first-order differential equation can be displayed by plotting several of the functions in the family on the same screen.

In Problems 1–4 plot the solution indicated for the specified choice of the parameter in the given viewing window.

1. $\dfrac{dy}{dx} - 2xy = 0$; $y = Ce^{x^2}$ for $C = -3 + 0.25j$, $j = 0, 1, 2, \cdots, 24$; in the viewing window $-2 \le x \le 2$, $-5 \le y \le 5$

2. $y' = 2\sqrt{|y|}$; $y = |x + C|(x + C)$ for $C = -2 + 0.2j$, $j = 0, 1, 2, \cdots, 20$; in the viewing window $-3 \le x \le 3$, $-6 \le y \le 6$

3. $\dfrac{dP}{dt} = P(0.25 - 10^{-4}P)$; $P = \dfrac{2500Ce^{0.25t}}{(2500 - C) + Ce^{0.25t}}$ for $C = 250j$, $j = 1, 2, \cdots, 14$; in the viewing window $0 \le t \le 25$, $0 \le P \le 3000$

4. $y' = -10(x - 1)y$; $y = Ce^{-5(x-1)^2}$ for $C = 0.5j$, $j = 1, 2, \cdots, 14$; in the viewing window $0 \le x \le 2$, $0 \le y \le 7$

It is more difficult to visualize a two-parameter or three-parameter family of solutions for a higher-order differential equation. One way is to vary each parameter separately. A second way is to look at the members of the family that satisfy an initial condition such as $y(x_0) = y_0$.

In Problems 5–7 plot the solution indicated for the specified choice of each parameter in the given viewing window.

5. $y'' + (y')^2 = 0$; $y = \ln|x + C_1| + C_2$ in the viewing window $-1 \le x \le 25$, $-2 \le y \le 16$

(a) $C_1 = 1$; $C_2 = j$ for $j = 1, 2, \ldots, 12$
(b) $C_1 = j$ for $j = -10, -9, \ldots, 0, 1, 2$; $C_2 = 4$
(c) $C_1 = e^{(4-j)} - 2$; $C_2 = j$ for $j = -2, -1, 0, \ldots, 6$

6. $y'' + 2y' + 10y = 0$; $y = C_1e^{-x}\cos(3x - C_2)$ in the viewing window $-1 \le x \le 2.5$, $-7 \le y \le 7$

(a) $C_1 = 5$; $C_2 = 0.5j$ for $j = -2, -1, 0, 1, 2, \ldots, 8$
(b) $C_1 = j$ for $j = -4, -3, \ldots, 5, 6, 7$; $C_2 = 0$
(c) $C_1 = 5\sec j$; $C_2 = j$ for $j = -2, -1, 0, \ldots, 6$

7. $y'' - 6y' + 9y = 0$; $y = e^{3x}(C_1x + C_2)$ in the viewing window $0 \le x \le 1.8$, $-10 \le y \le 25$

(a) $C_1 = 0.01$; $C_2 = 0.01j$ for $j = -2, -1, 0, 1, 2, \ldots, 6$
(b) $C_1 = 0.05j$ for $j = -4, -3, \ldots, 5, 6, 7$; $C_2 = 0$
(c) $C_1 = 0.5 - 0.2j$; $C_2 = 0.2j$ for $j = 1, 2, 3, \ldots, 8$

18.3 A solution to an initial-value problem:

$$\text{solve} \quad \frac{dy}{dx} = F(x, y) \quad \text{subject to} \quad y(x_0) = y_0$$

can be approximated by using a short calculator or computer program. Select a Δx near zero (positive or negative but not zero) and construct the sequence $x_j = x_0 + j\,\Delta x$ for $j = 1, 2, \ldots$. A tangent line approximation for y at x_0 gives

$$y \approx y_0 + F(x_0, y_0)(x - x_0)$$

for x near x_0. Use this approximation for $x = x_1$ to get

$$y(x_1) \approx y_1 = y_0 + F(x_0, y_0)\,\Delta x$$

Repeatedly using these small tangent line approximations gives

$$y(x_{j+1}) \approx y_{j+1} = y_j + F(x_j, y_j)\,\Delta x$$

This method for the approximate solution of a first-order initial-value problem is called the **Euler method**. Notice that in the special case where $F(x, y) = f(x)$ and $y(x) = y_0 + \int_{x_0}^{x} f(t)\,dt$, the Euler method is equivalent to approximating $y(x_{j+1})$ using Riemann sums with left endpoints over an equally spaced partition with spacing Δx.

In Problems 8–11 use the Euler method to approximate the solution of the given initial-value problem on the indicated interval.

8. $y' = 1 + x \sin(xy)$, $y(0) = 0$ on $0 \le x \le 2$ with $\Delta x = 0.1$

9. $y' = \sin x^2$, $y(0) = 0.4$ on $0 \le x \le 1$ with $\Delta x = 0.05$

10. $y' = -xy + \dfrac{4x}{y}$, $y(0) = 1$ on $0 \le x \le 1$ with $\Delta x = 0.025$

11. $y' = \dfrac{x - y}{x + y}$, $y(2) = 1$ on $1 \le x \le 2$ with $\Delta x = -0.05$

Just as there are methods for numerical integration that are much more efficient than using Riemann sums with left endpoints, there are better methods for approximating initial-value problems than using the Euler method. A simple example described here is commonly called the **modified Euler method**. Given (x_j, y_j) where $y_j \approx y(x_j)$, set

$$z_{j+1} = y_j + m_1\,\Delta x \quad \text{where } m_1 = F(x_j, y_j)$$

and

$$y_{j+1} = y_j + 0.5(m_1 + m_2)\,\Delta x \quad \text{where } m_2 = F(x_{j+1}, z_{j+1})$$

Again in the special case where $F(x, y) = f(x)$, this modified Euler method is equivalent to a numerical integration rule, in this case the Trapezoidal Rule.

In Problems 12–15 use the modified Euler method to approximate the solution of the given initial-value problem on the indicated interval.

12. $y' = 1 + x \sin(xy)$, $y(0) = 0$ on $0 \le x \le 2$ with $\Delta x = 0.2$

13. $y' = \sin x^2$, $y(0) = 0.4$ on $0 \le x \le 1$ with $\Delta x = 0.1$

14. $y' = -xy + \dfrac{4x}{y}$, $y(0) = 1$ on $0 \le x \le 1$ with $\Delta x = 0.05$

15. $y' = \dfrac{x - y}{x + y}$, $y(2) = 1$ on $1 \le x \le 2$ with $\Delta x = -0.1$

18.4 Consider now the special nonlinear second-order initial-value problems of the form

$$y'' = F(x, y) \qquad y(x_0) = y_0 \qquad y'(x_0) = y_0'$$

(Notice that this special form has no dependence on y'.) Again for numerical computation, we seek a sequence y_j consisting of approximations for $y(x_j)$ for $j = 1, 2, 3, \ldots$, where $x_j = x_0 + j\,\Delta x$ for some Δx near zero (positive or negative but not zero). Using the central difference approximation for the second derivative

$$y''(x_j) = \frac{y(x_{j+1}) - 2y(x_j) + y(x_{j-1})}{(\Delta x)^2}$$

we want

$$F(x_j, y_j) \approx \frac{y_{j+1} - 2y_j + y_{j-1}}{(\Delta x)^2}$$

We choose to use this relationship to select y_{j+1}, obtaining the **difference equation**

$$y_{j+1} = 2y_j - y_{j-1} + (\Delta x)^2 F(x_j, y_j) \qquad \text{for } j = 2, 3, \ldots$$

To begin using this formula, we need y_0 and y_1. This first quantity is known from the initial conditions. We approximate y_1 using the fact that we also know $y'(x_0) = y_0'$ and $y''(x_0) = F(x_0, y_0)$. Taking a quadratic Taylor polynomial approximation for y centered at x_0 and evaluating at x_1 give

$$y_1 = y_0 + y_0'\,\Delta x + F(x_0, y_0)(\Delta x)^2 \approx y(x_1)$$

Now we can use y_0 and y_1 to find y_2 and later terms using the selected relationship.

In Problems 16–18 use the difference equation method to approximate the solution of the given initial-value problem on the indicated interval.

16. $y'' = -y^3$, $y(0) = 0$, $y'(0) = 0.025$ on $0 \le x \le 4$ with $\Delta x = 0.1$

17. $y'' = e^{-x} - y^2$, $y(1) = 2$, $y'(1) = 0.4$ on $1 \le x \le 2$ with $\Delta x = 0.025$

18. $y'' = \dfrac{1}{x + y}$, $y(3) = 0$, $y'(3) = 0$ on $3 \le x \le 7$ with $\Delta x = 0.2$

18.6 For vibrational models, it is important to understand the changes in the solutions when one of the constants in the differential equation changes. For example, how does the solution for a free damped motion problem change when the damping constant is increased or the spring constant is decreased? Much can be learned about these models by plotting solutions for varying constants.

In Problems 19 and 20 plot the solutions for the free damped motion problem for the given constants when the weight on the spring is released from rest ($x'(0) = 0$) from a point 1 ft above the equilibrium position ($x(0) = -1$).

19. $m = 0.25$ slug, $k = 4$ lb/ft, $\beta = 1.5, 1.6, 1.7, 1.8, 1.9, 2.0, 2.1, 2.2$ lb-s/ft

20. $m = 0.25$ slug, $\beta = 2$ lb-s/ft, $k = 3.7, 3.8, 3.9, 4.0, 4.1, 4.2, 4.3, 4.4$ lb/ft

APPENDICES

Greek Alphabet

α	A	Alpha	ν	N	Nu
β	B	Beta	ξ	Ξ	Xi
γ	Γ	Gamma	o	O	Omicron
δ	Δ	Delta	π	Π	Pi
ε	E	Epsilon	ρ	P	Rho
ζ	Z	Zeta	σ	Σ	Sigma
η	H	Eta	τ	T	Tau
θ	Θ	Theta	υ	Υ	Upsilon
ι	I	Iota	ϕ	Φ	Phi
κ	K	Kappa	χ	X	Chi
λ	Λ	Lambda	ψ	Ψ	Psi
μ	M	Mu	ω	Ω	Omega

APPENDIX I REVIEW OF PRECALCULUS MATHEMATICS

I.1 SOME BASIC MATHEMATICS

Sets A set A is a collection of objects called the **elements** of A. The symbolism $a \in A$ is used to denote that a is an element of a set A. A set B is a **subset** of a set A, written $B \subset A$, if every element of B is also an element of A. Sets A and B are equal, $A = B$, if they contain the same elements. The **union** $A \cup B$ of two sets A and B is the set containing the elements that are either in A or in B. The **intersection** $A \cap B$ is the set of elements that are elements in A and in B. The intersection of sets A and B is the set of elements that are *common* to both A and B. The set $\varnothing$ with no elements is called the **empty set**. If $A \cap B = \varnothing$, then A and B are said to be **disjoint** sets.

Laws of Exponents If n is a positive integer,

$$x^{-n} = \frac{1}{x^n}, \qquad x \neq 0$$

Also,

$$x^0 = 1, \qquad x \neq 0$$

If m and n are integers, then

$$(i)\ x^m x^n = x^{m+n}$$

$$(ii)\ \frac{x^m}{x^n} = x^{m-n}, \qquad x \neq 0$$

$$(iii)\ (x^m)^n = x^{mn}$$

$$(iv)\ (xy)^n = x^n y^n$$

$$(v)\ \left(\frac{x}{y}\right)^n = \frac{x^n}{y^n}, \qquad y \neq 0$$

Rational Exponents and Radicals If m and n are positive integers,

$$x^{m/n} = (x^m)^{1/n} = (x^{1/n})^m$$

$$x^{1/n} = \sqrt[n]{x}$$

provided all expressions represent real numbers. The laws of exponents hold for rational numbers.

EXAMPLE 1 $16^{3/4} = (16^{1/4})^3 = 2^3 = 8$ □

Binomial Theorem If n is a positive integer, the **Binomial Theorem** is

$$(a+b)^n = a^n + \frac{n}{1!}a^{n-1}b + \frac{n(n-1)}{2!}a^{n-2}b^2 + \cdots$$
$$+ \frac{n(n-1)\cdots(n-r+1)}{r!}a^{n-r}b^r + \cdots + b^n$$

where $r! = 1 \cdot 2 \cdot 3 \cdots (r-1)r$. For example,

$$(a+b)^2 = a^2 + 2ab + b^2$$
$$(a+b)^3 = a^3 + 3a^2b + 3ab^2 + b^3$$

EXAMPLE 2 Expand $(2x+4)^3$.

Solution When we identify $a = 2x$ and $b = 4$, the Binomial Theorem gives

$$(2x+4)^3 = (2x)^3 + 3(2x)^2 4 + 3(2x)(4)^2 + (4)^3$$
$$= 8x^3 + 48x^2 + 96x + 64$$ □

Special Factors

Difference of two squares: $X^2 - Y^2 = (X+Y)(X-Y)$

Difference of two cubes: $X^3 - Y^3 = (X-Y)(X^2+XY+Y^2)$

Sum of two cubes: $X^3 + Y^3 = (X+Y)(X^2-XY+Y^2)$

EXAMPLE 3

(*a*) $x^3 - 1 = (x-1)(x^2+x+1)$

(*b*) $x^3 + 8 = (x+2)(x^2-2x+4)$ □

Equations The equation $ax + b = 0$ is called a **first-degree** or **linear equation**. A linear equation has the unique solution $x = -b/a$, $a \neq 0$. A **second-degree** or **quadratic equation** is $ax^2 + bx + c = 0$, $a \neq 0$. Two methods for finding the solutions of a quadratic equation are factoring and the quadratic formula. Recall, the latter is given by

$$x = \frac{-b \pm \sqrt{b^2 - 4ac}}{2a}$$

When $b^2 - 4ac > 0$, the equation has two distinct real solutions. For example, we see from the quadratic formula that the solutions of $x^2 + x - 1 = 0$ are the irrational numbers $(-1 - \sqrt{5})/2$ and $(-1 + \sqrt{5})/2$. When $b^2 - 4ac = 0$, the solutions of a quadratic equation are **equal** (or have *multiplicity* 2). When $b^2 - 4ac < 0$, the solutions of a quadratic equation are **complex numbers.**

Complex Numbers A **complex number** is any expression of the form

$$z = a + bi \quad \text{where} \quad i^2 = -1$$

The real numbers a and b are called the **real** and **imaginary** parts of z, respectively. In practice, the symbol i is written as $i = \sqrt{-1}$. The number $\bar{z} = a - bi$ is called the **complex conjugate** of z.

EXAMPLE 4 If $z_1 = 4 + 5i$ and $z_2 = 3 - 2i$ are complex numbers, then their complex conjugates are, respectively,

$$\bar{z}_1 = 4 - 5i \quad \text{and} \quad \bar{z}_2 = 3 - (-2)i = 3 + 2i$$ □

Sum, Difference, and Product The **sum, difference**, and **product** of two complex numbers $z_1 = a_1 + b_1 i$ and $z_2 = a_2 + b_2 i$ are defined as follows:

$$(i)\quad z_1 + z_2 = (a_1 + a_2) + (b_1 + b_2)i$$
$$(ii)\quad z_1 - z_2 = (a_1 - a_2) + (b_1 - b_2)i$$
$$(iii)\quad z_1 z_2 = (a_1 a_2 - b_1 b_2) + (a_1 b_2 + b_1 a_2)i$$

In other words, to add or subtract two complex numbers we simply add or subtract the corresponding real and imaginary parts. To multiply two complex numbers we use the distributive law and the fact that $i^2 = -1$.

EXAMPLE 5 If $z_1 = 4 + 5i$ and $z_2 = 3 - 2i$, then

$$\begin{aligned} z_1 + z_2 &= (4 + 3) + (5 - 2)i = 7 + 3i \\ z_1 - z_2 &= (4 - 3) + (5 - (-2))i = 1 + 7i \\ z_1 z_2 &= (4 + 5i)(3 - 2i) \\ &= (4 + 5i)3 + (4 + 5i)(-2i) \\ &= 12 + 15i - 8i - 10i^2 \\ &= (12 + 10) + (15 - 8)i = 22 + 7i \end{aligned}$$ □

The product of a complex number $z = a + bi$ and its conjugate $\bar{z} = a - bi$ is the *real number*

$$z\bar{z} = a^2 + b^2$$

Quotient The **quotient** of two complex numbers is found by multiplying the numerator and denominator of the expression by the conjugate of the denominator.

EXAMPLE 6 If z_1 and z_2 are the complex numbers in Example 5, then

$$\frac{z_1}{z_2} = \frac{4+5i}{3-2i} = \frac{(4+5i)(3+2i)}{(3-2i)(3+2i)}$$
$$= \frac{2+23i}{3^2+2^2} = \frac{2}{13} + \frac{23}{13}i$$ □

If $a > 0$, then the square root of the negative number $-a$ is defined by

$$\sqrt{-a} = \sqrt{a}\sqrt{-1} = \sqrt{a}i$$

For example, $$\sqrt{-25} = \sqrt{25}\sqrt{-1} = 5i.$$

EXAMPLE 7 By the quadratic formula, the solutions of

$$x^2 - 4x + 20 = 0$$

are $$x = \frac{4 \pm \sqrt{-64}}{2} = \frac{4 \pm \sqrt{64}i}{2}$$

or $$x_1 = 2 + 4i \qquad x_2 = 2 - 4i$$ □

As illustrated in Example 7, the complex solutions of a quadratic equation with real coefficients are conjugates.

Every real number can also be considered a complex number by simply taking $b = 0$ in $z = a + bi$. The relationship between real and complex numbers is shown in the following diagram:

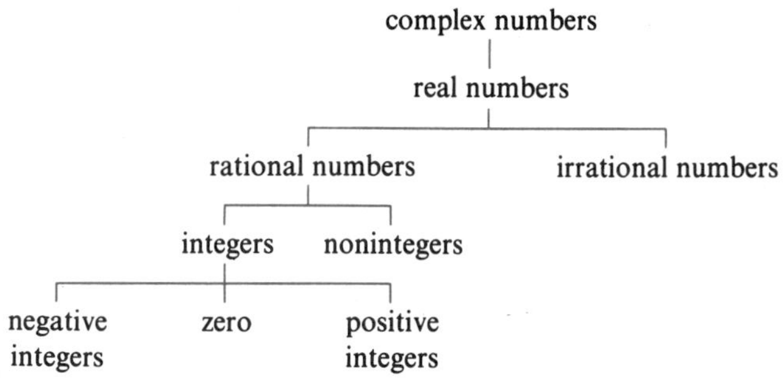

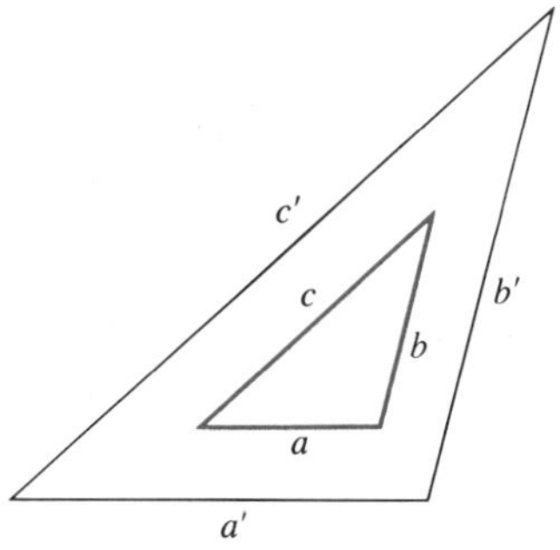

FIGURE I.1

Similar Triangles The lengths of corresponding sides in similar triangles are proportional. Thus, if Figure I.1 represents two similar triangles, the ratios of the lengths of the corresponding sides are equal:

$$\frac{a}{a'} = \frac{b}{b'} = \frac{c}{c'}$$

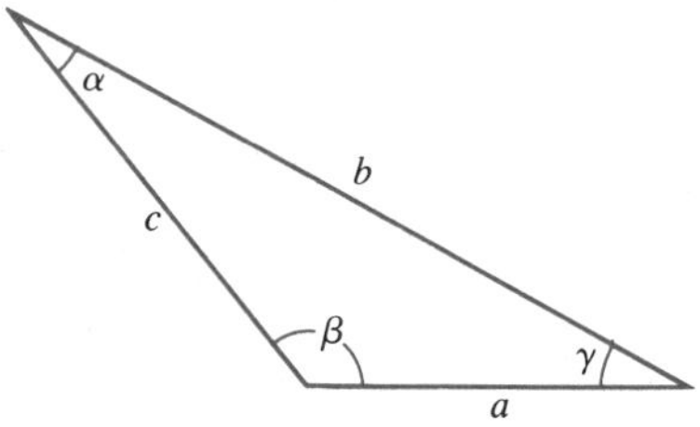

FIGURE I.2

Law of Sines For the triangle shown in Figure I.2, if we know (*a*) the length of one side and two angles or (*b*) the lengths of two sides and the angle opposite one of them, then the remaining parts of the triangle can be found from the **law of sines**:

$$\frac{\sin\alpha}{a} = \frac{\sin\beta}{b} = \frac{\sin\gamma}{c}$$

Law of Cosines Referring to the triangle in Figure I.2, if we know (*a*) the lengths of three sides or (*b*) the length of two sides and the angle between them, then the remaining parts of the triangle can be found from the **law of cosines**:

$$a^2 = b^2 + c^2 - 2bc \cos \alpha$$
$$b^2 = a^2 + c^2 - 2ac \cos \beta$$
$$c^2 = a^2 + b^2 - 2ab \cos \gamma$$

EXERCISES I.1 *Answers to odd-numbered problems begin on page A-111.*

In Problems 1–4 evaluate the given expression.

1. $8^{2/3}$ **2.** $16^{-5/4}$

3. $\sqrt{(-5)^2}$ **4.** $\sqrt[3]{8x^3y^9}$

In Problems 5–8 use the Binomial Theorem to expand the given expression.

5. $(3r - 4s)^2$ **6.** $(x + y - 1)^2$

7. $(2x - y)^3$ **8.** $(t^2 + r^3)^3$

In Problems 9–12 factor the given expression.

9. $4x^2y^2 - 9$ **10.** $x^4 - y^4$

11. $64a^3 - b^3$ **12.** $x^6 - 1$

In Problems 13–22 solve the given equation.

13. $2x - 3 = 6x + 9$ **14.** $3(x - 2) = -4(x - 1)$

15. $4x^2 = 9$ **16.** $36x^2 - 4 = 0$

17. $5x^2 = 2x$ **18.** $3x^2 + x = 0$

19. $x^2 + 2x - 15 = 0$ **20.** $x^2 + 6x + 9 = 0$

21. $2x^2 + 2x - 1 = 0$ **22.** $-x^2 + 3x + 2 = 0$

In Problems 23–32 perform the indicated operation. Write the answer in the form $a + bi$.

23. $(6 + 5i) - (-8 + 2i)$ **24.** $3i(2 + i) + 4(1 - 2i)$

25. $(4 + 6i)(-2 + 5i)$ **26.** $(1 - i)(1 + i)(3 + 7i)$

27. $\dfrac{1 + 3i}{3 - 4i}$ **28.** $\dfrac{i}{1 + i}$

29. $i\dfrac{2 + i}{1 - i}$ **30.** $\dfrac{1}{i(2 - i)}$

31. i^5 **32.** i^{-8}

In Problems 33–36 solve the given equation.

33. $x^2 + x + 1 = 0$ **34.** $3x^2 + 4 = 0$

35. $x^2 - 4x + 29 = 0$ **36.** $2x^2 - 2x + 1 = 0$

In Problems 37 and 38 determine x and y.

37.

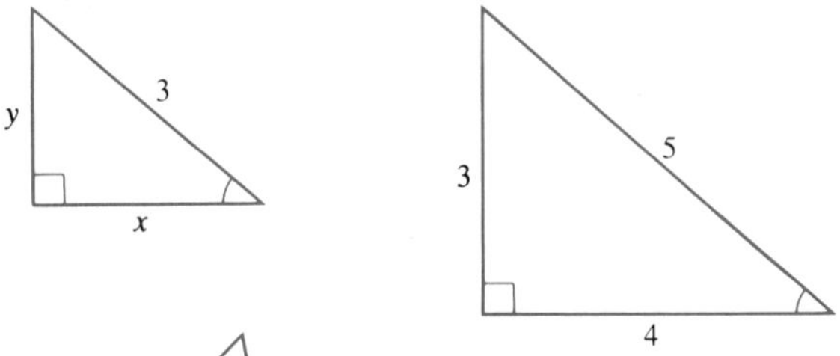

38.

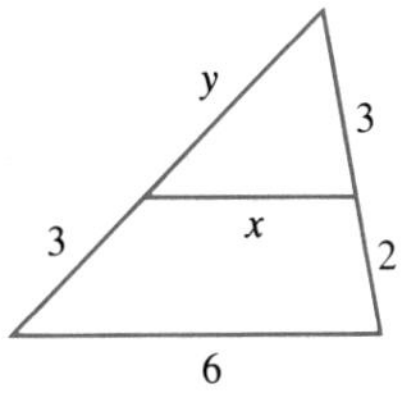

In Problems 39 and 40 find the indicated unknowns in the given triangle.

39.

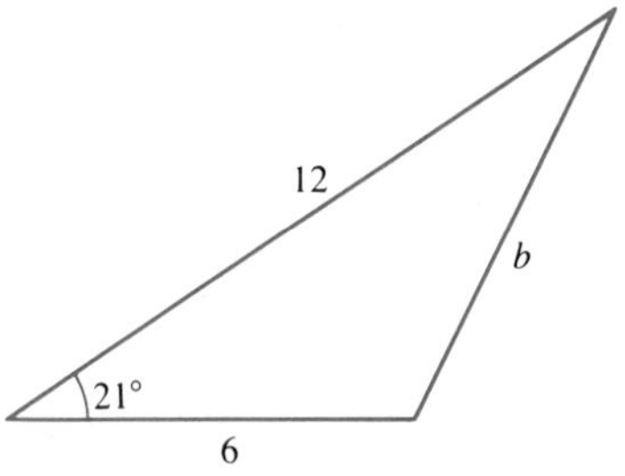

40.

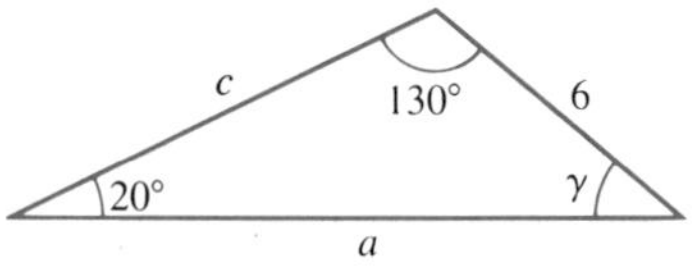

I.2 REAL NUMBERS AND INEQUALITIES

Real numbers are classified as either **rational** or **irrational**. A rational number can be expressed as a quotient a/b, where a and b are integers and $b \neq 0$. A number that is *not* rational is said to be irrational. For example, $\frac{5}{4}$, -6, $\frac{22}{7}$, $\sqrt{4} = 2$, and $1.32 = \frac{132}{100}$ are rational numbers, whereas $\sqrt{2}$, $\sqrt{3}/2$, and π are irrational. The sum, difference, and product of two real numbers are real numbers. The quotient of two real numbers is a real number provided the denominator is not zero.

The set of real numbers is denoted by the symbol R; the symbols Q and H are commonly used to denote the set of rational numbers and the set of irrational numbers, respectively. In terms of the union of two sets, we have $R = Q \cup H$. The fact that Q and H have no common elements is summarized by $Q \cap H = \varnothing$, where $\varnothing$ is the empty set.

Number Line The set of real numbers can be put into a one-to-one correspondence with the points on a horizontal line, which is called the **number line** or **real line**. A number a associated with a point P on the number line is called the **coordinate** of P. The point chosen to represent 0 is called the **origin**. As shown in Figure I.3, **positive numbers** are placed to the right of the origin and **negative numbers** are placed to the left of the origin. The number 0 is neither positive nor negative. The arrowhead on the number line points in the **positive direction**. As a rule, the words "point a" and "number a" are used interchangeably with "point P with coordinate a."

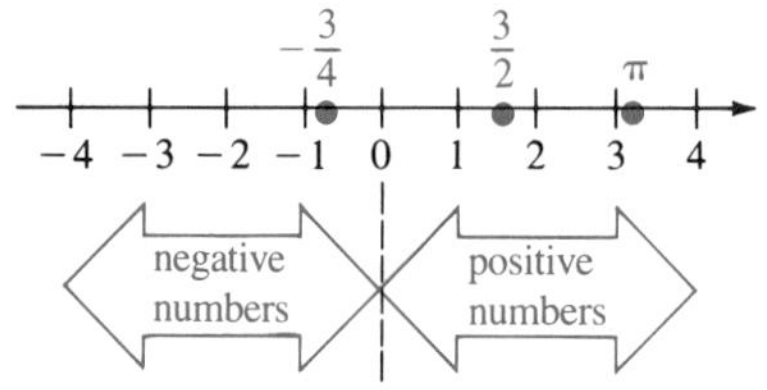

FIGURE I.3

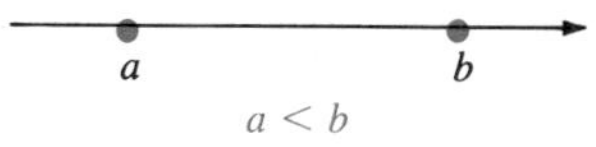

FIGURE I.4

Inequalities The set R of real numbers is an **ordered** set. A real number a is **less than** a real number b, written $a < b$, if the difference $b - a$ is positive. For example, $-2 < 5$, since $5 - (-2) = 7$ is a positive number. On the number line $a < b$ means that the number a lies to the left of the number b. See Figure I.4. The statement "a is less than b" is equivalent to saying "b is greater than a" and is written $b > a$. Thus, a number a is positive if $a > 0$ and negative if $a < 0$. Expressions such as $a < b$ and $b > a$ are called **inequalities.**

Properties of Inequalities The following is a list of some **properties of inequalities**:

(*i*) If $a < b$, then $a + c < b + c$ for any real number c.

(*ii*) If $a < b$ and $c > 0$, then $ac < bc$.

(*iii*) If $a < b$ and $c < 0$, then $ac > bc$.

(*iv*) If $a < b$ and $b < c$, then $a < c$.

Property (*iii*) indicates that if an inequality is multiplied by a negative number, then the sense of the inequality is *reversed.*

The symbolism $a \leq b$ is read "a is **less than or equal** to b" and means $a < b$ or $a = b$. The preceding four properties hold when $<$ and $>$ are replaced by $\leq$ and $\geq$, respectively. If $a \geq 0$, the number a is said to be **nonnegative**.

EXAMPLE 1 Solve the inequality $2x < 5x + 9$.

Solution First we use (*i*) to add $-5x$ to both sides of the inequality:

$$2x + (-5x) < 5x + 9 + (-5x)$$
$$-3x < 9$$

Then when we multiply both sides of the last inequality by $-\frac{1}{3}$, property (*iii*) implies $x > -3$. □

Intervals If $a < b$, the set of real numbers x that are *simultaneously* less than b and greater than a is written $\{x \mid a < x < b\}$ or simply $a < x < b$. This set is called an **open interval** and is denoted by (a, b). The numbers a and b are the **endpoints** of the interval.

Various kinds of intervals and their **graphs** on the number line are summarized in the following table.

Name	Symbol	Definition	Graph
Open interval	(a, b)	$\{x \mid a < x < b\}$	
Closed interval	$[a, b]$	$\{x \mid a \leq x \leq b\}$	
Half-open intervals	$(a, b]$	$\{x \mid a < x \leq b\}$	
	$[a, b)$	$\{x \mid a \leq x < b\}$	
Infinite intervals	(a, ∞)	$\{x \mid x > a\}$	
	$[a, \infty)$	$\{x \mid x \geq a\}$	
	$(-\infty, b)$	$\{x \mid x < b\}$	
	$(-\infty, b]$	$\{x \mid x \leq b\}$	
	$(-\infty, \infty)$	$\{x \mid -\infty < x < \infty\}$	

The symbols ∞ and $-\infty$ are read "infinity" and "negative infinity," respectively. These symbols do not represent real numbers. Their use is simply shorthand for writing an *unbounded* interval.

EXAMPLE 2

(*a*) The set $\{x \mid -3 < x < 8\}$ is $(-3, 8)$ in interval notation.

(*b*) The set $\{x \mid 5 < x \leq 6\}$ is $(5, 6]$ in interval notation.

(*c*) The set $\{x \mid x \leq -1\}$ is $(-\infty, -1]$ in interval notation. □

An interval that is a subset of another interval I is said to be a **subinterval** of I. For example, $[1, 2]$, $[2, \frac{5}{2}]$, and $(3, 6)$ are each subintervals of $[1, 6]$.

EXAMPLE 3 Solve the inequality $4 < 2x - 3 < 8$.

Solution From the properties of inequalities it follows that

$$4 < 2x - 3 < 8$$

$$4 + 3 < 2x < 8 + 3 \qquad [\text{by } (i)]$$

$$\frac{7}{2} < x < \frac{11}{2} \qquad [\text{by } (ii)]$$

In interval notation the solution is $(\frac{7}{2}, \frac{11}{2})$. □

The inequalities considered in Examples 1 and 3 are **first-degree** or **linear** inequalities. In the next example we solve a **second-degree** or **quadratic** inequality.

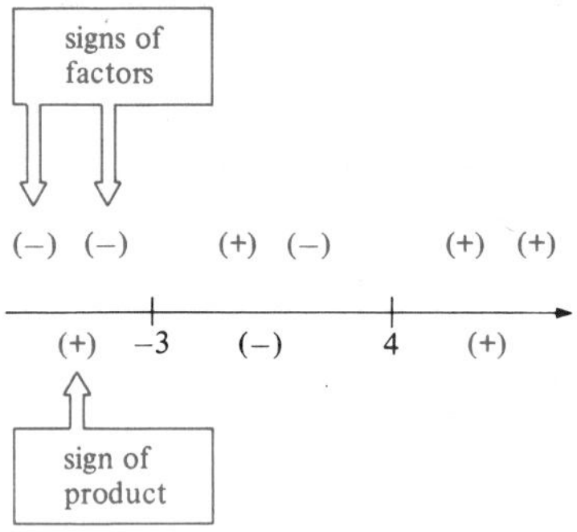

FIGURE I.5

EXAMPLE 4 Solve the inequality $(x + 3)(x - 4) < 0$.

Solution We begin by putting the two roots, -3 and 4, of the quadratic equation $(x + 3)(x - 4) = 0$ on the number line. By ascertaining the algebraic sign of each factor on the subintervals $(-\infty, -3)$, $(-3, 4)$, and $(4, \infty)$ of $(-\infty, \infty)$, we can determine the algebraic sign of the product. For example, by replacing x by *any* number in $(-\infty, -3)$, we see that both factors $x + 3$ and $x - 4$ are negative, and so the product $(x + 3)(x - 4)$ must be positive. In this case, the original inequality is not satisfied. Inspection of Figure I.5 indicates that the solution of the problem is $(-3, 4)$. □

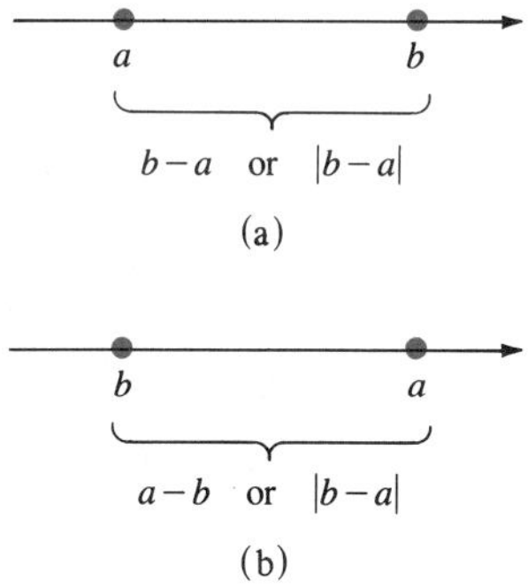

FIGURE I.6

Absolute Value

If a is a real number, then its **absolute value** is

$$|a| = \begin{cases} a, & a \geq 0 \\ -a & a < 0 \end{cases}$$

On the number line, $|a|$ is the distance between the origin and the number a. In general, the **distance between two numbers** a and b is either $b - a$ if $a < b$ or $a - b$ if $b < a$. In terms of absolute value, the distance between a and b is $|b - a|$. See Figure I.6. Note that $|b - a| = |a - b|$.

EXAMPLE 5

(*a*) $|-5| = -(-5) = 5$

(*b*) $|2x - 1| = \begin{cases} 2x - 1 & \text{if } 2x - 1 \geq 0 \\ -(2x - 1) & \text{if } 2x - 1 < 0 \end{cases}$

That is, $$|2x - 1| = \begin{cases} 2x - 1 & \text{if } x \geq \frac{1}{2} \\ -2x + 1 & \text{if } x < \frac{1}{2} \end{cases}$$ □

EXAMPLE 6 The distance between -5 and 7 is the same as the distance between 7 and -5.

$$|7 - (-5)| = |12| = 12 \text{ units} \quad \text{and} \quad |-5 - 7| = |-12| = 12 \text{ units}$$ □

Absolute Values and Inequalities Since $|x|$ gives the distance between a number x and the origin, the solution of the inequality $|x| < b$, $b > 0$, is the set of real numbers x that are *less* than b units from the origin. As shown in Figure I.7,

$$|x| < b \quad \text{if and only if} \quad -b < x < b \tag{1}$$

On the other hand, the solution of the inequality $|x| > b$ is the set of real numbers that are *more* than b units from the origin. Consequently,

$$|x| > b \quad \text{if and only if} \quad x > b \quad \text{or} \quad x < -b \tag{2}$$

See Figure I.8. The results in (1) and (2) hold when $<$ and $>$ are replaced with $\leq$ and $\geq$, respectively, and when x is replaced by $x - a$. For example, $|x - a| \leq b$, $b > 0$, represents the set of real numbers x such that the distance from x to a is less than or equal to b units. As illustrated in Figure I.9, $|x - a| \leq b$ if and only if x is a number in the closed interval $[a - b, a + b]$.

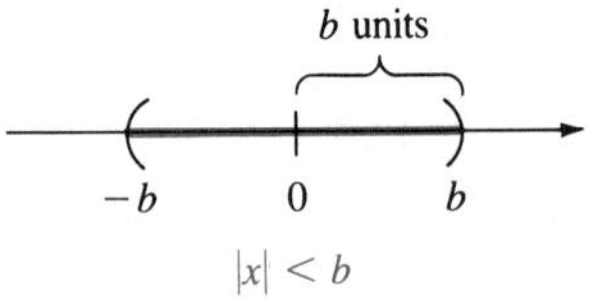

FIGURE I.7

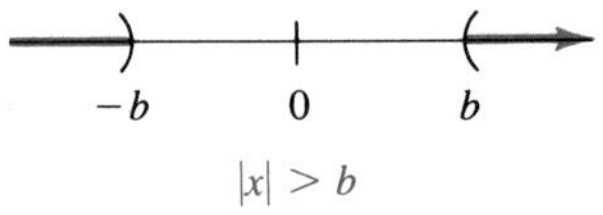

FIGURE I.8

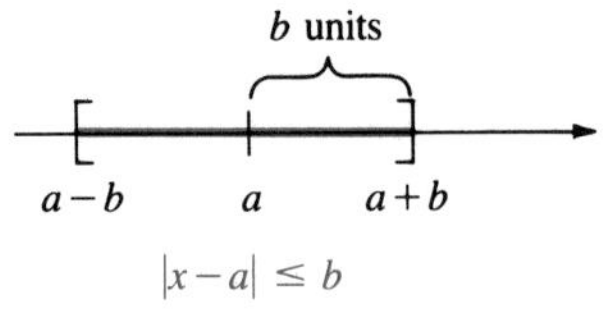

FIGURE I.9

EXAMPLE 7 Solve the inequality $|x - 1| < 3$.

Solution Using (1), we first rewrite the inequality as

$$-3 < x - 1 < 3$$

Therefore,

$$-3 + 1 < x - 1 + 1 < 3 + 1$$
$$-2 < x < 4$$

The solution is the open interval $(-2, 4)$. □

EXAMPLE 8 Solve the inequality $|x| > 2$.

Solution From (2) we have immediately

$$x > 2 \quad \text{or} \quad x < -2$$

The solution is a union of infinite intervals: $(-\infty, -2) \cup (2, \infty)$. □

EXAMPLE 9 Solve the inequality $0 < |x - 2| \leq 7$.

Solution The given inequality means $0 < |x - 2|$ *and* $|x - 2| \leq 7$. In the first case, $0 < |x - 2|$ is true for any real number except $x = 2$. In the second case, we have

$$-7 \leq x - 2 \leq 7 \quad \text{or} \quad -5 \leq x \leq 9$$

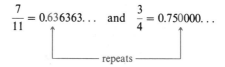

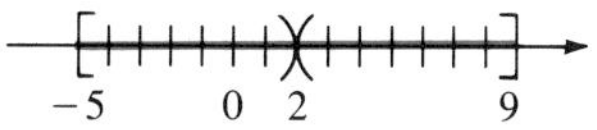

FIGURE I.10

or $[-5, 9]$. The set of real numbers x that satisfies both inequalities consists of all numbers in $[-5, 9]$ except 2. Written as a union of intervals, the solution is $[-5, 2) \cup (2, 9]$. See Figure I.10. □

Triangle Inequality In conclusion, the proof of the so-called **triangle inequality**

$$|a + b| \leq |a| + |b| \tag{3}$$

is left as an exercise.

▲ ***Remarks*** (*i*) Every real number has a nonterminating decimal representation. The rational numbers are characterized as *repeating* decimals; for example, by long division, we see

$$\frac{7}{11} = 0.636363\ldots \quad \text{and} \quad \frac{3}{4} = 0.750000\ldots$$

(repeats)

Hence, irrational numbers are *nonrepeating* decimal numbers. For example,

$$\pi = 3.141592\ldots \quad \text{and} \quad \sqrt{3} = 1.732050\ldots$$

(nonrepeating decimal)

(*ii*) In Example 8 there is no way of writing the solution of the inequality $|x| > 2$ as a single interval. The statement $x > 2$ or $x < -2$ is *not* equivalent to $2 < x < -2$. There are no real numbers that are simultaneously greater than positive 2 and less than negative 2.

(*iii*) Before moving on, a brief word about the square root of a nonnegative number x is in order:

The square root of $x \geq 0$, written $\sqrt{x}$, is always nonnegative.

For example, $\sqrt{64} = 8$, *not* ± 8. In fact, the square root is related to the absolute value:

$$\sqrt{x^2} = |x|$$

Thus, $\sqrt{x^2} = x$ if $x \geq 0$ and $\sqrt{x^2} = -x$ if $x < 0$. For example, $\sqrt{5^2} = 5$ since $5 > 0$ and $\sqrt{(-5)^2} = -(-5) = 5$ since $-5 < 0$.

EXERCISES I.2

Answers to odd-numbered problems begin on page A-111.

In Problems 1–4 write the given inequality using interval notation.

1. $-4 \leq x < 20$
2. $x \geq 5$
3. $x < -2$
4. $\frac{1}{2} < x < \frac{7}{4}$

In Problems 5–8 write the given interval as an inequality.

5. $(\frac{3}{2}, 6)$
6. $[-1, 5]$
7. $[20, \infty)$
8. $(-\infty, -7)$

In Problems 9–12 write the interval for the given graph.

9. 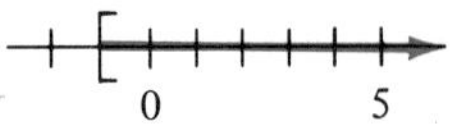

10. 

11. 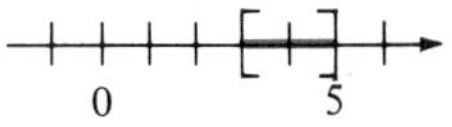

12. 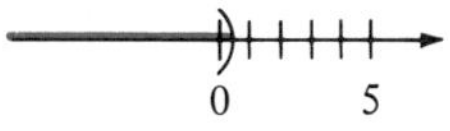

In Problems 13–28 solve the given inequality. Write the solution in interval notation.

13. $3x < -9$

14. $-2x > 8$

15. $4x + 1 > 10$

16. $-\frac{1}{2}x + 6 \leq 0$

17. $4x \geq 5x - 7$

18. $x + 12 \leq 5x$

19. $-4 < 1 - x \leq 3$

20. $1 \leq \dfrac{2x + 14}{3} < 2$

21. $x \leq 3x + 2 \leq x + 6$

22. $10 - x < 4x \leq 25 - x$

23. $(x - 1)(x - 9) < 0$

24. $x^2 + 2x - 3 \geq 0$

25. $10 - 3x^2 \geq 13x$

26. $-x^2 < 6x$

27. $4x^2 - 4x + 1 \geq 0$

28. $x^2 > 25$

In Problems 29–32 write the given expression without absolute value symbols.

29. $|4 - a|$, $4 - a$ is a negative number

30. $|-6a|$, a is a positive number

31. $|a + 10|$, a is greater than or equal to -10

32. $|a^2 - 1|$, a is a number in $(-1, 1)$

In Problems 33–36 solve for x.

33. $|4x| = 36$

34. $|-2x| = 16$

35. $|3 - 5x| = 22$

36. $|12 - \frac{1}{2}x| = x$

In Problems 37–46 solve the given inequality. Write the solution in interval notation.

37. $|x| < 4$

38. $|-\frac{1}{3}x| \leq 3$

39. $|1 - 2x| \leq 1$

40. $|5 + 4x| < 17$

41. $\left|\dfrac{x + 3}{-2}\right| < 1$

42. $0 < |x + 1| \leq 5$

43. $|x| > 6$

44. $|4 - x| > 0$

45. $|5 - 2x| > 7$

46. $|x + 9| \geq 8$

47. If $1/x < 4$, does it follow that $x > \frac{1}{4}$?

48. If $x^2 < 6x$, does it follow that $x < 6$?

In Problems 49–52 express the given statement as an inequality involving an absolute value. Express the solution of the given statement using interval notation.

49. The set of real numbers less than 4 units from 9

50. The set of real numbers less than or equal to $\frac{1}{2}$ unit from 1.5

51. The set of real numbers greater than 2 units from 2

52. The set of real numbers greater than -1 and less than 7

53. When equipment is depreciated linearly and loses all its initial worth of A dollars over a period of n years, its value V in x years $(0 \leq x \leq n)$ is given by $V = A(1 - x/n)$. If a computer costs \$100,000 initially and is depreciated over 20 years, determine the values of x such that $30{,}000 \leq V \leq 80{,}000$.

54. According to one theory, the most beneficial effect of exercise such as jogging is obtained when the pulse rate is maintained within a certain interval. The endpoints of the interval are obtained by multiplying the number (220 − age) by 0.70 and 0.85. Determine this pulse rate interval for a 30-year-old jogger. For a 40-year-old jogger.

In Problems 55–58 replace the comma between the given pair of real numbers with one of the symbols <, >, or =. Use a calculator.

55. $\pi, \dfrac{22}{7}$

56. $\dfrac{\pi}{2}, 1.5$

57. $\dfrac{180}{\pi}, 57.29$

58. $\dfrac{1}{\pi}, \dfrac{1}{3}$

The **midpoint** of an interval with endpoints a and b is the number $(a + b)/2$. In Problems 59–62 use this information to find an inequality $|x - c| < d$ whose solution is the given interval.

59. $(0, 8)$

60. $(-1, 6)$

61. $(-3, 4)$

62. $(-10, -2)$

63. Use $-|a| \leq a \leq |a|$ and $-|b| \leq b \leq |b|$ to prove the triangle inequality (3). [*Hint:* Add the inequalities.]

I.3 THE CARTESIAN PLANE

It is virtually impossible to pick up a text, journal, or news magazine without encountering some sort of graphical display of data, such as illustrated in Figure I.11, in a *coordinate plane* formed by the intersection of two perpendicular number lines. In mathematics such a coordinate plane is called a **Cartesian plane.***

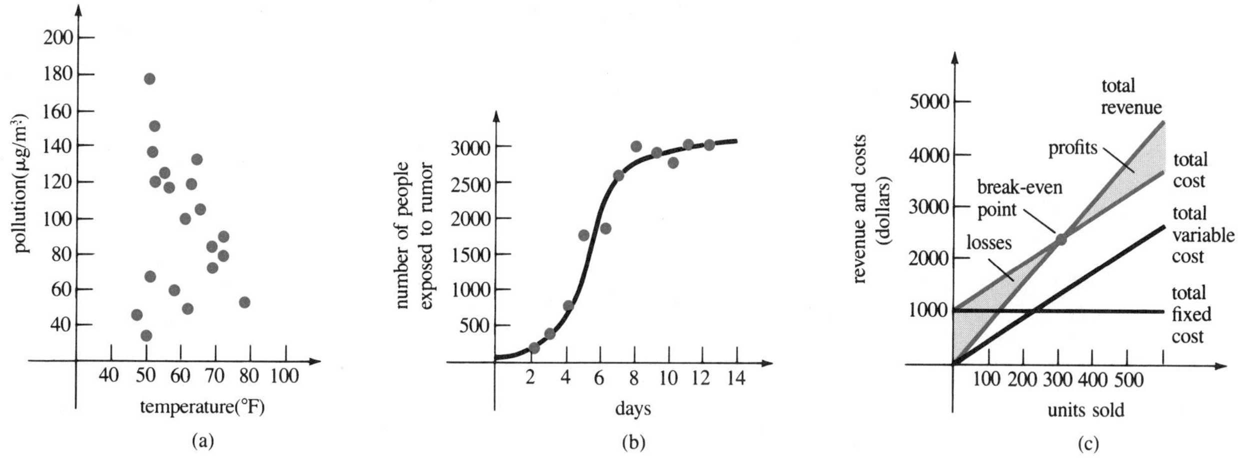

FIGURE I.11

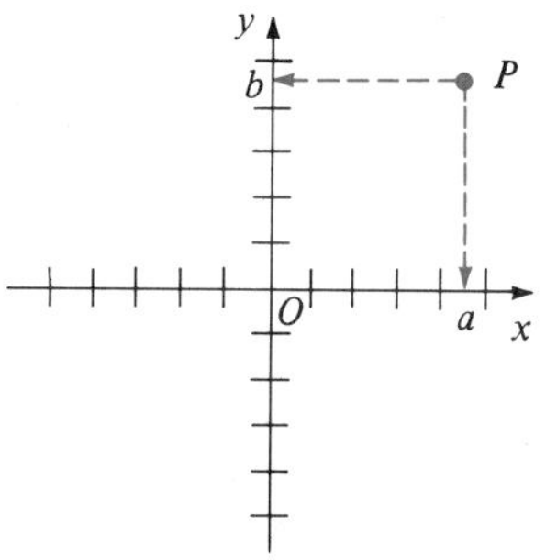

FIGURE I.12

In the following general discussion, we shall assume that the same scale has been used to mark off each number line. The point of intersection of these number lines, corresponding to the number 0 on both lines, is called the **origin** and is denoted by O. The horizontal number line is called the **x-axis** and the vertical line is called the **y-axis**. Numbers to the right of the origin on the x-axis are positive; numbers to the left of O are negative. On the y-axis numbers above the origin are positive; numbers below O are negative. Because the horizontal and vertical number lines are labeled with the letters x and y, a Cartesian plane is often simply called an **xy-plane**.

If P denotes a point in a Cartesian plane, we can draw perpendicular lines from P to both the x- and y-axes. As Figure I.12 shows, this determines a number a on the x-axis and a number b on the y-axis. Conversely, we see that specified numbers a and b on the x- and y-axes determine a unique point P in the plane. In this manner a one-to-one correspondence between points in a Cartesian plane and ordered pairs of real numbers

***René Descartes (1596–1650)** The Cartesian plane was named in honor of the French nobleman, lawyer, scientist, engineer, mathematician, gentleman soldier, playboy, philosopher, and prominent architect of the so-called Age of Reason, René Descartes. In mathematics Descartes is credited with being one of the principal "inventors" of the field of analytic geometry, which deals with the correspondence between equations in two variables and curves in a coordinate plane. Legend has it that the fundamental concepts of analytic geometry came to Descartes while he was lying in bed observing a fly crawling on the ceiling. Although his philosophical treatises *Rules for the Direction of the Mind, Discourse on Method,* and *Meditations* are still of some significance, Descartes probably achieved everlasting fame by penning the dictum: *Cogito ergo sum*—"I think therefore I am." Always of fragile health, Descartes died in 1650 at the age of 54 of influenza while serving as a philosophy tutor to Queen Christina of Sweden.

(a, b)* is established. We call a the **x-coordinate** or **abscissa** of the point P; b is called the **y-coordinate** or **ordinate** of the point. The axes are also called **coordinate axes** and P is said to have **coordinates** (a, b).

Quadrants The coordinate axes divide the Cartesian plane into four regions known as **quadrants**. Algebraic signs of the x-coordinate and y-coordinate of any point (a, b), located in each of the four quadrants, are indicated in Figure I.13(a). Points on a coordinate axis, such as $(2, 0)$ and $(0, -3)$ in Figure I.13(b), are considered to be not in any quadrant. This method of describing points in a plane is called a **rectangular** or **Cartesian coordinate system**. In this system, two points (a, b) and (c, d) are **equal** if and only if $a = c$ and $b = d$.

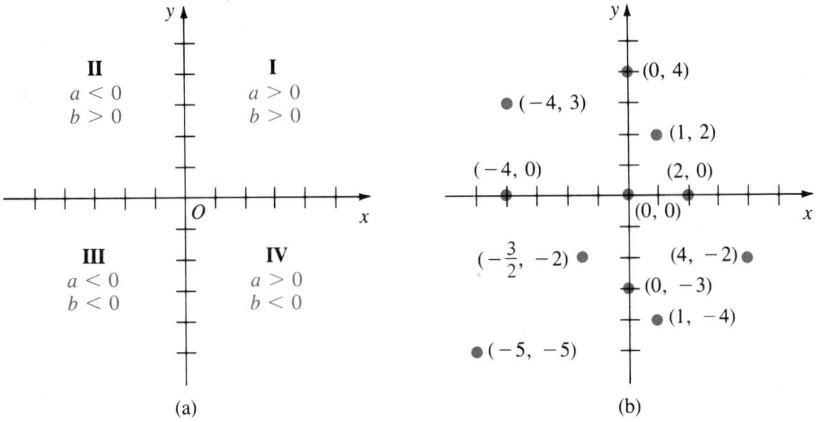

FIGURE I.13

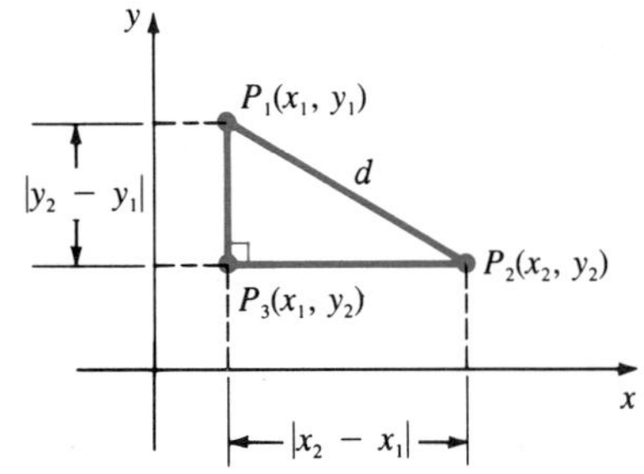

FIGURE I.14

Distance Formula We can obtain the distance $d(P_1, P_2)$ between two points $P_1(x_1, y_1)$ and $P_2(x_2, y_2)$ from the Pythagorean Theorem. As shown in Figure I.14, the three points P_1, P_2, and P_3 form a right triangle with a hypotenuse of length d and sides of lengths $|x_2 - x_1|$ and $|y_2 - y_1|$. Thus,

$$d^2 = |x_2 - x_1|^2 + |y_2 - y_1|^2$$

yields the **distance formula**

$$d(P_1, P_2) = \sqrt{(x_2 - x_1)^2 + (y_2 - y_1)^2} \tag{1}$$

EXAMPLE 1 Find the distance between the points $(-2, 3)$ and $(4, 5)$.

Solution By identifying P_1 as $(-2, 3)$ and P_2 as $(4, 5)$, we obtain from (1)

$$d(P_1, P_2) = \sqrt{(4 - (-2))^2 + (5 - 3)^2} = \sqrt{6^2 + 2^2} = \sqrt{40} = 2\sqrt{10} \qquad \square$$

Since $(x_2 - x_1)^2 = (x_1 - x_2)^2$ and $(y_2 - y_1)^2 = (y_1 - y_2)^2$, it does not matter which points are designated P_1 and P_2. In other words, $d(P_1, P_2) = d(P_2, P_1)$.

---

*This is the same notation used for an open interval. You should be able to tell from the context of the discussion whether this symbol refers to a point or an interval.

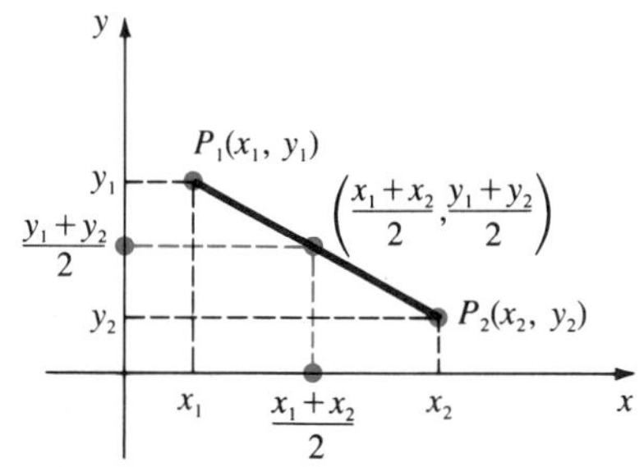

FIGURE I.15

Midpoint of a Line Segment The distance formula can be used to prove that the coordinates of the **midpoint of a line segment** from a point $P_1(x_1, y_1)$ to a point $P_2(x_2, y_2)$ are

$$\left(\frac{x_1 + x_2}{2}, \frac{y_1 + y_2}{2}\right) \tag{2}$$

See Figure I.15. If M denotes a point on the segment P_1P_2 with coordinates given in (2), then M is the midpoint of P_1P_2 provided

$$d(P_1, M) = d(M, P_2) \quad \text{and} \quad d(P_1, P_2) = d(P_1, M) + d(M, P_2) \tag{3}$$

In Problem 61 you will be asked to verify that the coordinates of M satisfy the equations in (3).

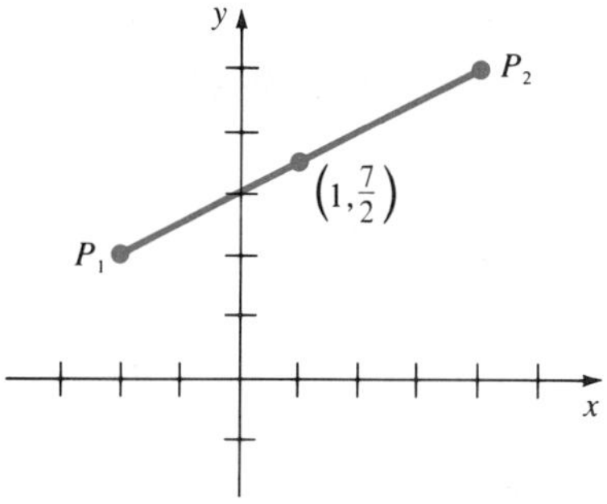

FIGURE I.16

EXAMPLE 2 Find the coordinates of the midpoint of the line segment from $P_1(-2, 2)$ to $P_2(4, 5)$.

Solution From (2) we have

$$x = \frac{-2+4}{2} = \frac{2}{2} = 1 \quad \text{and} \quad y = \frac{2+5}{2} = \frac{7}{2}$$

The midpoint $(1, \frac{7}{2})$ is shown in Figure I.16. □

Graphs A **graph** is any set of points (x, y) in the Cartesian plane. A graph can be an infinite set of points such as the points on the line segment P_1P_2 in Figure I.16 or simply a finite set of points as in Figure I.11(a) The **graph of an equation** is the set of points (x, y) in the Cartesian plane that are solutions of the equation. An ordered pair (x, y) is a **solution** of an equation if substitution of x and y into the equation reduces it to an identity.

EXAMPLE 3 The point $(-2, 2)$ is on the graph of $y = 1 - \frac{1}{8}x^3$, since

$$2 = 1 - \frac{1}{8}(-2)^3 \quad \text{is equivalent to} \quad 2 = 1 + \frac{8}{8} \quad \text{or} \quad 2 = 2$$ □

Point Plotting One way of sketching the graph of an equation, often done in elementary courses, is to **plot points** and then connect these points with a smooth curve. To obtain points on the graph we assign values to either x or y and then solve the equation for the corresponding values of y or x. Of course, we need to plot enough points so that the shape, or pattern, of the graph is evident.

EXAMPLE 4 Graph $y = 1 - \frac{1}{8}x^3$.

Solution By assigning values of x, we find the values of y given in the accompanying table. It seems reasonable, by connecting the points shown in Figure I.17(a) with a smooth curve, that the graph of the equation is that given in Figure I.17(b).

x	y
-3	$\frac{35}{8}$
-2	2
-1	$\frac{9}{8}$
0	1
1	$\frac{7}{8}$
2	0
3	$-\frac{19}{8}$

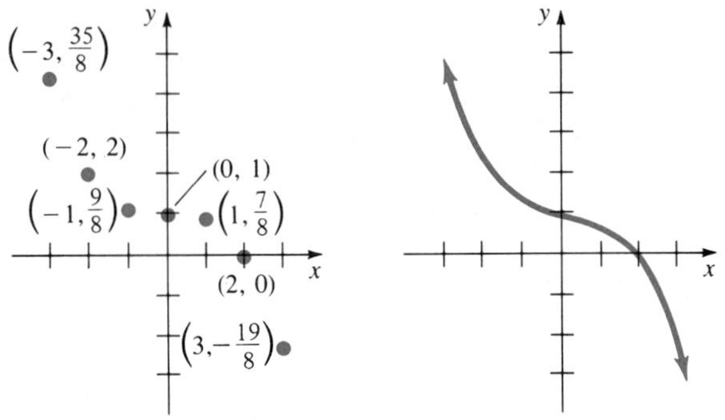

(a) plotting points

(b) graph of $y = 1 - \frac{1}{8}x^3$

FIGURE I.17 □

Symmetry Before plotting points, you can determine whether the graph of an equation possesses **symmetry**. Figure I.18 shows that a graph is

(*i*) **symmetric with respect to the *y*-axis** if, whenever (x, y) is a point on the graph, $(-x, y)$ is also a point on the graph;

(*ii*) **symmetric with respect to the *x*-axis** if, whenever (x, y) is a point on the graph, $(x, -y)$ is also a point on the graph; and

(*iii*) **symmetric with respect to the origin** if, whenever (x, y) is a point on the graph, $(-x, -y)$ is also a point on the graph.

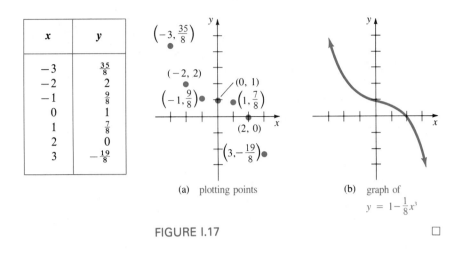

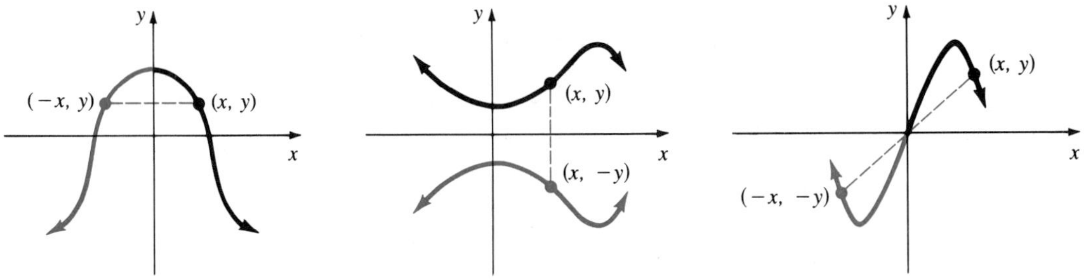

(a) symmetry with respect to y-axis

(b) symmetry with respect to x-axis

(c) symmetry with respect to origin

FIGURE I.18

Tests for Symmetry For an equation, (*i*), (*ii*), and (*iii*) yield the following three **tests for symmetry**. A graph of an equation is symmetric with respect to

(*i*) the **y-axis** if replacing x by $-x$ results in an equivalent equation;

(*ii*) the **x-axis** if replacing y by $-y$ results in an equivalent equation; and

(*iii*) the **origin** if replacing x by $-x$ and y by $-y$ results in an equivalent equation.

EXAMPLE 5 Determine whether the graph of $x = |y| - 2$ possesses any symmetry.

Solution

Test (i): $-x = |y| - 2$ or $x = -|y| + 2$ is not equivalent to the original equation.

Test (ii): $x = |-y| - 2$ is equivalent to the original equation, since $|-y| = |y|$.

Test (iii): $-x = |-y| - 2$ or $x = -|y| + 2$ is not equivalent to the original equation.

We conclude that the graph of $x = |y| - 2$ is symmetric with respect to the x-axis. □

Inspection of Figure I.17(b) shows pictorially that the graph of $y = 1 - \frac{1}{8}x^3$ has none of the symmetries we are considering.* You are encouraged to verify that each of the three tests for symmetry fails to yield an equivalent equation.

Detecting symmetry before plotting points can often save time and effort. For example, if the graph of an equation is shown to be symmetric with respect to the y-axis, then it is sufficient to plot points with x-coordinates that satisfy $x \geq 0$. As suggested in Figure I.18(a), we can find points in the second and third quadrants by taking the mirror images, through the y-axis, of the points in the first and fourth quadrants.

EXAMPLE 6 Graph $x = |y| - 2$.

Solution The entries in the accompanying table were obtained by assigning values to y and solving for x. We have taken $y \geq 0$, since we saw in Example 5 that the graph of the equation is symmetric with respect to the x-axis. In Figure I.19(a) color is used to indicate the points on the graph gained by symmetry. The graph of the equation, which seems to consist of two straight lines, is given in Figure I.19(b).

x	y
2	4
1	3
0	2
−1	1
−2	0

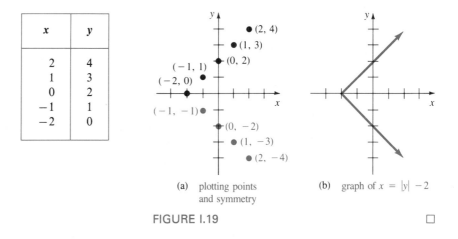

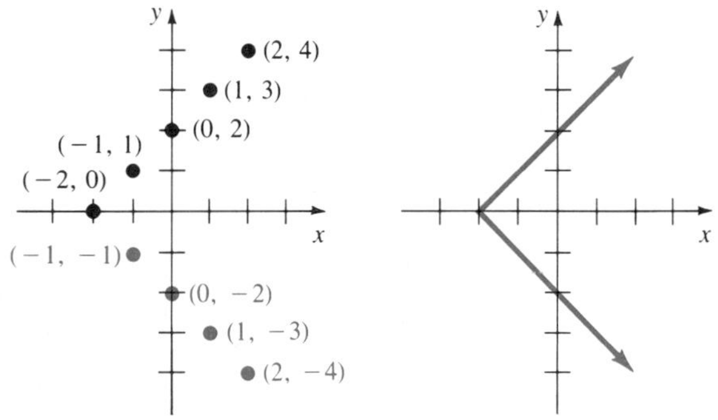

(a) plotting points and symmetry

(b) graph of $x = |y| - 2$

FIGURE I.19

□

*We are confining our attention to symmetry with respect to the two coordinate axes and the origin. A graph could, of course, possess other types of symmetries.

Intercepts When you sketch the graph of an equation, it is always a good idea to ascertain whether the graph has any **intercepts**. The x-coordinate of a point where the graph crosses the x-axis is called an ***x*-intercept**. The y-coordinate of a point where the graph crosses the y-axis is called a ***y*-intercept.** In Figure I.20 the x-intercepts of the graph are x_1, x_2, and x_3. The single y-intercept is y_1. Figure I.21 shows a graph that has no x- or y-intercepts. In Example 4 the x-intercept of the graph is 2; the y-intercept is 1. In Example 6 the x-intercept of the graph is -2; the y-intercepts are -2 and 2.

Since $y = 0$ for any point on the x-axis and $x = 0$ for any point on the y-axis, we can determine the intercepts of the graph of an equation in the following manner:

***x*-intercepts:** Set $y = 0$ in the equation and solve for x.
***y*-intercepts:** Set $x = 0$ in the equation and solve for y. (4)

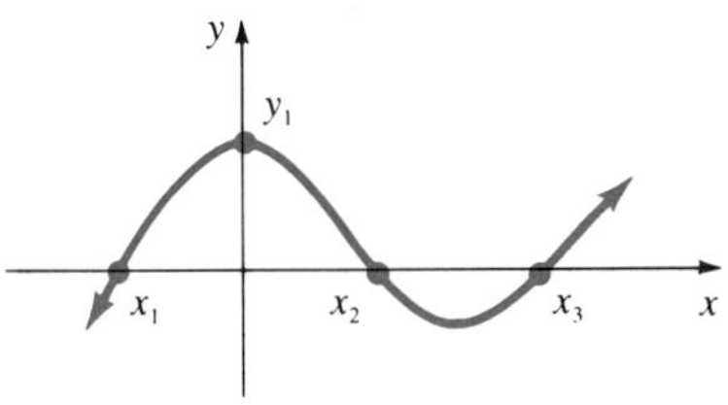

FIGURE I.20

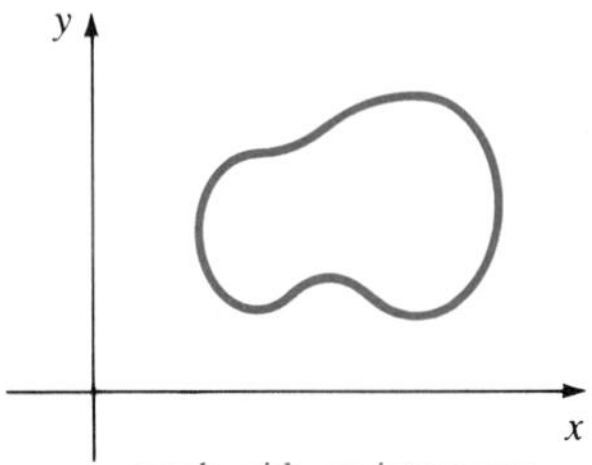

FIGURE I.21

EXAMPLE 7 Find the intercepts for the graph of (*a*) $y = 4x - 3$ and (*b*) $y = \dfrac{x^2 + 1}{x^2 + 5}$.

Solution

(*a*) Setting $y = 0$ yields $0 = 4x - 3$ or $x = \frac{3}{4}$. Setting $x = 0$ gives $y = -3$. The x- and y-intercepts are $\frac{3}{4}$ and -3, respectively.

(*b*) Here $y = 0$ if $x^2 + 1 = 0$ and $x^2 + 5 \neq 0$. Since $x^2 + 5 \neq 0$ for all real numbers, we have $x^2 + 1 = 0$ or $x^2 = -1$. But there are no real numbers that satisfy the last equation. Hence, the graph has no x-intercepts. Now, when we set $x = 0$, the equation gives $y = \frac{1}{5}$. The y-intercept is $\frac{1}{5}$. □

Circles The distance formula (1) enables us to find an equation for a very familiar plane curve. A **circle** is the set of all points (x, y) in the Cartesian plane that are equidistant from a fixed point $C(h, k)$. If r is the fixed distance, then a point $P(x, y)$ is on the circle if and only if

$$d(C, P) = \sqrt{(x - h)^2 + (y - k)^2} = r$$

Equivalently, we have the **standard form** for the equation of a circle with **center** $\boldsymbol{C(h, k)}$ and **radius** $\boldsymbol{r}$:

$$(x - h)^2 + (y - k)^2 = r^2 \tag{5}$$

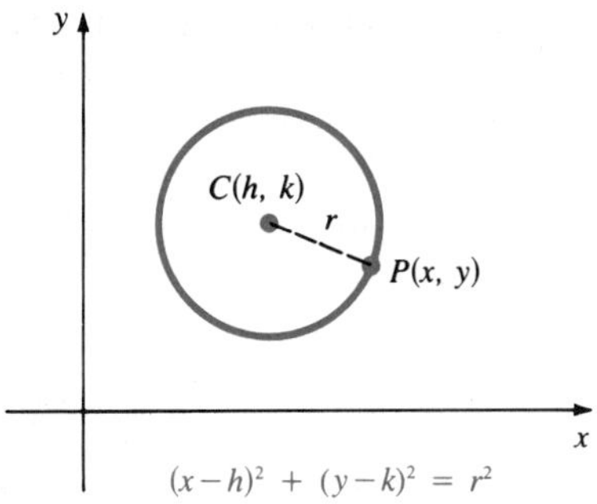

FIGURE I.22

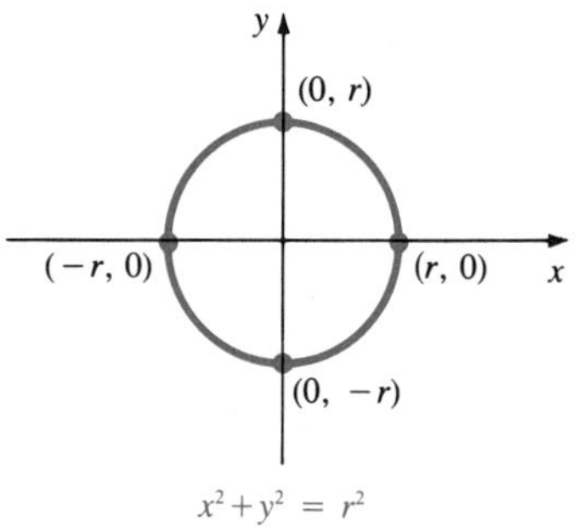

FIGURE I.23

See Figure I.22. If we set $h = 0$ and $k = 0$ in (5), we obtain the standard form for the equation of a circle with center at the origin

$$x^2 + y^2 = r^2$$

See Figure I.23.

EXAMPLE 8 Consider the line segment from $P_1(-2, 2)$ to $P_2(4, 5)$ in Example 2. Find an equation of the circle that passes through P_1 and P_2 with center at the midpoint M of P_1P_2.

Solution Since the center of the circle is the midpoint $M(1, \frac{7}{2})$, we must have $h = 1$ and $k = \frac{7}{2}$. The radius r of the circle is either $d(M, P_1)$ or $d(M, P_2)$. Using

$$d(M, P_1) = \sqrt{(-2 - 1)^2 + \left(2 - \frac{7}{2}\right)^2} = \sqrt{\frac{45}{4}} = \frac{3\sqrt{5}}{2} = r$$

we find from (5) that an equation for the circle is

$$(x - 1)^2 + \left(y - \frac{7}{2}\right)^2 = \frac{45}{4}$$

□

By squaring out the terms in (5), we see that every circle has an alternative equation of the form

$$Ax^2 + Ay^2 + Cx + Dy + E = 0, \qquad A \neq 0$$

However, the converse is not necessarily true; that is, not every equation of the form $Ax^2 + Ay^2 + Cx + Dy + E = 0$ is a circle. See Problems 43–48.

EXAMPLE 9 Show that $x^2 + y^2 - 4x + 8y + 2 = 0$ is an equation of a circle. Find its center and radius.

Solution We want to write the given equation in the form (5). To do this we must *complete the square* in both x and y. We begin by grouping the x-terms and the y-terms together and taking the constant term to the right side of the equality:

$$(x^2 - 4x \qquad) + (y^2 + 8y \qquad) = -2$$

Inside each set of parentheses we add the square of one-half the coefficient of the first-degree term; that is, we add $(-4/2)^2 = 4$ in the first and $(8/2)^2 = 16$ in the second. But to retain the equality we must also add these numbers to the right side:

$$(x^2 - 4x + 4) + (y^2 + 8y + 16) = -2 + 4 + 16$$

or

$$(x - 2)^2 + (y + 4)^2 = 18$$

The last equation is the standard form for the equation of a circle with center (h, k). Identifying $h = 2$, $k = -4$, and $r^2 = 18$, we conclude that the circle has center $(2, -4)$ and radius $3\sqrt{2}$. □

Conic Sections The circle is just one member of a class of curves known as **conic sections**. An equation of a conic section can always be expressed in the form $Ax^2 + By^2 + Cx + Dy + E = 0$, where A and B are not both zero.

EXAMPLE 10 Graph $9x^2 + 16y^2 = 25$.

Solution First, observe that replacing (x, y) in turn by $(-x, y)$, $(x, -y)$, and $(-x, -y)$ does not change the given equation. Hence, the graph is symmetric with respect to the y-axis, x-axis, and origin. Furthermore, $9(1)^2 + 16(1)^2 = 25$ indicates that $(1, 1)$ is a point on the graph. Finally,

$$y = 0 \quad \text{implies} \quad 9x^2 = 25 \quad \text{or} \quad x = \pm\frac{5}{3}$$

$$x = 0 \quad \text{implies} \quad 16y^2 = 25 \quad \text{or} \quad y = \pm\frac{5}{4}$$

The x-intercepts are $-\frac{5}{3}$ and $\frac{5}{3}$; the y-intercepts are $-\frac{5}{4}$ and $\frac{5}{4}$. By connecting the points in Figure I.24(a) with a smooth curve, we obtain the graph in Figure I.24(b). Observe that the graph is *not* a circle.

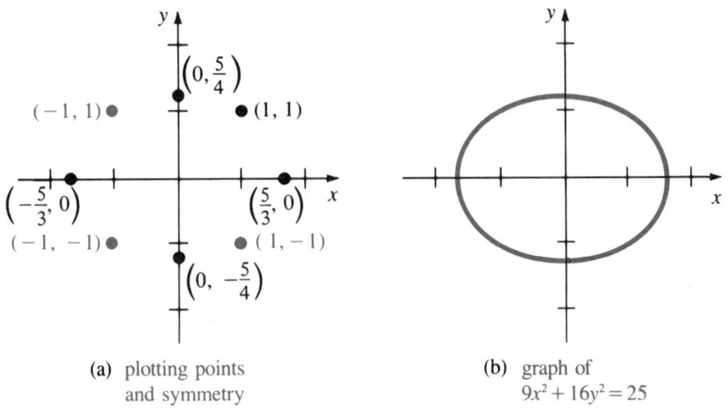

FIGURE I.24 □

The Ellipse When written in the equivalent form

$$\frac{x^2}{(5/3)^2} + \frac{y^2}{(5/4)^2} = 1$$

the equation in Example 10 is a special case of the standard form

$$\frac{(x-h)^2}{a^2} + \frac{(y-k)^2}{b^2} = 1 \tag{6}$$

For $a = b$, the graph of (6) is a circle of radius a. If $a \neq b$, the graph of (6) is called an **ellipse with center (h, k)**. Thus, the graph of $9x^2 + 16y^2 = 25$ is an ellipse with center at the origin. The ellipse, along with the other conic sections, is considered in detail in Chapter 11.

EXERCISES 1.3

Answers to odd-numbered problems begin on page A-112.

In Problems 1–6 the point (a, b) is in the first quadrant. Determine the quadrant of the given point.

1. $(a, -b)$ **2.** (b, a)

3. $(-b, -a)$ **4.** $(-a, a)$

5. $(-a, b)$ **6.** $(-b, a)$

In Problems 7–10 find the distance between the given points.

7. $P_1(3, -1), P_2(7, -3)$ **8.** $P_1(0, 5), P_2(-8, -2)$

9. $P_1(\sqrt{3}, 0), P_2(0, -\sqrt{6})$ **10.** $P_1(\frac{5}{2}, 5), P_2(-\frac{3}{2}, 5)$

In Problems 11 and 12 determine whether the given points are the vertices of a right triangle.

11. $(16, 2), (-6, -2), (20, 10)$

12. $(-2, -8), (0, 3), (-6, -5)$

In Problems 13 and 14 use the distance formula to determine whether the given points are collinear.

13. $(1, 3), (-2, -3), (4, 9)$

14. $(0, 2), (1, 1), (5, -2)$

In Problems 15 and 16 solve for x.

15. $P_1(x, 2), P_2(1, 1), d(P_1, P_2) = \sqrt{10}$

16. $P_1(x, 0), P_2(-4, 3x), d(P_1, P_2) = 4$

17. Find an equation that relates x and y if it is known that the distance from (x, y) to $(0, 1)$ is the same as the distance from (x, y) to $(x, -1)$.

18. Show that the point $(-1, 5)$ is on the perpendicular bisector of the line segment from $P_1(1, 1)$ to $P_2(3, 7)$.

In Problems 19 and 20 find the midpoint of the line segment from P_1 to P_2.

19. $P_1(4, 7), P_2(8, -3)$ **20.** $P_1(-3, 5), P_2\left(\frac{1}{2}, \frac{3}{4}\right)$

21. If the coordinates of the midpoint of the line segment from $P_1(1, 3)$ to $P_2(x_2, y_2)$ are $(3, 4)$, what are the coordinates of P_2?

22. Figure I.25 shows the midpoints of the sides of a triangle. Determine the coordinates of the vertices of the triangle.

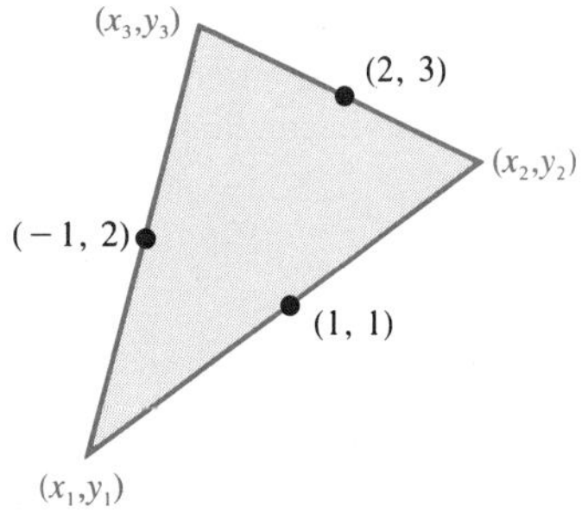

FIGURE I.25

23. If M is the midpoint of the line segment from $P_1(2, 3)$ to $P_2(6, -9)$, find the midpoint of the line segment from P_1 to M and the midpoint of the line segment from M to P_2.

24. The point $P(x, y)$ with coordinates defined by $x = x_1 + r(x_2 - x_1)$, $y = y_1 + r(y_2 - y_1)$ is on the line segment from $P_1(x_1, y_1)$ to $P_2(x_2, y_2)$. Use $P_1(2, 3)$, $P_2(6, -9)$, $r = \frac{1}{4}$, $r = \frac{1}{2}$, and $r = \frac{3}{4}$, and compare your results with those of Problem 23. Find a point two-thirds of the way from P_1 to P_2.

In Problems 25–32 graph the given equation. Determine any symmetry.

25. $y = 2x + 1$

26. $y = x - 5$

27. $x = y^2$

28. $y = \sqrt{x}$

29. $y = |x|$

30. $4y - x^3 = 0$

31. $x^2 = y^2$

32. $y = x^2 + x$

In Problems 33–36 complete the given graph using the indicated symmetry.

33. x-axis

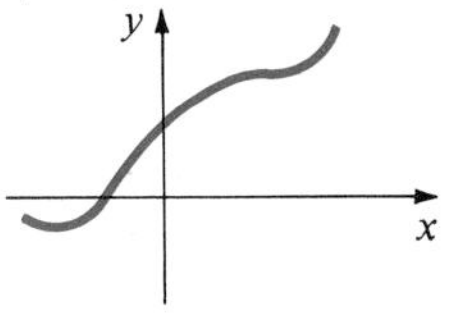

FIGURE I.26

34. y-axis

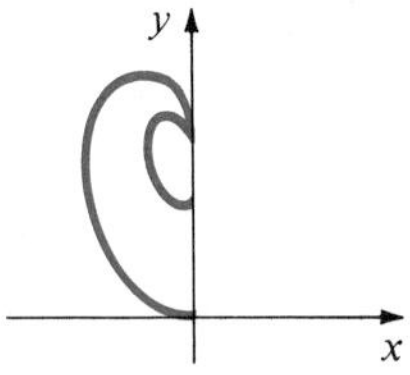

FIGURE I.27

35. Origin

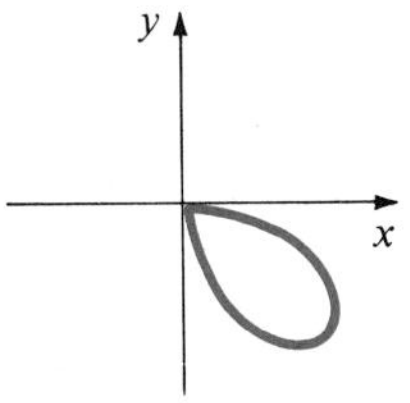

FIGURE I.28

36. x- and y-axes

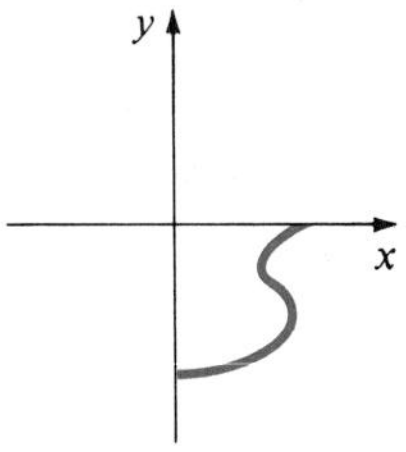

FIGURE I.29

In Problems 37–42 find an equation in standard form of the circle that satisfies the given conditions.

37. Center $(4, -6)$, radius 8

38. Center $(-\frac{5}{2}, \frac{1}{2})$, radius $\sqrt{2}$

39. Center $(3, -4)$, passing through the origin

40. Center $(6, 2)$, tangent to the x-axis

41. Center $(1, 1)$, passing through $(5, 2)$

42. Center at the midpoint M of the line segment from $P_1(3, 8)$ to $P_2(5, 2)$, radius $\frac{1}{2}d(M, P_2)$

In Problems 43–48 determine whether the given equation is an equation of a circle. If so, give its center and radius.

43. $x^2 + y^2 + 8x - 6y = 0$

44. $2x^2 + 2y^2 - 16x - 40y = 37$

45. $3x^2 + 3y^2 - 18x + 6y = -2$

46. $x^2 + y^2 + 10y + 26 = 0$

47. $x^2 + y^2 - 12x + 8y + 52 = 0$

48. $x^2 + y^2 + \frac{1}{2}x + y = 0$

In Problems 49–52 the given equation is an equation of an ellipse. Graph.

49. $25x^2 + 4y^2 = 100$

50. $x^2 + 4y^2 = 36$

51. $\dfrac{x^2}{4^2} + \dfrac{y^2}{2^2} = 1$

52. $x^2 + \dfrac{y^2}{7} = 1$

In Problems 53 and 54 the given equation is an equation of an ellipse. Put the equation in standard form and give the center of the ellipse.

53. $x^2 + 4y^2 - 4x + 40y + 88 = 0$

54. $5x^2 + 2y^2 + 60x - 8y + 178 = 0$

In Problems 55–58 graph the set of points (x, y) that satisfy the given equation or inequality.

55. $xy = 0$

56. $xy > 0$

57. $xy < 0$

58. $y < x$

59. Find an equation(s) for the circle(s) passing through $(1, 3)$ and $(-1, -3)$ that has radius 10.

60. Graph $y + |y| = x + |x|$.

61. Let M denote the point on the line segment P_1P_2 with coordinates given in (2). Verify that these coordinates satisfy both equations in (3).

I.4 LINES

The notion of a line plays an important role in the study of differential calculus. Before discussing lines, however, it is convenient to introduce two special symbols.

Increments If $P_1(x_1, y_1)$ and $P_2(x_2, y_2)$ are any two points in the plane, we define the **x-increment** to be the difference in x-coordinates

$$\Delta x = x_2 - x_1$$

and the **y-increment** to be the difference in y-coordinates

$$\Delta y = y_2 - y_1$$

We read the symbols "Δx" and "Δy" as "delta x" and "delta y," respectively.

EXAMPLE 1 Find the x- and y-increments for (*a*) $P_1(4, 6)$, $P_2(9, 7)$; (*b*) $P_1(10, -2)$, $P_2(3, 5)$; and (*c*) $P_1(8, 14)$, $P_2(8, -1)$.

Solution

(*a*) $\Delta x = 9 - 4 = 5$, $\Delta y = 7 - 6 = 1$

(*b*) $\Delta x = 3 - 10 = -7$, $\Delta y = 5 - (-2) = 7$

(*c*) $\Delta x = 8 - 8 = 0$, $\Delta y = -1 - 14 = -15$ □

Example 1 shows that an increment can be positive, negative, or zero.

Slope Suppose L denotes a nonvertical line in the Cartesian plane. Associated with such a line there is a number called the **slope** of the line. If $P_1(x_1, y_1)$ and $P_2(x_2, y_2)$ are distinct points on L, then the **slope** m of the line is defined to be the quotient

$$m = \frac{\Delta y}{\Delta x} = \frac{y_2 - y_1}{x_2 - x_1} \tag{1}$$

The increment $\Delta x = x_2 - x_1$ is said to be the *change in x* or **run** of the line; the corresponding *change in y* or **rise** of the line is $\Delta y = y_2 - y_1$. Thus,

$$m = \frac{\text{change in } y}{\text{change in } x} = \frac{\text{rise}}{\text{run}}$$

Assuming, for the sake of discussion, that $P_1(x_1, y_1)$ and $P_2(x_2, y_2)$ are chosen such that $x_1 < x_2$, we see that the slope of the line illustrated in Figure I.30(a) is positive, whereas the slope of the line in Figure I.30(b) is negative. If a line has positive slope, then the ordinates of points on the line increase as the abscissas increase; for lines with negative slope, ordinates decrease as abscissas increase. If a line is horizontal, $\Delta y = 0$ and therefore its slope is zero. See Figure I.30(c). The slope of a vertical line is undefined, since (1) has no meaning when $\Delta x = 0$. See Figure I.30(d).

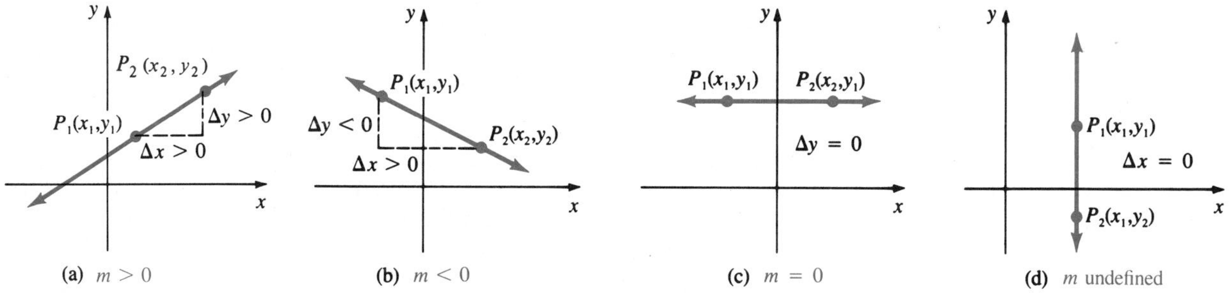

FIGURE I.30

EXAMPLE 2 Find the slope of the line that passes through the points (*a*) (6, 3), (−2, 5); (*b*) (7, −2), (4, −2); and (*c*) (1, 5), (1, −3).

Solution Using (1) in the first two cases results in

(*a*) $m = \dfrac{5-3}{-2-6} = \dfrac{2}{-8} = -\dfrac{1}{4}$

(*b*) $m = \dfrac{-2-(-2)}{4-7} = \dfrac{0}{-3} = 0$. The line through (7, −2) and (4, −2) is horizontal.

(*c*) Since $\Delta x = 1 - 1 = 0$, the slope of the line through (1, 5) and (1, −3) is undefined. Hence, the line is vertical. □

The slope of a line is *unique* in the sense that any pair of points on the line determines the same quotient m. This can be proved from the fact that ratios of corresponding sides of similar triangles are equal. If

$$\Delta x = x_2 - x_1, \quad \Delta y = y_2 - y_1 \quad \text{and} \quad \Delta x' = x'_2 - x'_1, \quad \Delta y' = y'_2 - y'_1$$

then we see from Figure I.31 that the right triangles $P_1P_2P_3$ and $P'_1P'_2P'_3$ are similar, and as a consequence

$$\frac{y_2 - y_1}{x_2 - x_1} = \frac{y'_2 - y'_1}{x'_2 - x'_1} \quad \text{or} \quad m = \frac{\Delta y}{\Delta x} = \frac{\Delta y'}{\Delta x'}$$

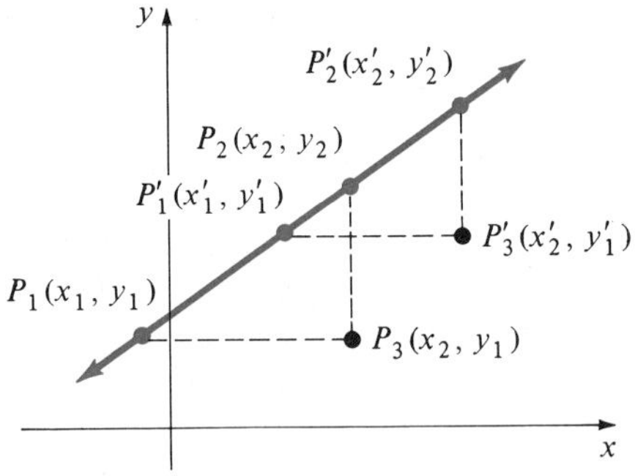

FIGURE I.31

Parallel Lines Two lines L_1 and L_2 with slopes m_1 and m_2 are **parallel** if and only if $m_1 = m_2$. Of course, two vertical lines—that is, lines parallel to the *y*-axis—are also parallel but have undefined slopes.

Perpendicular Lines Two lines L_1 and L_2 with slopes m_1 and m_2 are **perpendicular** if and only if $m_1m_2 = -1$. In other words, lines with slopes are perpendicular when one slope is the *negative reciprocal* of the other:

$$m_1 = -\frac{1}{m_2} \quad \text{and} \quad m_2 = -\frac{1}{m_1}$$

Equations of Lines The concept of slope enables us to find **equations of lines.** Through a point $P_1(x_1, y_1)$ there passes only one line L with specified slope m. To find an equation of L, let us suppose that $P(x, y)$ denotes any point on the line for which $x \neq x_1$. By equating slopes,

$$\frac{y - y_1}{x - x_1} = m \quad \text{we obtain} \quad y - y_1 = m(x - x_1) \tag{2}$$

Since the coordinates of all points $P(x, y)$ on the line, *including* $P_1(x_1, y_1)$, satisfy the last equation in (2), we conclude that it is an equation of L. This particular equation is called the **point–slope form**. We summarize:

*The **Point–Slope Form** for the Equation of a Line*

$$y - y_1 = m(x - x_1) \tag{3}$$

EXAMPLE 3 Find an equation for the line through (6, −2) with slope 4.

Solution From the point–slope form (3) we obtain

$$y - (-2) = 4(x - 6)$$

Equivalently, $$y + 2 = 4x - 24 \quad \text{or} \quad y = 4x - 26$$ □

The point–slope form (3) yields two other important forms for equations of lines with slope. If the line L passes through the y-axis at $(0, b)$ then (3) gives

$$y - b = mx \quad \text{or} \quad y = mx + b$$

The number b is called the ***y*-intercept** of the line. Furthermore, if L is a horizontal line through $P_1(x_1, y_1)$, then setting $m = 0$ in (3) yields $y - y_1 = 0(x - x_1)$ or $y = y_1$. In summary:

*The **Slope–Intercept Form** for the Equation of a Line*

$$y = mx + b \tag{4}$$

*Equation of a **Horizontal Line** Through* $P_1(x_1, y_1)$

$$y = y_1 \tag{5}$$

Two distinct points $P_1(x_1, y_1)$ and $P_2(x_2, y_2)$ also determine a unique line. If the line has slope, then an equation can be obtained from (3) by computing $m = (y_2 - y_1)/(x_2 - x_1)$ and using the coordinates of either P_1 or P_2. Finally, if the line through P_1 and P_2 is vertical, then any pair of points on the line have the same x-coordinate. Thus, if $P(x, y)$ is on the vertical line through $P_1(x_1, y_1)$, we must have $x = x_1$. We summarize this last case:

*Equation of a **Vertical Line** Through* $P_1(x_1, y_1)$ (6)

$$x = x_1$$

EXAMPLE 4 Find an equation for the line through $(2, -3)$ and $(-4, 1)$.

Solution By designating the first point as P_1, it follows from (1) that the slope of the line through the points is

$$\frac{1-(-3)}{-4-2} = -\frac{2}{3}$$

Using the point–slope form (3), we get

$$y-(-3) = -\frac{2}{3}(x-2) \quad \text{or} \quad y = -\frac{2}{3}x - \frac{5}{3}$$

Alternative Solution I We can, of course, choose $(-4, 1)$ as P_1:

$$y-1 = -\frac{2}{3}(x-(-4)) \quad \text{or} \quad y = -\frac{2}{3}x - \frac{5}{3}$$

Alternative Solution II From the slope–intercept form (4) we can write $y = -\frac{2}{3}x + b$. Substituting $x = 2$ and $y = -3$ in the last equation gives $-3 = -\frac{4}{3} + b$, so $b = -\frac{5}{3}$. As before, $y = -\frac{2}{3}x - \frac{5}{3}$. □

EXAMPLE 5

(*a*) An equation of a horizontal line through $(4, 9)$ is $y = 9$.

(*b*) An equation of a vertical line through $(-1, -2)$ is $x = -1$. □

Linear Equation Any equation

$$ax + by + c = 0 \tag{7}$$

in which both x and y appear to the first-power, and a, b, and c are constants, is a **linear equation**. As the name suggests, the graph of an equation of form (7) is a straight line. The following summarizes three special cases of (7):

(*i*) $a = 0$, $b \neq 0$, horizontal line: $y = -\dfrac{c}{b}$

(*ii*) $a \neq 0$, $b = 0$, vertical line: $x = -\dfrac{c}{a}$

(*iii*) $a \neq 0$, $b \neq 0$, (7) can be written as

$$y = -\frac{a}{b}x - \frac{c}{b} \tag{8}$$

Comparing (8) with (4), we see that when $b \neq 0$, the line given by (8) has slope $-a/b$ and y-intercept $-c/b$.

EXAMPLE 6 Find an equation for the line through $(-1, 5)$ perpendicular to the line with equation $2x + y + 4 = 0$.

Solution Writing the given equation as $y = -2x - 4$, we see that the slope is -2. Thus the line through $(-1, 5)$ perpendicular to $2x + y + 4 = 0$ has slope $\frac{1}{2}$. Hence, from (3) we obtain

$$y - 5 = \frac{1}{2}(x - (-1)) \quad \text{or} \quad y = \frac{1}{2}x + \frac{11}{2} \quad \text{or} \quad x - 2y + 11 = 0 \qquad \square$$

Graphs To graph an equation of a line, we need determine only two points whose coordinates satisfy the equation. When $a \neq 0$ and $b \neq 0$ in (7), the line must cross both coordinate axes. The ***x*-intercept** is the x-coordinate of the point where a line crosses the x-axes; the ***y*-intercept** is the y-coordinate of the point where a line crosses the y-axis. Since a point on the x-axis has coordinates $(x, 0)$, the x-intercept is found by setting $y = 0$ in the equation and solving for x. Similarly, we get the y-intercept by setting $x = 0$.

EXAMPLE 7 Graph the line with equation $2x - 3y + 12 = 0$.

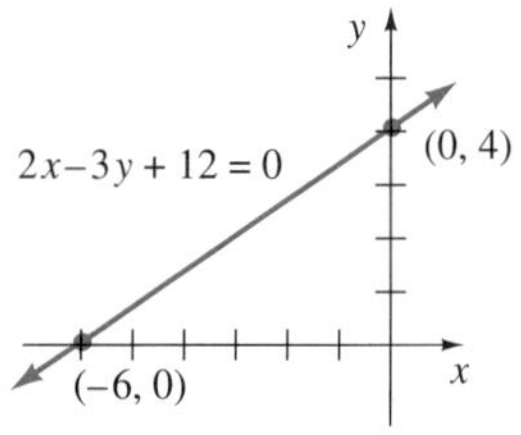

FIGURE I.32

Solution By setting $y = 0$ in the equation, we obtain

$$2x + 12 = 0 \quad \text{or} \quad x = -6$$

The x-intercept is -6. Now when $x = 0$,

$$-3y + 12 = 0 \quad \text{implies} \quad y = 4$$

The y-intercept is 4. As shown in Figure I.32, the line is drawn through the points $(-6, 0)$ and $(0, 4)$. $\square$

EXERCISES I.4

Answers to odd-numbered problems begin on page A-112.

In Problems 1–4 find the slope of the line through the given points.

1. $(4, 1), (6, -2)$
2. $(3, 0), (6, 9)$
3. $(\frac{1}{2}, 4), (-\frac{3}{2}, 10)$
4. $(-3, 2), (11, 2)$

In Problems 5 and 6 find $P_2(x_2, y_2)$ using the given information.

5. $P_1(3, -2)$, $\Delta x = 4$, $\Delta y = 5$
6. $P_1(0, 7)$, $\Delta x = 0$, $\Delta y = -3$

In Problems 7–24 find an equation for the line that satisfies the given conditions.

7. Through $(5, -2)$ and $(1, 2)$
8. Through $(0, 0)$, slope 8
9. Through $(1, 3)$, $\Delta x = 3$, $\Delta y = 9$
10. Through $(\frac{1}{4}, -\frac{1}{2})$, x-intercept $\frac{1}{8}$
11. y-intercept 8, slope 1
12. x-intercept -3, y-intercept $\frac{1}{2}$
13. Through $(10, -\frac{3}{2})$, slope 0
14. Through $(-\frac{7}{3}, \frac{3}{8})$, parallel to y-axis
15. Through $(1, 2)$, parallel to $4x + 2y = 1$
16. Through $(4, 4)$, parallel to $x - 3y = 0$
17. Through $(0, 0)$, perpendicular to $y = -\frac{1}{4}x + 7$
18. Through $(\frac{1}{2}, -1)$, perpendicular to $3x + 4y - 12 = 0$

19. Through (2, 3) and the point common to $x + y = 1$ and $2x + y = 5$

20. Through the point common to $2x + 3 = 0$ and $y + 6 = 0$, slope -2

21. Through the midpoint of the line segment from $(-1, 3)$ to $(4, 8)$, perpendicular to the line segment

22. Through the midpoints of the line segments between the x- and y-intercepts of $3x + 4y = 12$ and $x + y = -6$

23. Through (4, 2) parallel to the line through (5, 1) and $(-1, 7)$

24. Through (3, 9), slope undefined

In Problems 25–32 graph the line with the given equation. Give the slope and the x- and y-intercepts.

25. $y = 2x + 3$

26. $y = -3x$

27. $2y - 5 = 0$

28. $x = -4$

29. $5x + 3y = 15$

30. $x - y + 6 = 0$

31. $3x - 8y - 10 = 0$

32. $-\frac{1}{2}x + \frac{1}{4}y = 2$

In Problems 33 and 34 graph the line through (3, 2) with the given slope.

33. $\frac{3}{4}$

34. $-\frac{1}{2}$

In Problems 35–40 find the value of k so that the graph of the given linear equation satisfies the indicated condition.

35. $kx + 3y = 1$, passes through (5, 1)

36. $-x + 7y = k$, x-intercept $\frac{3}{2}$

37. $x - ky + 3 = 0$, y-intercept -4

38. $2x + ky + 1 = 0$, perpendicular to $-5x + 10y = 3$

39. $kx + y = 0$, parallel to $3x - 7y = 12$

40. $kx + \sqrt{3}y = k$, slope $\sqrt{3}$

In Problems 41 and 42 determine whether the given points are collinear.

41. $(0, -4), (1, -1), (3, 5)$

42. $(-2, 3), (1, \frac{3}{2}), (-1, \frac{1}{2})$

43. Find the coordinates of the point P in Figure I.33.

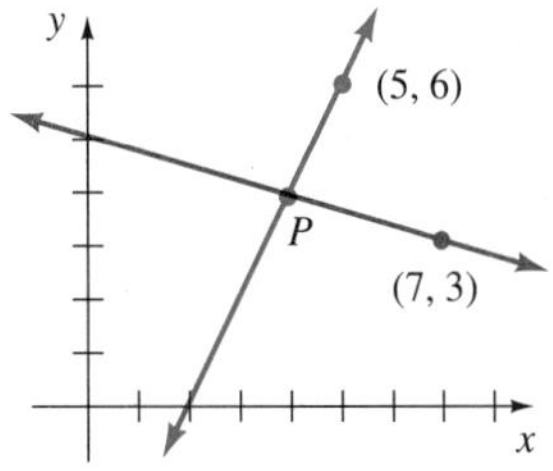

FIGURE I.33

44. A line through (2,4) has slope 8. Without finding an equation of the line, determine whether the point $(1, -5)$ is on the line.

In Problems 45 and 46 use slopes to verify that P_1, P_2, and P_3 are vertices of a right triangle.

45. $P_1(8, 2)$, $P_2(1, -11)$, $P_3(-2, -1)$

46. $P_1(8, 2)$, $P_2(-3, 0)$, $P_3(5, 6)$

47. Show that an equation of the line in Figure I.34 is

$$\frac{x}{a} + \frac{y}{b} = 1$$

This is called the **intercept form** for the equation of the line.

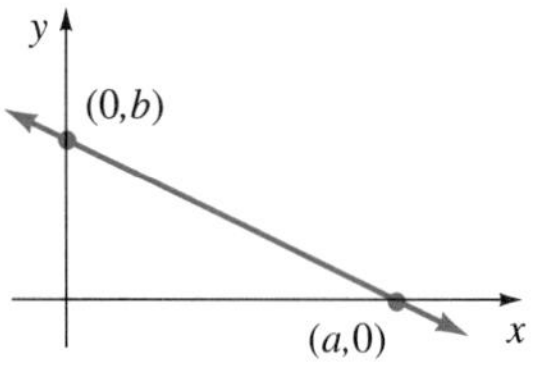

FIGURE I.34

In Problems 48 and 49 find the intercept form for the equation of the line with the given intercepts.

48. x-intercept 3, y-intercept $-\frac{1}{2}$

49. x-intercept 5, y-intercept 2

50. Show that the **two-point form** for the equation of the line through $P_1(x_1, y_1)$ and $P_2(x_2, y_2)$, $x_1 \neq x_2$, is

$$y - y_1 = \left(\frac{y_2 - y_1}{x_2 - x_1}\right)(x - x_1)$$

The **distance D from a point $P_1(x_1, y_1)$ to a line** $ax + by + c = 0$ is

$$D = \frac{|ax_1 + by_1 + c|}{\sqrt{a^2 + b^2}}$$

Use this result in Problems 51–54.

51. Find the distance from (2, 5) to the line $2x + 3y - 1 = 0$.

52. Find the distance from the origin to the line $y = \frac{3}{4}x - 10$.

53. Find the distance from the x-intercept of $-x + 3y + 4 = 0$ to the line through (1, 1) and $(-1, 13)$.

54. Find the distance between the parallel lines $x + y = 1$ and $2x + 2y = 5$.

I.5 MATRICES AND DETERMINANTS

Matrices A **matrix** is any rectangular array of numbers:

$$\begin{bmatrix} a_{11} & a_{12} & \cdots & a_{1n} \\ a_{21} & a_{22} & \cdots & a_{2n} \\ \vdots & & & \vdots \\ a_{m1} & a_{m2} & \cdots & a_{mn} \end{bmatrix}$$

The numbers in the array are called **entries** or **elements** of the matrix. If a matrix has m rows and n columns, we say that its **size** is m by n (written $m \times n$). An $n \times n$ matrix is called a **square** matrix or a matrix of **order *n***. A 1×1 matrix is simply a constant. A matrix is usually denoted by a capital boldface letter such as **A**, **B**, **C**, or **X**.

The entry in the ith row and jth column of an $m \times n$ matrix **A** is written a_{ij}. An $m \times n$ matrix **A** is then abbreviated as $\mathbf{A} = [a_{ij}]_{m \times n}$.

EXAMPLE 1 (*a*) $\mathbf{A} = \begin{bmatrix} 1 & 2 & 3 \\ 0 & 5 & -6 \end{bmatrix}$ is a 2×3 matrix.

(*b*) $\mathbf{B} = \begin{bmatrix} 9 & 7 & 0 & 8 \\ \frac{1}{2} & -2 & 6 & 1 \\ 0 & 0 & -1 & 6 \\ 5 & \sqrt{3} & \pi & -4 \end{bmatrix}$ is a 4×4 square matrix or a matrix of order 4. □

Vectors A matrix

$$\begin{bmatrix} a_1 \\ a_2 \\ \vdots \\ a_n \end{bmatrix}$$

with n rows and one column is called a **column vector**. A matrix

$$[a_1 \quad a_2 \quad \cdots \quad a_n]$$

with one row and n columns is called a **row vector**.

Equality Two $m \times n$ matrices **A** and **B** are **equal** if $a_{ij} = b_{ij}$ for each i and j.

Scalar Multiple In the discussion of matrices, real numbers are often referred to as **scalars**. If k is a real number, the **scalar multiple** of a matrix **A** is defined to be

$$k\mathbf{A} = \begin{bmatrix} ka_{11} & ka_{12} & \cdots & ka_{1n} \\ ka_{21} & ka_{22} & \cdots & ka_{2n} \\ \vdots & & & \vdots \\ ka_{m1} & ka_{m2} & \cdots & ka_{mn} \end{bmatrix} = [ka_{ij}]_{m \times n} \tag{1}$$

EXAMPLE 2 From (1) we have

$$5\begin{bmatrix} 2 & -3 \\ 4 & -1 \\ \frac{1}{5} & 6 \end{bmatrix} = \begin{bmatrix} 5(2) & 5(-3) \\ 5(4) & 5(-1) \\ 5(\frac{1}{5}) & 5(6) \end{bmatrix} = \begin{bmatrix} 10 & -15 \\ 20 & -5 \\ 1 & 30 \end{bmatrix} \qquad \square$$

Matrix Addition When two matrices **A** and **B** are the same size, we can add them by adding their corresponding entries. The **sum** of two $m \times n$ matrices is the matrix

$$\mathbf{A} + \mathbf{B} = [a_{ij} + b_{ij}]_{m \times n}. \tag{2}$$

EXAMPLE 3 It follows from (2) that the sum of

$$\mathbf{A} = \begin{bmatrix} 2 & -1 & 3 \\ 0 & 4 & 6 \\ -6 & 10 & -5 \end{bmatrix} \quad \text{and} \quad \mathbf{B} = \begin{bmatrix} 4 & 7 & -8 \\ 9 & 3 & 5 \\ 1 & -1 & 2 \end{bmatrix}$$

is

$$\mathbf{A} + \mathbf{B} = \begin{bmatrix} 2+4 & -1+7 & 3+(-8) \\ 0+9 & 4+3 & 6+5 \\ -6+1 & 10+(-1) & -5+2 \end{bmatrix} = \begin{bmatrix} 6 & 6 & -5 \\ 9 & 7 & 11 \\ -5 & 9 & -3 \end{bmatrix} \qquad \square$$

The **difference** of two $m \times n$ matrices is defined as

$$\mathbf{A} - \mathbf{B} = \mathbf{A} + (-\mathbf{B}) \quad \text{where} \quad -\mathbf{B} = (-1)\mathbf{B}$$

Matrix Multiplication Under some circumstances two matrices can be multiplied together. If **A** is a matrix with m rows and p columns, and **B** is a matrix with p rows and n columns, then the **product AB** is the $m \times n$ matrix

$$\mathbf{AB} = \begin{bmatrix} a_{11} & a_{12} & \cdots & a_{1p} \\ a_{21} & a_{22} & \cdots & a_{2p} \\ \vdots & & & \vdots \\ a_{m1} & a_{m2} & \cdots & a_{mp} \end{bmatrix} \begin{bmatrix} b_{11} & b_{12} & \cdots & b_{1n} \\ b_{21} & b_{22} & \cdots & b_{2n} \\ \vdots & & & \vdots \\ b_{p1} & b_{p2} & \cdots & b_{pn} \end{bmatrix}$$

$$= \begin{bmatrix} a_{11}b_{11} + a_{12}b_{21} + \cdots + a_{1p}b_{p1} & \cdots & a_{11}b_{1n} + a_{12}b_{2n} + \cdots + a_{1p}b_{pn} \\ a_{21}b_{11} + a_{22}b_{21} + \cdots + a_{2p}b_{p1} & \cdots & a_{21}b_{1n} + a_{22}b_{2n} + \cdots + a_{2p}b_{pn} \\ \vdots & & \vdots \\ a_{m1}b_{11} + a_{m2}b_{21} + \cdots + a_{mp}b_{p1} & \cdots & a_{m1}b_{1n} + a_{m2}b_{2n} + \cdots + a_{mp}b_{pn} \end{bmatrix} \tag{3}$$

Note carefully in (3) that the product $\mathbf{C} = \mathbf{AB}$ is defined only when the number of columns in the matrix **A** is the same as the number of rows in the matrix **B**. The size of the product can be determined from

$$\mathbf{A}_{m \times p}\mathbf{B}_{p \times n} = \mathbf{C}_{m \times n}$$

Also, you might recognize that the entries in, say, the ith row of the final matrix $\mathbf{C} = \mathbf{AB}$ are formed by using the component definition of the inner or dot product of the ith row (vector) of **A** with each of the columns (vectors) of **B**.

EXAMPLE 4 Find the product of the following matrices.

$$(a)\quad \mathbf{A} = \begin{bmatrix} 4 & 7 \\ 3 & 5 \end{bmatrix},\quad \mathbf{B} = \begin{bmatrix} 9 & -2 \\ 6 & 8 \end{bmatrix} \qquad (b)\quad \mathbf{A} = \begin{bmatrix} 5 & 8 \\ 1 & 0 \\ 2 & 7 \end{bmatrix},\quad \mathbf{B} = \begin{bmatrix} -4 & -3 \\ 2 & 0 \end{bmatrix}$$

Solution (*a*) From (3) we have

$$\mathbf{AB} = \begin{bmatrix} 4\cdot 9 + 7\cdot 6 & 4\cdot(-2) + 7\cdot 8 \\ 3\cdot 9 + 5\cdot 6 & 3\cdot(-2) + 5\cdot 8 \end{bmatrix} = \begin{bmatrix} 78 & 48 \\ 57 & 34 \end{bmatrix}$$

$$(b)\quad \mathbf{AB} = \begin{bmatrix} 5\cdot(-4) + 8\cdot 2 & 5\cdot(-3) + 8\cdot 0 \\ 1\cdot(-4) + 0\cdot 2 & 1\cdot(-3) + 0\cdot 0 \\ 2\cdot(-4) + 7\cdot 2 & 2\cdot(-3) + 7\cdot 0 \end{bmatrix} = \begin{bmatrix} -4 & -15 \\ -4 & -3 \\ 6 & -6 \end{bmatrix}$$ □

In general, matrix multiplication is not commutative; that is, $\mathbf{BA} \neq \mathbf{AB}$. Observe that in part (*a*) of Example 4, the product **BA** is

$$\mathbf{BA} = \begin{bmatrix} 30 & 53 \\ 48 & 82 \end{bmatrix}$$

whereas in part (*b*) the product **BA** is *not defined*, since the first matrix (in this case **B**) does not have the same number of columns as the second matrix has rows.

Associative Law Although we shall prove it, matrix multiplication is **associative**. If **A** is an $m \times p$ matrix, **B** a $p \times r$ matrix, and **C** an $r \times n$ matrix, then the product

$$\mathbf{A}(\mathbf{BC}) = (\mathbf{AB})\mathbf{C}$$

is an $m \times n$ matrix.

Distributive Law If **B** and **C** are $r \times n$ matrices and **A** is an $m \times r$ matrix, then the **distributive law** is

$$\mathbf{A}(\mathbf{B} + \mathbf{C}) = \mathbf{AB} + \mathbf{AC}$$

Furthermore, if the product $(\mathbf{B} + \mathbf{C})\mathbf{A}$ is defined, then

$$(\mathbf{B} + \mathbf{C})\mathbf{A} - \mathbf{BA} + \mathbf{CA}$$

Linear Equations Recall that any equation of the form $ax + by = c$, where a, b, and c are real numbers, is said to be a **linear equation** in the variables x and y. The graph of a linear equation in two variables is a straight line. For real numbers a, b, c, and d, $ax + by + cz = d$ is a linear equation in the variables x, y, and z and is the equation of a plane. In general, an equation of the form

$$a_1x_1 + a_2x_2 + \cdots + a_nx_n = b_n$$

where $a_1, a_2, \ldots, a_n$ and b_n are real numbers, is a linear equation in the n variables $x_1, x_2, \ldots, x_n$.

Matrices can be used to solve **systems of linear equations**. Systems of linear equations are also called **linear systems**. A general system of m linear equations in n

unknowns has the form

$$\begin{aligned} a_{11}x_1 + a_{12}x_2 + \cdots + a_{1n}x_n &= b_1 \\ a_{21}x_1 + a_{22}x_2 + \cdots + a_{2n}x_n &= b_2 \\ \vdots \qquad\qquad & \quad \vdots \\ a_{m1}x_1 + a_{m2}x_2 + \cdots + a_{mn}x_n &= b_m \end{aligned} \tag{4}$$

Augmented Matrix The solution of a linear system does not depend on what symbols are used as variables. Thus, the systems

$$\begin{aligned} 2x + 6y + z &= 7 \\ x + 2y - z &= -1 \\ 5x + 7y - 4z &= 9 \end{aligned} \quad \text{and} \quad \begin{aligned} 2u + 6v + w &= 7 \\ u + 2v - w &= -1 \\ 5u + 7v - 4w &= 9 \end{aligned}$$

have the same solution (10, -3, 5). In other words, when solving a linear system, the variables are immaterial; it is the coefficients of the variables and the constants that determine the solution of the system. In fact, we can solve a system of form (4) by dropping the variables entirely and performing operations on the rows of the array of coefficients and constants:

$$\left[\begin{array}{cccc|c} a_{11} & a_{12} & \cdots & a_{1n} & b_1 \\ a_{21} & a_{22} & \cdots & a_{2n} & b_2 \\ \vdots & & & & \vdots \\ a_{m1} & a_{m2} & \cdots & a_{mn} & b_m \end{array}\right]$$

This array is called the **augmented matrix** of the system (4).

EXAMPLE 5

(*a*) The augmented matrix $\left[\begin{array}{ccc|c} 1 & -3 & 5 & 2 \\ 4 & 7 & -1 & 8 \end{array}\right]$ represents the linear system

$$\begin{aligned} x_1 - 3x_2 + 5x_3 &= 2 \\ 4x_1 + 7x_2 - x_3 &= 8 \end{aligned}$$

(*b*) The linear system

$$\begin{aligned} x_1 - 5x_3 &= -1 \\ 2x_1 + 8x_2 &= 7 \\ x_2 + 9x_3 &= 1 \end{aligned} \quad \text{is the same as} \quad \begin{aligned} x_1 + 0x_2 - 5x_3 &= -1 \\ 2x_1 + 8x_2 + 0x_3 &= 7 \\ 0x_1 + x_2 + 9x_3 &= 1 \end{aligned}$$

Thus, the augmented matrix of the system is

$$\left[\begin{array}{ccc|c} 1 & 0 & -5 & -1 \\ 2 & 8 & 0 & 7 \\ 0 & 1 & 9 & 1 \end{array}\right]$$

□

Elementary Row Operations Two linear systems are **equivalent** if they have exactly the same solutions. Since the rows of an augmented matrix represent the equations in a linear system, we can obtain equivalent linear systems by performing the following **elementary row operations** on an augmented matrix:

(*i*) Multiply a row by a nonzero constant.

(*ii*) Interchange any two rows.

(*iii*) Add a nonzero multiple of one row to any other row.

Of course, when we add a multiple of one row to another, we add the corresponding entries in the rows.

Gauss–Jordan Elimination To solve a system such as (4) using an augmented matrix we shall use the **Gauss–Jordan elimination method**. This means we carry out a succession of elementary row operations until we arrive at an augmented matrix in **reduced row-echelon form**:

(*i*) The first nonzero entry in a nonzero row is a 1.

(*ii*) The remaining entries in a column containing a first entry 1 are all zeros.

(*iii*) In consecutive nonzero rows, the first entry 1 in the lower row appears to the right of the 1 in the higher row.

(*iv*) Rows consisting of all zeros are at the bottom of the matrix.

EXAMPLE 6 (*a*) The augmented matrices

$$\begin{bmatrix} 1 & 0 & 0 & 2 \\ 0 & 1 & 0 & 8 \\ 0 & 0 & 0 & 0 \end{bmatrix} \quad \text{and} \quad \begin{bmatrix} 0 & 0 & 1 & -6 & 0 & 2 \\ 0 & 0 & 0 & 0 & 1 & 4 \end{bmatrix}$$

are in reduced row-echelon form. You should verify that the criteria for this form are satisfied.

(*b*) The augmented matrices

$$\begin{bmatrix} 1 & -7 & 2 \\ 0 & 0 & 0 \\ 1 & 0 & 0 \end{bmatrix} \quad \text{and} \quad \begin{bmatrix} 2 & 6 & 0 & 5 & 0 \\ 0 & 1 & 0 & 0 & 0 \\ 0 & 0 & 7 & 3 & -2 \\ 0 & 0 & 0 & 1 & 4 \end{bmatrix}$$

are not in reduced row-echelon form. You should determine which of the four criteria listed above are not satisfied. □

Once an augmented matrix in reduced row-echelon form has been attained, the solution of the system will be apparent by inspection. In terms of the equations of the original system, our goal is simply to make the coefficient of x_1 in the first equation* equal to 1 and then use multiples of that equation to eliminate x_1 from the remaining equations. The process is repeated on the other variables.

---

*We can always interchange equations so that the first equation contains the variable x_1.

To keep track of the row operations on an augmented matrix we shall utilize the following notation:

Symbol	Meaning
R_{ij}	Interchange rows i and j
cR_i	Multiply the ith row by the nonzero constant c
$cR_i + R_j$	Multiply the ith row by c and add to the jth row

EXAMPLE 7 Solve the linear system

$$\begin{aligned} 2x_1 + 6x_2 + x_3 &= 7 \\ x_1 + 2x_2 - x_3 &= -1 \\ 5x_1 + 7x_2 - 4x_3 &= 9 \end{aligned}$$

using Gauss–Jordan elimination.

Solution Using row operations on the augmented matrix of the system, we obtain

$$\left[\begin{array}{ccc|c} 2 & 6 & 1 & 7 \\ 1 & 2 & -1 & -1 \\ 5 & 7 & -4 & 9 \end{array}\right] \xrightarrow{R_{12}} \left[\begin{array}{ccc|c} 1 & 2 & -1 & -1 \\ 2 & 6 & 1 & 7 \\ 5 & 7 & -4 & 9 \end{array}\right] \xrightarrow[-5R_1 + R_3]{-2R_1 + R_2} \left[\begin{array}{ccc|c} 1 & 2 & -1 & -1 \\ 0 & 2 & 3 & 9 \\ 0 & -3 & 1 & 14 \end{array}\right]$$

$$\xrightarrow{\frac{1}{2}R_2} \left[\begin{array}{ccc|c} 1 & 2 & -1 & -1 \\ 0 & 1 & \frac{3}{2} & \frac{9}{2} \\ 0 & -3 & 1 & 14 \end{array}\right] \xrightarrow[3R_2 + R_3]{-2R_2 + R_1} \left[\begin{array}{ccc|c} 1 & 0 & -4 & -10 \\ 0 & 1 & \frac{3}{2} & \frac{9}{2} \\ 0 & 0 & \frac{11}{2} & \frac{55}{2} \end{array}\right]$$

$$\xrightarrow{\frac{2}{11}R_3} \left[\begin{array}{ccc|c} 1 & 0 & -4 & -10 \\ 0 & 1 & \frac{3}{2} & \frac{9}{2} \\ 0 & 0 & 1 & 5 \end{array}\right] \xrightarrow[-\frac{3}{2}R_3 + R_2]{4R_3 + R_1} \left[\begin{array}{ccc|c} 1 & 0 & 0 & 10 \\ 0 & 1 & 0 & -3 \\ 0 & 0 & 1 & 5 \end{array}\right]$$

The last matrix is in reduced row-echelon form and represents the system

$$\begin{aligned} x_1 + 0x_2 + 0x_3 &= 10 \\ 0x_1 + x_2 + 0x_3 &= -3 \\ 0x_1 + 0x_2 + x_3 &= 5 \end{aligned}$$

Hence, it is evident that the solution of the system is $x_1 = 10$, $x_2 = -3$, $x_3 = 5$. □

Determinants Suppose **A** is an $n \times n$ matrix. Associated with **A** is a *number* called the **determinant of A**, which is denoted by det **A**. Symbolically we distinguish a matrix **A** and the determinant of **A** by writing

$$\mathbf{A} = \begin{bmatrix} a_{11} & a_{12} & \cdots & a_{1n} \\ a_{21} & a_{22} & \cdots & a_{2n} \\ \vdots & & & \vdots \\ a_{n1} & a_{n2} & \cdots & a_{nn} \end{bmatrix} \quad \text{and} \quad \det \mathbf{A} = \begin{vmatrix} a_{11} & a_{12} & \cdots & a_{1n} \\ a_{21} & a_{22} & \cdots & a_{2n} \\ \vdots & & & \vdots \\ a_{n1} & a_{n2} & \cdots & a_{nn} \end{vmatrix}$$

A **determinant of order 2** is the number

$$\begin{vmatrix} a_{11} & a_{12} \\ a_{21} & a_{22} \end{vmatrix} = a_{11}a_{22} - a_{12}a_{21} \tag{5}$$

EXAMPLE 8 By (5), $\begin{vmatrix} 4 & -3 \\ 2 & 9 \end{vmatrix} = 4 \cdot 9 - (-3)(2) = 42.$ □

A **determinant of order 3** can be evaluated by **expanding the determinant by cofactors of the first row**:

$$\begin{vmatrix} a_{11} & a_{12} & a_{13} \\ a_{21} & a_{22} & a_{23} \\ a_{31} & a_{32} & a_{33} \end{vmatrix} = a_{11}\begin{vmatrix} a_{22} & a_{23} \\ a_{32} & a_{33} \end{vmatrix} - a_{12}\begin{vmatrix} a_{21} & a_{23} \\ a_{31} & a_{33} \end{vmatrix} + a_{13}\begin{vmatrix} a_{21} & a_{22} \\ a_{31} & a_{32} \end{vmatrix} \tag{6}$$

EXAMPLE 9 By (6),

$$\begin{vmatrix} 8 & 5 & 4 \\ 2 & 4 & 6 \\ -1 & 2 & 3 \end{vmatrix} = 8\begin{vmatrix} 4 & 6 \\ 2 & 3 \end{vmatrix} - 5\begin{vmatrix} 2 & 6 \\ -1 & 3 \end{vmatrix} + 4\begin{vmatrix} 2 & 4 \\ -1 & 2 \end{vmatrix}$$

$$= 8(0) - 5(12) + 4(8) = -28$$ □

Cofactors In general, the **cofactor** of an entry in the ith row and jth column of a determinant is $(-1)^{i+j}$ times that determinant formed by deleting the ith row and jth column. Thus, the cofactors of a_{11}, a_{12}, and a_{13} are, respectively,

$$\begin{vmatrix} a_{22} & a_{23} \\ a_{32} & a_{33} \end{vmatrix}, \quad -\begin{vmatrix} a_{21} & a_{23} \\ a_{31} & a_{33} \end{vmatrix}, \quad \text{and} \quad \begin{vmatrix} a_{21} & a_{22} \\ a_{31} & a_{32} \end{vmatrix}$$

The cofactor of, say a_{23} is $-\begin{vmatrix} a_{11} & a_{12} \\ a_{31} & a_{32} \end{vmatrix}$.

Some Properties of Determinants

(*i*) If every entry in a row (or column) is zero, then the value of the determinant is zero.

(*ii*) If two rows (or columns) are equal, then the value of the determinant is zero.

(*iii*) Interchanging two rows (or columns) results in the negative of the value of the original determinant.

(*iv*) A determinant can be expanded by the cofactors of any row (or column).

EXAMPLE 10

(*a*) From (*i*), $\begin{vmatrix} 1 & 6 & 0 \\ 3 & 2 & 0 \\ 4 & 5 & 0 \end{vmatrix} = 0.$

(*b*) From (*ii*), $\begin{vmatrix} 1 & 6 & 6 \\ 3 & 2 & 2 \\ 4 & 5 & 5 \end{vmatrix} = 0.$

(*c*) Since $\begin{vmatrix} 2 & 3 \\ 4 & 5 \end{vmatrix} = -2$, it follows from (*iii*) that $\begin{vmatrix} 4 & 5 \\ 2 & 3 \end{vmatrix} = 2.$ □

EXAMPLE 11 Expand the determinant $\begin{vmatrix} 3 & 4 & 3 \\ 2 & 0 & 0 \\ 7 & 1 & 2 \end{vmatrix}$.

Solution Taking advantage of the zeros in the second row, we expand the determinant by the cofactors of the second row:

$$\begin{vmatrix} 3 & 4 & 3 \\ 2 & 0 & 0 \\ 7 & 1 & 2 \end{vmatrix} = -2\begin{vmatrix} 4 & 3 \\ 1 & 2 \end{vmatrix} + 0\begin{vmatrix} 3 & 3 \\ 7 & 2 \end{vmatrix} - 0\begin{vmatrix} 3 & 4 \\ 7 & 1 \end{vmatrix} = -2(8 - 3) = -10$$ □

Cramer's Rule Systems of n linear equations in n unknowns can sometimes be solved by means of determinants. The solution of the system

$$a_{11}x + a_{12}y = b_1$$
$$a_{21}x + a_{22}y = b_2$$

is

$$x = \frac{\begin{vmatrix} b_1 & a_{12} \\ b_2 & a_{22} \end{vmatrix}}{\begin{vmatrix} a_{11} & a_{12} \\ a_{21} & a_{22} \end{vmatrix}} \qquad y = \frac{\begin{vmatrix} a_{11} & b_1 \\ a_{21} & b_2 \end{vmatrix}}{\begin{vmatrix} a_{11} & a_{12} \\ a_{21} & a_{22} \end{vmatrix}}$$

provided that the determinant of the coefficient matrix $\begin{bmatrix} a_{11} & a_{12} \\ a_{21} & a_{22} \end{bmatrix}$ is not zero. This last condition also guarantees that there is only one solution. Similarly, the system of three equations in three unknowns

$$a_{11}x + a_{12}y + a_{13}z = b_1$$
$$a_{21}x + a_{22}y + a_{23}z = b_2$$
$$a_{31}x + a_{32}y + a_{33}z = b_3$$

has the unique solution

$$x = \frac{\begin{vmatrix} b_1 & a_{12} & a_{13} \\ b_2 & a_{22} & a_{23} \\ b_3 & a_{32} & a_{33} \end{vmatrix}}{\begin{vmatrix} a_{11} & a_{12} & a_{13} \\ a_{21} & a_{22} & a_{23} \\ a_{31} & a_{32} & a_{33} \end{vmatrix}} \qquad y = \frac{\begin{vmatrix} a_{11} & b_1 & a_{13} \\ a_{21} & b_2 & a_{23} \\ a_{31} & b_3 & a_{33} \end{vmatrix}}{\begin{vmatrix} a_{11} & a_{12} & a_{13} \\ a_{21} & a_{22} & a_{23} \\ a_{31} & a_{32} & a_{33} \end{vmatrix}} \qquad z = \frac{\begin{vmatrix} a_{11} & a_{12} & b_1 \\ a_{21} & a_{22} & b_2 \\ a_{31} & a_{32} & b_3 \end{vmatrix}}{\begin{vmatrix} a_{11} & a_{12} & a_{13} \\ a_{21} & a_{22} & a_{23} \\ a_{31} & a_{32} & a_{33} \end{vmatrix}} \tag{7}$$

provided that the determinant of the coefficient matrix is not zero. The foregoing procedure illustrates **Cramer's Rule.***

EXAMPLE 12 Solve:

$$\begin{aligned} x + 2y + \ \ z &= 1 \\ x \qquad + 3z &= 4 \\ 4x + 2y + \ \ z &= 5 \end{aligned}$$

Solution Since $\begin{vmatrix} 1 & 2 & 1 \\ 1 & 0 & 3 \\ 4 & 2 & 1 \end{vmatrix} = 18$, the system can be solved by Cramer's Rule (7):

$$x = \frac{\begin{vmatrix} 1 & 2 & 1 \\ 4 & 0 & 3 \\ 5 & 2 & 1 \end{vmatrix}}{\begin{vmatrix} 1 & 2 & 1 \\ 1 & 0 & 3 \\ 4 & 2 & 1 \end{vmatrix}} = \frac{24}{18} = \frac{4}{3} \qquad y = \frac{\begin{vmatrix} 1 & 1 & 1 \\ 1 & 4 & 3 \\ 4 & 5 & 1 \end{vmatrix}}{\begin{vmatrix} 1 & 2 & 1 \\ 1 & 0 & 3 \\ 4 & 2 & 1 \end{vmatrix}} = -\frac{11}{18} \qquad z = \frac{\begin{vmatrix} 1 & 2 & 1 \\ 1 & 0 & 4 \\ 4 & 2 & 5 \end{vmatrix}}{\begin{vmatrix} 1 & 2 & 1 \\ 1 & 0 & 3 \\ 4 & 2 & 1 \end{vmatrix}} = \frac{16}{18} = \frac{8}{9}$$ □

EXERCISES I.5

Answers to odd-numbered problems begin on page A-113.

In Problems 1–6 state the size of the given matrix.

1. $\begin{bmatrix} 1 & 2 & 3 & 9 \\ 5 & 6 & 0 & 1 \end{bmatrix}$

2. $\begin{bmatrix} 0 & 2 \\ 8 & 4 \\ 5 & 6 \end{bmatrix}$

3. $\begin{bmatrix} 1 & 2 & -1 \\ 0 & 7 & -2 \\ 0 & 0 & 5 \end{bmatrix}$

4. $[5 \quad 7 \quad -15]$

5. $\begin{bmatrix} 1 & 5 & -6 & 0 \\ 7 & -10 & 2 & 12 \\ 0 & 9 & 2 & -1 \end{bmatrix}$

6. $\begin{bmatrix} 1 \\ 5 \\ -6 \\ 0 \\ 7 \\ -10 \\ 2 \\ 12 \end{bmatrix}$

---

*Named after the Swiss mathematician Gabriel Cramer (1704–1752).

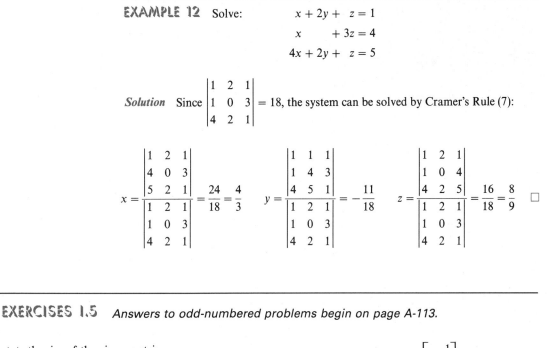

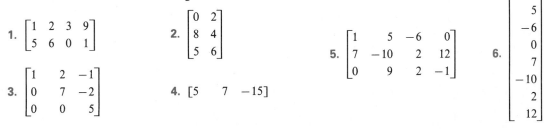

In Problems 7–10 determine whether the given matrices are equal.

7. $\begin{bmatrix} 1 & 2 & 3 \\ 4 & 5 & 6 \end{bmatrix}, \begin{bmatrix} 1 & 2 \\ 3 & 4 \\ 5 & 6 \end{bmatrix}$

8. $\begin{bmatrix} 1 & 2 \\ 0 & 1 \end{bmatrix}, \begin{bmatrix} 1 & 0 \\ 2 & 1 \end{bmatrix}$

9. $\begin{bmatrix} \sqrt{(-2)^2} & 1 \\ 2 & \frac{2}{8} \end{bmatrix}, \begin{bmatrix} -2 & 1 \\ 2 & \frac{1}{4} \end{bmatrix}$

10. $\begin{bmatrix} \frac{1}{8} & \frac{1}{5} \\ \sqrt{2} & 1 \end{bmatrix}, \begin{bmatrix} 0.125 & 0.2 \\ 1.414 & 1 \end{bmatrix}$

In Problems 11 and 12 determine the values of x and y for which the matrices are equal.

11. $\begin{bmatrix} 1 & x \\ y & -3 \end{bmatrix}, \begin{bmatrix} 1 & y-2 \\ 3x-2 & -3 \end{bmatrix}$

12. $\begin{bmatrix} x^2 & 1 \\ y & 5 \end{bmatrix}, \begin{bmatrix} 9 & 1 \\ 4x & 5 \end{bmatrix}$

In Problems 13 and 14 find c_{23} and c_{12} for the matrix $\mathbf{C} = 2\mathbf{A} - 3\mathbf{B}$.

13. $\mathbf{A} = \begin{bmatrix} 2 & 3 & -1 \\ -1 & 6 & 0 \end{bmatrix}, \mathbf{B} = \begin{bmatrix} 4 & -2 & 6 \\ 1 & 3 & -3 \end{bmatrix}$

14. $\mathbf{A} = \begin{bmatrix} 1 & -1 & 1 \\ 2 & 2 & 1 \\ 0 & -4 & 1 \end{bmatrix}, \mathbf{B} = \begin{bmatrix} 2 & 0 & 5 \\ 0 & 4 & 0 \\ 3 & 0 & 7 \end{bmatrix}$

15. If $\mathbf{A} = \begin{bmatrix} 4 & 5 \\ -6 & 9 \end{bmatrix}$ and $\mathbf{B} = \begin{bmatrix} -2 & 6 \\ 8 & -10 \end{bmatrix}$, find

(a) $\mathbf{A} + \mathbf{B}$ (b) $\mathbf{B} - \mathbf{A}$ (c) $2\mathbf{A} + 3\mathbf{B}$

16. If $\mathbf{A} = \begin{bmatrix} -2 & 0 \\ 4 & 1 \\ 7 & 3 \end{bmatrix}$ and $\mathbf{B} = \begin{bmatrix} 3 & -1 \\ 0 & 2 \\ -4 & -2 \end{bmatrix}$, find

(a) $\mathbf{A} - \mathbf{B}$ (b) $\mathbf{B} - \mathbf{A}$ (c) $2(\mathbf{A} + \mathbf{B})$

17. If $\mathbf{A} = \begin{bmatrix} 2 & -3 \\ -5 & 4 \end{bmatrix}$ and $\mathbf{B} = \begin{bmatrix} -1 & 6 \\ 3 & 2 \end{bmatrix}$, find

(a) $\mathbf{AB}$ (b) $\mathbf{BA}$ (c) $\mathbf{A}^2 = \mathbf{AA}$ (d) $\mathbf{B}^2 = \mathbf{BB}$

18. If $\mathbf{A} = \begin{bmatrix} 1 & 4 \\ 5 & 10 \\ 8 & 12 \end{bmatrix}$ and $\mathbf{B} = \begin{bmatrix} -4 & 6 & -3 \\ 1 & -3 & 2 \end{bmatrix}$, find

(a) $\mathbf{AB}$ (b) $\mathbf{BA}$

19. If $\mathbf{A} = \begin{bmatrix} 1 & -2 \\ -2 & 4 \end{bmatrix}, \mathbf{B} = \begin{bmatrix} 6 & 3 \\ 2 & 1 \end{bmatrix}$, and $\mathbf{C} = \begin{bmatrix} 0 & 2 \\ 3 & 4 \end{bmatrix}$, find

(a) $\mathbf{BC}$ (b) $\mathbf{A}(\mathbf{BC})$

(c) $\mathbf{C}(\mathbf{BA})$ (d) $\mathbf{A}(\mathbf{B} + \mathbf{C})$

20. If $\mathbf{A} = [5 \quad -6 \quad 7]$, $\mathbf{B} = \begin{bmatrix} 3 \\ 4 \\ -1 \end{bmatrix}$, and

$\mathbf{C} = \begin{bmatrix} 1 & 2 & 4 \\ 0 & 1 & -1 \\ 3 & 2 & 1 \end{bmatrix}$, find

(a) $\mathbf{AB}$ (b) $\mathbf{BA}$ (c) $(\mathbf{BA})\mathbf{C}$ (d) $(\mathbf{AB})\mathbf{C}$

In Problems 21–26 use Gauss–Jordan elimination to solve the given linear system.

21. $\begin{aligned} x_1 - x_2 &= 11 \\ 4x_1 + 3x_2 &= -5 \end{aligned}$

22. $\begin{aligned} 3x_1 - 2x_2 &= 4 \\ x_1 - x_2 &= -2 \end{aligned}$

23. $\begin{aligned} x_1 - x_2 - x_3 &= -3 \\ 2x_1 + 3x_2 + 5x_3 &= 7 \\ x_1 - 2x_2 + 3x_3 &= -11 \end{aligned}$

24. $\begin{aligned} x_1 + 2x_2 - x_3 &= 0 \\ 2x_1 + x_2 + 2x_3 &= 9 \\ x_1 - x_2 + x_3 &= 3 \end{aligned}$

25. $\begin{aligned} x_1 + 2x_2 + 2x_3 &= 2 \\ x_1 + x_2 + x_3 &= 0 \\ x_1 - 3x_2 - x_3 &= 0 \end{aligned}$

26. $\begin{aligned} x_1 - x_2 - 2x_3 &= 11 \\ 2x_1 + 4x_2 + 5x_3 &= -35 \\ 6x_1 - x_3 &= -1 \end{aligned}$

27. For the determinant $\begin{vmatrix} -5 & 6 & 0 \\ 3 & 4 & 2 \\ -2 & 1 & 5 \end{vmatrix}$ find **(a)** the value of the cofactor of the entry in the third row and second column, and **(b)** the value of the cofactor of the entry in the second row and second column.

28. If $\begin{vmatrix} a_{11} & a_{12} & a_{13} \\ a_{21} & a_{22} & a_{23} \\ a_{31} & a_{32} & a_{33} \end{vmatrix} = 3$, determine the values of the following:

(a) $\begin{vmatrix} a_{31} & a_{32} & a_{33} \\ a_{21} & a_{22} & a_{23} \\ a_{11} & a_{12} & a_{13} \end{vmatrix}$ (b) $\begin{vmatrix} a_{12} & a_{13} & a_{11} \\ a_{22} & a_{23} & a_{21} \\ a_{32} & a_{33} & a_{31} \end{vmatrix}$

(c) $\begin{vmatrix} a_{11} & a_{12} & a_{13} \\ a_{11} & a_{12} & a_{13} \\ a_{21} & a_{22} & a_{23} \end{vmatrix}$ (d) $\begin{vmatrix} a_{11} & a_{12} & 0 \\ a_{21} & a_{22} & 0 \\ a_{31} & a_{32} & 0 \end{vmatrix}$

In Problems 29 and 30 expand the given determinant by cofactors other than those of the first row.

29. $\begin{vmatrix} 2 & -3 & 1 \\ 0 & 2 & -4 \\ 0 & 6 & 5 \end{vmatrix}$

30. $\begin{vmatrix} 4 & 3 & -1 \\ 7 & 0 & -2 \\ -1 & 5 & 8 \end{vmatrix}$

In Problem 31 and 32 find the values of λ that satisfy the given equation.

31. $\begin{vmatrix} 3-\lambda & 5 \\ 1 & -1-\lambda \end{vmatrix} = 0$

32. $\begin{vmatrix} 1-\lambda & -4 & 4 \\ 0 & 3-\lambda & -2 \\ -2 & 0 & -3-\lambda \end{vmatrix} = 0$

In Problems 33–36 use Cramer's Rule to solve the given system of equations.

33. $\begin{aligned} -3x + y &= 3 \\ 2x - 4y &= -6 \end{aligned}$

34. $\begin{aligned} x + y &= 4 \\ 2x - y &= 2 \end{aligned}$

35. $\begin{aligned} x - 2y - 3z &= 3 \\ x + y - z &= 5 \\ 3x + 2y \quad &= -4 \end{aligned}$

36. $\begin{aligned} x - y + 6z &= -2 \\ -x + 2y + 4z &= 9 \\ 2x + 3y - z &= \tfrac{1}{2} \end{aligned}$

APPENDIX II SOME $\varepsilon-\delta$ PROOFS

Proof of Theorem 1.4(i) Let $\varepsilon > 0$ be given. To prove (*i*) we must find a $\delta > 0$ so that

$$|f(x) + g(x) - L_1 - L_2| < \varepsilon \quad \text{whenever} \quad 0 < |x - a| < \delta$$

Since $\lim_{x\to a} f(x) = L_1$ and $\lim_{x\to a} g(x) = L_2$, we know there exist numbers $\delta_1 > 0$ and $\delta_2 > 0$ for which

$$|f(x) - L_1| < \frac{\varepsilon}{2} \quad \text{whenever} \quad 0 < |x - a| < \delta_1 \tag{1}$$

$$|g(x) - L_2| < \frac{\varepsilon}{2} \quad \text{whenever} \quad 0 < |x - a| < \delta_2 \tag{2}$$

Now, if we choose δ to be the smaller of δ_1 and δ_2, then (1) and (2) *both* hold and so

$$\begin{aligned} |f(x) + g(x) - L_1 - L_2| &= |f(x) - L_1 + g(x) - L_2| \\ &\le |f(x) - L_1| + |g(x) - L_2| \\ &< \frac{\varepsilon}{2} + \frac{\varepsilon}{2} = \varepsilon \end{aligned}$$

whenever $0 < |x - a| < \delta$. ■

Proof of Theorem 1.4(ii)

$$\begin{aligned} |f(x)g(x) - L_1L_2| &= |f(x)g(x) - f(x)L_2 + f(x)L_2 - L_1L_2| \\ &\le |f(x)g(x) - f(x)L_2| + |f(x)L_2 - L_1L_2| \\ &= |f(x)||g(x) - L_2| + |L_2||f(x) - L_1| \\ &\le |f(x)||g(x) - L_2| + (1 + |L_2|)|f(x) - L_1| \end{aligned} \tag{3}$$

Since $\lim_{x\to a} f(x) = L_1$ and $\lim_{x\to a} g(x) = L_2$, we know there exist numbers $\delta_1 > 0$, $\delta_2 > 0$, $\delta_3 > 0$ such that $|f(x) - L_1| < 1$ or

$$|f(x)| < 1 + |L_1| \quad \text{whenever} \quad 0 < |x - a| < \delta_1 \tag{4}$$

$$|f(x) - L_1| < \frac{\varepsilon/2}{1 + |L_2|} \quad \text{whenever} \quad 0 < |x - a| < \delta_2 \tag{5}$$

$$|g(x) - L_2| < \frac{\varepsilon/2}{1 + |L_1|} \quad \text{whenever} \quad 0 < |x - a| < \delta_3 \tag{6}$$

Hence, if we choose δ to be the smaller of δ_1, δ_2, and δ_3, we have from (3), (4), (5), and (6),

$$|f(x)g(x) - L_1L_2| < (1 + |L_1|) \cdot \frac{\varepsilon/2}{1 + |L_1|} + (1 + |L_2|) \cdot \frac{\varepsilon/2}{1 + |L_2|} = \varepsilon$$ ■

Proof of Theorem 1.4(iii) We shall first prove that

$$\lim_{x\to a}\frac{1}{g(x)}=\frac{1}{L_2},\qquad L_2\neq 0$$

Consider

$$\left|\frac{1}{g(x)}-\frac{1}{L_2}\right|=\frac{|g(x)-L_2|}{|L_2||g(x)|}\tag{7}$$

Since $\lim_{x\to a} g(x)=L_2$, there exists a $\delta_1>0$ such that

$$|g(x)-L_2|<\frac{|L_2|}{2}\quad\text{whenever}\quad 0<|x-a|<\delta_1$$

For these values of x,

$$|L_2|=|g(x)-(g(x)-L_2)|\leq |g(x)|+|g(x)-L_2|<|g(x)|+\frac{|L_2|}{2}$$

gives

$$|g(x)|>\frac{|L_2|}{2}\quad\text{and}\quad\frac{1}{|g(x)|}<\frac{2}{|L_2|}$$

Thus, from (7),

$$\left|\frac{1}{g(x)}-\frac{1}{L_2}\right|<\frac{2}{|L_2|^2}|g(x)-L_2|\tag{8}$$

Now for $\varepsilon>0$ there exists a $\delta_2>0$ such that

$$|g(x)-L_2|<\frac{|L_2|^2}{2}\varepsilon\quad\text{whenever}\quad 0<|x-a|<\delta_2$$

By choosing δ to be the smaller of δ_1 and δ_2, it follows from (8) that

$$\left|\frac{1}{g(x)}-\frac{1}{L_2}\right|<\varepsilon\quad\text{whenever}\quad 0<|x-a|<\delta$$

We conclude the proof using Theorem 1.4(*ii*):

$$\lim_{x\to a}\frac{f(x)}{g(x)}=\lim_{x\to a}\frac{1}{g(x)}\cdot f(x)=\lim_{x\to a}\frac{1}{g(x)}\cdot\lim_{x\to a}f(x)=\frac{L_1}{L_2}$$ ■

EXERCISES

1. Use Definition 1.2 to prove Theorem 1.1(*ii*).
2. Use Definition 1.2 to prove Theorem 1.2.
3. If $\lim_{x\to a} f(x)=L_1$ and $\lim_{x\to a} g(x)=L_2$, use Definition 1.2 to prove $\lim_{x\to a}[f(x)-g(x)]=L_1-L_2$.
4. Use Definition 1.2 to prove the Squeeze Theorem. [*Hint:* For $\varepsilon>0$ there exist numbers $\delta_1>0$ and $\delta_2>0$ such that $L-\varepsilon<f(x)<L+\varepsilon$ whenever $0<|x-a|<\delta_1$ and $L-\varepsilon<h(x)<L+\varepsilon$ whenever $0<|x-a|<\delta_2$. Let δ be the smaller of δ_1 and δ_2.]

APPENDIX III PROOF OF TAYLOR'S THEOREM

Let x be a fixed number in $(a - r, a + r)$ and let the difference between $f(x)$ and the nth-degree Taylor polynomial of f at a be denoted by

$$R_n(x) = f(x) - P_n(x)$$

For any t in the interval $[a, x]$ we define

$$F(t) = f(x) - f(t) - \frac{f'(t)}{1!}(x - t) - \frac{f''(t)}{2!}(x - t)^2 - \cdots$$
$$- \frac{f^{(n)}(t)}{n!}(x - t)^n - \frac{R_n(x)}{(x - a)^{n+1}}(x - t)^{n+1} \qquad (1)$$

With x held constant we differentiate F with respect to t using the Product and Power Rules:

$$F'(t) = -f'(t) + \left[f'(t) - \frac{f''(t)}{1!}(x - t)\right]$$
$$+ \left[\frac{f''(t)}{1!}(x - t) - \frac{f'''(t)}{2!}(x - t)^2\right] + \cdots$$
$$+ \left[\frac{f^{(n)}(t)}{(n - 1)!}(x - t)^{n-1} - \frac{f^{(n+1)}(t)}{n!}(x - t)^n\right]$$
$$+ \frac{R_n(x)(n + 1)}{(x - a)^{n+1}}(x - t)^n$$

for all t in (a, x). Since the last sum telescopes, we obtain

$$F'(t) = -\frac{f^{(n+1)}(t)}{n!}(x - t)^n + \frac{R_n(x)(n + 1)}{(x - a)^{n+1}}(x - t)^n \qquad (2)$$

Now it is evident from (1) that F is continuous on $[a, x]$ and that

$$F(x) = f(x) - f(x) - 0 - \cdots - 0 = 0$$

Furthermore, $$F(a) = f(x) - P_n(x) - R_n(x) = 0$$

Thus, $F(t)$ satisfies the hypotheses of Rolle's Theorem (Theorem 3.4) and so there exists a number c between a and x for which $F'(c) = 0$. From (2) we obtain

$$R_n(x) = \frac{f^{(n+1)}(c)}{(n + 1)!}(x - a)^{n+1}$$

■

APPENDIX IV TABLE OF INTEGRALS

Basic Forms

1. $\int u\,dv = uv - \int v\,du$

2. $\int u^n\,du = \frac{1}{n+1}u^{n+1} + C, n \neq -1$

3. $\int \frac{du}{u} = \ln|u| + C$

4. $\int e^u\,du = e^u + C$

5. $\int a^u\,du = \frac{1}{\ln a}a^u + C$

6. $\int \sin u\,du = -\cos u + C$

7. $\int \cos u\,du = \sin u + C$

8. $\int \sec^2 u\,du = \tan u + C$

9. $\int \csc^2 u\,du = -\cot u + C$

10. $\int \sec u \tan u\,du = \sec u + C$

11. $\int \csc u \cot u\,du = -\csc u + C$

12. $\int \tan u\,du = -\ln|\cos u| + C$

13. $\int \cot u\,du = \ln|\sin u| + C$

14. $\int \sec u\,du = \ln|\sec u + \tan u| + C$

15. $\int \csc u\,du = \ln|\csc u - \cot u| + C$

16. $\int \frac{du}{\sqrt{a^2 - u^2}} = \sin^{-1}\frac{u}{a} + C$

17. $\int \frac{du}{a^2 + u^2} = \frac{1}{a}\tan^{-1}\frac{u}{a} + C$

18. $\int \frac{du}{u\sqrt{u^2 - a^2}} = \frac{1}{a}\sec^{-1}\frac{u}{a} + C$

19. $\int \frac{du}{a^2 - u^2} = \frac{1}{2a}\ln\left|\frac{u+a}{u-a}\right| + C$

20. $\int \frac{du}{u^2 - a^2} = \frac{1}{2a}\ln\left|\frac{u-a}{u+a}\right| + C$

Forms Involving $\sqrt{a^2 + u^2}$

21. $\int \sqrt{a^2 + u^2}\,du = \frac{u}{2}\sqrt{a^2 + u^2} + \frac{a^2}{2}\ln|u + \sqrt{a^2 + u^2}| + C$

22. $\int u^2\sqrt{a^2 + u^2}\,du = \frac{u}{8}(a^2 + 2u^2)\sqrt{a^2 + u^2} - \frac{a^4}{8}\ln|u + \sqrt{a^2 + u^2}| + C$

23. $\int \frac{\sqrt{a^2 + u^2}}{u}\,du = \sqrt{a^2 + u^2} - a\ln\left|\frac{a + \sqrt{a^2 + u^2}}{u}\right| + C$

24. $\int \frac{\sqrt{a^2 + u^2}}{u^2}\,du = -\frac{\sqrt{a^2 + u^2}}{u} + \ln|u + \sqrt{a^2 + u^2}| + C$

25. $\int \frac{du}{\sqrt{a^2 + u^2}} = \ln|u + \sqrt{a^2 + u^2}| + C$

26. $\int \frac{u^2\,du}{\sqrt{a^2 + u^2}} = \frac{u}{2}\sqrt{a^2 + u^2} - \frac{a^2}{2}\ln|u + \sqrt{a^2 + u^2}| + C$

27. $\displaystyle\int \frac{du}{u\sqrt{a^2+u^2}} = -\frac{1}{a}\ln\left|\frac{\sqrt{a^2+u^2}+a}{u}\right| + C$

28. $\displaystyle\int \frac{du}{u^2\sqrt{a^2+u^2}} = -\frac{\sqrt{a^2+u^2}}{a^2u} + C$

29. $\displaystyle\int \frac{du}{(a^2+u^2)^{3/2}} = \frac{u}{a^2\sqrt{a^2+u^2}} + C$

Forms Involving $\sqrt{a^2 - u^2}$

30. $\displaystyle\int \sqrt{a^2-u^2}\,du = \frac{u}{2}\sqrt{a^2-u^2} + \frac{a^2}{2}\sin^{-1}\frac{u}{a} + C$

31. $\displaystyle\int u^2\sqrt{a^2-u^2}\,du = \frac{u}{8}(2u^2-a^2)\sqrt{a^2-u^2} + \frac{a^4}{8}\sin^{-1}\frac{u}{a} + C$

32. $\displaystyle\int \frac{\sqrt{a^2-u^2}}{u}\,du = \sqrt{a^2-u^2} - a\ln\left|\frac{a+\sqrt{a^2-u^2}}{u}\right| + C$

33. $\displaystyle\int \frac{\sqrt{a^2-u^2}}{u^2}\,du = -\frac{1}{u}\sqrt{a^2-u^2} - \sin^{-1}\frac{u}{a} + C$

34. $\displaystyle\int \frac{u^2\,du}{\sqrt{a^2-u^2}} = -\frac{u}{2}\sqrt{a^2-u^2} + \frac{a^2}{2}\sin^{-1}\frac{u}{a} + C$

35. $\displaystyle\int \frac{du}{u\sqrt{a^2-u^2}} = -\frac{1}{a}\ln\left|\frac{a+\sqrt{a^2-u^2}}{u}\right| + C$

36. $\displaystyle\int \frac{du}{u^2\sqrt{a^2-u^2}} = -\frac{1}{a^2u}\sqrt{a^2-u^2} + C$

37. $\displaystyle\int (a^2-u^2)^{3/2}\,du = -\frac{u}{8}(2u^2-5a^2)\sqrt{a^2-u^2} + \frac{3a^4}{8}\sin^{-1}\frac{u}{a} + C$

38. $\displaystyle\int \frac{du}{(a^2-u^2)^{3/2}} = \frac{u}{a^2\sqrt{a^2-u^2}} + C$

Forms Involving $\sqrt{u^2 - a^2}$

39. $\displaystyle\int \sqrt{u^2-a^2}\,du = \frac{u}{2}\sqrt{u^2-a^2} - \frac{a^2}{2}\ln\left|u+\sqrt{u^2-a^2}\right| + C$

40. $\displaystyle\int u^2\sqrt{u^2-a^2}\,du = \frac{u}{8}(2u^2-a^2)\sqrt{u^2-a^2} - \frac{a^4}{8}\ln\left|u+\sqrt{u^2-a^2}\right| + C$

41. $\displaystyle\int \frac{\sqrt{u^2-a^2}}{u}\,du = \sqrt{u^2-a^2} - a\cos^{-1}\frac{a}{u} + C$

42. $\displaystyle\int \frac{\sqrt{u^2-a^2}}{u^2}\,du = -\frac{\sqrt{u^2-a^2}}{u} + \ln\left|u+\sqrt{u^2-a^2}\right| + C$

43. $\displaystyle\int \frac{du}{\sqrt{u^2-a^2}} = \ln\left|u+\sqrt{u^2-a^2}\right| + C$

44. $\displaystyle\int \frac{u^2\,du}{\sqrt{u^2-a^2}} = \frac{u}{2}\sqrt{u^2-a^2} + \frac{a^2}{2}\ln\left|u+\sqrt{u^2-a^2}\right| + C$

45. $\displaystyle\int \frac{du}{u^2\sqrt{u^2-a^2}} = \frac{\sqrt{u^2-a^2}}{a^2u} + C$

46. $\displaystyle\int \frac{du}{(u^2-a^2)^{3/2}} = -\frac{u}{a^2\sqrt{u^2-a^2}} + C$

Forms Involving $a + bu$

47. $\displaystyle\int \frac{u\,du}{a+bu} = \frac{1}{b^2}(a+bu-a\ln|a+bu|) + C$

48. $\displaystyle\int \frac{u^2\,du}{a+bu} = \frac{1}{2b^3}\left[(a+bu)^2 - 4a(a+bu) + 2a^2\ln|a+bu|\right] + C$

49. $\displaystyle\int \frac{du}{u(a+bu)} = \frac{1}{a}\ln\left|\frac{u}{a+bu}\right| + C$

50. $\displaystyle\int \frac{du}{u^2(a+bu)} = -\frac{1}{au} + \frac{b}{a^2}\ln\left|\frac{a+bu}{u}\right| + C$

51. $\displaystyle\int \frac{u\,du}{(a+bu)^2} = \frac{a}{b^2(a+bu)} + \frac{1}{b^2}\ln|a+bu| + C$

52. $\displaystyle\int \frac{du}{u(a+bu)^2} = \frac{1}{a(a+bu)} - \frac{1}{a^2}\ln\left|\frac{a+bu}{u}\right| + C$

53. $\displaystyle\int \frac{u^2\,du}{(a+bu)^2} = \frac{1}{b^3}\left(a + bu - \frac{a^2}{a+bu} - 2a\ln|a+bu|\right) + C$

54. $\displaystyle\int u\sqrt{a+bu}\,du = \frac{2}{15b^2}(3bu-2a)(a+bu)^{3/2} + C$

55. $\displaystyle\int \frac{u\,du}{\sqrt{a+bu}} = \frac{2}{3b^2}(bu-2a)\sqrt{a+bu} + C$

56. $\displaystyle\int \frac{u^2\,du}{\sqrt{a+bu}} = \frac{2}{15b^3}(8a^2+3b^2u^2-4abu)\sqrt{a+bu} + C$

57. $\displaystyle\int \frac{du}{u\sqrt{a+bu}} = \frac{1}{\sqrt{a}}\ln\left|\frac{\sqrt{a+bu}-\sqrt{a}}{\sqrt{a+bu}+\sqrt{a}}\right| + C, \quad \text{if } a>0$

$\displaystyle\qquad = \frac{2}{\sqrt{-a}}\tan^{-1}\sqrt{\frac{a+bu}{-a}} + C, \quad \text{if } a<0$

58. $\displaystyle\int \frac{\sqrt{a+bu}}{u}\,du = 2\sqrt{a+bu} + a\int\frac{du}{u\sqrt{a+bu}}$

59. $\displaystyle\int \frac{\sqrt{a+bu}}{u^2}\,du = -\frac{\sqrt{a+bu}}{u} + \frac{b}{2}\int\frac{du}{u\sqrt{a+bu}}$

60. $\displaystyle\int u^n\sqrt{a+bu}\,du = \frac{2u^n(a+bu)^{3/2}}{b(2n+3)} - \frac{2na}{b(2n+3)}\int u^{n-1}\sqrt{a+bu}\,du$

61. $\displaystyle\int \frac{u^n\,du}{\sqrt{a+bu}} = \frac{2u^n\sqrt{a+bu}}{b(2n+1)} - \frac{2na}{b(2n+1)}\int\frac{u^{n-1}\,du}{\sqrt{a+bu}}$

62. $\displaystyle\int \frac{du}{u^n\sqrt{a+bu}} = -\frac{\sqrt{a+bu}}{a(n-1)u^{n-1}} - \frac{b(2n-3)}{2a(n-1)}\int\frac{du}{u^{n-1}\sqrt{a+bu}}$

Trigonometric Forms

63. $\displaystyle\int \sin^2 u\,du = \tfrac{1}{2}u - \tfrac{1}{4}\sin 2u + C$

64. $\displaystyle\int \cos^2 u\,du = \tfrac{1}{2}u + \tfrac{1}{4}\sin 2u + C$

65. $\displaystyle\int \tan^2 u\,du = \tan u - u + C$

66. $\displaystyle\int \cot^2 u\,du = -\cot u - u + C$

67. $\displaystyle\int \sin^3 u\,du = -\tfrac{1}{3}(2+\sin^2 u)\cos u + C$

68. $\displaystyle\int \cos^3 u\,du = \tfrac{1}{3}(2+\cos^2 u)\sin u + C$

69. $\displaystyle\int \tan^3 u\,du = \tfrac{1}{2}\tan^2 u + \ln|\cos u| + C$

70. $\displaystyle\int \cot^2 u\,du = -\tfrac{1}{2}\cot^2 u - \ln|\sin u| + C$

71. $\displaystyle\int \sec^3 u\,du = \tfrac{1}{2}\sec u\tan u + \tfrac{1}{2}\ln|\sec u + \tan u| + C$

72. $\displaystyle\int \csc^3 u\,du = -\tfrac{1}{2}\csc u\cot u + \tfrac{1}{2}\ln|\csc u - \cot u| + C$

73. $\displaystyle\int \sin^n u\,du = -\frac{1}{n}\sin^{n-1}u\cos u + \frac{n-1}{n}\int\sin^{n-2}u\,du$

74. $\displaystyle\int \cos^n u\,du = \frac{1}{n}\cos^{n-1}u\sin u + \frac{n-1}{n}\int\cos^{n-2}u\,du$

75. $\displaystyle\int \tan^n u\,du = \frac{1}{n-1}\tan^{n-1}u - \int\tan^{n-2}u\,du$

76. $\displaystyle\int \cot^n u\,du = \frac{-1}{n-1}\cot^{n-1}u - \int\cot^{n-2}u\,du$

77. $\displaystyle\int \sec^n u\,du = \frac{1}{n-1}\tan u\sec^{n-2}u + \frac{n-2}{n-1}\int\sec^{n-2}u\,du$

78. $\displaystyle\int \csc^n u\,du = \frac{-1}{n-1}\cot u\csc^{n-2}u + \frac{n-2}{n-1}\int\csc^{n-2}u\,du$

79. $\displaystyle\int \sin au\sin bu\,du = \frac{\sin(a-b)u}{2(a-b)} - \frac{\sin(a+b)u}{2(a+b)} + C$

80. $\displaystyle\int \cos au\cos bu\,du = \frac{\sin(a-b)u}{2(a-b)} + \frac{\sin(a+b)u}{2(a+b)} + C$

81. $\displaystyle\int \sin au\cos bu\,du = -\frac{\cos(a-b)u}{2(a-b)} - \frac{\cos(a+b)u}{2(a+b)} + C$

82. $\displaystyle\int u\sin u\,du = \sin u - u\cos u + C$

83. $\displaystyle\int u\cos u\,du = \cos u + u\sin u + C$

84. $\displaystyle\int u^n\sin u\,du = -u^n\cos u + n\int u^{n-1}\cos u\,du$

85. $\int u^n \cos u\, du = u^n \sin u - n \int u^{n-1} \sin u\, du$

86. $\int \sin^n u \cos^m u\, du = -\dfrac{\sin^{n-1} u \cos^{m+1} u}{n+m} + \dfrac{n-1}{n+m} \int \sin^{n-2} u \cos^m u\, du$

$= \dfrac{\sin^{n+1} u \cos^{m-1} u}{n+m} + \dfrac{m-1}{n+m} \int \sin^n u \cos^{m-2} u\, du$

87. $\int \dfrac{du}{1 - \sin au} = \dfrac{1}{a} \tan\left(\dfrac{\pi}{4} + \dfrac{au}{2}\right) + C$

88. $\int \dfrac{du}{1 + \sin au} = -\dfrac{1}{a} \tan\left(\dfrac{\pi}{4} - \dfrac{au}{2}\right) + C$

89. $\int \dfrac{u\, du}{1 - \sin au} = \dfrac{u}{a} \tan\left(\dfrac{\pi}{4} + \dfrac{au}{2}\right) + \dfrac{2}{a^2} \ln\left|\sin\left(\dfrac{\pi}{4} - \dfrac{au}{2}\right)\right| + C$

Inverse Trigonometric Forms

90. $\int \sin^{-1} u\, du = u \sin^{-1} u + \sqrt{1 - u^2} + C$

91. $\int \cos^{-1} u\, du = u \cos^{-1} u - \sqrt{1 - u^2} + C$

92. $\int \tan^{-1} u\, du = u \tan^{-1} u - \tfrac{1}{2} \ln(1 + u^2) + C$

93. $\int u \sin^{-1} u\, du = \dfrac{2u^2 - 1}{4} \sin^{-1} u + \dfrac{u\sqrt{1 - u^2}}{4} + C$

94. $\int u \cos^{-1} u\, du = \dfrac{2u^2 - 1}{4} \cos^{-1} u - \dfrac{u\sqrt{1 - u^2}}{4} + C$

95. $\int u \tan^{-1} u\, du = \dfrac{u^2 + 1}{2} \tan^{-1} u - \dfrac{u}{2} + C$

96. $\int u^n \sin^{-1} u\, du = \dfrac{1}{n+1}\left[u^{n+1} \sin^{-1} u - \int \dfrac{u^{n+1}\, du}{\sqrt{1 - u^2}}\right], \quad n \neq -1$

97. $\int u^n \cos^{-1} u\, du = \dfrac{1}{n+1}\left[u^{n+1} \cos^{-1} u + \int \dfrac{u^{n+1}\, du}{\sqrt{1 - u^2}}\right], \quad n \neq -1$

98. $\int u^n \tan^{-1} u\, du = \dfrac{1}{n+1}\left[u^{n+1} \tan^{-1} u - \int \dfrac{u^{n+1}\, du}{1 + u^2}\right], \quad n \neq -1$

Exponential and Logarithmic Forms

99. $\int u e^{au}\, du = \dfrac{1}{a^2}(au - 1)e^{au} + C$

100. $\int u^n e^{au}\, du = \dfrac{1}{a} u^n e^{au} - \dfrac{n}{a} \int u^{n-1} e^{au}\, du$

101. $\int e^{au} \sin bu\, du = \dfrac{e^{au}}{a^2 + b^2}(a \sin bu - b \cos bu) + C$

102. $\int e^{au} \cos bu\, du = \dfrac{e^{au}}{a^2 + b^2}(a \cos bu + b \sin bu) + C$

103. $\int \ln u\, du = u \ln u - u + C$

104. $\int \dfrac{1}{u \ln u}\, du = \ln|\ln u| + C$

105. $\int u^n \ln u\, du = \dfrac{u^{n+1}}{(n+1)^2}[(n + 1)\ln u - 1] + C$

106. $\int u^m \ln^n u\, du = \dfrac{u^{m+1} \ln^n u}{m+1} - \dfrac{n}{m+1} \int u^m \ln^{n-1} u\, du, \quad m \neq -1$

107. $\int \ln(u^2 + a^2)\, du = u \ln(u^2 + a^2) - 2u + 2a \tan^{-1} \dfrac{u}{a} + C$

108. $\int \ln|u^2 - a^2|\, du = u \ln|u^2 - a^2| - 2u + a \ln\left|\dfrac{u + a}{u - a}\right| + C$

109. $\displaystyle\int \frac{du}{a + be^u} = \frac{u}{a} - \frac{1}{a}\ln|a + be^u| + C$

Hyperbolic Forms

110. $\displaystyle\int \sinh u\, du = \cosh u + C$

111. $\displaystyle\int \cosh u\, du = \sinh u + C$

112. $\displaystyle\int \tanh u\, du = \ln \cosh u + C$

113. $\displaystyle\int \coth u\, du = \ln|\sinh u| + C$

114. $\displaystyle\int \operatorname{sech} u\, du = \tan^{-1}(\sinh u) + C$

115. $\displaystyle\int \operatorname{csch} u\, du = \ln|\tanh \tfrac{1}{2}u| + C$

116. $\displaystyle\int \operatorname{sech}^2 u\, du = \tanh u + C$

117. $\displaystyle\int \operatorname{csch}^2 u\, du = -\coth u + C$

118. $\displaystyle\int \operatorname{sech} u \tanh u\, du = -\operatorname{sech} u + C$

119. $\displaystyle\int \operatorname{csch} u \coth u\, du = -\operatorname{csch} u + C$

Forms Involving $\sqrt{2au - u^2}$

120. $\displaystyle\int \sqrt{2au - u^2}\, du = \frac{u - a}{2}\sqrt{2au - u^2} + \frac{a^2}{2}\cos^{-1}\left(\frac{a - u}{a}\right) + C$

121. $\displaystyle\int u\sqrt{2au - u^2}\, du = \frac{2u^2 - au - 3a^2}{6}\sqrt{2au - u^2} + \frac{a^3}{2}\cos^{-1}\left(\frac{a - u}{a}\right) + C$

122. $\displaystyle\int \frac{\sqrt{2au - u^2}}{u}\, du = \sqrt{2au - u^2} + a\cos^{-1}\left(\frac{a - u}{a}\right) + C$

123. $\displaystyle\int \frac{\sqrt{2au - u^2}}{u^2}\, du = -\frac{2\sqrt{2au - u^2}}{u} - \cos^{-1}\left(\frac{a - u}{a}\right) + C$

124. $\displaystyle\int \frac{du}{\sqrt{2au - u^2}} = \cos^{-1}\left(\frac{a - u}{a}\right) + C$

125. $\displaystyle\int \frac{u\, du}{\sqrt{2au - u^2}} = -\sqrt{2au - u^2} + a\cos^{-1}\left(\frac{a - u}{a}\right) + C$

126. $\displaystyle\int \frac{u^2\, du}{\sqrt{2au - u^2}} = -\frac{(u + 3a)}{2}\sqrt{2au - u^2} + \frac{3a^2}{2}\cos^{-1}\left(\frac{a - u}{a}\right) + C$

127. $\displaystyle\int \frac{du}{u\sqrt{2ua - u^2}} = -\frac{\sqrt{2au - u^2}}{au} + C$

Some Definite Integrals

128. $\displaystyle\int_0^{\pi/2} \sin^{2n}x\, dx = \int_0^{\pi/2} \cos^{2n}x\, dx = \frac{\pi}{2}\,\frac{1 \cdot 3 \cdot 5 \cdots (2n - 1)}{2 \cdot 4 \cdot 6 \cdots 2n}, \quad n = 1, 2, 3, \ldots$

129. $\displaystyle\int_0^{\pi/2} \sin^{2n+1}x\, dx = \int_0^{\pi/2} \cos^{2n+1}x\, dx = \frac{2 \cdot 4 \cdot 6 \cdots 2n}{1 \cdot 3 \cdot 5 \cdots (2n + 1)}, \quad n = 1, 2, 3, \ldots$

APPENDIX V FORMULAS FROM GEOMETRY

Area A; circumference C; volume V; surface area S

RECTANGLE

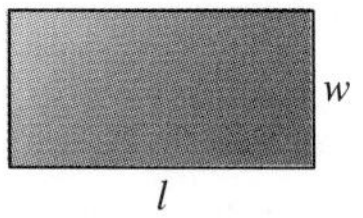

$A = lw \qquad C = 2l + 2w$

PARALLELOGRAM

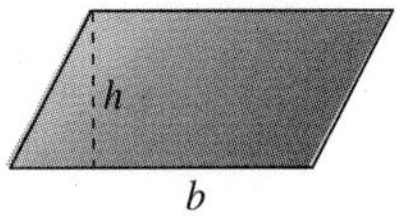

$A = bh$

TRAPEZOID

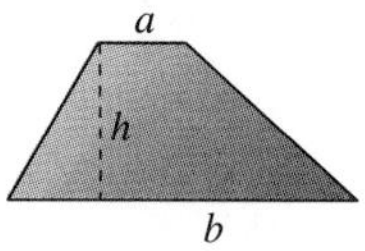

$A = \frac{1}{2}(a + b)h$

RIGHT TRIANGLE

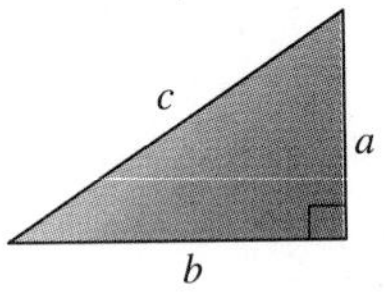

Pythagorean Theorem:
$c^2 = a^2 + b^2$

TRIANGLE

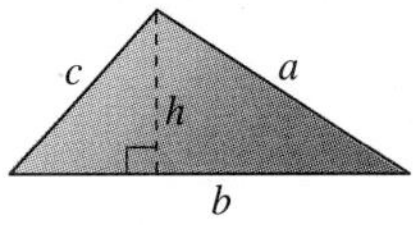

$A = \frac{1}{2}bh \qquad C = a + b + c$

EQUILATERAL TRIANGLE

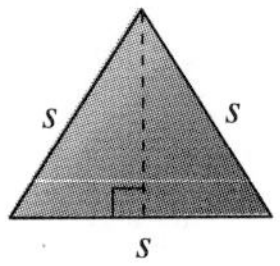

$h = \frac{\sqrt{3}}{2}s \qquad A = \frac{\sqrt{3}}{4}s^2$

CIRCLE

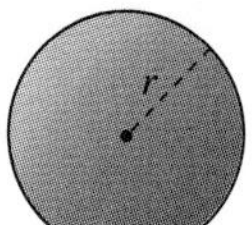

$A = \pi r^2 \qquad C = 2\pi r$

CIRCULAR RING

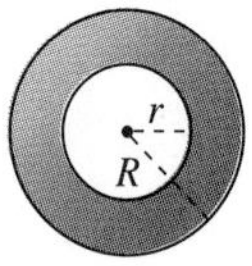

$A = \pi(R^2 - r^2)$

CIRCULAR SECTOR

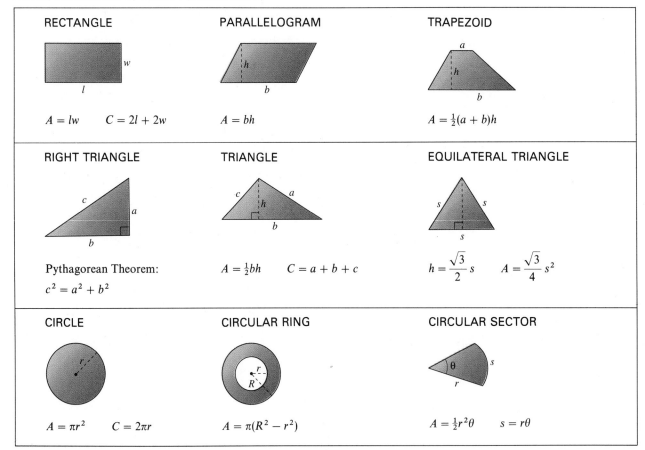

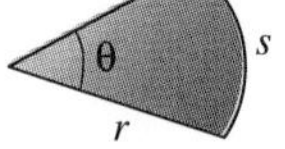

$A = \frac{1}{2}r^2\theta \qquad s = r\theta$

ELLIPSE

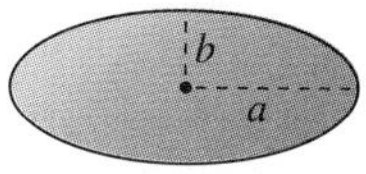

$A = \pi ab$

ELLIPSOID

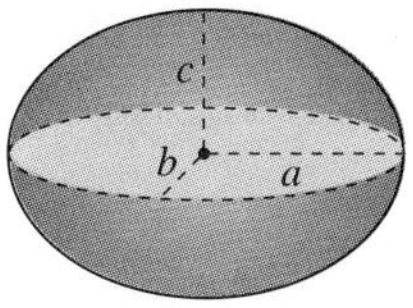

$V = \frac{4}{3}\pi abc$

SPHERE

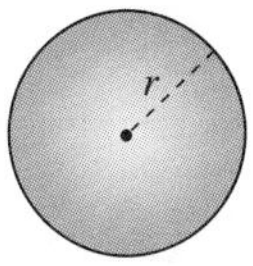

$V = \frac{4}{3}\pi r^3 \qquad S = 4\pi r^2$

RIGHT CYLINDER

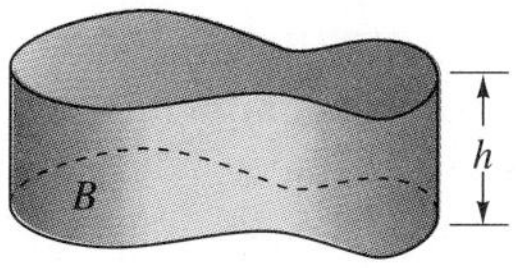

$V = Bh$, B area of base

RIGHT CIRCULAR CYLINDER

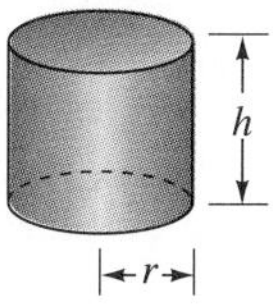

$V = \pi r^2 h \qquad S = 2\pi rh$

RECTANGULAR PARALLELEPIPED

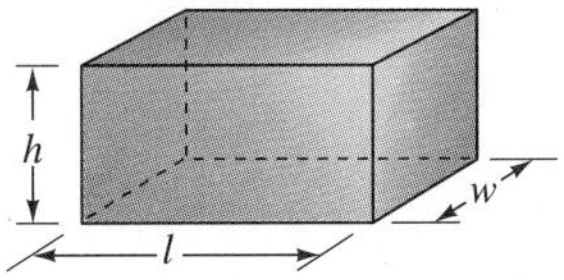

$V = lwh \qquad S = 2(hl + lw + hw)$

CONE

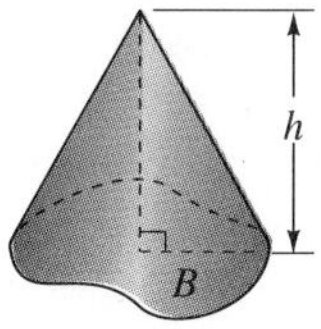

$V = \frac{1}{3}Bh$, B area of base

RIGHT CIRCULAR CONE

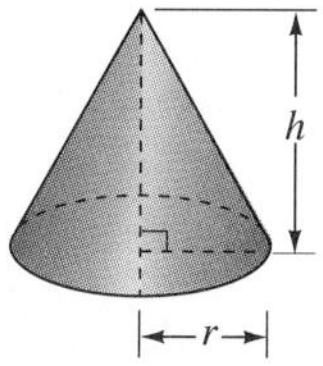

$V = \frac{1}{3}\pi r^2 h \qquad S = \pi r\sqrt{r^2 + h^2}$

FRUSTUM OF A CONE

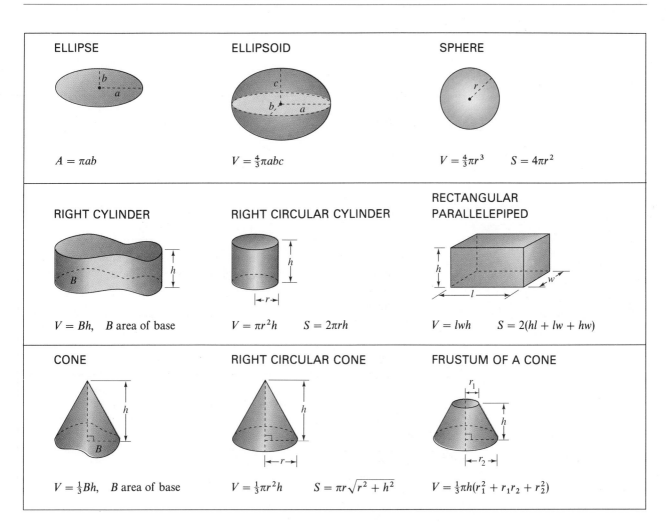

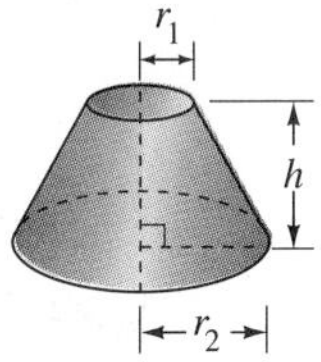

$V = \frac{1}{3}\pi h(r_1^2 + r_1 r_2 + r_2^2)$

APPENDIX VI FRACTALS

Introduction As you have seen in the preceding chapters of this text, the techniques of calculus effectively model a wide variety of phenomena. This is especially so when the relationship between the dependent variables and the independent variables of the phenomenon of interest can be modeled by a continuous and differentiable function. Within the domain of possibilities for the independent variables, the algorithms of calculus can evaluate specific local behaviors of the function as well as certain global characteristics.

Some phenomena follow a sort of evolutionary process in which the present conditions supply the raw material for the conditions of the next stage, and those conditions in turn become the raw material for the subsequent stage. For example, in economics, financial conditions today supply the basis for tomorrow's developments. Similarly, weather forecasters formulate daily and weekly weather predictions based on present conditions. Modeling such systems requires a formation rule or function that defines the features of each stage based on the conditions of the preceding stage. In order to observe the pattern of outcomes at each stage, we simply pass the outcomes generated by the formation rule at the preceding stage back into the rule. Such a simple feedback model not only monitors the evolution of the process but also generates the next domain value for the subsequent application of the formation rule.

The mathematical mechanism for defining such a feedback process involves the composition $f(g(x))$ of two functions f and g. The notation $f(g(x))$ implies that the function g first produces a range value $g(x)$ from a domain value x. Then the function f operates on the outcome of the function g to yield the result of the composition. Of course, more than two functions could be chained together in this way, and some might even be applied more than once in the composition. For example, $f(g(f(x)))$ suggests that first a domain value x is transformed by f, then g acts on the outcome, and finally f again operates on the result of function g's action. Now, the feedback model for the sort of evolutionary phenomena described here simply involves the repeated composition $f(f(f \cdots f(x)))$ of a function f with itself.

Constructing mathematical models for these feedback processes can lead to very striking visual representations that possess a number of features currently under investigation in the developing field of **fractal geometry**. Frequently generated through the type of iteration techniques characteristic of function composition, these features include self-similarity and a measure of complexity known as fractal dimension.

Iteration We use the term *iterative* to describe a repetitive process. Repeatedly composing a function with itself as discussed above is one example of iteration. More generally, iterative processes in mathematics include repetitive geometric processes as well as iterative algebraic and numerical patterns.

EXAMPLE 1 Let $f(x) = x/2$ and let a_n denote the composition of function f with itself n times at $x = 2$. Write a formula for a_n and compute $\lim_{n\to\infty} a_n$.

Solution From the pattern of the first three terms a_1, a_2, and a_3:

$$a_1 = f(2) = \frac{2}{2} = 1$$

$$a_2 = f(f(2)) = f(a_1) = f(1) = \frac{1}{2}$$

$$a_3 = f(f(f(2))) = f(a_2) = f\left(\frac{1}{2}\right) = \frac{1}{4}$$

we find that the general form for a_n must be $a_n = (\frac{1}{2})^{n-1}$ Accordingly, $\lim_{n\to\infty}(\frac{1}{2})^{n-1} = 0$. □

In the early chapters of this text we paid considerable attention to the behavior of functions and their graphs. However, in Example 1 our interest focused more on the sequence of outcomes than on the properties of function f. In fact, the object of attention in fractal geometry is characteristically the state of affairs after an iterative process is repeated many times rather than just the properties of the construction rule itself. The following example clarifies this fact and also exhibits an iterative geometric process.

EXAMPLE 2 In Problem 72 in Exercises 10.1 a construction was given which involved repeatedly replacing each segment ________________ in a polygon with four equal segments

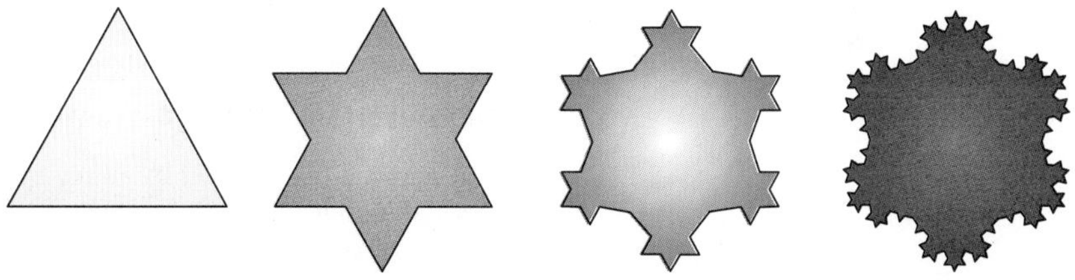

Stage 0 Stage 1 Stage 2 Stage 3

FIGURE VI.1

Beginning with the three segments that form an equilateral triangle and repeating the construction process ad infinitum result in the snowflake-like curve discovered in 1904 by Swedish mathematician Helge Von Koch. Figure VI.1 illustrates the first four stages of this process. A number of properties make this curve interesting:

(*i*) It is produced by an infinite process involving geometric iteration.

(*ii*) The perimeter is infinite (see Problem 72 in Exercises 10.1), but the enclosed area is finite (see Problem 54 in Exercises 10.3).

(*iii*) The curve is so irregular that it fails to have a tangent at every single point. In this sense an equation written to define the curve would be nowhere differentiable.

(*iv*) The curve is self-similar, a property discussed below. □

It seems obvious that the shape of the resulting figure depends in part on the figure initially selected for beginning the construction process. Accordingly, it should not come as a surprise that the same process can exhibit very different outcomes when that process is started from distinct initial domain values.

Attractors

EXAMPLE 3 After developing Newton's Method for recursively approximating a zero of a function f in Section 2.11, we concluded with a remark that under certain conditions the method could fail to converge. Specifically, instead of converging to a zero of f which acts as a sort of **attractor** for the sequence of outcomes as shown in Figure VI.2, the recursively obtained estimates might either oscillate through a finite set of points as exhibited in Figure VI.3 or actually approach infinity. As suggested in the preceding paragraph, the particular outcome depends completely on the selected initial point. □

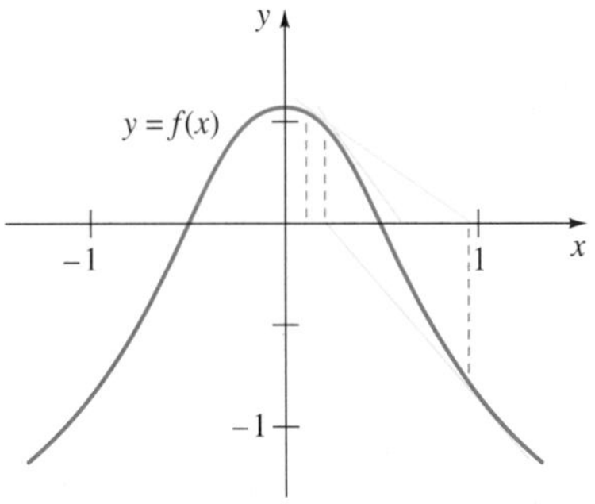

FIGURE VI.2

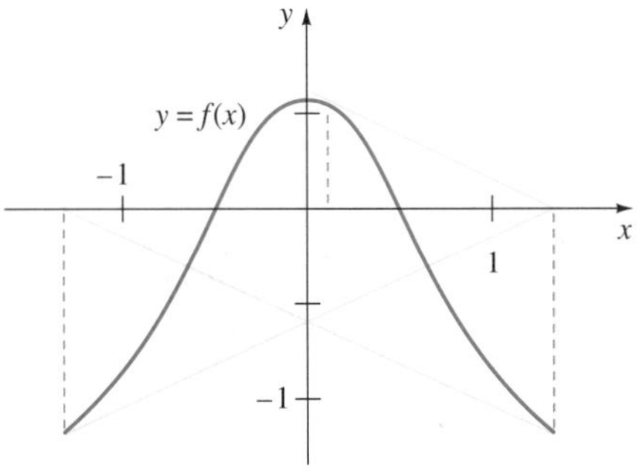

FIGURE VI.3

DEFINITION VI.1 Attractor

If a sequence is bounded by either approaching a finite limit or oscillating through a finite set of finite values, then the value, or the set of values, is called the **attractor** of the sequence.

If the sequence diverges to infinity, then infinity is called the **attractor** of the sequence.

In the case of geometric iteration such as that represented by the Koch curve construction, the attractor is the shape finally produced after infinitely many steps of the iteration process. Altering the shape of the replacement pattern (Problem 1) or the shape of the initial figure would ultimately result in a different attractor shape than that generated by the Koch conditions. Similarly, changing either the function $f(x)$ to which Newton's Method is applied or the initial point from which the process begins will result in a different set of attractors depending on how many zeros exist and the properties of the curve defined by the selected function $f(x)$.

Self-Similarity and Fractal Dimension An experiment commonly called the **Chaos Game** involves marking dots inside of an equilateral triangle according to

TI-81
GRAPHING
CALCULATOR

```
:ClrDraw
:0→Xmin
:1→Xmax
:0→Ymin
:1→Ymax
:X=.15
:Y=.1
:0→C
:Lbl 1
:C+1→C
:Pt-On(X,Y)
:If C>800
:End
:Rand→N
:If N<.3333
:Goto 2
:If N>.6666
:Goto 3
:.5(X+.15)→X
:.5(Y+.1)→Y
:Goto 1
:Lbl 2
:.5(X+.85)→X
:.5(Y+.1)→Y
:Goto 1
:Lbl 3
:.5(X+.5)→X
:.5(Y+1)→Y
:Goto 1
```

the outcomes on a sequence of random rolls of a die. Specifically, begin from an arbitrarily selected vertex and repeatedly use the die to select one of the three vertices at random. With each roll of the die, move to the midpoint between the last point and the next randomly selected vertex. At the midpoint place a dot before continuing to the next midpoint indicated by the die. The accompanying program for the Texas Instruments TI-81 Graphing Calculator uses the built-in random number generator to simulate the plotting of 800 points according to the rules of the Chaos Game.

While one might conjecture that the dots would randomly distribute themselves throughout the entire interior of the triangle, in fact a highly structured pattern of triangular regions appears. First defined in 1916 by the great Polish mathematician Waclaw Sierpinski (1882–1969), the gasket that bears his name exhibits a feature which we now refer to as **self-similarity** and is characteristic of many fractals. See Figure VI.4.

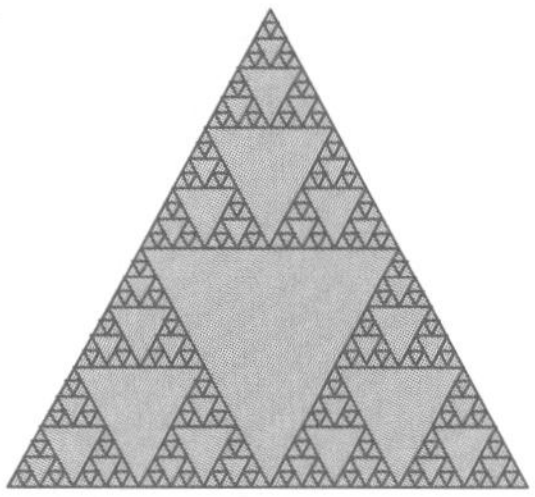

FIGURE VI.4

DEFINITION VI.2 Self-Similarity

A figure is called **self-similar** when each section of the figure contains a part that is an exact replica of the whole.

Any picture of the Sierpinski Gasket is an imperfect representation of the actual geometric object generated by infinitely many steps of the Chaos Game. However, even these visual approximations suggest that magnifying small sections of the completed gasket would reveal details with the same structure and level of complexity as that displayed by the whole figure. In fact, when a figure strictly possesses this property of self-similarity, we can use this feature to obtain a numerical measure of the level of complexity present.

According to our common notions of dimension, we understand a line to have dimension 1, a plane to have dimension 2, and a solid to have dimension 3. In fact, we can compute these values by reference to a simple growth process. Imagine a "cloning" operation that allows squares to reproduce themselves some given number of times r in one or more directions. For example, Figure VI.5 exhibits the result of allowing the square A to grow $r = 9$ times in each of two directions for a total array of $N = 81$ squares. Indeed, computing this total number N of squares follows from simple multiplication in which $81 = 9 \times 9$, or equivalently from a simple power rule in which $81 = 9^2$. In general, the formula $N = r^D$ supplies a rule for computing the dimension D of the space covered by this cloning operation. By solving this formula for D, we obtain $D = \ln N/\ln r$.

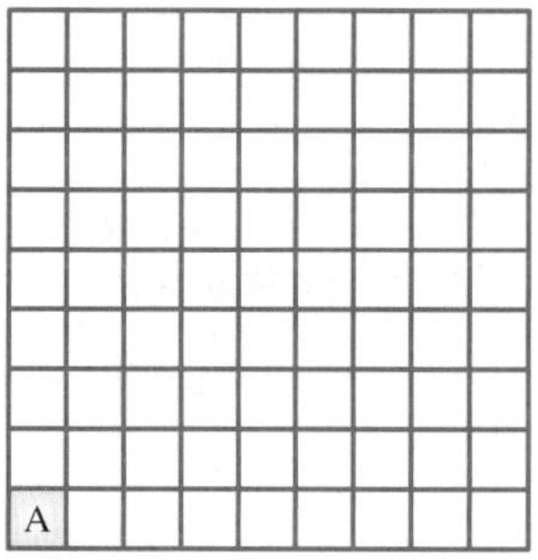

FIGURE VI.5

The similarity dimension extends our notion of dimension in order to provide a means for measuring the complexity of a fractal shape when the whole figure is a replica of its similar parts. As we have seen, we can generate objects like the Koch curve by a recursive construction process of scaling and substitution, in which the scale factor r and the number N of pieces substituted both remain constant. The replacement process is analogous to the manner in which the square clones itself as described above. The self-similarity dimension is the quotient obtained by dividing the logarithm of the number N of pieces in a replacement step by the logarithm of the scaling factor r. The resulting constant supplies a measure of complexity of the limit object obtained after the construction process is repeated ad infinitum.

THEOREM VI.1 Similarity Dimension

The **similarity dimension** of a self-similar fractal shape is given by

$$D = \frac{\ln N}{\ln r}$$

The similarity dimension yields one measure of a figure's complexity. In the case of the Koch curve, although the construction process must be continued ad infinitum, any particular segment at a given stage of the construction is $r = 3$ times longer than the $N = 4$ segments that will be drawn in its place at the next stage. The scaling ratio $r = 3$ represents a kind of magnification power. More important, the self-similarity fractal dimension is $D = \ln 4/\ln 3 \approx 1.26$.

Mandelbrot Set During 1979, while working at IBM's Thomas J. Watson Research Center, the French mathematician and geometer Benoit Mandelbrot extended a problem of iteration in the complex plane which was studied by another French mathematician Gaston Julia (1893–1978) during the early part of the century. In order to keep track of the separate results obtained by iterating different points, Mandelbrot used a computer to plot a black dot at the location of a point in the complex plane if its sequence of iterates remained bounded. If the point led to an unbounded sequence of points, then Mandelbrot did not plot a dot at the point's initial location. The resulting two-dimensional graph simultaneously exhibited such extraordinary symmetries and complexity that both the graph and the processes that generated it have since become objects of more intensive investigation.

In order to comprehend something of Mandelbrot's fascinating results, we must first understand how to operate with complex numbers. Each number in the set of complex numbers $\{z = x + yi \mid x, y \text{ real numbers and } i^2 = -1\}$ corresponds to a point (x, y) in a plane called the complex plane. We refer to x as the real part of the complex number and y as the imaginary part. You should review the definitions of addition, subtraction, multiplication, and division of complex numbers as well as the examples supplied in Appendix I. Keep in mind that when two complex numbers are added or multiplied subject to the definitions, the result always has a geometric representation as a point on the complex plane.

Representing complex numbers as points on a plane also permits a measurement of the distance $|z_1 - z_2|$ between two such numbers. More precisely, if $z_1 = a + bi$ and $z_2 = c + di$, then

$$\begin{aligned} |z_1 - z_2| &= |(a + bi) - (c + di)| \\ &= |(a - c) + (b - d)i| \\ &= \sqrt{(a - c)^2 + (b - d)^2} \end{aligned}$$

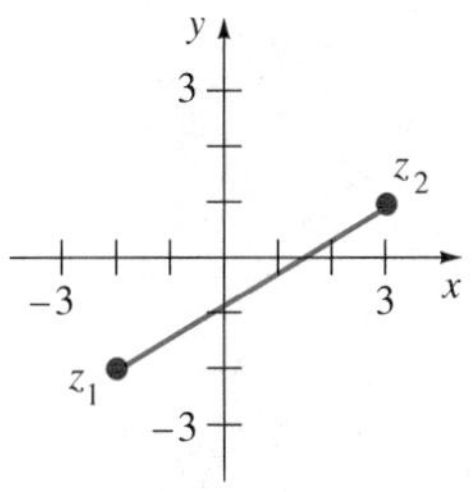

FIGURE VI.6

EXAMPLE 4 Compute the distance between the two points associated with $z_1 = -2 - 2i$ and $z_2 = 3 + i$.

Solution The distance between z_1 and z_2:

$$|z_1 - z_2| = |(-2 - 2i) - (3 + i)| = |(-2 - 3) + (-2 - 1)i|$$
$$= \sqrt{(-5)^2 + (-3)^2} = \sqrt{34}$$

is indicated in color in Figure VI.6. □

If a complex number $c = a + bi$ is iterated through a function of the form $f(z) = z^2 + c$, then the resulting sequence of outcomes plotted on the complex plane will either stay within a finite region of the complex plane or diverge to infinity. Beginning with any selected complex number $c = a + bi$, we need to determine the behavior of the following sequence of complex iterates:

$$c_0 = c, \quad c_1 = (c_0)^2 + c, \quad c_2 = (c_1)^2 + c, \quad c_3 = (c_2)^2 + c, \ldots$$

EXAMPLE 5

(*a*) Iterating the complex number $c = 0 - i$ through the function $f(z) = z^2 + c$ results in a *bounded* sequence:

$$\begin{aligned}
c_0 &= c &&= && 0 - i \\
c_1 &= (c_0)^2 + c &&= (0 - i)^2 + (0 - i) &&= -1 - i \\
c_2 &= (c_1)^2 + c &&= (-1 - i)^2 + (0 - i) &&= 0 + i \\
c_3 &= (c_2)^2 + c &&= (0 + i)^2 + (0 - i) &&= -1 - i \\
c_4 &= (c_3)^2 + c &&= (-1 - i)^2 + (0 - i) &&= 0 + i \\
&\vdots && && \vdots
\end{aligned}$$

(*b*) Iterating the complex number $c = 0.5 + 0i$ through the function $f(z) = z^2 + c$ results in an *unbounded* sequence:

$$\begin{aligned}
c_0 &= c &&= && 0.5 + 0i \\
c_1 &= (c_0)^2 + c &&= (0.5 + 0i)^2 + (0.5 + 0i) &&= 0.75 + 0i \\
c_2 &= (c_1)^2 + c &&= (0.75 + 0i)^2 + (0.5 + 0i) &&= 1.0625 + 0i \\
c_3 &= (c_2)^2 + c &&= (1.0625 + 0i)^2 + (0.5 + 0i) &&= 1.6289 + 0i \\
c_4 &= (c_3)^2 + c &&= (1.6289 + 0i)^2 + (0.5 + 0i) &&= 3.1533 + 0i \\
&\vdots && && \vdots
\end{aligned}$$

□

The **Mandelbrot Set** shown in Figure VI.7 consists of the collection of all points in the complex plane for which the sequence of iterates is bounded and therefore does not diverge to infinity. When plotted on the complex plane, such points are usually colored black. Points outside of the Mandelbrot Set all have sequences of iterates that diverge to infinity and these points are assigned colors according to their distance from the edge of the bounded set.

The exceedingly irregular boundary of the Mandelbrot Set forms a barrier between these two types of behavior, points that upon iteration remain within the set

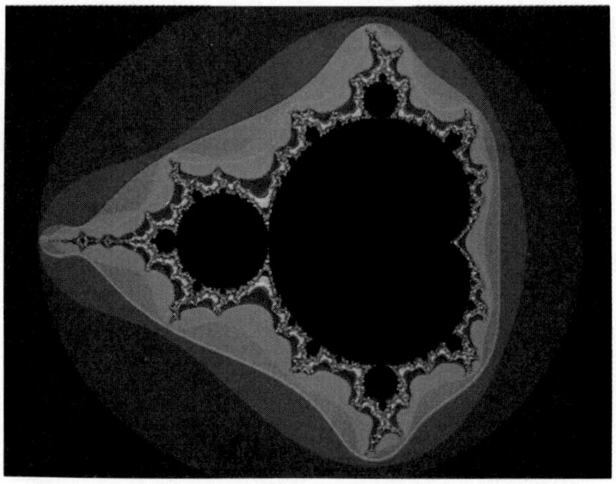

FIGURE VI.7 The Mandelbrot Set

versus those that diverge in their iteration sequences. Although the fractal dimension D of this very complex boundary is not currently known, mathematicians working in this field conjecture that the probable value is $D = 2$. If correct, this value would imply a measure of complexity significantly greater than that of the Koch curve, which as we have shown has fractal dimension equal to $\ln 4/\ln 3 \approx 1.26$.

It is, in fact, the extreme irregularity of the Mandelbrot boundary which complicates assigning colors to exterior points based on their distances from the boundary. After all, computing the minimum distance for a given exterior point would necessitate identifying the nearest boundary point, a task made nearly impossible by the boundary's complexity. However, an alternative approach to measuring this distance involves computing for each exterior point the rate at which its associated sequence of iterates diverges. Specifically, define a circle centered at the origin having some large fixed radius. The number of iterations required for a divergent sequence to pass out of the circle then provides a measure for the divergence rate. A small count suggests rapid divergence and implies that the point was already out toward the large circle, far from the boundary of the Mandelbrot Set. A high count indicates rather slow progress toward infinity and close proximity to the set. Accordingly, the indirect distance measure based on the divergence rate described above conveniently permits color assignments.

The color plates in Figures VI.8–VI.11 indicate the results of zooming in to selected portions of the boundary area. Incredible and unexpected patterns emerge, including small replicas of the larger Mandelbrot Set. Indeed, this intricate figure generated by iterative processes displays the classical features of fractal shapes including high complexity and a significant degree of self-similarity.

Julia Sets The Mandelbrot Set not only displays fascinating properties of its own but, as we shall see, also serves to organize and catalog another infinite collection of sets called **Julia Sets**, named for French mathematician Gaston Julia. Like the Mandelbrot Set, we generate each of these sets by iterating complex numbers z through a quadratic function having the form $f(z) = z^2 + c$. However, a comparison of the iteration rule used to form the Mandelbrot Set with that used in making Julia Sets will highlight differences in the construction processes.

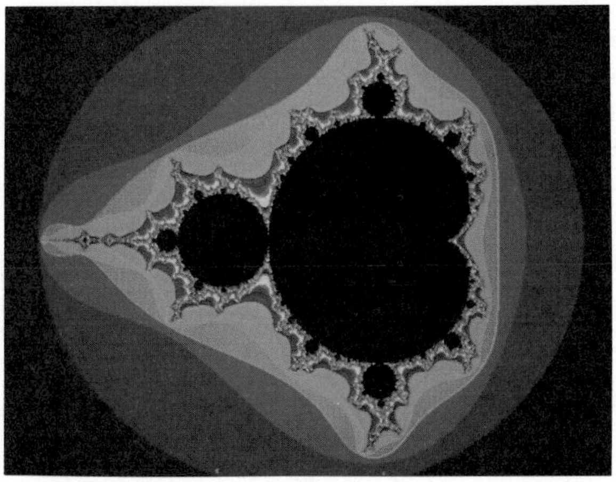

FIGURE VI.8 The Mandelbrot Set

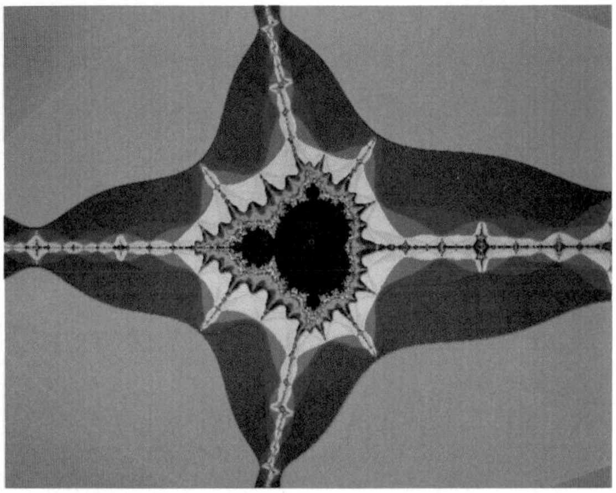

FIGURE VI.9

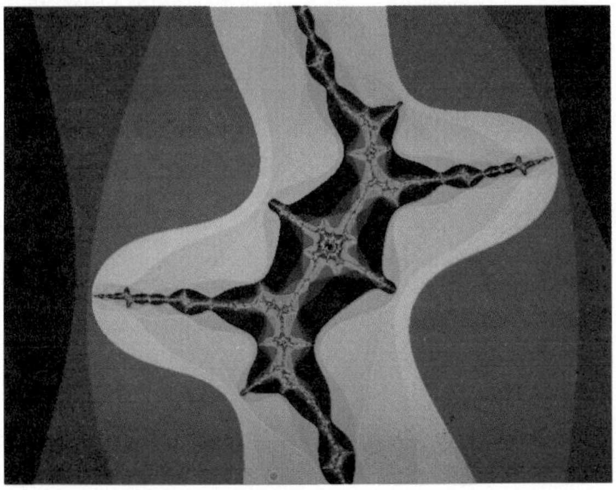

FIGURE VI.10

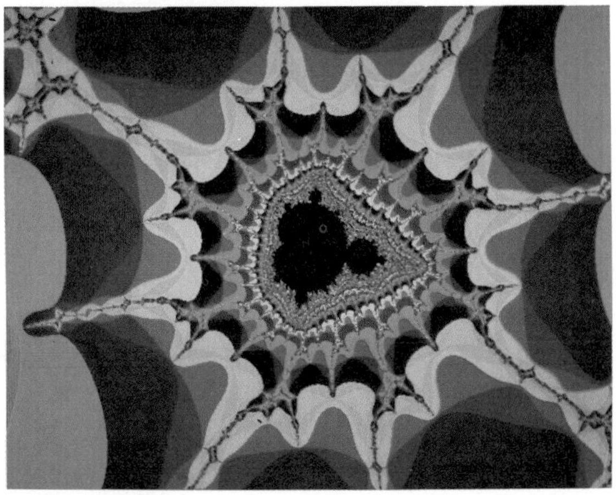

FIGURE VI.11

Figures VI.8–VI.11 display successive magnifications of the Mandelbrot Set. The enlarged section is visible to the left of the center in Figure VI.8.

Mandelbrot Set	Julia Set
Each complex number $c = a + bi$ is iterated through a function of the form $f(z) = z^2 + c$. Notice that the constant in the function f changes each time a distinct number c is selected for iteration.	Each complex number $z = x + yi$ is iterated through a function of the form $f(z) = z^2 + c$, where $c = a + bi$ is the same constant for all values of z. Notice that once a value for c is selected, the function f remains the same for all complex numbers iterated through it.
The Mandelbrot Set consists of those points that are associated with complex numbers for which the sequence of iterates remains bounded.	For a given value of c, the Julia Set consists of those points that lie on the boundary between points associated with bounded sequences of iterates and points associated with unbounded sequences.

EXAMPLE 6 Identify the Julia Set generated by iterating complex numbers $z = x + yi$ through the quadratic function $f(z) = z^2 + c$ with $c = 0 + 0i$.

Solution The Julia Set consists of those points (x, y) that lie on the unit circle $x^2 + y^2 = 1$.

Points on the unit circle are associated with complex numbers $z = x + yi$ whose iterates under $f(z)$ are associated with points also lying on the unit circle (see Problem 6(**a**)).

Points within the unit circle are associated with complex numbers $z = x + yi$ whose iterates under $f(z)$ are associated with points lying inside of the unit circle at ever-decreasing distances from the origin (0, 0). In this case the sequences of iterates are bounded (see Problem 6(**b**)).

Points outside of the unit circle are associated with complex numbers $z = x + yi$ whose iterates under $f(z)$ are associated with points lying outside of the unit circle at ever-increasing distances from the origin (0, 0). In this case the sequences of iterates are unbounded (see Problem 6(**c**)). □

We can now add some precision to our earlier comment that the Mandlebrot Set provides a kind of organizing catalog of the Julia Sets. Each complex number $c = a + bi$ uniquely identifies:

(*i*) Some point (a, b) on the complex plane which lies either within the Mandelbrot Set or outside of it.

(*ii*) A Julia Set formed by iterating all complex numbers z through $f(z) = z^2 + c$.

Remarkably, each Julia Set associated with a point within the Mandelbrot Set possesses the geometric property of connectedness, a feature which implies that you can trace a line from any point in the set to any other point in the set without lifting your pencil from the paper. In contrast, each Julia Set associated with a point outside of the Mandelbrot Set consists of points which are like a dust of disconnected dots. Accordingly, the Mandelbrot Set catalogs the set of points which represent connected Julia Sets as displayed in VI.12–VI.15 and VI.17. In contrast and as shown in VI.18 and VI.19, points outside of the Mandelbrot Set represent disconnected Julia Sets.

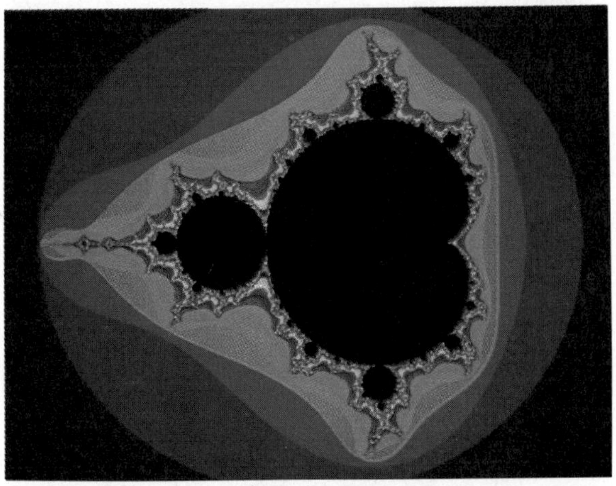

FIGURE VI.12 The Mandelbrot Set

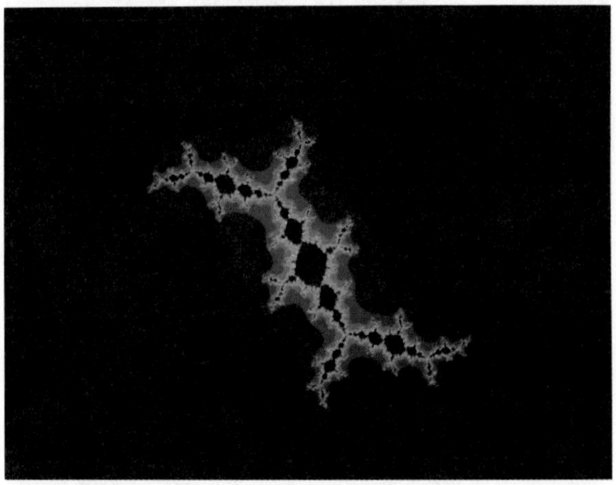

FIGURE VI.13

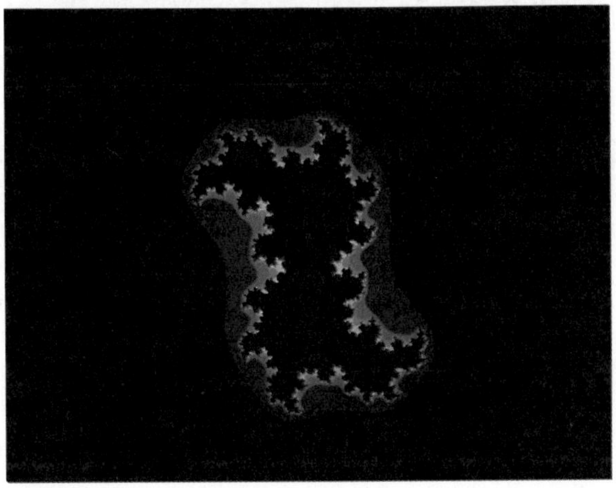

FIGURE VI.14

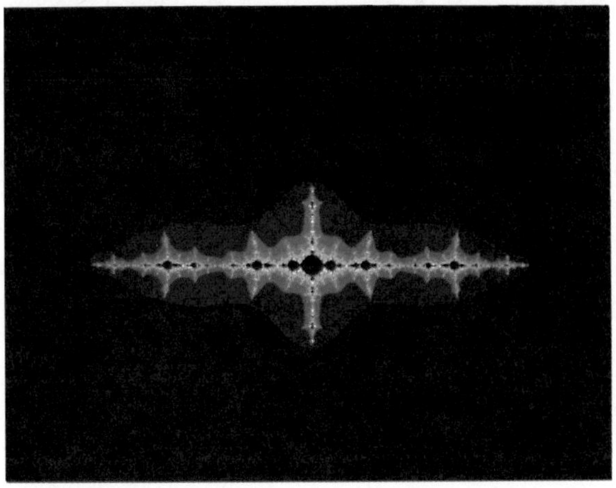

FIGURE VI.15

Figures VI.13–VI.15 display the Julia Sets associated with points within the interior of the Mandelbrot Set as shown in Figure VI.12.

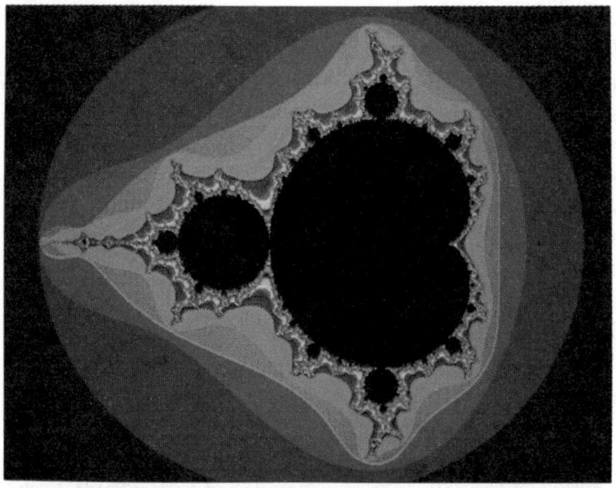

FIGURE VI.16 The Mandelbrot Set

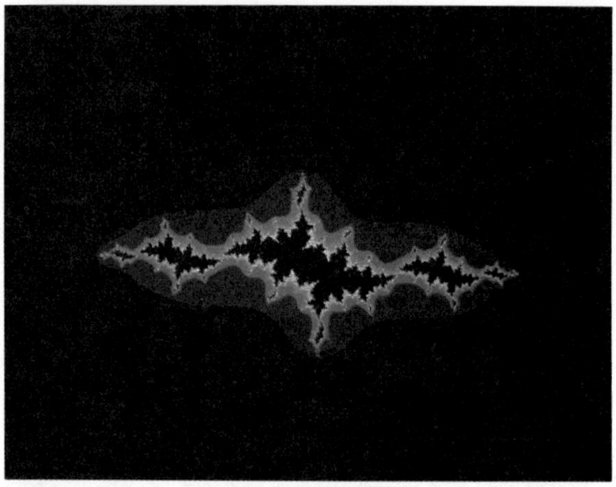

FIGURE VI.17

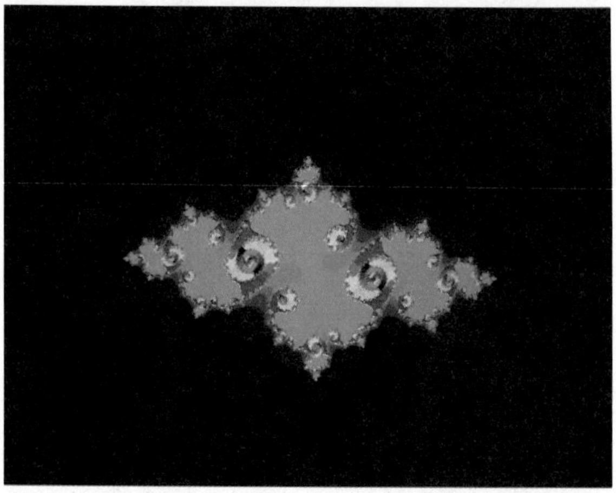

FIGURE VI.18

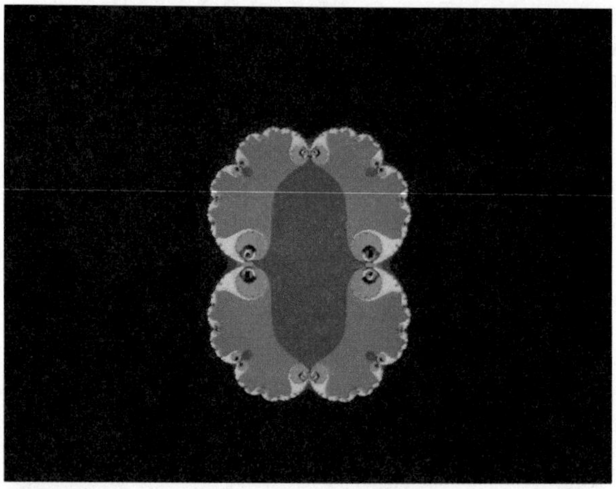

FIGURE VI.19

Figures VI.18–VI.19 display the Julia Sets associated with points exterior to the Mandelbrot Set as shown in Figure VI.16. These stand in marked contrast to the Julia Set of Figure VI.17 which is associated with a point within the Mandelbrot interior.

Concluding Comments The foregoing overview of fractals including the Mandelbrot Set and Julia Sets barely introduces the range of topics currently under study in this fascinating and integrative branch of modern mathematics. As suggested in the opening comments of this appendix, one impetus for the development of fractals is the promise they hold for modeling the behavior of those physical systems that essentially follow iterative patterns. Often such systems exhibit a kind of unpredictability which until recently was not well understood. Thus, while we have discussed certain properties of fractals including iteration, self-similarity, and fractal dimension, in this introduction we have not demonstrated the way in which fractal geometry addresses sources of unpredictability. However, we have sought to demonstrate in a limited way that, while the algorithms of the calculus offer one set of models we can use to analyze a portion of our world, fractal geometry offers new and powerful methods for comprehending the nature of the phenomena that surround us.

EXERCISES

1. Shown below each of the following grids is a line segment and a replacement shape formed from a number of line segments. Perform the replacement process three times. Determine the scaling factor and the number of replacement segments substituted at each stage for each segment in the prior stage. Compute the self-similarity dimension of the figure eventually produced by the replacement process.

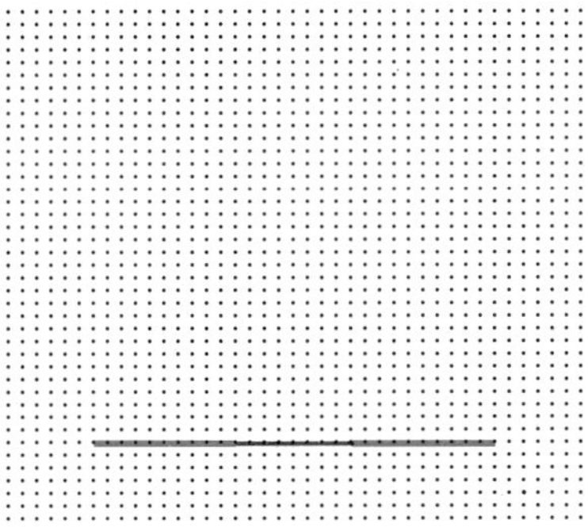

Replace every straight line segment

with the pattern:

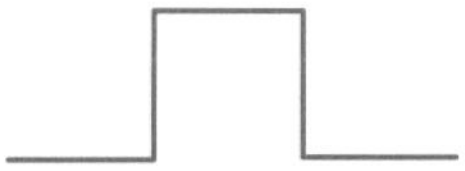

Replace every straight line segment

with the pattern:

2. Use the following process to construct the Sierpinski Gasket:

(*i*) Connect the midpoints on the sides of the level 0 figure as shown to form the level 1 figure in which the central shaded triangle is deleted.

(*ii*) In each of the remaining triangles join the midpoints in order to form the level 2 figure shown. Again, delete each of the central shaded triangles.

(*iii*) Continue the process initiated in steps (*i*) and (*ii*) ad infinitum.

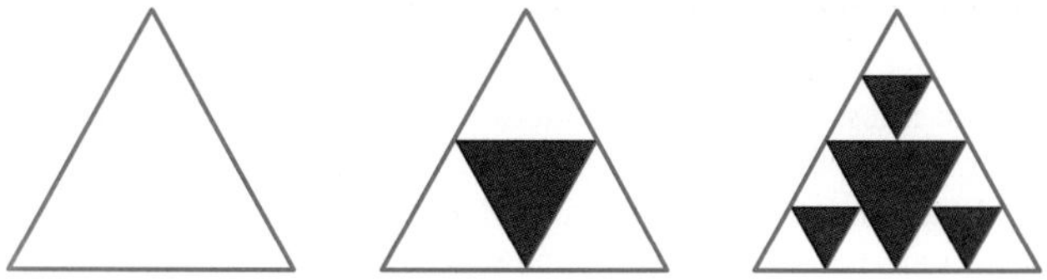

Let the area of the level 0 triangle be 1 square unit. Write an expression for the total area A_n of the undeleted sections remaining at level n. Compute the limit of A_n as n approaches infinity.

3. Apply Newton's Method to the function

$$f(x) = \frac{1}{(1 + x^2)^2} - \frac{25}{36}$$

starting at each of the initial points given. Which of the sequences of outcomes has an attractor at a zero of $f(x)$? What is the attractor for the remaining sequences?

(a) $x_0 = 0.05$ (b) $x_0 = 0.07$ (c) $x_0 = 0.09$

4. Determine whether or not each given sequence is bounded. Identify the attractor.

(a) $a_n = 3\left(\frac{1}{5}\right)^n$ (b) $a_n = \sin\left(n\frac{\pi}{2}\right)$

(c) $a_n = 3 + \left(\frac{1}{10}\right)n$

5. How many iterations of $c = -0.5 + 0.5i$ are required through the function $f(z) = z^2 + c$ before the point associated with the outcome z_n is more than 2 units from the origin? The z_n are defined by

$$z_1 = f(c) \qquad z_2 = f(f(c)) \qquad z_3 = f(f(f(c)))$$

6. Let $f(z) = z^2 + (0 + 0i)$.

(a) The point (0, 1) on the unit circle is associated with the complex number $z = 0 + i$. Show that by iterating $z = 0 + i$ through $f(z)$ each outcome is associated with a point also lying on the unit circle.

(b) The point (0.5, 0.5) lies within the unit circle. Show that by iterating the associated complex number $z = 0.5 + 0.5i$ through $f(z)$ each outcome is associated with a point closer to the origin than that point associated with the prior iterate.

(c) The point $(1, -2)$ lies outside of the unit circle. Show that by iterating the associated complex number $z = 1 - 2i$ through $f(z)$ each outcome is associated with a point farther from the origin than that point associated with the prior iterate.

ANSWERS

Test Yourself, Page xix

1. False 2. True 3. False 4. True
5. 12 6. -243 7. $\dfrac{3x^3 + 8x}{\sqrt{x^2 + 4}}$
8. $0, 7; -1 + \sqrt{6}, -1 - \sqrt{6}$ 9. $2(x + \frac{3}{2})^2 + \frac{1}{2}$
10. $x^2(x - 5)(x + 3); (x - 2)(x + 2)(x^2 + 4);$ $(x - 3)(x^2 + 3x + 9)$
11. False 12. False 13. True
14. $x = 6$ or $x = -6$ 15. $-a + 5$
16. **(a)**, **(b)**, **(d)**, **(e)**, **(g)**, **(h)**, **(i)**, **(l)**
17. (i) **(d)**; **(c)**; (iii) **(a)**; (iv) **(b)**
18. $-2 < x < 2; |x| < 2$
19.

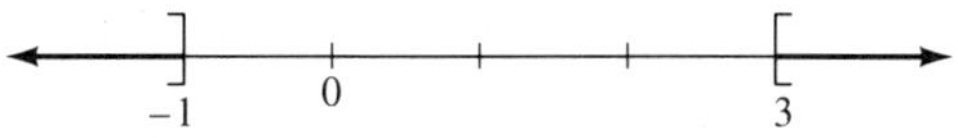

20. $(-\infty, -2) \cup (\frac{8}{3}, \infty)$ 21. Fourth
22. $(5, -7)$ 23. $x_1 = -12$ and $y_2 = 9$
24. $(1, -5); (-1, 5); (-1, -5)$
25. x-intercept -2, y-intercepts -4 and 4
26. Second and fourth 27. $x = 6$ or $x = -4$
28. $x^2 + y^2 = 25$ 29. $d(P_1, P_2) + d(P_2, P_3) = d(P_1, P_3)$
30. **(c)** 31. False 32. $k = -27$
33. 8 34. Slope $\frac{2}{3}$, x-intercept -9, y-intercept 6
35. $y = -5x + 3$ 36. $y = 2x - 14$
37. $y = -\frac{1}{3}x + 3$ 38. $y = -\frac{5}{8}x$
39. $x - \sqrt{3}y + 4\sqrt{3} - 7 = 0$
40. (i) **(g)**; (ii) **(e)**; (iii) **(h)**; (iv) **(a)**; (v) **(b)**; (vi) **(f)**; (vii) **(d)**; (viii) **(c)**

Exercises 1.1, Page 12

1. $12; 0; 3 - \sqrt{3}; a^2 + a$
3. $2; 1.5; 3; 1; 0; -a$ if $a < -1$ and $-2a + 1$ if $a \geq -1$
5. 6 7. $-10a - 5h + 9$
9. $3a^2 + 3ah + h^2$ 11. $[-1, \infty)$
13. $(0, \infty)$ 15. $[-5, 5]$
17. The set of real numbers except 4
19. The set of real numbers except 0 and $\frac{1}{2}$
21. $[0, \infty)$ 23. $(-\infty, 3)$ 25. $[1, \infty)$
27. $(-\infty, 4]$ 29. A function
31. Not a function 33. $[1, \infty), [-1, \infty)$
35. $(-2, 2], \{-2, -1, 1, 2\}$ 37. $5; -4$

39.

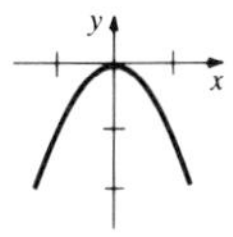

41.

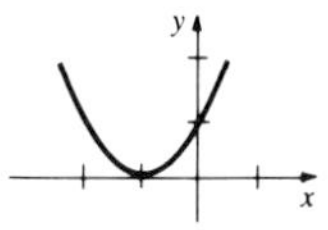

43.

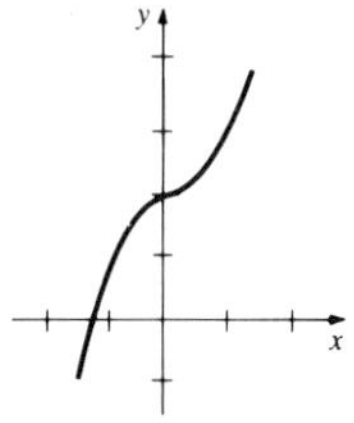

45.

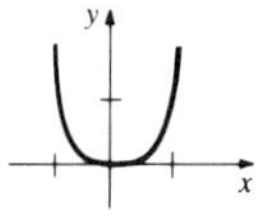

47.

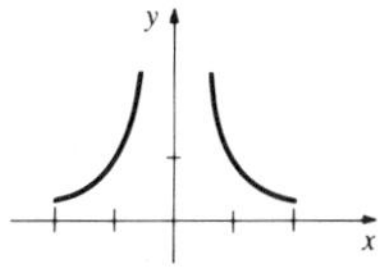

49.

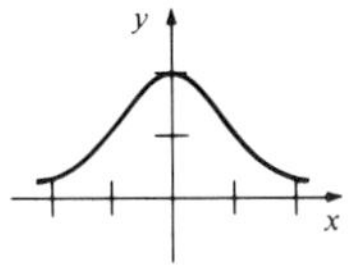

51.

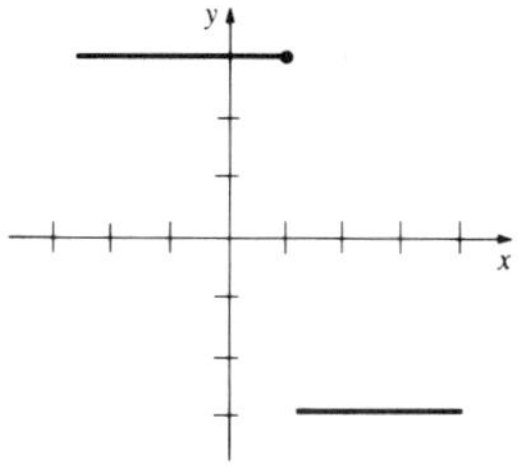

53.

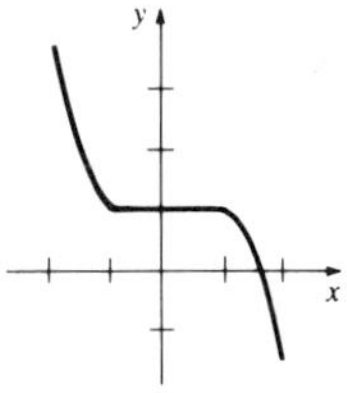

55. 123.9 long tons 57. $T_f = \frac{9}{5}T_c + 32$
59. $t = 0$ and $t = 6$

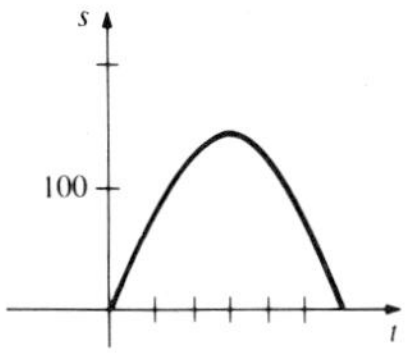

61. 50 g 63. $\Delta x = 3, \Delta y = 45$

65. $-2x + 13; 6x - 3; -8x^2 - 4x + 40;$

$\dfrac{2x + 5}{-4x + 8}, x \neq 2$

67. $3x^2 + 4x^3; 3x^2 - 4x^3; 12x^5; \dfrac{3}{4x}, x \neq 0$

69. $\dfrac{x^2 + x + 1}{x(x + 1)}; \dfrac{x^2 - x - 1}{x(x + 1)}; \dfrac{1}{x + 1}; \dfrac{x^2}{x + 1},$

$x \neq 0$ and $x \neq -1$

71. $2x^2 + 5x - 7; -x + 1; x^4 + 5x^3 - x^2 - 17x + 12;$

$\dfrac{x + 3}{x + 4}, x \neq 1$ and $x \neq -4$

73. $3x + 16; 3x + 4$ 75. $4x^2 + 1; 16x^2 + 8x + 1$

77. $\dfrac{3x + 3}{x}; \dfrac{3}{3 + x}$ 79. $x; x$

81. $128x^9; \dfrac{1}{4x^9}$

83. $10; \dfrac{353}{64}; 1153; 19$

85. $f(x) = 2x^2 - x, g(x) = x^2$

87. $36x^2 - 36x + 15$ 89. $-2x + 9$

91. False

93.

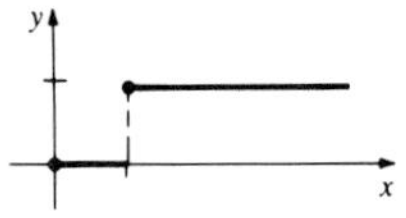

95. $y = 2 - 3U(x - 2) + U(x - 3)$

Exercises 1.2, Page 22

1.

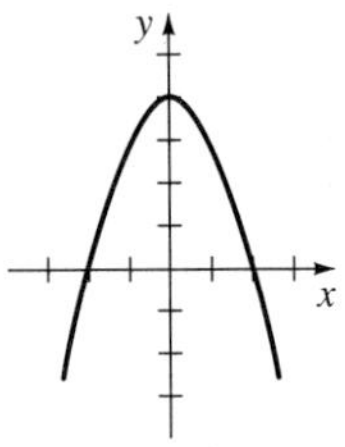

3.

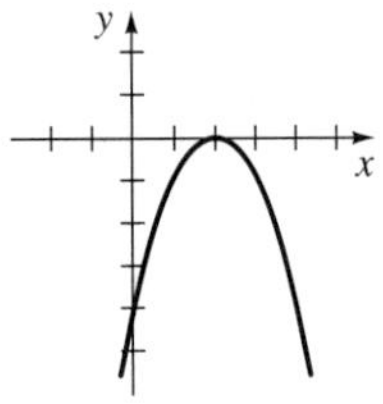

5.

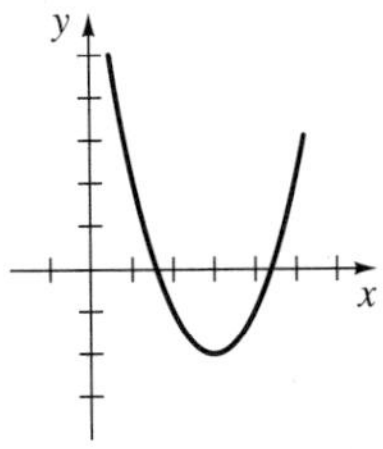

7.

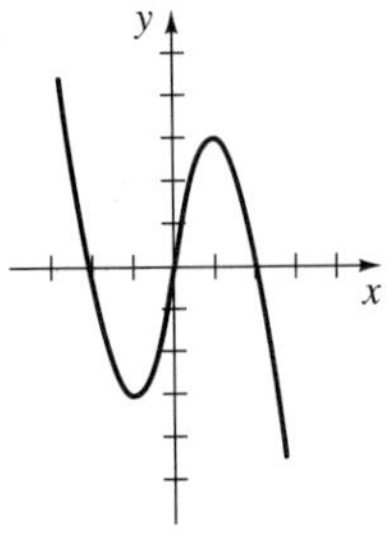

9.

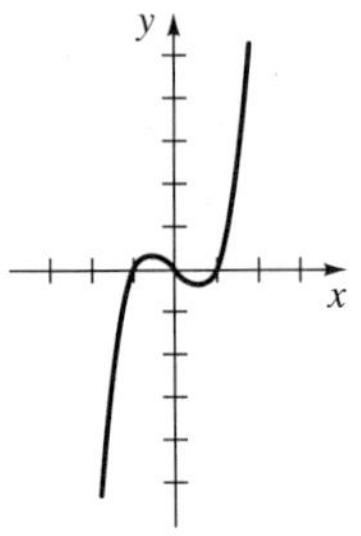

11.

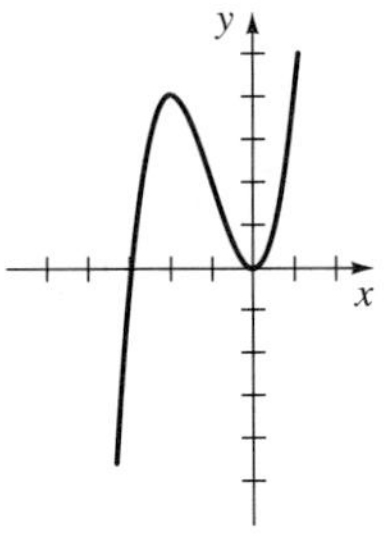

13.

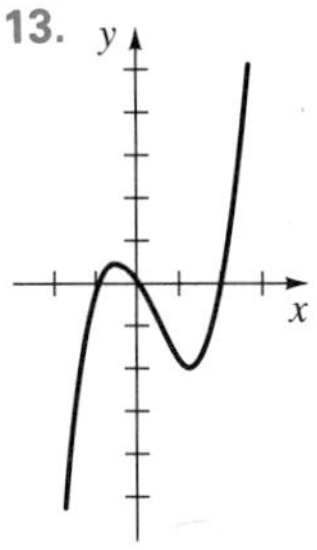

15.

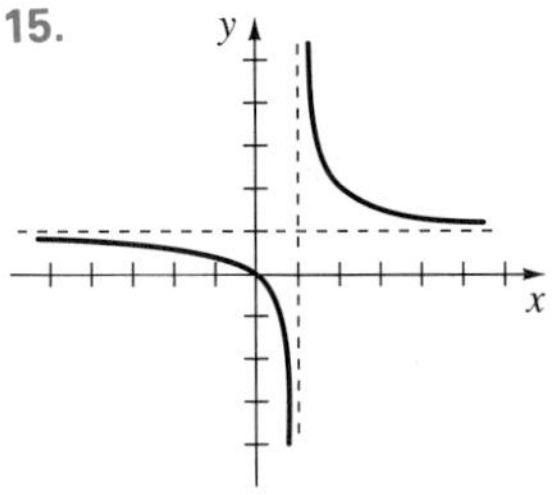

17.

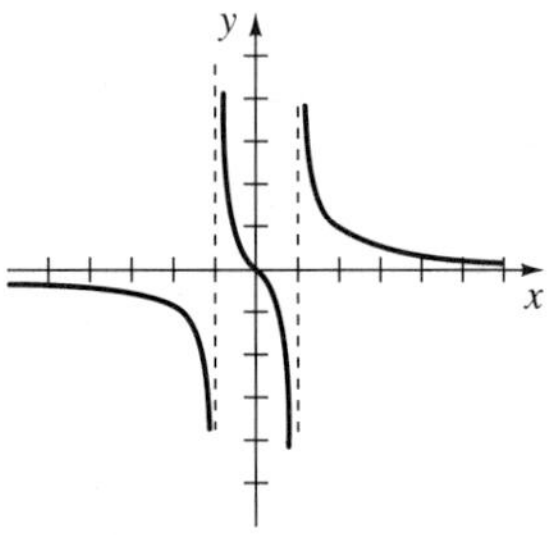

19.

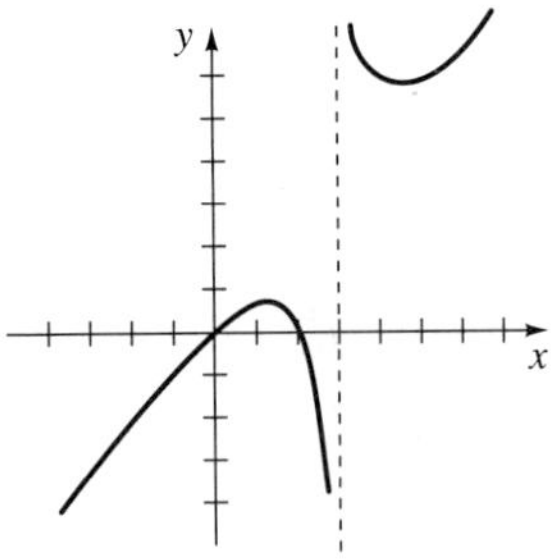

21.

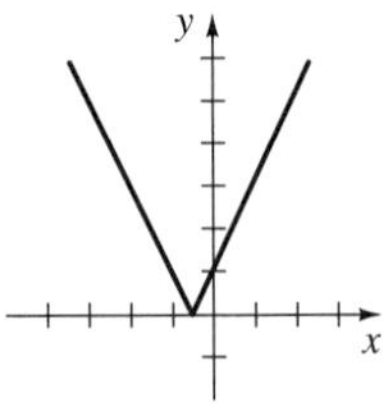

23.

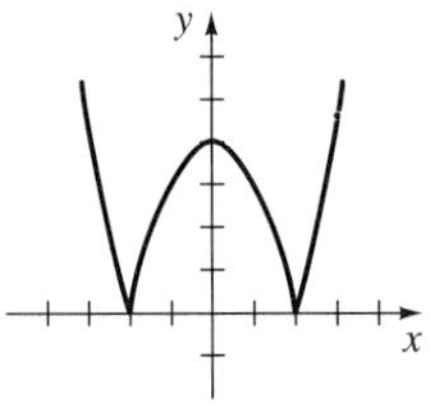

25.
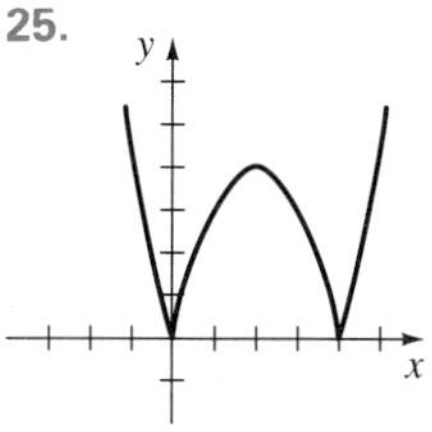

27.
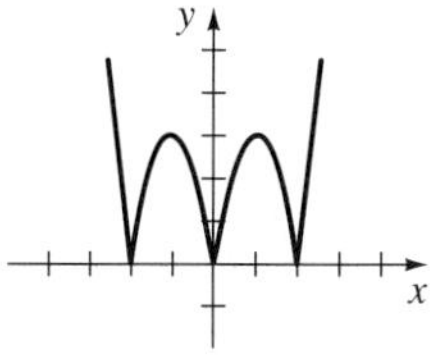

29.
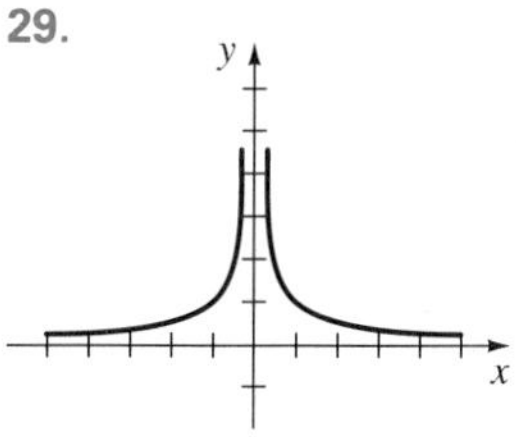

31.
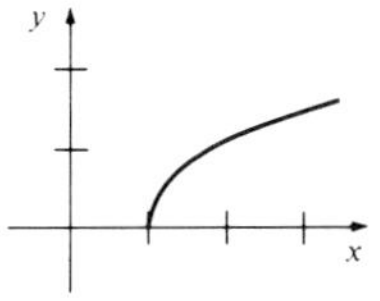

33.
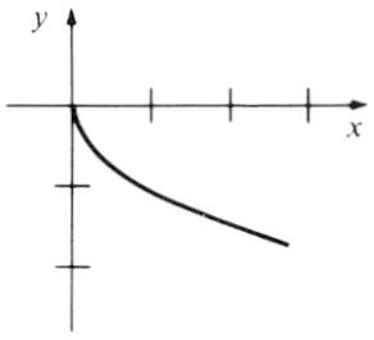

35.
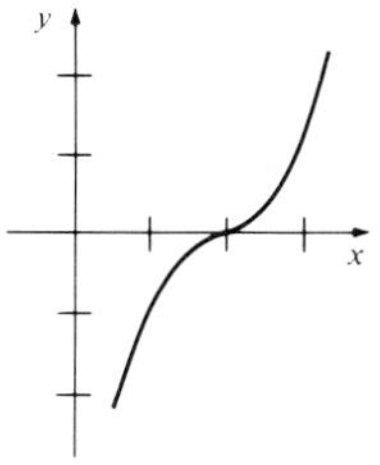

37.
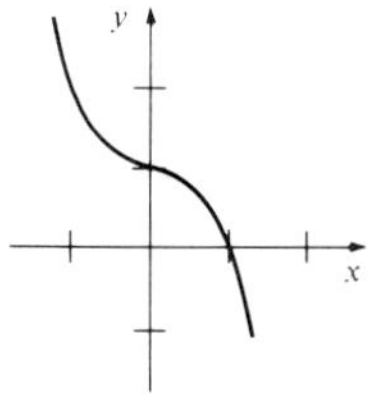

39.
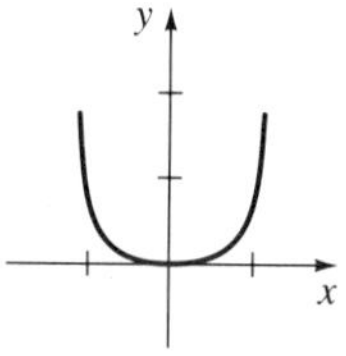

41. $y = |x - \frac{3}{2}|$ 43. $y = -(x - 2)^2 + 3$
45. $y = (x + 4)^2 + 5$

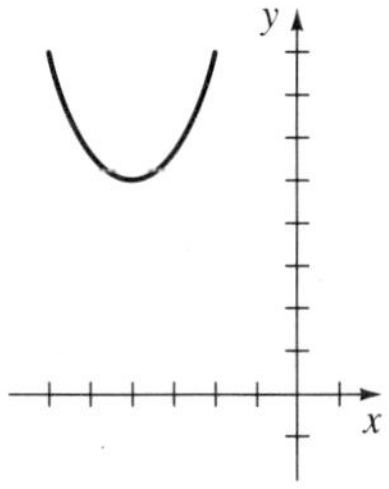

47. (e) 49. (b) 51. (c)

53.
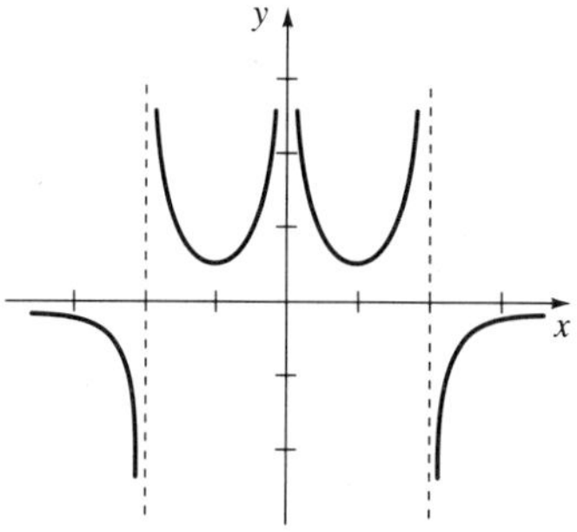

55.
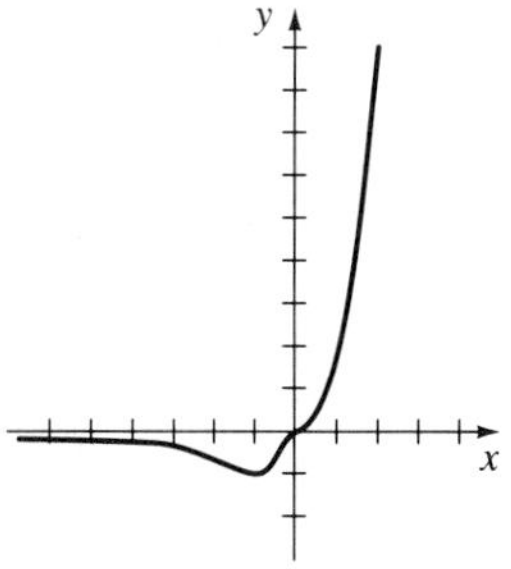

57.
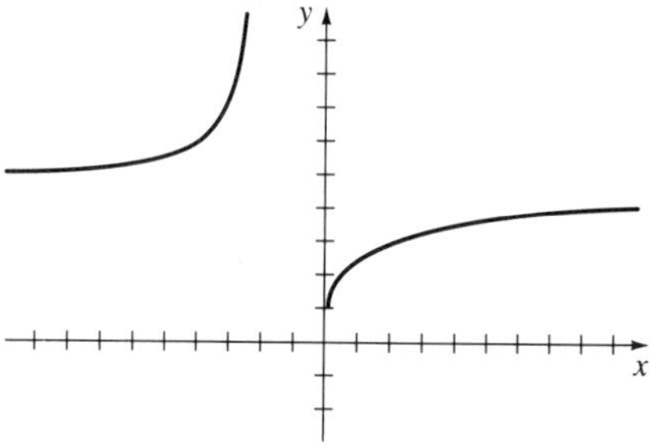

59.
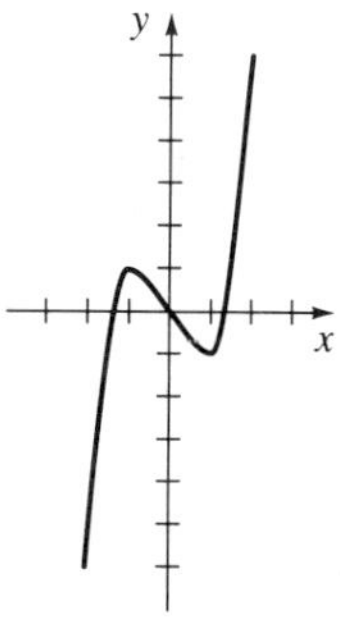

61.
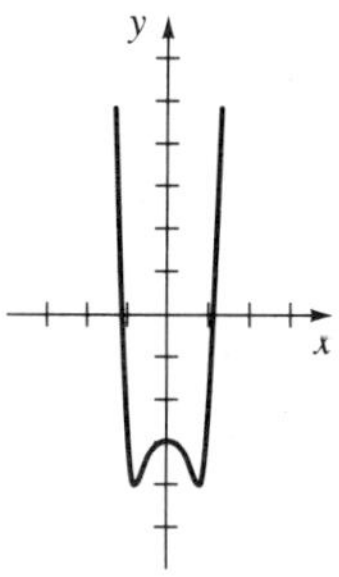

63. Two basic shapes:

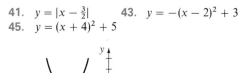

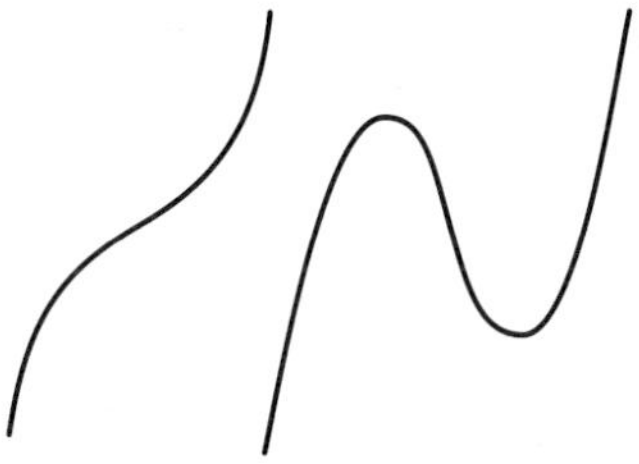

65. $f + g$ and $f - g$ are odd, whereas fg and f/g are even.

Exercises 1.3, Page 33

1. $9°$ 3. $330°$ 5. $-240°$ 7. $7\pi/6$
9. $5\pi/3$ 11. $-5\pi/6$ 13. $-\frac{1}{2}$
15. $-\sqrt{3}/2$ 17. $2/\sqrt{3}$ 19. $-1/\sqrt{3}$
21. $1/\sqrt{3}$ 23. 0

25. $\tan t = -2$
$\cot t = -\frac{1}{2}$
$\sec t = \sqrt{5}$
$\csc t = -\sqrt{5}/2$

27. $\cos t = \sqrt{15}/4$
$\tan t = -1/\sqrt{15}$
$\cot t = -\sqrt{15}$
$\sec t = 4/\sqrt{15}$
$\csc t = -4$

29. $t = 4\pi/3 + 2n\pi$, $t = 5\pi/3 + 2n\pi$, $n = 0, \pm 1, \pm 2, \ldots$
31. $\pi/4, \pi/2, 3\pi/2, 7\pi/4$ 33. $(\sqrt{6} - \sqrt{2})/4$
35. $-\sqrt{2 - \sqrt{2}}/2$
43. For all values of t, $|\sin t| \leq 1$ and $|\cos t| \leq 1$.

45.
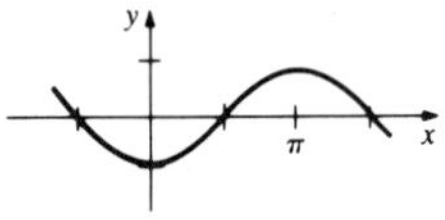

47.
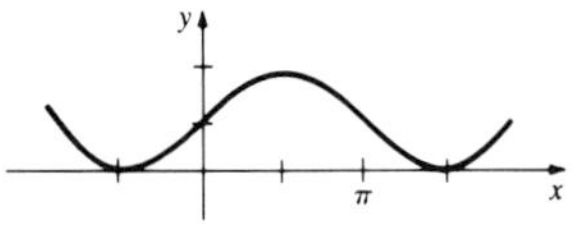

49.
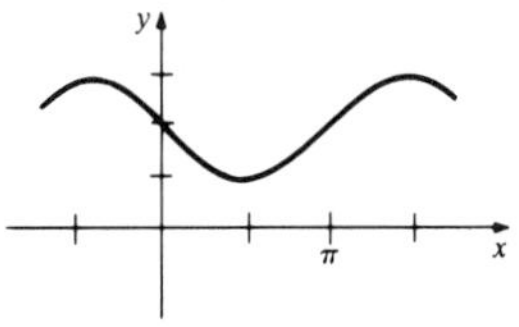

51.
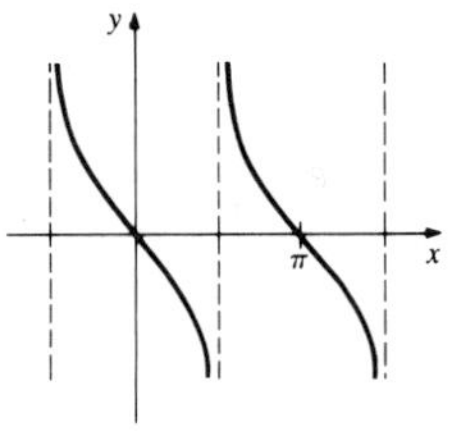

53.
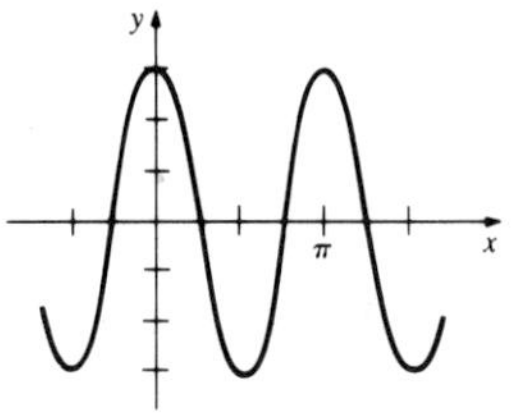

55.
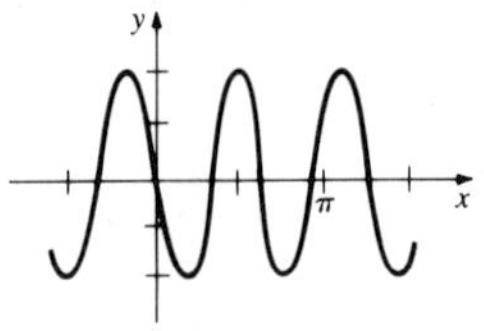

57.
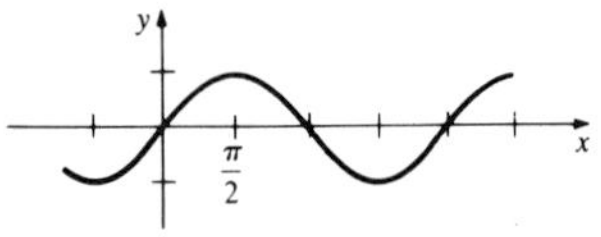

59.
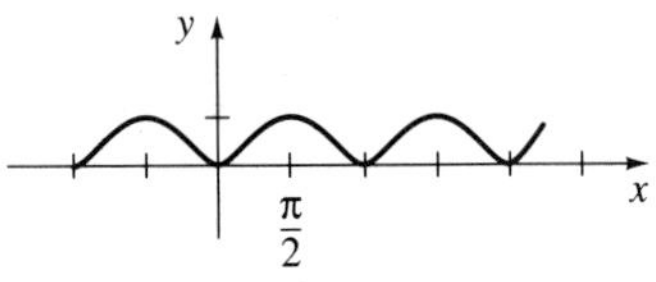

61. $y = 2 + \sin 4x$ 63. $y = \sin(x - \pi/6)$
65. $(1 + 4x)\cos x$; $(1 + 4x)/\cos x$; $1 + 4\cos x$; $\cos(1 + 4x)$
67. $\sin x \cos x$; $\tan x$; $\sin(\cos x)$; $\cos(\sin x)$
69. $\sqrt{x} + \sqrt{x}\cos 2x$; $(1 + \cos 2x)/\sqrt{x}$; $1 + \cos(2\sqrt{x})$; $\sqrt{1 + \cos 2x}$
71. $t = 9$ and $t = 21$; $80°$ occurs at $t = 15$; $60°$ occurs at $t = 3$
73. Even 75. Odd 77. Even

79.
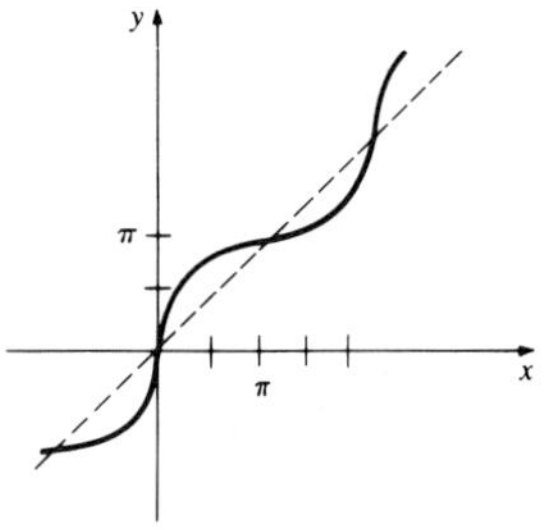

81.
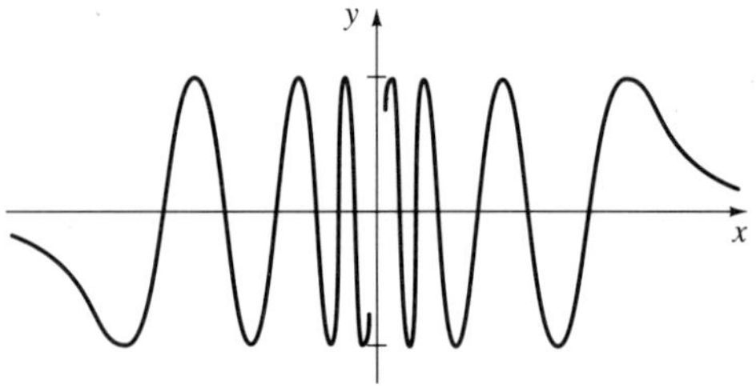

83.
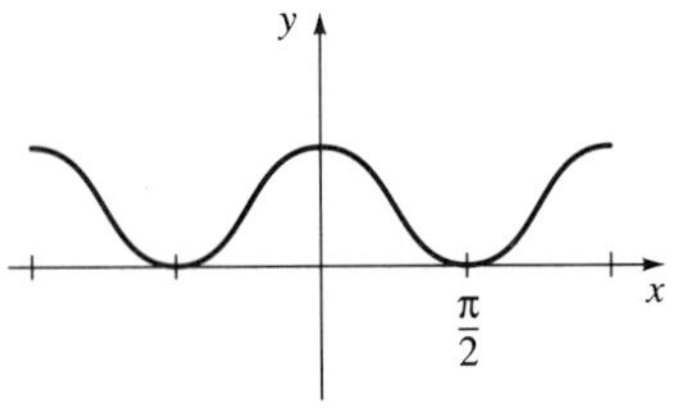

85. 0.8411; 1.8235 **87.** 1.2059; 0.3649

Exercises 1.4, Page 38

1. $P(A) = 4\sqrt{A}$ **3.** $A(s) = (\sqrt{3}/4)s^2$
5. $V(A) = A^{3/2}$ **7.** $d(x) = \sqrt{x^4 + x^2 - 6x + 9}$
9. $A(l) = l\sqrt{4r^2 - l^2}$
11. $A(x) = (-1 + \sqrt{5})x^2/2$
13. $A(r) = 8(\sqrt{2} - 1)r^2$ **15.** $A(x) = (40 - 2x)^2$
17. $V(r) = 40\pi r^2 + \frac{4}{3}\pi r^3$
19. $d(t) = 20\sqrt{13t^2 + 8t + 4}$
21. $V(h) = \begin{cases} 120h^2, & 0 \le h < 5 \\ 1200h - 3000, & 5 \le h \le 8 \end{cases}$
23. $V(h) = (\pi/27)h^3$ **25.** $A(r) = 2\pi r^2 + 2V_0/r$
27. $C(x) = \sqrt{5}k\sqrt{x^2 + \frac{1}{4}} + k(20 - x)$; k is cost per mile
29. $h(\theta) = 300 \tan\theta$ **31.** $L(\theta) = 3\csc\theta + 4\sec\theta$
33. $V(\theta) = 2\pi R^3 \sin^2\theta\cos\theta$,
$S(\theta) = 2\pi R^2(\sin^2\theta + 2\sin\theta\cos\theta)$
35. $T(\theta) = \frac{12}{5} - \frac{9}{5}\tan\theta + 3\sec\theta$
37. $S(\theta) = 2V/R + \pi R^2(\csc\theta - \frac{2}{3}\cot\theta)$

Exercises 1.5, Page 52

1. 8 **3.** Does not exist **5.** 2
7. Does not exist **9.** 0 **11.** 3 **13.** 0
15. Does not exist; 2; does not exist
17. 3; 3; 3 **19.** 0; 2; does not exist; 5
21. Does not exist; 0; 0; 1 **23.** Does not exist
25. The correct statement should be $\lim_{x\to 0^+} \sqrt{x} = 0$.
27. Since $\lim_{x\to 0^-} \sqrt[3]{x} = \lim_{x\to 0^+} \sqrt[3]{x} = 0$, the given statement is correct.
29. $-\frac{1}{4}$ **31.** $\frac{1}{3}$ **33.** $\frac{1}{4}$ **35.** 5
37. -8 **39.** Choose $\delta = \varepsilon$.
41. Choose $\delta = \varepsilon$. **43.** Choose $\delta = \varepsilon$.
45. Choose $\delta = \varepsilon/3$. **47.** Choose $\delta = 2\varepsilon$.
49. Choose $\delta = \varepsilon$. **51.** Choose $\delta = \varepsilon/8$.
53. Choose $\delta = \sqrt{\varepsilon}$. **55.** Choose $\delta = \varepsilon^2/5$.
57. Choose $\delta = \varepsilon/2$. **59.** Choose $\delta = \min\{1, \varepsilon/7\}$.
61. Choose $\delta = \sqrt{\varepsilon}$. **63.** Choose $\delta = \sqrt{a\varepsilon}$.

Exercises 1.6, Page 61

1. 15 **3.** -12 **5.** 4 **7.** 4
9. $-\frac{8}{5}$ **11.** 14 **13.** $\frac{28}{9}$ **15.** $\sqrt{7}$
17. Does not exist **19.** 1 **21.** -10
23. 3 **25.** 60 **27.** 14 **29.** $\frac{1}{5}$
31. $-\frac{1}{8}$ **33.** 3 **35.** Does not exist
37. 2 **39.** $\frac{128}{3}$ **41.** Does not exist
43. -2 **45.** $a^2 - 2ab + b^2$ **47.** $2x$
49. $-1/x^2$ **51.** $\frac{1}{2}$ **53.** $\frac{1}{5}$
55. Does not exist **57.** $12 + 3h$ **61.** 3

Exercises 1.7, Page 71

1.

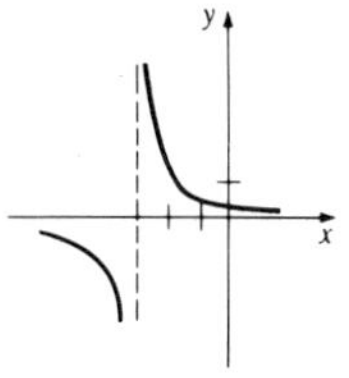

VA: $x = -3$
HA: $y = 0$

3.

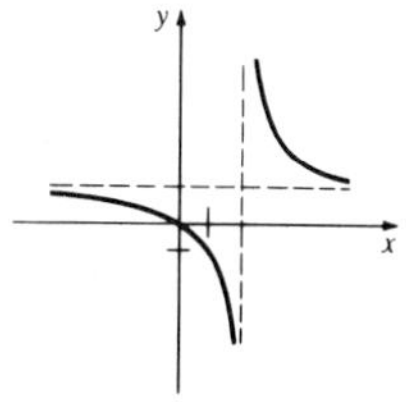

VA: $x = 2$
HA: $y = 1$

5.

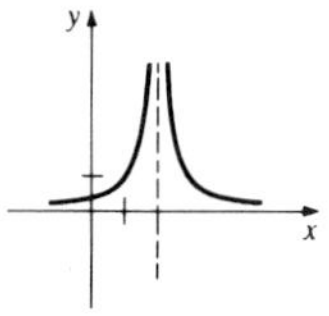

VA: $x = 2$
HA: $y = 0$

7.

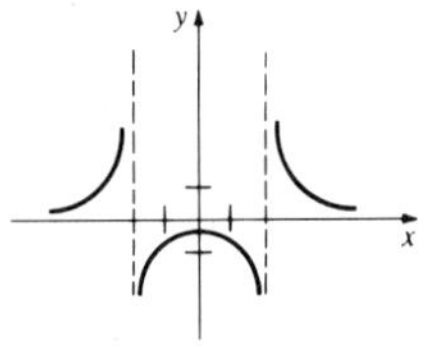

VA: $x = -2$ and $x = 2$
HA: $y = 0$

9.

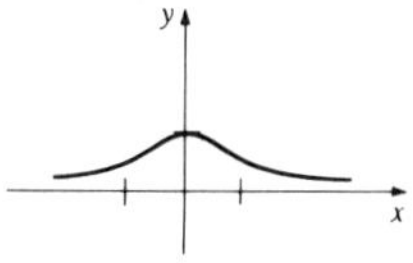

VA: None
HA: $y = 0$

11.

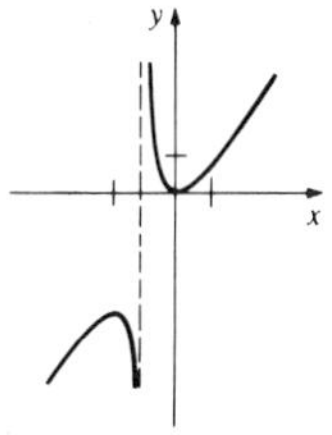

VA: $x = -1$
HA: None

13.

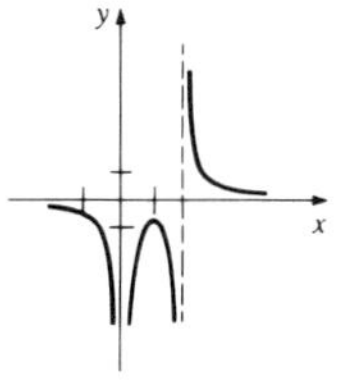

VA: $x = 0$ and $x = 2$
HA: $y = 0$

15.

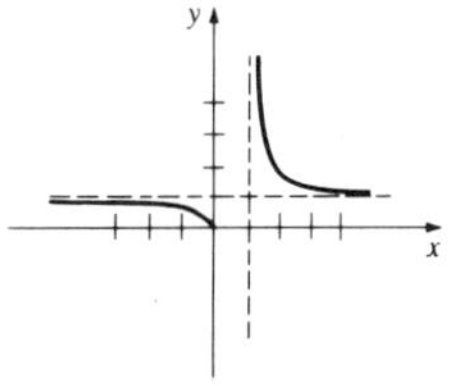

VA: $x = 1$
HA: $y = 1$

17.

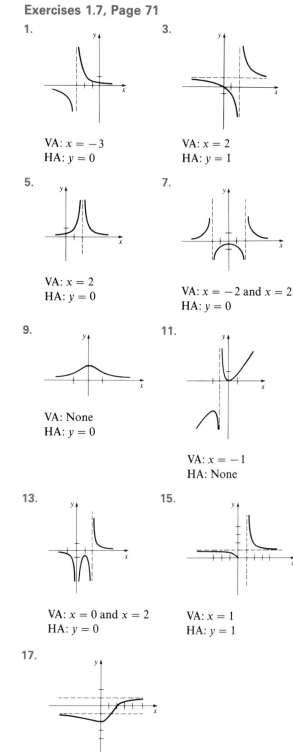

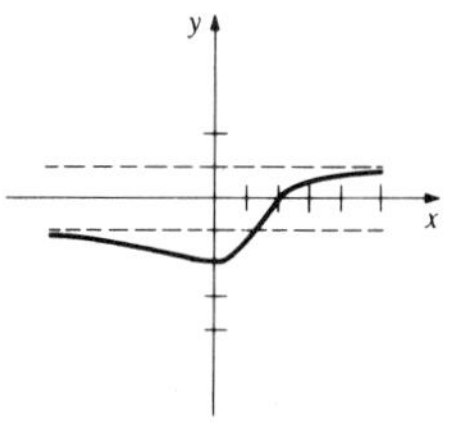

VA: None
HA: $y = 1$, $y = -1$

19. 1; 3; does not exist; 2; 1; 3   21. 2; $-\infty$; 0; 2
23. $-\infty$; $-\frac{3}{2}$; ∞; 0   25. $-\infty$ 27. ∞
29. ∞ 31. ∞ 33. $\frac{1}{4}$ 35. 5
37. 0   39. $-\frac{1}{4}$ 41. $\frac{5}{2}$ 43. 4
45. $-2/\sqrt{3}$ 47. $1/\sqrt{2}$ 49. 0
51. -1
53.

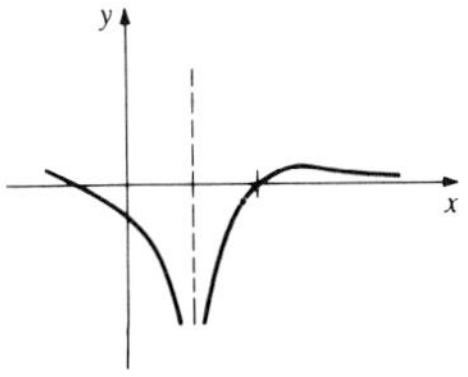

55.

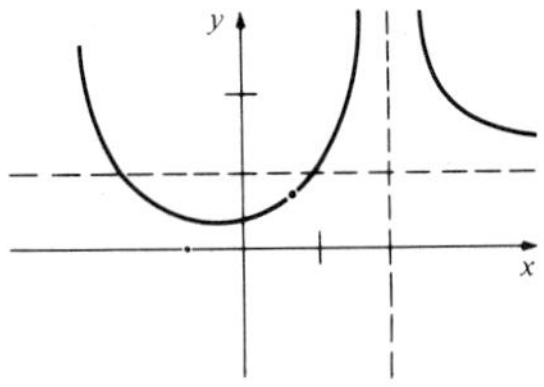

57. $m \to \infty$ 59. 0   61. ∞ 63. ∞
65. 2   67. 1
69. ∞; approximately 2.72; 1

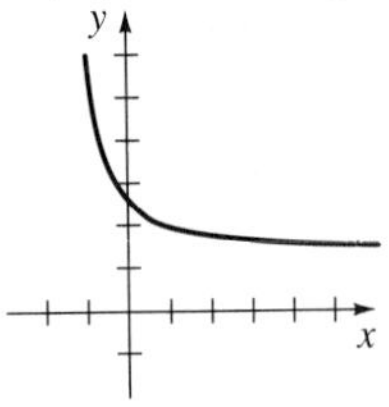

71. Choose $N = 7/4\varepsilon$.   73. Choose $N = -30/\varepsilon$.

Exercises 1.8, Page 80

1. None   3. 3 and 6
5. $n\pi/2, n = 0, \pm 1, \pm 2, \ldots$ 7. 2   9. 0
11. Continuous; continuous
13. Continuous; continuous
15. Not continuous; not continuous
17. Continuous; not continuous
19. Not continuous; not continuous
21. Not continuous; continuous   23. $m = 4$
25. $m = 1, n = 3$
27.

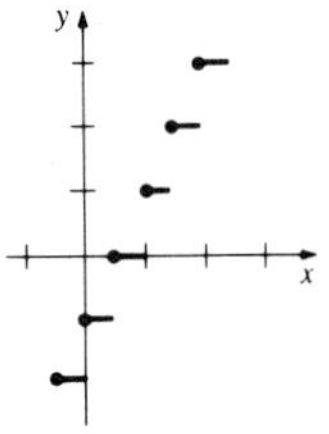

Discontinuous at $n/2$, where n is any integer
29. $\sqrt{3}/2$ 31. 0   33. 1   35. 1
37. $(-3, \infty)$ 39. $c = 4$
41. $c = -\sqrt{2}, c = 0, c = \sqrt{2}$
47. $0, \pm\pi/2, \pm 3\pi/2, \pm 2\pi, \ldots$
49. Define $f(9) = 6$.
57. $-1.21, -0.64, 1.33$

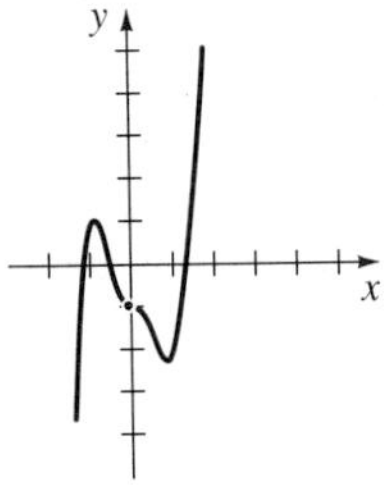

59. 2.21   61. 0.74

Chapter 1 Review Exercises, Page 82

1. True   2. False   3. False   4. False
5. True   6. False   7. False   8. True
9. False   10. True   11. True
12. False   13. False   14. True
15. True   16. False   17. False
18. False   19. False   20. False
21. True   22. True   23. False
24. True   25. False   26. True
27. True   28. False   29. True
30. False   31. $\frac{32}{5}$ 33. (0, 10]
35. -7 and 5   37. 4   39. $-\frac{1}{5}$
41. 3^- 43. $-\infty$ 45. -2 47. 10
49. $\frac{1}{6}$
53. The midpoint of the interval $[x_1, x_2]$ is mapped into the midpoint of the interval $[f(x_1), f(x_2)]$.
55. $V = 6l^3$; $V = 2w^3/9$; $V = 3h^3/4$
57.

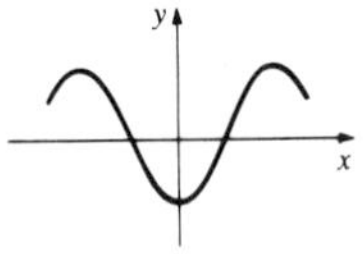

59. $f(-4) = 3, f(-3) = 0$
$f(-2) = -2, f(-1) = 0$
$f(0) = 2.5, f(1) = 2$
$f(1.5) = 1, f(2) = 0$
$f(3.5) = 3, f(4) = 4$
61. 26   63. $\frac{1}{2}$ 65. Does not exist
67. 4   69. $2a$

71.

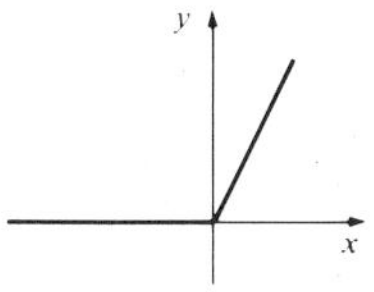

f is continuous at every real number.

73. $[-2, 1)$ and $(1, 2]$ 75. Yes

77. Vertical asymptotes: $x = -\frac{5}{2}$, $x = \frac{5}{2}$, $x = -9$; horizontal asymptote; $y = \frac{1}{2}$

79. (g) 81. (h) 83. (a) 85. (e)

87. (i) (a), (f); (ii) (b), (d), (e); (iii) (c), (d), (e); (iv) (d), (e); (v) (a), (b), (d), (e); (vi) (a), (d), (e); (vii) (f); (viii) (a), (b), (c)

Exercises 2.1, Page 98

1. -4.5 3. 7

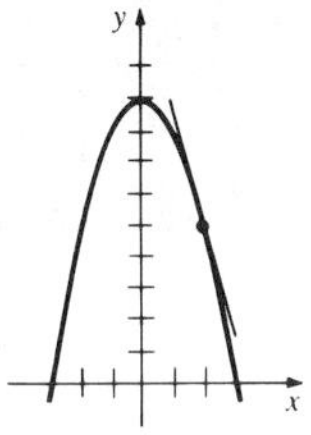

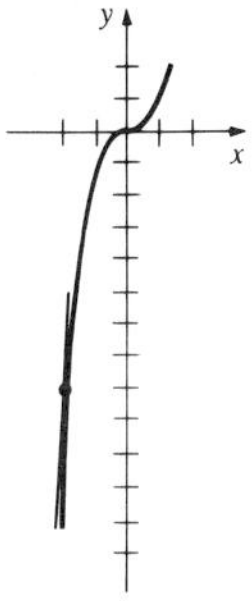

5. $(3\sqrt{3} - 6)/\pi$

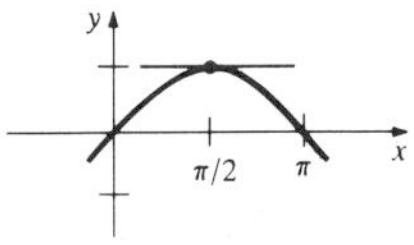

7. 2 9. 6 11. 8 13. 3

15. -9 17. $-2/x_0^3$; $\frac{1}{4}$; -54

19. $y = -10x + 26$ 21. $y = 6x$

23. $y = -2x + 1$ 25. 1

27. Possible horizontal tangents at $x = 1$ and $x = 3$

29. Possible horizontal tangents at $x = -2$ and $x = 1$; possible vertical tangent at $x = 3$; tangent does not exist at $x = 6$

31. The line is not tangent to the graph at the point.

33. $(\sqrt{3\,\Delta x + 1} - 1)/\Delta x$; ; 1.5

0.2649	1.3246
0.1402	1.4018
0.0149	1.4889
-0.3675	1.8377
-0.1633	1.6334
-0.0151	1.5114

37. 58 mph 39. 3.8 h 41. -14

43. -4.9 m/s; 5 s; -49 m/s

45. 448 ft, 960 ft, 1008 ft, 960 ft; 144 ft/s; hits the ground at $t = 16$ s; $-32t_1 + 256$ ft/s; -256 ft/s; 1024 ft

47. Approximately 1.3 m to the right (or above) O, approximately 2.7 m to the right (or above) O; 0.7 m/s; approximately 1.5 m/s; $t = 3$; $[0, 3]$; $[3, 7]$

Exercises 2.2, Page 107

1. 0 3. -3 5. $6x$ 7. $8x - 1$

9. $2x + 2$ 11. $3x^2 + 1$

13. $-3x^2 + 30x - 1$ 15. $-1/x^2$

17. $-1/(x-1)^2$ 19. $-1/x^2 - 2/x^3$

21. $y = -x - 4$ 23. $y = 2x - 2$

25. $(-4, -6)$

27. $1/(2\sqrt{x})$; $(0, \infty)$; by Theorem 1.7,

$$f'_+(0) = \lim_{\Delta x \to 0^+} \frac{1}{\sqrt{\Delta x}} \text{ does not exist}$$

29. $(2x + 1)^{-1/2}$; $(-\frac{1}{2}, \infty)$

31. $\frac{1}{3}(x - 4)^{-2/3}$; $(-\infty, 4) \cup (4, \infty)$

35. $f'_-(2) = -1$, whereas $f'_+(2) = 2$

37. $f'_-(1) = f'_+(1) = 4$

39. f is differentiable at $x = 0$ and $f'(0) = 0$.

41. The graph of a horizontal line $f(x) = k$ has zero slope ($f'(x) = 0$).

43.

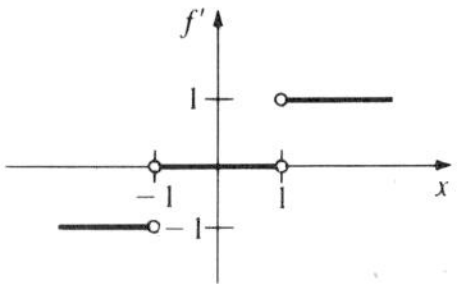

45.

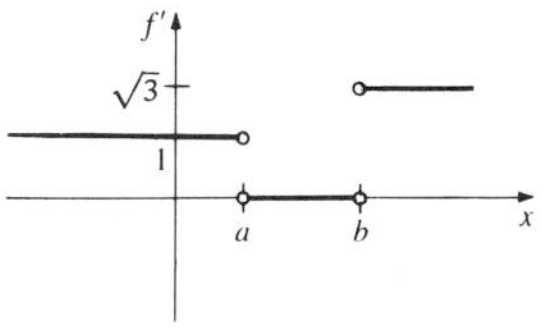

47.

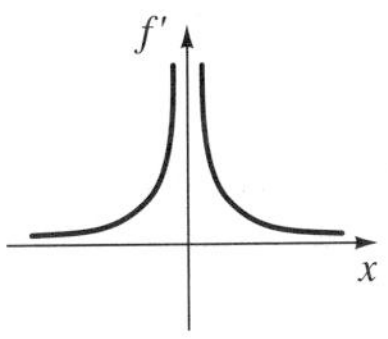

53. $\frac{3}{2}a^{1/2}$

Exercises 2.3, Page 114

1. $9x^8$ 3. 0 5. $14x - 4$

7. $x^4 - 12x^3 + 18x$

9. $20x^4 - 20x^3 - 18x^2$ 11. $2x + 2$

13. $192u^2$ 15. $18z^2 + 2az$
17. $-12\beta^3 + 21\beta^2 - 10\beta$ 19. $6x^5 + 40x^3 + 50x$
21. $y = 6x + 3$ 23. $y = 20x - 56$
25. $(4, -11)$ 27. $(3, -25), (-1, 7)$
29. $y = \frac{1}{4}x - \frac{7}{2}$ 31. $x = 4$
33. $(-4, \infty), (-\infty, -4)$
35.

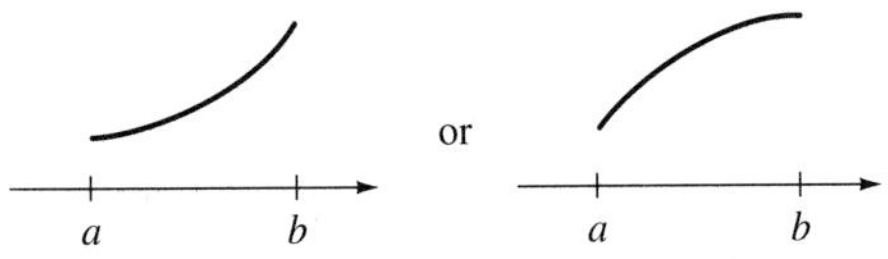

In each case the graph is rising on the interval; that is, $f(x)$ is increasing as x increases.
37. $(2, 8)$ 39. $(\frac{1}{4}, -\frac{3}{16})$
41. $(-1, -4), (-5, 20)$
43. $y = \frac{1}{3}x + \frac{2}{27}, y = \frac{1}{3}x - \frac{2}{27}$
45. On the graph of f, $(-\frac{3}{2}, \frac{3}{4})$; on the graph of g, $(-\frac{3}{2}, -\frac{1}{2})$
47. $b = 8, c = -9$
49. $-\frac{1}{4}, \frac{2+\sqrt{2}}{2}, \frac{2-\sqrt{2}}{2}$ 51. $\frac{5\sqrt{5}}{4}$
53. $S = 4\pi r^2$ 55. -15 N

Exercises 2.4, Page 120

1. $-1/x^2$ 3. $-2x^{-3}/25$ 5. $12x - 2x^{-3}$
7. $5x^4 - 9x^2 + 4x - 28$
9. $8 + \frac{8}{x^3} + \frac{3}{x^4}$ 11. $\frac{-20x}{(x^2+1)^2}$
13. $\frac{-17}{(2x-5)^2}$ 15. $72x - 12$
17. $\frac{t^2 + 2t}{(2t^2 + t + 1)^2}$ 19. $18z^2 + 22z + 6$
21. $\frac{6x^2 + 8x - 3}{(3x+2)^2}$ 23. $\frac{4x^3 + 9x^2 + 1}{(3x^3 + x)^2}$
25. $-\frac{1}{u^2} - \frac{2}{u^3} - \frac{3}{u^4} - \frac{4}{u^5}$
27. $\frac{2x^3 + 8x^2 - 6x - 8}{(x+3)^2}$
29. $\frac{-6}{(3x+1)^3}$ 31. $6, x \neq 0$
33. $y = -16x + 12$ 35. $y = -20x - 12$
37. $(0, 24), (-\sqrt{5}, -1), (\sqrt{5}, -1)$
39. $(-1, \frac{1}{2}), (0, 0), (1, \frac{1}{2})$ 41. $(-\frac{1}{4}, -16), (\frac{1}{4}, 16)$
43. $(3, \frac{3}{2}), (-5, \frac{1}{2})$ 45. -6 47. $\frac{11}{5}$
49. 18 51. $y = -8x + 8, y = -8x - 8$
53.

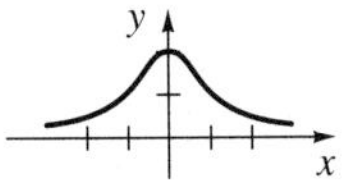

$(0, 2), (-1, 1), (1, 1)$

55. $x < 0$ 57. $x > -\frac{9}{4}$
59. $-16km_1m_2$
61. $-\frac{RT}{(V-b)^2} + \frac{2a}{V^3}$

Exercises 2.5, Page 127

1. $\frac{3}{2}$ 3. 0 5. 1 7. 4 9. 0
11. 36 13. $\frac{1}{2}$ 15. Does not exist
17. 3 19. $\frac{3}{7}$ 21. 0 23. -4
25. 4 27. $\frac{1}{2}$ 29.

0.0157	0.0167	0.0173	0.0174	0.0175	0.0175

;

0.0175; The limit appears to be $\pi/180$.
33. $2x + \sin x$ 35. $7\cos x - \sec^2 x$
37. $x\cos x + \sin x$ 39. $\cos 2x$
41. $x^2\sec x\tan x + 2x\sec x + \sec^2 x$ 43. $\cos x$
45. $-\frac{\csc^2 x + x\csc^2 x + \cot x}{(x+1)^2}$
47. $\frac{4x\tan x - 2x^2\sec^2 x + 2x}{(1 + 2\tan x)^2}$ 49. $\frac{1}{1+\cos\theta}$
51. $2\sin u\cos u$ 53. 0
55. $x^2\sin x\sec^2 x + x^2\sin x + 2x\sin x\tan x$
57. $(\pi/2, \pi/2)$ 59. $3\sqrt{3}x + 6y = \sqrt{3}\pi + 3$
61. $6\sqrt{3}x - 9\sqrt{3}y = \sqrt{3}\pi - 18$
63. $12x - 6y = 16\pi + 3\sqrt{3}$ 65. $y = x - 2\pi$
67. $2\cos 2x$
69. $-160/3$; As the angle of elevation increases, the length s of the shadow decreases.
71.

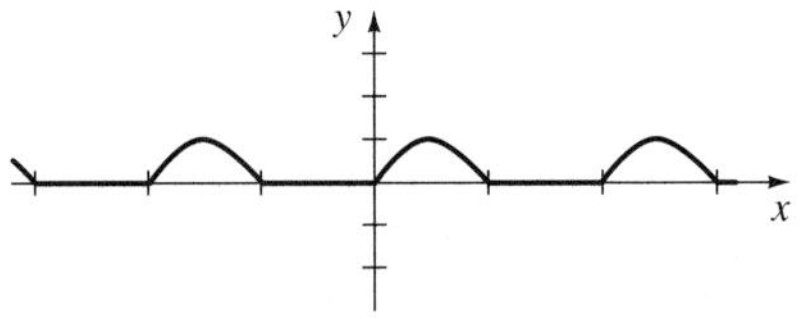

not differentiable at 0, $\pm\pi, \pm 2\pi, \ldots$
73.

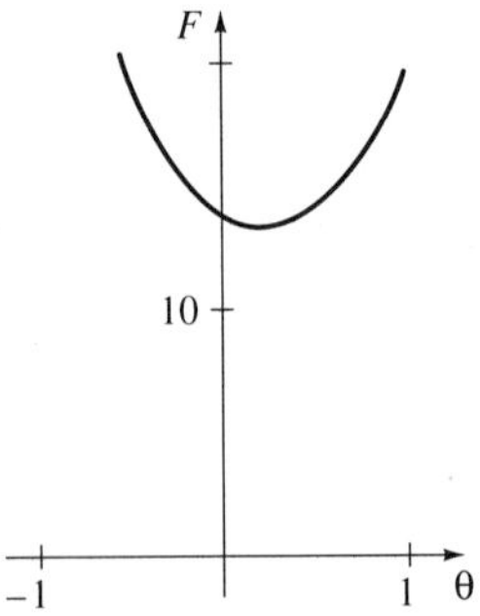

$\frac{14(0.2\cos\theta - \sin\theta)}{(0.2\sin\theta + \cos\theta)^2}$; 0.197 radian or 11.31°;

approximately 13.73 lb; From the figure it is seen that the boy exerts the smallest force (magnitude 13.73 lb) when he pulls the sled by holding the tow rope at the angle of 0.197 radian (11.31°) from the horizontal.

Exercises 2.6, Page 134

1. $-150(-5x)^{29}$ 3. $200(2x^2 + x)^{199}(4x + 1)$
5. $-4(x^3 - 2x^2 + 7)^{-5}(3x^2 - 4x)$
7. $-10(3x - 1)^4(-2x + 9)^4 + 12(-2x + 9)^5(3x - 1)^3 = (3x - 1)^3(-2x + 9)^4(-54x + 108)$
9. $3 \sin^2 x \cos x$ 11. $\dfrac{8x(x^2 - 1)}{(x^2 + 1)^3}$
13. $10[x + (x^2 - 4)^3]^9[1 + 6x(x^2 - 4)^2]$
15. $4(t^{-1} + t^{-2} + t^{-3})^{-5}(t^{-2} + 2t^{-3} + 3t^{-4})$
17. $-3(2 + u \sin u)^{-4}(u \cos u + \sin u)$
19. $\dfrac{(2v - 5)^3(-6v^2 + 45v - 5)}{(v + 1)^9}$
21. $\pi \cos(\pi x + 1)$ 23. $8 \sin 4x \cos 4x$
25. $-3x^5 \sin x^3 + 3x^2 \cos x^3$
27. $\left(\sec^2 \dfrac{1}{x}\right)\left(-\dfrac{1}{x^2}\right)$
29. $\dfrac{5 \cos 6\theta \cos 5\theta + 6 \sin 5\theta \sin 6\theta}{\cos^2 6\theta}$
31. $5(\sec 4x + \tan 2x)^4(4 \sec 4x \tan 4x + 2 \sec^2 2x)$
33. $\cos(\sin x)\cos x$
35. $24x \sin^2(4x^2 - 1)\cos(4x^2 - 1)$
37. $360t^2(1 + t^3)^3(1 + (1 + t^3)^4)^4(1 + (1 + (1 + t^3)^4)^5)^5$
39. -54 41. -7 43. $y = -8x - 3$
45. $12x - 2y = 3\pi + 2$
47. $y - \dfrac{\sqrt{6}}{4} = \dfrac{12}{(3\sqrt{6} + 2\sqrt{2})\pi}\left(x - \dfrac{1}{2}\right)$
49. $(-\sqrt{3}/3, -3\sqrt{3}/16), (\sqrt{3}/3, 3\sqrt{3}/16)$; no
51. $\theta = (2n + 1)\pi/4, n = 0, 1, 2, \ldots$
53. $5/8\pi$ in/min 55. $(3u^2 + 2)(9x^8 + 8x)$
57. $3F'(3x)$ 59. $-10/(-10x + 7)$
63. $f'(g(h(x)))g'(h(x))h'(x)$

Exercises 2.7, Page 138

1. -2 3. 32 5. $60x^{-4}$
7. $6x + 16 - 40x^{-6}$
9. $(3x - 4)(180x^2 - 192x + 32)$
11. $-128(4t + 2)^{-3}$ 13. $-100 \cos 10x$
15. $-x \sin x + 2 \cos x$
17. $\dfrac{4 \cos^2\theta + 6 \cos \theta + 8 \sin^2\theta}{(3 + 2 \cos \theta)^3}$
19. $\sec^3 x + \tan^2 x \sec x$ 21. $1440x^2 + 120x$
23. $-\pi^3 \cos \pi x$ 27. $n!$ 29. $(-4, 48)$
31. $(1, \infty), (-\infty, 1)$ 33. $y = -7x$ 35. $\frac{1}{18}$
37. $f(x) = -2x^2 + 3x - 6$
39. -32 41. 23
43. $q(t) = q_0 \cos \dfrac{1}{\sqrt{LC}}\, t$
45.

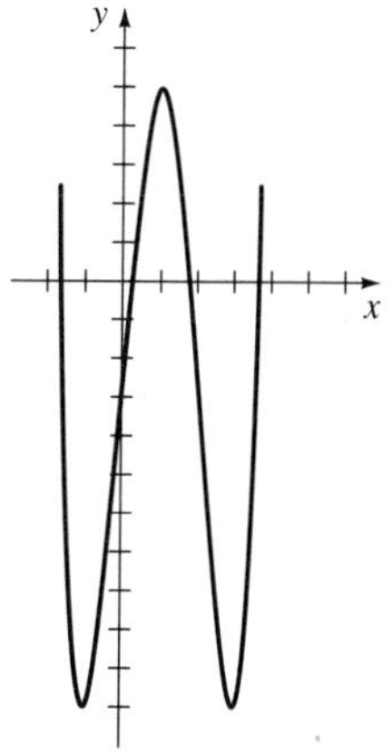

92, 32, −4, −16, −4, 32, 92;
The graph is concave upward when $f'' > 0$ and concave downward when $f'' < 0$.
49. 2, $\kappa \to 0$ as $x \to \infty$ implies that the graph becomes close to linear for very large values of x.

Exercises 2.8, Page 143

1. $1/(2y - 2)$ 3. $(2x - y^2)/(2xy)$
5. $(y + 1)/(2y - x)$ 7. $(4x - 3x^2y^2)/(2x^3y - 2y)$
9. $[x^2 - 4x(x^2 + y^2)^5]/[y^2 + 4y(x^2 + y^2)^5]$
11. $(2x^4y^4 + 3y^{10} - 6x^9y)/(6xy^9 - 3x^{10})$
13. $(1 - x)/(y + 4)$ 15. $3/[2y(x + 2)^2]$
17. $[\cos(x + y) - y]/[x - \cos(x + y)]$
19. $\cos y \cot y$ 21. $(\cos 2\theta)/r$ 23. $-\frac{2}{5}$
25. $-\frac{1}{3}$ and $-\frac{2}{3}$ 27. $-8x + 3y = 22$
29. $2x - 4y = 2 - \pi$ 31. $(1, 2), (-1, -2)$
33. $(-\sqrt{5}, 2\sqrt{5}), (\sqrt{5}, -2\sqrt{5})$
35. $(3 - x)/(y + 4)$; $(x - 3)^2 + (y + 4)^2 = -2$ does not describe a real locus of points.
37. $(y^3 - 2x^2)/y^5$ 39. $-25/y^3$
41. $-(\sin y)/(1 - \cos y)^3$ 43. $-2/(y - x)^3$
45. $y' = (1 - 3x^2y)/x^3$ is equivalent to $y' = -2x^{-3} - 3x^{-4}$
47. $y = 1 - \sqrt{x - 2}$
49. $y = \begin{cases} \sqrt{4 - x^2}, & -2 \le x < 0 \\ -\sqrt{4 - x^2}, & 0 \le x \le 2 \end{cases}$
53. $-(x/y)\, dx/dt$ 55. $y = -x + 3$; $(\sqrt[3]{2}, \sqrt[3]{4})$
57. Approximately (1, 0.68), (1, −0.68); approximately $y = 0.23x + 0.45$, $y = -0.23x - 0.45$; $(\sqrt{6}/2, \sqrt{2}/2), (\sqrt{6}/2, -\sqrt{2}/2), (-\sqrt{6}/2, \sqrt{2}/2), (-\sqrt{6}/2, -\sqrt{2}/2)$

Exercises 2.9, Page 148

1. $15x^{1/2}$ 3. $-\frac{4}{3}x^{-7/3}$ 5. $\frac{1}{2}x^{-1/2} - \frac{1}{2}x^{-3/2}$
7. $\frac{4}{3}(x^3 + x)(x^2 - 4)^{-1/3} + 2x(x^2 - 4)^{2/3}$
9. $\dfrac{33}{2}\dfrac{(x-9)^{1/2}}{(x+2)^{5/2}}$ 11. $\frac{1}{2}x^{-1/2}\cos\sqrt{x}$
13. $1 + x(x^2 + 1)^{-1/2}$
15. $\frac{1}{3}[(t^2 - 1)(t^3 + 4t)]^{-2/3}(5t^4 + 9t^2 - 4)$
17. $-\frac{1}{2}(\theta + \sin\theta)^{-3/2}(1 + \cos\theta)$
19. $\frac{8}{5}s^3(s^4 + 1)^{-3/5}$ 21. $-\frac{1}{4}x^{-3/2}$
23. $3x^{-1}(\sqrt{x} + 1)^2 - x^{-3/2}(\sqrt{x} + 1)^3$
25. $-2.4(\sin\theta)^{2.4} + 3.36(\sin\theta)^{0.4}\cos^2\theta$
27. $x - 12y = -16$ 29. $7x - 8y = 3$
31. $-(y + 2\sqrt{xy})/(x + 2\sqrt{xy})$
33. $y^{-1/2}(x^2 - 1)^{3/2}(6x^2 - 1)$
35. $8\sqrt{3}x + 8y = 4\sqrt{3}, 8\sqrt{3}x - 8y = 4\sqrt{3}$
37. Vertical tangent at (4, 0)
39. No vertical tangents
41. Vertical tangent at (0, 1) 43. $\pi/8$
45. $-ky^{-2}(2k/y - 2k/R + v_0^2)^{-1/2}$
49. $N_a = \frac{1}{2}(\sqrt{C^2 + 4k^2} - C); \frac{1}{2}(-1 + C/\sqrt{C^2 + 4k^2})$
53. Approximately 4.951; approximately 20.615

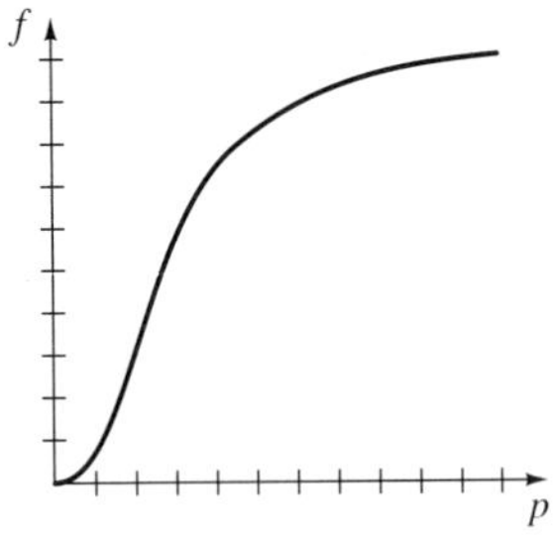

Exercises 2.10, Page 154

1. 0 3. $-(2x)^{-3/2}\,dx$
5. $16x^3(x^4 - 1)^{-2/3}\,dx$ 7. $\dfrac{4x}{(x^2 + 1)^2}\,dx$
9. $-x \sin x\,dx$
11. $\Delta y = 2x\,\Delta x + (\Delta x)^2, dy = 2x\,dx$
13. $\Delta y = 2(x + 1)\,\Delta x + (\Delta x)^2, dy = 2(x + 1)\,dx$
15. $\Delta y = -\dfrac{\Delta x}{x(x + \Delta x)}, dy = -\dfrac{1}{x^2}\,dx$
17. $\Delta y = \cos x \sin \Delta x + \sin x(\cos \Delta x - 1)$, $dy = \cos x\,dx$
19.

25	20	5
11.25	10	1.25
2.05	2	0.05
0.2005	0.2	0.0005

21. 6.0833 23. 16 25. 0.325

27. 0.4 29. $\dfrac{1}{2} + \dfrac{\sqrt{3}\pi}{120} \approx 0.5453$
31. 1.11; −2.9 33. 9π; 8π
35. Exact volume is $\Delta V = \frac{4}{3}\pi(3r^2t + 3rt^2 + t^3)$, approximate volume is $dV = 4\pi r^2 t$, where $t = \Delta r$; $(0.1024)\pi$ in^3
37. ± 6 cm^2, ± 0.06, $\pm 6\%$ 41. 2048 ft, 160 ft
43. Minimum at the equator ($\theta = 0°$), maximum at the North Pole ($\theta = 90°$N); 981.9169; approximately 0.07856 cm/s^2
45. 0.01 s

Exercises 2.11, Page 162

1. One real root 3. No real roots
5. 3.1623 7. 1.5874 9. 0.6823
11. −1.1414, 1.1414 13. 0, 0.8767
15. 2.4981 17. 1.6560 19. 0.7297
21. 0.0915 23. 0.0337, 44.494; 44.497
25. 1.8955
29.

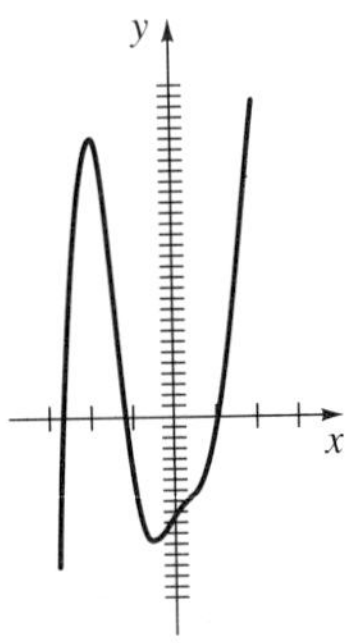

1.0000, −1.2494, −2.6638
31.

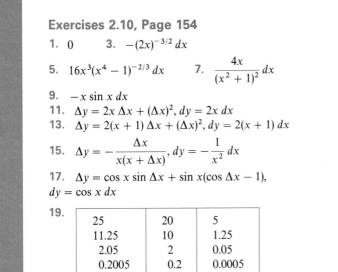

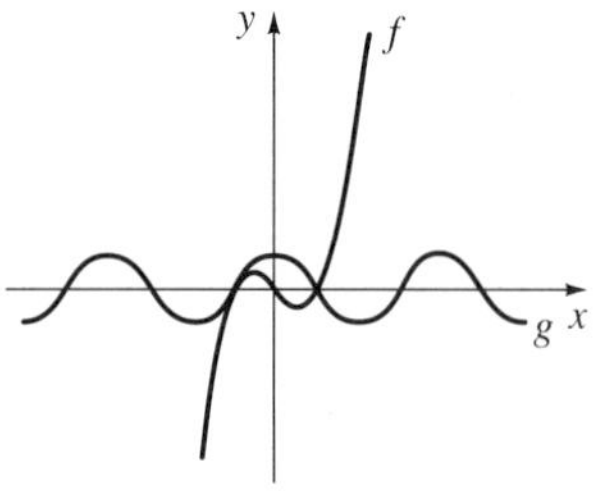

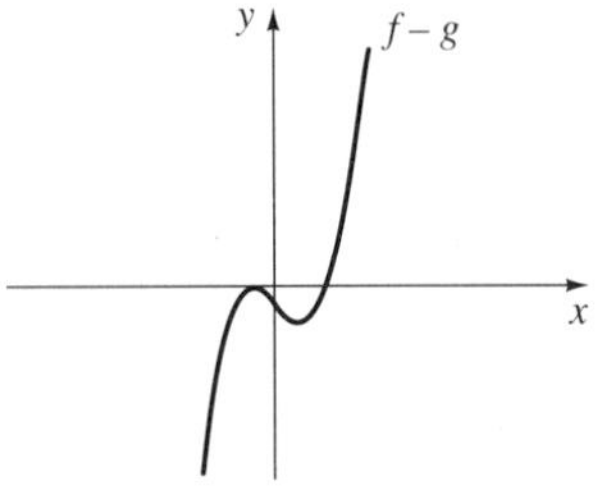

The graphs in parts (**a**) and (**b**) suggest that there is a negative and a positive root. A computer zoom-in of the graph of $f - g$ in a neighborhood of the suspected "negative root" shows that there is no root. Newton's Method gives the approximation 1.4645 for the positive root.

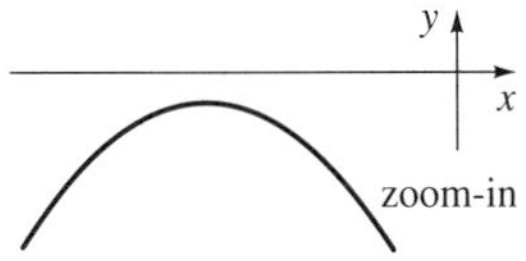

Chapter 2 Review Exercises, Page 165

1. True **2.** True **3.** False **4.** False
5. True **6.** True **7.** False **8.** True
9. False **10.** False **11.** True
12. True **13.** True **14.** True
15. False **16.** True **17.** 0 **19.** 3; 27
21. $(x^3)^2 \cdot 3x^2$ **23.** $\sin x$ **25.** -3
27. 0 **29.** 18 **31.** $x = \sqrt{n\pi}$, $n = 0, 1, 2, \ldots$
33. $y = 6x - 9$, $y = -6x - 9$ **35.** (4, 2)
37. Among others, $(\pi/2, 0)$ and $(3\pi/2, 0)$
39. $\pi/3$, $2\pi/3$
41. $(-16 \sin 4x)F'(\sin 4x) + (16 \cos^2 4x)F''(\sin 4x)$
43.

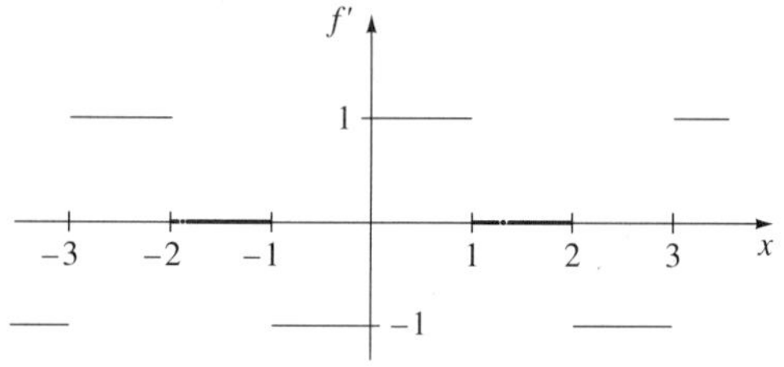

45. 0.2448 **47.** $2y^{1/3}/3x^{5/3}$
49. $-(16x \sin 4x + 4 \sin 4x + 4 \cos 4x)/(4x + 1)^2$
51. $2 - 2x^{-3} + 2x$
53. $10(t + \sqrt{t^2 + 1})^9(1 + t(t^2 + 1)^{-1/2})$
55. $0.08x^{-0.9}$
57. $x^2(x^4 + 16)^{1/4}(x^3 + 8)^{-2/3} + x^3(x^4 + 16)^{-3/4}(x^3 + 8)^{1/3}$
59. $\dfrac{2 + x^{1/3}}{3x^{1/3}(1 + x^{1/3})^2}$ **61.** $\frac{405}{8}(3x)^{-1/2}$
63. $2 + 6t^{-4}$ **65.** 1.6751

Exercises 3.1, Page 174

1. $-1, -2, 2, 8$; 19, 18, 18, 8
3. 18, -23, 23, 18; 6, 1, 1, -6
5. $-\frac{15}{4}$, 17, 17, -128; 0, 2, 2, -2
7. $1, 1 - \pi, \pi - 1, 0; \frac{1}{2}, 1, 1, \pi^2$
9. 6 and -6; 8 and -8
11. $6\sqrt{2}, -6\sqrt{2}$; 15; -4 and 8
13. $(-\infty, -3)$, $(0, 3)$
15. $v(t) = 2t$, $a(t) = 2$

17. $v(t) = 2t - 4$, $a(t) = 2$

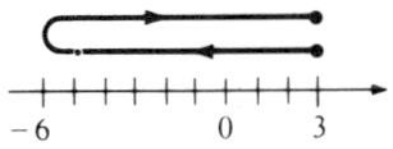

19. $v(t) = 6t^2 - 12t$, $a(t) = 12t - 12$

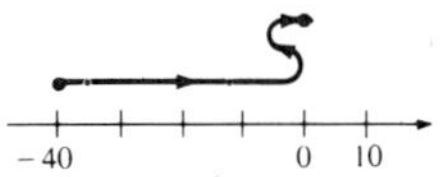

21. $v(t) = 12t^3 - 24t^2$, $a(t) = 36t^2 - 48t$

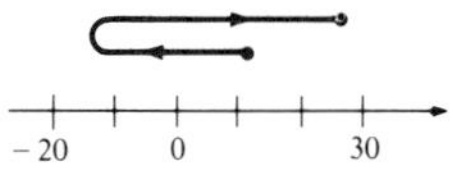

23. $v(t) = 1 - 2t^{-1/2}$, $a(t) = t^{-3/2}$

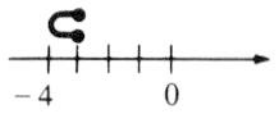

25. $v(t) = (\pi/2)\cos \frac{\pi}{2} t$, $a(t) = -(\pi/2)^2 \sin \frac{\pi}{2} t$

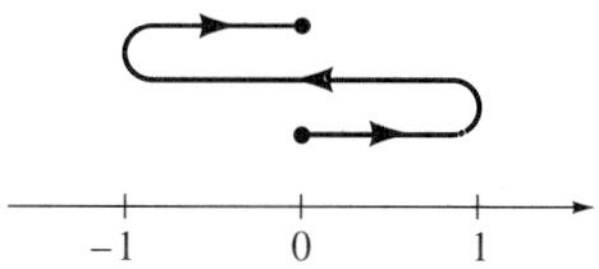

27.

positive	negative
zero	zero
positive	positive
positive	negative
negative	negative
negative	positive

29. $v > 0$ on $(-\infty, \frac{3}{2})$, $v < 0$ on $(\frac{3}{2}, \infty)$; 42 ft
31. $64\sqrt{2}$ ft/s; 16 ft/s^2
33. $-8\sqrt{\pi}$ ft/s; the y-coordinate is decreasing

Exercises 3.2, Page 180

1. $\dfrac{dV}{dt} = 3x^2 \dfrac{dx}{dt}$ **3.** $8\sqrt{3}$ cm^2/h

5. $\frac{4}{3}$ in/h 7. $s\cos\theta\dfrac{d\theta}{dt}+\sin\theta\dfrac{ds}{dt}=\dfrac{dx}{dt}$

9. 24 cm/min 11. -6 or 6 13. $\frac{4}{9}$ cm^2/h
15. $\frac{35}{4}$ lb/in^2/s 17. $5/(32\pi)$ m/min
19. $-\frac{1}{3}$ in^2/min
21. $\sqrt{3}/10$ ft/min; 71.45 min; 0.035 ft/min
23. $-1/(4\pi)$ ft/min; $-1/(12\pi)$ ft/min; approximately -0.0124 ft/min
25. $-\sqrt{2}/2$ ft/min 27. -360 mi/h
29. $8\pi/9$ km/min 31. $\pi/12$ km/s
35. 17 knots 37. $\dfrac{dh}{dt}=\dfrac{16}{\pi(12-h)^2}$
39. 668.7 ft/min
41. Increase; approximately 2.8% per day

Exercises 3.3, Page 190

1. Abs. max. $f(2)=-2$, abs. min. $f(-1)=-5$;
abs. max. $f(7)=3$, abs. min. $f(3)=-1$;
no extrema;
abs. max. $f(4)=0$, abs. min. $f(1)=-3$
3. Abs. max. $f(4)=0$, abs. min. $f(2)=-4$;
abs. max. $f(1)=f(3)=-3$, abs. min. $f(2)=-4$;
abs. min. $f(2)=-4$;
abs. max. $f(5)=5$
5. No extrema;
abs. max. $f(\pi/4)=1$, abs. min. $f(-\pi/4)=-1$;
abs. max. $f(\pi/3)=\sqrt{3}$, abs. min. $f(0)=0$;
no extrema
7. $\frac{3}{2}$ 9. $-1, 6$ 11. $2, \frac{4}{3}$ 13. 1
15. $\frac{3}{4}$ 17. $-2, -\frac{11}{7}, 1$
19. $2n\pi$, $n=0, \pm1, \pm2, \ldots$
21. Abs. max. $f(3)=9$,
abs. min. $f(1)=5$
23. Abs. max. $f(8)=4$,
abs. min. $f(0)=0$
25. Abs. max. $f(0)=2$,
abs. min. $f(-3)=-79$
27. Abs. max. $f(3)=8$,
abs. min. $f(-4)=-125$
29. Abs. max. $f(2)=16$,
abs. min. $f(0)=f(1)=0$
31. Abs. max. $f(\pi/6)=f(5\pi/6)=f(7\pi/6)$
$=f(11\pi/6)=\frac{3}{2}$,
abs. min. $f(\pi/2)=f(3\pi/2)=-3$
33. Abs. max. $f(\pi/8)=f(3\pi/8)=f(5\pi/8)$
$=f(7\pi/8)=5$,
abs. min. $f(0)=f(\pi/4)=f(\pi/2)=f(3\pi/4)$
$=f(\pi)=3$
35. Endpoint abs. max. $f(3)=3$,
rel. max. $f(0)=0$,
abs. min. $f(-1)=f(1)=-1$
37. $c_1, c_3, c_4, c_{10}; c_2, c_5, c_6, c_7, c_8, c_9$;
abs. min. $f(c_7)$, endpoint abs. max. $f(b)$;
rel. max. $f(c_3)$, $f(c_5)$, $f(c_9)$;
rel. min. $f(c_2)$, $f(c_4)$, $f(c_7)$, $f(c_{10})$
39.

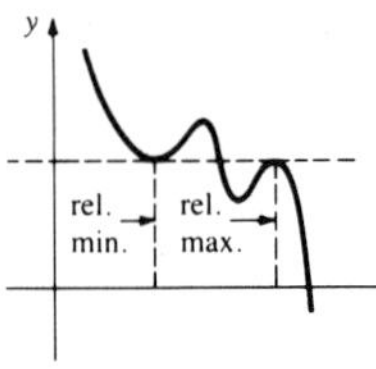

43. $s(t)\ge 0$ only for $0\le t\le 20$;
$s(10)=1600$
45. n an even positive integer
47. $f(-a)$ is a rel. min. 49. c; c; $c-k$; c/k
51.

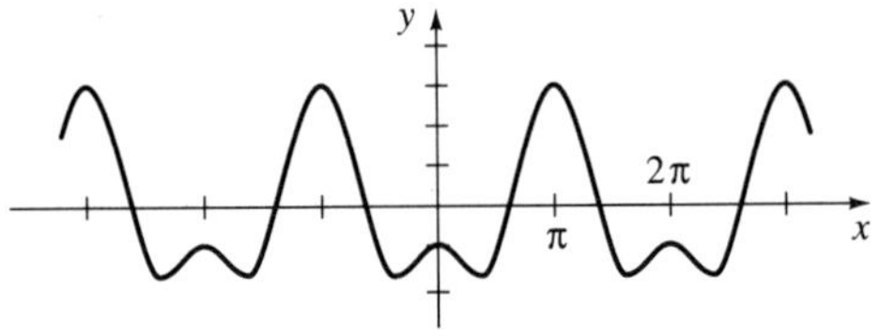

$0, \pi/3, \pi, 5\pi/3, 2\pi$;
abs. max. $f(\pi)=3$,
abs. min. $f(\pi/3)=f(-5\pi/3)=-\frac{3}{2}$
53.

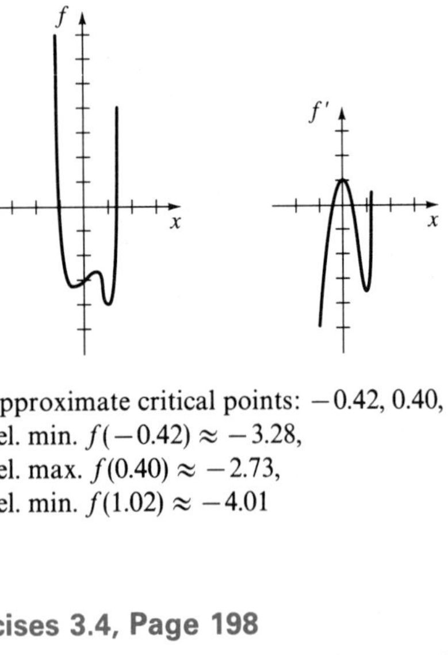

approximate critical points: $-0.42, 0.40, 1.02$;
rel. min. $f(-0.42)\approx -3.28$,
rel. max. $f(0.40)\approx -2.73$,
rel. min. $f(1.02)\approx -4.01$

Exercises 3.4, Page 198

1. $c=0$ 3. $f(-3)=0$ but $f(-2)\ne f(-3)$
5. $c=-\frac{2}{3}$ 7. $c=-\pi/2, \pi/2$, or $3\pi/2$
9. f is not differentiable on the interval.
11. $f(a)\ne 0$ and $f(b)=0$, so $f(a)\ne f(b)$
13. $c=3$ 15. $c=\sqrt{13}$
17. f is not continuous on the interval.
19. $c=\frac{9}{4}$ 21. $c=1-\sqrt{6}$
23. f is not continuous on $[a, b]$.

25. Increasing on $[0, \infty)$, decreasing on $(-\infty, 0]$
27. Increasing on $[-3, \infty)$, decreasing on $(-\infty, -3]$
29. Increasing on $(-\infty, 0]$ and $[2, \infty)$, decreasing on $[0, 2]$
31. Increasing on $[3, \infty)$, decreasing on $(-\infty, 0]$ and $[0, 3]$
33. Decreasing on $(-\infty, 0]$ and $[0, \infty)$
35. Increasing on $(-\infty, -1]$ and $[1, \infty)$, decreasing on $[-1, 0)$ and $(0, 1]$
37. Increasing on $[-2, 2]$, decreasing on $(-2\sqrt{2}, -2]$ and $[2, 2\sqrt{2})$
39. Increasing on $(-\infty, 0]$, decreasing on $[0, \infty)$
41. Increasing on $(-\infty, 1]$ and $[3, \infty)$, decreasing on $[1, 3]$
43. Increasing on $\left[-\dfrac{\pi}{2} + 2n\pi, \dfrac{\pi}{2} + 2n\pi\right]$, decreasing on $\left[\dfrac{\pi}{2} + 2n\pi, \dfrac{3\pi}{2} + 2n\pi\right]$, where n is an integer
45. f is always increasing, since $f'(x) > 0$ for all x.
47. Since the average speed on the time interval is 60 mph, the Mean Value Theorem implies there must be some time in the interval at which the speed is exactly equal to 60 mph. This exceeds the legal speed limit.
51. The derivative of fg will be positive on (a, b) if $f(x) > 0$ and $g(x) > 0$ for all x in (a, b).
55. $c \approx 0.3451$ radian

Exercises 3.5, Page 204

1. Rel. max. $f(1) = 2$

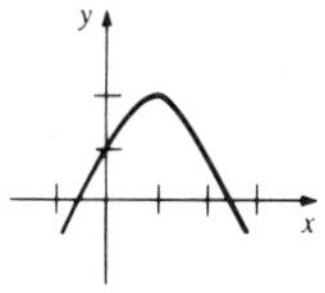

3. Rel. max. $f(-1) = 2$, rel. min. $f(1) = -2$

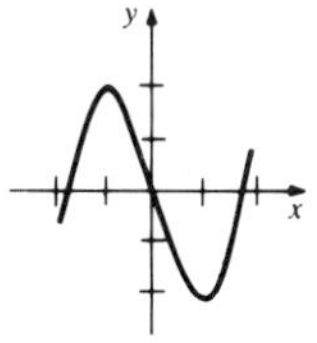

5. Rel. max. $f(\frac{2}{3}) = \frac{32}{27}$, rel. min. $f(2) = 0$

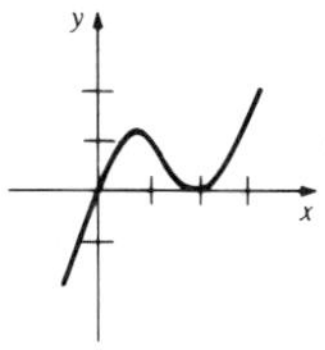

7. No extrema

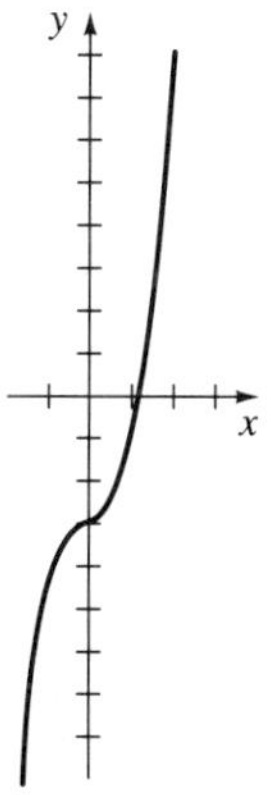

9. Rel. min. $f(-1) = -3$

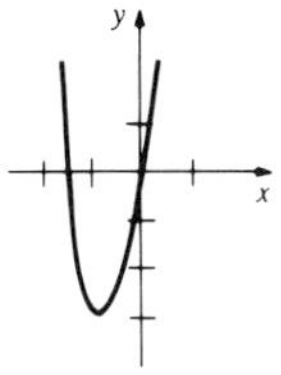

11. Rel. min. $f(0) = 0$

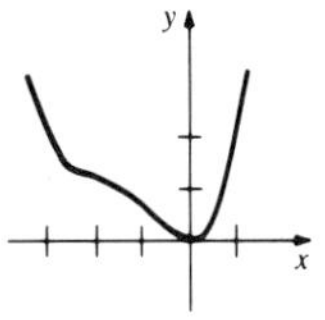

13. Rel. max. $f(0) = f(3) = 0$, rel. min. $f(\frac{3}{2}) = -\frac{81}{16}$

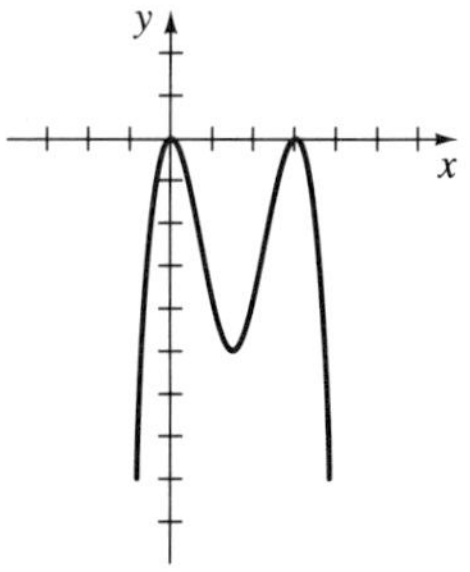

15. Rel. max. $f(0) = 0$, rel. min. $f(1) = -1$

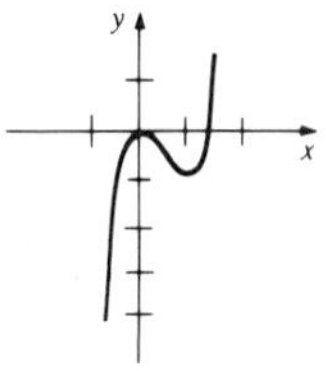

17. Rel. max. $f(-3) = -6$, rel. min. $f(1) = 2$

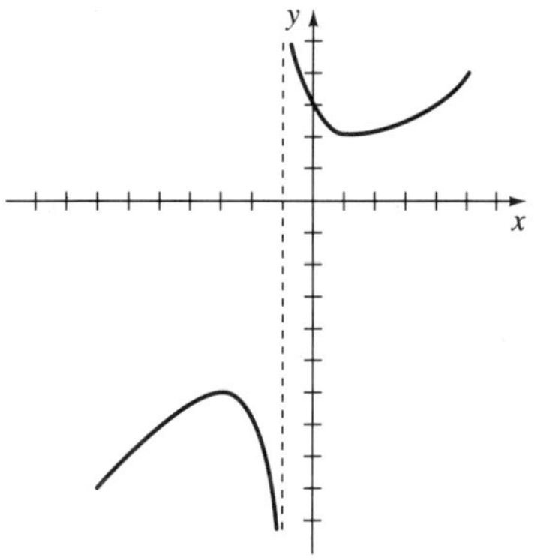

19. Rel. max. $f(\sqrt{3}) = 2/3\sqrt{3}$,
rel. min. $f(-\sqrt{3}) = -2/3\sqrt{3}$

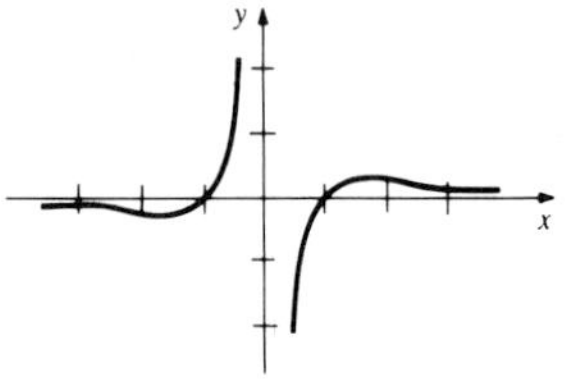

21. Rel. max. $f(0) = 10$

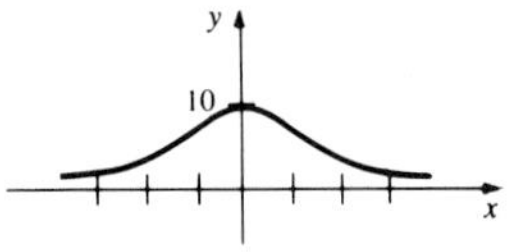

23. Rel. min. $f(-2) = f(2) = 0$,
rel. max. $f(0) = \sqrt[3]{16}$

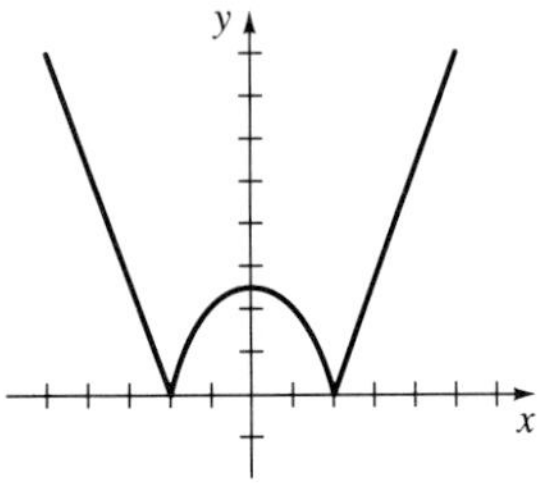

25. Rel. max. $f(\sqrt{2}/2) = \frac{1}{2}$,
rel. min. $f(-\sqrt{2}/2) = -\frac{1}{2}$

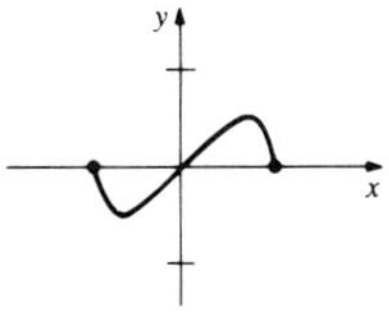

27. Rel. max. $f(-8) = 16$,
rel. min. $f(8) = -16$

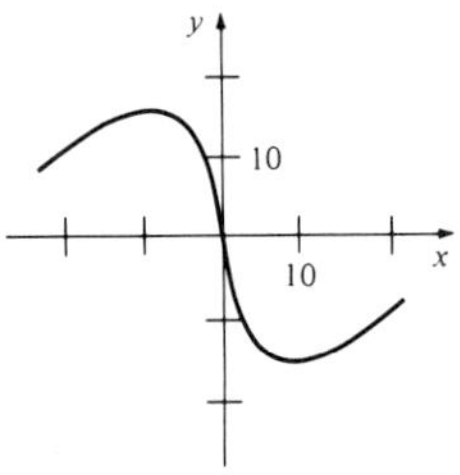

29. Rel. max. $f(0) = 0$,
rel. min. $f(-2) = -12(2)^{2/3}$,
rel. min. $f(2) = -12(2)^{2/3}$

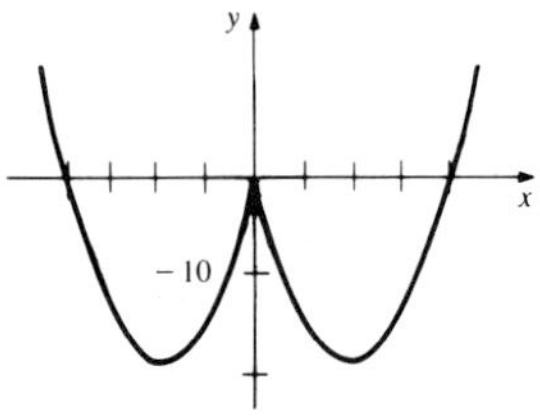

31.

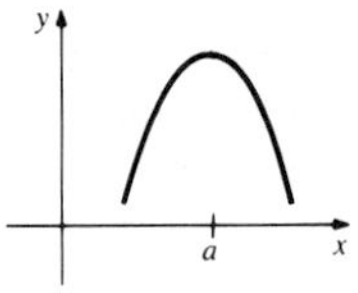

33.

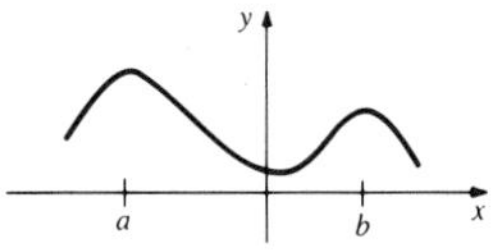

35.

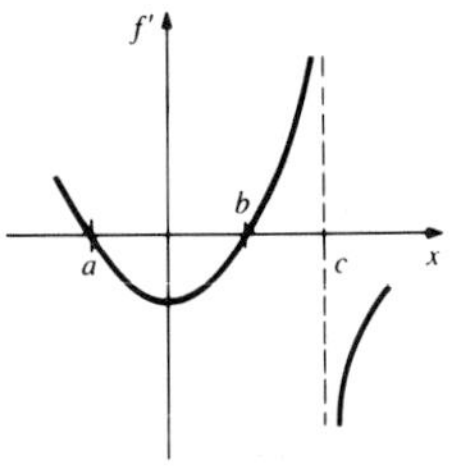

37.

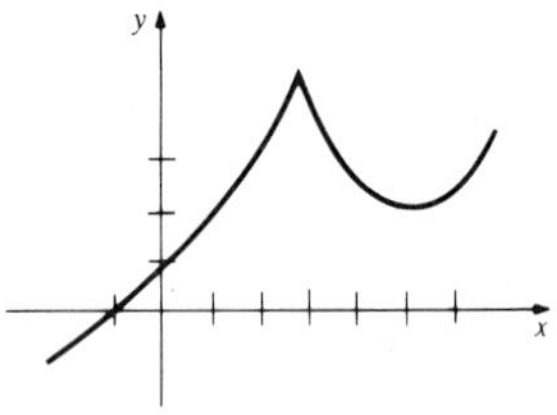

39.

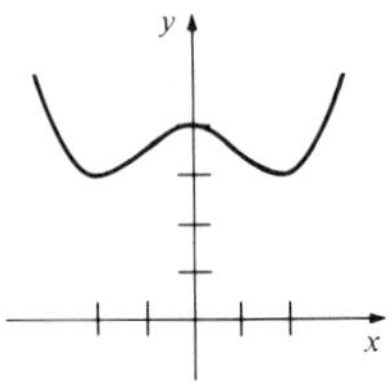

41. Rel. min. $f'(-2) = -13$

43. $(n\pi, \pi/2 + n\pi), (\pi/2 + n\pi, \pi + n\pi)$, $n\pi/2, n = 0, \pm 1, \pm 2, \ldots$;
rel. max. $f(-\pi/2) = f(\pi/2) = \cdots = 1$,
rel. min. $f(0) = f(\pi) = \cdots = 0$

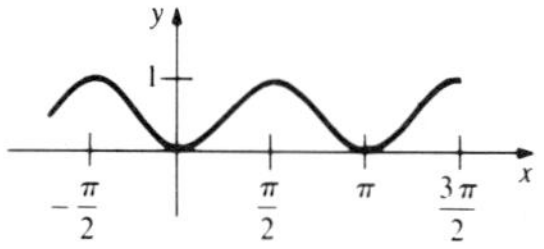

45. $f(x) = -\frac{1}{2}x^2 + 2x + 4$ 47. Yes
49. Only $f + g$ necessarily has a relative maximum at c.
51. Physically, this means that there is the same energy loss regardless of the direction in which the sound travels. $T(1) = 1$ is a relative maximum.

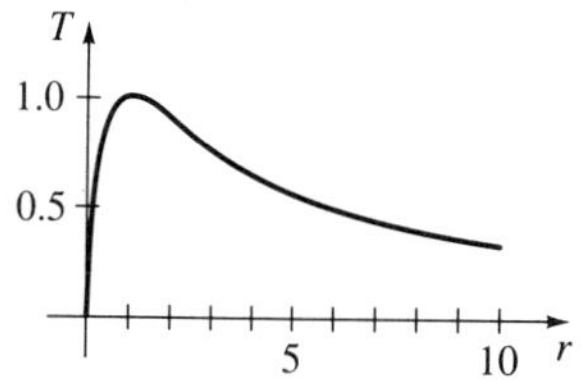

Exercises 3.6, Page 213

1. Concave downward on $(-\infty, \infty)$
3. Concave upward on $(-\infty, 2)$, concave downward on $(2, \infty)$
5. Concave upward on $(-\infty, 2)$ and on $(4, \infty)$, concave downward on $(2, 4)$
7. Concave upward on $(-\infty, 0)$, concave downward on $(0, \infty)$
9. Concave upward on $(0, \infty)$, concave downward on $(-\infty, 0)$
11. Concave upward on $(-\infty, -1)$ and on $(1, \infty)$, concave downward on $(-1, 1)$
13. f' decreasing on $(-\infty, -1]$, f' increasing on $[-1, 0]$, f' decreasing on $[0, 1]$, f' increasing on $[1, \infty)$
17. $(-\sqrt{2}, -21 - \sqrt{2}), (\sqrt{2}, -21 + \sqrt{2})$
19. $(0, 0), (\pm\pi, 0), (\pm 2\pi, 0), \ldots$
21. $(0, 0), (\pm\pi, \pm\pi), (\pm 2\pi, \pm 2\pi), \ldots$
23. Rel. max. $f(\frac{5}{2}) = 0$

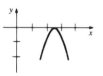

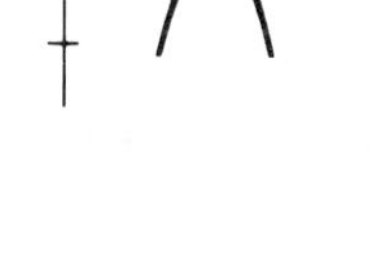

25. Point of inflection $(-1, 0)$

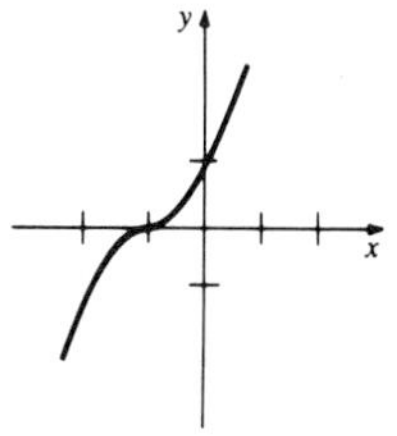

27. Rel. max. $f(-1) = 4$, rel. min. $f(1) = -4$, points of inflection $(0, 0)$, $(-\sqrt{2}/2, 7\sqrt{2}/4), (\sqrt{2}/2, -7\sqrt{2}/4)$

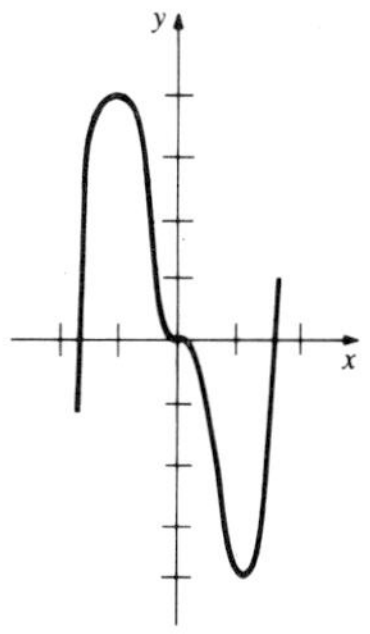

29. Rel. max. $f(\sqrt{2}) = \sqrt{2}/4$, rel. min. $f(-\sqrt{2}) = -\sqrt{2}/4$, points of inflection $(0, 0)$, $(-\sqrt{6}, -\sqrt{6}/8), (\sqrt{6}, \sqrt{6}/8)$

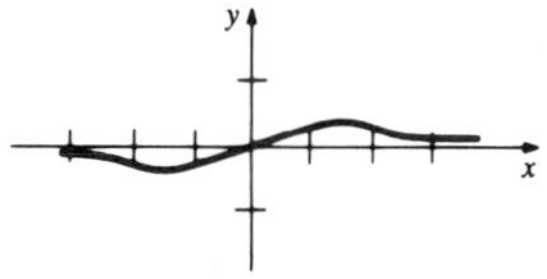

31. Rel. max. $f(0) = 3$

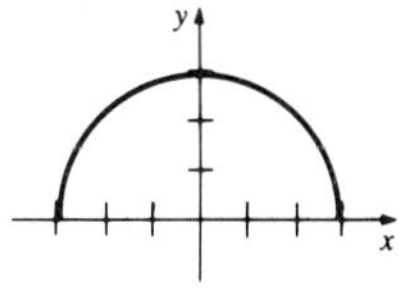

33. Rel. min. $f(-\frac{1}{4}) = -3/4^{4/3}$, points of inflection $(0, 0), (1/2, 3/2^{4/3})$

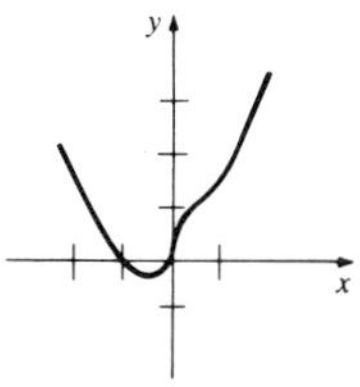

35. Rel. max. $f(2\pi/3) = f(4\pi/3) = 1$, rel. min. $f(\pi/3) = f(\pi) = f(5\pi/3) = -1$, points of inflection $(\pi/6, 0)$, $(\pi/2, 0)$, $(5\pi/6, 0)$, $(7\pi/6, 0)$, $(9\pi/6, 0)$, $(11\pi/6, 0)$

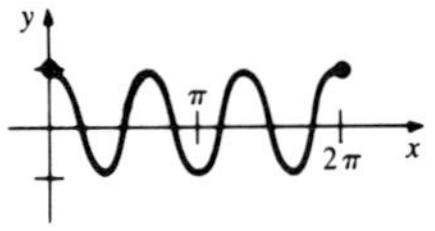

37. Rel. max. $f(\pi/4) = \sqrt{2}$, rel. min. $f(5\pi/4) = -\sqrt{2}$, points of inflection $(3\pi/4, 0)$, $(7\pi/4, 0)$

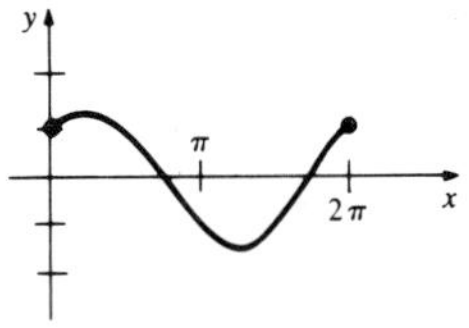

39. Rel. max. $f(\pi/4) = \frac{1}{2}$

41. Rel. min. $f(\pi) = 0$

43.

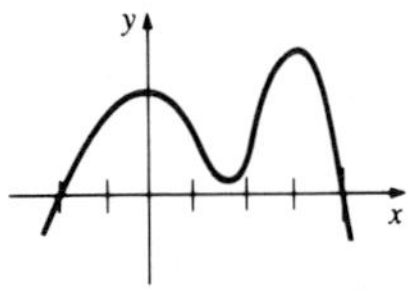

45.

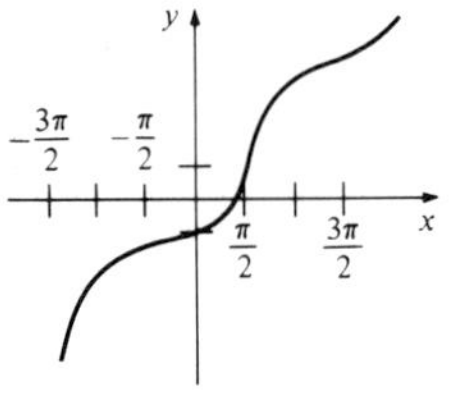

47. $f(x) = -\frac{1}{6}x^3 + \frac{1}{2}x^2 + \frac{2}{3}x$

49.

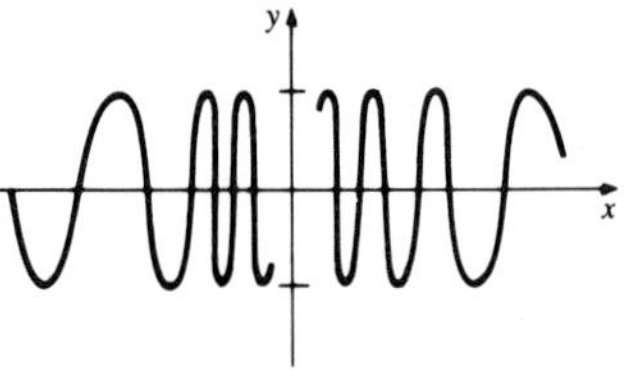

53. $(c, f(c))$ is a point of inflection.

Exercises 3.7, Page 221

1. 30, 30 3. $\frac{1}{2}$ 5. $\frac{1}{3}$ and $\frac{2}{3}$

7. $(2, 2\sqrt{3})$, $(2\ -2\sqrt{3})$; $(0, 0)$ 9. $(\frac{4}{3}, -\frac{128}{27})$

11. Base $\frac{3}{2}$, height 1

13. Base 6, top 3, height $3\sqrt{3}/2$

15. 750 ft by 750 ft 17. 2000 m by 1000 m

19. Base 40 cm by 40 cm, height 20 cm

21. Base $\frac{80}{3}$ cm by $\frac{80}{3}$ cm, height $\frac{20}{3}$ cm; maximum volume $128{,}000/27\ \text{cm}^3$

23. Radius $2/3\pi$ m, height $2/3$ m

25. Radius $2R/3$, height $H/3$

27. Square with length of side $(a + b)/\sqrt{2}$

29. Height $\frac{15}{2}$ cm, width 15 cm

31. Radius $\sqrt[3]{16/\pi}$, height $2\sqrt[3]{16/\pi}$

33. $\sqrt{\frac{2}{3}}R$; $(2\sqrt{3}/27)\pi R^3$; $2\pi(1 - \sqrt{2/3})$ radians

35. $\pi/12$; $32 - 16\sqrt{3}$

37. Radius of circular portion $10/(4 + \pi)$, width $20/(4 + \pi)$, height of rectangular portion $10/(4 + \pi)$

39. The wire should not be cut at all. Bend the entire portion into a circle of radius $1/(2\pi)$ m.

41. Length of cross-section $\sqrt{3}d/3$, width of cross-section $\sqrt{6}d/3$

43. Radius of cylinder (and hemisphere) $\sqrt[3]{9}$, length of cylinder $2\sqrt[3]{9}$

45. 35 47. $-\frac{1}{8}$

49. $y = h/2$; maximum distance h

51. $\frac{50}{11}$ m from I_1 53. $16\sqrt{2}$ ft

55. Minimum costs occur when $x = 4$.

59. Let y denote the difference between Young's rule and Cowling's rule. Then y_{max} occurs at $t = 12(\sqrt{2} - 1) \approx 5$ yr. At this age $y_{max} \approx 0.04D_a$.

63. $w_0L^4/384EI$

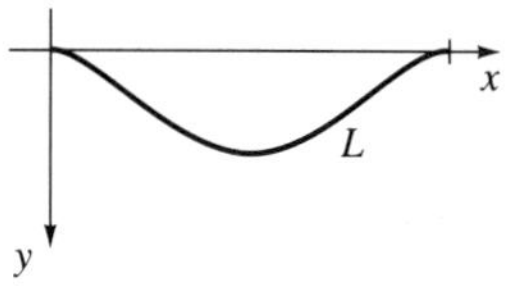

65. $L = x + 2\sqrt{4 + (4 - x)^2}$; $4 - 2/\sqrt{3}$; $L = x + \sqrt{1 + (4 - x)^2} + \sqrt{4 + (4 - x)^2}$; the length of AB is close to 3.2.

Exercises 3.8, Page 231

1. 40,000; 16,100; 60 3. 60; 59

5. 183; 180; 186 7. $R''(x) < 0$ for all $x > 0$

11. $\frac{2}{3}$, inelastic 13. $\frac{9}{8}$, elastic

15. $\frac{7}{16}$, inelastic

17. $\frac{18}{7}$, decreases by approximately 15.4%

19. 4.47

Chapter 3 Review Exercises, Page 232

1. False 2. False 3. False 4. True

5. True 6. False 7. False 8. True

9. True 10. False

11. f need not be differentiable at c.

13. Abs. max. $f(-3) = 348$, abs. min. $f(4) = -86$
15. Abs. max. $f(3) = \frac{9}{7}$, abs. min. $f(0) = 0$
17.

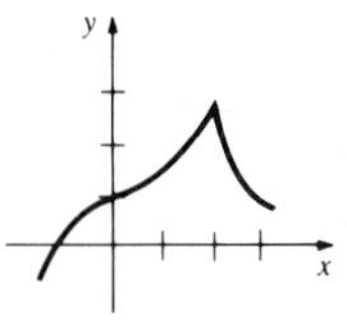

19. Maximum velocity is $v(2) = 12$, maximum speed is $|v(-1)| = |v(5)| = 15$

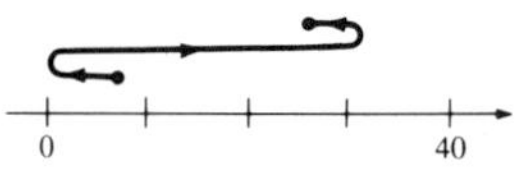

23. Zeros are a, b, and $(a + b)/2$.
25. Rel. max. $f(-3) = 81$, rel. min. $f(2) = -44$

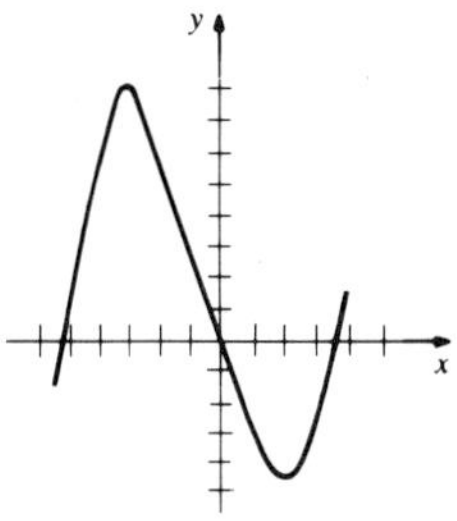

27. Rel. max. $f(0) = 2$, rel. min. $f(1) = 0$

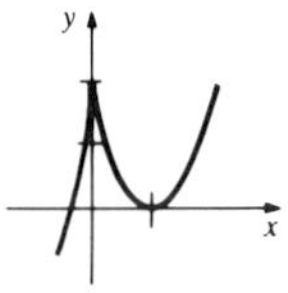

29. Rel. min. $f(0) = 0$, points of inflection $(-3, 27)$, $(-1, 11)$
31. Point of inflection $(3, 10)$ 33. $(a + b + c)/3$
35. $\frac{5}{4}$ ft/s 39. 2
43. Fly to point 17.75 km from the nest.
45. A 10-ft section of fence should be attached to the 40-ft-long side of the house; the other three sections should be 50 ft long.
47. $x = 195$, $y = 390$; 57,037.5 ft^2
49. 10 ft from the 10-ft mast
55. Approximate dimensions: $11.83 \times 8.66 \times 3.17$; approximate maximum volume: 324.76 in^3
57. 36,300; 13,056; 236; 600; 224; 400
59. Approximately 31.5 cm

Exercises 4.1, Page 247

1. $3x + C$ 3. $\frac{1}{6}x^6 + C$ 5. $\frac{3}{2}x^{2/3} + C$
7. $t - \frac{25}{12}t^{0.48} + C$ 9. $x^3 + x^2 - x + C$
11. $\frac{2}{7}x^{7/2} - \frac{4}{3}x^{3/2} + C$ 13. $\frac{16}{3}x^3 + 4x^2 + x + C$
15. $\frac{1}{3}x^3 - 4x + C$ 17. $-r^{-1} + 5r^{-2} + C$
19. $-\frac{1}{2}x^{-2} + \frac{1}{3}x^{-3} - \frac{1}{4}x^{-4} + C$
21. $16w^4 - 16w^3 + 6w^2 - w + C$
23. $-4\cos x + \frac{1}{4}x^{-4} + C$
25. $-\cot x + \csc x + C$ 27. $-2\cot x + 3x + C$
29. $\frac{1}{2}x^2 + C$ 37. $x^2 - 4x + 5$
39. $y = 2x^3 + 9x + C$ 41. $y = -1/x + C$
43. $y = x - x^2 - \cos x + C$ 45. $y = x^2 - x + 1$
47. $f'(x) = x^2 + C_1$, $f(x) = \frac{1}{3}x^3 + C_1x + C_2$
49. $f(x) = x^4 + x^2 - 3x + 2$
51. $y^{-2} = 2x^{-1} + C$
53. $y + y^2 + \frac{1}{3}y^3 = x + x^2 + \frac{1}{3}x^3 + C$
55. $\cos y = x^{-1} - 5x + C$
57. $y^3 = -3x^{-1} + 30$ 59. $y = \dfrac{\omega^2}{2g}x^2$
61. $h(t) = \dfrac{1}{2500}(-t + 100\sqrt{5})^2$; $100\sqrt{5}$ s
63. $V(t) = \left[k\left(\dfrac{4\pi}{3}\right)^{1/3} t + C\right]^3$

Exercises 4.2, Page 257

1. $-\frac{1}{6}(1 - 4x)^{3/2} + C$ 3. $-\frac{1}{10}(5x + 1)^{-2} + C$
5. $-\frac{5}{32}(3 - 4x)^{8/5} + C$ 7. $\frac{2}{3}(x^2 + 4)^{3/2} + C$
9. $\frac{3}{4}(z^2 + 9)^{2/3} + C$ 11. $\frac{1}{40}(4x^2 - 16x + 7)^5 + C$
13. $\frac{1}{2}(x^3 + 3x - 16)^{2/3} + C$
15. $\dfrac{1}{3}\left(3 - \dfrac{2}{v}\right)^{3/2} + C$ 17. $-\frac{9}{4}(1 - \sqrt[3]{x})^{4/3} + C$
19. $-\frac{1}{2}(4 + \sqrt{t})^{-4} + C$ 21. $\frac{1}{7}(x - 1)^7 + C$
23. $-\frac{1}{4}\cos 4x + C$ 25. $\frac{1}{3}(2t)^{3/2} - \frac{1}{6}\sin 6t + C$
27. $\frac{1}{2}\sin x^2 + C$ 29. $\frac{1}{5}\sin(5x + 1) + C$
31. $-2\csc\sqrt{x} + C$ 33. $-\frac{1}{2}\cot(z^2 + 2z) + C$
35. $\frac{1}{18}\sin^6 3x + C$ 37. $\frac{1}{6}\tan^3 2x + C$
39. $\frac{1}{7}\tan 7x - x + C$ 41. $-2\cot x - \csc x + C$
43. $\frac{1}{2}x - \frac{1}{4}\sin 2x + C$ 45. $\frac{1}{2}x + \frac{1}{16}\sin 8x + C$
47. $\frac{3}{2}x + \sin 2x + \frac{1}{8}\sin 4x + C$
49. $\sin x - \frac{1}{3}\sin^3 x + C$
51. $\frac{2}{3}(t + 2)^{3/2} - 4(t + 2)^{1/2} + C$
53. $-\frac{3}{4}(1 - x)^{4/3} + C$
55. $\tan y = \frac{1}{2}x - \frac{1}{4}\sin 2x + C$
57. $y = x + \cos x - \pi$

Exercises 4.3, Page 262

1. $3 + 6 + 9 + 12 + 15$ 3. $\dfrac{2}{1} + \dfrac{2^2}{2} + \dfrac{2^3}{3} + \dfrac{2^4}{4}$
5. $-\frac{1}{7} + \frac{1}{9} - \frac{1}{11} + \frac{1}{13} - \frac{1}{15} + \frac{1}{17} - \frac{1}{19} + \frac{1}{21} - \frac{1}{23} + \frac{1}{25}$
7. $(2^2 - 4) + (3^2 - 6) + (4^2 - 8) + (5^2 - 10)$

9. $-1+1-1+1-1$ 11. $\sum_{k=1}^{6}(2k)^2$

13. $\sum_{k=1}^{7}(2k+1)$ 15. $\sum_{k=0}^{12}(3k+1)$

17. $\sum_{k=1}^{5}(-1)^{k+1}\frac{1}{k}$ 19. $\sum_{k=1}^{8}6$

21. $\sum_{k=1}^{4}\frac{(-1)^{k+1}}{k^2}\cos\frac{k\pi}{p}x$ 23. 420 25. 65

27. 109 29. 3069 31. 101,262

33. 70,940 35. Yes; let $j = k + 3$.

37. $\sum_{k=0}^{6}(k+6)$; $\sum_{k=1}^{7}(k+5)$; $\sum_{k=2}^{8}(k+4)$

39. $\sum_{k=1}^{7}\frac{1}{10^k}$ 41. $\sum_{k=1}^{n}\left(3+\frac{2k}{n}\right)\frac{2}{n} = 8 + \frac{2}{n}$

43. 20

45. $-f(0) + f(n)$; $f(k) = \sqrt{k}$ and $f(k) = \frac{-1}{k+1}$

49. $n(n+1)$ 51. 252.28 kJ

53. $\bar{x} = \frac{1}{n}\sum_{k=1}^{n} x_k$

Exercises 4.4, Page 270

1. 12 3. 18 5. 28 7. $\frac{8}{3}$ 9. $\frac{4}{3}$

11. $\frac{16}{3}$ 13. $\frac{1}{4}$ 15. 4 17. $\frac{32}{5}$

19. 5 21. $\frac{25}{2}$ 23. $\frac{77}{60}$; $\frac{25}{12}$

25. $\lim_{n\to\infty}\sum_{k=1}^{n} f\left(a + (k-1)\frac{b-a}{n}\right)\frac{b-a}{n}$;

$\lim_{n\to\infty}\sum_{k=1}^{n} f\left(a + (2k-1)\frac{b-a}{2n}\right)\frac{b-a}{n}$

27. 9 29. $\frac{1}{6}$ 31.

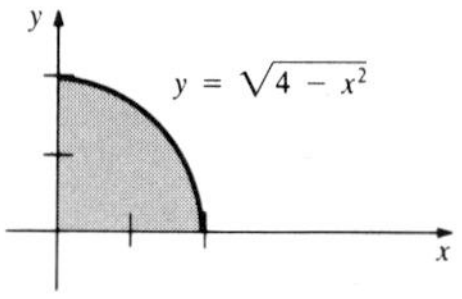

33. $\frac{2}{3}$ 35. 258

Exercises 4.5, Page 278

1. $\frac{19}{2}$; $\frac{5}{2}$ 3. $\frac{33}{2}$; 1 5. $\frac{189}{256}$; $\frac{3}{4}$

7. $(3-\sqrt{2})\pi/4$; π 9. 5 11. 24

13. -4 15. $\frac{2}{3}$ 17. $\frac{5}{6}$ 19. $-\frac{3}{4}$

25. $\frac{2}{3}$ 27. $\int_{-2}^{4}\sqrt{9+x^2}\,dx$

29. $\int_{0}^{2}(x+1)\,dx$ 33. 1.1118625, 1.1116552

Exercises 4.6, Page 282

1. 12 3. 10 5. $-\frac{3}{2}$ 7. -20

9. 56 11. -20 13. 24 15. 0

17. 20 19. -56 21. 64 23. 20

25. 8 27. 0 29. 3 31. $\frac{1}{2}$

33. $(\pi+2)/4$ 35. 0 37. 36 39. 10

43. $x^3 \le x^2$ for all x in $[0, 1]$.

45. The function $f(x) = (x^3+1)^{1/2}$ is increasing on the interval $[0, 1]$. Therefore, $f(0) \le f(x) \le f(1)$ or $1 \le f(x) \le 2^{1/2}$. The result follows from the comparison property with $b - a = 1$, $m = 1$, and $M = 2^{1/2}$.

Exercises 4.7, Page 294

1. $(3x^2-2x)^6$ 5. 4 7. 12 9. 46

11. 1 13. $-\frac{1}{3}-\frac{\sqrt{2}}{6}$ 15. $\frac{2}{3}$ 17. 0

19. $-\frac{2}{3}$ 21. $-\frac{28}{3}$ 23. $\frac{8}{3}$ 25. $\frac{9}{2}$

27. 21 29. $\frac{128}{3}$ 31. 1 33. $\frac{65}{4}$

35. $\sqrt{6}-\sqrt{3}$ 37. $\frac{1}{2}$ 39. 1 41. $\frac{2}{3}$

43. $\frac{4\pi+6}{(\pi+3)(\pi+2)}$ 45. 0 47. 0

49. 5 51. 22 53. 4 55. $\frac{19}{6}$

57. 9 59. $\frac{38}{3}$ 61. $\frac{116}{15}$ 63. 3

65. $\int_{a}^{x^2} f(t)\,dt$; $2xf(x^2)$ 67. $-6\sqrt{24x+5}$

69. 28

71. The limit of the sum is the same as $\int_0^{\pi}\sin x\,dx$.

73. $f(x) = x^{-2}$ is not continuous on $[-1, 1]$.

77.

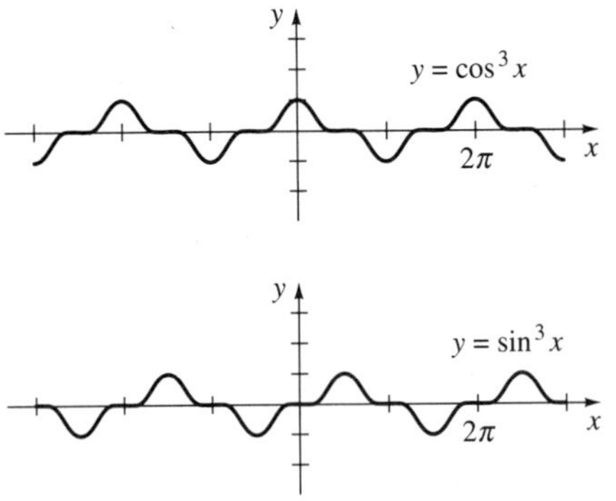

0; 0

Exercises 4.8, Page 306

1. 78, $M_3 = 77.25$ 3. 22, $T_3 = 22.5$

5. 1.7564, 1.8667 7. 1.1475, 1.1484

9. 0.4393, 0.4228 11. 0.4470, 0.4900

13. $\frac{26}{3}$, 8.6611 15. 1.6222 17. 0.7854

19. 0.4339 21. 11.1053 23. $n \ge 8$

25. 1.11

27. For Simpson's Rule: $n \ge 26$; for Trapezoidal Rule: $n \ge 366$

29. Trapezoidal Rule gives 1.10.

31. For $n = 2$ and $n = 4$, the Midpoint Rule gives 36, which is the exact value of the integral.

33. 2/3; $M_8 = 21/32$; $T_8 = 11/16$; $E_8 = 1/96$ for Midpoint Rule and $E_8 = 1/48$ for Trapezoidal Rule. The error for the Midpoint Rule is one-half the error for the Trapezoidal Rule.
37. Approximately 7.1
39. Approximately 4976 gallons
41.

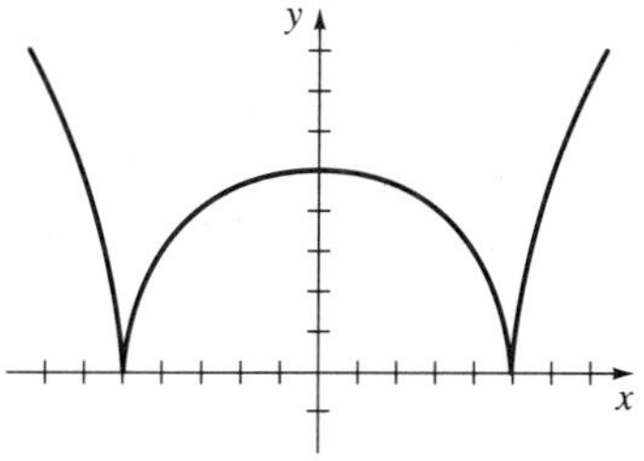

41.4028

Chapter 4 Review Exercises, Page 308

1. False 2. True 3. True 4. True
5. True 6. False 7. False 8. False
9. False 10. False 11. True
12. True 13. True 14. True
15. True 16. False 17. True
18. True 19. True 20. True
21. $f(x)$ 23. $\sqrt{6}$ 25. 3720 27. 250
29. Of the same length 31. -6
33. $\frac{1}{505}(5t+1)^{101} + C$ 35. $\frac{1}{2}$ 37. 0
39. $-\frac{1}{56}\cot^7 8x + C$ 41. -6 43. $\frac{11}{2}$
45. 0 47. π 49. $E_4 \approx 0.0833$
51. Midpoint Rule: 0.3074, Trapezoidal Rule: 0.3160, Simpson's Rule: 0.3099
53. 130 ft^2

Exercises 5.1, Page 322

1. $\frac{4}{3}$ 3. $\frac{81}{4}$ 5. $\frac{9}{2}$ 7. $\frac{11}{2}$ 9. $\frac{11}{4}$
11. $\frac{11}{6}$ 13. 2 15. $\frac{3}{4}(2^{4/3} + 3^{4/3})$
17. 4 19. 2π 21. $\frac{7}{3}$ 23. $\frac{27}{2}$
25. $\frac{32}{3}$ 27. $\frac{81}{4}$ 29. 4 31. $\frac{10}{3}$
33. $\frac{64}{3}$ 35. $\frac{128}{5}$ 37. $\frac{118}{3}$ 39. 22
41. $\frac{9}{2}$ 43. $\frac{8}{3}$ 45. 8 47. $2\sqrt{2} - 2$
49. $4\sqrt{3} - 4\pi/3$ 51. $\frac{19}{4}$ 53. $\frac{5}{2}$
55. $\pi a^2/4$ 57. $\frac{52}{3}$
61. Area is 2 square units.

Exercises 5.2, Page 326

1. $(5625)\sqrt{3}/16$ 3. $\frac{1024}{3}$ 5. 128
7. $10\pi/3$ 9. 9 11. $2a^3/3$
13. (d) $V = \dfrac{h\pi}{3}[r_1^2 + r_1 r_2 + r_2^2]$

Exercises 5.3, Page 335

1. $\pi/2$ 3. $4\pi/5$ 5. $\pi/6$ 7. $1296\pi/5$
9. $\pi/2$ 11. $32\pi/5$ 13. 32π 15. $7\pi/3$
17. $256\pi/15$ 19. $3\pi/5$ 21. 36π
23. $500\pi/3$ 25. $16\pi/105$ 27. π^2
29. $(4\pi - \pi^2)/4$ 31. $4\pi/5$ 33. $\pi/6$
35. $8\pi/15$ 37. $250\pi/3$ 39. $36\sqrt{3}\pi/5$
41. $3\pi/2$ 43. 16π 45. $8\pi/5$
47. $21\pi/10$ 49. $\pi/6$ 51. $248\pi/15$
53. 4π 55. $625\pi/6$ 57. $45\pi/2$
59. $(\pi^2 - 2\pi)/2$ 61. $(\pi/3)r^2h$ 63. $(4\pi/3)r^3$
65. $(4\pi/3)ab^2$ 67. $43\pi/2$
69. $V = 4[\frac{1}{63}a^2 + \frac{1}{35}b^2 + \frac{1}{15}c^2 + \frac{1}{3}d^2 + \frac{2}{35}ac + \frac{2}{15}bd]$; approximately 1.319 cubic units;

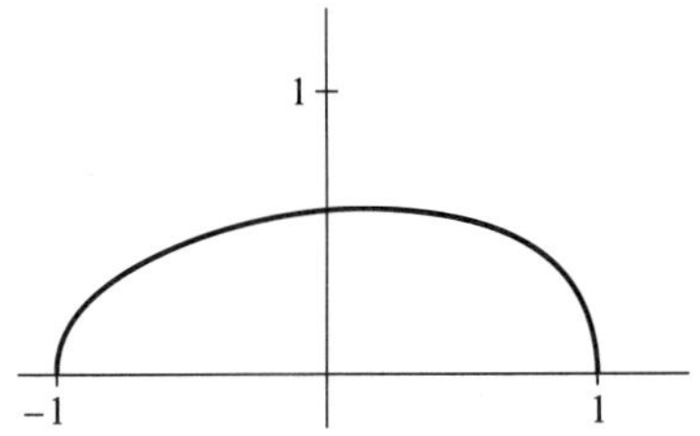

approximately 1.263 cubic units

Exercises 5.4, Page 340

1. $2\sqrt{2}$ 3. $\frac{1}{27}[13^{3/2} - 8] \approx 1.44$ 5. 45
7. $\frac{10}{3}$ 9. $\frac{4685}{288} \approx 16.27$ 11. 9
13. $\displaystyle\int_{-1}^{3} \sqrt{1 + 4x^2}\, dx$ 15. $\displaystyle\int_{0}^{\pi} \sqrt{1 + \cos^2 x}\, dx$
17. $\frac{1}{27}[40^{3/2} - 8] \approx 9.07$
19. The integrand of $\displaystyle\int_0^1 x^{-1/3}\, dx$ is discontinuous on [0, 1]; 6
21. $\pi/2$ 23. $(x_2 - x_1)\sqrt{1 + m^2}$
25. 1.4804 27. 1.222

Exercises 5.5, Page 345

1. $208\pi/3$ 3. $(\pi/27)[10^{3/2} - 1] \approx 3.56$
5. $(\pi/6)[37^{3/2} - 1] \approx 117.32$ 7. $100\sqrt{5}\pi$
9. $253\pi/20$
11. $(\pi r/6h^2)[(r^2 + 4h^2)^{3/2} - r^3]$; approximate percentage error is $0.99\% < 1\%$
13. $A_s/A_e = h/2(R + h)$; approximately 11.9%; 6380 km; $A_s/A_e \to \frac{1}{2}$ as $h \to \infty$; approximately 49%
15. $12\pi a^2/5$
21. $(2\pi/3)\displaystyle\int_1^8 (4x^{-1/3} - x^{1/3})\sqrt{9x^{2/3} + 4}\, dx$
23. 26.211

Exercises 5.6, Page 350

1. -4 3. $\frac{34}{3}$ 5. 3 7. 0 9. 2
11. $\frac{61}{9}$ 13. 24 15. $\frac{1}{12}$ 17. 0
19. $3\sqrt{3}/\pi$ 21. $-1 + \dfrac{2\sqrt{3}}{3} \approx 0.15$
23. 12 25. 103° 29. $2kt_1/3$
31. The car travels at a constant rate of 60 mph for 40 miles and at a constant rate of 40 mph for the next 30 miles; 51.4; 49.4
33. $[f(x+h) - f(x)]/h$

Exercises 5.7, Page 354

1. $s(t) = 6t - 7$ 3. $s(t) = \frac{1}{3}t^3 - 2t^2 + 15$
5. $s(t) = -\frac{5}{2}\sin(4t + \pi/6) + \frac{5}{2}$
7. $v(t) = -5t + 9$, $s(t) = -\frac{5}{2}t^2 + 9t - \frac{9}{2}$
9. $v(t) = t^3 - 2t^2 + 5t - 3$, $s(t) = \frac{1}{4}t^4 - \frac{2}{3}t^3 + \frac{5}{2}t^2 - 3t + 10$
11. $v(t) = \frac{21}{4}t^{4/3} - t - 26$, $s(t) = \frac{63}{28}t^{7/3} - \frac{1}{2}t^2 - 26t - 48$
13. $\frac{11}{225}$ km 15. 256 ft 17. 30.62 m
19. 400 ft; 6 s 21. -80 ft/s
23. Approximately 848.5 m/s
29. 17 cm 31. 34 cm 33. 24 cm

Exercises 5.8, Page 360

1. 3300 ft-lb 3. 222,750 ft-lb 5. $\frac{2}{5}$ ft
7. 10 N-m; 27.5 N-m 9. 7.5 ft-lb; 37.5 ft-lb
11. 453.1×10^8 joules 13. 127,030.9 ft-lb
15. 45,741.6 ft-lb 17. 57,408 ft-lb
19. 900,000 ft-lb 21. 24,960 ft-lb
23. $2{,}700{,}000 - 100x$; 2.65×10^9 ft-lb
27. 126 N-m 29. 5.5 N-m
31. $v_0 = \sqrt{2km_1/R}$; 11,183.1 m/s; 5086.4 m/s

Exercises 5.9, Page 366

1. 196,000 N/m^2, 4,900,000π N; 196,000 N/m^2, 784,000π N; 196,000 N/m^2, 19,600,000π N
3. 499.2 lb/ft^2, 244,640 lb; 59,904 lb, 29,952 lb
5. 121.59 lb 7. 1280 lb 9. 3660.8 lb
11. 13,977.6 lb 13. 9984π lb 15. 5990.4 lb
17. $30{,}420\sqrt{17}$ lb 19. $4{,}992{,}000\sqrt{2}$ lb

Exercises 5.10, Page 375

1. $-\frac{2}{7}$ 3. $-\frac{13}{30}$ 5. 1 7. $\frac{115}{36}$
9. $\frac{4}{7}$ 11. $\frac{19}{5}$ 13. $\frac{11}{10}$ 15. $\frac{15}{2}$
17. $\bar{x} = 3$, since $\rho(x)$ is symmetric about the line $x = 3$.
19. $\bar{x} = -\frac{2}{7}, \bar{y} = \frac{17}{7}$ 21. $\bar{x} = \frac{17}{11}, \bar{y} = -\frac{20}{11}$
23. $\bar{x} = \frac{10}{9}, \bar{y} = \frac{28}{9}$ 25. $\bar{x} = \frac{3}{4}, \bar{y} = \frac{3}{10}$
27. $\bar{x} = \frac{12}{5}, \bar{y} = \frac{54}{7}$ 29. $\bar{x} = \frac{93}{35}, \bar{y} = \frac{45}{56}$
31. $\bar{x} = \frac{1}{2}, \bar{y} = \frac{8}{5}$ 33. $\bar{x} = \frac{16}{35}, \bar{y} = \frac{16}{35}$
35. $\bar{x} = \frac{3}{2}, \bar{y} = \frac{121}{540}$ 37. $\bar{x} = -\frac{7}{10}, \bar{y} = \frac{7}{8}$
39. $\bar{x} = 0, \bar{y} = 2$ 41. $\bar{x} = 0, \bar{y} = (\pi + 8)/8$
45. $2\pi^2 a^2 b$ 47. $4\pi^2 ab$

Exercises 5.11, Page 381

1. 0.11 L/s $\approx$ 6.76 L/min 3. 1.85 mg/L
5. 94% 7. \$49,600 9. \$320/year
11. $CS = \$625$; $PS = \$625$
13. $CS = \$12.50$; $PS = \$2.33$
15. $CS = \$45$; $PS = \$31.50$
19. $\dfrac{1-k}{2-k}\,\dfrac{b^{2-k} - a^{2-k}}{b^{1-k} - a^{1-k}}$

Chapter 5 Review Exercises, Page 382

1. False 2. False 3. True 4. True
5. True 6. True 7. True 8. True
9. True 10. False 11. False
12. False 13. Joules 15. 2500 ft-lb
17. 6 19. $\displaystyle\int_a^b f(x)\,dx$
21. $\displaystyle\int_a^b f(x)\,dx - \int_b^c f(x)\,dx + \int_c^d f(x)\,dx$
23. $\displaystyle\int_a^b 2\,dx$ 25. $\displaystyle -\int_a^b 2f(x)\,dx + \int_b^c 2f(x)\,dx$
27. $\displaystyle -\int_a^0 \frac{1}{2}x\,dx + \int_0^{2b}\left(b - \frac{1}{2}x\right)dx$
29. $\bar{x} = \dfrac{\displaystyle\int_0^2 x(f(x) - g(x))\,dx}{\displaystyle\int_0^2 (f(x) - g(x))\,dx}$, $\bar{y} = \dfrac{\displaystyle\frac{1}{2}\int_0^2 ([f(x)]^2 - [g(x)]^2)\,dx}{\displaystyle\int_0^2 (f(x) - g(x))\,dx}$
31. $\displaystyle 2\pi\int_0^2 x(f(x) - g(x))\,dx$
33. $\displaystyle\int_0^2 (f(x) - g(x))^2\,dx$
35. $\dfrac{315\sqrt{41}}{16}\pi$ ft^2
37. $\frac{256}{45}$ 39. 37.5 joules 41. 624,000 ft-lb
43. 2040 ft-lb 45. 691,612.8 ft-lb
47. $(40^{3/2} - 8)/27$ 49. 17,066.7 N
51. $\frac{3}{4}$ m from the left along the 1-m bar and $\frac{6}{5}$ m from the left along the 2-m bar

Exercises 6.1, Page 393

1. 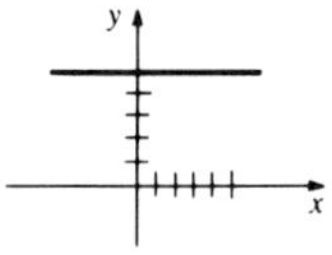

Not one-to-one

3. 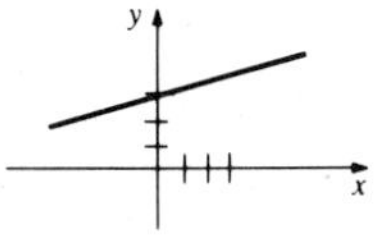

$f^{-1}(x) = 3x - 9$

5. 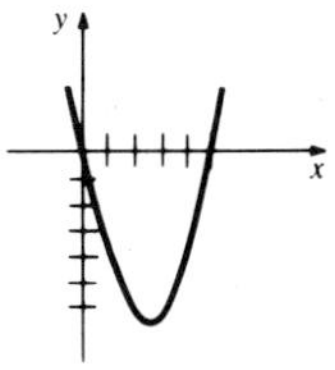

Not one-to-one

7. 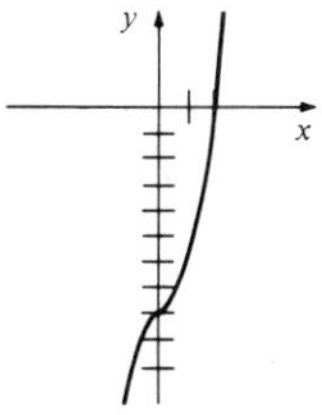

$f^{-1}(x) = \sqrt[3]{x + 8}$

9. 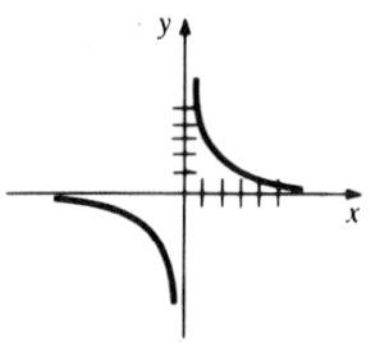

$f^{-1}(x) = 4/x$

11. 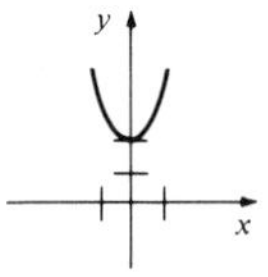

Not one-to-one

13. $f^{-1}(x) = x - 5$ 15. $f^{-1}(x) = \sqrt[3]{\dfrac{x - 7}{3}}$

17. $f^{-1}(x) = \frac{1}{2}x^3 + 2$

19. $f^{-1}(x) = \sqrt[3]{(1 - x)/8x}$

21. $f^{-1}(x) = \dfrac{2 - x}{1 - x}$

23. $f'(x) > 0$ for all x shows that f is increasing on $(-\infty, \infty)$. The result follows from Theorem 6.2.

25. Since f is a polynomial function, it is continuous. Now $f(-2)$ is a relative maximum and $f(1)$ is a relative minimum. Since these are the only extrema, any value of y chosen between $f(1)$ and $f(-2)$ must correspond to at least two values of x. Thus, f is not one-to-one.

27. Domain: $[0, \infty)$
Range: $[-2, \infty)$

29. Domain: all real numbers except 0
Range: all real numbers except -3

31. Domain: $[0, \infty)$
Range: $[5, \infty)$

33. 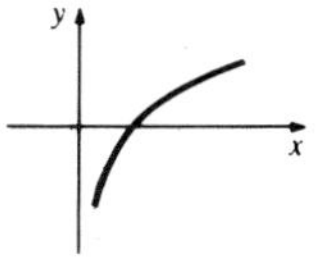

35. 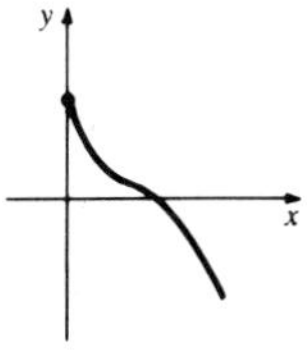

37. $\frac{2}{3}$ 39. $(5, 3)$; $y = \frac{1}{10}x + \frac{5}{2}$

41. $(8, 1)$; $y = \frac{1}{60}x + \frac{13}{15}$

43. $f'(x) = 1/(x + 1) > 0$ for $x > -1$. Hence, by Theorem 6.2, f^{-1} exists and so f is one-to-one; 2

45. $f^{-1}(x) = 1/(x - 2)$;
$(f^{-1})'(x) = -1/(x - 2)^2$

47. $F(x) = (5 - 2x)^2, x \geq 5/2$;
$F^{-1}(x) = \dfrac{5 - \sqrt{x}}{2}$

49. $F(x) = x^2 + 2x + 4, x \geq -1$;
$F^{-1}(x) = -1 + \sqrt{x - 3}$

51. $\frac{51}{4}$ 55. No 59. $f^{-1}(x) = (x + \sqrt[3]{x})^3$

Exercises 6.2, Page 401

1. $x > -1$ 3. $x \neq \pm 1$ 5. No

7. Yes 9. No 11. $10/x$ 13. $1/2x$

15. $(4x^3 + 6x)/(x^4 + 3x^2 + 1)$ 17. $3x + 6x \ln x$

19. $(1 - \ln x)/x^2$ 21. $1/x(x + 1)$

23. $\tan x$ 25. $-1/x(\ln x)^2$

27. $(1 + \ln x)/x \ln x$ 29. $1/[(x \ln x)\ln(\ln x)]$

31. $12(t^2 + 1)/t(3t^2 + 6)$

33. $(x^2 + 6x + 7)/(x + 1)(x + 2)(x + 3)$

35. $y/x(2y^2 - 1)$ 37. $y(1 - x)/x(2y^2 + 1)$

39. $(2x - x^2y - y^3)/(x^3 + xy^2 - 2y)$

41. $y = x - 1$ 43. $y = 4x - 8$

45. $(\frac{1}{4}, -\ln 2)$

47. 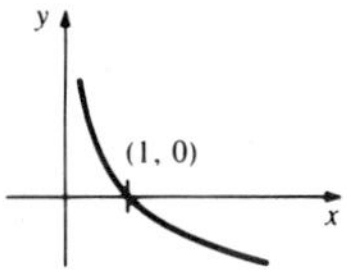

49. 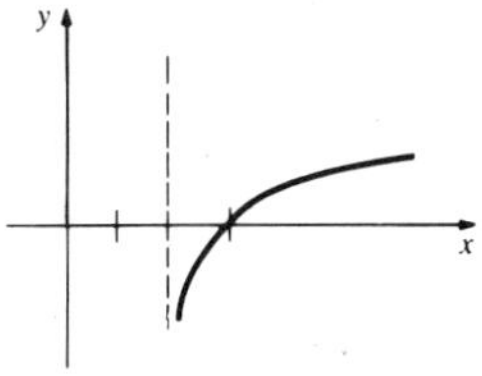

51. x- and y-intercepts are at the origin; symmetry with respect to the y-axis; no asymptotes; $f(0) = 0$ is a relative minimum; concave downward on $(-\infty, -1)$ and $(1, \infty)$, concave upward on $(-1, 1)$

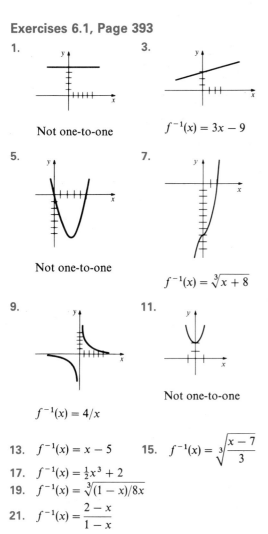

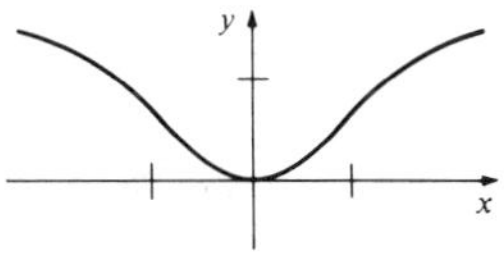

55.

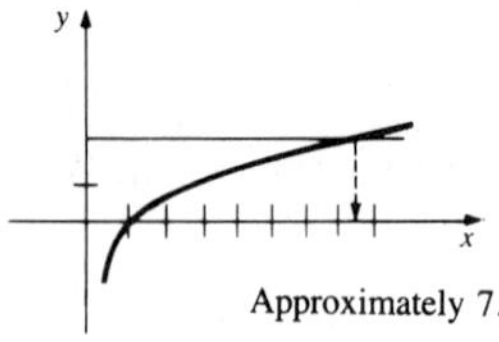

57.

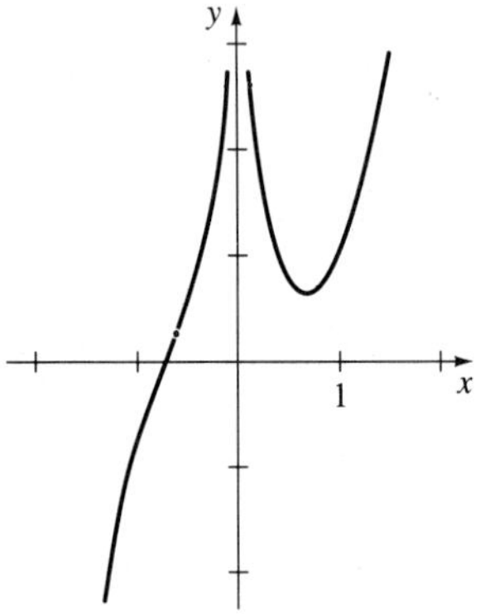

-0.7047; rel. min. $f\left(\frac{1}{\sqrt[3]{3}}\right)=\frac{1}{3}(1+\ln 3)$

Exercises 6.3, Page 407

1. $-e^{-x}$ 3. $\frac{1}{2}x^{-1/2}e^{\sqrt{x}}$
5. $-e^{-2x}(2x+1)/x^2$ 7. $-\frac{5}{2}(1+e^{-5x})^{-1/2}e^{-5x}$
9. $(4x^3+2xe^{x^2})/(x^4+e^{x^2})$
11. $-4/(e^x-e^{-x})^2$ 13. 1
15. $2xe^{3x}/(x^2+1)+3e^{3x}\ln(x^2+1)$
17. $-4e^{(x+2)/(x-2)}/(x-2)^2$ 19. $2xe^{x^2}e^{e^{x^2}}$
21. $9e^{9x}$ 23. $\frac{1}{3}t^{-2/3}e^{t^{1/3}}+\frac{1}{3}e^{t/3}$
25. $(2x+1)^2e^{-(1-x)^4}[4(2x+1)(1-x)^3+6]$
27. $(x^2e^x+e^{2x})/(x+e^x)^2$ 29. $2e^{2x}\sec^2 e^{2x}$
31. $e^{x+y}/(1-e^{x+y})$
33. $(-ye^{xy}\sin e^{xy})/(1+xe^{xy}\sin e^{xy})$
35. $(-y^2+ye^{x/y})/(2y^3+xe^{x/y})$ 37. $-64e^{-4x}$
39. $e^x/(e^x+1)^2$ 41. $y=ex$ 43. $-\frac{1}{2}$
45.

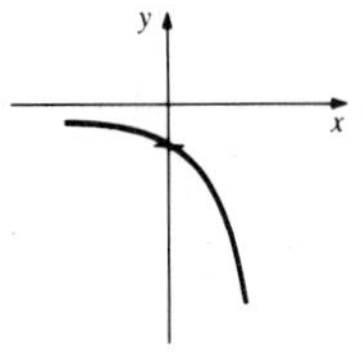

47.

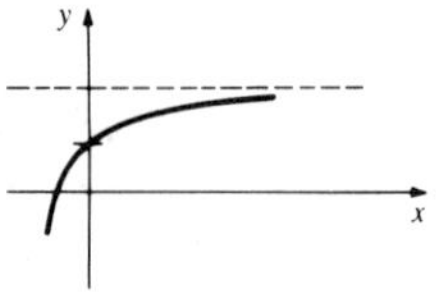

49.

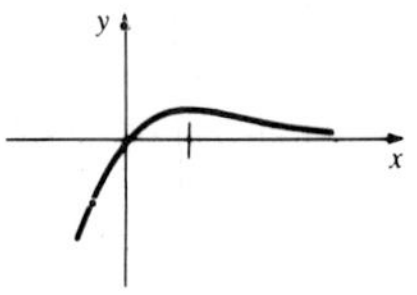

$f(1)=e^{-1}$ is an abs. max.

51.

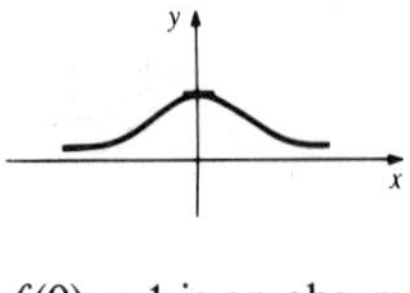

$f(0)=1$ is an abs. max.

53.

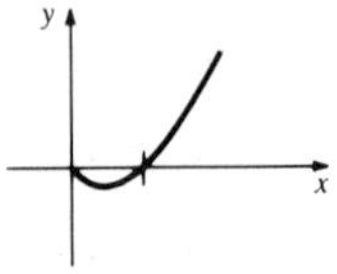

$f(e^{-1})=-e^{-1}$ is an abs. min.

55. Since $f'<0$ for all $x>0$, f is decreasing on $(0,\infty)$. By Theorem 6.2, f has an inverse.

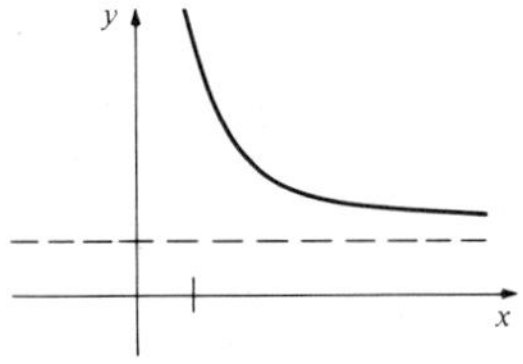

Range of $f=\{y\,|\,y>1\}$; $f^{-1}(x)=2/(\ln x)$; domain of $f^{-1}=\{x\,|\,x>1\}$

57.

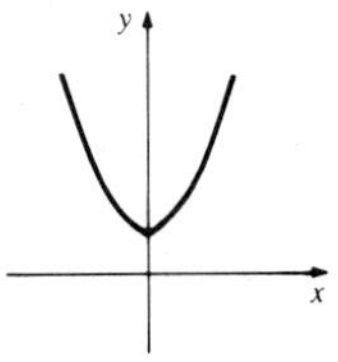

$$\frac{dy}{dx}=\begin{cases}e^x, & x>0\\ -e^{-x}, & x<0\end{cases};\ \text{no}$$

61. 0, 2

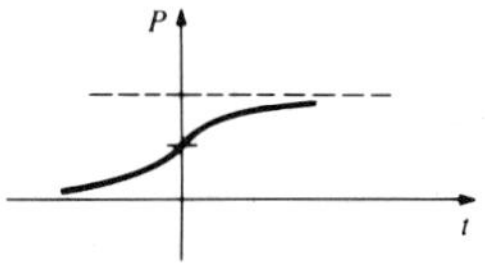

63. 88.22 cm; 9.95 cm/yr; $\frac{1}{4}$ yr

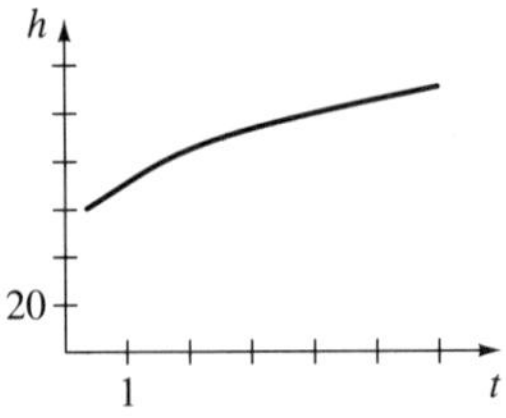

about 3.3 years old

65. Temperature is dropping at a rate of approximately 2°K per minute.

67. \$6.724.44, \$6.749.29

69. x-intercept 1.3863 (approximately); y-intercept -12

ANSWERS CHAPTER 6

Exercises 6.4, Page 414

1. $\frac{1}{3}\ln|x| + C$ 3. $\frac{1}{2}\ln|2x - 1| + C$
5. $\frac{1}{2}\ln(x^2 + 1) + C$ 7. $\frac{1}{2}\ln|x^2 + 4x - 3| + C$
9. $x - \ln|x + 1| + C$
11. $\frac{1}{2}x^2 + 4x + \ln|x + 2| + C$ 13. $\ln|\ln x| + C$
15. $\frac{1}{2}(\ln x)^2 + C$ 17. $\frac{3}{4}(1 + \ln x)^{4/3} + C$
19. $\ln|1 + \sin t| + C$ 21. $\frac{1}{10}e^{10x} + C$
23. $-\frac{1}{6}e^{-2x^3} + C$ 25. $2e^{\sqrt{x}} + C$
27. $-\frac{1}{4}e^{-2x^2+4} + C$
29. $x + 3e^x + \frac{3}{2}e^{2x} + \frac{1}{3}e^{3x} + C$
31. $-e^{-t} + t + C$ 33. $\frac{1}{3}e^{3x} + 2e^{-x} - \frac{1}{5}e^{-5x} + C$
35. $\ln(e^x + e^{-x}) + C$ 37. $e^x - \ln(e^x + 1) + C$
39. $2e^{x/2} + C$ 41. $-\frac{1}{5}\ln|\cos 5x| + C$
43. $\theta + 2\ln|\sec\theta + \tan\theta| + \tan\theta + C$
45. $\frac{1}{2}\ln 9$ 47. 0 49. $-2 + \ln 4$
51. $2\ln 8$ 53. $\frac{40}{9} - 2\ln 3$ 55. $\frac{1}{2}(1 - e^{-4})$
57. $e - e^{-1} - 2$ is not area; area is $e + e^{-1} - 2$.
59. $\frac{\pi}{2}(e^4 - 1)$ 61. $\pi(1 - e^{-1})$ 63. $\frac{3}{4}$
65. $15\pi/16 + \pi\ln 2$
67. C_1x^4 where C_1 is an arbitrary constant; $C_1(x + 1)^{-2}e^x$ where C_1 is an arbitrary constant
69. $\frac{1}{2}y^2 - y + \ln|y + 1| = -\frac{1}{x} + C$
71. $-\frac{1}{2}e^{-2y} = \frac{1}{3}e^{3x} + C$
73. $\frac{1}{2}\ln|\csc 2y - \cot 2y| = \sin x + C$
75. $y = 3e^{-(t-1)^2/2}$ 81. 1
85. $\frac{x^2}{2}\ln x - \frac{x^2}{4} + C$ 87. 0.740; 5, 41

Exercises 6.5, Page 425

1. $e^{\pi\ln 2}$; 8.8 3. $e^{e\ln 10}$; 522
5. $e^{-\sqrt{5}\ln 7}$; 0.01 7. $4^x\ln 4$
9. $-2(10^{-2x})\ln 10$ 11. $2x20^{x^2}\ln 20$ 13. 0
15. $x^2 2^x\ln 2 + 2x2^x$ 17. $\sqrt{5}x^{\sqrt{5}-1}$
19. $(\sin 2)x^{(\sin 2)-1}$
21. $(t^4 + 3t^2)^{(\ln 4)-1}(4t^3 + 6t)\ln 4$
23. $4^x(\ln 4 + e^x\ln 4 - e^x)/(1 + e^x)^2$
25. $\frac{1}{2}5^{x/2}\ln 5$ 27. $-(\ln x)^{-1-1/e}/ex$
29. $1/(x\ln 4)$ 31. $\ln x/(x\ln 10) + (\log_{10}x)/x$
33. $6x^\pi/(6x - 4)\ln 3 + \pi x^{\pi-1}\log_3|6x - 4|$
35. $3^{x^3+1}/x\ln 3 + x^2 3^{x^3+1}(\ln 3)(\log_3 x^3)$
37. $9^x(\ln 9)^2$ 39. $8e(e - 1)(e - 2)(2x)^{e-3}$
41. $y/(2^y\ln 2 - x)$ 43. $1/(x + xe^y)\ln 10$
45. $7^x/\ln 7 + C$ 47. $3(10^{x/3})/\ln 10 + C$
49. $(10^{-2} - 10^{-3})/\ln 10$
51. $e^x + x^{e+1}/(e + 1) + (e^e)x + C$
53. $(1 + 2^t)^{21}/(21\ln 2) + C$ 55. $(2^e - 2^{e^{-1}})/\ln 2$
57. $\tan\theta + 2^{1+\tan\theta}/\ln 2 + 2^{-1+2\tan\theta}/\ln 2 + C$
59. $[\ln(1 + 5^x)]/\ln 5 + C$ 61. $64\ln 4$
63. $(-1/\ln 4, -4^{-1/\ln 4}/\ln 4)$ 65. $225/(16\ln 2)$
67. $(x^2 + 4)^{2x}\left[\frac{4x^2}{x^2 + 4} + 2\ln(x^2 + 4)\right]$
69. $x^x 2^x[1 + \ln 2 + \ln x]$
71. $x(x - 1)^x\left[\frac{1}{x} + \frac{x}{x - 1} + \ln(x - 1)\right]$
73. $(\ln|x|)^x\left[\frac{1}{\ln|x|} + \ln(\ln|x|)\right]$
75. $x^{x^x}x^x\ln x\left[1 + \ln x + \frac{1}{x\ln x}\right]$
77. $x^x(1 + \ln x)\sec^2 x^x$
79. $\frac{1}{2}\sqrt{\frac{(2x + 1)(3x + 2)}{4x + 3}}\left[\frac{2}{2x + 1} + \frac{3}{3x + 2} - \frac{4}{4x + 3}\right]$
81. $\frac{(x^3 - 1)^5(x^4 + 3x^3)^4}{(7x + 5)^9}\left[\frac{15x^2}{x^3 - 1} + \frac{16x^3 + 36x^2}{x^4 + 3x^3} - \frac{63}{7x + 5}\right]$
83. $y = 3x - 2$ 85. $2x^{2x}\left[2(1 + \ln x)^2 + \frac{1}{x}\right]$
87.

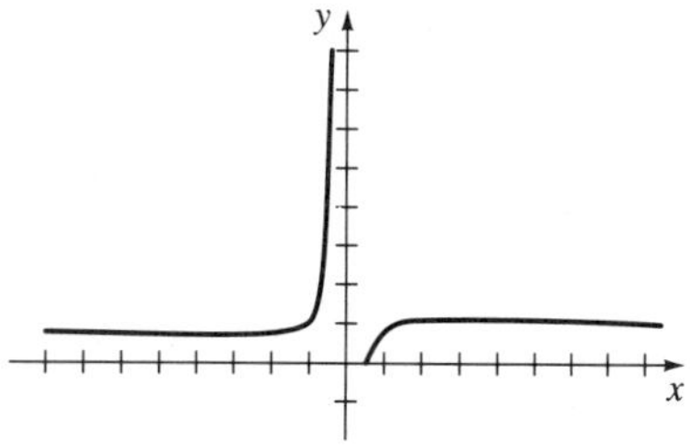

rel. max. $f(e) \approx 1.4447$, rel. min. $f(-e) \approx 0.6922$

89.

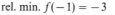

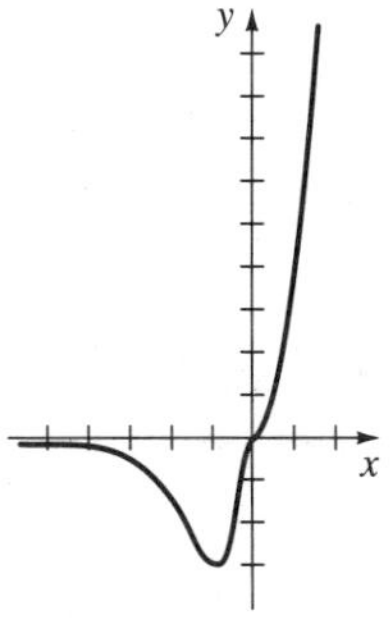

rel. min. $f(-1) = -3$

91. Approximately 251 times as strong; the San Fernando Valley earthquake was approximately 5 times as strong
93. 7.4; base

Exercises 6.6, Page 429

1.

$(e^{\Delta x}-1)/\Delta x$
1.0517092
1.0050167
1.0005000
1.0000500
0.9999500

Exercises 6.7, Page 436

1. $\sinh x = -\frac{1}{2}$, $\cosh x = \sqrt{5}/2$, $\tanh x = -\sqrt{5}/5$, $\coth x = -\sqrt{5}$, $\operatorname{sech} x = 2\sqrt{5}/5$, $\operatorname{csch} x = -2$
3. $10 \sinh 10x$ 5. $\frac{1}{2}x^{-1/2}\operatorname{sech}^2\sqrt{x}$
7. $-6(3x-1)\operatorname{sech}(3x-1)^2\tanh(3x-1)^2$
9. $-3 \sinh 3x \operatorname{csch}^2(\cosh 3x)$
11. $3 \sinh 2x \sinh 3x + 2 \cosh 2x \cosh 3x$
13. $2x^2\sinh x^2 + \cosh x^2$ 15. $3 \sinh^2 x \cosh x$
17. $\frac{2}{3}(x-\cosh x)^{-1/3}(1-\sinh x)$
19. $4 \tanh 4x$ 21. $(e^x+1)/(1+\cosh x)^2$
23. $e^{\sinh t}\cosh t$
25. $\dfrac{1+\operatorname{cost}\sinh 2t - 2\sin t \cosh 2t}{(1+\sinh 2t)^2}$
27. $y = 3x$ 29. $\frac{1}{8}\cosh 8x + C$
31. $\frac{1}{5}\sinh(5x-4)+C$ 33. $\frac{1}{3}\tanh x^3 + C$
35. $-3 \operatorname{csch} \sqrt[3]{x} + C$ 37. $\frac{1}{3}(1+\sinh 2x)^{3/2}+C$
39. $\frac{1}{5}\ln(7+\cosh 5x)+C$ 41. $-\frac{1}{3}e^{-\cosh 3x}+C$
43. $\frac{1}{7}\sinh^7 x + C$ 45. $\sinh e^x + C$
47. $2 \sinh 1$ 49. $\sinh 3 + \sinh 1 - 4$
51. $\pi(\cosh 3 - 1)$
53. $(\pi/2)(b-a)+(\pi/4)(\sinh 2b - \sinh 2a)$
55. $(x^2+1)/2x, x>0$ 57. $-2\operatorname{sech}^2 x \tanh x$
61.

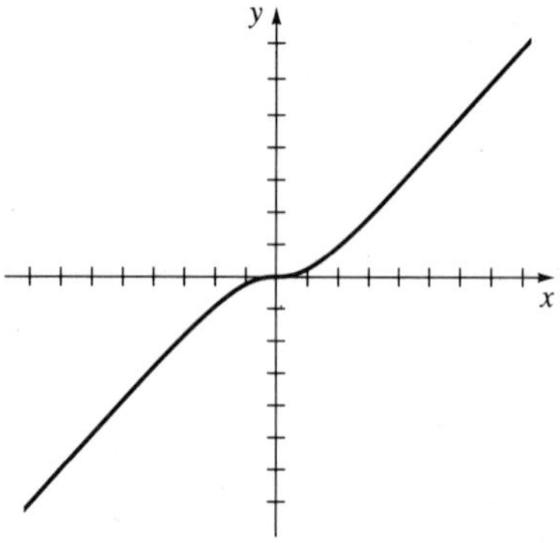

As $x \to \infty$, $\tanh x \to 1$, and so the graph becomes close to the graph of $y = x - 1$. As $x \to -\infty$, $\tanh x \to -1$, and so the graph becomes close to the graph of $y = x + 1$.

Exercises 6.8, Page 446

1. $P(t) = P_0e^{kt}$; approximately 7.9 years; 10 years
3. $A(t) = 100e^{-0.0051t}$; 88.5 mg; approximately 135.9 h
5. Approximately 24,180 years
7. \$6,665.45; 12; \$6,651.82
9. Approximately 64.5°
11. $X(t) = \dfrac{A}{B} - \dfrac{A}{B}e^{-Bt}$; as $t \to \infty$, $X(t) \to A/B$; $t = (\ln 2)/B$
13. $C(t) = C_s + (C_0 - C_s)e^{-kAt/V}$
15. $i(t) = \frac{6}{5} - \frac{6}{5}e^{-20t}$
17. $v(t) = \dfrac{mg}{k} + \left(v_0 - \dfrac{mg}{k}\right)e^{-kt/m}$; $v \to mg/k$ as $t \to \infty$;

$$s(t) = \frac{mg}{k}t - \frac{m}{k}\left(v_0 - \frac{mg}{k}\right)e^{-kt/m} + \frac{m}{k}\left(v_0 - \frac{mg}{k}\right) + s_0$$

19. $y = Ce^{4x}$ 21. $y = \frac{1}{10} + Ce^{-5x}$
23. $y = \frac{1}{4}e^{3x} + Ce^{-x}$ 25. $y = x^{-1}\ln x + Cx^{-1}$
27. $y = C/(e^x+1)$ 29. $y = \sin x + C\cos x$
31. $(\sec\theta + \tan\theta)r = \theta - \cos\theta + C$
33. $(x+1)y = 20 + \ln x$
35. $A(t) = 200 - 170e^{-t/50}$
37. $A(t) = 1000 - 1000e^{-t/100}$ 39. 64.38 lb
41. $y = \begin{cases} 1-e^{-x}, & 0 \le x \le 1 \\ (e-1)e^{-x}, & x > 1 \end{cases}$

Chapter 6 Review Exercises, Page 449

1. True 2. False 3. True 4. False
5. False 6. True 7. False 8. True
9. True 10. True 11. False
12. True 13. False 14. False
15. True 16. False 17. False
18. True 19. True 20. False 21. 3
23. $a - b$ 25. 2
27. Logarithmic differentiation
29. Catenary 31. $1/x + 2/(4x-1)$
33. $\cos(e^x\ln x)(e^x/x + e^x\ln x)$
35. $e^x + 2e^{2x}$ 37. $7x^6 + 7^x\ln 7$
39. $(x^2+e^{2x})^{\ln x}\left[\dfrac{(2x+2e^{2x})\ln x}{x^2+e^{2x}} + \dfrac{\ln(x^2+e^{2x})}{x}\right]$
41. $(e^x - y^2)/(2xy + e^y)$ 43. $3x^2e^{x^3}\cosh e^{x^3}$
45. $\sinh(\cosh x)\sinh x$ 47. $-\frac{1}{10}\ln|1-5x^2| + C$
49. $\frac{1}{2}\ln|x(x+2)| + C$ 51. $\dfrac{1}{8} + \dfrac{6}{\ln 7} + 7^\pi$
53. $\frac{1}{4}\ln|\sin 4x| + C$ 55. $\frac{1}{3}(e^{-6} - e^{-15})$
57. $(24)^t/\ln 24 + C$ 59. $-\cosh(1/x) + C$
61. $-\frac{1}{9}(\sinh 3x)^{-3} + C$
63. $f^{-1}(x) = \sqrt[3]{\dfrac{x-8}{x}}$, $(f^{-1})'(x) = \dfrac{8}{3x^2}\left(\dfrac{x-8}{x}\right)^{-2/3}$

65. $1225\pi/72$

67.

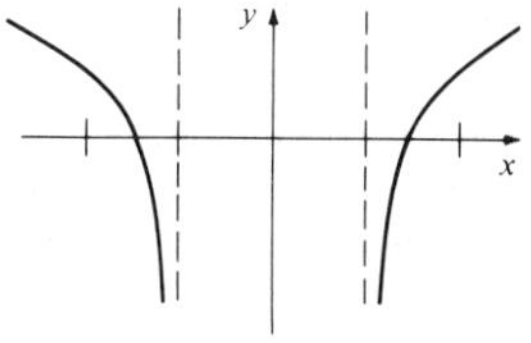

71. $A(t) = \dfrac{k_1 M}{k_1 + k_2}(1 - e^{-(k_1+k_2)t})$; as $t \to \infty$, $A \to k_1M/(k_1 + k_2)$. If $k_2 > 0$, the material will never be completely memorized;

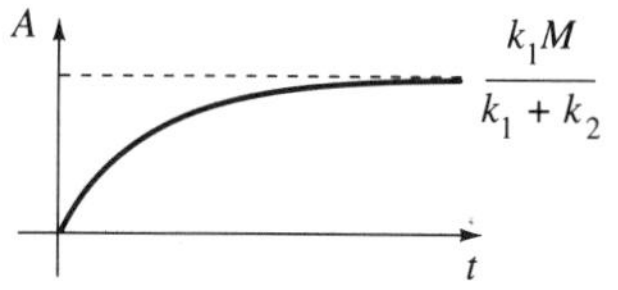

73. $P(t) = P_0 e^{k \sin t}$

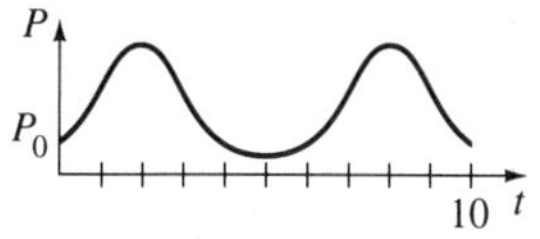

75. $y = C \csc x$ 77. $y = -\frac{1}{4}t + Ct^5$

79. $y = \frac{1}{4} + C(x^2 + 4)^{-4}$

81. $dW/dF < 1$ means that the animal is never gaining more weight than it is consuming in feed; approximately 287 lb

Exercises 7.1, Page 469

1. 0 3. $3\pi/4$ 5. $\pi/4$ 7. $3\pi/4$

9. $-\pi/3$ 11. $7\pi/6$ 13. $\sqrt{3}/2$

15. $-\pi/4$ 17. $\frac{4}{5}$ 19. 2 21. $\sqrt{2}$

23. $\pi/3$ 25. π 27. $-1/\sqrt{3}$ 29. -1

31. $4\sqrt{2}/9$ 33. $\sqrt{3}(2 + \sqrt{10})/9$

35. $13/\sqrt{170}$ 37. $\sqrt{1 - x^2}$ 39. $\sqrt{x^2 - 1}$

41. $2x/(x^2 + 1)$ 45. 3

47. $\cos t = \sqrt{5}/5$, $\tan t = -2$, $\cot t = -\frac{1}{2}$, $\sec t = \sqrt{5}$, $\csc t = -\sqrt{5}/2$

49. 5 is not in the interval $(-\pi/2, \pi/2)$.

51.

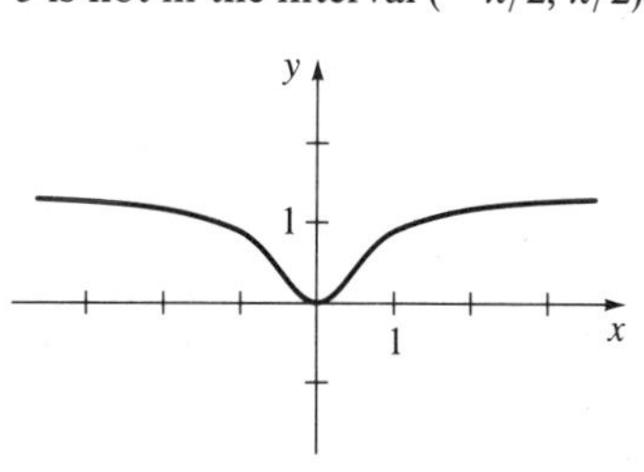

$f(x) \to \pi/2$ as $x \to \infty$

53.

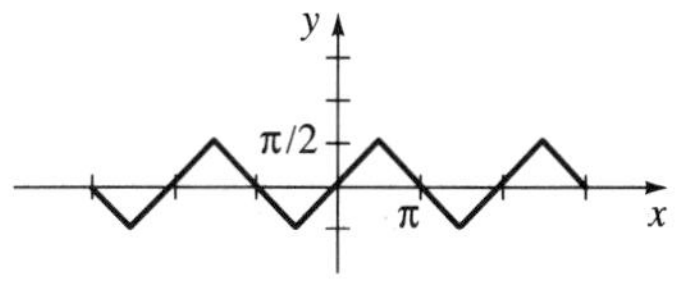

The function $\sin x$ is defined for every real number x. Since $-1 \le \sin x \le 1$ is the domain of $\sin^{-1}x$, we must then have for every real number x, $-\pi/2 \le \sin^{-1}(\sin x) \le \pi/2$. Note, for example, on the interval $-\pi/2 \le x \le \pi/2$ that the values of the function $f(x) = \sin^{-1}(\sin x)$ increase from $-\pi/2$ to $\pi/2$, on the interval $\pi/2 \le x \le 3\pi/2$ the values of f decrease from $\pi/2$ to $-\pi/2$, on the interval $3\pi/2 \le x \le 5\pi/2$ the values of f increase from $-\pi/2$ to $\pi/2$, and so on.

55. 0.28 radian (16.2°); 114.26 ft/s

57. 0.20 radian or 11.31° south of west

Exercises 7.2, Page 478

1. $\dfrac{5}{\sqrt{1 - (5x - 1)^2}}$ 3. $\dfrac{-8}{4 + x^2}$

5. $\dfrac{1}{1 + x} + x^{-1/2}\tan^{-1}\sqrt{x}$

7. $\dfrac{2(\cos^{-1}2x + \sin^{-1}2x)}{\sqrt{1 - 4x^2}(\cos^{-1}2x)^2}$

9. $\dfrac{-2x}{(1 + x^4)(\tan^{-1}x^2)^2}$ 11. $\dfrac{2 - x}{\sqrt{1 - x^2}} + \cos^{-1}x$

13. $3\left(x^2 - 9\tan^{-1}\dfrac{x}{3}\right)^2\left(2x - \dfrac{27}{9 + x^2}\right)$

15. $1/(t^2 + 1)$ 17. $-4 \sin 4x/|\sin 4x|$

19. $\dfrac{2x \sec^2(\sin^{-1}x^2)}{\sqrt{1 - x^4}}$

21. $2x(1 + y^2)/(1 - 2y - 2y^3)$

23. $\sin^{-1}x + \cos^{-1}x = \text{constant}$

25. $\frac{1}{5}\tan^{-1}5x + C$ 27. $\pi/12$

29. $-2(1 - x^2)^{1/2} - 3\sin^{-1}x + C$

31. $-x + 2\tan^{-1}x + C$

33. $-\frac{4}{3}(2 - 3t^2)^{1/2} + C$

35. $\dfrac{\sqrt{10}}{10}\tan^{-1}\dfrac{\sqrt{10}}{5}x + C$ 37. $\dfrac{2}{3}\left(\dfrac{\pi}{4}\right)^{3/2}$

39. $\sec^{-1}(x + 1) + C$

41. $\frac{1}{2}x^2 + \frac{1}{2}\tan^{-1}\left(\dfrac{x + 2}{2}\right) + C$

43. $\frac{1}{4}\tan^{-1}(2t^2 + 1) + C$ 45. $\pi/48$

47. $\tan^{-1}e^x + C$ 49. $\sqrt{3}/3$

51. $y = ((2 + \pi)/4)x - \frac{1}{2}$ 53. $\sin^{-1}(\frac{3}{5}) - \sin^{-1}(\frac{1}{5})$

55. $\frac{\pi}{4}[\sec^{-1}(4) - \sec^{-1}(\frac{25}{16})]$ 57. $\sqrt{2}/120$ rad/s

59. $10\sqrt{3}$ ft

61. Approximately 1.74 rad/s; π; approximately 14.51 rad/s

63. 5.5730 65. 0.4389

67. Rel max. $f(\sqrt{2}/2) = -\sqrt{2} + 3\tan^{-1}(\sqrt{2}/2)$; rel. min. $f(-\sqrt{2}/2) = \sqrt{2} + 3\tan^{-1}(-\sqrt{2}/2)$

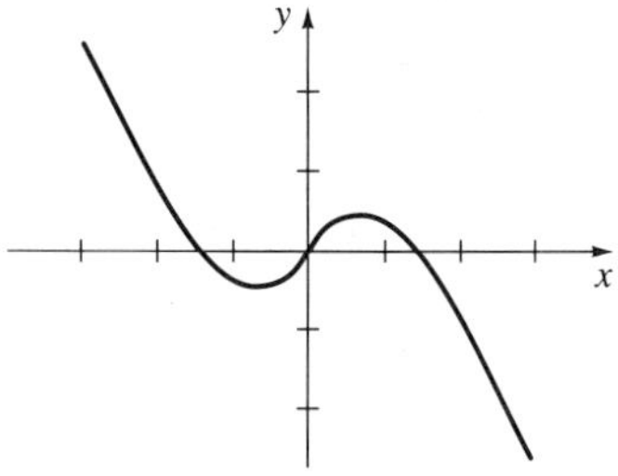

Exercises 7.3, Page 486

1. $3/\sqrt{9x^2+1}$ 3. $-2x/[1-(1-x^2)^2]$

5. $\sec x$ 7. $3x^3/\sqrt{x^6+1} + \sinh^{-1}x^3$

9. $-1/x^2\sqrt{1-x^2} - (\operatorname{sech}^{-1}x)/x^2$

11. $-1/(x\sqrt{1-x^2}\operatorname{sech}^{-1}x)$

13. $3(\cosh^{-1}6x)^{-1/2}\sqrt{36x^2-1}$

15. $(x+1)/\sqrt{x^2-1}$ 17. $\sinh x/\sqrt{\cosh^2 x+1}$

19. $\coth^{-1}x$ 21. $\frac{1}{2}\sinh^{-1}2x + C$

23. $\begin{cases}\frac{1}{2}\tanh^{-1}2x + C, & |x| < \frac{1}{2}\\ \frac{1}{2}\coth^{-1}2x + C, & |x| > \frac{1}{2}\end{cases}$

or $\frac{1}{4}\ln\left|\frac{1+2x}{1-2x}\right| + C, \quad 2x \neq 1$

25. $\frac{1}{3}\cosh^{-1}\frac{3}{4}x + C, 3x > 4$ 27. $\frac{1}{2}\sinh^{-1}x^2 + C$

29. $\begin{cases}-\tanh^{-1}(x+1) + C, & |x+1| < 1\\ -\coth^{-1}(x+1) + C, & |x+1| > 1\end{cases}$

or $\frac{1}{2}\ln\left|\frac{x+2}{-x}\right| + C, \quad x+1 \neq 1$

31. $\sinh^{-1}(x+2) + C$ 33. $\cosh^{-1}e^x + C$

35. $\sinh^{-1}(\sin\theta) + C$ 37. $(t^2+16)^{1/2} + C$

39. $x - \ln\left|\frac{1+x}{1-x}\right| + C$

41. Approximately 2.6998

43. Approximately 0.9730

45. $\pi/6$; $\pi(\sqrt{3}-1)$; $\frac{\pi}{2}\ln\left(\frac{2+\sqrt{3}}{6-3\sqrt{3}}\right)$

47. $y = (T/w)\cosh(wx/T) + 1 - T/w$

55. $\frac{1}{2}\ln|1+x| - \frac{1}{2}\ln|1-x| + C = \frac{1}{2}\ln\left|\frac{1+x}{1-x}\right| + C$

Chapter 7 Review Exercises, Page 488

1. True 2. False 3. False 4. False
5. False 6. False 7. True 8. True
9. True 10. True

11. $-\frac{3|x|}{x^2\sqrt{x^2-9}}$ 13. $\frac{(\cot^{-1}x)^{-2}}{1+x^2}$

15. $-4x^2/\sqrt{1-x^2}$

17. $\frac{(4x\cos x^2)(\tan^{-1}(\sin x^2))}{1+\sin^2 x^2}$

19. $\sqrt{1-y^2}/(\sqrt{1-y^2}-1)$

21. $1/(\sqrt{1-x^2}\sqrt{(\sin^{-1}x)^2+1})$

23. $e^{x\cosh^{-1}x}\left[\frac{x^2}{\sqrt{x^2-1}} + x\cosh^{-1}x + 1\right]$

25. $\frac{1}{2}\sin^{-1}\frac{2}{3}x + C$ 27. $\pi/6$

29. $\sin^{-1}\left(\frac{x+3}{4}\right) + C$ 31. $1/\pi$ 33. $\pi/18$

35. $\frac{1}{3}\sinh^{-1}3$ 37. $\frac{5}{13}$ 39. $\pi/6 - \tan^{-1}(\frac{1}{2})$

43. $k\left(\frac{L}{R}\right)^a\left(\tan^{-1}\left(\frac{H - R\tan\theta}{R}\right) - \tan^{-1}(-\tan\theta)\right)^a$

45. $\ln\left(1 + \frac{2}{\sqrt{3}}\right)$

Exercises 8.1, Page 495

1. $\frac{1}{5}(x+1)^5 - \frac{1}{4}(x+1)^4 + C$

3. $\frac{4}{5}(x-5)^{5/2} + \frac{22}{3}(x-5)^{3/2} + C$

5. $\frac{2}{3}(x-1)^{3/2} + 2(x-1)^{1/2} + C$

7. $\frac{2}{9}(3x-4)^{1/2} - \frac{26}{9}(3x-4)^{-1/2} + C$

9. $2\sqrt{x} - 2\tan^{-1}\sqrt{x} + C$

11. $t - 8\sqrt{t} + 8\ln(\sqrt{t}+1) + C$

13. $\frac{3}{10}(x^2+1)^{5/3} - \frac{3}{4}(x^2+1)^{2/3} + C$

15. $-(x-1)^{-1} - (x-1)^{-2} - \frac{1}{3}(x-1)^{-3} + C$

17. $2x^{1/2} + 3x^{1/3} + 6x^{1/6} + 6\ln|x^{1/6} - 1| + C$

19. $2\sqrt{e^x - 1} - 2\tan^{-1}\sqrt{e^x - 1} + C$

21. $-\frac{4}{3}(1-\sqrt{v})^{3/2} + \frac{4}{5}(1-\sqrt{v})^{5/2} + C$

23. $\frac{4}{3}(1+\sqrt{t})^{3/2} + C$

25. $\ln|x^2+2x+5| + \frac{5}{2}\tan^{-1}\left(\frac{x+1}{2}\right) + C$

27. $-2\sqrt{16-6x-x^2} - \sin^{-1}\left(\frac{x+3}{5}\right) + C$

29. $\frac{506}{375}$ 31. $6 + 20\ln\frac{11}{14}$ 33. $\frac{177}{2}$

35. $\frac{1}{1326}$ 37. $3 + 3\ln\frac{2}{3}$ 39. $\frac{1}{168}$

41. Let $u = t^2$.

43. $-\frac{3}{2} + 3\ln 2$ 45. $32\pi/3 - 4\pi\ln 3$

47. $\frac{232}{15}$

Exercises 8.2, Page 503

1. $\frac{2}{3}x(x+3)^{3/2} - \frac{4}{15}(x+3)^{5/2} + C$
3. $x \ln 4x - x + C$ 5. $(x^2/2)\ln 2x - x^2/4 + C$
7. $-x^{-1}\ln x - x^{-1} + C$
9. $t(\ln t)^2 - 2t \ln t + 2t + C$ 11. $\frac{3}{2}\ln 3$
13. $\pi/4 - \frac{1}{2}\ln 2$ 15. $x \sin^{-1}x + (1-x^2)^{1/2} + C$
17. $\frac{1}{3}xe^{3x} - \frac{1}{9}e^{3x} + C$
19. $-\frac{1}{4}x^3e^{-4x} - \frac{3}{16}x^2e^{-4x} - \frac{3}{32}xe^{-4x} - \frac{3}{128}e^{-4x} + C$
21. $\frac{1}{2}e^{x^2}(x^2-1) + C$ 23. $-12e^{-2} + 8e^{-1}$
25. $\frac{1}{8}t \sin 8t + \frac{1}{64}\cos 8t + C$
27. $-x^2\cos x + 2x \sin x + 2\cos x + C$
29. $\frac{1}{3}x^3\sin 3x + \frac{1}{3}x^2\cos 3x - \frac{2}{9}x \sin 3x - \frac{2}{27}\cos 3x + C$
31. $\frac{1}{17}e^x[\sin 4x - 4\cos 4x] + C$
33. $\theta \sec\theta - \ln|\sec\theta + \tan\theta| + C$
35. $\frac{1}{3}\cos x \cos 2x + \frac{2}{3}\sin x \sin 2x + C$
37. $\frac{1}{3}x^2(x^2+4)^{3/2} - \frac{2}{15}(x^2+4)^{5/2} + C$
39. $\frac{1}{2}x[\sin(\ln x) - \cos(\ln x)] + C$
41. $-\frac{1}{2}\csc x \cot x + \frac{1}{2}\ln|\csc x - \cot x| + C$
43. $\frac{1}{4}x^2 - \frac{1}{4}x \sin 2x - \frac{1}{8}\cos 2x + C$
45. $x(\sin^{-1}x)^2 + 2\sqrt{1-x^2}\,\sin^{-1}x - 2x + C$
47. $-\dfrac{x^2e^x}{x+2} + xe^x - e^x + C$
49. $-\frac{1}{2}xe^x\cos x + \frac{1}{2}xe^x\sin x + \frac{1}{2}e^x\cos x + C$
51. $5\tan^{-1}2 - 2$
53. $-2\sqrt{x+2}\cos\sqrt{x+2} + 2\sin\sqrt{x+2} + C$
55. $\frac{1}{2}(\sin^{-1}x)^2 + C$ 57. $3\ln 3 + e^{-1}$
59. $5\pi(\ln 5)^2 - 10\pi\ln 5 + 8\pi$ 61. $2\pi^2$
63. $s(t) = \frac{1}{2} - \frac{1}{2}e^{-t}\sin t - \frac{1}{2}e^{-t}\cos t$
65. Approximately 31.8 ft-lb 67. $\bar{x} = 1, \bar{y} = \pi/8$
73. $-\frac{1}{3}\sin^2x \cos x - \frac{2}{3}\cos x + C$
75. $\frac{1}{30}\cos^2 10x \sin 10x + \frac{1}{15}\sin 10x + C$
77.

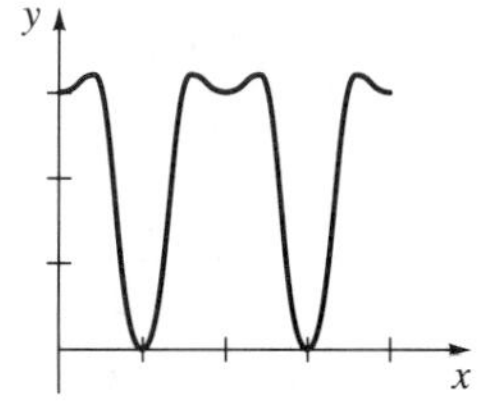

$\frac{17}{4}\pi$ square units

Exercises 8.3, Page 511

1. $\frac{2}{3}(\sin x)^{3/2} + C$ 3. $\sin x - \frac{1}{3}\sin^3x + C$
5. $\frac{1}{4}\sin^4x - \frac{1}{6}\sin^6x + C$ 7. $25\sqrt{2}/168$
9. $\frac{1}{60}$ 11. $\frac{3}{8}t - \frac{1}{4}\sin 2t + \frac{1}{32}\sin 4t + C$
13. $\frac{1}{16}x - \frac{1}{64}\sin 4x + \frac{1}{48}\sin^3 2x + C$
15. $\frac{3}{128}x - \frac{1}{128}\sin 4x + \frac{1}{1024}\sin 8x + C$
17. $\frac{1}{12}\tan^6 2t + \frac{1}{8}\tan^4 2t + C$
19. $\tan x + \frac{1}{3}\tan^3x + C$
21. $-\frac{1}{11}\cot^{11}x - \frac{1}{13}\cot^{13}x + C$
23. $\frac{2}{3}\sec^{3/2}x + 2\sec^{-1/2}x + C$
25. $\frac{1}{4}\sec^3x \tan x - \frac{1}{8}\sec x \tan x - \frac{1}{8}\ln|\sec x + \tan x| + C$
27. $\ln|\sin x| + \frac{1}{2}\cos^2x + C$
29. $\frac{1}{7}\tan^{-7}(1-t) + \frac{1}{5}\tan^{-5}(1-t) + C$
31. $\sqrt{3} - \pi/3$ 33. $\frac{1}{3}\tan^3x - \tan x + C$
35. $-\frac{1}{2}\cot^2t - \ln|\sin t| + C$
37. $\pi^2/12 + \sqrt{3}\pi/16$
39. $-\frac{1}{6}\cos 3x + \frac{1}{2}\cos x + C$
41. $\frac{1}{4}\sin 2x - \frac{1}{12}\sin 6x + C$ 43. $\frac{5}{12}$
47. $\pi/4$

Exercises 8.4, Page 517

1. $-\sin^{-1}x - \sqrt{1-x^2}/x + C$
3. $\ln|x/6 + \sqrt{x^2-36}/6| + C$
5. $\frac{1}{3}(x^2+7)^{3/2} + C$
7. $-\frac{1}{3}(1-x^2)^{3/2} + \frac{1}{5}(1-x^2)^{5/2} + C$
9. $-x/4\sqrt{x^2-4} + C$
11. $\frac{1}{2}x\sqrt{x^2+4} + 2\ln|\sqrt{x^2+4}/2 + x/2| + C$
13. $\sin^{-1}(x/5) + C$
15. $\frac{1}{4}\ln|4/x - \sqrt{16-x^2}/x| + C$
17. $\ln|\sqrt{x^2+1}/x - 1/x| + C$
19. $-(1-x^2)^{3/2}/3x^3 + C$
21. $x/\sqrt{9-x^2} - \sin^{-1}(x/3) + C$
23. $\frac{1}{2}\tan^{-1}x + x/[2(1+x^2)] + C$
25. $x/[16\sqrt{4+x^2}] - x^3/[48(4+x^2)^{3/2}] + C$
27. $\ln|\sqrt{x^2+2x+10}/3 + (x+1)/3| + C$
29. $\frac{1}{16}\tan^{-1}[(x+3)/2] + (x+3)/[8(x^2+6x+13)] + C$
31. $-(5x+1)/[9\sqrt{5-4x-x^2}] + C$
33. $\ln|x^2+4x+13| + C$
35. $x - 4\tan^{-1}(x/4) + C$
37. $\sqrt{3} + 2\pi/3$ 39. $\sqrt{2}/50$
41. $2\sqrt{3} - \frac{172}{81}$ 43. $\sec^{-1}(e^x) + C$
45. $(\sqrt{3}/3)\ln\left(\dfrac{\sqrt{2}-1}{2-\sqrt{3}}\right)$
49. $(\sqrt{3}\pi/9)(\sqrt{3} - 1 - \pi/12)$
51. $12\sqrt{2}\pi - 4\pi\ln(\sqrt{2}+1)$
53. $2 - \sqrt{2} - \ln(\sqrt{6} - \sqrt{3})$
55. The slope at (x, y) is $-y/\sqrt{s^2-y^2}$. This is the same as dy/dx; $x = 10\ln(10/y + \sqrt{100-y^2}/y) - \sqrt{100-y^2}$
57. Approximately 49.01 lb
59. $\frac{1}{3}x^3\sin^{-1}x + \frac{1}{3}\sqrt{1-x^2} - \frac{1}{9}(1-x^2)^{3/2} + C$
61. $-\sqrt{9+x^2}/9x + C$

Exercises 8.5, Page 528

1. $-\frac{1}{2}\ln|x| + \frac{1}{2}\ln|x-2| + C$
3. $-2\ln|x| + \frac{5}{2}\ln|2x-1| + C$

5. $-\frac{1}{6}\ln|x+3| + \frac{1}{6}\ln|x-3| + C$
7. $\frac{5}{8}\ln|x-4| + \frac{3}{8}\ln|x+4| + C$
9. $-\frac{1}{2}\ln|x+3| + \frac{1}{2}\ln|x+1| + C$
11. $-\frac{1}{6}\ln|2x+1| + \frac{2}{3}\ln|x+2| + C$
13. $6\ln|x| - \frac{7}{2}\ln|x+1| - \frac{3}{2}\ln|x-1| + C$
15. $\frac{1}{2}\ln|x+1| - \ln|x+2| + \frac{1}{2}\ln|x+3| + C$
17. $-2\ln|t| - t^{-1} + 6\ln|t-1| + C$
19. $\ln|x| - \ln|x+1| + (x+1)^{-1} + C$
21. $-\frac{1}{3}(x-3)^{-3} + C$
23. $-2(x+1)^{-1} + \frac{3}{2}(x+1)^{-2} + C$
25. $-\frac{1}{2}(x^2-1)^{-1} + C$
27. $-\frac{1}{32}\ln|x+1| - \frac{1}{16}(x+1)^{-1} + \frac{1}{32}\ln|x+5| - \frac{1}{16}(x+5)^{-1} + C$
29. $-\frac{19}{16}\ln|x| - \frac{19}{8}x^{-1} + \frac{11}{8}x^{-2} - \frac{3}{2}x^{-3} + \frac{35}{16}\ln|x+2| + C$
31. $\frac{1}{3}x^3 - x^2 + 6x - 10\ln|x+1| - 8(x+1)^{-1} + C$
33. $-\frac{1}{2}\ln 3$ 35. $2\ln\frac{5}{3} - \frac{14}{15}$
37. $\ln|1+\sin x| - \ln|2+\sin x| + C$
39. $\frac{1}{9}\ln\left|\frac{e^t-2}{e^t+1}\right| + \frac{1}{3(e^t+1)} + C$
41. $\frac{1}{4}\ln(1+\sqrt{1-x^2}) + \frac{1}{4}(1+\sqrt{1-x^2})^{-1} - \frac{1}{4}\ln|1-\sqrt{1-x^2}| - \frac{1}{4}(1-\sqrt{1-x^2})^{-1} + C$
43. $\frac{1}{4}\ln\frac{15}{7}$ 45. $8\pi\ln\frac{2}{3} + 11\pi/3$
47. $8\pi\ln 2 - 4\pi$ 49. 276 students
51. $\frac{1}{3}\tan^{-1}x - \frac{1}{6}\tan^{-1}(x/2) + C$
53. $\frac{1}{2}\ln|x^2+2x+5| - \frac{1}{2}\ln|x^2+6x+10| - 2\tan^{-1}(x+3) + C$
55. $-\ln|x| + \frac{1}{2}\ln(x^2+1) + \tan^{-1}x + C$
57. $\frac{1}{6}\ln|x-3| - \frac{1}{12}\ln(x^2+9) + \frac{1}{2}\tan^{-1}(x/3) + C$
59. $\frac{1}{2}(x+1)^{-1} + \frac{1}{2}\tan^{-1}x + C$
61 $-\frac{1}{4}\ln|t+1| + \frac{1}{4}\ln|t-1| - \frac{1}{2}\tan^{-1}t + C$
63. $\frac{1}{16}\tan^{-1}(x/2) + (x-8)/[8(x^2+4)] + C$
65. $5\ln|x+1| - \ln|x^2+2x+2| - 7\tan^{-1}(x+1) + C$
67. $\frac{1}{3}\ln|x-1| - \frac{1}{6}\ln|x^2+x+1| - (\sqrt{3}/3)\tan^{-1}[(2x+1)/\sqrt{3}] + C$
69. $-x^{-1} - \frac{3}{2}\tan^{-1}x - x/[2(x^2+1)] + C$
71. $\frac{1}{2}\ln(x^2+4) - \frac{11}{16}\tan^{-1}(x/2) + (12+5x)/[8(x^2+4)] + C$
73. $-\frac{1}{4}(4x^2+5)^{-1} + C$
75. $\frac{3}{4}\ln|x| - \frac{3}{8}\ln|x^2+2x+2| - 3\tan^{-1}(x+1) - (9x+8)/[4(x^2+2x+2)] + C$
77. $\frac{1}{6}\ln\frac{8}{3} + (\sqrt{2}/6)\tan^{-1}(\sqrt{2}/2)$ 79. 0
81. $3(x+1)^{1/3} + \ln|(x+1)^{1/3} - 1| - \frac{1}{2}\ln|(x+1)^{2/3} + (x+1)^{1/3} + 1| - \sqrt{3}\tan^{-1}\{[2(x+1)^{1/3}+1]/\sqrt{3}\} + C$
83. $\ln 9 - \frac{1}{2}\ln 17$ 85. $\pi^2/2 - \pi$
87. $\ln\left|\frac{T-T_s}{T+T_s}\right| - 2\tan^{-1}\left(\frac{T}{T_s}\right) = 4T_s^3kt + C$

Exercises 8.6, Page 533

1. $-\sqrt{9+x^2}/9x + C$
3. $\sqrt{x^2-5} - \sqrt{5}\cos^{-1}(\sqrt{5}/x) + C$
5. $1/(16+20x) - \frac{1}{16}\ln|(4+5x)/x| + C$
7. $\frac{1}{7}t^2(1+2t)^{3/2} - \frac{1}{105}(3t-1)(1+2t)^{3/2} + C$
9. $-(3-x) + 9/(3-x) + 6\ln|3-x| + C$
11. $\frac{1}{4}\tan^4\theta - \frac{1}{2}\tan^2\theta - \ln|\cos\theta| + C$
13. $x\ln(x^2+16) - 2x + 8\tan^{-1}(x/4) + C$
15. $-\frac{1}{2}\tan[(\pi/4)-x] + C$
17. $-\sqrt{2x-x^2} + \cos^{-1}(1-x) + C$
19. $63\pi/512$ 21. $\frac{3}{13}(e^{2\pi}+1)$
23. $\frac{1}{2}e^{-x}\cos x - \frac{1}{2}x^2e^{-x}\cos x + \frac{1}{2}e^{-x}\sin x + xe^{-x}\sin x + \frac{1}{2}x^2e^{-x}\sin x$
25. $2\sqrt{x} - \frac{14}{\sqrt{15}}\arctan\left(\frac{1+2\sqrt{x}}{\sqrt{15}}\right) - \ln(4+\sqrt{x}+x)$
27. $6\sqrt{1+x^{1/3}} - 4(1+x^{1/3})^{3/2} + \frac{6}{5}(1+x^{1/3})^{5/2}$
29. $\frac{2}{\sqrt{3}}\arctan\left(\frac{\sin x}{\sqrt{3}(1+\cos x)}\right)$
31. $x - \frac{\sin x}{1+\cos x}$
33. $\frac{1}{2}\ln\left(\frac{\sin x}{1+\cos x}\right) - \frac{\sin^2 x}{4(1+\cos x)^2}$
35. $\ln\left(\frac{1+\frac{\sin x}{1+\cos x}}{-1+\frac{\sin x}{1+\cos x}}\right) - \frac{\sin x}{1+\cos x}$
37. $\frac{1}{\sqrt{3}}\ln\left(-2-\sqrt{3}+\frac{\sin x}{1+\cos x}\right) - \frac{1}{\sqrt{3}}\ln\left(-2+\sqrt{3}+\frac{\sin x}{1+\cos x}\right)$
39. $-1+\sqrt{3}$ 41. $-\frac{1}{2}\ln(\frac{1}{5}) + \frac{1}{2}\ln(\frac{1}{3})$
43. $\frac{1}{2h}(2\mu r + 2hr^2)^{1/2} - \frac{2\mu}{(2h)^{3/2}} \cdot \ln(\sqrt{\mu+hr} + \sqrt{hr}) = t + C$

Chapter 8 Review Exercises, Page 534

1. True 2. False 3. True 4. True
5. True 6. False 7. False 8. False
9. False 10. True 11. $2\sqrt{x} - 18\ln(\sqrt{x}+9) + C$
13. $(x^2+4)^{1/2} + C$
15. $\frac{3}{256}\tan^{-1}(x/2) + x/[32(x^2+4)] + x/[32(x^2+4)^2] - x^3/[128(x^2+4)^2] + C$
17. $x - 4x^{-1} + C$
19. $\frac{1}{2}\ln(x^2+4) - \frac{5}{2}\tan^{-1}(x/2) + C$
21. $(\ln x)^{10}/10 + C$
23. $(t^2/2)\sin^{-1}t - \frac{1}{4}\sin^{-1}t + \frac{1}{4}t\sqrt{1-t^2} + C$
25. $\frac{1}{5}(x+1)^5 - \frac{3}{4}(x+1)^4 + C$
27. $x\ln(x^2+4) - 2x + 4\tan^{-1}(x/2) + C$

29. $-\frac{2}{125}\ln|x| - \frac{1}{25}x^{-1} + \frac{2}{125}\ln|x+5| - \frac{1}{25}(x+5)^{-1} + C$
31. $-\frac{1}{12}\ln|x+3| - \frac{1}{2}(x+3)^{-1} + \frac{1}{12}\ln|x-3| + C$
33. $\tan t - t + C$ 35. $\frac{1}{13}\tan^{13}x + \frac{1}{11}\tan^{11}x + C$
37. $y\sin y + \cos y + C$ 39. $\sin t - \frac{1}{5}\sin^5 t + C$
41. $\frac{1}{6}(1 + e^w)^6 + C$
43. $-\frac{1}{8}\csc^2 4x - \frac{1}{4}\ln|\sin 4x| + C$ 45. $\frac{1}{4}$
47. $\sec x - \tan x + x + C$ 49. $\frac{5}{2}\ln 2 - \frac{3}{2}\ln 3$
51. $(e^x/10)[\cos 3x + 3\sin 3x] + C$
53. $\frac{1}{2}t\cos(\ln t) + \frac{1}{2}t\sin(\ln t) + C$
55. $2\sqrt{x}\sin\sqrt{x} + 2\cos\sqrt{x} + C$
57. $-\frac{2}{3}\cos^3 x + C$
59. $\frac{1}{2}(x+1)\sqrt{x^2+2x+5} + 2\ln|\sqrt{x^2+2x+5}/2 + (x+1)/2| + C$
61. $\frac{1}{7}\sec^7 x - \frac{2}{5}\sec^5 x + \frac{1}{3}\sec^3 x + C$
63. $\frac{1}{4}t^4 - \frac{1}{2}t^2 + \frac{1}{2}\ln(1+t^2) + C$
65. $\frac{5}{2}\ln(x^2+1) + \tan^{-1}x - \frac{1}{2}(x^2+1)^{-1} + C$
67. $\frac{1}{4}x^2 - \frac{1}{4}x\sin 2x - \frac{1}{8}\cos 2x + C$
69. $\sin x e^{\sin x} - e^{\sin x} + C$ 71. $\sqrt{6} - 2$
73. $t\sinh^{-1}t - (t^2+1)^{1/2} + C$ 75. $\ln\frac{3}{2}$
77. $\frac{1}{39}\tan^{13}3u + \frac{1}{45}\tan^{15}3u + C$
79. $3\tan x + \sec x + C$
81. $\frac{1}{2}x^2(1+\ln x)^2 - \frac{1}{2}x^2(1+\ln x) + \frac{1}{4}x^2 + C$
83. $e^{e^x} + C$ 85. $t^2 - \ln(1 + e^{t^2}) + C$
87. $\frac{1}{5}\sin^{-1}(5x+2) + C$
89. $\sin x\ln|\sin x| - \sin x + C$
91. Approximately 47 generations

Exercises 9.1, Page 548

1. 0 3. 2 5. $\frac{1}{3}$ 7. 10 9. -6
11. $\frac{1}{2}$ 13. $\frac{7}{5}$ 15. $\frac{1}{6}$
17. Does not exist 19. $\frac{1}{2}$ 21. 0
23. 0 25. 5 27. Does not exist
29. -2 31. $-\frac{1}{8}$ 33. -1
35. Does not exist 37. $\frac{1}{9}$ 39. 3
41. $\frac{q}{p}a^{q-p}$ 43. $A = 25\left[\frac{\theta - \frac{1}{2}\sin 2\theta}{\theta^2}\right]; 0; \frac{50}{3}$
45.

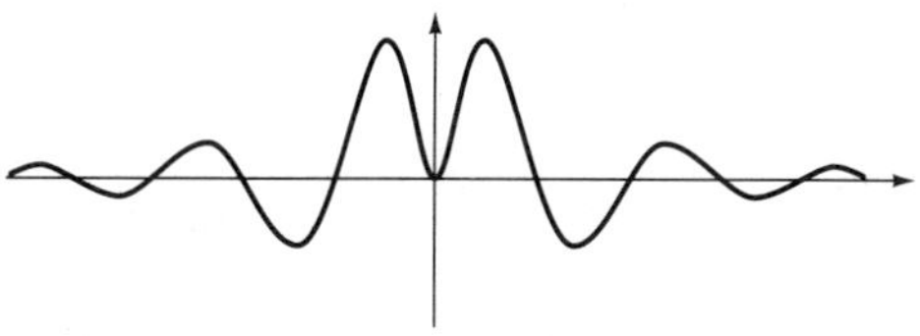

0; $\lim_{x\to\infty} x\sin x$ does not exist, but $x\sin x \not\to \infty$ as $x \to \infty$.
49. $-\frac{1}{2}$ 51. 1 53. 1 55. 0
57. -4 59. $\frac{1}{4}$ 61. 1 63. e^3
65. 1 67. $\frac{1}{4}$ 69. 0 71. $\frac{1}{5}$
73. 0 75. 1 77. $\frac{1}{24}$
79. Does not exist 81. $e^{-1/3}$ 83. 1
85. $\frac{1}{2}$
87.

0.1	13,781
0.01	1.6×10^{43}

89. Does not exist; 0
91. 1 if $n = 1$, does not exist if $n > 1$

Exercises 9.2, Page 558

1. $\frac{1}{81}$ 3. Diverges 5. $\frac{1}{2}e^6$
7. Diverges 9. $\frac{1}{2}$ 11. 0 13. $-\frac{1}{18}$
15. $3e^{-2}$ 17. 1 19. $\pi/2$ 21. $\frac{1}{2}$
23. 4 25. $\ln 2$ 27. $\frac{1}{4}\ln\frac{7}{3}$ 29. $\frac{1}{21}$
31. $\frac{1}{6}$ 33. 2
35. The integral $\int_1^\infty (1/x)\,dx$ diverges; $\pi\int_1^\infty (1/x^2)\,dx = \pi$; the integral $\int_1^\infty \left(\frac{\sqrt{x^4+1}}{x^3}\right)dx$ diverges.
37. 2.86×10^{10} joules 41. $1/s, s > 0$
43. $1/(s-1), s > 1$ 45. $1/(s^2+1), s > 0$
47. $e^{-s}/s, s > 0$ 49. Integrate by parts
51. $\bar{R} = r_0, \bar{V} = 6(\frac{4}{3}\pi\bar{R}^3)$
53. Diverges for $k \le 1$; converges for $k > 1$
55. Diverges for $k \ge 0$; converges for $k < 0$
57. Since $\sin^2 x \le 1$, use $g(x) = 1/x^2$; the integral $\int_1^\infty (1/x^2)\,dx$ converges.
59. On $[0, \infty)$, $1/(x + e^x) < 1/e^x = e^{-x}$; use $g(x) = e^{-x}$; the integral $\int_0^\infty e^{-x}\,dx$ converges.
63. Diverges 65. 100 67. $2\sqrt{2}$
69. Diverges 71. 6 73. $-\frac{1}{4}$
75. Diverges 77. Diverges 79. $-\frac{4}{3}$
81. $\pi/4$ 83. $\pi/2$ 85. 4
87. $2\sqrt{3}$; the integral $\pi\int_{-2}^1 dx/(x+2)$ diverges.
89. $\frac{1}{2}\ln 2$
91. Diverges for $k \ge 1$; converges for $k < 1$
93. $\pi/2$ 95. π
97. It is common practice to define $f(0) = 1$. Hence, it is not an improper integral.

Chapter 9 Review Exercises, Page 561

1. False 2. True 3. False 4. False
5. False 6. False 7. True 8. False
9. True 10. False 11. True
12. False 13. True 14. True
15. False 16. False 17. False
18. False 19. False 20. False

21. 0 23. $8\sqrt{3}\pi/9$ 25. Does not exist
27. $\frac{4}{25}$ 29. -2 31. 1 33. e^{-1}
35. Does not exist 37. $\frac{3}{2}\sqrt[3]{9}$ 39. 0
41. Diverges 43. 0 45. Diverges
47. $2 - 2e^{-1}$ 49. $\frac{2}{3}$
51. 2π; the areas are infinite. 55. $\frac{1}{2}$
57. $p_1 v_1 \ln(v_2/v_1)$ 59. $f''(x)$ 61. $1 + \ln\sqrt{2}$
63. 1404.8 to 1285.4; 955.5 to 945.4; 4.15 years; 3.77 years; approximately 8 years
65. $\text{Li}(100) \approx 29$, $x/\ln x \approx 22$ for $x = 100$; there are 25 primes less than 100.

Exercises 10.1, Page 575

1. $\frac{1}{3}, \frac{1}{5}, \frac{1}{7}, \frac{1}{9}, \ldots$ 3. $-1, \frac{1}{2}, -\frac{1}{3}, \frac{1}{4}, \ldots$
5. $3, -3, 3, -3, \ldots$ 7. 10, 100, 1000, 10,000, . . .
9. $1, 1 + \frac{1}{2}, 1 + \frac{1}{2} + \frac{1}{3}, 1 + \frac{1}{2} + \frac{1}{3} + \frac{1}{4}, \ldots$
11. 2, 4, 12, 48, . . . 13. Choose $N > 1/\varepsilon$.
15. Choose $N > (1 - \varepsilon)/\varepsilon$.
17. Choose $N > -(\ln \varepsilon)/\ln 10$. 19. 0
21. 0 23. $\frac{1}{2}$ 25. Sequence diverges
27. Sequence diverges 29. $\frac{1}{4}$ 31. 0
33. 0 35. Sequence diverges 37. 0
39. $\frac{5}{7}$ 41. 1 43. 6 45. 1
47. 1 49. $\ln\frac{4}{3}$ 51. 0
53. $\{2n/(2n - 1)\}$ 55. $\{(-1)^{n+1}(2n + 1)\}$
57. $\{2/3^{n-1}\}$ 59. $-\frac{1}{2}, -\frac{1}{4}, -\frac{1}{8}, -\frac{1}{16}, \ldots$
61. $3, 1, \frac{1}{3}, \frac{1}{3}, \ldots$ 63. $\frac{40}{9}$ ft; $15(\frac{2}{3})^n$
65. 15, 18, 18.6, 18.72, 18.74, 18.75, . . .
67. After first year: $1 + r$;
after second year: $(1 + r) + r(1 + r) = (1 + r)^2$;
after third year: $(1 + r)^2 + r(1 + r)^2 = (1 + r)^3$

69.

34	55	89	144	233
21	34	55	89	144
55	89	144	233	377

71. 2, 2.8, 4; 5.7, 8, 11.3
73. Use $a_n = 0$ and $c_n = 1/4^n$.
75. Use $a_n = 4$ and $c_n = 4 + 1/n$.

79. $$1 + \cfrac{1}{2 + \cfrac{1}{2 + \cfrac{1}{2 + \frac{1}{2}}}}, \quad 1 + \cfrac{1}{2 + \cfrac{1}{2 + \cfrac{1}{2 + \cfrac{1}{2 + \frac{1}{2}}}}};$$

1.5, 1.4, 1.4167, 1.4138, 1.4143, . . . ; the sequence appears to converge to $\sqrt{2}$.
81. 3
83. After n such bisections, there are $3 \cdot 2^n$ chords ($n \geq 1$);
$a_1 = 0.26794919243$
$a_2 = 0.06814834742$
$a_3 = 0.01711027725$
$a_4 = 0.00428215352$
$a_5 = 0.00107082505$
$a_6 = 0.00026772418$, $\sqrt{a_n}$ is the length of each chord ($n \geq 0$);
$b_0 = 3$
$b_1 = 3.10582854123$
$b_2 = 3.13262861325$
$b_3 = 3.13935020283$
$b_4 = 3.14103194988$
$b_5 = 3.14145247629$
$b_6 = 3.14155760277$, b_n is the length of $3 \cdot 2^n$ chords ($n \geq 0$). The chords taken together form a polygonal approximation to the length of the semicircumference, which is π. In this way we see that $b_n \to \pi$ as $n \to \infty$.

Exercises 10.2, Page 582

1. Increasing 3. Not monotonic
5. Increasing 7. Nonincreasing
9. Increasing 11. Not monotonic
13. Bounded and increasing
15. Bounded and increasing
17. Bounded and decreasing
19. Bounded and decreasing
21. Bounded and increasing
23. Bounded and decreasing 25. $\{(-1)^n\}$
27. The sequence is increasing. It is also bounded, since $0 \leq \int_1^n e^{-t^2}\,dt \leq \int_1^n e^{-t}\,dt < e^{-1}$.
29. Taking the limit as $n \to \infty$ of the given formula yields $L = bL/(a + L)$. Solving this last equation for L gives $L = 0$ or $L = b - a$.

Exercises 10.3, Page 590

1. $3 + \frac{5}{2} + \frac{7}{3} + \frac{9}{4} + \cdots$
3. $\frac{1}{2} - \frac{1}{6} + \frac{1}{12} - \frac{1}{20} + \cdots$
5. $1 + 2 + \frac{3}{2} + \frac{4}{6} + \cdots$
7. $2 + \frac{8}{3} + \frac{48}{15} + \frac{384}{105} + \cdots$
9. $-\frac{1}{7} + \frac{1}{9} - \frac{1}{11} + \frac{1}{13} - \cdots$
11. Converges to $\frac{15}{4}$ 13. Converges to $\frac{2}{3}$
15. Diverges 17. Converges to 9000
19. Diverges 21. $\frac{2}{9}$ 23. $\frac{61}{99}$
25. $\frac{1313}{999}$ 27. 1 29. $\frac{1}{2}$ 31. $\frac{17}{6}$
33. By Theorem 10.7 35. By Theorem 10.7
37. By Theorem 10.7
39. The series is a constant multiple of the divergent harmonic series and, hence, is divergent by Theorem 10.8.
41. By Theorem 10.10 43. $|x| < 2$
45. $-2 < x < 0$ 47. 75 ft 49. 1000
51. 18.75 mg
53. $L_n = d\dfrac{1 - (1 - p)^n}{p}$, since $0 < p < 1$, $L_n \to d/p$ as $n \to \infty$;

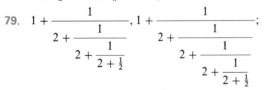

155.56 mg, rounding up in each case we find: parasthesia 20 days, ataxia 49 days, dysarthia 96 days, neither deafness nor death can occur at this daily dosage; 3.02 mg.

63. The argument is invalid, since $1 + 2 + 4 + 8 + \cdots$ is a divergent series.

Exercises 10.4, Page 597

1. Converges 3. Diverges 5. Diverges 7. Converges 9. Converges 11. Diverges 13. Converges 15. Converges 17. Converges 19. Converges 21. Converges 23. Converges 25. Diverges 27. Converges 29. Converges 31. Diverges 33. Converges 35. Converges 37. Converges 39. Diverges 41. No

Exercises 10.5, Page 601

1. Converges 3. Diverges 5. Converges 7. Converges 9. Diverges 11. Converges 13. Converges 15. Converges 17. Diverges 19. Diverges 21. $0 \le p \le 1$ 23. p any real number

Exercises 10.6, Page 608

1. Converges 3. Diverges 5. Converges 7. Converges 9. Converges 11. Converges 13. Diverges 15. Conditionally convergent 17. Absolutely convergent 19. Absolutely convergent 21. Absolutely convergent 23. Diverges 25. Conditionally convergent 27. Diverges 29. Conditionally convergent 31. Absolutely convergent 33. Diverges 35. 0.84147 37. 5 39. 0.9492 41. Less than $\frac{1}{101} \approx 0.009901$ 43. The series contains mixed algebraic signs but is not an alternating series; converges 45. The series is not an alternating series; converges 47. $a_{k+1} \le a_k$ is not satisfied for k sufficiently large; diverges 49. Diverges; converges; converges; diverges 51. Adding zeros to the series for $S/2$ does not affect its value.

Exercises 10.7, Page 612

1. $(-1, 1]$ 3. $[-\frac{1}{2}, \frac{1}{2})$ 5. $[2, 4]$ 7. $(-5, 15)$ 9. $\{0\}$ 11. $[0, \frac{2}{3}]$ 13. $[-1, 1)$ 15. $(-16, 2)$ 17. $(-\frac{75}{32}, \frac{75}{32})$ 19. $[\frac{2}{3}, \frac{4}{3}]$ 21. $(-\infty, \infty)$ 23. $(-3, 3)$ 25. $(-\infty, \infty)$ 27. $(-\frac{15}{4}, -\frac{9}{4})$ 29. 4 31. $x > 1$ or $x < -1$ 33. $x < -\frac{1}{2}$ 35. $(-2, 2)$ 37. $x < 0$ 39. $0 \le x < \pi/3, 2\pi/3 < x < 4\pi/3, 5\pi/3 < x \le 2\pi$

Exercises 10.8, Page 617

1. $\sum_{k=0}^{\infty} x^k; (-1, 1)$ 3. $\sum_{k=0}^{\infty} \frac{(-1)^k 3^k}{5^{k+1}} x^k; \left(-\frac{5}{3}, \frac{5}{3}\right)$ 5. $\sum_{k=2}^{\infty} \frac{k(k-1)}{2} x^{k-2}; (-1, 1)$ 7. $\ln 4 + \sum_{k=1}^{\infty} \frac{(-1)^{k+1}}{k4^k} x^k; (-4, 4]$ 9. $\sum_{k=0}^{\infty} (-1)^k x^{2k}; (-1, 1)$ 11. $\sum_{k=0}^{\infty} \frac{(-1)^k}{2k+1} x^{2k+1}; [-1, 1]$ 13. $\sum_{k=0}^{\infty} \frac{(-1)^k}{(k+1)(2k+3)} x^{2k+3}; [-1, 1]$ 15. $(-3, 3]$ 17. 0.0953 19. 0.4854 21. 0.0088 25. $\sum_{k=0}^{\infty} \frac{(-1)^k}{k!} x^k$ 27. 0.3678 29. 0.1973 31. Set $x = 1$.

Exercises 10.9, Page 625

1. $\sum_{k=0}^{\infty} \frac{x^k}{2^{k+1}}$ 3. $\sum_{k=0}^{\infty} \frac{(-1)^k}{k+1} x^{k+1}$ 5. $\sum_{k=0}^{\infty} \frac{(-1)^k}{(2k+1)!} x^{2k+1}$ 7. $\sum_{k=0}^{\infty} \frac{x^k}{k!}$ 9. $\sum_{k=0}^{\infty} \frac{x^{2k+1}}{(2k+1)!}$ 15. $\sum_{k=0}^{\infty} \frac{(-1)^k}{5^{k+1}} (x-4)^k$ 17. $(\sqrt{2}/2) + (\sqrt{2}/2)(x - \pi/4) - (\sqrt{2}/2 \cdot 2!)(x - \pi/4)^2 - (\sqrt{2}/2 \cdot 3!)(x - \pi/4)^3 + \cdots$ 19. $\frac{1}{2} - (\sqrt{3}/2)(x - \pi/3) - (1/2 \cdot 2!)(x - \pi/3)^2 + (\sqrt{3}/2 \cdot 3!)(x - \pi/3)^3 + \cdots$ 21. $\ln 2 + \sum_{k=1}^{\infty} \frac{(-1)^{k+1}}{k2^k} (x-2)^k$ 23. $\sum_{k=0}^{\infty} \frac{e}{k!} (x-1)^k$ 25. $x + \frac{1}{3}x^3 + \frac{2}{15}x^5 + \frac{17}{315}x^7 + \cdots$ 27. $\sum_{k=0}^{\infty} \frac{(-1)^k}{k!} x^{2k}$ 29. $\sum_{k=0}^{\infty} \frac{(-1)^k}{(2k)!} x^{2k+1}$ 31. $-\sum_{k=0}^{\infty} \frac{x^{k+1}}{k+1}$ 33. $1 + x^2 + \frac{2}{3}x^4 + \frac{17}{45}x^6 + \cdots$ 35. 0.71934; four decimal places 37. 1.34983; four decimal places 39. Equate $\sec x = (R + y)/R$ with $1 + x^2/2$ and solve for y. From $L = Rx$, $y = Rx^2/2 = (Rx)^2/2R$ becomes $y = L^2/2R$. For $L = 1$ mi, $y = 7.92$ in. 47. $0 + 0 + 0 + \cdots$

Exercises 10.10, Page 629

1. $1 + \frac{1}{3}x - \frac{2}{3^2 2!}x^2 + \frac{1 \cdot 2 \cdot 5}{3^3 3!}x^3 + \cdots; 1$

3. $3 - \frac{3}{2 \cdot 9}x - \frac{3}{2^2 2! 9^2}x^2 - \frac{3(1 \cdot 3)}{2^3 3! 9^3}x^3 + \cdots; 9$

5. $1 - \frac{1}{2}x^2 + \frac{1 \cdot 3}{2^2 2!}x^4 - \frac{1 \cdot 3 \cdot 5}{2^3 3!}x^6 + \cdots; 1$

7. $8 + \frac{8 \cdot 3}{2 \cdot 4}x + \frac{8 \cdot 3 \cdot 1}{2^2 2! 4^2}x^2 - \frac{8 \cdot 3 \cdot 1}{2^3 3! 4^3}x^3 + \cdots; 4$

9. $\frac{1}{4}x - \frac{2}{4 \cdot 2}x^2 + \frac{2 \cdot 3}{4 \cdot 2^2 2!}x^3 - \frac{2 \cdot 3 \cdot 4}{4 \cdot 2^3 3!}x^4 + \cdots; 2$

11. $|S_2 - S| < a_3 = \frac{x^2}{9}$

13. $x + \sum_{k=1}^{\infty} \frac{1 \cdot 3 \cdot 5 \cdots (2k-1)}{2^k k!(2k+1)} x^{2k+1}$

15. Expand the integrand of $2\int_0^{l/2} \sqrt{1 + (64d^2/l^4)x^2}\, dx$ in a binomial series and integrate term-by-term.

17. $P_0(x) = 1$, $P_1(x) = x$, $P_2(x) = \frac{1}{2}(3x^2 - 1)$

19. $\sqrt{2} + \frac{\sqrt{2}}{2^2}(x-1) - \frac{\sqrt{2}}{2^4 2!}(x-1)^2 + \frac{\sqrt{2}(1 \cdot 3)}{2^6 3!}(x-1)^3 + \cdots$

Chapter 10 Review Exercises, Page 631

1. False 2. False 3. True 4. True
5. True 6. False 7. False 8. False
9. True 10. True 11. False
12. True 13. True 14. False
15. False 16. False 17. False
18. True 19. False 20. False
21. False 22. False 23. True
24. True 25. True 26. False
27. 20, 9, $\frac{4}{5}$, 16 29. 12 31. $-\pi/4$
33. 4 35. Converges 37. Converges
39. Converges 41. Diverges 43. Diverges
45. $\frac{61{,}004}{201}$ 47. $[-\frac{1}{3}, \frac{1}{3}]$ 49. $\{-5\}$
51. $\frac{4}{3}$ 53. $1 - \frac{1}{3}x^5 + \frac{4}{3^2 2!}x^{10} + \cdots$
55. $x - \frac{2^2}{3!}x^3 + \frac{2^4}{5!}x^5 + \cdots$
57. $\sum_{k=0}^{\infty} \frac{(-1)^{k+1}}{(2k+1)!}(x - \pi/2)^{2k+1}$ 59. \$6 million

Exercises 11.1, Page 639

1. $(0, 0); (0, -\frac{1}{8}); y = \frac{1}{8}$ 3. $(0, 0); (3, 0); x = -3$
5. $(0, 0); (-\frac{5}{2}, 0); x = \frac{5}{2}$ 7. $(1, 0); (1, 1); y = -1$
9. $(4, 3); (\frac{9}{2}, 3); x = \frac{7}{2}$
11. $(4, -3); (4, -2); y = -4$
13. $(-2, 2); (-2, \frac{9}{4}); y = \frac{7}{4}$
15. $(-3, 3); (-2, 3); x = -4$
17. $(-2, -\frac{5}{8}); (-2, -\frac{7}{24}); y = -\frac{23}{24}$
19. $(\frac{1}{8}, -\frac{5}{2}); (-\frac{3}{8}, -\frac{5}{2}); x = \frac{5}{8}$
21. $y = 4x^2$ 23. $(y-1)^2 = -12(x-1)$
25. $(y-4)^2 = -12(x-1)$
27. $(y-4)^2 = 10(x + \frac{5}{2})$ 29. $y = x^2 + 2$
31. $y = (\frac{2}{9})x^2 - 2$; $y = (\frac{2}{9})(x+2)^2 - 2$
33. $-x + 8y = 8$ 37. $y = \frac{3}{500}x^2$; 5.4 m
43. $(-b/2a, (4ac - b^2)/4a)$, $(-b/2a, (1 + 4ac - b^2)/4a)$
45. $x^2 + y^2 + 2xy + 2x - 14y + 1 = 0$

Exercises 11.2, Page 646

1. $(\pm 4, 0), (0, \pm 5); (0, \pm 3)$
3. $(\pm 1, 0), (0, \pm\sqrt{10}); (0, \pm 3)$
5. $(\pm 4, 0), (0, \pm 2\sqrt{2}); (\pm 2\sqrt{2}, 0)$
7. $(\pm\sqrt{7}, 0), (0, \pm 2); (\pm\sqrt{3}, 0)$
9. $(-4, 3), (6, 3), (1, -3), (1, 9); (1, 3 - \sqrt{11}), (1, 3 + \sqrt{11})$
11. $(-6, 2), (10, 2), (2, -4), (2, 8); (2 - 2\sqrt{7}, 2), (2 + 2\sqrt{7}, 2)$
13. $(-\frac{1}{2} - \sqrt{2}, 1), (-\frac{1}{2} + \sqrt{2}, 1), (-\frac{1}{2}, -1), (-\frac{1}{2}, 3); (-\frac{1}{2}, 1 - \sqrt{2}), (-\frac{1}{2}, 1 + \sqrt{2})$
15. $(-\sqrt{3}, 3), (\sqrt{3}, 3)\ (0, 0), (0, 6); (0, 3 - \sqrt{6}), (0, 3 + \sqrt{6})$
17. $(-6, -1), (4, -1), (-1, -4), (-1, 2); (-5, -1), (3, -1)$
19. $\left(4 - \frac{\sqrt{5}}{5}, 2\right), \left(4 + \frac{\sqrt{5}}{5}, 2\right), (4, 1), (4, 3); \left(4, 2 - \frac{2\sqrt{5}}{5}\right), \left(4, 2 + \frac{2\sqrt{5}}{5}\right)$
21. $\frac{5}{13}$ 23. $\sqrt{2}/\sqrt{5}$ 25. $x^2/25 + y^2/4 = 1$
27. $x^2/12 + y^2/16 = 1$
29. $(x+1)^2/16 + (y+2)^2/25 = 1$
31. $4x^2/31 + 4y^2/3 = 1$ 33. $x^2/8 + y^2/2 = 1$
35. $(x-4)^2/64 + (y+2)^2/100 = 1$
37. One possibility is $x^2/16 + 9y^2/128 = 1$.
39. $-\frac{18}{5}, \frac{18}{5}$
43. $(4, 1); (2, 1), (6, 1), (4, 4), (4, -2); (4, 1 - \sqrt{5}), (4, 1 + \sqrt{5})$
45. Approximately 2.85×10^7 miles; approximately 4.35×10^7 miles
47. Approximately 0.97 49. $\frac{1}{3}$
51. Approximately 4667 km
53. $R = r + \sqrt{a^2 + b^2}$ 55. 25.5296

Exercises 11.3, Page 654

1. $(\pm 4, 0); (\pm 5, 0); y = \pm 3x/4$

3. $(0, \pm 3); (0, \pm 3\sqrt{2}); y = \pm x$
5. $(\pm 5, 0); (\pm\sqrt{29}, 0); y = \pm 2x/5$
7. $(0, \pm 4), (0, \pm 5); y = \pm 4x/3$
9. $(0, 2), (2, 2); (1 - \sqrt{2}, 2), (1 + \sqrt{2}, 2);$
$y = x + 1, y = -x + 3$
11. $(4, -7), (4, 5); (4, -1 - 2\sqrt{10}), (4, -1 + 2\sqrt{10});$
$y = 3x - 13, y = -3x + 11$
13. $(-\frac{3}{2}, -\frac{4}{3}), (\frac{5}{2}, -\frac{4}{3}); (\frac{1}{2} - 2\sqrt{2}, -\frac{4}{3}),$
$(\frac{1}{2} + 2\sqrt{2}, -\frac{4}{3}); 6x - 6y = 11, 6x + 6y = -5$
15. $(-2, 3), (-2, 7); (-2, 5 - \sqrt{29}), (-2, 5 + \sqrt{29});$
$-2x + 5y = 29, 2x + 5y = 21$
17. $(2 - \sqrt{3}, 1), (2 + \sqrt{3}, 1); (2 - \sqrt{6}, 1), (2 + \sqrt{6}, 1);$
$y = x - 1, y = -x + 3$
19. $(-9, -1), (1, -1); (-4 - 5\sqrt{2}, -1),$
$(-4 + 5\sqrt{2}, -1); y = x + 3, y = -x - 5$
21. 3 23. $\sqrt{3}$ 25. $x^2/9 - y^2/16 = 1$
27. $(x - 2)^2/25 - (y + 7)^2/25 = 1$
29. $(y - 2)^2/9 - (x - 1)^2/7 = 1$
31. $y^2/16 - x^2/2 = 1$ 33. $4x^2/9 - 4y^2/27 = 1$
35. $(x + 1)^2/16 - (y - 2)^2/20 = 1$
37. $y^2/36 - x^2/64 = 1$
39. $(2\sqrt{2}, \sqrt{2}), (-2\sqrt{2}, -\sqrt{2})$
41. $(-2\sqrt{26}, -8), (2\sqrt{26}, -8)$ 43. 12π
47. 3, 6, 9, 13, 16, 17, 19

Exercises 11.4, Page 659

1. $(2, -1)$ 3. $(-\frac{3}{2}, \frac{7}{2})$
5. The graph of $y = |x|$ is translated one unit to the right and up four units.
7. The graph of $y = \sin x$ has been translated $\pi/2$ units to the left.
9. The graph of $x^2 - y^2 = 4$ has been translated to the left one unit and up one unit.
11. $(4\sqrt{2}, -2\sqrt{2})$ 13. $(-(1 + \sqrt{3})/2, (\sqrt{3} - 1)/2)$
15. $(-1, -7)$
17. Ellipse rotated 45°, $3X^2 + Y^2 = 8$
19. Parabola rotated 45°, $Y^2 = 4\sqrt{2}X$
21. Ellipse rotated by $\theta \approx 26.6°$,
$(X + \sqrt{5})^2/8 + Y^2/3 = 1$
23. $Y = X^2$; XY: $(0, \frac{1}{4})$; xy: $(-\frac{1}{8}, \sqrt{3}/8)$;
$Y = -\frac{1}{4}$, $2x - 2\sqrt{3}\,y = 1$
27. Hyperbola 29. Parabola 31. Ellipse

Chapter 11 Review Exercises, Page 660

1. True 2. False 3. True 4. True
5. True 6. False 7. True 8. True
9. False 10. False 11. False
12. True 13. False 14. False
15. True 16. True 17. $(0, \frac{1}{8})$

19. $(0, 2)$ 21. $y = -5$ 23. 6
25. $(2, -1), (6, -1)$ 27. $(4, -3)$
29. $-\sqrt{5}, \sqrt{5}$ 33. 10^8 m; 9×10^8 m
35. $x + y + 4 = 0$ 37. $20\sqrt{2}$ ft

Exercises 12.1, Page 669

1.

x	−5	−3	−1	1	3	5	7
y	6	2	0	0	2	6	12

3.

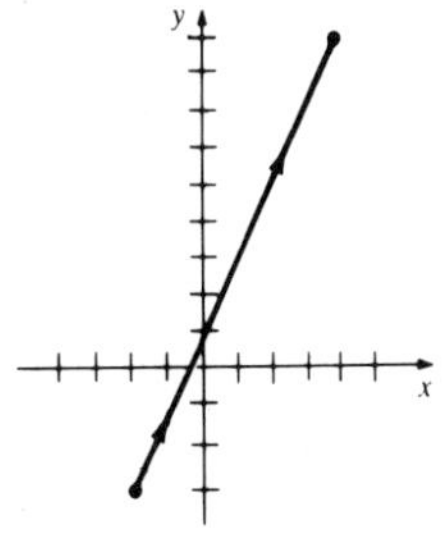

5.

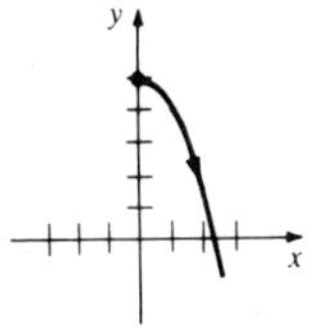

7.

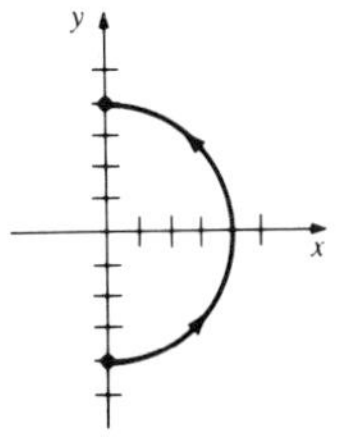

9.

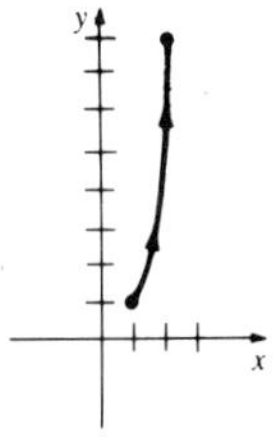

11. $y = x^2 + 3x - 1, x \geq 0$
13. $x = 1 - 2y^2, -1 \leq x \leq 1$
15. $y = \ln x, x > 0$
17.

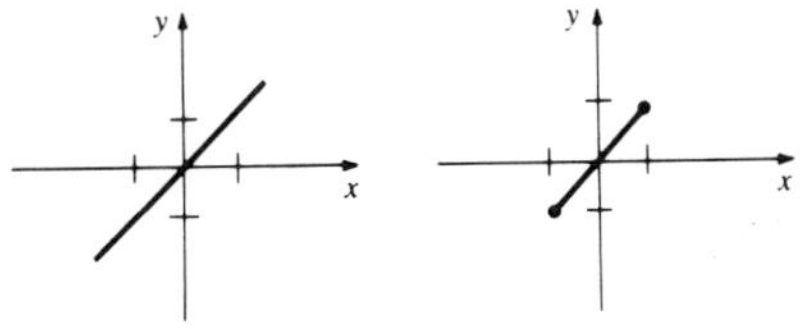

19.

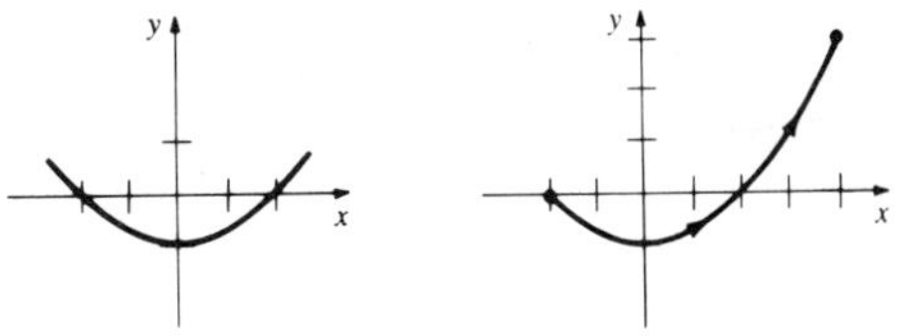

21.

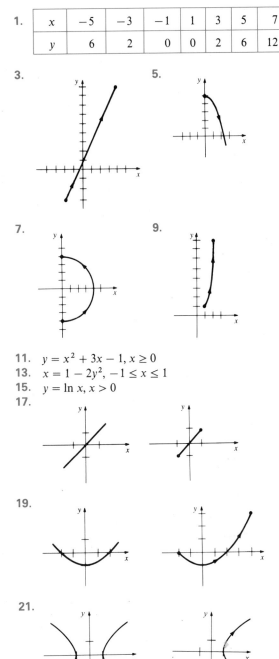

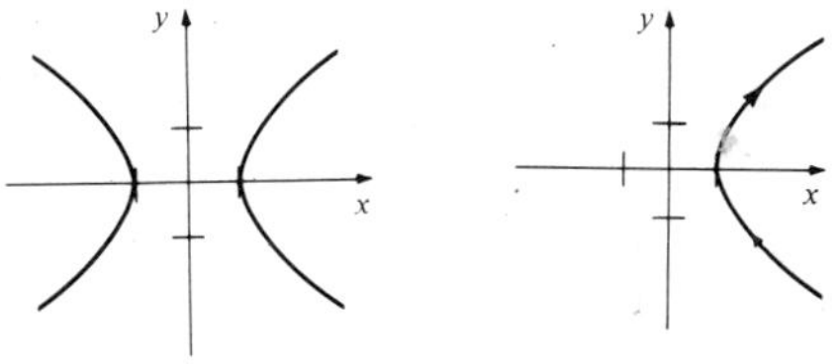

23.

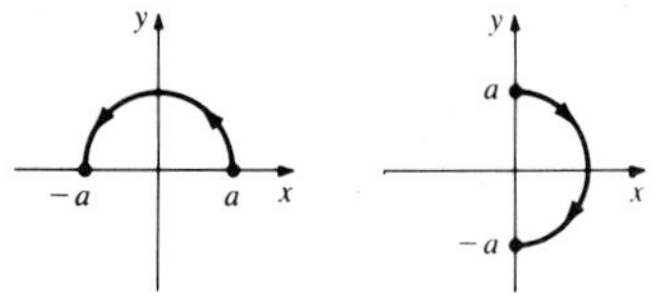

25.

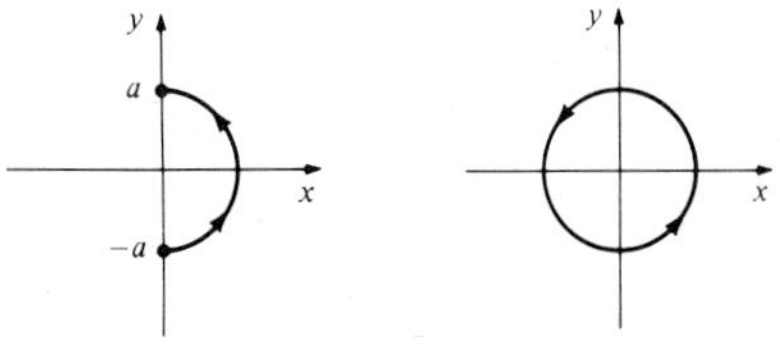

27.

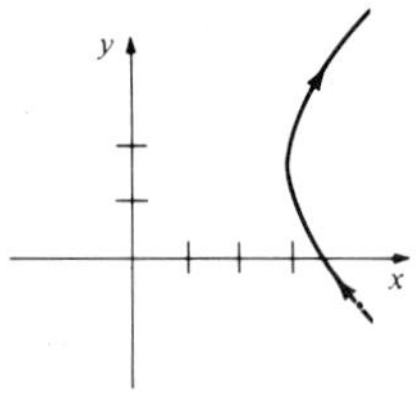

29. **(a)**, **(f)**

31. $x = \pm\sqrt{r^2 - L^2\sin^2\phi}$, $y = L\sin\phi$

33. $x = 2a\cot\theta$, $y = 2a\sin^2\theta$

35. $x = a(\cos\theta + \theta\sin\theta)$, $y = a(\sin\theta - \theta\cos\theta)$

37. $x^{2/3} + y^{2/3} = b^{2/3}$

39.

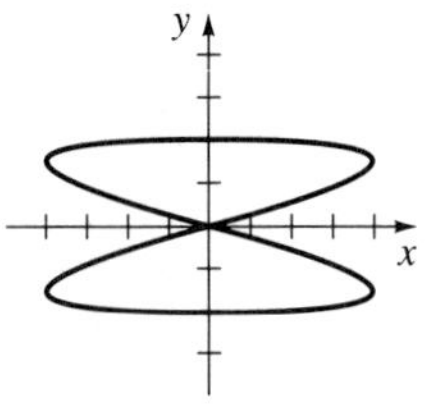

41.

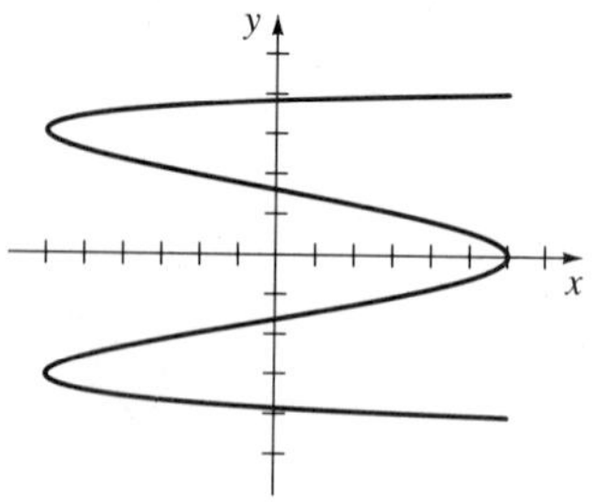

43.

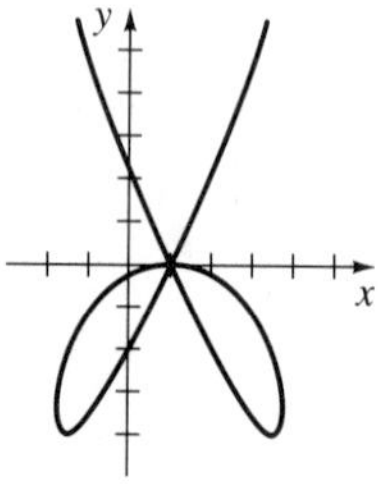

Exercises 12.2, Page 676

1. $\frac{3}{5}$ 3. 24 5. -1 7. $y = -2x - 1$

9. $4x - 3y = -4$ 11. $\sqrt{3}/4$

13. $y = 3x - 7$

15. Hor. tan. at (0, 0); ver. tan. at $(2/3\sqrt{3}, \frac{1}{3})$, $(-2/3\sqrt{3}, \frac{1}{3})$

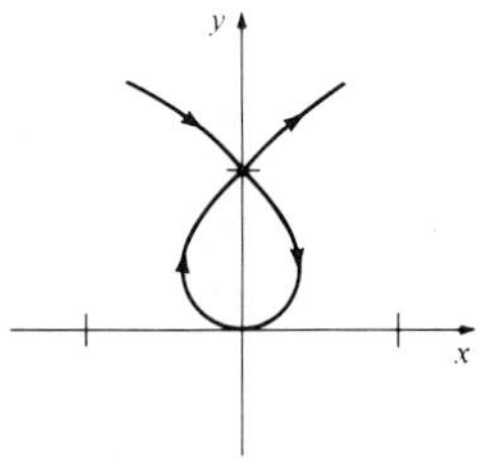

17. Hor. tan. at $(-1, 0)$ and $(1, -4)$; no ver. tan.

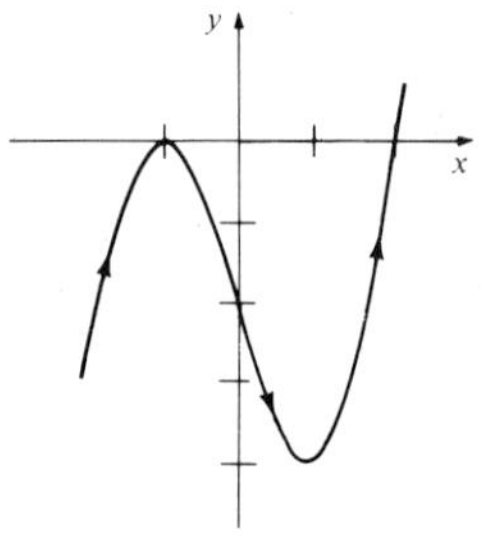

19. $3t$, $1/(2t)$, $-1/(12t^3)$

21. $-2e^{3t} - 3e^{4t}$, $6e^{4t} + 12e^{5t}$, $-24e^{5t}$, $-60e^{6t}$

23. Concave upward for $0 < t < 2$, concave downward for $t < 0$ and $t > 2$

25. $\frac{104}{3}$ 27. $\sqrt{2}(e^{\pi} - 1)$ 29. $3b/2$

31. -0.6551; -5.9991, 1.0446, 9.7361

Exercises 12.3, Page 681

1.

3.

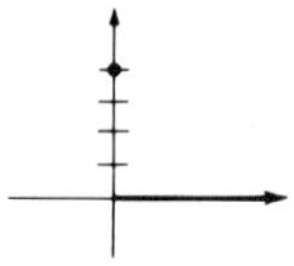

5. $(6, -5\pi/4)$; $(6, 11\pi/4)$; $(-6, 7\pi/4)$; $(-6, -\pi/4)$

7. $(2, -4\pi/3)$; $(2, 8\pi/3)$; $(-2, 5\pi/3)$; $(-2, -\pi/3)$

9. $(1, -11\pi/6)$; $(1, 13\pi/6)$; $(-1, 7\pi/6)$; $(-1, -5\pi/6)$

11. $(\frac{1}{2}, -\sqrt{3}/2)$ 13. $(-\frac{7}{2}, 7\sqrt{3}/2)$

15. $(-2\sqrt{2}, -2\sqrt{2})$

17. $(3\sqrt{2}, -3\pi/4)$; $(-3\sqrt{2}, \pi/4)$

19. $(2, -\pi/6)$; $(-2, 5\pi/6)$

21. $(5, -\pi/2)$; $(-5, \pi/2)$

23. $(5, 2.4981)$; $(-5, -0.6435)$

25. $r = 5 \csc \theta$ 27. $\theta = \tan^{-1} 7$
29. $r = 2/(1 + \cos \theta)$ 31. $r = 6$
33. $r = 1 - \cos \theta$ 37. $x = 2$
39. $(x^2 + y^2)^3 = 144x^2y^2$ 41. $(x^2 + y^2)^2 = 8xy$
43. $x^2 + y^2 + 5y = 0$
45. $8x^2 - 12x - y^2 + 4 = 0$
47. $3x + 8y = 5$ 51. $80/\sqrt{41}$ cm/s

Exercises 12.4, Page 689

1.

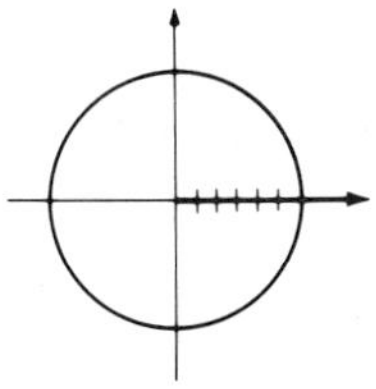

3.

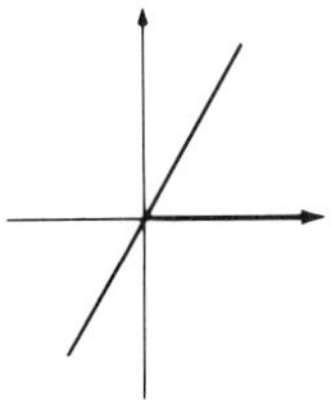

5.

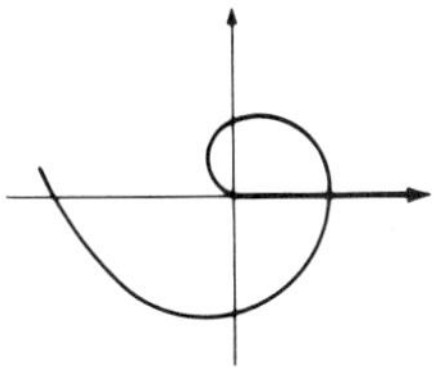

7.

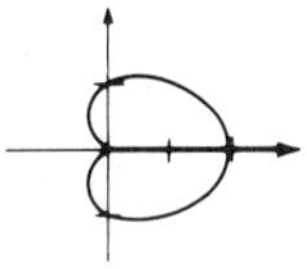

9.

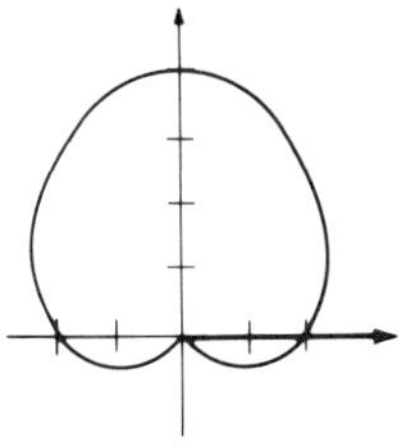

11.

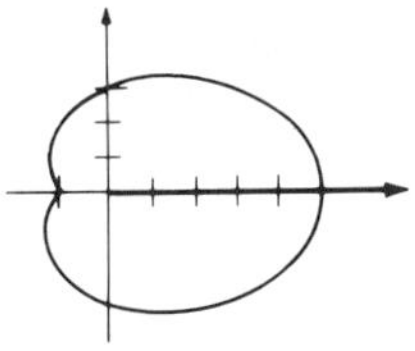

13.

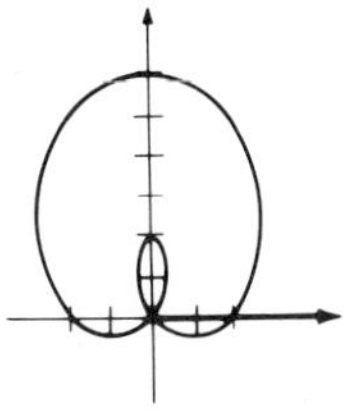

15.

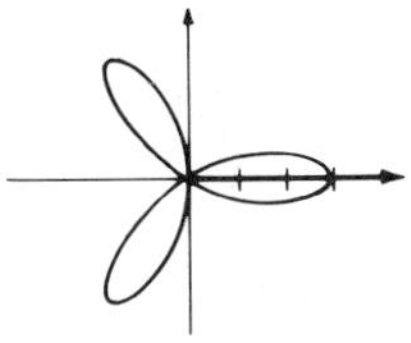

17.

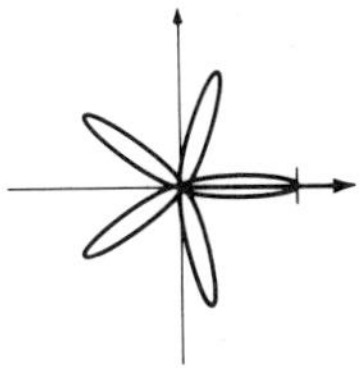

19.

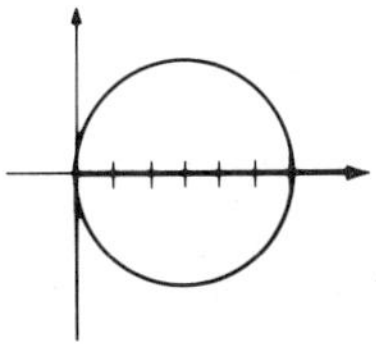

21.

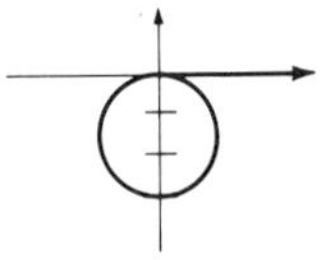

23.

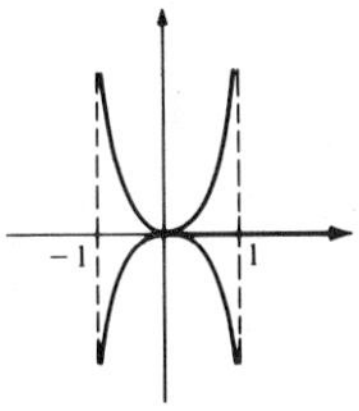

25.

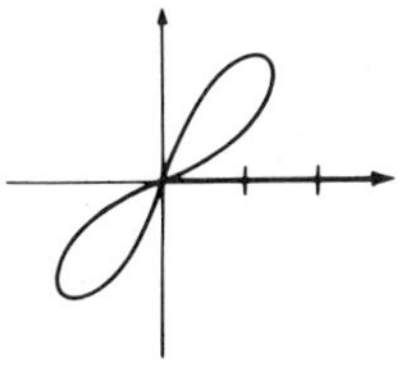

27.

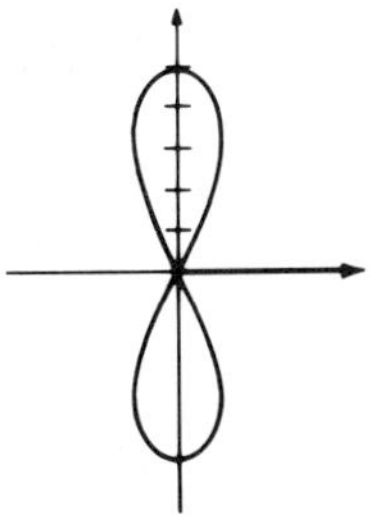

29.

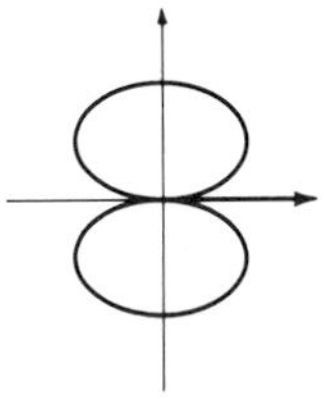

31.

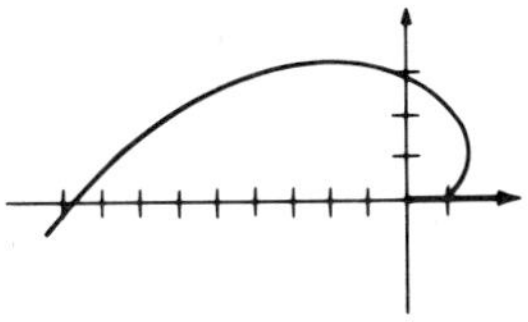

33.

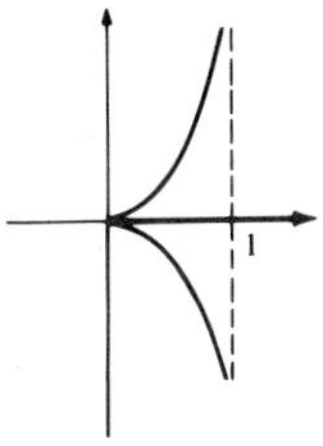

35. $-2/\pi$ 37. $\sqrt{3}/15$ 39. $y = -x + 1$
41. Hor. tan. at $(3, \pi/3)$ and $(3, 5\pi/3)$; ver. tan. at $(4, 0)$, $(1, 2\pi/3)$, and $(1, 4\pi/3)$
43. $(2, \pi/6)$, $(2, 5\pi/6)$
45. $(1, \pi/2)$, $(1, 3\pi/2)$, origin
47. $(3, \pi/12)$, $(3, 5\pi/12)$, $(3, 13\pi/12)$, $(3, 17\pi/12)$, $(3, -\pi/12)$, $(3, -5\pi/12)$, $(3, -13\pi/12)$, $(3, -17\pi/12)$
49. $(0, 0)$, $(\sqrt{3}/2, \pi/3)$, $(\sqrt{3}/2, 2\pi/3)$

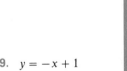

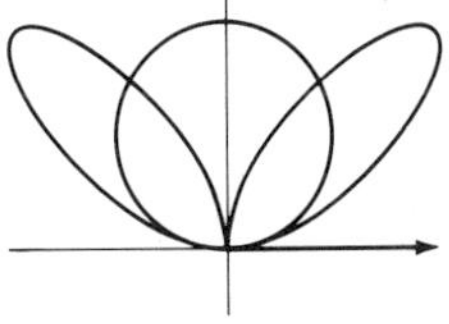

51. Symmetry with respect to the x-axis
53. Symmetry with respect to the origin
55. Symmetry with respect to the x-axis
59. **(e)** 61. **(f)** 63. **(b)**

Exercises 12.5, Page 695

1. π 3. 24π 5. 11π 7. $9\pi/2$
9. π
11. $r^2 = 9\cos 2\theta$ is not nonnegative on $[0, 2\pi]$; 9
13. $9\pi^3/4$ 15. $\frac{1}{4}(e^{2\pi} - 1)$ 17. $\frac{1}{8}(4 - \pi)$
19. $\frac{1}{6}(2\pi + 3\sqrt{3})$ 21. $\pi + 6\sqrt{3}$
23. $18\sqrt{3} - 4\pi$ 25. $\frac{1}{2}(4\pi + 3\sqrt{3})$
27. $\sqrt{15} - \cos^{-1}(\frac{1}{4})$ 29. $\sqrt{5}(e^2 - 1)$
31. $\frac{1}{2}[\sqrt{2} + \ln(\sqrt{2} + 1)]$
33. $A = \dfrac{1}{2}\displaystyle\int_{\theta_1}^{\theta_2} r^2\, d\theta = \dfrac{1}{2}\int_a^b r^2 \dfrac{d\theta}{dt}\, dt$
$= \dfrac{1}{2}\displaystyle\int_a^b (L/m)\, dt = L(b - a)/2m$

Exercises 12.6, Page 701

1. $e = 1$, parabola

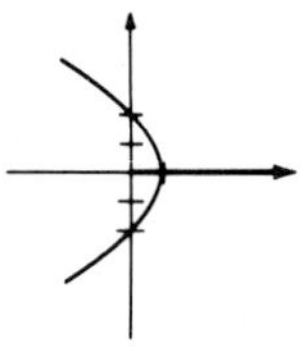

3. $e = \frac{1}{4}$, ellipse

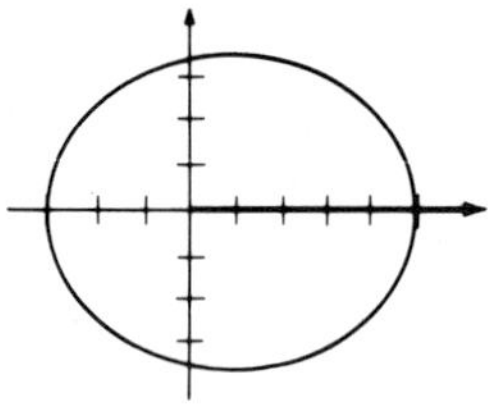

5. $e = 2$, hyperbola

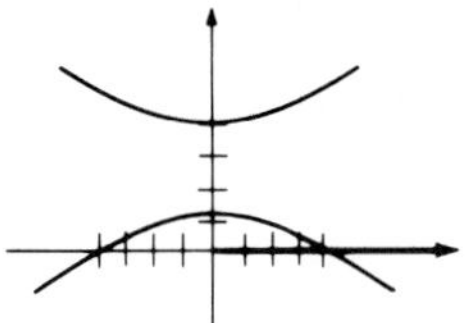

7. $e = 2$, hyperbola

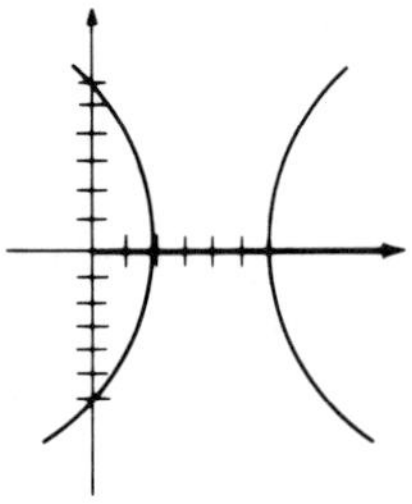

9. $e = \frac{4}{5}$, ellipse

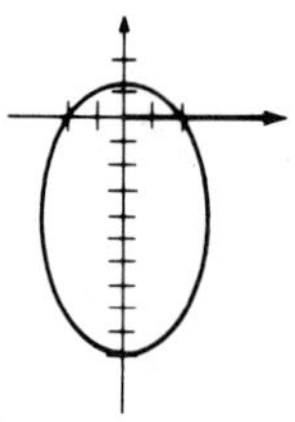

11. Center: $(-2, 0)$; foci: $(0, 0)$, $(-4, 0)$;
vertices: $(-6, 0)$, $(2, 0)$, $(-2, -2\sqrt{3})$, $(-2, 2\sqrt{3})$
13. $r = 3/(1 - \cos\theta)$ 15. $r = 30/(2 + 5\sin\theta)$
17. $r = 5/(1 + \sin\theta)$ 21. $r = 9600/(1 - 0.2\cos\theta)$
23. $r = (8.7 \times 10^8)/(1 - 0.97\cos\theta)$

Chapter 12 Review Exercises, Page 702

1. False 2. False 3. True 4. True
5. True 6. False 7. False 8. True
9. True 10. True 11. True
12. True 13. False 14. True
15. True 16. True 17. True
18. False 19. False 20. False
21. $3x + 3\sqrt{3}y = \pi$ 23. $(8, -26)$
25. $y^2 < 0$ for $|x| > 1$; $x = \sin t$, $y = \sin 2t$;
$(\sqrt{2}/2, 1)$, $(\sqrt{2}/2, -1)$, $(-\sqrt{2}/2, 1)$, $(-\sqrt{2}/2, -1)$

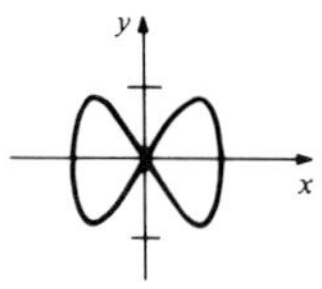

27. $5\pi/4$
29. $x + y = 2\sqrt{2}$; $r = 2\sqrt{2}/(\cos\theta + \sin\theta)$
31. $r = 3 + 2\cos\theta$ 33. $r = \dfrac{1}{1 - \cos\theta}$
35. $\sqrt{3} - \pi/3$
37. $x = 3at/(1 + t^3)$, $y = 3at^2/(1 + t^3)$
39. $r = \dfrac{3a\cos\theta\sin\theta}{\cos^3\theta + \sin^3\theta}$; $3a^2/2$ 41. $(2\pi - 3\sqrt{3})/2$

Exercises 13.1, Page 722

1. $6\mathbf{i} + 12\mathbf{j}$; $\mathbf{i} + 8\mathbf{j}$; $3\mathbf{i}$; $\sqrt{65}$; 3
3. $\langle 12, 0\rangle$; $\langle 4, -5\rangle$; $\langle 4, 5\rangle$; $\sqrt{41}$; $\sqrt{41}$
5. $-9\mathbf{i} + 6\mathbf{j}$; $-3\mathbf{i} + 9\mathbf{j}$; $-3\mathbf{i} - 5\mathbf{j}$; $3\sqrt{10}$; $\sqrt{34}$
7. $-6\mathbf{i} + 27\mathbf{j}$; $\mathbf{0}$; $-4\mathbf{i} + 18\mathbf{j}$; 0; $2\sqrt{85}$
9. $\langle 6, -14\rangle$; $\langle 2, 4\rangle$ 11. $10\mathbf{i} - 12\mathbf{j}$; $12\mathbf{i} - 17\mathbf{j}$
13. $\langle 20, 52\rangle$; $\langle -2, 0\rangle$
15. $2\mathbf{i} + 5\mathbf{j}$

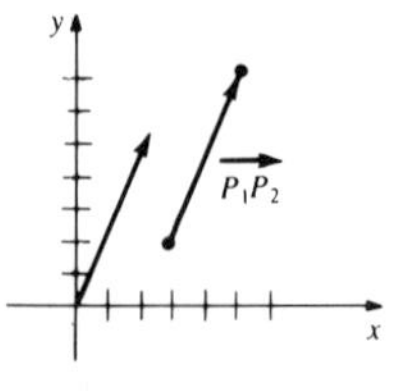

17. $2\mathbf{i} + 2\mathbf{j}$

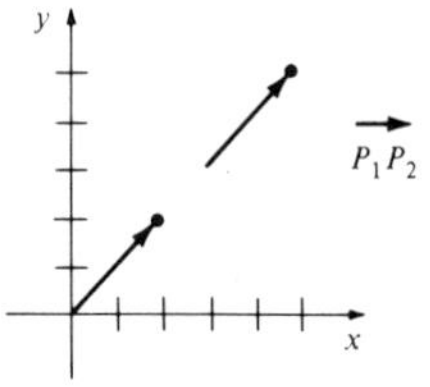

19. $(1, 18)$ 21. **(a)**, **(b)**, **(c)**, **(e)**, **(f)**

23. $\langle 6, 15\rangle$ 25. $\left\langle \frac{1}{\sqrt{2}}, \frac{1}{\sqrt{2}}\right\rangle; \left\langle -\frac{1}{\sqrt{2}}, -\frac{1}{\sqrt{2}}\right\rangle$

27. $\langle 0, -1\rangle; \langle 0, 1\rangle$ 29. $\left\langle \frac{5}{13}, \frac{12}{13}\right\rangle$

31. $\frac{6}{\sqrt{58}}\mathbf{i} + \frac{14}{\sqrt{58}}\mathbf{j}$ 33. $\left\langle -3, -\frac{15}{2}\right\rangle$

35.

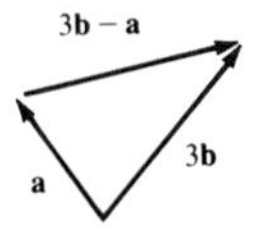

37. $-(\mathbf{a} + \mathbf{b})$

41. $\mathbf{a} = \frac{5}{2}\mathbf{b} - \frac{1}{2}\mathbf{c}$ 43. $\pm\frac{1}{\sqrt{2}}(\mathbf{i} + \mathbf{j})$

45. $|\mathbf{a} + \mathbf{b}| \leq |\mathbf{a}| + |\mathbf{b}|$; when P_1, P_2, and P_3 are collinear

49. Approximately 31° 51. Approximately 153 lb

Exercises 13.2, Page 729

1.–5.

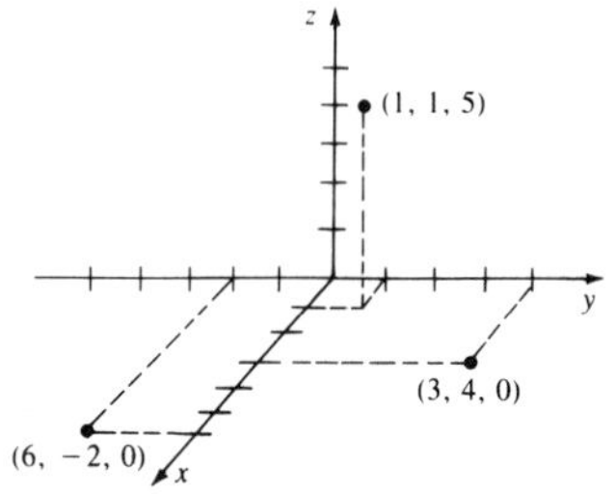

7. The set $\{(x, y, 5) \mid x, y \text{ real numbers}\}$ is a plane perpendicular to the z-axis, 5 units above the xy-plane.

9. The set $\{(2, 3, z) \mid z \text{ a real number}\}$ is a line perpendicular to the xy-plane at $(2, 3, 0)$.

11. $(0, 0, 0), (2, 0, 0), (2, 5, 0), (0, 5, 0), (0, 0, 8), (2, 0, 8), (2, 5, 8), (0, 5, 8)$

13. $(-2, 5, 0), (-2, 0, 4), (0, 5, 4); (-2, 5, -2); (3, 5, 4)$

15. The union of the coordinate planes

17. The point $(-1, 2, -3)$

19. The union of the planes $z = -5$ and $z = 5$

21. $\sqrt{70}$ 23. 7; 5 25. Right triangle

27. Isosceles

29. $d(P_1, P_2) + d(P_1, P_3) = d(P_2, P_3)$

31. 6 or -2 33. $(4, \frac{1}{2}, \frac{3}{2})$

35. $P_1(-4, -11, 10)$

37. Approximately $(1.407, 0.2948, 0.9659)$

39. $\langle -3, -6, 1\rangle$ 41. $\langle 2, 1, 1\rangle$ 43. $\langle 2, 4, 12\rangle$

45. $\langle -11, -41, -49\rangle$ 47. $\sqrt{139}$ 49. 6

51. $\langle -\frac{2}{3}, \frac{1}{3}, -\frac{2}{3}\rangle$ 53. $4\mathbf{i} - 4\mathbf{j} + 4\mathbf{k}$

55.

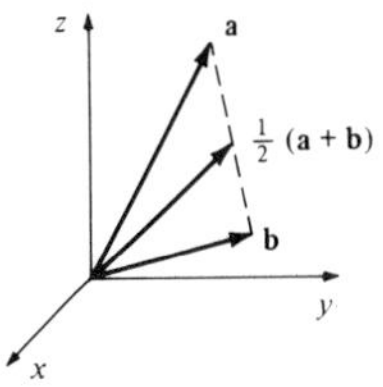

Exercises 13.3, Page 738

1. $25\sqrt{2}$ 3. 12 5. -16 7. 48

9. 29 11. 25 13. $\langle -\frac{2}{5}, \frac{4}{5}, 2\rangle$

15. **(a)** and **(f)**, **(c)** and **(d)**, **(b)** and **(e)**

17. $\langle \frac{4}{9}, -\frac{1}{3}, 1\rangle$ 21. 1.11 radians or 63.43°

23. 1.89 radians or 108.43°

25. $\cos\alpha = 1/\sqrt{14}$, $\cos\beta = 2/\sqrt{14}$, $\cos\gamma = 3/\sqrt{14}$; $\alpha = 74.5°$, $\beta = 57.69°$, $\gamma = 36.7°$

27. $\cos\alpha = \frac{1}{2}$, $\cos\beta = 0$, $\cos\gamma = -\sqrt{3}/2$; $\alpha = 60°$, $\beta = 90°$, $\gamma = 150°$

29. 0.9553 radian or 54.73°; 0.6155 radian or 35.27°

31. $\alpha = 58.19°$, $\beta = 42.45°$, $\gamma = 65.06°$ 33. $\frac{5}{7}$

35. $-6\sqrt{11}/11$ 37. $72\sqrt{109}/109$

39. $\langle -\frac{21}{5}, \frac{28}{5}\rangle; \langle -\frac{4}{5}, -\frac{3}{5}\rangle$

41. $\langle -\frac{12}{7}, \frac{6}{7}, \frac{4}{7}\rangle; \langle \frac{5}{7}, -\frac{20}{7}, \frac{45}{7}\rangle$ 43. $\langle \frac{72}{25}, \frac{96}{25}\rangle$

45. 1000 ft-lb 47. 0; 150 N-m

49. Approximately 1.80 angstroms

Exercises 13.4, Page 746

1. $-5\mathbf{i} - 5\mathbf{j} + 3\mathbf{k}$ 3. $\langle -12, -2, 6\rangle$

5. $-5\mathbf{i} + 5\mathbf{k}$ 7. $\langle -3, 2, 3\rangle$ 9. $\mathbf{0}$

11. $6\mathbf{i} + 14\mathbf{j} + 4\mathbf{k}$ 13. $-3\mathbf{i} - 2\mathbf{j} - 5\mathbf{k}$

17. $-\mathbf{i} + \mathbf{j} + \mathbf{k}$ 19. $2\mathbf{k}$ 21. $\mathbf{i} + 2\mathbf{j}$

23. $-24\mathbf{k}$ 25. $5\mathbf{i} - 5\mathbf{j} - \mathbf{k}$ 27. 0

29. $\sqrt{41}$ 31. $-\mathbf{j}$ 33. $\mathbf{0}$ 35. 6

37. $12\mathbf{i} - 9\mathbf{j} + 18\mathbf{k}$ 39. $-4\mathbf{i} + 3\mathbf{j} - 6\mathbf{k}$

41. $-21\mathbf{i} + 16\mathbf{j} + 22\mathbf{k}$ 43. -10

45. 14 square units 47. $\frac{1}{2}$ square unit

49. $\frac{7}{2}$ square units 51. 10 cubic units

53. Coplanar

55. 32; in the xy-plane, 30° from the positive x-axis in the direction of the negative y-axis; $16\sqrt{3}\mathbf{i} - 16\mathbf{j}$

63. $\mathbf{A} = \mathbf{i} - \mathbf{k}$, $\mathbf{B} = \mathbf{j} - \mathbf{k}$, $\mathbf{C} = 2\mathbf{k}$

Exercises 13.5, Page 753

1. $\langle x, y, z\rangle = \langle 1, 2, 1\rangle + t\langle 2, 3, -3\rangle$

3. $\langle x, y, z\rangle = \langle \frac{1}{2}, -\frac{1}{2}, 1\rangle + t\langle -2, 3, -\frac{3}{2}\rangle$

5. $\langle x, y, z\rangle = \langle 1, 1, -1\rangle + t\langle 5, 0, 0\rangle$

7. $x = 2 + 4t$, $y = 3 - 4t$, $z = 5 + 3t$

9. $x = 1 + 2t$, $y = -2t$, $z = -7t$

11. $x = 4 + 10t$, $y = \frac{1}{2} + \frac{3}{4}t$, $z = \frac{1}{3} + \frac{1}{6}t$

13. $\frac{x - 1}{9} = \frac{y - 4}{10} = \frac{z + 9}{7}$

15. $\dfrac{x+7}{11} = \dfrac{z-5}{-4}, y = 2$

17. $x = 5, \dfrac{y-10}{9} = \dfrac{z+2}{12}$

19. $x = 4 + 3t, y = 6 + \frac{1}{2}t, z = -7 - \frac{3}{2}t;$
$\dfrac{x-4}{3} = 2y - 12 = \dfrac{2z+14}{-3}$

21. $x = 5t, y = 9t, z = 4t; \dfrac{x}{5} = \dfrac{y}{9} = \dfrac{z}{4}$

23. $x = 6 + 2t, y = 4 - 3t, z = -2 + 6t$

25. $x = 2 + t, y = -2, z = 15$

27. Both lines pass through the origin and have parallel direction vectors.

29. $(0, 5, 15), (5, 0, \frac{15}{2}), (10, -5, 0)$

31. $(2, 3, -5)$ 33. Lines do not intersect.

35. Yes

37. $x = 2 + 4t, y = 5 - 6t, z = 9 - 6t, 0 \le t \le 1$

39. $40.37°$

41. $x = 4 - 6t, y = 1 + 3t, z = 6 + 3t$

43. The lines are not parallel and do not intersect.

Exercises 13.6, Page 759

1. $2x - 3y + 4z = 19$ 3. $5x - 3z = 51$

5. $6x + 8y - 4z = 11$ 7. $5x - 3y + z = 2$

9. $3x - 4y + z = 0$ 11. The points are collinear.

13. $x + y - 4z = 25$ 15. $z = 12$

17. $-3x + y + 10z = 18$ 19. $9x - 7y + 5z = 17$

21. $6x - 2y + z = 12$

23. Orthogonal: **(a)** and **(d)**, **(b)** and **(c)**, **(d)** and **(f)**, **(b)** and **(e)**; parallel: **(a)** and **(f)**, **(c)** and **(e)**

25. **(c)**, **(d)** 27. $x = 2 + t, y = \frac{1}{2} - t, z = t$

29. $x = \frac{1}{2} - \frac{1}{2}t, y = \frac{1}{2} - \frac{3}{2}t, z = t$

31. $(-5, 5, 9)$ 33. $(1, 2, -5)$

35. $x = 5 + t, y = 6 + 3t, z = -12 + t$

37. $3x - y - 2z = 10$ 39.

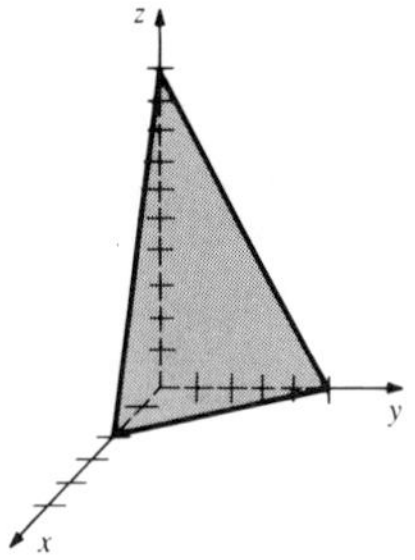

41.

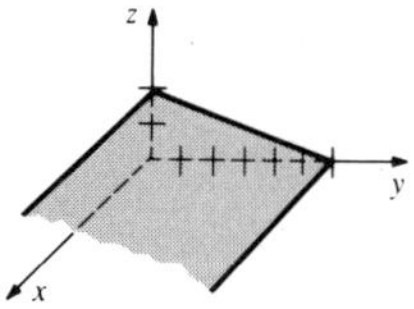

43.

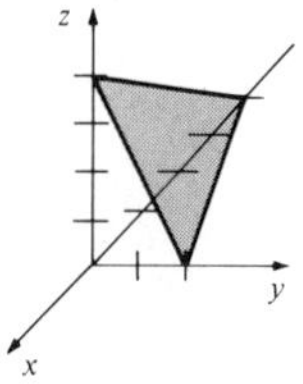

47. $3/\sqrt{11}$ 49. $107.98°$

Exercises 13.7, Page 764

1.

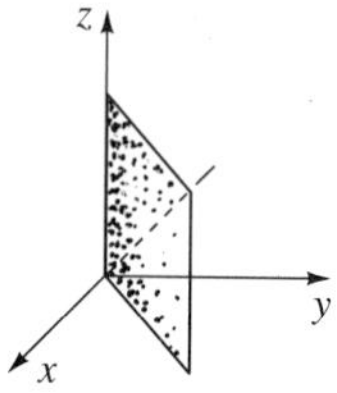

3.

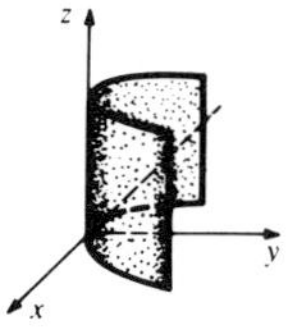

5.

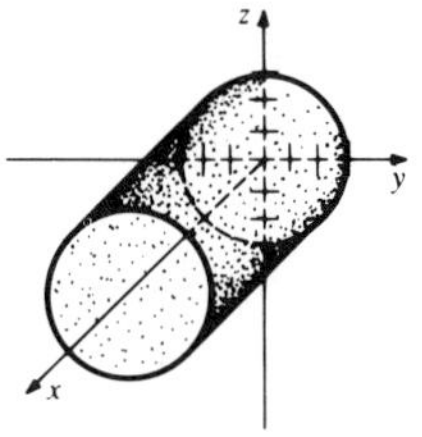

7.

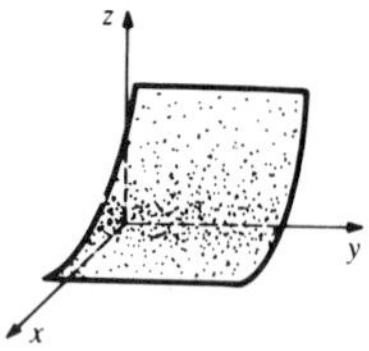

9.

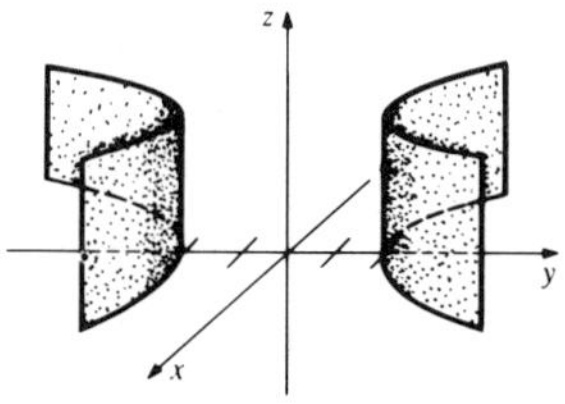

11.

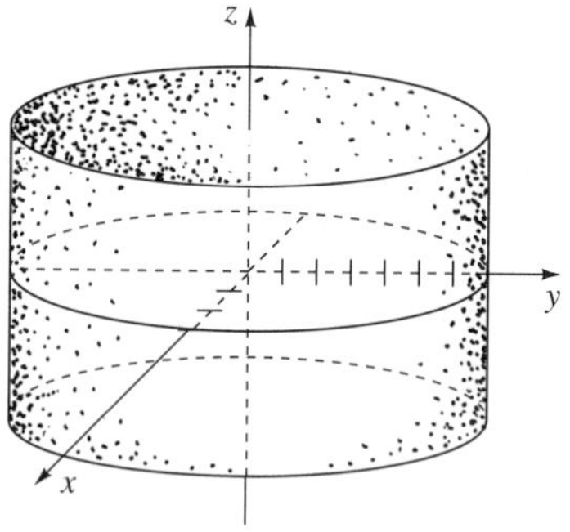

13.

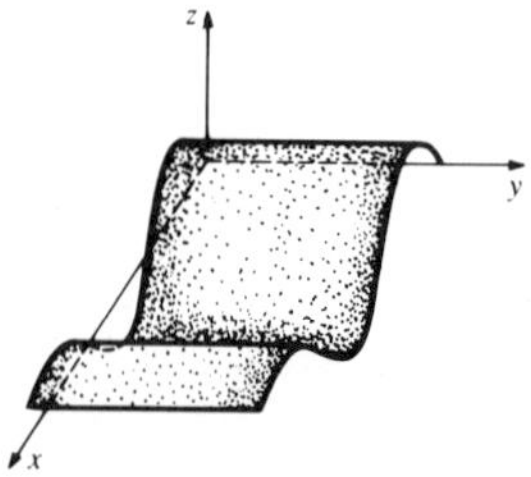

15.

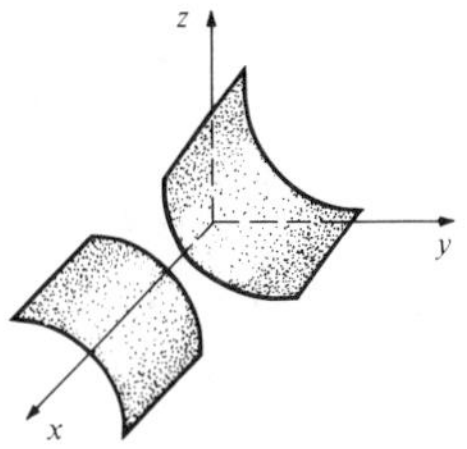

17.

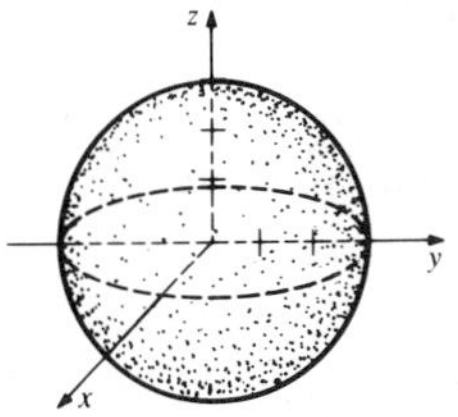

19.

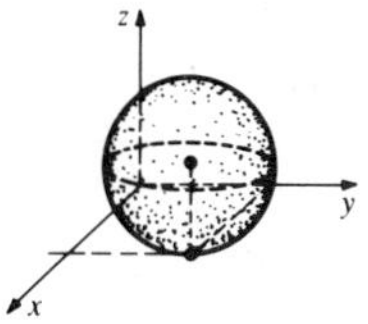

21. Center $(-4, 3, 2)$; radius 6
23. Center $(0, 0, 8)$; radius 8
25. $(x+1)^2 + (y-4)^2 + (z-6)^2 = 3$
27. $(x-1)^2 + (y-1)^2 + (z-4)^2 = 16$
29. $x^2 + (y-4)^2 + z^2 = 4$ or $x^2 + (y-8)^2 + z^2 = 4$
31. $(x-1)^2 + (y-4)^2 + (z-2)^2 = 90$
33. All points on the upper half of the sphere $x^2 + y^2 + (z-1)^2 = 4$
35. All points on and outside of the sphere $x^2 + y^2 + z^2 = 1$

Exercises 13.8, Page 722

1. Paraboloid

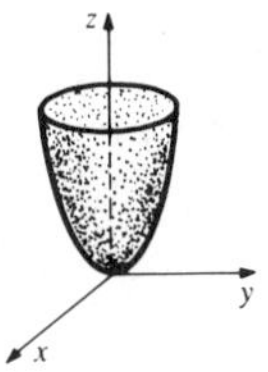

3. Ellipsoid

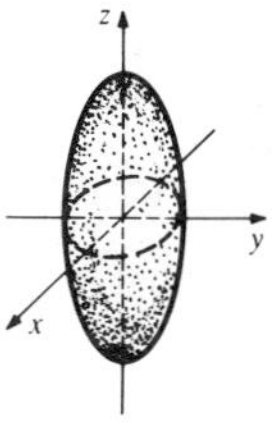

5. Hyperboloid of one sheet

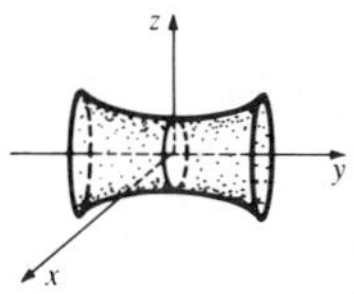

7. Elliptical cone

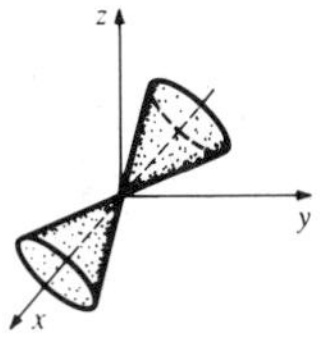

9. Hyperbolic paraboloid

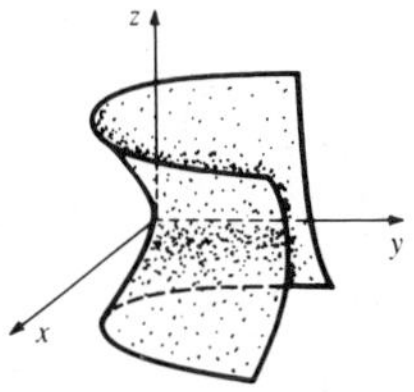

11. Hyperboloid of two sheets

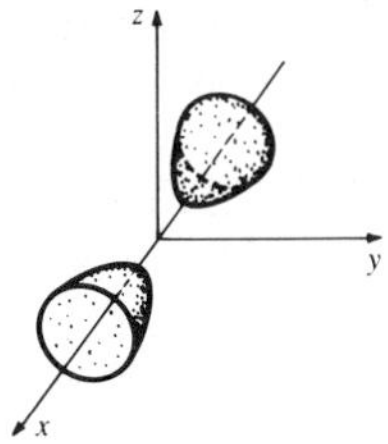

13. Ellipsoid

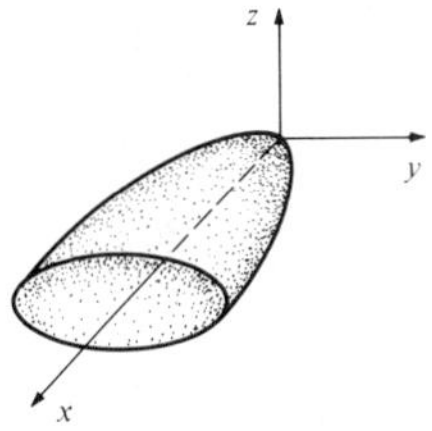

15.

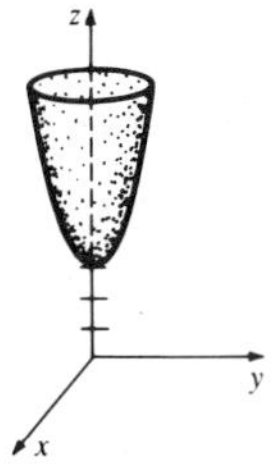

17.

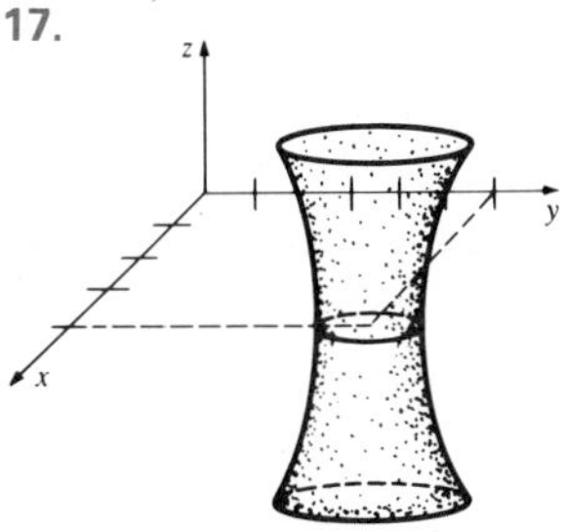

19. One possibility is $y^2 + z^2 = 1$; z-axis
21. One possibility is $y = e^{x^2}$; y-axis
23. $y^2 = 4(x^2 + z^2)$
25. $y^2 + z^2 = (9 - x^2)^2, x \geq 0$
27. $x^2 - y^2 - z^2 = 4$
29. $z = \ln \sqrt{x^2 + y^2}$
31. 1, z-axis; 2, y-axis; 4, z-axis; 6, z-axis; 10, z-axis; 11, x-axis; 14, z-axis

33.

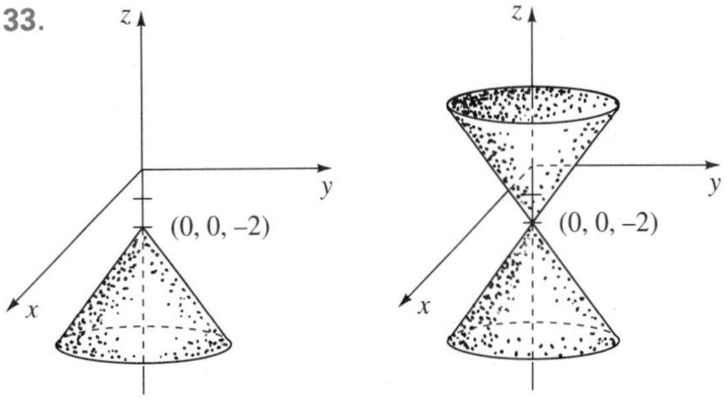

35. Area of cross-section is $\pi ab(c - z)$; $(\pi/2)abc^2$
37. $(2, -2, 6), (-2, 4, 3)$

Chapter 13 Review Exercises, Page 773

1. True 2. False 3. False 4. True
5. True 6. True 7. True 8. True
9. True 10. True 11. $9\mathbf{i} + 2\mathbf{j} + 2\mathbf{k}$
13. $5\mathbf{i}$ 15. 14 17. $-6\mathbf{i} + \mathbf{j} - 7\mathbf{k}$
19. (4, 7, 5) 21. (5, 6, 3) 23. $-36\sqrt{2}$
25. 12, −8, and 6 27. $3\sqrt{10}/2$
29. 2 units 31. $(\mathbf{i} - \mathbf{j} - 3\mathbf{k})/\sqrt{11}$ 33. 2
35. $\frac{26}{9}\mathbf{i} + \frac{7}{9}\mathbf{j} + \frac{20}{9}\mathbf{k}$ 37. Elliptic cylinder
39. Hyperboloid of two sheets
41. Hyperbolic paraboloid
43. $x^2 - y^2 + z^2 = 1$, hyperboloid of one sheet; $x^2 - y^2 - z^2 = 1$, hyperboloid of two sheets
45. Sphere; plane 47. $\dfrac{x-7}{4} = \dfrac{y-3}{-2} = \dfrac{z+5}{6}$
49. The direction vectors are orthogonal and the point of intersection is (3, −3, 0).
51. $14x - 5y - 3z = 0$ 53. $-6x - 3y + 4z = 5$
57. $-qvB\mathbf{k}$; $\mathbf{v} = (1/m|\mathbf{r}|^2)(\mathbf{L} \times \mathbf{r})$
59. Approximately 192.4 N-m

Exercises 14.1, Page 785

1.

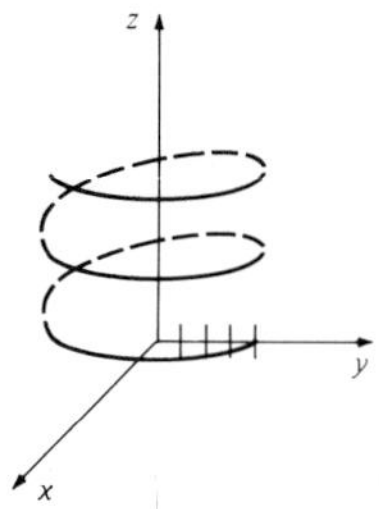

3.

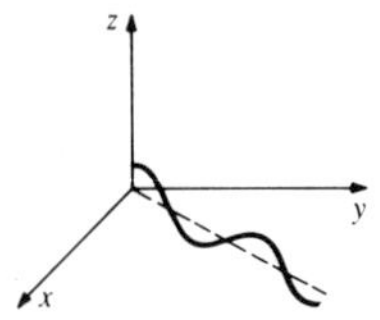

5.

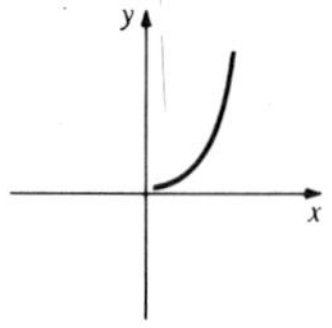

7.

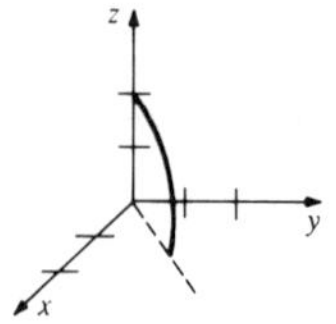

9.

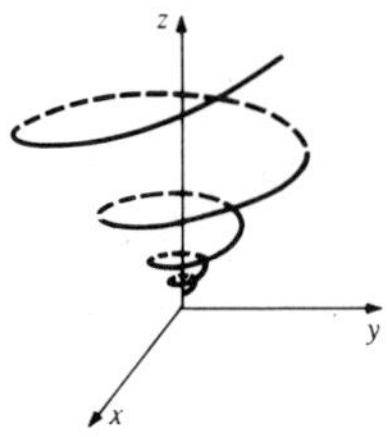

11.

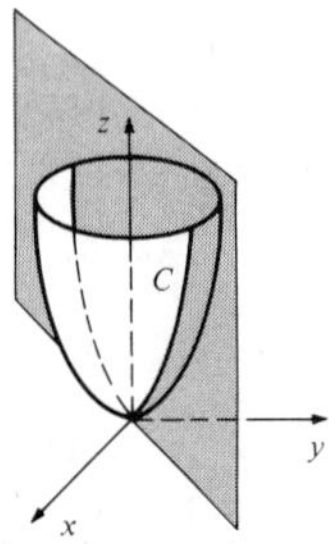

$\mathbf{r}(t) = t\mathbf{i} + t\mathbf{j} + 2t^2\mathbf{k}$

13.

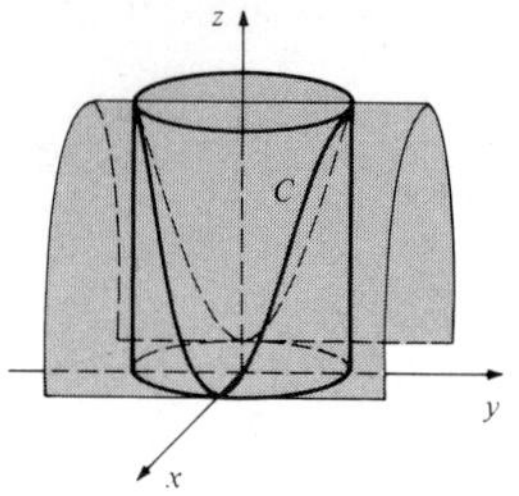

$\mathbf{r}(t) = 3\cos t\mathbf{i} + 3\sin t\mathbf{j} + 9\sin^2 t\mathbf{k}$

15. $2\mathbf{i} - 32\mathbf{j}$
17. $(1/t)\mathbf{i} - (1/t^2)\mathbf{j}$; $-(1/t^2)\mathbf{i} + (2/t^3)\mathbf{j}$
19. $\langle e^{2t}(2t+1), 3t^2, 8t-1\rangle$; $\langle 4e^{2t}(t+1), 6t, 8\rangle$

21.

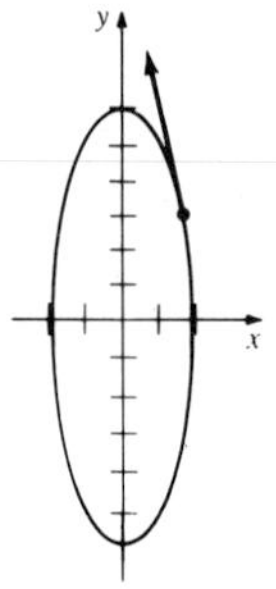

23.

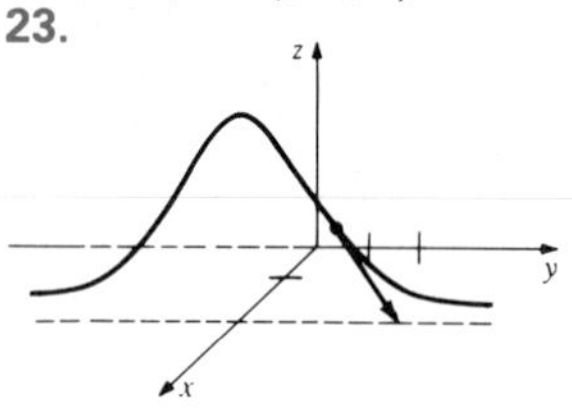

25. $x = 2 + t$, $y = 2 + 2t$, $z = \frac{8}{3} + 4t$
27. $\mathbf{r}(t) \times \mathbf{r}''(t)$ 29. $\mathbf{r}(t) \cdot [\mathbf{r}'(t) \times \mathbf{r}'''(t)]$
31. $2\mathbf{r}_1'(2t) - (1/t^2)\mathbf{r}_2'(1/t)$ 33. $\frac{3}{2}\mathbf{i} + 9\mathbf{j} + 15\mathbf{k}$
35. $e^t(t-1)\mathbf{i} + \frac{1}{2}e^{-2t}\mathbf{j} + \frac{1}{2}e^{t^2}\mathbf{k} + \mathbf{c}$
37. $(6t+1)\mathbf{i} + (3t^2 - 2)\mathbf{j} + (t^3 + 1)\mathbf{k}$
39. $(2t^3 - 6t + 6)\mathbf{i} + (7t - 4t^{3/2} - 3)\mathbf{j} + (t^2 - 2t)\mathbf{k}$
41. $2\sqrt{a^2 + c^2}\pi$ 43. $\sqrt{6}(e^{3\pi} - 1)$
45. $a\cos(s/a)\mathbf{i} + a\sin(s/a)\mathbf{j}$
47. Differentiate $\mathbf{r}(t) \cdot \mathbf{r}(t) = c^2$.

Exercises 14.2, Page 791

1. Speed is $\sqrt{5}$.

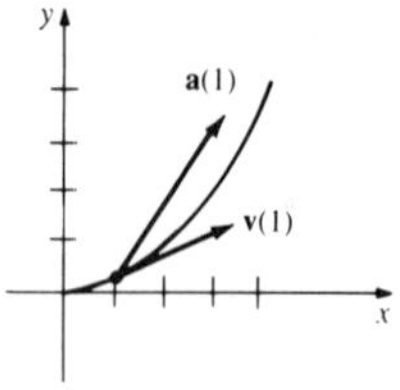

3. Speed is 2.

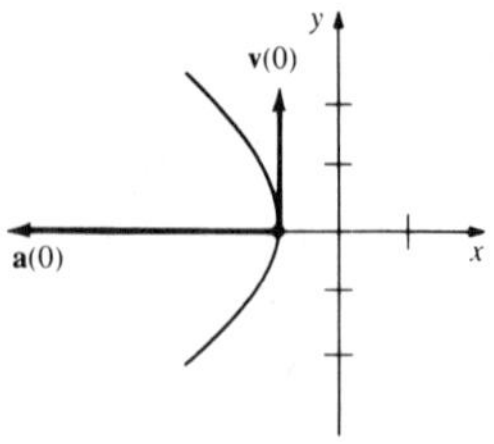

5. Speed is $\sqrt{5}$.

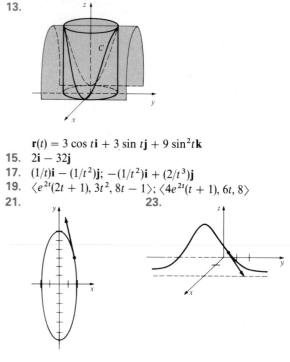

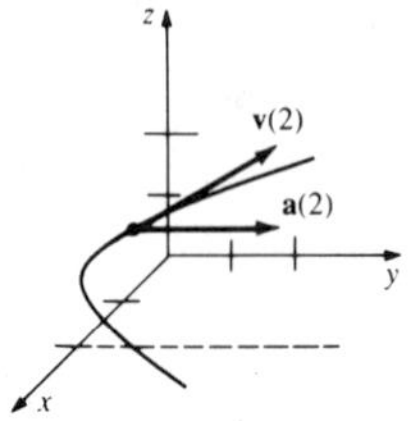

7. Speed is $\sqrt{14}$.

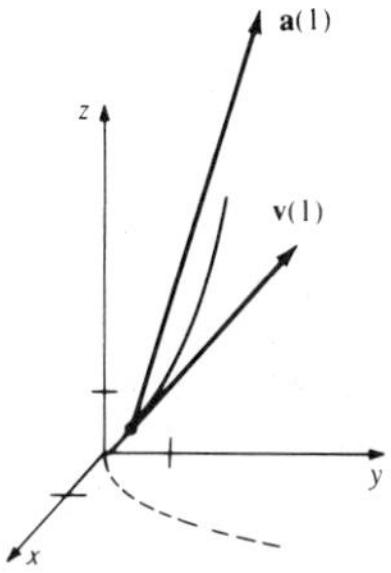

9. (0, 0, 0) and (25, 115, 0);
$\mathbf{v}(0) = -2\mathbf{j} - 5\mathbf{k}$, $\mathbf{a}(0) = 2\mathbf{i} + 2\mathbf{k}$,
$\mathbf{v}(5) = 10\mathbf{i} + 73\mathbf{j} + 5\mathbf{k}$, $\mathbf{a}(5) = 2\mathbf{i} + 30\mathbf{j} + 2\mathbf{k}$

11. $\mathbf{r}(t) = (-16t^2 + 240t)\mathbf{j} + 240\sqrt{3}t\mathbf{i}$ and $x(t) = 240\sqrt{3}t$, $y(t) = -16t^2 + 240t$; 900 ft; approximately 6235 ft; 480 ft/s

13. 72.11 ft/s 15. 97.98 ft/s

17. Assume that (x_0, y_0) are the coordinates of the center of the target at $t = 0$. Then $\mathbf{r}_p = \mathbf{r}_t$ when $t = x_0/(v_0\cos\theta) = y_0/(v_0\sin\theta)$. This implies $\tan\theta = y_0/x_0$. In other words, aim directly at the target at $t = 0$.

21. 191.33 lb

23. $y = -\dfrac{g}{2v_0^2\cos^2\theta}x^2 + (\tan\theta)x + s_0$ is an equation of a parabola.

25. $\mathbf{r}(t) = k_1e^{2t^3}\mathbf{i} + \dfrac{1}{2t^2 + k_2}\mathbf{j} + (k_3e^{t^2} - 1)\mathbf{k}$

27. Since $\mathbf{F}$ is directed along $\mathbf{r}$, we must have $\mathbf{F} = c\mathbf{r}$ for some constant c. Hence, $\boldsymbol{\tau} = \mathbf{r} \times (c\mathbf{r}) = c(\mathbf{r} \times \mathbf{r}) = \mathbf{0}$. If $\boldsymbol{\tau} = \mathbf{0}$, then $d\mathbf{L}/dt = \mathbf{0}$. This implies that $\mathbf{L}$ is a constant.

Exercises 14.3, Page 798

1. $\mathbf{T} = (\sqrt{5}/5)(-\sin t\mathbf{i} + \cos t\mathbf{j} + 2\mathbf{k})$
3. $\mathbf{T} = (a^2 + c^2)^{-1/2}(-a\sin t\mathbf{i} + a\cos t\mathbf{j} + c\mathbf{k})$,
   $\mathbf{N} = -\cos t\mathbf{i} - \sin t\mathbf{j}$,
   $\mathbf{B} = (a^2 + c^2)^{-1/2}(c\sin t\mathbf{i} - c\cos t\mathbf{j} + a\mathbf{k})$,
   $\kappa = a/(a^2 + c^2)$
5. $3\sqrt{2}x - 3\sqrt{2}y + 4z = 3\pi$
7. $4t/\sqrt{1 + 4t^2}$, $2/\sqrt{1 + 4t^2}$
9. $2\sqrt{6}, 0, t > 0$ 11. $2t/\sqrt{1 + t^2}$, $2/\sqrt{1 + t^2}$
13. 0, 5 15. $-\sqrt{3}e^{-t}, 0$
17. $\kappa = \dfrac{\sqrt{b^2c^2\sin^2t + a^2c^2\cos^2t + a^2b^2}}{(a^2\sin^2t + b^2\cos^2t + c^2)^{3/2}}$
23. $\kappa = 2$, $\rho = \frac{1}{2}$; $\kappa = 2/\sqrt{125} \approx 0.18$, $\rho = \sqrt{125}/2 \approx 5.59$; the curve is sharper at (0, 0).
25. κ is close to zero.

Chapter 14 Review Exercises, Page 798

1. True 2. True 3. True 4. False
5. True 6. False 7. True 8. True
9. False 10. True 11. $\sqrt{2}\pi$
13. $x = -27 - 18t$, $y = 8 + t$, $z = 1 + t$
15.

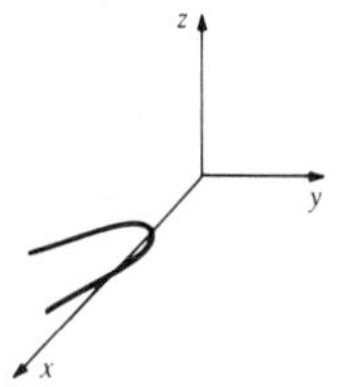

17. $-t^2\sin t + 2t\cos t - 2\sin t\cos t + 8t^3e^{2t} + 12t^2e^{2t}$

19. $(t + 1)\mathbf{i} + \left(\dfrac{1}{m}t^2 + t + 1\right)\mathbf{j} + t\mathbf{k}$;
$x = t + 1$, $y = \dfrac{1}{m}t^2 + t + 1$, $z = t$

21. $\mathbf{v}(1) = 6\mathbf{i} + \mathbf{j} + 2\mathbf{k}$, $\mathbf{v}(4) = 6\mathbf{i} + \mathbf{j} + 8\mathbf{k}$, $\mathbf{a}(t) = 2\mathbf{k}$ for all t

23. $\mathbf{i} + 4\mathbf{j} + (3\pi/4)\mathbf{k}$

25. $\mathbf{T} = (\tanh 1\mathbf{i} + \mathbf{j} + \operatorname{sech} 1\mathbf{k})/\sqrt{2}$,
   $\mathbf{N} = \operatorname{sech} 1\mathbf{i} - \tanh 1\mathbf{k}$,
   $\mathbf{B} = (-\tanh 1\mathbf{i} + \mathbf{j} - \operatorname{sech} 1\mathbf{k})/\sqrt{2}$,
   $\kappa = \frac{1}{2}\operatorname{sech}^2 1$

Exercises 15.1, Page 808

1. $\{(x, y) \mid (x, y) \neq (0, 0)\}$ 3. $\{(x, y) \mid y \neq x^2\}$
5. $\{(s, t) \mid s, t \text{ any real numbers}\}$ 7. $\{(r, s) \mid |s| \geq 1\}$
9. $\{(u, v, w) \mid u^2 + v^2 + w^2 \geq 16\}$
11. Figure 15.22 13. Figure 15.15
15. Figure 15.16 17. Figure 15.20
19.

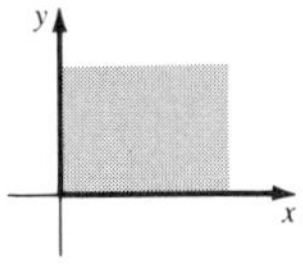

21.

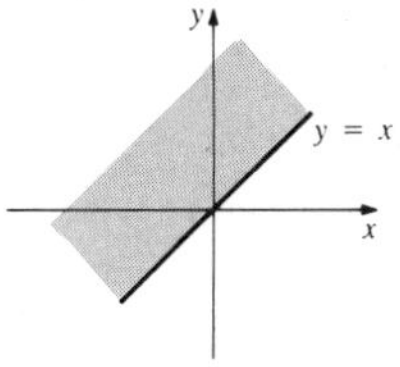

23. $\{z \mid z \geq 10\}$ 25. $\{w \mid |w| \leq 1\}$
27. 10, −2 29. 4, 4
31. A plane through the origin perpendicular to the xz-plane
33. Upper half of a cone
35. Upper half of an ellipsoid
37.

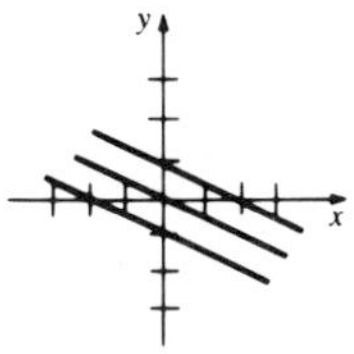

39.

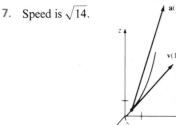

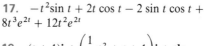

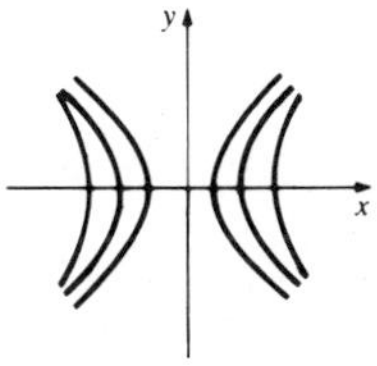

41.
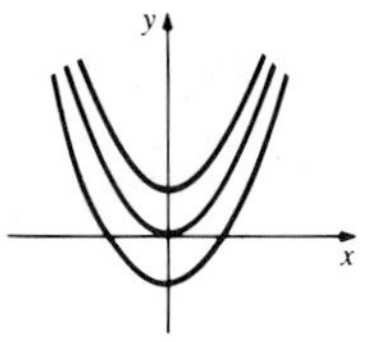

43. Elliptical cylinders 45. Ellipsoids

47.
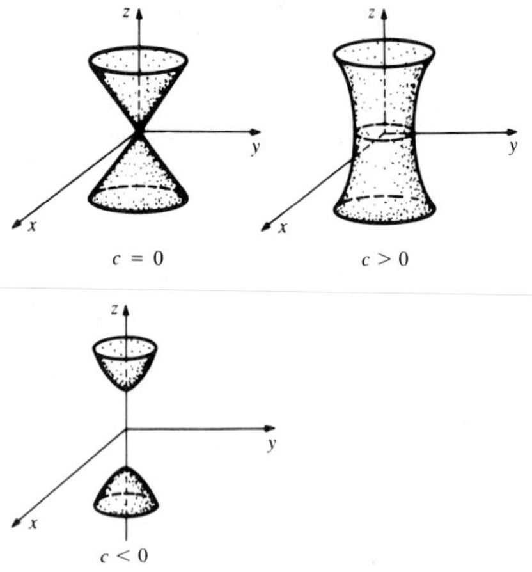

49.
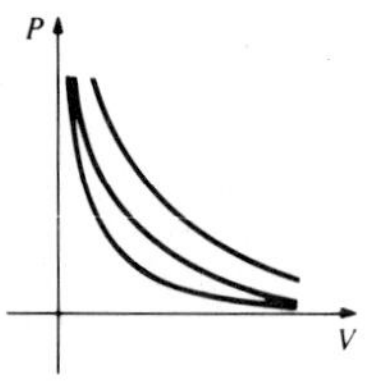

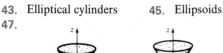

51. $C(r, h) = 2.8\pi r^2 + 4.6\pi rh$
53. $V(r, h) = \frac{11}{9}\pi r^2 h$ 55. 15,600 cm^2
57. 1397.2 59. 1086.4 cal/h

Exercises 15.2, Page 817

1. 26 3. Does not exist 5. 1
7. Does not exist 9. 108 11. 1
13. $\frac{1}{3}$ 15. $-\frac{1}{3}$ 17. 360
19. Does not exist 21. -3 23. 0
25. 0 27. 0 29. 0
31. $\{(x, y) \mid x \geq 0 \text{ and } y \geq -x\}$
33. $\{(x, y) \mid x/y \neq (2n + 1)\pi/2,$
$n = 0, \pm 1, \pm 2, \ldots, y \neq 0\}$
35. Continuous; not continuous; not continuous
37. f is continuous at (0, 0). 39. Choose $\delta = \varepsilon/3$.
41. Choose $\delta = \varepsilon$.

Exercises 15.3, Page 824

1. $\partial z/\partial x = 7, \partial z/\partial y = 16y$
3. $\partial z/\partial x = 6xy + 4y^2, \partial z/\partial y = 3x^2 + 8xy$
5. $\partial z/\partial x = 2x - y^2, \partial z/\partial y = -2xy + 20y^4$
7. $\partial z/\partial x = 20x^3y^3 - 2xy^6 + 30x^4$,
$\partial z/\partial y = 15x^4y^2 - 6x^2y^5 - 4$
9. $\partial z/\partial x = 2x^{-1/2}/(3y^2 + 1)$,
$\partial z/\partial y = -24\sqrt{x}y/(3y^2 + 1)^2$
11. $\partial z/\partial x = -3x^2(x^3 - y^2)^{-2}, \partial z/\partial y = 2y(x^3 - y^2)^{-2}$
13. $\partial z/\partial x = -10 \cos 5x \sin 5x$,
$\partial z/\partial y = 10 \sin 5y \cos 5y$
15. $f_x = e^{x^3y}(3x^3y + 1), f_y = x^4e^{x^3y}$
17. $f_x = 7y/(x + 2y)^2, f_y = -7x/(x + 2y)^2$
19. $g_u = 8u/(4u^2 + 5v^3), g_v = 15v^2/(4u^2 + 5v^3)$
21. $w_x = x^{-1/2}y, w_y = 2\sqrt{x} - (y/z)e^{y/z} - e^{y/z}$,
$w_z = (y^2/z^2)e^{y/z}$
23. $F_u = 2uw^2 - v^3 - vwt^2\sin(ut^2)$,
$F_v = -3uv^2 + w\cos(ut^2), F_x = 128x^7t^4$,
$F_t = -2uvwt\sin(ut^2) + 64x^8t^3$
25. -16 27. $x = -1, y = 4 + t, z = -24 + 2t$
29. -2 31. y^2e^{xy} 33. $20xy - 6y^2$
35. $18uv^2t^2$ 37. $-e^{r^2}(4r^2 + 2)\sin\theta$
39. $-60x^3y^2 + 8y$
43. $\partial z/\partial x = -x/z, \partial z/\partial y = -y/z$
45. $\partial z/\partial u = (vz - 2uv^3)/(2z - uv)$,
$\partial z/\partial v = (uz - 3u^2v^2)/(2z - uv)$
47. $A_x = y\sin\theta, A_y = x\sin\theta, A_\theta = xy\cos\theta$
59. $\dfrac{\partial u}{\partial t} = \begin{cases} -gx/a, & 0 \leq x \leq at \\ -gt, & x > at \end{cases}$; for $x > at$ the motion is that of a freely falling body;
$\dfrac{\partial u}{\partial x} = 0$ for $x > at$; the string is horizontal.
61.
$$\frac{\partial^2 z}{\partial x^2} = \lim_{\Delta x \to 0} \frac{f_x(x + \Delta x, y) - f_x(x, y)}{\Delta x}$$
$$\frac{\partial^2 z}{\partial y^2} = \lim_{\Delta y \to 0} \frac{f_y(x, y + \Delta y) - f_y(x, y)}{\Delta y}$$
$$\frac{\partial^2 z}{\partial x\, \partial y} = \lim_{\Delta x \to 0} \frac{f_y(x + \Delta x, y) - f_y(x, y)}{\Delta x}$$
63. 10

Exercises 15.4, Page 831

1. $\Delta z = 0.2, dz = 0.2$
3. $\Delta z = -0.79, dz = -0.8$
5. $\varepsilon_1 = 5\,\Delta x, \varepsilon_2 = -\Delta x$
7. $\varepsilon_1 = y^2\,\Delta x + 4xy\,\Delta y + 2y\,\Delta x\,\Delta y$,
$\varepsilon_2 = x^2\,\Delta y + 2x\,\Delta x\,\Delta y + (\Delta x)^2\,\Delta y$
9. $dz = 2x\sin 4y\,dx + 4x^2\cos 4y\,dy$
11. $dz = 2x(2x^2 - 4y^3)^{-1/2}\,dx - 6y^2(2x^2 - 4y^3)^{-1/2}\,dy$
13. $df = 7t\,ds/(s + 3t)^2 - 7s\,dt/(s + 3t)^2$
15. $dw = 2xy^4z^{-5}\,dx + 4x^2y^3z^{-5}\,dy - 5x^2y^4z^{-6}\,dz$
17. $dF = 3r^2\,dr - 2s^{-3}\,ds - 2t^{-1/2}\,dt$
19. $dw = du/u + dv/v - ds/s - dt/t$ 21. 0.9%
23. $-mg(0.009)$, decreases 25. 15%
27. 4.9%

29. $x_h = L\cos\theta - l\cos(\theta + \phi)$,
$y_h = L\sin\theta - l\sin(\theta + \phi)$;
in the first case $(x_e, y_e) = (L, 0)$ and in the second $(x_e, y_e) = (0, L)$; in the first case the approximate maximum error is $2\pi L/180$, while in the second it is $\pi L/180$.
31. 13.0907

Exercises 15.5, Page 836

1. $x^2 + 4x + \frac{3}{2}y^2 - y = C$
3. $\frac{5}{2}x^2 + 4xy - 2y^4 = C$
5. $x^2y^2 - 3x + 4y = C$ 7. Not exact
9. $xy^3 + y^2\cos x - \frac{1}{2}x^2 = C$ 11. Not exact
13. $xy - 2xe^x + 2e^x - 2x^3 = C$
15. $x + y + xy - 3\ln|xy| = C$
17. $x^3y^3 - \tan^{-1}(3x) = C$
19. $-\ln|\cos x| + \cos x \sin y = C$
21. $y - 2x^2y - y^2 - x^4 = C$
23. $x^4y - 5x^3 - xy + y^3 = C$
25. $xy^2 + x^2y - y + \frac{1}{3}x^3 = \frac{4}{3}$
27. $4xy + x^2 - 5x + 3y^2 - y = 8$
29. $P(x, y) = ye^{xy} + y^2 - y/x^2 + h(x)$
31. $k = 10$ 33. Exact 35. Not exact
37. Exact 39. Exact

Exercises 15.6, Page 841

1. $\partial z/\partial x = 3x^2v^2e^{uv^2} + 2uve^{uv^2}$, $\partial z/\partial y = -4yuve^{uv^2}$
3. $\partial z/\partial u = 16u^3 - 40y(2u - v)$,
$\partial z/\partial v = -96v^2 + 20y(2u - v)$
5. $\partial w/\partial t = -3u(u^2 + v^2)^{1/2}e^{-t}\sin\theta -$
$3v(u^2 + v^2)^{1/2}e^{-t}\cos\theta$,
$\partial w/\partial\theta = 3u(u^2 + v^2)^{1/2}e^{-t}\cos\theta - 3v(u^2 + v^2)^{1/2}e^{-t}\sin\theta$
7. $\partial R/\partial u = s^2t^4e^{v^2} - 4rst^4uve^{-u^2} + 8rs^2t^3uv^2e^{u^2v^2}$,
$\partial R/\partial v = 2s^2t^4uve^{v^2} + 2rst^4e^{-u^2} + 8rs^2t^3u^2ve^{u^2v^2}$
9. $\dfrac{\partial w}{\partial t} = \dfrac{xu}{(x^2 + y^2)^{1/2}(rs + tu)} + \dfrac{y\cosh rs}{u(x^2 + y^2)^{1/2}}$,
$\dfrac{\partial w}{\partial r} = \dfrac{xs}{(x^2 + y^2)^{1/2}(rs + tu)} + \dfrac{sty\sinh rs}{u(x^2 + y^2)^{1/2}}$,
$\dfrac{\partial w}{\partial u} = \dfrac{xt}{(x^2 + y^2)^{1/2}(rs + tu)} - \dfrac{ty\cosh rs}{u^2(x^2 + y^2)^{1/2}}$
11. $dz/dt = (4ut - 4vt^{-3})/(u^2 + v^2)$
13. $dw/dt\,|_{t=\pi} = -2$
15. $dp/du = 2u/(2s + t) + 4r/[u^3(2s + t)^2] -$
$r/[2u^{1/2}(2s + t)^2]$
17. $dy/dx = (4xy^2 - 3x^2)/(1 - 4x^2y)$
19. $dy/dx = y\cos xy/(1 - x\cos xy)$
21. $\partial z/\partial x = x/z$, $\partial z/\partial y = y/z$
23. $\partial z/\partial x = (2x + y^2z^3)/(10z - 3xy^2z^2)$,
$\partial z/\partial y = (2xyz^3 - 2y)/(10z - 3xy^2z^2)$
25. $\dfrac{\partial u}{\partial x} = -\dfrac{F_x(x, y, z, u)}{F_u(x, y, z, u)}, \dfrac{\partial u}{\partial y} = -\dfrac{F_y(x, y, z, u)}{F_u(x, y, z, u)}$,
$\dfrac{\partial u}{\partial z} = -\dfrac{F_z(x, y, z, u)}{F_u(x, y, z, u)}$

35. 5.31 cm^2/s 37. 0.5976 in^2/yr
41. Approximately 380 cycles per second; decreasing

Exercises 15.7, Page 850

1. $(2x - 3x^2y^2)\mathbf{i} + (-2x^3y + 4y^3)\mathbf{j}$
3. $(y^2/z^3)\mathbf{i} + (2xy/z^3)\mathbf{j} - (3xy^2/z^4)\mathbf{k}$
5. $4\mathbf{i} - 32\mathbf{j}$ 7. $2\sqrt{3}\mathbf{i} - 8\mathbf{j} - 4\sqrt{3}\mathbf{k}$
9. $\sqrt{3}x + y$ 11. $\frac{15}{2}(\sqrt{3} - 2)$
13. $-1/2\sqrt{10}$ 15. $98/\sqrt{5}$ 17. $-3\sqrt{2}$
19. -1 21. $-12/\sqrt{17}$
23. $\sqrt{2}\mathbf{i} + (\sqrt{2}/2)\mathbf{j}$, $\sqrt{5/2}$
25. $-2\mathbf{i} + 2\mathbf{j} - 4\mathbf{k}$, $2\sqrt{6}$
27. $-8\sqrt{\pi/6}\mathbf{i} - 8\sqrt{\pi/6}\mathbf{j}$, $-8\sqrt{\pi/3}$
29. $-\frac{3}{8}\mathbf{i} - 12\mathbf{j} - \frac{2}{3}\mathbf{k}$, $-\sqrt{83{,}281}/24$
31. $\pm 31/\sqrt{17}$
33. $\mathbf{u} = \frac{3}{5}\mathbf{i} - \frac{4}{5}\mathbf{j}$; $\mathbf{u} = \frac{4}{5}\mathbf{i} + \frac{3}{5}\mathbf{j}$; $\mathbf{u} = -\frac{4}{5}\mathbf{i} - \frac{3}{5}\mathbf{j}$
35. $D_{\mathbf{u}}f = (9x^2 + 3y^2 - 18xy^2 - 6x^2y)/\sqrt{10}$;
$D_{\mathbf{u}}F = (-6x^2 - 54y^2 + 54x + 6y - 72xy)/10$
37. $(2, 5), (-2, 5)$ 39. $-16\mathbf{i} - 4\mathbf{j}$
41. $x = 3e^{-4t}$, $y = 4e^{-2t}$
43. One possible function is $f(x, y) = x^3 - \frac{2}{3}y^3 + xy^3 + e^{xy}$.

Exercises 15.8, Page 855

1.
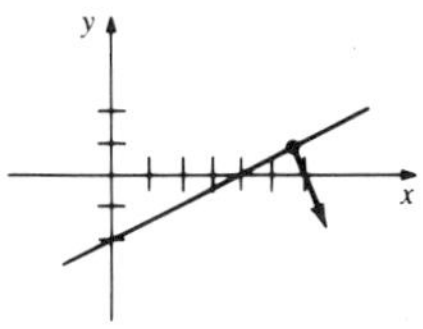

3.
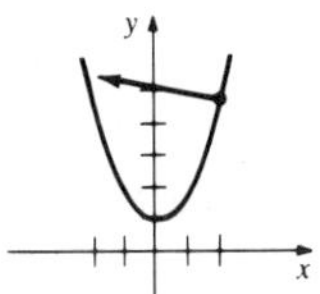

5.
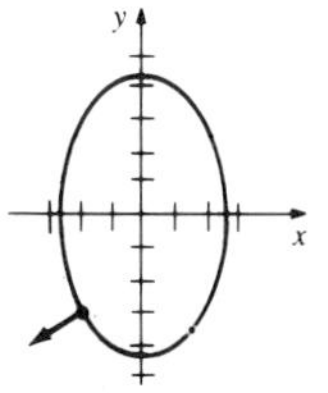

7.
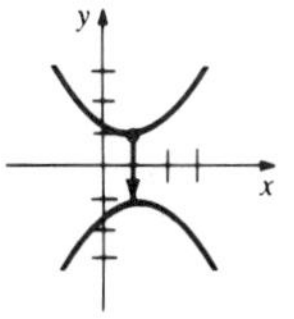

9.
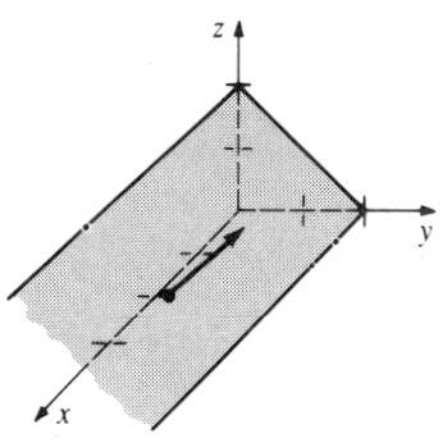

11.
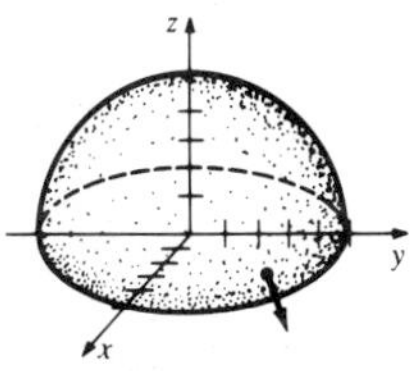

13. $(-4, -1, 17)$ 15. $-2x + 2y + z = 9$
17. $6x - 2y - 9z = 5$ 19. $6x - 8y + z = 50$
21. $2x + y - \sqrt{2}z = (4 + 5\pi)/4$
23. $\sqrt{2}x + \sqrt{2}y - z = 2$
25. $(1/\sqrt{2}, \sqrt{2}, 3/\sqrt{2}), (-1/\sqrt{2}, -\sqrt{2}, -3/\sqrt{2})$
27. $(-2, 0, 5), (-2, 0, -3)$
33. $x = 1 + 2t, y = -1 - 4t, z = 1 + 2t$
35. $(x - \frac{1}{2})/4 = (y - \frac{1}{3})/6 = -(z - 3)$

Exercises 15.9, Page 861

1. Rel. min. $f(0, 0) = 5$
3. Rel. max. $f(4, 3) = 25$
5. Rel. min. $f(-2, 1) = 15$
7. Rel. max. $f(-1, -1) = 10$, rel. min. $f(1, 1) = -10$
9. Rel. min. $f(3, 1) = -14$ 11. No extrema
13. Rel. max. $f(1, 1) = 12$
15. Rel. min. $f(-1, -2) = 14$
17. Rel. max. $f(-1, (2n + 1)\pi/2) = e^{-1}$, n odd; rel. min. $f(-1, (2n + 1)\pi/2) = -e^{-1}$, n even
19. Rel. max. $f((2m + 1)\pi/2, (2n + 1)\pi/2) = 2$, m and n even; rel. min. $f((2m + 1)\pi/2, (2n + 1)\pi/2) = -2$, m and n odd
21. $x = 7, y = 7, z = 7$ 23. $(\frac{1}{6}, \frac{1}{3}, \frac{1}{6})$
25. $(2, 2, 2), (2, -2, -2), (-2, 2, -2), (-2, -2, 2)$; at these points the least distance is $2\sqrt{3}$
27. $8\sqrt{3}abc/9$ 29. $x = 2, y = 2, z = 15$
31. $\theta = 30°, x = P/(4 + 2\sqrt{3}), y = P(\sqrt{3} - 1)/2\sqrt{3}$
33. Abs. max. $f(0, 0) = 16$
35. Abs. min. $f(0, 0) = -8$
37. Abs. max. $f(\frac{1}{2}, \sqrt{3}/2) = 2$, abs. min. $f(-\frac{1}{2}, -\sqrt{3}/2) = -2$
39. Abs. max. $f(\sqrt{2}/2, \sqrt{2}/2) = f(-\sqrt{2}/2, -\sqrt{2}/2) = \frac{3}{2}$, abs. min. $f(0, 0) = 0$
41. Abs. max. $f(\frac{8}{5}, -\frac{3}{5}) = 10$, abs. min. $f(-\frac{8}{3}, \frac{3}{5}) = -10$
43. $(0, 0)$ and all points $(x, \pi/2x)$ for $0 < x \le \pi$; abs. max. $f(x, \pi/2x) = 1$ for $0 < x \le \pi$, abs. min. $f(0, 0) = f(0, y) = f(x, 0) = f(\pi, 1) = 0$

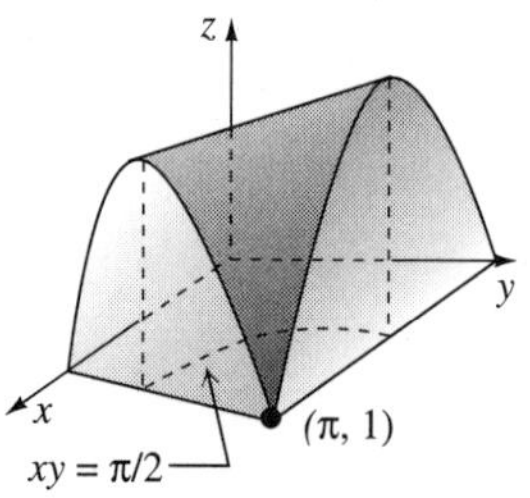

Exercises 15.10, Page 865

1. $y = 0.4x + 0.6$ 3. $y = 1.1x - 0.3$
5. $y = 1.3571x + 1.9286$
7. $v = -0.84T + 234$; 116.4, 99.6

Exercises 15.11, Page 872

1.

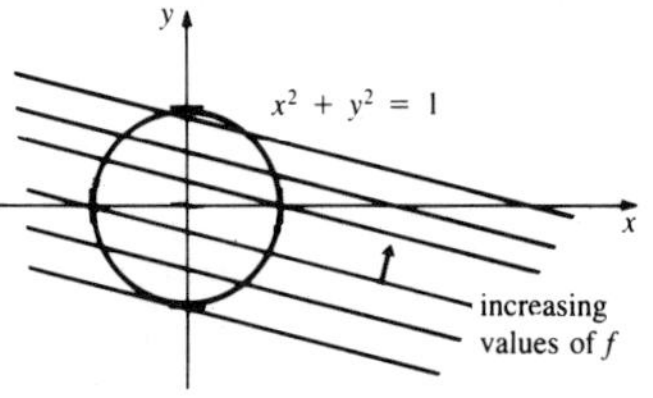

f appears to have a constrained maximum and a constrained minimum.
3. Max. $f(1/\sqrt{10}, 3/\sqrt{10}) = \sqrt{10}$; min. $f(-1/\sqrt{10}, -3/\sqrt{10}) = -\sqrt{10}$
5. Max. $f(1, 1) = f(-1, -1) = 1$; min. $f(1, -1) = f(-1, 1) = -1$
7. Min. $f(\frac{1}{2}, -\frac{1}{2}) = \frac{13}{2}$
9. Max. $f(1/\sqrt[4]{2}, 1/\sqrt[4]{2}) = f(-1/\sqrt[4]{2}, -1/\sqrt[4]{2})$
$= f(1/\sqrt[4]{2}, -1/\sqrt[4]{2})$
$= f(-1/\sqrt[4]{2}, 1/\sqrt[4]{2}) = \sqrt{2}$,
min. $f(0, 1) = f(0, -1) = f(1, 0) = f(-1, 0) = 1$
11. Max. $f(\frac{9}{16}, \frac{1}{16}) = \frac{729}{65{,}536}$; min. $f(0, 1) = f(1, 0) = 0$
13. Max. $F(\sqrt{5}, 2\sqrt{5}, \sqrt{5}) = 6\sqrt{5}$; min. $F(-\sqrt{5}, -2\sqrt{5}, -\sqrt{5}) = -6\sqrt{5}$
15. Max. $F(1/\sqrt{3}, 2/\sqrt{3}, \sqrt{3}) = 2/\sqrt{3}$
17. Min. $F(\frac{1}{3}, \frac{1}{3}, \frac{1}{3}) = \frac{1}{9}$
19. Min. $F(\frac{1}{3}, \frac{16}{15}, -\frac{11}{15}) = \frac{402}{225}$
21. f is the square of the distance from a point (x, y) on the graph of $x^4 + y^4 = 1$ to the origin. The points $(\pm 1, 0)$ and $(0, \pm 1)$ are closest to the origin, whereas the points $(\pm 1/\sqrt[4]{2}, \pm 1/\sqrt[4]{2})$ are the farthest from the origin.
23. F is the square of the distance from a point (x, y) on the line of intersection of the two given planes. The point $(\frac{1}{3}, \frac{16}{15}, -\frac{11}{15})$ is closest to the origin.
25. Max. $A(4/(2 + \sqrt{2}), 4/(2 + \sqrt{2})) = 4(3 + 2\sqrt{2})$
27. Max. $V(12 - 9/2\sqrt{5}, 6/\sqrt{5}) = (9\pi/2)(24 - \sqrt{5})$
29. Max. $z = P + 4(2 - \sqrt{4 + P\sqrt{27k}})/\sqrt{27k}$
31. Max. $F(k/3, k/3, k/3) = k/3$
33. $(-1/\sqrt{5}, 4 + \sqrt{5}, -2/\sqrt{5})$ is the point farthest from the xz-plane, whereas $(1/\sqrt{5}, 4 - \sqrt{5}, 2/\sqrt{5})$ is the point closest to the xz-plane.
35. $\lambda = 1/\gamma S(1 - e^{v_f/\gamma_n})$; $n = 2$: $M_0 = 600.33P$, $n = 3$: $M_0 = 89.42P$, $n = 4$: $M_0 = 63.48P$, $n = 5$: $M_0 = 54.70P$; minimum total mass of a two-stage rocket is 6.71 greater than the minimum total mass of a three-stage rocket.

Chapter 15 Review Exercises, Page 874

1. False 2. False 3. True 4. True
5. False 6. False 7. False 8. True
9. True 10. False 11. $-\frac{1}{4}$
13. The ellipse $3x^2 + y^2 = 28$
15. $\dfrac{\partial F}{\partial r} g'(w) + \dfrac{\partial F}{\partial s} h'(w)$ 17. f_{yyzx}
19. $-F(x)$
21. $f_x(x, y)g'(y)h'(z) + f_{xy}(x, y)g(y)h'(z)$
23. $e^{-x^3y}(-x^3y + 1)$ 25. $-\frac{3}{2}r^2\theta(r^3 + \theta^2)^{-3/2}$
27. $6x^2y \sinh(x^2y^3) + 9x^4y^4\cosh(x^2y^3)$
29. $-60s^2t^4v^{-5}$ 31. $\frac{1}{2}\mathbf{i} + \frac{1}{2}\mathbf{j}$
33. $(6x^2 - 2y^2 - 8xy)/\sqrt{40}$
35.

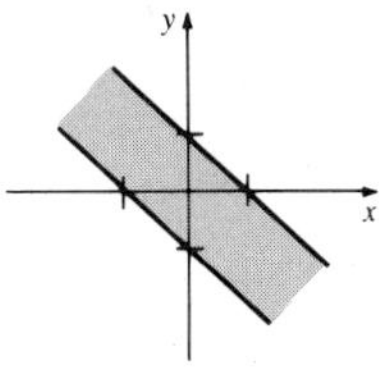

37. $2x\,\Delta y + 2y\,\Delta x + 2\,\Delta x\,\Delta y - 2y\,\Delta y - (\Delta y)^2$
39. $dz = 11y\,dx/(4x + 3y)^2 - 11x\,dy/(4x + 3y)^2$
41. $x^2\cos y^3 - y = C$
43. $x = -\sqrt{5}, z - 3 = \frac{4}{3}(y - 1)$
45. $2; -2/\sqrt{2}; 4$
47. $4\pi x + 3y - 12z = 4\pi - 6\sqrt{3}$
49. $3x + 4y = 25$ 51. $x = 2, y = 1, z = 2$
53. -8.77 cm/s 57. Not an extremum
59. Relative maximum 61. $A = \frac{1}{2}L^2\cos\theta\sin\theta$
63. $A = 2xz + 2yz - 5z^2$
65. $V = 16xy - 4xy\sqrt{x^2 + y^2}$

Exercises 16.1, Page 883

1. 52 3. 8; 8 5. 60 7. 10π
9. The integrand $f(x, y) = x + 5y$ is not nonnegative over the region.
11. 80 13. 34 15. 66 17. 18

Exercises 16.2, Page 887

1. $24y - 20e^y$ 3. $x^2e^{3x^2} - x^2e^x$ 5. $\dfrac{x}{2}\ln 5$
7. $2 - \sin y$ 9. 37 11. $-4\sqrt{2}/3$
13. $-\frac{4}{21}$ 15. $18 - e^3 + 3e$ 17. $\frac{10}{3}$
19. $(\pi/4)\ln 9$ 21. $\frac{1}{4}e^2 + \frac{1}{4}$ 23. π
25. e^{-1} 27. $2 - \pi$ 29. $(3\sqrt{3} - \pi)/6$
31.

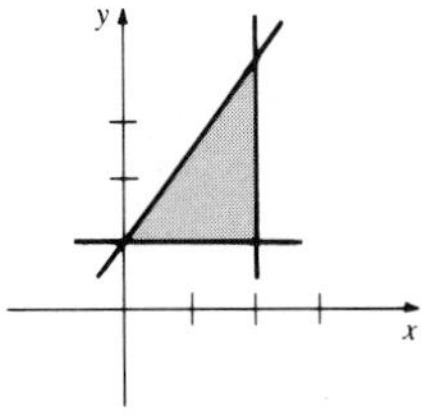

33.

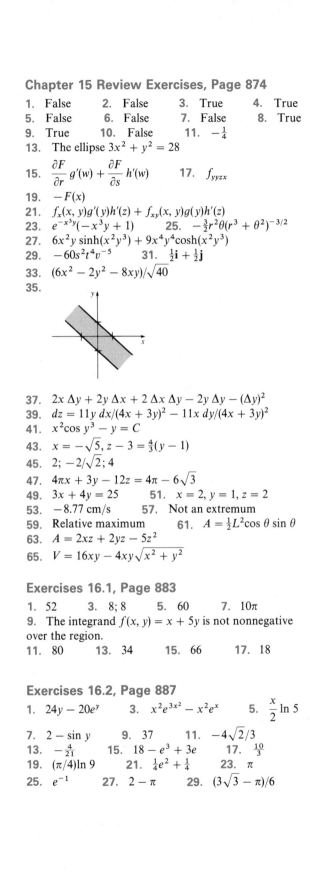

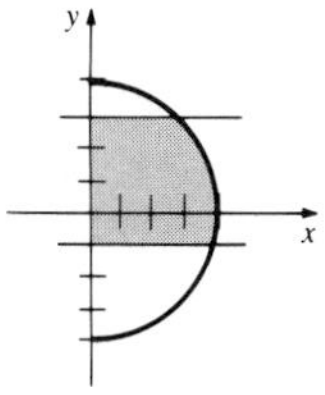

35. Both integrals equal 9.
37. Both integrals equal $13\pi^2 - 16$.
39. Both integrals equal $\frac{32}{5}$.

Exercises 16.3, Page 894

1. $\frac{1}{21}$ 3. $\frac{25}{84}$ 5. 96 7. $2\ln 2 - 1$
9. $\frac{14}{3}$ 11. 40 13. $\frac{9}{2}$
15. $e^4 - e + 3 - 4\ln 4$ 17. $\frac{63}{4}$
19. **(c)** 16π 21. 18 23. 2π 25. 4
27. $30\ln 6$ 29. $15\pi/4$ 31. $\frac{16}{9}$ 33. $\frac{35}{6}$
35. $\displaystyle\int_0^4 \int_{\sqrt{x}}^2 f(x, y)\,dy\,dx$
37. $\displaystyle\int_1^{e^3} \int_{\ln y}^3 f(x, y)\,dx\,dy$
39. $\displaystyle\int_0^1 \int_{y^3}^{2-y} f(x, y)\,dx\,dy$ 41. $(2^{3/2} - 1)/18$
43. $\frac{2}{3}\sin 8$ 45. $\pi/8$

Exercises 16.4, Page 900

1. $\bar{x} = \frac{8}{3}, \bar{y} = 2$ 3. $\bar{x} = 3, \bar{y} = \frac{3}{2}$
5. $\bar{x} = \frac{17}{21}, \bar{y} = \frac{55}{147}$ 7. $\bar{x} = 0, \bar{y} = \frac{4}{7}$
9. $\bar{x} = (3e^4 + 1)/[4(e^4 - 1)]$, $\bar{y} = 16(e^5 - 1)/[25(e^4 - 1)]$
11. $\frac{1}{105}$ 13. $4k/9$ 15. $\frac{256}{21}$ 17. $\frac{941}{10}$
19. $a\sqrt{10}/5$ 21. $ab^3\pi/4; a^3b\pi/4; b/2; a/2$
23. $ka^4/6$ 25. $16\sqrt{2}k/3$ 27. $a\sqrt{3}/3$

Exercises 16.5, Page 906

1. $27\pi/2$ 3. $(4\pi - 3\sqrt{3})/6$ 5. $25\pi/3$
7. $(2\pi/3)(15^{3/2} - 7^{3/2})$ 9. $\frac{5}{4}$
11. $\bar{x} = 13/3\pi, \bar{y} = 13/3\pi$
13. $\bar{x} = \frac{12}{5}, \bar{y} = 3\sqrt{3}/2$
15. $\bar{x} = (4 + 3\pi)/6, \bar{y} = \frac{4}{3}$ 17. $\pi a^4k/4$
19. $(ka/12)(15\sqrt{3} - 4\pi)$ 21. $\pi a^4k/2$
23. $4k$ 25. 9π 27. $(\pi/4)(e - 1)$
29. $3\pi/8$ 31. 250
33. Approximately 1450 m^3 35. $\sqrt{\pi}/2$
37. $2\pi dD_0[d - (R + d)e^{-R/d}]$; $\dfrac{2d^2 - (R^2 + 2dR + 2d^2)e^{-R/d}}{d - (R + d)e^{-R/d}}$; $2\pi d^2D_0, 2d$

Exercises 16.6, Page 911

1. $3\sqrt{29}$ 3. $10\pi/3$ 5. $(\pi/6)(17^{3/2}-1)$
7. $25\pi/6$ 9. $2a^2(\pi-2)$ 11. $8a^2$
13. $2\pi a(c_2-c_1)$ 15. π

Exercises 16.7, Page 918

1. 48 3. 36 5. $\pi-2$ 7. $\frac{1}{4}e^2-\frac{1}{2}e$
9. 50

11. $\int_0^4\int_0^{2-(x/2)}\int_{x+2y}^4 F(x,y,z)\,dz\,dy\,dx;$
$\int_0^2\int_{2y}^4\int_0^{z-2y} F(x,y,z)\,dx\,dz\,dy;$
$\int_0^4\int_0^{z/2}\int_0^{z-2y} F(x,y,z)\,dx\,dy\,dz;$
$\int_0^4\int_x^4\int_0^{(z-x)/2} F(x,y,z)\,dy\,dz\,dx;$
$\int_0^4\int_0^z\int_0^{(z-x)/2} F(x,y,z)\,dy\,dx\,dz$

13. $\int_0^2\int_{x^3}^8\int_0^4 dz\,dy\,dx;$
$\int_0^8\int_0^4\int_0^{\sqrt[3]{y}} dx\,dz\,dy;$
$\int_0^4\int_0^2\int_{x^3}^8 dy\,dx\,dz$

15.
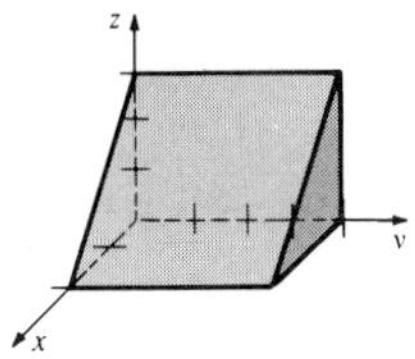

17.
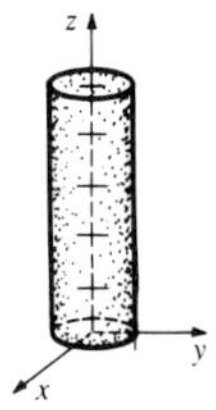

19.
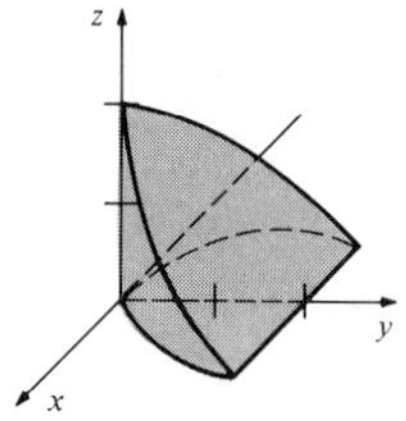

21. $16\sqrt{2}$ 23. 16π
25. $\bar{x}=\frac{4}{5}, \bar{y}=\frac{32}{7}, \bar{z}=\frac{8}{3}$
27. $\bar{x}=0, \bar{y}=2, \bar{z}=0$
29. $\int_{-1}^1\int_{-\sqrt{1-x^2}}^{\sqrt{1-x^2}}\int_{2y+2}^{8-y}(x+y+4)\,dz\,dy\,dx$
31. $2560k/3$; $\sqrt{80/9}$ 33. $k/30$
35. $k\int_{-5}^5\int_{-\sqrt{25-x^2}}^{\sqrt{25-x^2}}\int_{\sqrt{x^2+y^2}}^5 \times$
$(x^2+y^2)\sqrt{x^2+y^2+z^2}\,dz\,dy\,dx$

Exercises 16.8, Page 927

1. $(-10/\sqrt{2}, 10/\sqrt{2}, 5)$ 3. $(\sqrt{3}/2, \frac{3}{2}, -4)$
5. $(0, 5, 1)$ 7. $(\sqrt{2}, -\pi/4, -9)$
9. $(2\sqrt{2}, 2\pi/3, 2)$ 11. $(4, -\pi/2, 0)$
13. $r^2+z^2=25$ 15. $r^2-z^2=1$
17. $z=x^2+y^2$ 19. $x=5$
21. $(2\pi/3)(64-12^{3/2})$ 23. $625\pi/2$
25. $(0, 0, 3a/8)$ 27. $8\pi k/3$
29. $(\sqrt{3}/3, \frac{1}{3}, 0)$; $(\frac{2}{3}, \pi/6, 0)$
31. $(-4, 4, 4\sqrt{2})$; $(4\sqrt{2}, 3\pi/4, 4\sqrt{2})$
33. $(2\sqrt{2}, 0, -2\sqrt{2})$; $(2\sqrt{2}, 0, -2\sqrt{2})$
35. $(5\sqrt{2}, \pi/2, 5\pi/4)$ 37. $(\sqrt{2}, \pi/4, \pi/6)$
39. $(6, \pi/4, -\pi/4)$ 41. $\rho=8$
43. $\phi=\pi/6$ 45. $x^2+y^2+z^2=100$
47. $z=2$ 49. $9\pi(2-\sqrt{2})$
51. $2\pi/9$ 53. $(0, 0, \frac{7}{6})$ 55. πk

Exercises 16.9, Page 936

1. $(0, 0), (-2, 8), (16, 20), (14, 28)$

3.
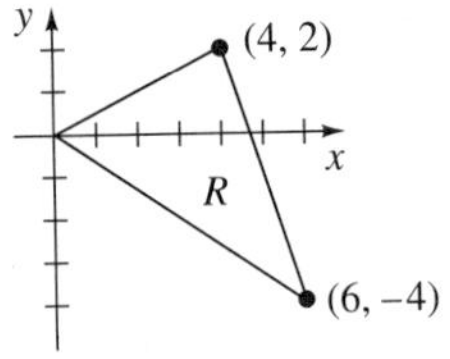

5.
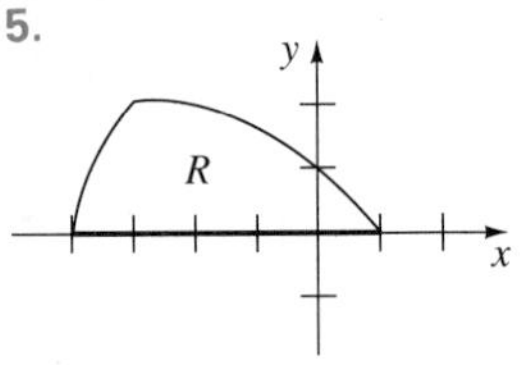

7. $-2v$ 9. $-1/3u^2$

11.
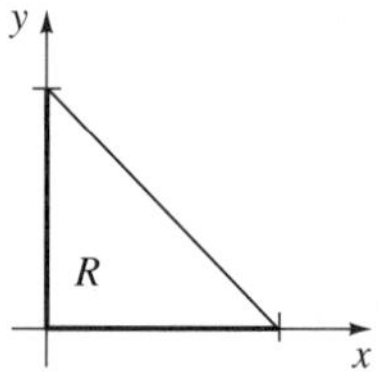

$(0, 0)$ is the image of every point on the boundary $u=0$.
13. 16 15. $\frac{1}{2}$ 17. $\frac{1}{4}(b-a)(c-d)$
19. $\frac{1}{2}(1-\ln 2)$ 21. $\frac{315}{4}$ 23. $\frac{1}{4}(e-e^{-1})$
25. 126 27. $\frac{5}{2}(b-a)\ln\frac{d}{c}$ 29. $15\pi/2$

Chapter 16 Review Exercises, Page 937

1. True 2. True 3. False 4. False
5. $32y^3-8y^5+5y\ln(y^2+1)-5y\ln 5$
7. Square region 9. $f(x,4)-f(x,2)$
11. $\int_0^4\int_{x/2}^{\sqrt{x}} f(x,y)\,dy\,dx$ 13. $(\sqrt{2}, 2\pi/3, \sqrt{2})$
15. $z=r^2$, $\rho=\csc\phi\cot\phi$
17. $-y\cos y^2+y\cos y^4$

19. $e^2 - e^{-2} + 4$ 21. $1 - \sin 1$ 23. $\frac{10}{3}$
25. 320π 27. $\frac{37}{60}$
29. $\int_0^{1/\sqrt{2}} \int_{\sqrt{1-x^2}}^{\sqrt{9-x^2}} \frac{1}{x^2+y^2}\,dy\,dx + \int_{1/\sqrt{2}}^{3/\sqrt{2}} \int_x^{\sqrt{9-x^2}} \frac{1}{x^2+y^2}\,dy\,dx$
31.

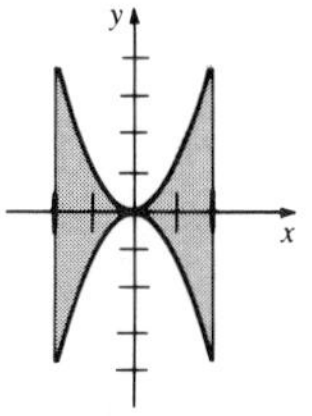

33. $(1 - \cos 1)/2$ 35. $5\pi/8$
37. $(2\pi/3)(2^{3/2} - 1)$
39. $\int_0^1 \int_x^{2x} \sqrt{1-x^2}\,dy\,dx$; $\int_0^1 \int_{y/2}^{y} \sqrt{1-x^2}\,dx\,dy + \int_1^2 \int_{y/2}^{1} \sqrt{1-x^2}\,dx\,dy$; $\frac{1}{3}$
41. $41k/1512$ 43. 8π 45. 0

Exercises 17.1, Page 952

1. $-125/3\sqrt{2}$; $-250(\sqrt{2} - 4)/12$; $\frac{125}{2}$
3. 3; 6; $3\sqrt{5}$
5. -1; $(\pi - 2)/2$; $\pi^2/8$; $\sqrt{2}\pi^2/8$ 7. 21
9. 30 11. 1 13. 1 15. 460
17. $\frac{26}{9}$ 19. $-\frac{64}{3}$ 21. $-\frac{8}{3}$ 23. 0
25. $\frac{123}{2}$ 27. 70
29.

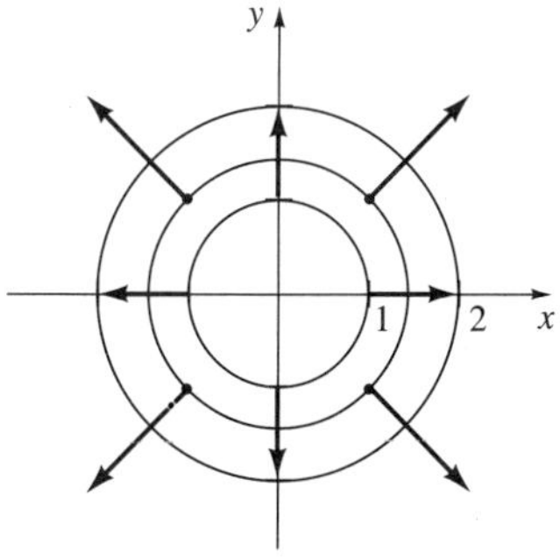

31.

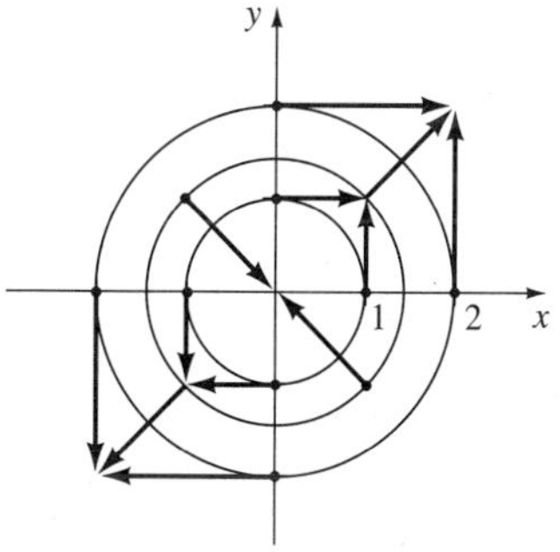

33.

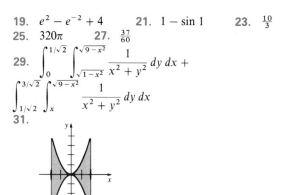

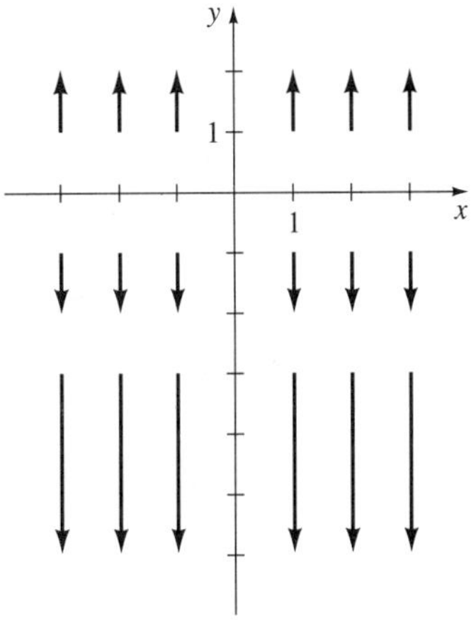

35. $-\frac{19}{8}$ 37. e 39. -4 41. 0
43. On each curve the line integral has the value $\frac{208}{3}$.
47. $\bar{x} = \frac{3}{2}$, $\bar{y} = 2/\pi$

Exercises 17.2, Page 961

1. $\frac{16}{3}$ 3. 14 5. 3 7. 330
9. 1096 11. $\phi = x^4y^3 + 3x + y$
13. Not a gradient field
15. $\phi = \frac{1}{4}x^4 + xy + \frac{1}{4}y^4$ 17. $3 + e^{-1}$
19. 63 21. $8 + 2e^3$ 23. 16
25. $\pi - 4$ 27. $\phi = (Gm_1m_2)/|\mathbf{r}|$

Exercises 17.3, Page 968

1. 3 3. 0 5. 75π 7. 48π
9. $\frac{56}{3}$ 11. $\frac{2}{3}$ 13. $\frac{1}{8}$
15. $(b - a) \times$ (area of the region bounded by C)
19. $3a^2\pi/8$ 23. $45\pi/2$ 25. π
27. $27\pi/2$ 29. $3\pi/2$ 33. 3π

Exercises 17.4, Page 975

1. $\frac{26}{3}$ 3. 0 5. 972π
7. $(3^{5/2} - 2^{7/2} + 1)/15$ 9. $9(17^{3/2} - 1)$
11. 36 13. $k\sqrt{3}/12$ 15. 18 17. 28π
19. 8π 21. $5\pi/2$ 23. $-8\pi a^3$
25. $4\pi kq$ 27. $(1, \frac{2}{3}, 2)$ 29. $(0, 0, \frac{2}{3})$; $8\sqrt{2}\pi k$

Exercises 17.5, Page 980

1. $(x - y)\mathbf{i} + (x - y)\mathbf{j}$; $2z$ 3. $\mathbf{0}$; $4y + 8z$
5. $(4y^3 - 6xz^2)\mathbf{i} + (2z^3 - 3x^2)\mathbf{k}$; $6xy$
7. $(3e^{-z} - 8yz)\mathbf{i} - xe^{-z}\mathbf{j}$; $e^{-z} + 4z^2 - 3ye^{-z}$
9. $(xy^2e^y + 2xye^y + x^3yze^z + x^3ye^z)\mathbf{i} - y^2e^y\mathbf{j} + (-3x^2yze^z - xe^x)\mathbf{k}$; $xye^x + ye^x - x^3ze^z$
29. $2\mathbf{i} + (1 - 8y)\mathbf{j} + 8z\mathbf{k}$ 37. 6

Exercises 17.6, Page 987

1. -40π 3. $\frac{45}{2}$ 5. $\frac{3}{2}$ 7. -3

9. $-3\pi/2$ 11. π 13. -152π
15. 112 17. Take the surface to be $z = 0$; $81\pi/4$

Exercises 17.7, Page 994

1. $\frac{3}{2}$ 3. $12a^5\pi/5$ 5. 256π 7. $62\pi/5$
9. $4\pi(b - a)$ 11. 128 13. $\pi/2$

Chapter 17 Review Exercises, Page 996

1. True 2. True 3. False 4. True
5. True 6. True 7. True 8. True
9. True 10. True 11. True
12. True

13. $\nabla\phi = -\dfrac{x}{(x^2 + y^2)^{3/2}}\mathbf{i} - \dfrac{y}{(x^2 + y^2)^{3/2}}\mathbf{j}$

15. $6xy$ 17. 0 19. 0 21. $56\sqrt{2}\pi^3/3$
23. 12 25. $2 + 2/3\pi$ 27. $\pi^2/2$
29. 5π 31. 180π
33. $(\ln 3)(17^{3/2} - 5^{3/2})/12$ 35. $6(e^{-3} - 1)$
37. $-4\pi c$ 39. 0 41. 125π 43. 3π
45. $\frac{5}{3}$

Exercises 18.1, Page 1003

1. Linear, second-order
3. Nonlinear, first-order
5. Linear, fourth-order
7. Nonlinear, second-order
9. Linear, third-order
35. $m_1 = 2, m_2 = 3$;
corresponding solutions are $y_1 = e^{2x}$ and $y_2 = e^{3x}$
37. $m_1 = 0, m_2 = 4$;
corresponding solutions are $y = 1$ and $y = x^4$

Exercises 18.2, Page 1007

1. $x\ln|x| + y = Cx$
3. $(x - y)\ln|x - y| = y + C(x - y)$
5. $x + y\ln|x| = Cy$
7. $\ln(x^2 + y^2) + 2\tan^{-1}(y/x) = C$
9. $\ln|y| = 2\sqrt{x/y} + C$
11. $y^9 = C(x^3 + y^3)^2$
13. $(y/x)^2 = 2\ln|x| + C$
15. $e^{2x/y} = 8\ln|y| + C$
17. $x\cos(y/x) = C$ 19. $y + x = Cx^2e^{y/x}$

Exercises 18.3, Page 1013

1. $y = C_1 + C_2e^{x/3}$
3. $y = C_1e^{-4x} + C_2e^{4x}$
5. $y = C_1\cos 3x + C_2\sin 3x$
7. $y = C_1e^{x} + C_2e^{2x}$
9. $y = C_1e^{-4x} + C_2xe^{-4x}$
11. $y = C_1e^{(-3+\sqrt{29})x/2} + C_2e^{(-3-\sqrt{29})x/2}$
13. $y = C_1e^{2x/3} + C_2e^{-x/4}$
15. $y = e^{2x}(C_1\cos x + C_2\sin x)$
17. $y = e^{-x/3}\left(C_1\cos\dfrac{\sqrt{2}}{3}x + C_2\sin\dfrac{\sqrt{2}}{3}x\right)$
19. $y = C_1e^{-x/3} + C_2xe^{-x/3}$
21. $y = 2\cos 4x - \frac{1}{2}\sin 4x$
23. $y = -\frac{3}{4}e^{-5x} + \frac{3}{4}e^{-x}$
25. $y = -e^{x/2}\cos(x/2) + e^{x/2}\sin(x/2)$
27. $y = 0$ 29. $y = e^{2(x-1)} - e^{x-1}$
31. $y'' + y' - 20y = 0$
33. $q(t) = q_0e^{-20t}(\cos 60t + \frac{1}{3}\sin 60t)$
35. $y = C_1 + C_2e^{-x} + C_3e^{5x}$
37. $y = C_1e^{x} + C_2e^{4x}\cos x + C_3e^{4x}\sin x$
39. The auxiliary equation must be $(m - m_1)^2 = m^2 - 2m_1m + m_1^2 = 0$. Thus, the differential equation is $y'' - 2m_1y' + m_1^2y = 0$.

Exercises 18.4, Page 1019

1. $y = C_1\cos x + C_2\sin x + x\sin x + \cos x\ln|\cos x|$
3. $y = C_1\cos x + C_2\sin x + \frac{1}{2}\sin x - \frac{1}{2}x\cos x$
$= C_1\cos x + C_3\sin x - \frac{1}{2}x\cos x$
5. $y = C_1\cos x + C_2\sin x + \frac{1}{2} - \frac{1}{6}\cos 2x$
7. $y = C_1e^{x} + C_2e^{-x} + \frac{1}{4}xe^{x} - \frac{1}{4}xe^{-x}$
$= C_1e^{x} + C_2e^{-x} + \frac{1}{2}x\sinh x$
9. $y = C_1e^{2x} + C_2e^{-2x}$
$+\dfrac{1}{4}\left(e^{2x}\ln|x| - e^{-2x}\displaystyle\int_{x_0}^{x}\frac{e^{4t}}{t}\,dt\right), x_0 > 0$
11. $y = C_1e^{-x} + C_2e^{-2x} + (e^{-x} + e^{-2x})\ln(1 + e^{x})$
13. $y = C_1e^{-2x} + C_2e^{-x} - e^{-2x}\sin e^{x}$
15. $y = C_1e^{x} + C_2xe^{x} - \frac{1}{2}e^{x}\ln(1 + x^2) + xe^{x}\tan^{-1}x$
17. $y = C_1e^{-x} + C_2xe^{-x} + \frac{1}{2}x^2e^{-x}\ln x - \frac{3}{4}x^2e^{-x}$
19. $y = C_1e^{x/2} + C_2xe^{x/2} + \frac{8}{9}e^{-x} + x + 4$
21. $y = \frac{3}{8}e^{-x} + \frac{5}{8}e^{x} + \frac{1}{4}x^2e^{x} - \frac{1}{4}xe^{x}$
23. $y = C_1e^{-3x} + C_2e^{3x} - 6$
25. $y = C_1e^{-2x} + C_2xe^{-2x} + \frac{1}{2}x + 1$
27. $y = C_1\cos 5x + C_2\sin 5x + \frac{1}{4}\sin x$
29. $y = C_1e^{-x} + C_2e^{3x} - \frac{4}{3}e^{2x} - \frac{2}{3}x^3 + \frac{4}{3}x^2$
$- \frac{28}{9}x + \frac{80}{27}$
31. $y = e^{4x}(C_1\cos 3x + C_2\sin 3x) + \frac{1}{10}e^{3x}$
$- \frac{126}{697}\cos 2x + \frac{96}{697}\sin 2x$
33. $y = \frac{5}{8}e^{-8x} + \frac{5}{8}e^{8x} - \frac{1}{4}$
35. $i_p(t) = -\frac{30}{13}\cos 60t + \frac{45}{13}\sin 60t$
$+ \frac{160}{17}\cos 40t - \frac{40}{17}\sin 40t$
37. $y = C_1x + C_2x\ln x + \frac{2}{3}x(\ln x)^3$

Exercises 18.5, Page 1024

1. $y_1(x) = c_0\displaystyle\sum_{n=0}^{\infty}\frac{(-1)^n}{(2n)!}x^{2n}$

$y_2(x) = c_1\displaystyle\sum_{n=0}^{\infty}\frac{(-1)^n}{(2n+1)!}x^{2n+1}$

3. $y_1(x) = c_0$

$y_2(x) = c_1 \sum_{n=1}^{\infty} \frac{x^n}{n!}$

5. $y_1(x) = c_0\left[1 + \frac{1}{3 \cdot 2}x^3 + \frac{1}{6 \cdot 5 \cdot 3 \cdot 2}x^6 + \frac{1}{9 \cdot 8 \cdot 6 \cdot 5 \cdot 3 \cdot 2}x^9 + \cdots\right]$

$y_2(x) = c_1\left[x + \frac{1}{4 \cdot 3}x^4 + \frac{1}{7 \cdot 6 \cdot 4 \cdot 3}x^7 + \frac{1}{10 \cdot 9 \cdot 7 \cdot 6 \cdot 4 \cdot 3}x^{10} + \cdots\right]$

7. $y_1(x) = c_0\left[1 - \frac{1}{2}x^2 - \frac{3}{4!}x^4 - \frac{21}{6!}x^6 - \cdots\right]$

$y_2(x) = c_1\left[x + \frac{1}{3!}x^3 + \frac{5}{5!}x^5 + \frac{45}{7!}x^7 + \cdots\right]$

9. $y_1(x) = c_0\left[1 - \frac{1}{3!}x^3 + \frac{4^2}{6!}x^6 - \frac{7^2 \cdot 4^2}{9!}x^9 + \cdots\right]$

$y_2(x) = c_1\left[x - \frac{2^2}{4!}x^4 + \frac{5^2 \cdot 2^2}{7!}x^7 - \frac{8^2 \cdot 5^2 \cdot 2^2}{10!}x^{10} + \cdots\right]$

11. $y_1(x) = c_0$; $y_2(x) = c_1 \sum_{n=1}^{\infty} \frac{1}{n} x^n$

13. $y_1(x) = c_0 \sum_{n=0}^{\infty} x^{2n}$; $y_2(x) = c_1 \sum_{n=0}^{\infty} x^{2n+1}$

15. $y_1(x) = c_0\left[1 + \frac{1}{4}x^2 - \frac{7}{4 \cdot 4!}x^4 + \frac{23 \cdot 7}{8 \cdot 6!}x^6 - \cdots\right]$

$y_2(x) = c_1\left[x - \frac{1}{6}x^3 + \frac{14}{2 \cdot 5!}x^5 - \frac{34 \cdot 14}{4 \cdot 7!}x^7 - \cdots\right]$

17. $y_1(x) = c_0[1 + \frac{1}{2}x^2 + \frac{1}{6}x^3 + \frac{1}{6}x^4 + \cdots]$
$y_2(x) = c_1[x + \frac{1}{2}x^2 + \frac{1}{2}x^3 + \frac{1}{4}x^4 + \cdots]$

19. $y = 6x - 2[1 + \frac{1}{2!}x^2 + \frac{1}{3!}x^3 + \frac{1}{4!}x^4 + \cdots]$

Exercises 18.6, Page 1035

1. A weight of 4 lb ($\frac{1}{8}$ slug) attached to a spring is released from a point 3 units above the equilibrium position with an initial upward velocity of 2 ft/s. The spring constant is 3 lb/ft.
3. $x(t) = \frac{1}{2}\cos 2t + \frac{3}{4}\sin 2t$
5. $x(t) = -5 \sin 2t$
7. A weight of 2 lb ($\frac{1}{16}$ slug) is attached to a spring whose constant is 1 lb/ft. The system is damped with a resisting force numerically equal to 2 times the instantaneous velocity. The weight starts from the equilibrium position with an upward velocity of 1.5 ft/s.
9. $\frac{1}{4}$ s, $\frac{1}{2}$ s, $x(\frac{1}{2}) = e^{-2} \approx 0.14$
11. $x(t) = e^{-2t}(\frac{1}{2}\cos 4t + \frac{1}{2}\sin 4t)$
13. Overdamped for $\beta > \frac{5}{2}$; critically damped for $\beta = \frac{5}{2}$; underdamped for $0 < \beta < \frac{5}{2}$
15. $x(t) = \frac{1}{625}e^{-4t}(24 + 100t) - \frac{1}{625}e^{-t}(24\cos 4t + 7\sin 4t)$; $x(t) \to 0$ as $t \to \infty$

Chapter 18 Review Exercises, Page 1036

1. True 2. True 3. False 4. True
5. True 6. False 7. True 8. True
9. $xy = C$ 11. $\ln|y| = e^{y/x} - 1$
13. $y = C_1 e^{(1-\sqrt{3})x} + C_2 e^{(1+\sqrt{3})x}$
15. $y = C_1 e^{-2x} + C_2 e^{5x}$
17. $y = C_1 \cos\frac{x}{3} + C_2 \sin\frac{x}{3}$
19. $y = -24\cos 6x + 3\sin 6x$
21. $y_1(x) = c_0\left[1 - \frac{1}{3 \cdot 2}x^3 + \frac{1}{6 \cdot 5 \cdot 3 \cdot 2}x^6 - \frac{1}{9 \cdot 8 \cdot 6 \cdot 5 \cdot 3 \cdot 2}x^9 + \cdots\right]$

$y_2(x) = c_1\left[x - \frac{1}{4 \cdot 3}x^4 + \frac{1}{7 \cdot 6 \cdot 4 \cdot 3}x^7 - \frac{1}{10 \cdot 9 \cdot 7 \cdot 6 \cdot 4 \cdot 3}x^{10} + \cdots\right]$

23. $y = e^x(C_1\cos x + C_2\sin x) - e^x\cos x \ln|\sec x + \tan x|$
25. $y = \frac{1}{2}\cos x + \frac{1}{2}\sin x + \frac{1}{2}\sec x$
27. $y = C_1 e^{-3x} + C_2 e^{4x} - \frac{1}{10}xe^{2x} - \frac{13}{100}e^{2x}$
29. $0 < m \le 2$
31. $x(t) = e^{-4t}\left(\frac{26}{17}\cos 2\sqrt{2}t + \frac{28\sqrt{2}}{17}\sin 2\sqrt{2}t\right) + \frac{8}{17}e^{-t}$

Exercises I.1, Page A-6

1. 4 3. 5 5. $9r^2 - 24rs + 16s^2$
7. $8x^3 - 12x^2y + 6xy^2 - y^3$
9. $(2xy - 3)(2xy + 3)$
11. $(4a - b)(16a^2 + 4ab + b^2)$
13. -3 15. $-\frac{3}{2}, \frac{3}{2}$ 17. $0, \frac{2}{5}$
19. $-5, 3$ 21. $-\frac{1}{2} - \frac{\sqrt{3}}{2}, -\frac{1}{2} + \frac{\sqrt{3}}{2}$
23. $14 + 3i$ 25. $-38 + 8i$ 27. $-\frac{9}{25} + \frac{13}{25}i$
29. $-\frac{3}{2} + \frac{1}{2}i$ 31. i
33. $-\frac{1}{2} - \frac{\sqrt{3}}{2}i, -\frac{1}{2} + \frac{\sqrt{3}}{2}i$ 35. $2 - 5i, 2 + 5i$
37. $x = \frac{12}{5}, y = \frac{9}{5}$ 39. $b = 6.75$

Exercises I.2, Page A-11

1. $[-4, 20)$ 3. $(-\infty, -2)$ 5. $\frac{3}{2} < x < 6$
7. $x \ge 20$ 9. $[-1, \infty)$ 11. $[3, 5]$
13. $(-\infty, -3)$ 15. $(\frac{9}{4}, \infty)$ 17. $(-\infty, 7]$
19. $[-2, 5)$ 21. $[-1, 2]$ 23. $(1, 9)$

25. $[-5, \frac{2}{3}]$ 27. $(-\infty, \infty)$ 29. $-4 + a$
31. $a + 10$ 33. -9 or 9 35. $-\frac{19}{5}$ or 5
37. $(-4, 4)$ 39. $[0, 1]$ 41. $(-5, -1)$
43. $(-\infty, -6) \cup (6, \infty)$ 45. $(-\infty, -1) \cup (6, \infty)$
47. No. The solution is $(-\infty, 0) \cup (\frac{1}{4}, \infty)$.
49. $|x - 9| < 4$; $(5, 13)$
51. $|x - 2| > 2$; $(-\infty, 0) \cup (4, \infty)$
53. $4 \le x \le 14$ 55. $<$ 57. $>$
59. $|x - 4| < 4$ 61. $|x - \frac{1}{2}| < \frac{7}{2}$

Exercises I.3, Page A-21

1. Fourth quadrant 3. Third quadrant
5. Second quadrant 7. $2\sqrt{5}$ 9. 3
11. Not a right triangle 13. Collinear
15. $x = 4$ or $x = -2$ 17. $x^2 = 4y$
19. $(6, 2)$ 21. $(5, 5)$ 23. $(3, 0), (5, -6)$
25. No symmetry

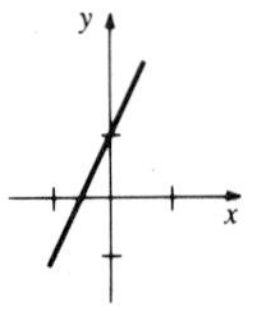

27. Symmetry with respect to x-axis

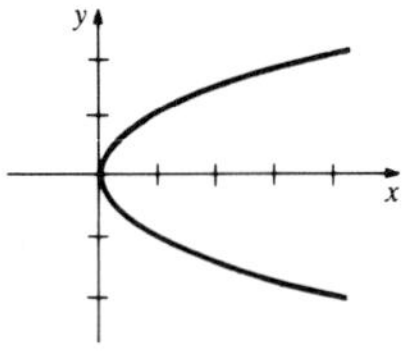

29. Symmetry with respect to y-axis

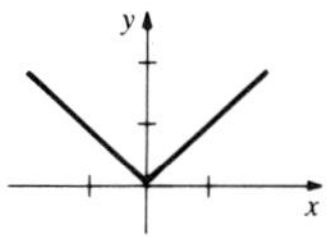

31. Symmetry with respect to x- and y-axes and the origin

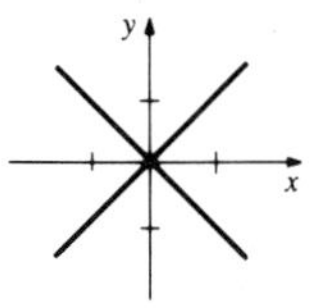

33.

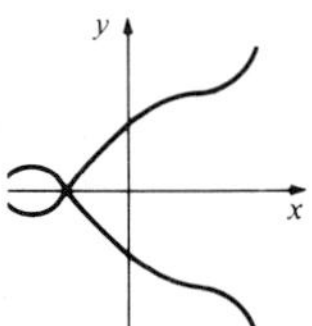

35.

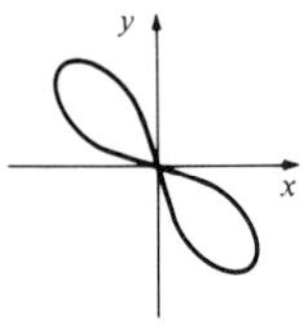

37. $(x - 4)^2 + (y + 6)^2 = 64$
39. $(x - 3)^2 + (y + 4)^2 = 25$
41. $(x - 1)^2 + (y - 1)^2 = 17$
43. Circle, $C(-4, 3)$, $r = 5$
45. Circle, $C(3, -1)$, $r = \sqrt{28/3}$
47. Equation does not describe a circle but rather the single point $(6, -4)$.
49.

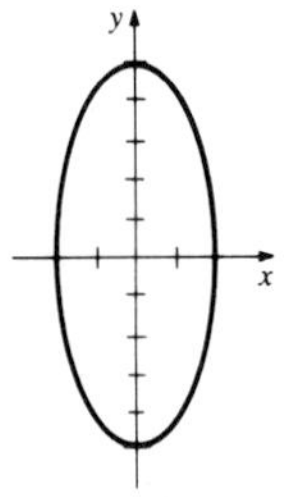

51.

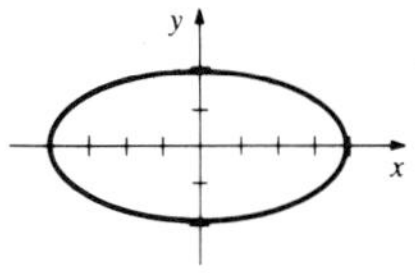

53. $\dfrac{(x - 2)^2}{4^2} + \dfrac{(y + 5)^2}{2^2} = 1$, $C(2, -5)$
55. The graph consists of the x- and y-axes.
57.

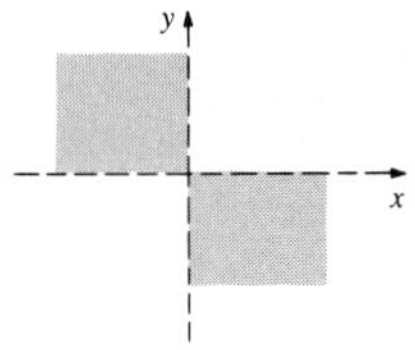

59. $(x + 9)^2 + (y - 3)^2 = 100$
$(x - 9)^2 + (y + 3)^2 = 100$

Exercises I.4, Page A-27

1. $-\frac{3}{2}$ 3. -3 5. $P_2(7, 3)$
7. $y = -x + 3$ 9. $y = 3x$ 11. $y = x + 8$
13. $y = -\frac{3}{2}$ 15. $y = -2x + 4$
17. $y = 4x$ 19. $y = -3x + 9$
21. $y = -x + 7$ 23. $y = -x + 6$
25. $2, -\frac{3}{2}, 3$ 27. 0, none, $\frac{5}{2}$

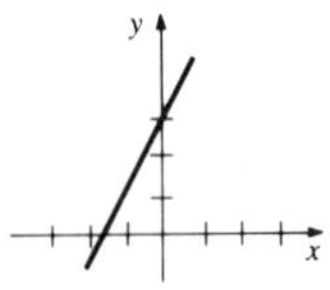

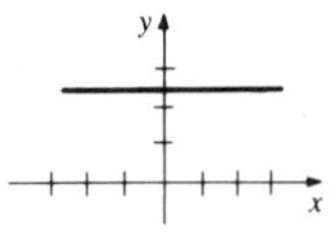

29. $-\frac{5}{3}, 3, 5$ 31. $\frac{3}{8}, \frac{10}{3}, -\frac{5}{4}$

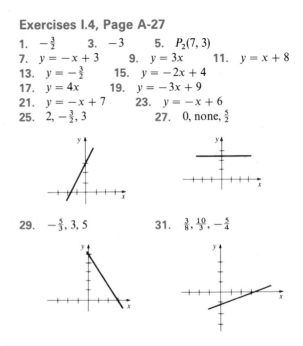

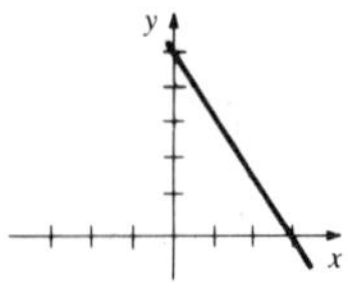

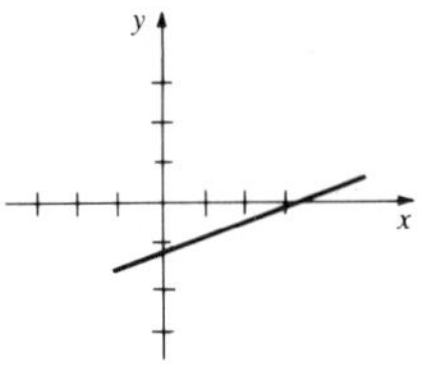

33.

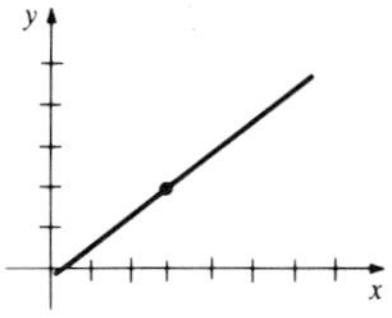

35. $-\frac{2}{5}$ 37. $-\frac{3}{4}$ 39. $-\frac{3}{7}$

41. Collinear 43. (63/16, 31/8)

49. $x/5 + y/2 = 1$ 51. $18\sqrt{13}/13$

53. $17\sqrt{37}/37$

Exercises I.5, Page A-37

1. 2×4 3. 3×3 5. 3×4

7. Not equal 9. Not equal

11. $x = 2, y = 4$ 13. $c_{23} = 9, c_{12} = 12$

15. $\begin{bmatrix} 2 & 11 \\ 2 & -1 \end{bmatrix}; \begin{bmatrix} -6 & 1 \\ 14 & -19 \end{bmatrix}; \begin{bmatrix} 2 & 28 \\ 12 & -12 \end{bmatrix}$

17. $\begin{bmatrix} -11 & 6 \\ 17 & -22 \end{bmatrix}; \begin{bmatrix} -32 & 27 \\ -4 & -1 \end{bmatrix};$ $\begin{bmatrix} 19 & -18 \\ -30 & 31 \end{bmatrix}; \begin{bmatrix} 19 & 6 \\ 3 & 22 \end{bmatrix}$

19. $\begin{bmatrix} 9 & 24 \\ 3 & 8 \end{bmatrix}; \begin{bmatrix} 3 & 8 \\ -6 & -16 \end{bmatrix}; \begin{bmatrix} 0 & 0 \\ 0 & 0 \end{bmatrix};$ $\begin{bmatrix} -4 & -5 \\ 8 & 10 \end{bmatrix}$

21. $x_1 = 4, x_2 = -7$

23. $x_1 = 0, x_2 = 4, x_3 = -1$

25. $x_1 = -2, x_2 = -2, x_3 = 4$ 27. 10; -25

29. 68 31. $-2, 4$ 33. $x = -\frac{3}{5}, y = \frac{6}{5}$

35. $x = -4, y = 4, z = -5$

ANSWERS TO CALCULATOR/COMPUTER ACTIVITIES

1.1, Page 453

1. -17031, -603.96875, 3.16832, 32649, 9763993
3. 7, 5.64, 1.25, 5, 5.24, 5, 4.99

1.2, Page 453

5.

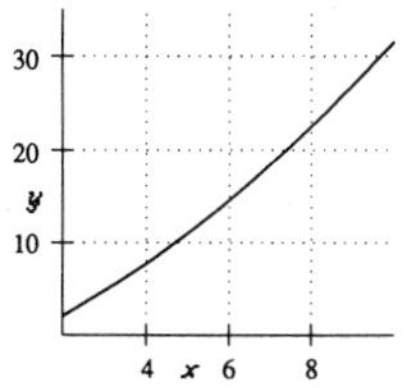

7.

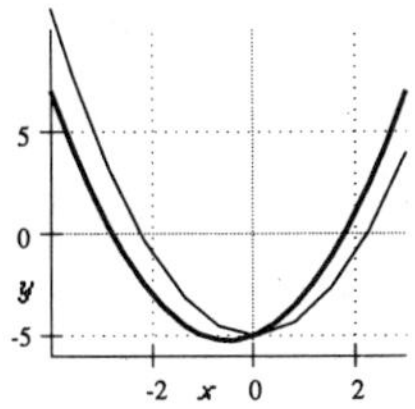

Using the SHARP EL-5200 (which is by far the slowest of the graphic calculators), we get the following times.

9. 46.5 s and 24.5 s

1.3, Page 453

A few sample graphs are shown below.

11.

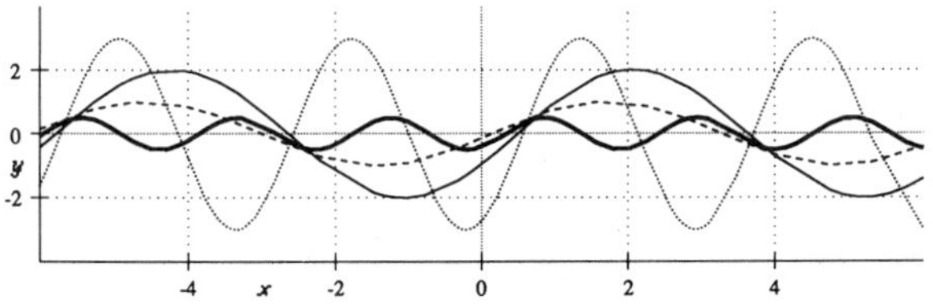

1.5, Page 454

13.

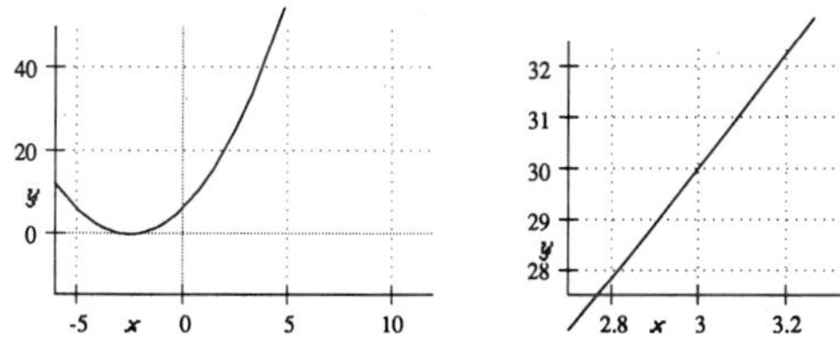

15.

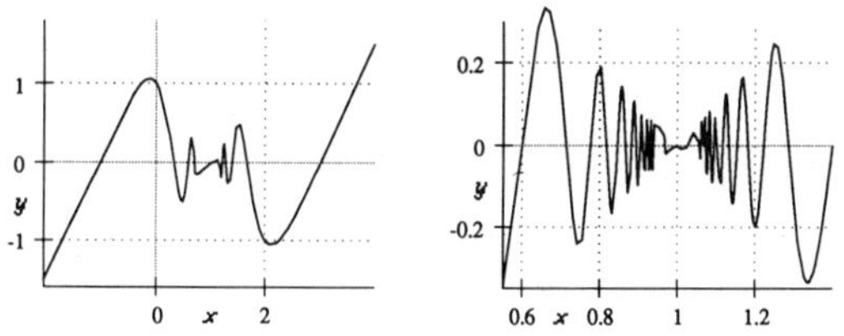

1.7, Page 454

17.

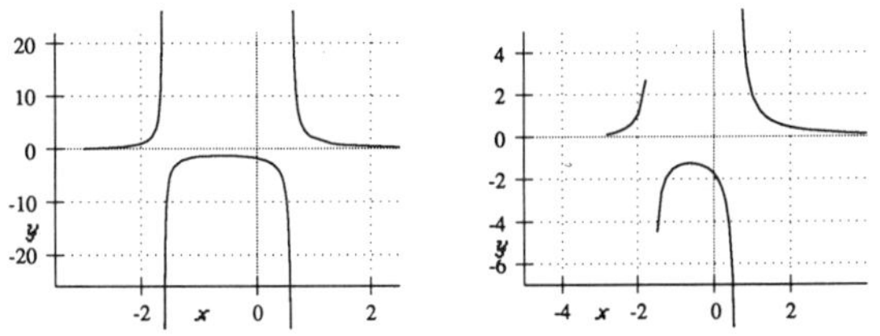

19.

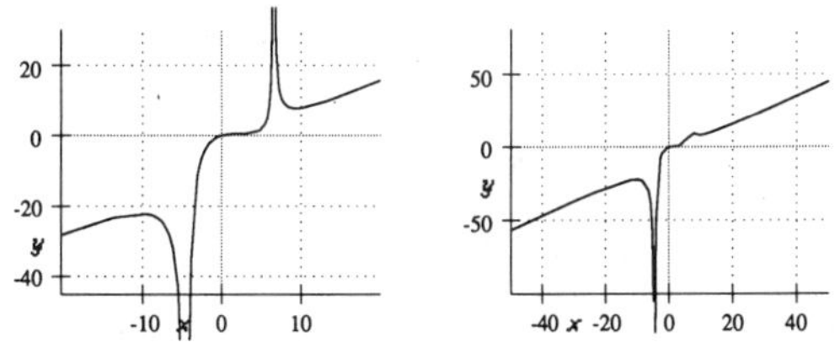

1.8, Page 454

21.

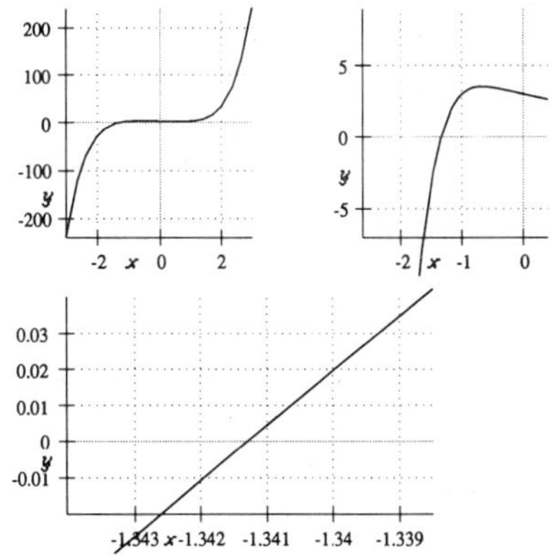

23.

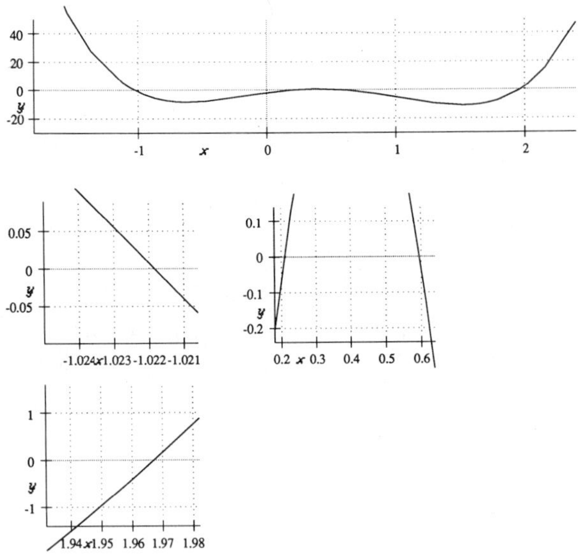

25.

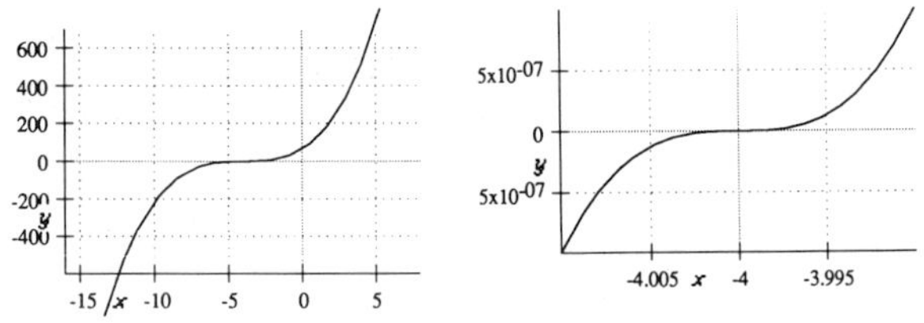

2.1, Page 455

1. 284.1875, 182.8046938, 191.0803141, 190.2358906, 190.320162, 190.31174, 190.3126, 190.312
3. 2.04356605, 1.5123419, 1.256726532, 1.25393519, 1.247767475, 1.247704083, 1.247677, 1.247675, 1.247833333

2.2, Page 455

5. 206, 190.4687563, 190.3140625, 190.3125156, 190.3125, 190.3125, 190.3125, 190.312, 190.32
7. 1.167188365, 1.240301564, 1.247668686, 1.247676157, 1.247676893, 1.2476769, 1.24767675, 1.247676667, 1.247625, 1.24775, 1.2525

2.3, Page 456

9. $20x^3 - 12x + 4$ **11.** $7\cos(7x + 2)$
13. $x^x(\ln x + 1)$
15. $\begin{cases} 2x + 1, & \text{if } |x + \frac{1}{2}| > \frac{\sqrt{13}}{2} \\ -2x - 1, & \text{if } |x + \frac{1}{2}| < \frac{\sqrt{13}}{2} \end{cases}$

2.6, Page 456

17.

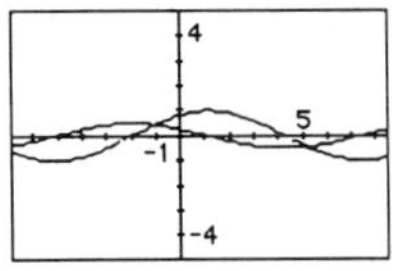

19.

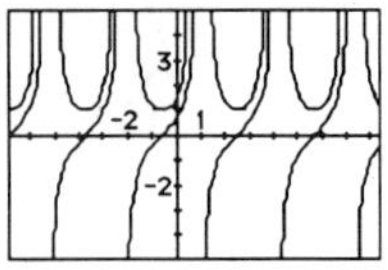

21.

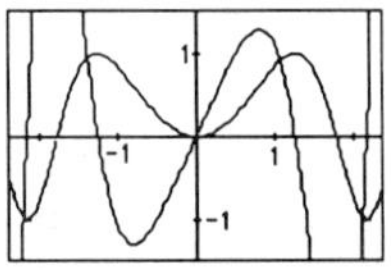

23.

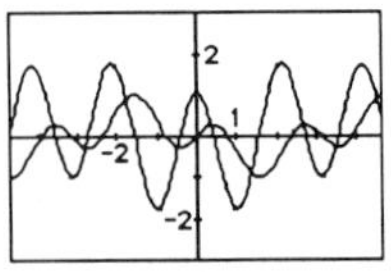

2.11, Page 456

Reworking the first root of Example 4 from Section 2.11, we compare the following Newton's Method result:
.9, .83827291, .8372947813, .8372938452
25. $\Delta x = 0.1$;
.9, .8322897837, .8375374402, .8372826492, .8372943611, .8372938214, .8372938463, .8372938452
$\Delta x = 0.02$;
.9, .8380620657, .8372931284, .8372938464, .8372938452
27. .8, .9, .8353654123, .8372158313, .8372939952, .8372938452

3.3, Page 457

1. $(-4, 273)$, $(0.7847, 102.9)$, $(2.546, 105.6)$, $(6.346, 12.69)$, $(10, 495)$
3. $(-2, 0.9048)$, $(-1.047, 1.298)$, $(1.047, 0.5614)$, $(5.236, 2.823)$, $(7.330, 2.569)$, $(11.52, 3.775)$, $(13.61, 3.589)$, $(15, 3.834)$

3.4, Page 457

5. 2.186787455
7. This function does not satisfy the conditions of the theorem, but we can still find $c = -3$.

3.5, Page 457

9. -1, -0.8854560257, -0.5680647467, -0.1205366803, 0.354604887, 0.7485107482, .9709418174
11. -0.959492736, -0.6548607339, -0.1423148383, 0.4154150130, 0.8412535328, 1

3.6, Page 457

13. Concave up on $(-\infty, 0.308654)$, $(0.504042, 0.614105)$, $(6.38590, 6.49596)$, $(6.7, \infty)$; concave down on $(0.308654, 0.4)$, $(0.5, 0.504042)$, $(0.61415, 6.38590)$, $(6.49596, 6.5)$
15. Concave up on $(0, 0.0664339)$, $(1.56873, 5)$; concave down on $(0.664339, 1.56873)$

4.3, Page 458

1. 2807606.971
3. 5.187377518
5. 85.63378028
7. 0.175742986

4.4, Page 458

9. 1.782371753, 1.784659108, 1.785459187
11. 4.223919088, 4.293125437, 4.329113166
13. 1.734240766, 1.803397687, 1.823488079

4.8, Page 458

15. $\Delta x = 0.1333333333$, $G(0.2666666666) = 0.2491505144$, $G(0.5333333332) = 0.4643013383$, $G(0.7999999998) = 0.646970503, \ldots,$ $G(4) = 0.804896038$
17. $\Delta x = 0.2$, $G(-0.6) = 0.2453145078$, $G(-0.2) = 0.6287354032$, $G(0.2) = 1.028734337, \ldots,$ $G(3.8) = 13.25796605$
19. $\Delta x = 0.1333333333$, $G(-1.733333333) = 0.2405090101$, $G(-1.4666666667) = 0.4574535541$, $G(-1.2) = 0.52415300122, \ldots, G(2) = 2.996235848$

5.1, Page 459

1. Intersection at $x = 0.04520146418$ and 3.056872776; area 14.77883398
3. Intersection at $x = -1.95231675$, 0, and 1.95231675; area 1.972848888
5. Intersection at $x = -3$ and 6.88210135; area 717.2203493
7. Intersection at $x = -6.099554395$ and 9.96086448; area 67.75283506

5.4, Page 459

9. 13.90379795
11. 3.879338259
13. 30.17282194
15. 15.28079116
17. 20.69672591

5.5, Page 459

19. 14.42359945
21. 10.07446712

5.6, Page 459

23. $A(f) = 4.981230886$; $c = 2.761176955$
25. $A(f) = 0.5118276717$; $c = 1.180152309$ or 1.961440345

6.2, Page 460

1. Special points: $(0.1, -2.304225)$, $(\pi/2, 0)$, $(3.14, -6.442354)$
3. Limit as $x \to -0.5$ is ∞; limit as $x \to \infty$ is 1; special points: $(0, 2)$, $(3, 4/3)$
5. Special points: $(0.1, -2.0403)$, $(0.574, 0)$, $(2.434, 1.5047)$, $(4.502, 0)$, $(4.669, -0.0638)$, $(4.831, 0)$, $(8.124, 2.8667)$, $(10.977, 0.7872)$, $(14.291, 3.4427)$, $(17.267, 1.2397)$, $(20, 3.7438)$

6.3, Page 460

7. Limit as $x \to 0^-$ is 3; limit as $x \to 0^+$ is 0; special points: $(-4, 2.1592048)$, $(4, 1.827026)$
9. Limit as $x \to \pm\infty$ is 0; special points: $(-3, 1.3925\ \mathrm{E} - 4)$, $(0, 1.128)$, $(3, 1.3925\ \mathrm{E} - 4)$
11. Special points: $(0, 2.5)$, $(0.0239, 2.5110)$, $(0.5524, 0)$, $(0.6862, -0.2490)$, $(0.8194, 0)$, $(1.6301, 4.2914)$, $(0.2370, 3.2602)$, $(3.1678, 5.4173)$, $(3.9231, 4.5898)$, $(4.5309, 4.9490)$, $(7.4831, 0)$, $(11.2737, -4.9932)$, $(14.9610, 0)$, $(18.7008, 5.0003)$, $(20, 4.2730)$

6.4, Page 460

13. 1.0000000
15. 127.2930003
17. 298.202937

6.5, Page 460

19. The only special point on each graph is the x-intercept at $x = 1$.
21. Limit as $x \to 0^+$ is $-\infty$; limit as $x \to \infty$ is 0; special points: $(1, 0)$, $(1.57905, 0.369156)$, $(4, 0.0390625)$

6.7, Page 461

23. 21.5818943
25. -0.02450142759 or -1.603099753
27. Special points: $(-5, -47.209949)$, $(-3.19196, 0)$, $(-2.1773, 2.272735)$, $(0, 1)$, $(2.1773, 2.272735)$, $(3.19196, 0)$, $(5, -47.209949)$
29. Limit as $x \to 0$ is 0; special points: $(-10, -0.9999092)$, (10.09999092)

7.1, Page 705

1. $(-1, 0)$, $(-0.771178, 0.330674)$, $(0, 0)$ $(0.771778, -0.330674)$, $(1, 0)$
3. $(-1, 0)$, $(-\sqrt{3}/2, \pi/4)$, $(-0.5, 0)$, $(0, -\pi/4)$, $(0.5, 0)$, $(\sqrt{3}/2, \pi/4)$, $(1, 0)$
5. $$y = \begin{cases} \pi - \tan^{-1}\left(\dfrac{1}{x}\right), & x < 0 \\ \dfrac{\pi}{2}, & x = 0 \\ \tan^{-1}\left(\dfrac{1}{x}\right), & x > 0 \end{cases}$$
7. $$y = \begin{cases} \pi - \sin^{-1}\left(\dfrac{1}{x}\right), & x \le -1 \\ \sin^{-1}\left(\dfrac{1}{x}\right), & x \ge 1 \end{cases}$$

9.

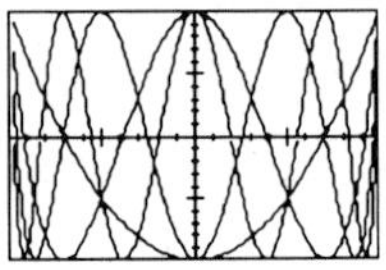

7.3, Page 705

11. $(-2\pi, 0)$, $(-3\pi/2, 0.881373587)$, $(-\pi, 0)$, $(-\pi/2, -0.881373587)$, $(0, 0)$, $(\pi/2, 0.881373587)$, $(\pi, 0)$, $(3\pi/2, -0.881373587)$, $(2\pi, 0)$

13. $(1.724715, 0.2100181456)$, $(6, 0.235415)$; limit as $x \to 1^+$ is $+\infty$

8.5, Page 706

1. $$\frac{3.799974306}{x + 1.600015221} + \frac{-4.100178898}{x - 2.900081213} + \frac{6.199887414}{x - 5.699856494} + \frac{1.500317179}{x - 7.200077513}$$

3. $$\frac{3}{x-6} + \frac{9x+2}{x^2+2x+9} + \frac{5}{2x^2+3x+11}$$

5. $$\frac{13}{x+29} + \frac{5}{x-35} + \frac{-7}{x-9} + \frac{x+6}{x^2+x+17}$$

7. $$\frac{6}{2x+1} + \frac{24}{x-19} + \frac{10}{x-18} + \frac{-5}{x-7}$$

9. $$\frac{23x+45}{5x^2+2x+106} + \frac{87}{2x+53} + \frac{-34x-61}{3x^2+6x+11}$$

9.2, Page 706

1. $x \in [0, 100]$; 0.8862269255
3. $x \in [0, 50]$; 3.323350970
5. $x \in [0, 1000]$; 4.722017777 E − 3
7. $\int_0^1 \sin\left(\frac{1}{t}\right) dt$; $t \in [0.001, 1]$, 0.504066497
9. $\int_0^1 \frac{1}{t^{(\ln t + 1)}} dt$; $t \in [0.0001, 1]$, 0.886226925

10.1, Page 707

1. 916 3. 7 5. 31,251

7. $x_1 = 0.9553364891$; $x_5 = 0.7844362247$; $x_{10} = 0.7326982079$

9. $x_1 = 6.782329983$; $x_5 = 3.410881529$; $x_{10} = 3.304837114$

11. $x_1 = 4.313604991$; $x_5 = 5.358793418$; $x_{10} = 5.351150341$

10.2, Page 707

13. The sequence is not monotonic because
$$\frac{2n}{2n+1} < 1 < \frac{2n+2}{2n+1}.$$

15. The sequence is decreasing because
$$a_{2n+1} = \frac{4n^2+4n}{4n^2+4n+1} a_{2n-1} < a_{2n-1}.$$

17. $N = 10$

21. On a TI-81 calculator: 0.2000000004, 0.2000000089, 0.1999998411, 0.1999797092

10.3, Page 708

23. 2.522977329 25. −0.8224545959

10.6, Page 708

29. $S_{800} = 0.6925225712$, error bound 1.248 E − 3; $A_{800} = 0.6931467909$, error bound 3.242 E − 4; $S_{900} = 0.6925919336$, error bound 1.110 E − 3; $A_{900} = 0.6931468726$, error bound 5.549 E − 4

31. $S_{80} = 0.8646647168$, error bound 1.3964 E − 119; $A_{80} = 0.8646647168$, error bound 6.987 E − 120; $S_{100} = 0.8646647168$, error bound 1.0603 E − 160; $A_{100} = 0.8646647168$, error bound 5.302 E − 161

10.7, Page 708

33. All visible in the viewing window $-1.6 \le x \le 1.6$, $-2.1441 \le y \le 19.9806$

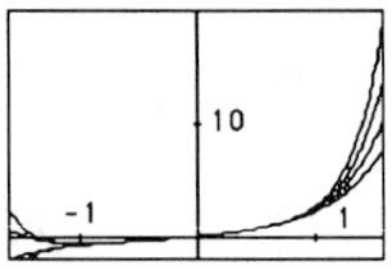

35. All visible in the viewing window $-10 \le x \le -1$, $-18.5833 \le y \le 47.8063$

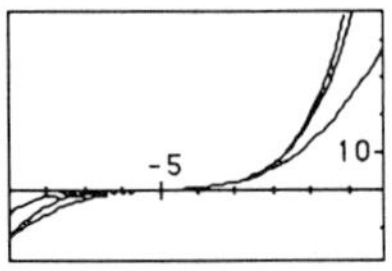

37. All visible in the viewing window $-1 \le x \le 9$, $-7.1597 \le y \le 2.5284$

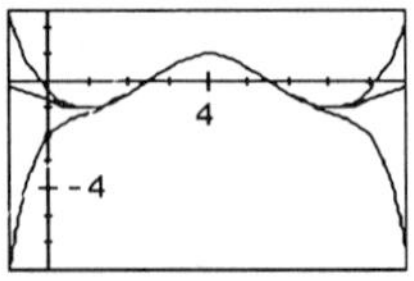

10.9, Page 709

39.

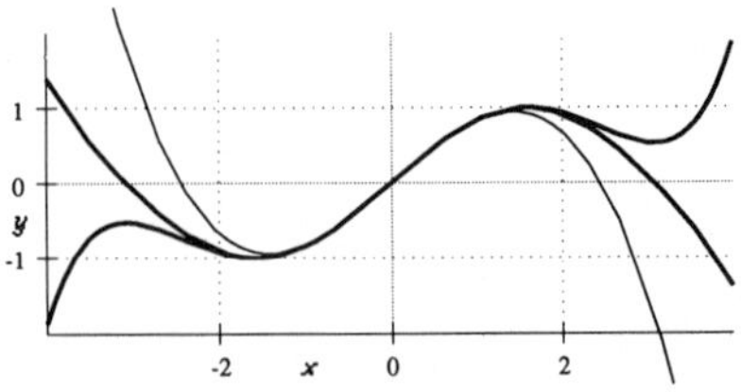

41.

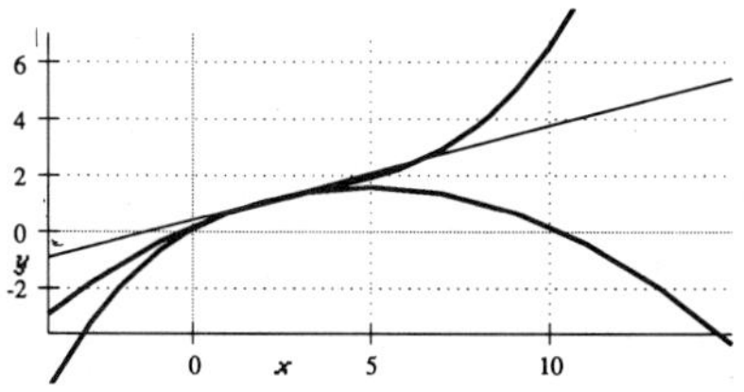

43. $P_3(-2) = -3.333333333$, $P_7(-2) = -3.625396825$, $P_{11}(-2) = -3.626859067$;
$P_3(0.2) = 0.2013333333$, $P_7(0.2) = 0.2013360025$, $P_{11}(0.2) = 0.2013360025$;
$P_3(3) = 7.5$, $P_7(3) = 9.958928571$, $P_{11}(3) = 10.01760755$

45. $P_3(0.5) = 0.8753817427$, $P_4(0.5) = 0.8772495641$, $P_5(0.5) = 0.8776036186$;
$P_3(1) = 0.5403022008$, $P_4(1) = 0.5403023042$, $P_5(1) = 0.5403023059$;
$P_3(3) = -1.06969908$, $P_4(3) = -0.766705224$, $P_5(3) = -0.9716519291$

11.1, Page 709

1.

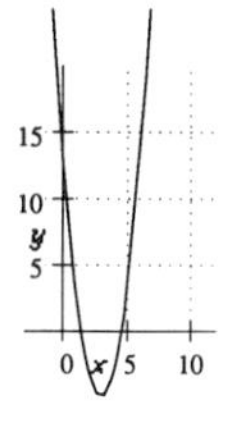

3.

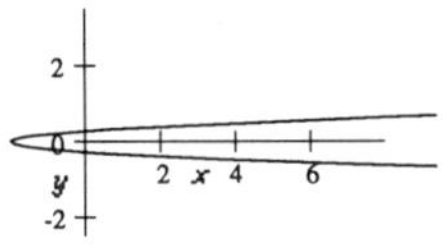

5.

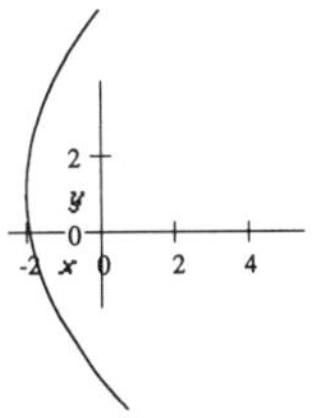

11.2, Page 710

7. 6.282746050 A.U. **9.** 59.88786939 A.U.

11.3, Page 710

11. Asymptotes $y = \pm 0.8x$; center (0, 0); vertices $(0, \pm 0.8\sqrt{2})$

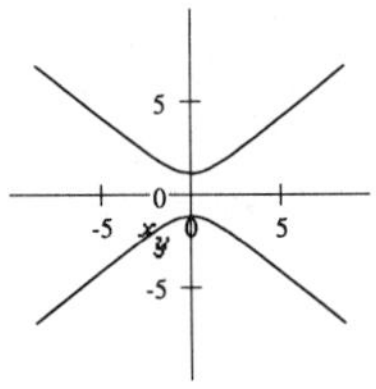

13. Asymptotes $y - 5 = \pm x$; center (0, 5); vertices $(0, 5 \pm \sqrt{20})$

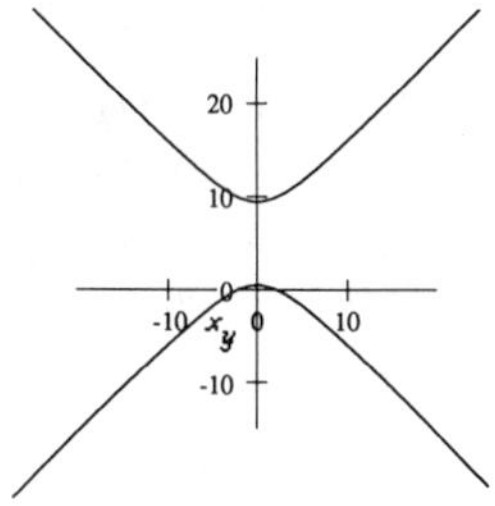

15. Asymptotes $y + 15 = \pm 2(x - 3.5)$; center (3.5; −15); vertices $(3.5, -15 \pm 2\sqrt{22.75})$

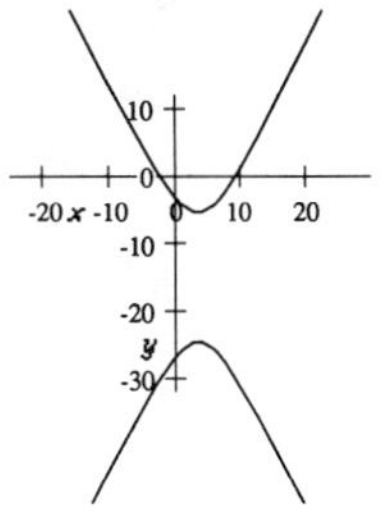

12.1, Page 710

1.

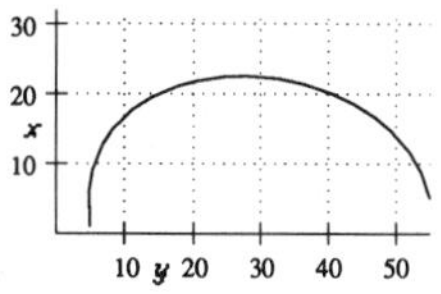

3.

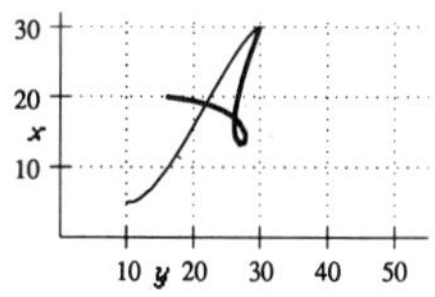

5.

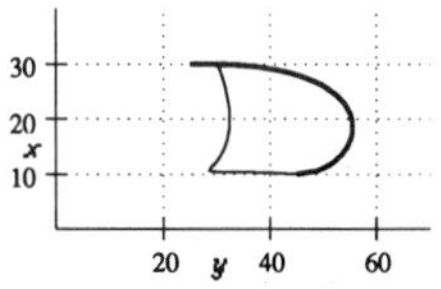

7. One example: $P_0(48, 42)$, $P_1(60, 60)$, $P_2(31,55)$, $P_3(30, 45)$, $P_4(29, 35)$, $P_5(51, 30)$, $P_6(50, 20)$, $P_7(49, 10)$, $P_8(20, 5)$, and $P_9(32, 23)$

9. One example: $P_0(55, 42)$, $P_1(49, 48)$, $P_2(35, 32)$, $P_3(50, 32)$, $P_4(24, 32)$, $P_5(43, 12)$, and $P_6(53, 22)$

12.2, Page 711

11. Tangent line at $t = 1$ is

$$y - q_3 = \frac{q_3 - q_2}{p_3 - p_2}(x - p_3).$$

13. The piecewise-defined curve has continuous slope.

15. Arc length: 67.13442705;
line segments: 189.7635957

12.3, Page 711

17. (0.2836621855, −0.9589242747), (4.890738004, 1.039558454), (1.449431018, 3.728156344), (−10.01199793, −22.90414842)

19. TI-81 has none of these problems; CASIO calculators sometimes cannot convert the origin to polar form; SHARP calculators sometimes cannot convert polar coordinates with negative r to rectangular form.

12.4, Page 712

21.

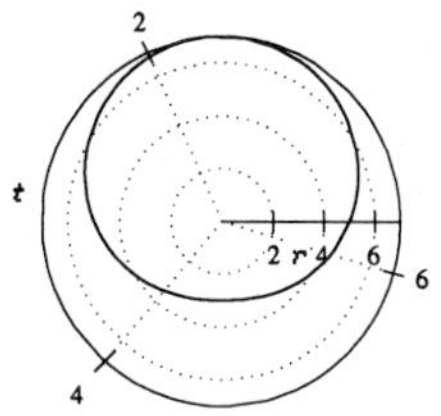

23.

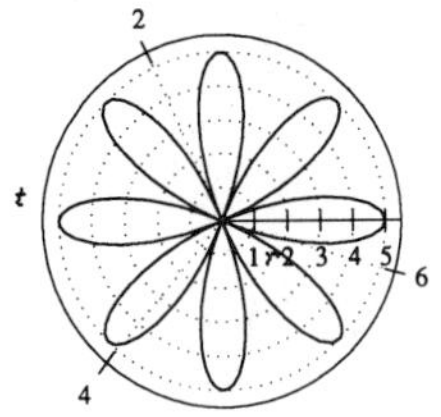

25.

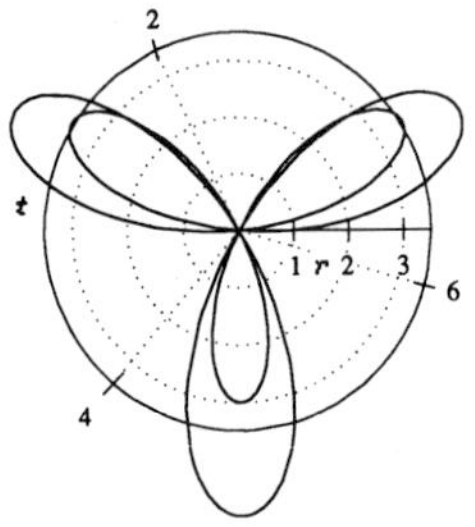

27.

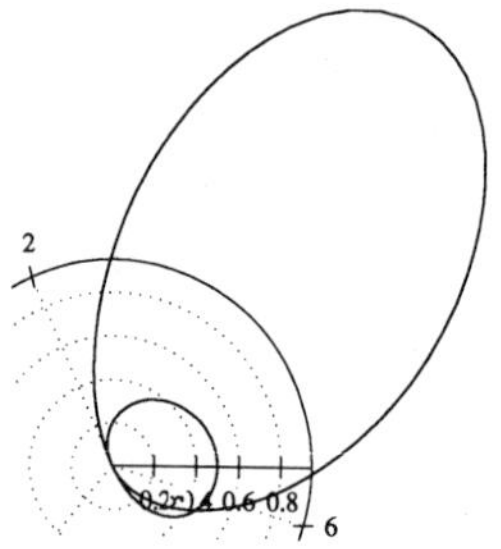

29.

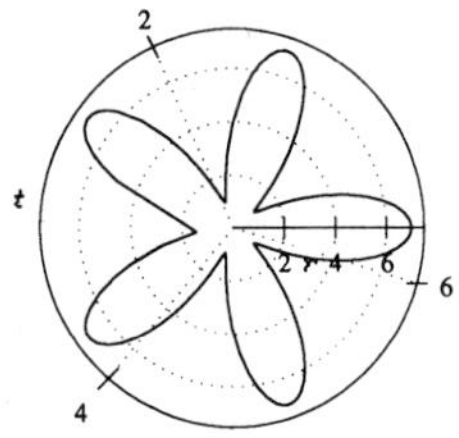

31. (0, 0), (0.8652873684, 1.306928869)
(0.7356576076, 1.813990757)
(48.37466268, 5.074964287)
(41.12759519, 5.582026175)

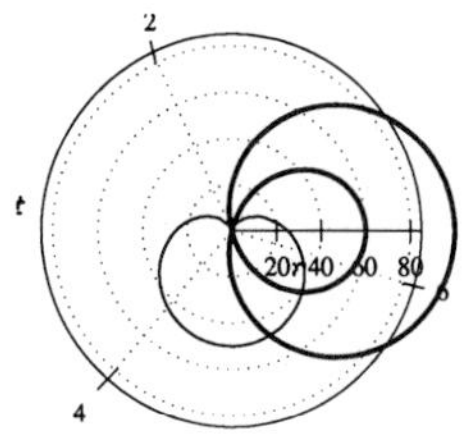

12.5, Page 712

33. 10.29855055 **35.** 12.11056028 **37.** 16

12.6, Page 712

39. $r = \dfrac{1+e}{1+e\cos\theta}$

41.

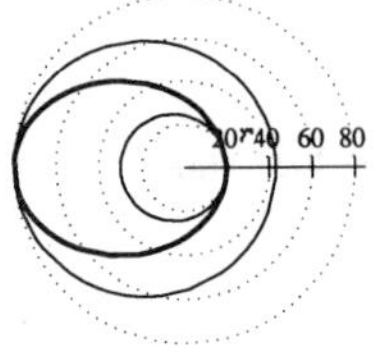

43.

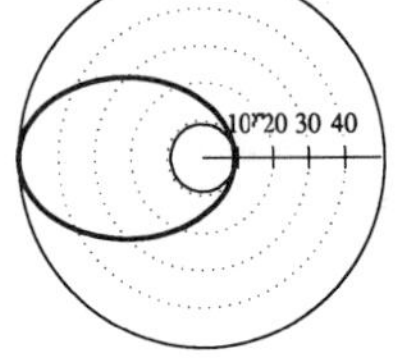

45. Axis: $y = \dfrac{\sqrt{3}}{2}x$

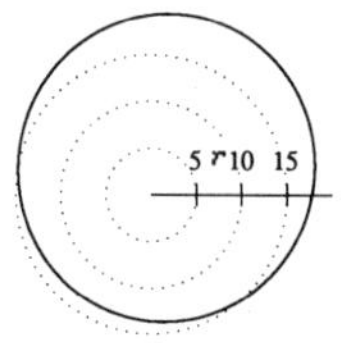

47. Axis: $y = -\dfrac{1}{\sqrt{3}}x$

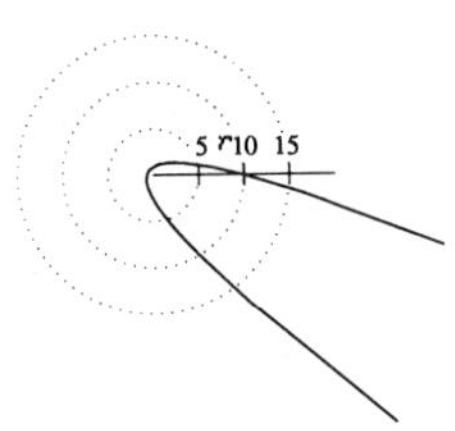

49. Axis: $y = -(\tan 5)x$

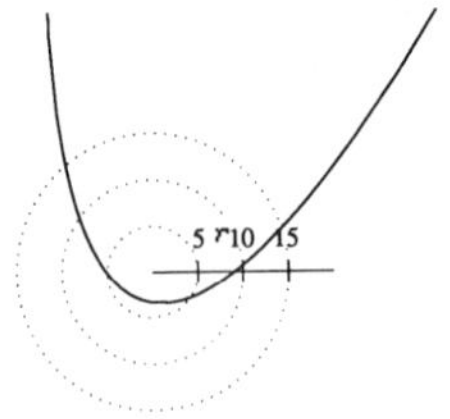

13.1, Page 1039

1. $\mathbf{c} = \mathbf{a} + \mathbf{b} = \langle -1.8, 5.09\rangle$

5. $\mathbf{c} = \mathbf{a} + \mathbf{b} = \langle 60.3, 30.6\rangle$, $\mathbf{w} = \mathbf{b} - \mathbf{a} = \langle -10.1, 66.4\rangle$

13.2, Page 1039

7. $\mathbf{a} + \mathbf{b} = \langle 6, 8, -6\rangle$, $\mathbf{b} - \mathbf{a} = \langle 2, 18, -12, 66.4\rangle$, $|\mathbf{a}| = 6.164414003$, $|\mathbf{b}| = 16.30950643$

9. $0.0571\mathbf{c} = \langle 13.2472, 52.9888, -36.544\rangle$, $\mathbf{b} - \mathbf{c} = \langle 340, -591, 1442\rangle$, $|\mathbf{a} + \mathbf{c}| = 1522.622409$, $|\mathbf{a} + \mathbf{b} + \mathbf{c}| = 1861.909235$
11. $\mathbf{c}/|\mathbf{c}| = \langle 0.4820971617, 0.1910196301, 0.8550402491\rangle$, $(\mathbf{b} + \mathbf{c})/|\mathbf{b} + \mathbf{c}| = \langle 0.1019883230, 0.9518910147, 0.2889669152\rangle$, $|18\mathbf{a} + 67\mathbf{c}| = 8149.981472$, $\mathbf{a} - \mathbf{b} + \mathbf{c} = \langle 181, 54, 147\rangle$

13.3, Page 1040

13. $\mathbf{a}\cdot\mathbf{b} = 5$, $\mathbf{b}\cdot\mathbf{a} = 5$, $|\mathbf{a}| = 7.681145748$, $|\mathbf{b}| = 19.10497317$
15. $\mathbf{a}\cdot\mathbf{c} = 383512$, $\mathbf{b}\cdot\mathbf{c} = -68480$, $|(\mathbf{a}\cdot\mathbf{c})\mathbf{a}| = 182620586.1$, $|(\mathbf{a} + \mathbf{b})\cdot\mathbf{c}| = 315032$
17. $|\mathbf{c}|^2 = 12086$, $(\mathbf{a} + \mathbf{c})\cdot\mathbf{b} = -8573$, $(\mathbf{a} + \mathbf{b})\cdot\mathbf{c} = -5529$, $\mathbf{a}\cdot(\mathbf{b} + \mathbf{c}) = 3886$

13.4, Page 1040

19. $\mathbf{a}\times\mathbf{c} = \langle 5708, 1440, -784\rangle$, $\mathbf{b}\times\mathbf{c} = \langle 10116, -544, -5840\rangle$, $\mathbf{a}\times\mathbf{b} = \langle 5708, 1440, -784\rangle$, $\mathbf{a}\cdot(\mathbf{b}\times\mathbf{c}) = -276128$
21. $\mathbf{a}\times\mathbf{c} = \langle 3794, -3374, 412\rangle$, $\mathbf{b}\times\mathbf{c} = \langle -4120, -3668, 1822\rangle$, $\mathbf{a}\times(\mathbf{b}\times\mathbf{c}) = \langle -302814, 329370, -21660\rangle$, $\mathbf{a}\cdot(\mathbf{b}\times\mathbf{c}) = -331158$

13.8, Page 1040

23.

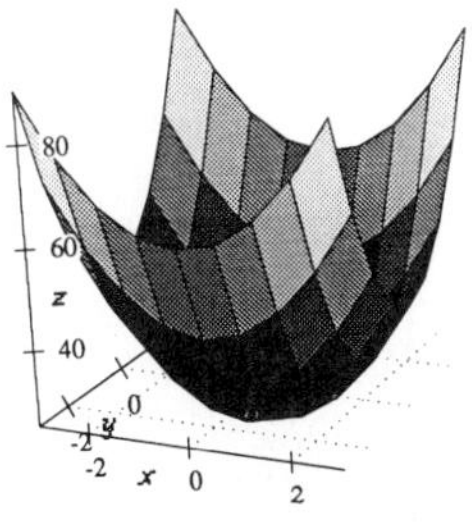

25.

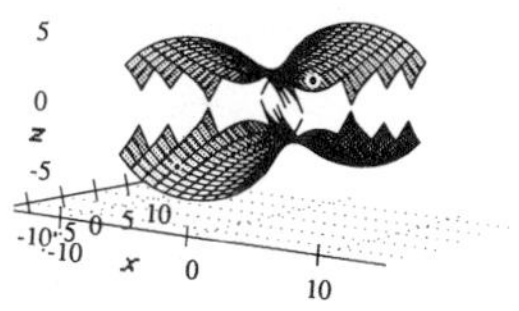

27.

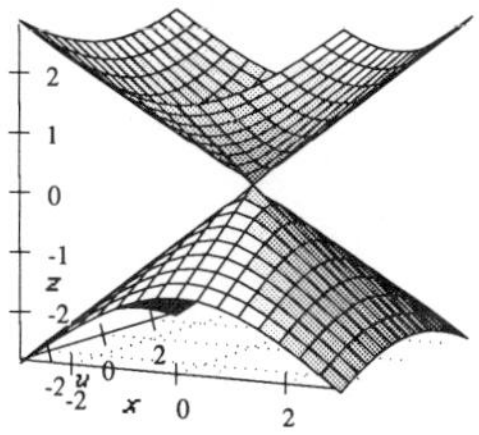

14.1, Page 1041

1.

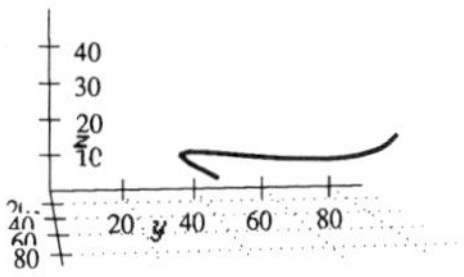

3.

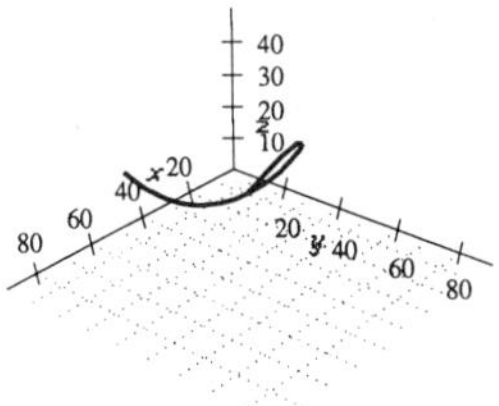

5.

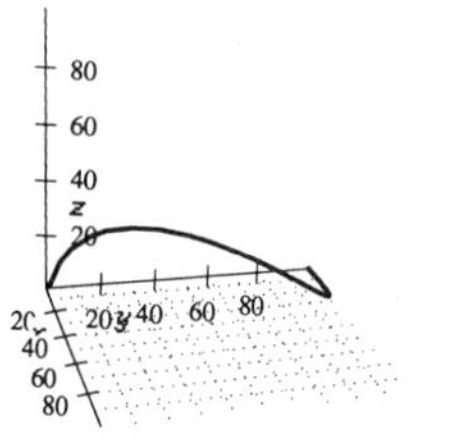

7. Tangent line at $t = 0$ on the curve is $x = 13 + 3(30 - 13)t$, $y = 45 + 3(5 - 45)t$, $z = 5 + 3(25 - 5)t$
9. Curve arc length 93.29246126; line segment arc length 157.4771217
11. $40 + 40 + 20 = 100$; $0 + 50 + 50 = 100$; $50 + 50 + 0 = 100$; $60 + 15 + 25 = 100$

14.2, Page 1042

13.

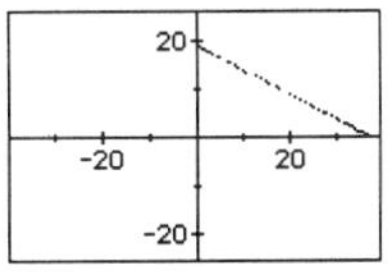

15.

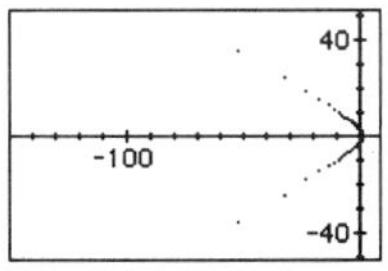

17.

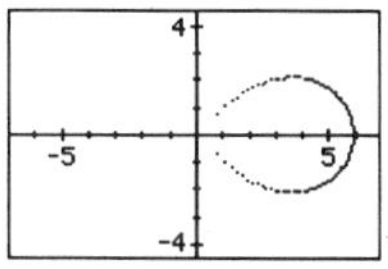

14.3, Page 1042

19. At $t = -2\pi/3$, $\rho = 4.630064794$, $(h, k) = (-5.12484016, 3)$
At $t = 0$, $\rho = 1$, $(h, k) = (0, 0)$
At $t = \pi/6$, $\rho = 1.613743061$,

$(h, k) = (-0.1980890609, -0.5773502692)$
At $t = 2$, $\rho = 5.933318809$,
$(h, k) = (5.9991678479, 3.97370255)$

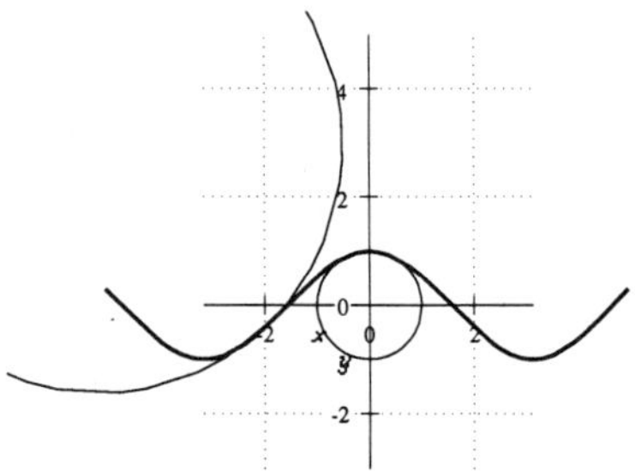

21. At $t = -\pi/4$, $\rho = 6$,
$(h, k) = (4.242640687, 4.242640687)$
At $t = 0$, $\rho = 0.6$, $(h, k) = (2.4, 0)$
At $t = \pi/6$, $\rho = 3.493689492$,
$(h, k) = (2.138195178, -2.641412932)$
At $t = \pi/4$, $\rho = 6$, $(h, k) = (4.242640687, -4.242640687)$
At $t = \pi/2$, $\rho = 0.6$, $(h, k) = (0, -2.4)$
At $t = 3\pi/4$, $\rho = 6$,
$(h, k) = (-4.242640687, -4.242640687)$
23. At $t = \pi/2$, $\rho = 9.8$, $(h, k) = (0, -4.8)$;
circle contains the ellipse
25. $b = \pm 20$ **27.** 0.6115517658

15.1, Page 1043

1. $z = \dfrac{x^3}{3} + x(y^2 - 1) - \dfrac{y}{2} + 1$

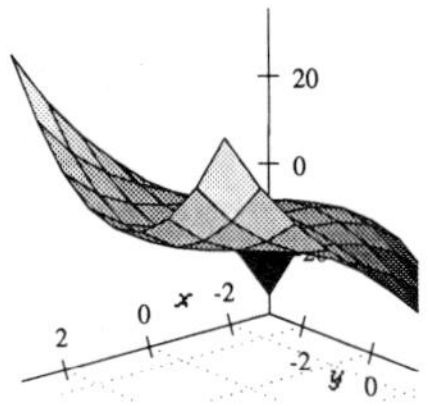

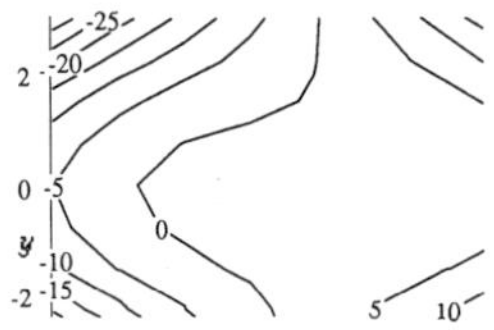

3. $z = x^4 - 3xy^2 + y^3 - x - y$

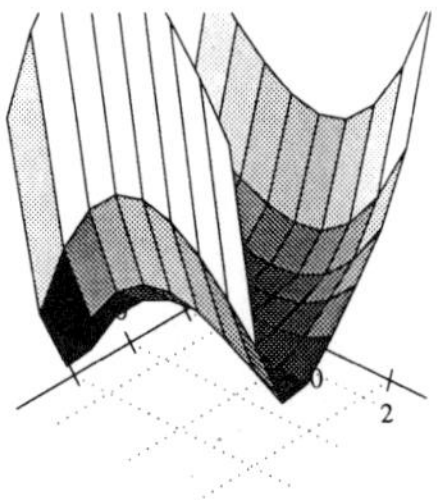

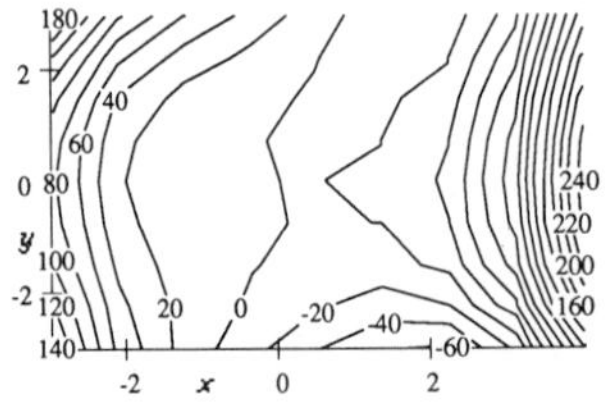

5. $z = |xy|$

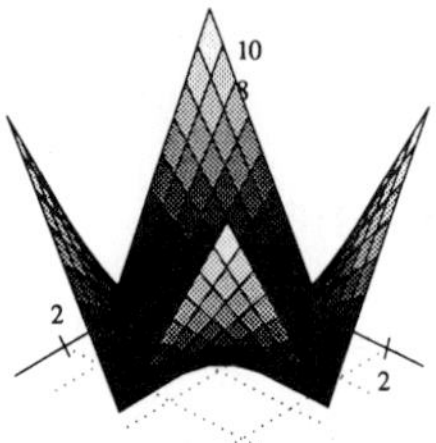

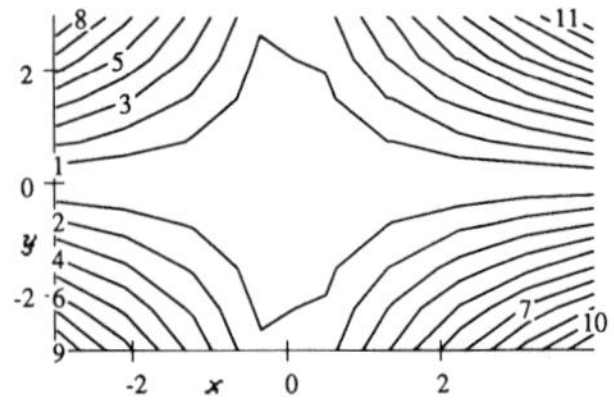

7. $z = \cos\left(\dfrac{x^2 + 2y^2}{4}\right)$

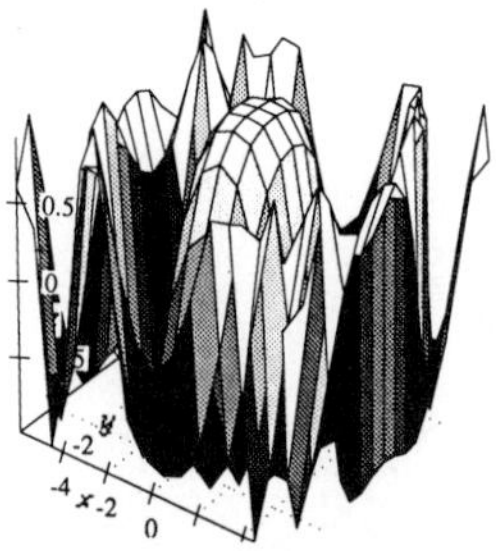

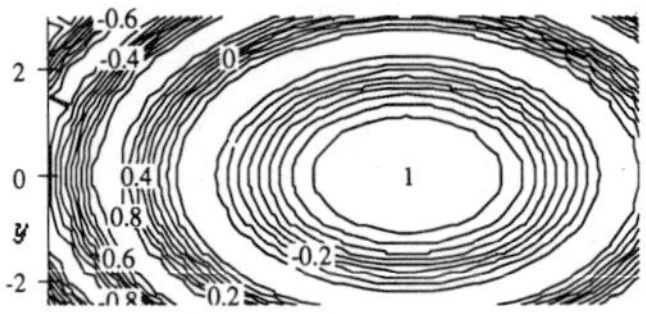

15.2, Page 1043

9.

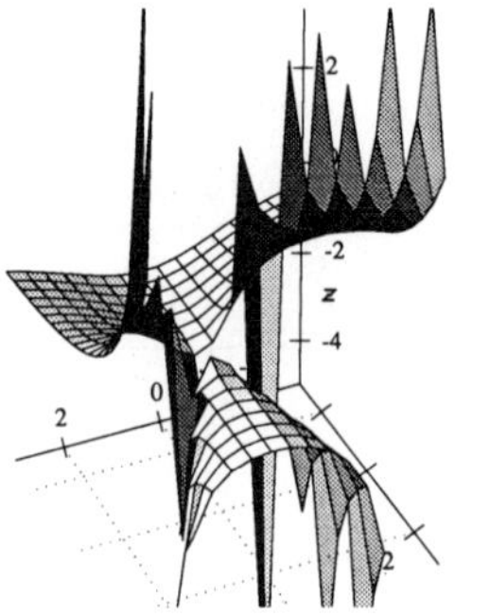

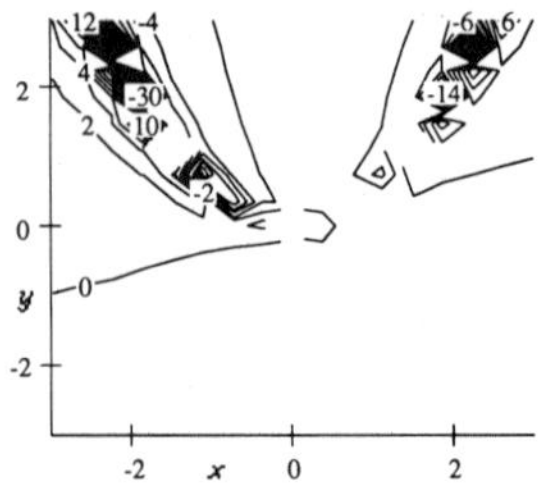

11.

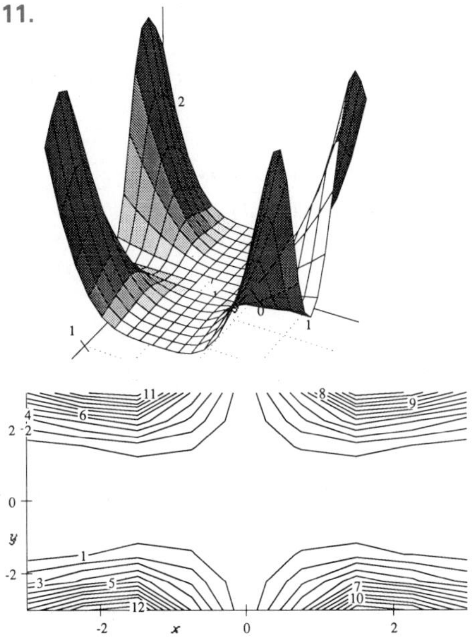

15.3, Page 1044

13.

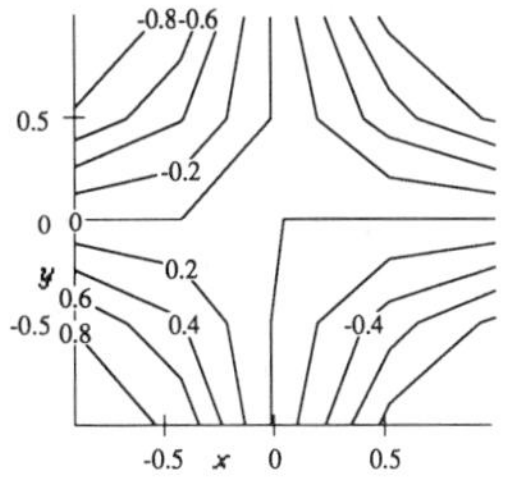

$f(x)$

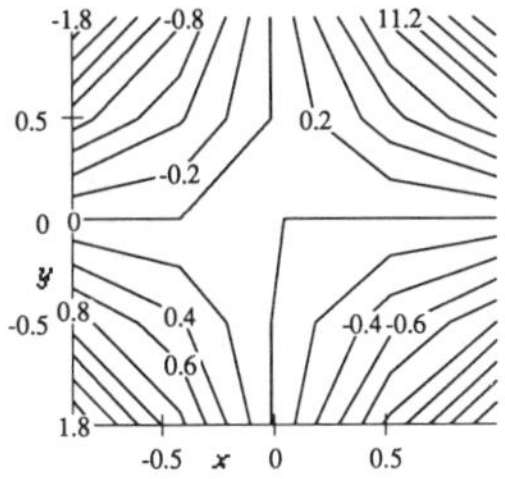

$p_2(x, y) = 2xy$

15.

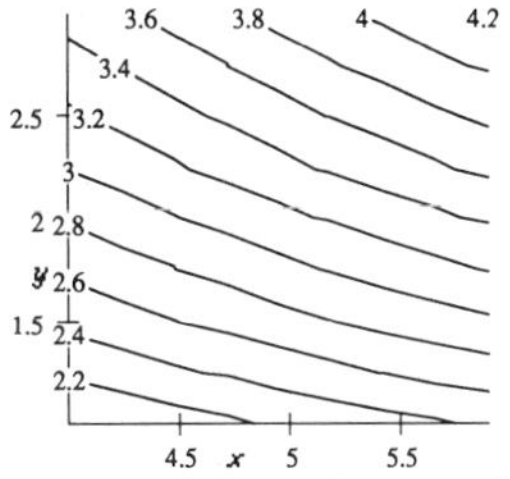

$f(x)$

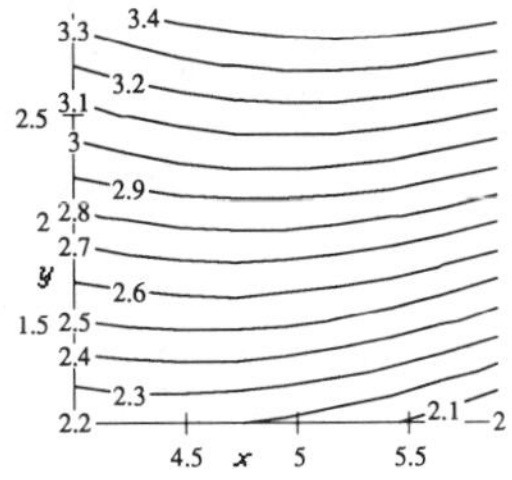

$$p_2(x) = \sqrt{10} + \frac{5}{2\sqrt{10}}(x-2) + \frac{1}{\sqrt{10}}(y-5) + \frac{1}{2}\left[-\frac{5}{8\sqrt{10}}(x-2)^2 + \frac{1}{2\sqrt{10}}(x-2)(y-5) - \frac{1}{10\sqrt{10}}(y-5)^2\right]$$

17.

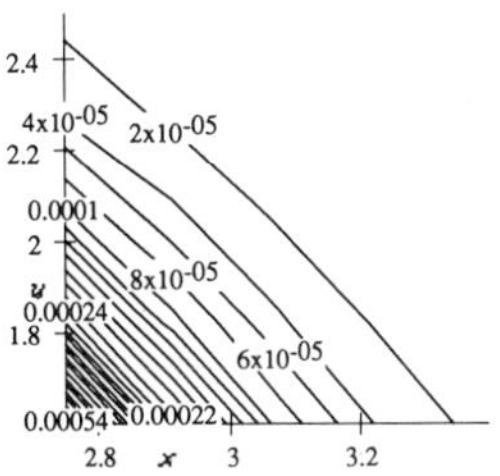

$f(x)$

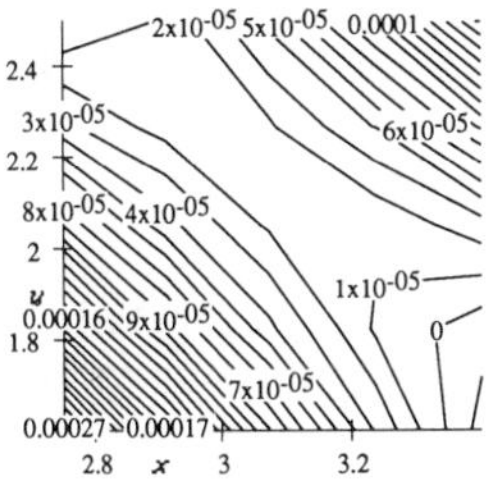

$$p_2(x, y) = [15 - 85(x-3) - 60(y-2) + 165(x-3)^2 + 340(x-3)(y-2) + 105(y-1)^2](e^{-13})$$

15.4, Page 1044

19. $f_y(0.8, 2.0) = -0.878$, $f_{yy}(0.8, 2.0) = 1.184$, $f_y(1.4, 2.0) = -2.014$, $f_{yy}(1.4, 2.0) = 2.008$

21. $f_x(1.2, 2.0) = 2.2975$, $f_y(1.2, 2.0) = -1.576$, $f_{xx}(1.2, 2.0) = 3.875$, $f_{xy}(1.2, 2.0) = -1.8425$, $f_{xy}(1.2, 2.0) = 1.88$

23. $p_2(x, y) = 6.698 + 2.2975(x-1) - 1.576(y-2) + 1.9375(x-1)^2 - 1.8425(x-1)(y-2) + 0.94(y-1)^2$

15.9, Page 1045

25.

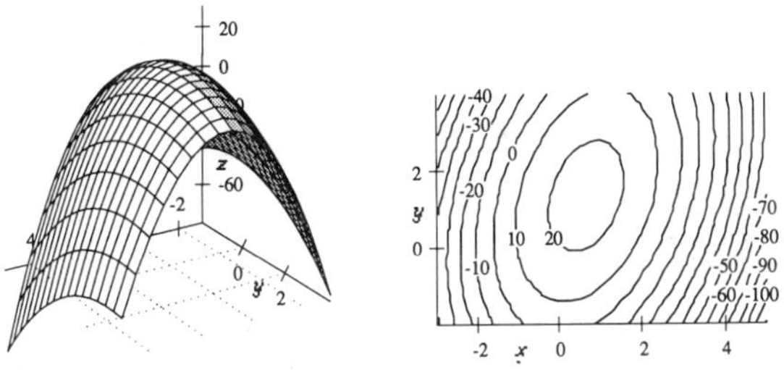

27.

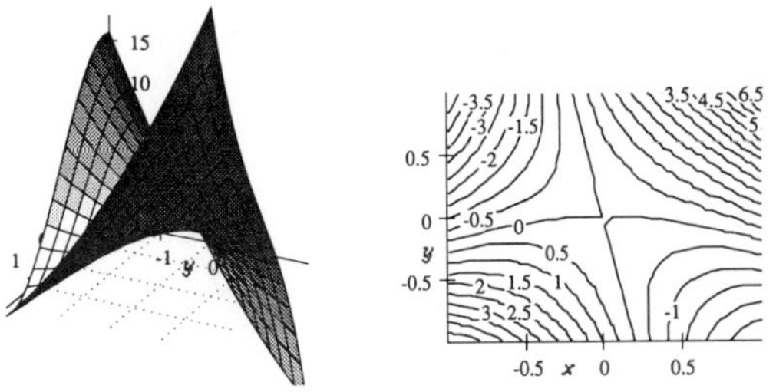

15.10, Page 1045

29. $y = -7.1285 + 0.137225x$, linear model
$y = 10.46426644 * e^{0.0027299409x}$, exponential model
$y = 0.050705054x^{1.135177185}$, power model

31. $y = 5388.776831 + 65.64381334x$, linear model
$y = 2.949868784\text{E}26 * e^{0.7001866463x}$, exponential model
The power model cannot be calculated for this data.

16.1, Page 1045

1. 1962.069855, 1251.233156, 1083.482697, 1010.021556
3. 52.19956281, 64.0887437, 67.89392742, 69.76272218

16.2, Page 1046

5. −0.071032573, −0.0822807142
7. 197.0448171, 197.0448171
9. 27.93310501, 27.93310661
11. 0.7853439999, 0.7853981633

16.3, Page 1047

13. 17932/945 or 18.97566138 **15.** 1890.197369

16.6, Page 1047

17. 2.667912471 **19.** 10.08224472

16.7, Page 1048

21. 0.2986955138 **23.** 11.03678176

17.1, Page 1048

1. −17.98257577, 0, 35.48288295
3. 1.461380487, −2.063418294, 2.53383162
5. 11.30840984, 3.9063554, 2.697088316, 2.07114214
7. 143.5995902 **9.** 113.3315603

17.2, Page 1049

11. 1.536229, 1.470218, 1.414097, 1.367246, 1.328964
13. 1.612598, 1.620343, 1.627831, 1.638447

17.3, Page 1049

15. 0.3678794412 **17.** 0

17.4, Page 1049

19. −2.53708663
21. 9.622398

17.5, Page 1049

23.

```
>
  curl(f, v);
```
array(1 . . . 3,

$[-3\sin(x^2y^3z^5)x^2y^2z^5 - 7\cos(x^3+7yz)y,$
$-\sin(3x+y^2z)y^2 + 2\sin(x^2y^3z^5)xy^3z^5,$
$3\cos(x^3+7yz)x^2 + 2\sin(3x+y^2z)yz])$

```
>
  diverge(f, v);
```
$-3\sin(3x+y^2z) + 7\cos(x^3+7yz)z$
$-5\sin(x^2y^3z^5)x^2y^3z^4$

25.

```
>
  grad((2+7*x*y-z^2)/sin(1+x+y^2+z^3),v);
```
array(1 . . . 3,

$$\left[7\frac{y}{\sin(1+x+y^2+z^3)} - \frac{(2+7xy-z^2)\cos(1+x+y^2+z^3)}{\sin(1+x+y^2+z^3)^2},\right.$$

$$7\frac{x}{\sin(1+x+y^2+z^3)} - 2\frac{(2+7xy-z^2)\cos(1+x+y^2+z^3)y}{\sin(1+x+y^2+z^3)^2},$$

$$\left.-2\frac{z}{\sin(1+x+y^2+z^3)} - 3\frac{(2+7xy-z^2)\cos(1+x+y^2+z^3)z^2}{\sin(1+x+y^2+z^3)^2}\right])$$

27.

$f := [5x^3 + 6x^2y + 12xy^2 + 7y^3, x^5y^8z^2, 2 + y - z]$

```
>
  curl(f, v);
```
array(1 . . . 3,

$[1 - 2x^5y^8z, 0, 5x^4y^8z^2 - 6x^2 - 24xy - 21y^2])$

```
>
  curl(", v);
```
array(1 . . . 3,

$[40x^4y^7z^2 - 24x - 42y,$
$-2x^5y^8 - 20x^3y^8z^2 + 12x + 24y, 16x^5y^7z])$

29.

```
  grad(diverge([sec(x*y+z^2),cot(x^4+5*y*z),
    csc(x^3*y^5*z^2)],v),v);
```
array(1 . . . 3,

$[\sec(xy+z^2)\tan(xy+z^2)^2y^2 + \sec(xy+z^2)^3y^2$
$+ 40\csc(x^4+5yz)^2z\cot(x^4+5yz)x^3$
$+ 6\csc(x^3y^5z^2)\cot(x^3y^5z^2)^2x^5y^{10}z^3$
$+ 6\csc(x^3y^5z^2)^3x^5y^{10}z^3$
$- 6\csc(x^3y^5z^2)\cot(x^3y^5z^2)x^2y^5z,$
$\sec(xy+z^2)\tan(xy+z^2)^2xy + \sec(xy+z^2)^3xy$
$+ \sec(xy+z^2)\tan(xy+z^2)$
$+ 50\csc(x^4+5yz)^2z^2\cot(x^4+5yz)$
$+ 10\csc(x^3y^5z^2)\cot(x^3y^5z^2)^2x^6y^9z^3$
$+ 10\csc(x^3y^5z^2)^3x^6y^9z^3$
$- 10\csc(x^3y^5z^2)\cot(x^3y^5z^2)x^3y^4z,$
$2\sec(xy+z^2)\tan(xy+z^2)^2zy + 2\sec(xy+z^2)^3zy$
$+ 50\csc(x^4+5yz)^2z\cot(x^4+5yz)y$
$- 5\csc(x^4+5yz)^2$
$+ 4\csc(x^3y^5z^2)\cot(x^3y^5z^2)^2x^6y^{10}z^2$
$+ 4\csc(x^3y^5z^2)^3x^6y^{10}z^2$
$- 2\csc(x^3y^5z^2)\cot(x^3y^5z^2)x^3y^5])$

18.1, Page 1050

1. 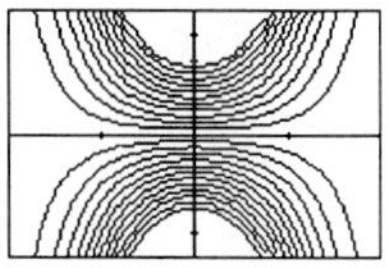

3. 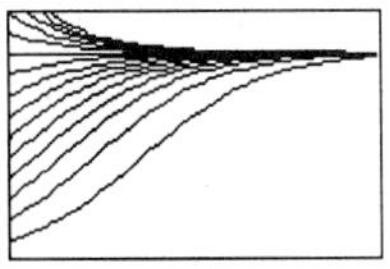

5. (a)

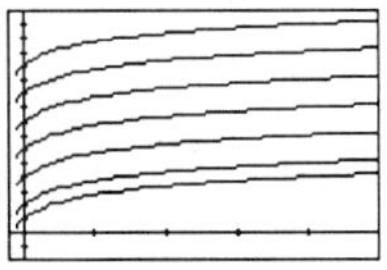

(b)

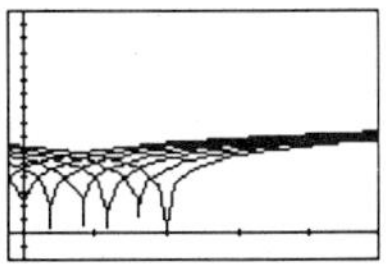

(c)

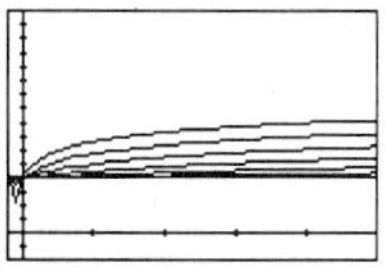

7. (a)

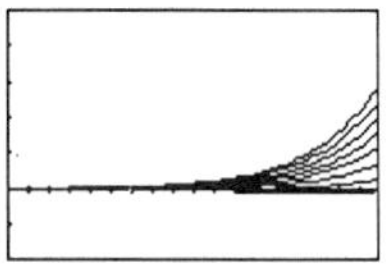

(b)

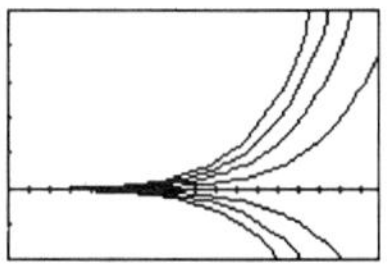

(c)

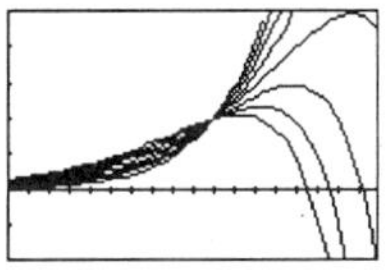

18.3, Page 1050

9. $y(1) \approx 0.6894567783$

11. $y(1) \approx 0.7266109233$

13. $y(1) \approx 0.7111708112$

15. $y(1) \approx 0.7320534678$

18.4, Page 1051

17. $y(2) \approx 0.7918102024$

18.6, Page 1051

19. 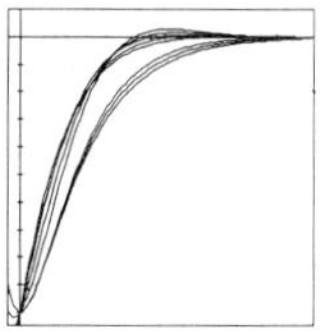

INDEX

FOURIER 1768–1830

Jean-Baptiste Fourier, a tailor's son, was orphaned at 8, ghost-writing sermons at 12, and teaching mathematics at 16. Elected to the infamous Committee of Surveillance during the French Revolution, he saved many from death. In 1807 his paper on heat propagation was rejected because it lacked proofs, but today he is honored by a bronze statue in Auxerre, by the Institute Fourier in Grenoble, and by the vocabulary of mathematics everywhere.

EULER 1707–1783

After first taking lessons from his pastor-father, Leonhard Euler became a student of John Bernoulli. He spent most of his life at the Berlin and St. Petersburg academies. Euler withheld his own work on calculus of variations so young Lagrange could publish first; and he showed similar generosity on many other occasions. Generations of mathematicians followed LaPlace's advice: "Read Euler, he is our master in all." Euler kept up his unparalleled output, although totally blind the last 17 years of his life.

GAUSS 1777–1855

Johann Friedrich Carl Gauss, the only son of a bricklayer and the grandson of a poor gardener, was not yet three when he corrected his father's computation of a payroll. Number theory and astronomy later profited from this computational facility—aided by a memory, it is said, that enabled him to dispense with a table of logarithms. The Gauss-Weber statue in Göttingen honors his invention of the electric telegraph. His mathematics is its own monument.

LAGRANGE 1736–1813

Joseph Louis Lagrange was the last of his mother's 11 children and the only one to live beyond infancy. In his teens he was already a professor at the Royal Artillery School in Turin. Invited there through the efforts of Euler and D'Alembert, he spent 20 productive years at the court of Frederick the Great, until the latter's death in 1786. Thereupon Louis XVI installed him in the Louvre, where, it is said, he was a favorite of Marie Antoinette. He deplored the excesses of the French Revolution, yet helped the new government establish the metric system. He was the first professor at the École Polytechnique, where calculus and number theory were his specialties.

LAPLACE 1749–1827

Pierre-Simon Laplace was successful in combining science with politics. Napoleon made him minister of the interior but dismissed him because he "searched for subtleties everywhere and carried into administration the spirit of the infinitely small"—meaning, the infinitesimal calculus. Yet Napoleon then made him a senator; and after Napoleon's downfall, Laplace became a marquis.

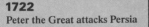

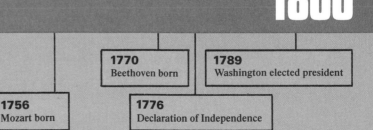

Karlskirche, Vienna, built by Johann Fischer von Erlach during the years 1716–1737

1700 — 1800

1722
Peter the Great attacks Persia

1756
Mozart born

1770
Beethoven born

1776
Declaration of Independence

1789
Washington elected president

LEJEUNE-DIRICHLET 1805–1859

Lejeune-Dirichlet was Riemann's teacher and Gauss's first successor. He married a sister of the composer Mendelssohn, and soon his home was a center of artistic society—even though he continued to sit for hours on end with his friend Jacobi "in mathematical silence." His contributions to the field of number theory were vast and far-reaching.

PICARD 1856–1941

Émile Picard excelled in classics, but he found geometry distasteful and simply memorized the theorems to avoid penalties. By 1928 he was a member of 40 learned societies. Picard's lecturing was incomparable; one had to hurry to get a seat. It has been said that every French mathematician over 30 was once a pupil of his. "You are the best thing I have done in mathematics," he told his last packed audience of students, old and new.

RIEMANN 1826–1866

Until the age of ten, Georg Friedrich Bernhard Riemann was educated chiefly by his father, a Lutheran pastor. Excused from high school mathematics because of his precocity, he flashed through Legendre's 859-page treatise on number theory in six days and through the other classics with equal ease. Although Riemann was befriended by Seyffer, Lejeune-Dirichlet, and Gauss, he had trouble taking care of himself and three dependents on $300 a year. Throughout his life he had an affinity for the study of calculus.

HILBERT 1862–1943

In nonmathematical subjects, David Hilbert's school record was undistinguished; and he did not especially concern himself with mathematics, because, he said, "I certainly knew I would do it later." Hilbert's method of research was to master a field, then leave it when his success seemed at its peak. Thus, by repeated specialization, he was considered a broad mathematician. In 1925 Hilbert contracted pernicious anemia and was among the first Europeans to be saved by the American liver preparation that later led to the discovery of vitamin B_{12}.

POINCARÉ 1854–1912

Until the age of eight, Jules-Henri Poincaré was educated at home, chiefly by his mother. He excelled both in classics and mathematics, winning first prize in two nationwide competitions. Even after reading a book at high speed, he always knew the page and line on which any subject was treated. Because he could not see the blackboard clearly, he followed mathematical lectures by just sitting back and listening. Honored as an author as well as for mathematics (particularly in analysis), he was a superb expositor of scientific subjects and wrote 30 books.

LIE 1842–1899

Marius Sophus Lie showed no marked interest in mathematics until the age of 26. He was responsible for initiating work on continuous groups. He befriended many well-known mathematicians, among them Darboux, who rescued him from arrest as a spy when Lie, an enthusiastic hiker, tried to walk from Paris to Italy during the war of 1870.

1800 — **1900**

1818 Marx born

1865 President Lincoln assassinated

1879 Einstein born

1898 Spanish-American War